REVIEWS in MINERALOGY
Volume 36

PLANETARY MATERIALS

J.J. PAPIKE, EDITOR

CONTENTS

Cover: Barred olivine chondrule in the Allende CV3 carbonaceous chondrite, set in a matrix of sub-micron-sized silicate, oxide, sulfide and carbonaceous material.

Transmitted light, crossed polarizers. Long dimension: 1 mm.

Series Editor: Paul H. Ribbe

MINERALOGICAL SOCIETY OF AMERICA

WASHINGTON, D.C.

REVIEWS IN MINERALOGY

(Formerly: SHORT COURSE NOTES)

ISSN 0275-0279

Volume 36

PLANETARY MATERIALS

ISBN 0-939950-46-4

ADDITIONAL COPIES of this volume as well as those listed on the following page may be obtained at moderate cost from:

THE MINERALOGICAL SOCIETY OF AMERICA
1015 EIGHTEENTH STREET, NW, SUITE 601
WASHINGTON, DC 20036 U.S.A.

List of volumes currently available in the *Reviews in Mineralogy* series

Vol.	Year	Pages	Editor(s)	Title
1-7	*out*	*of*	*print*	
8	1981	398	A.C. Lasaga R.J. Kirkpatrick	KINETICS OF GEOCHEMICAL PROCESSES
9A	1981	372	D.R. Veblen	AMPHIBOLES AND OTHER HYDROUS PYRIBOLES—MINERALOGY
9B	1982	390	D.R. Veblen, P.H. Ribbe	AMPHIBOLES: PETROLOGY AND EXPERIMENTAL PHASE RELATIONS
10	1982	397	J.M. Ferry	CHARACTERIZATION OF METAMORPHISM THROUGH MINERAL EQUILIBRIA
11	1983	394	R.J. Reeder	CARBONATES: MINERALOGY AND CHEMISTRY
12	1983	644	E. Roedder	FLUID INCLUSIONS (Monograph)
13	1984	584	S.W. Bailey	MICAS
14	1985	428	S.W. Kieffer A. Navrotsky	MICROSCOPIC TO MACROSCOPIC: ATOMIC ENVIRONMENTS TO MINERAL THERMODYNAMICS
15	1990	406	M.B. Boisen, Jr. G.V. Gibbs	MATHEMATICAL CRYSTALLOGRAPHY (Revised)
16	1986	570	J.W. Valley H.P. Taylor, Jr. J.R. O'Neil	STABLE ISOTOPES IN HIGH TEMPERATURE GEOLOGICAL PROCESSES
17	1987	500	H.P. Eugster I.S.E. Carmichael	THERMODYNAMIC MODELLING OF GEOLOGICAL MATERIALS: MINERALS, FLUIDS, MELTS
18	1988	698	F.C. Hawthorne	SPECTROSCOPIC METHODS IN MINERALOGY AND GEOLOGY
19	1988	698	S.W. Bailey	HYDROUS PHYLLOSILICATES (EXCLUSIVE OF MICAS)
20	1989	369	D.L. Bish, J.E. Post	MODERN POWDER DIFFRACTION
21	1989	348	B.R. Lipin G.A. McKay	GEOCHEMISTRY AND MINERALOGY OF RARE EARTH ELEMENTS
22	1990	406	D.M. Kerrick	THE Al_2SiO_5 POLYMORPHS (Monograph)
23	1990	603	M.F. Hochella, Jr. A.F. White	MINERAL-WATER INTERFACE GEOCHEMISTRY
24	1990	314	J. Nicholls J.K. Russell	MODERN METHODS OF IGNEOUS PETROLOGY—UNDERSTANDING MAGMATIC PROCESSES
25	1991	509	D.H. Lindsley	OXIDE MINERALS: PETROLOGIC AND MAGNETIC SIGNIFICANCE
26	1991	847	D.M. Kerrick	CONTACT METAMORPHISM
27	1992	508	P.R. Buseck	MINERALS AND REACTIONS AT THE ATOMIC SCALE: TRANSMISSION ELECTRON MICROSCOPY
28	1993	584	G.D. Guthrie B.T. Mossman	HEALTH EFFECTS OF MINERAL DUSTS
29	1994	606	P.J. Heaney, C.T. Prewitt, G.V. Gibbs	SILICA: PHYSICAL BEHAVIOR, GEOCHEMISTRY, AND MATERIALS APPLICATIONS
30	1994	517	M.R. Carroll J.R. Holloway	VOLATILES IN MAGMAS
31	1995	583	A.F. White S.L. Brantley	CHEMICAL WEATHERING RATES OF SILICATE MINERALS
32	1995	616	J. Stebbins P.F. McMillan D.B.Dingwell	STRUCTURE, DYNAMICS AND PROPERTIES OF SILICATE MELTS
33	1996	862	E.S. Grew L.M. Anovitz	BORON: MINERALOGY, PETROLOGY AND GEOCHEMISTRY
34	1996	438	P.C. Lichtner, C.I. Steefel, E.H. Oelkers	REACTIVE TRANSPORT IN POROUS MEDIA
35	1997	448	J.F. Banfield K.H. Nealson	GEOMICROBIOLOGY: INTERACTIONS BETWEEN MICROBES AND MINERALS

FOREWORD

'Planetary Materials' was the brain-child of James J. Papike, recent President of the Mineralogical Society of America and current Director of the Institute of Meteoritics at the University of New Mexico. It was probably not intentional that this volume be the largest ever produced in the *Reviews in Mineralogy* series, but this work has exceeded by nearly 200 pages the next smaller volume (33). Perhaps it is no coincidence that both books were produced apart from MSA short courses—the normal venue for most of the *RiM* volumes.

Given the sheer mass of material, this tour de force of Solar System mineralogy has been handled differently than most books in the series: pagination is by chapter, i.e. the pages in Chapter 1 are numbered 1-1, 1-2, 1-3, etc. and those in Chapter 2, 2-1, 2-2, 2-3, etc. This facilitated indexing and increased the speed of publication which would otherwise have suffered set-backs because of new life (the birth of Rhian Jones and Adrian Brearley's baby girl, Elena) and death and illness (in the family and person, respectively, of the Series Editor).

Indexes are luxuries in the *RiM* series because publication deadlines are usually too short for most volumes, but, thankfully, the 1039-page text of 'Planetary Materials' was indexed by the above-named parents—a heroic and absolutely essential undertaking for such an encyclopedic work.

I thank Dr. Jodi Rosso for an extensive array of assistance with both software and hardware (I was born in too early a generation!) and more particularly for her assistance with the copy editing of Chapter 2 and the 220+ pages of tables that arrived in this office in a variety of states of disorder. Without her help this volume would have lingered in the *RiM* Editorial Office another three months beyond the eight it took to complete.

Paul H. Ribbe
Blacksburg, Virginia
September 10, 1998

PREFACE

"Look at trees and see forests."

We seek to understand the timing and processes by which our solar system formed and evolved. There are many ways to gain this understanding including theoretical calculations and remotely sensing planetary bodies with a number of techniques. However, there are a number of measurements that can only be made with planetary samples in hand. These samples can be studied in laboratories on Earth with the full range of high-precision analytical instruments available now or available in the future. The precisions and accuracies for analytical measurements in modern Earth-based laboratories are phenomenal. However, despite the fact that certain types of measurements can only be done with samples in hand, these samples will always be small in number and not necessarily representative of an entire planetary surface. Therefore, it is necessary that the planetary material scientists work hand-in-hand with the remote sensing community to combine both types of data sets. This exercise is in fact now taking place through an initiative of NASA's Curation and Analysis Planning Team for Extraterrestrial Materials (CAPTEM). This initiative is named "New Views of the Moon: Integrated Remotely Sensed, Geophysical, and Sample Datasets." As preliminary results of the Lunar Prospector mission become available, and with the important results of the Galileo and Clementine missions now providing new global data sets of the Moon, it is imperative that the lunar science community synthesize these new data and integrate them with one another and with the lunar-sample database. Integrated approaches drawing upon multiple data sets can be used to address key problems of lunar origin, evolution, and resource definition and utilization.

The idea to produce this Reviews in Mineralogy (RIM) volume was inspired by the realization that many types of planetary scientists and, for that matter, Earth scientists will need access to data on the planetary sample suite. Therefore, we have attempted to put together, under one cover, a comprehensive coverage of the mineralogy and petrology of planetary materials. The book is organized with an introductory chapter that introduces the reader to the nature of the planetary sample suite and provides some insights into the diverse environments from which they come. Chapter 2 on Interplanetary Dust Particles (IDPs) and Chapter 3 on Chondritic Meteorites deal with the most primitive and unevolved materials we have to work with. It is these materials that hold the clues to the nature of the solar nebula and the processes that led to the initial stages of planetary formation. Chapter 4, 5, and 6 consider samples from evolved asteroids, the Moon and Mars respectively. Chapter 7 is a brief summary chapter that compares aspects of melt-derived minerals from differing planetary environments.

Many individuals worked very hard, for over a year, to make this book become a reality. First and foremost are the authors of the individual chapters. Acknowledgments appear at the end of each chapter thanking the reviewers of the chapters and the support staff that helped assemble the materials. However, I must single out a special person for special thanks, Paul Ribbe. Paul kept his cool while receiving many huge manuscripts with endless numbers of figures and tables. His 'can do' attitude and professionalism are largely responsible for this volume reaching completion.

Thank you, Paul.

Jim Papike
Albuquerque, NM
July 1, 1998

TABLE of CONTENTS

Chapter 3 CHONDRITIC METEORITES

Adrian J. Brearley and Rhian H. Jones

Chapter 4

NON-CHONDRITIC METEORITES FROM ASTEROIDAL BODIES

D.W. Mittlefehldt, T.J. McCoy, C.A. Goodrich & A. Kracher

Chapter 5 LUNAR SAMPLES

J.J. Papike, G. Ryder & C.K. Shearer

Chapter 6

MARTIAN METEORITES

H.Y. McSween, Jr & A.H. Treiman

Chapter 7

COMPARATIVE PLANETARY MINERALOGY: CHEMISTRY OF MELT-DERIVED PYROXENE, FELDSPAR, AND OLIVINE

James J. Papike

Chapter 1

THE PLANETARY SAMPLE SUITE AND ENVIRONMENTS OF ORIGIN

C.K. Shearer, J.J. Papike and F.J.M. Rietmeijer

Institute of Meteoritics
Department of Earth and Planetary Sciences
University of New Mexico
Albuquerque, New Mexico 87131

PLANETARY SAMPLE SUITE

Introduction

The planetary sample suite discussed in the text of this volume of *Reviews in Mineralogy* consists of materials from the Moon, Mars, and a wide variety of smaller bodies such as asteroids and comets. This suite of materials has been used to reconstruct the chronology and evolution of our solar system, define past and current processes in a wide range of planetary settings, and provide ground truth for remote sensing exploration. The first planetary samples in our collection were delivered to Earth as meteorites. These samples include materials from a large number of asteroids, the Moon and Mars. Although records of meteorite falls extend as far back as 1478 BC, E.F.F. Chladni, in 1794, was one of the first to argue that meteorites were extraterrestrial in origin and linked to atmospheric fireballs. In the second half of this century, humans have been far less passive and much more systematic in collecting planetary materials. Robotic or human sample return missions have been made to the Earth's Moon. Research teams, primarily from the United States and Japan, have recovered meteorites from Antarctica. Interplanetary dust particles (IDPs) are actively being collected in the stratosphere, from terrestrial polar ices, deep-sea sediments, and within impact features on spacecraft. Future sample return missions to Mars and selected asteroids are planned for the beginning of the 21st century.

A summary of the suite of documented planetary materials is presented in Table 1. This table lists the extent (number and/or mass of samples) of the sample suite from each planetary body that has been sampled either actively or passively It is obvious that such a summary is antiquated from the time it is put together. In particular, with a major effort to retrieve meteorites from Antarctica, totals for many suites of meteorites change a great deal from year to year. Notwithstanding, the intent of Table 1 is to give the reader a sample context in which to place the information presented in the subsequent chapters.

Moon

The suite of samples representing the Moon consists of over 2200 individual samples with a total weight of over 384 kg (Table 1). Most of these samples were collected by manned missions to the Moon (Apollo missions 11, 12, 14, 15, 16, 17). Additional samples were either collected by Soviet robotic missions (Luna 16, 20 and 24) or fell to Earth as lunar meteorites.

A total of 381.7 kg of samples were collected by the Apollo missions. These missions collected samples from the near-Earth side of the Moon. Sample sites represent a range of geological settings including mare sites (Apollo 11, 12), highlands sites (Apollo 14, 16) and mare-highland boundary sites (Apollo 15, 17). Samples were collec-

0275-0279/98/0036-0001$05.00

Table 1. Description of the planetary sample suite (Graham et al. 1985, Vaniman 1991, Grossman 1994).

Planetary Sample Type	Lithologies	Weight of Samples	Number of Samples
Moon			
Apollo Missions	regolith, basalt, highlands	381.7 kilograms	2196
Luna Missions	regolith, basalt, highlands	300 grams	
Lunar Meteorites	basalt, highlands	2.6 kilograms	13
Mars			
SNC Meteorites	basalt, basalt cumulates	≈ 70 kilograms	12
4 Vesta (HED parent body)			
Eucrites	basalts, basalt cumulates	>200 kilogram	122
Diogenites	basalt cumulates	≈70 kilograms	26
Howardites	"regolith"	≈36 kilograms	47
Asteroid Belt (not including 4 Vesta)			
Other Achondrites			
Non-Antarctic falls and finds		>1500 kilograms	38
U.S. Anarctic finds			79
Iron Meteorites			
Non-Antarctic falls		>1679 kilograms	42
Non-Antarctic finds			683
U.S. Anarctic finds		> 335 kilograms	68
Stony-iron			
Non-Antarctic falls		> 611 kilograms	10
Non-Antarctic finds			63
U.S. Anarctic finds		> 30 kilograms	23
Enstatite Chondrites			
Non-Antarctic falls			13
Non-Antarctic finds			11
U.S. Anarctic finds		> 4 kilograms	63
Ordinary Chondrites			
Non-Antarctic falls		>10,730 kilograms	736
Non-Antarctic finds			850
U.S. Anarctic finds		>1465 kilograms	5243
Carbonaceous Chondrites			
Non-Antarctic falls		> 462 kilograms	35
Non-Antarctic finds			32
U.S. Anarctic finds		> 15 kilograms	205
Interstellar Dust Particles			
Over 15 stratosphere missions have collected IDPs with a chondritic signature.		<< 1 gram	~10,000

ted by a variety of methods including direct collection of hand-sample-sized rocks and chips of boulders and scoop, rake, and core samples of regolith (Vaniman et al. 1991a). The Apollo 11 mission returned the fewest samples (58 samples and 21.6 kg), and the Apollo 17 mission collected the most (741 samples and 110.5 kg). The samples returned by the Apollo missions consist of a wide variety of lunar lithologies including mare basalts, highland lithologies (e.g. anorthosites, norites, gabbros, basalts), breccias, regoliths, and impact melts.

Three unmanned Luna missions (Luna 16, 20, and 24) returned slightly more than 300 grams of lunar regolith. This material consists of soil and small rock fragments collected by drilling 35, 27, and 160 cm into the lunar surface. Lithologies sampled by these missions include mare basalts and several highland lithologies.

Lunar meteorites are the third source of lunar samples (Table 1). A total of 13 lunar meteorites have been recovered and documented. All but two have been collected in Antarctica by U.S. and Japanese expeditions. The first lunar meteorite was collected in 1979 (Yamato 793274). The two lunar meteorites from outside Antarctica were recovered from western Australia (Hill et al. 1991) and the Sahara desert (Bischoff and Weber 1997). The largest lunar meteorite (Yamato 86032) has a total mass of 648.4 grams. The total mass of all the lunar meteorites is 2.6 kg. Within this suite of samples are basaltic breccias, anorthositic breccias and mare gabbros.

Mars

It was not realized until the late 1970s and early 1980s (McSween et al. 1979, Walker et al. 1979, Wasson and Wetherill 1979, Bogard and Johnson 1983, Becker and Pepin 1984) that differentiated meteorites referred to as the SNC (shergottites, nakhlites, and chassignites) group were our first samples of Mars. This conclusion was based on the young crystallization ages (<1.3 Ga.) of the few SNC meteorites available at that time (McSween et al. 1979, Walker et al. 1979, Wasson and Wetherill 1979), and the match of several geochemical-isotopic fingerprints between measurements made by the Viking missions and the SNCs (Bogard and Johnson 1983, Becker and Pepin 1984, Swindle et al. 1986). A more detailed discussion of these arguments is presented by McSween and Treiman (this volume, Chapter 6).

The suite of martian samples consists of 12 meteorites (Table 1). The first martian sample that was collected was the Chassigny meteorite. It was seen to fall in Haute-Marne, France on October 3, 1815 at 8:00 AM (Graham et al. 1985). A total of 6 martian samples were recovered between 1977 and 1994 in Antarctica. Sample size for the martian samples ranges from a known mass of approximately 18 kg (Zagami) to 12 gm (QUE94201). All samples are magmatic in origin and include 4 basalts, 3 lherzolites, an orthopyroxenite, 3 clinopyroxenites-wehrlite, and a dunite. The orthopyroxenite (ALH84001) has superimposed upon its igneous lithology a secondary mineralization assemblage consisting of carbonate and sulfide.

Howardites, eucrites, diogenites (HED)

The howardites, eucrites, and diogenites constitute a suite of meteorite lithologies (HED) that are known to be related through oxygen isotope systematics (Clayton and Mayeda 1983). They are believed to be remnants of an ancient (4.6 Ga.) and complex basaltic magma system on asteroid 4 Vesta (Consolmagno and Drake 1977, Binzel and Xu 1993, Binzel 1996). The eucrites are pigeonite- and plagioclase-bearing basalts that represent either surface or near-surface basaltic liquids or crystal accumulations from basaltic liquids. Diogenites are orthopyroxenites that represent very efficient subsurface accumulations of orthopyroxene from basaltic magmas. The howardites are breccias made up of mechanical mixtures of eucrites and diogenites.

HED make up approximately 3% of the meteorite population and approximately 2% of the U.S. Antarctic collection located at the Johnson Space Center. There are approximately 195 known samples of the HED parent body (presumably 4 Vesta). Of this total, 122 are eucrites, 26 diogenites, and 47 howardites. The U.S. Antarctic meteorite collection contains 83 eucrites, 14 diogenites, and 24 howardites.

Other achondrites

In addition to the HED, SNC, and lunar meteorites, there are a wide variety of other differentiated achondritic meteorites, representing parent bodies other than 4 Vesta. These meteorites include the aubrites, ureilites, acapulcoites-lodranites, and angrites.

Although referred to as achondritic meteorites, these groups are samples from parent bodies that either preserved nebular signatures or totally eradicated those signatures through melting. Models for the origin of ureilites and aubrites range from nebular condensates to igneous cumulates. Many recent researchers propose models in which the ureilites are condensates or impact melted condensates and the aubrites represent brecciated and reassembled igneous lithologies. Acapulcoites-lodranites are considered to be residua from which basaltic melt was partially or totally extracted. Angrites may represent either igneous cumulates or a quenched basaltic magma. Chapter 4 discusses these meteorite-types in much more detail.

These types of achondrites make up less than 1.5% of the non-Antarctic meteorite population and approximately 2% of the U.S. Antarctic collection located at the Johnson Space Center. There are approximately 117 samples of the these types of achondrites. Of this total, 79 are U.S. Antarctic finds, 17 are non-Antarctic falls, and 21 are non-Antarctic finds.

Iron meteorites

Iron meteorites are samples of cores of disrupted asteroids. Following disruption, there clearly was very little mixing between the asteroids' silicate mantles and iron cores. Therefore, the iron meteorites provide evidence for core formation on a small scale and provide us with some insight into core formation in general. The iron meteorites are derived from a wide variety of parent bodies. Over 60 groups have been identified based on geochemical characteristics (Ni, Ga, Ge, Ir), cooling rate estimates, and exposure ages. A more detailed discussion of these fingerprints is presented in Chapter 4 of this volume.

Based on a 1985 summary of meteorites (Graham et al. 1985), over 725 iron meteorites, weighing approximately 1670 kg, have been documented and classified. They make up approximately 5% of the total non-Antarctic meteorite falls (Graham et al. 1985). This number does not include the very large number of non-Antarctic iron meteorites that have been found without an observed fall. Iron meteorites make up approximately 1.2% of the U.S. Antarctic collection. This translates to over 68 samples with a total weight of greater than 335 kg (Grossman 1994).

Stony-iron meteorites

Stony-iron meteorites, such as pallasites and mesosiderites, are mixtures of roughly equal proportions of metal and silicate phases. Pallasites may represent samples from core-mantle boundaries of disrupted differentiated asteroids, whereas mesosiderites are surface-derived breccias. The metal in pallasites is similar to that found in fractionated iron meteorites, whereas the metal in mesosiderites appears to be similar to that found in unfractionated chondritic meteorites. Therefore, these samples provide us glimpses into two different types of processes: core formational processes (pallasites) and early surficial processes on asteroids (mesosiderites).

At least 96 samples of stony-iron meteorites have been documented and classified in Antarctic and non-Antarctic collections (Graham et al. 1985, Grossman 1994). The total weight of samples is greater than 640 kg. The ratio of mesosiderite samples to pallasite samples is approximately 1:1. The stony-iron meteorites make up approximately 1.2% of the non-Antarctic falls, 3.6% of the non-Antarctic finds (Graham et al. 1985), and 0.4% of the U.S. Antarctic meteorite collection (Grossman 1994).

Chondrites

Chondrites are agglomeritic rocks consisting of a wide variety of components (refractory inclusions, chondrules, matrix, sulfides, silicates, Fe-Ni metal) with very different formational histories. In unequilibrated chondrites, these individual components provide unique information about the astrophysical sources of interstellar dust (interstellar grains), solar nebular conditions and processes (CAIs, chondrules, matrix) and accretional events within the solar nebula. Many chondrites have experienced secondary processing, such as aqueous alteration and thermal metamorphism, probably within a parent body environment. The mineralogy of these chondrites provides insights into the conditions and durations of such secondary processes within planetesimals.

Based on the 1985 summary of meteorites (Graham et al. 1985), 1681 chondrites have been documented and classified. From this compilation, only 67 have been classified as carbonaceous chondrites. The chondrites make up approximately 85% of the total non-Antarctic meteorite falls (Graham et al. 1985) which is equal to approximately 11,200 kg of sample. This does not include the large number of non-Antarctic, chondritic meteorites that have been found without an observed fall. Chondrites make up approximately 95% of the U.S. Antarctic collection or over 5400 samples with a weight of greater than 1483 kg (Grossman 1994).

Interplanetary dust particles (IDPs)

Interplanetary dust particles (IDPs) are actively being collected from the stratosphere, from polar ices, and within impact features on spacecraft. Detailed descriptions of these various collections are presented by Brownlee (1978), Rietmeijer and Warren (1994), and Rietmeijer (this volume, Chapter 2). The stratosphere collection of IDPs are particularly important because they represent planetary materials that complement meteorites. Beginning in the 1950s, these particles have been collected in the atmosphere, first on filters and then, in the 1970s, on inertial impaction surfaces. Since May 1981, NASA has used aircraft to collect IDPs from the stratosphere at an altitude of 17 to 19 km. More than 15 stratosphere missions have collected IDPs that have chondritic signatures. An average mission that is flown for 35 hours will recover approximately 5 IDPs on a 30 cm^2 collector. Obviously, missions with larger collectors do recover higher numbers of particles. The average IDP size is approximately 10 μm. IDPs collected by this NASA program are curated at the Johnson Space Center.

ENVIRONMENTS

Introduction

The properties of the parent bodies dictate the chemical and mineralogical characteristics of the sample making up this planetary suite.. The location of the sample source in the nebula may be reflected in the extent of volatile loss, the availability of distinct reservoirs during planetary assembly and the extent of preservation of primitive, unprocessed planetary material. For example, a volatile loss episode in the early nebula resulted in substantial heterogeneities in the noble gases and volatile elements. This volatile loss, which appears to have been more prevalent in the inner nebula, is reflected in noble gas and volatile element depletions in the inner planets, and to a lesser extent, in achondrite meteorites.

Planet size plays a significant role in the duration of magmatism (McSween 1985), style of magmatism (Wilson and Head 1981, BVSP 1981), and the composition of magmas (BVSP 1981). The size of the planet/asteroid dictates both the thermal history of the body and the pressure regime under which melting will occur. O'Hara (1968), Kushiro (1972), and Presnall (1979) have documented the role of pressure on the composition of a partial melt derived from the Earth's mantle. The effects of pressure within the context of particular planetary settings have been illustrated by Delano (1980) and Longhi (1992) for the Moon, Longhi et al. (1992) for Mars, and Stolper (1977) and Grove and Bartels (1992) for the parent body of the HED meteorites. To summarize the basic elements of these papers, in dry planetary mantles, quartz-normative basaltic magmas are the first liquids produced during partial melting at low pressures. Increasing pressure will drive the cotectic melting reaction to increasingly olivine-normative compositions. In addition, a broad spectrum of mantle P-T environments should produce a variety of mantle mineral assemblages. Therefore, in a large planetary body with many P-T-volatile mantle regimes such as the Earth, a spectrum of melt compositions should be produced. On the other hand, magmatism on a parent body as small as 4 Vesta should produce a fairly limited range of magmas of distinct composition, produced at relatively low pressure (≈ 1 kb) and low volatile content. Clearly, planetary environment dictates many of the characteristics of planetary materials. The following sections summarize some of the fingerprints of contrasting environments and the nature of these environments. For a comparison of some orbital, physical, and petrologic properties of selected parent bodies, see Tables 2 and 3.

Fingerprints of planetary environments

There are numerous geochemical and mineralogical "fingerprints" that reflect both the reservoir from which the planetary sample was derived and the environment under

Table 2. Summary of planetary, Moon and asteroidal properties.

Orbital Properties	**Earth**	**Moon**	**Mars**	**4 Vesta**
Mean Distance from Sun				
AU	1.000		1.524	2.36
10^6 km	149.6		227.9	353.1
Orbital Period (days)	365.256		686.980	1325.88
Mean Orbital Velocity (km/s)	29.79	1.03	24.13	< 5
Eccentricity	0.0167	0.055	0.0934	
Inclination to Eclipitic (degrees)	--------	5° 09'	1.85°	
Physical Properties	**Earth**	**Moon**	**Mars**	**4 Vesta**
Mass				
Earth = 1	1.0000	0.0123	0.1074	0.00005
10^{23} kg	59.76	0.735	6.421	0.0027
Equatorial radius				
Earth = 1	1.0000	0.2725	0.532	0.045
km	6378	1738	3398	288
Mean density (gm/cm^3)	5.52	3.343	3.94	3.8
Zero pressure density (gm/cm^3)	4.0		3.75	
Moment of inertia (•Ma^2)	0.33078	0.392	0.365	0.3
Sidereal rotation period	23.9345 hr	27.32 days	24.6229 hr	5.34 hr
Equatorial surface gravity (m/s^2)	9.78	1.62	3.72	
Equatorial escape velocity (km/s)	11.2	2.37	5.02	0.36
Heat Flow (mK/m^2)	63	~29		
Petrologic Properties	**Earth**	**Moon**	**Mars**	**4 Vesta**
Duration of volcanism (b.y)	4.55 to 0.0	4.55 to 2.0 (?)	4.55 to 0.0	4..55
Conditions of magmatism				
pressure of origin	> 100 kbars	> 20 kbars	> 100 kbars	< 1 kbar
depth of origin	100-200 km	100 -1000 km	50-200 km	<288 km
oxygen fugacity	FMQ ± 1.5	IW - 1	FMQ to FMQ-1.2	IW - 2
Mg/Mg+Fe of bulk planet	.89 - .82	.82 - .78	.65 - .78	.64 - .80

Table 3. Selected properties of IDP parent bodies.

	ASTEROIDS		COMETS	
	C-type	P- and D-types	Oort cloud	Kuiper belt
Heliocentric distance (AU) [1]	2-4	~3-5	$>10^4$	~30-100
Original heliocentric distance (AU)	Same	Same	2 0-30	Same
Near-circular orbits	Yes	Yes	No	Yes and no
Geocentric velocity (km s^{-1}) [2]	~5.5	~5.5	5.5-11, <1.2 AU, and >11 for <1.2 AU	< 5
Diameter (km)	1-1000		1-10, up to 100 [3];	1-300, up to 1000 [3,4]
T accretion (K)	<200	<200	80-60 [3]	30-10 [3]
T storage (K)	120-180 [5]	<<200	<60; possibly <30 [6]	30-10
T alteration	300-425 [7]	Unknown	Unknown	Unknown
T perihelion (K)	Earth crossing asteroids and Near-Earth asteroids; unknown		T_c = 140-180 [8]; T_s = 350-400 [9]; crust on volatile depleted nuclei up to 775 [10]	Unknown
T of Zodiacal dust at 1 AU (K) [11]	255-300; maximum at 370 ('black bodies' 1 μm in diameter) or 500-550 ('gray' bodies); aggregate IDPs behave as 'black bodies'			
Maximum T of atmospheric entry (K)	(700-1300) [12] – (~1400) [13]			Their size <10μm will promote strong heating [14]

Sources: [1] Bell et al. (1989); [2] Flynn (1989); [3] Weissman (1985); [4] Flynn (1996); [5] Fanale and Salvail (1989); [6] Stern and Shull (1988); [7] Zolensky et al. (1993); [8] equilibrium central temperature (T_c), McSween, Jr. and Weissman (1989) and Klinger (1983); [9] surface temperature (T_s), Sagdeev et al. (1986); [10] Brownlee (1978); [11] Gustafson (1994); [12] Nier and Schlutter (1992, 1993); [13] Rietmeijer (1996); [14] Love and Brownlee (1991).

which the sample evolved. These planetary fingerprints can be attributed to a variety of processes, ranging from heterogeneities in the early solar nebula to the late-stage evolution of a planetary body. Some of the more useful fingerprints in identifying planetary origin and nature of a planetary environment include oxygen isotope compositions, noble gases signature, ages of magmatic activity, elemental signatures (K/U, Mn/Fe, Eu anomaly, element valance state) and mineral assemblages (silicates, hydrous silicates, carbonates, oxides, metals).

The variation in the abundance of the three oxygen isotopes indicates the existence of separate reservoirs of planetary material and provides an important geochemical fingerprint concerning the source of the sample. The oxygen isotope plot in Figure 1 clearly seperates various oxygen isotope reservoirs among planetary materials. Within this plot, the terrestrial fractionation line, with a slope of 0.5, exhibits the variability that can be attributed to simple mass fractionation. The Earth, Moon, and two subgroups of enstatite chondrites (EH, EL). This illustrates the point that identical oxygen isotopic signatures do not necessarily indicate derivation of samples from the same body, merely derivation from a part(s) of the solar nebula with similar oxygen isotopic characteristics. Mars and the ordinary chondrites plot at $\delta^{17}O$ values above the terrestrial fractionation line, whereas 4 Vesta (HED) and certain iron meteorites plot slightly below this line. The oxygen isotopic composition of minerals from Allende refractory inclusions define a line with a distinctly different slope (1.0) from the terrestrial isotopic fractionation line. The deviation of the slope of this line from that expected for simple fractionation indicates that this line represents a mixing line of two different reservoirs in the solar nebula. On this line also lie some of the carbonaceous chondrites and the ureilites.

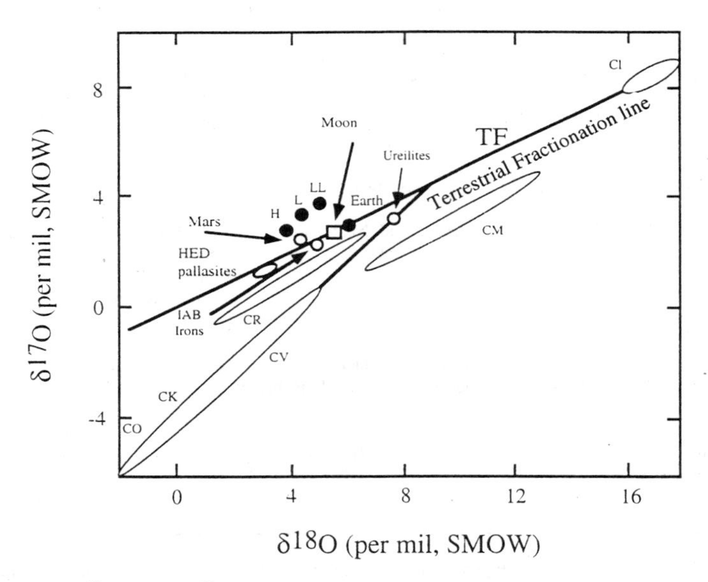

Figure 1. A plot of $\delta^{17}O$ against $\delta^{18}O$ for the Earth, Moon, Mars, and various meteorite classes. Modified after Clayton and Mayeda (1983, 1984), Sears and Dodd (1988), and Taylor (1988).

Two different arrays of noble gas compositions have been identified in planetary samples: "solar" abundances and "planetary" abundances. Planetary and meteoritic material is depleted in the light noble gases relative to solar wind abundances. A plot of $^{36}Ar/^{132}Xe$ versus $^{84}Kr/^{132}Xe$ (Fig. 2) also shows differences in noble gas compositions among planetary materials. The noble gas data from chondritic meteorites clearly define a distinct population array that may pass near the solar value. The terrestrial and martian atmospheres and Martian samples (SNC) define populations that deviate from the chondrite meteorites and solar abundances. Mars appears to be distinctly different from Earth. Compared to the Earth, the martian atmosphere is more depleted in both total noble gases and light noble gases. The differences among planetary materials in noble gas concentrations and noble gas ratios are not a product of solar nebula heterogeneities, but

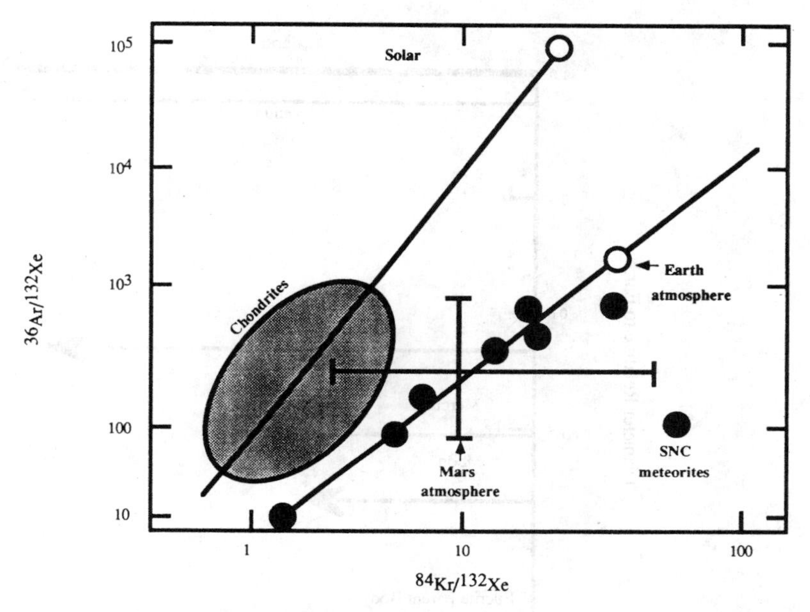

Figure 2. A plot of $^{36}Ar/^{132}Xe$ versus $^{84}K/^{132}Xe$ that illustrates differences among meteoritic and planetary noble gas data. Modified after Ott and Begemann (1985).

are a result of secondary processes involving atmospheric evolution following accretion (i.e. hydrodynamic blow-off).

Crystallization ages of magmatic assemblages also provide some clues to the planetary setting of a sample. McSween (1985) illustrated that there appears to be a relationship between planet size and the duration of igneous activity (Fig. 3). This relationship is related to the insulating capacity of large bodies that permits retention of heat produced by the decay of long-lived radionuclides. The result is that magmatism on asteroid-size bodies was arrested early in the evolution of the solar system (4.5 Ga), whereas in larger bodies the magmatic activity continued to 1.5 billion years after accretion (Moon, Mercury) through the present (Earth). The initial suggestions that the SNC meteorites were martian in origin were based on their relatively young crystallization ages (1.3 Ga to 180 Ma).

Certain elemental ratios in planetary bodies can be useful fingerprints that identify sources of materials and provide useful information concerning post-accretional planetary processes and conditions. Depletions in moderately volatile elements occur to varying degrees in the terrestrial planets and meteorites. This major volatile depletion event in the inner nebula probably occurred before or during planetesimal assembly. Ratios such as K/U (Fig. 4) and Mn/Fe (see Chapter 7) illustrate the elemental planetary signatures caused by this early volatile depletion episode. The K/U ratio reflects the behavior of potassium, a moderately volatile element, as compared to uranium, a refractory element. Compared to the K/U ratio in CI carbonaceous (≈ primordial abundance ratio) and ordinary chondrites, potassium is depleted to varying degrees. Measurements of K/U ratios for Mars and Earth (1 to 1.5×10^4) overlap and are a factor of 4 lower than the CI value. The Moon and the HED parent body exhibit an even more significant depletion (K/U = 2500).

In addition to the features that are tied to nebular heterogeneity, elemental ratios may

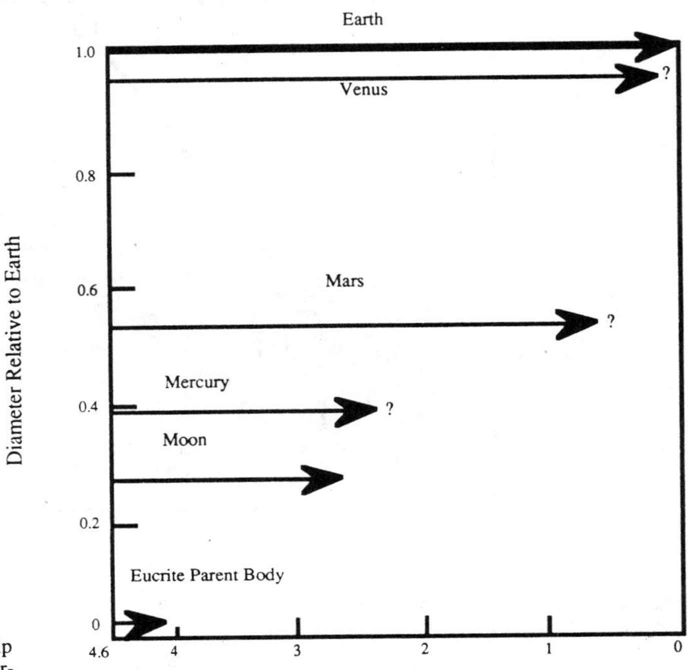

Figure 3. The relationship between planet size and duration of igneous activity. Modified after McSween (1985).

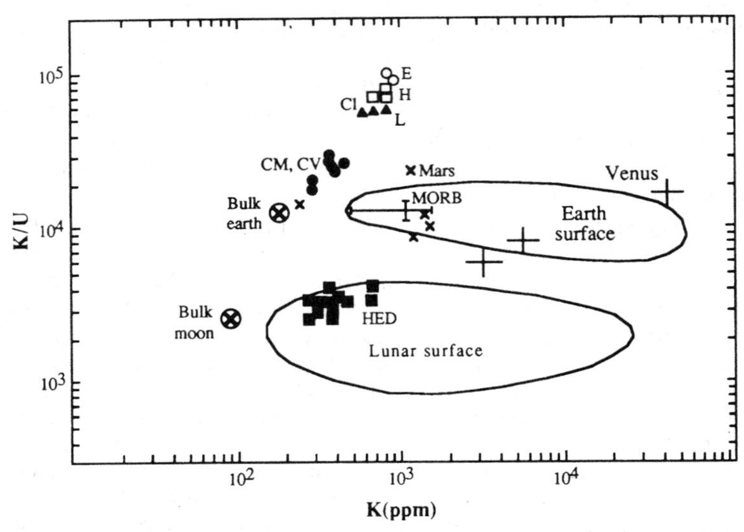

Figure 4. Plot of K (a moderately volatile element) versus U (a refractory element) that demonstrates the usefulness of an elemental fingerprint for distinguishing among sources of a variety of planetary materials. The CI value is taken to represent primordial abundances of U and K. Modified after BVSP (1981).

also reflect planetary conditions and processes following accretion. The depletions of siderophile elements in planetary samples have been used to estimate the extent of core formation and the composition of the core on the planetary body in question (Laul et al. 1986, Newsom and Drake 1982). The inferred valance states of iron (Fe^{3+}, Fe^{2+}, Fe metal), chromium (Cr^{3+}, Cr^{2+}), titanium (Ti^{4+}, Ti^{3+}), and europium (Eu^{3+}, Eu^{2+}) have been used to estimate fO_2 values for planetary bodies. For example, Crozaz and McKay (1989), using the experimental study of McKay et al. (1989), were able to estimate the fO_2 of crystallization for an achondrite using the Eu/Gd ratio in plagioclase and pyroxene. Similar calculations have been made for lunar, terrestrial, martian, and HED samples (Fig. 5).

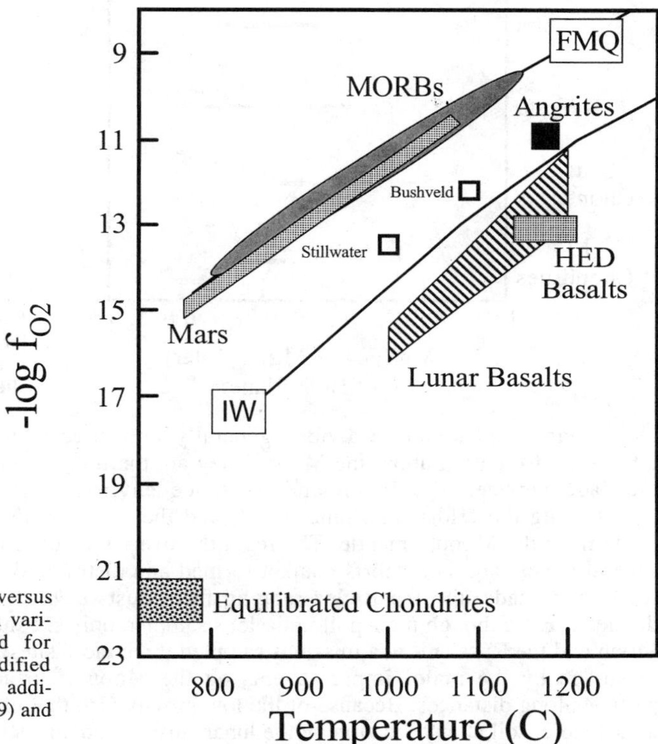

Figure 5. Oxygen fugacity versus temperature for basalts from various planetary settings and for equilibrated chondrites. Modified after BVSP (1981); includes additional data from McKay (1989) and Crozaz and McKay (1989).

Moon

The present suite of lunar samples represents the crust from an evolved planet. The chemistry, mineralogy, and petrology of these samples provide valuable information that allows us to reconstruct the evolutionary history of the Moon's crust and mantle.

The Moon orbits the Earth at a mean distance of 384,400 km. The density of the Moon is 3.34 g cm^{-3} compared to the Earth's density of 5.52 g cm^{-3}. The interior of the Moon (radius 1,738 km) comprises a feldspathic crust (thickness ~65 km frontside and ~100 km backside), a differentiated mantle and perhaps a small core (Taylor 1982). The surface of the Moon can be divided into two major regions: the relatively low, smooth, dark (low albedo) maria and the densely cratered, light-colored (high albedo), rugged highlands. Although the mare basalts cover ~17% of the surface area of the Moon, they make up only ~1% by volume of the lunar crust (Head 1976).

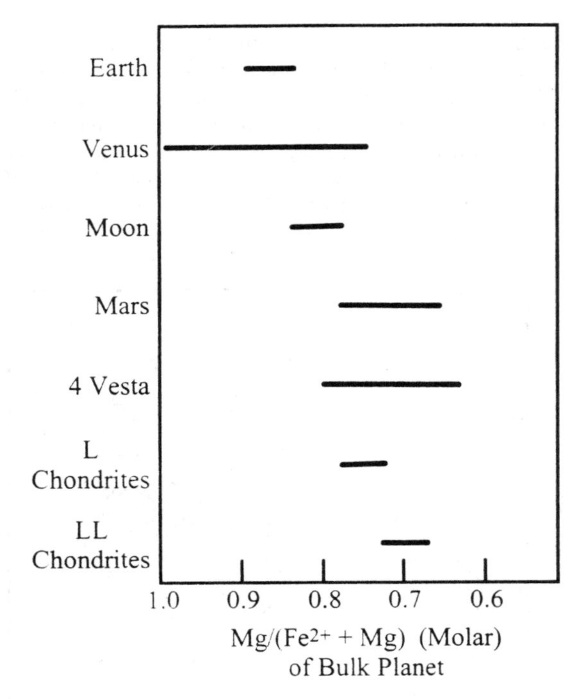

Figure 6. Range in Mg/(Mg+Fe) for the bulk compositions of different planetary bodies. Modified after BVSP (1981).

Lunar samples can be divided generally into three main groups, each giving us different information about the Moon. They are the regolith samples, which sample the interface between the Moon and its space environment; the highlands samples, representing the feldspathic lunar crust; and the mare basalts, which represent partial melts from the Moon's mantle. The regolith covers the entire lunar surface to depths of several meters and is a debris blanket formed and distributed by impact processes. The regolith is made up of particles ranging from dust-sized to blocks several meters in diameter. Even though the Apollo missions sampled only six sites on the front side of the Moon and the Russian Luna missions sampled three more sites, our sampling coverage is expanded by the major impact events on the Moon's surface that have transported particles long distances. Because of the low gravity (1/6 that of Earth) and the lack of an atmosphere, soil at any location on the lunar surface contains a small but important exotic component that increases the spatial significance of the lunar sample collection.

The mare basalt samples represent a period of igneous activity of about one billion years (~4.0 to 3.0 Ga). These samples provide a window into lunar mantle processes and are the best probes of the lunar mantle. Unlike the Earth, which has kimberlite occurrences that provide us with xenolithic mantle samples from great depths and earthquake activity which provides seismic signals to a heavily instrumented terrestrial surface, the Moon provides us with few ways to look into its interior other than the mare basalts. A comparison of the chemistry [Mg/(Mg+Fe)], mineralogy, and structure of the mantle of the Moon to that of the Earth, Mars, and 4 Vesta is shown in Figures 6 and 7.

The highland samples provide material from the feldspathic lunar crust. The early history of the Moon is thought to have been dominated by a global magma ocean (Taylor 1982). In this ocean of unknown depth, but probably >500 km, plagioclase floated, forming an anorthositic crust 65 to 100 km thick during the period ~4.6 to 4.4 Ga.

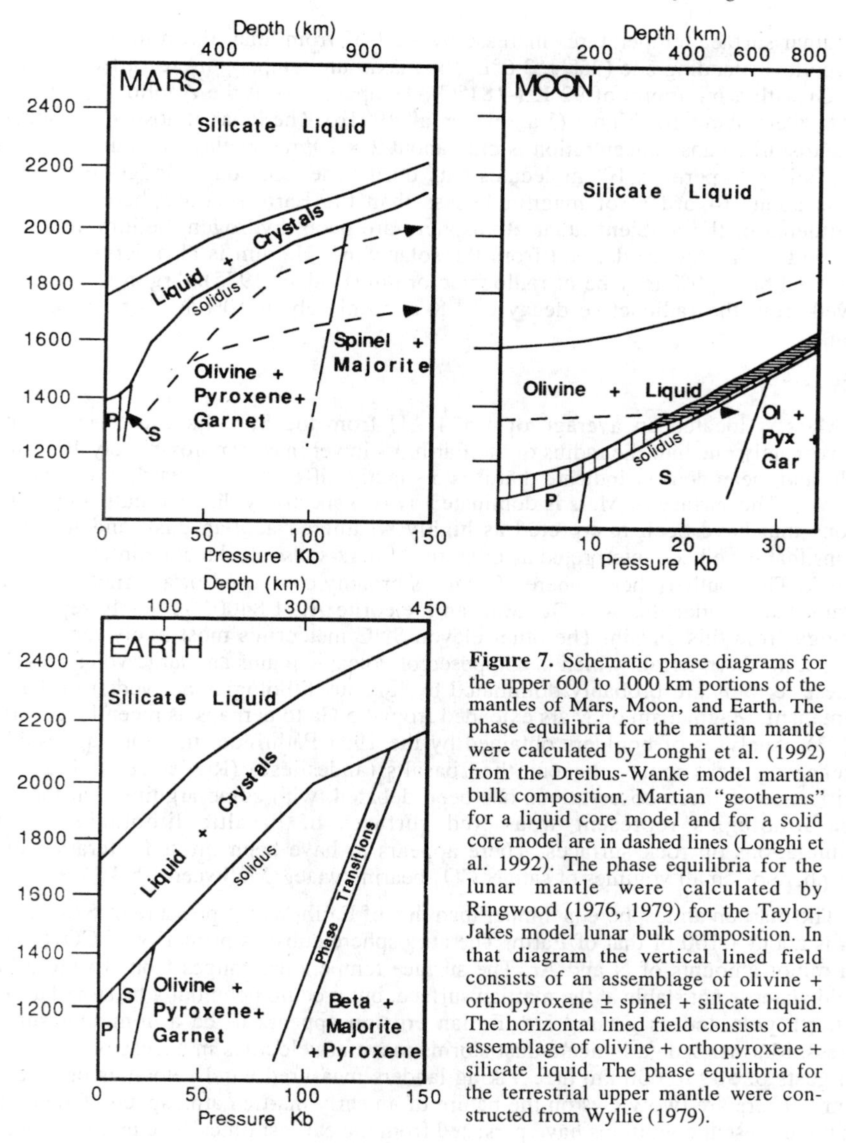

Figure 7. Schematic phase diagrams for the upper 600 to 1000 km portions of the mantles of Mars, Moon, and Earth. The phase equilibria for the martian mantle were calculated by Longhi et al. (1992) from the Dreibus-Wanke model martian bulk composition. Martian "geotherms" for a liquid core model and for a solid core model are in dashed lines (Longhi et al. 1992). The phase equilibria for the lunar mantle were calculated by Ringwood (1976, 1979) for the Taylor-Jakes model lunar bulk composition. In that diagram the vertical lined field consists of an assemblage of olivine + orthopyroxene ± spinel + silicate liquid. The horizontal lined field consists of an assemblage of olivine + orthopyroxene + silicate liquid. The phase equilibria for the terrestrial upper mantle were constructed from Wyllie (1979).

Anorthosites make up ~80% of the lunar crust. The other ~20% is composed of younger rocks (the Mg-suite) which intruded the crust at ~4.43 to 4.1 Ga, after most of the magma ocean had crystallized (Warren and Kallemeyn 1993).

The Moon lacks significant water and the lithologies making up its crust crystallized at a very low oxygen fugacity (Fig. 5). Therefore, all lunar lithologies thus far sampled are anhydrous and contain elements in reduced valance states (e.g., Fe^0, Fe^{2+}, Cr^{2+}, Ti^{3+}). Recent observations by the Lunar Prospecter orbital mission indicate water ice occurs at the poles of the Moon (Feldmann et al. 1998). The source of the water ice may be cometary in origin.

Lunar surface temperatures increase by ~280 K from lunar dawn to lunar noon. At the Apollo 15 landing site (26°N, 3.6°E) the maximum temperature recorded was 374 K (101°C) with a minimum of 92 K (-181°C). Temperatures at the Apollo 17 site (20°N, 30.6°E) were about 10° higher (Vaniman et al. 1991b). The lunar atmosphere is tenuous. The undisturbed gas concentration is only about 2×10^5 molecules cm^{-3} during the lunar night, falling to perhaps 10^4 molecules cm^{-3} during the lunar day (Hodges et al. 1974). This is about 14 orders of magnitude less than the Earth's atmosphere. The major constituents of the ambient lunar atmosphere are neon, hydrogen, helium, and argon. Neon and hydrogen are derived from the solar wind. Helium is also derived from the solar wind but ~10% may be of radiogenic origin (Hodges 1975). Argon is mostly ^{40}Ar derived from the radioactive decay of ^{40}K and only about 10% is ^{36}Ar of solar wind origin.

Mars

Mars is located an average of 1.524 AU from the Sun. Its equatorial radius is approximately one half the radius of the Earth. Its lower mass (approximately 10% that of Earth) and mean density indicate that it is distinctly different from Earth in composition (Table 2). The surface of Mars is dominated by two spectrally distinct units: high-albedo regions that have been interpreted as highly weathered aeolian dust, and low-albedo regions that have been interpreted as mixtures of dark soils, blocks of bedrock, and in situ bedrock. The southern hemisphere of Mars is broadly composed of an Ancient Cratered Terrain that is older than 4.0 Ga. Martian meteorite ALH 84001 probably represents a lithology from this terrain. The other eleven SNC meteorites most likely represent the younger, northern terrain which is composed of volcanic plains and large volcanoes. Both of these terrains are probably dominated by igneous lithologies derived from basaltic magmatism. Magmatism on Mars extended from 4.5 Ga to perhaps as recently as 180 Ma (Fig. 3). Analyses of boulders obtained by the 1997 Pathfinder mission suggested the presence of rocks more siliceous than basalts (andesites?) (Rieder et al. 1997). The significance of these observations has been debated, with some arguing that the high-silica lithologies represent weathered surfaces of basaltic lithologies or dust accumulations on rock surfaces. There appears to have been limited alteration of the lithosphere by small volumes of saline, CO_2-bearing water (McSween 1994).

The martian atmosphere is thinner than that of Earth, with a pressure of 6 to 10 mbar. This is about 1/100 of that of Earth. The atmosphere consists primarily of CO_2 (≈ 95%) with minor amounts of N and Ar. The surface temperature ranges from 150 to 300 K. Liquid water is not stable at the martian surface, but it could potentially be stored as ice at the north pole and as ground ice. Eolian erosion appears to be a dominant surficial process. Mariner 9 data indicate dust storms move at velocities in excess of 200 km hr^{-1} with gusts of 500 to 600 km hr^{-1}. Viking landers measured wind velocities of up to 100 km hr^{-1}. There are two views on the nature of an early martian atmosphere. One opinion is that the present conditions have persisted from the earliest times. The preferred point of view at this time calls for a more hospitable early Mars, with a thicker and wetter atmosphere.

Mars accreted from a different population of planetesimals than did Earth, based on the different oxygen isotope, noble gas, volatile element compositions of the SNC meteorites versus terrestrial samples and a different mean density from Earth. Based on the SNC meteorites, Dreibus and Wanke (1985) estimated the composition of the martian mantle. Compared to Earth, their model for the martian mantle has a lower Mg/(Mg + Fe) value (Fig. 6), a lower abundance of siderophile elements with chalcophile affinities (Co, Ni, Cu) and a higher abundance of volatile elements, alkali elements and siderophile elements with weak chalcophile affinities (Mn, Cr, W). The lower Mg/(Mg + Fe) and higher alkali element content of the martian mantle will result in higher olivine/pyroxene, pyroxene/garnet, and pyroxene/spinel ratios in mantle mineral assemblages relative to the terrestrial mantle. The siderophile element variability may reflect a high oxidation state of the martian mantle, resulting in the retention of Mn, Cr, and W in the mantle during core formation (Dreibus and Wanke 1985, McSween 1994). A high oxidation state for the martian mantle is also corroborated by mineral assemblages in the SNC meteorites. Using coexisting titanomagnetite and ilmenite solid solution members, several studies have calculated the fo_2 of crystallization for the SNC magmas to be between the FMQ buffer and 1.2 log fo_2 units below the FMQ buffer (Fig. 5) (Reid and Bunch 1975, McSween and Stolper 1978, Smith and Hervig 1979). Delano and Arculus (1980) calculated much more reducing conditions (fo_2 = FMQ-3) from the same mineral assemblages as Reid and Bunch (1975).

Unlike the Moon, there is abundant evidence for abundant volatiles, including water, in the early martian mantle (Dreibus and Wanke 1985). Wanke and Dreibus (1988) argued that much of the original water was lost during accretion and core formation. From trapped melt inclusions, Johnson et al. (1991) and McSween and Harvey (1993) calculated that martian magmas may contain ~1.4 wt % H_2O. However, the low H content of the amphibole in these inclusions (Popp and Bryndzia 1992, Watson et al. 1994, McSween 1994) indicate that this is probably an upper limit for martian magmas.

Longhi (1992) calculated the equilibrium mineral assemblages in the martian mantle using the Dreibus and Wanke (1985) mantle composition, phase equilibrium relations and estimated temperature-pressure profiles. According to his results, approximately half of the martian mantle consists of olivine + pyroxene (orthopyroxene and clinopyroxene) + an aluminous phase (Fig. 7). The aluminous phase changes with increasing pressure, from plagioclase → spinel → garnet. At pressures of greater than 12 to 14 GPa, the olivine is transformed to the spinel structure and pyroxene and garnet are converted to majorite. Near the core-mantle boundary at pressures of 22 to 24 Gpa, spinel and majorite are transformed to the perovskite structure plus magnesiowustite. The appearance of this high pressure assemblage is dependent upon the location of the core-mantle boundary (McSween 1994). The phase transitions are thought to occur at slightly lower pressures than on Earth (Fig. 7) because of the more Fe-rich nature of the martian mantle (Fig. 6).

Asteroid belt

General description. The main asteroid belt lies between the orbits of Mars and Jupiter at approximately 1.8 to 5.2 AU from the Sun. Several other clusters of asteroids occupy near-Earth, Mars and Jupiter orbits. As of 1989, 4044 asteroids have been named. A majority of the asteroids (≈ 85%) have diameters of less than 100 km. Only three asteroids (1 Ceres, 2 Pallas, and 4 Vesta) have diameters greater than 500 km. Approximately 700 asteroids that are larger than 1 km in diameter are in an orbit that crosses Earth's path. The total mass of asteroids in the asteroid belt is ~3.7×10^{21} kg. This is roughly 5% of the mass of the Moon. The favored explanation for this low distribution of mass is that the belt lost substantial material during the growth of Jupiter soon after T_0 (4.56 Ga).

There are variations both in distribution and composition in the asteroid belt. Major depletions in the abundance of asteroids (Kirkwood Gaps) occur at the $^2/_1$ and $^3/_1$ resonance with Jupiter and at $^3/_7$, $^2/_5$, and $^1/_4$ ratios of asteroid/jovian orbital periods. Asteroids that occupy these orbits experience large increases in orbital eccentricity. The resulting orbits may well become Earth-crossing and thus provide a delivery mechanism for asteroidal material to Earth. Gradie and Tedesco (1982) and numerous follow-on studies (i.e., Bell et al. 1989) clearly demonstrated that the asteroid belt was zoned with

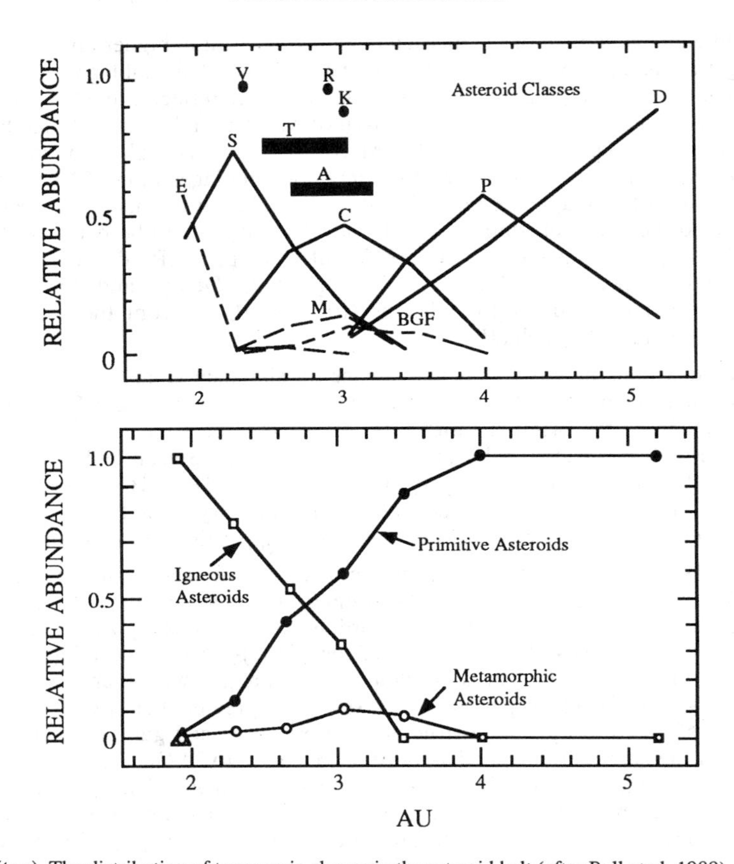

Figure 8 (top). The distribution of taxonomic classes in the asteroid belt (after Bell et al. 1989).

Figure 9 (bottom). The inferred distribution of differentiated, metamorphic, and primitive meteorites in the asteroid belt (after Bell et al. 1989).

regard to the extent of differentiation and composition. The main asteroid belt can be divided into three major groups based on differences in asteroidal evolution (Figs. 8 and 9). Those asteroids that had experienced an igneous history are dominant inward from 2.7 AU. These differentiated asteroids include classes A,E,M, R, S and V and are typified by moderate- to high-albedo (Table 4). These differentiated asteroids could potentially be parent bodies of iron meteorites, basaltic achondrites, pallasites, and enstatite chondrites (Bell et al. 1989). Bodies that experienced "metamorphism" predominantly occur between 2.8 and 3.8 AU. These bodies include the C-type asteroids and are typified by moderate- to low-albedo. The surfaces of these asteroids appear to have been affected by thermal metamorphism in the presence of liquid water. Asteroids of this central belt may be similar in composition to various types of chondritic meteorites (Bell et al. 1989). Primitive, or "unprocessed", asteroids occur outside 3.4 AU and make up approximately 100% of the asteroids outside 4.0 AU. P- and D-type asteroids dominate this region of the asteroid belt. These classes of asteroids have low albedo, contain water as ice, and do not contain hydrous minerals that are observed in the "middle" belt asteroids that experienced thermal metamorphism and hydration. They also may be carbon-rich. There are no meteorite analogs for the P-type asteroids, although P and D types may be represented by

Table 4. Asteroidal spectral taxonomy, surface mineralogy and meteorite analogues (after Lipschutz et al. 1989, Bell et al. 1989).

Type	No.	Surface Mineralogy	Suggested Meteorite Analogues
A	3	olivine, olivine + metal	olivine achondrites, pallasites
B	6		
C	41	hydrated silicates and carbon/	CI-CM assemblages with some
F	10	organics/opaques	additional alteration or
G	5		metamorphism
D	19	carbon/organic rich	organic-rich cosmic dust grains?
P	23	silicates (?) + carbon-rich	CI-CM assemblages plus organics (?)
E	8	iron-free enstatite, forsterite	enstatite achondrites
M	21	metal, metal + enstatite	irons
Q	1	olivine, pyroxene, metal	ordinary chondrites
R	1	pyroxene + olivine + metal	olivine-rich achondrites
S	73	metal + olivine + pyroxene	pallasites or olivine-rich stony-iron
V	1	pyroxene, feldspar	basaltic achondrites
T	1	hydrated silicates and carbon/ organics/opaques but more highly altered than BCFG	highly altered CI-CM assemblages

IDPs (Bradley et al. 1996). The T-type may be represented in the meteorite collection by highly altered carbonaceous chondrites (Bell et al. 1989).

4 Vesta. HED meteorites are thought to have been derived from asteroid 4 Vesta (McCord et al. 1970, Consolmagno and Drake 1977, Wetherill 1987, Binzel and Xu 1993, Binzel 1996) and 20 other small Vesta-like asteroids (5 to 10 km) in the inner asteroid belt. Several researchers, however, disagree that the source for the HED meteorites is 4 Vesta (Wasson and Chapman 1996). Although 4 Vesta is not in an orbital location capable of delivery to the Earth, the small Vesta-like asteroids extend from 4 Vesta to the 3:1 mean motion jovian resonance in the inner asteroid belt, thus providing a mechanism for Vesta material to reach Earth as meteorites (Binzel and Xu 1993, Binzel 1996). The orbital distribution of these smaller bodies is consistent with their being ejected from Vesta by one or more major impacts (Binzel 1996).

Based on Hubble Space Telescope (HST) images, 4 Vesta has a mean diameter of 530 kilometers and a shape that can be fit by an ellipsoid of radii 280, 272, and 227 (± 12) kilometers (P.C. Thomas et al. 1996). The mean density of Vesta calculated from the mass reported by Schubart and Matson (1979) is 3.8 ± 0.6 gm cm^{-3} (P.C. Thomas et al. 1996). Vesta's reflectance spectrum has been studied for a long period of time and is remarkably well-defined. Earliest studies of Vesta date back to 1929. High-precision spectrophotometry indicates that the average surface of Vesta is analogous to howardite and/or polymict eucrite assemblages (Gaffey 1996). HST observations also indicate that the Vesta surface is geologically diverse. The western hemisphere appears to be dominated by low-albedo units that may be analogous to surface eucritic basalts (Gaffey

1996). Assuming that volcanism occurred within 20 m.y. of T_0, much of the surface of this asteroid consists of lithologies that were formed approximately 4.5 Ga ago. This old surface has been disrupted by various impacts. The largest impact structure occurs near the south pole. This structure has a mean diameter of 460 kilometers and an average depth below the rim of 13 kilometers. Exposed within this structure is a high-calcium, coarse grained pyroxene-rich plutonic lithology (crustal intrusions?), or an olivine -rich lithology (mantle?) or both (P.C.Thomas et al. 1997). Thomas et al. (1997) estimated that approximately 1% of 4 Vesta was ejected during this impact event.

Vesta lacks significant water and has a very low oxygen fugacity. Available data on oxygen fugacity versus temperature relationships for basaltic magmatism on 4 Vesta indicate an oxygen fugacity of 1 to 2 log units below the iron-wüstite buffer (Fig. 6). This overlaps with conditions of lunar magmatism. Therefore, basalts that crystallized on 4 Vesta are anhydrous, contain fairly reduced mineral assemblages and have elements in reduced valance states (e.g., Fe^0, Fe^{2+}, Cr^{2+}, Ti^{3+}). Also see Chapter 7.

Other achondrites. The planetary environment for other basaltic achondites (excluding the SNC and lunar meteorites) is more poorly constrained than that of the HED meteorites. There are no direct observations of the orbits of these achondrites to directly determine if they are from the asteroid belt. Rather, indirect evidence comes from spectral studies and the conditions of formation as implied by mineral assemblages. Based on spectral data, most were probably derived from the inner asteroid belt (Bell et al. 1989). Although a magmatic origin for some of these "basaltic" lithologies is controversial, many others appear to represent vestiges of early solar system basaltic magma systems (approximately 4.5 Ga.). Those that most likely represent portions of asteroidal basaltic magma systems (acapulcoites-lodronites and angrites) appear to have crystallized at fairly reducing conditions, close to the iron-wüstite buffer (Crozaz and McKay 1989, McKay 1989) (Fig. 5). Miyamoto and Takeda (1994) estimated that many of the acapulcoite -lodronite lithologies reflect melting at peak temperatures of approximately 1000°C and cooling rates of 2° to 3°C yr^{-1}. They also concluded that reduction at the iron-wüstite buffer occurred after melting.

Iron meteorites. Iron meteorites represent samples of cores of differentiated asteroids and therefore provide us with snapshots of early core formation. The iron meteorites are derived from a diverse variety of parent bodies. Although at least 13 main groups and approximately 60 distinct, smaller groups have been documented (Buchwald 1975, Wasson and Wetherill 1979), it is questionable whether these different meteorite classes represent over 60 different, disrupted parent bodies or that the irons were segregated in numerous small parent bodies that were eventually accreted into a larger parent body. Possible sources for these meteorites are the M- and S-class asteroids. These include 6 Hebe, 8 Flora, and 18 Melpomene, which have diameters in the range of 50 to 200 kilometers.

Formation of fractionated, iron meteorite parent bodies through melting and core formation occurred early in the history of the solar system (approximately 4.555 Ga). Re-Os isotopic systematics have been interpreted as indicating slight differences in core formation ages (Smoliar et al. 1996). High degrees of melting of the parent body facilitated metal segregation and core formation (Taylor 1989). The rate of cooling and the depth of segregation has been a subject of considerable debate. This is attributed to the wide range of cooling rates that have been reported for the same iron meteorites. For example, the calculated cooling rates for Bristol, a type IVA meteorite, extend from less than 10°C per m.y. (Wood 1964) to 20,000°C per m.y. (Narayan and Goldstein 1985). More recently, differences in calculated cooling rates have been reconciled by taking into account minor and trace element components. It is most likely that the iron meteorites cooled at rates of 1 to 100°C per m.y. at depths of 10 to 300 km (Wasson 1985, Saikumar and Goldstein 1988).

The structure of iron meteorite parent bodies following differentiation and prior to disruption is also widely debated. Several parent body models have been proposed: asteroids with a central core, asteroids consisting of a fragmented and reaccreted core, asteroids with large metal pockets which failed to segregate into a central core, and asteroids composed of smaller differentiated bodies that had already experienced core segregation. The first model is consistent with cooling rate data only if many of the differentiated parent bodies that experienced efficient core formation are small. The latter three models are consistent with suggested cooling rates and larger parent bodies.

Chondrites. The primitive nature of chondrites requires that they originated on parent bodies that escaped the extensive post-accretional processes experienced by the parent bodies of differentiated meteorites. One line of evidence for their origin in the asteroid belt is based upon the reconstructed orbits of several chondritic meteorites using photographic and eye-witness documentation. Each of the calculated paths is highly elliptical and the greatest distance its orbit extends from the sun is in the asteroid belt. Spectral observations have been interpreted to indicate that a majority of chondrite parent bodies occur at distances greater than 2.8 AU in the asteroid belt (Table 4, Figs. 8 and 9).

Based upon O isotopes (Fig. 1) and mineral-chemical distinctions, chondritic meteorites were derived from numerous parent bodies. Clayton and coworkers (e.g. Clayton and Mayeda 1983, 1984) conservatively estimated that the number of parent bodies for the stony meteorites exceeds 20. It is estimated that the chondrites were derived from a minimum of 9 parent bodies, with the ordinary chondrites representing at least 3 parent bodies, the enstatite chondrites representing 2 or more parent bodies, and the carbonaceous chondrites representing at least 5 different parent bodies. Over 80 asteroids have been identified as having spectral characteristics consistent with a chondrite or near-chondrite surface composition. The O isotopic data and orbital reconstructions indicate that the chondrites were derived from a relatively small portion of the solar nebula.

As noted above, reflectance spectra for asteroids have been used to distinguish meteorite types, define compositional-differential zoning in the asteroid belt, and identify possible meteorite analogs among the asteroids (Table 4). Potential asteroidal analogs for the chondrites are C (and B, F, G subclasses) class (CI, CM chondrites), K class (CV and CO chondrites), T class (highly altered carbonaceous chondrites), and Q class (ordinary chondrites). A possible example of a CI-CM parent body is 1 Ceres, a C class asteroid. It is the largest discovered asteroid. Its spectral properties have been interpreted as indicating a surface mineralogy consisting of hydrous silicates, water ice, and NH_4-bearing minerals. It is located 2.77 AU from the Sun, has an effective diameter of 932.6 ± 5.2 km and a mean density of 2.3 g cm^{-3}. It contains 33% of the mass of the entire asteroid belt. Nine of the 13 most massive asteroids fall into the C taxonomy class.

The identification of potential parent bodies for the ordinary chondrites is much more paradoxical. Ordinary chondrites have been delivered to Earth fairly commonly, yet a spectrophotometric match with known asteroids is ambiguous. Several 1- to 2-km Q class asteroids which cross Earth's orbit have the appropriate spectral signature, but none have reflectance spectra that match ordinary chondrites. The S class asteroids are fairly common in the inner asteroid belt and have been suggested as potential parent bodies for the ordinary chondrites (Anders 1978; Greenberg and Chapman 1983, Wetherill 1985). Yet, spectral properties of S class asteroids are not similar to ordinary chondrites. This enigma may be attributed to surface effects that masks the actual mineral composition of the asteroids (Wetherill and Chapman 1988).

Comprehensive studies of the chondrites indicate the age of the parent bodies to be 4.555 Ga. This age is only a few million years younger than that of the oldest materials formed in the solar system (Ca-, Al-rich refractory inclusions, CAIs). The initial components of the chondrites consisted of chondrules and matrix, with anhydrous assemblages of olivine, pyroxene, glass, metallic iron, and troilite (FeS). It is still being debated whether these parent bodies accreted hot (>800°C) or cold (<400°-500°C) (Hutchison 1996, Rubin and Brearley 1996). Following accretion, these anhydrous mineral assemblages experienced varying degrees of aqueous alteration and metamorphism on their parent bodies. Metamorphic conditions on many of the ordinary chondrite parent bodies ranged from 300° to 950°C at pressures of less than 2 kbar (Dodd 1981, Sears and Hasan 1987, Scott et al. 1989). Similar P-T regimes may have occurred on the parent bodies of the carbonaceous chondrites and enstatite chondrites, but the restricted sample collection prohibits extending these P-T regimes to these parent bodies. The occurrence of lower temperature regimes on asteroids is implied by numerous interpretations of the aqueous alteration of carbonaceous chondrites. These data indicate parent body temperature regimes of 25° to 150°C (DuFresne and Anders 1962, Bunch and Chang 1980, Hayatsu and Anders 1981, Clayton and Mayeda 1984, Zolensky et al. 1989). Aqueous fluids involved in this low temperature alteration were characterized by values of Eh < 0 V and pH = 8 to 12 (DuFresne and Anders 1962, Bunch and Chang 1980, Hayatsu and Anders 1981, Clayton and Mayeda 1984). Either hydrocryogenic processes involving water ice that was accreted into asteroids (Gooding 1984, Rietmeijer 1985) or liquid water produced by asteroidal heating could have directly promoted hydrous alteration of silicates. Whether the hydrous alteration occurred within the parent body or in the surface regolith is unknown. Metamorphism and alteration may have occurred over a period of 100 to 200 million years. However, interpretation of the data obtained from Rb-Sr and I-Xe systems is controversial (Brannon et al. 1988). Cooling of chondritic material through 400°C at rates of 10^{1-3}°C per m.y. occurred after accretion had ended (Dodd 1981, Wasson 1985).

Asteroid surfaces and the meteorite record (breccias, silicate fracturing, silicate structure deformation) attest to the effects of impact events on various chondrite parent bodies from accretion to a significant time after the body was cool. The meteorite record indicates many parent bodies experienced shock pressures of a few tens of kilobars to almost 1 megabar (Dodd 1981, Stoffler et al. 1988) and post-shock cooling rates of 100°C/d to 1°C/100 yr (Taylor and Heymann 1971, Smith and Goldstein 1977). These cooling rates suggest burial depths of 0.5 m to 1 km (Taylor and Heymann 1971). The deeper burial depths indicate the formation of relatively large craters on the surfaces of these parent bodies (Scott et al. 1989). This is attested to by many recent images of asteroids such as Ida, Gaspra and Mathilda, in both near-Earth orbit and in the asteroid belt. Impacts on small asteroids will probably not produce abundant shocked materials (Scott et al. 1989). The presence of craters also raises the possibility that asteroids developed regoliths that probably are less mature than the lunar regolith, with its high agglutinate content (McKay et al. 1989). Asteroidal regoliths may be thin, considering the pronounced radius of curvature and weak gravity of asteroids. However, McKay et al (1989) suggested that the regolith on the surfaces of large asteroids (>100 km) will be much thicker than the lunar regolith, whereas smaller asteroids (>10 km) will have regolith slightly thicker than that on the Moon.

Our understanding of the structure of parent bodies from which the chondrites originated is based on metallographic cooling rates, radiometric age determinations, ^{244}Pu chronothermometry and thermal history as indicated by mineral assemblages. Most of these data indicate parent body size on the order of 100 km in diameter. Two end-member models for the internal structure of chondritic asteroids have been proposed:

a layered internal structure model (Pellas and Fieni 1988, Lipschutz et al. 1989), and a fragmentation and assembly model (Taylor et al. 1987). In the former model, parent bodies consist of concentric zones of different metamorphic grades. The highest metamorphic grades would be the most deeply buried in the asteroid interior. Mineral assemblages reflecting lower temperature and higher volatile content should reside on the outer portions of the asteroids. Data that show a correlation between Pu-fussion track densities in whitlockite and metamorphic grade in H chondrites (Pellas and Fieni 1988) and an inverse relation between petrographic type and metallographic cooling rate in unshocked H chondrites (Lipschutz et al. 1989) support the layered internal structure model. In the second model, parent bodies consist of disrupted fragments of concentrically zoned asteroids. Accretion of these disrupted fragments occurred after or during metamorphism at temperatures ≥ 500°C. This model was devised to explain the apparent lack of an inverse relationship between petrographic type and metallographic cooling rate for the ordinary chondrites (Taylor et al. 1987).

Interplanetary dust particles

Just after sunset and just before dawn, particularly in the tropics, there is a faint glow in the sky at the point where the sun has set or is about to rise over the horizon. This so-called Zodiacal light and 'Gegenschein', respectively, are due to sunlight scattered by dust particles in interplanetary space of the inner part of the solar system. This dust continuously falls into the sun. The Zodiacal dust cloud is, in principle, a cross-section of all dust-producing small bodies in the solar system (Brownlee 1994). The chemical and mineralogical properties of collected IDPs suggest that they arrived from a limited number of parent bodies. It is also possible that a wide range of parent bodies is represented by the collected IDPs but we have no frame of reference by which to recognize these origins.

There are two groups of IDPs: (1) micrometer-sized chondritic aggregates and (2) micrometer-sized non-chondritic particles. Chondritic materials are often attached to the surface of non-chondritic IDPs, which suggests that the groups have a common origin. It is not necessarily true that they are uniquely related to each other in terms of parent body sources. There are three observations that define the proper context of chondritic IDPs among other solar system materials. First, they differ significantly in form and texture from components of carbonaceous chondrites and contain some mineral assemblages that do not occur in any meteorite class (Mackinnon and Rietmeijer 1987, Rietmeijer 1992a). Second, chondritic IDPs are carbon-rich, with 1 to 46 wt % C and an average bulk carbon content that is ~2-3 times higher than that of the most carbon-rich CI carbonaceous chondrites (K.L.Thomas et al. 1996). Third, reflectance spectra (450-800 nm) of aggregate IDPs are similar to those of P- and D-class outer belt asteroids (Bradley et al. 1996).

The properties of chondritic aggregate IDPs firmly establish an association with the least-altered protoplanets in the solar system: C-, P-, and D-class asteroids, Kuiper Belt comets, and Oort Cloud comets (Table 3). The P- and D-class outer-belt asteroids and the Kuiper Belt comets are still in the regions of the solar system between 3-5 AU and >30AU where they accreted approximately 4.55 b.y. ago. Oort cloud comets accreted between 20-30 AU. This radial distribution is important because the temperatures of major modification processes during dust accretion, and proto-planet evolution, continuously decreased as a function of increasing heliocentric distance. Carbon-rich chondritic IDPs represent the least altered materials from the early system that for 4.55 b.y. were preserved in parent bodies with very little alteration.

The well-known breaking-up of comet nuclei by tidal forces, e.g., the comet

Shoemaker-Levy 9 'string-of-pearls' impact into Jupiter during 1994 July, already suggested that comet nuclei are texturally heterogeneous bodies. Yet, the collimated dust release seen on the nucleus of comet P/Halley was a surprise (Keller et al. 1986). This jetting activity confirmed that comet nuclei are either primordial rubble piles on a small scale (Weissman 1986) or consist of large (tens of centimeters to hundreds of meters) porous boulders in a matrix of dirty ice (Gombosi and Houpis 1986) (Fig. 10). This icy-glue model offers two petrologically distinct environments: dirty ice and the compacted boulders which resemble CI carbonaceous chondrites. Small-scale jetting activity is also related to the sudden 'blow-off' of parts of the refractory crust that built up during repeated perihelion passage of the comet (Stöffler 1989). This crust may eventually seal off the nucleus from sublimation at which point the comet becomes inactive. During a subsequent perihelion passage the water vapor pressure may forcefully disrupt a portion of this crust causing the collimated release of dust. Asteroids can also form rubble piles, such as Earth-crossing asteroid 4769 Castalia (Hudson and Ostro 1994), which may not contain the dirty-ice 'glue.' They probably accreted as mixtures of anhydrous minerals, organic materials and ice that may still be preserved in some of today's asteroids. Fanale and Salvail (1989) modeled the thermal regime due to solar heating as a function of time on the (large) C-class asteroid Ceres (470 km in diameter). They showed that primordial ice could still exist below 10 to 100 m at its equator and 1 to 10 m below the surface at latitudes >40°. In their model, ice is lost via diffusion and subsequent sublimation at the surface, or it is partially present as water of hydration.

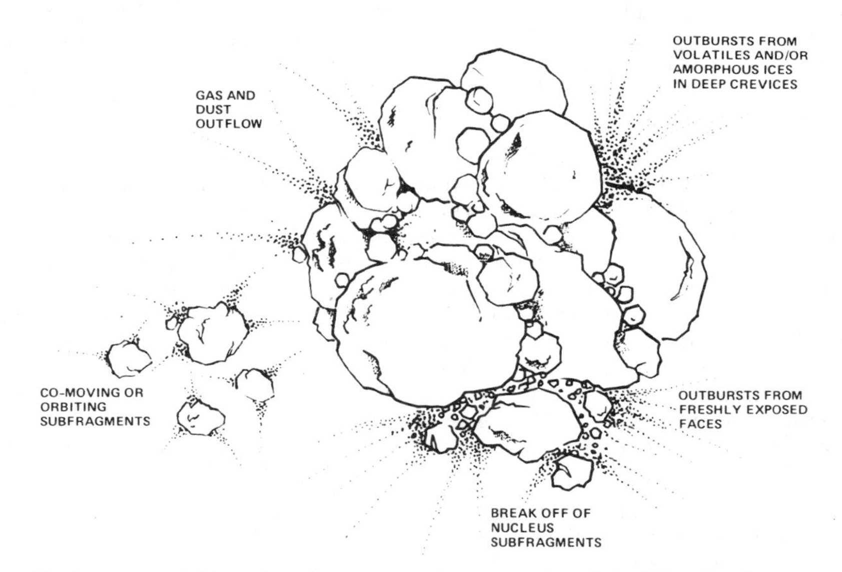

Figure 10. A conceptual illustration of a comet nucleus as a primordial rubble pile of many smaller fragments weakly bonded by local melting at the contact surfaces. In this model the collimated dust release is possible from between fragments while disruptions of the nucleus produce debris with a wide range of sizes. By courtesy of the author. [Used by permission of the editor of *Nature*, from Weissman (1986), Fig. 1, p. 243.]

Comets and many asteroids have irregular, non-hydrostatic shapes and low densities. The C-class asteroid Mathilde was recently found to have a density of 1.3 g cm^{-3} (EOS, 78:285-286, 1997). The extremely low mean density of comet nuclei, e.g. in the range of 0.1 to 0.2 g cm^{-3} for comet P/Halley (Rickman 1986), supports the idea that no widespread internal heating occurred in comets since their formation (Whipple 1987) and that they are texturally heterogeneous bodies that experienced only modest thermal regimes during their lifetime. The conditions were only sufficient to induce minor, and probably localized, alteration and compaction. These carbon-rich, dirty-ice bodies did not experience the electromagnetic induction heating during the Sun's T-Tauri phase that affected inner- and main-belt asteroids (McSween and Weissman 1989). They were also unlikely to have been heated by the decay of long-lived radionuclides or by impacts (McSween and Weissman 1989). The evidence from meteorites and infrared spectroscopy indicates that C-class bodies experienced temperatures that point to an internal heat source at one time during their evolution (Table 3).

Temperatures in icy protoplanets were too low to support the prolonged presence of liquid water, but periodic comets occupy a unique niche among the small solar system bodies (Table 3). As they approach perihelion, solar radiation induces thermal regimes conducive to aqueous alteration that exist for periods varying from weeks to months. Comet nucleus simulation experiments show that transient pockets of water could exist during perihelion, depending upon dust grain size, the dust ice ratio and porosity (Kömle et al. 1991) and that temperatures during perihelion are high enough for hydrocryogenic alteration analogous to that on Earth. In terrestrial permafrost, a thin, 1-8 nm (273 K) to 0.6 nm (193 K) layer of interfacial water occurs at the surfaces of grains that are embedded in ice (Anderson and Morgenstern 1973). This layer will transport dissolved chemical species and aid in grain comminution. The activity of this interfacial water layer significantly alleviates the thermal constraints on aqueous alteration in icy protoplanets and it extends the period of time available for alteration (Rietmeijer 1985).

Thermal alteration refers to modifications in ice- and water-free environments that existed locally (?) in a parent body. Periodic comets appear to be efficient environments for thermal alteration. During each perihelion passage a residue of dust and non-volatile organic materials builds up at the nucleus surface (Stöffler 1989). This anhydrous crust can reach high temperatures depending on the distance to the sun. The crust on comet P/Halley locally reached 400 K (Sagdeev et al. 1986). Under these conditions, dehydration of nanometer-sized layer silicates and nucleation of ultrafine-grained iron-magnesium silicates in amorphous materials is possible (Rietmeijer 1996). Carbon-rich materials in chondritic IDPs include organic matter, amorphous and poorly graphitized carbons (Rietmeijer 1992b), which could be related to each other as a function of increasing temperature (Rietmeijer and Mackinnon 1985). The extent of thermal alteration in this black crust remains uncertain. Also uncertain are the possible thermal effects in IDPs during solar system sojourn. When IDPs spiral toward the sun, they are heated by solar radiation to temperatures that conceivably could induce thermal metamorphism during the time (10^4 to 10^5 years) they require to reach 1 AU (Table 3). There are, however, no data to support this type thermal of alteration.

Flynn (1990) pointed out that gravitational focusing by the Earth's mass biases the IDPs that are able to enter the atmosphere. This process favors IDPs with low geocentric velocity in near-circular orbits (cf. Table 3). Although comets are the most prolific dust producers, the ratio of cometary to asteroidal dust in the atmosphere may be reversed from this ratio in near-Earth space due to selection effects related to atmospheric entry survival. The literature uses both "geocentric velocity," which refers to the particle velocity relative to the Earth but prior to the particle encountering the Earth's gravitational pull, and "atmospheric entry velocity." Atmospheric entry velocity is the geocentric velocity augmented by the effect of gravitational acceleration as the particle is attracted towards the Earth. The lowest possible entry velocity is 11.1 km s^{-1} (Flynn 1989, 1990). Asteroidal dust enters the atmosphere at ~12 km s^{-1}. Cometary debris has entry velocities that peak at ~20 km s^{-1} but as high as 65 km s^{-1} (comet P/Halley). The modal dust velocity is 14.5 km s^{-1}. Due to collisions with air molecules in the Earth's atmosphere between 100 and 80 km altitude, IDPs slow down until they reach a terminal velocity of 1-10 cm s^{-1} (brownlee, 1985). This process has two implications. First, the slow-down causes a 10^6-fold increase in the flux of IDPs in the lower stratosphere compared to the flux in near-Earth space, which makes it possible to collect IDPs efficiently. Second, each IDP is flash-heated for 5 to15 s as a function of its mass, size, entry velocity, and entry angle (Love and Brownlee 1991). Asteroidal dust will experience less heating than cometary debris of the same size and density. Knowing the maximum flash heating temperature allows a determination of the pre-entry velocity of IDPs, which allows a separation of cometary and asteroidal IDPs (Table 3).

ACKNOWLEDGMENTS

We thank Steve Simon and Hap McSween for constructive reviews and Adrian J. Brearley, George Flynn, Rhian Jones, Kase Klein, Aurora Pun and Joyce K. Shearer for useful discussions and editorial comments. CKS, JJP, and FJMR were supported by NASA grants and by the Institute of Meteoritics.

REFERENCES

Anders E (1978) Most stony meteorites come from the asteroid belt. *In* Asteroids: An exploration Assessment. (eds) D Morrison, WC Wells, NASA CP-2053:145-157

Anderson DM, Morgenstern NR (1973) Physics, chemistry, and mechanics of frozen ground. *In*: Permafrost: A Review, p 257-288, National Academy of Sciences, Washington, DC

Becker RH, Pepin RO (1984) The case for a martian origin of the shergottites: Nitrogen and noble gases in EETA79001. Earth Planet Sci Lett 69:225-242

Bell JF, Davis DR, Hartmann WK, Gaffey MJ (1989) Asteroids: The Big Picture. *In* Binzel RP, Gehrels R, Matthews MS (eds) Asteroids II:921-945. Univ of Arizona Press, Tucson

Binzel RP (1996) Astronomical evidence linking Vesta to the HED meteorites: A review. Workshop on evolution of igneous asteroids: Focus on Vesta and the HED meteorites. LPI Technical Report 96-02:2

Binzel RP, Xu S (1993) Chips off asteroid 4 Vesta: Evidence for the parent body of basaltic achondrite meteorites. Science 260:186-191

Bischoff A, Weber D (1997) Dar Al Gani 262: The first lunar meteorite from the Sahara. Meteoritics Planetary Science 32:A13-A14

Bogard DD, Johnson P (1983) Martian gases in an Antarctic meteorite. Science 221:651-654

Bradley JP, Keller LP, Brownlee DE, Thomas KL (1996) Reflectance spectroscopy of interplanetary dust particles. Meteoritics Planet Sci 31:394-402

Brannon JC, Podosek FA, Lugmair GW (1988) Initial $^{87}Sr/^{86}Sr$ and Sm-Nd chronology of chondritic meteorites. Proc Lunar Planet Sci Conf 18:555-564

Brownlee DE (1978) Interplanetary dust: Possible implications for comets and presolar interstellar grains. *In* Protostars & Planets, Gehrels T (ed) p 134-150, Univ of Arizona Press, Tucson

Brownlee DE (1985) Cosmic dust: collection and research. Ann Rev Earth Planet Sci 13:147-173

Brownlee DE (1994) The origin and role of dust in the early solar system. *In* Zolensky ME, Wilson TL, Rietmeijer FJM, Flynn GJ (ed) Analysis of Interplanetary Dust, AIP Conf Proc 310:127-143, Am Inst Physics, New York

Buchwald VF (1975) Handbook of Meteorites. Univ of California Press, Berkeley

Bunch TE, Chang S (1980) Carbonaceous chondrites-II. Carbonaceous chondrite phyllosilicates and light element geochemistry as indicators of parent body processes and surface conditions. Geochim Cosmochim Acta 44:1543-1577

BVSP (1981) Basaltic Volcanism on the Terrestrial Planets. Pergamon Press, New York

Clayton RN, Mayeda TK (1983) Oxygen isotopes in eucrites, shergottites, nakhlites, and chassignites. Earth Planet Sci Lett 62:1-6

Clayton RN, Mayeda TK (1984) The oxygen isotope record for Murchison and other carbonaceous chondrites. Earth Planet Sci Lett 67:151-161

Consolmagno GJ, Drake M J (1977) Composition and evolution of the eucrite parent body: Evidence from rare earth elements. Geochim Cosmochim Acta 41:1271-1282

Crozaz G, McKay G (1989) Minor and trace element microdistributions in Angra dos Reis and LEW 86010: Similarities and differences. Lunar Planet Sci XX:208-209

Delano JW (1980) Chemistry and liquidus phase relations of Apollo 15 red glass: Implications for the deep lunar interior. Proc Lunar Planet Sci Conf 11:251-288

Delano JW, Arculus RJ (1980) Nakhla: oxidation state and other constraints. Lunar Planet Sci XI:219-221

Dodd RT (1981) Meteorites, A Petrologic-Chemical Synthesis. Cambridge Univ Press, New York

Dreibus G, Wanke H (1985) Mars, a volatile-rich planet. Meteoritics 20:367-381

DuFresne ER, Anders E (1962) On the chemical evolution of the carbonaceous chondrites. Geochim Cosmochim Acta 26:1849-1862

Fanale FP, Salvail JR (1989) The water regime of asteroid (1) Ceres. Icarus 82:97-110

Feldman WC, Binder AB, Barraclough BL, Belian RD (1998) First results from the lunar prospector spectrometers. Lunar Planet Sci 29, Abstr #1936. Lunar and Planetary Institute, Houston

Flynn GJ (1989) Atmospheric entry heating of micrometeorites. Proc Lunar Planet Sci Conf 19:673-682

Flynn GJ (1990) The near-Earth enhancement of asteroidal over cometary dust. Proc Lunar Planet Sci Conf 20:363-371.

Flynn GJ (1996) Sources of 10 micron interplanetary dust: The contribution from the Kuiper belt. *In* Gustafson BÅS, Hanner MS (ed) Physics, Chemistry and Dynamics of Interplanetary Dust. Astron Soc Pacific Conf Series 104:171-175

Gaffey MJ (1996) Asteroid spectroscopy: Vesta, the basaltic achondrites, and other differentiated asteroids. LPI Technical Report 96-02:8

Gombosi TI, Houpis HLF (1986) An icy-glue model of cometary nuclei. Nature 324:43-44

Gooding JL (1984) Aqueous alteration on meteorite parent bodies: Possible role of "unfrozen" water and the Artarctic meteorite analog. Meteoritics 19:228-229

Gradre J, Tedesco E (1982) Compositional structure of the asteroid belt. Science 216:1405-1407

Graham AL, Bevan AWR, Hutchison R (1985) Catalogue of Meteorites, 4th Edn. British Museum, London, and Univ of Arizona Press, Tucson

Greenberg R, Chapman CR (1983) Asteroids and meteorites: Parent bodies and delivered samples. Icarus 55:455-481

Grossman JN (1994) The meteoritical bulletin, No. 76, 1994 January: The U.S. Antarctic meteorite collection. Meteoritics 29:100-143

Grove TL, Bartels KS (1992) The relation between diogenite cumulates and eucrite magmas. Proc Lunar Planet Sci 22:437-445

Gustafson BÅS (1994) Physics of zodiacal dust. Ann Rev Earth Planet Sci 22:553-595

Hayatsu R, Anders E (1981) Organic compounds in meteorites and their origins. *In* Cosmo and Geochemistry 99:1-37

Head JW (1976) Lunar volcanism in space and time. Rev Geophys Space Phys 14:265-300

Hill DH, Boynton WV, Haag RA (1991) A lunar meteorite found outside the Antarctic. Nature 352: 614-617

Hodges RR Jr (1975) Formation of the lunar atmosphere. The Moon 14:139-157

Hodges RR Jr, Hoffman JH, Johnson FS (1974) The lunar atmosphere. Icarus 21:415-426

Hudson RS, Ostro SJ (1994) Shape of Asteroid 4769 Castalia (1989PB) from inversion of radar images. Science 263:940-943

Hutchison R (1996) Hot accretion of the ordinary chondrites: The rocks don't lie. Lunar Planet Sci 27: 579-580

Johnson MC, Rutherford MJ, Hess PC (1991) Chassigny petrogenesis: Melt compositions, intensive parameters, and water contents of martian (?) magmas. Geochim Cosmochim Acta 55:349-366

Keller HU, Arpingy C, Barbieri C, Bonnet RM, Cazes S, Coradini M, Cosmovici CB, Delamere WA, Huebner WF, Hughes DW, Jamar C, Malaise D, Reitsema HJ, Schmidt HU, Schmidt WKH, Seige P, Whipple FL, Wilhelm K (1986) First Halley multicolour camera imaging results from Giotto. Nature 321:320-326

Klinger J (1983) Classification of cometary orbits based on the concept of orbital mean temperature. Icarus 55:169-176

Kömle NI, Steiner G, Baguhl M, Kohl H, Kochan H, Thiel K (1991) The effect of non-volatile porous layers on temperature and vapor pressure of underlying ice. Geophys Res Lett 18:265-268

Kushiro I (1972) The effects of water on the composition of magmas formed at high pressures. J Petrol 13:311-334

Laul JC, Smith MR, Wanke H, Jagoutz E, Dreibus G, Palme H, Spettel B, Burghele A, Lipschutz ME, Verkouteren RM (1986) Chemical systematics of the Shergotty meteorite and the composition of its parent body (Mars). Geochim Cosmochim Acta 50:909-926

Lipschutz ME Gaffey MJ, Pellas P (1989) Meteoritic parent bodies: nature, number, size and relation to present-day asteroids. *In* Binzel RP, Gehrels T, Matthews MS (eds) Asteroids II: p 740-777, Univ of Arizona Press, Tucson

Longhi J (1992) Experimental petrology and the petrogenesis of mare volcanics. Geochim Cosmochim Acta 56:2235-2252
Longhi J, Knittle E, Holloway JR, Wanke H (1992) The bulk composition, mineralogy, and internal structure of Mars. *In* Mars. HH Kieffer, BM Jakosky, CW Snyder, MS Matthews (eds) p 185-208, Univ of Arizona Press, Tucson
Love SG, Brownlee DE (1991) Heating and thermal transformation of micrometeoroids entering the Earth's atmosphere. Icarus 89:26-43
Mackinnon IDR, Rietmeijer FJM (1987) Mineralogy of chondritic interplanetary dust particles. Reviews Geophysics 25:1527-1553
McCord TB, Adams JB, Johnson TV (1970) Asteroid Vesta: spectral reflectivity and compositional implications. Science 168:1445
McKay DS, Swindle TD, Greenberg R (1989) Asteroidal regoliths: What we do not know. *In* Asteroids II. Binzel RP, Gehrels T, Matthews MS (eds) p 617-642, Univ of Arizona Press, Tucson
McKay GA (1989) Partitioning of rare earth elements between major silicate minerals and basaltic melts. *In* Geochemistry and Mineralogy of Rare Earth Elements. BR Lipin, GA McKay (eds) Rev Mineral 21:45-77
McSween, HY Jr (1985) SNC meteorites: Clues to martian petrologic evolution? Rev Geophys 23:391-416
McSween HY Jr (1994) What we have learned about Mars from the SNC meteorites. Meteoritics 29: 757-779
McSween HY Jr, Harvey RP (1993) Outgassed water on Mars: Constraints from melt inclusions in SNC meteorites. Science 259:1890-1892
McSween HY Jr, Stolper EM (1978) Shergottite meteorites I: mineralogy and petrography. Lunar Planet Sci IX:732-734
McSween HY Jr, Weissman PR (1989) Cosmochemical implications of the physical processing of cometary nuclei. Geochim Cosmochim Acta 53:3263-3271
McSween HY Jr, Taylor LA, Stolper EM (1979) Allan Hills 77005: A new meteorite type found in Antarctica. Science 204:1201-1203
Miyamoto M, Takeda H (1994) Thermal history of lodranites Yamato 74357 and MAC 88177 as inferred from the chemical zoning of pyroxene and olivine. J Geophys Res 99:5669-5677
Narayan C, Goldstein JI (1985) A major revision of iron meteorite cooling rates- an experimental study of the growth of the Widmanstatten pattern. Geochim Cosmochim Acta 49:397-410
Newsom HE, Drake MJ (1982) The metal content of the eucrite parent body: constraints from the partitioning behavior of tungsten. Geochim Cosmochim Acta 46:2483-2489
Nier AO, Schlutter DJ (1992) Extraction of helium from individual interplanetary dust particles by step-heating. Meteoritics 27:166-173
Nier AO, Schlutter DJ (1993) The thermal history of interplanetary dust particles collected in the Earth's stratosphere. Meteoritics 28:675-681
O'Hara MJ (1968) The bearing of phase equilibria studies in synthetic and natural systems on the origin and evolution of basic and ultrabasic rocks. Earth Sci Rev 4:69-133
Ott U, Begemann F (1985) Are all "martian" meteorites from Mars? Nature 317:509-512
Pellas P, Fieni C (1988) Thermal histories of ordinary chondrite parent asteroids. Lunar Planet Sci XIX:915-916
Popp RK, Bryndzia, LT (1992) Statistical analysis of Fe^{3+}, Ti, and OH in kaersutite from alkalic igneous rocks and mafic mantle xenoliths. Am Mineral 77:1250-1257
Presnall DC (1979) Fractional crystallization and partial fusion. *In* The Evolution of the Igneous Rocks: 50th Anniversary Perspectives. HS Yoder (ed) p 59-75, Princeton Univ Press, Princeton, NJ
Reid AM, Bunch TE (1975) The nakhlites, part II: where, when, and how. Meteoritics 10:317-324
Rickman K (1986) Masses and densities of comets Halley and Kopff. European Space Agency Spec Publ 249:195-205
Rieder R, Economou T, Wanke H, Turkevich A, Crisp J, Bruckner J, Dreibus G, McSween HY Jr (1997) The chemical composition of martian soil and rocks returned by the mobile proton x-ray spectrometer: preliminary results from the x-ray mode. Science 278:1771-1774
Rietmeijer FJM (1985) A model for diagenesis in proto-planetary bodies. Nature 313:293-294
Rietmeijer FJM (1992a) Mineralogy of primitive chondritic protoplanets in the early solar system. Trends Mineralogy 1:23-41
Rietmeijer FJM (1992b) Pregraphitic and poorly graphitised carbons in porous chondritic micrometeorites. Geochim Cosmochim Acta 56:1665-1671
Rietmeijer FJM (1996) The ultrafine mineralogy of a molten interplanetary dust particle as an example of the quench regime of atmospheric entry heating. Meteoritics Planet Sci 31:237-242
Rietmeijer FJM, Mackinnon IDR (1985) Poorly graphitized carbon as a new cosmothermometer for primitive extraterrestrial materials. Nature 316:733-736

Rietmeijer FJM, Warren JL (1994) Windows of opportunity in the NASA Johnson Space Center Cosmic Dust Collection. *In* Analysis of Interplanetary Dust. Zolensky ME, Wilson TL, Rietmeijer FJM, Flynn GJ (eds) Am Inst Physics Conf Proc 310: 255-275

Ringwood AE (1976) Limits on the bulk composition of the Moon. Icarus 28:325-349

Ringwood AE (1979) Origin of the Earth and Moon. Springer-Verlag, New York, 295 p

Rubin AE, Brearley AJ (1996) A critical evaluation of the evidence for hot accretion. Icarus 124:86-96

Sagdeev Z, Blamont J, Galeev AA, Moroz VI, Shapiro VD, Shevchenko VI, Szego K (1986) Vega spacecraft encounters with comet Halley. Nature 321:259-262

Saikumar V, Goldstein JI (1988) An evaluation of the methods to determine the cooling rates of iron meteorites. Geochim Cosmochim Acta 52:715-725

Schubart J, Matson, DL (1979) Masses and densities of asteroids. *In* Asteroids. T Gehrels (ed) p 84-97, Univ of Arizona Press, Tucson

Scott ERD Taylor GJ Newsom HE Herbert F Zolensky M, Kerridge JF (1989) Chemical, thermal and impact processing of asteroids. *In* Asteroids II. RP Binzel, T Gehrels, MS Matthews (eds) p 701-739, Univ of Arizona Press, Tucson

Sears DWG, Dodd RT (1988) Overview and classification of meteorites. *In* Meteorites and the Early Solar System. JF Kerridge, MS Matthews (eds) p 3-34, Univ of Arizona Press, Tucson

Sears DWG, Hasan FA (1987) The type 3 ordinary chondrites: A review. Surv Geophys 9:43-97

Smith BA, Goldstein JI (1977) The metallic microstructure and thermal histories of severely reheated chondrites. Geochim Cosmochim Acta 41:1061-1072

Smith JV, Hervig RL (1979) Shergotty meteorite: Mineralogy, petrology, and minor elements. Meteoritics 14:121-142

Smoliar MI, Walker RJ, Morgan JW (1996) Re-Os isotope systematics in the Cape York meteorite shower. Lunar Planet Sci XXVII:1225-1226

Stern AA, Shull JM (1988) The influence of supernovae and passing stars on comets in the Oort cloud. Nature 332:407-411

Stöffler D (1989) Petrographic working model of the ROSETTA sampling and modelling subgroup for a comet nucleus. European Space Agency Spec Publ 302:23-29

Stöffler D, Bischoff A, Buchwald V, Rubin AE (1988) Shock effects in meteorites. *In* Meteorites and the Early Solar System. JF Kerridge, MS Matthews (eds) p 165-202, Univ of Arizona Press, Tucson

Stolper EM (1977) Experimental petrology of eucritic meteorites. Geochim Cosmochim Acta 41:587-611

Swindle TD, Caffee MW, Hohenberg CM (1986) Xenon and other noble gases in shergottites. Geochim Cosmochim Acta 50:1001-1015

Taylor GJ (1989) Metal segregation in asteroids. Lunar Planet Sci XX:1109-1110

Taylor GJ, Heymann D (1971) Post shock thermal histories of reheated chondrites. J Geophy Res 76:1879-1893

Taylor GJ, Maggiore P Scott ERD Rubin AE, Keil K (1987) Original structures and fragmentation and reassembly histories of asteroids: Evidence from meteorites. Icarus 69:1-13

Taylor SR (1982) Planetary Science: A Lunar Perspective. Lunar and Planetary Institute, Houston

Taylor SR (1988) Planetary compositions. *In* Meteorites and the Early Solar System. JF Kerridge, MS Matthews (eds) p 512-534, Univ of Arizona Press, Tucson

Thomas KL, Keller LP, McKay DS (1996) A comprehensive study of major, minor, and light element abundances in over 100 interplanetary dust particles. *In* Physics, Chemistry and Dynamics of Interplanetary Dust. Gustafson BAS, Hanner MS (eds) Astron Soc Pacific Conf Series 104:283-286

Thomas PC, Binzel RP, Gaffey MJ, Zellner BH, Storrs AD, Wells E(1996) Vesta: Spin pole, size, and shape from HST images. LPI Technical Report 96-02:32

Thomas PC, Binzel RP, Gaffey MJ, Storrs AD, Wells E, Zellner BH (1997) Impact excavation on asteroid 4 Vesta: HST results. Science 277:1492-1495

Vaniman DT, Dietrich J, Taylor GJ, Heiken G (1991a) Exploration, samples, and recent concepts of the Moon. *In* Lunar Sourcebook. Heiken G, Vaniman DT, French B. (eds) p 5-26, Cambridge Univ Press, Cambridge, UK

Vaniman D, Reedy R, Heiken G, Olhoeft G, Mondell W (1991b) The lunar environment. *In* Lunar Sourcebook. Heiken G, Vaniman DT, French B (eds) p 27-60, Cambridge Univ Press, Cambridge, UK

Walker D, Stolper EM, Hayes JF (1979) Basaltic volcanism: The importance of planetary size. Proc Lunar Planet Sci Conf 10:1995-2015

Wanke H, Dreibus G (1988) Chemical composition and accretional history of terrestrial planets. Phil Trans Roy Soc London A235:545-557

Warren PH, Kallemeyn GW (1993) The ferroan-anorthositic suite, the extent of primordial lunar melting, and the bulk composition of the moon. J Geophy Res 98:5445-5455

Wasson JT (1985) Meteorites: Their record of early solar system history. WH Freemaan & Co, New York

Wasson JT, Chapman CR (1996) Space weathering of basalt-covered asteroids: Vesta an unlikely source of the HED meteorites. LPI Technical Report 96-02:38

Wasson JT, Wetherill GW (1979) Dynamical, chemical, and isotopic evidence regarding the formation locations of asteroids and meteorites. *In* Asteroids (ed) T Gehrels, p 926-974, Univ of Arizona Press, Tucson

Watson LL, Hutcheon ID, Epstein S, Stolper EM (1994) Water on Mars: Clues from deuterium/hydrogen and water contents of hydrous phases in SNC meteorites. Science 265:86-90

Weissman PR (1985) The origin of comets: Implications for planetary formation. *In* Protostars & Planets II. Black DC, Matthews MS (eds) p 895-919, Univ of Arizona Press, Tucson

Weissman PR (1986) Are cometary nuclei primordial rubble piles? Nature 320:242-244

Wetherill GW (1985) Asteroidal sources of ordinary chondrites. Meteoritics 20:1-22

Wetherill GW (1987) Dynamical relationship between asteroids, meteorites, and Apollo-Amor objects. Phil Trans Soc London A323: 323-337

Wetherill GW, Chapman CR (1988) Asteroids and Meteorites. *In* Meteorites and the Early Solar System. JF Kerridge, MS Matthews (eds) p 35-67, Univ of Arizona Press, Tucson

Whipple FL (1987) The cometary nucleus: Current concepts. Astron Astrophys 187:852-858

Wilson L, Head JW (1981) Ascent and eruption of basaltic magma on the Earth and Moon. J Geophys Res 86:2971-3001

Wood JA (1964) The cooling rates and parent planets of several iron meteorites. Icarus 3:429-459

Wyllie PJ (1979) Petrogenesis and the physics of the Earth. *In* The Evolution of the Igneous Rocks Fiftieth Anniversary Perspectives. HS Yoder (ed) p 483-520, Princeton Univ Press, Princeton, NJ

Zolensky ME, Bourcier WL, Gooding GL (1989) Aqueous alteration on hydrous asteroids: Results of EQ3/6 computer simulations. Icarus 78:411-425

Zolensky ME, Barrett R, Browning L (1993) Mineralogy and composition of matrix and chondrule rims in carbonaceous chondrites. Geochim Cosmochim Acta 57:3123-3148

Chapter 2

INTERPLANETARY DUST PARTICLES

Frans J.M. Rietmeijer

Institute of Meteoritics and
Department of Earth and Planetary Sciences
The University of New Mexico
Albuquerque, NM 87131

"In a dark wood wandering"

(Hella S. Haasse 1949, translation: L.C. Kaplan, edited by A. Miller 1991, Academy Chicago Publishers)

"But we may have missed an important finding!"

(Jessberger et al. 1988)

INTRODUCTION

Imagine picking random micrometer size objects from the vast reservoir of the Earth's stratosphere in order to study the evolution of protoplanets out there in the tremendous reaches of the solar system. That is exactly what interplanetary dust particle (IDP) research is all about. It is a vast forest of unknown truths to be discovered as part of a never-ending story that will not be finished here. New concepts emerge that are later adjusted to accommodate new findings. The IDPs are fossil presolar dust that offers a unique window to the materials and processes, e.g. dust accretion, in the evolving solar system 4500 Myrs ago. Dust was the building block of the terrestrial planets, the cores of giant planets, and asteroids and comets. This research is not as far-fetched as it sounds. A recent image of the Eagle Nebula shows the common birth of stars in the Galaxy.

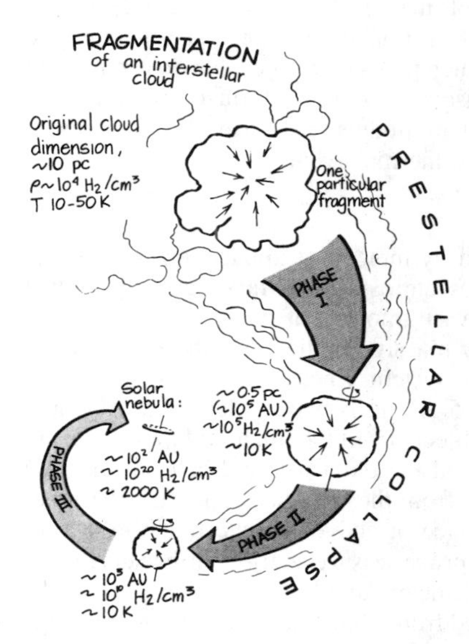

Figure 1. Cartoon depicting the sequence of events during the collapse of a fragment in a molecular cloud and the formation of a circumstellar (solar nebula) disk around a young star. Source: Wood JA, Chang S (eds) (1985) The cosmic history of the biogenic elements and compounds. NASA SP-476, 80 p. Courtesy of NASA.

The gravitational collapse of a portion of a dark molecular cloud with interstellar dust when triggered by some cause (e.g. a supernova shock wave) ultimately produces a central star surrounded by a flattened disc of dust and gas (Morfill et al. 1985). The solar nebula was the result of this normal star-forming process 4500 Myrs ago (Fig. 1). The dust in the disc is referred to as presolar dust. Once a central star (Sun) has formed during the collapse of the cloud, the dust is gravitationally bound to the star and they evolve together. Young stars evolve via a T-Tauri phase, which is a thermal event that heats up the surrounding disc. The temperatures due to this event decrease as a function of

0275-0279/98/0036-0002$10.00

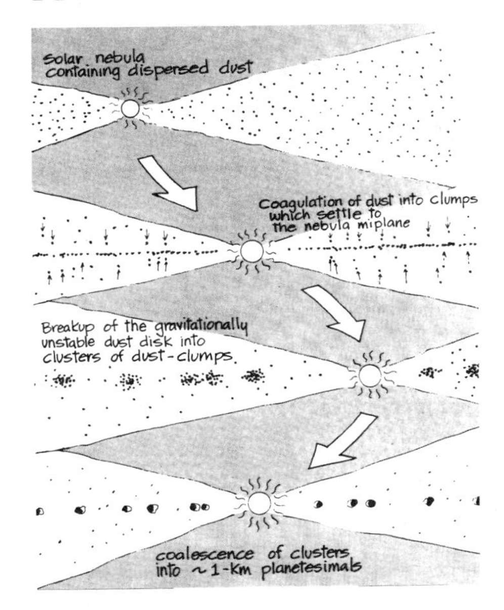

Figure 2. Cartoon showing the evolution of the dusty circumstellar disk wherein presolar dust and (condensed) solar nebula dust settle to the midplane of the disk followed by accretion into dust-clumps (or, protoplanets). Ultimately the planetesimals and planets grow in an increasingly less-dusty solar system. Source: Wood JA, Chang S (eds) (1985) The cosmic history of the biogenic elements and compounds. NASA SP-476, 80 p. Courtesy of NASA.

increased heliocentric distance. Presolar dust in the disc closest to the star evaporates at temperatures as high as ~1700°C. Some fraction of the resulting vapors condenses in cooler regions at larger heliocentric distance. There is a transition zone that separates the inner part of the disc with condensed dust and an outer part where presolar dust remained unaffected. Finally, all dust is cleared from the disc as a result of dust accretion into protoplanets (Fig. 2). According to current solar system models, chondritic IDPs represent protoplanets that were beyond the reach of the T-Tauri thermal event.

Dust in the Solar System

Dust in the solar system is continually produced by meteoroid impacts on moons and asteroids (airless bodies) and planets with a tenuous atmosphere (Mars), and by mutual collisions among asteroids. These processes release vast amounts of debris ranging in size from 'conventional' meteorites to IDPs. Sufficiently massive bodies feel the gravitational pull of the Sun and Jupiter. The region of the solar system wherein this 'choice' favors either one body is located in asteroid belt at ~2 to ~5 AU in between Mars (1.5 AU) and Jupiter (5.2 AU). Broadly speaking, meteorite size objects from the inner and main asteroid belts fall towards the sun. Objects of similar mass from the outer asteroid belt fall towards Jupiter and are ejected from the solar system but dust from these bodies can reach the inner solar system. The sublimation of water ice at the surface of active comets (much less from icy asteroids) liberates dust from these bodies. This process is very efficient because these icy protoplanets are small (typically 1 to 10 km in diameter) unconsolidated dirty ice balls. Interplanetary space typically contains IDPs generated from collisions and sublimation plus interstellar dust that passes freely through today's solar system. The typical mass of collected IDPs ranges from 10^{-12} to 10^{-9} grams.

Once the very low-mass IDPs are ejected from their parent bodies, they show orbital behavior that is fundamentally different from the chaotic paths followed by more massive (larger) meteoroids. While still subjected to the gravitational pull from the sun (99% of the mass in the solar system) and the giant planets (mostly Jupiter), the IDPs also feel the solar wind pressure. As a result of both forces the radius of their orbits slowly decay due to

Poynting-Robertson (PR) drag (Brownlee 1994). They slowly spiral towards the sun whereby their orbits are circularized which favors gravitational focussing by the Earth (Flynn 1989). A 10 μm size IDP from the main asteroid belt reaches 1 AU after $\sim 6 \times 10^4$ years while cometary dust requires 10^4 to 10^5 years under the influence of PR drag (Flynn 1996a). The cumulative micrometeoroid (or IDP) flux at 1 AU is $>10^{-5}$ m^2 s^{-1} for particles $<10^{-9}$ g and the peak in the mass flux is defined by IDP of 1.5×10^{-5} g which corresponds to dust ~200 μm in diameter (ρ = 0.28 g cm^{-3}) (Love and Brownlee 1993). The result of this particular behavior due to PR drag is a steady-state population of IDPs at 1 AU (i.e. an Earth-orbit crossing heliocentric distance) that is homogeneous with regard to all possible contributing sources, including short-period (SP) comets (Fig. 3). For example, comet P/Schwassmann-Wachmann 1 alone provides up to 6% of the mass of grains 5 μm to 2 cm in diameter in the Zodiacal dust cloud (Fulle 1992).

Figure 3. The nucleus of comet P/Halley photographed by the GIOTTO spacecraft during the comet's latest apparition. The nucleus measured 15 × 8 × 7.5 km. Its non-hydrostatic 'potato' shape shows this object is a primitive accretionary body. The collimated dust release (dust jets) is visible to the left of the nucleus and is caused by rapid dust ejection from between 'refractory boulders' in a primitive 'rubble pile.' Courtesy of H.U. Keller; copyright: Max-Planck-Institut für Aeronomie, Lindau, Germany.

Origins of IDPs

IDPs originate in active periodic comets from the Oort cloud and Kuiper belt, from carbon-rich, outer-belt (icy) asteroids), main-belt asteroids, and (ice-free) protoplanets that in today's solar system have evolved into near-Earth and Earth-crossing asteroids. These sources include protoplanets that do not yield 'conventional' meteorites. It is noted though that some meteoroid types will not deliver 'conventional' meteorites because they (1) are too fragile, (2) have high velocities that promote complete burn-up in the atmosphere, or (3) have not evolved into Earth-crossing orbits by purely gravitational effects. The in situ measurements at comet P/Halley during its apparition during 1985/86 were a technological milestone. This database remains the only direct observations of a major type of dust-producing body in the solar system. The Halley data showed, among other things, that the chemical compositions of its dust are similar to chondritic IDPs. The data could not constrain the mineralogy of comet P/Halley dust. We should remain cautious with interpretations of IDP origins because there is probably considerable diversity among comets and asteroids. The proportions of asteroid and cometary IDPs in the stratosphere are not constrained but it is unlikely that these IDPs are dominated by cometary debris. Some investigators studied the chemical and mineralogical links between chondritic IDPs and carbonaceous chondrites and unequilibrated ordinary chondrite meteorites (Brownlee 1978a; Klöck and Stadermann 1994). Others explored a continuum of these properties between chondritic IDPs and meteorites (Rietmeijer 1985a, Rietmeijer and McKay 1985). The various sources of collected IDPs causes variations in the delivery of dust types to the Earth's atmosphere but it requires a critical relationship between the timing of dust collection to capture these astronomical events.

A survey of the NASA/JSC Cosmic Dust Collection showed that chondritic IDPs in the lower stratosphere might display inter-annual variations (Rietmeijer and Warren 1994). It is conjectured that the variations are related to the dust bands associated with the Koronis and Themis asteroid families. The spectral S-class Koronis asteroids and spectral C-class Themis asteroids have two peaks in their annual Earth encounters (Dermott and Liou 1994). Collisions among near-Earth asteroids that are mostly impact debris from main-belt asteroids (Bottke et al. 1994), and Earth-crossing asteroids including extinct comet nuclei (Rabinowitz et al. 1994), also contribute dust to the Zodiacal cloud. These asteroids are C- and S-class objects but chondritic IDPs with petrological and chemical properties of ordinary (S-class) chondrite meteorites have not yet been found (Flynn 1996b). It remains possible that S-class IDPs are present among the collected IDPs but this observation alludes to temporal variations in the constituents of the Zodiacal cloud as a function of Earth-encounter opportunities along its orbit. Temporal variations in IDP types entering the atmosphere were already suggested by Crozier (1966) who observed increased abundances of black magnetic spherules 5 to 60 μm in diameter during the Fall.

Research goals

The primary research goals are (1) characterization of the mineral and chemical properties of collected IDPs, (2) definitions of the dust properties in the early solar system and (3) mineral alteration in protoplanets. Important are also the interactions of IDPs with the Earth's atmosphere, i.e. dynamic pyrometamorphism and contamination by stratospheric aerosols. If we are able to develop a complete picture of IDP histories from formation in volatile-rich parent bodies to collection, there is an opportunity to study the common types of presolar dust. In 'conventional' meteorites only the most refractory presolar dust (e.g. diamond and SiC) survived parent body alteration and preserved the distinct isotopic signatures of their sources (Anders and Zinner 1993). This situation is different for common types of interstellar dust. To explain the extinction features and polarization of light caused by interstellar dust, Mathis and Whiffen (1989) proposed a model of composite interstellar dust that is a mixture of 'silicate,' amorphous carbon and graphite grains. The composite dust can reach up to a few times 0.1 μm in size. These grains were selected on the basis of their physical but not their mineralogical properties. In fact, silicate can be olivine or pyroxene, or both, with unspecified $Mg/(Mg+Fe^{2+})$ (mg) ratios. They can be amorphous or crystalline solids. That is, the infrared spectrum of a Mg_2SiO_4 material that does not have long-range crystallographic ordering may already have bonding properties of forsterite. Graphite in this model may be substituted by amorphous carbons or polycyclic aromatic hydrocarbons (Zhang et al. 1996).

This chapter deals with extraterrestrial materials collected in the lower stratosphere that are known as interplanetary dust particles. The IDPs survived atmospheric entry without melting although they did not remain unheated. During their short (<2 months) residence time in the stratosphere they are not destroyed by erosion and weathering. Once extraterrestrial material reaches the Earth's surface it is subject to geological processes in their environment whereby small objects (including IDPs) are destroyed, particularly when they are fragile or glass-rich objects. Searching for micrometeorites at the surface is only viable in environments wherein concentration effects increased their time-integrated abundance, such as deep-sea sediments, placers, 'hard-grounds,' land-ice sheets and glaciers. Materials recovered from these deposits are known as micrometeorites. Micrometeorites are biased towards resistant iron-nickel spheres (± sulfur) and CM type spheres (Brownlee 1978b, 1985; Kurat et al. 1994). Micrometeorites are studied ever since the first magnetic spherules with inferred cosmic origin were recovered from deep-sea sediments by the HMS Challenger in the late 1880s (Murray and Renard, 1883). I will not discuss micrometeorites recovered at the Earth's surface.

I present the pre-1982 IDP research separately because it established the uniqueness of stratospheric IDPs. This research emphasized the properties that distinguish IDPs from natural terrestrial dust and anthropogenic dust in the stratosphere and to variations among aggregate IDPs that stand out among the collected particles. I offer a hierarchical accretion model for a context of the growing body of chemical data, and for comparisons of the compositions of IDPs, comet P/Halley dust and carbonaceous chondrite matrix.

PRE-1982 IDP RESEARCH

Upper stratospheric dust collections: The first lessons

The first lesson learned from stratospheric dust collection was the fact that the collection technique is indiscriminate with regard to the types and origins of collected particles. A sampling of the stratosphere will invariably include particles from different sources with relative contributions that vary as a function of time, altitude, and possibly geographic location. At any time the stratosphere contains extraterrestrial dust, dust from natural terrestrial sources (volcanic dust, wind-blown dust, biomass burning) and dust related to anthropogenic activities (Mackinnon et al. 1982, Zolensky and Mackinnon 1985, Zolensky et al. 1989a). During the late sixties balloons were the only means to carry dust collectors aloft to altitudes as high as about 37 km altitude. The early studies suffered from contamination by sulfate aerosols in the upper stratosphere (Bigg et al. 1970). These efforts did not sample adequate volumes of the upper stratosphere to ensure a reasonable chance of collected IDPs. Collected fluffy aggregates that were generally smaller than ~2 μm in size (up to ~6 μm) were composed of grains ranging from 5 to 150 nm in diameter (Bigg et al. 1971). Unambiguous identification of the aggregates was difficult because routine manipulation of (sub-) micrometer particles was not possible. Their morphology resembled the type of dust expected to exist in comet nuclei based on remote sensing data of active short-period comets but they also resembled condensed matter such as soot particles. Determination of mineralogical and chemical properties was impossible, as transmission electron microscopes [TEM] with high point-to-point resolution were not equipped with energy dispersive spectrometers. Recognition of extraterrestrial dust from the stratosphere was made credible based on scanning electron microscope [SEM] and electron microprobe analyses of debris in impact craters on space hardware (e.g. manned spacecraft windows) that flew in low-Earth orbit. For example, the Gemini S-010 experiment that had residues in three impact craters had chemical signatures consistent with meteoroid materials (Griffith et al. 1971). The C/Si (0.29) and Mg/Si (0.57) ratios in debris from a crater 290 μm in diameter are comparable to those measured in comet P/Halley dust sixteen years later (Jessberger et al. 1988). Brownlee et al. (1973) completed the first successful cosmic dust collection at 35 km altitude where ~80% of the dust consisted of transparent Al_2O_3 spheres mostly 2 to 3 μm in diameter. Their origin was a mystery. Only ~10% of collected particles was possibly extraterrestrial material that melted during atmospheric entry (Brownlee and Hodge 1973). They included rare spheres such as a large (12 μm in diameter) sphere with a chondritic composition. Also, opaque spheres of (a) Fe, Mg (Si was not detected due to interference from the substrate) plus minor amounts of Ni, Ti, Cr, and (b) Fe, Ni, S with Fe/Ni ≈ 20 (Brownlee et al. 1973).

Lower stratospheric dust collections: Early results

The collection of dust in the lower stratosphere began in March 1974 (Brownlee et al. 1976a). Brownlee's pioneering efforts of dust collection and analyses capitalized on the improved collection techniques, the availability of high-flying aircraft and advances of small-particle manipulation and analyses during the mid-1970s. A typical analyzed particle was ~10 μm in diameter but most collected particles were <2 μm. Most particles >10 μm had chondritic compositions (Brownlee et al. 1977a). Spherical or ovoid shapes indicated that some particles reached their melting temperature late during their lifetime. When fast-moving (micro-) meteoroids decelerate in the Earth's atmosphere their kinetic energy is dissipated as heat that can lead to melting and ablation. In the case of natural dust this particular shape alone supports an extraterrestrial origin. Spherical particles are also linked to high-energy anthropogenic, i.e. space-related, activities such as the use of solid-fuel propulsion systems ascending in the atmosphere, re-entered debris from low-Earth orbit due to catastrophic satellite failure, orbital decay and entry break-up, and solid fuel propulsion systems fired in orbit.

Anthropogenic dust

Anthropogenic dust usually stands out among natural dusts by its unique chemical compositions, or rather the combination of elements. From the earliest days of collection it was evident that anthropogenic dust, including 'space age' debris, persists in the lower stratosphere. Most particles have an obvious, easily traceable source. Others are elusive. The source of the mysterious Al_2O_3 spheres (AOS) is unique and easy to determine. The AOS range in size from ~2 μm in diameter (the collector cut-off size) to ~8 μm whereby 90% of all particles <4 μm are transparent AOS spheres (Brownlee et al. 1976b). The spheres are effluents of solid-fuel rockets such as the Titan III rocket. From 1974-1979 these spheres were the most abundant particle type in the stratosphere between 20 and 35 km altitude and their spatial density remained constant at 10^{-2} m^{-3} (>5 μm) (Brownlee et al. 1976b). Zolensky et al. (1989a) reported a ten-fold increase in AOS >1 μm in diameter in the lower stratosphere between 1976 and 1984. The US Space Transportation System (STS, Space Shuttle) is most likely the major source of AOS spheres. They predicted that the hiatus in the STS program from January 28, 1986 till September 29, 1988 would lead to depletion or even the disappearance of AOS spheres in the lower stratosphere. Rietmeijer and Flynn (1996) confirmed this prediction but did not find a direct link between the abundance of AOS >9 μm in diameter and the timing of STS launches during a period from May 1981 till July 1991. Rietmeijer and Flynn (1996) submitted that stratospheric residence times of these spheres are longer than predicted by the Stokes-Cunningham law. Thus, some AOS are hollow spheres, or massive spheres can be 'rafted' in nonspherical aggregates of anthropogenic debris.

Other forms of anthropogenic dust occur as irregular micrometer sized Fe,Cr,Ni-particles are readily recognized as stainless steel. Other elemental combinations are not as straightforward but include thermal paints from rockets and satellites. Brownlee (1978b) found (rare) aluminum spheres and irregular particles with small and variable amounts of Mg, Si, S, Ca, Ti, Fe, Ni and Cu. The origins of Al-prime spheres and particles that still persist in the stratosphere today are enigmatic. The hydrogen and magnesium isotopic compositions of Al-prime particles are generally consistent with a man-made origin and indigenous Mg isotope mass fractionation indicates an important role of vaporization and condensation reactions in their formation (McKeegan 1986). They are most likely solid-rocket effluents or re-entered space debris from low-Earth orbit. The abundances of several

Al-prime particles (~7 μm in size) with a fine-grained crust adhering to an aggregate core on the collectors showed an apparent correlation with the abundance of chondritic IDPs (Flynn et al. 1982). Flynn et al. (1982) inferred that close to chondritic Fe/Ni ratios support an extraterrestrial source of Al-prime particles. In another Al-prime particle, Zolensky (1987) found high-temperature refractory minerals with stable oxygen isotope ratios to prove their extraterrestrial origin (McKeegan 1987).

Interplanetary dust particles

The properties of IDPs (Table 1) show it is possible to group these particles on the basis of chemistry and/or morphology but it is not immediately apparent how to fit their mineralogical properties. A workable first-order classification scheme made a chemical subdivision into chondritic and non-chondritic IDPs. Spherical IDPs most likely form when non-spherical particles reach their melting point during atmospheric entry heating. Chondritic aggregates are probably precursors of chondritic spheres (Table 1) but the original properties of unmelted aggregates may have been modified by thermal alteration.

A particle is 'chondritic' when the elements Mg, Al, Si, S, Ca, Ti, Cr, Mn, Fe and Ni occur in relative proportions similar (within a factor 2) to their cosmic abundance, represented by the CI carbonaceous chondrite composition (Brownlee et al. 1976a). Several terrestrially common elements, such as Cu, Cl, Na and Cd, do not occur in cosmic proportions (Brownlee et al. 1976a). The Fe/Ni ratios of collected chondritic IDPs cluster around the cosmic (or, chondritic) ratio, Fe/Ni = 18.8 (Brownlee et al. 1977b). Trace element abundances in two aggregate IDPs (Table 2) are very close to the abundances in CI meteorites with the exception of gold (200 to 420 times CI) and zinc enrichments of 16 to

Table 1. Properties of stratospheric dust particles determined prior to 1982. Sources: Brownlee (1978a; -b); Brownlee et al. (1976b; 1977a; -b) and Flynn et al. (1978)

	CHONDRITIC PARTICLES		NON-CHONDRITIC PARTICLES	
	Aggregates	Spheres	Silicate particles	FSN particles
Optical properties	Black (opaque)	Transparent but usually opaque due to Fe_3O_4	Transparent and opaque	opaque
Chemical signature	Chondritic with bulk carbon and sulfur both >4 wt. %; carbon ranged from 2.2 to >15 wt. %	Chondritic but no sulfur	Fe,Mg-silicates with sulfur	S/Fe = 1 to 0 Fe/Ni = ∞ to 10
Grain sizes (μm)	0.001 to >1, mostly ~0.1-0.3, some are polycrystalline aggregates	No data	Ol: 4 x 5 –16 x 9 En (plate): 6 x 6 FeS: 0.7	2.4 x 3.8 - 6 x 6
Minerals (identified by XRD)	$Fe_{1-x}S$; Fe_3O_4	Ol, incl. Mg = 60; pyroxene, incl. Mg = 60; magnetite	Fo_{100-47}; En_{100-83}; magnetite	(Ni)-$Fe_{1-x}S$, x = 0-0.1; Magnetite
Phases (based on chemical signature)	Mg-pyroxene; Olivine, Fe/Si ratio variable; poorly-ordered or amorphous Fe,Mg,Si phase; Pentl.; (Ni) –Pyrrh.; High Si grain; a Si,Al,Ca,Ti grain	No data	Fe-sulfides: 1-5 wt. % nickel	$Fe_{1-x}S$ and Fe_3O_4

Table 2. Minor and trace elements in aggregate IDPs

ELEMENTS	ABUNDANCES	REFERENCES
^{4}He [cm^3 (STP) g^{-1}]	0.002-0.25	Rajan et al. 1977
^{20}Ne [cm^3 (STP) g^{-1}]	$4.4 \pm 1.6 \times 10^{-4}$	Hudson et al. 1981
^{36}Ar	$5.1 \pm 0.3 \times 10^{-13}$	
^{84}Kr	$< 4 \times 10^{-15}$	Note: Fe/Ni = 15.75
^{132}Xe	$\sim 10^{-15}$	
Uranium (10^{-9} ppb)	<15 Orgueil (CI) : 6-221 Allende (CV): 5-17 Crust (Terra) : 2300	Flynn et al. 1978
Fe (g)	1.20 - 2.28 (x 10^{-8})	Ganapathy and Brownlee 1979
Ni	9.18 – 7.32 (x 10^{-10})	
Na	4.86 – 3.20 (x 10^{-10})	
Cr	1.76 – 3.11 (x 10^{-10})	Note: Fe/Ni = 13.1 - 31.1
Zn	4.89 – 1.07 (x 10^{-10})	
Co	3.82 – 2.83 (x 10^{-11})	
Sc	2.90 – 5.68 (x 10^{-13})	
Au	3.03 – 6.39 (x 10^{-13})	
Ir	3.42 – 6.11 (x 10^{-14})	
$\Delta(^{25}Mg/^{24}Mg \pm 2\sigma$ (per mil)/ $\delta^{26}Mg \pm 2\sigma$ (per mil)	$-0.8 \pm 0.6/3.2 \pm 1.1$ $2.2 \pm 1.3/1.0 \pm 2.8$ $-1.9 \pm 0.5/4.0 \pm 0.8$ $0.2 \pm 0.7/3.5 \pm 1.1$	Esat et al. 1979 Note: Fe/Ni = 15.6

35 times CI (Ganapathy and Brownlee 1979). It is apparent that major, minor and trace element abundances of the aggregate IDPs are close to, or even higher (zinc), than the chondritic (or solar) abundances of the elements. It is unlikely that natural aggregates of (known) fine-grained terrestrial or lunar materials can match the proportions for six, cosmically abundant, elements (C, Mg, Si, S, Fe, Ni). Flynn et al. (1978) found that one or two major elements in any given IDP could be undetectable by energy dispersive spectrometry, and that there is sub-micrometer chemical heterogeneity for some minor elements, e.g. titanium. The chondritic signatures are reliable indicators for an extraterrestrial origin for chondritic IDPs (4 μm to 25 μm in size) collected in the lower stratosphere. This chemical fingerprinting is supported by the chondritic composition of micrometeoroid residue in impact craters on the surface of Skylab IV (Brownlee et al. 1974). This impact residue and similar ones in impact craters on other satellites proved that chondritic micrometeoroids occur as small bodies in space.

Chondritic aggregate IDPs. Aggregates make up ~90% of chondritic IDPs with a mass ranging from 5×10^{-11} to 10^{-8} g. They are compact aggregates with little pore space (Brownlee 1978a; Brownlee et al. 1976a, 1977a) (Fig. 4). In rare chondritic aggregates the loosely bound components form a highly porous structure (Fig. 5a) and occasionally with rare whiskers (Fig. 5b). Aggregate IDP densities are between 2 g cm^{-3} and 3.5 g cm^{-3} (Brownlee et al. 1977b). In one of the first TEM studies of aggregate IDPs, Fraundorf and Shirck (1979) reported irregular clumps (up to ~500 nm) forming the matrix of these particles. The matrix clumps consist of tiny crystallites (<10 nm to ~100 nm). In some particles they are closely packed at scales of 2 to 4 μm (Fig. 6). The correlated Fe/Mg weight ratios of these clumps are scattered around a line from (Fe/Si = 0.1)/(Mg/Si = 1.4) to (Fe/Si = 2)/(Mg/Si = 0.7) (Brownlee et al. 1980). These values are close to the CI element ratio Mg/Si = 0.9 but the Fe/Si element ratios are much lower (CI: Fe/Si = 1.8). An

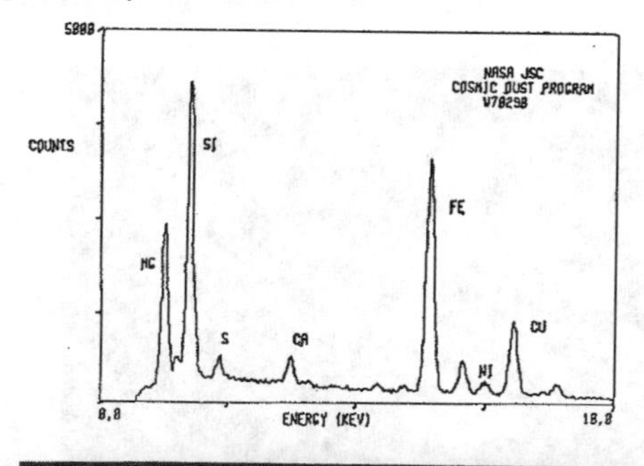

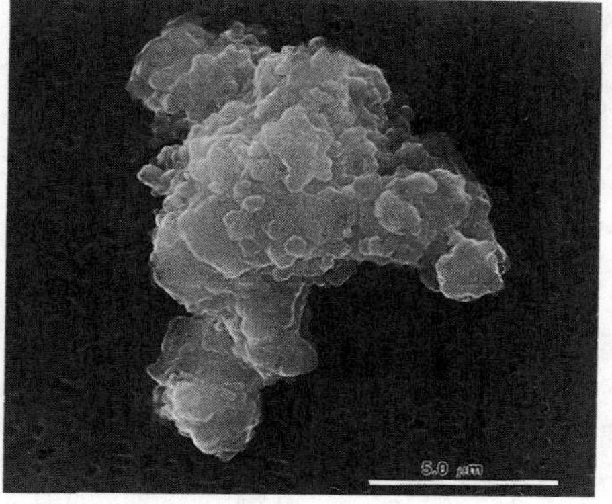

Figure 4. Scanning electron micrograph of chondritic filled aggregate IDP W7029B5. The EDS spectrum shows the Mg, Al, Si, Ca, S and Fe in chondritic proportions. The chemical signature plus morphology support, but do not prove, an extraterrestrial origin. The copper peak is an instrumental artifact and results from spurious x-rays generated within the microscope. NASA photo S82-27567.

amorphous carbonaceous coating softens the contours of 'clumps.' It is also the embedding medium for the tiny crystallites in the 'clumps' (Fraundorf and Shirck 1979). Larger crystals (up to 200 nm) occur within the matrix. They include platy Ca-poor, Mg-rich pyroxene and subhedral Fe-sulfides (Fraundorf and Shirck 1979). Fraundorf (1981) made a TEM study of eight chondritic aggregates (4 μm to 15 μm) and three coarse-grained IDPs (4 μm to 10 μm). The coarse-grained particles included an IDP of coarse-grained Fe-sulfides with fine-grained material (13-05-01C), one IDP with fragments of radial grown pyroxene (13-05-03C) and an olivine aggregate IDP (14-06-09A). Fraundorf (1981) found that

- The average composition of five aggregate IDPs (excluding 14-03-09A) show Mg/Si = 0.71 ± 0.115, Fe/Si = 0.59 ± 0.17, and S/Si = 0.49 (three IDPs, only). Sulfur was depleted (S/Si <0.15) in the two other chondritic IDPs and in 14-03-09A.
- All particles are mixtures of (1) non-crystalline clumps with non-stoichiometric (silicate) compositions, (2) crystals <10 to 100 nm in size, and (3) crystals >100 nm. The matrix clumps have tiny crystals of 'silicate, iron sulfide and metallic iron' in a carbonaceous material (clumps were also described as 'polycrystalline'). The variable volume fractions of constituents defined two groupings:

	Matrix clumps	*Crystals <10-100 nm*	*Crystals >100nm*
Mean ±1σ	0.45 ± 0.15	0.32 ± 0.08	0.23±0.14
Range (N = 6)	0.2-0.6	0.2-0.40	0.1-0.5
and, N = 2	0.95	zero	0.05

- Crystals >100 nm have 'defect' structures. They were identified by a combination of electron diffraction, dark field TEM imaging and chemical composition (Table 3).

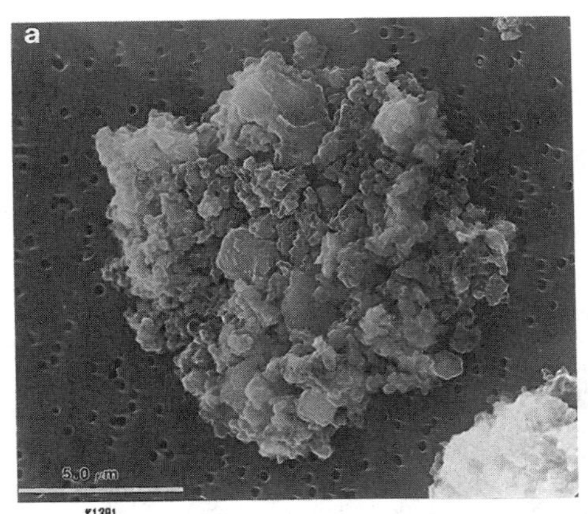

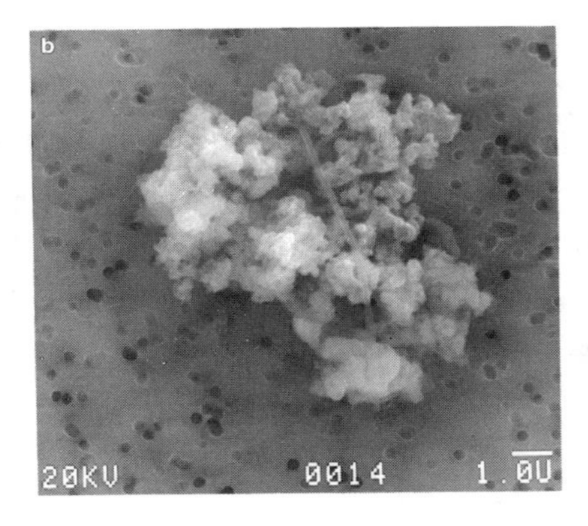

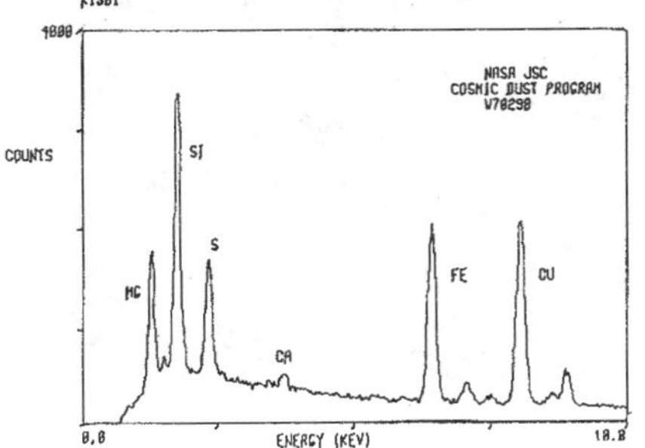

Figure 5. Scanning electron micrograph of two highly porous chondritic aggregate IDPs in the NASA/JSC Cosmic Dust Collection. (a) W7029B23 showing a fine-grained matrix with embedded platy grains. Its EDS spectrum shows the Mg, Al, Si, Ca, S and Fe in chondritic proportions. The high bremsstrahlung background indicates the presence of low atomic number elements (e.g. carbon). The chemical signature plus morphology support an extraterrestrial origin. The copper peak is an instrumental artifact due to spurious x-rays (NASA photo S82-27575) and (b) L2005E5 with a long whisker embedded in its matrix. NASA photo S90-38116.

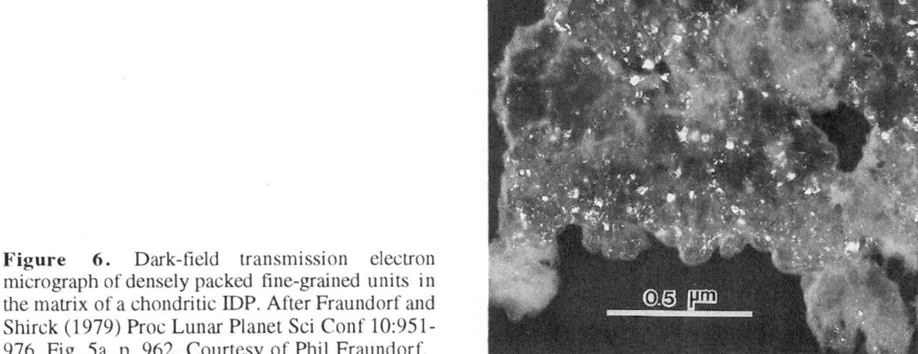

Figure 6. Dark-field transmission electron micrograph of densely packed fine-grained units in the matrix of a chondritic IDP. After Fraundorf and Shirck (1979) Proc Lunar Planet Sci Conf 10:951-976, Fig. 5a, p. 962. Courtesy of Phil Fraundorf.

Table 3. Large mineral grains in chondritic aggregate IDPs (source: Fraundorf 1981)

	SIZE (μm)	COMPOSITION
Olivine	0.2 x 1.0	Fo97
	0.3 x 0.3	Fo99
	0.6 x 1.0	Fo70
Pyroxene	>1.0 (euhedral laths elongated along [100]);	enstatite
	0.2	enstatite
Fe-sulfide	0.2 (polycrystalline aggregate)	pyrrhotite
	0.2 (single-crystal grain)	pyrrhotite; S/Fe = 1.2 (at. ratio)
Low-Ni Fe-metal	0.2 (droplet)	teanite, Ni ~6%
Fe-carbide	Polycrystalline rim (15 nm) on teanite droplet	Cohenite: $(Fe,Ni)_3C$

- Decorations of magnetite nanocrystals on silicates formed during atmospheric entry. This conclusion was based on the Fe/Si and S/Si atomic ratios in combination with TEM analyses. That is, these ratios are Fe/Si = 0.48 and S/Si = 0.67 in particle 13-06-05A containing pyrrhotite, while particle 14-03-09A with Fe/Si = 0.92 and S/Si < 0.06 is characterized by the highest abundance of magnetite nanocrystals.
- Some IDPs have a 'silicon-excess,' i.e. they contain more silica than is required to fit a stoichiometric pyroxene composition. This 'excess silicon' must be viewed with caution (Fraundorf 1981) but is not due to significant silicone oil contamination of collected IDPs (Fraundorf and Shirck 1979).

Proof that chondritic aggregates are extraterrestrial materials includes (1) large amounts of solar wind implanted ^{4}He and other noble gases (Table 2) and (2) the presence of amorphous coatings (40 nm thick) on some constituent grains due to radiation processes (Brownlee 1978a). Both properties indicate that the IDPs were exposed to the solar wind for 10 to100 years (Rajan et al. 1977). Considering the available time, it is unlikely that the measured helium content is due to uranium decay (Rajan et al. 1977). To account for the measured abundances, they calculated a required decay time of 4.6×10^9 years and uranium content of 200 ppm by wt which is not observed in IDPs (Table 1). Esat et al. (1979) measured the magnesium isotopic compositions of aggregate IDPs (Table 2), two transparent spheres, one black sphere, and six forsteritic monomineralic particles (Fe/Ni = 29.75). All measured IDPs had terrestrial magnesium isotopic compositions within 2 parts per thousand. The ^{42}Ca/^{44}Ca ratios (0.3098-0.3160) in one forsteritic IDP and a Ca-rich pyroxene grain were within ~1% of the terrestrial value. The isotopic compositions confirm that these IDPs are solar system materials, and possibly presolar dust that was unaffected by the sun's T-Tauri thermal event.

Chondritic Non-aggregate IDPs. The smooth surface of a subgroup of chondritic IDPs (Fig. 7) is morphologically distinction from aggregate particles. Brownlee (1978a) reported the first large chondritic IDP (15 μm) with this distinct smooth surface. This particle contained chamosite (a trioctahedral chlorite) or serpentine that was "identifiable by X-ray diffraction." These morphologically distinct IDPs were the only particles wherein layer silicates could be identified. It is not true that IDPs with a smooth surface are uniquely hydrated (Brownlee, pers. comm.). Lacking TEM characterizations, the interior structures of these chondritic smooth IDPs remained unknown.

Non-chondritic IDPs. They are mostly Mg-rich olivines and pyroxenes, and Fe,Ni-sulfides (Table 1). Euhedral crystals are present. Brownlee et al. (1977a) reported 'whisker-like rod.' The iron-nickel sulfides occur as irregular masses, aggregates, euhedral

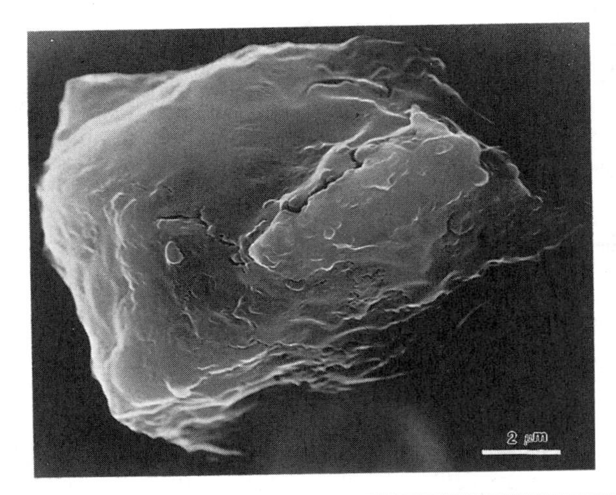

Figure 7. Scanning electron micrograph of chondritic IDP W7017B12 with a characteristic smooth surface. Its EDS spectrum resembles the spectrum shown in Figure 6 but with a much lower background. In other particles the smooth surface may show 'cracked' regions. Surface regions can also be platy, cracked, fibrous or vesicular (Schramm et al. 1989). NASA photo S81-39963.

Figure 8. Scanning electron micrograph of a coarse-grained Mg,Fe-silicate IDP with chon-dritic material attached to the surface. Its EDS spectrum shows minor peaks for Al, K, Ca and Cr. The particle rests on a nucleopore filter during scanning electron microscope analysis. The black tear is damage to the original negative. NASA photograph S-92-35376.

Figure 9. Scanning electron micrograph of a coarse-grained sulfide IDP with chondritic material attached to the surface. NASA photograph S92-35813

single-crystal grains and stacks of platelets (Brownlee 1978b). They often have chondritic material adhered to the surface (Figs. 8 and 9) which suggests a physical relationship between chondritic and non-chondritic IDPs. In fact, sub- to anhedral grains of silicates and sulfides were also found embedded in the matrix of chondritic aggregates.

Some non-chondritic IDPs occurs as compact silicate spheres. They are 'secondary micrometeorites' due to meteor ablation in the atmosphere. During melting and evaporation, elements with a high boiling point (refractory) are concentrated, such as calcium in a silicate sphere (cf. Esat et al. 1979). Quenched melt IDPs can have iron-nickel mounds on the surface as a result of centrifugal separation of immiscible liquids (Brownlee 1978b). The Fe-Ni-sulfur (FSN) particles are mostly spheres (Brownlee 1978b). They are also 'secondary micrometeorites'. Because the S/Fe ratios of FSN spheres show a continuous trend that approaches zero, it is possible that the iron-nickel spheres are FSN spheres that lost all sulfur during ablation (Brownlee 1978b).

Summary of IDP data

The pre-1982 research demonstrated that extraterrestrial dust survived deceleration in the Earth's atmosphere and can be collected from the lower stratosphere for laboratory studies. Chondritic IDPs that survived without apparent melting still show mineralogical and chemical signatures associated with atmospheric entry heating, such as magnetite nanocrystals (Table 1) and loss of volatile elements (e.g. sulfur). Minor heating is inferred when volatile elements, such as zinc, sodium and helium, have CI abundances, or higher (Ganapathy and Brownlee 1979). The relative abundances of unmelted IDPs appeared to be a function of particle size (Brownlee et al. 1977b):

Size	*Chondritic IDPs*	*Silicate IDPs*	*Sulfide IDPs*
5-10 μm	55%	12%	33%
>10 μm	Present	No data	Extremely rare
>30 μm	No data	Very abundant	No data

The early work established several basic facts of chondritic aggregate IDPs, viz.

1. Particles (3 to 25 μm) with a chondritic composition for the major rock-forming elements also have chondritic minor and measured trace element abundances. High volatile element contents (i.e. carbon and zinc) suggest they are different from carbonaceous chondrites, although the major element compositions of chondritic IDPs, CI and CM carbonaceous chondrite are similar (Table 4).
2. One chondritic aggregate broke apart on collector into >100 small 'fragments' (Brownlee 1978b; Brownlee et al. 1976a, 1977b), which showed they are mixtures of matrix clumps (~100 nm to ~500 nm), Fe,Ni-sulfide and Mg,Fe-silicate grains. The matrix clumps are polycrystalline aggregates with a carbonaceous matrix.

Table 4. Average composition of chondritic aggregate IDPs (atomic fractions) compared with CI matrix, CI and CM meteorite bulk compositions (sources: Brownlee 1978a and Hudson et al. 1981)

	Aggregate IDPs	*Bulk CI*	*CI matrix*	*Bulk CM*
Mg/Si	0.85 ± 0.15	1.06	0.92	1.04
Fe/Si	0.63 ± 0.26	0.90	0.54	0.84
S/Si	0.35 ± 0.13	0.46	0.13	0.23
Al/Si	0.063 ± 0.023	0.085	0.093	0.084
Ca/Si	0.048 ± 0.017	0.071	0.011	0.072
Na/Si	0.049 ± 0.048	0.060	0.016	0.035
Ni/Si	0.037 ± 0.020	0.051	0.047	0.046
Cr/Si	0.012 ± 0.0047	0.013	0.012	0.012
Mn/Si	0.015 ± 0.0017	0.009	0.005	0.006
Ti/Si	0.0022± 0.0006	0.002	0.001	0.002

Chondritic IDPs, CI and CM carbonaceous chondrites are (1) very fine-grained, (2) have carbon and sulfur contents >4 wt %, and (3) contain pyrrhotite, Fe-poor olivines and pyroxenes, and magnetite. As Flynn et al. (1978) noted the complex and fragile structure of aggregate IDPs is incompatible with a prior history of residence in regolith environments on a meteorite parent body. The chemical properties of IDPs merged into a coherent framework that showed distinct deviations from CI and CM bulk, and CI matrix, compositions. Calcium in IDPs is 'depleted' compared to the bulk meteorites but 'enriched' relative to the CI matrix, while the S/Si ratio in IDPs is intermediate between CI matrix and bulk CI (Table 4). The CI meteorites experienced aqueous alteration that, among other things, resulted in the formation of Ca-bearing sulfate and carbonate veins. It follows that aqueous alteration of the IDP parent bodies was nonexistent, or limited in scope, compared to the meteorite parent bodies.

The morphological and mineralogical data (Table 1) appear as a haphazard collection of facts lacking trends, yet, suggesting that systematic relationships are likely to exist among these chondritic IDP properties. The first model for IDPs postulated that the minerals and textures recorded two episodes of aggregation: (1) aggregation of basic' building blocks of close to chondritic compositions, and (2) aggregation of matrix and similar-sized, or larger, monomineralic grains (Fraundorf et al. 1982a). The explosive growth in stratospheric IDP collection and research after 1982 provided new insights into the unique chondritic aggregate and cluster IDPs.

Collection of stratospheric dust

Micrometeoroids encounter the Earth at velocities of >11 km s^{-1}. At 100-80 km altitude they decelerate to cm s^{-1}. This slow-down in velocity results in a 10^6-fold increase in IDPs in the stratosphere compared to the flux in near-Earth space. Hence, the stratosphere is an excellent location for the efficient collection of IDPs. An analysis of >150 particles showed that the flux of IDPs >10 μm was constant at 3×10^{-6} m^2 s^{-1} between 1974 and 1976 (Brownlee 1978b). There is no reason to assume that this flux has changed since. The implication is that a sampling of a relatively small volume of the lower stratosphere will bring a reasonable yield of IDPs. Stratospheric dust is collected using inertial impact collectors mounted underneath the wings of high-flying NASA U2 aircraft (Brownlee et al. 1976a). The combination of aircraft speed and collector size means that a sampling of 30-40 hours captures about ten IDPs >10 μm in size. On May 22, 1981, the NASA Johnson Space Center (JSC) Cosmic Dust Program made its first sampling of the stratosphere between 17 and 19 km altitude (Warren and Zolensky 1994). The flight paths were more-or-less along the pacific coastline from south of the equator to Alaska, and over the contiguous US from the east to the west coasts. This program has since sampled the lower stratosphere during distinct intervals (Rietmeijer and Warren 1994). Particles are collected on Lexan plates covered by a layer of high viscosity silicone oil to entrap collected particles that impact the collector surface. Two types of collectors are in use, viz. small area collectors (SACs: 30 cm^2) and large area collectors (LACs: 300 cm^2). Four collectors are housed in each of two pylons mounted underneath the wings of high-flying aircraft (Fig. 10). A pressure sensor opens and closes the pylons to ensure that collections take place only in the lower stratosphere. During pre- and post flight collector handling and curation in a class 100 clean room, contamination of collected dust with micrometer-size extraneous dust is low (Mackinnon et al. 1982). Warren and Zolensky (1994) reviewed the details of collection and curation. Collectors were flown haphazardly over the years (no collections during January and February due to aircraft maintenance), the length of collection period varies, and the actual hours of collection (30.7 to 65 hours) vary within these periods. This situation required a normalization procedure to extract information on particle abundances from this collection (Rietmeijer and Warren 1994).

Figure 10. Pylons (arrows) housing four small area collectors each mounted underneath the wings of a NASA WB-57F stratospheric aircraft. NASA photograph S81-31582.

Silicone oil

The particles are picked off the collectors by hand and rinsed with hexane to remove the silicone oil. This process is not 100% efficient. Residual silicone oil remains at surfaces in aggregate IDPs (Sandford and Walker 1985) and in fluffy poorly graphitized carbon particles (Rietmeijer 1987). A residual silicone oil layer is not considered a problem for mineralogical studies of the particles. Fraundorf and Shirck (1979) found this thin layer interfered with the identification of thin coatings of amorphous material on grains in IDPs. Schramm et al. (1989) reported experimental evidence that residual silicone oil is not a major problem during energy dispersive spectroscope analyses of major elements. Some residual oil invariably remains. In some cases even vigorous washing will not remove the oil layer resulting in severe contamination. These cases are recognized.

Curation

Collected (unmelted) chondritic IDPs in the NASA/JSC collection range from 2 μm to 41 μm have a Gaussian size distribution with mean = 12 μm (Rietmeijer and Warren 1994). Brownlee et al. (1980) reported chondritic IDPs that were 2 to 41 μm in size but up to 60 μm in size. Fraundorf et al. (1982b) described IDPs of 2 μm to 22 μm in size (98% of total amount) and three particles of 36 to 38 μm. The data suggest a collector cut-off size of ~2 μm, although smaller particles are collected (Zolensky and Mackinnon 1985, Zolensky et al. 1989a). This method of collection is biased against dust <2 μm that occurs as individual particles. Cluster IDPs break apart during collection whereby the fragments from the main particle cover a sizeable area on the collector in a recognizable spray pattern. When a fragmented cluster on the collector is exposed during subsequent hours of collection there is a possibility of contamination with another cluster IDP, or with an individual IDP. Vice versa, an IDP already on the collector may become contaminated with a cluster particle. The probability of overlapping impacts is low and it is not perceived as a serious problem (Flynn 1994) although this will not be true for very small particles in a larger cluster IDP.

Particles collected by the NASA program receive a preliminary identification. These data for each particle are listed in Cosmic Dust Catalogs [catalogs are available upon request from the Curator of Cosmic Dust at NASA/JSC, Houston, TX 77058, USA]. The data include a SEM image, an energy dispersive spectrum (EDS) representing the bulk composition, size, shape, color, luster and whether the particle is transparent, translucent, or opaque (in visible light). Particles are classified in to the following groups: (a) Cosmic (C), (b) TCN (terrestrial contamination, natural), (c) TCA (terrestrial contamination, artificial), (d) AOS (see above), and '?' ('Identification uncertain'). The particles listed in the Cosmic Dust Catalogs are available upon request to investigators. They are studied using a wide range of state-of-the-art experimental techniques (Sutton 1994).

CLASSIFICATION OF STRATOSPHERIC DUST

Simple, first-order, particle classification

A reliable first-order classification scheme employs simple and unambiguous criteria that are easy to determine. It can also be expanded without changing its nature. Fraundorf et al. (1982b) made the first subdivision using Mg/Si and Al/Si peak intensity ratios in the EDS spectrum of a particle. This approach is biased towards silicates but it separates silicate/chondritic, iron-nickel, and 'low-Z' particles (Z: atomic number that in this context refers to elements $Z < 11$). They found that 'detectable K' is associated with natural terrestrial particles (volcanic dust). Mackinnon et al. (1982) used the information from the NASA/JSC Cosmic Dust Catalogs for particles on collectors W7010, W7017 and W7029 that sampled the lower stratosphere during three periods between May 22 and December 15, 1981. The first subdivision was made according to morphology: (1) spheres, (2) aggregates, and (3) fragments (all particles that were neither a sphere nor an aggregate). Spheres, with mean diameters ranging from 5.1 μm to 7.6 μm in size, made up 45% of all collected particles, followed in abundance by fragments (35%), and aggregates (20%). Element peak ratios in the EDS spectra define the chemical particle types (Table 5).

To test classification using morphology and bulk particle chemistry, I compare the results obtained by Mackinnon at al. (1982) with the micrometeorite taxonomy developed by Brownlee et al. (1982) (Table 6). Ideally, all stratospheric dust types occur in different collections but not necessarily with the same abundances. If a particle type is apparently restricted to one particular collection it indicates investigator bias or temporal variations in its occurrence in the lower stratosphere. The result shows a remarkable similarity between both classifications with differences almost certainly due to investigator bias (Table 6). These simple classifications recognize all types of IDPs with a high level of confidence. Mackinnon et al. (1982) and Brownlee et al. (1982) independently recognized the same IDP types among the collected stratospheric particles, that is, chondritic particles, silicate spheres (incl. MMS), metal ± S spheres, metal ± S particles, mafic particles, Ca,Al,Si fragments, and CAS/Ti dust.

Refined particle classification

A refined classification combines particle morphology, chemistry and optical properties (Kordesh et al. 1983). They found that AOS spheres make up 56% of collected spheres 3 to 11.5 μm in diameter. Only 8% of the spheres are opaque. The others are transparent to translucent (73%) or translucent (19%). To account for a bimodal size distribution of AOS and their optical properties, Mackinnon and McKay (1986) suggested

Table 5. Stratospheric dust types defined by their chemical composition (source: Mackinnon et al. 1982)

TYPE	EDS SIGNATURE
Chondritic	Major elements in chondritic abundances (within a factor 3)
Silicate	High Mg/Si ratio, some Ca, Al; Si-only; high Si with Mg, Al and variable amounts of Mn, K, Cl, S
Metal + S	Fe, Ni, S
Metal – S	Fe,Ni
CAS/T	Ca, Al, Si, enriched in Al, Ca, Ti; depleted in Fe and S
Aluminum	Al only
Al-prime	high amounts of Fe, Ti, Cr in addition to Al; variable Ni, Si, S Very high Al, minor amounts of Al, Fe, Ti, Cr, Ni, Si Al with Mg, Si, Fe, and low to trace amounts of Ti and Cr
Mafic	'Chondritic'
Carbonate	No data
Other	High Si with large mounts of Mn, S, Ca, or Au

Table 6. Comparison of stratospheric dust types in two different collections listed in decreasing order of abundance reported by Mackinnon et al. (1982)

CHEMICAL TYPE	MACKINNON ET AL. (1982)	BROWNLEE ET AL. (1982)
SPHERES		
Aluminum	AOS spheres	AOS spheres
Silicate	Present	Stony spheres; also Metal Mound Spheres (MMS) covered with kamacite mounds; often carbonaceous material is present
Metal + S		FSN spheres, usually no metallic iron
Metal – S		Fe,Ni spheres, iron usually oxidized
Al prime	Present	No data
CAS/Ti	Present	Present
IRREGULAR SOLID PARTICLES		
Silicate	Present: silica, clay	No data
Mafic	Present	Mono- and polycrystalline pyroxene and olivine with chondritic material attached
Chondritic smooth	Present	Present
Metal ± S	Present	FSN with up to a few percent zinc; Fe, Ni with roughly cosmic Fe/Ni ratios
Aluminum	Present	Present
Al-prime	Present	Present
Carbonate	Present	Calcite with chondritic material attached
Ca,Al,Si	Present	Present
AGGREGATES		
Chondritic	Present	Present
Al prime	Include refractory IDPs	(Flynn et al. 1982; Zolensky 1987)
Aluminum	Possibly collapsed hollow AOS spheres	Many Al-prime particles appear to be aggregates
Silicate	Minor Fe, Mg, S, Al, Cr	No data

that these spheres were from different sources (e.g. tactical vs. orbiter rockets), or were partially oxidized fuel and spent solid rocket fuels. The later collections include a few broken hollow spheres, which raises the possibility that a bimodal size distribution reflects differential settling rates of hollow and massive spheres. About 70% of the particles in the NASA Cosmic Dust Collection are opaque (Yamakoshi 1994). The opaque particles include >90% of all chondritic IDPs and Al prime aggregates (Kordesh et al. 1983). The small (4 to 14 μm) silicate aggregates include IDPs and volcanic ash (Rietmeijer 1993a) but a definitive statement of their origin is difficult based on major element chemistry and morphology alone. Brownlee (1978b) suggested that iron-nickel spheres are FSN spheres that lost sulfur during ablation. Mass loss during ablation produces progressively smaller spheres. It is significant that the average size of FeS spheres was 7.4 μm and the average Fe sphere on the same collectors was 4.0 μm (Mackinnon and McKay 1986). Some metal ± S spheres may be hollow objects that grew on the fusion crust of conventional meteorites during ablation (Rietmeijer and Mackinnon 1984).

Collection and curation bias

Collection and curation bias is a concern when the data are used for statistical analyses or to make inferences of the sources of collected particles. A comparison of particle types in different collections due to catastrophic events offers an opportunity to investigate bias in collected particles. Particle type abundances in the stratosphere are often strongly size-dependent. For example, collected particles of ~1 μm in size are nearly all AOS spheres (Brownlee, personal comm.). Table 7 shows the fractions of volcanic dust, particles related to rocket launches and IDPs collected during a five-year period (Fraundorf et al. 1982b, Zolensky et al. 1989a). Fraundorf et al. (1982b) indicated that their data (U13-R18) were biased towards larger particles, and against transparent (mostly AOS) spheres. Their concern was justified when compared with results from studies on accurate stratospheric particle abundances on three collectors that included U2-9 and W7017 (Zolensky and Mackinnon 1985, Zolensky et al. 1989a). The Fraundorf et al. (1982b) data are best compared with particles >6 μm on the U2-9 and W7017 collectors. The fraction of IDPs was remarkably constant during this five-year period. The consistency of IDP abundances (Table 7) shows that different investigators use similar criteria. There is little bias against IDPs > 2 to 5 μm. The high volcanic dust fractions during 1976/77 and after late 1980 (Table 7) are real and show the global impact on the Northern Hemisphere of events that injected debris directly in the stratosphere. Volcanic eruptions from 1976 till September 1981 that probably contributed dust to the lower stratosphere during this period (McClelland et al. 1989) are listed in Table 8. There is a distinct but broad correlation between these eruptions and high fractions of volcanic dust during this 5-year period. Mackinnon et al. (1984) showed that the amount and size of dust in the lower stratosphere from one single point source (the El Chichón volcano) decays continuously in a predictable manner. The continuing increase that started after1980 (Table 7) points to more than one contributing point source.

Chemical classification

Sc/Fe vs. Co/Fe ratios. Well-chosen element ratios, especially when determined by non-destructive techniques, are useful to classify stratospheric dust. Lindstrom (1992) used Instrumental Neutron Activation Analysis and found that various chondrite meteorites and suites of igneous rocks from many sources in the solar system (lunar rocks, eucrites, Martian meteorites, terrestrial basalts) occupy discrete fields in a matrix of Sc/Fe and Co/Fe ratios. This matrix identified unmelted chondritic IDPs among the stratospheric dust particles although some IDPs had an 'igneous' Sc/Fe-Co/Fe signature. Unfortunately very little is known of the petrological properties of the IDPs analyzed by Lindstrom (1992)

Table 7. Percentage of IDPs, volcanic dust and AOS spheres among collected particles in the lower stratosphere between 1976 March and September 15, 1981

Collector	Time of collection	IDPs	Volcanic dust	AOS
U2-9	March, 1976	[42] (12)	[58] (34)	[29] (5)
U13	Nov. '76 –April '77	39	2	12
U14	July-August, 1977	39	5	17
U18	Late 1978	44	1	5
U21	Late 1980	32	22	6
U23	Early 1981	35	33	0
R18	July 7-Sept. 15, 1981	17	61	0
W7017	July 7-Sept. 15, 1981	[45] (22)	[25] (23)	[59] (23)

[x]: dust >6μm and (x): dust 1-5 μm.

Table 8. Major volcanic eruptions from 1976 till September 1981

VOLCANO	*ERUPTION TIME*
Augustine, Alaska	Jan. 22, 1976
Soufrière, St. Vincent Isl.	April 13, 1979
Sierra Negra, Galapagos	Nov. 13, 1979
Mt. St. Helens, Oregon, US	May 18, 1980
Gareloi, Alaska	September, 1980
Alaid, Kuril Islands	April 27-29, 1981
Pagan, Mariana Islands	May 22, July 6, 1981

which makes it somewhat difficult to evaluate the usefulness of Sc/Fe-Co/Fe signatures for dust classification. For example, U2015E10 is a very large (45 × 30 μm) chondritic particle with a smooth surface but its optical properties (Cosmic Dust Catalog 5(1) 1984) do not match those of chondritic IDPs (Mackinnon et al. 1982, Kordesh et al. 1983). Yet, it has an 'igneous' Sc/Fe-Co/Fe signature that is distinct from chondritic IDPs. It also has a very low density of 0.5 g cm^{-3} (Zolensky et al. 1989b). Assuming this particle is a unique IDP, its low density probably prevented melting during atmospheric entry. Lacking other data, the properties of U2015E10 are also consistent with volcanic ash.

Fe/Ni ratio. The cosmic Fe/Ni ratio is 18.8. The Fe/Ni ratios for IDPs with chondritic abundances for the major rock-forming elements are clustered around the cosmic value (Brownlee et al. 1977b). Sutton and Flynn (1988) and Flynn and Sutton (1990, 1991a) determined major and minor element abundances in sixteen stratospheric particles listed as 'comic' or 'C?'. Their Fe/Ni data (Table 9) plus data from Table 2 show a normal distribution with a mean of 17.1 (1σ = 5.8) for Fe/Ni ratios ranging from 9.1 to 31.1. Schramm et al. (1989) analyzed 200 IDPs and found average Fe/Ni ratios of 23.2 (chondritic smooth IDPs) and 29.4 (chondritic porous IDPs). Coarse-grained IDPs (mostly silicate particles) have an average Fe/Ni ratio of 30.8 (Schramm et al. 1989). These values are within the range of the Gaussian distribution, which shows that the Fe/Ni ratio is diagnostic tool for separation of chondritic IDPs from non-chondritic IDPs, including terrestrial dust.

Volatile elements. Volatile elements with condensation temperatures <750°C, were measured in >150 stratospheric particles using Synchrotron X-ray Fluorescence (SXRF), Proton-Induced X-ray emission (PIXE) and Secondary ion mass spectroscopy (Arndt et al. 1996, Flynn et al. 1996). The volatile element distributions subdivide 'cosmic' and

Table 9. Fe/Ni ratios of chondritic IDPs and low-Ni chondritic IDPs

IDPS	FE/NI RATIO	IDPS	FE/NI RATIO
Chondritic IDPs		L2001*D1	16.9
U2022G1	15.5	L2003*D1	15.9
W7029*A27	27.7	L2003*E1a	16.1
U2015G1	18.0 (main particle)	L2003*E1b	9.1
W7027C5	10.3	L2003*E1c	10.1
U2022G17	21.0	Low-Ni chondritic IDPs	
W7013H17	21.3	U2022B2	241
W7013A11	14.1	U2001B6	957
U2022G2	13.0	W7066*A4	350
U2022C18	23.1	L2002*C1	333

probably 'cosmic' particles in to (1) chondritic IDPs, (2) 'low-nickel' particles and (3) 'low-zinc' particles (Sutton and Flynn 1988). Among the chondritic IDPs some particles are enriched in zinc relative to the CI value (Sutton and Flynn 1988). Particles with deviations from the CI abundance in one or more rock-forming elements also have deviant abundances for minor and trace elements. Particles U2022G17 and W7027C5 have correlated enrichments in Cu, Ga, Ge and Se that were not seen in other chondritic IDPs (Flynn and Sutton 1990).

Low-Ni particles. The 'low-nickel' particles have non-chondritic Fe/Ni ratios (Table 9). The lithophile, siderophile and volatile elements in three particles (U2001B6, W7066*A4, L2002*C1) are similar to terrestrial basalts (Flynn and Sutton 1990, 1991a). Unlike chondritic aggregates they are differentiated ('igneous') materials that are conceivably from the moon, Mars, or the eucrite parent body in the asteroid belt. Particle W7066*A4 contains 900 ppm lead. The ratio Zn/Cu = 57 in L2002*C1 is reminiscent of Mt. St. Helens ejecta (Rose et al. 1982). The phases in U2001B6 include calcium alumino-silicates, a variety of 'Ti,Fe-Al' phases, iron sulfides and chromite (Sutton et al. 1990). Although there are no detailed TEM data available for particles of this group, the petrological properties of three particles from the same stratospheric sampling that included L2002*C1 suggest that they are probably terrestrial volcanic ash (Fig. 11; Rietmeijer 1992a). It remains possible that 'low-nickel' particles include bona fide IDPs but none of the particles studied so far by TEM shows features consistent with an extraterrestrial origin. They were a persistent and significant (>10%) component among collected particles between 1981May and 1991 July (Flynn and Sutton 1992a). Flynn et al. (1993a) discussed that 'low-nickel' particle W7066*A4 may be a contaminant in a bona fide cluster IDP. Uncertainties in the origin of 'low-nickel' particles stem in part from the emphasis on bulk chemistry without regard for particle petrology and size (Rietmeijer 1997a).

Low-Zn IDPs. The 'low-zinc' particles, with Zn is ~0.1 × CI, are C type IDPs with otherwise chondritic major element abundances and enrichments of volatile elements relative to CI (Flynn and Sutton 1992a). They argued that low zinc abundances are the result of volatile element loss during atmospheric entry heating. This interpretation assumes selective zinc loss because other volatile elements in these particles are enriched relative to the CI standard. Flynn and Sutton (1992a) assumed that zinc occurs in (iron-rich) sphalerite that is present in chondritic IDPs (Mackinnon and Rietmeijer 1987). Yet, the volatile element abundances in the low-zinc IDPs are similar to these distributions in CI and CM carbonaceous chondrites and ordinary chondrites. A few chondritic IDPs are petrologically similar to CI (Keller et al. 1992a) and CM (Bradley and Brownlee 1991, Rietmeijer 1996a) meteorite matrix.

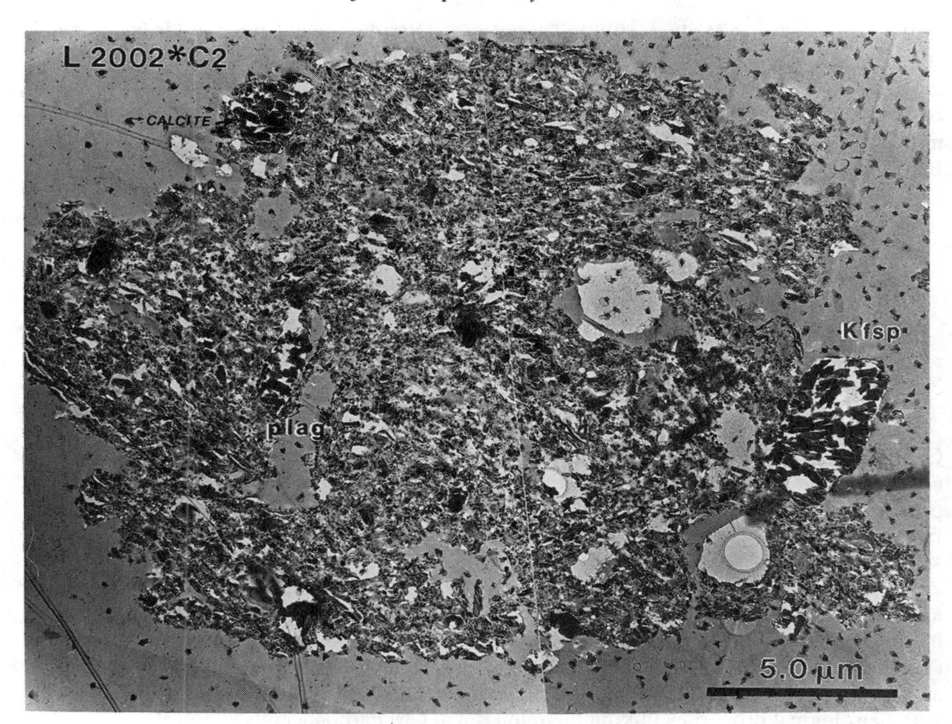

Figure 11. Transmission electron micrograph of an ultrathin (80-100 nm) section of stratospheric particle L2002*C2. This vesicular volcanic ash particle has a compact silica-rich matrix with smectite proto-phyllosilicates and large grains of alkali-feldspar (Kfsp), plagioclase (plag) and calcite. The gray background is the embedding epoxy on top of a holey carbon support film. The black dots are an artifact. [Used by permission of the editor of Proceedings of Lunar and Planetary Science 22:196, from Rietmeijer (1992b), Fig. 2.]

CHONDRITIC IDPs

Classification

Chondritic IDPs are readily recognizable among the collected particles by their major rock-forming element compositions (within a factor 2 or 3). They occur as aggregates and fragments (Mackinnon et al. 1982) whereby the latter include layer silicate-rich IDPs (Brownlee 1978a, Schramm et al. 1989). A subdivision of chondritic IDPs includes (1) particles that form highly porous and somewhat porous aggregates, and (2) smooth particles, i.e. (1a) chondritic porous (CP) IDPs, (1b) chondritic filled (CF) IDPs, and (2) chondritic smooth (CS) IDPs (cf. Mackinnon and Rietmeijer 1987). The CS IDPs are smooth on a 10 µm scale and are usually not porous (Brownlee, personal comm.).

Infrared classification

Infrared (IR) spectroscopy is a non-destructive technique and therefore of potentially great value for classification. The silicate mineral that dominates the IDP spectrum determines its IR classification. In some cases, a strong IR emitter that determines the IR spectrum may occur as a minor constituent of the particle. There are three distinct IR groups, viz. (1) olivine-rich particles, (2) pyroxene-rich particles, and (3) layer silicate-rich

IDPs (Sandford and Walker 1985). Chondritic IDPs dominated by anhydrous minerals (IR) tend to be aggregates (Bradley and Brownlee 1986) while phyllosilicate-rich (IR) IDPs are mostly 'single smooth grains' (Sandford and Walker 1985, Table 1). The layer silicate basal spacing is either 0.73 nm (serpentine-septechlorite) or 1.0 to 1.2 nm (smectite or mica) and they do not occur together in the same IDP (Bradley 1988, Bradley and Brownlee 1986). The modified chondritic IDP classification becomes

(I) (CP and CF) aggregate IDPs	(Ia)	olivine-rich particles
	(Ib)	pyroxene-rich particles
(II) Smooth (compact) IDPs	(IIa)	serpentine-rich particles
	(IIb)	smectite-rich particles

Mackinnon and Rietmeijer (1987) added 'plus carbons' to each of the four subtypes in recognition that these IDPs are generally more carbon-rich than CI carbonaceous chondrites.

Detailed TEM data are available for two IDPs in the 'layer silicate' IR class: (1) 'low-Ca' hydrated IDP (Tomeoka and Buseck 1984) and (2) hydrated IDP r21-M3-5A (Tomeoka and Buseck 1985) with high (+625.3 to +818.5) D/H ratios (Zinner et al. 1983) proving its extraterrestrial origin. This IDP contains large (2.5 μm) high-Al, low-iron fassaite (Ti-diopside) crystals with partially hydrated (smectite) amorphous material surrounding some pyroxenes. Tomeoka and Buseck (1985) linked this IDP with carbonaceous chondrites. The low-Ca IDP shares gross mineralogical similarities with CI and CM carbonaceous chondrite matrix. In detail there are differences, i.e. partially hydrated, amorphous ferromagnesiosilica materials with 'smectite-like' compositions (Tomeoka and Buseck 1984). It is a compact aggregate of pyrrhotite, low-Ni pentlandite and amorphous material. The IR properties place both IDPs among the hydrated IDPs but with petrological properties linking them to CI and CM carbonaceous chondrites.

Solar flare tracks and D/H anomalies in three IDPs of the 'olivine' IR class prove their extraterrestrial origin. They were studied by TEM (Christoffersen and Buseck 1986a). One IDP is an aggregate with matrix clumps plus subhedral single-crystal olivines (Mg = 0.93 to 0.98) (50 nm to 1 μm and 1 to 5 μm) and enstatite (0.1 μm). A second aggregate IDP has olivine grains (Mg = 0.39) (0.5 to 2.0 μm) embedded in a carbonaceous matrix with mostly sub-micrometer, rounded low-Ni pentlandite grains and lath-shaped Ca-clinopyroxene (Al_2O_3 = 3 to 4 wt %) (50-100 nm, and up to 200-300 nm). The third one is a compact non-chondritic IDP with chondritic material at the surface and consisting of (sub-) micrometer size olivines (Mg = 0.72), 5C pyrrhotite (0.1 to 0.5 wt % Ni) and rare sphalerite. One 'pyroxene' IR class particle is a highly porous aggregate of 'matrix clumps' and enstatite, diopside and fassaite (Christoffersen and Buseck 1986b) that are probably responsible for its IR signature.

Infrared classification: Limitations. Sandford and Walker (1985) found that some IR spectra indicate the presence of materials from both olivine and pyroxene classes. Secondary properties, e.g. an IR signature, conceivably impose apparent order among petrologically diverse chondritic IDPs. A drawback of IR classification is the seemingly exclusive nature of the major subgroups. As a result, the carbon-rich CP IDP W7029C1, wherein 11% of its constituents are poorly ordered, non-stoichiometric smectite and well-ordered kaolinite and talc (Rietmeijer and Mackinnon 1985a), was considered an anomalous IDP. Later TEM studies found that amorphous materials and small amounts of saponite occur in many 'anhydrous' chondritic IDPs (Zolensky and Lindstrom 1992) as well as non-stoichiometric plagioclase and alkali feldspars (Rietmeijer 1991). Gibson and Bustin (1994) reported the presence of water-bearing phases in carbon-rich aggregate IDP L2006B16 that on the basis of TEM analyses classified as an 'anhydrous' particle. These disparate classifications illustrate that textural and mineralogical heterogeneity affect IDP classification, particularly when a diagnostic mineral is present in small amounts. Classification became more confusing when Zolensky and Lindstrom (1992) decided that the mere presence of layer silicates is enough to classify an IDP as 'hydrous.' Small amounts of layer silicates are undetectable by IR spectroscopy of IDPs.

This practice exposed the weakness of classification that mixes properties of primary (olivine, pyroxene) and secondary (layer silicates) minerals. Another concern pertains to the very nature of the aggregate IDPs. Bradley et al. (1989) performed automated thin-film point count analyses of IDPs from the 'pyroxene' and 'olivine' IR classes. Cluster analyses confirmed the 'olivine' IR classification of IDP U220SP66, that is, 68% of data-points contributed to an olivine cluster. In IDP U222B22, ~40% of the data-points contributed to a primary cluster for enstatite. Pyroxene-rich aggregate IDPs contain both large single-crystal pyroxenes (up to ~1 μm) and smaller pyroxenes dispersed in the matrix (Bradley et al. 1989). The largest single-crystal grains may determine the IR classification of aggregate IDPs. Both IR and point-count classifications thus become an indicator of component mixing in chondritic aggregate IDPs, viz. single-crystal grains and matrix units. A similar situation determined classification of the IDP with the unique refractory minerals reported by Christoffersen and Buseck (1986b). IR spectroscopy is useful methods for a very broad subdivision of chondritic IDPs. It is also helpful to scan for phases with a strong and distinct infrared signature within IDPs such as carbonates (Sandford 1986a). An IDP with a strong carbonate absorption band at 6.8 μm in the IR spectrum was found by TEM analysis to contain large single-crystal carbonates. They were identified as rhombohedral grains (50 to 300 nm, in diameter) of breunnerite and magnesian siderite with minor calcium and manganese (Tomeoka and Buseck 1986).

What alternative? Particle classification is sophisticated enough to recognize the chondritic IDPs among collected particles but there is no satisfactory classification for these IDPs. The existing classification disconnects hydrated and anhydrous IDPs. It does not recognize the possibility of a continuum from hydrous to fully hydrated particles. The best subdivision is into CF and CP aggregate IDPs and chondritic smooth IDPs. These particles are typically carbon rich with ~2 to 3 times the CI abundance (Thomas et al. 1996). During October 3-10, 1989 and June/July, 1991, compact chondritic IDPs with a rough surface (CR IDPs) abounded in the lower stratosphere (Rietmeijer and Warren 1994). This morphology is distinct from silicate fragments with adhered chondritic material and CS IDPs. All three CM IDPs identified to data belong to the CR IDP subtype. Clearly, morphology and bulk composition of IDPs are not suitable for classification purposes. The type CI and CM IDPs do not show the aggregation textures that are so typical for chondritic aggregates. This observation suggests that other physical particle properties should be considered for chondritic IDP classification.

Porosity

The porosity was determined using image analyses techniques. The most comprehensive data set shows a range from 0 to 46% with two distinct peaks at 0% and 4% (Corrigan et al. 1997). A statistical treatment of the data in Table 10, excluding the mode at zero percent, shows a Gaussian distribution (at a 95% confidence limit) with a mean = 11.4% porosity. The distribution drops off sharply at 4% to ~50% of the modal abundance. Assuming a solid matter density of 3.5 g cm^{-3} for a chondritic particle (cf. Love and Brownlee 1994), the average porosity than corresponds to an average particle density of 3.1 g cm^{-3} and the modal value yields 3.4 g cm^{-3}. Some CP IDPs are fractal aggregates with a calculated density ranging from 0.08 to 0.14 g cm^{-3} (Rietmeijer 1993a), which corresponds to 98 to 96% porosity of these IDPs.

Table10. Porosity of chondritic IDPs

	IDP SUBTYPE	POROSITY (%)	SOURCE
U2-22-B28	Aggregate	98.7-82.6	Mackinnon et al. 1987
No data	Aggregate	91.63	
No data	CS	69.75	
Clu17 (#)	Aggregate	12	Strait et al. 1996;
U2022E17	Serpentine, CS	21	(#) Thomas et al. (1995a)
W7027E11	Serpentine, CS	15	
W7027A17	Smectite, CS	7, 14	
L2007-3	Smectite, CS	5	
L2001-8	No data	3	Corrigan et al. 1997
L2001-15		0	
L2001-16		0	
L2005-10		4	
L2005AA1		0	
L2005AA3		2	
L0225AA4		0	
L2005C32		27	
L2005C37		4	
L2005D32		7	
L2005E36		4	
L2005F39		4	
L2008J1		14	
L2005J4		4	
L2005K9		0	
L2005O10		3	
L2005V1		14	
L2005V13		9	
L2005Y6		1	
L2005Z3		9	
L2005Z6		10	
L2005Z8		5	
L2005Z9		3	
L2005Z13		0	
L2006-1		39	
L2006D13		7	
L2009J2		12	
L2011O10		6	
L2011R3		46	
LAC-1		2	

Density

Density (ρ) was obtained by several techniques. Direct weighing the particle on a fiber balance (Fraundorf et al. 1982c), synchrotron X-ray excitation (Flynn and Sutton 1991b), an EDS-calibrated neutron activation method (Zolensky et al. 1989b), and TEM determination of IDP mass relative to an iron standard and volume (Love et al. 1994). The measured densities range from 0.3 g cm^{-3} to 4.3 g cm^{-3} (Table 11). Sutton et al. (1991) mapped density variations within an IDP (average density = 2.4 ± 0.7 g cm^{-3}) and found sub-micrometer regions with $\rho \cong 5$ g cm^{-3} (probably sulfide- or metal(-oxide)-rich domains). Variations in bulk density among aggregate IDPs correspond, at least in part, to variations in the proportions of their constituents. The skewed Gaussian distribution for a large database obtained by Love and Brownlee (1994) is markedly different from the bimodal distribution reported by Flynn and Sutton (1991b) on a much smaller data set. The particles measured by Love and Brownlee (1994) were biased against highly porous IDPs but this bias causes only a small effect as only an estimated <15% of the total IDP population was missed. The mean density obtained by Love and Brownlee (1994) matches

Table 11. Mass, size and density of chondritic IDPs

IDP	SUBTYPE	SIZE (μm)	MASS (pg)	DENSITY (g cm^{-3})
U13-3-02	?	7	690	2.2
U13-3-25	?	7	620	1.7
U13-4-2a	?	11	2800	2.2
U21-5-2b3	?	4	92	1.2
U18-6-01	?	8	1300	2.2
U15-7-07	?	8	460	1.0
U23-4-4a	?	15	2500	0.7
W7029*A27	C	10	No data	1.6
U2015G1	C	17 x 20 x 20	No data	1.6
W7027C5	C	22 x 30 x 35	No data	>0.5
U2022G17	C	8 x 8 x 14	No data	0.4
W7013H17	C	12 x 12 x 15	No data	0.8
W7013A11	C	17 x 12 x 26	No data	1.7
U2022G2	C	15 x 7 x 15	No data	2.0
U2022B2	C (low-Ni)	12 x 8 x 15	No data	0.8
U2022C18	C	11 x 9 x 14	No data	0.9
U2001B6	C? (low-Ni)	Ø=25 x 15	No data	0.6
W7066*A4	C (low-Ni)	7 x 6 x 8	No data	1.7
U2022G1	C	30 x 20 x 15	No data	0.7
U2015F1	CS	No data	4550	2.0
U2015E10	CF	No data	16300	0.5
W7029*A2	CP	No data	209	0.5
W7029*A3	CP	No data	2160	0.5
W7013A8	CF	No data	2080	0.4
LAC1	CF	No data	45200	3.4
"Chondritic" (*)	not specified	5-15	No data	Mean = 2.0 Mode = 1.8 Range = 0.3-4.3

Sources: Fraundorf et al. 1982c, Flynn and Sutton 1991b, Zolensky et al. 1989b. (*) density of 81 black chondritic IDPs on collectors U2-30 and U2012; one chondritic particle (6.5 g cm^{-3}) was probably dominated by iron-nickel metal (Love et al. 1994). The Fe/Ni ratios of particles U2015G1 through U2022G1 are presented in Table 9.

the average of the population from Flynn and Sutton (1991b) with ρ = 1.4 to 2.2 g cm^{-3}. The densities <1.4 g cm^{-3} form a skewed Gaussian distribution with a mean of 0.7 g cm^{-3}, which corresponds to 80% porosity. The mean density, ρ = 2.0 g cm^{-3} (Table 11), is consistent with 43% porosity. The porosity support three groupings, (1) 0-53%, (2) 70-80%, and >90% porosity. The highly porous CP IDPs are indeed rare. For comparison, matrix porosity of hydrated CI and CM chondrites ranges from 1% to 13 % (Corrigan et al. 1997).

Aggregate and collapsed aggregate IDPs

The aggregate nature of chondritic IDPs is the key to appreciate the histories of these unique materials. Aggregate IDPs consist of matrix clumps of nanometer sized iron-magnesium silicates, Fe,Ni-sulfides and oxides in an amorphous matrix, and large silicate and sulfide grains (Table 3). The pyroxenes include distinct enstatite single-crystals (Fig. 12). Similar matrix clumps, enstatite (0.5 × 0.2 μm) and FeS-grains occur in a compact (porosity <30%) hydrated IDP along with patches of a Fe-rich, ferromagnesio-silica glass' and smectite, Mg = 0.45 (Fig. 13) (Bradley 1988). This compact particle is a collapsed aggregate. Not every compact IDP is a hydrated particle. The IDP L2005Z17 is a compact particle but it is an anhydrous IDP with large anhedral low-Ca pyroxene crystals. And, IDP L2011R3 is a porous (~22%) saponite-type IDP with abundant low-Ca pyroxene

Figure 12. Bright-filed (left) and dark-field (right) transmission electron micrographs of a bladed enstatite grain embedded in the matrix of an aggregate IDP. Scale bars are 0.1 micrometer. After Fraundorf and Shirck (1979) Proc Lunar Planet Sci Conf 10:951-976, Fig. 3a, p. 958. Courtesy of Phil Fraundorf.

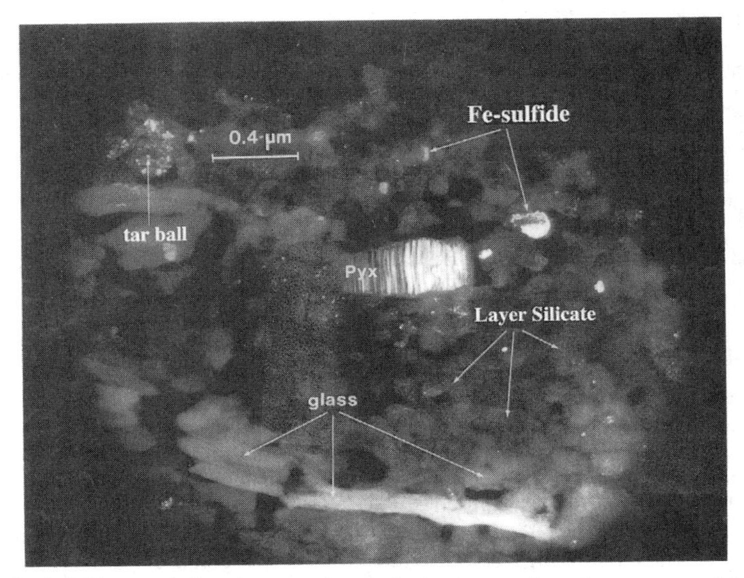

Figure 13. Dark-field transmission electron micrograph showing a collapsed aggregate particle. Rounded matrix units are still visible. Embedded in the matrix are GEMS (formerly tar balls), iron-rich glass, Fe-sulfides and a bladed enstatite grain. After Bradley (1988) Geochimica et Cosmochimica Acta 52, 889-900, Fig. 7, p. 896.

(Zolensky and Barrett 1994, Figs. 1d and 1f). These authors used the classification by Zolensky and Lindstrom (1992) that a hydrous IDP only requires the mere presence of a layer silicate grain. Still, both IDPs illustrate that secondary minerals are not unique to collapsed aggregates and embedded pyroxenes bear no relationship to the post-accretion particle history. The same matrix units occur in both anhydrous and hydrated IDPs (Zolensky and Barrett 1994), confirming similar observations made by others.

The grouping of chondritic IDPs according to porosity supports a model of gradual collapse of fluffy aggregates into compact particles. The data cannot confirm whether CP IDPs are truly unique particles, or whether they are precursors to the more common CF IDPs with the same matrix clumps of ultrafine-grained constituents (Fraundorf 1981, Bradley 1988, Rietmeijer 1989a). The aggregate texture is the result of mixing matrix clumps, large silicates and sulfides in variable proportions. In this model, mineral grain sizes could correspond to episodic aggregation (Fraundorf et al. 1982a, Bradley 1988). Rietmeijer (1989a, 1992b, 1996b) suggested that the ultrafine-grained silicates in the matrix clumps could be secondary phases due to dynamic pyrometamorphism.

In an attempt to reconcile various chondritic IDP classifications, Rietmeijer (1994a) proposed a classification that is based on the properties of identifiable textural entities in these particles. They can be classified on the basis of openness of their structure as (I) aggregate IDPs and (II) collapsed aggregate IDPs. In highly porous aggregates the constituents are loosely bound (CP IDPs) (Fig. 5). Most aggregates are more compact with little pore space (CF IDPs). The collapsed aggregate particles have a typically smooth surface and are known as CS IDPs but they show at least vestiges of the accretion aggregate texture (Fig 13). Changes in the properties of these entities trace the evolution of the particles from accretion through post-accretion alteration to parent body modification. This classification has its problems too but it predicts that the matrix units were initially amorphous materials. Accretion textures survive in chondritic aggregates. The porosity reflects the accretion process, post-accretion alteration at scales comparable to the size of individual aggregates, or both.

MODIFICATIONS OF IDPS

Solar System sojourn of IDPs

Once ejected from their parent body, IDPs are exposed to the space environment wherein they are subjected to the bombardment by energetic particles that causes chemical and mineralogical alteration (known as space weathering). The 50-nm thick polycrystalline rim of magnetite on an iron-nickel alloy grain in IDP U219C2 due to solar wind particles is an example of this type of alteration (Fig. 14; Bradley 1994a). Olivine, enstatite and diopside crystals in IDPs are damaged by particles of energies above 100 eV. Heavy particles (i.e. Fe) leave tracks at >0.5 MeV/nucleon. These interactions leave a track along which the crystal lattice is destroyed (Fig. 19) (Bradley et al. 1984a). The density of tracks in a grain depends on the time that the IDP is exposed to the space environment, its distance from the sun and the flux of solar flare nuclei. Track densities are on the order of 10^{10} to 10^{11} cm^{-2} (Thiel et al. 1991). The orbits and mean heliocentric distances of asteroids and comets are different and IDPs from these sources follow different trajectories due to PR drag towards the sun. As a consequence, the track density in grains from these sources are different and can constrain their origin (Sandford 1986b). In many IDPs crystalline grains have a rim of ~10 nm up to ~200 nm thick that can be amorphous or polycrystalline due to radiation damage (Bradley 1994a, Bradley and Brownlee 1986, Christoffersen and Buseck 1986a, Thiel et al. 1991). This rim may recrystallize during atmospheric entry heating (Rietmeijer 1996b). A rim >200 nm on a single-crystal grain with a very high solar flare

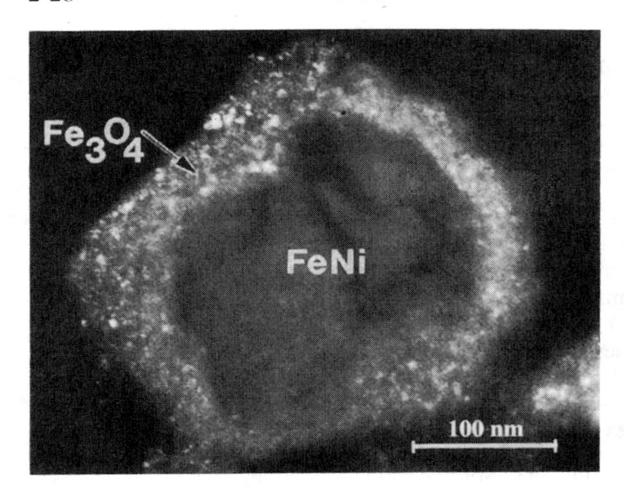

Figure 14. Dark-field transmission electron micrograph of a kamacite grain decorated by a magnetite rim in IDP U219C2 due to energetic particle bombardment in space. After Bradley (1994) Geochimica et Cosmochimica Acta 58, 2123-2134, Fig. 1d, p. 2124.

track density is believed to be diagnostic of comets from the Kuiper belt (Flynn 1996a). On their way to an Earth crossing orbit, IDPs experience solar heating but the induced temperatures are probably too low (<300°C) to cause significant thermal alteration although light hydrocarbon materials may suffer polymerization. In any case, the thermal effects incurred during solar system sojourn are likely to be overprinted during atmospheric entry heating.

Dynamic pyrometamorphism

Dynamic pyrometamorphism is thermal alteration that occurs at a very short (seconds) time-scale with very steep dT/dt gradients during heating and quenching. Lightning strike induced alteration is a terrestrial example of this process. The minimum velocity at which IDPs enter the Earth's atmosphere is 11.1 km s^{-1}. Due to collisions with the air molecules between 100-80 km altitude they decelerate whereby the kinetic energy is transformed into thermal energy. As a result they are flash-heated for 5 to15 s. A micrometeoroid (10 to 15 μm in diameter) will typically establish thermal equilibrium between its interior and its surface. A fraction of the incoming particles evaporates (shooting stars) and other particles survive but are melted with accompanying mass loss (they form quenched-liquid spheres). Still others survive apparently intact as recognizable chondritic aggregates and non-chondritic IDPs. Love and Brownlee (1991) reviewed the physical parameters that control flash heating and calculated the resulting time-temperature profiles in IDPs. Briefly, these properties are particle size, density, velocity, and entry angle. The average entry angle for IDPs is ~45°. Thus, to a first approximation, the major variables determining the thermal profiles in collected stratospheric particles <100 μm become particle mass, size and velocity. Love and Brownlee (1994) derived a simple correlation between the peak heating temperature and entry velocity of incoming IDPs as a function of particle mass (Fig. 15). Typical peak heating temperatures of cometary dust are ~300°C higher than for slower moving asteroidal particles of the same size and density (Love and Brownlee 1991, 1994). It is in principle possible to subdivide dust from these sources on the basis of the maximum entry temperature. Dynamic pyrometamorphism occurs with varying levels of intensity in all particles even though they show no outward sign of melting and ablation (sphere formation). This event may overprint or erase evidence of earlier mineral features.

Alteration. Dynamic pyrometamorphic alteration includes a number of reactions that are generally characterized by oxidation and volatile element loss. The chondritic IDPs

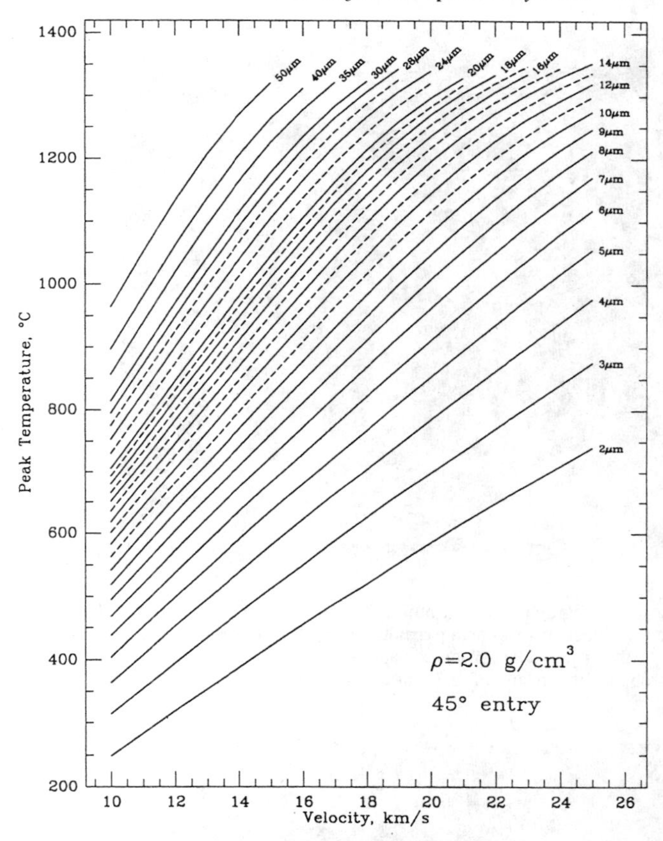

Figure 15. Calculated peak atmospheric entry temperatures for micrometeoroids as a function of velocity for particles with a density 2 g cm^{-3} entering the earth's atmosphere at a 45° angle. The size-density product is constant and the graph can be used for any reasonable density of particles. [Used by permission of the editor of Meteoritics, from Love and Brownlee (1994), Fig 1, p. 70.]. Courtesy of Stan Love.

typically contain carbonaceous materials that during flash heating can contribute to reducing conditions that may be very localized within a particle. Some forms of alteration take a readily recognizable form such as an iron oxide rim around an IDP. In general, this particular thermal alteration of IDPs is not well studied in detail. Changes that are believed to be dynamic pyrometamorphic alteration are,

- Oxidation of Mg,Fe-silicates that results in magnetite nanocrystal decorations of these grains (Fraundorf 1981, Germani et al. 1990). Ablation and oxidation of silicates and sulfides along the particle perimeter produce a generally discontinuous polycrystalline magnetite or maghemite rim of 50- to150-nm thick (Fig. 16) (Germani et al. 1990; Keller et al. 1992b, 1993). Maghemite rims are especially well-developed on pentlandite and pyrrhotite along the particle perimeter due to oxidation and sulfur loss resulting in the formation if vesicles underneath the rim (Rietmeijer 1992c). Small iron-oxide crescents develop on ferromagnesiosilica materials along the perimeter (Fig. 17) (Rietmeijer 1996b, Thomas et al. 1995a).
- Topotactic intergrowths of magnetite plates in, and on, olivine with $[111]_{mt} \parallel [010]_{ol}$ and $[110]_{mt} \parallel [100]_{ol}$, narrow coherent laihunite lamellae in Fe,Mg-olivine, $Fe^{2+}_{(2-3x)}Fe^{3+}_{2x}SiO_4$ with $0.24 < x < 0.37$ (Keller et al. 1992b), cellular intergrowths of maghemite in Fe-Mg-rich olivine, co-precipitation of andradite-rich garnet, $Mg_{0.9}Fe^{2+}_{1.1}Ca_{1.2}(Fe^{3+}_{1.4}Al_{0.6})(Al_{0.3}Si_{2.7})O_{12}$, and formation of Ca-bearing olivine, $Mg_{1.2}Fe_{0.7}Ca_{0.1}SiO_4$ (Rietmeijer 1996c).

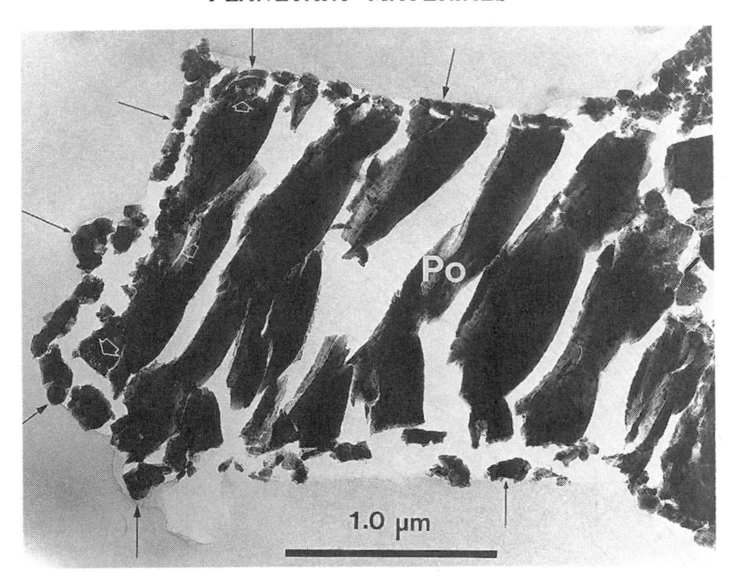

Figure 16. Transmission electron micrograph of an ultrathin section of stratospheric particle L2005E40 showing a continuous iron oxide rim (black arrows) on a pyrrhotite crystal with vesicles due to sulfur loss (open white arrows) underneath the rim. The shattered appearance of the section is typical in ultramicrotomed IDPs and is probably the result of thermal stresses induced during atmospheric entry. (Rietmeijer, unpublished data).

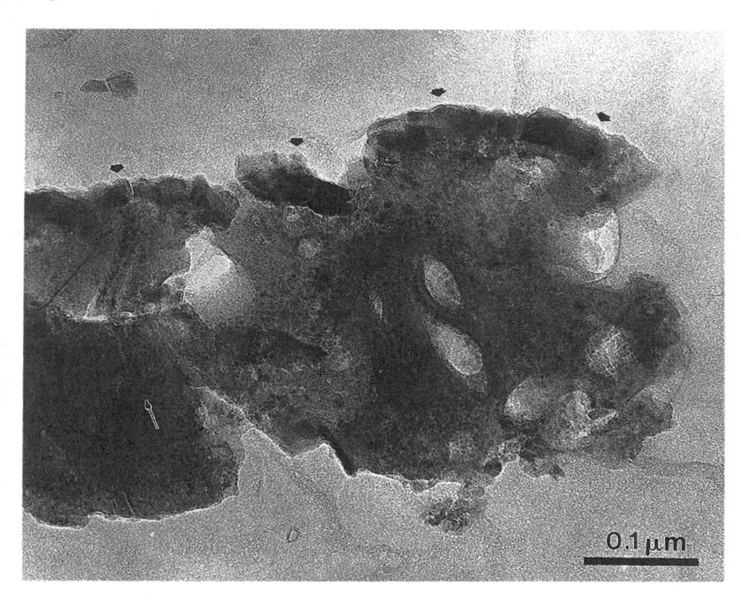

Figure 17. Transmission electron micrograph of an ultrathin section of stratospheric particle L2005B22 with large vesicles and a narrow, discontinuous iron oxide rim (black arrows) and flow structure (black and white arrow). The gray background is the embedding epoxy. Modified after Rietmeijer (1996) Meteoritics and Planetary Science 31:237-242.

- Oxidation of metallic bismuth (Mackinnon and Rietmeijer 1984) and decomposition of bruennerite into magnesiowüstite, periclase and CO_2 (Keller et al. 1996a).
- Carburization in the particle interior and formation of (ε-carbide, $(Fe,Ni)_3C$, an orthorhombic Fe,Ni-carbide with a superlattice structure indicating interstitial ordering of carbon, cohenite, and FCC carbide (austenite) (Fraundorf 1981, Christoffersen and Buseck 1983a, Bradley et al. 1984b). Rounded carbide grains have a rim of amorphous carbon with graphite lattice fringes (0.34 nm) parallel to the interface. This ordered zone is ~2.5 nm wide (Bradley et al. 1984b). The carbides (500 to 1000 nm in size) can be associated with kamacite (low-Ni, Fe,Ni alloy) grains. Bradley et al. (1984b) reported rare filamentous carbon forming tubules up to several micrometers in length CP IDPs. Some tubules were amorphous carbon, others also contained graphitic carbon. Most tubules contained at least one embedded iron-rich mineral grains, i.e. Fe,Ni-carbide, magnetite, or pyrrhotite. This particular unique carbon morphology is associated with Fisher-Tropsch reactions.
- Annealing and erasure of solar flare tracks in constituent iron-magnesium silicate crystals (Fraundorf et al. 1982d, Sandford and Bradley 1989).
- Degradation of lattice fringes in layer silicates (Zolensky and Lindstrom 1992). Atmospheric entry heating simulation experiments found that layer silicates in IDPs decomposed at 810°C into a material with an infrared signature suggestive of pyroxene (Sandford and Bradley 1989). Olivines and pyroxenes should be considered as possible products of severe heating in IDPs dominated by serpentine or smectite. Formation of vesicular lattice fringe textures and development of lattice fringes >0.7 nm in serpentine (Rietmeijer 1996a, Zolensky and Lindstrom 1992). These features define the so-called 'intermediate phase' that forms during thermal decomposition of serpentine in to olivine plus pyroxene (Akai 1988).
- Local melting, fusion and devolatilization of the matrix (Germani et al. 1990) such as diopside plus ferromagnesiosilica material reacting to a vesicular, amorphous Fe^{3+}-rich, Ca-pyroxene-like material (Rietmeijer 1996d).
- Loss of sulfur and zinc (Flynn and Sutton 1992a), magnesium from amorphous matrix materials (Rietmeijer 1996a) and helium (see below).
- Structural and morphological changes are subtler. Christoffersen and Buseck (1983b, 1986b) reported enstatite single-crystals with irregular strain contrast in bright-field TEM images, bending of (100) lattice fringes, broadening of diffraction maxima, and untwinned clinoenstatite with (100) stacking faults. Many TEM studies reported mottled textures in single-crystal olivines, pyrrhotite and pentlandite in heated IDPs. It remains unproven but it is possible that thermal stresses during flash-heating and ultra-rapid quenching are responsible for these structural features. The very nature of dynamic pyrometamorphism favors kinetically driven rather then thermodynamically controlled reactions. Hence, grain size (surface free energy) becomes a controlling parameter. Nanometer size iron-nickel sulfides in many IDPs occur as rounded grains while larger grains in the same particles have angular (sub- to anhedral) shapes (Christoffersen and Buseck 1986a,b; Rietmeijer 1996a). The carbide-coated teanite droplets (<50 nm) could be quenched melts (Fraundorf 1981, Fig. 5). If melting is accompanied by loss of sulfur, it may result in the ultrafine-grained polycrystalline 'spotted' magnetite grains (Fraundorf 1981, Blake et al. 1988) and dense masses of magnetite grains <50 nm (Christoffersen and Buseck 1986b) in the matrix of IDPs. Thin (<100nm) sections of IDPs for TEM studies are prepared by ultramicrotomy. Oftentimes portions of the constituent grains shatter in a glass-like manner during sectioning (e.g. Thomas et al. 1995a, Fig. 7a; Rietmeijer 1996b, Fig. 7; Klöck and Stadermann 1994, Fig. 11). This common behavior is probably caused by thermal

stresses during flash heating and quenching. Quench rates based on the time-temperature profiles calculated by Love and Brownlee (1991) are estimated at 10^5-10^6°C per hour. Despite these ultrahigh rates, Mg,Fe-silicates and iron-oxides (3-26 nm in size) nucleated and grew (almost no diffusion) in the quenched ferro-magnesiosilica matrix of a melted IDP in less than 10 s (Rietmeijer 1996b).

Flynn (1996a) submitted that mineralogical and chemical indicators of thermal alteration should show a correlated behavior in a particle. Unfortunately the mineralogical database is too small to investigate correlated behavior but it is not a priori evident that these correlations should exist. For example, distinct iron oxide rims on IDPs are associated with Fe,Ni-sulfides whereas only thin rims develop on ferromagnesiosilica materials. The iron oxide nanocrystals are limited to Mg,Fe-silicates. The nature of the newly formed iron oxides is a function of the relative proportions of matrix and micrometer size pentlandite, pyrrhotite and Mg,Fe-silicate grains. The relative proportions of these constituents probably also control volatile element loss. This is a rather simple-minded explanation of a probably complex set of parameters that include particle size (individual IDPs or cluster IDP fragments) and pre-entry properties.

Zinc loss. The carrier phases of zinc in chondritic IDPs that may control the behavior of this element during dynamic pyrometamorphism are uncertain. Still, the low zinc contents ($<0.1 \times$ CI) of a subset of chondritic IDPs are probably a good indicator of thermal alteration (Fig. 18). All particles with zinc depletion have a (partial) magnetite rim to confirm atmospheric entry heating (Flynn et al. 1993b). This finding is qualitatively solid but the fact remains that the zinc host phases are poorly known. Sphalerite is a candidate but it is a rare phase in chondritic (Mackinnon and Rietmeijer 1987) and non-chondritic IDPs (Christoffersen and Buseck 1986a). Zinc occurs in matrix units of IDPs (Rietmeijer 1989a, Klöck and Staderman 1994) and in smectite proto-crystallites in the matrix of hydrated IDPs (Thomas et al. 1992a) and a CS IDP,

$$[Mg_{1.31}Fe_{1.30}Zn_{0.06}](Si_{3.42}Al_{0.58})O_{10.12} \text{ and } [Mg_{1.56}Fe_{1.02}Zn_{0.06}](Si_{3.45}Al_{0.55})O_{10.12}$$

(all Al in tetrahedral sites) (Blake et al. 1988). These observations offer the possibility that zinc is a persistent trace element (but generally below the EDS detection limit) in aggregate

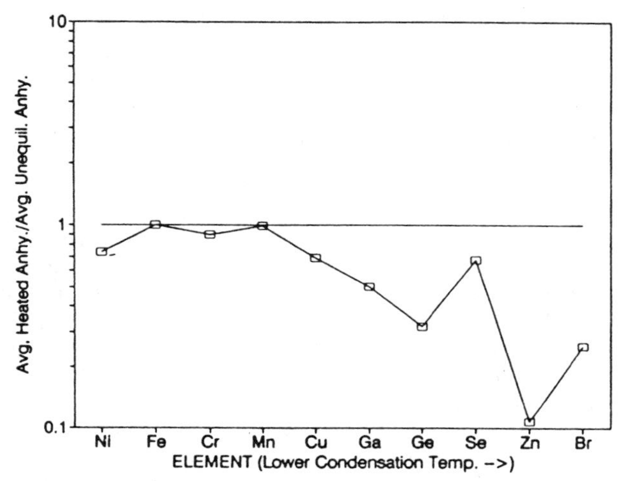

Figure 18. Average element abundances in eight severely heated IDPs normalized to these elements in five unequilibrated aggregate IDPs showing distinct depletions in the volatile element abundances due to atmospheric entry heating. [Used by permission of the series editor of the American Institute of Physics, from Flynn (1994), Fig 2, p. 139.]. Courtesy of George Flynn.

IDP matrix, which alleviates the restriction that zinc is uniquely present in sulfides. If it is a common matrix element, the abundance of this element in chondritic IDPs may be developed into a quantitative indicator of dynamic pyrometamorphism.

Solar flare tracks. Solar flare tracks occur as linear defects in micrometer size olivine and pyroxene grains in IDPs that were exposed to space during solar system sojourn (Fraundorf et al. 1980). Fraundorf et al. (1980, 1982d) argued on the basis of theory and experiment that silicates in IDPs should record observable solar flare tracks. Bradley et al. (1984a) successfully located the first solar flare tracks in olivine (Mg = 0.40-0.45) grains in two IDPs (Fig. 19) and they have since been found in many IDPs. The track density, 10^{10} to 10^{11} cm^{-2}, is consistent with exposure ages in interplanetary space of ~10^4 years, which are similar to calculated Poynting-Robertson lifetimes. Klöck and Stadermann (1994) and Zolensky and Lindstrom (1992) noted that in some IDPs only one or two silicate grains contain tracks. The penetration depth of solar wind ions is about one micrometer and the presence of solar flare tracks supports that the host phase was not a fragment of a larger particle. Still, many IDPs do not contain solar flare tracks in their constituent silicate minerals. The absence of tracks is no evidence that the particle was not in interplanetary space. It could have been shielded from space exposure within a very large particle, or it experienced thermal alteration. Flash-heating experiments show that tracks remain visible to TEM analysis up to ~600°C (Fraundorf et al. 1982d). The absence of these tracks could be due to dynamic pyrometamorphic alteration above ~600°C. Solar flare tracks in an IDP confirm its extraterrestrial origin and broadly constrain its maximum peak heating temperature <600°C.

Peak heating temperatures. Peak temperatures of atmospheric entry heating are recorded in mineral reactions and assemblages, and by measured helium release temperatures. Low, 200°-300°C, temperatures are indicated by metallic bismuth oxidation (Mackinnon and Rietmeijer 1984), layer silicate lattice fringe deterioration (Zolensky and Lindstrom 1992) and by ε-carbide formation (Christoffersen and Buseck 1983a). The

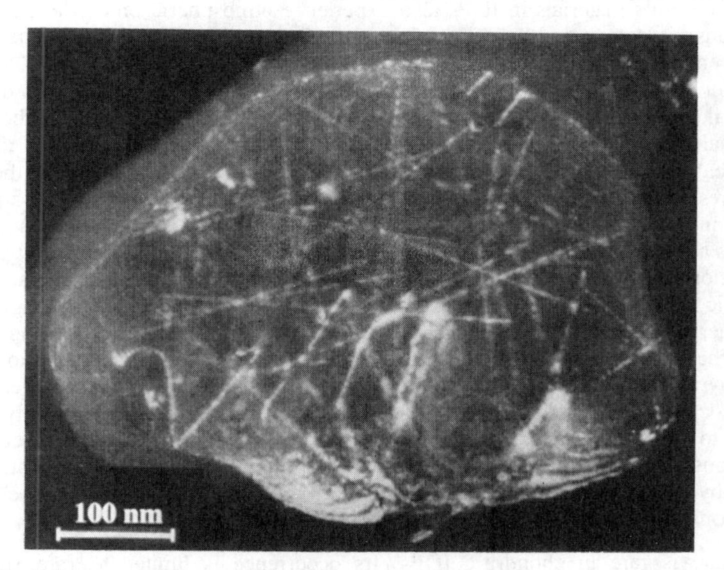

Figure 19. Dark-field transmission electron micrograph of solar flare tracks (white linear features) in an olivine grain ($Mg_{1.2}Fe_{0.8}SiO_4$) from IDP U220B11. After Bradley (1994) Geochimica et Cosmochimica Acta 58:2123-2134, Fig. 1b, p. 2124.

formation of amorphous carbon filaments occurs below 525°C, or above 525°C when graphitic carbon is present (Bradley et al. 1984b). A temperature of ~1150°C is inferred from cellular intergrowths of maghemite in olivine in a (partially) melted IDP (Rietmeijer 1996c). The He-release temperatures show that most unmelted IDPs reached a maximum temperature between 485°C and 1020°C (Nier and Schlutter 1992, 1993). The ^{4}He contents in fifteen chondritic IDPs ranged from 0.0008 and 0.28 cm^3 (STP) g^{-1} (average: 0.054 ± 0.02 cm^3 (STP) g^{-1}) (Nier and Schlutter 1990), which is similar to earlier measurements (Table 2). The measured helium content of an IDP is the helium that was retained after peak heating. Determining the temperature at which the residual helium is lost yields the maximum entry temperature (Nier and Schlutter 1992, 1993, Brownlee et al. 1995).

Endothermic reactions. Endothermic reactions in IDPs act as sinks of heat generated during atmospheric entry. The resulting dynamic pyrometamorphic temperatures will be lower than without the presence of a thermal sink. In rare cases, endothermic reactions protect minerals from thermal alteration. The loss of sulfur from iron-nickel sulfides in reactions producing iron oxides is endothermic till about 500°C (Rietmeijer 1993b). The role of endothermic mineral reactions in dynamic pyrometamorphic alteration of IDPs is still unappreciated, including in the physical models of atmospheric entry heating. Recently, Flynn (1995) showed that endothermic reactions could cause significant thermal gradients in a particle. Both sulfides and carbonaceous materials are common in chondritic IDPs. The latter includes polycyclic aromatic hydrocarbons, PAHs (Allamandola et al. 1987, Clemett et al. 1993), unsaturated carbons (Radicati di Brozolo and Fleming 1992), hydrogenated amorphous carbons with different structural disorder detected by Raman spectroscopy ('glassy' or turbostratic carbons) (Wopenka 1988). Also, amorphous, turbostratic and poorly graphitized carbons based on TEM studies (Rietmeijer 1992b,d; Rietmeijer and Mackinnon 1985b). The ordering detected by Raman spectroscopy is an indigenous feature due to intrinsic differences in chemical compositions of the amorphous carbons (Wopenka 1988).

Carbonaceous materials in IDPs also experience atmospheric entry flash heating but their response is determined by their unique thermodynamic properties. That is, the high activation energies of carbonization (293 kJ mole^{-1}) and graphitization (1087 kJ mole^{-1}) (Bonijoly et al. 1982) will ensure a large degree of survival of these materials during this thermal spike. Rietmeijer and Mackinnon (1985b) submitted that changes in the graphitic carbon basal spacing are a thermometer for thermal alteration at several hundred degrees on a 'geological time scale' (Fig. 20). Kinetic factors for this alteration increase the reaction temperatures from 2000° to 3000°C at a time scale of a few hours. Unless favorable hydrocarbon compositions or structures overcome the kinetic barriers, it is unlikely that indigenous amorphous and poorly graphitized carbons in IDPs show significantly dynamic pyrometamorphic alteration. The PAHs in IDPs are affected by thermal alteration during atmospheric entry. They undergo evaporation and in situ re-condensation, fragmentation, and polymerization to higher molecular weight, less volatile, components (Thomas et al. 1995a). The endothermic nature of these reactions favors the very conditions during dynamic pyrometamorphism and they are of considerable interest. For example, D/H 'hot spots' associated with carbonaceous materials in chondritic IDPs are highly enriched compared to D/H values in carbonaceous chondrites. The values in IDPs (cf. isotopic compositions of IDPs) could be indigenous but the possibility of mass fractionation during dynamic pyrometamorphism as a function of composition and structure of light hydrocarbons in IDPs deserves consideration.

Graphite is rare in chondritic IDPs. Its occurrence is limited to rare filamentous carbons (Bradley et al. 1984b) and to narrow (<10 nm) rims surrounding grains embedded in amorphous carbon material (Mackinnon and Rietmeijer 1987) that occur mostly on small

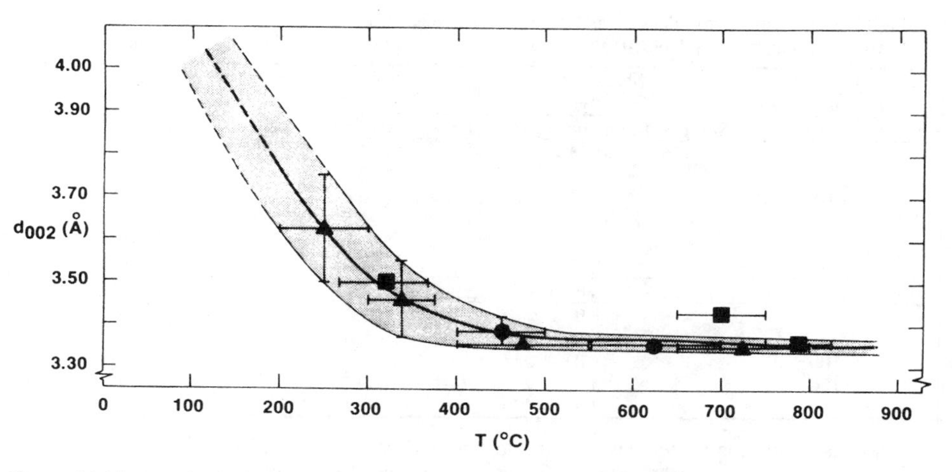

Figure 20. The poorly graphitized carbon cosmothermometer of average d_{002} interlayer spacings (Å) based on terrestrial metamorphic occurrences of these carbons (depicted by the symbols and error bars within the shaded area). [Used by permission of the editor of Nature, from Rietmeijer and Mackinnon (1985), Fig 2a, p. 735.]

small (<0.1 µm) embedded metal and metal oxide grains. The ~0.34 nm lattice fringes can be partially ordered (Bradley 1994a, Bradley et al. 1984b) or well ordered (Mackinnon et al. 1985) structures. The graphite rims are stabilized by the surface free energy of the enclosed grains.

Stratospheric contamination of IDPs

Residence time. Gravitational settling (cm s^{-1}) and circulation currents determine the residence time of IDPs from the time of deceleration at 100-80 km altitude till the time of collection at 17-19 km altitude. Most IDPs >2 µm in size are non-spherical particles (Fraundorf 1981, Flynn and Sutton 1991b). Their settling rate is a function of particle diameter, shape, density, and the physical properties of the atmosphere as a function of altitude. In general, IDPs fall rapidly (1 to 2 days) from ~80 km to 30-45 km altitude. In the upper stratosphere they slow down because they no longer behave as spherical bodies in an almost vacuum. The calculated residence times of IDPs, with size and density measured by Flynn and Sutton (1991b), range from 21 days to 224 days (Table 12). Also listed are calculated residence times for hypothetical aggregates to illustrate the effects of particle size, shape and density. Large aggregates rapidly fall through the stratosphere but they expose a considerable amount of surface for potential aerosol contamination.

Contamination mechanism. Contamination of IDPs with terrestrial particles >2 µm is probably not a serious problem (Brownlee et al. 1977b) but a similar statement for natural and anthropogenic aerosols <1.0 µm is most likely incorrect. During May 1985, the upper stratosphere over the southern US contained an unanticipated high abundance of volcanic dust <8.0 µm; mode <1.0 µm (Rietmeijer 1993c). Contamination occurs by sticking when IDPs collide with aerosols in the atmosphere. Zolensky and Mackinnon (1985) calculated that the collision frequency of a 10 µm particle with particles 1 µm in diameter is ~5×10^9 years. For particles 0.1 µm in diameter it is calculated at ~10^8 years. Thus, contamination with particles >0.1 µm is non-existent under ambient conditions. After a catastrophic dust injection, such as a major volcanic eruption, the enhanced particle number densities result in higher collision frequencies. Particle number densities typically

Table 12. Size, shape factor, density, and calculated residence time in the stratosphere for collected IDPs and hypothetical aggregates. Note: *a*, *b* and *c* are orthogonal particle dimensions whereby *a*>*b*>*c*.

Size (μm), Da = (*a*+*b*+*c*)/3	Shape factor, F = (*b*+*c*)/2*a*	Density, g cm^{-3}	Time (days)
Sources: Rietmeijer 1993c; Rietmeijer and Mackinnon 1997			
18	0.56	1.7	21
22	0.58	0.7	35
10	0.57	0.4	224
4	0.43	1.8	241
8	0.83	0.45	214
9	0.63	0.36	266
Hypothetical Aggregate IDPs			
60	1.0	0.01	267
60	1.0	0.1	27
100	1.0	0.1	8
100	0.5	0.1	14
100	0.9	0.01	86

increase as a function of decreasing particle size and contamination with aerosols <100 nm is an ever present concern. Also, the collector surfaces are often coated with sulfuric acid and huge numbers of nanometer-size stratospheric dust particles. When an IDP is collected it may impact an already contaminated surface. Chemical contamination is primarily a surface phenomenon but mineral aerosols may be more difficult to recognize. Arguments to appreciate their indigenous or extraneous origin in aggregate IDPs could be less straightforward. Mackinnon and Rietmeijer (1984) found platy single-crystal Bi_2O_3 and $Bi_2O_{2.75}$ grains (0.25 to 0.5 μm) in an aggregate IDP. A narrow amorphous carbon rim on these grains, the cosmic abundance of bismuth, and no documented volcanic sources for these oxides were the arguments to accept that they were dynamic pyrometamorphic oxidation products of indigenous metallic bismuth. Rietmeijer and Mackinnon (1997) noticed that similar bismuth oxides occurred in the upper stratosphere at ~35 km altitude during May 1985, and that their presence in three IDPs correlated with the times of major volcanic eruptions. They argued from chemical abundances associated with volcanic activity that nanometer size Bi_2O_3 and $Bi_2O_{2.75}$ grains might have a volcanic origin. Other data than those obtained by AEM and TEM analyses are required to decide the origin of these metal oxides in chondritic IDPs. This example serves to illustrate that a full understanding of dust contamination in the stratosphere is not yet possible because we lack details on the types and abundances of potential chemical and mineral contaminants in the stratosphere, and their sources.

Stratospheric contaminants. Stratospheric contaminants on collected IDPs are listed in Table 13. Small silica shards abound in volcanic eruption clouds. Their long residence times enhances the probability that they may contaminate dust settling in the lower stratosphere (Rietmeijer 1988). The other contaminants are species of volatile elements. Sulfur derives from the sulfate aerosol layer in the upper atmosphere and the generally high sulfur contents of the troposphere and lower stratosphere due to volcanic activity. High levels of chlorine and bromine-rich aerosols may mimic the sulfate aerosol behavior. Vis et al. (1987) reported the first bromine enrichments in IDPs. Sutton and Flynn (1988) found Br concentrations up to 28 times CI (Note: element enrichments or depletions are determined relative to the CI abundances of the elements).

Volatile element abundances. The abundances of volatile elements with low condensation temperatures (<750°C) increase as a function of heliocentric distance in the cooling solar nebula as temperatures monotonously decrease with greater distance from the sun (Palme et al 1988, Lipschutz and Woolum 1988). Accurate knowledge of indigenous volatile element abundances in chondritic IDPs would provide strong chemical support that

Table 13. Contaminant aerosols associated with IDPs collected in the lower stratosphere

CONTAMINANT SPECIES	POSSIBLE SOURCES	REFERENCES
Sulfuric acid	Sulfur-rich aerosol droplets	Mackinnon and Mogk 1985
Bromine (incl. $KBrO_3$)	Br: probably anthropogenic; K: volcanic or indigenous	Rietmeijer 1993d
Volatile elements: copper, potassium, sodium, gallium, germanium, selenium, zinc	Stratospheric aerosols of volcanic and meteoric origins	Jessberger et al. 1992; Stephan et al. 1994; Rietmeijer 1995
KCl	Volcanic (condensed) aerosol	Rietmeijer 1995
Na_2SO_3; Na_2S_2	S: sulfur-rich aerosol droplets Na: volcanic or indigenous, i.e. $2NaCl + H_2SO_4 = Na_2SO_4 + 2NaCl$	Rietmeijer 1993d
Fe (-oxide rim on IDPs)	Meteoric condensates	Jessberger et al. 1992
Bi_2O_3 and $Bi_2O_{2.75}$	Volcanic aerosol (?)	Rietmeijer and Mackinnon 1997
Silica	Volcanic ash (with adhered sulfuric acid aerosol droplets)	Rietmeijer 1988
SnO_2	Volcanic (?)	Fraundorf 1981; Rietmeijer 1989b
Cd-crystals	Anthropogenic: plating material	Germani et al. 1990
Al_2O_3 spheres	Anthropogenic: solid fuel effluents	Brownlee (1997, personal communication)
Al-prime particles	Probably anthropogenic (see text)	
Carbon soot	Anthropogenic: aircraft exhaust	

the morphologically and petrologically unique IDPs are a new type of extraterrestrial material. The review article by Arndt et al. (1996) presents the data for 89 stratospheric particles culled from the literature (Fig. 21). Flynn et al. (1996) summarized data for >70 particles (Fig. 22). The abundances are normalized to Fe = 1 and CI and presented as a function of 'condensation temperature of the elements' which is common practice among meteoriticists. All volatile elements are enriched but the bromine enrichments stand out at an average $50 \times$ CI abundance (Flynn et al. 1996). Flynn et al. (1996) and Arndt et al. (1996) agreed that bromine enrichments result from Br-rich stratospheric aerosol contamination. Flynn et al. (1996) pointed out that in two large (>35 μm) IDPs 50% of bromine was very weakly bound. In another IDP halogens were concentrated at the surface (Arndt et al. 1996). The observations support both a stratospheric and an indigenous component for bromine. It raises the possibility that volatile elements have dual origins. In IDPs the measured volatile element abundance is the sum of its indigenous concentration, minus the amount lost by evaporation during atmospheric entry, plus the stratospheric contaminant aerosol abundance (Rietmeijer 1995). This simple formula underscores the difficulties of this issue because there are no data on the indigenous volatile element abundances in IDPs. There are also no data on their stratospheric abundances as a function of time, altitude, geographic location, and the relative proportions of contributing sources. Even the details of the mechanism of aerosol-IDP interaction are unknown (Jessberger et al. 1992).

Origins of volatile elements. The origin of volatile elements in chondritic IDPs is a contentious and important issue. All IDPs are heated during atmospheric entry. During this thermal event an unknown fraction of indigenous volatile elements is probably lost by evaporation. The issue is compounded by the fact the host phases of most volatile elements in Table 13 are unknown. Their geochemical affinity suggests that they are associated with Fe,Ni-sulfides (Rietmeijer 1992b, 1995). The abundances of tin and bismuth, which occur as cassiterite and bismuth oxides, fit the general trend of volatile element enrichments (Fig. 23) (Rietmeijer 1989b). Jessberger et al. (1992) argued that, independent of their geochemical affinity, all volatile elements in IDPs are stratospheric contaminants. In their model, these aerosols resulted from evaporation of IDPs that lost all, or a fraction, of their

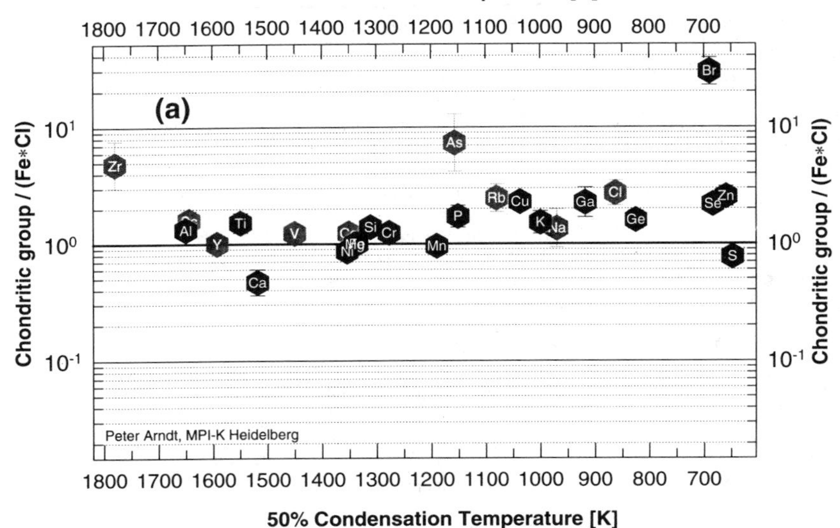

Figure 21. Geometric means (symbols) and standard errors (error bars) of the elements normalized to Fe = 1 and CI abundances in the group of 'chondritic' IDPs. Gray symbols indicate that an element is detected in <50% of IDPs in the database. Higher percentages are indicated by the black symbols. [Used by permission of the editor of Meteoritics and Planetary Science, from Arndt et al. (1996), Fig 10a, p. 827.] Courtesy of Peter Arndt.

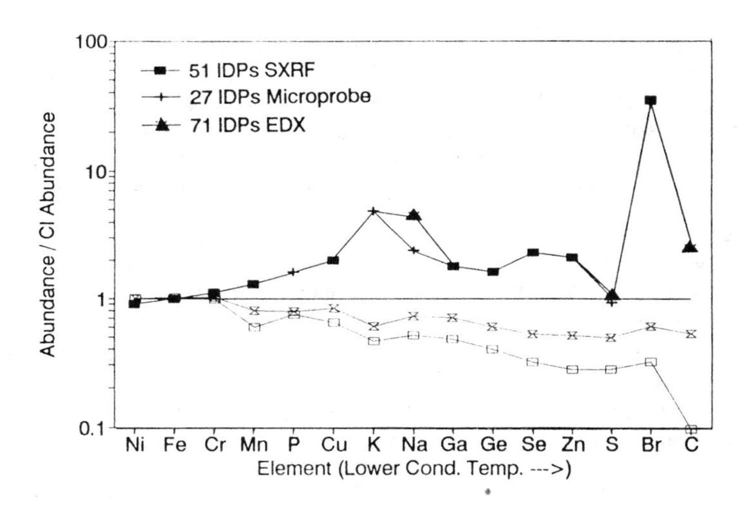

Figure 22. Average Fe and CI normalized compositions of 51 chondritic IDPs by synchrotron x-ray fluorescence, 27 IDPs by electron microprobe and 71 IDPs by energy dispersive spectrometry as a function of element condensation temperature (closed symbols). They include some of the same elements shown in Fig. 21). Open butterflies are CM carbonaceous chondrite abundances. Open squares are CV meteorite abundances [Used by permission of the editor of 1996 Astronomical Society of the Pacific, from Flynn et al (1996), Fig. 1., p. 293]. Courtesy of George Flynn.

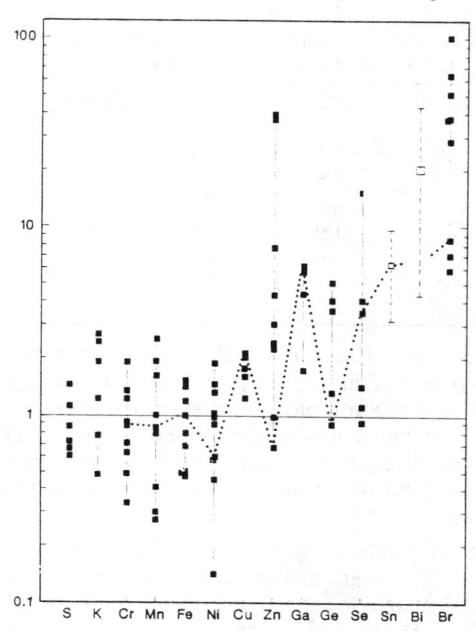

Figure 23. A summary of volatile element abundances in IDPs normalized to CI culled from published sources (solid squares) and calculated abundances for bismuth and tin in IDP W7029*A. Their abundances are based on estimates of the total mass of Bi_2O_3 and SnO_2 grains and the IDP. [Used by permission of the editor of Meteoritics, from Rietmeijer (1989), Fig 2, p. 46.]

volatile elements. Subsequently, these vapors condense as meteoric vapors that in turn contaminate the IDPs settling in the stratosphere. Using the sulfur, bromine and chlorine abundances in IDPs and micrometeorite fluxes, Sutton and Flynn (1990) calculated the amounts of sulfur and halogens that are contributed to the Earth's atmosphere by meteor evaporation. They concluded that the contributions of extraterrestrial sulfur and halogens are small compared to the influx of these elements from the troposphere.

Volatile elements from the troposphere. Anthropogenic and volcanic activities generate large amounts of volatile elements in the troposphere. Depending on chemical reactivity and the presence of sinks to scavenge these gas species and condensed aerosols, some fraction of these elements diffuses across the tropopause into the stratosphere (Yung et al. 1980, Symonds et al. 1988). Volatile elements associated with the Merapi volcano (Symonds et al 1987), Kilauea volcano (Naughton et al. 1976) and in volcanic eruption clouds (Smith et al. 1982, Sedlacek et al. 1982) show tremendous (average) enrichments (normalized to Fe and CI chondrites), viz. Cu: 40, Se: 187, Zn: 50-510 and Br: 10^4. Zinc, copper, selenium, germanium and gallium occurred at ppm levels in ejecta of the 1980 Mt. St. Helens and 1982 El Chichón eruptions (cf. Rietmeijer 1995). Small aerosol particles may remain in the stratosphere for many years (Rietmeijer 1993c).

If volatile elements in IDPs have volcanic origins, their abundances as a function of time should correlate with the times of volcanic eruptions. Unfortunately, the presentations of normalized volatile element abundances as a function of their condensation temperature (Flynn et al. 1996, Arndt et al. 1996) hide variations as a function of collection time. The average abundances for the 'chondritic group' of IDPs (Arndt et al. 1996) are listed as a function of their collection period (Table 14). The largest samples are L2005 (8 IDPs) and five particles on W7027/7029 and L2001/2/3. The remainders are samples of only two IDPs, or only one particle. Particle U2015G1 is a large aggregate that is very rich in zinc and copper (Zn/Cu = 37.5). This ratio is reminiscent of zincite and tenorite associated with the May 1980 and September 22, 1981 activity of Mt. St Helens (Rose et al. 1982). This

Table 14. Element/Fe normalized to CI abundances of zinc, copper, gallium, germanium and selenium in IDPs from the chondritic group arranged in chronological order of collection.

Collector	Collection period	Zinc	Copper	Gallium	Germanium	Selenium
W7013	05/22-07/06/81	2.7	2.1	3.4	2.4	3.6
W7017	07/07-09/15/81	0.9	1.6	1.6	1.5	3.2
W7027/29	09/15-12/15/81	1.2	2.3	2.6	1.4	2.7
U2015	06/22-08/18/83	45.0	1.2	–	–	–
U2022	04/09-06/26/84	2.0	3.2	1.9	1.3	1.5
L2001/2/3	Fall 1988	1.4	0.8	0.5	0.7	1.1
L2005	10/03-10/13/89	1.0	2.3	2.5	3.3	2.6
L2011	June-07/11/91	1.1	0.7	0.9	0.86	0.87

sample is excluded from the following argument. The very small database prohibits statistical analysis but there is a remarkable covariance among elemental abundances as a function of time. The oscillations in the data may be a response to short-term changes in stratospheric aerosol abundances. At this time, an indigenous origin of these elements can not be excluded based on the available data but it requires a detailed analysis of particle morphology, the stratospheric residence time, and the actual sites of volatile elements (i.e. at the surface or inclusions within the IDP).

Iron oxide rims. Iron oxide rims on stratospheric dust are most likely dynamic pyrometamorphic alteration. Jessberger et al. (1992) suggested that magnetite rims are the result of unspecified 'atmospheric processes.' For example, contamination by iron from the E-layer in the Earth's atmosphere between 80-110 km altitude (altitudes of IDP deceleration) or by passing directly through the vapor trail of an ablating 10 µm-size IDP. Flynn (1994) argued that the probability of contamination of an incoming IDP with iron and sodium from these sources is negligible. In fact, many IDPs do not have this iron oxide rim. The conclusion assumed vapor densities due to sporadic meteors and meteor showers. During meteor storms (e.g. the 1998 and 1999 Leonid storms) the possibility of IDP contamination by meteoric vapors, or (condensed) meteoric dust, increases but it is doubtful that it will be a cause for concern.

Not yet a new type of IDPs. Whether volatile element abundances support a new compositional type of chondritic IDPs (Flynn et al. 1996), or not, remains to be determined. The volatile elements include unknown fractions of indigenous and contaminant species, including volcanic, anthropogenic and meteoric aerosols. Presently the evidence is inconclusive because the arguments brought forward to support this notion involve the volatile elements themselves and are not fully convincing. Flynn et al. (1992b) found a linear correlation between copper (a moderately volatile siderophile element) and selenium (a more volatile chalcophile element) in chondritic IDPs and carbonaceous chondrites (Fig. 24). The normalized average ratios Cu/Fe = 2 and Se/Fe = 2.5 correspond to 5% of the (CI-) normalized volcanic Cu/Fe ratio and 1% of the Se/Fe ratio. In further support of an indigenous origin, Flynn et al. (1996) added carbon to the enrichment trend (Fig. 22). This argument is fallacious because carbon occurs in mineral constituents of aggregate IDPs. It is present in carbonates, carbides, lonsdaleite, amorphous carbons and poorly graphitized carbons (Rietmeijer 1992b). These modes of occurrence rule out any possibility that carbon is a contaminant species. The fact remains that volatile elements are systematically enriched as a function of their volatility. It is not yet clear that these trends are evidence for indigenous origins of volatile elements in chondritic IDPs.

Isotopic compositions in chondritic IDPs

Isotopic compositions are diagnostic for specific processes of nucleosynthesis. Isotopic compositions in natural materials have been critical to identify interstellar materials

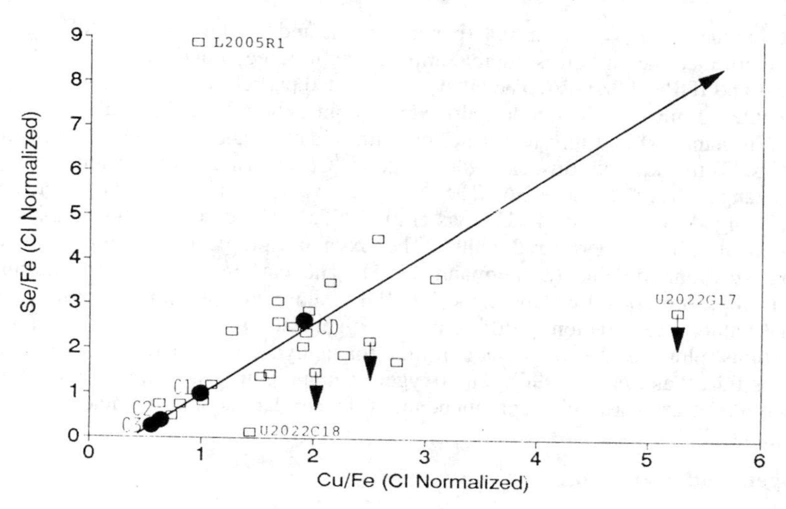

Figure 24. Se/Fe and Cu/Fe normalized to CI in 24 IDPs (open squares) and C1, C2 and C3 carbonaceous chondrites. [Reproduced with permission of the authors from Flynn and Sutton (1992) Lunar and Planetary Science XXIII: 373-374 Lunar an Planetary Institute, Houston.]

in meteorites and are potentially useful in this regard to understand the origins of IDPs, or individual components in chondritic aggregate IDPs.

Lithium, beryllium, boron and noble gases

In space, IDPs are bombarded by solar wind particles, solar flare particles, including so-called solar energetic particles (SEPs). Nier and Schlutter (1990) measured the helium and neon isotopes in chondritic IDPs (8-40 μm in size). The results, $^3He/^4He = 2.4 \times 10^{-4}$ (average), $^4He/^{20}Ne = 33 \pm 7$ and $^{20}Ne/^{22}Ne = 12 \pm 0.5$, matched earlier data (Table 2). Xu et al. (1991) found $^{20}Ne/^{22}Ne$ ratios of 33 and 50 in chondritic aggregates U2022B1 and U2034F10 and $^{21}Ne/^{20}Ne = 4.0\text{-}4.1 \times 10^{-3}$. These noble gases were implanted solar wind particles. These data support that IDPs arrived from different sources but with $^{20}Ne/^{22}Ne$ ratios in the range expected for solar wind neon (Nier 1994). The data support some SEP gasses, that are commonly seen in lunar soils, in IDPs (Nier and Schlutter, 1994). Xu et al. (1991) determined the lithium and boron isotopic compositions of two IDPs, $^7Li/^6Li = 8$ and 11.84 and $^{11}B/^{10}B = 3.6$ to 4.0. They are consistent with solar values. The ratios Li/Be/B at 46/0.3/23 and 148/0.2/24 are consistent with the solar values, 59/0.8/24.

Carbon, oxygen, magnesium and silicon

The ion microprobe made it possible to measure the isotopic compositions in IDPs. McKeegan et al. (1985) determined the Mg and Si isotopic compositions in fragments of three chondritic IDPs including hydrated IDP r3-M3-5A with a possible link to carbonaceous chondrites (Tomeoka and Buseck 1985) and a refractory CP IDP (Christoffersen and Buseck 1986b). The $\delta^{26}Mg$ values ranged from -6.6 to +3.6 and $\delta^{30}Si$ values from -21.7 to +17.2 (McKeegan et al. 1985). These values, which confirmed earlier magnesium isotope data (Table 2), show no deviations from the terrestrial value. The magnesium isotopic compositions in so-called GEMS in chondritic IDPs show a similar result (cf. section on GEMS). Esat et al. (1979) initially reported terrestrial magnesium isotopic compositions in two transparent spheres. Re-analysis of one sphere (W7029I18) showed a value $\delta^{26}Mg = 1.92 \pm 0.4$ (per mil) (Esat and Taylor 1987). McKeegan et al. (1985) determined the carbon isotopic compositions in three chondritic IDPs that included r3-M3-5A and IDP r21-M1-9A. The latter is a hydrated particle with iron- and magnesium-rich smectite (or mica) as the major hydrated phase plus abundant pentlandite and pyrrhotite (100 nm in diameter) and rare forsterite, enstatite platelets and rhombohedral Mg,Fe,Ca-carbonates (Tomeoka and Buseck 1986). The $\delta^{13}C$(°/∞) values for fragments of these particles ranged from -50 to +2.6. The $\delta^{13}C$(°/∞) values for fragments of cluster IDP L2008#5 ranged from -33 to -4 (Thomas et al. 1995a). The carbon isotopic ratios are not enriched relative to the terrestrial value. The oxygen isotopic compositions are on the terrestrial fractionation line (Stadermann 1991). The carbon, oxygen, magnesium and silicon isotopic compositions in chondritic IDPs show no enrichments relative to the terrestrial value. The situation is different for oxygen isotopes in refractory IDPs. These compositions plot on the fractionation line for anhydrous minerals in carbonaceous chondrites (Greshake et al. 1996). The oxygen isotopic compositions of a refractory IDP with enriched rare earth element abundances is similar to the particle analyzed by McKeegan (1987).

Hydrogen and nitrogen

Both hydrogen (Table 15) and nitrogen (Table 16) isotopic compositions show enrichments that could support an interstellar origin for constituents in aggregate IDPs. The D/H enrichments are associated with carbonaceous materials wherein they occur as 'hot spots' (McKeegan et al. 1985, 1987). McKeegan et al. (1985, 1987) did not find D/H enrichments in IDPs of the 'olivine' IR class although solar flare tracks in some of these particles support an extraterrestrial origin. Thomas et al. (1995a) explored the mineralogical and chemical variations of cluster IDPs (see next section) that were conjectured to be an overlooked type of chondritic IDP. The D/H ratios in cluster fragments, 5.6×10^{-4} to 1.9×10^{-3} (L2009; L2011) and 1.2×10^{-3} to 3.8×10^{-3} (L2005#31) (Table 15), are among the highest values measured in solar system materials (NOTE: the terrestrial δD(°/∞) values range from -200 to +75). Deuterium enrichments in extraterrestrial materials are the results of ion-molecule reactions that are only efficient at very low temperatures. Messenger and Walker (1996) calculated that the value δD(°/∞) = 24,800 puts an upper limit of -203°C (or,

Table 15. Hydrogen isotopic compositions in IDPs and cluster IDPs

IDP type	δD (‰)	Comments	References
chondritic	+498 to +1133	Carbon-rich (17.5 x 17.5 μm); CP IDP, 22.5 x 25 μm, linked with carbonaceous chondrites (Tomeoka and Buseck, 1985)	Zinner et al. 1983
chondritic	-386 to +2534; "hot spots" with D/H >9000	Carbon-bearing (Wopenka 1988) chondritic IDPs from all infrared classes (cf. infrared classification)	McKeegan et al. 1985
Nearly chondritic	-274 to +151 (1) -182 to +494 (2) -299 to –30 (3)	Individual IDPs of the 'LLS' (1), 'pyroxene' (2), and 'olivine' (3) infrared classes	McKeegan et al. 1987
C-rich chondritic	-55 to +822	Fragments of cluster IDP L2008#5; 4-29 wt. % C plus one 'magnetite' dominated fragment	Thomas et al. 1995a
Not specified	+1,147; +3 to +12,007; 0 to +5,570	Fragments of three cluster IDPs: L2008, L2009 and L2011	Messenger et al. 1996
C-rich	+800	Carbonaceous IDP (15 x 10 μm) with Ti-rich grains	Blake et al. 1989

70K) to the formation temperature. The nitrogen enrichments occur in fragments with strong signals for the presence of PAHs (Thomas et al. 1995a) but they were measured in a number of other IDPs. The nitrogen isotopic ratio $\delta^{15}N(‰) = +480$ supports $T < -203°C$ (Messenger and Walker 1996). These temperatures are consistent with materials from cold molecular clouds that were preserved among carbonaceous materials in IDPs (Messenger and Walker 1996). Similar low temperatures existed in the outer fringes of the solar system. These interpretations of hydrogen and nitrogen isotopic compositions in IDPs are based on comparison with H and N isotopic compositions in meteorites that have not experienced major thermal alteration. Thomas et al. (1995a) and Keller et al. (1996b) indicated that thermal alteration of PAHs and other carbonaceous materials occurs during atmospheric entry flash heating. It raises the possibility of mass fractionation that could modify the pre-entry isotopic composition of carbonaceous materials in IDPs.

PETROLOGY AND MINERALOGY OF AGGREGATE IDPS

Working hypothesis

Aggregate IDPs and Cluster IDPs. Aggregate IDPs occur as individual particles scattered on a collector. Some collectors also show dark splashes that are fragments resulting from the breakup of a larger particle (Fig. 25). During inertial impact collection cluster IDPs break into many fragments plus 'several hundred sub-fragments' (called 'fines') that are <5 μm in diameter (Thomas et al. 1995a). A survey of NASA/JSC Cosmic Dust Catalogs shows that cluster IDPs (typically >60 μm) consist of aggregate IDPs (81%), silicate particles (12%) and iron-nickel sulfide particles (7%) that are all <10 μm in size. Cluster fragments between 10 and 20 μm are mostly aggregates and rare silicate IDPs (Rietmeijer 1997b). The bulk composition of 53 fragments from cluster IDP L2008#5 (Thomas et al. 1995a) yield three groups, viz. chondritic and silicate fragments and sulfide and magnetite fragments, with the following size distributions:

- chondritic and silicate fragments range from 5 × 5 μm to 25 × 15 μm and have a Gaussian distribution with mean = 12.8 μm; 1σ = 4.7 μm [Note: mean and standard deviation are expressed as the root-mean-square sizes, r.m.s. = $(a^2 + b^2)^{0.5}$] with 12% of the fragments >20 μm (r.m.s.), and
- sulfide and magnetite fragments range from 5 × 5 μm to 17 × 15 μm and also have a Gaussian distribution with mean = 13.7 μm; 1σ = 4.6 μm, with ~20% of the fragments >20 μm (r.m.s.).

The student's t-test shows that these two distributions cannot be accepted as different populations. There is an important caveat when attempting to interpret the results from a statistical analysis of IDPs listed in the Cosmic Dust Catalogs or elsewhere. There does not

Table 16. Nitrogen isotopic compositions in IDPs and cluster IDPs

IDP type	$\delta^{15}N$(‰)	Reference
carbonate-rich IDP r21-M1-9A; IDPs of the olivine and pyroxene IR classes	zero to +200, extremes from about +350 to +500	Stadermann et al. 1989
fragments of cluster IDP L2005#31	about +260	Messenger et al. 1995
fragments of cluster IDP L2008#5	+56 to +260	Thomas et al. 1995a
sulfur-rich chondritic IDPU47-M1-3a; non-chondritic particle U47-M2-3, Mg,Fe-olivine (?) with high Na and P	411 ± 20; 306 ± 21	Stadermann et al. 1990

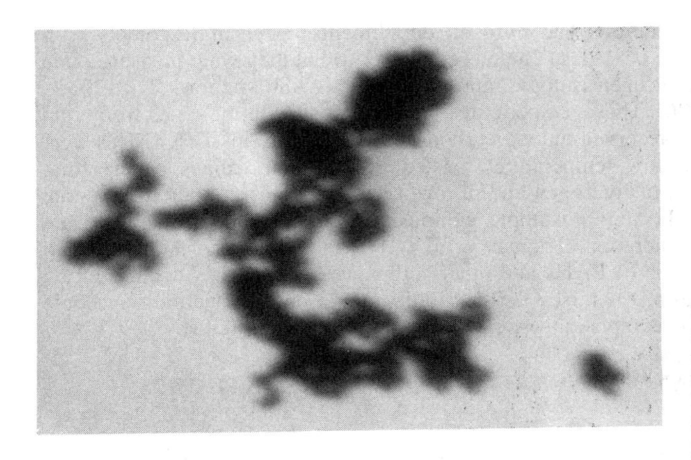

Figure 25. Photomicrograph of a cluster IDP and the collector surface. The out-of-focus appearance is due to the layer of silicone oil in which the cluster is embedded. NASA photo S82-26487.

appear to be a curation bias with regard to particles >2 to 5 μm. Whether this analysis reveals a fundamental property of IDPs related to their modes of formation, or their behavior during collection, is not clear. There are many sub-micrometer fragments on a collector surface that are not picked-off and cataloged, and that are therefore not considered in a statistical analysis of size distributions. Collected chondritic IDPs and magnesium-iron silicate and iron-nickel sulfide IDPs occur both as individual (or, unbound) particles and as fragments of cluster IDPs. Chondritic aggregate IDPs <10 μm occur with equal abundance as individual IDPs and as cluster fragments. About 80% of carbon-rich aggregates <10 μm size are cluster IDP fragments (Rietmeijer 1997b). Collapsed chondritic aggregates are rare among unbound IDPs but could be more common among cluster IDPs (Messenger et al. 1996).

Cluster IDPs with fragments up to 60 to 100 μm in size could be a previously under-appreciated resource. They offer an explanation for the apparent lack of correlation among various dynamic pyrometamorphic features and the apparent scarcity of solar flare tracks in some aggregate IDPs. Cluster IDPs have a lower density than individual fragments and its fragments may experience less thermal alteration than similar fragments that entered the atmosphere as individual particles. Cluster IDPs and aggregate IDPs were not subject to mechanical stresses $>10^4$ dyne cm^{-2} during deceleration in the upper atmosphere (Brownlee 1985) but the mechanical strength of aggregate and cluster IDPs must be different. Stratospheric dust particles include chondritic IDPs and non-chondritic particles with chondritic material attached to the surface. Subdivision based on particle porosity (Table 10) includes aggregates and collapsed aggregates (or CS IDPs). Chondritic IDPs without an accretion texture often resemble matrix material of CI and CM carbonaceous chondrites. This review is devoted to mineralogy and chemistry of chondritic aggregate IDPs and the relationships with non-chondritic IDPs. I will not consider minerals that are dynamic pyrometamorphic alteration or stratospheric contamination. I recognize that the origin of a mineral is often ambiguous, e.g. the filamentous carbons (Bradley et al. 1984b) and carbides (Christoffersen and Buseck 1983a) could be solar nebula dust or the most recent thermal alteration of IDPs.

Matrix units

The matrix of ultrafine-grained chondritic aggregate IDPs consists of 'principal components' that form discrete textural units (Rietmeijer 1992b). The nomenclature of the matrix units has reached a point of major confusion due to investigator preference and the steep learning curve of IDP-lore. Fraundorf 's descriptive term 'clumps' survived only in the early studies. Tomeoka and Buseck (1984) introduced a chemically rooted term 'FMA' for 'fluffy silicate that contains Fe, Mg and Al' in an effort to distinguish their findings from the carbon-bearing clumps. The nomenclature of units with properties defined by the original authors is summarized in Table 17. Upon reconsideration of the original data for 'tar balls', Bradley (1994b) re-christened these units as 'glass with embedded metals and sulfides' (GEMS). Its matrix, initially misidentified as carbonaceous material, i.e. tar balls (Bradley 1988), is a carbon-free, amorphous magnesiosilica material. Thomas et al. (1995a) reported unidentified 'fine-grained aggregates' (FGAs) in carbonaceous material with properties that resemble polyphase units and/or GEMS. Other texturally defined terms are 'micro-crystalline aggregates' and 'polycrystalline aggregates' (Fraundorf and Shirck 1979), and 'granular clusters'.

Table 17. Nomenclature of matrix units in aggregate IDPs

MATRIX UNITS				REFERENCES
Clumps with carbonaceous matrix				Fraundorf 1981
	Tar balls			Bradley 1988
Granular units				Rietmeijer 1989a
Granular units		Polyphase units (PUs)		Rietmeijer 1992b
	Reduced aggregates	Equilibrated aggregates	Unequilibrated aggregates	Bradley 1994a
			GEMS	Bradley 1994b
Granular units		PUs		Rietmeijer 1994a
Granular units		Ultrafine-grained PUs and GEMS are similar?		Rietmeijer 1996e
Granular units		Ultrafine-grained PUs	Coarse-grained PUs	Rietmeijer 1997c

Granular units (GUs). The descriptive term 'granular units' (GUs) is used for the 'non-crystalline clumps with non-stoichiometric (silicate) compositions' (cf. pre-1982 research) in the matrix of aggregate IDPs. The distinct (sub-) circular units can be loosely bound or closely packed into micrometer size patches up to 2 to 4 μm (Fraundorf and Shirck 1979, Rietmeijer 1989a, 1993a) (Figs. 6 and 26). They contain tiny, 2 nm to ~50 nm, crystals of silicate, iron sulfide and metallic iron in a carbonaceous material. The typically platy grains are disc-shaped when <30 nm in size and larger grains have sub- to euhedral shapes (Rietmeijer 1989a). The carbonaceous matrix is hydrocarbon and/or amorphous carbon (Fraundorf 1981, Christoffersen and Buseck 1986a, Wopenka 1988, Bradley 1988, Klöck and Staderman 1994).

Granular units are carbon bearing, ferromagnesiosilica materials with sulfur and minor aluminum, calcium, chromium, manganese and nickel (Christoffersen and Buseck 1986a, Rietmeijer 1989a), and traces of phosphorous (Rietmeijer 1989a, IDPs W7010*A2, U2015B9). A small amount of zinc occurs in these units in U2015B9 (Rietmeijer 1989a). Zinc up to 10 × CI occurs in some GUs (Klöck and Staderman 1994). Carbon-rich units containing a Mg, Si, Ca-silicate glass also exist (Bradley 1994a, IDP U222B42). These 'reduced aggregates' with a partial magnetite rim contain Fe,Ni metal (kamacite) and Fe-rich sulfide nano-grains embedded in a non-crystalline matrix. The metal grains can have a

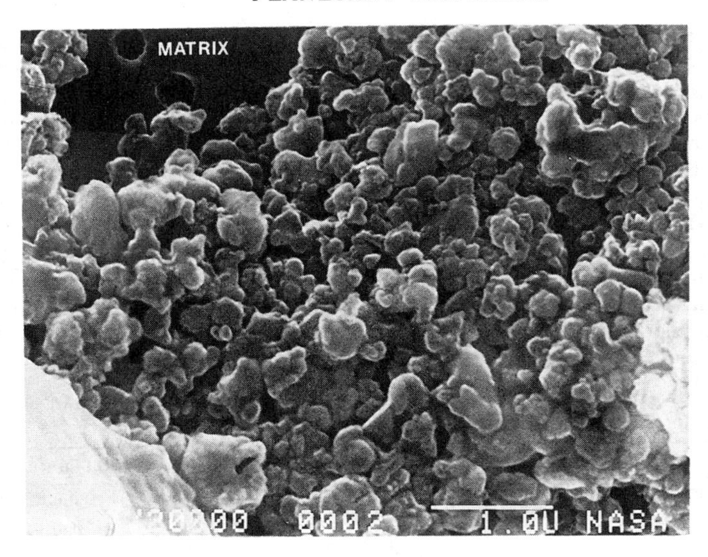

Figure 26. Scanning electron micrograph showing the matrix of granular units in IDP W7010*A2. (Rietmeijer, unpublished data).

thin (<10 nm) rim of Fe,Ni carbides or graphite (Bradley 1994a). The GUs are typically 0.1 μm in diameter (Rietmeijer 1993a). Larger (~0.3 μm) units are present (Bradley 1994a). Larger units result from fusion of the original individual units that form the fractal matrix of aggregate IDPs (Fig. 27) (Rietmeijer 1993a). Fraundorf (1981) and Rietmeijer (1989a) combined bulk composition and characteristic diffraction maxima in polycrystalline ('ring') electron diffraction patterns to identify the ultrafine minerals (Table 18).

Aggregate IDP u44-m1-7 consists mainly of matrix units with major and trace element abundances that are uniformly chondritic within a factor of 2 (Fig. 28) (Klöck and Staderman 1994). The magnesium and iron abundances obtained by spot (bulk) analyses within dispersed GUs from other IDPs are within 2 × CI abundance (Rietmeijer 1989a). The distinct Mg-ratio distribution in GUs (Figs 29) is similar to this distribution in comet Halley dust (Jessberger et al. 1988). It clearly differs from the Mg-distributions in hydrated solar system materials such as CI carbonaceous chondrites and hydrated IDPs (Lawler et al. 1989) (Fig. 30). The distributions of the S/Si and (Fe+Ni)/Si ratios suggest 'excess sulfur' that is possibly associated with carbonaceous materials in these units (Rietmeijer 1989a) but this association was not proven (Fig. 31).

Polyphase units (PUs). Polyphase units are (sub)-circular, carbon-free ferromagnesiosilica entities in the matrix of aggregate IDPs. There are three distinct types based on grain size and the iron oxidation states. They are polyphase units and GEMS.

Ultrafine-grained units with nanometer size grains of iron-magnesium olivines and pyroxenes, Fe,Ni-sulfides and magnetite in an amorphous ferromagnesiosilica matrix (Fig. 32) (Rietmeijer 1994a 1996e). The bulk Mg-ratios of these sub-micrometer ultrafine-grained PUs range from 0.7 to 0.17 (Table 19; Rietmeijer, unpublished data) with (Mg+Fe)/Si ratios of ~1.5 (Rietmeijer 1996e, 1997c). The bulk compositions are reducible to a stoichiometric serpentine dehydroxylate structural formula, $(Mg,Fe)_3Si_2O_7$. Within each unit (Mg+Fe)/Si ratios vary from 1 to 2, consistent with the presence of

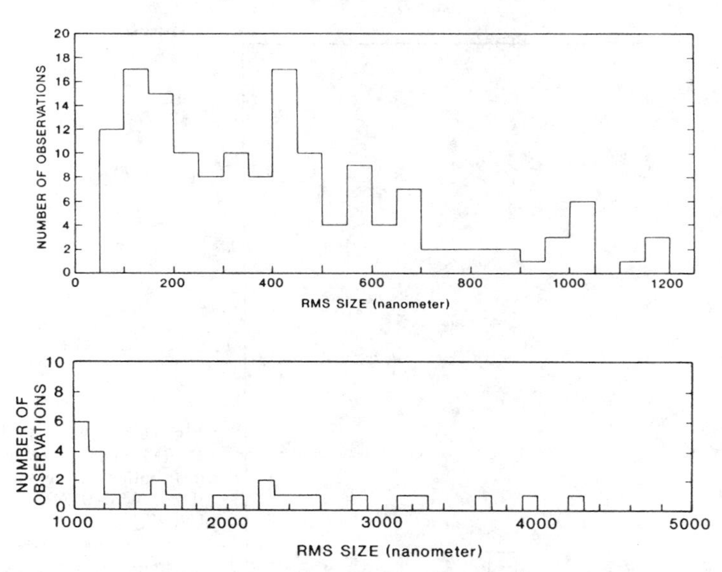

Figure 27. Histograms showing the number of granular units in IDP W7010*A2 as a function of the root-mean-square size (nm), excluding a few very large units between 5057 and 7580 nm. [Used by permission of the editor of Earth and Planetary Science Letters, from Rietmeijer (1993), Fig 2, p. 611.].

Table 18. Ultrafine-grained minerals in Granular Units

Silicates	Forsterite and Fe-rich olivine with traces of Mn and Ca; Enstatite – hypersthene, and possibly rare ferrosilite
Sulfides	Low-Ni pyrrhotite and low-Ni pentlandite, possibly troilite
Oxides (minor)	Magnetite, hercynite, chromite
Metals and carbides	(rare) taenite (~6 at% Ni) with {110} cohenite (Fe_3C) platelets decorations; (rare) kamacite (< 10 at% Ni)

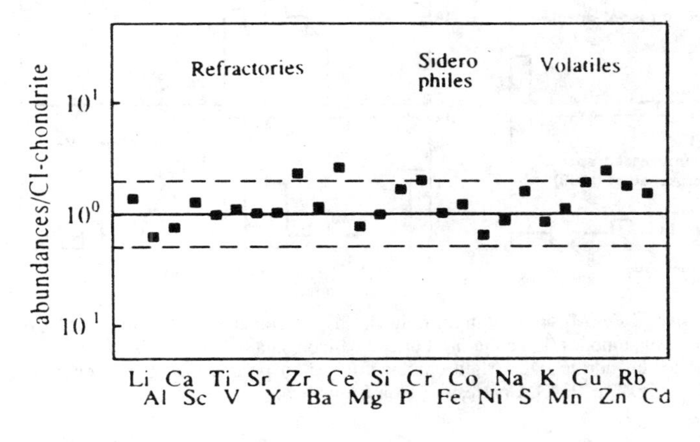

Figure 28. Major and trace element and REE, abundances normalized to CI in an IDP mostly consisting of matrix units. Rare embedded crystallites are <100 nm. The abundances are within a factor 2 of CI abundances. [Used by permission of the series editor of the American Institute of Physics, from Klöck and Stadermann (1994), Fig 4, p. 57.]. Courtesy of Wolfgang Klöck

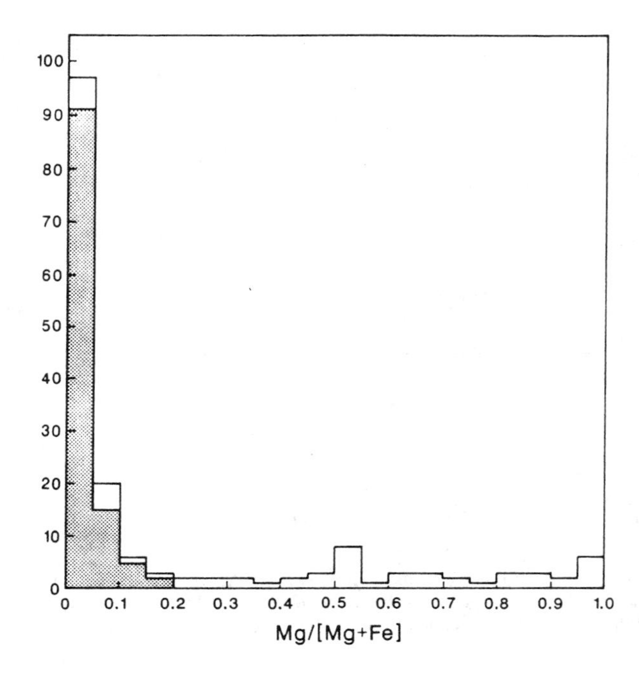

Figure 29. Histogram of Mg/(Mg+Fe^{2+}) (element ratio) distributions in granular units of IDPs W7010*A2, U2015*B and U2022C26 (shaded) augmented by data for aggregate IDPs culled from the literature. [Used by permission of the editor of Proceedings of the 19th Lunar and Planetary Science Conference, from Rietmeijer (1989), Fig. 4, p. 516.]

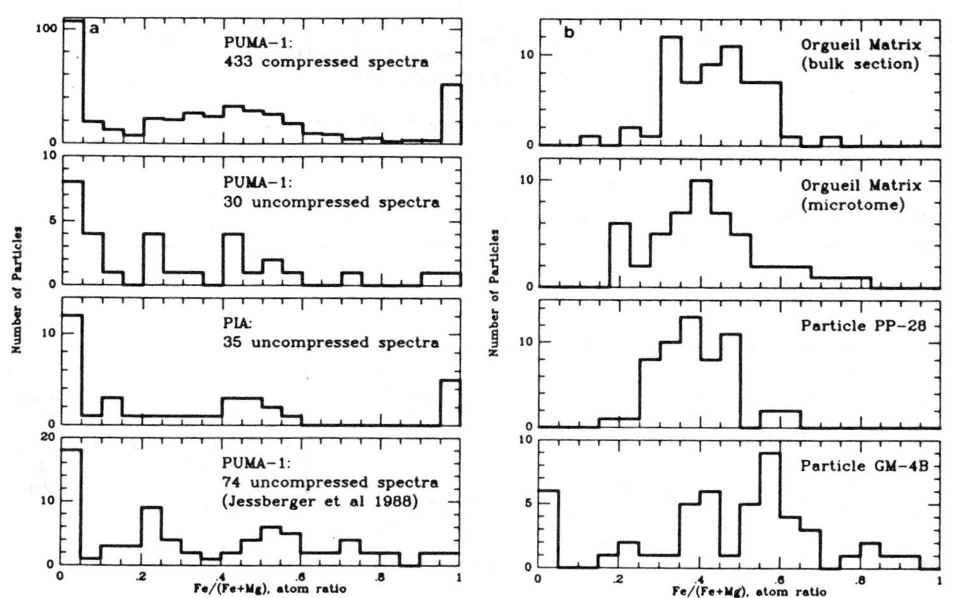

Figure 30. Histograms of corrected Fe/(Mg+Fe) (atomic) ratio in dust of comet P/Halley measured by two instruments in different operational modes (left column). For comparison data for the matrix of hydrated meteorite Orgueil and two hydrated chondritic IDPs (right column). [Used by permission of the editor of Icarus, from Lawler et al. (1989), Fig. 4, p. 234.]. Courtesy of Don Brownlee .

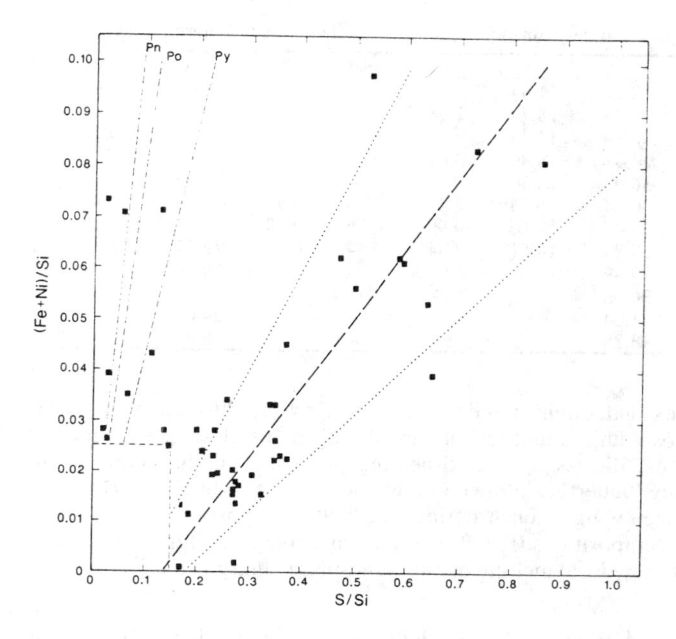

Figure 31. (Fe+Ni)/Si and S/Si ratios within granular units in the matrix of IDPs W7010*A2, U2015*B and U2022C26 compared with pentlandite, pyrrhotite and pyrite. The (heavy) dashed line delineates 'excess' sulfur. The blank box encloses 68% of the data. [Used by permission of the editor of Proceedings of the 19th Lunar and Planetary Science Conference, from Rietmeijer (1989), Fig. 5, p. 517.].

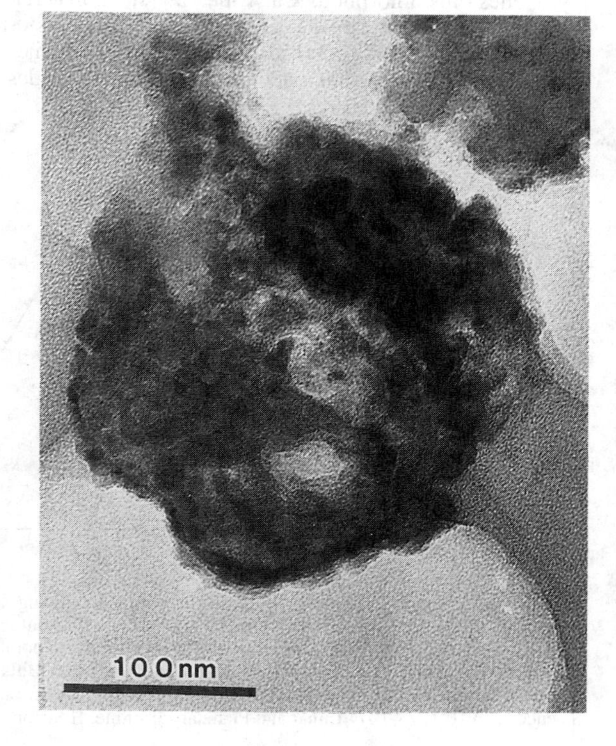

Figure 32. Transmission electron micrograph of an ultrafine-grained polyphase unit in chondritic aggregate IDP L2011A9 with nanometer size silicates and iron oxides in an amorphous ferro-magnesiosilica matrix. The sample is embedded in epoxy (gray) and is supported on a holey carbon film (dark gray bands) (Rietmeijer, unpublished data).

Table 19. Compositions of ultrafine-grained polyphase units (element wt. % normalized)

	Mg	Al	Si	S	Ca	Cr	Mn	Fe	Ni
1	9.99	0.36	28.21	11.34	-	0.27	-	48.03	1.79
2	10.88	0.59	26.35	18.79	1.18	0.27	0.02	40.41	1.50
3	19.19	-	30.30	14.44	-	-	-	34.36	1.42
4	15.18	0.39	24.44	18.89	0.61	0.12	-	38.33	2.04
5	22.44	-	36.95	10.30	-	-	-	29.93	0.36
6	26.37	0.44	30.84	9.61	2.41	0.39	0.46	28.02	1.44
7	34.96	1.56	36.47	4.03	0.90	0.16	0.29	21.12	0.52
8	22.26	0.72	28.89	14.51	2.54	0.34	0.13	29.27	1.33
9	18.55	1.10	29.49	-	-	-	-	50.86	-
10	28.82	0.90	36.06	-	-	-	-	34.21	-
11	34.94	-	41.04	-	-	-	-	24.02	-
12	39.39	-	43.88	-	-	-	-	16.73	-

ultrafine-grained pyroxenes and olivines (confirmed by electron diffraction analyses). Variations in Mg/Fe^{2+} ratios within a unit reflect variable Mg-ratios of silicates, variations in the 'local' proportions of silicates and Fe,Ni-bearing phases, or both. These internal variations are constrained by 'butterflies' drawn in a ternary diagram Mg-Fe-Si (Rietmeijer 1996e) The point where the 'wings' touch defines the bulk composition, which match exactly the measured bulk compositions (Fig. 33). Some units contain sulfur while they are sulfur-free in particles that show local melting due to atmospheric entry heating (Rietmeijer 1996e, 1997c).

Coarse-grained (10 to 410 nm in size) units of iron-magnesium olivines and pyroxenes plus amorphous Ca,Al-bearing silica material ('restite') with minor magnesium and iron (Rietmeijer 1997c). These three constituents often have equilibrium (triple-point) grain boundaries (Fig. 34) (Bradley 1988, Rietmeijer 1997c, Klöck and Stadermann 1994). The units are often associated with Fe,Ni-sulfides that have a size similar to that of

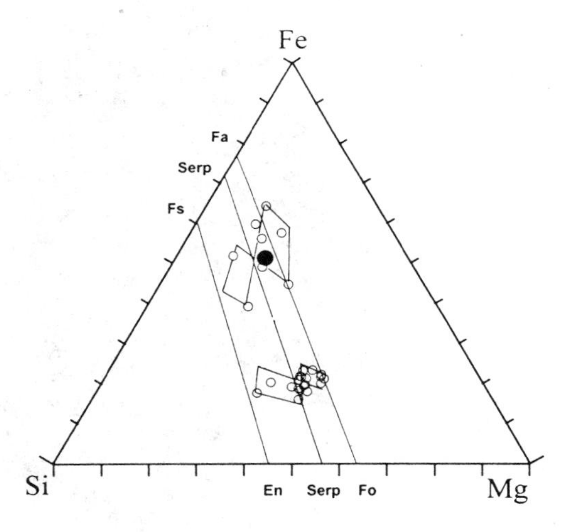

Figure 33. Ternary diagram Mg-Fe-Si (el. wt %) showing butterflies for the most iron-rich and magnesium rich ultrafine-grained PUs in IDP L2011K7. The butterflies enclose individual analyses of the units (open circles). The bulk composition of a unit is at the point where the wings meet which matches measured bulk composition and shown here for one unit (dot). Bulk compositions of ultrafine-grained PUs plot on a line delineating serpentine compositions. Reproduced from Rietmeijer (1996) Lunar and Planetary Science XXVII: 1073-1074 Lunar and Planetary Institute, Houston.

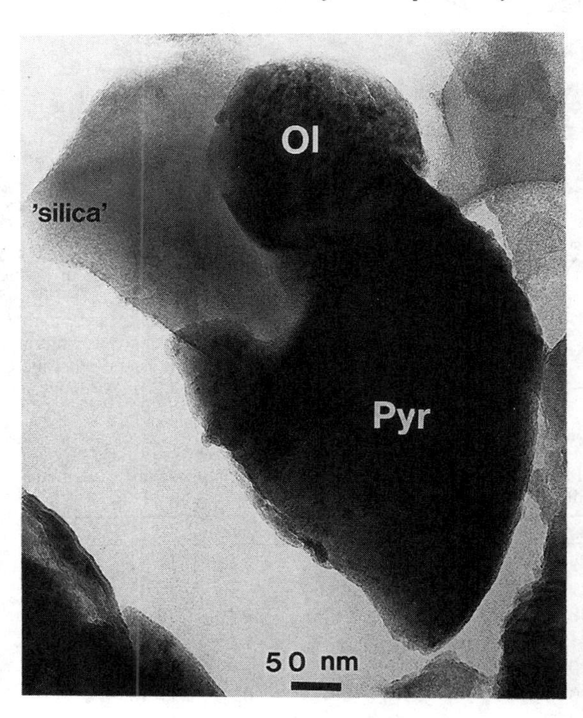

Figure 34. Transmission electron micrograph showing a coarse-grained polyphase unit in the matrix of chondritic aggregate IDP L2011A9. Notice the typical association of olivine, pyroxene and amorphous aluminosilica material with variable amounts of calcium, alkalis, magnesium and iron. Also notice the shattered appearance of the sample (Rietmeijer, unpublished data).

the unit constituents (Rietmeijer 1997c). Bradley (1994a) described similar units as 'equilibrated aggregates'. They were also reported by Tomeoka and Buseck (1984, Fig. 9). The olivine and pyroxene have equilibrated compositions (Klöck and Stadermann 1994) (Table 20). The Mg-ratios of the coarse-grained PUs are between 0.67 and 1.00 (average Mg = 0.77). Cluster analyses reveal the presence of olivine, pyroxene and a silica-rich phase (Bradley 1994a), which was confirmed by TEM identifications (Rietmeijer 1997c).

Table 20. Mg-ratios of equilibrated olivines and pyroxenes in (anhydrous) aggregate IDPs (source: Klöck and Stadermann 1994)

IDPs	Olivine	Pyroxene
W7013E17	97.6	97.8
W7027H10	95.9	96.2
U2015D16	95.7	95.8
U2015C16	93.1	94.4
W7027E6	87.4	88.8
L2005D18	86.8	87.7
W7028*C2	84.8	87.2
W7013E9	81.9	86.9

Coarse-grained PUs have composition that are restricted to the Mg-rich portion of a ternary diagram Si-Mg-Fe (el wt %) (Fig. 35). A line through the compositions of olivine, pyroxene and 'restite' of each unit intersects the Si-corner of the diagram. Rietmeijer (1997c) interpreted this arrangement as (equilibrium?) fractionation among these phases. The unit bulk compositions can be reduced to a stoichiometric smectite (dehydroxylate?) structural formula, $(Mg,Fe)_6Si_8O_{22}$. The 'structural formula' for the round glass unit (Table 21) also matches 'smectite' (assuming 22 O),

$$Mg_{2.4}Fe_{0.5}Mn_{0.1}Ca_{1.2}Cr_{0.1}Al_{1.3}[Al_{0.7}Si_{7.3}]O_{22}.$$

In IDP L2005K10 Zolensky and Lindstrom (1992, Fig 6) reported an angular unit (690 × 540 nm) with euhedral diopside, enstatite and forsterite grains (<100 nm) embedded in an

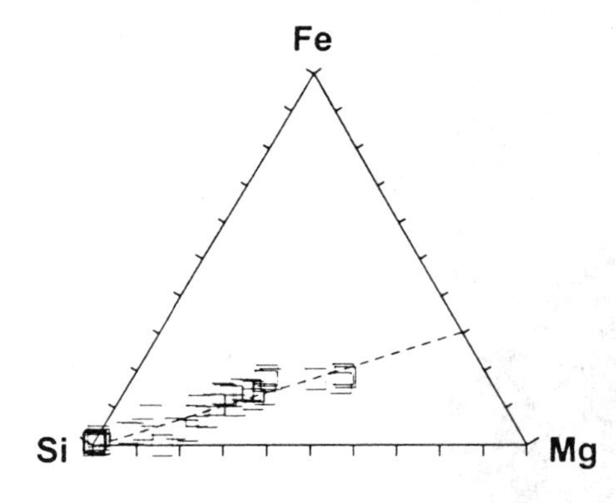

Figure 35. Ternary diagram Mg-Fe-Si (el. wt %) showing the compositional range in a single large coarse-grained PU ('Big Guy:' see Fig. 45). The data cluster at olivine and pyroxene compositions with the remainder as amorphous material ('restite') This unit is the most Fe-rich coarse-grained PU in IDP L2011A9 (Rietmeijer, unpublished data).

Table 21. Compositions of equilibrated aggregates and coarse-grained polyphase units (element wt. % normalized). Sources: Bradley (1994a; Table1) and Rietmeijer (1997c; unpubl. data)

	Mg	Al	Si	S	Ca	Cr	Mn	Fe	Ni
1	35.78	3.5	40.89	0.83	0.97	1.15	1.03	14.62	1.22
2	42.66	0.93	55.29	0.22	0.44	0.04	-	0.03	0.4
3	61.72	0.43	36.48	0.09	0.13	0.16	0.49	0.21	0.29
4	14.59	13.47	50.91	-	11.86	0.92	1.06	7.18	-
5	29.90	2.2	57.50	-	-	-	-	10.40	-
6	29.20	3.0	55.30	-	0.90	-	0.30	11.30	-
7	34.10	4.0	49.20	-	1.40	-	-	11.20	0.1
8	16.50	9.9	60.50	-	4.40	-	0.50	8.00	0.2

(1) bulk composition of equilibrated aggregate, (2) enstatite and (3) forsterite in a compact unit (0.3 x 0.2 μm), (4) round glass unit (1.0 μm in diameter), 5-8 bulk compositions.

amorphous matrix. Bradley (1988) reported a unit (500 × 400 nm) wherein olivine grains intimately associated with fassaite and 'glass' showed equilibrium grain boundaries. The observations suggest a scenario of continued crystallization of amorphous to equilibrated (holocrystalline) coarse-grained PUs.

Coarse-grained PUs have a bimodal size distribution with mean diameters of 530 nm and 1140 nm. In IDP L2011A9 one unit is 3.0 × 1.0 μm in size (Rietmeijer 1994a) This distribution pattern supports coagulation of smaller units. In some IDPs, coarse-grained PUs broke-up into individual minerals. Klöck and Staderman (1994) measured the compositions of pyroxenes and olivines, Mg = 1.0 to 0.68 with a marked abundance of pure enstatite and forsterite (Fig. 36). It is possible that they are fragments of coarse-grained PUs rather than support an unequilibrated nature of these analyzed IDPs. The olivines include a Mn-rich olivine, $Fo_{98.8}Fa_{0.4}Te_{0.8}$. In general, the petrographic relationships make it possible to recognize a desegregated PU but in other instances it is not clear whether a single-crystal grain in the matrix was part of a PU or has another origin.

GEMS. This acronym stands for 'glass with embedded metals and sulfides' (Bradley 1994b). They may contain magnetite that is presumably due to terrestrial oxidation (Brownlee, pers. comm.). The GEMS were previously also described as 'unequilibrated aggregates' (Table 17) and 'fine-grained' components (Bradley et al. 1989). The term 'glass' is potentially misleading because it could imply a molten precursor. Round objects

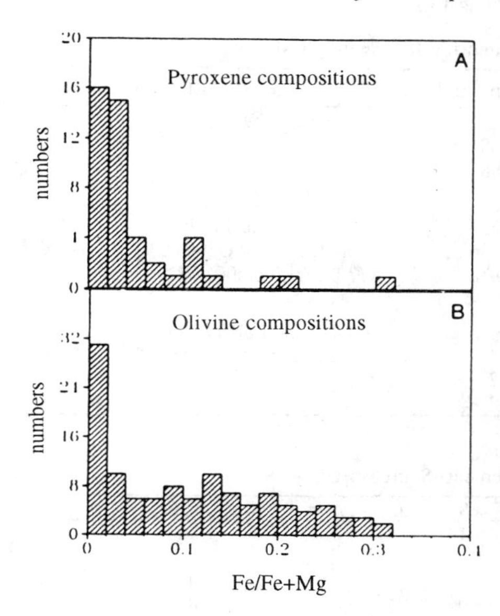

Figure 36. Histograms showing Fe/(Fe+Mg) distributions of pyroxenes and olivines from eight aggregate IDPs. The Mg ratios in the matrix units range from 1.0 to 0.68 and represent the full range for these units. The grains range from 0.1 to 1.0 μm in size. [Used by permission of the series editor of the American Institute of Physics, from Klöck and Stadermann (1994), Fig 2, p. 56]. Courtesy of Wolfgang Klöck.

as small as ~50 nm in diameter (Fraundorf 1981) and up to 1.0 μm in diameter (Bradley 1994a) in IDPs have been likened to quenched melt droplets. Yet, their spherical shape could reflect solid state re-adjustment to lower the surface free energy. In the case of GEMS, the term refers to a metastable material with properties that are intermediate between the ordered crystal state and the disordered state of a gas.

The GEMS are spherical units generally 0.1 to 0.5 μm in diameter with some units up to 1.0 μm that have an amorphous magnesiosilica matrix. The magnesium isotopic compositions are similar to chondritic IDPs values (Table 22). Bradley (1988) suggested that GEMS represent early solar system accretion. Bradley (1994b) argued that GEMS are preaccretionally irradiated grains due to the destruction of FeS grains embedded in a matrix of silicate glass due to continued irradiation by energetic H and He nuclei. As this process continues they ultimately become more rounded objects of an amorphous material with finely dispersed iron-nickel metal (kamacite) and iron sulfide nanocrystals <5 nm in size (Bradley 1994a,b). This proposed formation mechanism is unique but it relies only on similarities to the experimentally verified formation of polycrystalline rims on olivines and pyroxenes by irradiation processes. This process induces loss of magnesium and/or silicon towards the surface with resulting oxygen excess from ~20% up to 183% (Bradley 1994b, Bradley and Ireland 1996). Bulk compositions for GEMS and internal zoning are listed in Tables 23 and 24. The compositions for unequilibrated aggregates (or, GEMS) and their constituents are listed in Table 25. Aggregate IDPs that are dominated by GEMS (plus rare layer silicates) have an atmospheric entry velocity that suggests they have a cometary origin (Brownlee et al. 1995).

NOTE: The IDP research is interdisciplinary and attracts investigators from the Earth Sciences, Meteoritics, Chemistry, Physics, and Astrophysics. Thus comparisons of IDPs have to be made with data from different sources. Earth Scientists and Meteoriticists prefer presentations of chemical data as weight percent of the oxides and reduction in the form of element weight percents. Chemical data for ‘astronomical sources’ obtained by remote sensing methods yield chemical data in the form of ion or atomic abundances. Depending

Table 22. Magnesium isotopic compositions of chondritic IDPs with GEMS and of GEMS

$\Delta(^{25}Mg/^{24}Mg \pm 2\sigma$ (per mil)	$\delta^{26}Mg \pm 2\sigma$ (per mil)	REFERENCES
-0.8 ± 0.6	3.2 ± 1.1	Esat et al. 1979; chondritic aggregate IDPs (cf. Table 2)
2.2 ± 1.3	1.0 ± 2.8	
-1.9 ± 0.5	4.0 ± 0.8	
0.2 ± 0.7	3.5 ± 1.1	
1.9 ± 0.1	-0.15 ± 0.39	Esat and Taylor 1987; chondritic IDPs U2001B10, W7027A13, W7027A15, U2001A8
1.4 ± 0.1	0.19 ± 0.36	
1.0 ± 0.1	0.96 ± 0.38	
0.1 ± 0.2	2.86 ± 0.54	
-12 ± 10	9 ± 16	Bradley and Ireland 1994; GEMS in fragments from cluster IDP L2008#5, cf. Thomas et al. 1995a
-5 ± 7	-9 ± 14	
-1 ± 7	-5 ± 12	
14 ± 7	10 ± 11	

Table 23. Compositions of GEMS (atomic percent) Source Bradley 1994b

	O	Mg	Al	Si	S	Ca	Cr	Mn	Fe	Ni
bulk	61.9	2.9	0.8	16.9	6.1	-	0.3	-	11.1	-
bulk	56.2	22.3	0.6	13.3	3.2	-	0.1	-	4.2	0.1
center	52.9	24.6	1.2	12.1	3.9	0.2	0.3	0.2	4.4	0.2
edge	61.2	11.4	1.2	17.5	3.3	0.3	0.2	0.2	4.5	0.2
bulk	75.3	1.2	0.5	19.1	1.2	-	0.2	0.1	2.2	0.2
center	68.1	1.3	2.0	24.5	2.1	-	0.3	0.2	1.1	0.4
edge	70.1	1.5	-	11.6	3.8	-	-	-	12.7	0.3

Table 24. Bulk compositions of GEMS (atomic abundances normalized to Si)

	O	Mg	Al	S	Ca	Cr	Mn	Fe	Ni
1	3.76	0.38	0.12	0.14	0.01	0.02	0.005	0.42	0.03
2	3.7	0.24	0.11	0.08	0.02	0.04	-	0.38	0.03
3	3.96	1.11	0.08	0.10	0.08	0.03	0.008	0.70	0.05
4	3.51	0.65	0.11	0.05	0.03	0.02	0.004	0.12	0.006
5	4.99	0.82	0.12	0.12	0.04	0.01	0.01	0.43	0.03
6	-	0.83	0.09	0.07	0.04	0.015	0.01	0.49	0.05
7	-	0.58	0.13	0.14	0.02	0.013	0.011	0.42	0.03
8	-	0.66	0.11	0.11	0.037	0.013	0.005	0.69	0.06
9	-	0.49	0.13	0.39	0.027	0.02	0.017	0.72	0.07
10	-	0.959	0.064	0.54	0.045	0.017	0.01	0.656	0.027

Sources: 1-5 (Bradley and Ireland 1996); 6-9 'tar balls' in IDPs from the pyroxene IR class, and 10 bulk IDP (with 0.14 Na) for comparison (Bradley 1988, Table 2).

on the IDP investigator's preference or the target audience the IDP data are presented in both forms that can not be directly correlated.

Comparison of matrix units. The compositions of matrix units and their constituents are summarized in Figure 37. GEMS are petrologically distinct from PUs by the presence of iron-magnesium silicates in PUs and their absence in GEMS. In a ternary diagram Si-Mg-Fe (at %) GEMS and UAs occupy a low magnesium field consistent with their Mg,Fe-silicate-free nature. In this diagram, a line from the Si-corner to the Mg,Fe-join at Mg = 0.18 depicts the oxidation state of iron in this diagram separating reduced (Fe-FeO buffer) and oxidized ($FeO\text{-}Fe_2O_3$ buffer) phases. The compositions of ultrafine-grained PUs with iron-magnesium silicates and oxides plot below this line. The metal and sulfides

Table 25. Compositions of unequilibrated aggregates (element wt % normalized) (source: Bradley 1994a). 1,2: bulk compositions; 3-7: metal grains; 8-12: sulfides and 13-17: glass matrix.

	Mg	Al	Si	S	Ca	Cr	Mn	Fe	Ni
1	15.81	2.48	32.72	3.7	2.21	1.55	1.6	35.8	4.8
2	18.12	1.35	31.06	4.77	2.28	0.99	0.43	37.23	3.75
3	4.23	1.32	23.19	0.41	0.2	0.7	0.72	65.32	3.93
4	6.87	0.69	23.71	0.5	1.04	0.99	0.98	61.86	3.37
5	6.41	1.87	26.98	0.93	0.22	1.12	0.3	60.1	2.09
6	4.12	1.28	23.84	1.18	0.2	0.96	0.29	64.53	3.61
7	6.16	2.72	25.76	2.01	0.48	0.21	-	53.29	9.36
8	5.96	2.98	8.36	20.21	0.42	-	0.05	61.32	0.71
9	9.72	2.56	21.76	14.05	0.3	0.71	2.8	29.07	19.03
10	11.4	3.85	17.02	17.84	0.65	2.61	1.6	42.45	2.59
11	0.38	-	3.39	38.03	0.45	0.92	0.83	52.79	3.19
12	-	-	8.7	22.94	1.67	-	0.31	62.43	3.95
13	31.5	5.93	47.78	1.56	0.53	2.94	1.2	7.11	1.46
14	11.83	2.67	55.40	1.04	1.03	0.71	0.86	21.60	4.85
15	3.4	0.71	89.63	0.63	0.27	-	0.97	3.48	0.9
16	9.31	5.4	75.88	1.36	0.99	1.73	-	4.64	0.69
17	6.85	5.05	67.52	1.56	1.09	0.4	-	15.71	1.82

in unequilibrated aggregates (Table 25) plot in the field for reduced iron. The Mg ratios of both unequilibrated aggregates are 0.31 and 0.33 (Table 25). In fifteen GEMS the average value is Mg = 0.33 (Joswiak et al. 1996). The normalized (CI) Mg/Si, Al/Si and Si/O profiles in GEMS with gradients at a scale of 10 nm show no trends (Joswiak et al. 1996). Together these relationships point to different histories for GEMS and other types of PUs. This difference also shows in the Fe/Ni ratios. That is, the average Fe/Ni ratio for ultrafine-grained PUs is 23 (Table 19), (average) Fe/Ni = 13.3 in GEMS (Tables 23, 24) and Fe/Ni = 5.7 in the glass matrix of unequilibrated aggregates (see Table 25, analyses 1 and 2; just like the Fe/Ni ratios, the data in this table may be used to calculate the Mg-ratios).

Carbonaceous Units (CUs). Carbonaceous units are free of embedded silicates and sulfides (Rietmeijer 1994b). This unit was originally introduced to account for micrometer size clumps of poorly graphitized carbon in IDPs (Fig. 38) (Rietmeijer and Mackinnon 1985b). Irregularly shaped sub-micrometer units of vesicular carbonaceous materials occur in other particles (Fig. 39) (Klöck and Stadermann 1994, Thomas et al. 1994; Thomas et al. 1995a, Fig. 6a). The amorphous carbons are often fused into the 'cement' holding 'FGAs' together in some IDPs (Thomas et al. 1995). In other carbon-rich aggregate IDPs this 'cement' holds iron-magnesium silicates, sulfides and 'glass units' (or, GEMS) (Thomas et al. 1993a, Fig. 2; Thomas et al. 1994, Fig. 1; Bradley et al. 1996, Fig. 10). Vesicular carbons probably result from thermal devolatilization of organic materials. An apparent correlation between the presence of vesicular carbon and magnetite rims suggests volatile loss from carbon materials during atmospheric entry heating (Keller et al. 1996b). Vesicular carbon is probably not an experimental artifact due to interactions of the incident electron beam with the sample during TEM analysis. The response to thermal alteration of graphitizable carbons is a function of their structure and chemistry. Among other things, it could yield fluffy tangled networks of graphitic loops and tubules in pre-graphitic carbons (Fig. 40) (Rietmeijer 1992d). The properties of CUs remain poorly defined but the dust in the coma of comet P/Halley included an unexpected richness of discrete particles of light-elements with high carbon content (Jessberger et al. 1988). Fomenkova et al. (1992a) determined the total mass and number of the so-called CHON particles. Using a solid matter density of 2.0 g cm^{-3}, I calculate that an 'average' CHON particle was 0.6 μm in diameter. A type of CHON-like material in CUs is likely to fuse

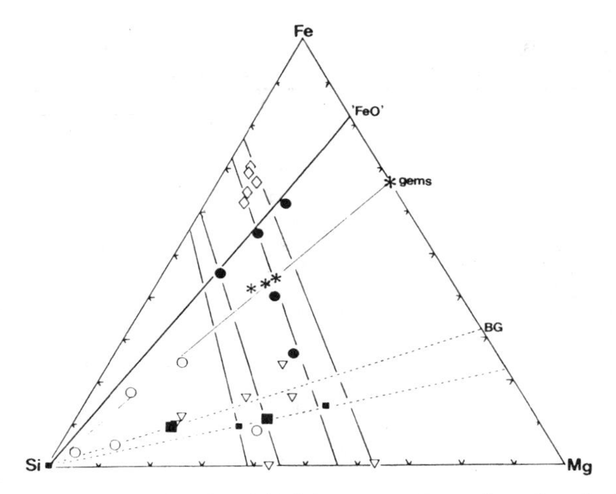

Figure 37. Ternary diagram Mg-Fe-Si (el. ratio) of the compositions of matrix units in aggregate IDPs. Unequilibrated and reduced aggregates (Bradley 1994a), and the average of 15 GEMS (star labeled 'gems') after (Joswiak et al. 1996). The matrix of unequilibrated aggregates (open circles) and the enclosed FeNi-metal (diamonds) (Bradley 1994a). Compositions of equilibrated aggregates (Bradley 1994a, Bradley et al. 1989) (triangles). Cluster compositions in coarse-grained PU 'Big Guy' (BG, see Fig. 45) (big and small squares) and its bulk composition (Rietmeijer, unpublished data) and the compositions of sulfur-free ultrafine-grained PUs (dots) (Rietmeijer, unpublished data, cf. Rietmeijer 1997c). The dashed lines are 'fractionation' lines for two coarse-grained PUs. The light solid line represents the 'average' mg-ratio of GEMS, although they show a much wider range (see text). The solid line labeled 'FeO' separates reduced (Fe-FeO) and oxidized ($FeO-Fe_2O_3$) fields. Composition lines are indicated for olivine, serpentine, pyroxene and smectite.

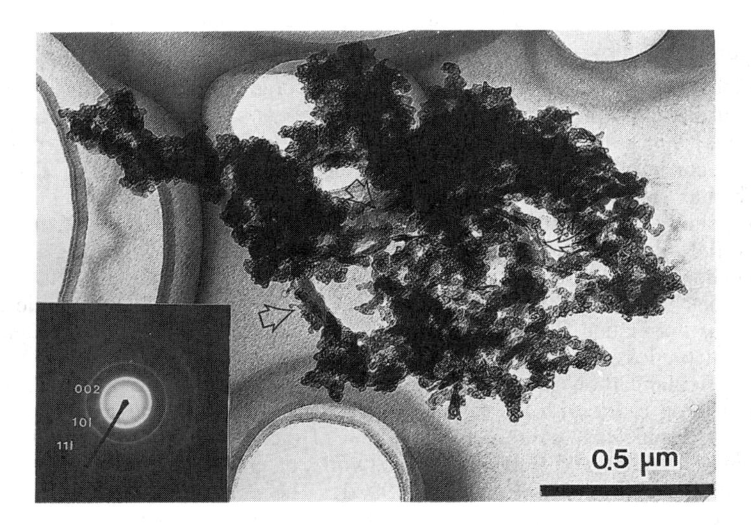

Figure 38. A poorly graphitized carbon grain in aggregate IDP W7029C1 dispersed on a holey carbon film. The open arrows indicate the concentric-circular structure that is common in this material. The polycrystalline (or 'ring') electron diffraction pattern is consistent with poorly graphitized carbon. After Rietmeijer and Mackinnon (1985) Nature 316 733-736, Fig.1d, p. 734.

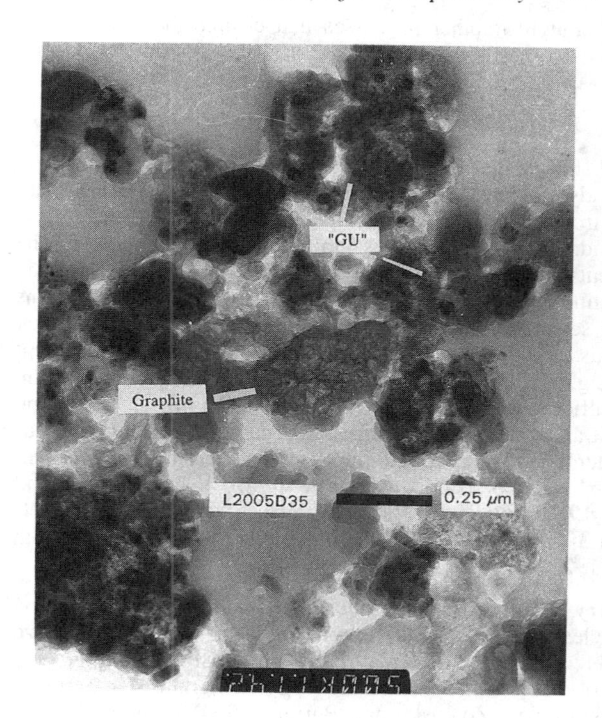

Figure 39. Bright-field transmission electron micrograph of chondritic aggregate IDP L2005D35 showing ultrafine-grained PUs (originally labeled 'GU') and a carbonaceous unit (labeled graphite) but which is probably a poorly graphitized carbon. After Klöck and Stadermann (1994), Fig. 1, p. 55. Courtesy of Wolfgang Klöck.

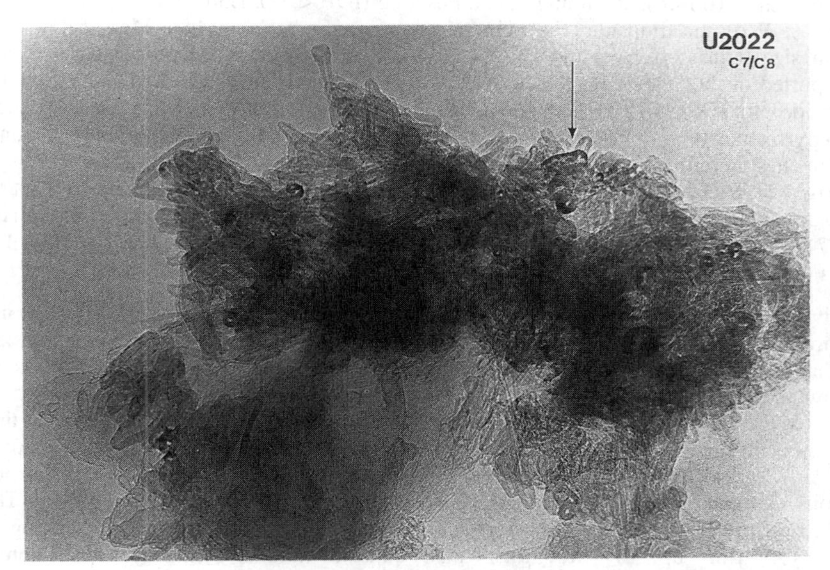

Figure 40. A poorly graphitized carbon grain in chondritic aggregate IDP U2022C7/C8 with buckminster-fullerene structures, i.e. open 'bucky-onions' and 'bucky-tubes'. [Used by permission of the editor of Geochimica et Cosmochimica Acta, from Rietmeijer (1992), Fig.2b, p. 1668.]

easily together during thermal alteration or other processes that change the composition and/or structure of CUs. Contiguous areas of fused CUs are probably common in C-rich chondritic aggregate IDPs.

Single-crystal grains

Single-crystal grains <10 to 100 nm in size and grains >100 nm among the matrix units of aggregate IDPs were already recognized in the earliest studies (Table 3). The platy single-crystal grains are mostly iron-magnesium silicates between 10 and 30 nm thick, anhedral to hexagonal Fe,Ni-sulfide grains, amorphous grains and rare plagioclase. The database does not permit statements about the size distributions. It is also not clear if the 100-nm size marks a critical distinction. There is some indication that embedded grains have a bimodal size distribution separated at about 100 nm (Christoffersen and Buseck 1986a, Germani et al. 1990). This bimodal distribution may indicate genetically different grains. The petrographical data (e.g. Bradley 1994a) are too few for a definitive evaluation but they hint that silicate and sulfide grains <100 nm are associated with at least some GEMS. The grains > 500 nm that are embedded in the matrix of aggregate and collapsed-aggregate IDPs include quadrilateral pyroxenes (including Ti-diopside), iron-magnesium olivines, Fe,Ni-sulfides and 'glass' (Tomeoka and Buseck 1984, Fig 3; Christoffersen and Buseck 1986a, Fig. 2; Bradley 1988, Fig.7; Rietmeijer 1989a, Fig.1, 1994a, Fig. 1; Bradley et al. 1989, Figs. 1 and 3; Germani et al. 1990, Fig. 6; Bradley 1994a, Fig. 1a; Zolensky and Barrett 1994, Fig. 1; Zolensky and Thomas 1995, Fig. 1).

Pyroxenes and refractory minerals. The silicate grains range from 'a few hundreds of nm (for most particles) (Zolensky and Barrett 1994) up to ~3 μm in size (Rietmeijer 1992b), although there is probably no real upper limit. Brownlee (personal comm.) observed grains up to 6 to 10 μm in IDPs. Enstatite (Mg >0.9) grains can be up to 5 μm (Christoffersen and Buseck 1986a). Zolensky and Barrett (1994) reported Ca-free pyroxenes (En = 100-46), including zoned enstatite (IDP <2005E36), Ca-poor pyroxenes (<25 mol % Wo) and diopside (Fig. 41). A pyroxene, $Fe_{0.06}Ca_{0.05}Mn_{0.03}Mg_{0.81}SiO_3$, was ~250 nm size (Christoffersen and Buseck 1986b). These pyroxene compositions match those reported in those studies (Klöck et al. 1990). Germani et al. (1990) found fassaite (Ti-diopside) in IDP W7029J10. Klöck et al. (1989) reported low-Fe, Mn-enriched (LIME) pyroxenes (Ca-poor) with up to 5.1 MnO wt % (Mg >0.95) (Table 26). Calcium-rich pyroxenes including diopside and augite, e.g. $En_{48}Fs_{14}Wo_{39}$ occur in some aggregate IDPs (Fig. 42) (Rietmeijer 1989a, 1994a). Some Ca-rich clinopyroxene has up to 4 wt % Al_2O_3. An unusual assemblage of diopside, multiple twinned fassaite, enstatite, olivine (Fo = 99–77), anorthite, perovskite ($CaTiO_3$) and spinel ($MgAl_2O_4$) occurs in a CP IDP that was part of a large cluster particle (Table 27) (Christoffersen and Buseck 1986b).

Olivine and enstatite, including untwinned clinoenstatite grains with (100) stacking faults, are abundant. Diopside and spinel are the most abundant of these minerals and are present as discrete anhedral crystals. More commonly, they are polycrystalline grains or intergrowths of a pair of these crystals along straight boundaries, or complex textures of several grains 50-100 nm in diameter. Similar intergrowths occur in IDP L2005K10 that also contains olivine (Fo = 100-76), enstatite (En = 98-81), pigeonite and anorthite, An < 94 mol % (Zolensky and Lindstrom 1992). Plagioclase and melilite co-occur in porous anhydrous aggregate IDP of the pyroxene (IR) subclass (Brownlee et al. 1994). The Ca,Al,Ti-rich minerals occur within aggregate particles. In addition, there are aggregate particles (~5 μm up to ~15 μm) that consist (almost) entirely of hibonite [$CaAl_{11}Ti_{0.5}Mg_{0.5}O_{19}$], gehlenite and perovskite in variable proportions along with diopside and Ti-diopside, and several of these IDPs contain corundum (Zolensky 1987, Greshake et al. 1996). These grains are between 50 nm and 500 nm. The oxygen isotopic composition

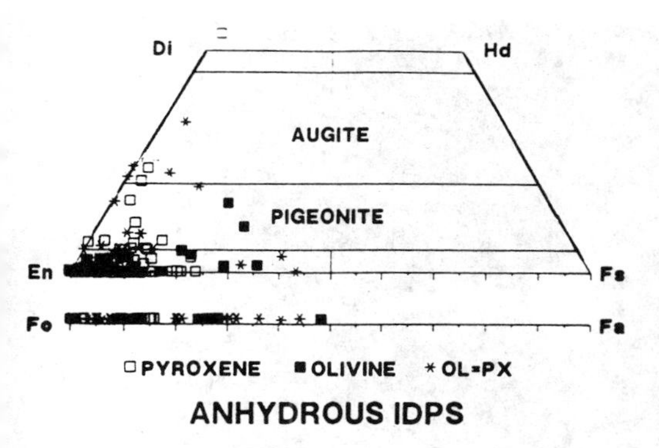

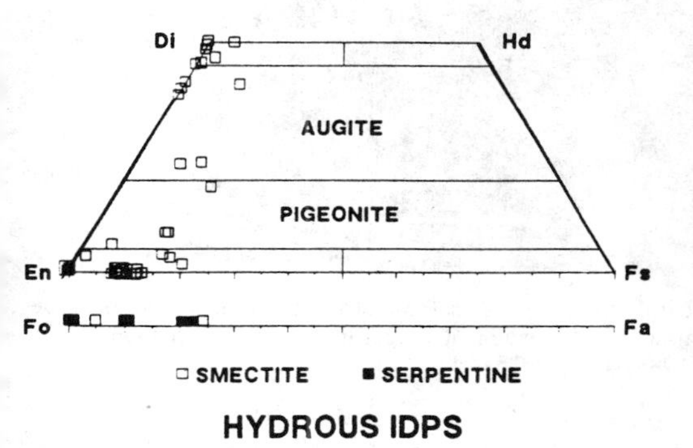

Figure 41. The pyroxene quadrilateral showing the compositions of pyroxenes and olivines in chondritic aggregate IDPs. Hydrated IDPs are identified on the basis of the presence of layer silicates but irrespective of their abundance. For anhydrous IDPs the symbols refer to the most abundant iron-magnesium silicate, i.e. pyroxene (open squares), olivine (solid squares), or about equal proportions of both (stars). Hydrated IDPs that are dominated by smectite are shown by open squares. Solid squares indicate that serpentine is the dominant layer silicate. [Used by permission of the editor of Meteoritics and Planetary Science, from Zolensky and Barrett (1994), Fig.3, p. 619.]. Courtesy of Mike Zolensky.

Table 26: Mn-bearing pyroxenes in chondritic IDPs (source: Klock et al., 1989)

	W7029*A28	U2015*B2	W7013E17	U2015D21	W7027H14	W7013D12	U2022G7	W7013C16	W7013G1
SiO_2	56.76	54.85	57.89	61.24	59.20	56.15	55.81	54.68	56.07
Al_2O_3	1.69	4.01	0.77	0.19	0.27	2.39	5.06	5.40	2.62
MgO	32.9	37.00	36.95	36.46	39.56	30.16	33.63	33.83	32.60
CaO	1.27	0.57	1.43	0.05	0.08	1.90	1.40	1.00	2.02
TiO_2	0.08	0.00	0.16	0.07	0.05	0.03	0.17	0.00	0.18
Cr_2O_3	1.55	0.93	0.68	0.78	0.49	1.52	1.81	1.60	1.47
MnO	1.72	1.78	0.95	0.87	0.26	5.12	1.04	2.03	4.14
FeO	4.03	0.87	1.17	0.34	0.09	2.73	1.08	1.46	0.90
En	0.91	0.98	0.95	0.99	1.00	0.91	0.95	0.95	0.94
Fs	0.06	0.01	0.02	0.01	0.00	0.05	0.02	0.03	0.02
Wo	0.03	0.01	0.03	0.00	0.00	0.04	0.03	0.02	0.04

of their host IDPs indicates an extraterrestrial origin with $\delta^{17}O_{smow}$ (°/oo rel.) to $\delta^{18}O_{smow}$ (°/oo rel.) ratios along the oxygen isotope fractionation line for anhydrous minerals in carbonaceous chondrites (McKeegan 1987, Greshake et al. 1996).

Distinct enstatite whiskers, ribbons and rods differ in their aspect ratio from the common enstatite platelets in chondritic aggregates with aspect ratios >20 (Bradley et al. 1983, Rietmeijer 1989a). They occur as thin blades (up to ~10 mm long and 200-300 nm wide; cf. IDPs L2005E4 and –E5 in NASA/JSC Cosmic Dust Catalog 11(1), 1990; Bradley 1994a, Fig. 1a) and as rods (Bradley 1994a, Rietmeijer 1989a). They are elongated along the crystallographic [100] direction and in some whiskers the displacement

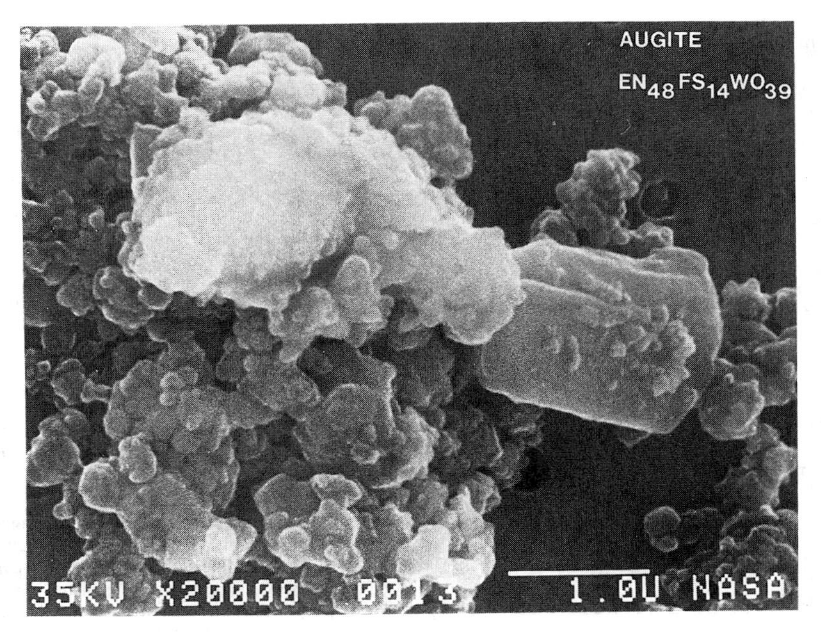

Figure 42. Scanning electron micrograph of an euhedral augite crystal in the matrix of granular units in chondritic aggregate IDP W7010*A2. [Used by permission of the editor of Proceedings of the 19th Lunar and Planetary Science Conference, from Rietmeijer (1989), Fig. 1, p. 514.]

Table 27. Compositions of refractory minerals in a CP IDP (source: Christoffersen and Buseck 1986b)

	CaO wt %	Al_2O_3 wt %	Cr_2O_3 wt %	TiO_2 wt %
Diopside	19.0-26.6	0.7-3.8	4.6,8.1	
Fassaite	22.56	19.89	0.4	5.5
Anorthite	18.0-19.8	34.8-35.9		
Perovskite	41.2			58.8
Spinel		67.7-73.1	1.7-2.7	

of (100) stacking faults is due to a screw dislocation. Only pure enstatite displays this particular habit (Bradley et al. 1983). Whiskers are rare in (micro)-meteorites. The only whiskers reported in 'conventional' meteorites are wollastonite whiskers growing inside vugs in the Allende meteorite wherein they have a possibly secondary origin (Miyamoto et al. 1979). Some enstatite platelets consist of fine intergrowths of ortho- and clino-enstatite lamellae typical of the proto-enstatite inversion (Bradley et al. 1983).

Olivines. The olivine grains are also 'a few hundreds of nm (for most particles)' (Zolensky and Barrett 1994) with compositions ranging from pure forsterite to Fo = 52 (Fig. 41). Previous studies also found values as low as Fo = 0.5. Minor elements in olivines are CaO = 0.1-0.2, MnO = 0.7-0.4 and Cr_2O_3 = 0.3 (wt %) (Christoffersen and Buseck 1986a). The Mn-rich pyroxenes in chondritic IDPs are accompanied by Mn-rich olivines with up to 2.6 wt % MnO, Mg >0.95 (Klöck et al. 1989) (Table 28). The manganese content shows a linear increase from MnO is 0.65 wt % in pure forsterite to 2.6 wt % in Fo = 0.95. Aggregate IDP W7029*A27 is the only particle wherein the intra-particle variations among 'silicates' are documented (Stephan et al. 1994) (Table 29). The

Table 28. Mn-bearing olivines in chondritic IDPs (source: Klock et al., 1989)

	W7029*A28	U2015*B2	W7013E17	W7017A14	U2034C10	U2015D21	U2011*B2	W7013G1	W7027H14	W7029B12	W7013D12
SiO_2	42.95	41.89	41.35	41.71	42.65	42.87	43.10	39.95	41.25	41.50	41.22
Al_2O_3	0.00	0.22	0.00	0.00	0.00	0.00	0.15	0.00	0.00	0.00	0.00
MgO	55.84	57.22	55.29	56.60	56.00	55.61	56.14	51.61	53.11	56.18	54.18
CaO	0.08	0.11	0.00	0.00	0.00	0.05	0.00	0.06	0.05	0.06	0.14
TiO_2	0.00	0.00	0.00	0.00	0.00	0.00	0.00	0.06	0.00	0.00	0.00
Cr_2O_3	0.14	0.17	0.18	0.32	0.32	0.36	0.18	0.76	0.17	0.49	0.52
MnO	0.67	0.32	0.92	0.93	0.82	0.95	0.16	2.63	1.88	1.01	1.52
FeO	0.32	0.07	2.26	0.44	0.21	0.16	0.27	4.93	3.54	0.75	2.42
Fo	0.995	0.995	0.97	0.99	0.99	0.99	0.10	0.93	0.95	0.98	0.96
Fa	0.000	0.000	0.02	0.00	0.00	0.00	0.00	0.05	0.03	0.01	0.03
Te	0.005	0.005	0.01	0.01	0.01	0.01	0.00	0.02	0.02	0.01	0.01

Table 29. Compositions of grains in the matrix of aggregate IDP W7029*A27 (source: Stephan et al. 1994)

	PYROXENES				OLIVINES		FELDSPAR		GLASS		SILICA
MgO	39.2	33.3	31.9	7.0	55.8	44.1	-	-	13.8	5.5	-
Al_2O_3	1.7	4.0	0.9	29.2	-	-	36.6	33.4	4.2	6.0	-
SiO_2	58.4	54.2	57.0	35.8	42.9	39.1	45.8	52.4	69.4	67.5	96.7
K_2O	-	-	-	-	-	-	-	-	-	-	0.7
CaO	-	1.8	1.2	21.0	-	-	17.7	14.2	0.2	5.1	-
TiO_2	-	0.2	0.2	2.3	-	-	-	-	-	-	-
Cr_2O_3	0.7	1.6	1.4	0.5	-	-	-	-	-	-	-
MnO	-	1.1	3.7	-	0.7	0.8	-	-	-	-	-
FeO	-	3.7	3.6	4.0	0.3	16.0	-	-	12.4	15.8	2.6
En, Fo	1.00	0.91	0.91	0.29	1.00	0.83					
Fa, Fs	-	0.06	0.06	0.09	-	0.17					
Wo	-	0.03	0.03	0.62							
An							1.00	1.00			

pyroxenes include $Mg_{0.85}Fe_{0.06}Mn_{0.06}Ca_{0.03}Si_2O_6$. There are no petrographic data available. Olivine and silica may be from coarse-grained PUs (cf. W7029*B1, Klöck and Stadermann 1994) (Fig. 43) but rare euhedral silica grains are present in the matrix of aggregate IDPs (Rietmeijer 1989a).

Sulfides. Iron-nickel sulfide grains are mostly embedded in the particle matrix. Thomas et al (1991) reported coarse-grained sulfides with up to 30 wt % Ni that are mostly ~3 to 4 μm in size but grains up to 8 μm were present. In general, sulfides embedded in chondritic IDP matrix are <8 μm in diameter (Rietmeijer 1992b). Grain morphologies range from rounded anhedral masses to euhedral pseudohexagonal plates (Zolensky and Thomas 1995). Zolensky and Thomas (1995) surveyed the compositions of sulfides that range from 200 nm to 15 μm in diameter. They overwhelmingly support pyrrhotite and stoichiometric troilite with Ni <5 at % Ni. The nickel content ranges from zero to within the range of pentlandite (Fig. 44). The database supports that pyrrhotite (0.1 to 0.5 wt % Ni) and pentlandite (Ni/Fe (at %) = 0.5 to 0.7) are the common sulfides in chondritic aggregate IDPs. Zolensky and Thomas (1995) reported that Ni-rich iron-sulfides are abundant in IDPs that contain any (but unspecified) quantity of layer silicates. In one IDP, Tomeoka and Buseck (1984) identified low-Ni sulfide as pyrrhotite with a hexagonal superstructure (Table 30). The superstructure dimensions are a = 0.69 nm = 2A and c = MC whereby A and C are dimensions of the pyrrhotite subcell (A = 0.345 nm and C = 0.575 nm), and M the multiplicity of a series of pyrrhotite superstructures. The cubic, high nickel (Table 31), sulfides are pentlandite. These sulfides occur individually but most commonly they are in clusters. Low-Ni pentlandite (Ni <3 at %) (Table 31) is unique to IDPs. It is present as thin

Figure 43. Dark-field transmission electron micro-graph of chondritic aggregate IDP W7029*B1 with ultrafine-grained matrix units (labeled 'GU') and a coarse-grained polyphase unit of forsterite, $Fo_{98.8}Fs_{0.4}Te_{0.8}$, and silica. [Used by permission of the series editor of the American Institute of Physics, from Klöck and Stadermann (1994), Fig 7, p. 60.]. Courtesy of Wolfgang Klöck.

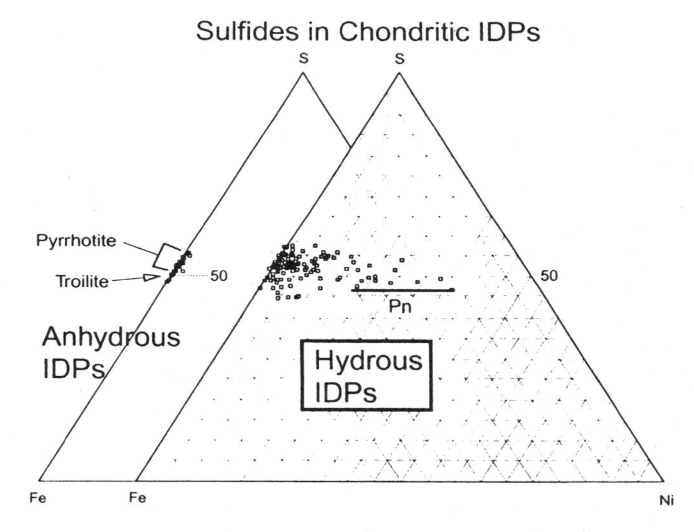

Figure 44. Compositions of micrometer sized iron nickel sulfides in chondritic aggregate IDPs. Hydrated IDPs are identified on the basis of the presence of layer silicates but irrespective of their abundance. [Used by permission of the editor of Geochimica et Cosmochimica Acta, from Zolensky and Thomas (1995), Fig. 2, p. 4709.] Courtesy of Mike Zolensky.

Table 30. Pyrrhotite compositions in a Low-Ca hydrated aggregate IDP (source: Tomeoka and Buseck 1984) (atomic percentages normalized)

Fe	46.1	48.5	48.0	43.9	48.7	48.0	45.5	46.0
Ni	0.4	0.9	0.4	3.5	0.9	0.4	1.7	2.2
S	53.5	50.6	51.6	52.6	50.4	51.6	52.8	51.8
Ni/Fe	0.01	0.02	0.01	0.08	0.02	0.01	0.04	0.05
(Fe+Ni)/S	0.87	0.98	0.94	0.90	0.98	0.94	0.89	0.93

(10 to 30 nm) euhedral plates up to ~800 nm in size (Tomeoka and Buseck 1984). These grains have mosaic textures. Germani et al. (1990) identified troilite, pyrrhotite and pentlandite in several particles. Rare Fe-rich, Ni-bearing sphalerite occurs in aggregate IDPs (Rietmeijer 1989a) and in a non-aggregate IDP of the 'olivine' IR class (Christoffersen and Buseck 1986a).

Table 31. Compositions of pentlandite and low-nickel pentlandite in a Low-Ca hydrated aggregate IDP (source: Tomeoka and Buseck 1984) (atomic percentages normalized)

	PENTLANDITE				LOW-NICKEL PENTLANDITE			
Fe	24.0	29.1	32.1	32.0	52.8	53.0	49.8	52.6
Ni	28.1	24.1	20.8	20.8	1.3	0.1	2.2	1.4
S	47.9	46.8	47.1	47.2	45.9	46.9	48.0	46.0
Ni/Fe	1.2	0.8	0.65	0.65	0.02	<0.01	0.04	0.03
(Fe+Ni)/S	1.1	1.1	1.1	1.1	1.2	1.1	1.1	1.2

Amorphous grains. The word 'grains' is perhaps misleading for the amorphous materials that occur as discrete matrix units. They can be fused into large contiguous patches (Fig. 45). The compositions range from Ca,Al-bearing ferromagnesiosilica materials with variable amounts of calcium and aluminum (Bradley 1988; Rietmeijer 1994a, 1997c) and variable Mg/Fe ratio ('iron-rich silicate glass,' Germani et al. 1990, Figs 5 and 6) to Fe,Mg-bearing aluminosilica materials (Fig. 46). Rare silica glass occurs also (Thomas et al. 1995b, Klöck and Stadermann 1994). An iron-rich 'pyroxene glass' is present in collapsed aggregate IDP U230A34 (Bradley 1988). The similarity of the glass and Fe-smectite compositions in the matrix (Table 32) implies that layer silicates formed during in situ hydration of the glass (Germani et al. 1990). The anorthite compositions in aggregate IDP W7029*A27 (Table 29) indicates a non-stoichiometric phase with excess silica similar to those in other IDPs (Rietmeijer 1991). Both glass compositions in W7029*A27 are within the range of coarse-grained PUs (Rietmeijer 1997c). The compositions of non-stoichiometric plagioclase, Ca/(Ca+Na) = 1.0-0.0, and alkali-feldspars, K/(K+Na) = 1.0-0.06 (Table 33) and layers silicates (Table 34) in hydrated IDPs are similar to those of the amorphous materials (Fig. 46) (Rietmeijer 1991). Both observations link the formation of layer silicates to the presence of amorphous materials with the appropriate composition.

Miscellaneous. Grains in this category are rare inclusions in the matrix of chondritic aggregates. They include Mg-rich carbonate (Germani et al. 1990, Fig. 1a) and euhedral breunnerite and magnesian siderite grains of ~1 μm (Table 35) (Tomeoka and Buseck 1986). Also, euhedral magnetite up to 300 nm in CP IDP W7029B1 (Christoffersen and Buseck 1983b) that based on grain size alone is not a dynamic pyrometamorphic alteration product. In addition, nanometer sized grains of spinel, chromite, tin oxide (Fraundorf 1981, Rietmeijer 1989b); chromium oxide (~100 nm in size) (Germani et al. 1990), bismuth metal (Mackinnon and Rietmeijer 1984), and titanium metal and Magnéli phases, Ti_nO_{2n-1}, n = 4,5,6 (Mackinnon and Rietmeijer 1987, Fig. 6; Rietmeijer and Mackinnon 1990). Rare low-Ni iron phosphide (?) and Ca-phosphate (?) is present (Rietmeijer 1989a, Zolensky and Lindstrom 1992).

Non-chondritic IDPs

Coarse-grained, non-chondritic, IDPs among the collected particles are recognized by chondritic material at the surface. They occur in the NASA/JSC Cosmic Dust Collection as eu- and subhedral iron-nickel sulfide and subhedral magnesium-iron silicate grains that may be single-crystal particles. Schramm et al. (1989) defined a coarse-grained IDP as a particle wherein mineral grains >3 μm comprise more than 50% of the particles mass. Petrological

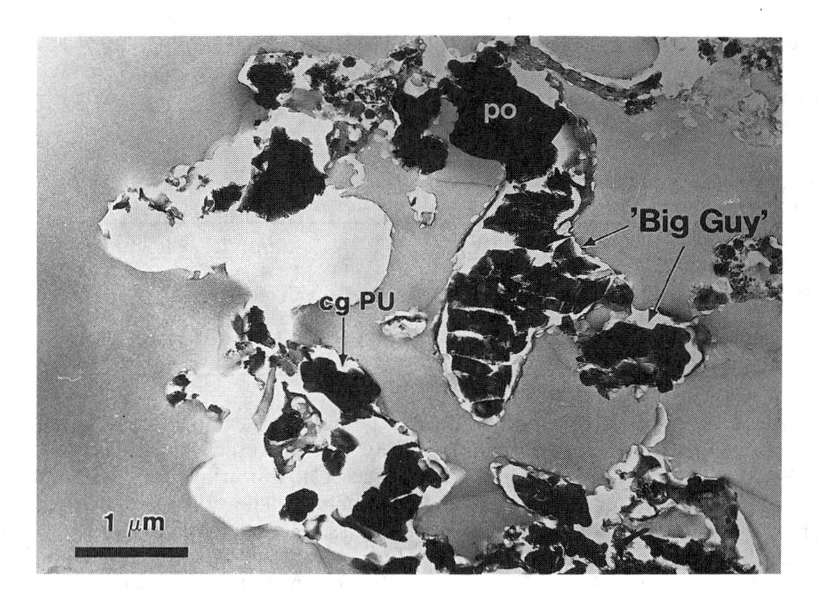

Figure 45. Transmission electron micrograph of an ultrathin section of aggregate IDP L2011A9 showing the coarse-grained PU 'Big Guy' associated with pyrrhotite (po). Its compositions are shown in Fig. 33. A coarse-grained polyphase unit of more common size is also indicated. The shattered appearance of the section and loss of sample (white areas) from the embedding epoxy (gray background) is typical for most IDPs. After Rietmeijer (1994) AIP Conf Proc 310: 231-240, Fig. 1, p. 234.

Table 32. Pyroxene glass and matrix compositions in IDP U230A34 (element wt % normalized).

	Pyroxene glass	Smectite matrix
Mg	12.73	17.99
Al	2.96	5.27
Si	39.24	52.58
S	0.36	1.15
Ca	0.10	-
Cr	1.14	0.64
Mn	0.46	0.17
Fe	42.63	21.67
Ni	0.34	0.63

Table 33. Compositions of non-stoichiometric plagioclase and alkali-feldspars in chondritic aggregate IDPs (Source: Rietmeijer, 1991)

	#1	#2	#3	#4	#5	#6	#7	#8	#9	#10
SiO_2	50.6	38.5	42.7	54.6	64.6	74.7	73.2	66.4	32.0	35.9
Al_2O_3	23.5	44.6	41.6	36.9	27.5	15.7	18.4	25.9	53.6	56.7
Na2O	0.0	4.5	9.7	4.5	2.0	3.7	2.2	1.7	2.0	0.8
K_2O	1.2	1.4	0.2	0.2	5.2	5.1	5.7	5.6	11.1	5.5
CaO	22.9	8.2	5.3	3.2	0.1	0.0	0.0	0.1	0.0	0.0
MgO	0.6	0.7	0.5	0.6	0.3	0.8	0.4	0.1	1.1	1.1
FeO	1.1	1.2	0.1	0.1	0.2	0.0	0.2	0.1	0.2	0.0
An	94.1	45.3	23.0	27.6	0.5	0.0	0.0	0.6	0.0	0.0
Ab	0.0	45.3	76.1	70.5	36.6	52.7	36.7	31.1	21.4	19.4
Or	5.9	9.4	0.9	1.8	62.9	47.3	63.3	68.3	78.6	80.6

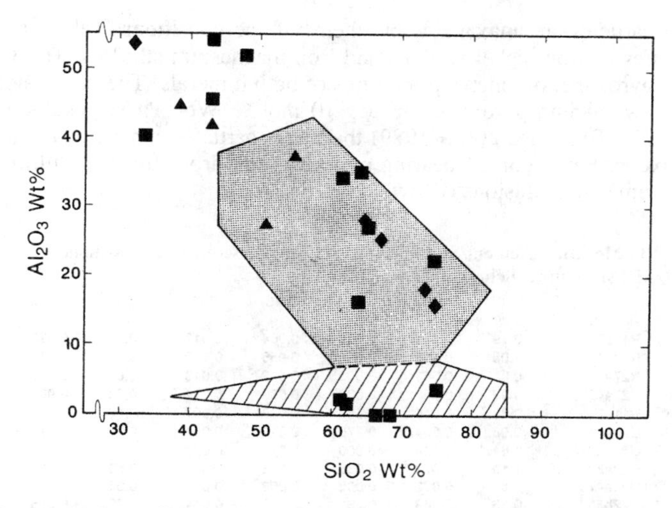

Figure 46. Diagram Al_2O_3 vs. SiO_2 (wt %) showing the compositional range of amorphous materials in chondritic aggregate IDPs after Thomas et al. (1989). They recognized distinct low-Al_2O_3 (hatched) and high-Al_2O_3 (shaded) compositions. Superimposed on these fields are the compositions of non-stoichiometric plagioclase (triangles), alkali-feldspar (diamonds) and layer silicates (squares) in three aggregate IDPs. [Used by permission of the editor of Trends in Mineralogy (Council of Scientific Research Integration, India) From Rietmeijer (1992) Fig. 8, p. 32.]

Table 34. Compositions of layer silicates in chondritic aggregate IDPs (source: Rietmeijer, 1991)

	#1	#2	#3	#4	#5	#6	#7
SiO_2	43.1	33.4	74.9	62.0	61.1	43.2	74.4
Al_2O_3	45.5	40.0	3.7	1.9	2.2	55.0	22.5
Na2O	1.1	2.0	5.1	2.8	4.0	0.6	1.9
K_2O	7.7	1.2	0.0	0.0	0.0	0.0	0.3
CaO	0.0	0.0	0.0	0.0	0.0	0.1	0.0
MgO	1.5	9.5	16.2	30.6	32.1	0.0	0.8
FeO	1.0	13.8	0.1	2.7	0.5	0.2	0.1
total	99.9	99.9	100.0	100.0	99.9	99.1	100.0
	#8	#9	#10	#11	#12	#13	#14
SiO_2	63.9	61.0	47.7	66.3	68.1	64.9	67.1
Al_2O_3	35.1	34.4	51.8	0.0	0.0	27.4	16.9
Na2O	0.5	0.8	0.0	0.0	0.0	0.0	0.0
K_2O	0.0	0.5	0.0	0.0	0.0	2.7	3.0
CaO	0.0	0.1	0.2	0.0	0.0	0.0	0.8
MgO	0.2	0.6	0.0	31.7	28.9	3.8	3.7
FeO	0.4	2.6	0.2	1.9	2.7	0.8	8.5
total	100.1	100.0	99.9	100.0	99.7	99.6	100.0

Table 35. Euhedral carbonates in IDP r21-M1-9a (source: Tomeoka and Buseck 1986)

$MgCO_3$	$FeCO_3$	$CaCO_3$	$MnCO_3$
91.0	6.8	1.5	0.7
72.7	26.2	0.3	0.8
59.2	38.7	1.4	0.7
47.1	49.6	2.7	0.6
36.2	57.3	5.0	1.5
20.9	70.2	7.1	1.8

data for these particles are unavailable but their bulk compositions (Table 36) support that they are mixtures of iron-nickel sulfides and iron-magnesium silicates. The (Mg + Fe)/Si ratios support pyroxene, olivine, and mixtures of both minerals. The bulk calcium contents suggest very low-calcium pyroxenes up to ~10 mol % Wo. These results are consistent with the report by Brownlee et al. (1989) that these particles consist of forsteritic olivine, Mg-rich pyroxene, Fe and/or Ca-bearing pyroxene, and iron sulfides. Sulfide and silicate IDPs contain kamacite inclusions (Brownlee et al. 1989).

Table 36. Element/Si (atom ratios) for coarse-grained non-chondritic IDPs. (Source: Schramm et al. 1989)

	Mg/Si	Al/Si	S/Si	Ca/Si	Cr/Si	Fe/Si	Ni/Si
R12a7	0.79	0.041	0.160	0.064	0.017	0.35	0.023
R27a1	1.05	0.105	0.000	0.056	0.012	0.55	0.000
R27a14	0.91	0.041	0.009	0.032	0.016	0.07	0.000
R27a18	0.98	0.000	0.000	0.003	0.014	0.33	0.000
U11a2	0.68	0.046	0.542	0.033	0.007	1.18	0.091
U20a14	0.98	0.064	0.270	0.023	0.017	0.74	0.012
U20a32	1.61	0.024	0.000	0.013	0.022	0.38	0.000
U22a32	1.19	0.077	0.036	0.033	0.049	0.63	0.000
U22a47a	0.89	0.072	0.000	0.062	0.013	0.04	0.011
U22b33	0.83	0.049	0.261	0.046	0.014	0.79	0.018
U22c2	0.68	0.259	0.009	0.316	0.000	0.02	0.000
U22c9	1.08	0.047	0.783	0.023	0.000	0.64	0.031
U22c10	0.77	0.047	0.058	0.056	0.011	0.13	0.008
U22c11	0.87	0.086	0.000	0.042	0.009	0.08	0.000
U22c14	0.93	0.073	0.016	0.053	0.024	0.53	0.022
U22c45	0.86	0.060	1.216	0.140	0.000	1.84	0.041
U24b9	0.96	0.093	0.078	0.059	0.016	0.23	0.000
U24b17	0.87	0.127	0.140	0.140	0.021	0.67	0.000
U24b19	1.15	0.060	0.350	0.046	0.013	0.57	0.023
U24b23	0.85	0.023	0.068	0.000	0.000	0.17	0.000
U24b24	1.03	0.065	0.395	0.122	0.025	0.53	0.018
U24b25	0.98	0.096	0.123	0.112	0.016	0.32	0.000
U24b45	1.10	0.077	0.000	0.000	0.000	0.58	0.000
U30a15	1.18	0.110	0.231	0.008	0.009	0.79	0.000
U30a21	0.75	0.103	0.160	0.437	0.017	1.13	0.000
U30a24	1.21	0.000	0.041	0.010	0.020	0.52	0.000
U30a34	0.87	0.055	0.146	0.027	0.014	0.27	0.014
U30a36	1.59	0.047	0.132	0.022	0.000	0.40	0.000
U30a38	0.99	0.045	0.259	0.027	0.000	0.38	0.000
U30b5	1.45	0.065	1.051	0.049	0.000	1.08	0.263
U30b31	1.01	0.124	0.000	0.079	0.015	0.04	0.000
U30b41	1.06	0.069	0.605	0.116	0.014	0.96	0.000
U35a15	1.3	0.049	0.056	0.039	0.017	0.24	0.000
U35a2	4.17	0.133	0.118	1.602	0.000	1.91	0.000
U35a3	4.98	0.194	0.956	0.381	0.000	1.87	0.121
U35a30	0.71	0.059	0.032	0.225	0.030	0.12	0.000

The sizes of coarse-grained particles analyzed by Schramm et al. (1989) probably define a bimodal distribution, viz. (1) mean = 8.2 μm (1σ = 2.1 μm; range 4-13 μm; N = 25) and (2) mean = 15.6 μm (1σ = 2.6 μm; range 12-22 μm; N = 11). Based on size alone, the grains of the first population are similar to those that are embedded in aggregate IDP matrix (see above). The second population could represent a distinct group of larger particles. The olivine- and pentlandite-rich IDP (Christof-fersen and Buseck 1986a) probably has an affinity with coarse-grained IDPs. Another particle consists of a sulfide grain attached to a forsterite grain (Rietmeijer 1996c). Compact IDP L2005Z17 has two firmly embedded Mg-rich, low-Ca clinopyroxene grains of 22 × 15 μm and 10.5 × 8 μm in size (Zolensky and Barrett 1994). Two coarse-grained silicate IDPs, viz. RB-27A19 (20 × 10 μm) and U2-11A19A (60 × 40 μm), are encrusted by Fe-sulfide and chondritic materials (Steele 1990). The former is nearly entirely forsterite (Table 37: #1, #2). The latter is an enstatite IDP (Table 37: #3, #4) with a partially included forsterite (15 × 15 μm) grain (Steele 1990).

Table 37. Coarse-grained silicate in non-chondritic IDPs (oxide wt %) (source: Steele 1990)

	#1	#2	#3	#4
SiO_2	40.90	40.8	42.0	42.3
Al_2O_3	0.07	0.14	0.11	0.14
Cr_2O_3	0.39	0.46	0.58	0.58
MgO	52.8	51.6	54.6	55.2
FeO	2.48	2.22	2.02	1.90
CaO	0.16	0.33	0.14	0.13
MnO	0.82	0.81	0.44	0.42
NiO	0.0	0.02	0.02	0.01
total	97.62	96.38	99.91	100.68

Seventeen sulfide (FSN) IDPs (5-15 µm) (Love et al. 1994) have an average density of 3.6 ±1.0 g cm^{-3} (my calculation). This value is close to the density of hypothetical FeS, 3.5-3.6 g cm^{-3} (Robie et al. 1967) [pyrrhotite: 4.625-4.79 g cm^{-3}, troilite: 4.83 g cm^{-3}]. The largest sulfides analyzed by Zolensky and Thomas (1995) are attached to chondritic IDPs such as a grain 17 × 8 µm in size attached to compact IDP L2007-15 (Zolensky and Lindstrom 1992). The compositions of these large grains are unknown but they are probably mostly pyrrhotite (Zolensky and Thomas 1995). They include pyrrhotite (~9 µm in size) with a mean sulfur content of 38.7 wt % and an average Ni content of 1.8 wt % (Schramm and Brownlee 1990). Unmelted Fe,Ni-sulfides often contain many small inclusions and embayed minerals of Na,Al,Si-rich material, Ca,Si,Mg-rich material, olivine, pyroxene, sphalerite and Cr-rich grains (Schramm and Brownlee 1990).

Secondary minerals in chondritic IDPs

The data support that hydration in chondritic aggregate IDPs is restricted to amorphous materials of the appropriate composition. Layer silicates formed with the least possible amount of cation diffusion. This statement is made realizing the ambiguity involving interpretations of petrographic textures and the small number of IDPs studied in full petrographic detail. The IR classification includes smectite- and serpentine-rich IDPs. It appears that smectite phases are more prevalent among these IDPs then serpentine minerals. This apparent relative abundance could be an artifact of sample selection. There are no obvious assemblages for the secondary minerals.

Smectite

Smectite proto-phyllosilicates occur as randomly orientated sinuous grains with a basal spacing of 1.0-1.2 nm in amorphous material. They are 10 to 50 nm long and up to ~5 nm thick. Their compositions are $[Mg_{1.9}Fe_{1.0}Al_{0.1}](Si_{3.8}Al_{0.2})O_{10}(OH)_2$ (Mg = 0.65; range: 0.69-0.60) (Tomeoka and Buseck 1984) and $[Mg_{1.81}Fe_{1.18}Al_{0.64}Zn_{0.07}](Si_4)O_{9.9}$ calculated on a water-free basis (Mg = 0.605) (Blake et al. 1988). Saponite with 1.0-1.4 nm basal spacing and Mg = 0.9-0.7 occurs in amorphous material often associated with embedded pyrrhotite and pentlandite, which contain up to 30 wt % Ni (Zolensky and Lindstrom 1992, Keller et al. 1993). Thomas et al. (1991) reported saponite compositions ranging from Mg = 1.0–0.7. Tomeoka and Buseck (1984) found Fe-rich smectite (or mica),

$$(Fe^{2+},Fe^{3+})_{2.8}Mg_{1.9}Al_{0.7}Si_4O_{10}(OH)_2 \text{ (Mg = 0.4)}.$$

Iron-rich smectite (1.0 to 1.2 nm basal spacing) (Table 32) is present in a collapsed aggregate IDP wherein round matrix units (100 to 200 nm in diameter) are still visible (Bradley 1988, Fig. 7). Poorly-ordered (Mg,Fe,Ca),Al-rich, non-stoichiometric members of the smectite group occur in CP IDP W7029C1 along with well ordered Mg-poor talc and kaolinite crystals (Rietmeijer and Mackinnon 1985a). Illite occurs in IDPs W7029E5, U2011C2 and U2022C7/C8 and rare pyrophyllite and mica co-occur with feldspars (Rietmeijer 1991).

Serpentine

Serpentine is described in a limited number of IDPs. The first reported IDP was a smooth particle with a 0.7 nm diffraction maximum in its XRD pattern (Brownlee 1978a). The IDP U230A11 consists of well-crystallized layer silicate with a ~0.7 nm basal spacing (cation/Si ratio ~1.5), and Fe-rich glass, pentlandite, pyrrhotite, (presumably) magnetite and carbonaceous material (Bradley 1988). Dense masses of coarse-grained polygonal-type serpentine occur in IDP L2005P17 (Zolensky and Lindstrom 1992). IDP W7017A14 with LIME olivines contains serpentine (Klöck et al. 1990). Other particles wherein serpentine, Mg = 0.85 to 0.65, comprises >50% of the particle contain small (<200 nm) embedded grains of manganese-rich olivine and iron-nickel sulfides with Ni up to 30 wt % (Thomas et al. 1991, 1992a, 1995b).

Salts

Salts in smectite-bearing IDPs commonly form scattered disc-shaped grains (Rietmeijer 1992b, Zolensky and Lindstrom 1992). Saponite, magnesite, pure and $Mg_{0.91}Ca_{0.09}CO_3$ to $Mg_{0.64}Fe_{0.36}CO_3$, and siderite co-occur (Radicati di Brozolo and Fleming 1992, Zolensky and Lindstrom 1992). Rare sulfates include barite ($BaSO_4$) (Rietmeijer 1990) and $Ba_{0.76}Ca_{0.22}Sr_{0.02}SO_4$, $MgSO_4$ ($\bullet nH_2O$; unknown), and basanite, $Ca_{0.66}Na_{0.34}SO_4\bullet(1/2H_2O)$ (Zolensky and Lindstrom 1992). Rare porous masses (up to 200 nm) of periclase occur in IDP L2005K10 (Zolensky and Lindstrom 1992).

Carbon phases

The only evidence that carbon was involved in aqueous alteration is the presence of carbonates (cf. salts) and platy grains of carbon-2H (or lonsdaleite) that formed by hydrous pyrolysis of hydrocarbons below 315°C (Rietmeijer and Mackinnon 1987). The carbon-2H grains are up to 0.6×0.2 μm in size.

ACCRETION

The mineralogy and chemistry of chondritic aggregate and cluster IDPs reflect their accretion history. Accretion that was determined by the availability of matrix, or principal, components, silicate and sulfide particles in proportions that varied as a function of time and location in the early solar system. The anhydrous and 'hydrated' (or, CS) IDPs may not be fundamentally different in their pre-alteration constituents.

Matrix aggregates

The matrix units were interpreted as aggregates of nanometer size silicates and sulfides held together by carbonaceous (Fraundorf et al. 1982a, Bradley 1988) or amorphous ferromagnesio-silica materials (Bradley 1994a). The implication of this interpretation is that accretion in the solar system started with dust grains much smaller than ~50 nm that become incorporated into distinct units. In terms of condensation models for the formation of solid materials in the solar system, this notion has a certain appeal but computer simulations of solar system accretion starting with these small grains fail to produce planets. Rietmeijer (1996e, 1997c) favored in situ thermal alteration of amorphous matrix units to produce the grains <50 nm while Bradley (1994b) suggested that GEMS were due to radiation induced amorphization and (partial) re-crystallization. In both models the smallest accreting grains were ~100 up to ~500 nm (Fig. 47), viz.

- carbonaceous units,
- granular units of ultrafine-grained silicates and sulfides embedded in a carbonaceous matrix, including 'reduced aggregates',

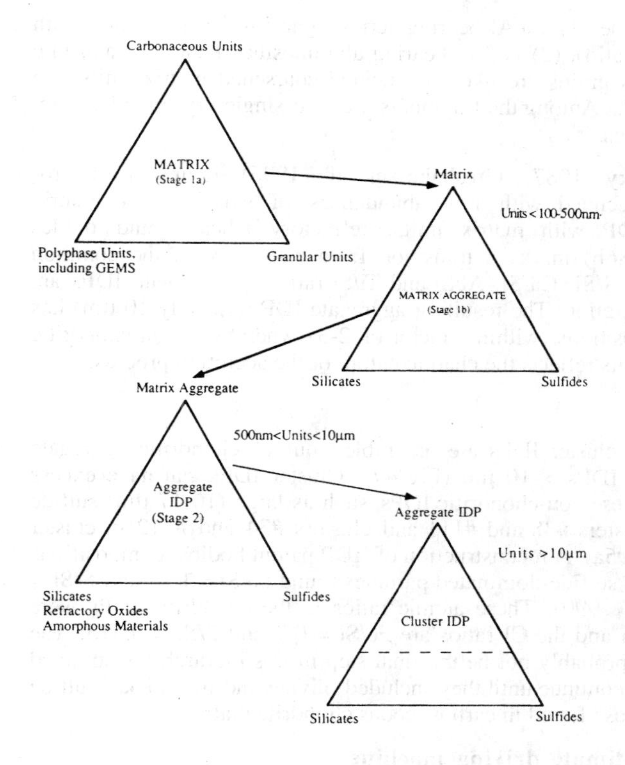

Figure 47. Ternary diagrams depicting the accretion hierarchy of 'matrix aggregates' (1a and 1b), aggregate IDPs (2) and to cluster IDPs (3). The second stage offers an explanation for the interrelationship among refractory IDPs and common chondritic aggregates. The dashed line in the third diagram delineates the relationship among chondritic aggregate and non-chondritic IDPs with chondritic material at the surface.

- carbon-free, polyphase units: (1) serpentine dehydroxylate units that include (1a) ultrafine-grained PUs and (1b) GEMS, and (2) smectite 'dehydroxylate' units, or coarse-grained PUs,
- iron-magnesium silicates and iron-rich sulfides.

Aggregate IDPs

Accretion of 'matrix aggregates' was characterized by the relative availability of these units in the accretion regions as a function of time and location in the early solar system (i.e. turbulent transport). These variations have two important consequences, (1) variable proportions of matrix units among aggregate IDPs are the rule rather than the exception and unique 'glass-rich' chondritic IDPs U219C11 and W7027A11 (Bradley et al. 1992) reflect transient micro-environments, and (2) chemical variations among IDPs are an acquired accretion property. That is, C/Si abundances are a function of the (local) ratios of CUs (to a lesser extend GUs) and polyphase units during accretion (Fig. 47). The relative proportions of grains >500 nm vary among aggregate IDPs (Blake et al. 1988, Bradley 1988, Rietmeijer 1994b, Thomas et al. 1993a, Fig. 2; Klöck and Stadermann 1994, Figs. 1, 7; Rietmeijer 1994a, Fig. 1; Thomas et al. 1994, Figs 1, 2; Thomas et al. 1995a, Fig. 14). These grains include (Fig. 47)

- single-crystal calcium-iron-magnesium silicates and Ca,Al-rich silicates,
- Ca,Ti,Al-oxides,
- pyrrhotite, troilite and pentlandite, and

- amorphous grains that include (1) Ca,Al-bearing ferromagnesiosilica materials (with variable Ca, Al and Mg/Fe ratio), (2) Fe,Mg-bearing aluminosilica materials, and rare silica glass. The amorphous grains are likely to include coarsened matrix units and fragments of fractionated PUs. Among the fragments are also single-crystals of mostly Mg-rich olivine and pyroxenes.

Refractory IDPs (Zolensky 1987, Greshake et al. 1996) represent 'micro-environments' wherein they accreted with low abundances of sulfides and 'matrix aggregates'. The aggregate IDP with matrix units, refractory silicates and oxides (Christoffersen and Buseck 1986b) marks a transition from refractory to the common aggregate IDPs. Variable Fe/Si, S/Si, Ca/Si, Al/Si and Ti/Si ratios in aggregate IDPs are determined during continued accretion. The resulting aggregate IDP (typically 10 μm) has approximately chondritic compositions (within a factor of 2-3) whereby a non-chondritic abundance for one or two elements reflects the chaotic nature of the accretion process.

Cluster IDPs

When accretion continues, cluster IDPs are assembled out of chondritic aggregate IDPs, silicate IDPs and sulfide IDPs > 10 μm (Fig 47). Cluster IDPs can have excess element abundances related to these non-chondritic IDPs, such as large (10-17 μm) sulfide IDPs and Fe-rich IDPs, i.e. clusters #38 and #111 and clusters #34 and #112 in cluster IDP L2008#5 (Thomas et al. 1995a). A reconstruction of 'IDP parent bodies' compositions using chondritic and silicate-and sulfide-dominated particles found Fe/Si = 1.14 and S/Si = 0.776 (Schramm and Brownlee 1990). These atomic ratios in the chondritic IDPs were Fe/Si = 0.697 and S/Si = 0.356 and the CI ratios are Fe/Si = 0.9 and S/Si = 0.515. The stage depicted in Figure 47 will probably not be the final step in this hierarchy. Continued accretion of larger entities could continue until they included olivine and iron-nickel sulfide grains up to ~200 μm such as those found in carbonaceous chondrite matrix.

Aggregate IDPs: The penultimate driving machine

To understand the information in IDPs requires a non-traditional approach to geological and mineralogical phenomena. Still, geological processes must remain fundamentally unchanged and apply to the processes in IDP parent bodies. The unique properties of chondritic aggregate IDPs, i.e. mixtures of ultrafine grains, amorphous materials and high-temperature minerals (i.e. non-chondritic IDPs), will allow alteration reactions to proceed under geologically extreme conditions and at short time scales. The very nature of the formation processes of constituents in chondritic aggregate IDPs, and the nature of their formation history by sedimentation and accretion in solar system environments, produced the penultimate nonequilibrium assemblages. Early alteration is a kinetically controlled, therefore unpredictable, free ride in these aggregates loaded with potential free energy. This situation that arises from innate particle properties (e.g. surface free energy) is compounded by external environmental conditions, e.g. water-ice-rock ratios. Small environmental changes trigger rapid, random, and probably small-scale responses. The typical lack of thermodynamic equilibrium in aggregate IDPs means that they carry no geochemical information on the alteration conditions. Kinetically controlled alteration will persist until the aggregates have equilibrated with the ambient thermodynamic conditions, which, to a first approximation, mostly likely happened in ice-free protoplanets, or at least ice-free domains therein. Once aggregate alteration has reached thermodynamic equilibrium will it be necessary to add energy for continued aqueous and/or thermal modification.

Aqueous alteration

The matrix units of chondritic aggregate IDPs were initially amorphous materials, possibly non-equilibrium condensates (Rietmeijer et al. 1997), or they became amorphous by subsequent alteration. Layer silicate formation depends critically on the presence of materials with appropriate composition, i.e. the serpentine dehydroxylate and smectite dehydroxylate units. Indeed, the Mg-ratios of serpentine mimic those of the ultrafine-grained PUs and GEMS (Fig. 48). The bulk compositions of coarse-grained PUs and smectite also show identical ranges, viz. Mg = 1.0-0.67 and Mg = 1.0-0.605, respectively. The similarities may be coincidence, or offer a clue to the relationships between anhydrous amorphous grains and layer silicates. That is, alteration involved limited diffusion in parent bodies wherein liquid water was a scarce commodity.

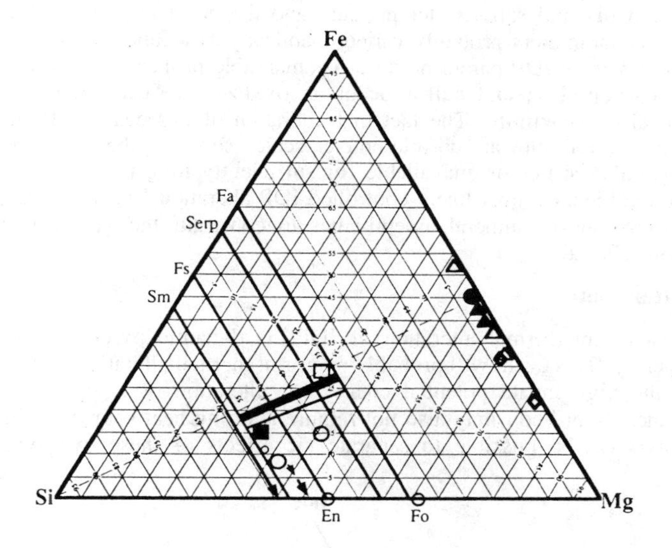

Figure 48. Ternary diagram (atom ratios) showing Mg ratios for bulk IDPs and matrix units. The average composition of anhydrous IDPs (Thomas et al. 1993a, Schramm et al. 1989) (solid triangles) and the same but corrected for FeS (Schramm et al. 1989) (open diamonds). The average composition of aggregate IDP subtypes (Thomas et al. 1993a) (open triangles). The solid line defines the lowest Mg-ratio determined for coarse-grained PUs and equilibrated aggregates. Large open circles and black arrows symbolize the range of Mg-ratios in these matrix units up to Mg = 1.0. The small open circle is the most Mg-rich 'restite' found in coarse-grained PUs. The heavy bar depicts Mg ratios of bulk IDP compositions. The arrow indicates the range of smectite Mg ratios from mostly 0.6-1.0 but as low as Mg = 0.4. The CI (open square), solar (dot) values (Anders and Grevesse 1989) and the average composition of comet Halley (Jessberger et al. 1988) (solid square), and the compositional lines for olivine, serpentine, pyroxene and smectite are presented.

Hydration is a gradual and continuous process that proceeds as a function of the availability of water which is in part determined by the accretion history, i.e. the dust to ice ratio, the relative proportions of matrix units, and the parent body thermal regime. Layer silicate and salt occurrences in aggregate IDPs reflect a response to barely perceptible alteration producing a few proto-phyllosilicate grains to substantial alteration of amorphous precursors. These conditions are reminiscent of hydro-cryogenic alteration in terrestrial permafrost environments wherein aqueous alteration proceeds via an interfacial water layer. This thin layer, 1-8 nm (0°C) to 0.6 nm (-80°C) of interfacial water occurs at the surface of grains embedded in ice (Anderson and Morgenstern 1973). Under these conditions migration and precipitation of chemical species occurs efficiently (Gibson et al. 1983). Rietmeijer (1985b) suggested that similar conditions of hydro-cryogenic alteration existed in icy protoplanets that are the IDP parent bodies. Under marginal conditions protophyllosilicate formation favors amorphous materials. The conditions in the majority of IDPs could not support continued alteration. Higher temperatures are possible under ice-free conditions. These conditions are reflected by the relationships of layer silicates in aggregate IDP W7029C1 wherein well-ordered Mg-poor talc and kaolinite are the products of closed-system decomposition of non-stoichiometric smectite (Fig. 49). Higher temperatures are required for aqueous alteration of non-stoichiometric plagioclase (An = 0-0.94) and alkali-feldspars (Or = 0.47-0.81). Assuming (almost certainly incorrect) that these reactions occurred at thermodynamic equilibrium, this aqueous alteration occurred at 175°-195°C (Rietmeijer 1991). The extent of aqueous alteration is a function of the water to rock ratio, fluid pH and salinity, temperature and the efficiency of fluid transport (i.e. porosity). These parameters probably varied chaotically as a function of time, temperature and space in aggregate IDP parent bodies. A remarkable property of chondritic aggregate IDPs is that carbon, iron and sulfur occur as oxidized, reduced and metallic species, oftentimes in close proximity. The fact that alteration of aggregate IDPs hardly reached thermodynamic equilibrium at sub-micrometer scales supports the view that energy for alteration was either scarce or unavailable for sufficiently long periods of time, or both. Because reaction kinetics most likely controlled IDP alteration I have not attempted to use mineral compositions or mineral assemblages to constrain the geochemical conditions during aqueous alteration.

Thermal alteration

The most recent thermal alteration of IDPs is dynamic pyrometamorphism during atmospheric entry. This event will probably overprint thermal alteration in the parent body. Because of the high innate potential energy for alteration of aggregate IDPs, thermal alteration in the parent body may also not require a strong heat source. Using the mineral grain sizes it might be possible to separate the effects of these two types of thermal

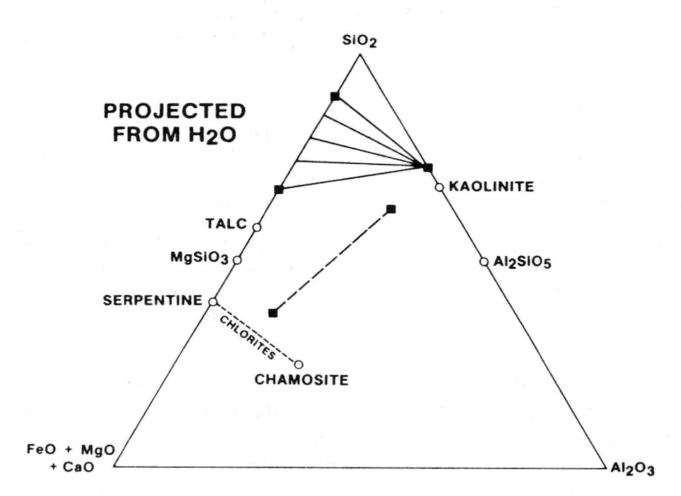

Figure 49. The compositions of poorly ordered layer silicates (dashed line) and single-crystal layer silicates (solid lines) in chondritic aggregate IDP W7029C1 plotted in the ternary diagram Al_2O_3-SiO_2-(FeO+MgO+CaO) (wt %). Although not a completely closed system, these layer silicates are related by the loss of divalent cations that resulted in the single-crystal phases. This interpretation is consistent with magnesite present in this IDP. (Rietmeijer and Mackinnon 1985, unpublished data)

alteration. For example, the ultrafine iron-magnesium silicates in ultrafine-grained PUs could have formed during atmospheric entry but olivines and pyroxenes in the coarse-grained PUs probably did not form in this manner. They required substantial diffusion and reorganization which seems to favor a thermal environment with a duration in excess of 10-15 s. Similar arguments apply to the proposition that turbostratic and poorly graphitized carbons are parent-body alteration, but there is no proof for this conjecture.

Chemical properties of aggregate IDPs

Neither matrix units nor accreting grains have chondritic compositions. These building blocks were assembled into a chondritic material during accretion. The aggregate IDP is the smallest volume of a chondritic material with deviations for one or two elements related to the accretion history, post-accretion alteration, or both. The major element compositions in these particles are mostly obtained by energy dispersive spectroscopy and the reported abundances are determined by the EDS detection limits. The abundances of minor elements may be a reason for caution in the absence of mineralogical data for the particles. All data for bulk compositions of aggregate IDPs, cluster fragments, and (reconstructed) cluster IDP compositions are listed in Table 38, except for the data on chondritic IDPs obtained by Schramm et al. (1989) which are listed in their paper.

Sodium content in bulk aggregate IDPs and fragments from cluster IDPs scatter widely from zero to Na/Si = 0.75 element ratio (Fig. 50). Two extreme values, viz. Na/Si = 1.7 (IDP W7029*A) and Na/Si = 4.9 (IDP L2005P9) were reported by Stephan et al. (1994) who indicated that L2005P9 was contaminated with terrestrial material containing Na, K, Ca and chlorine. I submit that Na/Si ratios > 0.1 (the bulk CI ratio) support contamination with stratospheric aerosols. The same origin can not be excluded for IDPs and fragments with Na/Si < 0.1 but the amount of indigenous sodium that survived atmospheric entry heating of the particles is unknown. This interpretation is consistent with the fact that Na-bearing phases are rare in aggregate and cluster IDPs. For phosphorous the story is different. The P/Si ratios in aggregate and cluster IDPs range from zero to 0.063 with a mean = 0.0315 (Bradley et al. 1996; Thomas et al. 1993a, 1995a). The cosmic ratio P/Si = 0.0086 suggests that phosphor-bearing minerals are present. They do but they are rare.

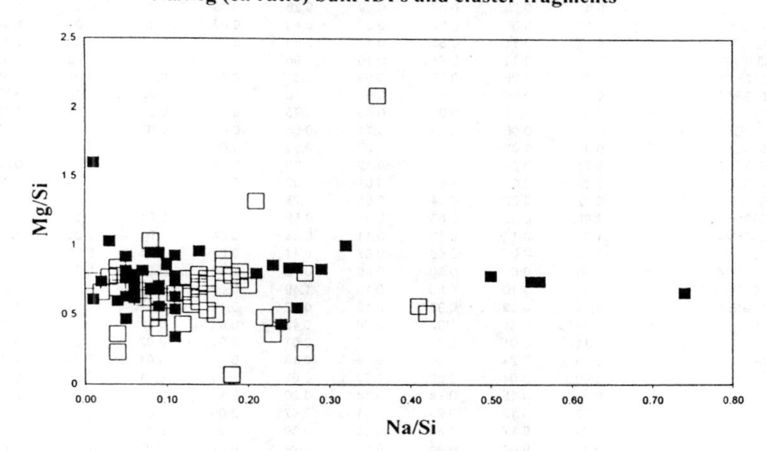

Figure 50. Mg/Si vs. Na/Si (el. ratio) in bulk aggregate IDPs (solid squares) and cluster fragments (open squares), except for two unusually high-sodium IDPs (see text). The CI Na/Si = 0.1.

Table 38. Elements/Si (wt %) abundances in chondritic aggregate IDPs and fragments of cluster IDPs.

		C/Si	Na/Si	Mg/Si	Al/Si	S/Si	Ca/Si	Cr/Si	Mn/Si	Fe/Si	Ni/Si
1	18-2-02			0.76	0.20	0.35	0.08			0.48	0.03
2	18-2-06			0.81	0.20	0.50	0.04			0.53	0.03
3	18-2-07			0.76	0.01	0.47	0.07			0.35	0.03
4	18-3-01			0.62	0.18	0.23	0.05			0.52	0.06
5	18-3-02			0.81	0.13	0.25	0.03			0.44	0.03
6	18-3-07			0.67	0.10	0.57	0.02			0.51	0.06
7	18-4-01B			0.91	0.15	0.14	0.04			0.41	0.04
8	18-4-04			1.15	0.13	0.08	0.02			0.15	0.01
9	18-4-11			1.20	0.05	0.07	0.01			0.38	0.01
10	18-5-03			0.53	0.20	0.14	0.03			0.59	0.02
11	18-5-08			1.10	0.13	0.04	0.05			0.16	0.01
12	18-5-17B			1.05	0.10	0.01	0.03			0.33	0.03
13	18-5-20			0.24	0.01	0.01	0.01			0.60	0.03
14	U2-13A1		0.25	0.84	0.11	0.86	0.11	0.02	0.03	2.97	0.15
15	U2-13A3		0.10	0.87	0.12	0.28	0.10	0.03	0.02	2.12	0.11
16	U2-20B11		0.14	0.96	0.06	0.54	0.05	0.02	0.01	0.66	0.03
17	U2-20B37		0.11	0.93	0.04	0.62	0.06	0.01	0.02	0.79	0.04
18	U2070A-1B			1.09	0.79	0.38	0.10			1.34	0.10
19	U2070A-2H			1.02	0.06	1.55	0.04	0.06		2.64	1.47
20	U2070A-3D			0.67	0.08	0.47	0.00			2.17	0.10
21	U2070A-6A			0.73	0.08	0.28	0.00		0.03	1.41	0.14
22	U20703A-5B			1.02	0.07	0.00	0.08			1.72	0.09
23	U2070A-2B			0.79	0.08	0.76	0.07		0.09	1.30	0.09
24	U2070A-8F			0.82	0.04	0.77	0.07	0.03	0.05	1.48	0.08
25	U2073A-5J			1.20	0.21	0.10	0.04			0.85	0.05
26	U2073A-7F			0.81	0.07	0.99	0.06			2.01	0.14
27	hydrated		0.05	0.82	0.08	0.34	0.02			0.74	0.03
28	L2005L6	1.00		0.85	0.08	0.26	0.12			2.25	0.21
29	L2005P9	2.00		0.80	0.13	0.21	0.01			2.80	0.07
30	L2005P13	0.79		0.81	0.07	0.23	0.17			1.79	0.01
31	L2006H5	0.47		0.50	0.07	0.43	0.12			1.06	0.06
32	L2005R7	0.47		0.63	0.07	0.15	0.01			1.00	0.04
33	L2006C12	0.68		0.70	0.14	0.36	0.01			4.55	0.03
34	L2006E10	0.73		0.73	0.09	0.14	0.03			1.20	0.11
35	L2006F10	1.07		0.71	0.07	0.28	0.01			0.93	0.06
36	L2006F12	0.54		0.69	0.10	0.18	0.03			2.08	0.08
37	L2006G1	1.54		0.74	0.06	0.31	0.05			0.92	0.06
38	L2006J14	1.69		0.92	0.08	0.12	0.02			0.77	0.05
39	W7029*A27	2.56	1.70	0.54	1.32		2.43	1.36	0.97	1.21	0.99
40	L2005P9	0.48	4.91	0.31	1.26		1.94	0.47	0.60	0.74	0.37
41	U2015E7	0.03	0.05	0.47	0.12	0.01	0.07			1.09	0.10
42	W7028*C2	0.44	0.05	0.76	0.08	0.23	0.06			0.60	0.04
43	U2022G7	0.44	0.01	0.61	0.05	0.42	0.05			0.30	0.02
44	U2017A2	0.85	0.03	1.03	0.04	0.35	0.04			0.36	0.02
45	U2015D21	0.63	0.07	0.82	0.07	0.36	0.07	0.02	0.01	1.27	0.05
46	U2015D22	0.61	0.06	0.66	0.09	0.28	0.05	0.01	0.01	1.31	0.05
47	U2015E3	0.51	0.06	0.62	0.07	0.40	0.07	0.02	0.01	1.67	0.05
48	U2034E3	1.13	0.11	0.54	0.13	0.63	0.04	0.01	0.01	1.54	0.08
49	U2034F27	0.68	0.11	0.74	0.10	0.96	0.10	0.06	0.05	2.85	0.18
50	W7013B13	0.51	0.08	0.68	0.09	0.31	0.06	0.02	0.01	0.68	0.03
51	W7013B17	0.41	0.06	0.79	0.06	0.10	0.04	0.02	0.01	0.82	0.02
52	W7013C16	0.47	0.05	0.92	0.06	0.75	0.04	0.01	0.01	1.12	0.06
53	W7013D12	0.84	0.06	0.63	0.11	0.58	0.05	0.02	0.08	1.65	0.06
54	W7013E9	0.99	0.09	0.68	0.07	0.34	0.06	0.01	0.01	0.81	0.03
55	W7013E17	0.74	0.24	0.43	0.32	0.30	0.07	0.01	0.01	0.57	0.02
56	W7013G1	0.29	0.01	1.60	0.01	0.02	0.02	0.02	0.05	0.31	0.01
57	W7013G6	0.42	0.02	0.74	0.05	0.25	0.09	0.03	0.03	0.92	0.05
58	W7013H5	1.88	0.05	0.63	0.07	0.69	0.09	0.01	0.01	1.17	0.05
59	W7013H24	1.97	0.11	0.63	0.11	0.98	0.04	0.01	0.01	1.10	0.04
60	W7027E6	0.69	0.09	0.56	0.07	0.41	0.05	0.01	0.01	0.83	0.04
61	W7027H14	1.56	0.04	0.60	0.06	0.50	0.05	0.01	0.01	0.97	0.03
62	W7029*A28	1.63	0.10	0.86	0.10	1.49	0.08	0.03	0.01	3.87	0.12
63	W7029*B9	1.70	0.09	0.95	0.12	0.72	0.10	0.01	0.01	1.52	0.04
64	clu11	1.39	0.21	1.32	0.08	0.43	0.97	0.01	0.02	0.78	0.06
65	clu13	0.11	0.01	0.74	0.02	0.01	0.01	0.02	0.01	0.10	0.00
66	clu14	0.16	0.24	0.50	0.24	0.13	0.13	0.01	0.01	2.22	0.07
67	clu15	0.09	0.04	0.69	0.07	0.02	0.00	0.01	0.01	0.57	0.00
68	clu16	0.32	0.15	0.76	0.14	0.90	0.15	0.01	0.01	2.05	0.13
69	clu17	0.83	0.12	0.76	0.01	0.47	0.06	0.01	0.01	1.15	0.08
70	clu18	0.28	0.17	0.85	0.08	0.30	0.06	0.01	0.01	1.68	0.04
71	clu19	0.59	0.04	0.36	0.04	0.24	0.02	0.00	0.00	0.45	0.03

Table 38 continued. Elements/Si (wt %) abundances in chondritic aggregate IDPs and fragments of cluster IDPs.

72	clu110	0.49	0.08	0.47	0.04	0.20	0.05	0.02	0.01	0.94	0.06
73	clu111	0.88	0.09	0.40	0.04	3.72	0.00	0.00	0.00	5.43	0.51
74	clu112	0.36	0.27	0.23	0.07	0.70	0.02	0.16	0.00	9.25	0.61
75	clu113	0.31	0.04	0.78	0.03	0.12	0.09	0.00	0.00	0.28	0.01
76	clu114	0.55	0.08	1.03	0.06	1.20	0.55	0.02	0.04	2.44	0.11
77	clu115	0.26	0.08	0.75	0.08	0.44	0.07	0.01	0.02	1.51	0.08
78	clu116	0.16	0.18	0.78	0.07	0.42	0.07	0.02	0.02	2.48	0.08
79	clu117	0.10	0.16	0.50	0.06	0.76	0.04	0.05	0.01	3.65	0.24
80	clu118	0.37	0.04	0.23	0.08	0.02	0.64	0.00	0.00	0.06	0.00
81	clu119	0.51	0.14	0.70	0.08	1.14	0.05	0.00	0.01	2.14	0.13
82	clu120	0.08	0.03	0.77	0.03	0.02	0.01	0.01	0.00	0.13	0.00
83	clu121	0.09	0.04	0.84	0.02	0.09	0.00	0.01	0.01	0.61	0.00
84	clu21	0.51	0.41	0.56	0.08	0.90	0.03	0.01	0.13	4.82	0.51
85	clu23	0.53	0.14	0.73	0.11	0.71	0.04	0.02	0.01	1.76	0.11
86	clu24	0.34	0.19	0.81	0.12	0.19	0.06	0.01	0.02	1.21	0.06
87	clu25	0.46	0.15	0.76	0.10	0.35	0.06	0.01	0.01	1.20	0.07
88	clu26	0.27	0.10	0.60	0.10	0.34	0.01	0.01	0.01	1.02	0.10
89	clu27	0.63	0.20	0.71	0.33	0.16	0.06	0.01	0.01	1.52	0.09
90	clu28	0.64	0.19	0.75	0.10	0.29	0.06	0.01	0.01	1.36	0.09
91	clu31	0.55	0.13	0.63	0.09	0.20	0.06	0.01	0.01	0.96	0.04
92	clu32	1.04	0.15	0.53	0.07	0.27	0.04	0.01	0.01	0.84	0.04
93	clu33	1.40	0.12	0.43	0.06	0.40	0.04	0.01	0.01	0.71	0.04
94	clu34	0.52	0.07	0.62	0.03	2.14	0.03	0.02	0.01	7.93	0.51
95	clu35	0.18	0.27	0.80	0.07	0.13	0.06	0.02	0.03	3.05	0.08
96	clu36	0.25	0.14	0.79	0.09	0.75	0.08	0.02	0.02	2.09	0.21
97	clu37	0.23	0.08	0.67	0.06	0.08	0.08	0.02	0.02	2.93	0.01
98	clu38	1.00	0.17	0.90	0.07	8.53	0.00	0.00	0.00	15.90	0.02
99	clu39	0.16	0.23	0.36	0.26	0.06	0.69	0.02	0.01	0.19	0.01
100	clu310	2.36	0.12	0.66	0.08	0.48	0.04	0.01	0.01	0.85	0.05
101	clu311	0.37	0.15	0.72	0.06	0.75	0.07	0.03	0.02	3.07	0.17
102	clu312	0.51	0.09	0.51	0.08	0.34	0.02	0.01	0.01	1.06	0.04
103	clu313	0.92	0.13	0.57	0.08	0.28	0.04	0.01	0.01	0.77	0.04
104	clu314	0.96	0.02	0.66	0.01	0.25	0.04	0.01	0.00	1.08	0.05
105	clu315	2.03	0.36	2.09	0.12	0.84	0.07	0.00	0.01	1.64	0.04
106	clu317	0.10	0.22	0.48	0.19	0.03	0.01	0.00	0.01	0.86	0.00
107	clu318	0.29	0.42	0.51	0.77	0.18	0.29	0.01	0.01	0.63	0.04
108	clu319	0.26	0.10	0.74	0.08	0.44	0.07	0.02	0.01	1.34	0.08
109	clu320	0.40	0.18	0.07	0.02	0.01	0.00	0.00	0.01	0.03	0.00
110	clu51	0.30	0.17	0.69	0.06	0.80	0.01	0.01	0.01	1.74	0.17
111	clu52	0.71	0.09	0.52	0.11	0.27	0.12	0.01	0.05	0.62	0.03
112	clu53	0.32	0.11	0.65	0.06	0.45	0.06	0.02	0.01	1.66	0.07
113	clu54	0.73	0.08	0.62	0.06	0.57	0.05	0.02	0.02	1.04	0.05
114	clu55	0.25	0.14	0.59	0.11	0.36	0.03	0.01	0.01	1.21	0.27
115	clu56	0.23	0.07	0.67	0.07	0.54	0.05	0.02	0.01	0.94	0.05
116	L2005#17l	0.52	0.32	1.00	0.09	0.52	0.08			2.15	0.06
117	L2005#17d	0.37	0.56	0.74	0.13	0.79	0.07			2.75	0.11
118	L2005#19l	0.72	0.23	0.86	0.12	0.77	0.08			2.18	0.11
119	L2005#19d	0.48	0.29	0.83	0.08	0.46	0.06			1.26	0.06
120	L2005#26l	0.22	0.11	0.34	0.03	0.17	0.11			1.56	0.07
121	L2005#26d	0.54	0.21	0.80	0.08	0.41	0.07			1.27	0.07
122	L2006#28l	1.62	0.50	0.78	0.12	0.68	0.05			1.23	0.04
123	L2006#28d	1.82	0.55	0.74	0.12	0.62	0.05			1.06	0.06
124	L2006#4l	0.27	0.09	0.71	0.10	0.17	0.10			0.75	0.03
125	L2006#4d	0.57	0.26	0.84	0.09	0.29	0.06			1.27	0.06
126	W7029*Al	0.19	0.08	0.95	0.05	0.14	0.02			1.50	0.04
127	W7029*Ad	0.46	0.11	0.78	0.08	0.57	0.07			2.06	0.11
128	L2006#14d	7.58	0.26	0.55	0.00	0.42	0.03			0.55	0.05
129	L2006#14d	9.78	0.74	0.66	0.11	0.53	0.11			0.79	0.06
130	L2006#14d	0.89	0.06	0.73	0.05	0.14	0.03			0.74	0.03

1-3 : Hudson et al. (1981)
14,15 : Ganapathy and Brownlee (1979)
16,17 : Bradley et al. (1984b)
18-26 : Bradley et al. (1996)
27 : Thomas et al. (1991)
28-38 : Keller et al. (1993)
39,40 : Stephan et al. (1994)
41-44 : Blanford et al. (1988)
45-63 : Thomas et al. (1993)
64-115 : Thomas et al. (1995a)
113-130: Thomas et al. (1993b)

Aggregate size

Particle size is an important parameter for comparisons of the chemical properties among IDPs and other fine-grained solar system materials. Most studies of aggregate IDPs find aluminum, calcium and sulfur abundances that are below CI levels (Fraundorf et al. 1982a, Schramm et al. 1989, Zolensky and Barrett 1994). The aggregate IDP sizes, using the data for porous and smooth particles from Schramm et al. (1989), show a skewed (S = 0.68) Gaussian distribution: mean = 10.5 μm, mode = 7.1 μm and 1σ = 5.2 μm (N = 164). Aggregate IDPs culled from different sources (Hudson, et al. 1981, Ganapathy and Brownlee 1979, Fraundorf 1981, Blanford et al. 1988, Thomas et al. 1993a) show a similar size distribution. The fragments of cluster IDP L2008#5 (Thomas et al. 1995a) show a bimodal distribution, viz. (1) mean = 9.2 μm with 1σ = 1.6, μm and (2) mean = 15.9 μm with 1σ = 4.4 μm. The first population matches the aggregate IDP distribution. The second population indicates a population of larger IDPs that include iron-nickel sulfide IDPs, iron-rich (magnetite) IDPs (and compositions that appear intermediate between sulfides and oxides) and silicate IDPs, which include Mg,Fe-olivine and pyroxene and Ca-rich clinopyroxene (#clu39; #clu118): $En_{22}Fe_{13}Wo_{65}$ and $En_{18.5}Fe_{5.5}Wo_{76}$ (my calculations).

Element distributions. The diagrams of the element (ratio) distributions for bulk IDPs and cluster fragments clearly show the effects of non-chondritic IDPs and thus the compositional shifts from aggregate IDPs to cluster IDPs. These extreme values are 'end-members' of a mixing line between chondritic and non-chondritic sulfide, magnetite or almost pure carbon IDPs. The effects of including large non-chondritic IDPs in the database are shown in the distributions of the Fe/S (Fig.51) and Mg/Fe ratios (Fig. 52) for aggregate and cluster IDPs. The extreme data points (Fig.51a) are for cluster fragments #111 (FeS, 12 ×12 μm), #34 (high Fe grain with sulfur, 15 × 17 μm) and #112 (high Fe grain, 11 × 12 μm). The data for #38 (FeS, 12 × 12 μm; Fe/Si = 15.9) is not shown. Otherwise, the aggregate IDPs and cluster fragments have overlapping compositions (Fig. 51b). Particles with Fe/S larger than the CI value (dot) reflect variations in the relative proportions of sulfides and matrix material. Sulfur loss due to atmospheric entry heating is likely to be responsible for the S/Si values less than CI.

The distribution of Mg/Fe ratios also shows a pattern with overlapping compositions for aggregate IDPs and cluster fragments with a halo of data for non-chondritic cluster fragments (Fig. 52a). These fragments include the same coarse-grained sulfide and magnetite IDPs (cf. Fig. 51a) and a few silicate IDPs. The extreme Mg/Si ratio represents carbon-rich fragments Clu315 (Thomas et al. 1993a) but mostly the compositions of aggregate IDPs and fragments overlap (Fig. 51b). Remarkably, most 10-μm-sized particles have Fe/Si and Fe/Mg ratios smaller than the CI values (Fig. 51b).

The range of the C/S ratios is dominated by a few sulfide cluster fragments (probably with chondritic and/or carbon material at the surface) and two very high carbon fragments from cluster IDP L2006#41 (Thomas et al. 1993b) (Fig. 53a). The C/S ratios in aggregate IDPs and cluster fragments show considerable overlap whereby most of the high S/Si particles are cluster fragments (Fig. 53b). These S/Si ratios (and similarly other Si-normalized ratios for these particles) could be due to lower than CI silicon abundances in these particles by variable proportions of chondritic material and non-chondritic (i.e. sulfide) particles in the analyzed sample.

Ternary presentations

The data are generally presented in a ternary Fe-Mg-Si diagram that is adequate for comparisons with other extraterrestrial materials and for the evaluation of relationships among IDPs. Schramm et al. (1989) determined bulk compositions of chondritic IDPs that

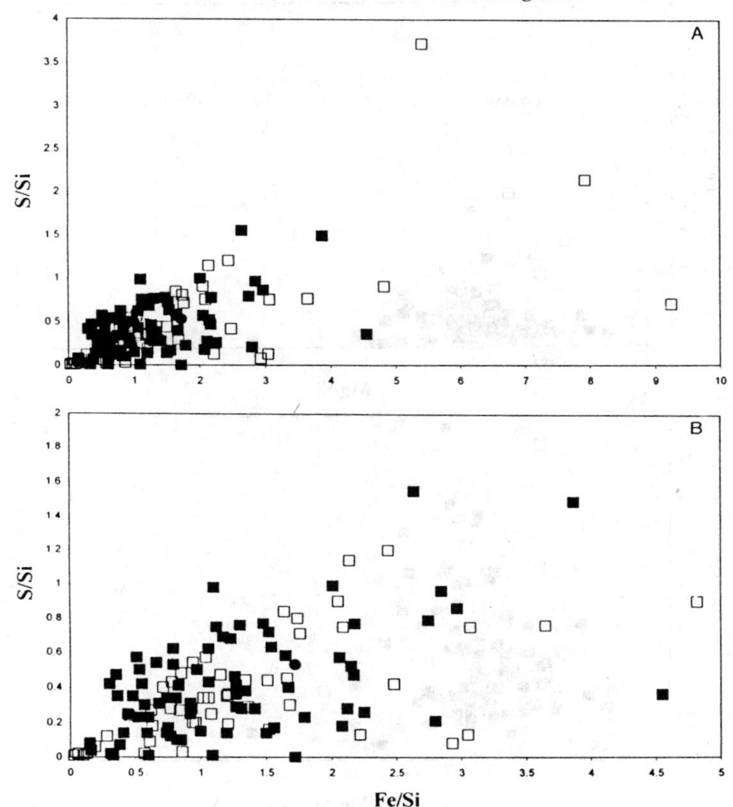

Figure 51. Fe/Si vs. S/Si (el. ratio) in bulk aggregate IDPs (solid squares) and cluster fragments (open squares). (A) full range of Fe/S including non-chondritic sulfide IDPs. (B) considerable overlap of the compositions for aggregate and cluster IDPs. The dot represents the CI ratios.

are mostly 6 to 15 μm in size but with a full range from 4 μm to 40 μm. These IDPs were subdivided into porous and smooth IDPs. In their classification a porous particle is an anhydrous aggregate and a smooth particle is composed of layer silicates. Their results reveal no substantial differences between these subgroups calculated on a sulfide-free basis when plotted in this ternary diagram (atomic ratios) (Fig. 54). This adjustment means that the resulting distributions show variations only among matrix aggregates plus embedded silicates, refractory oxides and amorphous grains in chondritic particles (Fig. 47). This result is consistent with the notion that both types of IDPs are aggregates and that hydration is not related to the aggregate nature of IDPs. The Ca/Si and Al/Si ratios in these porous and collapsed aggregate particles show the same accretion features. That is, variable fractions of fassaite, plagioclase and Ca-bearing aluminosilica materials (PUs) characterize the matrix aggregates (Fig. 55).

Studies that used automated thin-film point-count analyses initially claimed that the distribution patterns of data points from anhydrous aggregates in this ternary diagram were systematically different from hydrated IDPs (Bradley 1988). Subsequently, Bradley et al.

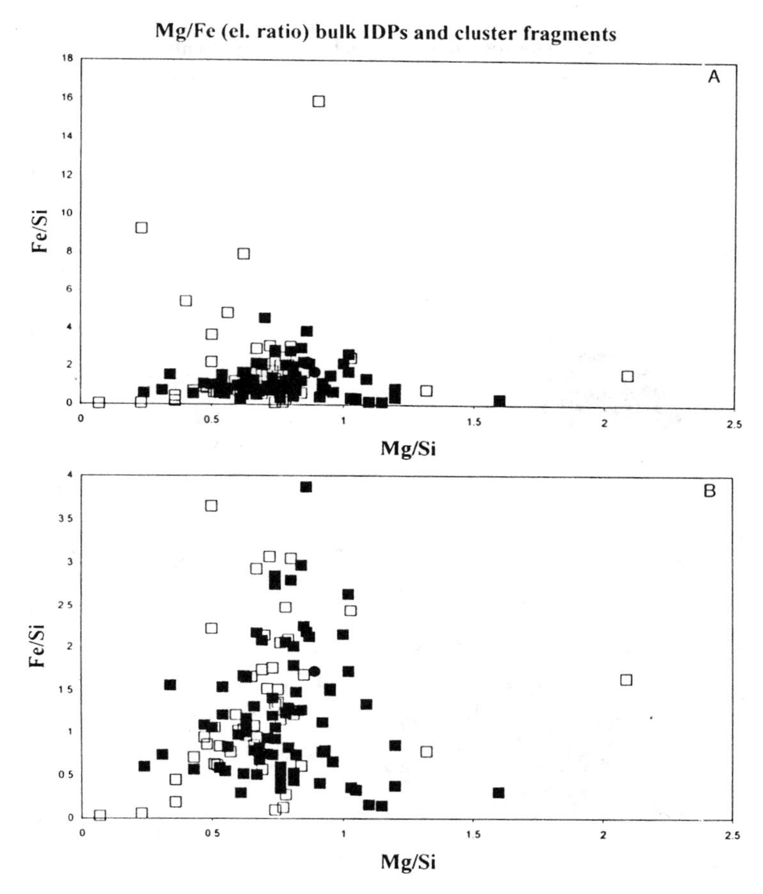

Figure 52. Mg/Si vs. Fe/Si (el. ratio) in bulk aggregate IDPs (solid squares) and cluster fragments (open squares). (A) cluster fragments in a halo around a tightly packed cluster of overlapping compositions (B) for aggregate and cluster IDPs. The black dot is the CI Mg/Fe ratio.

(1989) and Germani et al. (1990) showed that both subgroups have very similar distribution patterns. Similar distributions characterize mostly amorphous chondritic aggregate IDPs (Bradley et al. 1992). Yet, the individuality of some particles was evident, e.g. Mg-rich compositions for IDP U230A43 are due to carbonate grains embedded in its matrix (Bradley et al. 1992). These point count analyses reveal the chemical distributions in the matrix and in matrix aggregates. That is, chemical variations at a smaller scale than the data from Schramm et al. (1989) but entirely consistent with the aggregate nature of these particles. Very similar distributions occur indeed in olivine- and pyroxene-rich IDPs (Bradley et al. 1989, Fig. 2). An exception is IDP U2-13-M10#4, which does not have the low-Mg data points along the Fe-Si join of the ternary diagram. This distribution could support that its contains mostly coarse- and ultrafine-grained polyphase units. At this spatial resolution, however, the distributions are distinctly different from those of the bulk compositions of aggregate IDP. Data points for matrix aggregates fill the part of the diagram on the Si-rich side of the pyroxene line with data points reaching along the Fe-Si join into the Fe-apex. The different distribution patterns show the effect of sample size thus

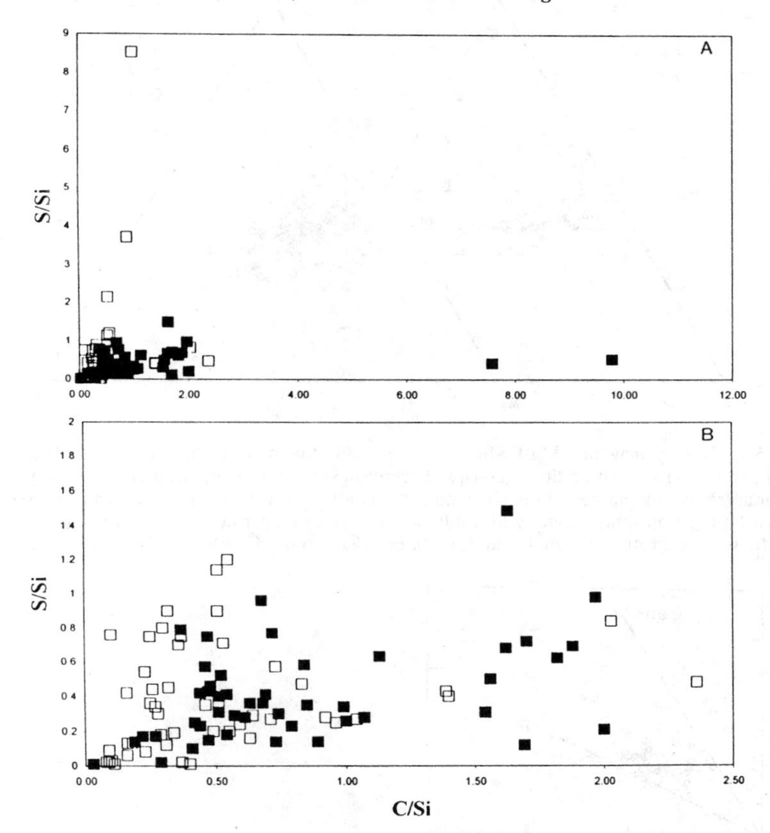

Figure 53. C/Si vs. S/Si (el. ratio) in bulk aggregate IDPs (solid squares) and cluster fragments (open squares). (A) the full range defined by non-chondritic sulfide IDPs and carbon-rich fragments with chondritic material attached to the surface, and (B) mostly overlapping compositions for both. The S/Si ratios trending upwards are due to sulfides in the particles. Otherwise there may be a weak correlation reminiscent of 'excess' sulfur in granular units (cf. Fig. 31).

illustrating that bulk compositions are dominated by variable proportions of silicates, refractory oxides, amorphous grains and sulfides embedded in a matrix that contributes much less mass to the bulk particle.

An aggregate and cluster IDP-meteorite connection

It seems only logical that there should be a connection between the unique chondritic aggregate and cluster IDPs and matrices of undifferentiated meteorites. The database is still too small to make judicious comparisons but there are tantalizing similarities to justify the exploration of a connection. For example, the grain sizes and compositions of forsterite, low-Ca and high-Ca (20 to 45 mol % Wo) and sulfides in an amorphous Fe-rich matrix of the unique Acfer 094 carbonaceous chondrite (Greshake 1997) are reminiscent of the properties of coarse-grained and ultrafine-grained PUs. Also, the typical compositional trends of coarse-grained PUs (cf. Fig. 35) resemble parts of the matrix in the Bells CM

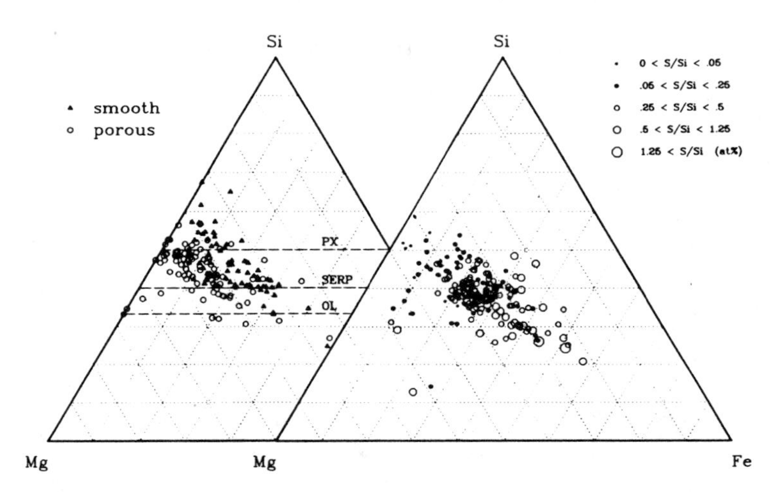

Figure 54. Ternary diagrams Mg-Fe-Si (at. ratio) with the bulk compositions of chondritic IDPs, including porous and smooth particles (collapsed aggregates) shown in the diagram on the right. The left-hand diagram shows the aggregate matrix compositions after correction for the presence of iron sulfides (sulfur loss during atmospheric entry will result in incomplete iron correction). [Used by permission of the editor of Meteoritics, from Schramm et al. (1989), Fig 12, p. 108.]. Courtesy of Don Brownlee.

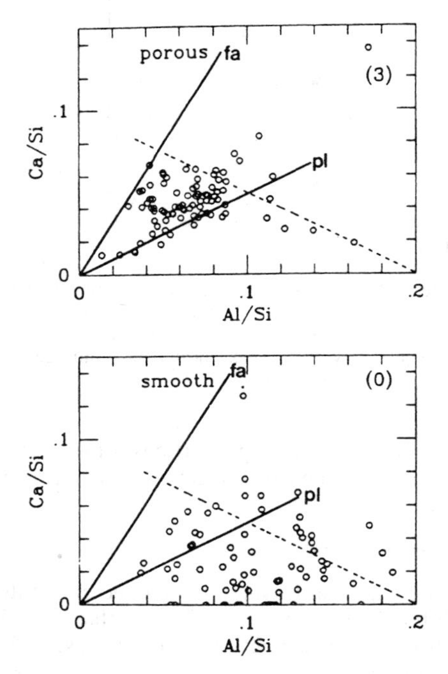

Figure 55. Al/Si vs. Ca/Si (at. ratios) for porous and smooth chondritic IDPs showing that these compositions are constrained by variations in the proportions of fassaite (fa) and plagioclase (pl) (either crystalline or amorphous) (solid lines), and amorphous 'restite' materials (dashed line). After Schramm et al. (1989) Meteoritics 24: 99-112, Fig. 8. Courtesy of Don Brownlee.

chondrite (Brearley 1995). These similarities may be coincidental and not reveal a grander picture. Many IDP investigators explore the possible links between collected IDPs, asteroidal (meteorites), comets and other astrophysical sources. I refer the reader to the review articles by Bradley et al. (1988) and Bradley (1994c), which discuss the IDP properties that could be relevant in this effort.

Chondritic IDPs and comet P/Halley

The IDP community is blessed (or cursed) by the fact that there is data from one dust-producing proto-planet. The semi-quantitative compositions, and sizes, of dust in comet P/Halley were measured by space probes during its 1985/86 perihelion. When Jessberger et al. (1988) reduced the comet data, the database available for comparison consisted of chondritic aggregate IDPs and matrices of CI and CM carbonaceous chondrites. The accretion hierarchy of chondritic IDPs implies that measured compositions are a function of particle size and a comparison with dust from this comet should occur at comparable levels of sample size. The dust compositions in this comet, solar system abundances, which are similar to CI abundances except for volatile elements (Anders and Grevesse 1989) and the composition of solar photosphere (Ross and Aller 1979) are presented in atomic abundances. The database expressed in atomic abundances is much smaller (Table 39) than the database used in Figures 51-53 (Table 38) (Note: Table 39 includes the average chondritic IDP compositions determined by Schramm et al. 1989).

Table 39. Element/Si (at.) for aggregate and cluster IDPs.

		C/Si	Na/Si	Mg/Si	Al/Si	S/Si	Ca/Si	Cr/Si	Mn/Si	Fe/Si	Ni/Si
1	13-05-01C			0.71	0.10	0.29	0.06	0.06	0.00	0.87	0.07
2	13-06-05A			0.83	0.09	0.67	0.05	0.04	0.00	0.38	0.05
3	13-07-08A			0.59	0.10	0.35	0.05	0.05	0.00	0.88	0.10
4	13-08-09			0.60	0.11	0.45	0.07	0.07	0.00	0.45	0.08
5	13-08-16			0.82	0.10	0.15	0.06	0.04	0.00	0.62	0.04
6	13-10-03			1.17	0.28	1.24	0.16	0.14	0.00	2.68	0.25
7	14-03-09A			0.59	0.14	0.06	0.05	0.05	0.00	0.92	0.05
8	14-06-09A			1.51	0.80	0.50	0.04	0.04	0.00	0.41	0.04
9	14-07-2A5			0.71	0.14	0.07	0.04	0.03	0.00	0.52	0.04
10	L2005R7	1.10		0.70		0.10	0.01			0.50	0.02
11	L2005L6	2.30		1.00		0.20	0.08			1.10	0.10
12	L2005P9	4.50		0.90		0.20	0.01			1.40	0.03
13	L2005R1	2.30		0.90		1.50	0.07			1.80	0.05
14	L2005P13	1.90		0.95		0.20	0.12			0.90	0.05
15	L2005P2	1.40		0.90		0.70	0.04			4.00	0.08
16	bulk IDP	2.00	0.100	0.80	0.10	0.40	0.04			0.70	0.02
17	Ol-IDP	1.00	0.050	1.20	0.05	0.30	0.03	0.01	0.02	0.60	0.02
18	Mixed-IDP	1.60	0.120	0.80	0.13	0.30	0.05	0.01	0.01	0.50	0.02
19	Pyrox-IDP	3.10	0.080	0.80	0.09	0.60	0.04	0.01	0.02	0.90	0.03
20	L2008#5	1.10	0.200	0.70	0.10	0.40	0.07	0.01	0.01	0.70	0.04
21	CP (bulk)	2.39	0.056	1.015	0.07	0.417	0.047	0.016		0.705	0.024
22	CS (bulk)	1.32	0.051	0.824	0.082	0.341	0.021	0.014		0.742	0.032
23	CI	0.72	0.057	1.075	0.085	0.515	0.061	0.013		0.900	0.049
24	Halley	4.40	0.054	0.540	0.037	0.389	0.034	0.005	0.003	0.281	0.022

1-9 : Fraundorf (1981)
10-15: Thomas et al. (1992a)
16 : Thomas et al. (1992b)
17-19: Thomas et al. (1993a)
20 : Thomas et al. (1995a)
21-22: Schramm et al. (1989)
23: Anders and Grevesse (1989)
24: Jessberger et al. (1988)

Fe/Si vs. S/Si

The outliers in the diagram Fe/Si vs. S/Si (Fig. 56) are non-chondritic IDPs dominated by iron-nickel sulfides, viz. 13-10-03 (Fraundorf 1981) and L2005R1 (Thomas et al. 1992a), and magnetite-dominate IDP L2005P2 (Thomas et al. 1992a), which is probably a sulfide particle altered during dynamic pyrometamorphism. In general, the Fe/Si and S/Si (atomic) ratios in aggregate IDPs are smaller than the CI (or, solar system) abundance. The solar system abundance represents a larger sample than cluster IDPs. The low S/Si values reflect sulfur loss during atmospheric entry. The Fe/Si ratios in aggregate IDPs are higher

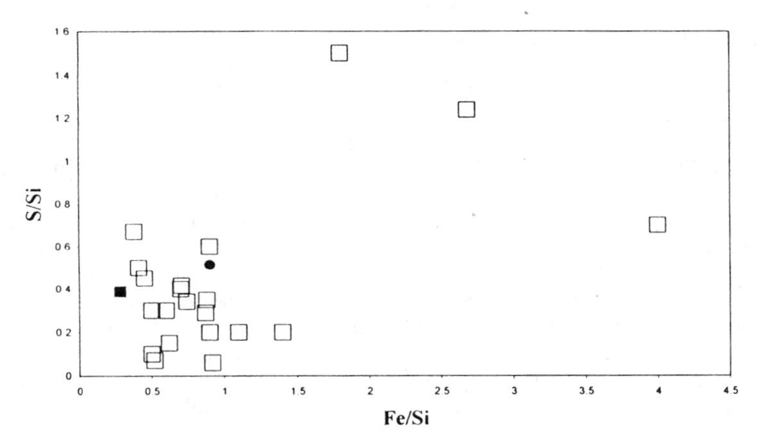

Figure 56. Fe/Si vs. S/Si (at. ratios) of aggregate IDPs (open squares). The outliers are defined a high amount of sulfides in the aggregates. The Fe/S ratios in comet Halley dust (solid square) and the CI value (dot) are shown.

than this ratio in comet Halley's dust. These higher ratios, as well as slightly higher S/Si ratios, indicate that the comparison of the Halley data should not be made at the level of aggregate IDPs that include sulfides and silicates between 0.5 and 10 µm (cf. Fig. 47). Indeed, Jessberger et al. (1988) found only a few FeS dominated grains.

Mg/Si vs. Fe/Si

The outliers in the Mg/Si vs. Fe/Si diagram include the non-chondritic IDPs 13-10-03 (Fraundorf 1981) and L2005P2 (Thomas et al. 1992a), as well as the olivine aggregate IDP 14-06-09 (Fraundorf 1981). The Mg/Si ratios in aggregate IDPs are intermediate between the CI and Halley abundances (Fig. 57). The higher Mg/Si ratios in aggregate IDPs compared to the comet data suggest that only a few dust particles in comet Halley were (non-chondritic) silicate-dominated grains.

C/Si vs. S/Si

Chondritic IDPs are carbon-rich compared to CI meteorites (Thomas et al. 1996). The outliers are again the non-chondritic IDPs L2005R1 and L2005P2 (Thomas et al. 1992a) that both have some small amounts of carbon materials at the surface. Pyroxene dominated IDPs (Thomas et al. 1992a, 1993a) have a high FeS content and they may not be a good match with the Halley dust data. The C/Si ratios in aggregate IDPs plot in between the CI and Halley dust abundances (Fig. 58). The diagram shows that matrix aggregates with nominal amounts of iron-nickel sulfides have S/Si ratios that match the comet dust (S/Si = 0.4) and the solar photosphere, S/Si = 0.36 (Ross and Aller 1976).

Matrix units and comet Halley dust

The mass of the dust particles for which chemical data were obtained in comet Halley ranges from 10^{-12} to 10^{-16} g (McDonnell et al. 1987, Fomenkova et al. 1992a). These measured particles masses compare with those of the matrix units in aggregate IDPs (Rietmeijer 1989a) and comparisons with IDPs have to be made at the level of the matrix and matrix aggregate (cf. Fig. 47). For comparison, the mass of a 10-µm-sized aggregate

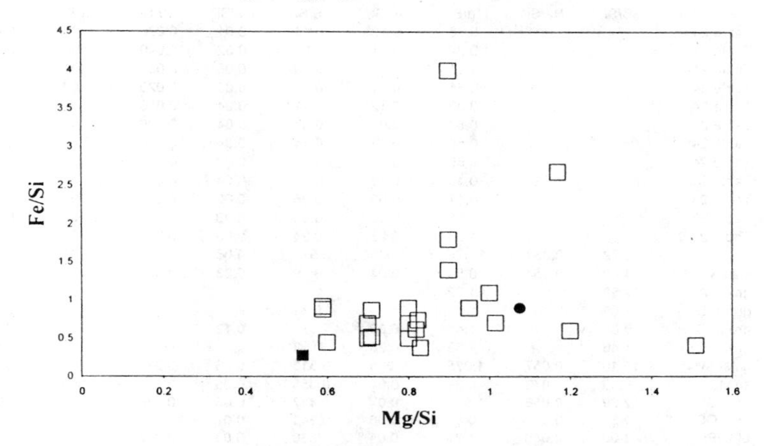

Figure 57. Mg/Si vs. Fe/Si (at. ratios) of aggregate IDPs (open squares) whereby sulfide-rich aggregates define the outliers. The Mg/Fe ratio in comet Halley dust (solid square) and the CI value (dot), are shown.

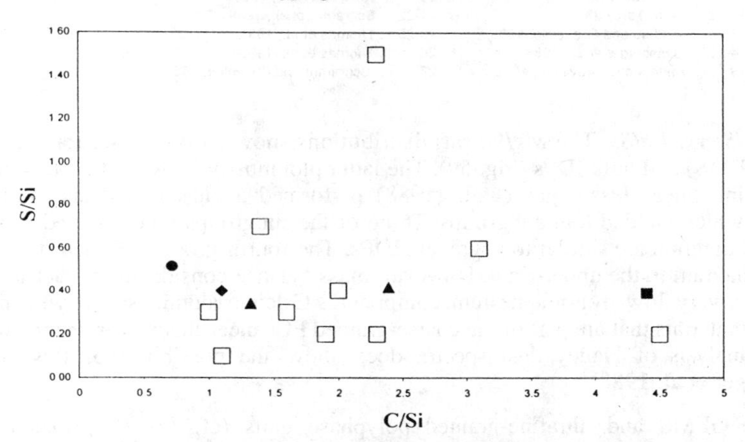

Figure 58. C/Si vs. S/Si (at. ratios) of aggregate IDPs (open squares), comet Halley dust (solid square), the CI value (dot), the bulk composition of cluster IDP L2008#5 (diamond) after Thomas et al (1995a) and CP and CS IDP bulk compositions (triangles) after Schramm et al. (1989).

IDP with an assumed density of 1.55 g cm^{-3} (cf. Table 11) is 8×10^{-10} g. The atomic ratios for GEMS and aggregate IDP bulk compositions are presented in Table 40.

Fe/Ni ratio. The Fe/Ni ratio in Halley's dust ($<10^{-15}$ to $>10^{-13}$ g) shows a peak at Fe/Ni = 20 (Fomenkova et al. 1992b) which is similar to Fe/Ni = 17.1 for chondritic IDPs (Table 9). This ratio is close to Fe/Ni = 23 for the ultrafine-grained PUs (Table 19) and to Fe/Ni = 13.3 in GEMS (Tables 23, 24). This result shows that chondritic (or, cosmic) Fe/Ni ratios are apparently determined by the earliest accreting dust in the solar system.

Table 40. Element/Si (at.) abundances in GEMS, comet P/Halley, the CI, solar system and solar photosphere abundances, and chondritic IDPs.

		C/Si	Na/Si	Mg/Si	Al/Si	S/Si	Ca/Si	Cr/Si	Mn/Si	Fe/Si	Ni/Si
1	Table 24			0.38	0.12	0.14	0.01	0.020	0.01	0.42	0.030
2	Table 24			0.24	0.11	0.08	0.02	0.040	0.00	0.38	0.030
3	Table 24			1.11	0.08	0.10	0.08	0.030	0.01	0.70	0.050
4	Table 24			0.65	0.11	0.05	0.03	0.020	0.00	0.12	0.010
5	Table 24			0.82	0.12	0.12	0.04	0.010	0.01	0.43	0.030
6	Table 24			0.83	0.09	0.07	0.04	0.020	0.01	0.49	0.050
7	Table 24			0.58	0.13	0.14	0.04	0.010	0.01	0.42	0.030
8	Table 24			0.66	0.11	0.11	0.02	0.010	0.01	0.69	0.060
9	Table 24			0.49	0.13	0.39	0.04	0.020	0.02	0.72	0.070
10	Table 23			0.17	0.05	0.36	0.00	0.020	0.00	0.66	0.000
11	Table 23			0.06	0.03	0.60	0.00	0.010	0.00	0.12	0.010
12	Table 23			1.68	0.05	0.24	0.00	0.010	0.00	0.32	0.010
13	CI	0.72	0.057	1.075	0.085	0.515	0.06	0.013	0.01	0.90	0.049
14	Halley	4.40	0.054	0.54	0.04	0.39	0.03	0.004	0.00	0.28	0.022
15	group A	17.50		1.32		0.77				0.84	
16	group B	1.56		1.08		0.31				0.35	
17	group C	0.06		1.45	0.08		0.12			0.18	
18	group D	1.46		1.59	0.28	0.63				6.06	
19	solar syst	10.10	0.057	1.075	0.085	0.515	0.06	0.010	0.010	0.90	0.049
20	sun	9.30	0.043	0.89	0.07	0.355	0.05	0.110	0.006	0.71	0.043
21	bulk CP	2.39	0.056	1.015	0.07	0.417	0.05	0.016		0.705	0.024
22	bulk CS	1.32	0.051	0.82	0.08	0.341	0.02	0.014		0.74	0.032
23	ol-IDPs	1.00	0.050	1.20	0.05	0.30	0.03	0.008	0.180	0.60	0.020
24	mix. IDPs	1.60	0.120	0.80	0.13	0.30	0.05	0.008	0.006	0.50	0.020
25	Pyr. IDPs	3.10	0.080	0.80	0.09	0.60	0.04	0.009	0.019	0.90	0.030
26	L2008#5	1.10	0.200	0.70	0.10	0.40	0.07	0.008	0.008	0.70	0.040
27	calc. Bulk		0.055	0.98	0.08	0.78	0.05	0.016		1.14	0.056

1 - 9 : from Table 24
10-12 : from Table 23
13 : Anders and Grevesse, 1989
14-18 : Jessberger et al., 1988
19 : Anders and Grevesse, 1989
20 : Ross and Aller, 1976
21-22 : Schramm et al., 1989
23-25 : Thomas et al., 1993a
26 : Thomas et al., 1995a
27 : Schramm and Brownlee, 1990

Mg/Si vs. Fe/Si. The Mg/Fe (at) distributions show a distinct separation for matrix units (GEMS) and bulk IDPs (Fig. 59). The latter plot mostly between the serpentine (1.5) and olivine lines. Jessberger et al. (1988) performed a cluster analysis of the Halley spectra, which yielded four subgroups. Three of the subgroups when plotted in the Mg/Si vs. Fe/Si diagram are similar to aggregate IDPs. The fourth group is C-rich dust and plots off the diagram to the upper-right. However, mass balance considerations dictate that there must be a very low iron-magnesium component. Calcium-aluminosilica (with minor Mg and Fe) materials that are part of the coarse-grained PUs meet this requirement. In fact, the cluster analysis of Halley dust spectra does show the possibility of this component (Jessberger et al. 1988).

The GEMS and ultrafine-grained polyphase units (cf. Fig. 37) match the bulk composition of comet Halley. This match suggests that most dust detected in comet Halley compares with the matrix units of chondritic aggregate IDPs. At the level at which comet Halley was sampled its dust was very low in iron. It appears therefore that the observed dust distribution (Jessberger et al. 1988, Fig. 4a) is correct. The adjustment of the Halley data to match chondritic Mg:Fe:Si ratios that Jessberger et al. (1988) preferred seems unjustified by the data on matrix units in chondritic aggregate IDPs.

Matrix units. The Mg/Si vs. Fe/Si diagram for the matrix units is presented in element instead of atomic ratios (Fig. 60). The relationship among matrix units is an open issue at this time. The ultrafine- and coarse-grained PUs are distinct textural units with bulk compositions supporting that they could be non-equilibrium condensates (Rietmeijer et al. 1997). GEMS, or unequilibrated aggregates, are believed to have formed by radiation with energetic H and He nuclei that was accompanied by element loss. The Mg/Fe ratios in the

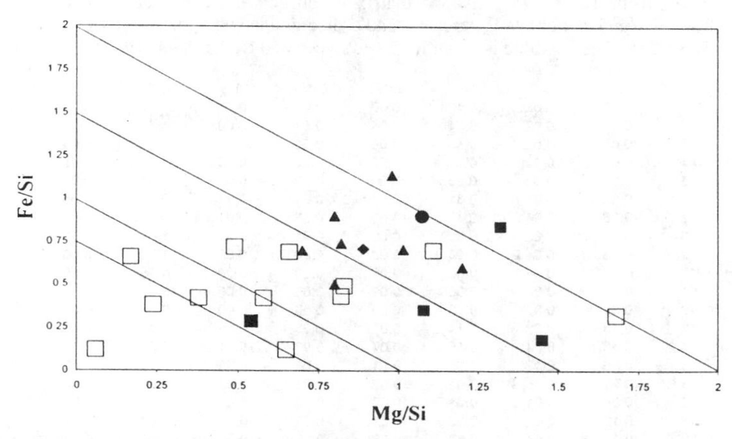

Figure 59. Mg/Si vs. Fe/Si (at. ratios) of GEMS (open squares), bulk compositions of aggregate IDPs (solid triangles), bulk comet Halley (large solid square), comet Halley subgroups (small solid squares), the CI and solar system values (dot) and the solar photosphere (diamond). The lines delineate the (Mg,Fe)/Si ratios for olivine, serpentine, pyroxene and smectite. GEMS and ultrafine-grained PUs with smectite-like ratios provide the best match with the bulk composition of comet Halley dust.

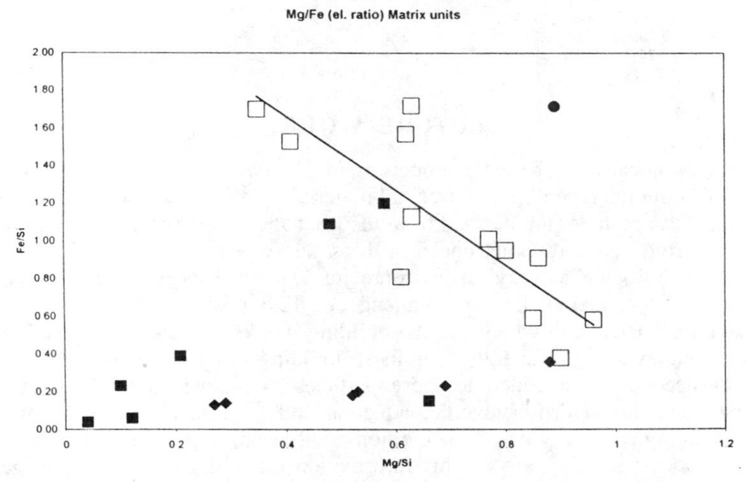

Figure 60. Mg/Si vs. Fe/Si (el. ratios) of matrix units in chondritic aggregate IDPs, viz. the anti-correlated compositions for ultrafine-grained PUs (open squares), the matrix of unequilibrated aggregates (solid squares), coarse-grained PUs (diamonds) and the CI ratio (dot).

matrix of unequilibrated aggregates and coarse-grained PUs define a linear trend at low Fe/Si ratios (Fig. 60, Table 41). For comparison, the CI value (McSween and Richardson 1977) has a much higher Mg/Fe ratio. The Mg/Fe ratios of ultrafine-grained PUs are scattered around a line defined by serpentine compositions. These relationships are identical to the ones shown in Mg-Fe-Si ternary diagram (Fig. 37) and confirm similar compositions for matrix units of chondritic IDPs due to irradiation and condensation.

Table 41. Element/Si (wt %) abundances in ultrafine-grained polyphase units (from Table 19), bulk and matrix of unequilibrated aggregates (from Table 25) and equilibrated aggregates and coarse-grained polyphase units (from Table 21). The CI value is taken from McSween and Richardson (1977).

	Mg/Si	Al/Si	S/Si	Ca/Si	Cr/Si	Mn/Si	Fe/Si	Ni/Si
1	0.35	0.01	0.40	0.00	0.01	0.00	1.70	0.06
2	0.41	0.02	0.71	0.04	0.01	0.00	1.53	0.06
3	0.63	0.00	0.48	0.00	0.00	0.00	1.13	0.05
4	0.62	0.02	0.77	0.02	0.00	0.00	1.57	0.08
5	0.61	0.00	0.28	0.00	0.00	0.00	0.81	0.01
6	0.86	0.01	0.31	0.08	0.01	0.01	0.91	0.05
7	0.96	0.04	0.11	0.02	0.00	0.01	0.58	0.01
8	0.77	0.02	0.50	0.09	0.01	0.00	1.01	0.05
9	0.63	0.04	0.00	0.00	0.00	0.00	1.72	0.00
10	0.80	0.02	0.00	0.00	0.00	0.00	0.95	0.00
11	0.85	0.00	0.00	0.00	0.00	0.00	0.59	0.00
12	0.90	0.00	0.00	0.00	0.00	0.00	0.38	0.00
13	0.48	0.08	0.11	0.07	0.05	0.05	1.09	0.15
14	0.58	0.04	0.15	0.07	0.03	0.01	1.20	0.12
15	0.66	0.12	0.03	0.01	0.06	0.03	0.15	0.03
16	0.21	0.05	0.02	0.02	0.01	0.02	0.39	0.09
17	0.04	0.01	0.01	0.00	0.00	0.01	0.04	0.01
18	0.12	0.07	0.02	0.01	0.02	0.00	0.06	0.01
19	0.10	0.07	0.02	0.02	0.06	0.00	0.23	0.03
20	0.88	0.01	0.02	0.02	0.03	0.03	0.36	0.03
21	0.29	0.26	0.00	0.23	0.02	0.02	0.14	0.00
22	0.52	0.04	0.00	0.00	0.00	0.00	0.18	0.00
23	0.53	0.05	0.00	0.02	0.00	0.01	0.2	0.00
24	0.69	0.08	0.00	0.03	0.00	0.00	0.23	0.00
25	0.27	0.16	0.00	0.07	0.00	0.08	0.13	0.03
CI	0.89	0.099	0.529	0.101	0.019	0.017	1.716	0.091

1 -12 : from Table 19
13-19: from Table 25
20-25: from Table 21

FUTURE WORK

The mineralogical and chemical properties of chondritic aggregate and cluster IDPs indicate that this unique type of extraterrestrial material preserves the accretion history in the solar nebula. These particles are derived from the least altered carbon-rich, icy protoplanets in the outer asteroid belt and comet nuclei in the solar system. They provide a window to the 'onset of mineralogical activity' in environments wherein energy to sustain petrological alteration was scarce. Herein lies the challenge of IDP research. We need answers to questions such as 'How and where did the building blocks of these particles form?' and 'How, when and where did alteration of these building blocks take place?' Also 'How much energy needed to be added to these particles to initiate parent body alteration?' Reaction rate data for hydro-cryogenic, aqueous and thermal alteration seem to be a prerequisite although the very nature of the non-equilibrium aggregates and clusters could make this a moot issue. To rely for this information on mineral formation in geological environments is almost certainly flawed. Maybe the very nature of the chaotic mineralogical properties and the high levels of potential energy for alteration in the chondritic aggregate and cluster IDPs are the defining properties of the state-of-affairs during the early solar system history. The chemical properties of these particles are well established but without a rigorous mineralogical framework it remains difficult to assess these data. Unexpected complications arise from incomplete knowledge of the interactions of IDPs with the atmosphere, i.e. dynamic pyrometamorphism. Collections of IDPs in the lower stratosphere avoid the periods immediately following a major volcanic eruption such as Mount St. Helens (1980), El Chichón (1982) and Mount Pinatubo (1991). Still, all stratospheric dust collections show volcanic signatures. An important issue for future research is the origin of volatile elements in IDPs, i.e. indigenous or contamination.

Mineralogical and petrological studies using electron microscope techniques have a high priority. In particular of the matrix units that include fossil presolar dust that is at least 4500 Myrs old, and who knows how much older still. In 25 years of interdisciplinary research on IDPs we learned a great deal. We know we can collect and recognize extraterrestrial fossils. There is a long road ahead that wanders 'through a dark wood.' This research is challenging and it is fun too.

ACKNOWLEDGMENTS

FJMR was supported by NASA Grants NAGW-3626 and NAG5-4441. I am grateful for constructive reviews by Don Brownlee, Gary Huss, Kase Klein and Ian Mackinnon, but errors and omissions are my responsibility. I thank my colleagues for permission to reproduce their work. Fleur Rietmeijer-Engelsman provided technical support at UNM.

REFERENCES

Allamandola LJ, Sandford SA, Wopenka B (1987) Interstellar polycyclic aromatic hydrocarbons and carbon in interplanetary dust particles and meteorites. Science 237:56-59

Akai J (1988) Incompletely transformed serpentine-type phyllosilicates in the matrix of Antarctic CM chondrites. Geochim Cosmochim Acta 52:1593-1599

Anders E, Grevesse N (1989) Abundances of the elements: Meteoritic and solar. Geochim Cosmochim Acta 53:197-214

Anders E, Zinner E (1993) Interstellar grains in primitive meteorites: Diamond, silicon carbide, and graphite. Meteoritics 28:490-514

Anderson DM, Morgenstern NR (1973) Physics, chemistry, and mechanics of frozen Ground p 257-288 In Permafrost: A Review. Nat'l Acad Sci, Washington, DC

Arndt P, Bohsung J, Maetz M, Jessberger EK (1996) The elemental abundances in interplanetary dust particles. Meteoritics Planet Sci 31:817-833

Bigg EK, Ono A, Thompson WJ (1970) Aerosols at altitudes between 20 and 37 km. Tellus XXII:550-563

Bigg EK, Kviz Z, Thompson WJ (1971) Electron microscope photographs of extraterrestrial particles. Tellus XXIII:247-260

Blake DF, Mardinly AJ, Echer CJ, Bunch TE (1988) Analytical electron microscopy of a hydrated interplanetary dust particle. Proc Lunar Planet Sci Conf 18:615-622

Blake D, Fleming RH, Bunch TE (1989) Identification and characterization of a carbonaceous, titanium containing interplanetary dust particle. Lunar Planet Sci XX:84-85

Blanford GE, Thomas KL, McKay DS (1988) Microbeam analysis of four chondritic interplanetary dust particles for major elements, carbon and oxygen. Meteoritics 23:113-121

Bonijoly M, Oberlin M, Oberlin A (1982) A possible mechanism for natural graphite formation. Int'l J Coal Geol 1:283-312

Bottke, Jr. WF, Nolan NC, Greenberg R, Kolvoord RA (1994) Collisional lifetimes and impact statistics of near-Earth asteroids. In Gehrels T (ed) Hazards due to Comets & Asteroids, p 337-357. Univ Arizona Press, Tucson, AZ

Bradley JP (1988) Analysis of chondritic interplanetary dust thin-sections. Geochim Cosmochim Acta 52:889-900

Bradley JP (1994a) Nanometer-scale mineralogy and petrography of fine-grained aggregates in anhydrous interplanetary dust particles. Geochim Cosmochim Acta 58:2123-2134

Bradley JP (1994b) Chemically anomalous, pre-accretionally irradiated grains in interplanetary dust from comets. Science 265:925-929.

Bradley J (1994c) Mechanisms of grain formation, post-accretional alteration, and likely parent body environments of interplanetary dust particles (IDPs). In Zolensky ME, Wilson TL, Rietmeijer FJM, Flynn GJ (eds) Analysis of Interplanetary Dust. AIP Conf Proc 310:89-104 Am Inst Phys, New York

Bradley JP, Brownlee DE (1986) Cometary particles: Thin sectioning and electron beam analysis. Science 231:1542-1544

Bradley JP, Brownlee DE (1991) An interplanetary dust particle linked directly to type CM meteorites and an asteroidal origin. Science 251:549-552

Bradley JP, Ireland T (1996) The search for interstellar components in interplanetary dust particles. In Gustafson BS, Hanner MS (eds) Physics, Chemistry and Dynamics of Interplanetary Dust, Astron Soc Pacific Conf Series 104:275-282

Bradley JP, Brownlee DE, Veblen DR (1983) Pyroxene whiskers and platelets in interplanetary dust: evidence of vapour phase growth. Nature 301:473-477

Bradley JP, Brownlee DE, Fraundorf P (1984a) Discovery of nuclear tracks in interplanetary dust. Science 226:1432-1434

Bradley JP, Brownlee DE, Fraundorf P (1984b) Carbon compounds in interplanetary dust: Evidence for formation by heterogeneous catalysis. Science 223:56-58

Bradley JP, Germani MS, Brownlee DE (1989) Automated thin-film analyses of anhydrous interplanetary dust particles in the analytical electron microscope. Earth Planet Sci Lett 93:1-13

Bradley JP, Humecki HJ, Germani MS (1992) Combined infrared and analytical electron microscope studies of interplanetary dust particles. Astrophys J 394:643-651

Bradley JP, Sandford SA, Walker RM (1988) Interplanetary Dust Particles. In Kerridge JF, Matthews MS (eds) Meteorites and the Early Solar System, p 861-898, Univ Arizona Press, Tucson, AZ

Bradley JP, Keller LP, Brownlee DE, Thomas KL (1996) Reflectance spectroscopy of interplanetary dust particles. Meteoritics Planet Sci 31:394-402

Brearley AJ (1995) Aqueous alteration and brecciation in Bells, an unusual, saponite-bearing, CM chondrite. Geochim Cosmochim Acta 59:2291-2317

Brownlee DE (1978a) Interplanetary dust: Possible implications for comets and presolar interstellar grains. In Gehrels T (ed) Protostars & Planets, p 134-150 Univ Arizona Press, Tucson, AZ

Brownlee DE (1978b) Microparticle studies by sampling techniques. In McDonnell AM (ed) Cosmic Dust, p 295-336. Wiley Interscience, New York

Brownlee DE (1985) Cosmic dust: Collection and research. Ann Rev Earth Planet Sci 13:147-173

Brownlee DE (1994) The origin and role of dust in the early solar system. In Zolensky ME, Wilson TL, Rietmeijer FJM, Flynn GJ (eds) Analysis of Interplanetary Dust, AIP Conf Proc.310:127-143 Am Inst Phys, New York

Brownlee DE, Hodge PW (1973) Ablation debris and primary micrometeoroids in the stratosphere. Space Res 13:1139-1151

Brownlee DE, Hodge PW, Bucher W (1973) The physical nature of interplanetary dust as inferred by particles collected at 35 km. In Hemenway CL, Millman PM, Cook AF (eds) Evolutionary and Physical Properties of Meteoroids. NASA SP-319:291-295

Brownlee DE, Tomandl DA, Hodge PW, Hörz F (1974) Elemental abundances in interplanetary dust. Nature 252:667-669

Brownlee DE, Tomandl D, Hodge PW (1976a) Extraterrestrial particles in the stratosphere. In Elsasser H, Fechtig H (eds) Interplanetary dust and the Zodiacal light, p 279-284 Springer-Verlag, New York

Brownlee DE, Ferry GV, Tomandl D (1976b) Stratospheric aluminum oxide. Science 191:1270-1271

Brownlee DE, Rajan RS, Tomandl DA (1977a) A chemical and textural comparison between carbonaceous chondrites and interplanetary dust. In Delsemme AH (ed) Comets, Asteroids, Meteorites interrelations, evolutions and origins, p 137-141 University of Toledo, Toledo, OH

Brownlee DE, Tomandl DA, Olszewski E (1977b) Interplanetary dust: A new source of extraterrestrial material for laboratory studies. Proc Lunar Sci Conf 8:149-160

Brownlee DE, Pilachowski L, Olzewski E, Hodge PW (1980) Analysis of interplanetary dust collections. In Halliday I, McIntosh BA (eds) Solid Particles in the Solar System p 333-342, D Reidel, Dordrecht, Holland

Brownlee DE, Olszewski E, Wheelock M (1982) A working taxonomy for micrometeorites. Lunar Planet Sci XIII:71-72

Brownlee DE, Schramm LS, Wheelock MM, Maurette M (1989) Large mineral grains in interplanetary dust. Lunar Planet Sci XX:121-122

Brownlee DE, Joswiak DJ, Love SG, Bradley JP, Nier OA, Schlutter DJ (1994) Identification and analysis of cometary IDPs. Lunar Planet Sci XXV:185-186

Brownlee DE, Joswiak DJ, Schlutter DJ, Pepin RO, Bradley JP, Love SG (1995) Identification of individual cometary IDP's by thermally stepped He release. Lunar Planet Sci XXVI:183-184

Christoffersen R, Buseck PR (1983a) Epsilon Carbide: A low-temperature component of interplanetary dust particles. Science 222:1327-1329

Christoffersen R, Buseck PR (1983b) Mineralogy and microstructure of some C-type interplanetary dust particles as determined by analytical electron microscopy. Lunar Planet Sci XIV:111-112

Christoffersen R, Buseck PR (1986a) Mineralogy of interplanetary dust particles from the 'olivine' infrared class. Earth Planet Sci Lett 78:53-66

Christoffersen R, Buseck PR (1986b) Refractory minerals in interplanetary dust. Science 234:590-592

Clemett SJ, Maechling CR, Zare RN, Swan PD, Walker RM (1993) Identification of complex aromatic molecules in individual interplanetary dust particles. Science 262:721-725

Corrigan CM, Zolensky ME, Dahl J, Long M, Weir J, Sapp C, Burkett PJ (1997) The porosity and permeability of chondritic meteorites and interplanetary dust particles. Meteoritics Planet Sci 32:509-515

Crozier WD (1966) Nine years of continuous collection of black, magnetic spherules from the atmosphere. J Geophys Res 71:603-611

Dermott SF, Liou JC (1994) Detection of asteroidal dust particles from known families in near-Earth orbit. In Zolensky ME, Wilson TL, Rietmeijer FJM, Flynn GJ (eds) Analysis of Interplanetary Dust, AIP Conf Proc 310:11-21 Am Inst Physics, New York

Esat TM, Taylor SR (1987) Mg Isotopic composition of some interplanetary dust particles. Lunar Planet Sci XVIII:269-270

Esat TM, Brownlee DE, Papanastasiou DE, Wasserburg GJ (1979) The Mg isotopic composition of interplanetary dust particles. Science 206:190-192

Flynn GJ (1989) Atmospheric entry heating of micrometeorites. Proc Lunar Planet Sci Conf 19:673-682.

Flynn GJ (1994) Changes to the composition and mineralogy of interplanetary dust particles by terrestrial encounters. In Zolensky ME, Wilson TL, Rietmeijer FJM, Flynn GJ (eds) Analysis of Interplanetary Dust. AIP Conf Proc 310:127-143 Am Inst Phys, New York

Flynn GJ (1995) Thermal gradients in interplanetary dust particles: The effect of an endothermic phase transition. Lunar Planet Sci XXVI:405-406

Flynn GJ (1996a) Sources of 10 micron interplanetary dust: The contribution from the Kuiper belt. In Gustafson BS, Hanner MS (eds) Physics, Chemistry and Dynamics of Interplanetary Dust, Astron Soc Pacific Conf Series 104:171-175

Flynn GJ (1996b) Are the S-Type asteroids the parent bodies of ordinary chondrite meteorites?: Evidence from the interplanetary dust recovered from the Earth's stratosphere. Lunar Planet Sci XXVII:365-366

Flynn GJ, Sutton SR (1990) Synchrotron X-ray fluorescence analyses of stratospheric cosmic dust: New results for chondritic and low-nickel particles. Proc Lunar Planet Sci Conf 20:335-342

Flynn GJ, Sutton SR (1991a) Chemical characterization of seven Large Area Collector particles by SXRF. Proc Lunar Planet Sci 21:549-556

Flynn GJ, Sutton SR (1991b) Cosmic dust particle densities: Evidence for two populations of stony micrometeorites. Proc Lunar Planet Sci 21:541-547

Flynn GJ, Sutton SR (1992a) Trace elements in chondritic stratospheric particles: Zinc depletions as a possible indicator of atmospheric entry heating. Proc Lunar Planet Sci 22:171-184

Flynn GJ, Sutton SR (1992b) Element abundances in stratospheric cosmic dust: Indications for a new chemical type of chondritic material. Lunar Planet Sci XXIII:373-374

Flynn GJ, Fraundorf P, Shirck J, Walker RM, Zinner E (1978) Chemical and structural studies of 'Brownlee' particles. Proc Lunar Sci Conf 9:1187-1208

Flynn GJ, Fraundorf P, Keefe G, Swan P (1982) Aluminum-rich spinel aggregates in the stratosphere: From Earth, or not? Lunar Planet Sci XIII:223-224

Flynn GJ, Sutton SR, Bajt S, Klöck W (1993a) New low-Ni (igneous?) particles among the C and C? types of cosmic dust. Lunar Planet Sci XXIV:499-500

Flynn GJ, Sutton SR, Bajt S, Klöck W, Thomas KL, Keller LP (1993b) Depletions of sulfur and/or zinc in IDPs: Are they reliable indicators of atmospheric entry heating? Lunar Planet Sci XXIV:497-498

Flynn GJ, Bajt S, Sutton SR, Zolensky M, Thomas KL, Keller LP (1996) The abundance pattern of elements having low nebular condensation temperatures in interplanetary dust particles: Evidence for a new type of chondritic material. In Gustafson BS, Hanner MS (eds) Physics, Chemistry and Dynamics of Interplanetary Dust, Astron Soc Pacific Conf Series 104:291-294

Fomenkova MN, Kerridge JF, Marti K, McFadden L-A (1992a) Compositional trends in rock-forming elements of comet Halley dust. Science 258:266-268

Fomenkova M, Kerridge J, Marti K, McFadden L (1992b) Iron-rich particles in comet Halley dust. Lunar Planet Sci XXIII:381-382

Fraundorf P (1981) Interplanetary dust in the transmission electron microscope: diverse materials from the early solar system. Geochim Cosmochim Acta 45:915-943

Fraundorf P, Shirck J (1979) Microcharacterization of 'Brownlee' particles: Features which distinguish interplanetary dust from meteorites? Proc Lunar Planet Sci Conf 10:951-976

Fraundorf P, Flynn GJ, Shirck J, Walker RM (1980) Interplanetary dust collected in the earth's stratosphere: The question of solar flare tracks. Proc Lunar Planet Sci Conf 11:1235-1249.

Fraundorf P, Brownlee DE, Walker RM (1982a) Laboratory studies of interplanetary dust. In Wilkening LL (ed) Comets, p 383-409, Univ Arizona Press, Tucson, AZ

Fraundorf P, McKeegan KD, Sandford SA, Swan P, Walker RM (1982b) An inventory of particles from stratospheric collectors: Extraterrestrial and otherwise. Proc Lunar Planet Sci Conf 13, J Geophys Res 87 Suppl:A403-A408

Fraundorf P, Hintz C, Lowry O, McKeegan KD, Sandford SA (1982c) Determination of the mass, surface density, and volume density of individual interplanetary dust particles. Lunar Planet Sci XIII:225-226

Fraundorf P, Lyons T, Schubert P (1982b) The survival of solar flare tracks in interplanetary dust silicates on deceleration in the earth's atmosphere. Proc Lunar Planet Sci Conf 13, J Geophys Res 87 Suppl:A409-A412

Fulle M (1992) Dust from short-period comet P/Schwassmann-Wachmann 1 and the replenishment of the interplanetary dust cloud. Nature 359:42-44

Ganapathy R, Brownlee DE (1979) Interplanetary dust: Trace element analysis of individual particles by neutron activation. Science 206:1075-1076

Germani MS, Bradley JP, Brownlee DE (1990) Automated thin-film analyses of hydrated interplanetary dust particles in the analytical electron microscope. Earth Planet Sci Lett 101:162-179

Gibson EK, Jr, Bustin R (1994) Volatiles in interplanetary dust particles: A comparison with volatile-rich meteorites. In Zolensky ME, Wilson TL, Rietmeijer FJM, Flynn GJ (eds) Analysis of Interplanetary Dust. AIP Conf Proc 310:173-184 Am Inst Phys, New York

Gibson EK, Wentworth SJ, Mckay DS (1983) Chemical weathering and diagenesis of cold desert soil from Wright Valley, Antarctica: An analog of martian weathering processes. Proc Lunar Planet Sci Conf 13 Part 2, J Geophys Res 88 Suppl:A912-A928

Greshake A (1997) The primitive matrix components of the unique carbonaceous chondrite Acfer 094: A TEM study. Geochim Cosmochim Acta 61:437-452

Greshake A, Hoppe P, Bischoff A (1996) Mineralogy, chemistry, and oxygen isotopes of refractory inclusions from stratospheric interplanetary dust particles and micrometeorites. Meteoritics Planet Sci 31:739-748

Griffith OK, Renzema TS, Hallgren DS, Hemenway CL (1971) Electron microprobe studies of cosmic dust impact craters. Space Res XI:383-392

Hudson B, Flynn GJ, Fraundorf P, Hohenberg CM, Shirck J (1981) Noble gases in stratospheric dust particles: Confirmation of extraterrestrial origin. Science 211:383-386

Jessberger EK, Christoforidis A, Kissel J (1988) Aspects of major element composition of Halley's dust. Nature 332:691-695

Jessberger EK, Bohsung J, Chakaveh S, Traxel K (1992) The volatile element enrichment of chondritic interplanetary dust particles. Earth Planet Sci Lett 112:91-99

Joswiak DJ, Brownlee DE, Bradley JP, Schlutter DJ, Pepin RO (1996) Systematic analyses of major element distributions in GEMS from high speed IDPs. Lunar Planet Sci XXVII:625-626

Keller LP, Thomas KL, McKay DS (1992a) An interplanetary dust particle with links to CI chondrites. Geochim Cosmochim Acta 56:1409-1412

Keller LP, Thomas KL, McKay DS (1992b) Thermal processing of cosmic dust: Atmospheric heating and parent body metamorphism. Lunar Planet Sci XXIII:675-676

Keller LP, Thomas KL, McKay DS (1993) Carbon abundances, major element chemistry, and mineralogy of hydrated interplanetary dust particles. Lunar Planet Sci XXIV:785-786

Keller LP, Thomas KL, McKay DS (1996a) Mineralogical changes in IDPS resulting from atmospheric entry heating. In Gustafson BS, Hanner MS (eds) Physics, Chemistry and Dynamics of Interplanetary Dust, Astron Soc Pacific Conf Series 104:295-298

Keller LP, Thomas KL, McKay DS (1996b) Carbon petrography and the chemical state of carbon and nitrogen in IDPs. Lunar Planet Sci XXVII:659-660

Klöck W, Stadermann FJ (1994) Mineralogical and chemical relationships of interplanetary dust particles, micrometeorites and meteorites. In Zolensky ME, Wilson TL, Rietmeijer FJM, Flynn GJ (eds) Analysis of Interplanetary Dust, AIP Conf Proc 310:51-87 Am Inst Physics, New York

Klöck W, Thomas KL, McKay DS, Palme H (1989) Unusual olivine and pyroxene composition in interplanetary dust and unequilibrated ordinary chondrites. Nature 339:126-128

Klöck W, Thomas KL, McKay DS, Zolensky ME (1990) Olivine compositions in anhydrous and hydrated IDPs compared to olivines in matrices of primitive meteorites. Lunar Planet Sci XXI:637-638

Kordesh KM, Mackinnon IDR, McKay DS (1983) A new classification and database for stratospheric dust particles. Lunar Planet Sci XIV:389-390

Kurat G. Koeberl C, Presper T, Brandstätter F, Maurette M (1994) Petrology and geochemistry of Antarctic micrometeorites. Geochim Cosmochim Acta 58:3879-3904

Lawler ME, Brownlee DE, Temple S, Wheelock MM (1989) Iron, magnesium, and silicon in dust from comet Halley. Icarus 80:225-242

Lindstrom DJ (1992) Scandium/iron and cobalt/iron ratios as indicators of the sources of stratospheric dust particles. Lunar Planet Sci XXIII:779-780

Lipschutz ME, Woolum DS (1988) Highly labile elements. In Kerridge JF, Matthews MS (eds) Meteorites and the Early Solar System, p 462-487, Univ Arizona Press, Tucson, AZ

Love SG, Brownlee DE (1991) Heating and thermal transformation of micrometeoroids entering the Earth's atmosphere. Icarus 89:26-43

Love SG, Brownlee DE (1993) A direct measurement of the terrestrial mass accretion rate of cosmic dust. Science 262:550-553

Love SG, Brownlee DE (1994) Peak atmospheric entry temperatures of micrometeorites. Meteoritics 29:69-70

Love SG, Joswiak D, Brownlee DE (1994) Densities of stratospheric micrometeorites. Icarus 111:227-236

Mackinnon IDR, McKay DS (1986) Refinements and developments on the stratospheric dust database and classification scheme. Lunar Planet Sci XVII:510-511

Mackinnon IDR, Mogk DW (1985) Surface sulfur measurements on stratospheric particles. Geophys Res Lett 12:93-96

Mackinnon IDR, Rietmeijer FJM (1984) Bismuth in interplanetary dust. Nature 311:135-138

Mackinnon IDR, Rietmeijer FJM (1987) Mineralogy of chondritic interplanetary dust particles. Reviews Geophys 25:1527-1553

Mackinnon IDR, McKay DS, Nace G, Isaacs AM (1982) Classification of the Johnson Space Center Stratospheric Dust Collection. Proc Lunar Planet Sci Conf 13, J Geophys Res 87 Suppl:A413-A421

Mackinnon IDR, Gooding JL, McKay DS, Clanton US (1984) The El Chichón stratospheric cloud: Solid particulates and settling rates. J Volc Geothermal Res 23:125-146

Mackinnon IDR, Rietmeijer FJM, McKay DS, Zolensky ME (1985) Microbeam analyses of stratospheric particles. In Armstrong JT (ed) Microbeam Analysis-1985, p 291-297 San Francisco Press, San Francisco, CA

Mackinnon IDR, Lindsay C, Bradley JP, Yatchmenoff B (1987) Porosity of serially sectioned interplanetary dust particles. Meteoritics 22:450-451

Mathis JS, Whiffen G (1989) Composite interstellar grains. Astrophys J 341:808-822

Morfill GE, Tscharnuter W, Völk HJ (1985) Dynamical and chemical evolution of the protoplanetary nebula. In Black DC, Matthews MS (eds) Protostars and Planets II, p 493-533. Univ Arizona Press, Tucson, AZ

Murray J, Renard AF (1883) 'On the measurement characters of volcanic ashes and cosmic dust, and their origin in deep-sea sediment deposits'. Proc Roy Soc Edinburgh 12:474-495

McClelland L, Simkin T, Summers M, Nielsen E, Stein TC (1989) Global Volcanism 1975-1985, Smithsonian Institution Scientific event alert network (SEAN) 665 p. Prentice Hall, Englewood Cliffs, NJ, Am Geophys Union Washington, DC

McDonnell JAM et al. (1987) The dust distribution within the inner coma of comet P/Halley 1982i: encounter by Giotto's impact detectors. Astron Astrophys 187:719-741

McKeegan KD (1986) Hydrogen and magnesium isotopic abundances in aluminum-rich stratospheric dust particles. Lunar Planet Sci XVII:539-540

McKeegan KD (1987) Oxygen isotopes in refractory stratospheric dust particles: Proof of extraterrestrial origin. Science 237:1468-1471

McKeegan KD, Walker RM, Zinner E (1985) Ion microprobe isotopic measurements of individual interplanetary dust particles. Geochim Cosmochim Acta 49:1971-1987

McKeegan KD, Swan P, Walker RM, Wopenka B, Zinner E (1987) Hydrogen isotopic variations in interplanetary dust particles. Lunar Planet Sci XVIII:627-628

McSween HY jr, Richardson SM (1977) The composition of carbonaceous chondrite matrix. Geochim Cosmochim Acta 41:1145-1161

Messenger S, Walker RM (1996) Isotopic anomalies in interplanetary dust particles. In Gustafson BS, Hanner MS (eds) Physics, Chemistry and Dynamics of Interplanetary Dust, Astron Soc Pacific Conf Series 104:287-290

Messenger S, Clemett SJ, Keller LP, Thomas KL, Chillier XDF, Zare RN (1995) Chemical and mineralogical studies of an extremely deuterium-rich IDP. [abstr] Meteoritics 30:546-547

Messenger S, Walker RM, Clemett SJ, Zare RN (1996) Deuterium enrichments in cluster IDPs. Lunar Planet Sci XXVII:867-868

Miyamoto M, Onuma N, Takeda H (1979) Wollastonite whiskers in the Allende meteorite and their bearing on a possible post-condensation process. Geochem J 13:1-15

Naughton JJ, Greenberg VA, Goguel R (1976) Incrustations and fumarolic condensates at Kilauea Volcano, Hawaii: Field, drill-hole and laboratory observations. J Volc Geothermal Res 1:149-165

Nier AO (1994) Helium and neon in interplanetary dust particles. In Zolensky ME, Wilson TL, Rietmeijer FJM, Flynn GJ (eds) Analysis of Interplanetary Dust, AIP Conf Proc 310:115-126, Am Inst Physics, New York

Nier AO, Schlutter DJ (1990) Helium and neon isotopes in stratospheric particles. Meteoritics 25:263-267

Nier AO, Schlutter DJ (1992) Extraction of helium from individual interplanetary dust particles by step-heating. Meteoritics 27:166-173

Nier AO, Schlutter DJ (1993) The thermal history of interplanetary dust particles collected in the Earth's stratosphere. Meteoritics 28:675-681
Nier AO, Schlutter DJ (1994) A search for solar energetic particle helium in interplanetary dust particles. Lunar Planet Sci XXV:999-1000
Palme H, Larimer JW, Lipschutz ME (1988) In Kerridge JF, Matthews MS (eds) Meteorites and the Early Solar System, p 436-461 Univ Arizona Press, Tucson, AZ
Rabinowitz DR, Bowell E, Shoemaker E, Muinonen K (1994) The population of Earth crossing asteroids. In Gehrels T (ed) Hazards due to Comets & Asteroids, p 285-312 Univ Arizona Press, Tucson, AZ
Radicati di Brozolo F, Fleming RH (1992) Mass spectrometric observation of organic species in a single IDP thin section. Lunar Planet Sci XXIII:1123-1124
Rajan RS, Brownlee DE, Tomandl D, Hodge PW, Farrar H, Britten RA (1977) Detection of ^{4}He in stratospheric particles gives evidence of extraterrestrial origin. Nature 267:33-134
Rietmeijer FJM (1985a) On the continuum between chondritic interplanetary dust and CI and CM carbonaceous chondrites: A petrological approach. Lunar Planet Sci XVI:698-699
Rietmeijer FJM (1985b) A model for diagenesis in proto-planetary bodies. Nature 313:293-294
Rietmeijer FJM (1987) Silicone oil: A persistent contaminant in chemical and spectral microanalyses of interplanetary dust particles. Lunar Planet Sci XVIII:836-837
Rietmeijer FJM (1988) Enhanced residence of submicron Si-rich volcanic particles in the lower stratosphere. J Volc Geothermal Res 34:173-184
Rietmeijer FJM (1989a) Ultrafine-grained mineralogy and matrix chemistry of olivine-rich chondritic interplanetary dust particles. Proc Lunar Planet Sci Conf 19:513-521
Rietmeijer FJM (1989b) Tin in a chondritic interplanetary dust particle. Meteoritics 24:43-47
Rietmeijer FJM (1990) Salts in two chondritic porous interplanetary dust particles. Meteoritics 25:209-214
Rietmeijer FJM (1991) Aqueous alteration in five chondritic porous interplanetary dust particles. Earth Planet Sci Lett 102:148-157
Rietmeijer FJM (1992a) A detailed petrological analysis of hydrated, low-nickel, nonchondritic stratospheric dust particles. Proc Lunar Planet Sci 22:195-201
Rietmeijer FJM (1992b) Mineralogy of primitive chondritic protoplanets in the early solar system. Trends Mineralogy1:23-41
Rietmeijer FJM (1992c) Endothermic reactions constrain dynamic pyrometamorphic temperatures in two iron-rich interplanetary dust particles. Lunar Planet Sci XXIII:1151-1152
Rietmeijer FJM (1992d) Pregraphitic and poorly graphitised carbons in porous chondritic micrometeorites. Geochim Cosmochim Acta 56:1665-1671
Rietmeijer FJM (1993a) Size distributions in two porous chondritic micrometeorites. Earth Planet Sci Lett 117:609-617
Rietmeijer FJM (1993b) Micrometeorite dynamic pyrometamorphism: Observation of a thermal gradient in iron-nickel sulfide. Lunar Planet Sci XXIV:1201-1202
Rietmeijer FJM (1993c) Volcanic dust in the stratosphere between 34 and 36 km altitude during May 1985. J Volc Geothermal Res 55:69-83
Rietmeijer FJM (1993d) The bromine content of micrometeorites: Arguments for stratospheric contamination. J Geophys Res 98(E4):7409-7414
Rietmeijer FJM (1994a) A proposal for a petrological classification scheme of carbonaceous chondritic micrometeorites. In Zolensky ME, Wilson TL, Rietmeijer FJM, Flynn GJ (eds) Analysis of Interplanetary Dust, AIP Conf Proc 310:231-240 Am Inst Physics, New York.
Rietmeijer FJM (1994b) Searching for a principal component mixing model for chondritic interplanetary dust particles: The use of size analyses. Lunar Planet Sci XXV:1129-1130
Rietmeijer FJM (1995) Post-entry and volcanic contaminant abundances of zinc, copper, selenium, germanium and gallium in stratospheric micrometeorites. Meteoritics 30:33-41
Rietmeijer FJM (1996a) CM-like interplanetary dust particles in the lower stratosphere during July 1989 October and 1991 June. [abstr] Meteoritics Planet Sci 31:278-288
Rietmeijer FJM (1996b) The ultrafine mineralogy of a molten interplanetary dust particle as an example of the quench regime of atmospheric entry heating. Meteoritics Planet Sci 31:237-242
Rietmeijer FJM (1996c) Cellular precipitates of iron-oxide in extraterrestrial olivine in a stratospheric interplanetary dust particle. Mineral Mag 60:877-885
Rietmeijer FJM (1996d) Principal components constrain dynamic pyrometamorphism in a partially melted interplanetary dust particle. Lunar Planet Sci XXVII:1071-1072
Rietmeijer FJM (1996e) The butterflies of principal components: A case of ultrafine-grained polyphase units. Lunar Planet Sci XXVII:1073-1074
Rietmeijer FJM (1997a) Not all cluster particles in the NASA/JSC Cosmic Dust Collection are extraterrestrial. Lunar Planet Sci XXVIII:1171-1172

Rietmeijer FJM (1997b) First-order properties of chondritic cluster IDPs based on data from the NASA/JSC Cosmic Dust Catalogs. Lunar Planet Sci XXVIII:1169-1170

Rietmeijer FJM (1997c) Principal components: Petrology and chemistry of polyphase units in chondritic porous interplanetary dust particles. Lunar Planet Sci XXVIII:1173-1174

Rietmeijer FJM, Flynn GJ (1996) Lower stratospheric abundances of aluminum oxide and Al'-spheres > 9 micrometers from May 22, 1981, to July 1991. Meteoritics Planet Sci 31:A114-A115

Rietmeijer FJM, Mackinnon IDR (1984) Melting, ablation and vapor phase condensation during atmospheric passage of the Bjurböle meteorite. J Geophys Res 87 Suppl:B597-B604

Rietmeijer FJM, Mackinnon IDR (1985a) Layer silicates in a chondritic porous interplanetary dust particle. Proc Lunar Planet Sci Conf 16, J Geophys Res 90 Suppl:D149-D155

Rietmeijer FJM, Mackinnon IDR(1985b) Poorly graphitized carbon as a new cosmothermometer for primitive extraterrestrial materials. Nature 316:733-736

Rietmeijer FJM, Mackinnon IDR (1987) Metastable carbon in two chondritic porous interplanetary dust particles. Nature 326:162-165

Rietmeijer FJM, Mackinnon IDR (1990) Titanium oxide Magnéli phases in four chondritic interplanetary dust particles. Proc Lunar Planet Sci Conf 20:323-333

Rietmeijer FJM, Mackinnon IDR (1997) Bismuth oxide nanoparticles in the stratosphere J Geophys Res102(E3):6621-6627

Rietmeijer FJM, McKay DS (1985) An interplanetary dust particle analog to matrices of CO/CV carbonaceous chondrites and unmetamorphosed unequilibrated ordinary chondrites. [abstr] Meteoritics 20:743-744

Rietmeijer FJM, Warren JL (1994) Windows of opportunity in the NASA Johnson Space Center Cosmic Dust Collection. In Zolensky ME, Wilson TL, Rietmeijer FJM, Flynn GJ (eds) Analysis of Interplanetary Dust, AIP Conf Proc 310:255-275 Am Inst Physics, New York

Rietmeijer FJM, Fu G, Karner JM (1997) Alteration of presolar dust based of transmission electron microscope/analytical electron microscope studies of chondritic interplanetary dust particles and nonequilibrium simulation experiments. In Zolensky ME, Krot AN, Scott ERD (eds) Workshop on Parent-body and Nebular Modification of Chondritic Materials.LPI Tech Rpt 97-02, Part 1:51-53

Robie RA, Bethke PM, Beardsley KM (1967) Selected X-ray crystallographic data molar volumes, and desmiry of minerals. US Geol Survey Bull 1248:1-87

Rose WI, Chuan RL, Woods DC (1982) Small particles in plumes of Mount St. Helens. J Geophys Res 87(C7):4956-4962

Ross JE, Aller LH (1976) The chemical composition of the sun. Science 191:1223-1229

Sandford SA (1986a) Acid dissolution experiments: Carbonates and the 6.8-micrometer bands in interplanetary dust particles. Science 231:1540-1541

Sandford SA (1986b) Solar flare track densities in interplanetary dust particles: The determination of an asteroidal versus cometary source of the zodiacal dust cloud. Icarus 68:377-394

Sandford SA, Bradley JP (1989) Interplanetary dust particles collected in the stratosphere: Observations of atmospheric entry heating and constraints on their interrelationships and sources. Icarus 82:146-166

Sandford SA, Walker RM (1985) Laboratory infrared transmission spectra of individual interplanetary dust particles from 2.5 to 25 microns. Astrophys J 291:838-851

Schramm LS, Brownlee DE (1990) Iron-nickel sulfides in interplanetary dust. Lunar Planet Sci XXI:1093-1094

Schramm LS, Brownlee DE, Wheelock MM (1989) Major element composition of stratospheric micrometeorites. Meteoritics 24:99-112

Sedlacek WA, Heiken G, Zoller WH, Germani MS (1982) Aerosols from the Soufriere eruption plume of 17 April 1979. Science 216:1119-1121

Smith DB, Zielinski RA, Rose Jr. WI, Huebert BJ (1982) Water-soluble material on aerosols collected within volcanic eruption clouds. J Geophys Res 87(C7):4963-4972

Stadermann FJ (1991) Rare earth and trace element abundances in individual IDPs. Lunar Planet Sci XXII:1311-1312

Stadermann FJ, Walker RM, Zinner E (1989) Ion microprobe measurements of nitrogen and carbon isotopic variations in individual IDPs. [abstr] Meteoritics 24:327

Stadermann FJ, Walker RM, Zinner E (1990) Stratospheric dust collection: An isotopic survey of terrestrial and cosmic particles. Lunar Planet Sci XXI:1190-1191

Steele IM (1990) Minor elements in forsterites of Orgueil (Cl), Alais (Cl) and two interplanetary dust particles compared to C2-C3-UOC forsterites. Meteoritics 25:301-307

Stephan T, Jessberger EK, Klöck W, Rulle H, Zehnpfenning J (1994) TOF-SIMS analysis of interplanetary dust. Earth Planet Sci Lett 128:453-467

Strait MM, Thomas KL, McKay DS (1996) Porosity of interplanetary dust particles. Lunar Planet Sci XXVII, 1285-1286

Sutton SR (1994) Chemical compositions of primitive solar system particles. In Zolensky ME, Wilson TL, Rietmeijer FJM, Flynn GJ (eds) Analysis of Interplanetary Dust, AIP Conf Proc 310:145-157 Am Inst Physics, New York

Sutton SR, Flynn GJ (1988) Stratospheric particles: Synchrotron X-ray fluorescence determination of trace element contents. Proc Lunar Planet Sci Conf 18:607-614 Cambridge University Press

Sutton SR, Flynn GJ (1990) Extraterrestrial halogen and sulfur contents in the stratosphere. Proc Lunar Planet Sci Conf 20:357-361

Sutton SR, Bradley JP, Flynn GJ (1990) Trace element compositions and mineralogy of low-nickel stratospheric particles. Lunar Planet Sci XXI:1225-1226

Sutton SR, Cholewa M, Bench G, Saint A, Legge JGF, Weirup D, Flynn GJ (1991) Scanning transmission ion microscopy (STIM): A new technique for density mapping of micrometeorites. Lunar Planet Sci XXII:1363-1364

Symonds RB, Rose WI, Reed MH, Lichte FE, Finnegan DL (1987) Volatilization, transport and sublimation of metallic and non-metallic elements in high temperature gases at Merapi Volcano, Indonesia. Geochim Cosmochim Acta 51:2083-2101

Symonds RB, Rose WI, Reed MH (1988) Contribution of Cl- and F-bearing gases to the atmosphere by volcanoes. Nature 334:415-418

Thiel K, Bradley JP, Spohr R (1991) Investigation of solar flare tracks in IDPs: Some recent results. Nucl Tracks Radiat Meas 19:709-716 Int'l J Radiat Appl Instrum Part D.

Thomas KL, Klöck W, McKay DS (1989) Compositional comparison of IDP glasses and UOC chondrule glasses. [abstr] Meteoritics 24:332

Thomas KL, Keller LP, Klöck W, McKay DS (1991) Mineralogical and chemical constraints on parent bodies for hydrated interplanetary dust particles. Lunar Planet Sci XXII:1395-1396

Thomas KL, Keller LP, Flynn GJ, Sutton SR, Takatori K, McKay DS (1992a) Bulk compositions, mineralogy, and trace element abundances of six interplanetary dust particles. Lunar Planet Sci XXIII:1427-1428

Thomas KL, Keller LP, Blanford GE, Klöck W, McKay DS (1992b) Carbon in anhydrous interplanetary dust particles: Correlations with silicate mineralogy and sources of anhydrous IDPS. Lunar Planet Sci XXIII:1425-1426

Thomas KL, Blanford GE, Keller LP, Klöck W, McKay DS (1993a) Carbon abundance and silicate mineralogy of anhydrous interplanetary dust particles. Geochim Cosmochim Acta 57:1551-1566

Thomas KL, Klöck W, Keller LP, Blanford GE, McKay DS (1993b) Analysis of fragments from cluster particles: Carbon abundances, bulk chemistry, and mineralogy. Meteoritics 28:448-449

Thomas KL, Keller LP, Blanford GE, McKay DS (1994) Quantitative analyses of carbon in anhydrous and hydrated interplanetary dust particles. In Zolensky ME, Wilson TL, Rietmeijer FJM, Flynn GJ (eds) Analysis of Interplanetary Dust, AIP Conf Proc 310:165-172 Am Inst Physics, New York

Thomas KL, Blanford GE, Clemett SJ, Flynn GJ, Keller LP, Klöck W, Maechling CR, McKay DS, Messenger S, Nier AO, Schlutter DJ, Sutton SR, Warren JL, Zare RN (1995a) An asteroidal breccia: The anatomy of a cluster IDP. Geochim Cosmochim Acta 59:2797-2815

Thomas KL, Keller LP, Clemett SJ, McKay DS, Messenger S, Zhare RN (1995b) Hydrated cluster particles: Chemical and mineralogical analyses of fragments from two interplanetary dust particles. Lunar Planet Sci XXVI:1407-1408

Thomas KL, Keller LP, McKay DS (1996) A comprehensive study of major, minor, and light element abundances in over 100 interplanetary dust particles. In Gustafson BS, Hanner MS (eds) Physics, Chemistry and Dynamics of Interplanetary Dust, Astron Soc Pacific Conf Series 104:283-286

Tomeoka K, Buseck PR (1984) Transmission electron microscopy of the 'LOW-CA' hydrated interplanetary dust particle. Earth Planet Sci Lett 69:243-254

Tomeoka K, Buseck PR (1985) Hydrated interplanetary dust particle linked with carbonaceous chondrites? Nature 314:338-340.

Tomeoka K, Buseck PR (1986) A carbonate-rich, hydrated, interplanetary dust particle: Possible residue from protostellar clouds. Science 231:1544-1546

Vis RD, van der Stap CCAH, Heymann D (1987) On the use of a nuclear microprobe for trace element analysis in meteorites and cosmic dust. Nuclear Instr Methods Physics Res B22:380-385

Warren JL, Zolensky ME (1994) Collection and curation of interplanetary dust particles recovered from the stratosphere. In Zolensky ME, Wilson TL, Rietmeijer FJM, Flynn GJ (eds) Analysis of Interplanetary Dust, AIP Conf Proc 310:105-114, Am Inst Physics, New York

Wopenka B (1988) Raman observations on individual interplanetary dust particles. Earth Planet Sci Lett 88:221-231

Xu Yin-lin, Xie P, Fan C-Y, Cao Y-M (1991) Discovery of new compositions—Li and B in interplanetary dust particles. Sci China Series A 34:209-213

Yamakoshi K (1994) Extraterrestrial Dust, Laboratory studies of interplanetary dust. Terra Scientific Publ, Tokyo, Kluwer Academic Publ, Dordrecht, Netherlands 213 p

Yung Y, Pinto JP, Watson RT, Sander SP (1980) Atmospheric bromine and ozone perturbations in the lower stratosphere. J Atmos Sci 37:339-353

Zhang K, Guo B, Colarusso P, Bernath PF (1996) Far-infrared emission spectra of selected gas-phase PAHs: Spectroscopic fingerprints. Science 274:582-583

Zinner E, McKeegan KD, Walker RM (1983) Laboratory measurements of D/H ratios in interplanetary dust. Nature 305:119-121

Zolensky ME (1987) Refractory interplanetary dust particles. Science 237:1466-1468

Zolensky M, Barrett R (1994) Compositional variations of olivines and pyroxenes in chondritic interplanetary dust particles. Meteoritics 29:616-620

Zolensky ME, Lindstrom D (1992) Mineralogy of 12 large 'chondritic' interplanetary dust particles. Proc Lunar Planet Sci 22:161-169

Zolensky ME, Mackinnon IDR (1985) Accurate stratospheric particle size distributions from a flat plate collection surface. J Geophys Res 90(D3):5801-5808

Zolensky ME, Thomas KL (1995) Iron- and iron-nickel sulfides in chondritic interplanetary dust particles. Geochim Cosmochim Acta 59:4707-4712

Zolensky ME, McKay DS, Kaczor LA (1989a) A tenfold increase in the abundance of large solid particles in the stratosphere, as measured over the period 1976-1984. J Geophys Res 94 (D1):1047-1056

Zolensky ME, Lindstrom DJ, Thomas KL, Lindstrom RM, Lindstrom MM (1989b) Trace element compositions of six 'chondritic' stratospheric dust particles. Lunar Planet Sci XX:1255-1256

ADDENDUM (2001)

Since 1997 when the original text was prepared, Brownlee et al. (1999) have shown that the amorphous matrix of GEMS (see p. 2-53) has a ferromagnesiosilica composition and not a pure magnesiosilica composition as was initially reported by Bradley (1994, op. cit.). This matrix is therefore indistinguishable from the matrix of the ultrafine-grained (ufg) PUs (see p. 2-46). These units appear to be identical but whether GEMS and ufg PUs have a similar origin or not remains a subject of debate.

There is now a better appreciation of the origin of coarse-grained ferromagnesiosilica PUs (see p. 2-50). Based on results of laboratory condensation experiments to simulate dust formation in circumstellar discs, Rietmeijer et al. (1999) have shown that the compositions of these units are defined by mixing of metastable eutectic "FeSiO" and "MgSiO" dust condensates along well-constrained mixing lines in a ternary system MgO-'Fe-oxides'-SiO_2.

Brownlee DE, Joswiak DJ, Bradley JP (1999) High spatial resolution analyses of GEMS and other ultrafine grained IDP components (abstr). Lunar Planet Sci XXX: CD ROM #2031

Rietmeijer FJM, Nuth III JA, Karner JM (1999) Metastable eutectic condensation in a Mg-Fe-SiO-H_2-O_2 vapor: Analogs to circumstellar dust. Astrophysical J 527: 395-404

Chapter 3

CHONDRITIC METEORITES

Adrian J. Brearley and Rhian H. Jones

Institute of Meteoritics
Department of Earth and Planetary Sciences
University of New Mexico
Albuquerque, New Mexico 87131

In Nature's infinite book of secrecy
A little I can read.

William Shakespeare, Anthony and Cleopatra, *Act 1, sc. 2, l. 9-10*

INTRODUCTION

Chondritic meteorites are the oldest and most primitive rocks in the solar system. They formed as conglomerates of particles, many of which record individual, diverse nebular histories. In addition to providing considerable insights into the earliest history of the solar nebula, chondrites are hosts for interstellar grains that predate solar system formation and survived processing in the protoplanetary disk (solar nebula) environment. Chondrites display a wide diversity of isotopic compositions, including large variations in oxygen isotopes and evidence for the existence of short-lived radionuclides in the early solar system. Chondrites therefore offer a unique opportunity to study the earliest stages of formation of our planetary system. Our efforts to understand these meteorites are complicated considerably by the fact that most chondrites have suffered secondary processing to varying degrees, either in the nebula after formation of individual components, or on asteroids after accretion. Studies of these secondary processes provide insights into the geological evolution of asteroids. One of the greatest challenges in chondrite research is to disentangle the effects of primary and secondary processes, in order to gain a fuller understanding of the separate processes.

In this chapter we present an overview of the mineralogy of chondrites. We do not attempt to enter into detailed discussions about the complexities of isotopic systems or the cosmochemical evolution of chondritic bulk compositions, which are essential to a complete understanding of the origins of this group of meteorites. In the Introduction, we present an overview of chondrite properties, including classification and terminology, and a brief description of the processes that they record. A complete compilation of the mineralogy of meteorites has been published recently by Rubin (1997a,b).

Processes affecting material in chondrites: overview

Chondrites are complicated rocks. They represent the culmination of a sequence of complex events, including primary processes that affected individual components during their formation, and processes that have affected the whole rock after accretion. Here we summarize the main types of processing that chondrites have undergone. Further discussion is provided in following sections, where components are described individually. For more details and further reading the reader is referred to several excellent summaries in Kerridge and Matthews (1988).

Primitive chondrites are mixtures of diverse components that experienced a wide range of pre-accretionary histories (primary processes) in the solar nebula. The components occur in varying proportions in the different chondrite groups (Fig. 1).

0275-0279/98/0036-0003$30.00

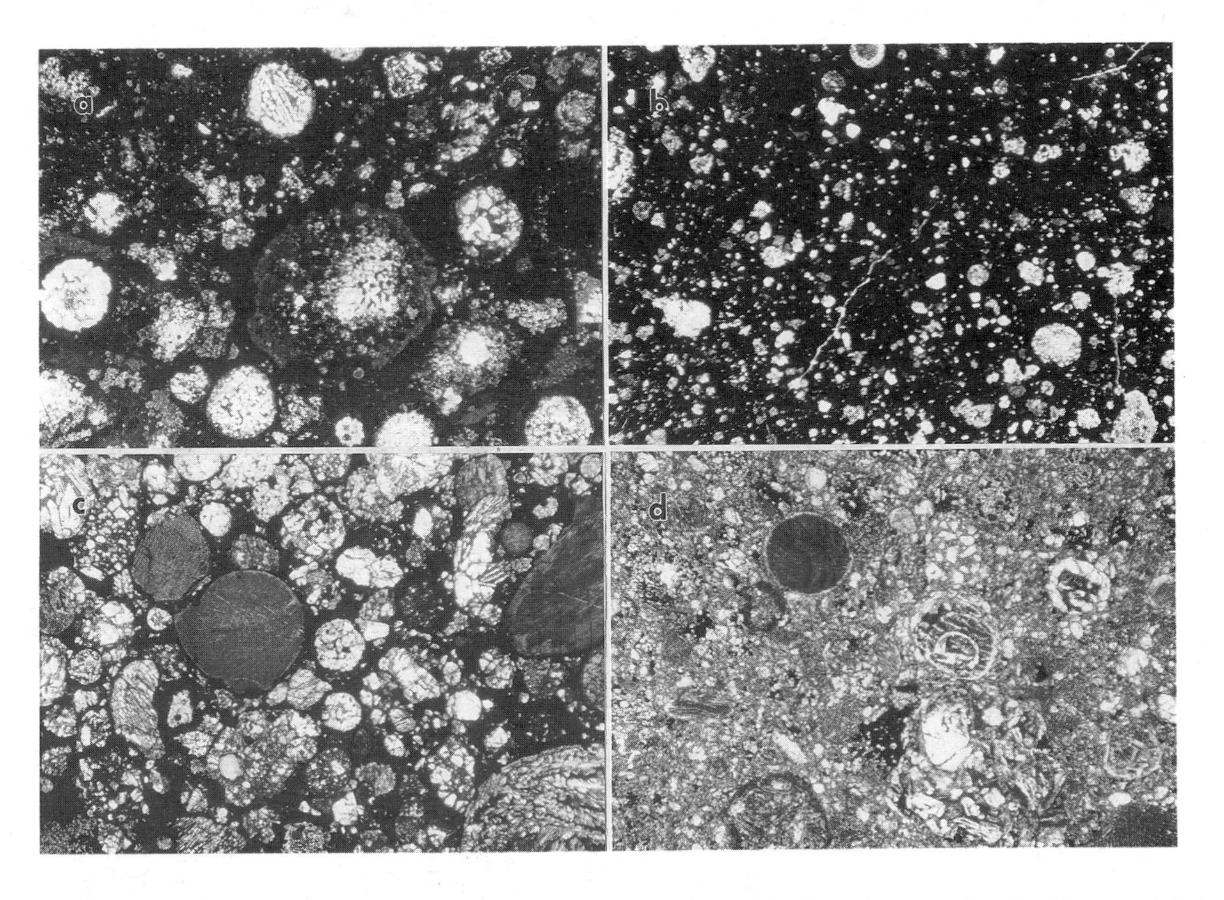

Figure 1. Optical photomicrographs of chondrite textures, showing variations in chondrule sizes and abundances between various chondrite groups (a,b,c), and the effect of metamorphism in ordinary chondrites (c and d). All photos are to the same scale, 7.25 mm across. (a) CV3 chondrite, Allende; (b) CM2 chondrite, Murchison; (c) Type 3 ordinary chondrite, Semarkona (LL3); (d) Type 5 ordinary chondrite, Tuxtuac (LL5).

Chondrules, which are the most common component of most chondrite groups, are submillimeter-sized ferromagnesian silicate objects that show evidence for an igneous origin, and are believed to have formed in flash-heating events in the solar nebula. The next most abundant component, matrix, consists of a fine-grained, disequilibrium mixture of silicates, oxides, metal, sulfides and organic constituents that represent a low-temperature fraction of nebular material. Interstellar grains are found embedded in matrix. Refractory and mafic inclusions, including calcium-, aluminum-rich inclusions (CAI), amoeboid olivine aggregates (AOA) and other refractory objects, are the products of high-temperature processes including evaporation, condensation, and melting, and contain a variety of exotic isotopic anomalies. Coarse-grained opaque (metal and sulfide) particles are distributed throughout the matrix in many chondrite groups.

Secondary processes affecting chondrites include aqueous alteration, thermal metamorphism, and shock metamorphism. Aqueous alteration results in the production of hydrous phases from essentially anhydrous primary precursor assemblages (Fig. 1b), as well as oxidation effects (e.g. Zolensky and McSween 1988). Aqueous alteration probably occurred on asteroids, after accretion, when fluids were mobilized in a low-temperature heating event. The source of water is probably ice that accreted in the original parent body. Aqueous alteration has also been suggested to be a nebular process in some cases, occurring through the interaction of low-temperature nebular gases with solid phases prior to accretion.

Thermal metamorphism is most commonly assumed to have taken place on asteroids, within a few tens of millions of years after accretion (e.g. McSween et al. 1988). It is manifested by increasing degrees of chemical and textural equilibration of primary components (Fig. 1c, d). The highest degree of metamorphism recorded by rocks that can still be properly called chondrites is at temperatures below the appearance of partial melts, specifically in the metal-sulfide system Fe-Ni-S, so that peak temperatures are below 950°C. Partially melted chondritic material (primitive achondrites) is described in Chapter 4. Metamorphic overpressures were low and of little significance since most chondrite parent asteroids are small, less than 100 km in diameter. The general view is that individual components of chondrites were accreted cold, and that metamorphism occurred during subsequent heating of the body. The heat source for metamorphism, as well as for melting of ice for aqueous alteration, is hard to identify: it is commonly attributed to the decay of short-lived radioisotopes such as ^{26}Al, although electromagnetic induction by a massive solar wind is also a possibility. An alternative scenario, retrograde metamorphism, invokes accretion of hot material so that no separate heat source is necessary. The thermal structure of the parent body would be an "onion shell" model for a body heated by radioactive decay. However, for ordinary chondrites, cooling rates do not appear to be a function of peak metamorphic temperatures as would be expected for an onion shell body, and a "rubble pile" model has been suggested in which metamorphism occurs in smaller planetesimals, and cooling rates are controlled by the burial depth of these planetesimals after accretion of the larger parent body.

Many chondrites have suffered impact processing to varying degrees (e.g. Bunch and Rajan 1988). The effects are independent of the nature and degree of aqueous alteration and thermal metamorphism, and occurred prior to, during, and/or after these processes took place. Many chondrites are breccias, attesting to the common impact processing that took place in the asteroid belt during the early history of the solar system (Fig. 2). A wide variety of breccia types is recognized, including primitive, regolith, fragmental, impact-melt, and granulitic breccias and regolith agglomerates. The dominant clasts are typically fragments of the same compositional group as the host chondrite, including cognate clasts of related compositions. Clasts of chondrites of other groups are less common. Dark

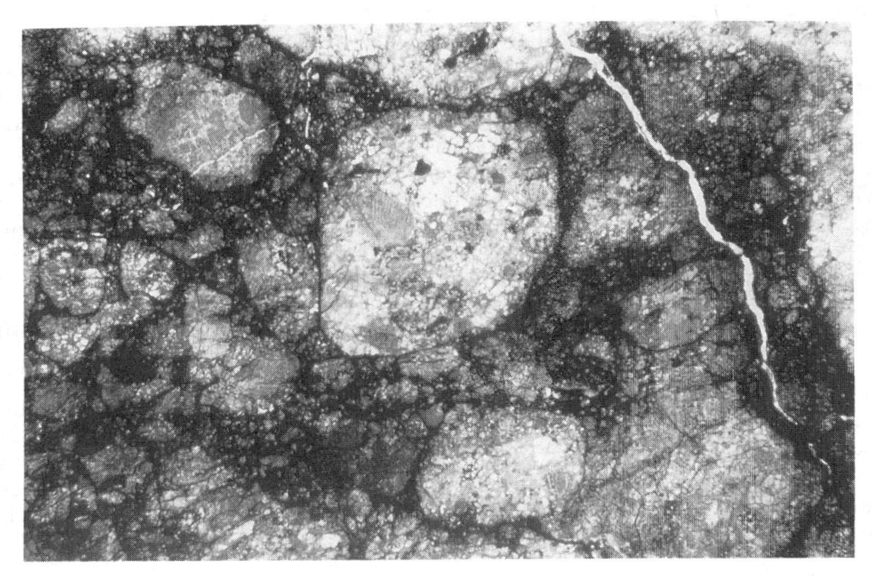

Figure 2. Transmitted light photomicrograph of the Naryilco LL ordinary chondrite, showing light-dark regolith breccia texture. The sample is 2 cm across.

portions of light-dark structured regolith breccias contain high abundances of noble gases that were probably implanted during irradiation by an ancient solar wind, in a regolith environment. The effects of impact processing are recorded in unbrecciated as well as brecciated chondrites in the form of shock metamorphism of the constituent minerals and occurrence of impact melts. Shock pressures up to about 90 GPa are recorded in chondrites (e.g. Stöffler et al. 1988).

Ages of chondrites and their components

The primitive nature of chondrites allows us to determine the ages of some of the earliest solar nebular events, and hence to estimate the age of the solar system. In addition, dating of metamorphic and shock events helps to interpret early geological processes that occurred on asteroids. Age determination for chondritic components as well as whole rocks is accomplished by long-lived radioisotope geochronology (e.g. Rb-Sr, U-Pb, Pb-Pb, Ar-Ar), similar to that applied to dating of terrestrial rocks. A second important method for dating early solar system events uses evidence from short-lived, extinct radionuclides that were present in the solar nebula epoch, following their generation by stellar nucleosynthesis shortly before formation of the solar nebula. Evidence from such systems provides relative ages of chondritic components and places constraints on timescales for the duration of the solar nebula. Three of the most valuable short-lived radioisotopes used for such dating are ^{26}Al, ^{53}Mn and ^{129}I, which decay to daughter isotopes ^{26}Mg, ^{53}Cr and ^{129}Xe with half lives of 0.75, 3.7 and 15.7 Ma, respectively. Relative ages of processes recorded in chondrites are summarized in Figure 3.

Some of the oldest material in the solar system is that in CAIs. These objects consistently give the oldest ages, both from long-lived and short-lived radioisotope chronometry (e.g. Tilton 1988). CAIs in the Allende CV chondrite have been studied extensively, using Rb-Sr and Pb-Pb techniques. Pb/Pb model ages for Allende inclusions of close to 4.56 Ga have been determined by several groups: the best estimate of the age

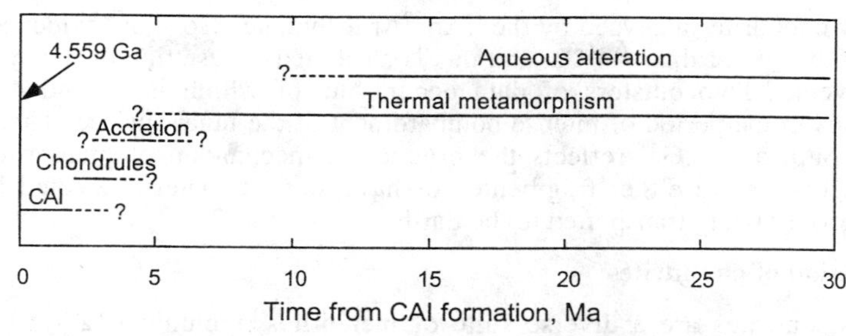

Figure 3. Timescales of chondrite formation. Times are referenced to the onset of CAI formation, 4.56 Ga ago. Times that have been determined are shown with solid lines; times that are unknown are shown as dashed lines.

of Allende CAIs, and hence the solar system, is 4.559±0.004 Ga (Chen and Wasserburg 1981). Evidence from the extinct radionuclide systems ^{26}Al, ^{53}Mn and ^{129}I shows that many CAIs are significantly older than chondrules. Many CAIs yield an initial ^{26}Al/^{27}Al ratio of ~5×10^{-5}, whereas typical values for typical chondrules are $<1 \times 10^{-5}$ (see MacPherson et al. 1995). If this difference is attributable solely to decay of ^{26}Al, it implies an age difference between the formation of CAIs and chondrules of at least 2 million years. Similar age differences are determined from the ^{53}Mn and ^{129}I chronometers (e.g. Swindle et al. 1996). This conclusion supports models for solar nebula evolution over a period of several million years (Podosek and Cassen 1994). However, it also raises a problem because it requires extended storage of CAIs within the region of chondrite formation.

Age determinations for whole-rock chondrites have been made using the Sr, Nd and Pb geochronometers, although these are hampered by uncertainties in analytical data, lack of precise knowledge of the ^{87}Rb decay constant, and the influence of terrestrial Pb contamination. Most values are consistent with ages around 4.555 Ga, a few Ma younger than the age of CAIs (e.g. Tilton 1988, Taylor 1992). Three chondrites yield high-precision model Pb-Pb ages around 4.552 Ga (Chen and Wasserburg 1981, Manhès et al. 1987). Minster et al. (1982) summarized Rb-Sr data for H, LL and E chondrite whole rocks, and found an age of 4.498±0.015 Ga, although this would change to 4.555 Ga with a change in the ^{87}Rb decay constant from 1.420 to 1.402×10^{-11} /a (Nyquist et al. 1986).

The timing of secondary alteration processes on asteroids has been inferred using several methods. Timing of the onset of aqueous alteration in CI chondrites, determined from Rb-Sr techniques, was estimated as occurring ~50 Ma after accretion (Macdougall et al. 1984), whereas evidence from the ^{53}Mn system gives an earlier age of <20 Ma after CAI formation (Endress et al. 1996, Hutcheon and Phinney 1996). The I-Xe chronometer suggests that aqueous alteration in the Semarkona ordinary chondrite occurred ~10 Ma after formation of chondrules (Swindle et al. 1991). The presence of radiogenic ^{26}Mg, derived from the decay of ^{26}Al, in secondary feldspar in an ordinary chondrite indicates that metamorphism also occurred early, within ~6 Ma of CAI formation (Zinner and Göpel 1992). Data from the ^{129}I and Pb-Pb systems also support an early onset of metamorphism, with metamorphic activity continuing for tens of millions of years (Swindle and Podosek 1988, Göpel et al. 1994). ^{40}Ar-^{39}Ar ages of chondrites of several different classes and petrologic types show a peak at around 4.5 Ga (e.g. Turner 1988), although there is a significant spread in the data of the order of 100 Ma. This is interpreted as evidence for the extended duration of parent body metamorphism, and is consistent with observed disturbances in internal Rb-Sr and Pb-Pb isochrons for H, L and E chondrites in the age range of 4.3 to 4.45 Ga.

Many chondrites analyzed by the ^{40}Ar-^{39}Ar technique also show evidence of Ar loss at times more recent than 4.4 Ga, and this is attributed to resetting of the system during impact events. Two clusters of data occur, one of which is around 4 Ga, which corresponds to the period of intense bombardment of the lunar surface. The other, more recent, group, at <2 Ga, reflects the production mechanism of meteorites whereby asteroidal bodies are either fragmented or have surface material ejected in energetic impacts prior to being transported to the Earth.

Classification of chondrites

The chondrites are a diverse suite of meteorites, including 12 well established groups, plus several other grouplets and unique chondrites whose properties differ significantly, warranting separate classifications. The different groups are defined by properties including their bulk chemistries, isotopic compositions, oxidation states, and proportions of individual components (see below). The primary divisions of chondrite classification are the carbonaceous (C), ordinary (O), and enstatite (E) classes (Fig. 4), each of which contains distinct groups that are closely related. The C class consists of 6 groups with distinct bulk compositional and oxygen isotopic characteristics: CI, CM, CR, CV, CO, and CK. The letter designating the group refers to the typical chondrite fall of the group, for example Ivuna is the typical CI chondrite. In addition to these groups, carbonaceous chondrites with properties similar to ALH85085 have recently been termed the CH group, although this name is not widely accepted. The O class is divided into three groups, H, L and LL. H chondrites have high total Fe contents, L chondrites have low total Fe contents, and LL chondrites have low metallic Fe relative to total Fe, as well as low total Fe contents. E chondrites also contain a high iron (EH) and a low iron (EL) group. Classification of two further chondrite groups, R and K, is not agreed on at present. Rumuruti-like (R) and Kakangari-like (K) chondrites are sufficiently different from all other chondrites that they have been suggested to represent additional classes. Since their classification is uncertain, we treat them individually in this chapter and do not attempt to suggest any affinities with other classes or groups.

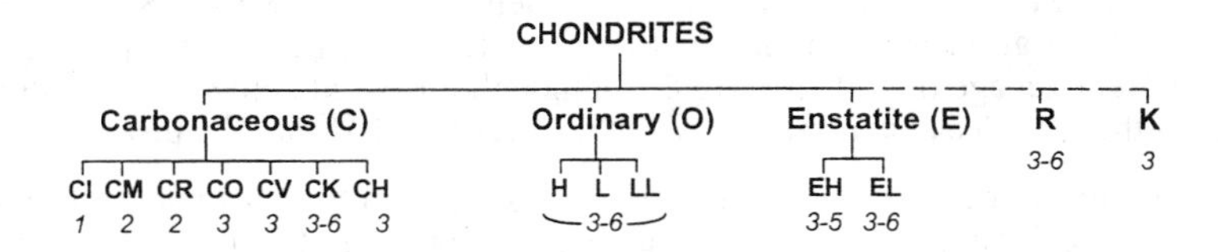

Figure 4. Classification of chondrite classes and groups. Petrologic types within each group are in italics.

A classification scheme that identifies the degree and nature of secondary alteration (petrologic type) experienced by a chondrite was introduced by Van Schmus and Wood (1967). In the current usage of this scheme, type 3 is the least altered material, types 2 to 1 represent increasing degrees of aqueous alteration, and types 4 to 6 represent increasing degrees of thermal metamorphism. More recently, several chondrites have been designated as type 7. This term has been used ambiguously and is discussed further in a section below (Type 4-6 chondrites: non-opaque material). The type 3 group has been subdivided for the O and CO chondrites into decimal subtypes, 3.0-3.9, of which 3.0 is the most primitive material that has suffered minimal thermal processing. The O and E chondrites are all of petrologic types 3-6. Most C chondrites are of petrologic types 1-3, except for the CK group (types 3-6). Table 1 provides a summary of the criteria used to define petrologic types.

The degree of shock metamorphism recorded in a chondrite is determined from a variety of mineralogical and textural parameters. Stöffler et al. (1991) defined increasing shock stages of S1 to S6 for ordinary chondrites. Assignment of shock stage is based on shock effects observed in olivine and plagioclase (Table 2). Since olivine is rare in enstatite chondrites, Rubin et al. (1997) extended the shock classification scheme to orthopyroxene and used the same S1 to S6 shock stages for these chondrites. For carbonaceous chondrites, the same scheme can be applied (Scott et al. 1992), although it must be based largely on observations of shock effects in olivine. The higher porosity of many C chondrites relative to O and E chondrites means that C chondrites record a lower equilibration temperature and a higher post-shock temperature increase for the same incident shock wave.

An additional classification parameter is used to identify the degree of terrestrial weathering that a chondrite has experienced. Two schemes are in use: one for hand specimens of Antarctic meteorites, and the other for polished sections. Weathering categories for hand specimens of Antarctic meteorites are: A = minor rustiness; B = moderate rustiness; C = severe rustiness; e = evaporite minerals visible to the naked eye (e.g. Grossman 1994). Wlotzka (1993) suggested the following progressive alteration stages for polished sections: W0 = no visible oxidation of metal or sulfide, some limonite staining; W1 = minor oxide rims around metal and troilite, minor oxide veins; W2 = moderate oxidation of about 20-60% of metal; W3 = heavy oxidation of metal and troilite, 60-95% being replaced; W4 = complete oxidation of metal and troilite, but no oxidation of silicates; W5 = beginning alteration of mafic silicates, mainly along cracks; W6 = massive replacement of silicates by clay minerals and oxides. Massive veining with iron oxides also develops independently of weathering grade. The degree of weathering appears to be correlated with terrestrial age within specific terrestrial climatic conditions.

Bulk compositions, O isotopes, oxidation states and other chondrite properties: comparisons between different groups

The various chondrite groups are each distinct and well defined, and each has a unique set of properties. Properties that characterize each chondrite group include the abundances of individual components, sizes of chondrules and CAIs, proportions of different types of chondrules, bulk compositions, O isotopic compositions, and oxidation state. The narrow range of properties within each group implies that each group is derived from a localized region, possibly a single asteroid. It is generally not possible to interpret variations between groups in terms of systematic trends that can be related to smoothly changing variables of the solar nebula. Although certain properties may be interpreted as varying with heliocentric distance (Rubin and Wasson 1995), other properties, or the properties of additional chondrite groups, cannot be fitted into the same trends (Scott and Newsom 1989, Weisberg et al. 1996). The variation between groups must therefore be regarded as representing localized heterogeneity of the nebular environment within the chondrite source region.

Petrographic characteristics of the chondrite groups are summarized in Table 3. Variations in the proportions (vol %) of various components are considerable. The abundance of chondrules varies from 80% in O and E chondrites to less than 1% in the CI group (Fig. 1). This property may indicate differences in the efficiency of chondrule formation from one region to another, or it may reflect differences in accretion properties. Matrix abundance also shows a wide range, complementary to chondrule abundances. Refractory inclusions are a significant component (>10%) of CO and CV chondrites, less abundant in CM and CK, and only a minor component (<1%) in all other groups. Variations in metal abundances reflect variation in oxidation states (see below), as well as

Table 1: Summary of criteria for petrologic types.

Criterion	Petrologic type 1	2	3	4	5	6
Homogeneity of olivine and low-Ca pyroxene compositions	-	>5% mean deviations		≤5%	homogeneous	
Structural state of low-Ca pyroxene	-	predominantly monoclinic		>20% monoclinic	<20% monoclinic	orthorhombic
Feldspar	-	minor primary grains only		secondary, <2 μm grains	secondary, 2-50 μm grains	secondary, >50 μm grains
Chondrule glass	-	altered, mostly absent[1]	clear, isotropic, variable abundance	devitrified, absent		
Metal: maximum bulk Ni, wt%	-	<20%; taenite minor or absent	kamacite and taenite in exsolution relationship >20%			
Sulfides: mean Ni content	-	>0.5 wt%	<0.5 wt%			
Matrix	all fine-grained, opaque	mostly fine, opaque	clastic and minor opaque	transparent, recrystallized, coarsening from 4 to 6		
Chondrule-matrix integration	no chondrules	chondrules very sharply defined		chondrules well defined	chondrules readily delineated	chondrules poorly defined
Carbon, wt%	3-5	0.8-2.6	0.2-1	<0.2		
Water, wt%	18-22	2-16	0.3-3	<1.5		

After Van Schmus and Wood (1967), with modifications including some from Sears and Dodd (1988).
[1]Chondrule glass is rare in CM2 chondrites, but is preserved in many CR2 meteorites.

Table 2: Classification scheme for shock metamorphism in chondrites.

Shock stage	Description	Effect resulting from equilibration peak shock pressure: OLIVINE	PLAGIOCLASE	ORTHOPYROXENE	Shock pressure (GPa)*
S1	unshocked	*sharp optical extinction*, irregular fractures			<4-5
S2	very weakly shocked	*undulatory extinction,* irregular fractures	*undulatory extinction,* irregular fractures	*undulatory extinction,* irregular fractures, some planar fractures	5-10
S3	weakly shocked	*planar fractures,* undulatory extinction, irregular fractures	undulatory extinction	*clinoenstatite lamellae on (100),* undulatory extinction, planar fractures, irregular fractures	15-20
S4	moderately shocked	*weak mosaicism,* planar fractures	undulatory extinction, partially isotropic, planar deformation features	*weak mosaicism,* twinning on (100), planar fractures	30-35
S5	strongly shocked	strong mosaicism, planar fractures, planar deformation features	*maskelynite*	strong mosaicism, planar fractures	45-55
S6	very strongly shocked	*solid state recrystallization* and staining, ringwoodite, melting	shock melted (normal glass)	*majorite, melting*	75-90
	shock melted	whole rock melting (impact melt rocks and melt breccias)			

From Stöffler et al. (1991) and Rubin et al. (1997). Shock levels in O chondrites are characterized by effects in olivine and plagioclase; shock levels in C chondrites are characterized by effects mostly in olivine (Scott et al., 1992); shock levels in E chondrites are characterized by effects mostly in orthopyroxene. The prime shock criteria for each shock stage are in italics.
* Shock pressures for O chondrites only.

Table 3. Petrographic characteristics of the chondrite group.

	Chondrule abundance[1] (vol%)	Matrix abundance (vol%)	Refractory inclusion abundance[2] (vol%)	Metal abundance[3] (vol%)	Chondrule mean diameter (mm)
CI	<<1	>99	<<1	0	-
CM	20	70	5	0.1	0.3
CR	50-60	30-50	0.5	5-8	0.7
CO	48	34	13	1-5	0.15
CV	45	40	10	0-5	1.0
CK	15	75	4	<0.01	0.7
CH	~70	5	0.1	20	0.02
H	60-80	10-15	0.1-1?	10	0.3
L	60-80	10-15	0.1-1?	5	0.7
LL	60-80	10-15	0.1-1?	2	0.9
EH	60-80	<2-15?	0.1-1?	8	0.2
EL	60-80	<2-15?	0.1-1?	15	0.6
R	>40	36	0	0.1	0.4
K	27	73[4]	<0.1	0[4]	0.6

[1]Chondrule abundance includes mineral fragments.
[2]Refractory inclusion abundance includes CAI + AOI.
[3]Metal abundance is for metal outside chondrules.
[4]Matrix abundance includes metal.
Abundance sums of less than 100 vol% are because of significant sulfide components in most cases.
Sources: Scott et al. (1996) except for CO data (McSween, 1977a), K (Kakangari) chondrite data (Weisberg et al., 1996), and "CH" (ALH85085) chondrule mean diameter from Grossman et al. (1988a) and Scott (1988).

the effects of aqueous alteration. ALH85085-like chondrites are extremely metal-rich C chondrites. There is a wide range of mean chondrule sizes, from 1.0 mm in CV chondrites down to 0.15 mm in CO chondrites. Chondrules in ALH85085-like chondrites are unusually small compared with all other groups, with a mean diameter of 20 μm. Chondrule sizes within most chondrite groups commonly show log-normal size-frequency distributions, and mean diameters are well defined.

The bulk compositions of CI chondrites are a very close match to the composition of the solar photosphere, excluding a few very volatile elements (Fig. 5). Hence, elemental abundances in bulk CI chondrites are viewed as a measure of average solar system abundances and are used as a reference composition for many types of solar system materials. CI chondrites are the most enriched of all chondrite groups in volatile elements such as H, C, and N which are present in organic materials and hydrous minerals in these meteorites. Bulk compositions of the other chondrite groups are similar to, but distinct from, CI chondrites, with most non-volatile elements varying within a factor of <2 relative to CI abundances (Fig. 6). Each chondrite group has a unique and well defined elemental abundance pattern. In general, carbonaceous chondrites are enriched in refractory lithophile elements relative to CI, and depleted in volatile lithophile elements. Siderophile and chalcophile elements show volatility-controlled abundance patterns. Elemental abundance patterns for O, R and K chondrites are fairly flat and refractory lithophile elements are depleted relative to CI. E chondrites have the lowest refractory lithophile element abundances. In addition to the abundance patterns shown, other bulk composition parameters that resolve the chondrite groups include

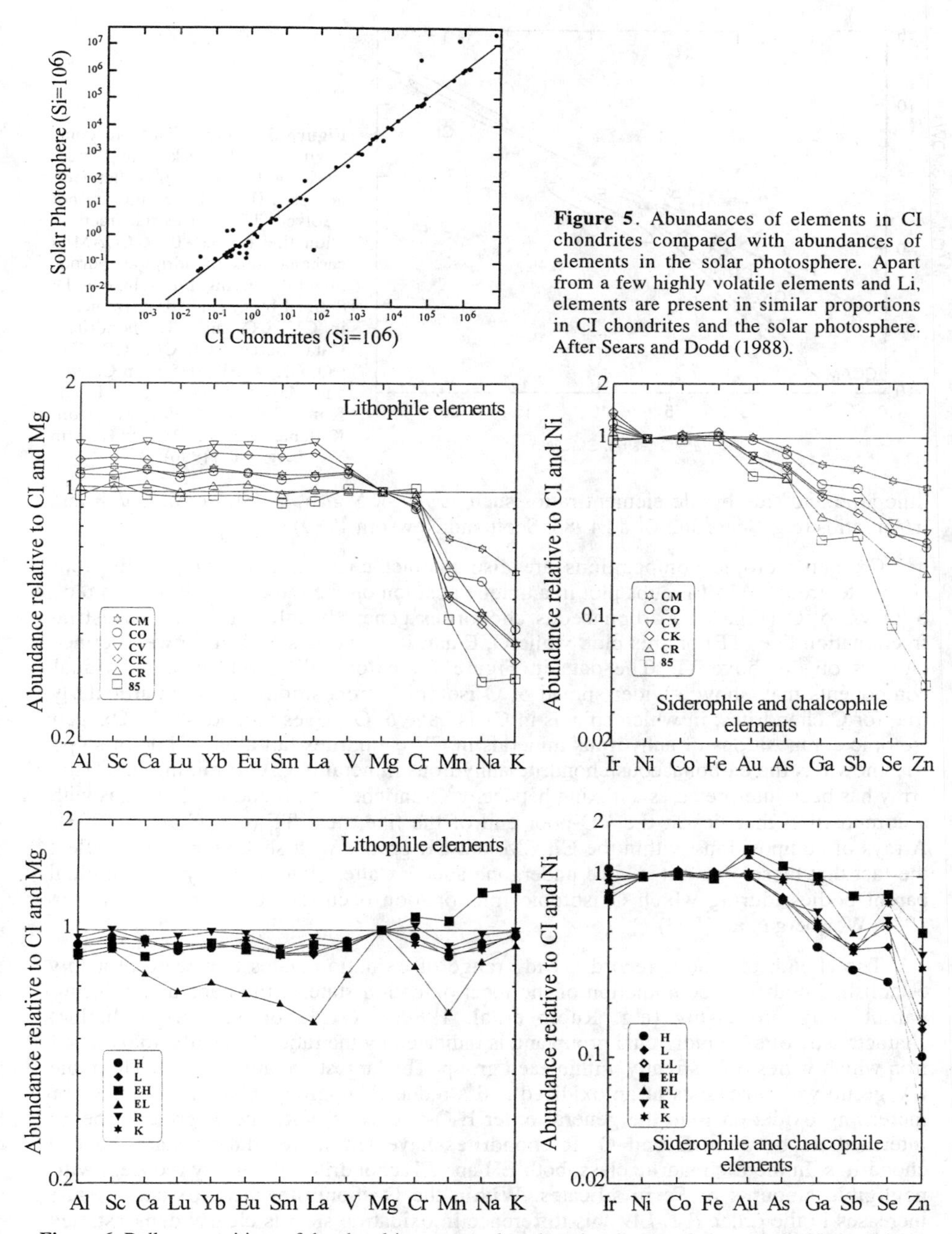

Figure 5. Abundances of elements in CI chondrites compared with abundances of elements in the solar photosphere. Apart from a few highly volatile elements and Li, elements are present in similar proportions in CI chondrites and the solar photosphere. After Sears and Dodd (1988).

Figure 6. Bulk compositions of the chondrite groups, plotted as abundances relative to bulk CI chondrites. Elements are arranged in order of increasing volatility. LL chondrite lithophile elements are not resolvable from L chondrites. Data sources: CI, CM, CO, CV, H, L, EH, EL from Wasson and Kallemeyn (1988); CK from Kallemeyn et al. (1991); CR from Kallemeyn et al. (1994); ALH85085 from Wasson and Kallemeyn (1990); R from Kallemeyn et al. (1996); K from Weisberg et al. (1996).

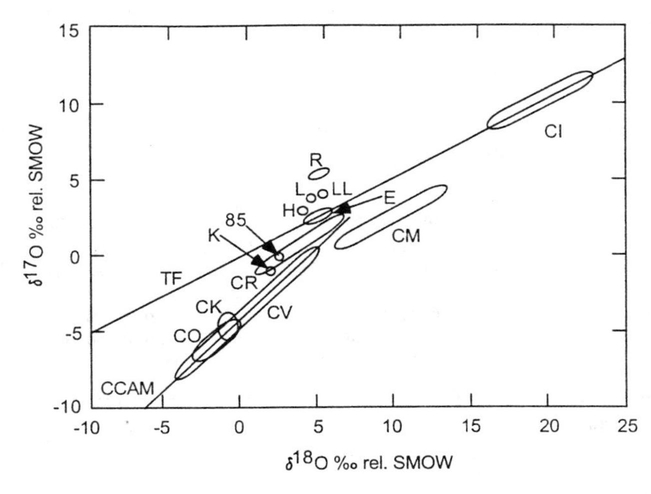

Figure 7. Oxygen isotopic compositions of bulk chondrites, showing the variability between groups. EH and EL groups are not resolved. TF = terrestrial fractionation line (slope ~0.5). CCAM = carbonaceous chondrite anhydrous minerals mixing line (slope ~ 1), defined by the anhydrous minerals in CV, CO and CK chondrites. Data sources: CI, CM, CR, CV, CO, CK, ALH85085 from Clayton and Mayeda (1989); H, L, LL, E from Clayton (1993); R from Kallemeyn et al. (1996); K from Weisberg et al. (1996).

lithophile and siderophile element ratios such as Al/Si, Mg/Si, Ca/Si, Fe/Si, Ga/Ni and Ir/Au ratios (e.g. Sears and Dodd 1988, Scott and Newsom 1989).

Oxygen isotopic compositions are also distinct characteristics of the different chondrite groups. Most groups plot in a unique position on the three-isotope oxygen plot, $\delta^{17}O$ vs. $\delta^{18}O$ (Fig. 7). Carbonaceous chondrites generally fall below the terrestrial fractionation line (TF) on this plot, while O, E and R chondrites each form well defined clusters on or above TF. The data in Figure 7 are for bulk chondrites: individual components may show a wider spread of O isotopic compositions. This is particularly true for C chondrites, in which spinels in CAIs have $\delta^{18}O$ values around -40‰. Oxygen isotopic compositions of anhydrous minerals in CV chondrites fall along a line of slope ~1, known as the carbonaceous chondrite anhydrous mineral (CCAM) mixing line. This array has been interpreted as a mixing between ^{16}O-enriched solids and a nebular gas with a composition that lies at the ^{16}O-poor end of the line, near TF (e.g. Clayton 1993). Arrays of compositions within the CI, CM and CR groups with shallower slopes reflect the fact that these chondrites have undergone aqueous alteration, probably on asteroidal parent bodies, during which O isotopic fractionation occurred (Clayton and Mayeda 1984, Weisberg et al. 1993).

The chondrite groups record a wide range of oxidation states that were probably established both as a combination of the local oxidation state of the nebula and during parent body processing (e.g. Rubin et al. 1988a). Oxidation state is a distinct characteristic of each individual group and is indicated by the ratio of metallic to oxidized iron which varies only slightly within each group. The largest variation occurs within the CV group which consists of an oxidized and a reduced subgroup. Overall, the order of increasing oxidation is in the general order E-O-C classes, with the K grouplet being intermediate between E and O. R chondrites have similar oxidation states to CO chondrites. In the E chondrite class, both EH and EL chondrites are highly reduced, with negligible amounts of Fe in silicates. Within the O chondrite class, oxidation state increases in the order H-L-LL: this difference in oxidation state is clearly demonstrated by the Fe/(Fe+Mg) ratios of olivine and pyroxene which increase in the same sequence (see section below, Type 4-6 chondrites: non-opaque material), as well as increasing Co contents of kamacite. For C chondrites, ALH85085-like chondrites are highly reduced, and CR-CO-CV-CK-CM-CI represents a trend of increasing oxidation states. The main

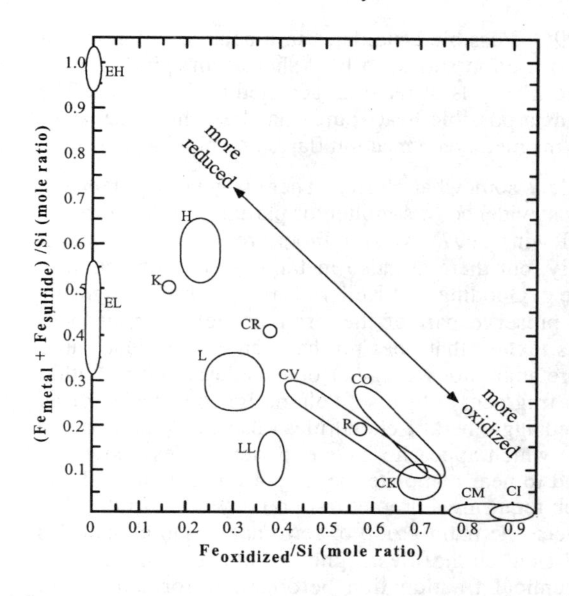

Figure 8. Urey-Craig diagram showing relative iron contents and oxidation states of the chondrite groups. Iron present in metal and sulfide phases is plotted vs. iron present in silicate and oxide phases, for bulk chondrite compositions. Compositions with the same bulk iron content but varying oxidation state would fall on a line of slope -1. CM and CI chondrites do not contain iron metal or iron sulfides: Fe present in other phases in these two groups does not appear on the diagram. For all other groups, all iron in the chondrite is included. After Larimer and Wasson (1988) with additional data for three CK chondrites and Efremovka (oxidized CV) from Jarosewich (1990), Karoonda (CK) from Mason and Wiik (1962a), Renazzo (CR) from Mason and Wiik (1962b), and Y-793575 (R chondrite) from Yanai (1992).

trends of differences in oxidation state, coupled with bulk iron contents, are summarized in a diagram known as a Urey-Craig plot (Fig. 8). In this plot, compositions with a constant bulk Fe/Si ratio, but varying oxidation states, would plot along a line of slope -1. The order of increasing oxidation state is represented by the order of increasing oxidized Fe (the Fe(ox) axis).

CHONDRULES: PRIMARY PROPERTIES

Introduction

Chondrules are the most abundant components of chondrites, comprising up to 80 vol % of O and E chondrites and >15 vol % of most other chondrite groups, except for the CI group in which chondrules are absent. The majority of chondrules are submillimeter-sized, igneous spheres, consisting predominantly of ferromagnesian silicate material (olivine, pyroxene, and a feldspathic glass). Their droplet shapes and igneous textures lead to an obvious interpretation that they are formed from molten droplets (e.g. Sorby, 1877). However, the environment in which chondrules formed is a matter of considerable debate and little consensus. At present, models in which chondrules are formed in the solar nebula are most popular. Planetary origins, such as during volcanic and impact events, have not been in favor for some time. Important arguments against such origins are summarized by Taylor et al. (1983) and include an absence of fractionation of REE in bulk chondrules, inconsistent with a volcanic origin, and differences in abundances of various components between chondrites and known regoliths such as that on the Moon. Chondrules are considered to have formed from precursor dustballs that were entrained in the nebular gas, rather than by condensation of liquid droplets directly from a gas. A transient heating event melted the dustballs, and they were subsequently cooled, initially at rates around hundreds of degrees per hour or faster, before accreting into their chondritic parent bodies (e.g. Wasson 1993, Hewins 1996). Temperature, time and other important constraints on chondrule-forming processes are largely based on mineralogical and textural studies of chondrules from unequilibrated chondrites as well as experimental analogs. The nature of the heating event is not understood (e.g. Boss 1996). Possible candidates in a nebular environment include shock waves, possibly in the accretion shock, in bow shocks, in spiral density waves, or shock waves propagating from parcels of material accreted to the nebula. The feasibility of lightning in the nebula as a possible heat source has been hotly debated. Other suggested environments include magnetic reconnection flares and bipolar outflows.

The exact definition of a chondrule is somewhat blurred. There is general agreement that a chondrule is an object that shows evidence of a molten or partially molten droplet origin (e.g. Grossman et al. 1988b, Hewins 1997). Most authors prefer to use the term chondrule for silicate-rich objects only, but there are also metallic droplet objects that have been referred to as chondrules (e.g. Gooding and Keil 1981). Chondrule fragments are usually easily identified if they preserve part of their rounded exterior surface. However, a fragment with an igneous texture that does not have an easily identifiable droplet origin becomes somewhat more ambiguous. Angular objects described as "lithic clasts" have been interpreted both as fragments of larger volume igneous rocks, or as chondrules. Fine-grained objects termed agglomeratic chondrules (also called dark-zoned chondrules and coarse-grained lumps) which appear to have experienced less extensive heating, and mild sintering as opposed to near complete melting, are irregular in shape and would not fit into a rigid definition requiring a droplet morphology. Chondrules that are intermediate in composition between the main group of ferromagnesian chondrules and CAI are generally termed Al-rich chondrules. Igneous clasts, whose bulk compositions show evidence for chemical fractionation before their formation, are discussed in Chapter 4.

In this section we discuss the mineralogy and petrologic characteristics of ferromagnesian and Al-rich chondrules. Summaries of chondrule properties not discussed here, such as detailed bulk chemistries, and oxygen isotopic data may be found in Grossman et al. (1988b). We describe the primary mineralogy of chondrules in unequilibrated chondrites of petrologic types 2 and 3, as well as anhydrous silicate grains in CI chondrites (type 1) that may be derived from chondrules. The effects of mild metamorphism in petrologic subtypes 3.0-3.9, and aqueous alteration of chondrule phases in petrologic types 2 and 3 are also discussed.

The majority of chondrules are ferromagnesian in composition. Their bulk compositions are generally close to CI chondrite compositions for refractory and moderately volatile elements, and they are generally depleted in volatile and siderophile elements relative to bulk CI chondrites (e.g. Grossman et al. 1988a). Bulk compositions of chondrules vary within the ferromagnesian suite. Correlations between the Fe^0/Fe^{II} ratio and volatile element depletions, have led to suggestions that reduction and volatile loss from common precursor assemblages may account for a large part of the variation in bulk compositions (Sears et al. 1996). Alternatively, individual chondrule precursor assemblages (dustballs) may vary in their inventories of volatile-rich and reducing agent material (Grossman 1996a).

Chondrule compositions also vary between the chondrite groups, although the range of textures observed is similar in most groups. In ferromagnesian chondrules, olivine and pyroxene are the dominant phenocrysts (Fig. 9). Porphyritic chondrules may have predominantly olivine phenocrysts (PO: porphyritic olivine), pyroxene (PP: porphyritic pyroxene), or a mixture of both minerals (POP: porphyritic olivine-pyroxene). Granular olivine-pyroxene (GOP) textures have a uniformly smaller grain size than porphyritic textures. Barred olivine (BO) textures usually consist of one or more sets of strongly elongated, prismatic olivine crystals that occur in parallel orientation and most of which exhibit coincidental extinction in cross-polarized light. Radial pyroxene (RP) textures consist of thin needles of pyroxene radiating from a single point on the edge of the

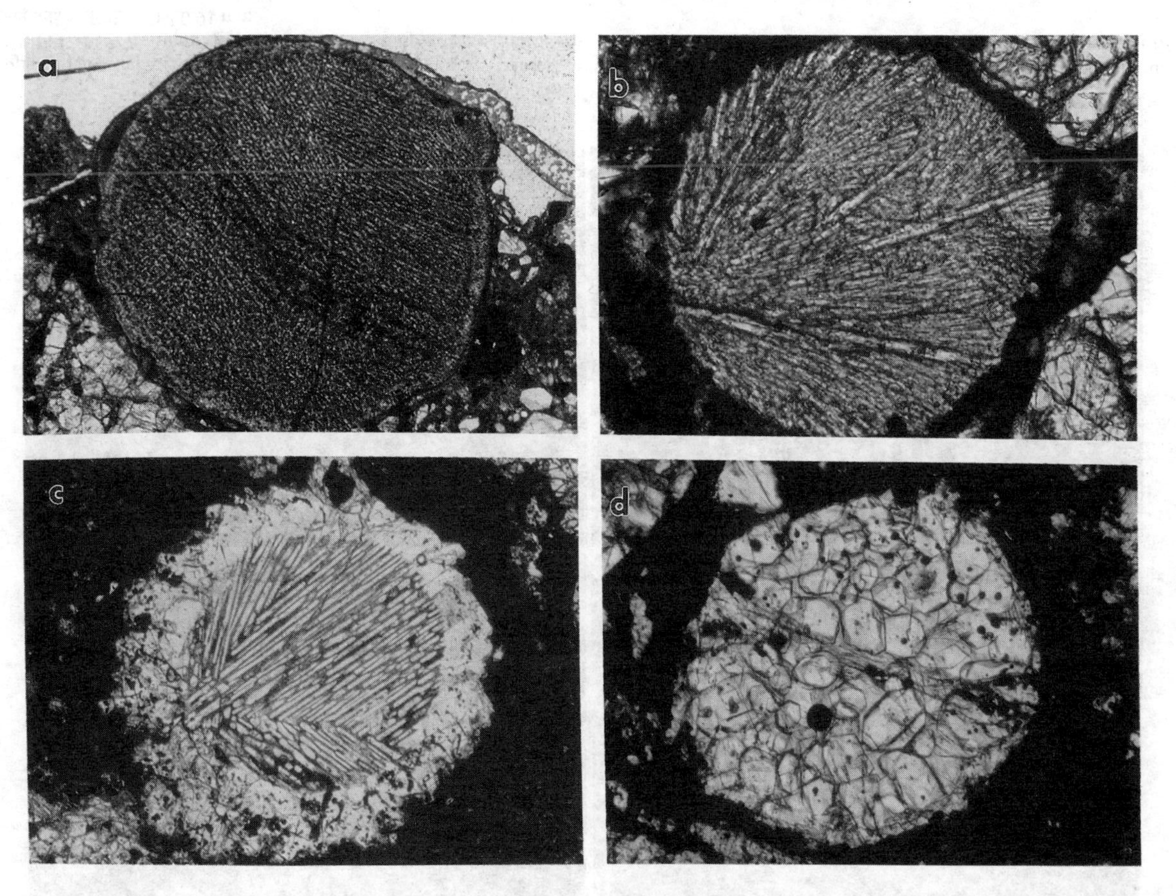

Figure 9. Optical photomicrographs showing variation in chondrule textures: (a) cryptocrystalline (C), PPL, field of view (FOV) 2.1 mm; (b) radial pyroxene (RP), PPL, FOV 0.4 mm; (c) barred olivine (BO), PPL, FOV 1.7 mm; (d) porphyritic olivine (PO), type IA, PPL, FOV 0.7 mm.

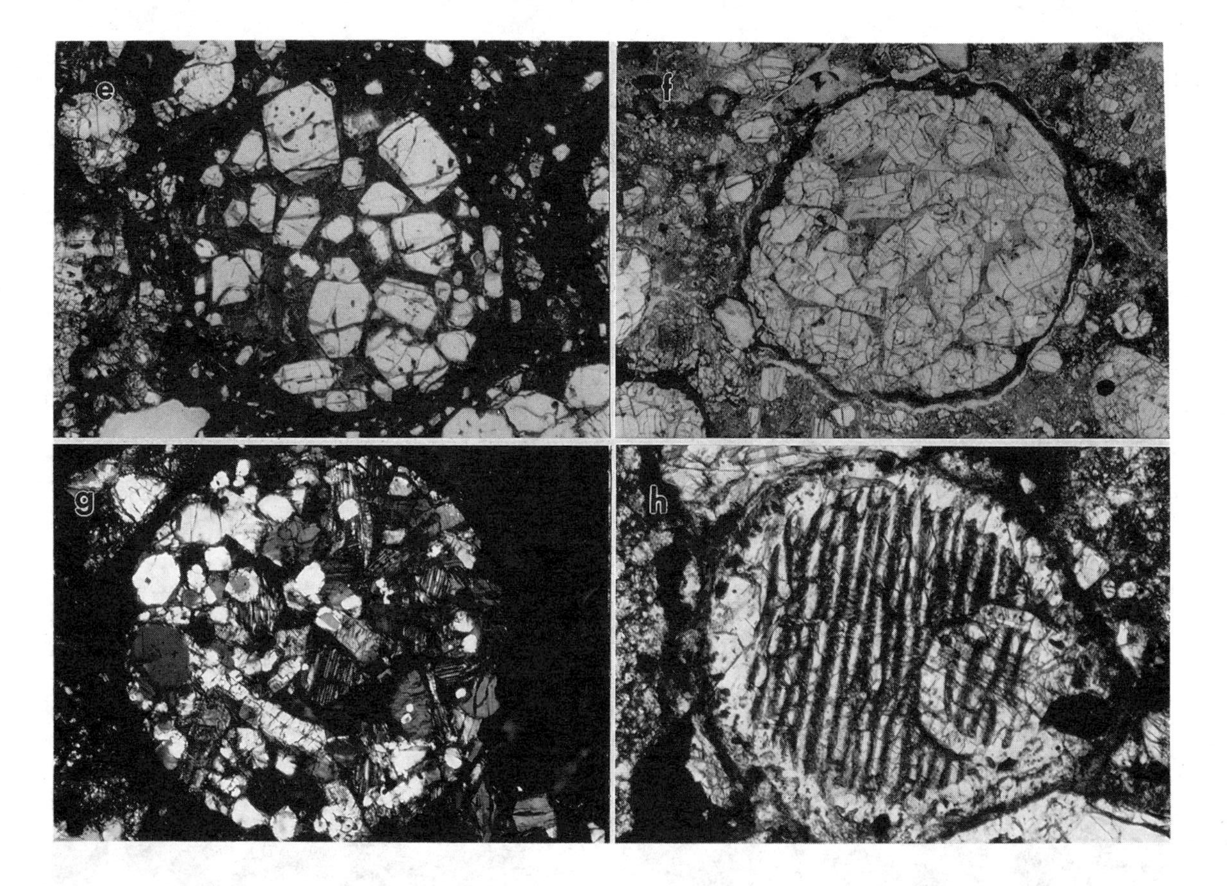

Figure 9. Optical photomicrographs showing variation in chondrule textures: (e) porphyritic olivine (PO), type IIA, PPL, FOV 2.1 mm; (f) porphyritic pyroxene (PP), type IB, PPL, FOV 1.7 mm; (g) porphyritic olivine / pyroxene (POP), type IAB, XPL, FOV 0.9 mm; (h) compound chondrule consisting of a BO primary and a BO secondary, PPL, FOV 0.91 mm.

chondrule. Cryptocrystalline chondrules (C) consist of extremely fine-grained intergrowths (grain size ≤2 μm) of pyroxene and glass and exhibit multiple optical extinction domains under crossed nicols. Interstitial material in most chondrule textural types, commonly termed mesostasis, is frequently glassy in unequilibrated chondrites. In many cases, mesostasis glass contains quench microcrystallites, commonly Ca-rich pyroxene. Completely glassy chondrules are extremely rare. Al-rich chondrules, chromite-rich chondrules and silica-bearing chondrules have different mineralogies that are described further below.

Several classification schemes have been proposed for chondrules. The textural descriptions above were defined mainly by Gooding and Keil (1981) and are used extensively in the literature. However, compositional classifications are also applied, especially for porphyritic chondrules, to distinguish between FeO-poor and FeO-rich suites of chondrules in unequilibrated chondrites. McSween (1977a) introduced the terms type I (FeO-poor, reduced chondrules) and type II (FeO-rich, oxidized chondrules). McSween also used type III for RP chondrules, but this term is used infrequently. Scott and Taylor (1983), Jones (1994), and Hewins (1997) have developed the textural scheme to include all types of chondrules, with subdivisions A and B referring to olivine-rich and pyroxene-rich chondrules, e.g. type IAB chondrules are initially FeO-poor and contain both olivine and pyroxene phenocrysts, type IIA chondrules are initially FeO-rich and contain predominantly olivine phenocrysts. The distinction between type I and type II is somewhat arbitrary, but refers to chondrules in which the mg# of olivine and low-Ca pyroxene is >90 and <90 for type I and type II chondrules, respectively. The textural definitions have been used to identify chondrules that have known initial compositions in metamorphosed or aqueously altered chondrites, so that the extent of alteration can be assessed. A different classification scheme, introduced by Sears et al. (1992), defines chondrule types according to the compositions and cathodoluminescence properties of olivine and mesostasis glass. In unequilibrated chondrites, groups A and B of this scheme are approximately the same as types I and II of the textural scheme.

Porphyritic chondrules are considered to have formed from droplets that were extensively melted, but in which abundant heterogeneous nucleation sites were preserved during the chondrule melting event (e.g. Lofgren 1996). The size of the relict grains that acted as nucleation sites is disputed. According to some authors, the cores of grains, possibly up to tens of microns across, were commonly preserved (e.g. Wasson 1993, Weisberg and Prinz 1996). In the other extreme, nucleation sites could have been microscopic and would therefore remain unidentifiable in the present chondrules (Lofgren 1996). Barred and radial textures develop when melts are heated to superliquidus temperatures, eliminating most viable nuclei. Nucleation can also occur by seeding of a molten chondrule with impinging fine-grained dust (e.g. Connolly and Hewins 1995). Clear examples of relict grains, with compositions that are foreign to their host chondrules, can be recognized. These occur as the cores of grains that are overgrown with compositions that crystallized from the host chondrules. They may be primitive condensates that constituted a part of the nebular dust inventory, or they may be derived from previous generations of chondrules (Steele 1988, Jones 1996a). Many chondrites also contain "compound chondrules" (Fig. 9h), which consist of two or more chondrules that were conjoined during chondrule formation (Wasson et al. 1995).

Semarkona: A type 3.0 ordinary chondrite

The most extensively studied unequilibrated ordinary chondrite (UOC) is Semarkona, LL3.0. Because this is one of the least equilibrated chondrites, chondrule mineral compositions are considered to be primary and hence provide important information about chondrule precursors and chondrule formation. The most significant secondary processing that chondrules in Semarkona have experienced is the late introduction of Na into mesostasis after solidification, either in the nebula as volatile elements condensed (Sears et al. 1996), or during mild metasomatism on the parent body (Grossman 1996b). Fine-grained matrix in Semarkona has undergone mild aqueous alteration (Alexander et al. 1989a), but this alteration does not appear to have modified chondrule mineral compositions to an extent discernible with electron microprobe techniques. All textural types of chondrules are found in Semarkona. Porphyritic chondrules, which constitute ~80% of all chondrules in ordinary chondrites (Gooding and Keil 1981), have been studied extensively and most of the following description is focused on the porphyritic suite.

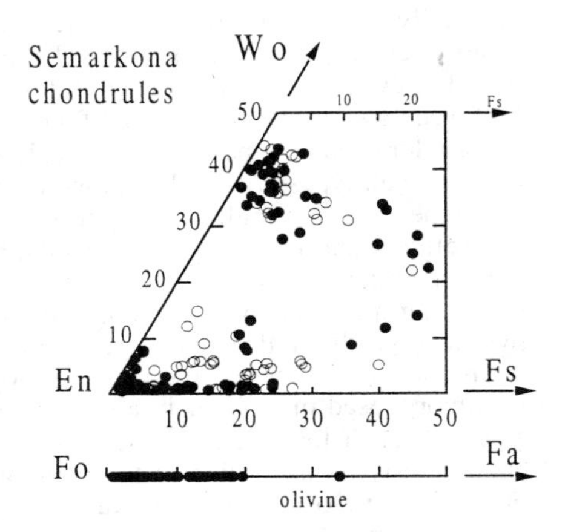

Figure 10. Pyroxene and olivine compositions from chondrules in the Semarkona (LL3.0) chondrite. Open circles are single electron microprobe analyses (Noguchi 1989), filled circles are means from individual chondrules (Jones and Scott 1989; Jones 1990, 1994, 1996b; Huang et al. 1996).

Silicate and oxide minerals. Mean Fa contents of *olivine* in individual chondrules in Semarkona vary from $Fa_{<1}$ to $Fa_{\sim 35}$, with mean Fa contents of most porphyritic chondrules falling below Fa_{20} (Fig. 10). In porphyritic chondrules, olivine occurs both as phenocrysts and as small, rounded grains (chadacrysts) that are poikilitically enclosed in clinoenstatite. Olivine chadacrysts are usually very similar in composition to phenocryst olivines in the same chondrules and are considered to have crystallized in situ during chondrule cooling (Jones 1994). The composition of chondrule olivine varies as a function of chondrule textural type (Scott and Taylor 1983). PO chondrules lie in two well defined fields: olivine in type IA chondrules has low Fa (<2 mol %) and olivine in type IIA chondrules has higher Fa (10-25 mol %). Poikilitic pyroxene chondrules (types IAB and IB) have Fa = 1-10 mol %, and POP chondrules span a wide range of Fa contents (3-25 mol %). Olivine from BO chondrules falls in both the type IA and IIA compositional fields with no intermediate compositions.

Minor element contents of olivine grains are also very variable (Fig. 11; Table A3.1). In type IIA chondrules, P_2O_5 contents of olivine up to 0.17 wt % have been reported (Jones 1990). Extremely forsteritic olivine, FeO < 2 wt %, is significantly different from more FeO-rich olivine, being enriched in incompatible elements (Ca, Al, Ti) and depleted in Mn and Cr. Incompatible trace elements such as Sc, V, Y, and Lu are also enriched in these olivines (Alexander 1994), and they also usually show blue cathodoluminescence (CL). "Blue" forsterites from a variety of unequilibrated chondrites were described by Steele (1986) who suggested that they are primitive condensates. Alternatively, these forsterites may have crystallized in situ from the host chondrule, and the high incompatible element contents may be related to disequilibrium and/or the very

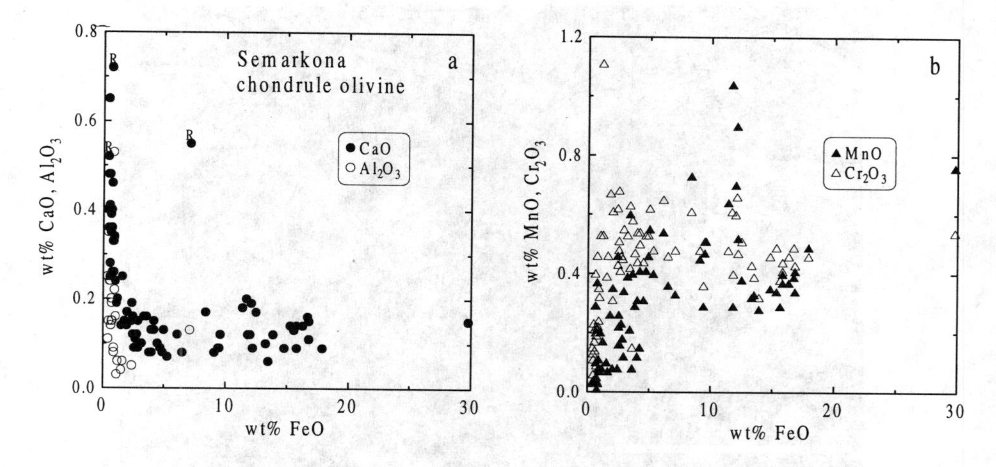

Figure 11. Minor element variation plots for olivine from chondrules in the Semarkona (LL3.0) chondrite. Al_2O_3 contents of olivines with FeO >3 wt % are generally below electron microprobe detection limits. R = relict forsterite grains in type II chondrules. Data sources: Jones and Scott (1989), Jones (1990, 1994, 1996b), Huang et al. (1996).

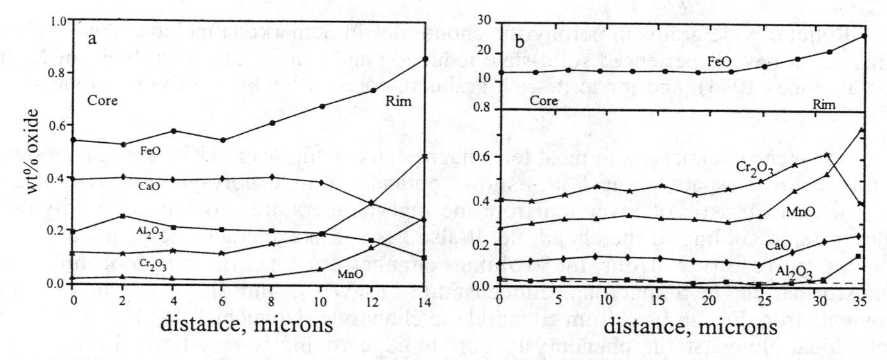

Figure 12. Olivine zoning in chondrules in the Semarkona (LL3.0) chondrite. (a) Type IA chondrule (Jones and Scott 1989) and (b) Type IIA chondrule (Jones 1990).

reducing conditions of crystallization (Jones and Scott 1989, Alexander 1994).

Olivine phenocrysts are typically zoned. Zoning from cores to edges of grains in type I chondrules may vary by only 0.5 wt % in FeO, whereas in type II chondrules, zoning is more pronounced (Figs. 12 and 13). Entrainment of hosts of small melt pockets in FeO-rich cores of some olivine grains in type IIA chondrules (Jones 1990) may be the result of supercooling. Minor element zoning in chondrule olivine phenocrysts is generally consistent with an origin by crystallization from chondrule melts (Jones 1990 1994 1996b). In the most reduced chondrules with incompatible-element rich forsterites, CaO and Al_2O_3 contents often decrease at the edges of grains while MnO and Cr_2O_3 increase, and patches with high concentrations of incompatible elements (corresponding to patches with varying CL colors) may occur within olivine grains. This has been interpreted as either partial resorption of a condensate relict grain (Steele 1986), or crystallization under strongly disequilibrium conditions (Jones and Scott 1989).

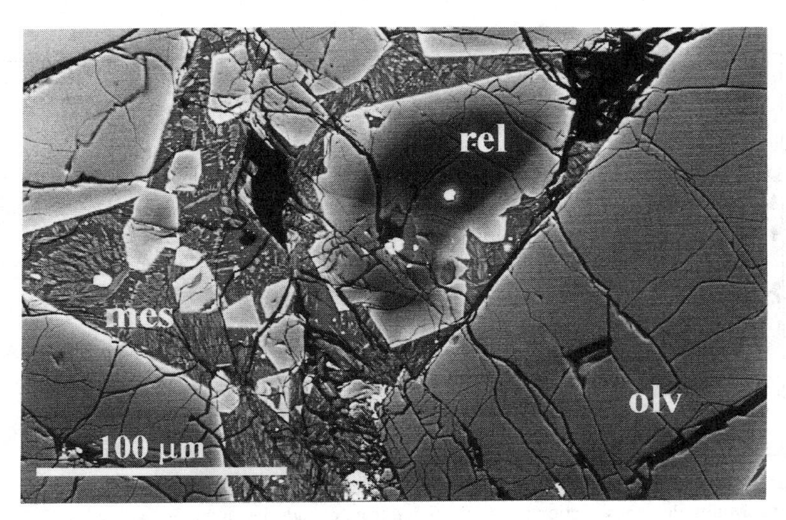

Figure 13. BSE image showing the texture of a type IIA chondrule in Semarkona. Olivine phenocrysts (olv) are normally zoned, and mesostasis (mes) consists of glass plus quench crystallites. A relict forsteritic olivine grain (rel) occurs at the core of one phenocryst and contains a small inclusion of Fe,Ni metal (white).

Relict olivine grains in porphyritic chondrules in Semarkona include "dusty" olivine grains that have experienced solid-state reduction and contain abundant blebs of Ni-free metal (Jones 1994), and forsterite-rich grains in cores of FeO-rich olivines (Jones 1990; Fig. 13).

Pyroxene phenocrysts in most ferromagnesian chondrules in UOC are clinoenstatite. *Clinoenstatite* (space group $P2_1/c$) shows optically visible polysynthetic twinning on (100), characteristic of inversion from the high-temperature protoenstatite polymorph during rapid cooling of the chondrule. It also has shrinkage cracks perpendicular to the twinning, resulting from the volume change during this polymorphic phase transformation. In Semarkona, clinoenstatite has $Wo_{<2}$, and shows a range of FeO contents from $Fs_{<1}$ to $Fs_{\sim 30}$ from chondrule to chondrule (Noguchi 1989; Fig. 10). Within individual clinoenstatite phenocrysts, core-to-edge zoning is very limited for FeO and minor elements. Minor element contents of clinoenstatite are plotted in Figure 14 and analyses are given in Table A3.2. As for olivine, the most En-rich clinoenstatites (FeO < 1 wt %) appear to be more enriched in incompatible elements (Al, Ti) than those with higher Fs contents. Trace element concentrations in clinoenstatite from Semarkona determined by Alexander (1994) and Jones and Layne (1997) show that they are extremely depleted in REE. Reported concentrations of La are 2-4 ppb (Jones and Layne 1997) and 15-85 ppb (Alexander 1994) and concentrations of Yb are 10-20 ppb (Jones and Layne 1997) and 17-53 ppb (Alexander 1994). In some of the more FeO-rich chondrules, clinoenstatite shows a faint lamellar zoning in back-scattered electron images, parallel to the twinning (Jones 1994; Fig. 15a). This zoning results from minor compositional zoning of FeO, MnO, CaO, Al_2O_3 and Cr_2O_3, with variation in FeO of less than 0.5 wt %.

Augite seldom occurs as discrete crystals, but is commonly present as thin (<20 µm) overgrowths on clinoenstatite phenocrysts. Augite shows a similar range of Fs contents to clinoenstatite (Fig. 10), and the Fs contents of clinoenstatite and augite in individual chondrules are very similar. Wo contents decrease with increasing FeO, from around

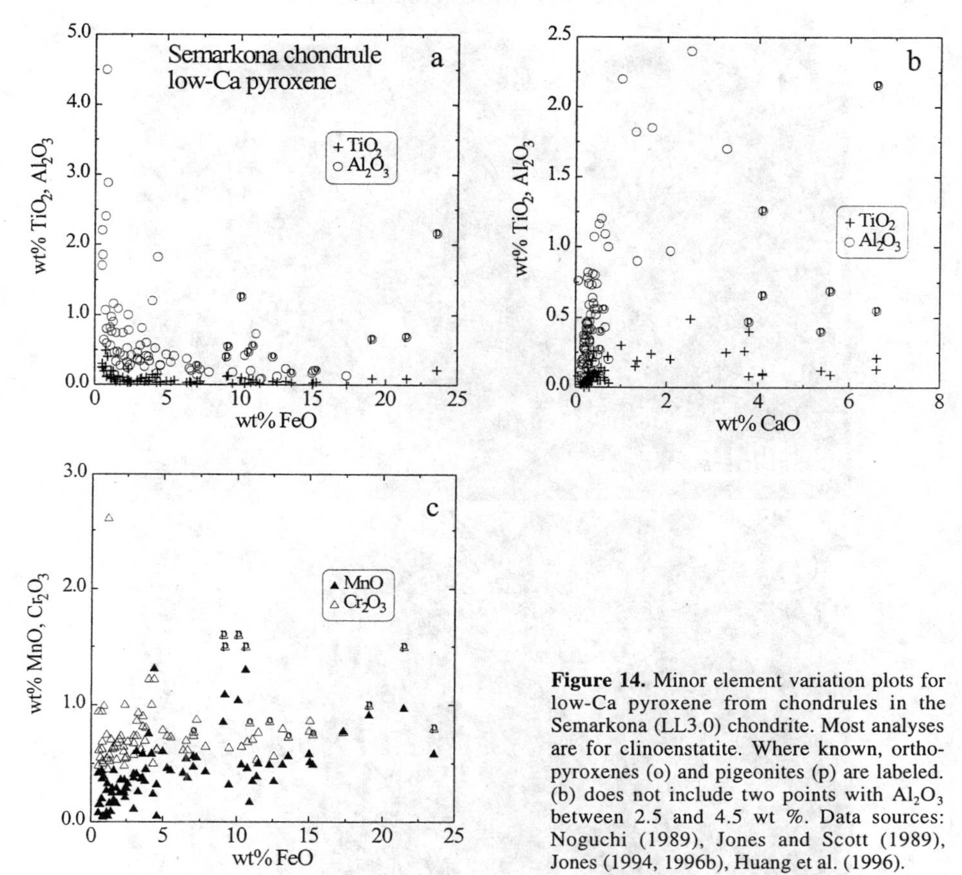

Figure 14. Minor element variation plots for low-Ca pyroxene from chondrules in the Semarkona (LL3.0) chondrite. Most analyses are for clinoenstatite. Where known, orthopyroxenes (o) and pigeonites (p) are labeled. (b) does not include two points with Al_2O_3 between 2.5 and 4.5 wt %. Data sources: Noguchi (1989), Jones and Scott (1989), Jones (1994, 1996b), Huang et al. (1996).

Wo_{40} for $<Fs_5$ to about Wo_{25} for $Fs_{>30}$. Augites are quite strongly zoned, with increasing CaO, Al_2O_3, TiO_2 and Na_2O from the clinoenstatite contact towards the mesostasis (Jones 1994; Fig. 16). Minor and trace element concentrations in augite are the highest of all the pyroxene structural types encountered in chondrules (Fig. 17; Table A3.3). Positive correlations are observed between TiO_2 and Al_2O_3, Al_2O_3 and CaO, and Cr_2O_3 and MnO (Noguchi 1989). REE patterns from augites are generally fairly flat with small negative Eu anomalies: concentrations are 0.1-1 ppm La, 0.4-1.3 ppm Yb (Alexander 1994, Jones and Layne 1997). Ca-rich pyroxenes also occur as small quench crystallites in chondrule mesostasis. Most of these are probably augite, but one chondrule contains fassaite with 3 wt % TiO_2 and 13 wt % Al_2O_3 (Jones and Scott 1989).

In some porphyritic chondrules, particularly the more FeO-rich ones, rare orthorhombic *enstatite* phenocrysts are present. These pyroxenes are significantly more CaO-rich than clinoenstatite (Wo_{2-5}), and do not show polysynthetic twinning. In some instances, orthopyroxene shows oscillatory zoning (Watanabe et al. 1986, Noguchi 1989, Jones 1996b; Fig. 15b) which is interpreted as forming either because of fluctuating redox conditions in the nebular gas (Watanabe et al. 1986), or during crystallization of the chondrule by disequilibrium kinetic effects (Jones 1996b). Orthopyroxene is also present as wide intergrowths over clinoenstatite cores in some of the most FeO-rich chondrules in which pyroxenes have a coarse barred texture. *Pigeonite*, Wo_{5-15}, occurs

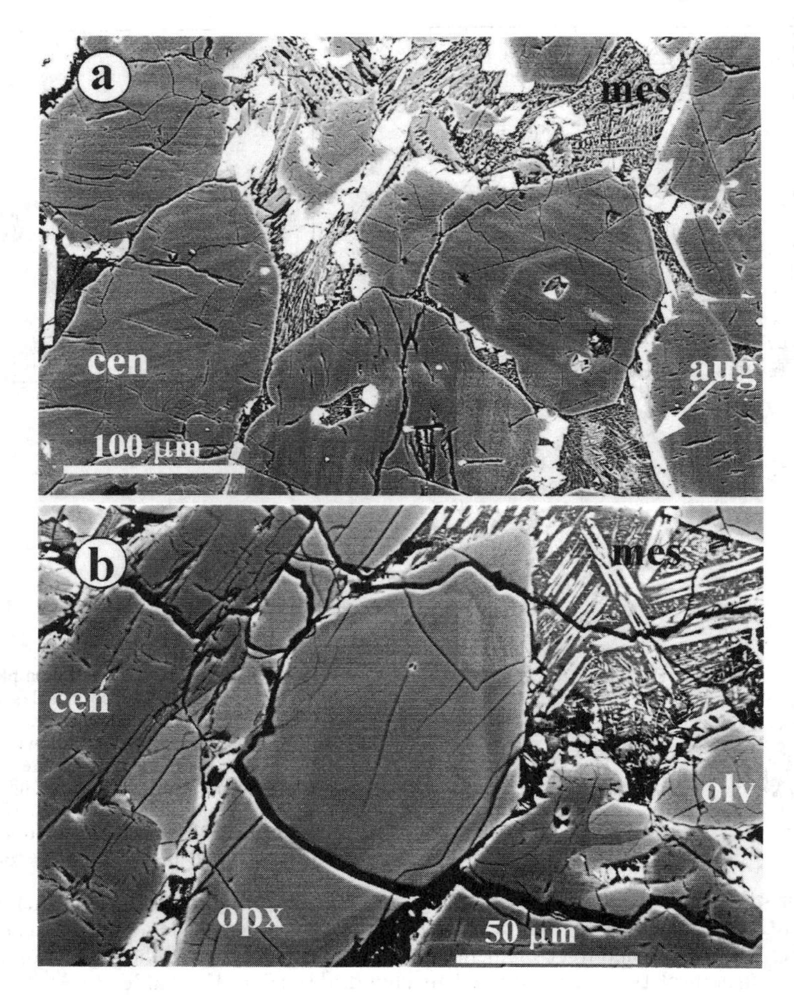

Figure 15. Pyroxene zoning in chondrules in the Semarkona (LL3.0) chondrite, BSE images. (a) Clinoenstatite phenocrysts (cen) show faint lamellar zoning and have narrow rims of augite (aug). (b) Oscillatory zoned orthopyroxene (opx) in a POP chondrule that also contains intergrown olivine (olv) and clinoenstatite (cen) phenocrysts and a mesostasis (mes) consisting of glass plus quench crystallites.

rarely as individual crystals, but occurs as a discrete, intermediate layer between low-Ca pyroxene (clinoenstatite or orthopyroxene) and augite overgrowths, particularly in more FeO-rich chondrules (Fig. 7 of Noguchi 1989). Minor and trace element concentrations of enstatite and pigeonite are shown in Figure 14 (see also Table A3.2). Figure 10 includes many analyses of intermediate, pigeonitic, compositions, exceeding the observed abundance of pigeonite that has been described. It is not clear whether all these analyses are truly pigeonite, or whether the microprobe analyses overlapped or included low-Ca pyroxene and augite phases.

Type IIA chondrules contain primary *chromite* that crystallized at a late stage and is present as small, euhedral crystals in mesostasis. This chromite has Cr/(Cr+Al) ratios around 0.94, and mean atomic compositions, based on 32 oxygen atoms, of around

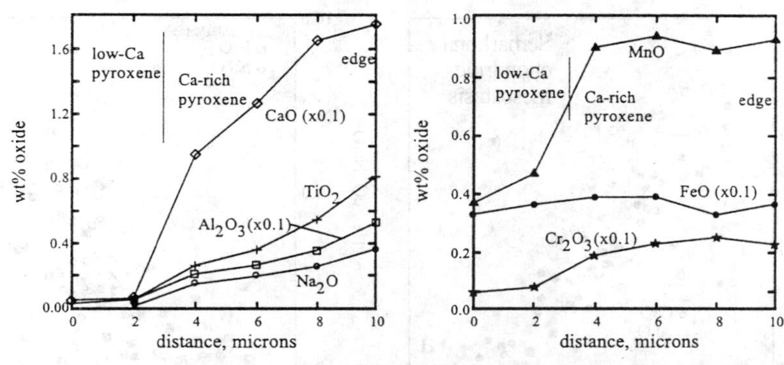

Figure 16. Zoning in pyroxene in a typical type IAB chondrule in the Semarkona (LL3.0) chondrite (Jones 1994). Electron microprobe traverse is from the outer part of a clinoenstatite grain towards the edge of its augite overgrowth.

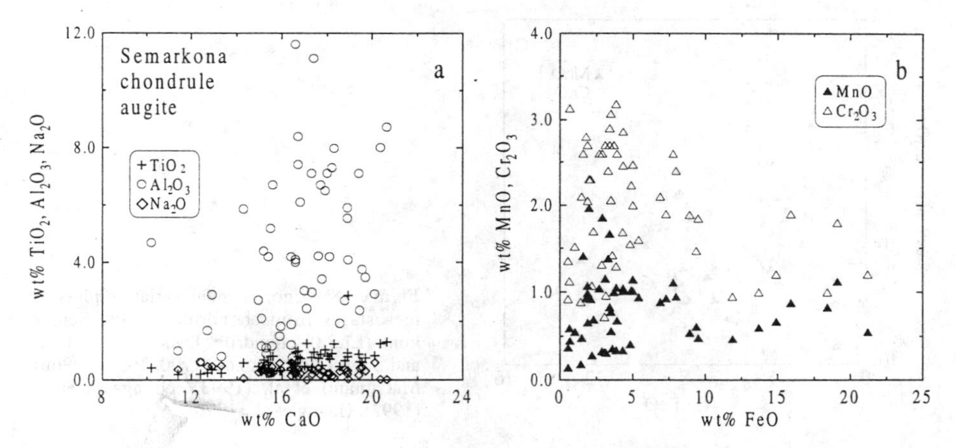

Figure 17. Minor element variation plots for augite from chondrules in the Semarkona (LL3.0) chondrite. Data sources: Noguchi (1989), Jones and Scott (1989), Jones (1994, 1996b), Huang et al. (1996).

0.2 for Si, 0.3 for Ti, 0.2 for V, 0.2 for Mn, 1.9 for Mg, and 0.01 for Zn (Johnson and Prinz 1991). Olivine-spinel geothermometry gives temperatures of around 1400°C for chromite crystallization in Semarkona chondrules (Johnson and Prinz 1991).

Chondrule *mesostasis* compositions reported in the literature are either compositions of glass only, or average compositions of glass plus quench microcrystallites. For the porphyritic chondrules in Semarkona, mesostasis compositions cover a wide range, from 45-73 wt % SiO_2. CaO and Al_2O_3 contents of mesostases in the entire suite of these chondrules are negatively correlated with SiO_2 and also show wide ranges (Fig. 18; see also Table A3.4). TiO_2 is also negatively correlated with SiO_2. FeO and Na_2O are positively correlated with SiO_2, and Na_2O is negatively correlated with CaO. These elements all lie on extremely well defined trends that cover the entire chondrule suite. Cr_2O_3 and MnO contents are low and show no strong trends. In many chondrules, mesostasis is zoned with increasing concentrations of Na and other volatile elements towards the edge of the chondrule (Matsunami et al. 1993, Grossman 1996b). This is interpreted as late-stage addition of Na either during chondrule formation (Sears et al. 1996), or on the parent body, after accretion (Grossman 1996b). Significant

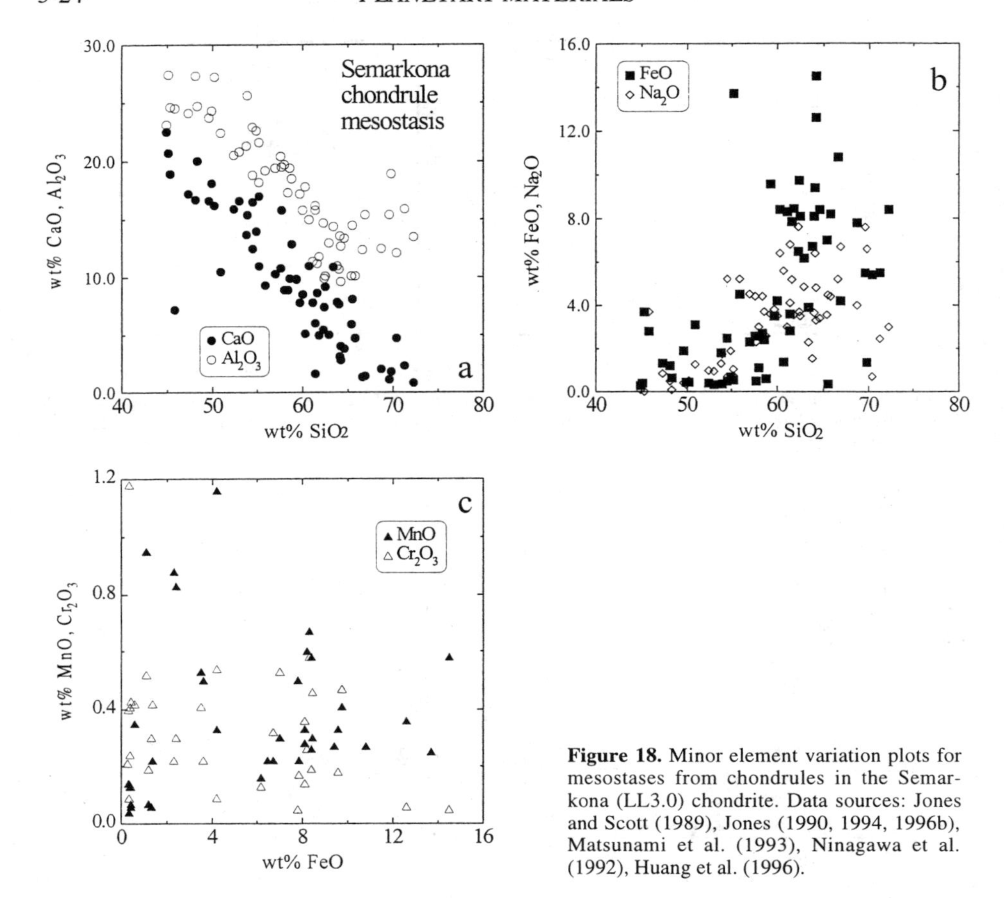

Figure 18. Minor element variation plots for mesostases from chondrules in the Semarkona (LL3.0) chondrite. Data sources: Jones and Scott (1989), Jones (1990, 1994, 1996b), Matsunami et al. (1993), Ninagawa et al. (1992), Huang et al. (1996).

concentrations of P_2O_5 are present in the mesostases of type IIA chondrules, up to 3.5 wt % (Jones 1990). REE abundances in the mesostases of porphyritic chondrules are about 10xCI abundances and show fairly flat patterns (Alexander 1994, Jones and Layne 1997).

Metal, sulfide and carbide minerals. Rounded grains of Fe,Ni metal and sulfides are common in Semarkona chondrules, distributed throughout the silicates and mesostases. In type IA chondrules, metal is common and sulfide is low in abundance (Jones and Scott 1989, Jones 1994). Troilite is more abundant than metal in type II chondrules (Jones 1990, 1996b), and pentlandite may also be present. The amounts of metal and sulfide vary enormously in individual chondrules, from trace amounts to droplets with diameters >30% that of the host chondrule. In addition, some metallic chondrules have been described that have only minor amounts of silicates (Gooding and Keil 1981, Grossman and Wasson 1985).

Metal in Semarkona chondrules shows a range of compositions, from ~1 to 62 wt % Ni (Affiatalab and Wasson 1980, Zanda et al. 1994). Most metal is kamacite; taenite is less abundant. Compositional variation within type II chondrules is small, but metal in type I chondrules is more heterogeneous (Jones and Scott 1989, Zanda et al. 1994). Co contents of kamacite are 0.2-0.4 wt %. Cr and P contents vary in the range <0.02 to 0.7 wt % and <0.02 to 1 wt %, respectively, and Si is generally below electron microprobe detection limits (Jones 1994, Zanda et al. 1994). Mean Ni and Cr contents of metal in individual chondrules are positively correlated with the Fa content of olivine in the same chondrules, indicating that the composition of metal in each chondrule was established by the oxidation state of the host chondrule during chondrule formation (Zanda et al. 1994). Cr and P may be present either in solid solution in the metal, or as tiny (<1 μm), reduced inclusions such as sulfides, phosphides and silicides. Rare metal grains in the most reduced chondrules contain spherical inclusions of silica glass, ≤5 μm across, that are inferred to have precipitated from the metal in the liquid state. Siderophile element abundances in individual chondrules, determined by Grossman and Wasson (1985), may be used to infer trace element concentrations of metallic phases.

A metallic chondrule described by Grossman and Wasson (1985) actually contains no metal. It consists of several large, polycrystalline cohenite grains, each surrounded by and partly intergrown with magnetite. Around the magnetite layers are layers of troilite. This type of carbide-magnetite assemblage (CMA) is common in Semarkona (Taylor et al. 1981, Krot et al. 1997a), occurring both in chondrules and matrix. CMAs consist of magnetite, troilite, Ni-rich sulfide, carbides, and Ni-rich taenite. The original assemblages of metal and troilite have been replaced by carbides and magnetite, either by gas-solid reactions in the nebula (Taylor et al. 1981), or in the parent body by low-temperature reactions with a C-H-O-bearing fluid (Krot et al. 1997a). Carbides show a wide range in composition, from 1-8 wt % Ni and 0.15-1.3 wt % Co. Cohenite, $(Fe,Ni)_3C$, contains ~1.5 wt % Ni and haxonite, $(Fe,Ni)_{23}C_6$, contains ~4.5 wt % Ni. The Ni-rich sulfides could be either an intergrowth of troilite and pentlandite, or a poorly characterized sulfide phase that can have a large range of Ni contents. Taenite grains have ~60 wt % Ni and 1-2 wt % Co.

Type 3.1-3.9 ordinary chondrites

The type 3 division of the petrologic type classification covers a wide range of the effects of progressive metamorphism, warranting further subdivision into subtypes 3.0-3.9. Chondrule minerals respond to this mild metamorphism by progressive equilibration of silicate mineral compositions which takes place by diffusional exchange throughout the chondrite. The glass of chondrule mesostasis also becomes progressively devitrified, resulting in crystallization of feldspar and Ca-rich pyroxene. Crystallization of feldspar has dramatic effects on the thermoluminescence (TL) properties of the bulk chondrite: TL sensitivity, corresponding to the amount of feldspar present, varies by about 3 orders of magnitude from subtype 3.0 to subtype 3.9 (Sears et al. 1980). Sears et al. (1995a) interpret differences in the temperature and width of the induced TL peak between subtypes 3.3-3.5 and subtypes 3.5-3.9 to reflect a change from low-temperature, ordered feldspar to high-temperature, disordered feldspar, corresponding to a peak metamorphic temperature of 500°-600°C for subtype 3.5.

The most significant changes in olivine compositions with metamorphism take place in the type 3 series. Olivine in type 3.9 chondrites is essentially equilibrated from chondrule to chondrule and throughout the chondrite, including equilibration with matrix olivine. Since diffusion rates are generally considerably slower in pyroxene than in olivine (e.g. Freer 1981), pyroxene equilibration lags behind olivine and the corresponding equilibration of pyroxene only occurs in higher petrologic types.

Silicate and oxide minerals. *Olivines* with low primary FeO contents (type I chondrules) also have limited zoning initially (Fig. 12a). With increasing petrologic subtype these grains first become more strongly zoned, as FeO diffuses into the grains, then they become equilibrated (McCoy et al. 1991a). Olivines from type II chondrules, that are initially more FeO-rich with marked zoning, acquire relatively less FeO and

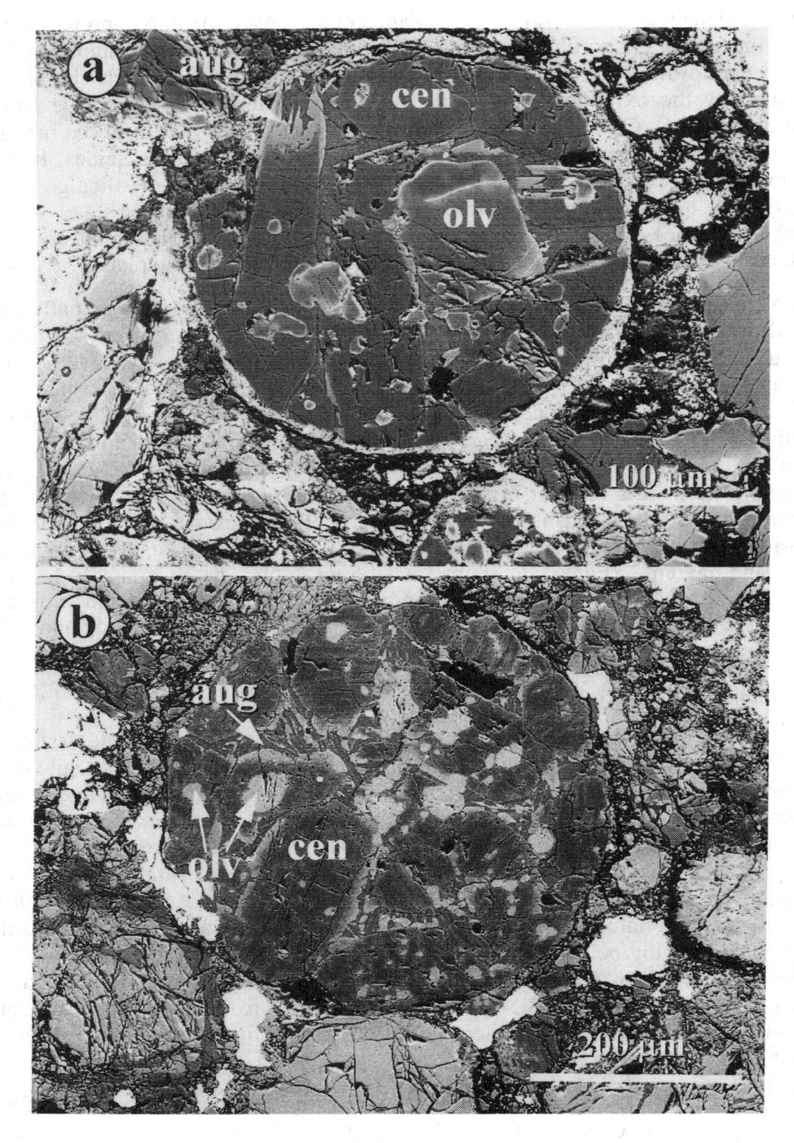

Figure 19. BSE images showing changes in olivine and low-Ca pyroxene compositions in type IAB chondrules through the petrologic subtype 3 series. (a) Tieschitz, H3.6. Zoning in olivine (olv) results from diffusion of Fe into the grains along grain boundaries and cracks. Clinoenstatite (cen) does not show strong zoning. It is rimmed with augite (aug). (b) Dhajala, H3.8. Olivine is essentially homogeneous (~Fa_{17}) and FeO-rich compared with low-Ca pyroxene. Clinoenstatite shows pronounced lamellar zoning resulting from diffusion of Fe into the grains.

approach the equilibrium composition faster. Diffusion of Fe into olivine can be observed in BSE images (Fig. 19). In chondrites of subtype 3.6-3.9, chondrule olivine compositions are relatively homogeneous and converge on compositions observed in type 4-6 OC (see section below, Type 4-6 chondrites: non-opaque material). For H, L and LL groups, some of the best defined, equilibrated compositions in high subtypes are as follows: Dhajala, H3.8, $Fa_{\sim17}$ (Sears et al. 1984a); ALHA-788084, H3.9, $Fa_{\sim18}$ (Tsuchiyama et al. 1988); Bremevörde, H3.9, $Fa_{\sim16}$ (Lux et al. 1980); ALHA-77216, L3.8, $Fa_{\sim26}$ (Matsunami et al. 1990a); ALHA-77304, LL3.8, $Fa_{\sim25}$ (Tsuchiyama et al. 1988). In intermediate petrologic subtypes, olivine compositions show a wide range in FeO content, as illustrated in histograms such as those in Figure 20. Representative olivine compositions in subtype 3.6 chondrites are given in Table A3.1.

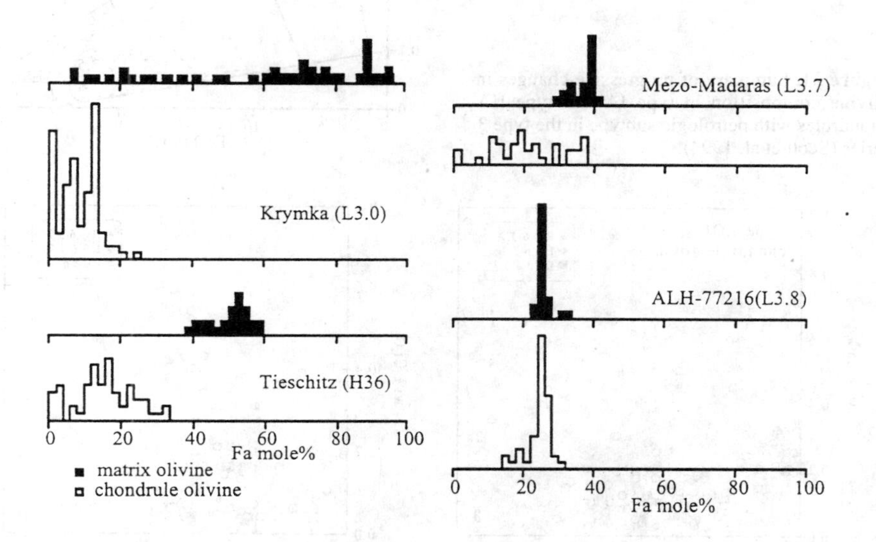

Figure 20. Histograms showing the distribution of olivine compositions in chondrules and matrix from type 3 OC (Matsunami 1990a). Krymka (LL3.1), Tieschitz (H3.6) and ALHA-77216 (L3.8) represent chondrites with increasing degrees of equilibration.

Three type 3 chondrites, Willaroy (H3.6), Suwahib (Buwah) (L3.7) and Moorabie (L3.8), are notably reduced compared with the other members of their groups (Wasson et al. 1993). Mean compositions of olivine in these three chondrites are Fa_{15}, Fa_{14}, and Fa_{16}, respectively. Reduction is thought to have taken place during parent body metamorphism.

Minor element contents of olivines in the type 3 series also show progressive changes towards equilibrated compositions (Figs. 21 and 22). On plots of CaO vs. FeO, chondrule olivine compositions move progressively to lower CaO contents. Changes in olivine compositions in type IA and type IIA chondrules are summarized by Scott et al. 1994 (Fig. 21). In the compositional classification scheme (Sears et al. 1992), these metamorphic paths on the CaO / Fa plot are defined as the trends A1→A3→A4→A5, A2→A4→A5 and B1→B2→B3→A5. MnO contents of chondrule olivines show a positive correlation with FeO (Fig. 22b). Different textural types of chondrules maintain differences in their primary olivine compositions, at least up to subtype 3.6. For example, in Parnallee (LL3.6), type IA and IIA chondrule olivines contain 0.23 wt % and 0.44 wt % MnO, respectively (McCoy et al. 1991a). Cr_2O_3 contents of olivines in chondrules decrease rapidly in the type 3 series (McCoy et al. 1991a, DeHart et al. 1992; Fig. 22c). From subtypes 3.0 to 3.3 to 3.6 they decrease from 0.37 to 0.17 to 0.04 in type IA chondrules, and 0.43 to 0.16 to 0.03 in type IIA chondrules (McCoy et al. 1991a). Rapid diffusion of Cr in olivines suggests that Cr is predominantly in the 2+ oxidation state

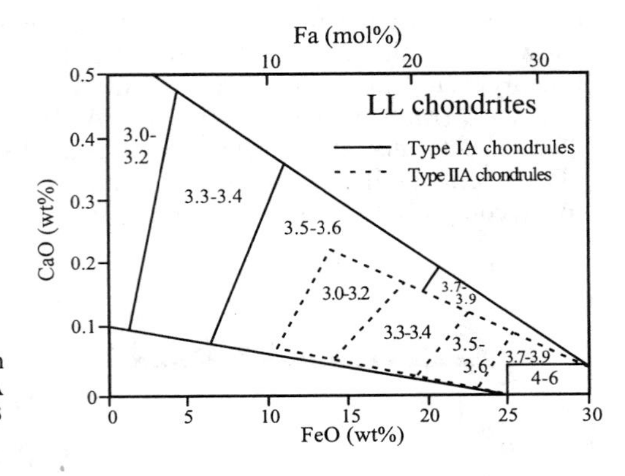

Figure 21. Summary of progressive changes in olivine composition in type IA and type IIA chondrules with petrologic subtype in the type 3 series (Scott et al. 1994).

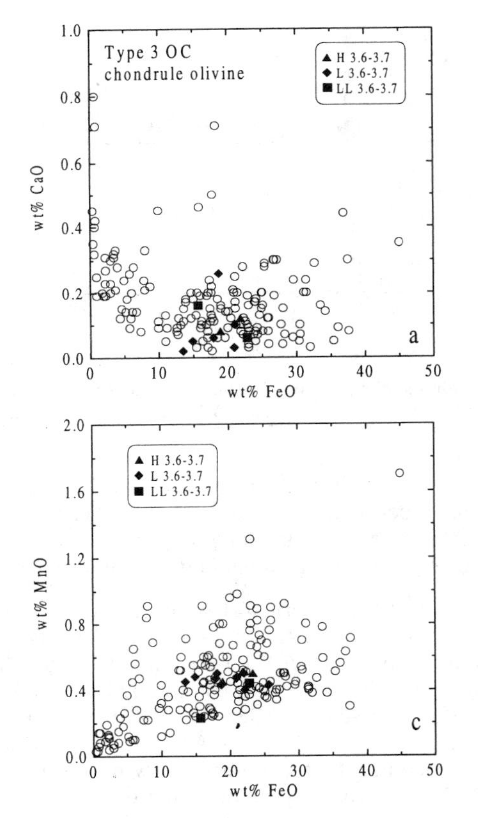

Figure 22. Minor element variation plots for olivine from chondrules in type 3 chondrites. Subtype 3.1-3.5 chondrites are open circles, subtype 3.6 and 3.7 are solid symbols. No data for 3.8 or 3.9 chondrites are available. Data sources: Dodd (1973, 1978), Fujimaki et al. (1981), Gooding et al. (1983), Rubin and Pernicka (1989), McCoy et al. (1991a), Krot and Wasson (1995), Wasson et al. (1995), Weisberg and Prinz (1996), Jones (1996b).

(Jones and Lofgren 1993). Al_2O_3 contents of olivines in type IA chondrules also decrease through the type 3 series. Incompatible trace element concentrations in olivine from chondrules in Bishunpur (LL3.1) and Chainpur (L/LL3.4) do not differ significantly from those in Semarkona (Alexander 1994).

Oscillatory zoning in a forsteritic olivine grain ($Fa_{0.2}$) in the ALHA76004 (LL3.2/3.4) chondrite was described by Steele (1995a). The olivine occurs in an inclusion that would generally be described as a type I chondrule. Oscillations on a scale of 5-10 μm observed in CL images correspond to cyclic variations in Al and Ti, while Mg, Fe, Ca and Cr vary monotonically across the zoned region. This zoning in incompatible elements is interpreted as taking place during crystallization from a melt.

Dusty, metal-bearing relict olivine grains are common in type 3 ordinary chondrites (Nagahara 1981a, Rambaldi 1981, Rambaldi and Wasson 1981,1982; Kracher et al. 1984, Jones and Danielson 1997). These grains are produced by solid-state reduction of more FeO-rich olivine and contain abundant micron-sized blebs of Ni-poor metal. Typical compositions are Fa_{1-10} and minor element contents are similar to those in the broad population of chondrule olivines (Jones and Danielson 1997). Dusty, chromite-bearing relict olivine grains have also been described in type 3 ordinary chondrite chondrules (Watanabe et al. 1984, Ruzicka 1990). Dusty, metal- and chromite- bearing grains all have an appreciable number of dislocations. Dislocations may arise from impact deformation (Watanabe et al. 1984), or they may be produced by stresses generated during exsolution of metal (Ruzicka 1990).

In a TEM study, Töpel-Schadt and Müller (1985) reported no dislocations in all chondrule olivines in Chainpur (L/LL3.4), and most olivine in Tieschitz (H3.6). Ruzicka (1990) reported a very low density of dislocations ($\leq 10^7$ cm^{-2}) in Chainpur chondrule olivine. However, some chondrules contain heavily deformed olivine with higher dislocation densities (10^8 to 10^9 cm^{-2}). This is probably the result of shock deformation. Undeformed olivine in chondrules in Parnallee (LL3.6) has isolated, often curved sub-grain boundaries that probably formed during initial grain growth (Ashworth and Barber 1977).

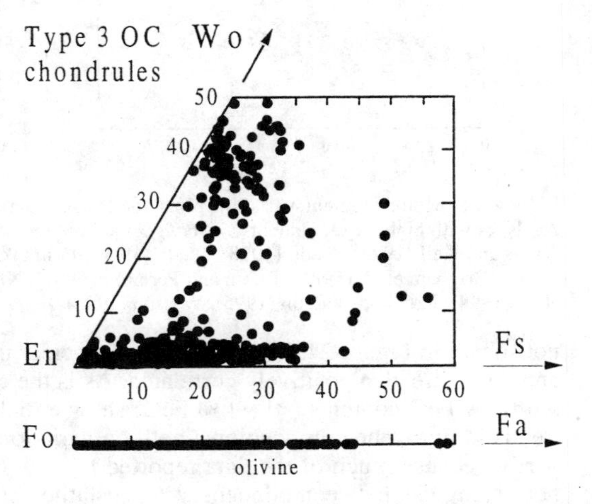

Figure 23. Pyroxene and olivine compositions in chondrules from type 3 ordinary chondrites. Data sources: Kurat (1967), Dodd (1973, 1978), Fodor and Keil (1975), Ikeda (1980), Fujimaki et al. (1981), Planner (1983), Gooding et al. (1983), Brigham et al. (1986), Rubin and Pernicka (1989), Noguchi (1989), McCoy et al. (1991a), Ehlmann et al. (1994), Krot and Wasson (1995), Wasson (1995), Weisberg and Prinz (1996), Jones (1996b).

In the petrologic type 3 series, changes in *pyroxene* compositions are less dramatic than changes in olivine compositions. In BSE images, Fe diffusion into clinoenstatite is observed along grain boundaries and shrinkage cracks, and lamellar zoning becomes more marked with increasing petrologic subtype (Tsuchiyama et al. 1988; Fig. 19b). Oscillatory zoning in enstatite is preserved at least up to petrologic type 3.6 (Watanabe et al. 1986, Jones 1996b). There is little difference overall in the ranges of *low-Ca pyroxene* compositions observed in individual chondrites throughout the type 3 series, generally $Fs_{<1}$ to $Fs_{\sim 35}$ mol % (Fig. 23). Oxide variation diagrams (Fig. 24) show a wide range of compositions for most minor elements. Representative analyses are given in Table A3.2. Several aluminous compositions, up to 16 wt % Al_2O_3, (Ikeda 1980, Noguchi 1989) are

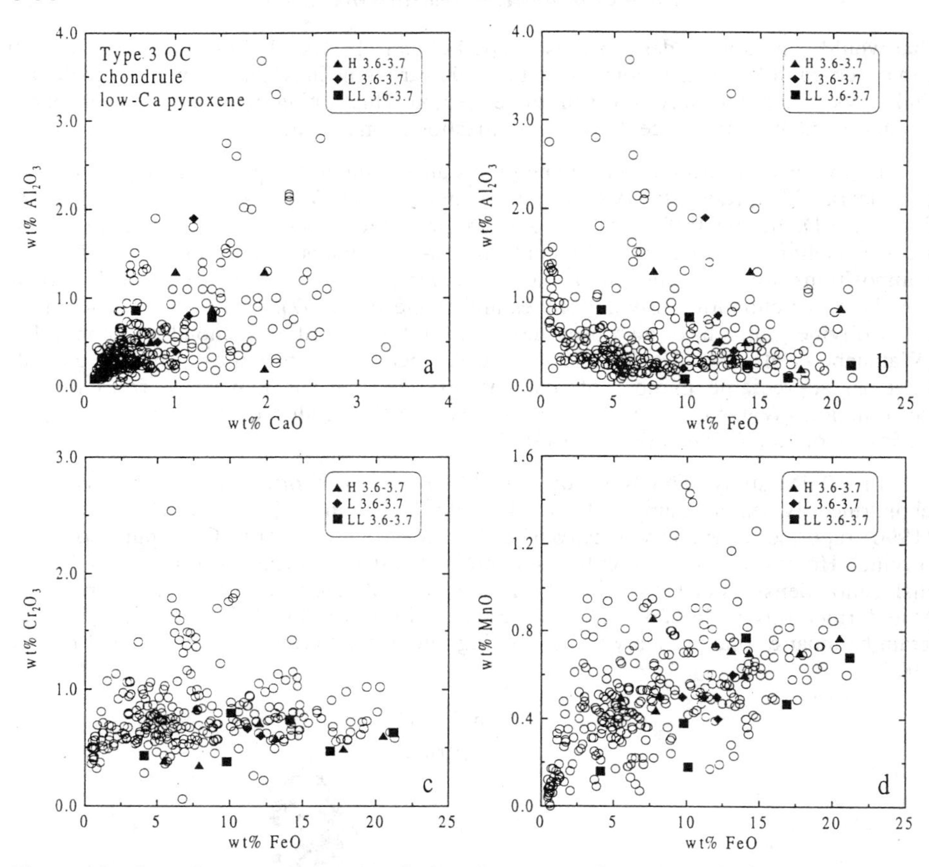

Figure 24. Minor element variation plots for low-Ca pyroxene from chondrules in type 3 chondrites. Analyses with high Al_2O_3 contents (7-16 wt %) are not included on the Al_2O_3 plots (see text). Data sources: Fodor and Keil (1975), Dodd (1978), Ikeda (1980), Fujimaki et al. (1981), Planner (1983), Gooding et al. (1983), Brigham et al. (1986), Rubin and Pernicka (1989), Noguchi (1989), McCoy et al. (1991a), Ehlmann et al. (1994), Krot and Wasson (1995), Wasson (1995), Jones (1996b).

not shown in Figure 24. MnO shows quite a clear positive correlation with FeO, and CaO and Al_2O_3 are also positively correlated. As is the case for Semarkona, low-Ca pyroxenes with low FeO contents (<1 wt % FeO) show enrichments in Al_2O_3. The absence of clear trends in pyroxene compositions, both on variation diagrams and with petrologic type, is partly because much of the data reported for low-Ca pyroxenes in these chondrites does not distinguish between clinoenstatite, enstatite and pigeonite. Many analyses for low-Ca pyroxene lie between 2 and 5 mol % Wo, but it is not clear what proportions of these are clinoenstatite and orthorhombic enstatite. Some individual analyses may overlap low-Ca pyroxene and augite, and others may be means that include both these phases. In a well constrained study, McCoy et al. (1991a) show a clear trend of decreasing Cr_2O_3 content with increasing Fs content in low-Ca pyroxene in type IA chondrules from LL chondrites, with increasing petrologic type. The loss of Cr_2O_3 from low-Ca pyroxene is consistent with loss of Cr_2O_3 from olivine, although it occurs more slowly in pyroxene as metamorphism proceeds, consistent with slower diffusion in pyroxene than in olivine. Concentrations of other elements do not vary significantly in the same series. Trace element concentrations in low-Ca pyroxene in Bishunpur (LL3.1) are indistinguishable from those in Semarkona (Alexander 1994).

TEM studies (Ashworth and Barber 1977, Töpel-Schadt and Müller 1985, Ruzicka 1990) show that the most common defects in low-Ca pyroxenes are (100) microstructures that represent stacking faults and microtwins. Low-Ca pyroxene grains also contain unit dislocations, partial dislocations, and both open and healed microcracks, which indicate the presence of strain. Unit dislocations attain local maximum densities of $\sim 5 \times 10^8$ cm^{-2}. Strain in low-Ca pyroxene is probably derived from the martensitic inversion of protoenstatite to clinoenstatite during cooling. Some microcracks may have been filled with residual chondrule liquids, indicating that protoenstatite inversion occurred above the solidus (Ruzicka 1990). In Mezö-Madaras and Tieschitz, low-Ca pyroxenes frequently consist of an intimate intergrowth of orthopyroxene and clinopyroxene parallel to (100). Orthopyroxene occupies about half of the volume in Mezö-Madaras. Occasional lattice fringes with a spacing of 13.5 Å were observed in Tieschitz (Töpel-Schadt and Müller 1985).

Pigeonite in a Parnallee (LL3.6) chondrule shows exsolution features in the form of a set of well defined lamellae of C2/c pyroxene, usually approximately parallel to (001) (Ashworth and Barber 1977). Regularities in the lamella width and spacing are consistent with an origin by coarsening after spinodal decomposition. A chondrule pigeonite in Chainpur (L/LL3.4) contains antiphase domain structures resulting from inversion from high temperatures. Both these examples indicate qualitatively that the chondrules containing the pigeonite were not simply quenched from the temperature of crystallization.

Pigeonite and *augite* compositions from the type 3 series are illustrated in Figure 23 (see also Tables A3.2 and A3.3). In comparison with Semarkona (Fig. 10), augite compositions tend to be more Wo-rich for equivalent Fs contents, and several are diopsidic ($Wo_{>45}$). These diopsides appear to be restricted to low FeO contents, $Fs_{<15}$. The highest Wo contents are diopsides from a spinel-bearing chondrule in Mezö-Madaras (Kurat 1967). Minor element variations in augite are shown in Figure 25. Concentration ranges are similar to those in Semarkona (Fig. 17), although some interelement correlations are different. Trace element analyses for two augites from Bishunpur (LL3.1) and Chainpur (L/LL3.4) are reported by Alexander (1994): REE patterns are fairly flat with ~0.5 ppm La, and ~0.7 ppm Yb.

Mesostases in chondrules in the type 3 ordinary chondrites show a similar range of compositions to those in Semarkona. Analyses in the literature are generally mean compositions of mesostasis, including crystalline phases. Although mesostasis glass becomes progressively more devitrified with increasing petrologic subtype through the type 3 series, detailed descriptions of changes in mineralogy as devitrification proceeds do not exist. Changes in TL sensitivity are interpreted in terms of feldspar crystallization, as discussed above. In a TEM study, Töpel-Schadt and Müller (1985) showed that mesostases of chondrules in Mezö-Madaras (L3.7) consisted of glass and small (<2 µm) idiomorphic olivine crystals that are relatively FeO-rich (Fa_{30-40}). Small amounts of plagioclase (An_{75-90}) were also observed. Tiny chromite crystals, <0.5 µm, were found embedded in glass in one olivine-rich chondrule. In a porphyritic chondrule in Tieschitz (H3.6), plagioclase with composition An_{45} to An_{90} occurred in the mesostasis, along with olivine and pyroxene crystals a few microns in size. Ikeda (1980) reports compositions of feldspar, An_{65-82}, in chondrules in ALHA76004 (L/LL3.4), but it is not clear whether this feldspar is primary or secondary (Table A3.5).

For all chondrules in the type 3 series, mesostasis compositions show general trends

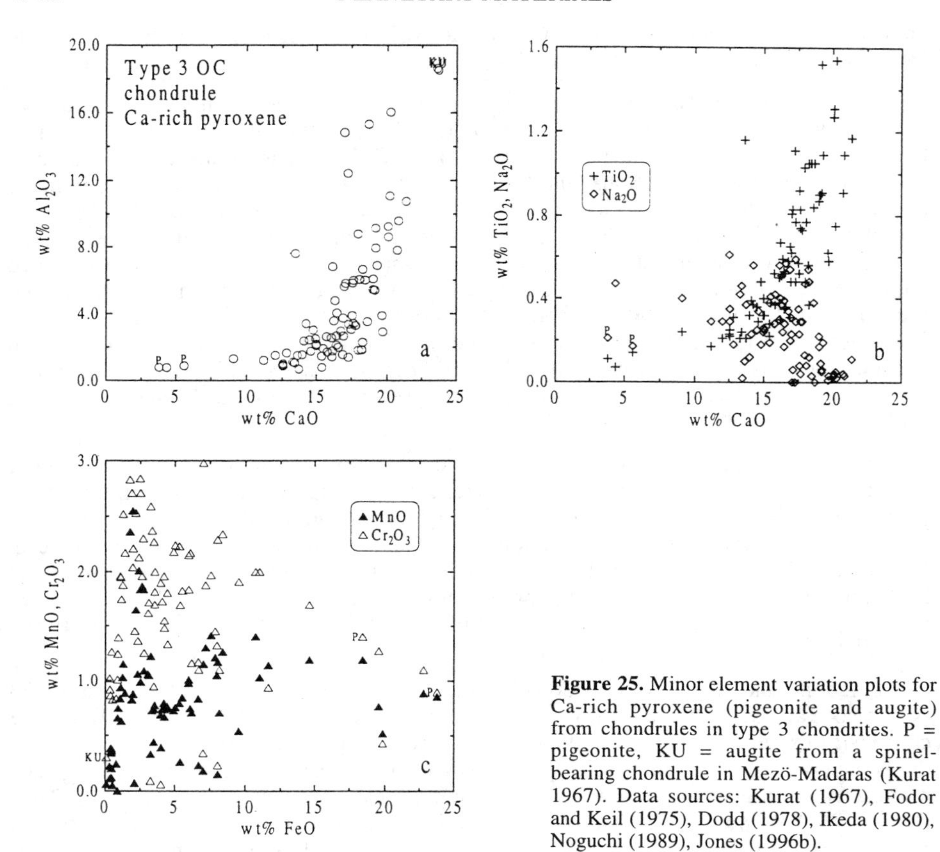

Figure 25. Minor element variation plots for Ca-rich pyroxene (pigeonite and augite) from chondrules in type 3 chondrites. P = pigeonite, KU = augite from a spinel-bearing chondrule in Mezö-Madaras (Kurat 1967). Data sources: Kurat (1967), Fodor and Keil (1975), Dodd (1978), Ikeda (1980), Noguchi (1989), Jones (1996b).

of decreasing CaO and Al_2O_3 with increasing SiO_2, as in Semarkona (Fig. 26; Table A3.4). There is a group of analyses with 55-65 wt % SiO_2, for which CaO contents are very low (<1 wt %). These chondrules also tend to have low FeO contents, but they are not limited to any obvious group of chondrules with similar textures or in any specific meteorites. FeO and Na_2O contents are very variable and FeO does not show a strong correlation with SiO_2. Several chondrules have Na_2O contents in the range 8-15 wt %, higher than any in Semarkona. MnO contents show a stronger correlation with FeO than in Semarkona. McCoy et al. (1991a) noted some changes in mesostasis compositions with petrologic type in specific textural types of chondrules: in type IA chondrules, FeO contents increase and TiO_2 contents decrease with petrologic type, and in type IIA chondrules, FeO and MnO contents decrease with increasing petrologic type. The chondrules in Figure 26 with mesostasis SiO_2 contents >90 wt % are radial pyroxene and cryptocrystalline chondrules in Sharps (H3.4), analyzed by Rubin and Pernicka (1989).

Spinel group minerals show a wide range of compositions in type 3 OC and are commonly zoned (Bunch et al. 1967). *Chromite* occurring in type II chondrules has Cr/(Cr+Al) ratios around 0.9, and mean atomic compositions, based on 32 oxygen atoms, of around 0.15 for Si, 0.4 for Ti, 0.2 for V, 0.14 for Mn, 1.3 for Mg, and 0.06 for Zn (Johnson and Prinz 1991). Bunch et al. (1967) also determined chromite compositions for type 3 OC (Table A3.6). Although these chromites were not necessarily in chondrules,

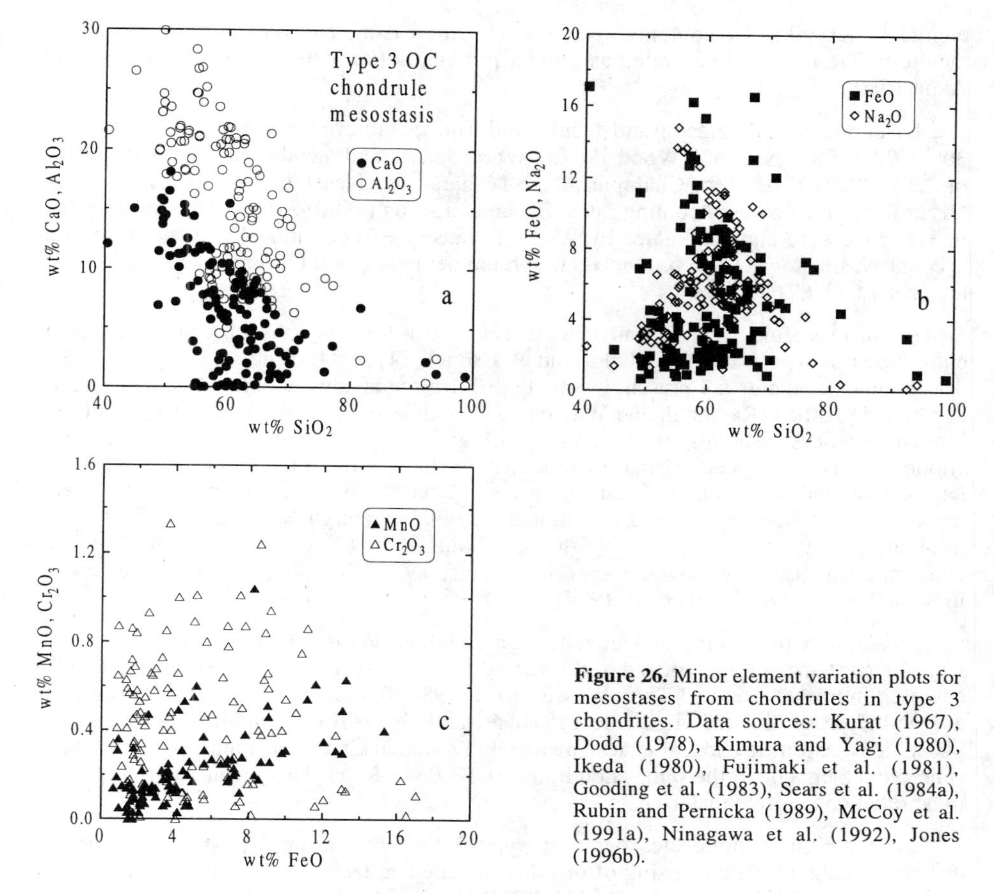

Figure 26. Minor element variation plots for mesostases from chondrules in type 3 chondrites. Data sources: Kurat (1967), Dodd (1978), Kimura and Yagi (1980), Ikeda (1980), Fujimaki et al. (1981), Gooding et al. (1983), Sears et al. (1984a), Rubin and Pernicka (1989), McCoy et al. (1991a), Ninagawa et al. (1992), Jones (1996b).

their compositions are similar to those determined by Johnson and Prinz (1991). Minor element contents in wt % are 0.2-3.7 TiO_2, 0.4-1.1 MnO, 0.4-0.9 V_2O_3. Some chondrites contain two compositionally distinct chromites. Calculated Fe_2O_3 contents for the chromites analyzed by Bunch et al. are 0-3.4 wt % for H3, 1.7-4.5 wt % for L3 and 0.7-5.0 wt % for LL3. Ikeda (1980) gives compositions of Al-rich spinels and chromites from chondrules in ALH76004.

Metal, sulfide and carbide minerals. Metal grains and metal-troilite assemblages are common in chondrules of type 3 ordinary chondrites, many occurring as rounded grains that are frequently concentrated towards the edges of the chondrule. Most metal in chondrules is kamacite which typically occurs as spheroidal grains, contains small amounts of taenite and schreibersite, and may be surrounded by a sulfide rim. Some spheroidal objects in chondrules consist entirely of sulfides. *Metal* compositions in chondrules differ from those in other occurrences, for example metal within chondrules has lower Co contents than metal on chondrule surfaces or in matrix (Affiatalab and Wasson 1980, Rambaldi and Wasson 1981, 1984; Nagahara 1982). Kamacite in chondrules shows considerable heterogeneity, for example compositions in Chainpur chondrules range from 3-5.5 wt % Ni and 0.1-2 wt % Co. Taenite has variable compositions, e.g. from 35-50 wt % Ni in Chainpur, and low Co contents, ~0.2 wt %. In taenite in Krymka, Cr contents decrease with increasing Ni content. Polycrystalline taenite in Tieschitz is interpreted as a relict primary solidification structure (Bevan and Axon 1980).

Ni contents of kamacite and taenite indicate metamorphic temperatures between 300°-500°C for type 3 OC (Wood 1967a). Wood determined metallographic cooling rates of 0.2° - 1°C/10^6 years for Chainpur, Mezö-Madaras and Tieschitz, but metal from other OC did not give coherent cooling rates. Bevan and Axon (1980) argued that cooling rates in Tieschitz were underestimated by Wood, because the Wood model does not take into account Ni heterogeneities produced at high temperatures, and that more realistic values are closer to 1°C/10^3 years.

Metal containing Cr, Si and P at levels of 0.1-1 wt % may be found in type I chondrules in type 3 OC (Rambaldi and Wasson 1981, 1984; Scott and Taylor 1983). These minor elements are present in small (<10 μm) inclusions of chromites, phosphates and euhedral silica (Rambaldi and Wasson 1981, Zanda et al. 1994). The grain size of the inclusions increases with increasing petrologic subtype. Schreibersite in Krymka chondrules is associated with iron oxide, phosphate and minor amounts of Co-rich tetrataenite, and is Co- and Cr-rich (up to 1 and 2 wt % Co and Cr, respectively). Some metal grains in Bishunpur contain schreibersite with very high Ni contents (~50 wt %), implying growth at temperatures of ~400°C (Rambaldi and Wasson 1981). Metal in rare chondrules in Bishunpur contains spherical inclusions of silica similar to those described in Semarkona above (Zanda et al. 1994).

In addition to spheres of kamacite, some chondrules in Chainpur, Bishunpur and Krymka contain a highly corroded, Si-bearing kamacite (0.1-0.4 wt % Si) that is totally enclosed in troilite (Rambaldi and Wasson 1981, 1984; Rambaldi et al. 1980). Phosphates are abundant at the metal-sulfide interface, and the sulfide contains inclusions of phosphates, chromite and a silicate. The metal is zoned in Cr, and contains higher Cr than spheroidal kamacite in the same chondrules (0.2-1.0 wt % Cr). This metal was interpreted to be relict condensate material.

Rare metallic chondrules occur in type 3 OC. For example, Nagahara (1982) described a chondrule consisting of metal (kamacite and taenite) and troilite exhibiting a lamellar intergrowth texture, in ALHA-77278. A metallic chondrule in Bishunpur has abundant small metal grains in its interior, with interstitial silicate glass (Rambaldi and Wasson 1981).

Carbide-magnetite assemblages (CMAs) occur in several type 3 OC (Taylor et al. 1981, Scott et al. 1982, Fredriksson et al. 1989, Krot et al. 1997a), and are present both in chondrules and matrix. They consist of magnetite, troilite, Ni-rich sulfide, carbides, Ni-rich taenite and Co-rich metal (Fig. 27). The original assemblages of metal and troilite have been replaced by carbides and magnetite, either by gas-solid reactions in the nebula (Taylor et al. 1981), or in the parent body by low-temperature reactions with a C-H-O-bearing fluid (Krot et al. 1997). The abundance of CMAs relative to unaltered metal-troilite assemblages varies widely in different chondrites, and the relative abundances of carbides and magnetite within the assemblages are also highly variable. Within individual chondrules, CMAs may occur in the outer portions while unaltered metal-troilite assemblages are present in the interior. Carbides show a wide range in composition, from 1-14 wt % Ni and 0-1.3 wt % Co. Cohenite, $(Fe,Ni)_3C$, contains 1.5-2 wt % Ni and haxonite $(Fe,Ni)_{23}C_6$, contains 4-5 wt % Ni. Ni-rich sulfides could be either an intergrowth of troilite and pentlandite, or a poorly characterized sulfide phase. Taenite grains have a wide range of compositions, 25-70 wt % Ni and 0-3 wt % Co. Kamacite is Co-rich: compositions up to 35 wt % Co are reported in Ngawi, although in most other

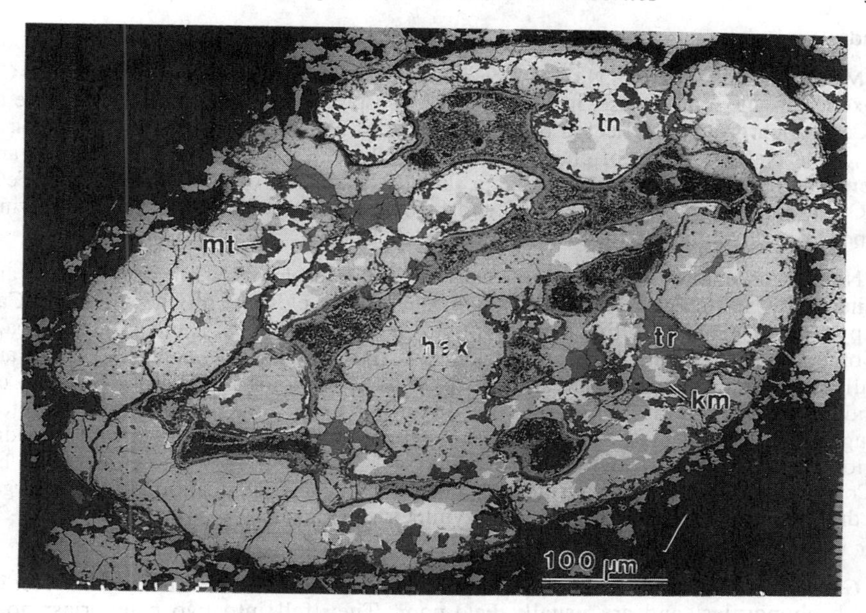

Figure 27. BSE image of a carbide-magnetite assemblage (CMA) in the L3.6 chondrite, LEW87284, consisting of coarse-grained intergrowths of Ni-rich taenite (tn) and haxonite (hax) and minor magnetite (mt), troilite (tr), and Co-rich kamacite (km). Black areas in the center of the nodule consist of a fine-grained mixture of silicate and carbide grains. After Krot et al. (1997a).

type 3 chondrites compositions are 2-8 wt % Ni and 0-10 wt % Co. Co contents of metals are much higher than those of unaltered metal in the same chondrites.

Al-rich chondrules. Al-rich chondrules are widespread, albeit rare constituents of ordinary chondrites and examples from many type 3 and type 4 chondrites have been described (Noonan 1975, Nagahara and Kushiro 1982, Bischoff and Keil 1983a,b, 1984; Bischoff et al. 1989, McCoy et al. 1991b, Russell et al. 1996). These objects, arbitrarily defined as having >10 wt % Al_2O_3 in the bulk chondrule, are part of a continuum of chondrule bulk compositions that lie between CAI and ferromagnesian chondrules. Within the group of Al-rich chondrules, bulk compositions vary considerably in their CaO/Na_2O ratios and Cr_2O_3 contents.

Al-rich chondrules show a variety of mineralogies and textures (Bischoff and Keil 1984). Some Ca-Al-rich chondrules consist of elongate, partly skeletal crystals of fassaite embedded in a fine-grained, microcrystalline to glassy mesostasis. Others are essentially BO chondrules, with a high proportion of mostly glassy, Ca-Al-rich mesostasis. Others contain laths of calcic plagioclase embedded in an extremely fine-grained mesostasis. Another type is fine-grained throughout, with small, skeletal crystals of fassaite and/or olivine and a groundmass consisting of olivine, plagioclase and fassaite. Most plagioclase compositions are anorthitic, An_{70-90}. Fassaite varies widely in composition, with 12-25 wt % CaO, 6-24 wt % MgO, 6-24 wt % Al_2O_3, and 0.6-4.3 wt % TiO_2. Olivine compositions fall into two groups, Fa_{13-26} and $Fa_{0.2-5}$, the more FeO-rich group being similar in composition to equilibrated olivine compositions throughout the host chondrites. Euhedral spinels occur at the rims of chondrules and are sometimes zoned: these are high in FeO (12-20 wt %) and Cr_2O_3 is usually <4 wt % although some more chromian spinels are also observed. Fe,Ni metal and troilite occur rarely in these chondrules.

Minerals found in (Ca,Na)-Al-rich chondrules include spinel, fassaite, low-Ca pyroxene, olivine, plagioclase, nepheline and mesostasis. The higher bulk Na relative to the Ca-Al-rich group is reflected by the presence of nepheline, more sodic plagioclase ($An_{<70}$) and Na_2O contents of glasses as high as 10 wt %. Most of these chondrules are extremely fine-grained, and some contain skeletal crystals of olivine. Spinels have FeO and Cr_2O_3 contents varying from 15-17 and 1-10 wt %, respectively. Fassaite and olivine compositions are similar to those in the Ca-Al-rich group.

Na-Al-rich chondrules have bulk Na_2O contents as high as 15 wt %. They are dominated by Na-rich glass, and also contain skeletal and elongate olivine and minor Ca-rich pyroxene and spinel. Olivine compositions in chondrules from H chondrites are Fa_{15-20}. Some chondrules with high bulk Na_2O also have high Cr_2O_3 contents. These are usually extremely fine grained and nearly opaque in transmitted light because of abundant small chromite, chromian-hercynian spinel or ilmenite grains distributed throughout the chondrules. Some small skeletal olivines are observed, as well as sodic plagioclase (An_{18-25}) and low-Ca pyroxene. Two Cr-Al-rich chondrules described by McCoy et al. (1991b) are porphyritic and contain olivine and plagioclase as well as large, euhedral grains of Cr-Al-spinels that are asymmetrically zoned (39-54 wt % Al_2O_3, 18-22 wt % FeO, 10-25 wt % Cr_2O_3).

Glass-rich chondrules. Glass-rich chondrules (Krot and Rubin 1994) are a subset of Al-rich chondrules, and are usually FeO-poor. They fall into two categories: non-porphyritic types that contain 90-99 vol % glass, and porphyritic types that contain 55-85 vol % glass. The non-porphyritic glass-rich chondrules contain skeletal crystals of olivine and fassaitic pyroxene that extend from the chondrule edge towards the center. Porphyritic types contain skeletal, barred or euhedral phenocrysts of olivine, low-Ca pyroxene and/or fassaite. A few chondrules also contain accessory spinel, troilite, and metallic Fe,Ni, and one contains euhedral crystals of merrillite. Glass in these chondrules is Al-rich (15-33 wt % Al_2O_3). Two groups of glasses contain high and low Ca contents, 0.1-3 and 8-15 wt % CaO. All the glasses are FeO-poor (0.6-3 wt % FeO). Glass compositions overlap completely with the main trends of mesostasis compositions in ferromagnesian chondrules (Fig. 26), although they are not included in those plots. Olivines have variable Fa contents, reflecting different degrees of metamorphism of the host chondrite, and CaO contents up to 0.6 wt %. Low-Ca pyroxene, $Wo_{0.5-6.0}$, is less equilibrated and has low FeO contents ($Fa_{0.6-2}$). It is commonly significantly more Al-rich (2-9 wt % Al_2O_3) than low-Ca pyroxene in typical ferromagnesian chondrules (cf. Fig. 24a), and contains 0.1-1.1 wt % TiO_2. Ca-rich pyroxene ($Fs_{<4}$, Wo_{35-50}) is commonly fassaitic, and contains 5-20 wt % Al_2O_3 and 1-4 wt % TiO_2. Most spinels are Al- and Mg-rich and contain ~0.3 wt % TiO_2 and 0.3-0.5 wt % Cr_2O_3. They have variable FeO contents, from 0.4-15 wt %.

Chromite-rich chondrules. Rare chondrules and inclusions in OC are enriched in Cr, and contain >13 wt % Cr_2O_3 in their bulk compositions (Ramdohr 1967, Noonan and Nelen 1976, Krot et al. 1993). They are typically small, only 100-300 μm in diameter, and are more common in H than in L and LL chondrites. They have granular, porphyritic and complex textures and consist of chromite grains embedded in a plagioclase-rich mesostasis. Accessory ilmenite, Ca-rich pyroxene and phosphates (merrillite and apatite) also occur. Some chondrules are chemically and/or mineralogically zoned. Chromite / spinel compositions vary widely from one chondrule to another, as well as being zoned in some cases. Cr/(Cr+Al) ratios vary from 0.2-0.9 (Krot et al. 1993). FeO contents are in the range 20-30 wt % for chromites and 12-25 wt % for spinels, and TiO_2 contents are as high as 2 wt % in the chromites. ZnO contents are up to 0.8 wt %. Plagioclase compositions in the mesostasis are also variable, $Ab_{40\text{-}90}$ and $Or_{1\text{-}55}$ mol %. Ilmenite in Cr-rich chondrules has high MgO (6 wt %) and low MnO (1 wt %) in comparison with matrix ilmenite in ordinary chondrites (3.5 wt % MgO, 2.2 wt % MnO).

Silica-bearing chondrules. Ferromagnesian chondrules are typically olivine-normative. However, rare (<2% of chondrules), silica-bearing chondrules and clasts have been observed in several type 3 OC (Fujimaki et al. 1981, Planner 1983, Matsunami et al. 1992, Brigham et al. 1986, Wasson and Krot 1994, Krot and Wasson 1994, Wood and Holmberg 1994). These chondrules consist almost entirely of silica and low-Ca pyroxene, in which Na, Ca and Al are minor elements. Many silica-bearing chondrules (silica-pyroxene-fayalite chondrules and clasts) also contain secondary, extremely ferroan fayalite. The origin of silica-rich chondrule melts has been attributed to a variety of nebular processes (Brigham et al. 1986, Wasson and Krot 1994, Krot and Wasson 1994, Wood and Holmberg 1994).

Silica-pyroxene chondrules have C, RP and porphyritic textures, with silica constituting up to 40 vol % of the chondrule. Silica is a liquidus phase and crystallized from the chondrule melt. It is very pure, commonly >99 wt % SiO_2 with the only impurity being <1 wt % FeO, and is inferred to be cristobalite in most cases. Low-Ca pyroxene compositions are variable, both within and between chondrules. This is partly because the chondrules are in partially equilibrated chondrites, although some compositional heterogeneity is also primary. Accessory phases include Ca-rich pyroxene, Fe,Ni metal, troilite, merrillite, and glass.

Silica-pyroxene-fayalite chondrules and clasts have been described by Brigham et al. (1986) and Wasson and Krot (1994). They consist of intergrown primary silica and low-Ca pyroxene, with secondary, highly fayalitic olivine occurring as a lacy network throughout the silica. The proportions of these three minerals are highly variable. Fayalites may be as FeO-rich as Fa_{99}, and typically contain about 1.5 wt % MnO and no detectable NiO. Fayalite contents often decrease towards the edges of the clasts, probably as a result of equilibration with the host chondrite. Low-Ca pyroxene has variable compositions, from ~7-35 wt % FeO. Some clasts contain abundant Ca-rich pyroxene, which has very variable compositions. FeO-rich, Ca-rich pyroxenes are inferred to be secondary. Minor chromite and troilite may be present in these clasts. Fayalite is considered to have formed by reaction of silica with FeO which was produced by oxidation of Fe,Ni metal or troilite (Brigham et al. 1986, Wasson and Krot 1994), or in a region of the nebula enriched in Fe and alkalis (Wood and Holmberg 1994).

Merrihueite and roedderite are minerals of the solid solution series, $(K,Na)_2(Fe,Mg)_5Si_{12}O_{30}$. They have been observed in chondrules in a limited number of type 3 ordinary chondrites, including Mezö-Madaras, ALHA77011, ALHA77115, and ALHA77278 (Dodd et al. 1965, 1966; Wood and Holmberg 1994, Krot and Wasson 1994). Merrihueite and roedderite occur in association with low-Ca pyroxene, fayalitic olivine, ± silica, and ± feldspathic mesostasis (Fig. 28), and are interpreted as being a reaction product of silica with alkali-rich vapors, either in the nebula (Wood and Holmberg 1994) or on a parent body (Krot and Wasson 1994). The minerals occur as thin veins within silica, or as compact grains that partially or almost completely replace silica phenocrysts. Fayalitic olivine, which occurs as veins similar to those in silica-bearing chondrules described above, postdates merrihueite formation. Merrihueite contains 4-8 wt % MgO, 19-25 wt % FeO, 0.5-3 wt % Na_2O, and 3-7 wt % K_2O. Roedderite contains 11-17 wt % MgO, 5-14 wt % FeO, 2-3 wt % Na_2O, and 4-5 wt % K_2O, and roedderite analyses span the compositional gap between merrihueite in OC and roedderite in enstatite chondrites and silicate inclusions in IAB irons.

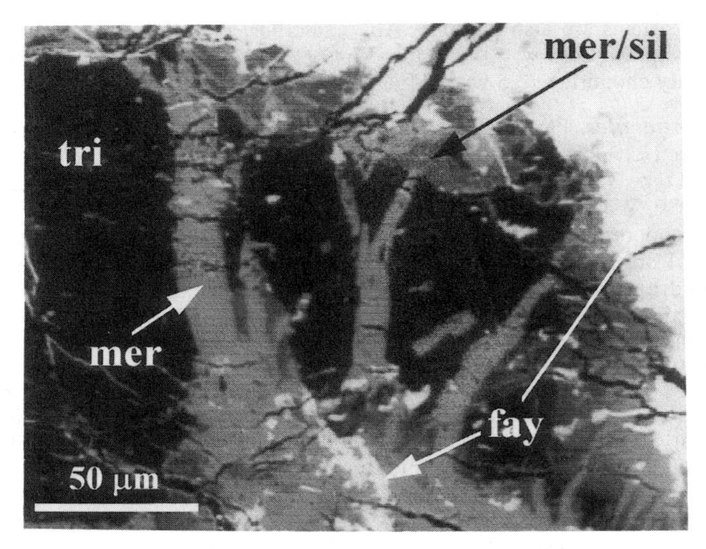

Figure 28. BSE image showing merrihuite (mer) - silica (sil) - fayalite (fay) association in a chondrule in the Mezö-Madaras L3 chondrite (after Wood and Holmberg 1994). Silica phenocrysts are probably tridymite (tri). Merrihuite and fayalite are secondary minerals that formed along cracks, probably by reactions between silica and an iron- and alkali-rich vapor.

Fujimaki et al. (1981) described a unique chondrule in the L3 chondrite, ALHA-77015, that contains a myrmekitic intergrowth of albite and a silica mineral, interstitial glass, and minor euhedral crystals of ferropseudobrookite. Silica contains minor amounts of Al_2O_3, Na_2O, TiO_2 and FeO, and is suggested to be either tridymite or cristobalite. Ferropseudobrookite crystals are zoned from 66 wt % TiO_2 at the center to 79 wt % at the edge. They contain minor amounts of CaO, Cr_2O_3 and MnO and no measurable MgO.

CO chondrites

The CO chondrites are all of petrologic type 3. They show a sequence of increasing degree of metamorphism that has resulted in subdivision of the type 3 series in a manner analogous to the OC type 3 series (Keck and Sears 1987, Scott and Jones 1990, Sears et al. 1991a). In practice, the highest petrologic subtype that has been described for CO chondrites is 3.7. It is important to note that the subtype divisions only indicate a relative scale of metamorphism within each chondrite group, and that the same subtype number does not necessarily indicate equivalent conditions of metamorphism for the ordinary and CO chondrites. Two CO3.0 chondrites have been identified, ALHA77307 and Colony. These represent the least equilibrated members of the series, in which chondrule silicate compositions are considered to be primary. However, chondrule mesostases in ALHA77307 have undergone aqueous alteration (Ikeda 1983). There is some question as to whether ALHA77307 belongs to the main CO chondrite group (e.g. Kallemeyn and Wasson 1982) but as its chondrule mineral compositions are very similar to those in the other CO chondrites we include it as part of the CO group here.

In CO3 chondrites, a large number of isolated olivine and pyroxene grains occur in the matrix, about 8 vol % of the whole chondrite, compared with about 40 vol % for chondrules (McSween 1977a). Several authors have argued that isolated olivine grains are derived from chondrules by fragmentation (McSween 1977b, Nagahara and Kushiro 1982, Rubin et al. 1985, Jones 1992). However, there is some discussion about whether incompatible-element rich forsterites are also derived from chondrules or whether they are primitive condensates (Steele 1989). We do not describe the compositions of isolated grains separately here. The population of isolated pyroxene grains in these chondrites has not been studied in detail.

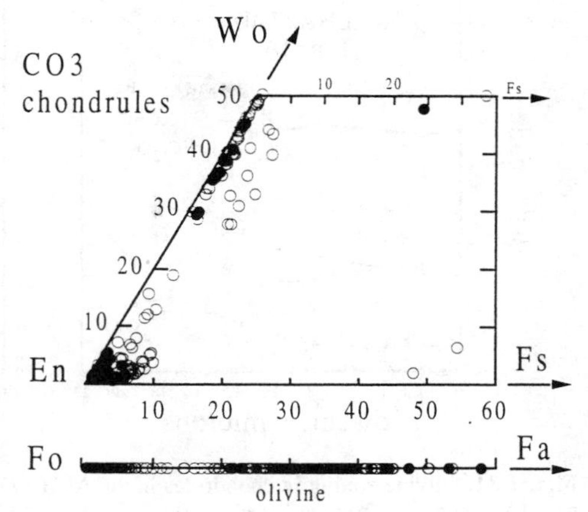

Figure 29. Pyroxene and olivine compositions in chondrules from CO chondrites. Solid circles are ALHA77307 (subtype 3.0), open circles are for all other CO chondrites (subtypes 3.1-3.7). Data sources: Kurat (1975), Ikeda (1982), Rubin and Wasson (1988), Noguchi (1989), Scott and Jones (1990), Jones (1992, 1993 and unpublished).

Silicate and oxide minerals. The range of *olivine* compositions in porphyritic chondrules in CO chondrites is from <Fa_1 to Fa_{60} (Fig. 29). This range is considerably wider than in porphyritic chondrules in the least equilibrated OC. Two distinct types of PO chondrules occur (Kurat 1975, McSween 1977a). Type IA chondrules are commonly metal-rich and, in the least equilibrated CO chondrites, FeO-poor. Type IIA chondrules are FeO-rich and commonly contain chromite: olivine is strongly zoned in these chondrules in the least equilibrated chondrites. Type IAB chondrules are intermediate in FeO content. BO chondrules are rare and are both FeO-poor and FeO-rich (e.g. Scott and Taylor 1983). FeO-rich, pyroxene-rich porphyritic chondrules are also rare, and few pyroxene compositions above Fs_{10} have been reported in all the CO chondrites.

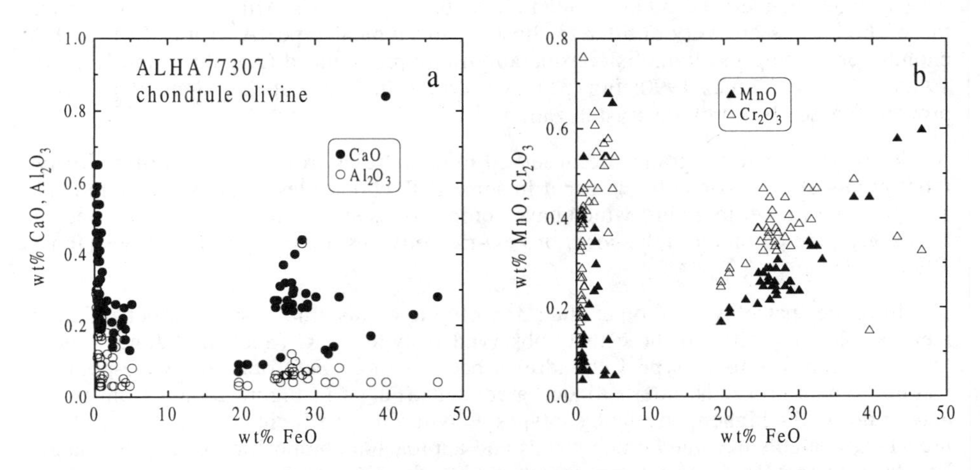

Figure 30. Minor element variation plots for olivine from chondrules in ALHA77307 (CO3.0). Data sources: Scott and Jones (1990), Jones (1992 and unpublished).

Minor element compositions in olivine from chondrules in ALHA77307 are shown in Figure 30 (see also Table A3.1). The distinct bimodal distribution in FeO contents in these plots is partly an artifact, because minor element data have only been reported for olivine in type IA and IIA chondrules. These chondrule types were selected specifically

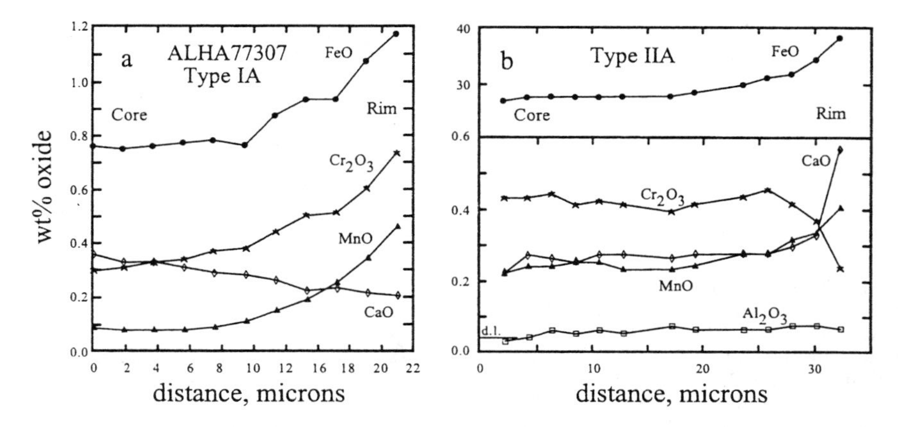

Figure 31. Olivine zoning in chondrules in the ALHA77307 (CO3.0) chondrite (Scott and Jones 1990). (a) type IA chondrules (b) type IIA chondrules.

for the purpose of using the olivine compositions as an indicator of petrologic subtype (Scott and Jones 1990). The plot of CaO vs. FeO contents of olivines in ALHA77307 chondrules shows a similar pattern to olivines in Semarkona chondrules (Fig. 11). Forsteritic olivine with $Fa_{<1}$ has a wide range of CaO contents, up to 0.65 wt % CaO, is rich in other incompatible elements, including Al_2O_3 and TiO_2, and has low Cr_2O_3 and MnO contents. MnO and FeO are strongly correlated in FeO-rich olivines.

Olivine in type IA chondrules in ALHA77307 shows very limited zoning of FeO and minor elements (Nagahara and Kushiro 1982a, Scott and Jones 1990; Fig. 31). Jones (1992) described "macroporphyritic" type I chondrules that contain a limited number of large olivine phenocrysts and only minor amounts of mesostasis. Mineral compositions in these chondrules are very similar to those of the typical type IA group. In type IIA chondrules, zoning in olivine is considerably more pronounced (Nagahara and Kushiro 1982a, Scott and Jones 1990; Fig. 31). A decrease in Cr_2O_3 at the edges of grains is probably caused by chromite crystallization.

Two types of relict grains are observed in chondrules from CO chondrites. Dusty, relict grains of metal-rich olivine occur in some POP chondrules in Colony (Rubin et al. 1985). Relict forsterite grains, which have compositions similar to olivines from type IA chondrules, are common in the cores of FeO-rich olivines in type IIA chondrules (Jones 1992 1993).

In higher petrologic subtypes than 3.0, olivine compositions show effects of mild metamorphism similar to the effects observed in type 3 OC (Scott and Jones 1990). Olivine compositions in type I chondrules become more FeO-rich, and CaO-poor, as metamorphism proceeds. Intermediate Fa contents (Fa_{5-20}) in Figure 28 are from type I chondrules in the higher petrologic subtypes. Olivines in type II chondrules lose CaO and their FeO contents become homogeneous and approach a composition of approximately Fa_{36} in subtype 3.7. Minor element variation plots for olivines from subtypes 3.1-3.7 are shown in Figure 32 (see also Table A3.1). MnO contents are well correlated with FeO. The slope of the MnO/FeO correlation is significantly shallower than the slope for ordinary chondrites (Fig. 22). Cr_2O_3 contents of most chondrule olivines in these chondrites are below 0.2 wt %, suggesting that Cr diffuses rapidly during metamorphism. In contrast, CaO and Al_2O_3 contents do not differ significantly from those in ALHA77307.

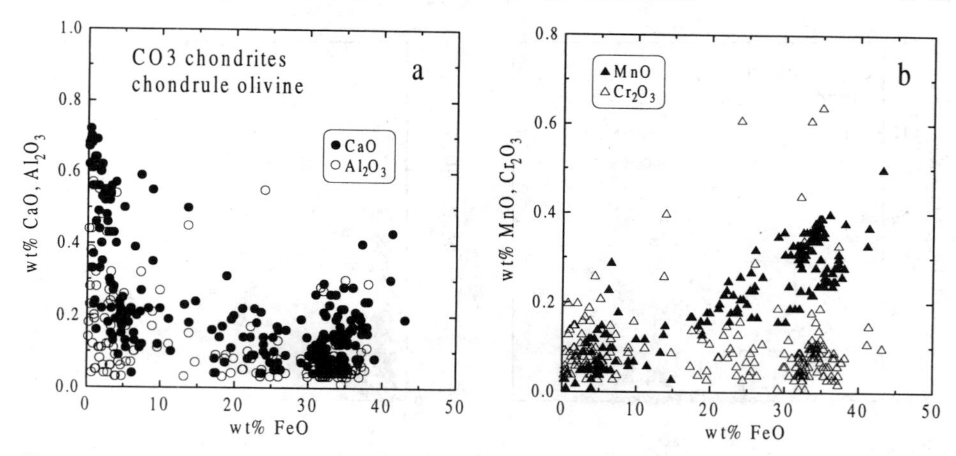

Figure 32. Minor element variation plots for olivine from chondrules in CO3 chondrites, subtypes 3.1-3.7. Three points with 1.0-1.2 wt % Cr_2O_3 are not included in (b). Data sources: Kurat (1975), Steele (1986), Rubin and Wasson (1988), Scott and Jones (1990), Jones (1993 and unpublished).

Pyroxene in chondrules in CO chondrites is usually polysynthetically twinned clinoenstatite, with augite overgrowths. Ikeda (1982) reports twinned low-Ca clinopyroxene and orthopyroxene in ALHA77003. Several chondrules in ALHA77307 contain pigeonite (Noguchi 1989). Minor element contents of *low-Ca pyroxenes*, including pigeonites, in CO3 chondrites are shown in Figure 33 (see also Table A3.2). The restricted range of FeO contents does not appear to be an artifact, as very few analyses of pyroxene with higher FeO contents have been reported in any CO chondrite study. Two FeO-rich compositions ($Fs_{>50}$) have been determined in ALHA77003 (subtype 3.5). Compositional zoning from cores to edges of low-Ca pyroxene grains is generally very limited (Noguchi 1989), but CaO and Cr_2O_3 are zoned from cores to rims of orthopyroxene grains in ALHA77003 (Ikeda 1982). Low-Ca pyroxene compositions in ALHA77307 are generally indistinguishable from those in CO chondrites of higher petrologic subtype, reflecting the limited effects of metamorphism.

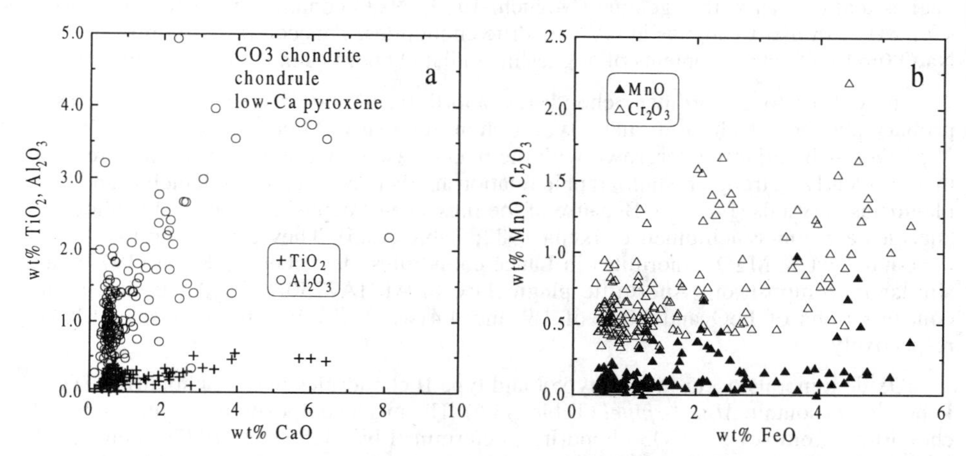

Figure 33. Minor element variation plots for low-Ca pyroxene from chondrules in CO3 chondrites. Data sources: Kurat (1975), Ikeda (1982), Rubin and Wasson (1988), Noguchi (1989), Scott and Jones (1990), Jones (unpublished).

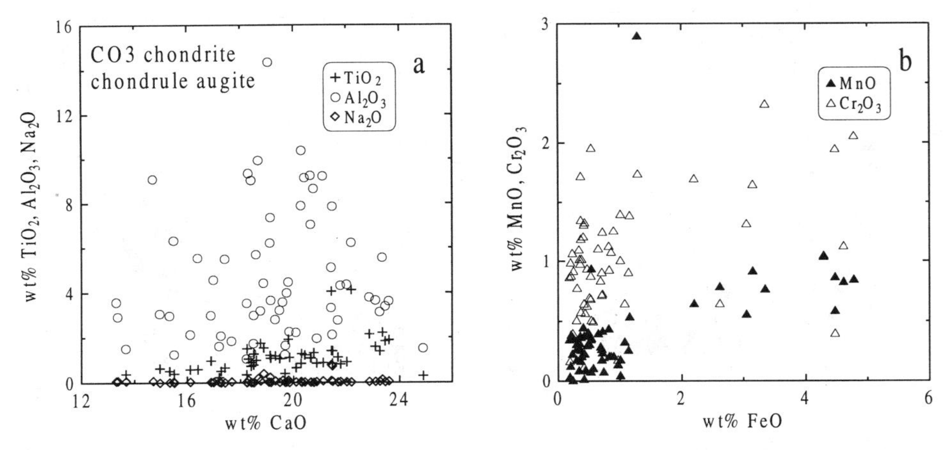

Figure 34. Minor element variation plots for augite from chondrules in CO3 chondrites. Data sources: Kurat (1975), Ikeda (1982), Noguchi (1989), Jones (1992 and unpublished).

Augite compositions in CO chondrite chondrules are summarized in Figures 29 and 34 (see also Table A3.3). Most of these data are from two papers: Ikeda (1982: ALHA77003), and Noguchi (1989: ALHA77307 and ALHA77003). Wo contents of augites show a wide range which probably reflects zoning within individual grains as well as interchondrule variation, but these variations are not well documented. One chondrule in ALHA77003 consists of euhedral diopside phenocrysts (Wo_{48-50}) plus glass. In ALHA77307, the range of Fs contents in augite is extremely limited: in type I chondrules all the analyses reported are $<Fs_{1.5}$. For ALHA77003 (subtype 3.5), a wider range of Fs contents in augite in type I chondrules is reported, up to Fs_8. It seems unlikely that this difference could be caused by the mild metamorphism that ALHA77003 has experienced, because diffusion in augite is very slow. Since the difference occurs within one data set (Noguchi 1989), this suggests a real difference in chondrule augite compositions between the two chondrites. The highest TiO_2 contents are observed in augites that coexist with pigeonite (Noguchi 1989). Na_2O contents are very low, mostly <0.1 wt %, similar to augites in CV chondrite chondrules. This contrasts sharply with the Na_2O (up to 0.5 wt %) contents of augites in similar chondrules in the OC Semarkona.

In contrast to the ordinary chondrites, anorthitic *plagioclase* (An_{60-90}) occurs as a primary phenocryst phase in many type I chondrules in CO3 chondrites (Ikeda 1982). *Nepheline* is invariably intergrown with the plagioclase in an interfingering texture, and there is clearly a strong crystallographic relationship between nepheline lamellae and host plagioclase crystals (Fig. 35). Because of the presence of nepheline, reported plagioclase analyses are non-stoichiometric (Ikeda 1982; Table A3.5). They contain up to 1 wt % FeO and 1 wt % MgO. Anorthites in Lancé chondrules, described by Kurat (1975), are similar in composition. Anorthitic plagioclase in ALHA77307 (An_{69}) has maximum concentrations of FeO and MgO of 1.9 and 4.4 wt % (Miúra and Tomisaka 1984), respectively.

Oxide minerals occur in both type I and type II chondrules in CO3 chondrites. Type I chondrules contain *Mg,Al-spinel* (Table A3.6). The mean composition of spinel in type I chondrules from several CO3 chondrites, determined by McSween (1977b) contains 4 wt % FeO, 0.3 wt % TiO_2 and 0.2 wt % Cr_2O_3. A macroporphyritic type I chondrule in ALHA77307 contains several Mg,Al-spinel grains enclosed in forsterite. Spinel has a low FeO content of 0.45 wt %, 0.2 wt % TiO_2 and 1.4 wt % Cr_2O_3 (Jones 1992). One

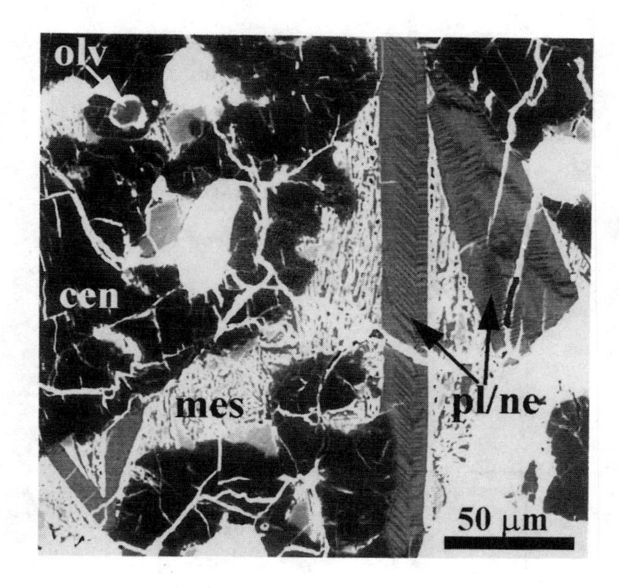

Figure 35. BSE image of a type I chondrule from the CO chondrite, Kainsaz. The chondrule contains clinoenstatite phenocrysts (cen), which enclose olivine phenocrysts (olv) and metal and sulfide grains (white) in a poikilitic texture, as well as elongate laths of plagioclase (pl: lighter gray) with intergrown nepheline (ne: darker gray) and fine-grained mesostasis (mes).

chondrule in ALHA77003 contains primary Mg,Al-spinel phenocrysts in association with olivine, augite, and plagioclase (Ikeda 1982). These spinels are zoned with increasing FeO contents towards the edges of the grains. Olivine grains are also zoned, but augite is highly magnesian ($Fs_{<1.5}$). The Fe-Mg zoning in spinel and olivine is likely to be secondary and to result from metamorphism. Type II chondrules contain small, euhedral chromian *hercynite* crystals (Kurat 1975, McSween 1977b), the average composition of which is given in Table A3.6. Geothermometry applied to olivine-chromite pairs from type II chondrules in CO chondrites shows progressively lower temperatures with increasing petrologic subtype. This is interpreted as progressive equilibration from initial chondrule crystallization temperatures (Johnson and Prinz 1991).

Only limited data are available for chondrule *mesostases* in CO3 chondrites. Most of the data are for Ornans (subtype 3.3: Rubin and Wasson 1988), and ALHA77003 (subtype 3.5: Ikeda 1982). Mesostases show a range of SiO_2 contents, from 40-60 wt % (Fig. 36; Table A3.4). Mesostases from chondrules in Ornans are significantly richer in Al_2O_3 and depleted in FeO, Cr_2O_3 and MnO (which is mostly below detection limits of the electron microprobe), compared with all analyses from ALHA77003. The reasons for this are not clear, although the chondrules studied by Rubin and Wasson (1980) were disaggregated from the whole rock, and are biased towards a larger size range than the mean chondrule size in CO chondrites. Interelement correlations for mesostases from all chondrules available are not very evident in this rather limited data set. In the plot of FeO vs. SiO_2, a bimodality in FeO contents for low SiO_2 contents reflects the bimodality in bulk FeO contents of olivine-rich chondrules (type IA and type IIA). Higher SiO_2 contents and intermediate FeO contents in the mesostases are from pyroxene-rich type I chondrules. Glass also occurs as inclusions in olivine grains, and these inclusions may also contain tiny euhedral spinel crystals and/or a vapor bubble (McSween 1977b).

Mesostasis compositions vary within chondrules as well as from one chondrule to another. One chondrule in ALHA77003, which contains diopside phenocrysts, has a strongly zoned mesostasis which increases in Na_2O (6 to 10 wt %), K_2O (0.1 to 2 wt %), and FeO (3.5 to 10 wt %), and decreases in SiO_2 (57 to 44 wt %) from core to edge, while

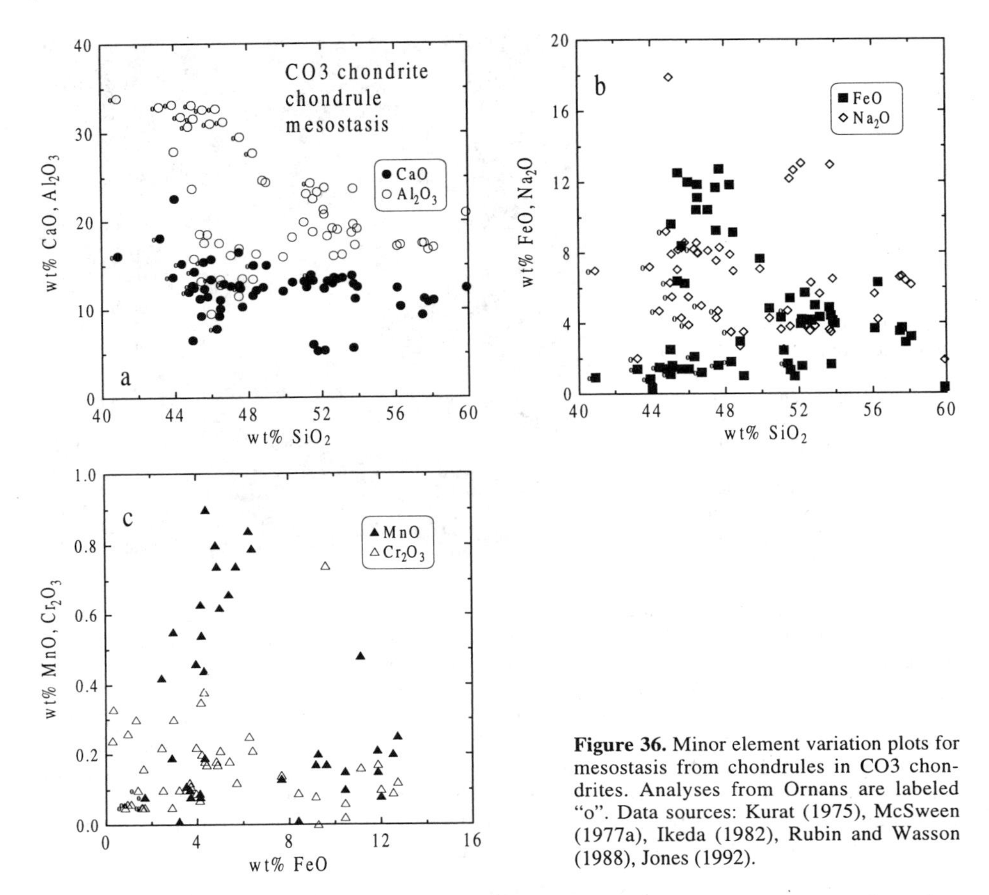

Figure 36. Minor element variation plots for mesostasis from chondrules in CO3 chondrites. Analyses from Ornans are labeled "o". Data sources: Kurat (1975), McSween (1977a), Ikeda (1982), Rubin and Wasson (1988), Jones (1992).

Al_2O_3 and CaO remain roughly constant. This zoning is interpreted as occurring after solidification of the chondrule, but prior to accretion into the parent body (Ikeda 1982). Two glass inclusions within a single large forsterite grain in ALHA77307 have very different SiO_2 contents of 45 and 60 wt % (Jones 1992).

In ALHA77307, mesostases of many chondrules are altered to *phyllosilicates* (Ikeda 1983). Phyllosilicate compositions have fairly uniform (Na+K)/Al ratios. Most of them plot in the range between chlorite and smectite, and the main phyllosilicate phases are interpreted to be chlorite and/or berthierine whose compositions range from $(Mg,Fe)_{11.5}AlSi_{7.5}O_{20}(OH)_{16}$ to $(Mg,Fe)_{10}Al_4Si_6O_{20}(OH)_{16}$. The compositions suggest addition of FeO and MgO to, and loss of CaO and alkalies from, the original chondrule mesostasis glass compositions.

Metal, sulfide and associated minerals. Type I chondrules in CO3 chondrites commonly contain abundant spherical or irregular assemblages of metal (kamacite and taenite) and sulfide, whereas type II chondrules are usually devoid of opaque minerals. Opaque minerals also occur throughout the matrix of CO chondrites. In Colony (3.0), most type I chondrules are metal-rich, containing 10-25 vol % metallic Fe,Ni blebs and additional sulfide (Rubin et al. 1985). In ALHA77307 (3.0), type I chondrules contain Fe,Ni metal, troilite and/or magnetite and pentlandite (Nagahara and Kushiro 1982a,

Ikeda 1983, Scott and Jones 1990). Magnetite appears to replace metal, and pentlandite appears to replace troilite. Magnetite is also common in Ornans (3.3), in which opaque assemblages consist of troilite, magnetite-troilite and/or magnetite-metal (Rubin and Wasson 1988). Metal in Ornans is all taenite, with kamacite being absent (Wood 1967a, McSween 1977a). Magnetite in Ornans contains significant Cr_2O_3 (1-2 wt %), Al_2O_3 (0.5-2 wt %) and MgO (0.4-0.7 wt %) (Rubin and Wasson 1988). Many magnetite grains are commonly intergrown with tiny grains of Ca-phosphate. Troilite in Lancé (3.4) is mostly very porous and is finely intergrown with silicates and phosphates (Kurat 1975). The carbides haxonite and cohenite have been observed in ALHA77307 (Scott and Jones 1990). The carbides are small (<30 μm) and irregularly distributed in kamacite. Haxonite contains ~5 wt % Ni, and cohenite 2-4 wt % Ni; both contain 0.2-0.4 wt % Co, and up to 0.9 wt % Cr, 0.1 wt % Si and 0.2 wt % P.

Metal compositions in CO3 chondrites change as a function of petrologic subtype. The Ni content of kamacite increases, and the Ni content of taenite decreases, with increasing degree of metamorphism (McSween 1977a, Scott and Jones 1990). Metal grains are zoned: kamacite has Ni-poor rims and taenite has Ni-rich rims of tetrataenite. Relatively homogeneous tetrataenite grains also occur. Taenite and tetrataenite contain 0.1-1.3 wt % Co (Scott and Jones 1990). In three CO chondrites, ALHA77307(3.0), ALHA77029 (3.4) and ALHA77003 (3.5), taenite in chondrules is richer in Cr (0.1-0.3 wt %) than taenite outside chondrules (<0.06 wt %) (Scott and Jones 1990). Taenite in Lancé contains 0.2-1.0 wt % Cr (Kurat 1975). In ALHA77307, a single plessite grain with ~24 wt % Ni, and two Co-rich taenite grains with 2.0-2.5 wt % Co and 60-64 wt % Ni were all observed outside chondrules (Scott and Jones 1990).

Mean Ni contents of kamacite in individual chondrites range from 4-7 wt %. In kamacite, Co contents increase and Cr contents decrease progressively with increasing petrologic subtype (McSween 1977a, Scott and Jones 1990; Fig. 37). Co concentrations in kamacite from chondrules and matrix are similar in ALHA77029 and ALHA77003. However, in ALHA77307, the Co content of metal in chondrules tends to be lower than for metal in matrix (Scott and Jones 1990). Kamacite in ALHA77307 contains 0.1-0.8 wt % of both Cr and P (means are 0.6 wt % Cr and 0.2 wt % P) and 0.03-0.11 wt % Si: Cr and P contents of chondrule kamacites are higher than those in matrix. Si and P contents of kamacites in ALHA77029 and ALHA77003 are <0.02 wt %, whereas kamacite in type I chondrules in Lancé contains 0.0-1.1 wt % Si, 0.03-0.9 wt % Cr, and 0.02-0.3 wt % P (Kurat 1975). High Cr, Si and P contents of kamacite are also characteristic of metal in the least metamorphosed LL chondrites as well as CM chondrites.

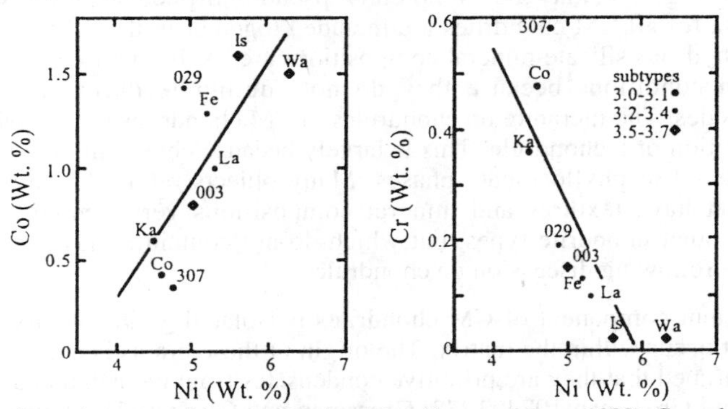

Figure 37. Mean kamacite compositions in CO chondrites (Scott and Jones 1990). Co and Ni increase, and Cr decreases, with increasing petrologic subtype. 307 = ALHA77307, 029 = ALHA77029, 003 = ALHA77003, Co = Colony, Fe = Felix, Is = Isna, Ka = Kainsaz, La = Lancé, Wa = Warrenton.

In ALHA77307, opaque clots, a few tens to several hundred microns in diameter, occur in the matrix (Ikeda 1983). Their shapes are sometimes spherical and sometimes irregular. They consist of aggregates of PCP with troilite and magnetite. Fe,Ni metals are observed as relic minerals in the central parts of some opaque clots, and this suggests that they formed by the alteration of materials consisting mainly of Fe,Ni metals. The Ni contents of sulfides are <0.5 wt %.

Plagioclase-rich chondrules. CO3 chondrites contain a suite of reduced, plagioclase-rich chondrules with complex mineralogies. These chondrules are fairly abundant in Lancé, but rare in the other CO chondrites: they have been observed in Ornans, Kainsaz, ALHA77003, and ALH83108 (Kurat and Kracher 1980, Palme and Wlotzka 1981, Jones 1997). Kurat and Kracher (1980) originally described these objects as basaltic lithic fragments, but more recent work has defined them as chondrules. They are characterized by the presence of abundant laths of primary anorthitic plagioclase (An_{70-90}) and phenocrysts of orthopyroxene and augite, in an ophitic texture. Many of the chondrules contain discrete units of clinoenstatite with olivine enclosed in a poikilitic texture, as well as associated Fe,Ni metal and troilite. A second, rarer, type of discrete unit consists of plagioclase, nepheline, and spinel. The chondrules have undergone secondary metasomatism, resulting in alteration of plagioclase to nepheline (Fig. 35). Fe-Mg exchange has resulted in zoning of originally FeO-poor olivine and spinel. In addition, mesostasis contains a ferrosalite pyroxene ($\sim Fs_{37}Wo_{50}$) that is probably an alteration product of groundmass diopside, plagioclase and silica. Clinoenstatites, orthopyroxenes and augites are all FeO-poor ($<Fs_2$). Spinels contain significant TiO_2 (~1 wt %), Cr_2O_3 (~1 wt %), V_2O_3 (~0.5 wt %), and ZnO (~0.3 wt %).

CM chondrites

CM chondrites are all of petrologic type 2. They have been affected by aqueous alteration but not by thermal metamorphism. Well-defined chondrules are low in abundance in CM chondrites, comprising only about 12 vol % of each rock (McSween 1979). The degree of alteration experienced by different members of the CM group varies, but no classification scheme has been generally adopted that describes a progressive alteration sequence. Various schemes have been proposed by Ikeda (1983), Kojima et al. (1984), and Browning et al. (1996). Progressive replacement of low-Ca pyroxene and olivine by phyllosilicates occurs to differing degrees in recognizable chondrules in CM chondrites. Murchison, which has been studied extensively, is one of the least altered CM chondrites. At advanced stages of alteration, such as in Yamato(Y)-791824, olivine grains may be completely pseudomorphed (Kojima et al. 1984). Chondrule data for all CM chondrites are included together in the following discussion. In general, anhydrous silicate mineral compositions are likely to reflect initial chondrule crystallization conditions because they do not equilibrate during low-temperature alteration episodes. The literature on chondrules in CM chondrites is somewhat confusing as to the definition of a chondrule. This is largely because chondrule mesostasis glass is commonly altered to phyllosilicate phases. Many objects referred to as inclusions or aggregates that have textures and mineral compositions very similar to porphyritic chondrules in other chondrite types, but which do not contain a glassy mesostasis, are included in the following discussion on chondrules.

An important component of CM chondrites is isolated grains of olivine, pyroxene and spinel that occur within the matrix. The origin of these grains is controversial. Some authors have argued that they are primitive condensates from the nebula gas (Fuchs et al. 1973, Olsen and Grossman 1974, 1978; Grossman and Olsen 1974, Onuma et al. 1976). Others have shown textural and compositional similarities between these isolated grains

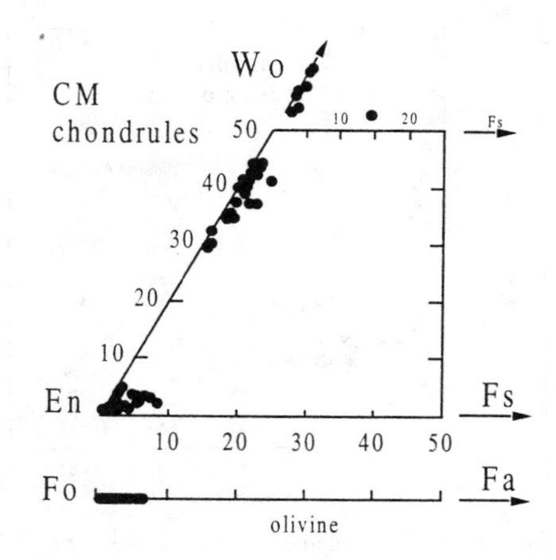

Figure 38. Pyroxene and olivine compositions in chondrules from CM chondrites. Data sources: Richardson and McSween (1978), Olsen (1983), Rubin and Wasson (1986), Steele (1986), Noguchi (1989), Simon et al. (1994), Brearley (1995).

and grains that are clearly in chondrules, and argued that the isolated grains have disaggregated from chondrules (McSween 1977b, Richardson and McSween 1978, Desnoyers 1980, Roedder 1981, Brearley 1995). Aqueous alteration and brecciation are important processes that tend to disaggregate chondrules. Since isolated grains are mineralogically similar to chondrule phases, we do not discuss them separately. In addition to the coarse grains discussed in these papers, isolated grains down to 0.1 μm in size also bear compositional similarities to the chondrule grain population and are not discussed further.

Silicate and oxide minerals. Most chondrules in CM chondrites are porphyritic, including porphyritic olivine and poikilitic pyroxene textural types. The range of *olivine* compositions is from Fa_1-Fa_{60}, with the majority of chondrule olivine being $<Fa_5$ and many with $<Fa_1$. On the pyroxene and olivine plot of Figure 38, as well as minor element plots presented below, only analyses presented in tables in the literature are included, and the range of olivine compositions is limited. Examples of histograms showing the full range of chondrule olivine compositions are given by Fuchs et al. (1973), Desnoyers (1980), and Brearley (1995) (Fig. 39). Scott and Taylor (1983) found that BO chondrules in Murray were more FeO-rich than any of the porphyritic chondrules analyzed in the same study. However, BO chondrules may also be FeO-poor (e.g. Richardson and McSween 1978). Olivine in PP chondrules tends to be low in FeO.

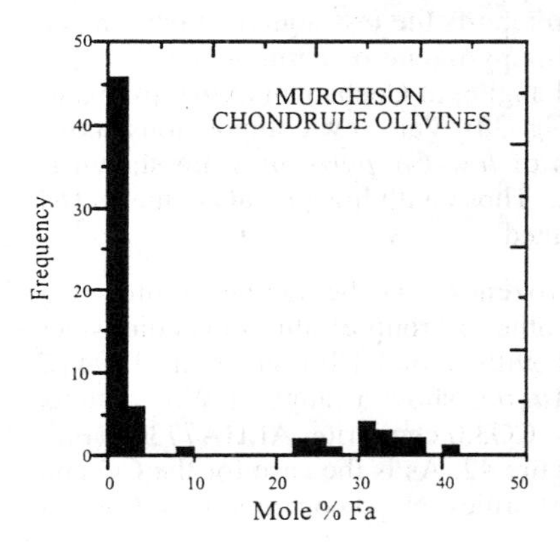

Figure 39. Histogram showing the distribution of olivine compositions in chondrules in the Murchison CM chondrite (Brearley 1995).

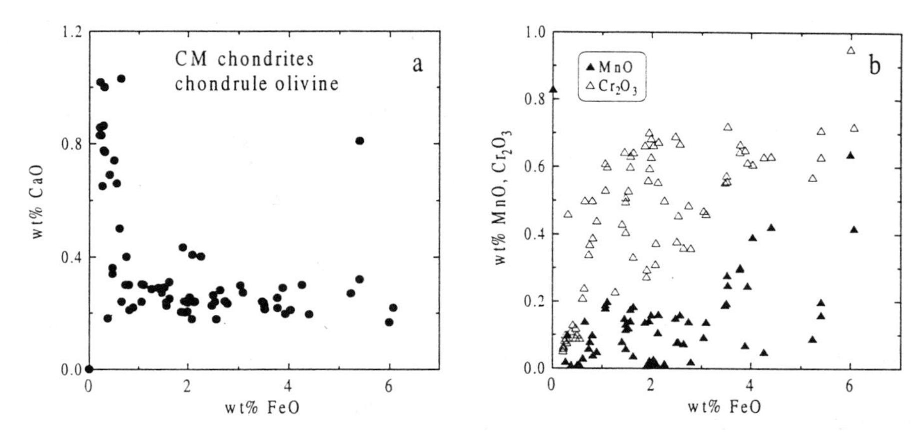

Figure 40. Minor element variation plots for olivine from chondrules in CM chondrites. Data sources: Richardson and McSween (1978), Olsen (1983), Rubin and Wasson (1986), Steele (1986), Simon et al. (1994), Brearley (1995).

Forsteritic olivines from type I chondrules show only limited zoning from core to rim, while more fayalitic olivine grains from FeO-rich chondrules are quite strongly zoned, e.g. from 20-32 or 40-55 mol % Fa (Desnoyers 1980, Akai and Kanno 1986). Minor element contents of chondrule olivine grains show similar patterns to those of other chondrite groups (Fig. 40: see also Desnoyers 1980). CaO contents reach high values of up to 1 wt % in the most forsteritic compositions, and Al_2O_3 contents also range up to 1 wt %. Scott and Taylor (1983) show CaO increasing with FeO for FeO-rich chondrules in Murray. Cr_2O_3 and MnO contents are both positively correlated with FeO in the forsterites as well as in the type II series (Desnoyers 1980). Figure 40 includes one olivine composition with an FeO content of 0.0 wt %, from a silica-bearing chondrule in Murchison (Olsen 1983). Representative olivine analyses are given in Table A3.1.

Rare dusty, relict olivine grains have been described in Murchison (Kracher et al. 1984, Jones and Danielson 1997). Relict grains of forsterite in the cores of FeO-rich olivine grains in type II chondrules have not been described specifically. However, they are clearly present in published BSE images (e.g. Fig. 2d of Metzler et al. 1992) and appear to be similar to those in CO chondrites.

Pyroxene in CM chondrite chondrules is usually low-Ca clinoenstatite, and augite occurs as rims on these grains. Noguchi (1989) also reports the less abundant occurrence of orthopyroxene. Müller et al. (1979) found a clinopyroxene of composition Wo_{20} to Wo_{30}, consisting of an intergrowth of pigeonite and augite, in Cochabamba. All pyroxene is very low in FeO, the mean being less than Fs_2 (Fig. 38). Very few compositions above Fs_{10} have been reported. Minor element contents of *low-Ca pyroxenes* are shown in Figure 41 and compositions are given in Table A3.2. Those with higher CaO contents (>1 wt %) may be orthoenstatite, but this is not documented.

Figure 38 shows two groups of Ca-rich pyroxenes, one being the augites that typically rim low-Ca pyroxenes (Wo_{30-45}), and the other a group of aluminous diopsides from disaggregated samples that are associated with spinel (Simon et al. 1994a). Analyses of both types are given in Table A3.3. *Augites* show a range of Wo contents similar to the range observed in chondrules from the CO3.0 chondrite, ALHA77307 (Fig. 29). Minor element variation plots are shown in Figure 42. As is the case for the CO and CV chondrite chondrules, and in contrast to OC chondrules, Na_2O contents of all Ca-rich

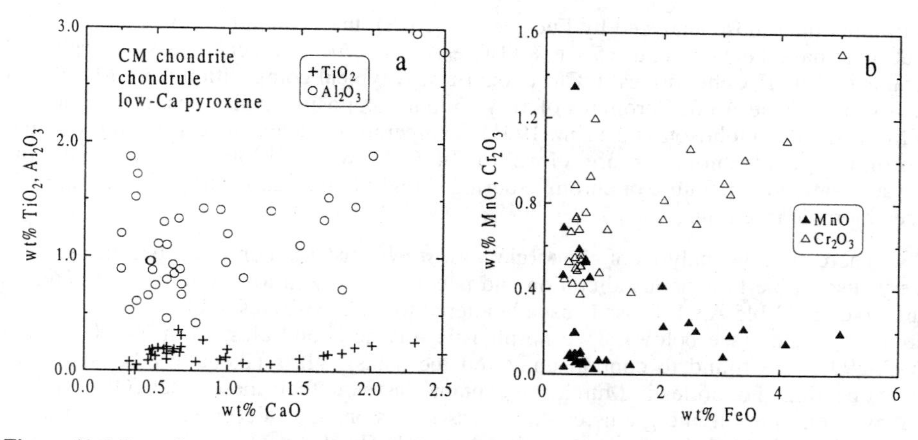

Figure 41. Minor element variation plots for low-Ca pyroxene from chondrules in CM chondrites. Data sources: Olsen (1983), Rubin and Wasson (1986), Noguchi (1989), Brearley (1995).

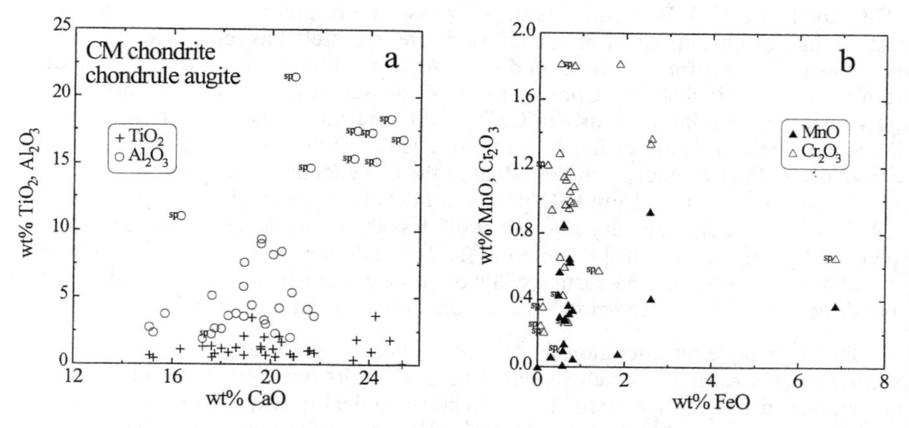

Figure 42. Minor element variation plots for augite from chondrules in CM chondrites. sp = augites that occur with spinel (Simon et al. 1994). Data sources: Olsen (1983), Noguchi (1989), Simon et al. (1994), Brearley (1995).

pyroxenes are extremely low. All Al_2O_3 contents above 10 wt % are for the diopsides associated with spinel. These diopsides are markedly enriched in CaO as well as Al_2O_3, and depleted in MnO (which is below detection limits in most analyses) compared with the augites. Some of them occur in chondrules from Murchison in which the primary mineralogies are forsteritic olivine, aluminous diopside, and spinel, with phyllosilicate replacing mesostases.

Spinel is present as inclusions in forsteritic olivine, sometimes associated with glass. These spinels are Al- and Mg-rich and commonly have very low FeO (<1 wt %) and Cr_2O_3 (<3 wt %) contents (Fuchs et al. 1973; Table A3.6). Spinel associated with more FeO-rich olivine is more FeO- and Cr_2O_3-rich (up to 11 wt % FeO and 25 wt % Cr_2O_3). Aluminous spinels typically contain ~0.2 wt % TiO_2. CaO contents are generally very low, but one analysis contains 1.6 wt % CaO (Fuchs et al. 1973). Spinels in disaggregated Murchison samples, described by Simon et al. (1994a: see above), are significantly more chromian than those analyzed by Fuchs et al. (1973): they contain up to 24 wt % Cr_2O_3 but most have FeO contents <5 wt % (Table A3.6). *Chromite* occurs as small euhedral grains in type IIA chondrules in CM chondrites. A typical composition from Murchison is given in Table A3.6. Chromites of very similar compositions also occur in a range of CM chondrites (Johnson and Prinz 1991). Temperatures determined from the olivine-chromite geothermometer are close to 1400°C, which probably represents the crystallization temperature of chromite during chondrule formation and indicates minimal metamorphic resetting.

There are few analyses of chondrule *mesostasis* for CM chondrites, because glass is very susceptible to aqueous alteration and is consequently in low abundance. Examples are given in Table A3.4. Glass is usually altered to phyllosilicates with a texture referred to as "spinach" (see below). Two porphyritic olivine chondrules, from Y-790123 and Y-75293, were found to contain unaltered mesostasis glass (Ikeda 1983). Also, two silica-bearing chondrules in Murchison contain glasses in their mesostases (Olsen 1983). Many chondrule olivine grains contain glass inclusions that are preserved because the olivine has shielded them from the altering fluids (Fuchs et al. 1973, Roedder 1981). Glass inclusions also occur in spinel grains (Fuchs et al. 1973, Simon et al. 1994a). Glass inclusions may contain Mg,Al-spinel, metal, troilite, and/or void bubbles (Fuchs et al. 1973, Roedder 1981). FeO contents of the glasses correspond to the FeO content of the host olivine. Compositions of glass inclusions are interpreted as representing the original mesostasis compositions of the chondrules. Analyses plotted in Figure 43 include data from these glass inclusions. Compositions of glasses in chondrules in CM chondrites show negative correlations between CaO, Al_2O_3 and SiO_2, similar to those observed in the Semarkona (LL3.0) ordinary chondrite (Fig. 18). The CM glass data show a bimodality in FeO vs. SiO_2, similar to that in the CO chondrites, and the result of the fact that the data are obtained from FeO-poor and FeO-rich PO chondrules. Na_2O contents of CM chondrule glasses are very low, generally less than 2 wt %. Glass compositions from spinel and olivine inclusions lie in similar fields and define similar trends to compositions of unaltered interstitial mesostases. Two glass compositions from a silica-bearing chondrule (Olsen 1983: see below) are notable outliers from the general trends.

The main aqueous alteration product of chondrules, "spinach", consists of a mixture of *phyllosilicates* and is given its name because of its deep green to yellowish brown appearance in transmitted light. In its most recognizable textural association, spinach commonly replaces chondrule mesostasis. Electron microprobe analyses of spinach have low totals, usually 80-90%, indicating that it consists of hydrated phases. The composition of spinach varies from point to point, but it is always markedly Ca,Al,Si-poor and Mg,Fe-rich relative to glass compositions, and Ca,Al-rich and (Mg+Fe)-poor relative to olivine. On an atomic ternary plot of Fe-Mg-Si, spinach analyses from a single meteorite lie within a narrow, nearly linear, field (Richardson and McSween 1978). Spinach compositions also vary from one chondrite to another. Progressive reaction of low-Ca pyroxene and olivine occur with increasing degrees of aqueous alteration. With progressive alteration, spinach changes from ferrous chlorite (and/or berthierine) with high Al_2O_3 contents and minor smectite components, to intermediate chlorite (and/or berthierine), and finally to magnesian serpentine containing less Al_2O_3 (Ikeda 1983). For high degrees of aqueous alteration, the compositions of chondrule spinach approach the compositions of the matrix of the chondrite (Kojima et al. 1984). In an HRTEM study of the spinach groundmass in chondrules of Y-791198, Akai and Kanno (1986) observed a phyllosilicate having a 14-Å interlayer spacing. This phyllosilicate is thought to be a 14-Å chlorite with an Fe/(Fe+Mg) ratio of 0.4-0.7 and ^{IV}Al of 1.8-3.0 on the basis of $O_{20}(OH_8)$. The presence of 7-Å platy phyllosilicates, probably serpentine, was also noted.

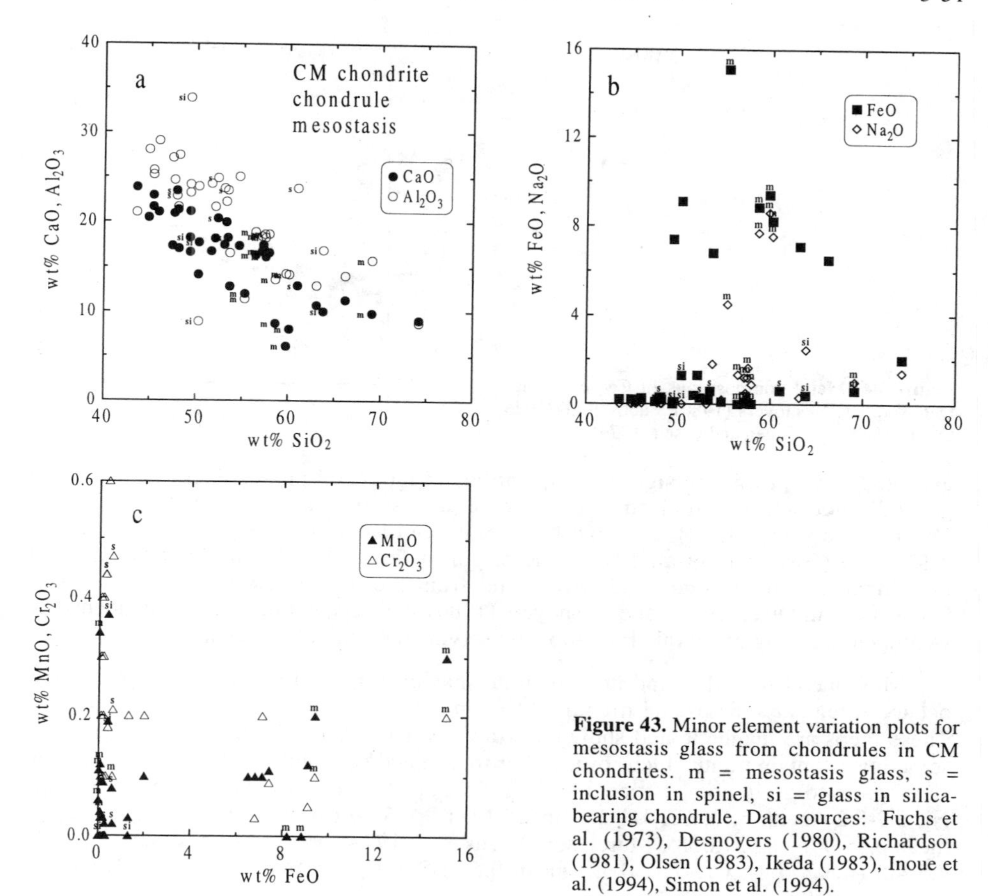

Figure 43. Minor element variation plots for mesostasis glass from chondrules in CM chondrites. m = mesostasis glass, s = inclusion in spinel, si = glass in silica-bearing chondrule. Data sources: Fuchs et al. (1973), Desnoyers (1980), Richardson (1981), Olsen (1983), Ikeda (1983), Inoue et al. (1994), Simon et al. (1994).

Alteration of mesostasis glass to spinach requires considerable mass transport. Fuchs et al. (1973) and Olsen and Grossman (1978) suggested that spinach was formed before accretion, as a product of late-stage, low-temperature gas phase reactions between chondrules and the cooling primordial nebula. Other authors argue that alteration occurred after accretion, and that metal and troilite within chondrules and the matrix of the meteorite are reservoirs for mass exchange (Richardson and McSween 1978, Richardson 1981).

Metal and sulfide minerals. Metal in CM chondrites is rare, but small rounded beads are often observed within forsterite grains. Rare isolated metal grains are also found in matrix. Metal enclosed in forsterite is commonly surrounded by a thin crescent of glass (Desnoyers 1980). Some metallic spherules are also observed in glass inclusions in olivine (Fuchs et al. 1973, Roedder 1981). Metal in CM chondrites is probably mostly kamacite with Ni contents of 4-13 wt % (Fuchs et al. 1973, Zanda et al. 1994). Co contents (0.2-0.7 wt %) are very clearly correlated with Ni (Fig. 44). Metal also has a very unusual composition compared with other chondrite groups, being rich in Cr (up to 1 wt %), and P (up to 3 wt %). High enrichments of Cr and P would normally be expected to indicate highly reducing conditions. However, under the reducing conditions necessary for such concentrations of these elements to occur, Mn and Si would also be

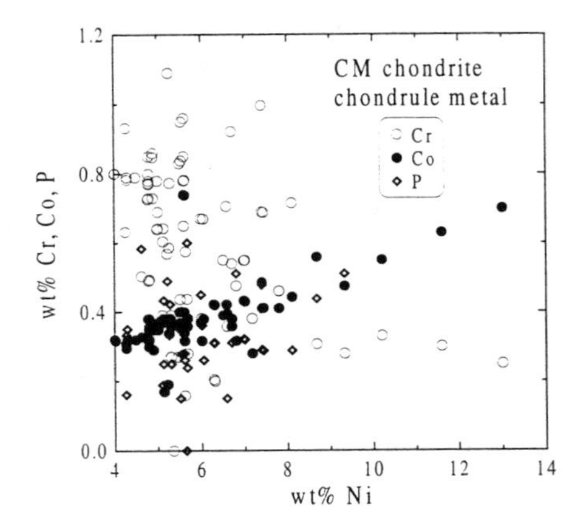

Figure 44. Metal compositions in CM chondrites. Data sources: Desnoyers (1980), Fuchs et al. (1973), Olsen (1983), Grossman and Olsen (1974).

expected to be present in significant quantities in the metal phase. These elements are generally not detected in electron microprobe analyses (Fuchs et al. 1973, Olsen et al. 1973a, Zanda et al. 1994), although the average Si content of one large metal grain is 0.12 wt % (Grossman et al. 1979). The origin of the Cr and P enrichments has been interpreted as direct condensation of metal from the nebula gas (Olsen et al. 1973, Grossman and Olsen 1974), and Desnoyers (1980) suggests that the original condensates were melted during chondrule formation, preserving their initial compositions.

Most metal in CM chondrite chondrules has been altered to an Fe-S-O phase that occurs in regular spheres and irregular blebs in the interstitial phyllosilicate mesostasis. These blebs are magnetic and show no x-ray powder pattern (Fuchs et al. 1973). They sometimes contain minute blebs of metal or troilite and pentlandite. Microprobe analyses show grain-to-grain compositional variabilities, given as (wt %): Fe, 37-45 (mean 42); Ni, 5.9-8.2 (mean 6.1); Cr, 1.4-2.5 (mean 1.6); S, 16-19 (mean 17.6); C, mean 0.2; P, mean 0.3; O, mean 31.0, for Murchison (Fuchs et al. 1973), and Fe, 40-57 (mean 44); Ni, 7.5-4.5 (mean 5.8); Cr, 0.81-0.42 (mean 0.55); S, 18.1-3.5 (mean 14.5) for Niger (I) (Desnoyers 1980).

Silica-bearing chondrules. Two silica-bearing chondrules were observed in Murchison by Olsen (1983). The first is a porphyritic pyroxene chondrule which contains rounded grains of unaltered metal, and has a mesostasis consisting of glass plus microlites of FeO-free augite. Numerous round to elliptical pods of essentially pure silica, 5-15 μm in longest dimension, are observed embedded in the glass in the outer region of the chondrule. The second chondrule contains two large, rounded crystals of FeO-free enstatite and minor metal, but mostly consists of a fine-scale intergrowth of augite and plagioclase (An_{89-93}) with interstitial lenses of glass of three distinct compositions. One consists essentially of SiO_2, Al_2O_3 and CaO, one has variable amounts of Ti, Mg, and Fe, and the third is almost pure silica with minor Al_2O_3 (1.7 wt %) and a trace of TiO_2 (0.05 wt %). The former two compositions are included in Figure 43. An unusual Ti,Cr oxide is associated with an augite crystal, but no analysis is given.

CV chondrites: Primary mineralogy of chondrules

The CV3 chondrites are divided into oxidized and reduced subgroups, based on the relative abundances of metal and magnetite, and Ni contents of sulfide minerals (McSween 1977c). The proportions of matrix and chondrules also differ between these groups. Here we discuss the silicate mineralogy of chondrules in the entire CV group without making a distinction between the reduced and oxidized subgroups. This is mostly because few analyses of chondrules in CV chondrites have been reported, and the majority of these are for the extensively studied chondrite, Allende. Allende is a member of the oxidized subgroup, but it and Mokoia differ from some other members of this subgroup (e.g. Kaba, Bali, and Grosnaja) in having a lower matrix / chondrule ratio (0.7 vs. 1.2) and a higher abundance of opaque mineral bearing chondrules. The matrix/chondrule ratio in the reduced subgroup (Arch, Efremovka, Leoville, and Vigarano) is similar to that in Allende (0.6: McSween 1977c). CV chondrites are generally of low petrologic subtypes. Guimon et al. (1995) assigned petrologic subtypes to the CV chondrites, based on TL data and other petrologic parameters. All are types 3.3 and lower, and Allende is assigned subtype 3.2. A discussion of the degree of metamorphism experienced by members of the CV group is summarized by Krot et al. (1995). The CV3.0 chondrites identified by Guimon et al. (1995) include ALHA81003, Arch, Axtell, Bali, Kaba, and Leoville, and there are very few data for chondrules in these least equilibrated CV chondrites. Thus, much of the data discussed below may not represent the least equilibrated chondrule data for the CV group. Evidence for metasomatism and aqueous alteration of chondrules in the CV chondrites has been discussed extensively recently (Krot et al. 1995), and is described in the following section.

Chondrules in CV chondrites are dominated by type I, porphyritic, olivine- and pyroxene-rich (POP) textural types (McSween 1977c, Simon and Haggerty 1979, 1980; Scott and Taylor 1983, Rubin and Wasson 1987). Both type IA and type IIA PO chondrules are observed, but type I chondrules are by far the more abundant. BO chondrules are low in abundance and RP chondrules are extremely rare.

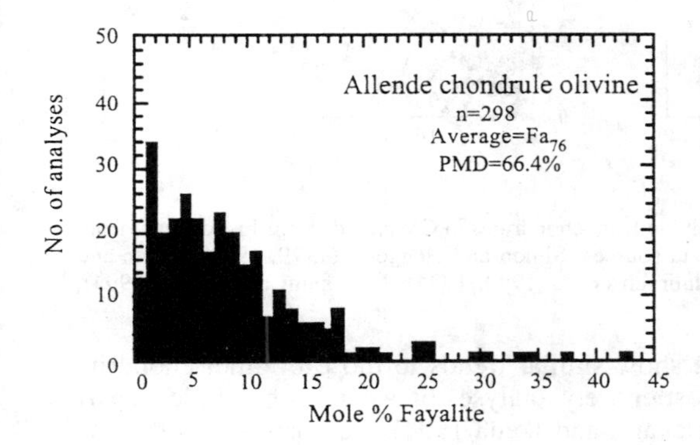

Figure 45. Histogram showing the distribution of olivine compositions in chondrules in the CV chondrite, Allende (Simon et al. 1995).

Silicate and oxide minerals. In most chondrules, *olivine* is forsteritic: histograms of Fa contents of chondrule olivines from several different chondrites show that the majority of olivine compositions are below Fa_{10}, with a tail of compositions spreading to Fa_{45} (Simon et al. 1995, Kracher et al. 1985, Fitzgerald and Jaques 1982, Murakami and Ikeda 1994, Cohen et al. 1983; Fig. 45). Mean chondrule olivine compositions for Allende and Axtell are Fa_8 and Fa_7, respectively (Simon et al. 1995), whereas the means for Leoville and Tibooburra chondrules are lower, around Fa_4 (Kracher et al. 1985, Fitzgerald and Jaques 1982). In Y-86751, there are two populations of chondrules. In the first, chondrules have fine-grained rims, their forsterite grains are not zoned, and all olivine compositions are below Fa_{10}. The second population consists of chondrules without fine-grained rims, in which olivine grains are strongly zoned and the distribution of olivine compositions is much more ferroan than that in Allende (Murakami and Ikeda 1994).

Olivine compositions in BO chondrules in Allende are concentrated at about Fa_4, with a few compositions up to Fa_{18} (Simon and Haggerty 1979, Lumpkin 1980). Type IIA chondrules in Vigarano have mean olivine compositions in the range Fa_{35-45} (Scott and Taylor 1983). Rare, dusty relict olivine has been described in an FeO-poor chondrule from Vigarano (Kracher et al. 1984).

Olivine in type I chondrules in Vigarano and Mokoia shows very limited zoning in FeO, whereas olivine in type IIA chondrules is more strongly zoned (Scott and Taylor 1983, Cohen et al. 1983). In Allende, olivine in type I chondrules shows complicated zoning. Some zoning is relatively strong (e.g. Fa_{3-20}, core to edge of grains), and appears to be diffusion controlled (e.g. Ikeda and Kimura 1995). This is similar to the zoning observed in low petrologic subtype ordinary and CO chondrites. A second type of zoning consists of Fa-rich rims on forsterite grains with steep compositional gradients between the two types of olivine, the origin of which is controversial and is discussed in more detail below.

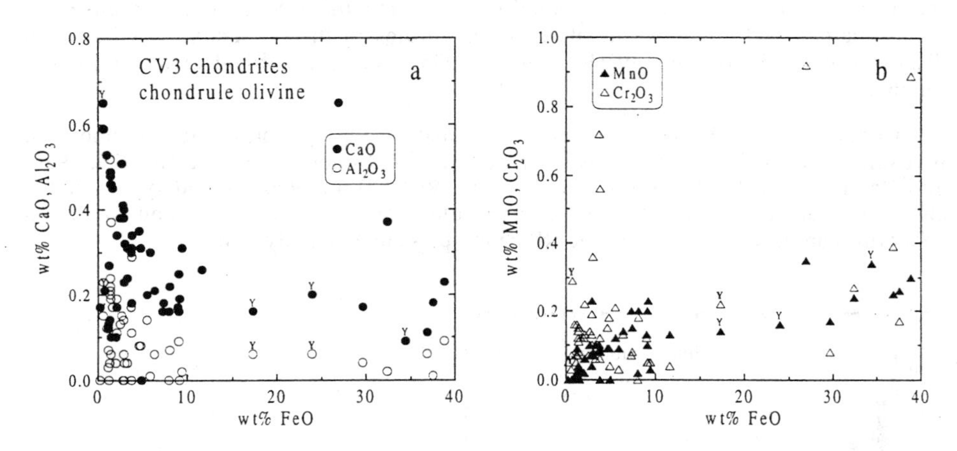

Figure 46. Minor element variation plots for olivine from chondrules in CV chondrites. Most data are from Allende; data from Y-86751 are labeled Y. Data sources: Simon and Haggerty (1979, 1980), Rubin and Wasson (1987), Peck and Wood (1987), Weinbruch et al. (1990, 1994), Murakami and Ikeda (1994), Kimura and Ikeda (1995).

Minor element contents of olivine show similar trends to those in other chondrite groups. The plots in Figure 46 are almost entirely analyses of Allende chondrules, apart from five analyses from Y-86751 (Murakami and Ikeda 1994). Compared with the CO chondrites, distributions of CaO vs. FeO and Cr_2O_3 vs. MnO are more similar to those of subtype 3.1-3.7 CO chondrites (Fig. 32) than to the distribution in the CO3.0 chondrite, ALHA77307 (Fig. 30). This suggests that olivine in type I chondrules in Allende also shows evidence for metamorphism. The data for chondrules in Y-86751 do not appear to show this effect. However, this chondrite contains many type I chondrules (the unrimmed population described above) in which olivine is strongly zoned. The olivine analysis with 27% FeO and 0.65% CaO is from a POP chondrule in Allende (Rubin and Wasson 1987). In Mokoia, CaO contents show a similar pattern to those in Allende (Cohen et al. 1983). Representative olivine analyses are given in Table A3.1.

Pyroxene assemblages in Allende are quite diverse compared with other chondrite groups (Noguchi 1989). The typical association of twinned clinoenstatite overgrown with augite is common. Clinoenstatite in Allende is commonly altered to ferrous olivine, as

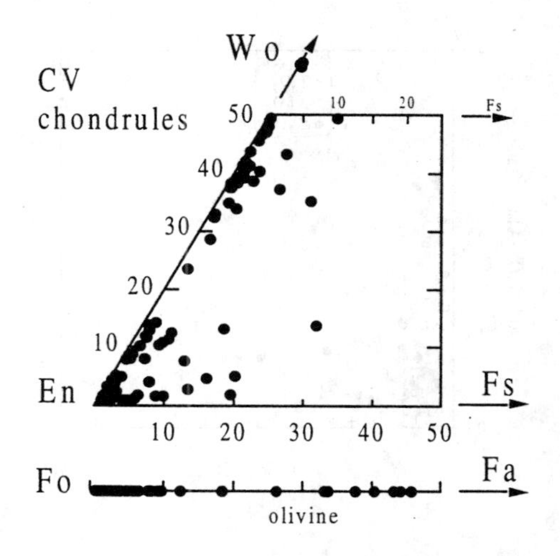

Figure 47. Pyroxene and olivine compositions in chondrules from CV chondrites. Data sources: Simon and Haggerty (1979, 1980), Lumpkin (1980), Cohen et al. (1983), Rubin and Wasson (1987), Peck and Wood (1987), Noguchi (1989), Weinbruch et al. (1990, 1994), Murakami and Ikeda (1994), Kimura and Ikeda (1995).

discussed below. However, orthopyroxene and pigeonite are also common, and these may occur as an intermediate layer between clinoenstatite and its augite overgrowth. In addition, pigeonite phenocrysts, with or without augite overgrowths, were noted by Noguchi (1989), and the dominant pyroxene in three out of 20 chondrules studied by Rubin and Wasson (1987) was pigeonite. In one POP chondrule studied by Rubin and Wasson (1987), the only pyroxene is diopside. As was the case for olivine, the pyroxene data used in Figures 47, 48, and 49 are mostly for Allende: few analyses from other CV chondrites are given in the literature.

Figure 47 shows the range of compositions of Allende chondrule pyroxenes. All the pyroxenes analyzed by Noguchi (1989) were in low-FeO chondrules, with Fs < 10. However, more FeO-rich pyroxenes have been observed in Allende in other studies (Rubin and Wasson 1987, Simon and Haggerty 1980, Kimura and Ikeda 1995, Lumpkin 1980), as well as in Mokoia (Cohen et al. 1983) and Y-86751 (Murakami and Ikeda 1994). Three compositions of $Wo_{58}Fs_{<1}$ were reported by Noguchi. These fassaitic pyroxenes contain >13 wt % Al_2O_3 and up to 3.4 wt % TiO_2, and occur in an olivine-clinopyroxene chondrule.

Minor element compositions of *low-Ca pyroxenes* ($Wo_{<5}$) are plotted in Figure 48. Figure 48b shows trends similar to other chondrite groups, with a wide range and high values of CaO and Al_2O_3 contents for the most enstatitic pyroxenes. *Ca-rich pyroxene* compositions are shown in Figure 49. Al_2O_3 contents of the most Wo-rich pyroxenes show a bimodal distribution. Most are below 6 wt % Al_2O_3, but a second group has Al_2O_3>10 wt %. This group includes the fassaitic compositions from a chondrule in Allende described above (~25 wt % CaO; Wo ~ 58), but the others, all from Allende (Noguchi 1989, Kimura and Ikeda 1995), are not described specifically. Augites that coexist with pigeonite have high TiO_2/Al_2O_3 ratios (Noguchi 1989). Kimura and Ikeda (1995) report even higher ranges of Al_2O_3 (up to 22.5 wt %) and MnO (up to 1.4 wt %) for augite in Allende chondrules. Na_2O contents of augites are consistently low (<0.1 wt %), being below or close to microprobe detection limits in most analyses. This is similar to other carbonaceous chondrite chondrules, and in marked contrast to ordinary chondrites. Representative pyroxene compositions are given in Tables A3.2 and A3.3.

Pyroxenes in type I chondrules from Allende have been studied by TEM (Müller 1991a, Müller et al. 1995). Clinoenstatite in the same chondrules is twinned on (100) and

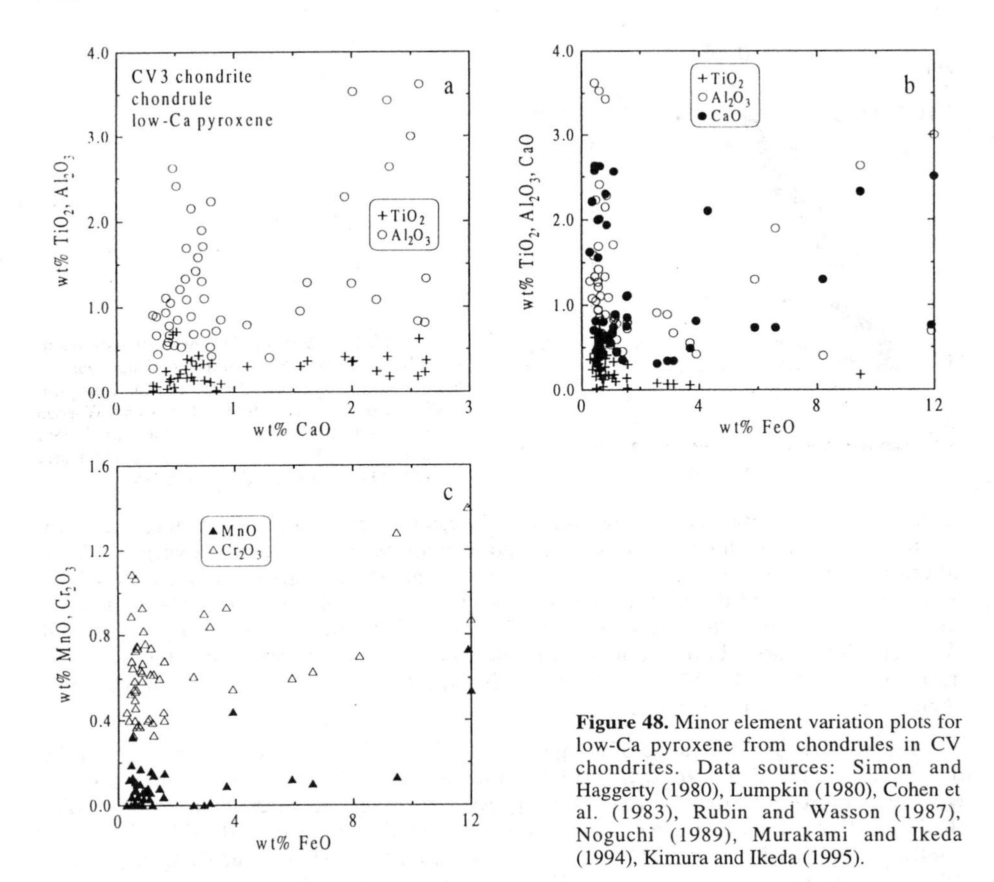

Figure 48. Minor element variation plots for low-Ca pyroxene from chondrules in CV chondrites. Data sources: Simon and Haggerty (1980), Lumpkin (1980), Cohen et al. (1983), Rubin and Wasson (1987), Noguchi (1989), Murakami and Ikeda (1994), Kimura and Ikeda (1995).

the widths of twin lamellae vary between 0.15 μm and unit cell scale. No orthorhombic enstatite intergrowths were detected. Orthopyroxene with higher CaO contents was observed in two chondrules. Pyroxenes with bulk compositions of $<Fs_3$, $Wo_{11\text{-}33}$ consist of lamellar intergrowths of pigeonite and diopside on (001), with the lamellae 5-30 nm in width. An average wavelength of between 25 and 33 nm corresponds to cooling rates on the order of 2-10°C/h (Weinbruch and Müller 1995). Lamellae of diopside on (100) are rare. Pigeonite lamellae are characterized by antiphase domains with the displacement vector 1/2(**a**+**b**). Regions of homogeneous pigeonite and diopside were observed in some regions of the grains with intergrowths. These pigeonites also contain antiphase domains, and some display modulated structures that indicate incipient exsolution.

Chromite occurs in type IIA chondrules in Allende (Weinbruch et al. 1994), as well as in an FeO-rich BO chondrule (Lumpkin 1980). Compositions in these two settings are markedly different: Cr/(Cr+Al) ratios are lower in the BO than the PO chondrules (0.68 and 0.78, respectively). Also, TiO_2 and MgO contents are higher in the BO chondrule than in the PO chondrules (Table A3.6). Temperatures derived from olivine-chromite pairs in type IIA chondrules are higher for ALH84037 and Leoville than for Vigarano and Allende, suggesting mild reheating in Vigarano and Allende (Johnson and Prinz 1991). The mean Cr/(Cr+Al) ratio of chromites from CV chondrites in the Johnson and

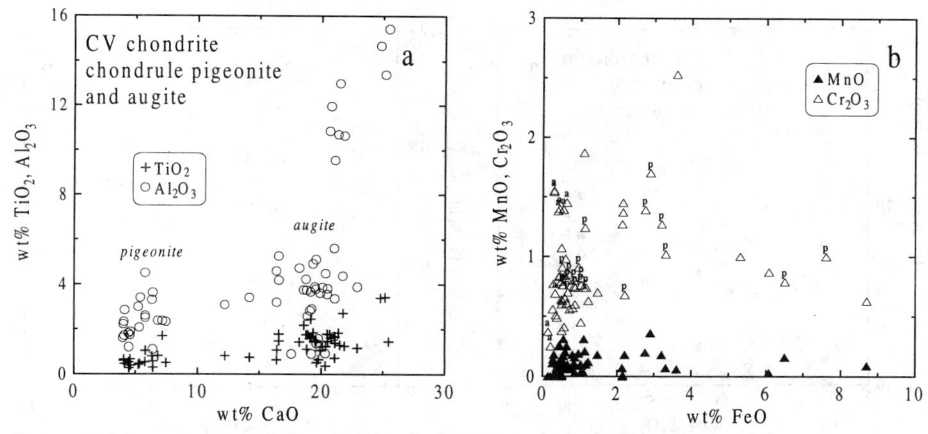

Figure 49. Minor element variation plots for pigeonite and augite from chondrules in CV chondrites. In (b), p = pigeonite and a = aluminous augite (Al_2O_3 >8 wt %). Data sources: Lumpkin (1980), Cohen et al. (1983), Rubin and Wasson (1987), Noguchi (1989), Murakami and Ikeda (1994), Kimura and Ikeda (1995).

Prinz study is 0.79. Chromian *spinel* (Cr/(Cr+Al) = 0.3-0.5) occurs in two BO chondrules in Allende, included in olivine or interstitial plagioclase (Ikeda and Kimura 1995). Occurrences of primary Mg,Al-spinel have been noted in Allende and Y-86751 (Ikeda and Kimura 1995, Murakami and Ikeda 1994; Table A3.6).

Plagioclase has been observed in several Allende chondrules. Noguchi (1989) reported plagioclase that appears to be primary, with compositions >An_{80}, in some orthopyroxene and/or pigeonite-bearing chondrules. Simon and Haggerty (1980) observed plagioclase with average compositions An_{84} in two type I chondrules. In Allende and Y-86751, small anhedral grains of plagioclase ($An_{80\text{-}90}$) that are partly replaced by feldspathoids (see below) occur in the mesostases of many chondrules (Murakami and Ikeda 1994, Ikeda and Kimura 1995). Anorthitic plagioclase in Allende (An_{91}) has maximum concentrations of FeO and MgO of 0.4 and 5.0 wt % (Miúra and Tomisaka 1984), respectively. A representative plagioclase analysis is given in Table A3.5. Plagioclase was a minor constituent of two type I chondrules studied by TEM (Müller 1991a, Müller et al. 1995). In one chondrule, plagioclase is An_{90} and contains numerous magnetite crystallites, mostly 10-20 nm in size, at least some of which are randomly oriented. In the other chondrule, only one plagioclase grain was observed, ~An_{80}. Small (10-30 nm) type **b** antiphase domains were observed in plagioclase in both chondrules. Cooling rates were estimated to be 3-30°C/h (Weinbruch and Müller 1995).

Chondrule *mesostasis* compositions in CV chondrites are sparse in the literature. This is partly because many chondrules do not contain glassy mesostases, but consist of holocrystalline assemblages, predominantly of plagioclase and Ca-rich pyroxene, that are commonly altered (e.g. Ikeda and Kimura 1995). Glass compositions that are reported are shown in Figure 50 (see also Table A3.4). Most elements are present in concentrations very similar to those in CO chondrites (Fig. 36), but the trends between different elements observed in the other chondrite groups are not evident in this limited data set. This may partly be because five of the points are average mesostasis compositions from different chondrule types described by Simon and Haggerty (1980), and the variation in composition within each chondrule type is obscured. One chondrule with a high Na_2O content (14.5 wt %) in the mesostasis is from one of the volatile-rich chondrules in Allende described by Isobe et al. (1986) which has augite and olivine phenocrysts. Simon and Haggerty (1980) also describe two glass-rich chondrules, one that is 100% glassy,

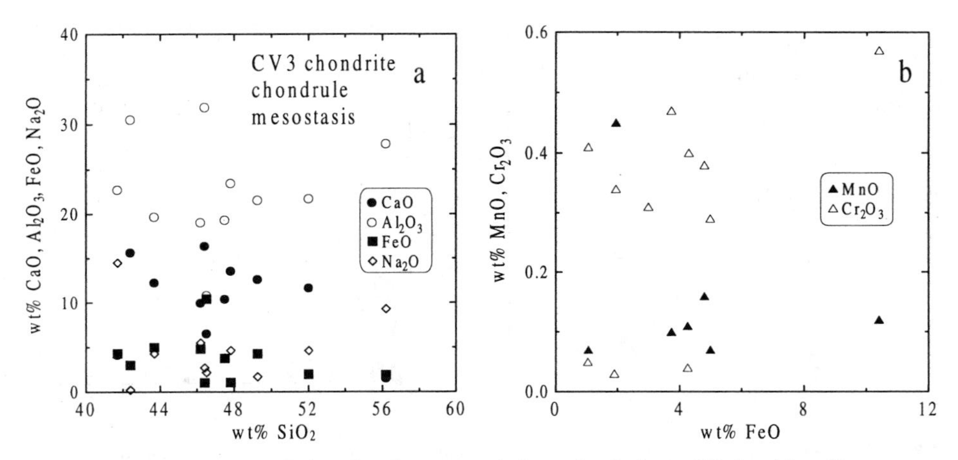

Figure 50. Minor element variation plots for mesostasis from chondrules in CV chondrites. Data sources: Simon and Haggerty (1980), Kracher et al. (1985), Isobe et al. (1986), Murakami and Ikeda (1994), Kimura and Ikeda (1995).

and one with 67% glassy mesostasis and 33% fine-grained, anhedral, untwinned pyroxene. The glass in one of these chondrules, presumably the pyroxene-bearing one, is FeO-rich (26 wt %) and the pyroxene is $Fs_{14}Wo_4$. Several of the chondrule mesostases in Allende analyzed by Rubin and Wasson (1987) have compositions close to stoichiometric plagioclase (Ab_5Or_1), with an average composition (wt %) of 43% SiO_2, 38% Al_2O_3 19% CaO, 0.5% Na_2O and <0.04% K_2O.

Three small grains of *corundum*, 1-1.5 μm in size, were observed in a type I chondrule in Allende (Müller et al. 1995). They occur in association with an amorphous material rich in Mg, Al, Si, and Fe. The corundum contains a high density of dislocations and polysynthetic twin lamellae parallel to (0001) which are interpreted as being derived by shock.

Metal, sulfide and magnetite assemblages. Type I chondrules in CV chondrites commonly contain opaque mineral assemblages that were described as opaque nodules by Haggerty and McMahon (1979). Some chondrules contain no opaque minerals, but most are riddled with subrounded grains of metal, magnetite and/or sulfide minerals, some of which may be quite porous. Grains of opaque minerals sometimes armor chondrules or are concentrated inside the chondrule boundaries (McSween 1977c). Opaque minerals in CV chondrites are described in detail in a section below.

Plagioclase-olivine inclusions (POI). A group of objects in CV chondrites termed plagioclase-olivine inclusions (POI) has been described by Sheng et al. (1991). These objects have similar characteristics to chondrules, with bulk compositions that are transitional between CAIs and ferromagnesian chondrules, and are related to Al-rich chondrules in ordinary chondrites. Mineral assemblages are dominated by plagioclase and olivine with variable amounts of spinel, fassaite, enstatite, and diopside. Textures are commonly sub-ophitic to intersertal with plagioclase and olivine phenocrysts in a fine-grained matrix. Porphyritic, granular, R and BO textures are also observed.

Plagioclase compositions range from An_{82} to An_{98}, with An_{95} the most frequently observed composition. Plagioclase contains small amounts of MgO (0.3-0.9 wt %) and FeO (0-0.4 wt %), but no K_2O. There is a positive correlation between Mg and Na contents, similar to that found in Type B CAIs. *Olivine* compositions are variable between different POIs and range from Fa_1 to Fa_{37}. Minor zoning is observed in individual grains. Olivine clusters near the edges of the POIs are more FeO-rich and more strongly zoned. TiO_2 contents in some olivines are high, up to 0.3 wt %, but high TiO_2 contents are not correlated with high Al_2O_3 or CaO contents. CaO varies up to 0.4 wt %. *Spinels* are Mg,Al-rich, generally containing <4 wt % FeO. Individual spinel grains are zoned in FeO. Cr_2O_3 contents vary between 0.2 and 10 wt %, and one POI contains spinel with 20-30 wt % Cr_2O_3. TiO_2 contents range up to 3 wt % and V_2O_3 up to 0.7 wt %. Although much of the spinel appears to have crystallized from the molten POIs, some spinel grains are relict as indicated by resorbed textures in some POIs, cores of spinel grains with compositions that differ from their overgrowths, and isotopic differences between spinel grains within individual POIs.

Several different *pyroxenes* occur in POIs, including fassaite, diopside, Al-rich enstatite and Fe-poor pigeonite. All pyroxenes are FeO-poor (<1 wt % FeO). Most fassaites are zoned with respect to Ti and Al. Two groups with different Al_2O_3/TiO_2 ratios are present: in POIs that contain Ti-rich oxide minerals, the ratio is close to 1:1, whereas fassaites with a higher Al_2O_3/TiO_2 ratio coexist with Al-rich enstatite, pigeonite, and diopside. TiO_2 contents of fassaites range up to 13 wt %, Al_2O_3 up to 15 wt %, and MnO up to 1 wt %. Approximately 60-80% of the total Ti is inferred to be trivalent, indicating a very low fo_2 of $\sim 10^{-17}$ at 1300°C. Enstatites are also Al,Ti-rich, e.g. one enstatite contains 8 wt % Al_2O_3 and 2 wt % TiO_2. Appreciable V_2O_3, up to 0.3 wt %, is present in some low-Ca pyroxenes as well as fassaite.

Many POIs contain glassy or fine-grained *mesostases*. Mesostases have high Na_2O contents, up to 10 wt %, high SiO_2 and low CaO contents. Several POIs contain accessory armalcolite, accompanied by the Ti-rich oxides rutile, ilmenite, perovskite, and zirconolite ($CaZrTi_2O_7$). REE contents of zirconolite are high, e.g. 4.4 wt % Y_2O_3 and 0.9 wt % CeO in one grain. Sapphirine, which occurs in some POIs in Allende, has high TiO_2 and Cr_2O_3 contents (1-2 wt %) compared with terrestrial examples. Secondary phases containing Na and Fe, such as nepheline, sodalite and hedenbergite, are quite restricted in POIs. They occur only near the edges or along cracks and interstitial regions within the inclusions.

CV chondrites: Secondary mineralogy of chondrules

Chondrules in oxidized CV chondrites show evidence for a variety of secondary processes. The environments in which most of these processes occur are hotly debated (e.g. Krot et al. 1995). Here we summarize the main mineralogical effects of a variety of alteration types, as observed in chondrules. The effects of thermal metamorphism have not been separated clearly from general chondrule descriptions, and are included in the preceding section. The reduced CV subgroup appears to have largely escaped most secondary alteration, although chondrules in Vigarano and Efremovka have recently been shown to contain evidence for metasomatism (Kimura and Ikeda 1997).

Metasomatism. The mesostases of most chondrules in Allende and Y-86751 have experienced various degrees of iron-alkali-halogen metasomatic alteration that resulted in replacement of amorphous or partly crystalline mesostasis and plagioclase by nepheline, sodalite, and minor amounts of grossular, wollastonite, andradite, kirschsteinite, and hedenbergite (Murakami and Ikeda 1994, Ikeda and Kimura 1995, Kimura and Ikeda 1995). The minor Ca-rich phases only occur with sodalite. Mineral compositions are given by Kimura and Ikeda (1995). In Y-86751, merrillite is abundant in the mesostases of some chondrules (Murakami and Ikeda 1994). Low-Ca pyroxene crystallites and phenocrysts within altered mesostasis have been partly replaced by ferroan olivine. Secondary Ca-rich pyroxenes, which are significantly more ferroan than primary Ca-rich pyroxenes, occur at the rims of primary Ca-rich pyroxene, between enstatite and ferroan olivine replacing enstatite, and as overgrowths on enstatite along cracks in chondrules. Comparisons of the chemistries of glassy mesostases and mesostases of altered chondrules indicate that there must have been a net loss of Ca, while Na, K, Fe and Cl were gained, during metasomatic alteration. There is no correlation between the degree of alteration and chondrule textures, sizes, shapes, or primary chemistry (Kimura and Ikeda 1995), although in Y-86751, altered chondrules lack dark rims.

Metasomatism of chondrules has also been observed in several other oxidized CV3 chondrites, including ALH84028 and LEW86006 (Krot et al. 1995). In Mokoia, one type of PO chondrules, termed "recrystallized" chondrules, contains no mesostasis and abundant nepheline (Cohen et al. 1983). Ikeda and Kimura (1995) and Kimura and Ikeda (1995) suggested that the Fe-alkali-halogen metasomatism took place in the solar nebula by reactions with a hot Na- and Cl-rich vapor, whereas Krot et al. (1995) prefer an asteroidal setting for this process.

Alteration of low-Ca pyroxene. Low-Ca pyroxene in Allende chondrules is commonly altered to fayalitic olivine, Fa_{25-50} (Housley and Cirlin 1983, Ikeda and Kimura 1995). On the outer edge of chondrules, clinoenstatite makes sharp, continuous, but irregular contacts with Fe-rich olivine which frequently grades into aggregates of olivine blades indistinguishable from the matrix. Fe-rich olivine is also distributed throughout the clinoenstatite crystals along grain boundaries and internal fractures. The environment in which this alteration occurred has been suggested to be either reaction of enstatite with Fe metal on the parent body (Housley and Cirlin 1983), or reactions with the nebular gas (Palme and Fegley 1990, Ikeda and Kimura 1995).

Fayalitic olivine. In addition to the association with low-Ca pyroxene, fayalitic olivine occurs as rims, up to 15 μm in width, around type I chondrule phenocrysts and isolated forsterite grains, and along cracks and veins in these forsterites. Fayalite-rich haloes also occur around metallic Fe,Ni, magnetite and sulfide inclusions within forsterite (Hua et al. 1988). Fayalitic rims are abundant in Allende and ALHA81258, and also occur in Bali, Grosnaja, Mokoia, ALH84028, LEW86006, and Y-86751 (Peck and Wood 1987, Hua et al. 1988, Weinbruch et al. 1990, 1994; Murakami and Ikeda 1994, Krot et al. 1995). Those in Allende have been described in most detail. The FeO content of olivine in the rims is relatively constant across each rim, and has compositions that vary between Fa_{35} and Fa_{50} (Table A3.1). Compositions vary from chondrule to chondrule as well as within a chondrule, whereas the rims surrounding forsterites in the interiors of the chondrules are usually less ferrous than those at the periphery. The boundaries between fayalite rims and forsterite cores are abrupt, and occur over a distance of a few micrometers (Fig. 51). MnO contents in the fayalitic rims are around 0.2-0.3 wt %, and CaO contents are 0.1-0.5 wt %. Fayalitic rims contain high contents of Al_2O_3 and Cr_2O_3 that are mainly concentrated in submicron sized inclusions of various spinels along the forsterite-fayalite boundary, as well as in veins. The spinels are mostly hercynites and chromites, with some magnetite (Weinbruch et al. 1990, Müller 1991b, Müller et al. 1995). Some spinel grains contain significant V (V/Cr atomic ratio ~0.1). In one chondrule, chromites are oriented with respect to olivine (Müller 1991b), but in other chondrules, chromites are largely randomly oriented (Müller et al. 1995). The crystallographic orientation of olivine usually remains unchanged across fayalite-rich veins. Olivine is mostly free of dislocations, but one very FeO-rich grain (Fa_{38}) contains dislocations with a Burgers vector [001], in which the dislocation microstructure is dominated by long screw components parallel to [001] (Müller 1991b). A second grain, Fa_3, which contains a swarm of chromite crystallites, contains dislocations and stacking faults parallel to (010).

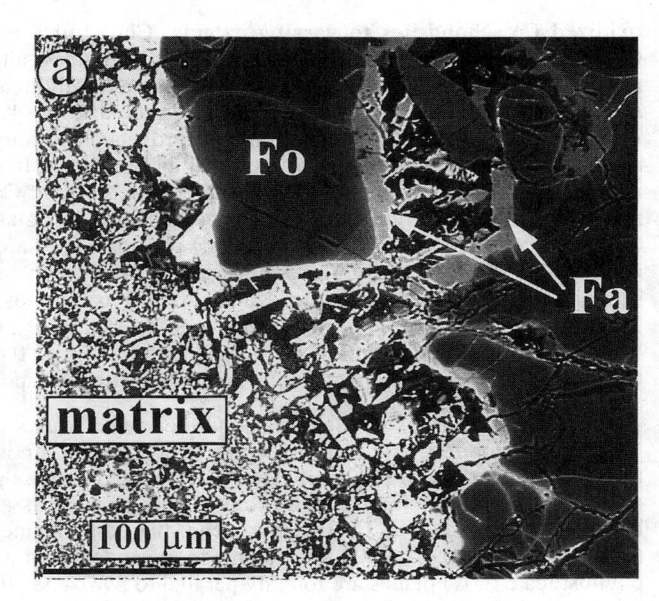

Figure 51. Fayalite-rich rims on forsteritic olivine in CV chondrites. (a) BSE image showing the edge of a type I chondrule in Allende, with fayalitic rims (Fa) surrounding forsterite (Fo) phenocrysts. (b; below) The zoning profile across a fayalite-rich rim on a forsterite grain in ALHA81258 (Krot et al. 1995).

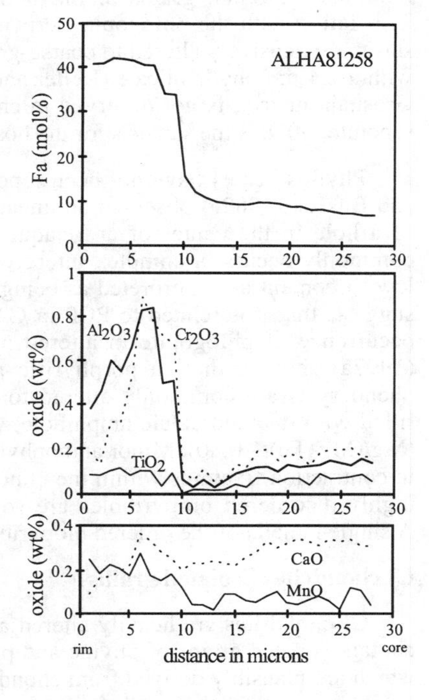

The origin of fayalite in these associations is controversial. Peck and Wood (1987), Hua et al. (1988) and Weinbruch et al. (1990, 1994) support models in which the fayalitic olivine condensed from an oxidized solar nebular gas. However, Krot et al. (1997b) proposed that they are products of a two-stage process, in which aqueous alteration of forsteritic olivine occurs first, producing intermediate phyllosilicates, and this stage is followed by dehydration to fayalitic olivine during thermal metamorphism.

Pure fayalite. In Kaba and Mokoia, pure fayalite grains, up to 100 μm in size, occur in association with magnetite, troilite and pentlandite in chondrules of different textural types (Hua and Buseck 1995). They also occur in matrix, forsterite-enstatite aggregates, and rims around CAI, chondrules and aggregates. These grains are clearly secondary and are suggested to form via reactions of SiO gas with magnetite and sulfide in an oxidized region of the solar nebula, at temperatures <900°C. Large fayalite grains (>30 μm) tend to have higher fayalite contents (>Fa_{97}) than smaller grains (Fa_{88-97}). Fayalite contains up to 1.0 wt % MnO. NiO contents are generally very low (<0.05 wt %).

Aqueous alteration. Aqueous alteration has affected chondrules in several of the oxidized CV chondrites to varying extents. Chondrules in Bali, Kaba, Mokoia and Grosnaja contain appreciable amounts of phyllosilicates, although the mineral species and proportions vary significantly even within an individual meteorite (Tomeoka and Buseck 1990, Keller and Buseck 1990a, Keller et al. 1994, Keller and McKay 1993). In Bali, the degree of aqueous alteration is heterogeneous throughout the meteorite (Keller et al. 1994), from chondrules that are fresh and unaltered to chondrules that are largely replaced by Mg-rich saponite. Framboidal magnetites and Ca-phosphates are accessory minerals. In Kaba, phyllosilicates replace enstatite and glass in the outer portions of chondrules (Keller and Buseck 1990a). The phyllosilicates are mostly coarse-grained Fe-bearing saponite and show a crystallographic orientation relationship such that c^* of saponite parallels a^* of enstatite. Kaba saponite is compositionally similar to that in Mokoia, with a mean composition (wt %): SiO_2, 45.2; TiO_2, 0.11; Al_2O_3, 5.0; FeO, 6.4; MgO, 22.1; CaO, 0.28; Na_2O, 1.6; K_2O, 0.30; Ni, 0.51; S, 0.47 [total: 82.0]. Submicron blebs of sulfides, possibly pentlandite, are mixed with the saponite.

The phyllosilicates in Mokoia chondrules are mixtures of Fe-bearing saponite and Na-rich phlogopite in coherent intergrowth with small amounts of Al-rich serpentine, termed high-aluminum phyllosilicate (HAP) by Cohen et al. (1983). Saponite and the phlogopite-serpentine intergrowths coexist with mesostasis glass on a submicron scale, between narrow laths of Ca-rich pyroxene that contain fine-scale exsolution lamellae (Tomeoka and Buseck 1990). Where phlogopite and serpentine replace this pyroxene, the phyllosilicate (001) planes are roughly parallel to pyroxene (100), suggesting a topotactic relationship. Small grains of magnetite, chromite, ilmenite and Fe-sulfides occur in association with the phlogopite-serpentine intergrowths. In Grosnaja, chondrule mesostasis is extensively altered to coarse-grained Na-saponite that is coherently interstratified with a 1.4-nm phyllosilicate (Keller and McKay 1993). Serpentine, which is observed in Grosnaja matrix, is not observed in chondrules. The Mg/(Mg+Fe) atomic ratio for the saponite, ~0.9, is the same as for the host chondrule olivines.

Phyllosilicates have not been reported in Allende chondrules. However, Tomeoka and Buseck (1982a) observed an unusual Fe,Ni,O-rich layered material along the edges of a hole in the center of an opaque nodule. This material is poorly crystalline and commonly occurs in complex intergrowths with small silicate grains. Its structure and low Si content are interpreted as being related to cronstedtite. Its chemical composition suggests that it is related to PCP in C2 chondrites. Kimura and Ikeda (1996) report the occurrence of phlogopite in altered mesostases of two Allende chondrules. Brearley (1997a) showed that in porphyritic pyroxene chondrules in Allende, clinoenstatite phenocrysts are commonly altered to hydrous phases. Veins within clinoenstatite are filled with talc and calcic amphibole, with a composition close to magnesiohornblende (Mg/(Mg+Fe) ~ 0.80). Minor anthophyllite is also present. Amphibole and talc also occur in contraction fractures within the clinoenstatite and as inclusions in olivine. Regions of highly disordered biopyriboles are sometimes present associated with grains of talc. A single lamella of the ordered biopyribole, jimthompsonite, was observed.

CI chondrites (isolated grains)

CI chondrites are heavily altered and do not contain chondrules. However, they do contain isolated grains of olivine and pyroxene in extremely low abundance (<1 vol %), which are plausibly derived from chondrule sources based on their mineral compositions, zoning characteristics, and the presence of melt inclusions (Leshin et al. 1997). We describe these grains here so that their similarities to chondrules in other carbonaceous chondrite groups are emphasized. Most of the analyses available are for the CI chondrite, Orgueil.

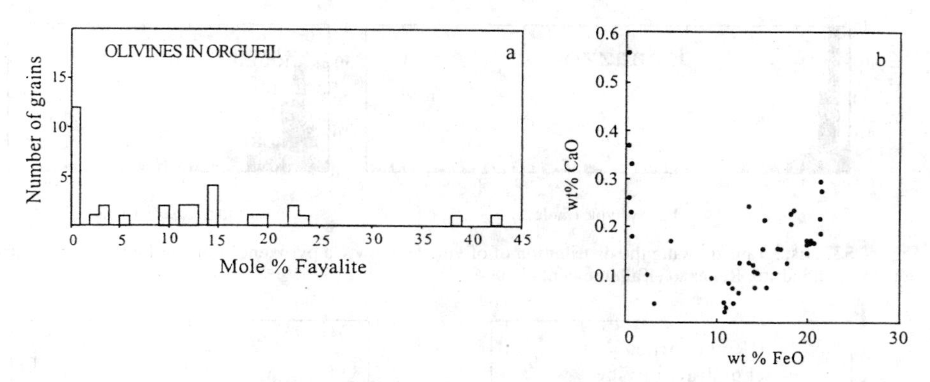

Figure 52. Olivine compositions in the Orgueil CI chondrite (Kerridge and Macdougall 1976). (a) Histogram showing the distribution of olivine compositions. (b) CaO vs. FeO contents.

Olivine grains in CI chondrites occur as fresh angular clasts up to 400 μm long, bounded by conchoidal fractures and poor to moderately developed cleavages (Reid et al. 1970, Kerridge and Macdougall 1976, Leshin et al. 1997). Several olivine grains from Orgueil contain 2-5 μm spherical blebs of metallic Fe,Ni, and some contain silicate melt inclusions with shrinkage bubbles (Leshin et al. 1997). Olivine shows a range in composition, up to Fa_{43}, although a high proportion of grains is forsteritic, $<Fa_2$ (Reid et al. 1970, Kerridge and Macdougall 1976; Fig. 52a; Table A3.1). Figure 52b illustrates the variation in CaO content with FeO content which shows a similar pattern to those in chondrule olivines from other carbonaceous chondrite groups as well as the most unequilibrated OC. MnO and Cr_2O_3 contents of olivine grains both vary in the range 0.1-0.6 wt % and these elements do not appear to be correlated with each other or with FeO (Leshin et al. 1997). Many olivine grains from Orgueil are zoned, with Fa contents increasing towards grain edges (Kerridge and Macdougall 1976, Leshin et al. 1997).

Low-Ca pyroxene grains in CI chondrites occur as angular, unaltered clasts and show parallel extinction, indicating that they are orthopyroxene (Reid et al. 1970, Kerridge and Macdougall 1976, Leshin et al. 1997). A polycrystalline aggregate of olivine and low-Ca pyroxene was separated from Orgueil (Leshin et al. 1997). Low-Ca pyroxene compositions range in FeO content from 1.4 to 15.5 wt %, and are Wo_{0-3}. *Ca-rich pyroxenes*, Wo_{41-46}, occur as angular fragments in Orgueil (Kerridge and Macdougall 1976, Leshin et al. 1997). Grains of composition Fs_{5-6} have significantly lower TiO_2 and Al_2O_3 contents than a grain that is significantly more FeO-rich (Fs_{15}). Representative pyroxene compositions are given in Tables A3.2 and A3.3.

CR chondrites

The chondrites Renazzo and Al Rais have been recognized as displaying properties different from other carbonaceous chondrite groups for some time (e.g. Wood 1967, McSween 1979). However, a formal definition of the CR chondrite group has only been accomplished more recently, since the discovery of several Antarctic and Algerian meteorites with similar properties (Weisberg et al. 1993, Kallemeyn et al. 1994). Members of this group are relatively metal-rich, containing ~7 vol % metal (Weisberg et al. 1995). CR chondrites show varying degrees of aqueous alteration, as well as evidence for mild thermal metamorphism that preceded the hydrothermal event. All CR chondrites are classified as petrologic type 2 (CR2). Al Rais is an anomalous member, in that it has higher abundances of matrix and dark inclusions, a higher volatile content, and higher degree of hydration than other members of the group.

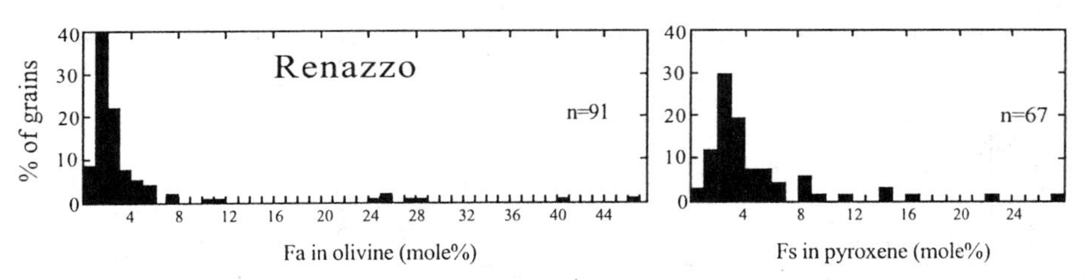

Figure 53. Histogram showing the distribution of olivine and low-Ca pyroxene compositions in chondrules in the CR chondrite, Renazzo (Kallemeyn et al 1994).

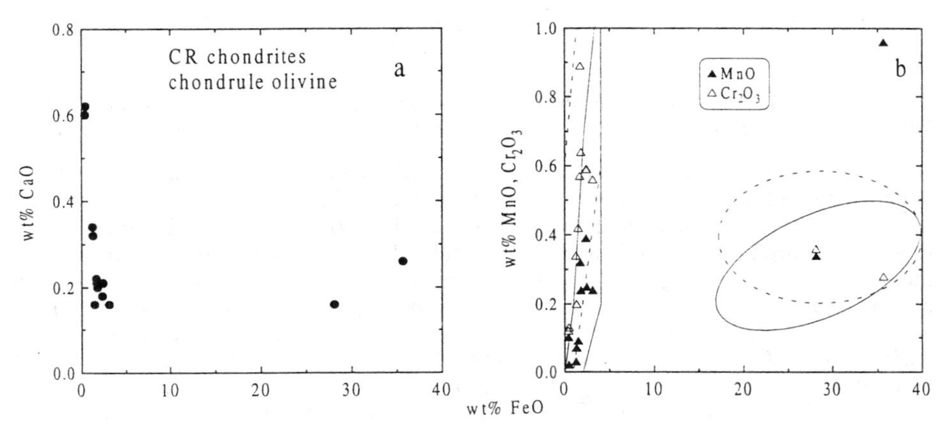

Figure 54. Minor element variation plots for olivine from chondrules in CR chondrites. In (b), areas outlined represent fields of data determined by Ichikawa and Ikeda (1995) for MnO vs. FeO (solid lines) and Cr_2O_3 vs. FeO (dashed lines). Data sources: Weisberg et al. (1993), Ichikawa and Ikeda (1995), Noguchi (1995).

Chondrules in CR chondrites are mostly porphyritic. BO chondrules are rare, although they are common in El Djouf 001. Most porphyritic chondrules are type I. Many type I chondrules are concentrically layered, having olivine- and/or pyroxene-rich cores surrounded by a continuous or discontinuous Fe,Ni metal layer ± a finer grained olivine-, pyroxene-rich layer ± a phyllosilicate-rich layer ± a carbonate-rich layer (Weisberg et al. 1993). CR chondrites contain fragmented olivine and pyroxene grains (e.g. Ichikawa and Ikeda 1995) which we describe here together with chondrules.

Silicate and oxide minerals. Chondrule *olivine* compositions vary from Fa_{0-70}. Compositional distributions are characterized by peaks between Fa_1 and Fa_3 within a broader cluster ranging from $<Fa_1$ to Fa_{4-6} (Fig. 53). The peak in FeO-rich compositions varies somewhat from one CR chondrite to another (Kallemeyn et al. 1994). For olivine in type I chondrules, CaO contents decrease with increasing FeO, and Cr_2O_3 and MnO are positively correlated with FeO (Fig. 54). MnO contents of FeO-rich olivines are also positively correlated with FeO, although the MnO/FeO ratio is lower for FeO-rich olivines than for forsteritic olivines. Forsteritic olivine generally shows no zoning, whereas FeO-rich olivine grains show normal zoning from core to rim (Noguchi 1995). Representative analyses are given in Table A3.1. Lee et al. (1992) suggested that Renazzo, MAC87320 and Al Rais represent petrologic subtypes 3.1-3.4, 3.4-3.5, and 3.6-3.7, respectively before aqueous alteration occurred. In comparison with type 3 chondrites from other chondrite groups (OC and CO), high Cr_2O_3 contents of olivines suggest that lower subtypes are more appropriate. Chromite-olivine pairs from type II

chondrules in Renazzo show no evidence for metamorphic resetting, whereas one chondrule from MAC87320 shows evidence for resetting at temperatures appropriate to subtype 3.1 (Johnson and Prinz 1991).

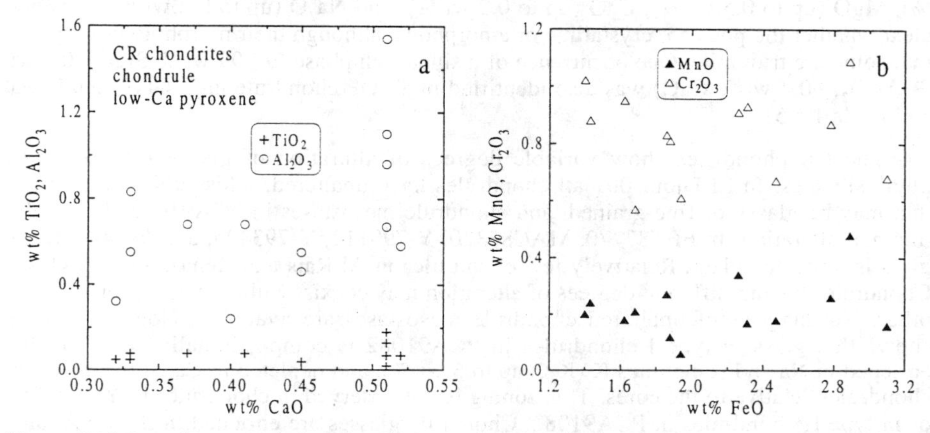

Figure 55. Minor element variation plots for low-Ca pyroxene from chondrules in CR chondrites. Data sources: Weisberg et al. (1993), Ichikawa and Ikeda (1995), Noguchi (1995).

Low-Ca pyroxene in CR chondrules is mostly clinoenstatite; orthopyroxene is rare. Like olivine, low-Ca pyroxene compositions are mostly FeO-poor, Fs_{1-7}. A few more FeO-rich grains have variable compositions as high as Fs_{60} in PCA91082 (Fig. 53). Wo contents vary from 0.5-1.0 mol %. Minor element contents of a limited number of low-FeO grains are shown in Figure 55. Cr_2O_3 and MnO contents may also vary up to 2.4 and 4 wt %, respectively (Weisberg et al. 1993, Ichikawa and Ikeda 1995, Noguchi 1995). *Pigeonite* appears to be present in CR chondrules, based on random analyses of pyroxenes (e.g. Noguchi 1995). Ca-rich pyroxene (*augite*) occurs in chondrules as small crystals both in the mesostases and on the margins of low-Ca pyroxene (Weisberg et al. 1993, Ichikawa and Ikeda 1995). Compositions are variable, from Fs_{1-9} and Wo_{25-50} in type I chondrules (Bischoff et al. 1993a, Noguchi 1995). A composition of ~$Fs_{38}Wo_{31}$ was analyzed in a type II chondrule (Noguchi 1995). Many Ca-rich pyroxenes have relatively high concentrations of Al_2O_3 (>5 wt %), Cr_2O_3 (up to 3.4 wt %), and MnO (up to 1.2 wt %). Compositions with up to 16 wt % Al_2O_3 and 5 wt % MnO have been reported (Noguchi 1995). Some type I chondrules also contain Ca-rich pyroxene with higher TiO_2/Al_2O_3 ratios, similar to the ratio in a Ca,Al-rich chondrule (Noguchi 1995). Pyroxene analyses are given in Tables A3.2 and A3.3.

Ca,Al-rich chondrules are rare in CR chondrites. A Ca,Al-rich chondrule observed in El Djouf 001 has a BO texture, with a coarse-grained intergrowth of fassaite (11 wt % Al_2O_3; 3 wt % TiO_2) and essentially end-member anorthite present between the forsteritic olivine laths (Bischoff et al. 1993a). A Ca,Al-rich chondrule with fassaitic pyroxene was also observed in PCA91082 (Noguchi 1995).

Calcic *plagioclase* is observed commonly in type I chondrules in PCA91082, occurring as laths and sometimes microphenocrysts. Plagioclase occurs in the groundmass of some chondrules in Y-8449 (Ichikawa and Ikeda 1995). Its composition varies from An_{28-99}, with negligible Or. Plagioclase in PCA91082 contains about 0.9 wt % MgO and 0.3 wt % FeO. Plagioclase and Ca-rich pyroxene also occur as irregular inclusions in magnesian olivine fragments (Ichikawa and Ikeda 1995).

Silica pods are observed at the edges of several chondrules in PCA91082 (Noguchi 1995). They are up to 10 μm in diameter, and are associated with plagioclase, pyroxene and mesostasis. They contain minor amounts of Al_2O_3 (0.2-0.6 wt %), FeO (0.5-1.3 wt %), MgO (up to 0.5 wt %), CaO (up to 0.2 wt %), and Na_2O (up to 0.2 wt %). It is not clear whether the pods are crystalline or amorphous, although their morphology suggests that some are tridymite. One occurrence of a silica-rich phase (97-99 wt % SiO_2, ~0.8 wt % Al_2O_3, ~0.4 wt % CaO) was also identified in a POP chondrule in Y-8449 (Ichikawa and Ikeda 1995).

The CR chondrites show variable degrees of alteration of glassy *mesostases* to phyllosilicates. In El Djouf 001 all chondrules have unaltered, feldspathic mesostases, that may be glassy or fine-grained, and chondrule mesostases in PCA91082 show only minimal alteration. In EET87770, MAC87320, Y-790112, Y-793495, and Y-8449, fresh glass is quite abundant. Relatively few chondrules in Al Rais and Renazzo retain glass. Chondrules having different degrees of alteration may coexist within micrometers of each other. No analyses of unaltered chondrule mesostases are available. Noguchi (1995) shows that glass in type I chondrules in PCA91082 is compositionally zoned, being enriched in Na and sometimes K (K_2O up to 3 wt %), and depleted in Ca, at the edges of chondrules relative to the cores. This zoning is not observed in chondrules in Y-793495, or in type II chondrules in PCA91082. Chondrule glasses are enriched in Si and Al and have lower concentrations of (Mg+Fe) than the phyllosilicates that replace them.

Chlorite- and serpentine-rich *phyllosilicates* are encountered in different chondrules and in different regions of the same chondrule (Weisberg et al. 1993). Chlorites in chondrule cores in Al Rais, Renazzo, and Y-790112 typically contain about 30 wt % SiO_2, 15 wt % Al_2O_3, 21-33 wt % FeO, and 11-18 wt % MgO. Phyllosilicates in chondrule rims are mixtures of Fe-rich serpentines and smectites (saponite) and/or chlorites. They typically contain 35-42 wt % SiO_2, 2-3 wt % Al_2O_3, 22-34 wt % FeO, and 17-21 wt % MgO. Phyllosilicates in Y-8449 and PCA91082 are richer in FeO than those in the other CR chondrites (Ichikawa and Ikeda 1995, Noguchi 1995). Phyllosilicates in PCA91082 are also considerably enriched in Na_2O compared with the other CR chondrites, consistent with their locations at the edges of chondrules where enhanced alkali element concentrations are observed in mesostases. Phyllosilicates in CR chondrites are more FeO-rich than those in CI chondrites and have a narrower range of compositions than those in CM chondrites.

Chromite occurs in type IIA chondrules (Johnson and Prinz 1991, Noguchi 1995) and in a ferroan olivine fragment in Y-8449 (Ichikawa and Ikeda 1995). Its composition is variable, and includes ~57 wt % Cr_2O_3, 25 wt % FeO, 6 wt % Al_2O_3, 7 wt % MgO, 1 wt % TiO_2, 0.8 wt % V_2O_5, and 0.3 wt % MnO.

Calcium carbonate, possibly calcite, was identified in the center of a POP chondrule in Y-8449 (Ichikawa and Ikeda 1995). The carbonate contains 52 wt % CaO and 1 wt % FeO, and has an MnO content <0.1 wt %.

Metal and sulfide minerals. Fe,Ni *metal* in CR chondrites is concentrated mainly in chondrules. The compositional range of chondrule metal in different CR chondrites is similar: 4-14 wt % Ni, 0.1-0.6 wt % Co, 0.01-2 wt % Cr, and 0.01-1 wt % P (Wood 1967a, Lee et al. 1992, Weisberg et al. 1993, Zanda et al. 1994, Noguchi 1995, Ichikawa and Ikeda 1995). Rare metal grains contain up to 22 wt % Ni. Rare metal grains in Renazzo contain small, spherical inclusions of silica glass that are inferred to have precipitated from molten metal (Zanda et al. 1994). Some metal in chondrule interiors in Y-8449 has narrow exsolution lamellae of taenite (Ichikawa and Ikeda 1995). Metal in chondrule rims is compositionally different from metal in chondrule cores. It also has complex zoning properties, with Ni and Co contents ~20% lower at grain edges relative to grain centers (Lee et al. 1992). Metal from all occurrences in CR chondrites shows a clear positive correlation between Ni and Co that corresponds to the CI abundance ratio of these two elements. The *sulfide* minerals, pyrrhotite and pentlandite, occur in association with metal (Kallemeyn et al. 1994). Pyrrhotite contains ~2 wt % Ni and <0.05 wt % Co, and pentlandite contains ~35 wt % Ni and ~0.9 wt % Co.

CK chondrites

Most members of the CK chondrite group are metamorphosed and are described in a section below (Type 4-6 chondrites: non-opaque material). One member of the group, Ningqiang, is a type 3 chondrite and its chondrules are described here. Ningqiang was originally described as an anomalous CV chondrite, but Kallemeyn et al. (1991) have shown that it is more closely related to the CK group. It differs sufficiently from the type 4-6 CK chondrites to be described as an anomalous member of the group (CK-an) (Kallemeyn et al. 1991). Chondrule characteristics described here are largely based on a study by Rubin et al. (1988b).

Most chondrules in Ningqiang are porphyritic (PO and POP); BO chondrules are rare. *Olivine* in porphyritic chondrules has a restricted range of Fa contents (Fa_{1-12}) and an average CaO content of 0.23 wt %. Olivine in microgranular chondrules has a wider range of compositions, Fa_{1-36}. The compositional distribution of random olivine compositions throughout the whole chondrite shows that most olivine is MgO-poor (<Fa_6), with a peak at Fa_{1-2}. *Low-Ca pyroxene* (clinoenstatite) occurs as chondrule phenocrysts; compositions throughout the whole chondrite peak at Fs_{0-1} and extend only up to Fs_5. *Mesostases* are microcrystalline: none of the chondrules contain isotropic glass. Mesostasis compositions contain about 50 wt % SiO_2, 20-33 wt % Al_2O_3, 15 wt % CaO, 1-7 wt % MgO, 3 wt % FeO, and 3 wt % Na_2O. Chondrules also contain up to ~25 vol % *opaque minerals*, including troilite grains and magnetite-sulfide ± metal assemblages up to 200 μm in size.

Ungrouped carbonaceous chondrites and carbonaceous chondrite grouplets

ALHA85085 and related chondrites. ALH85085 and a group of five other Antarctic meteorites with similar properties, which we refer to collectively here as ALH85085, have compositional and O and N isotopic properties that may relate them to the CR chondrite group (Weisberg et al. 1995). Acfer 182 also shows many similarities to ALH85085 (Bischoff et al. 1993b). The term "CH" chondrites has been used for this group but is not widely accepted. These chondrites have many extraordinary properties compared with other carbonaceous chondrite groups (see Introduction section). Most chondrules in ALH85085 are cryptocrystalline. Porphyritic chondrules comprise only ~20% of all chondrules and BO chondrules are rare (Scott 1988, Grossman et al. 1988a, Weisberg et al. 1988). Bulk compositions of cryptocrystalline chondrules approximate low-Ca pyroxene with a minor feldspathic component. Bulk compositions of BO and some porphyritic chondrules are Al-rich, commonly containing >10 wt % Al_2O_3. Chondrules themselves are low in abundance (~10 vol %), but silicate fragments that are probably derived from chondrules constitute ~60 vol % of the chondrite. Here we discuss both chondrules and silicate fragments together as chondrule material.

Most chondrule *olivine* is MgO-rich, with a peak in the olivine composition histogram at Fa_{1-2}, although rare chondrules with Fa contents up to 35 mol % are encountered (Scott 1988, Grossman et al. 1988a). For MgO-rich chondrule olivines, CaO contents are in the range 0.1-0.5 wt %, and MnO and Cr_2O_3 contents are up to 0.3 and 0.7 wt %, respectively (see Table A3.1). Some large olivine fragments are reversely zoned, ranging from Fa_{18} in the center to Fa_{12} on the edge, suggesting that reduction occurred after initial formation (Weisberg et al. 1988). *Low-Ca pyroxene* in chondrules is also MgO-rich, with a peak in the pyroxene composition histogram at Fs_2, although rare chondrules with Fs contents up to 33 mol % are encountered (Scott 1988, Bischoff et al. 1993b). Low-Ca pyroxenes ($Wo_{0.1-0.7}$) contain up to 0.9 wt % Al_2O_3, 0.1 wt % TiO_2, 0.8 wt % Cr_2O_3, and 0.3 wt % MnO (Scott 1988; Table A3.2). The mean pyroxene composition reported by Grossman et al. (1988a) contains 1.8 wt % CaO ($Wo_{3.4}$), 2.0 wt % Al_2O_3, and 0.2 wt % TiO_2. Chondrule *mesostases* generally consist of clear, nearly isotropic, feldspathic glass with high CaO contents. Rare isolated grains of Mg-Al *spinel* and ferroan chromian spinel occur in ALH85085 (Grossman et al. 1988a). Mg-Al spinel contains 69 wt % Al_2O_3, 28 wt % MgO, 1 wt % SiO_2, and 0.6 wt % FeO. Ferroan chromian spinel has a high Al_2O_3 content (48 wt %), 18 wt % FeO and 9 wt % Cr_2O_3.

Chondrules in Acfer 182 are larger than those in ALH85085: the mean chondrule diameter is 90 μm (Bischoff et al. 1993b). Smaller chondrules are dominated by cryptocrystalline textures, and porphyritic chondrules become increasingly abundant in the larger size ranges. The olivine composition distribution shows a broad peak between Fa_1 and Fa_4, and a range of olivine compositions up to Fa_{62}. Most low-Ca pyroxene compositions in Acfer 182 are $<Fs_5$, but rare compositions up to Fs_{57} have been analyzed.

LEW85332. LEW85332 is a unique, highly unequilibrated, carbonaceous chondrite (Rubin and Kallemeyn 1990). It is metal-rich (estimated 6-7 vol % metal) and Weisberg et al. (1995) have suggested that it is related to CR chondrites. It contains abundant chondrules that are predominantly porphyritic, many of which contain Fe,Ni metal. Isolated silicate grains, mostly olivine and clinoenstatite, constitute 8 vol % of the chondrite.

Olivine in LEW85332 is heterogeneous. The compositional distribution has a broad peak at Fa_{1-5}, and a continuous distribution of compositions up to Fa_{37} (Rubin and Kallemeyn 1990). The mean olivine composition contains 0.5 wt % Cr_2O_3, 9 wt % FeO, 0.2 wt % MnO, and 0.3 wt % CaO. Chondrule olivine phenocrysts are normally zoned. *Low-Ca pyroxene* is also very heterogeneous. The compositional distribution has a broad peak at Fs_{2-6}, and a continuous distribution of compositions up to Fs_{30}. The mean composition of low-Ca pyroxene ($Fs_{7.6}Wo_{1.7}$) contains 0.1 wt % TiO_2, 1.0 wt % Al_2O_3, 0.9 wt % Cr_2O_3, 5 wt % FeO, and 0.2 wt % MnO. Rare grains of *diopside* have compositions such as $Fs_{0.4}Wo_{50}$, with 1 wt % TiO_2 and 8 wt % Al_2O_3. Mesostases of some chondrules consist of transparent feldspathic glass, with an average composition of 60 wt % SiO_2, 19 wt % Al_2O_3, 8 wt % CaO, 6 wt % Na_2O, 6 wt % FeO, and 1 wt % MgO.

Bencubbin and Weatherford. Bencubbin and Weatherford are two very similar and unusual meteorites that have been variously described as polymict breccias containing iron and achondrite material (Simpson and Murray 1932, Beck and LaPaz 1949, Lovering 1962, Mason and Nelen 1968, McCall 1968) and chondritic breccias (Kallemeyn and Wasson 1978, Weisberg et al. 1990). Weisberg et al. (1995) suggested that they have affinities to the CR group. They consist of centimeter-sized metal clasts (~60%) and silicate clasts (~40%), with xenolithic clasts of various types of chondritic material (McCall 1968, Clayton and Mayeda 1978, Barber and Hutchison 1991). One of the reasons these meteorites have been described as chondrites is that textures and compositions of the silicate clasts are interpreted as being similar to barred olivine chondrules or chondrule fragments, with textures ranging from coarse-barred to cryptocrystalline (Weisberg et al. 1990). Olivine bars may be up to centimeter sized, considerably longer than those in chondrules from the other chondrite classes.

Most *olivine* compositions fall within an extremely narrow range, Fa_{2-4} (Mason and

Nelen 1968, Newsom and Drake 1979, Weisberg et al. 1990, 1995). A representative olivine composition contains 3.5 wt % FeO, 0.6 wt % Cr_2O_3, 0.2 wt % MnO, and 0.2 wt % CaO (Weisberg et al. 1990). *Low-Ca pyroxene*, both monoclinic and orthorhombic varieties, also occurs as large (millimeter-sized) crystals which fill the interstices between olivine bars and appear to have grown at the expense of earlier crystallized olivine. Low-Ca pyroxene compositions are typically $Fs_{2.3}Wo_{0.8}$, and contain ~0.1 wt % TiO_2, 0.7 wt % Al_2O_3 and Cr_2O_3, and 0.1 wt % MnO (Weisberg et al. 1990). *Ca-rich pyroxene* (Fs_2Wo_{43}) is also present, and contains 1 wt % TiO_2, 9 wt % Al_2O_3, 2 wt % Cr_2O_3, and 0.1 wt % MnO. Material interstitial to the olivine bars is a feldspathic, glassy material which contains fine needles of low-Ca pyroxene. *Glass* compositions contain 55 wt % SiO_2, 22-24 wt % Al_2O_3, 1-2 wt % FeO, 4-5 wt % MgO, and 14-15 wt % CaO (Newsom and Drake 1979, Weisberg et al. 1990).

Metal and associated phases in Bencubbin and Weatherford are described in a section below. Bencubbin also contains matrix material consisting of silicate glass containing fused droplets of metal, which acts as the cementing agent of the breccia (Mason and Nelen 1968, Ramdohr 1973, Newsom and Drake 1979). Matrix silicate contains ~22 wt % FeO, 35 wt % SiO_2, and 7 wt % Al_2O_3, and matrix metal has compositions similar to metal clasts.

Acfer 094. Acfer 094 is a unique, primitive carbonaceous chondrite with properties related to the CO and CM groups (Newton et al. 1995, Greshake 1997). The most abundant chondrule types are porphyritic. The composition of *olivine* in chondrules and chondrule fragments is highly variable, with a pronounced peak at Fa_{1-2} but a spread of olivine compositions up to Fa_{88}. Most *low-Ca pyroxenes* in chondrules are enstatites with $<Fs_2$, but compositions with higher FeO contents also occur. *Ca-rich pyroxene* has variable compositions, $Fs_{0.4-8}$ and Wo_{35-46}. Several Ca-rich pyroxenes have Al_2O_3 contents up to 9 wt %. Anorthitic *plagioclase* (An_{70-98}) occurs in chondrules. The most abundant metal phase is kamacite, 4-7 wt % Ni and 0.3-0.55 wt % Co. Ni-poor sulfides have a composition close to stoichiometric FeS, and pentlandites contain 17-26 wt % Ni.

Acfer 094 contains two unusual chondrules (Newton et al. 1995). One has an RP texture, with pyroxene compositions $Fs_{53}Wo_6$. Two different types of ferroan olivine occur in this chondrule: elongate crystals of Fa_{56}, and small grains of Fa_{73} enclosed in pyroxene. Tiny, Ni-free metal grains, some pentlandite, and a Cr-rich spinel are also present. The second chondrule consists of large olivine grains of composition Fa_{88} intergrown with laths of a silica phase, probably tridymite, and a SiO_2- and FeO-rich mesostasis.

Adelaide. Adelaide is a unique carbonaceous chondrite with affinities to the CO and CV chondrites (Fitzgerald and Jones 1977, Davy et al. 1978, Kallemeyn and Wasson 1982). Fitzgerald and Jones (1977) suggested that the lunar meteorite, Bench Crater, is related to Adelaide, although detailed petrographic comparisons have not been made. Chondrules in Adelaide include porphyritic and BO types, as well as plagioclase-rich chondrules (Davy et al. 1978, Sheng et al. 1991).

Olivine compositions throughout the meteorite range from 0.5-47.5 wt % FeO (Davy et al. 1978). *Low-Ca pyroxene*, mostly clinoenstatite, is mostly Mg-rich, $<Fs_3$, but grains up to Fs_{11} are also present. Al_2O_3 contents of low-Ca pyroxenes are typically around 1 wt %. Chondrule *mesostases* consist of isotropic or devitrified glass. Euhedral *chromite* crystals are observed at the peripheries of some olivine-rich chondrules. *Feldspar* in plagioclase-Ca-pyroxene chondrules is close to endmember anorthite.

Metal occurs as small globules included in some olivine crystals or occasionally within chondrule mesostasis (Davy et al. 1978). It is mostly kamacite, 5-10 wt % Ni. Sulfide minerals occur throughout the chondrite and have been weathered extensively. Most of the sulfide was originally troilite, with minor pentlandite. Magnetite is also a major phase throughout the chondrite, and mostly occurs within chondrules.

Belgica-7904. Belgica-7904 is a unique carbonaceous chondrite that has affinities to the CI and CM groups (Kojima et al. 1984, Skirius et al. 1986, Akai 1988, Mayeda et al. 1987, Tomeoka 1990a). It has suffered extensive aqueous alteration (e.g. Ikeda and Prinz 1993) followed by dehydration of phyllosilicates which probably occurred in an intense heating event, possibly shock heating (Kimura and Ikeda 1992). Chondrules, which are mostly porphyritic, are abundant in B-7904.

Olivine in chondrules varies widely in composition. Most olivine is $<Fa_2$ (Bischoff and Metzler 1991), but compositions as high as Fa_{66} occur in some type II chondrules. Olivine contains up to 0.9 wt % Cr_2O_3, 0.3 wt % NiO, 0.9 wt % MnO, and 0.7 wt % CaO (Skirius et al. 1986, Kimura and Ikeda 1992). Strong correlations exist between MnO and Cr_2O_3, and between Al_2O_3 and CaO in type I chondrules, and between MnO and FeO in type II chondrules. Olivine in type II chondrules is normally zoned, and relict forsterites are observed at the cores of some olivine grains. *Low-Ca pyroxene*, $Fs_{0-5}Wo_{0-14}$, occurs in large type I chondrules. It contains up to 0.9 wt % TiO_2, 4 wt % Al_2O_3, 1.3 wt % Cr_2O_3, and 0.3 wt % MnO (Kimura and Ikeda 1992). Grains with higher Wo content are probably pigeonite. *Ca-rich pyroxene* ($Fs_{0-3}Wo_{35-46}$) also occurs in type I chondrules. Minor element contents are variable, and include up to 2 wt % TiO_2, 9 wt % Al_2O_3, and 1.7 wt % Cr_2O_3. Na_2O is below electron microprobe detection limits. *Plagioclase* occurs as irregularly shaped grains in the cores of a few large type I chondrules. It is usually anorthitic (An_{90-99}), with the K_2O content below the detection limit. *Glass* is present in type I and type II chondrules, and compositions differ between the two chondrule types. Glasses in type II chondrules contain up to 2 wt % P_2O_5, and small phosphate grains, probably merrillite, also occur. Mg-Al-*spinel* occurs in forsterite phenocrysts in type I chondrules (Kimura and Ikeda 1992). Fine-grained, Al_2O_3-poor chromites are also present in type I chondrules, in Cr-rich ovoids described below. Type II chondrules contain Al_2O_3-rich chromites that are usually zoned, with FeO and Al_2O_3 increasing towards the edges of grains. Johnson and Prinz (1991) showed that chromite-olivine pairs in B-7904 retain their magmatic temperatures of about 1400°C.

Three types of *metal* occur in B-7904: kamacite, taenite, and Co-rich metal (Kimura and Ikeda 1992). A Ni-Co-rich type of kamacite occurs in forsterite phenocrysts in type I chondrules. Co and P contents are up to 3 wt % and 0.7 wt %, respectively. Ni-Co-poor kamacite (~1 wt % Ni, 0.02 wt % Co) occurs in clasts and matrix and is depleted in P (<0.1 wt %). Taenite has variable Ni and Co contents, up to ~3 wt % Co. Grains that are highly enriched in Ni (>50 wt %) may be fine-grained mixtures of awaruite and tetrataenite. Co-rich metal contains 11-16 wt % Co and 1.7-3.2 wt % Ni. *Troilite* occurs in all components of B-7904, and pentlandite is also common. *Pentlandite* contains 13-20 wt % Ni and 0.1-0.9 wt % Co. *Schreibersite* occurs rarely in forsterite phenocrysts and Cr-rich ovoids of type I chondrules. Ni contents vary from 10-17 wt %. Schreibersite occurring with taenite in Cr-rich ovoids (see below) and Mn-rich olivine-bearing clasts is more enriched in Ni (33-44 wt %). *Magnetite* occurs in type I chondrules, in association with kamacite and taenite.

Chondrules contain several secondary phases resulting from the aqueous alteration episode. *Phyllosilicates* occur in most components. Compositions of phyllosilicates in chondrules are mainly distributed between serpentine (chlorite) and Na-Al-talc. Three types of phyllosilicates occur (Kimura and Ikeda 1992). Low-Al, magnesian phyllosilicate occurs in small type I and type II chondrules and in the mantles of large type I chondrules. High-Al, magnesian phyllosilicates occur only in the central portions of large chondrules. High-Al, ferroan phyllosilicate occurs with opaque minerals in the cores of some large magnesian chondrules. Phyllosilicates in type II chondrules have higher CaO and P_2O_5 contents than those in type I chondrules, up to 16 wt % CaO and 14 wt % P_2O_5. Fine-grained merrillite may be mixed with the phyllosilicates. Fine-grained chromite may be intergrown with phyllosilicate in type I chondrules, which has an average Cr_2O_3 content of 1.0 wt %. *Magnesiowüstite* with an average chemical formula of $(Fe_{0.54}Mg_{0.39})_{0.93}O_{1.00}$ occurs within a troilite-taenite aggregate in the core of a type I chondrule (Kimura and Ikeda 1992).

Type I chondrules often have unusual spheroidal to ellipsoidal inclusions within the groundmass that are 20-200 μm in size and brown to black in color. These have been described as isotropic grains (Skirius et al. 1986), egg-shaped particles (Bischoff and Metzler 1991) and as Cr-rich ovoids (Kimura and Ikeda 1992). The brown inclusions consist mainly of phyllosilicates, and the black ones contain abundant opaque minerals such as troilite, taenite, and schreibersite. Phyllosilicates are similar in composition to low-Al phyllosilicates in chondrules, although they are enriched in Cr_2O_3 (0.3-9.9 wt %). They also have high contents of FeO (13-55 wt %), NiO (0-10 wt %), and P_2O_5 (0-7 wt %), suggesting that the analyses include fine-grained opaque minerals such as chromite, taenite, schreibersite, and troilite. Fine-grained chromite and a mineral suggested to be eskolaite (Kimura and Ikeda 1992) also occur in and around the Cr-rich ovoids. Eskolaite contains 81-91 wt % Cr_2O_3 and ~1 wt % FeO and V_2O_3, with minor ZnO (up to 0.4 wt %).

Coolidge and Loongana 001. Coolidge and Loongana 001 were defined as a grouplet by Kallemeyn and Rubin (1995), based on similarities in their chemical and petrographic properties. They appear to be related to the CV chondrites, but are distinct in many respects from the main CV group. They are both of similar petrologic type, 3.8-4, and shock stage, S2. Chondrules and chondrule fragments constitute ~65 vol % of these chondrites. Most chondrules are PO and POP types.

Compositional distributions of *olivine* in Coolidge and Loongana 001 are sharply peaked at Fa_{14} and Fa_{12}, respectively (Van Schmus 1969, McSween 1977c, Scott and Taylor 1985, Noguchi 1994, Kallemeyn and Rubin 1995). The average CaO and Cr_2O_3 contents of olivine are very low, both 0.02 wt %, and the MnO content is ~0.2 wt %. *Low-Ca pyroxene* in both chondrites is commonly clinoenstatite which shows lamellar zoning in BSE images typical of petrologic subtype 3.8-4 chondrites (Noguchi 1994). Low-Ca pyroxene compositions (Fs_{3-15}) are relatively unequilibrated compared with olivine with no peaks in the compositional distribution (Noguchi 1994, Kallemeyn and Rubin 1995). An average Wo content of 2 mol % for low-Ca pyroxene reported by Noguchi (1994) probably includes some orthopyroxene (~Wo_4) and pigeonite (~Wo_{10}) as well as clinoenstatite ($<Wo_2$). *Ca-rich pyroxenes* in Coolidge chondrules show a range of Wo contents from Wo_{20} to $>Wo_{50}$, but a restricted range of FeO contents, mostly $<Fs_{10}$ (Noguchi 1994). TiO_2, Al_2O_3, Cr_2O_3, and Na_2O contents of Ca-rich pyroxene in chondrules vary from 0-3, 0-10, 0-3 and 0-0.7 wt %, respectively. One BO chondrule contains both low-Ca and Ca-rich pyroxenes with extremely high Al_2O_3, 13-17 wt % in low-Ca pyroxene and 16-18 wt % in Ca-rich pyroxene. *Mesostasis* in many chondrules is recrystallized and contains no glass (Kallemeyn and Rubin 1995). Some chondrules in Coolidge contain glass with Ca-rich pyroxene needles (Noguchi 1994). Many chondrules in Coolidge contain *plagioclase* that varies in composition from An_{45-95} (Noguchi 1994). It is considered to be a primary crystallization product of the chondrule melt, rather than a devitrification product. Small aluminous *spinels* are observed in some chondrules in

Coolidge (Noguchi 1994). They are commonly enclosed in olivine crystals, and contain about 0.2 wt % V_2O_3, 0.1 wt % MnO, and 0.5 wt % ZnO. Chromian spinels are also observed, in association with metal and/or troilite grains, within chondrules or in chondrule rims. These contain about 0.6 wt % TiO_2 and V_2O_3, 0.2 wt % NiO, 0.5 wt % MnO and <0.1 wt % ZnO.

Fe,Ni *metal* and *troilite* grains occur commonly as spherules in chondrules. Most of the metal grains are kamacite (~7 wt % Ni, 0.35 wt % Co) with small amounts of taenite occurring as thin exsolution lamellae (Noguchi 1994). Chromian spinel and Ca-phosphate sometimes occur as inclusions in the metal and troilite grains. In comparison, metal in the matrix is highly altered.

In addition to the normal ferromagnesian chondrules, some chondrules in Coolidge consist of augite, plagioclase and silica, plus or minus small amounts of olivine (Noguchi 1994). These appear to differ from the Al-rich chondrules in OC because they are silica-saturated.

Type 3 enstatite chondrites

Enstatite chondrites are the most reduced group of chondritic meteorites and are characterized by complex assemblages of unusual metallic and sulfide phases. Here we describe ferromagnesian chondrules in EH3 chondrites. Chondrules in EL3 chondrites have not been described in detail. For the EH3 chondrites, Zhang et al. (1995) suggest petrologic subtypes of 3.0-3.3 for Y-691, Parsa, and ALH84206, 3.5 for Qingzhen, and 3.8 for Kota-Kota. A fragment of EH3-4 chondrite material, the Hadley Rille meteorite, has been found in an Apollo 15 soil sample (Haggerty 1972, Rubin 1997c). This millimeter-sized chondrite suffered impact melting during accretion on to the Moon.

Chondrules from two EH3 chondrites, Y-691 and Qingzhen, have been described in some detail (Ikeda 1988a,b, 1989a,b,c; Grossman et al. 1985). RP chondrules are the most abundant textural type. Porphyritic, granular and barred olivine-pyroxene chondrules are common, and cryptocrystalline and glassy types are rare. Pyroxene is the dominant phase in most chondrules: the amount of olivine in POP chondrules never exceeds 40 vol %, and PO chondrules do not occur. Olivine constitutes less than 2 vol % of the rock in E3 chondrites. Ikeda (1988b) described a suite of unusual silicate inclusions in Y-691. Many of these appear to be chondrules that were originally FeO-rich, but have undergone significant reduction after solidification. These objects are included in the following description of chondrules.

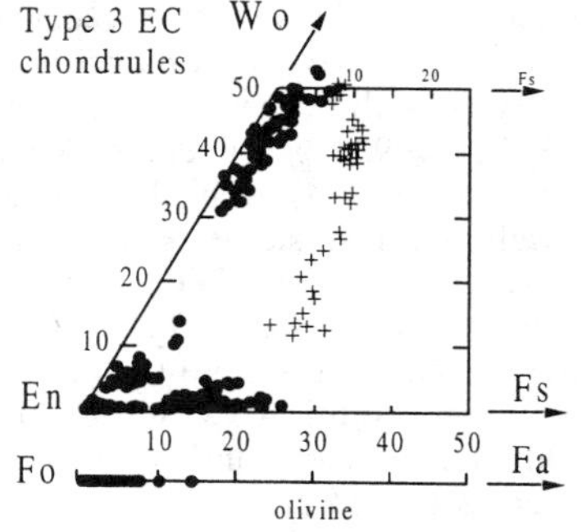

Figure 56. Pyroxene and olivine compositions in chondrules from type 3 enstatite chondrites. Crosses are analyses from a single chondrule, chondrule 152 of Ikeda (1988b). Data sources: Grossman et al. (1985), Lusby et al. (1987), Ikeda (1988b, 1989a), Rambaldi et al. (1983, 1984), Weisberg et al. (1994).

Silicate and oxide minerals. Chondrule *olivine* compositions summarized here vary from Fa_{0-15}, with most olivine being less than Fa_1 in Qingzhen (Fig. 56). FeO-rich olivines containing up to 20 wt % FeO were observed in ALHA81189 (EH3) by Lusby et al. (1987). Primary FeO-rich olivines in Y-691 are commonly reduced to a "decomposed

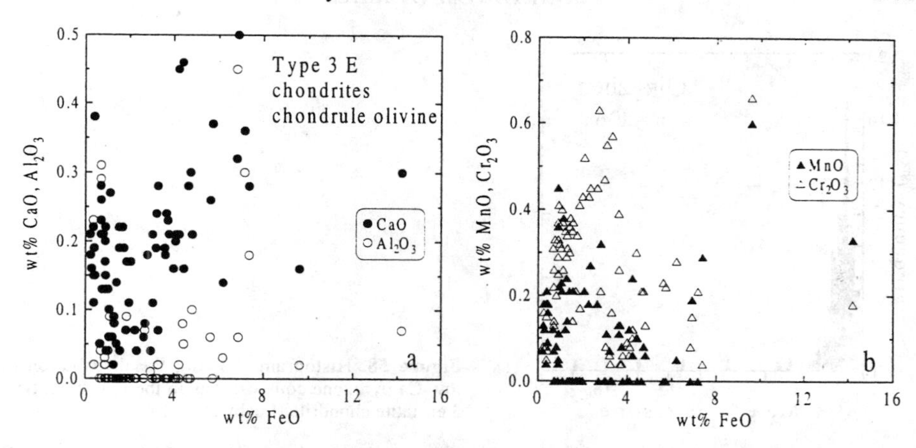

Figure 57. Minor element variation plots for olivine from chondrules in type 3 enstatite chondrites. Data sources: Grossman et al. (1985), Lusby et al. (1987), Ikeda (1989a), Rambaldi et al. (1983).

olivine" texture (Ikeda 1988b, 1989a; see below). Variations in CaO, Cr_2O_3 and MnO contents are illustrated in Figure 57 (see also Table A3.1). Many olivines contain MnO at extremely low levels, below electron microprobe detection limits. Lusby et al. (1987) described two types of silicates: "impure" forsterite and enstatite with <3 wt % FeO and detectable Cr_2O_3 and MnO, which show red luminescence, and "pure" forsterite and enstatite in which Cr_2O_3 and MnO are below detection limits, and which show bright blue luminescence. However, olivine analyses compiled here from several sources (Fig. 57) do not show such clear relationships: olivine grains with very low MnO contents generally have measurable and variable Cr_2O_3 contents. Forsteritic and FeO-rich olivine grains commonly show limited normal zoning (Binns 1967a, Rambaldi et al. 1983, Ikeda 1989a, Lusby et al. 1987). In addition, Lusby et al. (1987) described a PO chondrule fragment in ALHA81189 that has a typical type IIA texture, and in which euhedral olivine grains show concentric, oscillatory zoning from $Fa_{3\text{-}9}$ in magnesian zones to Fa_{25} in Fe-rich zones. The grains contain abundant metal blebs and narrow, parallel lamellae with Mg-rich compositions, indicating that they were reduced after crystallization.

Many, but not all, FeO-rich olivines in EH3 chondrites show varying degrees of reduction (Rambaldi et al. 1983, 1984; Lusby et al. 1987, Kitamura et al. 1988, Ikeda 1989a, Weisberg et al. 1994). Dusty, relict grains of olivine have been observed in FeO-poor chondrules in Qingzhen and Y-691 (Rambaldi et al. 1983, Kitamura et al. 1988). Metal in the dusty regions is Cr-rich and Ni-poor, unlike the metal in the rest of the chondrules that is Cr-poor and Ni-rich. In Y-691, olivine in several barred chondrules shows a reduced texture, termed "decomposed olivine", in which stripes of metal and Mg-rich olivine and/or enstatite occur approximately perpendicular to the long axis of the olivine bars (Ikeda 1988b, 1989a).

Pyroxene in EH3 chondrites varies widely in composition (Tables A3.2 and A3.3). Clinoenstatite, Ca-rich pyroxene (augite and diopside), pigeonite, and orthopyroxene (~Wo_5) are all present in chondrules in Y-691 (Ikeda 1989a); one chondrule contains pyroxene with five distinct compositions (Kitamura et al. 1987). Clinoenstatite and diopside are present in chondrules in Qingzhen (Grossman et al. 1985). Most low-Ca pyroxene is clinoenstatite (Mason 1966, Zhang et al. 1996), but in barred olivine-pyroxene chondrules it is commonly orthopyroxene. Most pyroxene is very magnesian, <Fs_4, with peaks in pyroxene histograms at $Fs_{1\text{-}2}$ (Fig. 58). However, enstatite

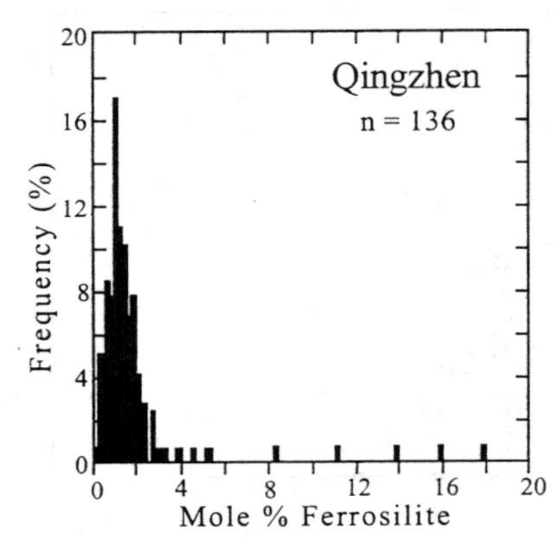

Figure 58. Histogram showing the distribution of low-Ca pyroxene compositions in the Qingzhen type 3 enstatite chondrite (Lusby et al. 1987).

compositions up to Fs_{25} are also fairly common (e.g. Rambaldi et al. 1984, Lusby et al. 1987, Ikeda 1988b, Kitamura et al. 1988, Weisberg et al. 1994; Fig. 56). Most low-Ca pyroxenes in Qingzhen and Y-691 have higher molar Fe/(Fe+Mg) ratios than coexisting olivine. Ca-rich pyroxenes show a more limited range of Fs contents, generally $<Fs_5$. Ca-rich pyroxenes (pigeonite, augite and diopside) with $>Fs_5$ are rare (Ikeda 1988b, Kitamura et al. 1988). Ikeda (1989a) identified two types of enstatite: primary enstatite, ~0.5-17 wt % FeO, which crystallized directly from chondrule melts, and secondary enstatite, <0.5 wt % FeO, which formed by reduction of ferroan olivine.

Low-Ca pyroxene compositions fall in several well defined fields in the pyroxene quadrilateral (Fig. 56: see also Fig. 59). Orthopyroxenes show a very wide range of Al_2O_3 contents, up to extremely high values of 14 wt %, although their TiO_2 contents are not unusually high (Fig. 59). Cr_2O_3 contents of all low-Ca pyroxenes are positively correlated with FeO, but MnO and FeO are not well correlated. Secondary enstatites, which are included in the minor element plots, have very low contents of FeO (<0.5 wt %), Al_2O_3 (<0.2 wt %), TiO_2, Cr_2O_3, MnO and CaO (all <0.1 wt %). Clinoenstatites with low FeO contents (<5 wt %) in enstatite chondrites show cathodoluminescence (CL) with a range of colors (Keil 1968a, Leitch and Smith 1982, McKinley et al. 1984, Weisberg et al. 1994, Zhang et al. 1996). Enstatites with very low MnO and Cr_2O_3 contents (generally below electron microprobe detection limits) show bright blue CL, while those with progressively higher MnO, Cr_2O_3, and FeO contents vary in color through blues, reds and purples.

Rare, intermediate *pigeonite* compositions between Wo_{10} and Wo_{30} have been reported by Kitamura et al. (1987 1988) and Ikeda (1988b). The FeO-rich ($>Fs_{15}$) pigeonite compositions in Figure 56 are from one chondrule in Y-691, described as an unusual silicate inclusion (Ikeda 1988b). Ikeda (1988b) suggests that augite and pigeonite in this chondrule are secondary phases produced by reaction of fassaitic diopside with an SiO_2-rich nebular gas.

Augites and *diopsides* show a wide range of compositions from Wo_{30-50}, but, except for the chondrules mentioned above, a limited range of Fs contents. Ca-rich pyroxenes occur as rims on low-Ca pyroxenes, individual phenocrysts, interstitial phases in barred chondrules, and as groundmass phases in mesostasis (Rambaldi et al. 1983, Kitamura et al. 1987, Lusby et al. 1987, Ikeda 1989a). Al-poor diopsides associated with plagioclase in the mesostases of some chondrules may have formed by decomposition of more Fe-rich, aluminous diopsides under reducing conditions (Ikeda 1989a). Analyses from all these occurrences are combined in Figure 60. Two diopside analyses contain ~3.4 wt %

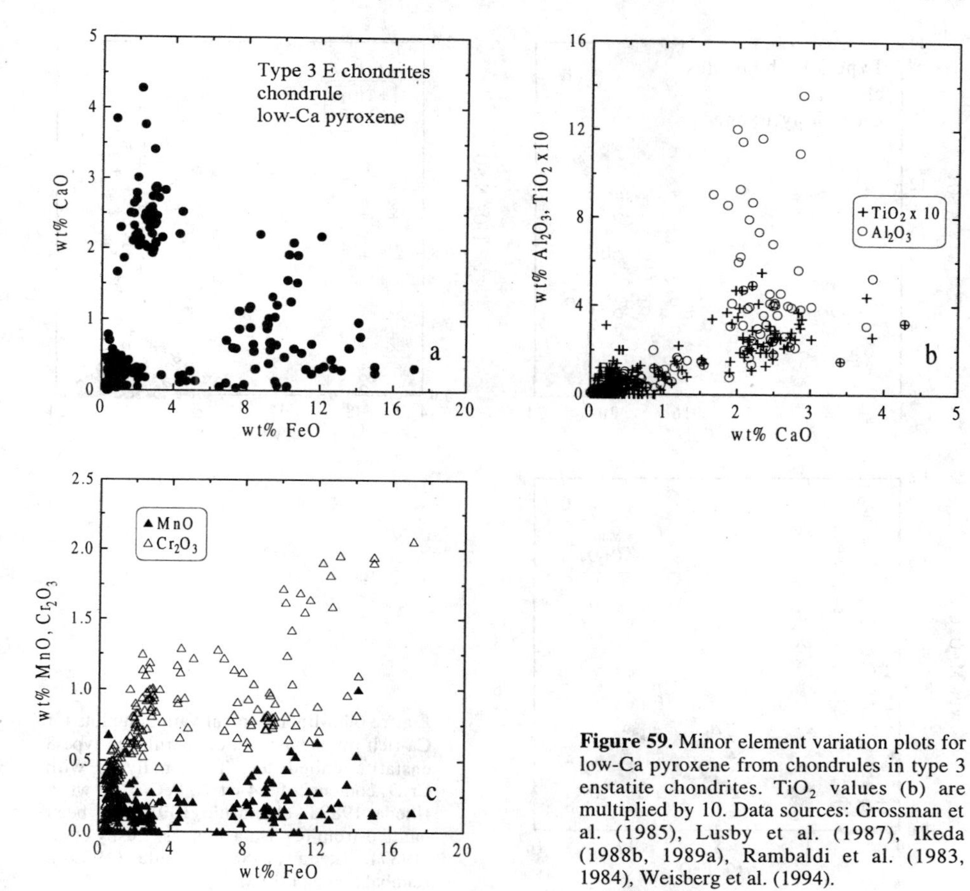

Figure 59. Minor element variation plots for low-Ca pyroxene from chondrules in type 3 enstatite chondrites. TiO_2 values (b) are multiplied by 10. Data sources: Grossman et al. (1985), Lusby et al. (1987), Ikeda (1988b, 1989a), Rambaldi et al. (1983, 1984), Weisberg et al. (1994).

Cr_2O_3. These are from a chondrule in Y-691, in which diopside occurs with silica in a glassy mesostasis (Ikeda 1989a). Many augites have no detectable Na_2O, but concentrations up to 0.5 wt % are not unusual and the highest value reported is 1.3 wt %.

Trace element concentrations in pyroxenes in several unequilibrated E chondrites have been measured by ion microprobe (Weisberg et al. 1994). Ti and V concentrations are 7-91 and 9-11 ppm, respectively for EH3 chondrites, but higher (266 and 59, respectively) in the EL3 chondrite, MAC88136. Sr was below detection. Cr, Mn, Ti, V and Sr are all significantly higher in more FeO-rich pyroxenes. For magnesian enstatites, REE abundance patterns relative to CI chondrite are either nearly flat or slightly enriched in LREE (La/Yb = 1.3-1.6), at about 0.5-4 × CI. In more FeO-rich enstatites, REE abundance patterns are LREE-depleted with La/Yb = 0.2-0.6.

FeO-rich pyroxenes are often reduced. Many of the Fe-rich, low-Ca pyroxenes in several EH3 chondrites (Kota-Kota, Parsa, Qingzhen, EET83322, LEW87223, Y-691) and one EL3 chondrite (MAC88136) contain blebs of Ni-poor Fe metal, as well as glass and/or silica, associated with more magnesian enstatite ($<Fs_3$) (Rambaldi et al. 1984, Lusby et al. 1987, Kitamura et al. 1988, Weisberg et al. 1994). The metal blebs and magnesian enstatite are generally aligned parallel to structural features in the Fe-rich host grain, such as lamellae, fractures, and grain boundaries. Reported Cr contents of metal

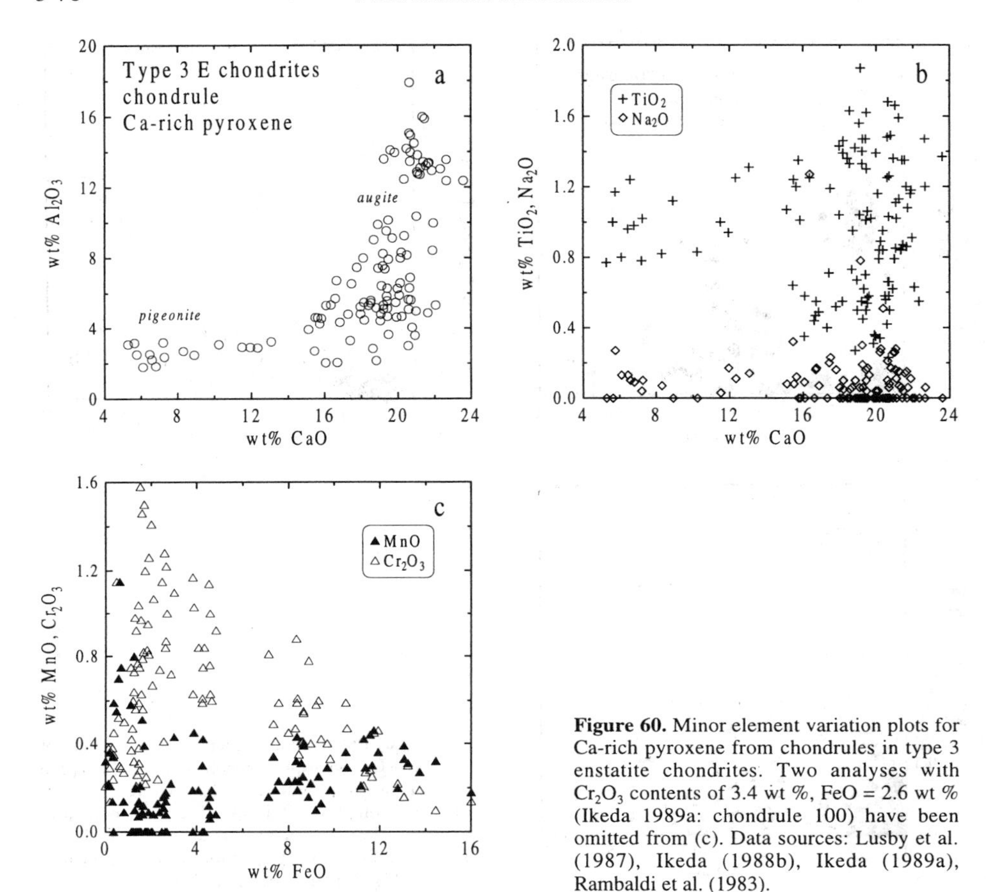

Figure 60. Minor element variation plots for Ca-rich pyroxene from chondrules in type 3 enstatite chondrites. Two analyses with Cr_2O_3 contents of 3.4 wt %, FeO = 2.6 wt % (Ikeda 1989a: chondrule 100) have been omitted from (c). Data sources: Lusby et al. (1987), Ikeda (1988b), Ikeda (1989a), Rambaldi et al. (1983).

blebs are up to 4.4 wt % (Rambaldi et al. 1984), although high Cr in microprobe analyses may be the result of overlap with Cr-rich spinels (Weisberg et al. 1994). Many of the Fe-rich pyroxene grains have an overgrowth of magnesian enstatite ($<Fs_2$) that is compositionally similar to enstatite crystals in the host chondrules.

Spinel occurs in some barred olivine-pyroxene chondrules in Y-691, in association with aluminous diopside or plagioclase (Ikeda 1988b, 1989a). It appears to be a late-stage crystallization product of the chondrule melts. Most spinels are Mg,Al-spinels with Cr/(Cr+Al) ratios of 0.006-0.02 (Cr_2O_3 < 2 wt %) and very low TiO_2 contents (<0.1 wt %). In some chondrules, the spinel composition is significantly more chromian (17-28 wt % Cr_2O_3), with low FeO contents (<3 wt %). TiO_2 contents of these spinels vary up to 1 wt %. The high Cr and low Fe contents may be the result of crystallization with higher FeO contents, followed by reduction. Spinel analyses are given in Table A3.6.

Feldspar occurs as primary phenocrysts in some chondrules, usually as laths in the mesostasis (Kitamura et al. 1987, Ikeda 1988b, 1989a). Compositions are variable (An_{48-88}) and maximum FeO and MgO contents are 1.6 and 3.3 wt %, respectively (Table A3.5). Plagioclase is sometimes partially replaced by nepheline (Ikeda 1988b). Al-rich chondrules in EH3 chondrites also commonly contain plagioclase (Bischoff et al. 1985).

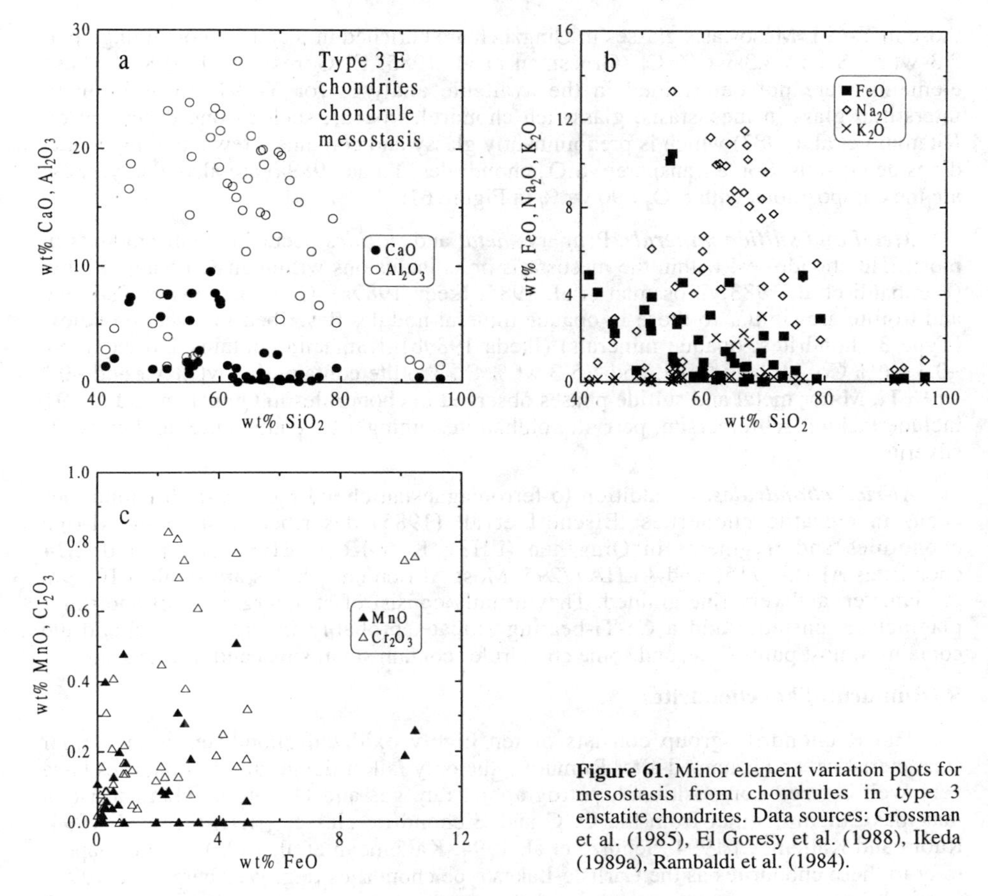

Figure 61. Minor element variation plots for mesostasis from chondrules in type 3 enstatite chondrites. Data sources: Grossman et al. (1985), El Goresy et al. (1988), Ikeda (1989a), Rambaldi et al. (1984).

Secondary plagioclase, close to pure albite in composition, occurs as a product of decomposition of aluminous diopside in some chondrules (Ikeda 1989a).

Silica is a common primary mineral in EH3 chondrules (Grossman et al. 1985, Kitamura et al. 1987, Ikeda 1989a). Silica minerals, probably tridymite (Binns 1967b), occur in the groundmass of most RP chondrules. In some cryptocrystalline chondrules, silica minerals occur as microphenocrysts and appear to have been a liquidus phase. One chondrule in Y-691 contains silica and diopside phenocrysts only (Kitamura et al. 1987). A group of chondrules described as transparent-SiO_2 chondrules by Ikeda (1989a) consist mainly of silica minerals that appear to be cristobalite and/or tridymite; in some cases they consist of silica glass. Analyses of silica minerals contain up to 1 wt % Al_2O_3, 1 wt % FeO, 0.4 wt % MgO, and 1 wt % Na_2O (Ikeda 1989a).

Mesostases of porphyritic and RP chondrules in Y-691 and Qingzhen are commonly glassy, but may also consist of a fine-grained assemblage of enstatite, diopside, plagioclase, silica and/or glass (Ikeda 1989a). Glass compositions are summarized in Figure 61 (see also Table A3.4). Analyses with low Al_2O_3 contents, <5 wt %, are analyses of the groundmasses of cryptocrystalline chondrules in Y-691 (Ikeda 1989a), plus one unusual PP chondrule (Ikeda 1988b). Alkali element contents of glasses are high, although K_2O contents of glasses in Qingzhen are systematically lower (<0.3 wt %) than those in Y-691. Mesostasis glasses in Qingzhen are enriched in S and Cl, containing up to 2.3 wt % S and 4.3 wt % Cl (Grossman et al. 1985, El Goresy et al. 1988). These elements were not determined in the available analyses for Y-691. In addition to interstitial glass in mesostasis, glass-rich chondrules occur, such as one described by Kitamura et al. (1987) which is predominantly glassy and contains a few small pigeonite-diopside crystals. Some transparent SiO_2 chondrules (Ikeda 1989a) are also glassy: these are the compositions with SiO_2 >90 wt % in Figure 61.

Metal and sulfide minerals. Primary *metal* and *sulfides* occur in small amounts in most EH3 chondrules, within the mesostasis or as inclusions within enstatite and olivine (Rambaldi et al. 1983, Grossman et al. 1985, Ikeda 1989a). Compositions of Fe-metal and troilite are similar to those in opaque mineral nodules described in a section below (Type 3 chondrites: opaque minerals) (Ikeda 1989b). Kamacite contains 2-6 wt % Ni, ~0.4 wt % Co, <0.1 wt % Cr, and 1.5-3 wt % Si. Troilite contains 1-7 wt % Cr and ~0.3 wt % Ti. Minor metal and sulfide phases observed in chondrules in Qingzhen and Y-691 include taenite, schreibersite, perryite, oldhamite, niningerite, daubreelite, and caswell-silverite.

Al-rich chondrules. In addition to ferromagnesian chondrules, Al-rich chondrules occur in enstatite chondrites. Bischoff et al. (1985) described a suite of Al-rich chondrules and fragments in Qingzhen (EH3), Kota-Kota (EH3), and paired EH4 chondrites ALHA77156 and ALHA77295. Most Al-rich chondrules are small, <100 μm in diameter, and very fine grained. They usually consist of an intergrowth of anorthitic plagioclase, enstatite and a Ca-Ti-bearing silicate, probably fassaite. One chondrule contains almost pure albite, and some chondrules contain small silica and troilite grains.

R (Rumuruti-like) chondrites

The R chondrite group consists of ten highly oxidized chondrites with similar characteristics, and is named after Rumuruti, the only fall of the group. R chondrites have very well defined compositional, petrographic, rare-gas and O-isotopic characteristics that make them distinct from the O, C and E chondrite classes (Bischoff et al. 1994, Rubin and Kallemeyn 1994, Schulze et al. 1994, Kallemeyn et al. 1996). Earlier papers refer to these chondrites as the Carlisle-Lakes type chondrites (e.g. Weisberg et al. 1991). Carlisle Lakes is the only member of the group that is unbrecciated. All other R chondrites are metamorphosed as well as brecciated, and are best described as R3-5 or R3-6 regolith breccias. The lowest petrologic subtype of material encountered outside equilibrated clasts is subtype 3.6 in ALH85151, and 3.8-3.9 in all the other R chondrites. Hence, primary chondrule mineralogy, particularly for olivine, is not observed. Most papers describing R chondrites do not make clear distinctions between chondrule mineral compositions and the coarse-grained component of matrix, because they are generally very similar. All the common textural types of chondrules are observed, although non-porphyritic types (RP, C, GOP) are usually rare. Fine-grained matrix has not been studied in detail. The following description is mostly for the mineralogy occurring throughout the R chondrites and combines a description of chondrules and coarse-grained matrix.

Silicate and oxide minerals. *Olivine* is the most abundant mineral in R chondrites, constituting 58-75 vol % of the whole rocks. Olivine compositions are essentially equilibrated in most R chondrites, with peaks in olivine composition histograms at Fa_{37-40} (Fig. 62; Table A3.1). Broader composition distributions are observed in several R chondrites: compositions of <Fa_1 are observed in ALH85151 (Score and Mason 1987), Rumuruti (Schulze et al. 1994), and Acfer 217 (Bischoff et al. 1994). In Y-82002, chondrule olivine compositions are significantly lower than matrix olivine compositions (Nakamura et al. 1993).

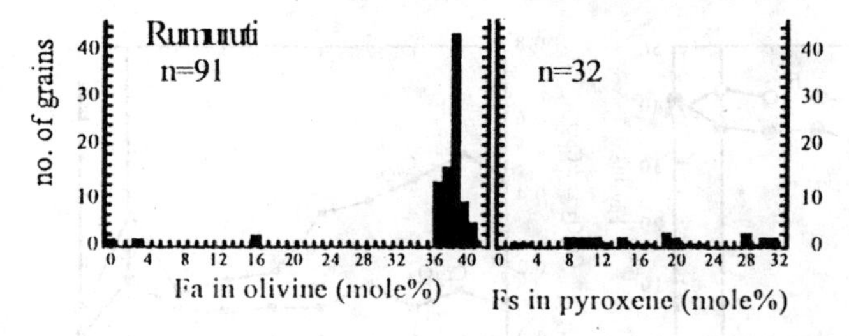

Figure 62. Histograms showing the distribution of olivine and low-Ca pyroxene compositions in R chondrites (Schulze et al. 1994, Kallemeyn et al. 1996).

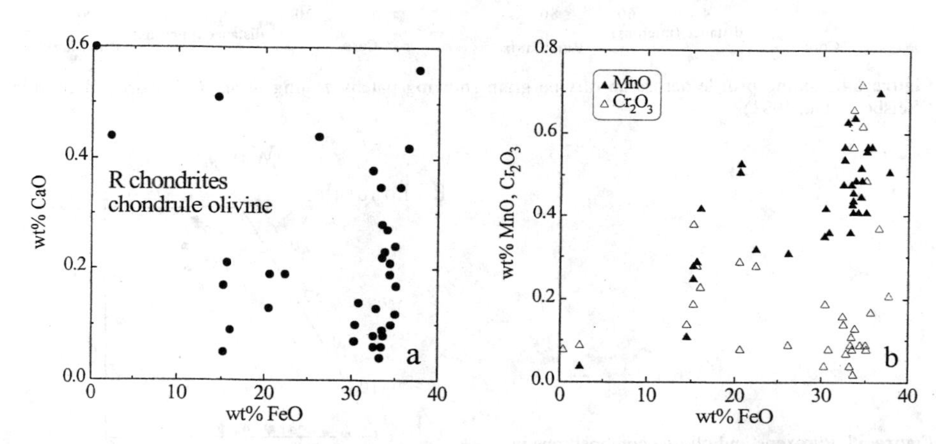

Figure 63. Minor element variation plots for olivine from chondrules in R chondrites. Data sources: Rubin and Kallemeyn (1989, 1994), Schulze et al. (1994), Bischoff et al. (1994), Weisberg et al. (1991).

Mean CaO contents of chondrule olivines in Y-82002, Acfer 217 and PCA91002 are low, <0.1 wt % (Nakamura et al. 1993, Bischoff et al. 1994, Rubin and Kallemeyn 1994). However, CaO contents range up to 0.6 wt % (Fig. 63). Olivines with high CaO contents and high FeO could have originated in type I chondrules with low FeO contents. MnO and Cr_2O_3 contents also show wide ranges (Fig. 63). R chondrite olivine generally contains significant amounts of NiO, consistent with the high oxidation state, with mean values between 0.1 and 0.2 wt %.

Some chondrule olivines in R chondrites show strong zoning. Weisberg et al. (1991) described two types of zoning in olivine in ALH85151 and Y-75302, patchy zoning and oscillatory zoning. Olivine in chondrules commonly shows patchy zoning, in the sense that zoning is not concentric about a single core. In many cases, zoning is related to fractures, and is clearly not the product of igneous crystallization. Zoning profiles show increasing FeO from core to rim, e.g. Fa_{2-43}, and MnO correlates closely with FeO, while CaO decreases and Cr_2O_3 remains essentially constant (Fig. 64). At the edges of grains showing patchy zoning, a narrow porous zone containing Ca-pyroxene and chromite is commonly present. In a chondrule in Acfer 217, forsteritic olivine grains ($Fa_{0.5}$) have patchy zoning with rims that are zoned up to Fa_{39} (Bischoff et al. 1994). Olivine fragments in the matrix of Rumuruti also show similar zoning, with core compositions as low as $Fa_{0.4}$ (Schulze et al. 1994). In one chondrule in ALH85151, all of the olivine exhibits oscillatory zoning in FeO and MgO (Weisberg et al. 1991). Cores of these olivines are Fa_{38}, and the cores are overgrown by a thin (2-3 μm) layer of magnesian olivine ($\leq Fa_{16}$), followed by a thinner (1 μm) layer of $\geq Fa_{44}$. MnO is correlated with FeO, CaO increases as FeO decreases, and Cr_2O_3 increases sharply at the edge of the fayalitic

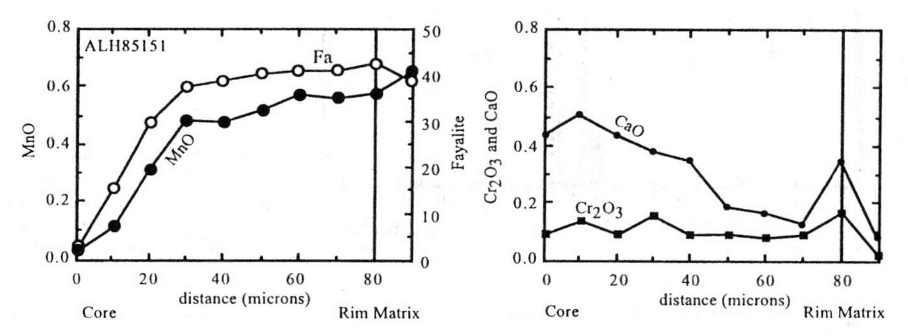

Figure 64. Zoning profile across an olivine grain showing patchy zoning in an R chondrite chondrule (Weisberg et al. 1991).

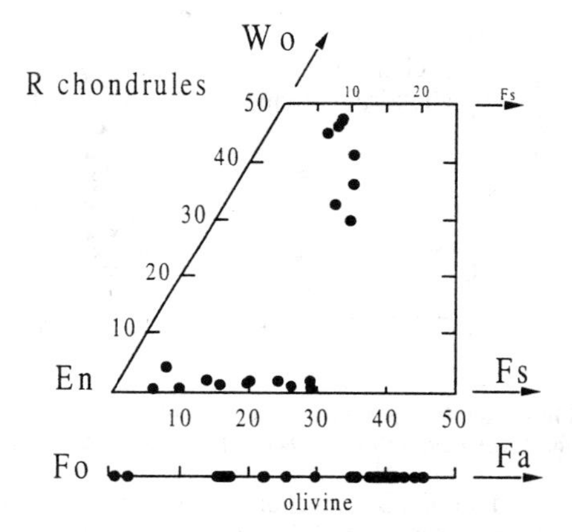

Figure 65. Pyroxene and olivine compositions in chondrules from R chondrites. Data sources: Rubin and Kallemeyn (1989, 1994), Schulze et al. (1994), Bischoff et al. (1994), Weisberg et al. (1991).

rim. In addition to these zoning types, zoning observed in olivine phenocrysts in a type IIA chondrule in PCA91002 may be igneous zoning (Rubin and Kallemeyn 1994).

Three *pyroxene* phases have been identified in R chondrites: low-Ca pyroxene which often shows polysynthetic twinning, augite and diopside (Fig. 65). In addition, intermediate compositions that appear to be pigeonite (up to Wo_{11}) are reported by Nakamura et al. (1993) in Y-82002 and Schulze et al. (1994) in Rumuruti. Ca-rich pyroxenes are more abundant than low-Ca pyroxene. Typical compositions are given in Tables A3.2 and A3.3.

Low-Ca pyroxene compositions show a wide spread of Fs contents, e.g. $Fs_{0\text{-}30}$ in PCA91002 and Rumuruti, and tend not to show peaks in histograms (Fig. 62). Wo contents are generally <2 mol % (CaO <1 wt %). Minor element contents are shown in Figure 66. One intermediate pyroxene composition, Wo_4, is notably more aluminous (7 wt % Al_2O_3) than any other pyroxene compositions (Schulze et al. 1994). Low-Ca pyroxene in Carlisle Lakes and ALH85151 commonly shows patchy zoning (Weisberg et al. 1991). Pyroxene grains are zoned in irregular patches, or the zoning is parallel to twin lamellae. Fayalite-rich olivine may occur as veins in the pyroxene and/or rims on it. An example of this zoning from one chondrule is from a composition of $Wo_{1.3}Fs_{19}$ to $Wo_{1.7}Fs_{28}$.

Fs contents of *Ca-rich pyroxenes* range from Fs_9 to Fs_{20}. In Acfer 217 and Rumuruti,

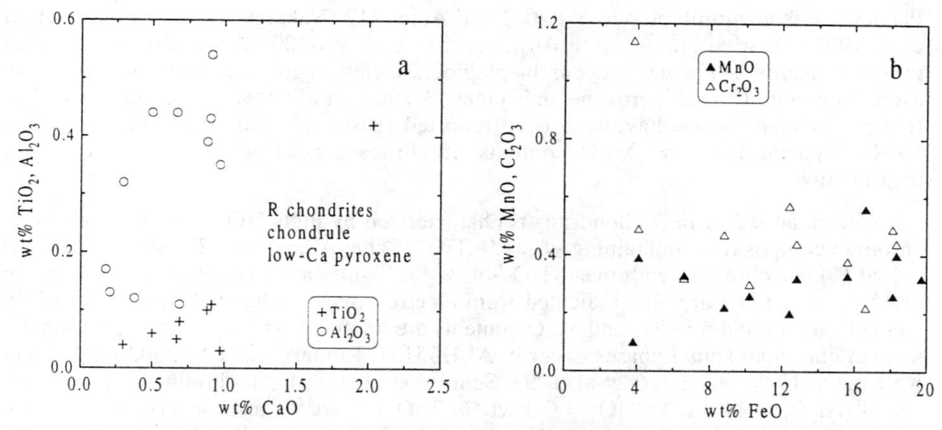

Figure 66. Minor element variation plots for low-Ca pyroxene from chondrules in R chondrites. One intermediate pyroxene composition with 2 wt % CaO (Wo_4) and 0.4 wt % TiO_2 also contains 7 wt % Al_2O_3 (Schulze et al. 1994). Data sources: Rubin and Kallemeyn (1989, 1994), Schulze et al. (1994), Bischoff et al. (1994), Weisberg et al. (1991).

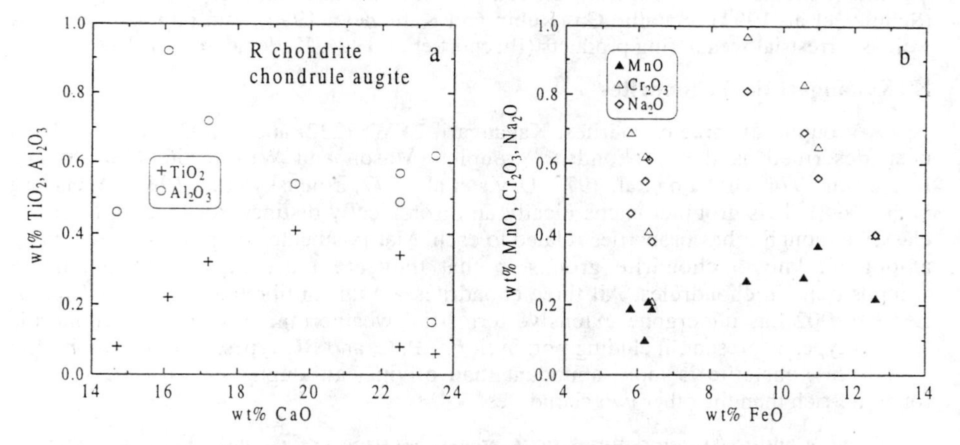

Figure 67. Minor element variation plots for Ca-rich pyroxene from chondrules in R chondrites. Data sources: Rubin and Kallemeyn (1989, 1994), Schulze et al. (1994), Bischoff et al. (1994).

most Ca-rich pyroxenes in chondrules and matrix are diopsides with a composition of $Fs_{11}Wo_{46-47}$ (Bischoff et al. 1994, Schulze et al. 1994). Al_2O_3 and TiO_2 contents are considerably lower than in other chondrite groups (Fig. 67). Na_2O contents are comparable to those in ordinary chondrites and significantly higher than Na_2O contents of augites in carbonaceous chondrites. Ca-rich pyroxenes contain minor NiO, ~0.12 wt % in PCA91002. In one type II chondrule in PCA91002, mesostasis consists of elongated skeletal and dendritic diopside grains embedded in a fine-grained to glassy diopsidic groundmass (Rubin and Kallemeyn 1994).

A range of *plagioclase* compositions is observed in R chondrites (Table A3.5). The mesostases of most chondrules are devitrified and consist predominantly of albite. Average compositions are $Ab_{87}Or_4$ in PCA91002 and ALH85151, $Ab_{82}Or_4$ in Carlisle Lakes, and $Ab_{86}Or_4$ in Rumuruti (Rubin and Kallemeyn 1989, 1994; Schulze et al. 1994).

Plagioclase is unequilibrated in Y-82002 and Acfer 217 (Nakamura et al. 1993, Bischoff et al. 1994). An-rich grains, up to An_{97}, occur in both Y-82002 and Acfer 217. An-rich grains in Acfer 217 usually occur in plagioclase-rich aggregates and chondrules, in association with Ca-rich pyroxene and spinel (Bischoff et al. 1994). Rare grains of alkali feldspar and orthoclase have also been reported (Bischoff et al. 1994, Schulze et al. 1994). Typical FeO and MgO contents of albites are <1 wt % and <0.1 wt %, respectively.

Chromian *spinel* in R chondrites is characterized by high TiO_2 contents, with mean chromite compositions containing ~6 wt % TiO_2 (Table A3.6). Only PCA91002 (TiO_2 = 3.4 wt %) has chromite with mean TiO_2 <4 wt %. Significant amounts of ferric iron (up to 15 wt % Fe_2O_3) are also indicated from electron microprobe analyses. Mean Al_2O_3 contents are around 5 wt %, and MgO contents are around 1 wt %. Two compositionally distinct chromian spinel phases occur in ALH85151, Rumuruti and Y82002 (Rubin and Kallemeyn 1989, Nakamura et al. 1993, Schulze et al. 1994). An aluminous "pleonaste" spinel (Al_2O_3 = 44 wt %; TiO_2 = 0.4 wt %, ZnO = 1 wt %) also occurs in Rumuruti. Magnetite is conspicuously absent from R chondrites except for its occurrence in one clast in PCA91241 (Rubin and Kallemeyn 1994, Kallemeyn et al. 1996).

Minor and accessory phases in R chondrites include chlorapatite, merrillite and ilmenite (Schulze et al. 1994, Bischoff et al. 1994, Rubin and Kallemeyn 1994), tridymite (Schulze et al. 1994), metallic Cu (Rubin and Kallemeyn 1994), and phyllosilicates, as well as terrestrial weathering products (Bischoff et al. 1994, Kallemeyn et al. 1996).

K (Kakangari-like) chondrites

A grouplet of three chondrites, Kakangari, LEW87232, and Lea County 002, have been described as the K chondrite grouplet (Mason and Wiik 1966, Graham and Hutchison 1974, Graham et al. 1977, Davis et al. 1977, Zolensky et al. 1989a, Weisberg et al. 1996). This grouplet is chemically and isotopically distinct from the O, E, and C classes although it has properties related to each. Matrix silicate compositions are unique among all known chondrite groups in that they are more Mg-rich than silicate compositions in chondrules. All three chondrites are unequilibrated, petrologic type 3. Lea Co. 002 has undergone extensive terrestrial weathering. A variety of chondrule textural types is present, including porphyritic, RP, C, and BO types. In most porphyritic chondrules, enstatite is more abundant than olivine, although Lea Co. 002 is more forsterite-rich than the other two chondrites.

Olivine and *pyroxene* compositions are all Mg-rich (<Fa_{10} and <Fs_{10}), and show a more restricted range in LEW87232 and Lea Co. 002 than in Kakangari (Zolensky et al. 1989a, Weisberg et al. 1996). Type II chondrules are absent. All three chondrites contain *low-Ca pyroxene* (clinoenstatite) and *Ca-rich pyroxenes* (Wo_{30-50}), and intermediate compositions up to Wo_{10} have been reported for Kakangari. Chondrule *mesostases* are generally clear to devitrified glass or are cryptocrystalline (Zolensky et al. 1989a, Weisberg et al. 1996).

In Kakangari, *metal* commonly occurs within chondrules, whereas in LEW87232 metal occurs as isolated objects outside the chondrules. Metal in LEW87232 is rimmed by enstatite laths similar to matrix enstatite. Kamacite compositions range from 5-8 wt % Ni in all three meteorites, and Co contents range from 0.2-0.5 wt % (Weisberg et al. 1996). Cr and P are generally <0.1 wt %. Kakangari and LEW 87232 have two types of taenite, one with ~24 wt % Ni and the other with 30-34 wt % Ni. Troilite is present in all three chondrites.

CALCIUM-ALUMINUM-RICH INCLUSIONS (CAI) OR REFRACTORY INCLUSIONS, FREMDLINGE AND AMOEBOID OLIVINE AGGREGATES

Introduction

Refractory inclusions or CAIs (Calcium-Aluminum-rich Inclusions) are a mineralogically and chemically diverse group of objects which occur mainly in carbonaceous chondrites, but are also found rarely in ordinary and enstatite chondrites. CAIs appear to have experienced complex evolutionary histories, sometimes involving several episodes of processing within the solar nebula and later after accretion within parent bodies. They have been the subject of intense study in a concerted effort to understand the physical and chemical conditions that were prevalent in the very earliest stages of the formation of our solar system. These studies have provided significant insights into very high temperature processes in the solar nebula, such as evaporation, condensation and melting, but even now some aspects of CAIs remain enigmatic and unresolved. As a consequence CAIs continue to be an exciting and fertile area of research. In this section, we review the mineralogy of CAIs in some detail, but we recognize, however, that in concentrating exclusively on the mineralogy of CAIs, we can present only a partial picture of their history. Many of the constraints on the origin of CAIs have come from detailed isotopic and bulk compositional studies which are beyond the scope of this review. These aspects, as well as the broader implications of CAIs for solar nebular models, have been discussed elsewhere in several excellent reviews (e.g. Grossman 1980, MacPherson et al. 1988, Podosek and Cassen 1994, MacPherson et al. 1995).

The mineralogy of CAIs is dominated by oxides and silicates, rich in Ca and Al, in addition to minor Ti and Mg. These phases all have extremely high vaporization temperatures, typically in excess of 1300K and, in many cases, are the phases predicted to condense from a solar nebular gas, based on equilibrium thermodynamic calculations (e.g. Grossman 1972). Among the refractory phases present in CAIs are corundum, hibonite, grossite, perovskite, anorthite and spinel, as well as melilite and fassaite. In the following sections we discuss the classification of CAIs, and the mineralogical and petrological characteristics of well characterized types of CAIs in the different chondrite groups.

Classification of CAI

The classification of CAIs is dominated by work on CV chondrites, especially Allende, mainly because of the large amount of material which became available after this meteorite fell in 1969. CV chondrites contain the highest abundance of CAIs of all the chondrite groups and contain a variety of different inclusion types. The number of different types of inclusions which occur in other chondrite groups is more restricted. Although CAIs appear to have a bewildering array of different mineralogical, chemical and isotopic characteristics, there are probably a rather limited number of basic inclusion types. The boundaries between different inclusion types are, in some cases, somewhat arbitrary and many cases gradations exist between one inclusion type and another.

Initial classifications of CAIs were based on their bulk compositional characteristics. Mason and coworkers (Martin and Mason 1974, Mason and Martin 1977, Taylor and Mason 1978, Mason and Taylor 1982) found that CAIs in Allende could be divided into 6 different groups (Fig. 68), based on the shapes of their rare earth element (REE) abundance patterns, one of which (Group IV) consists of ferromagnesian objects such as chondrules. The behavior of the REE in CAIs is strongly controlled by their relative volatilities during processes such as condensation and evaporation (e.g. Boynton 1989).

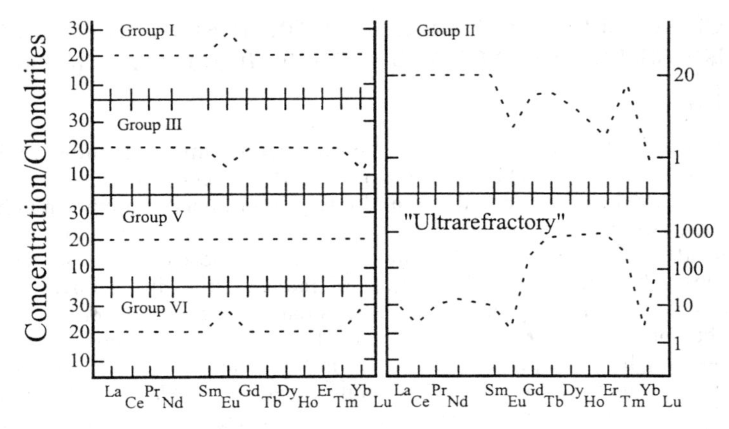

Figure 68. Schematic diagram showing generalized REE patterns observed in refractory inclusions from different compositional groups. After MacPherson et al. (1988) from data of Mason and Martin (1977) and others.

Hence the characteristic shapes of the REE patterns provide important constraints on the origins of individual CAIs. For example, Type II REE patterns appear to be formed solely during condensation.

Most workers have used classification schemes based on the petrographic properties of CAIs, rather than their chemistry. Grossman (1975) divided CAIs in Allende into 'coarse-grained' and 'fine-grained' inclusions, based on whether the components in the CAIs were too fine-grained to be adequately studied using a petrographic microscope. Two further types of 'coarse-grained' CAIs were recognized, termed Type A and Type B based on their mineralogical characteristics. More recently, an additional group of coarse-grained CAI, Type C has also been recognized (Wark 1987). Figure 69 illustrates the mineralogical differences between Type A, B and C inclusions in terms of the modal abundances of the two major mineralogical components, fassaite and melilite, and a third component consisting of all other phases (e.g. hibonite, spinel, anorthite and perovskite). Grossman and Ganapathy (1976) further divided 'fine-grained' inclusions into spinel-rich, fine-grained inclusions and amoeboid olivine aggregates (AOA) (Grossman and Steele 1976). An alternative petrographic scheme has also been proposed by Kornacki and Wood (1984a). This scheme is based on the relative abundances of three different components in CAIs in CV chondrites, termed rimmed or concentric objects, chaotic material (material that occurs interstitially between individual rimmed objects within aggregates) and finally, matrix-like material.

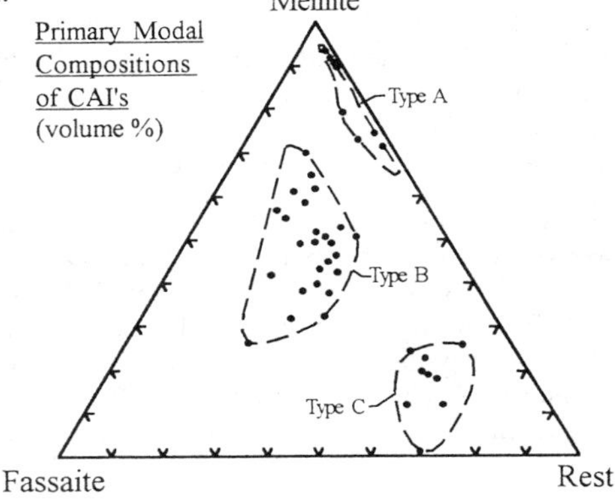

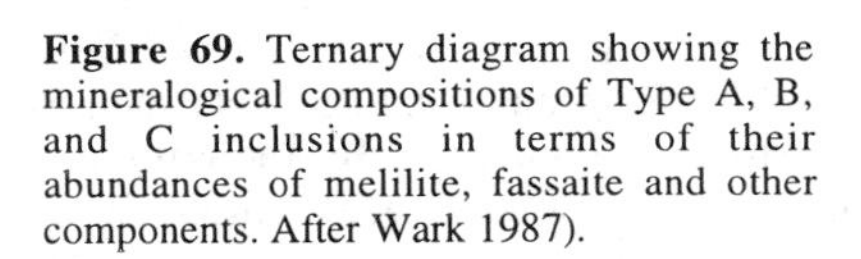
Figure 69. Ternary diagram showing the mineralogical compositions of Type A, B, and C inclusions in terms of their abundances of melilite, fassaite and other components. After Wark 1987).

As discussed by MacPherson et al. (1988), any classification scheme based solely on one particular type of property has a number of deficiencies. For example, Mason and Taylor (1982) pointed out that most CAIs do not have the mineralogical composition of Type A and B inclusions, as defined by Grossman (1975) and there exists a compositional continuum between the two types. Compositionally, CAIs of petrographic Type A, Type B, fine-grained and coarse-grained inclusions can all have Type II REE patterns.

Despite its deficiencies, the petrographic scheme, originally proposed by Grossman (1975), has gained widespread acceptance in the literature. Nevertheless, the compositional classification scheme of Mason and Taylor (1982) is an extremely important concept and has become a cornerstone of understanding the formation of CAIs.

Although the classification of CAIs is largely based on studies of CV chondrites, especially Allende, it can be extended to other chondrite groups, because essentially all the CAI types that are found in other carbonaceous chondrites, as well as the enstatite and ordinary chondrites, occur in CV chondrites. Type B inclusions, however, appear to occur exclusively in the CV chondrites and in this respect are somewhat anomalous.

Type A Inclusions. Type A CAIs can range up to 2 cm or more in size, but most typically are smaller than Type B inclusions. Type A inclusions are far more widespread than Type B inclusions, occurring in CV, CO, CR, CM, and ordinary chondrites and, unlike Type B inclusions, contain clinopyroxene as a minor phase. Based on studies of Allende, two types of Type A inclusion, designated "Fluffy" and "Compact" Type A inclusions have been recognized (MacPherson and Grossman 1979) on the basis of morphology, degree of alteration and melilite composition. However, because CAIs in Allende are much more altered than in other CV chondrites, it seems probable that this classification is somewhat artificial and that a gradation between these two types of inclusion exists.

Several examples of *"Fluffy" Type A* CAIs from Allende have been described (MacPherson and Grossman 1984, Kornacki and Wood 1984a), but they also occur in other CV chondrites (e.g. Vigarano, Sylvester et al. 1992). They show considerable textural and mineralogical diversity, but are characterized by their contorted, irregular shapes and melilite with åkermanite contents ranging between $Åk_0$ and ~$Åk_{35}$. Many inclusions consist of multiple nodules, each of which is surrounded by a Wark-Lovering rim sequence (see below), sometimes with each nodule separated by fine-grained, matrix-like material. The primary mineral assemblage of these inclusions is melilite, V-rich spinel, perovskite and often hibonite. Some inclusions contain coarse-grained melilite (up to 1.5 mm in size) that can be reversely zoned and contains kink bands. Other inclusions contain much finer-grained (<50 μm), undeformed melilite and coarse-grained crystals may be replaced by fine-grained melilite. The textural characteristics of unaltered "Fluffy" Type A inclusions in Vigarano indicate that these inclusions have experienced significant recrystallization and annealing.

Few *"Compact" Type A* inclusions have been studied in detail. According to Grossman (1975 1980) "Compact" Type A inclusions in Allende are coarse-grained (Fig. 70a), often consisting of only 2 or 3 mm-size crystals of blocky melilite, which enclose euhedral spinels (up to 50 μm) (Fig. 70b), perovskite, fassaite, and rhönite. The perovskite occurs as cubes (1-6 μm) and anhedral, purple grains (~70 μm). Prisms of fassaite (1-10 μm) are present as isolated grains within the melilite and are oriented in three directions, one direction being parallel to a set of cube edges in the perovskites (see also Simon et al. 1994b, 1995).

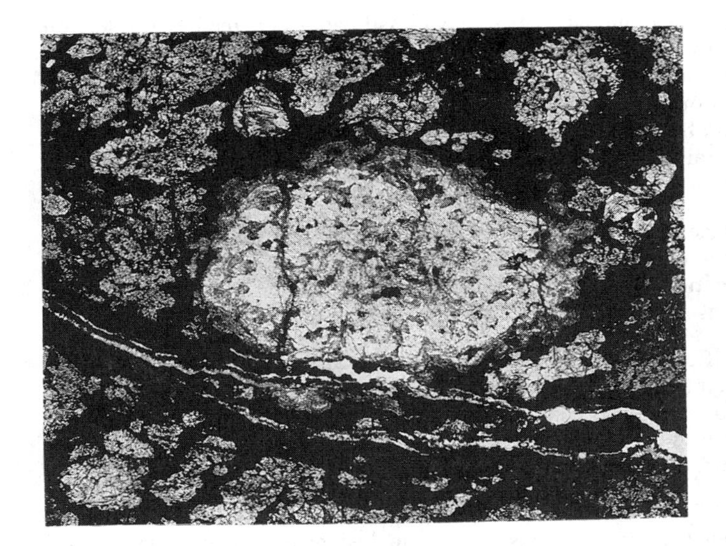

Figure 70. (a) Photomicrograph (plane polarized light) of an essentially unaltered "Compact" Type A inclusion from Efremovka. The inclusion is ~3.5 mm in maximum length and consists of melilite, with sub-ordinate spinel, very high Ti fassaite and minor rhönite. The inclusion is surrounded by a Wark-Lovering rim. Photo courtesy of G.J. MacPherson of the Smithsonian Institution.

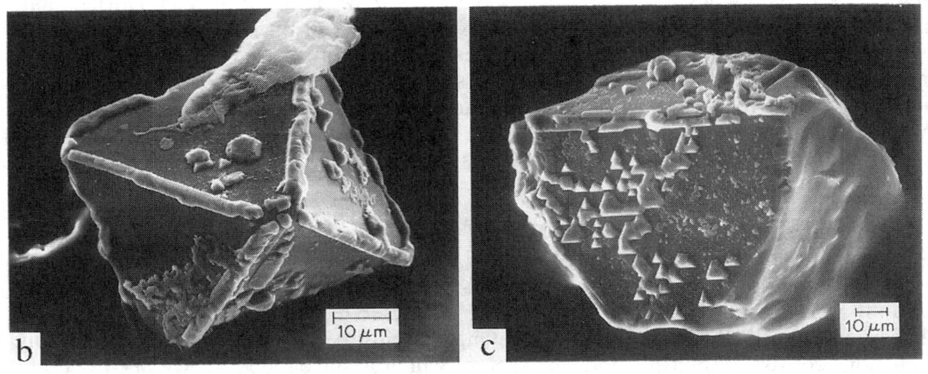

Figure 70 (cont'd). SEM images of spinel grains from Allende CAI. (b) Spinel from a "Compact" Type A inclusion (A1 1-16) with its apices decorated with perovskite rods. (c) spinel grain from same inclusion showing numerous epitaxial overgrowths of perovskite. Used by permission of the editor of *Geochimica et Cosmochimica Acta* from Barber et al. (1984), Figure 1.

Type B inclusions. Type B inclusions are unique to CV chondrites and have been intensively studied in several different meteorites (e.g. Allende, Efremovka, Leoville, Bali, Vigarano). Type B inclusions contain abundant fassaite (30-60 vol %), coarse ('primary') anorthite (5-25 vol %), spinel (15-30 vol %), and melilite (5-20 vol %) (Grossman 1980). Individual inclusions range in size from ~5 to ~25 mm. Wark and Lovering (1977, 1982a) further divided Type Bs into B1s which have melilite mantles and B2s which do not. A continuous sequence exists between these two endmembers (Wark and Lovering 1982a). The general model for the origin of Type B inclusions is that they crystallized from partially or completely molten droplets (e.g. MacPherson and Grossman 1981, Stolper 1982, Wark and Lovering 1982a,b; MacPherson et al. 1988). However, many Type B inclusions appear to have had complex histories (e.g. Podosek et al. 1991), possibly involving the capture of additional solid xenolithic material by molten droplets (e.g. El Goresy et al. 1985), multiple episodes of alteration and melting

(MacPherson and Davis 1993) or hot accretion into a parent body (e.g. Caillet et al. 1993).

Figure 71. (a) Photomicrograph (plane polarized light) of a Type B1 inclusion from Efremovka, about 6 mm at its maximum dimension. The thin section edge has truncated about 2/3 of the inclusion. A melilite mantle encloses a core consisting of melilite, pyroxene, spinel and anorthite and a Wark-Lovering rim is present on the margin of the inclusion, but the inclusion interior is virtually unaltered. Photo courtesy of GJ MacPherson, Smithsonian Institution.

Type B1 Inclusions have melilite-rich mantles, 0.5-3 mm wide, and fassaite-rich cores (Fig. 71a). They are typically coarser-grained (0.5-2 mm) than Type B2 inclusions (0.1-0.5 mm) and primary anorthite is usually concentrated in their cores. Although the mantle can be monomineralic, fassaite, anorthite and spinel may also be present (e.g. MacPherson and Grossman 1981, Wark and Lovering 1982a, Meeker et al. 1983, Podosek et al. 1991). Some inclusions contain little or no melilite in their cores (e.g. Allende Egg-3; Meeker et al. 1983). The mantles exhibit a range of textures from coarse-grained (0.5-1.5 mm), radially oriented laths to mosaics of fine-grained (<0.1 mm) equidimensional, triple-junctioned melilite. Both types of texture may be present in the same rim, sometimes with fine-grained material in the outer part of the rim. Type B1 inclusions usually contain more melilite and less fassaite than B2s. Melilite in the core is typically $Åk_{30-65}$, whereas in the rim the range is $Åk_{20-35}$ (but see below). Type B1s with unusual characteristics have also been reported (e.g. Simon et al. 1994b, 1995 in Axtell).

Type B2 inclusions have a uniform distribution of melilite, pyroxene, anorthite and spinel throughout their mass and unlike Type B1s show no zonation (Fig. 71b). Melilite ranges in composition from $Åk_{45-90}$ (Wark and Lovering 1982a).

Type B3 inclusions or *forsterite-bearing inclusions* are a relatively rare group of CAIs and only a small number have been described in detail (Blander and Fuchs 1975, Clayton et al. 1977, Dominik et al. 1978, MacPherson et al. 1981a, Clayton et al. 1984, Wark et al 1987, Davis et al. 1991). They include a higher proportion of the isotopically anomalous FUN inclusions than any other group of CAIs (e.g. Clayton et al. 1984, Davis et al. 1991). Wark et al. (1987) argued that this group of inclusions form a continuum with Type B1 and B2 CAIs and should be termed Type B3 inclusions. Their bulk compositions are less refractory than either the Type B1 or B2 inclusions, but their textures suggest that they formed by the same mechanism, i.e. crystallization from a liquid or a partially molten evaporative residue (Wark et al. 1987).

Type B3 inclusions are relatively large (>6-15 mm in size), vary in shape from

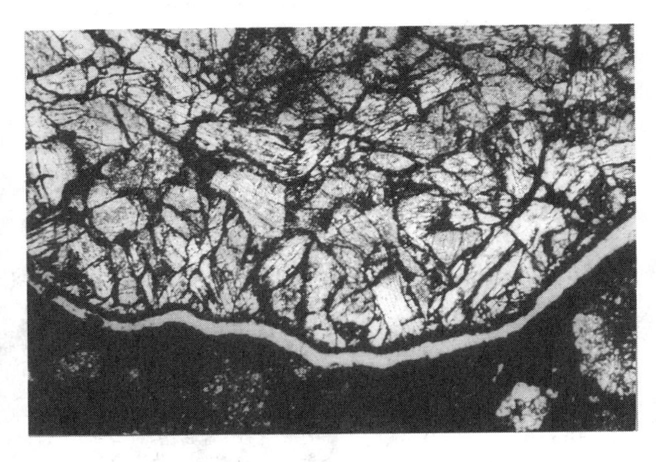

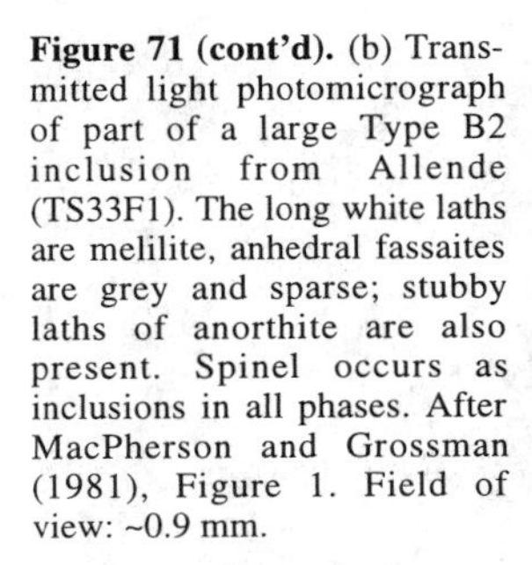

Figure 71 (cont'd). (b) Transmitted light photomicrograph of part of a large Type B2 inclusion from Allende (TS33F1). The long white laths are melilite, anhedral fassaites are grey and sparse; stubby laths of anorthite are also present. Spinel occurs as inclusions in all phases. After MacPherson and Grossman (1981), Figure 1. Field of view: ~0.9 mm.

spherical to subrounded to irregular and can contain vugs or vesicles. Mineralogically, they form a continuum from those dominated by fassaitic pyroxene (Ti-Al-bearing pyroxene) (50-63 vol %) with minor forsteritic olivine (1 vol %) to those in which forsterite is the most abundant phase (45 vol %) (Wark et al. 1987) and fassaite is subordinate (~31 vol %). Other primary phases are spinel (16-20 vol %), melilite (0-26 vol %) and minor anorthite (0-4 vol %). Some inclusions (e.g. Al6S3 - Clayton et al. 1977; TE - Dominik et al. 1978) have a concentrically zoned structure with olivine and fassaite in the core and spinel and fassaite in the surrounding mantle. In Allende, Type B3 inclusions have suffered secondary alteration in which melilite has been replaced by grossular, nepheline, hedenbergite, Al-Fe-diopside and monticellite (Clayton et al. 1984). However, Davis et al. (1991) have described one forsterite-bearing FUN inclusion from Vigarano which has suffered only minor secondary alteration.

Type C or plagioclase-rich inclusions. Grossman (1975) described a plagioclase-rich CAI in Allende and classified it as Intermediate (Type I) between Type A and Type B CAIs. Several other plagioclase-rich inclusions have now been recognized (e.g. Blander and Fuchs 1975, Lorin et al. 1978, Macdougall et al. 1981, Mason and Taylor. 1982, Wark 1987) in CV chondrites and one inclusion in the ungrouped carbonaceous chondrite, Adelaide. Wark (1987) argued that these CAIs constitute a compositionally and mineralogically distinct group of inclusions which he termed Type C. This designation is the one that has been adopted by subsequent workers (e.g. Beckett and Grossman 1988).

Type C inclusions are characterized by Group I or Group II trace element patterns, have plagioclase which is essentially pure anorthite, and are rich in fassaite. They are typically quite large (>2 mm) (Fig. 71c), coarse-grained and have somewhat variable textures. Wark (1987) recognized three textural subgroups, the first of which is characterized by coarse anorthite laths. The second group contains equigranular anorthite and pyroxene with an ophitic texture (Lorin et al. 1978) and the third group has a fine-grained groundmass and fassaite with a texture described by Wark (1987) as "lacy". Wark (1987) argued that the compositions and textures of Type C CAIs are consistent with crystallization from melt droplets which may have been liquid condensates. However, because of the difficulties of condensing liquids under low nebular pressures, Beckett and Grossman (1988) have presented arguments that the Type Cs could have formed as solid condensates (under special circumstances) which were later melted.

Fine-grained, spinel-rich inclusions. Fine-grained, spinel-rich inclusions were first recognized in Allende (Grossman 1975) and are characterized by their very fine-grained constituent phases (<1-~20 μm). Several different types of fine-grained inclusion have been recognized and a number of different classification schemes proposed (e.g. Wark 1979, Cohen et al. 1983, Kornacki et al. 1983, Kornacki and Wood 1984a), but none of these classification schemes appears to have been embraced by workers in the field. They have been variously termed fine-grained, alkali-rich, spinel aggregates (Type F) (Wark 1979), fine-grained, alkali-olivine aggregates (FAO), and Type Cc (concentric) inclusions (Cohen et al. 1983, Kornacki and Wood 1984a). Most commonly these inclusions are referred to as fine-grained, spinel-rich inclusions (e.g. Hashimoto and Grossman 1985, MacPherson et al. 1988, McGuire and Hashimoto 1989).

Most of the work on these objects has been carried out on inclusions from the Allende CV chondrite (e.g. Blander and Fuchs 1975, Wark 1978, 1979, 1981; Kornacki and Wood 1984a, McGuire and Hashimoto 1989), but they occur widely in other CV chondrites, e.g. Mokoia (Cohen et al. 1983), Y-86751 (Murakami and Ikeda 1994), and TIL 91722 (Birjukov and Ulyanov 1996). In Allende, the inclusions have irregular or lensoid shapes and have sizes (the longest dimension) up to 10 mm. Many of these inclusions consist of separate nodules of spinel, each of which is rimmed by a Wark-Lovering rim sequence and they are often mantled by fine-grained accretionary rims (see MacPherson et al. 1985). One of the widely recognized characteristics of fine-grained inclusions in Allende and other CV chondrites is that they often have an onion (Wark 1979) or concentric structure (Cohen et al. 1983).

Hashimoto and Grossman (1985) and McGuire and Hashimoto (1989) described the detailed mineralogical characteristic of zoned, fine-grained inclusions in Allende. In Allende, these inclusions are heavily altered and much of the mineralogy is of secondary origin. They consist of three mineralogically distinct, concentric layers (A-C) with zone C at the exterior. The central zone A consists of spinel (10-40%) + nepheline (20-40%) + Al-diopside + salite (10-35%), with minor olivine, sodalite ± grossular ± anorthite ± andradite ± hedenbergite ± perovskite ± ilmenite ± hibonite. The minerals have anhedral to subhedral morphologies and are loosely packed. Zone A sometimes contains 10-30 μm-sized concentric objects (see also Wark 1979, Cohen et al. 1983) with a core of spinel rimmed by nepheline, anorthite or grossular, which is mantled by clinopyroxene.

Zone B is texturally identically to zone A, but contains at least 10 vol % hedenbergite and/or andradite + spinel + clinopyroxene + nepheline ± minor sodalite ± olivine ± grossular ± anorthite ± perovskite ± ilmenite. Two mineral assemblages can exist in zone C. One is characterized by anhedral spinel, mantled by a layers of nepheline or anorthite, which are themselves rimmed by clinopyroxene. Nepheline + sodalite + olivine are interstitial phases with spinel ± clinopyroxene ± sodalite ± perovskite ± ilmenite. The second type consists of intergrown olivine and nepheline (sometimes associated with spinel) with interstitial clinopyroxene and sodalite.

Hibonite, analyzed by ion microprobe, from an Allende inclusion (Davis and MacPherson 1988) has a Type II REE pattern, suggesting that the inclusion may originally have consisted of fine-grained condensates. This is consistent with the view that these inclusions represent aggregates of independently formed grains (e.g. Wark and Lovering 1977, MacPherson and Grossman 1982, Brigham et al. 1985).

In Vigarano, fine-grained inclusions (e.g. 1623-14, 1623-16; Sylvester et al. 1992) are less altered and sometimes contain melilite. They consist of a core and a mantle rather than three distinct zones. The core zone is mineralogically similar to zone A of Allende inclusions and consists of spinel, fassaite, feldspathoids, anorthite and minor forsterite,

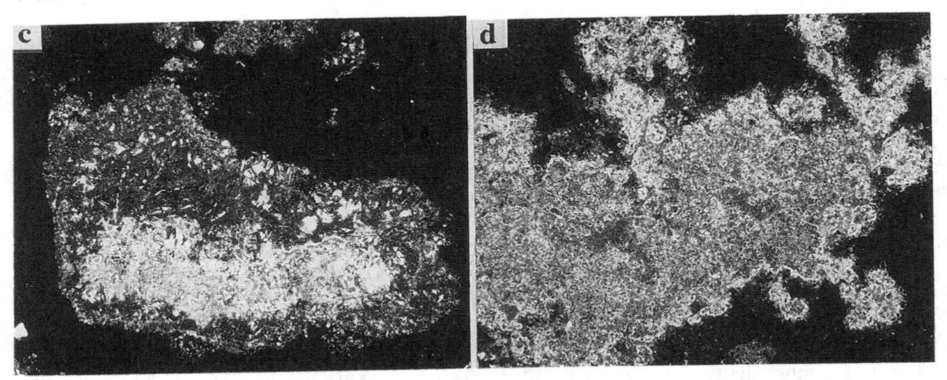

Figure 71 (cont'd). Transmitted light photomicrograph (plane polarized light) of (c) an Allende Type C inclusion (~2 mm in maximum dimension), consisting mostly of pyroxene and anorthite with subordinate spinel. The darker material around the perimeter consists of typical Allende fine-grained alteration products, probably nepheline, wollastonite, aluminous diopside, and grossular. No Wark-Lovering rim is present. (d) Similar photo of an amoeboid olivine inclusion from Allende consisting of fine-grained olivine, with spinel, aluminous diopside and some secondary alteration minerals such as nepheline. The field of view is 2.3 mm. Photos courtesy of G.J. MacPherson, Smithsonian Institution).

perovskite, diopside, ilmenite and an Fe-rich garnet and kirschteinite-rich monticellite. Fassaite, rather than Al-diopside, and salite are present in the core and grossular, andradite and hedenbergite are absent. The mantle consists of diopsidic augite, spinel, melilite and minor anorthite and feldspathoids.

Amoeboid olivine aggregates. Amoeboid olivine aggregates (AOA) have been recognized and described by several workers (e.g. Grossman and Steele 1976, McSween 1977c, Wark 1979, Cohen et al. 1983) in CV chondrites, including Allende and Mokoia. Strictly speaking, AOAs are not really refractory inclusions, but represent the least refractory endmember of a family of fine-grained inclusions. They are irregularly shaped, mm- to cm-sized objects (Fig. 71c) dominated by fine-grained olivine with accessory nepheline, sulfides, spinel, diopside, hedenbergite and fassaite. Olivine can occur as clumps of fine-grained (10 μm) crystals with other phases dispersed between the clumps. In some inclusions, fine-grained nepheline occurs in large patches. AOAs can enclose refractory nodules (<0.1 mm) consisting of a compact assemblage of spinel ± perovskite ± melilite (e.g. Hashimoto and Grossman 1987).

Unusual inclusions. Some inclusions have been recognized in CV chondrites that do not readily fit into the categories of inclusions described above. In Allende, these include the isotopically unusual, hibonite-bearing CAI HAL (J.M. Allen et al. 1980) and an unusual ultrarefractory inclusion (Wark 1986). Caillet (1994) also reported an unusual CAI from Axtell which lacks silicates and has a ferroan spinel in its core, intergrown with TiO_2-rich hibonite.

Wark-Lovering rims. Many inclusions in CV chondrites have distinct rims (20-50 μm in thickness) consisting of thin, multilayered bands (each ~5-10 μm in thickness), each with a different mineralogy (Fig. 72). The rims are continuous and maintain a remarkably constant width around the entire periphery of the inclusion. These rims were recognized by Christophe Michel-Lévy (1968) on inclusions in Vigarano, but were first studied in detail by Wark and Lovering (1977) and are now commonly referred to as Wark-Lovering rims. The mineralogy of the individual layers shows some variation and few rims show exactly the same sequence. However, an idealized mineralogy for a rim sequence (e.g. MacPherson et al. 1988) is as follows:

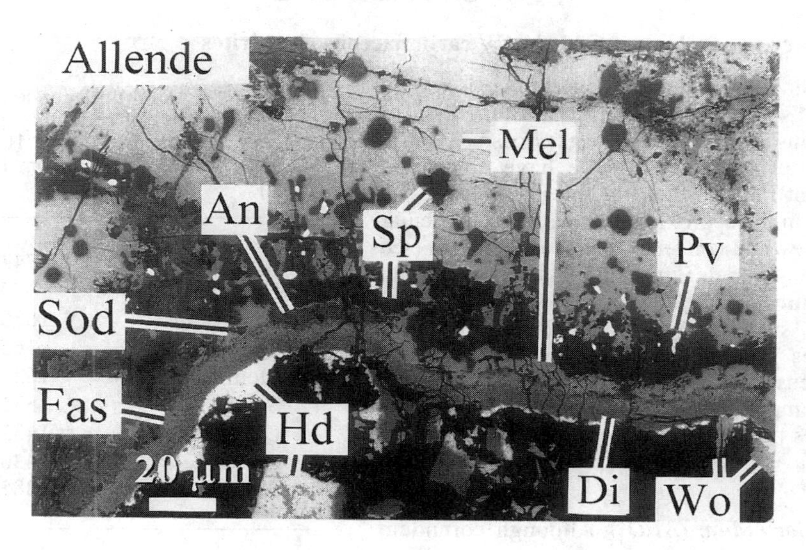

Figure 72. BSE image of part of a Wark-Lovering rim sequence on an Allende Type A inclusion. The layers which sequentially overlie the interior melilite consist of (1) spinel (Sp) + perovskite (Pv); (2) melilite (Mel) that is partially replaced with anorthite (An) and sodalite (Sod); (3) pyroxene that grades outward from fassaite (Fas; lighter gray) to diopside (Di; darker gray); (4) hedenbergite (Hd; bright); and (5) wollastonite (Wo). The dark region at the bottom is a cavity. After MacPherson et al. (1988).

- Innermost: spinel + perovskite ± hibonite ± fassaite
- Melilite or its alteration products + anorthite ± nepheline ± sodalite
- Pyroxene that changes in composition outwards through the rim from fassaite to Al-diopside
- Outermost: hedenbergite ± andradite or olivine ± spinel

The rim sequences on Type A inclusions in Allende appear to be somewhat different from those on Type Bs (Wark and Lovering 1977). In addition, there are differences between rims on CAIs in different CV chondrites, which probably reflect different processing after their formation (MacPherson et al. 1988). For example, alteration in the rims of oxidized CV chondrites such as Allende is considerable, whereas in the reduced CV chondrites, Vigarano and Leoville, the rims are relatively unaltered. It seems probable that the rims on CAIs in oxidized CV chondrites were all rather similar when they formed, but were subsequently altered to different degrees. This appears to be consistent with the observation that the CAIs in the different chondrites have been affected to similar degrees as the rims which surround them.

The origin of these rims is not completely resolved, but Wark and Lovering (1977) suggested that the different layers represent successive layers of condensation onto the exterior of CAIs and thus preserve a stratigraphy of condensation events in the solar nebula. The presence of refractory phases in Wark-Lovering rims that also occur in the interior of the CAIs which they enclose would suggest that there were successive episodes of evaporation and/or condensation. On the other hand, MacPherson et al. (1981) have suggested that different rim sequences on individual CAIs are the result of variable degrees of alteration of the host inclusion. A number of different possible scenarios for rim formation have also been discussed in some detail by Murrell and Burnett (1987) (see also Keller and Buseck 1994).

Primary mineralogy of CAIs in CV carbonaceous chondrites

Since the interpretation of CAIs is widely based on the calculated equilibrium condensation sequence (e.g. Grossman 1972, Lattimer et al. 1978), we have chosen to present mineral data for the major phases in CAIs in the approximate order that a particular phase would condense from a nebular gas (see Table 4) under typical solar nebular gas pressures (10^{-3}-10^{-5} atm). However, in many CAIs, such as Type B and C inclusions, the phases present did not necessarily form by condensation processes. In addition, because the temperature at which many of the minor phases present in CAIs condense is uncertain, we have treated these phases in a separate section.

Table 4. Equilibrium condensation temperatures (K) of refractory phases at a solar nebula pressure of 10^{-3} atm (Latimer et al. 1978; * = from Fegley 1991).

Phase	T (K)
Corundum	1742
Hibonite	~1735*
Perovskite	1680
Grossite	~1650*
Melilite	1625
Spinel	1535
Diopside	1435
Forsterite	1430
Anorthite	1385

Corundum (Al_2O_3). Although corundum is theoretically the first phase to condense from a gas of solar composition (e.g. Grossman 1972), it is remarkably rare in CAIs in CV and other chondrite groups. The dearth of corundum in CAIs has been attributed to the fact that during equilibrium condensation it reacts with the solar gas to form melilite (Grossman 1972). Wark (1986) reported an alumina phase, presumably corundum, as micron to submicron grains in the core of an unusual CAI (3643) from Allende (see Table A3.7). This phase contains ~1.7 wt % MgO, but low FeO (0.1 wt %). Corundum containing 1.6 wt % MgO and <0.04 wt % FeO also occurs in an Allende Type B inclusion (Bischoff and Palme 1987). Using TEM techniques, Greshake et al. (1996) found corundum inclusions, <0.2 μm in size, embedded in spinel in a refractory spherule consisting of spinel and perovskite from Acfer 094.

Hibonite ($CaAl_{12}O_{19}$). Keil and Fuchs (1971) first identified meteoritic hibonite in CAIs (their 'achondritic inclusions') in Leoville and Allende and it is now a widely recognized phase in CAIs from all the carbonaceous chondrite groups. Using powder x-ray diffraction, Keil and Fuchs (1971) determined hexagonal cell constants of a = 5.57±0.01 Å and c = 22.01±0.03 Å from a least squares analysis of 15 reflections for Leoville hibonites, comparable with cell constants for hibonite from Evisa Madagascar and published data.

Ideal hibonite has the simple composition $CaAl_{12}O_{19}$, but hibonites in CAIs are compositionally complex and exhibit coupled substitutions leading to significant concentrations of Ti, Mg, V, Fe, Si, Cr and Sc. Transition elements can also substitute into hibonite in more than one oxidation state. Based on electron microprobe analyses, there is a 1:1 correlation between Mg:Ti and a 1:-1 correlation of (Ti+Mg):(Al+V+Cr+Sc) in CAI hibonite indicating that Mg+Ti substitutes for two trivalent cations. This substitution indicates that the bulk of Ti in hibonite is Ti^{4+} which is charge balanced by Mg (e.g. Allen et al. 1978). However, a small component of Ti in hibonite is present as Ti^{3+} (e.g. Burns and Burns 1984, Beckett et al. 1988).

Hibonites exhibit cathodoluminescence under the electron beam which appears to be related to their minor element contents, but the exact relationship between minor element content and luminescence properties is poorly understood. Keil and Fuchs (1971) found two varieties of hibonite in Allende and Leoville CAIs, based on their compositions and

CL characteristics. Hibonite with bright-red orange CL is low in MgO (0.65-0.8 wt %) and TiO_2 (0.68-0.8 wt %), whereas bright blue luminescing hibonite has high MgO (3.3-3.7 wt %) and TiO_2 (6.5-7.9 wt %). Variations in hibonite CL which are related to composition have also been reported by Christophe Michel-Lévy et al. (1982).

Hibonite in CAIs contains up to ~8 wt % Ti, present as both Ti^{3+} and Ti^{4+} that can potentially be used to determine the fO_2 conditions at which the host CAI equilibrated. Using electron spin resonance (ESR) Beckett et al. (1988) measured Ti^{3+}/Ti^{4+} ratios in hibonite from a CAI from the Murchison CM chondrite, as well as several synthetic Ti-bearing hibonites with different Ti^{3+}/Ti^{4+} ratios. The results indicate substitution of Ti^{3+} and some Ti^{4+} into the 5-coordinated Al-site, the largest available crystallographic site in hibonite. Although some Ti^{4+} will substitute into the 5-coordinated Al-site, in order to maintain local charge balance, most Ti^{4+} probably substitutes into octahedral sites adjacent to tetrahedral Al sites in which substitution by Mg has occurred.

The concentrations of Ti^{3+} measured by ESR in Murchison hibonites range from 0.35 to 0.44 wt %, representing 23% of the total Ti present in the grain. Determination of the Ti^{3+}/Ti^{4+} ratio of the hibonite is based on the bulk Ti measured by electron microprobe, which carries some uncertainty. Using these data, Beckett et al. (1988) determined that the Murchison hibonite could have equilibrated with a gas of solar composition, i.e. under highly reducing conditions. However, because fO_2 is a function of temperature, there is some uncertainty in this value.

Electron microprobe analyses of hibonite can, at least, theoretically, be used to calculate Ti^{3+}, based on stoichiometric considerations. However, Beckett et al. (1988) found that, based on 81 analyses of hibonite with >1 wt % TiO_2, the calculated percentage of Ti^{3+} varies between -25 and +28%, suggesting that electron microprobe analyses cannot be reliably used to determine Ti^{3+} in hibonite.

Meteoritic hibonites vary in color from the most commonly observed blue (e.g. MacPherson et al. 1983) to green (Paque, pers. comm. in Ihinger and Stolper 1986) to orange (Allen et al. 1978). Occasionally, colorless crystals are also observed (J.M. Allen et al. 1980). Optical investigations (Ihinger and Stolper 1986) on synthetic hibonites show that its color is strongly controlled by oxygen fugacity. Hibonites change progressively in color from blue to green to orange to colorless as oxygen fugacity conditions are changed from below the iron-wüstite buffer up to air. The blue coloration is correlated with an absorption band at 715 nm in the optical spectra of both the synthetic samples and hibonite from the Blue Angel CAI in Murchison. These observations indicate that the Blue Angel hibonite equilibrated at an fO_2 of $\sim 10^{-(11-12)}$ assuming that it formed at 1430°C, oxygen fugacity conditions 4-5 orders of magnitude more oxidizing than canonical solar nebular conditions. Ihinger and Stolper (1986) found that orange hibonite from an Allende "Fluffy" Type A inclusion has a rather featureless, absorption spectra and equilibrated under much more oxidizing conditions ($fO_2 = \sim 10^{-6}$), perhaps indicating later processing in a more oxidizing environment

The presence of vanadium (Armstrong et al. 1982) and Fe^{2+}-Ti^{4+} charge transfer reactions (Burns and Burns 1984) have been proposed for the origin of the coloration in hibonite. However, Ihinger and Stolper (1986) have attributed the blue coloration to an increase in Ti^{3+} as conditions become more reducing. Synthetic hibonites which contain only Ti are all blue and display similar optical spectra, whereas synthetic hibonites which contain only V never developed a blue coloration under any of the oxygen fugacity conditions studied by Ihinger and Stolper (1986). However, the presence of V appears to be essential in producing the orange coloration found in complex hibonite compositions, because this color is not observed in synthetic samples which contain only Ti.

Representative electron microprobe analyses of hibonite from CAI in CV chondrites are reported in Table A3.8 and trace element data measured by ion microprobe are reported in Table A3.9.

Hibonite occurs as a rare, subordinate phase in *"Compact" Type A inclusions* in CV chondrites (Grossman 1975 1980). It occurs exclusively as acicular crystals, typically within 100 μm of the innermost inclusion rim (e.g. Allende; Grossman 1975, Leoville L1, Sylvester et al. 1993), associated with spinel, melilite and perovskite. Hibonite in an Allende "Compact" Type A inclusion (3898) (Podosek et al 1991) has a relatively narrow range in composition being TiO_2- and MgO-rich (8.5-9.2 wt % and 4.1-4.8 wt %, respectively). SiO_2 contents are low (0.34-0.74 wt %).

Grossman (1975), Blander and Fuchs (1975), Allen et al. (1978), Christophe Michel-Lévy et al. (1982), MacPherson and Grossman (1984) and Kornacki and Wood (1984a) have all reported hibonite in *"Fluffy" Type A inclusions* in Allende. The dominant hibonite type is the blue luminescent, orange pleochroic type, which is rich in Mg and Ti. Mg,Ti-poor hibonite is rare and occurs as patches within the Mg,Ti-rich crystals (e.g. Sylvester et al. 1992). This latter hibonite is colorless, but shows orange luminescence under an electron beam. Hibonite in "Fluffy" Type A inclusions (e.g. Grossman 1975, Allen et al. 1978, Christophe Michel-Lévy et al. 1982, MacPherson and Grossman 1984) spans nearly the entire compositional range (Fig. 73a) for TiO_2 and MgO reported by Keil and Fuchs (1971), although the range within individual inclusions is usually more restricted (e.g. Leoville inclusion L2; 1.99-4.4 wt % TiO_2; 1.37-2.23 wt % MgO; Christophe Michel-Lévy et al. 1982). In Allende CAIs, V_2O_3 contents can be variable (e.g. 0.3-2.5 wt %; Allen et al. 1978, MacPherson and Grossman 1984; 0.2-3.3 wt %; El Goresy et al 1980). More typically, however, the V_2O_3 contents of hibonites are relatively restricted, e.g. 0.32-0.41 wt % V_2O_3 in inclusion CG11 from Allende (Allen et al. 1978). Vanadium in hibonite in "Fluffy" Type A inclusions in Allende appears to show a positive correlation with Mg and Ti and is probably substituting for Al^{3+} in the hibonite structure (Fig. 73b). Allen et al. (1978) found Sc_2O_3 concentrations between 0.11-0.16 wt % for hibonite in Allende inclusion CG11.

Hibonite is extremely rare in *Type B inclusions* and appears to occur typically towards the exterior of the CAIs, close to the Wark-Lovering rim (e.g. Blander and Fuchs1975, Fuchs 1978). Blander and Fuchs (1975) observed 1-25 μm laths of hibonite

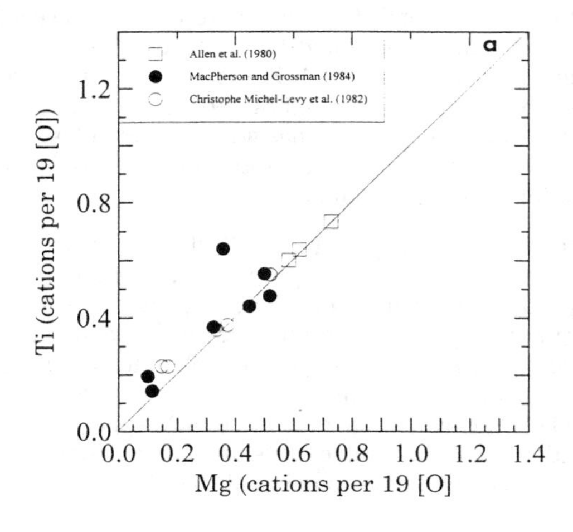

Figure 73. (a) Ti cations vs. Mg cations (per 19 oxygen atoms) in hibonite from "Fluffy" Type A inclusions in the CV chondrites, Allende and Leoville showing the strong positive correlation between these two elements. Data from Allen et al. (1980) and MacPherson and Grossman (1984) for Allende and Christophe Michel-Lévy et al. (1982) for Leoville.

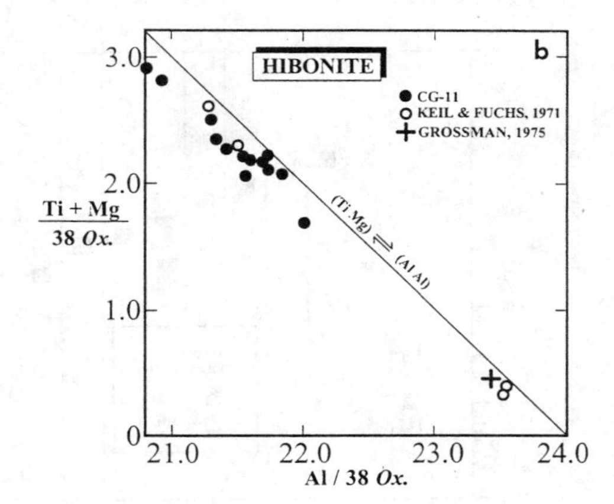

Figure 73 (cont'd). (b) Compositional plot of Ti+Mg vs. Al per 38 oxygens in hibonites from a "Fluffy" Type A inclusion (CG 11) from Allende consistent with the substitutional relationship $R^{2+} + R^{4+} = 2R^{3+}$. After Allen et al. (1978).

in the outer part of the mantle of a Type B1 inclusion (10/5) in Allende, which intrude toward the interior into a narrow band of Na-enriched glass. Similar observations were made by Fuchs (1978) in a Type B2 Allende CAI. (See also Bischoff and Palme 1987, Murrell and Burnett 1987, Podosek et al. 1991). The compositional range of these hibonites appears to be relatively restricted. Fuchs (1978) reported 2.5 wt % MgO and 4.3 wt % TiO_2., cf. 4 wt % MgO and 5.1-7.8 wt % TiO_2 (Bischoff and Palme 1987) and 3.6-5.8 wt % TiO_2 and 2.5-3.4 wt % MgO (3658; Podosek et al. 1991) in other Type B inclusions .

A Type B3 Vigarano CAI (1623-5) described by Davis et al. (1991) has a mantle that contains hibonite in association with melilite, spinel and perovskite. The hibonite has a restricted range of compositions with 4.15-5.88 wt % TiO_2 and 2.24-3.16 wt % MgO. Other detectable minor elements are FeO (0.26-0.57 wt %), V_2O_3 (0.21-0.28 wt %) and Cr_2O_3 (0.02-0.07 wt %).

Hibonite occurs rarely in *fine-grained, spinel-rich inclusions* (Kornacki and Wood 1984a, 1985a; Davis and MacPherson 1988, McGuire and Hashimoto 1989). Kornacki and Wood (1985a) found euhedral hibonites as tiny needles or coarse blades in the cores of concentric objects within fine-grained Allende inclusions (the so-called Type 2 inclusions of Kornacki and Wood 1984a). These hibonites have variable compositions with ~1-6.5 wt % TiO_2 and ~0.8-3.5 wt % MgO and up to ~0.45 wt % V_2O_3.

In CAIs with a large scale concentric zonation, hibonite occurs exclusively in zone A. McGuire and Hashimoto (1989) reported hibonite in one such CAI with 7 to 7.5 wt % TiO_2 and 3.7-3.9 wt % MgO but in a second inclusion, hibonite has a much more variable compostion with 0.2-5.3 wt % TiO_2 and 0.2 to 6.2 wt % MgO. SiO_2, Cr_2O_3, V_2O_3 and FeO all occur at concentrations levels <1 wt %. Hibonite in an unaltered, fine-grained inclusion in Vigarano ranges in composition from 4-5 wt % TiO_2 and ~0.2 wt % SiO_2 (Davis et al. 1987, Mao et al. 1990).

Hibonite occurs in several *unusual inclusions*. The core of the unusual Allende CAI, HAL consists almost exclusively of three, large (up to 1 mm), euhedral grains of hibonite (J.M. Allen et al. 1980). Two of the grains have a hexagonal outline with a frosty core and a colorless, transparent rim. The core region contains fine needles of a TiO_2-rich phase, possibly rutile, which are oriented with their long axes at 120° to one another in the (0001) plane of the hibonite. The hibonite is almost pure $CaAl_{12}O_{19}$, with very low

TiO_2 (0.06-1.06 wt %) and MgO (<0.01 wt %). Other elements (e.g. Si, V, Cr, Y) are

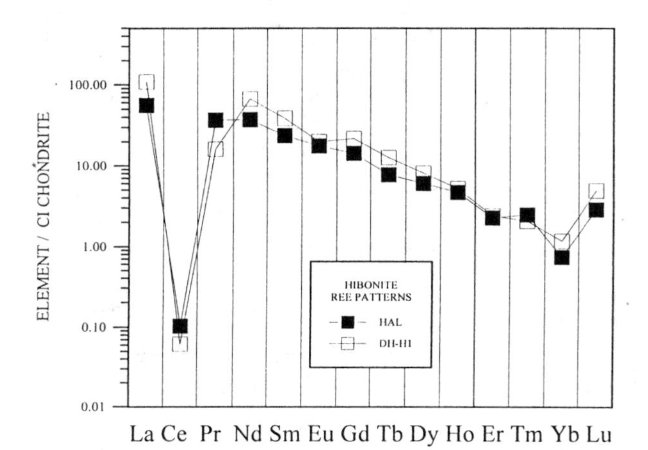

Figure 74. REE abundance diagram (normalized to CI chondrites) for hibonite from the unusual CAI HAL and hibonite from inclusion DH-H1 from the ordinary chondrite Dhajala. Ion microprobe analyses; data from Hinton et al. (1988).

below detection limits for EPMA. Compositional profiles from core to rim of one grain show some zoning in TiO_2, FeO and Sc_2O_3, with Ti and Sc highest in the core and Fe showing the reverse trend (~0.3 to 0.4 wt %). Fine-grained hibonites also occur in a layer which surrounds the inclusion and are richer in Zr, Sc, Ti, Fe and Si, but have very low MgO contents (0.25-0.59 wt %) (J.M. Allen et al. 1980).

Hibonite in HAL has a very unusual, highly fractionated REE pattern (Hinton et al. 1988), similar only to that found in a hibonite from inclusion DH-H1 in the ordinary chondrite, Dhajala. The pattern shows smoothly decreasing abundances from the LREE (La ~60 × CI) to the HREE (Yb, 1 × CI) (Fig. 74), with a superimposed depletion in Ce (~0.2 × CI). Hinton et al. (1988) suggested that the general depletion from LREE to HREE was produced by partitioning between hibonite and another solid phase or liquid.

Hibonite in the core of a complex, ultrarefractory inclusion (3643) from Allende (Wark 1986) shows partial pseudomorphic replacement by melilite and grossular. Hibonite crystals have variable compositions with 0.06-4.4 wt % TiO_2, 0.4-2.8 wt % MgO and 0.1-1.3 wt % V_2O_3. In the mantle of the inclusion, hibonites have higher MgO (2.6-3.8 wt %) and TiO_2 (4.1-6.1 wt %), but lower V_2O_3 (0.18-0.28 wt %) than the core.

Steele (1995b) described an unusual hibonite-bearing inclusion from Allende that has a central cluster of subparallel, tabular hibonites with low TiO_2 (~1.2-1.7 wt %) and MgO (~0.5-0.8 wt %) contents, surrounded by a region of zoned spinel. Hibonite also occurs within spinel and at the edge of the inclusion. Steele (1995b) interpreted the core hibonite as a relict on which later spinel and hibonite crystallized. The latter two occurrences of hibonite have much higher TiO_2 (~4.6-8.2 wt %) and MgO (2.4-4.4 wt %) contents than the hibonite core and FeO and V in the core hibonite are ~0.1 wt % and 240 ppm, compared with 0.29-0.70 wt % and 430 ppm, in the rim. The central hibonite grains also contain oriented, submicron lamellae of an exsolved phase, probably perovskite.

Hibonite also occurs in *other inclusions*. Wark and Lovering (1977) and Kornacki and Wood (1985a) reported hibonite in the cores of spinel-pyroxene inclusions from Allende. These inclusions (termed Type 3B in the classification scheme of Kornacki and Wood 1984a) are similar in morphology to the nodular or banded spinel-pyroxene inclusions found in CM chondrites (see Figs. 114a,b, below) (MacPherson et al. 1983,

1984; MacPherson and Davis 1994). These hibonites are compositionally variable with 2.82-7.3 wt % TiO_2, 1.7-3.75 wt % MgO, and ~0.05-0.22 wt % V_2O_3. Hibonite (8-9 wt % TiO_2; ~0.2 wt % SiO_2) also occurs in a melilite-rich Vigarano inclusion (1623-2) (Mao et al. 1990).

Grossite ($CaAl_4O_7$). The unnamed calcium aluminate phase, $CaAl_4O_7$ (now termed grossite; Weber and Bischoff 1994a) was first reported by Christophe Michel-Lévy et al. (1982) in a CAI from Leoville. The origin of $CaAl_4O_7$ in CAIs is uncertain. Some workers have argued that grossite is a stable condensate phase from a gas of solar composition, forming shortly after corundum and hibonite, but before perovskite, gehlenite and spinel (Blander and Fuchs 1975, Fegley 1991). However, in contrast, Geiger et al. (1988) have concluded that there is no stability field for grossite between 10^{-3} and 10^{-5} atm, based on a thermodynamic analysis. Its rarity in CAIs has been interpreted as evidence that it is not in fact a condensate, but may have crystallized from a melt (e.g. Christophe Michel-Lévy et al. 1982, Kimura et al. 1993) or have formed by sintering of preexisting refractory solids (Kimura et al. 1993). Despite these observations, Weber and Bischoff (1994b) and Weber et al. (1995) have presented textural and trace element evidence from inclusions in the CH chondrite Acfer 182 (see below) that grossite was probably a condensate phase in at least some cases.

Although rare, there are several reported occurrences of grossite in CV and other carbonaceous chondrites (see Weber and Bischoff 1994b for a review). The occurrence reported by Christophe Michel-Lévy et al. (1982) is in a large (6 mm) refractory inclusion (L2) in Leoville, probably a "Fluffy" Type A inclusion, where grossite coexists with gehlenitic melilite, spinel, perovskite and hibonite. In this inclusion, $CaAl_4O_7$ is a minor phase, occurring as highly birifringent blebs, 5-10 μm wide, enclosed in melilite, as well as associated with clusters of vermicular blebs of fine-grained (5-10 μm) perovskite. Grossite has also been reported in an Allende "Compact" Type A CAI by Paque (1987) and in two Type A CAI from Vigarano (Davis et al. 1987, Greenwood et al. 1993). In the inclusion, Victa, described by Greenwood et al. (1993), the grossite appears to have been partially replaced by pure hercynite. Electron microprobe analyses of $CaAl_4O_7$ (Christophe Michel-Lévy et al. 1982, Paque 1987; see Table A3.7) show that it is essentially stoichiometric, but contains significant amounts of minor elements such as SiO_2, (<~0.7 wt %) TiO_2 (<0.1 wt %), FeO (<~0.6 wt %) and MgO (<~0.13 wt %). FeO, in particular, appears to be most variable, ranging between 0.1 and 2.17 wt %.

Perovskite. Perovskite is widespread in CAIs in CV chondrites, but typically occurs as an accessory phase, often as small inclusions in phases such as melilite and fassaite. Marvin et al. (1970) were among the first authors to report perovskite in CAIs in Allende. Kornacki and Wood (1985b) have reported electron microprobe analyses of perovskite from several different types of CAIs in Allende and have shown that the perovskite minor element chemistry could be used to determine whether the inclusion has a Group II REE pattern (Mason and Taylor 1982). Inclusions with Group II REE patterns contain perovskites with <0.10 ZrO_2, whereas perovskites in inclusions with Group I, III, V and VI patterns typically contain >0.25 wt % ZrO_2. Electron microprobe analyses of perovskite are reported in Table A3.11 and trace element data in Table A3.12

Keller and Buseck (1994) reported TEM observations of perovskites from refractory (Wark-Lovering) rims on Type A and B CAIs in Allende and a Type B CAI in Vigarano and found that fine-scale twinning is commonly present. Twinning according to three twin laws is observed: (a) 90° rotation around [101], (b) 180° rotation around [101] and (c) a 180° rotation around [121], with the dominant being twinning on (121). These observations indicate that the perovskite was heated to above the cubic-orthorhombic phase transition (~1573K) and, based on studies of synthetic samples, cooled rapidly.

Perovskite is a common accessory phase in *"Compact" Type A inclusions* (e.g. Grossman 1975, 1980; Ulyanov et al. 1982, Kornacki and Wood 1985a, Fahey et al. 1987a, Podosek et al. 1991, Sylvester et al. 1993) and has a composition close to pure $CaTiO_3$. Grossman (1975) reported microprobe data for perovskites from a "Compact" Type A inclusion in Allende, which contain minor amounts of MgO (~0.15 wt %), Al_2O_3 (~0.31 wt %), Cr_2O_3 (<0.01 wt %) and FeO (<0.03 wt %) and detectable Y_2O_3, ZrO_2 and Nb_2O_5 (<0.3 wt %). Perovskites from the core and Wark-Lovering rim of an unaltered Efremovka inclusion (E2) (Ulyanov et al. 1982, Fahey et al. 1987a) also contain minor SiO_2, Al_2O_3, MgO and CaO. Rim perovskites tend to be slightly lower in SiO_2 (0.32 wt %, cf. 0.42 wt %), Cr_2O_3 (<0.02 wt % cf. 0.05 wt %) and MgO (0.14 wt % cf. <0.01 wt %) and higher in Al_2O_3 (0.63 wt % cf. 0.30 wt %) than those in the core. Ion microprobe analyses of REE in the core and rim perovskites (Fahey et al. 1987a) show that they have very similar, unfractionated, flat REE patterns (Fig. 75) at ~1000 × CI with negative Eu anomalies (~200 × CI), although there is some uncertainty (up to ~50%) in absolute concentrations, due to uncertainties in the sensitivity factors used for the analyses (Fahey et al. 1987a). Some of these grains may show zoning with REE enriched at their edges. There is also an enrichment in the refractory trace elements Zr, Nb and Y in the rim perovskites; Zr in the core perovskites ranges from 3100 - 3840 ppm, whereas in the rim the range is 5428-5813 ppm (Fig. 75). U (~11,000 ppb) and Th (45,000 ppb) concentrations in perovskites in the interior of an Allende "Compact" Type A inclusion (84C) have been reported by Murrell and Burnett (1987).

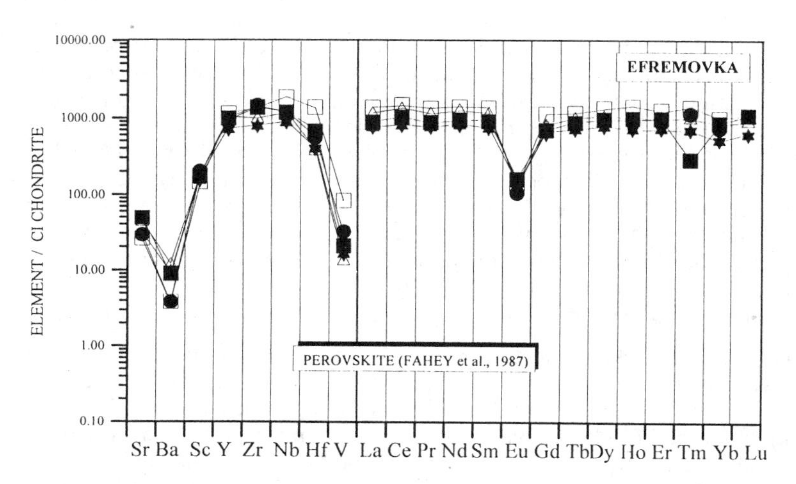

Figure 75. REE and refractory trace element abundance patterns (normalized to CI chondrites) for perovskites from a "Compact" Type A inclusion (E2) from Efremovka (ion microprobe analyses; data from Fahey et al. 1987). The data are from core and rim perovskites and show no systematic difference in REE patterns between the two different locations. Spot 1 = core; Spot 2 = intermediate; Spots 3-5 = rim.

Perovskite occurs associated with melilite, hibonite and spinel in *"Fluffy" Type A inclusions* in Allende and other CV chondrites (Grossman 1975, Allen et al. 1978, MacPherson and Grossman 1984, Meeker et al. 1983, Kornacki and Wood 1984a, 1985a,b; Kracher et al. 1985, Sylvester et al. 1992). In several Allende inclusions, perovskite contains minor amounts of MgO (<~0.23 wt %), Al_2O_3 (<~1.0 wt %), Cr_2O_3 (<0.2 wt %), FeO (<0.03 wt %), Y_2O_3 (<~0.21 wt %), ZrO_2 (<~0.68 wt %) and Nb_2O_5 (<0.04 wt %) (Grossman 1975). Similar minor element concentrations have been reported for perovskites in Allende inclusion CG11 (Allen et al. 1978) (0.06-0.39 wt % Sc_2O_3,

0.03-0.19 wt % FeO, 0.11-0.43 wt % ZrO_2, 0.06-0.35 wt % Y_2O_3, <0.07-0.14 wt % Cr_2O_3 and ~0.12 wt % V_2O_3). Kornacki and Wood (1985a,b) analyzed perovskites from two "Fluffy" Type A inclusions (the Type 3A inclusions of Kornacki and Wood 1984a) in Allende and found systematic differences in their minor element chemistries. One inclusion contains perovskites with higher Al_2O_3 (~0.75 wt %), V_2O_3 (0.13-0.15 wt %) and ZrO_2 (0.60 wt %) contents than the second (0.29-0.37 wt % Al_2O_3; <0.05 V_2O_3; 0.34 wt % ZrO_2). These perovskites are enriched in super-refractory elements (SRE), i.e. ZrO_2 and Y_2O_3, and occur in inclusions of different types which have Group III and VI REE patterns (e.g. Martin and Mason 1977).

U and Th concentrations in perovskites from an Allende "Fluffy" Type A inclusion (3529,44) have been reported by Ireland et al. (1990).

Perovskite is rare in *Type B inclusions* (Grossman 1980). However, Fuchs (1978) described several different occurrences of perovskite in a rhönite-bearing Type B2 inclusion from Allende. Perovskites occur as: (1) fine-grained, euhedral to subhedral crystals outlining larger, euhedral spinels within melilite, (2) vermicular perovskite enclosed in glass, sometimes at the contacts between rhönite and enclosing melilite, (3) vermicular perovskite in glass at grain boundaries between spinel and silicates, and (4) curved veins of perovskite which crosscut rhönite. Bischoff and Palme (1987) reported 0.17-0.33 wt % Al_2O_3 in perovskite from a Type B inclusion in Allende.

Perovskite, enriched in Nb and REE, occurs rimming spinel grains included in forsterite in the *Type B3* Allende inclusion TE (Dominik et al. 1978). Perovskite is also present in the mantle of a forsterite-bearing FUN inclusion from Vigarano (Davis et al. 1991) and may be present in inclusion CG-14 (Clayton et al. 1984).

Perovskite is a minor phase in *fine-grained inclusions* in Allende (e.g. Wark 1979, Cohen et al. 1983, Fegley and Post 1985, Kornacki and Wood 1985a,b; McGuire and Hashimoto 1989), Vigarano and Leoville (Davis et al. 1987, Mao et al. 1990, Sylvester et al. 1992). Several authors (e.g. Wark 1979, Cohen et al. 1983, Fegley and Post 1985) have all found fine-grained (~1 μm, but up to 8 μm in Kaba) perovskites inside spinels in the cores of concentric objects in CAIs in Allende, Mokoia and Kaba.

Kornacki and Wood (1985a,b) reported perovskite analyses from several spinel-rich inclusions in Allende. In one set of inclusions, ZrO_2 (0.16-1.4 wt %) and Y_2O_3 (0.01-0.39 wt %) are present in higher concentrations than a second group (<0.3 wt % ZrO_2; <0.2 wt % Y_2O_3). Al_2O_3 concentrations in these perovskites are ~1 wt % and other elements such as FeO, V_2O_3, and Ce_2O_3 are also detectable (0.1-0.79 wt %), but have variable concentrations. Perovskite analyses have also been reported by Murakami and Ikeda (1994) from a spinel-rich inclusion in Y-86751.

Perovskite occurs in layer II of the unusual hibonite-bearing Allende inclusion HAL (J.M. Allen et al. 1980). It contains minor amounts of Zr_2O_3 (0.58-0.83 wt %), Y_2O_3 (~0.1 wt %) and Sc_2O_3 (~0.1 wt %). Perovskite has been reported as an accessory phase in a melilite-rich inclusion from Vigarano (Mao et al. 1990).

Steele (1995b) found relatively large (15 μm) perovskite grains within the central hibonite core of an unusual hibonite-bearing Allende inclusion and smaller μm-sized grains in the surrounding spinel. The large grains contain fine, oriented lamellae of an Al_2O_3 phase, probably corundum, which has exsolved during cooling. Electron microprobe analyses of the perovskite, including the exsolution lamellae, show high Al_2O_3 contents (~1.9 wt %) whereas regions free of exsolution lamellae have lower Al_2O_3 (0.43 wt %), FeO (0-0.23 wt %), MgO (0-0.02 wt %) and SiO_2, suggesting that these elements are concentrated in the exsolved Al_2O_3. The bulk perovskite contains 0.09-0.33

wt % FeO and 0.03-0.17 wt % MgO. The central hibonite grain also appears to contain oriented, submicron lamellae of exsolved perovskite.

Semiquantitative PIXE analyses of perovskite from the mantle and crust region of an unusual refractory inclusion (3643) in Allende (Wark 1986) show that the heavy REEs, Y and Zr are much lower than in perovskite from typical Group I inclusions and are more typical of Group II inclusions (Martin and Mason 1977).

Melilite (gehlenite $Ca_2Al_2SiO_7$-åkermanite $Ca_2MgSi_2O_7$). Marvin et al. (1970) first reported melilite in CAIs in Allende and it was subsequently reported in Leoville by Keil and Fuchs (1971). Unlike terrestrial melilites, melilite in CAIs is low in Na_2O and can be consider to be a simple solid solution between gehlenite and åkermanite. The total compositional range of melilite in all CAIs in CV chondrites spans almost the complete range of the solid solution series ($Åk_{0-80}$) (Grossman 1980), but the range within individual inclusion types is somewhat more restricted (see below). Representative electron microprobe analyses of melilite are reported in Table A3.13 and trace element data, measured by ion microprobe, are reported in Table A3.14.

Melilite often shows strong evidence of deformation in the form of undulose extinction and/or well developed kink banding parallel to (001) (e.g. Grossman 1975, Allen et al. 1978, Christophe Michel-Lévy et al. 1982, Meeker et al. 1983, Barber et al. 1984, Podosek et al. 1991, MacPherson and Davis 1993). In some inclusions, brecciated regions within the melilite also occur (e.g. Podosek et al. 1991).

Grossman (1975) first described melilite in a *"Compact" Type A inclusion* (TS2F1) in Allende. In "Compact" Type As, coarse-grained melilite crystals ($Åk_{29-58}$) (2-3 mm in size) enclose crystals of spinel, perovskite, fassaite and rhönite. Melilite in "Compact" Type A inclusions has a very wide compositional range ($Åk_{1-80}$; Grossman 1980) and commonly shows reverse zoning rather than the normal or oscillatory zoning that is often found in Type B melilites. For example, melilite in Allende Type A inclusion 3898 (Podosek et al. 1991) has a range of compositions from $<Åk_1$-~$Åk_{45}$. Individual grains show monotonic reverse zoning with Al-rich margins ($Åk_{10-15}$), with core compositions in the range $Åk_{20-35}$ (Fig. 76). The most Mg-rich melilite compositions ($>Åk_{35}$) occur in narrow zones around fassaite inclusions in large melilite grains and Al-rich compositions occur only at the outer margins of the inclusion. Na_2O contents in melilites in this inclusion are below detection limits.

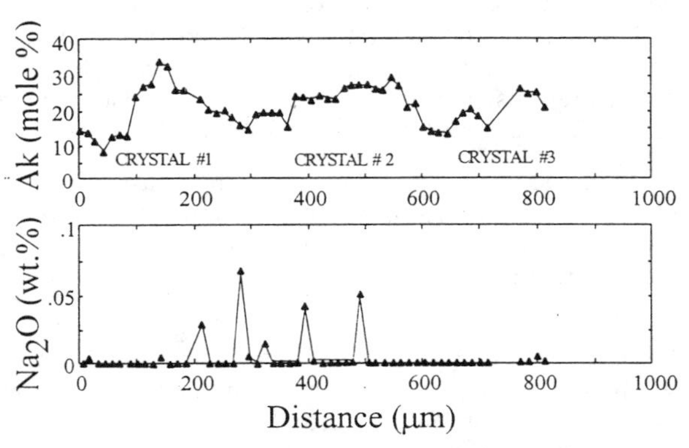

Figure 76. Compositional zoning profile across several adjacent melilite grains in a "Compact" Type A Allende inclusion (3898) showing the variation in åkermanite and Na_2O contents. Individual grains show monotonic reverse zoning in the Åk content of the melilite. After Podosek et al. (1991).

Petrographic descriptions and compositional data for melilite in "Compact" Type A inclusions in Vigarano and Leoville have been reported by Sylvester et al. (1993). Ulyanov et al. (1982) also described an unaltered "Compact" Type A inclusion from

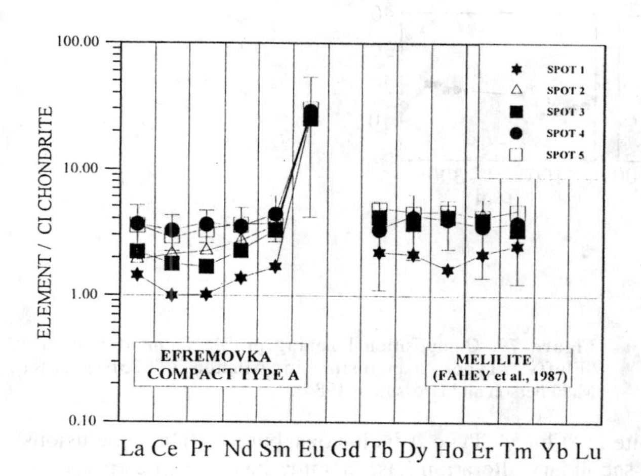

Figure 77. REE abundance diagram (normalized to CI chondrites) for melilites from core to rim of a "Compact" Type A inclusion from Efremovka (ion microprobe analyses: data from Fahey et al. 1987).

Efremovka which has a coarse-grained, melilite-rich core which is slightly richer in MgO (~$Åk_{24}$) than the rim melilites (~$Åk_{20}$) (Fahey et al. 1987a). However, ion microprobe analyses of the melilite show no compositional variation in trace element concentrations from core to rim (Fahey et al. 1987a). The melilites have smooth REE patterns (Fig. 77) with abundances 1-4 × CI and a large positive Eu anomaly (~11 × CI). Murrell and Burnett (1987) have reported concentrations of 2 ppm U in melilite from an inclusion (84C) in Allende.

In a TEM study of a "Compact" Type A inclusion from Allende, Barber et al. (1984) observed that melilite has a fragmented, but compact, non-porous texture in which cracks separate fragments which were once single grains. Overall the grains have a brecciated appearance and are highly strained with very high dislocation densities (>10^{10} cm^{-2}).

Melilite is one of the dominant phases in *"Fluffy" Type A inclusions* (e.g. Grossman 1980, MacPherson and Grossman 1984, Kornacki and Wood 1984a) and ranges from $Åk_{0-36}$ in these inclusions taken as a whole (Fig. 78), based on data from Allende, Vigarano, Leoville and Grosnaja (Fuchs 1971, Keil and Fuchs 1971, Blander and Fuchs 1975, Grossman 1975, Allen et al. 1978, Christophe Michel-Lévy et al. 1982, Meeker et al. 1983, MacPherson and Grossman 1984, Kracher et al. 1985). However, most inclusions contain melilite with a much more restricted range (<20 mol % Åk).

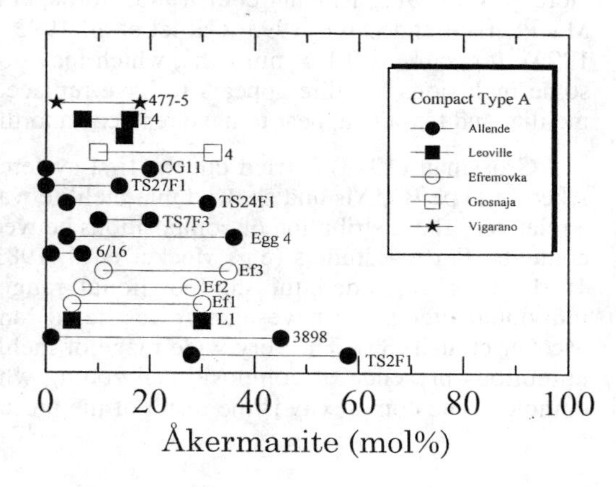

Figure 78. Compositional range of melilite in several "Compact" and "Fluffy" Type A inclusions from the CV chondrites Allende Vigarano, Leoville and Grosnaja (data from Fuchs 1971, Keil and Fuchs 1971, Blander and Fuchs 1975, Grossman 1975, Allen et al. 1978, Christophe Michel-Lévy et al. 1982, Meeker et al. 1983, MacPherson and Grossman 1984, Kracher et al. 1985).

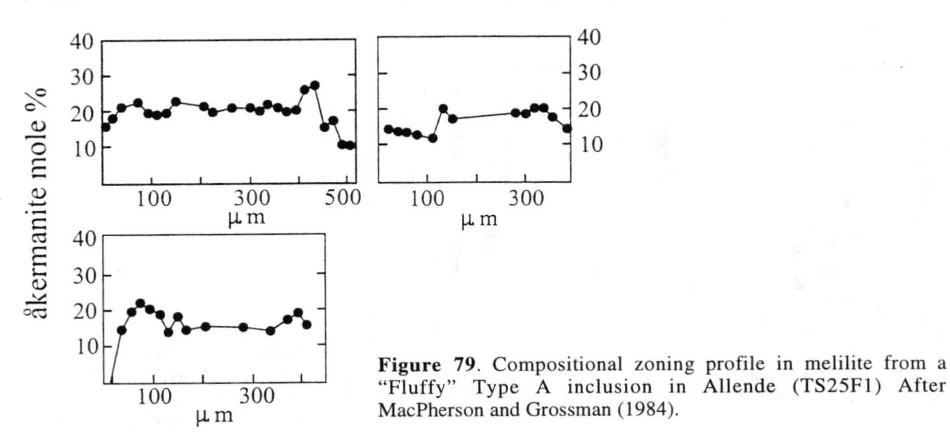

Figure 79. Compositional zoning profile in melilite from a "Fluffy" Type A inclusion in Allende (TS25F1) After MacPherson and Grossman (1984).

Zoning is present in melilite in "Fluffy" Type A inclusions, but in Allende inclusions is hard to determine, because secondary alteration has, in some instances, destroyed the original grain rim. Some inclusions contain subpopulations of melilites with reverse zoning (Allen et al. 1978, MacPherson and Grossman 1984) with an increase in gehlenite content from core to rim, e.g. $Åk_{17}$ in the core to $Åk_3$ in the rim (Fig. 79) (Allen et al. 1978). Sylvester et al. (1992) reported melilites in a much less altered "Fluffy" Type A inclusion from Vigarano which are zoned from $Åk_{8\text{-}18}$ in the cores to $Åk_{2\text{-}4}$ at the rims, set in a matrix of melilite grains of $Åk_{10\text{-}15}$. This range of melilite compositions thus falls within the range observed in Allende "Fluffy" Type As, suggesting that melilite rim compositions can still be obtained from altered inclusions in Allende.

Minor element contents in melilites from "Fluffy" Type A inclusions are all uniformly low, typically below detection limits for the electron microprobe with K_2O <0.02 wt %, Na_2O <0.03 wt %, FeO < 0.14 wt %, TiO_2 < 0.19 wt %, and Cr_2O_3 <0.02 wt % (Grossman 1975, Allen et al. 1978, Christophe Michel-Lévy et al. 1982, Meeker et al. 1983, MacPherson and Grossman 1984).

In inclusion CG11 from Allende, Allen et al. (1978) described needles of a unidentified prismatic phase, often oriented parallel to the (001) cleavage of melilite, which may be an exsolution product of some kind.

Melilite in *Type B inclusions* has been studied in considerable detail (e.g. Fuchs 1971, Gray et al. 1973, Blander and Fuchs 1975, Grossman 1975, Kurat et al. 1975, El Goresy et al. 1985, Kracher et al. 1985, Kornacki and Wood 1984a, Podosek et al. 1991, MacPherson and Davis 1993, Caillet et al. 1993, Sylvester et al. 1993, Goswami et al. 1994). It occurs as 0.1-5 mm laths, which may poikilitically enclose grains of spinel. In some inclusions melilite appears to have replaced anorthite and fassaite and in others, melilite and fassaite appear to have replaced anorthite.

Grossman (1975) carried out the first systematic study of melilite compositions in Allende Type B CAIs and showed that melilite was compositionally highly variable with a relatively flat distribution of compositions between $Åk_{15}$ and $Åk_{50}$ with a tail at high Åk contents. Further studies (e.g. Meeker et al. 1983, Kracher et al. 1985, Podosek et al. 1991) have expanded the compositional range somewhat ($Åk_{1\text{-}74}$) (Fig. 80), but individual inclusions have a more restricted range of melilite compositions (but see Meeker et al. 1983). The very wide range of melilite compositions is due, in part, to the ubiquitous presence of compositional zoning which, in some inclusions, is extremely complex. The complexity is the result of the presence of primary, igneous zoning, which

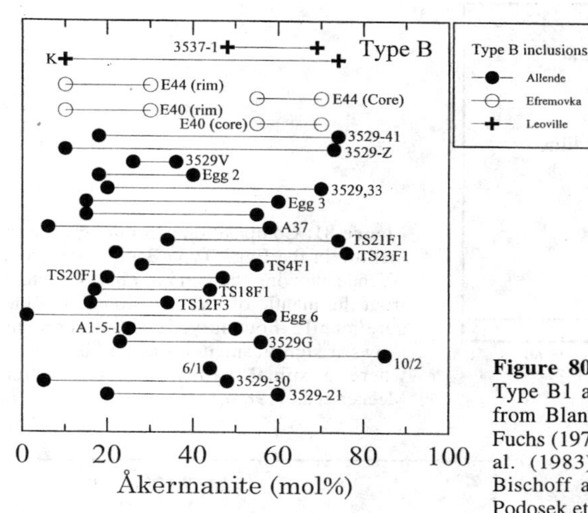

Figure 80. Compositional range of melilite from Type B1 and B2 inclusions in CV chondrites. Data from Blander and Fuchs (1975), Grossman (1975), Fuchs (1978), Wark and Lovering (1982b), Meeker et al. (1983), MacPherson and Grossman (1984), Bischoff and Palme (1987), Kracher et al. (1985), Podosek et al. (1991) and Goswami et al. (1994).

has, in many cases, been overprinted by zoning produced by episodes of recrystallization and possibly later melting (e.g. Podosek et al. 1991, MacPherson and Davis 1993).

One of the earliest forms of compositional zoning that was recognized in melilite in *Type B1 inclusions* is a progressive increase in Åk content from the rim into the core of the inclusion. This type of feature has been reported in several Type B1 inclusions from different CV chondrites including Bali (Kurat et al. 1975), Allende (Wark and Lovering 1982a, Meeker et al. 1983) and Efremovka (Goswami et al. 1994). Kurat et al. (1975) found that in an inclusion from Bali, the increase in Åk content from rim to core was well correlated with an increase in the Na_2O content of the melilite from 0.06 wt % at the exterior to 0.25 wt % in the core. In detail, in many inclusions the core to rim zoning can be complex, especially in inclusions which contain palisades of spinel (see below) (Wark and Lovering 1982b, Meeker et al. 1983) which often occur between the core and the mantle. In these inclusions, there can be an abrupt change in composition through the core/mantle boundary. For example, in the Allende Type B1 inclusion, Egg3, core melilite ranges in composition from $Åk_{55-64}$ (Wark and Lovering 1982b, Meeker et al. 1983), but becomes more gehlenitic outward, reaching $Åk_{31}$ at a spinel palisade where an abrupt discontinuity occurs (Fig. 81). Just inside the palisade the melilite composition is highly variable ($Åk_{31-54}$), but outside the palisade, in the mantle of the inclusion, the melilite is recrystallized, finer grained and more gehlenitic in composition ($Åk_{27-31}$) and progressively decreases to $Åk_{21-23}$ at the outer edge of the inclusion in contact with matrix (see also Wark and Lovering 1982b, Sylvester et al. 1993, Goswami et al. 1994 for data on inclusions from Allende, Vigarano and Efremovka).

Individual melilite grains in most Type B1 inclusions are zoned (see below), but Sylvester et al. (1993) and Caillet et al. (1993) have described a typical Type B1 CAI (3537-2) from Leoville which contains melilite with no evidence of normal or reverse zoning. However, the melilite does show an overall increase in Åk content from core to rim, with a well defined positive correlation between Åk content and Na_2O (Fig. 82). In melilite with $<Åk_{40}$, Na_2O is below detection (<~0.05 wt %), but above $Åk_{40}$, the Na_2O content increases progressively and systematically with Åk content (up to 0.22 wt % Na_2O at $Åk_{74}$).

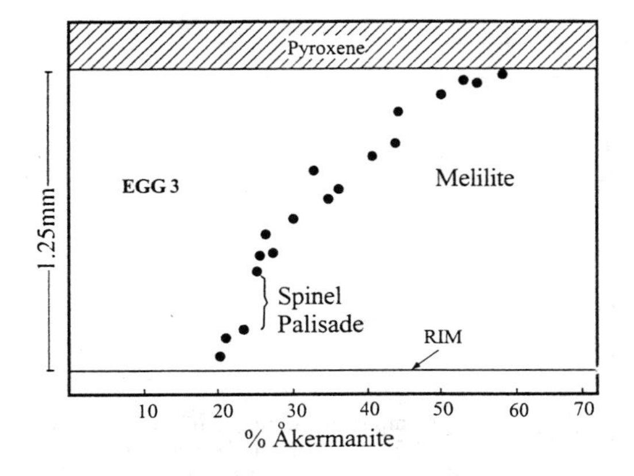

Figure 81. Compositional zoning profile in melilite in the Egg3 Type B inclusion from Allende showing the variation in Åk content from the mantle of the inclusion across the core/mantle boundary. The Åk content shows a significant jump at this boundary where a spinel palisade occurs. After Meeker et al. (1983).

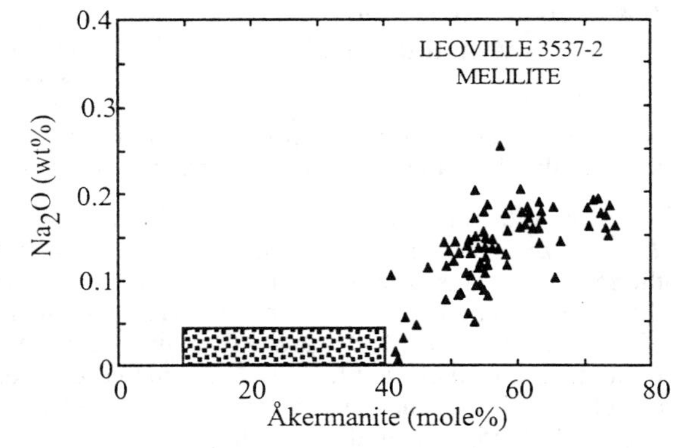

Figure 82. Plot of the Åk content of melilite vs. Na_2O in a Type B1 inclusion (3537-2) from Leoville. Na_2O is positively correlated with Åk content in melilite with Åk > 40, but is essentially below detection limits at Åk contents <40 (shaded box). After Caillet et al. (1993).

Individual melilite crystals in the core and mantle of typical Type B1 inclusions can be normally zoned, but reverse zoning can also be present (see below) (e.g. Meeker et al. 1983, Podosek et al. 1991), with Åk contents increasing towards the rim of individual crystals. For example, in the core of Allende inclusion 3529Z (Podosek et al. 1991) individual melilites have cores of ~$Åk_{30}$ with a smooth increase to $Åk_{58\text{-}73}$ in their rims (Fig. 83). The Na_2O contents of these crystals are typically very low (<0.03 wt %), but show random fluctuations up to ~0.10 wt % in their interiors, with a systematic increase towards the rims, up to ~0.14 wt % Na_2O. Melilite crystals in the inclusion mantle, that are oriented with their long axis normal to the margin of the inclusion, are zoned along their length, becoming more Åk-rich (up to ~$Åk_{58}$) into the interior of the inclusion, coupled with an increase in Na_2O from <0.03 wt % to 0.13-0.17 wt % at the core/mantle boundary.

In many Type B inclusions, melilite can also show reverse zoning (e.g. El Goresy et al. 1979, Meeker et al. 1983, MacPherson et al. 1984a). In a Type B1 inclusion (TS23F1) from Allende, MacPherson et al. (1984a) found that the melilite grains have extremely complex zoning patterns (Fig. 84). The cores are Åk-poor (up to $Åk_{27}$) and progressively increase in Åk content toward the rim (up to ~$Åk_{72}$). The Åk content then abruptly decreases reaching $Åk_{52}$ and then shows fluctuations up to $Åk_{60}$, before decreasing rapidly at the rim to $Åk_{68}$. Several other examples of similar zoning patterns were reported from the same inclusion. Based on an experimental study, MacPherson et al. (1984a) were able

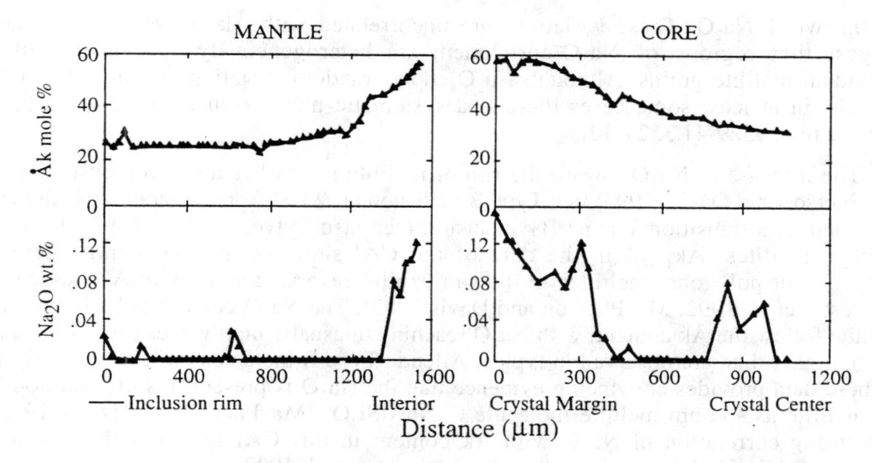

Figure 83. Compositional zoning profile in melilite in the Type B1 Allende inclusion 3529Z. The melilite shows typical normal zoning with an increase in Åk content toward the rim. Na_2O in melilite shows random fluctuations in the core, but increases systematically at the rims. After Podosek et al. (1993).

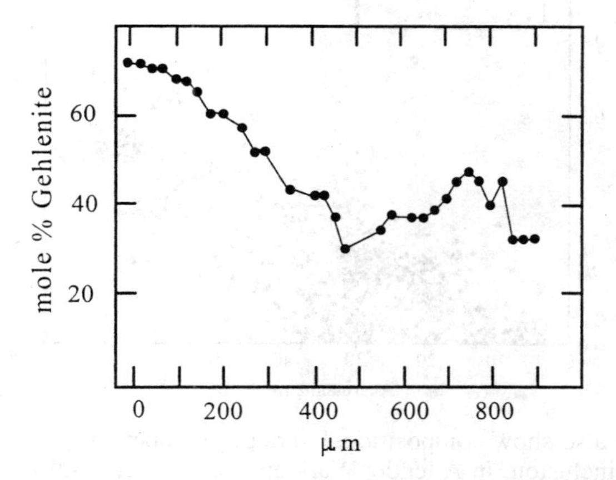

Figure 84. Complex compositional zoning profile in melilite in an Allende Type B1 inclusion (TS23F1) with normal zoning in the interior of the grains which changes to reverse zoning in the rim region. After MacPherson et al. (1984a).

to interpret this zoning pattern in terms of the crystallization sequence of anorthite, melilite and fassaite. For normal zoning, anorthite crystallizes before melilite or fassaite. However, at cooling rates between 0.5 and 50°C/hour, anorthite nucleation is suppressed and melilite and clinopyroxene coprecipitate. The crystallization of fassaite causes the Al/Mg ratio of the melt to increase, driving the crystallizing melilite to more gehlenitic compositions, causing the reverse zoning.

As noted earlier, some Type B inclusions have been affected by later processes, such as recrystallization which have produced additional complexities in melilite zoning (Podosek et al. 1991, MacPherson and Davis 1993). These complexities will be not be examined in detail here, but are sometimes related to different growth zones, indicated by the presence or absence of inclusions of spinels within the melilite. In several inclusions in Allende (e.g. 3529-41, 3529-30 and 3658; Podosek et al. 1991), random fluctuations in Na_2O content are present which range from below detection limits to regions with up to

0.1-0.2 wt % Na_2O. These variations are uncorrelated with Åk or FeO contents and suggest that regions of Na_2O-enrichment are heterogeneously distributed within individual melilite grains. Although Na_2O shows random variations in the interior of crystals, in at least some cases there is a systematic increase in Na_2O towards their margins (e.g. 3529-41,3529-30).

The increase in Na_2O toward the rim of melilite grains has also been observed by MacPherson and Davis (1993) in a Type B inclusion (1623-8) from Vigarano, which they classified as a transitional Type B1 inclusion (see also Sylvester et al. 1992). Coarse-grained melilites ($Åk_{12-72}$) in the core of the CAI show normal, concentric zoning ($Åk_{31-68}$), but polygonal melilites in the rim exhibit reverse zoning with Åk-rich cores (Sylvester et al. 1992, MacPherson and Davis 1993). The Na_2O content of both types of melilite follow the Åk content, with Na_2O reaching unusually high values (~0.46 wt % at $Åk_{75}$), higher than are observed in typical Allende Type B inclusions (Fig. 85). Analysis of these data provides convincing evidence that the Na_2O is present in solid solution in the melilite as sodium melilite molecule $CaNaAlSi_2O_7$ (MacPherson and Davis 1993). The strong correlation of Na_2O with Åk content in this CAI indicates that Na was incorporated during igneous crystallization (Sylvester et al. 1992).

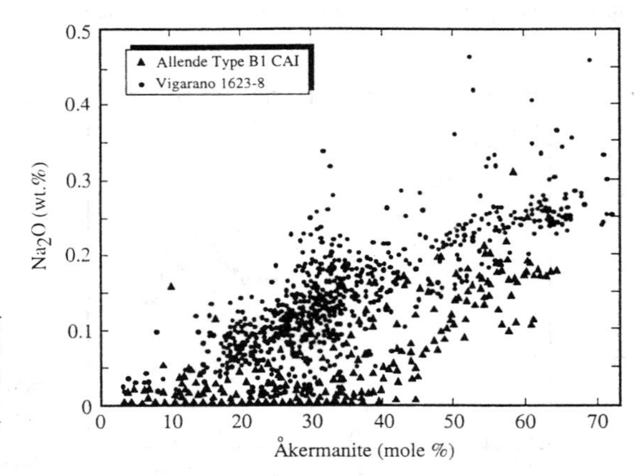

Figure 85. Plot of the Åk content of melilite vs. Na_2O for three Type B inclusions in Allende in comparison with data for melilite from a Type B1 inclusion (1623-8) from Vigarano. After MacPherson and Davis (1993).

Melilite in Type B inclusions can also show compositional variability depending on its specific spatial location within the inclusion. In Allende, Wark and Lovering (1982b) noted a significant difference in the composition of melilite within spinel framboids ($Åk_{65}$) in comparison with the host melilite ($Åk_{47}$) in a Type B1 CAI (Egg2) (see also Sylvester et al. 1993). In addition, some Type B inclusions in Allende and Vigarano contain texturally distinct regions which contain no spinel (spinel-free islands) (e.g. El Goresy et al. 1985, MacPherson et al. 1989a, Sylvester et al. 1993, Meeker 1995). In Type B1 Allende inclusion 5241, the spinel-free islands consist of melilite and fassaite with minor anorthite and are surrounded by areas which are rich in spinel associated with fassaite and melilite (El Goresy et al. 1985, MacPherson et al. 1989a, Meeker 1995). Melilite in the spinel-free regions is unzoned and within individual crystals has a compositional range $Åk_{52-68}$, with up to 2.7 mol % Na melilite in solid solution. The Na melilite content is positively correlated with increasing Åk content (El Goresy et al. 1985, Meeker 1995). This contrasts with melilites in the spinel-rich region which have low Na contents (<0.2 wt %) (e.g. Meeker 1995), show complex reverse zoning, and have a wider range in composition ($Åk_{43-62}$). The Na_2O content of melilites appears to be largely controlled by their proximity to late Na-rich alteration veins. The mantle melilite shows

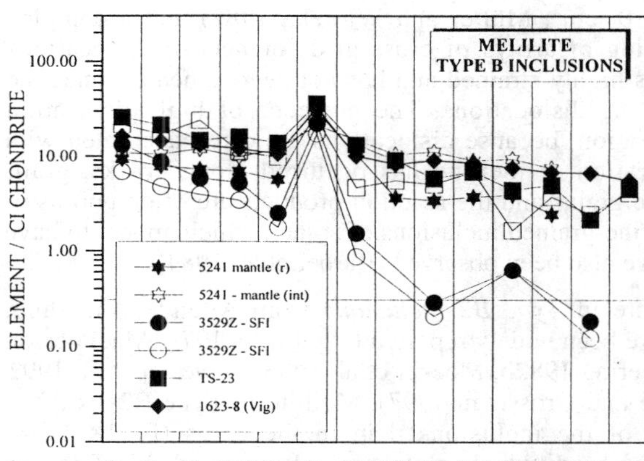

Figure 86. REE abundance diagrams (normalized to CI chondrites) for melilite from the mantle (r = rim, int = interior) of Type B1 Allende inclusion 5241 (ion microprobe analyses; data from MacPherson et al. 1989), from the spinel-free island (SFI) in Type B1 Allende inclusion 3529Z (data from Meeker 1995), from melilite which encloses fassaite and spinel in Allende Type B2 inclusion TS23 (data from Kuehner et al. 1989), and melilite from Vigarano Type B inclusion 1623-8 (data from MacPherson and Davis 1993).

oscillatory zoning, but overall decreases in Åk content (reverse zoning) from $Åk_{\sim33-\sim28}$ at the rim of the inclusion.

MacPherson et al. (1989a) measured REE abundances in melilite through the mantle of inclusion 5241 and found that they decrease progressively from the rim to the mantle-core boundary as a function of increasing Åk content. This is consistent with the inverse correlation between the partition coefficients (D) for REE and the Åk content of melilite (e.g. Beckett et al. 1990). The melilites in the mantle have relatively flat REE patterns (Fig. 86) (6-18 × CI) with positive Eu anomalies. However, in comparison, the REE patterns of melilite in the core of the inclusion are significantly different with enriched LREE relative to the HREE, but with a positive Eu anomaly. In the interiors of the core melilites, the REE are higher (La 8 × CI) than in their margins (La 2 × CI). Melilite in the spinel-free islands has a relatively flat REE pattern with a positive Eu anomaly, but lower REE abundances (1 × CI).

A spinel-free island has also been described in Allende B1 inclusion 3529Z (Podosek et al. 1991, Meeker. 1995). Melilite from this island has enriched LREE (La ~ 10 × CI), decreasing smoothly to the heavy REEs (Yb ~ 0.2 × CI) with a large positive Eu anomaly (~30 × CI) (Fig. 86) (Meeker 1995). It has been suggested that these spinel-free islands represent xenolithic material assimilated into the Type B inclusions when they were molten droplets (e.g. El Goresy et al. 1985, MacPherson et al. 1989a). However, based on zoning and textural arguments, Meeker (1985) has suggested that the spinel-free islands may not be xenoliths, but result from partial melting of material which contained either vesicles or spinel-free grains.

Type B1 inclusion 3537-1 from Leoville is also unusual (Caillet et al. 1993) and consists of two distinct regions, the most dominant having a relatively fine-grained, poikiloblastic texture consisting of fassaite, melilite, anorthite and spinel. Set within this poikiloblastic material are coarser-grained, relict islands with the same mineralogy. The melilite in the relict islands is $Åk_{48-69}$ whereas in the poikiloblastic region it is more aluminous ($Åk_{18-53}$). Na_2O in the melilite shows a positive correlation with increasing Åk content, with melilite in the relict islands typically having higher Na_2O contents (>0.1 wt %, cf. <0.1 wt %). TiO_2 contents are also lower in the relict islands (<0.05-0.07 wt % cf. <0.05-0.34 wt %), but K_2O in melilites in both regions is extremely low (≤0.04 wt %).

TEM studies of melilites from Type B1 inclusions in Allende show that they have very high dislocation densities (~10^9 cm^2; Müller and Wlotzka 1982) and a complex dislocation microstructure consisting of arrays of cross grids or nets of dislocations (Barber et al. 1984). The melilite is highly strained and has not been annealed since the deformation event which formed the dislocations. The presence of high dislocation densities has aided secondary alteration, because dislocations are often decorated with alteration phases. Secondary alteration of melilite has occurred along cracks, grain boundaries and at the surface of the grains and the reaction products are often porous in character. Fine scale twinning and fine-grained inclusions of spinel, which appear to have formed by exsolution processes, have also been observed (Barber et al. 1984).

Compositional data for melilite in *Type B2 inclusions* from Allende, Vigarano, Leoville, Efremovka and Axtell have been widely reported (e.g. Fuchs 1978, MacPherson and Grossman 1981, Wark and Lovering 1982b, Meeker et al. 1983, Sylvester et al. 1992 1993, Goswami et al. 1994, Simon and Grossman 1997). Melilite in Type B2s is often more gehlenitic at the periphery of the inclusions than in the cores (Fuchs 1978, MacPherson and Grosman 1981), and individual crystals are often zoned. MacPherson and Grossman (1981) found that melilites toward the margin of the inclusions are zoned both from core to rim ($Åk_{10-14}$) and along their length ($Åk_{11-20}$). Melilites in the interior have core to rim zoning ($Åk_{\sim15-60}$). Normal and reverse concentrically zoned melilites ($Åk_{24-49}$) have also been reported in some inclusions (e.g. Meeker et al. 1983).

Wark and Lovering (1982b) and Simon and Grossman (1997) observed that melilite in spinel palisade bodies in Type B2 CAIs in Allende, Vigarano and Axtell is compositionally distinct from melilite in the main bodies of the inclusions. For example, Wark and Lovering (1982b) found zoned melilite $Åk_{82-85}$ within spinel palisade bodies, cf. $Åk_{84-87}$ in the rest of the inclusions. Simon and Grossman (1997) reported that melilite in palisade bodies was normally zoned in Axtell and Vigarano inclusions and the compositional ranges lie completely within or extend to slightly more åkermanitic compositions than the range within the host inclusion.

REE and other trace elements for melilite from Type B inclusions have been reported by Mason and Martin (1974) and Nagasawa et al. (1977) using mineral separates. These REE patterns are similar and in good agreement with studies carried out by ion microprobe (e.g. MacPherson et al. 1989a (see above); Kuehner et al. 1989) (Fig. 86). Other reported trace element analyses for melilite in Type B inclusions include data for Li, Be and B (Phinney et al. 1979) and U and Th (Murrell and Burnett 1987).

Melilite in *Type B3* Allende inclusions varies in abundance from a trace phase (Clayton et al. 1984) to 26 vol % (Wark et al. 1987) and has suffered variable degrees of alteration to secondary phases such as grossular, nepheline, Al-diopside, and monticellite. Melilite in these inclusions spans the range from $Åk_{12}$ to $Åk_{88}$, but large compositional variations are usually only found in complex inclusions with a core and mantle (e.g. Clayton et al. 1984, Wark et al. 1987). The melilite contains no detectable Na melilite component in solid solution and no zoning is evident, in part because alteration is often extensive.

In inclusion CG14 (Clayton et al. 1984), melilite in the core region has a restricted range of compositions (~$Åk_{80}$) whereas melilite in the mantle region, just inside the Wark-Lovering rim, is highly variable in composition and has a lower åkermanite content ($Åk_{20-43}$) (see also Wark et al. 1987). Melilites in most other inclusions (e.g. Dominik et al. 1978, Wark et al. 1987) are Åk-rich and have very restricted compositional ranges ($Åk_{82-88}$) (but see Blander and Fuchs 1975).

Davis et al. (1991) have described a complex Type B3 inclusion (1623-5) in Vigarano, which has experienced only minor alteration. The inclusion consists of a core and a mantle, composed largely of melilite. Melilite compositions range from $Åk_{16}$ to $Åk_{100}$ with a distinct trimodal distribution and, although FeO and Na_2O contents are generally low (both <~0.66 and 0.40 wt %, respectively), they tend to increase toward the margin of the inclusion and adjacent to fractures, probably due to secondary alteration (e.g. Krot et al. 1995). Core melilite is essentially unzoned ($Åk_{89}$), except for regions of forsterite-free melilite which are zoned ($Åk_{65-87}$) adjacent to spinel grains and adjacent to forsterite ($Åk_{87-95}$). In the mantle, melilite ($Åk_{60-100}$) has locally pseudomorphed olivine and complex zoning is present.

Melilite is a relatively abundant phase in *Type C inclusions* (0-26 vol %), but in Allende inclusions has been extensively altered. Wark (1987) reported mean melilite compositions for several inclusions of $Åk_{25-58}$, but melilite in individual inclusions can range from $Åk_{15-40}$ but $Åk_{40-50}$ is more typical (Wark 1987). The concentrations of minor elements such as Cr, Ti, Fe, Na and K in melilites in Type C inclusions are typically near or below detection limits (<0.04 wt %).

Most studies of *fine-grained inclusions* in Allende have failed to find melilite (e.g. Grossman and Ganapathy 1976, Kornacki and Wood 1985a, McGuire and Hashimoto 1989), but it has been detected by x-ray diffraction and SEM studies (e.g. Clarke et al. 1970, Grossman et al. 1975). In Allende, these inclusions are highly altered and melilite appears to have been largely replaced by secondary phases. However, Davis et al. (1987) and Mao et al. (1990) have described a much less altered fine-grained inclusion (1623-3) from Vigarano which contains unaltered melilite ($Åk_{0-17}$) in the core and pure gehlenite (~$Åk_0$) in the mantle. Other occurrences of unaltered melilite have been reported in fine-grained inclusions in Vigarano ($Åk_{27-36}$; Sylvester et al. 1992), Leoville ($Åk_{4-13}$ and $Åk_{20-25}$; Mao et al. 1990), Mokoia (Cohen et al. 1983), Kaba ($Åk_{10-30}$, Fegley and Post 1985), and Y-86751 ($Åk_{7-36}$; Murakami and Ikeda 1994).

Mao et al (1990) reported reversely zoned melilite ($Åk_{7-16}$) in the core and mantle of an *unclassified melilite-rich inclusion* from Vigarano. Melilite also occurs in the core ($Åk_{1.9-6.6}$), mantle ($Åk_{1-12}$) and crust ($Åk_{1.5-5.2}$) of a complex Allende inclusion (3643; Wark 1986). Melilite in the core of this inclusion is a fine-grained, interstitial phase, but in the mantle melilite occurs as coarse-grained laths (100-200 μm), which often show kink banding, en echelon deformation twins and undulose extinction. The entire inclusion is rimmed by a complex "crust" consisting of three fine-grained refractory layers, one of which is dominated by melilite.

Spinel. Spinel is an extremely widespread phase in CAIs in CV chondrites. It typically occurs as small octahedra, often poikilitically enclosed in other phases, such as melilite and fassaite in Type A and B inclusions (Grossman 1980). Spinel also occurs in some Type A and B inclusions in two additional and distinct occurrences, termed 'framboids' and 'palisades' (El Goresy et al. 1979, Wark and Lovering 1982b, Simon and Grossman 1997). Framboids consist of small (10-200 μm), raspberry-shaped clusters of loosely or tightly packed, spinel crystals (El Goresy et al. 1979, Wark and Lovering 1982b) enclosed within phases such as fassaite, melilite and anorthite (Fig. 87a). Many framboids contain a central core of silicate grains, such as anorthite, melilite, fassaite or a combination of all three. Palisades, so named for their resemblance to stockade walls (Wark and Lovering 1977), range from chains or narrow bands of separate spinel crystals to continuous spinel ribbons (generally one spinel thick) (Fig. 87b). These chains appear as long arcs or rings in thin section and probably represent cross-sections of shells of spinel, which are 50 μm up to 2 cms in diameter. Framboids usually only appear in Type B CAIs (Wark and Lovering 1982b), but have also been observed in Type As (Simon et al. 1994b, 1995). Whilst most workers have favored a foreign origin for the palisades and framboids (e.g. Wark and Lovering 1982b), Simon and Grossman (1997) have argued

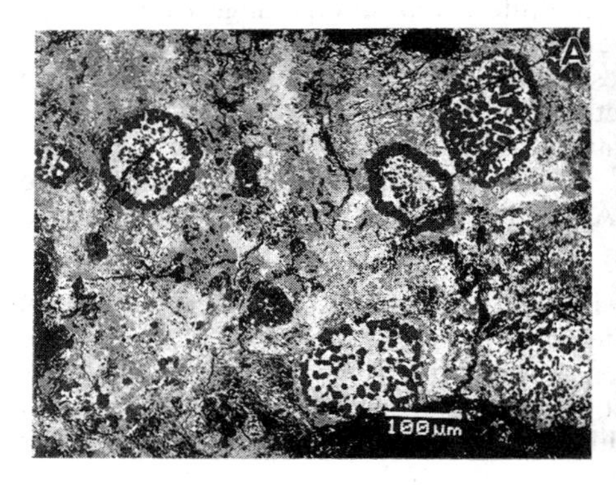

Figure 87. (a) BSE image of several spinel palisades in a Type B2 inclusion (AX2) from Axtell, enclosing anorthite and fassaite. Photo courtesy of Steve Simon (University of Chicago). (b) BSE image of a palisade body of spinel in a Type B inclusion in Axtell. An = anorthite, Sp = spinel, Fass = fassaite, Mel = melilite. After Simon and Grossman (1997).

that they form in situ in the CAIs as a result of crystallization of spinel around the periphery of vesicles. The layer of spinel later breaks and the vesicles fill with melt which then crystallizes.

Compositionally, the vast majority of spinels in CAIs are pure $MgAl_2O_4$, but more Fe and Cr-rich varieties also occur (see below). Representative electron microprobe analyses of spinel from a range of inclusions are reported in Table A3.15 and the occurrence and main compositional characteristics of spinels in CAI in CV carbonaceous chondrites are summarized in Table 5.

Barber et al. (1984) reported TEM observations of different types of spinels from a Type A and a Type B1 inclusion from Allende. The spinels all have remarkably similar microstructures with low dislocation densities ($<10^8$ cm^{-2}) and a characteristic mottled contrast that is probably due to the presence of very small defect clusters (2-3 nm in size).

Spinel in *"Compact" Type A CAIs* usually occurs as inclusions, < 50 μm in size, within melilite (e.g. Grossman 1975). The spinels can have a classic octahedral morphology and frequently have apices which are decorated with perovskite rods (e.g. Fig. 70d) (Barber et al. 1984). Perovskite can also occur as numerous triangular overgrowths on the [111] faces of the spinel inclusions. There are remarkably few compositional data for spinel in "Compact" Type As, although its occurrence is well documented in Allende (Grossman 1975, Podosek et al. 1991; Efremovka (Ulyanov et al. 1982, Fahey et al. 1987a) and Leoville (Sylvester et al. 1993). Spinels in Type A inclusions (both "Compact" and "Fluffy") in Allende generally contain < 2 wt % FeO, <0.5 wt % CaO, <0.7 wt % TiO_2 and <0.3 wt % Cr_2O_3 (Grossman 1975) (Fig. 88a,b). Podosek et al. (1991) reported 0.44-0.79 wt % V_2O_3 in spinel from Allende inclusion 3898. See also Fahey et al. (1987a) for spinel compositions in Efremovka Type A inclusions.

Spinel ($MgAl_2O_4$) is a common phase (5-20%) in *"Fluffy" Type A inclusions* (Grossman 1975, MacPherson and Grossman 1984, Kornacki and Wood 1985a). Grossman (1975) measured spinel compositions in 4 "Fluffy" Type A inclusions in Allende and found that they contain uniformly low concentrations of CaO (< 1 wt %) and Cr_2O_3 (<0.68 wt %), but FeO and TiO_2 show rather variable concentrations and, although typically low, can range up to 5 wt % and 2.2 wt %, respectively (see also Fuchs 1971,

Table 5. Summary of occurrence and minor element compositions (in wt%) of spinels in CAIs in CV carbonceous chondrites

Meteorite	Name	Type	TiO_2	Cr_2O_3	V_2O_3	FeO	CaO	ZnO	Reference
Allende									
	A	FTA	0.7-2.2	0.2	-	0.1	0.1	-	Fuchs (1971)
	B	FTA	0.2	-	-	-	-	-	Fuchs (1971)
	3898	CTA	0.15-0.72	0.06-0.34	0.44-0.79	-	-	-	Podosek et al. (1991)
	CG11(rim)	FTA	0.36-0.42	<0.15	0.37-0.43	6.34-6.97	-	-	Allen et al. (1978)
	CG11(core)	FTA	0.07-0.17	0.19-0.36	0.05-0.34	0.25-1.20	-	-	Allen et al. (1978)
	H126.78,1	Type B	0.17	0.19	0.22	0.20-0.35	0.3	-	El Goresy et al (1979)
		Type B	0.2-0.8	0.16-0.68	-	-	-	-	Grossman (1975)
	Egg 2	Type B1	?	0.1-0.3	0.3-0.6	-	-	-	Wark and Lovering (1982)
		Type B	0.93	-	-	<0.06	-	-	Bischoff and Palme (1987)
		Type B1	0.09-0.41	-	0.24-0.66	-	-	-	Podosek et al. (1991)
	3529-41	Type B	0.22-0.68	-	0.3-0.49	-	-	<0.08	Podosek et al. (1991)
	3529-30	Type B	0.2-0.73	-	0.2-1.07	-	-	-	Podosek et al. (1991)
	3529-21	Type B	0.13-0.39	-	0.09-0.25	-	-	n.d.	Podosek et al. (1991)
	3658	Type B	0.26-0.60	-	-	-	-	0.08	Podosek et al. (1991)
		Type B	0.25	0.25	0.19	<0.31	-	-	El Goresy et al. (1985)
		Type B1	0.24-0.29	-	0.22-0.59	up to 3.32‡	0.14-0.32	-	MacPherson and Davis (1993)
	110,A	Type B2	0.2-0.3	0.2-0.3	0.3-0.5	0.1-0.2	-	-	Wark and Lovering (1982)
		Type B2	0.2	-	-	up to 4.9††	0.1	-	Fuchs (1978)
Bali									
		Type B	0.15-0.33*	0.13-0.50†		0.05-0.24*			Kurat et al. (1975)
Leoville									
		Type B1	0.3	0.26	-	0.04	-	-	Kracher et al. (1985)
	3537-1	Type B1	0.11-0.33	-	0.3-0.51	0.05-0.45	0.08-0.77	-	Caillet et al. (1993)
	3537-2	Type B1	0.17-0.56	-	0.15-0.59	<0.2**	0.04-0.44	-	Caillet et al. (1993)
Vigarano									
	477-B	Type B	-	0.1-0.4 max 8.8	-	0.1-0.8 max 12.2	-	-	Sylvester et al. (1992)
Allende	1623-11	Type B	<0.44	-	<0.23		~0.18	-	Sylvester et al. (1992)
	CG 14	Type B3	~0.11	0.18-0.38	0.15-0.30(c) 0.32-0.42(m)	0.32-3.2(c) 0.08-2.9(m)	-	-	Clayton et al. (1984)
	TE	Type B3	0.1-0.3	0.1-0.3	-	<0.01	-	-	Dominik et al. (1978)
	1623-6	Type B3	0.01-0.66	0.02-0.34	0.11-0.29	-	-	-	Davis et al. (1991)

* - increases from rim to core; † - decreases from core to rim; ‡ - near alteration zones; †† - near rim of inclusion;
** - FeO can reach 1 wt% near the rim of the inclusion; (c) = core of inclusion; (m) = mantle of inclusion

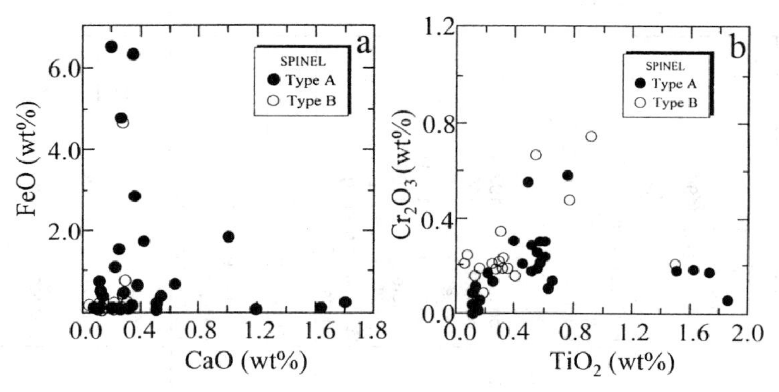

Figure 88. (a) CaO vs. FeO plot for spinels from Type A and Type B inclusions in CV chondrites. Data from Fuchs (1971), Allen et al. (1978), MacPherson and Grossman (1984), Grossman (1975), Christophe Michel-Lévy (1982), Kornacki and Wood (1985b). (b) Cr_2O_3 vs. TiO_2 plot for spinels from Type A and B inclusions in CV chondrites. Data from Fuchs (1971), Allen et al. (1978), MacPherson and Grossman (1984), Grossman (1975), Christophe Michel-Lévy (1982), Kornacki and Wood (1985b), Wark (1986), McGuire and Hashimoto (1989).

Allen et al. 1978, MacPherson and Grossman 1984). Compositionally variable, orange colored, V-rich spinels, sometimes with V_2O_3 contents >2 wt % and FeO up to 20 wt % can occur in some inclusions (Grossman 1980, Allen et al. 1978, MacPherson and Grossman 1984). In Allende inclusions, spinels located closer to the rims, in the Wark-Lovering rims or adjacent to zones of intense alteration have higher FeO (up to 20 wt % FeO) than in the cores (e.g. MacPherson and Grossman 1984). For example, in Allende inclusion CG11, Allen et al. (1978) found higher FeO and TiO_2 and lower Cr_2O_3 and V_2O_3 contents (6.34-6.97 wt % FeO; 0.36-0.42 wt % TiO_2; <0.15 wt % Cr_2O_3 and 0.37-0.43 wt % V_2O_3) in spinels in the rim of the inclusion, than those in the core (0.07-0.17 wt %; 0.19-0.36 wt %; 0.05-0.34 wt % and 0.25-1.20 wt %, respectively). Although the most V-rich spinels are also typically rich in Fe, there is no consistent correlation between these two elements, suggesting that the variation in V_2O_3 contents is probably a primary feature (MacPherson and Grossman 1984).

Spinels in *Type B1* and *B2 inclusions* have essentially identical compositions and are treated together. Spinels in Type B inclusions range in size from < 1 μm to > 100 μm and are usually poikilitically enclosed in laths of plagioclase, melilite and anhedral grains of fassaite (Grossman 1980). Framboids and palisades of spinel also occur in some inclusions (e.g. El Goresy et al. 1978, Wark and Lovering 1982a,b; Sylvester et al. 1993, Simon and Grossman 1997). Compositions of spinel (see Table A3.15) from a number of Type B inclusions from Allende, Bali, Leoville, Mokoia, and Vigarano have been reported in the literature (e.g. Gray et al. 1973, Grossman 1975, Kurat et al. 1975, Wark and Lovering 1982b, Meeker et al. 1983, Cohen et al. 1983, El Goresy et al. 1985, Kracher et al. 1985, Bischoff and Palme 1987, Podosek et al. 1991, Sylvester et al. 1992, MacPherson and Davis 1993, Caillet et al. 1993). Grossman (1975) found that spinels in two Type B Allende inclusions have concentrations of CaO, Cr_2O_3, FeO and TiO_2 always less than 1 wt % and mostly less than 0.5 wt % (Fig. 88a,b). The variation in minor element concentrations in spinel within individual inclusions is usually small and clear substitutional relationships between different elements have not been widely recognized. However, Podosek et al. (1991) reported that ZnO (up to 0.08 wt %) in Allende inclusion 3658 was correlated with higher FeO contents (>1 wt %).

In general, spinel does not vary in composition as a function of its spatial location within an inclusion. However, Kurat et al. (1975) reported that spinels in a Type B CAI from Bali show an increase in their mean Cr_2O_3 contents (0.13 to 0.50 wt %) and decrease in mean TiO_2 (0.33-0.15 wt %) and FeO (0.24 to 0.05 wt %) contents towards the core of the inclusion. MacPherson and Grossman (1981) also found slightly higher V_2O_3 contents in spinel from the margins of a Type B2 inclusion in Allende. Also in Allende, the FeO content of spinel increases closer to the rims of Type B inclusions (e.g. Fuchs 1978, MacPherson and Grossman 1981, MacPherson and Davis 1993, Caillet et al. 1993) and adjacent to alteration zones (e.g. MacPherson and Davis 1993). This increase in FeO is variable (2-5 wt %), but up to 12.2 wt % FeO accompanied by high V_2O_3 (15 wt %) and Cr_2O_3 (8.8 wt %) contents have been reported (Sylvester et al. 1992) in Vigarano inclusion 477-B.

In two Type B1 Allende CAIs (Egg6; Egg3), the Ti contents of spinel and enclosing fassaite are positively correlated (Fig. 89) which Meeker et al. (1983) interpreted as indicating equilibration between the spinel and pyroxene on a local scale. Using TEM, Doukhan et al. (1991) found that spinels in fassaite from Egg6 are euhedral and dislocation free. There are no topotactic orientation relationships between the spinel and fassaite, but a zone of unidentified microcrystalline alteration products occurs at the interface between the two phases. Also in Egg6, V-rich spinel (up to 3 wt % V_2O_3) occurs in the rim of a Type 3 fremdling (Meeker et al. 1983) and highly zoned spinels occur adjacent to a large fremdling (Zelda) (Armstrong et al. 1987). These spinels have Mg-rich cores, with very FeO- and V_2O_3-rich rims (30 wt % and 19 wt %, respectively).

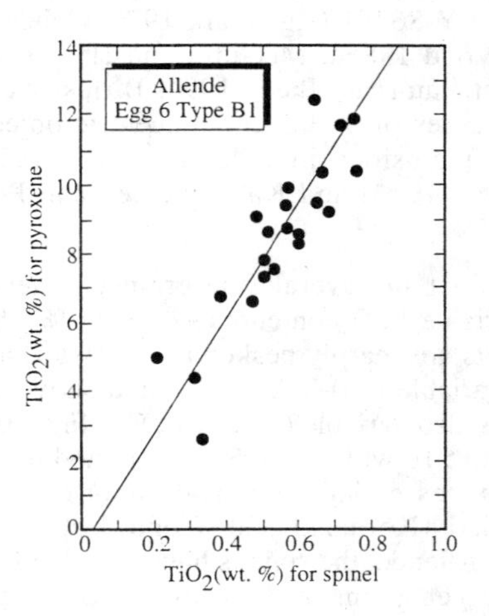

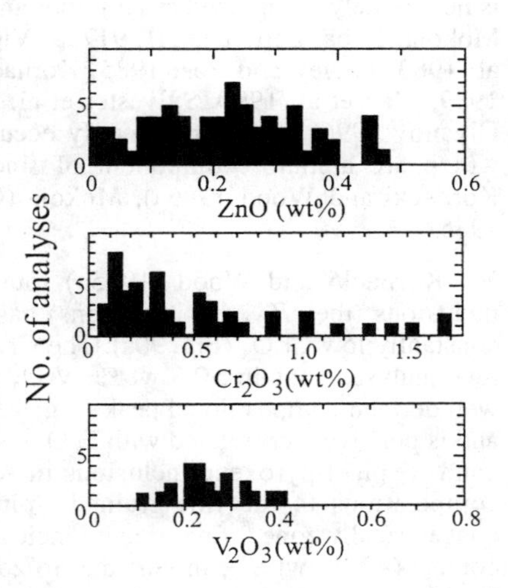

Figure 89 (left). Plot of Ti content of fassaite vs. the Ti content of coexisting spinel in the Type B1 inclusion Egg6 from Allende. After Meeker et al. (1983).

Figure 90 (right). Histograms showing the distribution of ZnO, Cr_2O_3, and V_2O_3 contents in spinel from spinel-rich inclusions in Allende. After Kornacki and Wood (1985b).

Trace element data (Li, Be and B) for spinels from one Type B inclusion have been reported by Phinney et al. (1979).

Spinel in *Type B3 CAIs* exhibits significant variations in FeO content, especially near the rims of inclusions or close to alteration zones. Such zones are especially common in Allende inclusions, but also occur in Vigarano. Spinel in the complex inclusion CG-14 (Clayton et al. 1984, Wark et al. 1987) has FeO contents between 0.34-3.2 wt % in the core and 0.08-2.9 wt % in the mantle. The minor element contents of both occurrences of spinel are very similar (~0.11 wt % TiO_2; 0.18-0.38 wt % Cr_2O_3), with the exception of V_2O_3 which is higher in the mantle (0.15-0.30 wt %, cf. 0.32-0.42 wt %) (Wark et al. 1987). Clayton et al. (1984) reported up to 13 wt % FeO, 0.15 wt % TiO_2 and 0.09 wt % V_2O_3 in spinel where alteration is most evident. Spinel in inclusion TE (Dominik et al. 1978) is compositionally similar but consistently has lower FeO contents (0.01 wt %). Spinel in the core region of Vigarano inclusion 1623-5 is also very low in FeO (<0.01 wt %) (Davis et al. 1991), but can reach 15 wt % towards the margins of the inclusion.

Spinel in *Type C inclusions* has a white or pink cathodoluminescence (Wark 1987), variable FeO contents (mean FeO - 0.13-12.23 wt %) and low concentrations of SiO_2, TiO_2, V_2O_3, Cr_2O_3 and CaO (<1 wt %). Mean TiO_2 and Cr_2O_3 contents of spinel from 7 different inclusions (Wark 1987) range from 0.18-0.39 wt % and 0.11-0.31 wt %, respectively with individual analyses varying by ±25% of the means. Higher Cr_2O_3 concentrations (1-2 wt %) were measured in spinel in Allende inclusion ABC (Macdougall et al. 1981) and Leoville inclusion LV2C (Lorin et al. 1978). In LV2C, the Cr_2O_3 and FeO contents of the spinel increase towards the rim of the inclusion. Mean V_2O_3 concentrations (0.18-0.77 wt %) in Type C inclusions are typically higher than either TiO_2 or Cr_2O_3, but exhibit a similar compositional variability within individual CAI (Wark 1987).

Clarke et al. (1970), Blander and Fuchs (1975) and Grossman and Ganapathy (1976) all found spinel in *fine-grained* CAIs from Allende using x-ray diffraction methods and it is now widely recognized as an important component of this type of inclusion in Allende, Mokoia, Kaba, Leoville, TIL 91722, Vigarano and Y-86751 (e.g. Wark 1979, Cohen et al. 1983, Fegley and Post 1985, Kornacki and Wood 1985a, McGuire and Hashimoto 1989, Mao et al. 1990, Sylvester et al. 1992, Murakami and Ikeda 1994, Birjukov and Ulyanov 1996). Spinel commonly occurs as the cores of 5-100 µm concentric objects which are a major component of fine-grained inclusions in Allende (Wark 1979, Kornacki and Wood 1985a), Mokoia (Cohen et al. 1983) and Kaba (Fegley and Post 1985).

Kornacki and Wood (1985a) found that spinel in several fine-grained Allende inclusions (their Type 2 inclusions) has very variable FeO contents (~2-23 wt %), but constantly low Cr_2O_3 (Fig. 90a). The Cr_2O_3 contents are sharply peaked at ~0.2 wt % and few analyses contain >0.3 wt %. V_2O_3 is quite variable (0.0-0.72 wt %) and shows no well defined compositional peak (Fig. 90b). ZnO is also variable (0-0.48 wt %) (Fig. 90c) and is positively correlated with FeO in spinels with >10 wt % FeO. Spinels in nodular or banded spinel-pyroxene inclusions in Allende have essentially identical minor element compositions to the fine-grained, spinel-rich CAI (Kornacki and Wood 1985a). In concentrically zoned, fine-grained inclusions from Allende, the spinels have variable FeO contents (2-25 wt %), in part due to zoning in larger grains at the rims (McGuire and Hashimoto 1989). However, in most inclusions the variation in FeO in spinel is only 2 to 3 wt %. Minor elements (e.g. TiO_2, Cr_2O_3, V_2O_3) are usually present in concentrations <0.5 wt % and Cr_2O_3 is negatively correlated with FeO.

Zoned spinels with FeO-poor cores (<1 wt % FeO) and FeO-rich rims (15 wt %) also occur in a zoned Vigarano inclusion (1623-14) (Sylvester et al. 1992). In inclusion 1623-16, spinel in the core contains up to 10.6 wt % FeO, whereas in the compact mantle the spinel is FeO poor (<0.5 wt %; Cr_2O_3 ~0.6 wt %). In Y-86751, spinels are all very FeO-rich (15-22 wt %; Mg/(Mg+Fe) = 0.4-0.85) (Murakami and Ikeda 1994) and although most grains are homogeneous, rare, large (~300 µm) spinel grains are zoned. Additional compositional data for spinels in fine-grained inclusions from Mokoia (Cohen et al.

1983) and Kaba (Fegley and Post 1985) all show low FeO and low concentrations of minor elements (<~0.5 wt %).

Spinel has been reported in a number of *unusual CAIs*, mostly from Allende (e.g. Fuchs. 1969, Wark 1986, Steele 1995b), but also Vigarano (e.g. Mao et al. 1990). Most spinels show variable FeO and low minor element contents (TiO_2, V_2O_3, Cr_2O_3<0.5 wt %), which vary somewhat depending on their location in the inclusions. For example, Steele (1995b) found that spinels surrounding a central core of hibonite in an unusual Allende inclusion are FeO-poor (1.9 wt %) but increase in FeO content towards the rim, reaching 9.3 wt %, with a total range of FeO contents of 0.98-11.5 wt %. In these spinels, TiO_2, ZnO and MnO are all positively correlated with increasing FeO. MnO in particular shows a very good correlation, increasing from ~100 ppm in FeO-poor spinels to ~200 ppm in FeO-rich spinels.

Pyroxene - fassaite. Dark green, highly pleochroic, Al-Ti-rich pyroxenes were first reported in CAIs in Allende by Marvin et al. (1970), Clark et al. (1970) and Fuchs (1971) and in Vigarano by Christophe Michel-Lévy et al. (1970). The term fassaite for this pyroxene was not generally adopted in the literature until the crystal structure refinement of Dowty and Clark (1973a) which showed that fassaite from Allende is structurally similar to terrestrial fassaite. Dowty and Clark (1973a) used the term titanian fassaite to distinguish this pyroxene from terrestrial fassaites which also have high Al_2O_3, but low TiO_2 contents (e.g. Mason 1974). The prefix titanian has generally been dropped in the meteoritic literature and the term fassaite has become synonymous with Al_2O_3-rich, TiO_2-bearing pyroxene. The occurrences and main compositional characteristics of fassaite in a range of inclusions in CV carbonaceous chondrites are summarized in Table 6 and representative electron microprobe analyses and trace element data are reported in Tables A3.16 and A3.17, respectively.

Electron microprobe analyses of fassaites from CAIs commonly show a cation deficiency in the structural formulae when Ti is calculated as Ti^{4+}, providing indirect evidence that Ti^{3+} is present. Dowty and Clark (1973a) measured polarized visible-region absorption spectra for fassaites from an Allende CAI and found two absorptions bands at 16,500 cm^{-1} and 21,000 cm^{-1}. The first of these bands was interpreted as the crystal field absorption for Ti^{3+} and the second as resulting from charge transfer between Ti^{3+} and Ti^{4+} in the M1 sites. Burns and Huggins (1973), however, reinterpreted both absorption bands as resulting from crystal field absorption due to Ti^{3+}. This reinterpretation has been disputed by Dowty and Clark (1973b). Nevertheless, the observations provide direct evidence for the presence of Ti^{3+} in fassaite and, because these fassaites are low in FeO, show that the intense color and pleochroism (dark-green to brown) is due to the presence of Ti^{3+}.

The structural formulae recalculated by Dowty and Clark (1973a), assuming 4 cations per 6 oxygens, was found to contain about 70% of the total Ti as Ti^{3+}. Grossman (1975) noted that as the Al_2O_3 content of fassaite in CAIs increases, the tetrahedral Si and octahedral Mg cations decrease when Ti is calculated as TiO_2. Grossman (1975) suggested that this was probably due to the presence of Ti^{3+} as argued by Dowty and Clark (1973a). Based on stoichiometric considerations, between one to two thirds of the Ti must be Ti^{3+}. Grossman (1975) suggested the coupled substitution Ti^{4+} and Al^{3+} for Si^{4+} and Mg^{2+} with Al^{3+} and Ti^{4+} substituting into the octahedral sites. More recent studies have revealed a wide range of Ti^{3+} contents in fassaite, based on stoichiometric constraints (e.g. Haggerty 1978a, 24-87% Ti as Ti^{3+}; MacPherson and Davis 1991, ~67% Ti^{3+}; Simon et al. 1991, 9-84% Ti^{3+}; see also Podosek et al. 1991, Davis et al. 1993). Mason (1974) and Grossman (1975) also noted that there is an increase in the cation

Table 6. Summary of occurrences and major element compositions of fassaite in CAIs in CV carbonaceous chondrites.

Meteorite	Name	Type	TiO_2	Al_2O_3	Reference
Allende					
	3898	Compact Type A	14-19	20-26	Podosek et al. (1991)
	A	Fluffy Type A	16.9	17.2	Fuchs (1971)
	CG 11†	Fluffy Type A	10-15	19-24	Allen et al. (1978)
	Egg 4	Fluffy Type A	?	?	Meeker et al. (1983)
	6/1	Type B	~9	18	Blander and Fuchs (1975)
	10/2		8	20	Blander and Fuchs (1975)
		Type B	7.74 (c) 3.06 (r)	19.6 (c) 19.7 20.46 (r)	Grossman (1975)
	Egg 3	Type B	9-12 (c) 2-5 (r)	nr?	Meeker et al. (1995)
	H126.78,1	Type B		11 (c) 18 (r)	El Goresy et al. (1978)
		Type B	~18	~19.5	Bischoff and Palme (1987)
	3529-21	Type B?	6-11 (c), <3 (r)	14-22 (c) 15-22 (r)	Podosek et al. (1991)
	3529-41	Type B1	3-10	13-22	Podosek et al. (1991)
	3529-30[1]	Type B1	12-Jul up to 17	nr	Podosek et al. (1991)
Bali	3658	Type B1	4.5-13	13-22.5	Podosek et al. (1991)
	5171,18	Type B2	5-6	16-20	Wark and Lovering (1982a)
Efremovka					
		Type B	4.2-7.5	19.4-17.2	Kurat et al. (1975)
Leoville					
	Ef3	Compact Type A	18	nr	Sylvester et al. (1993)
	3536-2	Compact Type A	7-8 (c) 11-12 (r)	nr	Mao et al. (1990)
	L4	Fluffy Type A	10-15	nr	Sylvester et al. (1993)
		Type B	5.0-8.1	14.3-20.3	Kracher et al. (1985)
	3537-1	Type B1	5.9-9.7 3.8-7.2	16.6-22.9 14.4-21.9	Caillet et al. (1993)
	3527-2	Type B1	2-11	13-25	Sylvester et al. (1992) Caillet et al. (1993)
Vigarano					
		Fluffy Type A	5.8-21.9	nr	Sylvester et al. (1991)
	477-B	Type B	6.5-14.5	16-21	Sylvester et al. (1991)
	1623-11	Type B	6-12	16-23	Sylvester et al. 1992
	Vig 2	Type B1	6-15 3-12 (c)	nr	Sylvester et al. (1993)
		Type B1	4-11	14-23	MacPherson and Davis (1993)
	1623-8	Type B2	5-11	17-23	Sylvester et al. 1992

(c) and (r) denote core and rim compositions of fassaite grains, respectively; nr = not reported
*Small fassaites in melilite in outer margins of inclusion; †· Fassaite from interior of inclusion
[2]· This inclusion has two compositional distinct types of fassaite

deficiency with increasing TiO_2 content when Ti is calculated as Ti^{4+}, suggesting that the Ti^{3+} content of the fassaite increases as total Ti increases. This suggestion has been confirmed for zoned fassaites in Allende Type B1 inclusions (Simon et al. 1991). Ti^{3+} decreases toward the rims of the fassaite grains, but Ti^{4+} remains essentially constant. This behavior has been attributed to the fact that Ti is a compatible phase in fassaite during igneous crystallization, but the larger Ti^{3+} cation fits more readily into the pyroxene M1 site and consequently is enriched in fassaite that grows during the earliest stages of crystallization (Simon et al. 1991).

Haggerty (1978a) suggested that the Ti^{3+} content of fassaite and coexisting phases could be used as an indicator of their condensation temperature. However, the use of this cosmothermometer has been questioned on the basis of textural criteria (e.g. El Goresy et al. 1977, Fuchs 1978) and the fact that the phases in Type B inclusions probably crystallized from a melt and are not, in fact, condensates (Grossman 1980).

Fassaite is a rare phase in *Type A CAIs* (Grossman 1975 1980), but a number of occurrences and compositional data have been reported for CAIs in Allende (Grossman 1975, Mao et al. 1990, Podosek et al. 1991), Efremovka (Ulyanov et al. 1982, Sylvester et al. 1993) and Leoville (Mao et al. 1990, Sylvester et al. 1993). In one Allende inclusion (3989), Podosek et al. (1991) found very Ti,Al-rich (14-19 wt % TiO_2; 20-26 wt % Al_2O_3), zoned fassaite with high V_2O_3 contents (~0.6-1.6 wt %). Zoned fassaites also occur in Leoville inclusions (Mao et al. 1990). Ulyanov et al. (1982) reported Sc-rich fassaite as a rare phase in the core of a "Compact" Type A inclusion in Efremovka.

Fassaite is a rare accessory phase in *"Fluffy" Type A inclusions*. Fuchs (1971) reported an analysis of a pyroxene from a "Fluffy" Type A CAI in Allende (inclusion A), with TiO_2 and Al_2O_3 contents of 16.9 and 17.2 wt %, respectively, which is clearly a fassaite. Ti and Al-rich fassaite (10-15 wt % TiO_2; 19-24 wt % Al_2O_3) occurs in the core of Allende inclusion CG11 (Allen et al. 1978), but fassaite in the rim regions of the inclusion has lower TiO_2 (~3-10 wt %) and Al_2O_3 (~4-18 wt %) contents. Compositionally distinct fassaites occur in the core of the inclusion depending on their textural relationship to other phases, e.g. fassaites with high Al_2O_3 (~24 wt %) and V_2O_3 (0.3-1.7 wt %) rim perovskite grains, whereas fassaite inclusions in melilite are richer in Ti and poorer in Al. Other minor occurrences of fassaite have been reported in Type A CAIs in Allende (Meeker et al. 1983), Leoville (Sylvester et al. 1993) and Vigarano (Sylvester et al. 1992).

Fassaite, as the dominant phase in *Type B inclusions*, has been studied extensively and compositional data for fassaite from inclusions in Allende, Bali, Efremovka, Leoville and Vigarano have been reported (e.g. Gray et al. 1973, Blander and Fuchs 1975, Kurat et al. 1975, Fuchs 1978, El Goresy et al. 1979, Meeker et al. 1983, Cohen et al. 1983, Kracher et al. 1985, El Goresy et al. 1985, Bischoff and Palme 1987, Podosek et al. 1991, Simon et al. 1991, Sylvester et al. 1992, 1993; Caillet et al. 1993, MacPherson and Davis 1993, Goswami et al. 1994). According to Simon et al. (1991) fassaite is present in four distinct, textural occurrences in Type B inclusions, the last two being minor. These are: (a) large blocky crystals, (b) irregularly shaped grains, poikilitically enclosed in melilite in the interior of the inclusions, (c) rounded grains in the melilite mantle, and (d) intergranular and poikilitic fassaite in the mantle melilite. Several authors (e.g. MacPherson et al. 1984a, Kuehner et al. 1989) have also noted thin (<10 μm) rims of fassaite which surround inclusions of spinel within the normally zoned cores of coarse-grained melilites. The rims can enclose 3 or 4 spinel grains or occur as meniscus-like infillings between two spinel crystals.

Fassaites in Type B1 inclusions show considerable compositional variability, most notably in their TiO_2 and Al_2O_3 contents (see Grossman 1975, Kurat et al. 1975). Grossman (1975) reported that the compositional ranges 5-10 wt % TiO_2 and 15-21 wt % Al_2O_3 are typical for Type B1 inclusions in Allende. However, the range of fassaite compositions exhibited from one inclusion to another can be quite different. Figure 91 shows examples of the compositional variation of fassaite in Type B Allende CAIs. These large variations are due mainly to the widely recognized concentric and sector compositional zoning that is present in fassaites in relatively unprocessed Type B1 inclusions (e.g. Grossman 1975, El Goresy et al. 1978, Grossman 1980, Wark and Lovering 1982a, Meeker et al. 1983, El Goresy et al. 1985, Podosek et al. 1991, Simon et al. 1991). Fassaites commonly have cores that are enriched in Ti and Al relative to the rims, but the variation in Al is usually much less than that for Ti. Core regions are highly

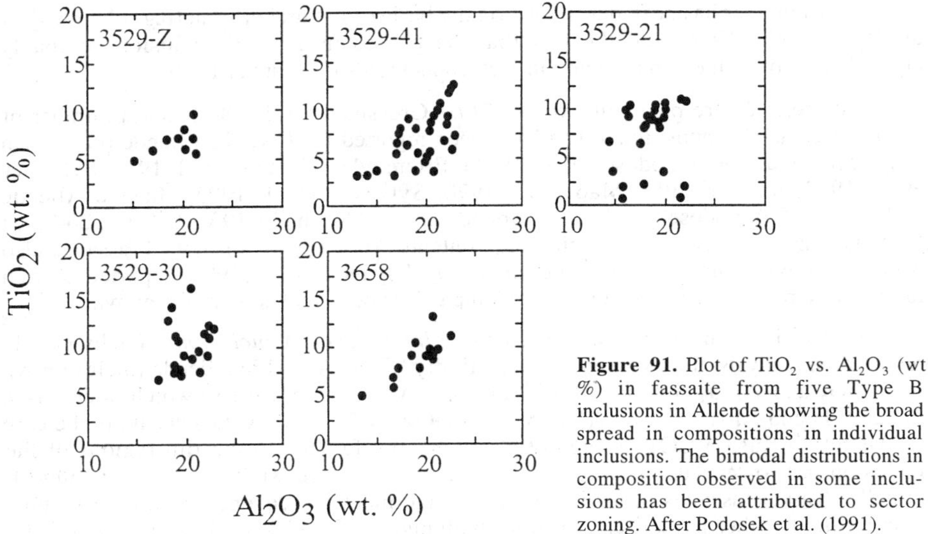

Figure 91. Plot of TiO_2 vs. Al_2O_3 (wt %) in fassaite from five Type B inclusions in Allende showing the broad spread in compositions in individual inclusions. The bimodal distributions in composition observed in some inclusions has been attributed to sector zoning. After Podosek et al. (1991).

pleochroic (Grossman 1980) and the rims may range from slightly pleochroic to non-pleochroic depending on the concentration of Ti (or rather Ti^{3+}, see above).

Simon et al. (1991) studied normal concentric zoning in fassaites in four Type B inclusions from Allende and found that some variations in zoning patterns occur from one inclusion to another, but overall very similar compositional trends are present. The cores of the fassaites tend to be Sc,Ti-rich and Mg,Si-poor relative to the rims (Fig. 92). In fassaites from one of the inclusions studied (TS34), the zoning profiles show monotonic changes in the elements from core to rim whereas two other inclusions (TS22 and TS23) have compositionally relatively flat interiors, but have sharply zoned rims. The enrichment in MgO in zoned fassaites was also noted by El Goresy et al. (1979) who also reported zoning in V_2O_3 and FeO in Allende inclusion H126.78,1.

Simon et al. (1991) found that MgO and (TiO_2+Ti_2O_3) in fassaites in Type B1 inclusions are negatively correlated which is attributable to the two substitutions Ti^{3+} + $^{IV}Al^{3+}$ = Mg^{2+} + Si^{4+} and Ti^{4+} + 2 $^{IV}Al^{3+}$ = Mg^{2+} + 2 Si^{4+}. These substitutions result in a positive correlation between Al_2O_3 and (TiO_2+Ti_2O_3) (Fig. 93), although there is some scatter due to the substitution of Al as Tschermak's molecule in the pyroxene, i.e. $^{IV}Al^{3+}$ + $^{VI}Al^{3+}$ = Mg^{2+} + Si^{4+}. However, at high (TiO_2+Ti_2O_3) contents (above ~12 wt % TiO_2), which typically occur in the cores of the very earliest fassaite to crystallize, there is no concomitant increase in Al_2O_3. Simon et al. (1991) suggested that at high Ti contents, ^{VI}Al begins to decrease while ^{IV}Al continues to increase, resulting in an approximately constant bulk Al. Fassaites with the highest TiO_2 contents (up to ~19 wt %) occur in the mantle of one inclusion (TS34) (Simon et al. 1991) and also have cores with high Sc contents (0.2-1.2 wt % Sc_2O_3). All other fassaites (termed subliquidus fassaites by Simon et al. 1991) have lower TiO_2 contents (<12 wt %).

El Goresy et al. (1985) and Meeker (1995) found some differences in zoning patterns between fassaite in spinel-free islands and in the main body of a complex inclusion from Allende (5241). In the spinel-free region, the fassaite has symmetric, concentric zoning with Ti,Al-rich cores (9 wt % TiO_2; ~18 wt % Al_2O_3) and Ti,Al- poor rims (~2.5 wt % TiO_2; ~16 wt % Al_2O_3). In spinel-rich regions, fassaite is compositionally similar but less highly zoned, and may be richer in Ti^{3+}. Podosek et al. (1991) and Meeker (1995) have

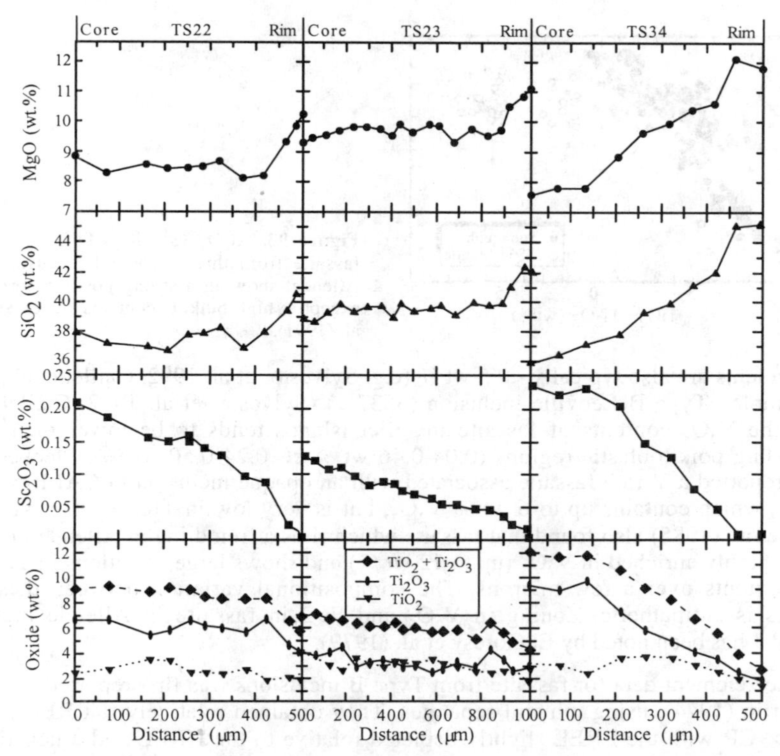

Figure 92. Core to rim compositional zoning profiles in fassaite from three different Type B inclusions in Allende, showing the different types of zoning present. After Simon et al. (1991).

also described concentric and sector zoning in fassaite in spinel-free islands in Allende Type B1 inclusion 3529Z.

Fassaites from Allende CAI 3529-1 are unzoned (Podosek et al 1991), but instead have distinct Ti, Al-rich cores (6-11 wt % TiO_2; 14-22 wt % Al_2O_3) with Ti-poor overgrowths (<3 wt % TiO_2; 15-22 wt % Al_2O_3) (Fig. 91).

Variations in fassaite composition can also occur depending on their textural location within inclusions. Kurat et al. (1975) analyzed fassaites (then termed aluminous titanaugite) in a Bali Type B1 CAI and observed a decrease in average TiO_2 and Al_2O_3 contents from the rim to the core of the inclusion (from 7.5 to 4.2 wt % TiO_2 and 19.4 to 17.2 wt % Al_2O_3). Similar observations have been made by Sylvester et al. (1992) and Caillet et al. (1993) in Leoville Type B1 inclusion 3537-2. MacPherson and Davis (1993) also reported that small fassaite grains enclosed within melilite in the outer regions of another Type B1 Allende CAI have higher TiO_2 (~17 wt %) and V_2O_3 (up to ~0.82 wt %, cf. <0.4 wt %), than the coarse-grained, zoned fassaite in the same inclusion (4-11 wt % TiO_2 and 14-23 wt % Al_2O_3). Other examples of compositionally and texturally distinct fassaite have been reported in Type B inclusions from Allende (Podosek et al. 1991; 3529-30; Meeker 1995; 3529Z), Vigarano (Sylvester et al. 1992; 1623-8), and Leoville (Caillet et al. 1993; 3537-1).

The minor element contents of fassaite in Type B inclusions are usually rather low. FeO contents are typically <0.6 wt % (Grossman 1975) and commonly < 0.02-0.09 wt %.

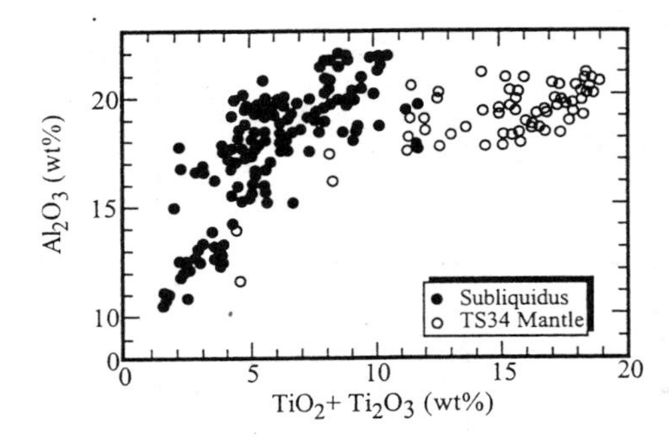

Figure 93. Al_2O_3 vs. TiO_2+Ti_2O_3 (wt %) for fassaite from three Type B1 inclusions from Allende, showing a strong positive correlation, except at high bulk Ti contents. After Simon et al. (1991).

V_2O_3 contents are also typically <0.5 wt % (e.g. Sylvester et al. 1992, Caillet et al. 1993). In a complex Type B Leoville inclusion (3537-1) (Sylvester et al. 1992, Caillet et al. 1993), the V_2O_3 contents of fassaite in relict islands tends to be lower than in the surrounding poikiloblastic regions (0.04-0.46 wt % cf. 0.24-0.50 wt %). Meeker et al. (1983) reported a V-rich fassaite associated with an opaque inclusion in CAI Egg6 from Allende, which contains up to 6 wt % V_2O_3, but is very low in TiO_2 (<~0.5 wt %). El Goresy et al. (1985) also found that fassaite which rims a fremdling in Allende inclusion 5241, is highly enriched in V_2O_3 (up to 12 wt %) and shows large variations in TiO_2 and V_2O_3 contents over a few microns. The compositional variation between these two elements is antipathetic. Zoning in V_2O_3 and FeO in fassaite in Allende inclusion H126.78,1 has been noted by El Goresy et al. (1979).

Trace element data for fassaite from Type B inclusions was first reported by Mason and Martin (1974), using mineral separates. They obtained relatively flat REE patterns (10-20 × CI) with the LREE slightly depleted relative to the HREE and a negative Eu anomaly. In general, these patterns are remarkably consistent with later ion microprobe analyses of fassaites in Type B inclusions (e.g. MacPherson et al. 1989a, Kuehner et al. 1989, Simon et al. 1991, Meeker 1995).

Simon et al. (1991) reported REE and other trace element concentrations (see Table A3.17) in zoned fassaite from three Type B inclusions from Allende (TS 23 (B1); TS 33 (B1); TS 34 (B1-B2)) and found that most fassaites have REE patterns with a positive slope from La (10-30 × CI) to Sm (~20-60 × CI) and relatively flat HREE (~30-100 × CI), with a large negative Eu anomaly (3-15 × CI) (Fig. 94a). In individual grains, the core (early) fassaites have the lowest REE and LIL element (Ta, Nb, Th, U) concentrations and the REE increase towards the rims. Some fassaite grains in one inclusion (TS23) have extremely high REE concentrations (LREE ~1000 × CI, HREE ~2000 × CI) (Fig. 94b).

Zoning in REE is present in zoned fassaites in spinel-free regions of Allende inclusion 5241 (MacPherson et al. 1989a) with HREE abundances of 30 and 100 × CI in the core and rim, respectively. In the spinel-rich region, fassaite cores and rims typically have depleted LREE (Ce ~9-15 × CI) and flat HREE (~20-40 × CI) with large negative Eu anomalies. In Allende inclusion TS23 (B), REE and other trace elements (Sc, V, Rb, Sr, Y, Zr, Nb, Ba) are highly enriched in the thin rims of fassaite which surround spinel inclusions in melilite (Kuehner et al. 1989). These grains have LREE 100-200 × CI and HREE 200-350 × CI with large negative Eu anomalies (~5 × CI) (Fig. 94c). In comparison, the normal subliquidus fassaites have LREE abundances of 10-30 × CI and HREE of ~40 × CI. Kuehner et al. (1989) have argued that the fassaite inclusions represent relict precursor grains which escaped melting.

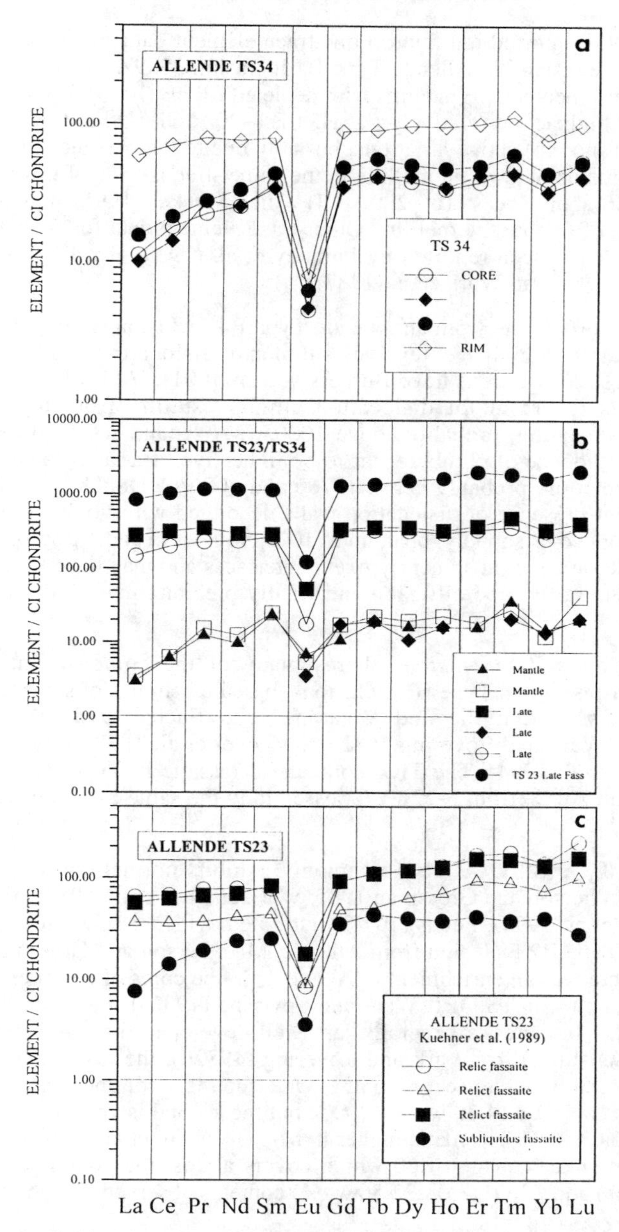

Figure 94. REE abundance patterns (normalized to CI chondrites) for: (a) Fassaites from Type B inclusion TS-34 from Allende. Cores of fassaites have low REE concentrations, but REE concentrations increase towards the rims (ion microprobe analyses; data from Simon et al. 1991). (b) Mantle and late crystallizing fassaites from Type B1 inclusion TS 34 from Allende. Core fassaites have low REE concentrations, but fassaites which texturally appear to have crystallized late in the sequence have extremely high REE abundances. Data from Simon et al. (1991). (c) Fassaites from inclusion TS23 in Allende (data from Kuehner et al. 1989). The analyses are from fassaite inclusions which occur as rims on spinel, inside melilite and from normal subliquidus fassaite.

Meeker (1995) reported ion microprobe trace element data for fassaites, inside and outside a spinel-free island in Allende Type B1 inclusion 3529Z. Both types of fassaites show very similar igneous REE patterns with depleted LREE (La ~4-7.5 × CI), a negative Eu anomaly and higher, but relatively flat HREE (~20 × CI). Other trace elements (Ba, Sc, V, Cr, Sr, Zr and Nb) show variations, possibly due to the presence of sector zoning. Low-Ti fassaite inclusions within melilite in the spinel-free islands of this inclusion have much higher REE abundances (100-200 × CI) which Meeker (1995) suggested could be due to resorption of fassaite by melilite. Other trace element data for fassaites in Type B CAIs in Allende have been reported by Phinney et al. (1979) (Li, Be, B), Murrell and Burnett (1987) (U, Th) and Wark et al. (1987) (U).

TEM studies of fassaite from an Allende Type B1 inclusion (Egg6) (Doukhan et al. 1991) show that it contains extensive, subplanar dislocation walls which are at equilibrium. These dislocations have Burgers vectors [001], $1/2[110]$ and $1/2[1\bar{1}0]$. This microstructure is the result of dislocation climb, resulting from high temperature deformation or annealing, probably above 1000°C. A second set of glide dislocations which show no evidence of climb appear to be indicative of a later, lower temperature. episode of deformation, probably between 700 and 1000°C. Doukhan et al. (1991) also noted the frequent presence of dislocation walls decorated with voids or bubbles. These voids may be rounded or subpolygonal, up to 1000 Å in width and 1 μm long. They occur adjacent to alteration zones at spinel-pyroxene interfaces and may have formed as a result of a fluid circulating by pipe diffusion and finally precipitating within the dislocation cores.

Fassaite in *Type B2 inclusions* shares many of the compositional and zoning characteristics of fassaite in Type B1 inclusions, based on studies of several inclusions in Allende, Efremovka, Leoville and Vigarano (e.g. Fuchs 1978, MacPherson and Grossman 1981, Wark and Lovering 1982a,b; Meeker et al. 1983, Sylvester et al. 1992 1993, Goswami et al. 1994). The TiO_2 contents of fassaite in Type B2s can vary from ~0.5-17 wt % and Al_2O_3 from 3-22 wt %, essentially the same as that observed in Type B1s.

Fassaite in Type B2 CAIs also commonly exhibits normal concentric and sector zoning (e.g. MacPherson and Grossman 1981, Wark and Lovering 1982a,b; Meeker et al. 1983, Goswami et al. 1994). Fuchs (1978) first noted evidence of zoning in fassaite in a rhönite-bearing Type B2 inclusion from Allende. MacPherson and Grossman (1981) also observed that there was an enrichment in Al and Ti in the cores of fassaites relative to the rims in Allende inclusion TS33F1. Wark and Lovering (1982a) suggested that the zoning found in Type B2 fassaites is generally less well developed than in Type B1s. In one inclusion (110,A) studied by Wark and Lovering (1982b), the fassaite has two distinct compositions, possibly due to sector zoning. One (the most common) is typical for Type B CAIs (3.81 wt % TiO_2; 14.30 wt % Al_2O_3), but the second is very low in Al (9.0 wt % Al_2O_3) and Ti (2.03 wt % TiO_2) and higher in Mg. Goswami et al. (1994) reported zoned fassaite in Efremovka inclusion E60 which covers almost the complete range of TiO_2 (0.19-12.21 wt %) and Al_2O_3 (3.08-22.56 wt %) contents observed in Type B2 inclusions.

Simon and Grossman (1997) investigated fassaite compositions from inside and outside palisade bodies in a Type B2 inclusion in Axtell. Both occurrences have the same composition, but are somewhat different from typical Type B inclusions in Allende and Axtell. The TiO_2 contents extend to higher values (16 wt %) and have higher V_2O_3 (0.2-1.6 wt %), but lower Sc_2O_3 (<0.1 wt %). Fassaite grains within palisade bodies can also be zoned with decreasing TiO_2 (7.0 to 5.9 wt %) and V_2O_3 (0.32 to 0.1 wt %) from core to rim (Simon and Grossman 1997).

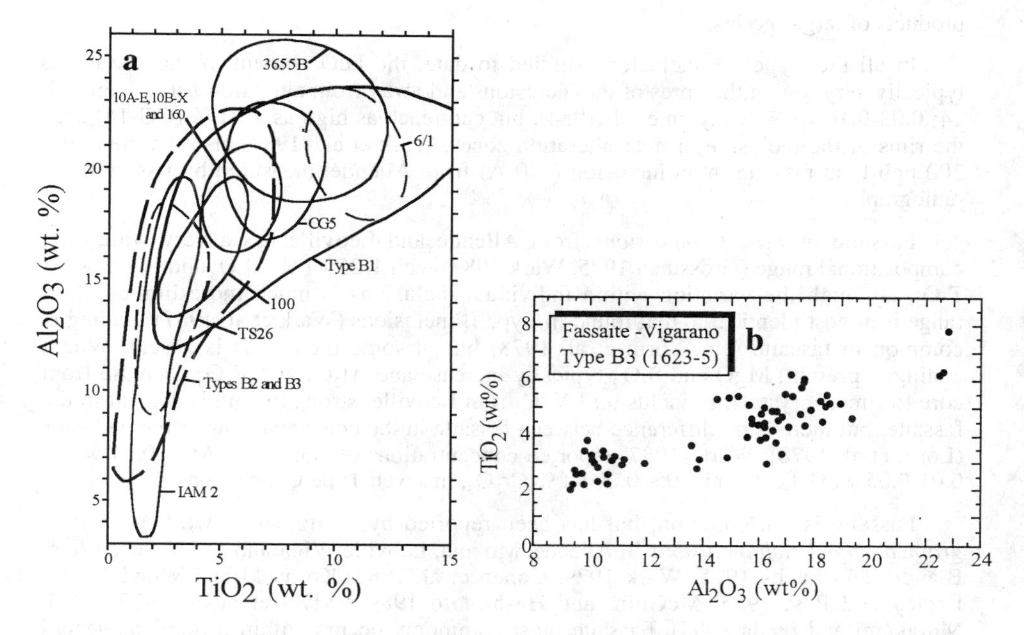

Figure 95. Al_2O_3 vs. TiO_2 (wt %) plot showing: (a) The compositional range of fassaites in several different Type B3 inclusions. After Wark et al. (1987). (b) The bimodal distribution of fassaite compositions in the Type B3 inclusion 1623-5 from Vigarano. After Davis et al. (1991).

MacPherson and Grossman (1981) found that minor elements such as FeO, MnO and NiO in fassaite are all below detection limits, but Cr_2O_3 is present in measurable concentrations (<0.06 wt %). U and Th data for fassaite in Allende inclusion 3529-41 have been reported by Murrell and Burnett (1987).

Fassaitic pyroxene is a major component of *Type B3 inclusions*, but typically contains lower concentrations of TiO_2 and Al_2O_3 than fassaite in Type B1 or B2 CAIs (Fig. 95a). Fassaites in most Type B3 inclusions have a similar compositional range (e.g. Blander and Fuchs 1975, Dominik et al. 1978, Clayton et al. 1977) and often show sector zoning (e.g. Allende inclusion, CG-14; Clayton et al. 1984). In Allende inclusion TE, Dominik et al. (1978) reported bimodal fassaite compositions which are probably due to sector zoning with one group low in TiO_2 and Al_2O_3 (2.24 wt % and 10.03 wt %, respectively) and a second group with higher TiO_2 and Al_2O_3 contents (4.29 wt % and 17.75 wt %, respectively). Similarly, Davis et al. (1991) reported bimodal fassaite compositions due to sector zoning in a Type B3 inclusion (1623-5) from Vigarano (Fig. 95b). Davis et al. (1991) calculated that about one third of the Ti in the core fassaites was trivalent (based on stoichiometric constraints), whereas adjacent to the mantle of the inclusion, Ti is mostly tetravalent.

In inclusion CG14, a complex Allende Type B3 inclusion with a core and a mantle, the fassaites show highly variable compositions (Clayton et al. 1984). Fassaite in the core of the inclusion has an unusual faint blue cathodoluminescence and shows both normal and reverse zoning. Normally zoned crystals have cores that are richer in Ti and Al than the edges and Mg and Si show an inverse relationship with Al and Ti in all grains. In comparison, fassaites in the mantle are finer-grained and have much lower Ti and Al contents, although some very Ti-Al-rich grains are present which may be alteration products of larger grains.

In all the Type B3 inclusions studied to date, the FeO content of the fassaite is typically very low in the cores of the inclusions and away from alteration zones (e.g. CG-14; 0.03-0.16 wt %; Clayton et al. 1984), but can reach as high as 4 wt % (CG-14) near the rims of the inclusion or near alteration zones. Wark et al. (1987) also reported 100-200 ppb U in fassaite in an inclusion (110-A) from Allende, measured by fission track radiography.

Fassaite in *Type C inclusions* from Allende and Leoville has a very wide total compositional range (Grossman 1975, Wark 1987) with 1-25 wt % Al_2O_3 and ~1-13 wt % TiO_2, although the variation within individual inclusions is much more limited. This range is almost identical to that found in Type B inclusions (Wark et al. 1987). Zoning is common in fassaite (e.g. Lorin et al. 1978) but in some inclusions is absent. Where zoning is present, MgO and SiO_2 typically increase and Al_2O_3 and V_2O_3 decrease from core to rim of crystals. In inclusion LV2C from Leoville, strong zoning is present in the fassaite, but there is no difference between fassaite in the cores and rims of the inclusion (Lorin et al. 1978). Wark (1987) reported concentrations of Na, K and Mn <0.04 wt %, 0.01-0.05 wt % FeO, and 0.05- 0.09 wt % Cr_2O_3, in seven Type C inclusions.

Fassaite is not common, but has been reported by a number of workers in *fine-grained, spinel-rich inclusions* in Allende, Mokoia, Leoville, Vigarano and Y-86751 (e.g. Blander and Fuchs 1975, Wark 1979, Cohen et al. 1983, Kornacki and Wood 1985a, Fegley and Post 1985, McGuire and Hashimoto 1989, Sylvester et al. 1992, 1993; Murakami and Ikeda 1994). Fassaite most commonly occurs within layered sequences around spinel grains (or nodules) in inclusions in Allende, Kaba and Vigarano (Wark 1979, Kornacki and Wood 1985a, Fegley and Post 1985, McGuire and Hashimoto 1989, Sylvester et al. 1992). The fassaite rims commonly zone into Al-diopside (see below) from the interior to the exterior of the nodule, but can sometimes be reversed (e.g. in Leoville inclusion 3526-2; Mao et al. 1990). Fassaite also occurs as rims around a phyllosilicate phase in Mokoia fine-grained inclusions (Cohen et al. 1983) and as rims (2-6 wt % TiO_2) around perovskite in Leoville inclusion, L3 (Sylvester et al. 1993). Ti-rich fassaite (>10 wt % TiO_2) has also been reported in a coarse-grained rim on a fine-grained inclusion in Allende (McGuire and Hashimoto 1989).

Fassaite in these inclusions shows considerable compositional variations in TiO_2 and Al_2O_3. Kornacki and Wood (1985a) reported that most fine-grained, spinel-rich inclusions contain fassaite with <11 wt % Al_2O_3 , <3 wt % TiO_2, ~0.2-1.6 wt % FeO, <0.1 wt % V_2O_3 and ~0.1-0.5 wt % Cr_2O_3. Fassaites in inclusions in Y-86751 (Murakami et al. 1994) and Kaba (Fegley and Post 1985) contain among the highest Al_2O_3 contents (25 and 30.3 wt %, respectively) from any CAI yet studied. In Y-86751, two groups of fassaites are present in fine-grained inclusions (Murakami and Ikeda 1994), one with high Al_2O_3 (22-25 wt %) and TiO_2 (10-18 wt %) and the second with lower Al_2O_3 (3-7 wt %) and TiO_2 (0-1 wt %).

Pyroxene - rhönite. Rhönite was first discovered in a meteorite by Fuchs (1971) who reported an occurrence in a CAI (probably a "Fluffy" Type A; Grossman 1980) in Allende. The rhönite contains much higher TiO_2, Al_2O_3, CaO and MgO contents and lower FeO and SiO_2 than terrestrial rhönite. The Allende rhönite (Fuchs 1971) is compositionally variable (e.g. 14.1-18.4 wt % TiO_2, 27.6-29.6 wt % Al_2O_3; 13.9-17.8 MgO; 17.0-19.6 wt % CaO; 0.3-4.2 wt % FeO; 18.6-20.1 wt % SiO_2) and also lacks Fe^{3+} and minor Na_2O and K_2O, which are present in terrestrial rhönite (see Table A3.21). In Allende, rhönite may occur in both monoclinic and triclinic forms (Fuchs (1971), whereas the terrestrial occurrences appear to be exclusively triclinic. Other minor occurrences of rhönite in Allende have been reported by Fuchs (1978) in a Type B2, Allen et al. (1978) in a "Fluffy" Type A and Podosek et al. (1991) in a "Compact" Type A. Fuchs (1978) reported V_2O_3 concentrations of ~1.0 wt %, but low FeO (0.5 wt %).

Pyroxene - diopside. Diopside is not an especially abundant phase in CAIs in CV chondrites, and in some highly altered inclusions (e.g. in Allende) there is also some question as to whether it is of primary or secondary origin (See Secondary Mineralogy of CAIs in CV carbonaceous chondrites). In this section we present data for occurrences of diopside whose origin appears to be primary.

Diopside is rare in *Type A inclusions* (~5 %) and typically occurs lining cavities or within the Wark-Lovering rims (e.g. Blander and Fuchs 1975, Grossman 1975). It can also occur as ribbons in the interior of inclusions, e.g. in a Leoville "Fluffy" Type A (Christophe Michel-Lévy et al. 1982). Diopside compositions in Allende "Fluffy" Type A inclusions have been reported by Fuchs (1971), Grossman (1975) and Blander and Fuchs (1975) and in Leoville by Sylvester et al. (1993). Sylvester et al. (1993) found compositions ranging from almost pure diopside with FeO contents <0.1 wt % and <0.03 wt % TiO_2 to Al-rich diopside with up to 8.99 wt % Al_2O_3 and 1 to 2 wt % TiO_2.

Diopside has been most commonly reported in *fine-grained, spinel-rich inclusions* in Allende, Mokoia and Kaba, where it occurs in rim sequences surrounding spinel grains or nodules (e.g. Wark 1979, Cohen et al. 1983, Fegley and Post 1985, Kornacki and Wood 1985a). As noted earlier the interior of these pyroxene rims is typically fassaitic in composition grading into aluminous diopside towards the exterior. Diopside has also been reported in fine-grained inclusions in Leoville (3536-1; Mao et al. 1990), Vigarano (1623-16; Sylvester et al. 1992), Y-86751 (Murakami and Ikeda 1994) and TIL 91722 (Birjukov and Ulyanov 1996). Diopside also occurs as rims around a phyllosilicate phase in Mokoia (Cohen et al. 1983) and as rims around perovskite in one inclusion in Leoville (Sylvester et al. 1993).

The diopsides commonly have Al_2O_3 contents that range from 0.6-11.2 wt % with TiO_2 from 0-~0.6 wt %. However, in a detailed study of zoned, fine-grained inclusions in Allende, McGuire and Hashimoto (1989) found that the clinopyroxenes are Al-bearing diopside and salite (probably secondary in origin) with up to 14 wt % Al_2O_3 and 7 wt % TiO_2, although TiO_2 contents <2 wt % are most typical. The diopside compositions are highly variable over distances of <20 μm and from one inclusion to another.

The most common occurrence of diopside in CAIs in CV carbonaceous chondrites is within the *Wark-Lovering rim* sequences (Wark and Lovering 1977) which occur on many CAIs. Diopside has been reported within rims on essentially all types of inclusion and typically occurs in one of the outer layers, although its position within the rim sequences varies from one inclusion to another.

Plagioclase. Marvin et al. (1970) first reported anorthite in a CAI from Allende and it is now recognized as a major phase in Type B and C inclusions. In many inclusions in Allende, two distinct generations of anorthite are present; coarse-grained, primary anorthite and a fine-grained, secondary alteration product associated with phases such as grossular, nepheline, wollastonite, etc. Occurrences and major compositional characteristics of anorthite in CAIs in CV chondrites are presented in Table 7. Representative electron microprobe analyses of anorthite from the different types of CAI discussed below are reported in Table A3.18 and trace element data, measured by ion microprobe, are reported in Table A3.12.

Anorthite in CAIs commonly exhibits a range of characteristic cathodoluminescence (CL) properties. Hutcheon et al. (1978) studied the CL properties of plagioclase in three

Table 7. Summary of occurrences, major and minor element compositions (in wt%) of anorthite in CAIs in CV carbonceous chondrites.

Meteorite	Name	Type	An	MgO	Na_2O	K_2O	TiO_2	FeO	Reference
Allende									
	3658	Type A		0.07	0.04-0.08				Podosek et al. (1991)
	TS33F1	Type B	99.0-99.3	0.13-0.14 (r)		<0.02		<0.02	Grossman (1975)?
				0.15-0.27 (c)					
	3529-Z	B1	100	0-0.16	0.02-0.09	<0.03			Podosek et al. (1991)
	3529-21	Type B	100	0.07	0.03-0.09	<0.03			Podosek et al. (1991)
	3529-30	Type B	100	0.13	0.05-0.12	<0.03			Podosek et al. (1991)
	3529-41	B1	100	~0.15*	b.d. to 0.11	<0.03			Podosek et al. (1991)
Bali									
		B1	100	<0.33	<0.11	<0.05	< 0.08	<0.01	
Leoville		B1	100	n.r.	n.r.	n.r.			Kracher et al. (1985)
	3537-1	Type B		0.25†	0.1†	£ 0.02			Caillet et al. (1993)
				<0.2 (p)	<0.06 (p)				
	3537-2	B1	99-100	~0.3	<0.15	£ 0.08		<1	Caillet et al. (1993)
Mokoia									
		Type B		0.04	n.d.		0.09		Cohen et al. (1983)
Vigarano									
	1623-11	B2	99-100	£0.21	£0.03		£0.1	£0.08	Sylvester et al. (1992)

(r) and (c) refer to rim and core compositions, respectively.* - rare points with up to 0.4 wt% Mgo were found in anorthite;
† - anorthite from relict parts of inclusions; (p) - anorthite from poikilitic part of inclusion

Type B inclusions (TS34, TS23 and TS21) from Allende and correlated the CL with minor element concentrations, specifically MgO and Na_2O (see also Steele et al. 1997). The plagioclase exhibits CL color variations from brilliant blue to dark blue, that are present in 4 distinct types of textural occurrences: (1) bands of alternating light and dark blue at twin boundaries due to the different crystallographic orientations of the twins, (2) sharp, angular, boundaries that appear to be crystallographically controlled and transect twinning, (3) irregular patches, 10-~100 μm in size, randomly distributed through the crystals and (4) gradual and irregular color changes in polycrystalline plagioclase aggregates, typically near the rims of the inclusions, which often show shock features. One or more of these CL textures typically occurs in each grain. Some variations in CL may be due to sector zoning (see Steele et al. 1997). Similar CL characteristics in anorthite have been reported extensively in other Type B inclusions (e.g. El Goresy et al. 1979, El Goresy et al. 1985, Podosek et al. 1991, MacPherson and Davis 1993).

From microprobe data, the bright blue CL regions have the lowest Mg and Na contents (<500 and ~600 ppm, respectively; Hutcheon et al. 1978) and the dark blue regions are low in Mg and Na (see Fig. 96a). Ti concentrations up to ~350 ppm are present, but are not correlated with either Mg or luminescence color (Hutcheon et al. 1978).

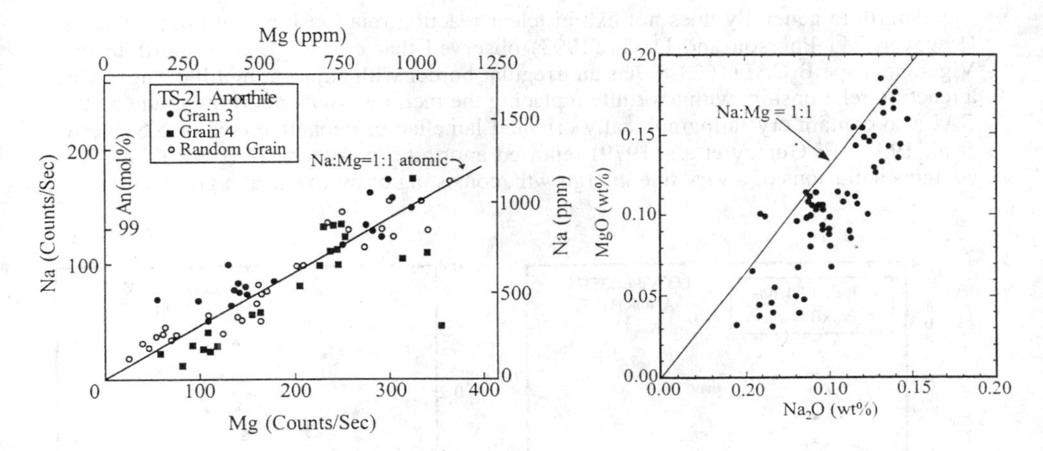

Figure 96. (a) Mg vs. Na (ppm) for plagioclase grains from a Type B inclusion (TS 21) in Allende. After Hutcheon et al. (1978). (b) Na_2O vs. MgO (wt %) contents of anorthite in a Type B Vigarano inclusion (1623-8) which show a strong positive correlation at higher Na_2O contents than is typically observed for most Type B inclusions. After MacPherson and Davis (1993).

The occurrence and compositional data for plagioclase in both *Type B1 and B2 inclusions* in Allende, Bali, Mokoia, Leoville and Vigarano have been reported by a number of authors (e.g. Gray et al. 1973, Blander and Fuchs 1975, Grossman 1975, Kurat et al. 1975, MacPherson and Grossman 1981, Cohen et al. 1983, Meeker et al. 1983, Kracher et al. 1985, El Goresy et al. 1985, Podosek et al. 1991, Sylvester et al. 1992, Caillet et al. 1993, MacPherson and Davis 1993). Grossman (1975) found that anorthites in Type B inclusions have a very restricted range of compositions (An_{100}-An_{99}) with 75% of the analyses being An_{100}. Further studies (see above) have confirmed this compositional range and also show that anorthite has consistently low minor element contents (summarized in Table 7), even in inclusions which have experienced complex histories (e.g. Podosek et al. 1991). K_2O is normally below detection for EPMA (<0.01

wt %), but some analyses contain up to 0.05 wt %. In two Allende Type B CAIs, Hutcheon et al. (1978) found a strong positive correlation between MgO and Na_2O in anorthite that lies on the 1:1 atomic trend line for one inclusion (TS21), whereas for a second (TS23) the data lie on a correlation line of less than one on an Na-Mg plot (Fig. 96a). Na and Mg concentrations for these plagioclases lie between ~70-1100 ppm for both elements and the range is similar in different grains. In a Type B Vigarano inclusion (1623-8), MgO is also correlated with Na_2O in coarse-grained anorthite, but at a higher Na_2O content than other Type B1 CAIs (MacPherson and Davis 1993) (Fig. 96b), indicating that it crystallized from a melt with higher Na_2O contents (see also Caillet et al. 1993).

It has been suggested that the correlation between Mg and Na is the result of a positive correlation between the Mg partition coefficient and liquid Si content in the late stages of crystallization (Peters et al. 1995). Alternatively, Steele at al. (1997) have shown that sector zoning will also produce a similar correlation between Mg and Na and also observed that Mg and Ti spikes occur at twin planes. This phenomenon may be due to the exsolution of clinopyroxene along the twin planes, as observed by Barber et al. (1984). However, Steele et al. (1997) favored the explanation that the Mg is simply substituting into the anorthite.

Anorthite generally does not exhibit clear reaction relationships with other phases. However, MacPherson and Davis (1993) observed that coarse-grained anorthite in a Vigarano Type B CAI (1623-8) has an irregular border with adjacent melilite, suggesting a reaction relationship, with anorthite replacing the melilite. Many anorthite grains in this CAI also contain crystallographically oriented lamellae of nepheline (see also Sylvester et al. 1992). El Goresy et al. (1979) reported anorthite in Allende CAI H126.78,1 that contains inclusions of a very fine intergrowth, consisting of pyroxene and grossular.

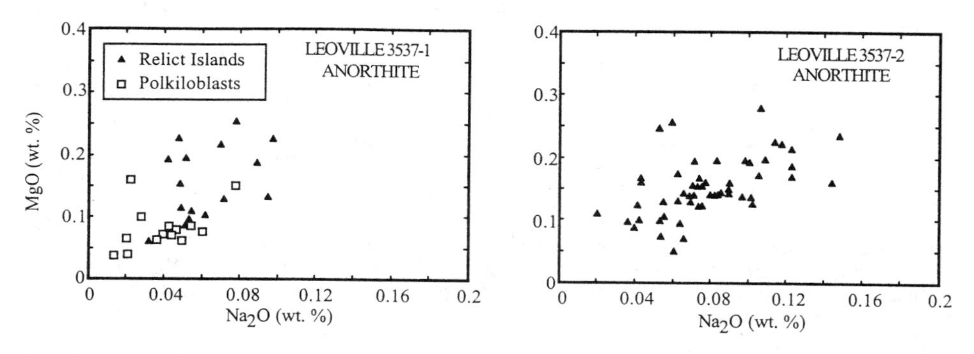

Figure 96 (cont'd). (c) Na_2O vs. MgO (wt %) contents in anorthite from Leoville Type B inclusions 3737-1 and 3527-2. After Caillet et al. (1993).

Anorthite grain compositions do not appear to be related to their spatial location within a CAI. However, Caillet et al. (1993) found that anorthite in relict islands in Leoville Type B CAI 3537-1 has slightly higher MgO (up to 0.25 wt %) and Na_2O (0.1 wt %) contents than in poikiloblastic regions (<0.2 wt % and <0.06 wt %, respectively) (Fig. 96c). Also, in Allende CAI TS33F1 (B2), MgO in anorthite is lower near the rim of the inclusion (0.13-0.14 wt %) than in the core (0.15-0.27 wt %) (MacPherson and Grossman 1981).

MacPherson et al. (1989a) determined REE concentrations in anorthite from a spinel-free island in Allende CAI 5241 by ion microprobe. REE patterns slope from the LREE (La - 6 × CI) to the HREE (~1 × CI) with a large positive Eu anomaly (50 × CI) (Fig. 97). An almost identical REE pattern was obtained from anorthite in a Vigarano CAI (1623-8; MacPherson and Davis 1993). Meeker (1995) reported REE and other trace element (Ba, Sc, Ti, V, Cr, Zr and Nb) data for anorthite from a spinel-free island in Allende Type B inclusion 3529Z. REE abundances are lower (0.15-2 × CI) with the LREE enriched relative to the HREE and a large positive Eu anomaly (30 × CI) consistent with element partitioning during igneous crystallization (Fig. 97). Phinney et al. (1979) reported Li, Be and B concentrations in anorthite from an Allende Type B1 CAI (EK-1-07).

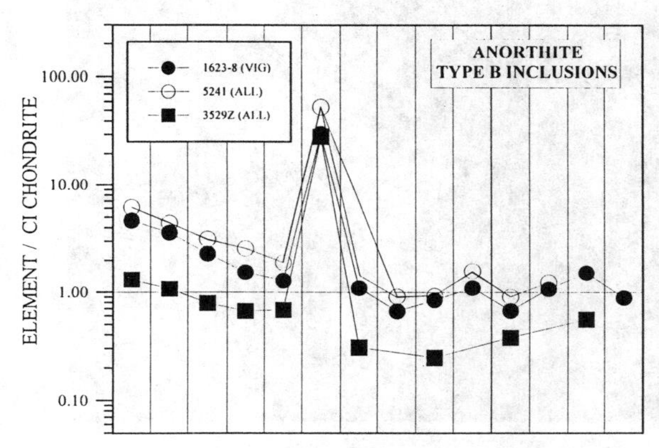

Figure 97. REE abundance patterns (normalized to CI chondrites) for anorthite from Type B inclusions in Allende (5241, 3529Z) and Vigarano (1623-8). (Ion microprobe analyses; data from MacPherson et al. (1989), MacPherson and Davis (1993) and Meeker (1995).

TEM observations of anorthite in Allende Type B1 CAIs (TS23, TS34) show that pericline twins are common, but only a few dislocations are present associated with voids and fine-grained inclusions (Barber et al. 1984). The anorthite contains two types of antiphase boundaries (Fig. 98); gently curved or straight **b** antiphase boundaries (APB) formed due to Si-Al ordering during the $C\bar{1}$ to $I\bar{1}$ phase transformation and irregular jagged **c** antiphase domains resulting from the $I\bar{1}$ to $P\bar{1}$ transformation that occurs below 250°C (Brown et al. 1963, Aldhart et al. 1980). These observations suggest formation of the plagioclase at high temperature with significant low temperature annealing. Barber et al. (1984) suggested that the c APBs may record the thermal history above 500°C. The plagioclase also contains numerous precipitates, probably fassaite, with a platelet morphology (0.1 μm long) (Barber et al. 1984). They occur along twins and twin boundaries and may have exsolved from an Mg-bearing plagioclase.

Primary anorthite is relatively rare in *Type B3 inclusions* but occurs in several inclusions including CAI 818a, CAI 110-A (Wark et al. 1987) and TE (Dominik et al. 1978).

Anorthite constitutes between 38 and 60 vol % of *Type C inclusions* (Wark 1987) and is essentially pure with Na_2O contents between 0.03-0.09 wt % ($An_{99.2-99.7}$) (Grossman 1975, Macdougall et al. 1981, Wark 1987). Wark (1987) also reported very low concentrations of MgO (<0.10 wt %), FeO (0.02-0.12 wt %), and TiO_2 (0.02-0.06 wt %), with K_2O, MnO, BaO, Cr_2O_3 and V_2O_3 all <0.04 wt %.

Anorthite is present in low abundances in *fine-grained inclusions* from Allende, Kaba, Leoville, Mokoia and Vigarano (e.g. Wark 1979, Cohen et al. 1983, Fegley and

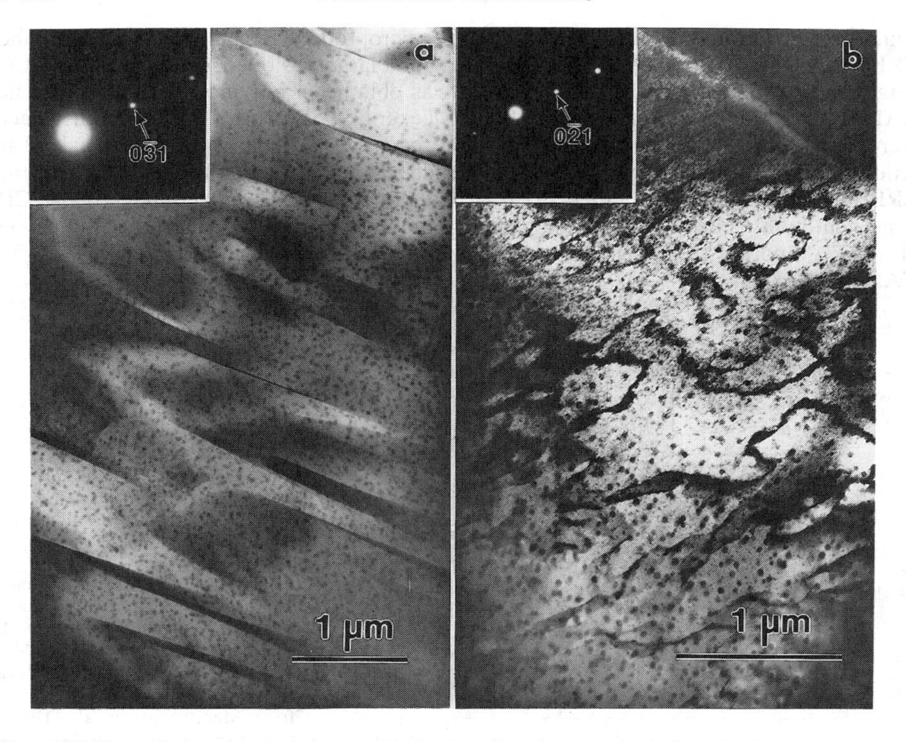

Figure 98. Transmission electron micrographs showing the microstructures of plagioclase in Allende. (a) Dark field image using the 031 reflection. Type **b** antiphase domains boundaries with a morphology that suggests a high temperature origin for the plagioclase and equilibration during slow cooling. (b) Dark field image using 021 reflection. Type **c** APBs in a nearby region of the same grain as in 98a, with a size and shape consistent with a well-equilibrated plagioclase. Used by permission of the editor of *Geochimica et Cosmochimica Acta* from Barber et al. (1984), Figure 6.

Post 1985, Mao et al. 1990, Sylvester et al. 1992), but there is some question as to whether it is primary or secondary in origin. In Allende, anorthite is sometimes the alteration product of melilite in coarse-grained CAIs, but McGuire and Hashimoto (1989) suggested that in fine-grained inclusions it may be primary. The anorthite in these inclusions is compositionally quite variable, ranging from An_{90} to An_{100} with a mean of An_{97} and Or < 0.5 mol %. Minor element concentrations such as Fe_2O_3 and MgO in all the anorthites studied are < 0.5 wt %.

Minor anorthite has also been observed rimming spinel in fine-grained inclusions in Allende (Wark 1979) and Mokoia (Cohen et al. 1983). In other fine-grained inclusions in Leoville (Mao et al. 1990) and Kaba (Fegley and Post 1985), the anorthite is essentially pure with only minor Na_2O (e.g. Leoville, <0.15 wt % Na_2O), but in Kaba can contain ~1.5 wt % MgO. Fegley and Post (1985) also reported one anomalous grain with a composition $An_{75}Ab_{25}$.

Minor anorthite (An_{90}) occurs in an *unusual cordierite-bearing CAI* in Allende (Fuchs 1969).

Olivine. Representative analyses of olivine in CAIs in CV chondrites are reported in Table A3.19. The only occurrence of olivine in a *Type B inclusion* was reported by Armstrong et al. (1987) who noted the presence of fine-grained olivine (<10 μm) adjacent to a large fremdling (Zelda) in the Allende Egg6 (B1) inclusion.

Forsteritic olivine is the dominant phase in *Type B3 CAIs* in Allende and Vigarano and has high concentrations of CaO (typically 1.0-2.07 wt %) and low FeO contents (typically 0-0.3 wt %) (Blander and Fuchs 1975, Dominik et al. 1978, Clayton et al. 1984, Wark et al. 1987). However, Dominik et al. (1978) reported CaO contents up to ~11 wt % in inclusion TE from Allende. The maximum CaO content of forsterite appears to increase as the abundance of olivine increases in the inclusion (Wark et al. 1987). Dominik et al. (1978) and Clayton et al. (1984) reported Al_2O_3 contents between 0.09-0.16 wt % in Allende inclusions TE and CG-14. TiO_2 and Cr_2O_3 concentrations are <0.1 wt % (Dominik et al. 1978, Wark et al. 1987). In Allende Type B3 CAIs, FeO in olivine can reach 3 wt % close to alteration zones within the CAI interior and 27 wt % near the margins of the inclusions (e.g. CG-14; Clayton et al. 1984) (see also Davis et al. 1991 for Vigarano CAI 1623-5). Davis et al. (1991) reported REE and other trace element data for olivine from the core of CAI 1623-5. The olivine has a fractionated REE pattern with depleted LREE (La = <0.1 × CI) relative to the HREE (Lu = ~4 × CI).

Olivine is present in *fine-grained inclusions* (e.g Wark 1979, Cohen et al. 1983, McGuire and Hashimoto 1989) and has a highly variable composition. In Mokoia (Cohen et al. 1983) and Allende inclusions (McGuire and Hashimoto 1989), olivine has the compositional range Fa_{10-43} and can have very high Al_2O_3 (<10 wt %) and Na_2O contents (<1.3 wt %), which have been attributed to the presence of submicron inclusions of spinel and nepheline. TiO_2, Cr_2O_3 and CaO are present in low concentrations, typically <0.5 wt %. In Vigarano and Kaba inclusions, the olivine contains very little FeO (e.g. Fo_{98}; Sylvester et al. 1992; $Fa_{98.35}$; Fegley and Post 1985).

Olivine (Fa_{22-34}) occurs in the rim sequence of the isotopically *unusual inclusion* HAL from Allende (J.M. Allen et al. 1980). Olivine Fo_{88}-Fo_{97} has also been reported in an armalcolite-bearing, unclassified CAI from Allende (Haggerty 1978b).

Accessory phases. A large number of minor and rare phases have been reported in CAIs in CV chondrites and are summarized in Table 8. Data on selected accessory phases that have been studied in detail are summarized below.

Paque et al. (1994) found a new Ca-Al-Ti silicate as an accessory phase in several Type A and B1 CAIs from Allende. Several to many grains, <10 µm in size, typically occur only in melilite crystals. The phase, termed temporarily as *UNK*, has a restricted compositional range with 29-32 wt % CaO, 24-28 wt % SiO_2, 20-27 wt % TiO_2, 16-20 wt % Al_2O_3 and <1 wt % MgO. Zoning may be present with Ti enriched in the cores. Paque et al. (1994) suggested a structural formula of $Ca_3Ti(Al,Ti)_2(Si,Al)_3O_{14}$ assuming between 7-13 wt % Ti^{4+}. From electron diffraction patterns, Barber et al. (1994) determined that UNK was either primitive hexagonal or triclinic and has cell parameters of a = 0.790±0.002 nm and c = 0.492±0.002 nm.

Marvin et al (1970) reported *hercynite* in CAIs in Allende that contain perovskite, and hercynite-rich spinels (up to 30 wt % Fe_2O_3) occur adjacent to a fremdling in the Egg6 Type B1 inclusion from Allende (Armstrong et al. 1987). Murakami and Ikeda (1994) also found hercynite-rich spinels (15-22 wt % FeO) in fine-grained inclusions in Y-86751. Hercynite may also be present in the unusual hibonite-bearing inclusion (HAL) from Allende (J.M. Allen et al. 1980).

Coulsonite containing 0.13 wt % Sc_2O_3 is a rare accessory phase as inclusions in melilite in a "Compact" Type A CAI in Efremovka (Ulyanov et al. (1982). In Allende, coulsonite-rich spinels (up to 19 wt % V_2O_3) occur adjacent to a large fremdling in the Egg6 Type B1 inclusion (Armstrong et al. 1987) and as inclusions (20-27 wt % V_2O_3;

Table 8. Primary accessory phases in CAIs in CV carbonaceous chondrites.

Mineral	Composition	CAI Type	Reference
Silicates			
Cordierite (high)	<0.7 wt% FeO	Unusual	Fuchs (1969)
	2.4-6.5 Na_2O		
Orthopyroxene	7.5 wt% Al_2O_3		Fuchs (1969)
Calcio-celsian	$(Ba,Ca)_4Al_8Si_6O_{32}$	B	Lovering et al. (1979)
Glass	8 wt% Na_2O	B1 (10/5)	Blander and Fuchs (1975)
UNK -	$Ca_3Ti(Al,Ti)_2(Si,Al)_3O_{14}$		Paque et al. (1994)
Hydroxides			
Pyrochlore	$(Ca,REE,Th)_{1.82}(Nb,Ti,Zr,Al)_{1.98}O_6(OH,F)$		Lovering et al. (1979).
Oxides			
Armalcolite	$FeTi_2O_5$-$MgTi_2O_5$	?	Haggerty (1978b)
Baddelyite	ZrO_2	Type B	Lovering et al. (1979)
Beckellite	$(Ce,Ca)_5(SiO4)_3(OH,F)$		Lovering et al. (1979)
Ca-Al-Mg titanate	$(CaMgFeMn)Ti_3O_7$ - $(TiAlCr)_2Ti_2O_7$		Haggerty (1978)
CaO	CaO	Type A	Greshake et al. (1996)
Chromite	$FeCr_2O_4$	Unusual (3643)	Wark (1986)
Coulsonite	FeV_2O_4	Compact Type A	Ulyanov et al. (1982)
		Egg6 Type B1	Armstrong et al. (1987)
		Type B	Fuchs and Blander (1977)
			Fuchs (1978)
		B1 (Egg6)	Armstrong et al. (1987)
Geikielo-ilmenite	$MgFeTiO_3$	Type B	Fuchs (1978)
Hercynite	$FeAl_2O_4$	?	Marvin et al (1970)
Magnetite	Fe_3O_4	Type B	see text
Magnesiowüstite	(Mg,Fe)O	Type B	Caillet et al. (1993)
Niobio-perovskite	$(Ca(Ti,Nb)O_3$	Type B	Lovering et al. (1979)
Rutile	TiO_2	Type A	Greshake et al. (1996)
Sc-Ti-Zr oxide		Unusual (HAL)	Allen et al. (1980)
Thorianite	ThO_2	Type B	Lovering et al. (1979)
Sulfides			
Millerite	NiS	Type B1, B2	Wark and Lovering (1980)
Molybdenite	MoS_2	Type B	Blander and Fuchs (1975)
		melilite-rich	Fuchs and Blander (1977)
Troilite	FeS	Type A and B	see text
Pentlandite	$(Fe,Ni)_9S_8$		see text
			Wark and Lovering (1982)
Metals			
Kamacite	Fe,Ni	Type A, B	see text
Taenite	Fe,Ni	Type A, B	see text
Awaruite	Fe,Ni		Casanova and Simon (1994)
Cu,Ni,Zn		ultrarefractory	Wark (1986)
Pt-group metals	Pt, Os, Ir, Ru, Rh	Type A, B	see text

60-68 wt % FeO) within metal grains in a rhönite-bearing Type B inclusion (Fuchs and Blander 1977, Fuchs 1978).

Magnetite has been recognized as a rare accessory phase in several CAIs including Type B CAIs in Allende (e.g. Blander and Fuchs 1975, Fuchs and Blander 1977, Fuchs 1978) and also in Vigarano. V-free magnetite occurs in veins adjacent to a large fremdling (Zelda) in the Egg6 Type B1 inclusion (Armstrong et al. 1987). In Vigarano,

V-rich magnetite sometimes occurs surrounding metal grains (MacPherson and Davis 1993) in Type B1 inclusion 1623-8. Magnetite spherules (2-5 μm) occur in olivine and fassaite in forsterite-bearing inclusion 9/16 from Allende (Blander and Fuchs 1975).

Haggerty (1978b) reported several grains of *armalcolite* ($FeTi_2O_5$-$MgTi_2O_5$) (10-25 μm) in two CAIs (ALL-3, ALL-6) in Allende whose primary mineralogy consists of anorthite, fassaite and spinel with olivine. The grains are compositionally variable with between 69.34-75.69 wt % TiO_2 and 8.02-21.24 wt % FeO. The armalcolite also contains variable amounts of MgO (2.31-14.61 wt %), CaO (0.01-3.40 wt %), Cr_2O_3 (0.19-10.31 wt %), Al_2O_3 (0.39-2.07 wt %) and ZrO_2 (<0.05-6.01 wt %).

In an Allende CAI, Haggerty (1978b) described several grains of a previously unrecognized *Ca-Al-Mg titanate* whose compositions lie along a solid solution series $(CaMgFeMn)Ti_3O_7$ - $(TiAlCr)_2Ti_2O_7$, with Ti calculated as TiO_2 varying between 78.81 and 85.73 wt %. Like armalcolite in the same inclusion, the phase contains variable concentrations of MgO (1.34-8.95 wt %), CaO (2.28-10.74 wt %), FeO (0.37-4.82 wt %), Cr_2O_3 (0.22-3.50 wt %) and Al_2O_3 (1.04-4.05 wt %) with minor MnO (<0.25 wt %). The phase is compositionally most closely related to zirconalite ($CaTiZr_2O_7$), but is dominated by Ti^{3+} rather than Zr. It may also be related to the loveringite mineral series.

With the exception of their occurrence in fremdlinge, *sulfides* are rather rare phases in CAIs. A number of minor occurrences have been reported, but in many cases the sulfide minerals have not been fully characterized.

In Allende, MacPherson and Grossman (1984) found that rare, very fine-grained *troilite* inclusions (<1 μm) occur in some "Fluffy" Type A CAI and Fuchs and Blander (1977) and Fuchs (1978) reported minor troilite associated with metal grains in a rhönite-bearing CAI (see also Haggerty 1978b). Troilite has been observed in Type B inclusions in Allende (Blander and Fuchs 1975, Wark and Lovering 1982b) and in Leoville (Caillet et al. 1993). Also, in Allende, troilite veins occur adjacent to a large fremdling (Zelda) in the Egg6 Type B1 inclusion (Armstrong et al. 1987). Troilite has also been reported in CAIs from Axtell (Simon et al. 1995).

Although apparently rather rare, *pentlandite* has been reported in a variety of CAIs in CV chondrites. Fuchs and Blander (1977) described pentlandite associated with high Ni metal in a Type B inclusion (10/5) and Cohen et al. (1983) reported small pentlandite grains within regions of what they describe as chaotic material within fine-grained CAIs in Mokoia. Other occurrences of Ni-bearing sulfides, probably pentlandite, have been reported in several Allende CAIs (Haggerty 1978b, El Goresy et al. 1979, Wark 1986) and in Kaba (Fegley and Post 1985).

In Allende, rare *molybdenite* (*MoS_2*) grains occur in a melilite-rich CAI (Fuchs and Blander 1977) and in a Type B CAI (Blander and Fuchs 1975). The molybdenite occurs as elongated blades, 1-3 μm wide and 16-24 μm long, embedded in Ni,Fe metal. The molybdenite is probably hexagonal. Fegley and Post (1985) also tentatively identified MoS_2 in a fine-grained inclusion from Kaba.

Metal. The main occurrence of metal in CAIs is in fremdlinge (see section on the Mineralogy of Fremdlinge), but metal also occurs as a minor accessory phase in many CAIs. One of the characteristics of metal inclusions in CAIs is that they can contain measurable amounts of refractory siderophile elements such as Mo, Re, W and platinum group elements, such as Pt, Os, Ir, Ru, Rh.

Metal has not been widely reported in "Compact" Type A inclusions, but Ulyanov et al. (1982) reported approximately 0.4 wt % Fe,Ni metal in a "Compact" Type A inclusion from Efremovka. Sylvester et al. (1993) also found rare metal grains in some Leoville and Efremovka "Compact" Type A inclusions. Grossman (1975) and MacPherson and Grossman (1984) have reported Fe,Ni grains (<10μm) as inclusions in other phases such as spinel in "Fluffy" Type A inclusions in Allende. Sylvester et al. (1993) also found nuggets of Os,Ir,Pt-bearing Fe,Ni grains in "Fluffy" Type A inclusion, L4 in Leoville.

Fe,Ni metal has been reported by several authors in Type B inclusions (Grossman 1975, Blander and Fuchs 1977, Fuchs 1978, El Goresy et al. 1979, Wark and Lovering 1982a,b, Meeker et al. 1983, Sylvester et al. 1992, Caillet et al. 1993). The metal occurs as small blebs or nuggets with rounded morphologies, up to several microns in size (e.g. Grossman 1975). In essentially all the Type B inclusions in Allende, the metal is Ni-rich but has a variable composition. Blander and Fuchs (1977) and Fuchs (1978) documented metal grains, 10-30 μm in size, but up to 125 μm, in a rhönite-bearing, Type B2 inclusion. The metal is Ni-rich (61-69 wt %) with minor V (0.4-0.6 wt %), Co (1.7-2.1 wt %), Cr (0.1-0.3 wt %), Si (0-0.1 wt %) and P (0-0.1 wt %). The metal occurs enclosed in all the primary phases (perovskite, fassaite, melilite and mesostases) of the inclusion. In a second Type B inclusion, Blander and Fuchs (1977) observed Ni-rich (~62 wt %) metal grains which contain inclusions of molybdenite. El Goresy et al. (1979) found metal grains with variable compositions in a Type B Allende inclusion, with a typical composition of ~60 wt % Ni and 30 wt % Fe. These grains are also relatively enriched in Pt (up to 2.5 wt %) and Rh, but have low concentrations of Ru, Os, Ir, W and Mo. Similarly, Wark and Lovering (1982b) found micron-sized nuggets of Ni-rich (55-95%) metal grains in the outer mantles of two Type B1 inclusions in Allende (Egg3 and 3529,33) and also in inclusion 5171,18 (Wark and Lovering 1982b). However, in Type B2 Allende inclusion 110,A, the metal grains are Fe-rich (70-100 wt % Fe; 30-0 wt % Ni) (Wark and Lovering 1982b) except within a spinel palisade body where rare, Ni-rich metal (60-70% wt %) occurs. In Axtell, Casanova and Simon (1994) found that the dominant metal phase in four CAIs they studied was awaruite.

In Type B CAIs from the reduced CV chondrites, Leoville and Vigarano, the metal typically contains less Ni than in Allende. For example, Ni,Fe metal grains that occur in a Type B1 inclusion (3537-2) in Leoville (Sylvester et al. 1992, Caillet et al. 1993) consist of both kamacite and taenite. The taenite is zoned from 31 to 41% Ni and 0.13 to 0.45% Co and kamacite contains 5.2-7.5 wt % Ni and 0.86-1.33 wt % Co. Kamacite and taenite also occur as blebs (up to 50 μm) associated with unidentified oxides and troilite grains in relict islands in Leoville Type B inclusion 3537-1 (Caillet et al. 1993) and as inclusions in fassaite and anorthite. Similarly, MacPherson and Davis (1993) found mostly taenite grains with ~35 wt % Ni and ~1 wt % Pt and Ru, in a Type B1 inclusion (1623-8) from Vigarano, but kamacite is also present with ~5 wt % Ni. Occurrences of Fe,Ni metal have also been reported in other Type B1 and B2 inclusions in Vigarano (e.g. Mao et al. 1990; 477-B, Sylvester et al. 1992 1993), but compositional data are not available. Goswami et al. (1994) have also reported metal grains in the core and mantle of Type B1 inclusions (E40, E65, E44) from Efremovka.

There are few reports of Fe,Ni metal in Type B3 inclusions, with the exception of examples documented by Blander and Fuchs (1975) and Wark et al. (1987). Blander and Fuchs (1975) reported metal spherules (2-5 μm) in olivine and fassaite in inclusion 9/16 from Allende and metal grains up to 20 μm in size are present in the core of another Allende inclusion (818a) (Wark et al. 1987). The metal in this inclusion ranges in composition between 90-97 wt % Fe and 3-8 wt % Ni. Blander et al. (1980) also found one grain of a Ni,Fe alloy containing uniformly alloyed Pt in a Type C Allende CAI (6/1) and rare Fe,Ni beads occur in a fine-grained, spinel-rich inclusion from Vigarano (Davis et al. 1987, Mao et al. 1990).

Wark (1986) found several Fe,Ni grains in the crust and mantle of an unusual, ultrarefractory inclusion (3643) from Allende. Although most metal grains in both occurrences are Ni-rich (crust—65-78 wt %; mantle—73-75 wt %), one grain with lower Ni contents was found in the crust (33 wt % Ni). Mao et al. (1990) reported Fe,Ni metal beads that sometimes contain Os, Ir and Ru in a melilite-rich inclusion in Vigarano (1623-3).

Wark (1986) found one grain of a highly unusual *CuNiZn alloy* in the mantle of an ultrarefractory inclusion (3643) from Allende. SEM EDS analysis showed that this phase has an approximate composition $Cu_{65}Ni_{18}Zn_{16}Fe_1$.

In addition to inclusions of Fe,Ni metal in CAIs, many inclusions also contain micron-sized nuggets of alloys rich in refractory (Mo, Re, W) siderophile elements and platinum group (Pt, Os, Ir, Ru, Rh) metals. For this reason they are sometimes referred to as *Refractory/Platinum group metals* or RPM (e.g. Wark and Lovering 1982b). These metal alloys occur in both Type A and Type B inclusions and occasionally occur in the same inclusions together with fremdlinge. Although they are probably related to fremdlinge in some way, they are distinct objects and are treated separately here from fremdlinge themselves.

RPM-rich nuggets have been reported in several *Type A inclusions* in Allende and Leoville (e.g. Wark and Lovering 1978, El Goresy et al. 1978, Allen et al. 1978, MacPherson and Grossman 1984, Sylvester et al. 1993). In general, the inclusions are extremely small and difficult to analyze. For example, Wark and Lovering (1978) and El Goresy et al. (1978) reported that the nuggets were <2 μm in size and enclosed within primary CAI phases. All contain Pt group metals (Re, Os, Ir, Pt, Ru, Rh), but some inclusions also contain high Fe (~10 wt %) but low Ni (~ 1 wt %). Similar observations were made by J.M. Allen et al. (1978) who reported several RPM nuggets, up to 0.3 μm in size, in "Fluffy" Type A inclusion CG11 from Allende. Semiquantitative analyses indicate that they consist of a Pt, Fe, Ni, Os Ir, Ru, Rh alloy which may contain smaller grains of a Ru, Os, Ir alloy. MacPherson and Grossman (1984) also found RPM nuggets in several "Fluffy" Type A inclusions from Allende, but reported no analytical data. Simon and Grossman (1992) found an unusual occurrence of crystallographically oriented lamellae of the alloy phase εRu-Fe in γNi,Fe in a "Compact" Type A inclusion from Leoville (L1). This phase probably formed by exsolution from taenite and equilibrated at ~873K.

El Goresy et al. (1978) reported RPM nuggets (<2 μm) enclosed in primary phases in *Type B inclusions* in Allende and Leoville. All contain Pt metals (Re, Os, Ir, Pt, Ru, Rh) but in comparison with similar nuggets in Type A inclusions, Fe and Ni are very low or absent. El Goresy et al. (1978) also found that the inclusions could consist of a single, complex Pt-metal alloy or multiple, submicron grains of different Pt alloys. Some nuggets are compositionally zoned with Os, Ir and Ru-enriched cores and Pt, Rh-rich rims. High concentrations of Mo, Nb and Ta are also present in the Pt metal nuggets as well as Zr in some rare grains. El Goresy et al. (1978) suggested that the Mo, Nb, Ta and Zr could be present as discrete oxide grains or in the metallic state in solid solution in the Pt metal. Some nuggets contain relatively high amounts of Cr (<4 wt %), very different from the Pt-metal nuggets which occur in fremdlinge (see below).

Also in Allende, Wark and Lovering (1982b) found μm-sized Pt-metal rich nuggets containing Mo, Re, W, Pt, Os, Ir, Ru, and Rh in a Type B2 inclusion (5171,18) and Hutcheon et al. (1987) reported a small (~3-6 μm) refractory metal nugget in a Type B CAI (C1) which contains Os, Ir and Ru with minor Cr. Pt-rich, Fe-bearing metal nuggets, < 1 μm in size, have also been reported in the outer portions of a Type B1 CAI (447B) from Allende (Zinner et al. 1991) which are depleted in the refractory siderophiles Re, Os, Ir, Ru. The same CAI also contains veins of a Pt-rich, Fe-bearing alloy with >15 wt % Pt but depleted in Os, Ir and Ru. Finally, in a Vigarano Type B1 inclusion (1623-8), MacPherson and Davis (1993) reported one metal grain with approximately 10 wt % Ni, 6 wt % Ru, 11 wt % Os, and 12 wt % Ir.

Dominik et al. (1978) found small metal beads enriched in Os, Ir, Pt, Rh, and Ru in *Type B3 inclusion* TE in Allende. Pt and Rh appear to be enriched at the edges of these beads and some grains also contain Ta and Nb.

Blander et al. (1980) were the first to report Pt-rich nuggets in a *Type C CAI* (6/1) from Allende. They found a distinct region of the inclusion which contained only ultrarefractory nuggets whereas the remainder of the inclusion only contains nuggets that are much less refractory in composition. The ultrarefractory nuggets are enriched in Os (~37-47.4 wt %), Mo (10.6-15.2 wt %), Ir (17.1-27.4 wt %) with lower concentrations of Re (3-4 wt %), W (6-8.5 wt %), Ru (3-8 wt %), Fe (3.5-6.8 wt %), and Ni (0.5-1.2 wt %). In a comprehensive study, Wark (1987) observed refractory metal nuggets in several Type C inclusions from Allende. In all cases they are extremely small (<5 μm) and three different types of nuggets exist: single-phase alloy nuggets, two phase nuggets and 'fremdlinge' of intergrown metal and silicates which are smaller and simpler than those found in Type B CAIs (El Goresy et al. 1978). The compositions of these different occurrences of alloys in individual inclusions are highly variable, but the single alloys are dominantly Mo, Pt, Ru and Ir-rich with significant concentrations of Fe, Ni, Rh, and Os. For example, RPM nuggets in CAI 100 studied by Wark (1987) have Mo between <0.5-35 wt % and Ru, Pt and Os range between 7.3-26.1 wt %, 13.5-45.1 wt %, and 2.2-12.5 wt %, respectively. Similar compositional ranges are exhibited by alloys within the small 'fremdlinge'. Two-phase nuggets have also been observed consisting of an Os,Ir,Ru alloy associated with Fe,Ni,Pt,Mo alloy.

The elemental correlations within these alloys are complex, but Wark (1987) found that positive correlations typically exist between element pairs which have similar volatilities (e.g. ReW, IrRu, NiFe, RuFe) whereas negative correlations exist when elements with very different volatilities are paired (e.g. OsPt, OsRh, IrRh). These observations suggest that the compositions of the refractory metal nuggets are largely controlled by fractional condensation from a gas phase (e.g. Fuchs and Blander 1980).

Cohen et al. (1983) found a μm-sized, Pt-rich metal grain included in spinel in just one out of 224 fine-grained, spinel-rich CAIs that they studied in Mokoia. Fegley and Post (1985) also found rare Pt-metal nuggets in an inclusion from Kaba.

Other inclusions. Wark (1986) analyzed numerous μm to submicron refractory metal nuggets from the core of an unusual, ultrarefractory inclusion (3643) in Allende. The metal nuggets are alloys of Re, W and Mo with the Pt group elements, Pt, Os, Ir, Ru and Rh and occur as inclusions within melilite, grossular and hibonite. Most are single phase metal, but one fremdlinge was found. Two groups of nuggets were found; one group, consisting of relatively large grains, is enriched in Os (18.5-57.0 wt % Os), whereas a second group consists of smaller grains which are Os-poor, and Pt-rich (20.5-63.9 wt % Pt).

Secondary mineralogy of CAIs in CV chondrites

Evidence of secondary alteration is commonly encountered in CAIs in some CV chondrites, but not in others. The most altered CAIs are found in CV chondrites from the oxidized subgroup, such as Allende and Axtell, whereas inclusions from reduced CV chondrites, such as Leoville, Vigarano and Efremovka, show much less secondary alteration. This secondary alteration has resulted in the replacement of primary phases by a plethora of new phases, in addition to the formation of veins. Based on studies of highly altered, fine-grained Allende CAIs (e.g. McGuire and Hashimoto 1989), the resistance of primary phases to alteration increases in the order: melilite, perovskite, Al,Ti-rich diopside, hibonite, spinel and aluminous diopside. Although many studies have noted the presence of alteration phases, detailed studies of their occurrence and composition in many inclusions is lacking, perhaps because they are typically very fine-grained. Unlike the alteration which has affected CAIs in CM chondrites, the bulk of the secondary phases in CAIs are anhydrous minerals, although hydrous phases are common in CAIs in Mokoia (e.g. Cohen et al. 1983).

The timing and location of this alteration is the subject of considerable discussion. Some workers favor alteration of the inclusions within a nebular environment (e.g. Allen et al. 1978, Hashimoto and Grossman 1987, McGuire and Hashimoto 1989), but it has also been argued that alteration could have occurred on a parent body (e.g. Housley and Cirlin 1983, Krot et al. 1995). Certainly there appear to be problems for the interpretation of these secondary phases as solar nebular condensates, as discussed by Grossman (1980). For a review of the different types of secondary alteration in CAIs in CV chondrites, as well as chondrules and matrix, see Krot et al. (1995).

Nepheline. Marvin et al. (1970) were the first to report nepheline in a CAI in Allende, associated with melilite. Nepheline is now recognized as one of the most widespread secondary phases in all types of CAIs in Allende and has also been reported to a lesser extent in CAIs in Leoville, Efremovka and Vigarano. These occurrences are summarized below and analyses are reported in Table A3.20.

Allen et al. (1978) and MacPherson and Grossman (1984) observed that nepheline was a common, fine-grained phase in *"Fluffy" Type A CAIs* in Allende (see also Murrell and Burnett 1987). It commonly occurs associated with anorthite, grossular and sodalite, typically replacing melilite; individual nodules show variable degrees of replacement of melilite. In one case, striated nepheline needles have been observed on a hibonite grain in a cavity in inclusion CG11 (Allen et al. 1978). In this inclusion, nepheline is Ca-rich (3.7 wt % CaO) and contains ~1.7 wt % K_2O. Murrell and Burnett (1987) also reported nepheline in rims within cavities on the interior of an Allende "Compact" Type A inclusion (84C).

Armstrong et al. (1987) noted the presence of nepheline adjacent to a large fremdling (Zelda) in the Egg6 *Type B1* Allende inclusion. Podosek et al. (1991) found that nepheline was rare in a suite of Allende Type B CAIs, but observed nepheline replacing melilite in Type B inclusion 3529-21. Nepheline is also rare in Vigarano inclusions, but MacPherson and Davis (1993) reported crystallographically oriented lamellae of nepheline replacing some anorthite grains in a Type B1 CAI (1623-8).

Nepheline is a common replacement product of melilite in the core and mantle of two *Type B3 inclusions* in Allende (Clayton et al. 1984, Wark et al. 1987). In inclusion CG-14 (Clayton et al. 1984), nepheline commonly occurs as thin rims surrounding spinel grains and in alteration zones, associated with small, iron-rich olivine grains. Nepheline in CAI 818a (Wark et al. 1987) is generally sodic in composition with 2-6 wt % CaO, 1.2-2.0 wt % K_2O and 0.1-2.0 wt % FeO, and is compositionally more variable in the mantle of the inclusion than the core (~4.4 wt % CaO).

Nepheline is a rare alteration phase of melilite in *Type C inclusions* (Wark 1987), but no analyses have been reported in the literature.

Clarke et al. (1970) suggested that nepheline was present in a *fine-grained CAI* from

Allende and Grossman et al. (1975) tentatively identified euhedral nepheline crystals in voids in another fine-grained inclusion using SEM and qualitative EDS analysis. Further studies, using x-ray diffraction, have confirmed the widespread presence of nepheline in these inclusions (Blander and Fuchs 1975, Grossman and Ganapathy 1976) (see also Wark 1979). Kornacki and Wood (1985a) found nepheline with ~1.7-3.4 wt % CaO, ~1.0-1.9 wt % K_2O and ~0.2-1.2 wt % FeO in fine-grained Allende inclusions. McGuire and Hashimoto (1989) also found that nepheline is a common replacement product of melilite in zoned inclusions in Allende. The nepheline in these inclusions is compositionally variable with 0.5-2.0 wt % K_2O and 1.2-6.5 wt % CaO, with concentrations of FeO and MgO <2.5 wt %. Nepheline is also present in a zoned, spinel-rich inclusion (1623-14) from Vigarano (Sylvester et al. 1992).

J.M. Allen et al. (1980) reported nepheline in the rim sequence of the *unusual, hibonite-bearing inclusion* HAL from Allende. The nepheline is Ca-rich (2.3-2.8 wt % CaO) and contains 1.4-1.7 wt % K_2O. Wark (1986) also observed nepheline replacing melilite in an ultrarefractory inclusion (3643) in Allende. The nepheline is relatively enriched in CaO (~4.3 wt %) with ~1.6 wt % K_2O. Rare nepheline occurs as a secondary phase in a melilite-rich inclusion from Vigarano (1623-2) (Mao et al. 1990).

Sodalite. Allen et al. (1978) suggested that sodalite may be present in *"Fluffy" Type A inclusion* CG11 from Allende, based on the presence of high Cl in analyses of very fine-grained material. MacPherson and Grossman (1984) confirmed the presence of sodalite, replacing melilite along with nepheline, grossular and anorthite, in several "Fluffy" Type A inclusions in Allende. Murrell and Burnett (1987) reported sodalite in rims within cavities on the interior of a "Compact" Type A inclusion (84C) in Allende.

Remarkably, there appear to be no confirmed occurrences of sodalite in Allende *Type B inclusions*, but Sylvester et al. (1993) found sodalite as a secondary alteration product in a Vigarano Type B2 inclusion (Vig 1).

The occurrence of sodalite in *fine-grained inclusions* in Allende was first suggested by Clarke et al. (1970) and Grossman et al. (1975) found euhedral grains of a silicate phase containing Na, Ca, Mg, Al and Cl in voids, using SEM techniques. They suggested that this phase was probably sodalite. Subsequent x-ray diffraction studies of fine-grained inclusions in Allende (Blander and Fuchs 1975, Grossman and Ganapathy 1976) have confirmed the widespread presence of sodalite. Sodalite has also been reported by Wark (1979) and Kornacki and Wood (1985a) in Allende inclusions. Analyses reported by Kornacki and Wood (1985a) contain ~0.1-0.3 wt % CaO, <0.05 wt % K_2O and 0.4-1.5 wt % FeO. McGuire and Hashimoto (1989) found fine-grained Cl and SO_2-bearing sodalite with <0.5 wt % FeO and MgO as a replacement product of melilite in all three zones of inclusions in Allende. Rare sodalite has also been reported in one zoned, spinel-rich inclusion from Vigarano (1623-14) and may possibly be present in a second (1623-16) (Sylvester et al. 1992).

Sodalite has been reported by Fuchs (1969b) in an *unusual cordierite-bearing CAI* in Allende, associated with orthopyroxene, spinel and anorthite. Wark (1986) also observed sodalite replacing melilite in an unusual ultrarefractory inclusion (3643) in Allende.

Grossular. Grossular was first identified as an accessory phase in CAIs in Allende based on limited x-ray diffraction data (Clarke et al. 1970, Fuchs 1969b,1971). Fuchs (1974) confirmed the presence of grossular in several inclusions from Allende using x-ray diffraction and electron microprobe. Grossular typically occurs as irregular masses (100 μm) or stringers, 5 μm wide and up to 200 μm long, associated with melilite and/or anorthite. The grossular in these inclusions has variable MgO (1.6-4.2 wt %) and FeO (0-3.4 wt %) (Fuchs 1974) but grossular in individual inclusions shows a more restricted range of compositions. Minor Na_2O (up to 0.4 wt %) is also present in grossular in most inclusions. Representative analyses are reported in Table A3.21.

Grossman (1975) presented x-ray diffraction evidence for the presence of grossular in *Type A inclusions* in Allende and Murrell and Burnett (1987) and Podosek et al. (1991) have also reported grossular. In inclusion 3898 (Podosek et al. 1991) the degree of alteration is very limited in comparison with other Allende CAIs and grossular is rare. Very rare grossular also occurs in veins within large crystals of wollastonite in a "Compact" Type A inclusion from Leoville (L1) (Sylvester et al. 1993).

Grossular occurs with nepheline, anorthite and sodalite, replacing melilite in *"Fluffy" Type A inclusions* in Allende (Allen et al. 1978, MacPherson and Grossman 1984, Murrell and Burnett 1987, MacPherson et al. 1988) (Fig. 99a). In some cases the grossular occurs as euhedral crystals lining cavities in this inclusion, associated with wollastonite needles (Fig. 99b). In inclusion CG11, Allen et al. (1978) found less than 6 mol % pyrope and andradite in solid solution in the grossular.

Figure 99. (a) Scanning electron micrograph of euhedral grossular crystals lining a cavity in a "Fluffy" Type A inclusion from Allende (CG11). Used by permission of Pergamon Press from Allen et al. (1978), Figure 1. (b) BSE image showing grossular replacing melilite in an Allende Type A inclusion. After MacPherson et al. (1988).

Fuchs and Blander (1975) reported grossular with compositions $Gr_{92}Py_8$ and $Gr_{95}Py_5$, respectively, in two Allende *Type B inclusions* (6/1, 10/2). Bischoff and Palme (1987) also reported grossular, which contains minor MgO (~1.2 wt %) and FeO (<0.5 wt %) in an Allende Type B inclusion. Other occurrences of grossular within melilite in Type B Allende CAIs have been reported by Fuchs (1978), Hutcheon et al. (1978) and Podosek et al. (1991).

Grossular occurs replacing melilite in the core and mantle of a *Type B3* Allende inclusion (818a; Wark et al. 1987). The core grossular has a more limited compositional range than that in the mantle, with MgO and FeO between 0.8-3.1 wt % and 0.3-1.5 wt %, respectively in the rim, compared with 0.0-3.8 wt % and 0.3-4.0 wt % in the mantle. Other elements, such as TiO_2 and Na_2O, are present in very low concentrations (<0.2 wt %), but MnO can reach 0.9 wt % in the mantle grossular, higher than in the core.

Wark (1987) described grossular (coexisting with nepheline, monticellite and an unidentified Mg-bearing, Ca-Al-silicate) as the most common replacement phase of melilite in *Type C inclusions* and observed some compositional variability, mostly in the

SiO_2, Al_2O_3, CaO and MgO contents. For individual analyses from all seven inclusions studied by Wark (1987), MgO and CaO exhibit the widest variation with a maximum range of 0.3-8.4 wt % and 31.9-38.1 wt %, respectively. Other elements such as Na_2O, TiO_2, FeO, V_2O_3 and Cr_2O_3 are generally present in low concentrations (<0.3 wt %), but TiO_2 values up 0.74 wt % and FeO up to 1.1 wt % have also been measured.

Clarke et al. (1970) deduced that grossular was present in one *fine-grained inclusion* in Allende, based on x-ray diffraction data and the bulk composition of the inclusion. Grossular has subsequently been reported in several fine-grained inclusions in Allende (Martin and Mason 1974, Grossman et al. 1975, Grossman and Ganapathy 1976, Wark 1979, Kornacki and Wood 1985a). Grossman et al. (1975) found euhedral crystals of grossular in voids in a fine-grained inclusion from Allende. Kornacki and Wood (1985a) found that grossular in Allende inclusions contains <5 wt % Fe_2O_3, <2 wt % MgO and <0.2 wt % MnO. However, McGuire and Hashimoto (1989) found that grossular that has replaced primary melilite in fine-grained CAIs in Allende has a much more variable composition than in any other type of inclusion and can be highly variable even over short distances within individual inclusions. Grossular in these CAIs often contains a significant component of andradite (0-25 mol %) in solid solution, in addition to minor amounts of TiO_2, Cr_2O_3, MnO, V_2O_3 and MgO (typically <0.5 wt %). Grossular with ~11.9 wt % FeO also occurs within regions of so called 'chaotic material' within fine-grained CAIs in Mokoia (Cohen et al. 1983).

J.M. Allen et al. (1980) reported grossular in the thick rim sequence of an isotopically *unusual, hibonite-bearing inclusion* (HAL) from Allende. The grossular contains a small pyrope component and up to 0.03 wt % Sc_2O_3. Wark (1986) described grossular with lilac fluorescence, coexisting with plagioclase, nepheline, sodalite and aenigmatite as a common replacement product of melilite in the core, mantle and crust of an unusual refractory inclusion (3643) from Allende. The grossular is almost pure with only minor amounts of TiO_2, MgO and FeO (<0.35 wt %).

Andradite. The first occurrence of andradite in a meteorite was described by Fuchs (1971) in two "Fluffy" Type A inclusions (Grossman 1980) from Allende. In both these inclusions (termed A and B), euhedral, equant crystals with a clear, greenish yellow color occur in cavities associated with wollastonite needles. Electron microprobe analyses of andradite show that it is essentially pure, with a composition very close to the theoretical composition (see Table A3.20). Andradite has since been recognized in other "Fluffy" Type A inclusions in Allende replacing melilite (Allen et al. 1978, MacPherson and Grossman 1984). Compositional data reported by Allen et al. (1978) for inclusion CG11 shows that andradite contains a small component of pyrope in solid solution, varying from 6 mol % in fine-grained andradite in the interior to <9 mol % in the Wark-Lovering rim. Andradite has not been widely reported in "Compact" Type A inclusions, with the exception of Murrell and Burnett (1987) who reported andradite in rims within cavities in the interior of an Allende inclusion (84C).

There are no reported occurrences of andradite in Type B1 and B2 inclusions, but Clayton et al. (1984) described andradite associated with diopside, hedenbergite and wollastonite needles in veins and cavities in a Type B3 inclusion (CG-14) from Allende. Essentially pure andradite also occurs replacing melilite in the mantle of a complex Type B3 inclusion (818a) described by Wark et al. (1987) and euhedral andradite is present at the interface between the porous melilite mantle of Vigarano inclusion 1623-5 and the surrounding matrix (Davis et al. 1991).

In contrast, andradite appears to be quite widespread in fine-grained CAIs in Allende (e.g. Wark 1979, Kornacki and Wood 1985a, McGuire and Hashimoto 1989), Mokoia

(Cohen et al. 1983) and Kaba (Fegley and Post 1985). Andradite occurs as a replacement product of melilite in fine-grained inclusions in Allende and has a composition which is close to pure endmember $Ca_3Fe_2^{3+}Si_3O_{12}$ (Kornacki and Wood 1985a, McGuire and Hashimoto 1989). Up to 12 mol % grossular component can be present in solid solution, but Al_2O_3 and MgO contents in andradite are generally < 1 wt %, but can range as high as 2.7 and 2.0 wt %, respectively. Other minor elements such as V_2O_3 and TiO_2 are present at concentrations between 0-0.2 wt %, but MnO contents may be higher and lie in the range 0-0.65 wt % (Kornacki and Wood 1985a, McGuire and Hashimoto 1989). In fine-grained CAIs in Mokoia, andradite occurs within regions of so called 'chaotic material' (Cohen et al. 1983) and andradite containing 1.47 wt % MgO has been reported in a fine-grained inclusion in Kaba (Fegley and Post 1985). It may also be present in inclusion 1623-16 from Vigarano, which contains an unspecified Fe-rich garnet (Sylvester et al. 1992).

Finally, J.M. Allen et al. (1980) have reported andradite ($And_{99}Gr_1$) in the rim sequence of an isotopically unusual hibonite-bearing inclusion (HAL) from Allende. The andradite contains a small pyrope component and up to 0.03 wt % Sc_2O_3.

Wollastonite. The first meteoritic occurrence of wollastonite was discovered in a CAI (inclusion B) in Allende by Fuchs (1971). Based on the petrographic description of this inclusion, Grossman (1980) concluded that it was a "Fluffy" Type A. Wollastonite occurs as acicular crystals, 2-5 μm in width and about 100 μm long, filling cavities within the inclusion. Fuchs (1971) described the wollastonite as occurring as felted masses or as a delicate open framework of randomly oriented needles (e.g. Figs. 99d,e). Electron microprobe analyses of this wollastonite show that it is essentially pure $CaSiO_3$ with minor MgO (0.05 wt %) and FeO (0.4 wt %). Representative analyses are reported in Table A3.21. Many similar occurrences of wollastonite have since been described from Allende inclusions, particularly by Grossman (1975), who identified wollastonite crystals, based on their acicular morphology, in cavities on slabs of both Type A and B inclusions. Based on this characteristic acicular morphology, wollastonite crystals have been widely interpreted as condensates (e.g. Allen et al. 1978), resulting from the alteration of primary melilite and/or other phases in the presence of a vapor phase. This conclusion appears to be supported by the association of wollastonite with other alteration phases such as grossular, anorthite and andradite, etc.

Wollastonite has been reported in *"Compact" Type A inclusions* from Allende (Murrell and Burnett 1987), Leoville (L1) and Efremovka (Ef2) (Sylvester et al. 1993). In inclusion L1, the wollastonite grains are coarse-grained (~100 μm across), whereas in Ef2 they are isolated, fine-grained, and occur interstitially.

Allen et al. (1978) found sprays and mats of wollastonite needles, 0.1-0.2 mm in length and as little as 1 μm in width, in cavities in *"Fluffy" Type A inclusion* CG11 in Allende (Fig. 99c). In many cases the wollastonite appears to have nucleated on euhedral grossular crystals. Barber et al. (1984) used TEM to examine needles of wollastonite from this inclusion and found that they are elongate parallel to a <100> zone axis and have a high density of stacking faults parallel to the elongation direction. The morphology of the grains appears to be that of a ribbon. Electron diffraction patterns from these grains show the presence of diffraction maxima characteristic of both wollastonite and $I\bar{1}$ wollastonite. Wollastonite, associated with nepheline, sodalite, and grossular, has also been reported by MacPherson and Grossman (1984) replacing melilite in some "Fluffy" Type A inclusions from Allende and has been interpreted as a secondary alteration product in these inclusions.

Figure 99 (cont'd). (c) SEM image of whiskers of wollastonite in cavities within a "Fluffy" Type A inclusion from Allende (CG11). Reproduced by permission of Pergamon Press: Allen et al. (1978), Figure 2a.

Wollastonite has not been widely reported in *Type B inclusions*, although this may in part be because most workers have emphasized the primary, rather than secondary mineralogical characteristics of this type of inclusion. However, Hutcheon et al. (1978) have reported wollastonite in fine-grained regions of alteration products (e.g. grossular, hedenbergite, etc.) in three Type B inclusions (TS 21, TS 23, TS 34) from Allende.

Dominik et al. (1978) reported a wollastonite-like phase, which occurs as acicular crystals filling voids in *Type B3 inclusion* TE from Allende. The composition of this phase, measured by electron microprobe, shows an excess in Ca and deficiencies in Si, compared with stoichiometric wollastonite. Clayton et al. (1984) described a similar occurrence of wollastonite in veins and small cavities in Allende inclusion CG-14.

Although *fine-grained inclusions* in Allende have been heavily altered, wollastonite is either rare (e.g. Kornacki and Wood 1985a) or appears to be essentially absent (e.g. McGuire and Hashimoto 1989). However, Cohen et al. (1983) found wollastonite within regions of so-called 'chaotic material' in fine-grained CAIs from Mokoia and it also occurs in a fine-grained inclusion in Kaba (Fegley and Post 1985). Kornacki and Wood (1985a) reported one analysis of wollastonite from a fine-grained inclusion which contains 0.88 wt % FeO and minor MnO (0.07 wt %) and MgO (0.25 wt %). In comparison, in Kaba (Fegley and Post 1985), the wollastonite is significantly richer in Fe (5.98 wt %) and Mg (5.23 wt % MgO).

Hedenbergite. Although hedenbergite is a rare phase in CAIs it has been reported in a number of inclusions. Representative analyses are reported in Table A3.22. Hedenbergite occurs in many of the Allende "Fluffy" Type A inclusions studied by MacPherson and Grossman (1984) where it occurs as 5-20 μm grains. Murrell and Burnett (1987) also observed hedenbergite in a "Compact" Type A inclusion (84C) from Allende. Hedenbergite does occur in Type B1 and B2 inclusions, but the only report is by Hutcheon et al. (1978) who found it in three Type B inclusions from Allende. Clayton et al. (1984) also described hedenbergite associated with diopside, andradite and wollastonite needles in veins and cavities in the Type B3 inclusion CG-14 from Allende.

One of the more common occurrences of hedenbergite is as an alteration product in fine-grained, spinel-rich inclusions (e.g. Wark 1979, Kornacki and Wood 1985a). In Allende inclusions, hedenbergite is nearly pure endmember $CaFeSi_2O_6$, with <1 wt % Al_2O_3 and MgO and between 0-1.6 wt % MnO (Kornacki and Wood 1985a, McGuire and Hashimoto 1989). Hedenbergite is also found in the so called 'chaotic material' within fine-grained CAIs in Mokoia (Cohen et al. 1983) and as rims around spinel grains in the same inclusions. Fegley and Post (1985) found compositionally variable hedenbergite ($Dp_{60}Hd_{40}$ - Hd_{100}) in a fine-grained inclusion from Kaba.

Finally, J.M. Allen et al. (1980) reported hedenbergite in the thick rim sequence of an isotopically unusual, hibonite-bearing inclusion (HAL) from Allende.

Diopside and salite. Although diopside can be a primary phase in CAIs (e.g. in Wark Lovering rim sequences), it may also be of secondary origin in some cases. Salite, on the other hand, is probably almost invariably secondary, based on its association with other phases which are clearly secondary. Representative analyses are reported in Table A3.22. SEM studies reported by MacPherson et al. (1981) and Hashimoto and Grossman (1985) show clearly that intense alteration can produce Al-Fe diopside from melilite. McGuire and Hashimoto (1989) also suggested that Al-diopside was primary, but the compositional variability observed was due to the introduction of FeO into primary Al-bearing diopside. Diopside of secondary origin appears to be a relatively rare phase in CAIs. Murrell and Burnett (1987) found an unspecified pyroxene (presumably diopside) associated with fine-grained nepheline, grossular and anorthite in a "Fluffy" Type A inclusion (G-1,0) from Allende. The only occurrences of secondary diopside in Type B inclusions are both from a CAI in Vigarano. MacPherson and Davis (1993) found interstitial diopside in a Type B inclusion (1623-8) and Sylvester et al. (1993) found Al-diopside as a secondary alteration product with sodalite and calcite in a Type B2 inclusions (Vig 1). Clayton et al. (1984) also described diopside associated with hedenbergite, andradite and wollastonite needles in veins and cavities in Type B3 inclusion CG-14 from Allende. Wark et al. (1987) also found diopside as an alteration product of melilite in the core and mantle of a complex Type B3 inclusion (818a) from Allende. The diopside in the interior this inclusion is Al-diopside with ~7 wt % Al_2O_3 and no detectable TiO_2 and in the mantle is also Al-diopside with 0.34-11.3 wt % Al_2O_3 and 0-1.6 wt % TiO_2. Minor elements such as FeO, V_2O_3 and Cr_2O_3 are typically present at concentration levels <0.5 wt %, although FeO can reach 2.8 wt % in diopside in the core. Wark et al. (1987) concluded that the diopside was the replacement product of melilite in regions where the replacement has been most intense. Salitic pyroxenes are also commonly encountered in the fine-grained, spinel-rich inclusions studied by McGuire and Hashimoto (1989) in Allende, and J.M. Allen et al. (1980) reported Fe-salite in the thick rim sequence on the *unusual, hibonite-bearing inclusion* (HAL) from Allende. The salite contains no detectable TiO_2, but up to 1.8 wt % Al_2O_3 in the most Mg-rich compositions.

Other phases. Secondary *anorthite* has been reported in several CAIs and is distinct from primary anorthite, because of its fine grain size and because it frequently displays a replacement relationship with melilite. For example, Podosek et al. (1991) observed some secondary anorthite, associated with melilite, in an Allende "Compact" Type A inclusion (3898). Fine-grained anorthite, often coexisting with other phases such as nepheline, grossular and sodalite, occurs replacing melilite in several "Fluffy" Type A inclusions in Allende (Allen et al. 1978, MacPherson and Grossman 1984, Murrell and Burnett 1987). Allen et al. (1978) reported up to 0.5 wt % Na_2O and $>An_{96}$ in anorthite in inclusion CG11.

Secondary anorthite is also commonly present in Type B1 and B2 inclusions. Hutcheon et al. (1978) reported secondary anorthite in three Type B inclusions (TS34, TS23 and TS21) from Allende and it is also a common replacement product of melilite in several inclusions studied by Podosek et al. (1991) (e.g. 3529Z, 3529-41, 3529-21, 3529-30, 3658). The alteration has often occurred along cleavage fractures, but the degree of alteration varies considerably from inclusion to inclusion in Allende. Wark et al. (1987) have also reported one Type B3 inclusion in which essentially pure anorthite, with blue cathodoluminescence, is present replacing melilite in the mantle of the inclusion. Kornacki and Wood (1985a) also reported fine-grained, anorthitic plagioclase ($>An_{99}$) in fine-grained inclusions from Allende which, in view of the intense alteration in these inclusions, is probably secondary in origin.

Other occurrences of secondary anorthite have been noted by Wark (1986) in an unusual ultrarefractory Allende inclusion (3643) and in two, zoned, fine-grained, spinel-rich inclusions in Vigarano (1623-14 and 1623-16) (Sylvester et al. 1992).

Monticellite ($CaMgSiO_4$) is a rare secondary phase in CAIs in CV chondrites. It occurs replacing melilite in the core of a complex Type B3 inclusion (818a) from Allende (Wark et al. 1987)(Table A3.20 and contain 1.1 wt % Al_2O_3 and 2.6 wt % FeO. In addition, Wark (1987) found rare monticellite as a replacement product of melilite in Type C inclusions in Allende. Based on energy dispersive electron microprobe analysis, only Al_2O_3, apart from SiO_2, MgO and CaO, is present in appreciable concentrations (2.8 wt %). Other minor elements, such as FeO, V_2O_3 and Cr_2O_3, are all <0.2 wt %. Sylvester et al. (1992) also observed rare monticellite in two zoned, spinel-rich, fine-grained inclusions in Vigarano. In inclusion 1623-14, the monticellite is kirschsteinite-rich with a compositional range $Kir_{41-49}Mont_{59-51}$, but has a much more restricted compositional range in inclusion 1623-16 ($Kir_{29}Mont_{71}$).

Wark (1986) reported a fine-grained alteration product of melilite in a refractory inclusion (3643) in Allende, which has a very variable composition, but has high SiO_2 (33 wt %), Al_2O_3 (35 wt %) and CaO (15-30% wt %) and also contains significant amounts of MgO, FeO and Na_2O. The best fit structural formula, based on 20 oxygen atoms, appears to be a substituted *aenigmatite*:

$$Na_2Fe_5(Ti,Si_6)O_{20} = Ca_2(Mg,Fe,Ca)_5(Al,Si)_7O_{20}.$$

McGuire and Hashimoto (1989) reported *sphene ($CaTiSiO_5$)* in zone A in a concentrically zoned, fine-grained inclusion in Allende (Table A3.20). This rare phase occurs as 5 μm, anhedral grains, partially enclosed in nepheline and contains ~4 wt % Al_2O_3 and minor (<1 wt %) V_2O_3, FeO, MnO, Na_2O and K_2O.

Ilmenite ($FeTiO_3$) is a rare phase in CAIs, but occurs most frequently in fine-graine, spinel-rich inclusions. Analyses are reported in Table A3.20. In Allende, ilmenite occurs replacing perovskite (Kornacki and Wood 1985a, McGuire and Hashimoto 1989) and is clearly secondary in origin. Ilmenite is actually quite common in the outer zone (C) of concentrically zoned inclusions in Allende (McGuire and Hashimoto (1989). It occurs as submicron grains often included within spinel, olivine and nepheline. Ilmenite has also been reported in fine-grained inclusions in Mokoia (Cohen et al. 1983), Kaba (Fegley and Post 1985) and Vigarano (Sylvester et al. 1992).

Steele (1995b) reported Mg-ilmenite, which appears to be secondary in origin, replacing perovskite along the boundary of an unusual hibonite-bearing inclusion in Allende. The ilmenite contains 2.97-7.81 wt % MgO and significant amounts of CaO and Al_2O_3 which may be contamination from adjacent perovskite.

Steele (1995b) reported an *Al_2O_3-rich phase* in an unusual, hibonite-bearing inclusion in Allende, which was interpreted as being the alteration product of primary hibonite. The phase is too fine-grained to obtain quantitative analyses, but contains variable amounts of Na and Si. Steele (1995b) suggested that the phase may be an intergrowth of corundum and nepheline or a modified Al_2O_3, β-alumina.

A variety of rare *phyllosilicate* phases have been observed in CAIs, mostly using TEM techniques. Tomeoka and Buseck (1982b) found a phyllosilicate phase with a 10-Å basal spacing in cavities within a fine-grained CAI from Allende. Based on compositional data, this phase was interpreted as a complex intergrowth of *montmorillonite* and *K-bearing mica.* In a fine-grained CAI in Bali, Keller et al. (1994) found rare, lath-shaped grains (200 nm long and ~20 nm wide) of Na-K bearing dioctahedral *mica*, closely intergrown with fine-grained *saponite*, sometimes in oriented intergrowths.

Keller and Buseck (1991) identified the brittle calcic mica, *clintonite* $Ca(MgAl)_3(Al_3Si_1)_{10}(OH)_2$, by TEM techniques in a Type A Allende inclusion. The clintonite occurs in an alteration vein as crystals 10-40 nm wide and 100-500 nm in length. The crystals show considerable strain contrast and contain layer terminations and other defects. Electron diffraction studies show that the clintonite is the 1M polytype and EDS analyses give a composition of

$$(Ca_{0.95}Na_{0.05})(Mg_{2.5}Al_{0.3}Fe_{0.05}Ti_{0.05}\square_{0.1})(Si_{1.6}Al_{2.4})_{10}(OH)_2.$$

In the outer region of the same inclusion in which clintonite occurs (see above), Keller and Buseck (1991) also found *margarite* in a lamellar intergrowth with anorthite. The margarite lamellae vary in width from 20-500 nm and are oriented with their c* axis parallel to $(11\bar{3})$ of anorthite. Electron diffraction and HRTEM images indicate that the margarite is a disordered intergrowth of two polytypes with 10- and 20-Å basal spacings. Based on EDS analysis the composition of the margarite is $(Ca_{0.95}Na_{0.05})(Mg_{0.15}Al_{1.65}Fe_{0.21})(Si_{1.8}Al_{2.2})_{10}(OH)_2$.

Cohen et al. (1983) first described a phyllosilicate phase which they termed a *High Al-Phyllosilicate* (HAP) in a fine-grained CAI from Mokoia. This phase occurs surrounding spinel grains where feldspathoids would typically occur in Allende fine-grained inclusions. X-ray diffraction and electron microprobe data suggest that this phase is a high-Al serpentine of some kind, but it is rich in Na_2O (2.75-4.44 wt %) and K_2O (0.48-0.96 wt %), suggesting some kind of mica. Cohen et al. (1983) suggested that it may be an intergrowth of two or more phases, possibly similar to the mica-montmorillonite intergrowths observed by Tomeoka and Buseck (1982b) in a fine-grained CAI in Allende. Fegley and Post (1985) and Keller and Buseck (1990a) also found a phase compositionally similar to HAP in a fine-grained CAI from Kaba. Although the HAP phase in CAIs has not itself been characterized in detail by TEM, Tomeoka and Buseck (1990a) found that HAP in chondrules and olivine-anorthite-rich aggregates in Mokoia consists of very fine-grained, randomly oriented crystallites, mostly with 10-Å basal spacings. However, some grains also have minor 7-Å interlayers. Tomeoka and Buseck (1990a) suggested that the 10-Å phase was probably *Na-rich phlogopite*, or possibly collapsed smectite, interlayered with serpentine (7-Å basal spacing).

Cohen et al. (1983) also reported another phyllosilicate phase which they termed *Low Al-Phyllosilicate* (LAP) in fine-grained CAIs from Mokoia. It occurs as aggregates <10 μm in size. The composition and texture of this phase (~34 wt % SiO_2; ~22 wt % MgO) suggest that it may be a *montmorillonite*-like phyllosilicate. Fegley and Post (1985) also found a compositionally similar phase in a fine-grained CAI from Kaba.

Wlotzka and Wark (1982) have described *zeolites* replacing melilite in CAIs from Leoville, but the identification is based solely on SEM EDS analyses and no XRD studies have been carried out to provide confirmation of their presence.

In Vigarano, Sylvester et al. (1993) reported *calcite* with Al-diopside and sodalite as fine-grained, secondary alteration products in a Type B2 inclusion (Vig1) and calcite also occurs in veins as a minor replacement product in a forsterite-bearing inclusion (1623-5) in Vigarano (Davis et al. 1991). Calcite has also been observed in fractures in a melilite-bearing inclusion (1623-3) for Vigarano and in a fine-grained inclusion from Leoville (3537-1) (Mao et al. 1990).

Wark (1987) reported a rare, *unidentified Mg-bearing Ca-Al silicate* in Type C inclusions in Allende. Based on EDS analyses, this phase contains low SiO_2 (~22.2 wt %), but high CaO (31.5 wt %) and Al_2O_3 contents (31.5 wt %). An unidentified silica-rich phase which has preferentially replaced melilite occurs in a forsterite-bearing inclusion (1623-5) in Vigarano (Davis et al. 1991). Type B3 inclusion TE (Dominik et al. 1978) contains a phase which has been described as veins of a potassium-rich, "layer structure" phase that replaces fassaite and is presumably a phyllosilicate phase of secondary origin. The potassium content in this phase reaches ~8 wt %, suggesting that it may be a mica.

Mineralogy of fremdlinge in CV carbonaceous chondrites

Fremdlinge (German for 'strangers') are rare, exotic, refractory siderophile element-rich objects that occur in CAIs in CV carbonaceous chondrites and are so-called because they were thought to be foreign objects that were incorporated into the host CAIs during their formation. This origin has, however, been questioned and many authors prefer the term 'opaque assemblages' to describe these objects (e.g. Blum et al. 1988, 1989; Simon and Grossman 1992). For the purposes of this review we use the term 'fremdlinge' with the caveat that an exotic or local origin for these objects is still not fully resolved (cf. El Goresy et al. 1978, Blum et al. 1988, 1989; Zinner et al. 1991).

Fremdlinge occur largely in coarse-grained Type B inclusions (Wark and Lovering 1978, El Goresy et al. 1978, Wark and Lovering 1982a, MacPherson and Grossman 1984, Simon and Grossman 1992, Sylvester et al. 1993, Caillet et al. 1993), and rarely in Type As. Fremdlinge are typically <50 μm in size, but very rarely can be larger, up to 1 mm (e.g. Zelda, a sulfide-rich fremdling from Allende; Armstrong et al. 1987). They are mineralogically complex and occur as inclusions within primary CAI phases. 70-80% of all fremdlinge occur within fassaite and the remaining 30-20% are distributed with decreasing frequency in spinel, melilite and anorthite (El Goresy et al. 1978). Fremdlinge also occur apparently captured between spinel grains and have embayed grain boundaries into the spinel (Fig 100). Fremdlinge are commonly surrounded by a thin rim of V-rich fassaite (e.g. Armstrong et al. 1985, Hutcheon et al 1987, Simon and Grossman 1992). Examples of fremdlinge from CAIs in CV chondrites are shown in Figures 101 and 102.

El Goresy et al. (1978) classified fremdlinge in CAIs in Allende and Leoville into three different types (1, 2 and 3), based on their abundance of Fe,Ni metal, textural occurrence and grain size of silicates and oxides. However, a number of fremdlinge which do not fit readily into these three categories have been recognized (e.g. Willy; Armstrong et al. 1985; Zelda, Armstrong et al. 1987).

The Type 1 fremdlinge contain 50-70% Fe,Ni metal, sulfides and Pt-metal grains, with the phosphates, oxides and silicates comprising the remaining 50-30%. Pt metal alloys typically occur as blebs (< 1 μm) within the Fe,Ni metal, silicates and phosphates. The different phases have interlocking grain boundaries and the silicate and sulfide components have a texture described as 'compact,' a feature which is unique to this group. Ca phosphates also appear to be more abundant than in other groups.

Type 2 fremdlinge contain 70 to 90% Fe-Ni metal and rare sulfides (El Goresy et al. 1978). Silicates and phosphates may also occur. Pt-metal alloys occur as small inclusions (often submicron) within the silicate phases and inside Fe,Ni metal. Many Type 2

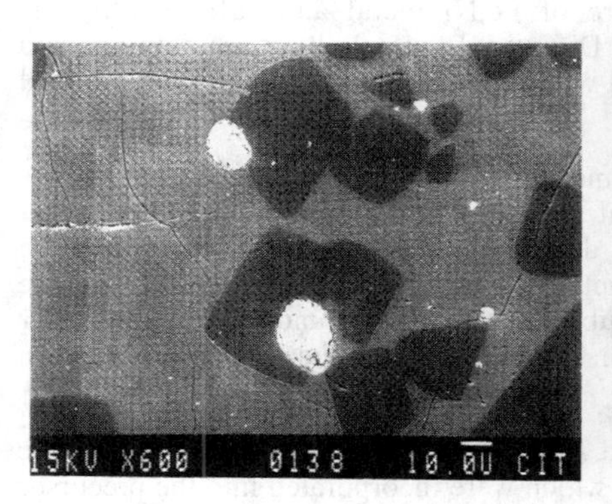

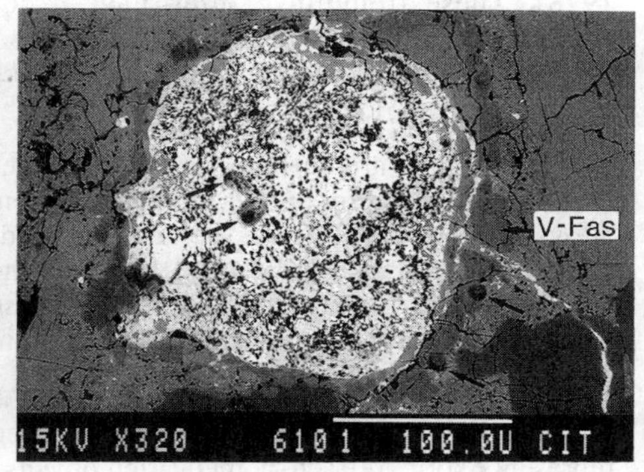

Figure 100 (left). BSE image showing two small fremdlinge associated with spinel and melilite in Type B inclusion TS 34 from Allende. The fremdlinge are commonly partially enclosed in embayed areas in spinel. Used by permission of the editor of *Geochimica et Cosmochimica Acta* from Armstrong et al. (1985), Figure 15.

Figure 101 (right). BSE image of a fremdling from the Allende Type B1 CAI Egg 6, which is a porous aggregate of mostly Ni,Fe metal, V-rich magnetite and pentlandite. Ca-rich pyroxene is also present and the whole fremdling is surrounded by a rim of V-rich fassaite. Used by permission of the editor of *Geochimica et Cosmochimica Acta* from Hutcheon et al. (1987), Figure 2.

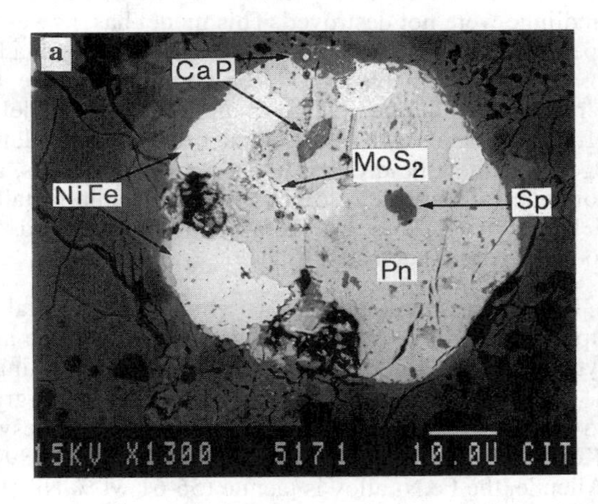

Figure 102. BSE image of a fremdling from CAI 5171. The fremdling consists of pentlandite, Ni,Fe metal with minor Fe-Cr-spinel (Sp), Ca-phosphate (CaP) and molybdenite (MoS_2). Os-Ru and Pt-Ir-Re nuggets are present within the metal. Used by permission of the editor of *Geochimica et Cosmochimica Acta* from Hutcheon et al. (1987), Figure 2.

fremdlinge contain numerous grains of one binary or one ternary alloy, a feature which appears to be unique to this group. Armstrong et al. (1985) observed many Type 2 inclusions (as well as others) which had a massive, 'welded,' or 'recrystallized' appearance. El Goresy et al. (1979) subsequently suggested two subgroups of Type 2 inclusions, one dominated by Ni- or Fe,Ni sulfide and a second by Fe,Ni metal.

In Type 3 fremdlinge, the metal content is highly variable (0-70%), and the inclusions resemble Type 1s in appearance. The silicate and oxide phases in Type 3s can have a spongy texture or occur as agglomerates of polygonal grains (El Goresy et al.

1978). These fremdlinge often have a core of Fe,Ni metal and sulfide which is surrounded by the silicates and oxide phases. Different Pt metal alloys are common and can coexist with each other or occur enclosed within other phases, typically Fe,Ni metal grains.

Fremdlinge are enriched in Pt-group elements (e.g. Re, Pt, Mo, Ru) and Pt metals occur as binary, ternary and multiple element alloy grains. The abundance of different alloys is variable (El Goresy et al.1978) and decreases in the following order (relative abundance in brackets: (a) Os, Ru and Rh metal grains (65%), (b) Pt with subordinate amounts of Rh, Ir and Os (25%), (c) Ir with subordinate Ru (5%), and (d) Re with minor Ru (5%). These metal grains can also contain volatile elements such as Zn, Ga, Ge, Sn and As.

According to El Goresy et al. (1978) and Armstrong et al. (1985 1987) fremdlinge represent exotic, refractory metal-rich objects which were incorporated into the precursor CAI material before the CAI formed. The formation of fremdlinge involves several different steps: (1) condensation of Ru-Os-rich nuggets at 1500-2000K, either in the nebula or in a supernova envelope, (2) condensation of Ni,Fe metal followed by oxidation of Fe and reaction with V in the nebular gas to produce V-rich magnetite, (3) aggregation of Ru,Os-rich nuggets, Ni,Fe metal and V-rich magnetite and associated phases at low T in the nebula, (4) mixing of these opaque assemblages with proto CAI material before or during a high temperature event which melted the silicate portion, followed by rapid cooling. The high temperature event was so brief that the delicate textures in the fremdlinge were not destroyed. This model has, however, been questioned based on low temperature (1073-1273K) oxidation experiments on a homogeneous alloy (Blum et al. 1988). These experiments reproduced, to a large degree, the mineralogy and compositions observed in fremdlinge and led to an alternative model for their origin. In this model the refractory metals alloys are condensates which, during melting of the CAIs, form homogeneous metallic droplets. At lower temperatures, after crystallization of the CAIs the homogeneous alloys exsolved into immiscible metallic phases (Ru, Os-rich nuggets and Ni,Fe metal) and were partially oxidized and sulfurized to form magnetite and associated phases such as sulfides.

Fe,Ni alloys. The dominant component of fremdlinge are metal alloys, specifically Fe,Ni and Pt group element alloys (Re, Os, Ir, Pt, Rh, Ru). Palme and Wlotzka (1976) first reported analytical data (EPMA) for *Fe,Ni metal* from a fremdling separated from an Allende inclusion. The metal contains significant quantities of Pt-group metals such as Ru, Rh, Os, Ir and Pt (10 wt %). Subsequent studies (e.g. El Goresy et al. 1978, 1979; Brandstätter and Kurat 1983, Armstrong et al 1985, Blum et al. 1989) show that in most fremdlinge from Allende, the Fe,Ni alloy is taenite (56-61 wt % Ni) (Fig. 103). In a large (150 μm), zoned fremdlinge ('Willy') from Allende (Armstrong et al. 1985) the Fe,Ni metal has a very constant composition (61 wt % Ni, 36 wt % Fe, 2 wt % Co, 0.4 wt % Pt) and smaller fremdlinge from the same CAI have very similar metal compositions (60 wt % Ni, 35 wt % Fe, <4% wt % Co, >0.5 wt % Pt). However, in some CAIs, the Ni content of metal in different fremdlinge can be quite variable. For example, in a Type B Allende inclusion, Bischoff and Palme (1987) found that the Fe and Co contents of metal are remarkably uniform (30.9-36.4 wt % Fe; 1.6-2.8 wt % Co) in several different fremdling, but the Ni content is variable (50.4-60.8 wt %). Rare fremdlinge containing metal with higher Ni contents have been observed, e.g. ~83 wt % Ni in a fremdling from the Type B Egg3 Allende inclusion (Blum et al. 1989). Bischoff and Palme also found one fremdling that contained metal with 61.3-67.3 wt % Ni, but a low Ir content (0.7-1.7 wt %). Other examples of Fe,Ni-rich fremdlinge in Allende have been reported by Wark and Lovering (1982a) and MacPherson and Grossman (1984).

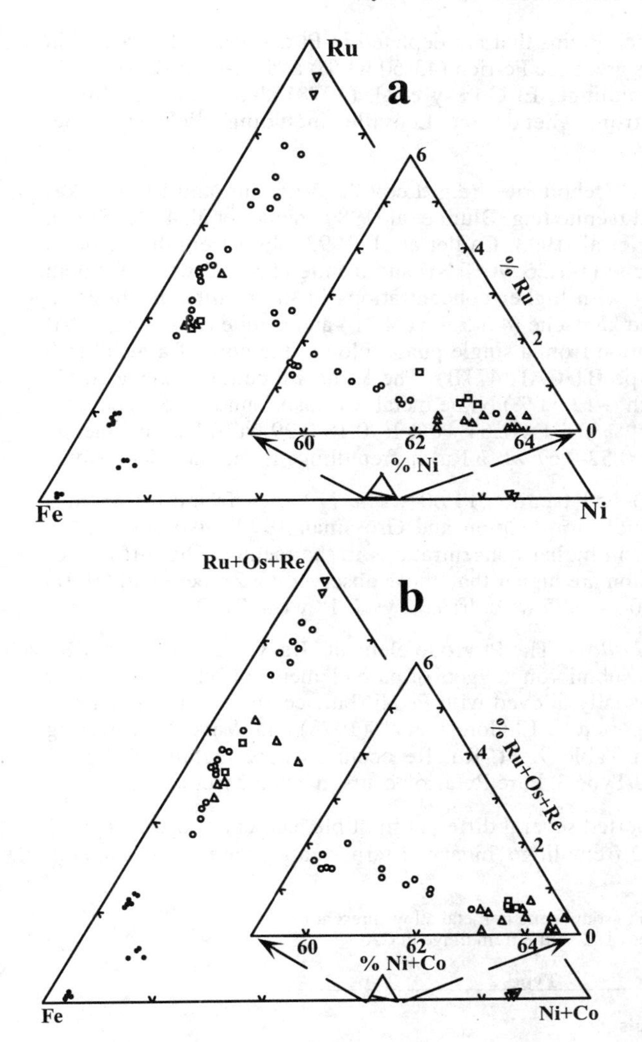

Figure 103. Compositions of metallic phases in opaque assemblages from five CAIs, nomalized and plotted in (a) the ternary system Ni-Fe-Ru and (b) the system (Ni+Co)-Fe-(Ru+Os+Re). Enlarged region shows the compositions of taenite (γFe-Ni) in detail. Open circles are for opaque assemblages in the Allende inclusion JIM, squares are Allende inclusion F3, triangles are Willy, inverted triangles are EGG-OA1 and filled circles are for Leoville inclusion LE0 575-OA1. After Blum et al. (1989).

Fe,Ni metal commonly contains Pt group elements in variable concentrations (e.g. in Type 1 and 3 fremdlinge, El Goresy et al. 1978). Armstrong et al. (1985) found that metal nuggets from a sulfide-rich fremdlinge in an Allende Type B1 CAI (5241) are often enriched in Pt (<15% wt %). In the sulfide-rich fremdling 'Zelda' (Armstrong et al. 1987) and other smaller fremdlinge in the same Allende Type B1 CAI (Egg6) (Palme et al. 1994), the metal contains 2.6 wt % Pt and 1.3 wt % Ir. Also in Allende, Bischoff and Palme (1987) reported metal in fremdlinge from a Type B CAI with a total range of Ir and Os contents of 0.7-8.3 wt % and 0.5-5.5 wt %, respectively. However, individual inclusions typically have more restricted ranges of Ir contents. Most metals are free of Mo and Pt, although examples containing up to 1.7 wt % Mo and 2.0 wt % Pt are present.

El Goresy et al. (1979) found Ni,Fe grains highly enriched in Pt (~40wt %), but with low Ir, Rh and W within Ni or Ni,Fe sulfides in a Type 2 fremdling in Allende Type B CAI H126.78. In the same CAI, fremdlinge that are dominated by Fe,Ni metal (~64.2%

Ni) contain Pt group element-rich grains that are depleted in Pt and Rh, but enriched in Os, Ru, Ir and Mo. These alloys are more Fe-rich (43-60 wt %) and Ni-poor (2-31 wt %) than in the sulfide-bearing fremdlinge. El Goresy et al. (1978) also found several Ni alloys in Type 3 fremdlinge from Allende and Leoville, including NiCoGe, NiGe, NiGaGe.

Fremdlinge from reduced CV chondrites (e.g. Leoville, Vigarano and Efremovka) often contain both kamacite and taenite (e.g. Blum et al. 1989, Zinner et al. 1991, Simon and Grossman 1992, Sylvester et al. 1993, Caillet et al. 1993). In a fremdling from a Leoville CAI (LEO575), kamacite (1-1.26 wt % Ni) and taenite (10-12.5 wt % Ni) both contain Pt group elements, but with higher concentrations in the taenite (Blum et al. 1989). Zinner et al. (1991) found kamacite (4.5-5.5 wt % Ni) and taenite (30-47 wt % Ni) that probably formed by exsolution from a single phase alloy in the core of a metal-rich fremdling from a Vigarano Type B1 CAI (477B). The kamacite contains elevated Co contents (0.9-2.4 wt %); Mo-rich (~12 wt %) Ni,Fe metal was also found. Pt elements are present in kamacite (1.5-1.74 wt % Os; 1.4-1.64 wt % Ir; 0.18-0.29 wt % Ru) and taenite (2-3 wt % Os; 1.5-2.13 wt % Ir; 0.52-0.67 wt % Ru) in fremdlinge in the same inclusion.

Kamacite (3.33 wt % Ni) and taenite (11.60 wt % Ni) in a fremdling from a "Compact" Type A Leoville inclusion (Simon and Grossman 1992) also contains Pt group metals which are present in higher concentrations in the taenite. The differences between the metal in this inclusion are higher than those observed by Zinner et al. (1991) (e.g. 5.67 vs. 0.99 wt % Os; 8.40 vs. 3.95 wt % Ir; 7.72 vs. 2.18 wt % Ru).

Pt-group metals and metal alloys. The Pt group elements Re, Os, Ir, Pt, Rh and Ru typically occur in fremdlinge as submicron nuggets of pure elements or binary, ternary or quaternary alloys and are not usually alloyed with Fe,Ni (but see above). The details of these occurrences have been reported by El Goresy et al. (1978) and Wark and Lovering (1978) and are summarized in Table 9. Pt, Ru, Re occur as pure metals in Type 1 fremdlinge and Pt in Type 2 and Type 3. Pure Re also occurs in Type 2 fremdlinge.

El Goresy et al. (1978) reported several different multiple element alloys in Type 1 fremdlinge (Table 9). In Type 2 fremdlinge, binary or ternary alloys are most common,

Table 9. Pt group element metal alloys present in Types 1, 2, and 3 fremdlinge in CAI.

Type 1	Type 2	Type 3
Pure metals		
Pt, Ru, Re	Re	Pt
Binary alloys		
Pt,Ir	PtIr,	OsRu
OsRu	OsRu	
OsPt		
Ternary alloys		
MoOsRu	OsRuIr	OsRhRu
OsReRu	OsRuPb	RuRhIr
MoOsRu		IrMoPt
		RuOsIr
Quaternary alloys		
PtWFeNi		OsRuRhIr
		OsIrRuOs
Multiple element alloys		
RuOsIrPtMo		PtRuIrSnGeOs
OsAsGePtNiFe		

but rare metal grains with highly unusual compositions have been found. For example, in a fremdling from a Type B Allende inclusion (H126.78), El Goresy et al. (1979) reported a very Nb-rich (22.8 wt %), Pt group metal-rich grain alloyed with Fe (28.2 wt %), Ni (9.22 wt %), Ru (24.2 wt %), and minor concentrations of Os, Ir and Mo.

In the large (150 μm) fremdlinge Willy from Allende (Armstrong et al. 1985), the Pt metal alloy is εRuOsFe containing minor Re, Ir, Co, Ni and sometimes W (~2 wt %). Refractory metal nuggets (~0.5-2 μm) consisting of OsRu and PtIrFe alloys are relatively abundant in the large fremdling Zelda from an Allende Type B1 CAI (Armstrong et al. 1987). The PtIr nuggets have a fairly constant composition of $Fe_{1.0}Pt_{0.4}Ir_{0.2}$, but the OsRu alloys are compositionally variable and fall into two groups with average compositions of $Fe_xOs_{1.0}Re_{0.09}Ru_{1.6}Ir_{0.2}$ and $Fe_xOs_{1.0}Re_{0.09}Ru_{0.9}Ir_{0.4}$, with Fe/Os = 0-0.9. Fremdlinge from the same CAI as Zelda also sometimes contain εRuOs nuggets distributed throughout the constituent phases (Palme et al. 1994). Wark (1986) also reported OsIr specks in Mo-sulfide, associated with a PtFeRu alloy in a fremdling in Allende inclusion 3643.

Compositionally heterogeneous εRu-Fe nuggets (9.71-51.60 wt % Fe; 32-53 wt % Ru) occur in taenite in JIM, a fremdling from Allende (Brandstätter and Kurat 1983, Blum et al. 1989). Two distinct groups of nuggets are present, one enriched in Os (20-30 wt %) and one with lower concentrations (~1-10 wt %) (Fig. 104). Blum et al. (1989) also found εRu-Fe with minor Fe (<4 wt %), but high Os (~28 wt %) in a fremdling from the Egg3 inclusion (Type B).

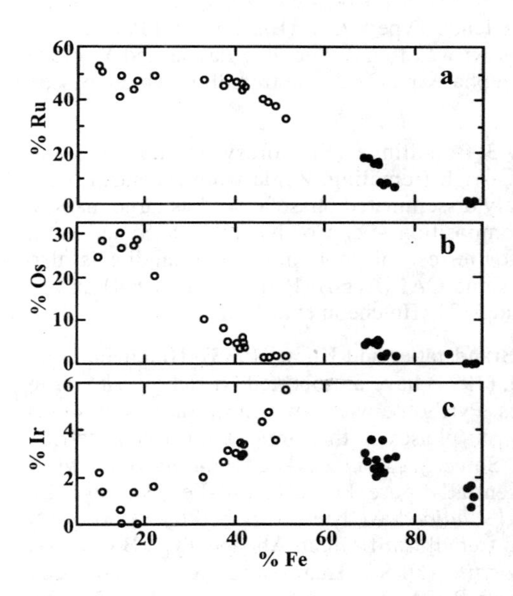

Figure 104. Compositional variation diagram showing Fe vs. (a) Ru, (b) Os and (c) Ir in metallic phases in opaque assemblages in two CAIs. Open circles are the compositions of εRu-Fe from an opaque assemblage in Allende CAI JIM, asterisks are the average composition of taenite in JIM and filled circles are the compositions of the three metallic phases from LE0575-OA1. After Blum et al. (1989).

Very small Pt-group metal particles, highly enriched in Os occur in fremdlinge from a Type B Allende inclusion (Bischoff and Palme 1987). The grains are too small to analyse precisely by electron microprobe, but appear to have Os contents between ~61 and ~83 wt %. Re and Ru are also present (3.2-8.5 wt % and 4.1-7.9 wt %, respectively) and Ir contents are variable (0-7.7 wt %, but typically around 1.5-2.5 wt %).

RuOs and RuFe-rich alloys have been reported in Ni,Fe metal in several fremdlinge from reduced CV chondrites. Blum et al. (1988) found a 2 × 10 μm RuOs-rich lamella in metal in a Leoville CAI and Zinner et al. (1991) also described exsolved RuOs-rich phases in both kamacite and taenite in a metal-rich fremdling from a Vigarano Type B1 CAI (477B). εRu-Fe with ~5 wt % Os occurs in taenite in a fremdling from Leoville inclusion LEO575 (Blum et al. 1989) (Fig. 104) and also occurs as thin (up to 1.5 μm), crystallographically oriented lamellae in taenite in two fremdlinge from a "Compact" Type A inclusion in Leoville (Simon and Grossman 1992). In these fremdlinge, the εRu-Fe exsolved from taenite during cooling. Lamellae in the different fremdlinge contain 13.59 and 16.23 wt % Ru, respectively. ~0.54 wt % Rh and Re and up to 7.94 wt % Os are also present. The different Ru-Fe ratios of the lamellae probably reflect a difference in bulk composition between the two blebs.

Other metals and alloys. *Mo metal* blebs occur rarely in Type 3 fremdlinge in Leoville and Allende CAIs (El Goresy et al. 1978). For example, El Goresy et al. (1979) reported a Mo metal grain with 91 wt % Mo in its core and a surrounding shell with 22.8 wt % Mo in an Allende Type B inclusion (H126.78). *Nb metal* and *WNbFeNi* alloys also occur as rare phases in Type 1 fremdlinge in CAIs in Allende and Leoville (El Goresy et al. 1978).

Sulfides. A wide range of different sulfide phases occur as both major and minor constituents of fremdlinge. *Troilite* occurs in a large (150 μm), zoned fremdlinge (Willy) in an Allende Type B inclusion (5241) (Armstrong et al. 1985). It contains minor V (~0.2 wt %), Cr (~0.25 wt %), Co (~0.5 wt %) and Ni (~1.2 wt %). Compositionally similar troilite (<0.3 wt % V, <0.3 wt % Cr, 0.2-1.1 wt % Ni, 0.1-1.4 wt % Ca) is also the main sulfide phase in fremdlinge from another Allende Type B CAI (Bischoff and Palme et al. 1987), although one grain with 13.9 wt % Ni was found. See also Palme and Wlotzka (1976) who found a Mo, Ru-bearing sulfide phase in an Allende fremdling which may be pyrrhotite or possibly troilite.

Pentlandite occurs in Types 1 and 3 fremdlinge (El Goresy et al. 1978) and constitutes 30% of a large (1 mm), sulfide-rich fremdling, Zelda from Allende (Armstrong et al. 1987) where it occurs as a finely disseminated phase as well as large masses. The pentlandite in Zelda has an average composition ~$Fe_{6.3}Co_{0.1}Ni_{2.2}(V,Cr)_{0.1}S_8$ and has a Ni concentration lower than is typical for terrestrial pentlandite. Pentlandite is also common in smaller fremdlinge from the same CAI (Egg6) (Palme et al. 1994) and in Allende Type B inclusions USNM 5241 and 5171 (Hutcheon et al. 1987)

In Allende, El Goresy et al. (1979), Brandstätter and Kurat (1983), Hutcheon et al. (1987), Blum et al. (1989) and Palme et al. (1994) have all reported Ni-rich *pyrrhotite* in fremdlinge, mostly from Type B inclusions. Pyrrhotite with low concentrations of V and Cr (~0.2 wt %) is one of the most abundant phases in the large (1 mm), sulfide-rich fremdling, Zelda (Armstrong et al. 1987). Some grains in Zelda contain up to 0.9 wt % Ru, but otherwise the pyrrhotite is essentially pure Fe sulfide (~$Fe_{0.96}S$). Sulfides belonging to the *FeS-NiS monosulfide solid solution* have been described by El Goresy et al. (1979) as a major component in several fremdlinge from an Allende Type B CAI and often occur in association with *heazlewoodite* (Ni_3S_2). Heazlewoodite has also been reported in a fremdling from the Egg3 Type B CAI in Allende (Blum et al. 1989). An additional Ni sulfide, *NiS* is also found in Type 3 fremdlinge in CAIs in Allende and Leoville (El Goresy et al. 1978). Other unspecified occurrences of Fe,Ni sulfides have also been reported in other fremdlinge (e.g. Wlotzka and Palme 1976, El Goresy et al. 1978, MacPherson and Grossman 1984).

El Goresy et al. (1978) described *MoS_2* grains, <~ 5 μm in size, that occur in Type 1 to 3 fremdlinge in CAIs in Allende and Leoville. Several other occurrences have also been reported in fremdlinge from Type A and B CAIs in Allende (e.g. Brandstätter and

Kurat 1983, Armstrong et al. 1985, Hutcheon et al. 1987, Blum et al. 1989, Palme et al. 1994). MoS_2 ($Mo_{0.96}Fe_{0.04}S_2$) is a minor phase in a large (1 mm), sulfide-rich fremdling, Zelda from Allende (Armstrong et al. 1987). Simon and Grossman (1992) also reported oriented, submicron lamellae of what appears to be MoS_2 in taenite in two fremdlinge from a "Compact" Type A CAI in Leoville.

Other minor occurrences of unusual sulfides such as WS_2 and unidentified V-rich and Fe, Ti sulfides have also been reported in fremdlinge. *WS_2* occurs in Type 1 and 3 fremdlinge and the V-rich and Fe,Ti sulfides are found only in Type 1 fremdlinge in CAIs in Allende and Leoville (El Goresy et al. 1978).

Phosphates. Unspecified Ca-phosphates occur in Types 1 and 3 fremdlinge in CAIs from Allende, Leoville, and Efremovka (El Goresy et al. 1978, Sylvester et al. 1993), but appear to be uncommon in Type 2s. Ca-phosphates have also been reported in a number of other Type A and B inclusions from Allende (e.g. MacPherson and Grossman 1984, Hutcheon et al. 1987, Blum et al. 1989, Palme et al. 1994).

Brandstätter and Kurat (1983), Armstrong et al. (1985) and Blum et al. (1989) have all reported *apatite* from a number of fremdlinge from Allende CAIs (typically Type B, e.g. 5241 and JIM). Apatite in the large (150 μm), fremdlinge Willy contains ~1.5 wt % Cl (Armstrong et al. 1985) and is often associated with scheelite or Mo-rich phases in the mantle of the inclusion. Armstrong et al. (1985) report a structural formulae of $Ca_{4.92}(Fe,Mg,Ni)_{0.08}(PO_4)Cl_{0.3}(OH,F)_{0.70}$ with minor Fe, Mg and Ni substituting for Ca. F and OH were also detected by ion microprobe.

Hutcheon et al. (1987) and Bischoff and Palme (1987) have both reported *whitlockite* in fremdlinge from Allende Type B CAIs. Armstrong et al. (1987) also reported 10-20 μm whitlockite grains distributed throughout a large (1 mm), sulfide-rich fremdling, Zelda from an Allende Type B1 inclusion. The whitlockite contains MgO (2.93 wt %), FeO (2.0 wt %) and Na_2O (2.95 wt %) and low concentrations of V_2O_3 (0.11 wt %) and Cr_2O_3 (0.16 wt %), similar to other meteoritic occurrences of whitlockite.

Tungstates-molybdates (scheelite $CaWO_4$ – powellite $CaMoO_4$). Armstrong et al. (1985) described the first meteoritic occurrence of scheelite in a large (150 μm), zoned fremdlinge (Willy) from Allende. It occurs as elongate laths often associated with apatite and contains a significant amount of Mo. Armstrong et al. (1985) have suggested that it may have formed by the breakdown of preexisting ferberite ($FeWo_4$). The scheelite is compositionally variable, ranging from $Ca(W_{0.87}Mo_{0.13})O_4$ in the center of the fremdling to $Ca(W_{0.96}Mo_{0.04})O_4$ in its outer regions. Bischoff and Palme (1987) also found a Ca,W,Mo-bearing phase in several fremdlinge from a Type B CAI in Allende. This phase has a composition consistent with a solid solution of scheelite and powellite. The compositions from several different fremdlinge span a range from $Ca(Wo_{0.56}Mo_{0.44})O_4$ to $Ca(W_{0.21}Mo_{0.79})O_4$.

Silicates. Various pyroxenes occur as minor phases in fremdlinge including fassaite, wollastonite and diopside. V-rich clinopyroxenes occur in Type 1 and 3 fremdlinge in CAIs in Allende and Leoville (El Goresy et al. 1978). Bischoff and Palme (1987) found rare V-rich *fassaite* grains with up to 6 wt % V_2O_3 in a fremdling from a Type B Allende CAI. Thin rims of V-rich fassaite (~2-30 μm) are commonly present surrounding small fremdlinge in Allende CAI TS34 (Armstrong et al. 1985). The rims are present irrespective of whether the fremdling occur enclosed within fassaite, melilite or between spinel grains. In one fremdling enclosed within fassaite, the rim has elevated V_2O_3 (~5 wt %), FeO (~1.5 wt %) and MgO (12 wt %) and low TiO_2 (~6.2 wt %) compared with the host (~14 wt % TiO_2; <1 wt % V_2O_3). Ti^{3+} is also lower (50%) than the host fassaite

(80%). Fine-grained fassaite also occurs in the rim of the large (150 μm), zoned fremdling Willy from Allende (Armstrong et al. 1985) and varies in composition from low V_2O_3 (0 %), high TiO_2 (16 wt %) to high V_2O_3 (16 wt %) and low TiO_2 (0 %). The fassaite has a Ti^{3+}/Ti ratio of 0.4-0.7.

Other occurrences of pyroxenes have been reported in fremdlinge in Allende and Leoville, e.g. *wollastonite,* occurs in Type 1 fremdlinge (El Goresy et al. 1978), *diopside* in the Allende fremdlinge Willy, (Armstrong et al. 1985) and *diopside* and an *Fe-rich clinopyroxene* in Allende Type B CAI USNM 5241 (Hutcheon et al. 1987).

Armstrong et al. (1987) reported common Fe-rich *olivine* (Fa_{55}) and rare Mg-rich olivine (Fa_{15}) in the rim of a large (1 mm), sulfide-rich fremdling, Zelda from Allende. The olivine contains excess Fe as well as minor V and Cr which Armstrong et al. (1987) interpreted as being due to the presence of fine-grained, intergrown V-rich magnetite.

El Goresy et al. (1978) found *melilite, plagioclase* (anorthite), and *sodalite* as minor accessory phases in Type 1 Fremdlinge in CAIs in Allende and Leoville and *nepheline* occurs in both Type 1 and 3 Fremdlinge.

Simon and Grossman (1992) reported the first meteoritic occurrence of *goldmanite* (vanadium-rich garnet) associated with a metal-rich fremdling from a Leoville "Compact" Type A inclusion. The composition of this garnet is not endmember goldmanite ($Ca_3V_2Si_3O_{12}$), but contains Al, Fe^{3+} and Ti with little or no Mg and Mn. The average structural formula determined by Simon and Grossman for this phase is $Ca_3(V_{1.22}Al_{0.46}Fe_{0.18}Ti_{0.13})Si_{2.87}Al_{0.13})O_{12}$, but there is some compositional variation in V_2O_3 (16.6-21.0 wt %), TiO_2 (0-7.1 wt %), Al_2O_3 (4.8-8.3 wt %), and Fe_2O_3 (0.6-6.8 wt %).

Oxides. Several different oxide phases have been reported as minor occurrences in fremdlinge including spinel, V-rich hercynite, V-rich magnetite, magnetite and magnesiowüstite. El Goresy et al. (1978) reported *spinel* in Type 1 and 3 fremdlinge in CAIs from Allende and Leoville. *V-rich spinel* occurs in the rim of a large (150 μm), zoned fremdlinge (Willy) from Allende (Armstrong et al. 1985). This latter spinel contains thin lamellae of a V-rich phase, possible FeV_2O_4, along the (100) planes that may have exsolved from a V-rich spinel-magnetite solid solution. A Cr-rich spinel has also been found in Type A inclusion USNM 5171 from Allende (Hutcheon et al. 1987). *Hercynite* and *V-rich hercynitic spinels* also occur in the rim of Willy. The latter contain up to 25 wt % FeO and 25 wt % V_2O_3 with Cr_2O_3 contents between 3.16-6.17 wt %, but low TiO_2 (<0.5 wt %). The highest V_2O_3contents occur where textural evidence indicates a reaction between V-rich magnetite and fassaite to form Al-V-rich spinels. *V-rich magnetite* is found in the same fremdling (Armstrong et al. 1985) with V_2O_3 contents in the core of the fremdling varying between 4 and 7 wt %, with 0.4-4 wt % Cr_2O_3, 0.7 to 1.3 wt % MgO and Al_2O_3 0.2-1.3 wt %. In a region between the mantle and rim of the fremdling the V_2O_3 contents reach 23 wt % with 5 wt % Cr_2O_3. *V-rich magnetite* also occurs in smaller fremdlinge from the same CAI (1.2-9.95 wt % V_2O_3), but has higher TiO_2 (0.58-1.26 wt %, cf. <0.03 wt %) (Armstrong et al. 1985, Hutcheon et al. 1987). Other occurrences of V-rich magnetite have been reported by MacPherson and Grossman (1984), Bischoff and Palme (1987), Armstrong et al. (1987), Blum et al. (1989), Simon and Grossman (1992) and Palme et al. (1994). Bischoff and Palme (1987) found that some fremdlinge contain magnetite with high V_2O_3 contents (e.g. 8.3-8.9 wt % and 12.0-14.4 wt % for two separate fremdlinge), but in others the V_2O_3 contents are much lower (0.9-1.3 wt %). Cr_2O_3 in magnetite ranges from ~1 up to 3.7 wt % in the same fremdlinge. V-Cr-poor magnetite has also been reported in several Allende CAIs including the Type B inclusions USNM 5241, Egg6 (Hutcheon et al. (1987) and JIM (Blum et al. 1989). See also Brandstätter and Kurat (1983).

Zinner et al. (1991) reported *magnesiowüstite* grains partly enclosed in metal-rich fremdlinge from a Type B1 CAI (477B) from Vigarano. This phase is a member of the periclase-magnesiowüstite solid solution series and has a composition $Per_{70}Wü_{30}$.

Cr-rich spinel occurs in fremdlinge in Type A CAI USNM 5171 from Allende (Hutcheon et al. 1987). Other minor oxide phases which have been observed in fremdlinge are unspecified V-rich oxides in Type 1 and 2 fremdlinge (El Goresy et al. 1978), a Mo-rich oxide in Willy, a large (150 μm), zoned fremdling from Allende (Armstrong et al. 1985) and baddelyite (ZrO_2) and Nb-rich ZrO_2 in Type 1 and 3 fremdlinge in Allende and Leoville (El Goresy et al. 1978).

Several other rare occurrences of unidentified phases have also been reported in Allende fremdlinge, including an unidentified *Mg,Fe molybdate* in the mantle of the fremdlinge, Willy (Armstrong et al. 1985) and small blebs of a V,Nb-rich phase included within apatite in a fremdling from CAI TS-34 in Allende (Armstrong et al. 1985). An unidentified Mo-rich phase which contains S, Ca, and Fe and could be a mixture of phases also occurs in a Type B1 CAI (Bischoff and Palme 1987).

Mineralogy of amoeboid olivine aggregates in CV carbonaceous chondrites

Amoeboid olivine aggregates (AOAs) have been widely recognized and described in CV carbonaceous chondrites (e.g. Grossman and Steele 1976, McSween 1977c, Simon and Haggerty 1979, Wark 1979, Cohen et al. 1983, Kornacki et al. 1983, Kornacki and Wood 1984a). AOAs have also been described in CO carbonaceous chondrites (McSween 1977a). They are irregularly shaped objects consisting largely of clumps (100-1000 μm in size) of polygonal olivine (1-20 μm in size) (Fig. 71d, p. 3-88) associated with other phases such as pyroxene, nepheline, sodalite and sometimes phyllosilicates (e.g. in Mokoia). The clumps are often surrounded by coarser-grained olivine. Grossman and Steele (1976) first described the mineralogy of AOAs in Allende in detail: olivine, as the dominant phase, shows a range of compositions from Fo_{64}-Fo_{99}, with a relatively flat peak in the distribution of olivine compositions between Fo_{73}-Fo_{90}. There appears to be little difference between olivine in the core and the rims of the aggregates. In Allende, nepheline and sodalite occur as interstitial phases and although they are extremely fine-grained, Grossman and Steele (1976) were able to obtain analyses which probably represent uncontaminated phases. Nepheline contains 2-5 wt % CaO and typically <2 wt % K_2O; sodalites appear to contain more Al and Si than the ideal structural formula and have CaO contents of 0-1.21 wt % with <0.1 wt % K_2O. Cl contents of between 5 and 6 wt % are typical. Pyroxene with a wide range of compositions occurs in AOAs in Allende (Grossman and Steele 1976). They are all Ca-rich (14-26.5 wt % CaO), but can be both Al-poor (<2 wt % Al_2O_3) with compositions from diopside to hedenbergite and Al-rich, sometimes with extremely high Al_2O_3 contents, up to 45 wt %. Other minor phases in AOAs in Allende are pure anorthite and spinel with variable Fe/(Fe+Mg) ratios.

Cohen et al. (1983) found that AOAs in Mokoia differed from those in Allende in having a lower abundance of interstitial feldspathoids. Instead, a phyllosilicate phase (LAP - Low Al-Phyllosilicate) is present between olivine grains (see Secondary Mineralogy of CAIs in CV chondrites). Fe-rich and Fe-poor olivines are both present, with the Fe-rich olivine typically occurring surrounding small pores. Cohen et al. (1983) found tiny veins and blebs of magnetite, pyrrhotite, and/or pentlandite in these AOAs.

A common characteristic of AOAs is that they frequently contain Al-rich refractory inclusions, usually ovoid or circular in appearance and sometimes up to 300 μm in size, in their interiors (Grossman and Steele 1976, Hashimoto and Grossman 1987). In Allende, lightly altered examples of these Al-rich inclusions contain many of the refractory phases which occur in CAIs, including melilite, fassaite, spinel and perovskite (Hashimoto and Grossman 1987) and sometimes have a zoned concentric structure often with a melilite core (Ak_4-Ak_{15}) usually associated with fassaite, spinel and perovskite. Petrographic studies of these Al-rich objects show that they exhibit a range of degrees of alteration and the alteration products are similar to those observed in large CAIs in Allende. In moderately altered inclusions, melilite has been replaced by fine-grained grossular, anorthite and feldspathoids and perovskite has been partially altered to ilmenite. Fassaite and spinel remain unaffected by the alteration. As the degree of alteration increases, fassaite is replaced by phyllosilicates and ilmenite, but primary spinel and perovskite persist, until at the highest level of replacement, spinel has been altered to form phyllosilicates or a mixture of olivine and feldspathoids. The phyllosilicates appear to a mixture of more than one phase of which Na-phlogopite, chlorite or aluminous lizardite all appear to be possible candidates (Hashimoto and Grossman 1987).

Primary mineralogy of CAIs in CM carbonaceous chondrites

CAIs in CM chondrites are much smaller and less common than those in CV carbonaceous chondrites, being typically <1mm, although rare examples up to 2 mm have been found (e.g. Hashimoto et al. 1986). As discussed below, however, CAIs in CM chondrites are of special significance, because many of them contain hibonite with a distinctive sky-blue color, a phase which is rare in CAIs in CV carbonaceous chondrites.

Corundum. Bar-Matthews et al. (1982) were the first to find corundum, associated with hibonite and rare perovskite, in a refractory inclusion (BB-5) in Murchison. The corundum appears to be completely surrounded by the hibonite grains, suggesting that corundum formation probably occurred before hibonite. The corundum is virtually pure Al_2O_3 with only minor CaO, TiO_2, SiO_2 and MgO (Table A3.7) and ion microprobe analysis (Hinton et al. 1988) shows that REE concentrations are close to or below detection limits (<0.05 ppm). Refractory trace elements such as Y and Zr are also extremely low (<3 ppm), but higher concentrations of Sc (289 ppm) and Ca (125 ppm) are present. In a second corundum-bearing inclusion from Murchison (MacPherson et al. 1984b), a mantle of corundum completely encloses a central core of hibonite. The corundum in this inclusion is also pure Al_2O_3 (99.96 wt %) with minor TiO_2, FeO, CaO and Sc_2O_3.

Hibonite ($CaAl_{12}O_{19}$). Although CAIs in CM chondrites are significantly less common and smaller than those in Allende, hibonite is much more common and has a larger grain size (>50 μm). In some inclusions in CM chondrites, it can be the major phase, typically associated with spinel and perovskite. Hibonite has been extensively studied, because it is postulated to be the primary condensate from a nebular gas or may have replaced corundum at lower temperatures (e.g. Grossman 1972, Fegley 1982). Consequently, its mineralogical, chemical and isotopic characteristics are likely to have recorded the highest temperature events in the solar nebula.

Fuchs et al. (1973) first reported hibonite-bearing CAIs in Murchison, which they described as white inclusions, 100-150 μm in diameter, with blue centers of micro-crystalline (3 × 10 μm) hibonite. Several different types of occurrence of hibonite have since been recognized in Murchison and other CM chondrites, based on their morphological and mineralogical characteristics. In several cases, these different occurrences of hibonite have been separated from the host meteorite by crushing and sieving, rather than being identified in situ. Representative electron microprobe analyses of hibonite are reported in Table A3.9 and trace element data, measured by ion microprobe, in Table A3.10.

Several *intact, hibonite-bearing* CAIs from CM chondrites have been described in detail. Armstrong et al. (1982) studied a large (1.5 mm) CAI from Murchison (named the Blue Angel) which has three, roughly concentric zones consisting of a hibonite-rich core, a calcite-rich mantle and a spinel-rich, layered rim. The hibonites are highly stoichiometric in all three zones (Fig. 105a) and conform to the stoichiometry Ca = 1 and Al+V+Fe+Cr+Mg+Ti = 12 for O = 19. However, they vary in composition from one zone to another and within single grains. The hibonites have high V_2O_3 contents with 0.77-1.80 wt % in the core, cf. 0.10- 0.57 wt % in the rims. Mg and Ti concentrations are also higher in rim hibonites (Fig. 105b). Cr_2O_3 and SiO_2 contents in Blue Angel hibonites are slightly higher than those reported from Allende and other meteorites, but Sc_2O_3, is lower (0.1 to 0.14 wt % in Allende, cf. ~0.05 wt % in the Blue Angel).

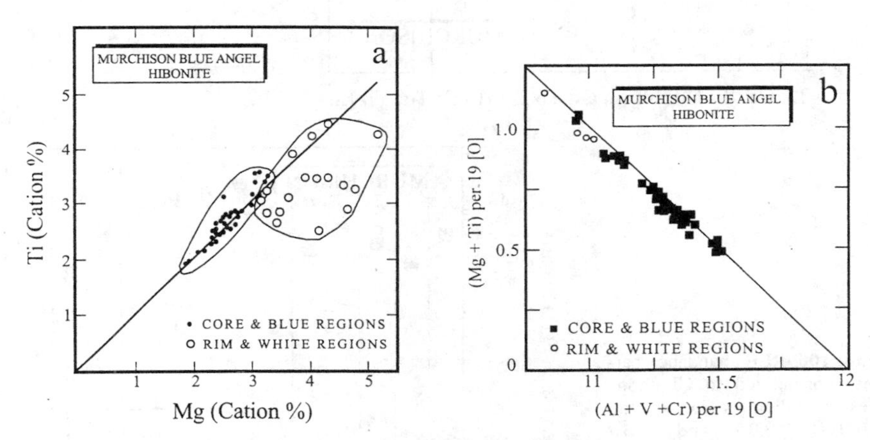

Figure 105. (a) Mg vs. Ti (cations per 19 O) plot for hibonites from the rim and core regions of the Blue Angel inclusion in Murchison. After Armstrong et al. (1980). (b) Mg+Ti vs. Al+V+Cr (cations per 19 O) in hibonites from the rim and core of the Blue Angel inclusion in Murchison. Points lie on or just below the line for the substitution Mg+Ti = 2(Al+V+Cr). Rim hibonites are significantly poorer in Mg+Ti and Al+V+Cr than core hibonites. After Armstrong et al. (1980).

In a second zoned, hibonite-bearing CAI (MUCH-1) from Murchison, fewer, but much larger, hibonite grains occur (MacPherson et al. 1983). These hibonites have uniformly low TiO_2 (~2.05 wt %), MgO (~1.07 wt %) and V_2O_3 below EPMA detection limits. REE patterns of MUCH-1 hibonite, obtained by ion microprobe, have flat LREE (40-50 × CI) with a progressive decrease in enrichment factor through the HREE and large negative Eu and Yb anomalies (Hinton et al. 1988) (Fig. 106a). This pattern is consistent with the Group III inclusions of Mason and Martin (1977). Hinton et al. (1988) argued that the REE pattern probably records a history of condensation in which hibonites passed through gases of changing composition as they grew and condensed with at least one other REE-bearing phase. Hibonite in a corundum-hibonite-bearing inclusion (BB-5) from Murchison (Bar-Matthews et al. 1982) has lower concentrations of MgO (0.65 wt %) and TiO_2 (2.01 wt %) than MUCH-1, but a very similar REE pattern (Fig. 106a) (Hinton et al. 1988).

El Goresy et al. (1984) found compositionally variable hibonite in a melilite-bearing CAI in Essebi. Hibonite that occurs in spinel + hibonite framboids in the melilite-rich core contains 1.4-9.0 wt % TiO_2 and 0.77-4.0 wt % MgO. However, in the rim sequence, MgO and TiO_2 in hibonites are lower (1.4-6.2 wt % TiO_2; 0.77-0.37 wt % MgO) and FeO is higher (<0.5 wt %, cf. <0.01 wt %). V_2O_3 contents in both occurrences are low (0.2-0.5

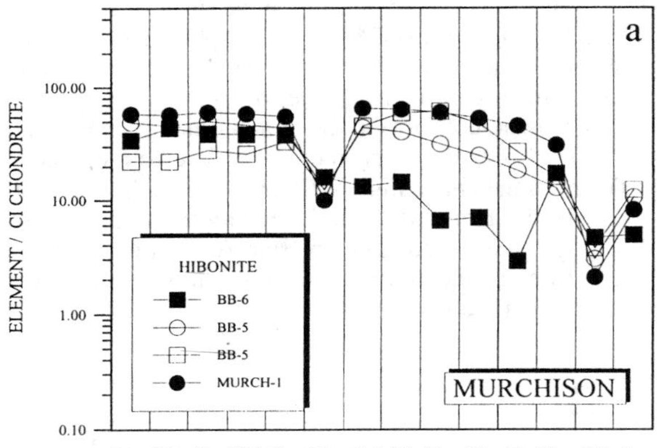

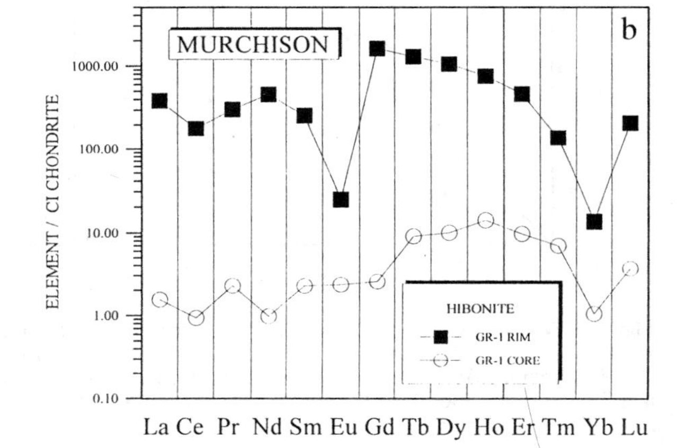

Figure 106. REE abundance patterns (normalized to CI chondrites) for hibonites from (a) the MUCH-1, BB5 and BB6 inclusions from Murchison, and (b) the core and rim of inclusion GR-1 from Murchison. Ion microprobe analyses. Data from Hinton et al. (1988).

wt %). An unusual feature of these hibonites is that they show significant departures from stoichiometry (El Goresy et al. 1984). Core hibonites deviate significantly from the 1:1 correlation typically observed for Ti^{4+} vs. Mg^{2+} and also appear to have excess Ca^{2+} (2.00 to 2.17 cations based on 38 oxygens). This non-stoichiometry is also observed for core hibonites in Al^{3+} vs. (Ti^{4+} +V^{3+} + Mg^{2+}) plots (Fig. 107). El Goresy et al. (1984) postulated that the observed deviation may be due to a complex substitution of the type 2 Ti^{3+} + Ca^{2+} + Mg^{2+} = 2 Al^{3+} + Ti^{4+}. An excess of Al_2O_3 in solid solution is observed in rim hibonites in this CAI with the total number of cations assigned to the Al^{3+} site (Ti + V + Al + Mg + Si) varying from 24.0 to 24.1 per 38 oxygen atoms. El Goresy et al. (1984) attributed the excess Al to the fact that non-stoichiometry can arise in Ca aluminates due to alteration of the stacking sequence of the three subunit cell slabs which lie on (001) (Schmidt and De Jonghe 1983).

MacPherson et al. (1984b) and Hinton et al. (1988) also found two compositionally distinct types of hibonite in a hibonite-corundum CAI (GR-1) from Murchison. This inclusion has a concentric structure with an inner, Mg-poor core and an outer, Mg-rich rim, which are separated by a corundum layer. Hibonite in both core and rim is very low in MgO and TiO_2 (<1.01 and <0.38 wt %), but the inner hibonite is enriched in Zr, Sc, Sr and Ba compared with rim hibonite. The REE patterns of rim and core are very different,

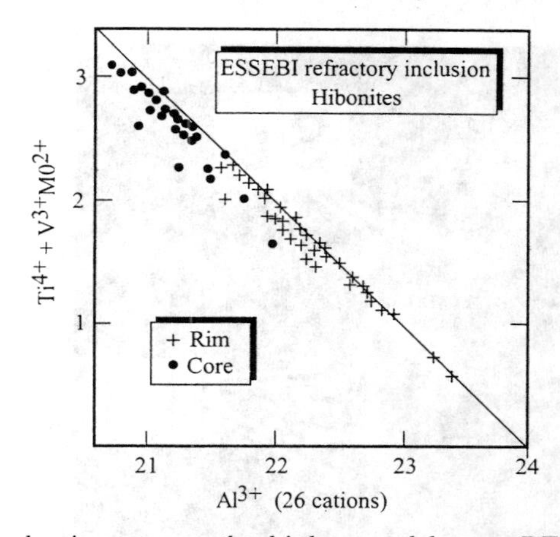

Figure 107. Ti^{4+}+V^{3+}+Mg^{2+} vs. Al^{3+} (cations) in core and rim hibonites from an unusual refractory inclusion in Essebi. The core hibonites with low Al contents show a marked departure from stoichiometry. After El Goresy et al. (1984).

having among the highest and lowest REE concentrations, respectively, measured in hibonites in CM chondrites (Fig. 106b). The core hibonite has a very irregular REE pattern with a marked step between the LREE (180-400 × CI; Gd ~1800 × CI) and HREE. Eu and Yb both show large, negative anomalies (20 and <10 × CI, respectively). In contrast, the rim hibonite has extremely low REE abundances (<1 to 10 × CI), with a general, but irregular increase from the LREE to Ho and then a decrease to Lu. Yb has a negative anomaly. Hinton et al. (1988) suggested that the depletion of LREE relative to HREE in the core hibonite could be explained by gas-solid partitioning during condensation, with the exception of the Eu anomaly, which would require the separate addition of volatile Eu prior to the formation of the rim hibonite. MacPherson et al. (1984b) suggested that the rim hibonite was formed by incongruent melting to corundum and a Ca-rich liquid, due to flash melting, followed by preferential volatilization of Ca. However, Hinton et al. (1988) rejected this idea based on the REE data for hibonite and suggested that the rim hibonite was formed by reaction of corundum with a solar nebular gas, before complete condensation of Ca.

Greenwood et al. (1994) found only one inclusion in situ in several thin sections of Cold Bokkeveld in which hibonite was the dominant phase. The hibonite in this inclusion is anhedral and has low TiO_2 (1.67 wt %) and MgO (0.8 wt %).

Macdougall (1981) first recognized a group of tiny (~70-310 μm) CAIs termed *spinel-hibonite spherules* in a hand-picked sample of ~500 chondrules from Murchison (Fig. 108). Further examples have been reported from Murchison by MacPherson et al. (1983, 1984b) and Simon et al. (1994c) and in Cold Bokkeveld by Greenwood et al. (1994). The spherules consist mostly of spinel and hibonite with lesser amounts of perovskite and rare iron sulfide and melilite. Although the hibonites in these spherules have somewhat variable textures they all contain crystals with a blade morphology, ranging in size from 5-20 μm.

The hibonites in the spherules have highly variable compositions (Macdougall 1981, MacPherson et al. 1983, 1984b; Simon et al. 1994c). One Murchison inclusion (BB-4) alone has hibonite compositions which span almost the complete range of Ti + Mg = 2 Al substitutions observed in meteoritic hibonites (MacPherson et al. 1983). REE patterns for hibonite from BB-6 (Fig. 106a), measured by ion microprobe (Hinton et al. 1988) are consistent with Group II rather than a Group III pattern (Mason and Martin 1974). The LREE are highly enriched, but decline in enrichment factor through the HREE and Tm is enriched relative to neighboring Er, a characteristic of Group II patterns. The hibonite is

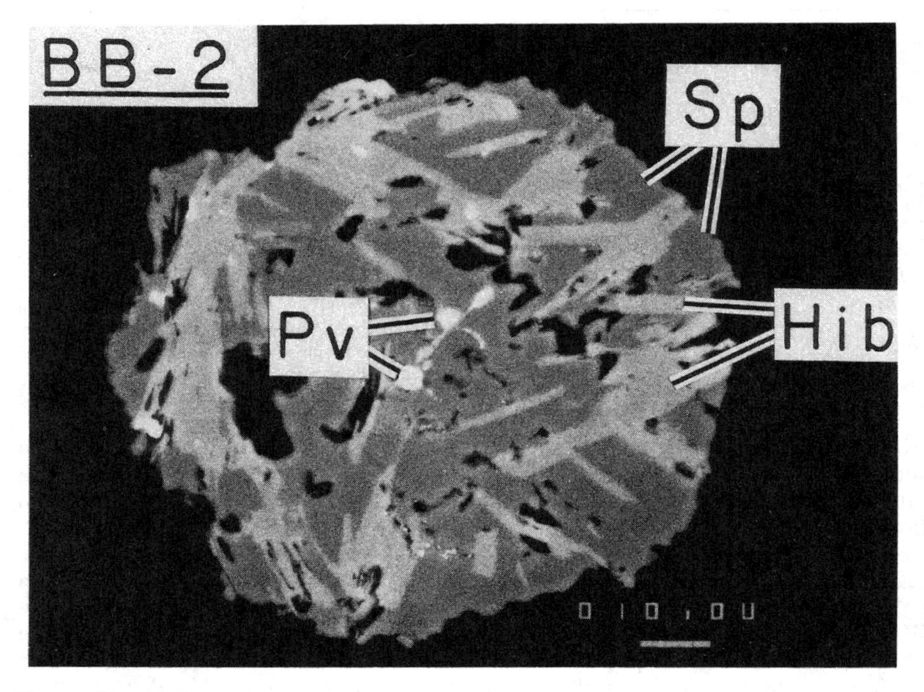

Figure 108. BSE image of a spinel-hibonite spherule (BB-1) from Murchison, in which hibonite (Hib) is concentrated toward the outer zone of the inclusion. The bulk of the inclusion consists of spinel (Sp) with minor perovskite (Pv) and melilite (Mel) and a mantle lacking hibonite. Scale is 10 μm. Used by permission of the editor of *Geochimica et Cosmochimica Acta* from MacPherson et al. (1983), Figure 2.

enriched in volatile elements (Na, K, Si, Cr, Mn and Fe) and depleted in refractory elements (Sc and Y), in comparison with hibonites which have Group III patterns. This type of pattern may be the result of partitioning of REE between hibonite and other condensed phases during condensation, but the volatile element enrichments may be secondary and were caused by equilibration with a volatile-rich gas or melt at lower temperatures.

Individual hibonite fragments or *Platy crystal fragments (PLACs)* have been decribed by MacPherson et al. (1983), Fahey et al. (1987b), Ireland (1988 1990) and Ireland et al. (1986, 1988). These are all isolated, individual, colorless to light-blue hibonite grains from Murchison, often with a platy morphology, some of which may be derived from larger CAIs. However, MacPherson and Davis (1994) found individual hibonites in situ in the matrix of Mighei. In Murchison, Murray, and Mighei, this type of hibonite is characterized by low TiO_2 contents (~ 2 wt %) (MacPherson et al. 1983, Fahey et al. 1987b, Ireland 1988, 1990), with rare exceptions with 5 and 6.5 wt % TiO_2 (Ireland et al. 1986). They also appear to be consistently low in MgO (<0.8 wt %; MacPherson et al. 1983).

Trace element analyses (ion microprobe) of Murchison and Murray hibonite fragments (Fahey et al. 1987b, Ireland 1988, 1990; Ireland et al. 1988, Hinton et al. 1988) show that they have rather similar REE patterns (Fig. 109a,b) which are typically relatively flat or show a progressive decrease to the HREE. They are all characterized by significant depletions in Eu and Yb, especially the PLACS, i.e. resemble Group III REE

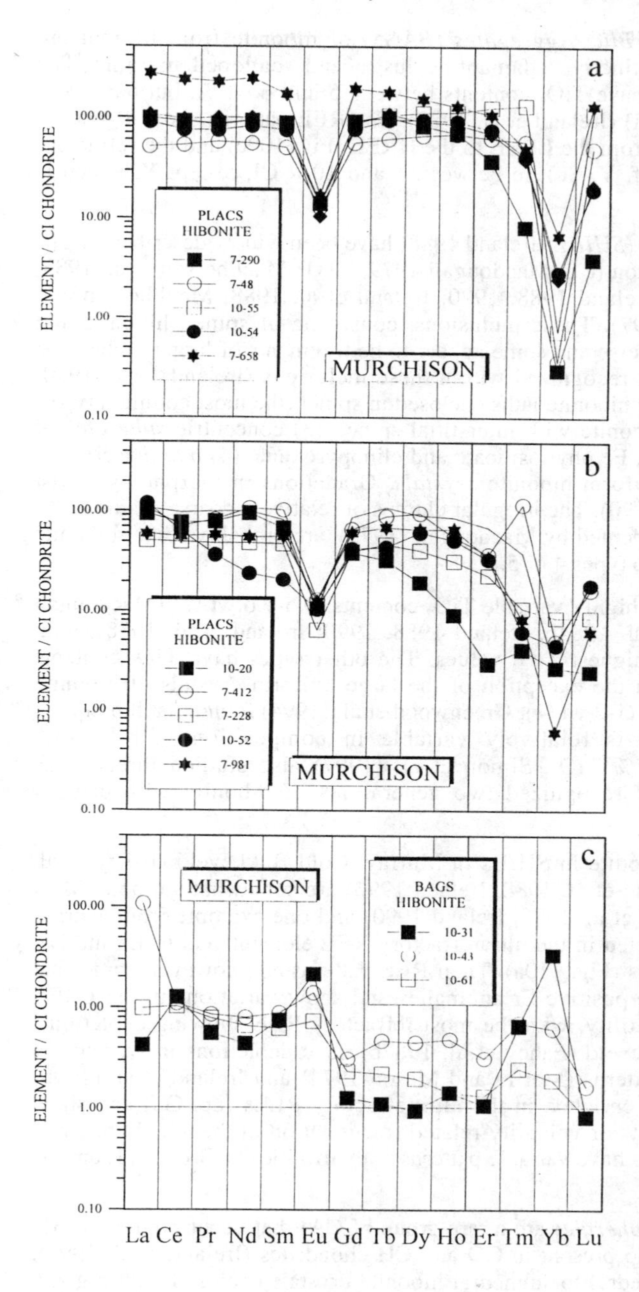

Figure 109. (a) Range of REE abundance patterns (normalized to CI chondrites) found in hibonite in PLACs from Murchison.
(b) Range of REE abundance patterns (normalized to CI chondrites) found in hibonite PLACs from Murchison.
(c) Representative REE patterns (normalized to CI chondrites) for hibonite from BAGS from Murchison. All data are from Ireland (1988, 1990) and Ireland et al. (1988b): ion microprobe analyses.

patterns of Martin and Mason (1974). The more volatile refractory trace elements, Ba, Nb and V are also depleted relative to elements such as Sc, Hf and Zr. The REE patterns are very similar to hibonite from the MUCH-1 CAI in Murchison (Hinton et al. 1988), indicating a similar formational history.

Ireland (1988) described *Blue Aggregates (BAGS)* of hibonite from Murchison. These aggregates have a distinctive adamantine lustre and scalloped margins. The hibonites in these aggregates have TiO_2 contents between 5 and 6 wt %, but have very low REE abundances (<1 × CI) (Ireland et al. 1988). The REE patterns show a general, but rather irregular, decrease from the LREE to the HREE (Fig. 109c) and the refractory trace elements (Er, Lu, Zr, Hf, Y, Sc) lie between 1 and 10 × CI, except Y, which is depleted (0.1 × CI).

Spinel-hibonite inclusions (SHIBs - Ireland 1988) have been widely described in CM chondrites, especially Murchison (e.g. Macdougall 1979, 1981; MacPherson et al. 1983, 1984b; Fahey et al. 1987b, Ireland 1988, 1990; Ireland et al. 1988, MacPherson and Davis 1994, Simon et al. 1997). These inclusions, consisting of spinel, hibonite and perovskite, are a texturally diverse and some appear to be fragments of larger inclusions. 5 types of hibonite have been recognized within these inclusions (Ireland 1988, 1990; Ireland et al. 1988): (1) *bladed* hibonite laths enclosed in spinel (the most common type), (2) *compact* aggregates of hibonite with interstitial spinel, (3) concentric *spherules* of hibonite surrounded by spinel, Fe phyllosilicate and clinopyroxene, (4) *massive* crystals of hibonite, and (5) large, uniform hibonite *crystals.* Gradations in morphology exist between the groups (Ireland 1990). The irregular clumps of relatively coarse-grained (20-50 μm) hibonite grains, documented by Macdougall (1979), are probably spinel-hibonite inclusions, possibly falling into types 4 or 5.

Hibonites in SHIBs have highly variable TiO_2 contents (0.5-9.0 wt %) (Macdougall 1979, 1981; MacPherson et al. 1984b, Ireland 1988, 1990; Ireland et al. 1988), with bladed hibonites having the highest TiO_2 values. The other types have TiO_2 contents between 1 and 7 wt %, with the exception of the large uniform crystals of hibonite which have low TiO_2 contents (1-2 wt %). Greenwood et al. (1994) found that hibonite in SHIBs in Cold Bokkeveld is relatively variable in composition with 2.1-3.8 wt % MgO and 4.2-6.0 wt % TiO_2. Simon et al. (1997) also studied three SHIB fragments in Murchison and recognized two generations of hibonite with distinct compositions.

The REE patterns of hibonite in SHIBs in Murray, Cold Bokkeveld (Fahey et al. 1987b) and Murchison (Ireland et al. 1988, Ireland 1990) are somewhat variable. Most SHIBS in Murchison (Ireland et al. 1988, Ireland 1990) and one example from Murray (Fahey et al. 1987b) are depleted in the ultrarefractory REE elements Gd to Er and Lu, i.e. similar to Group II patterns (Fig. 110a). Their REE patterns are, however, somewhat irregular and commonly have positive Er anomalies and show variations in the LREE which are attributable to volatility, with the most refractory REE showing depletions. Two SHIBs from Cold Bokkeveld (Fahey et al. 1987b) have depletions in Eu and Yb (Fig. 110b) like Group III patterns (Martin and Mason 1974) and Ireland et al. (1990) found two examples that are enriched in the ultrarefractory REEs (i.e. Gd) and show variable evidence of the effects of volatility-related fractionation in their REE patterns. The remaining group of SHIBs have variable patterns with affinities to Group I, II and III patterns.

Hibonite-bearing microspherules are a rare group of CAI that occur predominantly in CM chondrites, but are also present in CO and CH chondrites (Ireland et al. 1991). They typically consist of euhedral to subhedral hibonite crystals enclosed within glass, but one example consisting of a rounded hibonite core containing perovskite inclusions has also been reported in Murchison (Fig. 111a,b). Hibonites in these microspherules exhibit minor coupled substitution of Mg^{2+} and Ti^{4+} for $2Al^{3+}$, and contain between 1.8 and 2.3 wt % TiO_2 and 0.9 and 1.2 wt % MgO.

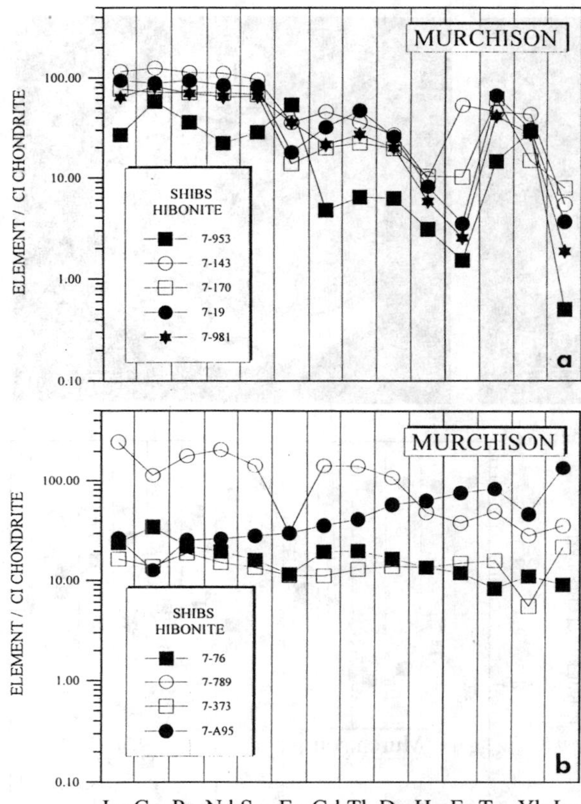

Figure 110. (a) Typical Type II REE abundance patterns (normalized to CI chondrites) for hibonite in SHIBS in Murchison and Murray. Ion microprobe analyses. Data from Ireland et al. (1988b) and Ireland (1990).
(b) REE abundance patterns (normalized to CI chondrites) for hibonite in SHIBS in Murchison and Murray which show similarities to Group III REE patterns. Ion microprobe analyses. Data from Ireland et al. (1988b), Ireland (1990) and Fahey et al. (1987).

Hibonites from two microspherules in Murchison (Ireland et al. 1991) have very different REE patterns. In one spherule, the hibonite is highly enriched in LREE (37-46 × CI) with depleted HREE and a negative Eu anomaly (Fig. 111c). The LREE decrease smoothly from La to Sm but the HREE show a steeper decrease from 45 × CI for Gd to 8 × CI for Yb and Lu. Refractory trace elements (Sc, Y, Zr and Hf) are all enriched by 15-30 × CI. The REE patterns in this inclusion appear to be consistent with equilibrium crystallization from a melt. Hibonite in the second inclusion has an extremely unusual and irregular REE pattern. The LREE are very low (3-6 × CI) with the HREEs being somewhat enriched (3-7 × CI) with positive Tm (18 × CI) and Yb (7 × CI) anomalies. No Eu anomaly is present. The refractory trace elements are also low (2-4 × CI) except Sc, Zr and Nb which range between 11-15 × CI. Hibonite in this inclusion shows no evidence of element partitioning and appears to have preserved evidence of an unequilibrated refractory precursor component (Ireland et al. 1991).

Several minor occurrences of hibonite have been reported in CM chondrites. In Mighei, hibonite generally occurs as an accessory phase in a variety of different types of CAIs (MacPherson and Davis 1994) which are not always easily fitted into the categories above. Compositionally, the Mighei hibonites show considerable variation and TiO_2 and MgO can vary by a factor of two in the same inclusion. V_2O_3 contents are always low, mostly <0.2 wt %, but the Mighei hibonites appear to contain more FeO (>0.3 wt %, cf. <0.2 wt % for Murchison). Lee and Greenwood (1994) have reported several occurrences

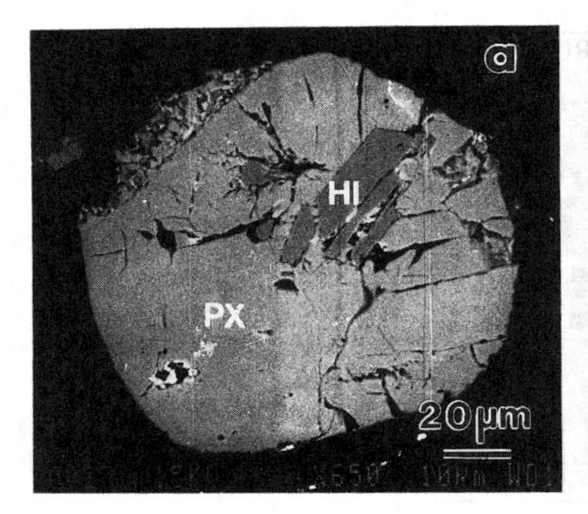

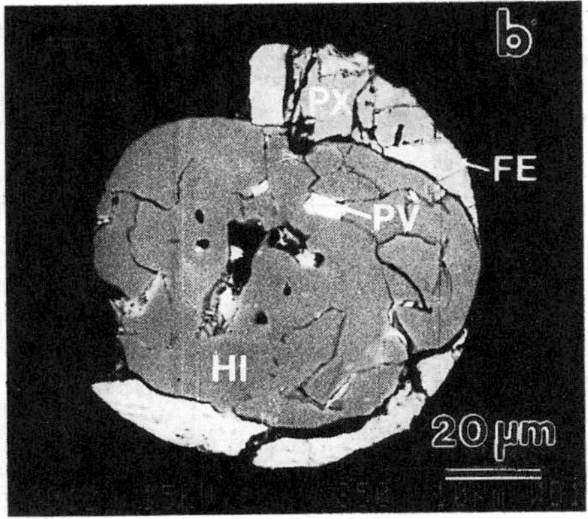

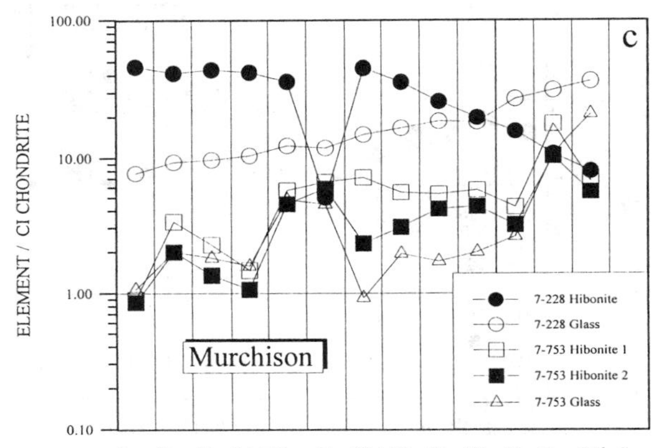

Figure 111. (a,b) BSE images of hibonite-bearing micro-spherules from Murchison (Ireland et al. 1991). Used by permission of the editor of *Geochimica et Cosmochimica Acta* from Ireland et al. (1991), Figures 1b,c.
(c) REE abundance patterns (normalized to CI chondrites) for hibonite in hibonite-bearing microspherules from Murchison. Ion microprobe analyses. Data from Ireland et al. (1991).

of hibonite as radiating laths in *spinel-pyroxene inclusions* in Murray. The hibonite appears to be compositionally similar to that found in other types of inclusion in Murchison, but shows a rather restricted range of compositions (MgO = ~2.7 wt %; TiO_2 of ~4-5 wt %).

Grossite ($CaAl_4O_7$). Grossite has only been reported in one CAI in a CM chondrite (Simon et al. 1994c) and occurs in a concentrically zoned refractory spherule from Murchison. Grossite occurs as anhedral, ~10 µm crystals in the spherule core, and is surrounded by concentric layers consisting of spinel + hibonite, melilite, anorthite with an outer layer of Al-diopside. This highly refractory assemblage probably crystallized from a melt at temperatures of ~2100°C. The grossite is compositionally homogeneous and pure with low concentrations of MgO, SiO_2, Sc_2O_3, TiO_2, V_2O_3 and FeO (<0.19 wt %). However, analyses reported by Simon et al. (1994c) indicate some non-stoichiometry with Ca cations generally greater than one and Al cations less than four cations per 7 oxygen atoms (Table A3.7).

Perovskite ($CaTiO_3$). Perovskite occurs widely as a minor phase in CAIs in CM chondrites (e.g. Fuchs et al. 1973, Macdougall 1979, 1981; Armstrong et al. 1982, MacPherson et al. 1983, Lee and Greenwood 1994, Greenwood et al. 1994). The perovskites (<10 μm in size) often occur as inclusions in phases such as melilite and spinel and are difficult to analyze, but appear to have compositions close to pure $CaTiO_3$, with minor amounts of V, Fe and Si. However, contamination from adjacent phases may be a problem (e.g. Al, Mg from spinel—Armstrong et al. 1982, Simon et al. 1994c) (Table A3.11). In a melilite-bearing inclusion from Essebi (El Goresy et al. 1984), perovskites in the rim sequence are somewhat enriched in ZrO_2 (0.20-0.52 wt %) and Y_2O_3 (0.10-0.26 wt %) in comparison with the core (ZrO_2 <0.05-0.32 wt %; Y_2O_3 <0.05-0.20 wt %). Some REE, determined by electron microprobe, are enriched in both core and rim perovskites; e.g. Ce_2O_3 0.13-0.50 wt %, but rim perovskites have higher concentrations of the HREE, Yb and Lu. Perovskite from ultrarefractory inclusion HIB-11 (Simon et al. 1996) is highly enriched in refractory elements, (e.g. 1.6 wt % Y_2O_3; 0.6 wt % Sc_2O_3) compared with typical CAI perovskite.

Ion microprobe analyses of three 10 μm perovskite inclusions in spinel from Murchison (Ireland et al. 1988; Fig. 112) show similar REE patterns with close affinities to Group II patterns. The LREE are enriched 40-400 × CI, with Ce and Sm having slightly higher abundances than the other REE. The HREE are depleted, except for Tm and Yb, which are enriched at about LREE levels. Refractory trace elements are rather low in abundance, but Nb is highly enriched (1,000-15,000 × CI). Ireland et al. (1990) also reported U (3.6-15 ppm) and Th (0.2-3.8 ppm) concentrations for perovskites in several different perovskite-bearing inclusions in Murchison.

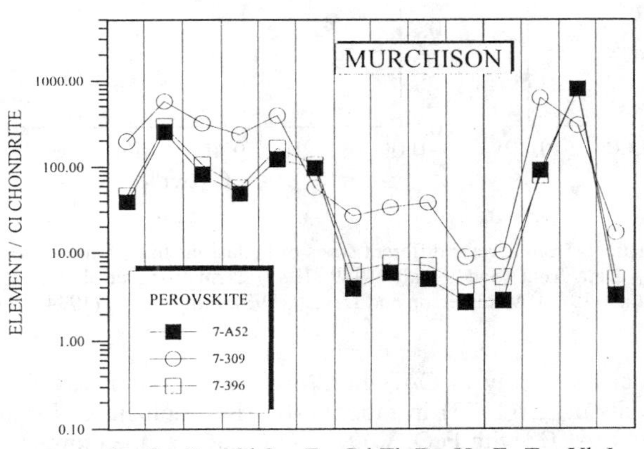

Figure 112. REE abundance patterns (normalized to CI chon-drites) from perovskite inclu-sions in spinel from a refractory inclusion in Murchi-son. All REE patterns are similar to Group II patterns from Allende. Ion microprobe analyses. Data from Ireland et al. (1988).

Melilite (gehlenite $Ca_2Al_2SiO_7$ - åkermanite $Ca_2MgSi_2O_7$). Melilite is much rarer in CAIs in CM chondrites than in CV chondrites. MacPherson et al. (1983) recovered three melilite-rich fragments from a CAI in Murchison, which resembles some "Compact" Type A inclusions from Allende. The CAI has a melilite-rich core, separated from a layered rim by a zone rich in spinel, perovskite, hibonite and small (10-15 μm) islands of melilite, which are partially replaced by calcite. The outer rim (25-30 μm thick) contains spinel, melilite, anorthite, and clinopyroxene. Melilites in the core and exterior of the inclusion are compositionally rather similar ($Åk_{2-23}$; cf. $Åk_{3-25}$) and are Na and Fe free in both occurrences.

No melilite-bearing inclusions were found by MacPherson and Davis (1994) in Mighei, but El Goresy et al. (1984) have described an unusual, melilite-bearing inclusion in Essebi. In this inclusion, melilite ($Åk_{10-26}$) is present as the dominant phase in the core and poikilitically encloses clusters of hibonite, perovskite and spinel. The Na melilite component in the core is negligible and the most gehlenitic compositions occur towards the rim region. The core is surrounded by a thin, complex rim sequence, consisting of 5 layers (designated A-E) in which the mineralogy becomes more refractory outwards. Melilite ($Åk_{0-10}$) with hibonite and perovskite inclusions is locally present in one rim layer (layer B) and has been partially replaced by nepheline and sodalite.

The rare melilite ($Åk_{0-25}$) found in *hibonite-bearing spherules* (e.g. BB-1) in Murchison is Fe and Na-free (MacPherson et al. 1983). Simon et al. (1994c) also reported a layer of melilite ($Åk_{5}$-$Åk_{6}$) in a highly refractory hibonite-bearing spherule from Murchison.

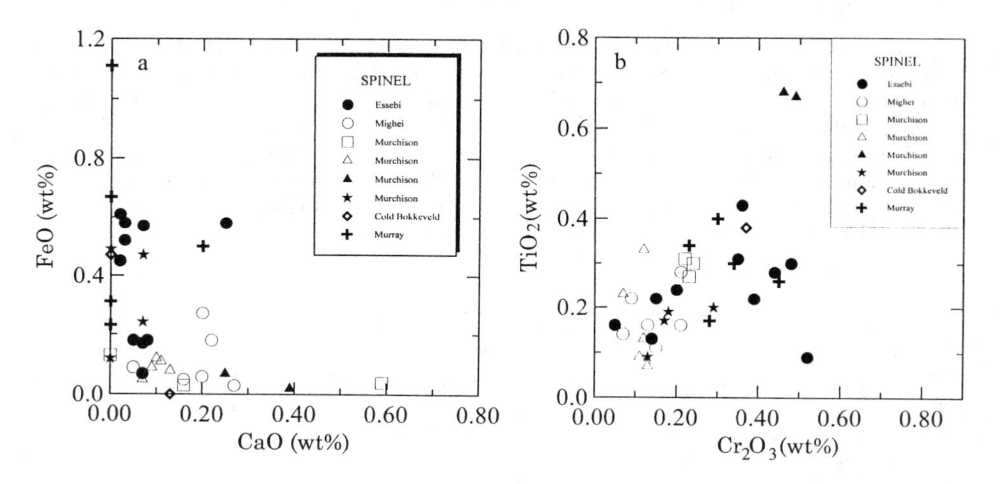

Figure 113. Oxide variation plots for spinels from several different types of inclusions in CM chondrites. (a) FeO vs. CaO. (b) TiO_2 vs. Cr_2O_3. Data from Macdougall (1979, 1981), Armstrong et al. (1982), MacPherson et al. (1983), El Goresy et al. (1984), MacPherson and Davis (1993), Simon et al. (1994) and Greenwood et al. (1994).

Spinel ($MgAl_2O_4$). Spinel occurs widely in CAIs in CM chondrites and was first recognized in Murchison by Fuchs et al. (1973) in hibonite-bearing inclusions. In all occurrences it is essentially pure $MgAl_2O_4$ with FeO, V_2O_3, TiO_2, Cr_2O_3 as the dominant minor elements (see Figs. 113a,b). Analyses of spinel in CAIs in CM chondrites are reported in Table A3.15.

Spinel from the Blue Angel inclusion in Murchison (Armstrong et al. 1982) shows some minor substitution of Fe, Cr and V for Al. Concentrations of these elements range between 0.24-0.49 wt % FeO, 0.10-0.27 wt % Cr_2O_3 and 0.13-0.29 wt % V_2O_3. Minor amounts of TiO_2 and SiO_2 are also present (<0.2 wt %). In comparison, MacPherson et al. (1983) found that spinels in several other, different types of CAI in Murchison have lower FeO (<0.13 wt %) and Cr_2O_3 (below detection) contents, but similar V_2O_3 (0.22-0.24 wt %). TiO_2 contents of the spinels also appear to be higher (>0.23 wt %). El Goresy et al. (1984) found that spinels in an Essebi inclusion are compositionally variable depending on their location. Spinels in the core have lower FeO contents (0.1 wt %) than those in the Blue Angel inclusion. In comparison, spinels in layer D of the rim are members of the spinel-corundum solid solution series (Baily and Russel 1969, Viertel and Siefert 1969). These spinels have excess Al_2O_3 which, expressed as the mole fraction of $Al_{2.67}O_4$ in solid solution, lies between 0.01 and 0.097. The presence of Al_2O_3 in solid solution indicates a very high temperature origin for these spinels. Their minor element contents are generally low (0.20 wt % TiO_2, 0.15 wt % V_2O_3), but FeO contents (0.05-0.5 wt %) are generally typically higher than the core spinel.

Spinels in *spinel-hibonite-bearing spherules* in Murchison (Macdougall 1981) contain ~0.27 wt % TiO_2, ~0.13 wt % Cr_2O_3, and ~0.21 wt % FeO. MacPherson et al. (1984b) also reported very similar concentrations of minor elements in spinel in this type of inclusion, but also analyzed V_2O_3 (~0.08 wt %). Simon et al. (1994c) found spinel in a $CaAl_4O_7$-bearing spinel-hibonite spherule from Murchison that contains excess Al_2O_3 in solid solution (see also El Goresy et al. 1984). Based on 4 oxygen atoms, the spinels have between 2.006 and 2.015 Al-site cations, equivalent to 1 to 2.3 mol % excess Al_2O_3. Minor element (Cr_2O_3, FeO, V_2O_3 and TiO_2) concentrations are <0.2 wt %.

Spinel in the *spinel-hibonite inclusions (SHIBs)* studied by MacPherson et al. (1984b) contain low concentrations of TiO_2 (0.88 wt %), V_2O_3 (~0.09 wt %), FeO (0.31 wt %), with Cr_2O_3 below detection limits.

Macdougall (1981) reported *spinel-perovskite spherules* from Murchison which are similar in size to the hibonite-containing spherules from the same meteorite, but contain only spinel and perovskite. However, no compositional data have been reported for these objects. A related group of objects are probably the *spinel spherules* containing only spinel described by Macdougall (1981) and Greenwood et al. (1994).

MacPherson et al. (1983) first recognized a group of inclusions in Murchison which are white in color and have a botryoidal appearance. These *spinel-pyroxene inclusions* often have either a nodular or banded internal structure (Fig. 114a,b) and consist largely of spinel, olivine and pyroxene, and typically lack hibonite and melilite. Further studies (e.g. MacPherson et al. 1984b, MacPherson and Davis 1994) on Murchison and Mighei show that spinel-rich objects with a whole range of textures transitional between the nodular and banded varieties exist. They also occur widely in Murray (Lee and Greenwood 1994) and Cold Bokkeveld (Greenwood et al. 1994).

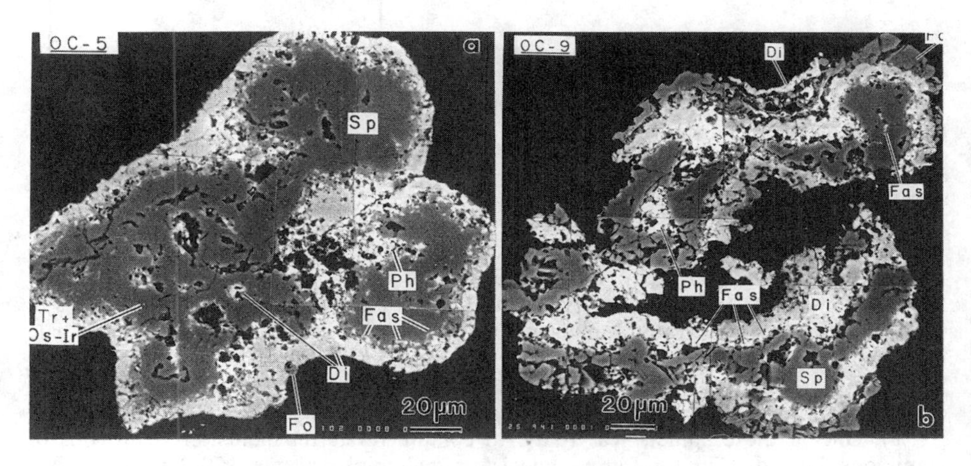

Figure 114. BSE images of (a) a nodular spinel-pyroxene inclusion from Murchison and (b) a banded spinel-pyroxene inclusion from Murchison. Sp = spinel, Ph = phyllosilicates, Fas = fassaite, Di = diopside, Fo = forsterite, Tr+Os–Ir = troilite enriched in Os and Ir. Used by permission of the editor of *Geochimica et Cosmochimica Acta* from MacPherson et al. (1983), Figures 8 and 9.

The nodular inclusions consist of rounded clumps of spinel that may be interconnected and are rimmed by pyroxene (Fig. 114a) which is zoned from fassaite to diopside at the outer edge (see below). A rim of Fe-rich phyllosilicate can also be present. Rare refractory metal-rich blebs and perovskite occur as inclusions within the spinel in Murchison whereas in Mighei and Murray, perovskite and hibonite can occur as accessory phases (MacPherson and Davis 1994, Lee and Greenwood 1994). In the banded inclusions, the spinel is arranged in elongate layers which are mantled by pyroxene (Fig. 114b). In Mighei, some extreme examples of this type of inclusion occur that consist of discrete pieces of spinel, rimmed by pyroxene, that are separated by up to 50 μm of matrix material.

In both types of inclusion, the spinel has an essentially identical composition to other occurrences in Murchison and Mighei (MacPherson et al. 1983, 1984b; MacPherson and Davis 1994). Spinels in Mighei show consistently good stoichiometry (MacPherson and Davis 1994) with well defined $R^{2+} + R^{4+} = 2\ R^{3+}$ substitutions (Fig. 115) and contain CaO and SiO_2 in low concentrations (<~0.08 wt %). Cr_2O_3 contents in the spinel are usually <0.6 wt %, but rare examples up to ~3 wt % Cr_2O_3 have been noted (MacPherson and Davis 1994). Spinel with high Cr_2O_3 contents also occurs in Cold Bokkeveld (Greenwood et al. 1994), but <1 wt % Cr_2O_3 is most typical. The FeO contents of spinels in spinel-pyroxene inclusions are typically >0.1 wt % (e.g. MacPherson and Davis 1994). Greenwood et al. (1994) found that 87% of the spinels they analyzed in Cold Bokkeveld contain <2 wt % FeO, although some spinels with up to 24.3 wt % FeO have been found in acid residues.

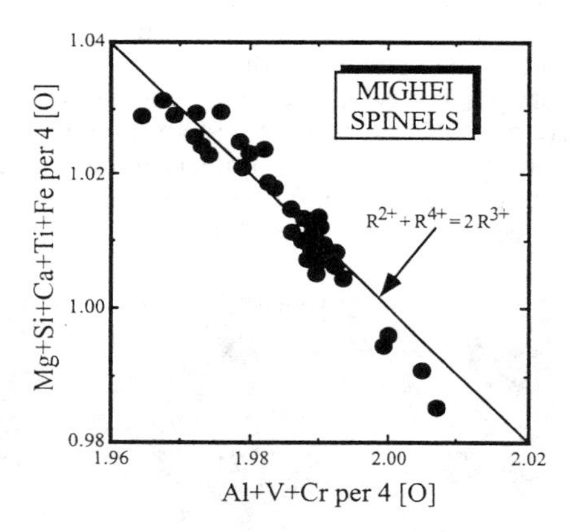

Figure 115. Spinel compositions in Mighei spinel-pyroxene inclusions showing that the spinels adhere to a well-defined $R^{2+} + R^{4+} = 2R^{3+}$ coupled substitution. After MacPherson and Davis (1994).

Macdougall (1979) observed single, euhedral crystals of pink *isolated spinel*, up to 400 μm in size, which are poor in FeO and Cr_2O_3, in separates from Murchison. MacPherson et al. (1983) also observed three distinct types of individual spinel grains in heavy mineral separates from Murchison. One type, consisting of just one grain, is a pale blue, perfect octahedron, 160 μm in size, which is pure $MgAl_2O_4$ with minor TiO_2 (~0.27 wt %), V_2O_3 (<0.23 wt %) and FeO (<0.04 wt %). The second type consists of dark red crystals, 130-200 μm in size, that are enriched in Cr_2O_3 (15-24 wt %) and FeO (6-7 wt %). One of these spinels was found to contain forsterite inclusions with rims of aluminous diopside. The third group consists of pale pink crystals, 150-250 μm in size, which have somewhat lower Cr_2O_3 (1.5-4 wt %) and FeO (<0.5 wt %) contents and may also contain inclusions of aluminous diopside. Some spinels may be zoned.

Simon et al. (1996) recognized an *ultrarefractory inclusion* from Murchison (HIB-

11) which is highly enriched in ultrarefractory elements and consists of spinel and Y-rich perovskite enclosed in Sc-fassaite. The spinel in this inclusion contains ~0.3 wt % Cr_2O_3 and 0.5 wt % V_2O_3. TiO_2 contents in the spinel are high (0.7 to 1.5 wt %) and may be due to contamination from adjacent perovskite.

A concentrically zoned, very fine-grained inclusion, similar to the fine-grained inclusions which occur in Allende, was reported by Macdougall (1979) with a core consisting dominantly of spinel with perovskite inclusions.

Fassaite. Fassaite is a rather rare phase in CAIs in CM chondrites. Macdougall (1981) reported an Al-Ti-rich pyroxene (presumably fassaite) as a minor phase in rims on some hibonite-bearing spherules from Murchison. Rare fassaite grains (26 wt % Al_2O_3; ~17 wt % TiO_2) also occur in a spinel-hibonite inclusion (MH10) from Murchison (Macdougall 1979). El Goresy et al. (1984) described fassaite with a narrow range of compositions (16.03-18.63 wt % TiO_2: 19.13-22.28 wt % Al_2O_3) in the core of a melilite-bearing inclusion in Essebi. This fassaite contains significant Sc_2O_3 (0.1-0.75 wt %) and V_2O_3 (0.31-0.56 wt %) and rare fassaites in layer E of the rim sequence around the inclusion have even higher concentrations (2.47-6.23 wt % Sc_2O_3; 0.38-0.99 wt % V_2O_3). In spinel-pyroxene inclusions in Mighei, fassaitic pyroxenes with TiO_2 contents >~2.5 wt % have cation totals significantly less than 4.0 cations per 6 oxygens (MacPherson and Davis 1994), due to the presence of significant Ti^{3+}. They have FeO contents <0.8 wt % and other minor elements (Cr_2O_3, MnO and V_2O_3) are all present at concentrations <~0.1 wt %. Fassaite with up to ~11 wt % TiO_2 and ~21 wt % Al_2O_3 is a common phase in the core of an unusual CAI found in Murray (Lee and Greenwood 1994). Lee and Greenwood (1994) reported a $Ti^{3+}/(Ti^{3+}+Ti^{4+})$ ratio of approximately 0.5 for these fassaites. Examples of fassaite which progressively zones into Al-diopside (see below) have been reported in a number of inclusions (e.g. MacPherson et al. 1983, Simon et al. 1994c, Greenwood et al. 1994).

Fassaitic pyroxene in *ultrarefractory inclusion* HIB-11 (Simon et al. 1996) is highly enriched in Sc_2O_3 (5.7-15 wt %) and is also Ti, Al-rich and Mg-poor compared with fassaite from Type B Allende CAIs. TiO_2 contents are 10.5-14.1 wt % and ZrO_2 is also relatively high (0.16-1.1 wt %). However, some fassaites have lower Sc_2O_3 and ZrO_2 and higher SiO_2 and MgO contents than is typical, a feature which is not correlated with any specific petrographic characteristic of the CAI. MgO and SiO_2 are positively correlated and both are negatively correlated with Sc_2O_3 (Fig. 116a). This trend is predicted by the fact that Sc_2O_3 is more compatible in fassaite than MgO during fractional crystallization from CaO-Al_2O_3-rich silicate melt (Simon et al. 1996). Sc_2O_3 is also negatively correlated with Y_2O_3 (Fig. 116b).

Diopside. Unlike CAIs in CV chondrites which contain very little diopside, CAIs in CM chondrites often contain diopside which is typically aluminous in character.

Rare Ca-rich pyroxenes occur in the Blue Angel inclusion from Murchison (Armstrong et al. 1982), associated with calcite or in the spinel-rich rim. Both occurrences are close to pure diopside with minor FeO (<1.3 wt %), Al_2O_3 (<3 wt %) and TiO_2 (<0.15 wt %), but pyroxene in the rim has higher FeO (~1.27 wt % cf. ~0.37 wt %) and Al_2O_3 (2.87 wt % cf. 0.81 wt %). Diopsides associated with calcite also contain detectable V, which is not apparent in the pyroxene in the rim. MacPherson et al. (1983) found that pyroxene in the rims of melilite-rich Murchison inclusions range in composition from Ti-poor, aluminous diopside to titaniferous fassaite. They are compositionally similar to pyroxenes in Allende Type B inclusions. Rare grains of Al-diopside (13.7 wt % Al_2O_3) with comparatively low TiO_2 contents (3.11 wt %) also occur with Sc-bearing fassaite in the rim of a melilite-bearing inclusion in Essebi (El Goresy et al. 1984).

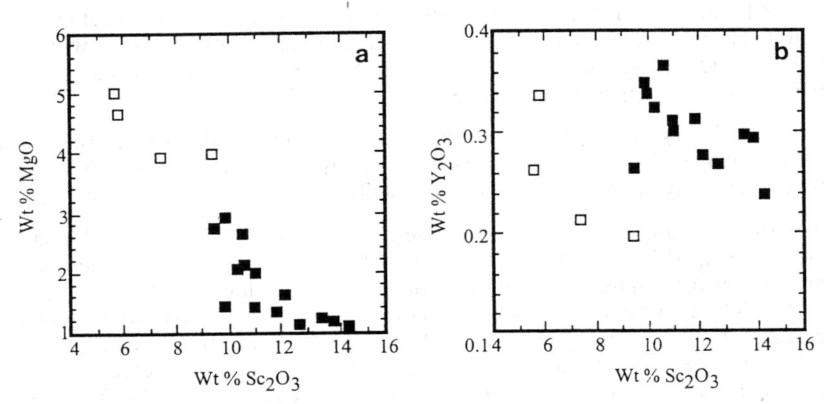

Figure 116. Oxide variation plots for fassaite from an ultrarefractory inclusion HIB 11 from Murchison, showing well-defined negative correlations between MgO and Sc_2O_3 (a) and Y_2O_3 and Sc_2O_3 (b). After Simon et al. (1996).

Macdougall (1979) reported very thin rims of diopside on essentially all intact *spinel-hibonite inclusions* in Murchison. The diopside is aluminous (~1.5 to 3.5 wt % Al_2O_3), but is low in FeO, TiO_2 and Cr_2O_3 (<1 wt %). However, such rims were not observed by MacPherson et al. (1984b), possibly due to their disruption during the disaggregation process used to isolate the inclusions.

Macdougall (1981) described thin (a few microns) rims of diopside on *spinel-hibonite-bearing spherules* from Murchison. Simon et al. (1994c) observed compositionally zoned pyroxene in a spinel-hibonite spherule in Murchison which ranges from pure diopside on the outer part of the spherule, through a low-Al, low-Ti diopside and finally to fassaite (~15 wt % Al_2O_3, 3-4 wt % TiO_2) toward the interior of the spherule.

Pyroxene also occurs in *spinel-pyroxene, olivine-spinel-pyroxene and olivine-pyroxene inclusions.* Pyroxene in pyroxene-bearing CAIs varies from nearly pure, Ti-free diopside to fassaite (MacPherson et al. 1983, 1984b; MacPherson and Davis 1994). Typically, the pyroxene closest to the spinel is fassaitic and becomes Al,Ti-poor with increasing distance from the spinel. Al_2O_3 can vary from ~2 wt % up to ~15 wt %. In one inclusion studied by MacPherson et al. (1984b) (OC-10), the Al_2O_3 and TiO_2 contents vary from 8.6-13.4 wt % and 0.4-2.5 wt % TiO_2, respectively. The diopside in these inclusions is typically Al-rich and Ti-poor, with low FeO contents (<0.16 wt %) (MacPherson et al. 1984b). Lee and Greenwood (1994) reported a similar variation in the Al_2O_3 contents (3 and ~13 wt %) of diopside in spinel-pyroxene inclusions from Murray (see also Greenwood et al. 1994).

In both Murchison and Mighei, pyroxene also occurs associated with forsteritic olivine in inclusions which lack spinel. In these inclusions, the pyroxene is compositionally much less variable and is diopsidic in composition with low Ti and Al contents (Greenwood et al. 1994).

In Murchison, a concentrically zoned, very fine-grained inclusion with an outer rim of diopside, was reported by Macdougall (1979) and MacPherson et al. (1983) reported an aluminous diopside rim on an isolated pink spinel crystal.

Minor phases. MacPherson et al. (1983) have reported the only occurrence of *anorthite* in a CM chondrite CAI. It occurs in rims on melilite-rich inclusions from Murchison and is pure endmember $CaAl_2Si_2O_8$ with only minor Na and no detectable K.

Olivine is a minor phase which occurs as small (<~5 μm) grains in spinel-pyroxene, olivine-pyroxene and spinel/olivine/pyroxene aggregates. In all occurrences in Murchison. Mighei, Murray and Cold Bokkeveld, the olivine is pure forsterite (MacPherson et al. 1983, 1984b; MacPherson and Davis 1994, Greenwood et al. 1994). Typical minor element concentrations are <0.18 wt % FeO, <0.3 wt % CaO and <0.1 wt % Al_2O_3.

Wark-Lovering rim sequences are rare on refractory inclusions in CM chondrites. However, MacPherson and Davis (1994) found essentially pure forsterite with a high CaO content (1.31 wt %) in one of the rim layers in a Wark-Lovering rim sequence on an inclusion fragment from Mighei. However, the high CaO content may be due to contamination from adjacent diopside.

Fe,Ni metal, refractory metal nuggets and *troilite* are all extremely rare in CM CAIs. In Murchison, MacPherson and Davis (1994) reported Fe,Ni metal inclusions inside olivine from spinel-pyroxene and olivine-pyroxene aggregates and Lee and Greenwood (1994) found metal grains in the diopside rim of an olivine-bearing, spinel-pyroxene inclusion in Murray. El Goresy et al. (1984) noted the presence of idiomorphic, refractory metal nuggets (up to 0.9 μm in size) enclosed in melilite or as grains interstitial to hibonite and spinel in the rim of an inclusion in Essebi. These grains are enriched in Pt-group refractory siderophile elements such as Ru, Re, Os, Ir and Pt (1-18 wt %) and also contain very high Mo contents (20-25 wt %). Troilite has been reported with pentlandite in the diopside rim of a spinel-pyroxene inclusion in Murray (Lee and Greenwood 1994).

Unidentified phases. MacPherson et al. (1984b) described tiny (<3 μm) grains of a *Ti,Zr,Y,Sc,Ca-rich phase* which occurs as inclusions in hibonite in a corundum-bearing inclusion (GR-1) from Murchison. Based on EDS analysis, the composition of this phase is ~29 wt % TiO_2, ~28 wt % ZrO_2, ~17 wt % Y_2O_3, ~13 wt % Sc_2O_3, and ~13 wt % CaO and is probably some type of oxide (see also CAI in CV chondrites). An unidentified *Ti-Al-bearing silicate* was found in an inclusion in Essebi (El Goresy et al. 1984), which contain several HREE which are above the detection levels for EPMA.

Secondary mineralogy of CAIs in CM chondrites

The CM chondrites have been widely affected by aqueous alteration which has resulted in the formation of a variety of secondary phases, especially hydrous minerals, in CAIs, chondrules and matrix (see Chondrule and Matrix sections). The location of alteration of CAIs in CM chondrites is the subject of some debate. MacPherson et al. (1983) argued that the alteration occurred before accretion into a parent body, but Greenwood et al. (1994) and Lee and Greenwood (1994) have provided evidence that it did, in fact, occur within an asteroidal environment (parent body alteration),

Phyllosilicates. Phyllosilicates are commonly present in CAIs in CM chondrites, but they have not been extensively studied. They typically occur interstitially to other phases in the CAIs, but in nodular spinel-pyroxene inclusions, a thin rim (<5 μm) of Fe-rich phyllosilicates often occurs between the layers of spinel and pyroxene (MacPherson and Davis 1994, Lee and Greenwood 1994, Greenwood et al. 1994). Outside the pyroxene rim a mantle of phyllosilicate is also present which is much poorer in Fe than the interior layer.

MacPherson et al. (1984b) reported EDS data for *phyllosilicates* from a spinel-hibonite spherule in Murchison. They are typically Fe-rich (30-44 wt % FeO) and Mg-poor (9-15 wt % MgO) and appear to have compositions consistent with Fe-rich *serpentines*. MacPherson and Davis (1994) also analyzed phyllosilicate phases in Mighei CAIs and found that they are compositionally variable in terms of their major elements,

Si, Mg and Fe. Like Murchison, most of the phyllosilicates have compositions consistent with ferroan Mg-serpentine, but tochilinite is also present in some CAIs. Some analyses suggest that a phase which is more aluminous than typical CM serpentines is also present.

Lee and Greenwood (1994) characterized the serpentines present in CAIs from Murray and Cold Bokkeveld, using TEM techniques. The phyllosilicate rims around spinel-pyroxene inclusions consist largely of micron to submicron serpentine crystals with a ~7-Å basal spacing. These serpentines are Fe-rich, but have variable Fe/Mg ratios between 4.8 and 8.6 and contain significant concentrations of Al_2O_3 (~8 wt %). Fe-rich and Al-bearing serpentines are also present intergrown with berthierine in an unusual CAI in Murray. In Cold Bokkeveld, compositionally variable serpentines occur in a variety of CAIs (Greenwood et al. 1994) (Fig. 117). In particular, Fe-rich serpentine, sometimes close to endmember cronstedtite in composition, occurs in a variety of spinel-bearing CAIs (e.g. spinel spherules, spinel and spinel-pyroxene inclusions). This phase occurs as laths (0.6-1 μm × 0.1-0.2 μm), and electron diffraction patterns frequently display streaking in diffraction rows with k ≠ 3, parallel to the c* direction, indicative of extensive stacking disorder. These serpentines contain minor MgO (~9 wt %), Al_2O_3 (3 wt %) and, occasionally, Ca, Ni and S.

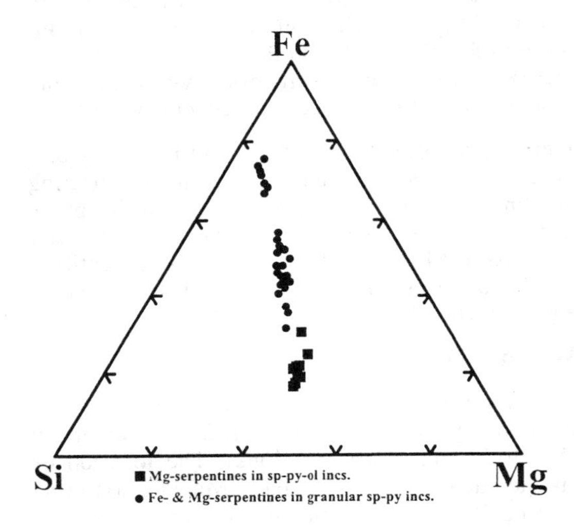

Figure 117. Compositions of phyllosilicate alteration products in CAIs from the Mighei CM chondrite, plotted in a ternary Si-Fe-Mg diagram. After Greenwood et al. (1994).

In spinel-pyroxene inclusions in Cold Bokkeveld, relatively coarse-grained, Fe-rich serpentine laths (700-800 nm × 75-100 nm) occur within a matrix of porous, poorly crystalline, Mg-rich serpentine. The Mg-rich serpentines are short, wavy crystals, a few tens of Ångströms in width, intergrown with crystallites with a circular morphology, similar to that of cross-sections of chrysotile (see CM matrix). Very Mg-rich serpentines with similar structural and morphological characteristics also occur surrounding high temperature phases in spinel-pyroxene-olivine inclusions in the same meteorite. These serpentines are often associated with serpentine fibers (20-150 Å in width), which may be straight, sinuous or wavy in character or have circular, chrysotile-like cross sections. Analytical data (EDS) for these Mg serpentines show they contain minor SO_3 (~3 wt %) and NiO (1.5 wt %), possibly due to the presence of localized intergrowths of tochilinite.

Other phases. Using TEM, Lee and Greenwood (1994) identified rare *chlorite* (14-Å basal spacing) and an Mg,Al-rich phyllosilicate phase with a 7-Å basal spacing in an unusual CAI in Murray. They interpreted the 7-Å phase as *berthierine*, which sometimes occurs intergrown with chlorite or, in most cases, with an Fe-rich serpentine and a Mg-rich phase such as Al-lizardite or amesite. Also in Murray, Lee and Greenwood (1994) found a phase high in Si, Al and Na in a CAI, using analytical TEM. Based solely on the chemical analysis they suggested that this phase was an Na-rich mica, probably *paragonite*.

Tochilinite or PCP (see CM matrix) appears to be a relatively common phase in altered CAIs in CM chondrites, but few details of its occurrence and composition are available. MacPherson and Davis (1994) found tochilinite in a number of inclusions in Mighei, but it is not clear what phase it has replaced. Lee and Greenwood (1994) also found tochilinite in a phyllosilicate rim around a spinel-pyroxene inclusion in Murray, which contains up to 37.0 wt % SO_3 and 4.9 wt % NiO. Greenwood et al. (1994) also reported TEM observations which indicate the presence of minor tochilinite intergrown with coarse and fine-grained serpentines in CAIs in Cold Bokkeveld.

Unlike inclusions in Allende, *feldspathoid* minerals are very rare in CAIs in CM chondrites. MacPherson et al. (1983) reported a feldspathoid mineral in a hibonite-bearing CAI (SH-4) in Murchison. Although the analyses contain high sodium, no Cl or S, which would be indicative of a sodalite group mineral were detected. El Goresy et al. (1984) reported that melilite in an Essebi CAI has been partially replaced by nepheline and sodalite.

Calcite is a common secondary phase in CAIs in CM chondrites. For example, calcite is a major secondary phase in the Blue Angel inclusion (Armstrong et al. 1982) and is essentially pure $CaCO_3$ with minor amounts of P and S. Many of the inclusions studied by MacPherson et al. (1983) in Murchison and MacPherson and Davis (1994) in Mighei also contain calcite, which also appears to be pure endmember $CaCO_3$, possibly intergrown on a microscopic scale with other minerals. In one inclusion in Mighei, calcite surrounds spinels which have a ragged appearance indicating that they may have been altered or replaced. Whether the replacement is directly by calcite or an intermediate phase which is subsequently replaced by calcite is not clear. Lee and Greenwood (1994) and Greenwood et al. (1994) also found calcite in CAIs in Cold Bokkeveld and El Goresy et al. (1984) described melilite in an inclusion in Essebi which shows evidence of minor replacement by calcite.

Gypsum has been reported as an alteration product of melilite in the mantle region of several inclusions in Murchison (Armstrong et al. 1982, MacPherson et al. 1983) and Essebi (El Goresy et al. 1984). Greenwood et al. (1994) reported rare calcium sulfate intergrown with calcite in a CAI from Cold Bokkeveld, which was identified by TEM techniques as an intergrowth of *hemihydrate* ($CaSO_4{\cdot}0.5H_2O$) and *anhydrite* ($CaSO_4$). A compositionally unusual amorphous phase which may be related to magnesian aluminocopiate ($(Mg,Al)(Fe,Al)_4(SO_4)_6(OH)_2{\cdot}20H_2O$) has been described by Lee and Greenwood (1994) in a CAI in Murray.

Whewellite ($Ca(C_2O_4){\cdot}H_2O$) was found by x-ray diffraction between olivine grains in a white inclusion in Murchison (Fuchs et al. 1973). This inclusion may be a CAI or an amoeboid olivine inclusion.

Secondary sulfides appear to very rare in CM chondrite inclusions. However, Lee and Greenwood (1994) and Greenwood et al. (1994) have reported *pentlandite* in single CAIs from Murray and Cold Bokkeveld.

Primary mineralogy of CAIs in CO carbonaceous chondrites

CAIs in CO chondrites have received much less attention than those in CV or CM chondrites, but brief general descriptions have been provided by several workers (e.g. Lancé, Kurat 1970; Isna, Methot et al. 1975; ALHA 77003, Ikeda 1982) and some ultrarefractory inclusions have also been studied (e.g. Fahey et al. 1994). The most detailed studies are those of Tomeoka et al. (1992) and Kojima et al. (1995) who studied CAIs in several Antarctic CO chondrites.

Most CAIs in CO chondrites range in size from 50 to 300 μm (Ikeda 1982, Tomeoka et al. 1992, Kojima et al. 1995), but inclusions up to 680 μm occur (e.g. Kojima et al. 1995). Tomeoka et al. (1992) and Kojima et al. (1995) found that most inclusions in Antarctic CO chondrites (Y-791717, Y-81020, Y-82050, Y-790992) are fine-grained, 'sinuous'-rimmed, complex CAIs (fine-grained, spinel-rich inclusions) comparable to those described by Kornacki et al. (1983) in Mokoia (CV) and similar to spinel-pyroxene inclusions in CM chondrites (MacPherson et al. 1983, 1984b; MacPherson and Davis 1994). Other rarer, unusual types of CAIs have also been observed by Tomeoka et al. (1992) and Kojima et al. (1995), but will not be discussed in detail here.

Like the ordinary chondrites, the CO chondrites display a well defined petrologic sequence which covers the range from petrologic 3.0 to 3.7 (Keck and Sears 1987, Scott and Jones 1990) and is characterized by the progressive equilibration of mineral compositions, probably due to parent body metamorphism (e.g. McSween 1977a, Scott and Jones 1990). This is reflected in an increase in the FeO content of chondrule olivines, and in CAIs by an increase in the FeO contents of spinel. Consequently, CAI in more metamorphosed CO chondrites contain spinel which is more hercynitic in composition than that commonly observed in other chondrite groups. Kojima et al. (1995) also observed an increase in the abundance of nepheline in CAIs as a function of petrologic type. The nepheline appears to be replacing melilite and anorthite, a reaction which may be the result of metamorphism.

Corundum. Kurat (1970) reported the first occurrence of corundum in a meteorite in a spinel-rich CAI from Lancé. The corundum occurs as red fluorescing crystals in the core of the inclusion and contains only minor SiO_2 (0.1 wt %) and TiO_2 (0.7 wt %).

Hibonite. Hibonite is a rare phase in CAIs in CO chondrites. Tomeoka et al. (1992) found hibonite in only 6 out of 54 CAIs examined in Y-791717 and Kojima et al. (1995) found one hibonite-bearing inclusion in Y-81020. In Y-791717, hibonite forms euhedral or platy grains, <50 μm in length, sometimes embedded within fassaite. These hibonites are compositionally variable, but contain <5.5 wt % TiO_2, <3.3 wt % MgO and are relatively enriched in FeO (<1.8 wt %) in comparison with hibonite from CAIs in CV and CM carbonaceous chondrites. In Y-81020, the hibonite occurs in a spherule, coexisting with fassaite.

Hibonite with low MgO (<1.0 wt %) and TiO_2 (<2.0 wt %) contents occurs in the core of an ultrarefractory CAI (HH-1) from Lancé (Fahey et al. 1994) and appears to have undergone partial pseudomorphic replacement by hercynite. The hibonite has ultrarefractory trace element abundance patterns with the HREE Gd and Tb enriched relative to the LREE (40 × CI, cf. 20 × CI). The HREE heavier than Tb show a progressive roll off with increasing mass, a feature that is also present in some platy hibonite crystals (PLACS) in Murchison (CM) (Hinton et al. 1988, Ireland et al. 1988).

Hibonite is also present in a very unusual CAI in Lancé (Kurat 1975). Two euhedral hibonite laths (20 × 10 μm) occur within a glass spherule (50 μm in diameter), similar to hibonite-bearing microspherules in Murchison (Ireland et al. 1991). The hibonite has low TiO_2 (1.5 wt %) and MgO (1.4 wt %) contents and has LREE-enriched (60-80 × CI) abundance patterns with a negative Eu anomaly (Fig. 118). The HREE decrease sharply in abundance to Er and Lu which are below detection limits (<1 × CI) (Ireland et al.

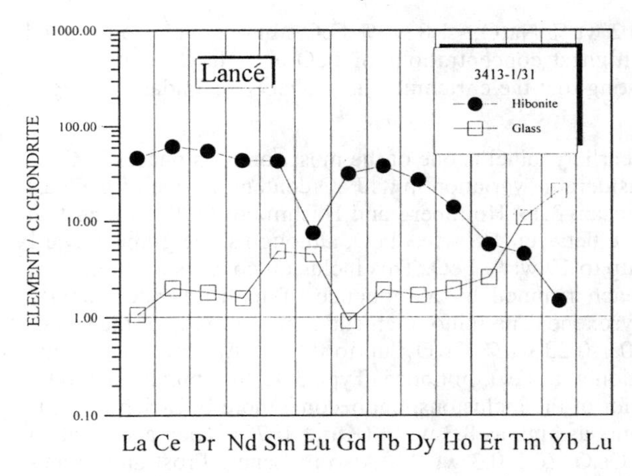

Figure 118. REE pattern for hibonite and coexisting glass (normalized to CI chondrites) in a glass spherule from Lancé. Ion microprobe analyses. Data from Ireland et al. (1991).

(1991). The coexisting glass has an essentially flat REE abundance pattern (20-30 × CI) with a negative Eu anomaly.

Grossite. Greenwood et al. (1992) reported two CAIs (out of a total of 81) from Colony which contain grossite. In both CAIs, grossite (<50 μm) occurs in the cores, rimmed by diopside (Greenwood, pers. comm. in Weber and Bischoff 1994b). Grossite with perovskite inclusions also occurs in the core of two CAIs in Y-81020 (Kojima et al. 1995).

Perovskite. Perovskite is a relatively common phase in a variety of CAIs in CO chondrites (e.g. Kurat 1970, Tomeoka et al. 1992, Fahey et al. 1994, Kojima et al. 1995). It frequently occurs as small inclusions (<10 μm) in spinel (e.g. Kurat 1970, Tomeoka et al. 1992, Kojima et al. 1995) and the proportion of perovskite can be as high as 10 vol % in some CAIs in Lancé (Kurat 1970). Perovskite from one inclusion in Lancé contains 1.5 wt % Al_2O_3 as well as significant MgO (1.5 wt %), ZrO_2 (2.2 wt %) and Y_2O_3 (1.1 wt %) and minor FeO (0.2 wt %). Tomeoka et al. (1992) found that perovskites in CAIs in Y-791717 are typically too small to analyze, but one analysis shows ~0.63 wt % Al_2O_3 and 0.37 wt % FeO. Frost and Symes (1970) also described a spherical CAI in Lancé, which they termed a chondrule, that contains a core of perovskite (1.52 wt % Al_2O_3; 0.87 wt % MgO) and spinel. Perovskite containing minor Al_2O_3 (~0.5 wt %) also occurs as inclusions within hibonite and hercynite in an ultrarefractory Lancé CAI (HH-1) (Fahey et al. 1994). Fahey et al. (1994) also reported ion microprobe trace element data for perovskite from this inclusion but the analyses usually consist of more than one phase due to overlap onto adjacent grains.

Melilite. Melilite is a relatively common phase in CAIs in CO chondrites (e.g. Kurat 1970, Methot et al. 1975, Holmberg and Hashimoto 1992, Tomeoka et al. 1992, Kojima et al. 1995). In Lancé Kurat (1970) found that melilite is compositionally variable and documented melilite from two CAIs with average compositions $Åk_{70}$ and $Åk_{10}$, respectively. The Na_2O content in the Åk-rich melilite is higher (0.2 wt % cf. <0.05), but K_2O in both examples is <0.05 wt %. Melilite ($Åk_{10-30}$) also occurs as thin (<1-5 μm) rims on spinel grains in a fine-grained inclusion from Kainsaz (Holmberg and Hashimoto 1992). The melilite contains minor Na_2O (0.1-0.2 wt %), but no detectable K_2O. Tomeoka et al. (1992) and Kojima et al. (1995) also found minor melilite in a variety of CAI types in several different Antarctic CO chondrites. The melilite in these CAIs has been partially to extensively replaced by nepheline and is gehlenitic in composition ($Åk_{1-}$

Na_2O and FeO contents (<0.1 to 2 wt % Na_2O; <1.8 wt % FeO). In one inclusion studied by Tomeoka et al. (1992), the highest concentrations of FeO and Na_2O occur on the margins of the inclusion, suggesting that the enrichment is probably secondary in origin (see also Kojima et al. 1995).

Spinel-hercynite. As noted earlier, spinel is one of the most common phases in CAIs in CO chondrites and shows considerable variation in its FeO content. In one of the least equilibrated CO chondrites (Kainsaz, 3.1), Holmberg and Hashimoto (1992) found that spinels in a fine-grained CAI have between 1-8 wt % FeO, although some grains toward the rim of the inclusion contain up to 19 wt % FeO. This inclusion consists of numerous, loosely packed spinel grains each rimmed by a sequence of melilite, occasionally anorthite and finally diopsidic pyroxene. The minor element concentrations of the spinel are low (0.08-0.49 wt % TiO_2; 0.1-0.23 wt % Cr_2O_3; up to 0.69 wt % ZnO), with ZnO showing a weak positive correlation with FeO. In Lancé (Type 3.4), the spinels tend to be FeO-poor (0.1 wt %) in the interior of the inclusions, but become more FeO-rich towards the exterior, reaching FeO contents as high as 8.3 wt % (Kurat 1970). They also contain minor TiO_2 (0.1-0.3 wt %) and Cr_2O_3 (0.2-0.3 wt %). Also in Lancé, Frost and Symes (1970) described a spherical CAI, which they termed a chondrule, that contains a core of spinel (1.35 wt % FeO) with perovskite. Spinel in fine to medium-grained inclusions in ALH 77003 (Type 3.5) (Ikeda 1982) are quite Fe-rich with MgO/(MgO+FeO) ratios of ~0.5. In Isna (Type 3.7), Methot et al. (1975) found that the spinels in all the inclusions that they studied showed little compositional variation and are Fe-rich (~22 wt % FeO; 13 wt % MgO), indicating that they have essentially equilibrated during metamorphism. Kojima et al. (1995) also found that the FeO content of spinel in CAIs from Y-81020, Y-790992 and Y-82050 increases as a function of petrologic type from <1 mol % hercynite in Y-81020 to 42-60 mol % hercynite in Y-82050.

Tomeoka et al. (1992) found that spinel in many CAIs in Y-791717 has been replaced by nepheline. The spinel shows a wide range of FeO contents (0.16 to 23.3 wt %) with the highest FeO contents occurring in CAIs in which spinel is most heavily replaced. Two of the inclusions have spinels with a very wide range of Fe/(Fe+Mg) ratios (0-0.576 and 0-0.42, respectively) and some spinels clearly show compositional zoning with FeO-rich rims. Tomeoka et al. (1992) also reported minor TiO_2 (<1.0 wt %), Cr_2O_3 (0.5 wt %) and V_2O_3 (<0.9 wt %).

Hercynite occurs mantling and apparently replacing hibonite that occurs in the core of an ultrarefractory CAI HH-1 from Lancé (Fahey et al. 1994). The hercynite is very Fe-rich with Mg/(Mg+Fe) ratios of 0.31-0.34. Minor element contents are typically low (<0.4 wt % TiO_2; <~0.03 wt % Cr_2O_3; 0.8-1.0 wt % ZnO).

Fassaite. Fassaite with 1-14 wt % TiO_2 and very high Al_2O_3 contents (8-33 wt %) occurs in CAIs in Y-791717, Y-81020, Y-790992, and Y-82050 (Tomeoka et al. 1992, Kojima et al. 1995). This fassaite is also unusual in having much higher FeO contents (<1-7 wt %) than fassaite in CAIs in other chondrite groups.

Diopside. Diopside, with a low FeO content, is the main pyroxene found in CAIs in CO chondrites (e.g. Kurat 1970, Methot et al. 1975, Ikeda 1982, Holmberg and Hashimoto 1992, Tomeoka et al. 1992). In melilite-spinel-bearing inclusions in Lancé, Kurat (1970) found endmember diopside with low FeO (~0.3 wt %) and TiO_2 (<0.1 wt %) contents but somewhat higher Al_2O_3 (~1.1 wt %). In the same meteorite, Fahey et al. (1994) also observed almost pure diopside (0.15 wt % Al_2O_3; <0.02 TiO_2; <0.02 Cr_2O_3) as a thin rim (3-10 μm) around an ultrarefractory CAI (HH-1). Other workers have found more aluminous diopsides in CO inclusions. Ikeda (1982) reported diopsides in CAIs in ALH 77003 with higher Al_2O_3 (14-15 wt %) and TiO_2 (0.7-2.7 wt %) contents.

Diopsides rimming these inclusions have similar compositions, but tend to be lower in Al_2O_3 (1.45-11.91) and slightly higher in FeO (0.39-0.86 wt %, cf. 0.27-0.35 wt %). Diopsides analyzed in CAIs in other Antarctic CO chondrites have a similar range of Al_2O_3 contents (1-9 wt %) and relatively low TiO_2 (<1-3 wt %) (Tomeoka et al. 1992, Kojima et al. 1995). Holmberg and Hashimoto (1992) also found very thin (<1 μm) rims of aluminous diopside (4-8 wt % Al_2O_3; 1-2 wt % TiO_2) on spinel grains in a fine-grained inclusion from Kainsaz.

Accessory phases. Very fine-grained *anorthite* (<0.1 wt % Na_2O) sometimes occurs in rims on spinel grains in a fine-grained CAI from Kainsaz (Holmberg and Hashimoto 1992). Kojima et al. (1995) also found anorthite in the rims around spinel grains in nodular and banded CAIs in Y-81020 and anorthite is also a major component in a small number of CAIs in Y-81020, Y-82050 and Y-790992 which consist of anorthite, fassaite and diopside. In CAIs in Y-790992, the anorthite has been extensively replaced by melilite.

Glass occurs coexisting with hibonite in a microspherule from Lancé (Kurat 1975, Ireland et al. 1991). The glass is highly calcic (25.5 wt % CaO) and aluminous (29.2 wt % Al_2O_3), but is low in SiO_2 (36.7 wt %) and MgO (6.8 wt %). The glass has an essentially flat REE abundance pattern (20-30 × CI) with a negative Eu anomaly (Fig. 118b) (Ireland et al. 1991).

Sulfides are both rare and poorly characterized in CAIs in CO chondrites. In Lancé, Kurat (1970) reported minor troilite in spinel-melilite-diopside-bearing CAIs and unidentified sulfides are also present in the ultrarefractory CAI HH-1 (Fahey et al. 1994). Holmberg and Hashimoto (1992) also noted rare Fe sulfides in a fine-grained inclusion in Kainsaz.

Kurat (1970) reported minor *Fe,Ni metal* in spinel-melilite-diopside-bearing CAIs in Lancé and rare metal occurs in a fine-grained inclusion in Kainsaz (Holmberg and Hashimoto 1992).

Kojima et al. (1995) found minor *ulvöspinel* associated with perovskite in a CAI from Y-790992.

Mineralogy of fremdlinge in CAIs in CO chondrites. Fremdlinge appear to be extremely rare in CO chondrites and only one example has been reported in the literature (Palme et al. 1982) from a CAI in Ornans. The fremdling was separated for analysis by INAA and studied by SEM. The fremdling has features similar to the Type 2 fremdlinge of El Goresy et al. (1978) (see section on fremdlinge in CV chondrites) and consists of an agglomerate of hundreds of submicron metal grains. Both taenite and Os,Ru,Re-rich particles are present. Normalized electron microprobe analyses show that the taenite contains ~43 wt % Ni and ~11.2 wt % Ir, as well as substantial Os, Ru and Re. The Os,Ru,Re-rich phase contains ~18 wt % Fe, 18 wt % Ni, 35-41 wt % Os and lower concentrations of Ir, Ru, and Re (<10 wt %).

Secondary mineralogy of CAIs in CO carbonaceous chondrites

Nepheline. Ikeda (1982) reported nepheline, low in K_2O and CaO (<0.42 wt %) in one medium-grained CAI in ALH 77003 and minor nepheline is also present in voids in the ultrarefractory CAI HH-1 from Lancé (Fahey et al. 1994). Holmberg and Hashimoto (1992) also tentatively identified minor nepheline as an interstitial phase in a fine-grained CAI from Kainsaz. In contrast, Tomeoka et al. (1992) and Kojima et al. (1995) found abundant nepheline as a replacement product of melilite and sometimes spinel in CAIs in several Antarctic CO chondrites (Fig. 119). The nepheline typically occurs as porous

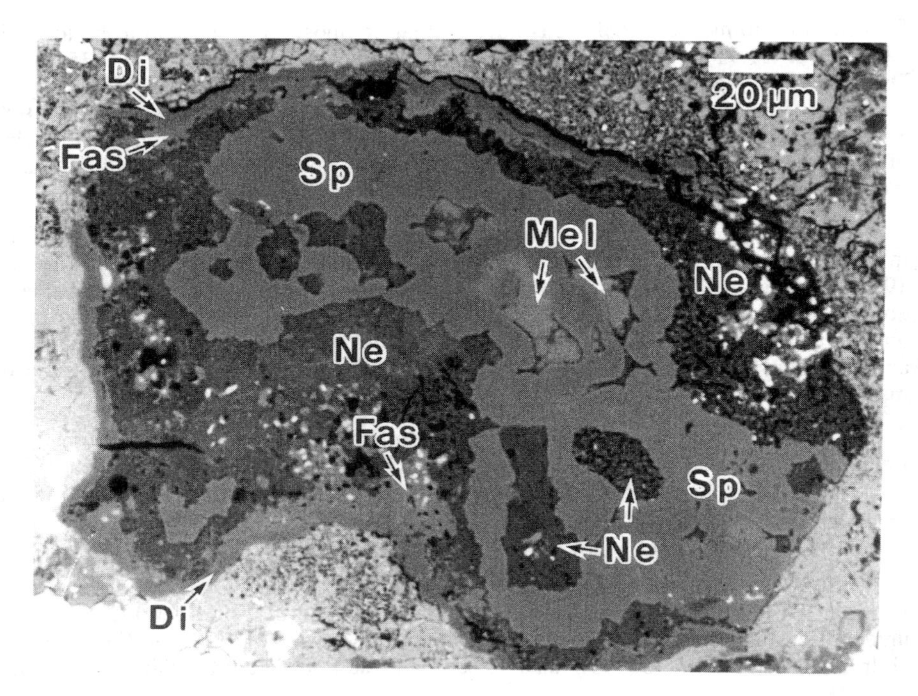

Figure 119. BSE image of a CAI from Yamato 791717. The spinel (Sp) core has been extensively replaced by nepheline (Ne), but a rim of unaltered diopside (Di) is still present, sometimes associated with unaltered fassaite (Fass). Melilite (Mel) grains are locally present in the interior. Used by permission of the editor of *Meteoritics* from Tomeoka et al. (1992), Figure 1a.

aggregates sometimes containing troilite, but in a few CAIs in Y-791717, occurs as large, homogeneous grains. These grains are essential pure nepheline with minor CaO (<1.32 wt %) and sometimes K_2O (typically <0.03-0.65, but sometimes up to 2 wt %). Nepheline in Y-82050 and Y-790992 CAIs contains 2-3 wt % CaO and <0.12 wt % K_2O.

Minor secondary phases. *Sodalite* may be present intergrown with nepheline in some altered inclusions in Y-791717 (Tomeoka et al. 1992), as indicated by the presence of Cl in some analyses, but has not been positively identified. Minor sodalite is also present with nepheline within voids in the ultrarefractory CAI HH-1 from Lancé (Fahey et al. 1994). Tomeoka et al. (1992) and Kojima et al. (1995) found *ilmenite* grains commonly occurring as small (<10 μm) inclusions in spinel in CAIs in Y-791717, Y-82050, and Y-790992. The ilmenite appears to have formed as a replacement product of perovskite and sometimes rims perovskite grains. Electron microprobe analyses of one grain show that the ilmenite is almost pure with minor MgO (1.78 wt %), Al_2O_3 (0.27 wt %) and MnO (0.47 wt %). Ilmenite with low MgO (<1.0) and TiO_2 (<2.0 wt %) is also present within a mantle of hercynite in the ultrarefractory CAI HH-1 from Lancé (Fahey et al. 1994). Other minor occurrences of secondary phases in CAIs in CO chondrites are Fe-rich *monticellite* ($Fe_{0.7}Mg_{0.3}CaSiO_4$) in Colony and Y-82050 (Greenwood et al. 1992, Kojima et al. 1995), and *troilite*, which may be secondary in origin, associated with nepheline in Y-82050 and Y-790992 (Kojima et al. 1995).

Primary mineralogy of CAIs in CR carbonaceous chondrites

CAIs are relatively common in CR chondrites (0.1-2.6 vol %) (e.g. McSween 1977c, Weisberg et al. 1993, Weber and Bischoff 1994b, Birjukov and Ulyanov 1996), but have not been studied in as much detail as CAIs in CV or CM chondrites. Weber and Bischoff (1994b) noted that CAIs in the CR chondrite Acfer 059-El Djouf 001 were typically small (<300 μm), but could occasionally reach up to 1 mm in apparent diameter. Most inclusions in this chondrite are fine-grained and irregularly shaped, commonly contain abundant melilite, spinel and Ca-pyroxene, with hibonite and perovskite as accessory phases in many inclusions. Weisberg et al. (1993) found "Compact" Type A and "Fluffy" Type A inclusions (see CV chondrites) as well as spinel-pyroxene aggregates (see CM chondrites) in the eight CR chondrites that they studied. The spinel-pyroxene aggregates (MacPherson et al. 1984b, MacPherson and Davis 1994), consist of several nodules, 1-200 μm in size, with spinel cores with minor perovskite, rimmed by a layer of diopsidic or fassaitic pyroxene.

Hibonite. Weber and Bischoff (1994b) found blue-green hibonite as a common accessory phase in CAIs in Acfer 059-El Djouf 001. In grossite-bearing inclusion 008/D, two morphological types of hibonite are present. Prismatic hibonite crystals have low MgO (0.55 wt %) and TiO_2 (0.52 wt %), compared with needle-like crystals with 4.6 wt % MgO and 6.3 wt % TiO_2. Both types of hibonite have low FeO contents (<0.40 wt %).

Grossite. Grossite in three inclusions in Acfer 059-El Djouf 001 contains minor SiO_2 and FeO (<0.4 wt %), with one exception with 0.76 wt % FeO that occurs in a CAI with FeO-bearing spinel and fassaite (Weber and Bischoff 1994b). Based on ion microprobe analyses the grossite in one inclusion (008/D) (Weber et al. 1995) has volatility-fractionated Group II abundance patterns (Mason and Taylor 1982) with enriched LREE (La to Sm) and Tm abundances (~110 × CI) relative to HREE (Gd-Lu) and Eu (~10 to 70 × CI) (Fig. 120).

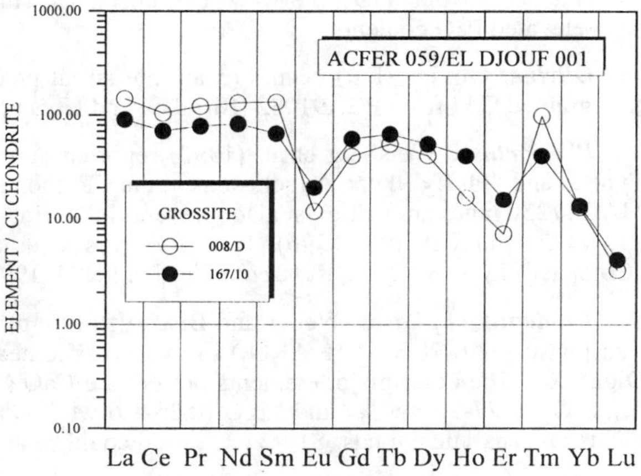

Figure 120. REE abundance pattern for grossite (normalized to CI chondrites) from two inclusions from CR chondrites. Ion microprobe analyses. Data from Weber et al. (1995).

Melilite. Weisberg et al. (1993) reported one melilite-rich, "Compact" Type A inclusion in both Renazzo and El Djouf 001. The Renazzo inclusion contains 88 vol % coarse-grained melilite of composition $Åk_{\sim 25}$. Melilite-rich, "Fluffy" Type A inclusions, up to 300 μm in size, with a nodular structure were also found in MAC87230 (1) and El Djouf 001 (5). These inclusions have a core of melilite with inclusions of spinel, surrounded by anorthite or spinel + perovskite and rimmed by diopside. The inclusions are surrounded by a Wark-Lovering rim sequence. Melilite (almost pure gehlenite) also occurs in nodular spinel-pyroxene aggregates in Renazzo and MAC87320.

Weber and Bischoff (1994b) found melilite in grossite-bearing inclusions in Acfer 059-El Djouf 001. The melilite is invariably gehlenitic (< $Åk_8$) and low in Na_2O and K_2O like melilite in similar inclusions from CH chondrites. Melilite ($Åk_{21}$) is also the dominant phase in a fine-grained CAI from PCA91082 (Birjukov and Ulyanov 1996).

Spinel. Spinel occurs in Compact and "Fluffy" Type A inclusions, as well as spinel-pyroxene aggregates in CR chondrites (Weisberg et al. 1993). Spinel in spinel-pyroxene aggregates contains some FeO (up to 0.44 wt %) and V_2O_3 (up to 0.65 wt %). Spinel is also present in grossite-bearing inclusions in Acfer 059-El Djouf 001 (Weber and Bischoff 1994b). In one inclusion, spinel contains <0.4 wt % SiO_2, TiO_2, Cr_2O_3, FeO and MgO. However, in another inclusion, the spinel has an FeO content of 3.0 wt %, correlating with a high FeO content in the coexisting grossite.

Perovskite. Weisberg et al. (1993) reported minor perovskite in Compact and "Fluffy" Type A inclusions, as well as spinel-pyroxene aggregates in CR chondrites. Weber and Bischoff (1994b) also found perovskite as an accessory phase in many CAIs in Acfer 059-El Djouf 001. Perovskite also occurs within the Wark-Lovering rim sequence on a "Compact" Type A inclusion in Renazzo (Weisberg et al. 1993).

Pyroxene. Diopside occurs in "Fluffy" Type A inclusions in CR chondrites and diopside and fassaite are also present rimming cores of spinel in spinel-pyroxene aggregates (Weisberg et al. 1993). Weisberg et al. (1993) also found fassaite within a Wark-Lovering rim sequence on a Renazzo "Compact" Type A inclusion. In grossite-bearing inclusions in Acfer 059-El Djouf 001 (Weber and Bischoff 1994b), pyroxene commonly occurs as layers or rims toward the periphery of the inclusions and ranges from Al-bearing diopside to fassaite in composition. The Al_2O_3 contents are highly variable (up to 12.9 wt %), but TiO_2 and FeO contents are usually low (<0.5 wt %). However, Weber and Bischoff (1994b) also found one inclusion (023/2) in which diopside (2.85 wt % TiO_2; 5.6 wt % FeO) coexists with spinel and grossite, which also have elevated FeO contents.

Olivine. Olivine (Fa_3) occurs as an interstitial phase to melilite and anorthite in a fine-grained CAI from PCA91082 (Birjukov and Ulyanov 1996).

Plagioclase. Weisberg et al. (1993) reported anorthitic plagioclase in "Compact" Type A and "Fluffy" Type A inclusions in the CR chondrites, Renazzo, El Djouf 001 and MAC87230 and anorthite is also present in a fine-grained CAI from PCA91082 (Birjukov and Ulyanov 1996). It also occurs within the Wark-Lovering rim on a "Compact" Type A CAI in Renazzo (Weisberg et al. 1993).

Unidentified phases. Weber and Bischoff (1994b) documented an unidentified Al-rich phase (58.6-78.3 wt % Al_2O_3) in two grossite-bearing inclusions in Acfer 059-El Djouf 001. The other major elements present are CaO (1.6-10.4 wt %) FeO (0.51-8.7 wt %), SiO_2 (0.77-9.8 wt %) and MgO (0.25-1.6 wt %). The phase may be hydrated as it yields low analytical totals (81-92 wt %) and could be an OH-bearing Al-oxide.

Mineralogy of CAIs in CH carbonaceous chondrites

Although CAIs only constitute ~1 vol. % of the components in CH chondrites (e.g. ALH 85085; Grossman et al. 1988a), they are abundant, but are extremely small. In ALH 85085, the inclusions are <107 μm in size (Grossman et al. 1988a), although different size ranges have been reported in different studies (25-75 μm, Weisberg et al. 1988, 5-70 μm, Kimura et al. 1993). Grossman et al. (1988a) found 50 CAIs in one thin section of ALH 85085 and estimated that another 50 >20 μm in size could also be present. In comparison, in Acfer 182, Weber and Bischoff (1994b) reported typical inclusion sizes of

<200 μm, with one large inclusion having an apparent diameter of ~450 μm.

Numerous inclusions from ALH 85085 and Acfer 182 have been studied and show that, in addition to being smaller, they differ in a number of important respects from CAIs in CV and CM chondrites. Most notably, grossite, a very rare phase in CV and CM inclusions, is common in CH chondrite CAIs and in many inclusions is often one of the major constituents (<30 vol %) (Kimura et al. 1993, Weber and Bischoff 1994b, Weber et al. 1995) (see below). In addition, all the primary phases have >~0.3 wt % FeO, higher than is normal in CAIs in CM and CV chondrites. CAIs in CH chondrites also contain no evidence of secondary alteration and phases such as feldspathoids and grossular are absent (see, however Grossman et al. 1988). These observations suggest that, in general, CAIs in CH chondrites have probably retained their original, primitive chemical and mineralogical characteristics.

In addition to grossite, CAIs in CH chondrites are dominated by melilite with spinel and perovskite as accessories in almost all inclusions (Grossman et al. 1988, Kimura et al. 1993). Some inclusions in both ALH 85085 and Acfer 182 also contain hibonite. Grossman et al. (1988a) classified inclusions in ALH 85085 into two groups: (a) melilite-rich inclusions, which have some affinities to Type A inclusions (Fig. 121), (b) spinel-hibonite-rich (SH) inclusions, and (c) spherules, which sometimes contain grossite. Each group of inclusions has igneous textures indicating crystallization from a liquid. Grossman et al. (1988a) also observed that almost all the inclusions in ALH 85085 have a rim consisting of aluminous diopside ± perovskite ± gehlenitic melilite.

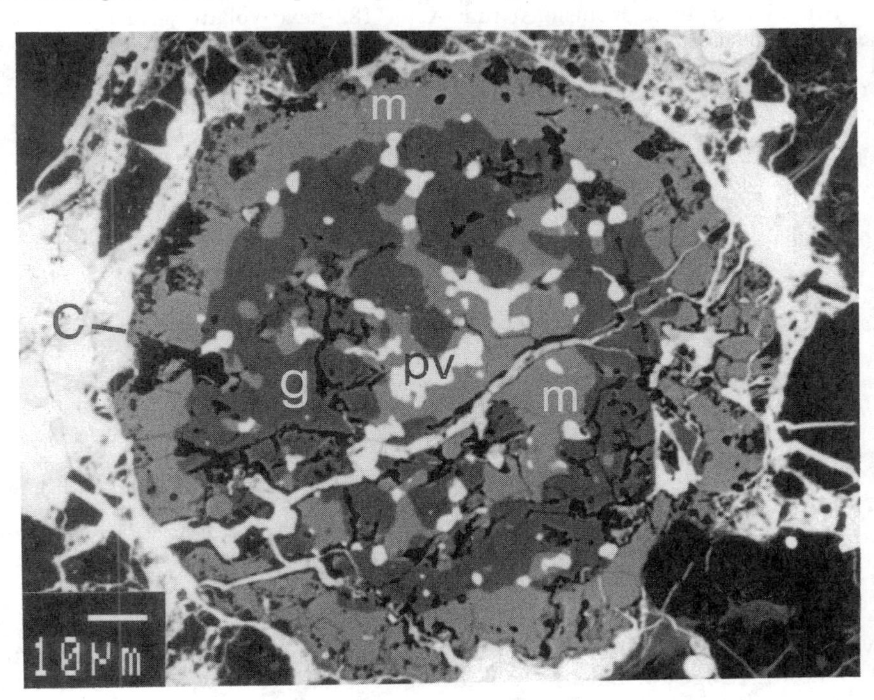

Figure 121. BSE image of a melilite-bearing inclusion from Acfer 182. Anhedral, polycrystalline grossite (g) complexes are enclosed in melilite (m). pv = perovskite. Used by permission of the editor of *Geochimica et Cosmochimica Acta* from Weber and Bischoff (1984b), Figure 2a.

Kimura et al. (1993) suggested a modification to this scheme to include only two groups, (a) spinel-(and melilite-)hibonite (SH) inclusions and (b) grossite-bearing inclusions. This modification was based on the observation that there is a smooth transition between Type A, melilite-rich inclusions and the SH inclusions in ALH 85085. The SH inclusions are similar to spinel-hibonite-rich inclusions in CM chondrites (Macdougall 1979, 1981; MacPherson et al. 1983, 1984b), but contain gehlenite-rich melilite and often grossite. They are typically round or ellipsoidal and many have a concentric texture, with either a hibonite or a spinel-rich core surrounded by a mantle of melilite and/or Ca-rich pyroxene. Only one inclusion in ALH 85085 has any resemblance to Type B inclusions in CV chondrites (Kimura et al. 1993). In addition to these groups, Weber and Bischoff (1994b) suggested an additional group, termed grossite-rich inclusions, based on a study of Acfer 182.

Hibonite. Hibonite has been found in several CAIs in ALH 85085 (Grossman et al. 1988a, Kimura et al. 1993). The hibonites are stoichiometric and have similar compositions to hibonite in spinel-hibonite CAIs from Murchison (CM) (MacPherson et al. 1983;1984b) with consistently low MgO (0.7-2.02 wt %) and TiO_2 (1.5-3.14 wt %) contents (Grossman et al. 1988, Kimura et al. 1993). Kimura et al. (1993) found hibonite in one inclusion (125) with significantly higher MgO (5.62 wt %) and TiO_2 (5.52 wt %) contents. Minor element concentrations in all hibonites are extremely low (Cr_2O_3, MnO, ZnO and V_2O_3 <0.1 wt % and typically <0.05 wt %), but FeO concentrations are somewhat higher (0.37-1.01 wt %).

Hibonites in three inclusions from Acfer 182 have volatility-fractionated trace element abundance patterns (i.e. Group II patterns of Mason and Taylor 1982) (Weber et al. 1995). REE abundances in hibonite in two inclusions are relatively low (LREE 15-40 × CI), but in a third are significantly higher (LREE 60-100 × CI) (Fig. 122).

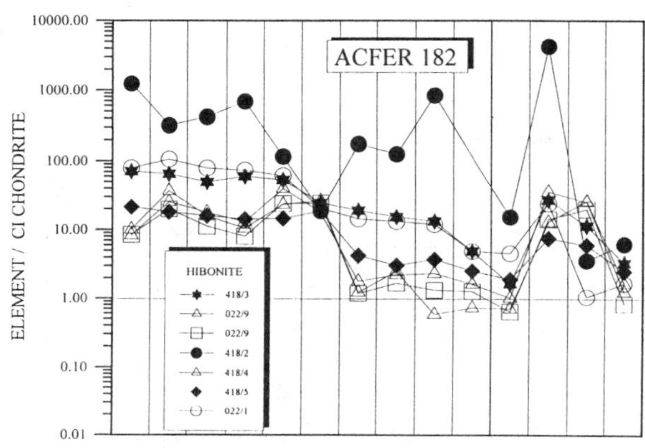

Figure 122. REE abundance pattern for hibonite (normalized to CI chondrites) from hibonite-spinel-bearing and grossite-bearing inclusions in Acfer 182. Ion microprobe analyses. Data from Weber et al. (1995).

Fine-grained hibonite is a typical accessory phase in *grossite-bearing inclusions* (Weber and Bischoff 1994b). In Acfer 182, the hibonite has highly variable compositions (0.55-4.6 wt % MgO; 0.52-6.3 wt % TiO_2) with typically <0.30 wt % FeO and SiO_2. In one inclusion (022/19) in Acfer 182, two morphologically and compositionally distinct types of hibonite occur. Mg,Ti-poor hibonites (0.87-1.09 wt % MgO; 1.85-2.69 wt % TiO_2) are present in the core, with lath-like hibonite (2.85 wt % MgO; 6.1 wt % TiO_2) occurring in a layer towards the rim of the inclusion.

Weber et al. (1995) found that hibonite in a few grossite-bearing inclusions has volatility-fractionated REE abundance patterns, with depleted highly refractory REE (Fig. 122) relative to less refractory REE (i.e. Group II patterns of Mason and Taylor 1982). The REE abundances and the shape of the patterns are typically identical to that of coexisting grossite (see below). One inclusion (418/2) contains hibonite with an unusual REE abundance pattern (Fig. 122), in which REE abundances progressively decrease from the LREE (La = 1000 × CI) to HREE (Lu = ~20 × CI), but with a negative Eu anomaly. This pattern appears to be a combination of Group II and III abundance patterns.

Grossite. Grossite is much more common in CAIs in CH chondrites than in any other chondrite group (Weber and Bischoff 1994b). In ALH 85085, Grossman et al. (1988a) reported grossite in 4 out of 50 inclusions and Kimura et al. (1993) found grossite in 17 out of 43 inclusions. In Acfer 182, Weber and Bischoff (1994b) found that 17 out of 31 inclusions studied were grossite-bearing including 13 which contain more than 30% grossite. Weber and Bischoff (1994a) have reported x-ray diffraction data for grossite in Acfer 182. Representative electron microprobe analyses of grossite are reported in Table A3.7 and trace element data, measured by ion microprobe are shown in Table A3.8.

Grossite occurs in many different textural and mineralogical associations in ALH 85085 (Grossman et al. 1988a, MacPherson et al. 1989a, Kimura et al. 1993), but grossite grains are rarely more than ~5 μm in size. Commonly, grossite-bearing CAIs occur which have a core dominated by euhedral grains of grossite intergrown with melilite and perovskite. The core is rimmed by layers of melilite and/or Ca-pyroxene (Kimura et al. 1993). In other inclusions, grossite grains occur enclosed within melilite crystals (Grossman et al. 1988, Kimura et al. 1993). Other rare and distinct occurrences of grossite have also been described which will not be discussed here.

In Acfer 182, Weber and Bischoff (1994b) identified several distinct groups of inclusions which contain grossite: (1) inclusions with a core of grossite (e.g. Fig. 123), (2) inclusions with grossite complexes in melilite, (3) inclusions with euhedral grossite and hibonite in melilite, (4) inclusions with a core of hibonite mantled by grossite, and (5) inclusions with grossite as an accessory phase. Inclusion types 1 to 4 are often grossite-rich. Grossite-bearing inclusions in Acfer 182 are generally similar to those in ALH 85085 (Grossman et al. 1988a, Kimura et al. 1993, Bischoff and Weber 1994b).

Grossite in most of the inclusions in ALH 85085 and Acfer 182 is pure stoichiometric $CaAl_4O_7$ (Grossman et al. 1988a, Kimura et al. 1993, Bischoff and Weber 1994). In comparison with ALH 85085, grossite in Acfer 182 has slightly lower SiO_2 (<0.30 wt %, but mostly <0.12 wt %; cf. 0.20-0.86 wt %) and FeO (<0.07-0.40 wt %; cf. 0.30-0.89 wt %). In both meteorites, grossite contains <0.10 wt % MgO, <0.30 wt % TiO_2 and <0.04 wt % V_2O_3.

Ion microprobe analyses of grossite from 16 inclusions in Acfer 182 (Weber et al. 1995) show low abundances of highly refractory trace elements compared with less refractory trace elements (i.e. volatility fractionated Group II patterns of Mason and Taylor (1982). Typical REE patterns for grossite show enrichments (~8-200 × CI) in the less refractory LREE (La to Sm) and Tm and the refractory HREE (Gd-Lu) and Eu are ~2 to 20 × CI (Fig. 124a). Lithophile trace elements generally show consistent behavior with the REE, i.e. Hf, Zr, Y are depleted (5-10 × CI) compared with 10-1000 × CI for less refractory elements such as Nb, V, Sr and Ba. In some inclusions, grossite has volatility-fractionated REE patterns, but the most volatile REEs, Eu and Yb, are enriched in comparison with a typical Group II pattern. The LREE in grossite in some CAIs can also be fractionated.

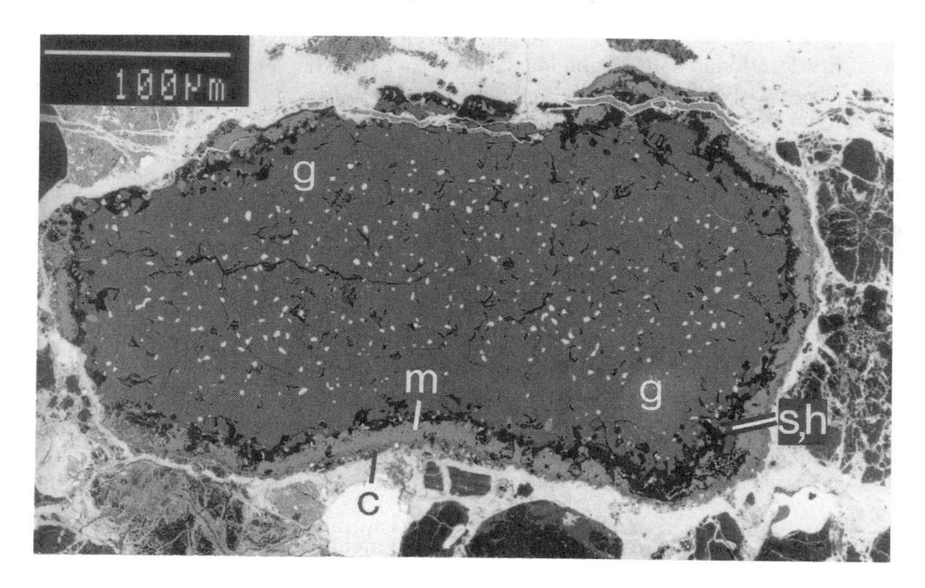

Figure 123. BSE image of a grossite-bearing inclusion in Acfer 182. The core of the inclusion consists of grossite (g) with inclusions of perovskite. The whole inclusion is rimmed by a sequence of hibonite (h) and spinel, followed by layers of melilite (m) and Ca-pyroxene (c). Used by permission of the editor of *Geochimica et Cosmochimica Acta* from Bischoff and Weber (1984b), Figure 1a.

Grossite REE abundance patterns in some inclusions are not dominated by volatility fractionation (Weber et al. 1995). Grossite in inclusion 418/8 has enriched highly refractory HREE relative to LREE (~20 × CI) (Fig. 124b). HREE abundances decrease smoothly from Gd (600 × CI) to 10-20 × CI for Yb and Lu. REE abundances in grossite in another CAI (418/2) decrease progressively from the LREE (La = 1000 × CI) to HREE (Lu = ~20 × CI) with a negative Eu anomaly. This pattern is a combination of Group II and III abundance patterns. Grossite associated with Zr,Y-rich perovskite and Sc-rich phases in inclusion 418/10 in Acfer 182 has a REE pattern similar to a Group III pattern, but has enriched HREE (60-70 × CI) and depleted LREE (~11 × CI). Another unusual feature of this pattern is a depletion in Ce. Only one inclusion (022/22) from Acfer 182 was found to contain grossite with a classic Group III pattern (Fig. 124b) with smoothly varying REE abundances (25-100 × CI) and negative Eu and Yb anomalies, which complement the positive anomalies in coexisting melilite.

Melilite. Melilite is common in *hibonite and spinel-bearing inclusions* in ALH 85085 (Grossman et al. 1988a, Kimura et al. 1993) and has compositions which are gehlenitic ($Åk_{1-29}$) with low Na_2O contents (<0.05 wt %) but comparatively high FeO contents (0.4 to 1.3 wt %). These compositions are similar to the range observed in Type A inclusions in Allende (Grossman 1975) as well as hibonite-rich inclusions in Murchison (MacPherson et al. 1983), with one inclusion (214) having a melilite composition ($Åk_{29}$) which approaches that of Type B inclusions in Allende.

Grossman et al. (1988a) measured melilite compositions in several *grossite-bearing inclusions* in ALH 85095, but the exact compositional range is not clear. Melilite occurs in 18 out of 21 inclusions studied by Weber and Bischoff (1994b) in Acfer 182 and has a very restricted range of compositions (<$Åk_8$). Melilite in the outer mantles of inclusions tends to be more Åkermanitic (<$Åk_5$) than melilite in the groundmass in the interior

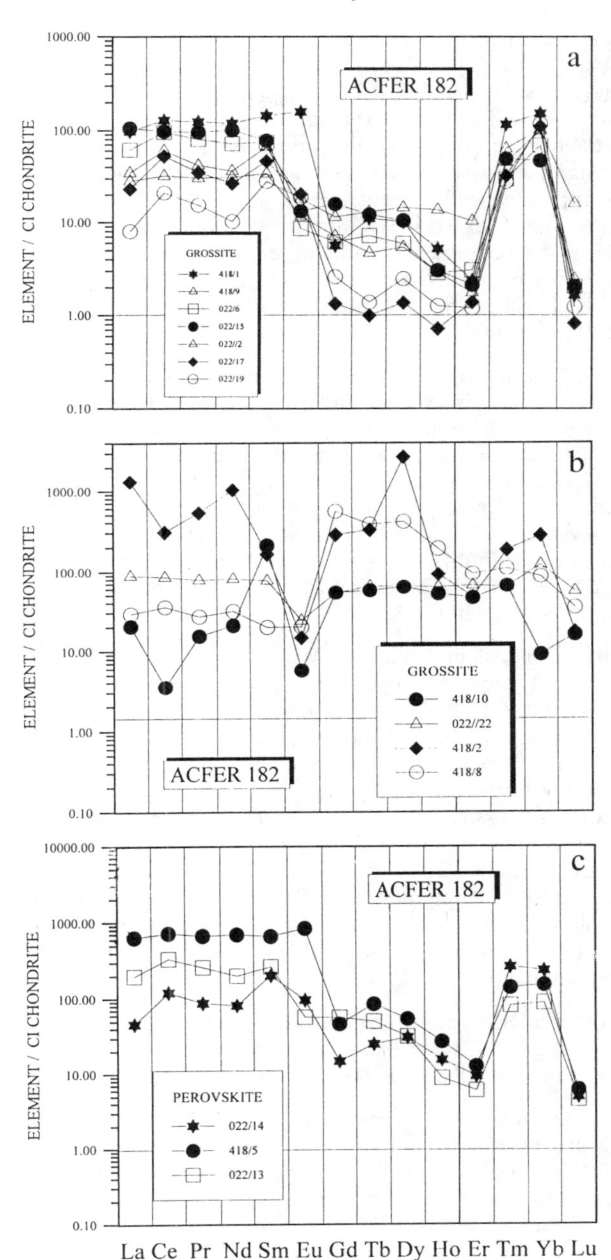

Figure 124. REE abundance patterns for grossite (a,b) and perovskite (c) (normalized to CI chondrites) from a variety of grossite-bearing inclusions in Acfer 182. Data from Weber et al. (1995).

(>$Åk_5$). Na_2O and K_2O contents are always low (<0.03 wt %) and FeO contents are typically <1 wt %, but most melilites contain <0.5 wt %.

In Acfer 182, Weber et al. (1995) found that melilite in grossite-bearing inclusions has volatility-fractionated REE patterns, with low abundances of HREE compared with LREE (i.e. Group II patterns of Mason and Taylor 1982). Trace element data for melilite in grossite-bearing inclusions are reported in Table A3.14. The shapes of the melilite REE patterns are typically identical to those of the coexisting grossite (e.g. Fig. 124a,b). However, the melilite REE abundances are either significantly lower or identical to the grossite. In two inclusions consisting of a grossite core with a melilite mantle, the grossite has a Group II pattern, but the melilite has positive anomalies in the most volatile REEs (Eu and Yb). This type of pattern is also observed in melilite in inclusion 418/8 in which coexisting grossite has an ultrarefractory pattern, but the melilite contains relatively low abundances of REE (0.3-10 × CI). Only one grossite-bearing inclusion (022/22) contains melilite with a classic Group III REE pattern with smoothly varying REE abundances (10-30 × CI) and positive Eu and Yb anomalies.

Perovskite. Grossman et al. (1988a) first observed perovskite in several inclusions in ALH 85085, but did not report any analyses. Weber and Bischoff (1994b) reported perovskite analyses from CAIs in Acfer 182 and found that they are pure endmember $CaTiO_3$, but contain 0.74-1.26 wt % Al_2O_3, which may be contamination from adjacent Al_2O_3-rich phases. FeO contents are invariably low (<0.4 wt %). Weber and Bischoff (1994b) also analyzed perovskite from a refractory element-rich inclusion 418/P which is dominated by a Y,Zr-rich perovskite (79 vol %) with a blue coloration. Two compositionally distinct perovskites occur in this inclusion; one large, complexly zoned perovskite grain with ~2 wt % Y_2O_3, ~4 wt % ZrO_2 and fine-grained perovskite with ~1 wt % Y_2O_3, ~9 wt % ZrO_2 and ~0.3 wt % HfO.

Perovskites in grossite-bearing inclusions in Acfer 182 (Weber et al. 1995) also have volatility-fractionated REE patterns. Trace element data for perovskites in grossite-bearing inclusions are reported in Table A3.12. The perovskite patterns typically have shapes identical to those of coexisting grossite and melilite, but the perovskite REE abundances are invariably higher, sometimes by an order of magnitude. In one grossite-bearing CAI, the perovskite has an ultrarefractory REE pattern with abundances 2 orders of magnitude higher than the coexisting grossite (Weber et al. 1995). The LREE are 600-1100 × CI and the HREE show a smooth decline from Gd at 100,000 × CI to ~10,000 × CI at Lu with a large, negative Eu anomaly (20 × CI) (Fig. 124b). The data for the refractory lithophile elements strongly complement the REE data. The Y,Zr-rich perovskite in inclusion 418/P (see above) has a similar ultrarefractory pattern with HREE enriched up to 10,000 × CI, with a progressive increase in REE abundance from Gd to Lu, with significant Eu, Tm and Yb anomalies (Fig. 124c).

Weber et al. (1995) also reported a REE pattern for perovskite from a grossite-free, hibonite-bearing inclusion in Acfer 182, which also has a volatility fractionated pattern, with enhanced LREE (1000 × CI) relative to HREE (6-100 × CI) (Fig. 124b).

Spinel. Spinel in *spinel-hibonite inclusions* in ALH 85085 is invariably fine-grained, stoichiometric and almost pure $MgAl_2O_4$ (Grossman et al 1988a, Kimura et al. 1993). FeO contents are invariably low (<0.50-0.66 wt %) and the concentrations of other minor elements such as V_2O_3 (0.01-0.9 wt %), Cr_2O_3 (0.1-0.4 wt %) and TiO_2 (0.1-0.4 wt %) are similar to those found in spinel in CM chondrites (MacPherson et al. 1983, 1984b; MacPherson and Davis 1994).

Spinel is typically fine-grained in *grossite-bearing inclusions* in Acfer 182. FeO contents are <0.50 wt % and TiO_2, Cr_2O_3 and CaO contents are typically in the ranges <~0.07-0.70 wt %, <0.06-0.76 wt %, and 0.1-0.20 wt %, respectively. In one inclusion from Acfer 182, the spinel has an unusually high FeO content (3.0 wt %), correlating with a high FeO content in the coexisting grossite.

Other silicate phases. *Pyroxene* in CAIs in ALH 85085 (Grossman et al. 1988a, Kimura et al. 1993) has highly variable compositions ranging from 1.4-19.0 wt % MgO, 2.4-40.3 wt % Al_2O_3 and 0.1-10.9 wt % TiO_2, i.e. diopside to fassaite. Kimura et al. (1993) found that fassaitic pyroxene (10.9 wt % TiO_2; 21.8 wt % Al_2O_3) is rare in spinel-hibonite inclusions in ALH 85085 and typically occurs in the cores of inclusions. Pyroxene commonly occurs at the outermost edge of grossite-bearing inclusions in Acfer 182 (Weber and Bischoff 1994b) and ranges from Al-bearing diopside to fassaite in composition. The Al_2O_3 contents are highly variable (1.4-6.8 wt %), but TiO_2 contents are usually low (<0.5 wt %). Fassaite, with a TiO_2 content of 8.2 wt % and extremely elevated Al_2O_3 (38.3 wt %) occurs intergrown with grossite in one inclusion in Acfer 182.

In ALH 85085, diopside typically occurs in the rims of the CAIs and has low Al_2O_3 (2.4 wt %) and TiO_2 (0.2 wt %) contents (Kimura et al. 1993). The FeO contents are typically low (<1.01 wt %) as are V_2O_3 (<0.30 wt %) and Cr_2O_3 (<1.1 wt %). However, Grossman et al. (1988a) reported aluminous diopside with 8.1 wt % Al_2O_3, 1.4 wt % TiO_2, and 1.2 wt % FeO in one CAI. Kimura et al. (1993) also found extremely aluminous diopside with Al_2O_3 contents up to 40.3 wt %, constituting more than 50 mol % $CaAl_2SiO_6$ (hypothetical Ca tschermakite) in ALH 85085 CAIs. This is the most aluminous composition ever measured for a natural pyroxene. Kimura et al. (1993) actually report one analysis whose composition is almost pure $CaAl_2SiO_6$. TiO_2 and Al_2O_3 are uncorrelated in these highly aluminous pyroxenes, an unusual feature for pyroxenes in CAIs. Since Ca tschermak-rich pyroxenes are only stable at high pressures, Kimura et al. (1993) suggested that this composition was metastable and produced by shock or quenched as a metastable phase after nonequilibrium condensation.

Ca-rich pyroxene (presumably diopside) in a grossite-free, hibonite-bearing inclusion in Acfer 182 has a volatility fractionated REE pattern with higher abundances of REE than the coexisting hibonite (Weber et al. 1995). The pattern shows a negative Eu anomaly, typical of Group II patterns. In contrast, Ca-pyroxene, associated with melilite and spinel in a grossite-bearing inclusion (418/10) has a very similar pattern to coexisting grossite, with enriched refractory HREE (60 × CI) relative to LREE (~20 × CI) but with negative Ce, Eu and Yb anomalies.

Both Grossman et al. (1988a) and Kimura et al. (1993) reported small amounts of *glass* in some CAIs in ALH 85085. Kimura et al. (1993) reported glass compositions from one inclusion and found that it has a variable composition with 17.7-19.7 wt % CaO and 36.5-39.5 wt % Al_2O_3. Minor elements such as TiO_2, V_2O_3 and Cr_2O_3 are all present at concentrations <0.11 wt %, but FeO concentrations are higher (~0.79 wt %).

Weber and Bischoff (1994b) reported very small grains of a *Sc,Zr-rich oxide* phase in a CAI (418/P) from Acfer 182, which is dominated (79 vol %) by a Y,Zr-rich perovskite. The phase could not be analyzed by electron microprobe without overlap onto adjacent phases, but contains ~34.7 wt % Sc_2O_3 and 58.6 wt % ZrO_2 and has a composition close to stoichiometric $Sc_2(Zr,Ti)_2O_7$.

Weber and Bischoff (1994b) described a *Sc,Al,Ca,Ti-rich phase* in an unusual refractory element-rich inclusion (418/P) in Acfer 182 which contains ~29.5 wt % Sc_2O_3, ~40 wt % Al_2O_3, ~12 wt % TiO_2, and ~16.5 wt % CaO. This phase has an ultrarefractory REE abundance pattern with highly enriched HREE, decreasing in abundance from Gd (110 × CI) to Lu (50 × CI), but with a negative Yb anomaly (Weber et al. 1995). The LREE are essentially flat at 6-20 × CI.

Mineralogy of CAIs in Kakangari, Lea Co 002 and LEW87232

Prinz et al. (1989) reported a number of CAIs in the unique carbonaceous chondrite,

Kakangari. Most of the CAIs are irregularly shaped aggregates (<50-400 μm), similar to spinel-pyroxene inclusions in CM chondrites (MacPherson et al. 1983, 1984b; MacPherson and Davis 1994). Most are spinel-rich ± perovskite and have rim sequences consisting of sodalite with an outer rim of aluminous diopside. Spinel is zoned with Al-rich cores (3-10 wt % Al_2O_3; 2-3 wt % FeO) and rims with up to 50 wt % Cr_2O_3 and 8-10 wt % FeO. The Cr-rich regions of the inclusions contain oriented needles of TiO_2 (<5 μm in size). Similar spinel-pyroxene inclusions also occur in the related chondrites Lea Co. 002 and LEW87232, but have not been studied in detail (Weisberg et al. 1996).

In Kakangari, Prinz et al (1989) also found an inclusion consisting of an irregular aggregate of hibonite grains (10-20 μm) with a restricted compositional range (3-4 wt % MgO; 5-5.5 wt % TiO_2; 0.9-1.0 wt % SiO_2; 0.65-0.86 wt % FeO). Each crystal is rimmed by a sequence of Cr,Al-rich spinel (10-20 wt % Cr_2O_3) and Al-diopside (1-2 wt % Al_2O_3, 2-4 wt % FeO).

Mineralogy of CAIs in the unique carbonaceous chondrites, Adelaide

Hutcheon and Steele (1982) and Huss and Hutcheon (1992) reported brief descriptions of several CAIs in the unique carbonaceous chondrite Adelaide. Hibonite occurs in two inclusions associated with melilite±spinel±perovskite. The melilite in one inclusion has a compositional range $Ak_{0.6-1.6}$. In this inclusion (HM-1), Huss and Hutcheon (1992) reported a 30 μm thick rim of a secondary, Ca-poor hydrous aluminosilicate. Two additional inclusions were also described, one consisting almost entirely of melilite and a second dominated by grossite with minor perovskite and rimmed by Fe-spinel.

Mineralogy of CAIs in the unique carbonaceous chondrites, Acfer 094

Periclase (MgO) has been reported from one spinel-perovskite spherule from the unique chondrite, Acfer 094 (Greshake et al. 1996). Rounded grains of periclase (100-150 nm) are extremely rare and occur enclosed in spinel or along spinel grain boundaries.

Mineralogy of CAIs in the unique carbonaceous chondrites, LEW 85332

Rubin and Kallemeyn (1990) described two "Fluffy" Type A CAIs in LEW 85332. Both inclusions contain melilite ($\mathring{A}k_{22}$) with low FeO (<0.2 wt %) and Na_2O (<0.04 wt %) and fine-grained perovskite. One inclusion (CAI1) contains Fe,Ni metal with minor Pt, Ru and Ir and the whole inclusion is surrounded by a continuous rim of Al-diopside (3.6 wt % Al_2O_3, 1.3 wt % TiO_2). The second inclusion (CAI2) contains small grains of spinel, which sometimes occur in clusters. This inclusion has a core consisting dominantly of Al-diopside (6.2 wt % Al_2O_3; 0.06 wt % TiO_2) and the whole inclusion is rimmed by diopside.

Mineralogy of CAIs in ordinary chondrites.

Bischoff and Keil (1983a 1984) surveyed a large number of ordinary chondrites and found a wide variety of Ca,Al-rich objects, but only a few of these have characteristics consistent with CAIs. The rest are chondrules or fragments which are closely related to chondrules. A total of 12 CAIs (the irregularly shaped inclusions of Bischoff and Keil 1984) were found in just 5 of 24 ordinary chondrites studied (Sharps, H3; Dhajala, H3; Weston, H4 (regolith breccia); ALHA77115, L3; Semarkona, LL3). Ca,Al-rich objects which appear to be CAIs have also been reported by Noonan (1975) from Clovis (No.1) (H3) and by Noonan and Nelen (1976) in Weston (H4) .

In general, the inclusions are small (100-250 μm), irregularly shaped and typically fine-grained (<10 μm), but are texturally and compositionally variable. They often have a layered appearance and have onion shell rims, similar to Wark-Lovering rims. Mineralogically, the inclusions typically contain perovskite, hibonite, spinel, ilmenite, fassaite, high- and low-Ca pyroxene, plagioclase (An>80), nepheline, sodalite, with minor olivine and unidentified secondary alteration products high in Na_2O, K_2O and FeO (Noonan 1975, Noonan and Nelen 1976, Bischoff and Keil 1983,1984).

Some inclusions described by Noonan and Nelen (1976) and Bischoff and Keil (1983a 1984) have a porous core of loose aggregates of hercynitic spinel and ilmenite surrounded by a series of layers. These rim sequences vary in complexity, but can consist of nepheline followed by fassaite+sodalite, Ca-rich pyroxene, fassaite and finally a possible layer of very fine-grained pyroxene. The porosity in the core of the inclusions is filled with an unidentified alkali and Fe-rich material.

Bischoff and Keil (1984) found one unusual, blue inclusion in Semarkona that shares similarities with the hibonite spherules found in the Murchison (CM) chondrite (Macdougall 1979, MacPherson et al. 1984b). This inclusion contains fine-grained hibonite, almost pure gehlenite, perovskite and Mg-spinel with some anorthite in the rim of the inclusion.

Hibonite. Bischoff and Keil (1983a) reported an analysis of hibonite with low TiO_2 (1.8 wt %), MgO (0.9 wt %) and FeO (0.9 wt %) from an inclusion in the Sharps H3 chondrite. In addition, Hinton and Bischoff (1984) described a hibonite clast (DH-H1) (70 μm long) in the H3 chondrite, Dhajala which is rimmed by Fe and Fe,Cr-rich spinel. The hibonite in DH-H1 it is almost pure, with MgO, TiO_2 and Na_2O contents below detection limits, but FeO contents of ~0.44 wt %. The extremely low Mg content was confirmed by ion microprobe analysis (17-200 ppm). Hinton et al. (1988) found that this hibonite has a highly fractionated REE pattern (Fig. 74), with smoothly decreasing abundances from the LREE (La = 100 × CI) to the HREE (Yb = 3 × CI), but with a very large depletion in Ce (~0.8 × CI). Hinton et al. (1988) suggested that the depletion from LREE to HREE was produced by partitioning between hibonite and another solid phase or liquid. The only hibonite with a comparable major and trace element composition is from the unusual Allende inclusion HAL (J.M. Allen et al. (1980) (see Fig. 74).

Spinel. Spinel in CAIs in ordinary chondrites are invariably much more hercynitic than spinels in CAIs in most carbonaceous chondrites, except the CK and metamorphosed CO chondrites. In ordinary chondrites, the spinels contain between 8.2-20.4 wt % FeO, 12.5 to 20.6 wt % MgO and have Cr_2O_3 < 0.42 wt %. Bischoff and Keil (1984) detected ZnO in spinel, but did not report concentrations. Only one inclusion, a blue hibonite-bearing inclusion from Semarkona, contains essentially pure $MgAl_2O_4$ with only 0.09-0.19 wt % FeO. Noonan (1975) reported Fe-spinel with ilmenite in CAIs in the H3 chondrite Clovis No.1 and several other H chondrites and Noonan and Nelen (1976) inferred the presence of Fe-bearing spinel in a CAI from Weston, based on x-ray maps.

Pyroxene. Fassaites in inclusions in ordinary chondrite CAIs have Al_2O_3 contents of 7.9-12.0 wt % and low TiO_2 contents (0.2-3.6 wt %) compared with fassaites in carbonaceous chondrites. The compositions within single inclusions show some variability. For example, in an inclusion from Weston (Noonan and Nelen 1976) the Al_2O_3 content varies between 9.6 and 11.9 wt % and TiO_2 between 3.1-3.6 wt %. FeO, MnO and Cr_2O_3 contents all appear to be low (<0.3 wt %). Noonan (1975) and Noonan and Nelen (1976) also reported diopside, rimming CAIs which consist of Fe-spinel and ilmenite, in Clovis No. 1 (H3) and Weston (H4, regolith breccia).

Mineralogy of CAIs in enstatite chondrites

Bischoff et al. (1985) searched several enstatite chondrites for CAIs and reported a total of 5 inclusions in Qingzhen, ALHA77295 and Kota-Kota. These inclusions range in size from 50-140 μm and contain various phases, including perovskite, hibonite, spinel, fassaite, anorthitic plagioclase, troilite and fine-grained material which is high in Na_2O, Cl and FeO which may be of secondary origin. Only one inclusion was found to be dominated by hibonite which contains 1.28 wt % MgO and 2.44 wt % TiO_2. The hibonite occurs in the center and at the edges of the inclusion. Most inclusions are spinel-rich and consist of a spinel-rich core in which the spinels are loosely packed and can contain small perovskite inclusions. The spinels are consistently low in FeO (<0.54 wt %) and also contain very low TiO_2 (<0.11 wt %) and Cr_2O_3 (<0.08 wt %).

Mineralogy of CAIs in metamorphosed carbonaceous chondrites: CK carbonaceous chondrites

Keller et al. (1992) and Noguchi (1993) have presented data on CAIs in CK chondrites. Rubin et al. (1988b) have also reported some data for CAIs in Ningqiang, a meteorite which was originally classified as an anomalous CV chondrite, but has since been reclassified as a CK (Kallemeyn et al. 1991). Keller et al. (1992) reported about 3 vol % CAIs in the anomalous CK chondrite, Maralinga, higher than in other CK chondrites. This observation is supported by Noguchi (1993) who found 6 CAIs in a single thin section of Maralinga, compared with two in Karoonda in a thin section of similar area. Keller et al. (1992) also noted two types of inclusions which they described as amoeboid inclusions and coarse-grained CAIs, but did not provide any detailed distinction between different types of CAIs. The mineralogy of these CAIs is dominated by plagioclase and clinopyroxene, with spinel and magnetite. In Ningqiang, CAIs which are similar to "Fluffy" Type A inclusions in Allende are present (Lin, pers. comm. in Rubin et al. 1988b) and Rubin et al. (1988b) also reported a plagioclase-rich CAI from Ningqiang which has some affinities to the Ca,Al-rich chondrules which occur in some carbonaceous chondrites (Wark 1987), but is more refractory in composition.

Spinel. Spinels in CAIs in Maralinga are emerald green, euhedral crystals of hercynite (27.4 wt % FeO; 11.4 wt % MgO), up to 200 μm in size, enclosed in plagioclase (Keller et al. 1992). They contain significant quantities of NiO (1-2 wt %) and average Cr_2O_3 contents of 0.6 wt %, but other minor elements (CaO, TiO_2 and MnO) are present in concentrations <~0.1 wt %. Hercynite-rich spinel (pleonaste) has also been described in CAIs in Karoonda, Maralinga, and EET 87507 (e.g. MacPherson and Delaney 1985, Noguchi 1993). In Karoonda, MacPherson and Delaney (1985) found hercynitic spinel in a large (~700 μm) CAI and in the core of a layered spinel-pyroxene inclusion that is similar to those found in the CM chondrite Murchison (e.g. MacPherson et al. 1983,1984b). In EET 87507, hercynitic spinel is also the dominant phase in a large (~600μm) CAI. Noguchi (1993) found that spinels in CAI in Karoonda, Maralinga and EET 87507 are compositionally similar, being rich in FeO (19-22 wt %) and relatively low in MgO (~12 wt %) as a result of metamorphic equilibration. Cr_2O_3 and ZnO contents are ~1 wt % or less, but TiO_2 is extremely low (<0.1 wt %). In an unusual plagioclase-rich CAI from Ningqiang, Rubin et al. (1988b) also reported spinel which is very low in FeO (0.26-0.66 wt %) and TiO_2 (0.21-0.27 wt %) but has somewhat elevated Cr_2O_3 contents (0.8-1.3 wt %).

Plagioclase. Polysynthetically twinned plagioclase, sometimes occurring as laths, is common in CAIs in the CK chondrite Maralinga and is nearly pure anorthite (Keller et al. 1992). However, Noguchi (1993) found considerably more compositional variation in plagioclase in CAIs in Maralinga with a range from An_{100} to An_{35}, with two peaks at An_{95-100} and An_{35-40}. In the very limited population of CAIs studied in Karoonda and EET 87507 (2 CAIs), Noguchi (1993) also found that plagioclase is a major phase with compositions An_{80-100} and An_{75-80}, respectively. Rubin et al. (1988b) also reported coarse-grained, polysynthetically twinned anorthite ($An_{93}Or_{0.3}$) containing 0.29-0.51 wt% MgO in an unusual plagioclase-rich CAI from Ningqiang.

Other silicate phases. In Ningqiang, *melilite* ($Åk_{10}$) occurs in two CAIs (Lin, pers. comm. in Rubin et al., 1988b) which may be related to Compact Type A inclusions in CV chondrites. Keller et al. (1992) observed zoned, coarse-grained fassaite in a CAI from Maralinga and *augite* occurs as a thin rim (<10 μm) on a spinel-pyroxene inclusion in Karoonda (Noguchi, 1993). It also occurs as a minor phase in CAIs from Maralinga and EET 87507 and in one CAI in the latter, the augite is compositionally zoned from core to rim with a decrease in Al_2O_3 from >15 wt % to ~10 wt %. Noguchi (1993) also found minor diopside in a large (~700 μm) CAI in Karoonda, associated with *grossular* and spinel. The grossular forms compact aggregates and is surrounded by thin augite rims. The grains are compositionally variable with 0.87-3.17 wt % Fe_2O_3 and 0-8.61 wt % MgO.

Magnetite. Noguchi (1993) described magnetite in a spinel-pyroxene CAI and spinel-rich CAI from Karoonda. In the spinel-rich CAI, magnetite occurs as an interstitial phase and as small inclusions in spinel. Euhedral magnetite also occurs in a spinel-rich inclusion in EET 87507, in some inclusions in Maralinga (Noguchi, 1993) and in a coarse-grained plagioclase-rich CAI in Ningqiang (Rubin et al. 1988b).

Accessory phases. Ca-phosphate occurs as a minor phase in CAIs in Maralinga (Noguchi, 1993) and unidentified sulfides have been reported by Rubin et al. (1988b) in a coarse-grained, plagioclase-rich CAI from Ningqiang.

Mineralogy of CAIs in metamorphosed carbonaceous chondrites: Coolidge/Loongana 001

Noguchi (1994) reported CAIs in Coolidge which range in size from 100-500 μm and are invariably fine-grained. Most of the CAIs consist of aggregates of subhedral to anhedral grains of spinel (a few tens of microns in size), associated with plagioclase (<10 μm). Many inclusions have diopside rims, a phase which is found rarely in the interiors of the inclusions. The diopside contains <5 wt% Al_2O_3, <0.8 wt% TiO_2 and significant amounts of Fe (Fs_{2-15}). Spinels are strongly zoned with aluminous, Mg-rich cores and Cr, Fe-rich rims, with Cr/(Cr+Al) ratios varying from 0.01 to 0.49 and Mg/(Mg+Fe) from 0.53 to 0.77. Plagioclase compositions are extremely heterogeneous, ranging from An_{95} to An_{27}, with a bimodal distribution. Noguchi (1994) also reported calcite, Ca-phosphate (probably both whitlockite and apatite) and unidentified opaque minerals in CAIs in Coolidge. It is unclear whether the calcite is a weathering product or not.

CHONDRITE MATRIX — PRIMARY AND SECONDARY MINERALOGY, DARK INCLUSIONS

Introduction

Matrix is a common constituent in all the carbonaceous chondrites (34-99 vol %), is less abundant in the unequilibrated ordinary chondrites (<5-15 vol %; Brearley 1996a) and is apparently absent in enstatite chondrites. Here we define matrix in primitive, unequilibrated chondrites as the "fine-grained, predominantly silicate material, interstitial to macroscopic, whole or fragmented, entities such as chondrules, inclusions and large isolated mineral (i.e. silicate, metal, sulfide and oxide) grains" (Scott et al. 1988a). This definition is distinct from the older definitions which considered matrix to be all components external to optically definable chondrules.

In recent years, the mineralogy of matrices has become increasingly well defined, as a result of electron microprobe and TEM studies. Matrix materials are remarkably diverse in character and vary from one chondrite group to another. In general, however, matrix is largely composed of a complex assemblage of unequilibrated material consisting of silicates (dominantly FeO-rich olivine), oxides, sulfides, sulfates, carbonates, Fe,Ni metal and carbonaceous materials. Many of these components appear to have experienced complex formational and thermal histories. Matrix in chondritic meteorites is also the host of interstellar grains, such as diamond, silicon carbide and graphite (see Interstellar Grains). For recent reviews of the properties and possible origins of chondrite matrices see Scott et al. (1988a), Buseck and Hua (1995) and Brearley (1996a).

The most common occurrence of matrix is an almost continuous groundmass interstitial to chondrules, but matrix also occurs as distinct lumps and clasts (e.g. Scott et al. 1984) and commonly as well defined, fine-grained rims around chondrules, CAIs etc. These rims have been variously called haloes (Bunch and Chang 1980), accretionary dark rims (King and King 1981), accretionary rims (MacPherson et al. 1985), accretionary dust mantles (Metzler et al. 1992), fine-grained rims (Brearley 1993), fine-grained dark rims (Hua et al. 1996). Here we use the term fine-grained rims (see also Zolensky et al. 1993) to describe these features.

Aqueous alteration of matrix

Almost all the carbonaceous chondrites groups (CI, CM, CO, CV, CR) have experienced aqueous alteration to different degrees and some unequilibrated ordinary chondrites have also been affected (see Barber 1985 and Zolensky and McSween 1988). In general, this alteration does not appear to have modified the bulk chemical compositions of the chondrites to any significant degree (e.g. the CI and CM chondrites), but it has overprinted the primary textural and mineralogical characteristics of matrices, chondrules and CAIs. The effects of aqueous alteration are especially prevalent in matrix materials, because of their high susceptibility to alteration by aqueous fluids as a result of their very fine grain size, high porosities and permeabilities. A full understanding of aqueous alteration processes is important for at least two reasons. First, the unraveling of the very earliest history of the components of carbonaceous chondrites requires an understanding of how aqueous alteration may have affected them, both mineralogically and chemically. Second, an understanding of the nature, mechanisms, conditions, locations and timing of aqueous alteration reactions is essential to a full appreciation of the behavior of volatiles, especially water, in the early stages of solar system formation and small bodies, in particular.

Mineralogy of CI carbonaceous chondrite matrices

One of the remarkable characteristics of the CI carbonaceous chondrites is that, despite their unfractionated bulk chemical composition in comparison with the solar photosphere (Anders and Grevesse 1989), they are the most highly hydrated of all the chondrite groups. Their matrices are dominated by very fine-grained phyllosilicate minerals (50-60 vol %; Boström and Fredriksson 1966, Kerridge 1967) which act as a groundmass for common magnetite crystals and a variety of accessory sulfides, carbonates, and sulfates. There is no evidence in these meteorites of primary components such as chondrules or CAIs, if they were ever present, except for rare grains of relatively coarse-grained Mg-rich olivine and low-Ca pyroxene. One of the most striking features of the CI chondrites is the widespread occurrence of crosscutting veins which contain carbonates and sulfates, providing unequivocal evidence that low temperature, hydrothermal alteration occurred within an asteroidal setting. Although CI chondrites are rare, clasts of material which are either fragments of CI chondrites themselves or have affinities to CIs have been observed in other chondrites (e.g. ureilites, Brearley and Prinz 1992; howardites, Reid et al. 1990).

Phyllosilicates. Phyllosilicates in CI chondrites occur within the very fine-grained matrix which is the dominant component of these meteorites. Several early x-ray and electron diffraction studies (e.g. DuFresne and Anders 1962, Kerridge 1964, 1969; Nagy et al. 1963, Nagy 1966, Mason 1962, Boström and Fredriksson 1966, Folinsbee et al. 1967, Bass 1971, Fuchs et al. 1973, Caillère and Rautureau 1974) showed that the dominant hydrous phases are phyllosilicates with basal spacings of ~7 Å and 14 Å, although the exact identity of these phases was the subject of some disagreement. Iddingsite (DuFresne and Anders 1962), chlorite and possibly montmorillonite (Nagy et al. 1963, Boström and Fredriksson 1966) or mica, chlorite or a mixed-layer type structure (Nagy 1966) have also been suggested to be present. No subsequent studies have confirmed the presence of 14-Å chlorite or mica, but x-ray diffraction analysis of glycolated phyllosilicates from Orgueil (Bass 1971) showed spacings of 7.3-7.5 Å and 17.5-18.9 Å, consistent with serpentine and a smectite-type mineral (e.g. montmorillonite), respectively. Electron diffraction studies, using TEM, have also provided evidence for the presence of serpentine (clinochrysotile; Kerridge 1962, 1964) and montmorillonite (Caillère and Rautureau 1974). Sepiolite and palygorskite may also be present in Orgueil (Kerridge 1962 1964). The phyllosilicate fraction of Revelstoke is also dominated by a phase with a 7-Å basal spacing which Folinsbee et al. (1967) suggested was a septechlorite, such as chamosite, based on limited electron diffraction data.

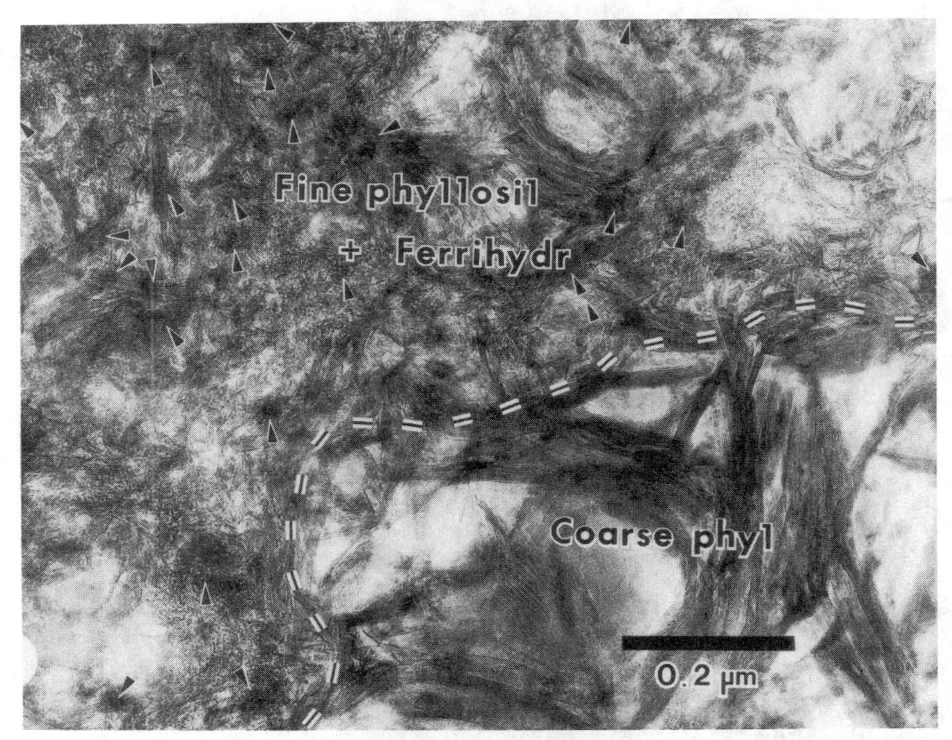

Figure 125. Transmission electron micrograph of the matrix of Orgueil showing the presence of regions of coarse and fine-grained phyllosilicates. Used by permission of the editor of *Geochimica et Cosmochimica Acta* from Tomeoka and Buseck (1988), Figure 6b.

Early TEM studies of mineral fragments from Orgueil (Nagy et al. 1963, Kerridge 1964, Boström and Fredriksson 1966, Nagy 1975, Caillère and Rautureau 1974) and Revelstoke (Folinsbee et al. 1967) showed that the phyllosilicates occur as flakes, curled flakes and rods or tubes down to <1 μm in size. HRTEM studies of the matrix of Orgueil using ion milled samples (Tomeoka and Buseck 1988) show that dominant phases in the matrix are Fe-bearing, Mg-rich serpentine and smectite (saponite) which are often poorly crystallized and very fine-grained. Two occurrences of these phyllosilicates (Fig. 125) are present (Tomeoka and Buseck 1988). The first, termed 'fine-grained' (in a relative sense), occur as randomly oriented packets, <100 to 300 Å in width, which yield only broad diffuse rings in electron diffraction patterns and are typically intimately intergrown with very fine-grained ferrihydrite. The second group of phyllosilicates is coarser-grained (300-3000 Å in width), and occurs in distinct, irregularly shaped clusters, 1 to 30 μm in size. The saponite shows variable d-spacings between ~10 and 15 Å, and the basal (001) layers are commonly wavy (Fig. 126). The commonly observed 10-Å d-spacing is lower than that expected for a typical smectite, but is probably the result of collapse of the interlayer site during electron beam irradiation or due to being under high vacuum in the TEM (Klimentidis and Mackinnon 1986). TEM observations from other CI chondrites such as Ivuna (Brearley 1992) show that the phyllosilicates present have essentially identical mineralogical characteristics to those in Orgueil. Interlayered serpentine/saponite and minor chlorite also occur in a CI chondrite clast from the Nilpena polymict ureilite (Brearley and Prinz 1992).

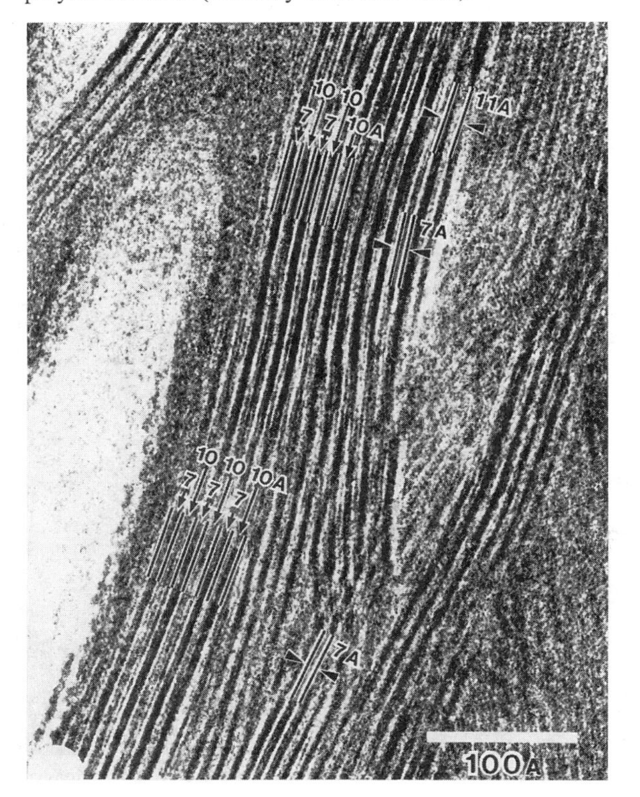

Figure 126. High resolution transmission electron micrograph of interlayered serpentine and saponite in the matrix of the CI chondrite, Orgueil. Used by permission of the editor of *Geochimica et Cosmochimica Acta* from Tomeoka and Buseck (1988), Figure 9b.

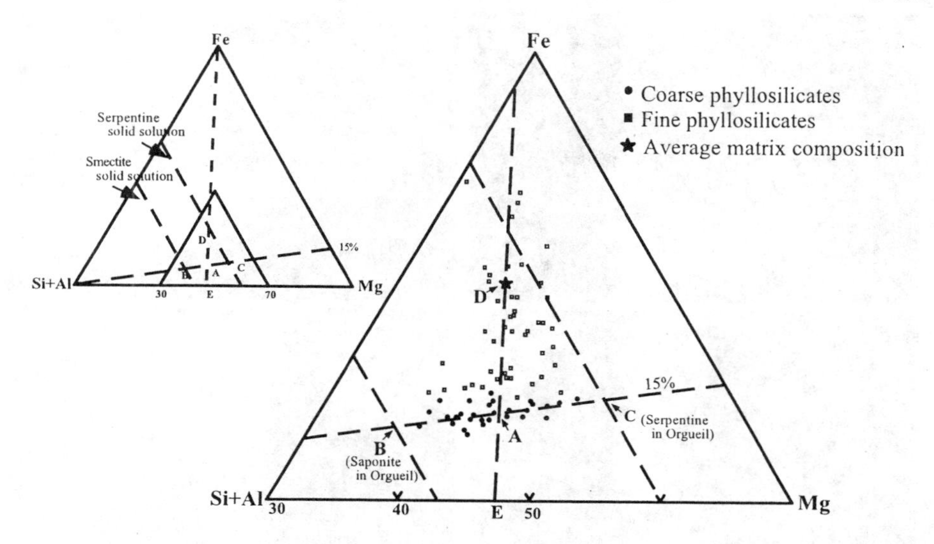

Figure 127. Ternary (Si+Al)-Fe-Mg (atom %) diagram showing the compositional range of matrix phyllosilicates in Orgueil. There are two distinct groupings, one consisting of relatively coarse-grained phyllosilicates which lie on a well-defined linear array and a second group of very fine-grained phyllosilicates which have variable compositions and extend towards the Fe apex, due to the fact that they are intimately mixed with ferrihydrite. After Tomeoka and Buseck (1988).

Compositional data for the fine-grained phyllosilicates in CI chondrites are rare and have largely been obtained by analytical electron microscopy (EDS analysis). Kerridge (1976) reported a large number of electron microprobe analyses of phyllosilicates in Orgueil. Although the exact phase present could not be identified, the data provided an upper limit of ~30% for the proportion of the smectite phase present (see also Kerridge 1977). Even using TEM techniques it is impossible to obtain analyses of individual serpentine and saponite grains, because of fine-scale intergrowths. On a (Si+Al)–Mg–Fe (atom %) diagram (Fig. 127) (Tomeoka and Buseck 1988), individual AEM analyses for coarse-grained phyllosilicates in Orgueil define a linear array which extends between the tie lines defining the solid solution ranges of serpentine and saponite. The compositions of very fine-grained phyllosilicates fall off this line and lie towards the Fe-rich apex, due to the fact that they are intimately intergrown with ferrihydrite (see below). Tomeoka and Buseck (1988) deduced that serpentine and saponite in Orgueil contain significant amounts of Al_2O_3 (3.3 wt % and 6.4 wt %, respectively). The Mg/(Mg+Fe) ratios of phyllosilicates are distinct from one CI chondrite (and CI clasts) to another, ranging from 0.85 in Orgueil (Tomeoka and Buseck 1988) to 0.72 in Nilpena (Brearley and Prinz 1992).

Tomeoka and Buseck (1988) proposed that the intimate intergrowth of serpentine/saponite and ferrihydrite was the result of dissolution of magnetite and Fe,Ni sulfides and the reprecipitation of small particles of ferrihydrite intimately associated with the phyllosilicates. Alternatively, Brearley and Prinz (1992), Zolensky et al. (1993) and Brearley (1997) have suggested that this intimate association is the result of the oxidation of ferrous saponite to Mg-rich saponite and ferrihydrite. Hence, the systematic variation in Mg/(Mg+Fe) ratios of phyllosilicates from one CI chondrite to another could be the result of different degrees of oxidation (e.g. Brearley and Prinz 1992, Brearley 1997).

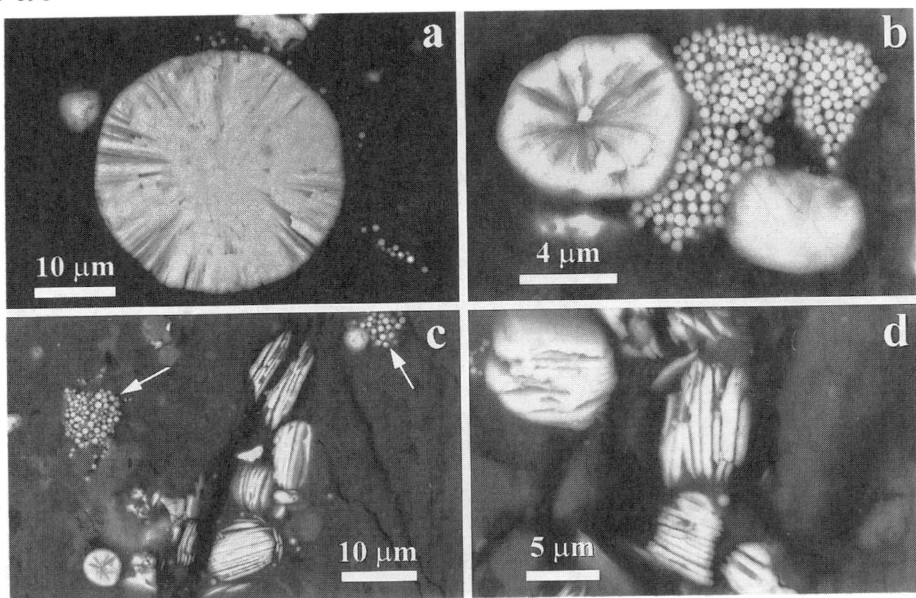

Figure 128. BSE images illustrating the range of magnetite morphologies present in the matrix of the CI carbonaceous chondrite Ivuna. (a) Spherulitic magnetite, (b) Framboidal magnetite associated with a spherulitic magnetites, (c) Concentration of magnetite plaquettes along a fracture. Framboidal magnetites are arrowed. and (d) close up of plaquette magnetites showing that they consist of stacks of platelets.

Oxides. *Magnetite* (Fe_3O_4) is the second most abundant phase in CI chondrites and was one of the first phases to be recognized in this group of meteorites (e.g. Clöez, 1864, Pisani, 1864, Cohen, 1894, Wiik 1956, Nagy and Claus 1962). X-ray diffraction studies (e.g. Mason 1962, Kerridge and Chatterji 1968, Kerridge 1970a, Bass 1971, Folinsbee et al. 1967) have shown that the unit cell edge of magnetite in Alais, Orgueil, Ivuna and Revelstoke is somewhat variable (8.393 to 8.43 Å) (see compilation in Nagy 1975), but is close to the value for terrestrial magnetite of 8.396 Å (Kerridge and Chatterji 1968).

A wide variety of magnetite morphologies including framboids, spherulites and plaquettes have been recognized in CI chondrites (e.g. Nagy and Claus 1962, Nagy et al. 1963, Boström and Fredriksson 1966, Jedwab 1965, 1967, 1968, 1971; Kerridge 1970a, Folinsbee et al. 1967, Nagy 1975, Kerridge et al. 1979a). Magnetite most commonly occurs as spherical to subspherical aggregates, 10-30 μm in size (Jedwab 1965, Kerridge 1970a), consisting of myriad μm to submicron grains of magnetite (Fig. 128a). The individual crystals are faceted (e.g. Kerridge 1970a). These magnetites have morphologies with a close resemblance to terrestrial framboidal pyrite that occurs in marine sediments (e.g. Jedwab 1965, Kerridge et al. 1979a). Spherical forms of magnetite (5 to 40 μm in diameter; Kerridge 1970a) are also common (Fig. 128b) and have spherulitic internal structures defined by individual radiating magnetite crystals (e.g. Boström and Fredriksson 1966, Kerridge 1970a, Kerridge et al. 1979a). An additional, unusual morphological group consists of disc-like platelets (Jedwab 1968) or 'plaquettes' of magnetite (Fig. 128c) which frequently occur within carbonate grains (e.g. Kerridge et al. 1979a). They also occur as individual grains and as spiral stacks of several platelets which show evidence of dislocation-induced growth spirals on the platelet surfaces (Jedwab 1971). The principle elongation plane of the platelets is probably {001}, rather than the preferred close-packed {111} planes (Kerridge et al. 1979a).

The origin of these unusual magnetite morphologies has been the subject of considerable debate. The framboidal and spherulitic morphologies resemble the morphologies of minerals which have grown terrestrially at low temperatures in the presence of aqueous solutions. Kerridge et al. (1979a) suggested that magnetites with these two morphologies could have grown from colloidal Fe hydroxide gels under similar conditions. However, Jedwab (1967) have argued that the plaquettes may have formed by condensation from a vapor phase (e.g. Donn and Sears,1963, Donn 1974), a view which has been challenged by Kerridge et al. (1979a). Several authors (e.g. Kerridge et al. 1979a) have argued there may, in fact, be several generations of magnetite present in CI chondrites which may be represented by the different morphological groups.

The magnetite content of different CI chondrites has been measured by a number of workers (e.g. Jeffrey and Anders 1970, DuFresne and Anders 1962, Boström and Fredriksson 1966, Kerridge and Chatterji 1968, Larson et al. 1974, Hyman et al. 1978, Hyman and Rowe 1983). For Orgueil, these determinations lie between 6 and 20 wt % Fe_3O_4, but numerous repeat determinations for several CI chondrites by Hyman et al. (1978) and Hyman and Rowe (1983) give values of: 11.0±0.4% – Orgueil, 8.9±0.9% –Alais, 11.1±0.3% – Ivuna, 9.4 % – Revelstoke, and 8.5% – Tonk.

Electron microprobe analyses of magnetites from CI chondrites are rare but show that they are essentially pure Fe_3O_4 with extremely low concentrations of minor elements (e.g. Boström and Fredriksson 1966, Folinsbee et al. 1967, Reid et al. 1990, Löhn and El Goresy 1992). Kerridge (1970a) measured the composition of spherulitic magnetites from Orgueil and found Ni contents up to 0.08±0.02 wt %.

Mössbauer spectroscopy of magnetite in Orgueil (Madsen et al. 1989) shows that the vacancy concentration is <0.0006, corresponding to a magnetite structural formula $Fe_{2.994}O_4$. Although magnetite is the most abundant Fe-bearing spinel phase in CI chondrites, Madsen et al. (1989) found that about 11% the magnetic fraction in Orgueil was probably *maghemite,* based on Mössbauer spectroscopy.

Ferrihydrite ($5Fe_2O_3.9H_2O$) was identified by Tomeoka and Buseck (1988) intimately mixed with matrix phyllosilicates (Fig. 129) in Orgueil. Ferrihydrite is well known to exhibit superparamagnetic behavior and is very abundant in Orgueil. It is probably the superparamagnetic phase, first recognized by Wdowiak and Agresti (1984), which constitutes about 30% of the total Fe in Orgueil (see also Madsen et al. 1986). Tomeoka and Buseck (1988) estimated that about 60% of Fe in the matrix of Orgueil is present as ferrihydrite. Electron diffraction studies of ferrihydrite in Orgueil show only two diffuse rings at ~2.58 Å and 1.50 Å, which are not consistent with any known crystalline ferrihydrite, but may be an amorphous hydrated gel, colloidal ferric hydroxide or hydrated ferric oxide polymer (Burns and Burns 1977). Ferrihydrite also occurs as granular masses in Orgueil which may have formed by oxidation of magnetite. The ferrihydrite in Orgueil appears to contain significant, but variable concentrations of Ni and S. In a CI chondrite clast from the Nilpena polymict ureilite, ferrihydrite also pseudomorphically replaces grains of pentlandite (e.g. Brearley and Prinz 1992).

Other minor oxide phases in CI chondrites include *limonite* (Nagy et al. 1963, Boström and Fredriksson 1966), *periclase* (Nagy and Andersen 1964) or *brucite* (Boström and Fredriksson 1966) and Ti_3O_5 (Brearley 1993a).

Sulfides. The sulfide minerals *troilite*, *pyrrhotite*, *pentlandite* and *cubanite* all occur in CI chondrites. *Troilite* has been reported (see below), but the identification of this phase is probably in doubt, since the major sulfide phase in CIs is now recognized as

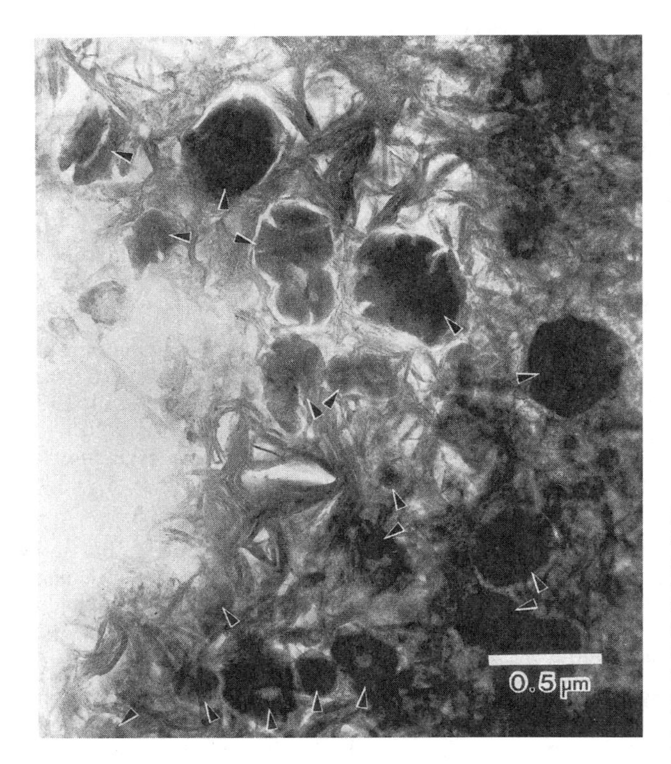

Figure 129. High resolution transmission electron micrograph showing relatively coarse-grained ferrihydrite, intimately mixed with phyllosilicates in the matrix of Orgueil. Used by permission of the editor of *Geochimica et Cosmochimica Acta* from Tomeoka and Buseck (1988), Figure 12.

pyrrhotite (see below). Troilite has very similar optical properties to pyrrhotite (e.g. Ramdohr 1973) and the two phases can easily be misidentified, based on optical methods alone. With this caveat in mind, Fitch et al. (1962) observed euhedral and corroded troilite crystals in Orgueil (see also Boström and Fredriksson 1966). Less corroded troilite grains were reported by Folinsbee et al. (1967) in Revelstoke (see also Leitch and Grossman 1977).

Pyrrhotite is now recognized as the most common sulfide in CI chondrites (Kerridge et al. 1979b) and occurs as laths and plates, 10-100 µm in diameter, which often have a hexagonal outline. It was first identified by x-ray diffraction in Revelstoke (Folinsbee et al. 1967) and subsequently found in situ in Orgueil by Kerridge (1970b). Pyrrhotites in Orgueil and Ivuna often show pronounced evidence of corrosion, which is not apparent in Alais (Kerridge et al. 1979b) or Revelstoke (Folinsbee et al. 1967). Small inclusions of pentlandite sometimes occur in Alais pyrrhotite. CI pyrrhotites contain 0.65 to ~1.3 wt % Ni, with pyrrhotite in Alais having a slightly lower mean Ni content than Ivuna or Orgueil (0.98 wt % vs. 1.15 wt %) (Kerridge et al. 1979b). Ni-bearing pyrrhotite also occurs in CI clasts in the Bholgati howardite (Reid et al. 1990) and the Nilpena polymict ureilite (with up to 7 atom% Ni) (Brearley and Prinz 1992).

Pentlandite is a rare phase in CI chondrites and has only been found in Revelstoke and Alais. In Revelstoke, Folinsbee et al. (1967) found rare, isolated pentlandite grains, ~20 µm in size, with an Fe:Ni ratio of ~1. In Alais, pentlandite sometimes occurs as small (~15 µm) grains included within hexagonal plates of pyrrhotite (Kerridge et al. 1979b) (see also Brearley and Prinz 1992).

Cubanite ($CuFe_2S_3$) occurs as a rare phase in Alais, Orgueil and Ivuna (Macdougall and Kerridge 1977, Kerridge et al. 1979b).

Carbonates. Carbonates occur in all the CI chondrites and constitute ~5 vol % of these meteorites, although their absolute modal abundance is variable. Fredriksson and Kerridge (1988) and Endress and Bischoff (1996) have reported lithic clasts with as high as 35% (Orgueil) and 10% (Ivuna) carbonate, respectively. Carbonates are present in veins, as irregular or subrounded isolated grains in the matrix and as elongate fragments resulting from brecciation (Richardson 1978, Fredriksson and Kerridge 1988, Endress and Bischoff 1996). Carbonate grain sizes vary from <10 μm up to several mm in length (Nagy and Andersen 1964, Fredriksson and Kerridge 1988, Johnson and Prinz 1993, Endress and Bischoff 1996). Based on compositional and textural relationships as well as isotopic data, two or more generations of carbonate may be present in CI chondrites (Richardson 1978, Macdougall et al. 1984, Macdougall and Lugmair 1989, Endress et al. 1996).

Carbonates can occur as pure phases or intergrown with other phases such as magnetite, phosphates, sulfates and sulfides (Richardson 1978, Fredriksson and Kerridge 1988, Johnson and Prinz 1993, Endress and Bischoff 1996).

Dolomite, breunnerite and calcite all occur in CI chondrites and Endress at al. (1996) reported a single occurrence of Mg,Ca-bearing siderite in Ivuna. Dolomite is the most abundant carbonate phase with breunnerite being relatively common, but much less so than dolomite. Some authors have referred to breunnerite as magnesite (e.g. Johnson and Prinz 1993), based on the observation that magnesite is the dominant endmember.

Dolomite was first reported by DuFresne and Anders (1962) in Orgueil and Ivuna, where it occurs as polycrystalline agglomerates. Fredriksson and Kerridge (1988) found that isolated dolomite aggregates are rarely larger than 100 μm in size and are much smaller than breunnerites. (See also Leitch and Grossman 1977). Dolomites in CI chondrites have a wide range of $MnCO_3$ contents (Fig. 130), but show a much more restricted range of $FeCO_3$ in solid solution (e.g. Fredriksson and Kerridge 1988, Johnson and Prinz 1993, Endress and Bischoff 1996). There do not appear to be any systematic compositional differences in dolomite from one CI chondrite to another (Fredriksson and Kerridge 1988, Johnson and Prinz 1993) (Fig. 130), although Endress and Bischoff (1996) found that dolomite in Alais and Tonk has a more restricted range of Ca/Mg ratios than in Orgueil and Ivuna. The total compositional variability of dolomites in CI chondrites can be expressed as $(Ca_{0.35-0.53}Mg_{0.34-0.51}Mn_{0.00-0.15}Fe_{0.02-0.13})CO_3$ (Endress and Bischoff 1996).

Different substitutional couples have operated during the formation of dolomite in different chondrites (Endress and Bischoff 1996). In Orgueil, Ivuna and Alais, $MnCO_3$ and $MgCO_3$ are negatively correlated (Fig. 131), but only in Orgueil is there a negative correlation between $MgCO_3$ and $CaCO_3$. In Alais and Ivuna, Mg and Ca are decoupled. These data suggest that in Orgueil Ca and Mg show a coupled substitution, but Mn only substitutes for Mg and not Ca in these dolomites. In contrast, in Ivuna and Alais Mn substitutes for both Mg and Ca and in Alais, Fe also substitutes for Mg, but apparently not for Ca. These observations have been interpreted as providing evidence that growth of the carbonates in different CI chondrites occurred in very different environments on the parent body (Endress and Bischoff 1996).

Riciputi et al. (1994) have reported major (Mg and Fe) and trace (B, Na, Mn, Sr, Ba) element data for dolomites from Orgueil, measured by ion microprobe. Calculated fluid compositions based on these data indicate that the dolomites crystallized from solutions

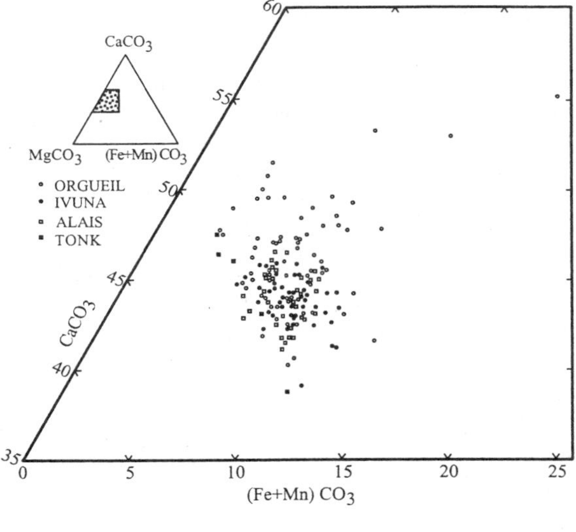

Figure 130 (left). Compositional range of dolomite in CI chondrites plotted in the ternary system $CaCO_3$-$MgCO_3$-$(Fe+Mn)CO_3$. After Kerridge and Fredriksson (1988).

Figure 131 (below). Substitutional relationships between different cations in dolomite in the CI chondrites Ivuna, Alais and Orgueil.

(a-c) $(Mn+Fe)CO_3$ vs. $MgCO_3$

(d-f) $MnCO_3$ vs. $MgCO_3$

(g) $(Mn+Fe)CO_3$ vs. $CaCO_3$

(h) $MnCO_3$ vs. $CaCO_3$

After Endress and Bischoff (1986).

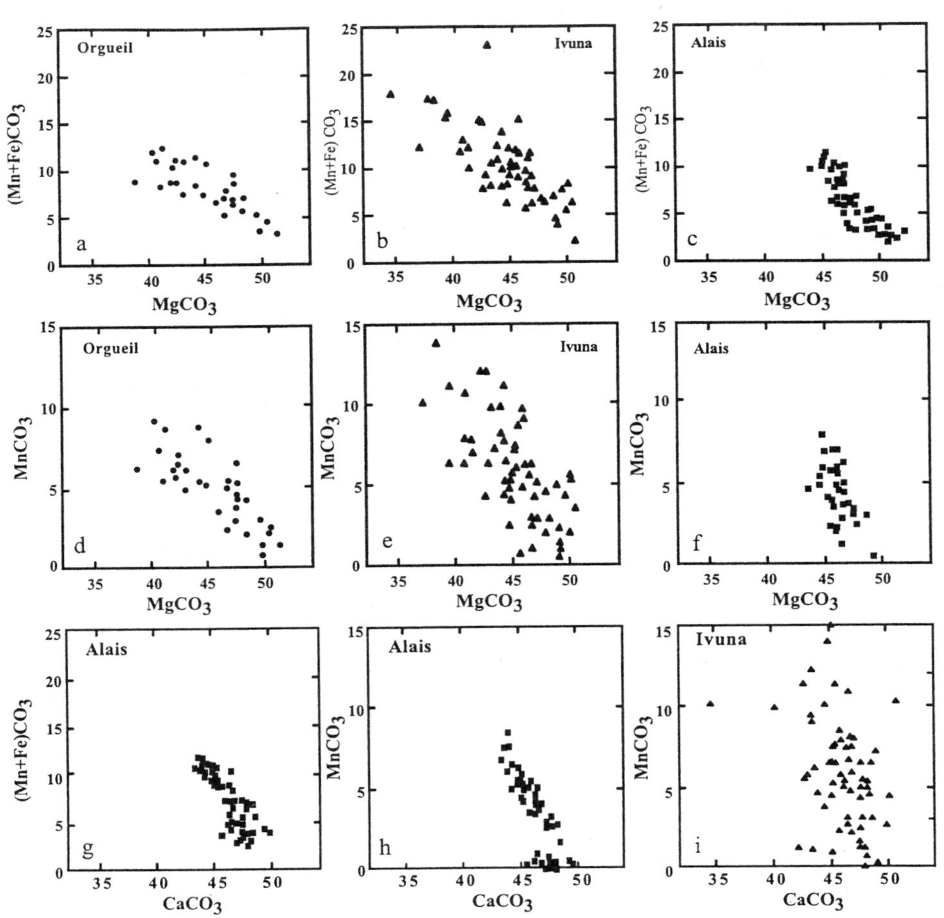

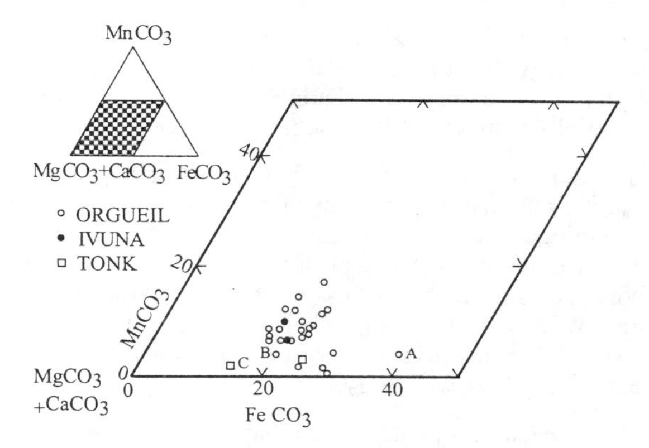

Figure 132. Compositional range of breunnerites in the three CI chondrites Orgueil, Ivuna and Tonk, plotted in the ternary $MnCO_3$-($MgCO_3$+$CaCO_3$)-$FeCO_3$. After Kerridge and Fredriksson (1988).

compositionally similar to terrestrial brines.

Breunnerite is common in CI chondrites (Pisani, 1864, DuFresne and Anders 1962, Folinsbee et al. 1967), but appears to be absent from Alais and is much less abundant in Ivuna than in Orgueil (Fredriksson and Kerridge 1988). Breunnerite tends to be coarser-grained (<1-3 mm) than dolomite or calcite (Fredriksson and Kerridge 1988) and is compositionally variable, with $MnCO_3$ contents up to 15 mol % and $FeCO_3$ up to 39 mol % (Fig. 132) (Fredriksson and Kerridge 1988, Johnson and Prinz 1993, Endress and Bischoff 1996). Zoning is absent (but see Nagy and Andersen 1964). Minor elements (e.g. Sr) are invariably below detection limits for EPMA. The full compositional range observed is $(Ca_{0.00-0.04}Mg_{0.53-0.78}Mn_{0.02-0.15}Fe_{0.15-0.39})CO_3$. Breunnerite in Orgueil and Ivuna contains higher $CaCO_3$ contents (~1-4 mol %) than in Alais (<0.8 wt %), but Fredriksson and Kerridge (1988) found one grain in Tonk with 5.5 mol % $CaCO_3$.

Calcite is extremely rare (Fredriksson and Kerridge 1988, Johnson and Prinz 1993, Endress and Bischoff 1996), but does occur in Ivuna, Alais and Orgueil. The typical occurrence of calcite is as clusters of several rounded grains, but spherulitic calcites have also been reported (Fredriksson and Kerridge 1988). Calcite is compositionally close to pure $CaCO_3$ (>99 mol %) with minor $FeCO_3$ (<0.6 mol %) and $MgCO_3$ (<0.2 mol %) (Fredriksson and Kerridge 1988, Johnson and Prinz 1993, Endress and Bischoff 1996).

Sulfates. Several sulfate phases, including gypsum, epsomite, bloedite and Ni-bloedite have been reported in CI chondrites. Berzelius (1834) reported 10.3 wt % of water soluble sulfates of Mg, Na and Ni in Alais. Subsequent studies (e.g. DuFresne and Anders 1962, Boström and Fredriksson 1966, Richardson 1978, Kerridge and Bunch 1979, Fredriksson and Kerridge 1988, Endress and Bischoff 1996) have all confirmed the presence of sulfates, which occur both as vein filling material (e.g. Richardson 1978) and as isolated grains within the matrix, sometimes associated with carbonate, magnetite or both (Endress and Bischoff 1996).

Gypsum ($CaSO_4 \cdot 2H_2O$) is a rare phase, but has been detected by x-ray diffraction (Bass 1971, Richardson 1978) and in situ (Nagy and Andersen 1964, Richardson 1978) in Orgueil as small, individual grains. Electron microprobe analysis of gypsum suggest variable Fe contents (3.1 wt % – Nagy and Andersen 1964; 0.7 wt % – Fredriksson and Kerridge 1988). Rare *anhydrite* may also be present (Bass 1971).

Epsomite ($(MgSO_4) \cdot 7H_2O$) was first reported by Cohen (1864) in Orgueil and DuFresne and Anders (1962) confirmed its presence using x-ray diffraction techniques. The modal abundace of epsomite in Orgueil is variable and values between 6.7 and 17% have been reported (e.g. DuFresne and Anders 1962, Boström and Fredriksson,1966,

Kerridge 1967). Boström and Fredriksson (1966) estimated that 2-3% epsomite occurs in veins and the rest is distributed as grains within the fine-grained matrix. In Ivuna, the epsomite appears to be mixed with the hexahydrate phase (DuFresne and Anders 1962), but this phase may have formed terrestrially from epsomite (Richardson 1978).

DuFresne and Anders (1962) handpicked one grain of *bloedite* ($MgSO_4 \cdot Na_2SO_4 \cdot 4H_2O$) from Ivuna. Na-bearing sulfate was also detected in Orgueil by Boström and Fredriksson (1966) using SIMS techniques, concentrated at the walls of veins with epsomite in the center. Fredriksson and Kerridge (1988) reported bloedites in Ivuna and Orgueil that contained significant concentrations of NiO (2.6-13.6 wt % NiO) and appear to have structural formulae consistent with *Ni-bloedites* (Nickel and Bridge 1977). The structural formula for the most Ni-rich compositions reported by Fredriksson and Kerridge (1988) is $Na_2Ni_{2.2}Mg_{4.5}(Fe,Mn,Ca)_{0.4}(SO_4)_{9.2} \cdot 2.8(H_2O)$.

Minor phases in CI chondrites include Ca-phosphate, which may be *merrillite* (Nagy and Andersen 1964) or *whitlockite* (Boström and Fredriksson 1966) (see also Endress and Bischoff 1996) and elemental *sulfur* (Nagy et al. 1961, DuFresne and Anders 1962, Boström and Fredriksson 1966).

Mineralogy of CM carbonaceous chondrite matrices

CM chondrites are one of the most abundant type of carbonaceous chondrite and have experienced extremely complex histories. Clasts of CM chondrites also occur widely in other meteorite types, including other carbonaceous chondrites (Wilkening 1978), howardites (Wilkening 1973, Bunch et al. 1979, Reid et al. 1990, Buchanan et al. 1993, Zolensky et al. 1996a) and possibly eucrites (Kozul and Hewins 1988, Zolensky et al. 1992, 1996a) and rarely in ordinary chondrites (Wilkening and Clayton 1974, Fodor and Keil 1976, Fodor et al. 1976) (see also Wilkening 1978). CM chondrites consist of primary high temperature minerals in chondrules, CAIs and amoeboid olivine aggregates, etc. set within an extremely fine-grained matrix of hydrous minerals. Chondrules and CAIs have been altered to different degrees by hydrous fluids in different CM chondrites.

In CM chondrites, fine-grained rims (Fig. 133) around chondrules, CAIs, PCP-rich objects, mineral aggregates and fragments are widespread (e.g. Bunch and Chang 1980, Metzler et al. 1992). In some CM chondrites (Metzler et al. 1992), the abundance of these rims is especially high such that essentially all the fine-grained material is present as rim material. These CM chondrites have been termed primary accretionary rocks (Metzler et al. 1992) and have not been modified to any extensive degree by brecciation. Most CM chondrites, however, are breccias consisting of fragments of primary accretionary rocks and surviving intact chondrules, CAIs, etc., embedded within a matrix of clastic fragments of chondrules, rims, etc., produced by brecciation (Metzler et al. 1992). The mineralogy of the fine-grained matrix and rims, which appear to be mineralogically essentially identical (Zolensky et al. 1993) is examined below, in addition to coarser-grained PCP-rich objects which occur associated with matrix material.

Phyllosilicates. Phyllosilicate minerals make up the major component of the matrix in CM chondrites. They are also widely developed in chondrules, CAIs and other mineral aggregates. Early x-ray diffraction studies (Kvasha 1948, DuFresne and Anders 1962, Mason 1963, Fuchs et al. 1973) showed that the major phase present has a ~7-Å basal spacing consistent with a serpentine-group mineral, although Kvasha (1948) and DuFresne and Anders (1962) suggested chlorite might also be present. Fuchs et al. (1973) found that their x-ray diffraction data for matrix material from Murchison, Mighei and Murray corresponded most closely to hexagonal and monoclinic septechlorites, such as

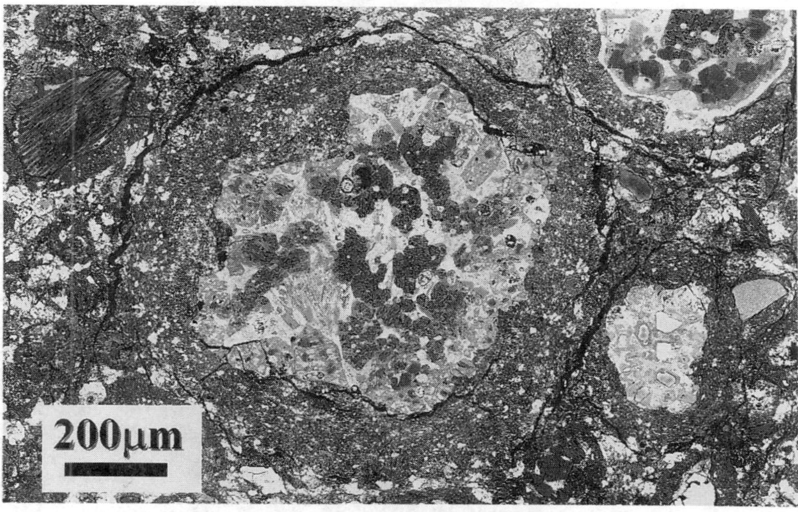

Figure 133. BSE image showing a typical fine-grained rim surrounding a chondrule in the ALH 81002 CM carbonaceous chondrite.

berthierine, rather than serpentine. X-ray diffraction analysis of four different groups of phyllosilicates, separated by density from Murray matrix, indicate that they are all either septechlorites or serpentines (Bunch and Chang 1980). However, Bunch and Chang (1980) preferred to classify them as Fe serpentines.

TEM studies have shown that a wide variety of different phyllosilicate types are present (dominantly serpentines), with very variable morphologies (summarized by Zolensky et al. 1993), degree of crystallinity and compositions. Moreover, phases with these different characteristics are commonly spatially associated or intergrown on a very fine-scale. Thus the challenge of unravelling their formational histories is extremely complex, especially so as many CM chondrites are brecciated.

The Fe^{3+}-rich serpentine, *cronstedtite,* with a distinctive platy or lath-like morphology has been shown by TEM studies to be one of the the major phases in the matrices of CM chondrites (Fig. 134). (e.g. Müller et al. 1977, 1979; Barber 1977, 1981; Mackinnon 1980, 1982; Akai 1980, 1982; Akai and Kanno 1986). Cronstedtite is often relatively coarse-grained. For example, Müller et al. (1979) described well crystallized cronstedtite grains up to 10 μm in length and 1 μm wide (parallel to c) in the matrix of Cochabamba. These crystals often show significant stacking disorder on (001) and consist of a disordered intergrowth of three different polymorphs, a feature which has been widely observed in other CM chondrites (e.g. Murchison, Mackinnon 1980; Y-791198; Akai and Kanno 1986). The faults associated with stacking disorder have a displacement of ±b/3 or ±2b/3 (Müller et al. 1979).

Electron microprobe analyses show that cronstedtite is not a pure endmember but always contains some Mg in solid solution. In Cochabamba, for example, Müller et al. (1979) found that cronstedtite had variable Mg/(Mg+Fe) ratios (mean 0.2), but with a relatively constant Al_2O_3 content (~3.3 wt %). Bunch and Chang (1980) found that a dark green phyllosilicate fraction separated by density from Murray was extremely FeO-rich (58.1 wt %) consistent with cronstedtite, with an Al_2O_3 content of 5.5 wt %.

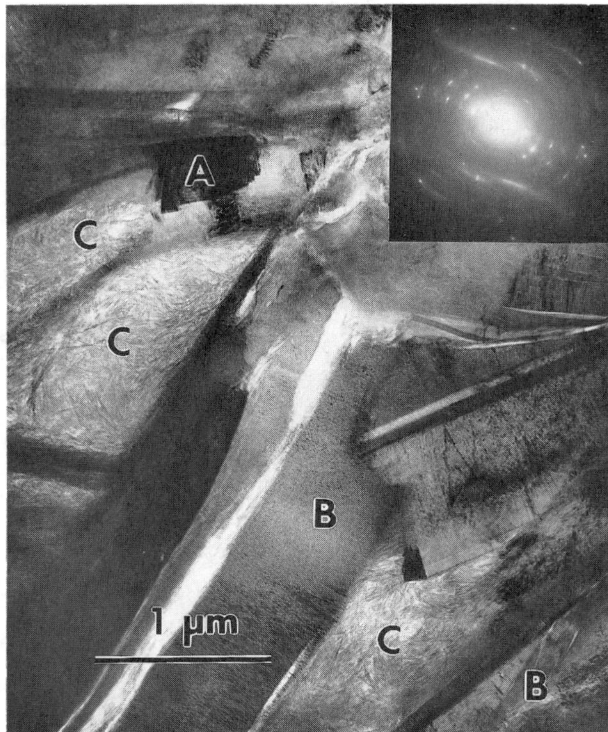

Figure 134. Low magnification transmission electron micrograph showing typical plate and ribbon-like crystals (e.g. A and B) of cronstedtite interspersed with fernlike growths (C) in the matrix of the CM carbonaceous chondrite, Murchison. After Barber (1981).

TEM studies (e.g. Barber 1981, Zolensky et al. 1993) have shown that *ferroan Mg-serpentines*, $(Fe,Mg)_6Si_4O_{10}(OH)_8$, are an important phase in CM chondrite matrices (e.g. Murchison, Murray, Nogoya, Mighei, Y-791198). These serpentines occur with a variety of morphologies and have remarkably similar characteristics from one CM chondrite to another. The identification of this serpentine in CM chondrite matrices, by TEM, is based on electron diffraction patterns for coarser-grained crystals, but for the very fine-grained materials is usually based solely on the presence of ~7-Å basal spacings in high resolution TEM images (e.g. McKee and Moore 1979, Mackinnon and Buseck 1979a, Barber 1981, Akai 1982), sometimes supplemented with compositional data (EDS x-ray analysis).

Most serpentines in CM chondrite matrices appear to be Fe-bearing chrysotile, rather than antigorite, based on electron diffraction and analytical electron microscope analyses (e.g. Barber 1981). The serpentines are mostly extremely fine-grained, sometimes poorly crystalline and can have fern-like or radiating fibrous morphologies (Fig. 135a-c) (McKee and Moore 1979, Mackinnon and Buseck 1979a, Barber 1981, Akai and Kanno 1986). Regions of poorly crystalline serpentine consist of aggregates of randomly oriented microcrystallites, just a few unit cells in thickness along the c direction (e.g. Mackinnon and Buseck 1979a, Akai 1980, 1982; Barber 1981). Individual microcrystallites often have a curved or wavy appearance (Fig. 135c) and tube-like morphologies (~10 nm in diameter) (Fig. 135d,e), a characteristic of terrestrial chrysotile, are also common (Mackinnon and Buseck 1979a, Barber 1981, Akai 1980, 1982; Akai and Kanno 1986, Brearley 1995). As well as very fine-grained serpentines, more coarsely crystalline varieties of chrysotile occur, including sectored cylindrical and spherulitic

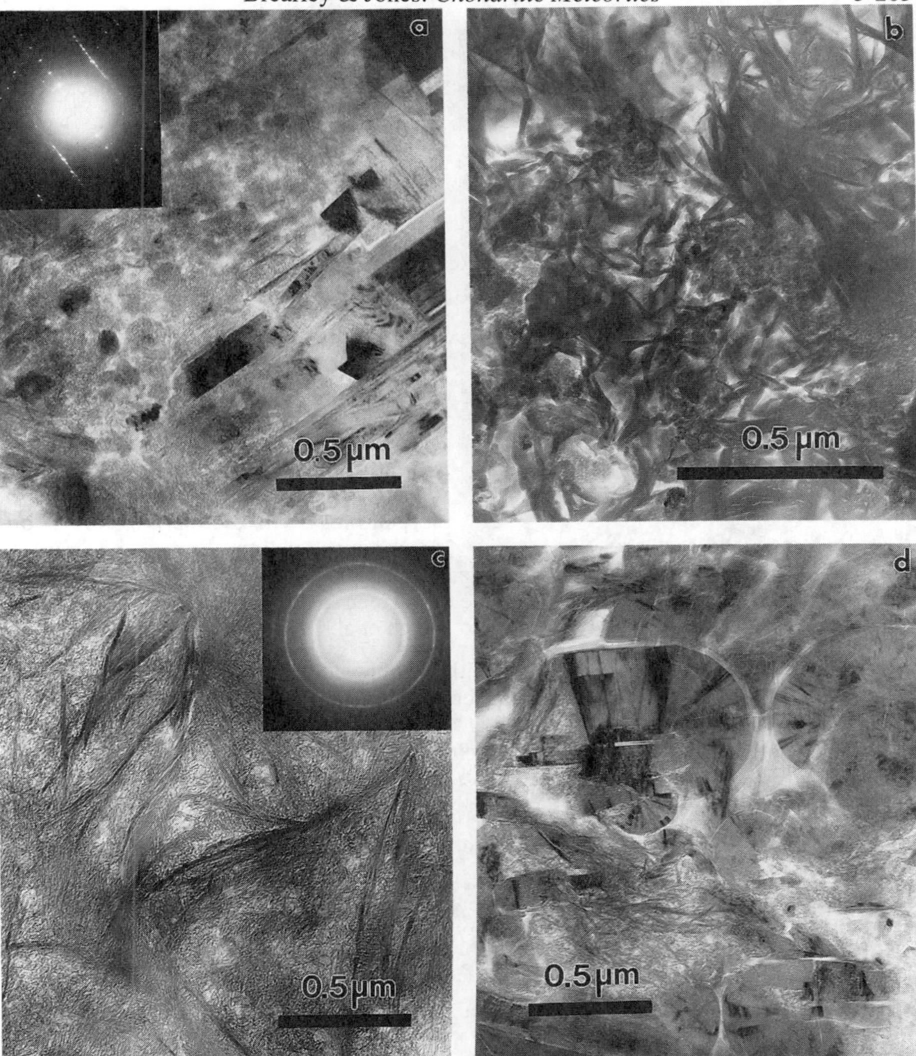

Figure 135. Transmission electron micrographs showing the range of serpentine textures present in the fine-grained matrix of CM carbonaceous chondrites: (a) coarse-grained platy serpentine, (b) fine-grained platy serpentine, (c) fern-like, microcrystalline serpentine, (d) Povlen textures in serpentine, (e—next page) Povlen textures in serpentine. After Barber (1981).

growth forms (McKee and Moore 1979, Mackinnon 1980, Barber 1981), similar to Povlen structures in terrestrial serpentines (e.g. Cressey and Zussman 1976, Cressey 1979). Well-crystallized, platy Mg-bearing, Fe-serpentines have also been observed in Cold Bokkeveld, Murchison (Barber 1981) and Y-791198 (Akai and Kanno 1986). Barber (1981) suggested that these serpentines may be closer to greenalite ($Fe_3SiO_5(OH)_4$) in composition, rather than cronstedtite, but Akai and Kanno (1986) argued that they may be related to lizardite.

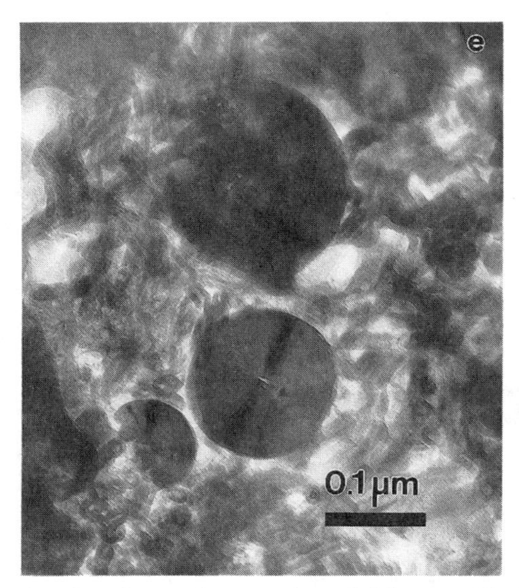

Ferroan serpentines in CM chondrites have compositions intermediate between end-member Mg serpentine ($Mg_3Si_2O_5(OH)_4$) and greenalite with a chemical formula approximated by $(Fe,Mg)_6Si_4O_{10}(OH)_8$ (e.g. Tomeoka et al. 1989a). Unlike terrestrial serpentines, the Mg-rich varieties in CM chondrites contain significant Al_2O_3 (2-5 wt %), S (1-2 wt %) and NiO (1-2 wt %). It is not yet clear whether the S or Ni are structurally bound or are present as interlayers of tochilinite within serpentine. The serpentines exhibit a very wide range of Mg-Fe substitution even within the same meteorite (e.g. Bunch and Chang 1980, Barber 1981, Akai and Kanno 1986). For example, Bunch and Chang (1980) separated three groups of serpentines with distinctive colorations (dark green, yellow, brown and gray) by density from the matrix of Murray. Each group of serpentines has a different Mg/(Mg+Fe) ratio which decreases from the dark green to the gray serpentines and Al_2O_3 contents between 2.4 and 5.3 wt %. (see also Bunch et al. 1979, Noro et al. 1980).

Zolensky et al. (1993) measured serpentine compositions in the matrices of a number of CM chondrites and found a a broad continuum of Fe/(Fe+Mg) ratios that differ somewhat from meteorite to meteorite (Fig. 136), an effect which has been attributed to variations in the degree of alteration of each meteorite (Zolensky et al. 1993). The total range of Fe/(Fe+Mg) ratios is ~0.3 to almost 1, but most of the data lie in the range from ~0.4-0.75 with the most Mg-rich serpentine compositions occurring in Nogoya (Bunch and Chang 1980, Zolensky et al. 1993).

The smectite, *saponite,* is extremely rare in CM chondrites, but occurs as the dominant phase, frequently containing coherent interlayers of *serpentine*, in the matrix of the unusual CM chondrite, Bells (Brearley 1995). The saponite/serpentine is Fe-rich (Fe/(Fe+Mg) = 0.38) and occurs as radiating or parallel to subparallel crystals (200 nm in length) and as ultrafine-grained, poorly crystalline material. Rare saponite also occurs in a fine-grained rim in Murray (Zolensky et al. 1993) and in a CM clast in the EET 87513 howardite (Buchanan et al. 1993).

Other rare phyllosilicates phases that have been reported in the matrices of CM chondrites are *chlorite* in Nogoya, ALH 84029 and Cochabamba (Zolensky et al. 1993) and Bells (Brearley 1995), *brucite* in Murchison (Mackinnon 1980) and *vermiculite* in Nogoya (Zolensky et al. 1993).

There appears to be some material in the matrices of CM chondrites which may be truly *amorphous* (e.g. Barber 1981). Many regions of matrix consist of domains of very poorly crystalline phyllosilicates, that occur in a groundmass which appears to be amorphous (e.g. Fig. 135c). According to Barber (1981), this material does not exceed 15 vol % of the matrix and this could be an overestimation caused by, for example, rapid beam damage to beam sensitive phyllosilicates.

Poorly characterized phases (PCP) and tochilinite. A widespread, unidentified phase rich in Fe, S, Ni and O, was first recognized by Fuchs et al. (1973) in Murchison

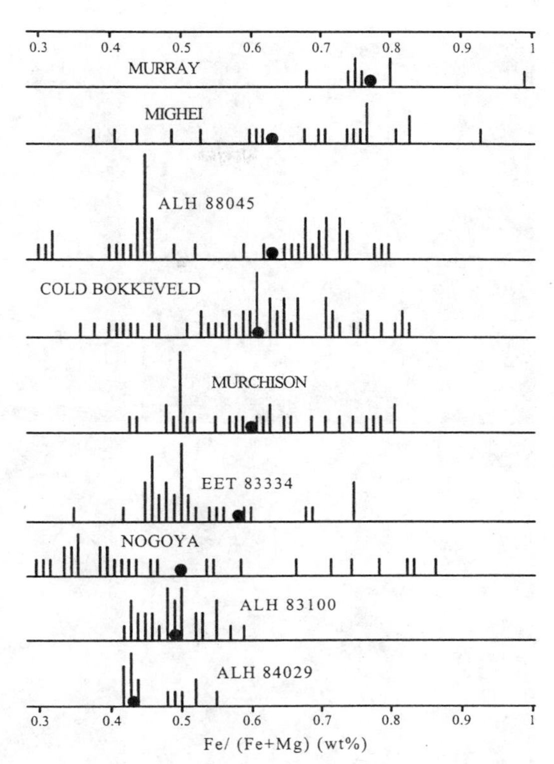

Figure 136. Histograms of the Fe/(Fe+Mg) ratios of serpentines in the matrix of several different CM carbonaceous chondrites, showing the extremely broad range of compositions. The range and average Fe/(Fe+Mg) ratios vary from one chondrite to another. After Zolensky et al. (1993).

and by Ramdohr (1973) (his Fe-C-O phase) in three other CM chondrites. Fuchs et al. (1973) termed this material 'poorly characterized phases' (PCP) and described several different occurrences including regular spheres, angular to irregular pieces with serated edges and irregular and rounded blebs. Examples of PCP in different CM chondrites are shown in Figure 137. These different occurrences of PCP vary from a few μm up to several hundred μm in size. The first two types occur most frequently within the matrix, whilst the blebs occur dominantly within chondrules and inclusions and probably replaced metal grains (see Chondrules). The term PCP-rich objects is now frequently applied to distinct occurrences of this material in the matrices of CM chondrites (e.g. Metzler et al. 1992) and other occurrences with distinct polygonal textures (Fig. 137d) have been commonly recognized (e.g. Bunch and Chang 1980).

PCP in Murray, Murchison and Nogoya has been studied in detail by Bunch and Chang (1980). They described PCP as being amorphous or poorly crystalline and high in Fe, Ni, O, C, S and, in some morphological forms, Si. Three types of PCP (Types I, II and III) have been distinguished: Type I PCP is massive and has high Ni and S contents, Type II is fibrous or needle-like with lower S, but higher Mg and Si and Type III (the most abundant) is granular to formless and has the highest Mg and Si contents, but low Ni. The carbon content of PCP increases from 2.2 wt % in Type I to 8.4 wt % in Type III (Bunch and Chang 1980). Si-bearing PCP aggregates sometimes consist of two phases, a PCP-rich outer portion and a core of an Fe-rich phyllosilicate phase.

PCP has been characterized in detail by high resolution TEM and electron diffraction techniques. Electron diffraction patterns from Type II PCP (fibrous) in Murchison and

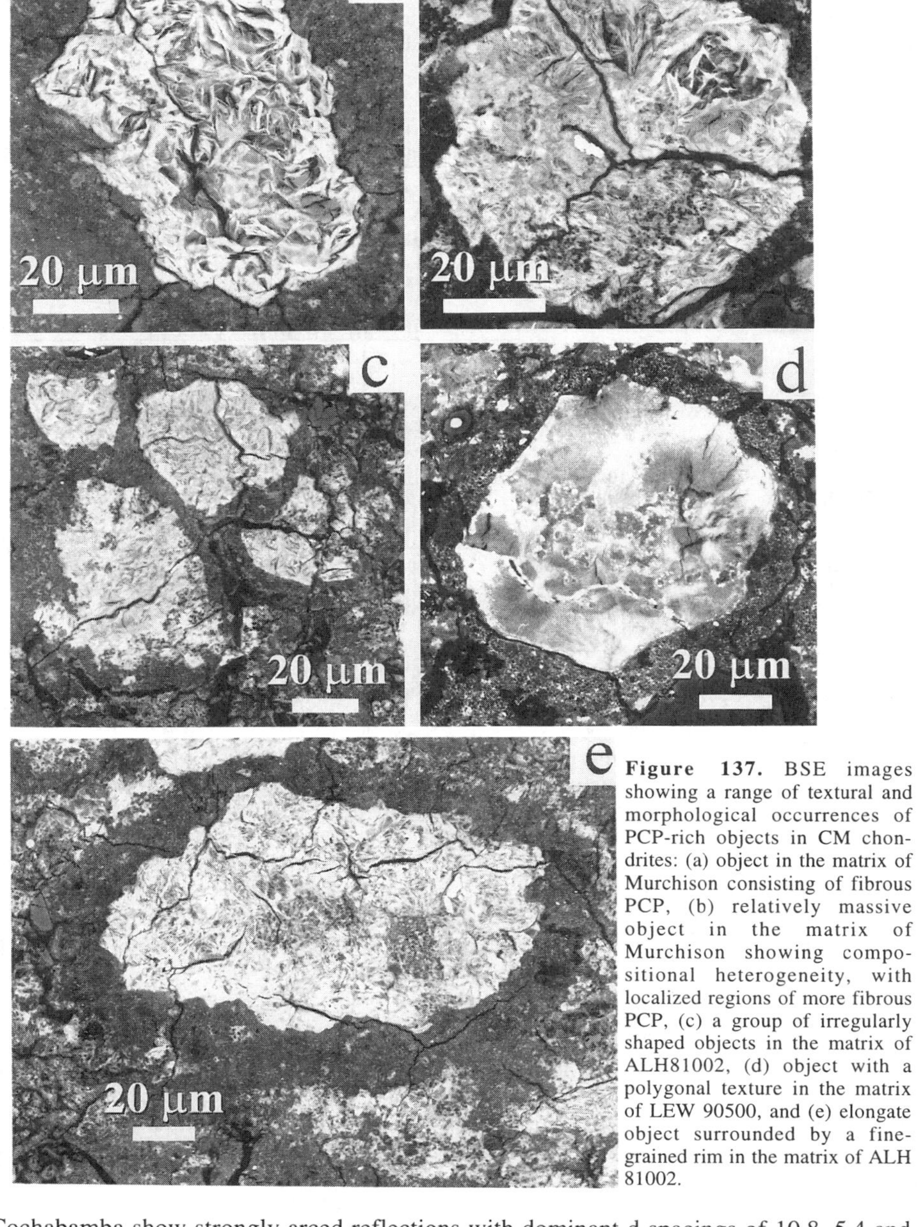

Figure 137. BSE images showing a range of textural and morphological occurrences of PCP-rich objects in CM chondrites: (a) object in the matrix of Murchison consisting of fibrous PCP, (b) relatively massive object in the matrix of Murchison showing compositional heterogeneity, with localized regions of more fibrous PCP, (c) a group of irregularly shaped objects in the matrix of ALH81002, (d) object with a polygonal texture in the matrix of LEW 90500, and (e) elongate object surrounded by a fine-grained rim in the matrix of ALH 81002.

Cochabamba show strongly arced reflections with dominant d spacings of 10.8, 5.4 and 2.7 Å (Barber 1981). The arcing is due to the undulation of the 10.8-Å layer planes along straight sections of PCP fibers (Fig. 138) (Barber et al. 1983). Using EDS analysis and electron energy loss spectroscopy (EELS), Barber (1981) and Barber et al. (1983) showed

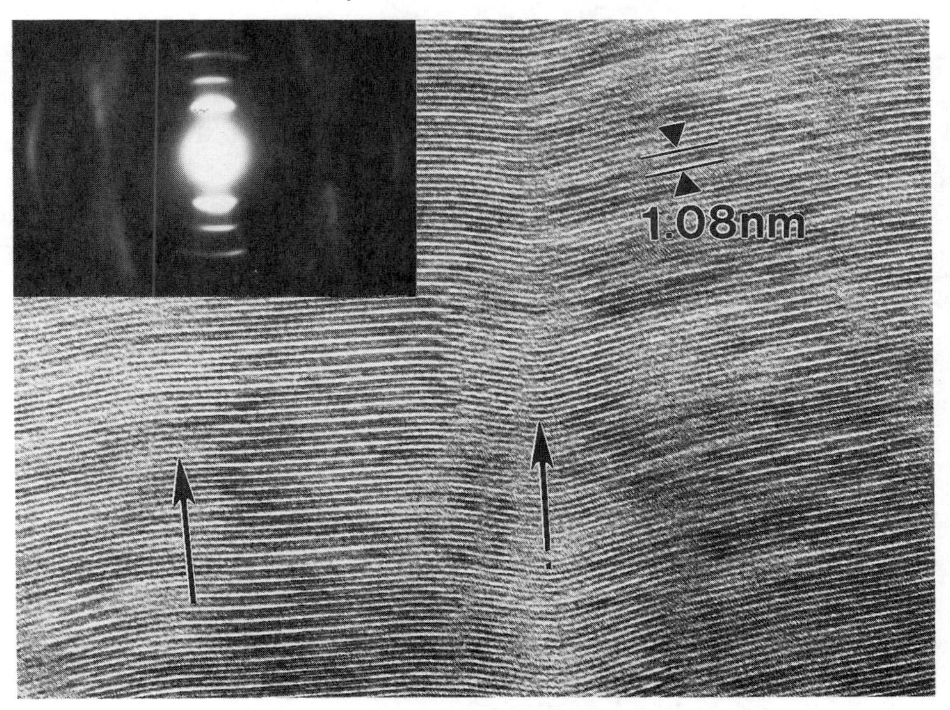

Figure 138. High resolution TEM image of undulatory PCP (tochilinite) from the matrix of the Murchison CM chondrite showing the presence of a dominant 10.8Å basal spacing and distinct bending in the lattice fringes at regular intervals (arrowed), presumably due to mismatch between the layers. After Barber et al. (1983).

that PCP is rich in Fe, Ni, S and O and is often intergrown coherently with serpentine (see also Mackinnon 1982). Using HRTEM, Akai (1982) also recognized a phase with an 11-Å d-spacing in the matrix of Y-74662, which is evidently the same phase.

HRTEM studies of the matrices of Murchison and Mighei (Mackinnon and Buseck 1979a,b; Akai 1980, Mackinnon 1980) have shown that crystallites with ordered and disordered interstratifications of 7-Å and 5-Å basal layers, with an overall basal periodicity of ~17 Å, are commonly present. These ordered sequences were interpreted as the regular interstratification of serpentine (S) and brucite (B) layers with the stacking sequence SBB (Mackinnon and Buseck 1979b, Akai 1980). Two distinct morphologies of this phase, tubes and sheets were recognized by Akai (1980) in Murchison and planar and corrugated interlayed structures have also been observed in Mighei (Mackinnon 1982, Tomeoka and Buseck 1983). Serpentine also occurs coherently interlayered with the SBB phase. Barber (1981) observed that very small crystallites (a few unit cells in width) of a phase consisting of interlayered 10-Å and 7-Å spacings also occurs in spongy or porous regions of Cold Bokkeveld and Murchison.

The resolution to the identity of the SBB phase and its relation to PCP was proposed by Mackinnon and Zolensky (1984). They suggested that the 10-Å Fe,Ni,S,O-rich phase observed by Barber et al. (1983) was the Fe-rich structural equivalent of the rare terrestrial phase, tochilinite, $6Fe_{0.9}S{\cdot}5(Fe,Mg)(OH)_2$ (see Jambor 1969 and Organova et al. 1972). It consists of coherently interstratified mackinawite-like sulfide sheets and

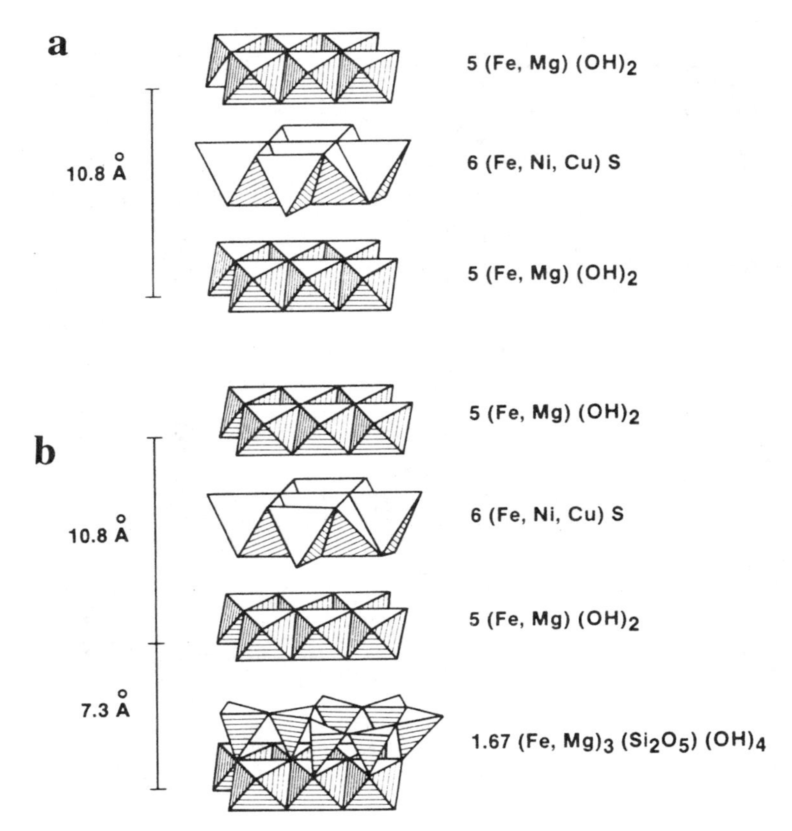

Figure 139. Schematic illustrations of the proposed structures for the two main types of PCP-phase in CM carbonaceous chondrites: (a) the tochilinite structure, (b) an interlayered serpentine structure. Used by permission of the editor of *Nature*, from Mackinnon and Zolensky (1984), Figure 1.

brucite sheets (Fig. 139). Mackinnon and Zolensky (1984) further proposed that the SBB-type mixed layered structures with a 17-Å basal layer spacing are ordered intergrowths of tochilinite (10-Å) and serpentine (7-Å). Zolensky et al. (1993) have since observed a phase with a 24-Å basal spacing in Nogoya, which appears to be a coherent, ordered intergrowth of two serpentine layers with a tochilinite layer.

The detailed microstructures of PCP in Mighei, Murchison and Murray have been studied by Tomeoka and Buseck (1985). They recognized two types of PCP: Type I which occurs in chondrules and aggregates, replacing metal and consists largely of tochilinite (their FESON phase, e.g. Tomeoka et al. 1989a) and a second, Type II, which occurs predominantly in the matrix and consists of intergrown tochilinite and cronstedtite with a variety of curved, rolled and cylindrical morphologies (Fig. 140a-f). The Type I and II PCPs of Tomeoka and Buseck (1985) appear to correspond to the Type I and II PCPs of Bunch and Chang (1980). Cronstedtite occurs in Type II PCP as individual, platy crystals and as ordered and disordered intergrowths with tochilinite, i.e. interlayered 10-Å and 7-Å phases (see also Nakamura and Nakamuta 1996).

Using x-ray diffraction, Nakamura and Nakamuta (1996) have shown that Type II PCP in the matrix of unbrecciated regions of Murchison consists of intergrown tochilinite

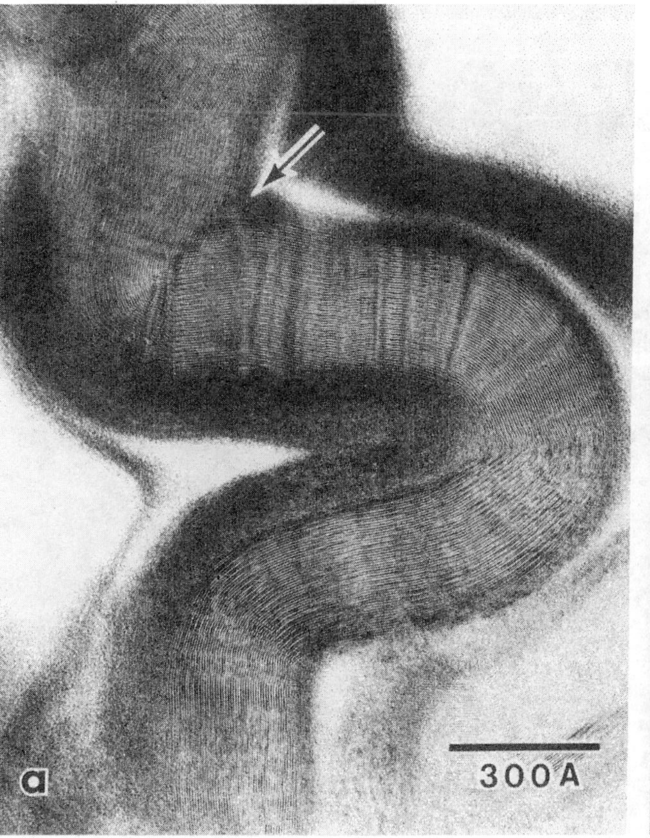

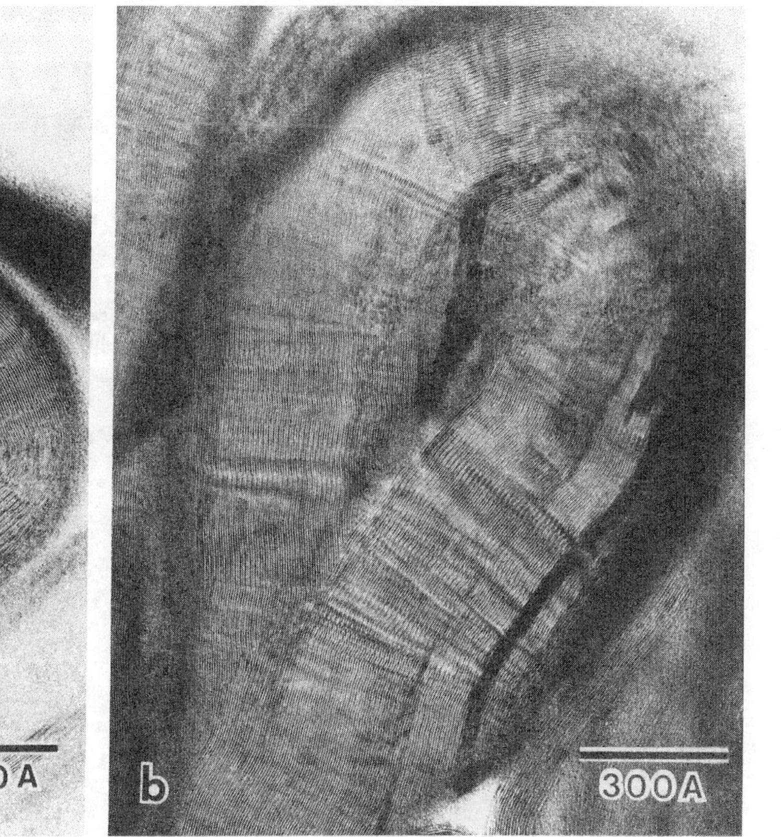

Figure 140a,b. High resolution transmission electron micrographs showing contorted structures in PCP (FESON) from the CM carbonaceous chondrite, Murchison. Used by permission of the editor of *Geochimica et Cosmochimica Acta*, from Tomeoka and Buseck (1985), Figures 17a,b.

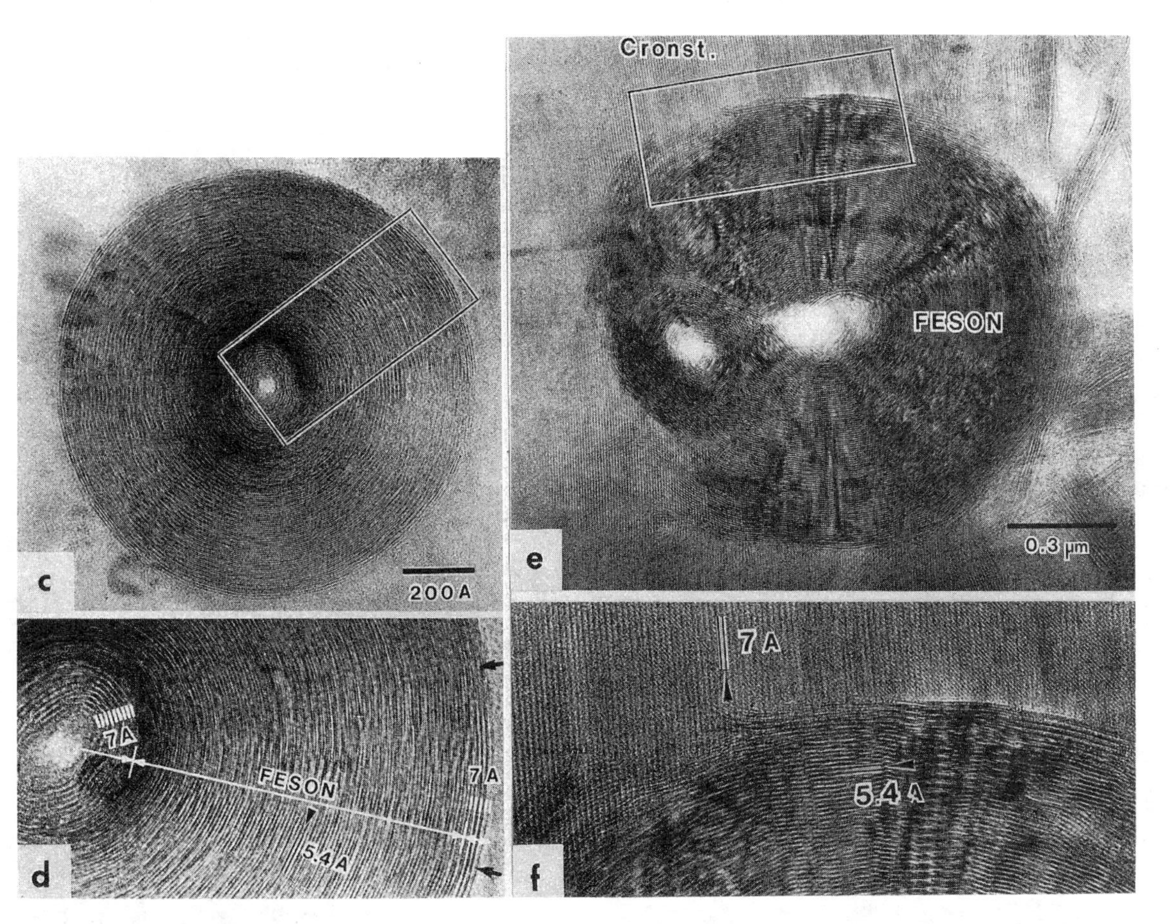

PFigure 140c-f. High resolution transmission electron micrographs showing PCP with cylindrical morphologies in the matrix of Murchison. Used by permission of the editor of *Geochimica et Cosmochimica Acta*, from Tomeoka and Buseck (1985), Figures 10a,b and Figures 12a,b.

and discrete cronstedtite. However, in brecciated regions Type II PCP consists of ordered and disordered mixed-layer cronstedtite and tochilinite. These observations have been interpreted as evidence that tochilinite was replaced by tochilinite/cronstedtite, during aqueous alteration, possibly aided by brecciation.

It is now evident that relatively coarse-grained Type I to III PCPs (Bunch and Chang 1980) and the SBB mixed layer structures in the matrix (Mackinnon and Buseck 1979b, Akai 1980) etc. are closely related types of material. The structural models proposed by Mackinnon and Zolensky (1984) appear to adequately explain the morphological and compositional variations observed in PCP in CM chondrites. The high Si and Mg and lower S and Ni contents of Type II and III PCPs analyzed by Bunch and Chang (1980) can be accounted for by increased interlayering of serpentine with tochilinite, at the scale of electron microprobe measurements. PCP is thus now widely accepted as being a coherently interlayered mixture of tochilinite with serpentine, where the degree of intergrowth varies. Electron microprobe data for PCP (Tomeoka and Buseck 1985) show the compositional range of PCP varies from 0-23.5 wt % SiO_2 and 0-21.5 wt % S (Fig. 141), with a negative correlation between these two elements. However, despite the fact that the identify of PCP is now known, the acronym is still widely used in the current literature and perhaps now can be considered to be a partially characterized phase (Tomeoka and Buseck 1985).

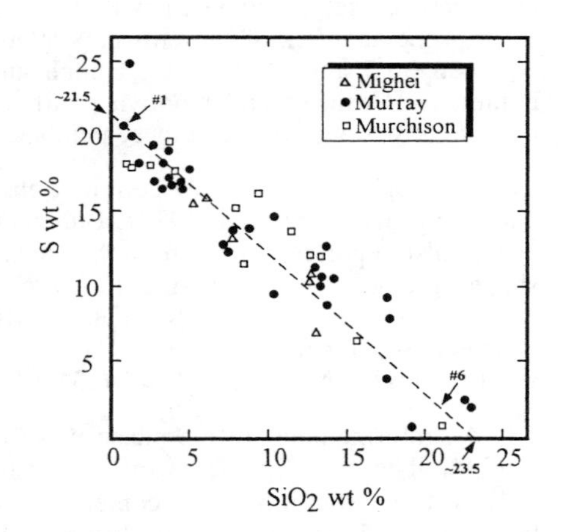

Figure 141. SiO_2 vs. S (wt %) plot showing the range of PCP compositions in Type II PCP in Mighei, Murchison and Murray. After Tomeoka and Buseck (1985).

PCP is often characterized by the undulatory character of tochilinite layers in platy and fibrous crystals (e.g. Barber et al. 1983), which is probably caused by a misfit between the mackinawite and brucite layers. This misfit may be, in part, controlled by composition because the brucite sheet can also contain Fe; Ni (and Cu) may substitute into the mackinawite. The high curvature observed in some grains, as well as the rolled morphologies (e.g. Akai 1982, Tomeoka and Buseck 1983, 1985), can be accounted for by extreme misfits between the different sheets.

Sulfides. Four main types of sulfide have been recognized in CM chondrites (Kerridge et al. 1979c): *troilite* (FeS), *pyrrhotite* ($Fe_{1-x}S$), *pentlandite* ($(FeNi)_9S_8$) and an intermediate sulfide which has compositions between troilite-pyrrhotite and pentlandite. All of these phases show considerable compositional heterogeneity and, with the exceptions of troilite and pyrrhotite, are all found intergrown with each other (Kerridge et al. 1979c, Brearley 1995). An additional Cr,P-rich sulfide phase has also been recognized in some CM chondrites (e.g. Bunch et al. 1979, Nazarov et al. 1997).

Fuchs et al. (1973) reported angular grains of *troilite* with a fragmented appearance in the matrix of Murchison and found that it was several times more common than pentlandite. However, they did not recognize pyrrhotite, which is also common in this meteorite (e.g. Kerridge et al. 1979c), so it is probable that some of the troilite grains were mistaken for pyrrhotite (Ramdohr 1973). Troilite grains, <50 μm in size, also occur in Y-791198 (Kojima et al. 1984, Metzler et al. 1992) and WIS91600 (Birjukov and Ulyanov 1996). Troilite often has a mottled appearance due to the presence of exsolved blebs of finely disseminated pentlandite (<1-5 μm in size) (Fuchs et al. 1973) and oriented lamellae of exsolved pentlandite can also occur (Ramdohr 1973). In Murchison, the Ni contents of troilite in pentlandite-free regions can reach as low as 0.2 wt %, but up to 1.7 wt % where pentlandite is common (Fuchs et al. 1973). Kerridge et al. (1979c) also reported low Ni contents (<0.5 wt %) in troilite from Murchison.

DuFresne and Anders (1962) found *pentlandite* in a heavy mineral separate from Mighei and it is now a widely recognized phase in CM chondrites (e.g. Fuchs et al. 1973, Ramdohr 1973, Bunch and Chang 1980, Metzler et al. 1992, Brearley 1995, Birjukov and Ulyanov 1996) and CM chondrite clasts (e.g. Bunch et al. 1979). Müller et al. (1979), Barber (1981) and Brearley (1995) have also unambiguously identified fine-grained pentlandite by TEM, commonly associated with very fine-grained phyllosilicates in the matrices of several CM chondrites. Ramdohr (1973) observed that pentlandite in Murray can occur closely associated with troilite, sometimes as exsolved lamellae. Electron microprobe analyses of Murchison pentlandite (Kerridge et al. 1979c) show Ni contents between 30.5 and 32.7 wt %. Bunch and Chang (1980) reported more variable Ni contents in Murray and Nogoya pentlandites; 20.7-34.0 wt % and 23.7-33.3 wt %, respectively, suggesting that these pentlandites may be intergrowths of two phases.

Pyrrhotite is a relatively common phase in the matrices of CM chondrites and CM clasts (e.g. Kerridge et al. 1979c, Bunch and Chang 1980, Bunch et al. 1979). Brearley (1995) also reported pyrrhotite in the unusual CM chondrite Bells that occurs as irregular shaped grains, 5-30 μm in size. In Murchison, pyrrhotite Ni contents range from 0.04-3.22 wt % (Kerridge et al. 1979c) and in Murray between 0.04 and 1.24 wt % (Bunch and Chang 1980). Pyrrhotites from a CM clast in the Jodzie howardite are sometimes zoned and have higher Ni contents (2.13-6.3 wt % Ni) (Bunch et al. 1979).

An intermediate sulfide phase was identified in CM chondrites by Kerridge et al. (1979c). The exact nature of this phase remains to be established, but Kerridge et al. (1979c) found that it has a Ni content between 0.5 and 24.5 wt %, suggesting that it may be a fine-grained intergrowth of different phases.

Bunch et al. (1979) first reported a new sulfide phase, rich in Cr and P, in a CM clast in the Jodzie howardite. This phase also occurs in other CM chondrites (Bunch and Chang 1980, Nazarov et al. 1997). Compositionally, this phase is an Fe, Ni sulfide with 3.1-23 wt % Cr, 1.2-4.0 wt % P and 3-10.4 wt % O (Bunch et al. 1979). TEM studies of this phase from Murchison and Mighei (Devouard and Buseck 1997) show that the sulfide actually consists of myriad crystalline domains and is probably a mixture of two, as yet unidentified phases, one of which may be a phosphide, based on electron energy loss spectroscopy (EELS).

Carbonates. Carbonates have been widely observed in CM chondrites (Kvasha 1948, Fuchs et al. 1973, Kerridge and Bunch 1979, Bunch and Chang 1980, Barber 1981, Armstrong et al. 1982) and CM clasts (e.g. Wilkening and Clayton 1974). Calcite is widespread, with dolomite being much rarer (Johnson and Prinz 1993). Rare aragonite also occurs (Müller et al. 1979, Barber 1981).

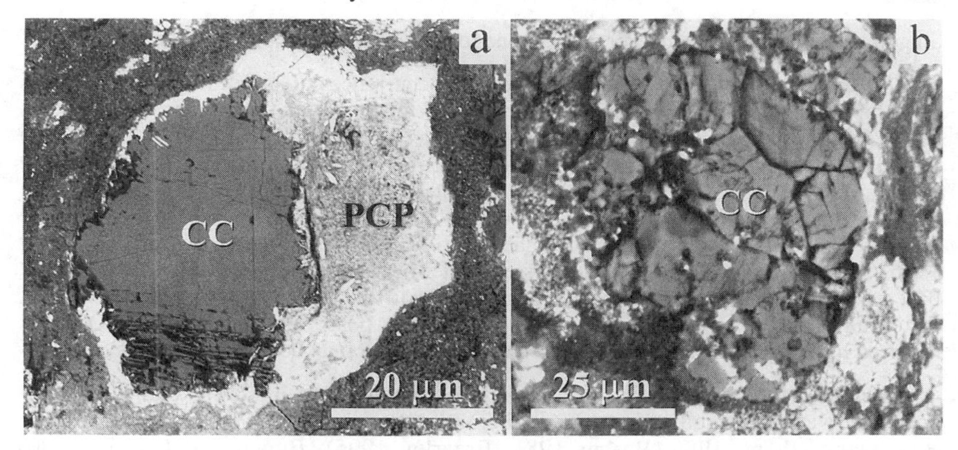

Figure 142. BSE images showing examples of the occurrence of calcite in the matrix of the CM chondrite Murchison. (a) irregularly-shaped calcite crystal completely enclosed within a PCP-rich object, (b) part of a partially fragmented spherular aggregate of calcite grains.

Calcite is the most common carbonate phase in CM chondrites (e.g Kvasha 1949, Fuchs et al. 1973, Bunch and Chang 1980, Barber 1981, Johnson and Prinz 1993, Lee 1993, Brearley 1995). In Murchison, it occurs as grains a few µm up to 80 µm in size (Fuchs et al. 1973) and occurs dominantly within the matrix, often associated with PCP-rich objects and inside altered CAIs and chondrules. Calcite grains occur as polycrystalline aggregates or as single grains (commonly with a rhombic form and often containing fine twins) (Fig. 142a,b) (e.g. Metzler et al. 1992) and as submicron grains in patches or in vein-shaped aggregates (Johnson and Prinz 1993). Barber (1981) found that calcite has low dislocation densities, but is often twinned on the $\{01\bar{1}8\}$ planes indicating that twin formation occurred by shock deformation. Calcite can sometimes occur associated with anhydrite (e.g. Bells, Brearley 1995) or hemihydrate and anhydrite (e.g. Cold Bokkeveld, Lee 1993).

Johnson and Prinz (1993) found that calcites in CM chondrites are pure $CaCO_3$, with minor FeO (mean = 0.35 wt %) (0.49 wt % $FeCO_3$). Mn and Mg were detectable in a few grains, never exceeding 0.51 wt % MgO and 0.83 wt % MnO. Trace element data (B, Na, Mg, Fe, Mn, Sr and Ba) for calcite, measured by ion microprobe, have been reported from several CM chondrites (Boriskino, ALH 83100, Nogoya, Murchison) (Riciputi et al. 1994).

Johnson and Prinz (1993) first identified *dolomite* in CM chondrites, although it appears to be relatively rare and certainly much less common than calcite. The dolomites contain variable, but significant FeO (1.48-4.27 wt %) and MnO contents (0.44-4.17 wt %). Riciputi et al. (1994) have reported B, Na, Mn, Sr, and Ba concentrations, measured by ion microprobe, for dolomites in the CM chondrites Boriskino and Nogoya. The trace elements do not follow their expected equilibrium partitioning behavior, consistent with the suggestion by Johnson and Prinz (1993) that the dolomites may have grown metastably.

Aragonite has been identified by electron diffraction in Cold Bokkeveld, Cochabamba and Murchison (Müller et al. 1979, Barber 1981). It is rare and occurs as much smaller grains than calcite, but the two phases may be associated. The aragonite is often very finely twinned and in Cochabamba has been observed as highly defective dendritic overgrowths on defect free calcite. Barber (1981) has suggested that the aragonite may have formed after calcite, due to a decrease in temperature.

Sulfates. Although not widespread, several occurrences of sulfates in CM chondrites have been reported. Rare *gypsum* has been reported in Mighei (DuFresne and Anders 1962), Murchison (Fuchs et al. 1973) and Nogoya (Bunch and Chang 1980). Lee (1993) described widespread, late stage, crosscutting veins containing *hemihydrate* ($CaSO_4 \cdot 0.5H_2O$) and *anhydrite* in the matrix of Cold Bokkeveld and inferred that the original phase present was gypsum, which had been dehydrated during sample preparation. *Anhydrite* with a fibrous habit also occurs in the matrix of Bells (Brearley 1995). Fuchs et al. (1973) also suggested that Mg sulfates are present in Murchison, based on the significant amounts of Mg and Na in the water soluble fraction of this meteorite. Finally, an occurrence of a Na-rich sulfate, possible *thenardite* (Na_2SO_4), was reported by King and King (1981) in a fine-grained rim on a chondrule in Murray.

Oxides. *Magnetite* is not an especially common phase in CM chondrites, but is present in variably amounts in many of them (e.g. Fuchs et al. 1973, Bunch et al. 1979, Bunch and Chang 1980, Barber 1981, Brearley 1995). Bunch and Chang (1980) described magnetite in Murray and Nogoya which occurs as small (<1-3 μm) spheres, equant grains and massive units. It also occurs in Cold Bokkeveld (Barber 1981) as isolated grains or as clusters or aggregates of small (<0.05 μm) grains. The unusual CM chondrite, Bells contains an extremely high abundance of magnetite for a CM chondrite (4.4 vol %; Hyman and Rowe 1986). It occurs distributed throughout the matrix as equant grains, 250 nm to 20 μm in size, and some grains have been extensively replaced by a well crystallized, unidentified Fe-oxide (Brearley 1995). Bunch and Chang (1980) also detected *hematite* and γFe_2O_3 by x-ray diffraction in Nogoya.

Other spinels which occur in CM chondrite matrices are *chromite* (e.g. Fuchs et al. 1973, Bunch et al. 1979, Barber 1981), $MgAl_2O_4$ *spinel* with minor Ti, V, Cr and Fe (Barber 1981) and *hercynite* ($(Mg,Fe)Al_2O_4$) with minor Ti and Cr (Barber 1981).

Minor phases in CM chondrite matrices. Several other phases such as olivines, pyroxenes, alkali halides and phosphates occur in CM chondrite matrices. Müller et al. (1979), Barber (1981), Akai and Kanno (1986) and Zolensky et al. (1993) have all observed common small grains of *olivine* (down to 0.1 μm in size) in the matrices of CM chondrites. Matrix olivines in CM chondrites are usually pure forsterites with no defects and some have well developed crystal shapes. However, Klöck et al. (1989) reported compositionally unusual forsteritic olivines in the matrices of Murchison and EET 83226 which contain elevated concentrations of MnO (1.4 wt %) (so-called LIME olivines—Low Iron, Manganese-Enriched olivines).

Orthoenstatite with lamellae on (001) has been reported in Cochabamba (Müller et al. 1979) and Barber (1981) also observed that rare, small clasts of Ca-rich pyroxene in the fine-grained matrices of several CM chondrites. Akai (1986) also reported Fe-Mg pyroxenes in the matrix of Y-791198 and Mg-rich (Fs < 2 mol %), low-Ca pyroxene also occurs in the matrix of the unusual CM chondrite Bells (Brearley 1995).

Fe,Ni metal is extremely rare in the matrices of CM chondrites, but Metzler et al. (1992) reported very fine-grained (<10 μm) blebs of Fe,Ni metal in fine-grained rims in Y-791198, surrounded by phyllosilicate phases. Rare, small grains (~0.2 μm) of the alkali halides, *halite* (NaCl) and *sylvite* (KCl) have been identified by EDS and some electron diffraction data (Barber 1981) in Murchison. *Cl-free apatite* has been found in Cochabamba (Müller et al. 1979) and Brearley (1993a) identified very rare, fine-grained (0.1-0.3 μm) crystals of the *Magnéli phases* Ti_5O_9 and Ti_8O_{15} in the matrix of the unusual CM chondrite Bells.

Primary mineralogy of CO carbonaceous chondrite matrices

Matrices in the CO3 chondrites typically constitute between 30-40 vol % of the meteorite (McSween 1977a, Scott and Jones 1990), although the least equilibrated of the group (ALH A77307, 3.0) has between 45 and 50 vol %. Matrix occurs as a groundmass to chondrules, isolated mineral grains (Fig. 143a) surrounding all the macroscopic components of the meteorite (Metzler et al. 1988, Brearley 1993b). The matrix is extremely fine-grained, essentially anhydrous and has a very low abundance of clastic mineral fragments above 1-2 μm in size.

The least equilibrated CO chondrite ALH A77307 (3.0) has a matrix which consists of a highly unequilibrated assemblage of Si and Fe-rich amorphous silicate material (Fig. 143b), in which fine-grained, submicron crystals of *olivine* (Fo_{100-3}), *low-Ca pyroxene*, *Fe,Ni metal*, *magnetite*, *pentlandite*, *pyrrhotite*, *anhydrite* and minor *mixed layer phyllosilicate* phases occur (Brearley 1993b). Three distinct occurrences of olivine occur: (1) submicron, irregularly shaped grains, with very variable compositions, closely associated with the amorphous component, (2) single grains of well crystallized, pure forsterite (200 nm - >4 μm) and single or clusters of LIME olivine with 1.5-2.5 wt % MnO (e.g. Klöck et al. 1989). Distinct aggregates or clusters of phases a few microns in size are present within the matrix and rims, based on their textural and mineralogical characteristics. Individual aggregates appear to have formed under very different conditions and experienced different thermal histories. In this meteorite, the mineralogy of rim materials appears to be essentially identical to other occurrences of matrix, suggesting that they represent material from the same reservoir.

The matrices of Kainsaz (3.1), Ornans (3.3), Lancé (3.4), and Warrenton (3.6) are dominated by fine-grained (<0.5μm) FeO-rich olivine (Christophe Michel-Lévy 1969, Keller and Buseck 1990b, Brearley 1994), which often appears to occur in closely intergrown clusters. Occasionally, more angular and coarser-grained crystals of olivine occur set within a groundmass of very fine-grained olivine (Fig. 143c). Minor *pyroxene* and accessory phases, *kamacite*, *taenite*, *awaruite* (in Warrenton) and *spinels* (hercynitic spinel and chromite) are also present (Keller and Buseck 1990b, Brearley 1994). A minor (amorphous?) feldspathic component, rich in Na, K and Al also occurs interstitially to the FeO-rich olivines. The olivines in the low petrologic type chondrites are compositionally heterogeneous, but by petrologic type 3.4 are completely equilibrated (Fig. 144) (Fa_{45-50}).

Secondary mineralogy of CO carbonaceous chondrite matrices

Some CO chondrites have experienced incipient aqueous alteration, indicated by the presence of phyllosilicates in their matrices and, in some cases within chondrules.

Phyllosilicates. Kerridge (1964) first observed fibrous phyllosilicates in Lancé and Ornans which were identified as clinochrysotile by electron diffraction. Keller and Buseck (1990b) confirmed the presence of fibrous phyllosilicates with a 7-Å basal spacing in Lancé matrix by HRTEM. The phyllosilicates occur interstitially to FeO-rich olivines and within veins within larger olivines. The fine-grained phyllosilicates occur as randomly oriented crystallites, often no more than a few unit cells in thickness, although grains up to 20 nm thick (measured parallel to c) also occur. Based on EDS analysis and the 7-Å basal spacing, Keller and Buseck (1990b) concluded that this phase was *Mg-rich serpentine* with an Mg/(Mg+Fe) ratio of ~0.73. Brearley (1993b) has also reported very fine-grained phyllosilicate phases, rarely more than 4 or 5 units cells in thickness with 7-, 10- and 14-Å basal spacings within amorphous material in the matrix of ALH A77307. The basal spacings of 7 Å and 14 Å indicate the presence of serpentine and chlorite. Brearley (1993b) also found clusters (0.3-1 μm in size) of an unusual mixed layer phase

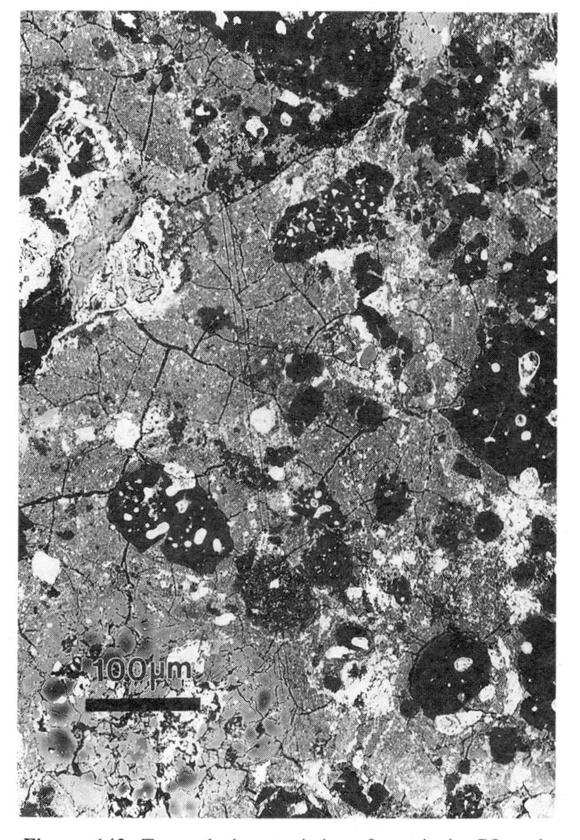

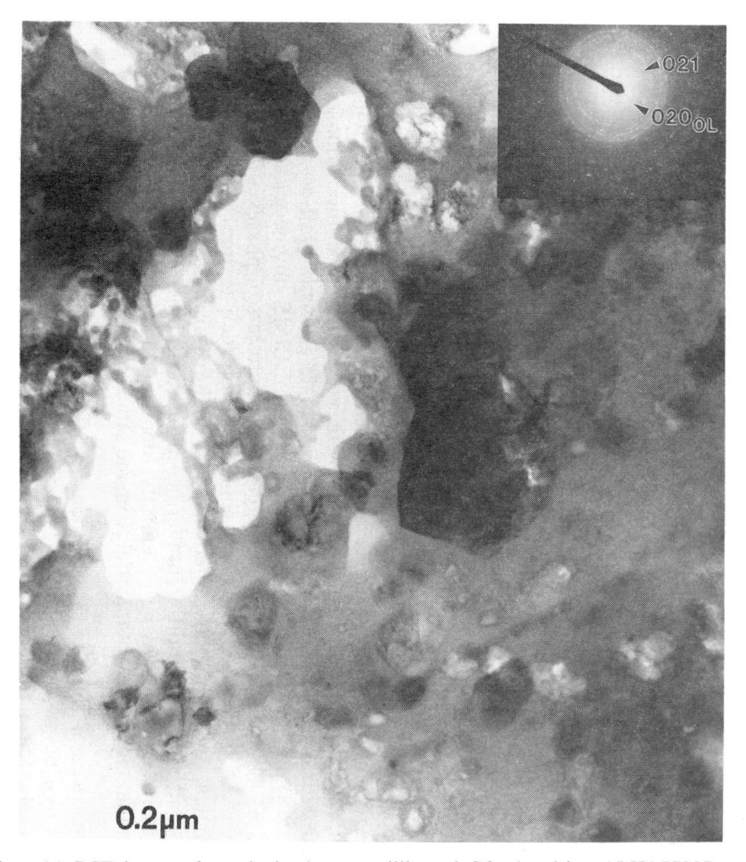

Figure 143. Textural characteristics of matrix in CO carbonaceous chondrites. (a) BSE image of matrix in the unequilibrated CO chondrite, ALHA77307. (b) Transmission electron micrograph of matrix in the unequilibrated CO chondrite, ALHA77307. The matrix consists of a highly unequilibrated assemblage of Si- and Fe-rich amorphous silicate material which acts as a groundmass to other fine-grained phases such as olivine and pyroxene. Used by permission of the editor of *Geochimica et Cosmochimica Acta* from Brearley (1993) Figures 1f and 6b.

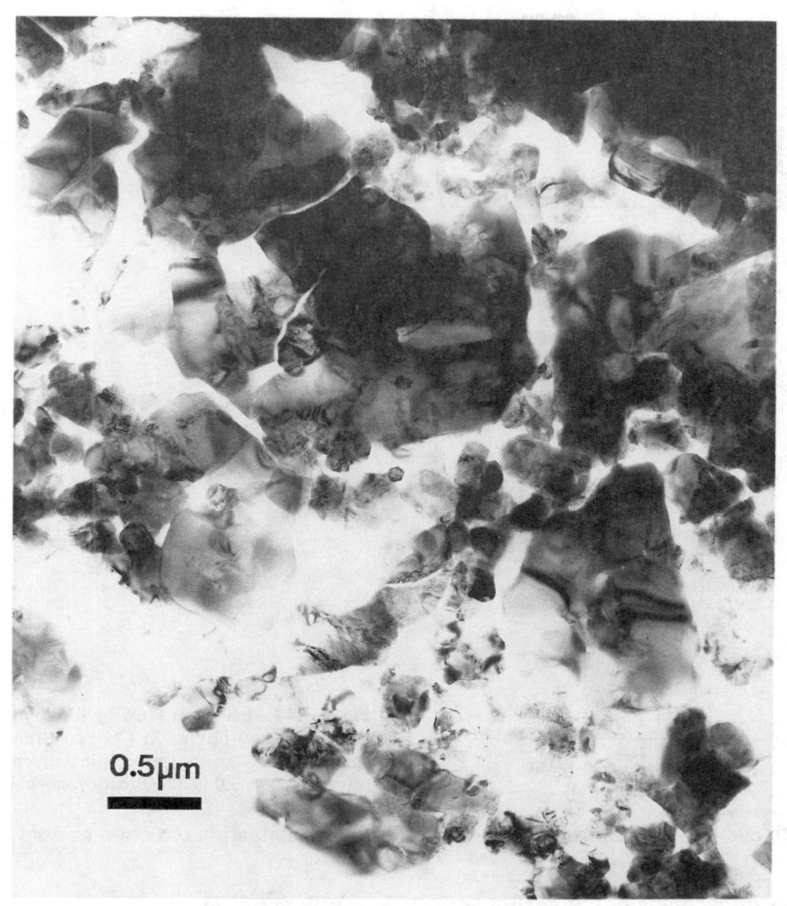

Figure 143 (cont'd). Textural characteristics of matrix in CO carbonaceous chondrites. (c) Transmission electron micrograph showing the characteristic texture of fine-grained matrix in CO chondrites >petrologic type 3.1 (Warrenton, 3.6). The matrix consists largely of fine-grained FeO-rich olivine with a variety of morphologies including lath shaped and subrounded to irregularly-shaped grains.

which consists of randomly interlayered phases with 7-, 8.5-, 9.6-, 1.08- and 1.2-Å basal spacings. Phyllosilicates are not present in the matrices of Kainsaz and Warrenton (Keller and Buseck 1990b).

Oxides. Keller and Buseck (1990b) found a poorly crystalline Fe^{3+} oxide dispersed between matrix olivines in Lancé, intimately associated with fine-grained Fe-bearing serpentine. It also occurs as thin coatings along crystal faces of euhedral olivine grains and along veins in olivine. Keller and Buseck (1990b) obtained two different electron diffraction patterns from these materials, one consistent with well crystallized ferrihydrite. The second phase may be any of several hydrous ferric oxides and also occurs locally along grain boundaries in Kainsaz and Warrenton, indicating extremely minor, incipient aqueous alteration.

Sulfates. Using TEM, Brearley (1993b) observed veins of fibrous *anhydrite* which crosscut regions of fine-grained matrix in ALH 77307. Individual crystallites are 1-2 μm

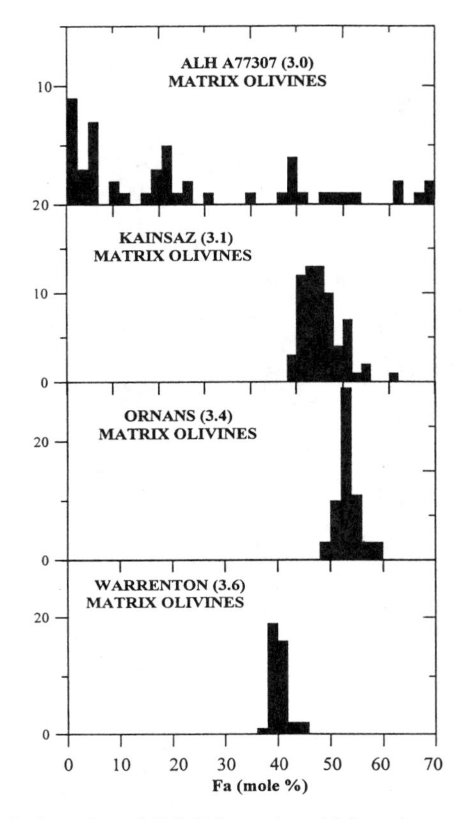

Figure 144. Histogram showing the compositions of fine-grained olivine in CO chondrites (AEM analyses) through the petrologic sequence from petrologic type 3.0 to 3.6 (Brearley, unpubl. data).

in length and 0.1-0.2 μm in width and occur with the elongation direction normal to vein walls.

Primary mineralogy of CV carbonaceous chondrite matrices

The matrices of CV chondrites have been quite extensively studied (e.g. Green et al. 1971, Ashworth and Barber 1975, Housley and Cirlin 1983, Peck 1984, 1985; Kornacki and Wood 1984, Toriumi 1989, Nakamura et al. 1992, Tomeoka and Buseck 1990, Keller and Buseck 1990a, Keller et al. 1994, Lee et al. 1996). Matrix material in the CV3 chondrites constitutes ~35-50 vol % and occurs as both chondrule rims (often layered) (e.g. Bunch and Chang 1980) and as interchondrule matrix. Fine-grained, mineralogically layered rims also commonly occur on CAIs in Allende which have been termed accretionary rims (MacPherson et al. 1985).

The exact primary mineralogy of CV chondrite matrix is the subject of considerable debate. Several authors have argued that FeO-rich olivines which are the dominant component are solar nebular condensates (e.g. Kornacki and Wood 1984b, MacPherson and Grossman 1985). However, Kojima and Tomeoka (1996) and Krot et al. (1995 1997b) have suggested alternatively that they may be the result of complex secondary processes involving aqueous alteration followed by metamorphism. Phases such as feldspathoids, wollastonite, andradite and grossular that are present in CAIs and chondrules and are regarded as secondary alteration phases are also present in the matrices of many CV chondrites. For simplicity, these phases are covered in this section because of the uncertainty in their origin.

Typical CV3 matrix material is somewhat coarser-grained than matrix in other chondrite groups, with an average grain size of ~5μm (Scott et al. 1988a, Toriumi 1989, Keller et al. 1994). However, there is considerable variation in matrix grain size and texture from one CV chondrite to another (Fig. 145). Krot et al. (1995) observed that the matrices of Allende and ALH 84028 (both oxidized CVs) were coarser-grained than matrix in the reduced CV chondrites, Vigarano and Leoville. Compaction of the matrix due to shock may also have occurred in both Leoville and Efremovka (e.g. Nakamura et al. 1992).

FeO-rich olivine is the dominant phase present in CV chondrite matrices (Green et al. 1971, Ashworth 1975, Kornacki and Wood 1984b, Peck 1984, Tomeoka and Buseck 1990, Keller and Buseck 1990, Keller et al. 1994, Krot et al. 1995) (Fig. 146a). Two morphologically distinct types of olivines have been recognized: (1) elongate or acicular, euhedral grains (e.g. Green et al. 1971) and (2) subhedral, rounded grains which are often finer grained (e.g. Toriumi 1989, Keller et al. 1994). The elongate grains can reach up to 20 μm in length and 1-3 μm in thickness (Fig. 146a), whereas the subhedral olivines are much finer grained (<100 nm to 3 μm) and occur in the interstices between the elongate olivines. The olivines show a range of microstructures which vary from one CV chondrite to another. Tomeoka and Buseck (1990), Keller and Buseck (1990a) and Keller et al. (1994) observed a high abundance of planar defects on (100) in olivine in the matrices of Mokoia, Kaba and Bali, respectively. Using TEM, Green et al. (1971) found that matrix olivines in Allende often contain voids (see also Ashworth and Barber 1975) and small inclusions and Peck (1984) observed submicron inclusions of sulfides, Fe,Ni metal and/or magnetite in olivine by SEM. These voids are often lined with thin rims of poorly graphitized carbon (Brearley (1996b, 1997c) which is sometimes associated with nanometer-sized inclusions of pentlandite (Fig. 146b).

The compositional range of matrix olivines varies from one CV chondrite to another indicating that they have been metamorphosed and equilibrated to different degrees (Fig. 147). Kaba has the largest range of olivine compositions (Fa_{10-85}; Peck 1984, Scott et al. 1988a) whilst Allende appears to be the most equilibrated with a well defined compositional peak between ~Fa_{45} to Fa_{56} (Peck 1984). Kaba contains areas of matrix which have been heavily affected by aqueous alteration. In these regions, olivine compositions are Fa_{16-48} with two distinct peaks at Fa_{20} and Fa_{38} (Fig. 148). Less altered regions of matrix contain olivines with a more restricted range of compositions (Fa_{38-54}). Scott et al. (1988a) reported data from Peck (1984), which shows that the average minor element contents for matrix olivines in 5 CV chondrites lie in the range 0.02-0.08 wt % TiO_2, 0.27-0.80 wt % Al_2O_3, 0.11-0.16 wt % Cr_2O_3, 0.27-0.38 wt % MnO, 0.08-0.22 wt % NiO and 0.10-0.33 wt % CaO.

In addition to olivine a wide variety of other phases have been identified in CV3 matrices and fine-grained rims, but they have not been investigated in detail. The assemblage of accessory phases varies from one chondrite to another and there are no clear mineralogical distinctions between their occurrences in the oxidized and reduced subgroups except that the former invariably contains magnetite rather than low Ni metal. Phases that have been reported in CV chondrite matrices are *enstatite*, *high-Ca pyroxene*, *fassaite*, *Fe,Ni metal*, *hedenbergite*, *nepheline*, *sodalite*, *andradite*, *awaruite*, *magnetite* (sometimes framboidal), *chromite*, *ilmenite*, *anorthite*, *spinel*, *troilite*, *Ca phosphate* and *pentlandite* (McSween 1977c, Peck 1983, Kornacki and Wood 1984, MacPherson et al. 1985, Scott et al. 1988a, Keller and Buseck 1990, Tomeoka and Buseck 1990, Nakamura et al. 1992, Keller et al. 1994, Murakami and Ikeda 1994, Lee et al. 1996, Krot et al.

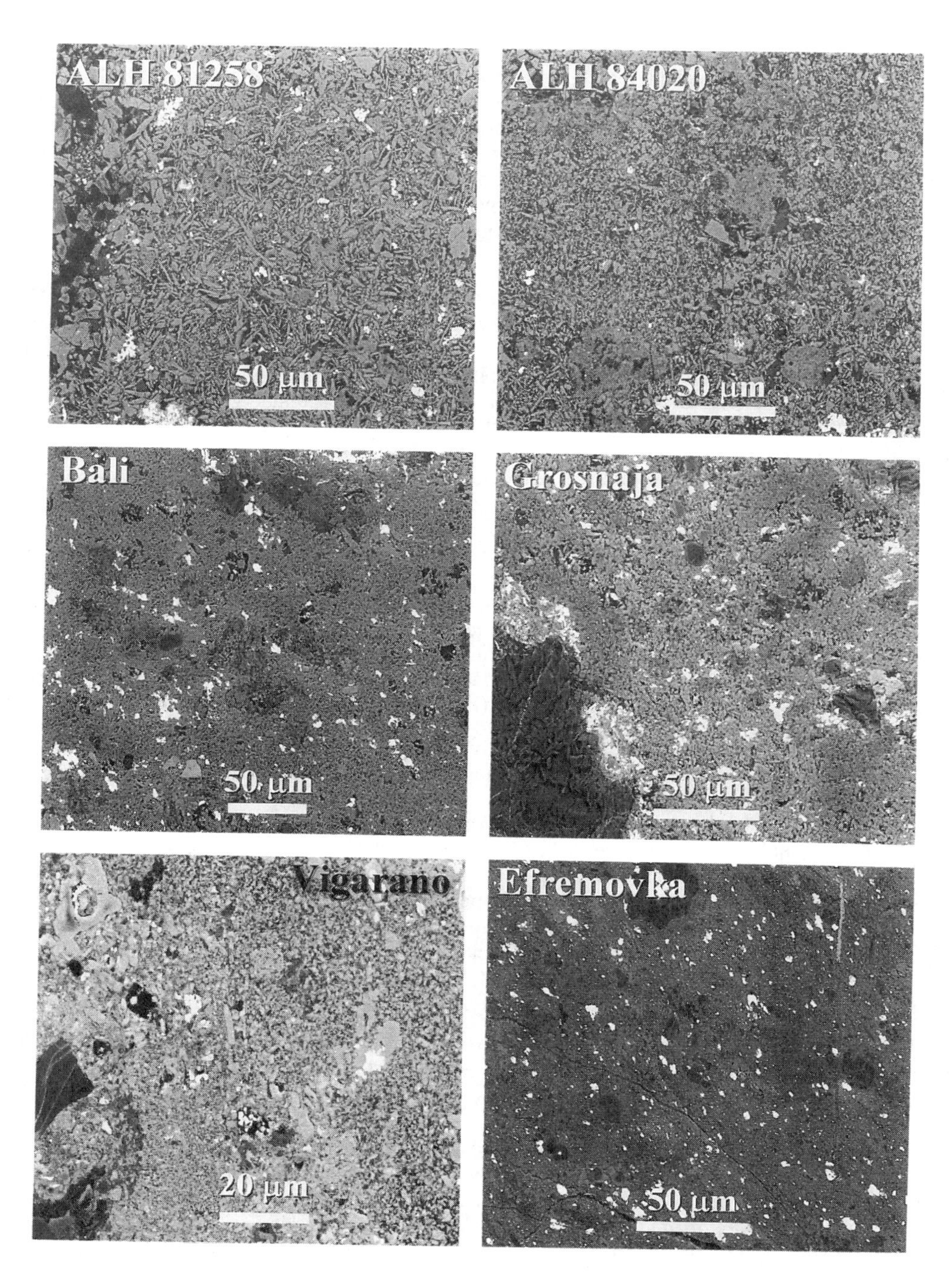

Figure 145. BSE images of the matrices of six different CV chondrites illustrating the variation in texture from one chondrite to another. (a) ALHA81258, (b) ALHA84028, (c) Bali, (d) Grosnaja, (e) Vigarano and (f) Efremovka. Used by permission of the editor of *Meteoritics* from Krot et al. (1995), Figure 7.

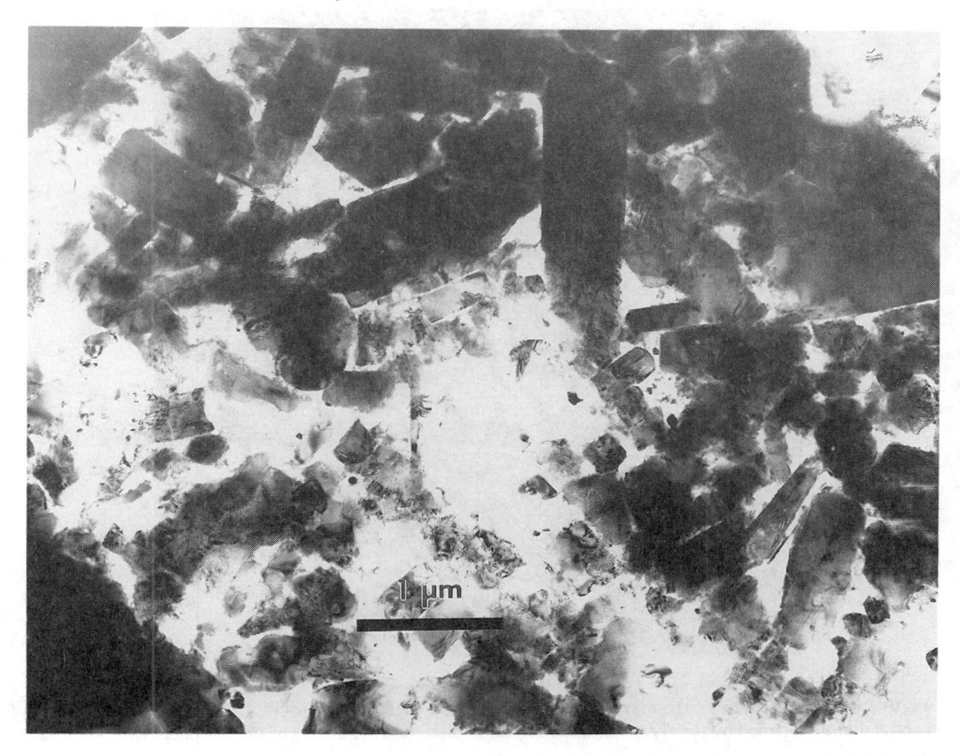

Figure 146. (a) Transmission electron micrograph showing platy and subrounded olivines grains in the matrix of the CV chondrite, Vigarano.

1997). Some of these minerals may have a secondary origin as noted earlier. Many of these phases are present in the distinct rim sequences which surround CAIs (e.g. Allende, MacPherson et al. 1985; Kaba, Hua et al. 1996).

In Allende, Kornacki and Wood (1984b) also described a matrix component, which they termed inclusion matrix, that occurs in olivine-rich aggregates (both rimmed and unrimmed). This inclusion matrix is similar in grain size and mineralogy to normal Allende matrix (i.e. contains platy laths of olivine (Fo_{50-65}), with nepheline (1.4-3.1 wt % CaO and 1.1-1.9 wt % K_2O), sodalite (up to 1 wt % FeO) and salitic pyroxene.

Secondary mineralogy of CV carbonaceous chondrite matrices

Several CV3 chondrites (Mokoia, Kaba, Bali, Grosnaja, Vigarano) have experienced aqueous alteration ranging from extremely minor (e.g. Vigarano; Lee et al. 1996) to highly altered with extensive development of phyllosilicates (e.g. Bali; Keller et al. 1994). The alteration is dominantly in the matrices, but is also evident in chondrules and CAIs in some CV chondrites (e.g. Mokoia and Bali) (Tomeoka and Buseck 1990, Keller et al. 1994).

Phyllosilicates. Tomeoka and Buseck (1990), Keller and Buseck (1990a), Keller et al. (1994) and Lee et al. (1996) have reported fine-grained saponite in Mokoia, Kaba, Bali and Vigarano, respectively. The occurrence of phyllosilicates in these meteorites is very similar although Bali and Kaba are more heavily altered and the alteration in

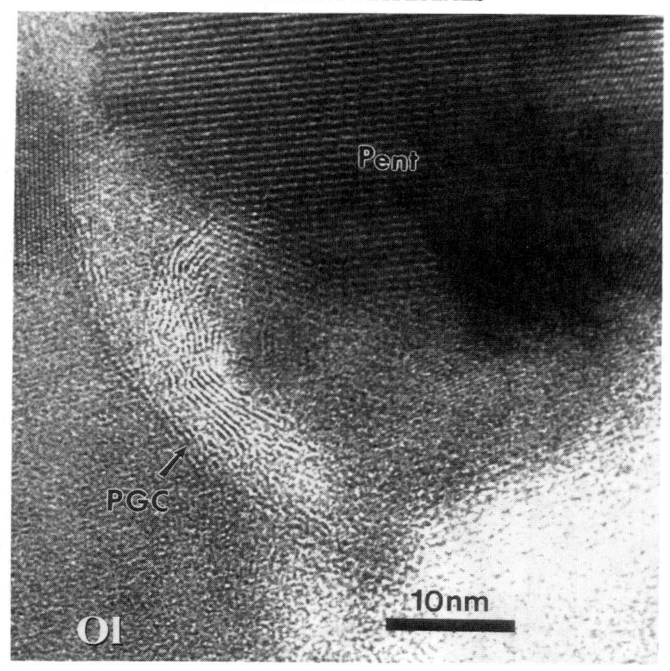

Figure 146 (cont'd). (b) High resolution transmission electron micrograph showing an inclusion of pentlandite surrounded by poorly graphitized carbon in olivine in the matrix of the CV chondrite, Allende.

Vigarano is only incipient. The saponites typically occur interstitially to grains of fine-grained FeO-rich olivine and have sometimes formed by replacement of olivine (Fig. 149a,b), forming oriented, lamellar intergrowths of the two phases (typically along [001] olivine; Lee et al. 1996). In Mokoia the saponite is typically fine-grained, up to 50 units cells in thickness, but can be very elongate and up to almost a micron in length (Tomeoka and Buseck 1990). In Kaba, micron-sized clusters of both coarse and fine-grained saponite occur (Keller and Buseck 1990a). The only CV chondrite which contains different phyllosilicates in its matrix is Grosnaja where serpentine and chlorite are present (Keller and McKay 1993).

Compositionally, the saponites in the matrices of Mokoia, Kaba and Bali are very similar, i.e. Fe-bearing Na saponite (Fe/(Fe+Mg) = 0-0.2), which contains ~1-2 wt % Na_2O (Tomeoka and Buseck 1990, Keller and Buseck,1990, Keller et al. 1994). The saponites in Bali extend to slightly more Mg-rich compositions with a range of Mg/(Mg+Fe) ratios between 0.8 and 1.0.

Oxides. Saponite in Mokoia is often associated within fine-grained *magnetite* (<0.5 μm), in addition to other Fe-rich grains which contain Cr, Mn, Al, Ti and may be spinels. Magnetite also occurs sparsely in Kaba and Vigarano matrices (Keller et al. 1994, Lee et al. 1996) but is extensively developed in heavily altered regions of Bali, where it often occurs as framboidal aggregates (up to 50 μm in size), that consist of numerous, individual magnetites with a grain size of 0.1-0.3 μm. Clusters (<100 μm) of grains of an unidentified, poorly crystalline Fe-oxide phase (possibly *ferrihydrite*) often occur in close association with saponite in Mokoia and Lee et al. (1996) also reported a finely fibrous Fe-oxide phase in Vigarano matrix, which they identified as ferrihydrite.

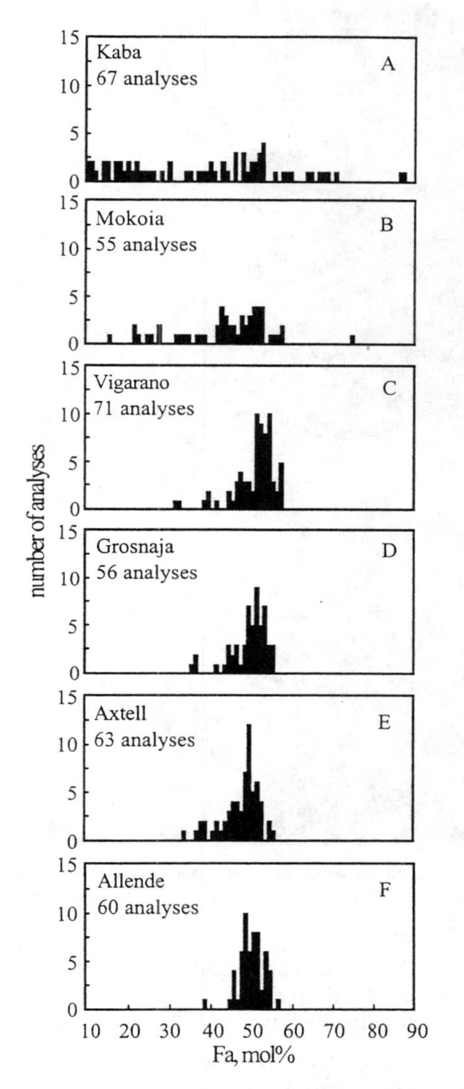

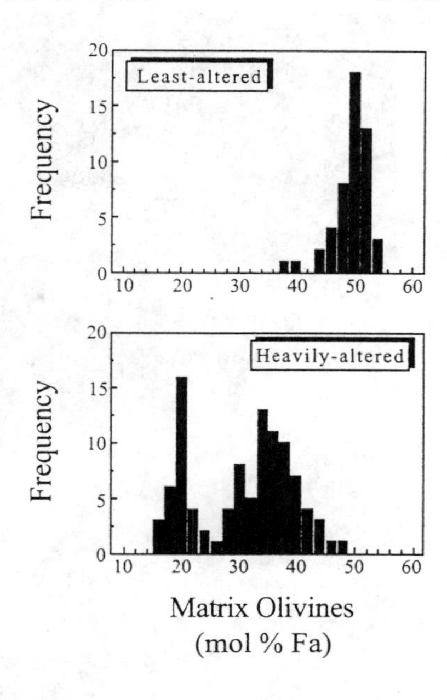

Figure 147 (left). Histograms showing the range in matrix olivine compositions in several CV chondrites:
(a) Kaba, (b) Mokoia, (c) Vigarano, (d) Grosnaja, (e) Axtell, (f) Allende.
After Krot et al. (1995).

Figure 148 (above). Histograms showing the compositional range of matrix olivines in (a) an unaltered region of Bali and (b) an altered region. After Keller et al. (1994).

Other minor secondary phases in CV chondrite matrices are *pentlandite* in Vigarano (Fuchs and Olsen 1973) and Bali (Keller et al. 1994), *pyrrhotite* in Vigarano (Zolensky et al. 1993) and Cl,F-free *apatite* also in Bali (Keller et al. 1994).

Mineralogy of dark inclusions in CV carbonaceous chondrites

The presence of unusual lithic clasts, often with a dark or fine-grained appearance, in CV chondrites has been recognized and studied by numerous authors (e.g. Clark et al. 1970, Fruland et al. 1978, Kurat et al. 1989, Johnson et al. 1990, Kojima and Tomeoka 1994, Krot et al. 1997b, Buchanan et al. 1997). These inclusions have come to be widely termed dark inclusions (DIs). DIs are angular clasts that range in size from <1mm up to several centimeters in size and have been observed in several CV chondrites, including

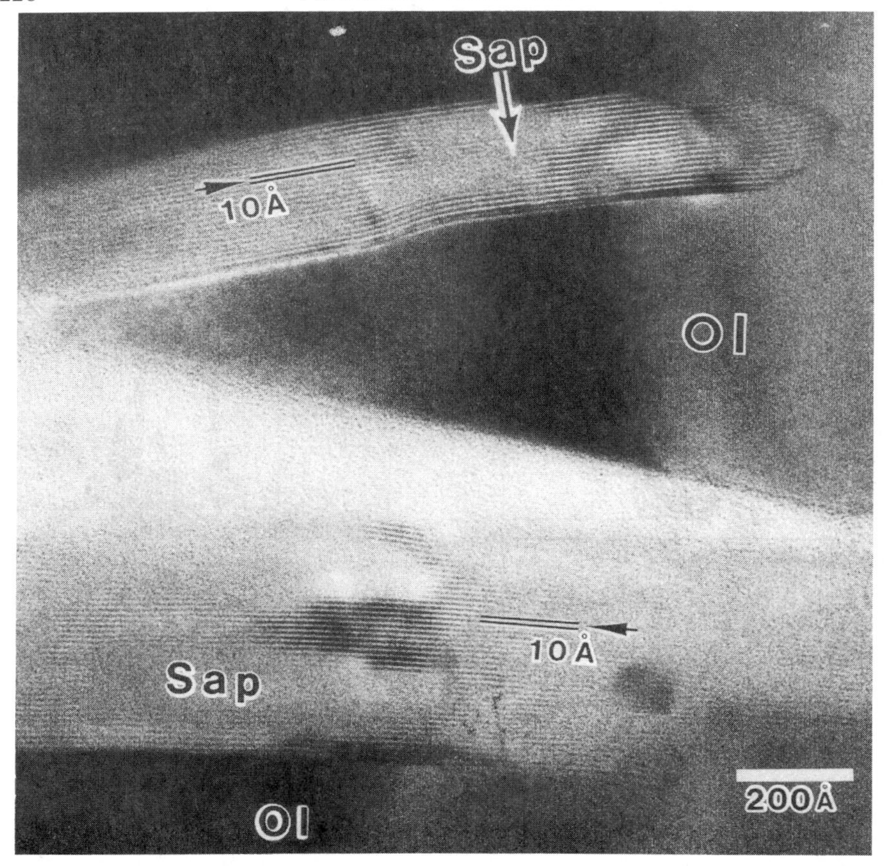

Figure 149. High resolution transmission electron micrograph showing saponite replacing olivine in the matrix of Mokoia (CV3). Used by permission of the editor of *Geochimica et Cosmochimica Acta*, from Tomeoka and Buseck (1990), Figure 4.

Allende, Leoville, Vigarano, Efre-movka and Mokoia. They exhibit considerable diversity in their textures, indicating that they may be derived from multiple sources and/or have experienced complex formational or alteration histories.

DIs have been classified into three different groups, Types A, B and C based largely on their petrographic characteristics (e.g. Johnson et al. 1990, Buchanan 1991, Krot et al. 1995). One endmember, Type A, resembles the host CV chondrites and consist of chondrules, CAIs, etc. embedded in a fine-grained matrix. In Allende, the chondrules and CAIs in the DIs are smaller (0.1-1 mm) than in the host chondrite (0.2-3.0 mm; Johnson et al. 1990). The second endmember, Type B, contains no chondrules and CAIs, but instead consists largely of distinct aggregates of fine-grained, FeO-rich olivine, usually embedded in a matrix of somewhat finer-grained olivine of similar composition. Most DIs, however, appear to be intermediate in character between the two endmembers (Type A/B) and contain porous aggregates of fine-grained olivine (Fig. 150a,b), as well as recognizable chondrules, that consist of regions of porous olivine as well as coarser-grained primary phases such as forsteritic olivine. A third type of dark inclusion, Type C, consists exclusively of fine-grained material with no chondrules, CAIs or mineral

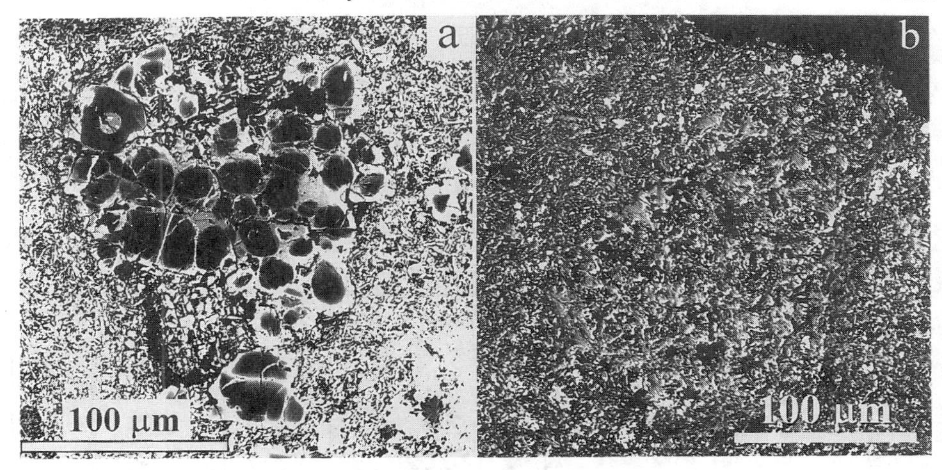

Figure 150. BSE images showing characteristics of Type A/B and Type B dark inclusions in Allende. (a) Image of a chondrule in dark inclusions 4314 showing extensive replacement of the primary chondrule forsterite by platy, FeO-rich olivine as well as the development of fayalitic rims on forsteritic olivine grains, (b) aggregate of platy fayalitic olivine in Allende dark inclusion 4301. The aggregate can be discerned by the fact that the olivines are coarser-grained than those in the surrounding matrix.

fragments and appears to represent a distinct group of inclusions which are not related in any obvious way to Type A or B DIs.

Several different hypotheses have been proposed for the origins of DIs. These include: (a) size sorted fractions of CV3 carbonaceous chondrite material (Fruland et al. 1978), (b) aggregates of primitive nebular condensates (Kurat et al. 1987), (c) samples of carbonaceous chondrite material that was melted and metamorphosed to different degrees within a regolith environment (Bunch and Chang 1983), (d) fragments of CV3 material that reacted with a solar nebular gas prior to their incorporation into the CV3 parent body (Johnson et al. 1990) and (e) fragments of CV3-like material that has been extensively aqueously altered and subsequently thermally metamorphosed within the CV parent body (Kojima and Tomeoka 1994, 1996; Krot et al. 1997).

Mineralogically, Type A, B and A/B DIs are dominated by fine-grained FeO-rich olivine which occurs as randomly oriented grains within the matrix and as distinct porous aggregates in Type B and Type A/B inclusions. The morphologies and compositions of the matrix and aggregate olivines are generally so similar that it is difficult to distinguish between them using BSE imaging techniques.

Type A and A/B dark inclusions. In Allende Type A and A/B DIs, chondrules are typically porphyritic or sometimes barred in texture and many have coarse-grained rims (Johnson et al. 1990). Chondrules in Type A and Type A/B DIs are typically dominated by forsteritic olivine (Fo_{98}), which in Type A/B DIs may be rimmed by fayalitic olivine with a porous appearance. Augitic pyroxene and rare enstatite also occur. Johnson et al. (1990) also found CAIs in Type A DIs, dominantly amoeboid olivine aggregates and fine-grained spinel-rich inclusions, which contain hibonite and secondary minerals such as nepheline and possibly andradite. In Type A Dis, the matrix olivines are fayalitic in composition and may be zoned with forsteritic cores. Minor *high calcium pyroxene*, *Ca Fe-pyroxene*, *low-Ca pyroxene*, *chromite*, *troilite*, *pentlandite*, *Fe,Ni metal*, *Ca phosphate*, *nepheline* and *andradite* have all been reported (Fruland 1978, Johnson et al. 1990). The Ca-Fe pyroxenes contain 0-9 wt % MgO (Fruland et al. 1978).

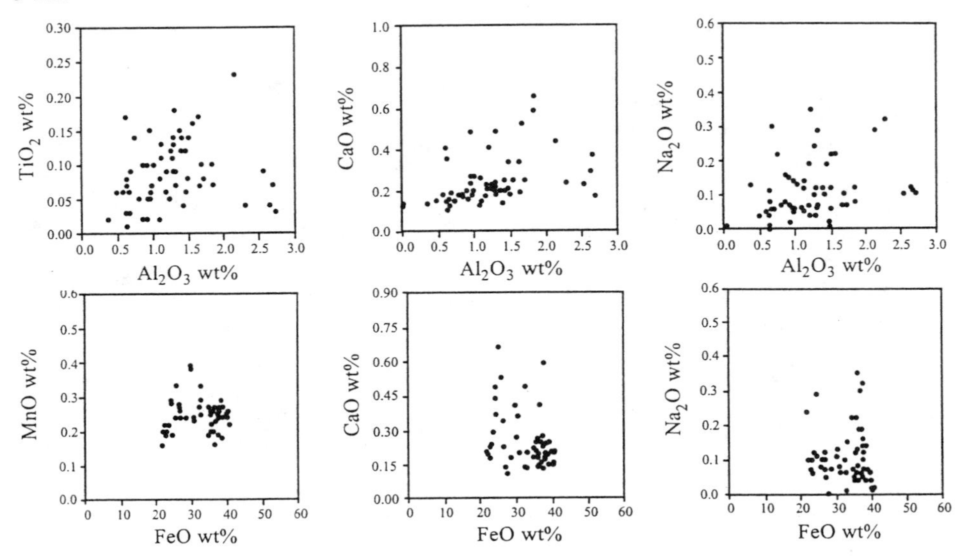

Figure 151. Variation diagrams showing the compositions of olivines in dark inclusion Allende AF. After Kurat et al. (1989).

A DI from Leoville (LV-2) (Kracher et al. 1985) contains chondrules which are significantly smaller than the host and abundant fragmented objects which may be broken up chondrules. Olivine compositions (taken from all occurrences) extend to much more fayalitic compositions ($Fa_{0\text{-}66}$) than in host Leoville ($Fa_{0\text{-}48}$). Low-Ca pyroxene is less abundant and more aluminous (mean ~2.8 wt % Al_2O_3; cf. ~1.9 wt %) than host Leoville and Ca-bearing pyroxene compositions that lie in the miscibility gap are more abundant than in the host. The matrix of LV2 is very fine-grained, contains rare polycrystalline metal, but euhedral magnetite, sometimes occurring as framboidal clusters, is abundant. A rim of extremely fine-grained material 1-2 mm in thickness is present surrounding this DI.

Type B dark inclusions. In Allende Type B DIs, such as Allende AF (Kurat et al. 1989, Kojima and Tomeoka 1996) several different types of porous aggregates occur that are set within a matrix of *fayalitic olivine*, *minor high-Ca pyroxene*, *Fe-sulfide*, Ca *phosphate*, *chromite*, *Fe,Ni metal*, *andradite* and *nepheline* (Johnson et al. 1990, Kojima and Tomeoka 1996). Olivine is the most abundant phase in the aggregates and occurs as stacks or as randomly oriented grains. Grains, <1 μm in size, are equant but coarser-grained olivines (up to 30 μm) have elongate or lath-like morphologies (Fruland et al. 1978, Kurat et al. 1989, Kojima and Tomeoka 1996), similar in character to Allende matrix. Olivine in the matrix is granular to anhedral and platy grains are relatively rare (Kojima and Tomeoka 1996). In DI Allende AF (Kurat et al. 1989, Kojima and Tomeoka 1996), the olivines vary from $Fa_{25\text{-}50}$ (22.0-40.5 wt % FeO), but have more restricted compositions (<4 wt % FeO) within individual porous aggregates (Kurat et al. 1989). These olivines have variable, but often high minor element contents; Al_2O_3 and Cr_2O_3 up to 3 wt %, < 0.25 wt % TiO_2, <0.8 wt % CaO; <0.4 wt % Na_2O, and <0.4 wt % MnO (Fig. 151). No clear correlations exist between any pair of elements although the refractory elements, Ca, Al and Ti, may show weak positive correlations. Kojima and Tomeoka (1996) found that matrix olivines in Allende AF also contain significant NiO (up to ~1.5 wt %) which is positively correlated with S (up to ~1.7 wt %). Kojima and

Tomeoka (1996) suggested these elements were present in microinclusions within the olivine. High minor element (Al, Ti, Cr) contents have also been reported in olivines in a second Type B Allende DI (A4301) (Johnson et al. 1990), but the fayalite content is much more restricted ($Fa_{39\pm3}$). The olivine in this DI contains submicron grains of pentlandite, chromite and poorly graphitized carbon (Brearley and Prinz 1996). Nepheline (1.5-2 wt % K_2O; 2-2.5 wt % CaO) is commonly present filling pore space within the porous aggregates and can sometimes be associated with rare sodalite. Within the aggregates, FeO-poor aluminous diopside (1-1.5 wt % FeO; 3-4 wt % Al_2O_3; ~1 wt % TiO_2) and salitic pyroxenes (11-13 wt % FeO; ~0.5 wt % Al_2O_3; 0.2-0.3 wt % FeO) also occur as minor phases.

Some aggregrates in Type B inclusions contain a significant component of Fe,Ni metal and sulfides in variable proportions associated with the fayalitic olivine. Pentlandite is the dominant sulfide with minor troilite and the metal phase is exclusively awaruite (~ 69 wt % Ni: 2.1 wt % Co; Kurat et al. 1989). Troilite also occurs in massive nodules intergrown with fayalitic olivine, Ca-phosphate and rimmed by a mixture of troilite and pentlandite. Troilite typically contains ~0.1 wt % Co and pentlandite is also Co-bearing (~ 1.6 wt % Co, 17-22 wt % Ni). Kurat et al. (1989) also observed a group of sulfide-bearing objects which have a massive troilite+pentlandite core with a rim of coarse-grained andradite, salitic pyroxene, olivine and nepheline. Inclusions of all these phases occur within the sulfide core, in addition to Ca phosphate, Ti-magnetite and rare native Cu. The clinopyroxene in this group of objects is very TiO_2- (10.2 wt %) and Al_2O_3- (20.0 wt %) rich, but has low FeO contents (~2 wt %). Ti-bearing augite also occurs as a minor phase.

In addition to the phases documented above, Kurat et al. (1989) also observed a large number of rare accessory phases in the Type B DI, Allende AF, including PGE nuggets, HgS, Ca-phosphate (whitlockite and apatite?), ilmenite, perovskite and calcite, associated with barite.

Some Type B Allende DIs have discontinuous, layered silicate rims (50-100 μm thick) which separate them from the host meteorite (Johnson et al. 1990). Up to three distinct layers can occur consisting of: 1) aluminous diopside (<3 wt % Al_2O_3) adjacent to the inclusion, 2) hedenbergitic pyroxene + andradite and/or kirschteinite and 3) an outer porous layer of hedenbergite which has a lower Fe content than in layer 2.

Two examples of Type B DIs from Vigarano have been described in detail by Kojima and Tomeoka (1993). Texturally they are essentially identical to Allende Type B DIs, and consist of aggregates and matrix comprised largely of platy fayalitic olivine (Fa_{40-50}). The aggregates and matrix contain tiny Fe,Ni metal grains ($Fe_{52}Ni_{48}$) to $Fe_{64}Ni_{36}$) and larger (2-15 μm), compositionally heterogeneous Ca-rich pyroxene grains (3.9-19.5 wt % CaO; 0.4-4.3 wt % Al_2O_3; 0.22-1.5 wt % TiO_2). Grains of low-Ca pyroxene, forsterite, Fe sulfide and chromite are minor phases in the matrix. Some exotic aggregates also occur which may be rich in diopside or Fe,Ni metal (Ni_{42-50}). Ca-phosphate also occurs as rare accessory phases. Vigarano DIs do not appear to contain feldspathoids.

Two examples of Leoville Type B DIs have been described briefly by Kracher et al. (1985) and Johnson et al. (1990), but their detailed mineralogy is not well characterized. However, they appear to differ in some respects from Type B DIs in Allende. They contain yellowish, translucent objects which Kracher et al. (1985) suggested were ghosts of severely altered chondrules, set within a fine-grained matrix (Fig. 152). A greenish brown material which has an olivine-like composition (Mg/(Mg+Fe) = 0.61-0.71) has replaced the chondrules but Johnson et al. (1990) have argued that this material is not olivine. This material has a massive appearance and differs markedly from the platy

Figure 152. BSE image showing a ghost chondrule in dark inclusion LV1 from Leoville. The chondrule has been completely pseudomorphed by massive, fine-grained, FeO-rich olivine.

olivine grains which constitute the equivalent porous aggregates in Allende. In the cores of some of these objects, forsteritic olivine is present. Very pure andradite, which may be formed by extensive alteration of CAIs, has also been reported by Kracher et al. (1985). The matrix of these inclusions is extremely fine-grained and contains coarser-grained opaque phases, mostly magnetite, embedded within it.

Mineralogy of CR carbonaceous chondrite matrices

Only the basic mineralogy of matrices in the CR chondrites is known. Weisberg et al. (1993) found that the matrices of several CR chondrites consist of a mixture of hydrous and anhydrous silicates too fine-grained to analyze by electron microprobe, as well as Ca carbonates, sulfides and magnetite framboids and platelets. Zolensky (1991) and Zolensky et al. (1993) found that the matrices of Renazzo, EET87770 and MAC87300 are dominated by fine-grained olivine with accessory pyrrhotite, pentlandite and phyllosilicates. Troilite occurs in Acfer 059/El Djouf 001 (Endress et al. 1994). In Renazzo, EET87770, MAC87300 (Zolensky et al. 1993) and Acfer 059/El Djouf 001 (Endress et al. 1994) the phyllosilicates are typically extremely fine-grained and are usually coherent intergrowths of serpentine and saponite (Zolensky et al. 1993, Endress et al. 1994—Acfer 059/El Djouf 001), identical to those observed in CI chondrites (e.g. Tomeoka and Buseck 1988). Analyses of carbonates in the matrices of CR chondrites show that they range from pure calcite to grains with up to 7.3 wt % FeO, 3.1 wt % MgO and 3.0 wt % MnO (Weisberg et al. 1993).

Mineralogy of dark inclusions in CR carbonaceous chondrites. Dark inclusions which have experienced hydrous alteration are abundant in CR chondrites and have been described by Weisberg et al. (1993), Bischoff et al. (1993b) and Endress et al. (1994) in Renazzo, Al Rais, MAC87320, Y-790112 and Acfer 059/El Djouf 001. According to Weisberg et al. (1993), the DIs are essentially identical to the matrices of CR chondrites themselves and can be considered to be the same material, although the sizes and abundances of carbonates, phyllosilicates etc., are somewhat variable from one dark inclusion to another. Most DIs appear to be closely related and have similar mineralogical affinities (e.g. Acfer 059/El Djouf 001, Endress et al. 1994). They consist of phyllosilicate fragments, chondrule and/or mineral fragments, *magnetite* and *sulfides*

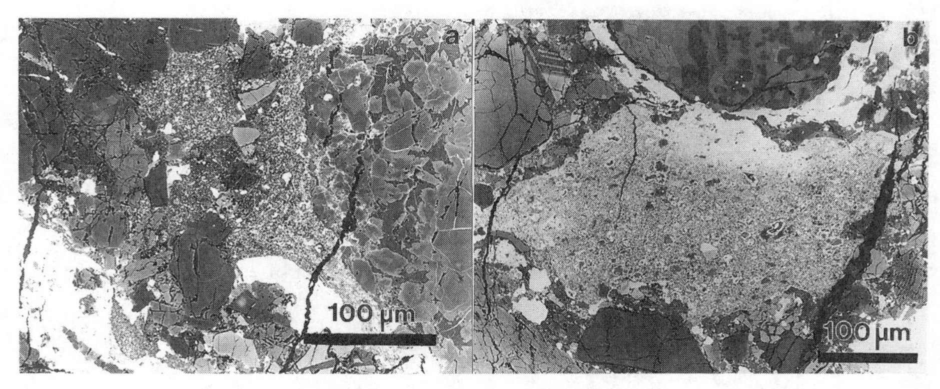

Figure 153. (a) BSE image illustrating a region of clastic interchondrule matrix in the unequilibrated ordinary chondrite, Bishunpur (LL3.1). (b) BSE image of a region of fine-grained matrix in the unequilibrated ordinary chondrite Bishunpur (LL3.1).

(*pyrrhotite*, $Fe_{0.93}S$ with 2-3 wt % Ni and *pentlandite*) set in a fine-grained phyllosilicate matrix consisting of smectite with subordinate serpentine. Magnetite grains, 1-25 µm in size, with framboidal and plaquette morphologies, are commonly present. *Cr-spinel, breunnerite, Ca-phosphate, ilmenite* and *schreibersite* also occur in the matrices and *Fe,Ni* occurs as inclusions within olivine grains (Bischoff et al. 1993b, Endress et al. 1994). *Refractory-rich nuggets* (Os,Mo,Ir-rich) were also reported in one inclusion in Acfer 059/El Djouf 001 (Endress et al. 1994).

Primary mineralogy of unequilibrated ordinary chondrite matrices

The mineralogy of matrix materials in the unequilibrated ordinary chondrites (UOCs) is now relatively well understood, and two distinct occurrences of matrix have been recognized (Ashworth 1977). The first type is a porous or clastic matrix that occurs interstitially to chondrules (interchondrule matrix) (Fig. 153a) and the second is finer-grained material that often rims chondrules and other macroscopic components (Fig. 153b) (e.g. Wlotzka 1983, Huss et al. 1981, Nagahara 1984, Alexander et al. 1989b, Matsunami et al. 1990b). In addition, several other, minor occurrences have been recognized, e.g. fine-grained rims, lumps or clasts (Scott et al. 1984). Fine-grained rims, in particular, have been extensively studied. (e.g. J.S. Allen et al. 1980, King and King 1981, Wilkening et al. 1984, Grossman and Wasson 1983, Matsunami 1984, Alexander et al. 1989b, Zolensky et al. 1993).

Interchondrule matrix consists of extremely fine grained material (<1µm), which acts as a groundmass to coarser-grained and often angular grains of low-Ca pyroxene and olivine (Ashworth 1977, Nagahara 1984, Matsunami 1984, Alexander et al. 1989a, Brearley et al. 1989, Matsunami et al. 1990b), whereas rim materials are typically more uniformly fine-grained (J.S. Allen et al. 1980, King and King 1981, Alexander et al. 1989b). The larger angular fragments have compositions consistent with them being fragments of chondrules (e.g. Alexander et al. 1989a).

The mineralogy of matrix and rims in UOCs is complex, especially that of the accessory phases which varies from one meteorite to another. Electron microprobe and TEM studies have identified (in decreasing order of abundance), *olivine* (Fo_{99} to Fo_{9}), *low-Ca pyroxenes* (En_{98-70}), *augite, albite, Fe,Ni* metal (*kamacite* and *taenite*), *troilite, magnetite, spinel, chromite, anorthite* (Ab_5An_{95}), *pentlandite, whitlockite* and *calcite* in various UOCs (Huss et al. 1981, Ikeda et al. 1981, Nagahara 1984, Matsunami 1984,

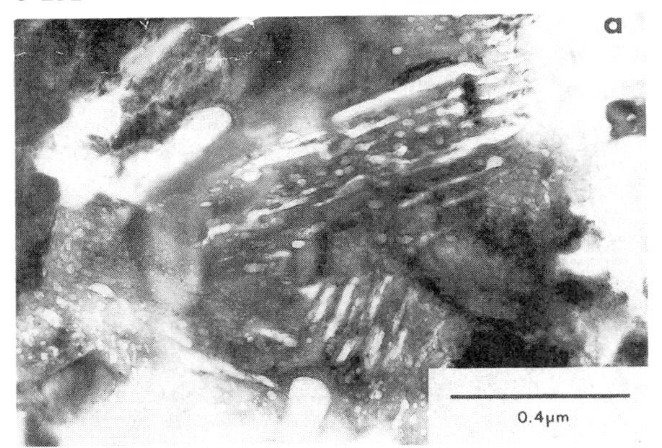

Figure 154. (a) Transmission electron micrograph of a highly porous Fe-silicate in the matrix of Semarkona which may have formed as a result of alteration of FeO-rich olivine. Used by permission of the editor of *Geochimica et Cosmochimica Acta* from Alexander et al. (1989a), Figure 3. (b—below) Transmission electron micrograph of a region of amorphous material in the matrix of Bishunpur. The amorphous phase has been partially replaced by phyllosilicates (smectite) and acts as a groundmass to small grains of Ni-rich metal.

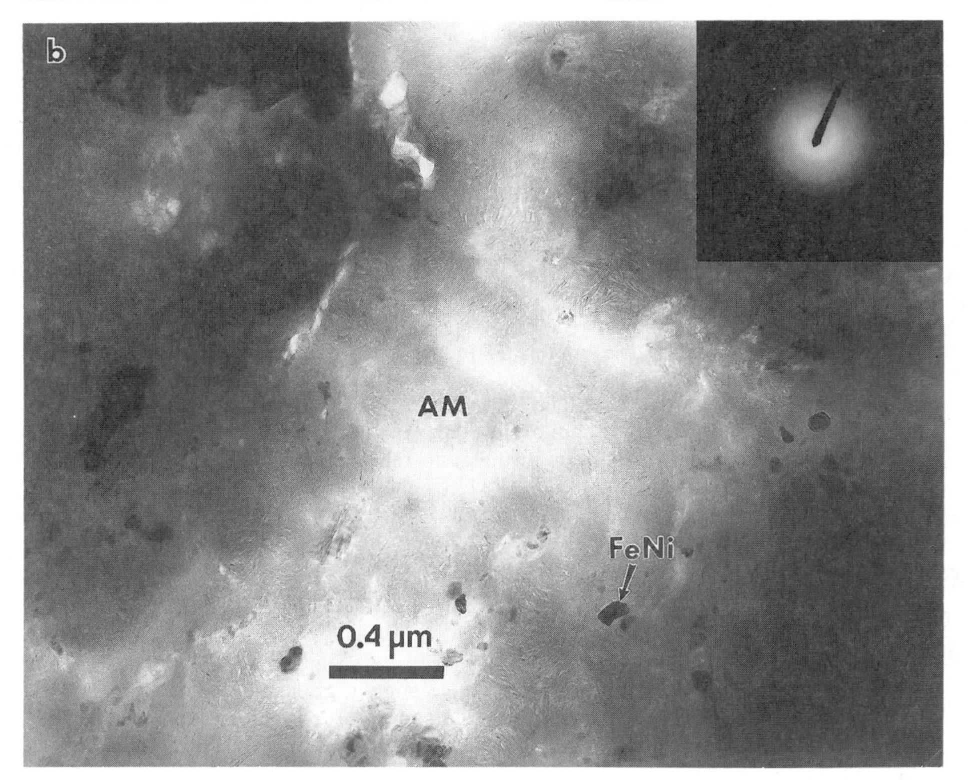

Matsunami et al. 1990b). Matsunami et al. (1990b) also reported rare, small (<5 μm) spherules of an unidentified SiO_2-rich (76-93 wt %) phase in chondrule rims in Chainpur and Sharps.

In the least equilibrated UOC, Semarkona (LL3.0), the matrix consists largely of phyllosilicates (Hutchison et al. 1987, Alexander et al. 1989a; see below), but fine-grained olivine and pyroxene grains are also present. These grains are invariably more Mg-rich than in other UOCs and have a more restricted range of compositions (e.g. ~Fa_5

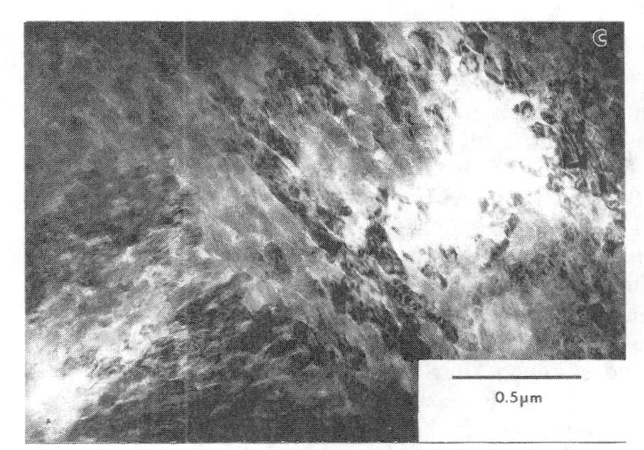

Figure 154 (cont'd) (c) Transmission electron micrograph of a typical region of matrix in Tieschitz which consists of a densely-packed groundmass of FeO-rich olivine. Used by permission of the editor of *Earth and Planetary Science Letters* from Alexander et al. (1989b), Figure 6b. (d) Transmission electron micrograph of matrix in the Sharps (H3.4) chondrite showing the platy, elongate morphology of fine-grained FeO-rich olivine.

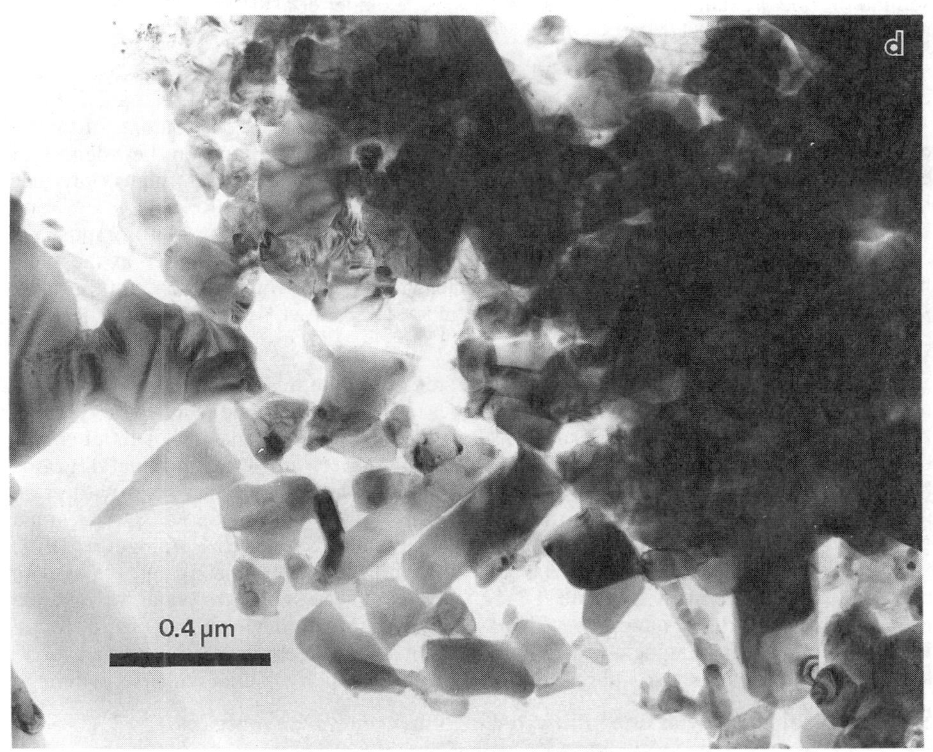

to Fa_{38} for olivine; Huss 1981, Alexander et al. 1989a,b). Both phases appear to have been altered by aqueous fluids; olivine has been replaced by an unidentified phase which contains numerous voids (Fig. 154a) and pyroxene also contains voids. Klöck et al. (1989) reported rare grains of forsteritic olivine which are unusually enriched in MnO (FeO<MnO; up to ~1.4 wt % MnO) in Semarkona matrix. These forsterites have been termed LIME (Low Iron, Manganese-Enriched) olivines, in contrast to normal matrix forsterites which have low MnO contents. Strained taenite with 65 wt % Ni is present in Semarkona matrix (Hutchison et al. 1987) and Alexander et al. (1989a,b) found troilite in fine-grained chondrule rims in Semarkona, whereas fine-grained, Ni-poor sulfides occur

in the matrix (see below).

TEM studies show that the very fine-grained component (<1μm) of matrix in Bishunpur (LL3.1) consists of an amorphous material (Fig. 154b), rich in normative albite, which acts as a groundmass to clastic olivines and pyroxenes (Ashworth 1977, Alexander et al. 1989a,b). Ikeda et al. (1981) also reported amorphous material in the matrices of two UOCS, ALH 764 (LL3) and ALH 77015 (L3), with a composition intermediate between albite and FeO-rich olivine. This amorphous phase is associated with FeO-rich olivine and contains spherical grains of submicron Fe,Ni metal. In other UOCs (e.g. Krymka, Chainpur, Tieschitz), the matrix consists largely of densely packed FeO-rich olivine (Fig. 154c) with a grain size <0.1μm (Alexander et al. 1989, Brearley et al. 1989). The groundmass olivines sometimes have a dendritic or elongate morphology, but more commonly are subrounded or equant. In Sharps, the matrix olivines are more loosely packed (Fig. 154d) and often have an elongate or platy morphology (Alexander et al. 1989a, Brearley, unpubl. data). Minor interstitial feldspathic material (amorphous to semiamorphous) occurs in Krymka and Sharps (Alexander et al. 1989a) (see also Huss et al. 1981, Nagahara 1984).

The compositions of fine-grained olivine in the UOCs become progressively more equilibrated as a function of increasing petrologic type (Huss et al. 1981, Nagahara 1984, Alexander 1989b, Matsunami et al. 1990b) (Fig. 155a). Minor element contents of matrix olivines are highly variable from one chondrite to another depending on their degree of equilibration (Matsunami et al. 1990b) (Fig. 155b). In Krymka (LL3.1) matrix olivines have ~0-0.23 wt % Al_2O_3, 0-~0.35 wt % Cr_2O_3, 0.03-0.8 wt % MnO, <0.2 wt % CaO and <0.6 wt % NiO over the compositional range Fa_{0-100}. MnO has a strong positive correlation with Fa content and NiO, Al_2O_3 and Cr_2O_3 tend to be higher in the most fayalitic olivines (Fig. 155b). In contrast, in Tieschitz (H3.6) which is more equilibrated, Al_2O_3, Cr_2O_3, CaO, and NiO contents in olivine are all <0.2 wt %, but MnO lies between 0.4 and ~0.65 wt % over the range Fa_{40-60} (Fig. 155b).

In Tieschitz, a distinctive matrix material, termed white matrix, rich in Na, K and Al (high in normative nepheline and albite) is present (e.g. Christophe Michel-Lévy 1976, Ashworth 1977, Hutchison et al. 1980, Nagahara 1984). This material fills channels between chondrules, etc. and contains olivine, augite, sulfides and metal grains (Ashworth 1981, Nagahara 1984). Hutchison (1987) also reported an overgrowth of an unusual Mn,Cr-enriched augite ($Wo_{30}En_{58}Fs_{12}$, <2.57 wt % MnO, <2.68 wt % Cr_2O_3) on a grain of low-Ca pyroxene in the white matrix of Tieschitz. Using TEM, Ashworth (1977) found plagioclase and nepheline as clastic grains within the white matrix and grains with compositions consistent with albite (~3-12 wt % Na_2O; Nagahara 1984) and nepheline (Alexander et al. 1989) have been found. However, Alexander et al. (1989) failed to identify nepheline or plagioclase using electron diffraction techniques, but instead found a layered intergrowth of two phases (Fig. 156a), one of which exhibits multiple twinning.

Secondary mineralogy of unequilibrated ordinary chondrite matrices

Kurat (1969) and Christophe-Michel-Lévy (1976) first suggested that the UOC Tieschitz had experienced aqueous alteration and Nagahara (1984) reported fluffy particles in Semarkona that may be phyllosilicates. TEM studies of the matrices of Semarkona (LL3.0) and Bishunpur (LL 3.1) (Hutchison et al. 1987, Alexander et al. 1989a,b) have shown that the hydrous alteration has occurred dominantly in the matrix, but chondrule mesostases have also been affected in some cases (Hutchison et al. 1987). Thin veins filled with carbides, which appear to be secondary in origin, are also present. In Semarkona and Bishunpur, the alteration assemblage consists of fine-grained smectite, maghemite and Ni-rich pyrrhotite (Hutchison et al. 1987, Alexander et al. 1989).

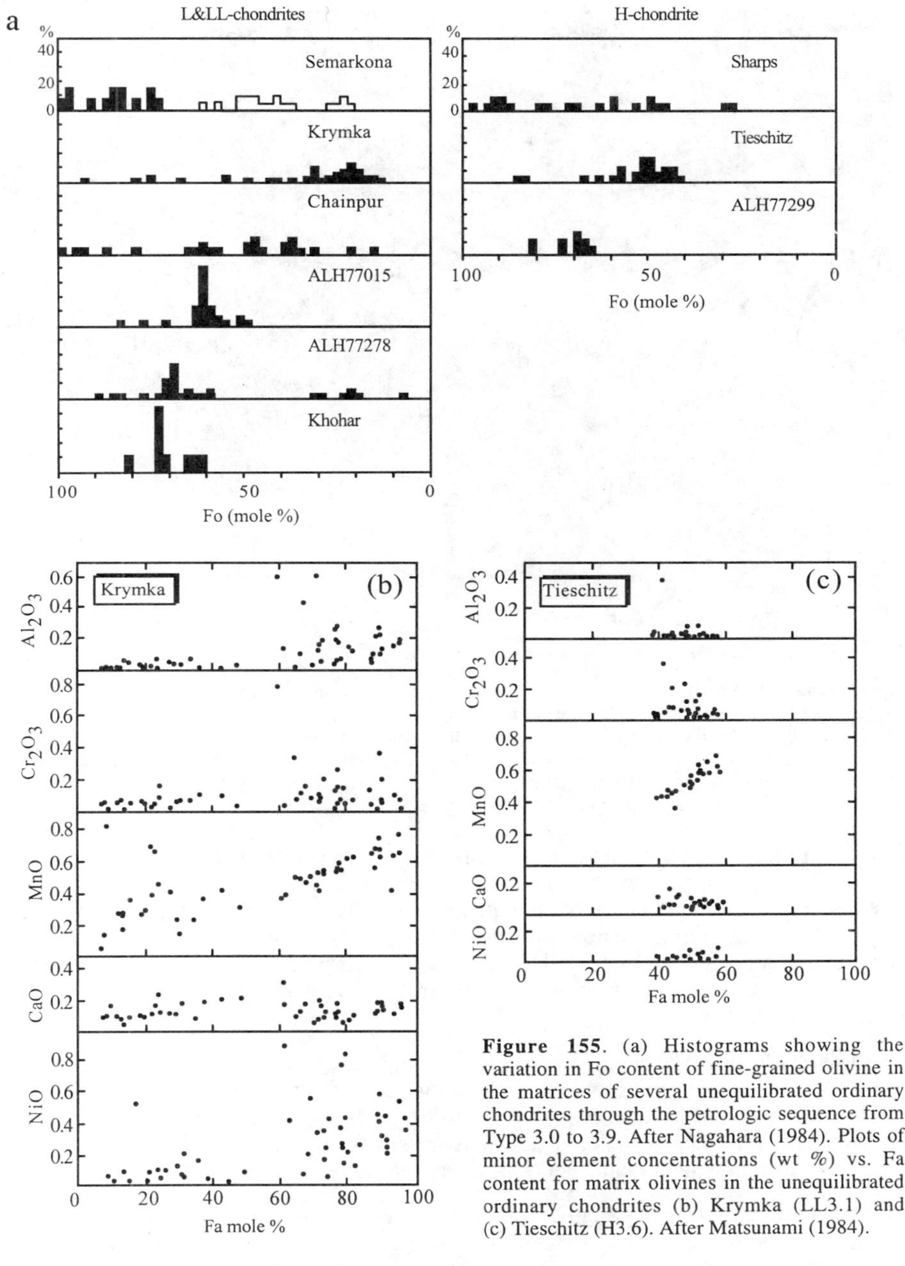

Figure 155. (a) Histograms showing the variation in Fo content of fine-grained olivine in the matrices of several unequilibrated ordinary chondrites through the petrologic sequence from Type 3.0 to 3.9. After Nagahara (1984). Plots of minor element concentrations (wt %) vs. Fa content for matrix olivines in the unequilibrated ordinary chondrites (b) Krymka (LL3.1) and (c) Tieschitz (H3.6). After Matsunami (1984).

Phyllosilicates. The matrix of Semarkona consists largely of microcrystalline, fibrous phyllosilicates with a basal spacing of 9.7-10 Å (Fig. 156b). This phase appears to be an Na-bearing, Fe-rich smectite, based on EDS analysis in the TEM (Hutchison et al. 1987, Alexander et al. 1989b). In some cases, the smectite has directly replaced matrix olivine and low-Ca pyroxene. Alexander et al. (1989b) also found distinct regions of

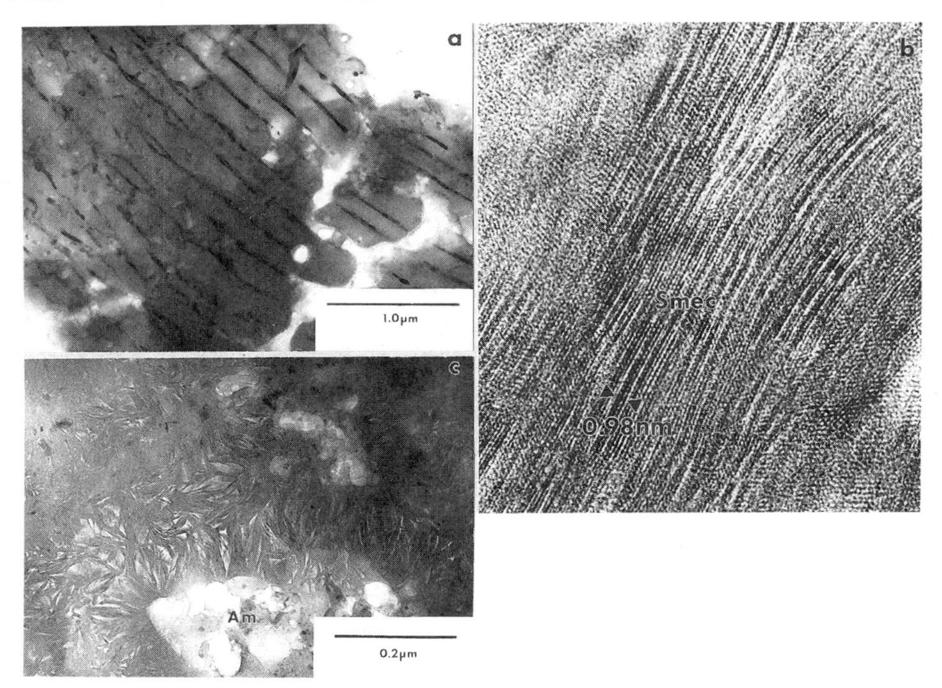

Figure 156a. (a) TEM image of nepheline-like phase showing twinning in the white matrix of Tieschitz. Used by permission of the editor of *Earth and Planetary Science Letters* from Alexander et al. (1989b), Figure 8c. (b) HRTEM image of smectite from the matrix of the unequilibrated ordinary chondrite, Semarkona. (c) TEM showing the development of fibrous phyllosilicates (smectite) in the matrix of the unequilibrated ordinary chondrite, Bishunpur. Used by permission of the editor of *Geochimica et Cosmochimica Acta* from Alexander et al. (1989a), Figure 6a.

coarser-grained smectite in Semarkona matrix. Smectite with both coarse and fine-grained morphologies, similar to those in Semarkona, also occurs in the matrix of Bishunpur (Alexander et al. 1989), but alteration in this meteorite is not as extensive (Fig. 156c). The smectite in Bishunpur is lower in Fe and higher in Ca than that in Semarkona and also contains up to 0.3 wt % Cl. In Semarkona, Alexander et al. (1989b) reported that some large sulfide phases were altering to an Fe-rich phyllosilicate phase, possibly a chlorite or berthierine.

Minor phases. Although *calcite* is not widely developed, it occurs in Semarkona as trains of twinned crystals which follow the general outlines of chondrules or as aggregates of crystals and isolated grains (e.g. Matsunami 1984, Nagahara 1984, Hutchison et al. 1987). It occurs rarely as a vein filling material. Compositionally, it appears to be pure calcite with (FeO+MgO) < 1 wt %.

Alexander et al. (1989a,b) found *troilite* in fine-grained chondrule rims in Semarkona and observed fine-grained, Ni-poor sulfides, possibly *pyrrhotite* in the matrix. A high Ni phase (18.4-26.2 wt %), possibly *pentlandite*, appears to have formed from alteration of troilite. Hutchison et al. (1987) reported fine-grained (<0.1 μm) crystals of Ni-poor *maghemite* in Semarkona matrix associated with smectite. *Magnetite* commonly occurs as strings within the matrix, sometimes associated with carbide, metal or sulfide (Taylor et al. 1981, Hutchison et al. 1987). Carbides (cohenite and haxonite) also occur in veins in Semarkona (e.g Taylor et al. 1981, Hutchison et al. 1987) and the cohenite has a composition intermediate between Fe_3C and Fe_2C (Hutchison et al. 1987). Lee et al. (1995) described the silicon nitride mineral, *nierite* in perchloric acid-resistant residues of Adrar 003 (LL3.2), Inman (L3.4) and Tieshitz (H3.6). The d-spacings are comparable to those of synthetic α-Si_3N_4, which has trigonal symmetry. It occurs as small (~2 × 0.4 μm) lath-shaped grains. A few nierite crystals in Inman and Tieschitz are intergrown with whiskers of a second silicon nitride, which is thought to be β-Si_3N_4. The β-Si_3N_4 whiskers may have acted as nucleation sites for the nierite. The nierite/β-Si_3N_4 ratio appears to be controlled by the metamorphic grade of the chondrite.

Mineralogy of matrices in unequilibrated unique chondrites

We provide here brief mineralogical data on the matrices of unusual chondrites or chondritic clasts which have been described in the literature.

Kakangari. Kakangari is an ungrouped, unequilibrated carbonaceous chondrite (Graham and Hutchison 1974, Davis et al. 1977), rich in fine-grained matrix material (30 vol %; Prinz et al. 1989). Based on TEM studies, Kakangari matrix has a primary mineralogy dominated by enstatite and olivine, with minor albite, anorthite, Cr-spinel, troilite and Fe,Ni metal (Brearley 1989). The matrix consists of an extremely fine-grained (<1μm), non-clastic component and a rarer, coarser-grained, clastic component consisting of angular mineral fragments. The matrix olivines and pyroxenes are Mg-rich ($Fo_{0\text{-}3}$ and Fs_{25}, respectively) (Brearley 1989). The non-clastic component of the matrix consists of irregular, distinct clusters or aggregates of crystals, 2-8 μm in diameter, distinguished by irregular boundaries and/or distinct mineralogies from adjacent mineralogical aggregates (Fig. 157). The enstatites are disordered intergrowths of ortho and clinoenstatite, with lamellae thicknesses that are both odd and even multiples of the 9-Å a-repeat, consistent with an origin by rapid cooling of protoenstatite from high temperature (Buseck et al. 1980). Some of the mineralogical aggregates show equilibrium textural features (e.g. 120° triple junctions), indicating that they experienced high temperature annealing.

Acfer 094. The highly unequilibrated, unique carbonaceous chondrites, Acfer 094 has a matrix mineralogy similar to ALH 77307 (CO) (Greshake 1997). Acfer 094 matrix consists of an unequilibrated assemblage of Si,Mg,Fe-rich amorphous material which is a groundmass to forsteritic ($Fa_{0\text{-}2}$) olivine grains (200-300 nm), low-Ca pyroxene ($Fs_{0\text{-}3}$; 300-400 nm), and Fe,Ni sulfides, mostly Ni-bearing pyrrhotites (up to 6.4 wt % Ni) with rare pentlandite (Greshake 1997). Larger (1-1.5 μm) grains of Ca-rich pyroxene ($Wo_{20\text{-}45}$) are also present. Minor aqueous alteration is indicated by the presence of fine-grained, poorly crystalline phyllosilicates and ferrihydrite (Greshake 1997). The phyllosilicates occur within regions of amorphous material and have 7- and 14-Å basal spacings, suggesting serpentine and chlorite. Serpentine also occurs locally in veins within olivine and probably formed as a result of terrestrial weathering.

Adelaide. The unique carbonaceous chondrite Adelaide has a matrix that is dominated by aggregates of fine-grained anhedral olivine (0.1-0.2 μm), but platy, euhedral olivines also occur (Brearley 1991). The olivines have a compositional range $Fa_{38\text{-}84}$. Clusters of anhedral low-Ca pyroxene (orthoenstatite) (0.1-0.5 μm in size) also occur, with elevated contents of Cr_2O_3 (0.93-1.32 wt %) and MnO (0.39-3.5 wt %). Other minor phases present in the matrix are magnetite, pentlandite and amorphous Fe-rich phase.

Bench Crater. Bench Crater is a small (1.5 × 3 mm) fragment of what appears to be an unusual carbonaceous chondrite returned from the Moon by Apollo 12 and occurs in one single grain mount (12037,188). It consists dominantly of fine-grained matrix which

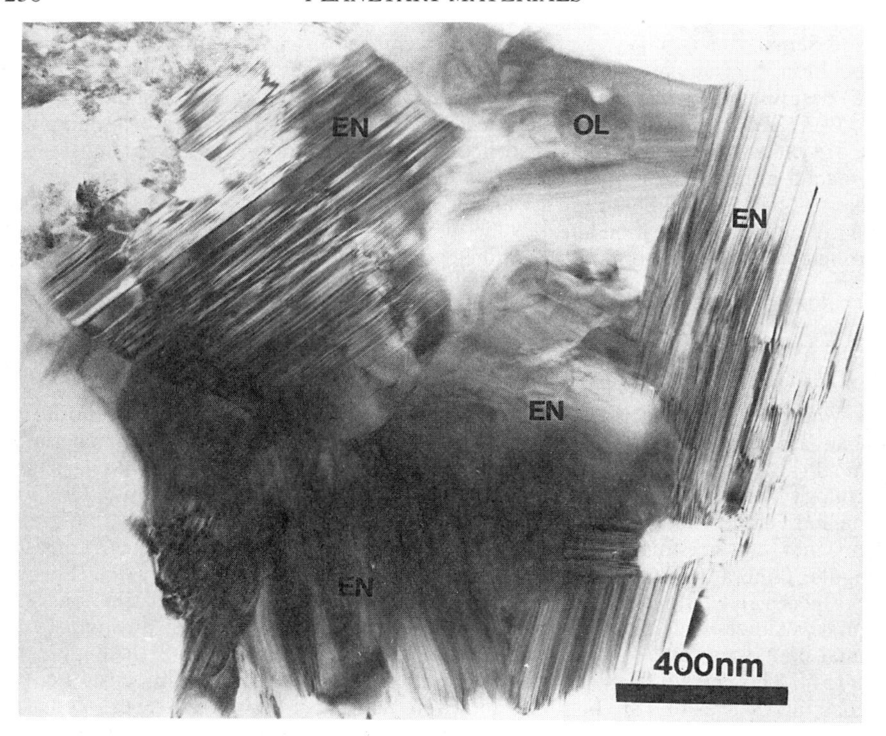

Figure 157. Transmission electron micrograph of a typical mineralogical aggregate in the matrix of the K chondrite, Kakangari. The aggregate consists of several grains of disordered clinoenstatite which contains inclusions of olivine. Used by permission of the editor of *Geochimica et Cosmochimica Acta*, from Brearley (1989), Figure 5a.

contains lath-shaped grains of pyrrhotite, framboids and platelets of magnetite, and minor pentlandite, calcite, dolomite, Mn-rich ilmenite and rare chalcopyrite (McSween 1976, Zolensky 1997). McSween (1976) also identified troilite, but Zolensky (1997) reported pyrrhotite. Aggregates of ferromagnesian silicates, possibly chondrules, occur set in the matrix. TEM studies of matrix in Bench Crater show that it contains fibrous flakes of saponite which are intimately mixed with a phase which has a composition intermediate between olivine and pyroxene (Zolensky 1997). This phase has been interpreted as being the result of the decomposition of serpentine, possibly during shock heating when the meteorite impacted the lunar surface.

LEW 85332. LEW 85332 is an unusual carbonaceous chondrite (Rubin and Kallemeyn 1990) with abundant chondrules and a hydrated matrix. The matrix consists of fine-grained interlayered saponite and serpentine (Brearley 1997b), associated with ferrihydrite, which may have formed during terrestrial weathering. The phyllosilicates have variable Mg/(Mg+Fe) ratios (0.1-0.75) but are typically more Fe-rich than in CI chondrites. Associated matrix minerals are troilite, magnetite and rare olivines and low-Ca pyroxenes.

CM1 clast in Kaidun. Zolensky et al. (1996b) provided a detailed description of an unusual, highly altered carbonaceous chondrite clast in the unique Kaidun breccia, which may be a a CM1 (i.e. completely altered CM) carbonaceous chondrite. Rounded objects

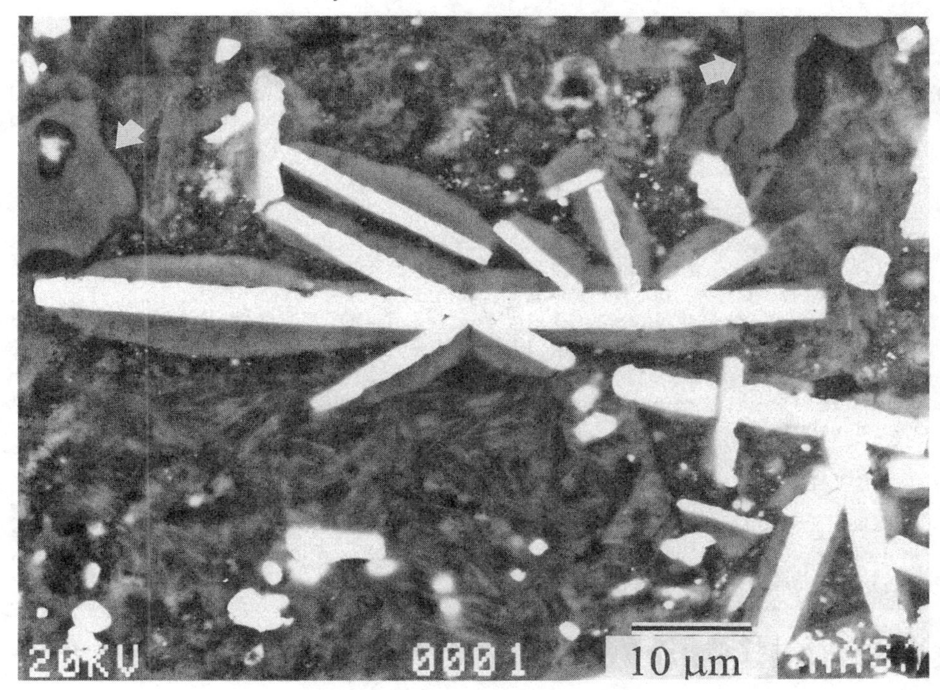

Figure 158. BSE image showing pyrrhotite in a CM1 clast in the Kaidun breccia. The pyrrhotite has a highly unusual morphology and is sheathed by phyllosilicates. Used by permission of the editor of *Meteoritics and Planetary Science* from Zolensky et al. (1996), Figure 6.

consisting dominantly of phyllosilicates are common in this clast and may be completely altered chondrules. The matrix phyllosilicates are dominantly Mg-serpentines sometimes intergrown with saponite and rarely chlorite. The serpentine displays fibrous, platy and cylindrical morphologies. Pyrrhotite with a very unusual acicular morphology (Fig. 158), sometimes coated by sheaths of phyllosilicate minerals, is common and has compositions ranging from $Fe_{0.90}Ni_{0.04}S$ to $Fe_{0.92}Ni_{0.02}S$. Pentlandite ($Fe_{4.6}Ni_{4.3}S_8$ to $Fe_{4.7}Ni_{4.4}S_8$) is less common than pyrrhotite. The other major opaque phase is framboidal magnetite with individual magnetites ranging from 1 to 5 μm. Small (<10 μm) apatite grains, sometimes with a circular appearance with a hollow central cavity, occur throughout the matrix of this meteorite. Rare diopside, andradite ($Ad_{75}Uv_{14}Sch_4Py_4Gr_3$) and melanite garnets have also been observed, in addition to fibrous pyroxene (~$En_{68}Wo_{32}$).

Metamorphosed carbonaceous chondrites

Mineralogy of CK carbonaceous chondrite matrices. The components of matrices in the CK chondrites are coarser-grained than in other chondrite groups as a result of the recrystallization of fine-grained matrix material during the metamorphism that has affected these meteorites. The matrices of CK chondrites are dominated by olivine grains (10-200 μm in size), plagioclase (20-100 μm), low-Ca pyroxene (Fs_{23-29}) and minor magnetite and sulfide grains (20-100 μm) (Scott and Taylor 1985, Kallemeyn et al. 1991). In Karoonda, finer grained material (<5 μm) may constitute ~10 vol % of the matrix (e.g. Brearley et al. 1987) and consists of augite (<2 μm), low-Ca pyroxene

($Fs_{\sim 27}$), anhedral olivine (<2 μm, ~Fa_{29}) and subhedral magnetite (<1 μm containing ~2 wt % Cr and 1 wt % Al). The smaller olivines often have high dislocation densities. Isolated grains of olivine and pyroxene from <20 to 1000 μm in size also occur distributed through the matrix. Magnetite and sulfide (pentlandite) inclusions are commonly found in matrix olivines, as well as in chondrule olivines (Scott and Taylor 1985, Keller et al. 1992, Rubin 1992).

The matrix olivines are unzoned and equilibrated with means between ~Fa_{29} and Fa_{33} (Scott and Taylor 1985, Kallemeyn et al. 1991) and Keller et al. (1992) reported 0.06 wt % NiO, 0.22 wt % MnO and <0.03 wt % CaO in olivine in Maralinga. Plagioclase in the matrices of PCA 82500, Karoonda (Scott and Taylor 1985) and Maralinga (Keller at al. 1992) have bimodal compositions (~An_{20-35} and ~An_{75-95}). Keller (1993) found that matrix plagioclase grains in Maralinga have sodic cores (An_{34-52}) with calcic plagioclase overgrowths (An_{98-76}). TEM studies show that the plagioclase has *e*- and *b*-type reflections in [001] zone axis electron diffraction patterns, indicative of the presence of Huttenlocher intergrowths.

Mineralogy of matrices in Coolidge and Loongana 001 carbonaceous chondrite grouplet. The matrix of Coolidge consist mostly of submicron materials with embedded crystals of micron-sized olivine (~Fa_{20-30}) and pyroxene which are probably fragments of chondrules (Noguchi 1994). The dominant phase in the fine-grained component of matrix is FeO-rich olivine with minor pyroxene and plagioclase. An Fe-rich phase, possibly a weathering product of metal, is also probably an important component.

Matrix mineralogy of aqueously altered and metamorphosed carbonaceous chondrites

Several carbonaceous chondrites have been recognized that do not readily fit into any of the existing carbonacous chondrite groups, but appear to have some affinities to both CI and CM carbonaceous chondrites, e.g. Y-82162, Y-86720 and Belgica 7904 (Ikeda 1992). It is generally agreed that these chondrites have experienced a complex history which probably involved aqueous alteration followed by a period of thermal metamorphism that resulted in the partial to complete dehydration of the phyllosilicate phases. Carbonaceous chondrite clasts that are probably related to these types of chondrites have been recognized in some howardites (e.g. Bholgati; Buchanan et al. 1993). The matrices of these meteorites now consist largely of dehydrated phyllosilicates, with minor amounts of opaque phases, carbonates and phosphates. They also contain a number of unusual clast types which have not been recognized in other carbonaceous chondrites. As a consequence, the mineralogy of these meteorites is extremely diverse, but will only be reviewed briefly here (see also Ikeda 1992 for a detailed review).

Yamato-82162. The mineralogy and petrology of Yamato (Y)-82162 has been described by Tomeoka et al. (1989b), Tomeoka (1990b), Akai (1990), Ikeda (1991,1992) and Bischoff and Metzler (1991). It appears to have the closest affinities to the CI chondrites. The matrix consists of dehydrated phyllosilicates, with minor amounts of pyrrhotite and Mg-Fe carbonates (breunnerites). The dehydrated phyllosilicates give diffraction rings consistent with olivine and probably formed by the thermal decomposition of serpentine and/or saponite (Tomeoka et al. 1989b, Akai 1990). In some cases the dehydration is incomplete and oriented intergrowths of olivine and partially transformed phyllosilicates are present (Akai 1990). Olivine (Fo_{65-83}; Zolensky et al. 1989b, 1991a) often contains irregularly shaped voids (Akai 1994). Minor enstatite and subcalcic pigeonite also occur. Coarser-grained phyllosilicates sometimes occur as isolated clusters, 50-300 μm in size, and in veins (1-200 μm wide). Using TEM, Tomeoka (1990) found that these phyllosilicates consist of a coherent, disordered intergrowth of two phases with d-spacings of ~10 Å and 13-15 Å, which may be Fe-bearing saponite and chlorite, respectively.

The matrix phyllosilicates have compositions close to smectite, yield higher totals and contain more Na_2O (1.4-4.6 wt %) than is typical for CI chondrites (Tomeoka et al. 1989b). An Na-rich trioctahedral mica such as Na phlogopite (Tomeoka et al. 1989b) may be present. Ikeda (1992) has suggested alternatively that they are actually a mixture of chlorite, serpentine and Na-bearing smectite. Mineralogically distinct clasts in Y-82162 often contain relatively coarse-grained phyllosilicates that have compositions that Ikeda (1992) described in terms of solid solutions between talc ($Mg_6Si_8O_{20}(OH)_4$) and sodium-aluminum talc, $NaMg_6(Si_7Al)O_{20}(OH)_4$. They appear to be mixtures of chlorite and sodian talc, with varying proportions in different clasts.

The carbonate mineralogy of Y-82162 differs from other CI chondrites in being dominated by breunnerite, with only minor dolomite and calcite (Tomeoka et al. 1989b, Ikeda 1992). Rare rhodocrosite (44.91 wt % MnO) also occurs in an unusual clast (Ikeda 1992). Minor oxide phases that occur in Y-82162 matrix and clasts (Ikeda 1992) include magnetite, usually pure, but sometimes with up to 5.49 wt % MnO and 3.86 wt % MgO, Fe,Mn-bearing periclase (MgO/(MgO+FeO) = 0.55-0.75; 6.2-7.2 wt % MnO) and zoned ilmenite. Sulfides present include pentlandite, chalcopyrite (Ikeda 1992) and a Cu-bearing sulfide with a composition $Cu_2Fe_3S_5$, intermediate between chalcopyrite and cubanite (Tomeoka et al. 1989b). Awaruite (67 wt % Ni) and apatite with variable F contents (0-3 wt %) have also been reported (Ikeda 1992).

Yamato-793321. Akai (1988, 1992) described an intermediate phase in the matrix of Y-793321 that consists of a disordered intergrowth of olivine and serpentine (Figs. 159a-c). The crystallographic orientation relationship between these phases is $c^*_{serp} || [100]_{ol}$, $[010]_{serp} || [100]_{ol}$ and $[100]_{serp} || [010]_{ol}$. Additional reflections in diffraction patterns from this phase have d-spacings of 9-13 Å which were interpreted as a transitional phase in the transformation of serpentine to olivine (Akai 1992) caused by thermal metamorphism.

Yamato-86720 and -86789. The mineralogy of Y-86720 (paired with Y-86789; Matsuoka et al. 1996) has been described in detail (Tomeoka et al. 1989c, Akai 1990, Bischoff and Metzler 1991, Zolensky et al. 1989c, Zolensky et al. 1991, Ikeda 1992). These two meteorites have properties intermediate between CM and CI chondrites (Ikeda 1992). In both meteorites, aqueous alteration was extremely advanced prior to metamorphism (Tomeoka et al. 1989c), because all the chondrules and clasts in this meteorite have been completely replaced by phases which have textures and compositions consistent with phyllosilicates. Tomeoka et al. (1989c) recognized a high-Al phase (19-25 wt % Al_2O_3) and a low-Al phase (2-5 wt % Al_2O_3) in altered chondrules in Y-86720 that may be mixtures of serpentine and a trioctahedral smectite. However, the totals of these phases are high, indicating that if phyllosilicates were present they have been dehydrated (e.g. Tomeoka 1989a) (see also Matsuoka et al. 1996 for Y-86789).

The matrix of Y-86720 consists of abundant, elongated, submicron olivine grains (Fo_{65-83}) and amorphous material (Fig. 159d,e) probably formed by the dehydration of precursor phyllosilicates (e.g. serpentine and saponite) (Tomeoka et al. 1989c, Akai 1990, Zolensky, et al. 1991a,b). The olivine often has a porous appearance and contains voids (Fig. 159e). In some cases, the dehydration appears to be incomplete and lattice spacings corresponding to saponite are present in HRTEM images (Akai 1990). The amorphous material is Fe-rich and may be ferrihydrite (Akai 1990). Fe sulfides, probably troilite (Tomeoka et al. 1989c) or pyrrhotite, are common (e.g. Zolensky 1989c, Bischoff and Metzler 1991, Ikeda et al. 1992). The pyrrhotite is only slightly nonstoichiometric ($Fe_{0.99}S$), so confusion with troilite is possible. The pyrrhotite occurs as large, euhedral

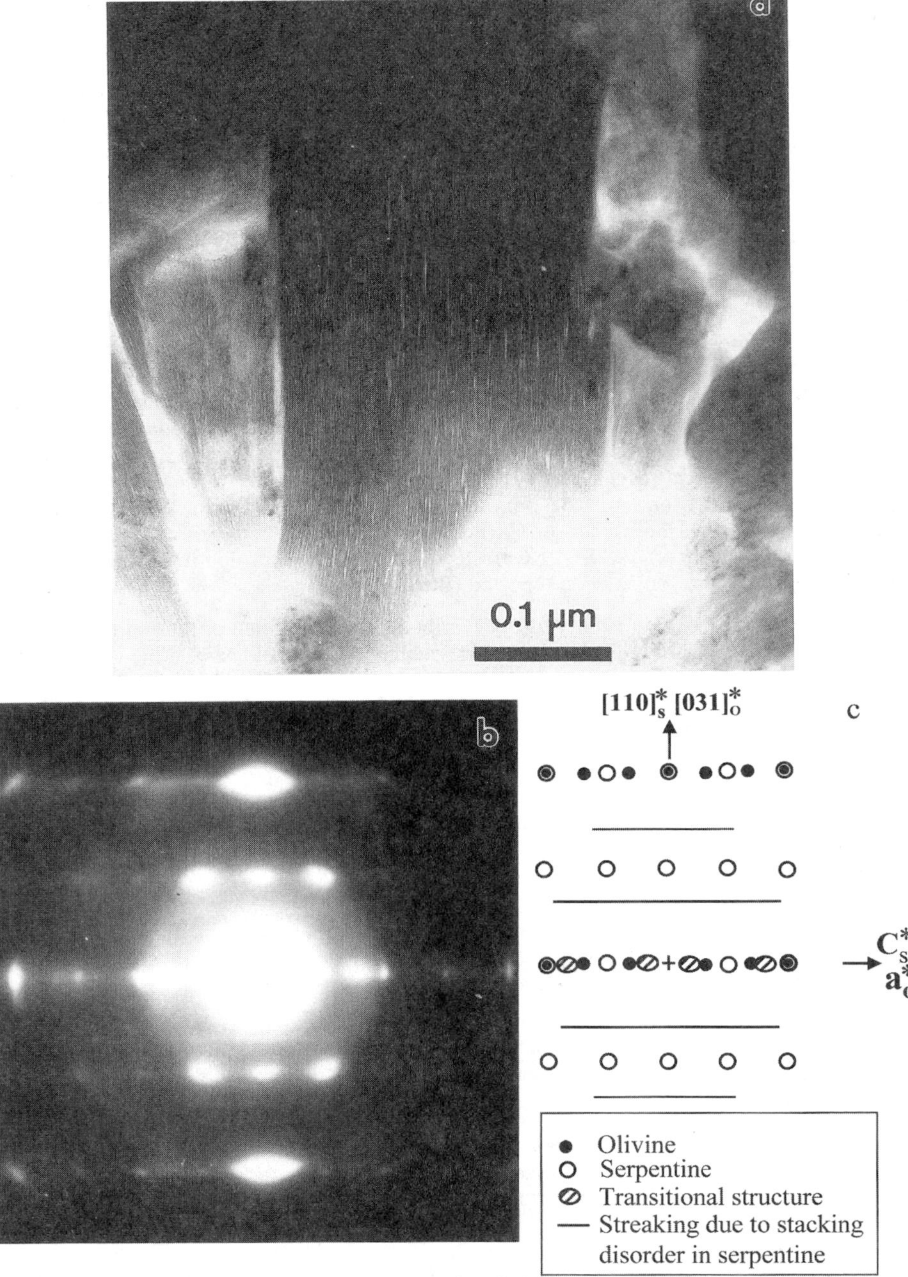

Figure 159. (a) Transmission electron micrographs of phyllosilicates from the matrix of the unusual carbonaceous chondrite Yamato 793321, which have been partially transformed to olivine. (b) Electron diffraction pattern for the mineral grain shown in (a). (c) Schematic diagram showing the crystallographic orientation relationship between olivine and serpentine. Key is shown in insert. Subscripts o and s refer to olivine and serpentine, respectively. After Akai (1988).

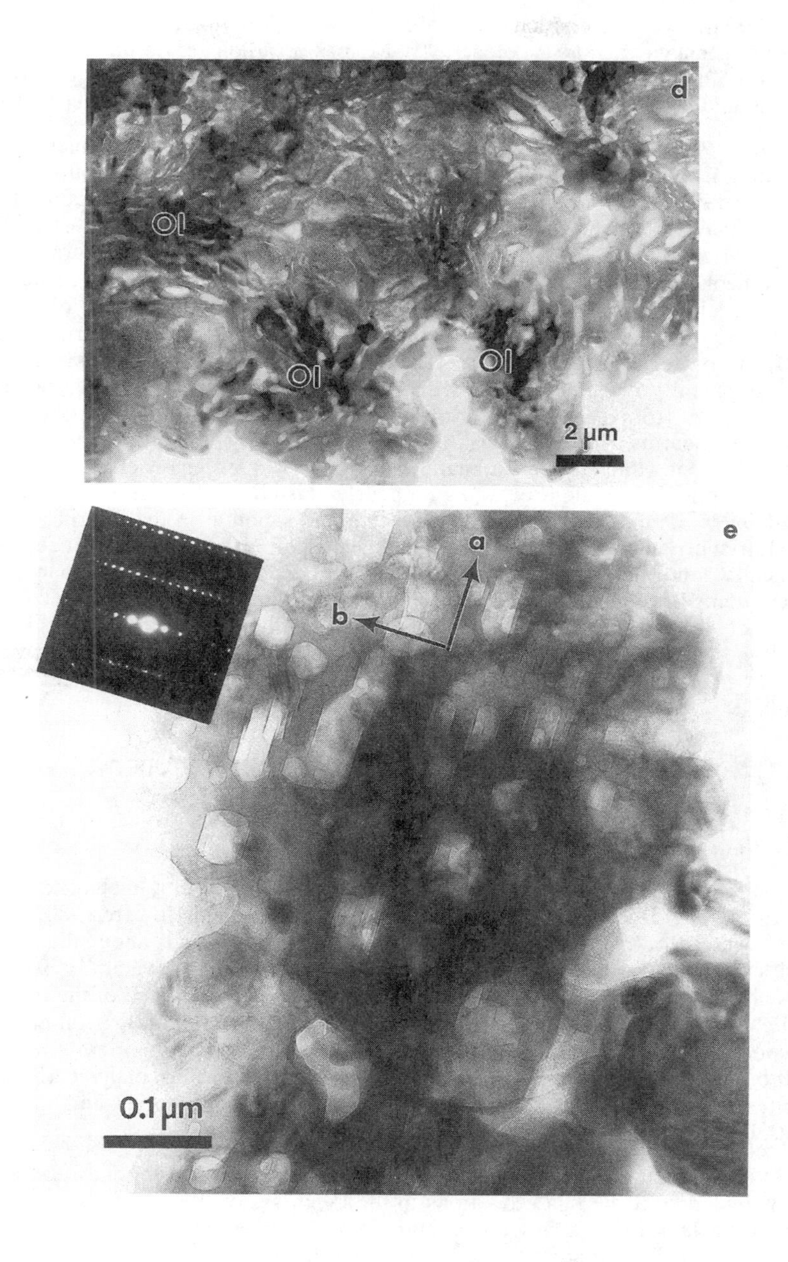

Figure 159 (cont'd). (d) TEM image of fine-grained platy olivine which has replaced phyllosilicates in the matrix of Yamato 82162. After Tomeoka et al. (1989b). (e) TEM image of olivines in the matrix of Yamato 86720 which have formed as a result of dehydration of phyllosilicate phases. The olivine has a porous appearance and contains many voids. After Akai (1994).

laths (50-200 μm long) and as small, irregularly shaped grains in the matrix (Tomeoka et al. 1989c, Zolensky et al. 1991, Ikeda 1992). Minor Ca carbonate, Ca-phosphate (apatite, whitlockite), ilmenite, magnetite, chromite, kamacite (Ni_5) and taenite (Ni_{42-62}) also occur in Y-86720 and 86789 (Tomeoka 1989c, Akai 1990, Ikeda 1992, Matsuoka et al. 1996).

Belgica 7904. The mineralogy of Belgica 7904 has been studied extensively but its classification is uncertain, although it has affinities to both CI and CM groups (Bischoff and Metzler 1991, Ikeda 1992. It has suffered less extensive aqueous alteration than Y-86720, but may have been thermally metamorphosed to a higher degree (Akai 1990). Belgica 7904 contains a variety of clasts with distinct mineralogies (Kimura and Ikeda 1992). Tomeoka et al. (1990b) reported submicron Fe,Ni metal particles, clusters of Fe sulfide and dehydrated phyllosilicates. Using electron diffraction and HRTEM, Akai (1988) found a phase with platy and curved morphologies in the matrix,. Electron diffraction patterns from this phase contain reflections consistent with crystallographically oriented intergrowths of olivine and serpentine (see above), that probably result from the partial transformation of serpentine to olivine during metamorphic heating (Akai 1988). Akai (1992) also observed thermally degraded saponite crystals in which no transitional phase is observed. Zolensky et al. (1989c, 1991) reported fine-grained olivine (Fo_{50-100}), with poorly crystalline phyllosilicates and a phase intermediate in composition between serpentine and saponite in the matrix of Belgica 7901. Matrix olivines contain abundant voids that often have well defined, crystallographically controlled outlines (Akai 1994). Sulfides, fine-grained phosphates and Ca carbonate veins have also been reported (Tomeoka 1990b, Bischoff and Metzler 1991, Zolensky et al. 1991). The dominant sulfide is troilite (<10 μm) (Tomeoka 1990, Bischoff and Metzler 1991) although Zolensky et al. (1989c) suggested that pyrrhotite ($Fe_{0.98}S$) was the major sulfide mineral. Rare pentlandite ($(Fe_{6.53}Ni_{2.44}Co_{0.07})_{9.04}S_8$) (Kimura and Ikeda 1992) and chromite have also been reported.

OPAQUE MINERALOGY OF UNEQUILIBRATED AND EQUILIBRATED CHONDRITES

Introduction: phase relations in the Fe,Ni system

Fe,Ni metal is one of the most widespread phases in chondritic meteorites. Chladni (1797) first recognized metallic metal grains in chondrites and Howard (1802) showed that they were alloys of Fe and Ni. However, little was known about the details of chondritic metals until the metallographic studies of Urey and Mayeda (1959), Kvasha (1961) and Knox (1963) which showed that chondritic metal consisted of the two phases kamacite and taenite with compositions similar to those found in iron meteorites (Ringwood 1961a, Stacey et al. 1961). Because Fe,Ni is such an important phase in unequilibrated and equilibrated ordinary chondrites, as well as many carbonaceous chondrites, we briefly review the characteristics of Fe,Ni minerals and their phase relations, before examining their detailed occurrences in chondrites.

An understanding of the behavior of Fe,Ni metal in ordinary chondrites has come from a knowledge of the phase relations in the Fe,Ni system. The details of this phase diagram, especially those at low temperature, have become increasingly well defined in recent years (e.g. Romig and Goldstein 1980, Reuter et al. 1989) (Fig. 160). The most important Fe,Ni phases in chondrites are the minerals kamacite (α – b.c.c), which contains up to 7.5 wt % Ni in solid solution, and taenite (γ – f.c.c.), the high Ni alloy (in the metallurgical literature, taenite is called austenite). The relationships between kamacite and taenite are largely controlled by the cooling history of their host chondrites. At temperatures above ~900°C, the stable phase is the homogeneous alloy taenite over a

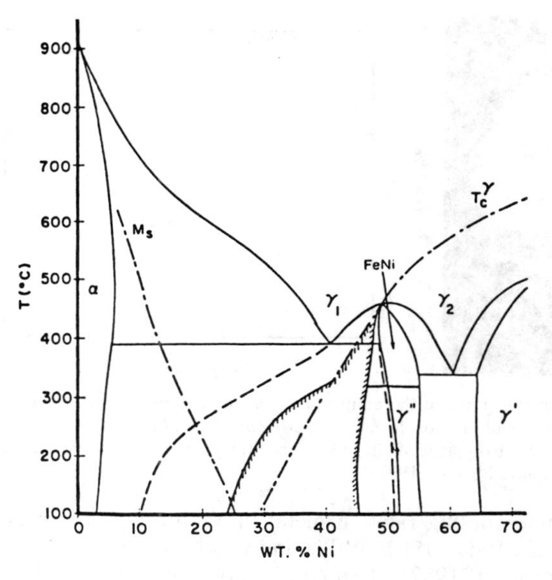

Figure 160. Low temperature phase diagram for the system Fe,Ni. (Reuter et al., 1989). Used by permission of the editor of *Meteoritics and Planetary Science* from Holland-Duffield et al. (1991), Figure 7.

wide range of Ni contents. During slow cooling a homogeneous alloy will enter the two phase field α+γ and kamacite (α) nucleates at the edge of taenite grains and grows inwards into the taenite (γ) grains. As cooling proceeds, the composition of the taenite in contact with kamacite will attempt to change its Ni content to follow the γ/α+γ phase boundary, i.e. will become more Ni-rich as temperature decreases. However, the interiors of the taenite grains have Ni compositions which are lower than the rims and a diffusion profile is established as the grains attempt to equilibrate. The magnitude of the difference between the Ni content of the core and the rim of the taenite grains is a function of their cooling rate. If cooling is relatively rapid, the core and rim are not able to equilibrate, because diffusion of Ni in the taenite is too slow and the grains develop a characteristic M-shaped Ni diffusion profile (e.g. Fig. 161) which was first described by Wood (1964). Under conditions where cooling is slower, however, the Ni diffusion profile is homogenized. At the interface between kamacite and taenite, a slight decrease in the Ni content of the kamacite is observed (the Agrell effect, Agrell et al. 1963). In rapidly cooled taenite grains, the Agrell effect disappears and the Ni profile through the center of the taenite grains is flattened.

With the determination of subsolidus phase relations in the Fe,Ni system, as well as diffusion coefficients for kamacite and taenite (Goldstein and Ogilvie 1965a, Goldstein et al. 1964), the basis was laid for using coexisting kamacite and taenite in chondrite metal as a cooling rate indicator for the equilibrated ordinary chondrites. A large number of workers have studied the detailed characteristics of the metal phases in ordinary chondrites starting with the work of Wood (1967). This study was the first to carry out detailed electron microprobe investigations of the compositional characteristics of kamacite and taenite in ordinary chondrites and use these data to constrain the cooling rates of chondrites. Wood (1967) developed a technique which allows the cooling rates of chondrites to be determined based on the relationship between the core Ni content of metal grains and their size. This model is based on the fact that in small metal grains, the distance over which Ni has to diffuse from the kamacite/taenite boundary to the interior of a grain is less than in a larger grain. Thus Ni contents of small grains should be higher than larger grains which have experienced the same cooling history. This technique

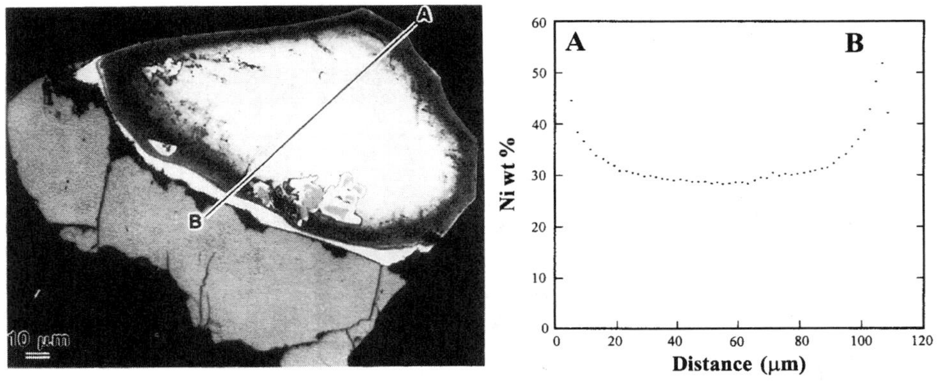

Figure 161. (a) BSE image of an etched metal grain in the equilibrated ordinary chondrite St. Séverin showing cloudy core and clear rim. Used by permission of the editor of *Meteoritics and Planetary Science* from Holland-Duffield et al. (1991), Figure 4a. (b) typical M-shaped compositional profile (A to B) for Ni across the same taenite grain. After Holland-Duffield et al. (1991).

has been widely applied to determine cooling rates in metal in equilibrated ordinary chondrites (e.g. Wood 1967, Scott and Rajan 1981, Willis and Goldstein 1981a, 1983; Holland-Duffield et al. 1991, Taylor et al. 1987). It should be noted that the absolute cooling rates determined using taenite compositions have been periodically subject to revision as a result of an improved understanding of the phase relations and diffusion coefficients in kamacite and taenite, especially the role of minor elements such as P (e.g. Romig and Goldstein 1980). However, typical measured cooling rates lie in the range 0.1-100°C/Ma.

In many equilibrated and rarely in unequilibrated ordinary chondrites, an intergrowth of kamacite and taenite occurs which is termed *plessite*. Plessite is not a mineral, but is a two-phase mixture of kamacite and taenite in which the two phases are very closely intergrown. Plessite usually forms at low temperatures from the decomposition of high temperature taenite compositions which have been retained during slow cooling (i.e. the low Ni cores of taenite grains). The grain size of kamacite and taenite in plessite can vary considerably depending on the temperature of decomposition and whether the metals have been annealed at a later stage in their histories, perhaps due to post-shock heating (see Shock Metamorphism). Although plessite can form by the direct decomposition of taenite (γ) to kamacite (α) and taenite (α), plessite can also form by a two stage reaction. This reaction involves the diffusionless transformation of taenite to the α_2-Fe,Ni phase, termed martensite (see below), which then decomposes isothermally to kamacite and taenite (see also Massalski et al. 1966).

Many different forms of plessite have been described in iron meteorites (see Buchwald 1975), but the most common type in ordinary chondrites is termed pearlitic plessite (or often just plessite in the literature). Pearlitic plessite is a plessite with specific textural characteristics, i.e. consists of a lamellar intergrowth of kamacite and taenite. The lamellae of taenite are extremely thin (< 2 µm).

Kamacite, taenite and plessite form during continuous slow cooling. However, many chondrites have experienced reheating as a result of shock events and cooled rapidly. Under such conditions modification of the metal can occur. If chondritic metal is heated into the γ (taenite) stability field it will compositionally homogenize. During subsequent rapid cooling, taenite transforms to the metastable cubic phase, α_2-Fe,Ni, termed *martensite*. Martensite is so-called because it it forms by a diffusionless martensitic transformation mechanism and the new phase forms parallel to the {111} planes in the taenite. Martensite is most commonly encountered in shock reheated chondrites (see Shock Metamorphism) and will decompose to plessite if reheating into the $\alpha+\gamma$ stability field occurs.

Opaque phases in unequilibrated ordinary chondrites

Several studies have examined the characteristics of the metallic phases in the unequilibrated ordinary chondrites (e.g. Ramdohr 1975, Bevan and Axon 1980, Scott and Rajan 1981, Afiattalab and Wasson 1980, Nagahara 1982b). Like the equilibrated ordinary chondrites, the main phases present are kamacite and taenite which are often associated with troilite. Kamacite and less commonly taenite are often polycrystalline (Affiatalab and Wasson 1980). Tetrataenite can also be present (Scott and Rajan 1981, Nagahara 1982b). In addition to the occurrence of metal as globules within chondrules (see Chondrule section), several other occurrences of metal have been recognized in unequilibrated ordinary chondrites, which appear to be broadly similar in the different chondrite groups (H, L and LL) (e.g. Bevan and Axon 1980, Afiattalab and Wasson 1980, Nagahara 1982). The occurrences are: (a) rare, large (up to 500 μm), rounded to subrounded clusters of metal and sulfide grains (and rarely chromite; Nagahara 1982) which are often referred to as 'metallic chondrules' (e.g. Bevan and Axon 1980, Nagahara 1982) (Fig. 162a), (b) isolated, angular to subrounded grains of variable size which occur externally to chondrules and are distributed throughout the chondrite (Fig. 162b), and (c) metal which occurs as irregular grains at the surface of chondrules or within fine-grained silicate rims on chondrules (e.g. Affiatalab and Wasson 1980) and (d) continuous chondrule rims that consist almost entirely of opaque phases (metal and troilite) and have a porous or 'sponge-like" appearance (e.g. Bevan and Axon 1980). Metals in occurrence (b) are most common. Metal particles in unequilibrated chondrites have median diameters of ~200 μm with only 10-25% having diameters <100 μm (Dodd 1974b, Afiattalab and Wasson 1980).

Trace element data for metals in unequilibrated ordinary chondrites, measured by INAA techniques, have been reported by Rambaldi (1976, 1977) and Kong and Ebihara (1996a,b).

Kamacite. Using electron microprobe techniques, Nagahara (1982b) showed that the mode of occurrence of kamacite and taenite varies in unequilibrated H, L and LL chondrites, but metal in all occurrences can be associated with troilite. In H3 chondrites, isolated grains of kamacite are most common, with composite grains of kamacite and taenite being much less abundant; discrete taenite grains are very rare. In L3 and LL3 chondrites, discrete taenite grains are much more common than composite grains, but single phase kamacite is still most abundant. These observations differ slightly from those of Bevan and Axon (1980) who found that most metal grains in Tieschitz (H3) are polycrystalline intergrowths of kamacite, taenite and troilite, although small (<50 μm) isolated grains of kamacite and zoned taenite can be present. Neumann bands are common and submicroscopic inclusions can occur along kamacite grain boundaries.

Kamacite is, however, clearly the most abundant phase in unequilibrated chondrites (e.g. Afiattalab and Wasson 1980) and shows compositional variations which are a function of both the degree of metamorphism and the chemical group (e.g. Afiattalab and Wasson 1980). In the least equilibrated chondrites (e.g. Bishunpur, Ngawi), matrix kamacite grains have a considerable range of both Co and Ni contents (Fig. 163) but in higher petrologic type 3 chondrites, such as Hallingeberg, Mezö-Madaras and Sharps, the matrix kamacite is largely equilibrated with a very narrow range of Co contents, but Ni

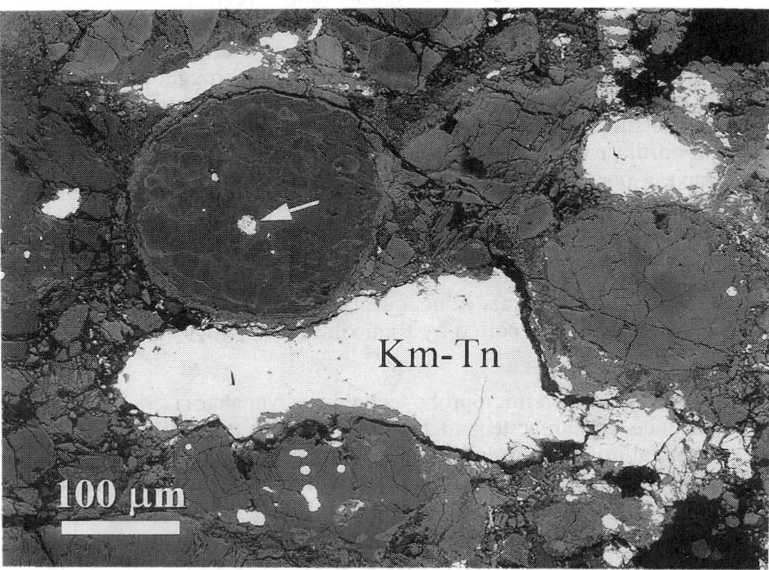

Figure 162. BSE image showing (a) a metal sulfide chondrule in the unequilibrated ordinary chondrite, Tieschitz (H3.6); Km = Kamacite, Tr = troilite; (b) the typical occurrences of metal in the matrix of the unequilibrated ordinary chondrite, Tieschitz (H3.6). A metal grain within a chondrule is arrowed.

can be variable in some cases (e.g. Hallingeberg). Afiattalab and Wasson (1980) found that metal in Chainpur showed an intermediate degree of equilibration with a compositional range of 2.7-6.1 wt % Ni and ~1 wt % Co.

Nagahara (1982b) showed that kamacite grains on chondrule surfaces and in the matrix were compositionally heterogeneous; some grains are unzoned, but others have Ni-rich cores and Fe-rich rims (see also Wood 1967, Lauretta et al. 1996) and different compositions are encountered in different grains. Such evidence suggests that these

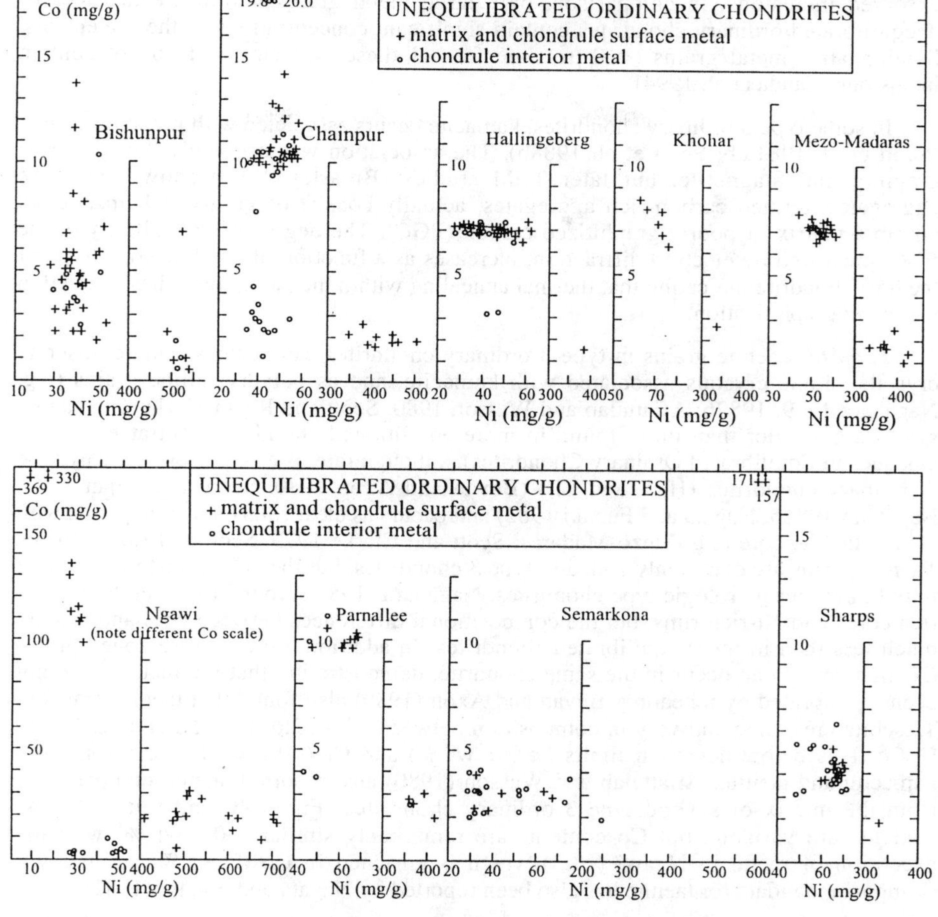

Figure 163. Ni vs. Co (mg/g) diagram for metal grains inside chondrules and those which occur in the matrix or on the exterior surface of chondrules in several unequilibrated ordinary chondrites. After Afiattalab and Wasson (1980).

chondrites were never heated above 500°C after they accreted. The zoning in the kamacite is believed to preserve the cooling history through and below ~450°C. Diffusion of Ni in kamacite is 2 orders of magnitude faster than in taenite so that metals in type 3 chondrites must have cooled rapidly in order to preserve primary compositional zoning.

Compositional data for kamacite (e.g. Afiattalab and Wasson 1980, Nagata and Funaki 1982, Smith et al. 1993) (Fig. 163) shows that the kamacite is typically richer in Co than taenite in type 3 chondrites. In ALHA 77270 (L3), low- and high-Co kamacite containing ~0.5 and 1.0 wt % Co occurs compared with <0.3 wt % for coexisting taenite (see also Smith et al. 1993 for ALHA 77214, L3).

Metal grains within chondrules in type 3 ordinary chondrites contain significant concentrations of Si, Cr and P (see Chondrules) either dissolved in the metal or as exsolved inclusions. In comparison, only large metal grains within the matrices of unequilibrated ordinary chondrites contain significant concentrations of these elements. Small matrix metal grains (<10 μm) are free of these elements and do not contain inclusions (Zanda et al. 1994).

In some type 3 ordinary chondrites, kamacite occurs associated with graphitic carbon (Scott et al. 1981a,b; Scott et al. 1988b). The association was originally thought to be graphite and magnetite, but later TEM studies (Brearley 1990) show that these aggregates, termed carbon-rich aggregates, actually consist of grains of kamacite set within a matrix of poorly graphitized carbon (PGC). The degree of crystallinity of the PGC, measured by electron diffraction, increases as a function of the petrologic type of the host chondrite indicating that thermal annealing within the parent body has resulted in increased graphitization.

Taenite. Taenite grains in type 3 ordinary chondrites can occur as single, discrete crystals, single crystals associated with kamacite, and as polycrystalline grains (e.g. Nagahara 1979, 1982b; Afiattalab and Wasson 1980, Scott and Rajan 1981). The taenite grains are smaller than those found in more equilibrated chondrites. Tetrataenite (see Opaques in Equilibrated Ordinary Chondrites), although rare, is also present in many type 3 ordinary chondrites (H, L and LL) (Bevan and Axon 1980, Scott and Rajan 1981, Nagahara 1982b, Nagata and Funaki 1982) and occurs as clear rims on taenite grains with cloudy taenite core (e.g. Mezö-Madaras; Scott and Clarke 1979, Scott and Rajan 1981). Taenite grains are commonly zoned in type 3 chondrites, but the zoning differs from that found in higher petrologic type chondrites. Nagahara (1982b) found that taenite had Fe-rich cores and Ni-rich rims, but the compositional differences between core and rim are much less than in more equilibrated chondrites. In addition, rare, reverse zoned grains (Ni-rich cores) can occur in the same chondrite, demonstrating that the metals have not been equilibrated by reheating. Bevan and Axon (1980) also found that taenite grains in Tieschitz are zoned and vary in composition between 33.5-52.6 wt % Ni and 0.2-0.4 wt % Co. P is below detection limits (<0.01 wt %) and Cr (<0.2 wt %) occurs in both kamacite and taenite. Afiattalab and Wasson (1980) also reported taenite compositions from the matrix of several type 3 ordinary chondrites (Fig. 163) and found that Ni contents are variable, but Co contents are remarkably similar (~0.2 wt %) with the exception of metal in Ngawi (LL3) which is enriched in Co (2 wt %, see below). Compositional data for taenite has also been reported by Nagata and Funaki (1982).

In some unequilibrated chondrites, such as Tieschitz, Krymka, Mezö-Madaras and Bishunpur, some etched grains of taenite consist of a mosaic of rounded dark areas (cloudy taenite) delineated by sharp, clear bands of tetrataenite (48-57 wt % Ni) (Wood 1967, Bevan and Axon 1980, Afiattalab and Wasson 1980, Scott and Rajan 1981). This texture has been termed polycrystalline and individual domains within the polycrystalline taenite shows typical compositional zoning with high Ni rims and lower Ni cores. Bevan and Axon (1980) also noted that clear, etched zones of tetrataenite which abut silicates or kamacite are broader than along mutual taenite grain boundaries and contain higher Ni contents (45-52 wt %). Bevan and Axon (1980) have attributed the formation of polycrystalline taenite to high temperature, rapid non-equilibrium crystallization from a metallic melt which established compositional heterogeneities in Ni content. During subsequent slower cooling at temperatures below ~700°C, kamacite probably nucleated and grew, resulting in an enhancement in the Ni zoning profiles, originally established at high temperatures. However, Scott and Rajan (1981) argued that this was not the case and that polycrystalline taenite grains are simply aggregates of small, individual metal grains which have not been annealed sufficiently to produce large grains.

Other metal phases. *Plessite* with 25-30 wt % Ni has also been reported by Nagata and Funaki (1982) in ALHA 77260 (L3) and rarely in Mezö-Madaras by Scott and Rajan (1981). Although comparatively rare, the high Ni metal phase *awaruite* (Ni_3Fe) has been reported in several LL3 chondrites (Taylor et al. 1981). A metal phase with high Co and low Ni (probably wairauite) was first reported in the LL3 chondrite Ngawi (Afiattalab and Wasson 1980, Rubin 1990). It occurs at the interface between a sulfide phase (troilite or pentlandite) and high Ni metal (tetrataenite or awaruite). In Ngawi, Rubin (1990) reported Co contents of 33 wt % in this phase.

Troilite. Although troilite is one of the major opaque phases in type 3 ordinary chondrites, remarkably its occurrences have not been studied in detail. It typically occurs associated with Fe,Ni metal, as anhedral metal and sulfide blebs in which the troilite sometimes rims the metal grains (e.g. Bevan and Axon 1980, Scott and Rajan 1981, Lauretta et al. 1996). Metal,troilite-rich rims around chondrules are also common (e.g. Grossman and Wasson 1987, Lauretta et al. 1996). Töpel-Schadt and Müller (1982) found antiphase domains, predominantly parallel to (001), with the displacement vector $1/3\langle\bar{1}10\rangle$ in troilite from the unequilibrated ordinary chondrites Chainpur (LL3), Mezö-Madaras (L3) and Tieschitz (H3).

Opaque phases in equilibrated ordinary chondrites

Fe,Ni metal. The petrography of metallic phases in equilibrated ordinary chondrites has been studied intensively by many workers, e.g. Ramdohr (1973), Sears and Axon (1976), Scott and Clarke (1979), Nagahara (1979), Clarke and Scott (1980), Willis and Goldstein (1983), Rubin (1990), Smith and Launspach (1991) and Smith et al. (1993). A large body of textural and compositional data now exists and the behavior of Fe,Ni metal during metamorphism and slow cooling is now well understood. The effects of shock heating on chondritic metal are examined under Shock Metamorphism.

The ordinary chondrites contain 8-20 wt % Fe,Ni metal, which occurs as grains, typically 100-200 μm in size, evenly distributed throughout the meteorites (e.g. Fig. 164), in addition to rounded blebs which occur inside chondrules (see Chondrules). Afiattalab

Figure 164. Optical micrograph (reflected light) showing the textures of metal in the equilibrated H chondrite Conquista (H4). Field of view = 2mm.

and Wasson (1980) showed that there are distinct changes in the morphology and grain shape of metal grains in ordinary chondrites in moving from petrologic type 3 to type 6. These are (a) the fraction of metal-bearing chondrules decreases and the amount of metal in chondrules decreases and (b) the fraction of fine-grained metal decreases and the fraction of coarse-grained metal increases. In addition, troilite and metal, which commonly occur as intergrowths in low petrologic type chondrites, tend to form separate grains with increasing petrologic type (Afiattalab and Wasson 1980), but Willis and Goldstein (1983) observed that taenite in Guareña (H6) and Colby (L6) occurs as both isolated grains or attached to sulfides or kamacite. A detailed analysis of the shape of metal grains in ordinary chondrites has been reported by Fujii et al. (1982).

Metal grains within chondrules in unequilibrated ordinary chondrites contain significant concentrations of Si, Cr and P (see Chondrules) either dissolved in the metal or as exsolved inclusions. In equilibrated chondrites, these elements were oxidized during metamorphism and have formed small inclusions of phases (<1-~20 μm) such as chromite, phosphate and silica which increase in size as a function of petrologic type (Zanda et al. 1994). However, in type 6 chondrites, complete diffusion of these elements out of the metal grains has occurred and no inclusions are present.

Trace element data for metals in equilibrated ordinary chondrites, measured by INAA techniques, have been reported by Rambaldi (1976, 1977), Kong et al. (1995a,b) and Kong and Ebihara (1996a,b). Widom et al. (1986) have also reported trace element data for metal nodules and veins in several ordinary chondrites, many of which appear to have formed by shock vaporization of chondritic material.

Kamacite (α-Fe,Ni; b.c.c) in most equilibrated chondrites occurs as grains which are single crystals, but can also be polycrystalline (e.g. in Guareña, Willis and Goldstein 1983). Grains consisting of both kamacite and taenite are also present (e.g. Urey and Mayeda 1959, Willis and Goldstein 1983). In many chondrites, shock-produced twin lamellae, termed Neumann bands, are common. Kamacite grains have heterogeneous compositions with higher Ni contents in their cores than at their edges (Wood 1967, Nagahara 1979). The actual Ni contents vary depending on the size of the grain, but in large grains are 6-7 wt % in the cores and 5-6 wt % within a few microns of the edges of the grains.

Sears and Axon (1975 1976) and Affiatalab and Wasson (1980) reported data on the Co and Ni contents of coexisting kamacite and taenite in equilibrated H, L and LL ordinary chondrites and showed that Co contents in kamacite progressively increase from H to L to LL chondrites. A more extensive survey of kamacite compositions in a large number of H, L and LL chondrites ranging from petrologic type 4 to type 6 (Rubin 1990), has fully defined the compositional ranges. Mean Co contents fall in distinct ranges within each of the different groups with an increase from H to LL: H – 0.44-0.51 wt %; L – 0.7-0.95 wt %; LL – 1.4-37 wt % (Fig. 165). Mean Ni contents in kamacite show the reverse trend: H – 6.9 wt %; L – 6.54 wt %; LL – 4.98 wt %). Data for Co and Ni in kamacite in equilibrated chondrites have also been reported by Nagahara (1979), Afiattalab and Wasson (1980), Smith and Launspach (1991) and Smith et al. (1993).

The increase in Co and decrease in Ni contents in kamacite in moving from the H to the LL group has been attributed to: (a) a decrease in the grain size of kamacite from H to LL chondrites which allows metal grains in LL chondrites to equilibrate faster and hence attain lower Ni contents and (b) the fact that Fe is more susceptible to oxidation than Co, such that the more oxidized LL chondrites contain metal with higher Co contents (Rubin 1990). Kamacite compositions also change as a function of petrologic type within individual groups, with type 3 and 4 chondrites having lower Ni and Co

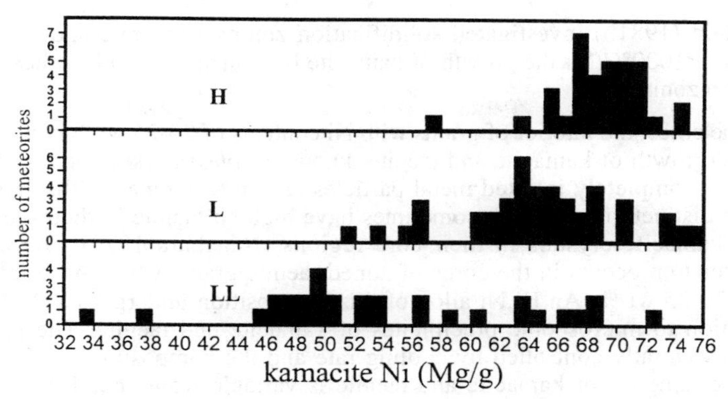

Figure 165. Histogram of kamacite Ni contents (mg/g) in H, L and LL chondrites. After Rubin (1990).

contents than type 6. In low petrologic type chondrites, this may be attributable to the presence of a primitive or pristine nebular metal component which had positively correlated Ni and Co (Rubin 1990).

Within many chondrites, more commonly L and LL, rather than H chondrites, kamacite grains with compositions which differ significantly from the typical mean have been reported which appear to be aberrant grains introduced as a result of brecciation processes. However, based on an extensive study of metal grains in Bruderheim (L6) and several Antarctic H, L and LL chondrites, Smith and Launspach (1991) and Smith et al. (1993), concluded that heterogeneities in the Ni and Co contents of kamacite and taenite were due to the presence of preexisting compositional heterogeneities which were not equilibrated during metamorphism (but see Rubin and Brearley 1996).

The abundance of *taenite* (γ Fe,Ni, f.c.c) is highest in the LL chondrites, consistent with their high bulk Ni content and decreases in abundance through the L to H chondrites where kamacite is most abundant. Taenite commonly occurs as single, anhedral crystals but polycrystalline grains also occur (e.g. Urey and Mayeda 1959, Wood 1967, Nagahara 1979, Hutchison et al. 1980a). Rare globular inclusions of troilite may also be present. Taenite grains invariably show core to rim zoning whose characteristics depend on the cooling rate of the grain. Taenite grains have M-shaped Ni zoning profiles (e.g. Wood 1967, Nagahara 1979, Scott and Clarke 1979, Willis and Goldstein 1983, Holland-Duffield et al. 1991) (e.g. Fig. 161) which form during slow cooling by diffusion controlled growth of kamacite and can be used to constrain cooling rates of the host meteorite (see above). Taenite commonly contains 25-35 wt % Ni at the center of crystals and 45-55 wt % Ni at the edges (Wood 1967, Nagahara 1979). For crystals whose centers contain <~34 wt % Ni, the phase present is actually martensite, which has formed at low temperatures from inversion of taenite with the appropriate composition. Many taenite grains also contain exsolution lamellae of kamacite which are superimposed on the M-shaped Ni diffusion profile (e.g. Nagahara 1979). The compositions of the two phases at their interface are variable even within a single grain.

Hutchison et al. (1980a) have argued that polycrystalline taenite in equilibrated chondrites is a relict solidification structure and that these chondrites have not been heated above ~700°C since they accreted (see also Bevan and Axon 1980). Such an origin would invalidate the use of the Wood (1967) model for determining metallographic cooling rates. This conclusion has, however, been challenged by Scott and Rajan (1981), who argued that polycrystalline taenite could be produced by hot working or accretion of grains which failed to coarsen during metamorphism. In addition, Willis and Goldstein (1981b) investigated solidification zoning in taenite and found that for cooling rates <1000°C/Ma the growth of kamacite from taenite would eradicate any relict solidification zoning.

In some cases, the interior of grains with Ni contents of 20-25 wt % consist of a fine-grained intergrowth of kamacite and taenite, known as *plessite* (see above). Plessite can also occur as completely isolated metal particles (e.g. in Guareña; Willis and Goldstein 1983) or as distinct grains which sometimes have high Ni taenite in their outer portions and many kamacite crystals in their core regions (Nagahara 1979). During cooling, plessite formation occurs in the cores of zoned taenite grains which have retained a Ni content of 20-25 wt %. An Fe,Ni alloy of this composition undergoes a Widmanstätten decomposition to microscopic precipitates of kamacite and taenite. The formation of plessite is essentially controlled by cooling rate and the composition of the alloy. The width of the lamellae of kamacite and taenite is variable from chondrite to chondrite. Wood (1967) showed that in slowly cooled chondrites (1-10°C/Ma) most zoned taenite grains are too small to have retained Ni contents appropriate for formation of plessite. However, mild reheating to 400°-500°C may have caused the decomposition of unstable martensite (see above) with ~30 wt % Ni to form plessite in such cases.

In addition to plessite formed by Widmanstätten decomposition, Axon and Grokhovsky (1982) and Grokhovsky and Bevan (1983) have reported plessite which results from the discontinuous precipitation reactions kamacite → kamacite + taenite and taenite → taenite + kamacite in the Richardton (H5) chondrite. The latter reaction is less extensive and more localized than the former and does not occur in all zoned taenite grains. The reactions occur by the growth of stable phases behind a grain boundary migrating into a supersaturated solid solution. The taenite to kamacite + taenite reaction occurs at the edges of zoned taenite grains in the matrix and involves a compositional change from taenite with 30-40 wt % Ni to kamacite (5 wt % Ni) and taenite (51 wt % Ni). The reaction takes place at 350°-400°C by nucleation and inward growth from high Ni taenite-silicate grain boundaries within zoned taenite grains, probably activated by strain within the taenite (Grokhovsky and Bevan 1983).

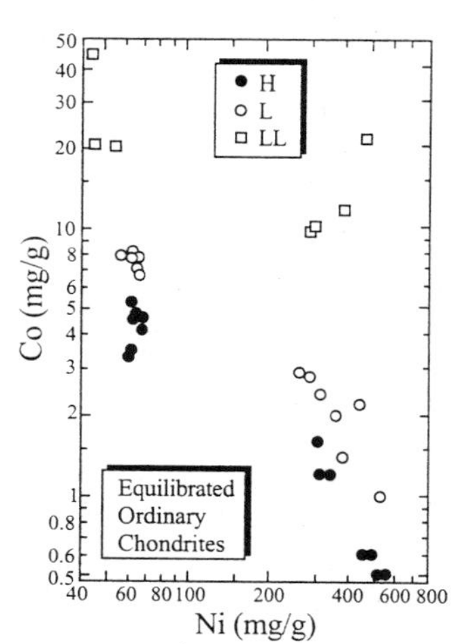

Figure 166. Co vs. Ni (mg/g) diagram for equilibrated H, L and LL chondrites. Kamacite compositions are on the left of the diagram and taenite on the right. After Afiattalab and Wasson (1980).

Sears and Axon (1976) and Afiattalab and Wasson (1980) reported Co contents in coexisting taenite and kamacite from several H, L and LL chondrites and showed that taenite always contains less Co then coexisting kamacite. Co contents in taenite are higher in LL chondrites (up to ~1.5 wt %) than L (~0.2-0.3 wt %) or H (~0.05-0.2 wt %)

chondrites (Fig. 166). In addition, both studies show that Co and Ni are positively correlated in taenite in the LL chondrites (Fig. 166) but are inversely correlated in the H and L groups. This phenomenon is caused by the fact the LL chondrites typically have low kamacite fractions, such that Co and most Ni are in taenite. The concentrations of these two elements vary inversely with the amount of metallic Fe present (Sears and Axon 1975, Afiattalab and Wasson 1981). The P content of taenite is typically < 0.01 wt % (Nagahara 1979).

In addition to kamacite and taenite, Taylor and Heymann (1971) found a Fe,Ni phase in ordinary chondrites which contains 49-57% Ni. They called this phase clear taenite, to differentiate it from the zoned variety of taenite which develops cloudy borders when etched with acid (often referred to as cloudy taenite). Scott and Clarke (1979) showed that clear taenite is optically anisotropic and is, in fact, ordered tetragonal Fe,Ni, now termed *tetrataenite,* that forms from taenite during cooling. Clear taenite occurs in H, L and LL chondrites and can occur as individual grains (10-60 μm size) (Taylor and Heymann 1971) or as well defined rims (1-5 μm wide) on normally zoned taenite grains (Scott and Clarke 1979) (Fig. 167a). Compositional zoning profiles show that the clear tetrataenite rims are relatively homogeneous, but a marked compositional discontinuity exists between the clear taenite and cloudy taenite in the interior of the grains (Fig. 167b).

In some chondrites (e.g. Olivenza, LL6, St. Séverin, LL6; Y-74160, LL7), tetrataenite can be as much as 40-50 wt % of the metal phase (Nagata et al. 1986, Nagata and Funaki 1982) and, based on SEM studies, may be very fine-grained (150-300 nm; Danon et al. 1979). Holland-Duffield et al. (1991) investigated the microstructure of metal grains from Kernouvé and St. Séverin by SEM. They found that in etched samples the central region of taenite grains has a honeycomb structure, which is coarser in St. Séverin than Kernouvé, due to a higher bulk Ni content of the grains and a slower cooling rate. The honeycomb structure is similar to the cloudy zone structure of iron meteorites and is a two-phase structure consisting of tetrataenite and martensite. In addition, three distinct structural zones were found to exist within the clear taenite rim. The grain size within these zones is essentially identical, but the degree of etching increases towards the interior of the grains. The origin of this microstructure is unclear, but is probably due to the presence of two phases, possibly γ''(FeNi) and γ'($FeNi_3$) which occur in high Ni alloys (<55 wt %) at low temperatures (<350°C) (Holland-Duffield et al. 1991). Willis and Goldstein (1983) and Holland-Duffield et al. (1991) have observed that the width of the outer clear tetrataenite rim is greater along grain edges in contact with troilite than edges in contact with silicates (e.g. in Guareña and St. Séverin). This feature is attributed to the fact that at high T (<700°C) Ni is soluble in troilite, but the solubility of Ni in troilite decreases at lower temperatures and diffuses out of the grains into adjacent taenite.

The magnetic properties of tetrataenite-rich ordinary chondrites have been measured by Nagata and Funaki (1982).

Although most metallic phases in ordinary chondrites are anhedral, Rubin (1994) reported a single grain of euhedral tetrataenite (48.3 wt % Ni; 2 wt % Co; 0.13 wt % Cu) in an impact melt clast in the LL6 chondrite, Jelica. The tetrataenite probably formed by primary crystallization from the impact melt.

Although tetrataenite with up to 57 wt % Ni is usually the highest Ni taenite found in ordinary chondrites, occurrences of very high Ni taenite with >57 wt % Ni have been reported. Christophe Michel-Lévy and Bourot-Denise (1992) found polycrystalline taenite grains containing 60 wt % Ni, 2.05 wt % Co and 0-0.45 wt % Cu, with little compositional variation, in the LL6 chondrite breccia, Dahmani, sometimes associated

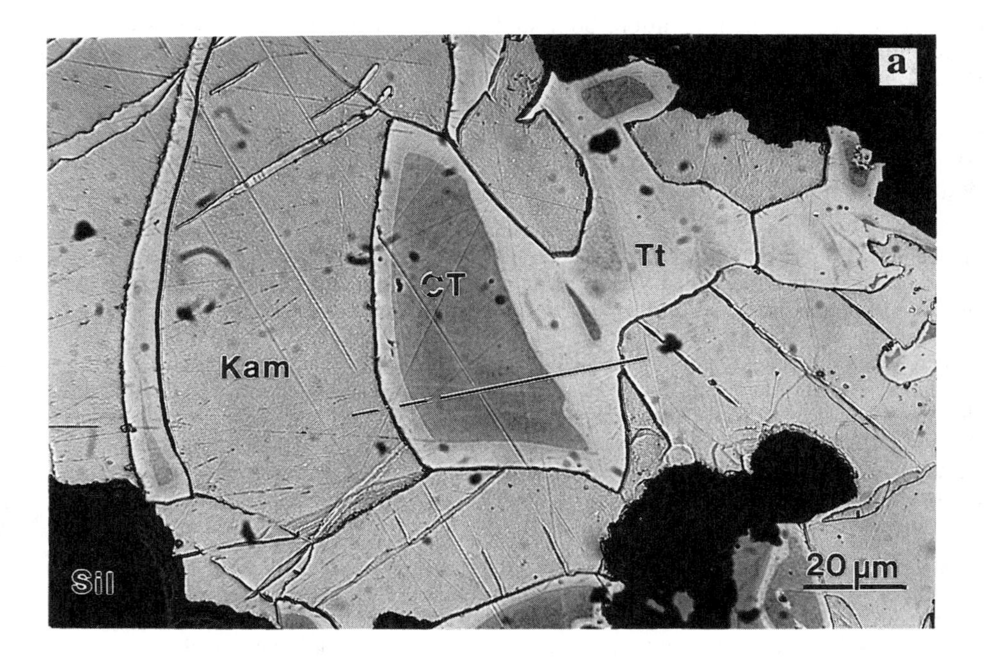

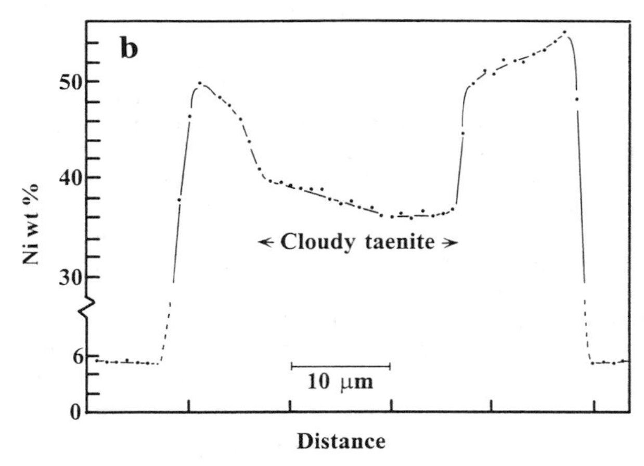

Figure 167. (a) Optical micrograph (reflected light) showing an etched metal grain in the Mezö-Madaras (L3) chondrite showing a cloudy core of taenite (CT) and a clear rim of tetrataenite (Tt): Kam = kamacite. (b) Electron microprobe profile across the metal grain showing the variation in Ni content across the grain shown in (a). There is a marked increase in the Ni content of the grain across the taenite-tetrataenite boundary. After Scott and Clarke (1979).

with troilite and minor pentlandite (10-15 wt % Ni, 0.2-0.7 wt % Co, 0-0.6 wt % Cu). Andrehs and Adam (1990) also reported a metallic grain in the LL6 chondrite Trebbin which contains coexisting alloys with 50 wt % Ni and 60-62.8 wt % Ni, respectively. These observations suggest that there may be a stable taenite phase intermediate in composition between tetrataenite (NiFe) and awaruite (Ni_3Fe). *Awaruite* has also been reported in a single LL5 chondrite Parambu (Danon et al. 1981, Rubin 1988), coexisting with troilite and pentlandite.

High Co metal (*wairauite*) has been reported in several equilibrated chondrites; EET87544 (LL4) (30.2 wt % Co) and the LL6 chondrites Appley Bridge (37.0 wt % Co),

Jelica (30.0 wt % Co), and Manbhoom (20.0 wt % Co) (Rubin 1990). Ni contents range from 1.4 -4.9 wt %.

Metallic copper. Copper is a rare ($1.4 \pm 2.1 \times 10^{-4}$ vol %; Rubin 1994), but widespread phase in ordinary chondrites, occurring in 205 out of 309 ordinary chondrites examined by Ramdohr (1973) and Rubin (1994). It was first reported by Quirke (1919) and subsequently by Duke and Brett (1965) and Olsen (1973). Rubin (1994) documented 9 different occurrences of copper in ordinary chondrites; the two most common are: (a) at kamacite-troilite interfaces and (b) adjacent to small troilite grains inside Ni-rich metal. In almost all cases the copper grains are irregularly shaped and have grain sizes between ~1-6 μm, although rare, unusually large grains have also been described (Olsen 1973). Limited electron microprobe data indicate that the copper contains ~1.5-2.0 wt % Ni in solid solution (Olsen 1973, Rubin 1994).

Opaque phases in unequilibrated and equilibrated and enstatite chondrites

Due to their formation under highly reducing conditions, the unequilibrated and equilibrated enstatite chondrites contain a highly unusual assemblage of opaque minerals, dominantly sulfides including such phases as niningerite, ferroan alabandite, oldhamite etc. Sulfide and metal constitute a major component of the enstatite chondrites (19-28% Fe,Ni metal, 7-15% troilite, Mason 1966, Keil 1968a). The phase relations between these different phases have provided many important insights into the genesis of these meteorites. Most notably they have enabled precise determinations of the equilibration temperatures (e.g. Fleet and MacRae 1987, Kissin 1989, Fogel et al. 1989, Zhang and Sears 1996) of the metamorphosed EH and EL chondrites which has not been possible for the equilibrated ordinary chondrites. In the unequilibrated enstatite chondrites, the opaque phases occur largely external to the chondrules in the form of clasts or distinct nodules (e.g. El Goresy et al. 1988, Kimura 1988, Ikeda 1989b, Zhang and Sears 1996). It is therefore appropriate to deal with these assemblages separately from chondrules. Chondrules themselves generally contain few opaques (see Chondrules), with niningerite apparently being the most common (e.g. Grossman et al. 1985, El Goresy et al. 1988).

Kamacite. Kamacite is by far the most abundant opaque phase in all enstatite chondrites (13.3-28.0 %) (Fig. 168). The anomalously low metal content of Blithfield (EL6; 1.6 wt %), determined by Keil (1968a) is because this meteorite is a breccia which contains centimeter-sized, sulfide-rich, metal-poor clasts (Rubin 1984), one of which was analyzed by Keil (1968a). The metal content of the matrix of Blithfield is actually extremely high (64%).

Ringwood (1961b) first observed that kamacite in enstatite chondrites was enriched in Si, based on x-ray diffraction measurements and chemical analyses of separated metal grains. Although the x-ray diffraction data of Ringwood (1961b) yielded consistently high Si contents, the agreement between the chemical analyses and subsequent electron microprobe analyses, performed by Keil (1968a), on the same meteorites is reasonable. Compositional data for kamacite in many enstatite chondrites including Abee (impact melt breccia), Indarch (EH4), Yilmia (EL6), Parsa (EH3), Kota-Kota (EH4), Adhi-Kot (EH4), South Oman (EH4), Y-691 (EH3) and Y-75261 have now been reported in the literature (e.g. Reed 1968, Wasson and Wai 1970, Buseck and Holdsworth 1972, Bhandari et al. 1980, Leitch and Smith 1980, Rubin and Keil. 1983, Rubin. 1983a,b,c; Kimura 1988, Ikeda 1989b, Nagahara 1991). Representative electron microprobe analyses of kamacite in enstatite chondrites are reported in Table A3.23.

The occurrences of kamacite in the EH3 chondrites, Y-691 (EH3) and Qingzhen have been described in considerable detail by Kimura (1988), El Goresy et al. (1988) and

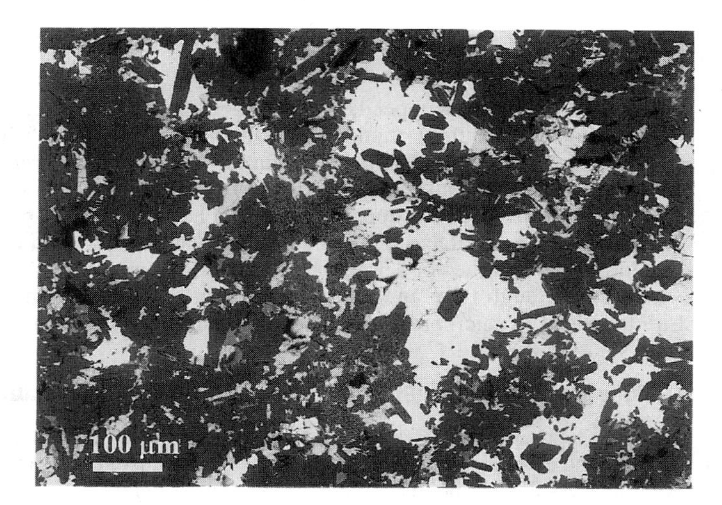

Figure 168. BSE image showing the typical distribution and morphology of metal (bright) and sulfide grains in the equilibrated enstatite chondrite, St Mark's (EH4).

Ikeda (1989b). Ikeda (1989b) recognized two different types, termed concentric nodules and massive nodules. The concentric nodules have a core of kamacite and troilite, surrounded by layers of silicate and a mantle of kamacite and troilite. Kamacite occurs in three types of massive nodule consisting of: (a) kamacite with minor amounts of several sulfide phases and graphite (the metal-graphite clasts of El Goresy et al. 1988), (b) kamacite and troilite with other minor phases (e.g. perryite), and (c) kamacite and niningerite as the main phases with several accessory sulfides. In an EH3-4 chondrite clast in Kaidun, Ivanov et al. (1996) also found kamacite occurring in three types of nodules—types I, II and III. Type I nodules have an irregular shape and consist of many small (<20 μm) grains of kamacite, Type II inclusions are more rounded and have a concentric structure, similar to those described by Ikeda (1989b) and Type III are rounded, but have no internal structure. All the inclusions are associated with other accessory phases such as troilite, schreibersite, etc. Ivanov et al. (1996) found that the nodules also contain very small (<5 μm) inclusions of SiO_2, glass, enstatite, roedderite and a mixture of SiO_2 and Na_2S_2, often contained within the metal.

El Goresy et al. (1988) showed that the metal grains in Qingzhen (EH3) and Y-691 (EH3) have quite distinct Si contents (Fig. 169). In Y-691, metal contains 1.2-2.4 wt % Si, whereas in Qingzhen the Si content is consistently higher (2.25-3.0 wt %). However, the Ni content in kamacite in both meteorites is very similar (1.5-4.0 wt %) with 0.25-0.35 wt % Co. In the EH3-4 clast studied by Ivanov et al. (1996), the metal contains ~3.33 wt % Si and 5.92 wt % Ni.

Electron microprobe analyses of kamacite in equilibrated enstatite chondrites have shown that there are clear compositional trends with increasing petrologic type in both the EH and EL series (e.g. Keil 1968a, Sears et al. 1982, Sears et al. 1984b, Zhang et al. 1995). Keil (1968a) initially showed that there were significant compositional differences between metal in EH4 and EH5 chondrites, compared with EL6. No E3, EL4, EL5 or EH6 chondrites were known at this time and the enstatite chondrites were divided into three groups, Type I (type 4), intermediate (type 5), and Type II (type 6) enstatite chondrites. It is now known that the Si and Ni contents of metals in the EH and EL groups are quite different and have different compositional trends as they become more equilibrated. Within the EL group, the Si content of kamacite increases from ~0.3-0.5 wt % in EL3 chondrites to 0.9-1.8 in EL5s and finally to 1.1-1.7 in EL6s (Keil 1968a, Sears

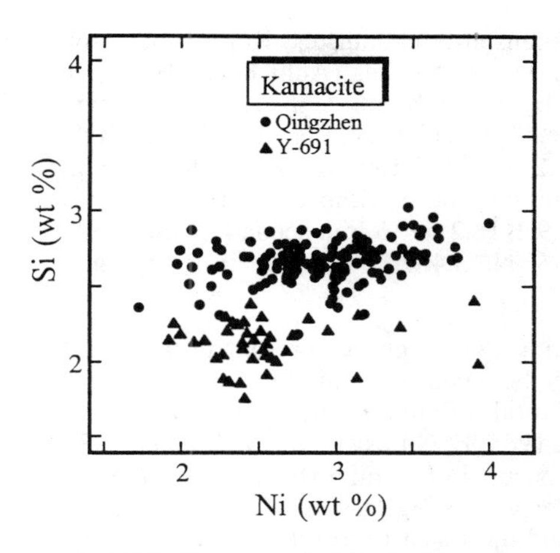

Figure 169. Si vs. Ni (wt %) contents of metal grains in the two EH3 chondrites Qingzhen and Yamato 691. The metal in the two chondrites has distinctly different Si contents, but a similar range of Ni contents. After El Goresy et al. (1988).

et al. 1982, Sears et al. 1984b, Zhang et al. 1995). In comparison, within the EH group the Si content of kamacite is around 2 wt % in EH3 chondrites (e.g. Ikeda 1989b), increasing to ~4 wt % in the EH6 chondrites, with EH4s having intermediate Si contents (2.6-3.5 wt %) (Keil 1968a, Sears et al. 1982, Zhang et al. 1995). Keil (1968a) found that the range of Si contents in kamacite, measured within individual meteorites, is remarkably restricted.

Keil (1968a) found that the average Ni contents of kamacite was variable from one chondrite to another, showing the highest variability in the EH4 chondrites. Subsequent work has shown that, like Si, the Ni content of metal within the EH and EL groups increases as function of petrologic type (e.g. Sears et al. 1982, Zhang et al. 1995). In EH3 chondrites, the kamacite contains 2.2-2.8 wt % Ni (e.g. El Goresy et al. 1988, Ikeda 1989b), but increases markedly to ~8 wt % in EH5-6 chondrites. A similar trend is observed in the EL chondrites, but the increase in Ni contents is much smaller and less well defined. However, EL3 chondrites contain, on average, <6 wt % Ni whereas types 5-6 contain >6 wt % (6.1-6.8 wt %) (Keil 1968a). Rambaldi et al. (1983) have suggested that the EH chondrites actually consist of two groups with high and low Ni contents in kamacite (2-4 wt % and 6-7 wt %, respectively).

The fact that the Ni contents of some kamacite grains in type 3 enstatite chondrites exceeds the approximate α-solubility limit (7.5 wt %) in the binary system Fe-Ni has been attributed to their high P contents (up to 0.7 wt %) (Keil 1968a). Keil (1968a) found that the mean P contents of kamacite within the EH4 chondrites show considerable variability (0.04-0.54 wt %), but in EL6 chondrites the mean values are much more restricted (0.09-0.13 wt %). Similarly, in EH4s the Co contents are more variable (0.46-0.77 wt % Co) than in EL6 chondrites (0.37-0.5 wt %).

Herndon and Rudee (1978) measured 0.52±0.05 wt % C in metal from Abee (impact melt breccia) by electron microprobe and also examined the microstructure of the metal using TEM techniques and found that the kamacite has a lamellar pearlite structure.

Taenite. Taenite is an extremely rare phase in enstatite chondrites (e.g. Ramdohr 1963, Keil 1968a). Buseck and Holdsworth (1972) found that taenite was more common in Yilmia (EL6) than other enstatite chondrites and has an average composition of 16.4 wt % Ni and 1.23 wt % Si. Taenite has also been reported by Ikeda (1989b), Lin et al. (1991) and Rubin (1997a) in enstatite chondrites.

Martensite. In an extensive study of the enstatite chondrites, Keil (1968a) found martensite (Keil termed this phase taenite) in two meteorites (Adhi-Kot, EH4; Hvittis, EL6). The identification of this phase was based on Ni content (9.8 and 9.1 wt %, respectively) measured by electron microprobe. The taenite is enriched in Si in both these chondrites (Adhi-Kot—0.8 wt %; Hvittis—3.2 wt %). Rubin and Keil (1983) reported heterogeneous martensite in two dark inclusions in the EH impact melt breccia, Abee, with 11.1-16.9 wt % Ni in one inclusion and 9.1-13.2 wt % in a second. Rubin (1983a) also reported martensite in Adhi Kot (8-11 wt % Ni, 3.4 wt % Si), sometimes intergrown with martensite (see also Rubin 1983b,c).

Troilite (FeS). Troilite is one of the major opaque phases present in the enstatite chondrites with modal abundances typically between 4.6 and 9.8 vol %. In type 3 enstatite chondrites, such as Y-691 (EH3), several different occurrences of troilite have been recognized (e.g. Nagahara 1985, Kimura 1988, El Goresy et al. 1988, Ikeda 1989b). According to Ikeda (1989b), troilite in Y-691 occurs in several different types of nodules with distinct mineral assemblages, i.e. nodules consisting largely of troilite, kamacite-troilite nodules, niningerite-troilite nodules and djerfisherite-troilite nodules. Troilite is also found in distinct clasts with daubreelite (El Goresy et al. 1988). Troilite in all these different occurrences is associated with or contains inclusions of additional sulfides, phosphides etc. Within the nodules, two types of troilite, termed porous (due to the presence of numerous pits) and massive, occur. The massive troilite often contains exsolution lamellae of daubreelite.

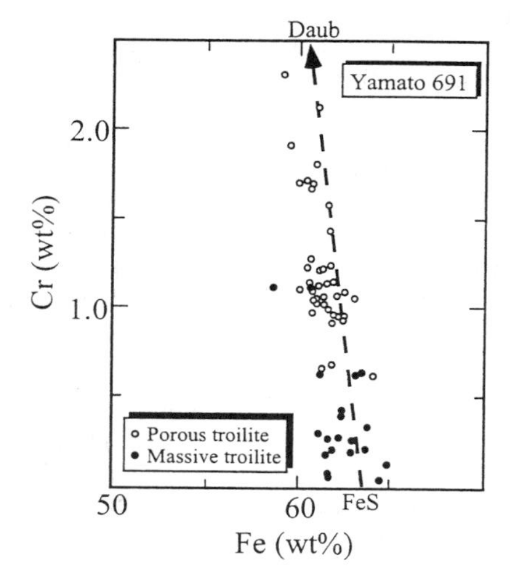

Figure 170. Cr vs. Fe (wt %) contents of massive and porous troilite in the EH3 chondrite Yamato 691. After Ikeda (1989).

In Y-691, Ikeda (1989b) found higher Cr contents in massive troilite (0.6-2.4 wt % Cr) than porous troilite (0.05-1.2 wt %) (Fig. 170). Porous troilite is also unusual in containing highly variable, but sometimes extremely high contents of Cl (0-8.2 wt %), which may be present in submicron inclusions. El Goresy et al. (1988) found that troilites in Qingzhen contain lower Ti contents (0.10-0.19 wt %) than those in Y-691 (0.20-0.33 wt %). Representative electron microprobe analyses of troilite in enstatite chondrites are reported in Table A3.24.

In equilibrated enstatite chondrites, troilite occurs as distinct grains, usually associated with Fe,Ni metal and other sulfides and can also occur as inclusions within SiO_2 (Keil 1968a). Troilite sometimes contains exsolution lamellae of ferroan alabandite and daubreelite (Keil and Andersen 1965). Troilite in equilibrated enstatite chondrites contains significant concentrations of Ti (0.2-1.0 wt %) (e.g. Keil 1968a, Leitch and Smith 1982). Sears et al. (1982) showed that the Ti contents of troilite in equilibrated EH and EL chondrites are distinctly different and Zhang et al. (1995) found that the Ti and Cr contents of troilite change as a function of increasing petrologic type. Ti and Cr increase from ~0.5 wt % and ~0.2 wt %, respectively, in EL3 chondrites to 0.8 and ~1.5 wt %, respectively in EL6 chondrites. For the EH group, the increase is from <0.25 wt % Ti and ~0.5 wt % Cr to 0.5 wt % Ti and ~3.7 wt % Cr in EH6s. Keil (1968a) also noted slightly lower Mn contents (0.03-0.21 wt %) in troilite in EL6 chondrites compared with EH4s (0.02-0.39 wt %). Analyses of troilite in equilibrated enstatite chondrites have also been reported by Rubin and Keil (1983), Rubin (1983a,b,c) and Sears et al. (1984b) (See Table A3.24).

Troilite in the unusual breccia, Y-75261, is more enriched in Cr (1.1-4.2 wt %) than any other enstatite chondrite (Nagahara 1991). However, Ti contents (0.2-0.8 wt %) lie in the typical range for enstatite chondrites and are negatively correlated with Cr and positively correlated with Fe contents. The high Cr and Ti concentrations may be due to the presence of submicron daubreelite lamellae.

Niningerite (FeMgMnS). Although Fredriksson and Wickman (1963) first reported an analysis of an unknown Fe-Mn-Mg sulfide phase in the St. Mark's (EH5) chondrite, niningerite was first fully recognized and described in several EH4 and 5 chondrites by Keil and Snetsinger (1967). Niningerite is a common accessory phase in unequilibrated EH chondrites (e.g. Rambaldi et al. 1986, El Goresy et al. 1988, Kimura 1988, Ehlers and El Goresy 1988, Ikeda 1989b). El Goresy et al. (1988) and Ehlers and El Goresy (1988) described six different occurrences of niningerite in the EH3 chondrites Qingzhen and Y-691. It occurs in: (1) chondrules as the most abundant sulfide, (2) in igneous clasts which consist dominantly of enstatite with interstitial SiO_2, (3) as niningerite-troilite-silicate clasts, (4) in multiphase metal-sulfide spherules, consisting of kamacite containing inclusions of niningerite, oldhamite and troilite plus accessory phases such as daubreelite and djerfisherite, (5) oldhamite-niningerite clasts, and (6) as inclusions in metal grains with oldhamite and troilite. Some of these occurrences are unique to particular chondrites, but some are found in more than one chondrite.

The grain sizes and textural characteristics of niningerite in different EH chondrites are highly variable. Grain sizes vary from a few microns up to 200 μm in some cases (e.g. Indarch, EH4) (Fig. 171). In equilibrated enstatite chondrites, Keil (1968a) reported abundances of niningerite of between 0.23 and 4.1 % with one outlier, Abee, which contains 11.2 vol %. The niningerite typically occurs intergrown with Fe,Ni metal and troilite and occasionally contains (111) exsolution lamellae of troilite of variable width (e.g. Abee, Kaidun III, South Oman, St. Mark's) or minute, rounded inclusions of metal. Niningerite can also occur as inclusions within SiO_2 (Keil 1968a). Analyses of niningerite are reported in Table A3.25.

Although niningerite contains Fe, Mn and Mg like ferroan alabandite, the two phases occupy distinct compositional fields within the FeS-MnS-MgS ternary system (Fig. 172). Keil and Snetsinger (1967) originally suggested that niningerite was only present in the less equilibrated Type I (EH4) and intermediate (EH5) enstatite chondrites, ferroan alabandite being the stable phase in Type IIs (EL6). However, after recognition of the EH and EL groups (Sears et al. 1982), it became clear that niningerite occurs only in the EH group in all petrologic types, including the EH3 chondrites (e.g. Y-691, Ehlers and

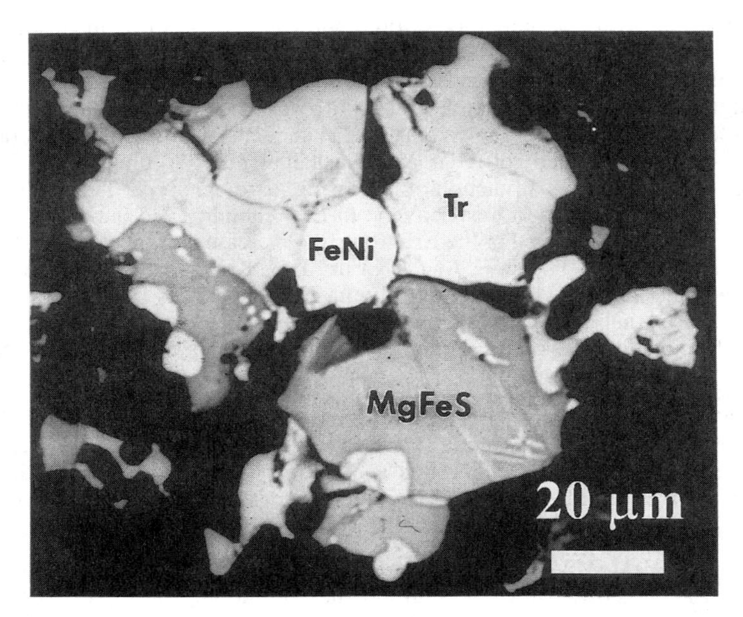

Figure 171. BSE image of a typical occurrence of niningerite (MgFeS) associated with Fe,Ni metal and troilite (FeS) in the EH4 chondrite, Kota-Kota.

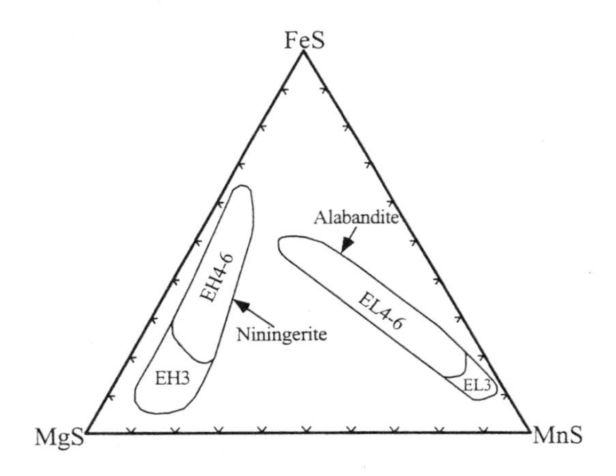

Figure 172. Ternary diagram MgS-FeS-MnS showing the compositional fields of alabandite and niningerite in enstatite chondrites. Data from Keil and Snetsinger (1967), Zhang et al. (1995).

El Goresy 1988, El Goresy and Ehlers 1989), whereas ferroan alabandite occurs in the EL group.

In a detailed study of niningerite in several EH chondrites (Abee, Kaidun III, South Oman (EH4), St. Mark's (EH5), Y-74730, Indarch (EH4), Y-691 (EH3) and Qingzhen (EH3), Ehlers and El Goresy (1988) found that some of the variability in the composition of niningerite was due to zoning within individual mineral grains. Two types of zoning occur: (a) normal zoning with a decrease in Fe from core to rim and reverse zoning which occurs only in Indarch, Y-691 and Qingzhen. The normal zoning is attributed to Fe diffusion into troilite during cooling and the reverse zoning is probably the result of

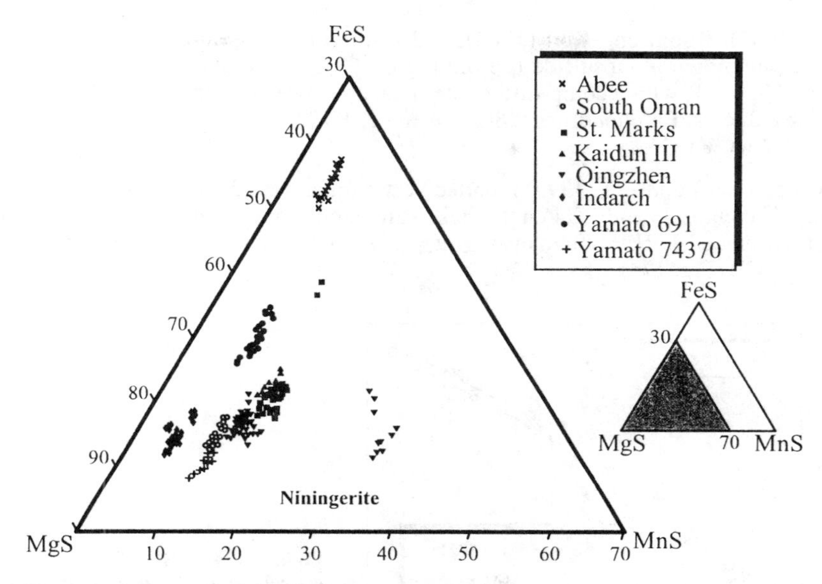

Figure 173. Ternary atom percent MgS-FeS-MnS diagram showing the compositional ranges of niningerite in several different equilibrated and unequilibrated enstatite chondrites. After Ehlers and El Goresy (1989).

thermal events which caused Fe to diffuse back into niningerite from troilite. El Goresy et al. (1988) and Ehlers and El Goresy (1988) also found that niningerite compositions in different EH chondrites occupy three distinct subgroups with different MnS contents in the MgS-MnS-FeS ternary system. The compositional variations within each group are essentially restricted to their MgS and FeS contents, which show considerable variability from one chondrite to another (Fig. 173). The three groups have compositional ranges 3.6-6.7 mol % MnS, 7.5 mol % MnS and 10-17 mol % MnS, respectively. Ehlers and El Goresy (1988) suggested that these different groups result from differences in Mn partitioning between enstatite and niningerite as a function of sulfur and oxygen fugacity during formation of the different chondrites.

Within the EH chondrites, there are clear trends in the composition of niningerite as a function of petrologic type. Skinner and Luce (1971) showed that variations in temperature result in considerable changes in the FeS and MgS contents of niningerite due to diffusion of FeS into coexisting troilite. In most EH3 chondrites, niningerite is enriched in MgS (>60 mol %) (e.g. Ehlers and El Goresy 1988, Ikeda 1989b), but the FeS content increases as a function of petrologic type to ~$(Mg_{0.4}Mn_{0.05}Fe_{0.55})S$. Keil (1968a) and Ehlers and El Goresy (1988) found, however, that the composition of niningerite in EH4 chondrites is highly variable, showing that it is not fully equilibrated. Mean Mg contents of niningerite in EH4 chondrites range from 10.1 to 23.5 wt % with Kota-Kota (EH4) showing a range in Mg contents from 18.7-27.1 wt %. Keil (1968a) also reported minor concentrations of Ca (0.39-3.03 wt %) and Cr (0.14-1.84 wt %) in niningerite in EH4 and EH5 chondrites and found that Cr and Ca are positively correlated and Mn and Mg negatively correlated with Fe (see also Ehlers and El Goresy 1988). In the impact melt breccia Abee, Ehlers and El Goresy (1988) found 1.8-2.2 wt % Cr and 2.2-2.9 wt % Ca in niningerite which also contains the highest FeS contents of all the EH chondrites they studied. Niningerite in EH3s is typically low in Ca (<1 wt %) and Cr (<0.61 wt %).

Compositions of niningerite in enstatite chondrites can also be found in Leitch and

Smith (1982), Rubin and Keil (1983), and Rubin (1983a). Notably, Nagahara (1991) found an unknown monosulfide (Fe,Mn,Mg,Ca,Cr)S in the unusual enstatite chondrite breccia, Y-75261, with a composition intermediate between niningerite in EH chondrites and alabandites in EL chondrites (28-23 wt % Fe; 12-22 wt % Mn; 6-11 wt % Mg; 2-3 wt % Mg; 1-2 wt % Cr).

Crozaz and Lundberg (1995) reported very similar REE patterns for niningerite from several different EH3 and EH4 chondrites. Although the abundances vary, all the patterns have fractionated HREE abundances (Lu = 1-7 × CI), decreasing rapidly to the LREE (0.1-0.7 × CI) (Fig. 174).

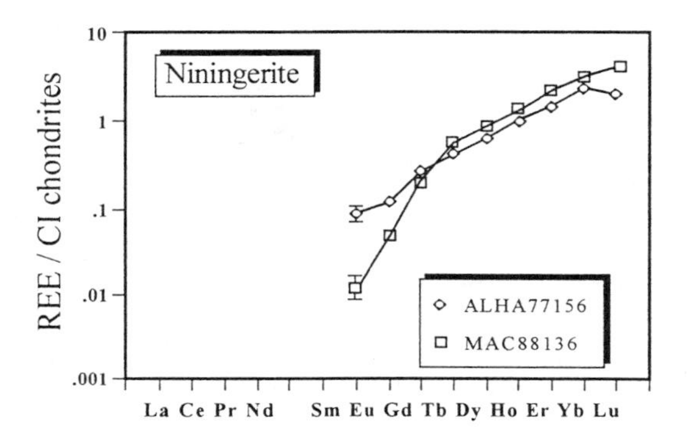

Figure 174. REE abundance diagram normalized to CI chondrites for niningerite grains from the enstatite chondrites ALH 77156 and MAC 88136. After Crozaz and Lundberg (1995).

Ferroan alabandite (MnFeS). Although ferroan alabandite was originally thought to occur only in Type II (type 6) enstatite chondrites (Keil 1968a), it is now known to occur in all EL chondrites, as an accessory phase (0.18-0.97 vol %) rather than niningerite. It occurs as discrete grains (5-100 μm) and as exsolution lamellae and cigar-shaped bodies in troilite (e.g. Keil 1968a, Buseck and Holdsworth 1972). Ferroan alabandite is richer in Mn and lower in Mg and Fe than niningerite (Fig. 172) and also shows a more restricted range of compositions from meteorite to meteorite than niningerite. In EL6 chondrites, Keil (1968a) found that ferroan alabandite ranges in composition from $(Mn_{0.53}Fe_{0.28}Mg_{0.16}Ca_{0.01}Cr_{0.009})S$ to $(Mn_{0.72}Fe_{0.18}Mg_{0.08}Ca_{0.006}Cr_{0.009})S$ and noted less well defined substitutional trends than for niningerite. Ca and S tend to increase with increasing Fe and Mn, decreases. Zhang et al. (1995) have shown that the composition of ferroan alabandite changes as function of petrologic type, becoming progressively enriched in Fe and depleted in Mn from EL3 to EL6 chondrites (Fig. 172). See Table A3.26 for representative electron microprobe analyses of alabandite. See also Rubin (1983b,c) for analyses of alabandite.

Crozaz and Lundberg (1995) measured REE abundances in ferroan alabandite from MAC88136 (EL3) and found fractionated REE patterns with significantly enriched HREE (Lu ~4 × CI) and depleted LREE which are typically below detection limits (Eu ~0.02 × CI).

Oldhamite (CaS). Oldhamite is a ubiquitous accessory phase in the enstatite chondrites (Ramdohr 1963, Mason 1966, Keil 1968a, Buseck and Holdsworth 1972, Leitch and Smith 1982, Rubin 1983c), although in some weathered chondrites it appears to be absent (e.g. Atlanta), due to its high susceptibility to terrestrial weathering. Unlike most other sulfides, oldhamite is not, in fact, an opaque phase, but has a distinct pink coloration in transmitted light. Keil (1968a) found that oldhamite commonly luminesces under electron bombardment with a variety of colors ranging from a common orange yellow to light green, dull grey and weak yellow.

In type 3 chondrites, such as Y-691 and Qingzhen, oldhamite is found as small grains within sulfide and metal nodules and as isolated mineral fragments (e.g. Rambaldi et al. 1986, Kimura 1988, El Goresy et al. 1988, Ikeda 1989b, Crozaz and Lundberg 1995). El Goresy et al. (1988) also found oldhamite as a minor phase in clasts dominated by niningerite. In more equilibrated chondrites, oldhamite typically occurs as individual grains, up to ~150 μm in size, within the enstatite matrix (e.g. Keil and Andersen 1965). Oldhamite is unusual in typically not being associated with other sulfides (e.g. Keil and Anderson 1965, Buseck and Holdsworth 1972).

Oldhamite is close to endmember CaS in composition, but contains minor Mg, Cr and Fe contents which are distinctly different in unequilibrated and equilibrated enstatite chondrites. In EH3 chondrites (e.g. Y-691), oldhamite contains <0.58 wt % Mg, <0.38 wt % Mn and <1.7 wt % Fe (El Goresy et al. 1988, Ikeda 1989b). Keil (1968a) reported higher average Mg and Fe in EH4 chondrites compared with EL6s (1.4 vs 0.52 wt % Mg; 0.60 vs 0.32 wt % Fe) and the reverse relationship for Mn (0.32 vs 0.96 wt % Mn). EH5 chondrites (originally termed intermediate) have Mn, Mg and Fe concentrations slightly lower than EH4s. Oldhamite in Kota-Kota (EH4) has anomalously high Fe contents (3.7-6.1 wt %) (Leitch and Smith 1982). Buseck and Holdsworth (1972) reported that Indarch (EH4) appeared to contain two compositionally distinct types of oldhamite, with ~35 mol % and 2 mol % MgS, respectively. For oldhamite compositions, see also Rubin and Keil (1983) and Rubin (1983a,c). Analyses of oldhamite are reported in Table A3.27.

Oldhamites are one of the major carriers of the REEs in the enstatite chondrites. Crozaz and Lundberg (1995) measured REE, Zr, Sc and Sr in unweathered and weathered oldhamite grains in three different petrologic settings in unequilibrated and equilibrated EH and EL chondrites. Weathering does not appear to affect the REE abundances. Seven different types of REE patterns (termed A to G), which show no correlation with petrologic setting or apparently petrologic type, were observed. The first 4 types of pattern form a continuum and are characterized by flat REE abundances at ~100 × CI in Type A, with decreasing REE abundances in moving to Type D patterns. However, Type B-D patterns have positive Eu and Yb anomalies, although the abundances of both these elements are the same as in Type A patterns (Fig. 175a-d). Type E patterns are flat at ~100 × CI with negative Eu anomalies and Type F is similar to Type A, but has slightly fractionated (elevated) HREE. Type G is unusual in having enrichments in both LREE and HREE up to ~600 × CI. Both Type F and G patterns were only observed in one weathered grain of oldhamite. Crozaz and Lundberg (1995) also found enrichments in Zr (13-14.5 × CI) in oldhamite in two EH chondrites, but no apparent enrichment in one EL3. Sr also appears to be enriched in oldhamite in EHs (14.5-21.6 × CI) in comparison with ELs, but Sc is similar in both groups (5.0-8.1 × CI).

Daubreelite ($FeCr_2S_4$). Borgström (1903) first recognized daubreelite in the enstatite chondrite Hvittis and the petrography and composition of this phase have subsequently been described in more detail by Ramdohr (1963), Mason (1966), Keil and Anderson (1965), Keil (1968a), Buseck and Holdsworth (1972) and El Goresy et al. (1988) (see also Rubin (1983a,b,c 1984). Daubreelite appears to occur in essentially all enstatite chondrites as a minor phase (<1.6 %, Keil 1968a), commonly as exsolution lamellae (a few microns to >100 μm wide) in troilite, parallel to {0001} or closely intergrown with troilite (e.g. Kimura 1988). It has also been observed as inclusions within SiO_2 (Keil 1968a). Electron microprobe analyses of daubreelite (Keil 1968a) show that it is compositionally close to $FeCr_2S_4$, but can contain minor Mn (~1-4 wt %), Mg (~<0.03-0.15 wt %), Ti (<0.03-0.15 wt %) and Zn (<0.03-0.55 wt %). The compositional ranges

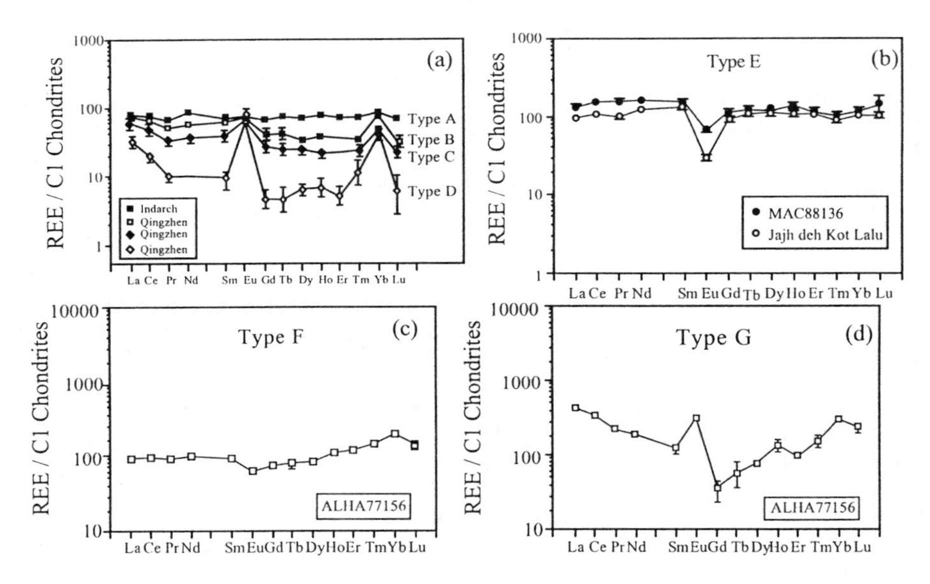

Figure 175. REE abundance diagrams normalized to CI chondrites, showing the seven different types of REE patterns encountered in oldhamite grains in enstatite chondrites: (a) Types A-D, (b) Type E, (c) Type F, and (d) Type G. After Crozaz and Lundberg (1995).

exhibited by minor elements are somewhat variable from one enstatite chondrite to another. Manganoan daubreelite appears to occur exclusively in metamorphosed enstatite chondrites. See Table A3.28 for electron microprobe analyses of daubreelite.

Zincian daubreelite. Keil (1968a,b) found that daubreelite in the Kota-Kota (EH4) and St. Mark's (EH5) chondrites is compositionally distinct from typical daubreelite, being enriched in Zn (4.2-5.5 Zn wt %), rather than Mn (<1.10 wt %). This variety of daubreelite has been referred to as zincian daubreelite and also typically occurs intergrown or exsolved from troilite. Keil (1968a,b) determined average structural formula for zincian daubreelite in Kota-Kota (EH4) and St. Mark's as $(Fe_{0.79}Zn_{0.21})(Cr_{1.96}Mn_{0.06}Zn_{0.03})S_4$ and $(Fe_{0.78}Zn_{0.19}Mn_{0.03})Cr_{1.98}Mn_{0.01}Ti_{0.01}S_2$, respectively. Zincian daubreelite also occurs in Yilmia (EL6) (Buseck and Holdsworth 1972). El Goresy et al. (1988) reported zincian daubreelites in Qingzhen (EH3) which contain considerable amounts of Na (0.20-0.45 wt %) and are the most Zn-enriched of any zincian daubreelites analysed in enstatite chondrites (5.7-8.11 wt % Zn).

Sphalerite (ZnS). Sphalerite is a relatively rare, but important accessory phase in enstatite chondrites which was first recognized by Ramdohr (1963). Descriptions of the occurrence and compositions of sphalerite in EH and EL chondrites have been reported by Rambaldi et al. (1986), Kimura (1988), El Goresy et al. (1988), El Goresy and Ehlers (1989), Kissin (1989) and Ikeda (1989b). Its presence, coexisting with troilite and kamacite, has been widely used as a geothermometer for the enstatite chondrites (e.g. Hutchison and Scott 1983, Kissin 1989, El Goresy and Ehlers 1989). The occurrence and composition of sphalerite in EH3-4 chondrites has been examined in some detail (e.g. El Goresy and Ehlers 1989, Kissin 1989, Ikeda 1989b). El Goresy and Ehlers (1989) found significant variations in the occurrence of sphalerite between different chondrites. In Qingzhen, sphalerites are ~30-150 μm in size and coexist with troilite, kamacite, perryite and schreibersite in the interiors of complex metal-sulfide clasts. Grains of sphalerite in contact with troilite are zoned with a progressive decrease in FeS and increase in MnS toward the troilite. Exsolution of troilite lamellae along (111) planes of sphalerite is evident adjacent to troilite. In Y-691, three types of sphalerite occur: (a) large, unzoned, primary sphalerite grains with inclusions of kamacite and troilite, (b) porous, zoned grains of sphalerite which decrease in FeS towards troilite grain boundaries, and (c) secondary, porous sphalerite which is associated with troilite, produced by the breakdown of djerfisherite. BSE images of sphalerite in Indarch (EH4) often show open pores and irregular dark and light zones. Inclusions of roedderite and oriented platelets of an unidentified phase may also be present.

Sphalerite in unequilibrated enstatite chondrites contains significant amounts of FeS, as well as minor MnS in solid solution. Other common minor elements are Mg (<1 wt %, Ni (0-~0.9 wt %), Cu (<0.05-1.23 wt %), Ga (<0.02-0.09 wt %) and sometimes Cr (<0.07 wt %). El Goresy and Ehlers (1989) reported compositional variability between the different types of sphalerites in Qingzhen (EH3), Y-691 (EH3), and Indarch (EH4), most notably in their FeS contents. Zoned crystals in Qingzhen range from core values of 49.6 to ~45 mol % FeS at the rims, in contact with kamacite and MnS is also zoned between 5.55 (core) and 6.48 mol % (rim). In Y-791, primary and secondary sphalerites are compositionally distinct. Secondary sphalerite has lower FeS contents (42.3-43.2 mol %, cf. 47.03-49.8 mol %), MnS (3.83 mol %, cf. 2.0 mol %) and Cu contents are higher (up to 1.23 wt %; cf. <0.67). In Indarch (EH4), sphalerite is difficult to analyze, but appears to have a much higher FeS content (53.5-56.0 mol % FeS), than either Qingzhen or Y-691. There also appear to be notable differences in the Ni contents of all types of sphalerite between Qingzhen (<~0.9 wt % Ni) and Y-691 (<0.07 wt % Ni). Rambaldi et al. (1986) have also reported small grains (<10 μm) of an unusual Ga-rich sphalerite in one metal sulfide nodule in Qingzhen which contains between ~2.1 and 3.7 wt % Ga. Representative analyses of sphalerites are reported in Table A3.29.

In more equilibrated enstatite chondrites, e.g. Pillistfer (EL6), sphalerite grains (~60 μm in size) are associated with kamacite and troilite and are compositionally homogeneous with ~47 mol % FeS and 20.6 mol % MnS (Kissin 1989).

Caswellsilverite ($NaCrS_2$). Caswellsilverite is a rare phase in enstatite chondrites, but has now been recognized in several meteorites. El Goresy et al. (1988) and Ikeda (1989b) reported caswellsilverite as a minor phase in kamacite-troilite nodules in Y-691 (EH3) and it also occurs in niningerite-oldhamite clasts and associated with an unknown Na-Cr-Cu-sulfide (El Goresy et al. 1988). Caswellsilverite in Y-691 is almost pure $NaCrS_2$, with minor Ca, Fe, Mn and K (El Goresy et al. (1988), but Ikeda (1989b) reported that the Na content of caswellsilverite varies from 3-15 wt %. The highest Na contents are consistent with caswellsilverite, but the phases with low Na may be weathering products. El Goresy et al. (1988) also found a Cu-bearing sulfide phase in Y-691 with up to 14.7 wt % Cu and 4.95 wt % Na which may be related to caswellsilverite. This phase has Cu negatively correlated with Na, suggesting a substitution of Na for Cu in the caswellsilverite structure. Analyses of caswellsilverite are reported in Table A3.30.

Djerfisherite ($K_3(Na,Cu)(Fe,Ni)_{12}S_{14}$). The highly unusual alkali, copper-iron sulfide, djerfisherite only occurs in the enstatite chondrites and achondrites and was discovered by Fuchs (1966) in the Kota-Kota (EH4) and St. Mark's (EH5) chondrites. Djerfisherite is probably the unknown mineral in St. Mark's that Ramdohr (1963) designated mineral C. It occurs in the EH3 chondrites Qingzhen and Y-691 (e.g. Rambaldi et al. 1986, El Goresy et al. 1988) and in some equilibrated enstatite chondrites, but has not been detected in others (Fuchs 1966, El Goresy et al. 1988). In Qingzhen, djerfisherite appears to occur exclusively outside chondrules (Grossman et al. 1985). In both Qingzhen and Y-691, textural evidence shows that djerfisherite is breaking down to porous (or frothy) secondary troilite and a mixture of fine-grained phases, including covellite, idaite and a K-bearing phase (El Goresy et al. 1988, El Goresy and Ehlers 1989, see also Rambaldi et al. 1986). Djerfisherite occurs as irregularly shaped inclusions in troilite in Y-691 (Kimura 1988) and as large grains (up to millimeters in size) as clasts, fragments, spherules and veins in the matrices of EH3 chondrites (El Goresy et al. 1988). It never occurs in chondrules. Ikeda (1989b) also described djerfisherite in kamacite-troilite nodules (see also El Goresy et al. 1988) and as the main phase in djerfisherite-troilite nodules in Y-691.

The composition of djerfisherite in Y-691 is highly variable, most notably in its Cu and Ni contents which appear to covary. Cu ranges between 1.43-4.20 wt % and Ni between 0.42-3.03 wt %. Na and K also show considerable variation and are negatively correlated in Y-691 and Qingzhen (Fig. 176). El Goresy et al. (1988) found two compositionally distinct types of djerfisherite in Y-691, one high in Cu (4.73 wt %) and low in Ni (0.49 wt %) and Na (1.14 wt %) and a second with lower Cu (2.26 wt %) and higher Ni (3.02 wt %) and Na (1.52 wt %). In comparison, djerfisherite in Qingzhen has a much more restricted range of compositions (El Goresy et al. 1988) and is distinct from djerfisherite in Y-691.

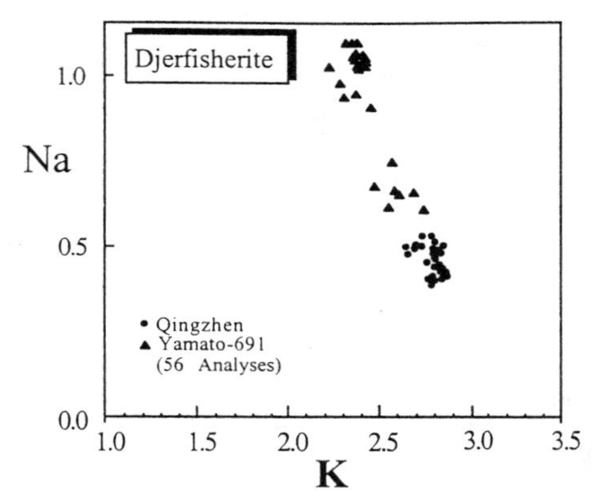

Figure 176. Na vs. K (cations) in djerfisherite from the EH3 chondrites Yamato 691 and Qingzhen. After El Goresy et al. (1988).

In Kota-Kota (EH4), djerfisherite occurs in rounded bodies, ~0.4 mm in size, associated with kamacite, schreibersite and troilite, but in St. Mark's (EH5) it occurs as 0.02-0.07 mm isolated grains in the silicate matrix (Fuchs 1966). Analyses of djerfisherite are reported in Table A3.31.

Other sulfide minerals. El Goresy et al. (1988) reported grains of *covellite (CuS)*, *idaite* (Cu_4FeS_6) and *bornite* (Cu_5FeS_4) as reaction products associated with the breakdown of djerfisherite in Y-691. In Y-691, El Goresy et al. (1988) found a *Zn,Cu-rich NaCr sulfide*, which probably formed from alteration of the Cu-bearing caswellsilverite phase. It contains 5.96-9.97 wt % Zn and 4.62-7.62 wt % Cu, with highly variable Na (1.86-12.9 wt %). El Goresy et al. (1988) found two unidentified *hydrated Na-Cr sulfides* (termed A and B) in distinct clasts and in troilite-daubreelite clasts in Qingzhen and Y-691. Phase A is confined to the daubreelite lamellae in troilite and is crystallographically oriented. In the clasts, the two sulfide phases are closely intergrown with each other and with daubreelite. Phase A gives totals of ~92 wt % with Na <1.14 wt % and phase B has totals of ~71 wt %. El Goresy et al. (1988) suggested that these phases may be preterrestrial in origin and are probably replacement products of caswellsilverite. However, they may be terrestrial weathering products (see Rubin 1997a). El Goresy et al.

(1988) also found a Zn,Cu-rich Na-Cr sulfide, which probably formed from alteration of the Cu-bearing caswellsiliverite phase.

Schreibersite ($FeNi_3P$). Schreibersite is a common accessory mineral in all the enstatite chondrites, occurring closely associated with or rimming kamacite grains (e.g. Keil and Anderson 1965, Keil 1968a, Buseck and Holdsworth 1972). In EH3 chondrites, it also occurs as rounded inclusions in kamacite, troilite or sphalerite (Kimura 1988, El Goresy et al. 1988, Ikeda 1989b). Keil and Anderson (1965) reported schreibersite grains as large as 300 μm in Jajh deh Kot Lalu. Keil (1968a) and Buseck and Holdsworth (1972) reported that schreibersite is compositionally variable, not only from grain to grain within a meteorite, but from meteorite to meteorite. The Ni contents of schreibersite in Y-691 (EH3) show a small compositional variation (15.22-18.35 wt % Ni) (Ikeda 1989b) and are indistinguishable from Qingzhen (El Goresy et al. 1988). Representative electron microprobe analyses of schreibersite are reported in Table A3.32.

In higher petrologic types, the Ni contents of schreibersite are highly variable. Keil (1968a) found that the range of average Ni contents in EH4 chondrites is 7.1-15.1 wt %, compared with 19.3-31.8 wt % in EL6 chondrites. Within individual meteorites, the Ni content can vary by as much as 26 wt % (e.g. Blithfield (EL6); Keil 1968a). Within the EH and EL groups, there is a progressive decrease in the Ni content of schreibersite as a function of petrologic type (e.g. Zhang et al. 1995). This decrease is largest in the EL chondrites which drop from ~40 wt % in EL3s to 12-30 wt % in EL6s. For EHs the decrease is from ~12-20 wt % Ni in EH3 chondrites to ~8-10 wt % in EH6s, although there is considerable scatter in the data for the EH5-6 chondrites. Si and Co contents in schreibersite in EH3 chondrites such as Y-691 range from ~0.14-0.35 wt % and 0-0.12 wt %, respectively (Ikeda 1989b, El Goresy et al. 1988). Keil (1968a) found that the average Co and Si contents of schreibersite in EH4 chondrites is higher than in EL6 chondrites (0.23-0.39 wt % Co vs 0.17-0.32; 0.20-0.46 wt % Si vs <0.03-0.10). Compositional data for schreibersite have also been reported by Leitch and Smith (1982) and Rubin (1983a,c 1984)

Perryite ($(Ni,Fe)_x(Si,P)$). The nickel silicide, perryite (sometimes termed the Henderson phase), was first described in enstatite chondrites by Fredriksson (1963), Fredriksson and Wickman (1963) and Ramdohr (1963) in St. Mark's (EH5) and Indarch (EH4). It is now recognized as a widespread accessory phase in the enstatite chondrites (e.g. Keil 1968a, Reed 1968), sometimes reaching 15% of the total opaques in the meteorite (e.g. Kota-Kota (EH4), Reed 1968). Perryite also occurs as a widespread minor accessory phase in metal and sulfide nodules in Y-691 (EH3) and in Qingzhen (EH3), commonly associated with kamacite and schreibersite (Kimura 1988, El Goresy et al. 1988, Ikeda 1989b) and sometimes graphite (El Goresy et al. 1988). In equilibrated enstatite chondrites, it typically occurs with kamacite, sometimes as small lath-shaped inclusions and also intergrown with troilite (e.g. Reed 1968, Keil, 1868a).

Compared with ideal $(Ni,Fe)_2(Si,P)$, perryite in enstatite chondrites is deficient in P. Reed (1968) reported the average structural formula of perryites in Kota-Kota (EH4) and South Oman as $(Ni_{1.9}Fe_{0.10})(Si_{0.61}P_{0.16})$ and $(Ni_{1.84}Fe_{0.16})Si_{0.76}P_{0.11})$, respectively. In the unequilibrated enstatite chondrites, such as Y-691, the Fe contents of perryite vary from grain to grain (6.76-10 wt %). The Ni contents of perryites in Y-691 and Qingzhen are distinctly different (El Goresy et al. 1988) (Fig. 177), with ~71-77 wt % Ni and ~76-82 wt %, respectively. However, the Si contents of perryite in both meteorites are comparable (11.8-13.9 wt %). In equilibrated enstatite chondrites, the perryite composition also varies from one chondrite to another. For example, Reed (1968) reported Si and Ni contents of 15.0 wt % and 6.3 wt %, respectively in Kota-Kota (EH4),

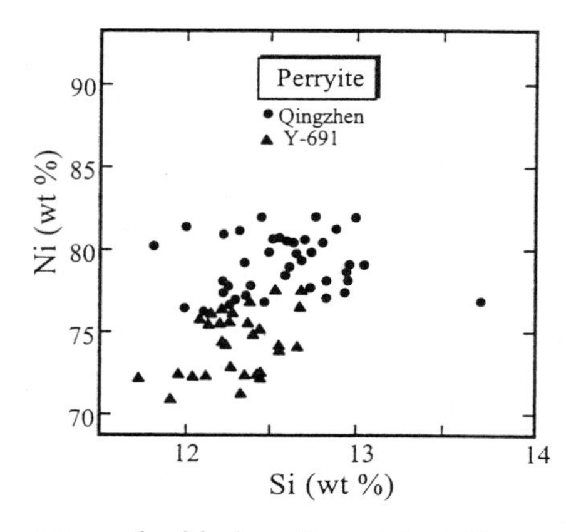

Figure 177. Ni vs. Si (wt %) diagram for perryites from the EH3 chondrites Yamato 691 and Qingzhen. After El Goresy et al. (1988).

compared with 4 wt % and 4 wt %, respectively in South Oman. See Table A3.33 for analyses of perryite.

Graphite. Graphite in most enstatite chondrites occurs associated with or as irregular or euhedral (~10-50 μm) inclusions in kamacite (e.g. Keil and Andersen 1965, Buseck and Holdsworth 1972, Leitch and Smith 1980). The graphite content of metal is lower in unequilibrated enstatite chondrites than higher petrologic types (Keil 1968a). This phenomenon probably results from the exsolution of graphite from carbon-bearing metal during metamorphism. In unequilibrated enstatite chondrites, El Goresy et al. (1988) found two distinct occurrences of graphite intergrown with metal (see also Kimura 1988). One type of graphite occurs as spherulites with radial oriented flakes, but the graphite can be fine-grained or amorphous. In the second type, graphite occurs as well oriented lamellae in kamacite. The spherulitic texture is abundant in Qingzhen, Y-691 and Y-74370, but is rare in Indarch (EH4), St. Mark's (EH5) and Kaidun III. The reverse is true for the oriented lamellae, which are most common in EH4 chondrites. Rare occurrences of graphite in enstatite chondrites have also been reported in Jajh deh Kot Lalu (Keil and Andersen 1965), Yilmia (EL6) (Buseck and Holdsworth 1972), Abee (Rubin and Keil 1983, Rubin 1997d, Rubin and Scott 1997) and Adhi Kot (Rubin 1983a) (see Shock Metamorphism) (see also Rubin et al. 1997).

Carbides. Cohenite was reported by Mason (1966) in Indarch (EH4), Abee and St Mark's (EH5). Cohenite occurs as thin pearlitic exsolution lamellae in kamacite in Abee (Herndon and Rudee 1978) and Rubin (1983a) found 0.5 to 2 μm elongate laths in kamacite in Adhi Kot.

Other minor phases. A number of other minor phases have been described in enstatite chondrites including *lawrencite* (Keil 1968a) in Indarch (EH4) (but see Rubin 1997a), *copper* metal (Ramdohr 1963, Fuchs 1966) in Hvittis and *osbornite* (TiN) in Yilmia (EL6) (Buseck and Holdsworth 1972). *Gold* was observed by Ramdohr (1963) in Hvittis, but Mason (1966) questioned this observation and suggested that the gold was probably osbornite (TiN). *Schöllhornite* also occurs as a minor phase in metal nodules in an EH3-4 clast in Kaidun (Ivanov et al. 1996), but may be a weathering product (see Rubin 1997). Several unidentified minerals have also been reported including a sulfide ($Fe_{0.537}Zn_{0.247}Mn_{0.160}Mg_{0.004}S$), associated with troilite, alabandite and daubreelite in Yilmia (EL6) (Buseck and Holdsworth 1972). A copper sulfide intermediate in composition between cubanite ($CuFe_2S_3$) and chalcopyrite ($CuFeS_2$) was described by Rambaldi et al. (1986) in a metal nodule in Qingzhen. The phase has a structural formula of $Cu_2Fe_3S_5$. Ikeda (1989b) reported an unidentified Cu-S-O phase in Y-691, which may be a weathering product of chalcocite or digenite.

Opaque mineralogy of CV carbonaceous chondrites

The opaque mineralogy of CV chondrites differs markedly depending on whether they belong to the reduced or oxidized subgroups. In reduced CV chondrites, such as Efremovka, Leoville, Vigarano and Arch, metal and pyrrhotite are the dominant opaque phases, whereas in the oxidized subgroup (e.g. Allende, Bali, Kaba, Grosnaja, Mokoia) metal is rare and magnetite is the dominant phase (McSween 1977c). According to McSween (1977), opaque phases in CV chondrites generally occur as subrounded grains inside chondrules, rather than independent phases in the matrix. However, Cohen et al. (1983) noted that in Mokoia, opaque assemblages also occur as discrete objects embedded in matrix (Cohen et al. 1983). Some CV chondrites appear to be breccias consisting of material from both oxidized and reduced subgroups. For example, in Leoville, centimeter-sized areas of the meteorite may contain either magnetite or metal almost exclusively, indicating postaccretional brecciation and mixing of material with varying oxidation states (McSween 1977c). Vigarano and Y-86751 also appear to be mixtures of oxidized and reduced material (Murakami and Ikeda 1994, Krot et al. 1995).

Diverse origins have been proposed for the opaque assemblages in the oxidized subgroup, including sulfidation and oxidation of Fe,Ni metal in an oxidizing environment in the solar nebula (Haggerty and McMahon 1979, Hua et al. 1988, Weinbruch et al. 1990, Murakami and Ikeda 1994), crystallization from metallic-sulfide-oxide liquids that formed immiscible droplets in crystallizing chondrule melts (Haggerty and McMahon 1979, McMahon and Haggerty 1980, Rubin 1991), and oxidation of Fe,Ni metal on an asteroid (Housley and Cirlin 1983, Blum et al. 1989, Krot et al. 1995). The dominant minerals in opaque nodules in the reduced subgroup are metal and troilite with minor magnetite and phosphates (McSween 1977c). Metal is mostly kamacite, but varying quantities of taenite (30-50% Ni) also occur (Fig. 178). Pentlandite in Allende and Vigarano contains 0.5 to 0.8 wt % Co (Fuchs and Olsen 1973).

Opaque nodules in CV chondrites of the oxidized group contain little or no metal, and magnetite predominates as the opaque phase. *Magnetites* in Allende, analyzed by Haggerty and McMahon (1979), have minor element variations in the following ranges: 0.5-0.7 wt % MgO, 0.8-1.9 wt % Al_2O_3, 1.5-2.1 wt % Cr_2O_3, ~0.1 wt % NiO, and <0.05 wt % TiO_2 and MnO. Rubin (1991) observed magnetites with lower abundances of MgO (0.07-0.5 wt %), Al_2O_3 (<0.04-0.74 wt %), and Cr_2O_3 (0.7-1.4 wt %) and determined CaO contents of <0.04-0.17 wt %. Magnetite in Axtell shows similar compositional ranges (Simon et al. 1995). In Y-86751, most magnetite has a Cr_2O_3 content of 1.0-1.5 wt % (Murakami and Ikeda 1994). In one chondrule in Y-86751, magnetite with a barred morphology also occurs in the mesostasis. This magnetite is compositionally distinct, and contains <0.5 wt % Cr_2O_3 and 0.6-1.0 wt % MgO. Magnetite grains in a type I chondrule in Allende that were studied by TEM sometimes contain small rounded inclusions of merrillite (Müller and Müller-Beneke 1992). Most magnetite grains contain stacking faults of the type 1/4 ⟨110⟩ {110} that probably formed during growth of the magnetite crystals.

Sulfide mineral compositions in the oxidized CV chondrites are variable, but consist mainly of pyrrhotite (described as troilite by McSween 1977c) and Ni-rich pentlandite (Haggerty and McMahon 1979, McSween 1977c; Fig. 178). Troilite and heazlewoodite (rhombohedral Ni_3S_2) are minor sulfide minerals. Pentlandite compositions in Allende are 17-23 wt % Ni, 0.8-1.5 wt % Co (Haggerty and McMahon 1979, Rubin 1991). Pyrrhotite averages 62.5 wt % Fe with <0.1 wt % Co+Ni (Haggerty and McMahon 1979). Magnetite and pentlandite are commonly finely intergrown.

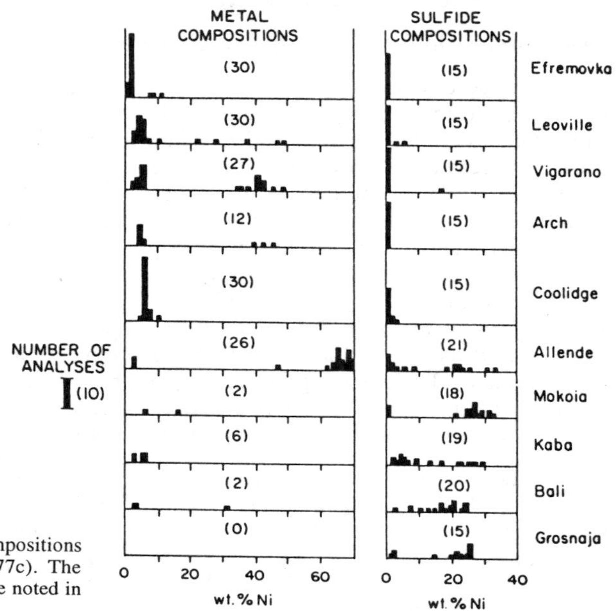

Figure 178. Metal and sulfide compositions in CV chondrites (McSween, 1977c). The numbers of microprobe analyses are noted in parentheses.

Metal compositions in opaque nodules in oxidized CV chondrites are very variable (Fig. 178), including kamacite and awaruite (Ni_3Fe), which is dominant in Allende (Clarke et al. 1970, Fuchs and Olsen 1973, McSween 1977c, Haggerty and McMahon 1979, McMahon and Haggerty 1980, Müller 1991a, Rubin 1991). Awaruite usually contains 65-68 wt % Ni and 1.7-2.6 wt % Co, although some compositions with 71-75 wt % Ni have also been reported. Euhedral, zoned awaruite grains were described by Rubin (1991). Ni-poor taenite (25 wt % Ni), Ni-rich taenite (37-63 wt % Ni) and a Co-rich metal (25-29 wt % Co, 1.4-2.6 wt % Ni) are present in Y-86751 (Murakami and Ikeda 1994).

Metal in Leoville contains tiny inclusions of phosphates, chromites, and silica that are inferred to have exsolved from metal during metamorphism (Zanda et al. 1994). Small amounts of merrillite, apatite, brianite, chromite, and fayalite may occur as accessory phases in opaque assemblages in Allende (Housley and Cirlin 1983, Rubin 1991, Kimura and Ikeda 1995).

Opaque mineralogy of CO carbonaceous chondrites

Opaque phases in CO chondrites have not been widely studied and only a general outline of their occurrence is available (e.g. Methot et al. 1975, McSween 1977a, Scott and Jones 1990, Shibata and Matsueda 1994, Shibata 1996). The most abundant phases present are kamacite, taenite and troilite and the metallic phases, sometimes associated with troilite, occur as (a) globules in the interior of chondrules, (b) grains attached to the chondrule surfaces but embedded within matrix or fine-grained chondrule rims, (c) grains within the matrix, (d) large metal chondrules. Awaruite may also be present (see Smith et al. 1993). Troilite is intimately associated with the metallic phases in the matrix, but is rare within chondrules. Details of opaque phases in CO chondrites are also discussed in the Chondrule section.

Shibata (1996) found that opaque minerals in the matrix of several CO chondrites typically occur as nodules ~100 μm in size, consisting of several phases, such as kamacite, troilite, taenite, tetrataenite and/or magnetite and cohenite. The identification of tetrataenite in this case is based only on composition (i.e. metal with >49 wt % Ni). In low petrologic type 3 chondrites, kamacite and tetrataenite are most abundant whereas in more metamorphosed CO chondrites, kamacite and taenite dominate and only minor tetrataenite is found. Ornans, however, appears to be unusual in containing only taenite (McSween 1977c).

In Y-82094, kamacite is the most abundant phase and ranges from submicron to several hundred microns in size. Taenite (and tetrataenite with Ni >49 wt %) is only a relatively minor phase and coexists with troilite. Inclusions of silicate and phosphates are abundant in metal within the matrix, but rare in chondrule metal. Shibata and Matsueda (1994) found Co contents in kamacite in Y-82094 between 0.4-0.6 wt %. Smith et al. (1993) found that kamacite in ALH 77307 (3.0) has low Co contents (0.2-0.5 wt %), but a high Ni phase, possibly awaruite, contains high Co (1.3-2.0 wt %). In most other CO chondrites, kamacite in metal nodules in the matrix contains few inclusion phases, whereas they are relatively common in chondrule kamacite.

McSween (1977a), in a survey of several CO chondrites, found that taenite has Ni-rich rims and kamacite has Ni-poor rims (see also Wood 1967). McSween (1977a) also found that the Ni, Cr and Co contents in kamacite in the CO chondrites vary as a function of increasing petrologic type. In Kainsaz (3.1), the average Ni and Co contents of kamacite are ~0.5 and ~4.5 wt %, respectively, but increase to ~1.6 wt % and ~6 wt % in Warrenton (3.6). Average Cr contents decrease from ~0.35 to almost zero from petrologic type 3.1 to 3.7. In Isna (3.7), Methot et al. (1975) found that most of the metal associated with troilite is taenite, but in metal chondrules is mostly kamacite. Taenite associated with troilite has Co contents of 0.1-0.6 wt %, whereas no Ni was detected in the associated kamacite. In comparison, isolated grains of kamacite in the matrix have much higher Co contents (1.7-2.4 wt %) than taenite (0.2-0.4 wt %).

In most CO chondrites, taenite has a narrow compositional range (Shibata and Matsueda 1994, Shibata 1996), and overall Co is negatively correlated with Ni. One unusual chondrite, Y-82094, has one group of taenite, low in Ni (16-24 wt %) and second group with high Ni contents (45 wt % Ni). Co contents in taenite are 0.1-0.4 wt % and Co is negatively correlated with Ni content in both groups. Taenite can also be polycrystalline in this chondrite and micron-sized intergrowths of kamacite and troilite occur at the interface between these two phases, which may have been produced as a result of shock metamorphism.

Cohenite has been recognized in several CO chondrites, including ALH 77307, Y-81020, Y-81025 and Y-74135 (Scott and Jones 1990, Shibata 1996), all low petrologic type chondrites. The cohenite occurs as isolated grains (<5 to 200 μm) or coexisting with tetrataenite, troilite, kamacite and magnetite. According to Shibata (1996), cohenite in Y-81025 is most abundant in the matrix, but does occur in chondrules as well. Cohenite can contain 4-5 wt % Ni (Shibata 1996, see also Scott and Jones 1990) and minor Co, Cr and P (<0.2 wt %). Shibata (1996) suggested that the cohenite may have formed by crystallization from a Fe-Ni-C melt.

Magnetite has only been reported in low petrologic type CO chondrites (ALH 77307, Y-81020, Y-81025 and Y-74135) and occurs as isolated grains and associated with Fe,Ni metal and/or troilite. It is most abundant in the matrix, but also occurs in chondrules and in some occurrences appears to have replaced Fe,Ni metal. In Y-81025, Shibata (1996) reported magnetite with a framboidal morphology, similar to that found in CI chondrites.

Whitlockite (<20 μm) is a common inclusion phase within kamacite and taenite within several CO chondrites studied by Shibata and Matsueda (1994). However, it only occurs in kamacite in Y-82094 and can occur along the external surfaces of the grains. Inclusions of the Na-rich phosphate minerals brianite and panethite occur only in kamacite in Y-82094. Whitlockite may have formed as result of reaction between P in the metal and Ca from the matrix as suggested by Rubin and Grossman (1985). The origin of the Na phosphate phases in Y-82094 is unclear, but reaction with Na-bearing phases in the meteorites such as glass or formation by reaction of P with a gas phase rich in Na, Mg and Ca have been suggested (Shibata and Matsueda 1994).

Opaque mineralogy of CM carbonaceous chondrites

Opaque phases in CM chondrites have been described in detail elsewhere (Chondrules and Matrix sections). In general, metallic iron exterior to chondrules is extremely rare in these meteorites due to aqueous alteration. However, rare occurrences of Fe,Ni metal grains in the matrix have been reported (e.g. Fuchs et al. 1973 – Murchison; Brearley 1995 – Bells). Grossman et al. (1979) reported a large grain of kamacite (~300 μm) isolated in the matrix of Murchison, with 6.82 wt % Ni, 0.39 wt % Co, 0.73 wt % Cr and 0.13 wt % P and, most notably, 0.12 wt % Si. The high concentrations of Ni, Co, Cr and P are typical of other occurrences of metal in CM chondrites (e.g. Olsen et al. 1973a), but are distinct from metal in other chondrites. The high concentration of Si in these metal grains is much higher than in any other carbonaceous or ordinary chondrite and appears to be the consequence of direct condensation of the metal from a solar nebular gas (e.g. Grossman and Olsen 1974, Grossman et al. 1979).

Opaque mineralogy of CR carbonaceous chondrites

Metal in CR chondrites is mostly concentrated in chondrules and is discussed in detail in the Chondrule section.

Opaque mineralogy of ALH85085

ALH85085 is one of the most metal-rich and sulfide-poor chondrites known (~20 vol % metal, 0.5 vol % sulfide minerals). Opaque minerals in ALH85085 include metal, minor troilite and magnetite, and trace quantities of taenite, schreibersite and osbornite (Scott 1988, Grossman et al. 1988a, Weisberg et al. 1988). *Metal* occurs both within chondrules and matrix, and sometimes wraps around chondrules and silicate fragments. Most metal is kamacite with 3-8 wt % Ni. A few grains with higher Ni contents (8-18 wt %) are composed of fine-grained plessite, and rare grains of tetrataenite are present. Concentrations of Co, Cr and P in kamacite and plessite are all in the range 0.0-0.5 wt %. Co and Ni are positively correlated. Many grains are zoned and have higher Ni contents at the cores than at the edges, but the zoning patterns of Ni, Cr, and Co vary from grain to grain. Although most metal grains have very low Si contents (<0.07 wt %), Weisberg et al. (1988) observed three grains of Si-bearing metal, containing 3.3-7.5 wt % Si.

Sulfide is predominantly troilite, with accessory pentlandite. Troilite contains up to 1.0 wt % Cr, 0.15 wt % Mn, 0.3 wt % Co, and 2 wt % Ni (Scott 1988, Grossman et al. 1988a, Weisberg et al. 1988). Grossman et al. (1988a) report one grain of a Ni-rich sulfide, possibly heazlewoodite. Cr-rich sulfide droplets 1-5 μm in size, occurring within a kamacite grain, are interpreted as troilite-daubreelite intergrowths (Scott 1988).

In Acfer 182, most *metal* is kamacite (5-7 wt % Ni), although Ni contents up to 21 wt % also occur (Bischoff et al. 1993b). Rare Si-bearing metal grains were reported by Prinz and Weisberg (1992). Sulfides occurring within dark, fine-grained inclusions are predominantly pyrrhotite containing about 1-1.5 wt % Ni. Sulfides in the matrix are close to stoichiometric troilite with lower Ni contents (<0.5 wt %). Rare pentlandite grains contain up to 32 wt % Ni.

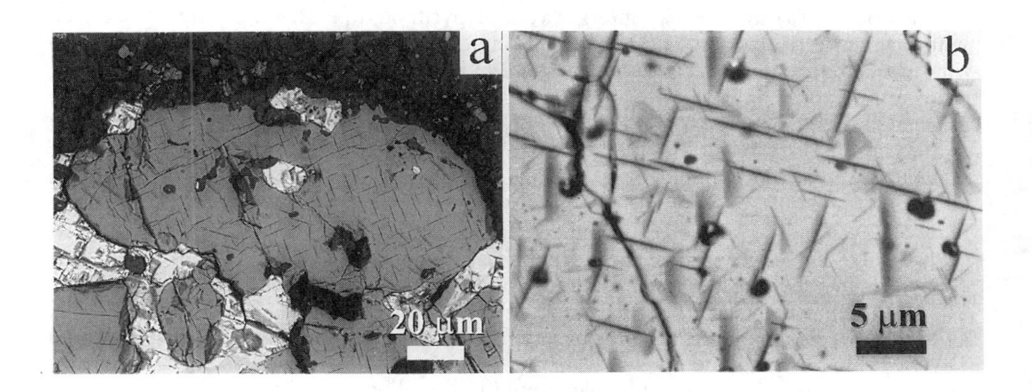

Figure 179. BSE images of magnetite in the CK chondrite, Karoonda. (a) Large magnetite grain associated with sulfides, (b) High magnification image of the grain shown in (a), illustrating the presence of numerous fine-scale exsolution lamellae of spinel on (111).

Opaque mineralogy of CK carbonaceous chondrites

The opaque mineralogy of CK chondrites has been studied in some detail by Geiger and Bischoff (1995) and additional data have been presented by Ramdohr (1973), Scott and Taylor (1985) and Rubin (1992 1993). *Magnetite* is the dominant opaque phase in CK chondrites constituting between 1.2 and 8.1 vol %. Two types of magnetite grains occur; large rounded grains up to 1 mm in size which often contain exsolution lamellae of ilmenite and spinel (Ramdohr 1973, Scott and Taylor 1985, Geiger and Bischoff 1995) (Fig. 179) and a second, much finer-grained type of magnetite (<10 µm) which is dispersed throughout the matrix. The larger magnetites often contain inclusions of sulfides, mostly *pentlandite* and OH and Cl-bearing *apatite*. The *ilmenite* exsolution lamellae in magnetite occur dominantly on {111} and vary in thickness as a function of metamorphic type, being absent in the type 3 CK chondrite, Watson, but increasing up to <5 µm in width in type 6 CKs. The lamellae are typically some 100s of microns in length. Spinel also occurs in magnetite as exsolution lamellae, <2 µm in width, on (100) or as small grains (<1 µm), homogeneously distributed throughout the large magnetite grains. Small spinel grains (spinel-hercynite solid solution, some with detectable gahnite) also occur decorating ilmenite exsolution lamellae in magnetite (Fig. 179). Curvilinear trails of very fine-grained magnetite and sulfide (dominantly pentlandite) are found throughout recrystallized silicates in CK chondrites and are thought to have formed as a result of shock mobilization of magnetite and sulfide (Rubin 1992).

Compositionally, magnetites in all the CK chondrites contain significant amounts of minor elements with mean compositions varying between 0.08-1.06 wt % MgO, 0.42-3.43 wt % Al_2O_3, 0.14-1.20 wt % TiO_2 and 2.8-5.5 wt % Cr_2O_3 (Rubin 1993, Geiger and Bischoff 1995). It is possible that at least some of these minor elements are present in fine-scale ilmenite or spinel lamellae.

Sulfides in the CK chondrites occur in low abundances (<0.1-1.3 wt %) but a large number of different phases have been reported. These include *troilite, pyrrhotite, pentlandite, pyrite, monosulfide, millerite, chalcopyrite* and sulfides rich in heavy siderophile elements such as Ru, Os, Ir, Au and Ag. Several paragenetic occurrences of sulfides occur: (a) myrmekitic intergrowths of pyrite and pentlandite, often associated with magnetite (Ramdohr 1973, Scott and Taylor 1985, Geiger and Bischoff 1995), (b) fine-grained, porous intergrowths of pentlandite and pyrite with pyrrhotite and/or monosulfide solid solution (Geiger and Bischoff 1995), (c) sulfide (largely pentlandite) associated with magnetite, often enclosed poikiloblastically within the magnetite (Geiger and Bischoff 1995), and (d) spherical to subspherical chondrules and nodules of magnetite and sulfide which occur in some CK chondrites (Rubin 1993). Pentlandites have variable compositions with 24.1-37.4 wt % Fe, and 27.2-41.2 wt % Ni; pentlandite in the CK3 chondrite Watson 002 contains 2.0 wt % Co. Pyrites are Ni-bearing with 0.2-4.5 wt % Ni and ~2 wt % Co. The sulfide phase containing ~7.0 wt % Ni in several CK chondrites may be either Ni-rich pyrrhotite or Fe-rich monosulfide solid solution (MSS). MSS compositions are variable with 27.0-32.4 wt % Ni and 25-32.7 wt % Fe. *Chalcopyrite* is a rare phase, only occurring in Karoonda and PCA 82500 (Scott and Taylor 1985, Kallemeyn et al. 1991, Geiger and Bischoff 1995) and *millerite* (<20 μm) grains with ~5 wt % Fe have only been found in Maralinga and Cook 003. Kallemeyn et al. (1991) and Rubin (1993) reported mackinawite in several CK chondrites, but Geiger and Bischoff (1995) have suggested that this phase may actually be a Ni-bearing pyrrhotite or Fe-rich monosulfide solid solution. Geiger and Bischoff (1995) also observed the sulfide phase $(Os,Ru,Ir)S_2$, a solid solution between laurite (RuS_2) and ehrlichmanite (OsS_2) in essentially all 19 CK chondrites that they studied. This phase also contains some Ir. Rare occurrences of cooperite (PtS), $(Au,Fe,Ag)_2S$ (a phase which deviates from the petrowskaite solid solution (Au,Ag)(S,Se) and the unidentified phase $(Fe,Au,Co)_2S_3$ have also been described by Geiger and Bischoff (1995).

Fe,Ni metal is extremely rare in CK chondrites, but trace amounts have been reported in Y-693 (Okada 1975), Y-82104 (Graham and Yanai 1986), Maralinga (Keller et al. 1992) and Ningqiang (Rubin et al. 1988). Very small (<1 μm) grains of Pt group metal-rich alloys have been observed in EET87514 and EET87526 (Geiger and Bischoff 1995), that are rich in Au and Fe with detecTable A3. Detectable amounts of Mo, Ru and Cr. Using electron diffraction, Hua et al. (1995) determined that Co-rich, Ni-poor metal (wairauite) (~39 wt % Co; 61 wt % Fe; <0.6 wt % Ni) in the anomalous CK chondrite Ningqiang has a Cs-Cl-type cubic structure with a = 0.286nm. The alloy is produced by slow cooling and is stable to below 500°C. The wairauite is associated with awaruite, magnetite and pentlandite indicating formation under oxidizing conditions.

Gieger and Bischoff (1995) found several rare telluride and arsenide phases in CK chondrites including chengbolite ($PtTe_2$) which occurs in several CK chondrites and a non-stocihiometric Au,Pt,Fe-telluride which may be a solid solution or a mixture of the $AuTe_2$ phases krennerite and calverite with chengbolite and frohbergite ($FeTe_2$). The arsenide phases appear to be solid solutions of the endmembers iridarsenite ($(Ir,Ru)As_2$), omeiite ($(O,Ru)As_2$), sperrylithe (PtS_2) and löllingite ($FeAs_2$).

Opaque mineralogy of the Coolidge and Loongana 001 carbonaceous chondrite grouplet

Noguchi (1994) reported Fe,Ni metal (mostly kamacite with a small amount of taenite) and troilite in chondrules in Coolidge. The taenite occurs in patches in the kamacite and inclusions of Cr-spinel and Ca-phosphate also occur as inclusions within the metal. Metal in the matrix is highly altered. Kamacite in Coolidge contains thin (<5 μm) exsolution lamellae of taenite (~23 wt % Ni, and 0.1 wt % Co). Ni and Co contents of kamacite are ~7 and 0.35 wt % with narrow ranges (~2.5 and ~0.2, respectively).

Opaque mineralogy of the unique carbonaceous chondrite, LEW 85332

LEW85332 has suffered extensive terrestrial alteration in which much of its metal has been oxidized to limonite. *Metal* occurs as small blebs within chondrules, but mostly occurs outside chondrules as angular to rounded grains (typically 2-50 μm in size). The unaltered metal consists predominantly of kamacite (6 wt % Ni), containing an average of 0.16 wt % Cr and 0.25 wt % Co. There is a moderately strong positive correlation between Co and Ni in kamacite, although unlike metal in the CR chondrites, the slope of the correlation lies below that of the CI Co/Ni ratio. Rare grains of martensite (8-21 wt % Ni) are present. *Troilite* occurs as rims that partially surround some kamacite grains, and also occurs as rounded grains and polycrystalline assemblages. Some grains contain patches or thin exsolution lamellae of pentlandite. Troilite contains significant Ni (0.7 wt %) and minor Co (0.1 wt %). *Pentlandite* is Ni-rich (30 wt % Ni) and contains 0.9 wt % Co.

Opaque mineralogy of K (Kakangari-like) chondrites

See Chondrule section for details of opaque phases in K chondrites.

Opaque mineralogy of Bencubbin

Metal in Bencubbin occurs both as micron-sized droplets within the silicate material and as large, sometimes polycrystalline, clasts up to 6 mm in diameter (Newsom and Drake 1979). Metal clasts have been deformed to varying degrees. Most metal is kamacite, with Ni below 8 wt %, although Ramdohr (1973) observed taenite in Bencubbin. A few Si-bearing grains with ~2 wt % Si also occur. Different metal clasts have different compositions. Ni and Co in kamacite are positively correlated and exhibit a trend line with a slope that is similar to the calculated condensation path of the solar elemental abundances. Ni and P are negatively correlated and also follow a calculated condensation path. Co, P, and Cr contents lie in the ranges 0.25-0.35, 0.16-0.35, and 0.07-0.33 wt %. Accessory minerals occurring at grain boundaries within mosaic kamacite grains include Na,Mg,Al-bearing silicates, a V-rich chromite mixed with silicate, and phosphate. Metal clasts in Bencubbin and Weatherford contain Cr-rich *troilite* globules (Mason and Nelen 1968, Ramdohr 1973, Newsom and Drake 1979). This troilite is apparently stoichiometric (Fe,Cr)S, and contains up to 30 wt % Cr. Ni contents are generally less than 1 wt %. In addition to troilite, Ramdohr (1973) lists the presence of daubreelite.

Opaque mineralogy of R (Rumuruti-like) chondrites

Metal and sulfide minerals. Metallic Fe,Ni is extremely rare to absent in most R chondrites. In Rumuruti and ALH85151, several grains of a high-Ni phase (67.5 and 64.0 wt % Ni), probably awaruite, have been observed (Schulze et al. 1994, Rubin and Kallemeyn 1989). Co contents are around 2 wt %. A grain of awaruite from Rumuruti with 73 wt % Ni and 1.5 wt % Ge was described by Schulze et al. (1994), and one grain of kamacite (~5.4 wt % Ni) was also observed. A metallic particle in ALH85151 consists of martensite (9.9 wt % Ni) surrounded by troilite, and several tiny grains of a Co-rich metal phase (1.4 wt % Ni, 37 wt % Co) situated at the boundaries between awaruite and troilite (Rubin and Kallemeyn 1989). Other rare metal grains in Acfer 217 and Rumuruti are rich in Au, Ag, Ir, and Pt (Bischoff et al. 1994, Schulze et al. 1994).

The major *sulfide* minerals in R chondrites are pyrrhotite and pentlandite. In ALH85151, troilite occurs rather than pyrrhotite. Several chondrules in Carlisle Lakes have large concentrations of small (5-20 μm) sulfide grains within 100 μm of the chondrule surface (Rubin and Kallemeyn 1989). Sulfide assemblages in Carlisle Lakes consist of polycrystalline aggregates of pyrrhotite and pentlandite, some of which contain chromite. Pyrrhotite contains minor amounts of Ni (<0.4 wt %). Pentlandite typically contains 28-33 wt % Ni and 1-2 wt % Co. One pentlandite grain in PCA91002 contains high amounts of Cu (6.8 wt %). Rare sulfide phases include chalcopyrite (Rubin and Kallemeyn 1989, Schulze et al. 1994) and pyrite, which may be somewhat variable in composition (Rubin and Kallemeyn 1994, Bischoff et al. 1994). One sulfide grain in Acfer 217 is rich in Os, Ru, and Fe (Bischoff et al. 1994).

INTERSTELLAR GRAINS

Interstellar grains are minute, mostly <20 μm across, and constitute only a minor component of chondrites, being present at levels less than hundreds of ppm. However, their significance far outweighs their low abundance, because their presolar nature provides a fascinating window into the astrophysical environment from which the solar system formed. The importance of interstellar grains in terms of understanding astrophysical models for stellar evolution and nucleosynthesis has led to a strengthening link between the astronomical and meteorite communities. Initial suspicions of an interstellar component in primitive chondrites were raised by observations of exotic noble gas (Xe and Ne) components (Reynolds and Turner 1964, Black and Pepin 1969). However, the study of interstellar grains has only burgeoned in the last decade, since they were successfully isolated in the laboratory. Analyses of larger individual grains have become possible with ion microprobe techniques. The presolar nature of the grains is evidenced by their exotic isotopic compositions that point to formation in late-stage stellar environments such as supernovae, AGB (asymptotic giant branch) stars, novae, and Wolf-Rayet stars. Mineralogically, interstellar grains are a simple suite consisting of diamond, silicon carbide, graphite, oxides (corundum and spinel), titanium carbide and silicon nitride. All these phases are refractory; the C-bearing phases formed under reducing conditions. Interstellar grains occur in all chondrite classes. They are embedded in chondrite matrix, and survived processing in the solar nebula environment. They are found in highest abundance in the most primitive chondrites, and are destroyed during metamorphism. The following section provides only a very brief review of the various interstellar minerals: for further information on this subject the reader is referred to reviews by Anders and Zinner (1993), Ott (1993) and Bernatowicz and Zinner (1997).

Diamond

Diamond is the most abundant type of interstellar grain, typically constituting ~400 ppm of CM2 chondrites. Detailed reviews of presolar diamonds have been written by Lambert (1992) and Huss and Lewis (1994a,b). Diamond is commonly described as microdiamond or nano-diamond because of its extremely fine-grained nature (Lewis et al. 1987, 1989; Fraundorf et al. 1989, Daulton et al. 1996). The mean crystallite size is ~20 Å, corresponding to only a few thousand atoms, and morphologies are generally euhedral. Single and multiple-twin boundaries are observed, as well as rare nanocrystals of lonsdaleite. The bulk density is only 2.22-2.33 g cm^{-3} compared with 3.51 g cm^{-3} for normal diamond. Possible formation mechanisms include chemical vapor deposition (CVD), grain-grain collisions of amorphous or graphitic C in interstellar shocks, and UV irradiation of graphite. The distribution of twin microstructures and an absence of dislocations suggest that most of the diamond formed by CVD (Daulton et al. 1996).

Interstellar diamonds are carriers of the presolar noble gas components HL, P3 and P6. HL consists of several different gases including Xe-HL, which is enriched in heavy

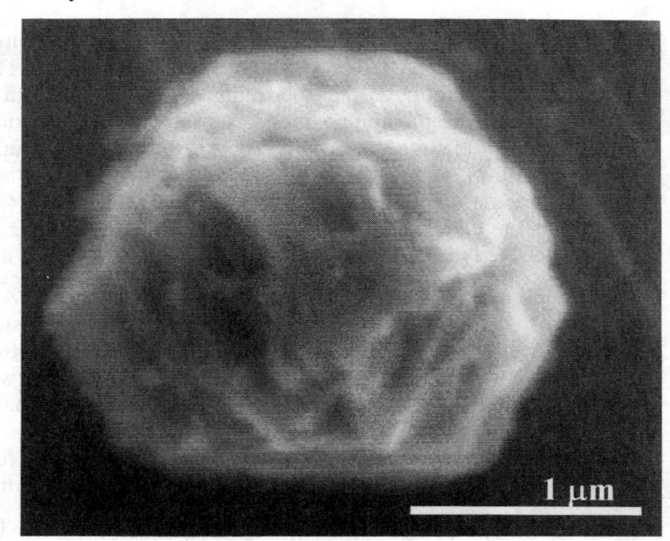

Figure 180. Interstellar SiC grain from the CI chondrite, Orgueil. Photo courtesy of Gary Huss.

and light isotopes of Xe. Such enrichments are expected from the *r*- and *p*-processes of nucleosynthesis and hence a supernova origin for the Xe, although not necessarily the diamond, is implied. The noble gases were probably emplaced by ion implantation. In addition to noble gases, isotopic anomalies for H, N, Ba and Sr are also observed in interstellar diamond. Carbon isotopic compositions are normal.

SiC

SiC abundances in the CI and CM chondrites are around 20 and 5 ppm, respectively. Grain sizes vary from <0.05 to 30 μm, with most grains between 0.3 and 3 μm (Fig. 180). Morphologies include irregular to round grains with euhedral platy surface features, and blocky, anhedral grains with flat, sometimes very smooth, surfaces. Although most grains have been observed in mineral separates, larger grains have also been observed in situ in CM chondrite matrix (Alexander et al. 1990). The isotopic properties of SiC grains show enormous diversity, and variations in key isotopic and elemental ratios are observed for different grain sizes. SiC grains are carriers of the Xe-S component, which is characteristic of *s*-process nucleosynthesis and links the grains to AGB stars (Lewis et al. 1994). Anomalous Kr compositions and Ne-E(H) components as well as *s*-process enrichments in other elements also indicate production in AGB stars. SiC grains are thought to condense in the outflowing stellar wind. Noble gases are incorporated by sorption during grain growth as well as by implantation from the wind. The relatively large sizes of SiC grains suggests that they form during the final, planetary nebula stage of AGB stars when the mass loss rate is high.

Numerous SiC grains >1 μm in size have been analyzed individually with the ion microprobe. Isotopic compositions of C, N, Si, Mg, Ca, Ti, and Zr have been measured (Anders and Zinner 1993, Hoppe et al. 1994 1996a). Four distinct populations occur: "mainstream" (>96% by mass), "large" (<2%), "X" (~1%) and "Y" (~1%). Isotopic ratios of C and N cover a vast range of about 3 orders of magnitude, consistent with the wide variation in $^{12}C/^{13}C$ ratios observed in carbon stars. C and N isotopic compositions of the majority of grains correspond to hydrostatic H-burning in the C-N-O cycle, but other grains correspond to He-burning as well as explosive H-burning events. Si isotopic compositions of SiC grains do not match solar compositions, and most do not show compositions expected from *s*-process nucleosynthesis. This latter observation is explained by the likelihood that the grains are a mixture from different stars of slightly non-solar initial isotopic compositions. Mg isotopic compositions are highly enriched in ^{26}Mg, apparently from the decay of ^{26}Al. Al occurs at levels up to several percent and is correlated with N, suggesting that it condensed as AlN, either in solid solution or as discrete inclusions. The initial ^{26}Al/^{27}Al ratio varies up to 2×10^{-2}, orders of magnitude higher than the canonical ratio for the solar nebula (5×10^{-5}) observed in CAIs. ^{26}Al probably decayed in SiC grains before they were incorporated into the solar nebula.

Ti is the most abundant minor element after N and Al. Many SiC grains have solar Ti/Si ratios, whereas others are depleted by up to 20x. Inclusions of TiC, 100-700 Å in size, have been observed with TEM in the interiors of some SiC grains (Bernatowicz et al. 1992). Some TiC grains have an epitaxial relationship with the surrounding SiC, suggesting either that both phases grew simultaneously or that TiC exsolved from SiC. Isotopic compositions for Ti are consistent with production by neutron exposure, although they deviate from calculations for AGB stars. As for Si, this is attributed to mixing of material from different stars of slightly non-solar initial isotopic compositions.

SiC grains of type X have very large anomalies in Si, Al, Ti and Ca isotopes and are probably derived from a supernova source (Hoppe et al. 1996b, Nittler et al. 1996). The largest SiC grains, ~5-20 μm, form another distinct population (Zinner et al. 1989, Virag et al. 1992). They contain much less Al and N than smaller grains, and only rarely have excesses of ^{26}Mg. They occur as distinct clusters with different properties, suggesting that they originated in a small number of stars.

Presolar ages of SiC grains have been calculated from cosmic-ray exposure ages, determined from excesses of ^{21}Ne produced by spallation of Si by cosmic rays (Tang and Anders 1988, Lewis et al. 1990, 1994). Presolar ages of >1000 Ma have been determined. Exposure ages of ~10^8 years are comparable to the lifetimes of molecular clouds and imply source regions of similar mass, 10^5-10^6 solar masses. The SiC population in chondrites is estimated to be derived from between 10 and 100 AGB stars in this source region (Alexander 1993).

Graphite

The abundance of interstellar graphite is highest in the CI chondrite Orgueil (~10 ppm), but very low in other chondrite groups (<1 ppm). Graphite occurs as grains up to ~10 μm in size, has a density around 2 g cm^{-3}, and has a fairly well ordered crystalline structure (Zinner et al. 1995, Bernatowicz et al. 1991, 1996). Graphite occurs in different morphologies, including aggregates of 0.1-0.3 μm spherules and compact grains of round and other morphologies. Grains with round or rounded morphologies occur as "onions", with smooth or shell-like platy surfaces and concentric shells, and "cauliflowers" that appear to be dense aggregates of smaller grains. Onions consist of layers of well crystallized graphite that frequently mantle a core of amorphous C. Cauliflowers consist of concentric layers of poorly graphitized C with turbostratic textures. Grains of both types sometimes contain small (20-50 nm) grains of TiC that are often rich in other refractory elements such as Zr and Mo. Other morphologies include some grains with euhedral, hexagonal shapes that are either platy or blocky, and anhedral, irregular grains with smooth surfaces. At least two distinct sources are implied for interstellar graphite: some grains come from the H-rich atmospheres of AGB stars, and others come from either Wolf-Rayet stars or the He-shell of Type II supernovae.

The ^{12}C/^{13}C ratio in compact grains covers a very wide range, from 8 to 1440, whereas that of spherulitic aggregates is close to solar. The $^{12}C/^{13}C$ ratio distribution varies amongst different density fractions, suggestive of four isotopic groups. Noble gases also show systematic trends with sample density, suggesting that more than one type of graphite is present. Graphite contains Ne-E(L), Kr-S and Xe-S. N, Si and H isotopic ratios of all grain types are mostly very close to solar ratios, although some grains contain large isotopic anomalies of O and Si. H, O, and probably also N may be present as organic compounds such as aromatic hydrocarbons. Si appears to be dispersed throughout the graphite. Al contents range from 300-30,000 ppm, and Al correlates with N, suggesting condensation as AlN. Many grains show clear evidence of extinct ^{26}Al.

Oxides

Interstellar oxide grains have been identified by O isotopic ion-imaging techniques on the ion microprobe (see summary by Nittler et al. 1997). Approximately 100 such grains with significant, presolar isotopic anomalies have been identified to date. These are predominantly corundum (Al_2O_3), although rare grains of spinel ($MgAl_2O_4$) have also been discovered. Many of the grains show well defined crystal surfaces, and sizes of the grains studied lie in the range ~0.5-5 μm. The range of O isotopic compositions is enormous, with the ratio $^{16}O/^{18}O$ varying from 100-40,000, and the ratio $^{16}O/^{17}O$ varying from ~300-5000. Many of the grains have large ^{26}Mg excesses that are probably attributable to the in situ decay of ^{26}Al. Inferred initial $^{26}Al/^{27}Al$ ratios are between 1.2×10^{-4} and 1.6×10^{-2}. Based on their isotopic characteristics, four groups of grains have been identified, each of which has a distinct source indicating formation in different stages of O-rich red giant and AGB star evolution.

Silicon nitride

Rare presolar grains of silicon nitride (Si_3N_4) have been identified (Nittler et al. 1995, Hoppe et al. 1996a). These grains have anomalous isotopic compositions similar to SiC grains X that point to an origin in Type II supernova ejecta.

Destruction of interstellar grains during metamorphism

Abundances of interstellar grains vary as a function of chondrite class and petrologic type (Huss 1990, Huss and Lewis 1995). Inferred pre-metamorphic abundances of diamond vary by a factor of about 2 between chondrite classes, although similar intial diamond/SiC ratios are observed in most classes. These observations suggest that all chondrite classes sampled the same reservoir of presolar grains, with variations in presolar grain abundances resulting from processing within the solar nebula. Abundances decrease progressively with increasing degrees of metamorphism, indicating that the grains are destroyed during metamorphic reheating. The pattern of destruction differs from one chondrite class to another. In type 3 OCs, graphite is most easily destroyed, followed by SiC then diamond. No interstellar grains remain above petrologic subtype 3.8. In CV3 and CO3 chondrites, diamond appears to be more stable than SiC. In EH chondrites, diamond and SiC survive much more severe metamorphism than in OCs, with SiC being present in Abee (EH5), and SiC is apparently more stable than diamond. Interstellar graphite has not been detected in EH, CV or CO chondrites.

TYPE 4-6 CHONDRITES: NON-OPAQUE MATERIAL

Introduction

In this section we describe chondrites that are classified as petrologic types 4-6. This includes the type 4-6 O chondrites (H, L, and LL), which are the most common type of meteorite falls and are thus at least volumetrically significant in terms of the chondrite record. For E chondrites, the EL group encompasses the whole type 3-6 metamorphic sequence, but only types 3-5 EH chondrites are known. Of the carbonaceous chondrites, only the CK group (types 4-6) contains material of petrologic type higher than type 3.

Type 4-6 chondrites have undergone significant equilibration, probably as a result of thermal metamorphism on chondrite parent bodies. The Van Schmus-Wood classification scheme (Van Schmus and Wood 1967) is used to classify ordinary chondrites into types 4-6 and has general features that are also applicable to E and CK chondrites (see Table 1). The most important textural changes common to all chondrite groups as metamorphism proceeds are as follows: (1) chondrule textures and outlines become increasingly blurred, and matrix becomes progressively more recrystallized and coarser grained; (2) chondrule mesostases become devitrified, which results in glass becoming increasingly rarer at the same time as feldspar and diopside become more abundant and coarser grained; (3) the diverse compositional ranges of most minerals in unequilibrated chondrites becomes progressively more equilibrated, (4) clinoenstatite, which is common in chondrules of type 3 chondrites, becomes progressively inverted to orthoenstatite. Usage of the term "equilibrated", which is traditionally applied to all type 4-6 chondrites, is not strictly correct because all minerals are not completely equilibrated, even in type 6 chondrites. This term is used loosely and describes the general homogeneity of major elements in olivine and pyroxene in type 4-6 chondrites. Peak temperatures of metamorphism and cooling rates for different petrologic types of ordinary and enstatite chondrites have been inferred from a variety of mineralogical thermometers, summaries of which may be found in Dodd (1981) and McSween et al. (1988) for the ordinary chondrites and Zhang et al. (1996) and Zhang and Sears (1996) for the enstatite chondrites.

The term petrologic type 7 has been used to describe several ordinary and one enstatite chondrite, and has been perceived as an extension of the petrologic type 3-6 sequence of metamorphism (Dodd 1981). This term was originally introduced to describe the L chondrite Shaw (Dodd et al. 1975), on the basis of high CaO contents of low-Ca pyroxene, the presence of large plagioclase grains (>100 μm), the absence of relict chondrule textures and evidence for anatexis. However, Shaw has since been shown to be a complex impact melt breccia (Taylor et al. 1979, Scott and Rajan 1979). Several other OC, as well as the E chondrite Happy Canyon, that have been classified as petrologic type 7 are also impact melt rocks (Rubin 1995, McCoy et al. 1995a). Use of the classification "type 7" for these impact-melt rocks and breccias is inappropriate. The term petrologic type 7 has also been used to describe a few unmelted ordinary chondrites that appear to have undergone more severe metamorphic heating than typical type 6 chondrites. Heyse (1978) classified the LL chondrite Uden and one clast in the LL breccia, St. Mesmin, as type 7, on the basis of high Wo contents in low-Ca pyroxene (see below). Preliminary classifications of type 7 have also been applied to some highly recrystallized chondrites in which there are no traces of melt, e.g. ALH84027 and EET92012 (Schwarz and MacPherson 1985, Satterwhite and Mason 1994). Although usage of the term type 7 may be appropriate for these chondrites, there is no clear description of how their properties differ from petrologic type 6 or the criteria for a type 7 classification. For this reason, usage of the type 7 classification is not generally accepted.

In the following discussion, we do not describe individual textural components of type 4-6 chondrites, such as chondrules and matrix, because textural distinctions become increasingly blurred as metamorphic grade increases, and many grains are recrystallized. Rather, we describe each mineral that may occur in any textural environment throughout the chondrite.

Ordinary chondrites

Type 4-6 OC have a simple mineralogy. Major minerals are olivine, low-Ca pyroxene and Fe,Ni metal (kamacite and taenite). Plagioclase, diopside and troilite are minor minerals, and accessory minerals include chromite, merrillite, and chlorapatite. Native copper, pigeonite, glass (in type 4 OC only), pentlandite, ilmenite, mackinawite, bravoite, and chalcopyrrhotite also occur (e.g. Gomes and Keil 1980). Olivine, pyroxene and metal compositions are used as diagnostic properties to determine the classification of ordinary chondrites into H, L, and LL groups (e.g. Mason 1963, Keil and Fredriksson 1964, Sears and Axon 1976, Afiattalab and Wasson 1980, Rubin 1990) as well as a suggested, intermediate L/LL group (Rubin 1990).

Many type 4-6 ordinary chondrites are breccias. Examples include St. Mesmin (Dodd 1974a), Plainview (Keil et al. 1980), Weston (Noonan and Nelen 1976), Adzi Bogdho (Bischoff et al. 1993c), Abbott (Fodor et al. 1976), and Kendleton (Ehlmann et al. 1988). We do not discuss breccias separately here, but include mineral compositions of the host rocks in the data we describe. The reader is referred to Fodor and Keil (1978), Keil (1982), Scott et al. (1985 1986), Bunch and Rajan (1988) and Sears et al. (1991b 1995b) for further discussions of OC breccias.

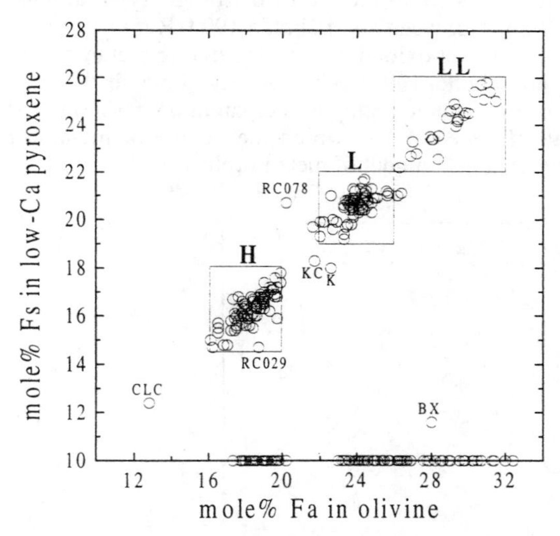

Figure 181. Fs vs. Fa contents of low-Ca pyroxenes and olivines in type 4-6 ordinary chondrites. Mineral compositions from H, L, and LL group chondrites fall in well-defined fields, with molar Fe/(Fe+Mg) ratios increasing with overall oxidation state of the chondrite. Outliers include the relatively unequilibrated LL4 chondrite, Bo Xian (BX), L4 chondrites Roosevelt County 078 (RC078), Kramer Creek (KC), and Kendleton (K), the H4 chondrite RC029 and the reduced H4 chondrite, Cerro los Calvos (CLC). Olivine data from Rubin (1990) are shown along the x axis. Data sources: Keil and Fredriksson (1964), Fodor et al. (1971a,b, 1972, 1976), Smith et al. (1972), Bunch et al. (1972), Lange et al. (1973, 1974), Graham and Nayak (1974), Jaques et al. (1975), Fudali and Noonan (1975), Gibson et al. (1977), Heyse (1978), Rubin et al. (1980), Gomes and Keil (1980), Klob et al. (1981), Scott et al. (1985, 1986), Ehlmann and Keil (1985), Ehlmann et al. (1988), Keil et al. (1980), McCoy and Keil (1988), Rubin (1990), Wasson et al. (1993), McCoy et al. (1991a, 1993, 1995b, 1995c).

Olivine and *low-Ca pyroxene* compositions in type 4-6 OC are generally equilibrated in their major element compositions. Within each chondrite, the standard deviation on the mean Fa content is <1 mol %, and commonly <0.5 mol % (e.g. Rubin 1990). Low-Ca pyroxene compositions may have a wider range of Fs contents. Within each chondrite group there is a range of mean Fa and Fs contents from individual chondrites (Fig. 181). The molar Fe/(Fe+Mg) ratio is higher in olivine than in coexisting low-Ca pyroxene. Compositional ranges in Fs vs. Fa for the H, L and LL groups were originally defined by Keil and Fredriksson (1964) and later updated by Gomes and Keil (1980). However, additional data since that time such as the compilation in Figure 181 expands the original fields. Although there is clearly a hiatus between the H and L groups, there is no hiatus between L and LL groups. Rubin (1990) tentatively defined an intermediate group of L/LL chondrites based on their olivine and kamacite compositions. Compositional ranges determined from Figure 181 are: H chondrites, Fa_{16-20}, $Fs_{14.5-18}$; L chondrites, Fa_{22-26}, Fs_{19-22}; L/LL chondrites, $Fa_{25.5-26.5}$ (after Rubin 1990); LL chondrites, Fa_{26-32}, Fs_{22-26}. There are several outliers from these fields, most of which are type 4 chondrites in which pyroxenes are likely to be incompletely equilibrated. The extremely low Fs content for Bo Xian (LL4) arises because this is a very unequilibrated type 4 chondrite, and the data are for type IA chondrules, initially the most FeO-poor material (McCoy et al. 1991a). Cerro los Calvos is one of a small group of reduced H chondrites (Whitlock et al. 1991, Wasson et al. 1993).

Fa contents of olivines and Fs contents of low-Ca pyroxenes within each chondrite group show slight increases as a function of petrologic type, although these effects are subtle (Scott et al. 1986, McSween and Labotka 1993, Rubin 1990). This observation has been used as an argument that oxidation occurred during metamorphism in the OC parent bodies (McSween and Labotka 1993). Alternatively, it may be the result of small changes in the composition of material accreting to the parent body (Scott et al. 1986), for a body that is concentrically layered and in which the degree of metamorphism is related to burial depth (the "onion shell" model of metamorphism).

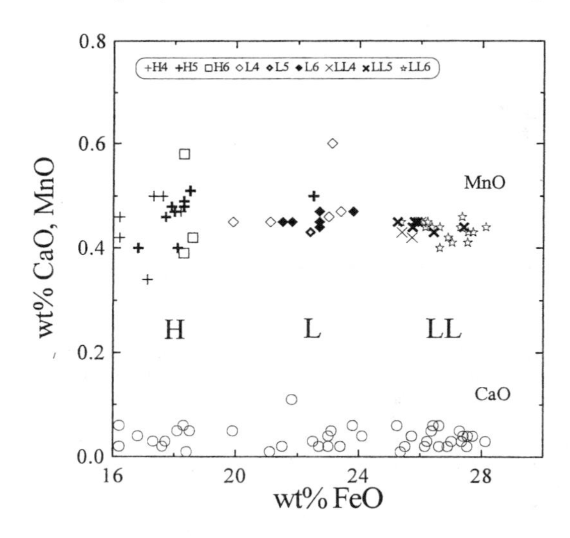

Figure 182. Minor element variation plots for olivine in type 4-6 ordinary chondrites. Data sources as for Figure 181.

Minor element contents of *olivines* in type 4-6 OC are very uniform. TiO_2, Al_2O_3 and Cr_2O_3 contents are all extremely low, <0.1 in type 4 OC and <0.05 in types 5 and 6. CaO contents are also very low, <0.1 wt % (Fig. 182; Table A3.34). MnO contents are between 0.4 and 0.5 wt %. Apart from some outliers which are mostly type 4 chondrites, there is a slight decrease in MnO content of equilibrated olivine from H to L to LL groups. Trace element concentrations in olivine from mineral separates from L6 and H6 chondrites have been determined by Allen and Mason (1973) and Curtis and Schmit

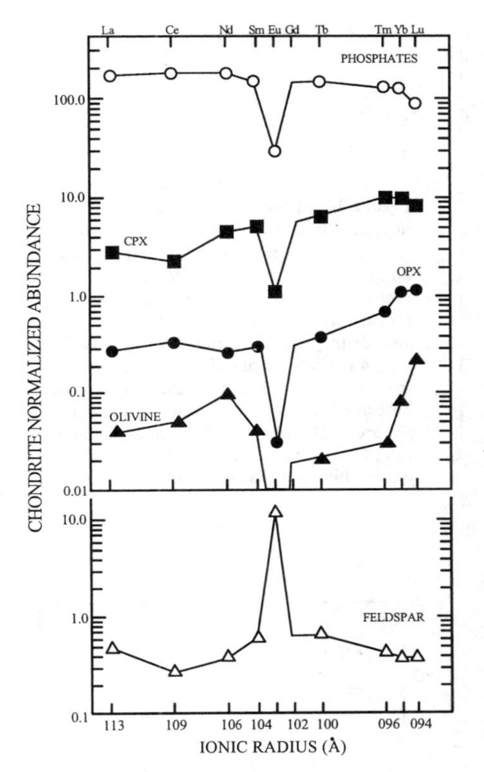

Figure 183. REE abundances in minerals from the L6 chondrite, Alfianello. Data from Curtis and Schmitt (1979).

(1979). REE abundances are low (Fig. 183). Dusty, metal-bearing and chromite-bearing olivine has been described in the H4 chondrites, Quenggouk and Guenie (Ashworth 1979, Christophe Michel-Lévy 1981). These grains probably originated as relict grains in chondrules.

The dominant pyroxene in type 4-6 OC chondrites is *low-Ca pyroxene*. This pyroxene is derived from the twinned clinoenstatite of type 3 chondrites, which inverts progressively to the orthorhombic form with increasing metamorphism. Weak lamellar zoning observed in BSE images, similar to that observed in type 3 OC, is preserved in some type 4 chondrites (Tsuchiyama et al. 1988, McCoy et al. 1991a). In type 5 chondrites, low-Ca pyroxene is predominantly orthopyroxene but it may be striated and may have extinction angles as high as 10° (e.g. Jobbins et al. 1966). Essentially all low-Ca pyroxene is orthorhombic in type 6. Disordered orthopyroxene occurs as single crystals in type 4-6 chondrites (Pollack 1966). Inversion of ortho-enstatite to clinoenstatite may also be induced by shock, and some highly shocked type 5 and 6 OC contain twinned grains of low-Ca pyroxene derived from this process. In addition to stacking disorder, chondritic pyroxenes display various degrees of disorder in the siting of Fe and Mg. This is common in type 3 and 4, and rare or absent in type 6 (Dundon and Walter 1967, Dodd 1969). Disordering of Fe and Mg in M1 and M2 sites of clino- and ortho-pyroxenes in an H5 chondrite suggest cooling rates of the order of degrees per day at temperatures of 500-600°C (Molin et al. 1994).

The Fs contents of low-Ca pyroxenes are discussed above. Wo contents show a significant range within each chondrite group, and this range is, to some extent, a function of the degree of metamorphism (Keil and Fredriksson 1964, Dodd 1969, Heyse 1978, Scott et al. 1986; Fig. 184; Table A3.35). Wo contents in two LL chondrites that have been classified as type 7 (Uden, and one clast from St. Mesmin: Heyse 1978) are consistently higher than Wo in most type 6 chondrites. However, in two LL chondrites that have been given preliminary classifications of type 7, low-Ca pyroxene compositions are Wo_2 (Schwarz and MacPherson 1985, Satterwhite and Mason 1994). Figure 184 appears to resolve clearly the H, L and LL chondrite groups for most type 4-6 OC. A few pyroxenes with $>Wo_3$ have been described in type 4 and 5 OC: Keil et al. (1978) described pyroxene of composition Wo_3 as pigeonite, and McSween and Patchen (1989) included several analyses of compositions $Wo_{3\text{-}5}$ in their orthopyroxene group. There is little discussion of the variability of low-Ca pyroxene compositions within individual chondrites, but there may be significant variation in the Wo component of up to 2 mol %,

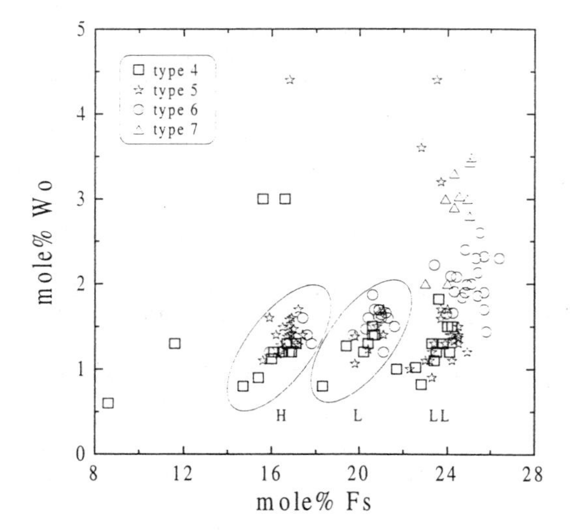

Figure 184. Low-Ca pyroxene compositions in type 4-6 ordinary chondrites. Fs contents increase with increasing oxidation state in the order H-L-LL. Fs and Wo contents show more or less well defined increases with increasing petrologic type within each chondrite group. Outliers are mostly from type 4 chondrites but also include pyroxenes from some type 5 chondrites that have been described as pigeonite. Data sources as for Figure 181, plus McSween and Patchen (1989), Schwarz and MacPherson (1985), Satterwhite and Mason (1994).

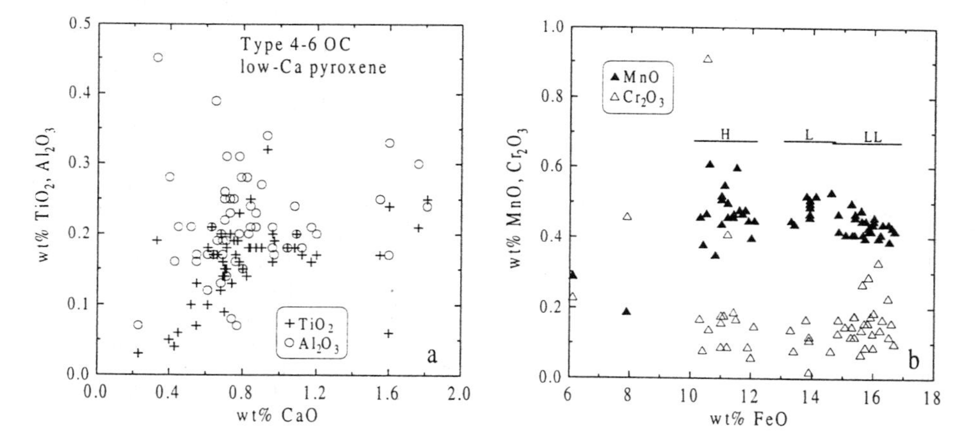

Figure 185. Minor element variation plots for low-Ca pyroxene in type 4-6 ordinary chondrites. One point with 1.1 wt % Al_2O_3 is not included in (a). Data sources as for Figure 181, plus McSween and Patchen (1989).

even in type 6 OC (e.g. Shervais et al. 1986, Koeberl et al. 1990).

Minor element contents of low-Ca pyroxenes show significant ranges from chondrite to chondrite (Fig. 185). Compositions are heterogeneous in type 4 chondrites, but there is also significant variation between type 6 chondrites, of 0.1-0.2 wt % in most minor elements. Within individual type 4 chondrites, there may be a significant range of minor element compositions (e.g. Fudali and Noonan 1975). The pigeonites mentioned above (Keil et al. 1978) have higher Cr_2O_3 than most low-Ca pyroxenes, 0.4-0.9 wt %. MnO contents of low-Ca pyroxenes are similar to those in olivine, 0.4-0.5 wt %, and MnO contents in LL chondrites appear to be slightly lower than those in H and L chondrites. Heyse (1978) found that in LL chondrites, Al and Na contents in low-Ca pyroxenes

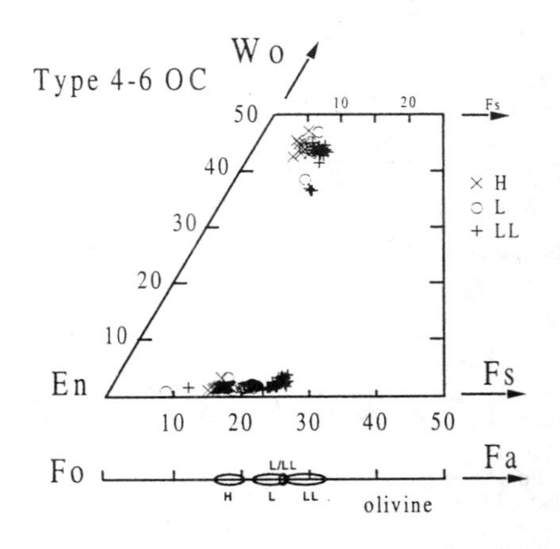

Figure 186. Pyroxene and olivine compositions from type 4-6 ordinary chondrites. Data sources as for Figure 181, plus McSween and Patchen (1989).

increase with Wo content. He interpreted this relationship as an indicator of increasing pressure with increasing metamorphic grade. However, McSween and Patchen (1989) were unable to reproduce these correlations. Trace element concentrations in low-Ca pyroxene from mineral separates from L6 and H6 chondrites have been determined by Allen and Mason (1973) and Curtis and Schmitt (1979). REE abundances are shown in Figure 183.

Ca-rich pyroxenes in type 4 chondrites range from augite to diopside in composition (Fig. 186). Diopsides coarsen with increasing metamorphic grade. In types 5 and 6 chondrites, all Ca-rich pyroxenes are diopsidic, with ~Wo_{45}. In type 7 chondrites, Wo contents are lower, as low as Wo_{36}. Mean Fs contents for each chondrite group increase in the series H, L, LL from means of approximately Fs_7 to $Fs_{8.5}$ to Fs_{10}. Two-pyroxene geothermometry has been applied to low-Ca pyroxene / diopside pairs in individual type 4-7 OC in an attempt to estimate peak temperatures of metamorphism for the different petrologic types (see Dodd (1981) and McSween and Patchen (1989) for summaries). This has met with only limited success. McSween and Patchen (1989) showed that the two pyroxenes do not give coherent temperatures, even in type 6 and 7 chondrites. Peak temperatures for type 6 and 7 chondrites obtained from Ca-rich pyroxenes were 900-960°C and 1150°C, respectively. The range in low-Ca pyroxene compositions ($Wo_{0.9\text{-}2.0}$) determined in one L6 chondrite gives a range in temperature of 650°-830°C (Shervais et al. 1986).

Minor element contents of diopsides are plotted in Figure 187 (see also Table A3.36). Trace element concentrations in Ca-rich pyroxene from mineral separates from L6 and H6 chondrites have been determined by Allen and Mason (1973) and Curtis and Schmitt (1979). REE abundances in diopside are 1-10× chondritic abundances, with an increase from LREE to HREE, and a significant negative Eu anomaly (Fig. 183). TEM studies of the microstructures of pyroxenes in type 4-6 OC have been carried out by Ashworth and Barber (1977), Ashworth (1980), Watanabe et al. (1985) and Langenhorst et al. (1995).

Plagioclase occurs in most type 4-6 ordinary chondrites. It is commonly the product of devitrification of chondrule glass, and grain size coarsens with increasing petrologic type. Many grains contain small subhedral to euhedral inclusions of olivine and pyroxene. Plagioclase is commonly albitic and shows little variation from chondrite to chondrite within each group. Mean compositions of data compiled here are $An_{82}Ab_{12}Or_6$ for H4-6 chondrites, $An_{84}Ab_{10}Or_6$ for L4-6 chondrites, and $An_{86}Ab_{10}Or_4$ for LL4-6

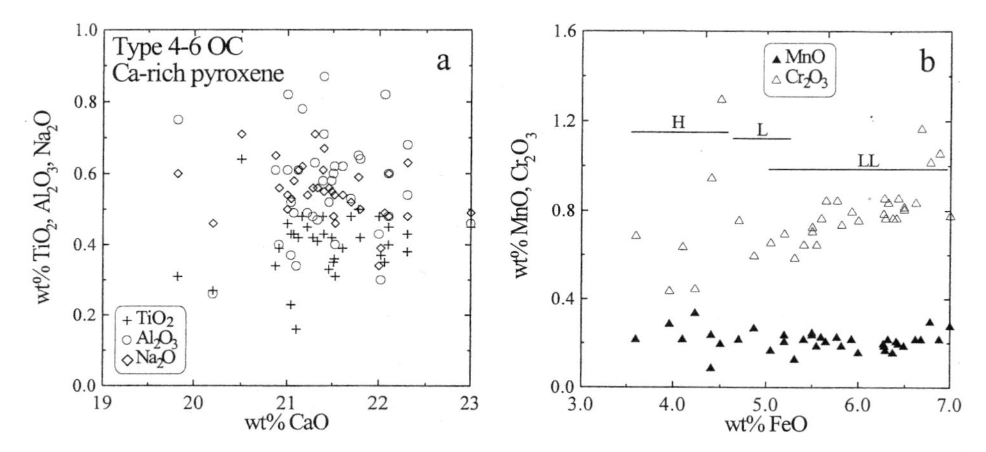

Figure 187. Minor element variation plots for augite in type 4-6 ordinary chondrites. One point with 1.9 wt % Al_2O_3 is not included in (a). Data sources: Fodor et al. (1971a,b, 1976), Smith et al. (1972), Bunch et al. (1972), Lange et al. (1973, 1974), Jaques et al. (1975), Heyse (1978), Gomes and Keil (1980), Klob et al. (1981), Ehlmann et al. (1988), Keil et al. (1988), McSween and Patchen (1989).

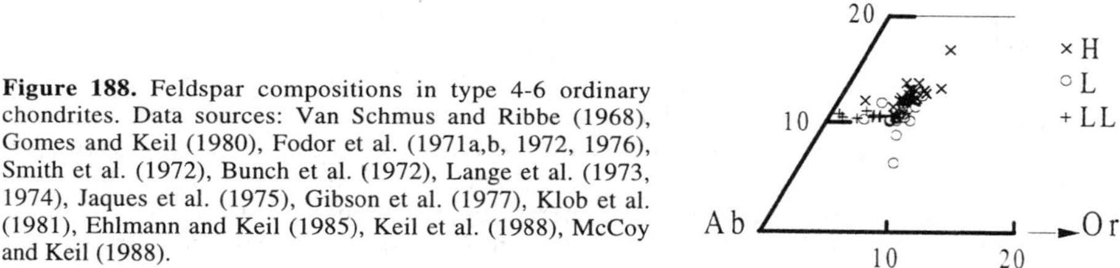

Figure 188. Feldspar compositions in type 4-6 ordinary chondrites. Data sources: Van Schmus and Ribbe (1968), Gomes and Keil (1980), Fodor et al. (1971a,b, 1972, 1976), Smith et al. (1972), Bunch et al. (1972), Lange et al. (1973, 1974), Jaques et al. (1975), Gibson et al. (1977), Klob et al. (1981), Ehlmann and Keil (1985), Keil et al. (1988), McCoy and Keil (1988).

chondrites (Fig. 188; Table A3.37). Van Schmus and Ribbe (1968) showed that feldspar in the LL chondrites has a range of compositions with a narrow range of An contents but varying Or contents. Individual plagioclase grains from type 6 chondrites may show patchy variation in K contents (Van Schmus and Ribbe 1968). Reported FeO and MgO contents vary from 0.1-1.2 and from 0.1-0.5 wt %, respectively. Miúra and Tomisaka (1984) discuss Fe and Mg substitutions in ordinary chondrite feldspars in some detail. P_2O_5 contents reported by Van Schmus and Ribbe are in the range 0.005-0.02 wt %. Secondary plagioclase is mostly in an intermediate to high structural state (Miyashiro 1962, Van Schmus and Ribbe 1968). Trace element concentrations in feldspar from mineral separates from L6 and H6 chondrites have been determined by Allen and Mason (1973) and Curtis and Schmitt (1979). REE abundance patterns are flat and show significant positive Eu anomalies (Fig. 183).

Anorthitic plagioclase, An_{70-90}, which is the typical primary feldspar composition in type 3 chondrites, is observed rarely as a relict phase in type 4-6 chondrites such as the L4 chondrite, Kramer Creek (Gibson et al. 1977) and the L6 chondrite, Leedey (Dodd 1969). MgO contents in the Kramer Creek anorthite are significantly lower than those in the secondary albitic feldspar (<0.1 wt %), while FeO contents are 0.5-1.0 wt %. An unusual inclusion in Los Martinez (L6) consists of a single crystal of highly zoned plagioclase ($Ab_{44}An_{55}Or_1$ in the core to $Ab_{79}An_{18}Or_3$ at the rim) intergrown with chromium-rich spinel (Brearley et al. 1991). The spinel has a well developed crystallographically controlled relationship with the host plagioclase, suggesting that it is the product of exsolution although the original phase is unknown.

Plagioclase of variable composition occurs in some highly shocked ordinary chondrites. In Paragould (LL5, shock stage S4-5), Rose City (H5, S6), and Stratford (L6, S4-5), plagioclase compositions vary from An_{9-45}, An_{9-24}, and An_{10-16}, respectively. This plagioclase is thought to have crystallized from shock-induced melts (Rubin 1992). Compositionally variable plagioclase (An_{12-28}) in Allegan (H5, S1) may also have been produced by shock melting. The low shock stage assigned to this chondrite may reflect the fact that shock occurred prior to thermal metamorphism (Rubin 1992). Nagahara (1980) reported variable plagioclase compositions in several ordinary chondrites. Compositionally variable plagioclase also occurs in the matrices of OC regolith breccias (Bischoff et al. 1983), where the range of compositions is from An_2Or_{93} to $An_{72}Or_{0.2}$.

Phosphate minerals typically constitute less than 2 vol % of ordinary chondrites. However, the distribution of phosphates may be heterogeneous within any chondrite: the dark fraction of St. Séverin (LL6) contains ~20 vol % chlorapatite (Crozaz and Zinner 1985). Phosphates were first discussed in detail by Shannon and Larsen (1925), who identified two phosphate mineral phases, chlorapatite and merrillite. There has been considerable discussion regarding whether the latter phase is truly merrillite, or whether it is whitlockite (e.g. see Fuchs 1969a, Van Schmus and Ribbe 1969, Mason 1971, Miúra and Matsumoto 1982, Rubin 1997b). Many authors use the name interchangeably. We refer to it here as merrillite, following Rubin (1997b). The relative proportions of the two phosphate minerals vary from chondrite to chondrite: merrillite is usually more abundant then chlorapatite. Ahrens (1970) suggested that a close spatial relationship with metal indicates that the phosphates form during metamorphism by oxidation of P-rich metal.

There is very little difference in composition in either phosphate phase between the H, L and LL groups, although there is some variation among reported analyses both within individual chondrites and between chondrites. Variations in CaO, MgO, FeO and Na_2O contents are illustrated in Figure 189 (see also Table A3.38). Chlorapatite essentially has the composition $Ca_5(PO_4)_3(Cl_{0.8}F_{0.1}OH_{0.1})$, typically containing 0.4-0.8 wt % F and 5-6 wt % Cl. Merrillite is essentially $Ca_3(PO_4)_2$, with significant substitutions of Na and Mg for Ca, so that its formula is close to

$2.55CaO{\cdot}0.28MgO{\cdot}0.15Na_2O{\cdot}0.02FeO{\cdot}P_2O_5$ (Van Schmus and Ribbe 1969).

In Ruhobobo (L6) there is a bimodal distribution of merrillite compositions (Klob et al. 1981), with differences in Na_2O and MgO contents (1 vs. 2.5 wt % Na_2O and 2 vs. 3.5 wt % MgO). Trace element analyses of phosphate separates were determined by Allen and Mason (1973).

Phosphates are major hosts for U, Pu and REE in ordinary chondrites. Typical U concentrations in merrillite and chlorapatite are 200 ppb and 3 ppm (e.g. Crozaz 1979, Pellas and Störzer 1981, Ebihara and Honda 1983, Crozaz et al. 1989). Individual grains from the same meteorite may vary in U abundances. There is a tendency for U concentrations to increase with petrologic type. In contrast to U, Pu and REE are more enriched in merrillite than in apatite (Reed et al. 1983, Reed and Smith 1985, Crozaz and Zinner 1985, Crozaz et al. 1989). Average REE abundances in a given phosphate phase, as well as the ratio of Ce abundances in merrillite and apatite, vary between meteorites by about a factor of two. Typical REE abundance patterns are shown in Figures 183 and 190. Rare grains of merrillite have REE patterns similar to patterns in chlorapatite (Crozaz et al. 1989). REE abundances in chlorapatite show more variability from meteorite to meteorite than those in merrillite. There do not appear to be any systematic

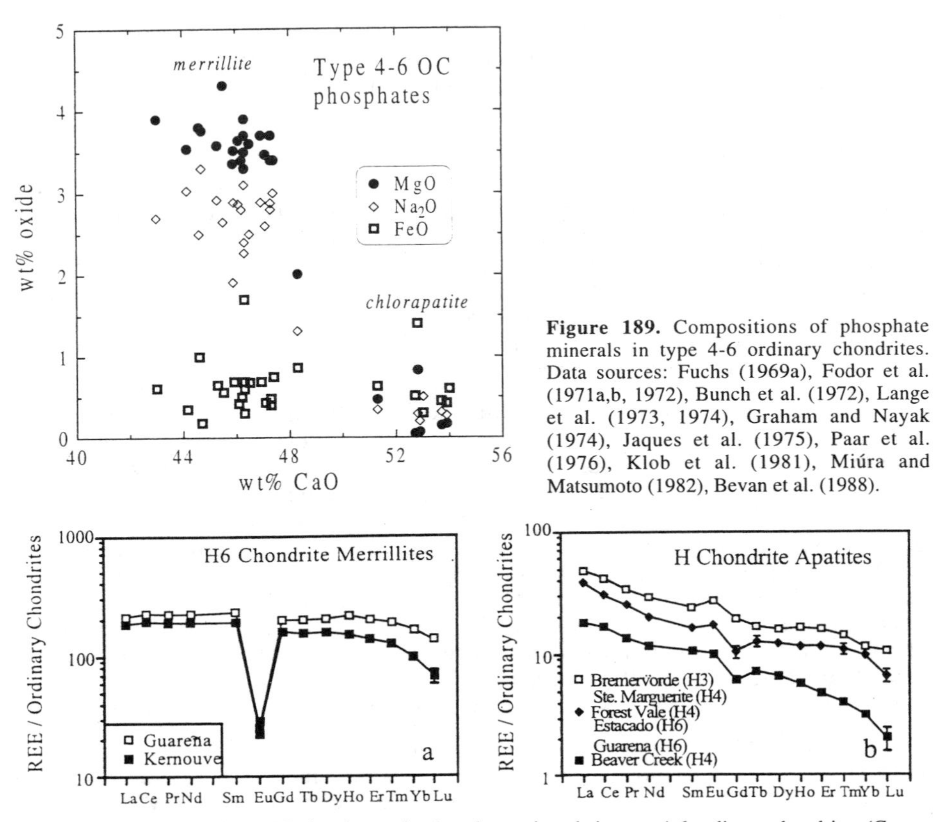

Figure 189. Compositions of phosphate minerals in type 4-6 ordinary chondrites. Data sources: Fuchs (1969a), Fodor et al. (1971a,b, 1972), Bunch et al. (1972), Lange et al. (1973, 1974), Graham and Nayak (1974), Jaques et al. (1975), Paar et al. (1976), Klob et al. (1981), Miúra and Matsumoto (1982), Bevan et al. (1988).

Figure 190. Representative REE abundances in phosphate minerals in type 4-6 ordinary chondrites (Crozaz et al. 1989). (a) Merrillite from two H6 chondrites, (b) chlorapatite from H3-H6 chondrites.

differences in REE concentrations in chlorapatite with petrologic type, although in Uden (LL7), U abundances and REE patterns for chlorapatite are similar to those in merrillite.

The abundance of *chromite* in the ordinary chondrites increases with increasing petrologic type. The mean composition also varies with petrologic type, and chromite compositions become more homogeneous as metamorphic grade increases. These observations are all consistent with chromite recrystallizing during metamorphism (Snetsinger et al. 1967, Bunch et al 1967, Dodd 1969). Chromite compositions are summarized in Figure 191 which illustrates trends within and between the H, L and LL groups (see also Table A3.39). FeO contents of chromites increase in the order H-L-LL, consistent with the overall degree of oxidation. TiO_2 and V_2O_3 contents also show an increase in the sequence H-L-LL, whereas Cr_2O_3, MgO, and MnO contents decrease, particularly if only the type 6 chondrites are considered. Chromite-olivine geothermometry applied to LL chondrites gives equilibration temperatures of <600°C for petrologic types 4-6, which probably reflects a closure temperature for cation diffusion in these minerals (Johnson and Prinz 1991). In addition to individual grains of chromite, chromite exsolution lamellae may be present in ilmenite (Buseck and Keil 1966). Trace element concentrations in chromite from mineral separates from L6 and H6 chondrites have been determined by Allen and Mason (1973) and Curtis and Schmitt (1979).

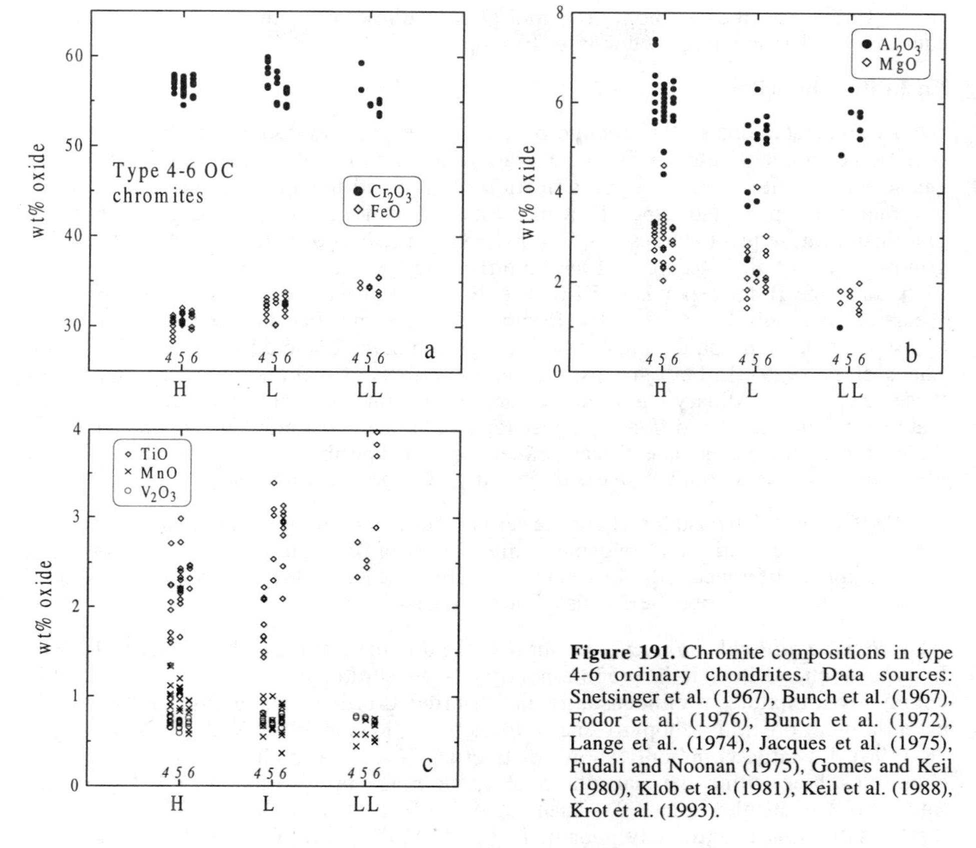

Figure 191. Chromite compositions in type 4-6 ordinary chondrites. Data sources: Snetsinger et al. (1967), Bunch et al. (1967), Fodor et al. (1976), Bunch et al. (1972), Lange et al. (1974), Jacques et al. (1975), Fudali and Noonan (1975), Gomes and Keil (1980), Klob et al. (1981), Keil et al. (1988), Krot et al. (1993).

Spinels with varying compositions are observed in type 4 and 5 OC (Table A3.39). These are mostly Al-rich spinels (Al_2O_3 in the range 20-60 wt %) that contain varying amounts of FeO (12-25 wt %) and MgO (6-20 wt %) (Fudali and Noonan 1975, McCoy et al. 1991b, Krot et al. 1993). Krot and Rubin (1994) report an Al-rich spinel composition with 7 wt % SiO_2 in a glass-rich chondrule in Dimmitt (H4). In an unusual Cr-rich inclusion in Los Martinez (L6), spinels enclosed in plagioclase show variable compositions from Cr-rich spinel to more chromitic compositions, with Cr substitution for Al and a positive correlation between Ti and Cr (Brearley et al. 1991).

Accessory minerals in type 4-6 OC include ilmenite, rutile, glass, silica and nepheline. Ilmenite occurs in association with chromite, metal, sulfide and rarely with aluminous spinels (Christophe Michel-Lévy 1981). It has the compositional range 51-54 wt % TiO_2, 35-45 wt % FeO, 1-7 wt % MgO, 0.6-10 wt % MnO, 0.03-1.4 wt % Cr_2O_3 (Snetsinger and Keil 1969). The Fe/Mg ratio increases in the series H-L-LL chondrites. Rutile occurs in association with metal, ilmenite and chromite, in some cases as fine exsolution lamellae (Buseck and Keil 1966, Fudali and Noonan 1975). Chondrule mesostasis glass is preserved in some type 4 OC (Gomes and Keil 1980). Glass preserved in an LL5 (Krahenberg) has a very low CaO content, 0.14 wt % (McCoy et al. 1991a). Rare grains of silica, usually cristobalite, have been reported in several OC (e.g. Grant 1969, Olsen et al. 1981). Nepheline is extremely rare (e.g. Fodor et al. 1977). Alexander et al. (1987) described scapolite (83 mol % marialite) containing 3.67 wt % Cl in an equilibrated clast in the LL3 chondrite Bishunpur.

Enstatite chondrites

An understanding of the metamorphic sequence in the enstatite chondrite group has become considerably clearer in recent years following the discovery of many Antarctic enstatite chondrites. An early classification of E chondrites divided them into type I, intermediate types, and type II (Keil 1968a), depending on their degree of metamorphism. Subsequent studies led to a division of the E chondrites into the EH and EL groups (Sears et al. 1982): type I and intermediate types correspond to EH3, EH4, and EH5, and type II corresponds to EL6. The discovery of EL3, EL4, and EL5 chondrites (Sears et al. 1984b, Lin et al. 1991, Rubin 1997e) has clarified this picture so that now two separate metamorphic sequences are recognized in the EH and EL groups (Keil 1989, Zhang et al. 1995). In both groups, the general effects of metamorphism are similar to those seen in the ordinary chondrites, although in detail there are important differences that result from the very different, highly reduced, mineral assemblages that characterize the enstatite chondrites. The silicate mineralogy of equilibrated enstatite chondrites is very simple, and essentially consists of enstatite, feldspar and minor silica.

Several type 4-6 enstatite chondrites are breccias, including impact melt breccias. In some of these, enstatite and feldspar compositions in host and clasts may show small compositional differences (Rubin 1983a,b,c; Rubin and Keil 1983, Rubin 1984, Nagahara 1991). We do not describe the details of such breccias here.

Silicate and oxide minerals. *Enstatite* is the dominant silicate phase in type 4-6 EC. In type 4 chondrites, it is predominantly clinoenstatite, in type 5 chondrites it is commonly a mixture of clinoenstatite and disordered orthoenstatite, and in type 6 it is essentially all ordered orthopyroxene (Pollack 1966, Mason 1966, Keil 1968a, Zhang et al. 1996). Enstatite commonly shows distinctive CL colors, both red and blue in EH4 chondrites, blue with some magenta in EH5, and magenta in EL5 and EL6 (Leitch and Smith 1982, McKinley et al 1984, Zhang et al. 1996). Leitch and Smith (1982) suggested that enstatite that is extremely poor in TiO_2, Al_2O_3, Cr_2O_3, MnO, and CaO shows blue CL, and enstatite containing higher concentrations of these minor elements shows red CL, but McKinley et al. (1982) argued that composition and CL color are not clearly related. Most enstatite is extremely FeO-poor, $<Fs_1$, but a few more FeO-rich grains, up to Fs_{11} (FeO ~ 7 wt %), have been reported in Abee (EH4) and St. Mark's (EH5) (Lusby et al. 1987). FeO contents of the majority of EL6 enstatites given in the literature vary from 0 to 1 wt %. However, Wasson et al. (1994) have suggested that most given FeO contents are too high. They suggested that FeO contents measured during electron microprobe analyses are influenced by secondary fluorescence effects from surrounding metal as well as the presence of small metal inclusions within many enstatite grains, and that the true values for FeO contents are likely to be less than 0.02 wt %.

Minor element contents of enstatites are summarized in Figure 192 (see also Table A3.35). The data available for this figure are limited, but similar Cr_2O_3 and MnO ranges were determined for the EH4 chondrite, ALHA77156 (McKinley et al. 1984). For the EL5 and EL6 chondrites, Cr_2O_3 and MnO contents of all enstatites are below electron microprobe detection limits. Ion microprobe analyses gave Cr contents of 399 and 178 ppm and Mn contents of 63 and 57 ppm for enstatites in St. Mark's (EH5) and Hvittis (EL6), respectively (Weisberg et al. 1994). Ti contents of 17-40 ppm and V contents of 4-17 ppm were determined for the same two chondrites. Trace element analyses of low-Ca pyroxene from Khairpur (EL6) were determined by Allen and Mason (1973).

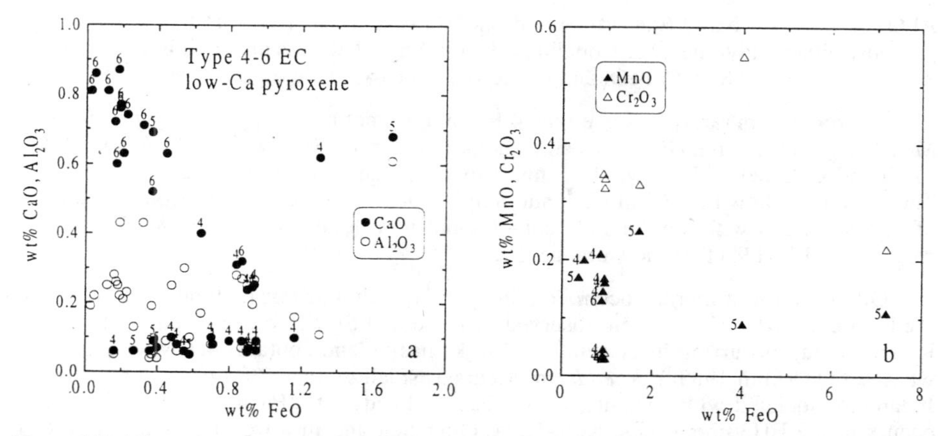

Figure 192. Minor element variation plots for low-Ca pyroxene in type 4-6 enstatite chondrites. All the type 6 data are for EL chondrites, and except for one analysis from an EL5 chondrite (0.7 wt % CaO), all type 4 and 5 data are for EH chondrites. Low-Ca pyroxenes in EH4 and EH5 chondrites with FeO between 1.5 and 8 wt % (not shown) have ~0.6 wt % CaO and ~0.5 wt % Al_2O_3. Data sources: Keil (1968), Buseck and Holdsworth (1972), Leitch and Smith (1982), Rubin and Keil (1983), Rubin (1983a,b,c), Sears et al. (1984), Lusby et al. (1987).

Only one occurrence of *pigeonite* has been described in type 4-6 E chondrites. It occurs in Jajh deh Kot Lalu (EL6) as thin rims on enstatites (Keil and Andersen 1965a). It is relatively FeO-rich (~1.7 wt %) and contains about 6 wt % CaO.

Cores of FeO-rich enstatites of compositions Fs_{1-10} in Abee (EH4) and St. Mark's (EH5) contain abundant, aligned, metal blebs (Lusby et al. 1987). The FeO-rich grains are completely surrounded by rims of FeO-poor (<1 wt %) enstatite, with very low minor element contents. The rims are suggested to have grown during planetary metamorphism.

Feldspar in type 4-6 EC is predominantly albitic. Feldspar grains coarsen during metamorphism: in type 4 and 5 chondrites feldspar grains are <2 μm (e.g. Rubin 1983a, Sears et al. 1984b), whereas in type 6 chondrites feldspar grains are up to 200 μm in size (e.g. Keil and Andersen 1965a). In EL6 chondrites, plagioclase may show prominent polysynthetic twinning. The mean feldspar composition for EH4 chondrites is $An_2Ab_{95}Or_3$, for EH5 chondrites is $An_1Ab_{93}Or_6$, and for EL6 chondrites is $An_{15}Ab_{81}Or_4$ (Keil 1968a; Table A3.37). Feldspar compositions in Eagle (EL6) vary from An_{15-17} and Or_{5-7} (Olsen et al. 1988). Feldspar in EH4 and EH5 chondrites typically contains 0.4-0.8 wt % FeO, and that in EL6 chondrites contains 0.1-0.6 wt % FeO. Trace element analyses of feldspar mineral separates from Khairpur (EL6) were determined by Allen and Mason (1973).

Silica is a minor but ubiquitous phase in type 4-6 EC. In EH4 chondrites, it typically contains 0.2-0.5 wt % Al_2O_3, 0.2-1.3 wt % FeO, and 0.1-0.2 wt % MgO, CaO and Na_2O. Keil (1968a) suggests that some of these minor elements may arise from overlap of the electron beam with small inclusions of troilite, niningerite, and daubreelite within silica grains. From the presence of significant impurities and x-ray powder diffraction data, most silica in EH4 chondrites was inferred to be α-cristobalite (Mason 1966, Rubin 1983a, Rubin and Keil 1983). Tridymite was identified in Indarch (EH4), and quartz in St. Mark's (EH5) by x-ray diffraction (Mason 1966). Silica in Yilmia (EL6) is widespread and occurs as anhedral crystals associated with enstatite and plagioclase, as well as with lacy-textured metal (Buseck and Holdsworth 1972). High concentrations of

Al_2O_3 (1.1 wt %), Na_2O (0.4 wt %) and K_2O (0.3 wt %) suggest that this silica is also probably either tridymite or cristobalite. Silica in Eagle (EL6) is also rich in Al_2O_3 (about 2 wt %), contains 0.8 wt % FeO, and is inferred to be tridymite (Olsen et al. 1988).

Olivine occurs rarely in some type 4 EC and has not been reported in types 5 and 6. Most olivine in Indarch (EH4) has a mean composition of $Fa_{0.5}$, with TiO_2, Al_2O_3, Cr_2O_3, MnO and CaO below 0.07 wt % (Binns 1967a, Leitch and Smith 1982; Table A3.34). These olivines show blue CL. Leitch and Smith (1982) also observed a second population of olivine grains with similar FeO contents but with higher Cr_2O_3 and MnO contents (0.24 and 0.14 wt %) that show orange CL.

Other minor minerals occurring in type 4,6 chondrites include richterite and roedderite. *Richterite* has been observed in Abee and St. Sauveur (Olsen et al. 1973b, Rubin 1983a), occurring in association with kamacite, and contains 4.1 wt % F and 6.5 wt % Na_2O (Rubin 1983a). *Roedderite* occurs as small grains (<30 μm) in the matrix of Indarch, associated with enstatite, plagioclase and tridymite (Fuchs et al. 1966). It also occurs in the EH3 chondrites, Kota-Kota, Qingzhen and in a troilite nodule in Y-691 (Fuchs et al. 1966, Rambaldi et al. 1986, Ikeda 1989b). Roedderites contain ~2 wt % FeO, ~0.4 wt % Al_2O_3, 3-4 wt % Na_2O, and 3-4 wt % K_2O. In perchloric acid-resistant residues of Indarch, Lee et al. (1995) report the presence of spinel, chromite, hibonite, rutile and a Na-Cr silicate, possibly cosmochlore, as well as nanodiamonds and SiC.

Nitrogen-bearing minerals. Several nitrogen-bearing minerals occur in enstatite chondrites. A silicon oxynitride mineral, named *sinoite* after its constituent elements, was discovered in the EL6 chondrite Jajh deh Kot Lalu (Andersen et al. 1964, Keil and Andersen 1965a). The mineral had been observed and described originally by Lacroix (1905) in Pillistfer and Hvittis. Its formula is close to Si_2N_2O, and its composition is approximately 56 wt % Si, 31 wt % N, 13 wt % O. It is orthorhombic and is distinguished in thin section by high-order birefringence. Sinoite occurs as irregular grains or lath-like crystals up to about 200 μm in length, the individual grains frequently being aggregated into larger patches. It has since been observed in several other EL6 chondrites (Keil and Andersen 1965b, Buseck and Holdsworth 1972, Rubin et al. 1997), within impact-melted portions of the EL4 chondrite QUE94368 (Rubin 1997e), and in a perchloric-acid resistant residue from Abee (Lee et al. 1995). In Yilmia, some of the crystals contain parallel lamellar bands that appear to be twinning rather than being due to compositional differences. Sinoite has been suggested to form either during metamorphism, with the N being released from the breakdown of osbornite and troilite (Fogel et al. 1989, Alexander et al. 1994), or from a liquid during impact melting (Rubin 1997e).

The silicon nitride mineral, *nierite* (α-Si_3N_4), has been observed in perchloric acid-resistant residues of Qingzhen (EH3) and Indarch (EH4) (Alexander et al. 1994, Lee et al. 1995, Russell et al. 1995). It has also been found in situ in Qingzhen (Alexander et al. 1994). The d-spacings of this mineral are comparable to those of synthetic α-Si_3N_4, which has trigonal symmetry (Lee et al. 1995). It occurs as small (~2 × 0.4 μm), euhedral laths and acicular grains within kamacite, perryite and schreibersite. A few nierite crystals are intergrown with whiskers of a second silicon nitride, which is thought to be β-Si_3N_4. Nierite is interpreted to have formed by exsolution of Si and N from N-bearing metal and sulfide phases during metamorphism. Thermodynamic calculations suggest that nierite is unstable with respect to sinoite during metamorphism. Some nierite is enriched in ^{15}N and may possibly be interstellar in origin (Russell et al. 1995).

CK chondrites

Most members of the CK group of carbonaceous chondrites are petrologic type 4-6.

This group includes Karoonda, Maralinga, and several Antarctic meteorites. Ningqiang, which is classified as an anomalous CK, is petrologic type 3 and is described in the Chondrule section above. Compositional, textural, and O-isotope data show that the CK group is closely related to CV and CO chondrites. Kallemeyn et al. (1991) made slight modifications to the criteria of Van Schmus and Wood (1967) in order to define petrologic types 4-6 for CK chondrites. Karoonda and Maralinga are both petrologic type 4. The most recrystallized member of the group is LEW87009, CK6. All CK chondrites contain regions of shock-induced, darkened silicates (olivine, pyroxene, and plagioclase) which contain abundant, tiny (<10 μm) grains of magnetite and pentlandite arranged in curvilinear trails.

Olivine is relatively homogeneous in all the CK chondrites, with a range of mean Fa from 29 to 33 (Kallemeyn et al. 1991). The mean Fa content of Karoonda has been given as 33.4 (Van Schmus (1969), 34.2 (Fitzgerald 1979) and 31.2 mol % (Kallemeyn et al. 1991). Scott and Taylor (1985) reported a range of Fa contents from 2-33 mol % for the same chondrite, with the lowest FeO contents occurring at the centers of large olivine grains, and Noguchi (1993) also showed a larger variation in mean Fa than earlier studies. CaO contents of olivine from remnant chondrules in Karoonda and PCA82500 (CK4/5) may range up to 0.3 wt % (Scott and Taylor 1985), although most olivine has <0.1 wt % CaO. Olivine contains significant amounts of NiO (0.4-0.6 wt %), and ~0.2 wt % MnO (Keller et al. 1992, Nakamura et al. 1993, Noguchi 1993). Typical compositions are given in Table A3.34. Rare aberrant olivine grains with significantly different compositions occur in many CK chondrites, e.g. a grain of composition Fa_{59} occurs in LEW86258 (CK4). Olivine in the matrix of Y-693 was shown to contain high dislocation densities ranging from 8×10^8 to 5×10^9, indicative of shock deformation (Nakamura et al. 1993).

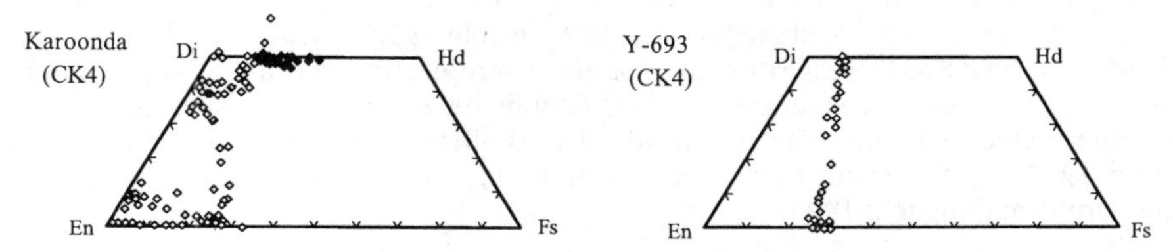

Figure 193. Pyroxene compositions in the CK chondrites, Karoonda (CK4) and Y-693 (CK4) (Noguchi 1993). Filled symbols are Fs-rich augites in matrix.

Low-Ca pyroxene and *Ca-rich pyroxene* (diopside and augite) occur as minor to accessory phases in CK chondrites. Figure 193 shows the distribution of pyroxene compositions in Karoonda and Y-693 (Noguchi 1993). Low-Ca pyroxene compositions in Karoonda and PCA82500 range from Fs_{1-30} and Fs_{6-25}, respectively (Scott and Taylor 1985, Noguchi 1993), whereas in Maralinga, Y-82104, Y-693, and EET87507 low-Ca pyroxene is more homogeneous, with compositions in the range Fs_{26-28} (Keller et al. 1992, Nakamura et al. 1993, Noguchi 1993). Low-Ca pyroxenes in Karoonda include one group with $<Wo_1$ and a second group with $\sim Wo_5$, and those in the more equilibrated examples also show a spread in Wo contents (Noguchi 1993). Low-Ca pyroxene in Karoonda shows patchy or lamellar zoning, with a range of compositions, e.g. Fs_{10-25} (Noguchi 1993). In the matrix of Y-693, and in Maralinga, low-Ca pyroxene is predominantly orthopyroxene (Keller et al. 1992, Nakamura et al. 1993). Noguchi (1993) observed a texture interpreted as olivine and Ca-rich pyroxene replacing low-Ca pyroxene in Maralinga. Mean TiO_2 and Cr_2O_3 contents of low-Ca pyroxenes are generally low in CK chondrites (<0.02 and <0.14 wt %, respectively), although they are higher in Karoonda (0.1 and 0.3 wt %, respectively). The low contents of these elements has been attributed to the ubiquitous presence in CK chondrites of Cr-bearing magnetite with exsolution lamellae of ilmenite (Noguchi 1993). Mean Al_2O_3 contents of low-Ca pyroxenes are high, 0.5-3.5 wt %, and MnO contents are ~0.2 wt % (Keller et al. 1992, Noguchi 1993). Representative analyses are given in Table A3.35.

Mean Ca-rich pyroxene compositions vary from Wo_{42-47} and are close to Fs_{10}. They contain 0.1-0.3 wt % TiO_2, 0.01-0.4 wt % Cr_2O_3, 1-2.6 wt % Al_2O_3, ~0.2 wt % Na_2O and ~0.1 wt % MnO (Keller et al. 1992, Noguchi 1993; Table A3.3.36). Fs-rich augites ($Fs_{10-25}Wo_{\sim50}$) occurring in the matrix of Karoonda show complex zoning (Noguchi 1993).

Low-Ca pyroxene and augite in matrices are commonly enclosed in plagioclase in an acicular texture (Nakamura et al. 1993, Noguchi 1993). Low-Ca pyroxene in this association has a high Al_2O_3/CaO ratio compared with low-Ca pyroxene in remnant chondrules. FeO contents are positively correlated with Al_2O_3 contents, suggesting that there is significant Fe_2O_3 present (Noguchi 1993).

Plagioclase compositions are very heterogeneous in all CK chondrites, with heterogeneity occurring within individual grains as well as between grains. Typical ranges of compositions are An_{18-100} in Karoonda and PCA82500 (Scott and Taylor 1985, Noguchi 1993) and An_{22-82} in ALH84038 (CK4) (Kallemeyn et al. 1991). Large grains (>several tens of micrometers) in Karoonda show reverse zoning, e.g. from An_{40} at the core to An_{75} at the edge (Noguchi 1993). Two distinct compositions occur in Maralinga: $An_{\sim20-40}$ at the cores of grains, and compositions around An_{80} occurring as overgrowths (Keller et al. 1992, Noguchi 1993; Table A3.37). Or contents vary from 1-8 mol % (Scott and Taylor 1985), although potassic feldspar also occurs rarely, e.g. a grain of composition $An_{19}Or_{34}$ was found in ALH82135 (CK4: Rubin 1992). Plagioclase is heterogeneous even in the CK6 chondrite, LEW87009 (An_{39-88}), where grains are typically 200 μm in size (Kallemeyn et al. 1991, Rubin 1992). Plagioclase in EET87860 (CK5/6) and EET83311 (CK5) exhibits partial isotropization and many grains contain numerous micrometer-size mafic crystallites (Rubin 1992). Plagioclase heterogeneity has been interpreted as being either the result of crystallization from shock-induced melts (Rubin 1992), or the result of mineral reactions in which augite is consumed during metamorphism (Noguchi 1993).

SHOCK METAMORPHISM

Introduction

The effects of shock metamorphism are widespread in chondritic meteorites and have been documented in detail by a number of authors. Shock metamorphism produces both textural changes within meteorites and, at the highest shock levels, produces mineralogical phase changes which sometimes result in the formation of new minerals, such as high pressure phases. The development of shock effects within meteorites is often heterogeneous, but nevertheless textural and mineralogical criteria have been developed which allow the degree of shock experienced by a given chondrite to be assigned a specific shock level. In this section we briefly review new mineral phases which are produced as a result of shock metamorphism and textural changes which occur. For reviews of shock metamorphism in meteorites see Stöffler et al. (1988) and Bischoff and Stöffler (1992). Detailed studies of shock effects in the different chondrite groups have been reported by Stöffler et al. (1991) (ordinary chondrites), Scott et al. (1992) (carbonaceous chondrites) and Rubin and Scott (1997) (enstatite chondrites).

In this section we concentrate on new shock minerals and shock effects in minerals. Other phenomena such as melt veins and pockets also occur widely in shocked chondrites, but will not be examined here. Details of the mineralogy and chemistry of these shock melt features have been reported by numerous authors (e.g. Dodd and Jarosewich 1979, 1982; Dodd et al. 1982, Ashworth 1985, Stöffler et al. 1991, Joreau et al. 1996).

Shock mineralogy and shock effects in ordinary chondrites

A sequence of shock stages based on mineralogical and textural criteria has been developed by Stöffler et al (1991) for the ordinary chondrites, which range from S1 (unshocked) to S6 (highly shocked). These shock stages are based on characteristic deformation microstructures in phases such as olivine and low-Ca pyroxene at low shock levels to melting and polymorphic phase transitions at higher shock levels (see Table 1).

Olivine. Shock effects in olivines in ordinary chondrites have been studied extensively and their evolution as a function of increasing shock is now well documented. These effects have been reviewed and summarized by Stöffler et al. (1991). As a function of increasing shock intensity, olivine shows the following shock features, with unshocked olivine showing only minor, irregular fractures (S1). At the lowest shock levels, olivine develops undulatory extinction and randomly oriented non-planar fractures (S2) (Carter et al. 1968). As the shock level increases, parallel planar fractures develop with a spacing of tens of microns in olivine with undulatory extinction (S3). The planar fractures have preferred crystallographic orientations (Müller and Hornemann 1969, Snee and Ahrens 1975, Reimold and Stöffler 1978, Bauer 1979, Langenhorst et al. 1995), which have been defined as {100}, {010}, {001}, {130} and {*hkl*} (bipyramids). The planar fractures on {130} and {*hkl*} appear to be diagnostic of shock deformation, based on shock experiments between 5 and 43 GPa (Müller and Hornemann 1969). Examples of planar fractures with these orientations have been reported in olivine in many ordinary chondrites (e.g. Müller and Hornemann 1969, Langenhorst et al. 1995).

At moderate levels of shock, olivine containing planar fractures develops a texture which has been termed mosaicism (S4). Individual grains have a mottled or mosaic appearance at extinction when viewed under a polarizing microscope. The individual domains are a few microns or less in size and are misoriented by 3 to 5°. The number and misorientation of the domains increases with increasing shock pressure. Olivine grains with highly developed mosaicism also contain planar deformation features (PDFs), 20-40 μm in length, in addition to parallel fractures (S5). However, the PDFs are submicroscopic lamellae, but have crystallographic orientations similar to those of the planar fractures.

As shock intensity increases to shock stage S6, olivine undergoes recrystallization in the solid state, usually adjacent to, or within, shock veins, resulting in extremely fine-grained polycrystalline aggregates. Unrecrystallized regions of olivine grains retain their planar fractures, PDFs and strong mosaicism. Total melting of olivine to form a melt containing blebs of troilite and Fe, Ni metal occurs adjacent to the recrystallized olivine.

Deformation microstructures in olivine that result from shock have been described by numerous authors (e.g. Ashworth and Barber 1975, 1977; Madon and Poirier 1983, Ashworth 1981, 1985; Langenhorst et al. 1995, Joreau et al. 1996). Planar fractures are commonly observed (e.g. Joreau et al. 1996). In shocked olivines, dislocation densities are often very high, particularly above shock stage S4. Values up to 2×10^{15} m^{-2} have been reported (e.g. Madon and Poirier 1983, Langenhorst et al. 1995) and the most common dislocation type has Burgers vector **b** = [001] and are [001] screw dislocations (e.g. Carter et al. 1968, Ashworth and Barber 1975, Ashworth 1981, Madon and Poirier 1983, Langenhorst et al. 1995). Such dislocations are diagnostic of deformation at relatively high strain rates and low temperatures (e.g. Carter 1971, Green and Radcliffe 1972, Blacic and Christie 1973). They often have long, straight segments parallel to [001] (e.g. Ashworth and Barber 1975, Ashworth 1985), but long dislocations which are not parallel to [001], but lie on {110} or (100) slip planes also occur (e.g. in Hedjaz, Ashworth and Barber 1975). In Tenham (L6), Price et al. (1979) reported dislocations with slip planes on {110} and (100) in olivine and Langenhorst et al. (1995) were able to determine glide planes for the screw dislocations of (100), {110} and {hk0}, typical for olivine (Poirier 1975) (see also Ashworth and Barber 1975, Joreau et al. 1996). Ashworth (1981 1985) reported a few dislocations with Burgers vector **b** = [010] in olivine in Butsura (H6), Alfianello (L6, S5) and Taiban (L6, S6) and **b** = [100] in olivine in Butsura.

In St. Séverin (LL6), Ashworth and Barber (1977) and Ashworth (1981) found evidence of dislocation recovery in olivine grains which they attributed to postshock annealing. The dislocations have Burgers vector **b** = [001] and have often regrouped into arrays which define subgrain boundaries. Dislocations also occur outside the dislocation arrays which are curved and often form small closed loops.

Raman spectra which provide an indication of the residual stress resulting from shock in olivines have been measured in olivines from two L6 chondrites (Y-7304 and 7305) (Miyamoto and Ohsumi 1995) (but see also Heymann 1990). X-ray diffraction data for shocked olivines in ordinary chondrites have been reported by Taylor and Heymann (1969).

Ringwoodite. Ringwoodite, the high pressure spinel polymorph of $(Mg,Fe)_2SiO_4$ olivine, was first reported by Binns et al. (1969) and Binns (1970) in black shock veins in the Tenham ordinary chondrite. Ringwoodite has now been widely recognized in many other highly shocked ordinary chondrites of shock stage S6 (e.g. Coleman 1977, Stöffler et al. 1991). Ringwoodite in ordinary chondrites commonly occurs as distinct purple crystals (isotropic in crossed polars), up to 100 μm in size, that occur as polycrystalline aggregates within or at the margins of recrystallized olivine grains. The grains are typically associated with melt pockets or thick melt veins. The microstructures of ringwoodite have been investigated in detail by a number of authors using TEM techniques (e.g. Putnis and Price 1979, Price et al. 1979, 1982; Price 1983, Madon and Poirier 1983). In Tenham, ringwoodite consists of partially inverted or faulted spinels (Putnis and Price 1979) and the faulted spinels occur commonly as fine-scale polycrystalline aggregates and as massive twinned crystals (Price et al. 1979). Spinels in the polycrystalline aggregates are usually rounded, have equilibrium grain boundaries and numerous planar defects (stacking faults) on {110}, as well as antiphase domains. These faults have been interpreted as the result of partial transformation from ringwoodite to wadsleyite (modified spinel or β-phase) during pressure release rather than forming by shock deformation (Putnis and Price 1979, Price et al. 1979). The massive crystals are twinned on {111} and {110} defects are also present. These grains have been interpreted as forming as a result of quasi-martensitic transformation of olivine to spinel (e.g. Poirier 1981). Stacking faults in ringwoodite have also been described by Madon and Poirier (1983), who determined a displacement vector of $(a/4)\langle 1\bar{1}0\rangle$. (see also Langenhorst et al. 1995, Chen et al. 1996, Sharp et al. 1997)

Wadsleyite (β-phase). Wadsleyite is the naturally occurring, orthorhombic, high pressure polymorph of $(Mg,Fe)_2SiO_4$, which has only been found in nature in shocked ordinary chondrites. In the Peace River meteorite, Price et al. (1982) found granular aggregates of well formed wadsleyite grains, as well as wadsleyite grains replacing ringwoodite. The two phases have the crystallographic orientation relationship

$[010]_\beta||[110]_\gamma$ and $[001]_\beta||[001_\gamma]$, the orientation in which the two phases have a continuous, near cubic close-packed oxygen sublattice. These textures were interpreted as evidence for post-shock inversion of ringwoodite to wadsleyite. Most wadsleyite grains contain stacking faults on (010) of the type 1/4[010](010). Intergrown lamellae of wadsleyite and ringwoodite have been documented from the Peace River (L6) chondrite by Price (1983) using high resolution TEM. Madon and Poirier (1983) also found small grains (<1 μm) of wadsleyite in shock veins in the Catherwood (L6) chondrite. The grains are often elongate; have low dislocations densities but frequently have stacking faults on (010) with displacement $1/2[\bar{1}01]$.

Enstatite. Shock effects in low-Ca pyroxenes are widespread in shocked ordinary chondrites, but have not been as well calibrated as a function of shock pressure as those in olivine (but see Stöffler 1991 and Rubin et al. 1997). Ashworth (1985) observed that the major shock effects in orthopyroxene in several shocked L chondrites (S5-S6) were the formation of shock lamellae of clinoenstatite and the development of dislocations. Both features are produced at extremely variable densities and some regions of orthopyroxene remain essentially undeformed (see also Ashworth 1981). In Tenham, enstatite exhibits intensive internal fracturing, dominantly parallel to (010) and (001), but also in other orientations (Langenhorst et al. 1995). A range of deformation microstructures are also present, such as high densities of dislocations and low angle subgrain boundaries that are probably the mosaic microstructure commonly observed by optical microscopy (Price et al. 1979, Langenhorst et al. 1995). Langenhorst et al. (1995) found that most dislocations occur in the glide plane (100) and the total dislocation density does not exceed 10^{12} m^{-2}, lower than in adjacent olivine and diopside. Within the orthopyroxene, clinoenstatite lamellae occur on (001) which dominantly appear to be even multiples of the 0.9 nm *a* repeat in width, suggesting a shock origin. (e.g. Buseck et al. 1980). Similar observations have been made by Joreau et al. (1996) on enstatite from Gaines County (H5), but in this case the enstatite also contains straight and narrow (<1 μm) fractures filled with amorphous material.

Small (~2 μm), rounded grains of clino and orthoenstatite ($En_{81.9}Fs_{17.1}Wo_{1.0}$) occur within melt pockets in Gaines County (H5) which have crystallized from the melt during cooling and are associated with plagioclase, diopside and glass (Joreau et al. 1996).

Diopside. Diopside is a minor phase in ordinary chondrites, but the microstructural effects of shock have been described by Ashworth (1980b 1985), Langenhorst et al. (1995) and Joreau et al. (1996). All the chondrites studied by Ashworth (1980) have been shocked to relatively low degrees and contain shock induced twins on (100) which cross-cut dislocations arrays, that may have formed during earlier crystal growth. The (100) twins may be associated with irregular cracks that can either be healed or filled with amorphous diaplectic glass. Partial dislocations form steps along the (100) twin boundaries and are consistent with the formation of twins by the generation and movement of partial dislocations with a Burgers vector parallel to [001]. Dislocations also occur between the twin planes which have the slip system (100)[001] (see also Joreau et al. 1996).

In more highly shocked chondrites, such as the L5-6 (S5-S6) chondrites studied by Ashworth (1985), mechanical twinning is present on (001) and (100) and the (001) twins are thicker than those on (100). In the highly shocked Tenham (L6, S6) chondrite, Langenhorst et al. (1995) have reported a variety of shock microstructures in diopside including mechanical twinning, high dislocation densities and planar deformation features. Mechanical twins occur parallel to (001) and (100) and the (100) twins are much thinner (<5 nm) than the (001) twins (70-260 nm). The number and thickness of (100) twins increases with increasing proximity to shock veins, indicating that formation of twins is enhanced where post-shock temperatures were higher. (001) twins appear to be diagnostic of shock, because experimentally deformed diopside only contains (100) twins.

Langenhorst et al. (1995) reported dislocation densities of 1×10^{14} cm^{-2} away from shock veins, but up to one order of magnitude lower adjacent to shock veins. The dislocation glide systems are dominantly (100)[001] and to a lesser degree {110}[001]. Diopside in Tenham also contains the first reported examples of amorphous lamellae (PDFs) in naturally shocked diopside. The lamellae are straight, narrow and are predominantly parallel to $(\bar{2}2\bar{1})$, $(2\bar{2}1)$ and $(2\bar{2}\bar{1})$ compared with experimentally shocked diopside in which the dominant lamellae are on $(33\bar{1})$ (Leroux et al. 1994). The PDFs in Tenham diopside have thicknesses of the order of 50 nm and appear to occur as pairs with variable interlamellar spacings (0.1-0.9 μm). Langenhorst et al. (1995) also reported that where the PDFs intersect twins, they are deflected and the twins displaced, indicating that the PDFs formed by shearing and that both features formed simultaneously.

Small (~2 μm), rounded grains of diopside ($En_{48.0}Fs_{5.2}Wo_{46.8}$) occur within melt pockets in Gaines County (H5) which have crystallized from the melt during cooling and are associated with enstatite, plagioclase and glass (Joreau et al. 1996).

Majorite. Smith and Mason (1970) reported the first natural occurrence of majorite in a veinlet in the Coorara (L6) meteorite, produced by high pressure transformation of low-Ca pyroxene to the higher density garnet structure. The majorite is extremely fine-grained and difficult to analyze, but has an Mg/(Mg+Fe) ratio of ~0.25. Coleman (1977) also described the occurrence of majorite in Catherwood (L6) and reported compositional data. Price et al. (1979) examined the microstructures of majorite in shock veins in Tenham. The majorite has space group *Ia*3*d* and occurs as equant and dendritic grains. The equant grains are 50-500 nm in size and have 120° grain junctions. An interstitial glassy phase is present and the textural evidence indicates that the majorite crystallized from the glass (Price et al. 1979). The dendritic majorite occurs as a single crystal intergrown with interstitial glass. The dendritic habit probably results from rapid growth from a supersaturated liquid. The two different habits of the majorite probably reflect differing thermal histories in different parts of the shock veins (see also Langenhorst et al. 1995).

Chen et al. (1996) described both low-Ca majorite and a majorite-pyrope solid solution in shock veins from Sixiangkou (L6, S6). Low-Ca majorite occurs as rounded polycrystalline grains, 15-300 μm in size, associated with ringwoodite and diaplectic plagioclase glass. The majorite contains a subgrain dislocation microstructure consistent with dislocation climb. The majorite has a composition essentially identical to unshocked low-Ca pyroxene in Sixiangkou and probably formed by solid state transformation of the pyroxene. The grains of majorite-pyrope garnet occur in the fine-grained matrix of the shock veins as idiomorphic crystals, 0.5-4 μm in size, and contain significant concentrations of Al_2O_3, CaO, Na_2O as well as Cr_2O_3. The compositions of these crystals indicate that they crystallized from a shock melt at very high pressures and temperatures.

$MgSiO_3$ ilmenite. Sharp et al. (1997) examined shock melt veins in the Acfer 040 (L5-6, S6) chondrite using TEM techniques and observed prismatic or plate-like grains with the composition $MgSiO_3$ with minor concentrations of FeO, Al_2O_3, Na_2O and Cr_2O_3. Based on electron diffraction studies, this phase was unambiguously identified as $MgSiO_3$ ilmenite with space group $R\bar{3}$, the first recorded occurrence of this phase in nature. The $MgSiO_3$ ilmenite occurs with ringwoodite and an amorphous phase which may have been $MgSiO_3$ perovskite, which became amorphous during depressurization.

Because $MgSiO_3$ ilmenite is not predicted as a stable liquidus phase at high pressure, Sharp et al. (1997) argued that it crystallized metastably during post shock decompression and heating.

$MgSiO_3$ perovskite. Crystalline $MgSiO_3$ perovskite has not been definitively identified in chondritic meteorites. However, tantalizing evidence to suggest that it was present in shock veins in the L5-6 chondrite, Acfer 040 has been presented by Sharp et al. (1997). The melt veins contain equant grains (~2 μm) of an amorphous material, surrounded by smaller mineral grains and interstitial glass. The grains of the amorphous phase are morphologically distinct from the glass and they have a composition similar to that of majorite garnet which occurs in the matrix of melt veins in other chondrites. The grains also contain significant Al_2O_3 (~5 wt %), CaO (3.95 wt %) and Na_2O (2 wt %) and are compositionally distinct from the interstitial glass which contains higher CaO and lower SiO_2 contents. Based on the fact that $MgSiO_3$ perovskite is unstable at atmospheric pressure and decomposes extremely rapidly during heating, Sharp et al. (1997) concluded that this phase was perovskite that crystallized at high pressure (~26 GPa and 2000°C) during shock, but has been amorphized, possibly during decompression or later heating.

Plagioclase. Stöffler et al. (1991) found a complete sequence of shock effects in plagioclase as a function of increasing shock intensity in ordinary chondrites. However, they argued that only shock effects which occur at high shock intensities should be used as indicators of shock intensity, because plagioclase of sufficient grain size for optical examination only occurs widely in type 5 and 6 ordinary chondrites. At low degrees of shock (S2-S3), plastic deformation of plagioclase results in undulatory extinction, whereas at moderate shock levels (S4), mosaicism, development of sets of PDFs and partial isotropization occur. Plagioclase completely transforms to diaplectic glass (maskelynite) at high shock levels (S5) and melts to form normal glass at the highest level of shock (S6). Maskelynite can be distinguished from normal glass by its refractive index (Stöffler et al. 1986).

Stöffler (1972) and Ostertag (1983) have described the optical appearance of these effects in considerable detail. PDFs have been shown to consist of isotropic lamellae with spacings of less than 2 μm within crystalline feldspar. Langenhorst et al. (1995) reported TEM observations of PDFs in oligoclase in the Tenham (L6) chondrite and found that they are 0.2-0.3 μm wide with a lamellae spacing of 0.5 μm. Up to four sets of PDFs, in different crystallographic orientations, can traverse individual feldspar grains. Peak pressures for the development of amorphous PDFs are between 24 and 34 GPa. Similar observations have been made by Joreau et al. (1996) on feldspar grains in Gaines County (H5). TEM observations on shocked plagioclase in ordinary chondrites have also been reported by Ashworth (1981, 1985).

In ordinary chondrites which have been shocked to shock level S5, maskelynite (diaplectic glass) is widespread (e.g. Stöffler et al. 1991). This clear, transparent glass retains the original shape and composition of the preexisting plagioclase grains, in comparison with normal glass that has a flow-deformed shape and occurs in melt veins and pockets in heavily shocked regions of the meteorites. For example, Ashworth (1985) found that plagioclase has been completely converted to glass in the L6 (S6) chondrite Taiban. In some melt pockets, plagioclase recrystallizes from the melt to form a matrix of polysynthetically twinned grains in which clinoenstatite and diopside may be embedded (e.g. Joreau et al. 1996).

Magnesiowüstite. In the fine-grained matrix of shock melt veins in Sixiangkou (L6, S6), Chen et al. (1996) found magnesiowüstite ($(Mg_{0.54}Fe_{0.46})O$) that occurs as irregular-shaped blebs, up to 5 μm long, interstitial to grains of idiomorphic majorite garnet. The blebs are segments of single crystals that fill channels between the majorite grains, suggesting that the two phases probably co-crystallized from a liquid at very high pressures and temperatures (e.g. 2050°-2300°C and 20-24 GPa). Submicroscopic magnetite crystals, sometimes as small as 3 nm in size, occur within the magnesiowüstite and probably exsolved during cooling.

Shock effects in metal. Metallic phases in ordinary chondrites are particularly susceptible to the effects of shock metamorphism, both in the solid state and as a result of melting. Many chondrites have been reheated as a result of shock which has affected the metal in a number of different ways. Several studies have described these effects in considerable detail (e.g. Buseck et al. 1966, Heymann 1967, Wood 1967, Begemann and Wlotzka 1969, Taylor and Heymann 1969, 1971; Smith and Goldstein 1977, Taylor et al. 1979, Scott and Rajan 1979, 1981; Scott 1982, Bennett and McSween 1996). They are reviewed only briefly here and include the formation of martensite (α_2Fe-Ni), disturbed M-shaped diffusion profiles, melting of metal (and troilite), the development of steep Ni gradients in martensite, the formation of secondary kamacite from taenite, chemical homogenization of metal grains and enrichment of phosphorus in metal.

Martensite is the metastable Fe-Ni phase, termed α_2Fe-Ni, that forms from taenite during rapid cooling. The development of martensite is extremely common in shock reheated chondrites (e.g. Heymann 1967, Taylor and Heymann 1971, Smith and Goldstein 1977). The formation of martensite is attributed to shock reheating of metal grains to temperatures within the taenite (γ) stability field, followed by rapid cooling. The temperature at which this occurs clearly depends on the Ni content of the alloy present, but can range from ~500°C for 30 wt % Ni to ~800°C for all compositions with Ni > 5 wt %. During rapid cooling, there is insufficient time for kamacite (α) to nucleate within the two phase field $\alpha + \gamma$ and the metastable taenite composition transforms to martensite, once the martensite transformation temperature is reached (see Opaques in Ordinary Chondrites). Taylor and Heymann (1971) found that martensite contained higher P contents (0.08 wt %) than metal in unreheated chondrites. Very fine-grained plessites (see below) are sometimes referred to as martensite in the literature (e.g. Taylor et al. 1979), sometimes in cases where martensite has not been unequivocally identified. In such cases the more general term plessite should probably be used (e.g. Scott and Rajan 1979).

Plessite is a fine-grained mixture of kamacite and taenite which occurs widely in unshocked and shocked equilibrated chondrites (see Opaques section). According to Massalski et al. (1966), plessite can form by a number of different mechanisms, either directly or indirectly from taenite, leading to plessites with differing microstructures. Type I plessite forms by precipitation of kamacite in a taenite host, Type II plessite results from the transformation of non-equilibrium taenite to α_2Fe-Ni (martensite) and Type III plessite forms from the decomposition of α_2Fe-Ni to kamacite (α) and taenite (γ) during reheating.

Plessite can also develop as a result of shock reheating (e.g. Scott and Rajan 1979). In weakly shocked L chondrites, lower than shock stage S4, Bennett and McSween (1996) described zoneless and fine-grained pearlitic plessite. The zoneless plessite consists of blocky interpenetrating laths of kamacite, consistent with formation by the decomposition of martensite during slow cooling between 500° and 300°C (Hutchison and Bevan 1983). The fine-grained pearlitic plessite is similar to that described by Grokhovsky and Bevan (1983) in Richardton (H5) that formed by discontinuous precipitation reactions (see Opaques section). In more highly shocked chondrites, coarser-grained plessite occurs. Taylor and Heymann (1970) found that in the shocked

Kingfisher (L5) chondrite, large (~1 mm) metal grains occur which are not martensite, but instead consist of an irregular array of kamacite and taenite grains, ranging from a few microns up to 100 μm in size (coarse plessite or coarse pearlitic plessite, Bennett and McSween 1996). The cores of the grains have bulk Ni contents of 14-15 wt % and kamacite contains 6.7-7.5 wt % Ni and taenite ~19-22 wt % Ni. P contents in kamacite are below detection (<35 ppm), but in taenite range from 60-130 ppm. No phosphides are present, but phosphates occur. Taylor and Heymann (1970) interpreted this microstructure as having resulted from the decomposition of preexisting α_2Fe-Ni, formed during the original parent body cooling, during later postshock heating. The postshock heating event did not reach temperatures within the taenite stability field, but stayed within the two phase $\alpha + \gamma$ field allowing breakdown of the martensite to produce intergrown kamacite and taenite (Type III plessite). Type III plessite has also been described by Taylor and Heymann (1971). Scott and Rajan (1979) also described fine-grained plessite in the highly shocked chondrite, Shaw with a composition 13.4-14.5 wt % Ni, 0.8 wt % Co.

In some shock reheated chondrites, the metal contains elevated concentrations of P (>0.10 wt %) (Taylor and Heymann 1971, Smith and Goldstein 1977), significantly higher than is typically present in unshocked equilibrated chondrites (e.g. <0.01 wt %). The increased P content has been attributed to local decreases in fO_2, possibly due to the presence of C or CO (Taylor and Heymann 1971) which act as reducing agents. Reduction of phosphates to P occurs which diffuses into the metal. This process would be facilitated by partial melting of the metal and the final amount of P in the metal is a function of the volume ratio of phosphate to metal and the maximum temperature attained after the shock event.

Several studies have shown that the Ni contents of individual metal grains (compositions measured at the center of grains) in reheated chondrites are often similar (Heymann 1967, Taylor and Heymann 1971, Begemann and Wlotzka 1969, Smith and Goldstein 1977, Bennett and McSween 1996). During shock reheating, melting, mixing and redistribution of metal occurs which results in a compositional equilibration of metal compared with metal in equilibrated ordinary chondrites which have not been reheated. Smith and Goldstein (1977) found that the degree of equilibration (i.e. compositional similarity of grain centers) increases as the degree of remelting increases. Solid state diffusion appears to play only a minor role in the equilibration process.

Cooling rates of shock reheated metal. An increase in Ni content is commonly observed at the edges of metal grains in shock reheated chondrites (Begemann and Wlotzka 1969, Taylor and Heymann 1971), which is due to Ni segregation during solidification of an Fe-Ni-S melt (e.g. Smith and Goldstein 1977). The steepness of the Ni gradient increases as a function of cooling rate and steep Ni gradients are indicative of relatively fast cooling (e.g. >15°C/day from temperatures >1050°C).

In some shock reheated chondrites, secondary kamacite (α) forms from taenite (γ) during cooling from maximum shock temperatures (Buseck et al. 1966, Wood 1967, Taylor and Heymann 1971, Smith and Goldstein 1977). The presence of secondary kamacite indicates that cooling rates were low enough to allow kamacite to nucleate and grow from taenite during cooling. Based on diffusion modeling of Ni concentration profiles, Smith and Goldstein (1977) estimated cooling rates for metal in several chondrites of between ~1°C/yr and 1°C/100 yr through the temperature interval ~700 to 500°C (see also Scott and Rajan 1979).

Shock melting of metal. Shock melting of metal and troilite is observed in ordinary chondrites that have been shocked to shock stage S4 and higher. Melting must have occurred at temperatures >900°C (e.g. Wood 1967, Taylor and Heymann 1971, Smith and Goldstein 1977) and the textural characteristics of the melted metal and sulfide change as the maximum shock temperature increases. In L chondrites of shock stage S4, metal melt droplets are very rare in comparison with troilite droplets and are also much smaller (<2 μm). Fine-grained mixtures of troilite and metal, termed 'fizzed troilite' occur at shock stages S4 and above. This texture was thought by Wood (1967) to be the result of eutectic melting, but Scott (1982) has argued that it is the result of rapid cooling (see troilite). In chondrites of shock stage S6, metal and sulfide melt droplets occur adjacent to most opaque grains and the droplets can reach 50 μm in size. Extensive melting results in the formation of ovoid metal grains (1-200 μm), agglomerated within troilite grains (e.g. Begemann and Wlotzka 1969, Taylor and Heymann 1971, Smith and Goldstein 1977, Bennett and McSween 1996). Several studies have shown that the globules are compositionally zoned with clear taenite rims and martensitic interiors (Begemann and Wlotzka 1969, Taylor and Heymann 1971). For example, in Orvino, Taylor and Heymann (1971) reported average Ni contents in the globule centers of 8.2 wt % and 12.3 wt % at their edges. P contents are also high and show zoning with a mean of 0.62 wt % in the cores and 0.51 wt % at the edges. In some examples, metal and troilite interfaces are intermixed in a eutectic structure (e.g. Taylor and Heymann 1971, Smith and Goldstein 1977). Injection of metal and troilite into fractures in silicates is also common and tiny blebs, as well as veins, are typically present. The metal globules appear to have crystallized from an Ni-Fe-S melt at temperatures between 1200 and 1350°C (Begemann and Wlotzka 1969).

In relatively highly shocked chondrites, metal-troilite spherules also form in addition to metal globules. For example, in several ordinary chondrites, Scott (1982) found a number of large (0.2-4 mm), troilite-metal globules with a dendritic or cellular texture. The metal dendrites consist of primary trunks with secondary branches and the spacings between these secondary branches are indicative of cooling rate (Flemings et al. 1970, Blau and Goldstein 1975). Scott (1982) estimated cooling rates of 10^{-7} to 10^{4}°C/sec over the temperature interval 1400-950°C for the globules. Chen et al. (1995) studied similar metal-troilite globules in shock-induced melt veins and melt pockets in the highly shocked Yanzhuang (H6, S6) chondrite. In globules in the melt pockets, metal dendrites have a crust-core structure with martensitic interiors (7.5-8.1 wt % Ni) and Ni-rich rims (12.5-23.3 wt %). Chen et al. (1995) showed that in melt veins the dendrites have three distinct microstructural areas consisting of a core (6.4-7.3 wt % Ni), a martensite zone between the core and rim (7.4-8.5 wt % Ni) and a Ni-rich rim (12.8-21.4 wt % Ni). These microstructures are produced by nonequilibrium crystallization under different cooling rates within the pockets (100-400°C/sec) and veins (6-30°C/sec), over the temperature interval 950-1400°C. Chen et al. (1995) also reported high concentrations of P in the dendrites (0.3-0.65 wt %), as a result of fast cooling.

In metal and troilite-rich melt pockets in Tenham, Joreau et al. (1996) found aggregates of very small grains (1 μm) of kamacite and taenite associated with troilite. The taenite grains are the smallest and often occur embedded within a polycrystalline troilite matrix. Dislocation densities are high in both kamacite and taenite.

Troilite. A large number of shock effects have been described in troilite in ordinary chondrites. Even at relatively low shock degrees (S2), shock melting of sulfides can occur and submicron to μm sized troilite melt droplets can be present (e.g. Stöffler et al. 1991, Bennett and McSween 1996). At higher shock stages (above S3), melting of troilite occurs along grain boundaries and unmelted troilite grain edges, adjacent to the melt, develop a bubbly or swiss cheese texture (Bennett and McSween 1996). Shock melting of troilite can also produce what has been termed a fine-grained 'fizz' of irregularly shaped metal grains set within a matrix of finely crystalline troilite (e.g. Scott 1982). The metal grains have grain sizes of 1-5 μm. Fizzed troilite has been reported in a number of ordinary chondrites (e.g. Wood 1967, Taylor and Heymann 1971, Smith and Goldstein 1977, Scott 1982, Bennett and McSween 1996). Scott (1982) has argued that 'fizzed' troilite forms by rapid solidification at cooling rates $>10^5$ °C/sec.

Shock experiments show that below 10 GPa troilite is monocrystalline, but becomes twinned between 10 and 20 GPa and polycrystalline at shock pressures between 35-60 GPa. Polycrystalline troilite is common in L chondrites which have been shocked to shock stage S4 (Bennett and McSween 1996). Silicate-troilite melt pockets become more abundant and incipient melting of large troilite grains along grain boundaries is apparent. At shock stage S5, troilite becomes strongly sheared and polycrystalline and troilite and Fe,Ni melt droplets occur adjacent to ~10% of opaque grains.

The microstructures of shock effects in troilite have not been widely studied with the exception of the TEM study of Joreau et al. (1996). Observations on troilite grains in Gaines County (H5) show that they contain high densities of dislocations, mosaicism and planar lamellae. The mosaicism is due to the presence of small subgrains or domains that are misoriented relative to one another. Joreau et al. (1996) observed two sets of extremely thin (≤10 nm), planar lamellae on the $\{11\bar{2}4\}$ planes which divide the subgrains into smaller, lozenge-shaped domains. The lamellae are amorphous and apparently have the same composition as the host troilite and can be considered to be PDFs produced by shock amorphization of troilite.

Smith and Goldstein (1977) found that the Ni contents of troilite in shock reheated chondrites was higher (<0.02 to 0.31 wt %) than in unreheated examples (<0.02 wt %). This observation was attributed to the fact that Ni solubility in troilite decreases as a function of temperature (Kullerud 1963). The high Ni contents in troilite are due to reequilibration of Ni with metal at elevated temperatures, followed by rapid cooling, which quenches in the high temperature composition. Smith and Goldstein (1977) estimated that cooling rates for troilite in shock reheated chondrites were slower than 25°C per day.

Shock effects in enstatite chondrites

Shock metamorphism in the enstatite chondrites has not been as extensively studied as shock effects in ordinary chondrites. However, Rubin et al. (1997) carried out a comprehensive study of shock effects in a large number of enstatite chondrites based on observations made by optical microscopy. Using shock effects in enstatite, Rubin et al. (1997) have proposed a shock classification for the enstatite chondrites which is comparable to the scheme of Stöffler et al. (1991) for the ordinary chondrites. Calibrations of shock using olivine as an indicator cannot be used for the enstatite chondrites because of its low abundance in EH3 and EL3 chondrites (~0.2-5 vol %) and absence in type 5 and 6 enstatite chondrites. However, shock effects in olivine, where it is present, can be used as further indicators of the shock stage (Rubin et al. 1997). No high pressure phases, such as majorite, which occur in ordinary chondrites, have been observed in the enstatite chondrites, but rare plagioclase is transformed to maskelynite at shock stage S5. In addition, several enstatite chondrites have been shocked so extensively that they contain impact melt clasts (e.g. Atlanta, Blithfield, Eagle, Hvittis, Khairpur; Rubin et al. 1997) or are impact melt rocks (e.g. Abee, Adhi Kot; Rubin and Scott 1997).

Enstatite in the enstatite chondrites displays the following sequence of shock features as a function of shock stage; shock stage S1 (unshocked), sharp optical extinction; S2 (very weakly shocked), undulose extinction; S3 (weakly shocked), development of clinoenstatite lamellae parallel to (100); S4 (moderately shocked), weak mosaicism; S5 (strongly shocked), strong mosaicism (Rubin et al. 1997). Based on these criteria, EL3 chondrites show a range of shock stages from S2 to S5, but EL5 and EL6 chondrites are only very weakly shocked (S2).

Shock melting and mobilization of metallic Fe-Ni and sulfides can produce a variety of effects in enstatite chondrites, including silicate darkening, the formation of opaque veins and rapidly solidified metal-sulfide mixtures. Silicate darkening is caused by the dispersion of submicron to micron-sized grains of metal and sulfide along grain boundaries and fractures in silicate minerals. Silicate darkening is observed in some EH and EL chondrites (Rubin et al. 1997), but is not well correlated with shock stage.

In many enstatite chondrites, opaque veins, consisting of kamacite and/or troilite, produced by shock melting, are present and in EH chondrites rare, thin veinlets of niningerite can be present. The veins occur in EH chondrites of shock stages S3 to S4 and EL chondrites S2 to S5. Rapidly solidified metal-sulfide mixtures consisting of fine-grained intergrowths of rounded to irregular shaped metal grains, surrounded by sulfide, occur in many enstatite chondrites. The textures are often cellular and can be mixtures of metal with troilite, niningerite, schreibersite, although metal/troilite is most common. These textures occur in S2 to S5 enstatite chondrites and also correlate poorly with the degree of shock.

Clasts or large opaque veins which probably have an origin by impact melting have been observed by Rubin et al. (1997) in a number of enstatite chondrites (Atlanta, Blithfield, Eagle, Hvittis and Khairpur). The clasts show a variety of textures: some are troilite-rich and others are distinguished by the absence of chondrules and/or a highly recrystallized appearance. Relict mineral grains indicative of incomplete melting may also occur (Rubin et al. 1997).

Several enstatite chondrites which have been interpreted as impact melt breccias have also been described. These include Abee, Happy Canyon and Adhi Kot (McCoy et al. 1995, Rubin and Keil 1983, Rubin and Scott 1997). In Abee, chondrules are almost 90% melted and the rock consists of angular clasts set in a matrix of crystallized impact melt. Graphite occurs as euhedral to subhedral laths (0.3-6 × 8-75 μm) throughout Abee and appears to have crystallized from the impact melt (Rubin 1997d, Rubin and Scott 1997). An example of a complete enstatite impact melt appears to be the unusual meteorite, Ilafegh 009, which probably formed by impact melting of EL chondrite material (McCoy et al. 1995).

Russell et al. (1992) reported a unique type of diamond isolated from acid residues of Abee. The diamonds are much coarser-grained (100 nm to 1 μm) than the typical nanometer-sized interstellar diamonds and have both platy and lath-shaped morphologies. The diamonds were originally interpreted by Russell et al. (1992) as having formed within the solar nebula, but Rubin and Scott (1997) have argued that because Abee is an impact melt breccia, the diamonds are produced by shock on the parent body. This appears to be supported by the presence of {111} twins, which occur in shocked synthetic diamond (Lee, pers. comm. 1995, reported in Rubin and Scott 1997).

Shock effects in carbonaceous chondrites

Scott et al. (1992) have examined shock effects in the different carbonaceous chondrite groups by optical microscopy and have found that, for the most part, these chondrites are less heavily shocked than the ordinary and enstatite chondrites. No high pressure phases, indicative of shock stages S5 or S6, have been reported in carbonaceous chondrites. Most CM2 and CO3 carbonaceous chondrites are unshocked, although a few contain olivine grains which indicate shock stages of S2 or S3, i.e. undulatory extinction in olivine grains. No shock veins or melt pockets were found in CM2 carbonaceous chondrites, but in the CO3 chondrite Lancé, micron-sized intergrowths of troilite and Fe,Ni occur at the interface between these two phases ('fizzed' troilite). This texture is indicative of local melting and rapid recrystallization (Scott 1982). Troilite grains are also polycrystalline or mosaicized. However, olivine grains indicate very low levels of shock (S1) in this chondrite.

Most CV3 chondrites are also unshocked, but several show shock effects in olivine consistent with shock stages S2-S4 (Scott et al. 1992). Ashworth and Barber (1975) reported TEM observations of olivine in Allende matrix and found that a few dislocations were present. They concluded that Allende was essentially unshocked. Efremovka is the most highly shocked of the CV chondrites studied by Scott et al. (1992) and chondrules in this meteorite are flattened and have a well defined alignment and elongation direction which has been attributed to shock. Efremovka also contains opaque, matrix-like regions which contain rounded metal-troilite particles which have fine-grained igneous textures and probably result from localized shock melting (Scott et al. 1992). These areas are associated with troilite-filled fractures in silicate grains and form veins 10-15 μm wide that can penetrate into chondrules. Leoville also shows extensive chondrule flattening and a well developed alignment is present. Shock microstructures in Leoville have been investigated by Nakamura et al. (1992). Relatively coarse-grained (0.1-2 μm) matrix olivines have high densities of microcracks and dislocations with Burgers vector b = [001], with dislocation densities of 3×10^9 cm^{-2} to 1×1^{10} cm^{-2}. Very fine-grained olivines (10-100 nm) also contain [001] screw dislocations, but the dislocation densities are typically lower. Enstatite in the matrix contains unit cell-scale intergrowths, which Nakamura et al. (1992) interpreted as being consistent with shock transformation.

Most CR chondrites contain olivines indicative of shock stages S1 or S2, but Renazzo and MAC87320 contain olivines indicative of shock stages S1, S2 and S3. These two meteorites may be breccias, such that the olivines were shocked before final assembly of the meteorite (Scott et al. 1992). Like the CR chondrites, the metamorphosed CK chondrites are all shock stage S1-S3, with the exception of EET 83311, which is strongly shocked. In two CK chondrites of shock stage S3 (EET 87860 and LEW 87009), olivines contain planar fractures and plagioclase is deformed but still anisotropic. EET 83311 is the most intensely shocked of all the carbonaceous chondrites studied by Scott et al. (1992). In situ melting has been extensive, locally up to 30%, and irregularly shaped regions of feldspathic melt are commonly present. Opaque melt veins are also present, which appear to have formed after the feldspathic melt.

ACKNOWLEDGMENTS

We are very grateful to Klaus Keil, Alan Rubin, Glenn MacPherson and Jeff Grossman for helpful reviews of this manuscript. We also thank Raymond Huskey and Kate Duke for technical assistance. This work was supported by the Institute of Meteoritics, University of New Mexico, and NASA grant NRA-97-282 to J.J. Papike.

APPENDIX

Representative mineral compositions in chondritic meteorites

The tables on the following pages contain electron and ion microprobe analyses of the following minerals: olivine in chondrules (A3.1), low-Ca pyroxene in chondrules (A3.2), Ca-rich pyroxene in chondrules (A3.3), chondrule mesostasis (A3.4), feldspar in type 3 ordinary chondrites (A3.5), oxides in type 3 ordinary chondrites (A3.6), corundum and grossite from CAIs in CV and CM chondrites (A3.7), trace element analyses of grossite from CAIs in CH and CM chondrites (A3.8), hibonite from CAIs in CV and CM chondrites (A3.9), hibonites from CAIs in CV, CM and ordinary chondrites (A3.10), perovskite from CAIs in CV and CM chondrites (A3.11), trace element analyses of perovskite from CAIs in CV and CH chondrites (A3.12), melilite from CAIs in CV, CM and CO chondrites (A3.13), trace element analyses of melilites from CAIs in CV and CH chondrites (A3.14), spinel from from CAIs in CV and CM chondrites (A3.15), fassaite from CAIs in CV chondrites (A3.16), trace element analyses of fassaites from CAIs in CV and CH chondrites (A3.17), anorthite from CAIs in CV chondrites (A3.18), olivine from CAIs in CV and CM chondrites (A3.19), wollastonite and grossular from CAIs in CV chondrites (A3.20), andradite, monticellite, nepheline, sphene and ilmenite from CAIs in CV chondrites (A3.21), diopside, hedenbergite and salite from CAIs in CV and CM chondrites (A3.22), kamacite in enstatite chondrites (A3.23), troilite in enstatite chondrites (A3.24), niningerite in enstatite chondrites (A3.25), alabandite in enstatite chondrites (A3.26), oldhamite in enstatite chondrites (A3.27), daubreelite in enstatite chondrites (A3.28), sphalerite in enstatite chondrites (A3.29), caswellsilverite in enstatite chondrites (A3.30), djerfisherite in enstatite chondrites (A3.31), schreibersite in enstatite chondrites (A3.32), perryite in the Yamato-691 (EH3) chondrite (A3.33), olivine in type 4-6 ordinary chondrites (A3.34), low-Ca pyroxene in type 4-6 ordinary chondrites (A3.35), Ca-rich pyroxene in type 4-6 ordinary chondrites (A3.36), feldspar in type 4-6 ordinary chondrites (A3.37), phosphates in type 4-6 ordinary chondrites (A3.38), and oxides in type 4-6 ordinary chondrites (A3.39).

Table A3.1. Olivine in chondrules (Type 1-3 Chondrites).

Meteorite	Semarkona	Semarkona	Semarkona	ALHA82110	Khohar	Parnallee	ALHA77307	ALHA77307	Isna	Isna	Murray	Bells
Type	LL3.0	LL3.0	LL3.0	H3.6	L3.6	LL3.6	CO3.0	CO3.0	CO3.7	CO3.7	CM2	CM
Ref	Jones and Scott (1989)	Jones (1994)	Jones (1990)	Wasson et al. (1995)	Wasson et al. (1995)	McCoy et al. (1991)	Jones (1992)	Jones (1992)	Jones (1993)	Jones (1993)	Rubin and Wasson (1986)	Brearley (1995)
Chemical composition (wt %)												
SiO_2	41.9	42.0	39.1	38.1	38.1	37.9	42.6	37.4	41.4	36.8	41.8	42.2
TiO_2	0.03	bd	bd	bd	nd	bd	0.05	0.03	0.08	0.02	bd	0.02
Al_2O_3	0.19	bd	bd	bd	nd	bd	0.12	0.07	0.25	0.08	0.05	0.01
Cr_2O_3	0.19	0.34	0.41	nd	nd	0.03	0.33	0.42	0.08	0.07	0.39	0.63
FeO	0.7	2.9	16.8	23.3	22.1	22.8	0.5	26.5	3.6	32.7	0.8	4.4
MnO	0.05	0.12	0.34	0.50	0.40	0.44	bd	0.25	0.06	0.33	0.04	0.42
MgO	55.6	54.6	42.9	38.2	38.8	39.2	55.8	34.8	53.8	30.3	55.7	52.9
CaO	0.40	0.09	0.15	bd	nd	0.06	0.40	0.25	0.43	0.07	0.21	0.19
NiO	nd	nd	nd	nd	nd	nd	nd	nd	nd	nd	nd	bd
Total	99.07	100.00	99.76	100.10	99.40	100.43	99.81	99.67	99.70	100.30	99.00	100.74
Cation formula based on 4 oxygens												
Si	0.996	0.998	0.995	0.996	0.997	0.986	1.003	0.998	0.990	1.003	0.995	1.003
Ti	0.001	0.000	0.000	0.000	0.000	0.000	0.001	0.001	0.001	0.000	0.000	0.000
Al	0.005	0.000	0.000	0.000	0.000	0.000	0.003	0.002	0.007	0.003	0.001	0.000
Cr	0.004	0.006	0.008	0.000	0.000	0.001	0.006	0.009	0.002	0.002	0.007	0.012
Fe	0.014	0.058	0.358	0.509	0.484	0.496	0.009	0.591	0.072	0.745	0.016	0.088
Mn	0.001	0.002	0.007	0.011	0.009	0.010	0.000	0.006	0.001	0.008	0.001	0.009
Mg	1.969	1.933	1.628	1.488	1.513	1.520	1.958	1.382	1.920	1.232	1.975	1.875
Ca	0.010	0.002	0.004	0.000	0.000	0.002	0.010	0.007	0.011	0.002	0.005	0.005
Ni	0.000	0.000	0.000	0.000	0.000	0.000	0.000	0.000	0.000	0.000	0.000	0.000
Total	2.999	2.999	3.000	3.004	3.003	3.014	2.991	2.996	3.004	2.995	3.001	2.991
Cation ratio 100 x Fe/(Fe+Mg)												
Fa	0.7	2.9	18.0	25.5	24.2	24.6	0.5	29.9	3.6	37.7	0.8	4.5

Table A3.1 (cont.). Olivine in chondrules (Type 1-3 Chondrites).

Meteorite	Allende	Allende[1]	Allende	Orgueil	Orgueil	PCA91082	PCA91082	ALH85085	ALH85085	Qingzhen	ALHA81189	ALH85151	Rumuruti
Type	CV3	CV3	CV3	CI1	CI1	CR2	CR2	CH	CH	EH3	EH3	R	R
Ref	Rubin and Wasson (1987)	Weinbruch et al. (1994)	Weinbruch et al. (1990)	Leshin et al. (1997)	Leshin et al. (1997)	Noguchi (1995)	Noguchi (1995)	Scott (1988)	Scott (1988)	Grossman et al. (1985)	Lusby et al. (1987)	Weisberg et al. (1991)	Schulze et al. (1994)
						Chemical composition (wt %)							
SiO_2	42.3	35.0	36.0	43.6	39.8	42.3	35.7	41.8	38.4	42.4	40.9	42.1	36.6
TiO_2	nd	0.01	0.01	nd	nd	bd	bd	bd	bd	bd	0.02	0.07	bd
Al_2O_3	0.13	0.06	0.02	0.12	bd	bd	bd	bd	bd	bd	0.02	bd	bd
Cr_2O_3	0.07	0.39	0.07	0.42	0.32	0.57	0.28	0.64	0.60	0.20	0.66	0.09	0.11
FeO	2.5	36.8	36.3	0.5	23.0	1.7	35.7	2.8	21.0	0.8	9.6	2.3	33.4
MnO	bd	0.25	0.27	0.08	0.40	bd	0.96	0.12	0.31	0.08	0.60	0.04	0.43
MgO	54.9	27.3	27.5	55.9	37.6	55.1	27.1	54.5	40.0	56.3	47.8	54.3	29.4
CaO	0.38	0.11	0.48	0.25	0.29	0.22	0.26	0.22	0.16	0.21	0.16	0.44	0.06
NiO	nd	0.03	0.04	nd	nd	nd	nd	nd	nd	nd	bd	nd	0.24
Total	100.28	99.96	100.68	100.87	101.41	99.80	99.91	100.08	100.47	100.00	99.76	99.34	100.24
						Cation formula based on 4 oxygens							
Si	1.000	0.984	0.998	1.014	1.021	1.001	0.999	0.994	0.990	0.998	1.006	1.004	1.006
Ti	0.000	0.000	0.000	0.000	0.000	0.000	0.000	0.000	0.000	0.000	0.000	0.001	0.000
Al	0.004	0.002	0.001	0.003	0.000	0.000	0.000	0.000	0.000	0.000	0.001	0.000	0.000
Cr	0.001	0.009	0.002	0.008	0.006	0.011	0.006	0.012	0.012	0.004	0.013	0.002	0.002
Fe	0.049	0.866	0.843	0.010	0.494	0.033	0.834	0.056	0.453	0.016	0.198	0.046	0.768
Mn	0.000	0.006	0.006	0.002	0.009	0.000	0.023	0.002	0.007	0.002	0.013	0.001	0.010
Mg	1.934	1.141	1.138	1.938	1.438	1.944	1.129	1.931	1.537	1.975	1.753	1.929	1.204
Ca	0.010	0.003	0.014	0.006	0.008	0.006	0.008	0.006	0.004	0.005	0.004	0.011	0.002
Ni	0.000	0.001	0.001	0.000	0.000	0.000	0.000	0.000	0.000	0.000	0.000	0.000	0.005
Total	2.998	3.011	3.001	2.980	2.976	2.994	2.998	3.000	3.004	3.000	2.987	2.994	2.993
						Cation ratio 100 x Fe/(Fe+Mg)							
Fa	2.5	43.1	42.6	0.5	25.6	1.7	42.5	2.8	22.8	0.8	10.1	2.3	38.9

[1] Fayalite-rich rim on forsteritic olivine

Table A3.2. Low-Ca pyroxene in chondrules (Type 1-3 Chondrites).

Meteorite	Semarkona	Semarkona	Semarkona	Semarkona	Semarkona	Khohar	Parnallee	Parnallee	Parnallee	ALH77003	ALH77003	ALH77003	Y-74662
Type	LL3.0	LL3.0	LL3.0	LL3.0	LL3.0	L3.6	LL3.6	LL3.6	LL3.6	CO3.5	CO3.5	CO3.5	CM2
Polymorph[1]	cen	cen	cen	opx	pig	cen	cen	opx	pig	cen	opx	pig	low-Ca
Ref	Jones and Scott (1989)	Jones (1994)	Jones (1996)	Jones (1996)	Jones (1996)	Wasson et al. (1995)	Jones (1996)	Jones (1996)	Jones (1996)	Ikeda (1982)	Ikeda (1982)	Ikeda (1982)	Noguchi (1989)
	Chemical composition (wt %)												
SiO_2	58.1	58.7	54.1	55.5	55.5	57.1	54.4	52.7	51.7	57.9	54.9	55.1	59.0
TiO_2	0.26	0.04	0.03	0.07	0.08	nd	bd	bd	0.11	nd	nd	0.55	0.19
Al_2O_3	1.07	0.32	0.20	0.56	0.47	0.40	0.10	0.24	0.78	0.91	2.77	3.54	0.85
Cr_2O_3	0.63	0.61	0.87	0.85	1.50	nd	0.47	0.63	0.90	0.53	1.55	1.67	0.50
FeO	0.76	2.30	15.10	10.90	10.60	8.20	16.90	21.20	23.80	1.13	4.28	2.40	0.53
MnO	0.21	0.27	0.52	0.18	1.31	0.50	0.47	0.68	0.86	0.00	0.42	0.45	0.06
MgO	37.9	37.7	27.2	30.7	27.4	33.0	27.0	23.2	17.5	39.0	33.1	32.5	39.2
CaO	0.41	0.21	0.74	0.63	3.80	0.30	0.20	0.54	3.80	0.49	1.89	3.94	0.63
Na_2O	bd	bd	bd	bd	0.13	0.10	bd	bd	0.21	0.01	0.00	0.13	0.00
K_2O	nd	nd	nd	nd	nd	nd	nd	nd	nd	0.05	0.00	0.04	nd
Total	99.34	100.20	98.80	99.40	100.80	99.60	99.50	99.20	99.70	100.01	98.90	100.32	100.98
	Cation formula based on 6 oxygens												
Si	1.970	1.986	1.979	1.973	1.973	1.994	1.984	1.976	1.976	1.956	1.917	1.894	1.967
Ti	0.007	0.001	0.001	0.002	0.002	0.000	0.000	0.000	0.003	0.000	0.000	0.014	0.005
Al	0.043	0.013	0.009	0.023	0.020	0.016	0.004	0.011	0.035	0.036	0.114	0.144	0.033
Cr	0.017	0.016	0.025	0.024	0.042	0.000	0.014	0.019	0.027	0.014	0.043	0.045	0.013
Fe	0.022	0.065	0.462	0.324	0.315	0.240	0.515	0.665	0.761	0.032	0.125	0.069	0.015
Mn	0.006	0.008	0.016	0.005	0.039	0.015	0.015	0.022	0.028	0.000	0.012	0.013	0.002
Mg	1.915	1.901	1.483	1.626	1.452	1.718	1.468	1.296	0.997	1.961	1.723	1.668	1.948
Ca	0.015	0.008	0.029	0.024	0.145	0.011	0.008	0.022	0.156	0.018	0.071	0.145	0.023
Na	0.000	0.000	0.000	0.000	0.009	0.007	0.000	0.000	0.016	0.001	0.000	0.009	0.000
K	0.000	0.000	0.000	0.000	0.000	0.000	0.000	0.000	0.000	0.002	0.000	0.002	0.000
Total	3.994	3.998	4.003	4.002	3.998	4.001	4.007	4.010	3.998	4.020	4.005	4.003	4.005
	Cation ratios Ca:Mg:Fe												
Fs	1.1	3.3	23.4	16.4	16.5	12.2	25.9	33.5	39.8	1.6	6.5	3.7	0.7
En	98.1	96.3	75.1	82.4	75.9	87.3	73.7	65.4	52.1	97.5	89.8	88.6	98.1
Wo	0.8	0.4	1.5	1.2	7.6	0.6	0.4	1.1	8.1	0.9	3.7	7.7	1.1

[1] cen = clinoenstatite, opx = orthoenstatite, pig = pigeonite

Table A3.2 (cont.). Low-Ca pyroxene in chondrules (Type 1-3 Chondrites).

Meteorite	Y-74662	Allende	Allende	Allende	Orgueil	Renazzo	ALH85085	Qingzhen	Y-691	Y-691	Y-691	Rumuruti	Rumuruti
Type	CM2	CV3	CV3	CV3	CI1	CR2	CH	EH3	EH3	EH3	EH3	R	R
Polymorph[1]	low-Ca	low-Ca	low-Ca	low-Ca (opx?)	opx	cen	low-Ca	low-Ca	low-Ca (cen?)	low-Ca (opx?)	pig	low-Ca	low-Ca
Ref	Noguchi (1989)	Noguchi (1989)	Rubin and Wasson (1987)	Noguchi (1989)	Leshin et al. (1997)	Weisberg et al. (1993)	Scott (1988)	Grossman et al. (1985)	Ikeda (1989c)	Ikeda (1989c)	Ikeda (1988b)	Schulze et al. (1994)	Schulze et al. (1994)
							Chemical composition (wt %)						
SiO_2	57.8	58.2	55.0	58.8	58.6	58.1	59.2	59.5	55.3	50.2	53.0	54.0	58.6
TiO_2	0.21	0.22	nd	0.23	nd	0.08	bd	0.04	0	0.47	1	bd	bd
Al_2O_3	1.44	1.21	0.69	0.81	0.44	0.68	0.40	0.27	0.30	11.46	3.15	0.12	0.17
Cr_2O_3	0.55	0.54	1.40	0.75	0.61	0.61	0.64	0.19	1.64	0.92	1.00	0.22	0.49
FeO	0.52	0.57	11.90	0.63	6.40	1.94	1.60	0.36	11.55	2.97	4.57	16.60	4.10
MnO	0.10	0.00	0.73	0.10	0.51	0.06	0.16	0.05	0.29	0.00	0.06	0.56	0.39
MgO	37.4	38.6	28.5	37.2	33.1	37.7	38.5	39.3	30.0	32.2	31.0	26.6	37.0
CaO	1.89	0.54	0.76	2.62	1.10	0.37	0.21	0.27	0.66	2.06	5.64	0.38	0.18
Na_2O	0.00	0.01	nd	0.01	nd	bd	nd	nd	0.00	0.00	0.00	0.05	0.06
K_2O	nd	nd	nd	nd	nd	nd	nd	nd	0.02	0.00	0.00	bd	bd
Total	99.92	99.94	98.98	101.12	100.76	99.54	100.71	99.98	99.77	100.30	99.40	98.53	100.99
							Cation formula based on 6 oxygens						
Si	1.954	1.960	1.979	1.969	2.009	1.975	1.985	1.995	1.971	1.720	1.866	1.989	1.984
Ti	0.005	0.006	0.000	0.006	0.000	0.002	0.000	0.001	0.000	0.012	0.026	0.000	0.000
Al	0.057	0.048	0.029	0.032	0.018	0.027	0.016	0.011	0.013	0.463	0.131	0.005	0.007
Cr	0.015	0.014	0.040	0.020	0.017	0.016	0.017	0.005	0.046	0.025	0.028	0.006	0.013
Fe	0.015	0.016	0.358	0.018	0.000	0.055	0.045	0.010	0.344	0.085	0.135	0.511	0.116
Mn	0.003	0.000	0.022	0.003	0.184	0.002	0.005	0.001	0.009	0.000	0.002	0.017	0.011
Mg	1.887	1.939	1.528	1.858	0.015	1.910	1.924	1.964	1.592	1.644	1.627	1.460	1.867
Ca	0.068	0.019	0.029	0.094	1.691	0.013	0.008	0.010	0.025	0.076	0.213	0.015	0.007
Na	0.000	0.001	0.000	0.001	0.040	0.000	0.000	0.000	0.000	0.000	0.000	0.004	0.004
K	0.000	0.000	0.000	0.000	0.000	0.000	0.000	0.000	0.001	0.000	0.000	0.000	0.000
Total	4.004	4.003	3.986	4.000	3.974	4.001	3.999	3.996	4.000	4.024	4.028	4.007	4.008
							Cation ratios Ca:Mg:Fe						
Fs	0.7	0.8	18.7	0.9	9.6	2.8	2.3	0.5	17.5	4.7	6.8	25.7	5.8
En	95.8	98.2	79.8	94.3	88.3	96.5	97.3	99.0	81.2	91.1	82.4	73.5	93.8
Wo	3.5	1.0	1.5	4.8	2.1	0.7	0.4	0.5	1.3	4.2	10.8	0.8	0.3

[1] cen = clinoenstatite, opx = orthoenstatite, pig = pigeonite

Table A3.3. Ca-rich pyroxene in chondrules (Type 1-3 Chondrites).

meteorite	Semarkona	Semarkona	Chainpur	Chainpur	ALHA77307	ALH77003	Y-74662 [1]	Murchison	Allende	Allende
type	LL3.0	LL3.0	LL3.4	LL3.4	CO3.0	CO3.5	CM2	CM2	CV3	CV3
ref	Jones (1994)	Jones (1996)	Noguchi (1989)	Noguchi (1989)	Noguchi (1989)	Noguchi (1989)	Noguchi (1989)	Simon et al. (1994)	Noguchi (1989)	Noguchi (1989)
					Chemical composition (wt %)					
SiO_2	52.3	49.1	48.5	46.2	52.5	51.0	51.7	42.2	51.9	47.0
TiO_2	0.34	0.41	1.17	1.11	0.66	1.75	2.05	1.84	1.46	1.29
Al_2O_3	2.1	4.7	10.7	12.4	5.5	3.2	3.3	18.3	3.8	13.0
Cr_2O_3	2.00	1.20	1.26	0.34	1.11	1.95	1.37	1.21	0.45	1.54
FeO	5.00	21.20	0.45	7.01	0.65	4.47	2.60	0.21	1.02	0.30
MnO	1.01	0.55	0.34	0.18	0.40	0.87	0.41	bd	0.03	0.12
MgO	20.4	13.6	15.9	14.7	21.7	17.9	18.9	11.4	20.5	15.7
CaO	15.1	10.2	21.4	17.3	17.4	18.8	19.7	24.8	20.5	21.5
Na_2O	0.30	bd	0.11	0.00	0.00	0.08	0.02	nd	0.02	0.04
K_2O	nd	nd	nd	nd	nd	nd	nd	nd	nd	nd
Total	98.60	101.00	99.75	99.23	99.98	99.93	100.06	100.06	99.69	100.40
					Cation formula based on 6 oxygens					
Si	1.924	1.865	1.746	1.703	1.865	1.868	1.870	1.537	1.868	1.684
Ti	0.009	0.012	0.032	0.031	0.018	0.048	0.056	0.050	0.040	0.035
Al	0.091	0.210	0.456	0.539	0.229	0.137	0.141	0.786	0.161	0.548
Cr	0.058	0.036	0.036	0.010	0.031	0.056	0.039	0.035	0.013	0.044
Fe	0.154	0.674	0.014	0.216	0.019	0.137	0.079	0.006	0.031	0.009
Mn	0.031	0.018	0.010	0.006	0.012	0.027	0.013	0.000	0.001	0.004
Mg	1.118	0.770	0.852	0.807	1.150	0.975	1.020	0.619	1.101	0.836
Ca	0.595	0.415	0.826	0.681	0.662	0.735	0.766	0.969	0.791	0.825
Na	0.021	0.000	0.008	0.000	0.000	0.006	0.001	0.000	0.000	0.000
K	0.000	0.000	0.000	0.000	0.000	0.000	0.000	0.000	0.000	0.000
Total	4.003	4.000	3.980	3.992	3.987	3.990	3.985	4.003	4.005	3.985
					Cation ratios Ca:Mg:Fe					
Fs	8.2	36.2	0.8	12.7	1.1	7.4	4.2	0.4	1.6	0.5
En	59.9	41.4	50.4	47.4	62.8	52.8	54.7	38.8	57.2	50.0
Wo	31.9	22.3	48.8	39.9	36.2	39.8	41.1	60.8	41.2	49.4

[1] V2O3 = 0.08

Table A3.3. (cont.) Ca-rich pyroxene in chondrules (Type 1-3 Chondrites)

Meteorite	Orgueil	Orgueil	Y-8449	Y-691	Y-691	Y-691	Y-691	Rumuruti	Rumuruti
Type	CI1	CI1	CR2	EH3	EH3	EH3	EH3	R	R
Ref	Leshin et al. (1997)	Kerridge and Mcdougall (1976)	Ichikawa and Ikeda (1995)	Ikeda (1989c)	Ikeda (1989c)	Ikeda (1989c)	Ikeda (1988b)	Schulze et al. (1994)	Schulze et al. (1994)
				Chemical composition (wt %)					
SiO_2	55.6	47.8	51.3	54.3	47.4	45.8	48.3	52.7	52.2
TiO_2	nd	1.47	0.52	0.55	0.91	0.66	1.33	0.06	0.32
Al_2O_3	1.7	7.2	4.6	2.0	10.0	15.0	5.6	0.6	0.7
Cr_2O_3	1.40	nd	2.93	1.46	0.72	0.31	0.42	0.41	0.83
FeO	3.20	8.77	2.71	1.59	2.86	1.46	9.43	6.30	10.60
MnO	0.35	nd	1.10	0.51	0.22	0.14	0.13	0.21	0.28
MgO	18.7	12.6	19.5	22.5	15.3	15.8	15.4	15.2	16.1
CaO	19.6	21.1	17.0	16.7	22.0	20.6	18.5	23.5	17.2
Na_2O	nd	0.27	0.17	0.17	0.00	0.00	0.00	0.61	0.69
K_2O	nd	nd	nd	0.02	0	0.02	0	bd	bd
Total	100.55	99.15	99.87	99.88	99.33	99.89	99.02	99.61	98.94
				Cation formula based on 6 oxygens					
Si	1.989	1.798	1.857	1.938	1.739	1.649	1.814	1.965	1.966
Ti	0.000	0.042	0.014	0.015	0.025	0.018	0.038	0.002	0.009
Al	0.072	0.320	0.197	0.085	0.431	0.639	0.247	0.027	0.032
Cr	0.040	0.000	0.084	0.041	0.021	0.009	0.012	0.012	0.025
Fe	0.000	0.000	0.082	0.047	0.088	0.044	0.296	0.196	0.334
Mn	0.096	0.276	0.034	0.015	0.007	0.004	0.004	0.007	0.009
Mg	0.011	0.000	1.054	1.197	0.838	0.850	0.860	0.845	0.904
Ca	0.997	0.707	0.660	0.639	0.863	0.796	0.746	0.939	0.694
Na	0.751	0.849	0.012	0.012	0.000	0.000	0.000	0.044	0.050
K	0.000	0.000	0.000	0.001	0.000	0.001	0.000	0.000	0.000
Total	3.955	4.010	3.994	3.991	4.010	4.010	4.019	4.036	4.022
				Cation ratios Ca:Mg:Fe					
Fs	5.2	15.1	4.6	2.5	4.9	2.6	15.6	9.9	17.3
En	54.1	38.6	58.7	63.6	46.8	50.3	45.2	42.7	46.8
Wo	40.7	46.3	36.8	33.9	48.3	47.1	39.2	47.4	35.9

Table A3.4. Mesostasis in chondrules (Type 2-3 Chondrites).

Meteorite	Semarkona	Semarkona	Sharps	Sharps	Ornans	ALH77003	Y-75293	Y-790123	Y-86751	Leoville	Qingzhen[1]	Y-691
Type	LL3.0	LL3.0	H3.4	H3.4	CO3.3	CO3.5	CM2	CM2	CV3	CV3	EH3	EH3
Ref	Jones and Scott (1989)	Jones (1990)	Rubin and Pernicka (1989)	Rubin and Pernicka (1989)	Rubin and Wasson (1988)	Ikeda (1982)	Ikeda (1983)	Ikeda (1983)	Murakami and Ikeda (1994)	Kracher et al. (1985)	Grossman et al. (1985)	Ikeda (1989c)
	Chemical composition (wt %)											
SiO_2	49.9	63.8	49.6	77.2	45.1	52.9	58.0	58.7	46.4	56.2	73.3	70.2
TiO_2	0.97	0.44	0.08	0.30	bd	nd	1.03	0.63	0.02	0.45	0.13	0.05
Al_2O_3	24.3	11.0	29.9	8.6	31.6	19.0	18.6	13.6	31.9	27.8	7.4	19.4
Cr_2O_3	0.43	0.32	bd	bd	bd	0.21	nd	nd	0.05	0.03	0.16	0
FeO	0.43	6.72	1.40	6.90	1.60	5.01	0.01	8.84	1.04	1.90	2.10	0.00
MnO	0.06	0.22	0.05	0.18	bd	0.62	0.34	0.00	bd	bd	bd	0.00
MgO	5.4	4.7	1.4	0.4	1.1	4.2	4.0	0.7	1.4	1.7	8.9	0.2
CaO	18.1	7.8	14.1	3.5	14.3	13.4	16.6	8.6	16.3	1.5	0.6	0.1
Na_2O	0.32	1.53	3.30	1.60	5.50	3.83	0.89	7.69	2.70	9.30	4.80	9.63
K_2O	bd	0.66	0.21	1.50	0.22	0.04	0.07	0.37	0.16	1.30	0.14	0.46
P_2O_5	nd	1.86	nd	nd	nd	nd	nd	nd	nd	nd	nd	nd
Total	99.90	99.11	100.04	100.13	99.42	99.17	99.46	99.07	99.96	100.18	97.57	100.11

[1] S = 1.17 wt %, Cl = 0.15 wt %

Table A3.5. Feldspars in chondrules (Type 3 Chondrites).

Meteorite **Type** **Ref**	ALH-76004 LL3.2/3.4 Ikeda 80	ALH77003 CO3.5 Ikeda 82	Allende CV3 Kimura and Ikeda (1995)	Y-691 EH3 Ikeda (1988b)	Y-691 EH3 Ikeda (1988b)	Rumuruti R Schulze et al. (1994)	Acfer 217 R Bischoff et al. (1994)
			Chemical composition (wt %)				
SiO_2	47.6	45.8	45.6	49.6	47.1	65.5	49.5
TiO_2	nd	nd	0.05	0.17	0.09	bd	nd
Al_2O_3	32.5	32.6	32.5	30.9	32.2	21.5	32.0
Cr_2O_3	0.12	0.03	0.00	0.00	0.07	bd	nd
FeO	0.47	0.21	0.88	0.84	0.78	0.91	0.33
MnO	0.15	0.03	0.07	0.00	0.00	bd	nd
MgO	0.20	0.72	0.41	2.34	2.50	0.11	0.05
CaO	16.4	16.5	18.2	13.0	15.8	2.3	14.1
Na_2O	2.03	2.94	1.75	3.86	2.34	9.2	3.5
K_2O	0.08	0.20	0.04	0.20	0.12	0.52	0.04
Total	99.58	99.02	99.42	100.92	100.96	100.01	99.52
			Cation formula based on 8 oxygens				
Si	2.197	2.139	2.131	2.254	2.152	2.885	2.270
Ti	0.000	0.000	0.002	0.006	0.003	0.000	0.000
Al	1.770	1.796	1.787	1.654	1.737	1.116	1.730
Cr	0.004	0.001	0.000	0.000	0.003	0.000	0.000
Fe	0.018	0.008	0.034	0.032	0.030	0.034	0.013
Mn	0.006	0.001	0.003	0.000	0.000	0.000	0.000
Mg	0.014	0.050	0.029	0.158	0.170	0.007	0.003
Ca	0.813	0.827	0.908	0.633	0.772	0.107	0.693
Na	0.182	0.266	0.158	0.340	0.207	0.786	0.311
K	0.005	0.012	0.002	0.012	0.007	0.029	0.002
Total	5.009	5.101	5.054	5.089	5.082	4.964	5.022
			Cation ratios Ca:Na:K				
An	81.4	74.8	85.0	64.3	78.3	11.6	68.8
Ab	18.2	24.1	14.8	34.5	21.0	85.2	30.9
Or	0.5	1.1	0.2	1.2	0.7	3.2	0.2

Table A3.6. Oxides in chondrules (Type 2-3 Chondrites): Chromite.

Meteorite	Bishunpur	Parnallee	Mean CO	Murchison	Allende	Allende	Rumuruti	Rumuruti
Type	LL3.1	LL3.6	CO3	CM2	CV3	CV3	R	R
Ref	Bunch et al. (1967)	Bunch et al. (1967)	McSween (1977a)	Fuchs et al. (1973)	Weinbruch et al. (1994)	Lumpkin (1980)	Schulze et al. (1994)	Schulze et al. (1994)
				Chemical composition (wt %)				
SiO_2	nd	nd	nd	nd	0.34	nd	nd	nd
TiO_2	0.43	1.66	0.90	1.20	0.85	4.00	5.10	4.00
Al_2O_3	0.10	3.32	51.30	14.00	6.87	15.40	5.20	5.40
Cr_2O_3	63.9	57.1	10.7	50.3	54.7	45.7	39.7	52.7
Fe_2O_3					2.9		12.3	1.5
FeO	34.3	35.8	32.3	27.2	31.7	27.3	34.2	32.6
MnO	0.58	0.44	0.40	0.30	0.20	nd	0.59	0.42
MgO	0.5	1.4	3.3	6.7	1.4	6.7	0.8	1.7
CaO	nd	nd	0.10	nd	0.05	bd	nd	nd
ZnO	nd	nd	nd	nd	0.07	nd	0.86	0.63
V_2O_3	0.72	0.45	nd	0.70	0.51	nd	nd	nd
NiO	nd	nd	nd	nd	0.05	nd	0.09	bd
Total	100.56	100.12	99.00	100.40	99.51	99.10	98..88	98.98
				Cation formula based on 32 oxygens				
Si	0.000	0.000	0.000	0.000	0.097	0.000	0.000	0.000
Ti	0.100	0.360	0.158	0.240	0.183	0.800	1.128	0.872
Al	0.030	1.140	14.112	4.400	2.328	4.831	1.803	1.846
Cr	15.050	13.170	1.974	10.600	12.375	9.613	9.228	12.079
Fe^{3+}					0.625		2.722	0.327
Fe^{2+}	8.540	8.930	6.382	6.070	7.587	6.075	8.410	7.905
Mn	0.150	0.100	1.148	0.060	0.048	0.000	0.147	0.103
Mg	0.240	0.620	0.025	2.660	0.597	2.657	0.351	0.735
Ca	0.000	0.000	0.000	0.000	0.015	0.000	0.000	0.000
Zn	0.000	0.000	0.000	0.000	0.015	0.000	0.187	0.135
V	0.170	0.110	0.000	0.140	0.117	0.000	0.000	0.000
Ni	0.000	0.000	0.000	0.000	0.012	0.000	0.021	0.000
Total	24.280	24.430	23.799	24.170	23.999	23.976	23.997	24.002
calculated Fe_2O_3	3.4	5.0						

Table A3.6 (cont.). Oxides in chondrules (Type 2-3 Chondrites): Spinels

Meteorite	ALHA77307	ALH77003	Murchison	Murchison	Y-86751	Y-691	Y-691
Type	CO3.0	CO3.5	CM2	CM2	CV3	EH3	EH3
Ref	Jones (1992)	Ikeda (1982)	Fuchs et al. (1973)	Simon et al. (1994)	Murakami et al. (1994)	Ikeda (1989c)	Ikeda (1988b)
				Chemical composition (wt %)			
SiO_2	nd	0.06	nd	0.10	0.11	nd	nd
TiO_2	0.16	0.34	0.20	0.27	0.33	0.08	0.34
Al_2O_3	70.8	68.2	69.6	50.4	63.3	71.8	48.3
Cr_2O_3	1.4	0.2	0.7	22.2	0.4	0.7	25.7
FeO	0.5	10.1	0.5	1.0	20.3	0.1	0.6
MnO	nd	0.07	nd	0.20	0.11	0.00	0.05
MgO	28.1	21.2	28.3	25.1	14.5	28.0	24.5
CaO	nd	0.00	0.00	0.01	bd	0.00	0.11
ZnO	nd	nd	nd	nd	nd	nd	nd
V_2O_3	nd	nd	nd	0.30	nd	nd	nd
NiO	nd	0.10	nd	nd	nd	nd	nd
Total	100.88	100.23	99.30	99.61	99.02	100.71	99.52
				Cation formula based on 32 oxygens			
Si	0.000	0.012	0.000	0.021	0.023	0.000	0.000
Ti	0.023	0.051	0.030	0.042	0.052	0.011	0.053
Al	15.767	15.916	15.730	12.279	15.730	15.959	11.865
Cr	0.209	0.031	0.120	3.627	0.067	0.104	4.233
Fe^{2+}	0.079	1.672	0.080	0.173	3.579	0.016	0.105
Mn	0.000	0.012	0.000	0.035	0.020	0.000	0.009
Mg	7.911	6.254	8.090	7.731	4.555	7.867	7.608
Ca	0.000	0.000	0.000	0.002	0.000	0.000	0.025
Zn	0.000	0.000	0.000	0.000	0.000	0.000	0.000
V	0.000	0.000	0.000	0.050	0.000	0.000	0.000
Ni	0.000	0.016	0.000	0.000	0.000	0.000	0.000
Total	23.989	23.964	24.050	23.960	24.026	23.957	23.898

Table A3.7. Representative electron microprobe analyses of corundum and grossite from CAIs in CV and CM carbonaceous chondrites.

	Corundum			Grossite				
Meteorite	Allende	Murchison	Murchison	Leoville	Allende	Murchison		
CAI Name	3643	BB-5	GR-1	L2	4691	B6		
CAI Type	UF	cor-hib	cor-hib	Type A	Type A	hib-gross spherule		
	1	2	3	4	5	6	7	8
SiO_2	0.40	0.04	bd	0.14	0.67	0.10	0.09	0.07
TiO_2	nd	0.32	0.19	0.10	0.09	0.17	0.19	0.17
Al_2O_3	97.60	98.53	99.96	78.10	78.85	77.76	77.60	77.25
Cr_2O_3	nd	nd	bd	<0.02	nd	nd	nd	nd
V_2O_3	nd	bd	bd	nd	nd	0.03	bd	bd
FeO	0.20	bd	0.02	0.63	bd	0.01	0.01	0.02
MgO	1.70	<0.02	bd	0.04	0.13	0.03	0.03	0.03
CaO	0.10	0.01	0.02	21.00	21.64	21.90	22.04	21.98
Sc_2O_3	nd	nd	0.07	nd	nd	nd	nd	nd
Total	100.00	98.90	100.26	100.01	99.33	100.00	99.96	99.52
	Cations Formula Based on 3 Oxygen Atoms			Cations Formula Based on 7 Oxygen Atoms				
Si	0.007	-	-	0.006	0.029	0.004	0.004	0.002
Ti	-	-	0.002	0.003	0.003	0.006	0.006	0.006
Al	1.959	1.990	1.996	3.988	3.961	3.972	3.968	3.968
Cr	-	-	-	-	-	-	-	-
V	-	-	-	-	-	0.001	-	-
Fe	0.003	-	-	0.023	-	-	-	-
Mg	0.043	-	-	0.003	0.008	0.002	0.002	0.002
Ca	0.002	-	-	0.975	0.988	1.017	1.025	1.027
Cations	2.014	1.990	1.999	4.997	4.989	5.002	5.005	5.005

Notes - UF = ultrarefractory inclusion; cor-hib = corundum-hibonite inclusion;
hib-gross spherule = hibonite-grossite bearing spherule.
Analysis 1 - Wark (1986); Analysis 2 - Bar-Matthews et al. (1992); Analysis 3 - MacPherson et al. (1984)
Analysis 4 - Christophe Michel-Lévy et al. (1982); Analyses 5 - Paque (1987)
Analyses 6-8 - Simon et al. (1994); bd = below detection; nd = not determined.

Table A3.8. Representative trace element analyses of grossite (concentrations in ppm—ion microprobe analyses) from CAIs in CH and CK chondrites (Weber et al. 1995).

CAI	022/2	022/6	022/9	022/15	418/1	418/9	022/17	022/14
Element								
Hf	<0.46	<0.22	<0.14	<0.48	<1.24	0.67±0.51	<0.18	<0.63
Zr	3.2	8.8	0.21±0.03	2.06	4.5	40	5.7	19
Y	1.39	2.48	1.64	0.99	2.16	12.6	1.19	4.9
Sc	33.6	12.8	224	32.7	136	26.1	10.7	36.7
Ti	12442	10610	1710	7831	19759	24499	27727	91786
Nb	20.8	10.8	0.03±0.01	2.04	32.2	3.7	5.7	113
V	147	192	121	403	357	418	663	1367
Sr	76.3	70	55.7	107	91.9	119	131	406
Ba	5.4	3.99	12.9	4.7	3.5	5.8	11.4	41.2
La	6.6	14.4	1	24.8	22.9	8.1	5.4	5
Ce	19.4	57.6	2.81	58.9	77.6	36.3	31.7	27.8
Pr	2.71	7.1	0.42	8.5	11	3.7	3.1	2.61
Nd	14	32.3	2.2	45.6	53.9	16.3	12	13.9
Sm	5	11	0.74	11.4	20.9	9.7	6.8	9.6
Eu	0.65	0.48	29	0.74	0.89	1.01	1.14	3.3
Gd	1.39±0.24	1.20±0.39	71	3.1	1.10±0.44	2.23±0.36	0.26±0.20	1.98±0.42
Tb	0.17±0.06	0.26±0.07	0.15±0.02	0.44±0.10	0.40±0.08	0.47±0.08	0.06±0.03	0.47±0.10
Dy	1.31	1.43	1.02	2.54	2.45	3.5	0.33±0.07	2.48
Ho	0.16±0.03	0.16±0.02	0.13±0.02	0.17±0.05	0.29±0.05	0.76±0.10	0.04±0.02	0.34±0.05
Er	0.28±0.06	0.49±0.05	0.25	0.34±0.09	0.38±0.08	1.64	0.22±0.05	0.54±0.11
Tm	1.1	0.72	0.28	1.19	2.82	1.54	0.78±0.05	2.12
Yb	2.4	1.67	0.27±0.04	1.14±0.15	3.7	2.49	2.69	13.4
Lu	0.06±0.02	0.05±0.01	0.03±0.01	<0.05	0.04±0.03	0.39±0.08	<0.02	<0.06

CAI	418/8	022/13	022/22	022/19	418/2	418/10	008/D	167/10
Element							CK	CK
Hf	<7.8	<0.36	2.13±0.93	<0.25	<4.1	0.55±0.46	<0.98	<1.43
Zr	186	2.92	220	4.5	36.5	39.7	21.3	41.3
Y	83.6	0.56±0.06	75.6	0.83	47.5	46.4	7.1	18.8
Sc	2698	31.4	202	5	543	267	97.9	70.8
Ti	20429	6814	15972	25439	23102	552	21541	21679
Nb	11.5	2.15	3.1	23.9	9.9	0.08±0.01	7.2	1.83
V	416	1594	884	609	2559	119	343	503
Sr	111	185	176	121	119	59.4	67.9	149
Ba	16.1	68.4	9.2	17.7	11.1	1.85	7.1	21.5
La	6.8	1.52	20.9	1.87	311	4.8	33.8	21.3
Ce	21.3	6	51.5	12.6	184	2.11	63.1	42.8
Pr	2.39±0.27	0.47±0.06	7	1.36	47	1.37	10.8	7
Nd	14.2	2.03	36.8	4.6	467	9.4	59.5	37.4
Sm	2.90±0.58	0.92±0.18	11.3	4.1	24.4	3.1	196	9.9
Eu	1.12±0.14	1.24	1.34	1.04	082±0.13	0.32	0.66±0.07	1.12
Gd	107	0.30±0.08	10.3	0.51±0.13	55.4	10.5	8.0±0.91	11.7
Tb	13.7	0.09±0.03	2.34	0.05±0.02	11.7	2.07	1.93	2.4
Dy	99.2	0.47±0.06	15.5	0.60±0.07	64.9	15.3	10	12.9
Ho	10.7	0.09±0.03	3.5	0.07±0.02	5.2	2.87	0.89	2.28
Er	14.9	0.06±0.03	10.5	0.19±0.05	7.5	7.3	1.14	2.43
Tm	2.65	0.33±0.05	1.54	0.64±0.07	4.5	1.59	2.38	0.99
Yb	<2.09	1.44	2.96	2.85	6.7±0.93	<0.22	3.1	3.26
Lu	<0.81	<0.03	1.34±0.12	<0.03	<0.42	0.39±0.05	0.08±0.04	0.10±0.04

Notes: All analyses are from CAI in the CH chondrite, Acfer 182, except analyses 008/D and 167/10 which are from Acter 059/El-Djouf 001 (CK)
Errors (1 στδ. δεϖ.) are given only of they exceed 10%; < upper limit (2 std. dev.)

Table A3.9. Representative electron microprobe analyses of hibonites from CAI in CV and CM carbonaceous chondrites.

Meteorite		Allende (CV)										Leoville (CV)				
CAI Type	Type A	Fluffy Type A							Fluffy Type A			Fluffy Type A				
CAI Name	TS12F1	TS28F1	TS28F1	TS25F1	TS29F1	TS27F1	TS24F1	TS52F1	CG11			L2				
Source	G (1975)	MacPherson and Grossman (1984)							Allen et al. (1978)			Christophe Michel-Lévy et al. (1982)				
	1	2	3	4	5	6	7	8	9	10	11	12	13	14	15	16
SiO_2	nd	0.36	0.15	1.39	0.26	0.72	0.23	1.15	0.43	0.29	0.35	0.44	0.20	0.37	0.26	0.77
TiO_2	1.59	1.37	5.25	6.09	3.80	4.18	5.82	1.19	6.78	7.19	8.30	1.77	6.00	4.40	4.00	1.99
Al_2O_3	86.58	89.25	83.14	79.84	84.40	81.70	80.29	87.44	79.21	78.91	75.75	88.40	80.00	83.90	86.00	85.50
Cr_2O_3	0.05	0.02	0.03	0.05	bd	nd	nd	nd	<0.07	<0.07	0.07	0.04	0.06	0.02	0.02	0.02
V_2O_3	nd	0.31	0.52	0.94	0.31	0.46	0.77	0.12	0.31	0.37	0.47	nd	nd	nd	nd	nd
FeO	0.60	0.04	0.02	0.08	bd	0.08	0.14	0.05	0.05	0.07	0.28	0.25	0.16	0.61	0.23	1.11
MnO	nd	nd	nd	nd	nd	nd	nd	nd	nd	nd	nd	nd	nd	0.02	0.02	0.02
MgO	0.53	0.86	2.60	2.82	2.17	3.78	3.25	1.17	3.53	3.73	4.23	1.39	3.20	2.23	2.16	1.37
CaO	8.13	8.12	7.36	8.86	8.22	8.55	8.64	8.60	8.82	8.25	8.26	8.40	8.40	8.60	8.40	9.70
Na_2O	nd	nd	nd	nd	nd	nd	nd	nd	nd	nd	nd	nd	nd	0.02	0.02	0.02
K_2O	nd	nd	nd	nd	nd	nd	nd	nd	nd	nd	nd	nd	nd	nd	nd	nd
Sc_2O_3	nd	<0.02	<0.02	nd	nd	0.04	0.05	0.04	0.14	0.12	0.10	nd	nd	nd	nd	nd
Y_2O_3	<0.02	nd	nd	nd	nd	nd	nd	nd	<0.03	<0.03	<0.03	nd	nd	nd	nd	nd
ZrO_2	0.07	nd	nd	nd	nd	nd	nd	nd	bd	bd	bd	nd	nd	nd	nd	nd
Nb_2O_4	<0.04	nd	nd	nd	nd	nd	nd	nd	nd	nd	nd	nd	nd	nd	nd	nd
Total	97.55	100.33	99.07	100.07	99.16	99.51	99.19	99.76	99.27	98.93	97.74	100.69	98.02	100.17	101.11	100.50
Cation Formula Based on 19 Oxygen Atoms																
Si	-	0.040	0.017	0.157	0.030	0.081	0.026	0.128	0.049	0.034	0.040	0.049	0.023	0.042	0.029	0.086
Ti	0.137	0.115	0.448	0.518	0.323	0.356	0.500	0.100	0.583	0.619	0.727	0.148	0.520	0.372	0.334	0.168
Al	1.709	11.695	11.113	10.640	11.259	10.918	10.812	11.533	10.663	10.649	10.397	11.566	10.869	11.131	11.259	11.322
Cr	0.005	0.002	0.003	0.004	-	-	-	-	-	-	-	0.004	0.005	0.002	0.002	0.002
V	0.000	0.028	0.047	0.085	0.028	0.042	0.070	0.011	0.029	0.034	0.044	-	-	-	-	-
Fe	0.058	0.004	0.002	0.008	-	0.007	0.013	0.005	0.004	0.007	0.027	0.023	0.015	0.057	0.021	0.104
Mn	-	-	-	-	-	-	-	-	-	-	-	-	-	0.002	0.002	0.002
Mg	0.091	0.143	0.440	0.475	0.367	0.640	0.553	0.194	0.600	0.637	0.735	0.230	0.550	0.374	0.357	0.229
Ca	0.999	0.968	0.895	1.073	0.997	1.038	1.058	1.032	1.079	1.013	1.031	0.999	1.037	1.037	1.000	1.167
Na	-	-	-	-	-	-	-	-	-	-	-	-	-	0.004	0.004	0.004
K	-	-	-	-	-	-	-	-	-	-	-	-	-	-	-	-
Sc	-	-	-	-	-	0.004	0.005	0.003	0.014	0.012	0.011	-	-	-	-	-
Y	-	-	-	-	-	-	-	-	-	-	-	-	-	-	-	-
Zr	0.004	-	-	-	-	-	-	-	-	-	-	-	-	-	-	-
Cations	13.002	12.995	12.965	12.960	13.004	13.086	13.037	13.006	13.019	13.003	13.010	13.018	13.020	13.022	13.009	13.086

Notes: G = Grossman (1975); bd = below detection; nd= not determined.

Table A3.9. (cont.) Representative electron microprobe analyses of hibonites from CAI in CV and CM carbonaceous chondrites.

Meteorite	Allende (CV)															
CAI Type	3643												HAL	HAL rim		
CAI Name	Ultrarefractory, unique				Spinel perovskite-hibonite inclusion								unique		FGSR* inclusions	
Source	Wark (1986)				Steele (1995)								Allen et al (1980)		M&G (1989)[1]	
	1	range	2	range	3	4	5	6	7	8	9	10	11	12	13	14
SiO_2	0.31	0.04-0.93	0.38	0.18-0.65	0.05	0.06	0.05	0.04	0.15	0.20	0.35	0.13	0.02	0.14	0.26	0.60
TiO_2	2.16	0.06-4.40	5.17	4.1-6.1	1.54	1.22	1.49	1.65	5.20	5.24	8.20	4.62	0.71	1.64	6.91	107.00
Al_2O_3	86.39	82.3-90.5	82.41	80.0-84.8	90.00	90.60	90.20	90.00	83.90	83.80	78.90	84.70	89.95	88.09	79.91	89.40
Cr_2O_3	0.05	0.00-0.12	0.03	0.00-0.06	0.01	0.00	0.00	0.00	0.04	0.04	0.07	0.03	0.02	0.02	0.28	0.04
V_2O_3	0.73	0.11-1.31	0.24	0.18-0.28	nd	nd	nd	nd	nd	nd	nd	nd	0.02	0.02	0.56	0.13
FeO	0.05	0.00-0.12	0.03	0.00-0.11	0.10	0.10	0.11	0.07	0.29	0.42	0.70	0.26	0.32	0.80	0.63	0.00
MnO	nd	nd	nd	nd	nd	nd	nd	nd	nd	nd	nd	nd	n.a	n.a	0.00	0.00
MgO	1.55	0.43-2.83	3.16	2.6-3.8	0.74	0.57	0.69	0.77	2.73	3.52	4.39	2.40	0.01	0.40	3.74	1.00
CaO	8.44	7.37-9.08	8.93	8.4-9.4	8.23	8.30	8.36	8.31	8.12	8.01	8.03	8.30	8.71	7.92	8.45	8.53
Na_2O	nd		nd		nd	nd	nd	nd	nd	nd	nd	nd	nd	nd	0.02	1.00
K_2O	nd		nd		nd	nd	nd	nd	nd	nd	nd	nd	nd	nd	bd	bd
Sc_2O_3	nd		nd		nd	nd	nd	nd	nd	nd	nd	nd	0.05	0.33	nd	nd
Y_2O_3	nd		nd		nd	nd	nd	nd	nd	nd	nd	nd	0.03	0.03	nd	nd
ZrO_2	nd		nd		nd	nd	nd	nd	nd	nd	nd	nd	0.02	0.08	nd	nd
Total	99.68		100.35		100.67	100.85	100.90	100.84	100.43	101.23	100.64	100.44	99.74	99.40	100.76	100.78
	Cation Formula Based on 19 Oxygen Atoms															
Si	0.035		0.043		0.006	0.007	0.006	0.004	0.017	0.022	0.039	0.015	-	0.016	0.030	0.067
Ti	0.183		0.437		0.128	0.101	0.124	0.137	0.438	0.439	0.696	0.389	0.060	0.139	0.587	0.089
Al	11.452		10.929		11.749	11.799	11.752	11.733	11.084	10.996	10.496	11.179	11.866	11.691	10.629	11.660
Cr	0.004		0.003		0.001	-	-	-	0.004	0.004	0.006	0.003	-	-	0.025	0.004
V	0.022		0.022		-	-	-	-	-	-	-	-	-	-	0.051	0.012
Fe	0.005		0.003		0.009	0.009	0.010	0.006	0.027	0.039	0.066	0.024	0.030	0.076	0.060	-
Mn	0.000		0.000		-	-	-	-	-	-	-	-	1.045	0.956	-	-
Mg	0.260		0.530		0.122	0.094	0.114	0.127	0.456	0.584	0.738	0.400	-	0.067	0.629	0.165
Ca	1.017		1.076		0.976	0.982	0.990	0.985	0.975	0.955	0.971	0.996	-	-	1.022	1.011
Na	-		-		-	-	-	-	-	-	-	-	-	-	0.005	0.002
K	-		-		-	-	-	-	-	-	-	-	-	-	-	-
Sc	-		-		-	-	-	-	-	-	-	-	0.005	-	-	-
Y	-		-		-	-	-	-	-	-	-	-	-	-	-	-
Zr	-		-		-	-	-	-	-	-	-	-	-	0.005	-	-
Cations	13.021		13.012		12.991	12.993	12.995	12.992	13.001	13.039	13.013	13.006	13.006	12.981	13.035	13.009

Notes: 1. McGuire andHashimoto (1989). *FGSR - fine-grained, spinel-rich inclusion. bd = below detection; nd= not determined.

Table A3.9. (cont.) Representative electron microprobe analyses of hibonites from CAI in CV and CM carbonaceous chondrites.

Meteorite	Murchison (CM)															
CAI Type	Cor+hib	sp-hib sp	sp-hib sp	sp-hib sp	sp-hib sp	sp-hib sp	hib-rich	hib-rich	hib	hib	PLAC	PLAC	PLAC	PLAC	PLAC	PLAC
CAI Name	BB-5	BB-1	BB-1	BB-2	BB-4	BB-4	MUCH-1	SH-4	DJ-3	DJ-6						
Source	BM et al*	MacPherson et al. (1983)									Ireland et al. (1988)					
	1	2	3	4	5	6	7	8	9	10	11	12	13	14	15	16
SiO_2	0.01	0.34	0.30	0.42	0.25	0.55	bd	0.33	<0.02	<0.02	1.06	bd	bd	bd	bd	bd
TiO_2	2.01	5.88	7.73	7.36	0.63	6.66	2.05	8.12	2.14	1.92	1.34	1.78	1.73	1.63	1.24	1.59
Al_2O_3	87.63	80.95	78.78	80.04	89.48	79.90	86.46	76.62	89.56	88.68	87.30	89.72	88.44	89.45	89.20	88.76
Cr_2O_3	nd	nd	nd	nd	nd	nd	nd	nd	nd	nd	nd	nd	nd	nd	nd	nd
V_2O_3	0.03	0.16	bd	bd	bd	bd	bd	bd	<0.02	<0.02	nd	nd	nd	nd	nd	nd
FeO	0.03	<0.03	bd	bd	bd	bd	bd	bd	<0.03	<0.03	bd	bd	bd	0.25	bd	0.15
MnO	nd	nd	nd	nd	nd	nd	nd	nd	nd	nd	nd	nd	nd	nd	nd	nd
MgO	0.63	3.12	4.47	4.45	1.13	4.62	1.07	4.41	0.77	0.69	1.06	98.00	1.00	1.20	0.95	1.06
CaO	8.34	7.84	8.52	8.21	8.32	8.20	8.70	8.58	7.90	7.84	8.33	8.54	8.25	8.56	8.48	8.44
Na_2O	nd	nd	nd	nd	nd	nd	nd	nd	nd	nd	nd	nd	nd	nd	nd	nd
K_2O	nd	nd	nd	nd	nd	nd	nd	nd	nd	nd	nd	nd	nd	nd	nd	nd
Sc_2O_3	nd	0.01	bd	bd	bd	bd	bd	bd	0.11	0.07	nd	nd	nd	nd	nd	nd
Y_2O_3	nd	nd	nd	nd	nd	nd	nd	nd	nd	nd	nd	nd	nd	nd	nd	nd
ZrO_2	nd	nd	nd	nd	nd	nd	nd	nd	nd	nd	nd	nd	nd	nd	nd	nd
Total	98.70	98.30	99.80	100.48	99.81	99.93	98.28	98.06	100.51	99.20	99.12	101.02	99.42	101.09	99.87	100.00

Cation Formula Based on 19 Oxygens

	1	2	3	4	5	6	7	8	9	10	11	12	13	14	15	16
Si	-	0.039	0.034	0.047	0.028	0.062	-	0.038	0.001	-	0.119	-	-	-	-	-
Ti	0.170	0.507	0.660	0.623	0.052	0.567	0.176	0.707	0.179	0.162	0.113	0.148	0.146	0.135	0.104	0.133
Al	11.690	10.933	10.544	10.615	11.771	10.648	11.605	10.457	11.710	11.744	11.566	11.679	11.673	11.651	11.736	11.688
Cr	-	-	-	-	-	-	-	-	-	-	-	-	-	-	-	-
V	-	0.015	-	-	-	-	-	-	0.001	-	-	-	-	-	-	-
Fe	0.000	-	-	-	-	-	-	-	0.001	-	-	-	-	0.023	-	0.014
Mn	-	-	-	-	-	-	-	-	-	-	-	-	-	-	-	-
Mg	0.110	0.553	0.757	0.746	0.188	0.778	0.181	0.761	0.128	0.115	0.183	0.162	0.166	0.198	0.158	0.177
Ca	1.010	0.962	1.036	0.990	0.995	0.994	1.062	1.064	0.940	0.944	1.003	1.010	0.990	1.014	1.015	1.010
Na	-	-	-	-	-	-	-	-	-	-	-	-	-	-	-	-
K	-	-	-	-	-	-	-	-	-	-	-	-	-	-	-	-
Sc	-	0.001	-	-	-	-	-	-	0.011	0.007	-	-	-	-	-	-
Y	-	-	-	-	-	-	-	-	-	-	-	-	-	-	-	-
Zr	-	-	-	-	-	-	-	-	-	-	-	-	-	-	-	-
Cations	12.980	12.990	13.031	13.021	13.034	13.049	13.024	13.027	12.961	12.972	12.984	12.999	12.975	13.021	13.013	13.022

Notes: cor+hib = corundum-hibonite inclusion; sp-hib sp = spinel-hibonite spherule; hib-rich - hibonite-rich inclusion; hib- hibonite fragment
PLAC - Platy crystal fragments; *Bar-Matthews et al. (1982)
bd = below detection; nd= not determined. Analyses 3-8 - Energy dispersive analyses.

Table A3.9 (cont.) Representative electron microprobe analyses of hibonites from CAI in CV and CM carbonaceous chondrites.

Meteorite	Murchison (CM)															
CAI Type	SHIB*	SHIB	SHIB	SHIB	SHIB	BAG*	BAG	BAG	NC							
Source	Ireland (1988)									Ireland (1986)						
	1	2	3	4	5	6	7	8	9	10	11	12	13	14	15	16
SiO_2	bd	bd	bd	bd	bd	bd	bd	bd	bd	bd	bd	bd	bd	bd	bd	bd
TiO_2	7.31	4.99	7.57	3.56	6.53	5.14	6.44	6.34	1.86	1.78	1.96	2.62	1.76	5.14	6.44	6.34
Al_2O_3	79.41	84.12	79.73	83.08	81.62	83.28	82.09	82.6	88.49	88.58	87.9	87.66	88.29	83.28	82.09	82.6
Cr_2O_3	nd	nd	nd	nd	nd	nd	nd	nd	nd	nd	nd	nd	nd	nd	nd	nd
V_2O_3	nd	nd	nd	nd	nd	nd	nd	nd	nd	nd	nd	nd	nd	nd	nd	nd
FeO	bd	0.21	bd	bd	0.23	bd	0.04	bd	bd	nd	nd	nd	nd	nd	nd	nd
MnO	nd	nd	nd	nd	nd	nd	nd	nd	nd	0.58	0.96	0.9	0.51	2.63	2.86	2.89
MgO	3.65	2.76	3.48	2.22	3.34	2.63	2.86	2.89	0.85	8.91	8.55	8.68	8.6	8.46	8.51	8.66
CaO	8.64	8.61	8.8	8.4	8.15	8.46	8.51	8.66	8.6	nd	nd	nd	nd	nd	nd	nd
Na_2O	nd	nd	nd	nd	nd	nd	nd	nd	nd	nd	nd	nd	nd	nd	nd	nd
K_2O	nd	nd	nd	nd	nd	nd	nd	nd	nd	nd	nd	nd	nd	nd	nd	nd
Sc_2O_3	nd	nd	nd	nd	nd	nd	nd	nd	nd	nd	nd	nd	nd	nd	nd	nd
Total	99.01	100.69	99.58	97.26	99.87	99.51	99.9	100.49	99.8	99.85	99.37	99.86	99.16	99.51	99.9	100.49
	Cation Formula Based on 19 Oxygen Atoms															
Si	-	-	-	-	-	-	-	-	-	-	-	-	-	-	-	-
Ti	0.628	0.420	0.648	0.308	0.555	0.437	0.547	0.536	0.156	0.150	0.170	0.220	0.150	0.440	0.850	0.540
Al	10.707	11.087	10.694	11.247	10.878	11.105	10.927	10.935	11.676	11.690	11.650	11.580	11.720	11.110	10.930	10.940
Cr	-	-	-	-	-	-	-	-	-	-	-	-	-	-	-	-
V	-	0.019	-	-	0.022	-	0.004	-	-	-	-	-	-	-	-	-
Fe	-	-	-	-	-	-	-	-	-	-	-	-	-	-	-	-
Mn	0.623	0.460	0.590	0.380	0.563	0.444	0.483	0.485	0.141	0.100	0.160	0.150	0.090	0.440	0.480	0.490
Mg	1.059	1.031	1.073	1.033	0.987	1.026	1.030	1.042	1.031	1.070	1.030	1.040	1.040	1.030	1.030	1.040
Ca	-	-	-	-	-	-	-	-	-	-	-	-	-	-	-	-
Na	-	-	-	-	-	-	-	-	-	-	-	-	-	-	-	-
K	-	-	-	-	-	-	-	-	-	-	-	-	-	-	-	-
Sc	-	-	-	-	-	-	-	-	-	-	-	-	-	-	-	-
Y	-	-	-	-	-	-	-	-	-	-	-	-	-	-	-	-
Zr																
Cations	13.017	13.017	13.005	12.968	13.006	13.012	12.991	12.998	13.004	13.010	13.010	12.990	13.000	13.020	12.990	13.100

Notes: bd = below detection; nd= not determined.
SHIB = Spinel-hibonite inclusion; BAG = Blue aggregate

Table A3.9 (cont.) Representative electron microprobe analyses of hibonites from CAI in CV and CM carbonaceous chondrites.

Meteorite	Murchison (CM)							Mighei (CM)				Murchison (CM)				
CAI Type	Spinel hibonite inclusion							1-3	1-4†	1-8‡	1-12†	Spinel-hibonite spherules			Grossite-bearing*	
Source	Simon et al (1997)							MacPherson andDavis (1993)				Macdougall (1981)			Simon et al. (1994)	
	1	2	3	4	5	6	7	8	9	10	11	12	13	14	15	16
SiO_2	1.43	0.21	0.37	0.08	0.67	0.53	0.46	0.42	0.32	0.05	0.23	0.06	0.18	0.12	0.31	0.22
TiO_2	0.07	0.70	8.61	1.83	10.14	1.35	7.03	2.36	5.04	1.33	5.01	2.20	5.05	6.73	4.60	5.58
Al_2O_3	89.49	89.96	78.96	88.66	77.30	88.41	79.68	85.64	82.42	88.53	82.70	88.50	78.60	80.36	82.53	82.75
Cr_2O_3	bd	bd	0.04	bd	0.13	bd	0.08	0.03	0.07	0.02	0.06	nd	nd	nd	nd	nd
V_2O_3	0.14	0.15	0.14	0.05	0.07	0.13	0.35	0.88	0.13	0.02	0.41	nd	nd	0.14	0.12	0.11
FeO	0.04	b.d.	b.d.	0.04	0.20	bd	bd	0.30	0.50	0.68	0.67	0.44	0.12	0.02	0.02	0.02
MnO	nd	nd	nd	nd	nd	nd	nd	nd	nd	nd	nd	nd	nd	nd	nd	nd
MgO	1.02	0.46	4.19	0.85	4.74	0.98	3.70	1.69	3.77	0.60	2.69	1.11	6.67	3.28	2.58	2.93
CaO	8.58	8.58	8.34	8.43	8.02	8.53	8.42	8.43	8.41	8.68	8.53	8.41	6.80	8.45	8.37	8.47
Na_2O	nd	nd	nd	nd	nd	nd	nd	nd	nd	nd	nd	nd	nd	nd	nd	nd
K_2O	nd	nd	nd	nd	nd	nd	nd	nd	nd	nd	nd	nd	nd	nd	nd	nd
Sc_2O_3	bd	bd	bd	0.12	0.07	bd	bd	nd	nd	nd	nd	nd	nd	nd	nd	nd
Y_2O_3	nd	nd	nd	nd	nd	nd	nd	nd	nd	nd	nd	nd	nd	nd	nd	nd
ZrO_2	nd	nd	nd	nd	nd	nd	nd	nd	nd	nd	nd	nd	nd	nd	nd	nd
TOTAL	100.77	100.06	100.65	100.06	101.34	99.93	99.72	99.75	100.66	99.91	100.30	100.72	97.42	99.10	98.53	100.08

Cation Formula Based on 19 Oxygen Atoms

	1	2	3	4	5	6	7	8	9	10	11	12	13	14	15	16
Si	0.158	0.024	0.042	0.009	0.075	0.059	0.052	0.048	0.036	0.006	0.026	0.007	0.007	0.014	0.035	0.025
Ti	0.006	0.059	0.730	0.154	0.856	0.114	0.600	0.200	0.426	0.112	0.425	0.184	0.184	0.578	0.395	0.473
Al	11.645	11.810	10.483	11.664	10.227	11.635	10.662	11.367	10.906	11.701	10.991	11.595	1.595	10.809	11.109	10.987
Cr	-	-	0.003	-	0.011	-	0.007	0.003	0.006	0.002	0.006	-	-	-	-	-
V	0.012	0.014	0.013	0.005	0.006	0.011	0.030	0.079	0.012	0.002	0.037	-	-	0.006	0.006	0.005
Fe	0.003	-	-	0.004	0.018	-	-	0.028	0.047	0.064	0.063	0.041	0.041	0.001	0.002	0.002
Mn	-	-	-	-	-	-	-	-	-	-	-	-	-	-	-	-
Mg	0.168	0.076	0.704	0.142	0.793	0.164	0.626	0.284	0.630	0.101	0.452	0.184	0.184	0.558	0.439	0.492
Ca	1.015	1.024 1.007?		1.008	0.964 1.020?		1.024	1.018	1.012	1.043	1.031	1.001	1.001	1.033	1.024	1.022
Na	-	-	-	-	-	-	-	-	-	-	-	-	-	-	-	-
K	-	-	-	-	-	-	-	-	-	-	-	-	-	-	-	-
Sc	-	-	-	0.012	0.007	-	-	-	-	-	-	-	-	-	-	-
Y	-	-	-	-	-	-	-	-	-	-	-	-	-	-	-	-
Zr	-	-	-	-	-	-	-	-	-	-	-	-	-	-	-	-
Cations	13.007	13.007	12.982	12.998	12.957	13.003	13.002	13.027	13.075	13.031	13.031	13.012	13.012	12.999	13.010	13.006

Notes: bd = below detection; nd= not determined.† Average of 2 analyses; ‡Average of 3 analyses. * Grossite-bearing spherule

Table A3.10. Selected trace element analyses of hibonites from CV, CM and ordinary chondrites (concentrations in ppm - ion microprobe measurements).

Meteorite	Allende (CV)	Dhajala (OC)	Murchison (CM)											
CAI Type	hib-rich	hib fragment	hib xstal	hib-rich	sp+hib sph*	Cor+hib CAI		hib+cor		PLAC	PLAC	PLAC	PLAC	PLAC
CAI Name	HAL	DH-H1	DJ-1	MUCH-1	BB-6	BB-5-1	BB-5-2	GR-1-C	GR-1-R*	7-29	7-51	7-228	13-13	13-25
Source	Hinton (1988)	Hinton et al. (1988)								Ireland et al. (1988)			Ireland (1990)	
	1	2	3	4	5	6	7	8	9	10	11	12	13	14
Sc	236	113	535	455	33.3±3.5	483	550	1336	449	371	368	112	247	401
V	0.4±0.1	<0.12	24±0.2	34	247	53	34	10	38	47	60	214	247	50
Sr	70	74	64	68	136	75	71	85	10	80	84	41	93	89
Y	4.0±0.2	3.6±0.5	41	76	<4.0	39	76	481	20	86	132	38	126	86
Zr	23	11.2±1.4	90	122	51±12	102	135	131	11	381	437	158	513	295
Nb	nd	nd	nd	nd	nd	nd	nd	nd	nd	0.16±0.05	0.27±0.08	1.2	0.77	0.35
Ba	7.49±0.16	0.48±0.06	13.92±0.29	10.41±0.41	68.2±1.9	13.78±0.49	7.17±0.74	19.6±0.6	0.92±0.19	5.6	5.7	2	9.1	11
La	13.1±0.18	25.54±0.38	11.79±0.22	13.62±0.4	8±0.6	11.53±0.38	5.23±0.54	90.7±1.2	0.36±0.09	23	15	12	32	13
Ce	0.062±0.014	0.037±0.016	29.01±0.41	34.57±0.73	26.4±1.2	27.54±0.67	13.4±1	108.7±1.5	0.56±0.14	55	37	30	76	33
Pr	3.28±0.09	1.44±0.09	4.89±0.15	5.36±0.26	3.5±0.4	4.48±0.24	2.48±0.37	27.1±0.6	0.2±0.07	7.5	4.8	4.3	10.5	4.7
Nd	16.89±0.25	30.21±0.52	23.85±0.4	26.44±0.69	17.4±1	21.05±0.64	11.7±1	207±2.1	0.44±0.14	42	27	22	53	23
Sm	3.49±0.13	5.68±0.26	7.7±0.26	8.13±0.44	5.6±0.7	6.44±0.41	4.88±0.74	38.0±1.1	0.33±0.14	13	8.4	7	14	6.5
Eu	0.99±0.04	1.12±0.07	0.67±0.04	0.56±0.06	0.9±0.16	0.69±0.07	0.7±0.17	1.36±0.12	0.13±0.05	0.80±0.05	0.83	0.36	0.8	0.64
Gd	2.8±0.28	4.27±0.52	9.71±0.65	12.79±1.29	2.6±1.1	8.67±1.02	8.9±2.8	316.1±5.7	0.5±0.5	18	11	7.9	27	9.2
Tb	0.28±0.05	0.46±0.1	1.39±0.12	2.32±0.26	0.53±0.24	1.46±0.21	2.16±0.55	46.9±1.2	0.33±0.14	3.6	2.8	1.7	5.4	2
Dy	1.47±0.13	1.97±0.24	8.49±0.36	14.56±0.76	1.6±0.6	7.68±0.57	14.9±2.7	256.4±2.9	2.43±0.47	20	20	10	30	12
Ho	0.26±0.03	0.29±0.05	1.39±0.09	2.96±0.22	0.39±0.15	1.39±0.15	2.66±0.49	42.1±0.8	0.78±0.15	3.9	5.7	1.8	5.4	2.5
Er	0.36±0.07	0.38±0.12	2.54±0.19	7.25±0.47	0.46±0.31	2.92±0.33	4.29±0.92	73.7±1.7	1.54±0.29	6	19	3.7	8.8	8.8
Tm	0.06±0.02	0.05±0.03	0.2±0.03	0.75±0.1	0.42±0.14	0.31±0.07	0.39±0.16	3.37±0.23	0.17±0.07	0.2	2.2	0.26	0.42	1.5
Yb	0.12±0.08	0.19±0.12	<0.24	0.34±0.25	0.76±0.3	0.49±0.23	0.61±0.6	<2.2	0.17±0.15	0.049±0.045	0.22±0.04	1.4	0.72±0.11	0.24±0.7
Lu	0.07±0.02	0.12±0.04	0.13±0.05	0.2±0.09	<0.12	0.26±0.08	0.3±0.21	5.12±0.48	0.09±0.07	0.096±0.025	0.94	0.21	0.26±0.06	1.8
Hf	nd	nd	nd	nd	nd	nd	nd	nd	nd	7.3±0.9	8.5	3.3±0.4	14	5.2±1.1

Notes: PLAC = Platy crystal fragments * spinel+hibonite spherule

Table A3.10. (cont.) Selected trace element analyses of hibonites from CV, CM and ordinary chondrites (concentrations in ppm - ion microprobe measurements).

Meteorite	Murchison (CM)												
CAI Type	10-31	10-43	10-61	13-33	13-60	13-61	7-19	7-143	7-170	7-426	H2-18	H2-12	H2-13
CAI Name	BAG	BAG	BAG	SHIB	SHIB	SHIB	SHIB	SHIB	SHIB	SHIB	spinel-hibonite		
Source	Ireland et al.(1988)					Ireland (1990)		Ireland et al. (1988)			Simon et al. (1997)		
Cr	nd	nd	nd	nd	nd	nd	nd	nd	nd	nd	103±6	148	778
Mn	nd	nd	nd	nd	nd	nd	nd	nd	nd	nd	<6.2	33.3±2.3	41.1±2.6
Sc	21	228	44	298	537	7.5	59	63	344	92	<18.8	597	<26
V	1958	3673	839	999	1999	88	1240	2460	2300	3000	736	62.2±8.5	396±56
Rb	nd	nd	nd	nd	nd	nd	nd	nd	nd	nd	2.30±0.47	1.05±0.31	0.86±0.36
Sr	100	72	40	169	153	45	219	119	54	33	42.4	85.3	33.3
Y	0.95	2.9	1.4	10.8	428	0.63	1.1	4.1	7.8	3.4	1.02±0.16	91.1	0.90±0.12
Zr	7.0±1.9	50±6	21±4	249	1937	2.3	14±3	18±5	72±11	33±10	4.59±0.54	346	3.43±0.39
Nb	23	1.3	0.77	5.7	19	26	19	38	3.5	15	11.5±0.8	2.42±0.38	3.92±0.38
Ba	21	46	18	127	51	34	64	29	16	5.4	41.1	11.4±0.9	9.61±0.66
La	1	2.5	2.3	4.9	63	1.08	15	27	18	22	0.394±0.067	53.6	17.8
Ce	7.6	6.6	6.2	10.7	45	7	49	75	43	53	2.92±0.29	106	43.1
Pr	0.5	0.76	0.7	1.4	10.7	0.51	6.2	10	6.3	8.3	0.73±0.19	18.9±1.0	5.84±0.45
Nd	2	3.6	3	7.2	80	1.5	29	50	32	38	0.63±0.29	86.3	20.9±1.3
Sm	1.1	1.3	0.92	1.4	9.3	0.98	9.6	14	10	12	2.25±0.23	20.9	5.64±0.31
Eu	1.2	0.78	0.41	1.3	1.3	0.39	2	2	0.77	1	0.411±0.077	0.70±0.11	0.414±0.062
Gd	0.25±0.04	0.61±0.07	0.46±0.07	2.2	117	0.15±0.03	4.2	9.0±1.0	3.9±0.6	6.3±1.0	<0.46	38.9±2.0	3.62±0.88
Tb	0.04±0.01	0.17±0.02	0.08±0.01	0.54	18	0.016±0.004	0.98	1.3±0.2	0.80±0.11	1.7±0.2	0.194±0.064	5.98±0.43	0.28±0.14
Dy	0.23	1.1	0.48	2.6	85	0.12	4.8	6.5	4.7	6.2	0.42±0.15	27.6±1.3	2.04±0.39
Ho	0.08±0.01	0.27	0.08±0.01	0.55	16	0.02±0.01	0.32	0.54±0.08	0.57	0.45±0.09	0.152±0.071	4.71±0.39	0.155±0.08
Er	0.17	0.5	0.21	1.3	28	0.05±0.01	0.4	0.84	1.6	0.55±0.07	0.33±0.12	7.73±0.56	<0.32
Tm	0.16	0.17	0.06±0.01	0.2	0.87	0.16	1	1.1±0.2	1.5	1.6	0.124±0.042	0.333±0.070	0.502±0.069
Yb	5.5	1	0.31	1.2	3.3±0.9	2.8	4.4	7	2.4	4.9	2.36±0.32	<1.2	0.50±0.25
Lu	0.02±0.00	0.04±0.01	0.02±0.00	0.21±0.03	1.2±0.3	0.017±0.003	0.045±0.008	0.13±0.04	0.19	<0.088	0.137±0.051	<0.40	0.114±0.069
Hf	0.12±0.11	0.87±0.11	0.22±0.05	2.1±0.07	30±8	0.05±0.03	0.19±0.08	0.50±0.27	0.51±0.24	0.57±0.35	nd	nd	nd
Th	nd	nd	nd	nd	nd	nd	nd	nd	nd	nd	0.315±0.095	6.45±0.44	0.81±0.13
U	nd	nd	nd	nd	nd	nd	nd	nd	nd	nd	0.275±0.087	0.117±0.058	0.140±0.053

Notes: Analyses of Simon et al. (1997) - errors are ±1s, based on counting statistics and are only reported where they exceed 5 % of the amount given; upper limites are <2s
BAG = Blue Aggregate; SHIB = spinel-hibonite inclusion.

Table A3.11. Representative electron microprobe analyses of perovskites from CAIs in CV and CM carbonaceous chondrites.

Meteorite	Allende (CV)											Leoville	Allende	
CAI Name	Type A							Fluffy Type A (FTA)				FTA	Type B	
CAI Type	TS2 FI	TS2 FI	T924 FI	T924 FI	T912 FI	TS12 FI	T926 FI	CG11				L2	A37	
Source	Grossman (1975)							Allen et al. (1980)				CML	B&P (1987)	
Notes								Interior	Interior	Rim	Interior			
	1	2	3	4	5	6	7	8	9	10	11	12	13	14
SiO_2	nd	nd	nd	nd	nd	nd	nd	0.38	0.22	0.24	0.12	0.28	0.17	0.28
TiO_2	57.07	57.17	56.97	56.16	57.38	57.52	54.85	58.15	58.23	57.56	57.91	56.80	57.70	57.90
Al_2O_3	0.31	0.28	0.68	0.39	0.28	0.24	1.45	0.34	0.38	0.48	0.35	0.50	0.17	0.33
Cr_2O_3	<0.01	<0.01	<0.01	0.03	<0.01	<0.01	0.02	<0.07	0.14	<0.07	<0.07	<0.02	bd	bd
V_2O_3	nd	nd	nd	nd	nd	nd	nd	bd	bd	bd	bd	nd	nd	nd
FeO	0.03	<0.02	<0.02	<0.02	0.15	0.18	0.07	0.06	0.09	0.16	0.19	0.05	nd	nd
MnO	nd	nd	nd	nd	nd	nd	nd	nd	nd	nd	nd	nd	nd	nd
MgO	0.15	0.13	0.08	0.06	0.13	0.13	0.23	0.05	0.04	0.05	0.06	<0.02	bd	bd
CaO	40.97	41.41	40.79	41.02	40.32	40.63	40.21	41.37	40.86	41.09	40.50	40.70	41.20	39.80
Sc_2O_3	nd	nd	nd	nd	nd	nd	nd	0.10	0.20	0.16	0.15	nd	nd	nd
Y_2O_3	<0.02	<0.02	0.10	0.03	0.21	0.12	0.09	0.17	0.35	0.14	0.28	nd	nd	nd
ZrO_2	0.09	0.08	0.12	0.17	0.30	0.47	0.68	0.26	0.43	0.43	0.11	nd	nd	nd
Nb_2O_4	<0.04	0.29	<0.04	<0.04	<0.04	<0.04	<0.04	nd	nd	nd	nd	nd	nd	nd
Total	98.62	99.36	98.74	97.86	98.77	99.29	97.60	100.88	100.94	100.31	99.67	98.33	99.24	98.31
	Cation Formula Based on 6 Oxygen Atoms													
Si	-	-	-	-	-	-	-	0.017	0.009	0.011	0.005	0.012	0.008	0.012
Ti	1.972	1.968	1.964	1.958	1.980	1.976	1.914	1.960	1.965	1.954	1.978	1.964	1.978	1.996
Al	0.016	0.016	0.036	0.022	0.016	0.012	0.080	0.018	0.020	0.026	0.018	0.028	0.010	0.018
Cr	-	-	-	0.002	-	-	-	-	0.005	-	-	-	-	-
V	-	-	-	-	-	-	-	-	-	-	-	-	-	-
Fe	0.002	-	-	-	0.006	0.006	0.002	0.002	0.003	0.006	0.007	0.002	-	-
Mn	-	-	-	-	-	-	-	-	-	-	-	-	-	-
Mg	0.010	0.008	0.006	0.004	0.008	0.008	0.016	0.003	0.003	0.003	0.004	-	-	-
Ca	2.016	2.030	2.004	2.038	1.982	1.988	2.000	1.987	1.964	1.988	1.971	2.004	2.014	1.956
Na	-	-	-	-	-	-	-	-	-	-	-	-	-	-
Sc	-	-	-	-	-	-	-	0.004	0.007	0.006	0.006	-	-	-
Y	-	-	0.002	-	0.006	0.002	0.002	0.004	0.008	0.003	0.006	-	-	-
Zr	0.002	0.002	0.002	0.004	0.006	0.010	0.016	0.006	0.009	0.009	0.002	-	-	-
Cations	4.018	4.024	4.014	4.026	4.004	4.006	4.030	4.001	3.993	4.006	3.997	4.010	4.008	3.982

Notes: bd = below detection; nd = not determined
CML = Christophe Michel-Lévy et al. (1982); B&P (1987) = Bischoff and Palme (1987)

Table A3.11. (cont.) Representative electron microprobe analyses of perovskites from CAIs in CV and CM carbonaceous chondrites.

Meteorite	Allende (CV)									Murch
CAI Name	3643									HIB-11
CAI Type	Unique	spinel-perovskite-hibonite inclusion								UF
Source	W (1986)	Steele (1995b)								S (1996)
	1	2	3	4	5	6	7	8	9	10
SiO_2	bd	0.29	0.23	0.24	0.18	0.55	0.14	0.23	0.21	0.08
TiO_2	56.8	56.9	57.5	57.1	57.8	57.3	58.2	57.8	58.2	55.97
Al_2O_3	bd	1.97	0.63	1.84	0.57	1.04	0.31	0.43	0.53	0.84
Cr_2O_3	0.2	nd	nd	nd	nd	nd	nd	nd	nd	0.4
V_2O_3	nd	nd	nd	nd	nd	nd	nd	nd	nd	bd
FeO	0.5	0.09	0.09	0.23	0.13	0.33	0.23	0.63	0.23	nd
MnO	nd	nd	nd	nd	nd	nd	nd	nd	nd	0.02
MgO	bd	0.03	0	0.07	0	0.15	0.02	0.02	0.03	38.63
CaO	42.5	40.9	41.7	40.3	41.8	40.6	41.4	40.9	41.3	0.56
Sc_2O_3	bd	nd	nd	nd	nd	nd	nd	nd	nd	nd
Y_2O_3	nd	nd	nd	nd	nd	nd	nd	nd	nd	1.6
ZrO_2	nd	nd	nd	nd	nd	nd	nd	nd	nd	0.21
Total	100	100.18	100.15	99.78	100.48	99.97	100.3	100.01	100.5	98.31

Cation Formula Based on 6 Oxygen Atoms

	1	2	3	4	5	6	7	8	9	10
Si	0.012	0.014	0.01	0.01	0.008	0.024	0.006	0.01	0.01	0.004
Ti	1.964	1.922	1.954	1.936	1.958	1.942	1.976	1.968	1.968	1.944
Al	0.028	0.104	0.034	0.098	0.03	0.056	0.016	0.022	0.028	0.046
Cr	-	-	-	-	-	-	-	-		-
V	-	-	-	-	-	-	-	-	-	0.012
Fe	0.002	0.004	0.004	0.008	0.004	0.012	0.008	0.024	0.008	0
Mn	-	-	-	-	-	-	-	-	-	-
Mg	2.004	0.002	-	0.004	-	0.01	0.002	0.002	0.002	0.002
Ca	-	1.968	2.018	1.946	2.018	1.96	2.002	1.984	1.99	1.912
Na	-	-	-	-	-	-	-	-	-	-
Sc	-	-	-	-	-	-	-	-	-	0.022
Y	-	-	-	-	-	-	-	-	-	0.04
Zr	-	-	-	-	-	-	-	-	-	0.004
Total	4.01	4.012	4.02	4.004	4.018	4.006	4.01	4.01	4.008	3.986

Notes: bd = below detection; nd = not determined. Murch = Murchison (CM); W (1986) = Wark (1986); S (1986) = Simon et al. (1996)
(2) In central hibonite, bulk analysis by scanning; (3) Area of (1) free of exsolution, (4) In central hibonite, bulk analysis by scanning,
(5) In central hibonite, bulk analysis by scanning, (6) Area of (4) free of exsolution; (7) In spinel zone, bulk analysis by scanning.
(7) Area of (6) free of exsolution.

Table A3.12. Trace element analyses (in ppm—ion microprobe analyses) of perovskite (pv) and anorthite (an) from CAIs in CV and CH chondrites.

Meteorite	Efremovka (CV)					Acfer 182 (CH)					Allende (CV)	Vigarano (CV)	Allende (CV)
CAI Name	E2					022/14	418/8	022/13	418/P	418/5t	5241	1623-8	3529Z
CAI Type	Compact Type A					Grossite-bearing inclusions					Type B	Type B	Type B
Source	Fahey et al.(1987)					Weber et al. (1995)					M (1989)	M&D (1993)	Meeker (1995)
Mineral	pv	pv	pv	pv	pv	pv	pv	pv	pv	pv	an	an	an
	1	2	3	4	5	6	7	8	9	10	11	12	13
Be	nd	nd	nd	nd	nd	nd	nd	nd	nd	nd	nd	0.268±0.034	nd
K	nd	nd	nd	nd	nd	nd	nd	nd	nd	nd	nd	82.5	nd
Mn	nd	nd	nd	nd	nd	nd	nd	nd	nd	nd	nd	<37	nd
Ti	nd	nd	nd	nd	nd	251047	254019	204650	294974	182246	nd	nd	356.75±107.5
V	916±5	737±5	1167±8	1815±11	4730±22	3573	1419	5692	723	4201	nd	3.70±0.29	6.35±1.9
Cr	nd	nd	nd	nd	nd	nd	nd	nd	nd	nd	nd	<6.0	1.42±0.5
Rb	nd	nd	nd	nd	nd	nd	nd	nd	nd	nd	nd	0.233±0.086	nd
Ba	22±1	27±2	21±1	9±1	9±2	45.8	7.2	21.8	32.6	30.8	nd	134	53.04±15.03
Sc	892±5	1136±8	1026±9	1165±8	854±9	52.8	13526	50.6	7726	24.9	nd	1.22±0.44	8.5±2.4
Y	1104±8	1493±10	1556±12	1535±14	1790±20	13.7	10736	10.1	21362	20.3	nd	0.748±0.058	nd
Zr	3100±26	3840±33	5525±47	5813±54	5428±68	53.3	4077	28.3	40102	42.9	nd	0.488±0.081	nd
Nb	218±5	216±6	292±8	279±9	489±15	367	251	239	752	305	nd	0.118±0.039	0.09±0.04
Sr	298±4	330±4	382±6	225±5	201±6	426	222	326	249	382	nd	392	456.81±122.35
La	178±3	267±4	207±5	195±5	326±8	10.8	192	46.4	135	149	1.44±0.08	1.09	0.31±0.01
Ce	485±6	798±8	628±9	605±10	908±16	72.9	341	202	230	436	2.65±0.12	2.17	0.65±0.18
Pr	68±2	101±3	79±3	76±4	121±6	7.8	58.7	23.2	57.1	59.8	0.28±0.017	0.205±0.033	0.07±0.02
Nd	368± 5	574±7	435±7	420±7	641±11	37.1	899	90.7	442	314	1.16±0.04	0.7±0.1	0.3±0.01
Sm	109±3	169±4	135±4	136±5	203±8	29.8	118	39	85	97	0.28±0.03	0.189±0.053	0.1±0.03
Eu	7±1	8±1	9±1	6±1	8±1	5.4	1.11±0.66	3.2	3.1±0.48	4.7	2.99±0.15	1.71	1.61±0.49
Gd	123±6	157±8	139±8	140±26	228±40	2.89±0.87	20211	11.1±1.68	925	9.1±3.9	nd	0.215±0.061	0.06±0.03
Tb	26±1	36±1	32±2	30±4	43±6	0.91±0.14	3529	1.82±0.30	242	3.1±0.77	0.033±0.007	0.024±0.011	nd
Dy	189±4	253±4	238±6	229±12	333±19	7.5	19619	7.7	1820	13.2±1.42	0.224±0.015	0.2042±0.038	0.06±0.02
Ho	40±1	57±1	56±2	54±3	80±4	0.86±0.12	1559	0.49±0.13	670	1.51±0.30	0.088±0.008	0.061±0.015	nd
Er	116±2	163±3	159±4	161±12	204±17	1.49±0.20	1545	0.96±0.18	2223	2.02±0.33	0.144±0.16	0.106±0.032	0.06±0.02
Tm	17±1	24±1	7±1	28±4	34±5	6.4	174	1.97±0.22	56.3	3.4	0.03±0.018	<0.026	nd
Yb	84±3	123±4	141±5	121±5	164±8	38.3	413	14.4	64	24.7	nd	0.248±0.042	0.09±0.03
Lu	15±1	22±2	27±2	27±2	27±2	0.12±0.05	144±23	<0.11	370	<0.15	nd	0.0215±0.0085	nd
Hf	41±14	38±20	70±25	85±31	145±41	<1.16	<466	0.98±0.48	406±47.7	<2.01	nd	nd	nd

Notes: M. (1989) = MacPherson et al. (1989); M&D = MacPherson and Davis (1993);

Table A3.13. Representative electron microprobe analyses of melilite from CAIs in CV, CM and CO carbonaceous chondrites.

Meteorite	Allende (CV)		Leoville (CV)					Allende (CV)			Efremovka (CV)					
CAI Type	Compact Type A		Fluffy Type A					Type B			Type B1					
CAI Name	3898		L2					A37			E40		E44		E65	
Source	P (1991)		Christophe Michel-Levy (1982)					Bischoff and Palme (1987)			Goswami et al. (1994)					
	1	2	3	4	5	6	7	8	9	10	11	12	13	14	15	16
SiO_2	31.74	22.21	22.60	23.90	25.00	27.30	28.60	21.80	27.30	31.10	29.83	38.59	26.66	38.55	30.17	35.00
TiO_2	0.07	0.15	0.19	0.07	0.07	0.05	0.04	<0.09	bd	2.62	0.06	0.02	bd	bd	0.02	0.01
Al_2O_3	20.22	35.89	37.90	36.10	34.80	29.40	26.40	35.00	26.40	15.80	23.79	10.85	29.01	10.80	25.14	17.09
Cr_2O_3	nd	nd	<0.02	<0.02	<0.02	<0.02	<0.02	nd	nd	nd	nd	nd	nd	nd	nd	nd
V_2O_3	0.02	<0.02	nd	nd	nd	nd	nd	nd	nd	nd	bd	bd	bd	bd	bd	bd
FeO	<0.02	<0.02	0.14	0.10	0.08	0.05	0.04	<0.09	bd	<0.07	0.03	0.02	0.01	bd	0.03	0.16
MnO	nd	nd	nd	nd	nd	nd	nd	nd	nd	nd	bd	bd	0.09	0.02	bd	bd
MgO	6.30	0.36	0.24	0.92	1.60	3.00	4.30	0.90	4.90	8.60	4.18	9.81	2.78	10.16	4.96	7.84
CaO	41.25	41.10	38.70	38.70	39.20	39.80	39.60	39.60	40.20	41.90	41.17	40.89	41.43	40.41	39.94	40.68
Na_2O	0.09	<0.03	nd	nd	nd	nd	nd	nd	nd	nd	0.02	0.10	0.01	0.05	0.07	0.19
K_2O	<0.03	<0.03	nd	nd	nd	nd	nd	nd	nd	nd	nd	nd	nd	nd	0.01	0.01
ZnO	nd	nd	nd	nd	nd	nd	nd	nd	nd	nd	nd	nd	nd	nd	nd	nd
Total	99.69	99.71	99.77	99.79	100.75	99.60	98.98	97.48	98.80	100.09	99.08	100.28	100.00	100.00	100.34	100.98
Ak	45.00	2.00														

Cations Formula Based on 14 Oxygen Atoms

	1	2	3	4	5	6	7	8	9	10	11	12	13	14	15	16
Si	2.904	2.036	2.049	2.163	2.242	2.485	2.620	2.045	2.520	2.868	2.744	3.487	2.435	3.489	2.723	3.144
Ti	0.005	0.010	0.013	0.005	0.005	0.003	0.003	-	-	0.182	0.004	0.001	-	-	0.001	0.001
Al	2.181	3.880	4.050	3.852	3.680	3.155	2.851	3.870	2.873	1.718	2.580	1.156	3.124	1.152	2.675	1.810
Cr	-	-	-	-	-	-	-	-	-	-	-	-	-	-	-	-
V	0.001	-	-	-	-	-	-	-	-	-	-	-	-	-	-	-
Fe	-	-	0.011	0.008	0.006	0.004	0.003	-	-	-	0.002	0.002	0.001	-	0.002	0.012
Mn	-	-	-	-	-	-	-	-	-	-	-	-	0.007	0.002	-	-
Mg	0.859	0.049	0.032	0.124	0.214	0.407	0.587	0.126	0.674	1.182	0.573	1.321	0.378	1.370	0.667	1.050
Ca	4.043	4.038	3.759	3.754	3.767	3.881	3.887	3.980	3.976	4.141	4.058	3.959	4.055	3.918	3.863	3.916
Na	0.016	-	-	-	-	-	-	-	-	-	0.004	0.018	0.002	0.009	0.012	0.033
K	-	-	-	-	-	-	-	-	-	-	-	-	-	-	0.001	0.001
Zn	-	-	-	-	-	-	-	-	-	-	-	-	-	-	-	-
Cations	10.009	10.013	9.913	9.906	9.913	9.935	9.952	10.020	10.043	10.091	9.964	9.943	10.003	9.940	9.945	9.967

Notes: bd = below detection; nd = not determined.
Analyses 11, 13 and 15 are from inclusion cores; analyses 12, 14, 16 are from inclusion mantles.

Table A3.13. (cont.) Representative electron microprobe analyses of melilite from CAIs in CV, CM and CO carbonaceous chondrites.

Meteorite	Allende (CV)															
CAI Type	B1				B1				B		B			B		
CAI Name	3529-Z				3529-41				3529-21		3529-30			3658		
Location	Mantle	Mantle	Core	Core	Core	Core	Mantle	Core	Rim	Core	Near	Core	Near Sp	W-L	Core	Core
	1	2	3	4	5	6	7	8	9	10	11	12	13	14	15	16
SiO_2	26.76	33.63	34.56	29.61	37.4	26.03	30.47	30.72	30.77	25.25	32.31	27.68	26.15	22.8	29.51	30.3
TiO_2	<0.02	0.03	0.03	<0.02	<0.02	0.02	0.15	<0.02	0.02	0.04	0.06	0.03	0.04	0.04	<0.02	0.04
Al_2O_3	28.43	17.2	15.55	24.92	10.76	30.72	22.99	22.81	21.89	31.13	17.56	25.99	28.15	35.02	23.29	22.31
Cr_2O_3	nd	nd	nd	nd	nd	nd	nd	nd	nd	nd	nd	nd	nd	nd	nd	nd
V_2O_3	nd	nd	nd	nd	nd	nd	nd	nd	nd	nd	0.02	<0.02	<0.02	nd	nd	nd
FeO	<0.02	<0.02	<0.02	<0.02	<0.02	<0.02	<0.02	<0.02	<0.02	<0.02	<0.02	<0.02	<0.02	0.04	<0.02	<0.02
MnO	nd	nd	nd	nd	nd	nd	nd	nd	nd	nd	nd	nd	nd	nd	nd	nd
MgO	3.19	7.27	7.91	4.52	9.92	2.43	5.25	5.4	5.62	2.28	7.3	4.14	3.11	0.78	4.99	5.57
CaO	41.61	41.78	41.31	41.53	41.25	41.23	41.22	41.41	41.26	41.36	40.58	40.83	40.69	41.16	40.93	41.15
Na_2O	<0.03	0.11	0.16	0.08	0.21	0.03	0.03	0.03	0.14	0.03	0.09	0.03	0.03	0.03	0.12	0.2
K_2O	<0.03	<0.03	<0.03	<0.03	<0.03	<0.03	<0.03	<0.03	<0.03	<0.03	<0.03	<0.03	<0.03	<0.03	<0.03	<0.03
ZnO	0.05	0.04	0.04	0.04	0.04	0.04	nd	nd	0.04	0.04	nd	nd	nd	0.04	0.04	0.04
Total	100.04	100.02	99.52	100.66	99.54	100.45	100.16	100.46	99.68	100.06	97.92	98.67	98.14	99.84	98.84	99.57
Ak	23	54	58	14	71	18	38	39	41	16	51	29	22	5	36	39
Cations Formula Based on 14 Oxygen Atoms																
Si	2.444	3.067	3.164	2.680	3.420	2.362	2.772	2.787	2.814	2.304	3.010	2.560	2.433	2.087	2.723	2.777
Ti	0.001	0.002	0.002	0.001	0.001	0.001	0.010	0.001	0.001	0.003	0.004	0.002	0.003	0.003	0.001	0.003
Al	3.062	1.849	1.678	2.659	1.160	3.286	2.466	2.440	2.360	3.349	1.929	2.834	3.088	3.779	2.534	2.410
Cr	-	-	-	-	-	-	-	-	-	-	-	-	-	-	-	-
V	-	-	-	-	-	-	-	-	-	-	0.001	0.001	0.001	-	-	-
Fe	0.002	0.002	0.002	0.002	0.002	0.002	0.002	0.002	0.002	0.002	0.002	0.002	0.002	0.003	0.002	0.002
Mn	-	-	-	-	-	-	-	-	-	-	-	-	-	-	-	-
Mg	0.434	0.988	1.079	0.610	1.352	0.329	0.712	0.730	0.766	0.310	1.014	0.571	0.431	0.106	0.686	0.761
Ca	4.073	4.083	4.052	4.027	4.041	4.008	4.018	4.026	4.044	4.044	4.051	4.046	4.057	4.036	4.047	4.041
Na	0.005	0.019	0.028	0.014	0.037	0.005	0.005	0.005	0.025	0.005	0.016	0.005	0.005	0.005	0.021	0.036
K	0.003	0.003	0.004	0.003	0.003	0.003	0.003	0.003	0.004	0.003	0.004	0.004	0.004	0.004	0.004	0.004
Zn	0.003	0.003	0.003	0.003	0.003	0.003	-	-	0.003	0.003	-	-	-	0.003	0.003	0.003
Cations	10.028	10.017	10.011	9.998	10.019	9.998	9.989	9.996	10.018	10.023	10.030	10.025	10.024	10.026	10.021	10.035

Notes: bd = below detection; nd = not determined. W-L = Wark-Lovering rim.
Analyses 1-16 - Podosek et al. (1991)

Table A3.13. (cont.) Representative electron microprobe analyses of melilite from CAIs in CV, CM and CO carbonaceous chondrites.

Meteorite	**Essebi (CM)**		**Murchison (CM)**							**Yamato 791717 (CO)**				
CAI Type	Unusual		Grossite spherule		sp-hib sp	melilite fragments				Unique melilite -rich				
CAI Name			B6		BB-1	MUM-2	MUM-2	MUM-1	MUM-1	Y17-21				
Source	EG et al. (1984)		S et al. (1994)			MacPherson et al. (1983)				Tomeoka et al. (1992)				
	1	2	3	4	5	6	7	8	9	10	11	12	13	14
SiO_2	27.40	22.10	22.95	23.07	21.44	25.40	22.68	22.06	27.66	22.90	23.70	24.30	23.60	24.50
TiO_2	0.10	0.11	0.07	0.02	nd	nd	nd	nd	nd	0.09	0.05	0.05	bd	0.03
Al_2O_3	28.00	36.60	35.24	34.89	36.47	30.62	35.71	37.02	27.68	36.20	34.10	35.00	33.50	32.40
Cr_2O_3	bd	bd	nd	nd	nd	nd	nd	nd	nd	bd	bd	bd	bd	bd
V_2O_3	bd	bd	bd	bd	nd	nd	nd	nd	nd	bd	0.03	0.11	bd	bd
FeO	bd	0.04	bd	bd	nd	nd	nd	nd	nd	0.05	0.52	1.31	1.83	0.84
MnO	0.02	0.07	nd	nd	nd	nd	nd	nd	nd	bd	bd	bd	0.10	bd
MgO	3.66	0.18	0.79	0.95	0.96	3.02	0.80	0.24	3.60	0.35	0.93	0.84	1.13	1.65
CaO	41.20	40.80	40.76	40.91	39.77	40.74	40.79	40.73	40.90	40.40	40.10	37.70	38.40	39.40
Na_2O	bd	bd	nd	nd	bd	0.23	0.25	bd	bd	0.05	0.20	0.81	0.59	0.17
K_2O	nd	nd	nd	nd	nd	nd	nd	nd	nd	0.03	0.03	0.14	0.13	0.03
Sc_2O_3	nd	nd	bd	bd	nd	nd	nd	nd	nd	nd	nd	nd	nd	nd
Total	100.38	99.90	99.81	99.84	98.64	100.01	100.23	100.05	99.85	100.10	99.70	100.30	99.30	99.00
	Cations Formula Based on 14 Oxygen Atoms													
Si	2.490	2.018	2.098	2.110	1.982	2.320	2.066	2.012	2.524	2.171	2.205	2.18	2.258 ?	
Ti	0.004	0.004	0.004	0.002	0.000	-	-	-	-	0.003	0.003	-	0.002	
Al	3.000	3.948	3.796	3.760	3.974	3.296	3.836	3.978	2.976	3.683	3.743	3.649	3.52	
Cr	-	-	-	-	-	-	-	-	-	-	-	-	-	
V	-	-	-	-	-	-	-	-	-	0.002	0.008	-	-	
Fe	-	0.002	-	-	-	-	-	-	-	0.04	0.099	0.141	0.065	
Mn	-	0.002	-	-	-	-	-	-	-	-	-	0.008	-	
Mg	0.492	0.020	0.108	0.130	0.132	0.412	0.110	0.032	0.490	0.127	0.114	0.156	0.227	
Ca	4.008	4.000	3.992	4.008	3.940	3.986	3.982	3.978	3.998	3.936	3.665	3.801	3.891	
Na	-	-	-	-	-	0.042	0.044	-	-	0.036	0.142	0.106	0.03	
K	-	-	-	-	-	-	-	-	-	0.004	0.016	0.015	0.004	
Sc	-	-	-	-	-	-	-	-	-	-	-	-	-	
Cations	9.994	9.994	9.998	10.010	10.028	10.056	10.038	10.000	9.988	10.002	9.995	10.056	9.997	
Ak	25	1	5	5.7	0.3	16.8	3.7	0.6	25.9					

Notes: bd = below detection; nd = not determined. EG et al. (1984) = El Goresy et al. (1984); S et al. (1994) = Simon et al. (1994)

Table A3.13. (cont.) Representative electron microprobe analyses of melilite from CAIs in CV, CM and CO carbonaceous chondrites.

Meteorite	Efremovka	Leoville (CV)					Vigarano (CV)					
CAI Type	B2	B		B1			Type B3					
CAI Name	E60	3537-1		3537-2			1623-5					
	G (1994)	Caillet et al. (1993)					Davis et al. (1991)					
	1	2	3	4	5	6	7	8	9	10	11	12
SiO_2	42.06	26.00	26.70	27.30	31.80	35.60	27.14	31.62	33.68	41.47	43.65	44.01
TiO_2	bd	<0.03	<0.03	<0.03	<0.03	0.03	0.08	0.11	0.30	0.10	0.02	0.06
Al_2O_3	4.76	29.50	28.40	27.20	20.20	14.50	28.56	19.92	17.43	4.15	0.74	0.32
Cr_2O_3	0.18	nd	nd	nd	nd	nd	0.02	0.04	0.04	0.05	bd	0.01
V_2O_3	0.11	<0.03	<0.03	0.02	<0.02	<0.02	0.01	nd	nd	nd	bd	bd
FeO	0.28	0.50	<0.04	<0.04	<0.04	<0.04	0.14	0.66	0.20	0.19	0.22	0.20
MnO	nd	nd	nd	nd	nd	nd	nd	nd	nd	nd	0.01	0.01
MgO	13.09	3.02	3.34	3.82	7.06	9.01	3.94	5.94	7.11	12.20	14.35	14.73
CaO	39.31	41.40	41.40	40.90	40.70	41.20	40.09	40.40	40.89	40.81	41.17	41.66
Na_2O	0.14	<0.02	<0.02	<0.02	<0.02	0.14	nd	0.40	0.29	0.14	0.06	0.04
K_2O	0.06	<0.02	<0.02	<0.02	<0.02	<0.02	0.02	0.01	0.01	0.01	nd	nd
Sc_2O_3	nd	nd	nd	nd	nd	nd	nd	nd	nd	nd	bd	0.02
Y_2O_3	nd	nd	nd	nd	nd	nd	nd	nd	nd	nd	bd	0.01
ZrO_2	nd	nd	nd	nd	nd	nd	nd	nd	nd	nd	bd	bd
Total	99.99	100.42	99.84	99.24	99.76	100.48	100.00	99.10	99.95	99.12	100.22	101.07
Ak		20	23	26	45	61	23	46	53	89	98	99

Cations Formula Based on 14 Oxygen Atoms

	1	2	3	4	5	6	7	8	9	10	11	12
Si	3.801	2.372	2.444	2.510	2.901	3.224	2.469	2.917	3.071	3.802	3.958	2.084
Ti	-	-	-	-	-	0.002	0.005	0.008	0.021	0.007	0.001	0.006
Al	0.507	3.173	3.064	2.949	2.173	1.548	3.063	2.166	1.874	0.449	0.079	3.883
Cr	0.013	-	-	-	-	-	0.001	0.003	0.003	0.004	-	-
V	0.008	-	-	0.001	-	-	0.001	-	-	-	-	-
Fe	0.021	0.038	-	0.000	-	-	0.011	0.051	0.015	0.015	0.017	0.004
Mn	-	-	-	0.000	-	-	-	-	-	-	0.001	-
Mg	1.763	0.411	0.456	0.524	0.960	1.216	0.534	0.817	0.966	1.667	1.939	0.047
Ca	3.806	4.047	4.060	4.030	3.979	3.998	3.908	3.993	3.995	4.009	4.000	3.939
Na	0.025	-	-	-	-	0.025	0.000	0.072	0.051	0.025	0.011	0.009
K	0.007	-	-	-	-	-	0.002	0.001	0.001	0.001	-	0.003
Sc	-	-	-	-	-	-	-	-	-	-	-	-
Y	-	-	-	-	-	-	-	-	-	-	-	-
Zr	-	-	-	-	-	-	-	-	-	-	-	-
Cations	9.951	10.041	10.024	10.014	10.012	10.012	9.994	10.027	9.997	9.978	10.006	9.975

Notes: bd = below detection; nd = not determined. G et al. (1984) = Goswami et al. (1994)
Analyses 2 and 3 are from poikiloblastic areas.

Table A3.14. Representative trace element analyses (in ppm—measured by ion microprobe) for melilites from CAI in CV and CH carbonaceous chondrites.

Meteorite	Efremovka					Allende		Allende					Vigarano	Allende	
CAI Name	E2					TS-23		5241					1623-8	3529Z	
CAI Type	Compact Type A					B1		B1					B	B	
Source	Fahey et al.(1987)					Kuehner et al (1989)		MacPherson et al. (1989)					M&D (1993)	Meeker (1995)	
	1	2	3	4	5	6	7	8	9	10	11	12	13	14	15
Ti	nd	nd	nd	nd	nd	nd	nd	nd	nd	nd	nd	nd	nd	39.4±10.33	34.41±8.35
Be	nd	nd	nd	nd	nd	nd	nd	nd	nd	nd	nd	nd	0.649±0.054	nd	nd
K	nd	nd	nd	nd	nd	nd	nd	nd	nd	nd	nd	nd	69.5	nd	nd
Sc	nd	nd	nd	nd	nd	<5.0	<5.2	nd	nd	nd	nd	nd	2.43± 0.81	4.14±1.07	3.52±0.8
V	nd	nd	nd	nd	nd	<13	18±7	nd	nd	nd	nd	nd	17.5±2.5	7.61±2.09	7.19±1.78
Cr	nd	nd	nd	nd	nd	14±9	37±10	nd	nd	nd	nd	nd	<6.0	22.7±6.11	18.66±4.47
Mn	nd	nd	nd	nd	nd	nd	nd	nd	nd	nd	nd	nd	15.9±8.7	nd	nd
Rb	nd	nd	nd	nd	nd	1.8±0.7	1.9±0.7	nd	nd	nd	nd	nd	0.71±0.21	nd	nd
Sr	nd	nd	nd	nd	nd	373±5	431±5	nd	nd	nd	nd	nd	249	692.4±155.37	569.43±117.4
Y	nd	nd	nd	nd	nd	5.3±0.5	3.0±0.4	nd	nd	nd	nd	nd	10.8	nd	nd
Zr	nd	nd	nd	nd	nd	0.34±0.24	0.54±0.31	nd	nd	nd	nd	nd	<1.6	0.46±0.13	0.38±0.09
Nb	nd	nd	nd	nd	nd	<0.31	2.11±0.59	nd	nd	nd	nd	nd	0.651±0.081	nd	0.06±0.04
Ba	nd	nd	nd	nd	nd	60.0±2.7	59.0±2.7	nd	nd	nd	nd	nd	39.5	81.43±18.6	90.31±19.36
La	0.34±0.03	0.45±0.04	0.52±0.05	0.85±0.05	0.83±0.05	6.09±0.28	5.07±0.26	3.9±0.11	2.52±0.07	1.97±0.09	2.16±0.09	1.37±0.02	3.84	3.1±0.7	1.6±0.34
Ce	0.60±0.06	1.28±0.10	1.07±0.08	1.95±0.12	1.77±0.08	12.92±0.64	10.32±0.59	9.65±0.24	6.31±0.15	4.78±0.15	4.73±0.15	2.83±0.04	8.69	5.27±1.15	2.87±0.59
Pr	0.09±0.01	0.20±0.02	0.15±0.02	0.32±0.02	0.29±0.02	1.31±0.27	2.14±0.35	1.57±0.06	1.06±0.04	0.75±0.03	0.69±0.03	0.365±0.011	1.17±0.08	0.52±0.12	0.35±0.07
Nd	0.62±0.03	1.22±0.05	1.02±0.06	1.57±0.06	1.61±0.06	7.3±1.0	5.05±0.88	7.63±0.17	4.45±0.11	4±0.09	3.41±0.09	1.57±0.02	5.43±0.27	2.37±0.54	1.55±0.31
Sm	0.25± 0.02	0.58±0.06	0.48±0.05	0.64±0.05	0.52±0.05	2.03±0.25	1.43±0.23	2.42±0.12	1.74±0.08	1.49±0.08	1.01±0.06	0.345±0.012	1.56±0.08	0.37±0.09	0.26±0.05
Eu	1.38±0.08	1.54±0.10	1.40±0.11	1.61±0.11	1.67±0.11	2.03±0.15	1.42±0.14	1.33±0.05	1.19±0.03	1.58±0.11	1.39±0.10	0.75±0.03	1.82	1.25±0.31	0.97±0.24
Gd	nd	nd	nd	nd	nd	2.60±0.63	0.91±0.69	nd	nd	nd	nd	nd	1.97±0.16	0.3±0.1	0.17±0.05
Tb	0.08±0.02	0.14±0.04	0.15±0.04	0.12±0.04	0.17±0.03	0.33±0.12	<0.20	0.55±0.03	0.38±0.02	0.24±0.03	0.2±0.02	0.046±0.003	0.299±0.032	nd	nd
Dy	0.52±007	1.03±0.11	0.90±0.13	0.99±0.11	1.06±0.12	1.41±0.34	1.22±0.33	3.59±0.1	2.7±0.06	1.39±0.09	0.96±0.06	0.288±0.009	2.19±0.11	0.07±0.03	0.05±0.02
Ho	0.09±0.02	0.24±0.03	0.23±0.03	0.22±0.03	0.25±0.03	0.38±0.13	0.38±0.13	0.75±0.05	0.5±0.03	0.3±0.02	0.19±0.02	0.059±0.002	0.471±0.041	nd	nd
Er	0.34±007	0.68±0.12	0.62±0.14	0.56±0.11	0.65±0.12	0.70±0.21	0.26±0.17	2.18±0.06	1.48±0.04	0.91±0.07	0.51±0.05	0.166±0.007	1.20±0.07	0.1±0.03	0.1±0.03
Tm	0.06±0.02	0.08±0.04	0.08±0.04	0.09±0.03	0.11±0.04	<0.12	<0.09	0.2±0.02	0.18±0.01	0.14±0.03	0.09±0.019	0.015±0.004	0.163±0.025	nd	nd
Yb	nd	nd	nd	nd	nd	0.35±0.23	<0.40	nd	nd	nd	nd	nd	1.08±0.09	0.03±0.03	0.02±0.02
Lu	nd	nd	nd	nd	nd	<0.11	<0.10	nd	nd	nd	nd	nd	0.135±0.021	nd	nd
Hf	nd	nd	nd	nd	nd	<0.36	<0.31	nd	nd	nd	nd	nd	bd	nd	nd
Ta	nd	nd	nd	nd	nd	<0.36	<0.28	nd	nd	nd	nd	nd	nd	nd	nd
Th	nd	nd	nd	nd	nd	0.43±0.12	0.15±0.08	nd	nd	nd	nd	nd	0.140±0.022	nd	nd
U	nd	nd	nd	nd	nd	0.07±0.06	<0.08	nd	nd	nd	nd	nd	bd	nd	nd

M&D (1993) = MacPherson and Davis (1993); Analyses 8-12 are a profile through the melilite mantle of the inclusion at distances of 0.09,0.27, 0.77,0.99 and 1.44 mm from the rim of the inclusion

Table A3.14. (cont.) Representative trace element analyses (in ppm—measured by ion microprobe) for melilites from CAI in CV and CH carbonaceous chondrites.

Meteorite	Vigarano (CV)					Acfer 182 (CH)								
CAI Name	1623-5					022/6	022/9	022/17	022/14	418/8	022/13	022/22	418/3	022/19
CAI Type	B3					Grossite-bearing inclusions								
Source	Davis et al. (1991)					Weber et al. (1995)								
	1	2	3	4	5	6	7	8	9	10	11	12	13	14
Ti	-	-	-	-	-	4478	6215	11467	29042	4071	4117	11841	9610	15870
Sc	5.63±0.61	7.92±0.53	10.8±1.3	6.73±0.74	59.2	42	21.8	13	23	1675	9.8	161	38.1	11
V	<11	6.8±5.1	10.9±5.7	<5.9	30.9±6.6	321	423	333	376	262	497	729	1289	586
Cr	51±13	63±15	55±14	67±15	29±10	nd	nd	nd	nd	nd	nd	nd	nd	nd
Mn	<33	81±15	21±17	354	46±22	nd	nd	nd	nd	nd	nd	nd	nd	nd
Rb	1.33±0.30	1.71±0.29	2.98±0.90	2.83±0.52	3.33±0.58	nd	nd	nd	nd	nd	nd	nd	nd	nd
Sr	271	384	321	379	226	134	205	146	386	344	262	222	258	173
Y	0.246±0.071	4.90±0.27	16.0±1.2	2.23 + 0.26	3.60±0.34	1.51±0.18	2.43	0.64	1.47	8.4	1.06±0.11	22.6	1.21±0.13	1.44
Zr	0.87±0.22	4.74±0.44	13.3±1.8	0.82±0.26	10.5±0.9	4.8	7.2	2.33	4.3	25	1.90±0.19	68.8	1.93	4.8
Nb	0.389±0.079	0.91±0.10	2.73±0.43	0.69±0.13	0.72±0.13	2.18	2.62	1.62	15.4	1.11±0.18	27.4	3.54±0.65	0.13±0.02	3.8
Ba	1.80±0.30	3.81±0.36	4.00±0.89	11.0±0.9	9.47±0.83	11.3	11.6	9.8	53.1	112	23.2	32.5	31.4	34.6
La	0.242±0.043	1.95±0.10	3.78±0.33	0.799±0.089	0.636±0.083	1.15±0.12	0.95	1.59	1.30±0.14	0.60±0.07	2.33	4.8±0.58	12.2	1.14
Ce	0.313±0.086	3.42±0.23	9.42±0.92	1.20±0.19	1.45±0.22	3.8	5.1	9	6.9	2.45±0.28	5.5	13.2±1.47	27.4	7.3
Pr	0.082±0.024	0.546±0.048	1.34±0.18	0.227±0.045	0.295±0.052	0.36±0.04	0.44±0.05	0.94	0.72±0.11	0.21±0.04	0.51±0.06	1.23±0.39	3.5	0.67±0.07
Nd	0.49±0.10	2.35±0.17	6.53±0.67	1.18±0.18	1.04±0.17	1.64	1.81	2.98	3.1	1.26	2.42	7.5±0.86	18.2	2.12
Sm	0.064±0.044	0.613±0.053	1.47±0.19	0.425±0.094	0.429±0.069	0.36±0.12	0.83±0.08	1.99±0.30	2.12±0.31	0.40±0.11	1.05±0.14	2.45±0.42	5.3±0.54	1.20±0.23
Eu	0.892	1.35	1.41±0.11	1.17±0.0	0.895±0.056	0.87	1.34	0.77	2.58	0.91±0.16	1.44±0.09	1.43±0.22	1.57	1.08
Gd	<0.14	0.90±0.12	2.24±0.41	0.32±0.13	0.47±0.14	<0.13	0.26±0.05	0.10±0.08	0.47±0.17	8.1	0.29±0.08	4.0±0.58	2.14±0.37	0.39±0.13
Tb	<0.028	0.139±0.021	0.322±0.074	0.042±0.023	0.127±0.028	<0.03	0.04±0.01	0.03±0.01	0.06±0.03	1.03±0.16	0.06±0.02	0.58±0.14	0.40±0.07	0.04±0.03
Dy	<0.11	0.73±0.10	2.08±0.39	0.178±0.095	0.52±0.13	0.21±0.04	0.31	0.10±0.03	0.49±0.07	7.3	0.42±0.05	4.1±0.49	2.2	0.21±0.06
Ho	0.072±0.020	0.246±0.028	0.290±0.073	0.052±0.024	0.141±0.032	0.03±0.01	0.10±0.01	<0.02	0.06±0.02	0.85±0.12	0.05±0.01	0.83±0.13	0.13±0.03	0.10±0.03
Er	<0.10	0.586±0.072	1.55±0.26	<0.14	0.223±0.079	0.10±0.04	0.27±0.03	0.04±0.03	0.14±0.06	1.17	0.09±0.02	2.45±0.27	0.29±0.06	0.19±0.06
Tm	<0.034	0.110±0.018	0.209±0.051	<0.046	0.023±0.021	0.05±0.02	0.07±0.02	0.13±0.02	0.36±0.09	0.23±0.05	0.27±0.05	0.36±0.08	0.44±0.06	0.19±0.05
Yb	<0.074	0.460±0.066	1.85±0.28	0.123±0.065	0.218±0.077	1.73	3.6	1.41	4.4	2.62	1.63	4.2±0.52	1.75	5.4
Lu	0.035±0.015	0.033±0.014	0.168±0.054	0.047±0.021	<0.04	<0.02	0.06±0.01	<0.01	<0.02	0.17±0.05	0.02±0.01	0.30±0.11	0.14±0.03	0.05±0.03
Hf	0.196±0.068	0.301±0.074	<0.37	0.59±0.12	0.26±0.11	0.32±0.23	<0.21	<0.16	0.69±0.26	<1.64	0.15±0.10	1.67±1.00	0.49±0.23	<0.36

Notes: For analyses 1-5, errors (1s) are only given of they exceed 5%. For analyses 6-14 errors (1s) are given only of they exceed 10%; < upper limit (2s) for all analyses.

Table A3.15. Representative electron microprobe analyses of spinel from CAI in CV and CM carbonaceous chondrites.

Meteorite	Allende (CV)							Leoville (CV)		Allende (CV)					
CAI Type	Type A			Fluffy Type A				Fluffy Type A		Fluffy Type A					
CAI Name	TS27 F1	TS24 F1	TS25 F1	CG11				L2		TS25F1	TS28F1	TS24F1	TS24F1	TS27F1	TS52F1
Source	Grossman (1975)			Allen et al. (1978)				CML et al (1982)		MacPherson and Grossman (1984)					
	1	2	3	4	5	6	7	8	9	10	11	12	13	14	15
SiO_2	nd	nd	nd	0.26	0.10	0.14	0.15	0.05	0.17	nd	nd	0.77	0.07	0.15	0.22
TiO_2	0.12	0.25	0.66	0.31	0.19	0.41	0.36	0.16	0.08	0.13	0.06	0.30	0.32	0.13	0.22
Al_2O_3	71.10	69.94	68.99	70.90	68.94	66.67	66.94	71.00	71.10	67.88	61.32	67.70	69.20	71.21	69.71
Cr_2O_3	<0.01	0.13	0.13	0.34	0.08	0.15	<0.05	0.18	0.24	0.15	0.20	0.21	0.18	nd	nd
V_2O_3	nd	nd	nd	0.30	1.20	0.37	0.43	nd	nd	1.63	4.91	2.28	0.79	0.93	0.60
FeO	0.06	0.17	1.87	0.07	0.09	6.50	6.34	0.29	3.70	4.01	20.67	0.06	0.07	1.72	0.20
MnO	nd	nd	nd	nd	nd	nd	nd	nd	nd	nd	nd	nd	nd	nd	nd
MgO	27.90	27.78	26.77	28.45	28.16	24.04	23.65	28.20	24.90	25.09	12.00	27.81	28.67	27.71	28.63
CaO	0.51	0.34	1.00	0.12	0.32	0.20	0.35	0.23	0.25	nd	nd	1.20	0.33	0.42	0.51
Sc_2O_3	nd	nd	nd	<0.02	<0.02	<0.02	<0.02	nd	nd	nd	nd	nd	nd	nd	nd
ZrO_2	0.04	<0 01	<0.01	<0.02	<0.02	<0.02	<0.02	nd	nd	nd	nd	nd	nd	nd	nd
Y_2O_3	<0.02	<0.02	<0.02	<0.02	<0.02	<0.02	<0.02	nd	nd	nd	nd	nd	nd	nd	nd
Nb_2O_5	0.04	<0.03	<0.03	nd	nd	nd	nd	nd	nd	nd	nd	nd	nd	nd	nd
Total	99.77	98.61	99.42	100.75	99.08	98.48	98.22	100.11	100.44	98.89	99.16	100.33	99.63	102.27	100.09

Cation Formula Based on 4 Oxygens Atoms

	1	2	3	4	5	6	7	8	9	10	11	12	13	14	15
Si	-	-	-	0.006	0.002	0.004	0.004	0.001	0.004	-	-	0.018	0.002	0.003	0.005
Ti	0.002	0.005	0.012	0.006	0.004	0.008	0.007	0.003	0.001	0.002	0.001	0.005	0.006	0.002	0.004
Al	1.994	1.985	1.964	1.970	1.955	1.955	1.965	1.986	2.008	1.963	1.930	1.907	1.951	1.968	1.956
Cr	-	0.002	0.002	0.006	0.002	0.003	-	0.003	0.005	0.003	0.004	0.004	0.003	-	-
V	-	-	-	0.006	0.023	0.007	0.009	-	-	0.032	0.105	0.044	0.015	0.017	0.011
Fe	0.001	0.003	0.038	0.002	0.002	0.135	0.132	0.006	0.074	0.082	0.462	0.001	0.001	0.034	0.004
Mn	-	-	-	-	-	-	-	-	-	-	-	-	-	-	-
Mg	0.989	0.997	0.963	1.000	1.010	0.891	0.878	0.997	0.889	0.918	0.478	0.990	1.022	0.968	1.016
Ca	0.013	0.009	0.026	0.003	0.008	0.005	0.009	0.006	0.006	-	-	0.031	0.008	0.010	0.013
Cations	3.000	3.002	3.005	2.998	3.006	3.008	3.004	3.001	2.988	3.000	2.980	3.000	3.008	3.002	3.009

Notes: bd = below detection; nd = not determined.

Table A3.15. (cont.) Representative electron microprobe analyses of spinel from CAI in CV and CM carbonaceous chondrites.

Meteorite	Allende (CV)			Bali (CV)		Leoville (CV)		Allende (CV)								
CAI Name	TS23 F1	TS23 F1	TS12 F3			3537-1	3537-2									
CAI Type	B	B	B	B	B	B	B	B3	B3	B3	B3	FGSR	FGSR	FGSR	FGSR	FGSR
Source	Grossman (1975)			Kurat et al. (1975)		Caillet et al. (1993)		Davis et al. (1991)				M&G (1989)			K&W (1985b)	
Notes				Means	Means											
	1	2	3	4	5	6	7	8	9	10	11	12	13	14	15	16
SiO_2	nd	nd	nd	0.77	0.42	0.10	0.11	0.04	0.03	0.14	0.15	0.06	0.18	0.34	<0.05	0.05
TiO_2	0.25	0.35	0.76	0.33	0.26	0.17	0.20	0.32	0.01	0.66	0.31	0.24	0.37	0.20	0.21	0.08
Al_2O_3	70.94	70.74	69.49	70.30	70.50	71.20	70.60	70.12	70.66	69.07	64.41	69.92	66.09	64.77	68.50	65.31
Cr_2O_3	0.20	0.18	0.58	0.29	0.37	nd	nd	0.31	0.02	0.28	0.09	0.15	49.00	0.00	0.09	<0.05
V_2O_3	nd	nd	nd	nd	nd	0.32	0.34	0.11	0.29	0.16	0.21	0.23	0.02	0.50	0.34	0.11
FeO	0.43	0.11	<0.02	0.24	0.08	0.13	nd	0.21	0.41	2.58	15.14	4.01	14.49	18.71	6.23	17.54
MnO	nd	nd	nd	nd	nd	nd	nd	nd	nd	nd	nd	0.00	0.20	0.00	<0.05	0.08
MgO	28.20	28.25	27.95	28.10	28.10	28.00	28.30	27.72	28.10	26.07	17.58	25.27	17.46	14.50	23.94	16.00
CaO	0.28	0.13	0.27	0.31	0.32	0.27	0.20	0.18	0.10	0.17	0.33	0.11	0.35	0.14	0.11	0.11
ZnO	nd	nd	nd	nd	nd	nd	nd	nd	nd	nd	nd	nd	nd	nd	<0.05	0.45
Y_2O_3	<0.02	<0.04	<0.02	nd	nd	nd	nd	nd	nd	nd	nd	nd	nd	nd	nd	nd
ZrO_2	<0 01	<0.01	<0.01	nd	nd	nd	nd	nd	nd	nd	nd	nd	nd	nd	nd	nd
Nb_2O_5	<0.03	<0.03	<0.03	nd	nd	nd	nd	nd	nd	nd	nd	nd	nd	nd	nd	nd
Total	100.30	99.76	99.05	100.34	100.05	100.19	99.75	99.01	99.62	99.13	98.22	99.99	99.64	99.16	99.42	99.75
Cation Formula Based on 4 Oxygen Atoms																
Si	-	-	-	0.018	0.010	-	-	0.001	0.001	0.003	0.004	0.001	0.005	0.009	-	0.001
Ti	0.004	0.006	0.014	0.006	0.005	0.002	0.003	0.006	0.000	0.012	0.006	0.004	0.007	0.004	0.004	0.002
Al	1.982	1.984	1.966	1.960	1.972	0.003	0.004	1.983	1.987	1.972	1.969	1.990	1.984	1.989	1.982	1.987
Cr	0.004	0.003	0.011	0.005	0.007	1.988	1.980	0.006	0.000	0.005	0.002	0.003	0.010	-	0.002	-
V	-	-	-	-	-	-	-	0.002	0.006	0.003	0.004	0.004	-	0.010	0.007	0.003
Fe	0.009	0.002	-	0.005	0.002	0.006	0.006	0.004	0.008	0.052	0.328	0.081	0.309	0.408	0.128	0.379
Mn	-	-	-	-	-	0.003	-	nd	nd	nd	nd	-	0.004	-	-	0.002
Mg	0.996	1.001	1.000	0.991	0.993	-	-	0.991	0.999	0.941	0.679	0.909	0.663	0.563	0.876	0.616
Ca	0.007	0.003	0.007	0.008	0.008	0.988	1.003	0.005	0.003	0.004	0.009	0.003	0.010	0.004	0.003	0.001
Na	-	-	-	-	-	0.007	0.005	-	-	-	-	-	-	-	-	-
Zn	-	-	-	-	-	-	-	-	-	-	-	-	-	-	-	0.008
Cations	3.002	3.000	2.998	2.993	2.996	2.997	3.001	2.998	3.003	2.994	3.002	2.995	2.992	2.987	3.000	3.001

Notes: bd = below detection; nd = not determined.
M&G (1989) = McGuire and Hashimoto (1989); K&W (1985b) = Kornacki and Wood (1985b)

Table A3.15. (cont.) Representative electron microprobe analyses of spinel from CAI in CV and CM carbonaceous chondrites.

Meteorite	Murchison (CM)			Mighei (CM)							Essebi (CM)			Murchison (CM)		
CAI Type														Grossite-bearing		
Source	MacPherson et al.(1983)			MacPherson and Davis (1993)							El Goresy et al. (1984)			Simon et al. (1994)		
	1	2	3	4	5	6	7	8	9	10	11	12	13	14	15	16
SiO_2	0.02	bd	0.67	0.02	0.03	0.04	0.03	0.57	0.04	0.08	0.11	0.48	0.26	0.11	0.02	0.15
TiO_2	0.30	0.23	0.27	0.16	0.30	0.22	0.31	0.13	0.22	0.28	0.28	0.22	0.16	0.13	0.33	0.23
Al_2O_3	71.44	71.10	68.65	71.20	70.67	70.93	69.95	68.32	70.28	70.38	70.80	71.90	73.20	70.99	70.96	70.19
Cr_2O_3	nd	nd	nd	0.24	0.13	0.16	0.23	2.34	0.26	0.19	0.21	0.09	0.13	0.12	0.12	0.07
V_2O_3	0.24	bd	0.23	0.05	0.48	0.39	0.35	0.14	0.15	0.44	0.30	0.15	0.06	0.09	0.38	0.66
FeO	0.13	bd	0.04	0.57	0.45	0.52	0.58	0.58	0.07	0.17	0.05	0.27	0.18	0.11	0.05	0.08
MnO	nd	nd	nd	nd	nd	nd	nd	nd	nd	nd	bd	0.02	bd	nd	nd	nd
MgO	27.52	28.40	27.11	28.09	28.28	27.96	28.02	27.78	28.17	28.39	27.60	25.60	25.70	27.82	28.20	27.92
CaO	bd	bd	0.59	0.07	0.02	0.03	0.03	0.25	0.07	0.07	0.16	0.20	0.22	0.11	0.07	0.13
Total	99.65	99.73	97.56	100.40	100.36	100.25	99.50	100.11	99.26	100.00	99.51	98.93	99.91	99.48	100.13	99.43
	Cation Formula Based on 4 Oxygen Atoms															
Si	-	-	0.016	0.000	0.001	0.001	0.001	0.014	0.001	0.002	0.001	0.011	0.006	0.003	0.000	0.004
Ti	0.005	0.004	0.005	0.003	0.005	0.004	0.006	0.002	0.004	0.005	0.005	0.004	0.003	0.002	0.006	0.004
Al	2.003	1.991	1.969	1.987	1.974	1.983	1.973	1.926	1.981	1.972	1.991	2.025	2.041	1.994	1.983	1.977
V	0.005	-	0.005	0.005	0.002	0.003	0.004	0.044	0.005	0.003	0.004	0.002	0.002	0.002	0.002	0.001
Cr	-	-	-	0.001	0.009	0.007	0.007	0.003	0.003	0.008	0.006	0.003	0.001	0.002	0.007	0.013
Fe	0.003	-	0.000	0.011	0.009	0.010	0.012	0.012	0.001	0.003	0.000	0.005	0.004	0.002	0.001	0.002
Mn	-	-	-	-	-	-	-	-	-	-	-	-	-	-	-	-
Mg	0.976	1.006	0.983	0.992	0.999	0.989	0.999	0.990	1.004	1.006	0.981	0.913	0.906	0.988	0.996	0.994
Ca	-	-	0.015	0.002	0.001	0.001	0.001	0.006	0.002	0.002	0.004	0.005	0.006	0.003	0.002	0.003
Cations	2.992	3.001	2.993	3.001	3.000	2.998	3.003	2.998	3.001	3.001	2.991	2.967	2.966	2.996	2.997	2.998

Notes: bd = below detection; nd = not determined.

Table A3.16. Representative electron microprobe analyses of fassaite from CAI in CV carbonaceous chondrites.

Meteorite	Allende											Kaba		Allende		
CAI Type	Type A					FTA					FTA	FGSR	FGSR	FGSR	FGSR	FGSR
CAI Name	TS24 F1	TS24 F1	TS24 FI	TS1 F1	TS24 F1	CG11										
Source	Grossman (1975)					Allen et al. (1978)					Fuchs*	F&G (1985)[1]		M&G (1989)[2]		
	1	2	3	4	5	6	7	8	9	10	11	12	13	14	15	16
SiO_2	55.01	53.61	55.32	51.59	49.75	28.12	30.76	33.34	37.65	46.43	33.50	32.39	35.44	32.23	44.27	42.79
TiO_2	<0.03	<0.03	<0.03	0.65	1.15	10.97	14.81	15.33	9.41	3.42	16.90	12.69	10.59	12.18	4.26	1.67
Al_2O_3	0.71	2.07	3.05	6.57	8.99	22.08	23.90	19.46	18.12	8.99	17.20	24.28	22.43	23.71	12.40	13.34
Cr_2O_3	nd	nd	nd	nd	nd	nd	nd	nd	nd	nd	bd	< 0.06	< 0.05	0.12	0.07	0.26
V_2O_3	nd	nd	nd	nd	nd	1.27	nd	nd	nd	nd	nd	0.27	0.16	0.58	0.13	0.28
FeO	<0.03	0.07	0.13	0.32	0.08	nd	nd	nd	nd	0.82	0.03	< 0.04	< 0.04	0.08	3.13	8.05
MnO	nd	nd	nd	nd	nd	nd	nd	nd	nd	nd	nd	< 0.03	bd	bd	0.05	bd
MgO	19.01	18.21	16.43	16.39	16.29	2.56	6.48	6.81	9.33	14.10	7.70	6.75	7.92	6.92	11.47	11.17
CaO	26.80	27.27	25.16	26.25	26.50	25.88	24.77	25.31	25.40	24.93	24.10	24.60	23.94	25.90	23.70	23.17
Na_2O	nd	nd	nd	nd	nd	nd	nd	nd	nd	nd	nd	0.02	0.03	0.04	0.38	0.09
K_2O	nd	nd	nd	nd	nd	nd	nd	nd	nd	nd	nd	nd	nd	0.02	bd	bd
Sum	101.53	101.13	100.09	101.77	101.76	90.88	100.72	100.25	99.91	98.69	99.40	101.00	100.51	101.78	99.86	100.82

Cations Formula Based on 6 Oxygen Atoms

	1	2	3	4	5	6	7	8	9	10	11	12	13	14	15	16
Si	1.960	1.920	1.980	1.840	1.770	1.170	1.140	1.240	1.390	1.720	1.255	1.187	1.294	1.181	1.633	1.598
Al	0.030	0.090	0.130	0.280	0.380	0.340	0.410	0.430	0.260	0.100	0.476	0.350	0.291	0.336	0.118	0.047
Ti	-	-	-	0.020	0.030	1.080	1.040	0.850	0.790	0.390	0.760	1.049	0.966	1.024	0.539	0.588
Cr	-	-	-	-	-	-	-	-	-	-	-	-	-	0.004	0.002	0.008
V	-	-	-	-	-	0.040	-	-	-	-	-	0.010	0.006	0.017	0.004	0.008
Fe	-	-	-	0.010	-	-	-	-	-	0.030	0.001	-	-	0.003	0.097	0.252
Mn	-	-	-	-	-	-	-	-	-	-	-	-	-	-	0.002	-
Mg	1.010	0.970	0.880	0.870	0.810	0.160	0.360	0.380	0.510	0.780	0.430	0.369	0.431	0.378	0.631	0.622
Ca	1.020	1.050	0.970	1.000	1.010	1.150	0.980	1.010	1.000	0.990	0.967	0.966	0.937	1.017	0.937	0.927
Na	-	-	-	-	-	-	-	-	-	-	-	0.002	0.002	0.003	0.027	0.006
K	-	-	-	-	-	-	-	-	-	-	-	-	-	0.001	-	-
Zn	-	-	-	-	-	-	-	-	-	-	-	-	-	-	-	-
Total	4.020	4.030	3.960	4.020	4.000	3.940	3.930	3.910	3.950	4.010	3.889	3.933	3.927	3.964	3.990	4.056

Notes: bd = below detection; nd = not determined. FGSR = fine-grained, spinel-rich inclusion
Analyses 6-7 are from inclusion interior; Analysis 8 from inner rim, Analysis 9 Outer rim.
*Fuchs (1971); [1] = Fegley and Post (1985); [2] = McGuire and Hashimoto (1989)

Table A3.16. (cont.) Representative electron microprobe analyses of fassaite from CAI in CV carbonaceous chondrites.

Meteorite	Allende															
CAI Type	Type B1									Type B		B1		B1		B
CAI Name												3529-Z		3529-41		3529-21
Source	El Goresy et al. (1985)									B&P (1987)[1]		Podosek et al. (1991)				
	1	2	3	4	5	6	7	8	9	10	11	12	13	14	15	16
SiO_2	37.30	37.40	43.10	38.40	37.50	37.00	37.30	41.30	40.50	29.70	29.00	43.58	36.82	43.55	36.10	46.20
TiO_2	9.30	10.40	2.20	9.24	9.56	10.20	10.60	2.02	0.80	17.20	18.80	4.86	9.69	117.00	10.28	0.46
Al_2O_3	18.10	17.30	15.80	17.70	18.40	18.90	17.70	18.20	20.10	19.70	20.00	14.83	20.98	16.72	21.69	15.66
Cr_2O_3	0.09	bd	0.10	0.06	0.11	bd	0.15	0.17	0.27	nd	nd	nd	nd	nd	nd	nd
V_2O_3	0.32	0.39	<0.05	0.37	0.33	0.28	0.47	2.05	3.15	nd	nd	nd	nd	nd	nd	nd
FeO	0.05	bd	0.06	bd	bd	bd	0.22	0.67	0.32	<0.05	bd	<0.02	<0.02	<0.02	0.02	0.06
MnO	nd	nd	nd	nd	nd	nd	nd	nd	nd	nd	nd	nd	nd	nd	nd	nd
MgO	8.78	8.56	11.60	9.07	8.74	6.52	8.69	10.10	9.50	7.40	7.00	11.87	7.94	11.34	7.42	12.46
CaO	24.60	23.60	24.30	24.50	24.20	24.10	24.40	25.10	24.90	24.70	25.10	24.86	24.41	24.92	24.38	25.82
Na_2O	nd	nd	nd	nd	nd	nd	nd	nd	nd	nd	nd	<0.03	<0.03	<0.03	<0.03	0.04
K_2O	nd	nd	nd	nd	nd	nd	nd	nd	nd	nd	nd	<0.03	<0.03	<0.03	<0.03	<0.03
ZnO	nd	nd	nd	nd	nd	nd	nd	nd	nd	nd	nd	0.04	0.07	<0.04	0.05	<0.04
Total	98.54	97.65	97.36	99.34	98.84	99.01	99.51	97.61	99.54	98.75	99.90	100.04	99.91	99.70	99.94	100.70
Cations Formula Based on 6 Oxygen Atoms																
Si	1.414	1.439	1.609	1.444	19.000	1.400	1.411	1.520	1.492	1.131	1.095	1.584	1.352	1.583	1.327	1.659
Ti^{3+}	0.181	0.278	0.044	0.208	0.218	0.235	0.229	0.007	-	0.492	0.534	0.133	0.268	0.087	0.284	0.012
Ti^{4+}	0.084	0.024	0.019	0.053	0.054	0.056	0.072	0.049	0.023	0.884	0.890	0.636	0.908	0.717	0.940	0.663
Al	0.811	0.784	0.696	0.784	0.821	0.843	0.787	0.798	0.780	-	-	-	-	-	-	-
V	0.010	0.012	-	0.007	0.011	0.010	0.014	0.061	0.093	-	-	-	-	-	-	-
Cr	0.003	-	0.003	0.002	0.003	-	0.004	0.005	0.008	-	-	-	-	-	0.001	0.002
Fe	0.002	-	0.002	-	-	-	0.007	0.021	0.010	-	-	-	-	-	-	-
Mg	0.497	0.490	0.645	0.509	0.493	0.480	0.490	0.556	0.523	0.420	0.394	0.643	0.435	0.614	0.406	0.667
Ca	1.000	0.973	0.982	0.989	0.981	0.976	0.986	0.992	0.982	1.008	1.015	0.968	0.961	0.971	0.960	0.993
Na	-	-	-	-	-	-	-	-	-	-	-	-	-	-	-	0.003
K	-	-	-	-	-	-	-	-	-	-	-	-	-	-	-	-
Zn	-	-	-	-	-	-	-	-	-	-	-	0.001	0.002	-	0.001	-
Cations	4.002	4.000	4.000	3.996	4.000	4.000	4.000	3.998	4.001	3.935	3.927	3.965	3.926	3.972	3.919	3.999
Ti^{3+}/Ti	68%	92%	69%	80%	80%	81%	76%	13%	0							

Notes: bd = below detection; nd = not determined. Analyses 4-9 are from spinel-rich areas

[1] B&P (1987) = Bischoff and Palme (1987)

Table A3.16. (cont.) Representative electron microprobe analyses of fassaite from CAI in CV carbonaceous chondrites.

Meteorite	Allende															
CAI Type	Type B															
CAI Name	TS23 F1	TS4 F1	TS21 FI	TS8 F3	TS4 F1	TS21 F1	TS8 F3	T812 F3	TS8 F3	TS23 FI	TS12 F3	T94 FI	TS4 FI	TSI2 F3	TS18 FI	TS18 FI
Source	Grossman (1975)															
	1	2	3	4	5	6	7	8	9	10	11	12	13	14	15	16
SiO_2	42.90	40.55	41.32	40.32	38.57	38.53	35.73	40.61	41.41	41.97	39.60	39.31	37.08	37.17	32.77	31.98
TiO_2	6.01	7.58	6.97	6.79	7.58	7.76	7.41	3.06	4.28	5.01	6.08	7.02	25.38	9.35	16.00	17.49
Al_2O_3	15.14	15.94	16.74	17.89	19.15	19.91	21.38	20.40	18.38	18.21	18.16	18.20	19.58	19.46	19.93	20.35
Cr_2O_3	nd	nd	nd	nd	nd	nd	nd	nd	nd	nd	nd	nd	nd	nd	nd	nd
V_2O_3	nd	nd	nd	nd	nd	nd	nd	nd	nd	nd	nd	nd	nd	nd	nd	nd
FeO	<0.03	0.49	0.04	0.27	0.57	0.36	<0.03	<0.03	0.13	<0.03	<0.03	0.49	8.07	<0.03	<0.03	<0.03
MnO	nd	nd	nd	nd	nd	nd	nd	nd	nd	nd	nd	nd	nd	nd	nd	nd
MgO	11.51	10.23	10.48	10.36	9.34	8.96	8.09	9.71	10.35	10.62	9.44	9.90	8.62	8.07	7.20	6.81
CaO	26.02	25.02	25.94	26.22	25.11	25.47	25.64	25.82	25.72	25.93	24.94	25.03	26.39	24.77	24.68	24.63
Na_2O	nd	nd	nd	nd	nd	nd	nd	nd	nd	nd	nd	nd	nd	nd	nd	nd
K_2O	nd	nd	nd	nd	nd	nd	nd	nd	nd	nd	nd	nd	nd	nd	nd	nd
Total	101.58	99.81	101.49	101.86	100.32	100.99	98.26	99.66	100.27	101.64	98.22	99.95	98.73	98.82	100.58	101.23
	Cations Formula Based on 6 Oxygen Atoms															
Si	1.540	1.490	1.490	1.460	1.410	1.400	1.340	1.480	1.510	1.510	1.470	1.440	1.380	1.380	1.210	1.180
Ti	0.160	0.210	0.190	0.180	0.210	0.210	0.210	0.080	0.120	0.140	0.170	0.190	0.230	0.260	0.450	0.480
Al	0.640	0.690	0.710	0.760	0.830	0.850	0.940	0.880	0.490	0.770	0.800	0.790	0.860	0.850	0.870	0.880
Cr	-	-	-	-	-	-	-	-	-	-	-	-	-	-	-	-
V	-	-	-	-	-	-	-	-	-	-	-	-	-	-	-	-
Fe	-	0.020	-	0.010	0.020	0.010	-	-	-	-	-	0.020	-	-	-	-
Mn	-	-	-	-	-	-	-	-	-	-	-	-	-	-	-	-
Mg	0.620	0.560	0.560	0.560	0.510	0.490	0.450	0.530	0.560	0.560	0.520	0.540	0.480	0.450	0.400	0.370
Ca	1.000	0.990	1.000	1.010	0.990	0.990	1.030	1.010	1.000	1.000	0.990	0.990	1.010	0.990	0.980	0.970
Na	-	-	-	-	-	-	-	-	-	-	-	-	-	-	-	-
K	-	-	-	-	-	-	-	-	-	-	-	-	-	-	-	-
	1.960	1.960	1.950	1.980	1.970	1.950	1.970	1.980	1.980	1.980	1.950	1.970	1.960	1.930	1.910	1.880

Table A3.16. (cont.) Representative electron microprobe analyses of fassaite from CAI in CV carbonaceous chondrites.

Meteorite	Allende							Bali			Allende					
CAI Type	Type B							Type B*			Type B			Type B		
CAI Name	TS12F3	TS8F3	TS23F1	TS12F3	TS12F3	TS18F1	TS18F1				TS-21			TS-23		
Source	Grossman (1975)							Kurat et al. (1975)			Hutcheon et al. (1978)					
	1	2	3	4	5	6	7	8	9	10	11	12	13	14	15	16
SiO_2	40.61	41.41	41.97	39.60	37.17	32.77	31.98	38.40	40.50	42.10	41.30	41.30	37.30	38.50	41.70	43.20
TiO_2	3.06	5.01	5.01	6.08	9.35	16.00	17.49	7.50	5.50	4.20	5.10	4.50	8.50	7.90	3.30	5.20
Al_2O_3	20.46	18.38	18.21	18.16	19.46	19.93	20.35	19.40	18.50	17.20	18.60	18.90	20.70	20.00	19.90	15.10
Cr_2O_3	nd	nd	nd	nd	nd	nd	nd	0.08	0.09	0.13	nd	nd	nd	nd	nd	nd
V_2O_3	nd	nd	nd	nd	nd	nd	nd	nd	nd	nd	nd	nd	nd	nd	nd	nd
FeO	<0.03	0.13	<0.03	<0.03	<0.03	<0.03	<0.03	0.20	0.07	0.07	nd	nd	nd	nd	nd	nd
MnO	nd	nd	nd	nd	nd	nd	nd	nd	nd	nd	nd	nd	nd	nd	nd	nd
MgO	9.71	10.35	10.52	9.44	8.07	7.20	6.81	8.70	10.00	10.80	10.30	10.20	8.10	9.00	9.90	11.10
CaO	25.82	25.72	25.93	24.94	24.77	24.68	24.63	25.50	25.80	25.50	25.40	25.50	25.00	25.10	25.40	25.20
Na_2O	nd	nd	nd	nd	nd	nd	nd	0.02	0.02	0.01	nd	nd	nd	nd	nd	nd
K_2O	nd	nd	nd	nd	nd	nd	nd	0.02	0.06	0.03	nd	nd	nd	nd	nd	nd
ZnO	nd	nd	nd	nd	nd	nd	nd	nd	nd	nd	nd	nd	nd	nd	nd	nd
Total	99.66	100.27	101.64	98.22	98.82	100.58	101.26	99.82	100.54	100.04	100.70	100.40	99.60	100.50	100.20	99.80
Cations Formula Based on 6 Oxygen Atoms																
Si	1.480	1.510	1.510	1.470	1.380	1.210	1.180	1.414	1.475	1.535	1.500	1.500	1.370	1.410	1.510	1.570
Ti	0.080	0.120	0.140	0.170	0.260	0.450	0.480	0.208	0.151	0.115	0.140	0.120	0.230	0.220	0.090	0.140
Al	0.880	0.790	0.770	0.800	0.850	0.870	0.880	0.842	0.794	0.739	0.790	0.810	0.900	0.860	0.850	0.650
Cr	-	-	-	-	-	-	-	0.002	0.003	0.004	-	-	-	-	-	-
V	-	-	-	-	-	-	-	-	-	-	-	-	-	-	-	-
Fe	-	-	-	-	-	-	-	0.006	0.002	0.002	-	-	-	-	-	-
Mn	-	-	-	-	-	-	-	-	-	-	-	-	-	-	-	-
Mg	0.530	0.560	0.560	0.520	0.450	0.400	0.370	0.477	0.543	0.587	0.560	0.550	0.440	0.490	0.530	0.600
Ca	1.010	1.000	1.000	0.990	0.990	0.980	0.970	1.006	1.007	0.996	0.990	0.990	0.980	0.980	0.990	1.000
Na	-	-	-	-	-	-	-	0.001	0.001	0.001	-	-	-	-	-	-
K	-	-	-	-	-	-	-	0.001	0.003	0.001	-	-	-	-	-	-
Cations	3.980	3.980	3.980	3.950	3.930	3.910	3.880	3.957	3.978	3.980	3.980	3.970	3.920	3.960	3.970	3.960

Notes: bd = below detection; nd = not determined.

*Average compositions. Analysis 8 - surface of inclusions (average of 4 analyses); Analysis 9 - intermediate center (average of 12 analyses); Analysis 10 - intermediate center (average of 8 analyses)

Table A3.16. (cont.) Representative electron microprobe analyses of fassaite from CAI in CV carbonaceous chondrites.

Meteorite	Allende															
CAI Type	B2					B2	B1	B1	B1	B1	B	B	B	FTA	B	B
CAI Name												Egg-3	Egg-3	Egg-4	Egg-3	Egg-6
Source	Wark and Lovering (1982b)											Meeker et al. (1983)				
	1	2	3	4	5	6	7	8	9	10	11	12	13	14	15	16
SiO_2	43.65	48.73	42.38	46.88	42.26	40.01	36.67	37.83	36.89	32.12	40.40	37.19	45.04	31.09	42.45	38.73
TiO_2	3.81	2.03	4.12	3.05	3.27	6.05	8.25	9.13	10.07	18.25	7.14	9.45	2.64	16.56	2.97	0.58
Al_2O_3	14.30	9.00	17.07	10.18	17.65	19.57	19.10	18.93	20.25	19.75	17.29	20.34	14.00	19.40	18.53	18.52
Cr_2O_3	0.06	0.04	0.02	0.04	0.02	0.11	0.00	0.05	0.01	0.05	bd	0.03	0.08	0.08	bd	0.28
V_2O_3	bd	bd	bd	0.07	0.10	bd	0.43	0.16	bd	bd	0.46	0.16	0.04	0.53	0.06	6.42
FeO	0.01	0.14	0.04	0.17	0.18	0.05	0.14	0.06	0.06	0.06	0.03	0.02	bd	0.03	0.02	0.49
MnO	nd	nd	nd	nd	nd	nd	nd	nd	nd	nd	nd	nd	nd	nd	nd	nd
MgO	12.62	15.26	11.24	14.06	10.99	9.93	8.69	8.43	8.13	5.85	9.40	8.56	12.59	6.78	10.12	9.83
CaO	24.90	25.18	25.25	25.71	25.79	26.00	26.28	24.85	25.08	25.29	24.84	25.06	25.26	24.60	25.50	24.45
Na_2O	nd	nd	nd	bd	bd	0.01	nd	nd	0.03	0.01	nd	nd	nd	nd	nd	nd
K_2O	nd	nd	nd	nd	nd	nd	nd	nd	nd	nd	nd	nd	nd	nd	nd	nd
Total	99.33	100.41	100.13	100.20	100.27	101.70	99.61	99.58	100.53	101.38	99.56	100.81	99.65	99.07	99.65	99.30
	Cations Formula Based on 6 Oxygen Atoms															
Si	1.598	1.758	1.541	1.703	1.537	1.441	1.364	1.398	1.351	1.186	1.510	1.380	1.650	1.200	1.560	1.430
Ti	0.105	0.055	0.113	0.083	0.089	0.164	0.231	0.254	0.277	0.507	0.200	0.260	0.072	0.480	0.082	0.016
Al	0.617	0.383	0.732	0.436	0.757	0.831	0.838	0.825	0.874	0.859	0.760	0.890	0.600	0.880	0.800	0.810
Cr	0.002	0.001	0.001	0.001	0.001	0.003	-	0.001	-	0.001	-	0.001	0.002	0.002	-	0.008
V	-	-	-	0.002	0.003	-	0.013	0.005	-	-	0.014	0.001	0.001	0.016	0.002	0.190
Fe	-	0.004	0.001	0.005	0.005	0.002	0.004	0.002	0.002	0.002	0.001	0.001	-	0.001	0.001	0.015
Mn	-	-	-	-	-	-	-	-	-	-	-	-	-	-	-	-
Mg	0.689	0.820	0.609	0.761	0.596	0.533	0.482	0.464	0.444	0.322	0.990	0.990	0.990	1.020	1.000	0.970
Ca	0.977	0.973	0.984	1.001	1.005	1.004	1.048	0.984	0.984	1.000	0.520	0.470	0.690	0.390	0.550	0.540
Na	-	-	-	-	-	0.001	-	-	0.002	0.001	-	-	-	-	-	-
K	-	-	-	-	-	-	-	-	-	-	-	-	-	-	-	-
Cations	3.988	3.995	3.980	3.994	3.993	3.978	3.980	3.933	3.935	3.878	3.995	3.993	4.005	3.989	3.995	3.979

Notes: bd = below detection; nd = not determined. FTA = "fluffy" Type A.

Table A3.16. (cont.) Representative electron microprobe analyses of fassaite from CAI in CV carbonaceous chondrites.

Meteorite	Efremovka		Allende											
CAI Type	B2		B3	B3	B3	B3	B3	B3	B3	B3	B3	B3	B3	B3
CAI Name	E60													
Source	Goswami et al.[1]		Wark et al. (1987)											
	1	2	3	4	5	6	7	8	9	10	11	12	13	14
SiO_2	52.24	33.99	45.30	46.20	50.10	46.80	51.00	45.80	49.40	47.32	51.20	42.20	45.30	45.00
TiO_2	0.19	12.21	2.50	0.80	1.00	2.20	0.30	1.60	bd	1.48	bd	6.40	1.55	1.45
Al_2O_3	3.08	22.56	12.10	11.70	9.60	13.70	7.50	13.50 S.1		12.01	6.90	16.60	16.50	16.20
Cr_2O_3	nd	nd	0.05	0.03	bd	0.10	0.10	bd	0.30	0.03	bd	bd	bd	bd
V_2O_3	bd	0.68	0.06	0.10	bd	bd	bd	bd	bd	0.10	bd	bd	bd	bd
FeO	0.81	bd	0.04	0.12	bd	bd	bd	bd	2.80	0.10	0.85	0.20	0.22	bd
MnO	0.08	0.02	13.30	13.50	14.60	12.80	15.80	13.00	14.80	13.72	13.80	10.00	11.20	11.90
MgO	18.95	7.05	25.10	25.90	24.80	24.40	25.30	25.40	25.40	25.32	27.20	24.50	25.40	25.10
CaO	22.65	23.39	nd	nd	nd	nd	nd	nd	nd	nd	nd	nd	nd	nd
Na_2O	bd	0.07	nd	nd	nd	nd	nd	nd	nd	nd	nd	nd	nd	nd
K_2O	bd	0.03	nd	nd	nd	nd	nd	nd	nd	nd	nd	nd	nd	nd
Total	98.00	100.00	98.40	98.30	100.0*	100.0*	100.0*	99.20	100.08	100.08	99.95	99.90	100.17	99.65
	Cations Formula Based on 6 Oxygen Atoms													
Si	1.917	1.253	1.671	1.707	1.8	1.687	1.838	1.671	1.966	1.711	1.862	1.537	1.635	1.632
Ti	0.005	0.339	0.069	0.022	0.027	0.06	0.008	0.044	-	0.04	-	0.175	0.042	0.04
AL	0.133	0.981	0.526	0.51	0.407	0.582	0.319	0.581	-	0.512	0.296	0.713	0.702	0.693
Cr	-	-	0.001	0.001	-	0.003	0.003	-	0.009	0.001	-	-	-	-
V	-	0.02	0.002	0.003	-	-	-	-	-	0.003	-	-	-	-
Fe	0.025	-	0.001	0.004	-	-	-	-	0.093	0.003	0.026	0.006	0.007	-
Mn	0.002	0.001	-	-	-	-	-	-	-	-	-	-	-	-
Mg	1.037	0.387	0.731	0.743	0.782	0.688	0.849	0.707	0.878	0.739	0.748	0.543	0.603	0.643
Ca	0.891	0.924	0.992	1.025	0.955	0.942	0.977	0.993	1.083	0.981	1.06	0.956	0.983	0.975
Na	-	0.005	-	-	-	-	-	-	-	-	-	-	-	-
K	-	0.001	-	-	-	-	-	-	-	-	-	-	-	-
Zn	-	-	-	-	-	-	-	-	-	-	-	-	-	-
Cations	4.01	3.91	4.00	4.01	3.97	3.961	3.993	3.995	4.029	3.991	3.991	3.931	3.971	3.982

Notes: bd = below detection; nd = not determined. Analyses 4-9 are from spinel-rich areas

[1]Goswami et al. (1994)

Analysis 3 — center of grain; analysis 4 * EDS analyses, normalized to 100 wt %

Analyses 10-14 — Zoning profile through grain from core to rim. Analyses 11-14 at distances of 5, 7, 10 and 10 microns from core.

Table A3.17. Trace element analyses of fassaites (concentrations in ppm, ion microprobe analyses) from CAIs in CV and CH chondrites.

Meteorite	Allende (CV)				Vigarano (CV)	Acfer 182 (CH)		Allende						
CAI Type	B				Type B			Type B						
CAI Name	TS 23				1623-8	418/10	022/lt	Core	TS34		Interior	Mantle fass	Late Fass	
Source	Kuehner et al. (1989)				M & D (1993)	Weber et al. (1995)		Simon et al. (1991)						
	1	2	3	4	5	6	7	8	9	10	11	12	13	14
K	9.5±1.0	12.0±1.2	27.1±1.8	2.3±0.5	nd	nd	nd	nd	nd	nd	nd	nd	nd	nd
Na	149±4	429±7	757±9	11±1	nd	nd	nd	9.2 ±0.6	12.34±0.70	43.59	51.56	13.8±1.1	195.2	7079
Sc	23.9±2.2	22.4±2.3	38.6±2.7	1012±7	723	62.9	282	1451	1307	586	124	5863	41.9±3.2	29±0.9
V	1104±16	738±15	600±11	1064±12	325	202	1233	3020	2536	838	1590	8769	489	48.4±8.3
Cr	530±16	465±15	210±14	329±14	<23	nd	nd	158±11	219±13	411	235±14	32.7±8.4	326	603
Ti	nd	nd	nd	nd	nd	681	72422	nd	nd	nd	nd	nd	nd	nd
Mn	nd	nd	nd	nd	<1.0	nd	nd	nd	nd	nd	nd	nd	nd	nd
Rb	1.7±0.6	1.7±0.6	1.6±0.6	3.2±0.7	39.2	nd	nd	nd	nd	nd	nd	nd	nd	nd
Sr	121±3	201±3	298±4	47±1	92.7	89.8	66.5	27.7±5.7	< 12.6	38.2±5.7	<15.4	22.1±9.8	153±19	411
Y	207.0±2.9	117.1±2.4	194.3±2.9	56.0±1.4	361	22.3	26.5	55	55.2	67.8	123	25.8	468	2693
Zr	359±7	125±4	355±7	414±7	4.03	62.8	191	429	360	346	272	1049	477	1659
Nb	17.6±1.5	20.1±1.7	25.6±1.9	2.40±0.50	0.46±0.10	0.56	3.8	2.26±0.31	2.52±0.35	2.29±0.34	8.54±0.66	5.0±0.7	117	138
Ba	23.3±1.5	45.1±2.2	47.1±2.2	<0.16	3.77	10.6	1.41	2.64	2.36	3.65	13.7	0.70±0.09	64.2	194.8
La	15.15±0.40	8.56±0.32	13.25±0.38	1.77±0.12	15.7	3.2	12.8	10.8	8.56	12.9	41.9	3.66±0.33	187.1	609.6
Ce	40.4±1.0	21.63±0.81	37.7±1.0	8.51±0.43	2.99±0.18	2.74	62.2	2.00±0.21	247±0.24	2.48±0.25	6.99±0.42	1.08 0.24	30.7	102.8
Pr	6.81±0.56	3.28±0.41	5.93±0.54	1.71±0.25	16.6	0.81±0.05	10.8	11.5±0.8	11.5±0.9	15.2±1.0	34.1	4.30±0.78	1309	516
Nd	35.3±2.1	18.6±1.6	31.5±2.0	10.8±1.0	6.82	6.3	46.1	5.37±0.31	5.06±0.32	6.44±0.37	11.59	3.42±0.30	41.7	165.9
Sm	11.39±0.51	6.41±0.42	12.2±0.6	3.72±0.27	0.366±0.029	1.83	197	0.25±0.03	0.26±0.03	0.35±0.04	0.45 ± 0.05	0.38±0.06	3.0±0.2	6.9
Eu	0.47±0.07	0.50±0.09	1.00±0.10	0.20±0.04	10.5	0.39	0.52±0.07	7.57±0.55	6.86±0.61	9.55±0.66	17.3±1.0	2.06±0.59	654±3.4	256.7
Gd	17.9±1.4	9.5±1.2	18.1±1.4	6.94±0.71	2.29	5	7	1.53±0.13	1.50±0.14	2.03±0.16	3.24±0.21	0.67±0.14	12.9±0.7	49.7
Tb	4.02±0.32	1.83±0.25	3.95±0.32	1.58±0.18	16	0.87	1.89	9.34	10.13±0.51	12.3	23.47	3.89±0.47	86.3±2.6	373.3
Dy	28.5±1.2	15.36±0.94	28.5±1.2	9.72±0.61	3.82	6.5	10.8	1.90±0.17	1.94±0.18	2.66±0.21	5.31±0.30	0.89±0.18	17.9±0.9	91.9
Ho	7.26±0.47	3.93±0.37	7.03±0.47	2.12±0.23	10.6	1.07	1.43	6.07	6.83	7.9	15.88	2.50±0.31	58.5	317.1
Er	26.40±0.88	14.78±0.71	24.32±0.87	6.62±0.40	1.48	2.83	3.4	1.15±0.08	1.06±0.08	1.46±0.10	2.73±0.14	0.86±0.11	11.5±0.5	44.9
Tm	4.23±0.23	2.13±0.18	3.62±0.22	0.92±0.10	10.2	0.68	2.39	5.71±0.38	5.27±0.39	7.18±0.46	12.37	2.39±0.39	53.4±2.2	266.8
Yb	22.4±1.1	11.95±0.85	21.2±1.1	6.52±0.53	1.72	0.67	5.2	1.31±0.10	0.99±0.10	1.27±0.11	2.37±0.15	0.48±0.11	9.7±0.5	49.6
Lu	5.34±0.30	2.37±0.23	3.77±0.26	0.68±0.12	7.35	0.23	038±0.05	0.18±0.10	0.37±0.14	3.16±0.41	8.44±0.68	25±0.5	111	1714
Hf	2.02±0.69	<1.1	4.39±0.74	8.71±0.60	0.808±0.093	0.38±0.35	0.85±0.59	16.29	12.79±0.51	9.17	4.81±0.44	34.7	4.9±1.3	13.8±12
Ta	3.92±0.68	1.55±0.58	2.94±0.63	0.32±0.28	nd	nd	nd	0.38±0.15	0.33±0.16	0.20±0.17	0.51±0.24	<0.39	7.9±1.0	13.6±09
Th	5.45±0.38	6.47±0.44	6.99±0.44	0.12±0.06	0.120±0.034	nd	nd	0.11±0.04	0.08±0.04	0.37±0.08	2.62±0.20	<0.11	12.8±0.8	183
U	0.69±0.13	0.74±0.14	0.91±0.15	<0.05	nd	nd	nd	0.03±0.0.2	0.04±0.03	0.04±0.03	0.35±0.08	<0.09	1.4±0.3	2.5±0.2

Notes; M & D (1993) = MacPherson and Davis (1993); Mantle fass = fassaite in the mantle of inclusion TS34; Late fass = late crystallizing fassaite in inclusion TS34 and TS23.
Analyses 1-4 uncertainties (±1 s) and upper limites (2 s) are based on counting statistics alone.
Analyses 8-12 - uncertainties are ±1 s and are only given where they exceed 5 % of the amount present; upper limits are < 2 s.

Table A3.18. Representative electron microprobe analyses of anorthite from CAIs in CV carbonaceous chondrites.

Meteorite	**Allende**	**Bali**		**Allende**						**Leoville**				
CAI Name	CG11						3529Z	7-AN	10-AN	3537-1	3537-1	3537-1	3537-2	3537-2
CAI Type	FTA	Type B		Type B1			B	B	B	B	B	B	B	B
Source	Allen[1]	Kurat et al. (1975)		El Goresy et al. (1985)			Meeker (1985)			Caillet et al. (1993)				
	1*	2	3	4	5	6	7	8	9	10	11	12	13	14
SiO_2	39.18	43.90	44.30	43.60	43.30	44.20	42.90	43.10	43.50	42.30	43.00	43.30	44.10	43.80
TiO_2	nd	0.06	0.08	0.04	0.06	bd	bd	bd	bd	0.07	0.03	<0.03	0.07	0.04
Al_2O_3	37.05	36.00	36.70	35.20	35.00	35.40	35.70	36.50	36.00	36.00	36.30	36.20	35.80	36.50
Cr_2O_3	nd	0.01	0.01	bd	0.06	<0.05	nd	nd	nd	nd	nd	nd	nd	nd
V_2O_3	nd	nd	nd	bd	bd	bd	bd	bd	bd	<0.02	0.10	<0.02	nd	nd
FeO	nd	0.02	0.01	bd	<0.05	bd	bd	bd	bd	0.08	0.16	0.09	0.25	0.24
MnO	nd	nd	nd	nd	nd	nd	nd	nd	nd	nd	nd	nd	<0.03	<0.03
MgO	bd	0.23	0.33	0.18	0.17	0.15	0.05	0.13	0.12	0.06	<0.03	<0.03	0.18	<0.03
CaO	19.68	19.80	19.20	19.20	19.60	18.70	20.50	20.70	20.40	19.70	20.00	20.20	20.60	20.70
Na_2O	0.46	0.09	0.11	0.14	0.10	0.03	0.09	0.07	0.13	0.03	0.17	0.03	0.13	0.05
K_2O	bd	0.04	0.05	nd	nd	nd	nd	nd	nd	<0.02	0.10	<0.02	<0.02	<0.02
Sc_2O_3	nd	nd	nd	nd	nd	nd	bd	bd	bd	nd	nd	nd	nd	nd
Total	96.37	100.15	100.79	98.36	98.29	98.48	99.20	100.50	100.10	98.24	99.86	99.82	101.13	101.33
						Cation Formula Based on 8 Oxygen Atoms								
Si	1.895	2.027	2.027	2.053	2.039	2.081	2.007	1.990	2.016	1.995	1.998	2.010	2.024	2.006
Ti	-	0.002	0.003	0.002	0.002	-	-	-	-	0.002	0.001	0.000	0.002	0.001
Al	2.113	1.960	1.980	1.954	1.945	1.962	1.967	1.988	1.962	2.001	1.989	1.981	1.937	1.971
V	-	-	-	-	-	-	-	-	-	-	-	-	-	-
Cr	-	-	-	-	-	-	-	-	-	-	0.004	-	-	-
Fe	-	0.001	-	-	-	-	-	-	-	0.003	0.006	0.003	0.010	0.009
Mn	-	-	-	-	-	-	-	-	-	-	-	-	-	-
Mg	-	0.016	0.023	0.013	0.012	0.010	0.004	0.009	0.009	0.004	-	-	0.012	-
Ca	1.02	0.980	0.941	0.966	0.988	0.944	1.026	1.024	1.009	0.995	0.996	1.005	1.013	1.016
Na	0.043	0.008	0.010	0.013	0.009	0.002	0.008	0.006	0.011	0.003	0.015	0.003	0.012	0.004
K	-	-	-	-	-	-	-	-	-	-	0.006	-	-	-
Total	5.071	4.996	4.987	5.001	4.997	4.999	5.012	5.017	5.007	5.004	5.015	5.001	5.011	5.009

Notes: bd = below detection; nd = not determined. FTA = "fluffy" Type A.

[1]Allen et al. (1978) * Energy dispersive analysis. 1 - Analyses 2 and 3 are each means of 8 analyses.

Analysis 10 is from a relict island in the inclusions and analyses 11 and 12 are from poikilitic regions.

Table A3.18. (cont.) Representative probe analyses of anorthite from CAIs in CV carbonaceous chondrites.

Meteorite	Allende										Kaba
CAI Name	3529-Z	3529-Z	3529-41	3529-41	3529-21	3529-21	3529-30				
CAI Type	B1	B1	B1	B1	B	B	B	B3	FGSR	FGSR	FGSR
Source	Podosek et al. (1991)							Wark[1]	M&G[2]	M&G[2]	F&G[3]
	1	2	3	4	5	6	7	8*	9	10	11
SiO_2	43.11	43.13	42.80	42.67	42.70	41.88	42.69	42.86	42.11	43.26	43.19
TiO_2	0.06	0.04	0.04	0.05	<0.02	<0.02	<0.02	nd	bd	0.08	0.15
Al_2O_3	36.52	36.56	36.22	36.89	35.87	36.21	35.68	36.37	37.21	36.38	34.73
Cr_2O_3	nd	nd	nd	nd	nd	nd	nd	nd	0.07	0.09	-
V_2O_3	nd	nd	nd	nd	nd	nd	nd	nd	0.01	0.06	< 0.06
FeO	<0.02	<0.02	0.03	<0.02	0.06	0.09	0.07	0.04	0.37	0.18	0.14
MnO	nd	nd	0.01	<0.01	<0.01	<0.01	<0.01	nd	0.09	bd	< 0.03
MgO	0.14	<0.03	0.09	<0.03	0.05	<0.03	0.03	0.01	0.50	0.02	1.47
CaO	20.46	20.22	20.26	19.98	20.04	19.93	19.63	20.06	19.84	19.88	19.48
Na_2O	0.06	0.04	0.09	0.04	0.09	0.03	0.06	0.04	0.19	0.36	0.08
K_2O	<0.03	<0.03	<0.03	<0.03	<0.03	<0.03	<0.03	bd	bd	bd	nd
ZnO	<0.04	<0.04	nd	nd	nd	nd	nd	nd	nd	nd	nd
TOTAL	100.35	99.99	99.54	99.63	98.81	98.14	98.16	99.49	100.39	100.31	99.24
An	100.00	100.00	100.00	100.00	99.00	100.00	99.00				

Cation Formula Based on 8 Oxygen Atoms

	1	2	3	4	5	6	7	8*	9	10	11
Si	1.993	1.998	1.994	1.984	2.003	1.980	2.013	1.998	1.949	2.000	2.016
Ti	0.002	0.001	0.001	0.002	-	-	-	-	-	0.003	0.005
Al	1.990	1.997	1.990	2.022	1.984	2.018	1.983	1.999	2.030	1.983	1.911
V	-	-	-	-	-	-	-	-	0.003	0.003	-
Cr	-	-	-	-	-	-	-	-	0.000	0.002	-
Fe	-	-	0.001	-	0.002	0.004	0.003	0.002	0.014	0.007	0.005
Mn	-	-	-	-	-	-	-	-	-	-	-
Mg	0.010	-	0.006	-	0.003	-	0.002	0.001	0.034	0.001	0.102
Ca	1.013	1.004	1.012	0.995	1.007	1.009	0.992	1.002	0.984	0.985	0.975
Na	0.005	0.004	0.008	0.004	0.008	0.003	0.005	0.004	0.017	0.032	0.007
Cations	5.013	5.004	5.012	5.007	5.007	5.014	4.998	5.006	5.031	5.016	5.021

Notes: bd = below detection; nd = not determined.
[1]Wark et al. (1987); [2]McGuire and Hashimoto (1989); [3]Fegley and Post (1985); *Mantle of the inclusion
FGSR = fine-grained spinel-rich inclusion

Table A3.19. Representative electron microprobe analyses of olivine from CAIs in CV and CM carbonaceous chondrites.

Meteorite	Leoville			Allende		Kaba	Allende					Vigarano		
CAI Name												1623-5		
CAI Type	Fluffy Type A			FGSR	FGSR	FGSR	Type B3					Type B3		
Source	CML (1982)			M&H (1989)		F&P	Wark et al. (1987)					Davis et al. (1991)		
Notes	W-L rim													
	1	2	3	4	5	6	7	8	9	10	11	12	13	14
SiO_2	42.40	41.00	41.10	38.14	38.12	43.19	42.62	42.90	39.60	37.20	36.60	42.40	42.76	41.81
TiO_2	0.03	0.05	0.02	0.25	0.18	bd	0.03	bd	0.09	0.01	bd	0.08	0.08	0.06
Al_2O_3	0.67	0.78	0.67	0.49	0.43	bd	0.09	nd	0.01	0.42	bd	0.06	0.07	0.06
Cr_2O_3	0.11	0.15	0.17	0.38	bd	0.23	0.01	bd	0.13	0.27	nd	bd	bd	bd
V_2O_3	nd	nd	nd	0.03	0.03	bd	nd	nd	nd	nd	nd	0.01	bd	0.01
FeO	0.85	3.00	5.60	23.41	26.32	1.66	0.15	6.20	10.90	20.80	37.00	0.05	0.05	0.02
MnO	<0.02	0.02	0.34	bd	bd	nd	bd	bd	0.11	0.11	0.30	nd	nd	nd
MgO	55.20	54.80	51.10	37.90	34.17	55.36	55.63	51.30	49.20	40.80	27.20	55.62	55.87	55.43
CaO	0.90	0.63	0.33	0.19	0.45	0.05	1.20	0.20	0.09	0.19	bd	1.69	1.72	1.69
Na_2O	bd	bd	bd	bd	0.09	0.02	nd	nd	nd	nd	nd	nd	nd	nd
K_2O	nd	nd	nd	nd	nd	nd	nd	nd	nd	nd	nd	nd	nd	nd
Total	100.16	100.43	99.46	100.99	99.81	99.17	99.74	100.60	100.04	99.80	101.10	99.91	100.55	99.08

Cation Formula Based on 4 Oxygen Atoms

	1	2	3	4	5	6	7	8	9	10	11	12	13	14
Si	0.997	0.973	0.996	0.989	1.010	0.996	1.004	1.024	0.977	0.966	1.010	0.999	1.001	0.994
Ti	0.001	0.001	-	0.005	0.004	-	0.001	-	0.002	-	-	0.001	0.001	0.001
Al	0.019	0.022	0.019	0.015	0.013	-	0.002	-	-	0.013	-	0.002	0.002	0.002
Cr	0.002	0.003	0.003	0.008	-	-	-	-	0.003	0.006	-	-	-	-
V	-	-	-	0.001	0.001	0.004	-	-	-	-	-	-	-	-
Fe	0.017	0.060	0.113	0.508	0.584	0.033	0.003	0.124	0.225	0.452	0.854	0.001	0.001	-
MnO	-	-	0.007	-	-	-	-	-	0.002	0.002	0.007	-	-	-
Mg	1.935	1.939	1.845	1.464	1.351	1.965	1.953	1.824	1.809	1.580	1.119	1.953	1.949	1.964
Ca	0.023	0.016	0.009	0.005	0.013	0.001	0.030	0.005	0.002	0.005	0.000	0.043	0.043	0.043
Na	-	-	-	-	0.005	0.001	-	-	-	-	-	-	-	-
K	-	-	-	-	-	-	-	-	-	-	-	-	-	-
Cations	2.992	3.014	2.993	2.995	2.981	3.000	2.994	2.976	3.020	3.024	2.990	2.999	2.997	3.004

Notes: bd = below detection; nd = not determined. W-L rim + Wark-Lovering rim.

CML (1982) = Christophe Michel-Lévy et al. (1982); M&G (1989) = McGuire and Hashimoto; F&P = Fegley and Post (1985)

Table A3.20. Representative electron microprobe analyses of accessory phases in CAIs in CV carbonaceous chondrites.

Meteorite	Allende	Allende	Kaba
CAI Type	FTA	FTA	FGSR
CAI Name		CG11	
Mineral	Wollastonite		
	1	2*	3
SiO_2	50.30	48.27	50.19
TiO_2	bd	nd	0.03
Al_2O_3	bd	nd	0.86
Cr_2O_3	bd	nd	bd
V_2O_3	nd	nd	< 0.07
Fe_2O_3	nd	nd	
FeO	0.40	0.48	5.98
MnO	nd	nd	0.27
MgO	0.05	nd	5.23
CaO	46.80	48.70	36.09
Na_2O	nd	nd	0.03
K_2O	nd	nd	nd
Total	97.60	97.45	98.82
Cation Formula Based on 6 Oxygens Atoms			
Si	1.997	1.940	0.980
Ti	-	-	-
Al	-	-	0.020
Cr	-	-	-
V	-	-	-
Fe^{3+}	-	-	-
Fe	0.013	0.020	0.098
Mn	-	-	0.004
Mg	0.003	-	0.152
Ca	1.991	2.100	0.755
Na	-	-	0.001
Cations	4.004	4.060	4.003

Meteorite	Allende										
CAI Type	FTA	Unusual	Unusual	B	B	FGSR	FGSR	B3		B3	
CAI Name	CG11										
Mineral	Grossular										
	4	5	6	7	8	9	10	11	range	12	range
SiO_2	38.62	39.37	38.90	37.50	36.40	38.87	39.09	39.20	37.9-41.4	39.30	37.1-41
TiO_2	nd	0.11	bd	<0.05	bd	0.33	0.13	0.13	0.0.36	0.01	0-0.08
Al_2O_3	24.00	23.35	23.20	25.20	28.00	20.12	22.76	22.80	21.5-25.3	23.60	21.6-25.9
Cr_2O_3	nd	0.01	bd	nd	nd	0.04	bd	nd		nd	
V_2O_3	nd	nd	nd	nd	nd	0.15	0.15	nd		nd	
Fe_2O_3	nd	nd	nd	nd	nd	3.71	0.40	nd		nd	
FeO	nd	0.39	0.20	0.39	0.46	bd	bd	1.00	0.3-1.5	1.20	0.3-4
MnO	nd	nd	nd	nd	nd	0.12	bd	0.02	0-0.15	0.16	0-0.9
MgO	nd	0.34	0.10	1.19	1.17	0.42	0.32	1.70	0.8-3.1	1.80	0-3.8
CaO	38.48	34.55	37.60	36.40	31.50	36.47	36.85	35.20	32.8-36.9	33.80	31.3-36.1
Na_2O	nd	nd	nd	nd	nd	0.00	0.09	0.06	0-0.4	0.12	0-0.21
K_2O	0.09	nd	bd	nd	nd	0.01	0.01	nd		nd	
Total	101.19	98.12	100.00	100.73	97.53	100.24	99.80	100.20		100.09	
Cation Formula Based on 12 Oxygens Atoms											
Si	2.880	1.993	1.952	1.867	1.841	2.954	2.945	1.959		1.959	
Ti	-	0.004	0.000	0.000	0.000	0.019	0.007	0.005		-	
Al	2.110	1.393	1.372	1.479	1.669	1.803	2.021	1.343		1.387	
Cr	-	-	-	-	-	0.002	-	-		-	
V	-	-	-	-	-	0.009	0.009	-		-	
Fe^{3+}	-	-	-	-	-	0.212	0.023	-		-	
Fe	-	0.017	0.008	0.016	0.019	-	-	0.042		0.050	
Mn	-	-	-	-	-	0.008	-	0.001		0.007	
Mg	-	0.026	0.007	0.088	0.088	0.048	0.036	0.127		0.134	
Ca	3.070	1.874	2.022	1.942	1.707	2.970	2.974	1.885		1.805	
Na	-	-	-	-	-	0.000	0.013	0.006		0.012	
Cations	0.010	5.306	5.362	5.393	5.325	0.001	0.001	5.367		5.353	

Notes: bd = below detection; nd = not determined. FTA = "fluffy" Type A.

Analysis 1 - Fuchs (1971), Analysis 2 and 4 - Allen et al. (1978), Analysis 3 - Fegley and Post (1985); Analyses 5 and 6 - Wark (1986);
Analyses 7 and 8 Bischoff and Palme (1987); Analyses 9 and 10 - McGuire and Hashimoto (1989); Analyses 11 and 12 - Wark et al. (1987)

Table A3.21. Representative electron microprobe analyses of accessory phases in CAIs in CV carbonaceous chondrites.

Meteorite	Allende	Allende	Allende	Allende	Kaba	Allende									
CAI Type	FTA	FTA	FGSR	FGSR	FGSR	B3	FTA	B3	FGSR	FGSR	hib-sp-pv inclusion				
CAI Name		CG11					CG11								
Mineral	andradite					Monticellite	Nepheline	Nepheline	Sphene	Sphene	Ilmenite				
	1	2*	3	4	5	6*	7	8	9	10	11	12	13	14	14
SiO_2	36.10	38.35	35.29	37.37	37.15	37.90	41.29	44.90	31.31	31.29	nd	nd	nd	nd	nd
TiO_2	nd	nd	bd	bd	0.07	bd	bd	bd	33.32	34.18	54.20	52.90	53.40	51.70	51.60
Al_2O_3	nd	bd	0.17	0.08	1.55	1.10	34.80	34.00	3.99	4.44	0.27	2.05	0.47	1.40	0.87
Cr_2O_3	nd	nd	bd	0.20	< 0.06	nd	nd	nd	bd	bd	nd	nd	nd	nd	nd
V_2O_3	nd	nd	0.04	0.04	< 0.07	nd	nd	nd	0.18	0.19	nd	nd	nd	nd	nd
Fe_2O_3	31.40	nd	29.96	28.76		nd	nd	nd	bd	bd	nd	nd	nd	nd	nd
FeO	nd	26.88	bd	bd	27.71	2.60	nd	bd	0.28	0.72	36.30	37.50	39.10	42.30	42.60
MnO	nd	nd	0.09	0.00	0.25	bd	nd	nd	bd	0.00	0.22	0.25	0.23	0.27	0.24
MgO	32.50	1.98	0.05	0.12	1.02	22.60	bd	0.20	0.33	0.07	7.81	5.88	5.22	3.11	2.97
CaO	nd	30.89	34.40	34.54	31.55	35.60	3.71	4.40	27.20	26.42	0.53	0.68	0.90	0.72	0.70
Na_2O	nd	bd	bd	0.03	0.05	bd	16.24	16.20	0.40	0.94	nd	nd	nd	nd	nd
K_2O	nd	bd	bd	bd	bd	bd	1.66	1.40	0.10	0.12	nd	nd	nd	nd	nd
Total	100.00	98.10	100.00	101.14	99.45	99.90	97.70	101.10	97.11	98.37	99.33	99.26	99.32	99.50	98.98

Cation Formula

No of [O]	12	12	12	12	12	4	16	16	5	5	6	6	6	6	6
Si	1.850	3.220	2.998	3.076	3.099	0.999	4.005	4.188	1.041	1.029	-	-	-	-	-
Ti	-	-	-	-	0.004	-	-	-	0.833	0.845	1.960	1.921	1.962	1.924	1.938
Al	-	-	0.017	0.008	0.152	0.034	3.980	3.739	0.156	0.172	0.015	0.117	0.027	0.082	0.051
Cr	-	-	0.000	0.013	-	-	-	-	-	-	-	-	-	-	-
V	-	-	0.003	0.003	-	-	-	-	0.005	0.005	-	-	-	-	-
Fe^{3+}	1.211	1.700	1.916	1.793	-	-	-	-	-	-	-	-	-	-	-
Fe	-	-	-	-	1.735	0.057	-	-	0.008	0.020	1.460	1.514	1.598	1.751	1.779
Mn	-	-	0.006	-	0.018	-	-	-	0.000	0.000	0.009	0.010	0.010	0.011	0.010
Mg	-	0.250	0.006	0.015	0.126	0.888	-	-	0.016	0.003	0.560	0.423	0.380	0.229	0.221
Ca	2.483	2.780	3.132	3.066	2.813	1.006	0.385	0.440	0.969	0.931	0.027	0.035	0.047	0.038	0.037
Na	0.000	-	-	0.005	0.008	-	3.055	2.930	0.026	0.060	-	-	-	-	-
K	-	-	-	-	-	-	0.205	0.167	0.004	0.005	-	-	-	-	-
Cations	5.544	7.950	8.078	7.979	7.955	2.984	11.630	11.464	3.058	3.070	4.032	4.021	4.024	4.035	4.037

Notes: bd = below detection; nd = not determined. FTA = "fluffy" Type A; FGSR - fine-grained spinel-rich inclusion. * Energy dispersive analysis
Analysis 1 - Fuchs (1971), Analysis 2 and 7 - Allen et al. (1978), Analyses 3, 4, 9 and 10 - McGuire and Hashimoto (1989); Analysis 5 - Fegley and Post (1985);
Analysis 6 - Wark et al. (1987); Analyses 11-15 - Steele (1995b)

Table A3.22. Representative electron microprobe analyses of diopside, hedenbergite and salite in CAIs in CV and CM chondrites.

Meteorite	Allende	Allende	Leoville	Allende	Kaba	Allende		Essebi†
CAI Type	FTA	FTA	FTA	FGSR	FGSR	FGSR		
CAI Name		CG11						
Mineral	Diopside	Diopside	Diopside	Diopside	Diopside	Rhönite		Diopside
	1	2*	3	4	5	6	range	7
SiO_2	53.50	52.38	53.20	54.56	53.89	19.10	18.6 - 20.1	44.30
TiO_2	0.70	0.98	0.26	0.02	0.52	16.80	14.1 - 18.4	3.11
Al_2O_3	3.10	3.98	2.87	1.64	2.61	28.90	27.6 - 29.6	13.70
Cr_2O_3	nd	nd	0.09	bd	< 0.05	nd	-	0.03
V_2O_3	nd	bd	nd	bd	0.07	nd	-	bd
FeO	0.10	0.50	0.50	0.14	0.23	1.90	0.3 - 4.2	0.98
MnO	nd	nd	<0.02	bd	bd	bd	-	bd
MgO	17.10	16.97	18.60	18.33	19.06	15.70	13.9 - 17.8	12.90
CaO	25.20	25.76	23.30	25.69	23.38	17.90	17.0 - 19.6	23.20
Na_2O	nd	nd	<0.02	0.03	0.07	bd	-	nd
K_2O	nd	nd	nd	nd	nd	bd	-	nd
Total	99.70	100.57	98.82	100.41	99.83	101.00		98.22
				Cations Formula Based on 6 Oxygens Atoms				
Si	1.933	1.890	1.934	1.960	1.936	0.722		1.645
Ti	0.019	0.030	0.007	0.001	0.014	0.478		0.086
Al	0.132	0.170	0.123	0.069	0.111	1.288		0.600
Cr	-	-	0.003	-	-	-		-
V	-	-	-	-	0.002	-		-
Fe	0.003	0.020	0.015	0.004	0.007	0.060		0.030
Mn	-	-	-	-				-
Mg	0.921	0.910	1.008	0.981	1.020	0.884		0.715
Ca	0.975	99.000	0.907	0.898	0.900	0.725		0.923
Na	-	-	-	0.002	0.005	-		-
Cations	3.982	4.010	3.996	3.915	3.995	4.157		3.999

Meteorite	Kaba	Allende				
CAI Type	FGSR	FGSR	FTA	FTA	FTA	FTA
CAI Name			CG11	CG11	CG11	CG11
Mineral	Hedenbergite				Salite	Fe-salite
	9	8	10*	11*	12*	13*
SiO_2	49.13	48.19	47.60	45.41	49.68	48.56
TiO_2	bd	bd	bd	nd	nd	nd
Al_2O_3	0.60	0.21	bd	bd	2.79	1.18
Cr_2O_3	< 0.05	bd	nd	nd	nd	nd
V_2O_3	< 0.06	bd	nd	nd	nd	nd
Fe_2O_3	bd	bd	nd	nd	nd	nd
FeO	25.36	27.41	25.77	27.90	10.61	20.55
MnO	0.51	0.46	nd	nd	nd	nd
MgO	2.18	0.07	0.32	bd	10.66	4.83
CaO	21.98	23.73	24.51	23.06	24.17	23.25
Na_2O	0.03	0.03	nd	nd	nd	nd
K_2O	bd	bd	nd	nd	nd	nd
Total	99.79	100.10	98.20	96.37	97.91	98.37
		Cations Formula Based on 6 Oxygens Atoms				
Si	1.993	1.986	1.990	1.960	1.920	1.960
Ti	-	-	-	-	-	-
Al	0.029	0.010	-	-	0.130	0.060
Cr	-	-	-	-	-	-
V	-	-	-	-	-	-
Fe^{3+}	0.861	-	-	-	-	-
Fe^{2+}	-	0.944	0.900	1.010	0.340	0.690
Mn	0.018	0.016	-	-	-	-
Mg	0.132	0.024	0.020	-	0.620	0.290
Ca	0.956	1.048	1.100	1.070	1.000	1.010
Na	0.003	0.002	-	-	-	-
Cations	3.992	4.030	4.010	4.040	4.010	4.010

Notes: bd = below detection; nd = not determined. FTA = "fluffy" Type A inclusion; FGSR = fine-grained spinel-rich inclusion; †CM chondrite

Analysis 1 - Fuchs (1971); Analysis 2 - Allen et al. (1978); Analysis 3 - McGuire and Hashimoto (1989), Analyses 4 and 5 - Fegley and Post (1985)

Analysis 6 - Fuchs? Analyis 7 - El Goresy et al. (1985); BaO Sc2O3, ZrO2, were also measured but below detection. Analysis 9 - Fegley and Post (1985).

Analysis 9 - McGuire and Hashimoto (1989); Analysis 10-13 - Allen et al. (1978)

Analysis 3 is from the Wark-Lovering rim on the inclusion.

Analysis 6 - total contains 0.7 wt % of elements which were described by Fuchs as the remainder. *Energy dispersive analysis

Table A3.23. Representative electron microprobe analyses of kamacite in EH and EL chondrites.

Meteorite	Yamato 691															
Type	EH3															
Source	Ikeda (1989)															
	1	2	3	4	5	6	7	8	9	10	11	12	13	14	15	16
Si	2.19	2.15	2.21	1.92	1.96	2.04	1.89	1.88	1.85	2.10	2.10	2.29	1.99	2.06	1.71	1.96
P	bd	bd	0.02	bd	bd	bd	bd	bd	0.03	bd	bd	0.03	0.04	0.02	0.01	0.03
Cr	0.01	bd	bd	0.01	bd	0.01	bd	0.01	0.02	bd	bd	0.02	bd	0.02	0.05	nd
Fe	94.35	94.06	94.01	95.55	94.37	95.69	94.87	94.40	94.74	94.65	95.34	94.51	94.25	95.49	95.34	94.43
Co	0.30	0.32	0.30	0.36	0.20	0.31	0.38	0.37	0.33	0.29	0.22	0.21	0.35	bd	0.34	0.32
Ni	3.35	3.49	3.25	2.64	3.30	2.00	3.19	3.04	3.44	3.44	2.52	2.44	3.39	2.60	2.46	3.28
Total	100.20	100.02	99.79	100.48	99.83	100.05	100.33	99.70	100.41	100.48	100.18	99.50	100.02	100.19	99.91	100.02

Meteorite	Indarch		Kota-Kota		Adhi Kot		Abee*		R80259	Blithfield			Yilmia	Blithfield		Yilmia	JdKL
Type	EH4		EH3,4		EH4				EL5	EL6			EL6	EL6		EL6	EL6
Source	Leitch and Smith (1982)								S (1984)	Rubin (1984)			B&H	Rubin (1984)		B&H	K&A
Notes										Clast 1	Clast 2	Matrix		Clast 1	Clast 2		
	mean	range	mean	range	mean	range	mean	range		mean	mean	mean		mean	mean	mean	
No. Anal.	79		32		52		45			45	15	61		2	2	5	
	1	2	3	4	5	6	7	8	9	10	11	12	13	14	15	16	17
Si	3.11	2.92-3.38	2.04	1.73-2.28	3	2.31-3.43	3.24	2.96-3.62	2.1	1.62	1.37	1.22	0.92	1.81	1.28	1.23	1.29
P	0.112	0.051-0.180	0.093	0.063-0.135	0.124	0.073-0.188	0.43	0.313-0.559	0.43	0.35	0.31	0.33	nd	< 0.05	< 0.05	0.01	bd
S	0.017	0-0.078	0.012	.006-0.035	0.009	00-0.027	0.012	0-0.026	nd	nd	nd	nd	nd	nd	nd	nd	nd
Fe	89.7	88.7-90.7	90.2	89.1-90.6	89.2	88.6-90.4	89.4	88.0-90.2	89.7	91.2	90.9	91.1	92.8	86	83.3	81	92
Ni	6.53	6.00-6.88	3.2	2.70-3.72	6.37	5.92-6.87	6.4	6.21-6.72	6.4	6.7	6.5	6.7	5.68	11.7	14.5	16.4	6.19
Co	0.38	0.302-0.432	0.433	0.379-0.477	0.362	0.284-0.407	0.474	0.41-0.54	0.64	0.09	0.13	0.13	0.43	0.23	0.23	0.28	0.37
Cu	nd	nd	nd	nd	nd	nd	nd		nd	nd	nd	nd	nd	nd	nd	0.06	nd
Total	99.849		95.811		99.265		99.954		99.27	99.96	99.21	99.48	99.83	99.74	39.31	98.98	99.85

bd = below detection; nd = not determined

JdKL = Jajh deh Kot Lalu; K&A (1965) = Keil and Anderson (1965); B&H (1972) = Buseck and Holdsworth (1972)

*Abee has been classified as an EH4, but Rubin and Scott (1997) have shown that Abee is an impact melt breccia.

Analyses 14-17 are martensite, although Keil and Andersen (1965) and Buseck and Holdsworth (1972) termed them taenite

Table A3.24. Representative electron microprobe analyses of troilite from EH and EL chondrites.

Meteorite	Yamato 691														
Source	Ikeda (1989)														
	m	m	m	p	p	m	m	p	p	p	p	m	p	p	m
	1	2	3	4	5	6	7	8	9	10	11	12	13	14	15
S	36.54	37.61	35.82	35.83	34.52	36.80	36.59	35.59	35.81	34.35	37.62	37.80	35.57	36.07	36.58
Cr	1.04	1.71	0.62	0.62	0.64	1.23	1.28	1.12	1.13	0.05	0.20	1.05	0.28	0.21	0.97
Fe	60.97	59.97	64.05	63.07	63.37	60.38	60.55	58.66	60.61	64.48	62.88	60.89	62.22	63.68	62.21
Ni	0.02	0.18	nd	nd	nd	bd	nd	nd	nd	nd	0.05	0.02	nd	nd	0.13
Cl	nd	nd	bd	0.33	1.28	bd	bd	4.54	0.66	0.84	bd	nd	1.12	0.10	0.00
Total	98.57	99.47	100.49	99.85	99.81	98.41	98.42	99.91	98.21	99.72	100.75	99.76	99.19	100.06	99.89

*m = massive troilite, p = porous troilite   bd = below detection; nd = not determined

Meteorite	Yamato 691						Kota-Kota		Indarch	Adhi-Kot		Abee	St.Mark's	RPA80259	JdKL	Blithfield		
Type	EH3						EH 3,4		EH4	EH4		EH4	EH5	EL5	EL6	EL6		
Source	El Goresy et al. (1989)						K (1968)	Leitch and Smith (1982)			R (1983)	L&S	K (1968)	S (1984	K&A	Rubin (1984)		
Notes	Primary troilite			Secondary troilite				Mean	Mean	Mean	Mean			Mean		Clast	Clast	Matrix
No. anal.								17	24	20	109			26		25	10	18
	1	2	3	4	5	6	7	8	9	10	11	13	14	15	16	17	18	19
Mg	nd	nd	nd	nd	nd	nd	nd	nd	nd	nd	0.05	nd	nd	0.07	nd	0.08	0.05	0.05
S	36.80	36.40	36.90	35.30	35.40	34.80	37.40	36.10	35.90	36.30	36.90	36.60	36.40	37.80	36.90	36.80	37.40	37.10
Ca	bd	bd	bd	bd	0.10	0.06	nd	nd	nd	nd	<0.05	nd	nd	<0.04	<0.02	nd	nd	nd
Ti	0.32	0.20	0.21	bd	bd	bd	0.27	0.29	0.26	0.35	0.40	0.38	0.38	0.47	0.67	0.71	0.55	0.31
Cr	0.91	1.04	1.13	0.52	0.45	0.42	0.60	0.43	1.52	1.73	2.50	1.62	0.70	3.10	0.65	0.93	0.93	0.87
Mn	nd	nd	nd	nd	nd	nd	0.05	0.82	1.01	0.93	0.40	0.71	0.09	0.48	<0.04	< 0.07	< 0.07	< 0.07
Fe	61.60	61.00	62.10	63.40	63.70	61.70	62.30	31.00	60.80	59.70	58.60	59.80	61.70	58.10	61.30	61.70	61.30	61.40
Co	nd	nd	nd	nd	nd	nd	nd	0.00	0.00	0.00	nd	0.00	nd	nd	nd	nd	nd	nd
Ni	nd	nd	nd	nd	nd	nd	nd	0.04	0.04	0.05	nd	0.05	nd	nd	nd	nd	nd	nd
Cu	nd	nd	nd	nd	nd	nd	nd	0.03	0.04	0.04	nd	0.04	nd	nd	nd	nd	nd	nd
Zn	nd	nd	nd	nd	nd	nd	nd	0.03	0.02	0.02	nd	0.01	0.12	0.10	nd	nd	nd	nd
Cl	bd	bd	bd	0.15	0.20	0.44	nd	nd	nd	nd	nd	nd	nd	nd	nd	nd	nd	nd
Total	99.63	98.64	100.34	99.47	99.81	97.45	100.74	98.74	99.58	99.12	98.85	99.22	99.38	100.12	99.52	100.22	100.23	99.73

bd = below detection; nd = not determined; Analysis 11 - mean of 109 analyses; Analysis 15, mean of 26 analyses; Analyses 17, 18, 19, mean of 25, 10 and 18 analyses, respectively
Sources: K (1968) = Keil (1968b), R = Rubin (1983); L&S = Leitch and Smith (1982), S (1984) = Sears et al. (1984); K&A = Keil and Anderson (1965)   JdKL = Jajh deh Kot Lalu

Table A3.25. Representative electron microprobe analyses of niningerites in EH and EL chondrites.

Meteorite	Yamato 691														
Type	EH3														
Source	Ikeda (1989)														
	1	2	3	4	5	6	7	8	9	10	11	12	13*	14*	15*
Mg	33.78	31.15	31.40	33.38	32.94	35.26	34.51	34.50	31.91	31.52	32.22	30.79	32.10	32.00	31.60
S	51.15	50.80	50.92	51.32	50.38	50.02	50.75	51.16	49.39	49.04	49.27	49.25	48.30	49.70	49.10
Ca	0.33	0.40	0.38	0.28	0.40	0.32	0.32	0.27	0.39	0.38	0.38	0.43	0.46	0.49	0.38
Cr	0.09	0.16	0.17	0.06	0.13	0.10	0.15	0.09	0.09	0.12	0.13	0.35	4.00	3.57	3.90
Mn	4.35	4.27	3.87	2.10	3.60	4.06	4.15	3.17	3.91	5.03	4.90	4.90	0.11	0.14	0.16
Fe	11.81	13.08	11.66	14.08	13.63	9.73	10.47	10.02	13.63	13.83	13.01	13.84	14.00	14.20	14.00
Ni	0.02	0.06	0.06	bd	bd	0.02	bd	0.07	0.12	0.16	0.05	0.08	0.48	1.30	0.02
Total	101.53	99.92	98.46	101.22	101.08	99.51	100.35	99.28	99.44	100.08	99.96	99.64	99.50	101.40	101.14

*Cu-bearing niningerites
bd = below detection; nd = not determined

Meteorite	Kota-Kota		Abee		Adhi-Kot			Indarch		Saint Sauveur		St. Mark's		R80259
Type	EH3,4		EH4		EH4			EH4		EH5		EH5		EL5
Source	Keil and Snetsinger (1967)				R (1983)	Keil and Snetsinger (1967)				Keil and Snetsinger (1967)				S (1984)
	1	2	3	4	5	6	7	8	9	10	11	12	13	14
Mg	23.50	18.4-27.1	10.10	8.9-11.6	12.90	11.30	10.9-12.3	18.30	17.5-19.3	13.20	10.9-14.6	22.70	21.6-24.9	11.20
S	46.90	45.4-47.8	41.00	3.43-4.54	42.30	42.60	41.7-43.1	43.40	41.8-43.5	42.70	41.5-43.3	47.40	47.3-49.4	41.50
Ca	0.39	0.28-0.57	3.03	2.34-3.66	12.90	1.96	1.57-2.28	1.28	1.19-1.42	2.55	2.22-2.91	0.53	0.42-0.69	3.00
Ti	nd		nd		<0.08	nd		nd						<0.08
Cr	0.14	0.10-0.17	1.84	1.64-2.04	1.70	1.97	1.82-2.16	1.66	1.44-2.09	1.77	1.69-1.96	0.40	0.32-0.51	1.30
Mn	11.60	9.8-12.4	4.02	3.61-4.26	6.90	7.10	6.3-7.6	6.50	6.1-7.0	3.93	3.43-4.54	11.80	10.5-12.9	12.80
Fe	15.60	12.8-19.9	37.10	35.6-38.2	34.40	34.20	33.4-35.4	27.00	24.5-28.5	35.20	34.0-36.8	16.60	15.5-18.3	32.10
Zn	nd		0.31	0.21-0.40	nd	nd		nd		nd		nd		nd
Total	98.13		97.40		99.90	99.13		98.14		99.35		99.43		101.90

bd = below detection; nd = not determined; Analysis 5 = mean of 71 analyses.
R (1983) = Rubin (1983); S (1984) = Sears et al. (1984)
R80259 = RKPA80259

Table A3.26. Representative electron microprobe analyses of alabandite from EL6 chondrites.

Meteorite					Yilmia	JdKL
Type	EL6				EL6	EL6
Source	Rubin (1984)				B&H	K&A
Notes	Clast 1	Clast 2	Matrix			
	mean (10)	mean (2)				
	1	2	3	4	5	6
Mg	7.20	6.20	4.30	12.80	4.35	2.15
S	38.90	39.40	39.40	40.40	39.20	37.40
Ca	0.26	0.35	0.24	0.54	0.20	0.20
Ti	< 0.08	< 0.08	< 0.08	< 0.08	nd	<0.02
Cr	0.14	0.23	0.13	0.23	0.10	0.55
Mn	40.20	39.10	44.60	28.00	42.50	45.80
Fe	13.50	14.80	12.10	18.20	14.60	11.70
Total	100.20	100.08	100.77	100.17	100.95	97.80

nd = not determined JdKL = Jajh deh Kot Lalu

K&A (1965) = Keil and Anderson (1965); B&H (1972) = Buseck and Holdsworth (1972)

Table A3.27. Representative electron microprobe analyses of oldhamite in EH and EL chondrites.

Meteorite	Yamato 691															
Type	EH3															
Source	Ikeda (1989)													El Goresy et al. (1988)		
	1	2	3	4	5	6	7	8	9	10	11	12	13	14	15	16
Na	nd	nd	nd	nd	nd	nd	nd	nd	nd	nd	nd	nd	nd	nd	nd	nd
Mg	0.35	0.5	0.4	0.43	0.32	0.49	0.42	0.4	0.37	0.33	0.37	0.41	0.44	0.41	0.39	0.39
S	44.05	43.9	42.76	45.18	45.33	44.98	44.07	44.99	44.56	44.97	45.11	44.45	43.37	43.30	42.80	43.30
Ca	53.34	55.07	53.8	53.49	54.72	52.5	54.23	53.12	53.07	53.46	54.63	53.31	54.63	54.30	53.50	54.00
Ti	nd	nd	nd	nd	nd	nd	nd	nd	nd	nd	nd	nd	nd	nd	nd	nd
Cr	nd	nd	nd	nd	nd	nd	nd	nd	nd	nd	nd	nd	nd	0.04	bd	bd
Mn	0.19	0.11	0.36	0.38	0.19	0.25	0.19	0.06	0.22	0.08	0.06	0.14	0.33	0.09	0.09	0.06
Fe	0.89	0.41	1.02	0.83	0.17	0.37	0.35	0.21	0.65	0.8	0.37	0.51	0.28	0.68	0.63	0.59
Ni	nd	nd	nd	nd	nd	nd	nd	nd	nd	nd	nd	nd	nd	0.10	0.10	0.09
Co	nd	nd	nd	nd	nd	nd	nd	nd	nd	nd	nd	nd	nd	bd	bd	bd
Cu	nd	nd	nd	nd	nd	nd	nd	nd	nd	nd	nd	nd	nd	0.18	0.21	0.18
Zn	nd	nd	nd	nd	nd	nd	nd	nd	nd	nd	nd	nd	nd	0.14	0.12	0.12
Total	98.82	99.99	98.34	100.31	100.73	98.59	99.26	98.78	98.87	99.64	100.54	98.82	99.05	99.24	97.84	98.73

bd = below detection; nd = not determined

Meteorite	Qingzhen			Adhi-Kot			Indarch				Kota-Kota		Abee		Yilmia	JdKL
Type	EH3			EH4			EH4				EH4		impact melt breccia		EL6	EL6
Source	El Goresy et al. (1988)			R (1983)	L&S (1982)		B&H		L&S (1982)		Leitch and Smith (1982)				B&H	K&A
Notes				Mean	mean	range	mean	mean	mean	range	mean	range	mean	range		
No. Anal.				17	18		5	5	20		11		19			
	1	2	3	4	5	6	7	8	9	10	11	12	13	14	15	16
Na	nd	nd	nd	nd	nd	-	nd	nd	nd	0.051-0.097	0.04	0.023-0.073	0.08	0.051-0.01	nd	nd
Mg	0.46	0.47	0.58	1.50	1.23	1.02-1.39	0.61	0.50	1.14	0.99-1.37	0.90	0.74-1.15	1.31	1.05-1.47	0.50	0.51
S	43.70	45.00	42.70	43.80	42.20	41.8-42.5	41.28	39.50	42.10	41.8-42.6	41.60	40.4-42.4	42.40	41.8-42.7	39.50	43.70
Ca	54.40	54.80	52.90	1.50	52.10	50.9-52.9	51.50	53.10	52.30	51.8-52.7	51.00	50.5-51.5	51.60	51.2-52.1	53.10	54.20
Ti	nd	nd	nd	<0.08	0.01	0-0.01	bd	bd	0.00	0-0.009	0.00	0-0.007	0.01	0-0.024	bd	<0.2
Cr	bd	bd	bd	<0.07	0.00	0-0.007	0.01	0.01	0.00	0-0.007	0.00	0-0.007	0.01	0-0.015	0.01	<0.04
Mn	0.20	0.13	0.16	0.50	0.28	0.22-0.32	0.29	1.88	0.21	0.112-0.317	0.17	0.15-0.19	0.23	0.194-0.27	1.88	1.01
Fe	0.08	0.81	1.70	1.10	0.71	0.64-0.79	0.41	0.55	0.61	0.43-0.90	5.50	3.71-6.10	0.53	0.39-0.63	0.55	0.44
Ni	bd	bd	0.14	nd	0.02	0.012-0.04	nd	nd	0.02	0.01-0.032	0.00	0-0.007	0.02	0.011-0.027	nd	nd
Co	bd	0.03	bd	nd	nd		nd	nd	nd		nd		nd		nd	nd
Cu	bd	0.03	bd	nd	0.03	0.015-0.05	nd	nd	0.04	0.024-0.051	0.02	0.012-0.027	0.03	0.022-0.047	nd	nd
Zn	0.07	0.03	bd	nd	nd		nd	nd	nd		nd		nd		nd	nd
Total	98.91	101.30	98.18	99.80	96.65		94.02	95.54	96.50		99.32		96.21		95.54	99.86

JdKL = Jajh deh Kot Lalu   bd = below detection; nd = not determined

R (1983) = Rubin (1983); L&S = Leitch and Smith (1982); K&A (1965) = Keil and Anderson (1965); B&H (1972) = Buseck and Holdsworth (1972)

Table A3.28. Representative electron microprobe analyses of daubreelites in EH and EL chondrites.

Meteorite	Qingzhen				Kota-Kota	St. Mark's	Yilmia		Blithfield			JdKL
Type	EH3				EH3,4	EH5	EL6		EL6			EL6
Source	El Goresy et al. (1988)				Keil (1968a)		B&H (1972)		Rubin (1984)			K&A
Notes									Clast 1	Clast 2	Matrix	
	1	2	3	4	5	6	7	8	9	10	11	12
Na	0.34	0.24	0.39	0.32	nd	nd	nd	nd	nd	nd	nd	nd
Mg	nd	nd	nd	nd	<0.03	<0.03	nd	nd	0.06	0.06	< 0.05	nd
S	43.00	44.20	43.20	45.10	43.00	43.90	44.40	43.60	44.30	46.00	45.80	44.40
Ca	bd	bd	bd	bd	nd	nd	nd	nd	nd	nd	nd	<0.02
Ti	bd	0.09	0.05	0.10	<0.02	0.12	0.08	bd	0.10	0.11	< 0.08	0.05
Cr	35.50	36.00	35.50	35.60	34.10	35.20	35.30	34.40	34.90	35.20	34.60	35.30
Mn	0.52	0.89	0.46	0.32	1.04	0.73	2.78	1.12	2.80	2.60	2.10	2.38
Fe	13.30	12.20	11.40	11.10	14.70	14.90	16.20	15.20	16.70	16.70	17.30	16.50
Cu	0.23	bd	0.23	0.19	nd	nd	nd	nd	nd	nd	nd	nd
Zn	5.70	6.61	7.53	8.11	5.20	4.26	0.14	3.94	nd	nd	nd	nd
Total	98.59	100.23	98.76	100.84	98.04	99.11	99.90	98.26	98.86	100.67	99.80	98.63

bd = below detection; nd = not determined

B&H (1972) = Buseck and Holdsworth (1972); K&A = Keil and Anderson (1965); JdKL = Jajh deh Kot Lalu

Analyses 8 is the means of 13 analyses; Analyses 9, 10 and 11 are means of 15, 5 and 6 analyses, respectively.

Table A3.29. Representative electron microprobe analyses of sphalerite from EH and EL chondrites.

Meteorite	Yamato 691									Qingzhen						
Type	Ikeda (1989)									El Goresy and Ehlers (1989)						
Source	EH3									EH3						
	1	2	3	4	6	7	8	9	10	11	12	13	14	15	16	17
Mg	0.48	0.71	0.71	0.99	0.61	1.04	0.80	0.67	1.00	0.64	0.71	0.70	0.70	0.61	0.63	0.64
S	35.22	35.24	34.84	36.05	35.42	35.76	34.69	34.64	35.54	34.70	34.40	34.30	34.20	34.20	34.40	33.90
Cr	0.24	0.07	bd	bd	0.28	bd	0.09	0.24	0.09	<0.05	<0.05	<0.05	bd	<0.05	<0.05	<0.05
Mn	6.18	0.98	1.07	1.05	1.03	1.22	1.03	1.14	2.23	3.85	3.49	3.44	3.45	3.45	3.45	3.45
Fe	32.64	29.85	27.75	29.65	31.24	28.16	29.81	32.49	27.19	27.80	28.00	27.60	27.80	28.50	29.20	28.90
Ni	0.44	0.07	0.01	0.10	0.13	0.31	0.18	0.31	bd	0.64	0.74	0.88	0.86	0.36	0.27	0.29
Cu	1.03	0.04	0.07	bd	0.05	0.21	0.17	0.14	0.16	0.25	0.14	0.77	0.46	0.35	0.41	0.40
Zn	21.86	32.60	33.92	31.75	29.79	33.14	32.28	29.42	32.38	30.90	31.80	29.20	29.20	29.20	29.60	29.30
Ga	nd	nd	nd	nd	nd	nd	nd	nd	nd	0.08	0.09	0.91	0.61	0.49	0.36	0.46
Total	98.09	99.56	98.37	99.59	98.55	99.84	99.05	99.05	98.59	98.86	99.37	97.80	97.28	97.16	98.32	97.34

Meteorite	ALHA 77295		Indarch								Pillistfer
Type	EH3,4		EH4								EL6
Source	Kissin (1989)							El Goresy and Ehlers (1989)			K (1989)
Notes	mean (2)		mean (2)								mean (5)
	2	3	4	5	6	7	8	9	10	11	16
S	33.18	35.86	35.26	35.14	36.13	34.02	35.37	34.30	34.30	34.60	34.86
Ca	nd	nd	bd	bd	bd	bd	0.80	nd	nd	nd	nd
Ti	nd	nd	nd	nd	nd	nd	nd	nd	nd	nd	nd
Cr	nd	nd	nd	nd	nd	nd	nd	0.07	bd	bd	nd
Mn	2.11	5.27	2.45	2.40	3.28	1.75	2.03	2.88	2.55	3.37	13.13
Fe	27.12	28.15	39.34	35.89	34.54	41.21	32.73	31.40	32.60	32.50	30.18
Ni	nd	nd	nd	nd	nd	nd	nd	nd	nd	nd	bd
Co	nd	nd	nd	nd	nd	nd	nd	nd	nd	nd	nd
Cu	0.16	0.27	bd	bd	bd	bd	0.75	<0.05	0.50	0.21	bd
Zn	33.62	28.93	23.02	25.10	22.46	20.34	26.19	23.50	27.00	24.20	20.39
Ga	0.43	0.38	0.00	0.38	0.61	0.81	0.71	<0.05	bd	0.09	bd
Total	96.62	98.86	100.07	98.91	97.02	98.10	98.57	92.15	96.65	94.88	98.56

K (1989) = Kissin (1989)   bd = below detection; nd = not determined

Table A3.30. Electron microprobe analyses of caswellsilverite.

Meteorite	Y-691	Qingzhen		
Type	EH3	EH3		
Source	I (1989)	El Goresy et al. (1989)		
	1	2	3	4
Na	15.31	15.80	15.70	15.90
S	45.49	46.90	46.60	46.90
K	0.07	bd	bd	bd
Ca	nd	0.17	0.20	bd
Cr	36.31	37.50	37.00	37.50
Mn	0.21	0.11	0.08	bd
Fe	0.77	0.85	0.72	0.77
Ni	0.07	bd	bd	bd
Cu	0.12	bd	bd	bd
Total	98.35	101.33	100.30	101.07

bd = below detection; nd = not determined
I = Ikeda (1989)

Table A3.31. Representative electron microprobe analyses of djerfisherite in EH and EL chondrites.

Meteorite	Yamato 691									Qingzhen			Y-691	St Mark's
Type	EH3									EH3			EH3	EH5
Source	Ikeda (1989)									El Goresy et al. (1989)				F (1966)
	1	2	3	4	5	6	7	8	9	10	11	12	13	14
Na	2.14	1.71	1.47	1.83	2.13	1.12	1.61	1.12	15.31	0.92	0.70	1.52	1.14	0.30
Source	33.27	34.68	35.73	35.53	35.30	33.67	33.23	33.79	45.49	30.30	29.60	31.20	32.20	33.80
K	8.03	6.74	6.56	6.58	6.46	7.76	7.95	7.76	0.07	8.08	8.40	7.07	7.94	8.70
Ca	nd	nd	nd	nd	nd	nd	nd	nd	nd	bd	bd	0.01	bd	bd
Cr	0.04	bd	bd	bd	0.03	0.04	bd	0.06	36.31	bd	bd	0.04	0.04	bd
Mn	bd	bd	bd	bd	bd	bd	bd	0.03	0.21	0.03	bd	0.13	0.01	bd
Fe	48.61	50.20	50.25	50.69	51.45	49.55	50.83	50.72	0.77	53.50	54.50	51.60	50.90	50.70
Ni	0.49	2.62	2.87	2.63	3.03	0.47	0.56	1.39	0.07	1.35	1.32	3.02	0.49	0.80
Cu	5.11	2.18	1.71	1.66	1.43	5.21	4.54	4.17	0.12	3.16	2.31	2.62	4.73	4.20
Zn	nd	nd	nd	nd	nd	nd	nd	nd	nd	bd	bd	bd	bd	bd
Cl	1.58	1.54	1.52	bd	bd	1.72	bd	bd	bd	1.48	1.48	1.49	1.50	1.00
Total	99.27	99.67	100.11	98.92	99.83	99.54	98.72	99.04	98.35	98.82	98.31	98.70	98.95	99.50

F (1966) = Fuchs (1966)   bd = below detection; nd = not determined

Table A3.32. Representative electron microprobe analyses of schreibersites in EH and EL chondrites.

	Yamato- 691 (EH3) Ikeda (1989)															
	1	2	3	4	5	6	7	8	9	10	11	12	13	14	15	16
Si	0.24	0.16	0.20	0.14	0.16	0.16	0.14	0.19	0.19	0.15	0.17	0.18	0.16	0.17	0.17	0.14
P	16.26	15.40	16.55	15.62	15.66	15.47	15.29	16.34	16.13	16.44	16.73	15.48	15.59	15.56	15.51	16.28
Cr	0.02	0.00	0.04	0.00	0.02	0.10	0.15	0.04	0.05	0.00	0.00	0.00	0.01	0.12	0.05	0.04
Fe	67.84	68.50	66.47	68.64	69.19	68.09	66.63	68.75	67.35	67.25	67.01	66.75	68.51	68.21	69.33	69.15
Co	0.09	0.05	0.00	0.04	0.12	bd	bd	bd	bd	0.06	0.00	0.01	0.07	bd	0.03	0.01
Ni	17.08	15.93	16.49	16.78	16.51	16.34	18.71	15.96	17.91	17.59	17.71	18.35	15.57	16.06	15.22	16.20
Total	101.53	100.04	99.75	101.22	101.66	100.16	100.92	101.28	101.63	101.49	100.62	101.77	99.91	100.12	100.31	101.82

Meteorite	Yamato - 691			Qingzhen			Indarch	Kota-Kota	Adhi-Kot		Abee	Yilmia	JdKL	Blithfield		
Type	EH3			EH3			EH4	EH4	EH4			EL6	EL6	EL6		
Source	El Goresy et al. (1988)			El Goresy et al. (1988)			Leitch and Smith (1982)			R (1983)	L&S	B&H	K&A	Rubin (1984)		
Notes														Clast 1	Clast 2	Matrix
	1	2	3	4	5	6	7	8	9	10	11	12	13	14	15	16
Si	0.14	0.14	0.16	0.25	0.26	0.16	0.20	0.19	0.21	0.40	0.24	0.24		0.15	0.06	0.10
S	nd	nd	nd	nd	nd	nd	0.01	0.01	0.01	nd	0.01	nd	nd	nd	nd	nd
P	14.20	13.90	13.60	15.10	15.10	15.00	15.40	15.10	15.00	15.80	15.30	15.50	15.40	14.60	15.00	14.70
Cr	bd	bd	bd	bd	bd	0.33	nd	nd	nd	nd	nd	nd	nd	nd	nd	nd
Fe	68.30	70.00	71.20	69.30	70.10	71.80	71.00	70.30	78.00	78.40	77.20	63.20	62.40	55.30	60.90	62.40
Ni	15.60	14.40	13.10	15.70	15.50	12.50	13.70	14.10	7.60	6.40	7.20	20.80	22.40	nd	nd	nd
Co	bd	bd	bd	bd	bd	bd	0.24	0.22	0.23	0.30	0.24	0.17	0.30	29.30	23.70	22.50
Cu	bd	bd	bd	bd	bd	bd	nd	nd	nd	nd	nd	0.09	nd	< 0.09	< 0.09	0.12
Total	98.24	98.44	98.06	100.35	100.96	99.79	100.55	99.92	101.05	101.20	100.19	100.00	100.50	99.35	99.66	99.82

*bd = below detection nd = not determined

R (1983) = Rubin (1983); L&S = Leitch and Smith (1982) B&H = Buseck and Holdsworth (1972); K&A = (1965); JdKL = Jajh deh Kot Lalu

Analyses 12 is mean of 10 analyses; Analyses 14, 15 and 16 are means of 8, 2 and 17 analyses, respectively.

Table A3.33. Representative electron microprobe analyses of perryite from Yamato 691 (Ikeda, 1989).

	1	2	3	4	5	6	7	8	9	10	11	12
Si	11.5	11.49	11.76	11.88	11.73	11.76	11.73	11.71	11.33	11.9	11.82	12.18
P	4.01	3.94	4.11	4.24	3.88	3.79	4.01	3.92	3.94	3.94	3.86	3.8
Cr	0.01	0.05	bd	bd	0.1	0.05	0.05	0.02	0.11	0.07	bd	bd
Fe	6.76	7.02	7.88	8.53	7.7	8.24	9.18	9.36	9.8	8.94	10	9.61
Co	0.05	0.02	bd	0.08	0.01	0.04	0.08	0.01	bd	bd	bd	0.08
Ni	79.18	78.08	77.4	75.96	77.97	76.61	75.65	76.16	75.7	76.11	74.7	75.71
Total	101.61	100.6	101.15	100.69	101.39	100.49	100.7	101.18	100.88	100.96	100.38	101.38

bd = below detection; nd = not determined

Table A3.34. Olivine in Type 4-6 chondrites.

Meteorite	Conquista	Santa Barbara	Nyirabrany	Uberaba	Timmersoi	Khanpur	Seoni	Willowbar	Mangwendi	Indarch[1]	Indarch[2]	Karoonda
Type	H4	L4	LL4	H5	L5	LL5	H6	L6	LL6	EH4	EH4	CK4
Ref	Keil (1978)	Berkley et al. (1978)	Heyse (1978)	Gomes et al. (1977a)	Smith et al. (1972)	Heyse (1978)	Bunch et al. (1972)	Lange et al. (1973)	Heyse (1978)	Leitch and Smith (1982)	Leitch and Smith (1982)	Noguchi (1993)
Chemical composition (wt %)												
SiO_2	39.6	38.3	37.4	39.1	38.1	37.1	39.8	37.1	37.0	42.2	42.3	37.5
TiO_2	0.02	0.02	0.02	nd	nd	0.02	nd	bd	0.01	0.04	0.03	0.00
Al_2O_3	0.07	nd	0.05	0.09	nd	0.04	nd	bd	0.02	0.09	0.09	0.02
Cr_2O_3	0.03	0.07	0.00	nd	0.02	0.02	nd	bd	0.01	0.24	0.05	0.00
FeO	16.2	23.1	25.7	18.3	22.4	25.7	18.3	22.7	26.4	0.6	0.6	26.4
MnO	0.46	0.60	0.42	0.48	0.43	0.44	0.58	0.47	0.44	0.14	0.04	0.18
MgO	43.7	39.7	37.1	42.5	38.9	37.1	41.9	39.7	36.3	56.3	56.3	35.1
CaO	0.02	0.05	0.04	bd	nd	0.04	0.06	0.02	0.05	0.19	0.14	0.04
NiO	nd	nd	nd	nd	nd	nd	nd	nd	nd	nd	nd	0.51
Total	100.10	101.84	100.75	100.47	99.85	100.50	100.64	99.99	100.18	99.75	99.54	99.78
Cation formula based on 4 oxygens												
Si	1.000	0.984	0.983	0.994	0.994	0.978	1.008	0.971	0.983	0.995	0.999	1.000
Ti	0.000	0.000	0.000	0.000	0.000	0.000	0.000	0.000	0.000	0.001	0.001	0.000
Al	0.002	0.000	0.002	0.003	0.000	0.001	0.000	0.000	0.001	0.002	0.003	0.001
Cr	0.001	0.001	0.000	0.000	0.000	0.000	0.000	0.000	0.000	0.004	0.001	0.000
Fe	0.342	0.496	0.566	0.389	0.489	0.568	0.388	0.497	0.586	0.011	0.012	0.589
Mn	0.010	0.013	0.009	0.010	0.010	0.010	0.012	0.010	0.010	0.003	0.001	0.004
Mg	1.644	1.519	1.454	1.610	1.513	1.461	1.582	1.549	1.435	1.979	1.981	1.395
Ca	0.001	0.001	0.001	0.000	0.000	0.001	0.002	0.001	0.001	0.005	0.003	0.001
Ni	0.000	0.000	0.000	0.000	0.000	0.000	0.000	0.000	0.000	0.000	0.000	0.011
Total	2.999	3.015	3.015	3.005	3.006	3.021	2.992	3.029	3.016	3.001	2.999	3.000
Cation ratio 100 x Fe/(Fe+Mg)												
Fa	17.2	24.6	28.0	19.5	24.4	28.0	19.7	24.3	29.0	0.6	0.6	29.7

[1] Olivine shows orange CL [2] Olivine shows blue CL

Table A3.35. Low-Ca pyroxene in Type 4-6 chondrites.

Meteorite	Conquista	Santa Barbara	Nyirabrany	Uberaba	Timmersoi	Khanpur	Seoni	Willowbar	Mangwendi	Uden
Type	H4	L4	LL4	H5	L5	LL5	H6	L6	LL6	LL7
Ref	Keil (1978)	Berkley et al. (1978)	Heyse (1978)	Gomes et al. (1977a)	Smith et al. (1972)	Heyse 78	Bunch et al. (1972)	Lange et al. (1973)	Heyse (1978)	Heyse (1978)
	Chemical composition (wt %)									
SiO_2	56.9	55.5	55.1	56.6	56.2	54.4	55.8	55.1	54.3	55.2
TiO_2	0.06	0.13	0.07	nd	0.17	0.14	0.21	0.15	0.16	0.17
Al_2O_3	0.21	0.08	0.16	0.25	0.17	0.17	0.21	0.15	0.18	0.25
Cr_2O_3	0.17	0.02	0.13	nd	0.08	0.12	0.14	0.08	0.19	0.27
FeO	10.30	13.90	14.83	11.60	13.40	15.38	10.60	14.60	16.01	15.63
MnO	0.46	0.51	0.47	0.48	0.44	0.46	0.61	0.53	0.46	0.41
MgO	31.30	29.50	29.04	31.00	28.80	28.36	31.10	29.40	27.54	26.60
CaO	0.45	0.74	0.55	0.7	0.64	0.69	0.63	0.8	0.96	1.55
Na_2O	0.03	nd	0.03	0.07	0.02	0.02	bd	bd	0.02	0.05
NiO	nd	nd	nd	nd	nd	nd	nd	nd	nd	nd
Total	99.88	100.38	100.36	100.70	99.92	99.72	99.30	100.81	99.82	100.15
	Cation formula based on 6 oxygens									
Si	2.001	1.980	1.974	1.989	2.004	1.969	1.983	1.966	1.971	1.993
Ti	0.002	0.003	0.002	0.000	0.005	0.004	0.006	0.004	0.004	0.005
Al	0.009	0.003	0.007	0.010	0.007	0.007	0.009	0.006	0.008	0.011
Cr	0.005	0.001	0.004	0.000	0.002	0.003	0.004	0.002	0.005	0.008
Fe	0.303	0.415	0.445	0.341	0.400	0.466	0.315	0.436	0.486	0.472
Mn	0.014	0.015	0.014	0.014	0.013	0.014	0.018	0.016	0.014	0.013
Mg	1.640	1.569	1.551	1.623	1.531	1.531	1.647	1.564	1.490	1.431
Ca	0.017	0.028	0.021	0.026	0.024	0.027	0.024	0.031	0.037	0.060
Na	0.002	0.000	0.002	0.005	0.001	0.001	0.000	0.000	0.001	0.003
Ni	0.000	0.000	0.000	0.000	0.000	0.000	0.000	0.000	0.000	0.000
Total	3.992	4.014	4.020	4.009	3.987	4.022	4.005	4.025	4.018	3.995
	Cation ratios Ca:Mg:Fe									
Fs	15.5	20.6	22.0	17.1	20.4	23.0	15.9	21.5	24.1	24.0
En	83.7	78.0	76.9	81.6	78.3	75.7	82.9	77.0	74.0	72.9
Wo	0.9	1.4	1.0	1.3	1.3	1.3	1.2	1.5	1.9	3.1

Table A3.35. (cont.) Low-Ca pyroxene in Type 4-6 chondrites.

Meteorite	Adhi-Kot[1]	Adhi-Kot[2]	St. Mark's	St. Mark's	RPA80259	Jajh deh Kot Lalu	Karoonda	EET87507
Type	EH4	EH4	EH5	EH5	EL5	EL6	CK4	CK5
Ref	Leitch and Smith (1982)	Leitch and Smith (1982)	Keil (1968)	Lusby et al. (1987)	Sears et al. (1984)	Keil (1968)	Noguchi (1993)	Noguchi (1993)
	Chemical composition (wt %)							
SiO_2	58.8	58.8	58.8	56.9	59.4	59.5	55.4	51.9
TiO_2	0.051	0.007	nd	bd	nd	nd	0.1	0.02
Al_2O_3	0.196	0.073	0.04	0.57	0.19	0.28	1.56	3.45
Cr_2O_3	0.32	0.04	nd	0.22	nd	nd	0.27	0.14
FeO	0.93	0.95	0.36	7.20	0.37	0.16	11.01	17.22
MnO	0.17	0.03	0.17	0.11	bd	bd	0.14	0.23
MgO	38.70	39.10	39.10	33.80	39.90	39.30	30.33	25.49
CaO	0.246	0.063	0.06	0.64	0.69	0.72	1.02	0.58
Na_2O	0.05	0.12	bd	bd	nd	bd	0.02	0.00
NiO	0.00	0.00	bd	nd	nd	bd	0.10	0.13
Total	99.47	99.18	98.53	99.44	100.55	99.96	99.90	99.11
	Cation formula based on 6 oxygens							
Si	1.990	1.993	1.999	1.982	1.984	1.995	1.958	1.906
Ti	0.001	0.000	0.000	0.000	0.000	0.000	0.003	0.001
Al	0.008	0.003	0.002	0.023	0.007	0.011	0.065	0.150
Cr	0.009	0.001	0.000	0.006	0.000	0.000	0.008	0.004
Fe	0.026	0.027	0.010	0.210	0.010	0.004	0.326	0.529
Mn	0.005	0.001	0.005	0.003	0.000	0.000	0.004	0.007
Mg	1.952	1.975	1.981	1.755	1.986	1.964	1.597	1.395
Ca	0.009	0.002	0.002	0.024	0.025	0.026	0.039	0.023
Na	0.003	0.008	0.000	0.000	0.000	0.000	0.001	0.000
Ni	0.000	0.000	0.000	0.000	0.000	0.000	0.003	0.004
Total	4.002	4.009	4.000	4.003	4.012	4.000	4.003	4.016
	Cation ratios Ca:Mg:Fe							
Fs	1.3	1.3	0.5	10.5	0.5	0.2	16.6	27.2
En	98.2	98.5	99.4	88.2	98.3	98.5	81.4	71.7
Wo	0.4	0.1	0.1	1.2	1.2	1.3	2.0	1.2

[1] Pyroxene shows red CL [2] Pyroxene shows blue CL

Table A3.36. Ca-rich pyroxene in Type 4-6 chondrites.

Meteorite	Avanhandava	Santa Barbara	Soko Banja	Itapicuru Mirim	Timmersoi	Khanpur	Seoni	Willowbar	Mangwendi	Uden	Karoonda	EET87507
Type	H4	L4	LL4	H5	L5	LL5	H6	L6	LL6	LL7	CK4	CK5
Ref	Paar et al. (1976)	Berkley et al. (1978)	McSween and Patchen (1989)	Gomes et al. (1977b)	Smith et al. (1972)	Heyse (1978)	Bunch et al. (1972)	Lange et al. (1973)	Heyse (1978)	Heyse (1978)	Noguchi (1993)	Noguchi (1993)
	Chemical composition (wt %)											
SiO_2	54.2	50.8	54.5	54.4	53.2	53.4	55.4	53.3	53.7	55.2	52.7	52.2
TiO_2	0.43	0.27	0.31	nd	0.45	0.38	0.48	0.4	0.43	0.17	0.34	0.27
Al_2O_3	0.68	0.26	0.40	0.71	0.48	0.54	0.43	0.60	0.52	0.25	1.66	2.58
Cr_2O_3	0.69	0.78	0.65	nd	0.60	0.66	0.44	0.70	0.74	0.27	0.35	0.13
FeO	3.60	7.00	5.41	4.40	4.87	5.05	3.96	5.20	5.82	15.63	6.65	6.74
MnO	0.22	0.28	0.22	0.24	0.27	0.17	0.29	0.24	0.19	0.41	0.13	0.11
MgO	17.3	19.5	16.6	17.3	16.6	16.7	17.3	16.6	16.6	26.6	15.5	16.9
CaO	22.3	20.2	21.5	21.4	22.1	22.3	22.0	22.1	21.0	1.6	22.3	20.6
Na_2O	0.63	0.46	0.46	0.67	0.48	0.48	0.34	0.60	0.53	0.05	0.17	0.18
K_2O	nd	nd	nd	nd	nd	nd	nd	nd	nd	nd	nd	nd
Total	100.05	99.55	100.10	99.12	99.05	99.67	100.64	99.74	99.62	100.15	99.85	99.80
	Cation formula based on 6 oxygens											
Si	1.976	1.897	1.995	1.999	1.973	1.968	2.002	1.966	1.980	1.993	1.951	1.925
Ti	0.012	0.008	0.009	0.000	0.013	0.011	0.013	0.011	0.012	0.005	0.009	0.007
Al	0.029	0.011	0.017	0.031	0.021	0.023	0.018	0.026	0.023	0.011	0.072	0.112
Cr	0.020	0.023	0.019	0.000	0.018	0.019	0.013	0.020	0.022	0.008	0.010	0.004
Fe	0.110	0.219	0.166	0.135	0.151	0.156	0.120	0.160	0.179	0.472	0.206	0.208
Mn	0.007	0.009	0.007	0.007	0.008	0.005	0.009	0.007	0.006	0.013	0.004	0.003
Mg	0.940	1.086	0.905	0.947	0.917	0.919	0.932	0.913	0.914	1.431	0.854	0.931
Ca	0.871	0.808	0.844	0.843	0.878	0.881	0.852	0.874	0.831	0.060	0.885	0.814
Na	0.045	0.033	0.033	0.048	0.035	0.034	0.024	0.043	0.038	0.003	0.012	0.013
K	0.000	0.000	0.000	0.000	0.000	0.000	0.000	0.000	0.000	0.000	0.000	0.000
Total	4.010	4.094	3.995	4.010	4.013	4.017	3.982	4.021	4.005	3.995	4.004	4.016
	Cation ratios Ca:Mg:Fe											
Fs	5.7	10.4	8.6	7.0	7.8	8.0	6.3	8.2	9.3	24.0	10.6	10.6
En	48.9	51.4	47.3	49.2	47.1	47.0	49.0	46.9	47.5	72.9	43.9	47.7
Wo	45.4	38.3	44.1	43.8	45.1	45.1	44.8	44.9	43.2	3.1	45.5	41.7

Table A3.37. Feldspar in Type 4-6 chondrites.

Meteorite	Avanhandava	Ipiranga	Seoni	Santa Barbara	Timmersoi	Ruhobobo	Indarch	St. Mark's	Jajh deh Kot Lalu	Maralinga	Maralinga
Type	H4	H5	H6	L4	L5	L6	EH4	EH5	EL6	CK4	CK4
Ref	Paar et al. (1976)	Gomes et al. (1978)	Bunch et al. (1972)	Berkley et al. (1978)	Smith et al. (1972)	Klob et al. (1981)	Leitch and Smith (1982)	Keil (1968)	Keil (1968)	Keller et al. (1992)	Keller et al. (1992)
Chemical composition (wt %)											
SiO_2	64.7	66.3	64.9	66.3	65.1	67.9	69.7	69.4	65.2	46.3	56.8
TiO_2	nd	nd	bd	0.02	nd	0.04	nd	nd	nd	0.03	nd
Al_2O_3	20.8	20.9	21.5	21.9	21.1	19.9	18.1	18.6	21.1	34.4	26.9
FeO	0.68	0.69	0.45	0.64	0.46	0.69	0.30	0.39	0.19	0.75	1.00
MgO	0.26	0.45	0.17	bd	0.06	nd	nd	nd	nd	0.09	0.07
CaO	2.2	2.4	2.2	1.9	2.4	2.2	0.1	0.2	3.3	16.7	8.2
Na_2O	9.9	8.96	9.7	9.1	9.87	8.4	11.5	10.4	9.5	2	6.6
K_2O	0.97	0.83	0.98	0.86	0.65	0.86	0.24	1.00	0.83	0.07	0.60
Total	99.51	100.52	99.85	100.76	99.68	99.94	99.92	100.00	100.08	100.34	100.17
Cation formula based on 8 oxygens											
Si	2.880	2.905	2.872	2.893	2.884	2.974	3.046	3.034	2.880	2.127	2.557
Ti	0.000	0.000	0.000	0.001	0.000	0.001	0.000	0.000	0.000	0.001	0.000
Al	1.092	1.080	1.122	1.127	1.102	1.028	0.933	0.959	1.099	1.863	1.428
Fe	0.025	0.025	0.017	0.023	0.017	0.025	0.011	0.014	0.007	0.029	0.038
Mg	0.017	0.029	0.011	0.000	0.004	0.000	0.000	0.000	0.000	0.006	0.005
Ca	0.105	0.112	0.102	0.091	0.116	0.101	0.004	0.010	0.154	0.822	0.396
Na	0.855	0.761	0.832	0.770	0.848	0.713	0.975	0.882	0.814	0.178	0.576
K	0.055	0.046	0.055	0.048	0.037	0.048	0.013	0.056	0.047	0.004	0.034
Total	5.029	4.959	5.011	4.952	5.007	4.891	4.981	4.955	5.001	5.031	5.034
Cation ratios Ca:Na:K											
An	10.3	12.2	10.3	10.0	11.6	11.7	0.4	1.0	15.2	81.9	39.3
Ab	84.2	82.8	84.1	84.7	84.7	82.7	98.3	93.1	80.2	17.7	57.3
Or	5.4	5.0	5.6	5.3	3.7	5.6	1.4	5.9	4.6	0.4	3.4

Table A3.38. PHOSPHATES IN TYPE 6 CHONDRITES

meteorite	Seoni	Willowbar	Seoni	Willowbar
type	H6	L6	H6	L6
ref	Bunch et al. (1972)	Lange et al. (1973)	Bunch et al. (1972)	Lange et al. (1973)
	chlorapatite	chlorapatite	merrillite	merrillite
	Chemical composition (weight percent)			
SiO_2	1.60	0.30	1.10	0.14
FeO	0.51	0.60	0.43	0.75
MnO	nd	0.10	nd	bd
MgO	0.05	bd	3.47	3.40
CaO	52.7	54	47.1	47.4
Na_2O	bd	0.60	2.60	3.00
P_2O_5	41.5	41.4	45.7	45.6
F	bd	nd	bd	nd
Cl	5.9	5.6	bd	nd
Total	102.26 [1]	102.6 [2]	100.4	100.29

[1] minus O = F,Cl, = 1.59, total = 100.67
[2] minus O = F,Cl, = 1.3, total = 101.3

Table A3.39. OXIDES IN TYPE 4-6 CHONDRITES

meteorite	Bath	Allegan	Butsura	Goodland	Lua	Walters	Soko Banja	Khanpur	Dhurmsala	Cullison	Gobabeb
type	H4	H5	H6	L4	L5	L6	LL4	LL5	LL6	H4	H4
ref	Bunch et al. (1967)	Bunch et al. (1967)	Bunch et al. (1967)	Bunch et al. (1967)	Bunch et al. (1967)	Bunch et al. (1967)	Bunch et al. (1967)	Bunch et al. (1967)	Bunch et al. (1967)	Krot et al. (1993)	Fudali and Noonan (1975)
oxide	chromite	chromite	chromite	chromite	chromite	chromite	chromite	chromite	chromite	spinel	spinel
	Chemical composition (weight percent)										
TiO_2	2.24	2.08	2.47	1.44	2.54	2.96	2.34	2.53	2.89	0.77	0.32
Al_2O_3	6.2	6.3	5.6	3.7	3.8	5.5	4.8	6.3	5.7	26.5	40.4
Cr_2O_3	56.4	56.5	56.8	59.8	57.6	56.0	56.3	54.7	55.2	39.1	26.3
FeO	30.5	32.0	31.4	32.8	33.0	32.8	34.3	34.3	33.5	23.2	19.7
MnO	0.93	1.05	0.95	0.80	1.00	0.91	0.45	0.58	0.75	0.35	0.23
MgO	2.5	2.3	2.8	1.6	2.2	2.1	1.6	1.7	2.0	8.6	12.3
ZnO	nd	nd	nd	nd	nd	nd	nd	nd	nd	0.80	nd
V_2O_3	0.70	0.71	0.66	0.81	0.71	0.72	0.78	0.75	0.69	nd	0.17
Total	98.75	100.24	100.06	100.18	100.13	100.22	99.78	100.11	100.03	99.32	99.25
	Cation formula based on 32 oxygens										
Ti	0.480	0.440	0.530	0.310	0.550	0.630	0.510	0.540	0.620	0.145	0.056
Al	2.090	2.100	1.870	1.250	1.280	1.840	1.640	2.110	1.910	7.841	11.019
Cr	12.740	12.630	12.710	13.620	13.080	12.570	12.810	12.310	12.430	7.758	4.810
Fe^{2+}	7.280	7.570	7.430	7.890	7.930	7.790	8.260	8.170	7.980	4.870	3.812
Mn	0.230	0.250	0.230	0.190	0.240	0.210	0.110	0.140	0.180	0.074	0.045
Mg	1.060	0.970	1.200	0.710	0.940	0.870	0.660	0.720	0.850	3.217	4.241
Zn	0.000	0.000	0.000	0.000	0.000	0.000	0.000	0.000	0.000	0.148	0.000
V	0.160	0.160	0.150	0.190	0.160	0.170	0.180	0.170	0.160	0.000	0.032
Total	24.040	24.120	24.120	24.160	24.180	24.080	24.170	24.160	24.130	24.053	24.015
calculated Fe_2O_3	0.56	1.67	1.11	1.67	2.24	1.11	2.22	2.22	1.67		

REFERENCES

Afiattalab F, Wasson JT (1980) Composition of the metal phases in ordinary chondrites: Implications regarding classification and metamorphism. Geochim Cosmochim Acta 44:431-446

Ahrens LH (1970) The composition of stony meteorites (VIII). Observations on fractionation between L and H chondrites. Earth Planet Sci Lett 9:345-347

Akai J (1980) Tubular form of interstratified mineral consisting of a serpentine-like layer plus two brucite-like sheets newly found in the Murchison (C2) meteorite. Mem NIPR Spec Issue 17:299-310

Akai J (1982) High resolution electron microscopic characterisation of phyllosilicates and finding of a new type with IIA structure in Yamato-74662. Mem NIPR Spec Issue 25:131-144

Akai J (1988) Incompletely transformed serpentine-type phyllosilicates in the matrix of Antarctic CM chondrites. Geochim Cosmochim Acta 48:1593-1599

Akai J (1990) Mineralogical evidence of heating events in Antarctic carbonaceous chondrites, Y-86720 and Y-82162. Proc NIPR Symp Antarct Met 3:55-68

Akai J (1992) T-T-T diagram of serpentine and saponite, and estimation of metamorphic heating degree of Antarctic carbonaceous chondrites. Proc NIPR Symp Antarct Met 5:120-135

Akai J (1994) Void structures in olivine grains in thermally metamorphosed Antarctic carbonaceous chondrites. Proc NIPR Symp Antarct Met 7:94-100

Akai J, Kanno J (1986) Mineralogical study of matrix- and groundmass phyllosilicates and isolated olivines in Yamato-791198 and -793321: with special reference to new finding of a 14-Å chlorite in groundmass. Mem NIPR Spec Issue 41:259-275

Akai J, Sekine T (1994) Shock effect experiments on serpentine and thermal metamorphic conditions in Antarctic carbonaceous chondrites. Proc NIPR Symp Antarct Met 7:101-109

Aldhardt W, Frey F, Jagodinski H (1980) X-ray and neutron investigation of the $P\bar{1}$-$I\bar{1}$transition in anorthite with low albite content. Acta Cryst A36:471-470

Alexander CMO'D (1990) In situ measurement of interstellar silicon carbide in two CM chondrites. Nature 348:715-717

Alexander CMO'D (1993) Presolar SiC in chondrites: How variable and how many sources? Geochim Cosmochim Acta 57:2869-2888

Alexander CMO'D (1994) Trace element distributions within ordinary chondrite chondrules: Implications for chondrule formation conditions and precursors. Geochim Cosmochim Acta 58:3451-3467

Alexander CMO'D, Hutchison RH, Graham AL, Yabuki H (1987) Discovery of scapolite in the Bishunpur (LL3) chondritic meteorite. Min Mag 51:733-735

Alexander CMO'D, Barber DJ, Hutchison R (1989a) The microstructure of Semarkona and Bishunpur. Geochim Cosmochim Acta 53:3045-3057

Alexander CMO'D, Hutchison R, Barber DJ (1989b) Origin of chondrule rims and interchondrule matrices in unequilibrated ordinary chondrites. Earth Planet Sci Lett 95:187-207

Alexander CMO'D, Swan P, Prombo CA (1994) Occurrence and implications of silicon nitride in enstatite chondrites. Meteoritics 29:79-85

Allen JM, Grossman L, Davis AM, Hutcheon ID (1978) Mineralogy, textures and mode of formation of a hibonite-bearing Allende inclusion. Proc Lunar Planet Sci Conf 9:1209-1233

Allen JM, Grossman L, Lee T, Wasserburg GJ (1980) Mineralogy and petrography of HAL, an isotopically unusual Allende inclusion. Geochim Cosmochim Acta 44:685-699

Allen JS, Nozette S, Wilkening LL (1980) A study of chondrule rims and chondrule irradiation records in unequilibrated ordinary chondrites. Geochim Cosmochim Acta 44:1161-1175

Allen RO, Mason B (1973) Minor and trace elements in some meteoritic minerals. Geochim Cosmochim Acta 37:1435-1456

Anders E, Grevesse N (1989) Abundances of the elements: Meteoritic and solar. Geochim Cosmochim Acta 53:197-214

Anders E, Zinner E (1993) Interstellar grains in primitive meteorites: diamond, silicon carbide and graphite. Meteoritics 28:490-514

Andersen CA, Keil K, Mason B (1964) Silicon oxynitride: A meteoritic mineral. Science 146:256-257

Andrehs G, Adams K (1990) The Trebbin chondrite. Meteoritics 25:319-321

Armstrong JT, Meeker GP, Huneke JC, Wasserburg GJ (1982) The Blue Angel: I. The mineralogy and petrogenesis of a hibonite inclusion from the Murchison meteorite. Geochim Cosmochim Acta 46:575-595

Armstrong JT, El Goresy A, Wasserburg GJ (1985) Willy: A prize noble Ur-Fremdling—Its history and implications for the formation of Fremdlinge and CAI. Geochim Cosmochim Acta 49:1001-1022

Armstrong JT, Hutcheon ID, Wasserburg GJ (1987) Zelda and company: Petrogenesis of sulfide-rich fremdlinge and constraints on solar nebula processes. Geochim Cosmochim Acta 51:3155-3173

Ashworth JR (1977) Matrix textures in unequilibrated ordinary chondrites. Earth Planet Sci Lett 35:25-34

Ashworth JR (1979) Two kinds of exsolution in chondritic olivine. Mineral Mag 43:535-538
Ashworth JR (1980a) Chondrite thermal histories: clues from electron microscopy of orthopyroxene. Earth Planet Sci Lett 46:167-177
Ashworth JR (1980b) Deformation mechanisms in mildly shocked chondritic diopside. Meteoritics 15: 105-115
Ashworth JR (1981) Fine structure in H-group chondrites. Proc R Soc London A374:179-194
Ashworth JR (1985) Transmission electron microscopy of L-group chondrites: 1. Natural shock effects. Earth Planet Sci Lett 73:17-32
Ashworth JR, Barber DJ (1975) Electron petrography of shock-deformed olivine in stony meteorites. Earth Planet Sci Lett 27:43-50
Ashworth JR, Barber DJ (1977) Electron microscopy of some stony meteorites. Phil Trans R Soc Lond A286:493-506
Axon HJ, Grokhovsky VJ (1982) Discontinuous precipitation in the metal of Richardton chondrite. Nature 296:835-837
Baily JT, Russel R Jr (1969) Preparation and properties of dense spinel ceramics in the $MgAl_2O_4$-Al_2O_3 system. Trans Brit Ceram Soc 68:159-164
Barber DJ (1977) The matrix of C2 and C3 carbonaceous chondrites (abstr). Meteoritics 12:172-173
Barber DJ (1981) Matrix phyllosilicates and associated minerals in C2M carbonaceous chondrites. Geochim Cosmochim Acta 45:945-970
Barber DJ (1985) Phyllosilicates and other layer-structured materials in stony meteorites. Clay Minerals 20:415-454
Barber DJ, Freeman LA, Bourdillon A (1983) Fe-Ni-S-O layer phase in C2M carbonaceous chondrites—a hydrous sulphide? Nature 305:295-297
Barber DJ, Hutchison R (1991) The Bencubbin stony-iron meteorite breccia: Electron petrography, shock history and affinities of a "carbonaceous chondrite" clast. Meteoritics 26:83-95
Barber DJ, Martin PM, Hutcheon ID (1984) The microstructures of minerals in coarse-grained Ca-Al-rich inclusions from the Allende meteorite. Geochim Cosmochim Acta 48:769-783
Barber DJ, Beckett JR, Paque JM, Stolper E (1994) A new titanium-bearing calcium aluminosilicate phase: II. Crystallography and crystal chemistry of grains formed in slowly cooled melts with bulk compositions of calcium-aluminum-rich inclusions. Meteoritics 29:682-690
Bar-Matthews M, Hutcheon ID, MacPherson GJ, Grossman L (1982) A corundum-rich inclusion in the Murchison carbonaceous chondrite. Geochim Cosmochim Acta 46:31-41
Bass MN (1971) Montmorillonite and serpentine in Orgueil meteorite. Geochim Cosmochim Acta 35:139-148
Bauer JF (1979) Experimental shock metamorphism of mono and polycrystalline olivine: A comparative study. Proc Lunar Planet Sci Conf 10:2573-2576
Beck CW, LaPaz L (1949) The Weatherford, Oklahoma, meteorite. Pop Astron 57:450-454
Beckett JR, Grossman L (1988) The origin of type C inclusions from carbonaceous chondrites. Earth Planet Sci Lett 89:1-14
Beckett JR, Live D, Tsay F-D, Grossman L, Stolper E (1988) Ti^{+3} in meteoritic and synthetic hibonite. Geochim Cosmochim Acta 52:1479-1495
Beckett JR, Spivack AJ, Hutcheon ID, Wasserburg GJ, Stolper EM (1990) Crystal chemical effects on the partitioning of trace elements between mineral and melt: An experimental study of melilite with applications to refractory inclusions from carbonaceous chondrites. Geochim Cosmochim Acta 54:1755-1774
Begemann F, Wlotzka F (1969) Shock induced thermal metamorphism and mechanical deformations in the Ramsdorf chondrite. Geochim Cosmochim Acta 33:1351-1370
Bennett MEI, McSween HY Jr (1996) Shock features in iron-nickel metal and troilite of L-group ordinary chondrites. Met Planet Sci 31:255-264
Berkley JL, Keil K, Gomes CB, Curvello WS (1978) Studies of Brazilian meteorites XII. Mineralogy and petrology of the Santa Bárbara, Rio Grande do Sul, chondrite. An Acad Brasil Ciênc 50:191-196
Bernatowicz TJ, Zinner EK (1997) Astrophysical implications of the laboratory study of presolar materials. AIP, New York, 750 pp
Bernatowicz TJ, Amari S, Zinner EK, Lewis RS (1991) Interstellar grains within interstellar grains. Astrophys J 373:L73-L76
Bernatowicz TJ, Amari S, Lewis RS (1992) TEM studies of a circumstellar rock. Lunar Planet Sci 23: 91-92
Bernatowicz TJ, Cowsik R, Gibbons PC, Lodders K, Fegley BJ, Amari S, Lewis RS (1996) Constraints on stellar grain formation from presolar graphite in the Murchison meteorite. Astrophys J 472:760-782
Berzelius JJ (1834) Uber Meteorsteine. Ann Phys Chem 33:113-123

Bevan AWR, Axon HJ (1980) Metallography and thermal history of the Tieschitz unequilibrated meteorite—metallic chondrules and the origin of polycrystalline taenite. Earth Planet Sci Lett 47: 353-360

Bevan AWR, McNamara KJ, Barton JC (1988) The Binningup H5 chondrite: A new fall from Western Australia. Meteoritics 23:29-33

Bhandari N, Shah VG, Wasson JT (1980) The Parsa enstatite chondrite. Meteoritics 15:225-234

Binns RA (1967a) Olivine in enstatite chondrites. Am Mineral 52: 1549-1554

Binns RA (1967b) Stony meteorites bearing maskelynite. Nature 213:1111-1112

Binns RA (1970) $(Mg,Fe)_2SiO_4$ spinel in a meteorite. Earth Planet Sci Lett 76:109-122

Binns RA, Davis RJ, Reed SJB (1969) Ringwoodite, natural $(Mg,Fe)_2SiO_4$ spinel in the Tenham meteorite. Nature 221:943-944

Birjukov V, Ulyanov A (1996) Petrology and classification of new Antarctic carbonaceous chondrites PCA91082, TIL91722 and WIS 91600. Proc NIPR Symp Antarct Met 9:8-19

Bischoff A, Keil K (1983a) Ca-Al-rich chondrules and inclusions in ordinary chondrites. Nature 303: 588-592

Bischoff A, Keil K (1983b) Catalog of Al-rich chondrules, inclusions and fragments in ordinary chondrites. Spec Pub No. 22, UNM Institute of Meteoritics, 33 pp

Bischoff A, Keil K (1984) Al-rich objects in ordinary chondrites: Related origin of carbonaceous and ordinary chondrites and their constituents. Geochim Cosmochim Acta 48:693-709

Bischoff A, Metzler K (1991) Mineralogy and petrography of the anomalous carbonaceous chondrites Yamato-86720, Yamato-82162 and Belgica-7904. Proc NIPR Symp Antarct Met 4:226-246

Bischoff A, Palme H (1987) Composition and mineralogy of refractory-metal-rich assemblages from a Ca-Al-rich inclusion in the Allende meteorite. Geochim Cosmochim Acta 51:2733-2748

Bischoff A, Stöffler D (1992) Shock metamorphism as a fundamental process in the evolution of planetary bodies: information from meteorites. Eur J Mineral 4:707-755

Bischoff A, Rubin AE, Keil K, Stöffler D (1983) Lithification of gas-rich chondrite regolith breccias by grain boundary and localized shock melting. Earth Planet Sci Lett 66:1-10

Bischoff A, Keil K, Stöffler D (1985) Perovskite-hibonite-spinel-bearing inclusions and Al-rich chondrules and fragments in enstatite chondrites. Chem Erde 44:97-106

Bischoff A, Palme H, Spettel B (1989) Al-rich chondrules from the Ybbsitz H4-chondrite: Evidence for formation by collision and splashing. Earth Planet Sci Lett 93:170-180

Bischoff A, Palme H, Ash RD, Clayton RN, Schultz L, Herpers U, Stöffler D, Grady MM, Pillinger CT, Spettel B, Weber H, Grund T, Endress M, Weber D (1993a) Paired Renazzo-type (CR) carbonaceous chondrites from the Sahara. Geochim Cosmochim Acta 57:1587-1603

Bischoff A, Palme H, Schultz L, Weber D, Weber HW, Spettel B (1993b) Acfer 182 and paired samples, an iron-rich carbonaceous chondrite: Similarities with ALH 85085 and relationship to CR chondrites. Geochim Cosmochim Acta 57:2631-2648

Bischoff A, Geiger T, Palme H, Spettel B, Schultz L, Scherer P, Schlüter J, Lkhamsuren J (1993c) Mineralogy, chemistry and noble gas contents of Adzhi-Bogdo—an LL3-6 chondritic breccia with L-chondritic and granitoidal clasts. Meteoritics 28:570-578

Bischoff A, Geiger T, Palme H, Spettel B, Schultz L, Scherer P, Bland P, Clayton RN, Mayeda TK, Herpers U, Michel R, Dittrich-Hannen B (1994) Acfer 217- a new member of the Rumuruti chondrite group (R). Meteoritics 29:264-274

Blacic JD, Christie JM (1973) Dislocation substructure of experimentally deformed olivine. Contrib Mineral Petrol 42:141-146

Black DC, Pepin RO (1969) Trapped neon in meteorites. II. Earth Planet Sci Lett 36:395-405

Blander M, Fuchs LH (1975) Calcium-aluminum-rich inclusions in the Allende meteorite: evidence for a liquid origin. Geochim Cosmochim Acta 39:1605-1619

Blander M, Fuchs LH, Horowitz C, Land R (1980) Primordial refractory metal particles in the Allende meteorite. Geochim Cosmochim Acta 44:217-223

Blau PJ, Goldstein JI (1975) Investigation and simulation of metallic spherules from lunar soils. Geochim Cosmochim Acta 39:305-324

Blum JD, Wasserburg GJ, Hutcheon ID, Beckett JR, Stolper EM (1988) Domestic origin of opaque assemblages in refractory inclusions in meteorites. Nature 331:405-409

Blum JD, Wasserburg GJ, Hutcheon ID, Beckett JR, Stolper EM (1989) Origin of opaque assemblages in C3V meteorites: Implications for nebular and planetary processes. Geochim Cosmochim Acta 53:543-556

Borgstom LH (1903) Die Meteoriten von Hvittis and Marjalahti. Bull Comm Geol Finlande 14:1-80

Boss AP (1996) A concise guide to chondrule formation models. *In* Chondrules and the Protoplanetary Disk (RH Hewins, RH Jones, ERD Scott eds) Cambridge Univ Press, 257-263

Böstrom K, Fredriksson K (1966) Surface conditions of the Orgueil meteorite parent body as indicated by mineral associations. Smithsonian Misc Coll 151:1-39

Boynton, WV (1989) Cosmochemistry of the rare earth elements: condensation and evaporation processes. *In* Geochemistry and Mineralogy of Rare Earth Elements (BR Lipin, GA McKay eds) Rev Mineral 21:1-24

Bradley JG, Huneke JC, Wasserburg GJ (1978) Ion microprobe evidence for the presence of excess ^{26}Mg in an Allende anorthite crystal. J Geophys Res 83:244-254

Brearley AJ (1989) Nature and origin of matrix in the unique type 3 chondrite, Kakangari. Geochim Cosmochim Acta 53:2395-2411

Brearley AJ (1990) Carbon-rich aggregates in type 3 ordinary chondrites: Characterization, origins, and thermal history. Geochim Cosmochim Acta 54:831-850

Brearley AJ (1991) Mineralogical and chemical studies of matrix in the Adelaide meteorite, a unique carbonaceous chondrite with affinities to ALH 77307 (CO3) (abstr) Lunar Planet Sci XII:133-134.

Brearley AJ (1992) Mineralogy of fine-grained matrix in the Ivuna CI carbonaceous chondrite (abstr) Lunar Planet Sci XXIII:153-154

Brearley AJ (1993a) Matrix and fine-grained rims in the unequilibrated CO3 chondrite, ALH A77307: Origins and evidence for diverse, primitive nebular dust components. Geochim Cosmochim Acta 57:1521-1550

Brearley AJ (1993b) Occurrence and possible significance of rare Ti-oxides (Magnéli phases) in carbonaceous chondrite matrices. Meteoritics 28:590-595

Brearley AJ (1994) Metamorphic effects in the matrices of CO3 chondrites: compositional and mineralogical variations (abstr) LPS XXV:165-166

Brearley AJ (1995) Aqueous alteration and brecciation in Bells, an unusual, saponite-bearing CM carbonaceous chondrite. Geochim Cosmochim Acta 59:2291-2317

Brearley AJ (1996a) The nature of matrix in unequilibrated chondritic meteorites and its possible relationship to chondrules. *In* Chondrules and the Protoplanetary Disk (RH Hewins, RH Jones, ERD Scott eds), Cambridge Univ Press, 137-152

Brearley AJ (1996b) A comparison of FeO-rich olivines in chondrules, matrix, and dark inclusions in Allende and the discovery of phyllosilicate veins in chondrule enstatite. Meteoritics 31:A21-22

Brearley AJ (1997a) Disordered biopyriboles, amphibole, and talc in the Allende meteorite: Products of nebular or parent body aqueous alteration? Science 276:1103-1105

Brearley AJ (1997b) Phyllosilicates in the matrix of the unique carbonaceous chondrite, LEW 85332 and possible implications for the aqueous alteration of CI chondrites. Met Planet Sci 32:377-388

Brearley AJ, Prinz M (1992) CI-like clasts in the Nilpena polymict ureilite: Implications for aqueous alteration processes in CI chondrites. Geochim Cosmochim Acta 56:1373-1386

Brearley AJ, Scott ERD, Mackinnon IDR (1987) Electron petrography of fine-grained matrix in the Karoonda C4 carbonaceous chondrite (abstr). Meteoritics 22:339-340

Brearley AJ, Scott ERD, Keil K, Clayton RN, Mayeda TK, Boynton WV, Hill DH (1989) Chemical, isotopic and mineralogical evidence for the origin of matrix in ordinary chondrites. Geochim Cosmochim Acta 53:2081-2098

Brearley AJ, Casanova I, Miller ML, Keil K (1991) Mineralogy and possible origin of an unusual Cr-rich inclusion in the Los Martinez (L6) chondrite. Meteoritics 26:287-300

Brigham CA, Papanastassiou DA, Wasserburg GJ (1985) Mg isotopic heterogeneities in fine-grained Ca-Al-rich inclusions (abstr). Lunar Planet Sci XVI:93-94

Brigham CA, Yabuki H, Ouyang Z, Murrell MT, El Goresy A, Burnett DS (1986) Silica-bearing chondrules and clasts in ordinary chondrites. Geochim Cosmochim Acta 50:1655-1666

Brown WL, Hoffman W, Laves F (1963) Über kontinuierliche und reversible Transformation des Anorthits ($CaAl_2Si_2O_8$) zwischen 25 und 350°C. Naturwiss 50:221

Browning LB, McSween HY Jr, Zolensky ME (1996) Correlated alteration effects in CM carbonaceous chondrites. Geochim Cosmochim Acta 60:2621-2633

Buchanan PC, Zolensky ME, Reid AM (1993) Carbonaceous chondrite clasts in the howardites Bholgati and EET87513. Meteoritics 28:659-682

Buchwald, VF (1975). Handbook of Iron Meteorites. Univ California Press, Berkeley, California 243 p

Bunch TE, Chang S (1980) Carbonaceous chondrites—II. Carbonaceous chondrite phyllosilicates and light element geochemistry as indicators of parent body processes and surface conditions. Geochim Cosmochim Acta 44:1543-1577

Bunch TE, Chang S (1983) Allende dark inclusions: Samples of primitive regoliths (abstr). Lunar Planet Sci 14:75-76

Bunch TE, Rajan RS (1988) Meteorite regolith breccias. *In* Meteorites and the Early Solar System (JF Kerridge, MS Matthews eds) University of Arizona Press, 144-164

Bunch TE, Keil K, Snetsinger KG (1967) Chromite composition in relation to chemistry and texture of ordinary chondrites. Geochim Cosmochim Acta 31:1568-1582

Bunch TE, Goles GG, Smith RH, Osborn TW (1972) The Seoni Chondrite. Meteoritics 7:87-95

Bunch TE, Chang S, Frick U, Neil J, Moreland G (1979) Carbonaceous chondrites-I. Characterization and significance of carbonaceous chondrite (CM) xenoliths in the Jodzie howardite. Geochim Cosmochim Acta 43:1727-1742

Burns RG, Burns VM (1977) Mineralogy of manganese nodules. *In* Marine Manganese Deposits (GP Glasby, ed) New York, Elsevier, 185-248

Burns RG, Burns VM (1984) Crystal chemistry of meteoritic hibonites. J Geophys Res 89:C313-C321

Burns RG, Huggins FE (1973) Visible region absorption spectra of a Ti^{3+} fassaite from the Allende meteorite. Am Mineral 58:955-961

Buseck PR, Holdsworth EF (1972) Mineralogy and petrology of the Yilmia enstatite chondrite. Meteoritics 7:429-447

Buseck PR, Hua X (1993) Matrices of carbonaceous chondrite meteorites. Ann Rev Earth Planet Sci 21:255-305

Buseck PR, Keil K (1966) Meteoritic rutile. Am Mineral 51:1506-1515

Buseck PR, Mason B, Wiik H-B (1966) The Farmington meteorite—mineralogy and petrology. Geochim Cosmochim Acta 30:1-8

Buseck PR, Nord GL, Veblen DR (1980) Subsolidus phenomena in pyroxenes. *In* Pyroxenes (CT Prewitt ed) Rev Mineral 7:117-211

Caillère S, Rautureau M (1974) Determination des silicates phylliteux des meteorites carbonées par microscopie et microdiffraction electroniques. Compte Rendus Acad Sci D279:539-542

Caillet CA (1994) A new type of Al-rich inclusion in Axtell CV3 chondrite (abstr). Meteoritics 29:453-454

Caillet C, MacPherson GJ, Zinner EK (1993) Petrologic and Al-Mg isotopic clues to the accretion of two refractory inclusions onto the Leoville parent body: One was hot, the other wasn't. Geochim Cosmochim Acta 57:4725-4643

Carter NL (1971) Static deformation of silica and silicates. J Geophys Res 76:5514-5540

Carter NL, Raleigh CB, DeCarli PS (1968) Deformation of olivine in stony meteorites. J Geophys Res 73:5439-5461

Casanova I, Simon SB (1994) Opaque minerals in CAIs, and classification of the Axtell (CV3) chondrite (abstr). Meteoritics 29:454-455.

Cassen P (1996) Overview of models of the solar nebula: Potential chondrule-forming environments. *In* Chondrules and the Protoplanetary Disk (RH Hewins, RH Jones, ERD Scott eds) Cambridge Univ Press, 21-28

Chen JH, Wasserburg GJ (1981) The isotopic composition of uranium and lead in Allende inclusions and meteoritic phosphates. Earth Planet Sci Lett 52:1-15

Chen M, Xie X, El Goresy A (1995) Nonequilibrium soldification and microstructures of metal phases in the shock-induced melt of the Yanzhuang (H6) chondrite. Meteoritics 30:28-32

Chen M, Sharp TG, El Goresy A, Wopenka B, Xie X (1996) The majorite-pyrope-magnesiowüstite assemblage: Constraints on the history of shock veins in chondrites. Science 271:1570-1573

Chladni EFF (1797) Fortsetzung der Bemerkungen über Feuerkugeln und niedergefallene Massen. Mag Neuest Zust Naturk (Jena) 1:17-30

Christophe Michel-Lévy M (1968) Un chondre exceptionnel dans la météorite de Vigarano. Bull Soc Franc Mineral 91:212-214

Christophe Michel-Lévy M (1976) La matrice noire et blanche de la chondrite de Tieschitz. Earth Planet Sci Lett 30:143-150

Christophe Michel-Lévy M (1981) Some clues to the history of the H-group chondrites. Earth Planet Sci Lett 54:67-80

Christophe Michel-Lévy M, Bourot Denise M (1992) Dahmani, a highly oxidised LL6 chondrite bearing Ni-rich taenite. Meteoritics 27:184-185

Christophe Michel-Lévy M, Caye R, Nelen J (1970) A new mineral in the Vigarano meteorite. Meteoritics 5:211

Christophe Michel-Lévy M, Kurat G, Brandstätter F (1982) A new calcium aluminate from a refractory inclusion in the Leoville carbonaceous chondrite. Earth Planet Sci Lett 61:13-22

Clarke RS, Scott ERD (1980) Tetrataenite—ordered FeNi, a new mineral in meteorites. Am Mineral 65:624-630

Clarke RS, Jarosewich E, Nelen J, Gomez M, Hyde JR (1970) The Allende, Mexico meteorite shower. Smithsonian Contrib Earth Sci 5:53

Clayton RN (1993) Oxygen isotopes in meteorites. Ann Rev Earth Planet Sci 21:115-149

Clayton RN, Mayeda T (1978) Multiple parent bodies of polymict brecciated meteorites. Geochim Cosmochim Acta 42:325-327

Clayton RN, Mayeda TK (1984) The oxygen isotope record in Murchison and other carbonaceous chondrites. Earth Planet Sci Lett 67:151-161

Clayton RN, Mayeda TK (1989) Oxygen isotope classification of carbonaceous chondrites (abstr). Lunar Planet Sci 20:169-170

Clayton RN, Onuma N, Grossman L, Mayeda TK (1977) Distribution of the pre-solar component in Allende and other carbonaceous chondrites. Earth Planet Sci Lett 34:209-224

Clayton RN, MacPherson GJ, Hutcheon ID, Davis AM, Grossman L, Mayeda TK, Molini-Velsko C, Allen JM, El Goresy A (1984) Two forsterite-bearing FUN inclusions in the Allende meteorite. Geochim Cosmochim Acta 48:535-548

Cloëz S (1864) Analyse chimique de la pierre météoritique d'Orgueil. Comp Rend Acad Sci Paris 59:37-38

Cohen E (1894) Meteoritikenkunde. Schweizerbart, Stuttgart, 340 p

Cohen RE, Kornacki AS, Wood JA (1983) Mineralogy and petrology of chondrules and inclusions in the Mokoia CV3 chondrite. Geochim Cosmochim Acta 47:1739-1757

Coleman LC (1977) Ringwoodite and majorite in the Catherwood meteorite. Can Mineral 15:97-101

Connolly HC, Hewins RH (1995) Chondrules as products of dust collisions with totally molten droplets within a dust-rich nebular environment: An experimental investigation. Geochim Cosmochim Acta 59:3231-3246

Cressey BA (1979) Electron microscopy of serpentine textures. Can Mineral 17:741-756

Cressey BA, Zussman J (1976) Electron microscopic studies of serpentinites. Can Mineral 14:307-313

Crozaz G (1979) Uranium and thorium microdistributions in stony meteorites. Geochim Cosmochim Acta 43:127-136

Crozaz G, Lundberg L (1995) The origin of oldhamite in unequilibrated enstatite chondrites. Geochim Cosmochim Acta 59:3817-3831

Crozaz G, Zinner E (1985) Ion probe determinations of the rare earth concentrations of individual meteoritic phosphate grains. Earth Planet Sci Lett 73:41-52

Crozaz G, Pellas P, Bourot-Denise M, de Chazal SM, Fieni C, Lundberg LL, Zinner E (1989) Plutonium, uranium and rare earths in the phosphates of ordinary chondrites—the quest for a chronometer. Earth Planet Sci Lett 93:157-169

Curtis DB, Schmitt RA (1979) The petrogenesis of L-6 chondrites: Insights from the chemistry of minerals. Geochim Cosmochim Acta 43:1091-1103

Danon J, Scorzelli RB, Souza-Azevedo I, Christophe Michel-Lévy M (1979) Iron nickel superstructure in metal particles of chondrites. Nature 281:469-471

Danon J, Christophe Michel-Lévy M, Jehanno C, Keil K, Gomes CB, Scorzelli RB, Souza Azevedo I (1981) Awaruite (Ni_3Fe) in the genomict LL chondrite Parambu: Formation under high fO_2. Meteoritics 16:305

Daulton TL, Eisenhour DD, Bernatowicz TJ, Lewis RS, Buseck PR (1996) Genesis of presolar diamonds: Comparative high-resolution transmission electron microscopy study of meteoritic and terrestrial nano-diamonds. Geochim Cosmochim Acta 60:4853-4872.

Davis AM, MacPherson GJ (1988) Rare earth elements in a hibonite-rich Allende fine-grained inclusion. Lunar Planet Sci XIX:249-250

Davis AM, Grossman L, Ganapathy R (1977) Yes, Kakangari is a unique chondrite. Nature 265:230-232

Davis AM, Grossman L, Allen JM (1978) Major and trace element chemistry of separated fragments from a hibonite-bearing Allende inclusion. Proc Lunar Planet Sci Conf 9:1235-1247

Davis AM, MacPherson GJ, Hinton RW, Laughlin JR (1987) An unaltered Group I fine-grained inclusion from the Vigarano carbonaceous chondrite (abstr). Lunar Planet Sci XVIII:223-224

Davis AM, MacPherson GJ, Clayton RN, Mayeda TK, Sylvester PJ, Grossman L, Hinton RW, Laughlin JR (1991) Melt solidification and late-stage evaporation in the evolution of a FUN inclusion from the Vigarano C3V chondrite. Geochim Cosmochim Acta 55:621-637

Davy R, Whitehead SG, Pitt G (1978) The Adelaide meteorite. Meteoritics 13:121-140

DeHart JM, Lofgren GE, Lu J, Benoit PH, Sears DWG (1992) Chemical and physical studies of chondrites: X. Cathodoluminescence and phase composition studies of metamorphism and nebular processes in chondrules of type 3 ordinary chondrites. Geochim Cosmochim Acta 56:3791-3807

Desnoyers C (1980) The Niger (I) carbonaceous chondrite and implications for the origin of aggregates and isolated olivine grains in C2 chondrites. Earth Planet Sci Lett 47:223-234

Devouard B, Buseck PR (1997) Phosphorus-rich iron, nickel sulfides in CM2 chondrites: condensation or alteration products (abstr). Met Planet Sci 32:A34-A34

Dodd RT (1969) Metamorphism of the ordinary chondrites: A review. Geochim Cosmochim Acta 33:161-203

Dodd RT (1973) Minor element abundances in olivines of the Sharps (H3) chondrite. Contrib Mineral Petrol 42:159-167

Dodd RT (1974a) The petrology of the St. Mesmin chondrite. Contrib Mineral Petrol 46:129-145

Dodd RT (1974b) The metal phase in unequilibrated ordinary chondrites and its implications for calculated accretion tempuratures. Geochim Cosmochim Acta 38:485-494.

Dodd RT (1978) The composition and origin of large microporphyritic chondrules in the Manych (L-3) chondrite. Earth Planetary Sci Lett 39:52-66

Dodd RT (1981) Meteorites: A Petrologic-Chemical Synthesis. Cambridge Univ Press, 368 p

Dodd RT, Jarosewich E (1979) Incipient melting in and shock classification of L-group chondrites. Earth Planet Sci Lett 44:335-340

Dodd RT, Van Schmus WR, Marvin UB (1965) Merrihueite, a new alkali-ferromagnesian silicate from the Mezö-Madaras chondrite. Science 149:972-974

Dodd RT, Van Schmus WR, Marvin UB (1966) Significance of iron-rich silicates in the Mezö-Madaras chondrite. Am Mineral 51:1177-1191

Dodd RT, Grover JE, Brown GE (1975) Pyroxenes in the Shaw (L-7) chondrite. Geochim Cosmochim Acta 39:1585-1594

Dodd RT, Jarosewich E, Hill B (1982) Petrogenesis of complex veins in the Chantonnay (L6) chondrite. Earth Planet Sci Lett 59:364-374

Dominik B, Jessberger EK, Staudacher T, Nagel K, El Goresy A (1978) A new type of white inclusion in Allende: petrography, mineral chemistry, $^{40}Ar/^{39}Ar$ ages and genetic implications. Proc Lunar Planet Sci Conf 9:1249-1266

Donn B (1964) The origin and nature of solid particles in space. Ann NY Acad Sci 119:5-16

Donn B, Sears GW (1963) Planets and comets: role of crystal growth in their formation. Science 140:1208-1211

Doukhan N, Doukhan JC, Poirier JP (1991) Transmission electron microscopy of a refractory inclusion from the Allende meteorite: Anatomy of a pyroxene. Meteoritics 26:105-109

Dowty E, Clark JR (1973a) Crystal structure refinement and optical properties of a Ti^{3+} fassaite from the Allende meteorite. Am Mineral 58:230-242

Dowty E, Clark JR (1973b) Crystal structure refinement and optical properties of a Ti^{3+} fassaite from the Allende meteorite: Reply. Am Mineral 58:962-964

Duke MB, Brett R (1965) Metallic copper in stony meteorites. US Geological Survey Professional Paper 525-B:B101-B103

Dundon RW, Walter LS (1967) Ferrous iron order-disorder in meteoritic pyroxenes and the metamorphic history of chondrites. Earth Planet Sci Lett 2:372-376

Ebihara M, Honda M (1983) Rare earth abundances in chondritic phosphates and their implications for early stage chronologies. Earth Planet Sci Lett 63:433-445

Ehlers K, El Goresy A (1988) Normal and reverse zoning in niningerite: A novel key parameter to the thermal histories of EH-chondrites. Geochim Cosmochim Acta 52:877-887

Ehlmann AJ, Keil K (1985) Classification of eight ordinary chondrites from Texas. Meteoritics 20:219-227

Ehlmann AJ, Keil K (1988) Classification of six ordinary chondrites from Texas. Meteoritics 23:361-364

Ehlmann AJ, Scott ERD, Keil K, Mayeda TK, Clayton RN, Weber HW, Schultz L (1988) Origin of fragmental and regolith meteorite breccias—evidence from the Kendleton L chondrite breccia. Proc Lunar Planet Sci Conf 18:545-554

Ehlmann AE, McCoy TJ, Keil K (1994) The Tatum, New Mexico, chondrite and its silica inclusion. Chem Erde 54:169-178

El Goresy A, Ehlers K (1989) Sphalerite in EH chondrites: I. Textural relations, compositions, diffusion profiles, and pressure temperature histories. Geochim Cosmochim Acta 53:1657-1668

El Goresy A, Nagel K, Ramdohr P (1977) Type A Ca,Al-rich inclusions in Allende meteorite: Origin of the perovskite-fassaite symplectite around rhönite and chemistry and assemblages of the refractory metals (Mo, W) and platinum metals (Ru, Os, Ir, Re, Rh, Pt) (abstr). Meteoritics 12:216

El Goresy A, Nagel K, Ramdohr P (1978) Fremdlinge and their noble relatives. Proc Lunar Planet Sci Conf 9:1279-1303

El Goresy A, Nagel K, Ramdohr P (1979) Spinel framboids and Fremdlinge in Allende inclusions: Possible sequential markers in the early history of the solar system. Proc Lunar Planet Sci Conf 10:833-850

El Goresy A, Ramdohr P, Nagel K (1980) A unique inclusion in the Allende meteorite: A conglomerate of hundreds of various fragments and inclusions (abstr). Meteoritics 15:286-287

El Goresy A, Palme H, Yabuki H, Nagel K, Herrwerth I, Ramdohr P (1984) A calcium-aluminum inclusion from the Essebi (CM2) chondrite: Evidence for captured spinel-hibonite spherules and for an ultrarefractory rimming sequence. Geochim Cosmochim Acta 48:2283-2298

El Goresy A, Armstrong JT, Wasserburg GJ (1985) Anatomy of an Allende coarse-grained inclusion. Geochim Cosmochim Acta 49:2433-2444

El Goresy A, Yabuki H, Ehlers K, Woolum D, Pernicka E (1988) Qingzhen and Yamato-691: A tentative alphabet for the EH chondrites. Proc NIPR Symp Ant Met 1:65-101

Endress M, Bischoff A (1996) Carbonates in CI chondrites: Clues to parent body evolution. Geochim Cosmochim Acta 60:489-507

Endress M, Keil K, Bischoff A, Spettel B, Clayton RN, Mayeda TK (1994) Origin of dark clasts in the Acfer 059/El Djouf 001 CR2 chondrite. Meteoritics 29:26-40

Endress M, Zinner EK, Bischoff A (1996) Early aqueous activity on primitive meteorite parent bodies. Nature 379:701-703

Fahey AJ, Zinner EK, Crozaz G, Kornacki A (1987a) Microdistributions of Mg isotopes and REE abundances in a Type A calcium-aluminum-rich inclusion from Efremovka. Geochim Cosmochim Acta 51:3215-3229

Fahey AJ, Goswami JN, McKeegan KD, Zinner EK (1987b) ^{26}Al, ^{244}Pu, ^{50}Ti, REE and trace element abundances in hibonite grains from CM and CV meteorites. Geochim Cosmochim Acta 51:329-350

Fahey AJ, Zinner E, Kurat K, Kracher A (1994) Hibonite-hercynite inclusion HH-1 from the Lancé (CO3) meteorite: The history of an ultrarefractory CAI. Geochim Cosmochim Acta 58:4779-4793

Fegley B, Jr. (1982) Hibonite condensation in the solar nebula. Lunar Planet Sci XIII:211-212

Fegley B Jr (1991) The stability of of calcium aluminate minerals in the solar nebula. Lunar Planet Sci XXII:367-368

Fegley B Jr, Palme H (1985) Evidence for oxidizing conditions in the solar nebula from Mo and W depletions in refractory inclusions in carbonaceous chondrites. Earth Planet Sci Lett 72:311-326

Fegley B Jr, Post JE (1985) A refractory inclusion in the Kaba CV3 chondrite: some implications for the origin of spinel-rich objects in chondrites. Earth Planet Sci Lett 75:297-310

Fitch F, Schwarcz HP, Anders E (1962) "Organized" elements in carbonaceous chondrites. Nature 193:1123-1125

Fitzgerald MJ (1979) The chemical composition and classification of the Karoonda meteorite. Meteoritics 14:109-116

Fitzgerald MJ, Jacques AL (1982) Tibooburra, A new Australian meteorite find, and other carbonaceous chondrites of high petrologic grade. Meteoritics 17:9-26

Fitzgerald MJ, Jones JB (1977) Adelaide and Bench Crater—members of a new subgroup of the carbonaceous chondrites. Meteoritics 12:443-458

Fleet ME, MacCrae ND (1987) Sulfidation of Mg-rich olivine and the stability of niningerite in enstatite chondrites. Geochim Cosmochim Acta 51:1511-1521

Flemings MC, Poirier DR, Barone RV, Brody HD (1970) Microsegregation in iron-based alloy. J Iron Steel Inst 208:371-381

Fodor RV, Keil K (1975) Implications of poikilitic textures in LL-group chondrites. Meteoritics 10:325-337

Fodor RV, Keil K (1976) Carbonaceous and non-carbonaceous lithic fragments in the Plainview, Texas, chondrite: origin and history. Geochim Cosmochim Acta 40:177-189

Fodor RV, Keil K (1978) Catalog of lithic fragments in LL-group chondrites. Spec Publ 19, UNM Institute of Meteoritics, 38 p

Fodor RV, Keil K, Jarosewich E, Huss GI (1971a) Mineralogy and chemistry of the Kyle, Texas, chondrite. Meteoritics 6:71-79

Fodor RV, Keil K, Jarosewich E, Huss GI (1971b) Mineralogy, petrology, and chemistry of the Burdett, Kansas, chondrite. Chem Erde 30:103-113

Fodor RV, Keil K, Jarosewich E (1972) The Oro Grande, New Mexico, chondrite and its lithic inclusions. Meteoritics 7:495-507

Fodor RV, Keil K, Wilkening LL, Bogard DD, Gibson EK (1976) Origin and history of a meteorite parent-body regolith breccia: Carbonaceous and noncarbonaceous lithic fragments in the Abbott, New Mexico, chondrite. New Mexico Geol Soc Spec Publ 6:206-218

Fodor RV, Keil K, Gomes CB (1977) Studies of Brazilian stone meteorites. IV Origin of a dark-colored unequilibrated lithic fragment in the Rio Negro chondrite. Revista Bras Geosciencias 7:45-57

Fogel RA, Hess PC, Rutherford MJ (1989) Intensive parameters of enstatite chondrite metamorphism. Geochim Cosmochim Acta 53:2735-2746

Folinsbee K, Douglas JAV, Maxwell JA (1967) Revelstoke, a new Type I carbonaceous chondrite. Geochim Cosmochim Acta 31:1625-1635

Fraundorf P, Fraundorf G, Bernatowicz T, Lewis R, Tang M (1989) Stardust in the TEM. Ultramicroscopy 27:401-412

Fredriksson K (1963) Chondrules and the meteorite parent bodies. Trans NY Acad Sci [II] 25:756-769

Fredriksson K, Kerridge JF (1988) Carbonates and sulfates in Cl chondrites: Formation by aqueous activity on the parent body. Meteoritics 23:35-44

Fredriksson K, Wickman FE (1963) Meteoriter. Sartryck ur Svensk Naturvetenskap 121-157

Fredriksson K, Jarosewich E, Wlotzka F (1989) Study Butte: A chaotic chondrite breccia with normal H-group chemistry. Z Naturforsch 44a:945-962

Freer R (1981) Diffusion in silicate minerals and glasses: a data digest and guide to the literature. Contrib Mineral Petrol 76:440-454

Frost MJ, Symes RF (1970) A zoned perovskite-bearing chondrule from the Lancé meteorite. Mineral Mag 37:724-726

Fruland RM, King EA, McKay DS (1978) Allende dark inclusions. Proc Lunar Planet Sci 9:1305-1329

Fuchs LH (1966) Djerfisherite, alkali copper-iron sulfide: a new mineral from enstatite chondrites. Science 153:166-167

Fuchs LH (1969a) The phosphate mineralogy of meteorites. *In* Meteorite Research (PM Millman ed) D Riedel, 683-695

Fuchs LH (1969b) Occurrence of cordierite and aluminous orthoenstatite in the Allende meteorite. Am Mineral 4:1645- 1653

Fuchs LH (1971) Occurrence of wollastonite, rhönite and andradite in the Allende meteorite. Am Mineral 56:2053-2068

Fuchs LH (1974) Grossular in the Allende (Type III carbonaceous) meteorite. Meteoritics 13:73-88

Fuchs LH (1978) The mineralogy of a rhönite-bearing calcium aluminum rich inclusion in the Allende meteorite. Meteoritics 13:73-88

Fuchs LH, Blander M. (1977) Molybdenite in calcium-aluminium-rich inclusions in the Allende meteorite. Geochim Cosmochim Acta 41:1170-1174

Fuchs LH, Blander M (1980) Refractory metal particles in refractory inclusions in the Allende meteorite. Proc Lunar Planet Sci Conf 11:929-944

Fuchs LH, Olsen E (1973) Composition of metal in Type III carbonaceous chondrites and its relevance to the source—assignment of Lunar metal. Earth Planet Sci Lett 18:379-384

Fuchs LH, Frondel C, Klein C (1966) Roedderite, a new mineral from the Indarch meteorite. Am Mineral 51:949-955

Fuchs LH, Jensen KJ, Olsen E (1970) Mineralogy and composition of the Murchison meteorite. Meteoritics 5:198-199

Fuchs LH, Olsen E, Jensen KJ (1973) Mineralogy, mineral-chemistry, and composition of the Murchison (C2) meteorite. Smithson Contrib Earth Sci 10:1-39

Fudali RF, Noonan AF (1975) Gobabeb, a new chondrite: coexistence of equilibrated silicates and unequilibrated spinels. Meteoritics 10:31-39

Fujii N, Miyamoto M, Kobayashi Y, Ito K (1982) On the shape of Fe-Ni grains among ordinary chondrites. Mem NIPR Spec Issue 25:319-330

Fujimaki H, Matsu-Ura M, Sunagawa I, Aoki K (1981) Chemical compositions of chondrules and matrices in the ALHA-77015 chondrite (L3). Mem NIPR Spec Issue 20:161-174

Fujimura A, Kato M, Kumazawa M (1982) Preferred orientation of phyllosilicates in Yamato-74642 and -74662 in relation to the deformation of C2 chondrites. Mem NIPR Spec Issue 25:207-215

Fujimura A, Kato M, Kumazawa M (1983) Preferred orientation of phyllosilicate [001] in matrix of Murchison meteorite and possible mechanisms of generating the oriented texture in chondrules. Earth Planet Sci Lett 66:25-32

Geiger T, Bischoff A (1995) Formation of opaque minerals in CK chondrites. Planet Space Sci 43:485-498

Geiger CA, Kleppa OJ, Mysen BO, Lattimer JM, Grossman L (1988) Ethalpies of formation of $CaAl_4O_7$ and $CaAl_{12}O_{19}$ (hibonite) by high temperature, alkali borate solution calorimetry. Geochim Cosmochim Acta 52:1729-1736

Gibson EK Jr, Lange DE, Keil K, Schmidt TE, Rhodes JM (1977) The Kramer Creek, Colorado meteorite: A new L4 chondrite. Meteoritics 12:95-107

Goldstein JI, Ogilvie RE (1965) A re-evaluation of the iron-rich portion of the Fe-Ni system. Trans AIME 233:2083-2087

Goldstein JI, Hanneman RE, Ogilvie RE (1964) Diffusion in the Fe-Ni system at 1 atm and 40 kbar pressure. Trans AIME 233:812-820

Gomes CB, Keil K (1980) Brazilian Stone Meteorites. University of New Mexico, 161 p

Gomes CB, Keil K, Jarosewich E (1977a) Studies of Brazilian meteorites VII. Mineralogy, petrology and chemistry of the Uberaba, Minas Gerais, chondrite. An Acad Brasil Ciênc 49:269-274

Gomes CB, Keil K, Jarosewich E, Curvello WS (1977b) Studies of Brazilian meteorites VIII. Mineralogy, petrology and chemistry of the Itapicuru Mirim, Maranhão, chondrite. An Acad Brasil Ciênc 49:407-412

Gomes CB, Keil K, Ruberti E, Jarosewich E, Silva JMLU (1978) Studies of Brazilian meteorites XVI. Mineralogy, petrology and chemistry of the Ipiranga, Paraná, chondrite. Chem Erde 37:265-270

Gooding JL, Keil K (1981) Relative abundances of chondrule primary textural types in ordinary chondrites and their bearing on conditions of chondrule formation. Meteoritics 16:17-43

Gooding JL, Mayeda TK, Clayton RN, Fukuoka T (1983) Oxygen isotopic heterogeneities, their petrological correlations, and implications for melt origins of chondrules in unequilibrated ordinary chondrites. Earth Planet Sci Lett 65:209-224

Göpel C, Manhés G, Allègre CJ (1994) U-Pb systematics of phosphates from equilibrated ordinary chondrites. Earth Planet Sci Lett 121:153-171

Goswami JN, Srinivasan G, Ulyanov AA (1994) Ion microprobe studies of Efremovka CAIs: I. Magnesium isotope composition. Geochim Cosmochim Acta 58:431-447

Graham AL, Hutchison R (1974) Is Kakangari a unique chondrite? Nature 251:128- 129

Graham AL, Nayak VK (1974) The Patora meteorite: an H6 fall. Meteoritics 9:137-139

Graham AL, Easton AJ, Hutchison R (1977) Forsterite chondrites; the meteorites Kakangari, Mount Morris (Wisconsin), Pontlyfni, and Winona. Mineral Mag 41:201-210

Grant RW (1969) The occurrence of silica minerals in meteorites. Meteoritics 4:180-181

Gray CM, Papanastassiou DA, Wasserburg GJ (1973) The identification of early condensates from the solar nebula. Icarus 20:213-239

Green HW II, Radcliffe SV (1972) Deformation processes in the Upper Mantle. *In* Flow and Fracture for Rocks (HC Heard, IY Borg, NL Carter, CB Raleigh eds) Am Geophys Union Monograph, Washington DC, 139-156

Green HW II, Radcliffe SV, Heuer AH (1971) Allende meteorite: A high voltage electron petrographic study. Science 172:936-939

Greenwood RC, Hutchison R, Huss GR, Hutcheon ID (1992) CAIs in CO3 meteorites: Parent body or nebular alteration? (abstr). Meteoritics 27:229

Greenwood RC, Morse A, Long JVP (1993) Petrography, mineralogy and Mg isotope composition of VICTA: A Vigarano $CaAl_4O_7$-bearing type A inclusion (abstr). Lunar Planet Sci XXIV:573-574

Greenwood RC, Lee MR, Hutchison R, Barber DJ (1994) Formation and alteration of CAIs in Cold Bokkeveld (CM2). Geochim Cosmochim Acta 58:1913-1035

Greshake A (1997) The primitive matrix components of the unique carbonaceous chondrite Acfer 094: A TEM study. Geochim Cosmochim Acta:437-452

Greshake A, Bischoff A, Putnis A, Palme H (1996) Corundum, rutile, periclase, and CaO in Ca, Al-rich inclusions from carbonaceous chondrites. Science 272:1316-1318

Grokhovsky VJ, Bevan AWR (1983) Plessite formation by discontinuous precipitation reaction from gamma-Fe,Ni in Richardton (H5) ordinary chondrite. Nature 301:322-324

Grossman JN (1994) The Meteoritical Bulletin, No. 76 1994 January: The U.S. Antarctic Meteorite Collection. Meteoritics 29:100-143

Grossman JN (1996a) Chemical fractionations of chondrites: signatures of events before chondrule formation. *In* Chondrules and the Protoplanetary Disk (RH Hewins, RH Jones, ERD Scott eds) Cambridge Univ Press, 243-253

Grossman JN (1996b) The redistribution of sodium in Semarkona chondrules by secondary processes (abstr). Lunar Planet Sci XXVII:467-468

Grossman JN, Wasson JT (1985) The origin and history of the metal and sulfide components of chondrules. Geochim Cosmochim Acta 49:925-939

Grossman JN, Wasson JT (1987) Compositional evidence regarding the origin of rims on Semarkona chondrules. Geochim Cosmochim Acta 51:3003-3011

Grossman JN, Rubin AE, Rambaldi ER, Rajan RS, Wasson JT (1985) Chondrules in the Qingzhen type-3 enstatite chondrite: Possible precursor components and comparison to ordinary chondrite chondrules. Geochim Cosmochim Acta 49:1781-1795

Grossman JN, Rubin AE, MacPherson GJ (1988a) ALH 85085: A unique volatile-poor carbonaceous chondrite with implications for nebular fractionation processes. Earth Planet Sci Lett 91:33-54

Grossman JN, Rubin AE, Nagahara H, King EA (1988b) Properties of chondrules. *In* Meteorites and the Early Solar System (JF Kerridge, MS Matthews eds) University of Arizona Press, 619-659

Grossman L (1972) Condensation in the primitive solar nebula. Geochim Cosmochim Acta 49:2433-2444

Grossman L (1973) Refractory trace elements in Ca-Al-rich inclusions in the Allende meteorite. Geochim Cosmochim Acta 37:1119-1140

Grossman L (1975) Petrography and mineral chemistry of Ca-rich inclusions in the Allende meteorite. Geochim Cosmochim Acta 39:433-453

Grossman L (1980) Refractory inclusions in the Allende meteorite. Ann Rev Earth Planet Sci 8:559-608

Grossman L, Ganapathy R (1976) Trace elements in the Allende meteorite-II. Fine-grained, Ca-rich inclusions. Geochim Cosmochim Acta 40:967-977

Grossman L, Olsen EJ (1974) Origin of the high-temperature fraction of C2 chondrites. Geochim Cosmochim Acta 38:173-187

Grossman L, Steele IM (1976) Amoeboid olivine aggregates in the Allende meteorite. Geochim Cosmochim Acta 40:149-155

Grossman L, Fruland RM, McKay DS (1975) Scanning electron microscopy of a pink inclusion from the Allende meteorite. Geophys Res Lett 2:37-40

Grossman L, Olsen E, Lattimer JM (1979) Silicon in carbonaceous chondrite metal: Relic of high-temperature condensation. Science 206:449-451

Guimon RK, Symes SJK, Sears DWG, Benoit PH (1995) Chemical and physical studies of chondrites. XII: The metamorphic history of CV chondrites and their components. Meteoritics 30:704-714

Haggerty SE (1972) An enstatite chondrite from Hadley Rille (abstr). *In* The Apollo 15 Lunar Samples (JW Chamberlain, C Watkins eds) Lunar Science Institute, 85-87

Haggerty SE (1978a) The Allende meteorite; Evidence for a new cosmothermometer based on Ti^{3+}/Ti^{4+}. Nature 276:221-225

Haggerty SE (1978b) The Allende meteorite: Solid solution characteristics and the significance of a new titanite mineral series in association with armalcolite. Proc Lunar Planet Sci Conf 9:1331-44

Haggerty SE, McMahon BM (1979) Magnetite-sulfide-metal complexes in the Allende meteorite. Proc Lunar Planet Sci Conf 10:851-870

Hashimoto A, Grossman L (1985) SEM-petrography of Allende fine-grained inclusions (abstr). Lunar Planet Sci XV1:323-324

Hashimoto A, Grossman L (1987) Alteration of Al-rich inclusions inside amoeboid olivine aggregates in the Allende meteorite. Geochim Cosmochim Acta 51:1685-1704

Hashimoto A, Hinton RW, Davis AM, Grossman L, Mayeda TK, Clayton RN (1986) A hibonite-rich Murchison inclusion with anomalous oxygen isotopic composition (abstr). Lunar Planet Sci XVII:317-318

Herndon JM, Rudee ML (1978) Thermal history of the Abee enstatite chondrite. Earth Planet Sci Lett 41:101-106

Herndon JM, Rowe MW, Larson EE, Watson DE (1975) Origin of magnetite and pyrrhotite in carbonaceous chondrites. Nature 253:516-518

Hewins RH (1996) Chondrules and the Protoplanetary Disk: An Overview. *In* Chondrules and the Protoplanetary Disk (RH Hewins, RH Jones, ERD Scott eds) Cambridge Univ Press, 3-9

Hewins RH (1997) Chondrules. Ann Rev Earth Planet Sci 25:61-83

Heymann D (1967) On the origin of hypersthene chondrites: Ages and shock effects of black meteorites. Icarus 6:189-221

Heymann D (1990) Raman study of olivines in 37 heavily and moderately shocked ordinary chondrites. Geochim Cosmochim Acta 54:2507-2510

Heyse JV (1978) The metamorphic history of LL-group ordinary chondrites. Earth Planet Sci Lett 40:365-381

Hinton RW, Bischoff A (1984) Ion microprobe magnesium isotopic analysis of plagioclase and hibonite from ordinary chondrites. Nature 308:169-172

Hinton RW, Davis AM, Scatena-Wachel DE, Grossman L, Draus RJ (1988) A chemical and isotopic study of hibonite-rich refractory inclusions in primitive meteorites. Geochim Cosmochim Acta 52:2573-2598

Holland-Duffield CE, Williams DB, Goldstein JI (1991) The structure and composition of metal particles in two type 6 ordinary chondrites. Meteoritics 26:97-103

Holmberg AA, Hashimoto A (1992) A unique, (almost) unaltered spinel-rich fine-grained inclusion in Kainsaz. Meteoritics 27:149-153

Hoppe P, Amari S, Zinner E, Ireland T, Lewis RS (1994) Carbon, nitrogen, magnesium, silicon and titanium isotopic compositions of single interstellar silicon carbide grains from the Murchison carbonaceous chondrite. Astrophys J 430:870-890

Hoppe P, Strebel R, Eberhardt P, Amari S, Lewis RS (1996a) Small SiC grains and a nitride grain of circumstellar origin from the Murchison meteorite: Implications for stellar evolution and nucleosynthesis. Geochim Cosmochim Acta 60:883-907

Hoppe P, Strebel R, Eberhardt P, Amari S, Lewis RS (1996b) Type II supernova matter in a silicon carbide grain from the Murchison meteorite. Science 272:1314-1316

Housley RM, Cirlin EH (1983) On the alteration of Allende chondrules and the formation of matrix. *In* Chondrules and their Origins. (EA King ed) Lunar and Planetary Institute. p 145-161

Howard E (1802) Experiments and observations on certain stony and metalline substances which at different times are said to have fallen on the earth: also on various kinds of native irons. Phil Trans Roy Soc London 92:168-212

Hua X, Buseck PR (1995) Fayalite in Kaba and Mokoia carbonaceous chondrites. Geochim Cosmochim Acta 59:563-578

Hua X, Adam J, Palme H, El Goresy A (1988) Fayalite-rich rims, veins, and halos around and in forsteritic olivines in CAIs and chondrules in carbonaceous chondrites: Types, compositional profiles and constraints on their formation. Geochim Cosmochim Acta 52:1389-1408

Hua X, Zinner EK, Buseck PR (1996) Petrography and chemistry of fine-grained dark rims in the Mokoia CV3 chondrite: Evidence for an accretionary origin. Geochim Cosmochim Acta 60:4265-4274
Huang S, Lu J, Prinz M, Weisberg MK, Benoit PH, Sears DWG (1996) Chondrules: Their diversity and the role of open-system processes during their formation. Icarus 122:316-346
Huss GR (1990) Ubiquitous interstellar diamond and SiC in primitive chondrites; abundances reflect metamorphism. Nature 437:159-162
Huss GR, Hutcheon ID (1992) Abundant $^{26}Mg^*$ in Adelaide refractory inclusions. Meteoritics 27:236
Huss GR, Lewis RS (1994a) Noble gases in presolar diamonds I: Three distinct components and their implications for diamond origins. Meteoritics 29:791-810
Huss GR, Lewis RS (1994b) Noble gases in presolar diamonds II: Component abundances reflect thermal processing. Meteoritics 29:811-829
Huss GR, Lewis RS (1995) Presolar diamond, SiC, and graphite in primitive chondrites: abundances as a function of meteorite class and petrologic type. Geochim Cosmochim Acta 59:115-160
Huss GR, Keil K, Taylor GJ (1981) The matrices of unequilibrated ordinary chondrites: Implications for the origin and history of chondrites. Geochim Cosmochim Acta 45:33-51
Hutcheon ID, Phinney DL (1996) Radiogenic $^{53}Cr^*$ in Orgueil carbonates: Chronology of aqueous activity on the CI parent body (abstr). Lunar Planet Sci XXVII:577-578
Hutcheon ID, Steele IM (1982) Refractory inclusions in the Adelaide carbonaceous chondrite (abstr). Lunar Planet Sci XIII:352-353
Hutcheon ID, Steele IM, Smith JV, Clayton RN (1978) Ion microprobe, electron microprobe and cathodoluminescence data for Allende inclusions with emphasis on plagioclase chemistry. Proc Lunar Planet Sci Conf 9:1345-1368
Hutcheon ID, Armstrong JT, Wasserburg GJ (1987) Isotopic studies of Mg, Fe, Mo, Ru and W in fremdlinge from Allende refractory inclusions. Geochim Cosmochim Acta 51:3175-3192
Hutchison MN, Scott SD (1983) Experimental calibration of the sphalerite cosmothermometer. Geochim Cosmochim Acta 47:101-108
Hutchison R (1987) Chromian-manganoan augite in the interchondrule matrix of the Tieschitz (H3) chondritic meteorite. Mineral Mag 360:311-316
Hutchison R, Bevan AWR (1983) Conditions and time of chondrule accretion. *In* Chondrules and Their Origins (EA King ed) Lunar and Planetary Institute, Houston, 162-179
Hutchison R, Bevan AWR, Agrell SO, Ashworth JR (1980a) Thermal history of the H-group chondritic meteorites. Nature 287:787-790
Hutchison R, Bevan AWR, Agrell SO, Ashworth JR (1980b) Accretion temperature of the Tieschitz, H3, chondritic meteorite. Nature 280:116-119
Hutchison R, Alexander CMO, Barber DJ (1987) The Semarkona meteorite: First recorded occurrence of smectite in an ordinary chondrite, and its implications. Geochim Cosmochim Acta 51:1875 -1882
Hyman M, Rowe MW (1983) Magnetite in CI chondrites. Proc Lunar Planet Sci Conf 13th; Pt 2, J Geophys Res Suppl 88:A736-A740
Hyman M, Rowe MW (1986) Saturization magnetization measurements of carbonaceous chondrites. Meteoritics 21:1-22
Hyman M, Rowe MW, Herndon JM (1978) Magnetite heterogeneity among CI chondrites. Geochem J 13:37-39
Ichikawa O, Ikeda Y (1995) Petrology of the Yamato-8449 CR chondrite. Proc NIPR Symp Antarc Met 8:63-78
Ihinger PD, Stolper E (1986) The color of meteoritic hibonite: An indicator of oxygen fugacity. Earth Planet Sci Lett 78:67-79
Ikeda Y (1980) Petrology of Allan Hills-764 chondrite (LL3). Mem NIPR Spec Issue 17:50-82
Ikeda Y (1982) Petrology of the ALH-77003 chondrite (C3). Mem NIPR Spec Issue 25:34-65
Ikeda Y (1983) Alteration of chondrules and matrices in the four Antarctic carbonaceous chondrites ALH 77307 (C3), Y-790123 (C2), Y-75293(C2) and Y-74662(C2). Mem NIPR Spec Issue 30:93-108
Ikeda Y (1988a) Petrochemical study of the Yamato-691 enstatite chondrite (EH3) I: Major element chemical compositions of chondrules and inclusions. Proc NIPR Symp Antarct Met 1:3-13
Ikeda Y (1988b) Petrochemical study of the Yamato-691 enstatite chondrite (EH3) II: Descriptions and mineral compositions of unusual silicate-inclusions. Proc NIPR Symp Antarct Met 1:14-37
Ikeda Y (1989a) Petrochemical study of the Yamato-691 enstatite chondrite (E3) III: Descriptions and mineral compositions of chondrules. Proc NIPR Symp Antarct Met 2:75-108
Ikeda Y (1989b) Petrochemical study of the Yamato-691 enstatite chondrite (E3) IV: Descriptions and mineral chemistry of opaque nodules. Proc NIPR Symp Antarct Met 2:109-146
Ikeda Y (1989c) Petrochemical study of the Yamato-691 enstatite chondrite (E3) V: Comparison of major element chemistries of chondrules and inclusions in Y-691 with those in ordinary and carbonaceous chondrites. Proc NIPR Symp Antarct Met 2:147-168

Ikeda Y (1991) Petrology and mineralogy of the Yamato-82162 chondrite (CI). Proc NIPR Symp Antarct Met 4:187-225

Ikeda Y (1992) An overview of the research consortium, "Antarctic carbonaceous chondrites with CI affinities", Yamato-86720, Yamato-82162, and Belgica-7904. Proc NIPR Symp Antarct Met 5:49-73

Ikeda Y, Kimura M (1995) Anhydrous alteration of Allende chondrules in the solar nebula I: Description and alteration of chondrules with known oxygen-isotopic compositions. Proc NIPR Symp Antarct Met 8:97-122

Ikeda Y, Prinz M (1993) Petrologic study of the Belgica 7904 carbonaceous chondrite: hydrous alteration, oxygen isotopes and relationship to CM and CI chondrites. Geochim Cosmochim Acta 57:439-452

Ikeda Y, Kimura M, Mori H, Takeda H (1981) Chemical compositions of matrices of unequilibrated ordinary chondrites. Mem NIPR Spec Issue 20:124-144

Ikeda Y, Noguchi T, Kimura M (1992) Petrology and mineralogy of the Yamato-86720 carbonaceous chondrite. Proc NIPR Symp Antarct Met 5:136-154

Inoue M, Nakamura N, Kojima H (1994) A preliminary study of REE abundances in chondrules, an inclusion and mineral fragments from Yamato-793321 (CM2) chondrite. Proc NIPR Symp Antarct Met 7:150-163

Ireland TR (1988) Correlated morphological, chemical and isotopic characteristics of hibonites for the Murchison carbonaceous chondrite. Geochim Cosmochim Acta 52:2827-2839

Ireland TR (1990) Presolar isotopic and chemical signatures in hibonite-bearing refractory inclusions from the Murchison carbonaceous chondrite. Geochim Cosmochim Acta 54:3219-3237

Ireland TR, Fahey AJ, Zinner EK (1988) Trace-element abundances in hibonites from the Murchison carbonaceous chondrite: Constraints on high temperature processes in the solar nebula. Geochim Cosmochim Acta 52:2841-2854

Ireland TR, Compston W, Williams IS, Wendt I (1990) U-Th-Pb systematics of individual perovskite grains from the Allende and Murchison carbonaceous chondrites. Earth Planet Sci Lett 101:379-387

Ireland TR, Fahey AJ, Zinner EK (1991) Hibonite-bearing microspherules: A new type of refractory inclusions with large isotopic anomalies. Geochim Cosmochim Acta 55:367-379

Isobe H, Kitamura M, Morimoto N (1986) Volatile-rich chondrules in the Allende meteorite. Mem NIPR Spec Issue 41:276-286

Ivanov I, MacPherson GJ, Zolensky ME., Kononkova NN, Migdisova LF (1996) The Kaidun meteorite: Composition and origin of inclusions in the metal of an enstatite chondrite clast. Met Planet Sci 31:621-626

Jaques AL, Lowenstein PL, Green DH, Kiss E, Nance WB, Taylor SR, Ware NG (1975) The Ijopega chondrite: a new H6 fall. Meteoritics 10:289-301

Jambor JL (1969) Coalignite from the Muskox Intrusion, North-West Territories. Am Mineral 54:437-447

Jarosewich E (1990) Chemical analyses of meteorites: A compilation of stony and iron meteorite analyses. Meteoritics 25:323-337

Jedwab J (1965) Structures framboidales dans la météorite d'Orgueil. Compt Rend Acad Sci Paris 261:440-444

Jedwab J (1967) La magnétite en plaquettes des météorites carbonées d'Alais, Ivuna, et Orgueil. Earth Planet Sci Lett 2:440-444

Jedwab J (1968) Variations morphologiques de la magnétite des météorites carbonées, d'Alais, Ivuna et Orgueil. *In* Origins and Distribution of the Elements (LH Ahrens ed) Pergamon 467-476

Jedwab J (1971) La magnétite de la météorite d'Orgueil vue au microscope électronique à abalayage. Icarus 15:319- 340

Jeffery PM, Anders E (1970) Primordial noble gases in separated meteoritic minerals. I. Geochim Cosmochim Acta 34:1175-1198

Jobbins EA, Dimes FG, Binns RA, Hey MH, Reed SJB (1966) The Barwell meteorite. Mineral Mag 35:881-902

Johnson CA, Prinz M (1991) Chromite and olivine in type II chondrules in carbonaceous and ordinary chondrites: Implications for thermal histories and group differences. Geochim Cosmochim Acta 55:893-904

Johnson CA, Prinz M (1993) Carbonate compositions in CM and CI chondrites and implications for aqueous alteration. Geochim Cosmochim Acta 57:2843-2852

Johnson CA, Prinz M, Weisberg MK, Clayton RN, Mayeda TK (1990) Dark inclusions in Allende, Leoville and Vigarano: Evidence for nebular oxidation of CV3 constituents. Geochim Cosmochim Acta 54:819-830

Jones RH (1990) Petrology and mineralogy of type II, FeO-rich, chondrules in Semarkona (LL3.0): Origin by closed-system fractional crystallization, with evidence for supercooling. Geochim Cosmochim Acta 54:1785-1802

Jones RH (1992) On the relationship between isolated and chondrule olivine grains in the carbonaceous chondrite ALHA77307. Geochim Cosmochim Acta 56:467-482

Jones RH (1993) Effect of metamorphism on isolated olivine grains in CO3 chondrites. Geochim Cosmochim Acta. 57:2853-2867

Jones RH (1994) Petrology of FeO-poor, porphyritic pyroxene chondrules in the Semarkona chondrite. Geochim Cosmochim Acta 58:5325-5340

Jones RH (1996a) Relict grains in chondrules: Evidence for chondrule recycling. *In* Chondrules and the Protoplanetary Disk (RH Hewins, RH Jones, ERD Scott eds) Cambridge Univ Press, Cambridge, 163-172

Jones RH (1996b) FeO-rich, porphyritic olivine chondrules in unequilibrated ordinary chondrites. Geochim Cosmochim Acta 60:3115-3138

Jones RH (1997) Alteration of plagioclase-rich chondrules in CO3 chondrites: Evidence for late-stage sodium and iron metasomatism in a nebular environment (abstr). *In* Workshop on Parent-body and Nebular Modification of Chondritic Materials. (ME Zolensky, AN Krot, ERD Scott eds). Lunar and Planetary Institute, 30-31

Jones RH, Danielson LR (1997) A chondrule origin for dusty relict olivine in unequilibrated chondrites. Met Planet Sci 32:753-760

Jones RH, Layne GD (1997) Minor and trace element partitioning between pyroxene and melt in rapidly cooled chondrules. Am Mineral 82:534-545

Jones RH, Lofgren GE (1993) A comparison of FeO-rich porphyritic olivine chondrules in unequilibrated chondrites and experimental analogues. Meteoritics 28:213-221

Jones RH, Scott ERD (1989) Petrology and thermal history of type IA chondrules in the Semarkona (LL3.0) chondrite. Proc Lunar Planet Sci Conf 19:523-536

Joreau P, Leroux H, Doukhan JC (1996) A transmission electron microscopy (TEM) investigation of opaque phases in shocked chondrites. Meteoritics 31:305-312

Kallemeyn GW, Rubin AE (1995) Coolidge and Loongana 001: A new carbonaceous chondrite grouplet. Meteoritics 30:20-27

Kallemeyn GW, Wasson JT (1982) The compositional classification of chondrites: III. Ungrouped carbonaceous chondrites. Geochim Cosmochim Acta 46:2217-2228

Kallemeyn GW, Boynton WV, Willis J, Wasson JT (1978) Formation of the Bencubbin polymict meteoritic breccia. Geochim Cosmochim Acta 42:507-515

Kallemeyn GW, Rubin AE, Wasson JT (1991) The compositional classification of chondrites: V. The Karoonda (CK) group of carbonaceous chondrites. Geochim Cosmochim Acta 55:881-892

Kallemeyn GW, Rubin AE, Wasson JT (1994) The compositional classification of chondrites: VI. The CR carbonaceous chondrite group. Geochim Cosmochim Acta 58:2873-2888

Kallemeyn GW, Rubin AE, Wasson JT (1996) The compositional classification of chondrites: VII. The R chondrite group. Geochim Cosmochim Acta 60:2243-2256

Keck BD, Sears DWG (1987) Chemical and physical studies of type 3 chondrites-VIII: Thermoluminescence and metamorphism in the CO chondrites. Geochim Cosmochim Acta 51:3013-3021

Keil K (1968a) Mineralogical and chemical relationships among enstatite chondrites. J Geophys Res 73:6945-6976

Keil K (1968b) Zincian daubreelite from the Kota-Kota and St. Mark's enstatite chondrites. Am Mineral 53:491-495

Keil K (1978) Studies of Brazilian meteorites XIV. Mineralogy, petrology, and chemistry of the Conquista, Minas Gerais, chondrite. Meteoritics 13:177-187

Keil K (1982) Composition and origin of chondritic breccias. *In* Workshop on Lunar Breccias and Soils and Their Meteoritic Analogs (GJ Taylor, LL Wilkening eds) Lunar and Planetary Institute, 97-109

Keil K (1989) Enstatite meteorites and their parent bodies. Meteoritics 24:195-208

Keil K, Andersen CA (1965a) Electron microprobe study of the Jajh deh Kot Lalu enstatite chondrite. Geochim Cosmochim Acta 29:621-632

Keil K, Andersen CA (1965b) Occurrences of sinoite, Si_2N_2O, in meteorites. Nature 207:745

Keil K, Fredriksson K (1964) The iron, magnesium and calcium distribution in coexisting olivines and rhombic pyroxenes in chondrites. J Geophys Res 64:3487-3515

Keil K, Fuchs LH (1971) Hibonite from the Leoville and Allende chondritic meteorites. Earth Planet Sci Lett 12:184-190

Keil K, Snetsinger KG (1967) Niningerite, a new meteoritic sulfide. Science 155:451-453

Keil K, Fodor RV, Starzyk PM, Schmitt RA, Bogard DD, Husain L (1980) A 3.6-b.y.-old impact melt rock fragment in the Plainview chondrite: Implications for the age of the H-group chondrite parent body and regolith formation. Earth Planet Sci Lett 51:235-247

Keller LP, Buseck PR (1990a) Aqueous alteration in the Kaba CV3 carbonaceous chondrite. Geochim Cosmochim Acta 54:2113-2120

Keller LP, Buseck PR (1990b) Matrix mineralogy of the Lancé CO3 carbonaceous chondrite: A transmission electron microscope study. Geochim Cosmochim Acta 54:1155-1163

Keller LP, Buseck PR (1991) Calcic micas in the Allende meteorite: Evidence for hydration reactions in the early solar nebula. Science 252:946-949

Keller LP, Buseck PR (1994) Twinning in meteoritic and synthetic perovskite. Am Mineral 79:73-79

Keller LP, McKay DS (1993) Aqueous alteration of the Grosnaja CV3 carbonaceous chondrite. Meteoritics 28:378

Keller LP, Clark JC, Lewis CF, Moore CB (1992) Maralinga, a metamorphosed carbonaceous chondrite found in Australia. Meteoritics 27:87-91

Keller LP, Thomas KL, Clayton RN, Mayeda TK, DeHart JM, McKay DS (1994) Aqueous alteration of the Bali CV3 chondrite: Evidence from mineralogy, mineral chemistry, and oxygen isotopic compositions. Geochim Cosmochim Acta 58:5589-5598

Kerridge JF (1962) Carbonaceous chondrites in the electron microscope. Fifth Intern. Congress Electron Microscopy, Academic Press, New York

Kerridge JF (1964) Low-temperature minerals from the fine-grained matrix of some carbonaceous meteorites. Ann N Acad Sci 119:41-53

Kerridge JF (1967) The mineralogy and genesis of the carbonaceous meteorites. *In* Mantles of the Earth and Terrestrial Planets (SK Runcorn, ed) Interscience, London, 35-47

Kerridge JF (1969) The use of selected area electron diffraction in meteorite mineralogy. *In* Meteorite Research (PM Millman, ed) Reidel, Dordrecht, 500-504

Kerridge JF (1970a) Some observations on the nature of magnetite in the Orgueil meteorite. Earth Planet Sci Lett 9:299-306

Kerridge JF (1970b) Meteoritic pyrrhotite. Meteoritics 5:149-152

Kerridge JF (1976) Major element compositions of phyllosilicates in the Orgueil carbonaceous meteorite. Earth Planet Sci Lett 29:194-200

Kerridge JF (1977) Correlation between nickel and sulfur abundances in Orgueil phyllosilicates. Geochim Cosmochim Acta 41:1163- 1164

Kerridge JF, Bunch TE (1979) Aqueous activity on asteroids: Evidence from carbonaceous chondrites. *In* Asteroids (T Gehrels ed) Univ of Arizona Press, 745-764

Kerridge JF, Chattterji S (1968) Magnetite content of a type I carbonaceous chondrite. Nature 220:775-776

Kerridge JF, Macdougall JD (1976) Mafic silicates in the Orgueil carbonaceous chondrite. Earth Planet Sci Lett 29:341-348

Kerridge JF and Matthews MS (1988) Meteorites and the Early Solar System. The University of Arizona Press, Tucson, 1269 pp

Kerridge JF, Mackay AL, Boynton WV (1979a) Magnetite in Cl carbonaceous chondrites: Origin by aqueous activity on a planetesimal surface. Science 205:395- 397

Kerridge JF, Macdougall JD, Marti K (1979b) Clues to the origin of sulfide minerals in Cl chondrites. Earth Planet Sci Lett 43:359-367

Kerridge JF, Macdougall JD, Carlson J (1979c) Iron-nickel sulfides in the Murchison meteorite and their relationship to phase Q1. Earth Planet Sci Lett 43:1-4

Kimura M (1988) Origin of opaque minerals in an unequilibrated enstatite chondrite, Yamato-691. Proc NIPR Symp Antarct Met 1:51-64

Kimura M, Ikeda Y (1992) Mineralogy and petrology of an unusual Belgica-7904 carbonaceous chondrite: Genetic relationships among the components. Proc NIPR Symp Antarct Met 5:74-119

Kimura M, Ikeda Y (1995) Anhydrous alteration of Allende chondrules in the solar nebula: Alkali-Ca exchange reactions and the formation of nepheline, sodalite and Ca-rich phases in chondrules. Proc NIPR Symp Antarct Met 8:123-138

Kimura M, Ikeda Y (1996) Comparative study on alteration processes of chondrules in oxidized and reduced CV3 chondrites (abstr) Met Planet Sci 31:A70-71

Kimura M, Ikeda Y (1997) Comparative study of anhydrous alteration of chondrules in reduced and oxidized CV chondrites. Proc NIPR Symp Antarct Met 10:191-202

Kimura M, Yagi K (1980) Crystallization of chondrules in ordinary chondrites. Geochim Cosmochim Acta 44:589-602

Kimura M, El Goresy A, Palme H, Zinner E (1993) Ca,Al-rich inclusions in the unique chondrite ALH85085: Petrology, chemistry, and isotopic compositions. Geochim Cosmochim Acta 57:2329-2359

King TVV, King EA (1981) Accretionary dark rims in unequilibrated ordinary chondrites. Icarus 48:460-472

Kissin SA (1989) Application of the sphalerite cosmobarometer to the enstatite chondrites. Geochim Cosmochim Acta 53:1649-1655

Kitamura M, Watanabe S, Isobe H, Morimoto N (1987) Diopside in chondrules of Yamato-691 (EH3). Mem NIPR Spec Issue 46:113-122

Klimentidis R., Mackinnon IDR (1986) High resolution imaging of ordered mixed-layer clays. Clays and Clay Minerals 34:155-164

Kitamura M, Isobe H, Watanabe S (1988) Relict minerals and their assemblages in Yamato-691 (EH3). Proc NIPR Symp Antarct Met 1:38-50

Klob H, Kracher A, Kurat G (1981) The Ruhobobo, Rwanda meteorite: a new L6 chondrite. Meteoritics 16:1-8

Klöck W, Thomas KL, McKay DS, Palme H (1989) Unusual olivine and pyroxene composition in interplanetary dust and unequilibrated ordinary chondrites. Nature 339:126-128

Knox R (1963) The microstructure of several stony meteorites. Geochim Cosmochim Acta 27:261-268

Koeberl C, Reimold WU, Horsch HE, Merkle RKW (1990) New mineralogical and chemical data on the Machinga (L6) chondrite, Malawi. Meteoritics 25:23-26

Kojima T, Tomeoka K (1993) Unusual dark clasts in the Vigarano CV3 carbonaceous chondrite: Record of parent body processes. Meteoritics 28:649-658

Kojima T, Tomeoka K (1996) Indicators of aqueous alteration and thermal metamorphism on the CV parent body: Microtextures of a dark inclusion from Allende. Geochim Cosmochim Acta 60:2651-2666

Kojima H, Ikeda Y, Yanai K (1984) The alteration of chondrules and matrices in new Antarctic carbonaceous chondrites. Proc 9th Symp Antarct Met. Mem NIPR Spec Issue 35:184-199

Kojima T, Yada S, Tomeoka K (1995) Ca-Al-rich inclusion in three Antarctic CO3 chondrites, Y-81020, Yamato-82050 and Yamato-790992: Record of low temperature alteration processes. Proc NIPR Symp Antarct Met 8:79-86

Kong P, Ebihara M (1996a) Distribution of W and Mo in ordinary chondrites and implications for nebular and parent body processes. Earth and Planetary Science Letters 137:83-94

Kong P, Ebihara M (1996b) Metal phases of L chondrites: Their formation and evolution in the nebula and in the parent body. Geochim Cosmochim Acta 60:2667-2680

Kong P, Ebihara M, Endo E, Nakahara H (1995a) Element distributions in metallic fractions of an Antarctic ordinary chondrite, ALH-77231 (L6). Proc 8th NIPR Symp Antarct Meteor 8:237-249

Kong P, Ebihara M, Nakahara H, Endo K (1995b) Chemical characteristics of metal phases of the Richardton H5 chondrite. Earth Planet Sci Lett 136:407-419

Kornacki AS, Wood JA (1984a) Petrography and classification of Ca, Al-rich and olivine-rich inclusions in the Allende CV3 chondrite. Proc Lunar Planet Sci Conf 14. J Geophys Res Suppl 89:B573-B587

Kornacki AS, Wood JA (1984b) The mineral chemistry and origin of inclusion matrix and meteorite matrix in the Allende CV3 chondrite. Geochim Cosmochim Acta 48:1663-1676

Kornacki AS, Wood JA (1985a) The identification of Group II inclusions in carbonaceous chondrites by electron microprobe analysis of perovskite. Earth Planet Sci Lett 72:74-86

Kornacki AS, Wood JA (1985b) Mineral chemistry and origin of spinel-rich inclusions in the Allende CV3 chondrite. Geochim Cosmochim Acta 49:1219-1237

Kornacki AS, Cohen RE, Wood JA (1983) Petrography and classification of refractory inclusions in the Allende and Mokoia CV3 chondrites. Mem NIPR Spec Issue 30:45-60

Kozul J, Hewins RH (1988) LEW 85300,02,03 polymict eucrites consortium II: Breccia clasts, CM inclusions, glassy matrix and assembly history. Lunar Planet Sci 19:647-648

Kracher A, Scott ERD, Keil K (1984) Relict and other anomalous grains in chondrules: Implications for chondrule formation. Proc Lunar Planet Sci Conf 14. (Part 2) J Geophys Res (Suppl) 89:B559-B566

Kracher A, Keil K, Kallemeyn GW, Wasson JT, Clayton RN, Huss GI (1985) The Leoville (CV3) accretionary breccia. Proc Lunar Planet Sci Conf 16:123-135

Krot AN, Rubin AE (1994) Glass-rich chondrules in ordinary chondrites. Meteoritics 29:697-707

Krot AN, Wasson JT (1994) Silica-merrihueite/roedderite-bearing chondrules and clasts in ordinary chondrites: New occurences and possible origin. Meteoritics 29:707-718

Krot AN, Wasson JT (1995) Igneous rims on low-FeO and high-FeO chondrules in ordinary chondrites. Geochim Cosmochim Acta 59:4951-4966

Krot AN, Ivanova MA, Wasson JT (1993) The origin of chromitic chondrules and the volatility of Cr under a range of nebular conditions. Earth Planet Sci Lett 119:569-584

Krot AN, Scott ERD, Zolensky ME (1995) Mineralogical and chemical modification of components in CV3 chondrites: Nebular or asteroidal processing? Meteoritics 30:748-775

Krot AN, Zolensky ME, Wasson JT, Scott ERD, Keil K, Ohsumi K (1997a) Carbide-magnetite assemblages in type-3 ordinary chondrites. Geochim Cosmochim Acta 61:219-237

Krot AN, Scott ERD, Zolensky ME (1997b) Origin of fayalitic olivine rims and lath-shaped matrix olivine in the CV3 chondrite Allende and its dark inclusions. Meteoritics 32:31-49

Kuehner SM, Davis AM, Grossman L (1989) Identification of relict phases in a once molten Allende inclusion. Geophys Res Lett 16:775-778

Kullerud G (1963) The Fe-Ni-S system. Carnegie Institute Yearbook 62:175-189

Kurat G (1967) Einige Chondren aus dem Meteoriten von Mezö-Madaras. Geochim Cosmochim Acta 31:1843-1858

Kurat G (1969) The formation of chondrules and chondrites and some observations on chondrules from the Tieschitz meteorite. *In* Meteorite Research (PM Millman ed) Reidel, Dordrecht, 185-190

Kurat G (1970) Zur genese der Ca-Al-reichen einschlusse im Chondriten von Lancé. Earth Planet Sci Lett 9:225-231

Kurat G (1975) Der kohlige Chondrit Lancé: Eine petrologische Analyse der komplexen Genese eines Chondriten. Tschermaks Min Petr Mitt 22:38-78

Kurat G, Kracher A (1980) Basalts in the Lancé carbonaceous chondrite. Z Naturforsch 35a:180-190

Kurat G, Hoinkes G, Fredriksson K (1975) Zoned Ca-Al-rich chondrule in Bali: New evidence against the primordial condensation model. Earth Planet Sci Lett 26:140-144

Kurat G, Palme H, Brandstätter F, Huth J (1989) Allende Xenolith AF: Undisturbed record of condensation and aggregation of matter in the Solar Nebula. Z Naturforsch 44a:988-1004

Kvasha LG (1948) Investigation of the stony meteorite Staroye Boriskino. Meteoritika 4:83-96

Kvasha LG (1961) Some recent observations on the textures of chondrites (in Russian). Meteoritika 20:124-136

Lacroix A (1905) Materiaux sur les météorites pierreuses: 1. Identité de composition des météorites de Pillistfer (1863) et de Hvittis (1901). Bull Soc Franc Mineral 28:70-76

Lambert DL (1992) The p-nuclei: Abundances and origins. Astron Astrophys Rev 3:201-256

Lange DE, Moore CB, Rhoton K (1973) The Willowbar meteorite. Meteoritics 8:263-275

Lange DE, Frost K, Sipiera PP, Moore CB (1974) The Canyonlands meteorite. Meteoritics 9:271-280

Langenhorst F, Joreau P, Doukhan JC (1995) Thermal and shock metamorphism of the Tenham chondrite: A TEM examination. Geochim Cosmochim Acta 59:1835-1845

Larimer JW, Wasson JT (1988) Siderophile element fractionation. *In* Meteorites and the Early Solar System (JF Kerridge, MS Matthews eds) The University of Arizona Press, Tucson, 416-435

Larson EE, Watson DE, Herndon JM, Rowe MW (1974) Thermomagnetic analysis of meteorites, 1. C1 chondrites. Earth Planet Sci Lett 21:345-350

Lattimer JM, Schramm DN, Grossman L (1978) Condensation in supernova ejecta and isotopic anomalies in meteorites. Astrophys J 219:230-249

Lauretta DS, Kremser DT, Fegley BJ (1996) A comparative study of experimental and meteoritic metal-sulfide assemblages. Proc NIPR Symp Antarct Met 9:97-110

Lee MR (1993) The petrogaphy, mineralogy and origins of calcium sulphate within the Cold Bokkeveld CM carbonaceous chondrite. Meteoritics 28:53-62

Lee MR, Greenwood RC (1994) Alteration of calcium- and aluminum-rich inclusions in the Murray (CM2) carbonaceous chondrite. Meteoritics 29:780-790

Lee MS, Rubin AE, Wasson JT (1992) Origin of metallic Fe-Ni in Renazzo and related chondrites. Geochim Cosmochim Acta 56:2521-2533

Lee MR, Russell SS, Arden JW, Pillinger CT (1995) Nierite (Si_3N_4), a new mineral from ordinary and enstatite chondrites. Meteoritics 30:387-398

Lee MR, Hutchison R, Graham AL (1996) Aqueous alteration in the matrix of the Vigarano (CV3) carbonaceous chondrite. Met Planet Sci 31:477-483

Leitch CA, Grossman L (1977) Lithic clasts in the Supahee chondrite. Meteoritics 12:125-140

Leitch CA, Smith JV (1982) Petrography, mineral chemistry and origin of Type 1 enstatite chondrites. Geochim Cosmochim Acta 46:2083-2097

Leroux H, Doukhan JC, Langenhorst F (1994) Microstructural defects in experimentally shocked diopside: A TEM characterization. Phys Chem Mineral 20:521-530

Leshin LA, Rubin AE, McKeegan KD (1997) The oxygen isotopic composition of olivine and pyroxene from CI chondrites. Geochim Cosmochim Acta 61:835-845

Lewis RS, Ming T, Wacker JF, Anders E, Steel E (1987) Interstellar diamonds in meteorites. Nature 326:360-362

Lewis RS, Anders E, Draine BT (1989) Properties, detectability and origin of interstellar diamonds in meteorites. Nature 339:117-121

Lewis RS, Amari S, Anders E (1990) Meteoritic silicon carbide: Pristine material from carbon stars. Nature 348:293-298

Lewis RS, Amari S, Anders E (1994) Interstellar grains in meteorites. II. SiC and its noble gases. Geochim Cosmochim Acta 58:471-494

Lin Y-T, Nagel H-J, Lundberg LL, El Goresy A (1991) MAC88136—The first EL3 chondrite (abstr). Lunar Planet Sci XXII:811-812

Lofgren GE (1996) A dynamic crystallization model for chondrule melts. *In* Chondrules and the Protoplanetary Disk (RH Hewins, RH Jones, ERD Scott eds) Cambridge Univ Press, 187-196

Löhn B, El Goresy A (1992) Morphologies and chemical composition of individual magnetite grains in CI and CM chondrites: A potential genetic link to their origin? Meteoritics 27:252

Lorin JC, Christophe Michel-Lévy M, Desnoyers C (1978) Ophitic Ca-Al inclusions in the Allende and Leoville meteorites: A petrographic and ion-microprobe study. Meteoritics 13:537-540

Lovering, JF (1962) The evolution of the meteorites—evidence for the coexistence of chondritic, achondritic and iron meteorites in a typical parent meteorite body. Researches on Meteorites (CB Moore, ed) Wiley, 179-197

Lovering JF, Wark DA, Sewell DKB (1979) Refractory oxide, titanate, niobate and silicate accessory mineralogy of some Type B Ca-Al inclusion in the Allende meteorite (abstr). Lunar Planet Sci IX:672-674

Lumpkin GR (1980) Nepheline and sodalite in a barred olivine chondrule from the Allende meteorite. Meteoritics 15:139-145

Lusby D, Scott ERD, Keil K (1987) Ubiquitous high-FeO silicates in enstatite chondrites. Proc Lunar Planet Sci Conf 17, J Geophys Res 92:E679-E695

Lux G, Keil K, Taylor GJ (1980) Metamorphism of the H-group chondrites: implications from the compositional and textural trends in chondrules. Geochim Cosmochim Acta 44:841-855

Macdougall JD (1979) Refractory-element-rich inclusions in C2 meteorites. Earth Planet Sci Lett 42:1-6

Macdougall JD (1981) Refractory spherules in the Murchison meteorite: Are they chondrules? Geophys Res Lett 8:966-969

Macdougall JD, Kerridge JF (1977) Cubanite: A new sulfide phase in CI meteorites. Science 197:561-562

Macdougall JD, Kerridge JF, Phinney D (1981) Refractory ABC (abstr). Lunar Planet Sci X:745-747

Macdougall JD, Lugmair GW, Kerridge JF (1984) Early solar system aqueous activity: Sr isotope evidence from the Orgueil CI meteorite. Nature 307:249-251

Mackinnon IDR (1980) Structures and textures of the Murchison and Mighei carbonaceous chondrite matrices. Proc Lunar Planet Sci Conf 11:839-852

Mackinnon IDR (1982) Ordered mixed-layer structures in the Mighei carbonaceous chondrite matrix. Geochim Cosmochim Acta 46:476-489

Mackinnon IDR, Buseck PR (1979a) High resolution transmission electron microscopy of two stony meteorites: Murchison and Kenna. Proc Lunar Planet Sci Conf 10:937-949

Mackinnon IDR, Buseck PR (1979b) New phyllosilicate types in a carbonaceous chondrite matrix. Nature 280:219-220

Mackinnon IDR, Zolensky ME (1984) Proposed structures for poorly characterized phases in C2M carbonaceous chondrite meteorites. Nature 309:240-242

MacPherson GJ, Davis AM (1993) A petrologic and ion microprobe study of a Vigarano Type B refractory inclusion: Evolution by multiple stages of alteration and melting. Geochim Cosmochim Acta 57:231-243

MacPherson GJ, Davis AM (1994) Refractory inclusions in the prototypical CM chondrite, Mighei. Geochim Cosmochim Acta 58:5599-5625

MacPherson GJ, Delaney JS (1985) A fassaite-two olivine-pleonaste-bearing refractory inclusion from Karoonda. Lunar Planetary Sci XVI:515-516

MacPherson GJ, Grossman L (1979) Melted and non-melted coarse-grained Ca,Al-rich inclusions in Allende (abstr). Meteoritics 14:479-480

MacPherson GJ, Grossman L (1981) A once molten, coarse-grained , Ca-rich inclusion in Allende. Earth Planet Sci Lett 52:16-24

MacPherson GJ, Grossman L (1982) Fine-grained spinel-rich and hibonite-rich Allende inclusions (abstr) Meteoritics 17:245-246

MacPherson GJ, Grossman L (1984) "Fluffy" Type A Ca,-Al-rich inclusions in the Allende meteorite. Geochim Cosmochim Acta 48:29-46

MacPherson GJ, Grossman L, Allen JM, Beckett JR (1981) Origin of rims on coarse-grained inclusions in the Allende meteorite. Proc Lunar Planet Sci Conf 12B:1079-1091

MacPherson GJ, Bar-Matthews M, Tanaka T, Olsen E, Grossman L (1983) Refractory inclusions in the Murchison meteorite. Geochim Cosmochim Acta 47:823-839

MacPherson GJ, Grossman L, Hashimoto A, Bar-Matthews M, Tanaka T (1984a) Petrographic studies of refractory inclusions from the Murchison meteorite. Proc Lunar Planet Sci Conf 15, J Geophys Res Suppl C299-C312

MacPherson GJ, Paque J, Stolper E, Grossman L (1984b) The origin and significance of reverse zoning in melilite from Allende Type B inclusions. J Geol 92:289-305

MacPherson GJ, Hashimoto A, Grossman L (1985) Accretionary rims on inclusions in the Allende meteorite. Geochim Cosmochim Acta 49:2267-2279

MacPherson GJ, Wark DA, Armstrong JT (1988) Primitive material surviving in chondrites: Refractory inclusions. *In* Meteorites and the Early Solar System (JF Kerridge, MS Matthews eds) University of Arizona Press, 746-807

MacPherson GJ, Crozaz G, Lundberg LL (1989a) The evolution of a complex type B Allende inclusion: An ion microprobe trace element study. Geochim Cosmochim Acta 53:2413-2427

MacPherson GJ, Davis AM, Grossman JN (1989b) Refractory inclusions in the unique chondrite ALH 85085. Meteoritics 24:297

MacPherson GJ, Davis AM, Zinner EK (1995) The distribution of aluminum-26 in the early solar system—A reappraisal. Meteoritics 30:365-386

Madon M, Poirier JP (1983) Transmission electron microscope observations of α, β and γ $(Mg,Fe)_2SiO_4$ in shocked meteorites: Planar defects and polymorphic transitions. Phys Earth Planet Int 86:69-83

Madsen MB, Morup S, Costa TVV, Knudsen JM, Olsen M (1986) Superparamagnetic component in the Orgueil meteorite and Mössbauer spectroscopy studies in applied magnetic fields. Nature 321: 501-503

Madsen MB, Morup S, Knudsen JM (1989) Extraterrestrial magnetite studied by Mössbauer spectroscopy. Hyperfine Interactions 50:659-666

Manhés G, Göpel C, Allègre CJ (1987) High resolution chronology of the early solar system given by lead isotopes (abstr). Terra Cognita 7:377)

Mao X-Y, Ward BJ, Grossman L, MacPherson GJ (1990) Chemical composition of refractory inclusions from the Vigarano and Leoville carbonaceous chondrites. Geochim Cosmochim Acta 54:2121-2132

Martin PM, Mason B (1974) Major and trace elements in the Allende meteorite. Nature 249:333-334

Mason B (1962) The carbonaceous chondrites. Space Science Rev 621-646.

Mason B (1963) Olivine composition in chondrites. Geochim Cosmochim Acta 27:1011-1023

Mason B (1966) The enstatite chondrites. Geochim Cosmochim Acta 30:23-39

Mason B (1971) Merrillite and whitlockite, or what's in a name. Mineral Record 2:277-279

Mason B (1974) Aluminum-titanium-rich pyroxenes with special reference to the Allende meteorite. Am Mineral 22:1198-1202.

Mason B, Martin PM (1974) Minor and trace element partitioning between melilite and pyroxene from the Allende meteorite. Earth Planet Sci Lett 22:141-144

Mason B, Martin PM (1977) Geochemical differences among components of the Allende meteorite. Smithson Contrib Earth Sci 19:84-95

Mason B, Nelen J (1968) The Weatherford meteorite. Geochim Cosmochim Acta 32:661-664

Mason B, Taylor SR (1982) Inclusions in the Allende meteorite. Smithsonian Contrib Earth Sci 25:1-30

Mason B, Wiik HB (1962a) Descriptions of two meteorites; Karoonda and Erakot. Am Mus Novitates 2115:1-10

Mason B, Wiik HB (1962b) The Renazzo meteorite. Am Mus Novitates 2106:1-11

Mason B, Wiik HB (1966) The composition of the Bath, Frankfort, Kakangari, Rose City and Tadjera meteorites. Am Mus Novitates 2272:1-23

Massalski TB, Park FR, Vassalmillet LF (1966) Speculations about plessite. Geochim Cosmochim Acta 30:649-662

Matsunami S (1984) The chemical compositions and textures of matrices and chondrule rims of eight unequilibrated ordinary chondrites. Proc NIPR Symp Antarct Met 35:126-148

Matsunami S, Nishimura H, Takeshi H (1990a) The chemical compositions and textures of matrices and chondrule rims of unequilibrated ordinary chondrites-II. Their constituents and implications for the formation of matrix olivine. Proc NIPR Symp Antarct Met 3:147-180

Matsunami S, Nishimura H, Takeshi H (1990b) Compositional heterogeneity of fine-grained rims in the Semarkona (LL3) chondrite. Proc NIPR Symp Antarct Met 3:181- 193

Matsunami S, Ninagawa K, Kubo H, Fujimara S, Yamamoto I, Wada T, Nishimura H (1992) Silica phase as a thermoluminescence phosphor in ALH-77214 (L3.4) chondrite. Proc NIPR Symp Antarct Met 5:270-280

Matsunami S, Ninagawa K, Nishimura S, Kubono N, Yamamoto I, Kohata M, Wada T, Yamashita Y, Lu J, Sears D, Nishimura H (1993) Thermoluminescence and compositional zoning in the mesostasis of a Semarkona group A1 chondrule and new insights into the chondrule-forming process. Geochim Cosmochim Acta 57:2101-2110

Matsuoka K, Nakamura T, Nakamuta Y (1996) Yamato-86789: A heated CM-like carbonaceous chondrite. Proc NIPR Symp Antarct Met 9:20-36

McCall GJH (1968) The Bencubbin meteorite: further details, including microscopic character of host material and two chondritic enclaves. Mineral Mag 36:726-739

McCoy TJ, Keil K (1988) Ingella Station, a new chondrite find from the Tenham strewnfield, Queensland, Australia. Meteoritics 23:381

McCoy TJ, Scott ERD, Jones RH, Keil K, Taylor GJ (1991a) Composition of chondrule silicates in LL3-5 chondrites and implications for their nebular history and parent body metamorphism. Geochim Cosmochim Acta 55:601-619

McCoy TJ, Pun A, Keil K (1991b) Spinel-bearing, Al-rich chondrules in two chondrite finds from Roosevelt County, New Mexico: Indicators of nebular and parent body processes. Meteoritics 26:301-309

McCoy TJ, Keil K, Wilson IE (1993) Classification of five new ordinary chondrites (RC 073, 074, 076-078) from Roosevelt County, New Mexico. Meteoritics 28:120-121

McCoy TJ, Keil K, Bogard DD, Garrison DH, Casanova I, Lindstrom M, Brearley AJ, Kehm K, Nichols RH, Hohenberg CM (1995a) Origin and history of impact-melt rocks of enstatite chondritic parentage. Geochim Cosmochim Acta 59:161-175

McCoy TJ, Ehlmann AJ, Keil K (1995b) The Fayette County, Texas, meteorites. Meteoritics 30:776-780

McCoy TJ, Ehlmann AJ, Keil K (1995c) The Travis County, Texas, meteorites. Meteoritics 30:348-351

McGuire AV, Hashimoto A (1989) Origin of zoned fine-grained inclusions in the Allende meteorite. Geochim Cosmochim Acta 53:1123-1133

McKee TR, Moore CB (1979) Characterization of submicron matrix phyllosilicates from Murray and Nogoya carbonaceous chondrites. Proc Lunar Planet Sci Conf 10:921-936

McKinley SG, Keil K, Scott ERD (1984) Composition and origin of enstatite in E-chondrites. Proc Lunar Planet Sci Conf 14, J Geophys Res Suppl 89:B567-B572

McMahon BM, Haggerty SE (1980) Experimental evidence bearing on the magnetite-alloy-sulfide association in the Allende meteorite: Constraints on the conditions of chondrule formation. Proc Lunar Planet Sci Conf 11:1003-1025

McSween HY Jr (1976) A new type of chondritic meteorite found in lunar soil. Earth Planet Sci Lett 31:193-199

McSween HY Jr (1977a) Carbonaceous chondrites of the Ornans type: a metamorphic sequence. Geochim Cosmochim Acta 44:477-491

McSween HY Jr (1977b) On the nature and origin of isolated olivine grains in carbonaceous chondrites. Geochim Cosmochim Acta 41:411-418

McSween HY Jr (1977c) Petrographic variations among carbonaceous chondrites of the Vigarano type. Geochim Cosmochim Acta 41:1777-1790

McSween HY Jr (1979) Alteration in CM carbonaceous chondrites inferred from modal and chemical variations in matrix. Geochim Cosmochim Acta 43:1761-1770

McSween HY Jr (1987) Aqueous alteration in carbonaceous chondrites: Mass balance constraints on matrix mineralogy. Geochim Cosmochim Acta 51:2469-2477

McSween HY Jr, Labotka TC (1993) Oxidation during metamorphism of the ordinary chondrites. Geochim Cosmochim Acta 57:1105-1114

McSween HY Jr, Patchen AD (1989) Pyroxene geothermometry in LL-group chondrites and implications for parent body metamorphism. Meteoritics 24:219-226

McSween HY Jr, Richardson SM (1977) The composition of carbonaceous chondrite matrix. Geochim Cosmochim Acta 41:1145-1161

McSween HY Jr, Sears DWG, Dodd RT (1988) Thermal metamorphism. *In* Meteorites and the Early Solar System (JF Kerridge, MS Matthews eds) University of Arizona Press, 102-113

Meeker GP (1995) Constraints on formation processes of two coarse-grained calcium-aluminum-rich inclusions: A study of mantles, islands and cores. Meteoritics 30:71-84

Meeker GP, Wasserburg GJ, Armstrong JT (1983) Replacement textures in CAI and implications for planetary metamorphism. Geochim Cosmochim Acta 47:707-721

Methot RL, Noonan AF, Jarosewich E, deGasparis AA, Al-Far DM (1975) Mineralogy, petrology and chemistry of the Isna (C3) meteorite. Meteoritics 10:121-132

Metzler K, Bischoff A, Stöffler D (1988) Characteristics of accretionary dark rims in carbonaceous chondrites. Lunar Planet Sci XIX:

Metzler K, Bischoff A, Stöffler D (1992) Accretionary dust mantles in CM chondrites: Evidence for solar nebula processes. Geochim Cosmochim Acta 56:2873-2897

Minster J-F, Birck J-L, Allègre CJ (1982) Absolute age of formation of chondrites by the ^{87}Rb-^{87}Sr method. Nature 300:414-419

Miúra Y, Matsumoto Y (1982) Merrillite, a whitlockite-group mineral in Yamato-75 chondrites. Mem NIPR Spec Issue 25 124-130

Miúra Y, Tomisaka T (1984) Composition and structural substitution of meteoritic plagioclases (I). Mem NIPR Spec Issue 9:210-225

Miyamoto M, Ohsumi K (1995) Micro Raman spectroscopy of olivines in L6 chondrites: evaluation of the degree of shock. Geophys Res Lett 22:437-440

Miyashiro A (1962) Common occurrence of high-temperature plagioclase in chondrites. Japan J Geol Geogr 33:235-237

Molin GM, Tribaudino M, Brizi E (1994) Zaoyang chondrite cooling history from Fe^{2+}-Mg intracrystalline ordering in pyroxenes. Mineral Mag 58:143-150

Müller WF (1991a) Microstructure of minerals in a chondrule from the Allende meteorite II. Thermal history deduced from clinopyroxenes and other minerals. Neues Jahrb Miner Abh 162:237-259

Müller WF (1991b) Microstructure of minerals in a chondrule from the Allende meteorite I. Olivine. Neues Jahrb Mineral, Abh 163:145-158

Müller WF, Hornemann U (1969) Shock-induced planar deformation structures in experimentally shock-loaded olivines and in olivines from chondritic meteorites. Earth Planet Sci Lett 7:251-264

Müller WF, Müller-Beneke G (1992) Stacking faults in magnetite from the Allende meteorite. Z Kristallogr 200:275-286

Müller WF, Wlotzka F (1982) Mineralogical study of the Leoville meteorite (CV3): Macroscopic texture and transmission electron microscope observations (abstr). Lunar Planet Sci XXIII:558-559

Müller WF, Kurat G, Kracher A (1977) Crystal structure amd composition of cronstedtite from the Cochabamba carbonaceous chondrite. Meteoritics 12:322

Müller WF, Kurat G, Kracher A (1979) Chemical and crystallographic study of cronstedtite in the matrix of the Cochabamba (CM2) carbonaceous chondrite. Tschermaks Min Petr Mitt 26:293-304

Müller WF, Weinbruch S, Walter R, Müller-Beneke G (1995) Transmission electron microscopy of chondrule minerals in the Allende meteorite: constraints on the thermal and deformational history of granular olivine-pyroxene chondrules. Planet Space Sci 43:469-483

Murakami T, Ikeda Y (1994) Petrology and mineralogy of the Yamato-86751 CV3 chondrite. Meteoritics 29:397-409

Murrell MT, Burnett DS (1987) Actinide chemistry in Allende Ca-Al-rich inclusions. Geochim Cosmochim Acta 51:985-999

Nagahara H (1979) Petrological study of Ni-Fe metal in some ordinary chondrites. Mem NIPR Spec Issue 15:111-122

Nagahara H (1980) Petrology of "equilibrated" chondrites. 2. Metamorphism and thermal history. Mem NIPR Spec Issue 17:32-49

Nagahara H (1981a) Evidence for secondary origin of chondrules. Nature 292:135-136

Nagahara H (1981b) Petrology of chondrules in ALH-77015 (L3) chondrite. Mem NIPR Spec Issue 20:145-160

Nagahara H (1982) Ni-Fe metals in the type 3 ordinary chondrites. Mem NIPR Spec Issue 25:86-96

Nagahara H (1983) Texture of chondrules. Mem NIPR Spec Issue 30:61-83

Nagahara H (1984) Matrices of type 3 ordinary chondrites-Primitive nebular records. Geochim Cosmochim Acta 48:2581-2595

Nagahara H (1991) Petrology of Yamato-75261 meteorite: An enstatite (EH) chondrite Breccia. Proc NIPR Symp Antarct Met 4:144-162

Nagahara H, Kushiro I (1982) Petrology of chondrules, inclusions and isolated olivine grains in ALH-77307 (CO3) chondrite. Mem NIPR Spec Issue 25:66-77

Nagasawa H, Blanchard DP, Jacobs JW, Brannon JC, Philpotts JA, Onuma N (1977) Trace element distributions in mineral separates of the Allende meteorite and their genetic implications. Geochim Cosmochim Acta 41:1587-1600

Nagata T, Funaki M (1982) Magnetic properties of tetrataenite-rich stony meteorites. Proc 7 Symp Antarct Met. Mem NIPR Spec Issue 25:222-250

Nagata T, Funaki M, Danon JA (1986) Magnetic properties of tetrataenite-rich meteorites; II. Mem NIPR Spec Issue 41:364-381

Nagy B (1966) Investigation of the the Orgueil carbonaceous meteorite. Geol Foren Stockholm Forh 88:235-272

Nagy B (1975) Carbonaceous chondrites. Elsevier, Amsterdam, 747 p

Nagy B, Andersen CA (1964) Electron probe microanalysis of some carbonate, sulfate and phosphate minerals in the Orgueil meteorite. Am Mineral 49:1730-1736

Nagy B, Claus G (1962) Notes on the petrography of the Orgueil meteorite. Proc. Intern. Meeting Milan 1962, Advances in Organic Geochemistry Pergamon, New York, 115-118

Nagy B, Meinschein WG, Hennessy DJ (1961) Mass spectroscopic analysis of the Orgueil meteorite: evidence for biogenic hydrocarbons. Ann NY Acad Sci 93:25-35

Nagy B, Menschein WG, Hennessy DJ (1963) Aqueous, low temperature environment of the Orgueil meteorite parent body. Am NY Acad Sci 108:534-552

Nakamura T, Nakamuta Y (1996) X-ray study of PCP from the Murchison CM carbonaceous chondrite. Proc NIPR Symp Antarct Met 9:37-50

Nakamura T, Tomoeka K, Takeda H (1992) Shock effects of the Leoville CV carbonaceous chondrite: a transmission electron microscope study. Earth Planet Sci Lett 114:159-170

Nakamura T, Tomeoka K, Takeda H (1993) Mineralogy and petrology of the CK chondrites Yamato-82104, -693 and a Carlisle Lakes-type chondrite Yamato-82002. Proc NIPR Symp Antarct Met 6:171-185

Nazarov MA, Brandstätter F, Kurat G (1997) Comparative chemistry of P-rich phases in CM chondrites (abstr). Lunar Planet Sci XXVIII:1003-1004

Newsom HE, Drake MJ (1979) The origin of metal clasts in the Bencubbin meteorite breccia. Geochim Cosmochim Acta 43:689-707

Newton J, Bischoff A, Arden JW, Franchi IA, Geiger T, Greshake A, Pillinger CT (1995) Acfer 094, a uniquely primitive carbonaceous chondrite from the Sahara. Meteoritics 30:47-56

Nickel EH, Bridge PJ (1977) Nickelbloedite, $Na_2Ni(SO_4)_2.4H_2O$, a new mineral from Western Australia. Mineral Mag 41:37-41

Ninagawa K, Yamamoto I, Wada T, Matsunami S, Nishimura H (1990) Thermoluminescence study of ordinary chondrites by TL spatial distribution readout system. Proc NIPR Symp Antarct Met 3: 244-253

Ninagawa K, Nishimura S, Kubono N, Yamamoto I, Kohata M, Wada T, Yamashita Y, Lu J, Sears DWG, Matsunami S, Nishimura H (1992) Thermoluminescence of chondrules in primitive ordinary chondrites, Semarkona and Bishunpur. Proc NIPR Symp Antarct Met 5:281-289

Nittler LR, Hoppe P, Alexander CMO'D, Amari S, Eberhardt P, Gao X, Lewis RS, Strebel R, Walker RM, Zinner E (1995) Silicon nitride from supernovae. Astrophys J 453:L25-L28

Nittler LR, Amari S, Zinner E, Woosley SE, Lewis RS (1996) Extinct ^{44}Ti in presolar graphite and SiC: Proof of a supernova origin. Astrophys J 462:L31-L34

Nittler LR, Alexander CMO'D, Gao X, Walker RM, Zinner E (1997) Stellar sapphires: The properties and origins of presolar Al_2O_3 in meteorites. Astrophys J 483:475-495

Noguchi T (1989) Texture and chemical composition of pyroxenes in chondrules in carbonaceous and unequilibrated ordinary chondrites. Proc NIPR Symp Antarct Met 2:169-199

Noguchi T (1993) Petrology and mineralogy of CK chondrites: Implications for the metamorphism of the CK chondrite parent body. Proc NIPR Symp Antarct Met 6:204-233

Noguchi T (1994) Petrology and mineralogy of the Coolidge meteorite (CV4). Proc NIPR Symp Antarct Met 7:42-72

Noguchi T (1995) Petrology and mineralogy of the PCA 91082 chondrite and its comparison with the Yamato-793495 (CR) chondrite. Proc NIPR Symp Antarct Met 8:33-62

Noonan AF (1975) The Clovis (no. 1) New Mexico, meteorite and Ca, Al and Ti-rich inclusions in ordinary chondrites. Meteoritics 10:51-59

Noonan AF, Nelen JA (1976) A petrographic and mineral chemistry study of the Weston, Connecticut, chondrite. Meteoritics 11:111-130

Noro H, Nagasawa K, Tokonami M (1980) Major element composition of clay minerals in the Murchison (C2) carbonaceous chondrite matrix. Proc 5 Symp Antarct Met. Mem NIPR Spec Issue 17 311-317

Nyquist LE, Takeda H, Bansal BM, Shih C-Y, Wiesmann H, Wooden JL (1986) Rb-Sr and Sm-Nd internal isochron ages of a subophitic basalt clast and a matrix sample from the Y75011 eucrite. J Geophys Res 91:8137-8150

Olsen EJ (1967) Amphibole: first occurrence in a meteorite. Science 156:61-62

Olsen EJ (1973) Copper-nickel alloy in the Blansko chondrite. Meteoritics 8:259-261

Olsen EJ (1983) SiO_2-bearing chondrules in the Murchison (C2) meteorite. *In* Chondrules and Their Origins (EA King ed) Lunar Planet Inst, 223-234

Olsen EJ, Grossman L (1974) A scanning electron microscopy study of olivine crystal surfaces. Meteoritics 9:243-254

Olsen E, Grossman L (1978) On the origin of isolated olivine grains in type 2 carbonaceous chondrites. Earth Planet Sci Lett 41:111- 127

Olsen E, Fuchs LH, Forbes WC (1973a) Chromium and phosphorus enrichment in the metal of type II (C2) carbonaceous chondrites. Geochim Cosmochim Acta 37:2037-2042

Olsen E, Huebner JS, Douglas JA, Plant AG (1973b) Meteoritic amphiboles. Am Mineral 62:869-872

Olsen EJ, Mayeda TK, Clayton RN (1981) Cristobalite-pyroxene in an L6 chondrite: Implications for metamorphism. Earth Planet Sci Lett 56:82-88

Olsen EJ, Huss GI, Jarosewich E (1988) The Eagle, Nebraska enstatite chondrite (EL6). Meteoritics 23:379-380

Onuma N, Nishida N, Okudera S, Ono Y (1976) Chemical inhomogeneity in a single euhedral olivine from the Murchison meteorite. Geochem J 10:205-210

Organova NI, Genkin AD, Drits VA, Dmitrik AL, Kuzimina O (1971) Tochilinite: a new sulfide hydroxide of iron and magnesium. Zap Mineral Obschestra 4:477-487

Ostertag R (1983) Shock experiments on feldspar crystals. Proc Lunar Planet Sci Conf 14, J Geophys Res 88:B-364-B376

Ott U (1993) Interstellar grains in meteorites. Nature 364:25-33

Paar W, Keil K, Gomes CB, Jarosewich E (1976) Studies of Brazilian meteorites II. The Avanhandava chondrite: mineralogy, petrology and chemistry. Rev Bras Geoc 6:201-210

Palme H, Fegley B Jr (1990) High-temperature condensation of iron-rich olivine in the solar nebula. Earth Planet Sci Lett 101:180-195

Palme H, Wlotzka F (1976) A metal particle from a Ca, Al-rich inclusion from the meteorite Allende, and the condensation of refractory siderophile elements. Earth Planet Sci Lett 33:45-60

Palme H, Wlotzka F (1981) Iridium-rich phases in Ornans (abstr). Meteoritics 16:373-374

Palme H, Wloztka F, Nagel K, El Goresy A (1982) An ultrarefractory inclusion from the Ornans carbonaceous chondrite. Earth Planet Sci Lett 61:1-12

Palme H, Hutcheon ID, Spettel B (1994) Composition and origin of a refractory-metal-rich assemblage in a Ca, Al-rich Allende inclusion. Geochim Cosmochim Acta 58:495-513

Paque JM (1987) $CaAl_4O_7$ from Allende type A inclusion NMNH 4691. Lunar Planet Sci XVIII:762-763

Paque JM, Beckett JR, Barber DJ, Stolper EM (1994) A new titanium-bearing calcium aluminosilicate phase: I. Meteoritic occurrences and formation in synthetic systems. Meteoritics 29:673-681

Peck JA (1983) Chemistry of CV3 matrix minerals and Allende chondrule olivine (abstr). Meteoritics 18:373-374

Peck JA (1984) Origin of the variation in properties of CV3 meteorite matrix and matrix clasts (abstr). Lunar Planet Sci XVI:635-636

Peck JA, Wood JA (1987) The origin of ferrous zoning in Allende chondrule olivine. Geochim Cosmochim Acta 51:1503-1510

Pellas P, Störzer D (1981) ^{244}Pu fission track thermometry and its application to stony meteorites. Proc R Soc London, Ser A 374:253-270

Peters MT, Shaffer EE, Burnett DS, Kim SS (1995) Magnesium and titanium partitioning between anorthite and Type B CAI liquid: dependence on oxygen fugacity and liquid composition. Geochim Cosmochim Acta 59:2785-2796

Phinney D, Whitehead B, Anderson D (1979) Li, Be and B in minerals of refractory-rich Allende inclusion. Proc Lunar Planet Sci Conf 10:885-905

Pisani F (1864) Etude chimique et analyse de l'aerolithe d'Orgueil. Compt. Rend 59:132-135

Planner HN (1983) Phase separation in a chondrule fragment from the Piancaldoli (LL3) chondrite. *In* Chondrules and their origins. (EA King ed) Lunar and Planetary Institute, Houston, 235-242

Podosek FA, Cassen P (1994) Theoretical, observational, and isotopic estimates of the lifetime of the solar nebula. Meteoritics 29:6-25

Podosek FA, Zinner EK, MacPherson GJ, Lundberg LL, Brannon JC, Fahey AJ (1991) Correlated study of initial $^{87}Sr/^{86}Sr$ and Al-Mg isotopic systematics and petrologic properties in a suite of refractory inclusions from the Allende meteorite. Geochim Cosmochim Acta 55:1083-1110

Poirier J-P (1975) On the slip systems of olivine. J Geophys Res 80:4059-4061

Poirier J-P (1981) Martensitic olivine-spinel transformation and the plasticity of the mantle zone. *In* Anelasticity of the Earth (FD Stacey, MS Patterson, A Nicolaus eds) Am Geophys Union, Washington DC, 113-117

Pollack SS (1966) Disordered orthopyroxene in meteorites. Am Mineral 51:1722-1726

Prinz M, Weisberg MK (1992) Acfer 182/207: A new ALH85085–type chondrite and its implications (abstr). Lunar Planet Sci XXIII:1109-1110

Prinz M, Weisberg MK, Nehru CE, MacPherson GJ, Clayton RN, Mayeda TK (1989) Petrologic and stable isotope study of the Kakangari (K-group) chondrite: Chondrules, matrix, CAI's (abstr.). Lunar Planet Sci. XX:870-871

Price GD (1983) The nature and significance of stacking faults in wadsleyite, natural beta $(Mg,Fe)_2SiO_4$ from the Peace River meteorite. Phys Earth Planet Int 33:137-147

Price GD, Putnis A, Agrell SO (1979) Electron petrography of shock-produced veins in the Tenham chondrite. Contrib Mineral Petrol 71:211-218

Price GD, Putnis A, Smith DGW (1982) A spinel to β-phase transformation in $(Mg,Fe)_2SiO_4$. Nature 296:729-731

Putnis A, Price GD (1979) High pressure $(Mg,Fe)_2SiO_4$ phases in the Tenham chondritic meteorite. Nature 280:217-218

Quirke TT (1919) The Richardton meteorite. J Geol 27:431-448.

Rambaldi ER (1977) Trace element content of metals from H- and LL-group chondrites. Earth Planetary Sci Lett 36:347-358

Rambaldi ER (1981) Relict grains in chondrules. Nature 293:558-561.
Rambaldi ER, Wasson JT (1981) Metal and associated phases in Bishunpur, a highly unequilibrated ordinary chondrite. Geochim Cosmochim Acta 45:1001-1015
Rambaldi ER, Wasson JT (1982) Fine, nickel-poor Fe-Ni grains in the olivine of unequilibrated ordinary chondrites. Geochim Cosmochim Acta 46:929-939
Rambaldi ER, Wasson JT (1984) Metal and associated phases in Krymka and Chainpur: Nebular formational processes. Geochim Cosmochim Acta 48:1885-1897
Rambaldi ER, Sears DW, Wasson JT (1980) Si-rich Fe-Ni grains in highly unequilibrated chondrites. Nature 287:817-820
Rambaldi ER, Rajan RS, Daode W, Housley RM (1983) Evidence for relict grains from chondrules of Qingzhen, an E3 enstatite chondrite. Earth Planet Sci Lett 66:11-24
Rambaldi ER, Housley RM, Rajan RS (1984) Occurrence of oxidized components in the Qingzhen enstatite chondrite. Nature 311:138-140
Rambaldi ER, Rajan RS, Housley RM, Wang D (1986) Gallium-bearing sphalerite in metal-sulfide nodules of the Qingzhen (EH3) chondrite. Meteoritics 21:23-32
Ramdohr P (1963) The opaque minerals in stony meteorites. J Geophys Res 68:2011-2036
Ramdohr P (1967) Chromite and chromite chondrules in meteorites. Geochim Cosmochim Acta 31:1961-1967
Ramdohr P (1973) The Opaque Minerals in Stony Meteorites. Elsevier, New York, 245 pp
Reed SJB (1968) Perryite in the Kota-Kota and South Oman enstatite chondrites. Mineral Mag 36:850-854
Reed SJB, Smith DGW (1985) Ion probe determination of rare earth elements in merrillite and apatite in chondrites. Earth Planet Sci Lett 72:238-244
Reed SJB, Smith DGW, Long JVP (1983) Rare earth elements in chondritic phosphates—implications for ^{244}Pu chronology. Nature 306:172-173
Reid AM, Bass MN, Fujita H, Kerridge JF, Fredriksson K (1970) Olivine and pyroxene in the Orgueil meteorite. Geochim Cosmochim Acta 34:1253-1255
Reid AM, Buchanan P, Zolensky ME, Barrett RA (1990) The Bholghati howardite: Petrography and mineral chemistry. Geochim Cosmochim Acta 54:2161-2166
Reimold WU, Stöffler D (1978) Experimental shock metamorphism of dunite. Proc Lunar Planet Sci Conf 9:2805-2824
Reuter KB, Williams DB, Goldstein JI (1989) Determination of the Fe-Ni phase diagram below 400°C. Metall Trans 20A:719-725
Reynolds JH, Turner G (1964) Rare gases in the chondrite Renazzo. J Geophys Res 69:3263-3281
Richardson SM (1978) Vein formation in the C1 carbonaceous chondrites. Meteoritics 13:141-159
Richardson SM (1981) Alteration of mesostasis in chondrules and aggregates from three C2 carbonaceous chondrites. Earth Planet Sci Lett 52:67-75
Richardson SM, McSween HY Jr (1978) Textural evidence bearing on the origin of isolated olivine crystals in C2 carbonaceous chondrites. Earth Planet Sci Lett 37:485-491.
Riciputi LR, McSween HY Jr, Johnson CA, Prinz M (1994) Minor and trace element concentrations in carbonates of carbonaceous chondrites, and implications for the compositions of coexisting fluids. Geochim Cosmochim Acta 58:1343-1351
Ringwood AE (1961a) Chemical and genetic relationships among meteorites. Geochim Cosmochim Acta 24:159-197
Ringwood AE (1961b) Silicon in the metal phase of enstatite chondrites and some geochemical implications. Geochim Cosmochim Acta 24:159-197
Roedder E (1981) Significance of Ca-Al-rich silicate melt inclusions in olivine crystals from the Murchison type II carbonaceous chondrite. Bull Minéral 104:339-353
Romig ADJ, Goldstein JI (1980) Determination of the Fe,Ni and Fe,Ni-P phase diagrams at low temperatures (700-300C). Met Trans 11A:1151-1159
Rubin AE (1983a) The Adhi Kot breccia and implications for the origin of chondrites and silica-rich clasts in enstatite chondrites. Earth Planet Sci Lett 64:201-212
Rubin AE (1983b) The Atlanta enstatite chondrite breccia. Meteoritics 18:113-121
Rubin AE (1983c) Impact melt-rock clasts in the Hvittis enstatite chondrite breccia: Implications for a genetic relationship between EL chondrites and aubrites. Proc Lunar Planet Sci Conf 14:B293-B300
Rubin AE (1984) The Blithfield meteorite and the origin of sulfide-rich, metal-poor clasts and inclusions in brecciated enstatite chondrites. Earth Planet Sci Lett 6:273-283
Rubin AE (1990) Kamacite and olivine in ordinary chondrites: Intergroup and intragroup relationships. Geochim Cosmochim Acta 54:1217-1232
Rubin AE (1991) Euhedral awaruite in the Allende meteorite: Implications for the origin of awaruite- and magnetite-bearing nodules in CV3 chondrites. Am Mineral 76:1356-1362

Rubin AE (1992) A shock-metamorphic model for silicate darkening and compositionally variable plagioclase in CK and ordinary chondrites. Geochim Cosmochim Acta 56:1705-1714
Rubin AE (1993) Magnetite-sulfide chondrules and nodules in CK carbonaceous chondrites: implications for the timing of CK oxidation. Meteoritics 28:130-135
Rubin AE (1994) Euhedral tetrataenite in the Jelica meteorite. Mineral Mag 58:215-221
Rubin AE (1995) Petrologic evidence for collisional heating of chondritic asteroids. Icarus 113:156-167
Rubin AE (1997a) Mineralogy of meteorite groups. Met Planet Sci 32:231-247
Rubin AE (1997b) Mineralogy of meteorite groups: An update. Met Planet Sci:32:733-734
Rubin AE (1997c) The Hadley Rille enstatite chondrite and its agglutinate-like rim: Impact melting during accretion to the Moon. Met Planet Sci 32, 135-141
Rubin AE (1997d) Igneous graphite in chondritic meteorites. Mineral Mag 61:699-703
Rubin AE (1997e) Sinoite (Si_2N_2O): Crystallization from EL chondrite impact melts. Am Mineral 82:1001-1006
Rubin AE, Brearley AJ (1996) A critical evaluation of the evidence for hot accretion. Icarus 124:86-96
Rubin AE, Grossman JN (1985) Phosphate-sulfide assemblages and Al/Ca ratios in type 3 chondrites. Meteoritics 20:479-489
Rubin AE, Kallemeyn GW (1989) Carlisle Lakes and Allan Hills 85151: Members of a new chondrite grouplet. Geochim Cosmochim Acta 53:3035-3044
Rubin AE, Kallemeyn GW (1990) Lewis Cliff 85332: A unique carbonaceous chondrite. Meteoritics 25:215-225
Rubin AE, Kallemeyn GW (1994) Pecora Escarpment 91002: A member of the new Rumuruti (R) chondrite group. Meteoritics 29:255-264
Rubin AE, Keil K (1983) Mineralogy and petrology of the Abee enstatite chondrite breccia and its dark inclusions. Earth Planet Sci Lett 64:118-131
Rubin AE, Pernicka E (1989) Chondrules in the Sharps H3 chondrite: Evidence for intergroup compositional differences among ordinary chondrite chondrules. Geochim Cosmochim Acta 53:187-195
Rubin AE, Scott ERD (1996) Abee and related EH chondrite impact-melt breccias. Geochim Cosmochim Acta 61:425-435
Rubin AE, Wasson JT (1986) Chondrules in the Murray CM2 meteorite and compositional differences between CM-CO and ordinary chondrite chondrules. Geochim Cosmochim Acta 50:307-315
Rubin AE, Wasson JT (1987) Chondrules, matrix and coarse-grained chondrule rims in the Allende meteorite . Geochim Cosmochim Acta 51:1923-1937
Rubin AE, Wasson JT (1988) Chondrules and matrix in the Ornans CO3 meteorite: Possible precursor components. Geochim Cosmochim Acta 52:425-432
Rubin AE, Wasson JT (1995) Variations of chondrite properties with heliocentric distance (abstr). Meteoritics 30:569
Rubin AE, Taylor GJ, Keil K, Nelson G (1980) The Correo and Suwanee Spring meteorites: Two new ordinary chondrite finds. Meteoritics 16:9-12.
Rubin AE, Scott ERD, Taylor GJ, Keil K, Allen JSB (1983) Nature of the H chondrite parent body regolith: Evidence from the Dimmitt breccia. Proc Lunar Planet Sci Conf 13:A741-A754
Rubin AE, James JA, Keck BD, Weeks KS, Sears DWG, Jarosewich E (1985) The Colony meteorite and variations in CO3 chondrite properties. Meteoritics 20:175-196
Rubin AE, Fegley B, Brett R (1988a) Oxidation state in chondrites. *In* Meteorites and the Early Solar System (JF Kerridge, MS Matthews eds) The University of Arizona Press, Tucson, 488-511
Rubin AE, Wang D, Kallemeyn GW, Wasson JT (1988b) The Ningqiang meteorite: Classification and petrology of an anomalous CV chondrite. Meteoritics 23:13-23
Rubin AE, Keil K, Scott ERD (1997) Shock metamorphism of enstatite chondrites. Geochim Cosmochim Acta 61:847-858
Russell SS, Lee MR, Arden JW, and Pillinger CT (1995) The isotopic composition and origins of silicon nitride in the ordinary and enstatite chondrites. Meteoritics 30:399-404
Russell SS, Srinivasan G, Huss GR, Wasserburg GJ, McPherson GJ (1996) Evidence for widespread ^{26}Al in the solar nebula and constraints for nebula time scales. Science 273:757-762
Ruzicka A (1990) Deformation and thermal histories of chondrules in the Chainpur (LL3.4) chondrite. Meteoritics 25:101-114
Satterwhite C, Mason B (1994) Descriptions of EET92012; EET92013; EET92016. Antarc Met Newsletter 17(1):12.
Schafer H, Müller WF, Hornemann U (1981) Electronmikroskopische analyse von spinell und melilith nach stosswellenbeanspruchung. Beitr Elektron Direktabb Oberfl 14: 275-277
Schmidt H, De Jonghe LC (1983) Structure and non-stoichiometry in calcium aluminates. Phil Mag 48:287-297

Schulze H, Bischoff A, Palme A, Spettel B, Dreibus G, Otto J (1994) Mineralogy and chemistry of Rumuruti: The first meteorite fall of the new R chondrite group. Meteoritics 29:275-286

Schwarz C, MacPherson G (1985) Description of ALH84027. Antarc Met Newsletter 8(2):9

Score R, Mason B (1987) Description of ALH85151. Antarc Met Newsletter 2:21

Scott ERD (1982) Origin of rapidly solidified metal-troilite grains in chondrites and iron meteorites. Geochim Cosmochim Acta 46:813-823

Scott ERD (1988) A new kind of primitive chondrite, Allan Hills 85085. Earth Planet Sci Lett 91:1-18

Scott ERD, Clarke RS Jr (1979) Identification of clear taenite in meteorites as ordered FeNi. Nature 281:113-124

Scott ERD, Jones RH (1990) Disentangling nebular and asteroidal features of CO3 carbonaceous chondrites. Geochim Cosmochim Acta 54:2485-2502

Scott ERD, Newsom HE (1989) Planetary compositions—clues from meteorites and asteroids. Z Naturforsch 44a:924-934.

Scott ERD, Rajan RS (1979) Thermal history of the Shaw chondrite. Proc Lunar Planet Sci Conf 10:1031-1043

Scott ERD, Rajan S (1981) Metallic minerals, thermal histories, and parent bodies of some xenolithic, ordinary chondrites. Geochim Cosmochim Acta 45:53-67

Scott ERD, Taylor GJ (1983) Chondrules and other components in C, O and E chondrites: similarities in their properties and origins. J Geophys Res 88:B275-B286

Scott ERD, Taylor GJ (1985) Petrology of types 4-6 carbonaceous chondrites. Proc Lunar Planet Sci Conf 15:C699-C709

Scott ERD, Taylor GJ, Rubin AE, Okada A, Keil K (1981a) Graphite-magnetite aggregates in ordinary chondritic meteorites. Nature 291:544-546

Scott ERD, Rubin AE, Taylor GJ, Keil K (1981b) New kind of type 3 chondrite with a graphite-magnetite matrix. Earth Planet Sci Lett 56:19-31

Scott ERD, Taylor GJ, Maggiore P (1982) A new LL3 chondrite, Allan Hills A79003, and observations on matrices in ordinary chondrites. Meteoritics 17:65-75

Scott ERD, Rubin AE, Taylor GJ, Keil K (1984) Matrix material in type 3 chondrites—Occurrence, heterogeneity and relationship with chondrules. Geochim Cosmochim Acta 48:1741-1757

Scott ERD, Lusby D, Keil K (1985) Ubiquitous brecciation after metamorphism in equilibrated meteorites. Proc Lunar Planet Sci Conf 16:D137-D148

Scott ERD, Taylor GJ, Keil K (1986) Accretion, metamorphism, and brecciation of ordinary chondrites: Evidence from petrologic studies of meteorites from Roosevelt County, New Mexico. Proc Lunar Planet Sci Conf 17, J Geophys Res 91:E115-E123

Scott ERD, Barber DJ, Alexander CMO, Hutchison R, Peck JA (1988a) Primitive material surviving in chondrites: matrix. *In* Meteorites and the Early Solar System (JF Kerridge, MS Matthews eds) University of Arizona Press, 718-745

Scott ERD, Brearley AJ, Keil K, Grady MM, Pillinger CT, Clayton RN, Mayeda TK, Wieler R, Signer P (1988b) Nature and origin of C-rich ordinary chondrites and chondritic clasts. Proc Lunar Planet Sci Conf 18:513-523

Scott ERD, Keil K, Stöffler D (1992) Shock metamorphism of carbonaceous chondrites. Geochim Cosmochim Acta 56:4281-4293

Scott ERD, Jones RH, Rubin AE (1994) Classification, metamorphic history, and pre-metamorphic composition of chondrules. Geochim Cosmochim Acta 58:1203-1209

Scott ERD, Love SG, Krot AN (1996) Formation of chondrules and chondrites in the protoplanetary nebula. *In* Chondrules and the Protoplanetary Disk (RH Hewins, RH Jones, ERD Scott eds) Cambridge Univ Press, 87-96

Sears DW, Axon HJ (1975) Metal of high Co content in LL chondrites. Meteoritics 11:97-100

Sears DW, Axon HJ (1976) Ni and Co content of chondritic metal. Nature 260:34-35

Sears DWG, Dodd RT (1988) Overview and classification of meteorites. *In* Meteorites and the Early Solar System (JF Kerridge, MS Matthews eds) Univ. Arizona Press, 3-31

Sears DWG, Grossman JN, Melcher CL, Ross LM, Mills AA (1980) Measuring metamorphic history of unequilibrated ordinary chondrites. Nature 287:791-795

Sears DW, Kallemeyn GW, Wasson JT (1982) The chemical classification of chondrites II: The enstatite chondrites. Geochim Cosmochim Acta 46:597-608

Sears DWG, Sparks MH, Rubin AE (1984a) Chemical and physical studies of type 3 chondrites-III. Chondrules from the Dhajala H3.8 chondrite. Geochim Cosmochim Acta 48:1189-1200

Sears DWG, Weeks K S, Rubin AE (1984b) First known EL5 chondrite—Evidence for dual genetic sequence for enstatite chondrites. Nature 308:257-259

Sears DWG, Batchelor DJ, Lu J, Keck BD (1991a) Metamorphism of CO and CO-like chondrites and comparisons with type 3 ordinary chondrites. Proc NIPR Symp Antarct Met 4:319-343

Sears DWG, Hasan FA, Batchelor JD, Lu J (1991b) Chemical and physical studies of type 3 chondrites—XI: metamorphism, pairing, and brecciation in ordinary chondrites. Proc Lunar Planet Sci Conf 21:493-512

Sears DWG, Lu J, Benoit PH, DeHart JM, Lofgren GE (1992) A compositional classification scheme for meteoritic chondrules. Nature 357:207-210

Sears DWG, Morse AD, Hutchison R, Guimon RK, Lu J, Alexander CMO'D, Benoit PH, Wright I, Pillinger C, Xie T, Lipschutz ME. (1995a) Metamorphism and aqueous alteration in low petrographic type ordinary chondrites. Meteoritics 30:169-181

Sears DWG, Huang S, Benoit PH (1995b) Chondrule formation, metamorphism, brecciation, an important new primary chondrule group, and the classification of chondrules. Earth Planet Sci Lett 131:27-39

Sears DWG, Huang S, Benoit PH (1996) Open-system behavior during chondrule formation. *In* Chondrules and the Protoplanetary Disk (RH Hewins, RH Jones, ERD Scott eds) Cambridge Univ Press, 221-231

Shannon EV, Larsen ES (1925) Merrillite and chlorapatite from stony meteorites. Amer J Sci 9:250-260

Sharp TG, Lingemann CM, Dupas C, Stöffler D (1997) Natural occurrences of $MgSiO_3$-ilmenite and evidence for $MgSiO_3$-perovskite in a shocked L chondrite. Science 277:352-355

Sheng YJ, Hutcheon ID, Wasserburg GJ (1991) Origin of plagioclase-olivine inclusions in carbonaceous chondrites. Geochim Cosmochim Acta 55:581-599

Shervais JW, Taylor LA, Cirlin E-H, Jarosewich E, Laul JC (1986) The Maryville meteorite: A 1983 fall of an L6 chondrite. Meteoritics 21:33-45

Shibata Y (1996) Opaque minerals in Antarctic CO3 carbonaceous chondrites, Yamato-74135, -790992, -791717, -81020, -81025, -82050 and Allan Hills-77307. Proc NIPR Symp Antarct Met 9:79-96

Shibata Y, Matsueda H (1994) Chemical composition of Fe-Ni metal and phosphate minerals in Yamato-82094 carbonaceous chondrite. Proc NIPR Symp Antarct Met 7:110-124

Simon SB, Grossman L (1992) Low temperature exsolution in refractory siderophile element-rich opaque assemblages from the Leoville carbonaceous chondrite. Earth Planet Sci Lett 110:67-75

Simon SB, Grossman L (1997) In situ formation of palisade bodies in calcium, aluminum-rich refractory inclusions. Meteoritics 32:61-70

Simon SB, Haggerty SE (1979) Petrography and olivine mineral chemistry of chondrules and inclusions in the Allende meteorite. Proc Lunar Planet Sci Conf 10:871-883

Simon SB, Haggerty SE (1980) Bulk compositions of chondrules in the Allende meteorite. Proc Lunar Planet Sci Conf 11:901-927

Simon SB, Grossman L, Davis AM (1991) Fassaite composition trends during crystallization of Allende Type B refractory inclusions. Geochim Cosmochim Acta 55:2635-2655

Simon SB, Grossman L, Podosek FA, Zinner E, Prombo CA (1994a) Petrography, composition, and origin of large, chromian spinels from the Murchison meteorite. Geochim Cosmochim Acta 58:1313-1334

Simon SB, Grossman L, Wacker JF (1994b) Unusual refractory inclusions from a CV3 chondrite found near Axtell, Texas (abstr). Lunar Planet Sci XXV:1275-1276

Simon SB, Yoneda S, Grossman L, Davis AM (1994c) A $CaAl_4O_7$-bearing refractory spherule from Murchison: Evidence for very high temperature melting in the solar nebula. Geochim Cosmochim Acta 58:1937-1949

Simon SB, Grossman L, Casanova I, Symes S, Benoit P, Sears DWG, Wacker JF (1995) Axtell, a new CV3 chondrite from Texas. Meteoritics 30:42-46

Simon SB, Davis AM, Grossman L (1996) A unique ultrarefractory inclusion from the Murchison meteorite. Meteoritics 31:106-115

Simon SB, Grossman L, Davis AM (1997) Multiple generations of hibonite in spinel-hibonite inclusions from Murchison. Met Planet Sci 32:259-269

Simpson ES, Murray DG (1932) A new siderolite from Bencubbin, Western Australia. Mineral Mag 23:33-37

Skinner BF, Luce FD (1971) Solid solution of the type (Ca,Mg,Mn,Fe)S and their use as geothermometers for the enstatite chondrites. Am Mineral 56:1269-1296

Skirius C, Steele IM, Smith JV (1986) Belgica-7904: A new carbonaceous chondrite from Antarctica: Minor element chemistry of olivine. Mem NIPR Spec Issue 41:243-258

Smith BA, Goldstein JI (1977) The metallic microstructures and thermal histories of severely reheated chondrites. Geochim Cosmochim Acta 41:1061-1072

Smith DGW, Launspach S (1991) The composition of metal phases in Bruderheim (L6) and implications for the thermal histories of ordinary chondrites. Earth Planet Sci Lett 102:79-93

Smith DGW, Bell JD, Frisch T (1972) The Timersoi, Niger, hypersthene chondrite. Meteoritics 7:1-16

Smith DGW, Miúra Y, Launspach S (1993) Fe, Ni and Co variations in the metals from some Antarctic chondrites. Earth Planet Sci Lett 120:487-498

Smith JV, Mason B (1970) Pyroxene-garnet transformation in Coorara meteorite. Science 168:832-833

Snee L, Ahrens T (1975) Shock-induced deformation features in terrestrial peridot and and lunar dunite. Proc Lunar Planet Sci Conf 6:833-842

Snetsinger KG, Keil K (1969) Ilmenite in ordinary chondrites. Amer Mineral 54:780-786

Snetsinger KG, Keil K, Bunch TE (1967) Chromite from "equilibrated" chondrites. Am Mineral 52:1322-1331

Sorby HC (1877) On the structure and origin of meteorites. Nature 15:495-498

Stacey FD, Lovering JF, Parry LG (1961) Thermomagnetic properties, natural magnetic moments, and magnetic anisotropies of some chondritic meteorites. J Geophys Res 66:1523-1534

Steele IM (1986) Compositions and textures of relic forsterite in carbonaceous and unequilibrated ordinary chondrites. Geochim Cosmochim Acta 50:1379-1395

Steele IM (1988) Primitive material surviving in chondrites: Mineral grains. *In* Meteorites and the Early Solar System (JF Kerridge, MS Matthews eds) The University of Arizona Press, 808-818

Steele IM (1989) Compositions of relic forsterites in Ornans (C3O). Geochim Cosmochim Acta 53:2069-2079

Steele IM (1995a) Oscillatory zoning in meteoritic forsterite. Am Mineral 80:823-832

Steele IM (1995b) Mineralogy of a refractory inclusion in the Allende (C3V) meteorite. Meteoritics 30: 9-14

Steele IM, Peters MT, Shaffer EE, Burnett DS (1997) Minor element partitioning and sector zoning in synthetic and meteoritic anorthite. Geochim Cosmochim Acta 61:415-423

Stöffler D (1972) Deformation and transformation of rock-forming minerals in natural and experimental shock processes: 1. Behavior of minerals under shock compression. Fortschr Mineral 49:50-113

Stöffler D, Ostertag R, Jammes C, Pfannschmidt G, Sen Gupta PR, Simons M, Papike JJ, Beauchamp RM (1986) Shock metamorphism and petrography of the Shergotty achondrite. Geochim Cosmochim Acta 50:889-903

Stöffler D, Bischoff A, Buchwald V, Rubin AE (1988) Shock effects in meteorites. *In* Meteorites and the Early Solar System (JF Kerridge, MS Matthews eds) p 165-202

Stöffler D, Keil K, Scott ERD (1991) Shock metamorphism of ordinary chondrites. Geochim Cosmochim Acta 55:3845-3867

Stolper E (1982) Crystallization sequences of Ca-Al-rich inclusions from Allende: An experimental study. Geochim Cosmochim Acta 46:2159-2180

Swindle TD, Podosek FA (1988) Iodine-xenon dating. *In* Meteorites and the Early Solar System (JF Kerridge, MS Matthews eds) The University of Arizona Press, Tucson, 1127-1146

Swindle TD, Grossman JN, Olinger CT, Garrison DH (1991) Iodine-xenon, chemical, and petrographic studies of Semarkona chondrules: Evidence for the timing of aqueous alteration. Geochim Cosmochim Acta 55:3723-3734.

Swindle TD, Davis AM, Hohenberg CM, MacPherson GJ, Nyquist LE (1996) Formation times of chondrules and Ca-Al-rich inclusions: Constraints from short-lived radionuclides. *In* Chondrules and the Protoplanetary Disk (RH Hewins, RH Jones, ERD Scott eds), Cambridge Univ Press, Cambridge, 77-86

Sylvester PJ, Grossman L, MacPherson GJ (1992) Refractory inclusions with unusual chemical compositions from the Vigarano carbonaceous chondrite. Geochim Cosmochim Acta 56:1343-1363

Sylvester PJ, Simon SB, Grossman L (1993) Refractory inclusions from the Leoville, Efremovka, and Vigarano C3V chondrites: Major element differences between Types A and B, and extraordinary refractory siderophile element compositions. Geochim Cosmochim Acta 57:3763-3784

Tang M, Anders E (1988) Interstellar silicon carbide: How much older than the solar system? Astrophys J 335:L31-L34

Taylor GJ, Heymann D (1969) Shock, reheating and the gas retention ages of chondrites. Earth Planet Sci Lett 7:151-161

Taylor GJ, Heymann D (1970) Electron microprobe study of metal particles in the Kingfisher meteorite. Geochim Cosmochim Acta 34:677-687

Taylor GJ, Heymann D (1971) Postshock thermal histories of reheated chondrites. J Geophys Res 76:1879-1893

Taylor GJ, Keil K, Berkley JL, Lange DE, Fodor RV, Fruland RM (1979) The Shaw meteorite: history of a chondrite consisting of impact-melted and metamorphic lithologies. Geochim Cosmochim Acta 43:323-337

Taylor GJ, Okada A, Scott ERD, Rubin AE, Huss GR, Keil K (1981) The occurrence and implications of carbide-magnetite assemblages in unequilibrated ordinary chondrites (abstr). Lunar Planet Sci XII:1076-1078

Taylor GJ, Scott ERD, and Keil K (1983) Cosmic setting for chondrule formation. *In* Chondrules and their Origins (EA King ed) Lunar and Planetary Institute, 262-278

Taylor GJ, Maggiore P, Scott ERD, Rubin AE, Keil K (1987) Original structures and fragmentation and reassembly histories of asteroids: Evidence from meteorites. Icarus 69:1-13

Taylor SR (1992) Solar System Evolution: a new perspective. Cambridge Univ Press, Cambridge, 307 pp

Taylor SR, Mason B (1978) Chemical characteristics of Ca-Al-rich inclusions in the Allende meteorite. Lunar Planet Sci IX:1158-1160

Tilton GR (1988) Age of the solar system. *In* Meteorites and the Early Solar System (JF Kerridge, MS Matthews eds) The University of Arizona Press, Tucson, 259-275

Tomeoka K (1990a) Phyllosilicate veins in a CI meteorite: Evidence for aqueous alteration on the parent body. Nature 345:138-140

Tomeoka K (1990b) Mineralogy and petrology of Belgica-7904: A new kind of carbonaceous chondrite from Antarctica. Proc NIPR Symp Antarct Met 3:40-54

Tomeoka K, Buseck PR (1982a) An unusual layered mineral in chondrules and aggregates of the Allende carbonaceous chondrite. Nature 299:327-329

Tomeoka K, Buseck PR (1982b) Intergrown mica and montmorillonite in the Allende carbonaceous chondrite. Nature 299:326-327

Tomeoka K, Buseck PR (1983) A new layered mineral from the Mighei carbonaceous chondrite. Nature 306:354-356

Tomeoka K, Buseck PR (1985) Indicators of aqueous alteration in CM carbonaceous chondrites: microtextures of a layered mineral containing Fe, S, O and Ni. Geochim Cosmochim Acta 49:2149-2163

Tomeoka K, Buseck PR (1988) Matrix mineralogy of the Orgueil CI carbonaceous chondrite. Geochim Cosmochim Acta 52:1627-1640

Tomeoka K, Buseck PR (1990) Phyllosilicates in the Mokoia CV carbonaceous chondrite: Evidence for aqueous alteration in an oxidizing condition. Geochim Cosmochim Acta 54:1787-1796

Tomeoka K, McSween HY Jr, Buseck PR (1989a) Mineralogical alteration of CM carbonaceous chondrites: a review. Proc NIPR Symp Antarct Met 2:221-234

Tomeoka K, Kojima H, Yanai K (1989b) Yamato-82162: A new kind of CI carbonaceous chondrite found in Antarctica. Proc NIPR Symp Antarct Met 2:36-54

Tomeoka K, Kojima H, Yanai K (1989c) Yamato-86720: A CM carbonaceous chondrite having experienced extensive aqueous alteration and thermal metamorphism. Proc NIPR Symp Antarct Met 2:55-74

Tomeoka K, Nomura K, Takeda H (1992) Na-bearing Ca-Al-rich inclusions in the Yamato-791717 CO carbonaceous chondrite. Meteoritics 27:136-143

Töpel-Schadt J, Müller WF (1982) Transmission electron microscopy on meteoritic troilite. Phys Chem Mineral 8:175-179

Töpel-Schadt J, Müller WF (1985) The submicroscopic structure of the unequilibrated ordinary chondrites, Chainpur, Mezö-Madaras, and Tieschitz: A transmission electron microscope study. Earth Planet Sci Lett 74:1-12

Toriumi M (1989) Grain size distribution of the matrix in the Allende chondrite. Earth Planet Sci Lett 92:265-273

Tsuchiyama A, Fujita T, Morimoto N (1988) FeMg heterogeneity in the low-Ca pyroxenes during metamorphism of the ordinary chondrites. Proc NIPR Symp Antarct Met 1:173-184

Turner G (1988) Dating of secondary events. *In* Meteorites and the Early Solar System (JF Kerridge, MS Matthews eds) The University of Arizona Press, Tucson, 276-288

Ulyanov AA, Korina MI, Nazarov MA, Sherbovsky EY (1982) Efremovka CAIs: Mineralogical and petrological data (abstr). Lunar Planet Sci XIII:813-814

Urey HC, Mayeda T (1959) The metallic particles of some chondrites. Geochim Cosmochim Acta 17:113-124

Van Schmus WR (1969) Mineralogy, petrology and classification of types 3 and 4 carbonaceous chondrites. *In* Meteorite Research (PM Millman, ed) Reidel, Dordrecht, 480-491

Van Schmus WR, Ribbe PH (1968) The composition and structural state of feldspar from chondritic meteorites. Geochim Cosmochim Acta 32:1327-1342

Van Schmus WR, Ribbe PH (1969) Composition of phosphate minerals in ordinary chondrites. Geochim Cosmochim Acta 33:637-640

Van Schmus WR, Wood JA (1967) A chemical-petrological classification for the chondritic meteorites. Geochim Cosmochim Acta 31:747-765

Viertel HJ, Seifert F (1969) Physical properties of defects spinels in the system $MgAl_2O_4$-Al_2O_3. N Jahrb Min Abh 134:167-182

Virag A, Wopenka B, Amari S, Zinner E, Anders E, Lewis RS (1992) Isotopic, optical and trace element properties of large single SiC grains from the Murchison meteorite. Geochim Cosmochim Acta 56:1715-1733

Wang D, Xie X (1981) Preliminary investigation of mineralogy, petrology and chemical composition of Qingzhen enstatite chondrite. Geochem Intl 1:68-81

Wark DA (1979) Birth of solar nebula: the sequence of condensation revealed in the Allende meteorite. Astrophys Space Sci 65:275-295

Wark DA (1986) Evidence for successive episodes of condensation at high temperatures in a part of the solar nebula. Earth Planet Sci Lett 77:129-148

Wark DA (1987) Plagioclase-rich inclusions in carbonaceous chondrite meteorites: Liquid condensates? Geochim Cosmochim Acta 51:221-242

Wark DA, Lovering JF (1977) Marker events in the early evolution of the solar system: Evidence from rims on Ca-Al-rich inclusions from carbonaceous chondrites. Proc Lunar Planet Sci Conf 8:95-112

Wark DA, Lovering JF (1978) Refractory/platinum metals and other opaque phases in Allende Ca-Al-rich inclusions (CAI's) (abstr). Lunar Planet Sci IX:1214-1216

Wark DA, Lovering JF (1982a) The nature and origin of type B1 and B2 inclusions in the Allende meteorite. Geochim Cosmochim Acta 46:2581-2594

Wark DA, Lovering JF (1982b) Evolution of Ca-Al-rich bodies in the earliest solar system: growth by incorporation. Geochim Cosmochim Acta 46:2595-2607

Wark DA, Boynton WV, Keays RR, Palme H (1987) Trace element and petrologic clues to the formation of forsterite-bearing Ca-Al-rich inclusions in the Allende meteorite. Geochim Cosmochim Acta 51:607-622

Wasson JT (1993) Constraints on chondrule origins. Meteoritics 28:14-28

Wasson JT, Kallemeyn GW (1988) Composition of chondrites. Phil Trans R Soc Lond A325:535-544

Wasson JT, Kallemeyn GW (1990) Allan Hills 85085: A subchondritic meteorite of mixed nebula and regolithic heritage. Earth Planet Sci Lett 101:148-161

Wasson JT, Krot AN (1994) Fayalite-silica association in unequilibrated ordinary chondrites: Evidence for aqueous alteration on a parent body. Earth Planet Sci Lett 122:403-416

Wasson JT, Wai CM (1970) Composition of metal, schreibersite and perryite of enstatite achondrites and the origin of enstatite chondrites and achondrites. Geochim Cosmochim Acta 34:169-184

Wasson JT, Rubin AE, Kallemeyn GW (1993) Reduction during metamorphism of four ordinary chondrites. Geochim Cosmochim Acta 57:1867-1878

Wasson JT, Kallemeyn GW, Rubin AE (1994) Equilibration temperatures of E chondrites: A major downward revision of the ferrosilite contents of enstatite. Meteoritics 29:658-662

Wasson JT, Krot AN, Lee MS, Rubin AE (1995) Compound chondrules. Geochim Cosmochim Acta 59:1847-1869

Watanabe S, Kitamura M, Morimoto N (1984) Analytical electron microscopy of a chondrule with relict olivine in the ALH-77015 chondrite (L3). Mem NIPR Spec Issue 35:200-209

Watanabe S, Kitamura M, Morimoto N (1985) A transmission electron microscope study of pyroxene chondrules in equilibrated L-group chondrites. Earth Planet Sci Lett 72:87-98

Watanabe S, Kitamura M, Morimoto N (1986) Oscillatory zoning of pyroxenes in ALH-77214 (L3) (abstr). Abstr Symp Antarct Met 11:74-75

Wdowiak TJ, Agresti DG (1984) Presence of a superparamagnetic component in the Orgueil meteorite. Nature 311:140-142

Weber D, Bischoff A (1994a) Grossite ($CaAl_4O_7$)- a rare phase in terrestrial rocks and meteorites. Eur J Mineral 6:591-594

Weber D, Bischoff A (1994b) The occurrence of grossite ($CaAl_4O_7$) in chondrites. Geochim Cosmochim Acta 58:3855-3877

Weber D, Zinner E, Bischoff A (1995) Trace element abundances and magnesium, calcium and titanium isotopic compositions of grossite containing inclusions from the carbonaceous chondrite, Acfer 182. Geochim Cosmochim Acta 59:803-823

Weinbruch S, Müller FW (1995) Constraints on the cooling rates of chondrules from the microstructure of clinopyroxene and plagioclase. Geochim Cosmochim Acta 59:3221-3230

Weinbruch S, Palme H, Müller WF, El Goresy A (1990) FeO-rich rims and veins in Allende forsterite: Evidence for high temperature condensation at oxidizing conditions. Meteoritics 25:115-125

Weinbruch S, Armstrong JT, Palme H (1994) Constraints on the thermal history of the Allende parent body as derived from olivine-spinel thermometry and Fe/Mg interdiffusion in olivine. Geochim Cosmochim Acta 58:1019-1030

Weisberg MK, Prinz M (1996) Agglomeratic chondrules, chondrule precursors and incomplete melting. *In* Chondrules and the Protoplanetary Disk (RH Hewins, RH Jones, ERD Scott eds) 119-127

Weisberg MK, Prinz M, Nehru CE (1988) Petrology of ALH85085: A chondrite with unique characteristics. Earth Planet Sci Lett 91:19-32

Weisberg MK, Prinz M, Nehru C (1990) The Bencubbin chondrite breccia and its relationship to CR chondrites and the ALH85085 chondrite. Meteoritics 25:269-279

Weisberg MK, Prinz M, Kojima H, Yanai K, Clayton RN, Mayeda TK (1991) The Carlisle Lakes-type chondrites: A new grouplet with high $\Delta^{17}O$ and evidence for nebular oxidation. Geochim Cosmochim Acta 55:2657-2669

Weisberg MK, Prinz M, Clayton RN, Mayeda TK (1993) The CR (Renazzo-type) carbonaceous chondrite group and its implications. Geochim Cosmochim Acta 57:1567-1586

Weisberg MK, Prinz M, Fogel RA (1994) The evolution of enstatite and chondrules in unequilibrated enstatite chondrites: Evidence from iron-rich pyroxene. Meteoritics 29:362-373

Weisberg MK, Prinz M, Clayton RN, Mayeda TK, Grady MM, Pillinger CT (1995) The CR chondrite clan. Proc NIPR Symp Antarct Met 8:11-32

Weisberg MK, Prinz M, Clayton RN, Mayeda TK, Grady MM, Franchi I, Pillinger CT, Kallemeyn GW (1996) The K (Kakangari) chondrite grouplet. Geochim Cosmochim Acta 60:4253-4263

Whitlock R, Lewis CF, Clark JC, Moore CB (1991) The Cerro los Calvos and La Bandera chondrites. Meteoritics 26:169-170

Wiik HB (1956) The chemical composition of some stony meteorites. Geochim Cosmochim Acta 9:279-289

Wilkening LL (1973) Foreign inclusions in stony meteorites-II. Carbonaceous chondritic xenoliths in the Kapoeta howardite. Geochim Cosmochim Acta 37:1985-1989.

Wilkening LL (1978) Carbonaceous chondritic material in the solar system. Naturwiss 65:73-79

Wilkening LL, Clayton RN (1974) Foreign inclusions in stony meteorites-II. Rare gases and oxygen isotopes in a carbonaceous chondritic xenolith in the Plainview gas-rich chondrite. Geochim Cosmochim Acta 38:937-945

Willis J, Goldstein JI (1981a) A revision of metallographic cooling rate curves for chondrites. Proc Lunar Planet Sci Conf 12B:1135-1143

Willis J, Goldstein JI (1981b) Solidification zoning and metallographic cooling rates of chondrites. Nature 293:126-127

Willis J, Goldstein JI (1983) A three dimensional study of metal grains in equilibrated, ordinary chondrites. Proc Lunar Planet Sci Conf 14:B287-B292

Wlotzka F (1983) Composition of chondrules, fragments and matrix in the unequilibrated ordinary chondrites Tieschitz and Sharps. *In* Chondrules and Their Origins (EA King, ed) Lunar and Planetary Inst., Houston, 296-318

Wlotzka F (1993) A weathering scale for the ordinary chondrites (abstr). Meteoritics 28:460

Wlotzka F, Wark DA (1982) The significance of zeolites and other hydrous alteration products in Leoville Ca-Al-rich inclusions. Lunar Planet Sci XIII:869-870

Wood JA (1964) The cooling rates and parent bodies of several iron meteorites. Icarus 3:429-459

Wood JA (1967a) Chondrites: Their metallic minerals, thermal histories, and parent planets. Icarus 6:1-49

Wood JA (1967b) Olivine and pyroxene compositions in type II carbonaceous chondrites. Geochim Cosmochim Acta 31:2095-2108

Wood JA, Holmberg BB (1994) Constraints placed on the chondrule-forming process by merrihueite in the Mezö-Madaras chondrite. Icarus 108:309-324

Yanai K (1992) Bulk composition of Yamato-793575 classified as Carlisle-Lakes type chondrite (abstr). Lunar Planet Sci XXIII:1559

Zanda B, Bourot-Denise M, Perron C, Hewins RH (1994) Origin and metamorphic redistribution of silicon, chromium, and phosphorus in the metal of chondrites. Science 265:1846-1849

Zhang Y, Sears DWG (1996) The thermometry of enstatite chondrites: A brief review and update. Met Planet Sci 31:647-655

Zhang Y, Benoit PH, Sears DWG (1995) The classification and complex thermal history of the enstatite chondrites. J Geophys Res100:9417-9438

Zhang Y, Huang S, Schneider D, Benoit PH, Sears DWG (1996) Pyroxene structures, cathodoluminescence and the thermal history of the enstatite chondrites. Met Planet Sci 31:87-96

Zinner E, Göpel C (1992) Evidence for ^{26}Al in feldspars from the H4 chondrite Ste. Marguerite (abstr). Meteoritics 27:311-312

Zinner E, Tang M, Anders E (1989) Interstellar SiC in the Murchison and Murray meteorites: Isotopic composition of Ne, Xe, Si, C. Geochim Cosmochim Acta 53:3273-3290

Zinner EK, Caillet C, El Goresy A (1991) Evidence for extraneous origin of magnesiowüstite-metal Fremdling from the Vigarano CV3 chondrite. Earth Planet Sci Lett 102:252-264

Zinner E, Amari S, Wopenka B, Lewis RS (1995) Interstellar graphite in meteorites: Isotopic compositions and structural properties of single graphite grains from Murchison. Meteoritics 30:209-226

Zolensky ME (1991) Mineralogy and matrix composition of "CR" chondrites Renazzo and EET 87770, and ungrouped chondrites Essebi and MAC87300 (abstr). Meteoritics 26:414

Zolensky ME (1997) Structural water in the Bench Crater chondrite returned from the Moon. Met Planet Sci 32:15-18

Zolensky ME and McSween, HY Jr (1988) Aqueous alteration. *In* Meteorites and the Early Solar System (JF Kerridge, MS Matthews eds) Univ. of Arizona, Tucson, 114-143

Zolensky ME, Score R, Schutt JW, Clayton RN, Mayeda TK (1989a) Lea County 001, an H5 chondrite, and Lea County 002, an ungrouped type 3 chondrite. Meteoritics 24:227-232

Zolensky ME, Barrett R, Prinz M (1989b) Petrography and mineralogy of Yamato-86720 and Belgica-7904. 14th Symp Ant Meteorites 24-26

Zolensky ME, Barrett R, Prinz M (1989c) Petrography, mineralogy and matrix composition of Yamato-82162, a new CI2 chondrite (abstr). Lunar Planet Sci XXVI:1253-1254

Zolensky ME, Prinz M, Lipschutz ME (1991) Mineralogy and thermal history of Y-82162, Y-86720, B-7904. Abstr Symp Antarct Met 16:195-196

Zolensky ME, Hewins RH, Mittlefehldt DW, Lindstrom MM, Xiao X, Lipschutz ME (1992) Mineralogy, petrology and geochemistry of carbonaceous chondritic clasts in the LEW 85300 polymict eucrite. Meteoritics 27:596-604

Zolensky ME, Barrett T, Browning L (1993) Mineralogy and composition of matrix and chondrule rims in carbonaceous chondrites. Geochim Cosmochim Acta 57:3123-3148

Zolensky ME, Weisberg MK, Buchanan PC, Mittlefehldt DW (1996a) Mineralogy of carbonaceous chondrite clasts in HED achondrites and the Moon. Met Planet Sci 31:518-537

Zolensky ME, Ivanov AV, Yang V, Mittlefehldt DW, Ohsumi K (1996b) The Kaidun meteorite: Mineralogy of an unusual CM1 lithology. Meteoritics 31:484-493

Chapter 4

NON-CHONDRITIC METEORITES FROM ASTEROIDAL BODIES

David W. Mittlefehldt

C23/Basic and Applied Research Departnment
Lockheed Martin SMSS Company
Houston, Texas 77058

Timothy J. McCoy

MRC 119, Department of Mineral Sciences
National Museum of Natural History
Smithsonian Institution
Washington, DC 20560

Cyrena A. Goodrich

Abteilung Kosmochemie
Max-Planck-Institute für Chemie
Postfasch 3060
55020 Mainz, Germany

Alfred Kracher

Department of Geological Sciences
Iowa State University
Ames, Iowa 50011

INTRODUCTION

This chapter considers all meteorites which are of asteroidal origin and which do not contain members displaying classic chondritic textures (i.e. chondrules in a fine-grained matrix). This chapter forms a logical bridge between the previous chapter on chondrites, and the following chapters on lunar and martian rocks, the products of planetary differentiation. We cover a wide range of petrologic types including undifferentiated, but highly metamorphosed chondritic materials, melt-depleted ultramafic rocks, ultramafic and mafic cumulates, mafic melts, metallic cumulates and melts, and stony-irons composed of various materials. In addition, we will touch on several unique non-chondritic meteorites, and on achondritic clasts from chondrites that do not appear to be impact melts. Table 1 summarizes the major meteorite types covered in this chapter, gives a synopsis of the characteristic mineralogies and petrologies, and probable modes of origin. The mineralogy of all meteorite groups was reviewed by Rubin (1997).

The meteorites and meteorite groups covered here have experienced a wide range of planetary processes. At one extreme, we have rocks such as the acapulcoites, which seem to be simply ultra-metamorphosed chondritic material. It is debatable whether these rocks ever reached their solidi, although the textures suggest some mobilization of minimum melts in the Fe,Ni-FeS system. At the other extreme, we have rocks such as the eucrites, which are planetary basalts far removed in composition from that of their bulk parent body. We also have materials, such as the IIE irons, in which impact mixing, and possibly melting, must have played an important role. We have rocks, the aubrites, that were formed under such low oxygen fugacities that elements normally strongly lithophile in character, Ca and the rare earth elements (REE), behaved as chalcophile elements. Finally, some of the meteorite types exhibit curious mixtures of nebular and planetary properties, such as the lodranites and ureilites.

0275-0279/98/0036-0004$20.00

Table 1. Synopsis of petrologic characteristics of meteorite groups and modes of origin.

GROUP	MINERALOGY	PETROLOGY	LIKELY ORIGIN
magmatic irons, and pallasite and mesosiderite metal phase	kamacite, tetrataenite, martensite, awaruite troilite, schreibersite ± graphite	coarse-grained metal, generally exsolved	fractional crystallization of core
non-magmatic irons IAB, IIE, IIICD metal phase	kamacite, tetrataenite, martensite, awaruite troilite, schreibersite ± graphite	coarse-grained metal, generally exsolved	impact melting (?)
IAB-IIICD silicate inclusions, winonaites	olivine, orthopyroxene, clinopyroxene, sodic plagioclase, ultrametamorphism, partial melting, impact mixing troilite, metal, daubreilite, schreibersite, graphite		fine-grained equigranular
IIE silicate inclusions	olivine, pyroxene, sodic plagioclase, metal, troilite; orthopyroxene, sodic plagioclase; orthopyroxene, clinopyroxene, sodic plagioclase	chondritic to gabbroic to silica-alkali-rich	metamorphism, partial melting, melt migration, shock
pallasites, silicates	olivine, chromite ± low-Ca pyroxene	coarse-grained, rounded to angular	cumulates from core-mantle boundary
IVA silicate inclusions	silica ± orthopyroxene	coarse-grained, igneous	partial melting (?)
brachinites	olivine, augite, Fe-sulfide, ± orthopyroxene ± sodic plagioclase	medium- to coarse-grained equigranular	metamorphosed chondrite, partial melting residue
acapulcoites-lodranites	olivine, orthopyroxene, clinopyroxene, metal ± troilite ± sodic plagioclase	fine- to coarse-grained equigranular	ultrametamorphism, partial melting ± melt migration
ureilites	olivine, pyroxene (some combination of pigeonite, orthopyroxene, augite)	coarse-grained granular	partial melting residue
aubrites	enstatite, forsterite, sodic plagioclase, metal, numerous sulfides	breccias of igneous lithologies	melted and crystallized residues from partial melting
diogenites, eucrites, howardites	orthopyroxene, olivine, chromite; pigeonite, calcic plagioclase, silica	coarse-grained orthopyroxenites, gabbros, fine- to coarse-grained basalts, brecciated	cumulates, crystallized melts, impact mixing
angrites	Al-Ti-diopside, anorthite, calcic olivine	fine- to coarse-grained igneous	crystallized melts, cumulates (?)
mesosiderites	orthopyroxene, olivine, chromite, pigeonite, calcic plagioclase, silica, phosphates	breccias of igneous lithologies, impact melts	impact melted crustal igneous lithologies

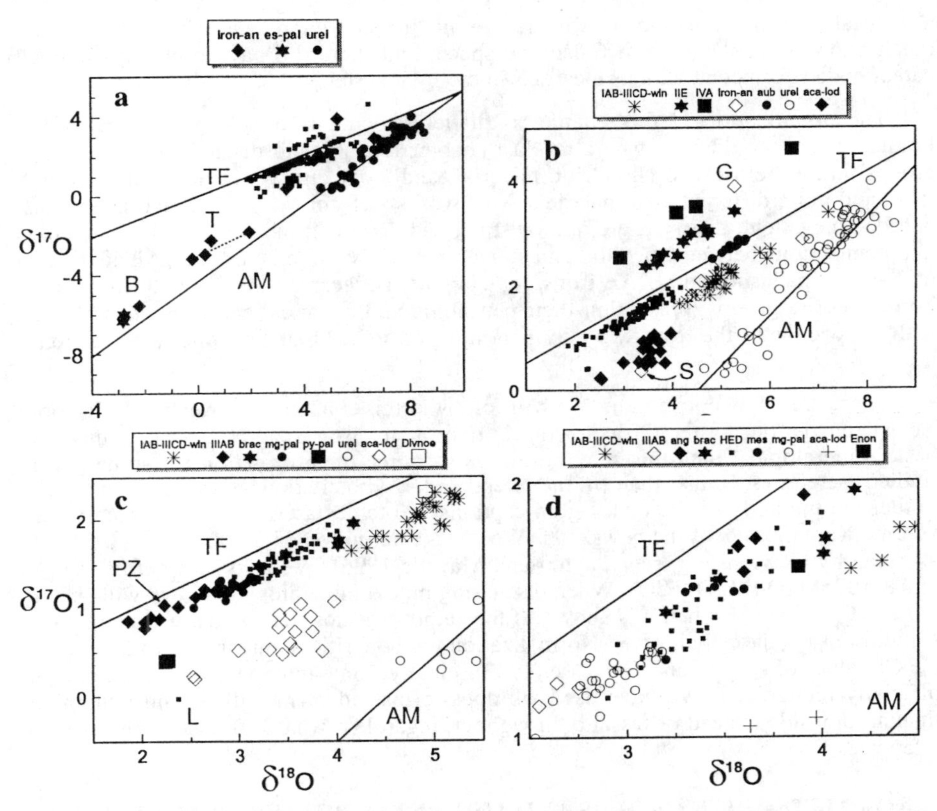

Figure 1. Oxygen-isotope plot for non-chondritic meteorites from asteroidal sources. In all parts of this figure, unlabeled symbols are for several different meteorite types, which will be separated and labeled in subsequent parts. The terrestrial fractionation (TF) and Allende mixing (AM) lines are shown for reference. (a) The entire range in O-isotopic compositions, showing the distributions of anomalous irons (iron-an), Eagle Station pallasites (es-pal), and ureilites (urei). The anomalous irons Bocaiuva (B) and Tucson (T) discussed in the chapter are labeled. Silicate and chromite samples of the anomalous iron Mbosi are connected with a tie line. (b) Metal-rich meteorites and winonaites plot along three mass fractionation lines; the IAB and IIICD irons and winonaites (win) plot on one below the terrestrial fractionation line, and the IIE and IVA irons on lines at higher $\Delta^{17}O$. The closely related acapulcoites and lodranites (aca-lod) form a cluster with significant dispersion in $\Delta^{17}O$. Aubrites (aub) are the only non-chondritic asteroidal meteorites that fall on the terrestrial fractionation line. The anomalous irons Guin (G) and Sombrerete (S) discussed in the chapter are labeled. (c) This portion of the figure clearly shows the much greater dispersion in $\Delta^{17}O$ for the acapulcoites and lodranites compared to many other non-chondritic meteorites, such as the IIIAB irons, main-group pallasites (mg-pal) and brachinites (brac). The two pyroxene pallasites (py-pal) plot near two members of the acapulcoite-lodranite clan. The unique achondrite Divnoe is similar in O-isotopic composition to the IAB and IIICD irons and winonaite meteorites, but distinct from them in $\Delta^{17}O$. The unique achondrite LEW 88763 (L) and the IIIAB iron containing chondritic silicates Puente del Zacate (PZ) discussed in the chapter are labeled. (d) An extreme close-up view shows the general O-isotopic similarity of the IIIAB irons, main-group pallasites, howardites, eucrites and diogenites (HED), mesosiderites (mes), angrites (ang) and brachinites. The unique meteorite Enon also occupies this region of O-isotope space. All data are from Clayton and Mayeda (1996).

Figure 1 places the meteorites discussed here in O-isotope space. Many of the meteorite groups, especially those that exhibit the clearest signature of planetary style igneous processes, occupy a fairly narrow region of O-isotope space just below the terrestrial fractionation line. However, overall, the non-chondritic meteorites occupy much of known solar system O-isotope space, and therefore have probably sampled parent bodies formed at various locations in the solar system.

The organization of this chapter is difficult to put in any single logical order. Logically, we should start where the last chapter left off, and discuss those meteorite groups that are closest to chondritic material, and work our way sequentially through more and more differentiated materials. This works well for the silicate-rich meteorites, but to some extent, the irons are quite distinct and do not fit in. Yet there are numerous interconnections between metallic and silicate materials; winonaites and IAB irons, and main-group pallasites and IIIAB irons, for example. Because of this, we will first discuss iron meteorite groups, presenting their petrologic and chemical characteristics. Armed with knowledge of the irons, we will then move on to tackle the achondrites and stony-irons.

There are a few conventions we will follow in this chapter. The Antarctic Meteorite Newsletter put out by the Office of the Curator at Johnson Space Center contains very basic mineralogical data that may be the only source of information for some poorly studied meteorites. Rather than fill the reference list with citations to these, we will refer to them in the body of the text, figure captions and tables as AMN X-Y, which means volume X, number Y of the newsletter. We use the definition of $\Delta^{17}O = \delta^{17}O - 0.52\ \delta^{18}O$ as given by Clayton (e.g. see Clayton and Mayeda 1996). We also here define mg# as molar 100*MgO/(MgO+FeO). When discussing modal mineralogy data, we will specify either wt % or vol % when it is known. If the author was not clear which was being used, we will opt for the nebulous %. Normalization to chondritic abundances are done using the CI values of Anders and Grevesse (1989). Finally, some mineral names are no longer officially recognized. We have used whitlockite instead of merrillite, and aluminian-titanian-diopside instead of fassaite throughout, regardless what was used in the original references.

IRON METEORITE GROUPS AND THE METAL PHASE OF STONY IRONS

General metallography and mineralogy

Metallic nickel-iron constitutes the largest mass proportion of extraterrestrial material available for study on earth. Among observed falls, however, iron meteorites account for only about 5%. Metallic meteorites are thus overrepresented in collections, because they are both durable and easy to recognize.

A typical iron meteorite consists mostly of metallic iron with 5 to 20% Ni, which is texturally continuous over the size of the meteorite, and contains various minor phases as inclusions. There are, however, individual exceptions to all of these features. Samples with up to 60% Ni are known, some meteorites are brecciated on a millimeter-to-centimeter scale, and a few objects are classified as metal-rich, even though they contain more sulfide than metal. However, no authenticated iron meteorite is known that contains less than 4% Ni. A thorough description of practically all iron meteorites known by the 1970s was presented by Buchwald (1975).

A number of meteorites consist partly of metallic iron and partly of silicates, and these are traditionally classified as stony-iron meteorites. In addition to the two major classes of stony-irons, pallasites and mesosiderites, some unusual individuals such as Lodran and Enon are often included in this category. However, the division between irons and stony-irons is somewhat arbitrary, since some iron meteorite breccias of group IAB contain as much silicate as mesosiderites, and some pallasites contain large silicate-

Figure 2. Photograph of a polished and etched slab of the Edmonton (Kentucky) IIICD iron meteorite, showing the Widmanstätten structure of a fine octahedrite (Of) with an average bandwidth of ~0.32 mm. The slab is 12 cm in its long dimension.

free areas of metal. Stony-irons and irons are sometimes collectively called "metal-rich meteorites" (Wasson 1974) to indicate that their metal content is higher than that of average solar system material.

The metallic phase of many iron meteorites shows a texture called the Widmanstätten pattern, an oriented intergrowth of body-centered cubic α-Fe,Ni, and high-Ni regions which consist of several phases (Fig. 2). The structure can be revealed by etching a well-polished sample with nital, a 2% solution of nitric acid in ethanol. The Widmanstätten pattern was discovered independently by G. Thompson and E.C. Howard in the first decade of the 19th century, some time prior to the eponymous Alois Beckh von Widmannstätten. Note that German literature, following the spelling of the proper name, uses "Widmannstätten" with "-nn-" whereas English literature always refers to "Widmanstätten structure" with a single "-n-."

The α-Fe,Ni phase, kamacite (known as ferrite in metallurgy), forms extensive lamellae with a Ni content <6%. A wide range of lamella widths is found among different meteorites, but within each one, the width of the kamacite lamellae is approximately constant. The more Ni-rich areas are generally referred to as taenite, but detailed study shows that these areas are complex mixtures of phases. A profile of the Ni content of taenite lamellae is generally M-shaped, with Ni contents close to 50% within a few micrometers of the kamacite interface, and much lower contents at the center of the lamellae. The 50% Ni region has subsequently been found to be an ordered, tetragonal phase of FeNi composition named tetrataenite (Clarke and Scott 1980). A sub-micron intergrowth of Ni-poor and Ni-rich phases called plessite is present within the taenite regions of more Ni-rich samples.

The formation of the Widmanstätten structure can be understood by considering the Fe-Ni phase diagram (Swartzendruber et al. 1991, Yang et al. 1996). Above about 800°C meteoritic metal forms γ-Fe,Ni regardless of composition, since at high temperatures there is no miscibility gap. During slow cooling α-Fe,Ni precipitates along the {111} planes of the γ-Fe,Ni parent. Depending on the bulk Ni content, nucleation of α-Fe,Ni occurs between 500° and 800°C, the lower temperatures corresponding to higher Ni contents. The continuity of the Widmanstätten structure in many large meteorite samples indicates that at the high temperature stage the γ-Fe,Ni crystals reached sizes of tens of centimeters to meters.

Taenite has traditionally been considered to be γ-Fe,Ni with varying Ni content, except for minor regions of tetrataenite. Recent work by Yang et al. (1997a,b) has cast doubt on this generally accepted picture. Based on detailed studies of the Ni-rich lamellae, they have concluded that if γ-Fe,Ni is present at all, it is only of minor importance. Most of the original γ-Fe,Ni undergoes complex decomposition reactions at temperatures below 400°C. Areas with Ni contents between 30% and 50% are actually intergrowths of tetrataenite and a body-centered cubic phase referred to as α_2-Fe,Ni. This phase is structurally identical to martensite, but unlike industrial martensite it did not originate by fast cooling. A very thin layer of awaruite, $FeNi_3$, separates the tetrataenite regions from kamacite. The bulk of the kamacite lamellae is apparently unaffected by the low-temperature reactions in the taenite region.

This reinvestigation of the Ni-rich areas of the Widmanstätten pattern represents a drastic revision of the traditional picture of iron meteorite mineralogy, and its implications have not yet become clear. Since diffusion in Fe-Ni alloys is too slow at low temperatures for an experimental determination of equilibrium reactions, most of the Fe-Ni phase diagram below 400°C is actually based on the phases observed in iron meteorites (Reuter et al. 1988, Yang et al. 1996, 1997b). Although meteorites as a natural long-term experiment have considerably increased our knowledge of Fe-Ni phase relations, the element of circularity that is thereby introduced invites caution when the meteorite-derived phase diagram is used to interpret meteorite structures.

The characteristic Widmanstätten pattern thus consists of lamellae of kamacite, separated by Ni-rich lamellae composed of several phases. Because kamacite lamellae are oriented along octahedral planes, meteorites that show this structure are known as octahedrites. According to the width of their kamacite lamellae, octahedrites are further subdivided into coarsest (Ogg, bandwidths >3.3 mm), coarse (Og, 1.3–3.3 mm), medium (Om, 0.5–1.3 mm), fine (Of, 0.2–0.5 mm), and finest octahedrites (Off, <0.2 mm). Plessitic octahedrites (Opl) are transitional to ataxites. The bandwidth boundaries, based on classifications developed in the 19th century, were redefined by Buchwald (1975) in such a way that most chemically related meteorites now fall into the same structural class. A few iron meteorites do not fit this textural classification scheme, and are regarded to have an anomalous texture. Some of these are specimens in which the Widmanstätten pattern has been partly or wholly obliterated by reheating in space (e.g. Juromenha; see Fig. 116 in Buchwald 1975, p. 95).

Samples with very high Ni content undergo the γ-α transformation at such low temperatures that only small kamacite spindles but no continuous lamellae form, and the texture is only microscopically visible. These meteorites are known as ataxites, designated D, and most have bulk Ni contents of ≥15%.

Some iron meteorites with Ni contents below 6% consist almost entirely of kamacite and show no Widmanstätten pattern; they are called hexahedrites and are designated H. The name refers to the cubic (hexahedral) cleavage of α-Fe,Ni single crystals. Most hexahedrites belong to chemical group IIA.

Etched kamacite in both hexahedrites and octahedrites usually shows features called Neumann bands, which are mechanical twins on the {211} planes of α-Fe,Ni. Twinning is induced by shock, either due to collisions in space, or during atmospheric passage (Buchwald 1975). Neumann bands can be best seen on the large, continuous kamacite crystals of hexahedrites, from which the usually more conspicuous Widmanstätten pattern is absent.

The ubiquity of Neumann bands indicates that they are formed easily at relatively low levels of shock. Most iron meteorites, however, show evidence of higher levels of shock as well. Shock pressures >13 GPa lead to a transient transformation of α-Fe,Ni to the high pressure ε modification, which manifests itself as "hatched kamacite." Even more severely shocked samples often show signs of eutectic melting at the edge of troilite nodules, secondary reactions of metal with carbide and phosphide, and deformation and even partial obliteration of the Widmanstätten pattern.

In addition to forming Widmanstätten lamellae or (in Ni-rich samples) kamacite spindles, α-Fe,Ni also commonly nucleates at the interface of non-metallic inclusions. This morphology, called swathing kamacite, forms around inclusions of troilite, schreibersite, silicates, etc., as a continuous layer of kamacite, which can be twice as wide as the average width of Widmanstätten lamellae (Buchwald 1975, p. 89).

Mineralogy of accessory phases

Except for a few transitional cases, there is a visually obvious difference between two types of mineral occurrences in metal-rich meteorites: (a) a major non-metallic lithology in meteorites such as pallasites or iron-silicate breccias such as Woodbine, and (b) minor occurrences of non-metallic minerals that are accessory phases within the metallic lithology. Separate sections of this chapter are dedicated to the silicate portions of stony-irons and iron-silicate breccias.

Table 2 gives an overview of minerals associated with the metal phase. By far the most common non-metallic minerals are troilite, schreibersite, and graphite. The carbides cohenite or, less frequently, haxonite, are common in some groups. In a small number of meteorites, notably the large Canyon Diablo iron associated with Meteor Crater, Arizona, some graphite has been transformed into diamond and lonsdaleite, a hexagonal high-pressure polymorph (Frondel and Marvin 1967), due to shock. Some graphite has formed by decomposition of cohenite, but it is also found as rims around non-metallic inclusions in a morphology that suggests subsolidus exsolution from the metal. "Cliftonite" is a carbon occurrence with cubic morphology, but is crystallographically identical to graphite. Its shape is imposed by the cubic structure of the host metal.

Troilite, the hexagonal, practically stoichiometric (Fe:S mole ratio usually ≥0.98) modification of FeS, differs structurally from terrestrial pyrrhotite. Schreibersite, $(Fe,Ni)_3P$, occurs either as coarse irregular or skeletal inclusions, or as very small euhedral crystals in the shape of rods or platelets. The latter were originally thought to be a different mineral, and named "rhabdite," a name still used to refer to this particular morphology. The Ni content of troilite is generally negligible, but both schreibersite and cohenite, $(Fe,Ni)_3C$, contain appreciable amounts of Ni. When they are in equilibrium, cohenite contains less and schreibersite more Ni than the host metal.

Platelet-shaped inclusions of chromite, troilite, or both, are relatively common. They can reach cm in size, but are usually only a few μm thick. Known as Reichenbach lamellae, they are mostly oriented along cubic planes of the parent γ crystal (Buchwald 1975), and enveloped by swathing kamacite. Their morphology suggests exsolution from the host metal.

Table 2. Mineralogy of the metal phase of iron and stony iron meteorites.

Elements and alloys			
kamacite	α-Fe,Ni (<6% Ni)	1	
martensite*	α_2-Fe,Ni	1	
taenite*	Fe,Ni (>30% Ni)	1	A
tetrataenite	FeNi	1	
awaruite	Ni_3Fe	1	
graphite	C (hexagonal)	2	
diamond	C (cubic)	5	
lonsdaleite*	C (hexagonal)	5	
copper	Cu	4	B
Sulfides			
troilite	FeS (hexagonal)	1	
mackinawite	FeS (tetragonal)	5	
sphalerite	(Zn,Fe)S	4	
alabandite	MnS	4	
pentlandite	$(Fe,Ni)_9S_8$	5	
chalcopyrite	$CuFeS_2$	3	
brezinaite*	Cr_3S_4	5	
daubréelite*	$FeCr_2S_4$	2	
djerfisherite	$K_3CuFe_{12}S_{14}S$	4	
Carbides, nitrides, phosphides			
cohenite	$(Fe,Ni)_3C$	2	
haxonite*	$(Fe,Ni)_{23}C_6$	3	
unnamed iron carbide*	$Fe_{2.5}C$	5	
carlsbergite*	CrN	3	
roaldite*	Fe_4N	4	
schreibersite*	$(Fe,Ni)_3P$	2	
Oxides			
chromite	$FeCr_2O_4$	3	C
silica	SiO_2	4	D
Phosphates			
sarcopside	$(Fe,Mn)_3(PO_4)_3$	4	
graftonite	$(Fe,Mn)_3(PO_4)_3$	4	
beusite	$Fe_3Mn_3(PO_4)_4$	5	E
buchwaldite*	$NaCaPO_4$	4	
galileiite*	$Na(Fe,Mn)_4(PO_4)_3$	4	
brianite*	$Na_2CaMg(PO_4)_2$	5	F
panethite*	$Na_2(Mg,Fe)_2(PO_4)_2$	5	F
farringtonite*	$(Mg,Fe)_3(PO_4)_2$	5	F
johnsomervilleite*	$Na_2Ca(Fe,Mg,Mn)_7(PO_4)_6$	5	
maricite*	$NaFePO_4$	5	

Column 3: Frequency of occurrence (see text for further explanation); 1 - ubiquitous, 2 - very common, 3 - common minor mineral, 4 - rare, 5 - only found in one or a few meteorites, * - not know as terrestrial minerals.
Column 4: notes; A - Taenite is apparently not a single phase (see text). Additional types of metal, such as ε-iron, can be produced by shock, B - The presence of small amounts of Cu-sulfides is suspected, but has never been unambiguously confirmed, C - Chromite associated with silicate inclusions contains Mg, Al, etc., but chromite associated with metal is usually close to pure $FeCr_2O_4$ in composition, D - Both tridymite and cristobalite have been found in iron meteorites, and other silica polymorphs may be present, E - Structurally continuous with graftonite; unlike all terrestrial beusite, the meteoritic variety is Ca-free, F - Brianite, panethite, and farringtonite may be part of a separate oxide/silicate lithology in the few meteorites where they have been found.

Most of the minerals listed in Table 2 have been discussed in greater detail by Buchwald (1977), but more have been discovered since then, for example roaldite, Fe_4N, (Nielsen and Buchwald 1981) and galileiite, $Na(Fe,Mn)_4(PO_4)_3$ (Olsen and Steele 1997). Other compounds known or suspected to be present are as yet insufficiently characterized to be either identified as a known mineral or described as a new one. Some phosphates, like brianite, panethite, and farringtonite, are probably more closely related to silicate inclusions, but are included in Table 2 because of their ambiguous status.

Several minerals in Table 2 are only known from meteorites, some others have been discovered in meteorites first, and only later found at a terrestrial location. Meteoritic minerals that have no naturally occurring terrestrial counterpart are identified with an asterisk in Table 2. Some of these are common in industrial products. For example, the mineral cohenite, Fe_3C, is a common inclusion in steel where it is known as cementite.

Column 3 of Table 2 denotes the frequency of occurrence by assigning minerals to five categories in order of decreasing abundance. By definition, all metal-rich meteorites contain one or more of the iron-nickel phases. A few other minerals, such as troilite and schreibersite, are also present in nearly all samples, and are designated "ubiquitous." A 2 denotes those minerals that are so common that they can usually be found in most large specimens. Rare minerals are designated 3 to 5 according to the following criteria: minerals that are present in many meteorites, but usually only in very small amounts, are assigned to category 3. An example is carlsbergite, CrN, which occurs in many members of several groups of irons, but in very small amounts (Buchwald and Scott 1971). Minerals that occur in only one or a few meteorites are assigned to categories 4 and 5. For minerals in category 5 there are usually theoretical reasons, in addition to observation, to assume that the mineral is confined to unusual specimens, such as in the case of brezinaite, Cr_3S_4, which occurs only in the highly reduced Tucson iron (Bunch and Fuchs 1969, Nehru et al. 1982). On the other hand, a mineral such as buchwaldite (Olsen et al. 1977), although rare, may well be present in many more samples than observed so far, and is therefore assigned to category 4.

Classification and chemical groups

Whereas the subdivision into hexahedrites, octahedrites, and ataxites is purely descriptive, a genetically more significant taxonomy is based on the trace element content of the metal phase. In a series of papers (Wasson 1967, 1969, 1970; Wasson and Kimberlin 1967, Wasson and Schaudy 1971, Schaudy et al. 1972, Scott et al. 1973, Scott and Wasson 1976, Kracher et al. 1980, Malvin et al. 1984, Wasson et al. 1989, 1998a,b) data on Ni, Ga, Ge, and Ir have been published on practically all known metal-rich meteorites, and several more elements have been determined in most others. When all likely pairings of samples are taken into account, the number of independent iron meteorites that had been analyzed as of 1990 was about 605 (Wasson 1990).

On plots of log(E) vs. log(Ni), where E denotes some well determined trace element, about 85% of iron meteorites are seen to fall into one of 13 clusters or chemical groups. With two exceptions these groups fall within restricted areas on a log(Ge) vs. log(Ni) plot (Fig. 3a), whereas on log(Ir) vs. log(Ni) plots (Fig. 3b) the same samples form narrow, elongated fields spanning a considerable range in Ir contents. It is customary to refer to these clusters as groups if they contain at least five members. Samples that fall outside defined chemical groups are called ungrouped (or sometimes anomalous) iron meteorites, and some of these form grouplets of two to four related individuals (Scott 1979a).

The fields outlined in Figure 3 are not equally populated, which is indicated by shading the more densely populated areas. For example, the shaded part of the IAB field

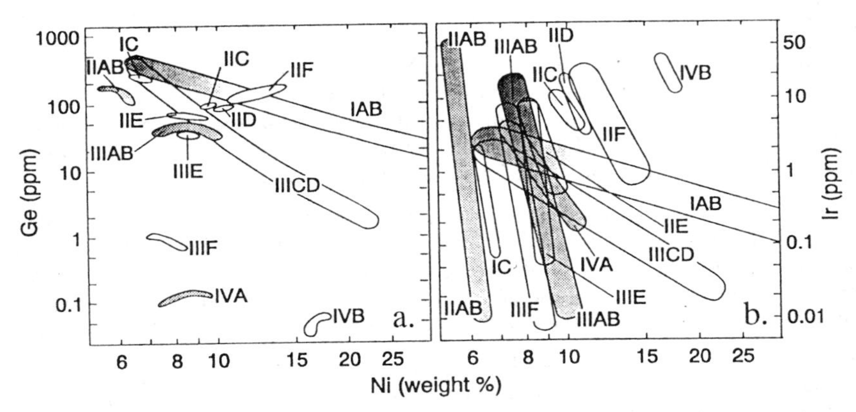

Figure 3. Plots of (a) Ge vs. Ni and (b) Ir vs. Ni showing the fields for the 13 iron meteorite groups. Germanium is one of the elements originally used to classify irons into groups I through IV based on decreasing Ge and Ga content. Note that most iron meteorite groups have limited ranges in Ge and Ni content; groups IAB and IIICD are unusual on this plot. The Ir content of most iron meteorite groups varies enormously from low to high Ni members, due to its strong fractionation between solid and liquid metal. This results in the observed steep trends on this plot for these so-called magmatic iron meteorite groups. Again, IAB and IIICD irons show an unusual distribution on this plot.

contains about 100 meteorites, whereas only 3 samples fall in the region ≥25% Ni. Likewise, in groups IIAB, IIIAB and IVA, the low-Ni, high-Ir regions contain more members than the opposite end of the outlines. The shaded areas in Figure 3 account for about half of all known iron meteorites.

The taxonomic significance of Ga and Ge derives from the fact that they are the most volatile siderophile elements (Wasson and Wai 1976, Wai and Wasson 1979), and in analogy to the situation in chondrites, more volatile elements tend to be more strongly fractionated between different groups (Wai and Wasson 1977). Thus both Ga and Ge show a narrow range within groups of similar iron meteorites, but large differences between different groups.

Early determinations of these elements (Lovering et al. 1957) resolved only four groups, designated by Roman numerals I through IV in order of decreasing content of Ga and Ge. Letters were added later, as in IVA and IVB, to distinguish additional groups. Not all of these groups proved to be independent, however, leading to the current nomenclature, in which some groups have a combination of two letters after their Roman numeral. For example, as more samples were analyzed, the hiatus between IIIA and IIIB was filled in, and the entire group is now designated IIIAB. On the other hand, no genetic relationship exists between groups IVA and IVB. A more detailed explanation of this nomenclature, together with a compilation of Ni, Ga, Ge, and Ir data, can be found in Wasson (1974).

Average compositions of the chemical groups calculated by Willis (1980) are given in Table 3. For elements like Ga, which do not show a large range within a given group, determining the average group content is fairly straightforward. However, in the case of an element like Ir, the "average" is model-dependent. As explained in more detail in the "Origin" section below, the modification of the Willis (1980) averages listed in Table 3 refer to models of the parent material from which the meteorites formed, rather than the average of actually analyzed samples.

Table 3. Average compositions of iron meteorite chemical groups.

Group	structure*	Ni wt%	Ga µg/g	Ge µg/g	Ir µg/g	C µg/g	P mg/g	S wt%	Cr µg/g	Co mg/g	Cu µg/g	Mo µg/g	W µg/g	Pd µg/g	Au µg/g	As µg/g	Sb ng/g	Group
IAB	see note 1	6.4	96	400	2.7	-	2.1	-	16	4.6	132	8.2	1.6	3.5	1.5	11	270	IAB
IC	var.	7.1	51.7	230	0.38	-	4.3	-	70	4.6	160	7.7	1.3	3.5	1.14	11	98	IC
IIAB	H (IIA)	6.15	58.3	173	10	2000[a]	6	0.20[a]	38	5.3	129	6.9	2.1	2.6	1.1	9.9	201	IIAB
	Ogg (IIB)	6.1[a]	-	64[b]	11a	-	20[a]	17.0[b]	-	-	-	-	-	-	1.0[a]	-	-	IIAB
		5.1[b]	-	-	1.3b	-	10[b]	-	-	-	-	-	-	-	-	-	-	IIAB
IIC	Opl	10.8	37.3	94.6	6.4	-	5.3	-	87	6.5	260	8.4	-	6	1.1	8.2	150	IIC
IID	Om	11.2	76.2	87	9.9	-	9.8	-	31	4.7	280	9.4	2.4	5.3	1.1	10	220	IID
IIE	anom.	9.13	24.4	68.3	4.1	-	-	-	-	4.7	416	6.8	0.78	5.3	1.6	16	300	IIE
IIF	var.	12.9	9.8	140	6.2	-	2.6	-	-	7.0	300	-	1	-	1.5	16	250	IIF
IIIAB	Om	8.49	19.7	38.9	4.1	140[a]	5.6	1.4[a,d]	40	5.0	160	7.2	1	3.5	1.2	10.5	265	IIIAB
		8.25[a]	18.5[c]	40[a]	5.0[a]	-	9.3[a]	5.0[b]	-	-	160	-	-	-	1.1[a]	-	-	IIIAB
		7.6[b]	-	35[b]	3.5[c]	-	-	6.0[c]	-	-	160	-	-	-	0.7[b]	-	-	IIIAB
		8.2[c]	-	36.25[c]	-	-	-	-	-	-	160	-	-	-	1.05[c]	-	-	IIIAB
IIICD	see note 2	-	-	-	-	-	-	-	-	-	-	-	-	-	-	-	-	IIICD
IIIE	Og	8.49	17.6	35.6	4.1	-	5.6	-	40	5.0	160	7.2	1	3.5	1.2	10.5	265	IIIE
IIIF	var.	7.95	6.82	0.91	3.2	-	2.2	-	210	3.6	170	7.2	1.2	4.4	0.91	11	86	IIIF
IVA	Of	8.0	2.15	0.124	1.9	200[a]	0.88	0.8[a]	140	4.0	150	5.9	0.6	4.6	1.5	7.6	9	IVA
		8.4[a]	-	0.12[a]	2.0[a]	-	3.3[a]	1.0[b]	-	-	-	-	-	-	1.6[b]	-	-	IVA
		-	-	0.15[b]	1.8[b]	-	2.4[b]	-	-	-	-	-	-	-	-	-	-	IVA
IVB	D	17.1	0.221	0.0526	22	40 [a]	1	0.03[a]	87.7	7.4	12	27	3	12	0.15	1.1	1.5	IVB
		17.0[b]	-	0.053[a]	20[a]	-	3.0[a]	0[b]	-	-	-	-	-	-	-	-	-	IVB
		-	-	-	-	-	5.7[b]	-	-	-	-	-	-	-	-	-	-	IVB

*The structural classification for typical group members. Larger groups often contain a few members with anomalous (reheated, brecciated, etc.) structures. No "typical" group structure exists for groups IC, IIE, IIF, and IIIF. **Note 1:** Structures in groups IAB and IIICD vary, in part because of the large spread in Ni values. Most low-Ni IAB members are Og, but group also includes finer structures and brecciated, silicate-rich members. **Note 2:** Because of the uncertainty of separating group IIICD from IAB, and reclassifications proposed after Willis (1980), no average composition is available for this group. The structure of IIICD members ranges from Om to D, depending on Ni content.

Data are from Willis (1980) except as noted:

[a]"Final" compositions from Table 5, Willis and Goldstein (1982:A444)

[b]Values of C_0 from Table 3, Jones and Drake (1983:1206)

[c]Model #6I from Table 4, Haack and Scott (1993)

[d]Esbensen et al. (1982)

The largest chemical group, IIIAB, comprises about one third of all known iron meteorites. On log(E) vs. log(Ni) plots, some groups (IIAB, IIC, IID, and IVA) occupy fields of similar shape, although in different regions of the plot. As Scott (1972) has demonstrated, the behavior of these groups is consistent with fractional crystallization from a melt, and these groups are therefore called "magmatic iron meteorite groups."

Groups IAB and IIICD, which account for about 20% of all iron meteorites, as well as the small group IIE, exhibit a different behavior from the magmatic groups. For example, their Ni and Ge contents span a larger range, and their Ir contents a smaller range, than those of the other groups. They also tend to be rich in inclusions, and some members contain abundant silicates or, in some cases, phosphates. Their formation mechanism was clearly different from that of the magmatic iron meteorite groups, but the nature of this process remains poorly understood. Kracher (1982) proposed formation of group IAB by cotectic crystallization of metal and sulfide, but Wasson et al. (1980) suggested that the non-magmatic groups originated in small impact melt pools.

Cooling rates

Two factors control the lamella width of the Widmanstätten pattern in octahedrites: the average Ni content of the metal and the cooling rate. The higher the Ni content, the lower the temperature at which the α phase begins to precipitate; the faster the cooling, the less time the α lamellae have to grow. Therefore higher Ni content and faster cooling lead to finer patterns. The growth process can be simulated by computer models of Ni diffusion during nucleation and growth of kamacite in the original taenite crystals. Wood (1964) and Goldstein and Ogilvie (1965) showed that cooling rates can be determined by comparing the computer models to actually observed textures and compositions. The calibration of these models requires accurately determined diffusion coefficients, which are sensitive to minor element content, and accurate phase diagrams. The original cooling rate estimates, which were generally in the range of 1-10°C/Ma, had to be revised upward by about a factor of 5 (Saikumar and Goldstein 1988) when the improved diffusion data of Dean and Goldstein (1986) became available.

Cooling rates determined from diffusion-controlled Ni concentration profiles resulting from phase transformations in the metal phase are collectively known as metallographic cooling rates. In the case of octahedrites and many stony-irons, these are generally derived from analyzing the Ni distribution in the Widmanstätten pattern. Quantification can be based on matching the M-shaped Ni profiles in taenite lamellae, or measuring width versus central Ni content of taenite regions. The two methods give similar results, but a third method, suggested by Narayan and Goldstein (1985) turned out to be invalid (Saikumar and Goldstein 1988). In irons that contain chondritic silicates it is possible to determine cooling rates from both the Widmanstätten patterns and isolated metal grains in the silicates. The two are generally in agreement (Herpfer et al. 1994). The most reliable cooling rate determinations for the major iron meteorite groups have been compiled in Table 4.

In the last decade, cooling rate determinations have undergone a number of refinements to take into account the presence of minor phases and elements, impingement of growing kamacite plates (Saikumar and Goldstein 1988), undercooling of kamacite nucleation, and local variations in Ni content (Rasmussen 1981, 1982). Nonetheless it is still uncertain whether kamacite requires the prior formation of phosphides to nucleate without undercooling (Saikumar and Goldstein 1988) or not (Rasmussen et al. 1995). Although most iron meteorites have a sufficiently high bulk P content for schreibersite to nucleate before kamacite, this is not the case for low-Ni members of group IVA. This is the same group that shows a significant and systematic difference in calculated cooling rates between its low-Ni and high-Ni members.

Table 4. Metallographic cooling rates of selected iron and stony iron meteorite groups.

meteorite group	cooling rate (°C/Ma)	ref.	parent body radius (km)
IAB	25	(a)	33
	310	(b)	10
IC	highly variable	(c)	
IIAB	0.8-3	(e)	165-90
	6-12	(d)	65-47
IID	5	(a)	70
IIIAB	3-75	(d)	90-20
	7.5-15	(a)	58-42
IIICD	10	(a)	51
	200	(b)	13
IVA	11-500	(d)	49-8
	40-325	(a)	27-10
IVB	30-260	(d)	30-11
	4300	(b)	3
pallasites	2.5-4	(a)	97-78
mesosiderites	0.5	(a)	206

References: (a) Yang et al (1997a), based on Goldstein (1969) as modified by Saikumar and Goldstein (1988), (b) Rasmussen (1989), (c) Scott (1977a), (d) Haack et al. (1990), (e) Randich and Goldstein (1978).

See text for method of calculation for parent body radius.

Since conventional metallographic cooling rates are based on the kamacite-taenite exsolution, both phases have to be present for a quantitative determination. Almost all meteorites contain sufficient kamacite to determine cooling rates by this method, even the members of group IVB, whose Ni content reaches 18% (Rasmussen et al. 1984). However, in hexahedrites no taenite is present. To circumvent this limitation, Randich and Goldstein (1978) developed a method to determine cooling rates from the growth of schreibersite. The metal-phosphide cooling rates of 0.8-3°C/Ma for group IIA are in agreement, within accepted error, with the kamacite-taenite cooling rates of 4-12°C/Ma for group IIB, which, based on chemical evidence, originated from the same parent melt. Cooling rates derived from two different reactions are thus in approximate agreement, and also consistent with an origin of group IIAB from a single melt reservoir.

A similar situation exists in group IIIAB, but in group IVA, which also shows a distinctly magmatic signature in terms of chemistry, cooling rates vary widely. In addition, four silicate-bearing irons, Bishop Canyon, Gibeon, São João Nepomuceno and Steinbach, are members of this group. These unusual features led Scott et al. (1996) and Haack et al. (1996) to postulate that the IVA parent body was fragmented while it was still hot, and cooled as a "rubble pile" rather than a body with a continuous core. A similar history has been proposed by Scott (1977a) for group IC, whose members, although chemically similar, show widely differing metallographic structures.

In principle, burial depths, and hence parent body sizes, can be inferred from metallographic cooling rates, but the calculations depend on the knowledge of the thermal properties of material of which we have no samples, viz., mantle and regolith of the iron meteorite parent bodies. Haack et al. (1990) have shown that the thickness of the regolith layer is a critical parameter, and since the accumulation of regolith and megaregolith is a stochastic process, there are fundamental limitations to such calculations. Considering these uncertainties, the complex formula relating cooling rate, thermal properties, and parent body radius can be simplified to give an estimate of parent body size:

$$R = g \times 149 \times CR^{-0.465}$$

where R is the parent body radius in km, g is a factor that is equal to 1 for a chondritic parent body, and the cooling rate CR is given in °C/Ma. Parent body radii based on this formula are shown in Table 4.

Ages

Most commonly used radionuclide clocks are all based on lithophile elements, such as K, Rb, Sm, Th and U, and are not applicable to iron meteorites. In fact, a considerable fraction of the light lithophile elements present in iron meteorites is due to cosmic-ray spallation rather than inherited from the parent body (Voshage 1967, Imamura et al. 1980, Honda 1988, Xue et al. 1995). However, the high content of noble metals, combined with their strong fractionation within groups, makes it feasible to use the β decay of ^{187}Re to ^{187}Os as a chronometer for iron meteorites. The half-life of ^{187}Re, which is about an order of magnitude longer than the age of the solar system, is not known accurately enough to make an absolute comparison with Rb-Sr or Sm-Nd ages of stony meteorites. Taking the value of 42.3±1.3 Ga (Lindner et al. 1989), Morgan et al. (1992, 1995) found that the ages of IIAB and IIIAB irons are identical (4.583 Ga), and indistinguishable from the age of chondrites, although high-Ni IIBs show some slight disturbance. The uncertainty in the half life places the lower limit of their age at 4.46 Ga. Shen et al. (1996) determined Re-Os isotopic systematics for a number of IAB, IIAB, IIIAB, IVA and IVB irons. Their data for IIAB irons yielded an age within uncertainty identical to that determined by Morgan et al. (1995). Further, Shen et al. (1996) determined an isochron age for IVA irons, which was 60±45 Ma older than that for IIAB irons. Horan et al. (1998) determined Re-Os isochron ages for groups IAB-IIICD, IIAB, IIIAB, IVA and IVB. They found that the IAB-IIICD, IIAB, IIIAB and IVB irons have the same age within uncertainties, although the IIIAB irons could be slightly older. Excluding the IVA irons, the ages obtained by Horan et al. (1998) and Shen et al. (1996) are identical within uncertainty, after recalculation to a common ^{187}Re half-life. Horan et al. (1998) found the IVA irons to be significantly younger than the other meteorite groups by about 80 Ma, in contrast to the conclusion of Shen et al. (1996).

Chronological information can also be derived from extinct radionuclide systems, if isotopic anomalies due to early radioactivity are preserved. Extinct radionuclides of potential significance for iron meteorite genesis are ^{107}Pd ($t_{1/2}$ 6.5 Ma), ^{182}Hf ($t_{1/2}$ 9±2 Ma) and ^{205}Pb ($t_{1/2}$ 14 Ma). Evidence for live ^{107}Pd in the form of radiogenic silver ($^{107}Ag^*$) has been found in iron meteorites of the magmatic groups IIAB, IIIAB, IVA, and IVB (Chen and Wasserburg 1990), as well as some ungrouped irons. The IVA irons have a higher $^{107}Ag^*/^{108}Pd$ ratio than other iron meteorite groups, indicating that the IVA irons are older. This agrees with the Re-Os results of Shen et al. (1996) but disagrees with those of Horan et al. (1998). The failure to find $^{107}Ag^*$ in members of groups IAB and IIICD may be due to the larger Ag content of these samples, and hence a higher detection limit, and not necessarily a later formation age.

The preservation of $^{107}Ag^*$ argues that fractional crystallization took place within a few million years of parent body accretion. Reconciling the rapid heat loss implied by this scenario with slow metallographic cooling rates probably requires more sophisticated models of parent body evolution than are currently available. Perhaps most small asteroids acquired the kind of insulating regolith envisioned by Haack et al. (1990) only after their interiors had solidified, but before they cooled through the temperature range at which metallographic cooling rates became established.

Refractory lithophile ^{182}Hf decays to refractory siderophile ^{182}W, and thus can be used to constrain the timing of metal-silicate separation (core formation) in differentiated bodies. Horan et al. (1998) have determined the W isotopic composition for the meteorites they studied by Re-Os chronometry. They found that all the iron meteorite groups have the same W isotopic composition within uncertainty, indicating that metal-silicate separation occurred on all these parent bodies within a limited time span (<5 Ma). Further, Horan et al. (1998) concluded that for all iron meteorite groups excluding the IIIAB irons, the time of crystallization as determined by Re-Os was measurably later than the time of metal-silicate separation as given by the W isotope systematics, suggesting that core crystallization may have been a protracted process.

A third kind of isotopic measurement that has attained particular importance for iron meteorites is the determination of cosmic-ray exposure ages. High energy cosmic-rays cause spallation in their targets, producing elements with characteristic isotopic signatures that are usually very different from average solar system isotope ratios. Cosmic-rays have a limited penetration depth. For the vast majority of matter in an asteroid, spallation only occurs after the material from which meteorites are formed has been fragmented to meter size or less. Exposure ages thus measure the time a meteorite has spent in space in essentially the same shape in which it fell.

In stony meteorites measurable spallation effects are usually limited to noble gases, but in iron meteorites with their high abundance of heavy target nuclei and low abundance of lithophile elements, it is possible to use the very precise $^{41}K/^{40}K$ method (Voshage 1967). Cosmic-ray exposure ages of iron meteorites are typically in the range of 200 to 1000 Ma, some 5 to 50 times longer than typical for stony meteorites. Tight clusters in exposure ages for groups IIIAB and IVA suggest discrete breakup events for their parent bodies 650±75 and 420±70 Ma ago, respectively (Voshage and Feldmann 1979). No other clusters have been observed. The highest $^{41}K/^{40}K$ exposure age measured for an iron meteorite is 2.3 Ga, or half the age of the solar system, for the ungrouped Deep Springs iron.

Exposure ages based on $^{26}Al/^{21}Ne$ ratios are systematically lower than $^{41}K/^{40}K$ ages (Aylmer et al. 1988). The two ages can be reconciled if it is assumed that the cosmic-ray flux increased by some 35% over the last 10^7 years, although the more mundane explanation of an error in spallation cross sections cannot yet be ruled out.

Origin of magmatic iron meteorite groups

Chemical evidence indicates that magmatic iron meteorite groups formed by fractional crystallization. The most plausible asteroidal setting for such a process would be the core of a differentiated asteroid. This would imply that each group comes from a separate asteroid, and that core compositions varied considerably between different parent bodies.

The partition coefficient of an element, E, is defined as $k_E = C_s/C_l$, where C_s is the concentration of E in the solid, and C_l is the concentration of E in the liquid. If the partition coefficient of an element is constant throughout the crystallization process, successive fractions of crystallizing solid follow the Scheil (or Rayleigh) equation,

$$C_s = k \times C_0 \times (1-g)^{k-1}$$

where C_0 is the concentration in the liquid before crystallization started, and g is the fraction that has already solidified. Mutually consistent partition coefficients can be derived from the fact that, given two elements A and B, the track of successive fractions of crystallizing solid on a plot of log(A) vs. log(B) is a straight line with slope $(k_A-1)/(k_B-1)$. This techniques was used by Scott (1972) to estimate partition coefficients for a number of elements based on log(E) vs. log(Ni) plots for magmatic iron meteorite groups.

In practice, the condition of constant partition coefficients is rarely fulfilled, particularly because k_E for most elements depends strongly on the concentration of non-metals (S, P, and C) in the parent melt (Willis and Goldstein 1982, Jones and Drake 1983, Malvin et al. 1986). Since the non-metals largely remain in the residual liquid, their influence on k_E becomes more pronounced as solidification progresses. Refined models have been proposed to take non-ideal behavior into account, such as trapping of small amounts of liquid (Scott 1979b), solid state diffusion (Narayan and Goldstein 1982), partition coefficients that change during crystallization (Sellamuthu and Goldstein 1985, Jones 1994), and liquid immiscibility (Ulff-Møller 1998).

For the largest meteorite groups, which are probably fairly representative samples of their asteroidal precursors, average group compositions can be estimated (Willis 1980). The average group composition, however, is not identical to the composition of the parent melt, since the last amount of residual material, which contains most of the non-metals, is apparently missing (Kracher and Wasson 1982). Average compositions based on actual samples as well as a fractional crystallization model with constant k_E values have been calculated by Willis (1980), and are given in Table 3.

Since P has a small but finite partition coefficient, plots of log(P) vs. log(Ni) can be used to infer initial P contents. Sulfur is almost entirely excluded from the crystallizing solid, and the S content of the parent melt can only be estimated indirectly. For example, the partitioning of Ge is very sensitive to the S content of the melt, $k_{Ge}<1$ for melts low in S, and >1 for S-rich melts. Phosphorous has a similar effect. The change in slope on log(Ge) vs. log(Ni) plots of group IIIAB (Fig. 3) can thus be modeled with a melt containing 4 to 5% S at the beginning of crystallization, even though there are no samples of the S-rich material in our meteorite collections.

Models that take the influence of S on trace element partitioning into account (Jones and Drake 1983, Malvin et al. 1986) have led to some revisions in the original group composition estimates of Willis (1980). It should be pointed out, however, that these revised figures, denoted with superscripted letters in Table 3, are intended to represent the parent melt from which meteorites crystallized. For example, the average Ni content of group IIAB is estimated to be 6.1%. Element distribution patterns suggest an initial S content of the IIAB parent melt of 17% (Jones and Drake 1983). Since S is effectively excluded from the crystallizing solid, this model leads to an original Ni content of $6.1\times(1/1.17)=5.1\%$. Both values are given in Table 3.

The largest discrepancies between the Willis (1980) values and the model melt compositions of Willis and Goldstein (1982) and Jones and Drake (1983) are for elements whose partition coefficients are highly sensitive to S content. This explains the large differences in Ge and Ir estimates for group IIAB. Since Ir concentrations in this group span five orders of magnitude, the average group value is poorly constrained by actual measurements. Meteorite collections are not unbiased samples of an entire core, and considering the large spread in values, even a very small bias can seriously affect the calculation of a meaningful average. An argument in favor of the lower Ir estimate is that it corresponds to an Ir/Ni ratio that is within a factor 2 of the chondritic value, in accordance with other roughly chondritic element/Ni ratios in groups IIAB.

Jones and Malvin (1990) presented a semi-empirical interaction model for the influence of non-metals on siderophile partitioning in the Fe-Ni-S-P system. The partition coefficients for siderophile element E are calculated from two parameters α and β_E, and the mole fraction X of the non-metal according to:

$$\log(k_E) = \log(k_E^*) + \beta_E \times \log(1\text{-}n\alpha X)$$

where k_E^* is the partition coefficient in the pure Fe-Ni-E system, and α and n depend only on the non-metal (n = 2, α = 1.09 for S; n = 4, α = 1.36 for P). The values of β_E depend on both E and the non-metal, and range from +0.5 (Cr, S) to -2.6 (Ir, S). Based on this model, Jones (1994) gave formulas for calculating slopes on log(E_1) vs. log(E_2) diagrams even if they are non-linear.

Little work has been done on the effect of C at low pressures, which also has a significant influence on trace element partitioning. Following Willis and Goldstein (1982), the C content of the original melt is assumed to be relatively small, and Jones and Malvin (1990) have confined their discussion to S and P.

The large sulfide nodules present in some magmatic iron meteorites are thought to be derived from trapped melt. If the amount of trapped melt could be determined, the modal sulfide content would allow an independent estimate of melt composition. Unfortunately there is no way to estimate the amount of trapped melt without resorting to parameters like partition coefficients, which themselves depend on S content. Esbensen et al. (1982) have tried to model melt trapping in members of the Cape York shower, which apparently represent mixtures of solid and trapped liquid in varying proportions. Estimating solid:liquid ratios between 93:7 to 36:64 in different samples on the basis of trace elements, they arrived at an S content of 2%. Taking into account the amount of solid that had already crystallized before the Cape York specimens formed, the calculated S content of the original liquid was 1.4% by weight.

Haack and Scott (1993) contend the assumptions of Esbensen et al. (1982) in calculating the solid:liquid ratios were oversimplified, and estimate an initial S content for the IIIAB parent melt of about 6%, close to the 5% calculated by Jones and Drake (1983). Recently, Ulff-Møller (1998) has estimated the S content of the IIIAB iron parent melt at 10.8 wt % by using the Ir-Au distribution of high-Ir IIIAB irons. The partition coefficient of Ir is highly sensitive and that of Au less sensitive to the S content of the melt (Jones and Malvin 1990).

To account for the low abundance of S estimated for iron meteorite parent melts, Keil and Wilson (1993) suggest that parent bodies lost considerable amounts of S into space by explosive volcanism during incipient differentiation. Eutectic Fe-S liquid is probably finely dispersed along grain boundaries and in the network of early silicate melt (Kracher 1985). A small amount of volatiles can cause liquid to be erupted, and lost into space due to the low gravity on small asteroids (Wilson and Keil 1991).

Anomalous iron meteorites

About 15% of all known iron meteorites do not belong to one of the defined chemical groups (Wasson 1990). Some ungrouped irons have unique features, others are in many respects similar to group members, except that their trace element contents do not fall near any clusters on log(E) vs. log(Ni) diagrams. Most iron meteorites with extreme compositions are ungrouped, although the most Ni-rich meteorite known, Oktibbeha County, appears to be related to group IAB (Kracher and Willis 1981). Among compositionally unique samples, Butler has a Ge concentration almost four times higher than any other iron meteorite, and is enriched in some other trace elements as well. A few ungrouped irons contain Si dissolved in the metallic phase, a feature characteristic of metal in enstatite chondrites. Tucson not only contains reduced Si, but highly reduced silicates and the sulfide brezinaite, but no troilite. The ungrouped meteorites Soroti and LEW 86211 contain more troilite than Fe,Ni by volume.

Most of the remaining ungrouped iron meteorites, however, are not so unusual, and it is likely that they are simply fragments of poorly sampled parent bodies (Scott 1979a). If each ungrouped individual or grouplet represents a separate parent body, then iron meteorites have sampled between 30 and 50 asteroids (Scott 1979a, Wasson 1990). In a few cases relationships to existing groups may have been overlooked, but even so it is obvious that iron meteorites sample many more differentiated parent bodies than the known classes of achondrites.

This is not unexpected. The most abundant types of iron meteorites are clearly related to stony meteorites: group IIIAB and main group pallasites to the howardite-eucrite-diogenite (HED) association, and the most common silicate inclusions, in group IAB as well as in group IIE, show affinities to winonaites and H chondrites, respectively. But the number of ungrouped iron meteorites alone (88 according to Wasson 1990) is greater than the number of all achondrite types, and the chances that a poorly sampled body is represented among them is correspondingly greater.

Some of the diversity of iron meteorites can be explained simply by the fact that they are easier to find than stony meteorites, and because of their greater number the probability of collecting rare types is higher. But even more important is the very long space lifetime of iron meteorites, as indicated by their long cosmic-ray exposure ages. When asteroids are broken up, the resulting meteorite-sized fragments have to undergo perturbations that bring them into earth-crossing orbits before they can be captured by earth. The longer the space lifetime of an object, the greater the probability of such a perturbation. Some of these perturbations are due to actual collisions, which may fragment a stony meteorite, but not an iron.

Most meteorites that arrive at earth come from particular locations in the asteroid belt where perturbations into earth-crossing orbits are likely. One such location is the 3:1 period resonance with Jupiter at 2.501 AU (Wasson 1990). The most common meteorite groups are probably fragments of asteroids that were at one time close to this location, or a similarly unstable one.

When objects some small distance away from the resonance are broken up by collisions, some fragments change their orbits to move closer to the resonance, increasing their probability of becoming earth-crossers. In general, the magnitude of change in orbital parameters depends on ejection velocity, which is in turn inversely correlated with mass. In other words, the debris close to the resonance can be expected to consist of a range of fragments, the larger ones predominantly from objects close to the resonance, the smaller ones sampling a more diverse population. This hypothesis is also consistent with the finding that ungrouped meteorites are more abundant among Antarctic meteorites than in Museum collections of iron meteorites. The Antarctic set is less biased toward larger sizes than collections elsewhere on earth that depend on chance discoveries. This is the explanation given by Wasson (1990) for the high abundance of ungrouped iron meteorites among the Antarctic collection.

The large number of iron meteorite parent bodies is also consistent with the observation that most asteroids in the inner asteroid belt are differentiated (Gaffey 1990). This indicates that the number of potential parent bodies for differentiated meteorites is relatively large. The reason why achondrites apparently sample only a small number of bodies is not a scarcity of differentiated asteroids, but rather the relatively short space lifetime of stony meteorites, which allows only samples from very few locations to reach earth as recoverable objects. This constraint is much relaxed for irons, hence a broader range of parent bodies is sampled by metallic meteorites.

SILICATE-BEARING IAB AND IIICD IRONS AND STONY WINONAITES

Of the 13 major iron meteorite groups, only the IAB and IIICD groups have broad ranges in Ni and some trace elements (e.g. Ga, Ge). These ranges are unlike those in the other 11 groups and difficult to explain by simple fractional crystallization of a metallic core. In addition, many members of these groups contain silicate inclusions, which are abundant in the IAB irons but are also found in a few IIICD irons. These silicate inclusions are roughly chondritic in mineralogy and chemical composition, but the textures are achondritic, exhibiting recrystallized textures. Silicate inclusions in IAB irons are linked through oxygen-isotopic (Clayton and Mayeda 1996) and mineral compositions to the stony winonaites and may be related to IIICD irons. Recently, comprehensive studies have been conducted of the winonaites (Kimura et al. 1992, Yugami et al. 1996, Benedix et al. 1998a), silicate inclusions in IAB irons (Yugami et al. 1997, Benedix et al. 1998b), silicate inclusions in IIICD irons (McCoy et al. 1993) and metal in the IAB and IIICD irons (Choi et al. 1995). The apparently contradictory features of these meteorites have led to a range of models to explain their formation, from those invoking impact melting (Choi et al. 1995) to those which suggest asteroid-wide partial melting (Kracher 1982, 1985). Yugami et al. (1997) suggested that partial melting and melt migration as a result of impact heating can produce the diversity of meteorite types observed on this parent body. Benedix et al. (1998b) proposed a model in which a partially-melted parent body is catastrophically disrupted and reassembled to explain the origins of these meteorites.

Table 5. Silicate-bearing IAB and IIICD irons, and winonaites.

Silicate-bearing IAB irons
Caddo County, Campo del Cielo, Canyon Diablo, EET 83333, EET 87505, EET 87504, EET 84300, EET 87506, Four Corners, Jenny's Creek, Kendall County, Landes, Leeds, Linwood, Lueders, Mertzon, Mundrabilla, Ocotillo, Odessa, Persimmon Creek, Pine River, Pitts, San Cristobal, Tacubaya, TIL 91725, Toluca, Udei Station, Woodbine, Youndegin, Zagora

Silicate-bearing IIICD irons
Maltahöhe, Carlton, Dayton

Winonaites
Winona, Mt. Morris (Wis.), Pontlyfni, Tierra Blanca, QUE 94535, Yamato 74025, Yamato 75261, Yamato 75300, Yamato 75305, Yamato 8005

Classification, petrology and mineralogy

Silicate-bearing IAB and IIICD iron meteorites and winonaites are listed in Table 5. As a group, the silicate-bearing IAB iron meteorites exhibit a number of common properties. The metallic textures of the IAB irons have been described in detail by several authors (e.g. Buchwald 1975). Overall, the silicate-bearing IAB iron meteorites span the range of structural classification from hexahedrites (e.g. Kendall County with 5.5% Ni) through fine octahedrites and coarsest octahedrites to ataxites (e.g. San Cristobal with 24.9% Ni). Silicate inclusions contain variable amounts of low-Ca pyroxene, olivine, plagioclase, calcic pyroxene, troilite, graphite, phosphates and Fe,Ni metal, and minor amounts of daubreelite and chromite. The abundance of silicates varies dramatically between different IAB irons, as well as between different sections of the same iron (Table 6). Plagioclase generally comprises ~10% of the total silicates present (Table 6), similar to the abundance found in ordinary chondrites (Van Schmus and Ribbe 1968, McSween et al. 1991). Overall, mafic mineral compositions are quite reduced (Table 7), with olivine compositions ranging from Fa_1 (Pine River) to Fa_8 (Udei Station) and low-Ca pyroxenes ranging from $Fs_{1.0}$ (Kendall County) to $Fs_{8.7}$ (Udei Station) (Bunch et al.

1970). Average anorthite contents of plagioclases are relatively narrow in range, from $An_{11.0}$ in Persimmon Creek to $An_{21.5}$ in Pine River, and calcic pyroxene is diopside varying from $Fs_{2.4}Wo_{44.7}$ in Mundrabilla to $Fs_{3.9}Wo_{43.9}$ in Persimmon Creek (Ramdohr et al. 1975, Bunch et al. 1970).

Table 6. Modal compositions (vol %) of silicate-bearing IAB and IIICD irons and winonaites. Data from McCoy et al. (1993) and Benedix et al. (1997a,b).

meteorite	section number	mafics*	plag	troilite	metal	W.P.	chr	daub	schreib	graph	Cohen
Troilite-rich IAB irons											
Mundrabilla	USNM 5914-1†	48.5	14.9	30.5	0.12	0.69	-	1.6	-	3.7	-
Persimmon Creek	USNM 2990	23.4	2.0	22.1	43.7	6.8	-	-	0.67	1.4	-
Pitts	USNM 1378	21.3	1.9	51.7	11.5	10.6	-	-	0.15	2.8	-
Zagora	USNM 6392	22.2	4.4	2.8	65.7	2.8	-	-	0.58	1.6	-
IAB irons with non-chondritic, silicate inclusions											
Caddo County	UNM 937†	37.2	44.7	4.7	5.9	7.3	-	-	0.23	-	-
Ocotillo	UH 226										
IAB irons with rounded inclusions											
Toluca	M8.53	5.1	1.4	23.8	43.1	9.9	0.08	-	14.9	1.7	-
	M8.79	0.7	-	1.4	56.7	22.6	-	-	9.4	4.4	4.9
	M8.150-1	-	-	17.9	68.4	0.55	-	-	4.3	-	8.9
	M8.150-2	8.5	1.7	28.4	50.9	0.50	-	0.07	4.7	5.4	-
Jenny's Creek	M217	-	-	67.9	-	13.7	-	0.92	14.2	-	3.2
Odessa	UH 256	-	-	3.3	85.0	1.4	-	0.14	0.79	8.4	1.1
Youndegin	M173	0.03	0.03	17.8	61.8	4.6	-	0.47	1.5	1.3	12.5
IAB irons with angular silicate inclusions and related winonaites											
Campo del Cielo	USNM 5615-2†	60.8	5.8	0.98	-	0.75	-	-	-	31.7	-
	USNM 5615-4†	72.1	10.2	6.4	1.6	1.3	0.24	-	0.06	8.1	-
	USNM 5615-6	44.4	8.4	4.0	31.7	0.83	-	-	2.9	7.7	-
	USNM 5615-8	20.6	12.0	3.9	41.2	3.9	-	-	1.4	16.9	-
	USNM 5615-9†	63.5	18.2	5.2	1.4	1.2	-	-	-	10.5	-
Copiapo	USNM 3204-2	17.9	2.3	0.43	71.0	4.6	0.11	-	1.2	2.5	-
EET 83333		13.8	2.6	0.91	78.4	1.3	-	-	1.3	1.8	-
EET 84300		5.8	1.4	1.5	84.9	1.6	-	-	2.8	1.9	-
EET 87504	,5	11.2	2.8	11.5	59.8	12.6	-	-	0.87	1.4	-
EET 87506		9.1	1.8	10.5	69.9	5.4	-	-	2.9	0.42	-
Four Corners	USNM 728	20.2	5.9	3.5	59.1	4.9	-	-	3.1	3.3	-
Landes	UH 147	35.7	2.9	1.3	50.9	2.4	-	-	0.49	6.3	-
Lueders	UH 255	25.4	3.7	1.1	45.3	21.2	-	-	1.6	1.8	-
Linwood	USNM 1416	8.3	0.4	6.9	46.7	10.0	-	-	5.7	21.9	-
Pine River	USNM 1421	24.8	4.6	1.1	65.4	0.70	-	-	2.2	1.3	-
Udei Station	USNM 2169	43.9	1.7	8.9	38.6	1.3	0.44	0.10	0.49	4.5	-
Woodbine	USNM 2169	36.6	7.0	7.0	3.4	9.8	-	-	6.8	1.4	-
Pontlyfni	M6.2	57.9	9.5	18.2	9.5	3.1	0.06	0.67	1.04	-	-
	M6.3	53.2	8.7	19.9	12.3	4.6	-	0.80	0.20	-	-
Winona	UH 133	66.1	13.3	5.1	1.8	13.4	0.10	-	-	-	-
	UH 195	68.2	12.2	6.1	2.3	11.2	0.10	-	-	-	-
	USNM 854	64.7	14.0	6.3	2.2	12.4	0.10	-	-	-	-
Mt. Morris (Wis.)	UH 157	65.5	9.4	5.8	3.5	15.1	-	0.20	-	-	-
	USNM 1198-2	65.7	11.5	8.7	0.2	13.2	-	0.30	-	-	-
Y-75300	,51-3 (coarse)	57.8	13.3	15.2	0.4	12.7	-	0.70	-	-	-
	,51-3 (fine)	75.2	12.2	7.5	-	5.1	-	-	-	-	-
Y-74025	,52-1	71.5	12.7	9.1	1.5	4.9	-	0.30	-	-	-
Tierra Blanca	M-3,1	63.1	8.8	1.4	-	26.2	-	-	-	0.56	-
	M-3,2	61.6	6.3	0.17	0.1	31.8	0.09	-	-	-	-
Y-8005	,51-3	52.9	8.1	8.3	12.5	16.8	-	0.7	0.6	-	-
IAB irons with phosphate-rich inclusions											
San Cristobal	UCLA	16.9‡	2.9	2.6	74.7	1.1	-	-	1.6	0.16	0.08

* Mafic silicates were not distinguished and include olivine, orthopyroxene, and clinopyroxene

† Thin section samples of separated silicate inclusions without metallic host.

‡ Includes ~ 0.9 vol.% brianite.

- = not observed.

Table 7. Representative analyses of silicates from IAB irons demonstrating the range of compositions, from Bunch et al. (1970).

	Olivine			Orthopyroxene			Clinopyroxene		Plagioclase	
	Pine River	Toluca	Udei Station	Pine River	Toluca	Udei Station	Pine River	Udei Station	Persimmon Creek	Pine River
	Chemical Composition (wt %)									
SiO_2	42.1	42.1	41.3	58.0	57.8	56.5	54.3	54.3	64.1	62.9
Al_2O_3	b.d.	b.d.	b.d.	0.51	0.17	0.85	0.96	1.72	21.6	23.0
TiO_2	b.d.	b.d.	b.d.	0.10	0.23	0.22	0.33	0.57	--	--
Cr_2O_3	b.d.	b.d.	0.09	0.19	0.06	0.37	0.38	1.16	--	--
FeO	1.17	4.5	7.5	2.55	4.5	5.6	1.00	1.84	0.68	0.29
MnO	0.17	0.24	0.41	0.30	0.22	0.45	0.18	0.28	--	--
MgO	57.0	53.4	51.0	38.4	35.6	34.7	19.4	18.6	--	--
CaO	b.d.	b.d.	b.d.	0.70	0.52	0.92	22.2	20.2	2.41	4.8
Na_2O	b.d.	b.d.	0.20	0.18	0.09	0.11	0.65	1.06	10.3	9.4
K_2O	b.d.	b.d.	b.d.	b.d.	b.d.	0.23	0.10	0.08	0.71	0.47
Total	100.44	100.24	100.50	100.93	99.19	99.95	99.50	99.81	99.80	100.86
	Cation Formula (O=4 for olivine, 6 for pyroxene, 8 for plagioclase)									
Si	0.990	1.004	0.995	1.967	1.993	1.955	1.967	1.962	2.849	2.777
Al^{IV}	b.d.	b.d.	0.004	0.021	0.007	0.035	0.033	0.038	1.132	1.191
Al^{VI}				b.d.	b.d.	b.d.	0.008	0.035		
Ti	b.d.	b.d.	b.d.	0.003	0.006	0.006	0.009	0.016	--	--
Cr	b.d.	b.d.	0.002	0.005	0.002	0.010	0.011	0.033	--	--
Fe	0.022	0.090	0.150	0.072	0.130	0.162	0.030	0.056	0.025	0.011
Mn	0.002	0.005	0.009	0.009	0.006	0.013	0.006	0.009	--	--
Mg	1.997	1.898	1.833	1.921	1.831	1.790	1.048	1.002	--	--
Ca	b.d.	b.d.	b.d.	0.025	0.006	0.034	0.862	0.782	0.115	0.228
Na	b.d.	b.d.	0.008	0.012	0.006	0.015	0.046	0.074	0.888	0.807
K	b.d.	b.d.	b.d.	b.d.	b.d.	0.005	0.005	0.004	0.040	0.027
Total Cations	3.011	2.997	3.001	4.035	4.000	4.025	4.025	4.011	5.049	5.041
	Molar Mineral End Members and mg# (100(MgO/(MgO+FeO)))*									
Wo	--	--	--	1.2	1.9	1.7	44.3	42.3	--	--
En	--	--	--	94.8	92.2	89.6	53.8	54.2	--	--
Fs	--	--	--	4.0	6.9	8.7	1.9	3.5	--	--
Or	--	--	--	--	--	--	--	--	3.9	2.5
Ab	--	--	--	--	--	--	--	--	85.1	76.0
An	--	--	--	--	--	--	--	--	11.0	21.5
mg#	98.9	95.5	92.4	96.4	93.4	91.7	97.2	94.7	--	--

Despite these gross similarities, extraordinary heterogeneity is found between individual samples and sometimes between different sections of the same sample. Bunch et al. (1970) developed a dual classification scheme (Odessa and Copiapo types) based on modes, mineralogies, mineral compositions, textures, and shapes of the inclusions. With more recent work, a five-fold classification scheme is needed. The five types are: (1) sulfide-rich IAB irons; (2) IAB irons with non-chondritic silicate inclusions; (3) IAB irons with rounded inclusions (Odessa type of Bunch et al. 1970); (4) IAB irons with angular silicate inclusions (Copiapo type of Bunch et al. 1970); and (5) IAB irons with phosphate-rich inclusions. We discuss one or two meteorites that illustrate the characteristics of each of these groups.

Sulfide-rich IAB irons. Several of the IAB irons have unusually high troilite contents, found as irregular masses (Pitts, Persimmon Creek), veins (Zagora) or large grains. The type member is Mundrabilla (Fig. 4). With 7.47 wt % Ni in the metallic host, Mundrabilla plots in the low-Ni cluster of IAB-IIICD irons on plots of Ni vs. Ir, Ga, and

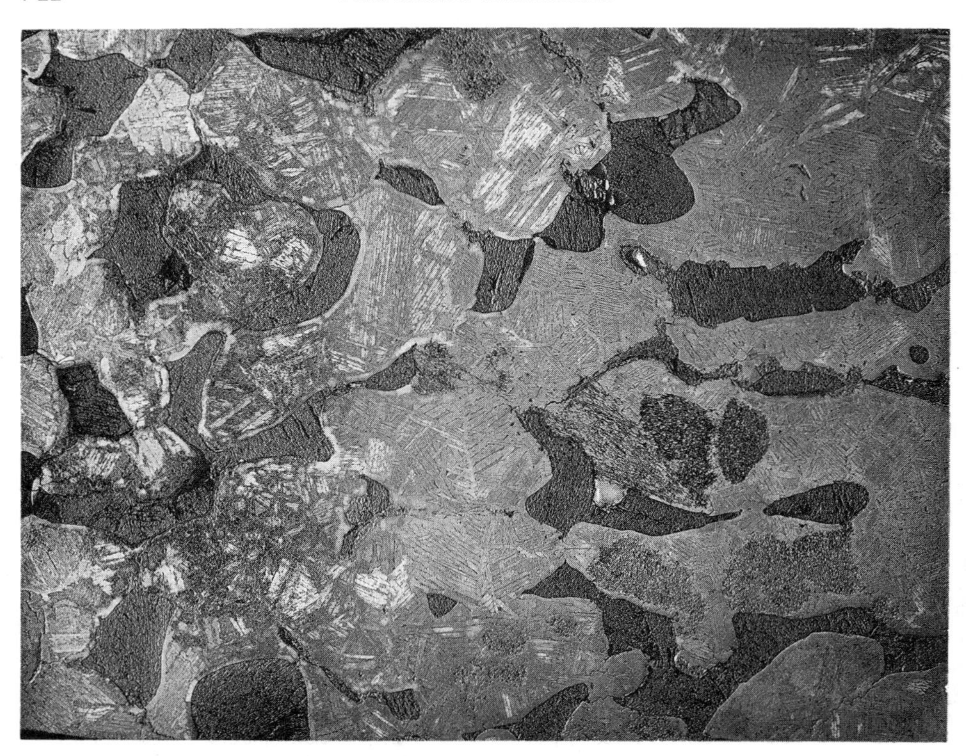

Figure 4. Photograph of the Mundrabilla iron meteorite showing metal-troilite intergrowths. White areas are Fe,Ni metal and gray areas are troilite. A cooling rate of ~5°C per annum during solidification is inferred from the weak dendritic texture. Field of view is 16 cm in maximum dimension.

Ge (Choi et al. 1995). Troilite comprises 25 to 35 vol % (Buchwald 1975) of Mundrabilla and occurs as mm-sized veins or lenses generally found along parent taenite grain boundaries (Buchwald 1975). Scott (1982) noted that the weak dendritic texture of the metal grains is a characteristic quench texture of metal-sulfide melts. In addition to abundant troilite, graphite, and minor amounts of schreibersite, these meteorites also contain silicate inclusions. Mundrabilla contains rare angular silicate inclusions, ranging in size from a few mm to a few cm, which are texturally similar to Winona (Benedix et al. 1998a) and silicates in Lueders (McCoy et al. 1996a).

IAB irons with non-chondritic silicate inclusions. IAB irons with non-chondritic silicates inclusions include Caddo County (basaltic inclusions) and Ocotillo (troctolitic inclusions). Coarse-grained, olivine-rich inclusions in some winonaites (e.g. Winona, Mt. Morris (Wis.)) also fit into this type. Caddo County, the type meteorite, is quite complicated in hand sample. Silicate inclusions are up to 7 cm in maximum dimension, but are truncated by the edge of the meteorite. Silicates comprise ~35 vol % (Fig. 5) of the slice. Metal occurs as the metallic host into which the clasts are embedded, as large grains within the silicate inclusions, and as veins which are clearly produced by post-solidification shock. In the hand sample, silicate grain sizes are observed to be highly variable and this is consistent with our thin section observations. Many inclusions, or parts of inclusions, consist of silicates with roughly chondritic modal proportions, but equigranular, recrystallized textures (Palme et al. 1991, Takeda et al. 1997a). These types

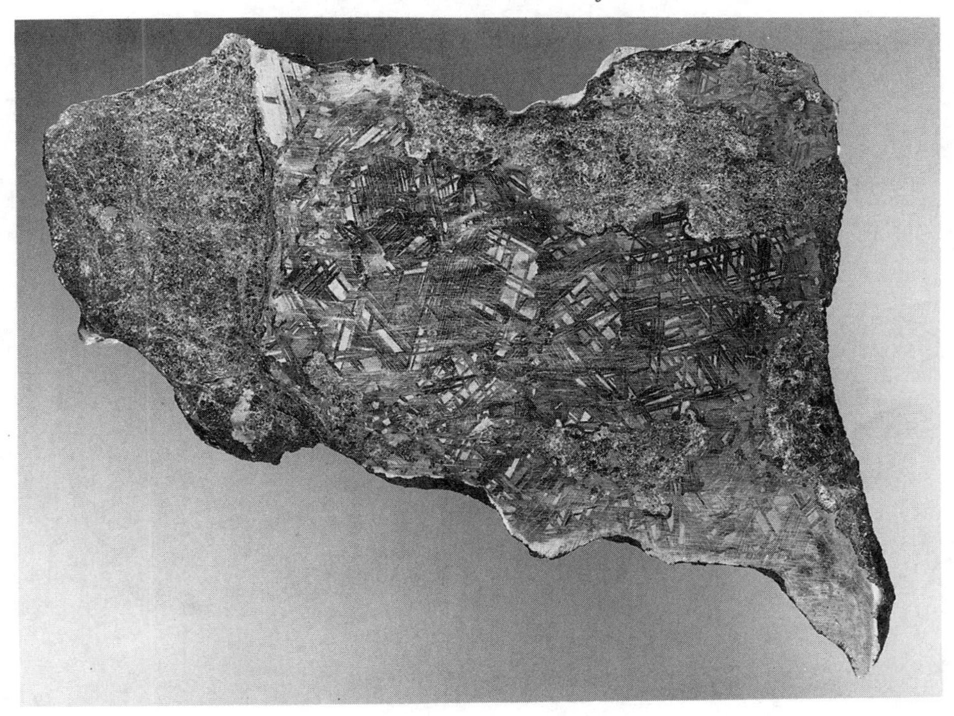

Figure 5. Photograph of Caddo County. Slab is 16.5 cm in maximum dimension. (Courtesy of the Smithsonian Institution).

of inclusions will be discussed in the section on IAB irons with angular silicate clasts. Rare inclusions of basaltic composition are also found in Caddo County. These are composed of major calcic pyroxene and sodic plagioclase and minor amounts of low-Ca pyroxene, olivine, troilite, and metal. Modal analyses by Takeda et al. (1993, 1994a), Yugami et al. (1997) and Benedix et al. (1998b; Table 6) indicate that these inclusions are broadly basaltic in composition, containing 45 to 55 vol % plagioclase, although texturally they are coarse-grained gabbros. Mafic silicates are relatively reduced (olivine, $Fa_{3.3}$; low-Ca pyroxene, $Fs_{6.5}$; calcic pyroxene, $Fs_{2.5}$), and plagioclase is $Ab_{75}An_{21}Or_3$ (Benedix et al. 1998b, Takeda et al. 1993, 1997a).

IAB irons with rounded inclusions. Several IAB irons contain inclusions which are typically rounded or ovoid and contain variable amounts of silicates, graphite and troilite, the latter two often being the dominant or sole constituents. These include Odessa, Toluca (Fig. 6) (and the paired Tacubaya), Canyon Diablo, Jenny's Creek and Youndegin. These meteorites have a relatively narrow range of Ni contents from 6.80 to 7.86 wt % Ni, within the low-Ni cluster of IAB irons. Bunch et al. (1970) referred to these as Odessa-types, while Benedix et al. (1998b) chose Toluca to represent this type, since a large number of specimens were readily available for study, allowing investigation of inter-inclusion heterogeneity.

Inclusions in Toluca consist of a core of either silicates or, more often, troilite surrounded by a sequence of swathing minerals such as graphite, schreibersite, and cohenite, although not all of these minerals are present in the sequence all of the time.

Figure 6. Photograph of a portion of the Toluca IAB meteorite (USNM 931) illustrating typical rounded to amoeboid graphite and/or troilite-bearing inclusions, sometimes containing angular silicate inclusions. (Courtesy of the Smithsonian Institution).

Inclusions which contain troilite and silicates are typically ovoid, while the graphite-troilite-rich inclusions tended to be rounded. The silicate inclusions, when present, are generally surrounded by cm-sized troilite areas. In this type, silicates do not appear to occur without either troilite or graphite-troilite surrounding them. Marshall and Keil (1965) and Buchwald (1975) reported similar inclusion properties for Odessa and Canyon Diablo, respectively. Mineral compositions of the rare silicates found in this type are reduced (Fa_{3-6}) and similar to those of other silicate inclusions in IAB irons.

A feature that seems to be common among the IAB irons is the heterogeneous distribution, morphology and mineralogy of the inclusions within a single meteorite. Silicate-bearing inclusions in Toluca are heterogeneously distributed and range in shape from nearly round to elongate and amoeboid (Benedix et al. 1998b). Modes of four Toluca inclusions (Table 6) showed ranges in abundance of mafic silicates (0-8.5 vol %), plagioclase (0-1.7 vol %), metal (43.1-68.4 vol %), troilite (1.4-28.4 vol %), schreibersite (4.3-14.9 vol %), cohenite (0-8.9 vol %) and graphite (0-5.4 vol %). This heterogeneity significantly complicates our ability to obtain representative samples.

IAB irons with angular silicate inclusions and winonaites. This type is characterized by the angular shapes of the inclusions and includes most of the IAB irons that Bunch et al. (1970) referred to as Copiapo type. Typical examples are Campo del

Figure 7. Photograph of large (~130 cm) polished slab of Campo del Cielo. Inclusions in this meteorite are more heterogeneously distributed and larger than those found in Lueders. (Courtesy of the Smithsonian Institution).

Cielo (Wlotzka and Jarosewich 1977, Bild 1977; Fig. 7) and Lueders (McCoy et al. 1996a; Fig. 8). The majority of IAB irons containing silicate inclusions fall into this category. Examples are Pitts, Persimmon Creek, and Zagora. The Ni contents of the members of this type span a wide range (6.58-13.78 wt %), extending beyond the low-Ni cluster of IAB irons. Of the inclusion types found in IAB irons, these angular inclusions most closely resemble the stony winonaites in texture, mineralogy, and mineral composition (Benedix et al. 1998a).

The members of this type are the most silicate-rich of the IAB iron meteorites, although silicate abundances are quite variable even in this group. The large El Taco mass of Campo del Cielo contains a few vol % of angular silicate inclusions, while other IAB irons such as Lueders, Landes, and Woodbine contain upwards of 40 vol % silicate inclusions (e.g. McCoy et al. 1996a). The inclusions are composed mainly of silicates of three basic morphologies: (1) fine-grained, recrystallized "chondritic" silicates (usually the more angular types); (2) medium-grained silicates; and (3) coarse-grained monomineralic crystals usually rounded and found individually in the metallic matrix, as first noted by Bunch et al. (1972). While some IAB irons of this type contain silicate inclusions which are apparently unrelated to each other (Campo del Cielo, Fig. 7), others (e.g. Lueders, Fig. 8) contain adjacent silicate inclusions which appear to have been invaded by molten metal. Adjacent silicate inclusions in these IAB irons, while slightly rotated, are obviously fragments of the same, larger parental inclusion. This relationship is not unlike that expected of a wall-rock invaded by melt from a dike.

The abundance of plagioclase relative to total silicate abundances is quite variable in these inclusions. Most of the inclusions have chondritic abundances of plagioclase

Figure 8. Photograph of polished slice of the Lueders IAB iron illustrating abundant angular silicate inclusions embedded in a metallic matrix.

(~10 vol %; Van Schmus and Ribbe 1968, McSween et al. 1991), although some exhibit enrichments (e.g. Campo del Cielo, Four Corners) or depletions (e.g. Udei Station) relative to chondrite modes. Given the small size and large grain size of some of these inclusions, at least part of this heterogeneity may be due to unrepresentative sampling, while others are clearly indicative of partial melting and melt migration. The depletion of plagioclase in Udei Station is quite striking, where areas up to 5 mm across are virtually devoid of plagioclase. In contrast, some inclusions in Campo del Cielo exhibit evidence for migration of melts in the form of veins of coarse-grained plagioclase and pyroxene with accompanying graphite, as first noted by Wlotzka and Jarosewich (1977).

The stony winonaites are most similar to these angular silicate inclusions. Indeed, they are so similar that Bevan and Grady (1988) have suggested that at least one, the Mount Morris (Wisconsin) winonaite, may be a separated silicate inclusion from the Pine River IAB iron. Some differences do exist (Benedix et al. 1998a); rare relict chondrules are found in Pontlyfni and, possibly, Mt. Morris (Wis.) (Benedix et al. 1998b), but these are apparently absent from silicate-bearing IAB irons. Both Winona and Mt. Morris (Wis.) contain mm-sized, coarse-grained areas which are dominated by olivine, in sharp contrast to the orthopyroxene-rich host. Finally, Tierra Blanca contains poikilitic calcic pyroxenes which reach 9 mm in maximum dimension (King et al. 1981). Neither of these textures is found in IAB silicates.

This group contains minerals with the most and least reduced compositions, which are found in Pine River ($Fa_{1.0}$), Kendall County ($Fs_{1.0}Wo_{0.8}$) and Udei Station ($Fa_{8.0}$; $Fs_{8.7}Wo_{1.7}$). Interestingly, Kendall County is probably the most reduced of the IAB irons, as Bunch et al. (1970) found no olivine grains they could analyze. Silicate inclusions in IAB irons generally have higher Fa and Fs values than the winonaites. However, as in the winonaites, Fa values are typically lower than Fs values in the IAB inclusions, indicating that reduction may have occurred. (The Fa content of olivine will be greater than the Fs content of orthopyroxene at equilibrium.) Supporting evidence for reduction comes from reverse zoning profiles in olivines in Campo del Cielo (Wlotzka and Jarosewich 1977).

Most of the existing bulk inclusion major and trace element compositions for IAB irons have been measured in this silicate-rich type (Table 8). Bulk major element compositions suggest that IAB silicate inclusions are similar to those of chondrites. Measured Si/Mg ratios for Landes (Kracher 1974), Woodbine (Jarosewich 1967), and Campo del Cielo (Wlotzka and Jarosewich 1977) are roughly similar to chondritic values as noted by Kracher (1974). Bild (1977) also found abundances of lithophile elements roughly similar to chondrites.

Bulk inclusion trace element analyses have been performed for Copiapo, Landes, Woodbine, Campo del Cielo (Bild 1977) and Udei Station (Kallemeyn and Wasson 1985). Measured rare earth element (REE) patterns and abundances for Copiapo and Landes are essentially chondritic. In contrast, Woodbine, Campo del Cielo and Udei Station exhibit fractionated patterns which can be roughly grouped into two types. The REE patterns measured in Campo del Cielo and Udei Station (Bild 1977, Kallemeyn and Wasson 1985) are negatively-bowed (V-shaped) with low REE^{3+} contents and a positive Eu anomaly. In contrast, Woodbine (Bild 1977) exhibits a positively-bowed pattern with high REE^{3+} contents and a negative Eu anomaly. Interestingly, the same two fractionated patterns have been widely observed in the winonaites (Benedix et al. 1998a and references therein). Kallemeyn and Wasson (1985) attribute these patterns to unrepresentative sampling of phosphates, which exhibit a pattern similar to that in Woodbine. Phosphates are concentrated near the edges of the inclusions in these meteorites and are often difficult to sample. While unrepresentative sampling may be important, Bild (1977) notes this range of patterns is consistent with heterogeneous distribution of plagioclase, diopside and phosphates, all of which occur in a low temperature melt and for which there is ample petrologic evidence of melt migration (Wlotzka and Jarosewich 1977, Benedix et al. 1998b).

IAB irons with phosphate-rich inclusions. Phosphates occur in many of the IAB irons as part of the rim sequence at the boundary between silicate inclusions and metal. However, a small number of IAB and IIICD irons are included in this type because they contain more abundant phosphates scattered throughout the silicate inclusions. In addition, these phosphates are often evolved Mg, Na-bearing phosphates (brianite, panethite, chladniite), rather than the more common Ca-bearing phosphates (whitlockite, apatite). Among the IAB irons, San Cristobal (24.97 wt % Ni; Choi et al. 1995) is the only member of this type (Scott and Bild 1974). We discuss here the IIICD irons Carlton (13.28 wt % Ni) and Dayton (17.03 wt % Ni), which contain abundant phosphates (McCoy et al. 1993), because of the possible relationship between IAB and IIICD irons (Kracher 1982, Choi et. al. 1995). These irons are all very high in Ni. Fuchs (1969) reported brianite in the low-Ni IAB Youndegin (6.80 wt % Ni), although compositional data were not given.

Silicate inclusions in San Cristobal examined by Benedix et al. (1998b) are subangular and incompletely rimmed by schreibersite. Veins of troilite and Fe,Ni metal cross-cut the inclusions. The silicates are equigranular and appear to be recrystallized. Silicates comprise ~20% of the sample and plagioclase comprises ~15% of the silicates. Minor graphite rims the inclusion and cohenite is observed within the metallic host. Brianite, first identified in San Cristobal by Scott and Bild (1974), is scattered

Table 8. Selected bulk chemical analyses of silicate-bearing IAB irons and winonaites.

	(1)	(2)	(2)	(2)	(3)			(4)	(5)	(6)	(6)	(7)	(8)	(9)
SiO_2	8.55	37.82	45.86	40.52	53.78	Na	mg/g	3.72	6.60	9.1	7.04	8.14	9.20	8.71
TiO_2	0.02	0.13	0.12	0.13	0.20	Mg	mg/g	146	181	158	208	169	149	163
Al_2O_3	0.53	1.36	3.41	1.64	4.13	Al	mg/g	7.0	12.3	14.2	12.4	13.2	14.0	14.2
Cr_2O_3	0.03	0.10	0.33	0.30	0.30	K	μg/g	337	627	750	606	733	--	--
FeO	0.49	3.95	4.94	3.33	3.17	Ca	mg/g	8.2	7.6	5.4	6.8	18	16	18
MnO	0.03	0.35	0.32	0.31	0.39	Sc	μg/g	8.10	10.4	5.8	8.3	12.3	12.1	11.8
MgO	5.19	28.63	28.27	31.94	31.58	V	μg/g	55	59	31	38	61	56	22
CaO	0.34	1.09	1.09	2.48	4.64	Cr	mg/g	1.95	2.68	2.55	1.46	2.09	1.86	0.95
Na_2O	0.29	0.52	1.23	0.47	1.74	Mn	mg/g	2.07	2.36	2.33	2.76	2.08	2.10	1.30
K_2O	0.02	0.04	0.11	0.06	0.07	Fe	mg/g	198	122	128	34.1	160	232	187
P_2O_5	--	0.39	0.26	0.03	n.d.	Co	μg/g	761	132	355	50.8	550	795	665
H_2O^+	--	1.13	0.85	0.76	n.d.	Ni	mg/g	12.1	3.3	4.23	0.73	12.9	13.0	18.7
H_2O^-	--	0.07	0.17	0.12	n.d.	Zn	μg/g	131	259	200	182	291	200	170
Fe (metal)	73.06	1.75	0.80	0.25	n.d.	Ga	μg/g	10.9	6.1	6.58	3.94	14.9	24.8	_7.84
Ni	9.08	0.18	0.27	0.10	n.d.	Ge	μg/g	37.5	5.6	30.7	3.46	32.2	90.7	15.7
Co	0.15	0.02	0.02	b.d.	n.d.	As	μg/g	2.57	0.45	n.d.	n.d.	n.d.	n.d.	n.d.
FeS	1.65	0.33	7.29	1.32	n.d.	Se	μg/g	11.2	15.9	n.d.	n.d.	n.d.	n.d.	n.d.
C	0.21	23.00	5.35	16.41	n.d.	Br	μg/g	0.27	0.56	n.d.	n.d.	n.d.	n.d.	n.d.
P	0.47	--	--	--	n.d.	Cd	ng/g	19	32	n.d.	n.d.	n.d.	n.d.	n.d.
						Sb	ng/g	121	47	n.d.	n.d.	n.d.	n.d.	n.d.
Total	100.11	100.86	100.69	100.17	100.00	La	ng/g	190	132	70	58	250	180	550
						Sm	ng/g	90	70	18	49	260	270	550
						Eu	ng/g	48	71	88	97	95	120	120
						Yb	ng/g	157	137	130	170	300	280	410
						Lu	ng/g	24	21	17	32	49	38	45
						Re	ng/g	99	4	n.d.	n.d.	n.d.	n.d.	n.d.
						Os	ng/g	1220	65	n.d.	n.d.	n.d.	n.d.	n.d.
						Ir	ng/g	1150	59	340	15.7	460	870	240
						Au	ng/g	265	44	86	12.9	185	310	210

(1) Woodbine whole rock, Jarosewich (1967), (2) Separated silicate inclusions from the El Taco mass of Campo del Cielo, Wlotzka and Jarosewich (1977), (3) Landes, Kracher (1974), (4) Tierra Blanca, Kallemeyn and Wasson (1985), (5) Udei Station, Kallemeyn and Wasson (1985), (6) Campo del Cielo silicate inclusions, Bild (1977), (7) Copiapo, Bild (1977), (8) Landes, Bild (1977), (9) Woodbine, Bild (1977)

throughout the silicates, but tends to be in contact with metal, either as veins within the silicates or at the edges of the inclusion.

McCoy et al. (1993) described inclusions in the Carlton and Dayton IIICD irons. These inclusions were extremely enriched in phosphates, in some cases comprising up to 70 vol % of the inclusion and forming the host into which the silicates were embedded. Chlorapatite is the dominant phosphate in Carlton, while Dayton contains abundant whitlockite, brianite and panethite. Silicates in these IIICD inclusions extend to more FeO-rich compositions than in IAB irons ($Fs_{11.6}$ in Dayton) and pyroxene compositions correlate with Ni concentration of the metallic host. In addition, plagioclase compositions are consistently less calcic in IIICD silicate-bearing inclusions ($An_{1.1-4.9}$) than in IAB inclusions ($An_{9.2-21.5}$).

Cooling rates

Cooling rates can constrain the physical setting of IAB irons during their solidification and subsolidus cooling. Scott (1982) argued that, in some cases, the cooling rate at the temperature of crystallization of an iron can be determined. If the parent taenite crystals have a dendritic texture, the distance between the dendrite limbs is proportional to the cooling rate at the time of solidification. The only IAB iron to which this method can be readily applied is Mundrabilla, which exhibits a weak dendritic pattern of the parent taenite crystals within the Fe,Ni-FeS intergrowth. Scott (1982) derived a cooling rate of ~5°C per annum, implying relatively rapid cooling at temperatures near the liquidus of Mundrabilla, about 1390°C as estimated using the Fe-S binary phase diagram from Ehlers (1972) and a S concentration of ~8 wt % given by Buchwald (1975). Using conventional methods of determining cooling rates, several authors (e.g. Herpfer et al. 1994, Yang et al. 1997a, Meibom, pers. comm. 1997) have found cooling rates of tens of °C/Ma for a number of IAB irons, implying a slow cooling environment at the time these irons cooled through ~500°C (see Table 4).

Table 9. Ages of silicate inclusions in IAB irons.

meteorite	K-^{40}Ar Ga (1)	^{39}Ar-^{40}Ar Ga (2)	I-Xe ΔMa* (3)	^{147}Sm-^{143}Nd Ga (4)
Caddo County				4.53 ± 0.02
Copiapo		4.45 ± 0.03	+1.36 ± 0.66	
Four Corners	4.49 ± 0.1			
Landes		4.43 ± 0.03	+2.61 ± 0.62	
Mundrabilla		4.52 ± 0.03	-0.68 ± 0.59	
Pitts		4.49 ± 0.03		
		4.52 ± 0.03		
Toluca	4.51 ± 0.1			
Woodbine		4.52 ± 0.03	-3.38 ± 0.30	

Sources: (1) Bogard et al. (1967); (2) Niemeyer (1979b); (3) Niemeyer (1979a), uncertain ages for Pitts silicates omitted; (4) Stewart et al. (1996). Ages from (1) and (2) corrected for new monitor age or decay constant by Herpfer et al. (1994).

*Age is in Ma relative to the Bjurböle chondrite with positive age indicating formation after Bjurböle.

Ages

Most ages for IAB irons are those of the silicate inclusions (Table 9), since it is these inclusions which can be readily dated by a number of isotopic systems (e.g. I-Xe, K-Ar, ^{39}Ar-^{40}Ar, ^{147}Sm-^{143}Nd). More recently, isotopic systems for direct dating of the metallic host have been developed (e.g. Re-Os), although the interpretation of internal isochrons (e.g. metal-schreibersite pairs, Shen et al. 1996) is not straightforward and may reflect a lengthy period of slow cooling. Absolute ages for the silicate inclusions range from 4.43 Ga to 4.53 Ga. The oldest age, measured in Caddo County, comes from the ^{147}Sm-^{143}Nd chronometer (Stewart et al. 1996), which closes at a relatively high temperature. Supporting evidence for early formation comes from I-Xe closure intervals relative to Bjurböle of -3.38 to +2.61 Ma (Niemeyer 1979a). Ages of 4.43 to 4.52 Ga are derived from the K-Ar and ^{39}Ar-^{40}Ar systems (Bogard et al. 1967, Niemeyer 1979b), which close at lower temperatures. These ages support the idea that partial melting, crystallization and metal-silicate mixing occurred very early in the history of the solar system, as suggested by a number of these authors.

Formation of the IAB and IIICD irons, and winonaites

In this section, we explore whether IAB and IIICD irons and the stony winonaites originate from a common parent body and, thus, reveal different aspects of the history of a common parent body. We then briefly review models for the origin of these meteorites.

A common parent body for IAB irons, IIICD irons and winonaites? It seems almost certain that the stony winonaites sample the same parent body as the IAB iron meteorites. Identical oxygen-isotopic compositions (Clayton and Mayeda 1996; and see Fig. 1) suggest a common oxygen-isotopic reservoir and, possibly, a common parent body. Mineralogies and mineral compositions are overlapping between winonaites and IAB silicate inclusions, particularly the angular silicate inclusions described above. In addition, textures are nearly identical between these two groups. While there are some differences (e.g. a number of silicate inclusion types are found in IAB irons which are not sampled as winonaites and vice versa), these differences probably result from unrepresentative sampling of these two populations, rather than real differences. Cosmic-ray exposure ages, which are commonly used to indicate sampling of meteorites by a common cratering event on a single parent body, are of little use given the considerable scatter in these ages in winonaites (Benedix et al. 1998a) and IAB irons (Voshage 1967).

It seems less clear whether IAB and IIICD irons sample a common parent body. Certainly, some features would seem to support a common parent body. Oxygen-isotopic compositions of silicate inclusions in IIICD irons (Clayton and Mayeda 1996) are essentially indistinguishable from those of the IAB irons and winonaites, again suggesting a common oxygen-isotopic reservoir, if not a common parent body. In addition, inclusions broadly similar in mineralogy to those in IIICD irons can be found among the IAB irons (McCoy et al. 1993, Yugami et al. 1997, Benedix et al. 1998b). However, important differences do exist. Most prominent among these are the differing trends on log-log plots of Ni vs. Ga, Ge and Ir for the metal phase, particularly at high Ni contents (see Fig. 3). Kracher (1982) suggested that these different trends could represent complementary fractional crystallization/partial melting trends on a common parent body. Choi et al (1995) suggested that no compositional hiatus exists between IAB and IIICD irons and, thus, they should be treated as a single group originating from a common parent body. While no hiatus exists at low Ni (<7.2 wt % Ni, Choi et al. 1995), at high Ni the groups are clearly distinguished, a surprising result if these indeed sample a common fractional crystallization/partial melting sequence. Differences exist within the silicate inclusions as well. Pyroxene compositions in IIICD silicate inclusions are correlated with Ni content in the host metal, a trend unlike that from IAB irons, and extend to higher Fs contents than do IAB irons (McCoy et al. 1993). In addition, plagioclase compositions are consistently more albitic than those in IAB irons (McCoy et al. 1993). In summary, the possibility that IAB and IIICD irons sample a common parent body cannot be excluded, but such a conclusion is probably premature. Further discovery of silicate-bearing IIICD irons may resolve this issue. It is also clear that even if IIICD irons sample a different parent body, they likely underwent many of the same processes as the IAB irons and winonaites.

Models for the formation of IAB irons and winonaites. While few theories have been postulated for the formation of the winonaites, several have been suggested for the IAB (and IIICD) irons. Wasson (1972) argued that elemental trends in the IAB irons were due to condensation of the metal directly from the nebula, giving rise to the term "non-magmatic." The major flaw of this hypothesis is that if the variation in elemental abundances is due to nebular processes, comparable ranges in the metal compositions should be found in chondritic material, but are not (Wasson et al. 1980). Another problem with this hypothesis is difficulty of forming parent taenite crystals tens of cms in size by condensation (Wasson et al. 1980).

Wasson et al. (1980) and Choi et al. (1995) have more recently championed a model which suggests IAB irons formed in localized impact melt pools within the chondritic megaregolith of an asteroid. These authors argue that impacts would selectively melt low-temperature fractions (Fe,Ni-FeS eutectic), which would migrate to form pools. Impacts occurred over a range of time and temperatures, producing the correlated variations between Ni and Ga, Ge and Ir. In an extreme view, each IAB iron represents an individual melt pool. While the pools would cool relatively quickly, trapping the unmelted, angular silicate inclusions, Choi et al. (1995) argued that both limited fractional crystallization and magma mixing would occur, producing both the high-Ni IAB irons and the scatter in Ga, Ge and Ir observed at high-Ni concentrations. Impact melting does provide a ready mechanism for mixing silicates and metal, but impact is probably incapable of producing copious quantities of Fe,Ni-FeS-rich melts by selective melting of these phases. Keil et al. (1997) summarize experimental and observational evidence that selective melting by impact occurs only locally and produces an extremely low percentage of melt. Any melt that is produced in this way quenches rapidly, making migration of these selective melts into larger pools difficult, if not impossible.

Several authors have proposed an alternative model in which IAB and IIICD irons formed as a result of partial melting and core formation (Kelly and Larimer 1977, Kracher 1982, 1985, McCoy et al. 1993). Kelly and Larimer (1977) argued that the composition of the metal was consistent with fractional melting of a single composition which was isolated from later melted material so that it would not reequilibrate. Wasson et al. (1980) marshaled several arguments against this model, the most compelling of which were the apparent contradictions between predicted and measured siderophile element partition coefficients and the unreasonably high temperatures required to form the last fractional melts. Kracher (1982, 1985) revived the partial melting model by suggesting that migration of the Fe,Ni-FeS eutectic melt could form a S-rich core at low temperatures. Thus, crystallization of both metal and sulfide determined the siderophile element trends observed. Choi et al. (1995) argued that crystallization of such a magma cannot produce the observed distribution of Ni concentrations. In fact, the system is probably far more complicated than envisioned by Kracher (1982, 1985), including large amounts of carbon and phosphorus, in addition to sulfur. Both McCoy et al. (1993) and Choi et al. (1995) pointed out that the partition coefficients of siderophile elements within such a complicated system remain essentially unknown. Finally, McCoy et al. (1993) argued that correlated trends between the properties of IIICD silicate-bearing inclusions (e.g. mineral compositions, modal mineralogies) and the Ni concentration of the metallic host are best explained by reaction between these two during a prolonged period of fractional crystallization of a common metallic magma. A serious flaw in these models is the inability to explain the mixing and retention of unmelted silicate clasts within the metallic core. This is particularly problematic in an asteroidal core, which crystallizes from the outside, thus armoring the inner core (Choi et al. 1995). Kracher (1982, 1985) argued that silicates may have been spalled into the core from the core-mantle boundary by impact-generated tectonic activity, although retention of the silicates remains problematic.

Takeda et al. (1994a) argued that IAB irons formed by partial melting and melt migration with heating by both ^{26}Al and impacts. Benedix et al. (1998b) extended this model. These authors argued against impact as a heat source for partial melting. Peak temperatures were insufficient to completely differentiate the body, although limited partial melting did occur. After the peak temperature was reached and cooling and crystallization had begun, a catastrophic impact occurred. The impact caused extensive mixing of metallic and silicate material of this body. Differing silicate lithologies were mixed to form winonaites and IAB irons were produced by co-mingling of silicate clasts and molten metal. After the impact, crystallization (creating high-Ni IAB irons), metamorphism and cooling occurred. Ages of 4.43 Ga to 4.53 Ga imply that this entire process occurred very early in the history of the solar system.

SILICATE-BEARING IIE IRONS

Despite including only eight silicate-bearing members, the IIE irons contain an enormous diversity of silicate inclusion types, ranging from silicate clasts with obvious chondrules, to elongate blebs of quenched basaltic melts. The existence of this stunning diversity within a single group has led to a range of models for the origin of IIE irons. Like the IAB irons discussed above, these models center on impact-generated melting and mixing (Wasson and Wang 1986, Olsen et al. 1994) and indigenous partial melting (McCoy 1995). IIE irons are also quite interesting because they are the only group of silicate-bearing irons for which young ages have been measured, thus lending additional support for the idea that impact played some role in the formation of these meteorites. In this section, we review the petrology and chronology of the silicate-bearing IIE irons and briefly discuss models for their origin.

Petrology and mineralogy

We divide the silicate-bearing IIE irons into five groups which adequately describe the range of inclusion types. Modal analyses and representative silicate compositions are gicen in Tables 10 and 11, respectively.

Netschaëvo. The host metal in Netschaëvo contains 8.6 wt % Ni, midway within the range for all IIE irons of 7.51 to 9.5 wt % Ni (Wasson and Wang 1986). Unfortunately, Netschaëvo was forged shortly after its recovery and this heating has altered much of its extraterrestrial record, although the exact extent of this alteration remains unknown. Netschaëvo is unique among silicate-bearing iron meteorites in containing angular clasts (Fig. 9) which are chondritic in the strictest sense. Bunch et al. (1970), Olsen and Jarosewich (1971), and Bild and Wasson (1977) noted that recrystallized chondrules typical of those found in type 6 ordinary chondrites are present within Netschaëvo. Mineralogically, these clasts are essentially identical to chondrites, containing olivine, orthopyroxene, sodic plagioclase and phosphates. Modally, Netschaëvo is richer in orthopyroxene and phosphates than typically found in ordinary chondrites (Table 10). Compositions of olivine ($Fa_{14.1}$) and orthopyroxene ($Fs_{13.6}$) are more reduced than found in H chondrites (Table 11) and the approximately equal Fa and Fs values suggest that reduction has played a role in the formation of these clasts. Rubin (1990) reported a metallographic cooling rate of ~3°C/Ma for metal within a silicate clast in Netschaëvo.

Table 10. Modal compositions (vol %) of silicate inclusions in IIE irons, excluding Fe,Ni metal and FeS.

	Colomera (1)	Elga (1)	Kodai-kanal (1)	Miles (2)*	Miles (2)**	Netsch-aëvo (3)	Techado (4)	Watson (5)	Weekeroo Station (1)
olivine	0	0	0	tr	0.6	26	81***	56	0
orthopyroxene	2-4	2-4	2-4	17.1	-	52		22	24
clinopyroxene	21-28	21-28	21-28	33.6	-	5	0	5	16
plag-trid-glass	67	68	73	49.3	91.1	14	19	15	59
phosphate	0.4-2	0.4-2	0.4-2	-	6.4	2	nd	1	0.4-2
yagiite	~3	-	-	-	-	-	-	-	-

*Average gabbroic inclusion
**Cryptocrystalline inclusion, also 0.6 vol% rutile and 0.3 vol% sodalite
***Includes both olivine and orthopyroxene
Sources of data: (1) Prinz et al. (1983b); (2) Ikeda and Prinz (1996); (3) Olsen and Jarosewich (1971); (4) McCoy, unpublished; (5) Olsen et al. (1994), the mode has been converted to vol% from the published wt % data.

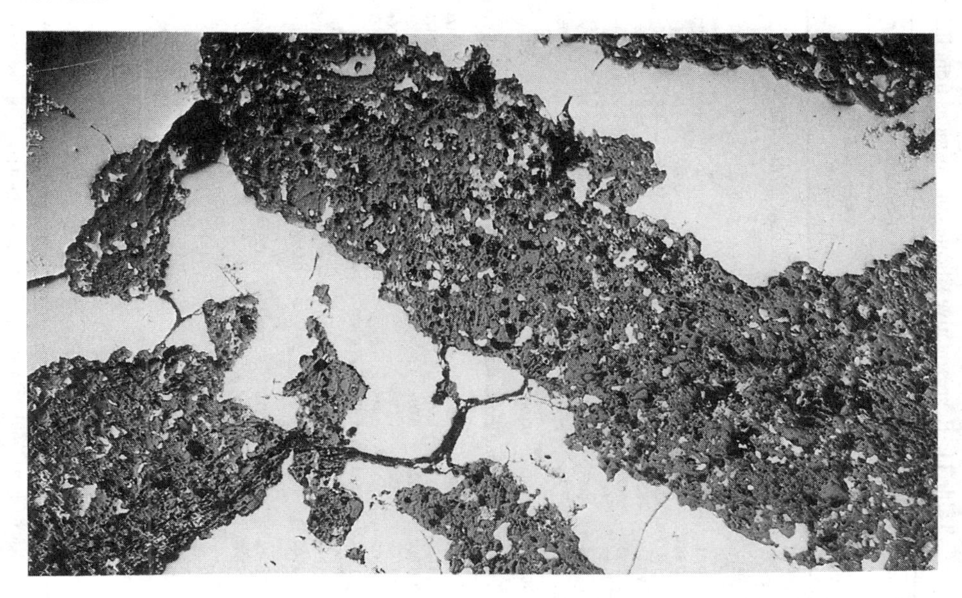

Figure 9. Photograph of angular, chondritic silicate inclusions in Netschaëvo. Field of view is 13 mm wide.

Techado. Techado contains 8.88 wt % Ni in its metallic host and, thus, also lies in the mid range of Ni concentrations for IIE irons. Casanova et al. (1995) described a single silicate inclusion ~1.5 cm in length from Techado. Like the inclusions in Netschaëvo, this inclusion is roughly chondritic in bulk composition, consisting of olivine, orthopyroxene, plagioclase, Fe,Ni metal and troilite. Although Casanova et al. (1995) suggested that the inclusion was unmelted, it has an unusual elongated shape (Fig. 10) that suggests that it might have been heated to temperatures sufficient to cause it to soften and become stretched. Unlike Netschaëvo, no relict chondrules were observed within the inclusion. Mineral compositions for olivine ($Fa_{16.4}$) and orthopyroxene

Table 11. Representative silicate compositions from IIE iron meteorites.

	OLIVINE		ORTHOPYROXENE				CLINOPYROXENE					PLAGIOCLASE						
	Netschaëvo	Watson	Netschaëvo	Watson	Weekeroo Station	Colomera	Netschaëvo	Watson	Weekeroo Station	Colomera		Netschaëvo	Watson		Weekeroo Station		Colormera	
	Chemical Composition (wt %)																	
SiO_2	40.0	39.3	56.3	56.4	55.1	55.2	55.3	54.2	52.7	52.1	51.0	65.5	67.1	67.1	67.9	67.4	67.1	65.3
Al_2O_3	b.d.	b.d.	0.22	0.51	0.30	0.06	0.70	0.74	0.96	1.18	4.1	20.8	19.0	18.3	19.6	18.9	19.9	18.4
Cr_2O_3	b.d.	0.06	0.24	0.71	0.25	0.24	n.d.	1.57	0.90	1.08	0.83	n.d.	b.d.	b.d.	n.d.	n.d.	n.d.	n.d.
TiO_2	b.d.	b.d.	0.22	0.16	0.30	1.33	n.d.	0.34	0.52	1.00	2.41	n.d.	b.d.	b.d.	n.d.	n.d.	n.d.	n.d.
FeO	13.5	18.7	9.1	11.2	13.7	14.8	3.5	5.58	11.7	8.2	4.6	0.40	0.61	0.29	0.27	0.23	0.17	0.17
MgO	46.0	40.4	31.8	28.6	28.5	27.1	17.0	16.4	15.7	15.6	14.6	n.d.	b.d.	b.d.	n.d.	n.d.	n.d.	n.d.
MnO	0.35	0.45	0.39	0.46	1.35	0.11	n.d.	0.30	0.75	0.38	0.39	n.d.	b.d.	b.d.	n.d.	n.d.	n.d.	n.d.
CaO	b.d.	0.04	0.75	1.88	1.47	1.07	21.9	19.1	17.2	19.3	21.0	2.96	0.43	0.28	2.33	1.91	0.36	0.05
Na_2O	b.d.	b.d.	0.06	0.09	b.d.	0.11	n.d.	0.84	b.d.	0.77	0.98	9.7	10.1	6.20	9.5	7.6	11.7	1.00
K_2O	n.d.	b.d.	b.d.	b.d.	b.d.	b.d.	n.d.	b.d.	b.d.	0.09	0.05	0.76	0.84	6.83	0.86	4.9	115	14.7
Total	99.85	98.95	99.08	100.01	100.97	100.02	98.40	99.07	99.43	99.70	99.96	100.12			100.46	100.94	100.38	99.62
	Atomic Formula (O = 4 for olivine, 6 for pyroxene, 8 for plagioclase)																	
Si	0.999	1.014	1.988	1.999	1.963	1.984	2.028	1.998	1.972	1.941	1.874	2.892	2.997	3.024	2.970	2.983	2.948	3.010
Al	-	-	0.008	0.021	0.012	0.003	0.030	0.032	0.042	0.054	0.176	1.083	1.000	0.972	1.010	0.994	1.031	0.997
Cr	-	0.001	0.008	0.020	0.009	0.007	-	0.046	0.027	0.037	0.026	-	-	-	-	-	-	-
Ti	-	-	0.006	0.004	0.008	0.036	-	0.009	0.015	0.024	0.066	-	-	-	-	-	-	-
Fe	0.282	0.404	0.269	0.332	0.408	0.445	0.107	0.172	0.335	0.255	0.141	0.016	0.023	0.011	0.010	0.005	0.006	0.005
Mg	1.712	1.555	1.674	1.511	1.519	1.452	0.930	0.901	0.876	0.866	0.799	-	-	-	-	-	-	-
Mn	0.007	0.010	0.013	0.014	0.041	0.003	-	0.009	0.024	0.011	0.013	-	-	-	-	-	-	-
Ca	-	0.001	0.028	0.071	0.056	0.041	0.861	0.755	0.690	0.770	0.825	0.141	0.021	0.014	0.109	0.090	0.017	0.003
Na	-	-	0.004	0.006	-	0.008	-	0.060	-	0.054	0.070	0.828	0.875	0.542	0.806	0.648	0.997	0.089
K	-	-	-	-	-	-	-	-	-	0.004	-	0.043	0.048	0.393	0.048	0.276	0.065	0.864
Total Cations	2.991	2.985	3.998	3.979	4.016	3.979	3.849	3.983	3.981	4.016	3.990	5.003	4.964	4.956	4.953	4.996	5.064	4.968
	Molar Mineral End Members and mg# (100(MgO/(MgO+FeO)))*																	
Fs	-	-	13.6	17.3	23.0	5.6	5.6	9.4	17.6	13.5	8.0	-	-	-	-	-	-	-
Wo	-	-	1.4	3.7	2.1	45.4	45.4	41.3	36.3	40.7	46.7	-	-	-	-	-	-	-
En	-	-	85.0	78.9	74.9	49.0	49.0	49.3	46.1	45.8	45.3	-	-	-	-	-	-	-
Ab	-	-	-	-	-	-	-	-	-	-	-	81.8	92.7	57.2	83.7	63.9	92.5	9.3
An	-	-	-	-	-	-	-	-	-	-	-	13.9	2.2	1.4	11.3	8.9	1.6	0.3
Or	-	-	-	-	-	-	-	-	-	-	-	4.3	5.1	41.4	5.0	27.2	5.9	90.4
mg#	85.9	79.4	86.2	82.0	78.8	76.5	89.6	84.0	70.5	77.2	85.0	-	-	-	-	-	-	

Data from Olsen et al. (1994) for Watson; Bunch and Olsen (1968) for Weekeroo Station and Colomera feldspars; all others from Bunch et al. (1970)

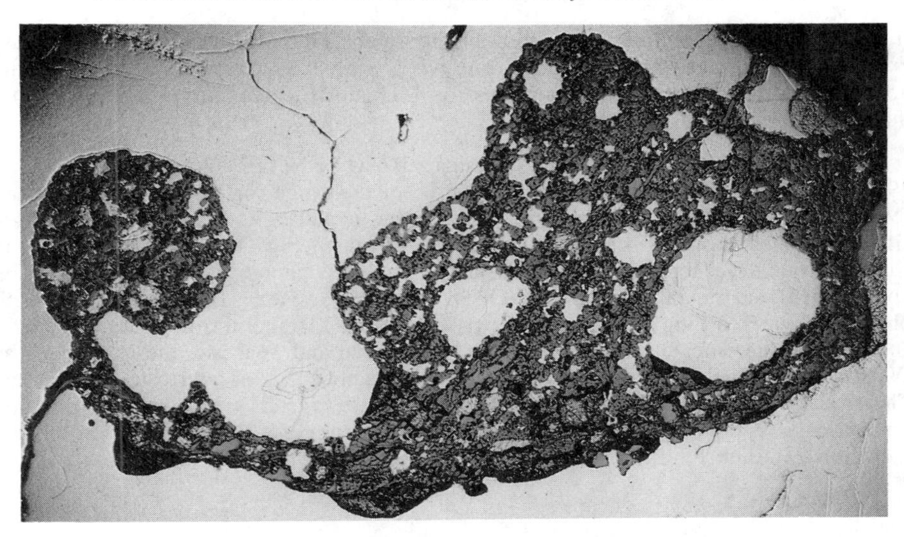

Figure 10. Photograph of elongate, curved silicate inclusion in Techado. The shape of this inclusions implies that it was heated to the point where it was ductile, but the bulk mineralogy is undifferentiated. Field of view is 13 mm wide.

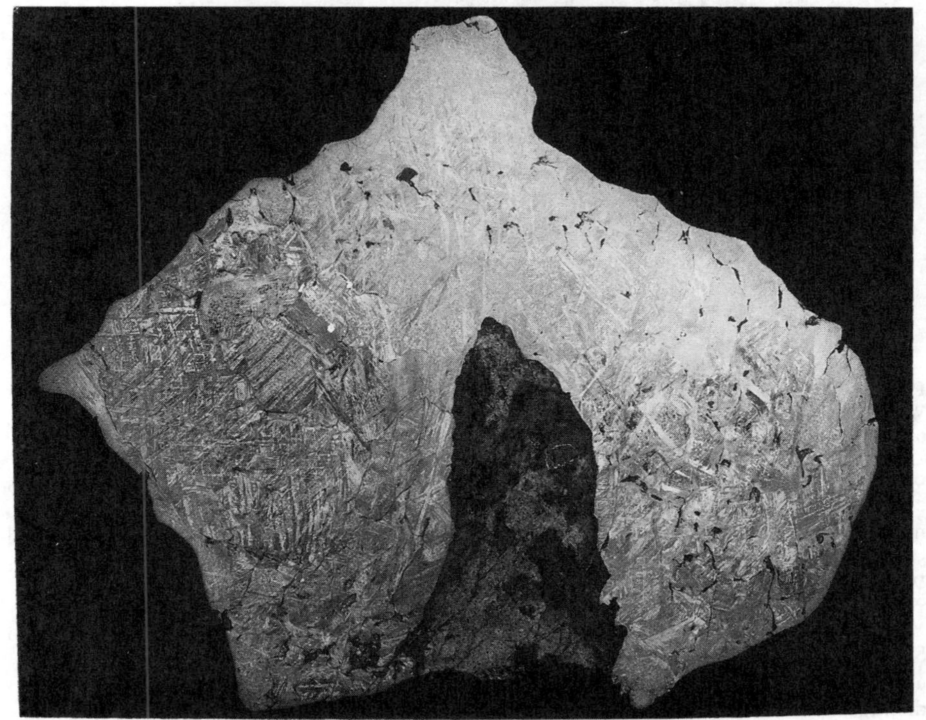

Figure 11. Photograph of a slab containing the large silicate inclusion in Watson. This inclusion is chondritic in silicate mineralogy, but essentially lacks metal and troilite. The silicate inclusion is ~10 cm in maximum dimension. (Photograph courtesy of the Smithsonian Institution).

($Fs_{15.3}Wo_{1.6}$) are slightly more FeO-poor than those observed in H chondrites. Plagioclase exhibits a small range of compositions from $An_{14.9-15.4}Or_{5.6-6.3}$. Casanova et al. (1995) did not report complete microprobe analyses for Techado silicates and, thus, they are not included in Table 11.

Watson. The Watson meteorite contains 8.07 wt % Ni (J.T. Wasson, pers. comm. 1995) within its metallic host, near the middle of the range for IIE irons. Olsen et al. (1994) reported a comprehensive study of Watson, including the discovery of a single silicate inclusion of ~30 cm^3 within the 93 kg mass (Fig. 11). Like inclusions in Netschaëvo and Techado, the Watson inclusion is roughly chondritic in bulk composition and a calculated modal mineralogy (Olsen et al. 1994, Table 10) contains 57 wt % olivine, 23 wt % orthopyroxene and 12 wt % feldspar. Metal and troilite are essentially absent from the silicate inclusion. Unlike Netschaëvo and Techado, the texture of the Watson silicate inclusion is decidedly igneous, not unlike that of a terrestrial peridotite. The silicate inclusion is dominated by orthopyroxene crystals which reach 1 mm and poikilitically enclose olivine crystals. This texture suggests that the Watson silicate inclusion experienced a high degree of melting. Mineral compositions (olivine $Fa_{20.6}$,

Figure 12. Photograph of a slab of Weekeroo Station exhibiting globular, silicate inclusions typical of Weekeroo Station, Miles, Colomera, Kodaikanal and Elga. Silicate inclusions are composed of plagioclase, orthopyroxene and clinopyroxene in a ratio of ~2:1:1. (Photograph courtesy of the Smithsonian Institution).

orthopyroxene $Fs_{17.6}Wo_{3.8}$) are within the range for H chondrites. Plagioclase is composed of an albitic host with K feldspar exsolution (antiperthite). Olsen et al. (1994) noted that shock-melted pockets are found in the metallic host and McCoy (1995) found shearing and shock deformation within the silicate inclusions, all of which indicate that Watson experienced a relatively severe post-formational shock event.

Weekeroo Station and Miles. The remaining silicate-bearing IIE irons (Weekeroo Station, Miles, Colomera, Kodaikanal and Elga) all contain what Prinz et al. (1983b) aptly described as globular silicate inclusions. Macroscopically, there is little to distinguish the silicate inclusions between these five meteorites and Figure 12 illustrates silicate inclusions in Weekeroo Station. Modal analyses (Table 10), however, indicate that Weekeroo Station and Miles mostly contain inclusions composed of orthopyroxene, clinopyroxene and plagioclase, while inclusions in Colomera, Kodaikanal and Elga consist almost exclusively of clinopyroxene and plagioclase with only minor orthopyroxene. For this reason, we discuss Weekeroo Station and Miles first.

Weekeroo Station (7.51 wt % Ni) is the lowest-Ni member of group IIE and Miles (7.96 wt % Ni) is one of the lowest-Ni members. Silicate inclusions in Weekeroo Station have been studied by Bunch and Olsen (1968), Bunch et al. (1970), Olsen and Jarosewich (1970), Prinz et al. (1983b), and McCoy (1995), and Miles silicate inclusions have been studied by McCoy (1995), Ikeda and Prinz (1996), Ikeda et al. (1997) and Ebihara et al. (1997). Both of these meteorites contain subequal amounts of plagioclase and pyroxene, with the pyroxene being comprised of subequal amounts of orthopyroxene and clinopyroxene. Most silicate inclusions in Miles are coarse-grained gabbros, but fine-grained, cryptocrystalline inclusions with a variey of mineral assemblages are also present (Ikeda and Prinz 1996, Ikeda et al. 1997). Weekeroo Station contains a mixture of coarse-grained pyroxene-plagioclase inclusions and inclusions with coarse pyroxene and radiating fine-grained plagioclase-tridymite mixtures. Bunch et al. (1970) reported a composition of Fs_{22} for orthopyroxene in Weekeroo Station. Miles also has a small range of orthopyroxene compositions from $Fs_{19.9\text{-}23.2}$ (McCoy, unpublished, see also Ikeda and Prinz 1996, Fig. 2 and Ikeda et al. 1997, Fig. 2). These compositions are considerably more FeO-rich than observed in the two IIE irons with "primitive" silicates, Netschaëvo and Techado. As is the case in most of the globular silicates, both plagioclase feldspar and potassium feldspar are present in both Weekeroo Station and Miles (Table 10). Pyroxene and plagioclase in Miles and Weekeroo Station exhibit shock deformation features indicative of mild shock.

Colomera, Kodaikanal and Elga. The other subgroup of globular silicate inclusions are those in Colomera (7.86 wt % Ni), Kodaikanal (8.71 wt % Ni) and Elga (8.25 wt % Ni). Prinz et al. (1980) examined silicate inclusions in all three meteorites, while Bunch and Olsen (1968) and McCoy (1995) examined Colomera and Kodaikanal. Elga was extensively studied by Osadchii et al. (1981), Bence and Burnett (1969) studied Kodaikanal, and Wasserburg et al. (1968) studied Colomera. Silicate inclusions in these meteorites are dominated by glass of plagioclase-tridymite composition and clinopyroxene in a ratio of ~2:1. Wasserburg et al. (1968) identified one sanidine crystal of 11 cm in length. Minor orthopyroxene, olivine, and phosphate are also found. Yagiite, $(K,Na)_2(Mg,Al)_5(Si,Al)_{12}O_{30}$, comprises ~3% of the silicate inclusions in Colomera (Table 10). Texturally, silicate inclusions in these meteorites are similar to, but more diverse, than those in Weekeroo Station and Miles. Buchwald (1975) and McCoy (1995) noted that in addition to the coarse-grained inclusions and those with radiating, fine-grained intergrowths of plagioclase and tridymite, glassy inclusions are also present and that these inclusions occur well beneath the heat-altered zone formed during atmospheric entry of the meteorite. McCoy (1995) observed coarse-grained and glassy silicate inclusions occurring within a few mm of one another. Mineral compositions within this group are somewhat more diverse, ranging from $Fa_{15\text{-}16}$ in Elga (Osadchii et al. 1981) and Fa_{23} in Colomera (Bunch et al. 1970). Clinopyroxene exhibits a range of compositions within both Colomera and Kodaikanal (Table 11; Bunch et al. 1970) and both plagioclase and potassium feldspar are present in these meteorites (Table 10).

Composition

Few analyses of silicate inclusions from IIE irons have been done. Bulk major element analysis of a composite of 12 Weekeroo Station inclusions was presented by Olsen and Jarosewich (1970), major and trace element data were given by Olsen et al. (1994) for an inclusion from Watson, and major and trace element data were given for 6 gabbroic and 3 cryptocrystalline inclusions from Miles by Ebihara et al. (1997). The composite of Weekeroo Station inclusions is unlike any chondrite in composition, although it is similar to bulk chondritic composition minus 50% olivine (Olsen and Jarosewich 1970). The Watson inclusion has a bulk composition like that of an H chondrite, minus most of the metal and troilite (Olsen et al. 1994). The Miles inclusions show fractionated lithophile element abundances, with the cryptocrystalline clasts showing the most extreme fractionations (Ebihara et al. 1997). The plagiophile elements Na, Al and K are enriched, as are the incompatible elements Ti and Hf, while Mg is depleted relative to H chondrites. The REE abundances are generally elevated over CI values, although some of the cryptocrystalline clasts have LREE depletions. These clasts clearly are not chondritic in composition, but neither are they what one might expect for a partial melt of a chondritic source (Ebihara et al. 1997).

Chronology

A particularly intriguing feature of the silicate-bearing IIE irons is the broad range of ages measured for different meteorites (Table 12). IIE irons can be divided into an "old" subgroup consisting of Techado, Miles, Weekeroo Station and Colomera and a "young" subgroup which includes Netschaëvo, Watson and Kodaikanal. Ages for a single meteorite from a variety of radiometric techniques generally agree with these broad groupings, although the spread in ages for Weekeroo Station ranges by as much as 200 Ma (Table 12). In addition to the ages given in Table 12, Burnett and Wasserburg (1967a) argued that the relatively low initial $^{87}Sr/^{86}Sr$ ratio of Kodaikanal for the very high Rb/Sr requires a late Rb/Sr fractionation, rather than simple metamorphic equilibration.

Table 12. Measured ages for silicate inclusions in IIE iron meteorites.

	^{39}Ar-^{40}Ar Ga	K-Ar Ga	Rb-Sr Ga	I-Xe ΔMa	Pb-Pb Ga
Colomera		4.24	4.51		
Kodaikanal		3.5±0.1	3.7±0.1		3.676±0.003
Miles	4.41±0.01				
Netschaëvo	3.75±0.03				
Techado	4.48±0.03				
Watson	3.656±0.005	~3.5			
Weekeroo Station	4.49±0.03		4.28, 4.39±0.07	+10 Ma	

Sources of ages: ^{39}Ar-^{40}Ar - Miles, Techado, Watson (Garrison and Bogard, 1995), Netschaëvo, Weekeroo Station (Niemeyer, 1980) (using corrected age for the St. Severin monitor); K-Ar - Colomera (D.D. Bogard, unpublished data), Kodaikanal (Bogard et al., 1969), Watson (Olsen et al., 1994); Rb-Sr - Colomera (Sanz et al., 1970), Kodaikanal (Burnett and Wasserburg, 1967b), Weekeroo Station (Burnett and Wasserburg, 1967b; Evensen et al., 1979) (all recalculated using $\lambda = 1.420 \times 10^{-11}$); I-Xe - (Niemeyer, 1980) (age in Ma relative to the Bjurböle chondrite, positive age indicates formation after Bjurböle); Pb-Pb - (Göpel et al., 1985).

Origin

A detailed discussion of the origin of IIE irons is beyond the scope of this review. Many of the arguments concerning a near-surface impact melt origin vs. a deep-seated indigenous origin are similar to those of other silicate-bearing irons. In particular, the siderophile element trends are difficult to explain by fractional crystallization of a largely metallic core and a core setting provides no ready mechanism for mixing silicates into the molten metal (Wasson et al. 1980). In support of a core origin is the difficulty of producing metallic metal melts from localized impact melt events and the slow metallographic cooling rates which appear to require deep burial (McCoy 1995). However, some features of silicate-bearing IIE irons are unique and it is these features which are worthy of further consideration.

Diversity of silicate inclusions types. Silicate inclusions in IIE irons are the most diverse in all silicate-bearing irons. These inclusion types can be arranged from most "primitive" to most "differentiated." Such a sequence would include five types: (1) Chondritic silicate inclusions in Netschaëvo, which are true chondritic clasts within a metallic matrix, (2) Partially melted, but undifferentiated clasts found in Techado, exhibiting evidence of partial melting in the overall outline of the inclusion, but maintaining an approximately chondritic mineralogy, (3) Totally melted inclusions which have lost metal and troilite, of the type observed in Watson, (4) Plagioclase–orthopyroxene–clinopyroxene "basaltic" partial melts of the type found in Weekeroo Station and Miles, which contain mineral abundances expected of a simple partial melt from a chondritic source, and (5) Plagioclase-clinopyroxene partial melts, such as those in Colomera, Kodaikanal, and Elga, which appear to be more differentiated than those in Weekeroo Station and Miles. There appears to be no correlation between the degree of evolution of the silicate inclusions and the Ni concentration in the host metal, which increases during fractional crystallization. Thus, the evolution of the silicates and metal does not appear to have progressed systematically. There are no clasts whose mineralogy would indicate that they are residues from partial melting of chondritic material.

Precursor chondritic material. A number of authors have argued that IIE irons represent impact melting at the surface of the H chondrite parent body. Olsen et al. (1994) pointed out the numerous similarities between H chondrites and Watson, including silicate mineral and bulk composition, oxygen-isotopic composition, and cosmic-ray exposure ages. Casanova et al. (1995) echoed this sentiment in their examination of a silicate inclusion in Techado. Clayton and Mayeda (1996) pointed out the similarities in oxygen-isotopic composition between H chondrites and IIE irons. The question of the link between IIE irons and H chondrites cannot be easily resolved, but IIE irons and H chondrites may not be as similar as previously suggested. In particular, oxygen-isotopic compositions differ between the two groups, $\Delta^{17}O = 0.73 \pm 0.09$ for H chondrites and 0.59 ± 0.05 for IIE irons, although there is some overlap. In addition, "primitive" IIE irons whose mineral compositions are unchanged by igneous processes are FeO-poor compared to H chondrites. However, the Sr isotopic results of Kodaikanal argue for late Rb/Sr fractionation (Burnett and Wasserburg 1967a), which could be due to shock melting and redistribution. The role of reduction in the formation of Netschaëvo and Techado is not fully established. These observations suggest that it is premature to declare a definitive link between IIE irons and H chondrites.

Age interpretation. Probably the most intriguing question in the genesis of IIE irons is the interpretation of the "young" ages for some silicate-bearing members. If these young ages record the formation of these rocks, this would provide definitive evidence for an impact-melt origin, since any indigenous heat source would have dissipated by 3.8 Ga in an asteroidal-size body.

We argue that the young ages do not represent the formation of these rocks, although some late melting and differentiation may have occurred. There is no correlation between the degree of differentiation experienced by the silicate inclusions and their ages. Thus, one cannot argue for a progression of increasing silicate differentiation with time. An alternative to the young formation scenario is that the ages were reset by shock events. Clearly, there is abundant evidence for post-formational shock in the form of deformation features within the silicate minerals. The real question is whether melting may have occurred at 3.8 Ga related to these shock events. We find the most compelling evidence for this in the form of the diversity of inclusion textures present within a single silicate-bearing IIE iron. Colomera contains coarse-grained igneous inclusions, fine-grained inclusions with radiating plagioclase-tridymite intergrowths, and glassy inclusions. This diversity clearly did not result from a single cooling event. Instead, we suggest that shock waves were focused into the inclusions, a feature commonly seen in the metallic hosts of some IIE irons (e.g. Watson; Olsen et al 1994). Intense, localized shock melting occurred within these inclusions and rapid cooling followed. Thus, silicate inclusions in some IIE irons record both early, slow cooling and later, rapid cooling following an intense shock event. An interesting consequence of this diversity is that different inclusions may have had dramatically different histories and yield different ages. Petrographic characterization of age-dated inclusions may prove absolutely essential in interpreting IIE iron ages and origins.

PALLASITES

Pallasites are stony irons composed of roughly equal amounts of silicate, dominated by olivine, and metal plus troilite. There are currently three separate pallasite types; the main-group pallasites, the Eagle Station grouplet, and the pyroxene-pallasite grouplet. The Eagle Station and pyroxene-pallasite grouplets contain 3 and 2 members, respectively, while the main-group contains approximately 41 members. Uncertainties in pairings makes precise census of main-group pallasites difficult. The three pallasite groups are distinguished from each other by differences in silicate mineralogy and composition, metal composition and O-isotopic composition. Figure 13 shows the O-isotopic compositions of the different pallasite groups, the most diagnostic characteristic, compared to meteorite groups in nearby O-isotope space. Note that the main-group pallasites are in the oxygen-isotope cluster occupied by IIIAB irons, mesosiderites, howardites, eucrites, diogenites, angrites and brachinites (Figs. 1d, 13).

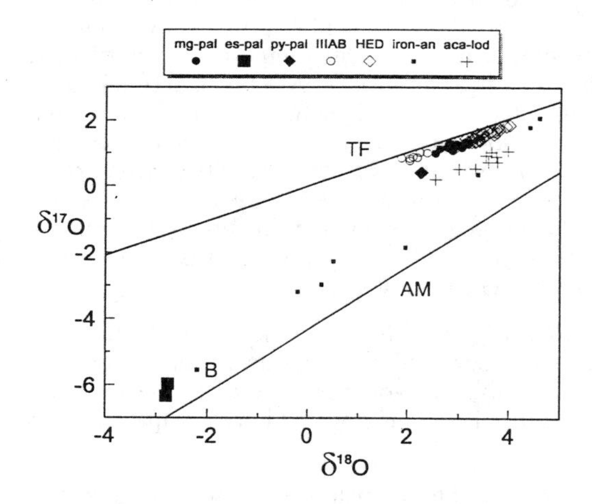

Figure 13. Oxygen-isotopic compositions of pallasites compared to other achondrites. Main-group pallasites (mg-pal) have O-isotopic compositions within the same O-isotope group as oxide phases in IIIAB irons, howardites, eucrites and diogenites (HED) and mesosiderites and angrites, which are not shown for clarity. The pyroxene-pallasites (py-pal) stand alone in O-isotope space, although an acapulcoite and a lodranite (aca-lod) plot nearby. The Eagle Station pallasites (es-pal) occupy a distinct region near the Allende mixing line (AM), and near silicates from the anomalous iron (iron-an) Bocaiuva (B). All data from Clayton and Mayeda (1996).

Individual olivine grains in pallasites can be on the order of a cm in size, and clusters of grains can be several cm in size (e.g. see Ulff-Møller et al. 1998). Olivine can be heterogeneously distributed. An extreme example is Brenham. Different specimens of Brenham can contain large regions of metal devoid of olivine with a few patches of typical pallasite material here and there (Fig. 14). This makes determination of modes

Figure 14. A polished and etched slab of the Brenham main-group pallasite display the very heterogeneous nature of some specimens of this pallasite. The specimen is 13 cm in maximum dimension. (Courtesy of the Smithsonian Institution).

difficult and fraught with uncertainties. The most reliable modal data obtained from polished slabs show that olivine varies from about 35-85 vol %, with chromite and phosphates generally <1 vol % each (Buseck 1977, Ulff-Møller et al. 1998). The olivine/metal weight ratio of pallasites ranges from 0.3-2.5. The macroscopic olivine-metal textures are also variable, with olivine grains varying from highly angular, fragmental shapes (Fig. 15) to well rounded grains (Fig. 16) (Buseck 1977, Scott 1977d). On a microscopic scale, even the most angular grains show evidence of rounding of the corners (Scott 1977d).

Main-group pallasites

The main-group pallasite silicates are composed dominantly of olivine, with minor amounts of low-Ca pyroxene, chromite and several different phosphates. Most main-group pallasites contain olivine of ~$Fo_{88\pm1}$ composition, but a few have anomalously ferroan olivines, down to Fo_{82} for Phillips County and Springwater (Fig. 17). Representative olivine compositions are given in Table 13. Few modern analyses of pallasite olivines are available in the literature, and most available data are from the partial analyses of Buseck and Goldstein (1969). Modern analyses for olivines from a few pallasite can be found in Davis and Olsen (1991), Mittlefehldt (1980), Righter et al. (1990) and Yanai and Kojima (1995). In addition to normal olivines, some pallasites also contain a minor amount of phosphoran olivine with 4 to 5 wt % P_2O_5 (Table 13) in zones continuous with normal olivine (Buseck 1977).

Main-group pallasite olivines contain very low concentrations of the trace transition elements (Table 14). The Sc contents vary from 0.5 to 2.4 μg/g, Cr from 160 to 600 μg/g, Mn from 1.5 to 3.2 mg/g and Zn from 5 to 8 μg/g (Davis 1977, Mittlefehldt 1980, and unpublished). The siderophile elements Co and Ni are generally at low concentrations. Nearly all modern Co analyses and most Ni analyses have been done by a bulk technique (INAA—Davis 1977, Mittlefehldt 1980 and unpublished), so contamination from the

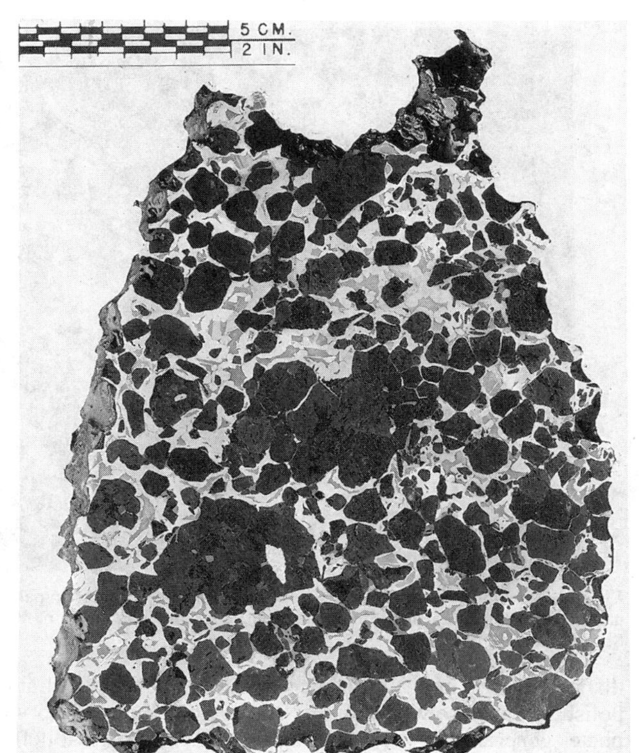

Figure 15. A polished and etched slab of the Salta main-group pallasite showing roughly cm-sized angular-subangular olivines dispersed in metal, and two larger olivine masses, one of which is veined by metal as though it was in the process of breaking up when the metal crystallized. (Photo courtesy of the Smithsonian Institution).

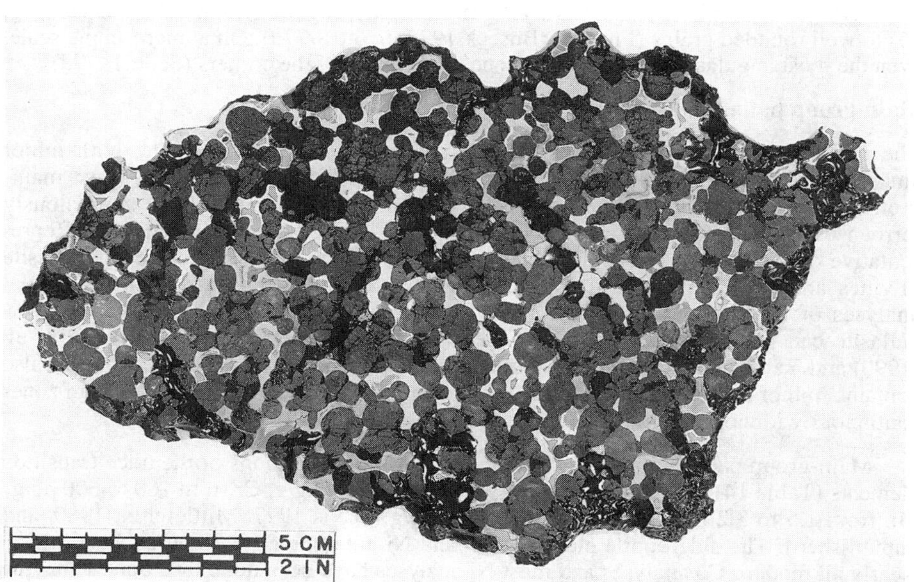

Figure 16. A polished and etched slab of the Thiel Mountains main-group pallasite showing roughly cm-sized, well rounded olivines dispersed in metal. (Photograph courtesy of the Smithsonian Institution).

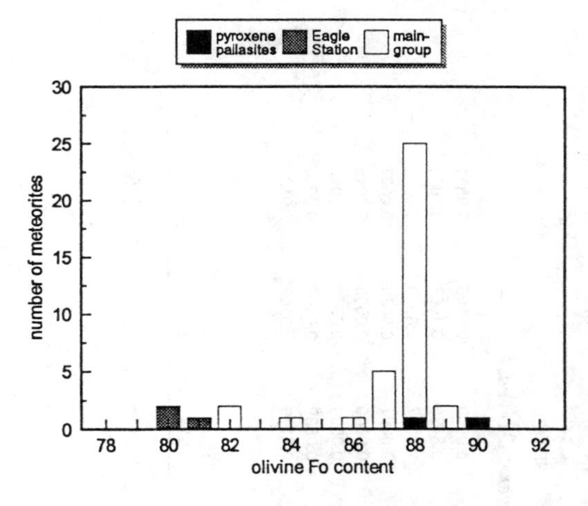

Figure 17. Histogram of Fo contents of pallasite olivines. Main-group pallasites have a narrow range in olivine compositions, except for three ferroan outliers. Data are from Boesenberg et al. (1995), Buseck (1977), Buseck and Goldstein (1969), Mittlefehldt (unpublished), Yanai and Kojima (1995).

metal phase is a worry. Main-group pallasite metal contains 660 to 900 times the Co and 4200 to 6400 times the Ni measured in most olivines. For a few pallasites, analyses of Ni in olivine have been done by ion microprobe (Reed et al. 1979) and these data agree well with the INAA bulk data (Mittlefehldt, unpublished), except that the latter are systematically lower than the ion probe data. Nickel in pallasite olivines shows a depletion in rims compared to cores, and in small compared to large grains (Reed et al. 1979). The systematically lower Ni content in the bulk analyses compared to the olivine core analyses reported by (Reed et al. 1979) are therefore understandable. The bulk analyses yield means of 7.4 μg/g Co and 18 μg/g Ni for main-group pallasite olivines.

Major and minor element contents are zoned in main-group pallasite olivines. Zhou and Steele (1993) showed that the count rates from electron microprobe analyses for Al, Cr and Ca decreased toward the rims of Springwater olivines, while that for Mn increased. Zhou and Steele (1993) did not report concentration profiles, only "normalized counts", but they estimated concentrations: Mn—2.75 mg/g; Cr—140 μg/g; Al—10 μg/g; Ca—64 μg/g. Presumably, these are for homogeneous cores. Miyamoto and Takeda (1994) similarly showed decreasing concentrations of CaO and Cr_2O_3, and Fa content for Esquel olivines toward the rim. The ion probe results of Hsu et al. (1997) also show that Cr decreases and Mn increases from the center to the rim for Brenham olivines.

Pyroxene occurs as a trace component in at least some main-group pallasites (Buseck 1977). The grains are typically only a few microns in size, and occur in symplectic intergrowths on the margins of olivine grains with one or more of the phases troilite, chromite, kamacite and phosphate. There are few analyses of pallasite pyroxenes; representative data are given in Table 15. The most remarkable feature of these pyroxenes is their very low Ca contents; 0.02 to 0.28 wt % CaO (Buseck 1977).

The phosphate mineralogy of main-group pallasites is complex. At least three phosphates are known to be native to the pallasites, farringtonite, stanfieldite and whitlockite, and a suite of poorly characterized phosphates believed to be terrestrial alteration produces are also present (Buseck and Holdsworth 1977). Representative farringtonite, stanfieldite and whitlockite analyses are given in Table 16. Phosphates must be the major repository for large ion lithophile elements in pallasites. The Springwater pallasite is an anomalous main-group pallasite based on olivine composition (Table 13), and the greater abundance of phosphates it contains (Buseck 1977). Ion probe analyses of all three phosphates from Springwater (Davis and Olsen 1991) show that farringtonite

Table 13. Compositions of olivine grains from representative pallasites.

	main-group							Eagle Station grouplet	pyroxene-pallasite grouplet
	Brahin	Giroux	Krasno-jarsk	Mount Vernon	Rawlinna	Spring-water	Spring-water	Eagle Station	Y-8451
Chemical Composition (wt %)									
SiO_2	36.5	40.8	40.0	40.4	35.9	39.0	35.5	38.7	40.2
FeO	13.7	10.7	11.8	12.8	14.3	17.2	16.2	19.0	10.2
MgO	43.2	48.9	48.0	47.5	44.8	43.9	43.0	42.2	49.1
MnO	0.30	0.25	0.19	0.29	0.26	0.33	0.45	0.18	0.37
Cr_2O_3	nd	0.052	0.041	0.049	nd	0.021	nd	0.031	0.08
CaO	nd	0.017	0.011	0.014	nd	0.004	nd	0.070	0.05
P_2O_5	3.82	nd	nd	nd	4.91	nd	4.14	nd	nd
Total	93.7$_{0}$	100.7$_{19}$	100.0$_{42}$	101.0$_{53}$	95.2$_{6}$	100.4$_{55}$	95.1$_{5}$	100.1$_{81}$	100.0$_{0}$
Cation Formula Based on 4 Oxygens (Ideal Olivine = 3 Cations per 4 Oxygens)									
Si	0.9289	0.9968	0.9899	0.9938	0.8912	0.9866	0.9000	0.9895	0.9891
Fe	0.2916	0.2186	0.2442	0.2633	0.2969	0.3639	0.3435	0.4063	0.2099
Mg	1.6384	1.7806	1.7704	1.7413	1.6574	1.6551	1.6247	1.6080	1.8005
Mn	0.0065	0.0052	0.0040	0.0060	0.0055	0.0071	0.0097	0.0039	0.0077
Cr	0.0000	0.0010	0.0008	0.0010	0.0000	0.0004	0.0000	0.0006	0.0016
Ca	0.0000	0.0004	0.0003	0.0004	0.0000	0.0001	0.0000	0.0019	0.0013
P	0.0823	0.0000	0.0000	0.0000	0.1032	0.0000	0.0889	0.0000	0.0000
Total Cations	2.8654	3.0026	3.0096	3.0058	2.8510	3.0132	2.8779	3.0102	3.0101
*Cation Ratios Fe/Mn and mg# (100*Mg/(Mg+Fe))*									
Fe/Mn	45	42	61	44	54	51	36	104	27
mg#	84.9	89.1	87.9	86.9	84.8	82.0	82.5	79.8	89.6

Phosphoran olivine analyses from Buseck (1977), Y-8451 from Yanai and Kojima (1995), all other analyses from Mittlefehldt (unpublished). nd = not determined.

Table 14. Trace element contents of olivine from representative pallasites.

	main-group						Eagle Station grouplet	
	Ahumada	Giroux	Glorieta Mountain	Krasno-jarsk	Mount Vernon	Spring-water	Cold Bay	Eagle Station
	Trace element contents in μg/g except Mn in mg/g							
Sc	1.86	1.09	2.29	1.23	1.65	0.54	2.55	2.66
Cr	595	312	280	228	306	169	289	404
Mn	2.2*	2.14	2.04	1.68	2.76	3.18	1.4*	1.34
Co	9.41	6.40	8.19	7.03	7.22	8.84	19.3	26.0
Ni	40.0	18.3	20.8	14.5	19.6	20.4	75.5	45.1
Zn	6.4	7.0	5.9	7.9	5.9	7.7	1.4	<2.4

Scandium, Cr, Co, Ni and Zn determined by INAA by Mittlefehldt (unpublished); Mn determined by INAA by Davis (1977), except * by electron microprobe analysis by Mittlefehldt (unpublished). Zinc in Eagle Station is a 2σ upper limit.

Table 15. Compositions of pyroxene grains from representative pallasites.

	main-group			pyroxene-pallasite grouplet Y-8451		
	Ahumada	Glorieta Mountain	Spring-water	(1)	(2)	(3)
	Chemical Composition (wt %)					
SiO_2	58.4	58.4	57.8	57.0	57.2	54.9
Al_2O_3	nd	nd	nd	0.13	0.26	0.29
TiO_2	nd	nd	nd	0.03	0.04	0.06
Cr_2O_3	nd	nd	nd	0.62	0.70	1.24
FeO	7.44	7.89	11.4	6.02	6.20	3.06
MnO	0.28	0.32	0.34	0.35	0.41	0.24
MgO	34.2	34.2	32.1	35.6	34.3	18.3
CaO	0.20	0.18	0.04	0.34	1.06	21.5
Na_2O	nd	nd	nd	0.04	0.05	0.63
Total	100.52	100.99	101.68	100.13	100.22	100.22
	Cation Formula Based on 6 Oxygens (Ideal Pyroxene = 4 Cations per 6 Oxygens)					
Si	2.0087	2.0042	2.0011	1.9678	1.9771	1.9896
^{IV}Al	0.0000	0.0000	0.0000	0.0053	0.0106	0.0104
Total tet*	2.0087	2.0042	2.0011	1.9731	1.9877	2.0000
Ti	0.0000	0.0000	0.0000	0.0008	0.0010	0.0016
^{VI}Al	0.0000	0.0000	0.0000	0.0000	0.0000	0.0020
Cr	0.0000	0.0000	0.0000	0.0169	0.0191	0.0355
Fe	0.2140	0.2265	0.3301	0.1738	0.1792	0.0927
Mn	0.0082	0.0093	0.0100	0.0102	0.0120	0.0074
Mg	1.7531	1.7492	1.6563	1.8316	1.7669	0.9884
Ca	0.0074	0.0066	0.0015	0.0126	0.0393	0.8349
Na	0.0000	0.0000	0.0000	0.0027	0.0034	0.0443
Total Cations	3.9914	3.9958	3.9990	4.0217	4.0086	4.0068
	*Cation Ratios Ca:Mg:Fe, Fe/Mn and mg# (100*Mg/(Mg+Fe))*					
Ca	0.4	0.3	0.1	0.6	2	43.6
Mg	88.8	88.2	83.3	90.8	89	51.6
Fe	10.8	11.4	16.6	8.6	9	4.8
Fe/Mn	26	24	33	17	15	13
mg#	89.1	88.5	83.4	91.3	90.8	91.4

Ahumada, Glorieta Mountain and Springwater analyses by Buseck (1977); Y-8451 analyses by Yanai and Kojima (1995). Y-8451 pyroxenes analyses are; (1) polysynthetically twinned coarse orthopyroxene, (2) non-polysynthetically twinned orthopyroxene inclusion in olivine, (3) clinopyroxene inclusion in orthopyroxene. nd = not determined.

Table 16. Compositions of phosphates from representative pallasites.

	WHITLOCKITE						STANFIELDITE				FARRINGTONITE	
	Ahumada	Ollague	Rawlinna	Spring-water	Cold Bay	Eagle Station	Ollague	Rawlinna	Spring-water	Eagle Station	Rawlinna	Spring-water
	Chemical Composition (wt %)											
CaO	48.2	49.0	43.8	47.30	45.4	48.15	26.4	24.4	25.58	25.61	0.07	0.077
P_2O_5	46.0	46.5	42.8	46.92	43.5	47.27	48.6	46.0	49.34	50.28	52.2	53.46
FeO	1.69	0.28	5.19	0.913	2.11	0.608	2.60	6.02	4.43	4.73	5.67	4.09
MgO	3.91	3.82	4.48	3.61	6.43	3.59	21.50	21.8	20.86	20.88	40.5	42.84
MnO	nd	0.10	nd	0.027	nd	0.025	0.41	0.32	0.488	0.301	0.14	0.225
Na_2O	0.20	0.90	2.96	2.50	1.08	1.55	nd	nd	nd	nd	nd	nd
K_2O	nd	0.06	0.11	0.070	nd	0.002	nd	nd	0.001	0.003	nd	0.002
SiO_2	0.21	0.10	0.88	0.026	1.94	0.022	nd	1.29	0.030	0.026	0.05	0.022
TiO_2	na	na	na	na	na	na	na	na	na	na	na	na
Al_2O_3	na	na	na	0.003	na	0.002	na	na	0.002	0.002	na	0.003
Cr_2O_3	na	na	na	0.011	na	0.006	na	na	0.023	0.006	na	0.012
Total	100.2_1	100.7_6	100.2_2	101.3_8	100.4_6	101.22_5	99.5_1	99.8_3	100.75_4	101.83_8	98.6_3	100.73_1
	Cation Formula Based on 8 Oxygens											
Ca	2.6578	2.6614	2.5214	2.5671	2.5070	2.6027	1.3837	1.3167	1.3416	1.3280	0.0035	0.0037
P	2.0043	1.9958	1.9469	2.0122	1.8981	2.0191	2.0128	1.9614	2.0448	2.0602	2.0660	2.0419
Fe	0.0727	0.0119	0.2332	0.0387	0.0909	0.0257	0.1064	0.2535	0.1813	0.1914	0.2217	0.1543
Mg	0.2999	0.2886	0.3587	0.2725	0.4938	0.2699	1.5673	1.6362	1.5217	1.5059	2.8214	2.8801
Mn	0.0000	0.0043	0.0000	0.0012	0.0000	0.0011	0.0170	0.0137	0.0202	0.0123	0.0055	0.0086
Na	0.0200	0.0885	0.3084	0.2455	0.1079	0.1516	0.0000	0.0000	0.0000	0.0000	0.0000	0.0000
K	0.0000	0.0039	0.0075	0.0045	0.0000	0.0001	0.0000	0.0000	0.0001	0.0002	0.0000	0.0001
Si	0.0108	0.0051	0.0473	0.0013	0.1000	0.0011	0.0000	0.0650	0.0015	0.0013	0.0023	0.0010
Ti	0.0000	0.0000	0.0000	0.0000	0.0000	0.0000	0.0000	0.0000	0.0000	0.0000	0.0000	0.0000
Al	0.0000	0.0000	0.0000	0.0002	0.0000	0.0001	0.0000	0.0000	0.0001	0.0001	0.0000	0.0002
Cr	0.0000	0.0000	0.0000	0.0004	0.0000	0.0002	0.0000	0.0000	0.0009	0.0002	0.0000	0.0004
Total Cations	5.0655	5.0595	5.4234	5.1436	5.1977	5.0716	5.0872	5.2465	5.1122	5.0996	5.1204	5.0903
	*Cation Ratios Fe/Mn and mg# (100*Mg/(Mg+Fe))*											
Fe/Mn	-	2.8	-	33	-	24	6.3	19	9.0	16	40	18
mg#	80.5	96.0	60.6	87.6	84.4	91.3	93.6	86.6	89.4	88.7	92.7	94.9

Springwater and Eagle Station data from Davis and Olsen (1991), remainder from Buseck and Holdsworth (1977).

Table 17. Analyses of chromite grains from representative pallasites.

	Glorieta Mountain	Brahin	Mount Vernon	Phillips County	Eagle Station
			Chemical Composition (wt %)		
TiO_2	0.32	0.23	0.27	0.16	nd
Al_2O_3	6.9	1.66	7.8	1.45	7.0
Cr_2O_3	62.1	69.1	61.3	68.5	62.1
V_2O_3	0.67	0.53	0.59	0.57	0.48
FeO	22.0	23.3	23.1	25.1	25.5
MgO	6.9	5.4	6.2	4.4	4.4
MnO	0.77	0.67	0.75	0.75	0.29
Total	99.66	100.89	100.01	100.93	99.77
			Cation Formula Based on 4 Oxygens (Ideal Spinel = 3 Cations per 4 Oxygens)		
Ti	0.0083	0.0061	0.0070	0.0043	0.0000
Al	0.2793	0.0687	0.3149	0.0605	0.2877
Cr	1.6862	1.9175	1.6598	1.9186	1.7118
V	0.0185	0.0149	0.0162	0.0162	0.0134
Fe	0.6319	0.6839	0.6616	0.7436	0.7435
Mg	0.3532	0.2824	0.3164	0.2323	0.2286
Mn	0.0224	0.0199	0.0218	0.0225	0.0086
Total Cations	2.9998	2.9934	2.9977	2.9980	2.9936
			*Cation Ratios Fe/Mn, mg# (100*Mg/(Mg+Fe)) and 100*Cr/(Cr+Al)*		
Fe/Mn	28	34	30	33	87
mg#	35.9	29.2	32.4	23.8	23.5
Cr/(Cr+Al)	85.8	96.5	84.1	96.9	85.6

All data taken from Bunch and Keil (1971).

contains very low abundances of the rare earth elements (REE) compared to either stanfieldite or whitlockite, while whitlockite is enriched in REE compared to stanfieldite from ~4 times for La to ~20 times for Lu.

Chromites are present in most, if not all, main-group pallasites; representative compositional data are given in Table 17. Pallasite chromites are fairly uniform in composition, except that Cr_2O_3 and Al_2O_3 vary and are anticorrelated; Cr_2O_3 from 69.1 to 60.5 wt %, Al_2O_3 from 1.66 to 9.1 wt %, maintaining a relatively constant (Cr+Al)/O atom ratio of ~0.5 (Bunch and Keil 1971).

Troilite and schreibersite are minor minerals in main-group pallasites. Troilite varies from ~0.1 to 7.3 vol %, and schreibersite varies from 0.3 to 2.8 vol % (Buseck 1977).

Eagle Station grouplet

In these pallasites, olivine is the dominant silicate phase, making up 75 to 80 vol % of the two meteorites measured by Buseck (1977). Minor clinopyroxene and orthopyroxene are reported in Eagle Station (Davis and Olsen 1991). Chromite has been reported by Bunch and Keil (1971) in Eagle Station, and whitlockite and stanfieldite, but not farringtonite, have been reported in the Eagle Station pallasites (Davis and Olsen 1991, Buseck and Holdsworth 1977).

The olivines in the Eagle Station grouplet are more ferroan than the main-group pallasites, have higher CaO contents, and slightly lower MnO contents leading to much higher FeO/MnO ratios (Table 13, Fig. 18). The trace transition elements in Eagle Station pallasites are also distinct. Their olivines contain higher Sc, Co and Ni than main-group pallasite olivines, but lower Zn contents (Table 14).

Only partial analyses of pyroxene are available. Davis and Olsen (1991) gave representative compositions of orthopyroxene (Wo_1Fs_{17}, 0.11 wt % Al_2O_3) and clinopyroxene ($Wo_{45}Fs_7$, 0.20 wt % Al_2O_3) in Eagle Station.

The major element compositions of stanfieldite and whitlockite in Eagle Station pallasites are very similar to those in the ferroan main-group pallasite, Springwater. The most noticeable difference is that whitlockite in the latter is richer in Na_2O than in the Eagle Station pallasites (Buseck and Holdsworth 1977, Davis and Olsen 1991; see Table 16). The REE in Eagle Station stanfieldite and whitlockite are LREE-depleted, a pattern quite distinct from that seen in Springwater phosphates (Davis and Olsen 1991).

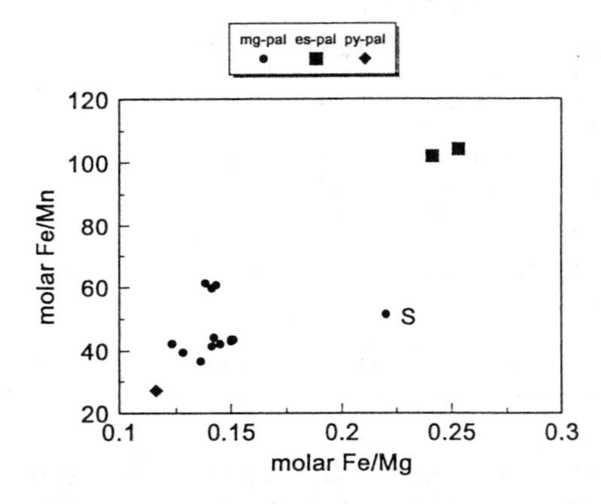

Figure 18. Molar Fe/Mn vs. Fe/Mg for pallasite olivines. The three pallasite types have distinct Fe/Mn ratios, but Fe/Mg ratios for the main-group and pyroxene-pallasites overlap (see Fig. 17). The three main-group pallasites with high Fe/Mn are Brenham, Krasnojarsk and Thiel Mountains. Springwater (S) is an anomalously ferroan main-group pallasite. All data from Mittlefehldt (unpublished).

Chromite in Eagle Station was analyzed by Bunch and Keil (1971). Eagle Station chromite grains are more ferroan when compared to the main-group pallasite chromites, except for those in Phillips County (Table 17), which also has anomalously ferroan olivines for a main-group pallasite (Buseck and Goldstein 1969).

Pyroxene-pallasite grouplet

The pyroxene-pallasite grouplet, consisting of Vermillion and Yamato 8451, has only recently been defined (Boesenberg et al. 1995) and very little data on them are yet in the literature. Hiroi et al. (1993) and Yanai and Kojima (1995a) gave brief descriptions of the first known member, Y-8451. Based on two modal analyses, this pallasite consists of about 55 to 63 vol % olivine, 30 to 43 vol % metal, 1 to 3 vol % pyroxene, 1 vol % troilite, plus minor whitlockite. Chromite was not mentioned. Olivine and pyroxene occur as 1 to 6 mm-sized rounded grains enclosed in a metal-troilite matrix. The occurrence of mm-sized pyroxenes distinguishes this pallasite from other types (compare with pyroxene textural descriptions for main-group and Eagle Station grouplet pallasites in Buseck 1977, Davis and Olsen 1991). The large pyroxenes poikilitically contain small olivine grains, and vice versa. Yanai and Kojima (1995a) list three distinct pyroxenes, polysynthetically twinned orthopyroxenes, which occur as large, rounded grains, non-polysynthetically twinned orthopyroxenes, which occur as inclusions in olivine, as aggregates, and as rims on polysynthetically twinned orthopyroxenes, and small clinopyroxenes. Hiroi et al. (1993) found only two distinct pyroxenes, orthopyroxene with ~2 mol % wollastonite and protopyroxene (inverted protoenstatite) with ~0.6 mol %. They also found that metal-olivine boundaries have slight dislocations.

Vermillion contains much less olivine, ~14 vol % (Boesenberg et al. 1995), than any other pallasite. The olivines occur as rounded to subrounded grains up to 1.5 cm in size in cm-sized bands interspersed with metal in bands of similar width. This texture is unique among pallisites. Minor amounts of orthopyroxene (~0.7 vol %), chromite (~0.2 vol %) and whitlockite (~0.07 vol %) are also present. The orthopyroxene occurs both along the rims of smaller olivine grains, and as large (mm-size) inclusions in olivine.

Pyroxene-pallasite olivines overlap the magnesian end of the range of main-group pallasites (Table 13, Fig. 17). Pyroxene-pallasite olivines have distinctly higher MnO contents, resulting in significantly lower FeO/MnO ratios than in the main-group pallasites (Table 13, Fig. 18).

There are three distinct pyroxene composition groups in Y-8451 that correlate with textures (Yanai and Kojima 1995a); the large, polysynthetically twinned orthopyroxenes have very low CaO contents, the non-polysynthetically twinned orthopyroxenes have intermediate CaO contents, and the clinopyroxenes are calcic (Table 15). The polysynthetically twinned orthopyroxenes also tend to be more magnesian than the non-polysynthetically twinned orthopyroxenes (Yanai and Kojima 1995a). Based on CaO content, the polysynthetically twinned orthopyroxene of Yanai and Kojima (1995a) appears to be equivalent to the protopyroxene of Hiroi et al. (1993). Boesenberg et al. (1995) described a similar range in orthopyroxene wollastonite contents for Vermillion, but do not give information of possible correlations between texture and composition.

Whitlockite in Vermillion and Y-8451 have LREE-depleted REE patterns similar to those determined for Eagle Station, but at higher concentrations (Boesenberg et al. 1995, Davis and Olsen 1991).

Metal phase

The metal phase of main-group pallasites is generally low in refractory siderophile elements such as Ir and high in Ni and moderately volatile siderophile elements such as As and Au (Fig. 19). Main-group pallasite metal is close in composition to Ni-rich IIIAB irons, but has slightly higher incompatible/compatible siderophile element ratios (e.g. Au/Ir) than IIIAB irons with similar Ir contents (Scott 1977b). There is also considerable scatter in main-group pallasite metal composition, particularly in the contents of compatible siderophile elements like Ir (Fig. 19a). The metal phase of Eagle Station grouplet pallasites is distinct in composition, having higher Ni and Ir, and lower moderately volatile siderophile element contents (Scott 1977b). Eagle Station grouplet metal is close in composition to IIF irons (Kracher et al. 1980). The two pyroxene-pallasites have widely different metal compositions (Wasson et al. 1998a,b). Y-8451 has Ni and Ir contents like those of the Eagle Station grouplet, but a Au content more like those of high Ni main-group pallasites, while Vermillion has Ni, Ir and Au contents similar to low Ni, high Ir main-group pallasites (Fig. 19). However, Vermillion has very different contents of moderately volatile siderophile elements (e.g. Ga, Ge) than any main-group pallasite (Wasson et al. 1998b). Wasson (pers. comm.) believes the metals in the two pyroxene-pallasites are too different in composition to support assignment to a single grouplet.

Ages

Very little is known of the formation ages of pallasites; older K-Ar ages are probably unreliable indicators of formation times. Recently, Mn-Cr dating has been done on several main-group pallasites (Hsu et al. 1997, Hutcheon and Olsen 1991, Shukolyukov and Lugmair 1997) and the Eagle Station pallasite (Birck and Allègre 1988). The data indicate that ^{53}Mn ($t_{1/2}$—3.7 Ma) was present at the time the pallasites were formed, and thus they formed very early in the history of the solar system. By tying the Mn-Cr data on the Omolon main-group pallasite to the Mn-Cr and Pb-Pb data on the LEW 86010 angrite, Shukolyukov and Lugmair (1997) determined an absolute age of 4.558±0.001 Ga for closure of the Mn-Cr system in Omolon. Eagle Station had a distinctly lower ^{53}Mn/^{55}Mn ratio at isotopic closure than did the main-group pallasites so far studied, suggesting Eagle Station reached closure ~10 to 15 Ma later (Hsu et al. 1997).

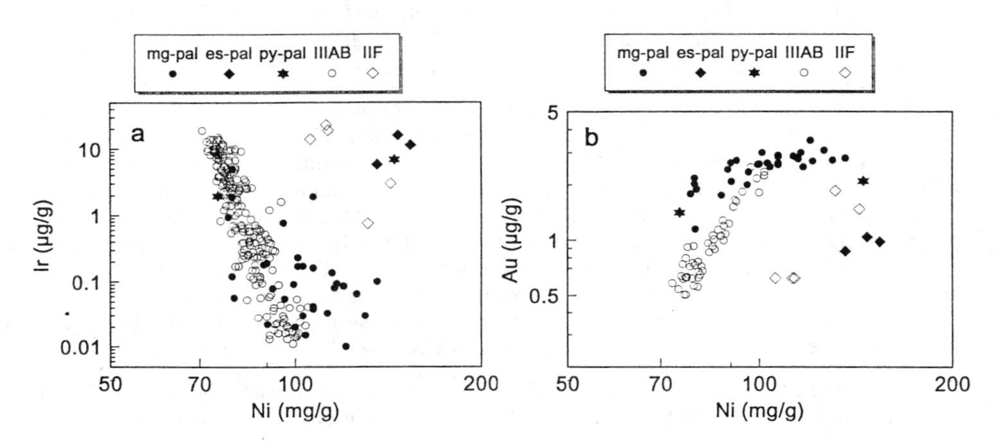

Figure 19. Diagrams of (a) Ir vs. Ni and (b) Au vs. Ni for metal from main-group pallasites, Eagle Station and pyroxene-pallasite grouplets compared to IIF and IIIAB irons. Main-group pallasites show wide ranges in Ir-Ni and Au-Ni, indicating that their metal is not simply quenched residual metallic melt. The high-Ni pyroxene-pallasite is Y-8451. Data are from Davis (1977), Esbensen et al. (1982), Malvin et al. (1984), Scott (1977), Scott and Wasson (1976), Ulff-Møller et al. (1998), Wasson (1974), Wasson et al. (1989, 1998a,b).

Cooling rates

Buseck and Goldstein (1969) determined cooling rates from the metal phase of pallasites. Because of changes in values of Ni diffusion parameters (Dean and Goldstein 1986), the cooling rate values given by Buseck and Goldstein (1969) need to be increased 5-fold (Saikumar and Goldstein 1988). This leads to cooling rates of 2.5-7.5°C/Ma for main-group pallasites. Cooling rates for three main-group pallasites determined by a different method agree with these values (Yang et al. 1997a, see Table 4). These cooling rates are lower than those of most irons (Table 4). Phosphorous is a moderately volatile element, and the formation of the Widmanstätten structure used in cooling rate determinations is strongly influenced by P content (Dean and Goldstein 1986). Because main-group pallasites contain higher moderately volatile siderophile element abundances than do Eagle Station grouplet pallasites (Scott 1977b), there could be a systematic difference in metallographic cooling rates calculated for the two types. Taken at face value, however, the cooling rate for Eagle Station, 10°C/Ma, is similar to those of main-group pallasites.

Miyamoto and Takeda (1994) have presented cooling rates based on fitting Fe-Mg and Ca zoning profiles in olivine for a main-group and a pyroxene-pallasite. The calculated cooling rates are very high, on the order of 10^6 times the metallographic cooling rates. It is difficult to evaluate the veracity of these cooling rates from the abstract, but the cooling rate based on Fe-Mg zoning is highly model dependent as it is a strong function of the assumed initial temperature.

Pallasite formation

The customary hypothesis for pallasite formation is that they were formed at the core-mantle boundary of their parent body. Metal will sink to form the core of a differentiated rocky body, while dunite is a plausible lowermost mantle material, either as partial melting residue or as cumulate from a crystallizing ultramafic melt. There are two possible problems with the simple core-mantle boundary model, both long discussed in the literature: (1) pallasites seem too abundant in the meteorite collection considering the small volume a core-mantle boundary will occupy in an asteroid, and (2) the large density difference between solid olivine and metal liquid will result in rapid separation of the two even in the low gravity field of asteroids. In spite of these problems, the core-mantle boundary origin still seems most plausible.

Main-group pallasite metal is close to IIIAB metal in composition (Fig. 19). Scott (1977c) showed that the mean composition of main-group pallasite metal is very similar to the composition of a model residual liquid after about 82% crystallization of a metallic melt like that parental to IIIAB irons. He suggested that main-group pallasites and IIIAB irons formed on the same parent body. This is compatible with a core-mantle boundary model for pallasite formation, assuming the core crystallized from the center out. However, Haack and Scott (1993) suggested that IIIAB irons were formed by inward dendritic growth from the core-mantle boundary. In order to explain the apparent close relationship between main-group pallasite metal and IIIAB irons, they suggested that some of the residual metallic melt was forced up to the core-mantle boundary by an unspecified process. The composition of metal in main-group pallasites is quite variable compared to Ni-rich IIIAB irons, which suggests the pallasite metal cannot be simply quenched residual melt. Ulff-Møller et al. (1997) suggest the wide range in Ir contents in main-group pallasite metal is caused by variable mixing of residual melt with earlier metal crystals at the time of metal-olivine mixing. As noted by Ulff-Moller et al. (1998), main-group pallasites have lower S contents than expected for trapped metallic melts. They thus concluded that the injected residual melt underwent a period of equilibrium crystallization, and that the last dregs of S-rich melt escaped the pallasite region.

The relationship between IIF irons and the metal phase of Eagle Station grouplet pallasites is not as close as that between IIIAB irons and the main-group. Kracher et al. (1980) suggested that IIF irons and Eagle Station grouplet pallasites were formed on similar parent bodies. The metal phase of the Bocaiuva silicate-bearing iron (discussed later) is also similar to IIF irons and the Eagle Station grouplet, and the oxygen-isotopic composition of Bocaiuva silicates is similar to those of the Eagle Station grouplet (Malvin et al. 1985, and see Fig 1a). All of these meteorites likely are from similar parent bodies formed in the same region of the solar system (Malvin et al. 1985).

Scott (1977d) addressed the problem of how the metal-silicate texture was formed through detailed study of the macrotextures of pallasite slabs. He found that pallasites display a range of textures, from those containing highly angular olivine grains with large olivine masses several cm in size, some of which are veined with metal, to pallasites with rounded olivines dispersed in the metal. The large olivine masses preserve textural evidence indicating that they were in the process of breaking up into smaller, angular olivine grains when the process was frozen by crystallization of the metal. Scott (1977d) favored a model in which solid olivine layers were invaded, broken up and dispersed by molten metal near its freezing point. Crystallization of the metal prevented density separation of the metal and olivine. Ullf-Møller et al. (1997) state that the most plausible mechanism to produce the textures seen in olivine masses by Scott (1977d) and by themselves in the Esquel main-group pallasite is by high pressure injection of metal, possibly by impact. Pallasites with rounded olivines were formed in a similar way, but annealing of the mixture with subsequent rounding of initially angular olivines occurred. Ohtani (1983) used thermodynamic, kinetic and physical models to constrain the time and temperature of formation of the olivine-metal textures. He suggested that the rounded olivine textures were formed at temperatures above the metal-troilite solidus, and that the time scale must have been <5 Ka or olivine and metal would have segregated. Pallasites with angular olivine were annealed at lower temperatures. In contrast, Buseck (1977) considered the rounded olivines to represent the primary shapes, while angular olivines were secondary. He suggested that pallasites are exotic cumulates; olivine and chromite grains represent cumulus minerals, while the metal is an intercumulus phase.

Davis and Olsen (1991) argued that the REE contents of pallasite phosphates are not compatible with simple models forming the silicates of all pallasites as either residual mantle or early cumulates during differentiation of their parent bodies. Many main-group and Eagle Station grouplet pallasites have REE patterns consistent with phosphate formation through redox reactions between the metal and olivine in which metallic P is oxidized. The REE in the phosphates are inherited from the olivine and show LREE depleted patterns consistent with either residual or cumulate dunites, and therefore core-mantle boundary models (Davis and Olsen 1991). However, the Springwater and Santa Rosalia main-group pallasites contain phosphates far too rich in REE for this scenario. Davis and Olsen (1991) suggested that the phosphates in these pallasites are consistent with crystallization at a late stage from a chondritic melt, and therefore indicate that some pallasites may have originated nearer the parent body surface. Note that both Santa Rosalia and Springwater yield cooling rates at the low end of the main-group pallasite range (Buseck and Goldstien, 1969), which may be inconsistent with the formation scenario inferred by Davis and Olsen(1991). Buseck and Holdsworth (1977) described textural evidence which they believed indicated that some phosphates crystallized from a residual melt. They did not specify which meteorites contained these textures, so one cannot correlate anomalous REE patterns with phosphate texture.

Minor and trace elements, and fayalite content show zoning profiles from core to rim in pallasite olivines. This, plus rounding of olivines suggestive of subsolidus annealing indicate that small scale element migration was occurring in pallasites at the time they cooled below the blocking temperature for these processes. The decreasing Fa and Ni contents at the rims, plus evidence for formation of at least some phosphates by oxidation of metallic P, suggest that redox reactions between metal and silicate were responsible for some element migration. However, the cause of zoning in Cr and Al remain obscure, and there is no consensus on the origin of all the phosphates and the pyroxenes.

Main-group pallasites appear to be closely related to the howardite, eucrite and diogenite (HED) meteorites. Oxygen-isotopic compositions show that all these meteorite types came from the same O-isotope reservoir (Clayton and Mayeda 1996, see Fig. 1d). Extreme differentiation can in principle produce olivines as magnesian as those in main-group pallasites and basalts as ferroan as basaltic eucrites on a single parent body (Mittlefehldt 1980). Hence, it is possible that all of these meteorite types, plus the IIIAB irons, represent fragments of a single parent asteroid. This will be discussed in more detail in the section on HED meteorites.

Clearly, time is ripe for renewed detailed studies of the pallasites. Recent ion microprobe studies of phosphate compositions, electron microprobe studies of zoning profiles, renewed attempts to date pallasite formation, and the characterization of new pallasite types all suggest that pallasite studies are being reinvigorated after their heyday in the mid-seventies.

SILICATE-BEARING IVA IRONS

The IVA iron meteorite group contains only four known silicate-bearing members. Unlike the other silicate-bearing iron groups discussed earlier in this chapter, silicates in IVA irons are distinctly non-chondritic in their mineralogy. Two of the IVA irons, Steinbach and São João Nepomuceno, contain abundant rounded pyroxene-tridymite inclusions (Reid et al. 1974, Bild 1976, Prinz et al. 1984, Ulff-Møller et al. 1995, Scott et al. 1996, Haack et al. 1996a). The IVA irons Bishop Canyon and Gibeon contain rare SiO_2 grains up to several cm in length (Schaudy et al. 1972, Buchwald 1975, Ulff-Møller et al. 1995, Scott et al. 1996, Marvin et al. 1997). Their linkage to IVA irons is confirmed by concentrations of Ni, Ga, Ge and Ir similar to those in IVA irons without silicates (Scott et al. 1996). Until recently, there existed a relative scarcity of basic data about the silicates in these meteorites. Their small number and unusual mineralogy make the origin of these meteorites probably the most enigmatic of all silicate-bearing irons.

Petrology and mineralogy

Pyroxene-tridymite-rich IVA irons. The pyroxene-tridymite-bearing IVA irons Steinbach and São João Nepomuceno have been widely studied. Figure 20a illustrates the main mass of Steinbach (from Scott et al. 1996). Metal and troilite occupy 29 vol % of this mass as irregularly shaped grains and two small, metal-rich areas are present. In this metal (Fig. 20b), metal grains that appear distinct actually have the same orientation of their Widmanstätten pattern over areas as large as 10 cm (McCall 1973), indicating that they were originally composed of single γ-Fe,Ni crystals of this size. Silicates comprise 19 vol % of São João Nepomuceno (Scott et al. 1996) but are heterogeneously distributed. Silicates are partly concentrated on grain boundaries of parent taenite crystals which reach 4 by 8 cm and silicate-free regions reach 10 by 20 cm. In areas where silicates are concentrated, the texture very closely resembles that of Steinbach.

Silicates in Steinbach and São João Nepomuceno are orthobronzite, clinobronzite and tridymite (Reid et al. 1974, Dollase 1967, Ulff-Møller et al. 1995, Scott et al. 1996). Modal data from Ulff-Møller et al. (1995) and Scott et al. (1996) show that Steinbach is composed of orthobronzite (37.3, 38.0 vol %, respectively), clinobronzite (4.5, 1.6), tridymite (20.2, 20.8), metal (32.4, 32.2), troilite (5.7, 6.6) and traces of chromite. The agreement between these values suggests that Steinbach is quite homogeneous on the scale of ~10 cm^2, although Scott et al. (1996) documented considerable heterogeneity in modes and pyroxene/tridymite ratios between individual polished thin sections. A smaller area of São João Nepomuceno (~3.5 cm^2) contains orthobronzite (37.8), clinobronzite (0.1), tridymite (5.1), metal and its weathering products (52.4), troilite (4.6) and traces of chromite (Scott et al. 1996). Compared to Steinbach, São João Nepomuceno is poorer in tridymite and clinobronzite.

A mixture of pyroxene and SiO_2 fills the spaces between the metallic host in both Steinbach and São João Nepomuceno. Silicate grain sizes are typically 1 to 5 mm (Ulff-Møller et al. 1995, Scott et al. 1996). Metal-silicate and silicate-silicate grain boundaries are typically complex. In many cases, metal adjacent to silicates forms cusps at triple junctions. Many adjacent pyroxene grains are irregular, indicative of silicate co-crystallization (Scott et al. 1996).

A remarkable feature of both Steinbach and São João Nepomuceno is the occurrence of twinned clinobronzite within orthobronzite. The clinobronzite islands are either totally enclosed in orthobronzite (Ulff-Møller et al. 1995, Scott et al. 1996) or occur at

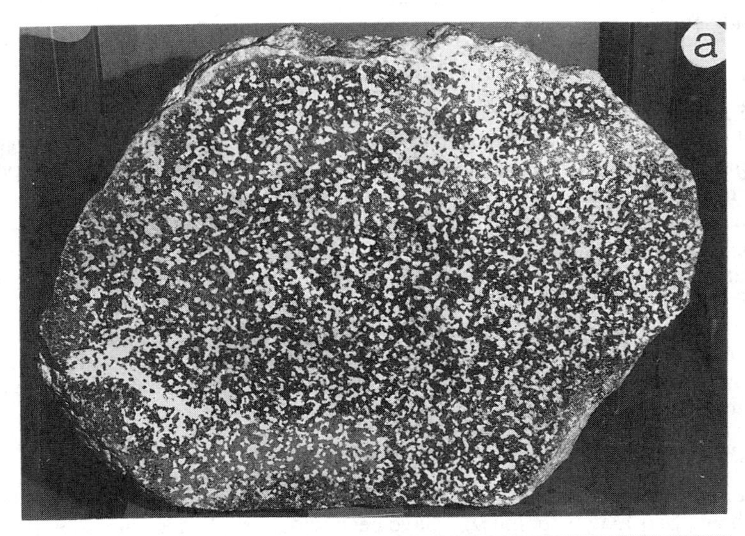

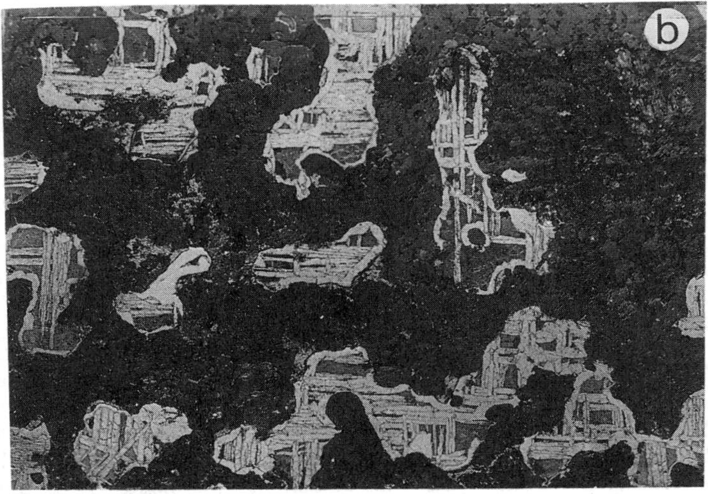

Figure 20. Photographs of the Steinbach silicate-bearing IVA iron. (a) A polished slice showing white metallic Fe,Ni and troilite in a dark silicate matrix. Longest dimension is 51 cm. (b) Part of an etched slice showing fine, oriented kamacite lamellae in metallic Fe,Ni. The continuity of the Widmanstätten pattern across the field shows that the grains are connected and once formed a single large skeletal taenite crystal. Width is 3.4 cm.

pyroxene-SiO_2 grain boundaries (Scott et al. 1996), but are not observed bordering metal. Reid et al. (1974) studied the pyroxenes by x-ray diffraction methods and Haack et al. (1996a) reported on a transmission electron microscope study of a pyroxene grain in Steinbach. These pyroxene grains contained a continuum of microstructures ranging from ordered clinobronzite to a highly striated microstructure consisting of a disordered intergrowth of clinopyroxene and orthopyroxene. Orthopyroxene lamellae are typically <10 unit cells in thickness, while clinopyroxene lamellae extend up to 50 unit cells in thickness. As noted earlier, São João Nepomuceno also contains a similar, but less abundant, clinobronzite, but this has not been examined by TEM.

Steinbach orthobronzite ($Fs_{15.0}$) is more FeO-rich than clinobronzite ($Fs_{14.5}$) (Reid et al. 1974, Scott et al. 1996). A similar relationship is seen in São João Nepomuceno (orthobronzite, $Fs_{14.2}$; clinobronzite, $Fs_{13.6}$), although São João Nepomuceno is slightly poorer in FeO than Steinbach (Scott et al. 1996). Ulff-Møller et al. (1995) also observed Ca-Cr-rich regions in orthobronzite and documented zoning within Steinbach pyroxenes and tridymite. Steinbach tridymite has been studied in detail by a number of authors (e.g. Tagai et al. 1977). Rare Al-rich regions may perhaps be related to contamination with Al polishing compound.

SiO_2-bearing IVA irons. Bishop Canyon and Gibeon contain large, but very rare, grains of SiO_2. Bishop Canyon contains two adjoining plate-like SiO_2 grains, probably tridymite (Buchwald 1975, Scott et al. 1996). Troilite with exsolved daubreelite occurs around the grains.

Silica grains from the widely-distributed Gibeon meteorite have been more widely-studied (Schaudy et al. 1972, Ulff-Møller et al. 1995, Scott et al. 1996, Marvin et al. 1997). Ulff-Møller et al. (1995) and Scott et al. (1996) studied the same sample, which contained a single crystal measuring 21 by 1 mm. This grain had been fractured into three pieces prior to complete solidification of the metallic host. Marvin et al. (1997) studied slices through a very large SiO_2 grain. The length of the original grain may have been up to 10 cm. These authors confirmed by x-ray diffraction studies and characteristic fluorescence that the grain was dominantly composed of tridymite. However, cracked areas 3-4 mm across in the center of the tridymite were quartz.

Ulff-Møller et al. (1995) and Scott et al. (1996) observed zoning in Gibeon tridymite, with FeO increasing from the core to the rim of the grain and Al_2O_3 decreasing.

Origin

The origin of the IVA irons remains enigmatic. Any favored interpretation depends critically on whether the origin of silicates in Steinbach and São João Nepomuceno are linked to that of the large SiO_2 grains in Bishop Canyon and Gibeon, or whether these two occurrences formed through independent processes. Certainly, recovery of more silicate-bearing IVA irons would substantially improve our understanding of their origin.

Prinz et al. (1984) suggested that silicates in pyroxene-tridymite-rich IVA irons formed by remelting of igneous cumulates. Several problems exist with this scenario. Prinz et al. (1984) suggested that the pyroxene/tridymite ratios are appropriate for eutectic mixtures, but new modal data of Ulff-Møller et al. (1995) and Scott et al. (1996) suggest that this is not the case. Cumulate remelting would also produce a melt with more Ca than Al and a strongly steepened REE pattern, neither of which is observed in Steinbach or São João Nepomuceno (Ulff-Møller et al. 1995).

A second model proposed by Prinz et al. (1984) and expanded upon by Scott et al. (1996) suggests oxidation of Si from metal and reaction with an olivine-bearing pallasite to produce pyroxene and tridymite observed in IVA irons. The gross texture of Steinbach is strikingly similar to some pallasites (e.g. Brenham; Nininger 1952). Oxidation of Si from metal could also form the silica observed in Bishop Canyon and Gibeon and may have formed silica in IIIAB irons (Kracher et al. 1977). Indeed, a pyroxene-bearing pallasite precursor (e.g. Yamato 8451, Yanai and Kojima 1995a) could provide a source of minor and trace elements observed in pyroxene of Steinbach and São João

Nepomuceno. A problem with this scenario is envisioning the presence of significant Si dissolved in metal, which requires very reducing conditions, and yet having relatively oxidized pyroxene is these meteorites (Fs_{14-15}).

Ulff-Møller et al. (1995) proposed that the silicates were derived by direct melting of a chondritic precursor with subsequent reduction. In this model, low degrees of partial melting would occur, with subsequent melt removal and reduction. Through this reduction, the Fe/Mn ratio of the melt was lowered and the SiO_2 content of the residual magma was increased. This process would occur most efficiently within the molten core of the asteroid, where sufficient reducing agents (e.g. P) exist. Ulff-Møller et al. (1995) argued that this reduction could have forced the melt to crystallize only tridymite.

It is unclear how the silicates were mixed with the metal in IVA irons. The oxidation of a pallasite-like precursor provides a ready explanation for mixing. Mechanisms which call for mixing of silicate melt into the molten core are more controversial. Ulff-Møller et al. (1995) proposed that large, metallic dendrites which formed at the core-mantle boundary could founder into the molten core, carrying pieces of the adhering mantle and, thus, explaining the mixing. Alternatively, a crystallizing core would undergo significant shrinkage and this might lead to injection of overlying, molten silicates into the cracks produced. Scott et al. (1996) criticized these passive mechanisms and suggested instead that silicate-bearing IVA irons formed during catastrophic breakup and reassembly of the IVA parent body, an idea explored in greater detail by Haack et al. (1996a). Evidence for such a process may be found in the complex cooling history required to explain the features of these meteorites. The orthobronzite-clinobronzite intergrowths clearly require very rapid cooling (Reid et al. 1974, Haack et al. 1996a) at high temperature, while the existence of the Widmanstätten pattern in these meteorites requires slow cooling at low temperature.

BRACHINITES

Brachina was originally thought to be a second chassignite (Johnson et al. 1977, Floran et al. 1978a, Ryder 1982), but was shown on the basis of chemistry, oxygen-isotopic and noble gas composition, and age to be unique (Nehru et al. 1983, Clayton and Mayeda 1983, Crozaz and Pellas 1983, Ott et al. 1983). Four other meteorites are now classified as brachinites: ALH 84025, Eagles Nest, Nova 003 (probably paired with Reid 013) and Hughes 026 (Table 18). Divnoe has been considered to be a brachinite on the basis of oxygen-isotopes (Clayton and Mayeda 1996), but is petrologically unique (Petaev et al. 1994). LEW 88763 has also been thought to be brachinite (AMN 14-2; Nehru et al. 1992, Kring and Boynton 1992), but its oxygen isotpoic composition is distinct from brachinites and falls near the trend defined by acapulcoites and lodranites (Fig. 1c), and is now considered to be unique (Swindle et al. 1998).

Petrography and mineral chemistry

Brachinites are dunitic wehrlites, consisting dominantly of olivine; reports for individual brachinites range from 79-93%, but Nehru et al. (1992) cite 74 to 98% for the group. They contain augite (4 to 15%), chromite (0.5 to 2%), Fe-sulfides (3 to 7%), minor phosphates, and minor Fe-Ni metal. Brachina contains 9.9% plagioclase, Nova 003 and Hughes 026 have <0.1%, and ALH 84025 and Eagles Nest have no plagioclase. Brachina, Nova 003 and Hughes 026 contain traces of orthopyroxene.

Brachina has an equigranular texture (subhedral to anhedral grains with 120° triple junctions) with average grain size 0.1 to 0.3 mm (Fig. 21a). Hughes 026 is also equigranular, with average grain size 0.65 mm. ALH 84025 is coarser-grained, with olivine

Table 18. Petrographic data for brachinites.

	olivine Fo *(modal %)*	**clinopyroxene** Wo/En *(modal %)*	**orthopyroxene** Wo/En *(modal %)*	**plagioclase** An/Ab *(modal %)*	**Fe sulfide** Ni wt % *(modal %)*	Texture	Ref.	Notes
Brachina	68-70 (80.4%)	38.7/48.4 (5.5%)	(trace)	22/76 (9.9%)	0.6-3.6 (2.9%)	equigranular (0.1-0.3 mm)	1	
ALH 84025	67-68 (79-90.2%)	44/46 (4.2-15%)	(none)	(none)	0.8-2.0 (3-4%)	prismatic-subhedral (up to 2.7 mm); equigranular	2-4	texture and grain size vary between thin sections
Nova 003	68.5	45/44	2.4/72.1	33.3			5	formerly named Window Butte
Reid 013	66.1	45.5/43.6	1.7/71.3	31.5/68.1			5	probably paired with Nova 003 (M. Prinz, pers. comm.)
Eagles Nest	68 (81%)	45/45 (6%)	(none)	(none)	(7%)	subequant, triple junctions (up to 1.5 mm)	6-8	
Hughes 026	65 (92.7%) (3.6%)	46-47/43-44	3/69 (1.6%)	32 (<0.1%)	(1.20%)	equigranular (0.65 mm)	9,10	previously, informally called Australia I

References: 1. Nehru et al. (1983); 2. Prinz et al. (1986a); 3. Warren and Kallemeyn (1989a); 4. AMN 8-2; 5. Wlotzka (1993); 6. Kring et al. (1991); 7. Kring and Boynton (1992); 8. Swindle et al. (1997); 9. M. Prinz (pers. comm.); 10. Nehru et al. (1996).

Figure 21. Photomicrographs of brachinites in transmitted light with partially crossed nicols. Scale bars are 0.2 mm and fields of view are ~2.9 × 1.7 mm. (a) Brachina. Equigranular texture of olivine, augite and plagioclase, with minor chromite and Fe,Ni metal and sulfides. (b) ALH 84025,7. Olivine and augite in equigranular texture. Interstitial troilite (black) left of center surrounds augite with fine twin lamellae.

up to 1.8 mm and augite up to 2.7 mm. Warren and Kallemeyn (1989a) observed that in two thin sections of ALH 84025 olivine is prismatic, with an aspect ratio of ~3:1, and shows a possible preferred orientation (Fig. 21b). However, Prinz et al. (1986a) reported an equigranular texture in two other thin sections; whereas yet another thin section is described as being polygonal-granular (AMN 8-2). Eagles Nest is also coarser-grained than Brachina. It contains olivine up to 1.5 mm, with subequant grain shapes and 120°

triple junctions (Kring et al. 1991). In all brachinites augite, chromite, and plagioclase appear to be interstitial to olivine. Sulfide is anhedral and interstitial, or occurs as veins with minor Fe-Ni metal cross-cutting the silicates and chromite. Brachina has 20-25 μm, rounded melt inclusions in olivine, similar to those in Chassigny (Floran et al. 1978a, Nehru et al. 1983). The inclusion-bearing olivine grains tend to be euhedral to subhedral (Floran et al. 1978a). Olivine in Eagles Nest has inclusions of chromite, apatite and troilite, but no melt inclusions (Swindle et al. 1998).

Olivine is homogeneous within each brachinite, and ranges from Fo_{65-70} among the brachinites. Augite is more magnesian: mg# = 79 to 82. In olivine MnO ranges from ~0.3 to 0.6%, Cr_2O_3 ranges from 0.01 to 0.08%, CaO ranges from 0.05 to 0.27% (Kring et al. 1991, Warren and Kallemeyn 1989a, Johnson et al. 1977, Nehru et al. 1983), and NiO is detectable—up to 0.05% (Nehru et al. 1992). In augite MnO ranges from 0.15 to 0.32%, Cr_2O_3 from 0.68 to 1.04%, Na_2O from 0.39 to 0.61%, TiO_2 from 0.1 to 0.4%, and Al_2O_3 from 0.55 to 0.98% (Kring et al. 1991, Warren and Kallemeyn 1989a, Johnson et al. 1977, Nehru et al. 1983).

Chromite has molar Cr/(Cr+Al) ratios of 0.81 to 0.83 in Brachina, ALH 84025, and Eagles Nest and 0.73 in Hughes 026, with mg# of 19 to 26 (Johnson et al. 1977, Warren and Kallemeyn 1989a, Nehru et al. 1983, Swindle et al. 1998, Nehru et al. 1996). Prinz et al. (1986a) reported that only minor Fe_2O_3 was required by stoichiometry for chromite in Brachina and Eagles Nest, but recalculation of published analyses does not confirm this. The TiO_2 contents of chromite range from 1.4% in Eagles Nest and ALH 84025 (Swindle et al. 1998, Warren and Kallemeyn 1989a) to ~2.9% in Brachina (Nehru et al. 1983).

The sulfide in brachinites is mainly Ni-rich (0.5 to 3%) troilite (Nehru et al. 1992), but pentlandite has also been observed in Brachina (Nehru et al. 1983). Metal contains 19 to 55% Ni (Nehru et al. 1992).

Chlorapatite is the dominant phosphate in brachinites (Nehru et al. 1992), but whitlockite also occurs in Hughes 026 (Nehru et al. 1996). Eagles Nest contains an Fe-phosphate, tentatively identified as being either ludlamite, vivianite or beraunite, but this is believed to be a terrestrial weathering product of apatite (Swindle et al. 1998).

Composition

Lithophile elements. Lithophile elements in Brachina are close to chondritic in abundance and are relatively unfractionated (Table 19, Fig. 22). Eagles Nest is slightly depleted in Al, Ca, K, Rb and more so in Na (0.07 × CI), and ALH 84025 is depleted in Al and Na (0.1 × CI), consistent with the lack of plagioclase in these two brachinites.

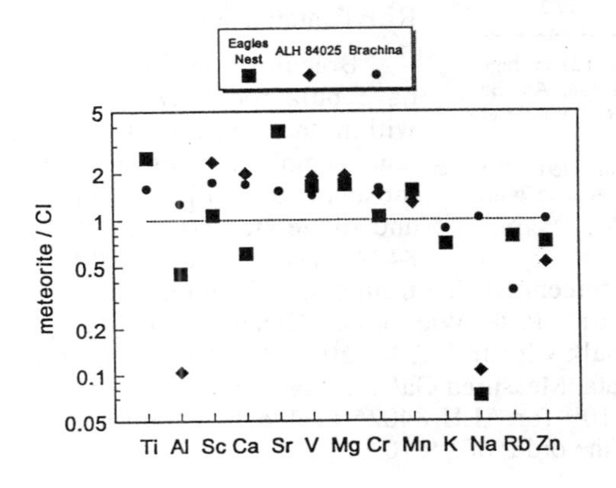

Figure 22. CI-normalized lithophile element abundances in brachinites. Data from Johnson et al. (1977), Nehru et al. (1983), Swindle et al. (1998), Warren and Kallemeyn (1989a).

Table 19. Bulk compositions of brachinites.

		ALH 84025 (1)	Brachina (2)	Eagles Nest* (3)
Na	mg/g	0.51	5.10	0.351
Mg	mg/g	184	162	161
Al	mg/g	0.90	11.0	3.91
Si	mg/g	170	178	
P	mg/g		1.2	
S	mg/g		15.2	
K	µg/g	_40	700	400
Ca	mg/g	18	15.5	5.5
Sc	µg/g	13.6	8.2	6.19
Ti	mg/g	_1.6	0.72	1.1
V	µg/g	108	77	94
Cr	mg/g	3.94	4.15	2.80
Mn	mg/g	2.53	2.6	3.04
Fe	mg/g	250	206	245
Co	µg/g	365	232	201
Ni	mg/g	5.10	3.09	2.20
Zn	µg/g	164	313	223
Ga	µg/g	2.1	7.6	1.5
As	µg/g	0.55	0.18	3.72
Se	µg/g	12.4	3.5	3.71
Br	µg/g	0.33	0.49	2.95
Rb	µg/g		2.0	1.8
Sr	µg/g		15	29.4
Zr	µg/g		2.7	8.4
Sb	ng/g	56		95
Cs	ng/g		200	104
Ba	µg/g		12	208
La	µg/g	0.065	0.665	2.66
Ce	µg/g		1.6	2.00
Nd	µg/g		0.86	2.51
Sm	µg/g	0.034	0.235	0.527
Eu	µg/g	0.033	0.08	0.131
Tb	µg/g		0.05	0.071
Yb	µg/g	0.098	0.21	0.186
Lu	µg/g	0.016	0.07	0.0255
Ir	ng/g	123	111	98
Au	ng/g	61	15	25.3
Th	ng/g		130	173

(1) whole rock, Warren and Kallemeyn (1989a); (2) average whole rock, Johnson et al. (1977), except Zn, Ga, As, Se and Br, Nehru et al. (1983); average whole rock, Swindle et al. (1998).

*Contamination and weathering have affected the composition of Eagles Nest. Data should be treated with caution (see Swindle et al. 1998).

Brachina has a nearly flat REE pattern, at ~1 × CI abundances (Fig. 23). ALH 84025 is LREE-depleted, with La = 0.28 × CI, consistent with its depletion in lithophile incompatible major elements. Eagles Nest, which is also depleted in lithophile incompatible major elements, is LREE-enriched, with La ~ 11 × CI, and shows a significant negative Ce anomaly. These features are most likely due to contamination by terrestrial REE (Swindle et al. 1998).

Incompatible lithophile element depletion in various meteorites can be gauged on a plot of CI-normalized Na/Sc vs. Sm/Sc ratios (Fig. 24). During melting or crystallization in ultramafic/mafic systems, Na and Sm will be more incompatible than Sc, and Na/Sc and Sm/Sc ratios will be low in partial melt residues or ultramafic cumulates. On this plot Brachina appears to be only slightly depleted and in this regard resembles acapulcoites and winonaites, while ALH 84025 is more depleted than most lodranites, and as depleted as some ureilites. Eagles Nest is not shown because of likely terrestrial REE contamination.

Brachina and Eagles Nest have bulk rock Fe/Mn ~ 80, within the range of those in lunar samples and distinct from those in howardites, eucrites and diogenites (HEDs). ALH 84025 has bulk rock Fe/Mn ~ 100. However, brachinites contain a few percent Ni-rich troilite and metal (e.g. Johnson et al. 1977, Nehru et al. 1992, Swindle et al. 1998, Warren and Kallemeyn 1989a), so bulk rock Fe/Mn ratios are higher than bulk silicate Fe/Mn ratios. The latter are more directly comparable to lunar or eucritic data. Measured Ga/Al ratios in brachinites are in the range 2×10^{-3} for Brachina to 7×10^{-4} for ALH 84025 (Table 19), significantly higher than those of eucrites which are on the order of 2×10^{-6}.

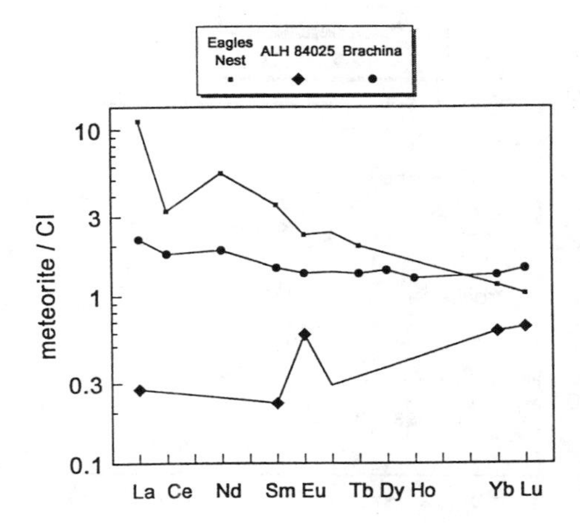

Figure 23. CI-normalized REE abundances in brachinites. Data from Johnson et al. (1977), Nehru et al. (1983), Swindle et al. (1998), Warren and Kallemeyn (1989a).

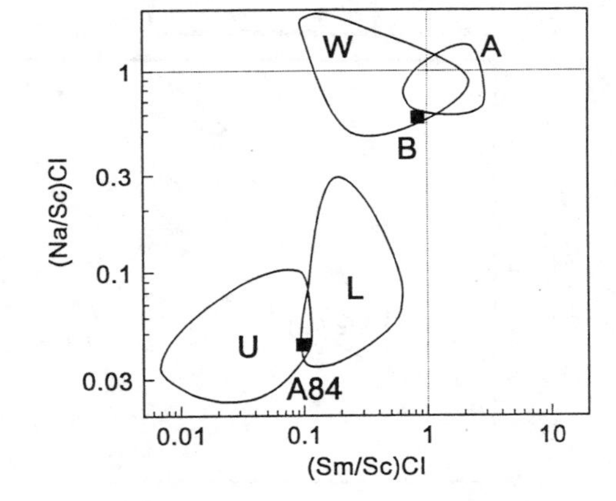

Figure 24. CI-normalized Na/Sc vs. Sm/Sc ratios for Brachina (B) and ALH 84025 (A84), compared to fields for acapulcoites (A), lodranites (L), ureilites (U) and winonaites-IAB silicates (W). Brachinite data are from Johnson et al. (1977), Nehru et al. (1983), Warren and Kallemeyn (1989a). Other data are from numerous sources.

Siderophile Elements. Siderophile element abundances in brachinites (Table 19) are near-chondritic (~0.1-1 × CI), similar to ureilites, but show a different pattern from ureilites (Fig. 25). Tungsten is enriched in Eagles Nest at ~4.6 × CI, but this may be due to terrestrial contamination. These abundances are much higher than in groups of rocks from differentiated bodies such as the Earth, Moon and Mars.

Noble Gases. Abundances of trapped noble gases in Brachina and ALH 84025 are high, and show a fractionated, or planetary-type pattern, similar to ureilites (Fig. 26). (A planetary-type noble gas pattern is one of increasing abundance, relative to solar elemental abundances, with increasing mass. This pattern is found for the terrestrial atmosphere, and in most chondrites.) Interelement ratios of Kr/Ar/Xe are similar in these two brachinites, and distinct from those in Chassigny (Ott et al. 1985a, 1987). In Eagles Nest, trapped noble gases are dominated by terrestrial contamination (Swindle et al. 1998).

Oxygen-isotopes. On an oxygen-isotope plot, the five brachinites and Divnoe plot along the line with slope ~1/2 defined by HED meteorites (Fig. 1c,d) and are

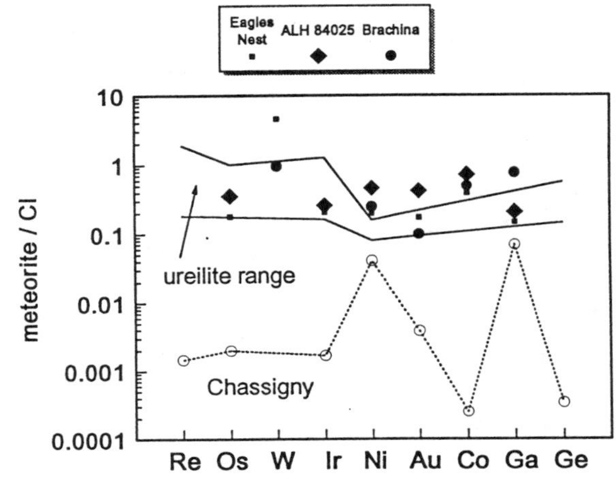

Figure 25. CI-normalized siderophile element abundances in brachinites, compared to the range in ureilites, and to Chassigny. Brachinite data are from Johnson et al. (1977), Nehru et al. (1983), Swindle et al. (1998) and Warren and Kallemeyn (1989a). Ureilite data are from numerous sources. Data for Chassigny from Treiman et al. (1986), Burghle et al. (1983), Warren (1987).

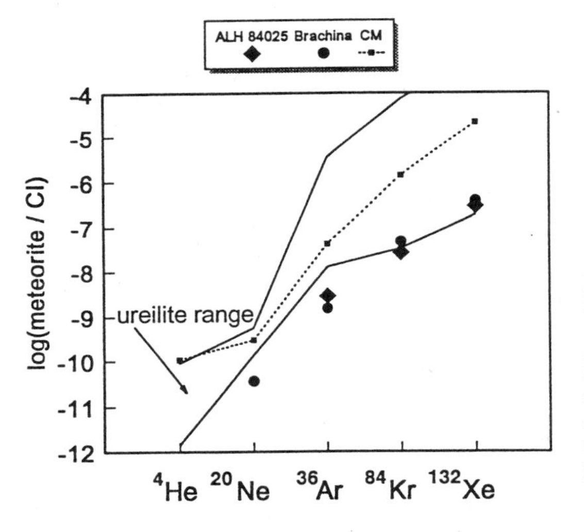

Figure 26. Noble gas abundances in Brachina (Bogard et al. 1983; Ott et al. 1983; Ott et al. 1985—1600° extraction) and ALH 84025 (Ott et al. 1987—1800° extraction) relative to cosmic abundances and compared to ureilites (for sources of ureilite data see Fig. 38). A typical CM chondrite (Haripura) as a representative planetary-type pattern is shown for reference (Mazor et al. 1970).

indistinguishable from the HED trend: $\Delta^{17}O$ = -0.26±0.08 for brachinites and -0.25±0.08 for HEDs (Clayton and Mayeda 1996).

Chronology

Chronological studies indicate that brachinites are extremely old. Brachina contains excess ^{53}Cr correlated with the Mn/Cr ratio (Wadhwa et al. 1998), which indicates that the short-lived radionuclide ^{53}Mn ($t_{1/2}$ ~ 3.7 Ma) was present at the time of last Cr isotopic equilibration. Combining the calculated $^{53}Mn/^{55}Mn$ ratio of Brachina with that ratio for the angrite LEW 86010 and its Pb-Pb age, yields an estimated formation age of 4.5637±0.0009 Ga for Brachina (Wadhwa et al. 1998).

In addition, Brachina, ALH 84025 and Eagles Nest all contain excess ^{129}Xe from in-situ decay of ^{129}I (Bogard et al. 1983, Ott et al. 1987, Swindle et al. 1993, 1998). In Brachina, a $^{128}Xe/^{129}Xe$ correlation indicates that retention of Xe began not later than ~4.4 Ga, and probably by 4.5 Ga ago (Bogard et al. 1983). In Eagles Nest, however, the lack of an I-Xe correlation indicates a disturbance resembling that in highly-shocked meteorites. The data indicate only that Eagles Nest began retaining Xe within ~50 Ma of primitive chondrites (Swindle et al. 1998). Fission track excesses due to decay of ^{244}Pu also indicate a formation age for Brachina of close to 4.5 Ga (Crozaz and Pellas 1983), and Sm-Nd isotopic data give a model age of 4.61 Ga (Bogard et al. 1983).

Other isotopic systems in brachinites indicate younger disturbances. The ^{39}Ar-^{40}Ar spectrum of Brachina shows a maximum age of 4.3 Ga and an average age of 4.1 Ga (Bogard et al. 1983), and its U-Th-He age is 400 Ma (Ott et al. 1985a). The ^{39}Ar-^{40}Ar spectrum for Eagles Nest yields a minimum age of 955 Ma and a maximum age of 1300 to 1500 Ma (Swindle et al. 1993). Rb-Sr isotopic data for a whole rock and a plagioclase-enriched sample of Brachina yield an apparent age of 2.5 Ga, with an initial $^{87}Sr/^{86}Sr$ ratio of 0.7070 (Bogard et al. 1983).

The cosmic-ray exposure age of Brachina is 2.0-3.5 Ma based on ^{3}He, ^{21}Ne and ^{38}Ar data (Bogard et al. 1983, Ott et al. 1983). ALH 84025 has a ^{21}Ne exposure age of 10.1 Ma (Ott et al. 1987), while for Eagles Nest the exposure ages are 44 Ma based on ^{21}Ne, and 49 Ma based on ^{38}Ar (Swindle et al. 1998). The $^{22}Ne/^{21}Ne$ ratios indicate that Brachina experienced very low shielding exposure (corresponding to ~3 cm depth), while Eagles Nest experienced higher shielding (Swindle et al. 1998).

Discussion

Nehru et al. (1983, 1992, 1996) consider brachinites to be a primitive achondrite group. Primitive achondrites are meteorites whose bulk compositions are approximately chondritic, but whose textures are igneous or metamorphic. The primitive achondrites are generally thought to be ultrametamorphosed chondrites or residues of very low degrees of partial melting on small parent bodies. Winonaites, silicates from IAB and IIICD irons, acapulcoites, lodranites, and recently even ureilites (Clayton and Mayeda 1996) have been considered to be primitive achondrites.

The near-chondritic abundances of lithophile elements, siderophile elements, and noble gases in Brachina support the interpretation that it has been only slightly modified by igneous processes, if at all. ALH 84025 also has siderophile elements and noble gases in near-chondritic abundances, but it is depleted in Al and Na, and has a fractionated REE pattern. Brachinites do not have a strictly chondritic mineralogy; their pyroxene is dominantly clinopyroxene rather than orthopyroxene, and only Brachina has a near-chondritic plagioclase content.

Nehru et al. (1992, 1996) suggest that brachinites evolved from CI-like material that was oxidized during planetary heating, converting orthopyroxene to olivine via the reaction $MgSiO_3 + Fe + 1/2O_2 \rightarrow (Mg,Fe)_2SiO_4$ (McSween and Labotka 1993). Partial melting occurred in some places, removing a basaltic component which is unsampled. Extensive thermal metamorphism produced equilibrated mineral compositions and textures. Equilibration temperatures calculated for brachinites range from 825° to 1070°C (orthopyroxene-clinopyroxene), 965° to 1246°C (Ca distribution between olivine and clinopyroxene), or 800° to 1080°C (olivine-chromite), supporting a high-temperature history (Nehru et al. 1996).

In contrast, Warren and Kallemeyn (1989a) have suggested that ALH 84025 is an olivine heteradcumulate, and that Brachina is an olivine orthocumulate. They point out that the coarse size, prismatic habit, and apparent preferred orientation of silicate grains in most of ALH 84025 are typical of cumulate rather than metamorphic rocks, and resemble similar features that have been argued to be the result of cumulus processes in ureilites (Berkley et al. 1980). In this model, the simple mineralogy, LREE-depletion, and low Al content of ALH 84025 indicate that it contains very little trapped liquid, while the presence of plagioclase, higher REE contents, and high Al content of Brachina indicate a far higher trapped liquid component. A cumulate model for Brachina was also discussed by Johnson et al. (1977) and Floran et al. (1978a).

A third model that has been discussed for Brachina is that it is an igneous rock that crystallized from a melt of its own composition (Johnson et al. 1977, Floran et al. 1978a). As a variant on this theme, Ryder (1982) modeled Brachina as an impact melt, but this model was largely contrived to accommodate Brachina and Chassigny on the same parent body.

There are enough petrologic and geochemical differences among the different brachinites to suggest that they may not have all formed by the same process. The question of whether brachinites are primitive achondrites, in the sense defined above, or are igneous cumulates, is one that applies to lodranites and ureilites as well. Trends in mineralogy, lithophile incompatible element abundances, and oxygen-isotopic systematics do not, however, show a coherent transition from the most chondritic to the least chondritic of the groups that have been considered to be primitive achondrites (Goodrich 1997c). Basaltic complements to brachinites are missing in the meteorite record, and this is also true for lodranites and ureilites. Whether the vagaries of meteorite delivery simply never sent them earthward, or whether they never existed remains unknown.

ACAPULCOITES AND LODRANITES

Acapulcoites and lodranites have evolved from being one or two "oddities" to a substantial group of meteorites that promise to greatly increase our understanding of partial melting and melt migration on asteroids early in the history of the solar system. This stems largely from the dramatic increase in the number of acapulcoites and lodranites recognized in our meteorite collections. We now recognize 9 acapulcoites which sample 5 distinct fall events and 14 lodranites representing 12 distinct fall events. The unique meteorite LEW 86220 is clearly related to this group, but is neither an acapulcoite or lodranite. With this increased sample suite, a number of recent, comprehensive studies have examined, and in some cases identified, this growing number of group members. Broader reviews of this group have been published by Nagahara (1992), Takeda et al. (1994a), Mittlefehldt et al. (1996) and McCoy et al. (1996b, 1997a,b).

Mineralogy and petrology

Acapulcoites and lodranites are composed of orthopyroxene, olivine, chromian diopside, sodic plagioclase, Fe,Ni metal, schreibersite, troilite, whitlockite, chlorapatite, chromite and graphite, a mineral assemblage broadly similar to that found in ordinary chondrites. However, mineral compositions, mineral abundances and textures of acapulcoites and lodranites differ from ordinary chondrites and, to a certain extent, from each other.

The distinction between acapulcoites and lodranites is not clear cut and disagreement exists as to the definitions and memberships of these groups. Acapulco is a fine-grained meteorite with approximately chondritic abundances of olivine, pyroxene, plagioclase, metal and troilite, while Lodran is a coarse-grained meteorite which is depleted in troilite and plagioclase. Comparison to these type meteorites suffices for classification of most of the members of this clan. However, meteorites like EET 84302 are problematic in being coarser-grained and depleted in troilite but not plagioclase. Most authors consider this meteorite intermediate between acapulcoites and lodranites (e.g. Takeda et al. 1994a, Mittlefehldt et al. 1996, McCoy et al. 1997a,b).

Modally, acapulcoites are similar to chondrites. In acapulcoites, orthopyroxene is more abundant than olivine (Yugami et al. 1997), as one might expect for chondritic silicates with relatively Mg-rich mafic mineral compositions. Plagioclase and troilite appear to occur in approximately the same abundances observed in ordinary chondrites (~10 wt % plagioclase and 5 to 6 wt % troilite), despite uncertainties due to sampling heterogeneity and weathering. In contrast, lodranites are sometimes enriched in olivines and often depleted in plagioclase and/or troilite relative to chondrites (McCoy et al. 1997b, Yugami et al. 1997). It should be emphasized that some lodranites contain relatively abundant plagioclase, but lack troilite, while others lack plagioclase but contain moderate abundances of troilite. In addition, plagioclase can be heterogeneously distributed, such as in EET 84302 (Takeda et al. 1994a).

Texturally, acapulcoites and lodranites are fine- to medium-grained equigranular rocks (Fig. 27a,c,e). Recrystallization is evident from abundant 120° triple junctions (Palme et al. 1981). Rare relict chondrules have been reported in the acapulcoites Monument Draw (McCoy et al. 1996b), Yamato 74063 (Yanai and Kojima 1991), and ALHA77081 (Schultz et al. 1982). In each case, only one or two relict chondrules have been observed. McCoy et al. (1996b) demonstrated that acapulcoites are significantly finer-grained (150 to 230 µm) than lodranites (540 to 700 µm), while EET 84302 is intermediate in grain size (340 µm). Both acapulcoites and lodranites generally exhibit only very minor shock effects (S1-S2), although a small number of lodranites (e.g. MAC 88177, Y-8307) are more heavily shocked, exhibiting shock melt veins which cross-cut the specimens. Substantial debate exists about the genesis of 1- to 5-µm blebs of Fe,Ni-FeS found in the cores of mafic silicates. Some authors (e.g. Takeda et al. 1994a) argue that these blebs decorate pre-existing planar fractures formed during an earlier shock event and that were subsequently annealed. Other authors (e.g. Schultz et al. 1982) suggest that these were trapped during chondrule formation and mark the outlines of relict chondrules. Zipfel et al. (1995) suggest that these were trapped during a period of extensive melting of acapulcoites. Acapulcoites and lodranites also exhibit a wide range of veins. Most acapulcoites contain mm-sized Fe,Ni-FeS veins which cross-cut all silicate phases. Monument Draw is unique in containing a cm-sized metal vein (Fig. 27b) with adjacent troilite-enriched areas. Acapulco and Monument Draw both contain phosphate veins hundreds of microns in length. Lodranites often exhibit plagioclase grains pinching between and partially enclosing mafic silicates (Fig. 27d).

Acapulcoites and lodranites have Mg-rich mafic silicate compositions compared to ordinary chondrites. Mafic silicate compositions of acapulcoites range from $Fa_{4.2-11.9}$ and $Fs_{6.5-12.6}$. In each case, the Fa content of olivine is approximately equal to or slightly less than the Fs content of orthopyroxene. McCoy et al. (1996b) reviewed evidence for zoning in mafic silicates in acapulcoites. Yugami et al. (1993) found orthopyroxene grains depleted in FeO in the rims relative to the core in low-FeO acapulcoite ALHA81187, although this feature is not observed in other acapulcoites. Lodranites exhibit similar ranges in composition of olivine ($Fa_{3.1-13.3}$) and orthopyroxene ($Fs_{3.7-13.8}$). Representative analyses that span this range are given in Table 20. Reverse zoning with rims lower in FeO than cores is common in both olivine and orthopyroxene of lodranites (McCoy et al. 1997a and references therein). Zoning is particularly prominent in olivines in high-FeO lodranites and in orthopyroxenes in lodranites in which the average Fs content of orthopyroxene is greater than the average Fa content of olivine. Chromian diopside exhibits a smaller range in FeO concentrations ($Fs_{4.3-5.2}$) and contains 0.74 to 1.74 wt % Cr_2O_3. Plagioclase compositions range from $An_{12.3-30.9}$ (McCoy et al. 1997a and

Figure 27. Photographs of acapulcoites and lodranites. (a) Photomicrograph of Acapulco showing its fine-grained, equigranular texture. Field of view is 5.2 mm. (b) Hand sample of Monument Draw with cm-sized metal veins. *On the following page:* (c) Photomicrograph of EET 84302, which exhibits a medium-grained, equigranular texture. Field of view is 5.2 mm. (d) Photomicrograph of Yamato 8002 with interstitial plagioclase, suggestive of melt migration. Field of view is 1.3 mm. (e) Photomicrograph of Lodran, which is coarse-grained and depleted in metal and troilite. Field of view is 5.2 mm.

references therein). Calculated two-pyroxene equilibration temperatures exhibit a substantial range from ~1000° to 1200°C.

One meteorite within the acapulcoite-lodranite clan is so unusual in both its texture and modal mineralogy as to deserve special attention. LEW 86220 (Fig. 27f) is linked to the other acapulcoites and lodranites through both its oxygen-isotopic and mineral

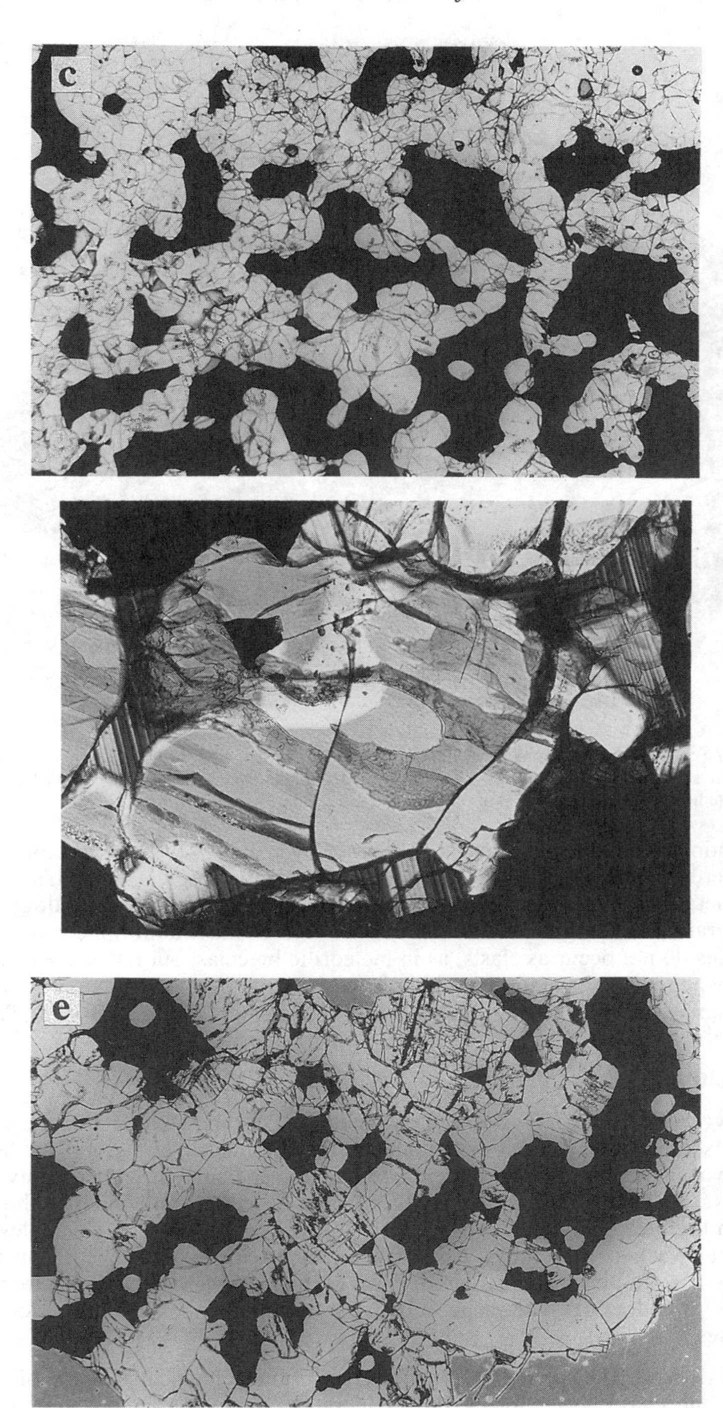
c
e

Figure 27 (cont'd). Photographs of acapulcoites and lodranites. (f) Photomosaic in crossed polars of an entire thin section of LEW 86220. Coarse-grained gabbroic, and Fe,Ni–FeS melts have intruded an acapulcoite host. Maximum dimension of specimen is 12 mm.

compositions (Clayton and Mayeda 1996, McCoy et al. 1997b). The meteorite consists of two different lithologies, one of acapulcoite modal mineralogy and grain size, the other of Fe,Ni metal, troilite and plagioclase-augite (basaltic-gabbroic) mineralogy and a very coarse grain size (plagioclase grains reach 9 mm in maximum dimension). These two lithologies do not occur as clasts, as in meteoritic breccias, but rather the coarse-grained lithology appears to have intruded into the acapulcoite lithology. A similar dichotomy of lithologies is found in silicate inclusions in the Caddo County IAB iron (Takeda et al. 1997a).

Composition

Oxygen-isotopic composition. Clayton et al. (1992) pointed out that the variability in oxygen-isotopic compositions of acapulcoites and lodranites, as a group, are beyond what could be attributed to mass-dependent fractionation and analytical uncertainty. Fayalite in olivine vs. $\Delta^{17}O$ (Fig. 28) are weakly correlated in lodranites, similar to the correlation found in ureilites (Clayton and Mayeda 1988). However, when only samples which had been acid-washed to remove terrestrial contamination (rust) were considered, the correlation strengthened considerably, suggesting primary nebular heterogeneity was preserved in lodranites. The range in both parameters is much smaller in lodranites than in ureilites.

Bulk chemical composition. A wide array of bulk chemical analyses of acapulcoites and lodranites have been published in the literature. Table 21 presents bulk major element

Table 20. Representative mineral compositions for acapulcoites and lodranites.

	OLIVINE					ORTHOPYROXENE					CLINOPYROXENE					PLAGIOCLASE		
	ALH A81187 (1)	Monument Draw (1)	Gibson (2)	EET 84302 (2)	LEW 88280 (2)	ALH A81187 (1)	Monument Draw (1)	Gibson (2)	EET 84302 (2)	LEW 88280 (2)	ALH A81187 (1)	Monument Draw (1)	Gibson (2)	EET 84302 (3)	LEW 88280 (2)	ALH A81187 (3)	Gibson (2)	EET 84302 (3)
	Chemical Composition (wt %)																	
SiO_2	41.5	42.1	42.0	40.4	39.8	58.0	58.3	57.9	57.2	55.6	54.4	54.5	53.6	54.0	52.1	63.5	64.3	62.4
Al_2O_3	n.d.	n.d.	b.d.	n.d.	b.d.	n.d.	0.28	0.42	n.d.	0.46	n.d.	0.75	1.00	1.18	1.07	23.1	22.6	23.9
Cr_2O_3	n.d.	n.d.	0.07	n.d.	b.d.	n.d.	0.25	0.47	n.d.	0.50	n.d.	1.18	1.74	1.56	1.60	n.d.	n.d.	n.d.
TiO_2	n.d.	n.d.	b.d.	n.d.	b.d.	n.d.	0.20	0.16	n.d.	0.12	n.d.	0.53	0.41	0.56	0.31	n.d.	n.d.	n.d.
FeO	4.27	9.45	3.10	8.17	12.2	4.79	6.92	3.86	5.67	7.88	2.06	2.63	2.62	2.15	2.98	0.16	0.28	0.13
MgO	54.2	47.1	54.8	50.2	46.3	34.8	32.0	34.4	34.5	32.9	19.0	16.8	17.4	18.0	17.7	0.03	b.d.	0.01
MnO	0.67	0.52	0.39	0.62	0.48	0.83	0.55	0.51	0.73	0.54	0.53	0.31	0.30	0.31	0.34	n.d.	n.d.	n.d.
CaO	b.d.	b.d.	b.d.	b.d.	b.d.	1.73	0.84	1.09	0.96	1.32	20.7	21.7	21.5	21.2	22.0	3.89	3.87	4.86
Na_2O	n.d.	n.d.	n.d.	n.d.	n.d.	n.d.	b.d.	b.d.	n.d.	b.d.	n.d.	0.74	0.65	0.70	0.71	9.27	9.27	8.71
K_2O	n.d.	n.d.	n.d.	n.d.	n.d.	n.d.	n.d.	n.d.	n.d.	n.d.	n.d.	n.d.	n.d.	n.d.	n.d.	0.29	0.66	0.30
Total	100.64	99.17	100.36	99.39	98.78	100.15	99.34	98.81	99.06	99.32	96.69	99.15	99.22	99.66	98.81	100.24	100.98	100.31
	Atomic Formula (O = 4 for olivine, 6 for pyroxene, 8 for plagioclase)																	
Si	0.988	1.035	0.996	0.991	1.000	1.995	2.027	2.003	1.993	1.959	2.021	1.994	1.965	1.963	1.931	2.801	2.821	2.758
Al	--	--	--	--	--	--	0.011	0.017	--	0.019	--	0.032	0.043	0.051	0.047	1.201	1.169	1.245
Cr	--	--	0.001	--	--	--	0.007	0.013	--	0.014	--	0.034	0.050	0.045	0.047	--	--	--
Ti	--	--	--	--	--	--	0.005	0.004	--	0.003	--	0.015	0.011	0.015	0.009	--	--	--
Fe	0.085	0.194	0.061	0.168	0.256	0.138	0.201	0.112	0.165	0.232	0.064	0.080	0.080	0.065	0.092	0.006	0.010	0.005
Mg	1.924	1.726	1.937	1.836	1.734	1.784	1.659	1.774	1.792	1.728	1.052	0.916	0.951	0.975	0.978	0.002	--	0.001
Mn	0.014	0.011	0.008	0.013	0.010	0.024	0.016	0.015	0.022	0.016	0.017	0.010	0.009	0.010	0.011	--	--	--
Ca	--	--	--	--	--	0.064	0.031	0.040	0.036	0.050	0.824	0.851	0.844	0.826	0.874	0.184	0.182	0.230
Na	--	--	--	--	--	--	--	--	--	--	--	0.052	0.046	0.049	0.051	0.793	0.789	0.746
K	--	--	--	--	--	--	--	--	--	--	--	--	--	--	--	0.016	0.037	0.017
Total Cations	3.012	2.965	3.003	3.009	3.000	4.005	3.958	3.978	4.007	4.021	3.977	3.984	4.000	3.999	4.039	5.003	5.007	5.002
	Molar Mineral End Members and mg# (100(MgO/(MgO+FeO)))*																	
Wo	--	--	--	--	--	3.2	1.7	2.1	1.8	2.5	42.5	46.1	45.0	44.2	44.9	--	--	--
En	--	--	--	--	--	89.9	87.7	92.1	89.9	86.0	54.2	49.6	50.7	52.3	50.3	--	--	--
Fs	--	--	--	--	--	6.9	10.6	5.8	8.3	11.6	3.3	4.4	4.3	3.5	4.8	--	--	--
Ab	--	--	--	--	--	--	--	--	--	--	--		--		--	79.8	78.3	75.1
An	--	--	--	--	--	--	--	--	--	--	--		--		--	18.5	18.1	23.2
Or	--	--	--	--	--	--	--	--	--	--	--		--		--	1.6	3.7	1.7
mg#	95.8	89.9	96.9	91.6	87.1	92.8	89.2	94.1	91.6	88.2	94.3	91.9	92.2	93.7	91.4	--	--	--

Data sources: (1) McCoy et al. (1996), (2)McCoy et al. (1997a), (3) Mittlefehldt et al. (1996).

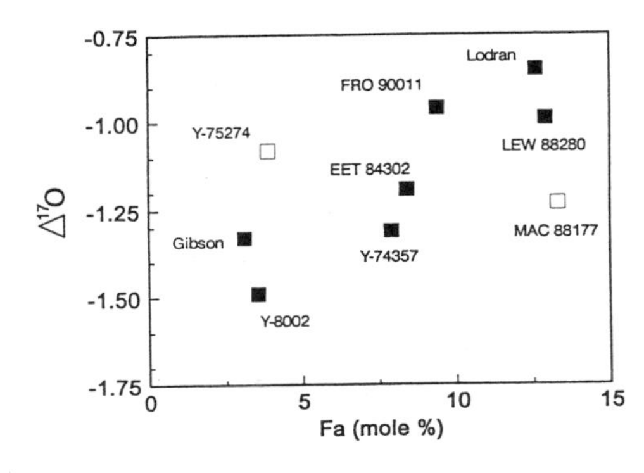

Figure 28. $\Delta^{17}O$ vs. Fa in olivine for lodranites. Non-Antarctic and acid-washed Antarctic lodranites are plotted as solid squares; non-acid-washed lodranites as open squares. The positive correlation for the former suggests nebular heterogeneity within the chondritic precursor to the lodranites.

Table 21. Bulk major element analyses (in wt %) of acapulcoites and lodranites.

	Acapulco (1)	ALHA 77081 (2)	Y-74063 (1)	Y-74357 (1)	Y-75274 (3)	Y-791493 (3)
SiO_2	37.8	40.9	39.0	37.7	24.1	34.9
TiO_2	0.07		0.08	0.09	0.02	0.05
Al_2O_3	2.06	2.27	2.96	0.20	0.18	0.90
Cr_2O_3	0.45	0.72	0.34	0.96	0.12	0.81
FeO	5.69	4.98	9.69	4.00	2.36	6.12
MnO	0.39	0.39	0.19	0.37	0.21	0.42
MgO	26.8	26.0	27.0	27.0	27.0	29.5
CaO	1.94	0.83	2.68	3.65	0.45	1.54
Na_2O	0.86	1.01	0.83	0.10	0.14	0.21
K_2O	0.06	0.08	0.07	0.02		
P_2O_5	0.37		0.46	0.26	0.27	0.49
Fe_2O_3	1.82		0.91	7.55	2.82	6.02
FeS	5.79	8.10	9.31	1.85	1.14	2.10
Fe	14.5	14.6	4.89	15.2	39.4	14.3
Ni	1.31	1.60	0.98	0.98	1.84	1.13
Co	0.11	0.08	0.03	0.08	0.14	0.07
sum	99.98	101.53	99.41	99.90	100.16	98.57

Data sources: (1) Yanai and Kojima (1991), (2) Nagahara and Ozawa (1986), (3) Haramura et al. (1983)

analyses, while Table 22 presents minor and trace elements. A complete discussion of this data is beyond the scope of this paper and the reader is referred to the original publications. However, it is clear that bulk chemical data can shed substantial light upon the processes which have effected these rocks. As noted by Mittlefehldt et al. (1996), a plot of Se/Co vs. Na/Sc (Fig. 29) is a proxy for tracking the abundances of the host minerals in which these elements are sighted. Thus, Se/Co is equivalent to troilite/metal, while Na/Sc equates to plagioclase/pyroxene. During partial melting, removal of the Fe,Ni-FeS eutectic should cause a large decrease in the Se/Co ratio. Continued partial melting at higher temperature to produce a basaltic melt, which is highly enriched in plagioclase, will cause a decrease in Na/Sc ratio. Thus, schematically, a meteorite should follow the track of decreasing Se/Co followed by a decrease in Na/Sc shown in Fig. 29. Note that many meteorites do not fall along this idealized path.

Table 22. Average bulk minor and trace element compositions of acapulcoites and lodranites.

		ACAPULCOITES							TRANSITIONAL	LODRANITES			
		Acapulco (1)	ALH 78230 (2)	ALHA 77081 (3)	ALHA 81187 (4)	ALHA 81261 (5)	Monument Draw (6)	Y-74063 (7)	EET 84302 (4)	FRO 90011 (6)	LEW 88280 (4)	Lodran (8)	MAC 88177 (5)
Na	mg/g	6.51	7.53	7.29	5.54	6.92	7.08	7.07	4.74	0.77	0.58	0.32	0.31
K	μg/g	510	540	660		620	390			70		14	6
Sc	μg/g	8.11	10.5	9.91	8.99	8.56	9.05	10.0	6.19	5.71	8.01	6.20	8.88
V	μg/g	83		93		92	96	91		82		72	80
Cr	mg/g	6.02	5.00	6.80	4.14	5.54	2.46	2.69	12.4	5.21	6.02	3.90	3.69
Mn	mg/g	3.07		2.91		2.70	2.60	2.64		2.90		2.60	3.19
Co	mg/g	0.88	0.77	0.77	0.73	0.76	0.70	0.46	1.76	0.71	0.47	1.23	0.17
Ni	mg/g	16.6	14.8	13.9	10.5	14.2	17.9	11.9	26.4	12.6	12.2	22.5	4.2
Zn	μg/g	220		280	150	240	100	90		130	170		60
Ga	μg/g	8.9		10.8		9.3	6.9			8.8			3.1
As	μg/g	2.2	1.8	2.1	1.8	2.3	2.7	1.7	4.5	2.3	1.8		0.5
Se	μg/g	9.0		9.4	2.6	8.6	11.5	22.5		5.0	9.2		4.8
Sb	ng/g	90		60	80	80					70		20
La	μg/g	0.70	0.27	0.26	0.27	0.35	0.32	0.28	0.05	0.06	0.09	0.02	0.15
Sm	μg/g	0.25	0.21	0.19	0.19	0.22	0.19	0.19	0.04	0.05	0.08	0.03	0.09
Eu	μg/g	0.11	0.08	0.09	0.09	0.09	0.09	0.10	0.01	0.01	0.01		0.06
Tb	μg/g	0.05		0.06	0.06				0.04	0.02	0.03	0.02	
Dy	μg/g		0.39			0.46	0.57	0.45		0.32	0.29		0.33
Yb	μg/g	0.23	0.27	0.28	0.24	0.25	0.20	0.23	0.12	0.14	0.21	0.13	0.21
Lu	μg/g	0.030	0.041	0.050	0.040	0.040	0.043	0.040	0.020	0.030	0.030	0.020	0.040
Hf	μg/g	0.13		0.17	0.20	0.22	0.24						0.08
Os	ng/g	960		950		690	220		1570				460
Ir	ng/g	870		780	810	600	310	2600	860	60	120	25	570
Au	ng/g	210		190	200	190	180	510	240	260	200	46	170

(1) average of analyses by Kallemeyn and Wasson (1985), Palme et al. (1981), Zipfel et al. (1995); (2) analysis by Fukuoka and Kimura (1990); (3) average of analyses by Kallemeyn and Wasson (1985), Schultz et al. (1982); (4) average of analyses by Mittlefehldt et al. (1996); (5) average of analyses by Mittlefehldt et al. (1996), Zipfel and Palme (1993); (6) analysis by Zipfel and Palme (1993); (7) average of analyses by Kimura et al. (1992); (8) average of analyses by Fukuoka et al. (1978).

Chronology

A variety of chronometers which close at relatively high temperature have been applied to Acapulco. Prinzhofer et al. (1992) reported a Sm-Nd isochron age of 4.60±0.03 Ga. This age is slightly older than the generally accepted age of the solar system (4.56 Ga) and has been questioned by a number of researchers (e.g. McCoy et al. 1996b). Göpel et al. (1992) reported a precise Pb-Pb isochron age of 4.557±0.002 Ga for U-enriched phosphates from Acapulco and the ^{129}I-^{129}Xe formation interval between Bjürbole and Acapulco is only 8 Ma (Nichols et al. 1994). Pellas et al. (1997) have shown that phosphates in Acapulco contained ^{244}Pu ($t_{1/2}$ 81.8 Ma) when they cooled through the fission track retention temperature, about 350-390 K. Thus, Acapulco must have formed only very shortly after the solar system.

A wider number of both acapulcoites and lodranites have had ^{39}Ar-^{40}Ar ages determined. These include Acapulco (4.503±0.011, 4.514±0.005 Ga), Monument Draw (4.517±0.011 Ga), ALHA81187 (4.507±0.024 Ga), ALHA81261 (4.511±0.007 Ga), EET 84302 (4.519±0.017 Ga) and Gibson (4.49±0.01 Ga) (McCoy et al. 1996b, 1997a; Mittlefehldt et al. 1996, Pellas et al. 1997). These ^{39}Ar-^{40}Ar ages are all substantially younger than the Pb-Pb age or I-Xe formation interval, suggesting that they reflect a prolonged period of cooling from high temperature (McCoy et al. 1996b). Pellas et al. (1997) showed that Acapulco initially cooled rapidly (~100°C/Ma) from high temperature, but cooled slowly at low temperature (~2°C/Ma).

Cosmic-ray exposure ages are between ~5.5 to 7 Ma for all acapulcoites and most lodranites examined to date (McCoy et al. 1996b, 1997a), possibly indicating sampling by a single impact event from a common parent body. This clumping between acapulcoites and lodranites strengthens arguments from similarities in oxygen-isotopic and mineral compositions and mineralogy, as well as the presence of transitional meteorites like EET 84302, that acapulcoites and lodranites are samples of a single parent body.

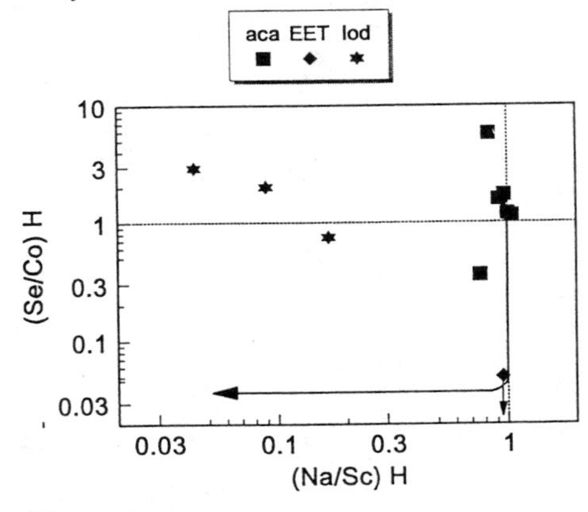

Figure 29. H-chondrite-normalized Se/Co vs. Na/Sc for acapulcoites (aca) and lodranites (lod).and the transitional meteorite EET 84302 (EET). See Mittlefehldt et al. (1996) for sources of data. These ratios serve as proxies for troilite/metal vs. plagioclase/pyroxene. Partial melting of a chondritic source should cause the residue to follow the arrowed line. High ratios for Se/Co in lodranites suggests that the Fe,Ni-FeS melts did not efficiently segregate from the residue when basaltic partial melts were removed, or that they were later re-introduced. The EET 84302 Se/Co ratio is an upper limit.

Discussion

Acapulcoites and lodranites formed from a chondritic precursor, based on the similarities in modal mineralogy between acapulcoites-lodranites and ordinary chondrites, the presence of rare relict chondrules in some acapulcoites, and the unfractionated bulk composition of acapulcoites (Fig. 29). This chondritic precursor was probably unlike known chondrites, differing in mineral, oxygen-isotopic and bulk compositions.

Acapulcoites and lodranites were extensively heated and experienced variable amounts of partial melting (Nagahara 1992, Takeda et al. 1994a, Mittlefehldt et al. 1996, McCoy et al. 1996b, 1997a,b). Disagreement exists over the extent of this partial melting in individual meteorites and in the efficiency with which melt migrated from their source regions. McCoy et al. (1996b, 1997a) argued that mm-sized Fe,Ni-FeS veins within acapulcoites represent melting of the Fe,Ni-FeS eutectic between ~950° and 1050°C. However, since these veins cross cut silicate phases, these authors argue that silicates must not have melted. Thus, acapulcoites experienced very low degrees of partial melting. Siderophile and chalcophile element evidence shows that these metallic-sulfide melts did not migrate out of many of the acapulcoites, although ALHA81187 lost and Y-74063 gained a small amount of these melts (Mittlefehldt et al. 1996). In contrast, lodranites exhibit depletions in troilite and plagioclase suggesting removal of the Fe,Ni-FeS and plagioclase-pyroxene partial melts and peak temperatures up to ~1250°C. In spite of the low modal troilite contents, some lodranites have Se/Co and Ir/Ni ratios indicating additions of a small amount of low temperature metallic-sulfide melt (Mittlefehldt et al. 1996). McCoy et al. (1997a) suggest that the coarser grain size of the lodranites reflects accelerated grain growth in the presence of a silicate melt.

Zipfel et al. (1995) hold a very different view of acapulcoite petrogenesis. These authors have suggested that Acapulco experienced extensive silicate partial melting and peak temperatures upwards of 1200°C. In part, these authors suggest that the inhomogeneous distribution of phosphates resulted from silicate melting. Mittlefehldt et al. (1996) have shown that acapulcoites have H chondritic ratios of Na/Sc, Sm/Sc and Eu/Yb, indicating that silicate partial melts, if they existed, did not migrate out of the acapulcoites. The presence of phosphate veins in some acapulcoites may suggest, instead, that phosphates are included in melting at the Fe,Ni-FeS eutectic (McCoy et al. 1996b), or that they participated in redox reactions with the metal at high temperatures (Mittlefehldt et al. 1996).

Takeda et al. (1994a) and Mittlefehldt et al. (1996) pointed out that while petrologic features might suggest partial melting of lodranites, melt migration was certainly complex. In particular, similar Se/Co ratios (Fig. 29) between acapulcoites and lodranites suggests that the Fe,Ni-FeS eutectic melt was not efficiently removed in at least some lodranites. Melt migration was addressed in considerable detail by McCoy et al. (1997b). In agreement with the work of Nagahara (1992) and Mittlefehldt et al. (1996), these authors argued that the migration of immiscible Fe,Ni-FeS and basaltic partial melts is complicated and could vary greatly depending on volatile contents, magma migration rates and conduit geometry. Indeed, Nagahara (1992) first pointed out that partial melts can migrate into, as well as out of, a source region, thus trapping migrating partial melts.

The migration of partial melts outside of the immediate source region was addressed by McCoy et al. (1997b), who argued that LEW 86220 sampled partial melts which migrated from a hotter lodranite source region and were trapped in a cooler acapulcoite host. At least some of these partial melts may have reached the surface where they may have been removed from the parent body by explosive volcanism (McCoy et al. 1997b). Recovery of an acapulcoite-lodranite regolith breccia would greatly elucidate whether basaltic partial melts formed thick deposits on the surface of the body.

Finally, all authors agree that these processes occurred very early in the history of the solar system, as reflected in the Pb-Pb, I-Xe and ^{244}Pu fission track retention ages. A substantial period of cooling or metamorphism (~50 Ma) occurred before Ar-Ar closure, leading to the younger ^{39}Ar-^{40}Ar ages, and another 100 Ma of cooling is required to explain the ^{244}Pu fission track ages. Finally, a small number of impacts, including one major impact at ~5.5 to 7 Ma, liberated most of the acapulcoites and lodranites from their common parent body.

UREILITES

Mineralogy, mineral chemistry and petrography

Olivine, pyroxene and chromite. Ureilites are essentially carbon-bearing ultramafic rocks. Their mineralogy is olivine + pyroxene, with <10% dark interstitial material, often referred to as matrix or vein material, consisting of varying amounts of carbon, metal, sulfides, and minor fine-grained silicates. None of the monomict ureilites contain plagioclase. The unusual ureilite LEW 88774 contains ~6% chromite. Modal pyroxene/(pyroxene+olivine) ratios range from 0 to ~0.9 (Table 23). In approximately 77% of the unpaired monomict ureilites, the sole pyroxene reported is pigeonite. In the remaining, a variety of pyroxene assemblages occur: pigeonite + augite, pigeonite + orthopyroxene, pigeonite + augite + orthopyroxene, orthopyroxene, and orthopyroxene + augite. The issue of the structural state of very low-Ca pyroxenes appears to be complex. For example, in LEW 88201, LEW 85440 and Y-791538, low-Ca pyroxene (~Wo_5) shows fine linear features under cross-polarized light (Takeda 1989, Takeda et al. 1992). Despite uniform composition, the Y-791538 pyroxene shows diffuse streaks in x-ray diffraction, which suggest that it was originally orthopyroxene but developed clinopyroxene lamellae during shock (Takeda 1989). In LEW 85440, TEM observations show an intimate alternation of ~10 nm-thick orthopyroxene and clinopyroxene lamellae

Table 23. Summary of petrographic information on ureilites. Ureilites are grouped by texture. Mineral compositions listed are for homogeneous cores.

	olivine Fo	pigeonite Wo/En	opx Wo/En	augite Wo/En	**py	texture	shock level	notes	references
Acfer 277	79	9.6/73			0.55, 0.33	typical	medium		1,2
ALH83014	82	8.0/77			0.13	typical	very low		AMN 8-1
ALHA77257	85	6.8/80			0.16	typical	low		13-15
ALHA78019	76	9.7/71			0.05	typical	very low		13-15
ALHA78262	78	8.4/72				typical	low		13-15
Asuka 87031	79	7.5/75			~0	typical	medium		3
Asuka 881931	76	8.8/73				typical	low-medium		3
EET 83225	87	10/78				typical	medium?		AMN 8-1
EET 87517	92	5.0/87			~0.5	typical	low		AMN 11-2
EET 90019	89	10/80				typical			AMN 15-1
FRO 90054	87		4.8/84	39/55	0.58,0.75	typical		1	22,23
GRA95205	79		4.0/77			typical			AMN 20-1
Hammadah al Hamra 126	79	8.4/74			0.3	typical	medium		4,35,36
*Havero	79	2.5/81			0.17	typical	high	2	5
Hughes 009	87		4.9/84	37/56	0.7	typical	medium		7,38
Jalanash	81	7.8/75			0.15	typical	medium		37
Kenna	79	9.8/72			0.25	typical	medium		5
*Lahrauli	79	7.7/74				typical	medium?		8,9
LEW 85328	80	9.0/74			0.33	typical	low		AMN 10-1
LEW85440	92		5.0/87	36/59	0.43,0.50	typical		3	24-26, AMN 13-2, AMN 13-3
LEW 88006	82	8.0/76				typical	low		AMN 13-2
LEW 88772	84	11/78				typical			AMN 16-1
*Novo Urei	79	10/72			0.31	typical	medium		5
Nullarbor 010	79	9.0/73	3.0/77		0.33	typical	very low	4	11
QUE 93336	77	10/72				typical	medium-high?	5	AMN 18-1
Roosevelt County 027	79	8.0/75			0.25	typical	low-medium		12
Y-74123	79	6.8/75			~0	typical			30-32
Y-74659	91	7.3/86	4.5/88		0.52	typical	low-medium		30,32,34
Y-790981	78	8.9/73			0.11	typical	medium	6	13,16,32
Y-791538	91	9.0/84	4.9/88		0.45	typical	low-medium		3,24,25
Y-791839	75	variable				typical			3
Y-82100	81	8.6/77			0.2	typical	low		3
Y-8448	78	variable				typical	low-medium		3

Table 23 (cont.). Summary of petrographic information on ureilites. Ureilites are grouped by texture. Mineral compositions listed are for homogeneous cores.

	olivine Fo	pigeonite Wo/En	opx Wo/En	augite Wo/En	**py	texture	shock level	notes	references
PCA 82506	78	6.0/76			0.48	typical/poikilitic	low		13,16
RKPA80239	83	6.7/78				typical/poikilitic	low		13,16
ALHA81101	79	8.1/78				mosaicized	high		16,17
Dar al Gani 084	79	5.0/81				mosaicized	high		4
*Dyalpur	84	11/76			0.41	mosaicized	high		5
FRO90036	80		3.0/81			mosaicized	high	13	1,22
Goalpara	79	4.4/76			0.33	mosaicized	high		5
LEW 86216	81		1.7/88		0.02	mosaicized	high		20
Y-74154	84	8.1/79				mosaicized	high	6	3
Dingo Pup Donga	84	5.0/81			~0.5	euhedral	medium		5
ALH 82106	95	4.9/90		37/60	0.35	bi-modal		7	13,18, AMN 9-3
EET 87511	85		4.5/82	36/55		bi-modal	low	8	20, AMN 11-2
Hammadah al Hamra 064	78		4.6/77	32/55	0.65	bi-modal	medium		4,37
META78008	77	16/66	4.5/76	32/55	0.05,0.37	bi-modal	medium		29
Y-74130	78	14/68	4.2/78	32/55	~0.5,~1	bi-modal			29,33
EET 83309	80-84	variable			0.27	polymict		9	19
EET 87720	79-87	8.0/83		35/52		polymict			21, AMN 12-3, AMN 13-1
Nilpena	80	9.0/75				polymict		10	10
North Haig	76-92	variable			0.26	polymict		11	5
Hajma	85	9.0/78				typical?	medium-high		6
LEW 87165	85	?Fs 13				?			AMN 12-1, AMN 13-1
LEW 88774	75		4.2/75	33/53	0.83-0.87	bi-modal?	medium?	12	27,28

*ureilite falls; **py = modal pyroxene/(pyroxene+olivine).

Notes: (1) paired with FRO93008; (2) pyroxene is clinobronzite; (3) paired with LEW 88201, LEW 882821 and LEW 88012; (4) paired with Nova 001; (5) paired with QUE 93341; (6) other highly variable pyroxene compositions are reported - these may be interstitial or shock melts; (7) paired with ALH 82130 and ALH 84136; (8) paired with EET 87523 and EET 87717; (9) compositions given are for monomict ureilitic clasts; (10) compositions given are for monomict ureilitic clasts; possibly paired with North Haig; (11) possibly paired with Nilpena; (12) contains ~6% chromite; (13) paired with FRO90168.

References: (1) Baba et al (1993); (2) Bland et al (1992); (3) Yanai and Kojima (1995b); (4) Weber and Bischoff (1996); (5) Berkley et al (1980); (6) Hutchison (1977); (7) Wlotzka (1994); (8) Malhotra (1962); (9) Bhandari et al (1981); (10) Jaques amd Fitzgerald (1982); (11) Treiman and Berkley (1994); (12) Goodrich et al (1987a); (13) Goodrich et al (1987b); (14) Goodrich and Berkley (1986); (15) Berkley and Jones (1982); (16) Berkley (1986); (17) Saito and Takeda (1990); (18) Berkley et al (1985); (19) Prinz et al. (1987); (20) Berkley (1990); (21) Warren and Kallemeyn (1991); (22) Folco (1992); (23) Fioretti and Molin (1996); (24) Takeda et al (1992); (25) Takeda (1989); (26) Takeda et al (1988b); (27) Warren and Kallemeyn (1994); (28) Prinz et al (1994); (29) Takeda et al (1989); (30) Takeda et al (1979a); (31) Takeda et al (1979b); (32) Takeda (1987); (33) Goodrich (1986b); (34) Takeda and Yanai (1978); (35) Sexton et al (1996); (36) Chikami et al. (1997b); (37) Weber and Bischoff (1998); Goodrich (1998).

with common (100) (Takeda et al. 1992). In Y-74130 and META78008, low-Ca pyroxene with $Wo_{\sim4.5}$ appears to be a complex mixture of clino- and ortho-pyroxene sequences in x-ray diffraction (Takeda 1987, Takeda et al. 1989). It is possible that in all ureilites low-Ca pyroxene with $Wo_{<5}$ was originally protoenstatite.

The typical ureilite texture is characterized by large, anhedral olivine and pyroxene grains averaging ~1 mm in size, which meet in triple junctions and have curved intergranular boundaries (Fig. 30a). Olivine grains may contain small, rounded inclusions of pyroxene; pigeonite or orthopyroxene may contain similar inclusions of olivine and/or augite. This poikilitic texture is especially common in PCA 82506, RKPA80239, and LEW 85440. In some ureilites the olivine and pyroxene grains show a pronounced elongation, with aspect ratios up to ~6:1 (Fig. 30a). Fabric analysis (Berkley et al. 1976, 1980; Goodrich et al. 1987a) shows that this elongation reflects both a foliation defined by the {100} crystal face of olivine, and a lineation defined by the crystallographic [001] **c**-axis of olivine. Several ureilites display a mosaicized texture with much smaller grain size, which is probably a result of shock deformation and/or recrystallization. These ureilites appear to have originally had the typical ureilite texture, as evidenced by dark matrix material outlining relict grain boundaries of large, elongate grains, and a common preferred orientation of grains within relict domains (Neuvonen et al. 1972). Dingo Pup Donga has a unique texture of euhedral, non-interlocking grains, and size-sorted layers (Berkley et al. 1980).

Four ureilites have an unusual bimodal texture (Fig. 30b). They are characterized by extremely large crystals of low-Ca pyroxene up to at least 15 mm in size which poikilitically enclose domains that have the typical ureilite texture. In ALH 82130 these are pigeonite, while in META78008, Y-74130, and EET 87511 they are orthopyroxene. In ALH 82130 and EET 87511 the typical-textured domains are devoid of pyroxene; in Y-74130 and META78008 they consist of olivine + augite, the latter of which contains small, rounded, poikilitic inclusions of pigeonite. These ureilites are sufficiently heterogeneous and coarse-grained that some thin sections show only the large pyroxene crystals and some show only the typical-textured domains.

Pyroxene in most ureilites is characterized by a lack of exsolution features. However, in ALH 82130, augite occurs as irregular lamellae and blebs in large poikilitic pigeonite crystals. This texture may be the result of unmixing of an original pigeonite at the pigeonite eutectoid reaction line at ~1246°C (Takeda et al. 1986, Mori et al. 1986), although if this is the case the pigeonite should be orthopyroxene. X-ray diffraction of the ALH 82130 pigeonite shows a weak orthopyroxene-like stacking sequence, which may be a remnant of orthopyroxene that later reinverted to pigeonite (Takeda et al. 1989). In LEW 88774, large pyroxene crystals have exsolved to roughly equal proportions of orthopyroxene and augite, in coarse lamellae ~50 μm wide (Warren and Kallemeyn 1994, Chikami et al. 1997a). This meteorite may have a bimodal texture, but this is not yet clear.

Olivine and pyroxene core compositions within each ureilite are very homogeneous in terms of mg#. Forsterite contents range from ~74 to 95 (Table 23, Fig. 31). There are no significant gaps in this range, but there is a spike at $\sim Fo_{79}$. Coexisting pyroxenes span a similar range (Table 23, Fig. 32), and indicate olivine-pyroxene equilibrium. There is no correlation between mg# and modal pyroxene/(pyroxene+olivine), nor between mg# and pyroxene type (Fig. 32).

Ureilite olivines are characterized by high CaO (~0.30-0.45 wt %) and Cr_2O_3 (~0.56 to 0.85 %), and pigeonite also has high Cr_2O_3 (up to ~1.26 %). Representative analyses of olivine and pyroxene cores are given in Tables 24 and 25. Examination of olivine and

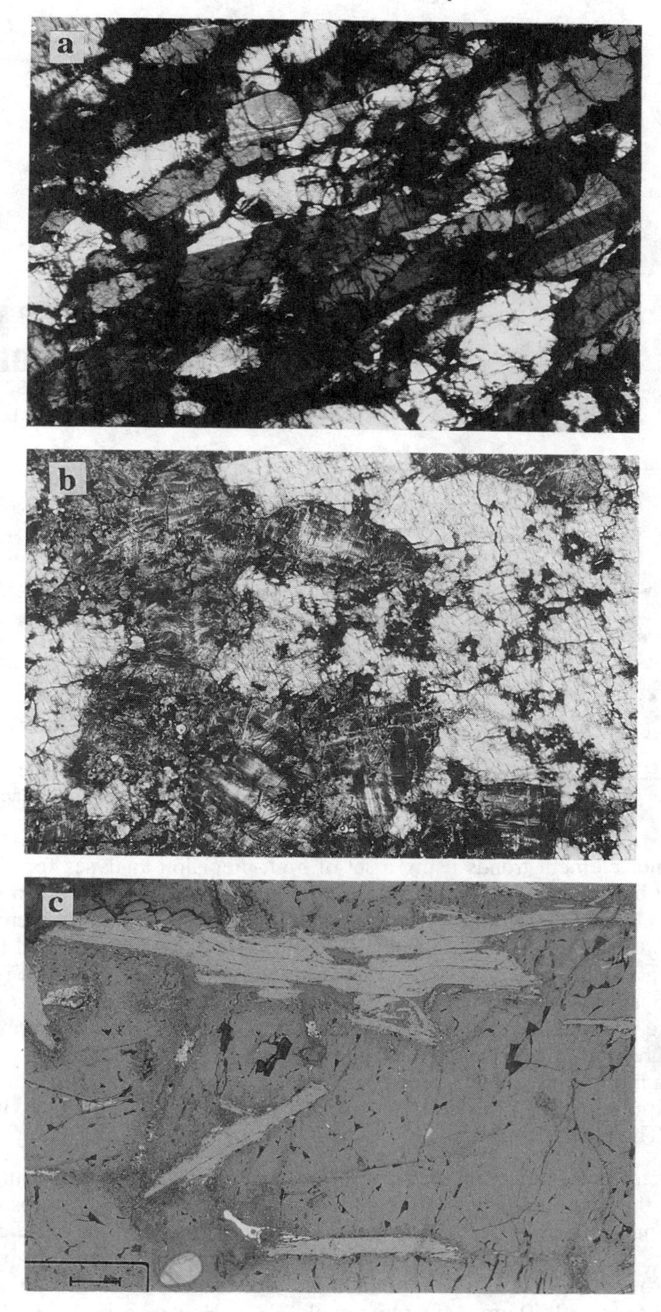

Figure 30. Photomicrographs of ureilites. (a) Kenna, showing typical ureilite texture and fabric. Field of view is 7 mm in width. (b) ALH 82130, illustrating bimodal texture of large low-Ca pyroxene crystal poikilitically enclosing areas that have the typical ureilite texture. Field of view is 5.8 mm in width. (c) Large euhedral graphite crystals in ALHA78019. Field of view is 1.5 mm in width.

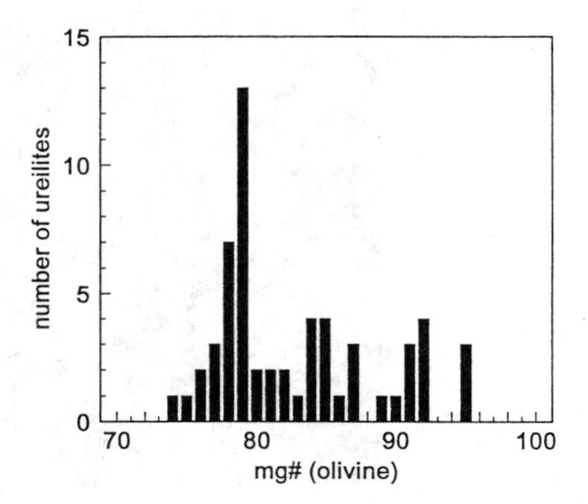

Figure 31. Histogram of mg# of olivine cores for monomict ureilites. Data from Table 23.

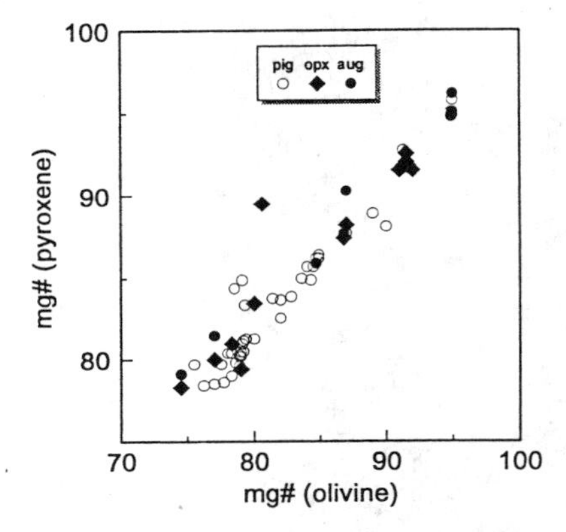

Figure 32. Correlation between mg# of olivine and pyroxene cores (pigeonite, orthopyroxene and augite) in monomict ureilites. Data from Table 23.

pyroxene minor element trends using a set of high-precision analyses for eight ureilites shows a very high degree of intragrain and intergrain homogeneity within each ureilite (Goodrich et al. 1987b). Olivine cores among ureilites have essentially identical Mg/Mn and Mg/Cr ratios (Fig. 33). Olivine in LEW 88774 plots off the Mg/Mn and Mg/Cr trends, with a low Fe/Mn ratio and an exceptionally high Fe/Cr ratio (Fig. 33). Both olivine and pyroxene show a negative correlation between FeO and MnO (Fig. 34), in contrast to the positive correlation with essentially constant FeO/MnO commonly shown by groups of related igneous rocks. Ratios of CaO/Al_2O_3 for olivine cores are high and vary by a factor of ~3 among ureilites. There is a significant correlation between CaO/Al_2O_3 in olivine and coexisting pyroxene cores, again indicating olivine-pyroxene equilibrium. There is no correlation of CaO/Al_2O_3 with mg#.

A characteristic feature of ureilites is the occurrence of reduced rims on olivine grains where they are in contact with carbonaceous matrix material or crosscut by veins of carbonaceous material (Berkley et al. 1980). These rims consist of nearly FeO-free olivine and/or enstatite, riddled with tiny inclusions of low-Ni metal. In most ureilites they are narrow (10-100 μm), and boundaries with cores are sharp; zonation can be detected only over ~30-40 μm (Miyamoto et al. 1985). However, in ALH 82130 and HH 126 the rims are considerably widened, to the point that the olivine grains have been almost completely reduced and only rare metal-free cores preserve the original composition (Berkley et al. 1985, Sexton et al. 1996). Pyroxene grains also sometimes

Table 24. Compositions of olivine grain cores from representative ureilites.

	Y-74659 (1)	ALHA77257 (1)	ALHA77257 (2)	Y-790981 (1)	Y-790981 (2)	Y-74130 (1)	ALHA78019 (1)	ALHA78019 (2)
Chemical Composition (wt %)								
SiO_2	40.3	39.3		38.2		37.6	37.5	
Al_2O_3	0.03	0.03	0.040	0.03	0.042	0.04	0.02	0.031
FeO	8.0	12.1	14.2	19.8	21.2	21.0	21.1	22.5
MgO	49.7	46.1	45.2	40.0	40.9	39.7	39.1	40.5
MnO	0.46	0.46	0.452	0.42	0.425	0.42	0.42	0.421
Cr_2O_3	0.58	0.72	0.743	0.55	0.606	0.39	0.72	0.698
CaO	0.30	0.34	0.340	0.28	0.326	0.24	0.41	0.453
Total	99.40	99.08		99.28		99.39	99.27	
Cation Formula Based on 4 Oxygens								
Si	0.9903	0.9883		0.9930		0.9830	0.9832	
Al	0.0009	0.0009		0.0009		0.0012	0.0006	
Fe	0.1650	0.2551		0.4304		0.4591	0.4627	
Mg	1.8201	1.7278		1.5496		1.5468	1.5278	
Mn	0.0096	0.0098		0.0092		0.0093	0.0093	
Cr	0.0113	0.0143		0.0113		0.0081	0.0149	
Ca	0.0079	0.0092		0.0078		0.0067	0.0115	
Total Cations	3.0051	3.0054		3.0022		3.0142	3.0100	
*Cation Ratios Fe/Mn and 100*Mg/(Mg+Fe) (mg#)*								
Fe/Mn	17	26	31	47	49	49	50	53
mg#	91.7	87.1	85.0	78.3	77.5	77.1	76.8	76.2

(1) Complete analyses by Takeda (1987); (2) Partial analyses by Goodrich et al. (1987b) are shown for comparison. These are high precision analyses for the minor elements.

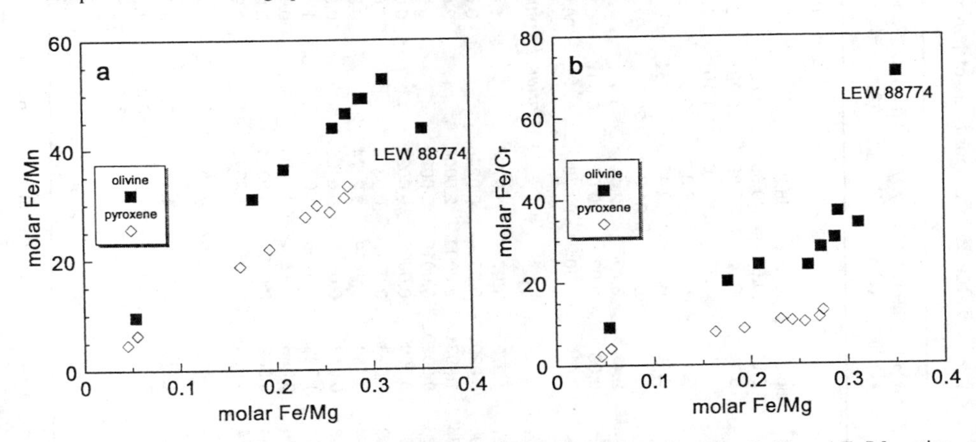

Figure 33. Correlations between (a) molar Fe/Mn and Fe/Mg ratios and (b) molar Fe/Cr and Fe/Mg ratios in olivine and pyroxene cores in 8 ureilites, indicating constant Mn/Mg and Cr/Mg ratios. Data from Goodrich et al. (1987b). The Fe/Mn-Fe/Mg trend has also been observed among other ureilites, based upon a less precise set of data (Mittlefehldt 1986). Olivine in LEW 88774 is off the trends shown by other ureilites. LEW 88774 data from M. Prinz (pers. comm.).

have reduced rims, consisting of enstatite with inclusions of low-Ni metal, but they are usually much narrower than the olivine reduction rims. In LEW 88774, chromite grains are also reduced along contacts with carbon, forming complex zones consisting mainly of

Table 25. Compositions of pyroxene grain cores from representative ureilites.

	Y-74659 (1)	Y-74659 (1)	ALHA77257 (1)	ALHA77257 (1)	ALHA77257 (2)	Y-790981 (1)	Y-790981 (1)	Y-790981 (2)	Y-74130 (1)	Y-74130 (1)	Y-74130 (1)	ALHA78019 (1)	ALHA78019 (2)
Chemical Composition (wt %)													
SiO_2	56.6	56.6	56.0	56.3		54.8	53.4		54.7	53.5	53.3	54.2	
Al_2O_3	0.48	0.51	0.61	0.66	0.65	1.13	0.54	1.39	1.92	1.96	2.82	0.53	0.54
TiO_2	0.10	0.11	0.06	0.09	0.081	0.06	0.30	0.093	0.13	0.14	0.25	0.05	0.042
Cr_2O_3	0.88	0.85	1.12	1.10	1.16	1.14	1.39	1.26	1.30	1.37	1.85	1.06	1.07
FeO	5.4	5.17	8.2	7.8	8.8	11.3	3.4	12.3	11.8	10.9	7.6	12.2	13.2
MnO	0.50	0.46	0.44	0.42	0.464	0.46	0.36	0.425	0.40	0.47	0.35	0.41	0.393
MgO	31.8	33.3	30.0	31.3	30.3	27.0	20.7	27.0	27.2	23.1	17.8	26.1	26.9
CaO	3.78	2.4	3.47	2.14	3.45	4.1	18.51	4.58	2.19	6.73	14.77	4.87	5.13
Na_2O	0.05	0.03	0.03	0.02	0.041	0.05	0.23	0.117	0.16	0.07	0.82	0.03	0.029
Total	99.54	99.41	99.90	99.78		100.11	98.79		99.79	98.25	99.55	99.48	
Cation Formula Based on 6 Oxygens													
Si	1.9787	1.9722	1.9746	1.9754		1.9615	1.9551		1.9563	1.9630	1.9505	1.9663	
^{IV}Al	0.0198	0.0209	0.0254	0.0246		0.0385	0.0233		0.0437	0.0370	0.0495	0.0227	
Total tet*	1.9985	1.9931	2.0000	2.0000		2.0000	1.9784		2.0000	2.0000	2.0000	1.9890	
Ti	0.0026	0.0029	0.0016	0.0024		0.0016	0.0083		0.0035	0.0039	0.0069	0.0014	
^{VI}Al	0.0000	0.0000	0.0000	0.0027		0.0092	0.0000		0.0372	0.0478	0.0721	0.0000	
Cr	0.0243	0.0234	0.0312	0.0305		0.0323	0.0402		0.0368	0.0397	0.0535	0.0304	
Fe	0.1564	0.1507	0.2409	0.2274		0.3395	0.1029		0.3526	0.3348	0.2320	0.3711	
Mn	0.0148	0.0136	0.0131	0.0125		0.0139	0.0112		0.0121	0.0146	0.0108	0.0126	
Mg	1.6568	1.7292	1.5765	1.6367		1.4403	1.1295		1.4497	1.2631	0.9713	1.4111	
Ca	0.1416	0.0889	0.1311	0.0805		0.1584	0.7262		0.0839	0.2646	0.5792	0.1893	
Na	0.0034	0.0020	0.0021	0.0014		0.0035	0.0163		0.0111	0.0050	0.0582	0.0021	
Total Cations	3.9984	4.0038	3.9965	3.9941		3.9987	4.0130		3.9869	3.9735	3.9840	4.0070	
*Cation Ratios Ca:Mg:Fe, Fe/Mn and 100*Mg/(Mg+Fe) (mg#)*													
Ca	7.2	4.5	6.7	4.1	6.6	8.2	37.1	8.9	4.4	14.2	32.5	9.6	9.7
Mg	84.8	87.8	80.9	84.2	80.3	74.3	57.7	72.6	76.9	67.8	54.5	71.6	70.8
Fe	8.0	7.7	12.4	11.7	13.1	17.5	5.3	18.6	18.7	18.0	13.0	18.8	19.5
Fe/Mn	11	11	18	18	19	24	9	29	29	23	21	29	33
mg#	91.4	92.0	86.7	87.8	86.0	80.9	91.7	79.6	80.4	79.0	80.7	79.2	78.4

(1) Complete analyses by Takeda (1987); (2) Partial analyses by Goodrich et al. (1987b) are shown for comparison. These are high precision analyses for the minor elements.

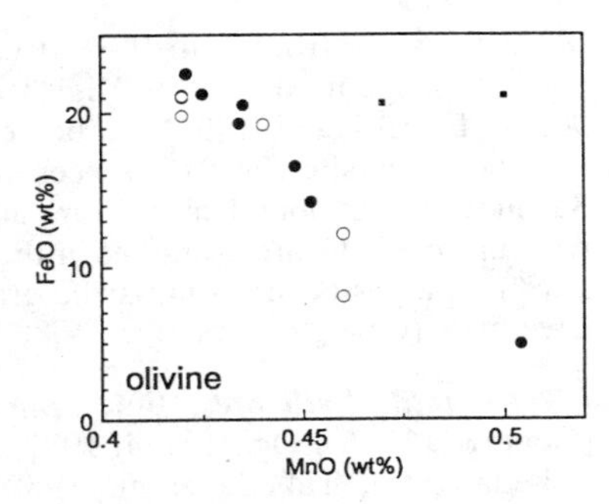

Figure 34. Negative correlation between MnO and FeO in olivine cores among ureilites. Data are from Goodrich et al. (1987b) (dots) and Takeda (1987) (circles). The data of Berkley and Jones (1982) (small squares) seem systematically high compared to other data for individual ureilites. The data of Berkley et al. (1980) are not shown because of problems with Mn standardization (Berkley, pers. comm.).

Cr-Fe-carbide, eskolaite (Cr_2O_3), and several other phases (Prinz et al. 1994, Warren and Kallemeyn 1994). Olivine rims show a negative correlation between FeO and MnO (Miyamoto et al. 1993), consistent with reduction. Chromite in LEW 88774 has 54 to 59% Cr_2O_3, 8 to 21% MgO, 14 to 19% Al_2O_3, 1 to 22% FeO and 0.6 to 2.8% TiO_2 (Warren and Kallemeyn 1994, Chikami et al. 1997a), and ranges from $FeCr_2O_4$-rich to $MgCr_2O_4$-rich to $MgAl_2O_4$-rich with increasing reduction (Prinz et al. 1994).

Carbon phases. The carbon-rich matrix of ureilites occurs mostly along silicate grain boundaries, but also intrudes the silicates along fractures and cleavage planes. In some ureilites, what appear to be isolated pockets of matrix are included in silicate grains, but these may be parts of veins intruding from grain boundaries in the unseen dimension. In rare cases matrix material surrounds poikilitic inclusions of olivine in low-Ca pyroxene.

Graphite is the most common polymorph of carbon in ureilites. Vdovykin (1970) found traces of chaoite and organic carbon compounds in Haverö, Novo Urei, Dyaplur and Goalpara. Diamond and lonsdaleite have been identified in Kenna (Gibson 1976, Berkley et al. 1976), Haverö and Goalpara (Marvin and Wood 1972) by electron diffraction and electron microprobe techniques. Diamond and lonsdaleite occur as small (<1 to 3 μm) anhedral to subhedral grains within a fine-grained matrix of graphite (Marvin and Wood 1972, Berkley et al. 1976). Diamond has also been identified in Dyalpur by TEM (Mori and Takeda 1988) and in Nilpena by x-ray diffraction of acid resistant residues (Ott et al. 1984), and is inferred to occur in other ureilites based upon the difficulty of cutting them.

In most ureilites the graphite is very fine-grained. However, in ALHA78019, ALHA78262, ALH 83014 and Nova 001 (paired with Nullarbor 010), it occurs as large, mm-sized, euhedral crystals (Fig. 30c), intergrown with metal, sulfide, or splotchy metal-sulfide intergrowths (Berkley and Jones 1982, Treiman and Berkley 1994, M. Prinz, pers. comm.). The elongate graphite blades commonly parallel the direction of silicate grain boundaries and have a strong tendency to align with their long axes parallel to the observed foliation. In places they intrude into silicate grains and, rarely, appear to be completely enclosed within silicate grains.

Carbon also occurs in the form of cohenite (Fe_3C) in metallic spherule inclusions in olivine and pigeonite in at least five ureilites (Goodrich and Berkley 1986). These spherules also contain C-bearing Fe-Ni metal, sulfide (predominantly troilite), and rare phosphorus-bearing minerals, and have textures typical of eutectic crystallization in the Fe-C system. Bulk compositions of the spherules range between 4 and 5 wt % C and 3 and 7 wt % Ni.

Metal, sulfides and phosphides. The composition of interstitial metal in ureilites is quite variable, both within individual specimens and among different ureilites. Nickel contents range from 1 to 9%, Si contents range from undetectable to ~3%, and Cr contents range up to ~1.5%. High P contents (1%) have been reported in metal in FRO 90054 (Fioretti et al. 1996). Carbon contents of the metal are unknown, but must be very low. Schreibersite (Fe_3P) has been reported in FRO 90054 and several other ureilites. Sulfides are not abundant, and are also quite variable in composition. High Cr varieties with up to ~34% are common in the interstitial material of Kenna, and low-Cr (0.5 to 2.8%) varieties occur as µm-sized grains within the silicate grains (Berkley et al. 1976). Brezinaite (Cr_3S_4) occurs in LEW 88774 (Prinz et al. 1994).

Interstitial silicates. Fine-grained interstitial silicates are common in ureilites (Goodrich 1986a, Ogata et al. 1987, 1988, 1991; Saito and Takeda 1989, Tomeoka and Takeda 1989, Takeda et al. 1988b, Mori and Takeda 1983, Prinz et al. 1994). Interpretations of these materials vary and unfortunately so does the nomenclature used in their description. They occur along grain boundaries in narrow zones, generally 10 to 20 µm wide, but up to 70 µm in Y-74123, and as veins intruding into olivine and pyroxene from the grain boundaries. They are often difficult to distinguish optically because they resemble the reduced margins of olivine and pyroxene core crystals and may be masked by interstitial metal and carbon. An exception is in Roosevelt County (RC) 027, in which they are up to 200 µm wide and optically prominent (Goodrich et al. 1987a). In Y-790981 and LEW 85328 these materials occur as pockets within pigeonite (Saito and Takeda 1989, Ogata et al. 1987), making the pigeonite cloudy.

The interstitial silicates consist of euhedral to subhedral crystals of low-Ca pyroxene and augite, surrounded by Si-Al-alkali glass. Low-Ca pyroxene ranges from Wo_{1-14} and augite ranges from Wo_{26-43}. These pyroxenes have much higher mg# than core pyroxene (up to mg# 99) and lower Mn/Mg, Cr/Mg, Na/Mg, Ti/Mg and Al/Mg ratios (Goodrich et al. 1987b). Augite contains up to 13% Al_2O_3 and 0.3% Na_2O. Aluminian-titanian-diopside with 5.05% TiO_2 occurs in an Al-rich vein in Y-74130 (Tomeoka and Takeda 1989). In LEW 88774 the interstitial pyroxenes are Cr-rich, and intergrown with µm-sized chromite and corundum (Prinz et al. 1994). The glass in ureilites contains up to 21% Al_2O_3, 4.5% Na_2O and 4.5% K_2O (Goodrich 1986a). Tiny grains of SiO_2 are also observed in the glass. All phases are highly variable in composition.

Polymict ureilites. Four ureilites are polymict; North Haig, Nilpena, EET 83309 and EET 87720. They are fragmental breccias containing lithic clasts of typical monomict ureilite material plus a variety of other lithic clasts, set in a matrix of smaller mineral fragments, carbon, suessite (Fe_3Si), sulfides, minor chromite and minor apatite (Jaques and Fitzgerald 1982, Prinz et al. 1983a, 1986b, 1987, 1988a; Keil et al. 1982, Warren and Kallemeyn 1989b, 1991). Olivine grains in the ureilite clasts range from ~Fo_{76-84}. Other lithic clasts include a porphyritic enstatite clast resembling material from enstatite chondrites or aubrites, clasts resembling chondritic material, feldspathic melt rocks, and clasts of a distinct Ca-Al-Ti-rich assemblage that resembles angrites. The angrite-like clasts contain anorthitic plagioclase (An_{96-98}), aluminian-titanian-diopside with up to 8% Al_2O_3 and 2% TiO_2, and olivine (Fo_{49-70}) containing up to 1.7% CaO (Prinz et al. 1986b 1987). The feldspathic melt clasts contain feldspathic glass with fine crystallites of olivine and sometimes low-Ca pyroxene or phosphate. Some contain larger angular crystals of olivine, pyroxene, or phosphate. The composition of the glass ranges from ~An_{5-50}. Bulk compositions of the clasts are high in SiO_2 (up to ~61%), Al_2O_3 (up to ~21%), and alkalis (up to ~7% Na_2O, 0.8% K_2O). Nilpena contains clasts of carbonaceous chondrite matrix-like material. This material has close affinities to CI-matrix, containing saponitic smectite clays, serpentine, magnetite, pentlandite, pyrrhotite and ferrihydrite, and differs from CM-matrix (Brearley and Prinz 1992).

The mineral fragments in polymict ureilites are mostly olivine and pyroxene having compositions consistent with derivation from monomict ureilites. However, in EET 83309 olivine compositions span a greater range (Fo_{62-98}) than in monomict ureilites. Some olivine fragments have significantly lower Cr_2O_3 and CaO contents than olivine in monomict ureilites. Plagioclase, which does not occur in monomict ureilites, is common among the mineral fragments in polymict ureilites, and spans the entire range of An_{0-100}. Plagioclase in the range An_{30-80} contains no detectable K_2O (Prinz et al. 1987). Modally, polymict ureilites are quite similar to monomict ureilites, with the significant addition of 1-2% plagioclase. They have modal pyroxene/(pyroxene+olivine) ratios of ~0.25.

Shock state. Monomict ureilites are classified in Table 23 as being of very low, low, medium, or high shock level. Very-low shock ureilites contain large, euhedral graphite crystals. In low-shock ureilites the silicates show only minor fracturing, undulatory extinction and kink bands, primarily in olivine. Diamonds are believed to be absent and graphite can sometimes be distinguished as small, euhedral crystals. These features indicate shock pressures <20 GPa (Carter et al. 1968). Medium shock ureilites show a greater extent of fracturing, undulatory extinction, and kink banding. Sub-grain boundaries may be prominent in olivine, and pyroxene may be cloudy due to glassy inclusions. Diamonds and/or lonsdaleite are present. In high shock ureilites the olivine is completely shattered or mosaicized (Lipschutz 1964) and pyroxene is mottled by melt glass. Diamonds and/or lonsdaleite are present. Shock pressures of at least 100 GPa are indicated by these features (Carter et al. 1968). Some of the criteria used in this classification scheme are open to question, however.

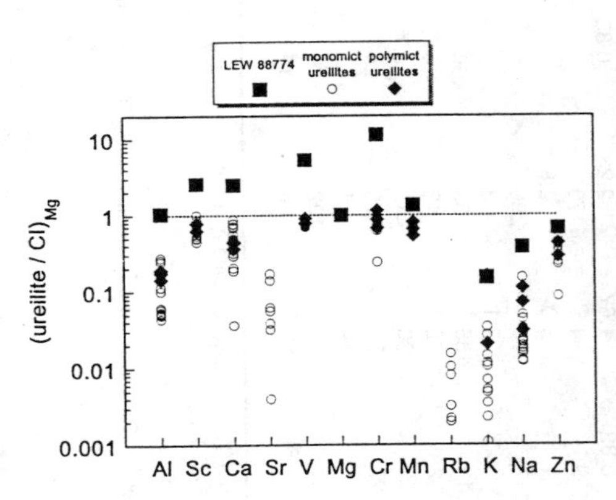

Figure 35. CI- and Mg-normalized lithophile element abundances in bulk monomict and polymict ureilites, and the unusual ureilite LEW 88774. Data plotted are averages of analyses from Bhandari et al. (1981), Binz et al. (1975), Boynton et al. (1976), Goodrich and Lugmair (1991, 1992, 1995), Goodrich et al. (1991, 1987a), Higuchi et al. (1976), Janssens et al. (1987), Jarosewich (1990), Jaques and Fitzgerald (1982), Kallemeyn and Warren (1994), Spitz and Boynton (1991), Takeda (1987), Wänke et al. (1972a), Warren, Kallemeyn (1989, 1992), Wasson et al. (1976), Wiik (1972), Yanai and Kojima (1995b). A few suspect values have not been plotted; Na and K values from Yanai and Kojima (1995b) and K from Wiik (1972). Data are plotted in order of increasing nebular volatility as estimated from calculated condensation temperatures compiled by Wasson (1985).

Chemistry

Lithophile elements. Lithophile element abundances in ureilites (Fig. 35, Table 26) reflect their ultramafic mineralogy. They are enriched in Sc, V, Mg, Cr and Mn, generally up to ~2.7 × CI. Chromite-rich LEW 88774 is unusually enriched in Sc, V and Cr (Table 26, Fig. 35). Calcium abundances are near-chondritic, varying mainly with pyroxene abundance and type. One analysis of Haverö (Wiik 1972) shows extremely low Ca (0.09 × CI) consistent with the upper limit reported by Wänke et al. (1972a), and LEW 88774 has exceptionally high Ca (3.7 × CI). Zinc is near-chondritic or slightly depleted. The plagiophile and incompatible elements Al, Sr, K, Na and Rb are moderately to strongly depleted (Fig. 35). Again, LEW 88774 is unusual, with high Al, K and Na (Table 26, Fig. 35). The polymict ureilite EET 83309 has higher abundances of Al, K and Na, and other polymict ureilites (EET 87720 and Nilpena) have abundances of these elements near the

Table 26. Select major*, minor and trace element contents of representative ureilites.

		ALH84136 *bi-modal* (1)	ALHA77257 *typical* (1)	ALHA81101 *mosaicized* (1)	EET 83309 *polymict* (2)	EET 87720 *polymict* (1)	Goalpara *mosaicized* (3)	Haverö** *typical* (4)	Kenna *typical* (3)	LEW 88774 *typical* (5)	Novo Urei *typical* (3)	PCA82506 *typical/poikilitic* (1)
Al_2O_3	wt %	0.23	0.25	0.16	0.70	0.57	0.21	0.21	0.23	2.46	0.35	0.11
Cr_2O_3	wt %	0.595	0.702	0.716	0.709	0.858	0.840	0.598	0.74	6.28	0.800	0.653
FeO	wt %	7.3	11.5	19.2	17.2	21.7	22.4	19.3	20.5	18.9	19.8	17.0
MnO	wt %	0.429	0.400	0.373	0.372	0.381	0.372	0.343	0.367	0.495	0.386	0.377
MgO	wt %	41.1	37.1	35.7	34.8	28.0	36.1	37.2	33.1	23.5	34.5	35
CaO	wt %	2.1	1.0	0.97	1.1	0.91	0.58	<0.14	1.3	4.67	1.1	0.52
Na_2O	wt %	0.0485	0.0247	0.0235	0.195	0.0485	0.018	0.033	0.033	0.364	0.047	0.0266
K	µg/g			8	238		42	35	1.3	121	17	
Sc	µg/g	10.7	7.7	7.2	7.8	8.9	6.3	4.9	8.5	21.5	8.1	5.8
V	µg/g	111	99	99	94	91	92		100	430	85	82
Co	µg/g	115	89	71	135	148	78	101	164	146	134	91
Ni	mg/g	1.55	0.89	0.82	1.67	2.23	0.85	0.92	1.18	1.69	1.30	0.82
Zn	µg/g	260	243	159	284	170		235		301		231
Ga	µg/g	2.6	1.84	1.44	3.0	2.8		1.13		28.7		1.8
As	ng/g	290	190	112	390	290				430		180
Se	µg/g	1.04		1.26	2.1	3				1.48		<0.9
La	ng/g	15	11	13	185	35	150	70	8.3	168	100	8.2
Sm	ng/g	25	2.8	2.6	62	12	6	14	1.6	146	12	<2
Eu	ng/g	4.5	<5	<6	30	<12	0.7	4.1	0.315	40	2.3	<2
Yb	ng/g	99	<40	27	76	25	24	25	39.5	380	45	27
Lu	ng/g	16	8.0	4.7	14	7		8.5	7.4	62		5.6
Ir	ng/g	149	184	37	249	102	80	240	560	480	440	198
Au	ng/g	32	20.0	11.6	32	28	19	24	40	34	70	17.5

* There are few modern analyses for SiO_2 and TiO_2. Based on data given in Jarosewich (1990), Takeda (1987) and Vdovykin (1970), SiO_2 contents are generally between 35-43 wt % and TiO_2 contents are <0.1 wt %.

** ureilite falls

References: (1) Warren and Kallemeyn (1992); (2) Warren and Kallemeyn (1989); (3) Boynton et al. (1976); (4) Wänke et al. (1972); (5) Kallemeyn and Warren (1994).

upper end of the ureilite ranges (Fig. 35). Y-74130 and Y-790981 have high Al and Na, probably due to a high augite content and a high content of Al-alkali-rich interstitial material, respectively.

The Ca/Al ratios of ureilites are superchondritic, ranging from ~2 to 14 × CI, with the exception of Y-74123, and the analysis of Haverö with low-Ca from Wiik (1972). Bulk ureilite compositions show a negative FeO-MnO correlation, similar to olivine and pyroxene (Fig. 33). The Mn/Mg ratios are nearly constant at 0.61±0.07 × CI, while Cr/Mg ratios are more variable: 0.79±0.17. (We suspect the unusually high Mn data of Spitz and Boynton (1991) are erroneous, and have not included them in this discussion.) Again, LEW 88774 is unusual, with Mn/Mg 1.3 × CI and Cr/Mg 11 × CI.

Rare earth elements. Rare earth element patterns have been determined for 28 ureilites. In general, REE abundances are low compared to other achondrites. This REE depletion is consistent with major incompatible element depletion. There is no correlation between degree of REE depletion and mg#. The classic CI-normalized REE pattern of ureilites is V-shaped at subchondritic values, with severe middle REE depletion (Fig. 36a). Europium ranges to as low as 0.005 × CI. Many ureilites, however, have LREE-depleted patterns, with La ranging from ~0.005-0.1 × CI, and negative Eu anomalies (Fig. 36b). LEW 88774 and FRO 90054 stand out with only slightly LREE-depleted patterns at near-chondritic values (Fig. 36b). For some ureilites, e.g. RC027 and Y-791538, one subsample may have a V-shaped REE pattern while another has a LREE-depleted pattern (compare Figs. 36a and 36b).

The nature and origin of the V-shaped REE patterns in ureilites has been a subject of considerable investigation and controversy. It has been shown (Boynton et al. 1976, Spitz and Boynton 1991, Goodrich and Lugmair 1995) that the LREE-enriched part of the pattern can be removed by leaching with weak acids, leaving a residue with a severely LREE-depleted pattern with a negative Eu anomaly, and yielding a leachate with a LREE-enriched pattern and positive Eu anomalies (Fig. 36d). It has been assumed that the LREE-enriched component is contained in an acid-soluble minor phase, which has extremely high LREE abundances. The LREE-enriched component also appears to be inhomogeneously distributed (Goodrich et al. 1987a, Goodrich et al. 1991), and it may be ubiquitous, since LREE-enriched leachates can be obtained even from samples that have bulk LREE-depleted patterns (Goodrich and Lugmair 1992).

The identity or host of the LREE-enriched component remains unknown. Electron microprobe and ion microprobe searches for LREE-enriched phases in interstitial areas have eliminated metal, sulfide, graphite, interstitial pyroxene and glass as LREE carriers (Spitz et al. 1988). However, ion microprobe analyses do show general LREE-enrichment in carbon-rich interstitial areas (Spitz et al. 1988, Guan and Crozaz 1995a). Some part of the LREE-enriched component is associated with olivine and is either on grain surfaces or internally equilibrated (Goodrich and Lugmair 1995). Guan and Crozaz (1995a) also observed LREE-enrichment in reduced rims on olivine grains in Novo Urei. The abundance of LREE-enriched component has been correlated only with the abundance of Sr and Ba (Goodrich and Lugmair 1995).

Polymict ureilites have higher REE concentration than most monomict ureilites and generally have distinct patterns (Fig. 36c). Ion microprobe analyses have shown that olivine and pyroxene fragments in EET 83309 have REE patterns typical of those in monomict ureilites (Guan and Crozaz 1995b). Feldspathic melt clasts in polymict ureilites have a variety of REE patterns unlike those of ureilites (Guan and Crozaz 1995b 1997). Aluminian-titanian-diopside in an angrite-like clast in North Haig has a REE pattern similar to that of aluminian-titanian-diopside in Angra dos Reis (Davis et al. 1988).

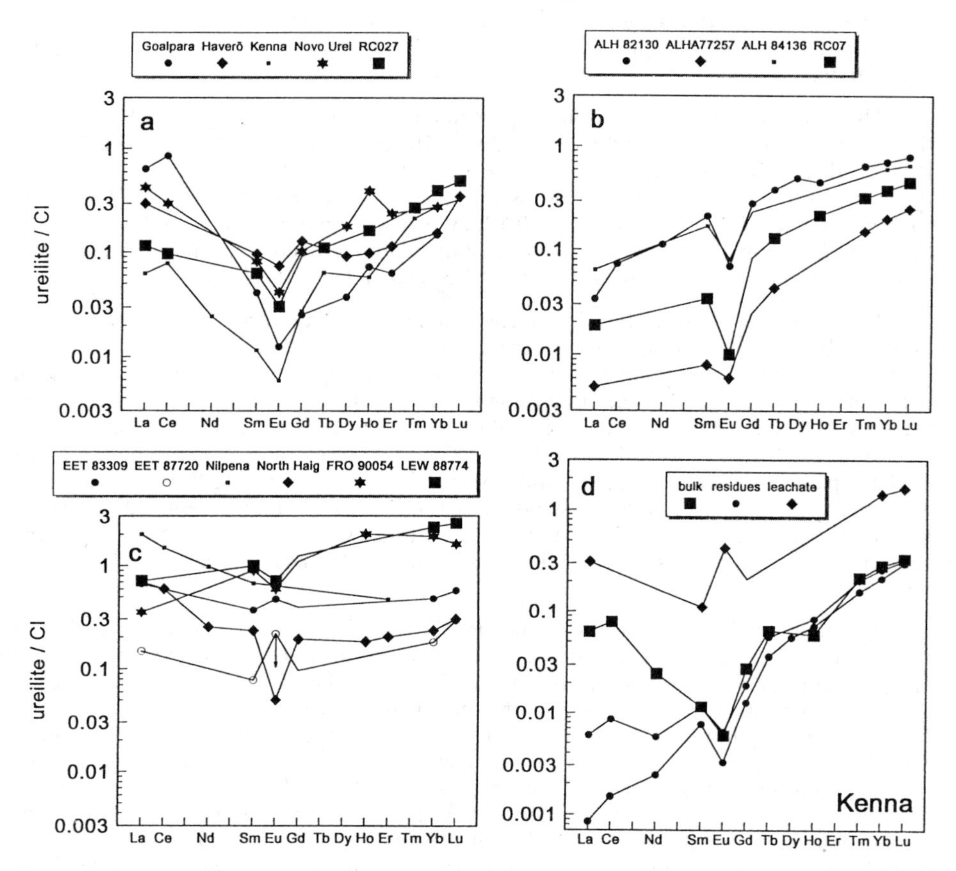

Figure 36. CI-normalized REE patterns for representative ureilites. (a) Many ureilites have V-shaped REE patterns. (b) Other ureilites have LREE-depleted patterns with negative Eu anomalies. Note that some ureilites (e.g. RC027) show V-shaped patterns in some subsamples and LREE-depleted patterns in other subsamples. (c) Polymict ureilites EET 83309, EET 87720, North Haig and Nilpena, unusual ureilite LEW 88774 and FRO 90054. Note that Eu in EET 87720 is an upper limit (arrow) and the positive Eu anomaly may not be real. (d) Leaching experiments, for example on Kenna, show that the LREE-enriched part of the V-shaped bulk ureilite REE pattern can be removed with weak acids. Leachates are LREE-enriched with positive Eu anomalies, and residues are extremely LREE-depleted with negative Eu anomalies. Data are from Boynton et al. (1976), Ebihara et al. (1990), Goodrich et al. (1987a), Jaques and Fitzgerald (1982), Spitz (1991, and pers. comm.), Spitz and Boynton (1991), Wänke et al. (1972a), Warren and Kallemeyn (1989, 1992).

Siderophile trace elements. Siderophile trace element data for bulk ureilites are summarized in **Figure** 37. Generally, abundances of the refractory siderophiles Re, Os, W and Ir are similar within each ureilite, and range from ~0.1-2 × CI among ureilites. Abundances of the more volatile siderophiles are generally lower, with Ni = 0.006 to 0.2 × CI, and tend to increase in the order Ni < Au < Co < Ga < Ge within each ureilite. EET 87511 has exceptionally low abundances of Ni, Au and Co, and an exceptionally steep Ni to Ge trend. LEW 88774 has very high Ga (2.8 × CI), considerably out of the range of other ureilites. Carbonaceous (metal-rich) vein separates from Kenna and Haverö show patterns similar to bulk ureilites, but at higher abundances (Fig. 37). There is no apparent correlation between siderophile element abundances and mg# among ureilites. Ureilite

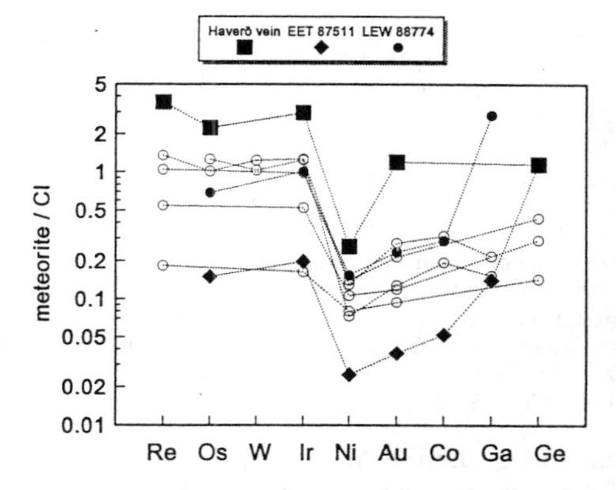

Figure 37. CI-normalized trace siderophile element abundances in representative ureilites (unlabeled). Bulk ureilites have higher abundances of the refractory elements Re, Os, W and Ir in approximately chondritic ratios, while abundances of the more volatile elements are lower and increase in the order Ni→Au→Co→Ga→Ge. Carbonaceous vein separates from Haverö show a similar pattern at higher abundances. LEW 88774 has exceptionally high Ga, while EET 87511 shows a highly fractionated pattern of volatile and moderately volatile siderophile elements. Data from Higuchi et al. (1976), Janssens et al. (1987), Kallemeyn and Warren (1994), Spitz and Boynton (1991), Warren and Kallemeyn (1992).

siderophile element abundances are considerably higher than those of ultramafic rocks from differentiated planets such as the Earth, Moon and Mars

There are good linear Re-Ir, Os-Ir, and to a lesser extent W-Ir correlations among bulk ureilites and carbonaceous vein separates, indicating that the refractory siderophiles are contained in a common component (e.g. see Goodrich 1992). The more volatile siderophiles also show linear correlations with Ir, suggesting that to a first approximation, ureilite siderophiles are a mixture of a refractory-rich and a refractory-poor component (Boynton et al. 1976, Higuchi et al. 1976, Janssens et al. 1987, Spitz and Boynton 1991). The refractory-rich component appears to be represented by the carbonaceous vein separates. Spitz (1992) observed two distinct Ga-Ir trends, suggesting distinct refractory-rich components for two groups of ureilites, but it seems likely this was spurious (Kallemeyn and Warren 1994). However, it has also been noted that ureilite siderophile element abundances are better correlated with solid metal/liquid metal partition coefficients than with volatility, suggesting that they are controlled by fractionation of metal in a magmatic system rather than mixing of two components (Goodrich et al. 1987a).

Carbon and nitrogen. Carbon contents measured for most ureilites range from ~2 to 6% (Bogard et al. 1973, Gibson 1976, Grady and Pillinger 1986, Grady et al. 1985a, Jarosewich 1990, McCall and Cleverly 1968, Wacker 1986, Wiik 1969, 1972). Measurements for individual ureilites are quite variable, as carbon is primarily contained in matrix material, which is inhomogeneously distributed. FRO 90054 has an unusually low carbon content of 0.24% (Grady and Pillinger 1993). There do not appear to be any correlations between carbon content and other geochemical or isotopic parameters. Nitrogen concentrations range from ~10 to 150 ppm (Grady et al. 1985a, Grady and Pillinger 1986, Murty 1994). The C/N ratios of ureilites are significantly lower than those in carbonaceous chondrites with similar C contents. Diamond is enriched in nitrogen; for example, diamond in Lahrauli contains 771 ppm N compared to 11.3 ppm for the whole rock (Murty 1994).

Noble gases. Ureilites contain trapped noble gases in chondritic abundances, with a fractionated or planetary type pattern as in CM chondrites (Fig. 38). Gas contents vary considerably, Xe contents vary by a factor of ~100 in bulk rocks, for example, but elemental and isotopic ratios are distinctive. Carbonaceous matrix material is enriched at least 600-fold in noble gases relative to the silicates, and the gases are largely contained in the carbon (Weber et al. 1971, 1976; Göbel et al. 1978). In diamond-bearing ureilites, diamond is the principal gas-carrier; graphite is virtually free of trapped gases (Weber et

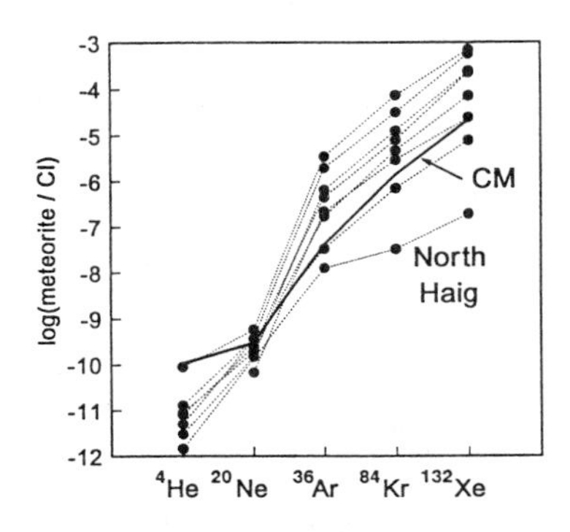

Figure 38. Elemental abundances of noble gases in representative ureilites relative to cosmic abundances. A typical CM (Haripura) pattern is shown for reference (Mazor et al. 1970). Ureilite data from Bogard et al. (1973), Goodrich et al. (1987b), Mazor et al. (1970), Müller and Zähringer (1969), Ott et al. (1993), Wacker (1986).

al. 1976, Göbel et al. 1978). However, the diamond-free ureilite ALHA78019 has trapped noble gas abundances comparable to those of diamond-bearing ureilites (Wacker 1986). In ALHA78019 the gases are contained in a fine-grained carbon whose structural state is unknown, while the coarse-grained graphite is gas-free. The fine-grained carbon has a near-chondritic Xe/C ratio.

Most of the ^{3}He in ureilites is cosmogenic in origin, while most of the ^{4}He appears to be trapped. The extremely low $^{4}He/^{3}He$ ratios are consistent with low U and Th contents. Ureilites have extremely low $^{40}Ar/^{36}Ar$ ratios, indicating very little radiogenic Ar and thus very early depletion of K. Neon in monomict ureilites appears to be a mixture of trapped and cosmogenic components. The composition of the trapped component has been estimated from Hajmah, which contains little cosmogenic Ne, and has been named Ne-U (Ott et al. 1985b). Solar wind implanted noble gases, notably Ne-B, are present in polymict ureilites (Ott et al. 1990, 1993), consistent with their being regolith breccias. Cosmic-ray exposure ages, based on ^{3}He and ^{21}Ne, range from 0.5 to 32 Ma for ten ureilites and show no apparent clustering (Bogard et al. 1973, Göbel et al. 1978, Goodrich et al. 1987a, Müller and Zähringer 1969, Ott et al. 1984, 1985b; Stauffer 1961, Weber et al. 1971).

Isotopic systematics

Oxygen-isotopes. On an oxygen-isotope plot, ureilites plot along the line with slope ~1 defined by C2-C3 matrix material and Allende CAIs (Fig. 1b), thought to be a nebular mixing line (Clayton et al. 1973, Clayton 1993). This pattern is unique among achondrites. Except for acapulcoites and lodranites, all other groups of achondrites and primitive achondrites plot along slope ~0.52 mass fractionation lines (e.g. see Fig. 1b,c). Ureilites show greater scatter about the CAI line than do the chondritic materials that define it, and it is possible to fit various slope ~0.52 lines to the data, possibly organizing some ureilites into groups related by mass-fractionation. Unique determination of such groups is difficult, and for those that have been postulated (Clayton and Mayeda 1988) there do not appear to be common petrographic features or chemical trends consistent with a fractionation relation.

There are no correlations between oxygen-isotopic composition and any petrographic or chemical parameters of ureilites except mg#. Ureilites show a good correlation (Fig. 39) between mg# (Fo in olivine) and $\Delta^{17}O$. Such a correlation is shown by some chondritic materials (e.g. Rubin et al. 1990), but amongst achondrite groups is otherwise shown only by lodranites (Fig. 39; see also Fig. 28).

Sm-Nd. Ureilites fall into 3 Sm-Nd groups: (1) A Kenna group including whole rock samples and a pyroxene separate from Kenna, a whole rock sample of Novo Urei, and a

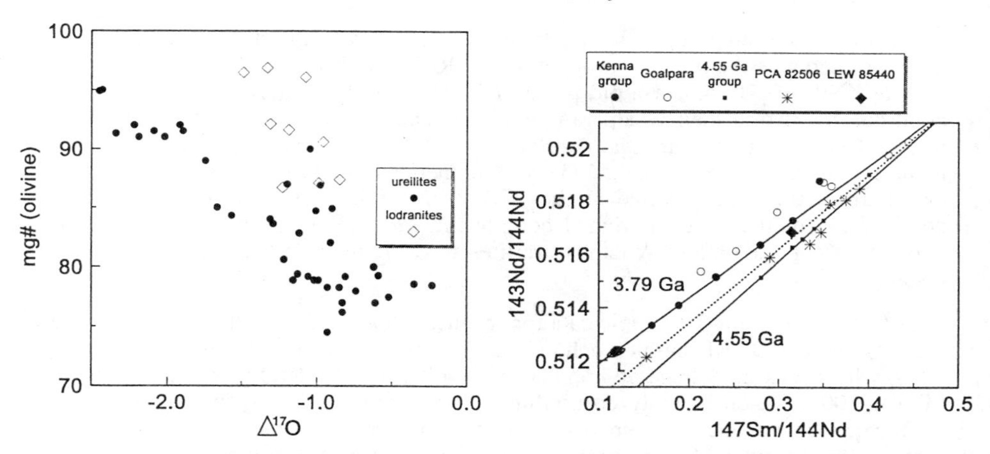

Figure 39 (left). Correlation between mg# of olivine and $\Delta^{17}O$ in ureilites. Data from Clayton and Mayeda (1996) and Table 23. Lodranites may show a similar correlation (see also Fig. 28). Data from McCoy et al. (1997a).

Figure 40 (right). Sm-Nd isotopic data for ureilites. Whole rock samples and a pyroxene separate from Kenna, and whole rock samples of Novo Urei and ALHA77257 define a line with a slope corresponding to an age of 3.79 Ga (Kenna group). Mineral separates and a whole rock sample of Goalpara are consistent with this line within error (analytical uncertainty not shown). Acid leachates of Kenna plot at the LREE-enriched end (low $^{147}Sm/^{144}Nd$) of the line (field labeled L). Samples of ALH 82130, META78008 and some samples PCA 82506 are consistent with the 4.55 Ga chondritic evolution line (4.55 Ga group). One whole rock sample of PCA 82506, and a leachate and residue generated from it, define a line with a slope corresponding to an age of 4.23 Ga (dashed line). A whole rock sample of LEW 85440 is consistent with this line. Note that the 3.79 Ga and the 4.23 Ga lines intersect the chondritic evolution line at the same point ($^{147}Sm/^{144}Nd$ ~ 0.51). Data from Goodrich and Lugmair (1991, 1992, 1995), Goodrich et al. (1991), Takahashi and Masuda (1990a), Torigoye-Kita et al. (1995a,b,c).

whole rock sample of ALHA77257. These define a $^{143}Nd/^{144}Nd$-$^{147}Sm/^{144}Nd$ line with a slope corresponding to an age of 3.79 Ga and an initial $^{143}Nd/^{144}Nd$ ratio of 0.50938. Mineral separates and a whole rock sample of Goalpara are consistent with this line (Fig. 40). Acid leachates of Kenna have $^{147}Sm/^{144}Nd$ and $^{143}Nd/^{144}Nd$ ratios very similar to those of the LREE-enriched whole rocks (Fig. 40), and are interpreted to represent the composition of the LREE-enriched component in Kenna (Goodrich and Lugmair 1995). (2) A 4.55 Ga group including some whole rock samples of PCA 82506, whole rock samples of ALH 82130 and META78008, and mineral fractions of META78008. These plot along the 4.55 Ga chondritic evolution (Fig. 40). (3) One whole rock sample of PCA 82506, and an acid leachate and residue generated from it, define a line with a slope corresponding to an age of 4.23 Ga. A whole rock sample of LEW 85440 is consistent with this line (Fig. 40). Both the 3.79 Ga line and the 4.23 Ga line intersect the chondritic evolution line at $^{147}Sm/^{144}Nd$ ~ 0.51. Interpretation of the significance of these data is controversial (Torigoye-Kita et al. 1995a,b; Goodrich et al. 1995), and clearly depends on understanding the LREE-enriched component in ureilites. Torigoye-Kita et al. (1995a,b) have argued that the 3.79 Ga line is a mixing line, and that the LREE-enriched component is a terrestrial contaminant, since its Sm-Nd isotopic composition is similar to that of average terrestrial crust. Soil from the site where Kenna was found has a similar, if not identical, composition (Goodrich and Lugmair 1995). However, Goodrich and Lugmair (1995) and Goodrich et al. (1995) argue that if the 3.79 Ga and 4.23 Ga lines are mixing lines, then samples devoid of LREE-enriched component, such as acid residues and leached mineral separates, should plot at the intersection of these lines with the chondritic evolution line, at $^{147}Sm/^{144}Nd \sim 0.51$. They suggest that these lines are isochrons and give the time of introduction of LREE-enriched material to highly-depleted ($^{147}Sm/^{144}Nd \sim 0.51$) 4.55 Ga-old proto-ureilite material. Torigoye-Kita et al. (1995a) point out that two of their Goalpara pyroxene separates (those with highest $^{147}Sm/^{144}Nd$ in Fig. 40a) have error bars that overlap the 4.55 Ga line, and argue that these points represent the uncontaminated, 4.55 Ga-old end member of a mixing line. However, a line drawn from the 4.55 Ga line at the $^{147}Sm/^{144}Nd$ ratios of these pyroxenes to the composition of the LREE-enriched component, falls below all but the most LREE-enriched Kenna samples. Whether ureilites have young Sm-Nd ages or not awaits resolution.

Rb-Sr. Rubidium-Sr isotopic data for ureilites scatter widely. Takahashi and Masuda (1990a) report a BABI (basaltic achondrite best initial, Papanastassiou and Wasserburg 1969) whole rock model age of 4.55 Ga, and a 4.01±0.06 Ga Rb-Sr internal isochron for META78008. All other analyzed ureilites have older model ages, indicating that they have experienced open system behavior at some younger time(s). Fifteen analyzed samples of Kenna including whole rocks, leachates and residues show essentially identical $^{87}Sr/^{86}Sr$ ratios of ~0.70866 (Goodrich and Lugmair 1995). Strontium in soil from near the Kenna recovery site has a much higher $^{87}Sr/^{86}Sr$ ratio (0.7199), and so does not appear to be a contaminant.

U-Th-Pb. U-Th-Pb systematics have been investigated for META78008 and Goalpara (Torigoye-Kita et al. 1995a,c). Extensive leaching procedures were used to remove the effects of contamination. The Pb, Th and U abundances in the residues were extremely low—for example, up to ~8 ng/g Pb, 3.2 ng/g Th, and 0.9 ng/g U in META78008. The leachates were significantly enriched in Pb—up to 250 ng/g in META78008 and 1478 ng/g in Goalpara. On a $^{207}Pb/^{206}Pb$-$^{204}Pb/^{206}Pb$ diagram, the META78008 leachates plot along a mixing line between the Pb isotopic composition of Canyon Diablo troilite and modern terrestrial Pb, indicating terrestrial Pb contamination. The least-contaminated residues yield a Pb-Pb isochron age of 4.563±0.021 Ga. All of the Goalpara samples showed evidence of mixing with modern terrestrial Pb, and in some cases another Pb component, probably old terrestrial Pb. On a U-Pb concordia diagram, the META78008 residues plot along a line connecting modern terrestrial Pb with the concordia point of 4.562 Ga, and the Goalpara residues define a general trend that intersects concordia at ~4.5 Ga. This old age for Goalpara, and the observation that leachable fractions are contaminated with terrestrial Pb, led Torigoye-Kita et al. (1995a) to suggest that REE in the leachable fractions are also terrestrial, and therefore that the 3.79 Ga Sm-Nd line for Goalpara and Kenna (Fig. 40a) is a mixing line.

Ar-Ar. ^{39}Ar-^{40}Ar determinations have been made on bulk samples of Kenna, Novo Urei, PCA 82506, and Haverö (Bogard and Garrison 1994). Interpretation of the Ar release spectra is uncertain due to possible contamination effects, but the most recent times of Ar degassing can be inferred. These times are ~3.3 Ga for Haverö, ~3.3 to 3.7 Ga for Novo Urei, ~4.1 Ga for Kenna and 4.5 to 4.6 Ga for PCA 82506. It is possible that the younger degassing ages record the same events that affected Sm-Nd ages.

Carbon and nitrogen. Carbon isotopic compositions of ureilites are in the range $\delta^{13}C$ = -11 to 0‰ (Grady et al. 1985a,b; Grady and Pillinger 1987a), and some correlation with mg# has been suggested. Graphite and diamond appear to have the same isotopic composition, but a small fraction of the carbon, which combusts at high temperature, is ^{13}C-depleted. This may be from carbide. FRO 90054, which has an unusually low carbon content, also has an exotic carbon isotopic composition: $\delta^{13}C$ = -24.3‰ (Grady and Pillinger 1993).

Nitrogen isotopic compositions in ureilites are variable, and there is evidence for distinct compositions in carbon phases ($\delta^{15}N$ = -110 to 60‰), silicates ($\delta^{15}N$ = 0±20‰), and metal (Grady et al. 1985a, Grady and Pillinger 1986, Murty 1994). Nitrogen in FRO 90054 is exotic: $\delta^{15}N$ = +64‰ (Grady and Pillinger 1993). Nitrogen isotopic compositions in polymict ureilites range up to $\delta^{15}N$ = +527‰ (Grady and Pillinger 1987b).

Discussion

Ureilites are clearly achondritic, and in terms of their mineralogy, textures, bulk lithophile element ratios, and depletion of lithophile incompatible elements, resemble groups of ultramafic rocks from evolved parent bodies such as the Earth, Moon and Mars. Thus, they have been widely believed to be either igneous cumulates or partial melt residues. However, they differ from such groups in several important ways. First, they show no evidence of being related to one another by igneous fractionation processes, and they lack complementary rock types (i.e. ureilitic basalts). Second, they have characteristics which are typical of chondritic materials and difficult to reconcile with extensive igneous processing, namely their oxygen-isotopic characteris-tics, high trace siderophile element abundances, and the presence of planetary-type noble gases. Ureilites remain incompletely understood in the context of the distinction between primitive and differentiated solar system materials. Discussions of ureilite genesis usually tacitly assume that they originated on one parent body, but this may not be the case.

Igneous processes among ureilites. Ureilites lack plagioclase, have superchondritic Ca/Al ratios, and are depleted in incompatible lithophile elements. These features indicate a high degree of igneous processing. Ureilites also show a high degree of internal textural and chemical equilibration, both within and between olivine and pyroxene, which suggests slow cooling at high temperatures. Estimates of pyroxene equilibration temperatures range from ~1200° to 1280°C (Takeda 1987, Takeda et al. 1989, Chikami et al. 1997a). In partial melting models, at least 15% melting is required to eliminate plagioclase, while ~20 to 30% melting is required to produce the more extreme REE depletions and negative Eu anomalies of ureilites (Spitz and Goodrich 1987, Warren and Kallemeyn 1992, Goodrich 1997a), assuming a chondritic source. In cumulate models, much higher degrees of processing are required, because the superchondritic Ca/Al ratios of olivine in ureilites could only be produced if their parent magmas were derived from previously plagioclase-depleted sources (Goodrich et al. 1987a).

The question of whether ureilites are residues or cumulates is one of the classic problems of ureilite research. Berkley et al. (1976, 1980) and Berkley and Jones (1982) argued that they are cumulates largely on the basis of their textures and fabrics. The multi-stage history required by the cumulate model, however, predicts the existence of numerous complementary rock types, particularly various basalts. Scott et al. (1993) and Warren and Kallemeyn (1992) argued that the existence of a large number of ureilites, and no apparently related basaltic meteorites, indicates that ureilites comprise the major rock type on their parent body and are therefore residues. Although some textural features of ureilites remain difficult to explain in a pure residue model (Treiman and Berkley 1994), few workers now advocate a cumulate model for ureilites. Warren and Kallemeyn (1989b) proposed a hybrid paracumulate model, in which ureilites form as "mushy, cumulate-like partial melt residues."

Regardless of whether they are residues or cumulates, ureilites show no correlations of mg# with bulk composition, mineral minor element composition, pyroxene type, pyroxene/(pyroxene+olivine) ratio, or REE depletion. Such correlations would be expected if ureilites were related to one another either by progressive partial melting of common source material, or by fractional crystallization of a common magma or similar magmas. Thus, ureilites do not seem to be a suite of related igneous rocks in the sense that other achondrite groups do.

This conclusion is consistent with the problem that large degrees of igneous processing and a high-temperature history are difficult to reconcile with the lack of oxygen-isotopic equilibration among ureilites (Clayton and Mayeda 1988). All other groups of achondrites and even most primitive achondrites show significant degrees of oxygen-isotopic equilibration (Fig. 39). Scott et al. (1993) suggested that individual ureilites formed in distinct zones that were sufficiently isolated from one another that they did not communicate isotopically. Warren and Kallemeyn (1992) also pointed out that kilometer-scale heterogeneities in oxygen-isotopic composition may have been preserved because oxygen diffusion is slow. However, the plausibility of deriving each (or at most a few) ureilite(s) from a distinct, kilometer-sized zone of a ureilite parent body (UPB), or a distinct UPB, decreases as the number of ureilites continues to grow. Alternatively, several models have suggested that the UPB accreted from material that was already depleted in the plagioclase component relative to chondrites (Takeda 1987, Kurat 1988). Such models require only a small degree of planetary igneous processing, which might not lead to oxygen-isotopic equilibration or readily distinguishable chemical trends. They also avoid the problem of locating the missing basaltic complements to ureilites, which in no way can be accounted for by the 1 to 2% plagioclase in polymict ureilites. Other suggestions for disposing of the basalts include blowing the crust off the UPB by volatile-enhanced, explosive volcanism (Warren and Kallemeyn 1992, Scott et al. 1993) or stripping it off via impacts.

Reduction among ureilites. The only apparent chemical trend observed among ureilites is the negative FeO-MnO (positive Fe/Mn-Fe/Mg with constant Mn/Mg) correlation shown by both bulk and mineral compositions (Figs. 33, 34), which indicates a relation by reduction, with no effects of fractional crystallization or partial melting (Mittlefehldt 1986, Goodrich et al. 1987b, Takeda 1987). Because ureilites contain carbon, it has been widely thought that carbon-silicate redox reactions must be responsible for this trend (Berkley and Jones 1982, Goodrich and Berkley 1986, Goodrich et al. 1987b, Walker and Grove 1993), particularly since there is also evidence in the form of reduced rims on primary minerals for a secondary, in-situ, carbon reduction event. Carbon redox reactions are strongly pressure dependent, and thus lead to estimates of the pressure range over which ureilites formed: experimental data indicate a pressure of ~2.5 MPa for the most magnesian ureilite and ~10 MPa for the most ferroan ureilite (Walker and Grove 1993).

However, several features of ureilites are inconsistent with a relation by progressive reduction of common material in a planetary setting: the lack of correlation between pyroxene/(pyroxene+olivine) and mg#; a lack of correlation between carbon content and mg#; and a lack of correlation between metal content or siderophile element abundances and mg#. Progressive reduction should lead to an increasing ratio of pyroxene/(pyroxene+olivine) through the reaction $(Mg,Fe)_2SiO_4 + C \rightarrow (Mg,Fe)SiO_3 + Fe + CO$ (McSween and Labotka 1993), decreasing carbon content through loss of CO, and increasing metal content, although metal may have segregated gravitationally. In addition, if reduction occurred in a magma or during partial melting, the Fe/Mn-Fe/Mg trend would show the effects of fractional crystallization or partial melting along with reduction, rather than pure reduction (Goodrich et al. 1987b).

The correlation between mg# and $\Delta^{17}O$ (Fig. 39) suggests that the reduction relation among ureilites may be a nebular, rather than planetary, feature. Some chondritic materials which show mg#-$\Delta^{17}O$ correlations also show negative FeO-MnO correlations (Takeda 1987, Goodrich 1997b), and it has been shown that a nebular mg#-$\Delta^{17}O$ correlation could survive 15 to 30% silicate partial melting, assuming that each ureilite melted in isolation from the others and did not experience any further (planetary) reduction due to carbon (Goodrich 1997a).

This raises another classic problem of ureilite research—the question of whether their carbon is primary. The fine grain size of graphite in most ureilites, the presence of diamonds, the restriction of carbon to interstitial areas and veins, and its obvious reaction relation with primary minerals, led to the idea that the carbon was shock-injected into ureilites late in their history, probably by a carbonaceous chondrite-like impactor (Wasson et al. 1976, Boynton et al. 1976). However, the discovery of large euhedral graphite crystals in apparently unshocked ureilites, and of iron-carbon alloy spherule inclusions within primary minerals, showed that graphite was a primary igneous phase, which had been disrupted and converted to diamond in most ureilites by late shock events (Berkley and Jones 1982, Goodrich and Berkley 1986). In this case, the reduced rims on primary silicate grains could have formed when ureilites were excavated by impact and a sudden drop in pressure caused a drop in carbon-controlled f_{O_2}. Sudden excavation is consistent with the lack of exsolution in pyroxene in most ureilites, high Ca contents in olivine, and cooling rate determinations on reduced rims (Toyoda et al. 1986). Several authors have proposed catastrophic impact disruption models for the UPB (Takeda 1987, Warren and Kallemeyn 1992). Polymict ureilites may place constraints on such models (Guan and Crozaz 1995b, 1997).

If carbon was primary in ureilites, then planetary reduction, which would have disrupted a nebular mg#-$\Delta^{17}O$ correlation, can be avoided only if all ureilites formed at pressures sufficiently high to prevent carbon redox reactions, that is ~7.5-10 Mpa (Warren and Kallemeyn 1992, Walker and Grove 1993). This implies a parent body with a minimum radius of ~100 km, which is large by the standard of present day asteroids. This is considered by some authors to be a problem, although it is likely that the average size of the asteroids has decreased (see Warren and Kallemeyn 1992). An alternative explanation for the mg#-$\Delta^{17}O$ correlation is that the UPB accreted with a radial gradient in $\Delta^{17}O$, upon which was superimposed a radial gradient in mg# created by pressure-dependent carbon-silicate redox reactions (Walker and Grove 1993, J.H. Jones, pers. comm.).

The question of the role of carbon in ureilite petrogenesis is closely associated with the issue of their noble gas contents, because the gases are largely contained in the carbon. The presence of trapped noble gases in near-chondritic abundances is difficult to explain in a scenario involving large-scale igneous events, because gases would be expected to be lost from open magmatic systems at high temperatures. No other achondrite group, except the lodranites, contains trapped noble gases in such high abundances. However, if melting occurred in a closed system, that is, at sufficiently high pressures, the gases might be largely retained (Berkley and Jones 1982, Wacker 1986). Berkley and Jones (1982) calculated that 10 MPa pressure would permit noble gas retention, and Wacker (1986) argued that under these conditions noble gases would actually diffuse into the interior of carrier carbon grains from surface sites, hence becoming more tightly trapped. Again, this model implies a large parent body. Alternatively, if the carbon was a late addition, the primitive gases may have been introduced after the igneous events occurred and the problem of their retention is avoided.

The high siderophile element abundances of ureilites are also unusual compared to other achondrites and rocks from evolved parent bodies, and bear on the question of whether the apparent reduction relation among them was established by nebular or planetary processes. If this relation was established in the nebula before accretion of the metal component, then there is no reason to expect a correlation between mg# and siderophile element abundances, as long as the degree of subsequent igneous processing was not high. If it was established by carbon reduction on the UPB, then ureilites should show either increasing metal content, if metal did not segregate or decreasing siderophile element abundances if metal did segregate, with increasing mg#. However, Goodrich et al. (1987b) proposed a complex magmatic model in which this is not the case. On the other hand, the siderophile element patterns shown by ureilites (Fig. 37) suggest the presence of two components, one refractory-rich and associated with metal-rich carbonaceous matrix material, and one refractory-poor (Boynton et al. 1976, Higuchi et al. 1976, Janssens et al. 1987, Spitz and Boynton 1991). This has led to the suggestion that the metal was a late addition, along with the carbon and noble gases (Wasson et al. 1976, Boynton et al. 1976), which again would be consistent with a lack of correlation of siderophile element abundances with mg#.

The relation of diamond to graphite remains unclear and this further complicates the picture. The idea that the diamonds were produced from original graphite by shock (Vdovykin 1970) is supported by the observation that apparently very low shock stage ureilites such as ALHA78019 do not appear to contain diamond (Wacker 1986). The high noble gas contents of diamonds relative to graphite can be explained because gas-free graphite has been shown to trap large amounts of noble gases upon shock transformation to diamond (Yajima and Matsuda 1989). The shock formation hypothesis for ureilitic diamond was originally taken as support for the carbon-gas-metal-injection hypothesis (Boynton et al. 1976, Wasson et al. 1976). However, studies of apparently low-shock ureilites have shown that there is no correlation of either noble gas abundance or carbon content with degree of shock. Matsuda et al. (1988, 1991) suggested that diamonds in ureilites were not produced by shock, but rather by chemical vapor deposition from an H_2-CH_4 gas onto refractory-rich early condensates in the solar nebula, thus explaining the association of refractory-rich siderophile elements with the carbonaceous vein material in ureilites. These authors found that fractionation of heavy noble gases during vapor deposition of diamond produced abundance patterns similar to those of ureilites. Conclusive evidence for diamonds in the very low shock stage (e.g. ALHA78019, Nullarbor 010) would support the vapor deposition hypothesis. If ureilitic diamonds did form by vapor deposition in the nebula, then at least part of their carbon may be primary.

How old are ureilites? Isotopic age data suggest that the ultramafic ureilite assemblage formed close to 4.55 Ga ago. Uranium-Th-Pb data clearly indicate ages of ~4.55 Ga for two ureilites, with no evidence for later disturbance aside from recent terrestrial Pb contamination (Torigoye-Kita et al. 1995a,c). Samarium-Nd data for several ureilites are also consistent with an ~4.55 Ga age, with no later disturbance (Fig. 40). These results may record the time of accretion of the UPB, consistent with the 4.56 Ga canonical age of meteorites. It is also possible that the ureilite assemblage was generated by planetary igneous processes within a few million years after accretion.

However, Sm-Nd data for several other ureilites (including one which has a 4.55 Ga U-Th-Pb age) suggest disturbances at 4.23 and 3.79 Ga, which are clearly associated with the poorly-understood LREE-enriched component in ureilites (Fig. 40). Whether this component is a terrestrial contaminant, in which case these young ages are meaningless, is controversial (Torigoye-Kita et al. 1995b, Goodrich et al. 1995). Because of the apparent pervasiveness of the LREE-enriched component in ureilites, and the importance that young Sm-Nd ages would have, this matter deserves further investigation. If it is shown that the LREE-enriched component is indigenous to ureilites and was introduced to the ultramafic assemblage at 3.79 and 4.23 Ga (Goodrich and Lugmair 1995), its nature may still be uncertain. It might have been derived from a foreign impactor, or it might represent a crustal component of the UPB. It could, therefore, indicate either impact events or internal heating events. The latter possibility would require a large UPB, which would have implications for some of the issues discussed above.

Ureilite precursor material. Because ureilites have high carbon contents, and because their oxygen-isotopic compositions are similar to those of carbonaceous chondrites, it has been suggested that ureilite precursor material was similar to carbon+metal-bearing carbonaceous chondrites (Vdovykin 1970, Clayton et al. 1976a, Higuchi et al. 1976, Wasson et al. 1976). Tomeoka and Takeda (1989) and Takeda (1987) suggested that Ca-Al-rich interstitial materials in ureilites are remnants of CV-like parent material. In general, many models for ureilite petrogenesis have assumed carbonaceous chondrite-like precursor material. In detail, however, the data do not show a clear link to any carbonaceous chondrite group. Siderophile and volatile element abundance patterns in the carbonaceous vein material in ureilites are inconsistent with derivation from CI chondrites (Boynton et al. 1976, Binz et al. 1975), and carbon isotopic compositions of ureilites are quite distinct from those of carbonaceous chondrites (Grady et al. 1985a). Nevertheless, the observation that chondrules and chondrule rims in Allende show an mg#-$\Delta^{17}O$ correlation similar to ureilites (Rubin et al. 1990) suggests the possibility that ureilite precursor material resembled some components of carbonaceous chondrites. Whether ureilite precursor material ever had a chondritic-type plagioclase component is, as discussed above, uncertain. It seems likely that ureilite precursor material is not represented in terrestrial museums.

LEW 88774. Chromite-bearing LEW 88774 is an unusual ureilite. Compared to other monomict ureilites it is exceptionally enriched in Cr, Ca, Al, Sc, V, Na, Ga, and to a lesser extent K and REE (Table 26; Figs. 35, 36c, 37) and it plots off the Fe/Mn-Fe/Mg and Fe/Cr-Fe/Mg trends shown by other ureilites (Fig. 33). It has the lowest mg# ($Fo_{\sim 74.5}$) of any ureilite, which is consistent with the presence of chromite in indicating that it formed at higher f_{O_2} than other ureilites (Prinz et al. 1994). However, its Fe/Mn ratio is low (Fig. 33a) and indicates that it is not part of the same reduction sequence as other ureilites. It also has the highest pyroxene/(pyroxene+olivine) ratio (0.9) among ureilites, which is inconsistent with its fitting into a reduction sequence. If this rock is a cumulate, extensive fractional crystallization of chromite is required. If it is a residue, its starting material must have been Cr-enriched. Most of the enriched elements in LEW 88774 are refractory, and so its precursor material may have been a Cr-rich refractory chondritic component, similar to chromite-rich chondrules (Krot et al. 1993), although it is not clear that Ga enrichment can be explained this way (Kallemeyn and Warren 1994). LEW 88774 may provide clues to distinguishing nebular from planetary features in ureilites.

AUBRITES

Among all of the achondrites, aubrites may be the most fascinating from a mineralogical perspective. These meteorites formed under highly reducing conditions and, thus, contain a variety of minerals unknown from earth. Aubrites are brecciated pyroxenites that consist primarily of FeO-free enstatite. Keil (1989) summarized the membership of the aubrites at that time. Included were Aubres, Bishopville, Bustee, Cumberland Falls, Khor Temiki, Mayo Belwa, Norton County, Peña Blanca Spring, Pesyanoe, ALHA78113, 20 Allan Hills meteorites within the ALH 83009 pairing group, 9 Lewis Cliff meteorites within the LEW 87007 pairing group, and Yamato 793592. Eight of the 9 non-Antarctic meteorites were observed falls. Also related are the non-brecciated meteorites Shallowater and Mt. Egerton, both of which differ substantially from the remainder of the group and are discussed separately later. The Si-bearing iron meteorite Horse Creek might be related to this clan. Aubrites are clearly related to enstatite chondrites, sharing similar mineralogy and mineral compositions (Keil 1968, Watters and Prinz 1979) and oxygen-isotopic compositions (Clayton et al. 1984), although the exact nature of this relationship is still controversial. In this section, we review the petrology and mineralogy of aubrites. We first focus on the silicate phases, which comprise the bulk of aubrites, and then turn our attention to the less abundant metallic phases, phosphides, nitrides and sulfides.

Table 27. Modal compositions of selected aubrites (Watters and Prinz, 1979).

	Bishopville	Cumberland Falls	Khor Temiki	Mayo Belwa	Norton County	Shallowater
enstatite	74.8	94.0	88.5	97.5	84.5	81.6
plagioclase	16.2	0.7	6.6	0.3	1.0	2.9
diopside	1.9	0.9	1.5	0.6	2.7	0.0
forsterite	6.7	1.5	3.6	1.6	10.0	4.7
kamacite	trace	0.7	0.1	trace	0.3	3.7
troilite	0.5	1.1	0.1	0.1	1.0	7.1

Mineralogy and petrology

Silicates. Enstatite, plagioclase, diopside, and forsterite are all found in aubrites. Bevan et al. (1977) reported rare fluor-amphibole in Mayo Belwa and Okada et al. (1988) reported silica in Norton County. Watters and Prinz (1979) reported the most comprehensive study of the mineralogy and mineral compositions of a suite of aubrites. Table 27 lists modal analyses for representative aubrites from this work.

Aubrites are dominated by pyroxenite clasts. Individual enstatite grains up to 10 cm in length have been reported from Peña Blanca Spring (Lonsdale 1947) and enstatite grains are commonly mm-sized in all aubrites. This extremely coarse grain size makes obtaining a representative sample for chemical analyses essentially impossible. The clasts exhibit igneous textures, with adjacent enstatite exhibiting irregular intergrown borders (Fig. 41) typical of co-crystallization (Okada et al. 1988). Post-crystallization shock has affected many of the enstatite grains, giving rise to a range of macroscopic appearances from white to colorless and transparent (Okada et al. 1988). Enstatite is essentially FeO-free (Table 28).

Diopside is also found in aubrites in abundances ranging from 0.2 to 8.1 vol % (Watters and Prinz 1979; Table 27). This diopside is essentially FeO-free and has $Wo_{40.1-46.1}$ (Table 28). Diopside occurs as both independent grains and as exsolution lamellae within enstatite (Reid and Cohen 1967, Watters and Prinz 1979, Okada et al. 1988). The abundance of diopside exsolution within the host grains ranges from ~9 to 25%, suggesting that the parent magma crystallized enstatite, pigeonite and diopside as distinct phases (Okada et al. 1988).

Plagioclase abundances are highly variable (0.3 to 16.2 vol %), although most aubrites are depleted in plagioclase relative to chondrites (~10 vol %, McSween et al. 1991). Plagioclase compositions range from $An_{1.8-8.2}$ (Table 28; Watters and Prinz 1979). Watters and Prinz (1979) observed one grain of $An_{23.8}$ in Khor Temiki, while other grains averaged $An_{2.0}$. Okada et al. (1988) reported a huge range of compositions from $An_{0-92.3}$, although these authors attributed much of this range to disequilibrium during rapid crystallization of impact-melt clasts within Norton County.

Figure 41. Photomicrograph in polarized light of Norton County illustrating the large enstatites which are mutually intergrown at their borders, suggesting co-crystallization. Field of view = 2.6 mm.

Forsterite shows a considerable range in abundance (0.3-10.0 vol %) and is essentially FeO-free (Table 28) (Watters and Prinz 1979). Grains up to 4 mm in size occur as distinct grains within the brecciated matrix and, less frequently, enclosed within enstatite grains.

Metal and associated phosphide and silicide. Metallic Fe,Ni is a minor but important constituent of most aubrites. It comprises from a trace to 0.7 vol % of aubrites (Watters and Prinz 1979), excluding the metal-rich Shallowater and Mt. Egerton. Aubritic metal is dominantly silicon-bearing kamacite with 3.7-6.8 wt % Ni and 0.12-2.44 wt % Si (Watters and Prinz 1979). Casanova et al. (1993a) observed a somewhat larger range in kamacite compositions, including the presence of Si-free kamacite in Norton County. Metal in aubrites occurs as inclusions in enstatite, interstitial grains up to several hundred μm in diameter, and cm-sized nodules. Okada et al. (1988) also observed rare taenite and tetrataenite within metal of Norton County. From the relationship between the composition and size of the taenite, these authors inferred a cooling rate of ≤1°C/Ma. A variety of phases occur intimately associated with aubritic metal and the metal itself is often associated with sulfides. Schreibersite is found associated with metal in most aubrites (Watters and Prinz 1979) and occasionally as isolated grains (Casanova et al. 1993a). Casanova et al. (1993a) observed the nickel silicide perryite ($(Ni,Fe)_5(Si,P)_2$) in the cm-sized metal nodules of Norton County and Mt. Egerton. Native copper has also been reported from aubrites (Ramdohr 1973).

Casanova et al. (1993a) analyzed the metal nodules from Norton County, ALH 84007, ALH 84008 and Mt. Egerton for the elements Cr, Co, Ni, Ga, As, W, Re, Ir and Au. With the exception of Cr, which behaves as a chalcophilic element at the oxygen fugacities experienced by aubrites, these elements were all present at approximately chondritic abundances. Casanova et al. (1993a) interpreted these results to indicate that

Table 28. Representative silicate compositions from aubrites (Watters and Prinz, 1979).

	enstatite		diopside		forsterite	plagioclase		
	Bustee	Norton County	Cumberland Falls	Bustee	Bustee	Khor Temiki	Khor Temiki	Norton County
	Chemical Composition (wt %)							
SiO_2	59.4	59.1	54.4	54.6	42.8	67.6	61.9	65.3
Al_2O_3	0.09	0.07	0.70	0.39	b.d.	20.3	24.2	21.8
TiO_2	0.02	<0.02	0.19	0.28	<0.02	<0.02	0.03	0.03
Cr_2O_3	0.02	<0.02	0.04	0.02	n.d.	n.d.	n.d.	n.d.
FeO	0.05	0.07	0.12	<0.02	<0.02	<0.02	<0.02	0.02
MnO	0.11	<0.02	0.05	0.04	<0.02	n.d.	n.d.	n.d.
MgO	40.3	39.6	22.8	20.2	56.9	0.08	0.14	0.03
CaO	0.42	0.64	21.3	24.0	0.05	0.43	5.1	1.77
Na_2O	0.02	0.02	0.33	0.29	n.d.	11.2	9.0	10.7
K_2O	n.d.	n.d.	n.d.	n.d.	n.d.	0.64	0.23	0.53
Total	100.4	99.5	100.0	99.8	99.99	100.4	100.6	100.1
	Cation Formula (O=4 for olivine, 6 for pyroxene, 8 for plagioclase)							
Si	1.983	1.991	1.947	1.968	1.004	2.956	2.733	2.873
Al	0.003	0.003	0.030	0.017	--	1.046	1.260	1.130
Ti	--	--	0.005	0.008	--	--	--	--
Cr	--	--	0.001	0.001	--	--	--	--
Fe	0.001	0.002	0.004	--	--	--	--	--
Mn	0.003	--	0.002	0.001	--	--	--	--
Mg	2.006	1.988	1.216	1.085	1.990	0.005	0.009	0.002
Ca	0.015	0.023	0.817	0.927	0.001	0.020	0.241	0.083
Na	0.001	0.001	0.022	0.020	--	0.950	0.771	0.913
K	--	--	--	--	--	0.036	0.013	0.030
Total	4.015	4.008	4.044	4.026	2.996	5.013	5.028	5.032
	Molar Mineral End Members							
Wo	0.7	1.1	40.1	46.1	--	--	--	--
En	99.2	98.8	59.7	53.9	--	--	--	--
Fs	0.1	0.1	0.2	0.0	--	--	--	--
Or	--	--	--	--	--	3.6	1.3	2.9
Ab	--	--	--	--	--	94.4	75.2	89.0
An	--	--	--	--	--	2.0	23.5	8.1

these cm-sized nodules had not been part of a large central core that experienced fractional crystallization, but rather were metal particles which never segregated to the core.

The unusual Si-bearing iron meteorite Horse Creek is likely related to the aubrites. Horse Creek was described by Buchwald (1975), who also summarized the previous literature. This meteorite contains 6.3 wt % Ni and 2.3 wt. Si in the bulk. Horse Creek is composed of dominant kamacite with an exsolved Ni silicide along the (111) planes of the kamacite. Thus, the meteorite's structure is similar to a Widmanstätten structure, but with nickel silicide in place of kamacite and kamacite in place of taenite, although Horse Creek is a hexahedrite. The Ni silicide has been described by some workers as perryite similar to that found in enstatite chondrites and achondrites, although Buchwald (1975) questions whether these are a single phase or a range of compositions in a solid solution. Schreibersite is also found in Horse Creek. Horse Creek is clearly related to metal in aubrites and may also represent metal which never segregated to the core of the aubrite parent body (Casanova et al. 1993a).

Table 29. Representative sulfide analyses from aubrites.

	troilite		oldhamite		daubreelite	alabandite		caswell.	heideite	niningerite
	Pesyanoe	Bishopville	Khor Temiki	Pena Blanca Spring	Pesyanoe	Mayo Belwa	Norton County	Norton County	Bustee	Bustee
	Chemical Compositions (wt %t)									
Ca	0.17	0.12	52.0	54.3	<0.02	0.72	0.29	b.d.	n.d.	0.45
Fe	62.5	57.1	<0.02	<0.02	17.8	23.5	17.1	b.d.	25.1	6.07
Ti	0.46	5.70	<0.02	0.12	0.17	0.16	<0.02	0.18	28.5	0.18
Mg	0.08	0.06	0.19	0.89	0.75	2.10	1.42	b.d.	<0.05	29.6
Mn	0.12	0.25	1.06	<0.02	1.27	33.7	42.9	0.08	0.02	12.8
Cr	0.21	0.52	<0.02	<0.02	33.9	1.50	0.13	37.4	2.9	0.13
Na	n.d.	n.d.	n.d.	n.d.	n.d.	n.d.	n.d.	15.7	n.d.	0.03
S	37.3	36.5	42.4	44.5	45.9	38.1	38.7	46.3	44.9	50.7
Total	100.84	100.25	95.65	99.81	99.79	99.78	100.54	99.66	101.42	99.96
	Atomic Formula (S=4)									
Ca	0.015	0.011	3.924	3.904	--	0.060	0.024	--	--	0.028
Fe	3.847	3.592	--	--	0.890	1.416	1.015	--	1.284	0.275
Ti	0.033	0.418	--	0.007	0.010	0.011	--	0.010	1.699	0.009
Mg	0.011	0.009	0.024	0.106	0.086	0.291	0.194	--	--	3.080
Mn	0.008	0.016	0.058	--	0.065	2.065	2.588	0.004	0.001	0.589
Cr	0.014	0.035	--	--	1.821	0.097	0.008	1.992	0.159	0.006
Na	--	--	--	--	--	--	--	1.891	--	0.003
Total	3.928	4.081	4.006	4.017	2.872	3.94	3.829	3.897	3.143	3.991

Data for troilite, oldhamite, daubreelite and alabandite from Watters and Prinz (1979); caswellsilverite from Okada and Keil (1982); heideite from Keil and Brett (1974); and niningerite from McCoy (1996).

Sulfides. A remarkable feature of aubrites is the occurrence of elements which are normally lithophile under oxidizing conditions becoming chalcophile under such highly reducing conditions. This gives rise to the formation of a variety of cubic sulfides, including troilite, FeS; oldhamite, CaS; alabandite, (Mn,Fe)S; niningerite, (Mg,Mn,Fe)S; daubreelite, $FeCr_2S_4$; heideite, $(Fe,Cr)_{1+x}(Ti,Fe)_2S_4$; djerfisherite, $K_3(Na,Cu)(Fe,Ni)_{12}S_{14}$; and caswellsilverite, $NaCrS_2$ (Fuchs 1966, Watters and Prinz 1979, Okada and Keil 1982, McCoy 1998). Representative analyses of these phases in aubrites are given in Table 29.

Among these sulfides, that which has garnered the most attention is oldhamite (CaS). Oldhamite typically is a minor phase in aubrites, occurring as grains on the order of a hundred microns in size within the brecciated matrix (Floss and Crozaz 1993). Rare clasts of oldhamite-rich lithologies have, however, been identified in Norton County (Wheelock et al. 1994) and Bustee (Kurat et al. 1992, McCoy 1998). These clasts contain abundant oldhamite, ~30% in the Bustee clasts, with co-existing enstatite, diopside, plagioclase and/or forsterite. Oldhamite in these clasts can contain inclusions of alabandite, niningerite, troilite, daubreelite, caswellsilverite, metal, heideite, forsterite and osbornite (TiN). In the Bustee clast, the Ti-rich sulfides titanian troilite containing 17.2 to 25.2 wt % Ti, osbornite and heideite are all found within oldhamite.

Unlike most other meteorites, where silicates or phosphates act as the major REE carriers, oldhamite is the host in aubrites (Kurat et al. 1992, Floss et al. 1990, Lodders et al. 1993, Wheelock et al. 1994). Rare earth element abundances in oldhamite grains are typically 100 to 1000 times those found in CI chondrites. Further, Floss and Crozaz (1993) recognized 10 distinct REE patterns in their study of 109 oldhamite grains from the aubrites Mayo Belwa, Bustee and Bishopville. Some aubritic oldhamite grains exhibit REE abundances and patterns similar to unequilibrated enstatite chondrites (e.g. Floss and Crozaz 1993; Crozaz and Lundberg 1995). Oldhamite also has an extraordinarily high melting temperature as a pure substance (2450-2525°C; Vogel and Heumann 1941; Chase et al. 1985).

A full discussion of the REE element abundances and patterns within aubrites is beyond the scope of this review. As stated earlier, representative bulk samples are virtually impossible to obtain, given the coarse grain size of the aubrites. Further, deriving petrogenetic information from such data requires extreme caution, since any "bulk" sample will consist of materials from a number of different lithologies. A small number of papers (e.g. Floss and Crozaz 1993, Lodders et al. 1993, Shimaoka et al. 1995) discuss mineral and bulk chemical data and interested readers should consult these references.

Unusual clasts and related meteorites. A variety of silicate-rich clasts and related meteorites are also found within aubrites and are briefly discussed here. With the exception of Shallowater and Mt. Egerton, all of the meteorites discussed here are either fragmental or regolith breccias (see Keil 1989 and references therein), consisting of a variety of rock clasts. Not surprisingly, impact-melt clasts are a fairly common component of aubrites. Several such clasts have been discussed by Okada et al. (1988). In the Cumberland Falls meteorite, clasts of chondritic material similar in composition to LL chondrites (Kallemeyn and Wasson 1985), but petrographically unique (Binns 1969), have been admixed during this surface residence. Since these lithologies do not speak directly to the origin or evolution of the aubrite parent body, they are not discussed further here.

The Shallowater aubrite has been recognized as unique since its original description by Foshag (1940). It is the only known unbrecciated aubrite. In addition, it contains significant quantities of both metal and troilite (Table 27). It is also the only aubrite to contain ordered orthopyroxene, rather than the disordered pyroxene common to other aubrites (Reid and Cohen 1967). Keil et al. (1989) noted that Shallowater consists of 80 vol % orthoenstatite crystals up to 4.5 cm in size which contain, as inclusions and in the interstices, xenoliths of an assemblage of twinned low-Ca clinoenstatite, forsterite, plagioclase, metallic Fe,Ni and troilite. These xenolithic phases can occur as polymineralic inclusions with all phases or as individual mineral grains and comprise ~20 vol % of the rock. The origin of Shallowater appears to be unrelated to that of the other aubrites. Keil et al. (1989) argue that Shallowater formed when a totally-molten enstatite-like asteroid was struck by an impactor of enstatite chondrite-like composition, but achondritic texture. The impactor was disrupted, forming the xenoliths, which were mixed into and quenched the enstatite-rich mantle of the molten target body. Thus, in this scenario Shallowater samples a separate parent body from the other aubrites with a dramatically different history. Several unanswered questions remain with this scenario, such as the lack of related lithologies which would have resulted from such an event.

Mt. Egerton is an unbrecciated meteorite composed of cm-sized enstatite crystals with about 21 wt % metallic Fe,Ni occurring in the interstices between the large enstatite laths. In addition to these phases, diopside, SiO_2, troilite, brezinaite and schreibersite are also found (Casanova et al. 1993b). Mineral and oxygen-isotopic compositions, as well as the unfractionated trace element composition of the metal, are all similar to other aubrites. In most respects, Mt. Egerton should be classified as an aubrite. It is, however, unusual in containing a far greater percentage of metal than other aubrites.

Fogel (1994, 1997) has found a small number of basaltic vitrophyre clasts within the Khor Temiki and LEW 87007 aubrites. The clasts are <1 cm in maximum dimension and consist of 60 to 90 vol % enstatite, minor diopside or olivine, troilite, metal and 10 to 40 vol % of glass which is highly enriched in Al_2O_3, Na_2O, K_2O and CaO relative to bulk aubrites or enstatite chondrites. In the case of the Khor Temiki vitrophyre, the glass also contains significant quantities of dissolved sulfur (0.82 wt %; Fogel 1994). It should be noted that not all dark clasts within aubrites are basaltic vitrophyres. Newsom et al. (1996) studied dark clasts in Khor Temiki and concluded that they are simply highly brecciated, containing a fine-grained matrix or enstatite with abundant inclusions.

Discussion

Although several authors have argued that aubrites formed by direct condensation from the solar nebula (Wasson and Wai 1970, Richter et al. 1979, Sears 1980), recent workers agree that the clasts within the breccias are of igneous parentage. Keil (1989) summarized the evidence for an igneous origin for aubrites, including textures of pyroxenitic clasts indicative of co-crystallization from a melt (Fig. 41), a range of clast types consistent with fractional crystallization (Okada et al. 1988), large crystal sizes inconsistent with condensation, and the presence of melt inclusions within enstatite grains (Fuchs 1974).

In a simplified model for the origin of aubrites (McCoy et al. 1997d), the direct precursor to the aubrites is an enstatite chondrite. Certainly, similarities in mineral and oxygen-isotopic compositions between enstatite chondrites and aubrites suggest some kind of link. Watters and Prinz (1979) argued that the precursor to aubrites may have been more similar to EL chondrites, while Fogel et al. (1988) favored an EH chondrite as the likely starting material. The enstatite chondrite is heated to the point of partial melting. The partial melts, both basaltic and Fe,Ni-FeS, are removed from the source region. The removal of these components is supported by the modal data of Watters and Prinz (1979) (Table 27) which demonstrate depletions relative to enstatite chondrites for most aubrites in plagioclase and for all aubrites in metal and troilite. Continued heating completely melts this ultramafic residual material, which subsequently crystallizes to form the coarse-grained, igneous aubrites. The pyroxene intergrowths typical of co-crystallization suggest that aubrites crystallized from a total or near-total melt, rather than being cumulates.

While this model may be broadly correct in concept, several questions and objections have arisen over the details of such a model. These questions include: (1) What happened to the basaltic partial melts? (2) Is aubritic oldhamite a nebular relict or an igneous crystallization product? (3) Can aubrites be derived by melting of known enstatite chondrites?

What happened to the basaltic partial melts? Differentiation of any likely chondritic precursor should produce a significant quantity of basaltic material. Despite this fact, brecciated aubrites contain a relative paucity of basaltic clasts. A number of theories have been proposed to explain the ultimate fate of these partial melts. As discussed earlier, Fogel (1994, 1997) has identified basaltic vitrophyres within aubrites which almost certainly sample these partial melts. Thus, partial melts are not completely absent from the aubrite parent body. However, it is clear that these rare basaltic clasts do not account for the abundant (10 to 20 vol %) basaltic material which would have been generated during partial melting of an enstatite chondrite-like source. One can, of course, speculate that we have an unrepresentative sample of the aubrite parent body, although the fact that aubritic regolith breccias do not contain abundant basaltic clasts suggests that such material does not occur as thick deposits on the surface of this body. Keil (1989) speculated that these materials could have been stripped from the parent body by impacts or, perhaps, the aubritic precursor was plagioclase-poor and abundant basalts were never formed. Wilson and Keil (1991) argued that basaltic material likely did exist and, owing to the presence of volatiles, may have been erupted at velocities exceeding the escape velocity of the aubrite parent body and, thus, were lost as small pyroclasts into space early in the history of the solar system. Finally, it is worth noting that both modal analyses (Watters and Prinz 1979) and bulk chemical analyses (Lodders et al. 1993) demonstrate that some aubritic samples are enriched in plagioclase and it is possible that much of the basaltic material was locally redistributed. The absence of basaltic melts is also a problem for understanding the genesis of ureilites and lodranites.

Is aubritic oldhamite a nebular relict or an igneous crystallization product? Although general consensus exists that aubrites formed from a melt, some authors have suggested that many oldhamite grains are nebular relicts which survived igneous processing (Floss and Crozaz 1993, Kurat et al. 1992, Lodders 1996). In particular, these authors cite as support the similarities in REE patterns between oldhamite in aubrites and unequilibrated enstatite chondrites, the inability to produce these patterns through fractional crystallization, and the remarkably high melting temperature of pure oldhamite. In contrast, the large size of some aubritic oldhamite grains (which are difficult to reconcile with the small size of oldhamites found in enstatite chondrites if aubritic oldhamite is a relict) (Wheelock et al. 1994, McCoy 1998), the presence of oldhamite partially enclosing silicates (Wheelock et al. 1994), REE patterns in some oldhamite grains consistent with igneous fractionation (Floss and Crozaz 1993, Wheelock et al. 1994) and the occurrence of phases as inclusions within oldhamite which have condensation temperatures which are lower than oldhamite (Wheelock et al. 1994, McCoy 1998) all suggest that at least some oldhamite grains formed by igneous processes.

A recent focus of research on the origin of aubrites has been experimental studies which both examine REE partitioning between oldhamite and silicate melts and replicate partial melting of enstatite chondrites in an effort to examine questions of aubrite petrogenesis. Partitioning of REE between oldhamite and silicate melts has been examined by Jones and Boynton (1983), Lodders (1996) and Dickinson and McCoy (1997). Partition coefficients measured in these works are generally around unity, while apparent partition coefficients determined by ratioing REE elements in oldhamite to those in bulk aubrites are on the order of 100 to 1000 (Wheelock et al. 1994). Dickinson and McCoy (1997) demonstrated that the strongest control on REE partitioning is the composition of the crystallizing oldhamite. Calcium-poor oldhamite (63 to 79% CaS) produces a LREE-depleted, HREE-enriched pattern with a negative Eu anomaly. In contrast, Ca-rich oldhamite (88-89% CaS) produces a bowed pattern with a positive Eu anomaly. Temperature also strongly influences partitioning. Partition coefficients increase with decreasing temperature. Experiments run for 2 days at 1200°C and annealed at 800°C for 9 days exhibit larger partition coefficients for Eu and Gd than those which were not annealed.

Measured partition coefficients are clearly inconsistent with a simple igneous origin in which oldhamite is the first crystallizing phase and REE abundances are determined solely by high temperature partitioning between the oldhamite and bulk liquid. Partition coefficients near unity are consistent with the nebular model. Oldhamite REE patterns may have been established by a complex process of partial melting, melt removal, fractional crystallization with oldhamite of varying compositions crystallizing from isolated melt pools, subsolidus annealing and exsolution of Mg, Mn, Cr, and Fe-rich phases, producing a wide range of REE patterns.

More direct evidence for the history of aubritic oldhamite comes from the work of Dickinson and Lofgren (1992), Fogel et al. (1996) and McCoy et al. (1997c), in which samples of the Indarch (EH4) chondrite were partially melted in the range of 1000° to 1500°C. Fogel et al. (1996) added synthetic CaS to their powdered Indarch to directly address the question of oldhamite melting. Each of these experiments produced melting of CaS at temperatures far below the melting temperature of the phase in isolation. McCoy et al. (1997c) found that at temperatures as low as 1200°C, no relict sulfides remained in the experimental charges and two sulfide melts, one Fe-rich and one Mg,Mn,Ca,Cr-rich, were found. Thus, it seems highly unlikely that oldhamite found within enstatite chondrites would survive as a relict phase within aubritic parent body. It is possible that oldhamite crystallizes from the Mg,Mn,Ca,Cr-rich sulfide melt or, as noted by Fogel et al. (1996), from the silicate melt, which contains weight-percent levels of dissolved S presumably combined with Ca or Mg. Thus, it appears likely that aubritic oldhamite is a product of igneous crystallization.

Can aubrites be derived by melting of known enstatite chondrites? Although aubrites clearly derived from a highly reduced precursor like enstatite chondrites, Keil (1989) presented several arguments that aubrites and enstatite chondrites formed on different parent bodies. These include the absence of enstatite chondrite clasts within aubritic breccias, the differences in cosmic-ray exposure and thus impact ejection ages, and the implausibility high thermal gradient necessary for a single enstatite chondrite-aubrite parent body. More relevant to this discussion, several differences exist between aubrites and known enstatite chondrites, e.g. Ti contents of troilite, (Keil 1989) and these features are not easily explained by igneous fractionation.

A unique oldhamite-pyroxene clast in Bustee contains the most Ti-rich troilite found in aubrites, as well as the Ti-rich phases osbornite and heideite. Thus, this occurrence may shed some light on the origin of the Ti enrichment in aubritic troilite relative to enstatite chondrite troilite. Previous workers have proposed that Ti could be enriched in aubritic troilite through immiscibility in the Fe-Ti-S system (Brett and Keil 1986), preferential fractionation of Ti-rich FeS (Fogel et al. 1988) or melting of osbornite (Casanova 1992). However, no immiscibility exists in the Fe-Ti-S system; temperature and density difference between Ti-rich and Ti-poor troilite are too small to produce fractionation; and the occurrence of Ti-rich troilite with osbornite in Bustee suggests that melting of osbornite is not producing the Ti-rich troilite. Instead, we suggest that co-crystallization with and/or exsolution from oldhamite provides such a mechanism. Indeed, oldhamite experiments by McCoy et al. (1997c) demonstrated immiscibility in the Fe-Ca-S system, in the form of one sulfide melt enriched in Fe and the other which contained all the Ca in the system. Oldhamite efficiently excludes Ti, which can be preferentially incorporated into troilite, thus providing a mechanism for enriching aubritic troilite in Ti and weakening arguments against a derivation of aubrites from known enstatite chondrites.

In addition, all of the Indarch melting experiments (Dickinson and Lofgren 1992, Fogel et al. 1996, McCoy et al. 1997c) demonstrated significant exchange of elements between different phases (e.g. Si between metal and silicate melt; S between sulfide and silicate melt). This movement of elements might produce compositions not readily predicted on the basis of equilibrium melting relations alone and could explain the origin of some features of aubrites not readily explicable by modeling of equilibrium or fractional partial melting.

HOWARDITES, EUCRITES AND DIOGENITES

The howardite, eucrite and diogenite (HED) meteorites make up the largest suite of crustal igneous rocks we have available from any solar system body other than the Earth and Moon. These rocks include basalts, cumulate gabbros and orthopyroxenites, plus brecciated mixtures of the igneous lithologies. Together, they provide an unmatched look at differentiation processes that occurred on asteroidal-sized bodies early in the history of the solar system. The HED meteorites have O-isotopic compositions similar to those of the IIIAB irons, main-group pallasites, mesosiderites, angrites and brachinites (Fig. 1c,d), and thus may have been formed in the same region of the solar nebula.

Mineralogy and petrology

Diogenites. The diogenites are coarse-grained orthopyroxenites, typically containing from ~84 to ~100 vol % orthopyroxene (Bowman et al. 1997, Hewins, pers. comm.). Most of them have been brecciated, so the original grain size is not well known, but it must have been at least 5 cm based on the largest reported clasts (Mason 1963). The typical diogenite, for example Johnstown, is composed of orthopyroxene clasts in a fine-grained fragmental matrix of orthopyroxene. Four exceptional diogenites are GRO 95555, LEW 88679, Tatahouine and the Yamato Type A, or Y-74013-type, diogenites. The initial description of GRO 95555 (AMN 19-2) states that this diogenite has a polygonal-granular texture of anhedral orthopyroxene grains from <1 to 2.4 mm in size. The initial description of LEW 88679 (AMN 15-1) states that this diogenite does not have a cataclastic texture, but rather is composed of anhedral grains of pyroxene up to 6 mm in size. Tatahouine apparently broke up during or after atmospheric passage as it is composed of numerous coarse-grained orthopyroxene fragments of cm size, generally without fusion crust on them. Unlike pyroxenes in most diogenites, those in Tatahouine show patchy extinction under crossed polars. The Yamato Type A diogenites have a granoblastic texture composed of equant, rounded orthopyroxene grains a few tens of μm to mm size, containing inclusions dominantly of chromite and troilite.

Chromite and olivine are common minor minerals, each making up between ~0 to 5 vol % (Bowman et al. 1997, Hewins, pers. comm.). EETA79002 may be unusually olivine-rich (Sack et al. 1991, 1994a), but olivine is very heterogeneously distributed in this diogenite (e.g. Bowman et al. 1997), and a mean abundance is currently unknown. Chromite occurs as coarse, mm-sized equant grains as well as in smaller grains a few tens to hundreds of μm in size. The smaller chromite grains are often poikilitically enclosed in orthopyroxene. The coarser chromites are often found as separate euhedral grains or fragments in the matrix, but some are preserved in their original igneous contact with orthopyroxene. Olivine grains are commonly a few hundred μm in size, but they are usually crystal fragments in the matrix. Olivine has been found in Roda in its original textural context with orthopyroxene, where igneous grain borders between the minerals are preserved. A third minor mineral in typical diogenites is anorthitic plagioclase (~0-5 vol %; Bowman et al. 1997, Hewins, pers. comm.). This mineral has not been well described in the literature, but where the textural setting is mentioned, it has been reported only as crystal fragments in the diogenite matrix.

Accessory minerals in the typical diogenite include diopside (~0 to 2 vol %), troilite (~0-3 vol %), metal (<1 vol %), a silica phase (~0 to 2 vol %), and rare phosphates (modal data from Bowman et al. 1997 and Hewins, pers. comm.). Diopside occurs as exsolution lamellae in the orthopyroxene, with thicknesses of at most a few μm. Troilite is highly variable in abundance between different diogenites and within a given diogenite. In some, it is a trace phase usually of μm-sized grains enclosed in orthopyroxene, often forming inclusion "curtains" within the pyroxene (Mori and Takeda 1981a). In other diogenites, troilite occurs as larger equant grains or polycrystalline aggregates in the matrix. Like troilite, metal frequently occurs as μm-sized grains in inclusion "curtains" in orthopyroxene. A silica phase is sometimes associated with metal and troilite in these inclusion "curtains" (Gooley and Moore 1976). A silica phase is also present as equant grains a few tens of µm in size in some diogenites, but the textural setting has not been well described. Phosphates are rare in diogenites, and not well described. In Roda, two rounded, 5-10 µm size light-rare-earth element-rich phosphates have been found enclosed in diopside exsolved from orthopyroxene (Mittlefehldt 1994).

The Yamato Type B, or Y-75032-type, diogenites are unusual. The many recovered specimens are undoubtedly all pieces of a single fall. These meteorites, described by Takeda et al, (1979a) and Takeda and Mori (1985) are brecciated rocks composed of subangular mineral and lithic clasts set in a black, glassy matrix containing fine-grained, clastic debris. The pyroxene in these meteorites is coarse-grained orthopyroxene and inverted pigeonite a few mm in size, which has exsolved augite lamellae and blebs. Enclosed in the rims of some pyroxene grains and interstitially at grain boundaries are irregular to triangular or lath-shaped plagioclase grains on the order of a few hundred µm in size. Plagioclase also is present as fragmental debris in the black glassy matrix. Chromite occurs as minute inclusions in the pyroxenes, and as clastic debris up to 300 µm in size in the black glassy matrix. Troilite has been reported in the black glassy matrix. Mittlefehldt and Lindstrom (1993) reported ilmenite and a silica polymorph in lithic clasts.

The compositions of pyroxenes in most diogenites are well known (Berkley and Boynton 1992, Fowler et al. 1994, Mittlefehldt 1994, Mittlefehldt and Lindstrom 1993, Sack et al. 1991, Takeda and Mori 1985, Takeda et al. 1979a, 1981). Table 30 gives representative pyroxene compositions for select diogenites, and Figure 42 shows a portion of the pyroxene quadrilateral displaying average orthopyroxene compositions for diogenites. Most diogenites contain orthopyroxene with a very uniform, common major

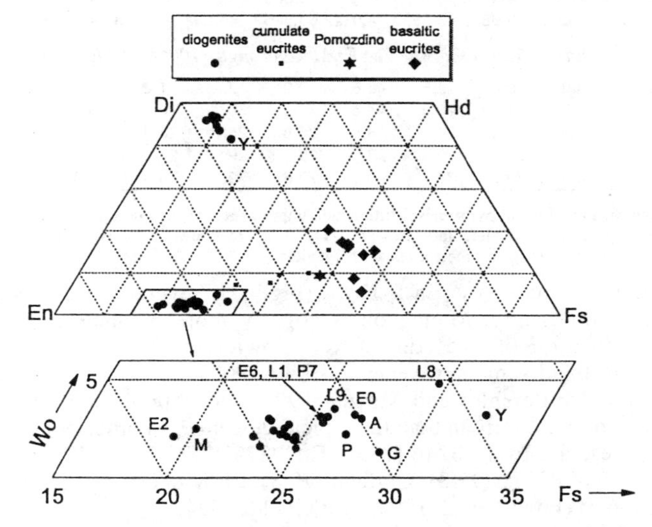

Figure 42. Pyroxene quadrilateral for HED pyroxenes, with detailed view of diogenite orthopyroxene compositions. Labeled diogenites are; E2 - most magnesian pyroxene in EETA79002, M - Manegaon, E6 - EET 83246, L1 - LAP 91900, P7 - PCA 91077, L9 - LEW 88679, P - Peckelsheim, E0 - EET 87530, A - ALH 85015, G - Garland, L8 - LEW 88008, Y - Yamato Type B diogenites. Diogenite data are from Fowler et al. (1994), Mittlefehldt (1994), Mittlefehldt and Lindstrom (1993), Mittlefehldt (unpublished). Bulk pigeonite compositions for cumulate eucrites, anomalous magnesian eucrite Pomozdino and representative basaltic eucrites are from the data in Table 30, plus Metzler et al. (1995). The dispersion of the basaltic eucrite data approximately parallel to lines of constant En content probably reflects the difficulty of determining bulk compositions in the exsolved pyroxenes.

Table 30. Compositions of pyroxene grains from representative diogenites and eucrites.

	EETA79002		PCA 91077	Johnstown		Peckels heim	EET 87530	Garland	LEW 88008	Y-75032		
	(1)	(2)	(3)	(4)	(5)	(6)	(7)	(8)	(9)	(10)	(11)	(12)
Chemical Composition (wt %t)												
SiO_2	55.6	54.5	54.1	54.1	52.6	54.8	54.1	54.5	53.2	53.4	53.6	53.4
Al_2O_3	0.59	0.85	0.69	0.99	1.11	0.40	0.79	0.32	1.18	0.69	0.77	0.72
TiO_2	0.10	0.10	0.07	0.11	0.20	0.04	0.10	0.02	0.14	0.16	0.35	0.19
Cr_2O_3	0.44	0.60	0.58	0.66	0.55	0.34	0.48	0.39	0.74	0.24	0.45	0.31
FeO	12.7	15.1	16.0	15.4	5.65	17.3	17.4	18.6	19.1	20.3	6.87	20.0
MnO	0.43	0.50	0.56	0.50	0.24	0.59	0.60	0.69	0.67	0.65	0.27	0.66
MgO	29.4	27.5	25.2	26.5	16.0	25.7	25.6	25.4	23.7	23.8	16.4	22.5
CaO	1.10	1.16	1.37	1.24	23.3	1.10	1.60	0.63	2.40	0.67	21.3	2.19
Na_2O	nd	nd	0.02	nd	nd	0.01	nd	nd	nd	nd	bd	nd
Total	100.36	100.31	98.59	99.50	99.65	100.28	100.67	100.55	101.13	99.91	100.01	99.97
Cation Formula Based on 6 Oxygens												
Si	1.9752	1.9623	1.9900	1.9670	1.9499	1.9900	1.9644	1.9851	1.9462	1.9746	1.9747	1.9791
^{iv}Al	0.0247	0.0361	0.0100	0.0330	0.0485	0.0100	0.0338	0.0137	0.0509	0.0254	0.0253	0.0209
Total tet*	1.9999	1.9984	2.0000	2.0000	1.9984	2.0000	1.9982	1.9988	1.9971	2.0000	2.0000	2.0000
Ti	0.0027	0.0027	0.0019	0.0030	0.0056	0.0011	0.0027	0.0005	0.0039	0.0044	0.0097	0.0053
^{vi}Al	--	--	0.0199	0.0094	--	0.0071	--	--	--	0.0047	0.0081	0.0106
Cr	0.0124	0.0171	0.0169	0.0190	0.0161	0.0098	0.0138	0.0112	0.0214	0.0070	0.0131	0.0091
Fe	0.3773	0.4547	0.4922	0.4683	0.1752	0.5254	0.5284	0.5666	0.5844	0.6278	0.2117	0.6199
Mn	0.0129	0.0152	0.0174	0.0154	0.0075	0.0181	0.0185	0.0213	0.0208	0.0204	0.0084	0.0207
Mg	1.5565	1.4756	1.3815	1.4359	0.8839	1.3909	1.3853	1.3788	1.2922	1.3116	0.9005	1.2428
Ca	0.0419	0.0448	0.0540	0.0483	0.9255	0.0428	0.0623	0.0246	0.0941	0.0265	0.8408	0.0870
Na	--	--	0.0014	--	--	0.0007	--	--	--	--	--	--
Total Cations	4.0036	4.0085	3.9852	3.9993	4.0122	3.9959	4.0092	4.0018	4.0139	4.0024	3.9923	3.9954
*Cation Ratios Ca:Mg:Fe, Fe/Mn and mg# (100*Mg/(Mg+Fe))*												
Ca	2.1	2.3	2.8	2.5	46.6	2.2	3.2	1.2	4.8	1.3	43.1	4.5
Mg	78.8	74.7	71.7	73.5	44.5	71.0	70.1	70.0	65.6	66.7	46.1	63.7
Fe	19.1	23.0	25.5	24.0	8.8	26.8	26.7	28.8	29.7	31.9	10.8	31.8
Fe/Mn	29	30	28	30	23	29	29	27	28	31	25	30
mg#	80.5	76.4	73.7	75.4	83.5	72.6	72.4	70.9	68.9	67.6	81.0	66.7

(1, 2) average magnesian and typical pyroxenes, Mittlefehldt, unpublished; (3, 6) average pyroxene, Fowler et al. (1994); (4, 5, 7, 8, 9) average pyroxenes, Mittlefehldt (1994); (10, 11, 12) average low-Ca pyroxene, high-Ca pyroxene and bulk pigeonite, Takeda and Mori (1985). All meteorites are diogenites.

element composition of ~$Wo_{2\pm1}En_{74\pm2}Fs_{24\pm1}$ (Fig. 42). Several diogenites are exceptions to this general uniformity. Manegaon is slightly more magnesian, with an average pyroxene composition of $Wo_{2.8}En_{76.8}Fs_{20.4}$ (Fowler et al. 1994, Fredriksson 1982, Mittlefehldt 1994) and some pyroxenes in EETA79002 are even more magnesian, up to $Wo_{1.8}En_{79.9}Fs_{18.3}$ (Mittlefehldt and Meyers 1991, and unpublished). There are several diogenites that are more ferroan than the typical diogenite. In order of decreasing mg# (in parentheses), these are EET 83246 (73.6), EET 87530 (72.7), Peckelsheim (72.6), ALH 85015 (71.9), LEW 88679 (71.3), Garland (70.8), LEW 88008 (69.6) and Yamato Type B diogenites (66.4) (Fowler et al. 1994, Mittlefehldt 1994).

Excluding low-Ca clinopyroxenes formed by shock deformation of orthopyroxene (Mori and Takeda 1981a), clinopyroxenes in diogenites are the product of subsolidus exsolution from the orthopyroxene or low-Ca pigeonite (for Yamato Type B diogenites) grains. Few analyses of these clinopyroxenes are available in the literature. For typical diogenites, many of the clinopyroxenes are too narrow to analyze with the electron probe. Mori and Takeda (1981a) determined an average lamella width of 77 nm for Ibbenbüren and 6 nm for Johnstown. They identified these as augite. Larger clinopyroxene grains are

Table 30 (cont'd). Compositions of pyroxene grains from representative diogenites and eucrites.

	Binda			Moama			Serra de Magé			Moore County		
	(13)	(14)	(15)	(16)	(17)	(18)	(19)	(20)	(21)	(22)	(23)	(24)
Chemical Composition (wt %)												
SiO_2	53.4	53.0	53.3	51.6	51.7	51.5	52.3	52.7	52.4	51.5	52.1	51.6
Al_2O_3	0.51	0.85	0.55	0.39	0.62	0.47	0.34	0.69	0.40	0.24	0.78	0.30
TiO_2	0.22	0.37	0.24	0.28	0.40	0.29	0.24	0.37	0.26	0.31	0.67	0.35
Cr_2O_3	0.24	0.40	0.26	0.27	0.37	0.27	0.12	0.23	0.14	0.15	0.32	0.17
FeO	21.7	8.48	20.1	24.8	11.3	23.9	27.4	10.5	24.4	29.2	13.9	27.3
MnO	0.71	0.37	0.67	0.89	0.47	0.80	0.95	0.42	0.85	0.97	0.48	0.91
MgO	22.2	14.7	21.3	19.2	13.1	18.6	18.6	13.4	17.6	15.5	12.2	15.1
CaO	1.05	21.6	3.52	1.65	22.3	3.71	0.75	22.3	4.62	2.65	19.7	4.70
Na_2O	bd	0.06	0.01	bd	0.05	bd	0.02	0.07	nd	bd	0.07	0.01
Total	100.03	99.83	99.95	99.08	100.31	99.54	100.72	100.68	100.67	100.52	100.22	100.44
Cation Formula Based on 6 Oxygens												
Si	1.9866	1.9745	1.9846	1.9782	1.9528	1.9688	1.9865	1.9699	1.9852	1.9903	1.9740	1.9896
^{iv}Al	0.0134	0.0255	0.0154	0.0176	0.0276	0.0212	0.0135	0.0301	0.0148	0.0097	0.0260	0.0104
Total tet*	2.0000	2.0000	2.0000	1.9958	1.9804	1.9900	2.0000	2.0000	2.0000	2.0000	2.0000	2.0000
Ti	0.0062	0.0104	0.0067	0.0081	0.0114	0.0083	0.0069	0.0104	0.0074	0.0090	0.0191	0.0101
^{vi}Al	0.0090	0.0118	0.0087	--	--	--	0.0017	0.0003	0.0031	0.0012	0.0088	0.0032
Cr	0.0071	0.0118	0.0077	0.0082	0.0110	0.0082	0.0036	0.0068	0.0042	0.0046	0.0096	0.0052
Fe	0.6752	0.2642	0.6259	0.7951	0.3570	0.7641	0.8704	0.3282	0.7731	0.9438	0.4405	0.8803
Mn	0.0224	0.0117	0.0211	0.0289	0.0150	0.0259	0.0306	0.0133	0.0273	0.0318	0.0154	0.0297
Mg	1.2309	0.8162	1.1819	1.0970	0.7374	1.0597	1.0529	0.7465	0.9937	0.8928	0.6889	0.8677
Ca	0.0419	0.8623	0.1404	0.0678	0.9025	0.1520	0.0305	0.8931	0.1875	0.1097	0.7998	0.1942
Na	--	0.0043	0.0007	--	0.0037	--	0.0015	0.0051	--	--	0.0051	0.0007
Total Cations	3.9927	3.9927	3.9931	4.0009	4.0184	4.0082	3.9981	4.0037	3.9963	3.9929	3.9872	3.9911
*Cation Ratios Ca:Mg:Fe, Fe/Mn and mg# (100*Mg/(Mg+Fe))*												
Ca	2.1	44.4	7.2	3.5	45.2	7.7	1.6	45.4	9.6	5.6	41.5	10.0
Mg	63.2	42.0	60.7	56.0	36.9	53.6	53.9	37.9	50.8	45.9	35.7	44.7
Fe	34.7	13.6	32.1	40.6	17.9	38.7	44.5	16.7	39.6	48.5	22.8	45.3
Fe/Mn	30	23	30	28	24	29	28	25	28	30	29	30
mg#	64.6	75.5	65.4	58.0	67.4	58.1	54.7	69.5	56.2	48.6	61.0	49.6

(13, 14, 15, 22, 23, 24) average low-Ca pyroxene, high-Ca pyroxene and bulk pigeonite, Pun and Papike (1995); (16, 17, 18) average low-Ca pyroxene, high-Ca pyroxene and bulk pigeonite Lovering (1975); (19, 20, 21) average low-Ca pyroxene, high-Ca pyroxene and bulk pigeonite Harlow et al. (1979). All meteorites are cumulate eucrites.

present in both of the meteorites, and have been analyzed by electron microprobe by Floran et al. (1981) and Mittlefehldt (1994). The larger clinopyroxene grains in diogenites are compositionally diopside. The clinopyroxene grains in the Yamato Type B diogenites are augites based on both compositional and crystallographic criteria (Mittlefehldt and Lindstrom 1993, Takeda and Mori 1985, Takeda et al. 1979a). Table 30 gives representative clinopyroxene analyses from selected diogenites.

Although the major elements in diogenite orthopyroxene grains are very uniform, the minor incompatible elements show considerable variation. Mittlefehldt (1994) showed that the contents of Al_2O_3 and TiO_2, for average orthopyroxenes vary by factors of ~3 and 4, respectively, and are positively correlated. Fowler et al. (1994) further showed that the minor incompatible element contents for orthopyroxene from individual diogenites can also be highly variable, with Al contents varying by a factor of 2 in pyroxenes from ALH 85015, for example. The variation in incompatible elements is decoupled from mg#, and orthopyroxenes with similar mg# can have quite different TiO_2 and Al_2O_3 contents (Mittlefehldt 1994). The compatible minor element Cr varies as well by factors of 2 to 3 (Fowler et al. 1994, Mittlefehldt 1994). Berkley and Boynton (1992) found a weak

Table 30 (cont'd). Compositions of pyroxene grains from representative diogenites and eucrites.

	Y-791195			Pomozdino			Sioux County		Bouvante	Y-75011 ,84C		Y-793164
	(25)	(26)	(27)	(28)	(29)	(30)	(31)	(32)	(33)	(34)	(35)	(36)
						Chemical Composition (wt %)						
SiO_2	49.9	51.1	50.3	50.6	51.4	50.7	49.1	50.7	49.8	53.4	47.4	48.5
Al_2O_3	0.27	0.68	0.33	0.29	0.87	0.47	0.23	0.47	0.15	0.84	0.96	0.59
TiO_2	0.16	0.40	0.23	0.30	0.65	0.36	0.25	0.32	0.13	0.10	0.82	0.33
Cr_2O_3	0.11	0.28	0.15	0.16	0.30	0.29	0.42	0.39	0.15	0.75	0.10	0.61
FeO	32.6	15.5	28.0	31.3	14.2	28.5	36.1	16.6	34.1	17.6	33.1	32.2
MnO	1.05	0.50	0.90	0.98	0.48	0.92	1.12	0.53	0.85	0.58	0.99	0.98
MgO	13.6	11.2	12.9	15.2	12.0	14.3	11.7	9.84	12.0	24.5	4.31	9.4
CaO	2.34	20.4	7.30	1.19	20.0	4.39	0.78	20.8	2.58	2.05	11.9	6.90
Na_2O	nd	nd	nd	0.04	0.10	0.05	nd	nd	nd	bd	0.03	nd
Total	100.03	100.06	100.11	100.06	100.00	99.98	99.70	99.65	99.76	99.82	99.61	99.51
						Cation Formula Based on 6 Oxygens						
Si	1.9752	1.9620	1.9734	1.9801	1.9598	1.9780	1.9786	1.9695	1.9903	1.9627	1.9537	1.9606
^{iv}Al	0.0126	0.0308	0.0153	0.0134	0.0391	0.0216	0.0109	0.0215	0.0071	0.0364	0.0463	0.0281
Total tet*	1.9878	1.9928	1.9887	1.9935	1.9989	1.9996	1.9895	1.9910	1.9974	1.9991	2.0000	1.9887
Ti	0.0048	0.0116	0.0068	0.0088	0.0186	0.0106	0.0076	0.0093	0.0039	0.0028	0.0254	0.0100
^{vi}Al	--	--	--	--	--	--	--	--	--	--	0.0003	--
Cr	0.0034	0.0085	0.0047	0.0050	0.0090	0.0089	0.0134	0.0120	0.0047	0.0218	0.0033	0.0195
Fe	1.0792	0.4977	0.9187	1.0244	0.4528	0.9299	1.2166	0.5393	1.1398	0.5410	1.1410	1.0886
Mn	0.0352	0.0163	0.0299	0.0325	0.0155	0.0304	0.0382	0.0174	0.0288	0.0181	0.0346	0.0336
Mg	0.8023	0.6409	0.7543	0.8865	0.6819	0.8314	0.7027	0.5697	0.7148	1.3420	0.2647	0.5663
Ca	0.0992	0.8393	0.3069	0.0499	0.8171	0.1835	0.0337	0.8658	0.1105	0.0807	0.5255	0.2989
Na	--	--	--	0.0030	0.0074	0.0038	--	--	--	--	0.0024	--
Total Cations	4.0119	4.0071	4.0100	4.0036	4.0012	3.9981	4.0017	4.0045	3.9999	4.0055	3.9972	4.0056
						*Cation Ratios Ca:Mg:Fe, Fe/Mn and mg# (100*Mg/(Mg+Fe))*						
Ca	5.0	42.4	15.5	2.5	41.9	9.4	1.7	43.8	5.6	4.1	27.2	15.3
Mg	40.5	32.4	38.1	45.2	34.9	42.8	36.0	28.8	36.4	68.3	13.7	29.0
Fe	54.5	25.2	46.4	52.2	23.2	47.8	62.3	27.3	58.0	27.5	59.1	55.7
Fe/Mn	31	31	31	32	29	31	32	31	40	30	33	32
mg#	42.6	56.3	45.1	46.4	60.1	47.2	36.6	51.4	38.5	71.3	18.8	34.2

(25, 26, 27) average low-Ca pyroxene, high-Ca pyroxene and bulk pigeonite, cumulate eucrite, Mittlefehldt and Lindstrom (1993); (28, 29, 30) average low-Ca pyroxene, high-Ca pyroxene and bulk pigeonite, anomalous, magnesian eucrite, Warren et al. (1990); (31, 32) average low-Ca pyroxene and high-Ca pyroxene, main-group eucrite, Mittlefehldt, unpublished; (33) average core, large pigeonite, Stannern-trend eucrite, Christophe Michel-Levy et al. (1987); (34, 35) core and rim pyroxene in least-metamorphosed basalt clast from polymict eucrite, Takeda et al. (1994c); (36) average bulk pigeonite, evolved, recrystalized basalt clast from polymict eucrite, Mittlefehldt and Lindstrom (1993).

positive correlation between orthopyroxene Cr_2O_3 content and mg# within individual diogenites.

There are relatively few analyses of diogenite olivine in the literature (Floran et al. 1981, Gooley 1972, Mittlefehldt 1994, Sack et al. 1991), and because many of them are either fragments in the matrix, or are of unknown textural context, it is difficult to know how they fit in diogenite petrogenesis. Representative olivine analyses are given in Table 31. Hewins (1981) found that orthopyroxene and olivine in 5 diogenites have an equilibrium distribution of Fe/Mn. Mittlefehldt (1994) presented data for olivine from Roda for grains in igneous contact with orthopyroxene, the only such case documented in the literature. The Fe/Mg data are compatible with subsolidus equilibration between olivine and orthopyroxene (Mittlefehldt 1994). Olivines in diogenites with the typical orthopyroxene composition are in the range Fo_{70-73} (Floran et al. 1981, Mittlefehldt 1994, Sack et al. 1991), whereas olivines from the slightly more ferroan diogenite Peckelsheim have an average composition of Fo_{65} (Gooley 1972), and those from the magnesian-

Table 31. Olivine analyses for representative diogenites, eucrites and howardites.

	Ellemeet (1)	Johnstown (2)	Peckelsheim (3)	Roda (4)	Y-793164 (5)	Washougal (6)	Y-7308 (7)	Y-82049 (8)	Y-82049 (9)	Yurtuk (10)
					Chemical Composition (wt %)					
SiO_2	37.1	37.5	35.89	36.9	31.0	40.2	36.70	36.2	35.0	37.0
FeO	26.9	25.8	30.69	24.9	60.4	12.1	26.55	33.6	37.6	30.7
MgO	35.1	35.9	32.65	36.9	6.9	47.3	36.17	30.0	26.5	32.4
MnO	0.55	0.51	0.44	0.52	1.26	0.28	0.50	0.63	0.72	0.57
Cr_2O_3	0.03	0.03	nd	0.02	nd	0.10	bd	nd	nd	0.02
CaO	0.07	0.07	0.02	0.04	0.08	0.07	0.03	0.06	0.05	0.04
Total	99.75	99.81	99.69	99.3	99.60	100.05	99.95	100.49	99.87	100.73
					Cation Formula Based on 4 Oxygens					
Si	0.9921	0.9959	0.9799	0.9832	1.0009	0.9960	0.9789	0.9931	0.9887	0.9966
Fe	0.6016	0.5730	0.7008	0.5549	1.6309	0.2507	0.5923	0.7709	0.8883	0.6915
Mg	1.3988	1.4208	1.3286	1.4653	0.3301	1.7466	1.4378	1.2265	1.1156	1.3006
Mn	0.0125	0.0115	0.0102	0.0117	0.0345	0.0059	0.0113	0.0146	0.0172	0.0130
Cr	0.0006	0.0006	--	0.0004	--	0.0020	--	--	--	0.0004
Ca	0.0020	0.0020	0.0006	0.0011	0.0028	0.0019	0.0009	0.0018	0.0015	0.0012
Total Cations	3.0076	3.0038	3.0201	3.0166	2.9992	3.0031	3.0212	3.0069	3.0113	3.0033
					*Cation Ratios Fe/Mn and mg# (100*Mg/(Mg+Fe))*					
Fe/Mn	48	50	69	-	47	43	52	53	52	53
mg#	69.9	71.3	65.5	72.5	16.8	87.4	70.8	61.4	55.7	65.3

(1,4) Mittlefehldt (1994); (2) Floran et al. (1981); (3) Gooley (1972); (5, 8, 9) Mittlefehldt and Lindstrom (1993); (6,10) Desnoyers (1982); (7) Miyamoto et al. (1978).

pyroxene bearing diogenite EETA79002 have an average composition of Fo_{76} (Sack et al. 1991, 1994a, and Mittlefehldt, unpublished).

Chromite analyses for diogenites have been presented by Berkley and Boynton (1992), Floran et al. (1981), Fredriksson (1982), Gooley (1972), Mittlefehldt (1994), Mittlefehldt and Lindstrom (1993) and Sack et al. (1991), and ilmenite analyses for Yamato Type B diogenites have been given by Mittlefehldt and Lindstrom (1993). Representative chromite and ilmenite analyses are given in Table 32. Unlike the orthopyroxenes and olivines, chromites in diogenites are variable in mg#. Sack et al. (1991, 1994a) show a range in mg# for EETA79002 chromites from ~14 to 32. This is essentially the same as the range found for the averages of chromites from all diogenites. The range in Al_2O_3 contents of average chromites is considerable, from ~6.8 to 21.6 wt %, excluding small chromites (Mittlefehldt 1994). Within individual diogenites, chromite grains can also be variable in Al_2O_3 composition. Gooley reported a range in Al_2O_3 content in Roda chromites from 8.6 to 18.6 wt %, although most of the grains he analyzed contained between 8.6 and 10.2 wt %. Mittlefehldt (1994) and Mittlefehldt and Lindstrom (1993) showed that small grains were of different composition than large grains, but the differences were not systematic; small grains in Shalka had lower Al_2O_3 contents and lower mg# than large grains, the opposite was true for ALHA77256, while for Yamato Type A diogenites small chromites had higher Al_2O_3 and lower mg#.

Few analyses of plagioclase from diogenites are available. Analyses are given for Garland, Johnstown, Manegaon, Peckelsheim, Roda and Yamato Type B diogenites by Floran et al. (1981), Fredriksson (1982), Gooley (1972), Mittlefehldt (1978, 1994) and Mittlefehldt and Lindstrom (1993). Most plagioclase in diogenites lies in the range An_{82}-An_{96}, although plagioclase as sodic as An_{73-76} is present in Peckelsheim, LEW 88679 and Yamato Type B diogenites (Gooley 1972, Takeda and Mori 1985, Mittlefehldt, unpublished). Table 33 gives representative plagioclase analyses from selected diogenites.

Table 32. Compositions of chromite and ilmenite grains from representative diogenites and eucrites.

	Shalka	Johns-town	Mane-gaon	ALHA 77256	Y-75032		Moama	Serra de Magé		Moore County		Y-791195		Pomozdino		Peters-burg
	(1)	(2)	(3)	(4)	(5)	(6)	(7)	(8)	(9)	(10)	(11)	(12)	(13)	(14)	(15)	(16)
Chemical Composition (wt %)																
TiO_2	0.50	0.42	0.22	0.78	1.96	54.1	5.39	3.2	52.9	10.1	52.9	11.9	51.6	4.52	54.3	3.68
Al_2O_3	6.84	9.37	16.1	21.6	7.90	0.02	7.72	8.3	0.08	5.7	bd	5.02	0.03	7.30	0.07	14.5
Cr_2O_3	60.6	56.3	48.7	43.8	55.0	0.37	50.1	53.0	0.27	42.3	0.19	39.1	1.75	49.8	0.22	43.9
V_2O_3	0.96	0.35	0.40	0.40	0.59	bd	0.67	0.73	bd	0.81	bd	0.76	bd	0.68	0.23	nd
FeO	27.2	30.1	27.8	27.0	31.8	42.4	32.7	33.1	42.8	39.5	42.9	42.2	44.7	34.2	43.5	34.3
MgO	3.90	2.54	4.33	6.24	1.95	3.15	2.38	1.30	2.35	1.31	1.92	0.98	1.36	1.15	1.56	1.63
MnO	0.51	0.51	0.49	0.43	0.55	0.89	0.81	0.58	0.86	0.66	0.82	0.60	0.90	0.67	1.10	0.51
Total	100.51	99.59	98.04	100.25	99.75	100.93	99.77	100.21	99.26	100.38	98.73	100.56	100.34	98.32	100.98	98.52
Cation Formula Based on 4 Oxygens (Ideal Spinel = 3 Cations per 4 Oxygens, Ideal Ilmenite = 2 Cations per 3 Oxygens)																
Ti	0.0131	0.0111	0.0056	0.0189	0.0523	0.9943	0.1428	0.0849	1.3255	0.2714	1.0010	0.3216	0.9709	0.1231	1.3392	0.0723
Al	0.2808	0.3883	0.6479	0.8211	0.3301	0.0006	0.3205	0.3453	0.0031	0.2400	--	0.2126	0.0009	0.3116	0.0027	0.4467
Cr	1.6686	1.5649	1.3147	1.1168	1.5418	0.0071	1.3953	1.4790	0.0071	1.1949	0.0038	1.1108	0.0346	1.4257	0.0057	0.9072
V	0.0268	0.0099	0.0110	0.0103	0.0168	--	0.0189	0.0207	--	0.0232	--	0.0219	--	0.0197	0.0060	--
Fe	0.7922	0.8850	0.7938	0.7282	0.9429	0.8666	0.9633	0.9770	1.1926	1.1803	0.9028	1.2681	0.9353	1.0357	1.1931	0.7497
Mg	0.2024	0.1331	0.2203	0.2999	0.1030	0.1147	0.1249	0.0684	0.1167	0.0698	0.0720	0.0525	0.0507	0.0621	0.0762	0.0635
Mn	0.0150	0.0152	0.0142	0.0117	0.0165	0.0184	0.0242	0.0173	0.0243	0.0200	0.0175	0.0183	0.0191	0.0205	0.0306	0.0113
Total Cations	2.9989	3.0075	3.0075	3.0069	3.0034	2.0017	2.9899	2.9926	2.6693	2.9996	1.9971	3.0058	2.0115	2.9984	2.6535	2.2507
*Cation Ratios Fe/Mn, mg# (100*Mg/(Mg+Fe)) and cr# (100*Cr/(Cr+Al))*																
Fe/Mn	53	58	56	62	57	47	40	56	49	59	52	69	49	50	39	66
mg#	20.4	13.1	21.7	29.2	9.9	11.7	11.5	6.5	8.9	5.6	7.4	4.0	5.1	5.7	6.0	7.8
cr#	85.6	80.1	67.0	57.6	82.4	92.5	81.3	81.1	69.4	83.3	100.0	83.9	97.5	82.1	67.8	67.0

(1-4) average typical chromites, diogenites, Mittlefehldt (1994); (5, 6) average chromite and ilmenite, diogenite, Mittlefehldt (unpublished); (7) average chromite, cumulate eucrite, Lovering (1975); (8-11) average chromite and ilmenite, cumulate eucrite, Bunch and Keil (1971); (12,13) average chromite and ilmenite, cumulate eucrite, Mittlefehldt and Lindstrom (1993); (14, 15) average chromite and ilmenite, anomalous magnesian eucrite, Warren et al. (1990); (16) average chromite, magnesian basalt clast, polymict eucrite, Buchanan and Reid (1996).

Table 32 (cont'd). Compositions of chromite and ilmenite grains from representative diogenites and eucrites.

	Ibitira		Bouvante		Stannern	Sioux County		Y-82066	Millbillillie		Juvinas		Pasamonte		Y-75011,84C	Y-793164
	(17)	(18)	(19)	(20)	(21)	(22)	(23)	(24)	(25)	(26)	(27)	(28)	(29)	(30)	(31)	(32)
Chemical Composition (Weight Percent)																
TiO_2	23.8	53.2	4.20	53.2	52.9	3.4	52.9	52.0	2.37	52.5	5.0	52.5	1.38	52.5	53.2	51.5
Al_2O_3	3.01	nd	8.15	bd	0.08	7.2	0.03	0.01	8.09	0.06	7.9	0.06	17.7	0.11	0.06	0.01
Cr_2O_3	19.8	nd	51.7	0.12	0.05	51.6	0.06	0.08	52.6	0.03	50.6	0.05	44.4	0.10	0.03	0.06
V_2O_3	nd	nd	nd	nd	bd	0.93	bd	bd	nd	nd	0.77	bd	0.61	bd	bd	bd
FeO	50.6	43.1	34.5	44.8	44.7	35.5	45.0	45.5	33.4	44.3	35.5	44.7	34.5	44.6	44.5	45.7
MgO	1.39	1.58	0.52	0.54	0.49	0.28	0.61	0.78	0.36	0.63	0.24	0.85	0.74	0.46	0.53	0.50
MnO	0.71	0.36	0.57	1.02	0.87	0.62	0.87	0.94	0.58	0.96	0.63	0.87	0.46	0.85	1.14	0.90
Total	99.31	98.24	99.64	99.68	99.09	99.53	99.47	99.31	97.40	98.48	100.64	99.03	99.79	98.62	99.46	98.67
Cation Formula Based on 4 Oxygens (Ideal Spinel = 3 Cations per 4 Oxygens, Ideal Ilmenite = 2 Cations per 3 Oxygens)																
Ti	0.6233	1.4066	0.1078	1.2903	1.4105	0.0625	1.4013	1.3802	0.0594	1.4107	0.1262	1.4187	0.0346	1.4297	1.3121	1.3497
Al	0.1236	--	0.3280	--	0.0033	0.2074	0.0012	0.0004	0.3177	0.0025	0.3124	0.0025	0.6959	0.0047	0.0023	0.0004
Cr	0.5452	--	1.3957	0.0031	0.0014	0.9970	0.0017	0.0022	1.3856	0.0008	1.3422	0.0014	1.1710	0.0029	0.0008	0.0017
V	--	--	--	--	--	0.0182	--	--	--	--	0.0207	--	0.0163	--	--	--
Fe	1.4737	1.2672	0.9851	1.2083	1.3254	0.7256	1.3256	1.3430	0.9307	1.3237	0.9961	1.3433	0.9625	1.3506	1.2205	1.3319
Mg	0.0721	0.0828	0.0265	0.0260	0.0259	0.0102	0.0320	0.0410	0.0179	0.0335	0.0120	0.0455	0.0368	0.0248	0.0259	0.0260
Mn	0.0209	0.0107	0.0165	0.0279	0.0261	0.0128	0.0260	0.0281	0.0164	0.0291	0.0179	0.0265	0.0130	0.0261	0.0317	0.0266
Total Cations	2.8588	2.7673	2.8596	2.5556	2.7926	2.0337	2.7878	2.7949	2.7277	2.8003	2.8275	2.8379	2.9301	2.8388	2.5933	2.7363
*Cation Ratios Fe/Mn, mg# (100*Mg/(Mg+Fe)) and cr# (100*Cr/(Cr+Al))*																
Fe/Mn	70	118	60	43	51	57	51	48	57	46	56	51	74	52	39	50
mg#	4.7	6.1	2.6	2.1	1.9	1.4	2.4	3.0	1.9	2.5	1.2	3.3	3.7	1.8	2.1	1.9
cr#	81.5	--	81.0	100.0	29.5	82.8	57.3	84.3	81.3	25.1	81.1	35.9	62.7	37.9	25.1	80.1

(17, 18) individual chromite and ilmenite grains, basaltic eucrite, Steele and Smith (1976); (19,20) average chromite and ilmenite, Stannern-trend eucrite, Christophe Michel-Levy et al. (1987); (21) average ilmenite, Stannern-trend eucrite, Bunch and Keil (1971); (22, 23) average chromite and ilmenite, main-group eucrite, Bunch and Keil (1971); (24) avearge ilmenite, main-group eucrite, Mittlefehldt and Lindstrom (1993); (25, 26) avearge chromite and ilmenite, main-group eucrite, Yamaguchi et al. (1994); (27-30) average chromite and ilmenite, main-group eucrites, Bunch and Keil (1971); (31) ilmenite, least-metamorphosed basalt clast, polymict eucrite, Takeda et al. (1994c); (32) average ilmenite, evolved basalt clast, polymict eucrite, Mittlefehldt and Lindstrom (1993).

Little is known of the composition of metal in diogenites. Gooley and Moore (1976) classified the metal as to textural setting depending on whether it occurred enclosed in orthopyroxene clasts or in the fragmental matrix. The mineralogy of the metal was not identified for 5 of the meteorites they studied, but for Ibbenbüren, Peckelsheim and Roda both kamacite and taenite occur. The metal in diogenites shows a wide range in Ni and Co contents, even within a single meteorite. For metal grains of unspecified mineralogy, median Ni contents range from ~0.07 wt % for that in the matrix of Shalka to 7.7 wt % for metal in the matrix of Ellemeet. Cobalt is variable too, from 0.11-1.7 wt % in these same grains, and the median Ni/Co ratio varies from 0.04-64.5. Roda contains some metal that is unusually rich in Co, with a median composition of 2.0 wt % Ni and 23.6 wt % Co. Kamacite varies in composition from 2.2-3.8 wt % Ni and from 0.5-7.6 wt % Co. Taenite varies in composition from 15.4-53.3 wt % Ni and from 1.1-2.7 wt % Co.

Cumulate eucrites. Cumulate eucrites are coarse-grained gabbros composed principally of low-Ca pyroxene and calcic plagioclase, with minor chromite and accessory silica polymorph, phosphate, ilmenite, metal, troilite (e.g. Delaney et al. 1984). Zircon has been listed for Serra de Magé, but the attribution is unclear (Gomes and Keil 1980, p. 77). Many cumulate eucrites are unbrecciated; ALH 85001, Binda, EET 87548 and Medanitos are exceptional in this. Binda, in fact, is a polymict breccia (Garcia and Prinz 1978) and some polymict eucrites contain abundant cumulate eucrite material (e.g. see Takeda 1991). The textures of unbrecciated cumulate eucrites are typically equigranular with subequal amounts of pyroxene and plagioclase grains 0.5-3 mm in diameter (Hess and Henderson 1949, Lovering 1975, Mittlefehldt and Lindstrom 1993). The cumulate eucrites have been subdivided into the feldspar- and orthopyroxene-cumulate eucrites based on modal abundances (Delaney et al. 1984).

The original igneous pyroxene of cumulate eudrites was pigeonite, which has subsequently undergone subsolidus exsolution of augite and, in some cases, inversion to orthopyroxene, with additional augite exsolution. Hess and Henderson (1949) determined that original pigeonite in Moore County had only partially inverted to orthopyroxene. They described four pyroxenes in Moore County; low-Ca pigeonite developed from the original igneous pigeonite by exsolution of coarse lamellae of augite, and hypersthene developed from the low-Ca pigeonite through inversion and exsolution of fine lamellae of salite. Mori and Takeda (1981b) found that the original pigeonite underwent a complex series of exsolution, decomposition and partial inversion to end as a mixture of seven pyroxene phases. Harlow et al. (1979) found that the original igneous pigeonite in Serra de Magé had inverted to hypersthene, with development of four types of augite exsolution during the subsolidus cooling history. Similarly, in Moama the original igneous pigeonite has exsolved augite and inverted to hypersthene (Lovering 1975, Takeda et al. 1976). Binda is the most magnesian of the cumulate eucrites, and the only one classified by Delaney et al. (1984) as an orthopyroxene-cumulate eucrite. Takeda et al. (1976) have shown that the original pyroxene in Binda was a low-Ca pigeonite that exsolved augite and inverted to hypersthene. Table 30 gives representative pyroxene analyses for selected cumulate eucrites, and estimated primary pigeonite compositions are plotted in Figure 42.

As is the case for all HED meteorites, plagioclase in cumulate eucrites is calcic, and is on average more calcic than that in basaltic eucrites. Basaltic eucrite plagioclase has compositions in the range of bytownite to anorthite, while in cumulate eucrites the plagioclase is anorthite, An_{91-95}. A-881394 contains unusually calcic plagioclase, An_{98} (Takeda et al. 1997b). The K_2O contents of cumulate eucrite plagioclases are very low, typically <0.1 wt %. Representative plagioclase analyses for selected cumulate eucrites are given in Table 33.

Table 33. Plagioclase analyses for representative diogenites, cumulate eucrites and basaltic eucrites.

	Johns-town (1)	LEW 88679 (2)	LEW 88679 (3)	Mane-gaon (4)	Peckels-heim (5)	Peckels-heim (6)	Roda (7)	Y-75032 (8)	Y-75032 (9)	Moama (10)	Serra dé Mage (11)	Moore County (12)
Chemical Composition (wt %)												
SiO_2	46.7	46.0	50.5	45.5	46.0	50.8	46.2	45.7	48.4	45.0	44.8	45.3
Al_2O_3	33.8	34.9	31.9	34.4	35.1	32.3	35.3	35.5	34.0	35.8	35.0	35.5
FeO	0.29	0.23	0.21	0.11	0.73	0.86	0.05	0.50	0.14	0.09	0.12	0.12
MgO	0.46	0.04	0.09	0.10	-	-	0.05	-	-	0.08	bd	nd
CaO	16.9	18.1	15.0	18.0	17.9	14.6	18.0	17.9	16.9	19.3	19.5	18.5
Na_2O	1.39	1.08	2.44	1.51	1.07	2.89	1.35	0.97	1.76	0.63	0.54	0.93
K_2O	0.07	0.03	0.18	0.10	0.02	0.14	0.10	0.03	0.22	0.02	0.02	0.06
Total	99.61	100.38	100.32	99.72	100.88	101.64	101.05	100.60	101.42	100.85	99.98	100.41
Cation Formula Based on 8 Oxygens (Ideal Plagioclase = 5 Cations per 8 Oxygens)												
Si	2.1522	2.1099	2.2920	2.1055	2.1030	2.2833	2.1049	2.0925	2.1877	2.0607	2.0715	2.0800
ivAl	1.8361	1.8868	1.7066	1.8764	1.8936	1.7109	1.8957	1.9159	1.8115	1.9307	1.9076	1.9213
Total tet*	3.9883	3.9967	3.9986	3.9819	3.9966	3.9942	4.0006	4.0084	3.9992	3.9914	3.9791	4.0013
Fe	0.0112	0.0088	0.0080	0.0043	0.0279	0.0323	0.0019	0.0191	0.0053	0.0034	0.0046	0.0046
Mg	0.0316	0.0027	0.0061	0.0069	0.0000	0.0000	0.0034	0.0000	0.0000	0.0055	0.0000	0.0000
Ca	0.8345	0.8896	0.7295	0.8925	0.8778	0.7049	0.8787	0.8782	0.8185	0.9450	0.9661	0.9102
Na	0.1242	0.0961	0.2147	0.1355	0.0949	0.2518	0.1193	0.0861	0.1543	0.0559	0.0484	0.0828
K	0.0041	0.0018	0.0104	0.0059	0.0012	0.0080	0.0058	0.0018	0.0127	0.0012	0.0012	0.0035
Total Cations	4.9939	4.9957	4.9673	5.0270	4.9984	4.9912	5.0097	4.9936	4.9900	5.0024	4.9994	5.0024
*Cation Ratios 100*Mg/(Mg+Fe) (mg#) and Molecular Proportion of Orthoclase (Or), Albite (Ab), Anorthite (An)*												
mg#	73.9	23.7	43.3	61.8	-	-	64.1	-	-	61.3	-	-
Or	0.4	0.2	1.1	0.6	0.1	0.8	0.6	0.2	1.3	0.1	0.1	0.4
Ab	12.9	9.7	22.5	13.1	9.7	26.1	11.9	8.9	15.7	5.6	4.8	8.3
An	86.7	90.1	76.4	86.3	90.2	73.1	87.5	90.9	83.0	94.3	95.1	91.3

(1) Floran et al. (1981); (2, 3) Mittlefehldt, unpublished; (4) Fredriksson (1982); (5, 6) Gooley (1972); (7) Mittlefehldt (1994); (8, 9) Mittlefehldt, unpublished data used in Mittlefehldt and Lindstrom (1993); (10) Lovering (1975); (11) Gomes and Keil (198x); (12) Mittlefehldt, unpublished data used in Mittlefehldt (1990).

Table 33 (cont'd). Plagioclase analyses for representative diogenites, cumulate eucrites and basaltic eucrites.

	Y-791195 (13)	Bou-vante (14)	Bou-vante (15)	Ibitira (16)	Juvinas (17)	Pasa-monte (18)	Peters-burg (19)	Pomoz-dino (20)	Y-82049 (21)	Y-82049 (22)	Y-793164 (23)	Y-793164 (24)
Chemical Composition (wt %)												
SiO_2	45.8	45.9	50.4	44.8	45.9	46.6	48.0	46.8	44.565	49.215	45.328	52.195
Al_2O_3	35.8	34.2	31.8	35.9	33.2	33.0	34.0	33.0	35.992	32.736	34.165	31.09
FeO	0.25	0.21	0.22	0.21	1.20	0.90	0.12	0.40	0.12	0.389	1.407	0.19
MgO	nd	0.01	0.01	nd	0.30	bd	0.73	0.01	nd	nd	nd	nd
CaO	18.6	18.0	15.2	18.7	18.0	17.4	17.4	18.4	19.008	16.187	17.671	13.758
Na_2O	1.00	1.23	2.42	0.50	1.00	1.80	1.27	1.62	0.706	1.845	0.664	2.669
K_2O	0.07	0.08	0.49	0.08	bd	bd	0.09	0.15	0.049	0.22	0.112	0.479
Total	101.52	99.64	100.49	100.11	99.60	99.70	101.61	100.38	100.44	100.59	99.35	100.38
Cation Formula Based on 8 Oxygens (Ideal Plagioclase = 5 Cations per 8 Oxygens)												
Si	2.0812	2.1243	2.2884	2.0623	2.1323	2.1588	2.1666	2.1557	2.0496	2.2382	2.1097	2.3571
^{iv}Al	1.9175	1.8618	1.7016	1.9482	1.8179	1.8020	1.8089	1.7917	1.9511	1.7548	1.8743	1.6549
Total tet*	3.9987	3.9861	3.9900	4.0105	3.9502	3.9608	3.9755	3.9474	4.0007	3.9930	3.9840	4.0120
Fe	0.0095	0.0081	0.0084	0.0081	0.0466	0.0349	0.0045	0.0154	0.0046	0.0148	0.0548	0.0072
Mg	0.0000	0.0007	0.0007	0.0000	0.0208	0.0000	0.0491	0.0007	0.0000	0.0000	0.0000	0.0000
Ca	0.9057	0.8923	0.7410	0.9203	0.8960	0.8637	0.8416	0.9082	0.9367	0.7888	0.8813	0.6657
Na	0.0881	0.1103	0.2132	0.0446	0.0901	0.1617	0.1112	0.1447	0.0630	0.1627	0.0599	0.2337
K	0.0041	0.0047	0.0284	0.0047	0.0000	0.0000	0.0052	0.0088	0.0029	0.0128	0.0067	0.0276
Total Cations	5.0061	5.0022	4.9817	4.9882	5.0037	5.0211	4.9871	5.0252	5.0079	4.9721	4.9867	4.9462
*Cation Ratios 100*Mg/(Mg+Fe) (mg#) and Molecular Proportion of Orthoclase (Or), Albite (Ab), Anorthite (An)*												
mg#	-	7.8	7.5	-	30.8	-	91.6	4.3	-	-	-	-
Or	0.4	0.5	2.9	0.5	-	-	0.5	0.8	0.3	1.3	0.7	3.0
Ab	8.8	10.9	21.7	4.6	9.1	15.8	11.6	13.6	6.3	16.9	6.3	25.2
An	90.8	88.6	75.4	94.9	90.9	84.2	87.9	85.6	93.4	81.8	93.0	71.8

(13) Mittlefehldt, unpublished data used in Mittlefehldt and Lindstrom (1993); (14, 15) Christophe Michel-Levy et al. (1987); (16) Wilkening and Anders (1975); (17, 18) Metzler et al. (1995); (19) Buchanan and Reid (1996); (20) Warren et al. (1990); (21, 22, 23, 24) Mittlefehldt, unpublished data used in Mittlefehldt and Lindstrom (1993).

Chromite is an ubiquitous minor mineral in cumulate eucrites, while ilmenite has been reported from many of them. Delaney et al. (1984) found ilmenite in all the cumulate eucrites they studied except Moama. The textures of chromites in cumulate eucrites are not well described in the literature. Hostetler and Drake (1978) mention that chromite and ilmenite occur as "well developed crystals" in Moore County, while Lovering (1975) describes chromites in Moama as occurring as equidimensional grains and elongate grains intergrown with tridymite. Details of chromite compositional variation are not described in the literature, but this mineral is apparently variable in composition in individual meteorites. The mean composition of ten chromite grains reported by Bunch and Keil (1971) for Moore County is significantly different than the mean of five grains analyzed by Hostetler and Drake (1978). Hostetler and Drake (1978) did not find zoning for the major elements in chromite grains, however. Mittlefehldt and Lindstrom (1993) show that chromite grains in ferroan cumulate eucrite Y-791195 vary from 8.6 to 15.7 wt % in TiO_2 and from 44.6 to 32.4 wt % in Cr_2O_3. Ilmenite occurs as individual grains, and composite grains with chromite and as exsolution lamellae in chromite (Bunch and Keil 1971, Hostetler and Drake 1978, Mittlefehldt and Lindstrom 1993). Representative chromite and ilmenite analyses are given in Table 32.

The composition of metal in the cumulate eucrites Binda, Moama, Moore County and Serra de Magé has received only cursory study (Duke 1965, Lovering 1964, 1975). For the unbrecciated cumulate eucrites, the metal is poor in Ni, with ≤0.5 wt % Ni. Cobalt has been determined only for Moama among these meteorites, and it is similarly low: 0.23 wt % (Lovering 1975). For Binda, one grain contained 0.44 wt % Ni and 1.38 wt % Co, while five other grains ranged from 1.9-2.1 wt % Ni and had ~2.2 wt % Co (Lovering 1975). The textural setting of the metal analyzed in Binda was not given, so it is difficult to compare these with the analyses of unbrecciated cumulate eucrites.

Basaltic eucrites. This subsection includes discussion of unbrecciated and monomict basaltic eucrites, as well as clasts from polymict HED samples that are similar to the basaltic eucrites. Some unusual igneous clasts will be dealt with separately, below. Basaltic eucrites are pigeonite-plagioclase rocks with fine to medium grain size. Textures, where preserved, are generally subophitic to ophitic. One exceptional eucrite is ALHA81001, which has a fine-grained quench texture consisting of pyroxene microphenocrysts set in a groundmass of glass or cryptocrystalline material (AMN 6-1; Warren and Jerde 1987). Almost all basaltic eucrites are brecciated, resulting in a rock composed of mineral and lithic fragments set in a fine-grained, generally fragmental matrix. Some eucrites have been highly recrystallized, resulting in a granular texture such as in Ibitira (Steele and Smith 1976). Vesicles are rare in eucrites; Ibitira is a spectacular exception. Vesicles have also been reported from PCA 91007 (Warren et al. 1996).

Minor and accessory phases in basaltic eucrites include the silica polymorphs tridymite and quartz, chromite, ilmenite, metal, troilite, whitlockite, apatite, olivine and rare zircon. The minor and accessory phases are generally found interstitial to pyroxene and plagioclase. However, cloudy pyroxene grains contain inclusions of chromite, ilmenite, metal and/or troilite, while some plagioclase grains contain inclusions of phosphates.

The original igneous pyroxene in basaltic eucrites was pigeonite, which subsequently underwent subsolidus exsolution of augite. Inversion of pigeonite to orthopyroxene is uncommon. Some of the original igneous pyroxene in Sioux County was orthopyroxene (Takeda et al. 1978a). Takeda and Graham (1991) described the mineralogic characteristics of basaltic eucrite pyroxenes, and related them to a metamorphic sequence of types from 1 to 6. In the least metamorphosed, type 1, basaltic eucrites, augite exsolution lamellae in the pigeonite are <0.1 μm thick, the pyroxenes are not cloudy, and the original igneous zoning is preserved. In the most metamorphosed, type 6, augite lamellae in pyroxenes are several μm thick, clouding of the pyroxenes is present, the pyroxenes are homogeneous in composition, and the pigeonite has partially inverted to orthopyroxene. Takeda and Graham (1991) used the amount of preservation of chemical zoning of the pyroxene as one of the criteria for classifying basaltic eucrites by metamorphic type. However, basaltic eucrites exhibit a range of grain sizes, and possibly not all had the same original igneous zoning.

Basaltic eucrites are ferroan, and their pyroxenes are iron-rich. Table 30 gives representative pyroxene analyses for selected basaltic eucrites, and estimated primary pigeonite compositions are shown in Figure 42. In the least metamorphosed basaltic eucrites, there are at least three types of zoning trends, as summarized by Takeda and Graham (1991). In one, the core of pyroxenes are magnesian pigeonites with low Ca contents. These are zoned through increasing Fe and Ca to ferroaugite or subcalcic ferroaugite compositions. Another type of zoning starts from low-Ca magnesium pigeonite cores, shows Fe/Mg zoning at roughly constant Ca content to low-Ca intermediate pigeonite compositions, and then is zoned through increasing Ca contents at roughly constant enstatite content to ferroaugite compositions. A third type of zoning is from relatively ferroan intermediate pigeonite compositions through increasing Ca contents at roughly constant enstatite content to ferroaugite compositions. In most cases, however, basaltic eucrite pyroxenes have undergone subsolidus Fe/Mg equilibration and augite exsolution, and are now composed of low-Ca, homogeneous intermediate or ferriferous pigeonite hosts containing lamellae of ferroaugite.

Plagioclase in basaltic eucrites is calcic, in the range of bytownite to anorthite (Table 33). Unlike cumulate eucrites, plagioclase compositions in individual basaltic eucrites can show considerable range in anorthite content. For example, Warren and Jerde (1987) reported that plagioclase in Nuevo Laredo spans the range ~An_{92-74}, which is most of the range observed for all basaltic eucrites. The K_2O contents are low, typically ≤0.2 to 0.3 wt %, although the more sodic plagioclase compositions can have K_2O contents of ~0.5 wt %.

Chromite and ilmenite (Table 32) are ubiquitous minor minerals in basaltic eucrites, but they have received little systematic study. Chromites in many basaltic eucrites have between ~3 to 6 wt % TiO_2 and ~7 to 9 wt % Al_2O_3 (Bunch and Keil 1971). Exceptionally titanian chromites are found in Ibitira, with ~15-24 wt % TiO_2, and these have low Al_2O_3 contents in the range ~3.0 to 5.5 wt % (Steele and Smith 1976). Chromites low in TiO_2 occur in Millbillillie (~2 wt % TiO_2; Yamaguchi et al. 1994) and Pasamonte (~1.4 wt %, Bunch and Keil 1971). The chromites in Millbillillie have Al_2O_3 contents within the typical range given above, while those in Pasamonte are unusually Al_2O_3-rich, 17.7 wt % (Bunch and Keil 1971). The titanian chromites in Ibitira appear to be clearly exceptional among basaltic eucrites, and the Al_2O_3-rich chromites from Pasamonte may also be unusual. The TiO_2-poor chromites in Millbillillie may simply represent the low TiO_2 end of the basaltic eucrite distribution.

Olivine in basaltic eucrites is rare, and is found in the more ferroan examples or in the mesostasis of the least-metamorphosed basaltic eucrites (e.g. Mittlefehldt and Lindstrom 1993, Takeda et al. 1994b). These olivines are very iron-rich (Table 31).

Partial cumulate eucrites. Some eucrites are partial cumulates, mixtures of cumulus crystals and a substantial amount of solidified melt. Warren et al. (1990) first suggested this based on their study of the unusual magnesian eucrite Pomozdino, and the following description is from this source. Pomozdino is a monomict breccia containing two types of mafic clasts in the comminuted matrix; coarser-grained ophitic-poikilophitic clasts and fine-grained, anhedral-granular clasts. Pomozdino originally contained two distinct primary pyroxenes, one, $Wo_{9.4}En_{42.7}Fs_{47.9}$ (Table 30, Fig. 42), compositionally similar to the primary pigeonite of cumulate eucrite Moore County, and one exhibiting Ca zoning, $Wo_{19}En_{40}Fs_{41}$ to $Wo_{39}En_{34}Fs_{27}$, more like that seen in pyroxenes of the basaltic eucrites Bouvante and Stannern. Both of these pyroxene types are magnesian, with mg# of ~47 and ~52, respectively. Compare these with bulk pyroxenes from basaltic eucrites such as Bouvante which have mg# ~ 38 (Christophe Michel-Levy et al. 1987, see Table 30). Chromite and ilmenite in Pomozdino are similar to those in Moore County and more magnesian than those of basaltic eucrites (see Table 32). Plagioclase, however, is more sodic, An_{81-87}, than those found in cumulate eucrites, An_{91-98}. Warren et al. (1990) considered Pomozdino to be a mixture of between 20 to 40% cumulus minerals with melt. The polymict eucrite Petersburg also contains basaltic clasts more magnesian than typical basaltic eucrites (Buchanan and Reid 1996, Mittlefehldt 1979). Although these authors did not favor a partial cumulate model for the formation of the Petersburg clasts, the general petrologic and compositional similarities with Pomozdino suggest that this might be a plausible scenario. Warren et al (1996) have also identified Y-791195, Y-791186 and RKPA80224 as possible partial cumulates.

Howardites, polymict eucrites and polymict diogenites. Howardites have long been known to be polymict breccias (Wahl 1952). More recently, numerous polymict breccias with bulk compositions like those of eucrites have been recovered from Antarctica, leading to recognition of polymict eucrites as a distinct rock type (e.g. Miyamoto et al. 1978, Olsen et al. 1978, Takeda et al. 1978b). Diogenites with basaltic eucritic clasts are also known (Lomena et al. 1976), and it has thus become obvious that, in terms of major components, howardites are intermediate members of an essentially continuous sequence of polymict breccias, including polymict eucrites and polymict diogenites (Delaney et al. 1983), that extends from the monomict eucrites to the monomict diogenites. Most of the material in the polymict breccias is identical or very similar to the basaltic eucrites, cumulate eucrites or diogenites (e.g. Delaney et al. 1984, Duke and Silver 1967), and the description of this material will not be repeated. Here we will describe the general features of the matrix plus brecciated and melted clasts of the polymict breccias, concentrating on the howardites.

The polymict breccias are composed of diverse mineral and lithic clasts set in a fine-grained fragmental to glassy matrix. Howardites also commonly contain glassy spheres and irregularly shaped particles. Lithic clasts in the polymict breccias tend to receive the most detailed study, so descriptions of the matrix are typically cursory, if given at all. There is no good census of matrix types in the polymict breccias. Many polymict breccias have fragmental matrixes that have been little modified by metamorphism; the howardites Bholghati, Kapoeta and Frankfort are examples (Mason and Wiik 1966a,b; Reid et al. 1990). Some polymict breccias have metamorphosed matrixes; polymict eucrite Y-792769 has a fine-grained, sintered matrix (Takeda et al. 1994b). Glassy or glass-rich matrixes are also common, for example in the paired polymict eucrites LEW 85300, LEW 85302 and LEW 85303 (Kozul and Hewins 1988), the howardite Monticello (Olsen et al. 1987) and the polymict breccia (polymict diogenite?) Y-791073 (Takeda 1986).

The polymict breccias, especially the howardites and polymict eucrites, contain breccia clasts, melt rocks and glass particles of various types. Duke and Silver (1967) first recognized breccia clasts in howardites. Bunch (1975) divided the breccia clasts into three types, and he restricted his definition to polymict clasts. These clasts consist of angular mineral and lithic clasts in either a fine-grained fragmental matrix (crystalline matrix breccias), glassy or devitrified matrix (glassy matrix breccias), or troilite matrix (sulfide matrix breccias). Labotka and Papike (1980) and Fuhrman and Papike (1981) also recognized sulfide matrix breccias as an important clast type, but lumped the crystalline and glassy matrix breccias, plus melt rocks, into their dark matrix breccia type. Melt rocks have a variety of textures. The matrix varies from glass to cryptocrystalline material to fine-grained quench-textured matrix; the melt rocks contain inclusions of rounded, partially melted minerals, subangular to rounded minerals, or euhedral plagioclase or pyroxene crystals (Bunch 1975, Delaney et al. 1984, Hewins and Klein 1978, Klein and Hewins 1979, Mittlefehldt and Lindstrom 1997, 1998). Glass particles vary from spherical to irregular in shape, from glass sensu stricto to devitrified glass; they may contain phenocrysts of olivine and/or low-Ca pyroxene (Bunch 1975, Hewins and Klein 1978, Labotka and Papike 1980, Mittlefehldt and Lindstrom 1998, Olsen et al. 1990).

Unusual igneous clasts. Most igneous clasts in polymict breccias are basically the same lithologies as occur as basaltic and cumulate eucrites and diogenites, and the descriptions of these meteorite types above serve to describe the clasts. There are some unusual clasts that deserve brief mention. Dymek et al. (1976) described a fine-grained porphyritic clast from the Kapoeta howardite composed of pyroxene microphenocrysts and a few chromite microphenocrysts in a holocrystalline, variolitic groundmass of acicular plagioclase and pyroxene. Mittlefehldt and Lindstrom (1997) described two similar clasts, a Kapoeta clast consisting of low-Ca pyroxene phenocrysts and microphenocrysts and ferroan olivine microphenocrysts in a cryptocrystalline granular textured groundmass, and a clast from the howardite EET 92014 consisting of skeletal pyroxene phenocrysts in a glassy groundmass containing microphenocrysts of pyroxene. The cores of the pyroxene phenocrysts in these clasts are magnesian, and can be more magnesian than core compositions of zoned basaltic eucrite pyroxenes (e.g. Mittlefehldt and Lindstrom 1997). Ikeda and Takeda (1985) briefly described magnesian olivine-orthopyroxene clasts in the Y-7308 howardite containing variable amounts of plagioclase, chromite and high-Ca clinopyroxene. Olivine is euhedral to anhedral and Fo_{65-73} in composition. Orthopyroxene is the major phase, and is similar in composition to pyroxenes in the ferroan diogenites; $Wo_{2-3}En_{67-72}Fs_{26-30}$. Ikeda and Takeda (1985) also describe unusually ferroan clasts composed of fayalitic olivine, Fo_{10-14}, hedenbergitic pyroxene, $Wo_{41}En_{16-19}Fs_{40-43}$, tridymite and plagioclase, An_{78-85}, with minor ilmenite, chromite, troilite, Fe-metal and whitlockite. Olivine is euhedral to subhedral a few tens of microns in size, and some is included within the larger pyroxene grains. Plagioclase and tridymite are several hundred microns in size.

Chondritic clasts. Chondritic clasts are a minor component of many HED meteorites, especially the polymict breccias. Zolensky et al. (1996) discussed the mineralogy and petrology of numerous chondritic clasts from HED meteorites and summarized previous work. They found that about 80% of the chondritic clasts are CM2 materials, while CR2 chondritic materials are less abundant. Some clasts appear to have CI chondrite parentage, but they have been heated, with consequent petrographic alteration (Buchanan et al. 1993). The polymict eucrite LEW 85300 contains a few chondritic clasts that are closest to CV3 chondrites in mineralogy and petrology, but have some unusual compositional characteristics (Zolensky et al. 1992).

Composition

The HED meteorites in general show limited ranges in major, minor and trace element composition within the different meteorite types. The polymict breccias, especially the howardites, are exceptional to this, and to a lesser extent, so are the cumulate eucrites. Bulk major and minor element compositions of selected HED meteorites are given in Table 34; trace element compositions are given in Table 35.

Table 34. Major element compositions of representative diogenite, eucrite and howardite whole rock samples.

	Johns-town (1)	Shalka (2)	Y-75032 (3)	Moama (4)	Serra de Magé (5)	Y-791195 (6)	Stannern (7)	Sioux County (8)	Juvinas (9)	Nuevo Laredo (10)	Y-74450 (11)	EETA 79004 (12)	Y-82049 (13)	Bhol-ghati* (14)	Kapoeta (15)	Y-7308 (16)
Chemical Composition (wt %)																
SiO_2	52.5	51.6	52.9	48.6	48.5	49.3	49.7	49.2	49.2	49.5	48.3	49.6	49.6	49.0	50.3	50.8
TiO_2	0.10	0.062	0.26	0.22	0.13	0.25	0.980	0.58	0.63	0.83	0.90	0.63	0.41	0.54	0.30	0.23
Al_2O_3	1.5	0.60	3.4	13.7	14.8	13.3	12.3	13.1	13	12.2	11.4	11.5	8.3	8.38	8.3	4.27
Cr_2O_3	0.82	2.41	0.69	0.61	0.626	0.34	0.34	0.316	0.305	0.282	0.396	0.413	0.79	0.733	0.694	1.02
FeO	15.9	16.3	18.6	14.8	14.4	17.3	17.8	18.4	18.7	19.6	18.6	17.8	17.8	18.1	17.5	16.7
MnO	0.484	0.553	0.61	0.50	0.476	0.58	0.525	0.551	0.515	0.584	0.531	0.504	0.54	0.522	0.495	0.523
MgO	25.5	25.8	20.9	11.9	10.7	7.7	6.97	6.88	6.6	5.55	7.58	8.67	15.2	16.7	15.8	21.4
CaO	1.83	0.73	3.61	9.47	9.75	10.5	10.7	10.4	11	10.3	9.95	8.95	6.6	7.19	5.20	3.83
Na_2O	0.02	0.04	0.10	0.22	0.249	0.40	0.62	0.406	0.377	0.508	0.51	0.398	0.22	0.278	0.276	0.13
K_2O	0.0011	0.0016	0.007	0.01	0.012	0.03	0.066	0.033	0.027	0.050	0.054	0.072	0.020	0.025	0.022	0.008
P_2O_5	0.041	0.002	bd	nd	0.030	bd	0.102	nd	nd	nd	nd	0.050	0.04	nd	nd	nd
S	0.22	nd	nd	nd	0.150	nd	nd	0.066	nd	nd	0.236	0.261	nd	nd	nd	0.110
Total	98.92	98.10	101.08	100.03	99.82	99.70	100.10	99.93	100.35	99.40	98.46	98.85	99.52	101.47	98.89	99.02
*Cation Ratios Fe/Mn, mg# (100*Mg/(Mg+Fe))*																
Fe/Mn	32	29	30	29	30	29	33	33	36	33	35	35	33	34	35	32
mg#	74.1	73.8	66.7	58.9	57.0	44.2	41.1	40.0	38.6	33.5	42.1	46.5	60.3	62.2	61.7	69.5

(1, 11, 16) Wänke et al. (1977); (2) McCarthy et al. (1972); (3, 6, 13) Mittlefehldt and Lindstrom (1993), (4) Lovering (1975); (5, 8) Palme et al. (1978); (7) McCarthy et al. (1973); (9, 15) Wänke et al. (1972b); (10) Warren and Jerde (1987); (12) Palme et al. (1983); (14) Laul and Gosselin (1990).

*average matrix

Table 35. Trace element compositions of representative diogenite, eucrite and howardite whole rock samples.

		Johns-town (1)	Shalka (2)	Y-75032 (3)	Moama (4)	Serra de Magé (5)	Y-791195 (6)	Stannern (7)	Sioux County (8)	Juvinas (9)	Nuevo Laredo (10)	Y-74450 (11)	EETA 79004 (12)	Y-82049 (13)	Bhol-ghati* (14)	Kapoeta (15)	Y-7308 (16)
Sc	μg/g	15.8	9.9	21.1	23.4	22.1	30.4	30.6	31.4	28.5	33.3	30.0	28.4	22.9	22.8	20.7	18.1
V	μg/g	115	--	--	114	111	--	--	--	--	61	67.5	--	--	109	--	124
Co	μg/g	38.1	18.0	16.2	8.6	9.50	6.2	7.18	5.29	5.8	2.15	12.0	7.30	21.4	29.0	28	20.5
Ni	μg/g	150	--	20	--	--	--	--	--	--	2.9	--	33	130	410	410	--
Ga	μg/g	0.18	--	--	--	--	--	2.17	--	2.16	1.2	1.46	1.9	--	--	1.04	0.51
Ba	μg/g	--	--	--	--	--	--	53	--	--	39	45.4	31	18	21	--	8.25
La	μg/g	0.063	0.009	0.40	0.18	0.58	0.49	5.08	2.39	2.58	3.83	4.79	1.53	1.55	1.61	1.39	0.68
Ce	μg/g	--	--	0.50	--	--	1.8	13.7	6.8	--	9.9	12.0	5.62	3.1	4.0	--	1.67
Sm	μg/g	0.059	0.004	0.367	0.16	0.35	0.407	3.15	1.42	1.48	2.32	2.78	0.88	0.99	1.01	0.54	0.42
Eu	μg/g	0.011	0.002	0.110	0.36	0.33	0.44	0.782	0.56	0.61	0.71	0.70	0.504	0.33	0.34	0.32	0.17
Gd	μg/g	0.24	--	--	--	--	--	--	--	2.3	--	3.4	1.35	--	1.3	1.25	0.54
Tb	μg/g	--	--	0.11	--	0.07	0.14	0.73	0.40	0.60	0.54	0.66	0.25	0.25	0.24	0.29	0.1
Dy	μg/g	0.21	--	--	--	0.6	--	5.05	2.6	3.2	3.7	4.2	1.75	--	--	1.26	0.66
Ho	μg/g	0.059	--	--	--	--	--	1.13	--	0.42	--	0.97	0.46	--	0.35	0.23	0.16
Yb	μg/g	0.17	0.030	0.64	0.42	0.39	0.58	2.90	1.55	1.72	2.40	2.61	1.32	0.98	1.00	0.89	0.49
Lu	μg/g	0.027	0.006	0.088	0.075	0.066	0.090	0.40	0.25	0.28	0.35	0.36	0.20	0.150	0.16	0.14	0.073
Hf	μg/g	--	--	0.31	--	0.14	0.57	2.34	1.17	1.3	1.61	2.08	1.33	0.68	0.69	0.6	0.33
Ta	ng/g	--	--	50	--	80	20	500	140	120	181	300	190	100	110	100	42
Ir	ng/g	6.4	--	--	--	--	--	--	--	27	0.083	--	--	3.8	21	20	--
Au	ng/g	1.7	--	--	--	--	--	--	2.5	7.9	1.5	0.27	1.4	--	8.6	6.8	0.45
Th	ng/g	--	--	--	--	--	110	600	--	--	440	--	320	110	230	--	--
U	ng/g	2.8	--	--	--	--	--	160	--	89	140	163	70	--	--	51	22

(1, 11, 16) Wänke et al. (1977); (2) Mittlefehldt (1994); (3, 6, 13) Mittlefehldt and Lindstrom (1993), (4) Mittlefehldt (1979); (5, 8) Palme et al. (1978); (7) Palme et al. (1988); (9, 15) Wänke et al. (1972b); (10) Warren and Jerde (1987); (12) Palme et al. (1983); (14) Laul and Gosselin (1990).
*average matrix

Major and minor element composition. Diogenites are very uniform in bulk major and minor element composition, with the exception of Cr_2O_3. Variation in Cr_2O_3 undoubtedly reflects variation in the amount of chromite in the meteorites. This partially reflects the mineralogical nature of the diogenites, and is not entirely a problem of sampling these coarse-grained rocks; analyses of some diogenites always contain more Cr_2O_3 than typical. For example, 10 analyses for Cr_2O_3 in Shalka whole-rock samples range from 1.1 to 2.4 wt %, while for Johnstown, 29 analyses range from 0.7 to 1.1 wt %. As mentioned above, some diogenites are slightly more magnesian or ferroan than the typical diogenite. Nevertheless, the compositions of these meteorites are not very different from the typical diogenite; for example compare magnesian diogenite Manegaon (MgO 26.5 wt %, FeO 14.3 wt %) and ferroan diogenite Garland (MgO 25.0 wt %, FeO 18.5 wt %) with the typical diogenites Johnstown and Shalka (MgO 25.5 to 25.8 wt %, FeO 15.9 to 16.3 wt %). The Yamato Type B diogenites are exceptional, being quite distinct in bulk composition from other diogenites, as expected from their distinct mineralogies (Table 34).

The cumulate eucrites show a wider range in bulk major and minor element composition. In part this reflects compositional differences among the rocks, but it is also complicated by difficulties in adequately sampling these coarse-grained, modally heterogeneous lithologies. Serra de Magé is particularly difficult to adequately sample; reliable measurements of Al_2O_3 in this rock vary from 12.7 to 20.9 wt %, reflecting differences in plagioclase content in the samples. The mg# of the cumulate eucrites will be less sensitive to sampling problems, as it is essentially defined by the pyroxene in the rock. The cumulate eucrites vary in mg# from ~65 for Binda to ~44 for Y-791195. The compatible element Cr generally reflects this variation in mg#, and is high in Binda and low in Y-791195. However, EET 87548 is exceptional in having the highest Cr content, yet a mg# in the mid-range of cumulate eucrites (Warren et al. 1996).

The basaltic eucrites are very uniform in composition. Most plot within a very limited range in mg#, CaO, Al_2O_3, TiO_2 or Cr_2O_3 contents. Figure 43 is a TiO_2 vs. mg# plot for basaltic eucrites, a variant of a diagram first used by Stolper (1977) to discuss their petrogenesis. Many basaltic eucrites plot within a small region of mg# ~ 38-40, TiO_2 ~ 0.6-0.7 wt %. These are frequently referred to as the main-group eucrites. Some eucrites plot within this same mg# band, but at higher TiO_2 contents. These are the Stannern-trend eucrites. Some eucrites have both higher TiO_2 contents and lower mg#, and are referred to as Nuevo Laredo-trend eucrites. Figure 44 shows histograms of CaO and Cr_2O_3 contents for basaltic eucrites, where the uniformity in bulk compositions of these samples becomes obvious.

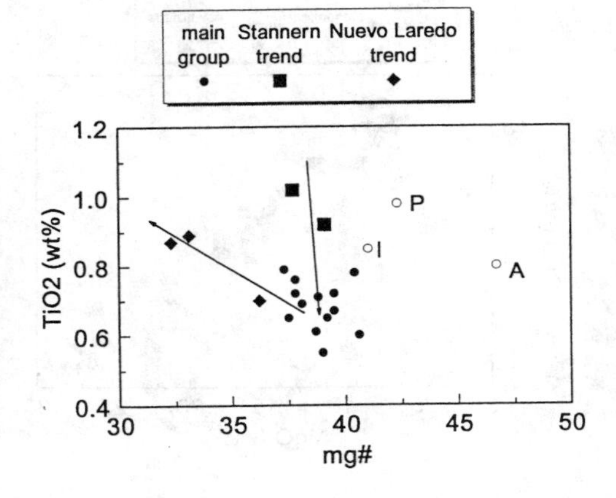

Figure 43. TiO_2 vs. mg# diagram for basaltic eucrites. The Stannern-trend (down arrow) is likely a trend of increasing partial melting. Most eucrites are members of the main-group. Their origin is controversial; they may either be primary partial melts or residual melts. The Nuevo Laredo-trend (left arrow) is a fractional crystallization trend. Anomalous eucrites are ALHA81001, Ibitira and Pomozdino. The data are averages of all available literature determinations.

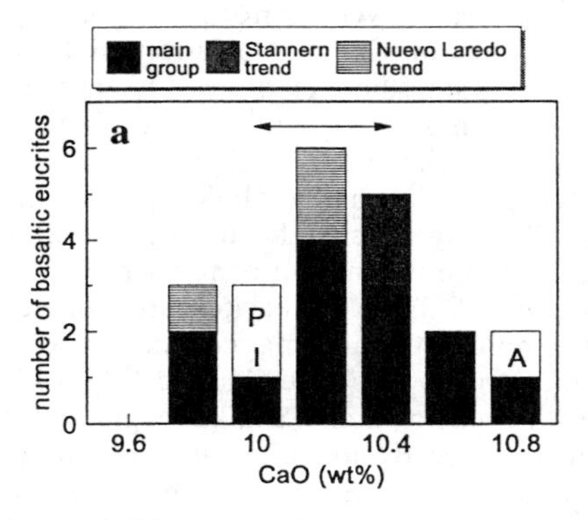

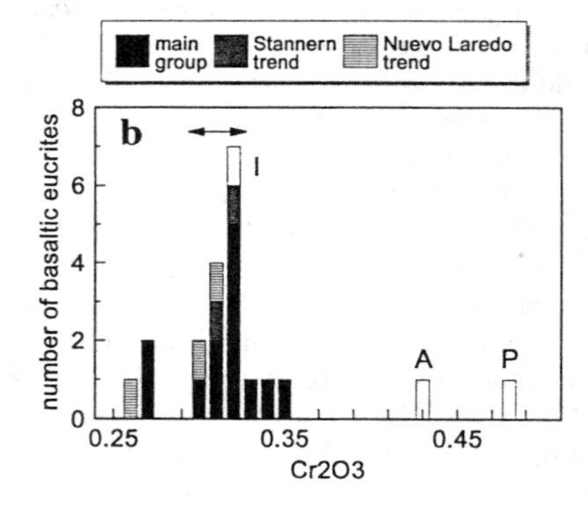

Figure 44. Histograms of (a) CaO and (b) Cr_2O_3 for basaltic eucrites. Both CaO and Cr_2O_3 have narrow ranges in basaltic eucrites, and the latter in particular has been taken as evidence for buffering by residual spinel in the eucrite source region, and hence a partial melting origin (Jurewicz et al. 1993). Arrows show the ranges in experimental, low temperature equilibrium partial melts of Murchison determined by these authors. Anomalous eucrites as in Figure 43.

The extreme heterogeneity in bulk major and minor element composition of the polymict breccias contrasts sharply with the uniformity seen for the igneous lithologies. For all intents and purposes, the polymict breccias are intermediate in composition between the diogenites and basaltic eucrites, and span wide ranges in composition. Figure 45 shows the relationship between CaO and MgO contents for polymict breccias compared to monomict basaltic eucrites and diogenites. The linear relationship displayed in this figure is consistent with the petrographic evidence that the polymict breccias are dominated by basalt clasts similar to the basaltic eucrites and orthopyroxene clasts similar to the diogenites. With few exceptions (see below), the polymict breccias would plot between basaltic eucrites and diogenites on any element-element plot (Fig. 46). Note that the single meteorite type, the howardites, spans a wide range in CaO and MgO contents, in contrast to what is observed for basaltic eucrites and diogenites (Fig. 45). This heterogeneity can also occur within individual howardites; seven analyses each of CaO and MgO for Kapoeta range from 3.1 to 9.9 and 8.1 to 22.2 wt %, respectively. This likely reflects differential sampling of basaltic and orthopyroxenitic clasts and matrix. Four matrix samples of Bholghati only vary from 6.7 to 7.2 wt % and 15.7 to 17.6 wt % for CaO and MgO, respectively (Laul and Gosselin 1990).

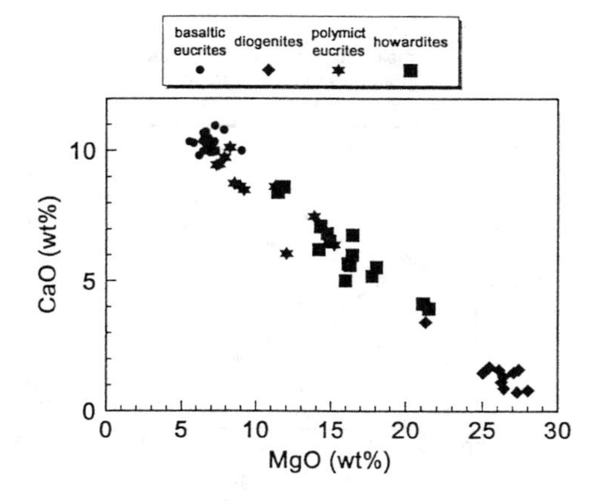

Figure 45. CaO vs. MgO for HED meteorites showing the mixing relationship between basaltic eucrites and diogenites to form the polymict eucrites and howardites. Although other components are present in the polymict breccias, they are dominantly composed of eucrite-like basalts and diogenite-like orthopyroxenes.

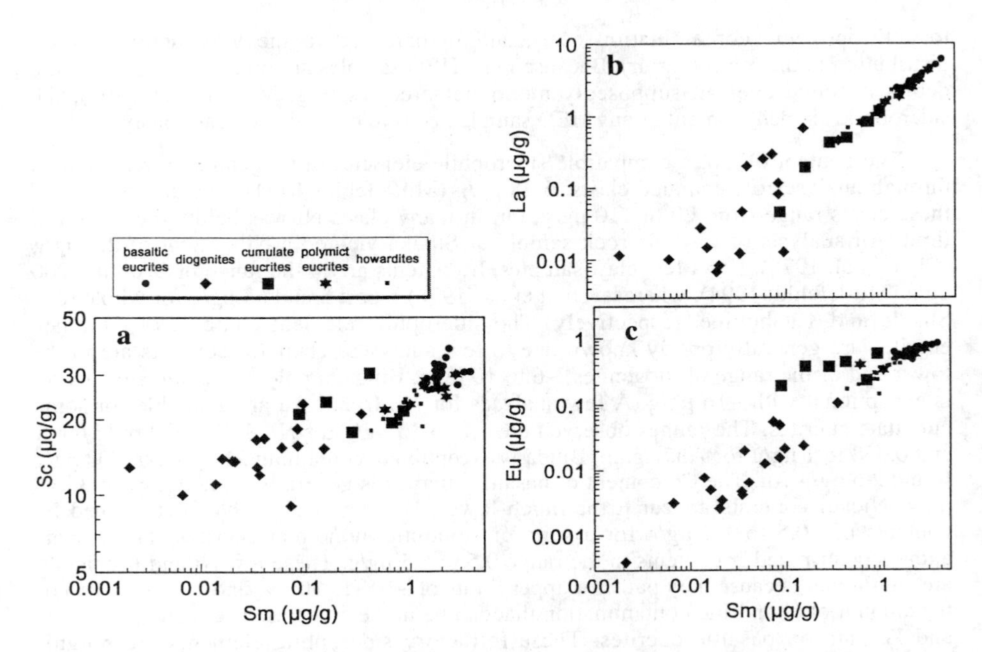

Figure 46. Scandium vs. Sm (a), La vs. Sm (b) and Eu vs. Sm diagrams for HED meteorites show the extreme variations in highly incompatible elements (La, Sm, Eu) in the HED suite compared to a moderately incompatible (bulk distribution coefficient near unity) element, Sc. The polymict breccias, howardites and polymict eucrites, are intermediate in trace element contents between basaltic eucrites and diogenites. The scatter in La and Sm content in diogenites and cumulate eucrites is due to the strong control of a trapped melt component on highly incompatible element contents in igneous cumulates. The "flattened" Eu-Sm trend between cumulate and basaltic eucrites is due to Eu becoming more compatible once plagioclase becomes a liquidus phase.

Trace lithophile element composition. Excluding the polymict breccias for the moment, the trace incompatible lithophile elements show systematic distributions related to rock type (Fig. 46). The contents of Sc, La, Sm and Eu, for example increase in the sequence diogenite, cumulate eucrite, basaltic eucrite. The La content varies by a factor of ~2000 in this suite, whereas Sc varies by only a factor of ~3. The trace incompatible elements are not uniform in the diogenites and the cumulate eucrites. For highly incompatible elements such as La and Sm, these meteorite types show wide ranges in content; the diogenites show a range of a factor of ~30 in La content, for example. The moderately incompatible element Sc exhibits much narrower ranges in content. For the basaltic eucrites, both the highly incompatible and moderately incompatible lithophile elements have narrow ranges, although the former are more variable than CaO, an incompatible major element. In some cases, the content of highly incompatible elements in whole-rock samples of diogenites are higher than observed in separated clasts. This suggests that even the monomict diogenites may contain incompatible element-rich debris in the matrix (Mittlefehldt 1994).

Siderophile element composition. Siderophile elements are strongly depleted in HED meteorites, and are generally higher in the polymict breccias than in the monomict breccias and unbrecciated samples. The Co and Ni contents of HED meteorites are fairly well known, but the contents of other siderophile elements are poorly known. In part this is due to there being relatively few analyses (e.g. Mo, W), in part due to the susceptibility to anthropogenic contamination (Au), and in part due to the very heterogeneous distribution in the rocks (e.g. Ir). Because most HED samples are breccias, and chondritic debris is found even in supposedly monomict breccias (e.g. Mittlefehldt 1994), the siderophile element content of any HED sample needs to be evaluated cautiously.

The content of Co, a compatible siderophile element, in diogenites is well known through analyses of separated clasts, 6-38 µg/g (Mittlefehldt 1994). The Ni contents of these clasts range from 20 to 110 µg/g, but in many clasts Ni was below the detection limit. An analysis of a whole-rock sample of Shalka yielded a Ni content of 2.1 µg/g (Chou et al. 1976). For a few clast samples, Ir contents are in the range of 2500 to 4000 pg/g (Mittlefehldt 1994), whereas Chou et al. (1976) report 8 and 63 pg/g for whole rock Shalka and Tatahouine, respectively. The siderophile element contents of cumulate eucrites are generally poorly known due to few analyses. Their Co contents are in the lower end of the range of diogenites, ~6 to 10 µg/g. Binda, a polymict cumulate eucrite, is exceptional with ~16 µg/g. A few analyses for Ni, Ir and Au are available for some cumulate eucrites. The ranges observed are 1.2 to 13 µg/g for Ni, 3.2 to 270 pg/g for Ir and 0.038 to 1 ng/g for Au. Again, Binda is exceptional, containing 21 µg/g Ni, 700 pg/g Ir and 2.5 ng/g Au. The Co content of basaltic eucrites is generally in the range of 3 to 8 µg/g. Nickel contents appear to be much lower. Warren et al. (1996) determined Ni contents of ~0.5 to 1.5 µg/g for a suite of Antarctic monomict eucrites. These same authors determined Ir contents in the range 0.5 to 6.5 pg/g. These low Ni and Ir contents are significant because they put firm upper limits of ~10^{-2} to 10^{-3} % and 10^{-3} to 10^{-4} % on the amount of chondritic contamination that can be in the breccias. There are limited Mo and W data on basaltic eucrites. These refractory siderophile elements are roughly correlated with refractory incompatible lithophile elements such as La in basaltic eucrites, but are depleted compared to CI abundances by factors of 30 for W/La and 570 for Mo/La (Newsom 1985, Newsom and Drake 1982).

Noble gases. Some howardites are enriched in noble gases compared to other HED meteorites. Excluding decay products and cosmogenic gases, these howardites contain two noble gas components, planetary- and solar-type gases (e.g Mazor and Anders 1967). Early in the study of noble gases in howardites it was believed that these gases were brought in by a carrier phase (Mazor and Anders 1967, Müller and Zähringer 1966). The characterization of carbonaceous chondrite fragments in howardites (Wilkening 1973) led to the identification of these clasts as the carriers of the planetary-type gases (Wilkening 1976). Mazor and Anders (1967) noted that the planetary- and solar-type gas contents were correlated, and thus believed that they were brought in by the same carrier. However, subsequent experiments have shown that the solar-type gases are a surface correlated component of mineral and glass fragments of the howardites themselves, not just of the chondritic materials (Black 1972, Padia and Rao 1989, Rao et al. 1991). These solar-type gases represent solar wind and solar flare gases implanted in the outer few µm of grains in the breccias, and show that these howardites were part of the regolith of their parent body.

Ages

It has long been known that the HED suite of meteorites is very old, not much younger than the chondrites. This was established by analyses of whole-rock suites (e.g. Birck and Allègre 1978, Papanastassiou and Wasserburg 1969), on individual meteorites (e.g. Allègre et al. 1975, Tera et al. 1997) and on igneous clasts from the polymict breccias (Nyquist et al. 1986) using the long-lived Rb-Sr, Sm-Nd and/or Pb-Pb chronometers, as well as by evidence for the decay products of the short-lived nuclides ^{129}I and ^{244}Pu (e.g. Rowe 1970). The best estimate of the age of magmatism for the

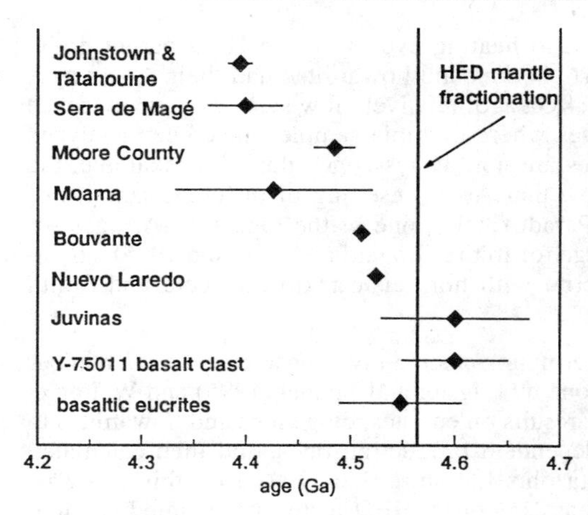

Figure 47. The age of magmatism on the HED parent body. The narrow band labeled HED mantle fractionation is the time of major parent body differentiation determined by Mn-Cr chronometry, and the width of the band includes the uncertainty on this age. The best estimate for the age of eucrite formation is given by the datum labeled basaltic eucrites. The ages of the brecciated basaltic eucrite Juvinas and the least metamorphosed basalt clast from Y-75011 are compatible with these age estimates. The brecciated basaltic eucrites Bouvante and Nuevo Laredo are significantly younger, but this could be due to impact metamorphism. The unbrecciated cumulate eucrites Moama, Moore County and Serra de Magé are significantly younger than the inferred HED mantle fractionation, and Serra de Magé is significantly younger than the basaltic eucrites best age estimate. The age of the Johnstown and Tatahouine diogenites are also significantly younger than the inferred HED differentiation time, but these rocks are brecciated and shocked, respectively, casting uncertainty on the meaning of their young age. The evidence shown here has been taken to indicate that the cumulate eucrites are ~100 Ma younger than the basaltic eucrites. Data are from Birck and Allègre (1981), Lugmair and Shukolyukov (1997), Nyquist et al. (1986), Smoliar (1993), Takahashi and Masuda (1990), Tera et al. (1997).

basaltic eucrites based on these data is in the range 4.51 Ga (Pb-Pb, Bouvante and Nuevo Laredo, Tera et al. 1997) to 4.60 Ga (Rb-Sr, Juvinas and Y-75011,84B least metamorphosed basalt clast, Allègre et al. 1975, Nyquist et al. 1986). Smoliar (1993) has examined all available Rb-Sr data on eucrites, and derived his best estimate for their formation age of 4.548±0.058 Ga. Recently, evidence for the presence of very short-lived ^{53}Mn ($t_{1/2}$ 3.7 Ma) and ^{60}Fe ($t_{1/2}$ 1.5 Ma) at the time of formation of basaltic eucrites has been found (Lugmair et al. 1994a,b; Shukolyukov and Lugmair 1993a,b). Lugmair and Shukolyukov (1997) have determined a Mn-Cr isochron for diogenite and basaltic eucrite samples, and combined this with a precise Pb-Pb age and Mn-Cr isotopic systematics for the angrite LEW 86010 to derive an estimate for the time of differentiation of the HED parent body of 4.5648±0.0009 Ga. Lugmair and Shukolyukov (1997) were careful to describe this time as "a time of HED mantle fractionation," rather than the time of formation of the diogenites and basaltic eucrites.

Age information on cumulate eucrites suggests that they are younger than basaltic eucrites. Tera et al. (1997) determined Pb-Pb ages for Moama, Moore County and Serra de Magé, all unbrecciated cumulate eucrites. These ages are, respectively, 4.426±0.094, 4.484±0.019 and 4.399±0.035 Ga, and are within the uncertainty of Sm-Nd ages determined on the same meteorites by Jacobson and Wasserburg (1984), Tera et al. (1997) and Lugmair et al. (1977). None of these ages overlaps the estimated time of HED parent-body differentiation given above within the stated 2σ uncertainty envelopes (Fig. 47). Takahashi and Masuda (1990) determined a Rb-Sr isochron for the diogenites Johnstown and Tatahouine of 4.394±0.011 Ga; again, this is outside the error envelope of the estimated HED parent-body differentiation time (Fig. 47). In this case, the difference is plausibly due to secondary processes as Johnstown is a breccia and Tatahouine has suffered shock damage.

Metamorphic and impact ages are best determined using the ^{39}Ar-^{40}Ar method, and Bogard (1995) has recently summarized these studies. He demonstrated that almost every eucrite and howardite studied by Ar-Ar chronometry contains evidence for significant Ar degassing, and argued that this was due to heating events on the HED parent body. Bogard (1995) further showed that most eucrites and howardites had their Ar-Ar ages reset within the time period of ~4.1 to 3.4 Ga ago; relatively few samples gave evidence for older or younger Ar-Ar ages. In cases where multiple samples have been analyzed from the same meteorite, the Ar-Ar ages are not always concordant. For example, the results compiled by Bogard (1995) show that Ar-Ar resetting of different samples of Kapoeta vary from ~3.44 to 4.48 Ga. Paradoxically, one of the oldest Ar-Ar ages for HED meteorites is the 4.495±0.015 Ga age for Ibitira (Bogard and Garrison 1995), an unbrecciated, highly metamorphosed eucrite with hornfelsic texture (Steele and Smith 1976).

Noble gas studies geared toward determining cosmic-ray exposure ages for a number of HED meteorites have recently been done by Eugster and Michel (1995) and Welton et al. (1997). The former authors used their results on eucrites, diogenites and howardites to determine composition and shielding dependent production rates, and then calculated cosmic-ray exposure ages for their data plus literature data. Based on this analysis, Eugster and Michel (1995) suggested that 81% of all HED meteorites studied fell into five exposure age clusters; 6±1, 12±2, 21±4, 38±8 and 73±3 Ma. Welton et al. (1997) determined the noble gas contents of diogenites, and calculated exposure ages for them. They found that 70% of the diogenites have ages consistent with the 21 and 38 Ma age clusters. Using these data and literature data on eucrites and howardites, Welton et al. (1997) performed statistical analysis to determine that the 21 and 38 Ma exposure age clusters are statistically significant, while the 12 Ma cluster is not. The 6 and 73 Ma clusters are the smallest identified by Eugster and Michel (1995) and so presumably they are also not statistically significant. Both the 21 and the 38 Ma age clusters contain the range of petrologic types; eucrites, polymict eucrites, diogenites and howardites.

HED meteorite petrogenesis

The earliest petrogenetic model for asteroidal differentiation was that of Mason (1962). He noted that the lithologies in the HED group could plausibly represent a fractional crystallization sequence from early orthopyroxenites, through increasing plagioclase content to the basaltic eucrites. Mason (1962) suggested that a chondritic body depleted in Na was totally melted, a metallic core segregated, and the silicate melt solidified via fractional crystallization. The eucrites were thus the residual liquid products of an extensive sequence of crystallization. In this model, the crust of the asteroid from top down was composed of eucrites, howardites and diogenites. Mason (1962) considered howardites a brecciated igneous rock type, not a mixture of lithologies.

Stolper (1977) presented an alternative model for eucrite petrogenesis in which the basaltic eucrites represent primary partial melts of their parent body. Stolper based his arguments on the results of melting experiments on basaltic eucrites in which he found conditions of temperature and oxygen fugacity at which the eucrites are multiply saturated with pyroxene, plagioclase, olivine, metal and spinel. Because this is a plausible mineral assemblage for a chondritic parent body, Stolper (1977) concluded that eucrites are primary partial melts of their parent body. In Stolper's model, the parent body undergoes minimal heating as total melting is not envisioned.

The Mason and Stolper models represent end member models, and several intermediate or hybrid models have been developed. Ikeda and Takeda (1985) developed a magma ocean model for HED petrogenesis. They invoke an accretional energy source to cause total melting of the outer portion of the parent asteroid, producing a thin magma ocean. The magma ocean undergoes fractional crystallization and volatile loss, producing first dunites and then orthopyroxenites (diogenites). At some point during crystallization, the surface of the magma ocean became a stable, crystalline "scum" layer composed of volatile-poor "Trend A" eucrites like Juvinas and Nuevo Laredo. The lower portions of the magma ocean suffered less volatile loss, and melts from this lower region were occasionally extruded as more volatile-rich "Trend B" magmas like Stannern.

Warren and Jerde (1987) presented a hybrid model of the Mason and Stolper petrogenetic schemes, in which members of the Stannern-trend are partial melts of the HED parent asteroid, while the Nuevo Laredo-trend eucrites and the main-group eucrites are residual liquids from melts that first produced the diogenite cumulates. Warren and Jerde (1987) also suggested that the main-group and Nuevo Laredo-trend eucrites might be mixtures of cumulus grains and residual liquid.

Hewins and Newsom (1988) recognized the two eucrite trends, A and B, of Ikeda and Takeda (1985) but suggested that these trends are the result of two distinct parent melts, rather than divergent evolution of a single magma ocean. Hewins and Newsom (1988) suggested that magnesian parent melts that were formed by high degrees of partial melting followed fractional crystallization paths to produce an incompatible-element-poor trend and an incompatible-element-rich trend (equivalent to trends A and B of Ikeda and Takeda 1985, respectively). The Hewins and Newsom (1988) scenario posits that magnesian parental melts were produced on the HED parent body, and they suggest that two clasts from polymict breccias are candidates for these melts. More recently, Mittlefehldt and Lindstrom (1997) studied two additional magnesian basalt clasts from howardites, and reviewed existing data on the other two. These authors concluded that all four magnesian basalt clasts so far identified are more likely impact melts of the parent body surface. In their opinion, there is no petrologic evidence for the existence of primary magnesian basalt clasts in HED meteorites. The reality of separate "A" and "B" trends in the basaltic eucrite suite is open to question.

The central issue in all these models is whether main-group eucrites are primary partial melts or residual liquids. Experimental studies of chondritic materials have been done to attempt to resolve this issue. Jurewicz et al. (1993, 1995a) performed melting experiments on CM, CV, H and LL chondrites. Volatile elements like Na were allowed to escape from the system in these experiments. Jurewicz et al. (1993, 1995a) showed that there is a close match between the low-temperature melting fraction of volatile-poor CM chondrites and the main-group eucrites suggested by Stolper (1977) to be primary partial melts. Consequently these authors favored Stolper's model for eucrite petrogenesis. One potential problem with these experimental studies is that a volatile-poor, CM chondrite-like source has insufficient pyroxene to produce orthopyroxene cumulates in abundance. This seems at odds with the HED meteorite suite, in which diogenites and orthopyroxene clasts in polymict breccias are modally significant. The problem is compounded by melting experiments on LL chondrites which do contain sufficient pyroxene to easily explain diogenites, but produce minimum melts unlike eucrites (Jurewicz et al. 1995a).

The partial melt vs. residual melt origin for main-group eucrites remains an unsettled problem in the meteoritical community. Indeed, a recent series of publications has resulted in a range of models for eucrite petrogenesis. Jones et al. (1996) consider that the eucrites Bouvante, Stannern, Chervony Kut, Cachari and Sioux County are primary partial melts. Ruzicka et al. (1997) believe basaltic eucrites, excluding Bouvante and Stannern, formed from a magma ocean that first fractionally crystallized dunites and orthopyroxenites, and then crystallized under conditions approaching equilibrium to produce the suite of basaltic eucrites. Righter and Drake (1997) suggested the eucrite parent body was totally molten, separated a core, and then underwent equilibrium crystallization to form dunites and orthopyroxenites. A basaltic residual liquid then underwent fractional crystallization to form the main-group basaltic eucrite suite, while

Bouvant and Stannern represent other residual melts. Note that Ruzicka et al. (1997) and Righter and Drake (1997) come to opposite conclusions as to which igneous rocks require equilibrium vs. fractional crystallization to form them. Warren (1997) presented an updated version of the Warren and Jerde (1987) model for eucrite petrogenesis. He suggested that the HED suite formed from a large magma system, probably a global magma ocean. Diogenite cumulates formed early while the system was rapidly crystallizing. The rate of cooling, and thus crystallization, slowed down and high mg# cumulate eucrites were produced. During this period of slow crystallization, the residual melt was repeatedly tapped to produce the basaltic eucrite suite. At the end of magmatism, low mg# cumulate eucrites are formed from the last dregs of the magma system. Clearly, we are not converging on a consensus opinion for eucrite petrogenesis.

The petrogenesis of diogenites is more clear cut, at least to a first order; these coarse-grained rocks are cumulates from a fractionally crystallizing magma. The nature of their parent magma is uncertain, however. Stolper (1977) suggested that they may have crystallized from a melt of essentially orthopyroxene composition. Recent study of minor element distributions in the pyroxenes, trace element study of bulk orthopyroxene separates and SIMS study of pyroxenes have tended to support this model (Fowler et al. 1994, 1995; Mittlefehldt 1994). Both bulk samples and pyroxene grains show wide ranges in incompatible element contents, suggesting a wide range in fractional crystallization, yet they have a very restricted range in mg# and mineralogy, suggesting a very narrow range in crystallization. The uniformity in mg# could be due to metamorphic equilibration (Fowler et al. 1994, 1995; Mittlefehldt 1994), but the nearly monomineralic nature of diogenites implies either a restricted range in crystallization or a parent melt very rich in orthopyroxene component (Mittlefehldt 1994). The former is at odds with the wide range in incompatible element abundances. Fowler et al. (1995) suggested that the diogenites may have been derived from a suite of parent mafic melts with different incompatible element contents, rather than a single orthopyroxenitic melt. Shearer et al. (1997) have concluded that diogenites did not form from highly orthopyroxene normative magmas. Rather, they suggest that the major and trace element systematics of the diogenite suite can be modeled as cumulates arising via small amounts (10-20%) of fractional crystallization of a suite of distinct basaltic magmas formed either by fractional melting of a homogeneous source, or equilibrium melting of heterogeneous sources.

Sack et al. (1991) suggested that two diogenites are unusually olivine-rich, and represent melting residues (restites) from the HED parent body rather than cumulates. Mittlefehldt (1994) argued that those two diogenites were not unusually olivine-rich, based on modal data provided by R. Hewins and on a bulk composition for one of them. Fowler et al. (1995) and Mittlefehldt (1994) showed that these two diogenites are not unusual in minor and trace element composition compared to other diogenites, and that a restite origin for them seems unlikely.

The cumulate eucrites are cumulates from a fractionally crystallizing mafic melt. Stolper (1977) demonstrated that the cumulate eucrites are too iron-rich to have crystallized from basaltic eucrites. Liquidus pyroxenes for Sioux County have an mg# of 71-72 (Stolper 1977), while the most magnesian cumulate eucrite Binda contains pyroxenes with a bulk mg# of 67 (Takeda et al. 1976). Using the data in Stolper (1977), one can infer that liquidus pyroxene in Nuevo Laredo or Lakangaon, the most ferroan basaltic eucrites, would have an mg# of ~59, compared with bulk pyroxene mg# of 58 and 56 for Moama and Serra de Magé, respectively (Harlow et al. 1979, Lovering 1975). Several researchers have suggested that the parent melt of the cumulate eucrites was unusual in that it had a LREE-enriched pattern. This was originally based on calculations using bulk analyses of mineral separates from Moore County and Serra de Magé, and mid-1970s vintage partition coefficients (Ma and Schmidt 1979, Ma et al. 1977), and has recently been reiterated by Hsu and Crozaz (1997) based on SIMS analyses of pyroxenes and plagioclase in Moama and Moore County and modern partition coefficients. These latter authors concluded that subsolidus redistribution did not affect the REE they measured, and that therefore the cumulate eucrites most likely were formed from melts with fractionated REE patterns which do not seem to be genetically related to the basaltic eucrites. Pun and Papike (1995) similarly found that nominal parent melts for the cumulate eucrites Binda, Moama, Moore County and Serra de Magé, calculated through inversion of mineral REE data obtained by SIMS, showed LREE-enriched patterns. These authors cautioned that unusual parent melts provided only one possible explanation of the result, and that subsolidus redistribution of the REE or inappropriate partition coefficients could be the cause of the unusual calculated patterns. Treiman (1997) showed that the cumulate eucrites Moore County and Serra de Magé can be modeled for major and trace elements as mixtures of cumulus minerals and trapped melt from parent melts similar to basaltic eucrites. He suggested that subsolidus redistribution of the REE causes parent melts calculated by inversion of SIMS data to have unusual REE patterns.

Thermal metamorphism of the HED parent body crust

As discussed above, almost all HED meteorites show textural evidence for extensive thermal annealing in the form of Fe-Mg equilibration and exsolution of pyroxenes and mineral grain recrystallization. Detailed study of the pyroxene textures led to the observation that the mg# and textural features were roughly correlated; more magnesian lithologies (diogenites, cumulate eucrites) are more annealed than are more ferroan lithologies (basaltic eucrites). As the former are intrusive rocks and the latter surficial rocks, this observation led to development of a layered-crust model for the HED parent body by H. Takeda and colleagues (e.g. Miyamoto and Takeda 1977, Takeda 1979, Takeda et al. 1976). In this model, the extent of pyroxene exsolution and inversion is related to the depth of formation of the rock, and the pyroxene textures are largely caused by slow cooling from the magmatic stage (Takeda 1979). It was recognized that some eucrites do not fit the overall trend, and later metamorphism was also noted as an important process in some cases (e.g. Takeda and Graham 1991). One possibility is that impact heating was an important agent of metamorphism, and this has been used to explain the textural differences between basalt clasts in polymict breccias and monomict eucrites (e.g. Nyquist et al. 1986, Takeda and Graham 1991). Heating by subsequent flows or intrusions is another possible metamorphism agent (Takeda and Graham 1991).

Other researchers have argued that impact heating (Keil et al. 1997) and contact metamorphism by later intrusions/extrusions (Yamaguchi et al. 1996) cannot possibly explain the great preponderance of metamorphosed lithologies in the HED suite. Yamaguchi et al. (1996, 1997) have addressed the issue of why thermal metamorphism was so prevalent among the basaltic eucrites, which presumably were extrusive flows on the HED parent body. These authors suggest that a high eruption rate on the parent body could bury early form basalts rapidly enough that heat conducted up from the mantle through the crust would be sufficient to anneal eucrites. Reestablishment of the parent body thermal gradient thus causes global thermal metamorphism in the crust. In this model, the earliest formed basalts would be the most highly metamorphosed, and the latest extrusions would be unmetamorphosed. This model can easily accommodate vesicular yet highly recrystallized eucrites such as Ibitira which must have been extruded onto the surface and yet suffered extensive metamorphism.

4 Vesta, the HED parent body?

The HED meteorite group is the only such group, excluding lunar and martian meteorites, for which we have a strong candidate for the parent body, the asteroid 4 Vesta. McCord et al. (1970) first showed that the reflectance spectrum of Vesta in the wavelength region from 0.3 to 1.1 μm was matched by the laboratory spectrum of a basaltic eucrite, Nuevo Laredo. Subsequently, Consolmagno and Drake (1977) argued that Vesta was indeed the parent body of the HED meteorites. Wasson and Wetherill (1979) discussed the dynamical problems of moving material from Vesta to Earth-crossing orbits where they could be sampled as meteorites. Basically, Vesta is far from orbital resonances such that only energetic events could propel material into regions where they would have a reasonable probability of evolving into Earth-crossing orbits. Such material is expected to be heavily shocked, in contrast to most HED meteorites. The yield of material from Vesta should be much smaller than from other, presumably differentiated asteroids more suitably placed in orbital-element space (Wasson and Wetherill 1979).

Hostetler and Drake (1978) proposed that the material ejected from Vesta was originally in fragments large enough to shield most of the mass from cosmic-ray bombardment. Some of these bodies evolved into Earth-crossing orbits, and collisions in space liberated the smaller pieces that now fall as meteorites, with cosmic-ray exposure ages of 5-75 Ma. Recently, Cruikshank et al. (1991) determined that three asteroids of ~1-3 km diameter in Earth-approaching orbits have spectral characteristics very similar to those of Vesta and the HED meteorites. These authors suggested that these three asteroids are the immediate sources of many of the HED meteorites. Cruikshank et al. (1991) did not favor Vesta as the ultimate source for the Earth-approaching asteroids. They were led by the oxygen-isotopic similarity of HED meteorites and main-group pallasites into a belief that the parent body of these meteorites was one and the same, and must therefore have been totally disrupted to liberate pallasites from the core-mantle boundary. Binzel and Xu (1993) surveyed numerous small asteroids in orbits similar to Vesta's and found 20 with diameters between 4 and 10 km that have spectra similar to that of Vesta. These small asteroids form a "trail" in orbital-element space from near Vesta to near the 3:1 resonance. Binzel and Xu (1993) proposed that these small asteroids are spalls from collisions on Vesta, and that some of their brethren reached the 3:1 resonance, had their orbits perturbed into Earth-crossing or Earth-approaching orbits, and ultimately delivered HED meteorites to the Earth.

Wasson and Chapman (1996) have argued that Vesta is not the HED parent body. They suggested that there must have been many asteroids with basaltic crusts, based on the number of differentiated parent bodies inferred from the meteorite record. They further suggested that space weathering has altered the reflectance spectra of most of these bodies to resemble S asteroids, and that one of these asteroids in a more favorable orbit is the parent body for HED meteorites. In their view, Vesta stands out as being like HEDs only because it was recently resurfaced, exposing fresh material. The controversy over whether Vesta is indeed the HED parent body will likely only be resolved by petrological and geochemical analysis of returned Vesta samples.

The geology of the surface of Vesta has come under increasing scrutiny through both ground- and space-based telescopic study. Based on this work, Vesta is known to harbor regions rich in orthopyroxene and regions more basalt-like (Binzel et al. 1997, Gaffey 1997). There is a general hemispheric dichotomy, with one hemisphere being richer in magnesian pyroxenes (more diogenite-like) than the other (Binzel et al. 1997). This hemisphere also gives spectral evidence for the presence of substantial olivine (Binzel et al. 1997, Gaffey 1997).

ANGRITES

The angrite grouplet consists of four meteorites, Angra dos Reis, LEW 86010, LEW 87051 and Asuka 881371. These rocks, while petrologically distinct, have identical oxygen-isotopic compositions (Clayton and Mayeda 1996), have similar unusual mineralogies (e.g. McKay et al. 1988a, 1990; Prinz et al. 1977, Yanai 1994) and share a number of geochemical characteristics (Mittlefehldt and Lindstrom 1990, Warren et al. 1995) that suggest that they originated on a common parent body. Although the angrites share a narrow region of O-isotope space with the HED meteorites, mesosiderites and brachinites (Fig. 1d), the unusual mineralogies and compositions of the angrites suggest that they are not closely related to any of these other meteorite groups. The angrites are crustal igneous rocks of generally basaltic composition, although Angra dos Reis is mineralogically and compositionally quite unusual.

Mineralogy and petrology

Angra dos Reis. The most detailed description of this rock was given by Prinz et al. (1977), and the synopsis given here is derived from this source unless otherwise noted. Angra dos Reis is a medium- to coarse-grained igneous rock composed dominantly of aluminian-titanian-diopside, with minor olivine, spinel and troilite, and accessory magnesian kirschsteinite, celsian, whitlockite, titanian magnetite, baddeleyite and metal. The pyroxenes occur in two textures; a groundmass of small xenomorphic grains on the order of 100 μm in size, and larger, poikilitic grains on the order of 1 mm in size. Pyroxene grains in the former join at triple junctures. The small groundmass grains are enclosed in the poikilitic grains, and occur along the margins of the large grains where they are partially enclosed. Olivine occurs as grains within the groundmass, or as aggregates of small equidimensional grains joining at triple junctures. Wasserburg et al. (1977) noted that olivine is heterogeneously distributed in the meteorite. Spinel occurs as xenomorphic grains "dispersed throughout" pyroxene. Whitlockite occurs as mm-sized grains dispersed throughout the rock, although again, Wasserburg et al. (1977) note that it is heterogeneously distributed. Magnesian kirschsteinite occurs as inclusions in olivine. Celsian and titanomagnetite are found in the groundmass. The compositions of olivine and kirschsteinite are given in Table 36, pyroxene in Table 37, spinel and titanomagnetite in Table 38, and celsian in Table 39.

Plagioclase has also been reported, but has not been found in thin section. Prinz et al. (1977) found a few grains of plagioclase in mineral separates prepared by Wasserburg et al. (1977). The composition of this plagioclase is given in Table 39. Lugmair and Marti (1977) report small amounts of feldspar in one subsample of Angra dos Reis prepared by Wasserburg et al. (1977), but did not characterize it. Störzer and Pellas (1977) found feldspar in subsamples they prepared for track work. These authors did not characterize the feldspar, but noted that two distinct types were present. The majority of the grains required etching conditions like those used to reveal tracks in bytownite, while one grain required etching conditions like those used to reveal tracks in albite-oligoclase (Störzer and Pellas 1977). The composition of plagioclase measured by Prinz et al. (1977) for Angra dos Reis is quite distinct from that in the other angrites (Table 39), but like those in HED meteorites (Table 33). Because plagioclase has not been found in thin section in Angra dos Reis, there is the possibility that the grains measured by Prinz et al. (1977) represent contaminants introduced during sample processing. Eucrites were being actively studied in Wasserburg's lab during the time period that Angra dos Reis was being prepared (Wasserburg et al. 1977). Both types of feldspar found by Störzer and Pellas (1977) have identical cosmic-ray track records, indicating identical exposure histories.

Table 36. Representative olivine and kirschsteinite analyses for angrites.

	Angra dos Reis (1)	Angra dos Reis (2)	LEW 86010 (3)	LEW 86010 (4)	Asuka 881371 (5)	Asuka 881371 (6)	Asuka 881371 (7)	Asuka 881371 (8)	Asuka 881371 (9)	Asuka 881371 (10)	Asuka 881371 (11)
Chemical Composition (wt %)											
SiO_2	36.3	34.6	32.5	32.4	39.9	33.3	39.6	34.2	37.6	30.6	31.0
Al_2O_3		0.33	0.03	0.01	0.04	0.02	0.06	0.06	0.04	0.04	-
FeO	38.3	26.2	50.6	32.2	10.3	43.1	13.5	43.5	28.6	53.4	43.3
MgO	24.3	8.90	13.4	4.93	47.5	20.5	45.4	19.6	32.8	0.35	2.30
MnO	0.60	0.42	0.62	0.39	0.18	0.48	0.16	0.50	0.34	0.77	0.64
Cr_2O_3	0.03	0.02	0.01	0.02	0.17	-	0.34	0.08	0.07	-	-
CaO	1.29	28.9	2.16	29.2	0.06	1.92	0.28	1.89	0.85	13.5	21.6
Total	100.82	99.37	99.32	99.15	98.15	99.32	99.34	99.83	100.30	98.66	98.84
Cation Formula Based on 4 Oxygens											
Si	1.0170	1.0154	0.9985	0.9913	0.9994	0.9816	0.9956	1.0012	1.0070	1.0068	0.9894
Al	0.0000	0.0114	0.0011	0.0004	0.0012	0.0007	0.0018	0.0021	0.0013	0.0016	0.0000
Fe	0.8974	0.6430	1.3002	0.8239	0.2158	1.0625	0.2839	1.0650	0.6406	1.4694	1.1558
Mg	1.0146	0.3892	0.6136	0.2248	1.7732	0.9006	1.7011	0.8551	1.3092	0.0172	0.1094
Mn	0.0142	0.0104	0.0161	0.0101	0.0038	0.0120	0.0034	0.0124	0.0077	0.0215	0.0173
Cr	0.0007	0.0005	0.0002	0.0005	0.0034	0.0000	0.0068	0.0019	0.0015	0.0000	0.0000
Ca	0.0387	0.9087	0.0711	0.9573	0.0016	0.0606	0.0075	0.0593	0.0244	0.4760	0.7387
Total Cations	2.9826	2.9786	3.0008	3.0083	2.9984	3.0180	3.0001	2.9970	2.9917	2.9925	3.0106
*Cation Ratios Larnite (La), Fosterite (Fo), Fayalite (Fa), Fe/Mn and mg# (100*Mg/(Mg+Fe))*											
La	2.0	46.8	3.6	47.7	0.1	3.0	0.4	3.0	1.2	24.3	36.9
Fo	52.0	20.1	30.9	11.2	89.1	44.5	85.4	43.2	66.3	0.9	5.5
Fa	46.0	33.1	65.5	41.1	10.8	52.5	14.2	53.8	32.4	74.9	57.7
Fe/Mn	63	62	81	82	56	89	83	86	83	68	67
mg#	53.1	37.7	32.1	21.4	89.2	45.9	85.7	44.5	67.1	1.2	8.6

(1) Average of olivine analyses (Prinz et al., 1977); (2) Average of kirschsteinite analyses (Prinz et al., 1977); (3) Average of olivine analyses (Crozaz and McKay, 1990); (4) Average of kirschsteinite analyses (Crozaz and McKay, 1990); (5, 6) Large, magnesian olivine core and rim, respectively (Yanai, 1994); (7, 8) Large, olivine xenocryst core and rim, respectively (Mikouchi et al., 1996); (9, 10) Groundmass olivine core and rim, respectively (Mikouchi et al., 1996); (11) Groundmass kirschsteinite (Yanai, 1994).

Table 37. Representative pyroxene analyses for angrites.

	Angra dos Reis (1)	LEW 86010 (2)	LEW 86010 (3)	LEW 86010 (4)	LEW 86010 (5)	LEW 86010 (6)	Asuka 881371 (7)	Asuka 881371 (8)	Asuka 881371 (9)	Asuka 881371 (10)	Asuka 881371 (11)
Chemical Composition (wt %)											
SiO_2	45.9	48.5	44.8	43.2	42.3	41.8	46.5	41.4	45.6	44.5	40.7
Al_2O_3	10.0	5.85	9.11	11.0	11.8	11.7	7.82	6.63	9.13	3.91	8.26
TiO_2	2.16	0.83	2.21	2.71	3.14	3.39	1.78	3.90	1.58	2.42	4.75
Cr_2O_3	0.21	0.67	0.35	0.17	0.13	0.16	0.38	bd	0.60	-	-
FeO	6.70	7.69	9.58	9.84	10.4	10.9	12.1	26.1	9.90	27.4	25.2
MnO	0.06	0.09	0.10	0.09	0.09	0.10	0.17	0.16	0.06	0.24	0.06
MgO	10.6	12.0	9.23	8.39	7.56	7.18	8.19	0.05	8.90	0.10	-
CaO	24.1	24.1	24.4	24.4	24.2	24.1	23.1	21.5	23.0	21.1	21.7
Na_2O	nd	nd	nd	nd	nd	nd	-	0.04	0.03	nd	0.15
Total	99.73	99.73	99.78	99.80	99.62	99.33	100.04	99.78	98.80	99.67	100.82
Cation Formula Based on 8 Oxygens (Ideal Plagioclase = 5 Cations per 8 Oxygens)											
Si	1.7185	1.8233	1.7073	1.6501	1.6237	1.6149	1.7751	1.7072	1.7460	1.8359	1.6537
^{iv}Al	0.2815	0.1767	0.2927	0.3499	0.3763	0.3851	0.2249	0.2928	0.2540	0.1641	0.3463
Total tet*	2.0000	2.0000	2.0000	2.0000	2.0000	2.0000	2.0000	2.0000	2.0000	2.0000	2.0000
Ti	0.0608	0.0235	0.0633	0.0778	0.0906	0.0985	0.0511	0.1209	0.0455	0.0751	0.1451
^{vi}Al	0.1598	0.0825	0.1165	0.1453	0.1576	0.1477	0.1270	0.0295	0.1581	0.0260	0.0493
Cr	0.0062	0.0199	0.0105	0.0051	0.0039	0.0049	0.0115	--	0.0182	--	--
Fe	0.2098	0.2418	0.3053	0.3143	0.3339	0.3522	0.3863	0.9001	0.3170	0.9454	0.8563
Mn	0.0019	0.0029	0.0032	0.0029	0.0029	0.0033	0.0055	0.0056	0.0019	0.0084	0.0021
Mg	0.5915	0.6723	0.5242	0.4776	0.4325	0.4134	0.4659	0.0031	0.5079	0.0061	--
Ca	0.9668	0.9708	0.9963	0.9986	0.9953	0.9977	0.9449	0.9500	0.9436	0.9328	0.9447
Na	--	--	--	--	--	--	--	0.0032	0.0022	--	0.0118
Total Cations	3.9968	4.0137	4.0193	4.0216	4.0167	4.0177	3.9922	4.0124	3.9944	3.9938	4.0093
*Cation Ratios Ca:Mg:Fe, Fe/Mn and mg# (100*Mg/(Mg+Fe))*											
Ca	54.7	51.5	54.6	55.8	56.5	56.6	52.6	51.3	53.4	49.5	52.5
Mg	33.5	35.7	28.7	26.7	24.6	23.4	25.9	0.2	28.7	0.3	0.0
Fe	11.9	12.8	16.7	17.6	19.0	20.0	21.5	48.6	17.9	50.2	47.5
Fe/Mn	110	83	95	108	115	107	70	161	167	113	408
mg#	73.8	73.5	63.2	60.3	56.4	54.0	54.7	0.3	61.6	0.6	0.0

(1) Average of analyses (Prinz et al., 1977); (2-6) Individual analyses (Crozaz and McKay, 1990); (7, 8) Mikouchi et al. (1996); (9-11) Individual analyses (Yanai, 1994).

Table 38. Representative oxide analyses for angrites.

	Angra dos Reis (1)	Angra dos Reis (2)	LEW 86010 (3)	LEW 86010 (4)	Asuka 881371 (5)	Asuka 881371 (6)
Chemical Composition (wt %)						
TiO_2	0.65	21.9	0.78	27.0	1.48	27.3
Al_2O_3	54.5	3.50	56.9	4.86	37.8	2.42
Cr_2O_3	3.30	1.02	1.21	1.11	20.9	0.01
V_2O_3	0.07	0.26	--	--	0.60	0.08
Fe_2O_3	3.40	18.5	1.86	40.2	--	--
FeO	28.4	48.8	32.2	25.8	31.4	68.4
MgO	8.00	1.18	5.87	1.16	6.98	0.02
MnO	0.18	1.17	0.17	0.36	0.26	0.22
CaO	0.76	1.21	0.07	0.06	0.13	0.09
SiO_2	0.51	1.23	0.05	0.06	0.25	0.10
Total	99.77	98.77	99.11	100.61	99.80	98.64
Cation Formula Based on 4 Oxygens (Ideal Spinel = 3 Cations per 4 Oxygens)						
Ti	0.0137	0.6052	0.0166	0.6786	0.0337	0.7960
Al	1.7990	0.1516	1.8985	0.1914	1.3492	0.1106
Cr	0.0731	0.0296	0.0271	0.0293	0.5004	0.0003
V	0.0016	0.0077	--	--	0.0146	0.0025
Fe^{3+}	0.0717	0.5116	0.0396	1.0110	--	--
Fe^{2+}	0.6652	1.4997	0.7623	0.7211	0.7952	2.2179
Mg	0.3339	0.0646	0.2476	0.0578	0.3150	0.0012
Mn	0.0043	0.0364	0.0041	0.0102	0.0067	0.0072
Ca	0.0228	0.0476	0.0021	0.0021	0.0042	0.0037
Si	0.0143	0.0452	0.0014	0.0020	0.0076	0.0039
Total Cations	2.9996	2.9992	2.9993	2.7035	3.0266	3.1433
*Cation Ratios Fe/Mn, mg# (100*Mg/(Mg+Fe)) and cr# (100*Cr/(Cr+Al))*						
Fe/Mn	156	41	187	71	119	307
mg#	33.4	4.1	24.5	7.4	28.4	0.1
cr#	3.9	16.4	1.4	13.3	27.1	0.3

(1) Average of spinel analyses (Prinz et al., 1977); (2) Average of titanomagnetite analyses (Prinz et al., 1977) (3) Average of hercynite analyses (McKay et al., 1988a); (4)) Average of titanomagnetite analyses (McKay et al., 1988a); (5) Spinel (Mikouchi et al., 1996); (6) Ulvöspinel (Mikouchi et al., 1996).

LEW 86010. Sadly, petrologic details for LEW 86010 are lacking, and brief descriptions only are available in numerous abstracts. LEW 86010 is composed primarily of aluminian-titanian-diopside, plagioclase and olivine, with minor kirschsteinite and troilite, and accessory whitlockite, hercynitic spinel, titanomagnetite and Fe,Ni metal (McKay et al. 1988a, Prinz et al. 1988b). Different authors describe the texture variously as "gabbroic or granular" (Delaney and Sutton 1988), "cumulate" (Goodrich 1988) or "subophitic to poikilitic" (Prinz et al. 1988b), but it can best be described as hypidiomorphic-granular (McKay et al. 1988a) (Fig. 48). Aluminian-titanian-diopsides are generally anhedral grains on the order of 2-3 mm in size and are zoned from mg# of 70-75 to about 50, with lower TiO_2 and Al_2O_3, and higher Cr_2O_3 in the cores (Delaney

Table 39. Representative plagioclase and celsian analyses for angrites.

	Angra dos Reis (1)	Angra dos Reis (2)	LEW 86010 (3)	Asuka 881371 (4)
Chemical Composition (wt %)				
SiO_2	46.2	33.6	43.3	43.8
Al_2O_3	33.6	28.0	36.7	35.4
FeO	0.21	0.43	0.23	0.37
MgO	--	--	0.03	0.19
CaO	17.2	1.20	20.5	20.2
Na_2O	1.47	0.15	0.03	0.04
K_2O	0.11	0.05	--	0.01
BaO	--	38.2	--	--
Total	98.79	101.63	100.79	100.01
Cation Formula Based on 8 Oxygens (Ideal Plagioclase = 5 Cations per 8 Oxygens)				
Si	2.1500	2.0135	1.9940	2.0316
^{iv}Al	1.8431	1.9778	1.9921	1.9354
Total tet*	3.9931	3.9913	3.9861	3.9670
Fe	0.0082	0.0216	0.0089	0.0144
Mg	--	--	0.0021	0.0131
Ca	0.8577	0.0771	1.0116	1.0040
Na	0.1326	0.0174	0.0027	0.0036
K	0.0065	0.0038	--	0.0006
Ba	--	0.8971	--	--
Total Cations	4.9981	5.0083	5.0114	5.0027
*Cation Ratios 100*Mg/(Mg+Fe) (mg#) and Molecular Proportion of Orthoclase (Or), Albite (Ab), Anorthite (An) and Celsian (Cs)*				
mg#	--	--	18.9	47.8
Or	0.7	0.4	--	0.1
Ab	13.3	1.8	0.3	0.4
An	86.0	7.7	99.7	99.5
Cs	--	90.1	--	--

(1) Average of plagioclase from mineral separates (Prinz et al., 1977); (2) Average of celsian analyses (Prinz et al., 1977); (3) Average of plagioclase analyses (Crozaz and McKay, 1990); (4) Plagioclase (Mikouchi et al., 1996).

and Sutton 1988, Goodrich 1988, McKay et al. 1988a, Prinz et al. 1988b). Plagioclase grains are subhedral to euhedral, on the order of 1 to 2.5 mm in size and are virtually end-member anorthite in composition (Goodrich 1988, McKay et al. 1988a). Olivine occurs as anhedral to subhedral grains roughly 1 to 2.5 mm in size and containing exsolution lamellae of kirschsteinite up to 20 μm in width (Delaney and Sutton 1988, Goodrich 1988, McKay et al. 1988a, Mikouchi et al. 1995). Olivine is uniform in composition with a mg# of 32-33 and a CaO content of 1.5 to 2.2 wt % (Crozaz and McKay 1990, McKay et al. 1988a, Prinz et al. 1988b). The compositions of olivine and kirschsteinite for LEW 86010 are given in Table 36, pyroxene in Table 37, spinel and titanomagnetite in Table 38, and plagioclase in Table 39.

A-881371 and LEW 87051. Asuka 881371 and Lewis Cliff 87051 are similar rocks. They are porphyritic-textured rocks containing large, subhedral to euhedral grains of magnesian olivine set in an ophitic textured groundmass of euhedral laths of anorthite intergrown with euhedral to subhedral, highly zoned aluminian-titanian-diopside and ferroan olivine (Fig. 49). Kirschsteinite and traces of titanomagnetite, spinel, whitlockite, troilite and Fe,Ni metal are also present (McKay et al. 1990, 1995; Mikouchi et al. 1996, Prinz and Weisberg 1995, Prinz et al. 1990, Yanai 1994).

The coarse olivines are variable in size and composition. Prinz et al. (1990) reported sizes from 0.15 to 1 mm for LEW 87051 with core compositions ranging from Fo_{73} to Fo_{90}, with the largest olivine being the most magnesian. The coarse olivines in A-881371 are up to about 2 mm in size, and core compositions have a narrower range of Fo_{81} to Fo_{89} (McKay et al. 1995, Mikouchi et al. 1996, Prinz and Weisberg 1995, Yanai 1994). The larger of the coarse olivines have very homogeneous cores with strongly zoned rims, while the smaller of the coarse olivines show zoning in the cores which is distinct from that in the rims (Mikouchi et al. 1996). Mikouchi et al. (1996) suggested that the zoning

Figure 48. Photomicrograph of LEW 86010 in transmitted light, partially crossed nichols, showing heterogeneous, hypidiomorphic texture. Major phases are anorthite (twinned), olivine (unzoned, exsolution lamellae) and Al-Ti-diopside (zoned). NASA/JSC photo S88 28784 courtesy of G. McKay.

observed in the smaller olivine cores was an artifact of off-center cuts through large olivine grains with zoned rims. The cores of the large olivine grains have high Cr_2O_3 contents, ~0.34 wt %, compared to the cores of groundmass olivine grains, 0.07 wt % (Mikouchi et al. 1996). The rims on the coarse olivines and the groundmass grains are strongly zoned to Fe- and Ca-rich compositions. Cores of groundmass olivines have compositions of about Fo_{66-70}, with roughly 1 mol % of the larnite (La) component, Ca_2SiO_4, and are zoned out to ~Fo_0, ~La_{20-25} (McKay et al. 1990, 1995; Mikouchi et al. 1996, Prinz and Weisberg 1995, Warren and Davis 1995, Yanai 1994). The larnite content of the rims of groundmass olivine grains is difficult to determine because of fine-scale intergrowth with kirschsteinite (e.g. McKay et al. 1990). Representative analyses of olivine and kirschsteinite grains for A-881371 are given in Table 36.

The aluminian-titanian-diopside grains are strongly zoned in Fe-Mg, but have nearly constant CaO content (McKay et al. 1990, Yanai 1994), while Al_2O_3 and TiO_2 are complexly zoned. Both Al_2O_3 and TiO_2 are high in the magnesian cores and initially drop with decreasing mg#, but as mg# continues to decrease, Al_2O_3 is roughly constant or increases slightly, while TiO_2 increases significantly (e.g. McKay et al. 1990, Warren and Davis 1995). Representative analyses of pyroxene grains for A-881371 are in Table 37.

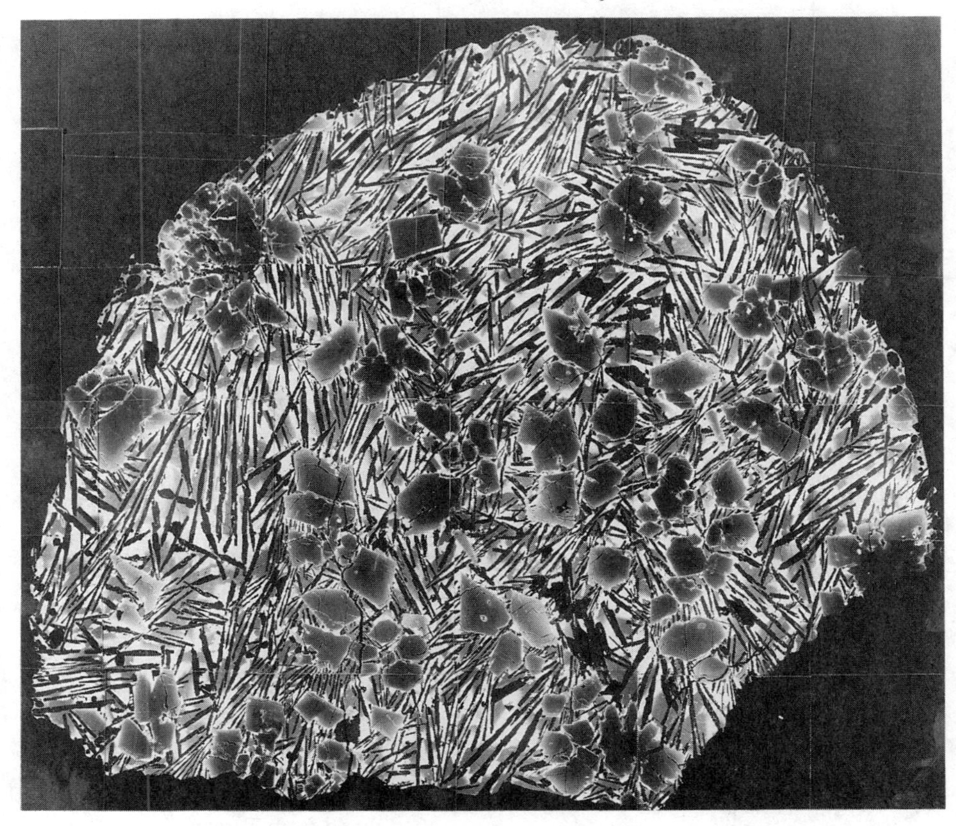

Figure 49. BSE image of LEW 87051 showing euhedral to anhedral olivine xenocrysts in a groundmass of anorthite (dark laths) and Al-Ti-diopside. Image is about 4.8 mm across. NASA/JSC photo S88 28784 courtesy of G. McKay.

Plagioclase in A-881371 and LEW 87051 are essentially pure anorthite. Representative analyses of plagioclase grains are given in Table 39. Representative analyses of spinel grains for A-881371 and LEW 87051 are given in Table 38.

Composition

The angrites are roughly basaltic in composition, although they are unlike any other known basalt in the solar system in that they are the most alkali-depleted. This is expressed by their very low Na_2O and K_2O contents (Table 40), and by their very low contents of the trace alkali elements, Li, Rb and Cs (e.g. Tera et al. 1970). Angrites also have low abundances of the moderately volatile element Ga, and they have lower Ga/Al ratios than any achondrite, lunar sample or martian meteorite (e.g. Warren et al. 1995). Paradoxically, however, angrites are not notably depleted in the highly volatile/mobile elements Br, Se, Zn, In and Cd compared to lunar basalts and eucrites (Warren et al. 1995). Figure 50 shows CI- and Sm-normalized moderately volatile and volatile element abundances for the angrites compared to the basaltic eucrites Sioux County, Juvinas and Ibitira. Also shown for comparison are the ranges in CI- and U-normalize volatile/mobile element abundances for 7 basaltic eucrite falls analyzed by Paul and Lipschutz (1990). (Both Sm and U are highly incompatible, refractory elements, and to first order, CI-

Table 40. Representative whole rock compositions of angrites.

		Angra dos Reis (1)	LEW 86010 (1)	LEW 87051 (1)	Asuka 881371 (2)
Bulk Major Element Composition					
SiO_2	wt%	43.7	39.6	40.4	37.3
TiO_2	wt%	2.05	1.15	0.73	0.88
Al_2O_3	wt%	9.35	14.1	9.19	10.1
Cr_2O_3	wt%	0.22	0.12	0.16	0.13
Fe_2O_3	wt%				0.63
FeO	wt%	9.8	18.2	19.4	23.4
MnO	wt%	0.10	0.20	0.24	0.20
MgO	wt%	10.8	7.0	19.4	14.8
CaO	wt%	23.1	18.4	10.4	12.5
Na_2O	wt%	0.0301	0.0211	0.0234	0.022
P_2O_5	wt%	0.13	0.13	0.08	0.17
S	wt%				0.59
Total		99.28	98.92	100.02	100.72
*Cation Ratios Fe/Mn and 100*Mg/(Mg+Fe) (mg#)*					
Fe/Mn		97	90	80	116
mg#		66.3	40.7	64.1	53.0
Whole Rock Trace Element Contents					
Sc	µg/g	49.6	56.7	36.0	32.9
Co	µg/g	21.3	21.8	27.4	
Ni	µg/g	58	39	45	
Zn	µg/g	1.73	1.8	0.89	
La	µg/g	6.14	3.38	2.32	2.34
Ce	µg/g	19.2	10.1	6.2	5.9
Sm	µg/g	5.76	2.68	1.48	1.39
Eu	µg/g	1.78	0.978	0.536	0.53
Tb	µg/g	1.39	0.66	0.36	
Yb	µg/g	4.82	2.64	1.52	1.38
Lu	µg/g	0.686	0.386	0.239	0.21
Hf	µg/g	2.79	2.21	1.17	1.03
Ta	ng/g	350	260	110	
Ir	ng/g	0.07	7.1	0.180	2.4
Th	ng/g	640	480	220	

(1) Data taken from Mittlefehldt and Lindstrom (1990), except Ni, Zn and Ir from Warren et al. (1995). All iron reported as FeO; (2) Major element data except Na_2O from Yanai (1994). Trace element data and Na_2O from Warren et al. (1995). Data identified by Warren et al. (1995) as having unusually high uncertainties not listed.

normalized element/Sm and element/U ratios are equivalent.) The angrites are depleted in the moderately volatile lithophile elements Li, Mn, Rb, K and Na, and lithophile/siderophile Ga compared to basaltic eucrites (Fig. 50), but within the ranges exhibited by basaltic eucrites for the more volatile/mobile elements, and moderately volatile siderophile elements. Note that volatile lithophile Cs in Angra dos Reis is more than an order of magnitude depleted relative to the narrow range exhibited by basaltic eucrites.

The Pb data displayed in Figure 50 are CI-normalized $1/\mu$ ($\equiv {}^{204}Pb/{}^{238}U$) as determined by chronologic studies of angrites and eucrites. For Angra dos Reis and Ibitira, the measured μ values of Wasserburg et al. (1977) were used. For LEW 86010, the conservatively low, inferred source region μ_1 of Lugmair and Galer (1992) was used. For the basaltic eucrite range, the inferred source μ_1 values for Bereba, Bouvante and Nuevo Laredo from Tera et al. (1997) were used. These data show that the angrite parent body was depleted in volatile/mobile Pb relative to the eucrite parent body (Lugmair and Galer 1992).

The angrites exhibit a range of fractionated refractory lithophile element patterns compared to basaltic eucrites (Fig. 51). A-881371, LEW 86010 and LEW 87051 show generally similar patterns, while Angra dos Reis is distinct. All angrites have superchondritic Ca/Al ratios, with Angra dos Reis showing extreme fractionation (Mittlefehldt and Lindstrom 1990), and compared to basaltic eucrites, have higher Ti/Hf ratios (Fig. 52a). The Antarctic angrites have generally flat REE patterns while Angra dos Reis has a fractionated pattern with depletions in the light and heavy REE compared to the middle REE, and only Angra dos Reis has a significant Eu anomaly (e.g. Ma et al. 1977, Mittlefehldt and Lindstrom 1990, Warren et al. 1995, see Fig. 51). The Antarctic angrites have higher Sc/Sm ratios than basaltic eucrites (Fig. 52b). Angra dos Reis has a

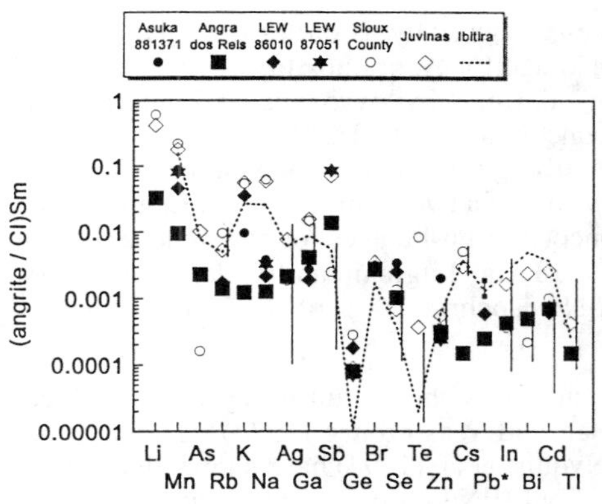

Figure 50. CI- and Sm-normalized moderately volatile and volatile elements in angrites plotted in order of decreasing nebular 50% condensation temperature. Data for the basaltic eucrites Sioux County, Juvinas and Ibitira are shown for comparison. Angrites are severely depleted in moderately volatile lithophile elements. The Pb data are CI-normalized $^{204}Pb/^{238}U$ ratios (see text). The K datum for LEW 86010 likely represents contamination acquired while in Antarctica. Angrite data are means of data from Laul et al. (1972), Lugmair and Galer (1992), Ma et al. (1977), Mittlefehldt and Lindstrom (1990), Nyquist et al. (1994), Schnetzler and Philpotts (1969), Tera et al. (1970), Warren et al. (1995), Wasserburg et al. (1977), Yanai (1994). The vertical line are ranges of CI-normalized volatile/U ratios for 7 basaltic eucrite falls from Paul and Lipschutz (1990).

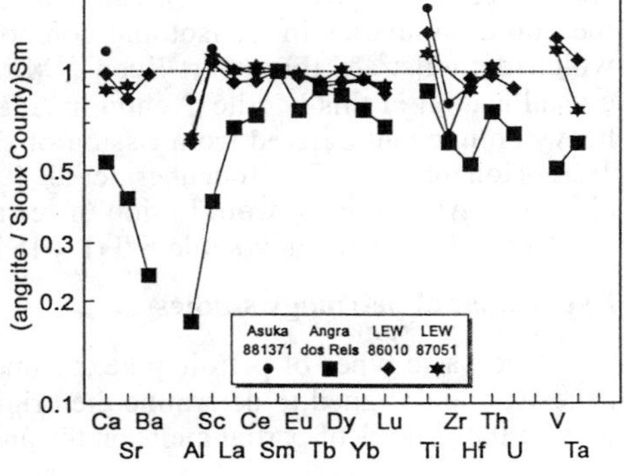

Figure 51. Sioux County- and Sm-normalized refractory lithophile elements in angrites plotted by increasing nominal valence and Z. Sioux County is used for normalization to emphasize the difference in angrite refractory element patterns from that of the most well studied asteroidal basalts, the eucrites. Note the strong depletions in Al and slight enrichment in Sc in antactic angrites, and the generally high Ti abundance relative to other tetravalent cations. Angra dos Reis has a very distinctive refractory lithophile element pattern compared to the other angrites. Data for this figure are from some of the sources listed in Figure 50, plus Morgan and Lovering (1964) and Tatsumoto et al. (1973).

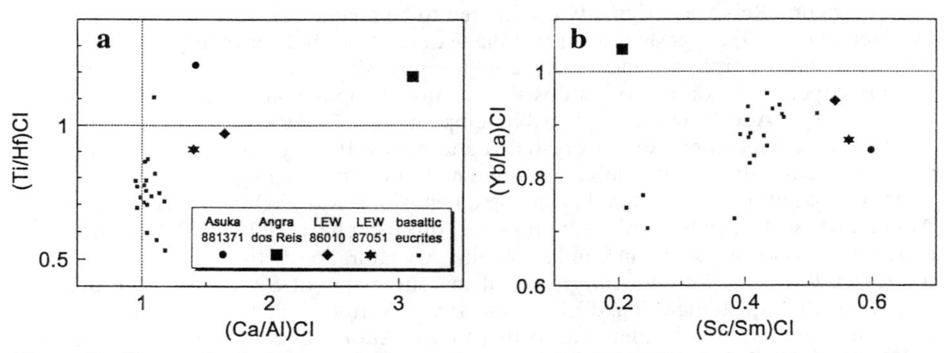

Figure 52. CI-normalized element ratio for angrites compared to basaltic eucrites. (a) Angrites have systematically higher Ca/Al and Ti/Hf than basaltic eucrites. The sole eucrite with $(Ti/Hf)_{CI} > 1$ is Ibitira, for which only single analyses of Ti and Hf are available. (b) The antarctic angrites have systematically higher Sc/Sm ratios than eucrites. Angra dos Reis is distinct from the other angrites in its refractory lithophile element ratios. Angrite data are from the sources listed in Figure 50.

lower Sc/Sm ratio, but this is because of its higher REE content. The Sc content of Angra dos Reis is substantially greater than that of basaltic eucrites (compare Table 40, Table 35). The pyroxene of Angra dos Reis contains ~200 ppb U (~25 times CI), roughly the same concentration as found in chondritic phosphates (Störzer and Pellas 1977).

Ages

Chronologic studies show that the angrites are virtually as old as the solar system. Lugmair and Marti (1977) determined a Sm-Nd age for Angra dos Reis of 4.55 Ga, but this was essentially a two-point "isochron" defined by pyroxene and whitlockite separates. Wasserburg et al. (1977) reported concordant U-Pb, Th-Pb and Pb-Pb ages for Angra dos Reis of 4.54 Ga. Jacobsen and Wasserburg (1984) determined a Sm-Nd age of 4.56 Ga for Angra dos Reis, but again, this is virtually a two-point isochron. Lugmair and Galer (1992) have determined very precise concordant Pb-Pb ages for pyroxene separates for Angra dos Reis and LEW 86010 of 4.5578 Ga. Lugmair and Galer (1992) and Nyquist et al. (1994) have determined Sm-Nd isochron ages of 4.55 and 4.53 Ga, respectively, for LEW 86010.

The angrites contain isotope anomalies caused by the in situ decay of short-lived nuclides, further indicating old ages. Jacobsen and Wasserburg (1984), Lugmair and Marti (1977), Lugmair and Galer (1992) and Nyquist et al. (1994) have shown that Angra dos Reis and LEW 86010 have anomalies in $^{142}Nd/^{144}Nd$ ratios consistent with decay of short-lived ^{146}Sm ($t_{1/2}$ 103 Ma). Lugmair et al. (1992) and Nyquist et al. (1994) have measured anomalies in Cr isotopic composition of mineral separates for LEW 86010 which demonstrate that short-lived ^{53}Mn ($t_{1/2}$ 3.7 Ma) was present at the time of crystallization of this angrite. Xenon in Angra dos Reis and LEW 86010 is enriched in a heavy component derived from fission of short-lived ^{244}Pu ($t_{1/2}$ 81.8 Ma) (Eugster et al 1991, Hohenberg 1970, Hohenberg et al. 1991, Lugmair and Marti 1977, Wasserburg et al. 1977). Although Xe from fission of refractory ^{244}Pu is present, no ^{129}Xe excess from the decay of short-lived, volatile ^{129}I ($t_{1/2}$ 16 Ma) has been measured.

Experimental petrology studies

Two basic types of petrologic experiments have been performed relevant to angrite genesis; one designed to determine the origin of Angra dos Reis, and the other designed to infer the genesis of partial melts on the angrite parent body.

Angra dos Reis was originally considered to be a cumulate (e.g. Prinz et al. 1977), but Treiman (1989) suggested rather that the texture of this meteorite indicated that it is a metamorphosed porphyry, and nearly a melt composition. Treiman (1989) performed melting experiments on an Angra dos Reis analog composition, AdoR1. He found that the liquidus of AdoR1 was at a plausible temperature, 1246°C, for a melt composition, and that the composition AdoR1 crystallized to nearly 100% pyroxene. Treiman (1989) could not demonstrate that AdoR1 was saturated in olivine, as might be expected for an asteroidal basalt. He argued that his analog composition was slightly richer in silica than Angra dos Reis which would suppress the olivine phase field, and that based on experiments done by others in similar systems, co-saturation with olivine was plausible for Angra dos Reis. Subsequently, several measurements of SiO_2 have been done on Angra dos Reis (Mittlefehldt and Lindstrom 1990, Warren et al. 1995, Yanai 1994), and the results are essentially identical to that of the AdoR1 analog. Lofgren and Lanier (1991) performed dynamic crystallization experiments on two Angra dos Reis analog compositions to see if the texture of this meteorite could be duplicated without recourse to mineral accumulation. Lofgren and Lanier (1991) found that under the appropriate crystallization conditions, textures like that of Angra dos Reis could be duplicated. However, they also found that olivine did not crystallize in amounts as high as that observed in Angra dos Reis.

McKay et al. (1988b) performed melting experiments on analog compositions of LEW 86010 in order to determine whether this rock could represent a melt composition.

They found that LEW 86010 contains slight excesses of olivine and plagioclase relative to a melt saturated in these phases plus calcic-pyroxene. The compositions of pyroxene produced in the experiments closely matched those measured in LEW 86010, and McKay et al. (1988b) therefore concluded that LEW 86010 crystallized from a melt close to the bulk rock composition. However, they suggested that this could mean either that LEW 86010 represents a melt composition, or that it was formed by accumulation of minerals in essentially cotectic proportions. Jurewicz et al. (1993) performed melting experiments on CM and CV chondrites at oxygen fugacity conditions relevant to angrite petrogenesis. They found that minimum melts in these chondrites were "angritic" in the sense that they resembled the composition of LEW 86010, although there were differences in detail.

Origin of the angrites

The genesis of the angrites is not well understood. In part this is because there are few members in the grouplet, so magmatic trends are not well represented. More important, though, are the differences between various members of the grouplet which makes it difficult to determine the relationships among them.

The petrogenesis of LEW 86010 seems relatively well understood. The bulk composition of this rock is similar to that produced by partial melting of CM or (especially) CV chondrites at oxygen fugacities of about 1 log unit above the iron-wüstite buffer (Jurewicz et al. 1993). The oxygen fugacity used for these experiments was that estimated by McKay et al. (1994) based on their experiments on the effects of f_{O_2} on Eu partitioning. Similarly, the results of melting experiments on synthetic LEW 86010 compositions show that this rock could represent a melt composition (McKay et al. 1988b, 1995). The refractory incompatible element pattern of LEW 86010 is relatively unfractionated (Fig. 51), and plausibly that of a partial melt extracted from an olivine-dominated residue. Finally, variations in trace element contents with major element zoning in minerals in LEW 86010 are consistent with closed crystallization of a melt with a composition like that of the bulk rock (Crozaz and McKay 1990). Minor fractionation of the Ca/Al ratio from chondritic may be due to minor hercynitic spinel in the residue, as suggested by Mittlefehldt and Lindstrom (1990) and demonstrated experimentally by Jurewicz et al. (1993).

The genesis of Asuka 881371 and LEW 87051 is less certain. Melting experiments on CM and CV chondrites suggest that LEW 87051 could be a high temperature partial melt (Jurewicz et al. 1993). However, olivine-melt equilibrium experiments and the compositions of the large olivines in A-881371 and LEW 87051 show that these olivines are xenocrysts (e.g. McKay et al. 1995, Mikouchi et al. 1996), as first suggested by Prinz et al. (1990). Thus, these meteorites are hybrid rocks. The measured groundmass compositions for LEW 87051 and A-881371 are similar to LEW 86010 in composition (Prinz and Weisberg 1995, Prinz et al. 1990). Hence, the groundmass fraction of these rocks could be a partial melt of the angrite parent body. The groundmass is still highly olivine normative, and Mikouchi et al. (1996) suggested that the groundmass contains a dissolved olivine xenocryst component. These meteorites may represent contaminated partial melts (e.g. Mittlefehldt and Lindstrom 1990), or possibly, they are impact melts (Jurewicz et al. 1993, Mikouchi et al. 1996).

The origin of Angra dos Reis is unknown. Although the texture and major element composition allow this rock to represent a melt composition (e.g. Treiman 1989), its fractionated trace element pattern is not easily understood as being that of a melt (Mittlefehldt and Lindstrom 1990). Jones (1982) suggested that Angra dos Reis was a mesocumulate, a cumulate with a significant fraction of trapped intercumulus melt. Mittlefehldt and Lindstrom (1990) calculated that a mixture of about 60% cumulus pyroxene and 40% melt could reproduce the REE pattern of Angra dos Reis. However, the melt would be very REE-rich, and they found no way to relate this melt to plausible partial melts of the angrite parent body, for example LEW 86010. The angrite parent body is extremely depleted in the moderately volatile elements Na, K, Rb and Ga compared to even the volatile-depleted eucrites (e.g. Mittlefehldt and Lindstrom 1990, Tera et al. 1970, Warren et al. 1995). Warren et al. (1995) have suggested that this does not extend to the most volatile elements, such as Cd and In. However, the estimated μ ($^{238}U/^{204}Pb$) for the angrite parent body does appear to be roughly a factor of 5 greater than that of the eucrite parent body (e.g. Lugmair and Galer 1992, Tera et al. 1997), suggesting a lower content of volatile Pb. It remains enigmatic how volatile depleted parent bodies were formed in the solar nebula. See discussion in Mittlefehldt (1987) and Humayun and Clayton (1995).

Angrites and eucrites represent the products of basaltic magmatism on asteroids, and it is instructive to compare them. Angrites are critically undersaturated in silica while eucrites are hypersthene normative, and thus mimic the dichotomy of alkaline and tholeiitic basalts on Earth (Mittlefehldt and Lindstrom 1990). Melting experiments on chondrites show that this difference can simply be due to differences in oxygen fugacity; melting under high f_{O_2} conditions yields melts critically undersaturated in silica, while melting under low f_{O_2} conditions yields melts that are hypersthene normative (Jurewicz et al. 1993, 1995b). The higher Ca/Al ratio for angrites compared to eucrites (Fig. 52a) is due to stabilization of hercynitic spinel in the residual source at high f_{O_2} causing a slight depletion in Al in the melt (Jurewicz et al. 1993, Mittlefehldt and Lindstrom 1990). Excluding Angra dos Reis, angrites also have slightly higher Sc/Sm ratios than eucrites (Fig. 52b) indicating that the residual source region for eucrites did contain some pyroxene. Based on melting experiments, the residual source region of LEW 86010 did not contain pyroxene (e.g. McKay et al. 1995).

MESOSIDERITES

Mesosiderites form one of the two major types of stony irons, and they are more enigmatic than the other group, the pallasites. Whereas pallasites are composed of materials logically expected for the core-mantle boundary region of a differentiated asteroid, the mesosiderites are composed of metal mixed in with mostly basaltic, gabbroic and pyroxenitic lithologies, and only minor olivine. Hence, mesosiderites seem to be mixtures of core material and crustal material, with little of the intervening mantle being present. This curious mix has fueled many imaginative scenarios for the formation of mesosiderites (see review by Hewins 1983) since the time of Prior (1918), who suggested that mesosiderites were formed by the invasion of a eucrite magma by a pallasite magma. Table 41 (below) gives a petrologic synopsis of all well-classified mesosiderites.

Bulk textures and classification

The silicate phase of mesosiderites is a brecciated mixture of igneous lithologies. The overall silicate texture is of mineral and lithic clasts in a fine-grained fragmental to igneous matrix (Floran 1978, Powell 1971). Large lithic clasts include basalts, gabbros, anorthosites, orthopyroxenites, and dunites. The basalts, gabbros and orthopyroxenites dominate, while dunites are subordinate and anorthosites are rare. Hence, the lithic clasts are dominantly from the crust of a differentiated asteroid, and are broadly similar to eucrites and diogenites. Mineral clasts consist mostly of coarse-grained orthopyroxene fragments up to at least 10 cm in size. Coarse-grained olivine clasts, up to at least 10 cm in size (McCall 1966), are less abundant, whereas coarse-grained plagioclase fragments a few mm in size are the least common. All-in-all, the silicate assemblage in mesosiderites is very similar to that in the howardites; with the exception of the lack of coarse-grained olivine in the latter (Prior 1918).

Mesosiderite petrologic classification has been organized similarly to that of chondrites, where compositional and textural features are used to form a grid of types (Van Schmus and Wood 1967). The mesosiderites are divided into three petrologic groups based on the orthopyroxene content (Hewins 1984, 1988; see Fig. 53). Compositional class A mesosiderite silicates are relatively basaltic in composition, containing more plagioclase and clinopyroxene. The compositional class B mesosiderite silicates contain a greater proportion of orthopyroxene, and are therefore more ultramafic. A unique mesosiderite from Antarctica, RKPA79015, contains almost exclusively orthopyroxene as its silicate phase (Prinz et al. 1982a), and is the sole member of compositional class C (Hewins 1988). There is no apparent relationship between the compositional classes of mesosiderites and the amount of metal and/or troilite they contain. The metal content varies from about 17-90 wt % and troilite from <1-14 wt % (Table 41). Many new mesosiderites have not yet been formally classified, so additional petrologic-compositional types may exist. Chaunskij, for example, is unusual in containing cordierite, a mineral not previously reported in mesosiderites, more metal and a higher level of shock damage in the metal than typical (Petaev et al. 1993), and a very different composition of metal (Wasson et al. 1998a).

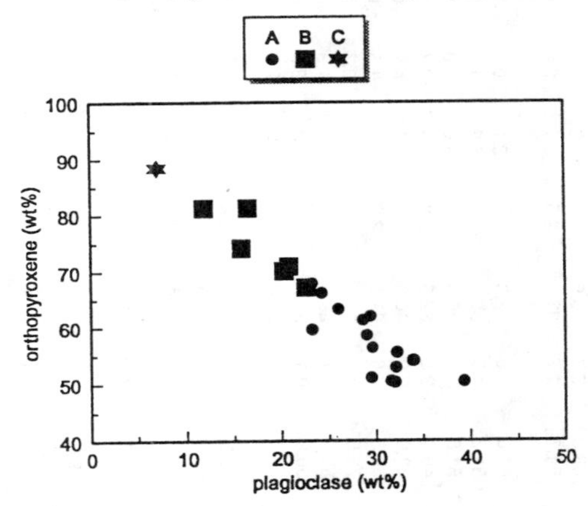

Figure 53. Orthopyroxene wt % vs. plagioclase wt % for mesosiderites showing the generally greater basaltic component, indicated by plagioclase, in the compositional class A mesosiderites compared to the more ultramafic-rich B mesosiderites. Data are from the summary in Table 41.

In addition to compositional distinctions based on the ratio of basaltic-gabbroic to orthopyroxenitic material in the breccias, mesosiderites have been further subdivided based on silicate textures. This is analogous to separation of chondrites into petrologic types 1 through 6. Like the chondrite sequence, the mesosiderite textural classification was originally thought to reflect increasing metamorphic equilibration of the silicates (Powell 1971). The lowest metamorphic grade (1) is characterized by a fine-grained fragmental matrix. Successively higher grades (2 and 3) are characterized by recrystallized matrixes, while the highest grade (4) mesosiderites are melt-matrix breccias (Floran et al. 1978b). Both compositional classes A and B mesosiderites contain members of textural grades 1 through 4 and the sole compositional class C mesosiderite is of textural grade 2.

This textural classification is somewhat too simplistic. Textures are variable in some mesosiderites; different samples of Estherville, for example, show textures indicating grades 3 and 4 (Hewins 1984). Also, Floran (1978) noted that even some textural grade 1 mesosiderites contain some igneous textured matrix material. One problem with the

Table 41. Petrologic characteristics of mesosiderites

Name	class	metal wt%	troil wt%	silicate wt%	norm/mode silicates to 100% opx	cpx	plag	oliv	trid	mer	cm	ilm	notes
ALHA77219	1B				76.0	1.4	12.9	4.1	2.9	1.3	1.4		mode
....					72.4	4.8	18.7		1.6	0.59	1.5	0.50	norm
....					70.0	6.0	18.8		3.4	1.8			matrix mode
....ALHA81059	1B				67.4	3.8	23.1		4.2	1.2	0.4		matrix mode, paired with ALHA77219
....ALHA81098	1B				72.8	2.2	22.1		0.3	0.2	2.3		matrix mode, paired with ALHA77219
ALHA81208	2B?				97.5					0.3	2.2		probably clast from one of above
Barea (f)	1A	56.6	1.6	41.8	63.6		27.3		6.2	2.9	1.2	0.7	norm
....					53.5	3.2	30.6	2.2	6.7	2.3	1.5		mode
....					53.9	2.9	34.4		7.4	1.4			matrix mode
Bondoc	3B(+)	43.7	0.68	55.62	81.3	0.5	16.5	0.5	0.2	0.2	0.8		mode, no corona
Budulan	3B(+)	70			70.2	2.6	20.3		4.1	1.8	1.0		mode
Chaunskij	?	90			46		34.2		5.9	4.7	1.2	0.9	mode, opx = total pyx, cordierite = 7.1%
....					46.8		34.7		7.6	5.2	0.8	0.8	norm, opx = total pyx, cordierite = 4.1%
Chinguetti	1B	80			81.3		11.8	2.1	2.1	1.7	1.0		mode
....					89.0		6.1		1.1	3.8			matrix mode
Clover Springs	2A	37			50.3	5.5	32.0	0.8	8.6	2.1	0.7		mode
....					47.9	1.1	37.5	0.3	10.6	2.3	0.3		matrix mode
Crab Orchard	1A	54.6	0.76	44.64	50.4	2.8	39.3	1.1	5.0	0.8	0.6		mode
....					49.3	2.7	33.5		11.3	2.9	0.3		matrix mode
Dalgaranga	A				54.1	1.7	33.9	2.5	4.1	3.4	0.3		mode
Donnybrook	3B(+)												
Dyarrl Island (f)	1A	17.5	2.3	80.2	61.6	9.0	22.8		2.4	0.7	1.0	2.1	norm
....					51.3	4.5	36.5		5.9	1.5	0.3		mode
Emery	3A	50.2	7.3	42.5	54.8		30.8		9.6	4.0	1.2	0.9	norm
....					46.2	2.5	32.5	3.8	9.8	3.7	1.5		mode
....					38.0	2.0	40.6		13.7	4.9	0.7	0.2	matrix mode
Estherville (f)	3/4A	56.4	3.40	40.2	59.3	3.6	27.1	2.2	5.5	1.5	0.8		mode
....					76.5		19.4		1.7	0.7	1.2	0.5	norm
....					53.2	3.7	29.5		11.9	1.3	0.4		matrix mode
Hainholz	4A	53.4	4.30	42.3	59.2	0.1	29.7	1.5	5.3	3.5	0.7		mode, no corona
....		53.3	2.00	44.7	63.4	1.7	27.4		3.9	2.0	1.4	0.19	norm
Harvard University	4A												
Lamont	3B				63.8		25.3	3.6	1.8	4.0	1.5	tr	matrix mode
Lowicz (f)	3A	59.6	0.61	39.79	63.3	4.0	26.0	0.7	5.6		0.4		mode
....					46.9	7.4	31.8	0.7	10.5	1.7	1.0		matrix mode
Mincy	3B(+)				73.6	2.1	17.8	0.7	2.6	1.4	1.8		mode
....					60.9	2.1	27.5		7.1	1.3	1.0		mode, coarse-grained
....					74.9	3.3	18.1		3.4	0.2	0.1		mode, fine-grained
....					79.5	3.1	15.6		1.4	0.3			matrix mode
Morristown	3A	49.8	0.66	49.54	51.1	4.2	29.5	6.0	6.5	1.7	1.0		mode
....					54.7	6.8	31.4		5.8	0.7	0.5		matrix mode

Table 41 (cont'd). Petrologic characteristics of mesosiderites

Name	class	metal	troil	silicate	norm/mode silicates to 100%								notes
		wt%	wt%	wt%	opx	cpx	plag	oliv	trid	mer	cm	ilm	
Mount Padbury	1A	49.8	7.6	42.6	59.6	2.6	23.2	1.9	8.5	3.3	0.9		mode
		57			44.7	4.3	35.0		13.0	2.9	0.2		matrix mode
Patwar (f)	1A	33.2	9.07	57.73	52.8	3.3	30.7	1.7	9.0	2.2	0.3		mode
		38.05	11.89	50.05	58.2	3.2	33.6		3.3	0.33	0.77	0.61	norm
					58.3	0.7	30.4		5.0	5.0	0.7		matrix mode
Pinnaroo	4A	64.08	7.76	28.16	50.9	1.4	35.4	1.8	6.5	3.6	0.2		mode, no corona
		54.6	13.9	31.5	57.3	3.5	32.5		3.1	2.0	0.69	0.93	norm
RKPA79015	2C(+)												
....RKPA80229	2C(+)												paired with RKPA79015
....RKPA80246	2C(+)												paired with RKPA79015
....RKPA80258	2C(+)				89.3	2.4	5.1	1.8		1.5			mode, paired with RKPA79015
....					87.7	2.8	9.0		0.2				matrix mode
....RKPA80263	2C(+)												paired with RKPA79015
Simondium	4A	42.7	8.8	48.5	63.8	1.2	27.7		3.7	1.2	1.8	0.48	norm
....					60.3	0.3	31.2	0.4	4.4	2.1	1.3		mode, no corona
Vaca Muerta	1A	47.3	12.63	40.07	52.9	2.0	32.1	1.4	8.0	3.0	0.6		mode
....					52.3	0.3	30.0		12.8	4.3	0.2		matrix mode
Veramin (f)	2B(+)	48.6	0.60	50.8	71.0	tr	20.8	5.7		1.0	1.5		mode, no corona
....					75.0	0.3	18.0	2.0	2.5	2.3			matrix mode
West Point	3A				66.2	3.5	24.2		4.3	1.8	tr		mode, Morristown?

Poorly characterized mesosiderites (paired meteorites listed in parentheses after the parent):

A-882023, Acfer 063, Acfer 265, EET 87500 (EET 87501, EET 92001), Eltanin (recovered in deep sea cores, USNS Eltanin), Ilafegh 002, LEW 86210, LEW 87006, MAC 88102, Murchison Downs, Pennyweight Point, QUE 86900 (QUE 93001, QUE 93002, QUE 93126, QUE 93150, QUE 93517, QUE 93575, QUE 93584, QUE 93586, QUE 94614, QUE 94639, QUE 94299), Um-Hadid, Weiyuan

References:

Classification: Floran (1978), Hewins (1984, 1988). Classifications indicated with (+) are considered petrologic grade 4 by Hewins (1984, 1988).

Wt % metal-trolite-silicate: Alderman (1940) - Pinnaroo; Floran (1978) - Hainholz, Pinnaroo, Simondium; Jarosewich and Mason (1969) - Patwar; Mason and Jarosewich (1973) - Barea, Budulan, Chinguetti, Clover Springs, Dyarrl Island, Emery, Mount Padbury; McCall (1966) - Mount Padbury; Powell (1971) - Bondoc, Crab Crchard, Estherville, Hainholz, Lowicz, Morristown, Patwar, Vaca Muerta, Veramin.

Modes: Hewins (1988) - ALHA81208, RKPA80258; Hewins and Harriott (1986) - Mincy; Petaev et al. (1993) - Chaunskij; Prinz et al. (1982) - RKPA79015, Prinz et al. (1980) - rest.

Matrix modes: Boesenberg et al. (1997) - Lamont; Hewins (1988) - ALHA77219, -81059, -81098; RKPA80258; Delaney et al. (1981) - rest.

Norms: Alderman (1940) - Pinnaroo; Floran (1978) - Hainholz, Pinnaroo, Simondium; Jarosewich and Mason (1969) - Patwar; Mason and Jarosewich (1973) - Barea, Budulan, Chinguetti, Clover Springs, Dyarrl Island, Emery, Mount Padbury; Nelen - Mason (1972) - Estherville; Marvin and Mason (1980) - ALHA77219; Petaev et al. (1993) - Chaunskij.

Figure 54. Polished and etched slab of Pinnaroo, a type 4A mesosiderite. Note the general coarse segregation of metal and silicate, the large silicate clast (upper center) and large metal clast, and the highly embayed borders between metal and silicate. This slab is 13 cm across. (Photograph courtesy of the Smithsonian Institution).

textural classification of mesosiderites is that it is frequently difficult to decide what is and what is not true matrix. Because of the pervasive, intervening metal plus troilite, it can be difficult to distinguish between true matrix material and small, fine-grained breccia or impact melt clasts. Finally, Hewins (1984, 1988) has shown that the type 2B, 2C and 3B mesosiderites have some igneous textured matrix, and so could be reclassified as petrologic grade 4. This is shown in Table 41 where their classification is given as, e.g. 3B(+), indicating possible classification as melt-matrix mesosiderites.

No systematic study of the metal-silicate textures at the macro-scale, similar to that done on pallasites, has been done for mesosiderites. There may be a correlation between silicate texture type and macro-scale metal-silicate textures. Both Pinnaroo (Fig. 54) and some parts of Estherville are classified as melt-matrix breccias (type 4A) and exhibit cm-scale separation of metal and silicate. The Patwar, Mount Padbury (Fig. 55) and Vaca Muerta fragmental matrix breccias (type 1A) exhibit fine-scale homogeneous distribution of metal and silicate. Some mesosiderites have anomalous metal-silicate textures, RKPA79015 (Fig. 56) being one example.

Mesosiderite matrix

Matrix textures. Mesosiderites exhibit essentially a continuum in particle sizes from large lithic and mineral clasts down to fine matrix grains, making the distinction between

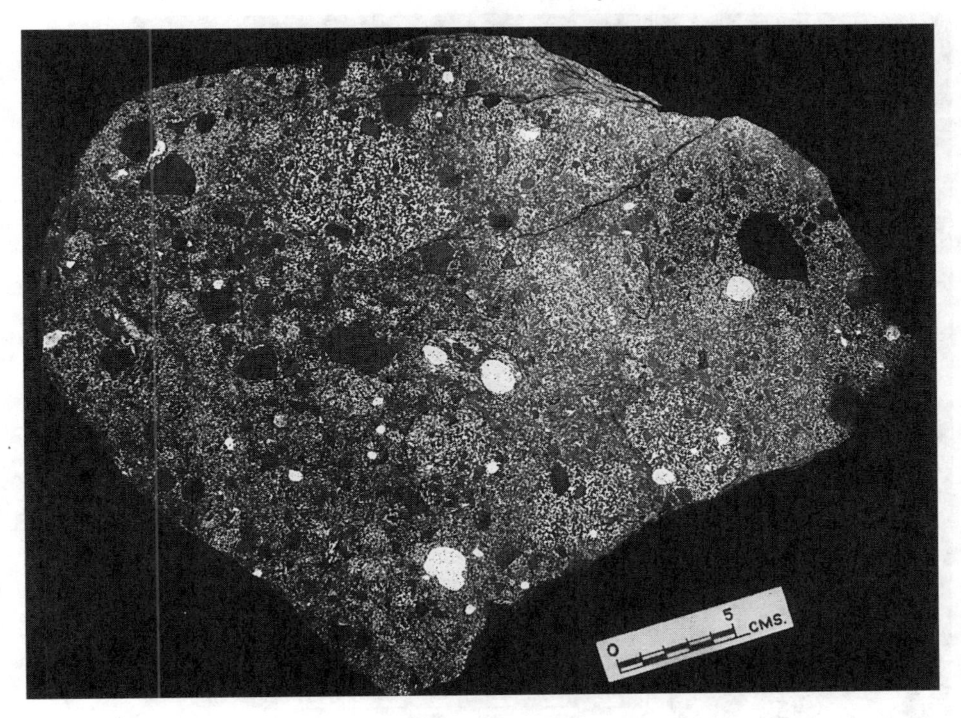

Figure 55. Polished slab of Mount Padbury, a type 1A mesosiderite. Numerous cm-sized silicate and metal clasts are dispersed in a finely divided metal-silicate matrix. Note metal-rich areas, some of which are metal-rich breccia clasts. (Photograph courtesy of the Smithsonian Institution).

matrix and clast arbitrary at some level. Because silicate textural domains are commonly bordered by troilite and/or metal, it is very difficult to distinguish matrix from fine-grained clasts in mesosiderites. Table 41 includes matrix modes for many of the mesosiderites, including members of all types except 4A and 4B. (A matrix mode is given for Estherville, but Delaney et al. (1981) did not indicate the textural type, 3A or 4A, corresponding to the Estherville thin section used for modal determination.)

According to Powell (1971) mesosiderites of textural grade 1 are characterized by clearly distinguishable cataclastic texture, with highly angular mineral and lithic fragments of all sizes. The fine-grained matrix contains much material of size <10 μm. Silicate-silicate grain boundaries in the matrix are not intergrown. These mesosiderites are texturally heterogeneous even on the microscopic scale.

Powell (1971) defined mesosiderites of textural grade 2 as still exhibiting a clearly cataclastic texture, but that the angularity is mostly confined to the larger lithic fragments. The matrix material is coarser-grained, and most grains are >10 μm in size. Contacts between small silicate grains may be sutured, and small grains may be intergrown with larger silicate fragments. Textural grade 2 mesosiderites are still texturally heterogeneous on a microscopic scale, but this is less pronounced than in grade 1.

Powell (1971) originally defined textural grade 3 as his highest metamorphic grade. Subsequent workers have shown that mesosiderites of textural class 3 display characteristic textures that are related to compositional type. Textural grade 3

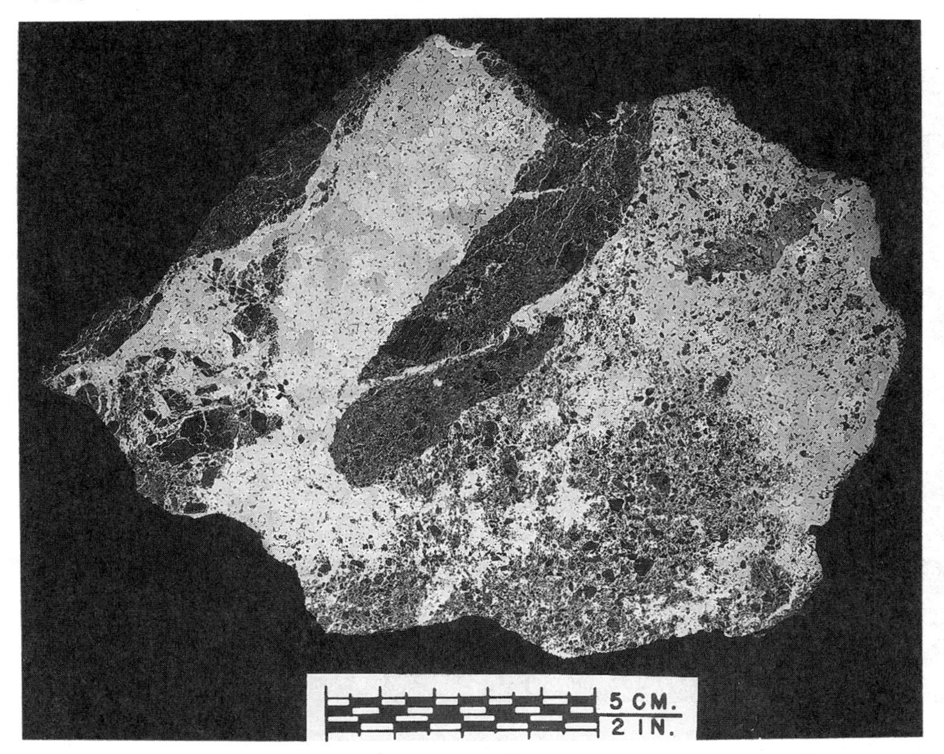

Figure 56. The type 2C mesosiderite RKPA79015 has an unusual texture, with silicate-free metal regions several cm in size. (Photograph courtesy of the Smithsonian Institution).

mesosiderites were originally split into the pyroxene poikiloblastic and plagioclase poikiloblastic textures for compositional class A and B, respectively (Floran 1978). Hewins (1984) reinterpreted the type 3B mesosiderite matrix to have an igneous texture, and therefore the texture of these mesosiderites will be referred to as plagioclase poikilitic. The plagioclase poikilitic matrix is characterized by plagioclase grains several mm in size enclosing numerous rounded to euhedral orthopyroxene grains on the order of 100 μm in size. As described by Floran (1978), the pyroxene poikiloblastic texture is characterized by large, inclusion-poor, low-Ca pyroxene grains, up to cm-size, with optically continuous overgrowths enclosing anhedral grains of plagioclase, tridymite, silica, troilite and metal. There are also optically continuous low-Ca pyroxene grains in the matrix enclosing a similar suite of minerals.

Mesosiderites of textural grade 4 exhibit igneous-textured matrixes. As originally defined by Floran et al. (1978b), these mesosiderites have fine- to medium-grained, intergranular textures composed of orthopyroxene granules between a network of subhedral, lath-shaped plagioclase grains. Minor to accessory minerals found between orthopyroxene and plagioclase grains include tridymite, augite, troilite, chromite, whitlockite and metal.

A unique texture found in mesosiderites is the corona texture developed around olivine grains (Delaney et al. 1981, Nehru et al. 1980, Powell 1971, Ruzicka et al. 1994).

Table 42. Compositions of pyroxene grains from representative mesosiderites.

	Mount Padbury		Chinguetti	Emery			Donnybrook					Mincy
	(1)	(2)	(3)	(4)	(5)	(6)	(7)	(8)	(9)	(10)	(11)	(12)
					Chemical Composition (wt %)							
SiO_2	53.1	52.6	54.7	52.9	53.2	52.6	54.8	54.7	54.6	54.0	54.0	53.9
Al_2O_3	0.58	0.97	1.33	0.68	0.66	1.14	0.44	0.62	0.92	1.35	1.20	1.36
TiO_2	0.40	0.64	0.09	0.50	0.40	0.66	0.08	0.10	0.16	0.21	0.28	0.22
Cr_2O_3	0.29	0.46	0.71	0.33	0.29	0.60	0.13	0.16	0.28	0.46	0.39	0.80
FeO	20.7	9.27	15.5	19.9	19.2	10.4	15.8	15.2	14.8	15.5	13.2	13.0
MnO	0.96	0.48	0.58	1.01	0.89	0.53	0.70	0.76	0.73	0.70	0.60	0.54
MgO	23.2	15.3	26.5	23.1	23.9	15.0	27.6	28.0	27.5	27.4	26.1	27.9
CaO	1.25	20.2	1.09	1.80	1.55	19.8	0.14	0.41	0.96	0.18	4.14	1.52
Total	100.48	99.92	100.50	100.22	100.09	100.73	99.69	99.95	99.95	99.80	99.91	99.24
					Cation Formula Based on 6 Oxygens							
Si	1.9641	1.9599	1.9671	1.9593	1.9631	1.9525	1.9838	1.9728	1.9687	1.9538	1.9531	1.9479
^{iv}Al	0.0253	0.0401	0.0329	0.0297	0.0287	0.0475	0.0162	0.0264	0.0313	0.0462	0.0469	0.0521
Total tet*	1.9894	2.0000	2.0000	1.9890	1.9918	2.0000	2.0000	1.9992	2.0000	2.0000	2.0000	2.0000
Ti	0.0111	0.0179	0.0024	0.0139	0.0111	0.0184	0.0022	0.0027	0.0043	0.0057	0.0076	0.0060
^{vi}Al	0.0000	0.0025	0.0235	0.0000	0.0000	0.0024	0.0026	0.0000	0.0078	0.0114	0.0043	0.0058
Cr	0.0085	0.0136	0.0202	0.0097	0.0085	0.0176	0.0037	0.0046	0.0080	0.0132	0.0112	0.0229
Fe	0.6403	0.2889	0.4662	0.6164	0.5925	0.3229	0.4784	0.4585	0.4463	0.4690	0.3993	0.3929
Mn	0.0301	0.0151	0.0177	0.0317	0.0278	0.0167	0.0215	0.0232	0.0223	0.0215	0.0184	0.0165
Mg	1.2789	0.8496	1.4203	1.2751	1.3143	0.8298	1.4890	1.5050	1.4777	1.4775	1.4069	1.5027
Ca	0.0495	0.8065	0.0420	0.0714	0.0613	0.7875	0.0054	0.0158	0.0371	0.0070	0.1604	0.0589
Total Cations	4.0078	3.9941	3.9923	4.0072	4.0073	3.9953	4.0028	4.0090	4.0035	4.0053	4.0081	4.0057
					*Cation Ratios Ca:Mg:Fe, Fe/Mn and mg# (100*Mg/(Mg+Fe))*							
Ca	2.5	41.5	2.2	3.6	3.1	40.6	0.3	0.8	1.9	0.4	8.2	3.0
Mg	65.0	43.7	73.6	65.0	66.8	42.8	75.5	76.0	75.4	75.6	71.5	76.9
Fe	32.5	14.9	24.2	31.4	30.1	16.6	24.2	23.2	22.8	24.0	20.3	20.1
Fe/Mn	21	19	26	19	21	19	22	20	20	22	22	24
mg#	66.6	74.6	75.3	67.4	68.9	72.0	75.7	76.6	76.8	75.9	77.9	79.3

(1, 2) impact melt matrix pyroxenes; (3) fragmental matrix pyroxene; (4, 6, 7, 8) matrix pyroxenes; (5) corona structure pyroxene; (9, 10, 11, 12) chadacrysts enclosed in poikilitic plagioclase; all data from Mittlefehldt, unpublished.

Many olivine grains in mesosiderites are surrounded by a zoned sequence containing, in order of abundance, orthopyroxene, plagioclase, whitlockite, chromite, clinopyroxene (type not identified) and ilmenite. Against the olivine, the corona mineralogy is dominantly orthopyroxene and chromite, with or without whitlockite. A middle zone still has orthopyroxene as the major mineral, but plagioclase makes up about 1/3 of the zone, whitlockite is variable in abundance from a few percent up to about 20%, and minor amounts of chromite are present. The outer zone is similar to the matrix in mineralogy, except that this zone is devoid of tridymite and metal.

Matrix mineral compositions. There are relatively few published analyses of matrix minerals in mesosiderites, and for many analyses in the literature, it is often difficult to determine what textural type of material has been analyzed. We will therefore rely heavily on unpublished data by one of us (DWM). We will not discuss the composition small olivine grains here, as the development of corona textures and the abundance of tridymite in the matrix indicates that these are basically small mineral clasts which could not have substantially reacted with the matrix. There are few, if any, analyses of matrix plagioclase available in the literature. Hence, we will focus on pyroxene in the matrix. We include in this category overgrowths on larger silicate grains.

Representative analyses of matrix pyroxene grains are given in Table 42, and Figure 57 shows the molar Fe/Mg vs. molar Fe/Mn relationships for mesosiderite pyroxenes. For the type 1A mesosiderite Mount Padbury, low-Ca pyroxene and calcic clinopyroxenes from impact melt matrix have low molar Fe/Mn ratios, 21 and 19, respectively, and are

Table 42 (cont'd). Compositions of pyroxene grains from representative mesosiderites.

	Mincy	Budulan	Pinnaroo					Chinguetti		Mincy		
	(13)	(14)	(15)	(16)	(17)	(18)	(19)	(20)	(21)	(22)	(23)	(24)
Chemical Composition (Weight Percent)												
SiO_2	53.5	53.1	53.1	52.5	51.8	53.9	52.4	55.0	54.2	53.7	53.5	53.2
Al_2O_3	0.95	0.77	0.63	0.57	0.95	0.81	0.92	0.60	0.80	0.73	1.53	0.99
TiO_2	0.34	0.24	0.27	0.20	0.35	0.13	0.29	0.09	0.12	0.13	0.36	0.39
Cr_2O_3	0.52	0.56	0.49	0.42	0.53	0.62	0.54	0.43	0.26	0.49	0.45	0.50
FeO	16.9	18.5	21.6	22.4	22.0	17.0	20.7	15.3	14.9	17.0	17.1	17.2
MnO	0.78	0.76	0.79	0.89	0.91	0.57	0.84	0.52	0.57	0.64	0.79	0.84
MgO	26.0	24.2	21.8	20.5	19.5	25.5	19.9	27.5	27.4	25.3	25.4	25.4
CaO	1.32	1.60	1.68	2.84	3.90	1.46	4.72	1.14	1.08	1.39	0.76	1.05
Total	100.31	99.73	100.36	100.32	99.94	99.99	100.31	100.58	99.33	99.38	99.89	99.57
Cation Formula Based on 6 Oxygens												
Si	1.9482	1.9608	1.9749	1.9701	1.9566	1.9668	1.9620	1.9743	1.9681	1.9716	1.9523	1.9534
^{iv}Al	0.0408	0.0335	0.0251	0.0252	0.0423	0.0332	0.0380	0.0254	0.0319	0.0284	0.0477	0.0428
Total tet*	1.9890	1.9943	2.0000	1.9953	1.9989	2.0000	2.0000	1.9997	2.0000	2.0000	2.0000	1.9962
Ti	0.0093	0.0067	0.0076	0.0056	0.0099	0.0036	0.0082	0.0024	0.0033	0.0036	0.0099	0.0108
^{vi}Al	0.0000	0.0000	0.0025	0.0000	0.0000	0.0016	0.0026	0.0000	0.0023	0.0032	0.0181	0.0000
Cr	0.0150	0.0164	0.0144	0.0125	0.0158	0.0179	0.0160	0.0122	0.0075	0.0142	0.0130	0.0145
Fe	0.5147	0.5713	0.6719	0.7030	0.6950	0.5188	0.6482	0.4593	0.4525	0.5220	0.5219	0.5282
Mn	0.0241	0.0238	0.0249	0.0283	0.0291	0.0176	0.0266	0.0158	0.0175	0.0199	0.0244	0.0261
Mg	1.4110	1.3318	1.2083	1.1465	1.0977	1.3867	1.1105	1.4712	1.4828	1.3843	1.3814	1.3900
Ca	0.0515	0.0633	0.0670	0.1142	0.1578	0.0571	0.1894	0.0438	0.0420	0.0547	0.0297	0.0413
Total Cations	4.0146	4.0076	3.9966	4.0054	4.0042	4.0033	4.0015	4.0044	4.0079	4.0019	3.9984	4.0071
*Cation Ratios Ca:Mg:Fe, Fe/Mn and mg# (100*Mg/(Mg+Fe))*												
Ca	2.6	3.2	3.4	5.8	8.1	2.9	9.7	2.2	2.1	2.8	1.5	2.1
Mg	71.4	67.7	62.1	58.4	56.3	70.7	57.0	74.5	75.0	70.6	71.5	70.9
Fe	26.0	29.1	34.5	35.8	35.6	26.4	33.3	23.3	22.9	26.6	27.0	27.0
Fe/Mn	21	24	27	25	24	29	24	29	26	26	21	20
mg#	73.3	70.0	64.3	62.0	61.2	72.8	63.1	76.2	76.6	72.6	72.6	72.5

(13, 14) chadacrysts enclosed in poikilitic plagioclase; (15, 16, 17, 18, 19) impact melt matrix pyroxenes; (20, 21) and (22, 23) core-rim pairs for large pyroxenes; (24) rim on large pigeonite; all data from Mittlefehldt, unpublished.

relatively magnesian, with mg# of 66.6 and 74.7, respectively. For the type 1B mesosiderite Chinguetti, a small fragmental low-Ca pyroxene grain from the matrix has a higher molar Fe/Mn of 26 than the Mount Padbury impact-melt matrix grains, and is quite magnesian with a mg# of 75.3.

For the type 3A, pyroxene poikiloblastic mesosiderite Emery, low-Ca pyroxene and calcic clinopyroxene analyses for matrix grains, plus low-Ca pyroxene from the corona around an olivine grain have low molar Fe/Mn, 19-21, and are magnesian, with mg# of 67.4 to 72.0. For type 3B mesosiderites, the plagioclase poikilitic textural grade, low-Ca pyroxene chadacrysts (mineral grains completely enclosed in a single host crystal) from poikilitic plagioclase show a range of compositions. Donnybrook pyroxene chadacrysts with similar mg# (75.9 to 77.9) can show a range of CaO contents from ~0.2 to ~4.1 wt % (Table 42). Mincy and Budulan have generally more ferroan chadacrysts with a narrower range in CaO contents. Note that for each of these mesosiderites, only one thin section was studied, so sample heterogeneity cannot be established. Hewins and Harriott (1986) found a limited range in CaO of chadacrysts in a thin section from a different specimen of Budulan than studied here. Also shown in Table 42 are representative analyses of matrix pyroxene grains from Donnybrook, and analyses of rims on large low-Ca pyroxene cores for Budulan and Mincy.

A range of pyroxene analyses are given for the type 4A mesosiderite Pinnaroo. These grains range from magnesian, low-Ca pyroxenes with molar Fe/Mn of ~27 to 30,

Table 42 (cont'd). Compositions of pyroxene grains from representative mesosiderites.

	Budulan		Patwar				Mount Padbury		Vaca Muerta			
	(25)	(26)	(27)	(28)	(29)	(30)	(31)	(32)	(33)	(34)	(35)	(36)
					Chemical Composition (wt %)							
SiO_2	53.2	53.6	50.9	52.2	51.4	51.8	50.3	52.4	52.5	50.6	51.6	53.2
Al_2O_3	0.92	0.60	1.15	0.99	0.62	0.69	0.66	0.75	0.56	0.30	1.24	0.21
TiO_2	0.26	0.13	0.83	0.48	0.40	0.42	0.44	0.46	0.35	0.23	0.60	0.18
Cr_2O_3	0.52	0.47	0.44	0.36	0.34	0.47	0.24	0.41	0.29	0.18	0.41	0.27
FeO	19.2	18.2	10.5	22.2	14.4	22.3	29.5	9.75	23.6	30.1	8.48	21.2
MnO	0.70	0.74	0.64	0.89	0.70	0.92	0.88	0.47	0.98	1.19	0.47	0.84
MgO	24.1	24.6	14.0	21.7	12.7	21.4	11.0	14.7	21.1	17.2	14.2	22.9
CaO	1.46	1.44	20.6	1.24	19.6	1.72	7.97	21.4	1.22	0.80	21.6	0.82
Total	100.36	99.78	99.06	100.06	100.16	99.72	100.99	100.34	100.60	100.60	98.60	99.62
					Cation Formula Based on 6 Oxygens							
Si	1.9563	1.9722	1.9346	1.9540	1.9585	1.9526	1.9717	1.9565	1.9665	1.9592	1.9523	1.9849
^{iv}Al	0.0399	0.0260	0.0515	0.0437	0.0278	0.0307	0.0283	0.0330	0.0247	0.0137	0.0477	0.0092
Total tet*	1.9962	1.9982	1.9861	1.9977	1.9863	1.9833	2.0000	1.9895	1.9912	1.9729	2.0000	1.9941
Ti	0.0072	0.0036	0.0237	0.0135	0.0115	0.0119	0.0130	0.0129	0.0099	0.0067	0.0171	0.0051
^{vi}Al	0.0000	0.0000	0.0000	0.0000	0.0000	0.0000	0.0022	0.0000	0.0000	0.0000	0.0076	0.0000
Cr	0.0151	0.0137	0.0132	0.0107	0.0102	0.0140	0.0074	0.0121	0.0086	0.0055	0.0123	0.0080
Fe	0.5905	0.5601	0.3338	0.6950	0.4589	0.7030	0.9671	0.3045	0.7393	0.9747	0.2683	0.6615
Mn	0.0218	0.0231	0.0206	0.0282	0.0226	0.0294	0.0292	0.0149	0.0311	0.0390	0.0151	0.0265
Mg	1.3207	1.3490	0.7930	1.2106	0.7212	1.2022	0.6426	0.8180	1.1779	0.9925	0.8007	1.2734
Ca	0.0575	0.0568	0.8389	0.0497	0.8002	0.0695	0.3348	0.8562	0.0490	0.0332	0.8757	0.0328
Total Cations	4.0090	4.0045	4.0093	4.0054	4.0109	4.0133	3.9963	4.0081	4.0070	4.0245	3.9968	4.0014
					*Cation Ratios Ca:Mg:Fe, Fe/Mn and mg# (100*Mg/(Mg+Fe))*							
Ca	2.9	2.9	42.7	2.5	40.4	3.5	17.2	43.3	2.5	1.7	45.0	1.7
Mg	67.1	68.6	40.3	61.9	36.4	60.9	33.0	41.3	59.9	49.6	41.2	64.7
Fe	30.0	28.5	17.0	35.5	23.2	35.6	49.7	15.4	37.6	48.7	13.8	33.6
Fe/Mn	27	24	16	25	20	24	33	20	24	25	18	25
mg#	69.1	70.7	70.4	63.5	61.1	63.1	39.9	72.9	61.4	50.5	74.9	65.8

(25, 26) core-rim pair for large pyroxene; (27, 28) impact melt clast pyroxenes; (29, 30) polygenic basalt clast RV-02 pyroxenes; (31) monogenic basalt clast RV-05 pyroxene; (32, 33, 34) gabbro clast RC-07 pyroxenes; (35, 36) pyroxene inclusions in tridymite, gabbro clast RC-07 ; all data from Mittlefehldt, unpublished except (31) from Mittlefehldt (1978).

to more calcic pyroxenes with molar Fe/Mn of 24 to 25. Some of these grains may not be true matrix grains. The low-Ca pyroxenes with the higher molar Fe/Mn are similar in composition to large low-Ca pyroxene clasts in the igneous matrix, suggesting that the smaller grains are simply smaller clasts.

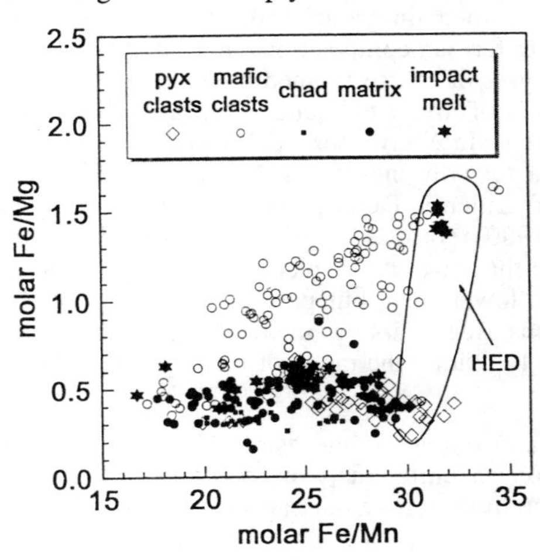

Figure 57. Molar Fe/Mg vs. Fe/Mn for mesosiderite pyroxenes compared to the field for HED meteorites. The HED field shows the effect of igneous processes; both partial melting and fractional crystallization result in relatively constant Fe/Mn but varying Fe/Mg. Mesosiderite mafic clasts show primarily the effects of FeO reduction; both Fe/Mn and Fe/Mg decrease during this process. Most mesosiderite matrix and impact melt pyroxenes, and pyroxene chadacrysts in plagioclase poikilitic mesosiderites form a band of roughly constant Fe/Mg but varying Fe/Mn. Data are from Mittlefehldt (1990 and unpublished).

Table 43. Compositions of olivine grains from representative mesosiderites.

	ALHA 77219	Crab Orchard	Emery			Hainholz			Mincy			Pinnaroo	Veramin
	(1)	(2)	(3)	(4)	(5)	(6)	(7)	(8)	(9)	(10)	(11)	(12)	(13)
	Chemical Composition (wt %)												
SiO_2	37.5	36.3	39.3	39.0	37.4	37.4	40.8	39.5	39.4	41.9	38.8	42.1	39.3
FeO	27.3	32.2	17.6	20.6	25.4	22.8	8.5	13.3	13.4	9.90	24.8	8.01	18.5
MgO	35.0	30.7	43.8	40.9	36.0	39.0	51.7	46.0	46.0	50.7	37.9	51.6	41.7
MnO	0.65	0.80	0.39	nd	0.64	0.63	0.21	0.32	0.32	nd	nd	0.20	0.50
Cr_2O_3	nd	0.04	nd	nd	nd	0.04	0.04	0.04	0.06	nd	nd	nd	0.03
CaO	0.05	0.08	nd	nd	nd	0.07	0.04	0.07	0.01	nd	nd	nd	0.03
Total	100.50	100.12	101.09	100.5	99.44	99.94	101.29	99.23	99.19	102.50	101.5	101.91	100.06
	Cation Formula Based on 4 Oxygens												
Si	0.9960	0.9939	0.9891	0.9988	0.9958	0.9800	0.9830	0.9933	0.9917	0.9998	1.0024	1.0026	1.0031
Fe	0.6064	0.7373	0.3705	0.4412	0.5656	0.4997	0.1713	0.2797	0.2821	0.1976	0.5359	0.1595	0.3949
Mg	1.3855	1.2527	1.6429	1.5611	1.4285	1.5231	1.8563	1.7239	1.7256	1.8029	1.4593	1.8313	1.5863
Mn	0.0146	0.0186	0.0083	--	0.0144	0.0140	0.0043	0.0068	0.0068	--	--	0.0040	0.0108
Cr	--	0.0009	--	--	--	0.0008	0.0008	0.0008	0.0012	--	--	--	0.0006
Ca	0.0014	0.0023	--	--	--	0.0020	0.0010	0.0019	0.0003	--	--	--	0.0008Total
Cations	3.0039	3.0057	3.0108	3.0011	3.0043	3.0196	3.0167	3.0064	3.0077	3.0003	2.9976	2.9974	2.9965
	*Cation Ratios Fe/Mn and mg# (100*Mg/(Mg+Fe))*												
Fe/Mn	41	40	45	--	39	36	40	41	41	--	--	40	37
mg#	69.6	62.9	81.6	78.0	71.6	75.3	91.6	86.0	85.9	90.1	73.1	92.0	80.1

(1) large olivine (Prinz et al., 1980); (2, 6, 7, 13) olivines from corona structures (Nehru et al. 1980); (3, 4, 5, 10, 12) large olivines (Mittlefehldt 1980); (11) small olivine (Mittlefehldt 1980); (8, 9) core and rim of large olivine (Mittlefehldt unpublished).

Mineral and lithic clasts

The igneous mineral and lithic clasts in mesosiderites are superficially similar to igneous clasts in howardites, eucrites and diogenites. However, many of them have subtle to stark differences with HED materials and are distinct igneous materials.

Ultramafic mineral clasts. Coarse-grained olivine clasts are commonly found in mesosiderites, but are absent in HED meteorites. These olivine clasts are typically rounded, single crystal fragments of cm size, although some brecciated clasts may have had polycrystalline precursors (McCall 1966, Mittlefehldt 1980). Texturally, many of them are very similar to the rounded olivine grains found in pallasites, and this led to the initial suggestion that mesosiderites were formed by mixing eucrite and pallasite magmas (Prior 1918). Table 43 presents representative analyses of mesosiderite olivine grains, and Figure 58 compares them to olivines from main-group pallasites. Mesosiderite olivines extend from more magnesian to more ferroan compositions than found in main-group pallasites (Fig. 58). No complete analyses of coarse-grained mesosiderite olivines are available in the literature, so the contents of minor elements, CaO and Cr_2O_3, are unknown, and only a few analyses containing MnO have been published. Nehru et al. (1980a) presented complete analyses for a few olivines from their study of corona textures, but these may be small mineral fragments. Delaney et al. (1981) presented partial analyses determined by Nehru et al. (1980a) for mesosiderite olivines, which again may be for small grains. Their data indicate that most mesosiderite olivine grains have FeO/MnO ratios in the range of ~30 to 42, lower than, but partially overlapping, the range for main-group pallasites. A few bona fide coarse-grained olivines studied by Mittlefehldt (1980) have FeO/MnO of 40 to 45, within the range of main-group pallasite olivines.

Coarse-grained low-Ca pyroxene clasts are common in mesosiderites. Many of these clasts are similar to diogenites and to orthopyroxenite clasts in howardites. Many are single crystal fragments, although polycrystalline pyroxenites are also present. These

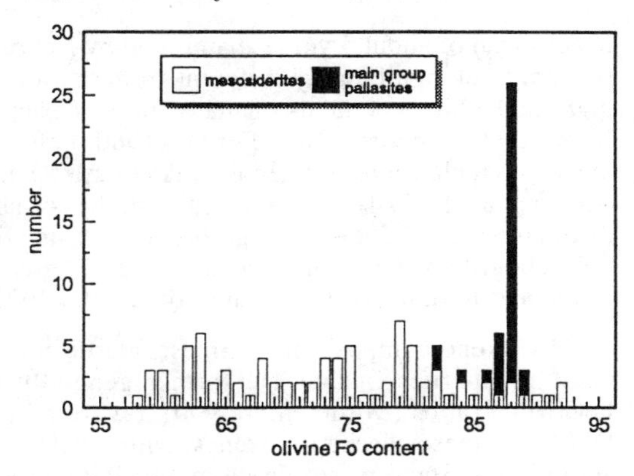

Figure 58. Histogram of olivine forsterite contents for individual mesosiderite olivine grains compared to average olivine compositions for main-group pallasites. Main-group pallasite olivines do not vary in composition within a given meteorite, except for minor zoning near the rims. Mesosiderite olivines show a wide range. Sources of pallasite data given in Figure 17. Mesosiderite data are from Agosto et al. (1980), Delaney et al. (1980), Mittlefehldt (1980 and unpublished), Nehru et al. (1980), Prinz et al. (1980).

clasts range in size up to about 10 cm. Coarse-grained orthopyroxenes are the most magnesian pyroxenes contained in mesosiderites, while coarse-grained pigeonites are among the most magnesian clinopyroxenes (Powell 1971). Systematics of pyroxene textural-compositional relationships are not well known. Powell (1971) determined that orthopyroxenes range from about Fs_{20} to Fs_{40}, and that in the mesosiderites Lowicz, Mincy and Morristown, orthopyroxenes more ferroan than about Fs_{30} were inverted from original pigeonite. This is similar to what is observed in the HED suite. Pyroxenes in typical diogenites ($Fs_{\sim25}$) crystallized as orthopyroxene, while that in the ferroan Yamato Type B diogenites ($Fs_{\sim32}$) crystallized as pigeonite. Table 42 gives representative coarse-grained low-Ca pyroxene analyses for mesosiderites.

Mafic lithic clasts. Mesosiderites contain basaltic and gabbroic clasts that are petrologically similar to basaltic and gabbroic eucrites and to mafic clasts in howardites (Ikeda et al. 1990, Kimura et al. 1991, McCall 1966, Mittlefehldt 1979, Rubin and Jerde 1987, 1988; Rubin and Mittlefehldt 1992). The original textures of these clasts vary from subophitic to ophitic for the finer-grained examples, to hypidiomorphic granular for the coarser-grained varieties. Many of them have been modified by shock metamorphism, which has caused granulation of minerals, recrystallization, formation of veins, impact melting and micro-faulting. These clasts are composed of original ferroan pigeonite and anorthitic plagioclase as the major phases, with minor to accessory silica (generally tridymite), whitlockite, augite, chromite and ilmenite. They typically contain metal and troilite, but it is difficult to determine whether these phases were added later by shock. The original pigeonite has undergone exsolution to form augite lamellae in either a lower Ca pigeonite or orthopyroxene host. Pyroxene textures in mesosiderites have not received the same level of detailed study accorded HED pyroxenes.

Detailed study of mesosiderite clasts has shown that many of the mafic clasts are distinguishable from similar HED meteorite materials. In general, basalts and gabbros from mesosiderites have higher modal proportions of tridymite and whitlockite than do similar HED lithologies (Nehru et al. 1980b, Rubin and Mittlefehldt 1992). Nehru et al. (1980b) presented summary data on basalt and gabbro clasts found in mesosiderite thin sections indicating a mean tridymite content of about 11 vol % for the basalt clasts, with a range of 5 to 14 vol %, while basaltic eucrites contain variable amounts of silica phases from ~1 to 8 vol % with a mean of about 4 vol % (Delaney et al. 1984). More recently, detailed petrologic study has been done on large basaltic and gabbroic clasts (up to several cm in size) mostly from the mesosiderite Vaca Muerta (Ikeda et al. 1990, Kimura et al. 1991, Mittlefehldt 1990, Rubin and Jerde 1987, Rubin and Mittlefehldt 1992). The basaltic clasts have a wider range of silica contents from about 0.4 to 13 vol % with a lower mean of about 5 vol % than found by Nehru et al. (1980b). This large clast mean is probably not significantly different from that of the basaltic eucrites. Nevertheless, the high modal silica contents of many mesosiderite basalt clasts serve to distinguish them from basaltic eucrites. Nehru et al. (1980b) also noted that mesosiderite basaltic clasts have modal chromite > ilmenite, unlike basaltic eucrites which have chromite < ilmenite (Delaney et al. 1984). But here again, studies of larger basaltic clasts remove some of the distinctiveness. Of the 19 large basaltic clasts with modal chromite and ilmenite data, only about half have chromite significantly greater than ilmenite (Kimura et al. 1991, Rubin and Jerde 1987, Rubin and Mittlefehldt 1992).

Pyroxene compositions in mesosiderite mafic clasts also tend to be distinct from those of eucrites in that the former generally have lower FeO contents and lower FeO/MnO ratios (Mittlefehldt 1990). Figure 57 is a diagram of molar Fe/Mg vs. molar Fe/Mn for mesosiderite pyroxenes, with a field for HED meteorite pyroxenes shown for comparison. Some mesosiderite mafic clasts have pyroxenes with Fe/Mg-Fe/Mn similar to those of HED mafic rocks, for example Mount Padbury clast RV-05 of Mittlefehldt (1979). However, pyroxenes in many mesosiderite mafic clasts have generally lower Fe/Mg and Fe/Mn, and these types of pyroxenes are not found among HED samples. Many coarse orthopyroxene and magnesian pigeonite clasts in mesosiderites have Fe/Mg-Fe/Mn characteristics indistinguishable from those of their HED meteorite counterparts. Representative pyroxene analyses for mesosiderite mafic clasts are presented in Table 42.

Plagioclase grains in mafic clasts are typically anorthitic, An_{95-88}, with orthoclase contents of about 0.1 to 0.7 mol %; very similar to that of plagioclase in eucrites and howardites. Table 44 gives representative plagioclase analyses for basalt and gabbro clasts.

Silicate compositions

Bulk silicate compositions. Mesosiderite silicates are broadly similar to howardites in bulk composition, as would be expected for breccias composed of similar minerals. Table 45 gives representative major element compositions for several mesosiderite bulk silicates, which can be compared with similar data for howardites in Table 34. One problem with mesosiderite silicate analyses is that it is impossible to remove all the metal, and even more so all the troilite, from the silicate prior to analysis. Hence, the abundance of "FeO" in mesosiderite silicates by techniques such as XRF are always too high and need adjustment. Because of this, we have favored the wet chemical analyses from a reputable analyst (E. Jarosewich), which give true FeO contents. We have also used some XRF analyses if S and Ni have been determined which allow us to make adjustments for both metal and troilite. One major difference between mesosiderite silicates and howardites is the generally lower FeO in the former, resulting in a systematically lower FeO/MnO ratio for mesosiderite silicates (Simpson and Ahrens 1977). Mesosiderite silicates are also quite enriched in P_2O_5 compared to howardites. There is a systematic variation in bulk silicate composition with mesosiderite type; compositional class A mesosiderites have higher CaO and Al_2O_3 and lower MgO and Cr_2O_3 than compositional class B mesosiderites, consistent with the differences in basaltic and ultramafic components between the compositional classes.

Selected lithophile minor and trace elements for representative mesosiderites are presented in Table 46, which can be compared with similar data for howardites in Table 35. Mesosiderites generally contain lower contents of the most incompatible trace elements, such as the light rare-earth-elements, than do howardites. There are also systematic differences between compositional classes A, B and C mesosiderites. Compositional class A mesosiderites generally contain higher contents of the most

Table 44. Compositions of plagioclases from representative mesosiderites.

	Pinnaroo	Patwar				Clover Springs	
	(1)	(2)	(3)	(4)	(5)	(6)	(7)
	Chemical Composition w (wt %)						
SiO_2	45.3	46.2	46.0	45.3	45.5	45.6	45.2
Al_2O_3	34.8	34.9	34.5	34.8	34.8	35.2	35.7
FeO	0.34	0.35	0.19	0.23	0.24	0.16	0.17
CaO	19.0	18.4	18.3	18.7	18.4	18.5	18.8
Na_2O	0.83	0.95	0.95	0.90	0.95	0.89	0.77
K_2O	0.07	0.09	0.07	0.12	0.11	0.07	0.03
Total	100.34	100.89	100.01	100.05	100.00	100.42	100.67
	Cation Formula Based on 8 Oxygens (Ideal Plagioclase = 5 Cations per 8 Oxygens)						
Si	2.0873	2.1109	2.1182	2.0911	2.0988	2.0930	2.0714
^{iv}Al	1.8900	1.8795	1.8726	1.8935	1.8921	1.9043	1.9284
Total tet	3.9773	3.9904	3.9908	3.9846	3.9909	3.9973	3.9998
Fe	0.0131	0.0134	0.0073	0.0089	0.0093	0.0061	0.0065
Ca	0.9381	0.9008	0.9029	0.9249	0.9094	0.9098	0.9231
Na	0.0742	0.0842	0.0848	0.0806	0.0850	0.0792	0.0684
K	0.0041	0.0052	0.0041	0.0071	0.0065	0.0041	0.0018
Total Cations	5.0068	4.9940	4.9899	5.0061	5.0011	4.9965	4.9996
	Molecular Proportion of Orthoclase (Or), Albite (Ab), Anorthite (An)						
Or	0.4	0.5	0.4	0.7	0.6	0.4	0.2
Ab	7.3	8.5	8.6	8.0	8.5	8.0	6.9
An	92.3	91.0	91.0	91.3	90.9	91.6	92.9

(1) impact melt matrix plagioclase; (2) impact melt clast RV-01 plagioclase; (3) plagioclase in monogenic basalt (?) clast contained in impact melt clast RV-01; (4); polygenic basalt clast RV-02 plagioclase (5); gabbro clast RC-05 plagioclase (6) gabbro clast RC-01 plagioclase; (7) gabbro clast RC-01 plagioclase inclusion in tridymite; all data from Mittlefehldt (1978) except (7) from Mittlefehldt (unpublished).

incompatible elements than do compositional class B mesosiderites, and the sole mesosiderite of compositional class C is poorest in these elements. There is overlap in trace element composition between compositional classes A and B, however, and howardites can show considerable variation in trace element content from one sample to another. Therefore, caution must be used in interpreting variations in trace element contents with compositional class among mesosiderites.

Ultramafic mineral clast compositions. There are relatively few bulk analyses for coarse-grained olivine and orthopyroxene clasts from mesosiderites. Mittlefehldt (1980) presented an analysis for a Fo_{82} olivine clast from Emery and a Fo_{92} olivine clast from Pinnaroo done by INAA, with comparative data for Brenham pallasite olivine (Fo_{88}). For both of the mesosiderite clasts, the INAA Cr content was about a factor of 10 higher than electron microprobe analyses, indicating that chromite inclusions are present. Cobalt and Ni contents of the mesosiderite olivines were substantially higher than in the Brenham olivine, suggesting metal inclusions are present in the mesosiderite olivines. The Sc and Sm contents of the Emery olivine are much higher than those of Brenham olivine (Mittlefehldt 1980), or pallasite olivines in general.

Orthopyroxene clast analyses were presented for Emery and Patwar by Mittlefehldt (1979) and for Bondoc by Rubin and Mittlefehldt (1992). In addition, Kimura et al. (1991) presented results for an orthopyroxene breccia clast and an olivine-orthopyroxene

Table 45. Bulk chemical analyses of representative mesosiderites.

	Barea 1A		Patwar 1A		ALHA 77219 1B		Lowicz 3A		Mincy 3B		Patwar 1A		Mount Padbury 1A
	(1)	(1*)	(1)	(1*)	(1)	(1*)	(2)	(2*)	(2)	(2*)	(3)	(4)	(5)
	Composition (wt %)												
SiO_2	21.71	51.9	25.76	51.5	32.27	52.5	50.12	53.6	47.46	52.7	50.3	48.8	48.4
TiO_2	0.11	0.26	0.16	0.32	0.16	0.26	0.342	0.37	0.263	0.29	0.58	0.48	0.72
Al_2O_3	4.01	9.58	5.86	11.71	3.93	6.40	10.58	11.32	6.69	7.43	13.5	13.4	12.5
Cr_2O_3	0.37	0.88	0.26	0.52	0.63	1.03	0.767	0.82	1.474	1.64	0.44	0.43	0.36
Fe_2O_3	--	--	--	--	8.87	--	--	--	--	--	--	--	--
FeO	5.61	13.4	6.86	13.7	7.86	12.8	13.36	8.2	16.78	9.6	16.5	18.2	20.4
MnO	0.24	0.57	0.31	0.62	0.41	0.67	0.533	0.57	0.530	0.59	0.53	0.59	0.56
MgO	6.62	15.8	6.88	13.7	12.56	20.4	13.85	14.8	18.27	20.3	6.97	7.32	6.54
CaO	2.54	6.07	3.63	7.25	3.14	5.11	7.97	8.53	5.29	5.87	10.5	9.47	9.53
Na_2O	0.12	0.29	0.19	0.38	0.15	0.24	0.31	0.33	0.22	0.24	0.36	0.33	0.44
K_2O	0.01	0.02	--	--	0.03	0.05	0.017	0.02	0.008	0.01	--	0.02	--
P_2O_5	0.52	1.24	0.14	0.28	0.30	0.49	1.325	1.42	1.243	1.38	--	--	--
Fe	50.24	--	33.70	--	23.63	--	--	--	--	--	--	--	--
Ni	6.19	--	4.20	--	2.86	--	0.35	--	0.47	--	--	--	--
Co	0.12	--	0.13	--	0.10	--	--	--	--	--	--	--	--
P	--	--	0.02	--	--	--	--	--	--	--	--	--	--
FeS	1.62	--	11.89	--	1.10	--	--	--	--	--	--	--	--
S	--	--	--	--	--	--	0.246	--	0.544	--	--	--	--
Total	100.03	--	99.99	--	98.00	--	99.770	--	99.242	--	99.68	99.04	99.45
	*Cation Ratios Fe/Mn and mg# (100*MgO/(MgO+FeO))*												
Fe/Mn	23	--	22	--	19	--	25	14	31	16	31	30	36
mg#	67.8	--	64.1	--	74.0	--	64.9	76.3	66.0	79.0	42.9	41.7	36.4

(1) Wet chemistry analyses of bulk meteorite (Jarosewich, 1990). The Fe_2O_3 is from rusted metal, and not the silicate fraction. (1*) Analyses renormalized on a metal- and trolite-free basis to 100%. (2) XRF analyses of bulk non-magnetic fraction (Simpson 1982). (2*) Analyses with measured FeO corrected for metal and troilite contamination using measured Ni and S, and renormalized on a metal- and trolite-free basis to 100%. (3) Electron microprobe analyses of fused-glass bead (Mittlefehldt 1979) on basaltic impact melt clast, (4,5) Average of metal-troilite corrected INAA on bulk sample and electron microprobe analyses of fused-glass bead (Mittlefehldt 1979) on polygenic basalt clast (4), and monogenic basalt clast (5).

breccia clast from Vaca Muerta. We will not consider these latter breccia clasts here because of the possibility of contamination by other silicate materials. In general, the mesosiderite orthopyroxenites are similar in composition to diogenites (Mittlefehldt 1979). Although Mittlefehldt (1979) pointed out some trace element compositional differences between the Patwar orthopyroxenite and diogenites, subsequent recoveries of diogenites from Antarctica have largely removed the distinction. The Bondoc orthopyroxenite is more magnesian than that in typical diogenites (mg# 81 vs. ~74-76) (Rubin and Mittlefehldt 1992), but is within the range of individual orthopyroxene grains found in the diogenite EETA79002, mg# 73-82 (Mittlefehldt, unpublished).

Mafic lithic clast compositions. Both the basalt and the gabbro clasts in mesosiderites show much wider ranges in trace element compositions than do petrographically similar materials in the HED suite, and some of them have slightly unusual major element compositions. Representative major element compositions for basalt clasts are given in Table 45, and selected trace-element analyses are given in Table 46. These should be compared to basaltic eucrite analyses given in Tables 34 and 35. All of the mesosiderite clasts have high Ni and Co contents indicating that metal is present. The Ni and Co contents of mesosiderite metal was used to estimate the true FeO contents of the samples. Some of the analyses also have high Se contents indicative of high modal troilite. In the absence of information on the troilite Se content, a correction to FeO for troilite was done assuming troilite has a cosmic S/Se ratio. The mg# and molar Fe/Mn ratios discussed below are based on these corrected FeO contents.

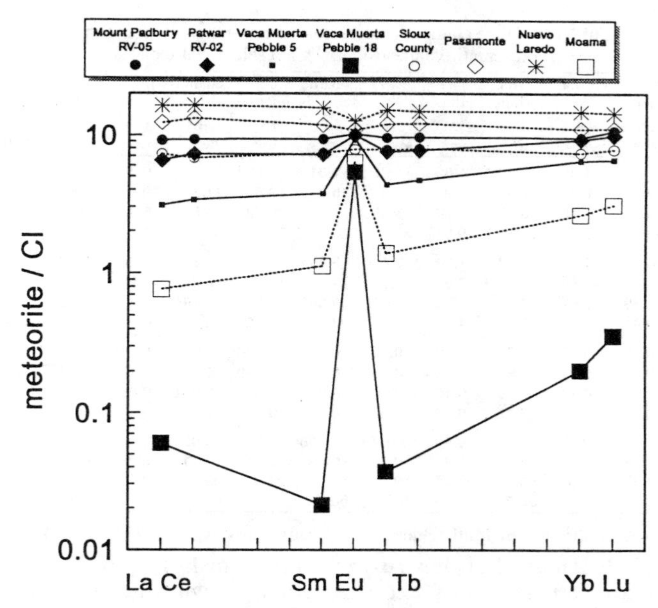

Figure 59. Rare earth element diagram for mesosiderite mafic clasts compared to select basaltic and cumulate eucrites. Mount Padbury RV-05 is a monogenic basalt, Patwar RV-02 a polygenic basalt, Vaca Muerta Pebble 5 an impact melt of cumulate gabbro(s), Vaca Muerta Pebble 18 an extremely fractionated cumulate gabbro (see Rubin and Mittlefehldt 1992 for classification). Data from Mittlefehldt (1979), Rubin and Jerde (1987), Rubin and Mittlefehldt (1992), Warren and Jerde (1987).

Some mesosiderite basalts are virtual clones of basaltic eucrites, for example, Mount Padbury clast RV-05 (Mittlefehldt 1979). This clast has a mg# of ~36, in the range of basaltic eucrites (33 to 41; see Table 34), a molar Fe/Mn of ~36, like that of basaltic eucrites, and a flat rare-earth-element (REE) pattern at about 9 to 10 × CI chondrite abundances. However, many of the basaltic clasts are distinct in major element composition. For example, Patwar basalt clast RV-02 (Mittlefehldt 1979) has a mg# of ~42, higher than that of basaltic eucrites, and a molar Fe/Mn of ~30, lower than that of basaltic eucrites. This clast also has a LREE-depleted pattern with a positive Eu anomaly, a pattern unknown among basaltic eucrite falls. Figure 59 shows REE patterns for mesosiderite basalt and gabbro clasts compared to basaltic and cumulate eucrites.

As is the case for the basalt clasts, some of the mesosiderite gabbro clasts are similar to cumulate eucrite gabbros in bulk major and/or trace element contents. Cumulate eucrites are more heterogeneous in major and trace element compositions than are basaltic eucrites, and there are fewer of them, so it is difficult to decide whether a given mesosiderite gabbro clast is distinct from the eucrite counterparts. The mesosiderite clast most similar to cumulate eucrites in major and trace element composition is Vaca Muerta Pebble 5, but this clast is texturally an impact melt (Rubin and Jerde 1987). Several mesosiderite gabbro clasts have trace element contents similar to those of cumulate eucrites (Mittlefehldt et al. 1992). However, many of the mesosiderite gabbroic clasts show extreme depletions in the most incompatible elements, such as the LREE, when compared to cumulate eucrites (Mittlefehldt 1979, Rubin and Mittlefehldt 1992). Extremes in this case are Vaca Muerta Pebbles 6 and 18, which have Sm abundances of 0.03 and 0.021 × CI chondrites, respectively (Rubin and Jerde 1987, Rubin and

Table 46. Trace element analyses of the non-magnetic fraction of representatitive mesosiderites and select mafic clasts.

	Barea	Mount Padbury	Patwar	Clover Springs	Esther-ville	Lowicz	Pinnaroo	Chin-guetti	Veramin	Mincy	RKPA 79015	Patwar		Mount Padbury
	1A	1A	1A	2A	3A	3A	4A	1B	2B	3B	2C	1A		1A
	wr	wr	wr	wr	wr	wr	wr	wr	wr	wr	wr	RV-01	RV-02	RV-05
Na mg/g	1.53	1.52	1.33	1.46	1.83	1.25	1.31	0.502	0.929	0.787	0.067	2.45	2.33	3.14
Sc μg/g	17.9	19.4	18.1	19.9	19.9	17.5	15.8	12.8	11.2	19.8	9.04	28.7	30.8	31.4
Cr mg/g	5.18	3.58	7.75	8.87	6.08	5.08	5	6.72	6.68	5.03	3.31	3.14	3.05	2.40
Co μg/g	52.6	113	43.8	72.1	39.5	51	320	79.4	108	138	95	48.0	63.0	22.2
Ni mg/g	1.29	4.2	1.8	2.52	1.04	2.1	5.44	16.4	2.8	3.28	7.49	4.00	2.18	0.60
Se μg/g	3.3	14.4	19.5	0.9	2.5	3.3	21.3	19.7	2.2	2.9	15.5	3.7	2.9	bd
La μg/g	5.86	1.39	0.82	2.16	1.78	0.49	0.58	1.05	0.16	0.42	0.32	1.98	1.53	2.15
Ce μg/g	12	3.4	2.3	10	4.5	1.6	bd	2.58	bd	bd	1.2	5.2	4.4	5.6
Sm μg/g	0.364	0.386	0.479	0.272	0.703	0.327	0.323	0.138	0.075	0.295	0.0431	1.28	1.06	1.38
Eu μg/g	0.23	0.23	0.24	0.23	0.32	0.18	0.2	0.068	0.044	0.15	0.011	0.54	0.56	0.57
Tb μg/g	0.098	0.086	0.11	0.08	0.15	0.094	0.07	0.042	0.031	0.058	bd	0.39	0.28	0.35
Yb μg/g	0.46	0.5	0.63	0.42	0.73	0.52	0.42	0.22	0.11	0.4	0.078	1.60	1.50	1.55
Lu μg/g	0.07	0.08	0.08	0.067	0.1	0.075	0.068	0.03	0.027	0.063	0.014	0.23	0.24	0.26
Hf μg/g	0.23	0.28	0.27	0.14	0.42	0.33	0.43	0.11	0.11	0.23	bd	1.1	0.87	0.88
Ir ng/g	20.6	60.6	14.6	66.2	bd	9.9	bd	24.8	bd	bd	3.8	27	bd	bd
Au ng/g	14.3	49.1	13.7	30.2	10.4	16	bd	33.6	bd	bd	57.6	21	17	bd

All analyses by INAA; whole rock by Mittlefehldt (unpublished), clasts by Mittlefehldt (1979).

Mittlefehldt 1992). These clasts have extreme Eu anomalies, with CI normalized Eu/Sm ratios of 220 to 260, the most extreme ratios known among solar system rocks (Mittlefehldt et al. 1992). In spite of these unusual trace element contents, these clasts have mg# and Fe/Mn ratios within the range of cumulate eucrites.

Mesosiderite metallic phase

Metallography. Powell (1969) provided the first comprehensive study of the metal phase of mesosiderites, although he primarily discussed the compositions and cooling rates of the metal. Like the silicates, metal occurs on a variety of scales and textural settings in mesosiderites from large, cm-sized clasts, to veins penetrating silicates, to matrix metal grains a few microns in size. An exceptional mesosiderite is RKPA79015 (Fig. 56), which contains large, several cm-sized areas of polycrystalline metal free of silicates (Clarke and Mason 1982). The boundaries between metal and silicate are generally irregular, with silicate grains commonly penetrating metal regions, and metal embays, penetrates, or veins, silicates (Powell 1969, 1971). Even the boundaries of the larger metal clasts exhibit fine-scale embayments with the silicate phase (Kulpecz and Hewins 1978, Powell 1969). Widmanstätten texture is developed in at least some of the larger metal clasts, but in general, kamacite occurs along metal-silicate and metal-metal boundaries, with taenite in the cores of the metal particles (Powell 1969). Tetrataenite has been observed rimming taenite cores in kamacite (Clarke and Mason 1982).

Troilite is common in mesosiderites, and makes up between <1 to about 14 wt % of the meteorites (Table 41). There are relatively few bulk sample modes or norms, so it is not possible to know how representative these values might be. In many cases, troilite occurs between silicate-rich and metal-rich regions in thin section. Schreibersite occurs included within kamacite, or at the metal-silicate grain boundaries of matrix metal particles (Kulpecz and Hewins 1978, Powell 1971).

Composition. The composition of the metal phase of mesosiderites was first extensively studied by Powell (1969) for bulk Ni content, by Wasson et al. (1974) for Ni, Ga, Ge and Ir, and by Begemann et al. (1976) for Fe, Ni and cosmogenic nuclides. Powell (1969) prepared his samples by grinding large surface areas of mesosiderite slabs, and therefore his analyses are of metal of all textural types; from clast to matrix metal.

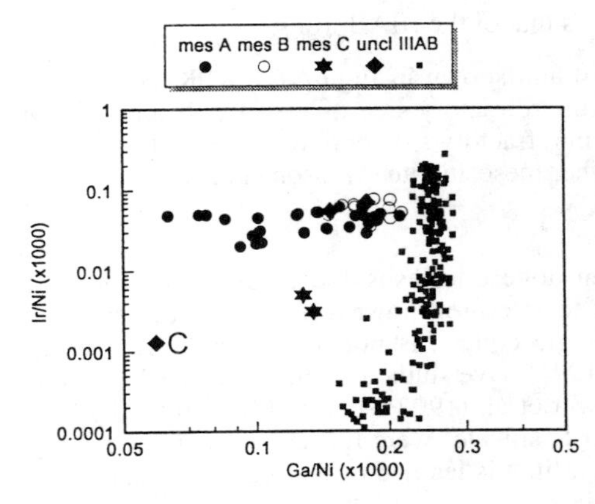

Figure 60. Comparison of mesosiderite metal and IIIAB irons on a plot of Ir/Ni vs. Ga/Ni. Mesosiderite metal does not exhibit the magmatic siderophile element trend shown by the IIIAB irons. Note that all low Ga/Ni mesosiderites are of compositional class A, and that duplicate analyses of the only compositional class C mesosiderite show its metal to be anomalous. The unclassified, anomalous mesosiderite Chaunsky (C) has a very unusual metal composition. Mesosiderite data are from Wasson et al. (1974, 1998a), Hassanzadeh et al. (1990); IIIAB iron data are from the compilation in Wasson (1974), plus more recent data from Esbensen et al. (1982), Kracher et al. (1980), Malvin et al. (1984), Scott and Wasson (1976), Wasson et al. (1989, 1998a).

Wasson et al. (1974) analyzed some of Powell's powders, powders they produced using the same technique, grains obtained by crushing and magnetic separation, and metal clasts. Begemann et al. (1976) prepared their metal separates by dry milling the samples, and they therefore included metal of all textural types. Hassanzadeh et al. (1990) analyzed large metal clasts from 15 mesosiderites for a suite of 11 siderophile elements.

As discussed by Wasson et al. (1974) and Hassanzadeh et al. (1990), and as is evident in the data of Begemann et al. (1976), the metal samples are invariably contaminated with silicates and oxides, and likely troilite as well. In addition, Wasson et al. (1974) and Hassanzadeh et al. (1990) have argued that the Ga and Fe, and possibly other siderophile element contents of the original metal may have been changed by interaction with the silicates. Hence, differences in the contents of Ni or other siderophile elements between different mesosiderites, or between different samples of the same mesosiderite, are difficult to interpret.

Using the original Powell (1971) classification for mesosiderites, Wasson et al. (1974) showed that textural grade 1 mesosiderites tended to be lower in Ga, Ge and Ir content than other mesosiderites. Using the current classification, it is clear that it is the compositional class A mesosiderites that tend to have the lower Ga, Ge and Ir contents. Figure 60 shows Ir/Ni vs. Ga/Ni for mesosiderite metal determined by Hassanzadeh et al. (1990) and Wasson et al. (1974, 1998a) compared with IIIAB irons. The compositional class B mesosiderites are all clustered at the high end of the Ga/Ni range, while the compositional class A mesosiderites span the range from the highest to lowest Ga/Ni ratios.

Hassanzadeh et al. (1990) divided mesosiderite metal compositions into three types; low-AuNi, intermediate-AuNi and high-AuNi on the basis of cluster analysis. They noted that all compositional class B mesosiderites except Chinguetti are in the low-AuNi group. They further argued that the classification of Chinguetti was not well known because only a small area was used for the modal analysis. One unusual feature of mesosiderite metal is the very small range in Ir contents compared to that found for igneous iron meteorite groups. Mesosiderite metal ranges in Ir from 1.7 to 6.3 μg/g, excluding the anomalous Chaunskij and RKPA79015 (Hassanzadeh et al. 1990, Wasson et al. 1974, 1998a). In contrast, IIIAB irons range in Ir from 0.011 to 19 μg/g (Wasson 1974). This difference is clearly shown in Figure 60, where IIIAB irons span three orders of magnitude in Ir/Ni whereas mesosiderite metal varies by only about a factor of four, excluding two anaomalous mesosiderites. The range in Ni content for mesosiderites, excluding Chaunskij is essentially the same as that of the IIIAB irons.

As noted by Hassanzadeh et al. (1990) and shown in Figure 60, RPKA79015 has anomalous metal composition. It is low in the refractory siderophiles Re, Ir and Pt, and slightly low in W. Ghaunskij is even lower in refractory siderophile elements than RPKA 79001 and has higher Ni and Cu than any other mesosiderite (Wasson et al. 1998a).

Ages

Modern age determinations have been done on mesosiderites using a variety of techniques. Some of the data are difficult to interpret, however, because the samples analyzed are polymict breccias and petrologic control is not provided. Murthy et al. (1977) and Brouxel and Tatsumoto (1990, 1991) have studied Estherville by Rb-Sr, Sm-Nd and U-Th-Pb chronometry, and Prinzhofer et al. (1992) studied Morristown by Sm-Nd. Other than indicating that the mesosiderite silicates were formed early in the history of the solar system and later metamorphosed, little is learned of the igneous evolution of the mesosiderite parent body from these studies.

A limited amount of chronometry has been done on igneous clasts from mesosiderites. Ireland and Wlotzka (1992) reported ion microprobe analyses for U-Pb on zircons. Zircon VM-2, which occurs in a "eucritic clast," is concordant in the U-Pb system and yielded a $^{207}Pb/^{206}Pb$ age of 4.563±0.015 Ga (Ireland and Wlotzka 1992). Petrologic details of this clast were not given, so it is not known whether it is a basalt that is petrologically equivalent to basaltic eucrites, or one of the magnesian, low Fe/Mn basalts with unusual REE patterns. Either way, this result indicates that magmatic processes occurred very early on the mesosiderite parent body. Stewart et al. (1994) performed Sm-Nd analyses of three igneous clasts from Vaca Muerta that had previously been studied by petrography and bulk composition (Rubin and Jerde 1987, Rubin and Mittlefehldt 1992), and a clast from Mount Padbury. The Sm-Nd isochron age of the Mount Padbury clast is 4.52±0.04 Ga. The ages of the Vaca Muerta clasts are 4.48±0.09 Ga for a basalt clast, 4.48±0.19 Ga for a gabbro clast and 4.42±0.02 Ga for an impact-melt clast.

Numerous mesosiderites have been studied using the ^{39}Ar-^{40}Ar technique in order to investigate the metamorphic and impact history of their parent body. Bogard and Garrison (1998) and Bogard et al. (1990) have determined Ar-Ar ages for 20 samples from 14 different mesosiderites, including whole rock and igneous clast samples. Most of the samples show stepwise release profiles with only modest increases in calculated age with extraction temperature. The average Ar-Ar age for these samples is about 3.9 Ga. (The published ages of Bogard et al. (1990) are in error due to a round-off error in the ^{40}K decay constant used (Bogard and Garrison 1998). The average age given here incorporates a correction for this error.)

Cosmic-ray exposure ages have been calculated from ^{36}Ar and ^{36}Cl measurements of the metal phase of several mesosiderites (Begemann et al. 1976). The calculated ages vary from 9 to 162 Ma, and there is no correlation with petrologic type. The calculated cosmic-ray exposure ages for type 1A mesosiderites vary from 9 to 133 Ma, for type 3B mesosiderites from 36 to 162 Ma, and for type 3A mesosiderites from 33 to 134 Ma.

Cooling rates

Cooling rates of mesosiderites have been an enigma ever since Powell (1969) demonstrated that they are among the lowest known. Cooling rates have been determined by a variety of techniques on both the silicate and metallic fractions. These cooling rate estimates are generally for different temperature regimes, so details of the thermal history of mesosiderites can be inferred.

Metallographic cooling rates. The larger metal clasts in mesosiderites have metallographic textures that allow cooling rates to be calculated. Powell (1969) originally calculated cooling rates for the temperature interval of ~500° to 350°C using the Wood method, and found rates of the order of 0.1°C/Ma, among the lowest cooling rates known. Subsequently, Ni diffusivity in Fe,Ni metal was revised (Saikumar and Goldstein 1988), resulting in recalculated cooling rates for mesosiderites of 1°C/Ma (see discussion in Bogard et al. 1990). Kulpecz and Hewins (1978) measured the Ni content in schreibersite and kamacite in the Emery mesosiderite, and used a non-isothermal, diffusion-controlled growth model for schreibersite in kamacite to calculate a cooling rate of 0.1°C/Ma. Here again, the more recent diffusivity data would require revision upward, probably of the same order as for the Wood method. However, in order for the method employed by Kulpecz and Hewins (1978) to be valid, schreibersite must have nucleated and grown solely from kamacite. Taenite and tetrataenite are typically present in mesosiderite metal particles, and even though they were not found in the grains studied by Kulpecz and Hewins (1978), it is likely that they were present out of the plane of the thin section, invalidating the calculated cooling rates (Agosto et al. 1980). Haack et al. (1996b) have reexamined the cooling rates of mesosiderites by modeling the Ni content of the centers of wide taenite lamellae and by modeling the central Ni content of kamacite using Powell's original data, and the phase diagram and diffusion constants of Saikumar and Goldstein (1988). Haack et al. (1996b) obtained very low cooling rates of ~0.01° to 0.03°C/Ma for several mesosiderites. Although there is uncertainty in the value of the low temperature cooling rate of mesosiderites, it seems clear that this value is very low, probably ≤1°C/Ma. Haack et al. (1996b) summarized metal phase compositional evidence that qualitatively argue that mesosiderite cooling rates are lower than those of other metal-bearing meteorites.

Silicate-based cooling rates. Estimates of cooling rates on the silicate fraction has been done by modeling element distributions in the minerals and by modeling Ar diffusion from bulk silicates. Delaney et al. (1981) measured steep Fe-Mg compositional profiles in pyroxene overgrowths on orthopyroxene cores for 4 mesosiderites, and suggested that these reflect cooling rates of 1° to 100°C per annum at temperatures in the range of 900° to 1150°C. Because diffusivity data for pyroxenes were lacking, they used data for olivines in this calculation. It is now known that Fe-Mg diffusion in pyroxenes is substantially slower than in olivine, and the cooling rates inferred for these pyroxene overgrowths should be much lower. Jones (1983) estimated values of ~2°C per annum for the highest rate given by Delaney et al. (1981), still ~10^6 to 10^8 times that estimated from the metal phase for low temperatures. Ganguly et al. (1994) reexamined the zoning profiles determined by Delaney et al. (1981) for Clover Springs and Lowicz overgrowths using updated diffusion data and more sophisticated cooling models. They calculated cooling rates of ~14°C/Ka at 1150°C and 5°C/Ka at 600°C. Ruzicka et al. (1994) modeled zoning profiles in overgrowths on plagioclase grains in corona structures, and estimated cooling rates of ≥100°C/Ka for temperatures of about 1100°C. Bogard et al. (1990) showed that the Ar-Ar release profiles and ages of mesosiderites were consistent with slow cooling at low temperature (500° to 350°C) as given by the metallographic cooling rates. Ganguly et al. (1994) estimated that the cooling rate at 250°C based on cation ordering in orthopyroxene was ≤1°C/Ma, similar to that determined on the metal phase for low temperatures.

Mesosiderite formation

As described above, the silicate fraction of mesosiderites is broadly similar to the

HED meteorites. Nevertheless, there are numerous differences in detail in their petrology and chemistry. Mesosiderites and HED meteorites show evidence for impact mixing on the surfaces of their parent bodies. Because mixing on asteroidal bodies is expected to be parent body-wide (e.g. Housen et al. 1979), Mittlefehldt (1979, 1990), Mittlefehldt et al. (1979) and Rubin and Mittlefehldt (1993) argued that mesosiderites and HEDs were formed on different parent bodies.

Because mesosiderites are composed of crustal silicates and iron metal presumed to represent core material of an asteroid, the formation of these meteorites requires complex processes. This complexity is compounded by estimates for cooling rates on the metal which indicate slow cooling, and hence deep burial (Powell 1969), while the textures and compositional zoning of the silicates give evidence for surficial processing and rapid cooling at high temperatures (Ganguly et al. 1994, Ruzicka et al. 1994). Indeed, the silicate textures and mineralogies of the mesosiderites indicate that these meteorites are polymict breccias formed by impact processes on the surface of a differentiated asteroid (Powell 1971). Rubin and Mittlefehldt (1993) presented a synopsis of the evolutionary history of mesosiderites, and what follows has been developed from this source and more recent work on mesosiderites.

Some of the igneous and meta-igneous clasts in mesosiderites are similar to the same materials found in the HED meteorite suite; they have essentially identical mg#, molar Fe/Mn, and trace element contents as eucrites and diogenites. Because of this, the earliest, post-accretion history of the mesosiderite parent body was likely similar to that of the HED parent body; parent body-wide igneous processes produced a mafic crust, a depleted ultramafic mantle and a metallic core. Just as there is active debate between partial melting vs. fractional crystallization models for formation of the HED parent body crust, this same uncertainty applies to the mesosiderite parent body crust. Age determinations on igneous and meta-igneous clasts have shown that the original mesosiderite parent body crust was formed about 4.56 Ga ago (e.g. Ireland and Wlotzka 1992). This is essentially the same as the time of formation of the HED meteorites, and shortly after the formation of chondritic meteorites (e.g. see Rubin and Mittlefehldt 1993). The presence of metal-poor breccia clasts in mesosiderites suggests that the parent body surface was gardened by impacts prior to metal-silicate mixing.

Metal-silicate mixing was the next major event (or events) on the mesosiderite parent body. A variety of models have been advanced to explain the metal-silicate mixing event including both internal and external mixing models (see review by Hewins 1983). Almost all models assume that the metallic and silicate fractions of mesosiderites originated in distinct regions, the core and crust, respectively, of either a single asteroid, or of two asteroids. The most plausible model is that of Wasson and Rubin (1985) and Hassanzadeh et al. (1990). These authors posit that an asteroidal core, stripped of its silicates by impacts except for some mantle olivine, accreted at low velocity to the mesosiderite parent body. Because the metal in mesosiderites appears to be undifferentiated, as indicated by the small range in Ir/Ni ratios (Fig. 60), these authors have suggested that the core was undifferentiated and therefore still largely molten at the time it was accreted. Hassanzadeh et al. (1990) further argued that metal-silicate mixing was responsible for the compositional distinction between compositional class A and B mesosiderites. They argued that all mesosiderites were originally of compositional class A, and that the region where the B mesosiderites were formed accreted more olivine with the metal. Through reduction of FeO and reaction with tridymite, this olivine was converted to orthopyroxene. The presence of metal-rich breccia clasts in mesosiderites suggests, but does not prove, that impacts gardened the surface of the parent body after metal-silicate mixing, and hence the proto-mesosiderite material remained on the parent body surface for some period of time.

Mittlefehldt (1979, 1990) and Rubin and Mittlefehldt (1992) have shown that many of the basalt and gabbro clasts in mesosiderites have undergone remelting after metal-silicate mixing. These are the clasts that are more magnesian than their eucrite counterparts, have low molar Fe/Mn ratios, LREE-depleted patterns with positive Eu anomalies, and high modal abundances of tridymite and/or whitlockite. This mineralogical and mineral compositional data shows that the mafic magmas were reduced by P from the metal phase (Mittlefehldt 1990). The LREE-depleted patterns and trace element signatures of some basaltic clasts indicate they are mixtures of basalts and cumulates (Mittlefehldt 1990, Rubin and Mittlefehldt 1992), and the extreme Eu/Sm fractionation of some gabbros indicate multiple melting and fractional crystallization steps (Mittlefehldt 1979, 1990; Mittlefehldt et al. 1992, Rubin and Mittlefehldt 1992). Because many of the mafic clasts in mesosiderites were remelted, it appears that a large fraction of the mesosiderite crust may have been remelted after metal-silicate mixing.

The overgrowths on pyroxene and plagioclase clasts and grains give evidence for rapid cooling at moderately high temperatures. The pyroxene overgrowths certainly must have formed at the time the mesosiderites were lithified into their current form, as these overgrowths can extend to cm size and poikilitically include all other minerals in the matrix (e.g. Floran 1978). Therefore, the evidence suggests that after the mesosiderites were assembled in their current form, they were still located near the surface of their parent body. Subsequently, the mesosiderites must have been put in a low-temperature environment where the cooling rate was many orders of magnitude lower, as indicated by the metallographic cooling rates and the cation ordering in orthopyroxene. Bogard et al. (1990) suggested that this was caused by a catastrophic disruption and reassembly of the mesosiderite parent body about 4.2 Ga ago, as indicated by the Ar-Ar ages. (We have adjusted the age quoted by Bogard et al. (1990) to correct for the computational error discovered by Bogard and Garrison (1998).) This event resulted in the mesosiderites being deeply buried in the parent body, where slow cooling at low temperatures caused Ar degassing. Haack et al. (1996b) have argued that break-up and reassembly did not occur on the mesosiderite parent body; instead they suggest that the cooling rates and petrology can be explained by essentially a single process: metal-silicate mixing by a large impact formed the mesosiderite breccias as we see them and led to the evidence for rapid cooling at high temperature. Continuing accretion of debris from the impact built up an insulating regolith. This gradually lowered the cooling rate at the mesosiderite location, leading to the low metallographic cooling rates at low temperature and Ar degassing. This model appears inconsistent with the petrologic and geochemical data on mesosiderite mafic clasts which suggest that melting and crystallization of much of the parent body crust occurred after metal-silicate mixing (Rubin and Mittlefehldt 1993).

MISCELLANEOUS NON-CHONDRITIC ASTEROIDAL METEORITES

A number of meteorites do not fit into well-defined groups on the basis of mineralogy, petrology or oxygen-isotopic compositions. In some cases, these meteorites formed by processes unknown from the main groups. In other cases, their origin was similar to that of a well-defined group, although probably representing another parent asteroid. Here we discuss several of these meteorites.

Bocaiuva

Bocaiuva is a unique silicate-bearing iron meteorite that was recovered as a single mass of 64 kg in 1947. The meteorite was studied by Desnoyers et al. (1985) and Malvin et al. (1985) and information given below is taken from these sources.

Bocaiuva is dominated by Fe,Ni metal (~85 vol %) which exhibits a Widmanstätten pattern in some portions. The width of the kamacite bands indicates the Bocaiuva is a finest octahedrite. Bocaiuva contains 8.49 wt % Ni in the metal, but a remarkably high Ge/Ga ratio (Malvin et al. 1985). The Ge/Ga ratio suggests that Bocaiuva may be related to IIF iron meteorites, Eagle Station pallasites, or a small number of ungrouped irons. Detailed comparison of the metal compositions suggests that none of these are closely related, however, and they may not have formed on the same parent body.

Silicates in Bocaiuva occur as nodules or aligned stringers up to 1 cm in maximum dimension. They are composed of olivine ($Fa_{7.7}$), orthopyroxene ($Fs_{7.6}$), minor plagioclase (An_{49}) and clinopyroxene ($Fs_{4.5}$, Wo_{42}) in approximately chondritic abundances. The mafic silicate compositions are similar to those in IAB irons, while the plagioclase is alkali-poor compared to plagioclase in IAB irons.

The oxygen-isotopic composition of Bocaiuva silicates is closest to that of the Eagle Station pallasites (Fig. 1a) and the CO and CV chondrites, suggesting that the parent bodies of all of these meteorites may have formed from a similar reservoir.

Malvin et al. (1985) argued for an origin by impact, largely because of the analogy with IAB irons (see section on silicate-bearing IAB irons) and the occurrence of apparently-aligned stringers of silicates indicative of flow. Desnoyers et al. (1985) noted that only a small part of the 64 kg mass had been sampled and suggested that more work was in order.

Divnoe

Divnoe is an ultramafic achondrite whose relationship to other groups has been debated. Divnoe has been studied by McCoy et al. (1992) and Weigel et al. (1996) who reported their work in abstracts and Petaev and Brearley (1994) and Petaev et al. (1994) who provided a more comprehensive study. The information given here is taken from those papers.

Divnoe consists predominantly of olivine (74.6 vol %, Fa_{20-28}) which occurs as mm-sized elongate grains with a strong preferred orientation. Olivine in Divnoe is unusual in containing exsolution lamellae of slightly more and less ferroan olivine, with differences in composition of 2 to 4 mol % Fa. Orthopyroxene (Fs_{20-28}) is also common throughout the meteorite, comprising 14.3 vol % of the bulk. Large clinopyroxene grains often poikilitically enclose other grains. The meteorite has been heavily shocked, dispersing melted Fe,Ni metal and troilite droplets into the surrounding rock. Plagioclase is depleted relative to chondrites, comprising 1.5 vol % of the bulk with a composition ranging from An_{32-45}. The oxygen-isotopic composition for Divnoe is similar to that for brachinites and HED meteorites in $\Delta^{17}O$, although Divnoe has higher $\delta^{18}O$ and $\delta^{17}O$ than these other meteorite groups (Fig. 1c). Divnoe is similar to the brachinites ALH 84025 and Brachina in its major element composition. However, Divnoe is strongly depleted in the highly incompatible lithophile elements compared to brachinites, and in this is more similar to lodranites. Refractory siderophile elements are enriched in Divnoe compared to brachinites. Divnoe and Brachina also contain the same trapped Xe isotopic pattern.

The relationship between Divnoe and other groups of meteorites is not clear. McCoy et al. (1992) argued that Divnoe might represent an extension of the correlation between Fa and $\Delta^{17}O$ observed in lodranites, although it is much more oxidized (Fa_{20-28}) than the most oxidized lodranite (Fa_{13}). Petaev et al. (1994) agreed that Divnoe and the lodranites shared a number of similarities and, at the minimum, formed through similar processes.

However, these authors concluded that Divnoe is a unique meteorite not genetically related to any other known meteorite. The more recent work of Weigel et al. (1996), however, provides evidence for a link between Divnoe and brachinites. Future recoveries of more brachinites and lodranites, both of which are small groups whose ranges are incompletely understood, will likely be necessary before this issue can be resolved.

Enon

Enon is a unique stony-iron meteorite find, but little is known about it. Its oxygen-isotopic composition is near the field defined by the differentiated meteorite groups (Clayton and Mayeda 1996), the IIIAB irons, main-group pallasites, HED meteorites, mesosiderites, angrites, and it plots near two of the four brachinites (Fig. 1d).

A brief description of Enon was given by Bunch et al. (1970), and what follows was taken from that source. Enon is composed of roughly equal amounts of Fe,Ni metal and silicates having a texture similar to mesosiderites. Silicates in Enon are olivine ($Fo_{91.1}$), orthopyroxene ($Wo_{1.9}En_{87.6}Fs_{10.5}$), clinopyroxene ($Wo_{43.7}En_{51.9}Fs_{4.4}$) and plagioclase ($Ab_{78.9}An_{14.7}Or_{6.4}$). The silicates in Enon are quite reduced as shown by the magnesian nature of the ferromagnesian minerals. The sodic plagioclase is similar to those in chondritic meteorites. Enon also contains chromite, whitlockite, troilite and schreibersite. Bunch et al. (1970) suggested that the silicate mineral assemblage of Enon is in, or near, equilibrium.

Major and trace element analysis by INAA/RNAA was reported by Kallemeyn and Wasson (1985), who also published an earlier analysis by Bild (1976). Niemeyer (1983) reported K, Cl, Te, I and U concentrations for an acid-washed silicate fraction of Enon. Enon has a chondritic lithophile element signature. With the exceptions of V, Cr and La, the Mg- and CI-normalized refractory and moderately volatile lithophile elements are at CI abundances. The slight depletions in V and Cr are plausibly due to undersampling of chromite. Refractory siderophile elements and the moderately volatile siderophile elements Au and As are enriched 10 to 20 $\times$ CI abundances relative to Mg, whereas the three most volatile siderophile elements analyzed, Ga, Sb and Ge, have lower abundances of 2 to 6 $\times$ CI. Enon has slightly low refractory siderophile element/Ni ratios, which Kallemeyn and Wasson (1985) attribute to shock mobilization of metal. The most volatile elements studied by Bild (1976) and Kallemeyn and Wasson (1985), Br, Se, Zn and Cd, show depletions relative to CI abundances. The compositional characteristics of Enon are thus those of a metal-rich chondrite. Kallemeyn and Wasson (1985) state that differential melt-solid transport resulting from shock melting is a plausible cause for the high metal and troilite content and texture of Enon.

Niemeyer (1983) did an Ar-Ar and I-Xe study of Enon. He determined an Ar-Ar age of 4.59 Ga for Enon, and found no evidence for loss of radiogenic Ar at low temperature. This suggests that Enon has not been disturbed by thermal events essentially since it was formed. No I-Xe age could be determined, but the Xe data suggested that short-lived ^{244}Pu was present in Enon at levels greater than typical for chondrites at the time of closure for Xe loss, indicating that Enon is an ancient object.

Enon appears to be a highly metamorphosed, metal-rich, reduced chondrite. Its silicates are distinct from those of the brachinites with similar O-isotopic composition. Enon is much more reduced with an olivine mg# 91 vs. 65-70 for brachinites, and molar FeO/MnO of 19 vs. 63-68 (Nehru et al. 1983, Warren and Kallemeyn 1989a).

Guin

The Guin iron meteorite was recovered as a single 34.5 kg mass in 1969. Guin is an ungrouped iron with similarities to IIE irons, but with enough significant differences to suggest that it samples yet another parent body. Guin has been studied by Rubin et al. (1986) and the description given here is taken from that work.

Guin is a coarse octahedrite with an unusually high Ni content. It is similar in composition to the iron meteorite groups IAB and IIICD, but is not a member of these groups. Guin contains ~6 vol % of silicate inclusions which are similar to silicate inclusions in the IIE irons Colomera, Kodaikanal and Elga. The largest inclusion consists of shock-melted plagioclase-rich matrix (~65 wt %, $Ab_{\sim87}$) surrounding large, partly-melted augite grains (~20 wt %, $Fs_{10.4}Wo_{41.4}$) with minor low-Ca pyroxene ($Fs_{20.9}Wo_{3.7}$). The oxygen-isotopic composition of Guin is similar to that of L or LL chondrites, and differs dramatically from that of IIE irons (Fig. 1b).

Rubin et al. (1986) discussed two possible origins for Guin (1) fractional crystallization in a deep-seated magma chamber and (2) shock-melting of chondritic material from near the parent body surface. These models have been discussed in the section on IIE irons and shall not be repeated here. However, Guin is an important meteorite since it suggests that whichever process formed Guin and the IIE irons operated on more than a single parent body.

LEW 88763

This small achondrite was originally classified as a brachinite (AMN 14-2). Nehru et al. (1992) presented a brief characterization of brachinites, and noted that LEW 88763 had an anomalous oxygen-isotopic composition compared to other brachinites (see Fig. 1c). Nakamura and Morikawa (1993) determined a number of lithophile element concentrations on LEW 88763, and noted that it has a composition very similar to that of Brachina, and a chondritic REE pattern. Swindle et al. (1998) did a detailed study of LEW 88763, and concluded that it is not a brachinite, but a unique achondrite. The description that follows is from Swindle et al. (1998).

LEW 88763 is an ultramafic rock composed of 71% olivine, 7% pigeonite and augite, 10% plagioclase, 6% opaque phases including taenite, troilite, chromite and ilmenite, 7% unidentified material and accessory whitlockite. It is a fine-grained rock with equigranular texture of mostly anhedral grains <1 mm in size. Compositions of the major minerals are: olivine Fo_{63-64}, pigeonite $Wo_4En_{67}Fs_{29}$, augite $Wo_{38}En_{46}Fs_{16}$ and plagioclase $Or_{2-7}Ab_{55-74}An_{19-44}$. Chromite has an mg# of 12 to 15. Fe-Mg have equilibrated in the mafic silicates; no Fe-Mg zoning was detected in augite, and olivine and pigeonite are close to Fe-Mg equilibrium. Compared to brachinites, olivine in LEW 88763 is slightly more ferroan, and at lower modal abundance (compare with Table 18).

LEW 88763 has a chondritic bulk composition. Refractory lithophile elements including both compatible and incompatible elements are in chondritic ratios, indicating that igneous processes have not altered the bulk composition. The REE patterns of three samples analyzed show small positive Eu anomalies and variable Ce anomalies, which Swindle et al. (1998) suggest may be due to terrestrial weathering. There is a slight depletion in moderately volatile and volatile elements consistent with nebular fractionation. Siderophile elements are in chondritic abundance in LEW 88763, with slight depletions in the most volatile siderophile elements measured. Swindle et al. (1998) suggest LEW 88763 shows a slight depletion in Se, which they attribute to loss of sulfide from the rock. However, Se is the most volatile element they measured and it seems the Se depletion could equally well be nebular in origin.

No ages have been determined for LEW 88763, but Swindle et al. (1998) infer that its K-Ar age is ≥ 4.5 Ga based on K and Ar measurements made on different splits. They found no evidence for ^{129}Xe excess from the decay of ^{129}I. The isotopic composition of Xe in LEW 88763 is identical to chondritic Xe.

Based on mineralogy, bulk composition and noble gas data, LEW 88763 is an ultra-metamorphosed chondrite of unique type.

Puente del Zacate

Puente del Zacate is a IIIAB iron with 8.2% Ni in its metallic host. This meteorite is unusual, however, in containing a silicate inclusion, which was studied by Olsen et al. (1996). The inclusion is 7 mm in diameter and contained within a troilite nodule. The inclusion itself is composed olivine (23 wt %, Fa_4), low-Ca pyroxene (14 wt %, Fs_6Wo_1), chromium diopside (15 wt %, Fs_3Wo_{47}), plagioclase (15 wt %, $An_{14}Or_4$), graphite (27 wt %), and traces of troilite, chromite, daubreelite and Fe,Ni metal. The bulk mineralogy is similar to chondrites. The bulk REE pattern calculated from mineral compositions and abundances is flat at about 2.5 × CI abundances. The bulk oxygen-isotopic composition of the clast is similar to phosphates and chromites in other IIIAB irons (Clayton and Mayeda 1996; see Fig. 1c). The petrography, mineralogy and mineral compositions within the silicate inclusion of Puente del Zacate are similar to those in IAB irons, although differences in oxygen-isotopic composition and composition of the metallic host preclude formation of these two groups on a common parent body.

The origin of Puente del Zacate is unclear. Unlike IAB irons, IIIAB irons display fractional crystallization trends consistent with formation in a single core of an asteroid. If this is, in fact, the case, the incorporation of an undifferentiated silicate inclusion is difficult to understand. This is especially true considering the link between IIIAB irons and main-group pallasites, which contain silicates formed by magmatic processes. Olsen et al. (1996) argued that IAB and IIIAB irons may represent a continuum in the degree of partial melting they experienced, with IAB irons forming from a parent body with a very small degree of partial melting and IIIABs representing a higher degree of partial melting, but with some undifferentiated regions remaining in the mantle. Again, this seems at odds with the IIIAB-main-group pallasite connection inferred from geochemical data, and the likely origins of these groups. The reader is referred to the sections iron meteorite groups, the IAB iron silicates and pallasites above.

Sombrerete

Further evidence supporting formation of IIE-like meteorites on other parent bodies comes from the Sombrerete meteorite. Sombrerete is an ungrouped iron with 10.0 wt % Ni in the metallic host (Malvin et al. 1984).

Sombrerete contains ~7 vol % of silicate inclusions, which were studied by Prinz et al. (1982b). The inclusions are 1 to 5 mm in size and rounded. Quench-textured albitic glass is the dominant phase and may contain needles of apatite and orthopyroxene, as well as microphenocrysts of orthopyroxene. The inclusions consist of orthopyroxene (14.7%, En_{68}), albitic glass (66.7%), quench plagioclase (9.0%), chlorapatite (8.0%) and traces of kaersutite, tridymite, chromite, ilmenite and rutile. The silicate inclusions are broadly similar to those in the IIE irons Weekeroo Station and Miles, although substantially enriched in Al_2O_3, Na_2O and P_2O_5. The oxygen-isotopic composition of Sombrerete silicates lies below the terrestrial mass fractionation line on a three-isotope oxygen plot (Fig. 1b), substantially different from IIE irons and Guin (Clayton and Mayeda 1996).

As in the case of the IIE irons and Guin, the origin of Sombrerete is not clear (see section on IIE irons). However, the existence of these meteorites suggests that at least three parent bodies experienced similar processes.

Tucson

The Tucson meteorite, with its striking ring shape, may be the most famous, and in many ways the most enigmatic, meteorite on Earth. The most complete petrologic study of Tucson was conducted by Nehru et al. (1982) and the information given here is from that work and sources referenced therein.

Tucson is an anomalous iron meteorite with 0.8 wt % Si and 2200 µg/g Cr in the metallic host and is chemically uniform. Small (0.1-2 mm) silicate inclusions comprise 8 vol % of the meteorite, are arranged in curved lines with flow orientation, and are elongate parallel to the direction of flow. The inclusions are composed of olivine (66.4 vol % of the silicate fraction; Fo_{99-100}), enstatite (30 vol %; 0.5-21 wt % Al_2O_3), diopside (3 vol %; 5-18 wt % Al_2O_3), minor plagioclase or glass and traces of spinel and brezinaite. Some of the pyroxenes are the most aluminous known in nature.

The compositions and textures of both the metal and silicate assemblages suggest that Tucson was essentially quenched from high temperature (1300-1500°C). The incorporation of aluminum into pyroxene resulted from the difficulty in nucleating feldspar. Nehru et al. (1982) favored a model for the origin of Tucson in which the silicate assemblage represented an impactor which mixed with the metal phase. The impact mixing caused melting of the silicates and rapid quenching of the metal and silicate, producing the quench textures of the silicates, the homogeneous composition of the metal and the flow alignment of silicates within metal.

It is interesting to note that the preferred scenario for the formation of Tucson bears a strong similarity to that suggested for the origin of Shallowater. However, the oxygen-isotopic composition of Tucson (Fig. 1a) differs dramatically from the enstatite meteorite clan which lie along the terrestrial fractionation line (Clayton et al. 1976b). Thus, Tucson may represent another example of a highly-reduced meteorite parent body which experienced processes similar to those postulated for formation of the Shallowater aubrite.

ACHONDRITIC CLASTS IN CHONDRITES

Some chondrites contain achondritic inclusions. In some cases, petrologic and geochemical study shows that they are obviously impact melts of the chondrite parent body (e.g. Keil et al. 1980). These are discussed in Chapter 3, not here. However, some achondritic clasts are obviously not impact melts of the parent body as they have distinct oxygen-isotopic compositions, mineralogies unlike bulk chondritic silicates and fractionated trace element abundances (e.g. Hutchison et al. 1988). These inclusions are of uncertain origin; they are not obviously nebular materials, nor are they easily understood in terms of parent-body igneous processes (e.g. Mittlefehldt et al. 1995) and hence are fascinating objects that deserve more study. Only a few of them have been studied in any detail, and surveys have been done on several others. Here we summarize what is known about these clasts. Bridges and Hutchison (1997) have done a survey of clasts in ordinary chondrites, and found that they occur in roughly 4% of ordinary chondrites. Of the 24 clasts they categorized, 10 are chemically fractionated and differ substantially from chondrules, and 8 could not be categorized.

Troctolites

Three clasts that can be classified as troctolites have been described from the L6 chondrites Barwell, Y-75097 and Y-793241. All of these inclusions have H chondrite O-isotopic compositions (Hutchison et al. 1988, Nakamura et al. 1994). The petrographic descriptions of these clasts are summarized from Hutchison et al. (1988), Sack et al. (1994), Yanai and Kojima (1993) and Yanai et al. (1983). These inclusions are dominated by olivine of varying grain sizes. The grains are subhedral to euhedral from <0.3 to >1.0 mm in size. Modal and normative data indicate that olivine makes up roughly 70 to 80% of the inclusions. Interstitial to the olivine is plagioclase or maskelynite, which makes up most of the remainder of the inclusion. Generally, fine-grained chromite is a minor component. Chlorapatite is a trace component in the Barwell and Y-793241 inclusions, whereas whitlockite makes up ~6% of the Y-75097 inclusion. Only a small surface area of the Y-793241 inclusion was available for study by Sack et al. (1994), so the modal data probably are not representative. Significantly, no pyroxene is reported from any of these inclusions. Yanai and Kojima (1993) report a barred olivine chondrule in the core of the Y-793241 inclusion, but petrographic details are not given.

Olivine compositions from all of these inclusions are within uncertainty the same as olivines from the host L chondrite (Hutchison et al. 1988, Sack et al. 1994). Plagioclase compositions are variable. In the Barwell inclusion, plagioclase in the core of the inclusion is very calcic, up to An_{74-70}, while near the margins of the inclusion they are more sodic, An_{20} (Hutchison et al. 1988). In the Y-75097 inclusion, plagioclase is only slightly more calcic than that in the host chondrite, and no mention is made of compositional variation within the inclusion (Sack et al. 1994). In the Y-793241 inclusion, plagioclase is much more calcic than that in the host chondrite, An_{61-44} vs. An_{19} (Sack et al. 1994).

Lithophile and siderophile element data are available for all of these inclusions. Two analyses have been done of the Barwell inclusion, one each of the interior and exterior (Hutchison et al. 1988). The LREE are slightly enriched and the HREE are depleted relative to CI chondrites, and there is a slight positive Eu anomaly. The patterns resemble that of plagioclase, which must be dominating the REE budget in the samples analyzed. The Barwell inclusion is depleted in siderophile elements, consistent with the low metal content. The Y-75097 inclusion has a highly fractionated lithophile element composition, but because of the heterogeneous nature of this inclusion (Nakamura et al. 1994, Sack et al. 1994), it is difficult to estimate a bulk inclusion composition. Many analyses of the inclusion display a common REE pattern with Eu at roughly chondritic abundance, moderate depletions in the LREE and HREE, and large depletions in the middle REE (e.g. Mittlefehldt et al. 1995, Nakamura et al. 1994, Warren and Kallemeyn 1989a). This type of pattern appears to represent that portion of the inclusion slightly interior of the margin with the host (Mittlefehldt et al. 1995). That part of the inclusion nearest the host has a REE pattern intermediate between this highly fractionated pattern and that of the host. The portion of the inclusion that may be the most interior (the original size and shape of the inclusion are unknown) is highly enriched in REE, with LREE at ~25 times CI abundances, HREE ~12 × CI and a negative Eu anomaly (Mittlefehldt et al. 1995). This portion of the inclusion has a high modal abundance of phosphates, which are controlling the REE content. Mittlefehldt et al. (1995) showed that all samples analyzed had Na contents higher than, and Sc contents lower than, those of H chondrites, and inferred therefore that the bulk inclusion could not have H chondrite composition. Analyses of the Y-793241 inclusion yield REE patterns similar to those of the intermediate regions of the Y-75097 inclusion (Mittlefehldt et al. 1995, Nakamura et al. 1994), but it is not known whether these analyses are representative.

The origin of these inclusions is unknown, but considering their general petrologic similarity, it is plausible they were all formed by the same process. Hutchison et al.

(1988) ascribed the origin of the Barwell clast to an unspecified planetary process, and they ruled out either nebular or impact processes. Mittlefehldt et al. (1995) pointed out the difficulties of producing pyroxene-free troctolites on asteroidal-sized bodies. Low pressure phase relationships do not yield olivine-plagioclase cumulates without recourse to ad hoc models; olivine-pyroxene, pyroxene and/or plagioclase-pyroxene rocks would be expected. Mittlefehldt et al. (1995) also discussed the difficulties of a nebular origin for these inclusions. No nebular materials, such as chondrules, proposed chondrule precursors, or matrix material, appear to be compositionally like the inclusions, and again, ad hoc models would be required to explain their genesis in the solar nebula. The origin of these clasts remains a conundrum.

"Norite"

Nakamura et al. (1990) and Misawa et al. (1992) describe a clast from the Hedjaz L3 chondrite which they classify as a norite. Based on the descriptions given and the photomicrographs, the major mafic mineral in this clast is olivine, and hence this is not a norite. The clast is small, only about 15 mg in mass, and yet the bulk lithophile element composition is not far removed from chondritic. Misawa et al. (1992) showed that Mg, Sc, Sr and Eu were slightly high compared to the range of L chondrites, while the moderately volatile Na, K and Rb were depleted. These authors suggested that impact melt processes could explain the unusual compositional features. Because of the small size of the sample, making representative sampling difficult, and its general similarity in bulk composition to L chondrites, we suspect that it is indeed impact melt material and will not discuss it further here.

Orthopyroxene-silica clasts

Ruzicka et al. (1995) described a silica-rich orthopyroxenite clast from the Bovedy L3 chondrite. The clast is composed of about 84% fine-grained orthopyroxene in equant to elongate grains showing normal igneous zoning with core mg# ~ 92 and rim mg# ~ 76. Tridymite and plagioclase make up about 6% each of the clast. Tridymite occurs as equant to elongate grains, generally interstitial to orthopyroxene. Plagioclase is interstitial to orthopyroxene and tridymite, and is $An_{77\text{-}66}$ in composition. Plagioclase is associated with sodic glass (about 3% of the clast). There is minor pigeonite (~1%) that occurs at the rims of orthopyroxene. Chromite, augite, metal and sulfide are accessory to trace phases.

This clast has a distinctive O-isotopic composition, plotting well away from the region of ordinary chondrites. It lies on a mass fractionation line from H chondrites, about 3.8 ‰ heavier in ^{18}O than the H chondrite mean. The bulk composition is distinct from average L chondrites for some elements. The moderately volatile alkali elements Na, K and Rb are all depleted relative to L chondrites, while volatile Cs is not. The refractory incompatible lithophile trace elements (Sc, Ti, Sr, REE, Hf, Th) are all depleted, but in general are not fractionated from each other. Aluminum is slightly depleted, Si slightly enriched and Mg, Ca, V, Cr and Mn are at roughly L chondrite abundances. Siderophile element abundances are depleted in the clast to ~0.2 × CI for refractory siderophile elements and to ~0.01 × CI for Ni, Co, Au and the chalcophile element Se. Iron is also depleted in the clast at ~0.5 × CI.

Ruzicka et al. (1995) suggested that this clast was formed by a complicated igneous process on its parent body involving melting and separation of low temperature metallic and silicate partial melts, melting at higher temperature, either as a continuation of magmatism or a second melting event, separating this high temperature melt from residual metal, olivine and orthopyroxene. Ruzicka et al. (1995) suggested that other silica-rich clasts and chondrules studied by other researchers (e.g. Bischoff et al. 1993,

Brigham et al. 1986, Wlotzka et al. 1983) may have had a magmatic origin. Some of these authors favored nebular, or at least non-igneous, origins for the clasts studied (Brigham et al. 1986, Wlotzka et al. 1983). Bischoff et al. (1993) could not choose between igneous processes, or thermal or impact metamorphism as the origin of the silica-alkali-rich clasts they studied. The very small sizes of the clasts studied by Bischoff et al. (1993), ~100-700 μm across, makes determining their petrogenesis difficult.

Nakamura et al. (1990) described a small (5 mg) orthopyroxene-silica clast from Hedjaz that may be similar to the Bovedy clast, but there is insufficient detailed information to make comparisons.

SUMMARY

The research discussed in this chapter summarizes past research and current thinking about the genesis of a wide diversity of meteorites which were affected by a common process—extensive heating on asteroidal bodies. An underlying theme of much of the current research on these rocks is the complexity of this process. Far from the simplistic ideas of whole body melting with lighter, basaltic components rising to the surface, a denser metallic component forming a core and depleted ultramafic components forming the mantle, modern meteorite research recognizes that a wide range of parameters can influence the final product. Clearly, non-chondritic meteorites were derived from a wide range of precursor chondrites, probably a wider range than is currently sampled by chondritic meteorites. Non-chondritic meteorite precursors had a wide range in volatile and moderately volatile element abundances (compare angrites, IVB irons with IAB iron metal and silicates, acapulcoites). Oxygen fugacities clearly differed by many orders of magnitude among the precursors, yielding the range from reduced aubrites to oxidized angrites. Possible variations in metal/silicate ratios and refractory lithophile/Mg ratios have been obscured by the extensive differentiation that produced many of these meteorites. The rocks were heated by a potentially diverse suite of heat sources, whose identification remains elusive (see however, Wood and Pellas 1991). The heat sources were apparently flexible enough to generate the wide range in heating observed. The degree of partial melting experienced by a parent body also varied dramatically, from very low degrees of partial melting producing acapulcoites and winonaites to very high degrees of partial melting to produce some iron meteorites and possibly the HED suite. Evidence from mesosiderites further suggests that differentiated crustal rocks were remelted and slowly crystallized on a scale larger than seems plausible for simple impact melting. Although the highly differentiated parent bodies may have had similar stratigraphies, the individual meteorites represent vastly different depths within the parent body, from metallic cores to the basaltic crust.

One interesting complexity, however, is that olivine-rich mantle rocks expected from differentiation of a chondritic precursor are surprisingly uncommon. Ureilites are potential mantle samples of their parent body, being basalt-depleted ultramafic rocks whose origin was probably as residues from partial melting. However, the O-isotopic heterogeneity, noble gas contents and siderophile element abundances of ureilites suggest that they may not have come from a parent body that underwent large-scale differentiation. The "dunite problem" has been discussed by several authors (e.g. Bell et al. 1989) and the solution likely has an explanation in asteroidal fragmentation and sampling, rather than in our understanding of melting and differentiation. Magmatic iron meteorites generally have very long cosmic-ray exposure histories compared to stony meteorites, suggesting that the mantle dunites stripped off iron meteorite cores could have been largely destroyed long ago. A second complexity is the lack of complementary basalts for the iron meteorites and the aubrite, lodranite and ureilite achondrites.

Collisional destruction may explain the lack of basalts to match the number of cores, as is generally invoked for the lack of dunites. However, the lack of basalts that can be associated with the depleted ultramafic achondrites is curious. This is especially true for the ureilites, as the polymict ureilites would be expected to contain a mixture of material from the parent body surface, including basalts.

Impact was a clear influence in the history of meteorite parent bodies. In some cases, impacts mixed lithologies from the same parent body (e.g. IAB and IIE irons) or from different parent bodies (e.g. mesosiderites). In many cases, stony achondrites experienced mixing of lithologies after solidification and cooling. These polymict brecciated achondrites vastly improve our sampling of these parent bodies, but also complicate our efforts to understand their petrogenesis. There are many unique irons, stony-irons or stones which are the only recognized representatives of their parent bodies. The full elucidation of the history of these parent bodies will almost certainly have to await further recoveries. It seems future research will likely serve to increasingly emphasize the complexity of asteroidal heating processes. Answers to many of our questions about the genesis of these rocks will be found through continued meteorite research. However, many of the most vexing problems will only be solved through eventual study of asteroidal materials in situ during planetary missions and on Earth using samples returned to our laboratories.

ACKNOWLEDGMENTS

We would like to thank several people for support and/or help in providing information used in this chapter: J.L. Berkley, G. Crozaz, C.R. Goodrich, J.N. Grossman, J.H. Jones, M. Prinz, T.D. Swindle, A.H. Treiman, D. Walker and J.T. Wasson. Reviews by the Pacific Rim review team of A.E. Rubin, H. Takeda and P.H. Warren resulted in great improvements in the manuscript, and we thank them for their Herculean efforts. The senior author rues the day he picked up the phone and heard J. Papike on the other end of the line, and vows never again to agree to produce a major manuscript under deadline pressure. He suspects the co-authors similarly wish they had not answered the phone when the senior author called. Support for D.W. Miitlefehldt came from NASA RTOP #152-13-40-21 to M.M. Lindstrom. Support for T.J. McCoy came from NASA grant NAG5-4490.

REFERENCES

Agosto WN, Hewins RH, Clarke RS Jr (1980) Allan Hills A77219, the first Antarctic mesosiderite. Proc Lunar Planet Sci Conf 11:1027-1045

Alserman AR (1940) A siderolite from Pinnaroo, South Australia. Trans Roy Soc S Aust 64:109-113

Allègre CJ, Birck JL, Fourcade S, Semet MP (1975) Rubidium-87/Strontium-87 age of Juvinas basaltic achondrite and early igneous activity in the solar system. Science 187:436-438

Anders E, Grevasse N (1989) Abundances of the elements: Meteoritic and solar. Geochim Cosmochim Acta 53:197-214

Aylmer D, Bonanno V, Herzog GF, Weber H, Klein J, Middleton R (1988) ^{26}Al and ^{10}Be production in iron meteorites. Earth Planet Sci Lett 88:107-118

Baba T, Takeda H, Saiki K (1993) Mineralogy of three EUROMET ureilites including an orthopyroxene-augite achondrite. Meteoritics 28:319

Begemann F, Weber HW, Vilcsek E, Hintenberger (1976) Rare gases and ^{36}Cl in stony-iron meteorites: cosmogenic elemental production rates, exposure ages, diffusion losses and thermal histories. Geochim Cosmochim Acta 40:353-368

Bell JF, Davis DR, Hartmann WK, Gaffey MJ (1989) Asteroids: The Big Picture. *In* Asteroids II, RP Binzel, T. Gehrels, MS Matthews (eds) Univ Arizona Press, 921-945

Bence AE, Burnett BS (1969) Chemistry and mineralogy of the silicates and metal of the Kodaikanal meteorite. Geochim Cosmochim Acta 33:387-407

Benedix GK, McCoy TJ, Keil K, Bogard DD, Garrison DH (1998a) A petrologic and isotopic study of winonaites: Evidence for early partial melting, brecciation, and metamorphism. Geochim Cosmochim Acta, in press

Benedix GK, McCoy TJ, Love SG, Keil K (1998b) A petrologic study of IAB irons: Constraints on the formation of the IAB-winonaite parent body. Geochim Cosmochim Acta (submitted)

Berkley JL (1986) Four Antarctic meteorites: Petrology and observations on ureilite petrogenesis. Meteoritics 21:169-189

Berkley JL (1990) Petrology of newly recovered orthopyroxene-bearing Antarctic meteorites: A new ureilite type? Lunar Planet Sci 21:69-70

Berkley JL, Boynton NJ (1992) Minor/major element variation within and among diogenite and howardite orthopyroxenite groups. Meteoritics 27:387-394

Berkley JL, Jones JH (1982) Primary igneous carbon in ureilites: Petrological implications. Proc Lunar Planet Sci Conf 13th, J Geophys Res 87:A353-A364

Berkley JL, Brown HG, Keil K, Carter NL, Mercier J-CC, Huss G (1976) The Kenna ureilite: An ultramafic rock with evidence for igneous, metamorphic and shock origin. Geochim Cosmochim Acta 40:1429-1437

Berkley JL, Taylor GJ, Keil K, Harlow GE, Prinz M (1980) The nature and origin of ureilites. Geochim Cosmochim Acta 44:1579-1597

Berkley JL, Goodrich CA, Keil K (1985) The unique ureilite, ALHA82106-82130: Evidence for progressive reduction during ureilite magmatic differentiation. Meteoritics 20:607-608

Bevan AWR, Grady MM (1988) Mount Morris (Wisconsin): A fragment of the IAB iron Pine River? Meteoritics 23:349-352

Bevan AWR, Bevan JC, Francis JG (1977) Amphibole in the Mayo Belwa meteorite: First occurrence in an enstatite achondrite. Mineral Mag 41:531-534

Bhandari N, Shah VG, Graham A (1981) The Lahrauli ureilite. Meteoritics 16:185-191

Bild RW (1976) A study of primitive and unusual meteorites. PhD thesis, Univ of California at Los Angeles, 234 p

Bild RW (1977) Silicate inclusions in group IAB irons and a relation to the anomalous stones Winona and Mt. Morris (Wis.). Geochim Cosmochim Acta 41:1439-1456

Bild RW, Wasson JT (1976) The Lodran meteorite and its relationship to the ureilites. Mineral Mag 40:721-735

Bild RW, Wasson JT (1977) Netschaëvo: A new class of chondritic meteorite. Science 197:58-62

Binns RA (1969) A chondritic inclusion of unique type in the Cumberland Falls meteorite. *In* Meteorite Research, PM Millman (ed) Reidel, Dordrecht, Netherlands, p 695-704

Binz CM, Ikramuddin M, Lipschutz ME (1975) Contents of eleven trace elements in ureilite achondrites. Geochim Cosmochim Acta 39:1576-1579

Binzel RP, Xu S (1993) Chips off of asteroid 4 Vesta: Evidence for the parent body of basaltic achondrite meteorites. Science 260:186-191

Binzel RP, Gaffey MJ, Thomas PC, Zellner BH, Storrs AD, Wells EN (1997) Geologic mapping of Vesta from 1994 Hubble Space Telescope images. Icarus 128:95-103

Birck JL, Allègre CJ (1978) Chronology and chemical history of the parent body of basaltic achondrites studied by the ^{87}Rb-^{87}Sr method. Earth Planet Sci Lett 39:37-51

Birck J-L, Allègre CJ (1988) Manganese-chromium isotope systematics and the development of the early solar system. Nature 331:579-584

Bischoff A, Geiger T, Palme H, Spettel B, Schultz L, Scherer P, Schlüter J, Lkhamsuren J (1993) Mineralogy, chemistry, and noble gas contents of Adzhi-Bogdo—an LL3-6 chondritic breccia with L-chondritic and granitoidal clasts. Meteoritics 28:570-578

Black DC (1972) On the origin of trapped helium, neon and argon isotopic variations in gas-rich meteorites, lunar soil and breccia. Geochim Cosmochim Acta 36:347-375

Bland P, Pillinger CT, Hutchison R (1992) A new ureilite from the Sahara—Acfer 277. Lunar Planet Sci 23:119

Boesenberg JS, Prinz M, Weisberg MK, Davis AM, Clayton RN, Mayeda TK, Wasson JT (1995) Pyroxene-pallasites: A new pallasite grouplet. Meteoritics 30:488-489

Boesenberg JS, Delaney JS, Prinz M (1997) Magnesian megacrysts and matrix in the mesosiderite Lamont. Lunar Planet Sci 28:125-126

Bogard DD (1995) Impact ages of meteorites: A synthesis. Meteoritics Planet Sci 30:244-268

Bogard DD, Garrison DH (1994) ^{39}Ar-^{40}Ar ages of four ureilites. Lunar Planet Sci 25:137-138

Bogard DD, Garrison DH (1995) ^{39}Ar-^{40}Ar age of the Ibitira eucrite and constraints on the time of pyroxene equilibration. Geochim Cosmochim Acta 59:4317-4322

Bogard DD, Garrison DH (1998) ^{39}Ar-^{40}Ar ages and thermal history of mesosiderites. Geochim Cosmochim Acta, in press

Bogard D, Burnett D, Eberhardt P, Wasserburg GJ (1967) ^{40}Ar-^{40}K ages of silicate inclusions in iron meteorites. Earth Planet Sci Lett 3:275-283
Bogard DD, Burnett DS, Wasserburg GJ (1969) Cosmogenic rare gases and the ^{40}K-^{40}Ar age of the Kodaikanal iron meteorite. Earth Planet Sci Lett 5:273-281
Bogard DD, Gibson EK, Moore DR, Turner NL, Wilken RB (1973) Noble gas and carbon abundances of the Haverö, Dingo Pup Donga, and North Haig ureilites. Geochim Cosmochim Acta 37:547-557
Bogard DD, Nyquist LE, Johnson P, Wooden J, Bansal B (1983) Chronology of Brachina. Meteoritics 18:269-270
Bogard DD, Garrison DH, Jordan JL, Mittlefehldt D (1990) ^{39}Ar-^{40}Ar dating of mesosiderites: Evidence for major parent body disruption <4 Ga ago. Geochim Cosmochim Acta 54:2549-2564
Bottke W F Jr., M C Nolan, R Greenberg, R A Kolvoord (1994) Velocity distributions among colliding asteroids. Icarus 107:255-268
Bowman LE, Spilde MN and Papike JJ (1997) Automated energy dispersive spectrometer modal analysis applied to the diogenites. Meteoritics Planet Sci 32:869-875
Boynton WV, Hill DH (1993) Trace-element abundances in several new ureilites. Lunar Planet Sci 24:167-168
Boynton WV, Starzyk PM, Schmitt RA (1976) Chemical evidence for the genesis of the ureilites, the achondrite Chassigny and the nakhlites. Geochim Cosmochim Acta 40:1439-1447
Brearley AJ, Prinz M (1992) CI chondrite-like clasts in the Nilpena polymict ureilite. Implications for aqueous alteration processes in CI chondrites. Geochim Cosmochim Acta 56:1373-1386
Brett R, Keil K (1986) Enstatite chondrites and enstatite achondrites (aubrites) were not derived from the same parent body. Earth Planet Sci Lett 81:1-6
Bridges JC, Hutchison R (1997) A survey of clasts and large chondrules in ordinary chondrites. Meteoritics Planet Sci 32:389-394
Brigham CA, Yabuki H, Ouyang Z, Murrell MT, El Goresy A, Burnett DS (1986) Silica-bearing chondrules and clasts in ordinary chondrites. Geochim Cosmochim Acta 50:1655-1666
Brouxel M, Tatsumoto M (1990) U-Th-Pb systematics of the Estherville mesosiderite. Proc Lunar Planet Sci Conf 20:309-319
Brouxel M, Tatsumoto M (1991) The Estherville mesosiderite: U-Pb, Rb-Sr, and Sm-Nd isotopic study of a polymict breccia. Geochim Cosmochim Acta 55:1121-1133
Buchanan PC, Reid AM (1996) Petrology of the polymict eucrite Petersburg. Geochim Cosmochim Acta 60:135-146
Buchanan PC, Zolensky ME, Reid AM (1993) Carbonaceous chondrite clasts in the howardites Bholghati and EET87513. Meteoritics 28:659-669
Buchwald VFB (1975) Handbook of iron meteorites. 3 vol. Berkeley: University of California Press
Buchwald VF (1977) The mineralogy of iron meteorites. Phil Trans Royal Soc London A286:453-491
Buchwald VF, Scott ERD (1971) First nitride (CrN) in iron meteorites. Nature Phys Sci 233:113-114
Bunch, TE (1975) Petrography and petrology of basaltic achondrite polymict breccias (howardites). Proc Lunar Sci Conf 6:469-492
Bunch TE, Fuchs LH (1969) A new mineral: brezinaite, Cr_3S_4, and the Tucson meteorite. Am Mineral 54:1509-1518
Bunch TE, Keil K (1971) Chromite and ilmenite in non-chondritic meteorites. Am Mineral 56:146-157
Bunch TE, Olsen E (1968) Potassium feldspar in Weekeroo Station, Kodaikanal, and Colomera iron meteorites. Science 160:1223-1225
Bunch TE, Keil K, Olsen E (1970) Mineralogy and petrology of silicate inclusions in iron meteorites. Contrib Mineral Petrol 25:297-240
Bunch TE, Keil K, Huss GI (1972) The Landes meteorite. Meteoritics 7:31-38
Burghle A, Dreibus G, Palme H, Rammensee W, Spettel B, Weckworth G, Wänke H (1983) Chemistry of shergottites and Shergottite Parent Body (SPB): Further evidence for the two component model for planet formation. Lunar Planet Sci 14:80-81
Burnett DS, Wasserburg GJ (1967a) Evidence for the formation of an iron meteorite at 3.8×10^9 years. Earth Planet Sci Lett 2:137-147
Burnett DS, Wasserburg GJ (1967b) ^{87}Rb-^{87}Sr ages of silicate inclusions in iron meteorites. Earth Planet Sci Lett 2:397-408
Buseck PR (1977) Pallasite meteorites-mineralogy, petrology, and geochemistry. Geochim Cosmochim Acta 41:711-740
Buseck PR, Goldstein JI (1969) Olivine compositions and cooling rates of pallasitic meteorites. Geol Soc Amer Bull 80:2141-2158
Buseck PR, Holdsworth E (1977) Phosphate minerals in pallasite meteorites. Mineral Mag 41:91-102
Carter NL, Raleigh CB, DeCarli P (1968) Deformation of olivine in stony meteorites. J Geophys Res 73:5439-5461

Casanova I (1992) Osbornite and the distribution of titanium in enstatite meteorites. Meteoritics 27: 208-209

Casanova I, Keil K, Newsom HE (1993a) Composition of metal in aubrites: Constraints on core formation. Geochim Cosmochim Acta 57:675-682

Casanova I, McCoy TJ, Keil K (1993b) Metal-rich meteorites from the aubrite parent body. Lunar Planet Sci 24:259-260

Casanova I, Graf T, Marti K (1995) Discovery of an unmelted H-chondrite inclusion in an iron meteorite. Science 268:469-608

Chase MW, Davies CA, Downing JR, Frurip DJ, McDonald RA, Syverud AN (1985) JANAF thermochemical tables. J Phys Chem Ref Data 14, Suppl. 1

Chen JH, Wasserburg GJ (1990) The isotopic composition of Ag in meteorites and the presence of ^{107}Pd in protoplanets. Geochim Cosmochim Acta 54:1729–1743

Chikami J, Mikouchi T, Takeda H, Miyamoto M (1997a) Mineralogy and cooling history of the calcium-aluminum-chromium enriched ureilite, Lewis Cliff 88774. Meteoritics Planet Sci 32:343-348

Chikami J, Mikouchi T, Miyamoto M, Takeda H (1997b) Mineralogical comparison of Hammadah Al Hamra 126 with some ureilites. Antarctic Meteorite Res 10:389-399

Choi B-G, Ouyang X, Wasson JT (1995) Classification and origin of IAB and IIICD iron meteorites. Geochim Cosmochim Acta 59:593-612

Chou C-L, Boynton WV, Bild RW, Kimberlin J, Wasson JT (1976) Trace element evidence regarding a chondritic component in howardite meteorites. Proc Lunar Sci Conf 7:3501-3518

Christophe Michel-Levy M, Bourot-Denise M, Palme H, Spettel B, Wänke H (1987) L'eucrite de Bouvante: chimie, pétrologie et minéralogie. Bull Minéral 110:449-458

Clarke, RS Jr., Mason B (1982). A new metal-rich mesosiderite from Antarctica, RKPA79015. Natl Inst Polar Res (Japan) Spec Issue 25:78-85

Clarke RS Jr, Scott ERD (1980) Tetrataenite—ordered FeNi, a new mineral in meteorites. Am Mineral 65:624-630

Clayton RN (1993) Oxygen-isotopes in meteorites. Ann Rev Earth Planet Sci 21:115-149

Clayton RN, Mayeda TK (1978) Genetic relations between iron and stony meteorites. Earth Planet Sci Lett 40:168-174

Clayton RN, Mayeda TK (1983) Oxygen-isotopes in eucrites, shergottites, nakhlites, and chassignites. Earth Planet Sci Lett 62:1-6

Clayton RN, Mayeda TK (1988) Formation of ureilites by nebular processes. Geochim Cosmochim Acta 52:1313-1318

Clayton RN, Mayeda TK (1996) Oxygen-isotope studies of achondrites. Geochim Cosmochim Acta 60:1999-2018

Clayton RN, Grossman L, Mayeda TK (1973) A component of primitive nuclear composition in carbonaceous chondrites. Science 181:485-487

Clayton RN, Mayeda TK, Onuma N, Shearer J (1976a) Oxygen-isotopic composition of minerals in the Kenna ureilite. Geochim Cosmochim Acta 40:1475-1476

Clayton RN, Onuma N, Mayeda TK (1976b) A classification of meteorites based on oxygen-isotopes. Earth Planet Sci Lett 30:10-18

Clayton RN, Mayeda TK, Rubin AE (1984) Oxygen-isotopic compositions of enstatite chondrites and aubrites. Proc Lunar Planet Sci Conf 15:C245-C249

Clayton RN, Mayeda TK, Nagahara H (1992) Oxygen-isotope relationships among primitive achondrites. Lunar Planet Sci 23:231-232

Consolmagno GJ, Drake MJ (1977) Composition and evolution of the eucrite parent body: evidence from rare earth elements. Geochim Cosmochim Acta 41:1271-1282

Crozaz G, Lundberg LL (1995) The origin of oldhamite in unequilibrated enstatite chondrites. Geochim Cosmochim Acta 59:3817-3831

Crozaz G, McKay G (1990) Rare earth elements in Angra dos Reis and Lewis Cliff 86010, two meteorites with similar but distinct magma evolutions. Earth Planet Sci Lett 97:369-381

Crozaz G, Pellas P (1983) Where does Brachina come from? Lunar Planet Sci 14:142-143

Cruikshank DP, Tholen DJ, Hartmann WK, Bell JF, Brown RH (1991) Three basaltic earth-approaching asteroids and the source of the basaltic meteorites. Icarus 89:1-13

Davis AM (1977) The cosmochemical history of the pallasites. Ph.D dissertation, Yale University, 285 p

Davis AM, Olsen EJ (1991) Phosphates in pallasite meteorites as probes of mantle processes in small planetary bodies. Nature 353:637-640

Davis AM, Prinz M, Laughlin JR (1988) An ion microprobe study of plagioclase-rich clasts in the North Haig polymict ureilite. Lunar Planet Sci 19:251-252

Dean DC, Goldstein JI (1986) Determination of interdiffusion coefficients in the Fe-Ni and Fe-Ni-P systems below 900°C. Metall Trans 17A:1131-1138

Delaney JS, Sutton SR (1988) Lewis Cliff 86010, an ADORable Antarctican. Lunar Planet Sci 19:265-266
Delaney JS, Nehru CE, Prinz M (1980) Olivine clasts from mesosiderites and howardites: Clues to the nature of achondritic parent bodies. Proc Lunar Planet Sci 11:1073-1087
Delaney JS, Nehru CE, Prinz M, Harlow GE (1981) Metamorphism in mesosiderites. Proc Lunar Planet Sci 12B:1315-1342
Delaney JS, Takeda H, Prinz M, Nehru CE, Harlow GE (1983) The nomenclature of polymict basaltic achondrites. Meteoritics 18:103-111
Delaney JS, Prinz M, Takeda H (1984) The polymict eucrites. Proc 15th Lunar Planet Sci Conf Part 1, J Geophys Res 89 Supl:C251-C288
Desnoyers C (1982) L'olivine dans les howardites: origine, et implications pour le corps parent de ces météorites achondritiques. Geochim Cosmochim Acta 46:667-680
Desnoyers C, Christophe Michel-Levy M, Azevedo IS, Scorzelli RB, Danon J, da Silva EG (1985) Mineralogy of the Bocaiuva iron meteorite: A preliminary study. Meteoritics 20:113-124
Dickinson TL, Lofgren GE (1992) Melting relations for Indarch (EH4) under reducing conditions. Lunar Planet Sci 23:307-308
Dickinson TL, McCoy TJ (1997) Experimental rare-earth-element partitioning in oldhamite: Implications for the igneous origin of aubritic oldhamite. Meteoritics Planet Sci 32:395-412
Dollase WA (1967) The crystal structure at 220°C of orthorhombic high tridymite from the Steinbach meteorite. Acta Cryst 23:617-623
Dreibus G, Palme H, Rammensee W, Spettel B, Weckworth G, Wänke H (1982) Composition of the Shergottite Parent Body: Further evidence of a two component model of planet formation. Lunar Planet Sci 13:186-187
Duke MB (1965) Metallic iron in basaltic achondrites. J Geophys Res 70:1523-1527
Duke MB, Silver LT (1967) Petrology of eucrites, howardites and mesosiderites. Geochim Cosmochim Acta 31:1637-1665
Dymek RF, Albee AL, Chodos AA, Wasserburg GJ (1976) Petrography of isotopically-dated clasts in the Kapoeta howardite and petrologic constraints on the evolution of its parent body. Geochim Cosmochim Acta 40:1115-1130
Ebihara M, Shinonaga T, Takeda H (1990) Trace element compositions of Antarctic ureilites and some implications to their origin. Meteoritics 25:359-360
Ebihara M, Ikeda Y, Prinz M (1997) Petrology and chemistry of the Miles IIE iron. II: Chemical characteristics of the Miles silicate inclusions. Antarctic Meteorite Res 10:373-388
Ehlers EC (1972) The Interpretation of Geological Phase Diagrams. San Francisco: WH Freeman and Co, 327 p
Esbensen KH, Buchwald VF, Malvin DJ, Wasson JT (1982) Systematic compositional variations in the Cape York iron meteorite. Geochim Cosmochim Acta 46:1913-1920
Eugster O, Michel T (1995) Common asteroid break-up events of eucrites, diogenites, and howardites and cosmic-ray production rates for noble gases in achondrites. Geochim Cosmochim Acta 59:177-199
Eugster O, Michel Th, Niedermann S (1991) ^{244}Pu-Xe formation and gas retention age, exposure history, and terrestrial age of angrites LEW86010 and LEW87051: Comparison with Angra dos Reis. Geochim Cosmochim Acta 55:2957-2964
Evensen NM, Hamilton PJ, Harlow GE, Klimentidis R, O'Nions RK, Prinz M (1979) Silicate inclusions in Weekeroo Station: Planetary differentiates in an iron meteorite. Lunar Planet Sci 10:376-378
Fioretti AM, Molin G (1996) Petrography and mineralogy of FRO93008 ureilite: Evidence for pairing with FRO90054 ureilite. Meteoritics 31:A43-A44
Fioretti AM, Molin G, Brandstätter F, Kurat G (1996) Schreibersite, metal, and troilite in ureilites FRO90054 and FRO93008. Meteoritics 31:A44
Floran RJ (1978) Silicate petrography, classification, and origin of the mesosiderites: Review and new observations. Proc Lunar Planet Sci Conf 9:1053-1081
Floran RJ, Prinz M, Hlava PF, Keil K, Nehru CE, Hinthorne JR (1978a) The Chassigny meteorite: a cumulate dunite with hydrous amphibole-bearing melt inclusions. Geochim Cosmochim Acta 42:1213-1229
Floran RJ, Caulfield JBD, Harlow GE, Prinz M (1978b) Impact origin for the Simondium, Pinnaroo, and Hainholz mesosiderites: Implications for impact processes beyond the earth-moon system. Proc Lunar Planet Sci Conf 9:1083-1114
Floran RJ, Prinz M, Hlava PF, Keil K, Spettel B, Wänke H (1981) Mineralogy, petrology, and trace element geochemistry of the Johnstown meteorite: a brecciated orthopyroxenite with siderophile and REE-rich components. Geochim Cosmochim Acta 45, 2385-2391
Floss C, Crozaz G (1993) Heterogeneous REE patterns in oldhamite from the aubrites: Their nature and origin. Geochim Cosmochim Acta 57:4039-4057
Floss C, Strait MM, Crozaz G (1990) Rare earth elements and the petrogenesis of aubrites. Geochim

Cosmochim Acta 57:4039-4057
Fogel RA (1994) Aubrite basalt vitrophyres: High sulfur silicate melts and a snapshot of aubrite formation. Meteoritics 29:466-467
Fogel RA (1997) A new aubrite basalt vitrophyre from the LEW 87007 aubrite. Lunar Planet Sci 28:369-370
Fogel RA, Hess PC, Rutherford MJ (1988) The enstatite chondrite-achondrite link. Lunar Planet Sci 19:342-343
Fogel RA, Weisberg MK, Prinz M (1996) The solubility of CaS in aubrite silicate melts. Lunar Planet Sci 27:371-372
Folco L (1992) Meteorites from the 1990/'91 expedition to the Frontier Mountains, Antarctica. Meteoritics 27:221-222
Foshag WF (1940) The Shallowater meteorite; a new aubrite. Am Mineral 25:279-286
Fowler GW, Papike JJ, Spilde MN, Shearer CK (1994) Diogenites as asteroidal cumulates: Insights from orthopyroxene major and minor element chemistry. Geochim Cosmochim Acta 58:3921-3929
Fowler GW, Shearer CK, Papike JJ, Layne GD (1995) Diogenites as asteroidal cumulates: Insights from orthopyroxene trace element chemistry. Geochim Cosmochim Acta 59:3071-3084
Fredriksson K (1982) The Manegaon diogenite. Meteoritics 17:141-144
Frondel C, Marvin UB (1967) Lonsdaleite, a hexagonal polymorph of diamond. Nature 214:587-589
Fuchs LH (1966) Djerfisherite, alkali copper-iron sulfide, a new mineral from the Kota-Kota and St. Mark's enstatite chondrites. Science 153:166-167
Fuchs LH (1969) The phosphate mineralogy of meteorites. *In* Meteorite Research (ed. PM Millman), p 683-695. Reidel
Fuchs LH (1974) Glass inclusions of granitic compositions in orthopyroxene from three enstatite achondrites. Meteoritics 9:342
Fuhrman M, Papike JJ (1981) Howardites and polymict eucrites: Regolith samples from the eucrite parent body. Petrology of Bholgati, Bununu, Kapoeta, and ALHA76005. Proc. Lunar Planet Sci 12B:1257-1279
Fukuoka T, Kimura M (1990) Chemistry of Y-74063, -74357 and ALH-78230 unique meteorites. Papers presented 15th Symp Antarctic Meteorites, NIPR, Tokyo, 99-100
Fukuoka T, Ma M-S, Wakita H, Schmitt RA (1978) Lodran: The residue of limited partial melting of matter like a hybrid between H and E chondrites. Lunar Planet Sci 9:356-358
Gaffey MJ (1990) Thermal history of the asteroid belt: Implications for accretion of the terrestrial planets. In Origin of the Earth (HE Newsom, JH Jones, eds) Oxford Univ Press p 17-28
Gaffey MJ (1997) Surface lithologic heterogeneity of asteroid 4 Vesta. Icarus 127:130-157
Ganguly J, Yang H, Ghose S (1994) Thermal history of mesosiderites: Quantitative constraints from compositional zoning and Fe-Mg ordering in orthopyroxenes. Geochim Cosmochim Acta 58: 2711-2723
Garcia DJ, Prinz M (1978) The Binda orthopyroxene cumulate eucrite: A slightly polymict brecciated melanorite. Meteoritics 13:473
Garrison DH, Bogard DD (1995) ^{39}Ar-^{40}Ar ages of silicates from IIE iron meteorites. Meteoritics 30:508
Gibson EK (1976) Nature of the carbon and sulfur phases and inorganic gases in the Kenna ureilite. Geochim Cosmochim Acta 40:1459-1464
Göbel R, Ott U, Begemann F (1978) On trapped noble gases in ureilites. J Geophys Res 83:855-867
Goldstein JI, Ogilvie RE (1965) The growth of the Widmannstätten pattern in metallic meteorites. Geochim Cosmochim Acta 29:893-920
Gomes CB and Keil K (1980) Brazilian Stone Meteorites. Univ New Mexico Press, 162 p
Goodrich CA (1986a) Trapped primary silicate liquid in ureilites. Lunar Planet Sci 17:273-274
Goodrich CA (1986b) Y73140: A ureilite with cumulus augite. Meteoritics 21:373-374
Goodrich CA (1988) Petrology of the unique achondrite LEW86010. Lunar Planet Sci 19:399-400
Goodrich CA (1992) Ureilites: a critical review. Meteoritics 27:327-252
Goodrich CA (1997a) Preservation of a nebular mg-$\Delta^{17}O$ correlation during partial melting of ureilites. Lunar Planet Sci 28:435-436
Goodrich CA (1997b) Iron-manganese-magnesium relations and mg-$\Delta^{17}O$ correlations in ureilites, lodranites, Allende chondrules, and ordinary chondrites: Nebular or planetary? Meteoritics Planet Sci 32:A49-A50
Goodrich CA (1997c) The chondrite-achondrite transition: Decoupling of oxygen-isotopic and geochemical changes. *In* Workshop on Parent-Body and Nebular Modification of Chondritic Materials. Zolensky ME, Krot AN, Scott ERD (eds) LPI Tech Report 97-02 Part 1, 15-17
Goodrich CA (1998) A ureilite (Hughes 009) with an unusual shock texture: Implications for the origin of metal in ureilites? Lunar Planet Sci 29:1123
Goodrich CA, Berkley JL (1986) Primary magmatic carbon in ureilites: Evidence from cohenite-bearing

spherules. Geochim Cosmochim Acta 50:681-691
Goodrich CA, Lugmair GW (1991) PCA82506: a ureilite with LREE-enriched component and a whole-rock Sm-Nd model age of 4.55 Ga. Lunar Planet Sci 22:467-468
Goodrich CA, Lugmair GW (1992) Addition of LREE-enriched material to a ureilite at 4.23 Ga: Evidence for episodic metasomatism? Lunar Planet Sci 23:429-430
Goodrich CA, Lugmair GW (1995) Stalking the LREE-enriched component in ureilites. Geochim Cosmochim Acta 59:2609-2620
Goodrich CA Keil K, Berkley JL, Laul JC, Smith MR, Wacker JF, Clayton RN, Mayeda TK (1987a) Roosevelt County 027: A low-shock ureilite with interstitial silicates and high noble-gas concentration. Meteoritics 22:191-218
Goodrich CA, Jones JH, Berkley JL (1987b) Origin and evolution of the ureilite parent magmas: Multi-stage igneous activity on a large parent body. Geochim Cosmochim Acta 51:2255-2273
Goodrich CA, Patchett PJ, Lugmair GW, Drake MJ (1991) Sm-Nd and Rb-Sr isotopic systematics of ureilites. Geochim Cosmochim Acta 55:829-848
Goodrich CA, Lugmair GW, Drake MJ, Patchett PJ (1995) Comment on "U-Th-Pb and Sm-Nd isotopic systematics of the Goalpara ureilite: Resolution of terrestrial contamination" by N Torigoye-Kita, K Misawa, and M Tatsumoto. Geochim Cosmochim Acta 59:4083-4085
Gooley R (1972) The chemistry and mineralogy of the diogenites. Ph.D dissertation, Arizona State University 198 p
Gooley R, Moore CB (1976) Native metal in diogenite meteorites. Am Mineral 61:373-378
Göpel C, Manhès G, Allègre CJ (1985) Concordant 3,676 Ma U-Pb formation age for the Kodaikanal iron meteorite. Nature 317:341-344
Göpel Ch., Manhès G, Allègre CJ (1992) U-Pb study of the Acapulco meteorite. Meteoritics 27:226
Grady MM, Pillinger CT (1986) The ALHA 82130 ureilite: Its light element stable isotope composition and relationship to other ureilites. Meteoritics 21:375-376
Grady MM, Pillinger CT (1987a) The EET83309 polymict ureilite: Its relationship to other ureilites on the basis of stable isotope measurements. Lunar Planet Sci 18:353-354
Grady MM, Pillinger CT (1987b) Unusual nitrogen isotopic composition of polymict ureilites. Meteoritics 22:394-395
Grady MM, Pillinger CT (1993) EUROMET ureilite consortium: A preliminary report on carbon and nitrogen geochemistry. Lunar Planet Sci 24:551-552
Grady MM, Wright IP, Swart PK, Pillinger CT (1985a) The carbon and nitrogen isotopic composition of ureilites: Implications for their genesis. Geochim Cosmochim Acta 49:903-915
Grady MM, Pillinger CT and Wacker J F (1985b) Carbon isotopes in HF/HCL residues of the unshocked ureilite ALHA78019. Meteoritics 20:652-653
Grossman JN (1994) Meteoritical Bulletin. Meteoritics 29:100-143
Guan Y, Crozaz G (1995a) The quest for the elusive LREE carrier in ureilites: An ion microprobe study. Lunar Planet Sci 26:527-528
Guan Y, Crozaz G (1995b) Rare-earth-element distribution in Elephant Morraine and the "missing" basaltic component associated with ureilites. Meteoritics 30:514-515
Guan Y, Crozaz G (1997) Rare earth elements in some lithic and mineral clasts of polymict ureilites and petrogenetic implications. Lunar Planet Sci 28:485-486
Haack H, Scott ERD (1993) Chemical fractionations in Group IIIAB iron meteorites: Origin by dendritic crystallization of an asteroidal core. Geochim Cosmochim Acta 57:3457-3472
Haack H, Rasmussen KL, Warren PH (1990) Effects of regolith/megaregolith insulation on the cooling histories of differentiated asteroids. J Geophys Res 95:5111-5124
Haack H, Scott ERD, Love SG, Brearley AJ, McCoy TJ (1996a) Thermal histories of IVA stony-iron and iron meteorites: Evidence for asteroid fragmentation and reaccretion. Geochim Cosmochim Acta 60:3103-3113
Haack H, Scott ERD, Rasmussen KL (1996b) Thermal and shock history of mesosiderites and their large parent asteroid. Geochim Cosmochim Acta 60:2609-2619
Haramura H, Kushiro I, Yanai K (1983) Chemical compositions of Antarctic meteorites I Proc Symp Antarctic Meteorites 8 109-121, NIPR, Tokyo
Harlow GE, Nehru CE, Prinz M, Taylor GJ, Keil K (1979) Pyroxenes in Serra de Magé: Cooling history in comparison with Moama and Moore County. Earth Planet Sci Lett 43:173-181
Haskin L, Warren P (1991) Chapter 8: Lunar Chemistry. *In* Lunar Sourcebook: A User's Guide to the Moon (G Heiken, D Vaniman, BM French, eds) Cambridge Univ Press, Cambridge, UK
Hassanzadeh J, Rubin AE, Wasson JT (1990) Compositions of large metal nodules in mesosiderites; links to iron meteorite group IIIAB and the origin of mesosiderite subgroups. Geochim Cosmochim Acta 54:3197-3208
Herpfer MA, Larimer JW, Goldstein JI (1994) A comparison of metallographic cooling rate methods used

in meteorites. Geochim Cosmochim Acta 58:1353-1366

Hess HH, Henderson EP (1949) The Moore County meteorite: A further study with comment on its primordial environment. Am Mineral 34:494-507

Hewins RH (1983) Impact versus internal origins for mesosiderites. Proc 14th Lunar Planet Sci Conf, Part 1: J Geophys Res 88:B257-B266

Hewins RH (1984) The case for a melt matrix in plagioclase-POIK mesosiderites. Proc 15th Lunar Planet Sci Conf, Part 1: J Geophys Res 89 Suppl, C289-C297

Hewins RH (1988) Petrology and pairing of mesosiderites from Victoria Land, Antarctica. Meteoritics 23:123-129

Hewins RH, Harriott TA (1986) Melt segregation in plagioclase-poikilitic mesosiderites. Proc 16th Lunar Planet Sci Conf, Part 2: J Geophys Res 91:D365-D372

Hewins RH, Klein LC (1978) Provenance of metal and melt rock textures in the Malvern howardite. Proc Lunar Planet Sci Conf 9:1137-1156

Hewins RH, Newsom HE (1988) Igneous activity in the early solar system. *In* Meteorites and the Early Solar System (JF Kerridge and MS Matthews, eds.) Univ Arizona Press, Tucson, AZ, 73-101

Higuchi H, Morgan JW, Ganapathy R, Anders E (1976) Chemical variation in meteorites—X Ureilites. Geochim Cosmochim Acta 40:1563-1571

Hiroi T, Bell JF, Takeda H, Pieters CM (1993) Spectral comparison between olivine-rich asteroids and pallasites. Proc NIPR Symp Antarctic Meteorites 6:234-245

Hohenberg CM (1970) Xenon from the Angra dos Reis meteorite. Geochim Cosmochim Acta 34:185-191

Hohenberg CM, Bernatowicz TJ, Podosek FA (1991) Comparative xenology of two angrites. Earth Planet Sci Lett 102:167-177

Honda M (1988) Statistical estimation of the production of cosmic-ray-induced nuclides in meteorites. Meteoritics. 23:3-12

Horan MF, Smoliar MI, Walker RJ (1998) ^{182}W and ^{187}Re-^{187}Os systematics of iron meteorites: Chronology for melting, differentiation, and crystallization in asteroids. Geochim Cosmochim Acta 62:545-554

Hostetler CJ, Drake MJ (1978) Quench temperatures of Moore County and other eucrites: residence time on eucrite parent body. Geochim Cosmochim Acta 42:517-522

Housen KR, Wilkening LL, Chapman CR, Greenberg RJ (1979) Asteroidal regoliths. Icarus 39:317-352

Hsu W, Crozaz G (1997) Mineral chemistry and the petrogenesis of eucrites: II. Cumulate eucrites. Geochim Cosmochim Acta 61:1293-1302

Hsu W, Huss GR, Wasserburg GJ (1997) Mn-Cr systematics of differentiated meteorites . Lunar Planet Sci 28:609-610

Humayun M, Clayton RN (1995) Potassium isotope cosmochemistry: Genetic implications of volatile element depletion. Geochim Cosmochim Acta 59:2131-2148

Hutcheon ID, Olsen E (1991) Cr isotopic composition of differentiated meteorites: A search for ^{53}Mn. Lunar Planet Sci 22:605-606

Hutchison R (1977) A crystalline ureilite from Oman. Meteoritics 12:263

Hutchison R, Williams CT, Din VK, Clayton RN, Kirschbaum C, Paul RL, Lipschutz ME (1988) A planetary, H-group pebble in the Barwell, L6, unshocked chondritic meteorite. Earth Planet Sci Conf 90:105-118

Ikeda Y, Prinz M (1996) Petrology of silicate inclusions in the Miles IIE iron. Proc NIPR Symp Antarctic Meteorites 9:143-173

Ikeda Y, Takeda H (1985) A model for the origin of basaltic achondrites based on the Yamato 7308 howardite. Proc Lunar Planet Sci Conf 15th; J Geophys Res 90:C649-C663

Ikeda Y, Ebihara M, Prinz M (1990) Enclaves in the Mt. Padbury and Vaca Muerta mesosiderites: Magmatic and residue (or cumulate) rock types. Proc NIPR Symp Antarctic Meteorites 3:99-131

Ikeda Y, Ebihara M, Prinz M (1997) Petrology of and chemistry of the Miles IIE iron. I: Description and petrology of twenty new silicate inclusions. Antarctic Meteorite Res 10:355-372

Imamura M, Shima M, Honda M (1980) Radial distribution of spallogenic K, Ca, Ti, V and Mn isotopes in iron meteorites. Z Naturforsch 35A:267-279

Ireland TR, Wlotzka F (1992) The oldest zircons in the solar system. Earth Planet Sci Lett 109:1-10

Jacobsen SB, Wasserburg GJ (1984) Sm-Nd isotopic evolution of chondrites and achondrites, II. Earth Planet Sci Lett 67:137-150

Janssens M-J, Hertogen J, Wolf R, Ebihara M, Anders E (1987) Ureilites: Trace element clues to their origin. Geochim Cosmochim Acta 51:2275-2283

Jaques AL, Fitzgerald MJ (1982) The Nilpena ureilite, an unusual polymict breccia: Implications for origin. Geochim Cosmochim Acta 46:893-900

Jarosewich E (1967) Chemical analysis of seven stony meteorites and one iron with silicate inclusions. Geochim Cosmochim Acta 31:1103-1106

Jarosewich E (1990) Chemical analyses of meteorites: A compilation of stony and iron meteorite analyses.

Meteoritics 25:323-337

Jarosewich E, Mason B (1966) Chemical analysis with notes on one mesosiderite and seven chondrites. Geochim Cosmochim Acta 33:411-416

Johnson JE, Scrymgour JM, Jarosewich E, Mason B (1977) Brachina meteorite—a chassignite from South Australia. Rec S Australia Mus 17:309-319

Jones JH (1982) The geochemical coherence of Pu and Nd and the $^{244}Pu/^{238}U$ ratio of the early solar system. Geochim Cosmochim Acta 46:1793-1804

Jones JH (1983) Mesosiderites: (1) reevaluation of cooling rates and (2) experimental results bearing on the origin of metal. Lunar Planet Sci 14:351-352

Jones JH (1994) Fractional crystallization of iron meteorites; constant versus changing partition coefficients. Meteoritics 29:423-426

Jones JH, Boynton WV (1983) Experimental geochemistry in very reducing systems: Extreme REE fractionation by immiscible sulfide liquids. Lunar Planet Sci 14:353-354

Jones JH, Drake MJ (1983) Experimental investigations of trace element fractionation in iron meteorites; II, The influence of sulfur. Geochim Cosmochim Acta 47:1199-1209

Jones JH, Malvin DJ (1990) A nonmetal interaction model for the segregation of trace metals during solidification of Fe-Ni-S, Fe-Ni-P, and Fe-Ni-S-P alloys. Met Trans B 21B:697-706

Jones JH, Treiman AH, Janssens M-J, Wolf R, Ebihara M (1988) Core formation on the eucrite parent body, the moon and the AdoR parent body. Meteoritics 23:276-277

Jones JH, Mittlefehldt DW, Jurewicz AJG, Lauer HV Jr, Hanson BZ, Paslick CR, McKay GA (1996) The origin of eucrites: An experimental perspective. *In* Workshop on Evolution of Igneous Asteroids: Focus on Vesta and the HED Meteorites, Mittlefehldt DW, Papike JJ, eds., LPI Tech Report 96-02, Part 1, 15

Jurewicz AJG, Mittlefehldt DW, Jones JH (1993) Experimental partial melting of the Allende (CV) and Murchison (CM) chondrites and the origin of asteroidal basalts. Geochim Cosmochim Acta 57:2123-2139

Jurewicz AJG, Mittlefehldt DW, Jones JH (1995a) Experimental partial melting of the St. Severin (LL) and Lost City (H) chondrites. Geochim Cosmochim Acta 59:391-408

Jurewicz AJG, Jones JH, Mittlefehldt DW, Longhi J(1995b) Making melts having 40%, 50% or 60% SiO_2 from chondritic materials: A synopsis of low-pressure, low-volatile, equilibrium melting relations. Lunar Planet Sci 26:707-708

Kallemeyn GW, Warren PH (1994) Geochemistry of LEW88774 and two other unusual ureilites. Lunar Planet Sci 25:663-664

Kallemeyn GW, Wasson JT (1985) The compositional classification of chondrites: IV Ungrouped chondritic meteorites and clasts. Geochim Cosmochim Acta 49:261-270

Keil K (1968) Mineralogical and chemical relationships among enstatite chondrites. J Geophys Res 73:6945-6976

Keil K (1989) Enstatite meteorites and their parent bodies. Meteoritics 24:195-208

Keil K, Brett R (1974) Heideite, $(Fe,Cr)_{1+x}(Ti,Fe)_2S_4$: A new mineral in the Bustee enstatite achondrite. Am Mineral 59:465-470

Keil K, Wilson L (1993) Explosive volcanism and the composition of cores of differentiated asteroids. Earth Planet Sci Lett 117:111-124

Keil K, Fodor RV, Starzyk PM, Schmitt RA, Bogard DD, Husain L (1980) A 3.6-b.y.-old impact-melt rock fragment in the Plainview chondrite: Implications for the age of the H-group chondrite parent body regolith formation. Earth Planet Sci Lett 51:235-247

Keil K, Berkley JL, Fuchs LH (1982) Suessite, Fe_3Si: A new mineral in the North Haig ureilite. Am Mineral 67:126-131

Keil K, Ntaflos Th., Taylor GJ, Brearley AJ, Newsom HE, Romig AD Jr. (1989) The Shallowater aubrite: Evidence for origin by planetesimal impact. Geochim Cosmochim Acta 53:3291-3307

Keil K, Stöffler D, Love SG, Scott ERD (1997) Constraints on the role of impact heating and melting in asteroids. Meteoritics Planet Sci 32:349-363

Kelly, W R, J W Larimer (1977) Chemical fractionation in meteorites-VII Iron meteorites and the cosmochemical history of the metal phase. Geochim Cosmochim Acta 41:93-111

Kimura M, Ikeda Y, Ebihara M, Prinz M (1991) New enclaves in the Vaca Muerta mesosiderite: Petrogenesis and comparison with HED meteorites. Proc NIPR Symp Antarctic Meteorites 4:263-306

Kimura M, Tsuchiyama A, Fukuoka T, Iimura Y (1992) Antarctic primitive achondrites, Yamato-74025, -75300, and -75305: Their mineralogy, thermal history, and the relevance to winonaite. Proc NIPR Symp Antarctic Meteorites 5:165-190

King EA, Jarosewich E, Daugherty FW (1981) Tierra Blanca: An unusual achondrite from West Texas. Meteoritics 16:229-237

Klein LC, Hewins RH (1979) Origin of impact melt rocks in the Bununu howardite. Proc Lunar Planet Sci

Conf 10:1127-1140

Kozul J, Hewins RH (1988) LEW 85300,02,03 polymict eucrites consortium—II: Breccia clasts, CM inclusion, glassy matrix and assembly history. Lunar Planet Sci 19:647-648

Kracher A (1974) Untersuchungen am Landes Meteorit. *In* Analyse extraterrestrischen Materials (W Kiesl and H Maliss, Jr., eds.), p 315-326. Springer

Kracher A (1982) Crystallization of a S-saturated Fe,Ni-melt, and the origin of the iron meteorite groups IAB and IIICD. Geophys Res Lett 9:412-415

Kracher A (1985) The evolution of partially differentiated planetesimal: evidence from iron meteorite groups IAB and IIICD. Proc Lunar Planet Sci Conf 15th, J Geophys Res 90:C689-C698

Kracher A, Wasson JT (1982) The role of S in the evolution of the parental cores of iron meteorites. Geochim Cosmochim Acta 46:2419-2426

Kracher A Willis J (1981) Composition and origin of the unusual Oktibbeha County iron meteorite. Meteoritics 16:239-246

Kracher A, Kurat G, Buchwald VF (1977) Cape York: The extraordinary mineralogy of an ordinary iron meteorite and its implications for the genesis of IIIAB irons. Geochem J 11:207-217

Kracher A, Willis J, Wasson JT (1980) Chemical classification of iron meteorites; IX, A new group (IIF), revision of IAB and IIICD, and data on 57 additional irons. Geochim Cosmochim Acta 44:773-787

Kring DA, Boynton WV (1992) The trace-element composition of Eagles Nest and its relationship to other ultramafic achondrites. Lunar Planet Sci 23:727-728

Kring DA, Boynton WV, Hill DH, Haag RA (1991) Petrologic description of Eagles Nest: A new olivine achondrite. Meteoritics 26:360

Krot A, Ivanova MA, Wasson JT (1993) The origin of chromatic chondrules and the volatility of Cr under a range of nebular conditions. Earth Planet Sci Lett 119:569-584

Kulpecz AA Jr, Hewins RH (1978) Cooling rate based on schreibersite growth for the Emery mesosiderite. Geochim Cosmochim Acta 42:1495-1500

Kurat G (1988) Primitive meteorites: An attempt towards unification. Phil Trans Royal Soc London A325:459-482

Kurat G, Zinner E, Brandstätter F (1992) An ion microprobe study of an unique oldhamite-pyroxenite fragment from the Bustee aubrite. Meteoritics 27:246-247

Labotka TC, Papike JJ (1980) Howardites: Samples of the regolith of the eucrite parent body: Petrology of Frankfort, Pavlovka, Yurtuk, Malvern, and ALHA 77302. Proc Lunar Planet Sci Conf 11:1103-1130

Laul J-C, Gosselin DC (1990) The Bholghati howardite: Chemical study. Geochim Cosmochim Acta 54:2167-2175

Lindner M, Leich RJ, Russ GP, Bazan JM, Borg RJ (1989) Direct determination of the half life of ^{187}Re. Geochim Cosmochim Acta 53:1597-1606

Lipschutz ME (1964) Origin of diamonds in the ureilites. Science 143:1431-1434

Lodders K (1996) An experimental and theoretical study of rare-earth-element partitioning between sulfides (FeS,CaS) and silicate and applications to enstatite achondrites. Meteoritics Planet Sci 31: 749-766

Lodders K, Palme H, Wlotzka F (1993) Trace elements in mineral separates of the Peña Blanca Spring aubrite: Implications for the evolution of the aubrite parent body. Meteoritics 28:538-551

Lofgren G, Lanier AB (1991) Dynamic crystallization experiments on the Angra dos Reis achondrite meteorite. Earth Planet Sci Lett 111:455-466

Lomena ISM, Touré F, Gibson EK Jr., Clanton US, Reid AM (1976) Aïoun el Atrouss: A new hypersthene achondrite with eucritic inclusions. Meteoritics 11:51-57

Lonsdale JT (1947) The Peña Blanca Spring meteorite, Brewster County, Texas. Am Mineral 32:354-364

Lovering JF (1964) Electron microprobe analysis of the metallic phase in basic achondrites. Nature 203:70

Lovering JF (1975) The Moama eucrite—A pyroxene-plagioclase adcumulate. Meteoritics 10:101-114

Lovering JF, Nichiporuk W, Chodos A, Brown H (1957) The distribution of gallium, germanium, cobalt, chromium, and copper in iron and stony-iron meteorites in relation to nickel content and structure. Geochim Cosmochim Acta 11:263-278

Lugmair GW, Galer SJG (1992) Age and isotopic relationships among the angrites Lewis Cliff 86010 and Angra dos Reis. Geochim Cosmochim Acta 56:1673-1694

Lugmair GW, Marti K (1977) Sm-Nd-Pu timepieces in the Angra dos Reis meteorite. Earth Planet Sci Lett 35:273-284

Lugmair GW, Shukolyukov A (1997) ^{53}Mn - ^{53}Cr isotope systematics of the HED parent body. Lunar Planet Sci 28:852-852

Lugmair GW, Scheinin NB, Carlson RW (1977) Sm-Nd systematics of the Serra de Magé eucrite. Meteoritics 12:300-301

Lugmair GW, MacIsaac C, Shukolyukov A (1994a) Small differences in differentiated meteorites recorded by the ^{53}Mn - ^{53}Cr chronometer. Lunar Planet Sci 24:813-814

Lugmair GW, MacIsaac C, Shukolyukov A (1994b) Small differences recorded in differentiated meteorites. Meteoritics 29:493-494

Ma M-S, Schmitt RA (1979) Genesis of the cumulate eucrites Serra de Magé and Moore County: A geochemical study. Meteoritics 14:81-89

Ma M-S, Murali AV, Schmitt RA (1977) Genesis of Angra dos Reis and other achondritic meteorites. Earth Planet Sci Lett 35:331-346

Malhotra PD (1962) A note on the composition of the Basti meteorite. Records Geol. Survey India 89: 479-481

Malvin DJ, Wang D, Wasson JT (1984) Chemical classification of iron meteorites; X, Multielement studies of 43 irons, resolution of group IIIE from IIIAB, and evaluation of Cu as a taxonomic parameter. Geochim Cosmochim Acta 48:785-804

Malvin DJ, Wasson JT, Clayton RN, Mayeda TK, Curvello WdS (1985) Bocaiuva—A silicate-inclusion bearing iron meteorite related to the Eagle-Station pallasites. Meteoritics 20:259-273

Malvin DJ, Jones JH, Drake MJ (1986) Experimental investigations of trace element fractionation in iron meteorites; III, Elemental partitioning in the system Fe-Ni-S-P. Geochim Cosmochim Acta 50:1221-1231

Marshall RR, Keil K (1965) Polymineralic inclusions in the Odessa iron meteorite. Icarus 4:461-479

Marvin U B, Wood JA (1972) The Haverö ureilite: Petrographic notes. Meteoritics 7:601-610

Marvin UB, Petaev MI, Croft WJ, Killgore M (1997) Silica minerals in the Gibeon IVA iron meteorite. Lunar Planet Sci 28:879-880

Mason B (1962) Meteorites. Wiley, NY 274 p

Mason B (1963) The hypersthene achondrites. Am Mus Novitates No. 2155, 13 p

Mason B, Jarosewich E (1973) The Barea, Dyarrl Island, and Emery meteorites, and a review of the mesosiderites. Mineral Mag 39:204-215

Mason B, Wiik HB (1966a) The composition of the Bath, Frankfort, Kakangari, Rose City, and Tadjera meteorites. Am Mus Novitates No. 2272, 24 p

Mason B, Wiik HB (1966b) The composition of the Barratta, Carraweena, Kapoeta, Mooresfort, and Ngawi meteorites. Am Mus Novitates No. 2273, 25 p

Matsuda J-I, Fukunaga K, Ito K (1988) On the vapor-growth diamonds formation in the solar nebula. Lunar Planet Sci 19:736-737

Matsuda J-I, Fukunaga K, Ito K (1991) Noble gas studies in vapor-growth diamonds: Comparison with shock-produced diamonds and the origin of diamonds in ureilites. Geochim Cosmochim Acta 55: 2011-2023

Matsuda J-I, Kusima A, Yajima H, Syono Y (1995) Noble gas studies in diamonds synthesized by shock loading in the laboratory and their implications on the origin of diamonds in ureilites. Geochim Cosmochim Acta 59:4939-4949

Mazor E, Anders E (1967) Primordial gases in Jodzie howardite and the origin of gas-rich meteorites. Geochim Cosmochim Acta 31:1441-1456

Mazor E, Heymann D, Anders E (1970) Nobles gases in carbonaceous chondrites. Geochim Cosmochim Acta 34:781-824

McCall GJH (1966) The petrology of the Mount Padbury mesosiderite and its achondrite enclaves. Mineral Mag 35:1029-1060

McCall GJH (1973) Meteorites and Their Origins. Wiley

McCall GJ, Cleverly WH (1968) New stony meteorite finds including two ureilites from the Nullarbor Plain, Western Australia. Mineral Mag 36:691-716

McCarthy TS, Ahrens LH, Erlank AJ (1972) Further evidence in support of the mixing model for howardite origin. Earth Planet Sci Lett 15:86-93

McCarthy TS, Erlank AJ, Willis JP (1973) On the origin of eucrites and diogenites. Earth Planet Sci Lett 18:433-442

McCarthy TS, Erlank AJ, Willis JP, Ahrens LH (1974) New chemical analyses of six achondrites and one chondrite. Meteoritics 9:215-221

McCord TB, Adams JB, Johnson TV (1970) Asteroid Vesta: Spectral reflectivity and compositional implications. Science 168:1445-1447

McCoy TJ (1995) Silicate-bearing IIE irons: Early mixing and differentiation in a core-mantle environment and shock resetting of ages. Meteoritics 30:542-543

McCoy TJ (1998) A pyroxene-oldhamite clast in Bustee: Igneous aubritic oldhamite and a mechanism for the Ti enrichment in aubritic troilite. Antarctic Meteorite Research 11:34-50

McCoy TJ, Scott ERD, Haack H (1993) Genesis of the IIICD iron meteorites: Evidence from silicate-bearing inclusions. Meteoritics 28:552-560

McCoy TJ, Keil K, Mayeda TK, Clayton RN (1992) Petrogenesis of the lodranite-acapulcoite parent body. Meteoritics 27:258-259

McCoy TJ, Ehlmann AJ, Benedix GK, Keil K, Wasson JT (1996a) The Lueders, Texas, IAB iron meteorite with silicate inclusions. Meteoritics Planet Sci 31:419-422

McCoy TJ, Keil K, Clayton RN, Mayeda TK, Bogard DD, Garrison DH, Huss GR, Hutcheon ID, Wieler R (1996b) A petrologic, chemical, and isotopic study of Monument Draw and comparison with other acapulcoites: Evidence for formation by incipient partial melting. Geochim Cosmochim Acta 60:2681-2708

McCoy TJ, Keil K, Clayton RN, Mayeda TK, Bogard DD, Garrison DH, Wieler R (1997a) A petrologic and isotopic study of lodranites: Evidence for early formation as partial melt residues from heterogeneous precursors. Geochim Cosmochim Acta 61:623-637

McCoy TJ, Keil K, Muenow DW, Wilson L (1997b) Partial melting and melt migration in the acapulcoite-lodranite parent body. Geochim Cosmochim Acta 61:639-650

McCoy TJ, Dickinson TL, Lofgren GE (1997c) Partial melting of Indarch (EH4) from 1000-1425°C: New insights into igneous processes in enstatite meteorites. Lunar Planet Sci 28:903-904

McCoy TJ, Dickinson TL, Lofgren GE (1997d) Experimental and petrologic studies bearing on the origin of aubrites. Papers presented to the 22nd Symp Antarctic Met, 103-105

McKay G, Lindstrom D, Yang S-R, Wagstaff J (1988a) Petrology of a unique achondrite Lewis Cliff 86010. Lunar Planet Sci 19:762-763

McKay G, Lindstrom D, Le L, Yang S-R (1988b) Experimental studies of synthetic LEW86010 analogs: Petrogenesis of a unique achondrite. Lunar Planet Sci 19:760-761

McKay G, Le L, Wagstaff J (1989) Is unique achondrite LEW86010 a crystallized melt? Lunar Planet Sci 20:675-676

McKay G, Crozaz G, Wagstaff J, Yang S-R, Lundberg L (1990) A petrographic, electron microprobe, and ion microprobe study of mini-angrite Lewis Cliff 87051. Lunar Planet Sci 21:771-772

McKay G, Le L, Wagstaff J, Crozaz G (1994) Experimental partitioning of rare earth elements and strontium: Constraints on petrogenesis and redox conditions during crystallization of Antarctic angrite Lewis Cliff 86010. Geochim Cosmochim Acta 58:2911-2919

McKay GA, Crozaz G, Mikouchi T, Miyamoto M (1995) Petrology of antarctic angrites LEW 86010, LEW 87051, and Asuka 881371. Antarc Meteorites 20:155-158

McSween HY, Labotka TC (1993) Oxidation during metamorphism of the ordinary chondrites. Geochim Cosmochim Acta 57:1105-1114

McSween HY Jr., Bennett ME III, Jarosewich E (1991) The mineralogy of ordinary chondrites and implications for asteroid spectrophotometry. Icarus 90:107-116

Metzler K, Bobe KD, Palme H, Spettel B, Stöffler D (1995) Thermal and impact metamorphism on the HED parent asteroid. Planet Space Sci 43:499-525

Mikouchi T, Takeda H, Miyamoto M, Ohsumi K, McKay GA (1995) Exsolution lamellae of kirschsteinite in magnesium-iron olivine from an angrite meteorite. Am Mineral 80, 585-592

Mikouchi T, Miyamoto M, McKay GA (1996) Mineralogical study of angrite Asuka-881371: Its possible relation to angrite LEW87051. Proc. NIPR Symp Antarctic Meteorites 9:174-188

Misawa K, Watanabe S, Kitamura M, Nakamura N, Yamamoto K, Masuda A (1992) A noritic clast from the Hedjaz chondritic breccia: implications for melting events in the early solar system. Geochem J 26:435-446

Mittlefehldt DW (1978) The differentiation history of small bodies in the solar system: The howardite and mesosiderite meteorite parent bodies. PhD dissertation, Univ California, Los Angeles, 208 p

Mittlefehldt DW (1979) Petrographic and chemical characterization of igneous lithic clasts from mesosiderites and howardites and comparison with eucrites and diogenites. Geochim Cosmochim Acta 43:1917-1935

Mittlefehldt DW (1980) The composition of mesosiderite olivine clasts and implications for the origin of pallasites. Earth Planet Sci Lett 51:29-40

Mittlefehldt D W (1986) Fe-Mg-Mn relations of ureilite olivines and pyroxenes and the genesis of ureilites. Geochim. Cosmochim. Acta 50:107-110

Mittlefehldt DW (1987) Volatile degassing of basaltic achondrite parent bodies: Evidence from alkali elements and phosphorus. Geochim Cosmochim Acta 51:267-278

Mittlefehldt DW (1990) Petrogenesis of mesosiderites: I Origin of mafic lithologies and comparison with basaltic achondrites. Geochim Cosmochim Acta 54:1165-1173

Mittlefehldt DW (1994) The genesis of diogenites and HED parent body petrogenesis. Geochim Cosmochim Acta 58:1537-1552

Mittlefehldt DW, Lindstrom MM (1990) Geochemistry and genesis of the angrites. Geochim Cosmochim Acta 54:3209-3218

Mittlefehldt DW, Lindstrom MM (1993) Geochemistry and petrology of a suite of ten Yamato HED meteorites. Proc NIPR Symp Antarctic Meteorites 6:268-292

Mittlefehldt DW, Lindstrom MM (1997) Magnesian basalt clasts from the EET 92014 and Kapoeta howardites and a discussion of alleged primary magnesian HED basalts. Geochim Cosmochim Acta 61:453-462

Mittlefehldt DW, Lindstrom MM (1998) Black clasts from howardite QUE 94200—Impact melts, not primary magnesian basalts. Lunar Planet Sci 29:1832

Mittlefehldt DW, Meyers B (1991) Petrology and geochemistry of the EETA79002 diogenite . Meteoritics 26:373

Mittlefehldt, DW, Chou, C-L, Wasson JT (1979) Mesosiderites and howardites: igneous formation and possible genetic relationships. Geochim Cosmochim Acta 43:673-688

Mittlefehldt DW, Rubin AE, Davis AM (1992) Mesosiderite clasts with the most extreme positive europium anomalies among solar system rocks. Science 257:1096-1099

Mittlefehldt DW, Lindstrom MM, M-S Wang, ME Lipschutz (1995) Geochemistry and origin of achondritic inclusions in Yamato-75097, -793241 and -794046 chondrites. Proc NIPR Symp Antarctic Meteorites 8:251-271

Mittlefehldt DW, Lindstrom MM, Bogard DD, Garrison DH, Field SW (1996) Acapulco- and Lodran-like achondrites: Petrology, geochemistry, chronology and origin. Geochim Cosmochim Acta 60:867-882

Miyamoto M, Takeda H (1977) Evaluation ofa crust model of eucrites from the width of exsolved pyroxene. Geochem J 11:161-169

Miyamoto M, Takeda H (1994) Chemical zoning of olivine in several pallasites suggestive of faster cooling . Lunar Planet Sci 25:921-922

Miyamoto M, Takeda H, Yanai (1978) Yamato achondrite polymict breccias. Mem Natl Inst Polar Res (Japan) Spec Issue No 8:185-197

Miyamoto M, Toyoda H, Takeda H (1985) Thermal history of ureilite as inferred from mineralogy of Pecora Escarpment 82506. Lunar Planet Sci 16:567-568

Miyamoto M, Furuta T, Fujii N, McKay DS, Lofgren GE, Duke MB (1993) The Mn-Fe negative correlation in olivines in ALHA77257 ureilite. J Geophys Res 98:5301-5307

Morgan JW, Walker RJ, Grossman JN (1992) Rhenium-osmium isotope systematics in meteorites I: Magmatic iron meteorite groups IIAB and IIIAB. Earth Planet Sci Lett 108:191-202

Morgan JW, Horn MF, Walker RJ, Grossman JN (1995) Rhenium-osmium concentration and isotope systematics in group IIAB iron meteorites. Geochim Cosmochim Acta 59:2331-2344

Mori H, Takeda H (1981a) Thermal and deformational histories of diogenites as inferred from their microtextures of orthopyroxene. Earth Planet Sci Lett 53:266-274

Mori H, Takeda H (1981b) Evolution of the Moore County pyroxenes as viewed by an analytical transmission electron microprobe (ATEM). Meteoritics 16:362-363

Mori H, Takeda H (1983) An electron petrographic study of ureilite pyroxenes. Meteoritics 18:358-359

Mori H, Takeda H (1988) TEM observations of carbonaceous material in the Dyalpur ureilite. Lunar Planet Sci 19:808

Mori H, Takeda H, Toyoda H (1986) Mineralogy of pigeonites from the Allan Hills 82106 ureilites. Meteoritics 21:466-467

Müller HW, Zähringer J (1966) Chemische Unterschiede bei Uredelgashaltigen Steinmeteoriten. Earth Planet Sci Lett 1:25-29

Müller HW, Zähringer J (1969) Rare gases in stony meteorites. *In* Meteorite Research (ed. P Millman), p. 845. D Reidel, Hingham, Massachusetts.

Murthy VR, Coscio MR Jr., Sabelin T (1977) Rb-Sr internal isochron and the initial $^{87}Sr/^{86}Sr$ for the Estherville mesosiderite. Proc Lunar Sci Conf 8:177-186

Murty SVS (1994) Interstellar vs. ureilitic diamonds: nitrogen and noble gas systematics. Meteoritics 29:507

Nagahara H (1992) Yamato-8002: Partial melting residue on the "unique" chondrite parent body. Proc NIPR Symp Antarctic Meteorites 5:191-223

Nagahara H, Ozawa K (1986) Petrology of Yamato-791493, "lodranite": Melting, crystallization, cooling history, and relationship to other meteorites. Mem NIPR Spec Issue No 41, 181-205

Nakamura N, Morikawa N (1993) REE and other trace lithophiles in MAC88177, LEW88280 and LEW88763. Lunar Planet Sci 24:1047-1048

Nakamura N, Misawa K, Kitamura M, Masuda A, Watanabe S, Yamamoto K (1990) Highly fractionated REE in the Hedjaz (L) chondrite: implications for nebular and planetary processes. Earth Planet Sci Lett 99:290-302

Nakamura N, Morikawa N, Hutchison R, Clayton RN, Mayeda TK, Nagao K, Misawa K, Okano O, Yamamoto K, Yanai K, Matsumoto Y (1994) Trace element and isotopic characteristics of inclusions in the Yamato ordinary chondrites Y-75097, Y-793241 and Y-794046. Proc NIPR Symp Antarctic Meteorites 7:125-143

Narayan C, Goldstein JI (1982) A dendritic solidification model to explain Ge-Ni variations in iron meteorite chemical groups. Geochim Cosmochim Acta 46:259-268

Narayan C, Goldstein JI (1985) A major revision of iron meteorite cooling rates; an experimental study of the growth of the Widmanstätten pattern. Geochim Cosmochim Acta 49:397-410

Nehru CE, Zucker SM, Harlow GE, Prinz M (1980a) Olivines and olivine coronas in mesosiderites. Geochim Cosmochim Acta 44:1103-1118

Nehru CE, Delaney JS, Harlow GE, Prinz M (1980b) Mesosiderite basalts and the eucrites . Meteoritics 15:337-338

Nehru CE, Prinz M, Delaney JS (1982) The Tucson iron and its relationship to enstatite meteorites. Proc 13th Lunar Planet Sci Conf, J Geophys Res B87, Suppl 1:A365-A373

Nehru CE, Prinz M, Delaney JS, Dreibus G, Palme H, Spettel B, Wänke H (1983) Brachina: A new type of meteorite, not a Chassignite. Proc Lunar Planet Sci Conf 14th J Geophys Res 88:B237-B244

Nehru CE, Prinz M, Weisberg MK, Ebihara ME, Clayton RN, Mayeda TK (1992) Brachinites: A new primitive achondrite group. Meteoritics 27:267

Nehru CE, Prinz M, Weisberg MK, Ebihara ME, Clayton RN, Mayeda TK (1996) A new brachinite and petrogenesis of the group. Lunar Planet Sci 27:943-944

Nelen J, Mason B (1972) The Estherville meteorite. Smithson Contrib Earth Sci 9:55-56

Neuvonen KJ, Ohlson B, Papuen TA, Hakli TA, Ramdohr P (1972) The Haverö ureilite. Meteoritics 7:515-531

Newsom HE (1985) Molybdenum in eucrites: Evidence for a metal core in the eucrite parent body. Proc 15th Lunar Planet Sci Conf, Part 2, J Geophys Res Suppl 90:C613-C617

Newsom HE, Drake MJ (1982) The metal content of the eucrite parent body: constraints from the partitioning behavior of tungsten. Geochim Cosmochim Acta 46:2483-2489

Newsom HE, Ntaflos Th., Keil K (1996) Dark clasts in the Khor Temiki aubrite: Not basalts. Meteoritics Planet Sci 31:146-151

Nichols RH, Jr., Hohenberg CM, Kehm K, Kim Y, Marti K (1994) I-Xe studies of the Acapulco meteorite: Absolute I-Xe ages of individual phosphate grains and the Bjurböle standard. Geochim Cosmochim Acta 58:2553-2561

Nielsen HP, Buchwald VF (1981) Roaldite, a new nitride in iron meteorites. Proc Lunar Planet Sci Conf 12, Geochim Cosmochim Acta, Suppl 16:1343-1348

Niemeyer S (1979a) I-Xe dating of silicate and troilite from IAB iron meteorites. Geochim Cosmochim Acta 43:843-860

Niemeyer S (1979b) ^{40}Ar-^{39}Ar dating of inclusions from IAB iron meteorites. Geochim Cosmochim Acta 43:1829-1840

Niemeyer S (1980) I-Xe and ^{40}Ar-^{39}Ar dating of silicate from Weekeroo Station and Netschaëvo iron meteorites. Geochim Cosmochim Acta 44:33-44

Niemeyer S (1983) I-Xe and ^{40}Ar-^{39}Ar analyses of silicate from the Eagle Station pallasite and the anomalous iron meteorite Enon. Geochim Cosmochim Acta 47:1007-1012

Nininger HH (1952) Out of the Sky. Dover

Nyquist LE, Takeda H, Bansal BM, Shih C-Y, Wiesmann H, Wooden JL (1986) Rb-Sr and Sm-Nd internal isochron ages of a subophitic basalt clast and a matrix sample from the Y75011 eucrite. J Geophys Res 91:8137-8150

Nyquist L, Bansal B, Wiesmann H, Shih C-Y (1994) Neodymium, strontium and chromium isotopic studies of the LEW86010 and Angra dos Reis meteorites and the chronology of the angrite parent body. Meteoritics 29:872-885

Ogata H, Takeda H, Ishii T (1987) Interstitial Ca-rich silicate materials in the Yamato ureilites with reference to their origin. Lunar Planet Sci 18:738-739

Ogata H, Mori H, Takeda H (1988) Mineralogy of interstitial rim materials of ureilites and their origin. Abstr Symp Antarctic Meteorites 13:138-141

Ogata H, Mori H, Takeda H (1991) Mineralogy of interstitial rim materials of the Y74123 and Y790981 ureilites and their origin. Meteoritics 16:195-201

Ohtani E (1983) Formation of olivine textures in pallasites and thermal history of pallasites in their parent body. Phys Earth Planet Interiors 32:182-192

Okada A, Keil K (1982) Caswellsilverite, $NaCrS_2$: A new mineral in the Norton County enstatite achondrite. Am Mineral 67:132-136

Okada A, Keil K, Taylor GJ, Newsom H (1988) Igneous history of the aubrite parent asteroid: Evidence from the Norton County enstatite achondrite. Meteoritics 23:59-74

Olsen E, Jarosewich E (1970) The chemical composition of the silicate inclusions in the Weekeroo Station iron meteorite. Earth Planet Sci Lett 8:261-266

Olsen E, Jarosewich E (1971) Chondrules: First occurrence in an iron meteorite. Science 174:583-585

Olsen EJ, Steele IA (1997) Galileiite: A new meteoritic phosphate mineral. Meteoritics Planet Sci 32: A155-A-156

Olsen E, Erlichman J, Bunch TE, Moore PB (1977) Buchwaldite, a new meteoritic phosphate mineral. Am

Mineral 62:362-364
Olsen E, Noonan A, Fredriksson K, Jarosewich E, Moreland G (1978) Eleven new meteorites from Antarctica, 1976-1977. Meteoritics 13:209-225
Olsen E, Dod BD, Schmitt RA, Sipiera PP (1987) Monticello: A glass-rich howardite. Meteoritics 22:81-96
Olsen EJ, Fredriksson K, Rajan S, Noonan A (1990) Chondrule-like objects and brown glasses in howardites. Meteoritics 25:187-194
Olsen E, Davis A, Clarke RS Jr., Schultz L, Weber HW, Clayton R, Mayeda T, Jarosewich E, Sylvester P, Grossman L, Wang M-S, Lipschutz ME, Steele IM, Schwade J (1994) Watson: A new link in the IIE iron chain. Meteoritics 29:200-213
Olsen EJ, Davis AM, Clayton RN, Mayeda TK, Moore CB and Steele IM (1996) A silicate inclusion in Puente del Zacate: A IIIA iron meteorite. Science 273:1365-1367
Osadchii Eu.G, Baryshnikova GV, Novikov GV (1981) The Elga meteorite: Silicate inclusions and shock metamorphism. Proc Lunar Planet Sci Conf 12B:1049-1068
Ott U, Begemann F, Löhr HP (1983) Noble gases in the Brachina and Chassigny meteorites. Lunar Planet Sci 14:586-587
Ott U, Löhr HP, Begemann F (1984) Ureilites: The case of the missing diamonds and a new neon component. Meteoritics 19:287-288
Ott U, Löhr HP, Begemann F (1985a) Noble gases and the classification of Brachina. Meteoritics 20:69-78
Ott U, Löhr HP, Begemann F (1985b) Trapped neon in ureilites—A new component. *In* Isotopic Ratios in the Solar System. (Ed. Centre National d'Etudes Spatiales), p 129-136. Cepadues-Editions; Toulouse, France
Ott U, Begemann F, Löhr HP (1987) Noble gases in ALH 84025: Like Brachina, unlike Chassigny. Meteoritics 22:476-477
Ott U, Löhr HP, Begemann F (1990) EET83309: A ureilite with solar noble gases. Meteoritics 25:396
Ott U, Löhr HP, Begemann F (1993) Solar noble gases in polymict ureilites and an update on ureilite noble gas data. Meteoritics 28:415-416
Padia JT, Rao MN (1989) Neon isotope studies of Fayetteville and Kapoeta meteorites and clues to ancient solar activity. Geochim Cosmochim Acta 53:1461-1467
Palme H, Baddenhausen H, Blum K, Cendales M, Dreibus G, Hofmeister H, Kruse H Palme C, Spettel B, Vilcsek E, Wänke H (1978) New data on lunar samples and achondrites and a comparison of the least fractionated samples from the earth, the moon and the eucrite parent body. Proc Lunar Planet Sci Conf 9:25-57
Palme H, Schultz L, Spettel B, Weber HW, Wänke H, Christophe Michel-Levy M, Lorin JC (1981) The Acapulco meteorite: Chemistry, mineralogy and irradiation effects. Geochim Cosmochim Acta 45:727-752
Palme H, Spettel B, Burghele A, Weckwerth G, Wänke H (1983) Elephant Moraine polymict eucrites: A eucrite-howardite compositional link. Lunar Planet Sci 14:590-591
Palme H, Wlotzka F, Spettel B, Dreibus G, Weber H (1988) Camel Donga: A eucrite with high metal content. Meteoritics 23:49-57
Palme H, Hutcheon ID, Kennedy AK, Sheng YJ, Spettel B (1991)Trace element distribution in minerals from a silicate inclusion in the Caddo IAB-iron meteorite. Lunar Planet Sci 22:1015-1016
Papanastassiou DA, Wasserburg GJ (1969) Initial strontium isotopic abundances and the resolution of small time differences in the formation of planetary objects. Earth Planet Sci Lett 5:361-376
Paul RL, Lipschutz ME (1990) Chemical studies of differentiated meteorites: I. Labile trace elements in Antarctic and non-Antarctic eucrites. Geochim Cosmochim Acta 54:3185-3196
Pellas P, Fiéni C, Trieloff M, Jessberger EK (1997) The cooling history of the Acapulco meteorite as recorded by the ^{244}Pu and ^{40}Ar-^{39}Ar chronometers. Geochim Cosmochim Acta 61:3477-3501
Petaev MI, Brearley AJ (1994) Exsolution in ferromagnesian olivine of the Divnoe meteorite. Science 266:1545-1547
Petaev MI, Clarke RS Jr, Olsen EJ, Jarosewich E, Davis AM, Steele IM, Lipschutz ME, Wang M-S, Clayton RN, Mayeda TK, Wood JA (1993) Chaunskij: The most highly metamorphosed, shock-modified and metal-rich mesosiderite. Lunar Planet Sci 24:1131-1132
Petaev MI, Barsukova LD, Lipschutz ME, Wang M-S, Arsikan AA, Clayton RN, Mayeda TK (1994) The Divnoe meteorite: Petrology, chemistry, oxygen-isotopes and origin. Meteoritics 29:182-199
Powell BN (1969) Petrology and chemistry of mesosiderites-I. Textures and composition of nickel-iron. Geochim Cosmochim Acta 33:789-810
Powell BN (1971) Petrology and chemistry of mesosiderites-II. Silicate textures and compositions and metal-silicate relationships. Geochim Cosmochim Acta 35:5-34
Prinz M, Weisberg MK (1995) Asuka 881371 and the angrites: Origin in a heterogeneous, CAI-enriched, differentiated, volatile-depleted body. Antarc Meteorites 20:207-210
Prinz M, Keil K, Hlava PF, Berkley JL, Gomes CB, Curvello WS (1977) Studies of Brazilian meteorites,

III. Origin and history of the Angra dos Reis achondrite. Earth Planet Sci Lett 35:317-330
Prinz M, Nehru CE, Delaney JS, Harlow GE, Bedell RL (1980) Modal studies of mesosiderites and related achondrites, including the new mesosiderite ALHA 77219. Proc Lunar Planet Sci Conf 11:1055-1071
Prinz M, Nehru CE, Delaney JS (1982a) Reckling Peak A79015: An unusual mesosiderite. Lunar Planet Sci 13:631
Prinz M, Nehru CE, Delaney JS (1982b) Sombrerete: An iron with highly fractionated amphibole-bearing Na-P-rich silicate inclusions. Lunar Planet Sci 13:634-635
Prinz. M, Delaney JS, Nehru CE, Weisberg MK (1983a) Enclaves in the Nilpena polymict ureilite. Meteoritics 18:376-377
Prinz M, Nehru CE, Delaney JS, Weisberg M, Olsen E (1983b) Globular silicate inclusions in IIE irons and Sombrerete: Highly fractionated minimum melts. Lunar Planet Sci 14:618-619
Prinz M, Nehru CE, Delaney JS, Fredriksson K, Palme H (1984) Silicate inclusions in IVA iron meteorites. Meteoritics 19:291-292
Prinz M, Weisberg MK, Nehru CE, Delaney JS (1986a) ALHA84025: A second Brachina-like meteorite. Lunar Planet Sci 17:679-680
Prinz. M, Weisberg MK, Nehru CE (1986b) North Haig and Nilpena: Paired polymict ureilites with Angra dos Reis-related and other clasts. Lunar Planet Sci 17:681-682
Prinz M, Weisberg MK, Nehru CE, Delaney JS (1987) EET83309: a polymict ureilite:Recognition of a new group. Lunar Planet Sci 18:802-803
Prinz M, Weisberg MK, Nehru CE (1988a) Feldspathic components in polymict ureilites. Lunar Planet Sci 19:947-948
Prinz M, Weisberg MK, Nehru CE (1988b) LEW86010, a second angrite: Relationship to CAI's and opaque matrix. Lunar Planet Sci 19:949-950
Prinz M, Weisberg MK, Nehru CE (1990) LEW 87051, a new angrite: Origin in a Ca-Al-enriched eucritic planetesimal? Lunar Planet Sci 21:979-980
Prinz M, Weisberg MK, Nehru CE (1994) LEW88774: A new type of Cr-rich ureilite. Lunar Planet Sci 25:1107-1108
Prinzhofer A, Papanastassiou DA, Wasserburg GJ (1992) Samarium-neodymium evolution of meteorites. Geochim Cosmochim Acta 56:797-815
Prior GT (1918) On the mesosiderite-grahamite group of meteorites: with analyses of Vaca Muerta, Hainholz, Simondium, Powder Mill Creek. Mineral Mag 18:151-172
Pun A, Papike JJ (1995) Ion microprobe investigation of exsolved pyroxenes in cumulate eucrites: Determination of selected trace-element partition coefficients. Geochim Cosmochim Acta 59:2279-2289
Ramdohr P (1973) The opaque minerals in stony meteorites. Akademie Verlag, Berlin
Ramdohr P, Prinz M, El Goresy A (1975) Silicate inclusions in the Mundrabilla meteorite. Meteoritics 10:477-479
Randich E, Goldstein JI (1978) Cooling rates of seven hexahedrites. Geochim Cosmochim Acta 42: 221-234
Rao MN, Garrison DH, Bogard DD, Badhwar G, Murali AV (1991) Composition of solar flare noble gases preserved in meteorite parent body regolith. J Geophys Res 96:19321-19330
Rasmussen KL (1981) The cooling rates of iron meteorites; a new approach. Icarus 45:564-576
Rasmussen KL (1982) Determination of the cooling rates and nucleation histories of eight group IVA iron meteorites using local bulk Ni and P variation. Icarus 52:444-453
Rasmussen KL (1989) Cooling rates and parent bodies of iron meteorites from group IIICD, IAB, and IVB. Physica Scripta 39:410-416
Rasmussen KL, Malvin DJ, Buchwald VF, Wasson JT (1984) Compositional trends and cooling rates of group IVB iron meteorites. Geochim Cosmochim Acta 48:805-813
Rasmussen KL, Ulff-Møller F, Haack H (1995) The thermal evolution of IVA iron meteorites; evidence from metallographic cooling rates. Geochim Cosmochim Acta 59:3049-3059
Reed SJB, Scott ERD, Long JVP (1979) Ion microprobe analyses of olivine in pallasite meteorites for nickel. Earth Planet Sci Lett 43:5-12
Reid AM, Cohen AJ (1967) Some characteristics of enstatite from enstatite achondrites. Geochim Cosmochim Acta 31:661-672
Reid AM, Williams RJ, Takeda H (1974) Coexisting bronzite and clinobronzite and the thermal evolution of the Steinbach meteorites. Earth Planet Sci Lett 22:67-74
Reid AM, Buchanan P, Zolensky ME, Barrett RA (1990) The Bholghati howardite: Petrography and mineral chemistry. Geochim Cosmochim Acta 54:2161-2166
Reuter KB, Williams DB, Goldstein JI (1988) Low temperature phase transformations in the metallic phases of iron and stony-iron meteorites. Geochim Cosmochim Acta 52:617-626
Richter GR, Wolf R, Anders E (1979) Aubrites: Are they direct nebular condensates? Lunar Planet Sci

10:1028-1030
Righter K, Drake MJ (1997) A magma ocean on Vesta: Core formation and petrogenesis of eucrites and diogenites. Meteoritics Planet Sci 32:929-944
Righter K, Arculus RJ, Delano JW, Paslick C (1990) Electrochemical measurements and thermodynamic calculations of redox equilibria in pallasite meteorites: Implications for the eucrite parent body. Geochim Cosmochim Acta 54:1803-1815
Robinson KL, Bild RW (1977) Silicate inclusions from the Mundrabilla iron. Meteoritics 12:354-355
Rowe MW (1970) Evidence for decay of extinct Pu^{244} and I^{129} in the Kapoeta meteorite. Geochim Cosmochim Acta 34:1019-1025
Rubin AE (1990) Kamacite and olivine in ordinary chondrites: Intergroup and intragroup relationships. Geochim Cosmochim Acta 54:1217-1232
Rubin AE (1997) Mineralogy of meteorite groups. Meteoritics Planet Sci 32:231-247
Rubin AE, Jerde EA (1987) Diverse eucritic pebbles in the Vaca Muerta mesosiderite. Earth Planet Sci Lett 84:1-14
Rubin AE, Jerde EA (1988) Compositional differences between basaltic and gabbroic clasts in mesosiderites. Earth Planet Sci Lett 87:485-490
Rubin AE, Mittlefehldt DW (1992) Classification of mafic clasts from mesosiderites: Implications for endogenous igneous processes. Geochim Cosmochim Acta 56:827-840
Rubin AE, Mittlefehldt DW (1993) Evolutionary history of the mesosiderite asteroid: A chronologic and petrologic synthesis. Icarus 101:201-212
Rubin AE, Jerde EA, Zong P, Wasson JT, Westcott JW, Mayeda TK, Clayton RN (1986) Properties of the Guin ungrouped iron meteorite: The origin of Guin and of group-IIE irons. Earth Planet Sci Lett 76:209-226
Rubin AE, Wasson JT, Clayton RN, Mayeda TK (1990) Oxygen-isotopes in chondrules and coarse-grained chondrule rims from the Allende meteorite. Earth Planet Sci Lett 96:247-255
Ruzicka A, Boynton WV, Ganguly J (1994) Olivine coronas, metamorphism, and the thermal history of the Morristown and Emery mesosiderites. Geochim Cosmochim Acta 58:2725-2741
Ruzicka A, Kring DA, Hill DH, Boynton WV, Clayton RN, Mayeda TK (1995) Silica-rich orthopyroxenite in the Bovedy chondrite. Meteoritics Planet Sci 30:57-70
Ruzicka A, Snyder GA, Taylor LA (1997) Vesta as the howardite, eucrite and diogenite parent body: Implications for the size of a core and for large-scale differentiation. Meteoritics Planet Sci 32:825-840
Ryder, G (1982) Siderophiles in the Brachina meteorite: impact melting? Nature 299:805-807
Sack RO, Azeredo WJ, Lipschutz ME (1991) Olivine diogenites: The mantle of the eucrite parent body. Geochim Cosmochim Acta 55:1111-1120
Sack RO, Azeredo WJ, Lipschutz ME (1994a) Erratum to R.O. Sack, W.J. Azeredo, and M.E. Lipschutz (1991) "Olivine diogenites: The mantle of the eucrite parent body." Geochim Cosmochim Acta 55:1111-1120. Geochim Cosmochim Acta 58:1044
Sack RO, Ghiorso MS, Wang M-S, Lipschutz ME (1994b) Igneous inclusions from ordinary chondrites: High temperature cumulates and a shock melt. J Geophys Res 99:26029-26044
Saikumar V, Goldstein JI (1988) An evaluation of the methods to determine the cooling rates of iron meteorites. Geochim Cosmochim Acta 52:715-726
Saito J, Takeda H (1989) Mineralogical study of LEW85328 ureilite. Lunar Planet Sci 20:938-939
Saito J, Takeda H (1990) Information of elemental distributions in heavily shocked ureilites as a guide to deduce the ureilite formation process. Lunar Planet Sci 21:1063-1064
Sanz HG, Burnett DS, Wasserburg GJ (1970) A precise $^{87}Rb/^{87}Sr$ age and initial $^{87}Sr/^{86}Sr$ for the Colomera iron meteorite. Geochim Cosmochim Acta 34:1227-1239
Schaudy R, Wasson JT, Buchwald VF (1972) The chemical classification of iron meteorites: VI. A reinvestigation of irons with Ge concentrations lower than 1 ppm. Icarus 17:174-192
Schmitt R, Laul JC (1973) A survey of the selenochemistry of major, minor, and trace elements. The Moon 8:182-209
Schnetzler CC, Philpotts JA (1969) Genesis of the calcium-rich achondrites in light of rare-earth and barium concentrations. *In* Meteorite Research (P.M. Millman, ed.), Reidel, Dordrecht, Holland, 206-216
Schultz L, Palme H, Spettel B, Weber HW, Wänke H, Christophe Michel-Levy M, Lorin JC (1982) Allan Hills A77081—An unusual stony meteorite. Earth Planet Sci Lett 61:23-31
Scott ERD (1972) Chemical fractionation in iron meteorites and its interpretation. Geochim Cosmochim Acta 36:1205-1236
Scott ERD (1977a) Composition, mineralogy and origin of Group IC iron meteorites. Earth Planet Sci Lett 37:273-284
Scott ERD (1977b) Pallasites—metal composition, classification and relationships with iron meteorites. Geochim Cosmochim Acta 41:349-360

Scott ERD (1977c) Geochemical relationship between some pallasites and iron meteorites. Mineral Mag 41:265-272
Scott ERD (1977d) Formation of olivine-metal textures in pallasite meteorites. Geochim Cosmochim Acta 41:693-710
Scott ERD (1979a) Origin of anomalous iron meteorites. Mineral Mag 43:415-421
Scott ERD (1979b) Origin of iron meteorites. *In* "Asteroids," editor T Gehrels. Tucson, AZ:Univ Ariz Press, p 892-921
Scott ERD (1982) Origin of rapidly solidified metal-troilite grains in chondrites and iron meteorites. Geochim Cosmochim Acta. 46:813-823
Scott ERD, Bild RW (1974) Structure and formation of the San Cristobal meteorite, other IB irons and group IIICD. Geochim Cosmochim Acta 38:1379-1391
Scott ERD, Clarke RS Jr (1979) Identification of clear taenite in meteorites as ordered FeNi. Nature 281:360-362
Scott ERD, Wasson JT (1975) Classification and properties of iron meteorites. Rev Geophys Space Phys 13:527-546
Scott ERD, Wasson JT (1976) Chemical classification of iron meteorites—VIII Groups IC, IIE, IIIF and 97 other irons. Geochim Cosmochim Acta 40:103-115
Scott ERD, Wasson JT, Buchwald VF (1973) The chemical classification of iron meteorites; VII, A reinvestigation of irons with Ge concentrations between 25 and 80 ppm. Geochim Cosmochim Acta 37:1957-1983
Scott ERD, Taylor GJ, Keil K (1993) Origin of ureilite meteorites and implications for planetary accretion. Geophys Res Lett 20:415-418
Scott ERD, Haack H, McCoy TJ (1996) Core crystallization and silicate-metal mixing in the parent body of the IVA iron and stony-iron meteorites. Geochim Cosmochim Acta 60:1615-1631
Sears DW (1980) Formation of E-chondrites and aubrites—A thermodynamic model. Icarus 43:184-202
Sellamuthu R, Goldstein JI (1985) Analysis of segregation trends observed in iron meteorites using measured distribution coefficients. Proc 15th Lunar Planet Sci Conf, J Geophys Res 90 (Suppl):C677-C688
Sexton A, Franchi IA, Pillinger CT (1996) Hammadah al Hamra 126—A new Saharan ureilite. Lunar Planet Sci 27, 1173-1174
Shearer CK, Fowler GW, Papike JJ (1997) Petrogenetic models for magmatism on the eucrite parent body: Evidence from orthopyroxene in diogenites. Meteoritics Planet Sci 32:877-889
Shen JJ, Papanastassiou DA, Wasserburg GJ (1996) Precise Re-Os determinations and systematics of iron meteorites. Geochim Cosmochim Acta 60:2887-2900
Sherman SB ,Treiman AH (1989) The olivine-fassaite liquidus: Experiments and implications for angrite achondrites and Ca-Al chondrules. Lunar Planet Sci 20:998-999
Shimaoka TK, Shinotsuka K, Ebihara M, Prinz M (1995) Whole rock compositions of aubritic meteorites: Implications for their origin. Lunar Planet Sci 26:1291-1292
Shukolyukov A, Lugmair GW (1993a) Live iron-60 in the early solar system. Science 259:1138-1142
Shukolyukov A, Lugmair GW (1993b) ^{60}Fe in eucrites. Earth Planet Sci Lett 119:159-166
Shukolyukov A, Lugmair GW (1997) The ^{53}Mn-^{53}Cr isotope system in the Omolon pallasite and the half-life of ^{187}Re . Lunar Planet Sci 28:1315-1316
Simpson AB (1982) Aspects of the composition and origin of achondrites and mesosiderites. PhD Dissertation, Univ Cape Town, 629 p
Simpson AB, Ahrens LH (1977) The chemical relationship between howardites and the silicate fraction of mesosiderites. *In* Comets, Asteroids, Meteorites—Interpretations, Evolution and Origins. AH Delsemme, ed., U. Toledo Press, 445-450
Smales AA, Mapper D, Webb MSW, Webster RK, Wilson J.D. (1970) Elemental composition of lunar surface material. Proc Apollo 11 Lunar Sci Conf:1575-1581
Smoliar MI (1993) A survey of Rb-Sr systematics of eucrites. Meteoritics 28:105-113
Spitz AH (1991) Trace element analysis of ureilite meteorites and implications for their petrogenesis. PhD dissertation, University of Arizona. 232 p
Spitz AH (1992) ICP-MS trace element analysis of ureilites: Evidence for mixing of distinct components. Lunar Planet Sci 23:1339-1340
Spitz AH, Boynton WV (1991) Trace element analysis of ureilites: New constraints on their petrogenesis. Geochim Cosmochim Acta 55:3417-3430
Spitz AH, Goodrich CA (1987) Rare earth element tests of ureilite petrogenesis models. Meteoritics 21:515-516
Spitz AH, Goodrich CA, Crozaz G, Lundberg L (1988) Ion microprobe search for the LREE host phase in ureilite meteorites. Lunar Planet Sci 19:1111-1112
Stauffer H (1961) Primordial argon and neon in carbonaceous chondrites and ureilites. Geochim

Cosmochim Acta 24:70-82

Steele IM, Smith JV (1976) Mineralogy of the Ibitira eucrite and comparison with other eucrites and lunar samples. Earth Planet Sci Lett 33:67-78

Steele IM, Olsen E, Pluth J, Davis- AM (1991) Occurrence and crystal structure of Ca-free beusite in the El Sampal IIIA iron meteorite. Am Mineral 76:1985-1989

Stewart BW, Papanastassiou DA, Wasserburg GJ (1994) Sm-Nd chronology and petrogenesis of mesosiderites. Geochim Cosmochim Acta 58:3487-3509

Stewart B, Papanastassiou DA, Wasserburg GJ (1996) Sm-Nd systematics of a silicate inclusion in the Caddo IAB iron meteorite. Earth Planet Sci Lett 143:1-12

Stolper E (1977) Experimental petrology of eucrite meteorites. Geochim Cosmochim Acta 41:587-611

Störzer D, Pellas P (1977) Angra dos Reis: Plutonium distribution and cooling history. Earth Planet Sci Lett 35:285-293

Swartzendruber LJ, Itkin VP, Alcock CB (1991) The Fe-Ni (iron-nickel) system. J Phase Equil 12:288-312

Swindle TD, Burkland MK, Kring DA (1993) Noble gases in the Brachinites Eagles Nest and LEW88763. Meteoritics 28:445-446

Swindle TD, Kring DA, Burkland MK, Hill DH, Boynton WV (1998) Noble gases, bulk chemistry, and petrography of olivine-rich achondrites Eagles Nest and Lewis Cliff 88763: Comparison to brachinites. Meteoritics Planet Sci 33:31-48

Tagai T, Sadanaga R, Takeuchi Y, Takeda H (1977) Twinning in tridymite from the Steinbach meteorite. Min. Journal 8:382-400

Takahashi K, Masuda A (1990a) The Rb-Sr and Sm-Nd dating and REE measurements of ureilites. Meteoritics 25:413

Takahashi K, Masuda A (1990b) Young ages of two diogenites and their genetic implications. Nature 343:540-542

Takeda H (1979) A layered-crust model of a howardite parent body. Icarus 40:455-470

Takeda H (1986) Mineralogy of Yamato 791073 with reference to crystal fractionation of the howardite parent body. Proc Lunar Planet Sci Conf 16th, Part 2; J Geophys Res 91:D355-D363

Takeda H (1987) Mineralogy of Antarctic ureilites and a working hypothesis for their origin and evolution. Earth Planet Sci Lett 81:358-370

Takeda H (1989) Mineralogy of coexisting pyroxenes in magnesian ureilites and their formation conditions. Earth Planet Sci Lett 93:181-194

Takeda H (1991) Comparisons of Antarctic and non-Antarctic achondrites and possible origin of the differences. Geochim Cosmochim Acta 55:35-47

Takeda H, Graham AL (1991) Degree of equilibration of eucritic pyroxenes and thermal metamorphism of the earliest planetary crust. Meteoritics 26:129-134

Takeda H, Mori H (1985) The diogenite-eucrite links and the crystallization history of a crust of their parent body. Proc Lunar Planet Sci Conf 15th, Part 2; J Geophys Res 90:C636-C648

Takeda H, Yanai K (1978) A thought on the ureilite parent body as inferred from pyroxenes in Yamato-Y74659. Proc 11:189-194, Lunar Planet Symp Tokyo, Inst Space Aeronaut Sci, Univ Tokyo

Takeda H, Miyamoto M, Ishii T, Reid AM (1976) Characterization of crust formation on a parent body of achondrites and the moon by pyroxene crystallography and chemistry. Proc Lunar Sci Conf 7:3535-3548

Takeda H, Miyamoto M, Duke MB, Ishii T (1978a) Crystallization of pyroxenes in lunar KREEP basalt 15386 and meteoritic basalts. Proc Lunar Planet Sci Conf 9:1157-1171

Takeda H, Miyamoto M, Yanai K, Haramura H (1978) A preliminary mineralogical examination of the Yamato-74 achondrites. Mem Natl Inst Polar Res (Japan) Spec Issue 8, 170-184

Takeda H, Miyamoto M, Ishii T, Yanai K, Matsumoto Y (1979a) Mineralogical examination of the Yamato-75 achondrites and their layered crust model. Mem Nat'l Inst Polar Res Spec Issue 12:82-108

Takeda H, Duke M, Ishii T, Haramura H, Yanai K (1979b) Some unique meteorites found in Antarctica and their relation to asteroids. Mem Nat'l Inst Polar Res Special Issue 15:54-73

Takeda H, Mori H, Yanai K (1981) Mineralogy of the Yamato diogenites as possible pieces of a single fall. Mem Natl Inst Polar Res Spec Issue No 20:81-99

Takeda H, Tachikawa O, Toyoda H (1986) Mineralogy of augite-bearing ureilites and some hypotheses on the origin of ureilites. Lunar Planet Sci 17:863-864

Takeda H, Mori H, Ogata H (1988a) On the pairing of Antarctic ureilites with reference to their parent body. Proc NIPR Symp Antarctic Meteorites 1:145-172

Takeda H, Mori H, Ogata H (1988b) Mineralogy of magnesian and calcic groups of ureilites and formation condition of ureilites. Lunar Planet Sci 19:1173-1174

Takeda H, Mori H, Ogata H (1989) Mineralogy of augite-bearing ureilites and the origin of their chemical trends. Meteoritics 24:73-81

Takeda H, Babi T, Mori H (1992) Mineralogy of a new orthopyroxene-bearing ureilite LEW88201 and the relationship between magnesian ureilites and lodranites. Lunar Planet Sci 23:1403-1404

Takeda H, Baba T, Saiki K, Otsuki M, Ebihara M (1993) A plagioclase-augite inclusion in Caddo County: Low-temperature melt of primitive achondrites. Meteoritics, 28:447

Takeda H, Mori H, Hiroi T, Saito J (1994a) Mineralogy of new Antarctic achondrites with affinity to Lodran and a model of their evolution in an asteroid. Meteoritics 29:830-842

Takeda H, Yamaguchi A, Nyquist LE, Bogard DD (1994b) A mineralogical study of the proposed paired eucrites Y-792769 and Y-793164 with reference to cratering events on their parent body. Proc NIPR Symp Antarctic Meteorites 7:73-93

Takeda H, Mori H, Bogard DD (1994c) Mineralogy and ^{39}Ar-^{40}Ar age of an old pristine basalt: Thermal history of the HED parent body. Earth Planet Sci Lett 122:183-194

Takeda H, Yugami K, Bogard D, Miyamoto M (1997a) Plagioclase-augite-rich gabbro in the Caddo County IAB Iron and the missing basalts associated with iron meteorites. Lunar Planet Sci 28:1409-1410

Takeda H, Ishii T, Arai T, Miyamoto M (1997b) Mineralogy of the Asuka 87 and 88 eucrites and the crustal evolution of the HED parent body. Antarctic Meteorite Res 10:401-413

Tera F, Eugster O, Burnett DS, Wasserburg GJ (1970) Comparative study of Li, Na, K, Rb, Cs, Ca, Sr and Ba abundances in achondrites and in Apollo 11 lunar samples. Proc. Apollo 11 Lunar Sci Conf: 1637-1657

Tera F, Carlson RW, Boctor NZ (1997) Radiometric ages of basaltic achondrites and their relation to the early history of the solar system. Geochim Cosmochim Acta 61:1713-1731

Tomeoka K, Takeda H (1989) Fe-S-Ca-Al-bearing carbonaceous veins in the Yamato-74130 ureilite: Evidence for the genetic link to carbonaceous chondrites. Lunar Planet Sci 20:1120-1121

Torigoye N, Yamamoto K, Misawa K, Nakamura N (1993) Compositions of REE, K, Rb, Sr, Ba, Mg, Ca, Fe and Sr isotopes in Antarctic "unique" meteorites. Proc NIPR Symp Antarctic Meteorites 6:100-119

Torigoye-Kita N, Misawa K, Tatsumoto M (1995a) U-Th-Pb and Sm-Nd isotopic systematics of the Goalpara ureilite: resolution of terrestrial contamination. Geochim Cosmochim Acta 59:381-390

Torigoye-Kita N, Misawa K, Tatsumoto M (1995b) Reply to the Comment by CA Goodrich, GW Lugmair, MJ Drake, and PJ Patchett on "U-Th-Pb and Sm-Nd isotopic systematics of the Goalpara ureilite: resolution of terrestrial contamination". Geochim Cosmochim Acta 59:4087-4091

Torigiye-Kita N, Tatsumoto M, Meeker GP, Yanai K (1995c) The 4.56 Ga age of the MET 78008 ureilite. Geochim Cosmochim Acta 59:2319-2329

Toyoda H, Haga N, Tachikawa O, Takeda H, Ishii T (1986) Thermal history of ureilite, Pecora Escarpment 82506 deduced from cation distribution and diffusion profile of minerals. Mem Nat'l Inst Polar Res Spec Issue 41:206-221

Treiman AH (1988) Angra dos Reis is not a cumulate igneous rock. Lunar Planet Sci 19:1203-1204

Treiman AH (1989) An alternate hypothesis for the origin of Angra dos Reis: Porphyry, not cumulate. Proc Lunar Planet Sci Conf 19:443-450

Treiman AH (1997) The parent magmas of the cumulate eucrites: A mass balance approach. Meteoritics Planet Sci 32:217-230

Treiman AH, Berkley JL (1994) Igneous petrology of the new ureilites Nova 001 and Nullarbor 010. Meteoritics 29:843-848

Treiman AH, Drake MJ, Janssens M-J, Wolf R, Ebihara M (1986) Core formation in the Earth and Shergottite Parent Body (SPB): Chemical evidence from basalts. Geochim Cosmochim Acta 50: 1071-1091

Treiman AH, Jones JH, Janssens M-J, Wolf R, Ebihara M (1988) Angra dos Reis: Complex silicate fractionations. Meteoritics 23:305-306

Ulff-Møller F, Rasmussen KL, Prinz M, Palme H, Spettel B, Kallemeyn GW (1995) Magmatic activity on the IVA parent body: Evidence from silicate-bearing iron meteorites. Geochim Cosmochim Acta 59:4713-4728

Ulff-Møller F, Tran J, Choi B-G, Haag R, Rubin AE, Wasson JT (1997) Esquel: Implications for pallasite formation processes based on the petrography of a large slab. Lunar Planet Sci 28:1465-1466

Ulff-Moller F (1998) Effects of liquid immiscibility on trace element fractionation in magmatic iron meteorites: A case study of group IIIAB. Meteoritics Planet Sci 33:207-220

Ulff-Moller F, Choi B-G, Rubin AE, Tran J, Wasson JT (1998) Paucity of sulfide in a large slab of Esquel: New perspectives on pallasite formation. Meteoritics Planet Sci 33:221-227

Van Schmus WR, Ribbe PH (1968) The composition and structural state of feldspar from chondritic meteorites. Geochim Cosmochim Acta 32:1327-1342

Van Schmus WR, Wood JA (1967) A chemical-petrologic classification for the chondritic meteorites. Geochim Cosmochim Acta 31:747-765

Vdovykin GP (1970) Ureilites. Space Sci Rev 10:483-510

Vogel R, Heumann T (1941) Das System Eisen-Eisensulfid-Kalziumsulfid. Archiv Eisenhüttenw 15:195-

199
Voshage H (1967) Bestrahlungsalter und Herkunft der Eisenmeteorite. Z Naturforsch 22a:477-506
Voshage H, Feldmann H (1979) Investigations on cosmic-ray-produced nuclides in iron meteorites; 3, Exposure ages, meteoroid sizes and sample depths determined by mass spectrometric analyses of potassium and rare gases. Earth Planet Sci Lett 45:293-308
Wacker J (1986) Noble gases in the diamond-free ureilite, ALHA78019: The roles of shock and nebular processes. Geochim Cosmochim Acta 50:633-642
Wadhwa M, Shukolyukov A, Lugmair GW (1998) ^{53}Mn-^{53}Cr systematics in Brachina: A record of one of the earliest phases of igneous activity on an asteroid. Lunar Planet Sci 29:1480
Wahl W (1952) The brecciated stony meteorites and meteorites containing foreign fragments. Geochim Cosmochim Acta 2:91.117
Wai CM, Wasson JT (1977) Nebular condensation of moderately volatile elements and their abundances in ordinary chondrites. Earth Planet Sci Lett 36:1-13
Wai CM, Wasson JT (1979) Nebular condensation of Ga, Ge and Sb and the chemical classification of iron meteorites. Nature 282:790-793
Walker D, Grove T (1993) Ureilite smelting. Meteoritics 28:629-636
Wänke H, Baddenhausen H, Spettel B, Teschke F, Quijano-Rico M, Dreibus G, Palme H (1972a) The chemistry of the Haverö ureilite. Meteoritics 7:579-590
Wänke H, Baddenhausen H, Balacescu A, Teschke F, Spettel B, Dreibus G, Palme H, Quijano-Rico M, Kruse H, Wlotzka F, Begemann F (1972b) Multielement analyses of lunar samples and some implications of the results. Proc Lunar Sci Conf 3rd:1251-1268
Wänke H, Baddenhausen H, Dreibus G, Jagoutz E, Kruse H, Palme H, Spettel B, Teschke F (1973) Multielement analyses of Apollo 15, 16 and 17 samples and the bulk composition of the moon. Proc 4th Lunar Sci Conf Geochim Cosmochim Acta Suppl 4:1461-1481
Wänke H, Baddenhausen H, Blum K, Cendales M, Dreibus G, Hofmeister H, Kruse H, Jagoutz E, Palme C, Spettel B, Thacker R, Vilcsek E (1977) On the chemistry of lunar samples and achondrites. Primary matter in the lunar highlands: A re-evaluation. Proc Lunar Sci Conf 8:2191-2213
Warren PH (1987) Mars regolith vs. SNC meteorites: Possible evidence for abundant crustal carbonates. Icarus 70:153-161
Warren PH (1997) Magnesium oxide-iron oxide mass balance constraints and a more detailed model for the relationship between eucrites and diogenites. Meteoritics Planet Sci 32:945-963
Warren PH, Davis AM (1995) Consortium investigation of the Asuka-881371 angrite: Petrographic, electron microprobe, and ion microprobe observations. Antarc Meteorites 20:257-260
Warren PH, Jerde EA (1987) Composition and origin of Nuevo Laredo trend eucrites. Geochim Cosmochim Acta 51:713-725
Warren PH, Kallemeyn GW (1989a) Allan Hills 84025: The second Brachinite, far more differentiated than Brachina, and an ultramafic achondritic clast from L chondrite Yamato 75097. Proc Lunar Planet Sci Conf 19:475-486
Warren PH, Kallemeyn GW (1989b) Geochemistry of polymict ureilite EET83309: and a partially-disruptive impact model for ureilite origin. Meteoritics 24:233-246
Warren PH, Kallemeyn GW (1991) Geochemistry of unique achondrite MAC88177: Comparison with polymict ureilite EET87720 and "normal" ureilites. Lunar Planet Sci 22:1467-1468
Warren PH, Kallemeyn GW (1992) Explosive volcanism and the graphite-oxygen fugacity buffer on the parent asteroid(s) of the ureilite meteorites. Icarus 100:110-126
Warren PH, Kallemeyn GW (1994) Petrology of LEW88774: An extremely Chromium-rich ureilite. Lunar Planet Sci 25:1465-1466
Warren PH, Jerde EA, Migdisova LF, Yaroshevsky AA (1990) Pomozdino: An anomalous, high-MgO/FeO, yet REE-rich eucrite. Proc Lunar Planet Sci Conf 20:281-297
Warren PH, Kallemeyn GW, Mayeda T (1995) Consortium investigation of the Asuka-881371 angrite: Bulk-rock geochemistry and oxygen-isotopes. Antarc Meteorites 20:261-264
Warren PH, Kallemeyn GW, Arai T, Kaneda K (1996) Compositional-petrologic investigations of eucrites and the QUE94201 shergottite. Antarctic Meteorites 21:195-197
Wasserburg GJ, Sanz HG, Bence AE (1968) Potassium-feldspar phenocrysts in the surface of Colomera, an iron meteorite. Science 161:684-687
Wasserburg GJ, Tera F, Papanastassiou DA, Huneke JC (1977) Isotopic and chemical investigations on Angra dos Reis. Earth Planet Sci Lett 35:294-316
Wasson JT (1967) The chemical classification of iron meteorites; I, A study of iron meteorites with low concentrations of gallium, germanium. Geochim Cosmochim Acta 31:161-180
Wasson JT (1969) The chemical classification of iron meteorites; III, Hexahedrites and other irons with germanium concentrations between 80 and 200 ppm. Geochim Cosmochim Acta 33:179-198
Wasson JT (1970) The chemical classification of iron meteorites; IV, Irons with Ge concentrations greater

than 190 ppm and other meteorites associated with group I. Icarus 12:407-423

Wasson JT (1972) Parent-body models for the formation of iron meteorites. Proc Intl Geol Cong 24: 161-168

Wasson JT (1974) Meteorites: Classification and Properties. Berlin and New York: Springer-Verlag, 316 p

Wasson JT (1985) Meteorites. Their Record of Early Solar-System History. New York: WH Freeman, 267 p

Wasson JT (1990) Ungrouped iron meteorites in Antarctica; origin of anomalously high abundance. Science 249:900-902

Wasson JT, Chapman CR (1996) Space weathering of basalt-covered asteroids: Vesta an unlikely source of the HED meteorites. *In* Workshop on Evolution of Igneous Asteroids: Focus on Vesta and the HED Meteorites, Mittlefehldt DW, Papike JJ, eds., LPI Tech Report 96-02, Part 1, 38-39

Wasson JT, Choi B-G, Jerde EA, Ulff-Moller F (1998a) Chemical classification of iron meteorites: XII. New members of the magmatic group. Geochim Cosmochim Acta 62:715-724

Wasson JT, Jerde E, Choi B-G, Ulff-Moller F (1998b) Chemical classification of iron meteorites: XIII. New ungrouped irons and members of the nonmagmatic groups. Geochim Cosmochim Acta 62, in press

Wasson JT, Kimberlin J (1967) The chemical classification of iron meteorites; II, Irons and pallasites with germanium concentrations between 8 and 100 ppm. Geochim Cosmochim Acta 31:149-178

Wasson JT, Rubin AE (1985) Formation of mesosiderites by low-velocity impacts as a natural consequence of planet formation. Nature 318:168-170

Wasson JT, Schaudy R (1971) The chemical classification of iron meteorites; V, Groups IIC and IIID and other irons with germanium concentrations between 1 and 25ppm. Icarus 14:59-70

Wasson JT, Wai CM (1970) Composition of the metal, schreibersite and perryite of enstatite achondrites and the origin of enstatite chondrites and achondrites. Geochim Cosmochim Acta 34:169-184

Wasson JT, Wai CM (1976) Explanation for the very low Ga and Ge concentrations in some iron meteorite groups. Nature 261:114-116

Wasson JT, Wang J (1986) A nonmagmatic origin of group-IIE iron meteorites. Geochim Cosmochim Acta 50:725-732

Wasson JT, Wetherill GW (1979) Dynamical, chemical and isotopic evidence regarding the formation locations of asteroids and meteorites. *In* Asteroids, T. Gehrels (ed) Univ Arizona Press, p 926-974

Wasson, JT, Schaudy R, Bild RW, Chou C-L (1974) Mesosiderites-I. Compositions of their metallic portions and possible relationship to other metal-rich meteorite groups. Geochim Cosmochim Acta 38:135-149

Wasson, JT, Chou C-L, Bild RW, Baedecker PA (1976) Classification of and elemental fractionation among ureilites. Geochim Cosmochim Acta 40:1449-1458

Wasson JT, Willis J, Wai CM, Kracher A (1980) Origin of iron meteorite groups IAB and IIICD. Z Naturforsch 35a:781-795

Wasson JT, Ouyang Xinwei, Jerde E (1989) Chemical classification of iron meteorites; XI, Multi-element studies of 38 new irons and the high abundance of ungrouped irons from Antarctica. Geochim Cosmochim Acta 53:735-744

Watters TR, Prinz M (1979) Aubrites: Their origin and relationship to enstatite chondrites. Proc Lunar Planet Sci Conf 10:1073-1093

Weber D, Bischoff A (1996) New meteorite finds from the Libyan Sahara. Lunar Planet Sci 27:1393-1394

Weber I, Bischoff A (1998) Mineralogy and chemistry of the ureilites Hammadah Al Hamra 064 and Jalanash. Lunar Planet Sci 29:1365

Weber HW, Hintenberger H, Begemann F (1971) Noble gases in the Haverö ureilite. Earth Planet Sci Lett 13:205-209

Weber, HW, Begemann F, Hintenberger H (1976) Primordial gases in graphite-diamond-kamacite inclusions from the Haverö ureilite. Earth Planet Sci Lett 29:81-90

Weigel A, Eugster O, Koeberl C and Krähenbühl U (1996) Primitive differentiated achondrite Divnoe and its relationship to brachinites. Lunar Planet Sci 27:1403-1404

Welten KC, Lindner L, van der Borg K, Loeken T, Scherer P, Schultz L (1997) Cosmic-ray exposure ages of diogenites and the recent collisional history of the howardite, eucrite and diogenite parent body/bodies. Meteoritics Planet Sci 32:891-902

Wheelock MM, Keil K, Floss C, Taylor GJ, Crozaz G (1994) REE geochemistry of oldhamite-dominated clasts from the Norton County aubrite: Igneous origin of oldhamite. Geochim Cosmochim Acta 58:449-458

Wiik HB (1969) On regular discontinuities in the composition of meteorites. Soc Fennica Commentationes Phys-Math 34:135-145

Wiik HB (1972) The chemical composition of the Haverö meteorite and the genesis of the ureilites. Meteoritics 7:553-557

Wilkening LL (1973) Foreign inclusions in stony meteorites-I. Carbonaceous chondrite xenoliths in the Kapoeta howardite. Geochim Cosmochim Acta 37:1985-1989

Wilkening LL (1976) Carbonaceous chondritic xenoliths and planetary-type noble gases in gas-rich meteorites. Proc Lunar Sci Conf 7 :3549-3559

Wilkening LL, Anders E (1975) Some studies of an unusual eucrite: Ibitira. Geochim Cosmochim Acta 39:1205-1210

Wilkening LL, Marti K (1976) Rare gases and fossil particle tracks in the Kenna ureilite. Geochim Cosmochim Acta 40:1465-1473

Willis J (1980) The bulk composition of iron meteorite parent bodies. PhD Dissertation, University of California at Los Angeles, 208 p

Willis J, Goldstein JI (1982) The effects of C, P, and S on trace element partitioning during solidification in Fe-Ni alloys. Proc 13th Lunar Planet Sci Conf, J Geophys Res 87 (Suppl):A435-A445

Wilson L, Keil K (1991) Consequences of explosive eruptions on small solar system bodies: The case of the missing basalts from the aubrite parent body. Earth Planet Sci Lett 104:505-512

Wlotzka F (1993) Meteoritical Bulletin, No. 75, 1993 December. Meteoritics 28:692-703

Wlotzka F (1994) Meteoritical Bulletin, No. 77, 1994 November. Meteoritics 29:891-897

Wlotzka F, Jarosewich E (1977) Mineralogical and chemical compositions of silicate inclusions in the El Taco, Camp del Cielo, iron meteorite. Smithsonian Contrib Earth Sci 19:104-125

Wlotzka F, Palme H, Spettel B, Wänke H, Jarosewich E, Noonan AF (1983) Alkali differentiation in LL-chondrites. Geochim Cosmochim Acta 47:743-757

Wood JA (1964) The cooling rates and parent planets of several iron meteorites. Icarus 3:429-459

Wood JA (1981) On the nature of the pallasite parent body: Midcourse corrections. Lunar Planet Sci 12:1200-1202

Wood JA, Pellas P (1991) What heated the meteorite planets? *In* The Sun in Time, CP Sonnet, MS Giampapa, MS Matthews (eds) Univ Arizona, 740-760

Xue S, Herzog GF, Souzis A, Ervin MH, Lareau RT, Middleton R, Klein J (1995) Stable magnesium isotopes, ^{26}Al, ^{10}Be, and ^{26}Mg/^{26}Al exposure ages of iron meteorites. Earth Planet Sci Lett 136:397-406

Yajima H, Matsuda J (1989) Further study on noble gases in shock-produced diamonds. Mass Spect 37:331-342

Yamaguchi A, Takeda H, Bogard DD, Garrison D (1994) Textural variations and impact history of the Millillillie eucrite. Meteoritics 29:237-245

Yamaguchi A, Taylor GJ, Keil K (1996) Global crustal metemorphism of the eucrite parent body. Icarus 124:97-112

Yamaguchi A, Taylor GJ, Keil K (1997) Metemorphic history of the eucritic crust of 4 Vesta. J Geophys Res 102:13381-13386

Yanai K (1994) Angrite Asuka-881371: Preliminary examination of a unique meteorite in the Japanese collection of antarctic meteorites. Proc NIPR Symp Antarc Meteorites 7:30-41

Yanai K, Kojima H (1984) Meteorites News. National Institute of Polar Research, Tokyo, Japan. 58 p

Yanai K, Kojima H (1987) Photographic Catalog of the Antarctic Meteorites. National Institute of Polar Research, Tokyo, Japan. 298 p

Yanai K, Kojima H (1991) Yamato-74063: Chondritic meteorite classified between E and H chondrite groups. Proc NIPR Symp Antarctic Meteorites 4:118-130

Yanai K, Kojima H (1993) General features of some unique inclusions in Yamato ordinary chondrites. Papers Presented to the 18th Symp Antarctic Meteorites, 57-60

Yanai K, Kojima H (1995a) Yamato-8451: A newly identified pyroxene-bearing pallasite. Proc NIPR Symp Antarctic Meteorites 8:1-10

Yanai K, Kojima H (1995b) Catalog of Antarctic Meteorites. National Institute of Polar Research Tokyo, Japan

Yanai K, Matsumoto Y, Kojima H (1983) A Brachina-like inclusion in the Yamato-75097 L6 chondrite: A preliminary examination. Mem Nat'l Inst Polar Spec Issue 30:29-35

Yang C-W, Williams DB, Goldstein JI (1996) A revision of the Fe-Ni phase diagram at low temperatures (<400°C). J Phase Equilibria 17:522-531

Yang C-W, Williams DB, Goldstein JI (1997a) A new empirical cooling rate indicator for meteorites based on the size of the cloudy zone of the metallic phases. Meteoritics Planet Sci 32:423-429

Yang C-W, Williams DB, Goldstein JI (1997b) Low-temperature phase decomposition in metal from iron, stony-iron and stony meteorites. Geochim Cosmochim Acta 61:2943-2956

Yugami K, Miyamoto M, Takeda H, Hiroi T (1993) Mineralogy of ALH81187: A partly reduced acapulcoite. Papers Presented to the 18th Symp Antarctic Meteorites, 34-37

Yugami K, Takeda H, Kojima H, Miyamoto M (1997) Modal abundances of primitive achondrites and the endmember mineral assemblage of the differentiation trend. Papers presented to the 22nd Symp Antarctic Meteorites, 220-222

Yugami K, Takeda H, Kojima H and Miyamoto M (1996) Mineralogy of the new primitive achondrites Y8005 and Y8307 and their differentiation from chondritic materials. Papers presented to the 21st Symp Antarctic Meteorites, 216-218

Zhou Y, Steele IM (1993) Chemical zoning and diffusion of Ca, Al, Mn, and Cr in olivine of Springwater pallasite . Lunar Planet Sci 24:1573-1574

Zipfel J, Palme H (1993) Chemical composition of new acapulcoites and lodranites. Lunar Planet Sci 24:1579-1580

Zipfel J, Palme H, Kennedy AK, Hutcheon ID (1995) Chemical composition and origin of the Acapulco meteorite. Geochim Cosmochim Acta 59:3607-3627

Zolensky ME, Hewins RH, Mittlefehldt DW, Lindstrom MM, Xiao X, Lipschutz ME (1992) Mineralogy, petrology and geochemistry of carbonaceous chondritic clasts in the LEW 85300 polymict eucrite. Meteoritics 27:596-604

Zolensky ME, Weisberg MK, Buchanan PC, Mittlefehldt DW (1996) Mineralogy of carbonaceous chondrite clasts in HED achondrites and the moon. Meteoritics Planet Sci 31:518-537

Chapter 5

LUNAR SAMPLES

James J. Papike[1], Graham Ryder[2] and Charles K. Shearer[1]

[1]*Institute of Meteoritics*
Department of Earth and Planetary Sciences
University of New Mexico
Albuquerque, New Mexico 87131

[2]*Lunar and Planetary Institute*
Center for Advanced Space Studies
3600 Bay Area Boulevard
Houston, Texas 77058

INTRODUCTION

This review of the lunar samples places major emphasis on mineralogy, petrology and geochemistry. The material presented in this chapter draws heavily from material covered in *Lunar Sourcebook: A User's Guide to the Moon* (Heiken et al. 1991) and readers desiring more complete information about the moon and its samples are referred to that comprehensive work. This chapter summarizes material covered in Chapter 5: Lunar Minerals; Chapter 6: Lunar Rocks, and Chapter 7: The Lunar Regolith, of the *Lunar Sourcebook.* Two of the authors of this chapter were also authors of the *Lunar Sourcebook* chapters mentioned above.

The material in this chapter is arranged as follows. First, a brief introduction to the lunar sampling sites and lunar evolution is given. Then the lunar regolith, which is a several meter-thick layer of unconsolidated debris and forms the interface between the Moon and its space environment, is reviewed. Next, we review the mineralogy of lunar samples, and extensive tabulated mineral data are reported in the appendix. We then review lunar basalts and volcanic glasses, and last we cover the rocks from the older, more feldspathic lunar highlands.

An important turning point in our understanding of the evolution of the Moon was made possible by the Apollo missions (1969-1972). The six Apollo manned missions and three Russian unmanned missions (1970-1976) returned over 380 kilograms of samples (SR Taylor 1982) from nine sampling sites located on the near side of the Moon, shown on Figure 1. Intense study of these Apollo and Luna samples and of several lunar meteorites (Warren 1994) have led to a basic understanding of the Moon's magmatic and impact history.

Perhaps the single most important concept resulting from the study of the lunar samples is the possible previous existence of a lunar magma ocean. A magma ocean was proposed early in the study of the Apollo 11 samples by Wood et al. (1970a,b) and Smith et al. (1970). Since these early pioneering papers, many variations of this model followed and several of these are summarized by Vaniman et al. (1991 *Lunar Sourcebook*). SR Taylor (1982) and Warren (1985) give detailed reviews of the magma ocean concept.

Why were we driven to the concept of a global magma ocean in the first place? The original suggestions by Wood et al. (1970a,b) and Smith et al. (1970) were based on the occurrence of small and scarce anorthosite samples among the dominantly basaltic sample suite recovered from Apollo 11. Based on this discovery and the previously mapped

0275-0279/98/0036-0005$20.00

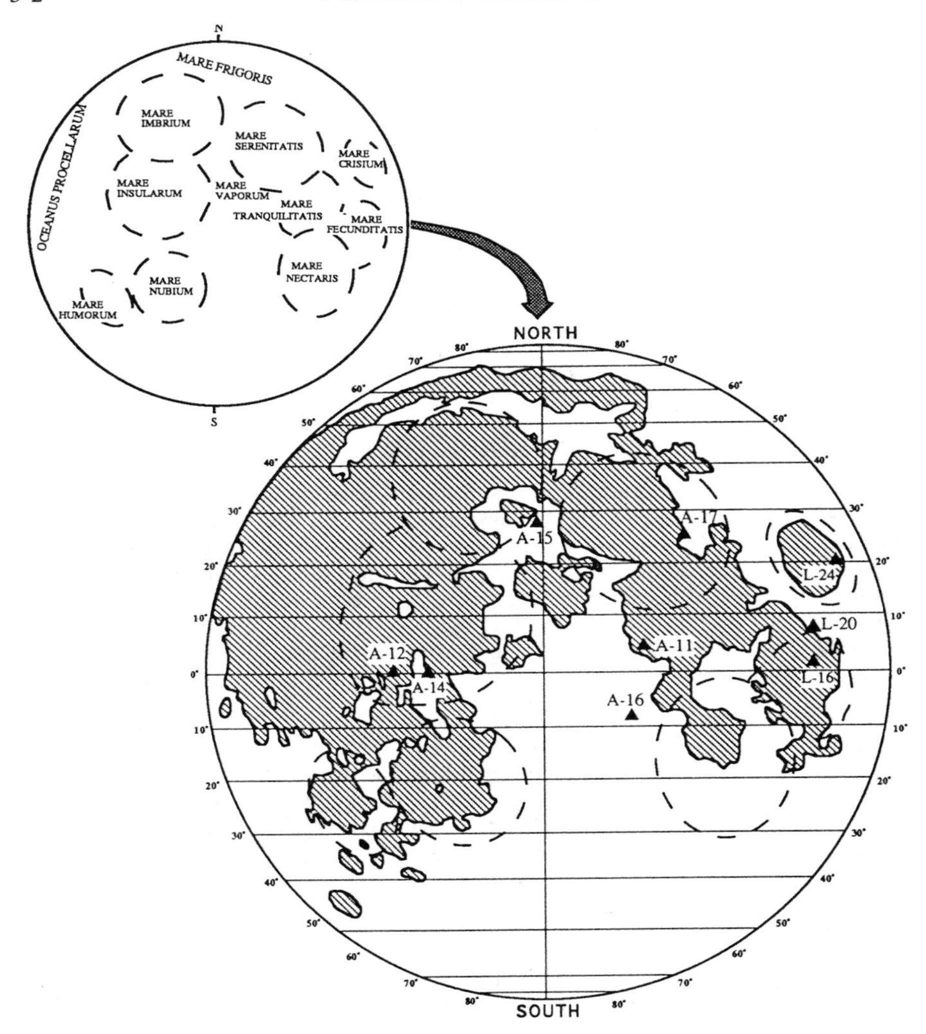

Figure 1. Schematic map of the Moon showing the distribution of mare basalts (pattern) and highland lithologies (white). The six Apollo landing sites (A11, A12, A14, A15, A16, and A17) and three Russian sampling sites (L16, L20, and L24) are indicated. Dashed lines indicate approximate boundaries of large basins.

geological units on the Moon, it was suggested that a significant portion of the lunar crust is dominated by anorthosite. Extraction of anorthosite from a body the size of the Moon (radius 1738 km) implies significant differentiation and feldspar separation of any reasonable lunar bulk composition. A simple model for accomplishing this planetary differentiation would be by deep melting to form a magma ocean, or magmasphere (Warren 1985), in which feldspar floats, forming the feldspathic outer crust.

SR Taylor (1982) summarized some of the observational facts about the Moon that led to the concept of an early, extensive differentiation. He inferred that the lunar crust,

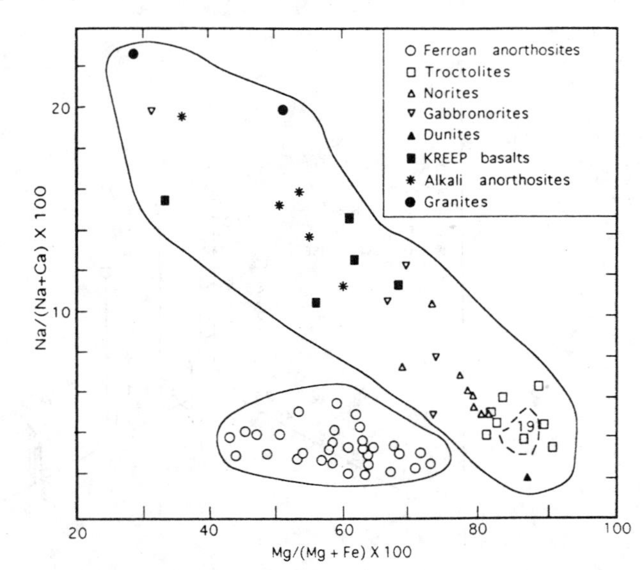

Figure 2. The atomic ratios Na/(Na+Ca) x Mg/(Mg+Fe) divide pristine highland lithologies into two groups: ferroan anorthosites (open circles) and the Mg-suite (other symbols).

~60 km thick, contains ~25 wt % Al_2O_3. Since most proposed bulk compositions contain less than 6 wt % Al_2O_3, the crust (~10% of the lunar volume) contains about 40 wt % of the total lunar Al_2O_3. Potassium, uranium, and thorium are also concentrated in the crust by two orders of magnitude in excess of their assumed lunar abundances. The early differentiation (4.6 to 4.4 Ga) could be accomplished either with a magma ocean or a closely spaced series of igneous events, but clearly a large fraction of the volume of the Moon was involved. Because we do not know with certainty the bulk composition of the Moon, we cannot specify with certainty the depth of the magma ocean, if one indeed existed. Estimates of the depth vary from 200 km to total Moon melting (SR Taylor 1982).

To unravel the complex history of the Moon, it is essential that we identify those rocks that formed from endogenous igneous activity as opposed to impact-generated melting. Warren and Wasson, (1977) presented criteria by which we can identify pristine igneous rocks which were neither formed by nor contaminated by impact. According to Warren (1985) the least ambiguous single means of identifying pristine rocks is by analysis of siderophile elements, because most meteorites have high siderophile contents, and rocks formed by impact can usually be identified by elevated concentrations of these elements. Warren (1993) presents a list of potential pristine rocks.

Pristine highlands rocks can be divided into two major groups based on the Na/(Na+Ca) molar versus the Mg/(Mg+Fe) molar content of their bulk rock compositions (Fig. 2). JL Warner et al. (1976) were the first to suggest the nature of these two highland rock groups. It is now generally agreed that the two major rock groups are the older ferroan anorthosites (>4.4 Ga) and the partly younger Mg-rich rocks (4.43 to 4.17 Ga) (SR Taylor et al. 1993). The ages overlap at the ancient end. These have commonly been referred to as Mg-suite or magnesian suite rocks, but in this Chapter we will call them Mg-rich rocks. They include dunites, troctolites, norites, and gabbronorites. SR Taylor et al. (1993) estimate that these make up ~20% of the lunar crust down to a depth of ~60 km, and ferroan anorthosite makes up most of the rest.

The pristine highland rocks and the breccias and regolith that are derived from them make up the lunar highlands (~83% of the lunar surface area). The remaining ~17% of the

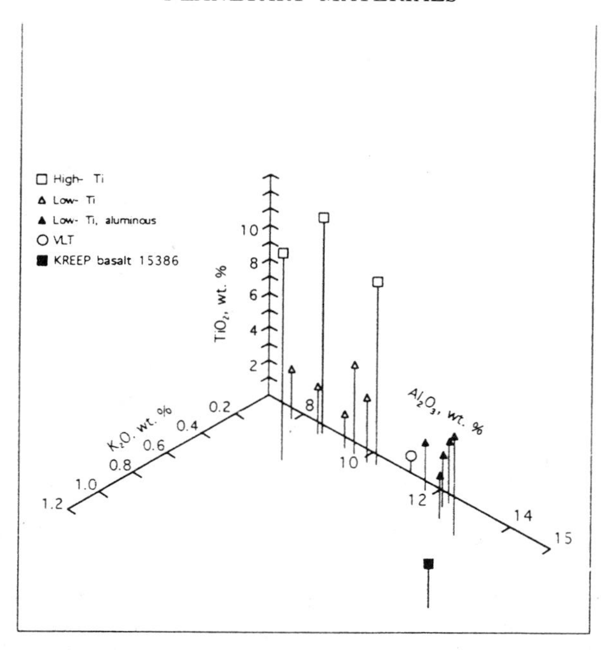

Figure 3. TiO_2 x Al_2O_3 x K_2O plot separates the major mare basalt types and KREEP basalts.

lunar surface area, and ~1% of the crustal volume, is composed of mare basalts (Head 1976). The mare basalts are enriched in FeO and TiO_2, and correspondingly depleted in Al_2O_3 and CaO, and have higher CaO/Al_2O_3 ratios than highland lithologies (GJ Taylor et al. 1991 *Lunar Sourcebook*). These chemical differences reflect the fact that the highland rocks are enriched in plagioclase and depleted in pyroxene relative to the mare terrain. Mare basalts are generally classified first on the basis of their TiO_2 contents. The three major groups of mare basalts are high-Ti (>9 wt % TiO_2), low-Ti (1.5 to 9 wt % TiO_2) and very low-Ti (VLT) basalts, with <1.5 TiO_2 wt % (GJ Taylor et al. 1991 *Lunar Sourcebook*). Figure 3 summarizes some of the chemical characteristics of mare basalts and also shows a KREEP (K-, REE-, and P-rich) basalt for comparison. The common model for the origin of mare basalt magmas is as remelting of mantle sources produced in the early differentiation and at least partly complementary to the feldspathic crust.

Although the magma ocean hypothesis has never been proven, it remains the leading model for the early differentiation of the Moon. One recent interpretation (Hess and Parmentier 1995) of the global consequences of a lunar magma ocean with a depth of 800 km is that plagioclase floated, forming a 60 km anorthosite crust near the lunar surface (Fig. 4). The denser olivine and pyroxene sank, forming a cumulate pile ~700 km thick with the later crystallizing, denser, low mg# [= Mg/(Mg+Fe) atomic] pyroxene assemblages near the top and the less dense, more magnesian, olivine and pyroxene assemblages near the bottom. Hess and Parmentier (1995) suggested that late-crystallizing melt was rich in trace incompatible elements forming a KREEP-rich residue that was also rich in ilmenite. If this model is correct, it leads to a gravitationally unstable cumulate pile, with dense cumulates overlying less dense cumulates. The predicted consequence of the instability (Spera 1992, Hess and Parmentier 1995) is convective overturn with the ilmenite layer sinking to deep mantle depths. Although this model is speculative, it potentially explains how KREEP (a late-crystallizing product of the magma ocean) can be brought to great depth in a zone of melting, which may include both Mg-rich cumulates and primitive mantle. This process could provide source regions for the pictritic lunar basalts (Shearer and Papike 1993) and Mg-, Al-rich magmas (the Mg-suite), which intruded the anorthositic crust during a period (4.43 to 4.17 Ga) after most of the magma ocean had crystallized (Warren and Kallemeyn 1993). The Mg-rich magmas, which apparently intruded the anorthositic crust, may have formed layered mafic complexes (James 1980) similar to the Stillwater Complex, Montana.

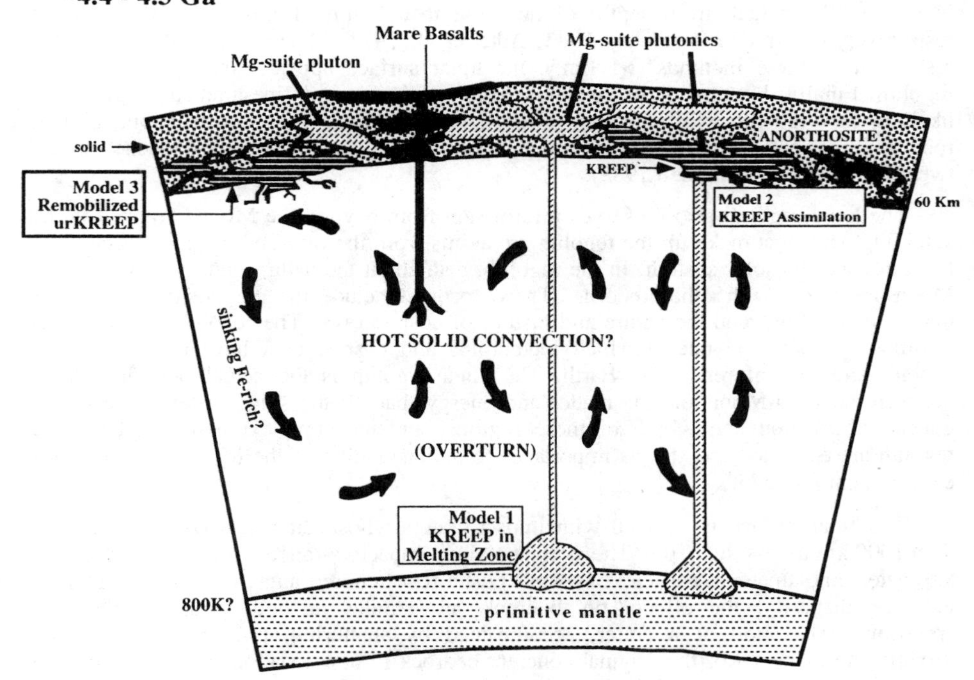

Figure 4. Schematic cross section of the Moon (radius 1738 km) illustrating the possible consequences of a magma ocean (depth 800 km). A 60 km-thick crust of anorthosite forms from floating plagioclase cumulates. Three models are illustrated for the possible origins of the younger Mg-suite which intrudes the anorthosite crust. Model 1 involves KREEP being delivered to depth by convective overturn and mixing with Mg-rich picritic melts. Model 2 is the same as model 1 but involves KREEP assimilation at the base of the crust. Model 3 involves mobilization of urKREEP (last dregs of the magma ocean) possibly caused by decompressional melting triggered by basin-forming impacts.

THE LUNAR REGOLITH

Introduction

All the lunar landings and all photographic investigations show that the entire lunar surface consists of a regolith layer that completely covers the underlying bedrock, except perhaps on some very steep slopes. On the airless, lifeless Moon, the regolith results from the continuous impact of large and small meteoroids and the steady bombardment of the lunar surface by charged atomic particles from the Sun and other stars.

The regolith is the source of virtually all of our information about the Moon. All direct measurements of physical and chemical properties of lunar material have been made on samples, both rocks and soils, collected from the regolith. Experiments, whether conducted by astronauts on the Moon or remotely monitored from the Earth, were done on or in the regolith layer. For example, heat flow measurements were made with sensors that were emplaced in the regolith; estimates of heat flow from the rocks below were made only by inference from the regolith data. Even seismic data were obtained from seismometers that were implanted on or in the regolith and were not coupled to bedrock. Remotely sensed x-ray fluorescence, infrared spectra, and gamma-ray signals come from the very top of the lunar regolith; in fact, from depths of no more than 20μm, 1 mm, and 10 to 20 cm respectively (Morris 1985, Pieters 1983, Adler et al. 1973, Metzger et al. 1973). At the resolution of these methods (>1 km^2), the lunar surface appears totally covered with regolith. Finally, because of its surficial, unconsolidated, and fine-grained nature, it is likely that the regolith will be the raw material used for lunar base construction, mining, road building, and resource extraction when permanent lunar bases are established in the twenty-first century (Mendell 1985).

The lunar regolith also preserves information from beyond the Moon. Trapped in the solid fragments that make up the regolith are atoms from the Sun and cosmic-ray particles from beyond the solar system. In the regolith, data about the nature and evolution of the Moon are mixed with other records. These records include the composition and early history of the Sun, and the nature and history of cosmic rays. The regolith also contains information about the rate at which meteoroids and cosmic dust have bombarded the Moon—and, by inference, the Earth. The lunar regolith is the actual boundary layer between the solid Moon and the matter and energy that fill the solar system. It contains critical information about both of these regions, and the complexities of studying the regolith are exceeded only by its importance to understanding of the Moon and the space environment around it.

The lunar surface is covered with impact craters whose diameters range from more than 1000 km to less than 1μm. The corresponding impactors range from asteroids tens of kilometers in diameter to particles of cosmic dust a few hundred angstroms across. Despite the size disparity, the effects on bedrock are similar at both ends of the size spectrum—excavation of a crater, accompanied by shattering, pulverization, melting, mixing, and dispersal of the original coherent bedrock to new locations in and around the crater. The moment that fresh bedrock is exposed on the Moon (e.g. by the eruption of a lava flow), meteoroid bombardment begins to destroy it. As the impacts continue, the original bedrock is covered by a fragmental layer of broken, melted, and otherwise altered debris from innumerable superimposed craters. This layer is the lunar regolith.

Studies of returned samples have shown that the bulk of this lunar regolith (informally called the lunar soil) consists of particles <1 cm in size, although larger cobbles and boulders, some as much as several meters across, are commonly found at the surface. Because the impact cratering events produce shock overpressures and heat, much of the pulverized material is melted and welded together to produce breccias (fragmental rocks) and impact melt rocks, which make up a significant portion of the regolith and add to its complexity.

The processes of regolith formation can be divided roughly into two stages. During the early stage, shortly after bedrock is first exposed and the regolith is still relatively thin (less than a few centimeters), both large and small impacts can penetrate the regolith and excavate fresh bedrock. The regolith layer builds up rapidly. As time goes on and the thickness increases (to a meter or more), only the larger (and far less frequent) impacts penetrate the regolith and bring up new bedrock. In this later stage, the smaller (and more numerous) impacts only disturb and mix (garden) the regolith layer already present, and the regolith thickness increases more slowly.

Since about 4 Ga, the impact flux on the lunar surface has been relatively low, and a regolith only a few meters thick is adequate to shield the underlying bedrock almost indefinitely. For this reason, the regolith thickness rarely exceeds 10 to 20 m. Regolith thicknesses on the maria are typically only a few meters (Langevin and Arnold 1977; SR Taylor 1982). Astronauts have drilled to a depth of approximately 3 m in the regoliths at Apollo sites, and estimates based on grain-size distributions suggest that the maximum thickness of the regolith may not exceed 20 m (McKay et al. 1974). Early estimates of regolith thicknesses by Oberbeck and Quaide (1968), based on crater-shape models, ranged from 3.3 m on Oceanus Procellarum to 16 m for the inner wall of the crater Hipparchus. At the four Surveyor mare sites, apparent regolith thicknesses range from 1 to 10 m (Shoemaker et al. 1968).

The current consensus is that the regolith is generally about 4 to 5 m thick in the mare areas but may average about 10 to 15 m in older highland regions. Beneath this true regolith is a complex zone that probably consists of large-scale ejecta and impact-fractured, brecciated bedrock (based on orbital radar data and modeling: Peeples et al. 1978, Langevin 1982). This layer of fractured bedrock has been called the megaregolith and may consist of large (>1 m) blocks. Some of the inferred properties of this megaregolith are different from those of the unconsolidated surficial material that has been sampled. However, the detailed properties of the megaregolith are essentially unknown.

The formation and evolution of the lunar regolith is a complex process. At any given spot, the nature and history of the regolith is determined by two completely random mechanisms. One is destructive—the excavation of existing regolith by impact craters. The other is constructive—the addition of layers of new material (either from bedrock or older regolith) that is excavated from either near (small) or distant (large) impact craters (McKay et al. 1974). Superimposed on these mechanical processes are the effects of solar and cosmic particles that strike the lunar surface.

These simultaneous processes combine to produce a regolith whose structure, stratigraphy, and history may vary widely, even between locations only a few meters apart. Surface layers can be buried and then reexposed. Single layers, or slabs containing multiple layers, can be transported, overturned, or buried and thus are rarely continuous over long lateral distances. Deciphering these complications is a major challenge that requires the application of a wide range of analytical techniques—petrologic studies, gas analyses, measurements of radioactivity, stable isotope studies, trace element geochemistry, magnetic measurements, and statistical modeling.

Lunar soil

The lunar soil is a somewhat cohesive, dark grey to light grey, very-fine grained, loose, clastic material derived primarily from the mechanical disintegration of lunar rocks. Figure 5 shows representative back scattered electron (BSE) images of lunar soil. The mean grain sizes of analyzed soils range from about 40 μm to about 800 μm and are typically between 60 and 80 μm. Individual lunar soil particles are mostly glass-bonded aggregates (agglutinates), as well as various rock, mineral, and glass fragments. The soils range in composition from basaltic to anorthositic, and they include a small (<2 wt %) meteoritic component. Although the chemical compositions of lunar soils show considerable variation, physical properties such as grain size, density, packing and compressibility are rather uniform.

Figure 5A. BSE images of thin sections from Apollo 17 core 70008,372 (A,B) and 70005,603 (C,D). Figures A,C have the same scale bar. Figures C,D are at a higher magnification as indicated by the scale bar in D. Core section 70008 has a coarse grain size and is less mature than core section 70005.

Upper: 70008, 372

(A) 1, High-Ti (devitrified) glass bead; 2, Ilmenite 3, Low-Ti glass bead.

(B) 4, plagioclase; 5, Agglutinate.

Lower: 70004,403

(C) 6, Agglutinate.

(D) 7, Glass bead, devitrified around perimeter; 8, Augite.

Petrographic studies of regolith samples make it possible to characterize the material in two complementary ways: by the relative proportions of different kinds of fragments and by the chemical and mineral compositions of individual rock and mineral fragments. Optical studies of regolith samples make it possible to conduct a census of the relative amounts (by volume) of different particles—rocks, mineral and glass fragments, and other components—in each sample. This determination of relative volumes of components, modal analysis, can be expanded to estimate the chemical composition of the soil as well if chemical data for fragments are determined directly with an electron microprobe. The bulk chemical composition, reconstructed in this way, can be used to interpret the sources of the particles, and the origin and evolution of the soil sample. Such studies show that five basic particle types make up the lunar soils: mineral fragments; pristine crystalline rock fragments; breccia fragments; glasses of various kinds; and the unique lunar constructional

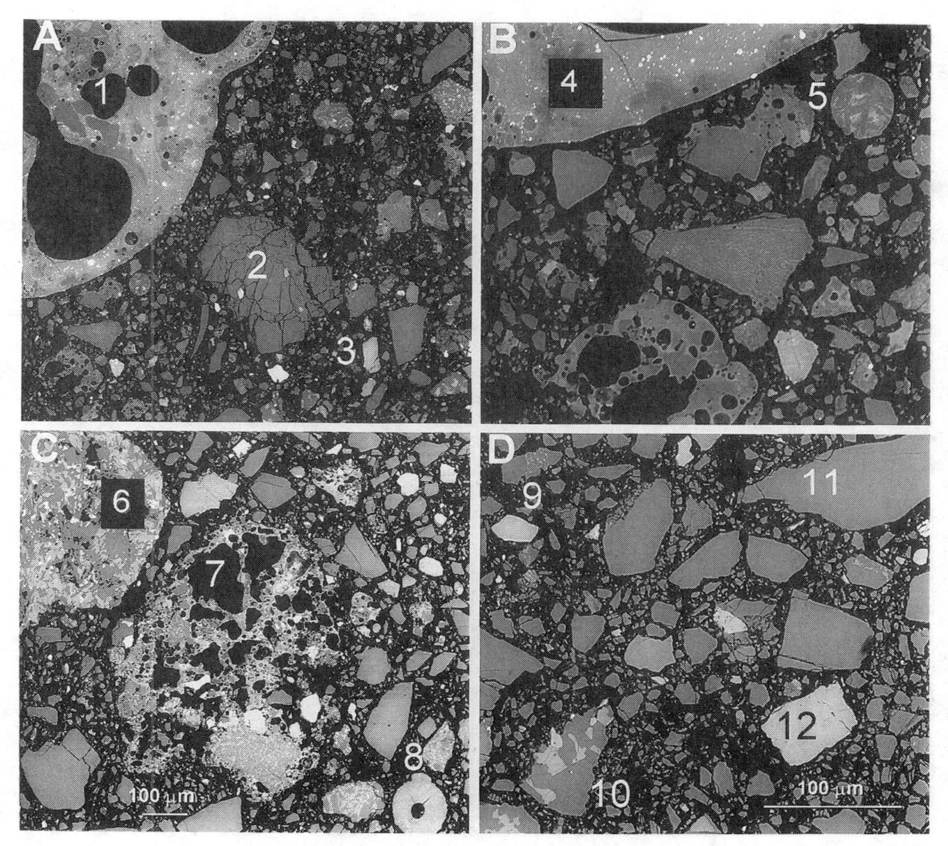

Figure 5B. BSE images of thin sections from Apollo 16 core 60010,6018 (top two) and 60009,60029 (bottom two). Figures A,B have the same scale bar. Figures C,D are at a higher magnification as indicated by the scale bar in D.

Upper: 60010,6018

(A) 1, Agglutinate; 2, Plagioclase; 3, Orthopyroxene.

(B) 4, Agglutinate with Fe-Ni metal; 5, low-Ti glass bead.

Lower: 60009,6029

(C) 6, Aggregate of anorthite and augite; 7, Agglutinate; 8, Glass bead (low-Ti).

(D) 9, Olivine; 10, Aggregate of anorthite, pyroxene and olivine; 11, Plagioclase; 12, Orthopyroxene.

particles called agglutinates. These diverse particles can also be divided into two groups: regolith-derived and bedrock-derived. All agglutinates, fragments of regolith breccias, and heterogeneous glasses have been formed by the action of meteorite impacts (chiefly impact melting) on regolith targets. These particles are sometimes called the fused soil component. The remaining fragments are pieces of igneous rocks, monomict breccias, and polymict breccias, which make up the bedrock-derived component.

Except for a few igneous rock fragments, the average size of rock clasts in the regolith is <250 µm. In general, it is difficult to identify unambiguously the rock type of particles this small; textures and mineral proportions may not be representative. For this reason, the compisitions of individual mineral grains in the fragments has been used instead as the basis for identifying the parent rock types.

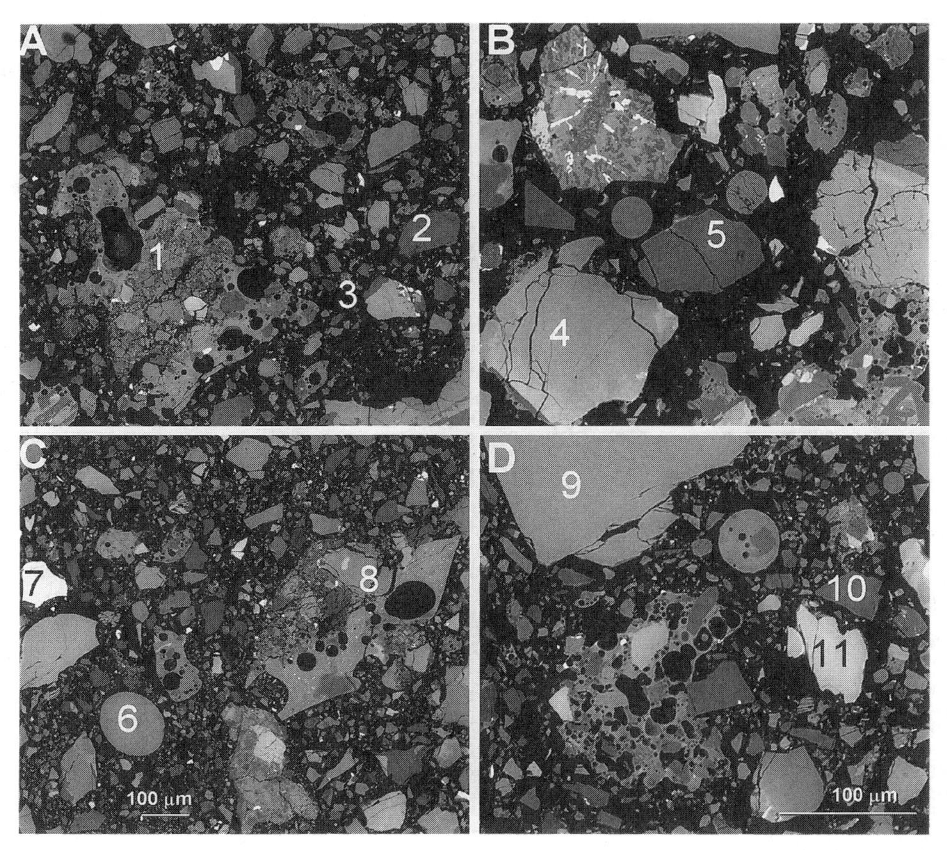

Figure 5C. BSE images of thin sections from Apollo 15 core 15011,6108 (top two) and 15010,6024 (bottom two). Figures A,C have the same scale bar. Figures C,D are at a higher magnification as indicated by the scale bar in D.

Upper: 15011,6108

(A) 1, Agglutinate; 2, Plagioclase; 3, Augite.

(B) 4, Augite; 5, Glass-bead.

Lower: 15010,6024

(C) 6, Glass bead (pyx composition); 7, Fayalite 8, Agglutinate.

(D) 9, Olivine; 10, Plagioclase; 11, Augite.

A further complication in studying the bedrock-derived component is that the relative abundance of particle types depends on the particle size. Polymineralic and lithic (rock) fragments dominate the coarser size fractions. In contrast, the finer soil fractions (<10 µm) are enriched in feldspars and glassy phases. The glasses come from two sources—the noncrystalline material in the groundmass of basaltic rocks, and the glassy bonding material originally present in breccias and agglutinates.

Modal analyses commonly provide information concerning the relations between the composition of the local bedrock and that of the regolith developed on it. All lunar soils have a minor exotic component derived from some distance away, but most soils appear to have been derived largely from bedrock in their immediate vicinity. The modal compositions of a regolith reference suite (Papike et al. 1982) illustrate the local influence well (Table 1, Fig. 6). The Apollo 11, Apollo 12, Luna 16, and Luna 24 missions landed

Table 1. Modal (vol %) abundance data for particles in the 1000-90 μm size fraction of representative soils from each mission (Simon et al. 1981).

	10084	12001	12033	14163	15221	15271	64501	67461	72501	76501	78221	21000	22001*	24999
Mineral Fragments														
Pyroxene + Olivine	4.2	18.3	26.3	2.6	16.1	13.5	1.0	0.5	5.2	17.3	9.8	6.4	8.9	40.2
Plagioclase	1.9	3.9	9.9	5.1	13.1	7.4	32.1	12.2	10.9	15.2	9.9	1.1	14.7	10.6
Opaque	1.1	0.2	1.3	-	0.1	0.3	-	1.1	0.1	2.8	0.4	-	0.1	0.2
Lithic Fragments														
Mare Basalts	24.0	12.9	7.5	2.2	3.1	3.2	0.3	0.5	2.9	9.2	5.7	18.3	1.7	6.9
ANT†	0.4	1.0	1.3	2.9	2.6	2.2	5.0	21.7	5.2	0.5	2.2	0.8	9.7	3.5
LMB¥	0.8	0.1	0.3	0.3	0.6	0.4	2.1	30.7	2.4	6.3	2.3	0.3	2.8	0.5
Feldspathic Basalt (KREEPY)	1.1	0.5	-	0.6	0.4	1.9	1.6	1.6	0.2	0.2	0.2	1.4	0.8	1.4
RNB/POIK§	-	2.3	3.7	10.9	2.7	2.8	8.3	7.9	9.7	8.1	4.4	2.8	10.9	2.2
Fused Soil Component														
DMB	7.5	9.5	11.9	19.3	13.3	12.9	13.9	11.1	22.6	4.2	12.0	15.0	15.0	10.6
Agglutinate	52.0	40.1	17.0	45.7	36.9	37.0	29.1	8.5	37.6	29.2	46.6	42.8	28.7	16.6
Glass Fragments														
Orange/Black	2.7	0.5	1.5	-	0.4	1.6	0.7	0.5	1.7	1.6	1.6	1.4	0.2	-
Yellow/Green	0.8	2.8	0.2	2.9	4.5	7.0	1.2	-	0.1	1.3	1.3	1.7	0.7	0.9
Brown	-	1.5	7.8	-	0.3	0.3	-	-	0.2	-	-	-	-	0.2
Clear	1.3	1.0	-	1.3	1.5	3.8	1.4	-	0.2	0.8	1.0	2.5	1.1	0.6
Miscellaneous														
Devitrified Glass	1.8	5.0	10.8	6.1	4.1	5.6	3.4	3.2	0.4	2.2	1.9	4.4	4.6	5.4
Others	0.3	0.5	0.5	-	0.3	0.2	-	0.5	0.4	1.1	0.7	1.1	0.1	0.3
Total	99.9	100.1	100.0	99.9	100.0	100.1	100.1	100.0	99.8	100.0	100.0	100.0	100.0	100.1
Number of points	625	823	666	311	1000	1008	942	189	801	820	1266	360	1333	634

* 500-90 μm fraction.
†ANT = anorthosite, norite, troctolite.
¥LMB = Light matrix breccia.
§RNB/POIK = Recrystallized noritic breccia/poikilitic breccia.

well inside mare regions. The soil samples from these sites contain abundant mare-derived, basaltic rock fragments and their mafic minerals, pyroxene, and olivine. The Apollo 16 and Luna 20 missions landed in highland regions, and Apollo 14 landed on a ridge apparently formed by material ejected from the Imbrium Basin. Soils from these missions show a preponderance of highland-derived lithic fragments (anorthositic rocks) and plagioclase feldspar. Apollo 15 and 17 landings were made in areas where highland hills meet the mare plains. Soil samples collected during these missions are intermediate in character between those from the mare and highland regions.

Agglutinates

Agglutinates are individual particles that are aggregates of small lunar soil particles (mineral grains, glasses, and even older agglutinates) bonded together by vesicular, flow-banded impact glass (Duke et al. 1970a, McKay et al. 1970, 1972; Heiken 1975). Agglutinates are small (usually <1 mm) and contain minute droplets of Fe metal and troilite (FeS). They form only at the lunar surface, by the melting and mixing produced by micrometeoritic bombardment of the regolith. Their abundance in a given soil increases with continued exposure of that soil at the lunar surface making them important indicators of soil evolution.

Agglutinates were one of the most interesting and surprising features observed in the first returned lunar soil samples. In some mature soils, agglutinates are the major constituent, and they may make up a much as 60% of the soil by volume. Agglutinates are unique to soils developed on terrestrial planets lacking an atmosphere, such as the Moon and probably Mercury. The formation of agglutinates requires (1) a rain of high-velocity micrometeoroids onto the surface of an airless planet, and (2) a target consisting of a regolith produced by a prior bombardment. Therefore, agglutinates are not found at all on Earth, even in association with terrestrial impact craters. Although some meteorites may

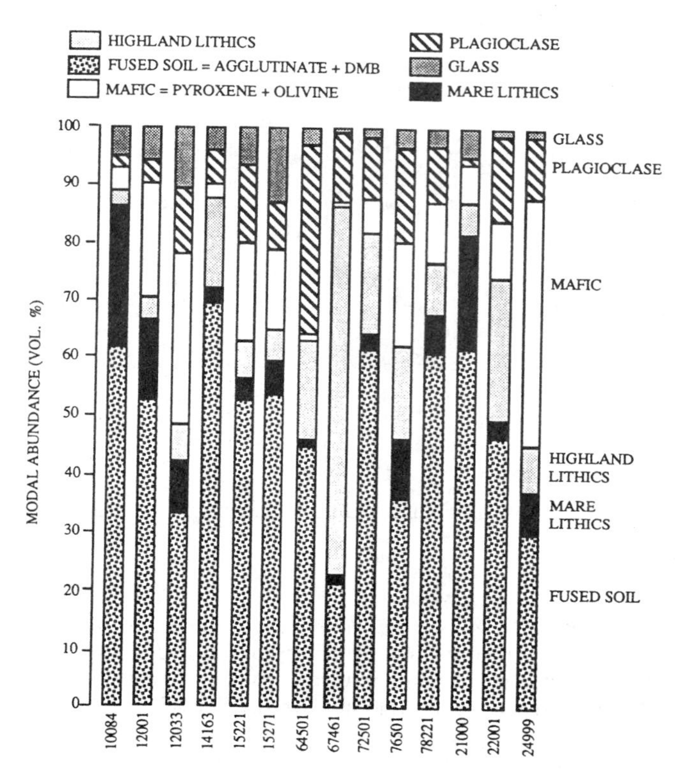

Figure 6. Bar graphs showing modal (volume) abundances of principal particle types in 14 lunar soil samples (Simon et al. 1981). This diagram distinguishes between rock fragments (mare lithics, highland lithics), single mineral and glass fragments (pyroxene and olivine, plagioclase, glass), and fused soil (agglutinates and dmb-Dark Matrix Breccia). Soil samples are from Apollo 11 (10084), Apollo 12 (12xxx), Apollo 14 (14163), Apollo 15 (15xxx), Apollo 16 (6xxxx), Apollo 17 (7xxxx), Luna 16 (21000), Luna 20 (22001), Luna 24 (24999).

have been derived from regolith-like deposits on some asteroids, no true agglutinates have been observed in them. So far, we have found true agglutinates only in lunar soils.

Agglutinates consist of soil grains bonded by glass, and the glass itself is produced by local impact melting of the lunar soil. For these reasons, agglutinates tend to mimic the composition of the soils from which they formed. However, an agglutinate can only incorporate particles smaller than itself. Therefore, agglutinates can only be representative of that fraction of a soil whose grain size is smaller than that of the agglutinates. However, agglutinates can range up to >1 mm in size, and it is reasonable to expect that the average composition of agglutinates found in a given soil will be generally similar to that of the bulk soil (Jolliff et al. 1991).

Although the chemical compositions of whole agglutinates may be soil-like, the compositions of the agglutinate glasses are not. The glass that forms the binder within agglutinates is, in contrast, a product of several complex processes. Primarily, the glass is a melt from the very small volume of soil that was directly impacted by a micrometeoroid.

This small volume may not have been representative of the total soil. Furthermore, small compositional changes may have occurred during melting, such as the escape of volatile elements, or the extraction of metallic iron from silicate melts, leaving the melt slightly depleted in iron. Additional effects could have been produced by the character of the mineral grains involved. Shocked grains with high internal strain energy and those with high surface-to-volume ratios (i.e. small grains), permit easier heat transfer and are likely to melt preferentially. This last possibility led Papike et al. (1981) to propose that fusion of the finest fraction of soil (F^3 model) is the chief process that determines the compositions of agglutinitic glass. The best test of this model has been provided by studies of Luna 24 soils, which are extremely immature, relatively pure mare soils (Basu et al. 1978). In these samples, the ratio of agglutinitic glass composition (Hu and Taylor 1978) to the bulk soil composition is quite similar to the ratio of the <10μm fraction composition (Laul et al. 1981) to that of the bulk soil. In particular, the enrichment and depletion patterns of specific elements in both agglutinates and in the <10μm fraction are quite similar (Walker and Papike 1981; Fig. 7). This study provides a large-scale demonstration that the chemical composition of agglutinitic glass is indeed biased toward that of the <10μm fraction of the original soil.

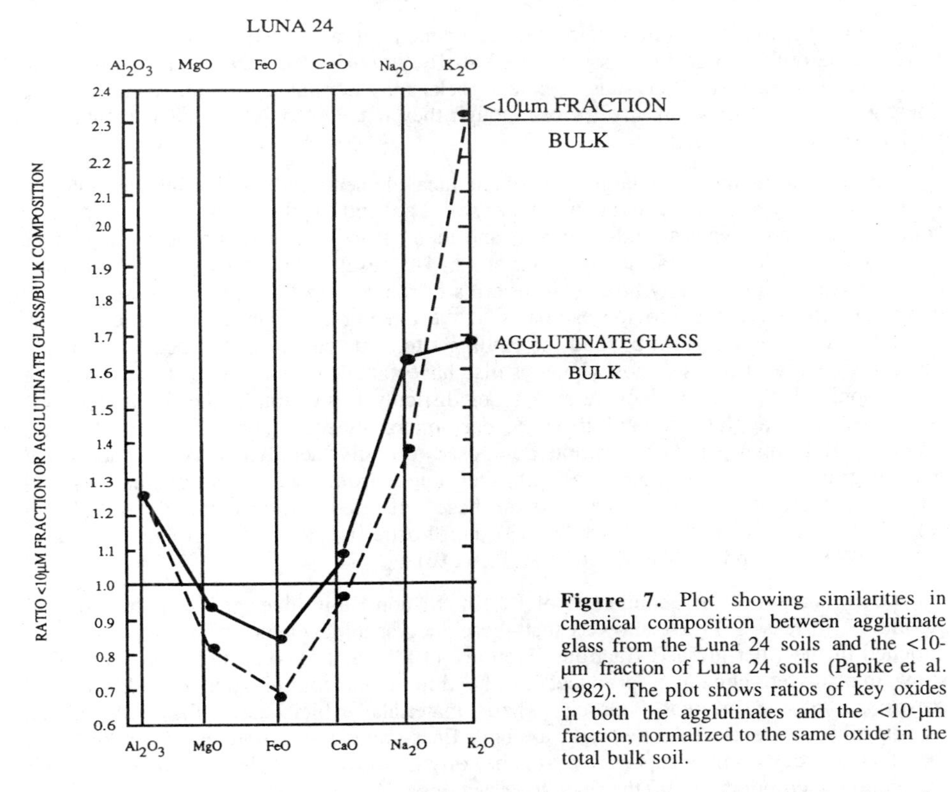

Figure 7. Plot showing similarities in chemical composition between agglutinate glass from the Luna 24 soils and the <10-μm fraction of Luna 24 soils (Papike et al. 1982). The plot shows ratios of key oxides in both the agglutinates and the <10-μm fraction, normalized to the same oxide in the total bulk soil.

Hörz and Cintala (1997) review the results of impact experiments related to the evolution of planetary regoliths. They report that the actual comminution of either lithic or monomineralic detritus is highly mineral-specific, with feldspar and mesostasis comminuting preferentially over pyroxene and olivine, thus resulting in mechanically fractionated fines, especially at grain sizes <20 μm. Such fractionated fines also participate preferentially in the shock melting of lunar soils, thus giving rise to agglutinate melts. As a result, agglutinate glasses are systematically enriched in feldspar components relative to the bulk composition of their respective host soils.

Chemical composition of lunar soils

The chemical compositions of lunar soils reflect their mixed origins. All regolith samples brought back from the Moon contain some components exotic to the collection site. Although Apollo 11 landed in the middle of a mare basalt plain, the soils do not have compositions equivalent to 100% mare basalt. There is chemical evidence for the presence of additional rock and mineral fragments from the anorthositic highlands, rare KREEP-bearing material, and a small meteoritic component.

Some minerals, including potassium feldspar, apatite, whitlockite, and zircon, are rare in the lunar soil, but nonetheless carry the bulk of such important trace elements as K, the rare earth elements (REE), P, and Zr. The abundance of major elements can be used to indicate roughly the proportions of highland anorthositic (Ca-, Al-rich) and mare basaltic (Fe-, Ti-rich) rocks that are present in the soil. The trace elements K, REE, and P can be used to estimate the percentage of KREEPy source rocks in a soil.

The proportions of different petrologic components in a soil vary with grain size, and therefore the bulk chemical composition will also vary with grain size. Minerals that were originally present as small crystals in parent rocks may dominate the composition of the finest-grained sizes of a lunar soil, even though they made up only a small percentage of the parent rock.

Multivariate analysis of a large array of chemical elements can provide insights into the variety of rock types represented in any lunar soil. Laul and Papike (1980a,b) chose several common lunar rock types as end-members, and then carried out mixing-model calculations based upon 35-element analyses of lunar soils. The results can be used to interpret the chemical composition of any lunar soil in terms of the principal rock types present in the immediate area, and can provide estimates of the amount and nature of exotic material introduced into the soil from beyond the landing site. This chemical-statistical approach is especially useful because most lunar soils have abundant fused material, making petrographic identification of source rock types difficult. For example, the Apollo 11 soils contain 60 vol % agglutinates and 20 vol % dark-matrix breccias (shock-compacted soils), leaving only a small part of the sample composed of easily identified rock fragments and mineral grains. Laul and Papike (1980a) determined, from their chemical analyses and mixing calculations, that the source rocks for these soils are (1) high-K mare basalt (10 wt %); (2) low-K mare basalt (62 wt %); (3) anorthosite, norite, and troctolite (highlands rocks) (14 wt %); and (4) low-K KREEP (13 wt %).

Determination of the abundances of K, Rb, Sr and Ba in nine grain-size fractions of Apollo 17 mare soil 71501 showed that these incompatible elements are systematically enriched in the finer-grained fractions. Korotev (1976) and Haskin and Korotev (1977) show that this enrichment is produced by the differential comminution of feldspar and mesostasis of the original bedrock. These materials, which carry the bulk of the incompatible elements in the bedrock, are both finer-grained and more easily broken than the other bedrock components (i.e. pyroxene, olivine, opaques), which accounts for their preferential incorporation into the finer size fractions.

Regolith evolution and maturity

The lunar regolith is produced by meteoroid impacts that shatter exposed lunar bedrock, which may consist of a wide variety of rock types. When this process begins, the resulting regolith material is both fresh and young. If the newly formed regolith is exposed at the lunar surface, it continues to be progressively modified by micrometeoroid impacts and by high-energy solar and cosmic charged particles. This modification process is called maturation, and regolith exposed to these processes for long periods of time becomes mature. If the fresh, young regolith soon becomes deeply buried (e.g. by ejecta from a nearby impact crater), it is not exposed to these surface processes, and it therefore remains fresh or immature. However, such buried regoliths may not remain buried indefinitely. Subsequent impacts may turn the regolith over (gardening) and bring young, buried regolith to the surface. At this point, it again becomes subjected to micrometeoroids and charged particles and the maturation process continues. This potential for alternate burial and exposure of lunar regolith gives rise to the concept of surface exposure age. This age is defined as the cumulative length of time that a given lunar soil has been exposed at (or near) the surface, as measured by the effects of some alteration process—micrometeoroid impact, pits, solar-wind implantation, or the effects of high-energy cosmic rays.

Bombardment by meteoroids, both large and small, produces the most significant changes in lunar soils:

1. Large particles are broken down into smaller ones.
2. Large impacts shock-indurate the regolith, producing coherent regolith breccias.
3. Micrometeoroid impacts produce small amounts of melt at the point of impact;the melt then cements nearby small grains to make agglutinates.
4. Impacts of all sizes may vaporize projectile and target material that is deposited on the exposed surfaces of regolith particles.
5. With moderate to large impacts that form craters more than a few meters in diameter, the ejected regolith forms layers that cover up neighboring soils. Such impacts may also excavate previously buried regolith, reexposing it to surface processes.
6. Continued bombardment produces a continuous churning of the regolith, mixing soils together at random.
7. Large impacts which penetrate the regolith layer also eject deep-seated bedrock fragments, often to significant distances, where they form the exotic component of the regolith at that site. These impacts may also generate a sheet of impact melt that covers regolith in and around the crater.
8. Finally, material ejected from a large crater may also produce secondary craters when it lands. These craters produce additional ejecta, and many other effects of the primary impacts, although at decreasing energy levels.

Therefore, meteoroid bombardment controls the maturation of the lunar soil in several different ways. It produces physical and chemical changes in the soil itself—comminution, vapor fractionation, the formation of aggregates (agglutinates and regolith breccias), and the addition of a meteoritic component. In addition, impacts produce other and more widespread changes—breaking up bedrock and adding it to the regolith, mixing together layers of soil that were originally discrete, introducing exotic rock components from distant sites, and turning over or reworking the exposed soil to depths that depend on the meteoroid size-frequency distribution and impact flux rate. Meteoroid impact also determines what volumes of regolith are brought to or near the lunar surface, where they can be affected by other processes.

Meteoroid impacts control the maturation of the regolith by regulating its excavation and exposure to near-surface processes, but they also produce their own maturation effects. Micrometeoroids comminute soil particles, a process that changes the distribution of particle sizes by producing smaller particles from larger ones. Therefore, the mean grain size of the <250 μm soil fraction is an index of maturity; finer-grained soils are generally more mature than coarser-grained ones, assuming the source rocks has similar grain sizes. Meteoroid impacts also produce agglutinates. Agglutinate content is one index of soil maturity; higher agglutinate contents imply more mature soils. This index conflicts to some extent with the previous one because agglutinate formation results from the welding of smaller particles into larger ones, and therefore acts to increase the overall grain size.

The intense and transient shock-wave heating produced by a meteoroid impact has several distinctive effects. First, volatile elements are vaporized from the targets (Ivanov and Florensky 1975, Naney et al. 1976, Keller and McKay 1992, Papike et al. 1997a). Because the gravity of the Moon is low, some of these vaporized elements can escape the Moon's gravitational field. As a result, the lunar soils can become progressively depleted in these elements with increasing maturity.

In addition, volatile elements that do not escape exhibit a measurable mass fractionation of their isotopes, probably produced during this vaporization process. Analyses of O and Si isotopic ratios in lunar rocks and soils indicate that lunar soils are enriched in ^{18}O and ^{30}Si relative to the crystalline rocks (Epstein and Taylor 1972). These workers also found that these large enrichments of the heavy isotopes of O and Si in lunar soils are directly related to the amount of solar-wind-derived H present. Successive partial fluorinations of soil grains, a technique that enables the extraction of Si and O from successively deeper layers in the grains, show that the enrichment of ^{18}O and ^{30}Si is essentially surface-correlated (i.e. restricted to the grain surface) (Epstein and Taylor 1972).

The concentrations of other volatile elements (e.g. Zn, Ga, Ge, Cd, Sb, Te, and Hg) in different grain-size fractions of lunar soil samples also show negative correlations with grain size, indicating that these elements are enriched in the finer fraction. Because small grains have higher surface-to-volume ratios than do larger ones, these results indicate that these elements have been deposited on grain surfaces (Krähenbühl et al. 1977).

Regolith turnover by meteoroid impacts does not simply take a volume of lunar soil with a given maturity and then expose or bury it; impacts also mix discrete volumes of soil together. This soil mixing is a complex process, but most measures of soil maturity make no distinction between soils matured *in situ* (i.e. which underwent very small scale mixing) and fresh, immature soils that had mature soils mixed into them. However, if soil mixing is sufficiently gross, and the mixed components have sufficiently different levels of maturity, then grain-size and track-density frequency distributions can demonstrate that a soil has been mixed.

Variation of soils with depth: The lunar core samples

Core samples of the lunar regolith have provided the most dependable information about the evolution of the regolith and about its textural and structural complexities. Identification of individual soil layers within the core samples has been based mostly on discrete changes in grain size and particle proportions from layer to layer. Depths of regolith samples by coring range from 10 cm (Apollo 11) to 298.6 cm (Apollo 17).

Core studies indicate that the lunar regolith is made up of discrete layers of material ejected from both large and small impact craters. Within a single layer, the observed grain size is related to the depth of cratering in the impact event, the ratio of bedrock to regolith in the excavated material, and the duration of subsequent exposure to micrometeoroid bombardment at the lunar surface. With a high percentage of bedrock in the ejecta layer, the material is coarse grained; conversely, the finest-grained regoliths have a smaller fresh bedrock component. The relationship between grain size and regolith thickness—that thicker regoliths have generally finer grain sizes—has been well-documented by McKay et al. (1980). Throughout a section of multilayered regolith, however, the grain size varies randomly from layer to layer. Exposure to micrometeoroid bombardment at the lunar surface further comminutes fragments in the regolith, but also builds up an increasing volume of agglutinates with time. Eventually, a steady-state grain size develops as the agglutinate content increases and balances comminution (Morris 1976).

The first cores from lunar regolith were collected at the Apollo 11 site, using hollow drive tubes that were hammered into the soil. These drive tubes were short (about 30 cm) and the incomplete sections they secured gave little indication of the textural complexity that is characteristic of the regolith sampled more deeply at other Apollo and Luna landing sites. A double drive-tube core sample (sample 12025/12028), collected at the Apollo 12 site on the rim of a 10-m-diameter crater, provided the first evidence that lunar regolith is layered. This 42-cm regolith sample is composed of ten discrete layers; the median grain sizes of the soils from the individual layers range from 61 μm to 595 μm.

A more complete view of the textural characteristics of the regolith, layer by layer, has been obtained from analyses of regoliths sampled by drilling at the Apollo 15, 16, and 17 sites. The 236-cm-deep section sampled by a core drill at the Apollo 15 site (15001 to 15006) consists of 42 distinct textural units, with median grain sizes from 44 μm to 89 μm; all the layers are poorly sorted (Heiken et al. 1973, 1976). Most of the samples from these layers form a tight grouping around a median grain size of 50 μm.

Also at the Apollo 15 landing site, a shorter drive tube, inserted only 20 m from the edge of Hadley Rille, penetrated the upper 60 cm of a 1-m-thick, immature regolith (lunar core sample 15010/11: McKay et al. 1980). Because of this sample's close proximity to bedrock, together with the process of continuous mass wasting of regolith at this location, it is coarser-grained than regolith samples collected elsewhere; it has an average mean grain size (average of several samples) of 85 μm (Fig. 8) (McKay et al. 1980).

At the Apollo 16 site, on the Cayley Plains, a 221-cm-thick section of highland regolith was sampled by drilling (Allton and Waltz 1980). In this core, three major subdivisions were recognized by Meyer and McCallister (1977) and 46 textural units by Duke and Nagle (1974). A systematic study of changes in grain size with depth (to 65cm) was made by McKay et al. (1977) for another Apollo 16 sample, core 60009/60010. In this core, there is a general increase in mean grain size with depth,. The upper 12 cm of the core shows the highest degree of *in situ* maturation and reworking, and it is the finest-grained unit sampled in this regolith section (Fig. 9).

Important recent geochemical studies of lunar cores are reported in Korotev (1991, 1997), Korotev and Morris (1993), and Korotev et al. (1997). Unfortunately, despite the detailed study of the Apollo cores, it has not been possible to identify any single layer as ejecta from a known impact crater at the landing site. A tenuous correlation has been made at the Apollo 17 site between Camelot Crater and a coarse-grained, immature layer in the drill core. Study of the lunar cores has confirmed that our general understanding of cratering mechanics and regolith formation seems to be correct. However, we still do not completely understand the interconnections of different stratigraphic features, such as grain size, sorting, the presence or absence of graded bedding, and the identification of former surface horizons.

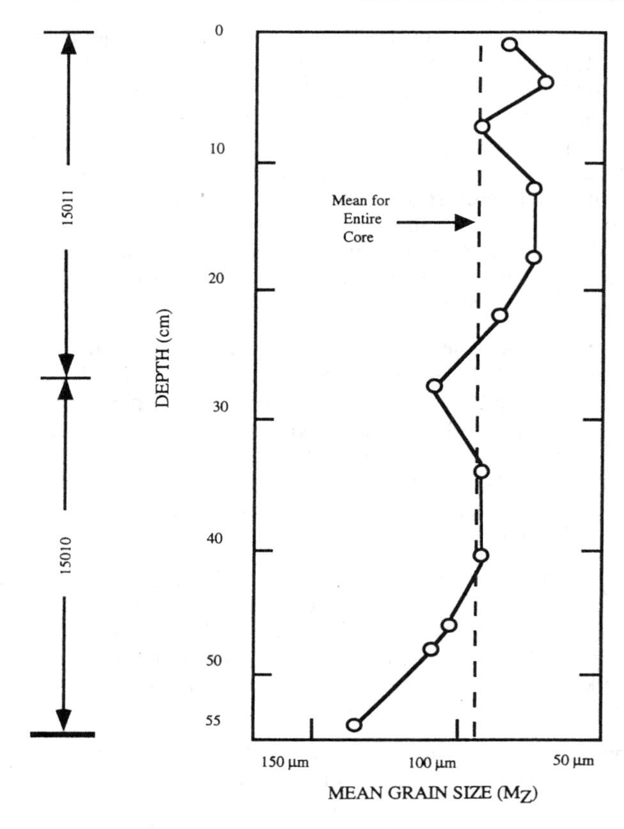

Figure 8. Plot of the data from the Apollo 15 drive-tube 15010/11, showing the sample depths and mean grain sizes for individual samples. This core was collected close to the rim of Hadley Rille, where regolith is thin and immature, as indicated by the relatively large mean grain size (about 100µm) for all portions of the core. (After McKay et al. 1980).

A wide range of stratigraphic data have been measured in the cores: visual color and texture, bulk chemistry, petrography, various maturity indices, the amounts of cosmogenic nuclides and particle tracks (which can be related to exposure ages), and their variations with depth. The most intensively studied cores are the Apollo 15 deep drill core (15001 to 15006), the Apollo 16 double drive tube 60009/60010, and the Apollo 17 deep drill core (70001 to 70009).

One of the simplest and most valuable techniques for understanding the nature and history of a section of regolith is to determine the variations, with depth, of the relative amounts of its different petrographic components i.e. the particles that can be identified using an optical (petrographic) microscope or a scanning electron microscope. These components are principally of two kinds: (1) those derived from bedrock, such as rock and mineral fragments; and (2) those produced as the result of exposure at the lunar surface-agglutinates, regolith breccias, and impact glasses. Petrographic variations with depth have helped to identify different layers in the lunar regolith at every station where a core has been collected. In particular, abrupt changes in the petrographic components have made it possible to locate the boundaries between discrete layers. Where such boundaries also coincide with breaks in the chemical and maturity profiles, a significant break in the history of regolith development may be inferred.

The variation of other petrographic constituents-rock and mineral fragments with depth in the lunar regolith depends essentially on the extent to which these bedrock-derived materials are introduced from both local and distant bedrock, and especially on the relative contributions of highland and mare sources. As shown above (see also Papike et al. 1982) it has not proven possible to correlate different stratigraphic layers, sampled by coring and identified on the basis of the variety of rock fragments, over any distance on the Moon. Therefore, the petrographic variations of the lunar regolith with depth can only be understood in terms of the mixing together of different lithologic components (e.g. Korotev 1991).

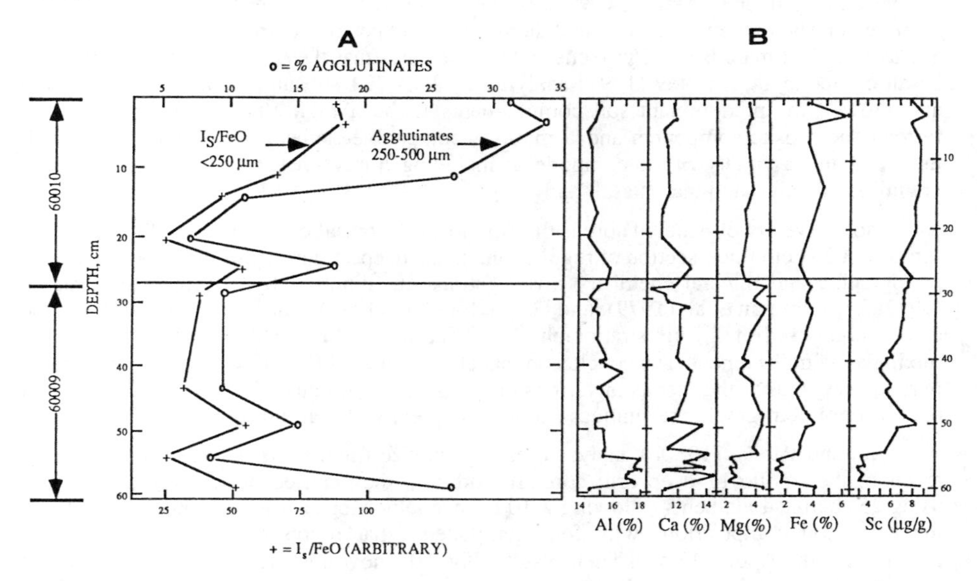

Figure 9. (A) Variation with depth (in centimeters) of two maturity indices: Is/FeO in the <250-µm size fraction, (for discussion of Is/FeO refer to Chapter 7 of Heiken et al. (1991*Lunar Sourcebook*) and agglutinate abundances in the 250- to 500-µm size fraction of the Apollo 16 drive core 60009/10 (McKay et al. 1977). The two indices correlate closely, despite the irregular variations in both throughout the core. Note the mature zone in the upper 15 cm of the core. (B) Variations in chemical composition, plotted for Al, Ca, Mg, Fe, and Sc as a function of depth in the Apollo 16 drive core 60009/10 (from Ali and Ehmann, 1977). The vertical scales give depth in centimeters. Significant chemical variations occur over depths of only a few centimeters, especially in the lower part of the core.

Because the lunar regolith is a mechanical mixture of several kinds of broken and melted rocks, variations in the chemical composition of the regolith with depth also provide a record of the influx and mixing of these materials through time. The chemical variations go in tandem with petrologic variations, but they have the advantage of including data from the fine-grained soil fractions that cannot be examined with petrographic methods. Variations in trace element abundances also record the input of different parent rocks, allowing investigation of the mixing process.

Two of the Apollo 16 cores have been studied in considerable detail. McKay et al. (1977) and Korotev (1991) showed that the content of mineral rock fragments in core 60009/60010 can be successfully modeled by mixtures of nearly pure anorthositic plagioclase feldspar and crystalline breccias (metamorphosed breccia plus poikilitic breccia), with modification by subsequent maturation. In a study of the core 64001/64002, Houck (1982) and Basu and McKay (1984) showed that a more complex situation existed. These cores contain a significant amount of mare basalt fragments. Although their abundance is low (mare basalts being exotic to the Apollo 16 site), there is a positive correlation between mare basalt fragments and monomineralic pyroxene fragments that may have been derived from the mare basalts.

Analyses of 36 soil samples from various depths of the Apollo 16 core 60009/60010 by Ali and Ehmann (1977), summarized in Figure 9, show that Al and Ca dominate over Mg and Fe, reflecting the dominance of plagioclase feldspar over such mafic minerals as olivine and pyroxene. Although Mg and Fe are lower in abundance, there is a weak positive correlation between them, and there is a strong positive correlation between Fe and Sc, as found in mare basalt Pyroxene, which indicates that they are both present in mare basalt components. Korotev (1991) analyzed by INA 121 subsamples from 60009/60010 and showed the predominant soil componments to be a highly-pure (99% plagioclase) ferroan-anorthosite component and a mafic, incompatible-element-rich Cayley soil which itself contains a variety of subcomponents including impact-melt breccia, fragmental and granulitic breccias, and mare basalt and glass.

Another very well-studied core is the Apollo 17 deep drill core 70001 to 70009, which sampled a 298 cm thick-section of regolith and is the deepest core from the Moon. Detailed petrographic and mineral chemistry investigations are reported by Vaniman and Papike (1977a,b), Vaniman et al. (1979), and GJ Taylor et al. (1979). These studies show that the core is composed of several stratigraphic units, each of which had a complex history of mixing and emplacement. The conclusion that at least three different rock types contributed pyroxene crystals to these soils in various proportions (Vaniman et al. 1979) indicates both the power of petrographic techniques and the complexity of regolith formation.

Laul and Papike (1980b) analyzed three grain-size fractions of soil samples from 30 levels of the Apollo 17 deep drill core. In addition, they carried out mixing calculations using 25 chemical elements (allowing a 10% uncertainty for each), attempting to duplicate the lunar soil compositions with four components that represent common rock types identified at the Apollo 17 site. Their results (Fig. 10) show that the chemical compositions of each of the 30 soils analyzed can be adequately expressed in terms of contributions from these four end members. These results are also consistent with the five stratigraphic units recognized on the basis of petrologic work (Vaniman et al. 1979), and this consistency between the two techniques strengthens the identification of these layers as discrete depositional units. Finally, these studies show that certain elements record the effects of soil maturation processes on the Moon. Comparison of the Is/FeO profile in the Apollo 17 deep drill core with the abundance of Zn, a volatile element, clearly shows a strong negative correlation (Fig. 11; Laul and Papike 1980b) implying that Zn has been progressively lost, probably by volatilization during impact, as the soil matured.

Comparison of soil chemistry with bedrock chemistry

Significant insights into the formation of the lunar regolith can be obtained by comparing the chemical compositions of soils from different landing sites (or from geologically distinct areas within a single site) with the compositions of common lunar rocks. These comparisons are one way to evaluate the degrees to which local, regional, and distant sources of different rock types have contributed to the lunar regolith at any single point. The complexity of regolith formation is indicated by the fact that no lunar soil, or any average composition of several soils from any region, is compositionally identical to any single lunar rock type. Average soil compositions from all lunar sample return missions are given in Tables 2 and 3.

The most common crustal rock type from the Moon, based on sample returns and chemical measurements made from orbit, is anorthositic gabbro, represented by samples 15455, 60335 and 66095 (SR Taylor 1982, Vaniman and Papike 1980). The Apollo 16

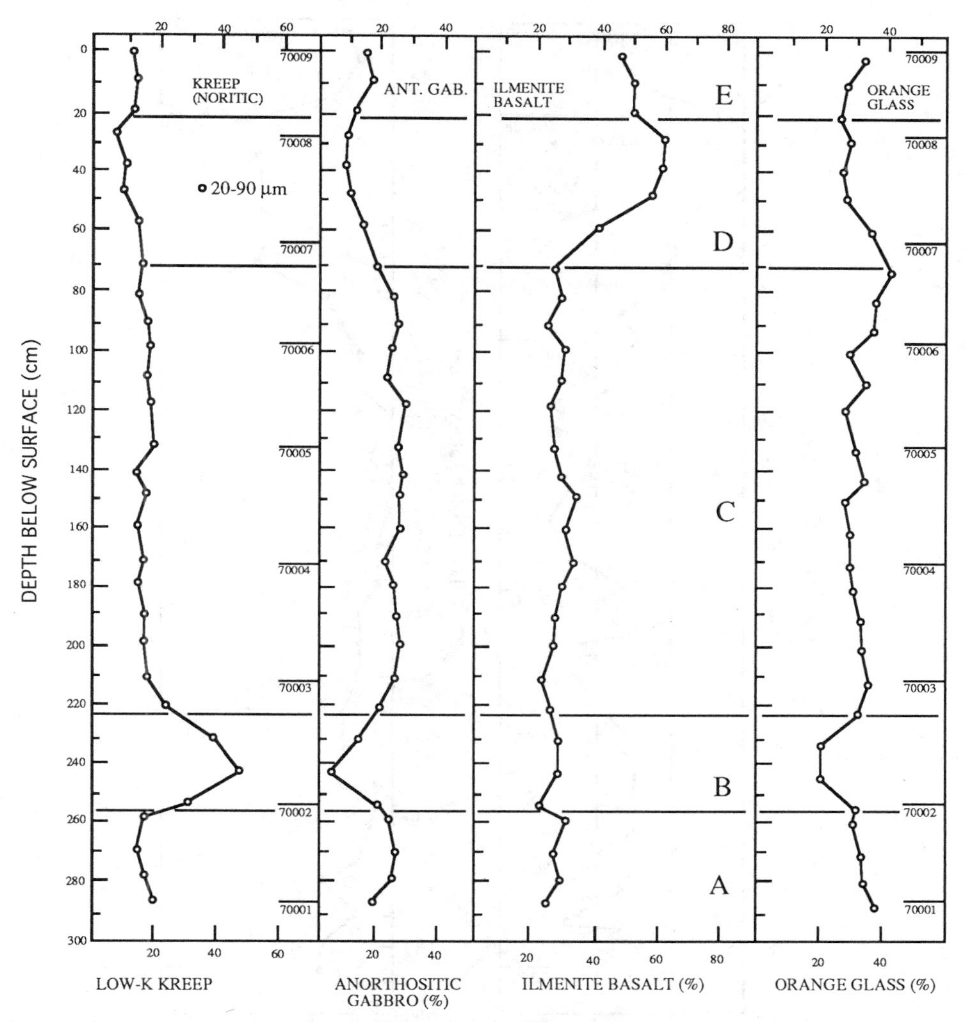

Figure 10. Calculated percentages of different lithologic components as a function of depth for soil samples from the Apollo 17 core 70001 to 70009. Each data point represents analysis for 25 chemical elements (Laul and Papike 1980b). Mixing models then use the chemical data to detemine the best match for the amounts of four rock types present: low-K noritic KREEP, anorthositic gabbro, ilmenite basalt, and orange glass. Vertical axis gives depth in core; horizontal axes give percentages for each component. Note that the boundaries (horizontal lines) of the five stratigraphic units (A-E) identified on the basis of petrographic studies coincide with significant variations in the abundances of different components.

and Luna 20 landing sites are well within highland areas, and these sites provide our best samples of the pre-mare lunar crust (e.g. Korotev 1997).

The Apollo 16 soils are the most similar to anorthositic gabbros. The average Apollo 16 soil is only slightly enriched in Al and Ca, but is substantially depleted in Mg in comparison with sample 60335, a typical anorthositic gabbro. It appears that some contribution from relatively pure anorthosite to the average Apollo 16 soil is responsible for these differences in major elements. Apollo 16 soils are also slightly depleted in

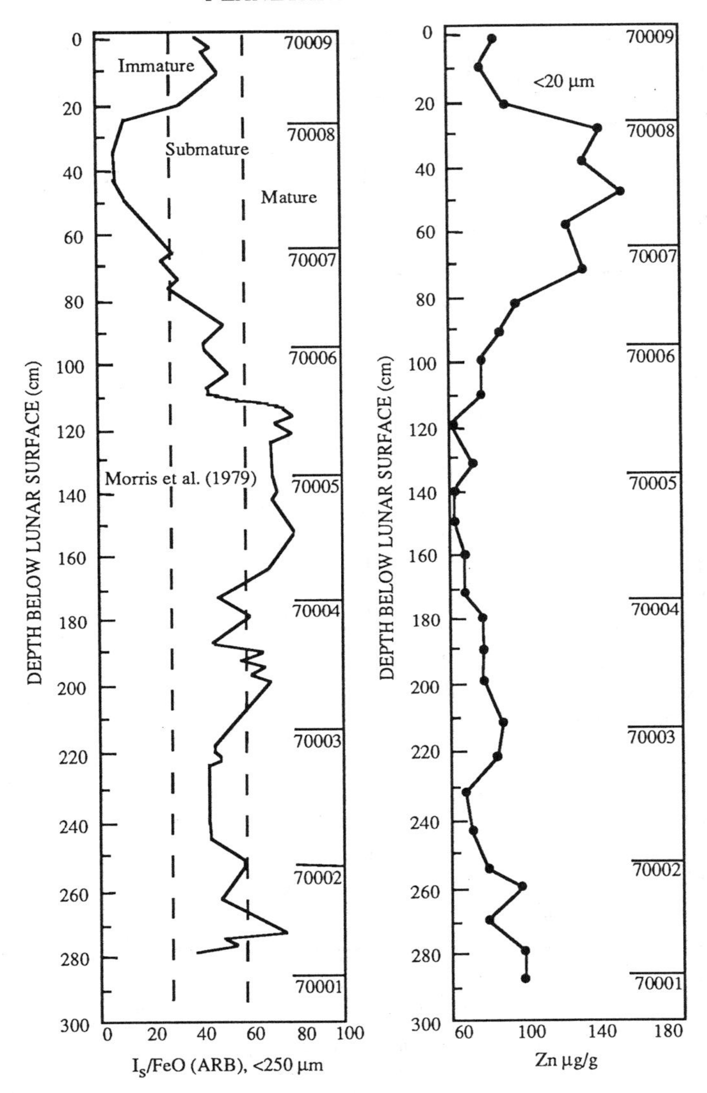

Figure 11. Depth profiles of the abundance of Zn, a volatile element, and Is/FeO, a maturity index, in the <20-µm size fraction of soil samples from the Apollo 17 drill core 70001 to 70009. These two quantities show a strong negative correlation, especially in the upper part of the core. This relationship suggests that Zn may be depleted in mature lunar soils, possibly having been expelled by vaporization during agglutinate production (Laul and Papike 1980b).

the KREEP component relative to sample 60335, having lower abundances of K, La, U, and Th. In contrast to the Apollo 16 soils, the highland soils from the Luna 20 are less similar to anorthositic gabbro, or for that matter, to any other lunar crustal rock.

In contrast, the Apollo 11, Luna 16, and Luna 24 soils have been collected from mare areas far from any exposed highland material. Despite their apparent derivation entirely from mare basalt lavas, the soils from these sites differ significantly from samples of mare

Table 2. Abundances (wt %) in bulk soils and separated size fractions of "reference suite" lunar soils and sample 15601 (Papike et al. 1982; LSPET 1972)

Sample	Size	SiO_2	TiO_2	Al_2O_3	FeO	MgO	CaO	Na_2O	K_2O	MnO	Cr_2O_3	Total
10084,1591	<1mm	41.3	7.5	13.7	15.8	8.0	12.5	0.41	0.14	0.213	0.290	99.8
12001,599	<1mm	46.0	2.8	12.5	17.2	10.4	10.9	0.48	0.26	0.220	0.410	101.1
12033,464	<1mm	46.9	2.3	14.2	15.4	9.2	11.1	0.67	0.41	0.195	0.387	100.8
14163,778	<1mm	47.3	1.6	17.8	10.5	9.6	11.4	0.70	0.55	0.135	0.200	99.8
15221,29	<1mm	46.0	1.1	18.0	11.3	10.7	12.3	0.43	0.16	0.154	0.325	100.5
15271,27	<1mm	46.0	1.5	16.4	12.8	10.8	11.7	0.49	0.22	0.162	0.350	100.4
64501,122	<1mm	45.3	0.37	27.7	4.2	4.9	17.2	0.44	0.10	0.056	0.090	100.3
67461,74	<1mm	45.0	0.29	29.2	4.2	3.9	17.6	0.43	0.055	0.055	0.075	100.8
72501,15	<1mm	45.2	1.4	20.1	9.50	10.0	12.5	0.44	0.17	0.120	0.230	99.7
76501,48	<1mm	43.4	3.2	18.1	10.8	12.0	12.8	0.38	0.10	0.145	0.270	101.2
78221,71	<1mm	43.0	4.2	17.0	12.6	11.0	12.4	0.37	0.10	0.163	0.350	101.1
Luna 16 21000,5	<1mm	-	3.5	15.5	16.5	8.1	11.8	0.38	0.11	0.230	0.300	-
Luna 20 22001,35	<1mm	-	0.48	23.5	7.27	9.7	14.1	0.35	0.068	0.100	0.180	-
Luna 24 24999,6	<1mm	-	1.0	11.7	20.2	9.7	11.1	0.27	0.027	0.270	0.467	-
Apollo 15 15601	<1mm	45.05	1.98	10.2	19.8	10.9	9.9	0.29	0.10	0.26	0.56	-

basalts returned as large rocks. In general, the soils are considerably enriched in Si relative to high-Ti basalts, depleted in Al relative to high-Al basalts, or enriched in Mg relative to low-Ti basalts. The Luna 24 soils and the Apollo 15 soil 15601 (from the edge of the Hadley Rille) are the two most basalt-rich soils from the Moon. Except for substantial Mg enrichment, Luna 24 soils are compositionally closest to the very low Ti basalts; they are, however, more enriched in the KREEP component.

The examples mentioned above, in which soil compositions can be easily related to a few local bedrock sources, actually represent the few exceptions among returned soil samples. Most other lunar soils are compositionally far removed from known lunar rocks.

Variation of soil chemistry among sites

Because the average soil compositions are primarily controlled by local bedrock, the variations of soil composition from site to site should reflect the variation of the average bedrock compositions at each site. If this is correct, then there are significant differences in lunar bedrock compositions, not only between maria and highland regions, but at different areas within these regions.

Both the Apollo 16 and Luna 20 sites are in similar-appearing highland regions, but the average composition of Luna 20 soils is much more mafic than those at the Apollo 16 site. The Luna 20 site is located in a highland region relatively close to Mare Fecunditatis,

Table 3. Trace element abundances (ppm) in bulk soils and separated size fractions of "reference suite" lunar soils and sample 15601 (from *Lunar Sourcebook* 1991).

Sample	Size**,μm	Sc	V	Co	Ba	Sr	La	Ce	Nd	Sm	Eu	Tb	Dy	Ho	Tm	Yb	Lu	Hf	Ta	Th	U	Ni*
10084,1591	Bulk	60.2	70	28.0	170	160	15.8	43	37	11.4	1.60	2.9	17	4.1	1.6	10.0	1.39	9.00	1.25	1.90	0.5	200
12001,599	Bulk	40.2	110	42.5	430	140	35.6	85	57	17.3	1.85	3.7	22	5.0	1.8	13.0	1.85	11.8	1.50	5.40	-	190
12033,464	Bulk	36.4	100	34.3	600	160	50.0	133	85	22.8	2.45	4.9	30	7.2	2.6	17.3	2.45	16.6	2.20	8.50	2.4	130
14163,778	Bulk	21.7	45	33.0	800	170	66.7	170	100	29.1	2.45	5.9	36	8.6	3.2	21.2	3.00	22.5	2.90	13.3	3.5	350
15221,29	Bulk	21.2	80	41.0	240	120	20.5	54	36	9.70	1.30	2.0	12	2.9	1.1	6.90	0.97	6.70	0.93	3.00	-	360
15271,27	Bulk	24.3	80	40.5	300	130	25.8	70	45	12.0	1.50	2.6	15	3.9	1.4	8.54	1.20	8.60	1.20	4.60	1.2	230
64501,122	Bulk	8.0	20	19.5	130	170	10.8	28	19	4.79	1.05	1.0	6.0	1.4	0.55	3.40	0.49	3.30	0.45	1.85	0.4	300
64761,74	Bulk	7.8	20	9.0	60	170	4.67	12	7.2	2.00	1.00	0.45	2.8	-	0.25	1.60	0.22	1.60	0.24	0.83		80
72501,15	Bulk	20.0	45	33.0	210	160	16.2	46	29	8.00	1.30	1.6	10.0	-	0.84	5.90	0.82	6.00	0.90	3.00	1.0	260
76501,48	Bulk	28.0	65	30.4	120	160	8.30	23	16	5.30	1.20	1.2	8.3	-	0.71	4.60	0.66	4.20	0.70	1.60	0.4	190
78221,71	Bulk	36.3	70	40.0	120	160	8.10	24	16	5.46	1.28	1.3	8.4	-	0.72	4.70	0.70	4.70	0.86	1.60	0.4	200
Luna 16 21000,5	Bulk	52.0	80	29.6	180	260	11.6	33	26	8.25	2.15	1.7	11	-	0.90	5.70	0.80	7.04	0.50	1.39	0.4	200
Luna 20 22001,35	>125	15.4	40	31.6	100	130	7.20	17	12	3.20	0.90	0.65	4.2	-	0.37	2.55	0.35	2.30	0.28	1.32	-	300
Luna 24 24999,6	Bulk	44.0	150	47.0	40	80	2.40	6.1	5.0	1.70	0.59	0.43	2.7	-	-	1.75	0.27	1.20	0.17	0.40	-	110
Apollo 15 15601	Bulk	35.1	200	48.9	135	-	11.3	29	-	6.3	1.01	1.33	9.7	-	-	5.2	0.9	4.9	0.60	1.52	0.46	170

*The Ni values are suspect of contamination from the Rh-plated Ni sieves.
**Bulk=<1 mm

and it is possible that there has been an appreciable contribution of material from the mare to the Luna 20 site. However, this explanation seems unlikely; the boundary between mare and highlands in this region is sharp, as indicated by both topography and albedo changes, implying that there has been little lateral migration of surface material. It is more likely that the composition of Luna 20 soils reflects the addition of an unknown mafic highland rock present near the site itself.

The Apollo 11, Apollo 12, Luna 16 and Luna 24 missions all landed on areas of mare basalt flows fairly far away from any highlands; despite this, the soil compositions from these missions are quite distinct from each other. The Apollo 11 soils are rich in Ti, Luna 16 soils have the highest Al of sampled mare soils, and Luna 24 soils are enriched in Fe. Apollo 11 soils have the highest abundance of Sc; Apollo 12 soils are significantly enriched in K, U, Th, and La; Luna 16 and Luna 24 soils are depleted in U, Th, and La. These variations and related differences have also been observed in Earth-based telescopic spectral studies and in measurements of gamma-ray and x-ray intensities made from lunar orbit during the later Apollo missions. In addition, the average compositions of mare soils from the Apollo 15 and Apollo 17 sites each have their own distinctive character.

The chemical analyses of lunar soils therefore suggest that the nine different regions of the Moon from which we have soil samples are composed of significantly different rock types.These rocks have been shattered, pulverized, differentially comminuted, and mixed to produce soils with a wide range of compositions. Even though small amounts of exotic materials are found in the soils from all missions, usually in the finer grain-size fractions, these occur in such small amounts that bulk soil compositions are not significantly affected.

Regolith breccias

Ballistic ejection from impact craters, together with gravitational creep, tend to emplace newer soils on top of older soils, burying them. However, even deeply buried soils may subsequently be ejected by meteoroid impacts and reexposed at the lunar surface. The irradiation histories deduced from discrete soil layers in some regolith cores reveal that such processes of burial and reexposure have, in fact, been operative at the surface of the Moon. It is therefore theoretically possible that some shallow cores can penetrate and sample layers of very old (i.e. >4.0 Ga) regolith which have been excavated and re-exposed at or near the lunar surface. However, a more definite source of samples of such ancient regolith has already been provided in the form of a group of samples called regolith breccias.

Regolith breccias are polymict breccias that contain rock, mineral and glass fragments in a glassy matrix. The presence of glass fragments and agglutinates within these breccias shows clearly that they have formed from earlier regolith deposits (Stöffler et al. 1980).

Regolith breccias are abundant among the samples returned by the Apollo 11 astronauts, and early studies showed that they were formed from the local regolith. There was considerable debate about the process that had indurated the originally loose regolith particles into a more or less coherent rock. Most investigators (e.g. King et al. 1970, Quaide and Bunch 1970) concluded that the breccias were shock-lithified, i.e. compacted and indurated by the action of shock waves passing through the regolith, generated by nearby meteoroid impacts. McKay et al. (1970) and Duke et al. (1970b) disagreed and said that the breccias were more analogous to hot welded deposits of volcanic ash. In particular, McKay et al. (1970) stated that the breccias were too porous and the materials in them not shocked enough to be consistent with shock lithification.

However, subsequent experimental studies have shown that the lunar regolith can be shock-lithified at pressures as low as 17 GPa (Schaal and Hörz 1980, Simon et al. 1986a). This result reflects the fact that the lunar soil is porous, and the pores collapse rapidly during passage of the compressive phase of the shock wave. During this collapse, the fine soil particles are efficiently melted by the resulting frictional heat. The small-scale melts then quickly cool to glass, bonding the unmelted soil particles to form a coherent rock. The materials used in shock-wave experiments did not retain any of their original porosity. It is therefore possible that some regolith breccias, which still retain considerable porosity, have formed at even lower shock pressures. This possibility has yet to be confirmed experimentally.

In a survey of Apollo 11 and Apollo 12 regolith breccias, Chao et al. (1971) distinguished four groups on the basis of texture: (1) porous and unshocked; (2) shock-compressed but still porous; (3) glass-welded; and (4) thermally metamorphosed and recrystallized. The coherence of these samples varies from extremely friable (crumbly) to very coherent. These characteristics reflect not only the intensity of the events that formed the breccias, but also their postformation histories. For example, porous breccias were probably formed at low shock pressure without suffering any later impacts, whereas the more compressed or welded breccias probably formed by more intense shock waves (closer to the impact) or were shocked again by later impacts.

Regolith breccias are important because they represent samples of old regolith that have been protected from the processes of maturation and mixing since the breccia formed. Some of the breccias are ancient, having formed between 500 and 1500 Ma. Over these long periods of time, further maturation and mixing of the original regolith will produce significant changes in the composition and properties between the regolith and the ancient material preserved in these breccias. Even over long periods of time, these differences are subtle, and detailed petrologic and chemical data are needed to identify them.

Results for Apollo 11 regolith breccias (Simon et al. 1984) show that they were formed locally; their compositions can be modeled closely by using appropriate proportions of local bedrock and soils. However, the breccias contain different populations of glass and plagioclase feldspar than the present-day soils, and they also have a higher ratio of high-K basalt to low-K basaltic components than do the soils. Although the exposed soil has been open to the addition of exotic materials much longer than have the breccias, both the present-day soil and the ancient breccias have similar amounts of highland component, indicating that little or no addition of highland material to the Apollo 11 site has occurred since the formation of the breccias.

By contrast, only five of the eleven Apollo 12 regolith breccias studied (Simon et al. 1985) could be modeled by using the composition of present-day Apollo 12 soils. One breccia (sample 12034) that could not be modeled was extremely KREEP-rich and clearly originated in a very different source regolith. Two other breccia samples were formed from anorthositic soils representing either a source exotic to the Apollo 12 site or perhaps the local premare regolith, present before eruption of the mare basalt lavas.

The Apollo 15 site is geologically diverse, with a highland-mare contact to the south and the deep Hadley Rille to the west (Heiken et al. 1991 *Lunar Sourcebook*). There are gradients with distance in the compositions of the regolith in both directions from the highland-mare contact (Basu and McKay 1979). Mare components become more abundant toward the mare, and highland components increase toward the highlands. The Apollo 15 regolith breccias appear to have formed from local materials at the site (Bogard et al. 1985, Korotev 1985, Simon et al. 1986b) and they can therefore provide information on local regolith dynamics and evolution. Simon et al. (1986b) showed that most of the breccias collected at the Apollo 15 site had not traveled far from where they were formed. Except for one anomalously KREEPy breccia (sample 15205) and one (sample 15306) with less green volcanic glass than the corresponding soil, each breccia is compositionally similar to the regolith at the spot (station) where it was collected. Comparison of the rock and mineral components of breccias and soils from the same station yields insights into regolith evolution over the whole site. For example, breccias collected at the edge of Hadley Rille contain slightly higher highland components than the soil there. This effect can be explained by the location; there is a continuous loss of regolith particles by gravitational creep into the rille, and the lost material is steadily replaced by material excavated from the mare bedrock under the thin regolith. Thus, any particles lost into Hadley Rille are replaced by local mare material. As time passed, the regolith became progressively richer in mare basalt components, while the breccias stayed the same, and the regolith has evolved to a point where the difference is detectable. On the other hand, soils from the highland stations at the Apollo 15 site are slightly richer in highland components than are the corresponding regolith breccia. This observation can also be explained by the morphology of the site. The highland samples were collected at the base of the Apennine Front, a mountain of highland rock that rises above the mare plain. With time, loose material from higher up on the mountain creeps or rolls down to collect at the base, enriching the soils there in highland materials while the breccias, which were formed much earlier, do not evolve. The fact that we can observe and explain such contrasts between breccias and soils demonstrates that both the geology and morphology of the Apollo 15 site have remained essentially the same since the mare basalts were erupted there at more than 3 Ga.

A multidisciplinary study of Apollo 16 regolith breccias that included petrography, chemistry, texture analysis, and attempts at age-dating the individual breccia components has been completed by McKay et al. (1986). They found, as did Simon et al. (1988a), that the breccias are generally similar to the soils, except that agglutinate contents are higher in the soils. Although the Apollo 16 regolith breccias apparently formed locally, their compositions do not correlate closely with the compositions of soils collected at the same stations. McKay et al. (1986) disaggregated two breccias, measured surface irradiation parameters, and found them to be very low, indicating that they had been buried at least a few meters deep in the regolith since they formed. They also found relatively high $^{40}Ar/^{36}Ar$ ratios in most of the breccias, indicating that the materials making up the breccias may have been exposed and irradiated at the lunar surface as early as 4 Ga. McKay et al. (1986) concluded that the Apollo 16 breccias contain ancient regolith that was only briefly exposed at the surface and is unlike any of the sampled Apollo 16 soils.

Finally, a characteristic of all regolith breccias studied thus far is a very low content of fused soil components (agglutinates and fragments of older regolith breccias) compared to present lunar soils. There are two possible explanations for this situation. Either the breccias were formed from extremely immature regolith (unlike any present-day regolith that was sampled) that contained very little fused soil, or agglutinates were present in the source material and were destroyed in the breccia-forming process. Because intact agglutinates and glass beads are observed in the Apollo 11 breccias, implying that they can survive moderate shock, Simon et al. (1984) preferred an origin from immature regolith. However, experimental studies (Simon et al. 1986a) have shown that agglutinates are easily destroyed by shock, although they can survive mild shock (<20 GPa). Simon et al. (1986a) then suggested that formation of regolith breccias from an immature regolith source can be inferred only for unshocked, porous regolith breccias. Agglutinates originally present in more intensely shock-compressed (or higher-grade) regolith breccias would probably be destroyed, and the absence of agglutinates in such breccias cannot be used as evidence for an immature source. Nonetheless, the virtual absence of agglutinates in unshocked regolith breccias is consistent with the idea that they are very ancient rocks that formed at a time of much more intense meteoroid bombardment, ~4 Ga (McKay et al. 1986). At that time, when both large and small impacts were frequent, agglutinates should have been rare due to the youth of the regolith, the higher turnover rates, and shorter exposure times. The results of McKay et al. (1986) suggest such an origin for Apollo 16 regolith breccias. Many other types of breccias seem to have formed at about that time as well.

LUNAR MINERALS

Introduction

Minerals have provided the keys to understanding lunar rocks because their compositions and atomic structures reflect the physical and chemical conditions under which the rocks formed. Analyses of lunar minerals, combined with the results of laboratory experiments, have enabled scientists to determine key parameters—temperature, pressure, cooling rate, and the partial pressures of such gases as oxygen, sulfur, and carbon monoxide—that existed during formation of the lunar rocks.

The most common silicate minerals are pyroxene, $(Ca,Fe,Mg)_2Si_2O_6$; plagioclase feldspar $(Ca,Na)(Al,Si)_4O_8$; and olivine $(Mg,Fe)_2SiO_4$. Potassium feldspar ($KAlSi_3O_8$) and the silica (SiO_2) minerals (e.g. quartz), although abundant on Earth, are notably rare on the Moon. Minerals containing oxidized iron (Fe^{3+} rather than Fe^{2+}) are absent on the Moon. The most striking aspect of lunar mineralogy, however, is the total lack of minerals that contain water, such as clays, micas and amphiboles.

Oxide minerals are next in abundance after silicate minerals. They are particularly concentrated in the mare basalts, and they may make up as much as 20 vol % of these rocks. The most abundant oxide mineral is ilmenite, $(Fe,Mg)TiO_3$, a black, opaque mineral that reflects the high TiO_2 contents of many mare basalts. The second most abundant oxide mineral, spinel, has a widely varying composition and actually consists of a complex series of solid solutions. Members of this series include: chromite, $FeCr_2O_4$; ulvöspinel, Fe_2TiO_4; hercynite, $FeAl_2O_4$; and spinel (sensu stricto), $MgAl_2O_4$. Another oxide phase, which is only abundant in titanium-rich lunar basalts, is armalcolite, $(Fe,Mg)Ti_2O_5$. As with the silicate minerals, no oxide minerals containing water (e.g. limonite) are native to the Moon. There was some debate about the origin of rare FeOOH found in some samples, but it is now generally believed that this material formed after contamination by terrestrial water.

Two additional minerals are noteworthy because, although they occur only in small amounts, they reflect the highly-reducing, low-oxygen environment under which the lunar rocks formed. Native iron (Fe) is ubiquitous in lunar rocks, and commonly contains small amounts of Ni and Co. Troilite, relatively pure FeS, is a common minor component; it holds most of the sulfur in lunar rocks.

Rare lunar minerals include apatite, $Ca_5(PO_4)_3(F,Cl)$, which contains only F or Cl and no OH, and the associated mineral whitlockite, $Ca_{16}(Mg,Fe)_2REE_2(PO_4)_{14}$. Rare sulfides, phosphides, and carbides occur in a variety of lunar rocks. Among these are a few that are largely of meteoritic origin and are very rare indeed: schreibersite, $(Fe,Ni)_3C$, and niningerite, (Mg,Fe,Mn)S. In detail, lunar mineralogy becomes quite complex when rare minerals are considered. A summary of known and suspect lunar minerals, compiled soon after the Apollo era, can be found in Frondel (1975) and Smith and Steele (1976). General reviews of silicate chemistry are available in Papike and Cameron (1976) and Papike (1987 1988).

Silicate minerals

The silicate minerals, especially pyroxene, plagioclase feldspar, and olivine, are the most abundant minerals in rocks of the lunar crust and mantle. These silicate minerals, along with other minerals and glasses, make up the various mare basalts and the more complex suite of highland rocks (melt rocks, breccias, and plutonic rocks). Meteoroid impacts over time have broken up and pulverized the lunar bedrock to produce a blanket of powdery regolith several meters thick, which forms the interface between the Moon and its space environment. The regolith therefore provides a useful sample of lunar minerals from a wide range of rocks, and Table 4 shows the average volume percentages of minerals in regolith collected at the Apollo and Luna sites. The data are for the 90 to 20 μm fraction, normalized so that the rock fragments are subtracted from the total. The resulting soil modes show the predominance of silicate minerals, especially pyroxene (8.5 to 61.1%), plagioclase feldspar (12.9 to 69.1%), and olivine (0.2 to 17.5%).

Table 4. Modal proportions (vol %) of minerals and glasses in soils from the Apollo (A) and Luna (L) sampling sites (90-20μm fraction, not including fused-soil and rock fragments). Revised from *Lunar Sourcebook* 1991.

	A-11	A-12	A-14	A-15 (H)	A-15 (M)	A-16	A-17 (H)	A-17 (M)	L-16	L-20	L-24
Plagioclase	21.4	23.2	31.8	34.1	12.9	69.1	39.3	34.1	14.2	52.1	20.9
Pyroxene	44.9	38.2	31.9	38.0	61.1	8.5	27.7	30.1	57.3	27.0	51.6
Olivine	2.1	5.4	6.7	5.9	5.3	3.9	11.6	0.2	10.0	6.6	17.5
Silica	0.7	1.1	0.7	0.9	-	0.0	0.1	-	0.0	0.5	1.7
Ilmenite	6.5	2.7	1.3	0.4	0.8	0.4	3.7	12.8	1.8	0.0	1.0
Mare Glass	16.0	15.1	2.6	15.9	6.7	0.9	9.0	17.2	5.5	0.9	3.4
Highland Glass	8.3	14.2	25.0	4.8	10.9	17.1	8.5	4.7	11.2	12.8	3.8
Others	-	-	-	-	2.3	-	-	0.7	-	-	
Total	99.9	99.9	100.0	100.0	100.0	99.9	99.9	99.8	100.0	99.9	99.9

(H) Denotes highland. (M) Denotes mare.

Pyroxene

Pyroxenes are the most chemically complex of the major silicate phases in lunar rocks. They are also informative recorders of the conditions of formation and the evolutionary history of these rocks. Pyroxenes are compositionally variable solid solutions, and they contain most of the major elements present in the host rocks.

For a review of pyroxene crystal chemistry, see Papike (1987). Briefly, in the pyroxene structure, the M1 and M2 sites provide a range of site volumes; as a result, pyroxenes can accommodate a wide variety of cations, and these cations reflect much of the chemistry and crystallization history of the rocks in which they occur. Ca, Na, Mn, Mg, and Fe^{2+} are accommodated in the large, distorted six- to eight-coordinated M2 site; Mn, Fe^{2+}, Mg, Cr^{3+}, Cr^{2+}, Ti^{4+}, Ti^{3+}, and Al occur in the six-coordinated M1 site; and Al and Si occupy the small, tetrahedral site. Potassium is too large to be accommodated in any of the pyroxene crystallographic sites at low pressure.

Pyroxene chemical analyses are listed in Table A5.2 for mare basalts, in Table A5.3 for highland clast-poor melt rocks and crystalline melt breccias as well as KREEP rocks, and in Table A5.4 for highland plutonic rocks (anorthosites and Mg-rich rocks). These tables show that Fe^{3+} (which would be listed as Fe_2O_3) does not occur in lunar pyroxenes and that sodium is low in abundance. Almost all mineral analyses reported at the end of the Chapter were taken from Heiken et al. (1991*Lunar Sourcebook*). The numbers above each analysis are the same as they appear in Heiken et al. (1991*Lunar Sourcebook*) in order to enable the reader to obtain the primary references to the data.

Although the major elements that define the pyroxene quadrilateral plot (Ca, Mg, and Fe) show important variations, the other, less abundant elements also show important trends. In lunar pyroxenes, these other elements include Al, Ti, and Cr. The very low oxygen fugacity (fO_2) during the crystallization of lunar rocks has resulted in reduced valence states for some Ti cations (Ti^{3+} rather than Ti^{4+}) and Cr cations (Cr^{2+} rather than Cr^{3+}). For example, Sung et al. (1974) suggest that as much as 30 to 40% of the Ti in Apollo 17 mare basalts occurs is trivalent and that most of this reduced Ti is in pyroxenes. These lower valence states are very rare in terrestrial pyroxenes. Pyroxene trace element data, determined by secondary ion mass spectrometry, is reported by Shearer et al. (1989), Papike et al. (1994), Papike (1996), and Papike et al. (1996).

Lunar pyroxenes also show evidence of substantial *subsolidus* reaction (i.e. recrystallization and other changes that take place below melting temperatures). Considerable work has been done to interpret the resulting features. It was discovered soon after the return of the Apollo 11 samples that subsolidus exsolution of two distinct pyroxenes, augite and pigeonite, had taken place within originally uniform pyroxene crystals (e.g. Ross et al. 1970). This process produced distinctive microscopic and submicroscopic exsolution lamellae, i.e. thin layers of pigeonite in augite, or vice versa. Papike et al. (1971) attempted to relate these exsolution features to the relative cooling histories of mare basalts. They pointed out that certain parameters of the pyroxene crystal unit cell (the length b and the angle ß) could also be used to estimate the composition of the intergrown augite and pigeonite. They also suggested that Δß (ß pigeonite-ß augite) could be used to indicate the degree of subsolidus exsolution and thus the relative annealing temperatures of the exsolved pyroxenes. Takeda et al. (1975) summarized similar exsolution data for pyroxene grains from Apollo 12 and 15 basalts. They compared the relative cooling rates (determined from exsolution studies) with absolute cooling rates determined from experimental studies. Ross et al. (1973) experimentally determined the 1-atmosphere augite-pigeonite stability relations for pyroxene grains from mare basalt 12021. Grove (1982) used exsolution lamellae in lunar clinopyroxenes as cooling rate indicators, and his results were calibrated experimentally. These studies all indicate that the cooling and subsolidus equilibration of igneous and metamorphic pyroxenes is a slow process; estimated cooling rates range from 1.5° to 0.2°C/hr for lava flows 6 m thick for Apollo 15 mare lavas (Takeda et al. 1975). McCallum and O'Brien (1996) studied the width and chemistry of exsolved pyroxenes in lunar highlands rocks to infer the stratigraphy of the highland crust.

Shock lamellae can be produced in pyroxenes by the shock waves due to meteoroid impact. However, these features are relatively rare, and they are much less well characterized than the analogous shock lamellae in plagioclase (Schaal and Hörz 1977).

Plagioclase feldspar

The silicate mineral feldspar has a framework structure of three-dimensionally linked SiO_4 and AlO_4 tetrahedra. For a review, see Papike (1988). The Si:Al ratio varies between 3:1 and 1:1. Ordering of Si and Al in specific tetrahedral sites can lead to complexities such as discontinuities in the crystal structure. Within this three-dimensional framework of tetrahedra containing Si and Al, much larger sites with 8 to 12 coordination occur that accommodate large cations (Ca, Na, K, Fe, Mg, Ba).

Aside from rare potassium- and barium-enriched feldspars, most lunar feldspars belong to the plagioclase series, which consists of solid solutions between albite ($NaAlSi_3O_8$) and anorthite ($CaAl_2Si_2O_8$). Because of the alkali-depleted nature of the Moon, lunar plagioclases are also depleted in Na (the albite component) relative to terrestrial plagioclases. Tables A5.5, A5.6, and A5.7 list plagioclase compositions for mare basalts, clast-poor melt rocks and crystalline melt breccias as well as KREEP rocks, and for highland plutonic rocks (anorthosites and Mg-rich rocks), respectively. The maximum chemical variation involves solid solution between albite and anorthite, a variability that can also be described as the coupled substitution between NaSi and CaAl, in which the CaAl component represents anorthite. The Ca abundance in the plagioclase, and therefore mol % anorthite correlate positively with the Ca/Na ratio in the host basalts (e.g. Papike et al. 1976, BVSP 1981). Plagioclase trace element data, determined by SIMS, are reported by Papike et al. (1996, 1997b).

Plagioclase from highland impact melts, breccias, and from KREEP rocks has more Na-rich compositions than that in highland plutonic rocks (i.e. less anorthite) along with enrichments in the geochemically similar elements potassium (K), rare earth elements (REE), and phosphorus (P), the association of minor elements referred to as KREEP. Plagioclase from coarse-crystalline igneous rocks has more restricted compositions; however, there is a positive correlation between the alkali content of the host rock and that of the plagioclase.

An interesting aspect of mare basalt plagioclases concerns the problem of where the observed Fe and Mg fit into the crystal structure. Smith (1974), in reviewing this problem, pointed out that (1) the Fe content in mare basalt plagioclase is higher than that in plagioclase from lunar highland rocks, and (2) the Fe content is positively correlated with albite (NaSi) content. Weill et al. (1971) noted peculiarities in lunar plagioclase cation ratios, and suggested that they might result from substitution of the theoretical "Schwantke molecule," where a vacant site normally occupied by a large cation (Na or Ca) is involved. Wenk and Wilde (1973) reviewed all available chemical data to accurately define the chemical components in lunar plagioclase. They concluded that (1) the deficiency of Al+Si in the tetrahedral sites (up to 0.06 atoms per eight-oxygen formula unit) is largely compensated by Fe and Mg substitution in the same tetrahedral sites; (2) the Ca/Na ratio in the octahedral large-cation sites increases correspondingly to maintain charge balance; and (3) the vacancy-coupled substitution of Ca + vacancy (the "Schwantke molecule") for 2 Na probably also occurs but is much less significant. Wenk and Wilde (1973) also made the interesting observation that the apparent number of vacancies increases with increasing Fe+Mg content. They suggested that the progressive substitution of larger Fe and Mg ions for the smaller Si in the tetrahedral sites decreases the volume of interstitial space available for the large Ca and Na atoms. Longhi et al. (1976) identified the same Fe and Mg components by using a slightly different approach. Schurmann and Hafner (1972) and Hafner et al. (1973), using Mössbauer and electron spin resonance techniques, showed that most of the iron in lunar plagioclase is present as Fe^{2+}. These studies all indicate that divalent Mg and Fe are important components in the tetrahedral structural sites in lunar plagioclase.

Plagioclase, like pyroxene, may undergo subsolidus reactions. Smith and Steele (1974) suggested that the grains of pyroxene and silica minerals observed as inclusions inside the plagioclase crystals of slowly-cooled highland plutonic rocks (e.g. anorthosites) may have formed by solid-state exsolution of the necessary elements (Ca, Fe, Mg, Si, etc.) from the original plagioclase grains. The plagioclase, when first crystallized, would have had high contents of Fe and Mg in its tetrahedral sites, perhaps comparable to the Fe and Mg contents of plagioclases in mare basalts. Smith and Steele (1974) proposed that, with falling temperature, the plagioclase component $CaSi_3(Mg,Fe^{2+})O_8$ broke down into pyroxene, $Ca(Mg,Fe^{2+})Si_2O_6$, and a silica mineral (SiO_2). However, this interesting suggestion was not supported by the subsequent results of Dixon and Papike (1975), who found that the volume of SiO_2 inclusions in plagioclases from anorthosites is much too low relative to the pyroxene inclusions for the two minerals to have formed by such a reaction. These authors preferred the explanation that the pyroxene inclusions and the plagioclase precipitated together from the original melt and later reacted during thermal annealing according to the exchange reaction: Ca(pyroxene) + Fe(plagioclase) = Fe(pyroxene) +

Ca(plagioclase). As a result, the plagioclase became more Ca-rich and the pyroxene more Fe-rich with time.

Olivine

The crystal structure of olivine, $(Mg,Fe)_2SiO_4$, consists of serrated chains formed of edge-sharing octahedra. These chains run parallel to the crystallographic c-axis. The octahedral chains are cross-linked by isolated SiO_4 tetrahedra. The major cations in the octahedral sites, Fe^{2+} and Mg, are distributed with a high degree of disorder (randomness) over both the M1 and M2 octahedral sites. However, the small amounts of Ca that may occur in olivine occupy only the M2 site (see Papike 1987).

Representative olivine analyses are listed in Tables A5.8 for mare basalts, A5.9 for highland clast-poor melt rocks and crystalline melt breccias and A5.10 for highland plutonic rocks (anorthosites and Mg rich rocks). The major compositional variation within olivines is caused by exchange of Fe and Mg; this exchange, and the resulting variations in composition, are represented by the ratio Fe/(Fe+Mg). The Fe end member, Fe_2SiO_4, is fayalite, and the Mg end member, Mg_2SiO_4, is forsterite. The most magnesian mare basalt olivine grains contain only 20 mol % fayalite (Fa), represented by the notation Fa_{20}. Most mare basalt olivines have compositions in the range Fa_{20}-Fa_{70}.

Very few olivines in mare basalts have compositions in the range Fa_{70-100}; however, a number of mare basalts do contain very Fe-rich olivine (Fa_{90-100}). These olivines are part of an equilibrium three-phase assemblage (Ca,Fe-pyroxene, Fe-olivine, silica) that crystallized stably from late-stage, Fe-enriched basalt melts. Some mare basalts, which cooled quickly during the late stages of crystallization, contain instead either Fe-rich pyroxene that crystallized metastably relative to the normal three-phase assemblage, or Fe rich pyroxenoid, pyroxferroite. The formation of extremely Fe-rich pyroxene violates a so-called "forbidden region" at the Fe-apex of the pyroxene quadrilateral.

Other significant elements in lunar olivines are Ca, Mn, Cr and Al. Calcium varies directly with the Fe content, and it may be an indicator of the cooling rate (Smith 1974). The experimental data of Donaldson et al. (1975) support this contention. Olivines in mare basalts are significantly enriched in Cr relative to olivines in terrestrial basalts. Cr_2O_3 values, which are commonly below detection limits (~0.1 wt %) in terrestrial olivines (Smith 1974) range up to 0.6 wt % in lunar olivines. Much or all of this Cr may be in the reduced Cr^{2+} valence state, and Haggerty et al. (1970) identified significant Cr^{2+} in lunar olivine using optical absorption techniques. Cr^{2+} is more readily accommodated into the olivine structure than is Cr^{3+}, which is the normal valence state for Cr in terrestrial olivines. The presence of Cr^{2+} is another result of the low oxygen fugacities that existed during mare basalt crystallization. Similarly, Cr^{2+} is much more abundant in lunar pyroxenes than in terrestrial pyroxenes (BVSP 1981).

Silica minerals: quartz, cristobalite, and tridymite

Silica minerals include several structurally different minerals, all of which have the simple formula SiO_2. These minerals are generally rare on the Moon. This rarity is one of the major mineralogic differences between the Moon and the Earth, where silica minerals are abundant in such common rocks as granite, sandstone, and chert. The near absence of silica minerals on the Moon is a result of several factors. For one thing, the Moon has apparently not evolved chemically beyond the formation of a low-silica, high-alumina anorthositic crust, so that high-silica granitic rocks are rare. For another, the Moon lacks hydrous and hydrothermal systems like those that can crystallize silica on Earth.

Despite their rarity on the Moon, the silica minerals are nevertheless important in classifying and unraveling the origin of some lunar rocks. Furthermore, lunar crustal rocks that contain silica minerals may be more abundant than their meager representation among the returned Apollo and Luna samples suggests. The silica minerals tend to concentrate along with incompatible elements, such as the KREEP component. For these reasons, the lunar silica minerals deserve greater consideration than their rarity would otherwise warrant.

The silica minerals found on the Moon are cristobalite, quartz, and tridymite. In spite of the intense impact cratering of the Moon, it is interesting that the high-pressure polymorphs of SiO_2, coesite and stishovite, which are known from young terrestrial impact craters, have not been found on the Moon. Explanations for their absence include the rarity of silica grains in the original target rocks and probable volatilization of silica during impact events in the high vacuum at the lunar surface (Papike et al. 1997a).

The crystal structures of the silica minerals are distinctly different from each other, but they all consist of frameworks of SiO_4 tetrahedra in which each tetrahedral corner is shared with another tetrahedron. A comparison of silica mineral structures, along with structure diagrams, can be found in Papike and Cameron (1976). All of the silica mineral structures contain little or no room for cations larger that Si^{4+}, hence the relatively pure SiO_2 composition of these minerals. The structures become more open in going from quartz to tridymite and cristobalite, and the general abundance of impurities increases accordingly.

Quartz occurs in a few granite-like (felsite) clasts as needle-shaped crystals that probably represent structural transformation (inversion) of original tridymite (Quick et al. 1981a). Some tridymite is preserved in these felsite clasts. The other rock type in which quartz is abundant is the rare fragments of coarse-grained lunar granites. The largest lunar granite clast yet found, from Apollo 14 breccia 14321, weighs 1.8 g and contains 40 vol % quartz (Warren et al. 1983a). A smaller granite clast, from sample 14303, was estimated to have 23 vol % quartz (Warren et al. 1983a). Based on their isotopic work on the large clast, Shih et al. (1985, 1993b) suggested that the sample crystallized in a deep-seated plutonic environment about 4.1 Ga. Consistent with the general absence of hydrous minerals on the Moon, the lunar granites do not contain mica or amphibole, as do granites on Earth.

The most common silica mineral in mare and KREEP basalts is not quartz but cristobalite, which can constitute up to 5 vol % of some basalts. This situation contrasts with the general absence of all silica minerals in terrestrial basalts. Lunar cristobalite commonly has twinning and curved fractures, indicating that it has inverted from a high-temperature crystal structure to a low-temperature one during cooling of the lavas (Dence et al. 1970, Champness et al. 1971). Other mare basalts contain crystals of the silica mineral tridymite that have incompletely inverted to cristobalite, producing rocks that contain both tridymite and cristobalite. In a study of Apollo 12 basalts, Sippel (1971) found that the coarser-grained samples contained cristobalite and quartz. Unfortunately, these mineral pairs are stable over fairly large temperature ranges and can also form metastably, outside of their equilibrium stability fields, so they are not useful for inferring the temperatures of lava crystallization.

Cristobalite tends to occur as irregular grains wedged between other crystals, while tridymite forms lathlike crystals. Klein et al. (1971) observed tridymite laths enclosed by pyroxene and plagioclase and suggested that tridymite was an early crystallizing phase.

Table A5.11a lists some representative analyses of cristobalite (analyses 13-15) and tridymite (analysis 16). Although the lunar silica minerals are nearly pure SiO_2, they typically contain contaminants such as Al_2O_3, TiO_2, CaO, FeO and Na_2O.

In those cases where silica minerals are observed in terrestrial basalts, they generally occur as rounded or embayed crystals of quartz that have been partly absorbed by the melt. These textures are ambiguous, and it is difficult to determine whether this quartz formed directly from the basaltic magma or whether it represents accidental inclusions picked up from other rocks through which the magma flowed. This is not the case in lunar mare basalts, where silica minerals have clearly crystallized *in situ*.

Other silicate minerals

Several other silicate minerals occur only rarely in lunar rocks. Some of these (e.g. tranquillityite and pyroxferroite) are unique to the Moon. Others, like zircon and potassium feldspar (K-feldspar), are rare on the Moon but common on Earth. These minerals occur in lunar basalts in small patches of high-silica residual melt formed during the last stages of crystallization of the mare lavas, and they are often accompanied by a silica mineral such as cristobalite. The same minerals also occur in unusual, high-silica, "KREEPy" highland rocks, which may have formed from a similar residual melt produced during large-scale crystallization of ancient highland igneous rocks.

Despite the rarity of these minerals on the Moon, they are important because (1) they act as recorders of the last stages of crystallization, and (2) they are commonly enriched in rare earth elements (REE) and in radioactive elements, some of which are useful in dating the samples.

As the basalt bedrock is gradually pulverized by meteoroid impacts, these rare minerals are released into the lunar soil. These minerals are fine-grained to begin with and, in the case of K-feldspar and cristobalite, are easily broken. As a result, they tend to become concentrated in the finer soil fractions. In mare terranes, the finest soil fractions have different compositions from the bulk soils; they are enriched in REE, radioactive elements, and in Al, K, and Na contained in feldspars.

Tranquillityite. Tranquillityite, $Fe_8(Y, Zr)_2Ti_3Si_3O_{24}$, is named for the Apollo 11 landing site in Mare Tranquillitatis. This mineral was first described in Apollo samples as "new mineral A" (Ramdohr and El Goresy 1970). Chemical analyses of tranquillityite, together with the structural formula, x-ray data, and density (4.7 g/cm^3) were first published by Lovering et al. (1971). Some of their analyses of tranquillityite from Apollo 11 and Apollo 12 basalts are listed in Table A5.11b.

Tranquillityite characteristically occurs in lunar mare basalts, where it forms small, (<100 μm) lath-shaped crystals associated with other rare minerals, such as pyroxferroite and apatite, in small pockets where the last minerals to crystallize are clustered. Tranquillityite is semiopaque; it is a nonpleochroic, and deep red in transmitted light because of its high TiO_2, and a gray to dark gray color in reflected light. Lovering et al. (1971) determined crystallographic dimensions for what appeared to be well-crystallized tranquillityite. However, further x-ray diffraction studies showed that some tranquillityites tend to be *metamict*, a state in which the original crystallographic order is partly or completely destroyed by radiation produced by decay of the relatively large amounts of U that they contain (Gatehouse et al. 1977, Lovering and Wark 1971). Gatehouse et al. (1977) were able to anneal this damage and recrystallize lunar tranquillityite grains by heating them to 800°C. Those heated in air formed a face-centered cubic crystal with a structure similar to fluorite (CaF_2), with an edge dimension of 4.85 Å. Grains reconstituted in vacuum formed with a slight rhombohedral distortion from cubic symmetry (edge dimension=4.743 Å). Heating above 900°C caused the tranquillityite structure to break down (Gatehouse et al. 1977). This result is consistent with the observation that tranquillityite is not found in metamorphosed (highly heated) basalt clasts (Lovering and

Wark 1975), in which it appears to have broken down, in part, to zircon.

Zircon. Zircon, ideally $ZrSiO_4$, is important not only because it tends to concentrate the heavy REE, but also because it is useful for age-dating the rocks in which it formed. Zircon is a refractory mineral that resists remelting and often incorporates Hf, Th, and U into its crystal structure, making it well-suited for U-Pb dating (e.g. Compston et al. 1984). Zircons also have a high retentivity for fission tracks, microscopic linear zones of damage produced by the recoil of U atoms that decay in the crystal by spontaneous fission. Zircon is especially suitable for age determinations based on the density of such tracks within the crystal (Braddy et al. 1975). Representative analyses of zircon are given in Table A5.11a.

The main source of lunar zircons appears to be the rare, high-silica granitic rocks (Lovering and Wark 1975). For example, rock 12013, a KREEPy granitic rock, contains 2200 μg/g Zr (LSPET 1970), which corresponds to approximately 0.1 vol % zircon (Drake et al. 1970). In this sample, zircons range from 4 to 80 μm in size. GJ Taylor et al. (1980) found 0.6 vol % zircon in a fragment of a similar rock type (quartz monzodiorite) from melt breccia 15405, and Keil et al. (1971) reported zircon in a "12013-like lithic fragment". Gay et al. (1972) found zircon in an anorthosite clast in breccia 14321. Representative analyses of these zircons are listed in Table A5.11a. Optical and physical properties of lunar zircons are described in Braddy et al. (1975). The full chemical composition and Raman spectra of a compositionally zoned zircon that shows partial metamictization was reported by Wopenka et al. (1996).

Because of the rarity of lunar granitic rocks and the durability of zircons, most zircons are found as isolated grains in soils and breccias. Zircons can also be found in metamorphosed basalt clasts, where they possibly form by the breakdown of tranquillityite (Lovering and Wark 1975). For their fission track study, Braddy et al. (1975) separated zircons from the "sawdust" left after breccia 14321 was cut for other studies. From 20 g of this sawdust they recovered 93 zircons, 70 of which were >100 μm in diameter. The zircons varied in U content from 15 to 400 μg/g, with a median content of 50 μg/g. Braddy et al. (1975), using methods of thermal annealing followed by etching, showed that zircons are excellent recorders of fission tracks.

Although they are rare, small, and difficult to work with, zircons are very important in dating lunar samples, especially very old (>4 Ga) highland rocks. U-Pb dating of lunar zircons, using ion microprobe analysis (Compston et al. 1984), has shown that zircons can survive the intense shock and heating of meteorite bombardment without serious disruption of their U-Pb systematics; they can therefore preserve the original rock ages. Compston et al. (1984) analyzed four zircons from a *clast* (a fragment of an older rock) within breccia 73217 and successfully determined a formation age for the clast of about 4356 Ma, a measurement that would otherwise not have been obtained.

Pyroxferroite. Although pyroxferroite had been synthesized in the laboratory (Lindsley 1967), the mineral was not observed in natural rocks until the return of the Apollo 11 samples. Pyroxferroite is an iron-rich pyroxenoid (a mineral structurally similar to pyroxene), whose formula is approximately $Ca_{1/7}Fe_{6/7}SiO_3$, with limited substitution of Mg for Fe. Some representative analyses are listed in Table A5.11a. Burnham (1971) determined that the pyroxferroite structure is based on a repeating pattern of seven SiO_4 tetrahedra; this structure is the same as that of a rare terrestrial mineral, pyroxmangite. Chao et al. (1970a) published some of the first x-ray, physical, and chemical data for pyroxferroite.

Pyroxferroite is found in Fe-rich basalts (ferrobasalts such as those at the Luna 24 site). As these rocks cooled, the compositions of the crystallizing pyroxenes changed, moving toward the CaFe side of the pyroxene quadrilateral. However, experimental studies by Lindsley and Burnham (1970) showed that pyroxferroite is only stable at approximately 1 GPa pressure, which on the Moon corresponds to a depth in the crust of several hundred kilometers (Chao et al. 1970a). Although the mare basalt magmas may have been derived from such depths, it is highly unlikely that they crystallized under such high pressures. Therefore, it is probable that pyroxferroite crystallized metastably (out of equilibrium) during rapid near-surface cooling of the basalts. To test this hypothesis, Lindsley et al. (1972) heated pyroxferroite crystals and found that those kept at 900°C for three days decomposed, forming the stable mineral assemblage Ca-,Fe-rich pyroxene + fayalite (Fe_2SiO_4) + tridymite (SiO_2). This result indicated that the lunar basalts cooled to below 900 °C within three days after crystallization of pyroxferroite. If they had remained above this temperature for a longer time, the pyroxferroite would have broken down to the three stable minerals.

Potassium feldspar. Another late-stage mineral found in lunar basalts is potassium feldspar, $KAlSi_3O_8$. Because the other basalt minerals (e.g. pyroxene, olivine, and plagioclase feldspar) accept very little of the relatively large K^+ ion into their structures, it becomes concentrated in the residual melt that remains after most minerals have crystallized. Early investigators were therefore not surprised to find minor amounts of late-formed potassium feldspar in Apollo 11 basalts (e.g. Agrell et al. 1970, Keil et al. 1970). Albee and Chodos (1970) observed a "K-rich phase" that approached potassium feldspar in composition but was nonstoichiometric.

In high-silica highland rocks, which have granitic or so-called "KREEPy" compositions, potassium feldspar is commonly neither minor nor fine-grained. In sample 12013, a KREEPy breccia that itself contains two different breccias (gray and black), there are abundant patches of fine-grained granitic material (felsite) that contain 50 vol % potassium feldspar, 40 vol % silica minerals, 5 vol % Fe-rich augite, and 5 vol % other phases (Quick et al. 1981a). Neither the black nor the gray breccias in 12013 contain any mare basalt components. Apparently 12013 was formed from a very unusual, SiO_2-rich, KREEPy, evolved terrane, probably somewhere in the highlands. Because 12013 appears to be an impact-produced breccia, it is not clear whether the variety of rock types it contains came from a single differentiated intrusion or from a number of unrelated sources (Quick et al. 1981a).

In addition to the occurrences of K-feldspar in highland samples (KREEPy and granitic rocks), and as a nonstoichiometric phase in some mare basalts, the mineral is also found in an unusual K-rich mare basalt type (very high-K basalt), which has been discovered as clasts in two Apollo 14 breccias (Shervais et al. 1983, 1984b, 1985b). These clasts have $K_2O > 0.5$ wt % and contain several percent potassium feldspar. Shervais et al. (1985b) concluded that these basalts were likely produced by the partial assimilation of granitic crust by a normal, low-Ti mare basalt magma on its way to the surface.

Some analyses of lunar potassium feldspars are listed in Table A5.11a. It is obvious that the feldspars can have a significant BaO content (also expressed as celsian feldspar). Quick et al. (1981a) reported that, in the potassium feldspar in the felsites of sample 12013, the celsian (Cn) content increases with the K:Na ratio in the potassium feldspar, and the composition ranges from 0.9 mol % Cn at K:Na = 1:1 to 2.9 mol % Cn at K:Na = 3.1.

Oxide minerals

The silicates, such as pyroxene, olivine, and feldspar, are the most abundant minerals in lunar rocks. With minor chemical differences, the common lunar silicate minerals are essentially the same as found on Earth. However, the nonsilicate minerals—especially the oxides—are far more distinctive in lunar rocks. These oxide minerals are potential ores for resource extraction at a lunar base, and they are particularly abundant in some mare basalts.

As mentioned earlier, several oxide minerals are important constituents of lunar rocks: ilmenite, $FeTiO_3$; spinels—with extensive chemical variations, $(Fe,Mg)(Cr,Al,Fe,Ti)_2O_4$; and armalcolite,$(Fe,Mg)Ti_2O_5$. The less abundant lunar oxide minerals include rutile, TiO_2, baddeleyite, ZrO_2, and zirconolite, $(Ca,Fe)(Zr,REE)(Ti,Nb)_2O_7$.

The major differences between the oxide minerals in lunar and terrestrial rocks arise from fundamental differences between both the surfaces and the interiors of these two planets. On the Moon, meteoroid impact and shock-metamorphic processes play a major role in altering rocks. These effects are not the same for all minerals. Shock damage and the formation of shock glasses from minerals, e.g. maskelynite from plagioclase feldspar, are observed chiefly in silicate minerals. Oxide minerals also record shock damage, but another effect of impact on oxide (and sulfide) minerals is to produce small amounts of chemical reduction.

In addition to the different surface environments of the Earth and Moon, the original conditions of formation of lunar rocks, most notably the volcanic ones, are different from those on Earth in three main aspects: (1) higher temperatures of formation; (2) lower oxygen fugacities during formation; and (3) the complete absence of water. These factors combine to produce oxide minerals on the Moon that are very different from those found on Earth. The temperatures at which igneous rocks melt and crystallize are much higher on the Moon than on Earth because of the absence of water, a chemical species that is efficient in lowering the melting temperatures of these melts. This temperature differential is not large, generally about 100° to 150°C for melting temperatures of about 1200°C, and this difference alone would not produce any major changes between terrestrial and lunar magmas.

In contrast, the differences in oxygen fugacity between the Earth and Moon led to pronounced differences in mineralogy (Sato et al. 1973, Sato 1978, Haggerty 1978b). As in the case of the silicate minerals, the low oxygen fugacity prevents any completely oxidized iron (Fe^{3+}) from forming. Indeed, oxygen fugacities on the Moon are so low that native iron (Fe°) is stable with FeO. As a result, metallic iron is ubiquitous in lunar samples of all kinds. Other elements in lunar minerals are also present in unusually reduced oxidation states (e.g. Ti^{3+}, Cr^{2+}) in comparison with terrestrial minerals.

The oxide minerals, although less abundant than silicates in lunar rocks, are of great significance because they retain signatures of critical conditions of formation (e.g. limited availability of oxygen) of the rocks in which they occur. Whereas most of the silicate minerals differ little from those on Earth, the opaque oxide phases reflect the reducing, anhydrous conditions that prevailed during their formation. By combining analyses of lunar oxide minerals with the results of laboratory experiments on their synthetic equivalents, the temperature and oxygen pressure conditions during formation of lunar rocks can be estimated (see Sato et al. 1973, Usselman and Lofgren 1976).

Because their oxygen is more weakly bonded than that in silicate minerals, oxide minerals are obvious and important potential feedstocks for any future production of lunar oxygen and metals. On Earth, similar oxide minerals commonly occur in economically recoverable quantities called ore deposits. However, most of these deposits have formed from hydrothermal waters (100° to 300°C or hotter). The Moon has little if any water and the presence of similar hydrothermal ore deposits on the Moon is not probable. However, there are other means of concentrating oxide minerals into exploitable ores. Crystal settling of dense minerals (e.g. chromite, ilmenite, and minerals containing the platinum-group elements) is possible within silicate magmas, if the magma remains liquid for a long enough time. On Earth, such accumulations are normally found in layered intrusions. These bodies form from large masses of magma that have been emplaced into crustal rocks without reaching the Earth's surface. Under such conditions, cooling is slower, and physical separation processes-which may include convection as well as settling-have time to act. Well-known examples of ore deposits resulting from these processes occur in the Stillwater Anorthosite Complex (Montana) and the Bushveld Igneous Complex (South Africa).

Ilmenite

Ilmenite, with the ideal formula $FeTiO_3$, is the most abundant oxide mineral in lunar rocks. The amount of ilmenite in a rock is a function of the bulk composition of the magma from which the rock crystallized (Rutherford et al. 1980, Norman and Ryder 1980a, Campbell et al. 1978); the higher the TiO_2 content of the original magma, the higher the ilmenite content of the rock. Ilmenite forms as much as 15 to 24 vol % of many Apollo 11 and 17 mare basalts (McKay and Williams 1979). However, the volume percentages of ilmenite in mare basalts vary widely across the Moon, as indicated by the range of TiO_2 contents in samples from different lunar missions.

The ilmenite crystal structure is hexagonal and consists of alternating layers of Ti- and Fe-containing octahedra. Most lunar ilmenite contains some Mg substituting for Fe (Table A5.12.), which arises from the solid solution that exists between ilmenite ($FeTiO_3$) and $MgTiO_3$, the mineral geikielite. Other elements are present only in minor to trace amounts (i.e. <1 wt %); these include Cr, Mn, Al, and V. In addition, ZrO_2 contents of up to 0.6 wt % have been reported from ilmenite in Apollo 14 and 15 basalts (El Goresy et al. 1971a,b; LA Taylor et al. 1973b). In fact, the partitioning of ZrO_2 between ilmenite and coexisting ulvöspinel (Fe_2TiO_4) has been experimentally determined (LA Taylor and McCallister 1972) and has been used as both a geothermometer (to deduce temperatures during crystallization) and as a cooling-rate indicator (LA Taylor et al. 1975, 1978; Uhlmann et al. 1979). Although terrestrial ilmenite almost always contains some Fe^{3+}, lunar ilmenite contains none. Table A5.12 lists representative ilmenite compositions from a wide range of lunar rock types.

Ilmenite commonly occurs in mare basalts as bladed crystals up to a few millimeters long. It typically forms near the middle of the crystallization sequence, where it is closely associated with pyroxene. It also forms later in the sequence and at lower temperatures, where it is associated with native Fe and troilite. In Apollo 17 rocks, ilmenite is frequently associated with armalcolite and occurs as mantles on armalcolite crystals (e.g. Haggerty 1973a, Williams and Taylor 1974). In these instances, ilmenite has possibly formed by the reaction of earlier armalcolite with the melt during crystallization.

The compositions of lunar ilmenites plot along the $FeTiO_3$-$MgTiO_3$ join; variation from $FeTiO_3$ is often expressed in wt % MgO. The appreciable MgO contents of many lunar ilmenites (>3 wt % MgO) are similar to terrestrial ilmenites that formed at high pressure in rocks called kimberlites (kimberlites come from deep in the Earth's mantle and are the igneous rocks that often contain diamonds). It was originally thought that the high Mg content in some lunar ilmenites indicated, as in terrestrial kimberlites, high pressures of formation. In general, the ilmenite with the highest Mg contents tends to come from relatively high-Mg rocks; ilmenite composition correlates with the bulk composition of the rock and therefore reflects magmatic chemistry rather than pressure. In detail, the distribution of Mg between ilmenite and coexisting silicate minerals in a magma is related to the timing of ilmenite crystallization relative to the crystallization of the other minerals. This crystallization sequence is itself a function of cooling rate and other factors, most notably the oxygen fugacity (Usselman et al. 1975). However, it is doubtful that the Mg contents in ilmenite all represent equilibrium conditions, because ilmenite compositions can vary significantly, even within distances of a few millimeters, within a single rock.

The stability curve of pure ilmenite as a function of temperature and fO_2 is significantly different from that of ulvöspinel (LA Taylor et al. 1973b), the spinel phase with which it is commonly associated, implying that the two minerals did not crystallize together. The data suggest that, in these mineral assemblages, ilmenite has formed by solid-state reduction of this high-Ti spinel at temperatures below their melting points. Rare grains of ilmenite also contain evidence for subsolidus reduction of ilmenite to rutile (TiO_2) + native Fe; other grains show reduction to chromite ($FeCr_2O_4$) + rutile + native Fe.

Spinels

Spinel is the name for a group of oxide minerals, all with cubic crystal symmetry, that have extensive solid solution within the group. Spinels are the second most abundant opaque mineral on the Moon, second only to ilmenite, and they can make up as much as 10 vol % of certain basalts, most notably those from the Apollo 12 and 15 sites. The general structural formula for these minerals is $^{IV}A^{VI}B_2O_4$, where ^{IV} and ^{VI} refer to cations with tetrahedral and octahedral coordination, respectively.

The basic spinel structure is a cubic array of oxygen atoms. Within the array, the tetrahedral A-sites are occupied by one-third of the cations, and the octahedral B-sites are occupied by the remaining two-thirds of the cations. In a normal spinel structure, the divalent cation, such as Fe^{2+}, occupies only the tetrahedral sites, and the two different sites each contain only one type of cation (e.g. $FeCr_2O_4$). If the divalent cation occurs in one-half of the B-sites, the mineral is referred to as an inverse spinel [e.g. $Fe(Fe,Ti)_2O_4$]. In lunar spinels, the divalent cations (usually Fe^{2+} or Mg^{2+}) occupy either the A- or both A- and B-sites (i.e. there are both normal and inverse lunar spinels), and higher-charge cations (such as Cr^{3+},Al^{3+},Ti^{4+}) are restricted to the B-sites.

The relations of the various members of the spinel group can be displayed in a diagram known as the Johnston compositional prism (e.g. Haggerty 1978a). The end-members represented include chromite, $FeCr_2O_4$; ulvöspinel, $FeFeTiO_4$ (commonly written as Fe_2TiO_4, but this is an inverse spinel with Fe^{2+} in both A- and B-sites); hercynite, $FeAl_2O_4$; and spinel (sensu stricto), $MgAl_2O_4$. Intermediate compositions among these end-members are designated by using appropriate modifiers (e.g. chromian ulvöpsinel or titanian chromite).

Most lunar spinels have compositions generally represented within the three-component system: $FeCr_2O_4$–$FeFeTiO_4$–$FeAl_2O_4$, and their compositions can be represented on a simple triangular plot. The addition of Mg as another major component provides a third dimension to this system; the compositions are then represented as points within a limited Johnston compositional prism in which the Mg-rich half (Mg > Fe) is deleted because most lunar spinels are Fe-rich (e.g. Agrell et al. 1970, Busche et al. 1972, Dalton et al. 1974, El Goresy et al. 1971b, 1976; Haggerty 1971a, 1972b,c, 1973b, 1978a; Nehru et al. 1974, 1976; LA Taylor et al. 1971). Most lunar spinel compositions fall between chromite and ulvöspinel. The principal cation substitutions in these lunar spinels can be represented by $Fe^{2+} + Ti^{4+} \Leftrightarrow 2(Cr,Al)^{3+}$. Other cations commonly present include V, Mn, and Zr. Representative compositions are given in Table A5.13.

Spinels are ubiquitous in mare basalts, where they occur in various textures and associations. These spinels are invariably zoned chemically. Such zoning occurs particularly in Apollo 12 and 15 rocks, in which chromite is usually the first mineral to crystallize from the melt. As the chromite crystals continue to grow, their TiO_2 and FeO contents increase, and their Al_2O_3, MgO and Cr_2O_3 contents decrease, with the overall composition moving toward that of ulvöspinel (Table A5.13). In most of the basalts that contain both titanian chromites and chromian ulvöspinels, the latter phase occurs as overgrowths and rims surrounding the chromite crystals. (Some individual ulvöspinel grains also occur as intermediate to late-stage crystallization products).

When observed using a reflected-light microscope, the ulvöspinel in these composite crystals appears as tan to brown rims about the bluish chromite. The contact between the two is commonly sharp, indicating a discontinuity in the compositional trend from core to rim. This break probably records a cessation in growth, followed later by renewed crystallization in which the earlier chromite grains acted as nuclei for continued growth of ulvöspinel (Cameron 1971). Some rocks (e.g. basalt 12018) contain spinel grains with diffuse contacts that reflect gradational changes in the composition of the solid solution. These textures could result from continuous crystallization of the spinel or from later solid-state diffusion within the crystal (LA Taylor et al. 1971).

Although most abundant in mare basalts, spinels also occur in highland rocks such as anorthosites, anorthositic gabbros, troctolites, and impact mixtures of these rock types (e.g. Haselton and Nash 1975). The spinels in anorthositic (plagioclase-rich) highland rocks tend to be chromite with lesser amounts of MgO, Al_2O_3, and TiO_2 (Table A5.13). However, certain highland rocks, notably the olivine-feldspar types (troctolites), contain pleonaste spinel. The composition of this spinel is slightly more Fe- and Cr-rich than an ideal composition precisely between the end members $MgAl_2O_4$ and $FeAl_2O_4$. This spinel is not opaque; under the microscope, it stands out because of its pink color, high index of refraction, and isotropic character in cross-polarized light.

Subsolidus reduction. Crystals that originally form from a melt may continue to change while the rock is solid but still hot. Such subsolidus reactions can occur at temperatures significantly below the melting point. In terrestrial rocks, the oxygen fugacity is relatively high during such changes, and terrestrial subsolidus reactions generally involve oxidation. However, in lunar rocks and soils, evidence for subsolidus *reduction* is extremely common. Lunar ulvöspinel grains are often reduced to ilmenite + native Fe; more rarely, ilmenite is reduced to rutile + native Fe or to chromite + rutile + native Fe (El Goresy et al. 1971a, 1972; Haggerty 1971b, 1972a,d, 1977; McCallister and Taylor 1973, LA Taylor et al. 1971). The causes of this late-stage reduction of ulvöspinel are speculative, but the effects have been quite pervasive (Brett 1975b, Haggerty 1978b, Sato 1978). In a few rocks (e.g. 14053, 14072), the Ti-rich ulvöspinel is reduced to a mixture of Ti-poor spinel + titanian chromite + ilmenite + native Fe (El Goresy et al. 1972, Haggerty 1972a).

Compositional changes in spinel during later subsolidus reduction are the opposite of those observed during primary crystallization (El Goresy et al. 1972, Haggerty 1972c, LA Taylor et al. 1971). During normal crystallization of spinel from a melt, the spinel typically begins as chromite and changes its composition toward ulvöspinel as growth continues. The net effect of later subsolidus reduction on the ulvöspinel is to form ilmenite + native Fe; the residual components enrich the remaining spinel so that its composition moves back toward chromite. The secondary generation of native Fe during these subsolidus reactions is of some importance. It provides evidence for the reducing nature of the reaction. It also increases the metal content of the rock involved. Spinel grains in the lunar soil also readily undergo reduction when shock metamorphosed by impacting micrometeoroids. This reduction is possibly caused by the presence in the soil of implanted solar-wind particles, notably the elements hydrogen and carbon, which create a reducing environment when heated to high temperatures during impact.

Armalcolite

Armalcolite is named after the Apollo 11 astronauts (ARMstrong, ALdrin, and COLlins). It was first recognized as a new mineral in samples from the Apollo 11 site, where it is a minor constituent in Ti-rich basalts (Anderson et al. 1970). Its composition is strictly defined as $(Fe_{0.5}Mg_{0.5})Ti_2O_5$, but the name is also used in a broader sense to describe solid solutions whose compositions vary between $FeTi_2O_5$ and $MgTi_2O_5$.

Armalcolite has a crystal structure like that of the mineral ferropseudobrookite, $FeTi_2O_5$. Titanium is restricted to the M2 site, and Mg, Al, and Fe occupy the M1 site. Detailed chemical analyses of armalcolite, with careful consideration of the ionic charge balance required within the crystal structure, have shown that appreciable Ti is present as Ti^{3+} rather than Ti^{4+} (Wechsler et al. 1976). Kesson and Lindsley (1975) examined the effects of Ti^{3+}, Al^{3+}, and Cr^{3+} on the stability of armalcolite, and later work showed that the Ti^{3+} content can be used to deduce the fO_2 at the time of crystallization (Stanin and Taylor 1979a,b, 1980). The presence of Ti^{3+} as well as Ti^{4+} in lunar armalcolites, due to the strongly reducing lunar environment, serves to distinguish the lunar variety from the armalcolites subsequently identified on Earth (Cameron and Cameron 1973), in which all Ti occurs as Ti^{4+}.

The occurrence of armalcolite is restricted to rocks with high TiO_2 contents that have also cooled rapidly. This rapid cooling (quenching) is essential to prevent early-formed armalcolite from reacting with the remaining liquid to form magnesian ilmenite. There are three distinct compositional types of armalcolite in lunar samples (Haggerty 1973a). The first and most abundant type is Fe-Mg armalcolite, and it is represented by intermediate compositions in the solid solution series $FeTi_2O_5$–$MgTi_2O_5$ (Table A5.14). This variety is the typical armalcolite observed in the high-Ti Apollo 11 and 17 basalts, although it is also found in basalt samples from all missions. Two varieties of this type have been characterized by their appearance in reflected-light microscopy as gray- vs. tan-colored; Haggerty (1973b) referred to these as ortho- and para-armalcolite, respectively. They have overlapping compositional ranges (Papike et al. 1974, Williams and Taylor 1974) and appear to be present in a variety of different textures. The most common type is the gray variety, which occurs with rims of high-Mg ilmenite, especially in Apollo 17 samples. There were suggestions that these two varieties had different crystal structures, but the crystal structures have since been shown to be identical (Smyth 1974).

The second compositional type of armalcolite is characterized by high contents of ZrO_2 (3.8-6.2 wt %), Cr_2O_3 (4.3-11.5 wt %), and CaO (3.0-3.5 wt %). This has been called Cr-Zr-Ca-armalcolite. The third type is intermediate in composition between the first type, Fe-Mg armalcolite, and the second type, Cr-Zr-Ca armalcolite. It has been called Zr-armalcolite and has distinctive amounts of ZrO_2 (2.0-4.4 wt %) Y_2O_3 (0.15-0.53 wt %), and Nb_2O_5 (0.26-0.65 wt %). Analyses of these various types of armalcolite are given in Table A5.14, and they are described in detail by Haggerty (1973a).

Other oxides

The only other oxide minerals of significant abundance in lunar rocks and soils are rutile (TiO_2) and baddeleyite (ZrO_2); representative compositions of these phases are given in Table A5.15. Rutile is generally associated with ilmenite, and it occurs most commonly as a reaction product from the reduction of ilmenite and/or armalcolite. Primary rutile occurs as discrete euhedral grains, also typically associated with ilmenite. Rutile in this association often contains Nb, Cr, Ta and REE (Marvin 1971, Hlava et al. 1972, El

Goresy and Ramdohr 1975; Table A5.15). Baddeleyite is common in certain Apollo 14 clast-poor impact melt rocks, e.g. 14310, 14073), where it is associated with schreibersite [$(Fe,Ni,Co)_3P$] (El Goresy et al. 1971a). Although these two minerals were originally thought to be indigenous to the Moon, it is now thought probable that the high Zr and P contents of these baddeleyite- and schreibersite-bearing rocks arise from meteoritic contamination that was incorporated into the original melts, which were produced by large meteoroid impact events.

Although rare on the Moon and not truly an oxide, the compound FeOOH has been found and often described as "rust" in lunar rocks from every mission, particularly those from Apollo 16 (El Goresy et al. 1973b, LA Taylor et al. 1973a). It has been conclusively shown that this phase, the mineral akaganeite (ß-FeOOH), is the product of contamination of the lunar rocks by terrestrial water vapor, which caused the oxyhydration of indigenous lawrencite, $FeCl_2$, to form this water-bearing phase (LA Taylor et al. 1973a, 1974). Therefore, to date, no evidence for indigenous water has been found in any lunar minerals.

Sulfide minerals

Sulfur is a relatively volatile element that plays a dual role on the Moon: it is important in the gases that drove lunar pyroclastic eruptions, and in the gases released during impact heating. For a planet with a surface otherwise poor in volatile elements, the Moon has a fair amount of sulfur. Lunar mare basalts, for instance, have about twice as much sulfur as do typical terrestrial basalts. On the Moon, this sulfur is present in sulfide (S) minerals; the low oxygen fugacities in the lunar environment apparently do not permit the formation of sulfate (SO_4) minerals.

Troilite

Troilite, FeS, is the most common sulfide mineral in lunar rocks. Although it almost always forms less than 1 vol % of any lunar rock, troilite is ubiquitous. It is commonly associated with native Fe, ilmenite, and spinel. The chemical composition of troilite is essentially that of stoichiometric FeS with less than 1 wt % of all other components (Table A5.16).

Based on a study of a small number of early Apollo samples, Skinner (1970) proposed that lunar troilite was always associated with native Fe in textures that result from crystallization at the 988°C eutectic point where both FeS and Fe form simultaneously. The formation of an immiscible sulfide melt late in the crystallization of a silicate magma preceded this eutectic crystallization. Some lunar troilite has undoubtedly formed in this way. However, other troilite occurrences are void of native Fe and require precipitation directly from a S-saturated silicate melt.

The most common occurrence of troilite is as an accessory phase in mare basalt, where it is usually a late-stage crystallization product. Secondary troilite forms later, in the solid rocks, in cases where the partial pressure of sulfur increases rapidly and sulfurizes native Fe during the high-temperature shock metamorphism produced by meteoroid impacts. Some Apollo 16 rocks, notably 66095, contain troilite that most likely formed as a direct result of this remobilization of sulfur during meteoroid impact.

Other sulfides

Other sulfide minerals positively identified in lunar rocks include chalcopyrite, $CuFeS_2$; cubanite, $CuFe_2S_3$; pentlandite, $(Fe,Ni)_9S_8$; mackinawite, $Fe_{1+x}S$; and sphalerite, (Zn,Fe)S. All these minerals are so rare as to be only geologic curiosities, and they have only minor applications in determining the origins of the rocks that contain them.

The Cu-bearing phases have only been found as small grains (<10 μm to 15 μm) in some Apollo 12 basalts (LA Taylor and Williams 1973) and in small cavities (vugs) in two Apollo 17 breccias, where chalcopyrite is associated with pentlandite (Carter et al. 1975). Pentlandite has also been reported from an Apollo 14 breccia (Ramdohr 1972). Mackinawite was identified as small (<5 μm) grains in certain Apollo 12 basalts (El Goresy et al. 1971b; LA Taylor et al. 1972). Sphalerite (with 28 mol % FeS in solution with ZnS) was observed in some Apollo 16 breccias, notably 66095 (Table A5.16; El Goresy et al. 1973a; LA Taylor et al. 1973a), where it was probably formed as a result of the mobilization of Zn and S during impact-produced shock metamorphism. It is only present as small grains (<20 μm) and in minor quantities (<0.01 vol %).

Native iron

Native iron metal, Fe^0, is only rarely found in terrestrial rocks. However, one of the surprising findings from the first-returned lunar samples, Apollo 11 mare basalts, was the presence of native Fe metal grains in every sample (e.g. Reid et al. 1970). These metal grains were produced by the crystallization of normal igneous melts under reducing conditions. Subsequently, native Fe metal was identified in all returned lunar rocks. In lunar rocks it is a ubiquitous minor phase, largely because of the low oxygen fugacities during original crystallization of lunar magmas and during subsequent meteoroid impacts.

Native Fe occurs in lunar rocks as three different minerals with different crystal structures. These minerals occur in various proportions and form intricate textures, either from exsolution during cooling or from later subsolidus reequilibration. These three minerals also have different chemical compositions, involving varying amounts of solid solution between Fe and Ni. Kamacite (α-Fe) has a body-centered cubic crystal structure and contains 0 to 6 wt % Ni. Taenite (γ-Fe) has a face-centered cubic crystal structure and contains 6 to 50 wt % Ni. Tetrataenite has a tetragonal crystal structure and is essentially FeNi with 50±2 wt % Ni.

In lunar samples, kamacite is the most abundant metal phase and taenite the second most abundant. Tetrataenite is only rarely observed, and that which occurs is most likely due to meteoritic contamination. These minerals are apparently formed by four different processes: (1) normal igneous crystallization; (2) subsolidus reduction of oxides, or breakdown of troilite and olivine (this process has occurred in rocks 14053 and 14072); (3) reduction of FeO in impact-produced silicate melts in the soil; and (4) meteoroid contamination. Effects of the first and second processes are readily apparent and were easily recognized in the first returned samples. The meteoroid metal component is to be expected because of the large flux of impacting bodies onto the lunar surface. However, the discovery that abundant native Fe could be produced in the soils during impact melting was unexpected.

Native Fe in lunar rocks

The amount of metal varies between samples as well as between sites, but is always less than 1 vol %. Representative analyses of metals are listed in Table A5.17. In mare basalts, native Fe makes up only a small portion of the opaque mineral content, which is mostly formed of the oxide minerals ilmenite and spinel. However, in some highland rocks such as the Apollo 16 breccias, native Fe is virtually the only opaque mineral present (Misra and Taylor 1975a,b). Some cubic Fe crystals have grown in the open spaces of breccias, apparently from Fe-rich vapor (Clanton et al. 1973).

Although there are substantial variations in the compositions of native Fe among the samples from a single landing site, some differences between sites can still be recognized.

Native Fe from Apollo 11 basalts is usually low in Ni (<1 wt %) and Co (<0.5 wt %); metals from the Apollo 17 samples are similar but may have higher Co contents. Metals from Apollo 12 and 15 basalts have compositions with 0 to 30 wt % Ni and 0 to 6 wt % Co. However, the high-Ni and high-Co metals in these basalts are rare. Native Fe enclosed within early-formed olivine crystals can contain 30 wt % Ni, whereas that in the later-crystallized parts of the rock can be virtually pure Fe (i.e. <0.2 wt % Ni). The metal in breccias from the Apollo 14 and 16 sites is difficult to distinguish from meteoritic metal and has wide-ranging compositions similar to those reported for meteorites (i.e. 0 to more than 50 wt % Ni, 0 to 8 wt % Co).

Native Fe in lunar soil

The lunar soil consists of comminuted rocks and minerals. Therefore, any mineral occurring in the rocks can become part of the soil (LA Taylor 1988). In particular, the native Fe metal grains now observed in the soil could have formed originally by normal crystallization of a silicate melt or by later subsolidus reduction of other minerals (e.g. oxides in the cooling magma). In addition, a large amount of metal can be contributed by impacting meteoroids. As a result of these diverse sources, the compositions of lunar Fe metal vary from essentially pure Fe to virtually any composition in the range reported for meteoritic metal. The compositions of metal grains larger than a few micrometers (i.e. those large enough for electron microprobe analysis), when taken as a whole for a given soil, tend to center around a bulk composition of Ni = 5 to 6 wt % and Co = 0.3 to 0.5 wt %; these values are the same as the average compositions of metal in chondrites.

However, magnetic studies of the lunar soil indicate that a significant amount of Fe metal is present as much smaller particles, well below 0.1 μm in size. Tsay et al. (1971) noticed that the ferromagnetic resonance (FMR) signal for single-domain Fe (i.e. from grains 40 to 330 Å in size) in the lunar soil was an order of magnitude greater than that from associated rock samples (see review article by LA Taylor and Cirlin 1986). These studies indicated that there is considerably more Fe metal in the lunar soils than in the rocks from which they were derived, that much of it is very fine-grained (<300 Å), and that it is not meteoritic metal. Since the soils are composed of disaggregated rocks, what is inherently different between the lunar rocks and soils? That is, where did this additional Fe metal come from? It must be produced by a process involved in the formation of the lunar soil itself.

There are two principal processes at work in forming lunar soils: (1) simple disaggregation, or the breaking of rocks and their minerals into smaller particles; and (2) agglutination, the welding together of rock and mineral fragments by the glass produced by melting due to small meteoroid impacts. These two processes compete to decrease and increase, respectively, the grain size of soil particles (Morris 1977, 1980).

The agglutinates (see section on the Regolith) contain much of the fine-grained, single-domain Fe metal particles in the soil. The majority of the metal grains in the agglutinates are from 100 to 200 Å in size, well within the single-domain size range of 40 to 330 Å for metallic Fe. The composition of most of these minute Fe particles is >99% Fe with only trace amounts of Ni and Co (Mehta and Goldstein 1979). By contrast, metal grains that are larger than a few micrometers across have higher Ni, Co, and P contents; these may be finely disseminated particles of meteoritic metal. Further details are contained in numerous studies characterizing the nature of the native Fe in lunar soils from various missions (e.g. Axon and Goldstein 1973, Goldstein and Axon 1973, Goldstein and Blau 1973, Hewins and Goldstein 1974, 1975; Ivanov et al. 1973, Mehta and Goldstein 1979, Misra and Taylor 1975a,b).

What is the origin of this abundant single-domain metallic iron? As originally discussed by Tsay et al. (1971), and later by others (e.g. Housley et al. 1973), the soil particles at the surface are exposed to the solar wind, and they are effectively saturated with solar-wind implanted protons (hydrogen nuclei) and carbon atoms in the outer 0.1 μm of their surface. When the soil is melted by a small impact, these elements produce an extremely reducing environment, which causes reduction of the Fe^{2+} in the agglutinate melt to Fe^0. This Fe metal then precipitates as myriad submicroscopic Fe^0 spheres disseminated throughout the quenched melt, i.e. in the agglutinate glass. This autoreduction process is responsible for producing the additional Fe^0 that occurs in agglutinate particles. Unfortunately, although this additional Fe metal is abundant in the soil, it is extremely fine-grained and may be too fine for easy concentration and beneficiation to produce iron metal as a resource.

Phosphate minerals

Lunar rocks and soils generally contain about 0.5 wt % P_2O_5, most of which is contained in the phosphate minerals whitlockite, $Ca_{16}(REE)_2(Mg,Fe)_2(PO_4)_{14}$, and apatite, ideally $Ca_5(PO_4)_3(OH,F,Cl)$ (Friel and Goldstein 1977), which occur in very minor amounts in most lunar rocks. Monazite (REE PO_4, enriched in LREE and Th) has been identified by Jolliff (1993). Whitlockite and apatite generally form crystals with hexagonal cross-sections. Apatite crystals with two well-shaped pyramidal ends (doubly-terminated), thought to be vapor-deposited, have been found in gas-formed cavities within lunar rocks.

As is the case for most minor lunar minerals, the phosphates occur as late-stage crystallization products in mare basalts. However, they are more abundant, and crystallize earlier, in certain KREEPy highland samples. Phosphates are also commonly found in association with metal particles (Friel and Goldstein 1977); in such occurrences they have probably formed by the oxidation of phosphorus out of the metal (Friel and Goldstein 1976, 1977).

Of the two major lunar phosphates, whitlockite is more abundant and has higher rare earth element (REE) contents. As Table A5.18 shows, lunar whitlockite can also contain significant amounts of FeO, MgO, Na_2O, and Y_2O_3, as well as REEs (especially the light REEs, La_2O_3, Ce_2O_3, Nd_2O_3). It is also U-enriched; crystals from the Apollo 12 basalts contain ~100 ppm U (Lovering and Wark 1971). These compositions contrast sharply with terrestrial and meteoritic whitlockites, which contain only trace amounts of rare earth elements (Albee and Chodos 1970).

As shown by the formula given above, apatite contains one OH, F, or Cl position per formula unit, and, in terrestrial apatites, the OH content may be quite high relative to F and Cl. However, the general lack of water in lunar samples causes lunar apatites to have low OH contents, and analyses indicate that they are Cl- and F-rich. Some nearly pure fluorapatites have been found on the Moon (e.g. Fuchs 1970, Shervais et al. 1984b), containing over 3 wt % F out of a possible maximum of 3.8 wt %. The ionic exchanges involved in producing these compositions are largely Si^{4+} for P^{5+}, balanced by REE^{3+} and Y^{3+} for Ca^{2+} (Fuchs 1970).

At least some of the difference between apatite and whitlockite can be attributed to different crystal/liquid partition coefficients. Experimental work by Dickinson and Hess (1983) shows Ds of about 9.5 for the distribution of light REE (Ce to Sm) in the system whitlockite/melt, compared to Ds of about 2.5 to 5.0 for apatite/melt. The apatite partition coefficients can approach and even exceed those of whitlockite in Si- and Na-rich liquids (Watson and Green 1981, Dickinson and Hess 1983), but sufficiently Si-and Na-rich liquids are rare on the Moon.

A comprehensive study of the chemistry of whitlockite and apatite was reported by Jolliff et al. (1993), and all analyses reported in table A5.18 were taken from this work. The analyses were carefully obtained by both EMP and SIMS techniques. The major findings of the Jolliff et al. (1993) study are the following. Total concentrations of REE oxides in whitlockites range from 9 to 13 wt %, and those in apatites range from 0.15 to 1 wt %. Ratios of REE concentrations in whitlockite to those in coexisting apatite range from ~10 to 60. The distribution of Mg and Fe between apatite and whitlockite is correlated with that of coexisting mafic silicates: Magnesium is strongly preferred by whitlockite, and Fe is preferred by apatite. Incorporation of REEs in whitlockite is dominated by the coupled substitution of $2REE^{3+}$ in Ca(B) sites + vacancy in Ca (IIA) for $2Ca^{2+}$ in Ca(B) sites and (Ca^{2+}, Na^{+}) in Ca (IIA). Other substitutions account for only a small portion of the REEs in whitlockite over the observed concentration range; thus, REE concentrations become partially saturated as the primary substitution approaches its stoichiometric limit of two REEs per fifty-six oxygens, leading to reduced whitlockite/melt partition coefficients, e.g. decreasing from 25 to 10 for Nd. The REE concentrations of lunar residual melts are not strongly depleted by whitlockite crystallization in assemblages consisting mainly of other minerals in typical proportions. Partition coefficients for the REEs in lunar apatite appear to be low and variable, e.g. ~0.2 to 0.8 for Nd.

INTRODUCTION TO MARE BASALTS

Mare basalts are exposed over 17% of the lunar surface area and are thought to make up less than 1% of the lunar crustal volume (Head 1976, Head and Wilson 1992) (Fig. 1). They primarily fill multi-ringed basins and irregular depressions on the earth-facing hemisphere of the Moon. They are especially profuse in Nectarian and post-Nectarian age basins such as Nectaris, Imbrium, Crisium, Humorum, and Serenitatis. Thinner deposits occur in older basins such as Tranquillititatis, Fecunditatis, and Procellarum. On the lunar far side, limited patches of mare basalts occur in younger craters and basins such as Planck, Apollo, Moscoviense, Tsiolkovsky. This preponderance of mare basalts on the Earth-facing hemisphere of the Moon has been attributed to the thicker highland crust on the far side of the Moon and the gravitational attraction of the Earth (Kaula et al. 1972; 1974, Toksoz et al. 1974; Solomon 1975). The thickness of these flood basalts ranges from 0.5 km to 1.3 km in irregular basins and up to 4.5 km in the central portions of younger basins (Bratt et al. 1985). The general style of volcanic activity was the eruption of large volumes of magma from relatively deep sources (not shallow crustal reservoirs) with very high effusion rates (Head and Wilson 1992; 1997). Evidence of lunar fire-fountaining is preserved in the form of spherical glass beads (Reid et al. 1973, Meyer et al. 1975, Delano 1979, 1980, 1986, Shearer and Papike 1993) and dark mantling deposits (Wilhelms and McCauley 1971, Head 1974, 1976; Head and Wilson 1992). Kirk and Stevenson (1989) have estimated the ratio of melt generated in the lunar mantle to melt erupted to range from 30:1 to 100:1 over the period of extensive lunar basalt volcanism.

In addition to the flood basalts associated with basin formation, there is abundant evidence of basaltic volcanism that pre-dated formation of the younger basins (i.e. pre-Nectarian). Although the petrologic record has been obscured by the early catastrophic impact history of the Moon, this record is retained in highland soils and breccias as clasts (Dickinson et al. 1985, Shervais et al. 1984b, 1985a,b, Neal et al. 1988, 1989a,b) or identified through remote sensing (Schultz and Spudis 1979, Bell and Hawke 1984, Hawke et al. 1990, Metzger and Parker 1979, Davis and Spudis 1985, 1987, Head and Wilson 1992, 1997, Antonenko et al. 1997). At least three types of ancient basaltic volcanism have thus far been identified: "KREEP" basalts, high-potassium basalts, and high-alumina basalts. Unlike younger episodes of basaltic magmatism (mare basaltic magmatism), the pre-mare KREEP basalts and high-potassium basalts have relatively high abundances of Al_2O_3 and incompatible trace elements. On the other hand, the pre-mare, high-alumina basalts are more mare-basalt like in composition and texture (Shervais et al. 1984b, 1985a,b; Dickinson et al. 1985). Numerous lines of evidence such as igneous textures, low siderophile trace-element abundances, and non-meteoritic siderophile-element ratios indicate that clasts representing these pre-mare basalt types are volcanic and not impact-derived (James 1980, Warren et al. 1997). The distribution and abundance of these lithologies are difficult to quantify. However, numerous studies imply that large volumes of these basaltic magmas ("cryptomaria") were erupted (Metzger and Parker 1979, Davis and Spudis 1985, 1987; Head and Wilson 1992, Antonenko et al. 1997). Head and Wilson (1992) have suggested that perhaps up to a third of the erupted basalts at the lunar surface was "cryptomaria".

PETROLOGY OF THE CRYSTALLINE MARE BASALTS

Classification

Two problems have complicated the classification of lunar basalts. First, the initial studies and classification of many of the mare basalts were conducted immediately following a sample return mission and prior to subsequent missions. This resulted in site-specific classification. For example, high-Ti, low-K basalts from the Apollo 11 and 17 landing sites have been referred to as Apollo 11 low-K basalts (Tera et al. 1970), Apollo 17 low-K basalts basalts (Tera et al. 1970), ophitic ilmenite basalts (Chao et al. 1970b, James and Jackson 1970), very high-Ti basalts (LSPET 1973, Brown et al. 1974, 1975,; Papike et al. 1974, 1976), olivine-porphyritic ilmenite basalt (Papike et al. 1974), mare basalt 1 (Brown et al. 1974), mare basalt 1A (Brown et al. 1975), poikilitic ilmenite basalt (RD Warner et al. 1975), microporphyritic ilmenite basalt (RD Warner et al. 1975), mare basalt 2 (Brown et al. 1974, 1975), and low-Mg ilmenite basalt (RD Warner et al. 1975). Certainly, there are many more names for similar rock types in the literature. As enumerated by Neal and Taylor (1992), even once some of these redundancies were corrected, this still resulted in the establishment of at least twenty-one different crystalline mare basalt types. Many of these different basalt types are similar in mineralogy and whole rock chemistry (Schuraytz and Ryder 1991). A second problem that complicates the classification of the mare basalts is that a wide range of modal and chemical compositions exists for single samples. This is partially a result of the study of small, non-representative samples of the different basalt lithologies. Numerous studies have demonstrated the problems in obtaining representative lunar samples (Rhodes et al. 1976, Papike et al. 1976, Haskin et al. 1977, Lindstrom and Haskin 1978, 1981; Schuraytz and Ryder 1988, 1991). These studies concluded that the mineralogical and chemical variations found within individual basaltic lithologies were a result of both magmatic differentiation and unrepresentative sampling. To alleviate some aspects of this sampling problem, Rhodes and Hubbard (1973) and Papike et al. (1976) averaged replicate superior analyses. It is apparent from these previous studies that sampling compromises to at least some extent any classification of mare basalt lithologies and interpretation of compositional variability within individual lithologies. It is not the intent of this summary paper to either drown the reader in extensive lunar terminology or baptize terrestrial and lunar petrologists with new classification schemes.

For the available data set of mare basalt bulk compositions, it is apparent that the most useful lithologic discriminant is TiO_2. The variation in TiO_2 is most clearly shown when TiO_2 is plotted against an index of differentiation such as mg#. Neal and Taylor (1992) illustrated the variability observed in mare basalts by plotting all available (>500) mare basalt analyses (see their Fig. 2, plot of TiO_2 versus mg#). Papike et al. (1976) and Papike and Vaniman (1978) defined three compositional groups based on TiO_2: very low-Ti (VLT) basalts (<1 wt % TiO_2), low-Ti basalts (1 to 5 wt % TiO_2), and high Ti basalts (9 to 14 wt % TiO_2). Intermediate-Ti basalts are rare in the lunar sample collection, but have been identified in regolith breccias from the Apollo 14 sample site (Ridley 1975, Papike et al. 1976, Neal and Taylor 1992). Neal and Taylor (1992) suggested the elimination of this intermediate-Ti field by having the low-Ti basalts extending from a TiO_2 of 1 to 6 wt % and the high-Ti basalts extending from 6 to 14 wt % TiO_2.

Our discussion of crystalline mare basalts will be organized into a description of different petrographic lithologies within the different TiO_2 chemical groups:

- *high-Ti basalts*: high-Ti, low-K basalts (Apollo 11 and 17) and high-Ti, high-K basalts (Apollo 11).
- *low-Ti basalts*: olivine basalts (Apollo 12 and 15), pigeonite basalts (Apollo 12 and 15), ilmenite basalts (Apollo 12), feldspathic basalts (?) (Apollo 12) high alumina basalts (Apollo 14, Luna 16), and very high potassium basalts (Apollo 14).
- *very low-Ti (VLT) basalts*: VLT (Luna 24 and Apollo 17).

The various mare basalt lithologies and picritic glass types are indexed to collection sites in Table 5.

High-Ti basalts

High-Ti, low-K basalts (Apollo 11 and 17 sites). Although there are textural, mineralogical and chemical differences among these basalts (Papike et al. 1976, Rhodes et al. 1976, BVSP 1981) that may be related by conditions of crystallization, shallow fractional crystallization processes (during transport to the lunar surface or on the surface) or differences in mantle sources, they may be generally grouped into low-potassium and high-potassium lithologies (Tera et al. 1970, Chao et al. 1970b, James and Jackson 1970, Papike et al. 1976). The Apollo 17 high-Ti, low-K basalts have been subdivided into two distinct types: those similar to the Apollo 11 low-K basalts, and those with higher abundances of olivine and Fe-Ti oxides (ilmenite and armalcolite) (Papike et al. 1976). The latter was referred to as very high-Ti basalt. Neal and Taylor (1992) classified all of these high-Ti basalts as either a high-Ti/low-Al/low-K group or a high-Ti/low-Al/high-K group. These groups have also been subdivided based on mineralogical and chemical characteristics into at least 11 petrologic-petrogenetic lithologies by various studies (Apollo 11 types A, B1, B2, B3, D and Apollo 17 types A, B1, B2, C, D, U).

The Apollo 11 and 17 high-titanium, low-K basalts are fine- to medium-grained and vesicular. Most exhibit subophitic to ophitic textures (Figs. 12 and 13). Porphyritic to microporhyritic textures have been documented in the Apollo 17 high-Ti lithologies (Fig. 14). The average grain sizes range between 0.25 to 1.5 mm, with the largest grain size ~3 mm. The ophitic textures are defined by the spatial relationship between plagioclase and pyroxene. These textures range from the subhedral plagioclase (An_{94-77}) being partially enclosed by pyroxene (subophitic) to being completely enclosed by pyroxene (ophitic). The textures change from subophitic to ophitic with increasing grain size (Papike et al. 1976). Pyroxenes are strongly zoned, with some having discontinuous rims of pyroxferroite (Chao et al. 1970a). The pyroxferroite rims on the pyroxene have decomposed to a fine-grained intergrowth of hedenbergite + silica + fayalite (Ware and Lovering 1970, Lindsley et al. 1972). In the slower-cooled, coarser-grained lithologies, pyroxenes tend to be euhedral and exhibit well-developed sector zoning. Large, early-formed crystals of olivine (Fo_{80-60}) occur in the finer-grained lithologies, whereas olivine is sparse to absent in the

Table 5. Basalt lithoilogies at sample collection sites.

Site	Lithologies

Apollo 11	High-Ti Basalts
	  High-K
	  Low-K
	Very Low-Ti Picritic Glass
	  Green
	High-Ti Picritic Glass
	  Orange

Apollo 12	Pigeonite Basalts Olivine Basalts Ilmenite Basalts High-Ti Picritic Glass Red

Apollo 14	Aluminous Basalts
	Very Low-Ti Picritic Glass
	  Green A, Green B, VLT
	Intermediate-Ti Picritic Glass
	  Yellow
	High-Ti Picritic Glass
	  Orange, Black

Apollo 15	Olivine Basalts
	Pigeonite Basalts
	KREEP Basalts
	Feldspathic Basalts*
	Very Low-Ti Picritic Glass
	  Green A, Green B, Green C, Green D, Green E
	Intermediate-Ti Picritic Glass
	  Yellow
	High-Ti Picritic Glasses
	  Orange, Red

Apollo 16	Very Low-Ti Picritic Glass Green

Apollo 17	Very Low-Ti Basalt
	High-Ti Basalts
	Very Low-Ti Picritic Glass
	  VLT
	High-Ti Picritic Glasses
	  74220 Orange, Orange I, Orange II

Luna 16	Aluminous Basalts
Luna 24	Very Low-Ti Basalts
Lunar Meteorites	These lunar meteorites are commonly breccias with a significant mare basalt component such as basalt clasts or picritic glasses. Elephant Moraine 87521, Yamato 793169, Asuka 881757, Asuka 31, Calcalong Creek, Yamato 793274 QUE 94281.

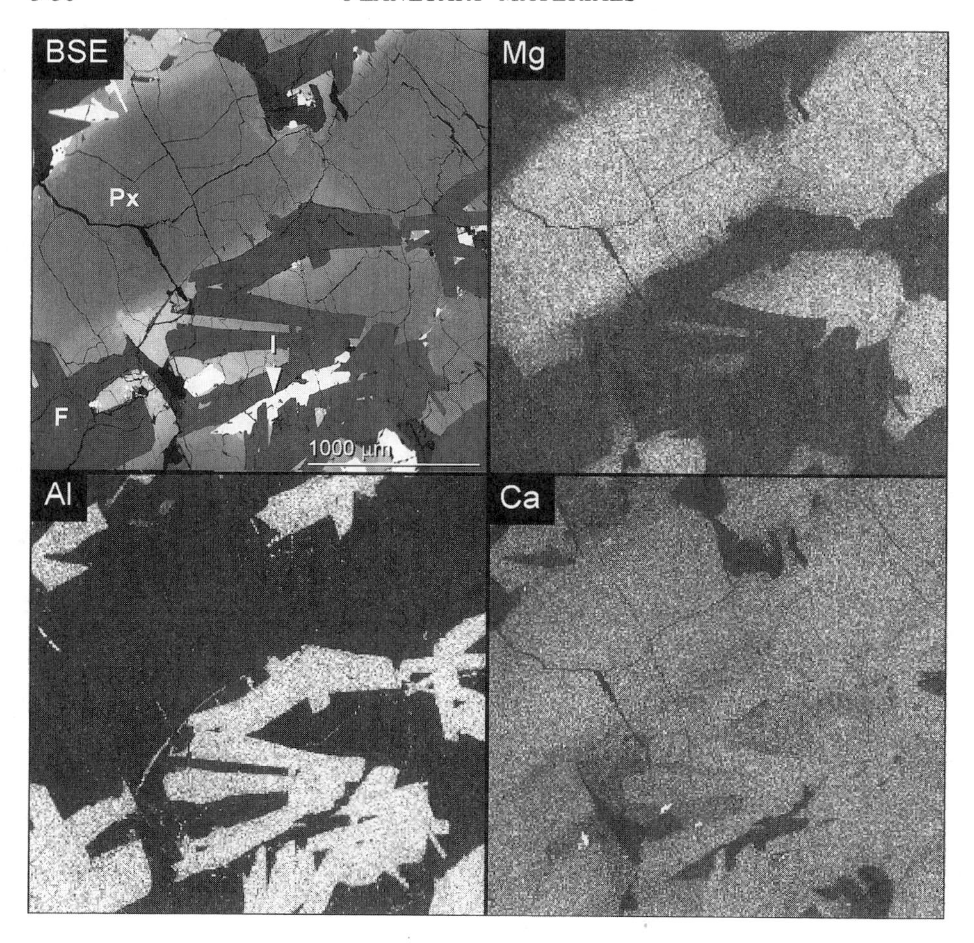

Figure 12. Backscatter electron image (BSE) and chemical distribution maps (Mg, Al, Ca) of subophitic texture exhibited by a high-Ti, low K mare basalt (10044). As with all the BSE images presented, the brighter the mineral phase, the higher the abundance of Fe. Compositional maps illustrate that the pyroxenes exhibit a decrease in Mg (and perhaps Ca) and an increase in Fe at rims. Plagioclase laths are well illustrated in the Al compositional map. Abbreviations: I = ilmenite, F = plagioclase, Px = pyroxene.

coarser grained basalts. The olivine is often mantled by Ti-rich pyroxene (Papike et al. 1976). Ilmenite forms elongate, bladed grains in the subophitic samples, grading into equant, anhedral grains in the ophitic samples. Armalcolite mantled by ilmenite occurs only in the Apollo 17 fine-grained, olivine-rich lithologies. Cr-Ti spinel occurs as inclusions within olivine grains and is mantled by ilmenite within the mesostasis. The mesostasis contains a variety of late-stage minerals including cristobalite, tridymite, ulvöspinel, apatite, whitlockite, pyroxferroite, fayalite, metallic iron, troilite, and tranquillityite (Lovering et al. 1971). The metallic iron is thought to have been produced through S loss during subsolidus reduction (Brett 1975b). The globules of troilite probably represent immiscible sulfide melt (Skinner 1970). Roedder and Weiblen (1970) and Roedder (1984) have also documented silicate melt immiscibility in these interstitial areas. This takes the form of coexisting potassium-, silica-rich glass and a feathery, iron-rich basaltic "glass". The

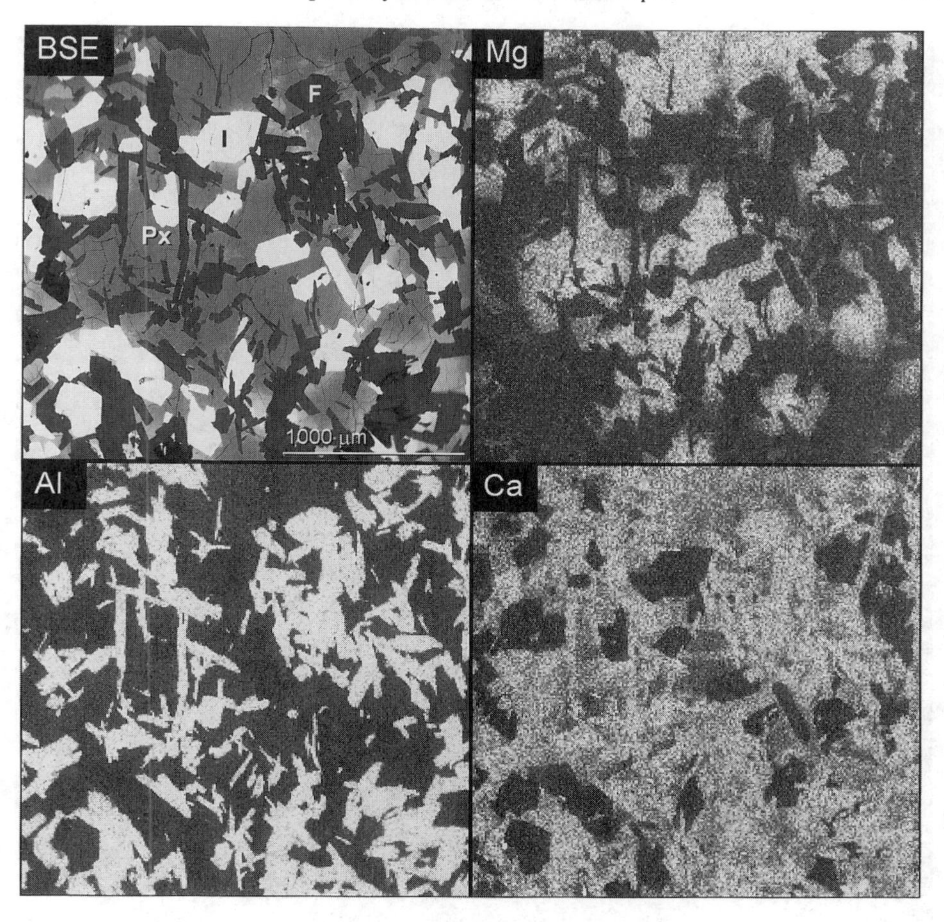

Figure 13. BSE and chemical distribution maps (Mg, Al, Ca) of ophitic texture exhibited by a high-Ti, low K mare basalt (10003,73). As well illustrated in both the BSE and Al map, feldspar and ilmenite are fully enclosed by pyroxene. Abbreviations: I = ilmenite, F = plagioclase, Px = pyroxene.

apparent crystallization sequence of these high-Ti magmas is olivine ⇒ spinel ⇒ ilmenite ⇒ pyroxene + plagioclase (Papike et al. 1976, Longhi 1987, 1992).

Modal data for these high-Ti, low-K basalts are presented in Table A5.19. The average modal mineralogy for major mineral phases in the lithologies from the Apollo 11 site is 49% clinopyroxene, 2.2% olivine, 31% plagioclase, and 14% ilmenite. These lithologies at the Apollo 17 site have average modes of major phases of 48% clinopyroxene, 31% plagioclase, and 14% ilmenite (Papike et al. 1976). Textural, modal, and chemical data are consistent with the Apollo 11 high-Ti, low-K basalts being part of one sill or lava flow that underwent near surface fractionation of olivine and ilmenite (Papike et al. 1976).

Some of the more Ti-rich and olivine-bearing lithologies collected from the Apollo 17 site are mineralogically and texturally distinct from the other high-Ti, low-K basalts. They are referred to as very high-Ti basalts by Papike et al. (1976) and BVSP (1981).

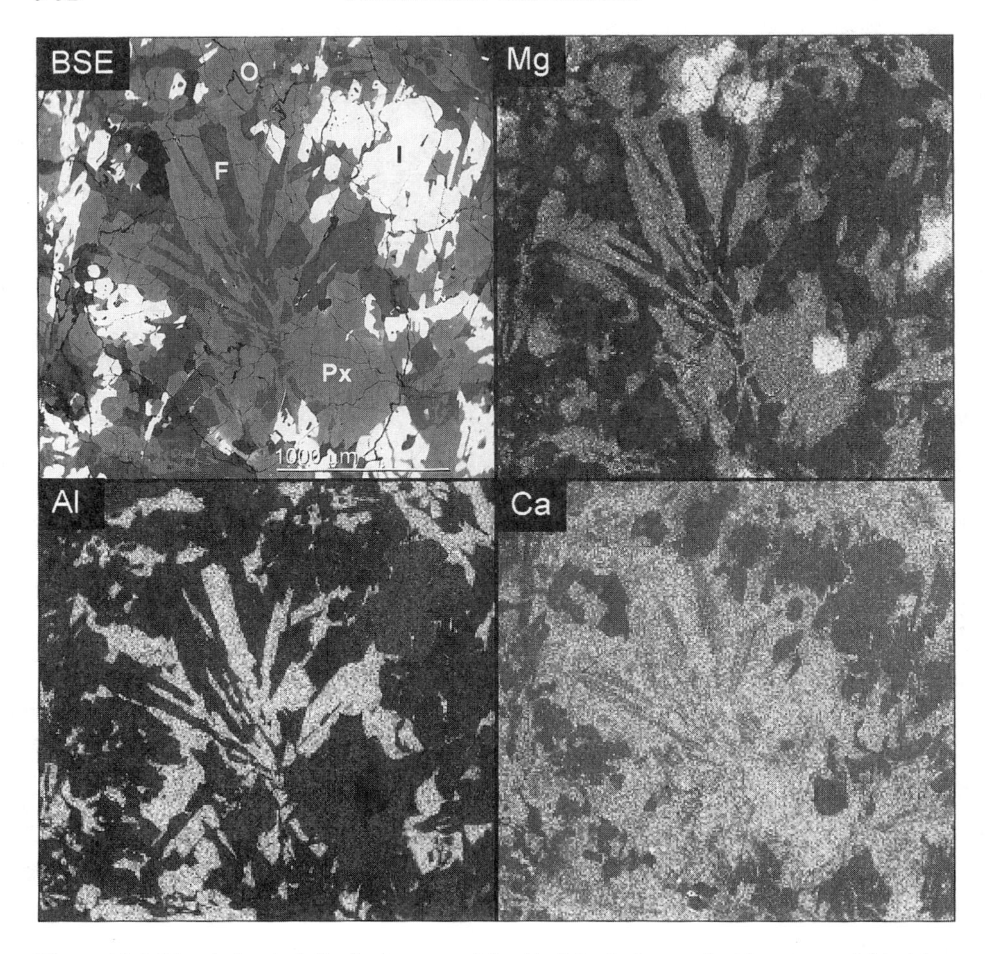

Figure 14. BSE and chemical distribution maps (Mg, Al, Ca) of microporphyritic texture exhibited by a high-Ti, low-K mare basalt (71035,28). Abbreviations: I = ilmenite, O = olivine, F = plagioclase, Px = pyroxene.

Descriptions of these basalts were given by Kridelbaugh and Weill (1973), Ridley and Brett (1973), Weigand (1973), Hodges and Kushiro (1974), Longhi et al. (1974), Meyer and Boctor (1974), Papike et al. (1974, 1976), and Dymek et al. (1975a). These lithologies vary from vitrophyres to microgabbros.In the vitrophyres, microphenocrysts of olivine, armalcolite, and spinel are set in a fine-grained, spherulitic, quenched intergrowth of clinopyroxene, plagioclase, ilmenite, tridymite, glass, troilite, and metallic iron. The olivine is commonly skeletal and mantled by titanaugite. In coarser-grained samples, the titanaugite occurs as discrete grains that contain inclusions of olivine. The titanaugite may exhibit well-developed sector zoning and is commonly mantled by pigeonite (Papike et al. 1976). The armalcolite is mantled by ilmenite in the finer-grained samples and occurs as subhedral inclusions within pyroxene and rarely within plagioclase. In addition to mantling the armalcolite, the ilmenite occurs as blocky, subhedral grains with residuum matrix embayments and inclusions in the coarser grained lithologies. The residuum matrix of the microgabbro lithologies is similar to those of other high-Ti basalts except that pyroxferroite or its breakdown products have not been identified in the microgabbro. The crystallization sequences in these lithologies are more complex than in the subophitic-ophitic lithologies.

Papike et al. (1976) documented a complex crystallization history of these basalts that was significantly affected by cooling rate, mg#, and oxygen fugacity. In coarser-grained lithologies, the apparent crystallization sequence was olivine + spinel ⇒ armalcolite ⇒ pyroxene ± plagioclase ⇒ plagioclase ⇒ ilmenite. In the fine-grained samples with high mg# (e.g. 74275, mg# = 0.50), plagioclase is the last major phase to crystallize. The suppression of plagioclase crystallization is typical of many quickly cooled lunar basalts. The interval over which Fe-Ti oxides occur in the crystallization sequence is dependent upon the FeO and TiO_2 content of the melt (Papike et al. 1976). In more iron-rich, fine-grained samples (e.g. 70215, mg# = 0.43), crystallization of the Fe-Ti oxides overlapped with part of the olivine crystallization interval.

In addition to textural differences, the modal mineralogy of this group of high-Ti, low-K basalts contains higher abundances of olivine and ilmenite than the other high-Ti, low-K lithologies (Table A5.19). The average modal abundances of major mineral phases in this Apollo 17 lithology are 47% clinopyroxene, 4.5% olivine, 23% plagioclase, and 24% ilmenite. Based on the modal mineralogy, textures, major element characteristics, and experimental studies, Kesson (1975) and Walker et al. (1975) concluded that high-Ti magmas similar to these were parental to the other high-Ti, low-K basalts at the Apollo 11 and 17 sites. Low-pressure fractional crystallization of armalcolite and olivine may link the lithologies.

High-Ti, high-K basalts (Apollo 11 site). This lithology contrasts with the other high-Ti basalts both in texture and chemistry. These "high-potassium" basalts (K_2O = 0.24 to 0.40 wt % compared to < 0.12 wt % for the other high-Ti basalts) tend to be finer-grained than the other high-Ti basalts with granular to intersertal textures. Several initial studies of these basalts classified them as intersertal ilmenite basalts (Chao et al. 1970b, James and Jackson 1970). The average grain sizes of these basalts range from 0.2 to 0.7 mm. In coarser grained samples pyroxene grains are up to 2.5 mm in length. The textures in this basalt type vary from anhedral laths of plagioclase (An_{82-75}) that poikilitically enclose subhedral grains of pyroxene and ilmenite in the coarser-grained samples (e.g. 10072, 10017), shown in Figure 15, to irregular fan shaped spherulitic intergrowths of plagioclase (An_{81-73}) and pyroxene in fine-grained samples (e.g. 10049). In the coarser-grained samples, pyroxene is zoned from colorless pigeonite cores to pink rims of Ti- and Fe-rich augite. Armalcolite is mantled by laths and blocky grains of ilmenite. Olivine occurs as anhedral to subhedral grains included within pyroxene or mantled by Ti-rich pyroxene. Cr and Cr-Ti spinel has been documented to occur as inclusions in the olivine. The mineralogy of the interstitial matrix is similar to those of the other high-Ti basalts and consists of plagioclase, pyroxene, residual glasses (immiscible K-rich glass and Fe-rich glass), metallic iron, troilite, cristobalite, tridymite, apatite, and whitlockite. The apparent crystallization sequence derived from textures is olivine ⇒ armalcolite ⇒ pyroxene (reaction between olivine and melt) ⇒ ilmenite (reaction between armalcolite and melt) ⇒ plagioclase (Papike et al. 1976).

In addition to textural differences, the modal mineralogy of this group of high-Ti, high-K basalts contains lower abundances of plagioclase than found in the high-Ti, low-K lithologies (Table A5.19). The average modal abundances of major mineral phases in this Apollo 11 lithology are 53% clinopyroxene, 0.1% olivine, 20% plagioclase, and 19% ilmenite. Variation of the modal mineralogy is not as great as that observed in the high-Ti, low-K basalts. Textural, modal, and chemical differences suggest that this lithology is not related to the high-Ti, low-K basalts.

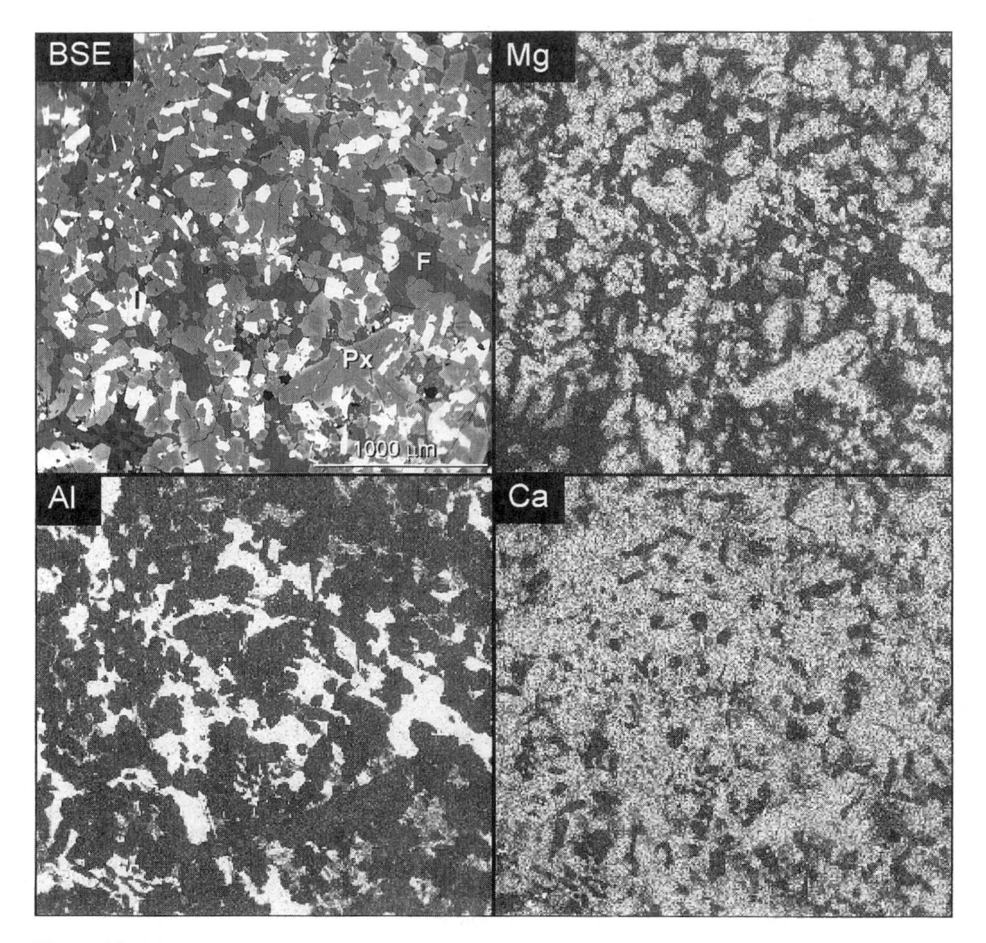

Figure 15. BSE and chemical distribution maps (Mg, Al, Ca) of anti-ophitic texture exhibited by high-Ti, high-K mare basalt (10017,20). Abbreviations: I = ilmenite, F = plagioclase, Px = pyroxene.

Low-Ti basalts

A number of low-Ti basalt lithologies were returned by the Apollo missions. These include the olivine basalts and pigeonite basalts collected by the Apollo 12 and 15 missions, ilmenite basalts from the Apollo 12 mission, high-alumina basalts from the Apollo 12 (?), Apollo 14, and Luna 16 missions, and the very-high-potassium basalts discovered as clasts in the Apollo 14 regolith.

Pigeonite basalt. These basalts exhibit an extraordinarily wide range of textures from vitrophyric to gabbroic. These highly porphyritic textures are typified by phenocrysts of highly zoned pyroxene. The average phenocryst size ranges from 1 mm in the vitrophric textured lithologies (12009, 15597) to several centimeters in the coarser gained samples (e.g. 12021, 15058). These textural variations reflect differences in cooling rate. The grain size of the mesostasis ranges from less than 0.5 mm to 6 mm. Textural variability is illustrated in Figures 16, 17, and 18. The sequence in these three figures from 15499 ⇒ 12052 ⇒ 15058 reflects decreasing cooling rate.

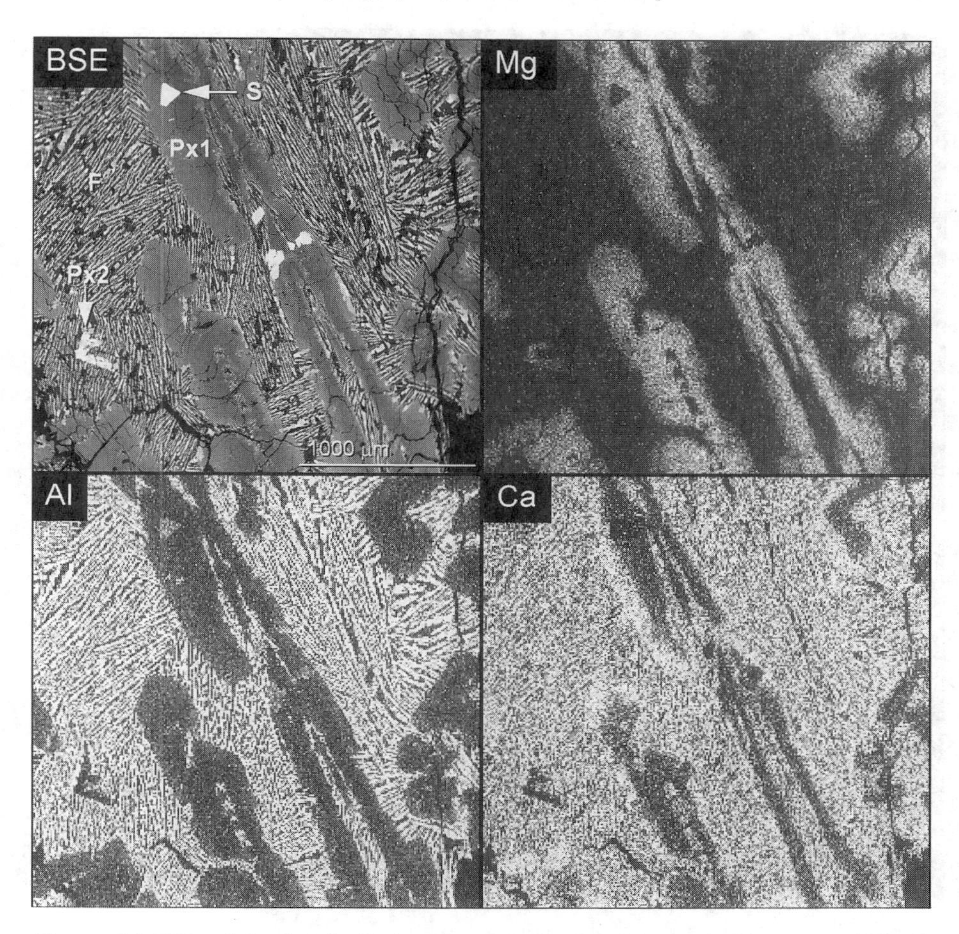

Figure 16. BSE and chemical distribution maps (Mg, Al, Ca) of the porphyritic texture exhibited by a rapidly cooled pigeonite basalt (15499). Large skeletal pyroxenes are zoned from high-Mg pigeonite to Fe-, Ca-rich clinopyroxene rims (see Mg and Ca maps). The Fe-, Ca-rich clinopyroxene occurs as rims on both the outside and inner hollows of the large pyroxene (Px1) and as small, individual grains in the surrounding matrix (Px2). Abbreviations: S = sulfide, F = plagioclase, Px1 and Px2 = pyroxene.

Pyroxene occurs as subhedral laths with cores of pigeonite that are rimmed by augite that is zoned to pyroxferroite. In samples that represent rapidly cooled lavas, the pyroxenes exhibit a "soda straw" morphology in which pyroxenes have hollow cores and pigeonite "channelways" parallel to c (Bence et al. 1971). In these textures, the early pigeonite is mantled with augite both outward to the crystal rim and inward toward the hollow cores (Figs. 16, 17). The compositional variability of the pyroxenes is represented by the analyses in Table A5.3. Opaque inclusions in the zoned pyroxene range from spinel (with ulvöspinel rims) associated with pigeonite cores to ilmenite associated with the augite rims. Plagioclase (An_{96-87}) occurs primarily as anhedral to subhedral laths intergrown in the mesostasis with pyroxene and ilmenite (Fig. 16). In the coarser grained samples it forms larger laths that are parallel to the long dimension of the pyroxene phenocrysts (Fig. 17). In some cases, the plagioclase radiates perpendicular to and penetrates to varying depths the outer augitic zones of the pyroxene phenocrysts. Anhedral to subhedral olivine (Fo_{75-30}) is

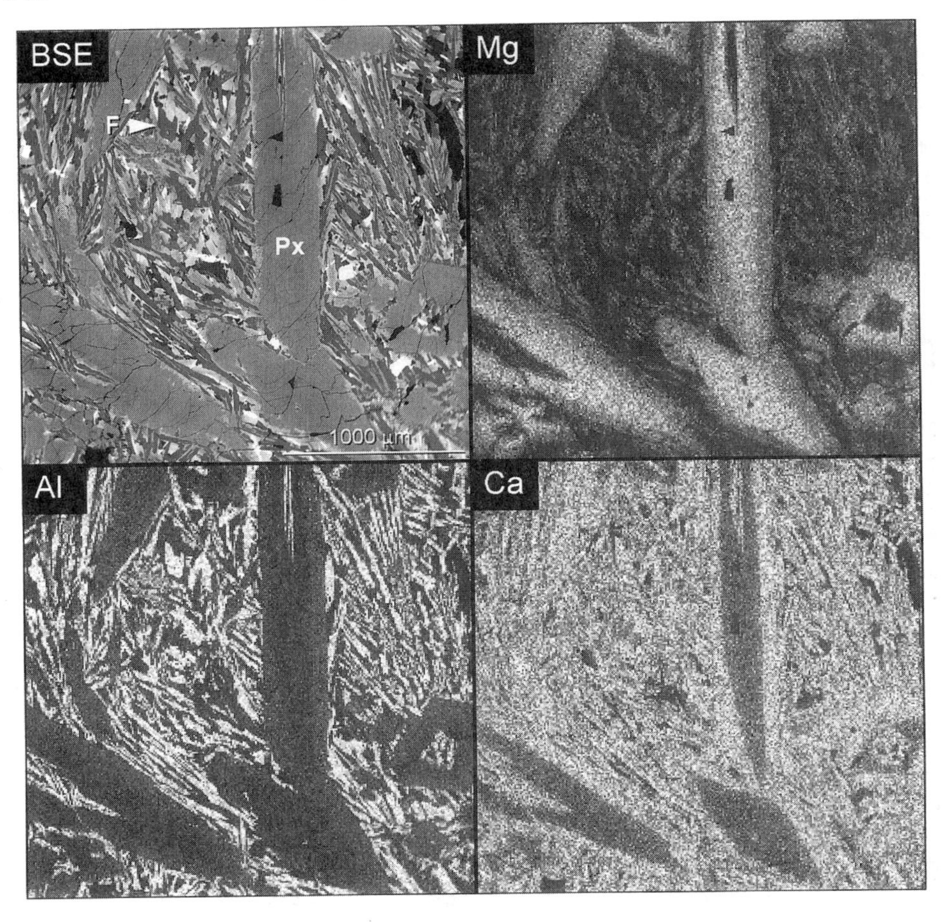

Figure 17. BSE and chemical distribution maps (Mg, Al, Ca) of the porphyritic texture exhibited by a pigeonite basalt (12052) that probably cooled at a rate slightly slower than 15499 (Fig. 16). The soda-straw pyroxene texture viewed do the c axis is evident. As in Figure 16, the large skeletal pyroxenes are zoned from high-Mg pigeonite to Fe-, Ca-rich clinopyroxene rims (see Mg and Ca maps). Plagioclase is significantly coarser than in sample 15499. Minor ilmenite and troilite in the fine-grained matrix are represented by bright grains in the BSE image. Abbreviations: F = plagioclase, Px = pyroxene.

in low abundance (< 4 vol %) to absent in these basalts. Olivine is more common in the Apollo 12 pigeonite basalts, in which it is embayed by pigeonite. Chrome spinel with thin rims of ulvöspinel occurs as minute inclusions in the olivine. Olivine is rare in the coarser-grained members of this rock type. Olivine and melt react to produce pyroxene and the relationship between olivine abundance and grain size is attributed to the reaction kinetics (GJ Taylor et al. 1991 *Lunar Sourcebook*). Ilmenite occurs as anhedral laths intergrown with the mesostasis. Additional accessory phases in the mesostasis include "glasses" representing late-stage immiscible liquids (Roedder and Weiblen 1970, Roedder 1984), cristobalite, tridymite, apatite, whitlockite, troilite, metallic iron, and alkali feldspar (Papike et al. 1976, Baldridge et al. 1979).

The pigeonite basalts probably represent samples of thin units in which cooling rate,

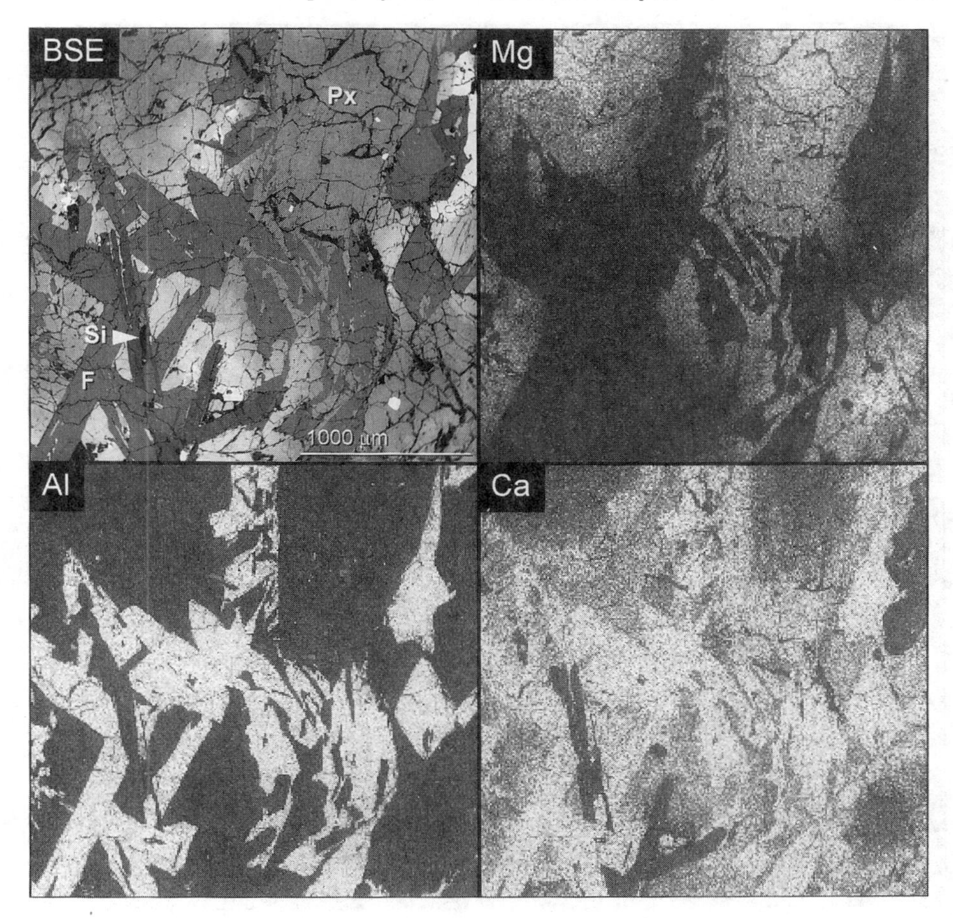

Figure 18. BSE and chemical distribution maps (Mg, Al, Ca) of the porphyritic texture exhibited by a slowly cooled pigeonite basalt (15058). Unlike 12052 and 15499, pyroxene is not skeletal in nature and the plagioclase does not form "feathery laths". The coarse pyroxenes are zoned from high-Mg pigeonite to Fe-, Ca-rich clinopyroxene rims (see Mg and Ca maps). Abbreviations: Si = tridymite, F = plagioclase, Px = pyroxene.

rather than near-surface fractionation, affected textures, crystallization sequence, and modal mineralogy. The general crystallization sequence of the pigeonite basalts is chrome spinel (ulvöspinel ⇒ olivine ⇒ pigeonite ⇒ augite ⇒ plagioclase ⇒ ilmenite. In some of the Apollo 15 pigeonite basalts, olivine may have crystallized prior to the spinel. Bence and Papike (1972) and Papike et al. (1976) interpreted deviations from this crystallization sequence and textural relations between plagioclase and pyroxene as reflecting retardation of plagioclase nucleation and crystallization due to different cooling rates.

The ranges of modes among the pigeonite basalts are 46 to 71% clinopyroxene, 0 to 4% olivine, 17 to 48% plagioclase, and 3 to 12% opaque mineral (ilmenite, spinel, metal, sulfides) (Table A5.20). The Apollo 12 pigeonite basalts show a limited relationship between grain size, modal mineralogy, and major element chemistry. This has been attributed to sampling. On the other hand, the Apollo 15 pigeonite basalts exhibit a decrease

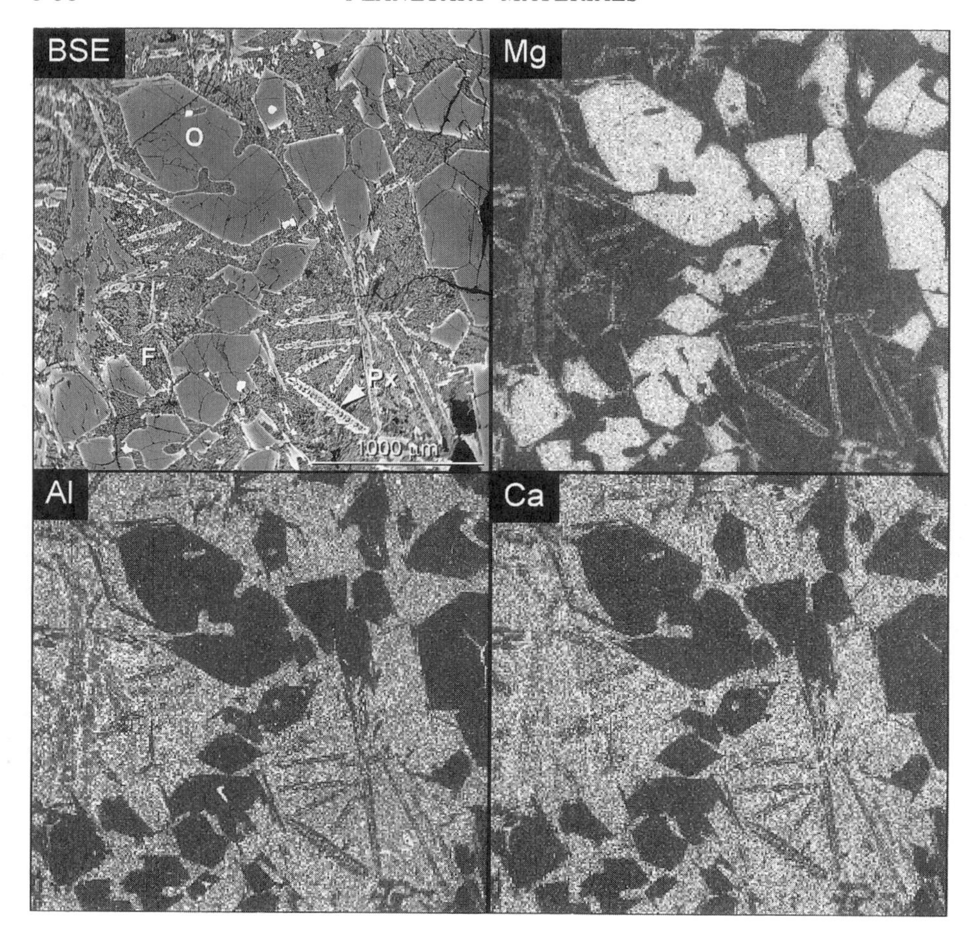

Figure 19. BSE and chemical distribution maps (Mg, Al, Ca) of the porphyritic texture exhibited by a rapidly cooled olivine basalt (12009). Some of the large euhedral olivine grains have thin pyroxene rims. Small skeletal pyroxenes are immersed in a fine grained matrix dominated by plagioclase (see Al and Ca maps). Abbreviations: O = olivine, Px = pyroxene.

in pyroxene and an increase in plagioclase with increasing grain size. This apparent relationship reflects the coarsening of plagioclase in the mesostasis of the slower-cooled samples (Bence and Papike 1972, Papike et al. 1976).

Olivine basalt. These olivine-rich lithologies are fine- to medium-grained basalts ranging from olivine vitrophyres to microgabbros or olivine cumulates (Figs. 19, 20, 21). The Apollo 15 olivine basalts are vesicular to scoriaceous. The average grain size ranges from 0.5 mm for the olivine phenocrysts in the vitrophyres to 3 to 4 mm in the coarser samples. Olivine ($Fo_{75\text{-}30}$) exhibits many morphologies. In the quickly cooled samples olivine is skeletal (12009). The olivine is subhedral in samples with intermediate-grain sizes, whereas in the coarser-grained samples olivine is either partially mantled by augite or forms rounded inclusions within pyroxene. The olivine commonly contains inclusions of metallic iron and chrome spinel. Within the vitrophyres and finer-grained samples, pyroxene occurs only in the matrix, in a radial, skeletal morphology intergrown with

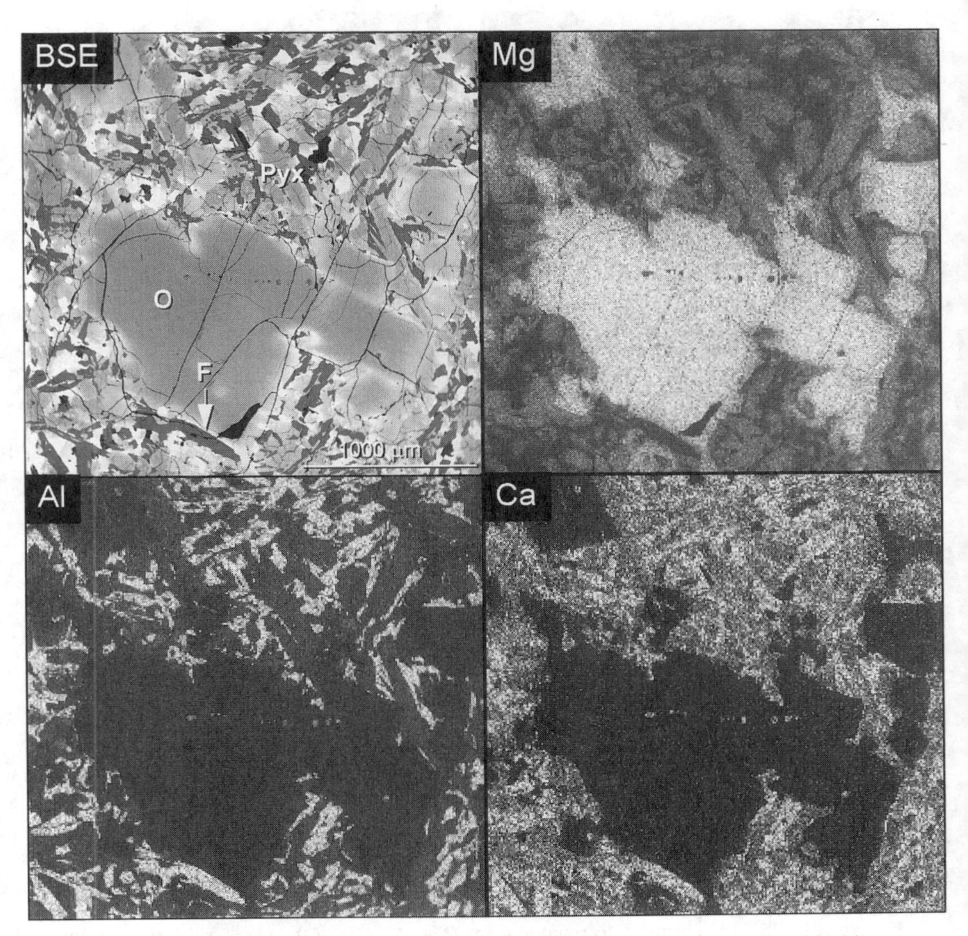

Figure 20. BSE and chemical distribution maps (Mg, Al, Ca) of the porphyritic texture exhibited by an olivine basalt (12004,55) that cooled at a rate slightly slower than 12009 (Fig. 19). Olivine is subhedral. Plagioclase and pyroxene is significantly coarser than in sample 12009. Abbreviations: O = olivine, F = plagioclase, Pyx = pyroxene.

plagioclase (An_{92-86}) (Fig. 20). With increasing grain size, pyroxene occurs as subhedral laths with pigeonite cores and augite mantles. The augite is zoned outward to pyroxferroite (Papike et al. 1976) (Figs. 21, 22). Plagioclase may partially to totally enclose pyroxene and olivine, be intersitial to pyroxene grains, or be partially enclosed by pyroxene. Opaque phases include Cr-spinel and metallic iron enclosed in olivine and Cr-Ti spinel, metallic iron, troilite and blocky ilmenite within the matrix. Other mesostasis phases include K-rich rhyolitic glass, fayalite, alkali feldspar, cristobalite, tridymite, apatite, and whitlockite (Papike et al. 1976). The general crystallization sequence for major phases in the olivine basalts was olivine + spinel ⇒ pigeonite ⇒ augite ⇒ plagioclase ⇒ ulvöspinel ⇒ ilmenite.

The average modal mineralogy (Table A5.21) for the Apollo 15 olivine basalts is 63% pyroxene, 7% olivine, 24% plagioclase and 5.5% opaque minerals. The Apollo 12 olivine basalts are much more enriched in olivine (Table A5.21) with an average modal mineralogy consisting of 53% clinopyroxene, 20% olivine, 19% plagioclase, and 7% opaque minerals

Figure 21. BSE and chemical distribution maps (Mg, Al, Ca) of the porphyritic texture exhibited by a slowly cooled olivine basalt (12035). Unlike 12004 and 12009, olivine is subhedral to anhedral; pyroxene and plagioclase are coarse-grained. Abbreviations: O = olivine, I = ilmenite, F = plagioclase, Px = pyroxene

(Papike et al. 1976). Modal mineralogy and major element chemistry have been interpreted as indicating that the Apollo 15 olivine basalts are products of limited amounts of olivine fractionation and that the samples in the collection are not cumulates (Kesson 1975). On the other hand, the suite of Apollo 12 olivine basalts may represent melt compositions to cumulates formed within a single thick lava flow. The vitrophyric samples (12002, 12009) appear to approach liquid compositions (Walker et al. 1976, Green et al. 1971a, Donaldson et al. 1975). Numerous lines of evidence indicate that the coarse-grained samples with high abundances of olivine and clinopyroxene represent cumulates in which the olivine may have been partially replaced by pyroxene. These lines of evidence include the positive correlation between plagioclase grain size and normative abundance of olivine (Walker et al. 1976), the observation that melts with olivine contents similar to the coarse-grained lithologies should have olivine with higher mg# (Green et al. 1971a, Newton et al. 1971), and the sympathetic variation between the abundance of olivine and pyroxene (assuming calculated melt densities of 12002) suggests some of the pyroxene formed from a reaction involving cumulate olivine and intercumulus melt (Walker et al. 1976).

Ilmenite basalt. The ilmenite basalts were first described in detail by Brett et al. (1971), Brown et al. (1971) and Klein et al. (1971). They are set apart from the other low-Ti basalts by their overall higher modal abundance of ilmenite, but they still qualify as low-Ti basalts with 2.7 to 5.0 wt % TiO_2. They are fine- to medium grained and exhibit subophitic to granular textures. The fine-grained samples (e.g. 12022) consist of abundant olivine (16 vol %, Fo_{80-50}) mantled by titanaugite and set in a quenched groundmass of pyroxene, ilmenite, and plagioclase (An_{94-87}). Laths of ilmenite are oriented parallel to (100) of olivine (Brett et al. 1971). In coarser-grained samples (e.g. 12051, 12064) olivine is far less common (<1.6 vol %). Pyroxene in these samples is highly zoned with pigeonite cores that are mantled by augite. Blades of cristobalite occur as individual grains and in the mesostasis. Other accessory minerals in the mesostasis include fayalite and phosphates. The general crystallization sequence for major phases in the ilmenite basalts was olivine + Cr spinel ⇒ pigeonite ⇒ augite + plagioclase + ulvöspinel + ilmenite. The average mode of the ilmenite basalt is 59% clinopyroxene, 3.5% olivine, 25% plagioclase, and 9% ilmenite. Table A5.22 illustrates the variation of the modal mineralogy among the samples.

High-alumina (or feldspathic) basalt. High alumina basalts are a collection of miscellaneous samples from Apollo (12 and 14) and Luna (16) sampling sites. They are related by their relatively high modal abundances of plagioclase (>38.5%) and high concentrations of Al_2O_3 (>11%). There is still some debate concerning the classification of some basalts in this group because they may be unrepresentative samples of basalts containing higher abundances of plagioclase (e.g. Apollo 12 feldspathic basalts) or are fractional crystallization products of melts that originally contained lower abundances of Al_2O_3 (e.g. Apollo 12 feldspathic basalts, evolved Luna 24 VLT basalts) (Nyquist et al. 1979, Neal et al. 1992, Neal and Taylor 1992). We will not include these lithologies in this description. Samples considered here are 14053 (Gancarz et al. 1971), 14072 (Longhi et al. 1972), B-1 from the Luna 16 mission (Albee et al. 1972) and clasts described by Shervais et al. (1984a,b; 1985a), Dickinson et al. (1985), Neal et al. (1989a,b) and Neal and Taylor (1992) in Apollo 14 regolith. It is also a point of debate whether the high-alumina basalts represent non-mare basaltic magmatism (Hubbard and Gast 1972) or high-Al mare basaltic magmatism (Ridley 1975). Here we consider all of these together.

Most of the high-alumina, low-Ti basalts exhibit subophitic to ophitic textures. Some of the clasts described by Shervais et al. (1985a,b) are vitrophyric. In the subophitic-ophitic samples, early olivine (Fo_{70-65}) commonly occurs as anhedral, rounded phenocrysts. Fe-rich, late-stage olivine (Fo_{20-0}) also occurs in the groundmass. This fine-grained olivine has been reduced to metallic iron + SiO_2 in sample 14053 (El Goresy et al. 1972). Pyroxene occurs in a fine-grained matrix with plagioclase and ilmenite or as large (5 mm), strongly-zoned laths enclosing subhedral laths of plagioclase. Minerals in the mesostasis include spinel, cristobalite, troilite, and residual, K-rich glass.

Shervais et al. (1984a,b; 1985a) and Dickinson et al. (1985) described over 45 high-alumina basalt clasts from Apollo 14 polymict breccia 14320. These clasts exhibit a wide range of textures, from coarse-grained subophitic to vitrophyres. Most are typified by olivine (Fo_{78-65}) phenocrysts that may exhibit resorbtion textures and are commonly mantled by pyroxene. In the medium- to coarse-grained lithologies, pyroxene is zoned from magnesian pigeonite cores to subcalcic augite rims. Plagioclase is blocky and ranges from An_{95} to An_{77}. The major opaque is ilmenite, which occurs as subhedral blades throughout the samples and as irregular grains in the mesostasis. Accessory phases include iron metal, troilite, chromite, cristobalite, and residual glass. The modal abundances determined by Dickinson et al. (1985) are 50% pyroxene, 25 to 35% plagioclase, 5 to 10% olivine, and 5 to 15% accessory phases. They have been classified into 5 groups based on mineral chemistries and whole rock chemistries (Dickinson et al. 1985). It is still debated whether these groups are real or a product of short-range unmixing. Selected modal data for the high alumina basalts are presented in Table A5.23.

Very high-potassium basalt. Shervais et al. (1985b) described clasts of this type of basalt in polymict lunar breccias from the Apollo 14 site. These basalts range in texture from medium- to coarse-grained ophitic to subophitic. They contain two generations of olivine. The early olivine phenocrysts (Fo_{73-61}) are commonly embayed or partially reabsorbed and are commonly mantled by magnesian pigeonite. Groundmass olivine (Fo_{50-36}) is euhedral to subhedral and exhibits a skeletal morphology. Pyroxene phenocrysts are zoned from magnesian pigeonite cores to augite mantles. Plagioclase forms small laths that range in composition from An_{95} to $An_{78.6}$. Blades of ilmenite are always interstitial. Spinel occurs as inclusions in olivine and pyroxene and as a phase in the groundmass. Potassium feldspar (Or_{84-95}) and K-, Si- rich glass are ubiquitous in the mesostasis. Other accessory phases include phosphates and iron metal. Modes of this lithology have not been presented in detail in the literature. Shervais et al. (1985b) indicated that these samples contain 0 to 22 vol % olivine and have a relatively high ratio of potassium feldspar to plagioclase. They deduced a crystallization sequence of olivine + spinel ⇒ pigeonite ⇒ plagioclase ⇒ augite.

Very low-Ti (VLT) basalts

The VLT crystalline basalts were first discovered in the Taurus-Littrow region in Apollo 17 deep drill core (Vaniman and Papike 1977a,b,c, GJ Taylor et al. 1977). Subsequently, VLT basalts were found among the Apollo 17 samples as clasts in impact melt rocks and breccias and among rock fragments in soils (GJ Taylor et al. 1978, RD Warner et al. 1978a, James and McGee 1980, Jolliff et al. 1996) and as the dominant basaltic lithology returned by the Luna 24 mission (Ma et al. 1978, Ryder and Marvin 1978). Most of the VLT fragments returned by the Apollo 17 and Luna 24 missions are extremely small (larger fragments have maximum dimensions of 2 to 4 mm).

The VLT basalts are vitrophyric to coarse-grained subophitic (Apollo 17)-ophitic (Luna 24) rocks. The vitrophyres consist of chromium spinel, olivine (Fo_{75}) and commonly pyroxene in a pale yellow/green to colorless glass. Olivine and pyroxene form skeletal crystals. Pyroxene is zoned from pigeonite cores to thin mantles of augite (Vaniman and Papike 1977c). In the coarser-grained lithologies, zoned and partially resorbed olivine (Fo_{76-05}) is in contact with pyroxene. The olivine in the Luna 24 VLT basalts exhibits more iron enrichment (Fo_{58-05}) than the Apollo 17 VLT basalts (Fo_{76-53}). Cr spinel occurs as inclusions in the olivine. Like the vitrophyres, the pyroxene is zoned from pigeonite cores to mantles of augite. The Fe enrichment in the Ca-rich pyroxenes is much more extensive in these coarser-grained samples. The pyroxene mantles are intergrown with plagioclase (An_{96-86}) in the Apollo 17 VLT, whereas in the Luna 24 VLT, the plagioclase is intergrown with both pyroxene and olivine. Groundmass contains iron metal, ilmenite, silica, troilite, ulvöspinel, and glass (Vaniman and Papike 1977c, GJ Taylor et al. 1978, RD Warner et al. 1978a, Papike and Vaniman 1978, James and McGee 1980, Ma et al. 1978, Ryder and Marvin 1978). Modes for VLT basalts are presented in Table A5.24. Although fragments are very small, Vaniman and Papike (1977c) determined the modes of two of the larger medium-grained fragments: Pyroxene 60 to 61%, plagioclase 28 to 31%, olivine 4.3 to 5.3%, ilmenite 0.2%. They also reconstructed the crystallization history of the Apollo 17 VLT basalts: olivine + Cr spinel ⇒ pigeonite ⇒ plagioclase + augite ⇒ ilmenite + ulvöspinel.

GEOCHEMISTRY OF THE CRYSTALLINE MARE BASALTS

Iajor elements

Major element characteristics of lunar basalts have been reviewed by Papike et al. 1976), BVSP (1981), Neal and Taylor (1992), and GJ Taylor et al. (1991). Selected ıajor element analyses of the mare basalts are presented in Tables A5.19 to A5.24. Within ıe context of terrestrial basalt classification (i.e. the basalt tetrahedron; Yoder and Tilley 962, Yoder 1976), the normative mineral assemblages of lunar basalts range from ıypersthene + quartz to hypersthene + olivine. Lunar basalts that are critically silica ındersaturated (with normative nepheline + olivine) have not yet been identified. The ıbsence of these types of magmas on the Moon attest to the volatile poor characteristics of he lunar mantle as well as the P-T regimes under which lunar magmas were derived. Relative to terrestrial basalts, lunar basalts exhibit (1) a spectacular range in TiO_2 contents (0.3 to 14 wt %, compared to less than 4 wt %); (2) reduced valance states for Fe, Ti, and Cr reflecting low magmatic oxygen fugacities; (3) depletions of alkali (K, Rb), siderophile, and volatile elements; and (4) absence of water (Papike et al. 1976, BVSP 1981, SR Taylor 1982, Newsom 1986).

The compositional diversity of the lunar basalts is partially illustrated in a plot of TiO_2 wt % versus mg# (Fig. 22). As an approximation, the basalts from the Apollo and Luna sampling sites form three distinct chemical clusters: A high-Ti cluster (9 to 14 wt % TiO_2), a low- to intermediate-Ti cluster (1 to 5 wt % TiO_2) and a very low-Ti cluster (<1 wt % TiO_2). The scarcity of basalts in the sample suite with 5 to 8 wt % TiO_2 may be a result of incomplete sampling; remote sensing studies suggest that only one third of all mare basalt types may be represented in the sample suite (Pieters 1978, 1990). One the other hand, the absence of basalts in this compositional range may reflect the composition of the lunar mantle and processes contributing to its evolution.

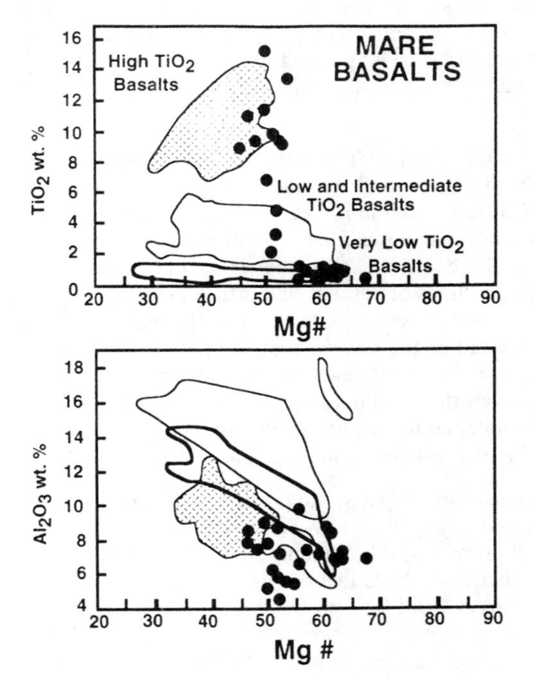

Figure 22. Plots of mg# versus TiO_2 and mg# versus Al_2O_3 for the crystalline mare basalts and picritic glasses. The fields for high-Ti, low-Ti, and very low-Ti basalts were defined from over 500 individual analyses compiled by Neal and Taylor (1992). They certainly do not all represent melt compositions or representative samples of basaltic lithologies. The filled circles represent picritic glass compositions that are thought to represent melt compositions.

Papike et al. (1976) pointed out that the mg# of the least fractionated high-Ti and low-Ti basalt suites are ~0.50. Basalts in these suites with mg# greater than 0.50 have been confirmed texturally to be cumulates. In contrast, the VLT basalts have mg# as high as 0.60. According to Papike et al. (1976) and BVSP (1981) this observation implies that compared to other basalts, either the VLT source was more magnesian or that the VLT basalts were produced by higher degrees of partial melting. Either scenario cannot be proven with the available data on the crystalline mare basalts. As pointed out by Delano (1986), Longhi (1987), and Shearer and Papike (1993), the crystalline mare basalts do not represent primary melts and therefore such implications are probably not valid.

The major element variation in the high-Ti basalts is a result of the fractionation of olivine and olivine + armalcolite ± chrome spinel. An increase in TiO_2 with decreasing MgO is exhibited in the more Mg-rich basalts (70017 and 74275) that have only olivine as a liquidus phase (Green et al. 1975, Walker et al. 1975, Longhi 1992). In less magnesian basalts, TiO_2 contents decrease with decreasing MgO (Fig. 23A). This reflects armalcolite and spinel joining olivine as a liquidus phase. Increases in Al_2O_3 (Fig. 22, 23B) and CaO with decreasing MgO are consistent with this fractional crystallization sequence. Based on the dispersion of the major and minor element data and differences in minor element ratios, these major element trajectories most likely represent the crystallization of distinct batches of high-Ti basalts (Rhodes et al. 1976, Papike et al. 1976, Neal and Taylor 1992). The apollo 11 high-K basalts are distinguished from the other high-Ti basalts by their higher K_2O and TiO_2 content and lower CaO and Al_2O_3 contents (Table A5.19). Papike et al. (1976) suggested that the limited major element variation in the high-K basalts indicates that the samples represent one thin flow or sill. The major and minor element characteristics of the Apollo 11 and 17 low-K basalts are similar to one another. They also show considerably more chemical variation than the high-K basalts. Based on textures, major and trace element characteristics (Ti, K, La), and ages, the low-K basalts have been subdivided into distinct chemical groups, which may represent distinct batches of magma and more than 5 separate igneous events (Rhodes et al. 1976, Papike et al. 1976, Geiss et al. 1977, Papanastassiou et al. 1977, Beaty and Albee, 1978, Guggisberg et al. 1979, Beaty et al. 1979a,b, Rhodes and Blanchard 1980, RD Warner et al. 1979, BVSP 1981, Neal and Taylor 1992).

The relationship between MgO and TiO_2 in the low-Ti basalts is the opposite to that found in a majority of the high-Ti basalts. In the low-Ti basalts, TiO_2 contents increase with decreasing MgO (Figs. 22 and 23C,D). In addition, CaO and Al_2O_3 increase with decreasing MgO (Figs. 22 and 23E,F). These major-element variations are a result of olivine and pigeonite being the liquidus phases rather than, Fe-Ti oxides. Individual lithologies show different major-element liquid lines of descent, indicating that they are not related by shallow-level fractional crystallization. For example, major- element variations within the Apollo 15 pigeonite basalts do not fall on the olivine control lines defined by the Apollo 15 olivine basalts. The same is true for the liquid line of descent relationship between the olivine and pigeonite basalts from the Apollo 12 sample site. Therefore, it is apparent from major-element data that pigeonite basalts are not related by fractional crystallization to the spatially associated olivine basalts from the same site (Papike et al. 1976).

The high-alumina and very high-potassium basalts have distinct major element characteristics that contrast sharply with those of other low-Ti basalts (Shervais et al. 1985a,b, Dickinson et al. 1985). Almost all of the high-alumina basalts contain greater than 11 wt % Al_2O_3. They also have K_2O and Na_2O contents that are greater than those of the other low-Ti basalts. The mg# commonly ranges from 0.57 to 0.34 and CaO/Al_2O_3 ranges from 0.81 to 1.0. Major-element variations are not highly correlated. For example, in the suite of samples occurring in Apollo 14 breccias, although there is an eight-fold increase in

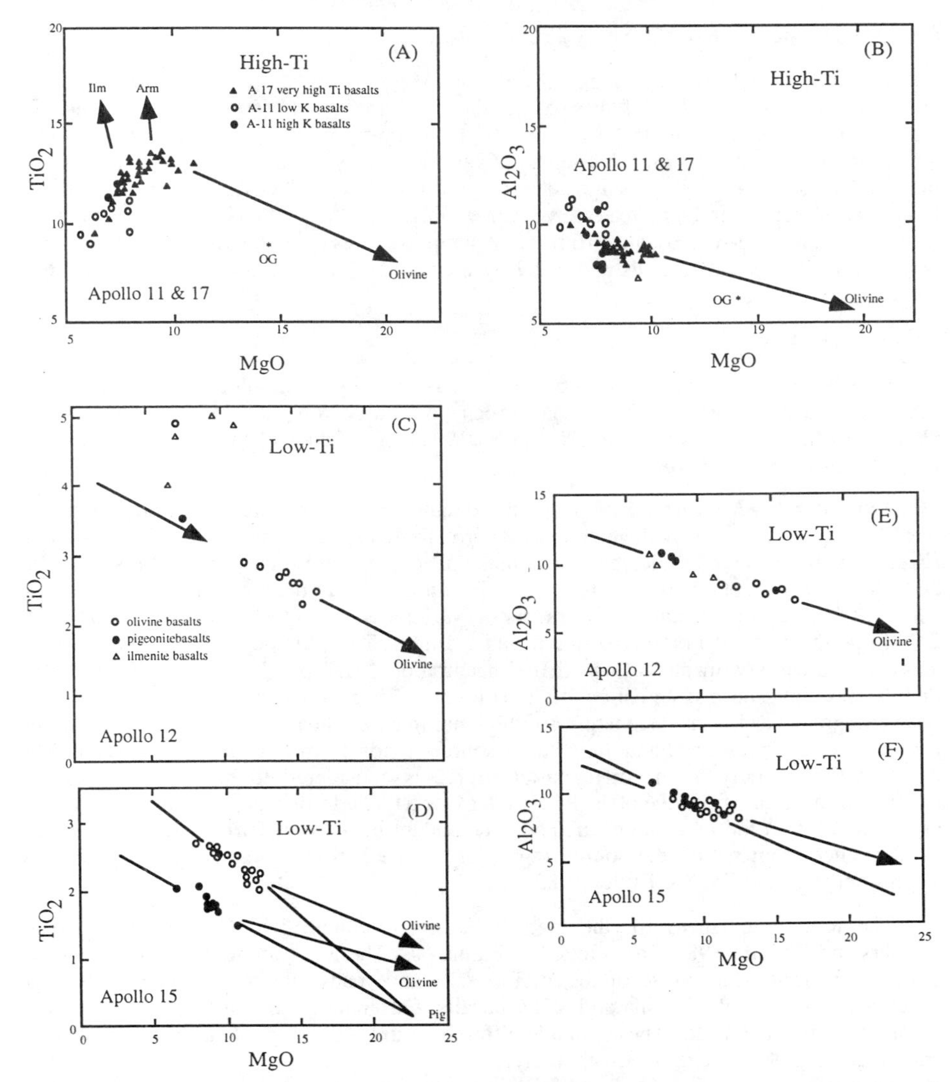

Figure 23. Chemical variation diagrams showing MgO-TiO_2 and MgO-Al_2O_3 relations for high-Ti (A,B) and low-Ti (C-F) mare basalts (from Papike et al. 1976). Arrows point toward the compositions of specific minerals that crystallize from these basaltic magmas. During crystallization of a magma, the evolving melt composition will be driven in the opposite direction of the individual mineral or combination of minerals that are crystallizing. Minerals represented in these diagrams are armalcolite (Arm), ilmenite (Ilm), pigeonite (Pig), and olivine.

some incompatible trace elements (Dickinson et al. 1985), there appears to be no correlation between MgO and TiO_2, CaO, or Al_2O_3 that would define a liquid line of descent. The very-high-potassium basalts have high Al_2O_3 (9.9 to 13.4 wt %) and Na_2O contents similar to the high-alumina basalts, but also have extraordinarily high K_2O contents (0.72 to 1.4 wt %). The mg# commonly ranges from 0.60 to 0.49 and CaO/Al_2O_3 ranges from 0.81 to 0.99. Although there appears to be no correlation between MgO and TiO_2, CaO and Al_2O_3

are anti-correlated with MgO (Shervais et al. 1985b).

The very low-Ti basalts exhibit a range of MgO contents, from the relatively high values of the Apollo 17 VLT basalts to those of the much more evolved Luna 24 ferrobasalts. Like the low-Ti basalts, with decreasing MgO, the VLT basalts exhibit increases in TiO_2, Al_2O_3 (Fig. 23), and CaO and a decrease in Cr_2O_3 The variation of the major elements in these basalts indicates compositional control primarily through olivine fractionation. Grove and Vaniman (1978) have shown that the Luna 24 ferrobasalts are not related to more Mg-rich basalts from other sampling sites. Compared to the other mare basalts, the VLT basalts are distinguished by their higher Al_2O_3 and lower K_2O and Na_2O contents.

Trace elements

Selected trace-element analyses of the various mare basalt lithologies are presented in Tables A5.25 to A5.30. These data have been previously compiled by Wentworth et al. (1979), BVSP (1981), Lofgren and Lofgren (1981), SR Taylor (1982), and GJ Taylor et al. (1991*Lunar Sourcebook*).

Rare earth elements. Several of the diagnostic geochemical characteristics of the mare basalts are shown in their chondrite-normalized rare earth element (REE) patterns. Essentially all the mare basalts thus far sampled have negative Eu anomalies (Figs. 24, 25, 26). The only exception to this ubiquitous Eu anomaly has been documented in several small (perhaps unrepresentative) fragments of VLT basalt from the Luna 24 mission (Fig. 24). In addition to the ubiquity of the negative Eu anomaly, the depth of the negative Eu anomaly increases with increasing REE concentration (VLT basalts ⇒ low-Ti basalts ⇒ high-Ti basalts). These systematic variations in the REE patterns for the mare basalts have been recognized as intrinsic signatures of the lunar mantle. They have been attributed to the formation of the lunar crust and mare basalt source region from a lunar magma ocean. The negative Eu anamoly in the mare basalt source is considered to be the result of the fractionation of Eu^{2+} from the other REE under low fO_2 conditions during plagioclase (lunar crust) flotation in the magma ocean. The concomitant increase in REE abundance reflects the sequence from which the source region crystallized from a lunar magma ocean (SR Taylor and Jakes 1974, SR Taylor 1982).

The REE abundances for the high-Ti basalts, normalized to C1 chondrite values (Anders and Ebihara 1982), are plotted in Figure 24. The REE abundances in the high-Ti basalts are higher than those of the VLT basalts and many of the low-Ti basalts. The exceptions are the high-alumina and VHK basalts. Differences exist in REE characteristics among the high-Ti basalts. These include differences in LREE slope, total REE abundance, and magnitude of the negative Eu anomaly.

The Apollo 11 high-Ti basalts have REE patterns possessing positive LREE slopes and negative HREE slopes. The high-K lithologies are more enriched in REE than the low-K lithologies. In the high-K samples, the chondrite normalized Ce abundance ($[Ce]_N$) and $[Yb]_N$ values are approximately 100 and 95, respectively (Fig. 24). The low-K basalts have $[Ce]_N$ and $[Yb]_N$ values that are approximately 60 (Fig. 24). The negative Eu anomaly is relatively large in both high-Ti basalts from the Apollo 11 site with $[Sm/Eu]_N$ between 2.6 and 3.8. The high-K samples have slightly higher $[Sm/Eu]_N$ values.

The Apollo 17 high-Ti basalts generally have similar patterns to each other, with negative Eu anomalies, positive LREE slopes and negative HREE slopes. Compared to the Apollo 11 high-Ti basalts, the Apollo 17 basalts have slightly steeper LREE slopes, flatter HREE slopes, lower overall abundances of REE ($[Ce]_N$ = 12 to 30 and $[Yb]_N$ = 30 to 50),

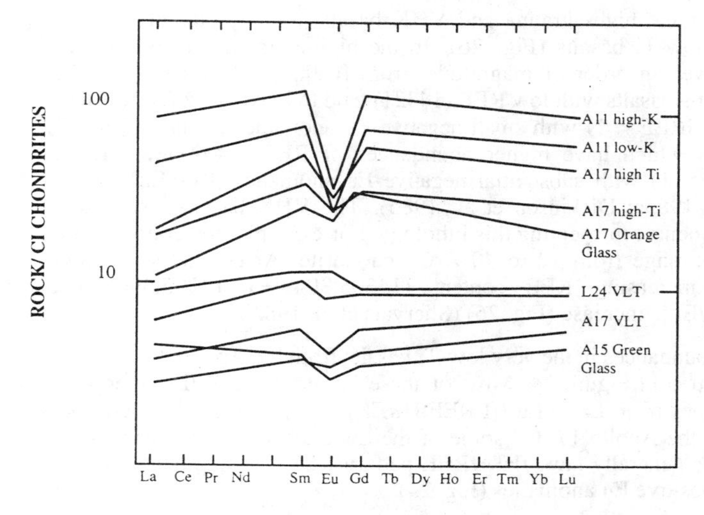

Figure 24. The REE contents in high-Ti and very low-Ti mare basalts, normalized to REE contents in chondritic meteorites (modified from GJ Taylor et al. 1991 *Lunar Sourcebook*).

and smaller negative Eu anomalies ($[Sm/Eu]_N$ = 1.3 to 1.7) (Fig. 24). The various types of Apollo 17 high-Ti basalts show slightly different to overlapping REE patterns. As shown in Figure 24, Group A basalt (75055), Group U basalt (70017), and a high-Ti picritic glass (74220) have diagnostic patterns.

Chondrite normalized rare earth element abundances in the low-Ti basalts (olivine basalts, pigeonite basalts, ilmenite basalts, high-alumina basalts and VHK basalts) are plotted in Figures 5 and 26. The olivine, pigeonite, and ilmenite basalts from the Apollo 12 and 15 sites exhibit similar REE patterns (Fig. 25). These basalts have convex-upward patterns with small negative Eu anomalies. The HREE slope for the Apollo 15 low-Ti basalts is somewhat steeper than that for the Apollo 12 basalts. The Apollo 15 low-Ti basalts have $[Ce]_N$ which ranges from 12 to 20, $[Yb]_N$ of approximately 10 and $[Sm/Eu]_N$ which ranges from 1.3 to 1.7. In the Apollo 12 low-Ti basalts, the $[Ce]_N$ ranges from 17 to 23, the $[Yb]_N$ ranges from 15 to 25, and the $[Sm/Eu]_N$ ranges from 1.6 to 1.9. The REE abundances for the Apollo 12 low-Ti basalts overlap considerably, although the ilmenite basalts generally appear to be slightly more enriched in the REE.

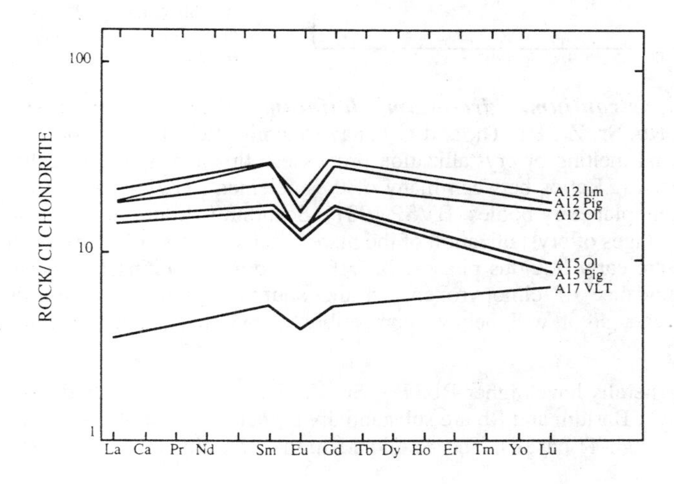

Figure 25. The REE contents in low-Ti mare basalts, normalized to REE contents in chondritic meteorites (modified from GJ Taylor et al. 1991 *Lunar Sourcebook*).

The REE patterns for the high-alumina and VHK basalts are substantially different from those of the other low-Ti basalts (Fig. 26). In the high-alumina basalts, the REE concentrations vary by over an order of magnitude, from $[Ce]_N$ = 10 to 100. The REE pattern for the high-alumina basalts with low REE (14321) tend to have fairly flat to LREE-depleted patterns (LREE/HREE < 1) with small negative Eu anomalies ($[Sm/Eu]_N \approx 1.2$). The high-alumina basalts which have higher abundances of REE (14321) are LREE-enriched (LREE/HREE >> 1) with substantial negative Eu anomalies ($[Sm/Eu]_N \approx 3.3$) (Fig. 26) (Shervais et al. 1985a; Dickinson et al. 1985). The VHK basalts also show a similar range among fragments representing this lithology. For example, the concentrations of Ce in the VHK basalts range from 12 to 40 × C1 chondrite. Analogous to the high-alumina basalts, with increasing REE content (14305,304 ⇒ 14305,390) both $[LREE/HREE]_N$ and $[Sm/Eu]_N$ increase (Fig. 26) (Shervais et al. 1985b).

Rare earth element abundances in the very low-Ti basalts from the Apollo 17 and Luna 24 sampling sites are plotted in Figure 24. Most of these basalts have REE patterns that have slightly positive slopes from La to Lu ($[LREE/HREE]_N < 1$) with $[Ce]_N$ abundances between 4 and 10. All of the Apollo 17 and some of the Luna 24 VLTs have negative Eu anomalies. Luna 24 VLTs generally have flat HREE patterns. Several fragments from the Luna 24 site have small positive Eu anomalies (Fig. 24).

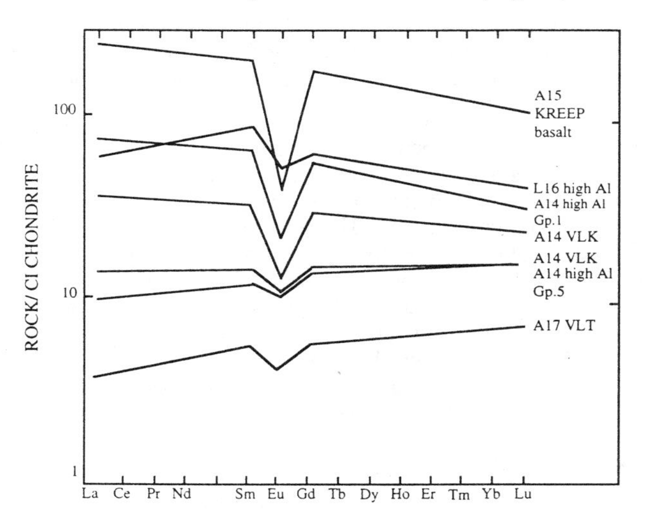

Figure 26. The REE contents in KREEP, high-alumina and very high potassium basalts, normalized to REE contents in chondritic meteorites (modified from GJ Taylor et al. 1991 *Lunar Sourcebook*).

Barium, rubidium, strontium, zirconium, hafnium, thorium, uranium, and scandium. Barium, Rb, Sr, Zr, Hf, Th, and U behave incompatibly in these basaltic systems during either partial melting or crystallization processes. Ilmenite and armalcolite should fractionate Hf from Zr. Ratios between many of these elements (K/U, Rb/Ba) are useful discriminators among planetary bodies (BVSP 1981). Scandium should also behave incompatibly during early stages of crystallization of the mare basalts because olivine and in some cases Fe-Ti oxides are early liquidus phases. Its behavior during melting is dictated by the presence and abundance of clinopyroxene in the source region. Scandium is compatible in clinopyroxene, so it will behave compatibly if pyroxene dominates the crystallization assemblage.

The high-Ti basalts generally have higher Rb, Ba, Sr, Zr, Hf, Th, and U abundances than basalts with lower TiO_2. Barium and Rb are substantially higher in the Apollo 11 high-K basalts than in the other high-Ti basalts. The Apollo 11 high-K basalts have an average

Ba content of 330 ppm and average Rb content of 6.2 ppm. Among the other high-Ti basalts, the Apollo 11 low-K basalts have average Ba and Rb concentrations of 108 ppm and 0.62 ppm, respectively, and the Apollo 17 high-Ti basalts average 60 ppm and 0.36 ppm (Papike et al. 1976, BVSP 1981, GJ Taylor et al. 1991 *Lunar Sourcebook*). Strontium in the high-Ti basalts ranges from approximately 120 to 195 ppm, with overlap among rock types. The abundance of Zr ranges from 150 to 500 ppm in the Apollo 11 low-K and Apollo 17 basalts to approximately 800 ppm in the Apollo 11 high-K basalts. The Zr/Hf ratio in the basalts averages about 33:1 (Papike et al. 1976). The high-K subgroup of basalts is also enriched in U (0.8 ppm) and Th (3 to 5 ppm) relative to the other high-Ti mare basalts (U = 0.06 to 0.22 ppm; Th = 0.20 to 0.72 ppm). The enrichment of highly incompatible elements in the Apollo 11 high-K basalts compared to other high-Ti basalts cannot be a result of either fractional crystallization or partial melting. These enrichments can only be attributed to differences in the characteristics of their mantle sources. Scandium concentrations among the high-Ti basalts are fairly uniform, ranging from 70 to 100 ppm. However, these Sc values are distinctly different from those of the low-Ti basalts (25 to 60 ppm). This difference has been attributed to the different roles played by clinopyroxene in the source regions of these basalts (SR Taylor and Bence 1975a, Papike et al. 1976).

The low-Ti basalts also exhibit significant differences in the abundance of highly incompatible elements between the Apollo 12 and 15 suite of basalts, high alumina basalts, and the VHK basalts. The low-Ti basalts from Apollo 12 and 15 (olivine, pigeonite, ilmenite basalts) have overlapping highly incompatible element abundances, with Rb = 0.45-2.0 ppm, Ba = 32-75 ppm, Zr = 55-200 ppm, Hf = 1-4 ppm, Th = 0.4-1.3 ppm, and U = 0.10-0.4. Samples making up the high alumina and VHK basalt suites exhibit considerably more variability in the highly incompatible elements. The high-alumina basalts with the higher abundance of incompatible elements (Group 1) are substantially enriched in Rb (10-13 ppm), Ba (110-230 ppm), Zr (210-400 ppm), Hf (~9 ppm), Th (1.8-2.7 ppm), and U (0.3-0.52 ppm) (Shervais et al. 1984a,b, Dickinson et al. 1985) compared to the low-Ti basalts from Apollo 12 and 15 sites. The incompatible element-depleted, high-alumina basalts have incompatible element abundances that overlap with the Apollo 12 and 15 samples (Shervais et al. 1984a,b, Dickinson et al. 1985). The Apollo 14 VHK basalts also exhibit a considerable enrichment and range in highly incompatible element abundances. In particular, Ba abundances in these basalts range from 100 to 840 ppm (Shervais et al. 1985b). Many of the other incompatible elements overlap with the high-alumina basalts. Strontium exhibits far smaller differences among the various low-Ti lithologies. The average Sr abundances for the Apollo 12 and 15 low-Ti basalts range from 101 to 148 ppm (GJ Taylor et al. 1991 *Lunar Sourcebook*). The Sr concentrations for the high-alumina and VHK basalts are somewhat lower (10-130 ppm) (Shervais et al. 1984a,b, 1985b). Scandium concentrations in the low-Ti basalts (34-66 ppm) show considerable overlap regardless of highly incompatible element concentrations. For example, in the Apollo 14 high-alumina basalts, Dickinson et al. (1985) observed an increase in average La from 3.4 ppm to 25 ppm with a accompanying change in average Sc from 62 to 59 ppm.

The very low-Ti basalts are considerably depleted in the highly incompatible elements compared to the other crystalline mare basalts. For example, the compilation of VLT basalt compositions show that Hf ranges from 0.6 to 1.4 ppm and that both Ba and Zr are approximately 50 ppm (BVSP 1981, GJ Taylor et al. 1991 *Lunar Sourcebook*). Scandium concentrations in the VLT basalts (47-57 ppm) are lower than in the high-Ti basalts and overlap with the low-Ti basalts. The data are too sparse and the samples are too small to conclude if real incompatible element differences exist between Luna 24 and Apollo 17 VLT basalts.

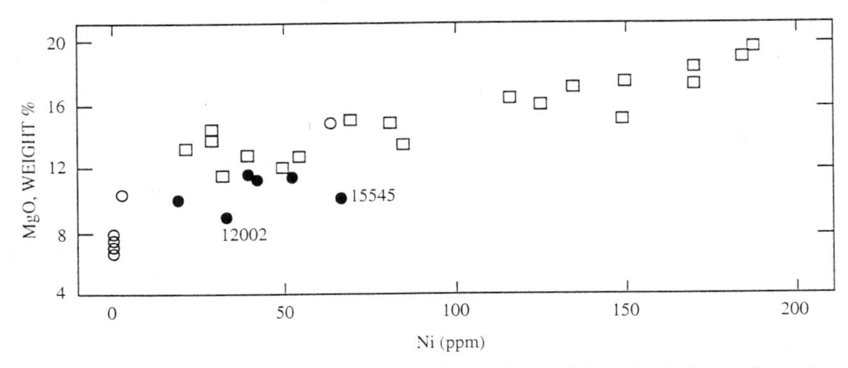

Figure 27. Ni (ppm) versus MgO (wt %) in mare basalts (circles) and in volcanic lunar glasses (squares). Open circles represent high-Ti basalts and solid circles represent low-Ti basalts (modified from Delano, 1986).

Nickel, cobalt, chromium, vanadium. Under extremely low fO_2 conditions, Ni and Co partition preferentially into a metallic phase. However, there is no evidence to suggest metallic iron is present in the source region for the mare basalts. In most documented cases, metallic iron crystallizes fairly late in the mare basalt crystallization sequence. Therefore, during melting and fractional crystallization, the fractionation of divalent Ni and, to a far lesser extent, divalent Co is controlled by olivine. The difference in Ni abundances between high-Ti and low-Ti basalts was controlled by olivine during the formation of their mantle sources (Lunar Magma Ocean [LMO] crystallization) and crystallization of the basalt following extraction from the mantle. Chromium in lunar basaltic magmas can be trivalent and divalent (Haggerty et al. 1970). Divalent Cr partitions into olivine and trivalent Cr forms chrome spinel. In most cases, the stability of chrome spinel on the liquidus will affect the behavior of Cr along the liquid line of descent.

As with the highly incompatible elements, Ni and Co contents in the crystalline mare basalts vary systematically with TiO_2 (Delano 1986, Shearer et al. 1991). In individual suites of basalts, Ni content is also correlated with mg# (Fig. 27). With increasing TiO_2, the Ni content decreases. In the high-Ti basalts, Ni and Co are depleted compared to most other mare basalts, with abundances of 2-10 ppm and 15-50 ppm, respectively. The low-Ti basalts show considerably more variability. The Ni contents of the olivine basalts range from 60 to 70 ppm. The pigeonite and ilmenite basalts have Ni abundances between 5 and 30 ppm. Cobalt abundances in the low-Ti basalts show more limited variation, with Co between 50 and 75 ppm. The high-alumina and VHK basalts also show considerable variability, with Ni contents between 10 and 120 ppm and Co between 19 and 41 ppm. Because many of the high-alumina and VHK lithologies are clasts in breccias, it is highly likely that their siderophile element signatures are not entirely magmatic. The very low-Ti basalts exhibit a range in Ni content from 30 to 80 ppm.

There is some overlap in chromium abundances among the different mare basalt lithologies. Differences in Cr between groups tends to be more of a function of mg# than TiO_2 content. For example, the Cr_2O_3 concentrations in the VLT basalts from the Luna 24 site (mg# $\approx$ 0.36) range from 0.15 to 0.29 wt %, whereas the Apollo 17 VLT basalts (mg# $\approx$ 0.54) have between 0.60 and 0.75 wt % Cr_2O_3. Overall, for the mare basalts there is a positive correlation between Cr_2O_3 and mg# (BVSP 1981). Within the Apollo 12 olivine basalts, a general decrease in Cr_2O_3 with decreasing mg# has been attributed to fractionation of olivine and minor amounts of chrome spinel (Haggerty et al. 1970, Papike et al. 1976).

Limited V data exist for the mare basalts. Unlike Cr, V abundances in the mare basalts tend to be related to TiO_2 contents. High-Ti basalts have lower V (≈ 50 ppm) and the Apollo 17 (≈ 200) and Luna 24 (140-180 ppm) VLT basalts. The V abundance in the olivine basalts from the Apollo 12 and 15 sites (~170 ppm) is also higher abundances than in other low-Ti basalts (~60-120 ppm) (Papike et al. 1976, BVSP 1981, Shervais et al. 1984a,b, 1985b, GJ Taylor et al. 1991 *Lunar Sourcebook*).

AGES OF THE MARE BASALTS

Formation ages of numerous mare basalt lithologies have been determined by radiometric dating. Estimated ages of large areas of mare terranes have been made by photogeology, using crater density-preservation, and age extrapolation from landing sites. The radiometeric ages for the mare basalts range from 3.8 to 3.16 Ga (summarized by BVSP 1981, Ryder and Spudis 1980, SR Taylor 1982, Nyquist and Shih 1992). Basaltic volcanism prior to basin formation has been documented to be as old as 4.2 Ga (LA Taylor et al. 1983). Photogeologic observations of dark-halo craters which have been interpreted as pre-basin basalt flows provide additional support for basaltic volcanism prior to 3.9 Ga (Schultz and Spudis 1983, Bell and Hawke 1984, Hawke et al. 1990, Head and Wilson 1992, 1997). The ages of the last gasps of mare volcanism have been estimated by photogeologic methods. Schaber (1973) estimated the ages of three flows to be 3.0±0.04, 2.7±0.03 and 2.5±0.03 Ga. Crater density statistics have been interpreted as indicating that the youngest basaltic units on the Moon may be between 0.9 and 2.0 Ga (Schultz and Spudis 1983).

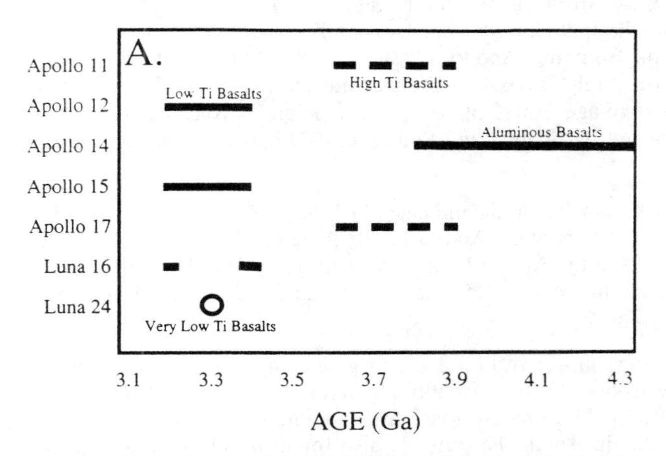

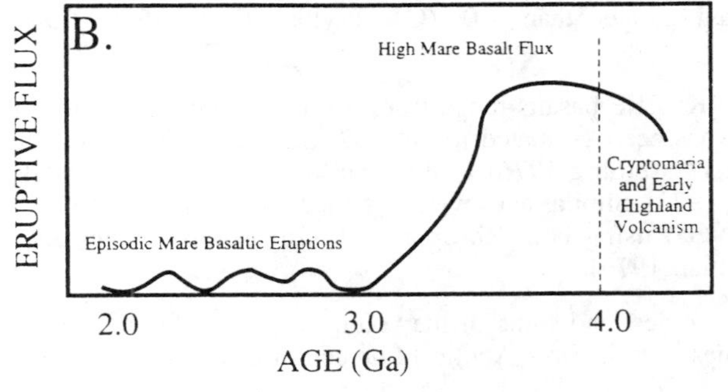

Figure 28. (A) Ages of mare basalts from various sampling sites (from Nyquist and Shih 1992). (B) Estimate of mare basalt eruptive flux (modified from Head and Wilson 1992, 1997).

The eruptive flux was not constant and a majority of mare basalts were emplaced between 3.8 and 3.2 Ga (Fig. 28). The 3.8 Ga age roughly corresponds with the rapid fall-off of meteorite bombardment and the excavation of the Imbrium basin. Head and Wilson (1992; 1997) suggested that the peak eruptive flux of mare basalts may have been a continuation of "cryptomare" volcanism (KREEP basalts, high-alumina basalts) (Fig. 28). Some eruptions during this peak eruptive flux may have lasted on the order of a year and emplaced 103 km^3 of basaltic lava (Head and Wilson 1992; 1997). Following this extensive eruptive pulse, eruptions were low volume and episodic and account for less than 5% of the total volume of mare basalts (Schaber, 1973, Schultz and Spudis 1983, Head and Wilson 1992, 1997).

Samples returned by Apollo and Luna missions present an ambiguous picture concerning the relation between magma composition and time (Papike et al. 1976). Head and Wilson (1992; 1997) concluded that although a variety of magma types were being erupted in nearside mare basins during the period 3.8 to 3.2 Ga, the early and intermediate phases of eruption within individual basins were dominated by high-Ti basalts. Later periods of eruption within a basin were predominately low-Ti basalts. They suggested that this may be a result of increasing depth of the mantle sources for various magma types. This latter conclusion is not consistent with high-pressure experimental results for the mare basalts (Stolper 1974, Green et al. 1975, Delano 1980, 1986, Chen et al. 1982, Chen and Lindsley 1983, Longhi 1987, 1992). We discuss below the relationship between magma type and depth of partial melting.

The high-Ti basalts from the Apollo 11 and 17 sampling sites are generally older than 3.5 Ga. The high-Ti picritic glass from the Apollo 17 site (Apollo 17 orange glass), is slightly younger (3.48 Ga) than the spatially associated, crystalline high-Ti basalts (3.59 to 3.79 Ga). The high-Ti basalts from the Apollo 11 site can be divided into two age-composition groups. The high-K, high-Ti basalts from that site are about 3.55 Ga, whereas the low-K, high-Ti basalts range in age from 3.59 to 3.79 Ga. High-Ti volcanic activity as young as 1 Ga has been suggested by Schultz and Spudis (1983) based on photogeologic observations.

Radioactive age-dating of the low-Ti basalts indicates that they formed at 3.08 to 3.37 Ga. The low-Ti basaltic volcanic activity at the Apollo 12 site is generally younger (3.08 to 3.29 Ga) than the volcanic activity at the Apollo 15 site (3.21 to 3.37 Ga). The ages of the different low-Ti lithologies from the Apollo 15 site (olivine and pigeonite basalts) are indistinguishable from one another.

The very low-Ti basalts range in age from 3.3 Ga to greater than 4.0 Ga. The very low-Ti picritic glass from the Apollo 15 site (Apollo 15 Green or Emerald glass) and a small basaltic fragment collected by the Luna 24 mission have radiometric ages of 3.3 Ga. The very low-Ti basalt fragments in Apollo 17 soils are also found in a breccia that has a melt-rich matrix that crystallized greater than 4.0 (GJ Taylor et al. 1991 *Lunar Sourcebook*).

The radiometric ages for the KREEP basalts range from 3.8 to 4.0 Ga. The age for Apollo 15 KREEP basalt 15382 has been estimated to be 3.82 Ga using Rb-Sr isotopic systematics (Papanastassiou and Wasserburg 1976) to 3.91 Ga using ^{40}Ar-^{39}Ar methods (Stettler et al. 1973, Turner et al. 1973). Similar ages were obtained for Apollo 15 KREEP basalt 15386 (3.85 Ga and 3.86 Ga) using both Rb-Sr and Nd-Sm isotopic systematics (Nyquist 1977, Carlson and Lugmair 1979).

The aluminous basalts are the oldest and some of the youngest basalts that have been radiometrically dated. Clasts of these basalts from Apollo 14 breccias range in age from 3.9

to 4.2 Ga (LA Taylor et al. 1983, Neal et al. 1989b, Snyder et al. 1995a). Aluminous basalts from the Apollo 12 and Luna 16 sites are much younger. A fragment of aluminous basalt from the Apollo 12 site has a crystallization age of 3.1 Ga, whereas the aluminous basalt collected from the Luna 16 site crystallized at 3.4 Ga (LA Taylor et al. 1983, Neal et al. 1989b, GJ Taylor et al. 1991 *Lunar Sourcebook*).

EXPERIMENTAL PHASE PETROLOGY OF CRYSTALLINE MARE BASALTS

Experimental studies on mare basalts have been summarized in numerous papers. Some of the best and most complete summaries are by Kesson and Lindsley (1974), Kesson (1975), BVSP (1981), GJ Taylor et al. (1991 *Lunar Sourcebook*), and Longhi (1992).

Dynamic crystallization experiments

As illustrated in the backscatter electron images and compositional maps of the various mare basalts (Figures 12-21), mare basalts have experienced a range of cooling histories that have influenced their textures (Bence and Papike 1972, Papike et al. 1976, GJ Taylor et al. 1991 *Lunar Sourcebook*), mineral morphologies and chemistries (Bence and Papike 1972, Shearer et al. 1989), and crystallization sequences (Bence and Papike 1972, Walker et al. 1976). Both dynamic crystallization experiments (e.g. Lofgren et al. 1974, 1979; Donaldson et al. 1975, Walker et al. 1976, Usselman et al. 1975, Grove and Walker 1977) and computational methods (e.g. Onorato et al. 1978) indicate that most mare basalts cooled at rates of 0.1° to 30°C per hour. This range in cooling rate can be produced by cooling in lava flows only a few meters thick (Brett 1975a, GJ Taylor et al. 1991 *Lunar Sourcebook*).

The dynamic crystallization experiments were undertaken by numerous investigators for the purpose of placing constraints on the thickness of lava flows, locating samples within flows and determining the relative importance of various factors (cooling rate, melt composition, number of nuclei prior to cooling) in producing the textures observed in lunar basalts. In rather early dynamic crystallization experiments, Lofgren et al. (1974) and Lofgren (1974) concluded that the porphyritic textures observed in many of the mare basalts could develop during crystallization induced at linear cooling rates. These porphyritic textures were more likely to develop if the density of nuclei in the melt were low and the basalt was saturated with only one phase (i.e. olivine) over a rather large temperature interval. Walker et al. (1976) demonstrated that for porphyritic olivine basalts, the cooling rate decreased between the time of the crystallization of the first phase and growth of other phases in the groundmass. They estimated that in olivine basalt 12002, the olivine initially crystallized at a cooling rate of approximately 1 °C per hour and slowed to a rate of between 0.2 to 0.1 °C during the crystallization of the groundmass. Comparison of experimental studies with textural variations in high-Ti basalts led Usselman et al. (1975) to conclude that two-stage cooling histories were required to produce textures in Apollo 17 high-Ti basalts.

Walker et al. (1975) demonstrated in their set of experiments that cooling rate will also affect the crystallization sequence. They showed that plagioclase crystallized prior to ilmenite during equilibrium crystallization and after ilmenite at rapid cooling rates (> 1°C per hour). The effect of cooling rate on delaying plagioclase nucleation and crystallization and its influence on minor and trace element partitioning has been demonstrated in synthetic and natural basalts by Bence and Papike (1972), Walker et al. (1975), Grove and Bence (1977) and Shearer et al. (1989). The cooling rate at which the plagioclase crystallized may be estimated using the observations of Grove and Walker (1977). They demonstrated that that the logarithm of the width of plagioclase grains in the groundmass of Apollo 15 pigeonite basalts was a linear function of the logarithm of the cooling rate.

In experiments designed to evaluate the effect of cooling rate on the textural and chemical characteristics of mesostasis in lunar basalts, Rutherford et al. (1974) and Hess et al. (1975) documented the role of cooling rate on liquid immiscibility. They observed that liquid immiscibility develops in the mesostasis of basalts at cooling rates of less than 2°C per hour and at temperatures below 1000°C. Longhi (1990, 1992) pointed out that the Fe-rich basaltic and Si-, K-rich rhyolitic melts from these experimental runs did not fractionate chemically as extensively as natural, immiscible melts in mineral inclusions and mesostasis.

Low pressure experiments

Low pressure experiments were performed on mare basalt compositions at lunar conditions to better understand their crystallization on the surface of the Moon. Results of low-pressure experiments for selected mare basalt types are summarized in Tables 6 through 8. The liquidus temperatures for this broad spectrum of compositions range from 1150°C to 1350°C (Kesson and Lindsley 1974, BVSP 1981, GJ Taylor et al. 1991 *Lunar Sourcebook*, Longhi 1992) and tend to be partially dependent on the mg# of the bulk composition. For example, the average Luna 24 very low-Ti basalt (mg# ≈ 0.4) has a liquidus temperature of approximately 1182°C (Grove and Vaniman 1978), whereas very

Table 6. Low-pressure (0-1 bar), equilibrium experimental results for high-Ti mare basalts. Table modified after Taylor et al. (1991).

Sample	Rock Type	Temp (°C)	Phases Present*
10072	high-Ti	1195	liq
	Mg/Mg+Fe=0.42	1159	liq+Ol+armal+ilm
	Walker et al. (1975)	1145	liq+Ol+ilm
		1140	liq+loCapx+ilm
		1130	liq+loCapx+hiCapx+ilm
		1122	liq+loCapx+hiCapx+ilm+pag
10020	low-K	1170	liq
	Mg/Mg+Fe=0.42	1160	liq+Ol+sp
	O'Hara et al. (1974)	1153	liq+Ol+sp
		1151	liq+Ol+sp+ilm
		1145	liq+Ol+sp+ilm+px+pag
		1133	liq+sp+ilm+pig+aug+plag
		1108	liq+sp+ilm+pig+aug+plag
71569	very high-Ti	1192	liq
	Mg/Mg+Fe=0.42	1156	liq+Ol+armal+sp
	Walker et al. (1975)	1146	liq+Ol+armal+sp
		1142	liq+Ol+hiCapx+armal+ilm
		1138	liq+Ol+hiCapx+ilm+plag
		1137	liq+hiCapx+loCapx+ilm+plag
		1127	liq+hiCapx+loCapx+ilm+plag
		1119	liq+hiCapx+loCapx+ilm+plag
74220	orange glass	1398	liq
	Mg/Mg+Fe=0.54	1302	liq+Ol
	Walker et al. (1975)	1251	liq+Ol
		1195	liq+Ol+sp
		1159	liq+Ol+ilm+sp
		1149	liq+Ol+ilm+sp
		1145	liq+Ol+ilm
		1140	liq+Ol+ilm+sp
		1130	liq+Ol+pig+ilm

*liq=liquid, Ol=olivine; armal=armalcolite; ilm=ilmenite; sp=spinel; loCapx=low-Ca pyroxene; hiCapx=high-Ca pyroxene; pig=pigeonite; aug=augite; plag=plagioclase.

Table 7. Low-pressure (0-1 bar), equilibrium experimental results for low-Ti mare basalts. Table modified after Taylor et al. (1991).

Sample	Rock Type	Temp(°C)	Phases Present*
12002	olivine basalt	1350	liq
	Mg/Mg+Fe=0.55	1328	liq+Ol
	Walker et al. (1976)	1266	liq+Ol
		1232	liq+Ol+sp
		1201	liq+Ol+sp
		1176	liq+Ol+sp+px
		1164	liq+Ol+sp+px
		1150	liq+Ol+sp+px
15065	pigeonite basalt	1273	liq+Ol
	Mg/Mg+Fe=0.50	1249	liq+Ol
	Walker et al. (1977)	1225	liq+Ol+sp+px
		1205	liq+Ol+sp+px
		1172	liq+Ol+sp+px
		1148	liq+Ol+sp+px
		1129	liq+sp+px+plag
		1125	liq+Ol+sp+px+plag
		1101	liq+sp+px+plag
14072	high alumina basalt	1285	liq
	Mg/Mg+Fe=0.55	1262	liq+Ol
	Walker et al. (1972)	1212	liq+Ol
		1190	liq+Ol+sp
		1175	liq+Ol+sp+pig+plag
		1170	liq+Ol+sp+pig+plag
		1140	liq+Ol+sp+pig+aug

*liq=liquid; Ol=olivine; sp=spinel; px=pyroxene; pig=pigeonite; aug=augite; plag=plagioclase.

Table 8. Low-pressure (0-1 bar), equilibrium experimental results for very low-Ti mare basalts. Table modified after Taylor et al. (1991).

	Rock Type	Temp(°C)	Phases Present*
Luna 24	average	1192	liq
	Mg/Mg+Fe=0.35-0.40	1189	liq
	Grove and Vaniman (1978)	1182	liq+plag
		1155	liq+plag+Ol+px
		1143	liq+plag+Ol+px
		1071	liq+plag+Ol+px
Apollo 15	green glass	1408	liq
	Mg/Mg+Fe=0.59	1392	liq+Ol
	Grove and Vaniman (1978)	1359	liq+Ol
		1221	liq+Ol
		1192	liq+Ol+px
		1163	liq+Ol+px
		1155	liq+Ol+px+plag
		1143	liq+Ol+px+plag

*liq=liquid; Ol=olivine; px=pyroxene; plag=plagioclase.

low-Ti pyroclastic glasses (mg# ≈ 0.6) have liquidus temperatures as high as 1408°C (Grove and Vaniman 1978). In many of the basalts, olivine (± spinel) is the first phase on the liquidus. This generally explains the dominance of olivine control lines in producing the variations in major element chemistry (Fig. 24). However, several bulk compositions investigated in the experiments summarized in Tables 6 and 8 do not have olivine as the first liquidus phase. The average Luna 24 basalt has plagioclase as the first phase on the liquidus (Grove and Vaniman 1978), whereas several of the high-Ti basalts are multi-saturated with olivine + armalcolite ± spinel ± ilmenite (Walker et al. 1975). Again, the mg# of the bulk composition generally controls whether olivine appears as the sole liquidus phase and the temperature interval over which olivine remains the sole liquidus phase. A comparison of the low pressure phase relations of Apollo 17 high-Ti basalt (mg# = 0.42) and a high-Ti pyroclastic glass (mg# = 0.54) illustrates this point. The high-Ti basalt is multi-saturated with olivine + armalcolite + spinel at the near-liquidus temperature of 1156°C. Olivine is never the only near-liquidus phase. In high-Ti pyroclastic glass, olivine is the only near-liquidus phase from 1302°C to 1195°C, where it is joined by spinel. An olivine + ilmenite + spinel assemblage does not appear until 1159°C. This relationship between mg# and olivine is illustrated for several other compositions by Longhi (1992).

The presence of magnesium-rich vitrophyres (Brett et al. 1971, BVSP 1981) and calculations of hypothetical melt compositions using Fe-Mg olivine/melt exchange coefficients (Walker et al. 1977, BVSP 1981, Longhi 1987, 1992) demonstrate the existence of low-Ti, olivine-saturated magmas with 11 to 13 wt % MgO and high-Ti, olivine + armalcolite-saturated magmas with 8 to 10 wt % MgO. Whether these parental crystalline mare basalts are primary compositions is questionable. Using a suite of pyroclastic Mg-rich glasses, Longhi (1987, 1992) illustrated crystallization sequences superimposed on liquid lines of descent for melts that approach primary compositions that could be parental to the more primitive crystalline mare basalts. Based on the assumption that primary magmas similar to these are parental to the crystalline mare basalts, the MgO contents for parental magmas for the high-Ti basalts should be 12 to 15 wt % and 15 to 19 wt % for the low-Ti and VLT mare basalts. A generalized low-pressure crystallization sequence for primary high-Ti mare basalts is olivine ⇒ olivine + spinel ⇒ olivine + spinel + armalcolite ⇒ armalcolite reacting with melt to form ilmenite ⇒ pigeonite ⇒ augite ± plagioclase ⇒ plagioclase. A generalized low-pressure crystallization sequence for primary low-Ti mare basalts is olivine ⇒ olivine + spinel ⇒ pigeonite ⇒ plagioclase ± augite ⇒ augite ± ilmenite. In both crystallization scenarios, olivine crystallizes alone over a fairly extensive temperature interval (100° to 200°C) (Longhi 1987, 1992). In addition, plagioclase is unstable near the liquidus and will become saturated in the melt only after extensive crystallization (>40 volume percent) (Longhi 1987).

High pressure experiments

Laboratory experiments on phase relations of mare basalts at high pressure are intended to provide insights into the formation of mare basalts within the lunar mantle. Ideally, these experiments should provide information about the pressure at which mare basalts were generated and the mineral assemblage with which the basaltic magma was in equilibrium. Numerous high pressure experiments have been run on a wide range of basalt compositions. A list of high pressure experiments is presented in Table 9. Interpretive problems concerning the depth of melting exist because of several assumptions. The first assumption is that during melting at least two of the original mantle minerals were in equilibrium with the basaltic melt before it was extracted. The depth of melting and segregation is assumed to be equivalent to the depth of multiple saturation determined experimentally. This may not necessarily be true. For example, extensive melting may exhaust all but one phase (i.e. olivine), resulting in the multiple saturation depth being meaningless. This appears to not be the case for many lunar basalts. Hess (1991, 1993) and Hess and Finnila (1997) demonstrated that most of the picritic (parental) magmas represented by pyroclastic glasses were multiply saturated at high pressure. The second assumption is that melting occurred at a single depth. Longhi (1992) has proposed models in which melting is polybaric in nature and, therefore, the multiple saturation depth represents an average depth of melting. The third assumption is that the melt compositions used in the high pressure experiments represent primary melts. Basalts that have undergone olivine fractionation since melting will give shallow multiple saturation depths. For most of the crystalline mare basalts, this is not a valid assumption, and therefore the estimated depth of melting and segregation is a minimum depth.

Despite these ambiguities, high-pressure phase relations of Mg-rich olivine normative mare basalts still provide valuable information. The most important aspect of the high-

Table 9. Temperature and pressure of multip[le saturation in high-pressure experiments on mare basalts and glasses (summary of data from Longhi 1992).

Sample	T(°C)	P(kbar)	phases	container	reference
			---low Ti---		
12002(b)	1380	12.5	ol, opx	Fe(h),Mo	1
12009(b)	1270	5-7	ol, pig	Fe(l)	2
"B" (s)	1300	10	ol, pig	G	3
12022(b)*	>1360	14-15	ol, pig	Fe(l)	4
12040(b)*	>1400	20-25	ol, opx	Fe(l)	4
14072(b)	1310	10	ol, opx, sp	Mo	5
14 VLT(sp)	1490	19	ol, opx	Fe(h)	6
	>1500	22	ol, pig	(Fe,Pt)	7
15016(b)	1310	11	ol, pig, sp	G	8
15016(s)	1350	12	ol, pig	Fe(h)	9
15065	>1270	<5	ol, pig, sp	Fe(h),Mo	10
15555(b)	1300	8.5	ol, pig	Fe(h)	10
15 Green (sp)	1450	17-18	ol, opx	Mo	11
17 VLT(sp)	1500	18	ol, opx(?)	Fe(h)	12
LUNA 24(s)	1200	5	aug, pl	Fe(h)	12
			---high Ti---		
10017	1220	10	aug, ilm	Fe(l)	13
avg. Ap11	<1200	<5	ol, aug, arm(?)	Fe(l)	14
15 Red(sp)	1460	25	ol, opx	Fe(h)	15
70017(b)	1230	5-7	ol, aug, sp	Mo	16
	1240	7.5	ol, opx, sp	G	16
70215(b)	1210	5-6	ol, aug, ilm, sp	Mo	16
	1230	6	ol, aug, sp	G	16
	1250	7-9	ol, aug	Fe(h)	17
70215(s)	1240	8	ol, aug, ilm	Fe(h)	9
74220(pg)	1490	20	ol, opx	Fe(h)	18
	1410	20	ol, opx	Mo	18
74275(b)	1320	12	ol, aug	Fe(h)	17

* = olivine-enriched composition; b = natural basalt; s = synthetic basalt; pg = natural picritic glass; s = synthetic picrite; Fe(l) = low-purity iron or mild steel; Fe(h) = high-purity iron; G = graphite; Mo = molybdenum; (Fe,Pt) = iron-platinum alloy

1: Walker et al. (1976); 2: Green et al. (1971b); 3: Kushiro (1972); 4: Green et al. (1971a); 5: Walker et al. (1972); 6: Chen et al. (1982); Chen and Lindsley (1983); 8: Hodges and Kushiro (1972); 9: Kesson (1975); 10: Walker et al. (1977); 11: Stolper (1974); 12: Grove and Vaniman (1978); 13: O'Hara et al. (1970); 14: Ringwood and Essene (1970); 15: Delano (1980); 16: Longhi et al. (1974); 17: Green et al. (1975); 18: Walker et al. (1975).

pressure studies is the pressures at which olivine + pyroxene ± spinel are stable. Multiple saturation pressures for high-Ti, low-Ti and VLT basalts overlap between 0.5 and 1.2 GPa (Table 9). This should be interpreted as a minimum pressure of melting. It also indicates that the sources for the different mare basalts are at similar depths. This has profound implications concerning the crystallization of a lunar magma ocean and the disruption of the resulting cumulate sequence (e.g. Ryder 1991; Shearer et al. 1991, Shearer and Papike 1993, Spera 1992, Hess and Parmentier 1995). An additional observation is that the mare basalts and the picritic glasses have different pressures of multiple saturation (Table 9). This may be interpreted in two ways: (1) the parental magmas for the picritic glasses and mare basalts are unrelated and in fact produced by melting at two different pressure regimes in the lunar mantle (less than 350 km and greater than 350 km); or (2) the multiple saturation depths for the mare basalts are an underestimation of the true depth of melting and the parental magmas for both the mare basalts and picritic glasses were generated at depths greater than 350 km. Based on the above assumptions, the latter appears to be more likely.

A second important observation from the high-pressure studies is that plagioclase is not stable along any high-pressure liquidus of a "primitive" mare basalt (Table 9). The absence of plagioclase as a residual phase in the source region for mare basalts requires that the ubiquitous negative Eu anomaly observed in lunar basalts is a characteristic of the mantle source. This requires plagioclase depletion to have occurred prior to melting and probably during the formation of the mantle source. Such a depletion has been used as an argument for the existence of a global lunar magma ocean early in the history of the Moon.

ISOTOPIC SIGNATURES OF THE CRYSTALLINE MARE BASALTS

Notable summaries of the isotopic systematics of lunar basalts include those by Nyquist (1977), BVSP (1981), SR Taylor (1982), and Nyquist and Shih (1992). The Nd-Sm, Rb-Sr, Lu-Hf, U-Pb and Hf-W isotopic systematics have provided fundamental and essential evidence for reconstructing the nature and evolution of the source region of lunar basalts. Relationships exhibited by these isotopic systems in mare basalts are summarized in Figure 29.

The εNd and initial Sr for lunar basalts indexed to age, landing site, and composition are illustrated in Figures 29A,B (Nyquist 1977, Nyquist and Shih 1992). Superimposed on the εNd versus time plot (Fig. 29A) are the chondrite evolution line and a growth curve for $^{147}Sm/^{144}Nd$ ~0.25. Most of the mare basalts plot at positive values of εNd and between these two curves. There appears to be little correlation between εNd and TiO_2 content of the basalts. Both high-Ti and low-Ti basalts from the Apollo 12, 15, and 17 collection sites lie along the 0.25 growth curve. This curve therefore represents the maximum observed depletion for LREE for the sources of the mare basalts. More recent measurements by Snyder et al. (1997) confirm that the Nd isotopic compositions of the ilmenite basalts are among the most radiogenic from the Moon (εNd = +10.5 to +11.2 at 3.2 Ga). The wide range in localities for which this maximum is observed indicates that LREE depletion relative to chondrites is a characteristic of the lunar mantle and is additional evidence of mantle-wide differentiation (Nyquist and Shih 1992). In their study of five Apollo 12 basalts, Snyder et al. (1997) interpreted the Sr and Nd isotopic systematics as indicating two distinct mantle sources for the Apollo 12 basaltic lithologies. Both of these sources were LREE-depleted for extended periods of time: up to 600 million years for the ilmenite basalt source and 900 million years for the pigeonite and olivine basalt source. In contrast, the high-alumina and very high-potassium basalts tend to fall at εNd values equal to and even slightly lower than the chondrite evolution line. Nyquist and Shih (1992) argued for this signature being imprinted on the high- alumina and very high-potassium basalts through either assimilation of a KREEP or highland material or mantle source hybridization. The lack of this signature in most other lunar basalts either implies different transport processes through the crust, different crustal structure during latter eruptive periods (i.e. KREEP preferentially stripped away during basin formation), or different mantle source regions. Figure 29B illustrates the variability in I_{Sr} for the mare basalts. The I_{Sr} values for the mare basalts are extremely low (slightly above BABI) and range from 0.69906±0.00004 for Luna 16 basalts to 0.69957±0.00005 for Apollo 12 sample 12065. Again, there appears to be a weak relationship between I_{Sr} and basalt composition.

Unruh et al. (1984) reported Lu-Hf data for a wide range of low- and high-Ti basalts. Figure 29C illustrates the $^{147}Sm/^{144}Nd$ and $^{176}Lu/^{177}Hf$ characteristics of the basalt sources compiled by Nyquist and Shih (1992) from the Nd and Hf isotopic data reported by Unruh et al. (1984). Figure 29C illustrates that the source for the lunar basalts had both non-chondritic Sm/Nd and Lu/Hf ratios. The source for the Apollo 12 feldspathic basalts plot the closest to the CHUR (Chondritic Uniform Reservoir). The source for the Apollo 12 ilmenite basalts and Apollo 17 high-Ti basalts plot the furthest away from CHUR. The sources for the high-Ti and low-Ti basalts define separate isotopic fields, with the sources for the low Ti-basalts showing a positive correlation between Lu/Hf and Nd/Sm and the

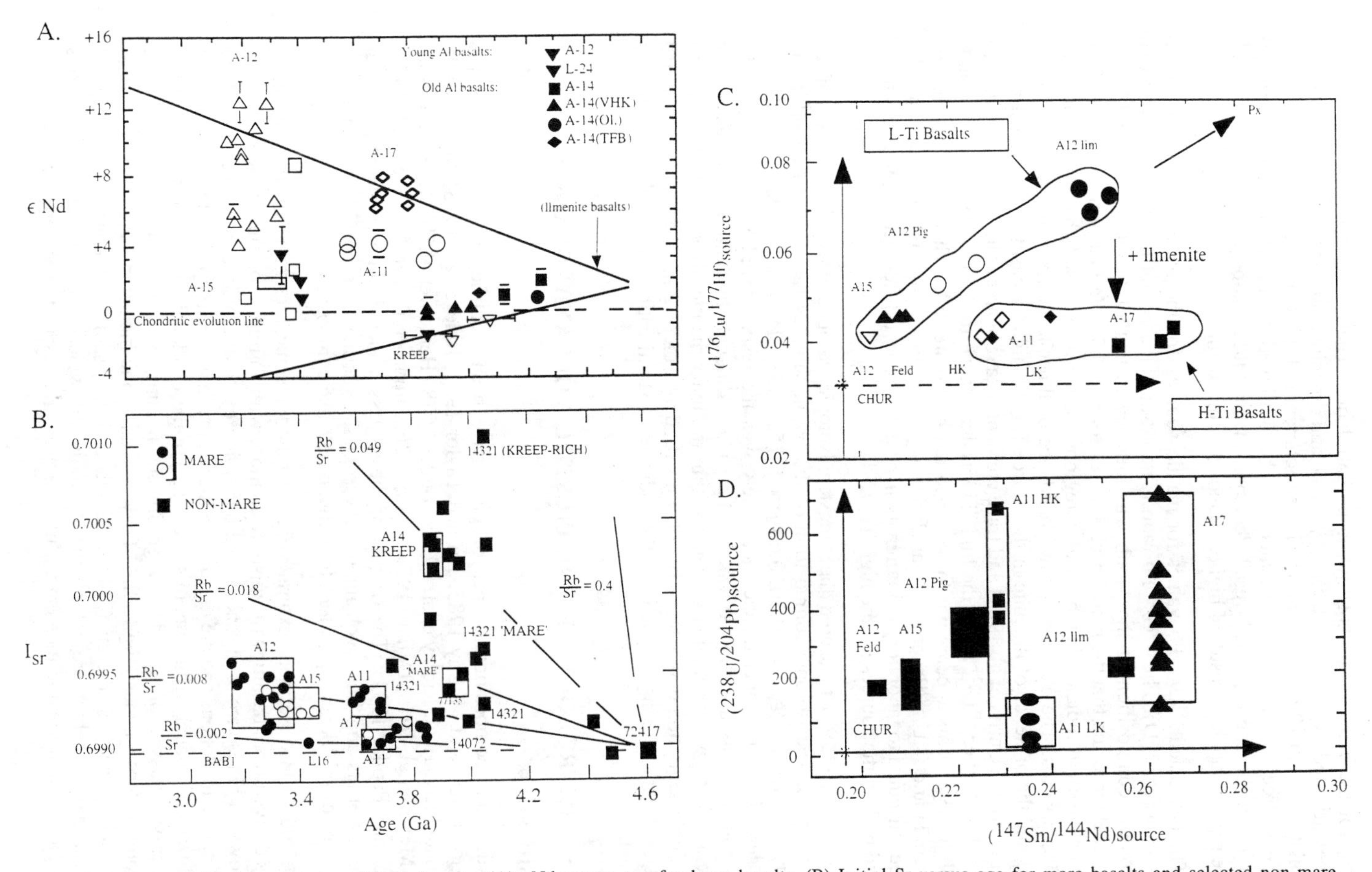

Figure 29. Isotopic characteristics of lunar basalts. (A) εNd versus age for lunar basalts. (B) Initial Sr versus age for mare basalts and selected non-mare lithologies. (C) $^{147}Sm/^{144}Nd$ versus $^{176}Hf/^{177}Hf$ calculated for mare basalt source regions assumed to have formed 4.56 Ga. (D) $^{147}Sm/^{144}Nd$ versus $^{238}U/^{204}Pb$ (= μ) calculated for mare basalt source regions assumed to have formed 4.56 Ga. [A, C, D modified from Nyquist and Shih (1992); B modified from Nyquist (1977).]

sources for the high-Ti basalts showing no correlation. These differences among sources may be attributed to variations in clinopyroxene and ilmenite abundances in the lunar mantle (Nyquist and Shih 1992). The progressive enrichment in Lu/Hf and Nd/Sm ratios observed for the sources of the low-Ti basalts may be attributed to increasing abundances of pyroxene in the source. The overall depletion in the Lu/Hf ratios of the high-Ti basalt source region may be a result of higher abundances of ilmenite.

The calculated $^{238}U/^{204}Pb$ values for mantle sources for the mare basalts range from 0 to ~650 (Fig. 29D). The $^{238}U/^{204}Pb$ ratio is based on the assumptions that the primordial Pb isotopic composition of the Moon was identical to that of troilite from the Canyon Diablo meteorite and that the mare basalt sources formed 4.56 Ga (Tatsumoto et al. 1973). These calculations indicate that the source regions for the mare basalts had non-chondritic, high U/Pb ratios. The absence or low abundance of ilmenite in the source for the low-Ti basalts and the presence of ilmenite in the source for the high-Ti basalts, as implied by the Lu-Hf isotopic (Unruh et al. 1984) and trace element data (Shearer et al. 1996b), should be reflected by higher $^{238}U/^{204}Pb$ in the high-Ti basalts relative to the low-Ti basalts (Nyquist and Shih 1992). This appears not to be supported by Figure 29D. This lack of correlation between Ti and $^{238}U/^{204}Pb$ and the variability of $^{238}U/^{204}Pb$ among high-Ti basalt samples may be attributed to laboratory contamination, or could reflect differences in the amount of trapped melt in the sources for the high-Ti basalt.

Lee et al. (1997) measured the tungsten isotopic compositions (^{182}Hf-^{182}W) of 21 lunar samples. The isotopic composition of the basalt samples ranges from chondritic to slightly radiogenic (ε_W = -0.50±0.60 to 6.75±0.42). They concluded that the ε_W heterogeneities in the lunar mantle most likely was the result of radioactive decay in the Moon following its accretion at 4.52 to 4.50 Ga. Although the authors suggested that differences in ε_W was a function of depth within the lunar mantle, they mistakenly equated the multiple-saturation depth for a basalt with its depth of origin. Therefore, the limited extent of mantle mixing implied by the radiogenic stratification of the lunar mantle is incorrect.

PETROLOGY OF THE PICRITIC VOLCANIC GLASSES

Classification

Numerous compositional types of glass have been identified in the lunar sample collection (Basu and McKay 1985, Hu and Taylor 1977, Reid et al. 1973, Meyer 1978, Delano and Livi 1981, Ridley et al. 1973a, Butler 1978, Heiken et al. 1974, McKay et al. 1973, Meyer et al. 1975, Delano 1979, 1980, 1986, Simon et al. 1989, Steele et al. 1992, Shearer and Papike 1993). The origins of these glasses include impact melting of lunar surface materials and lunar volcanism. These glasses define a compositional continuum ranging from soil-like compositions to mafic, basalt-like compositions.

Delano (1986) defined textural (Fig. 30) and chemical criteria for distinguishing glasses derived from basaltic volcanism from those with an impact origin. These criteria are as follows: (1) Presence or absence of schlieren and exotic inclusions. volcanic glasses are free of exotic inclusions and are uniform. Minerals that may be trapped in these glasses are crystallization products (ilmenite, metal, olivine) of these volcanic liquids (Figs. 30A,B,C). In contrast, glasses produced through impact melting of lunar surface materials are commonly heterogeneous (Fig. 30D). They contain exotic inclusions and streaks of fine-grained debris, called schlieren. (2) Intra-sample and inter-sample homogeneity. Compared to impact glasses, which are heterogeneous, individual volcanic glasses are homogeneous with regard to non-volatile elements. Unlike impact glasses that were produced by rapid

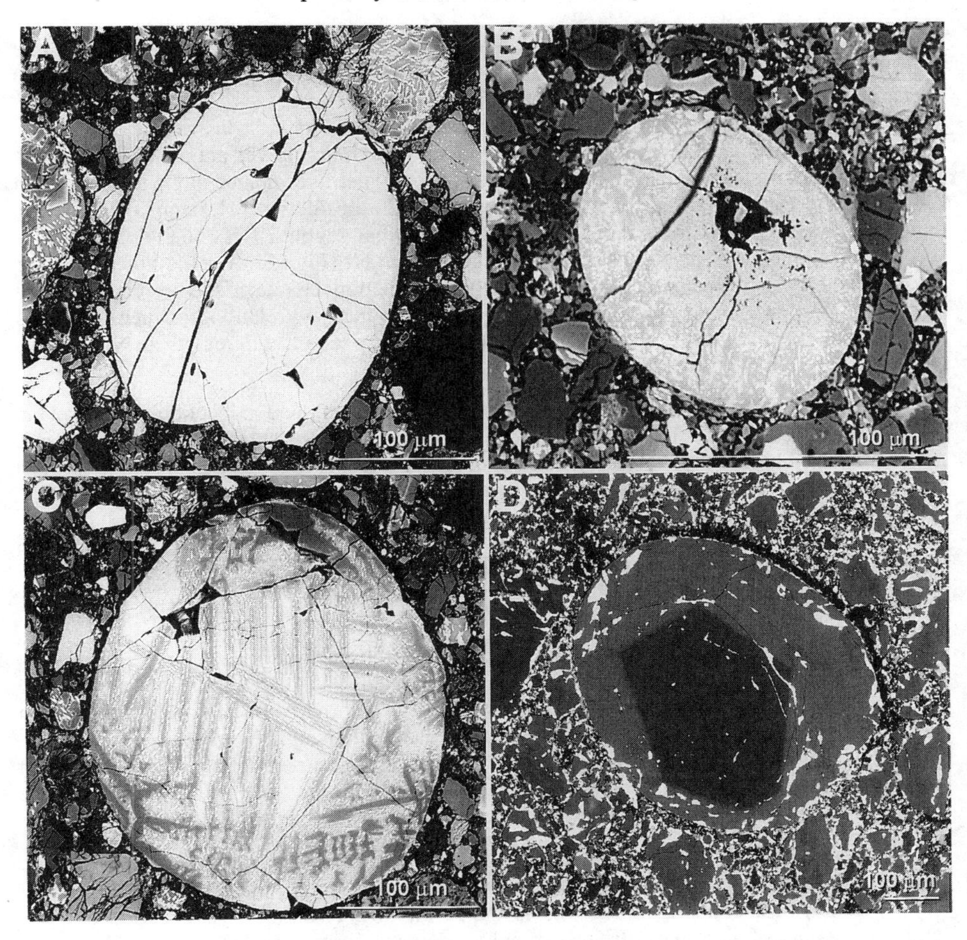

Figure 30. BSE images of glass beads within the lunar regolith. (A) Apollo 15 Green glass which represents a very low-Ti picritic magma that was erupted at the lunar surface through fire-fountaining. (B) Apollo 17 Orange glass which contains minor amounts of ilmenite. Like the Apollo 15 Green glass, this Mg-, Ti-rich glass is volcanic in origin. (C) Apollo 17 Orange glass which contains substantial blades of ilmenite due to slower cooling than B. (D) Heterogeneous impact glass containing large inclusion of olivine.

heating and cooling of heterogeneous lunar surface material, the volcanic glasses represent quenched products of a magma that was probably fairly homogeneous on a small scale. The individual volcanic glass beads show some compositional variability of volatile elements (S, Li) (Delano et al. 1994, Rutherford and Vogel 1995, Shearer et al. 1997). (3) Clustering of chemical compositions. The compositional arrays exhibited by impact glasses were primarily a function of the impact target and preferential volatilization of volatile elements. On the other hand, volcanic glasses from a single eruption tend to be chemically uniform and show chemical trends that were generated by crystal/liquid fractionation. (4) Surface-correlated and trapped volatiles. The best known examples of lunar volcanic glasses (Apollo 15 Green and Apollo 17 Orange glasses), have surfaces enriched in volatile elements compared to most other lunar samples. The glasses commonly have volatile concentrations that are equal to or less than those of mare basalts. The volatile loss from the glass and volatile-rich coatings are both products of fire-fountaining.(5) Ferromagnetic resonance intensity. During impact-melting processes, single- domain iron particles are formed by Fe reduction. This results in the impact glasses having a high ferromagnetic resonance intensity. Volcanic glasses have much lower abundances of single domain iron particles and therefore have lower ferromagnetic resonance intensities.(6) High Mg/Al ratios. Samples of volcanic glasses have Mg/Al ratios between 1.7 and 3.3, whereas impact glasses have ratios similar to those of the lunar regolith (less than 1.5).(7) Mg-correlated abundances of Ni. In a plot of MgO versus Ni (Fig. 28) for the volcanic lunar glasses, these two elements exhibit a positive correlation characteristic of crystal/liquid fractionation processes. In contrast, impact glasses show no correlation between MgO and Ni. Their MgO contents are probably related to the impact target, whereas their Ni contents may be related to the projectile.

Based on these criteria, major element characteristics, and collection site, Delano (1986) identified twenty-five distinct varieties of picritic volcanic glasses. Their major and trace element characteristics are presented in Table A5.31. Shearer and Papike (1993) presented trace element evidence for the existence of two additional volcanic glass types. Other potential volcanic glass types have been identified in studies of regolith returned by the Luna missions (GJ Taylor et al. 1977, Norman et al. 1978, Grove and Vaniman 1978) and in lunar meteorites (Arai and Warren 1996, Arai et al. 1997).

Textures

Lunar pyroclastic deposits are not composed entirely of homogeneous spherical glass beads. Beads are mixed with lunar regolith and are coated with both sublimates of volcanic origin and fine regolith fragments. The volatile coats are enriched in F, Cl, S, Zn, and Pb compared to other lunar materials. Some of the volatile elements (Au, Sb, and Ge) on the glass surfaces occur in the same ratios as are found in C1 chondritic meteorites (Morgan and Wandless 1984, Delano 1980). Some of these elements were transported in the eruptive gas phase as fluorides, chlorides and sulfides (Wasson et al. 1976).

Glass beads also can be fractured and substantially crystalline. Some samples of the high-Ti glasses that are commonly orange or red appear black when substatial crystallization occurred (Fig. 30). This is the result of substantial crystallization of fine-grained crystals of ilmenite, olivine, spinel, and metal from the glass. Although not as noticeable, the very low-Ti glasses may also contain minute crystals of olivine, spinel, and metal.

There is limited information concerning the straitgraphy of lunar pyroclastic deposits. The pyroclastic deposit sampled on the rim of Shorty Crater by the Apollo 17 mission indicates sub-units in the deposit stratigraphy that are related to the extent of glass crystallization and shape. This certainly is related to the evolution of fire-fountaining dynamics that produced the Apollo 17 deposit. Experiments by Uhlmann et al. (1974), radiative cooling calculations by Arndt et al. (1979, 1984), and observations concerning bead shape (McKay et al. 1973, Heiken and McKay 1977, Clanton et al. 1978) have shed some light on the nature of the fire-fountaining. Cooling rate experiments indicate that homogeneous glass beads must have cooled faster than 10° to 60°C/second. The spheres containing crystallites must have cooled slower. This cooling rate, which is much slower than that expected in a volatile-absent eruption into a lunar vacuum, and observed bead shapes suggest that magma droplets were colliding repeatedly in a hot, turbulent plume of volcanic gas for as long as 10 minutes (Arndt et al. 1984). Changes in glass characteristics in the pyroclastic deposit suggest that there were changing fire-fountaining conditions during the eruptions.

GEOCHEMISTRY OF THE PICRITIC VOLCANIC GLASSES

Major elements

Within the context of terrestrial basalt classification (i.e. basalt tetrahedron; Yoder and Tilly 1962, Yoder 1976), the normative mineral assemblages of the picritic glasses are hypersthene + olivine (Table 5.A31). Like the mare basalts, the volcanic glasses exhibit an extremely wide range in TiO_2 content from 0.20 to 17.0 wt % (Fig. 23). The glasses thus far documented have TiO_2 contents that correspond to mare basalts with very low TiO_2 (VLT; 0.2-1.0 wt %), intermediate TiO_2 (3.4-6.9 wt %), and high TiO_2 (8.6-17 wt %). At several sampling sites (Apollo 12, 14, 15, and 16), mare basalts with 1 to 3% TiO_2 are a dominant lithology. However, only two potential volcanic glass types thus far documented (Apollo 14 LAP, Apollo 15 anomalous yellow) have TiO_2 contents within that range (Shearer and Papike 1993). In addition to TiO_2, the picritic glasses show other chemical characteristics similar to the mare basalts. The picritic glasses have mineralogical (minor iron metal) and chemical (reduced valence states for Fe, Cr, and Ti) signatures indicative of formation at oxygen fugacities at or below the iron-wustite buffer. They are depleted in alkali, volatile, and siderophile elements. Water is absent from the glass beads.

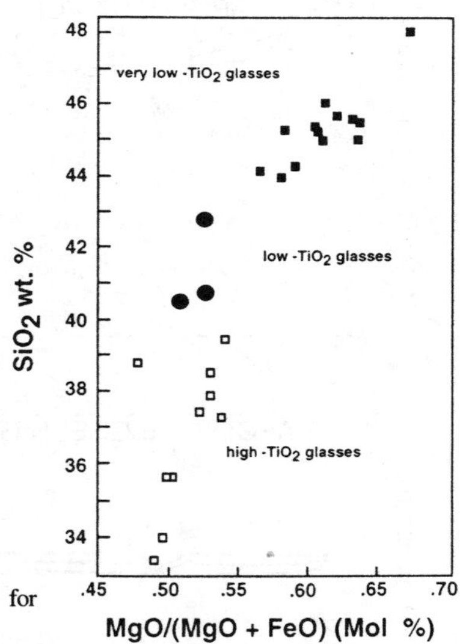

Figure 31. mg# = MgO/(MgO+FeO) (mol %) versus SiO_2 for the more Mg-rich members of each suite of picritic glasses.

Although a range of geochemical signatures exhibited by the glasses are similar to those of the mare basalts, other characteristics, such as mg# and MgO, CaO, and Al_2O_3 contents (Fig. 23) are consistently different. The picritic glasses are consistently higher in MgO and mg# and lower in Al_2O_3 and CaO than most fine-grained non-cumulate mare basalts. These differences are attributed to olivine fractionation. Among the most primitive members of each glass group, there appears to be a general positive correlation between mg# and SiO_2 (Fig. 31). This relationship has been attributed to the nature of the source, a residuum buffering effect on the mg#, or a pressure effect on melting (Delano 1986, Grove and Lindsley 1978, Steele et al. 1992, Shearer and Papike 1993). The high mg# values of the picritic glasses make them the best candidates for primary melts. However, experimental modeling of liquid lines of descent (Longhi 1987) indicates that picritic magmas represented by the glass beads are not related to the crystalline mare basalts by shallow crystal/melt fractionation processes. The major element characteristics of the volcanic glasses do indicate that a range of primary melts are responsible for the diversity of lunar basalts. In addition, they warn of the care that one must take in selecting the appropriate basalt compositions in evaluating both shallow and deep basalt-generating processes.

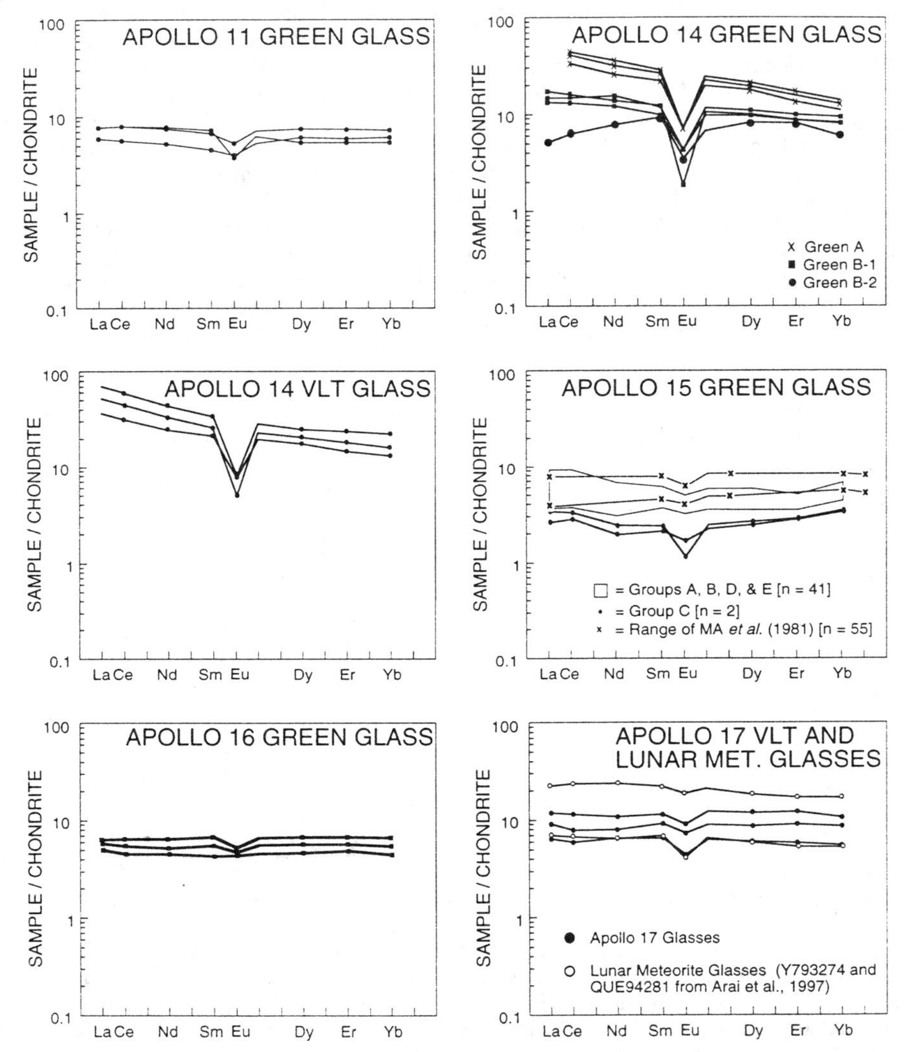

Figure 32A. REE contents in the volcanic, picritic glasses, normalized to REE contents in chondritic meteorites: Very Low-Ti glasses.

Trace elements

Rare earth elements. Rare earth element abundances in the very low-TiO_2 glasses, normalized to C1 chondrite values (Anders and Ebihara 1982) are plotted in Figure 32A (Ma et al. 1981a, Shearer et al. 1990, 1991, Shearer and Papike 1993, Steele et al. 1992; Arai et al. 1997). The glasses from Apollo 11, Apollo 15, Apollo 16, Apollo 17 and lunar meteorites have REE patterns that are essentially flat with abundances between 1.5 and 10 × C1 chondrite and a small negative Eu anomaly. In the Apollo 15 green C glass, Ma et al. (1981a) and Steele et al. (1992) observed a slightly positive slope from La to Lu. The REE abundances of Apollo 15 green A, B, D, and E glasses, Apollo 11 green glass, Apollo 16 green glass and Apollo 17 VLT glass overlap with [Ce]N ranging from 3.3 to 11 × C1 chondrite. The Apollo 17 VLT and the Apollo 15 Group B glasses show the widest variation in REE abundances ($[Ce]_N$ = 6 to 11 and 4 to 8.5, respectively). All the other glasses show a much more limited compositional dispersion. The two Apollo 15 group C glasses exhibit lower REE abundances than the previously noted groups with $[Ce]_N$ equal to 2.7 and 3.1 (Shearer and Papike 1993).

In contrast to these relatively flat patterns with minor Eu anomalies ($[Sm/Eu]_N < 2.5$), the very low-TiO_2 glasses from the Apollo 14 site (VLT, Green A, Green B type 1) are typically LREE-enriched with slightly larger negative Eu anomalies ($[Sm/Eu]_N$ = 6 to 14; Fig. 32A) (Shearer and Papike 1993). The total REE abundances of the Green A and VLT glasses overlap, with $[Ce]_N$ ranging from 30 to 70. The REE abundances of the more MgO-rich Green B type 1 glass are substantially lower ($[Ce]_N$ = 8 to 20). One Green B glass bead (referred to as type 2) contrasts with the other Green B glasses. Although identical in major element characteristics, duplicate analyses of this single glass bead show a LREE-depleted pattern (Fig. 32A).

Rare earth element abundances in the low- and intermediate-TiO_2 glasses (Apollo 14 yellow, Apollo 14 LAP (low alkali-picrite), Apollo 15 yellow, and Apollo 17 yellow) are plotted in Figure 32B. The low- to intermediate-TiO_2 glasses from the three sites shown (Apollo 14, Apollo 15, and Apollo 17) exhibit contrasting REE patterns (Fig. 32B). The Apollo 14 and Apollo 15 glasses have a pattern shape that is similar to that of the very low-TiO_2 green glasses from the Apollo 15 site. The pattern is flat with a small negative Eu anomaly. In contrast to the very low-TiO_2 glasses, the REE abundance is higher in the yellow glasses and the negative Eu anomaly is slightly deeper (Fig. 32B). The Apollo 14 yellow glasses have $[Ce]_N$ which range from 40 to 50, $[Yb]_N$ which ranges from 30 to 40, and $[Sm/Eu]_N$ which ranges from 2.1 to 3.5. Most of the ion microprobe analysis of the Apollo 15 yellow glasses (Shearer and Papike 1993) produced REE patterns that fell within the range observed by Ma et al. (1981a) and Hughes et al. (1988) (Fig. 32B). In these glasses, the $[Ce]_N$ ranges from 22 to 40, the $[Yb]_N$ ranges from 20 to 30, and the $[Sm/Eu]_N$ ranges from 1.25 to 2.1.

The Apollo 17 yellow glass (TiO_2 content = 6.5 to 6.9 wt %) has a REE pattern similar to many of the high-TiO_2 glasses (Hughes et al. 1988, 1989, BVSP 1981, Shearer et al. 1990; 1991, Shearer and Papike 1993). This yellow glass is LREE-depleted ($[La/Sm]_N < 1$), with $[Ce]_N$ from 15 to 35, and a $[Sm/Eu]_N$ from 1.1 to 1.5. The HREE content of this glass type is similar to that of the other yellow glasses ($[Yb]_N$ = 20 to 42; Fig. 32B).

The REE abundances for the high-TiO_2 glasses, normalized to C1 chondrite values, are plotted in Figures 32C,D. Figure 32D also compares the ion microprobe analyses for the Apollo 17 orange glass (Shearer and Papike 1993) with INA analyses of glass clods and individual glass beads (Hughes et al. 1988, 1989, BVSP 1981). The REE abundances in the high-TiO_2 glasses are generally higher than those of the very low- to intermediate-TiO_2 glasses. Differences exist in REE characteristics among the high-TiO_2 glasses. These include differences in LREE slope, total REE abundance and size of the negative Eu anomaly.

The Apollo 11 orange glasses have a REE pattern possessing a positive LREE slope and a negative HREE slope. The $[Ce]_N$ and $[Yb]_N$ values are approximately 20 and 25 to

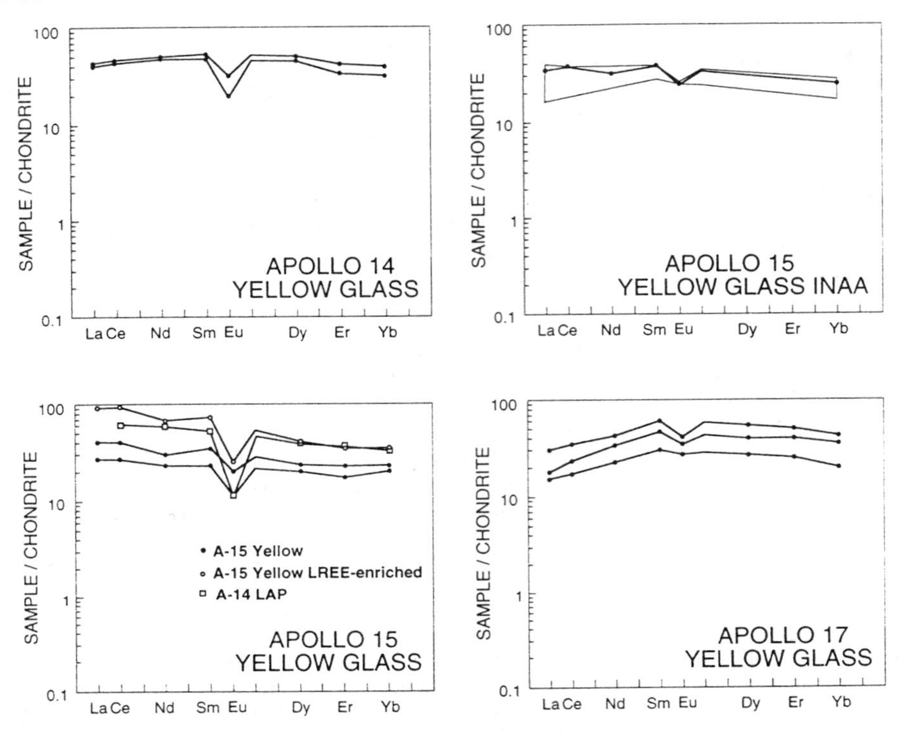

Figure 32B. REE contents in the volcanic, picritic glasses, normalized to REE contents in chondritic meteorites: Low- to intermediate-Ti glasses.

30, respectively (Fig. 32C). The negative Eu anomaly is slight with $[Sm/Eu]_N$ between 1.3 and 2. The Apollo 15 orange glass and the Apollo 17 orange glasses types 1, 2 and 74220 have similar patterns. Compared to the Apollo 11 orange glass, the orange glass from the Apollo 15 site has only slightly higher abundances of REE ($[Ce]_N$ = 31 and $[Yb]_N$ = 31) and a nearly parallel pattern except for Eu ($[Sm/Eu]_N$ = 1.2). The various types of Apollo 17 orange glass (Fig. 32D) show overlapping REE concentrations that cannot be distinguished from each other. Relative to the Apollo 11 and Apollo 15 orange glasses, the Apollo 17 glasses are slightly more enriched in LREE ($[Ce]_N$ = 30 to 50). The slope of the LREE is still positive. The HREE also have slightly higher abundances ($[Yb]_N$ = 25 to 50).

The high-TiO_2 glasses from the Apollo 12 (red), Apollo 14 (orange, red-black), and the Apollo 15 (red) sites have REE patterns with slightly negative (Apollo 12 red, Apollo 15 red) to slightly positive (Apollo 14 orange and red-black) LREE slopes. All have negative HREE slopes. The Apollo 12 red glass has REE patterns with $[Ce]_N$ that range from 60 to 80, $[Yb]_N$ values that range from 40 to 45, and negative Eu anomalies with $[Sm/Eu]_N$ between 1.5 and 2.3. The Apollo orange glass has higher total REE abundances ($[Ce]_N$ between 85 and 120), and a slightly larger negative Eu anomaly ($[Sm/Eu]_N$ = 2.5). The Apollo red-black glass has REE patterns with $[Ce]_N$ between 55 and 90 and negative Eu anomalies with $[Sm/Eu]_N$ between 1.3 and 1.9.

Barium, strontium, zirconium, and scandium. Within the very low-TiO_2 glasses, there are substantial differences among Ba, Sr, and Zr contents. A plot of Sr

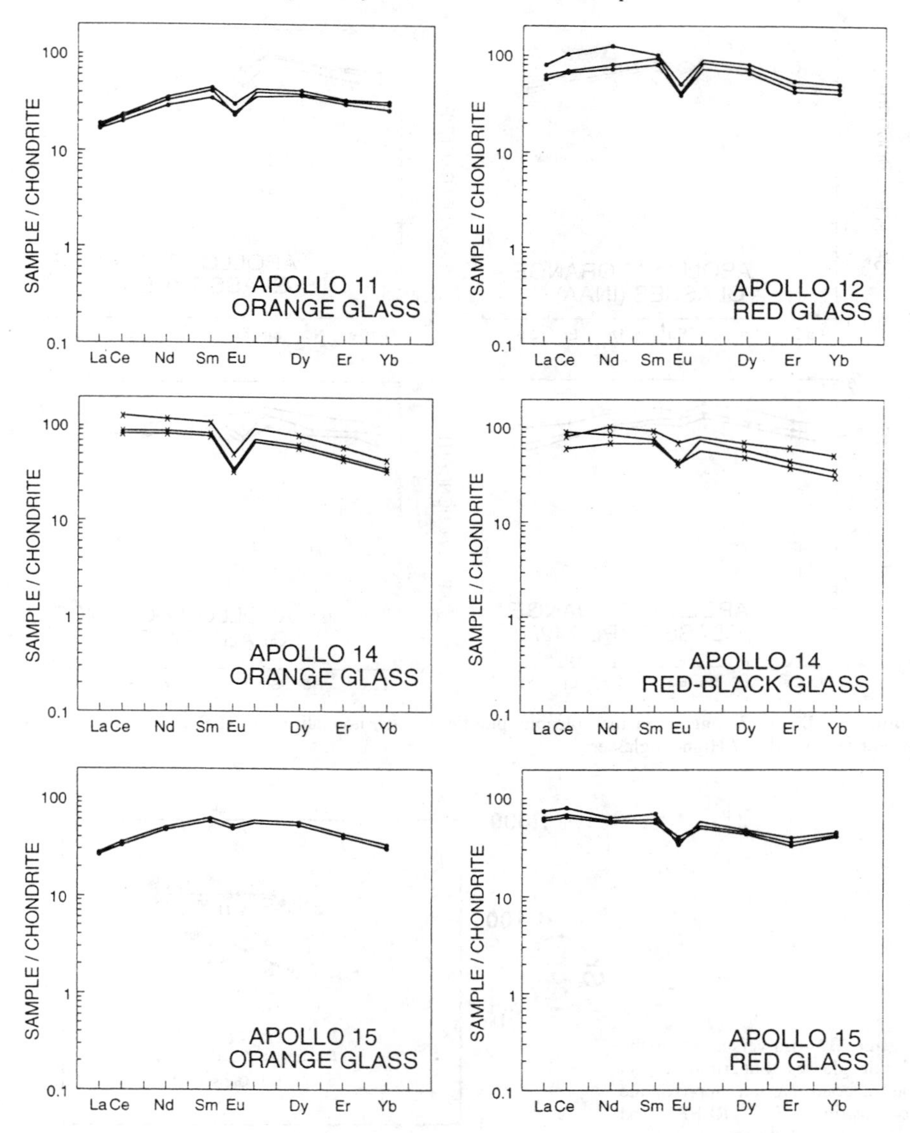

Figure 32C. REE contents in the volcanic, picritic glasses, normalized to REE contents in chondritic meteorites: High-Ti glasses.

versus Ba illustrates some of these differences (Fig. 33). The Apollo 14 glasses tend to be the most enriched in Ba, Sr, and Zr. The Apollo 15 green C glass has the lowest abundances of these incompatible elements. Ion microprobe analyses shows that scandium exhibits very little variation (Table A5.31) (Shearer and Papike 1993). INAA analyses of individual glass beads from the Apollo 15 site suggests there are real and subtle differences among the glasses in Sc (Steele et al. 1992). As noted by Delano (1990), there are substantial differences in Sc abundance between the picritic glasses and many of the crystalline mare basalts. These differences are very apparent in comparisons between

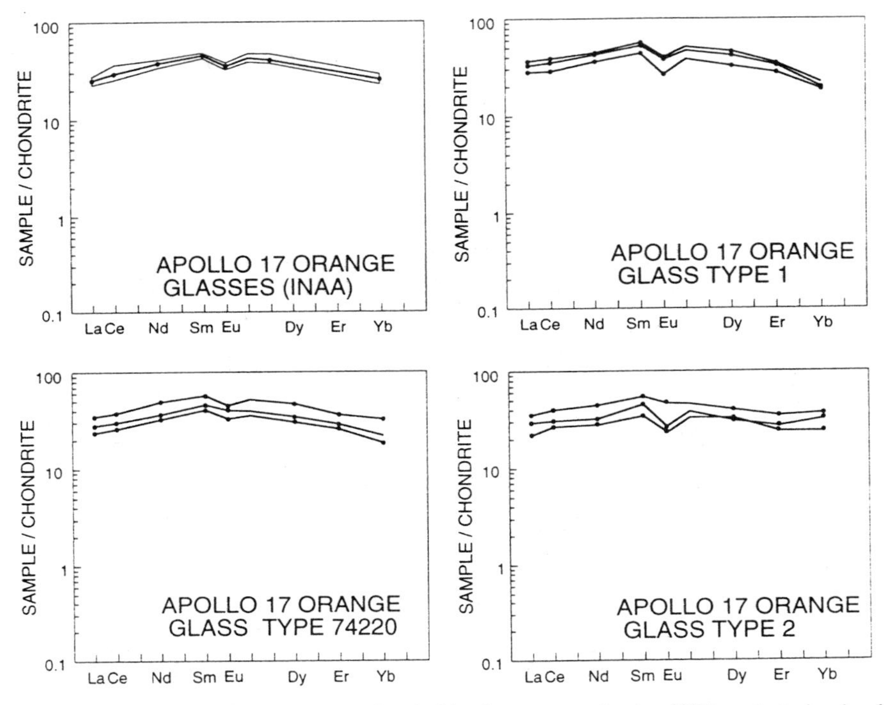

Figure 32D. REE contents in the volcanic, picritic glasses, normalized to REE contents in chondriti meteorites: Apollo 17 High-Ti glasses.

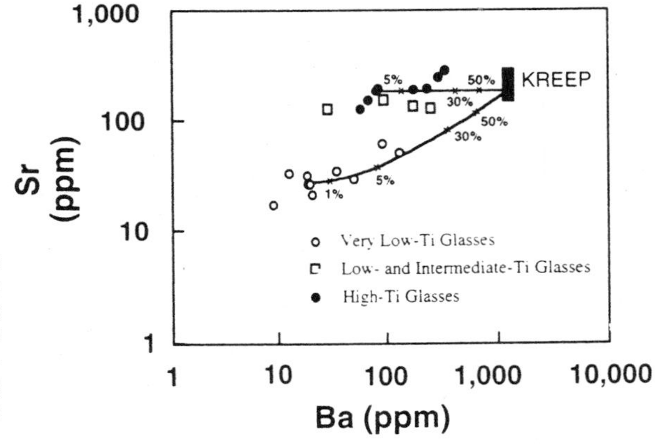

Figure 33. Ba versus Sr in the picritic glasses. Superimposed on the diagram are an approximate composition of "KREEP" and mixing lines between glass compositions and "KREEP". Mixing proportions along mixing lines are indicated.

basalts and glasses with similar TiO_2 contents.

The intermediate TiO_2 glasses also exhibit differences in Zr, Ba, and Sr among the various chemical types (Fig. 33). These incompatible elements decrease in the order Apollo 14 to Apollo 15 to Apollo 17. The LAP glass from Apollo 14 and the anomalous yellow glass from Apollo 15 (analysis 2b in Table A5.31) generally show incompatible element enrichments similar to the Apollo 14 intermediate-TiO_2 glasses (Fig. 33). The Apollo 15 glasses do show a measurable difference in Sc compared to the intermediate-TiO_2 glasses from the Apollo 14 and Apollo 17 sites (Table A5.31). In these three glass groups, Sc appears to be correlated with the abundance of TiO_2.

The high-TiO_2 glasses generally have higher Sr and Zr contents than glasses with lower TiO_2 (Fig. 33). They have overlapping abundances of Ba (Fig. 33) and Sc (Table A5.31). Among the high-TiO_2 glasses, the Apollo 12 red, the Apollo 14 orange and red-black, and the Apollo 15 red glasses have higher incompatible element concentrations than the high-TiO_2 glasses from the other sample sites.

Nickel, cobalt, chromium, vanadium. Microprobe studies of glass beads indicate that the Ni content in the very low-TiO_2 glasses is generally higher than that in the high-TiO_2 glasses (Delano 1986, Shearer et al. 1991). Beside this general observation, there is not a consistent relationship between Ni and either MgO or TiO_2 among the glass types. Within each glass type, the Ni content shows considerable variation (e.g. Apollo 14 VLT, 90 to 140 ppm) that may show a negative or positive correlation with MgO (Chen et al. 1982, Delano 1986, Shearer et al. 1991, 1996a; Steele et al. 1992). The Ni contents of the volcanic glasses is higher than that of mare basalts with similar TiO_2 content (SR Taylor 1982).

Unlike Ni, Co shows a very limited correlation with TiO_2. Although the very low-TiO_2 glasses may have some of the highest Co concentrations (Apollo 14 VLT and Green A) and the high-TiO_2 glasses some of the lowest Co concentrations (Apollo 14 red-black), there is overlap among the glasses (Fig. 34). Data from Shearer et al. (1996a) show that Ni and Co exhibit a positive correlation and that the Ni/Co ratio in the volcanic glasses exhibit limited fractionation. Ma et al. (1981a), Galbreath et al. (1990), Steele et al. (1992) and Shearer et al. (1996b) demonstrated that within the Apollo 15 green glasses (combining all types) there is a positive correlation between Co and incompatible elements such as Zr, REE, and Ba. Figure 34 illustrates that if the glass bead population is considered in its entirety, the incompatible element content (assuming Zr is representative of all of these incompatible elements) is not highly correlated with Co. In the case illustrated, a suite of samples with a given Co content may have a range of incompatible element concentrations.

The V contents of the glasses as determined by ion microprobe overlap over the full range of TiO_2 abundances. The high-TiO_2 glasses generally tend to have more Cr than the other volcanic glasses (Table A5.31).

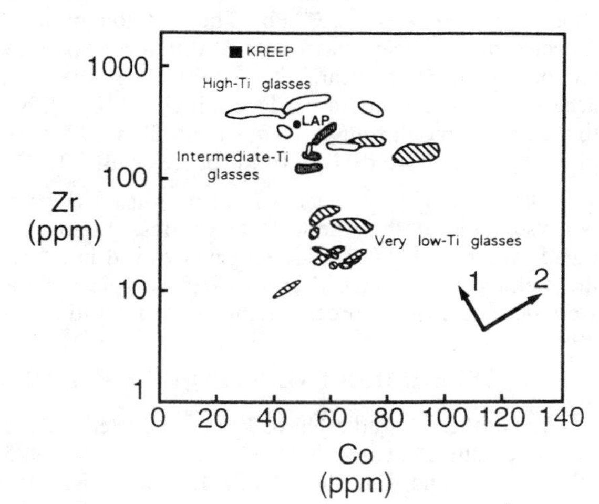

Figure 34. Concentrations of Zr versus Co in lunar picritic glasses.

Light lithophile elements (B, Be, Li). The insights offered by these elements in understanding terrestrial basaltic and rhyolitic processes (Ryan and Langmuir 1987, 1988) demonstrate their potential importance in understanding lunar basaltic magmatism. These elements have not been analyzed in any great detail in picritic glasses until recently (Shearer et al. 1994). The light lithophile elements (LLE) show a wide range of variability with Li ranging from 1.2 to 23.8 ppm, Be ranging from 0.06 to 3.09 ppm and B ranging from 0.11 to 3.87 ppm. At individual Apollo sampling sites, the LLE content is generally correlated to TiO_2. LLE abundances also parallel the enrichments of other lithophile elements such as Ba, Zr, Sr, and the REE. The volcanic glasses are distinct from the mare basalts in their abundance of LLE and LLE ratios.

ISOTOPIC SIGNATURES OF THE VOLCANIC GLASSES

The measured $^{238}U/^{204}Pb$ for the glasses and the calculated $^{238}U/^{204}Pb$ (μ) for the picritic magma sources contrast with similar measurements and calculations for the crystalline mare basalts. The $^{238}U/^{204}Pb$ ratios for the picritic glasses are equal to or slightly higher than the spatially associated crystalline mare basalts of similar composition (Tatsumoto et al. 1987). The calculated μ for the Apollo 17 orange glass is ~35 (Tera and Wasserburg 1976). For the Apollo 15 green glass it ranges from 19 to 55 (Tatsumoto et al. 1987). These are considerably less than the values of μ for the crystalline mare basalts (>200) (Nyquist and Shih 1992). In the Apollo 17 orange glass, the surface-correlated Pb appears to be in isotopic equilibrium with the glass (and magma) (Tera and Wasserburg 1976). However, in the Apollo 15 green glass, the isotopic composition of Pb on the glass bead surfaces was distinctly different from that in the glass bead interior (Tatsumoto et al. 1987). Tatsumoto et al. (1987) attributed this difference in the green glass to post-eruptive, lunar surface contamination.

The contrast between calculated $^{238}U/^{204}Pb$ for the picritic and "crystalline mare basalt" sources poses some interesting problems concerning U/Pb evolution in the lunar mantle and the nature of lunar basalt sources. Tera and Wasserburg (1974, 1976) and Tatsumoto et al. (1987) interpreted this difference as indicating that the picritic magmas, represented by the volcanic glasses, were derived from a source with considerably lower values of $^{238}U/^{204}Pb$. These differences between mare basalt and picritic glass sources in $^{238}U/^{204}Pb$ may reflect differences in the degree of volatile loss during early lunar mantle evolution. The differences do not reflect variable amounts of trapped residual liquid or differences in accessory mineralogy. Greater amounts of residual liquids with a KREEP-like signature should increase the $^{238}U/^{204}Pb$. The addition of a KREEP component should also be reflected in significant trace element differences between spatially associated mare basalts and picritic glasses (Shearer et al. 1990, 1991, Shearer and Papike 1993). The presence of accessory ilmenite should produce a higher $^{238}U/^{204}Pb$ in the high-Ti basalt source relative to the low-Ti basalt source (Nyquist and Shih 1992). However, both the low- and high-Ti picritic glasses were derived from sources with $^{238}U/^{204}Pb$ lower than the mare basalts.

Other isotopic systematics have been used with limited frequency. Based on the Sm-Nd systematics of the Apollo 15 green glass, Lugmair and Marti (1978) concluded that the mantle source for this picritic magma evolved in a "chondritic" Sm/Nd environment until the melting event. Sm-Nd and Lu-Hf isotopic systematics indicate that the mare basalts were derived from sources having non-chondritic Sm/Nd and Lu/Hf (Nyquist and Shih 1992).

EXPERIMENTAL STUDIES OF THE VOLCANIC GLASSES

Experimental studies have been conducted on a number of lunar glasses and their synthetic equivalents (Stolper 1974, Green et al. 1975, Delano 1980, 1986; Chen et al. 1982, Chen and Lindsley 1983, Longhi 1987, 1992). These studies have included determinations of phase relations and liquid lines of descent at low pressure and high pressure experiments to evaluate depths of melting, melting processes and melt properties (i.e. density).

Low-pressure experiments attempt to reconstruct conditions of crystallization experienced by the picritic magmas at the lunar surface. Results of selected low-pressure crystallization experiments on picritic glass compositions are presented in Tables 6 and 8. The picritic glasses have much higher liquidus temperatures than the crystalline mare basalts (Longhi 1992). Both the low-Ti and high-Ti glasses have olivine on the liquidus (Stolper 1974, Grove and Vaniman 1978, Longhi 1992). In the low-Ti glasses, pigeonite is the second major phase to crystallize, at approximately 200°C after olivine starts to precipitate (Longhi 1987, 1992). Plagioclase and augite crystallize after pigeonite. The calculated fractionation paths of the low-Ti glasses are sub-parallel to natural trends. In the two high-Ti picritic glasses that have been investigated, armalcolite is the second major phase to appear and develops a reaction relation with the liquid when ilmenite appears. The high-Ti glass compositions do not exhibit as an extensive interval of olivine crystallization as the low-Ti glasses. Pyroxene and plagioclase follow the crystallization of the Fe-Ti oxides. Both experiments and calculations (Longhi 1987, 1992) show that during crystallization of high-Ti picrites, residual liquids will follow paths of increasing TiO_2 and decreasing MgO during olivine crystallization, but TiO_2 and MgO will both decrease once olivine, armalcolite and ilmenite co-crystallize. Longhi (1987, 1992) and Shearer and Papike (1993) demonstrated that the liquid lines of descent for the picritic magmas generally do not overlap compositional fields for mare basalt magmas.

Results of high pressure experiments are presented in Table 9. These high pressure experiments indicate that the temperature and pressure of multiple saturation of the picritic glasses (1410-1500°C, and 1.7-2.5 GPa) exceed those of the primitive mare basalts (1200-1380°C, and 0.6-1.2 GPa). The high-Ti and very low-Ti picritic glasses have similar temperatures and pressures of multiple saturation and identical multiple saturation phases (olivine and orthopyroxene) (Table 9 and Fig. 35). These data may be interpreted in several ways. First, the multiple saturation depth may be interpreted as the minimum depth of melting and melt segregation. This interpretation assumes that both olivine and orthopyroxene were stable in the residua during melting. Based on the observations and calculations of Hess (1993) and Hess and Finnila (1997), this appears to be a reasonable assumption. If this is true, all picritic magmas represented by the volcanic glasses were generated deep in the lunar mantle (> 400 km). Melting that generated the mare basalts may have initiated at similar or shallower depths. Alternatively, the multiple saturation depth may represent an average depth of melting in a polybaric melting process (Longhi 1992). In this scenario, melting may have initiated at depths of approximately 1000 km in the lunar mantle and continued to 100 to 200 km. A third interpretation of the multi-saturation depth is that it is meaningless. High degrees of partial melting will result in the picritic magmas being saturated only with olivine in the source. Binder (1978, 1982, 1985) suggested that the picritic magmas were produced at fairly shallow depths (100-300 km) by high degrees of partial melting. Again, the work of Hess (1993) and Hess and Finnila (1997) strongly suggests that these melts were multiply saturated with olivine and orthopyroxene.

The origin of the exotic (relative to terrestrial magmas) high-Ti mare basalts and glasses has long been debated. Hubbard and Minear (1975) proposed that these high-Ti basalts were produced through ilmenite assimilation by low-Ti magmas at fairly shallow mantle depths. Ringwood and Kesson (1976) rejected this model based on the assimilation reaction stoichiometry and the thermal energy constraints. Several novel experimental approaches have been used to elucidate the origin of these magmas. Wagner and Grove (1997) evaluated the claim made by Ringwood and Kesson (1976) by investigating the dissolution rates of ilmenite in melts equivalent in composition to the Apollo 15 very low-Ti

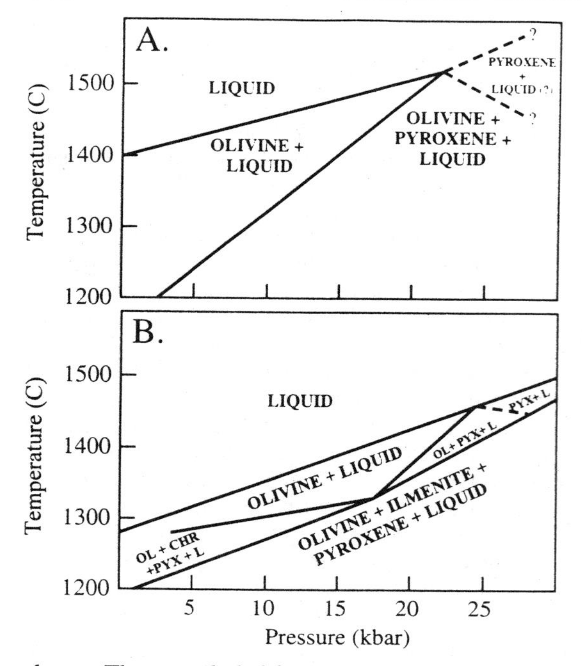

Figure 35. Simplified pressure-temperature phase diagram for (A) an Apollo 14 very low-Ti glass (Chen et al. 1982, Chen and Lindsley 1983) and (B) an Apollo 15 high-Ti glass (Delano 1980).

glasses. They concluded from their experimental results that the rate of ilmenite dissolution and the thermal requirements for dissolution did not eliminate the assimilation models.

An experimental study by Circone and Agee (1996) attempted to evaluate the prediction by Delano (1990) that the high density of the high-Ti magmas may potentially hinder segregation and migration of these melts in the lunar mantle. In this series of experiments, they performed static compressional density measurements on melt compositions equivalent to Apollo 14 black pyroclastic glass at pressures between 1.0 and 11.5 GPa. They concluded that the very high-Ti magmas will be denser than equilibrium olivine and orthopyroxene at pressures greater than 2.0 GPa (410 km) and 0.5 GPa (100 km), respectively. This indicates that the maximum depth of buoyant rise for magmas of this composition is approximately 400 km. If the high-Ti basalts are produced from great depths in the lunar mantle, a diapiric mechanism may be required to transport these magmas to the lunar surface (Hess 1991).

BASALT TYPES IDENTIFIED BY REMOTE SPECTRAL DATA

Our sampling of the Moon is represented by nine sample return missions and several meteorites of lunar origin. This limited collection of lunar materials compromises our ability to reconstruct the geology and history of the lunar crust and basaltic magmatism. Fortunately, remote spectral observations made by the Apollo, Galileo, Clementine Lunar Prospector missions, and Earth-based telescopic observations have extended our knowledge concerning the chemistry and distribution of mare basalts to unsampled regions of the Moon. The source of these data include γ-ray and x-ray spectrometry performed by the Apollo 15 and 16 missions, decades of Earth-based spectral observations in visible to near-infrared part of the electromagnetic spectrum, and recent (February 19 to May 3, 1994) visible and infrared images produced by the Clementine mission. Using Clementine images of lunar sample-return stations, several recent studies have attempted to improve extended-visible reflectance spectrometry mapping of the TiO_2 and FeO content of the lunar surface (i.e. Blewett et al. 1997a,b; Jolliff 1997). Detailed γ-ray spectroscopy is being undertaken by the Lunar Prospector mission during 1998.

The spectral reflectance character of lunar basalts tends to be dominated by two strong absorption bands of pyroxene near 0.97 μm and 1.0 μm. Plagioclase produces a weaker absorption band superimposed upon the clinopyroxene spectrum near 1.3 μm. The type and abundance of mafic minerals can be independently estimated from variations in the absorption features near 1.0 μm and 2.0 μm. Glasses of basalt composition similar to the picritic pyroclastic glasses have spectral reflectance characteristics that are essentially distinct from crystalline mare basalts. In the glasses and crystalline mare basalts, the combined abundance of iron and titanium strongly influences the extent of absorption in the ultraviolet and blue portions of their spectra. Plates of ilmenite in the high-Ti glasses (i.e. Apollo 17 black glass) produce reflectance spectrum of ilmenite. Therefore, reflectance spectra can be utilized to identify mare basalts, shed some light on their composition, distinguish mare basalts from pyroclastic deposits (volcanic glasses), and distinguish to some degree among the different glass compositions. High resolution multispectral images of selected lunar regions that were obtained in the 1970s (McCord et al. 1976, Johnson et al. 1977) provided some of the basic data used to define basalt types on the near side of the Moon. The elemental abundances of Th, U, K, and Fe at the lunar surface were mapped by γ-ray spectrometers on the Apollo 15 and 16 orbiting spacecraft. x-ray spectrometers on the same spacecraft mapped the variation of Al/Si and Mg/Si.

The x-ray and γ-ray data indicate that the mare are fundamentally mafic in composition (low Al/Si and high Fe) and that essentially all of the mare regions are composed of basalt flows. UV/VIS color ratios, albedo, 1 μm band strength, and 2 μm band strength have been used to distinguish different basalt types in unsampled regions of the Moon. In these studies, Pieters and McCord (1975) and Pieters (1978) recognized 13 distinct mare basalt types and three additional volcanic groups. Only four of these basalt types occur in regions that were sampled. It has been estimated that about two thirds to one half of the distinct mare basalt types on the near side of the Moon have not been sampled (Pieters 1978). The remote spectral data also illustrate that the maria are heterogeneous with regard to basalt type and that individual basins are filled with wide ranges of both sampled and unsampled basalt types. For example, unsampled high-Ti basalts from Oceanus Procellarum are superimposed on extensive, older low-Ti basalts. These high-Ti basalts are some of the youngest on the Moon. They are also distinct from high-Ti basalts from the Moon's eastern hemisphere (Pieters 1978, Pieters et al. 1980).

In addition to distinguishing among crystalline mare basalt types, remote spectral data have been used to identify possible pyroclastic deposits (i.e. dark mantle deposits) and to ascertain their distribution (Bell and Hawke 1984, Head and Wilson 1992). These deposits are fairly abundant and occur along rilles or lineaments, on the edges of the maria and overlap onto the adjacent highland regions (Hörz et al. 1991 *Lunar Sourcebook*, Head and Wilson 1992).

LUNAR HIGHLANDS ROCKS

The lunar highlands crust

The lunar *highlands* crust is virtually synonymous with lunar crust, for the volcanic plains that conspicuously occupy part of the surface compose far less than 1% of its total (Head and Wilson 1992). Nearly 300 years ago Galileo distinguished the terra from the mare according to topographic expression and albedo. Terra rise high above the mare plains, are more rugged and cratered, and are much brighter or more reflective. We now define the highland crust as that outer lower-density part of the Moon that is chemically and physically distinct from the underlying mantle and from any overlying mare plains. So in fact the highlands also underlie the lowlands. The crust-mantle boundary seems to be reasonably abrupt and defined on the basis of gravity and seismic studies at about 60-km depth in the central frontside and thickening to perhaps twice that depth on the far-side. However, the distribution is non-uniform, with dramatic crustal thinning beneath basins (Neumann et al. 1996).

Prior to Apollo, virtually nothing was known about the chemistry and petrology of the lunar highlands crust, although some physical properties of the actual surface had been inferred. That it had undergone intense impacting prior to the formation of the mare plains was rarely questioned by the time of the first sample collection in 1969, although previously the origin of the craters had been intensely debated. The work of Baldwin (1949) had been influential in this regard. But what was it that had been impacted? Some, assuming that the physical features of the Moon demanded that it had originated and remained cold, inferred a primitive, undifferentiated composition for the crust. In contrast, many topographic features were interpreted as consistent with deposition similar to terrestrial ignimbrites, leading others to infer a rhyolitic-dacitic or essentially granitic crust, even that tektites were of lunar origin. Only the Surveyor VII lander on the flanks of Tycho obtained any pre-Apollo chemical data for the highlands, an imprecise suggestion of some equivalent of terrestrial high-alumina basalt for the regolith at that location (Jackson and Wilshire 1968). (Lunar terminology having arisen in an uncoordinated, even chaotic, fashion following the acquisition of samples, such a composition would no longer be termed "high-alumina" in highland rock nomenclature.) Only with the landing of Apollo 11 and the acquisition of samples was there an unexpected recognition that the highlands crust was highly feldspathic and the inference made that it had formed by large-scale flotation of plagioclase from a global magma system immediately after the accretion of the Moon (Wood et al. 1970a,b, Anderson et al. 1970). The general context and significance of this concept are described in the Introduction to this Chapter.

First described as anorthosites (Wood et al. 1970a,b), the highlands particles found in the Apollo 11 regolith were actually fine-grained anorthositic norites that were highly metamorphosed breccias, and not the true igneous anorthosites later discovered on the Apollo 15 and Apollo 16 missions. It has since been recognized that many more types of igneous rocks with completely separate origins over a long period of crustal formation and modification contribute to the highlands crust (e.g. James 1980, Norman and Ryder 1979,1980b, Shervais and Taylor 1986, Warren 1993). Most of these igneous rocks appear to have been plutonic or hypabyssal in origin, but some incompatible element-rich volcanic rocks exist and others are inferred to have been extruded. Even most of the plutonic rocks such as anorthosites and norites have grain sizes smaller than terrestrial equivalents. (Rare fragments of ancient mare basalts do exist in the highlands terrain, but these are more appropriately described in the section on mare basalts). The idea that many rocks at the surface in the highlands are volcanic persisted even through sample site selection and planning for the Apollo 16 mission, after which it was recognized that at least most of the landforms of the highlands, even very flat plains, are of impact origin (Wilhelms 1987).

Like mare basalts, lunar highlands rocks are devoid of water in any form and thus lack hydrous minerals, and are low in other volatile components such as sodium and lead. They bear iron metal reflecting their low oxidation state, although the igneous rocks generally have low siderophile element abundances reflecting their low abundance within the bulk Moon. Oxygen isotopes demonstrate the links between highlands rocks and mare rocks, and between the Moon and the Earth. Other elemental ratios (especially among volatile and refractory incompatible elements, e.g. K/U) demonstrate the overwhelmingly lunar origin of the crustal rocks; the lunar crust is not an exotic plaster (for explanations, see SR Taylor 1982, Heiken et al. 1991 *Lunar Sourcebook*).

To a first order, the lunar highlands crust consists of rocks that were produced in igneous events over the period from 4.45 Ga to about 3.9 Ga. In addition, and dominating the sample collections, are derivative mixed rocks produced in impact events ranging in age from the first crustal production to the rapid decline in impacting after ~3.8 Ga. Lunar highlands rocks are commonly described as either pristine (igneous) rocks or as polymict (impact-mixed) rocks, and this is the first order classification of highlands rocks (see Classification of Highlands Rocks section below). In reality the distinction depends on the scale of observation: At least in the battered upper crust, there is no evidence of the preservation of igneous rock units at the scale of kilometers or even tens of meters. Instead, there is a megaregolith consisting of various-sized pieces of preserved igneous rocks jumbled in with brecciated rocks and polymict rocks. Remote observations of basin ejecta and structure and sample observations suggest that the lower half or two-thirds of the front-side crust may preserve igneous structure, but the details are obscure. At a contrasting scale, fragments of olivines only hundreds of microns across in breccias include some derived directly from igneous rocks. In practice, the distinction of pristine igneous from polymict brecciated rocks is made at the field observation through hand sample to roughly millimeter scale for lithic fragments. Several collected handsamples, such as troctolite 76535 (Gooley et al. 1974) are igneous rocks. A few boulders a meter or so across (e.g. norite 78235: Jackson et al. 1975), or large clasts of similar size within breccias (e.g. norite 77215: Minkin et al. 1978), similarly are little-modified igneous rocks. At the other end of the scale, some tiny fragments in breccias observed only in thin sections are recognized as fragments of igneous rocks (e.g. spinel troctolite clast in 67435: Prinz et al. 1973; granitic clasts in Boulder 1, Station 2, Apollo 17: Ryder et al. 1975a).

In addition to the question of scale, there is actually a continuum from pristine igneous rocks through thoroughly polymict breccias. Very few lunar highland samples are entirely unmodified in both chemistry and texture. Igneous rocks are modified by impact events. Mild fracturing to severe cataclasis can modify the rock fabric without changing (at the hand sample scale) the mineralogy, mode, or bulk chemistry, resulting in a monomict breccia. More extensive movement and melting creates a spectrum through genomict breccias (mixtures of related igneous rocks) through polymict fragmental breccias to impact melts of thoroughly mixed components. Through this process small amounts of meteoritic material can be added to the resulting rock. To understand the crust requires an understanding of the types of igneous rocks that are present, in part directly through those inferred to be pristine igneous rocks, and in part through constraints imposed on the components of polymict breccias that ultimately are derived from a mix of lunar igneous rocks and up to a few per cent extra-lunar contamination. The distinction of pristine igneous from polymict rocks has thus received considerable attention in lunar highlands studies and will be discussed in the following section.

Understanding the origins of individual rocks, their relationship to other rocks, their parental magmas, and their role and significance in lunar history are prime objectives of lunar sample studies. It requires not only sample descriptions including petrography, chemistry, and isotopic information, but as much information as possible about their vertical and lateral variations within the lunar crust. The Apollo lunar sample collection represents only six locations on the lunar surface, and two of these were strictly mare plains sites. This collection is supplemented by the three Luna regolith samples, of which only one (Luna 20, outside the Crisium basin) was from a highlands site, and by the meteorites from the Moon whose locations can only be restricted but not defined by remotely sensed chemical data. These sample collections cannot by themselves address igneous stratigraphies at any scale, nor relative abundances of lunar rock types at anything other than a local scale. Thus remotely sensed mineral and chemical data have been useful in extrapolating information obtained from the rock samples to a more general understanding of the structure and the production of the lunar crust (e.g. Spudis and Davis 1986, Lucey et al. 1995).

Distinction of pristine igneous from polymict rocks

The approach to using the chemistry and petrography of a lunar highlands rock varies according to whether that rock is a pristine igneous one or a polymict one, and—as noted above—there is a spectrum between these end-members. The bulk chemical composition of a polymict breccia does not correspond with any primary rock, though it closely corresponds with some average of a particular volume of crust. If not recognized as mixtures, analyses of polymict breccias can convey misleading impressions that there is less diversity in the crust than there really is and concerning igneous processes.

A polymict rock is easily recognized if it consists of fragments of different origins. Indeed, nothing better demonstrates the intensity of bombardment of the lunar surface through an extended period better than the fact that most samples returned by the Apollo missions to the highlands are breccias (consolidated fragmental rocks), consisting of discrete rock, mineral, and glass fragments set in a finer-grained matrix. The individual clasts represent a wide range of components, compositions, and ages. Many of these rocks and clasts contain shock-produced features such as maskelynite. However, in many cases the origin of a texturally homogeneous crystalline rock—which might be an igneous rock, an impact melt breccia, or a metamorphosed breccia—can be more ambiguous, requiring more detailed data. On the other hand, even a texturally homogeneous fragmental rock could be a monomict breccia faithfully representing in mineralogy and chemistry an igneous rock.

Although even the earliest authors had described some highlands rocks as igneous and others as polymict breccias, in many discussions, particularly those based on chemistry, the distinction was commonly ignored. The methods of discriminating origins were inadequately reviewed until Warren and Wasson (1977) assessed the varied criteria and made some carefully considered recommendations. Their criteria include abundances of meteoritic siderophile elements, petrographic textures, mineral compositional homogeneity, incompatible element abundances and relative abundances, ages, isotopic data, and plausibility. The methodology has not changed much since then. None of the criteria can be taken as absolute when used alone, i.e. they are neither totally inclusive nor totally exclusive. In most cases an absolute decision cannot be made, but a likelihood of pristinity can be assessed (Warren 1993).

The most reliable evidence of pristinity has usually been accepted as a lack of measureable meteoritic siderophile element contamination. Very obvious lunar igneous rocks such as the anorthosites and the troctolite 76535 have very low abundances of these siderophile elements, a characteristic shared with mare basalts. Polymict rocks including impact melts (and regoliths) have high abundances of siderophiles with relative abundances indicating a meteoritic origin. Thus Warren and Wasson (1977) suggested a cut-off value for siderophile elements ($<3 \times 10^{-4}$ times C1 chondrites) below which a rock could be considered to be uncontaminated and therefore not mixed. However, several terrestrial impact melts have no measureable meteoritic contamination but are nonetheless complex target mixes, so other criteria must be used in parallel. Further, while it is conceivable that a rock could have meteoritic contamination but not be mixed with other rocks, the near-surface crust contains mixed rocks through which a projectile must pass and so this seems unlikely. A few igneous rocks, such as norite 78235 and the dunite 72415, do contain slightly higher abundances of siderophile elements, and these rocks are recognized as igneous on other criteria, including non-meteoritic *relative* abundances of siderophile elements.

Textural and mineral phase homogeneity are less objective criteria, and more difficult to summarize. Clast-bearing, texturally heterogeneous samples are clearly impact produced, and even essentially basaltic looking (clast-poor, subophitic) impact melt samples such as 14310 can be recognized as polymict on the basis of meteoritic contamination, mildly heterogeneous textures, rare clasts, and non-cotectic/eutectic compositions in the absence of cumulate textures. However, a large-scale impact melt (as hypothesized for the Sudbury Igneous Complex; Grieve et al. 1991) could produce textures very similar to plutonic igneous rocks, and if a rock were produced from an achondritic impactor or settled out Fe-metal grains it would contain no measurable siderophile contamination. Nonetheless, most coarse-grained, homogeneous-textured igneous-looking rocks free of siderophile contamination are accepted as igneous, especially those that show pyroxene exsolution, accumulation, or other features suggesting slow cooling. Warren and Wasson (1978) and Norman and Ryder (1979) have argued why most of these rocks must be considered *bona fide* igneous rather than impact melt products.

Many lunar regoliths and polymict breccias contain elevated abundances of incompatible elements derived from a chemical component called KREEP (a name derived from its comparatively high abundances of potassium, rare earth elements, and phosphorus). The relative abundances of these incompatible elements in KREEP are fairly constant (Warren and Wasson 1977, 1979). KREEP is inferred to be ultimately derived in various ways from the late-stage dregs of the primordial magma system (urKREEP of Warren and Wasson 1979b), and its common inclusion as a cryptic component in most polymict materials leads to the suspicion that most rocks that contain KREEP are polymict. However, samples such as the Apollo 15 KREEP basalts contain significant KREEP abundances, yet lack siderophiles and clast-bearing examples and have all the characteristics of a volcanic sequence (Ryder 1987). A few others are plutonic. Thus the KREEP criteria if used alone can exclude genuine igneous rocks from the pristine igneous rock category.

Samples with ages greater than 4.2 Ga and radiogenic isotope ratios that are primitive, such as the ferroan anorthosites, are commonly accepted as igneous, in part because of the great difficulty in preserving old ages and primitive isotope ratios (which certainly indicate a lack of KREEP mixing) through any mixing in an impact event. Nonetheless, pristine igneous rocks can be comparatively young; the Apollo 15 KREEP basalts post-date even the Imbrium basin-forming event (Spudis 1978, Ryder 1994).

Plausibility as a criterion is perhaps subjective, but a rock consisting of almost pure plagioclase such as an anorthosite or of almost pure olivine such as a dunite, or an extreme composition such as granite, must surely be igneous. It would be implausible as a mixture. Conversely, if a rock appears to be of fairly average local crust composition, then suspicion will obviously arise that it might be polymict, and more stringent tests for, or scepticisim of, an igneous origin must be applied.

Warren (1993) using these criteria has selected 260 lunar samples as likely to be pristine igneous lithologies. He developed seven categories of confidence in pristinity ranging from 3 to 9; all numerical allusion to pristinity or pristinity index in this Chapter refer to Warren's (1993) index. Most of these 260 samples are very small. For those samples reasonably safely assumed to be igneous (pristinity categories 7 to 9), only 46 are larger than 1 g, only 33 larger than 5 g. Of these latter, more than half are ferroan anorthosites, and these include nearly all of the largest samples.

CLASSIFICATION OF LUNAR HIGHLANDS ROCKS

The samples collected from the lunar highlands have a great diversity in texture and in chemical composition. Classification—or at least nomenclature—is necessary to allow reasoned description, interpretation, and research planning. In this Chapter we partly follow the classification used in Heiken et al. (1991 *Lunar Sourcebook*) which conforms with many older ones or usage and reflects largely obvious groupings of rocks based on composition and origin. The essential distinction is between *pristine igneous rocks* and *polymict breccias*, which is genetic but uses the observational criteria described in the previous section. However, the subdivisions among igneous rocks are somewhat different from those in Heiken et al. (1991 *Lunar Sourcebook*), and are more similar to those used by Warren (1993). For the polymict breccias, we follow the classification of Stöffler et al. (1980), so the groups are mainly texturally rather than compositionally defined. The classification is not entirely unambiguous as it is neither totally inclusive nor totally exclusive, and varied criteria are used. Thus it is as much nomenclature as classification, summarized in Table 10.

Unlike mare basalts, for which related suites can be confidently established on the basis of sample location, petrography, chemistry, and isotopic data, the highlands igneous rocks are individual samples with only tenuously inferential relationships among them (except perhaps for the Apollo 16 ferroan anorthosites and the group of Apollo 15 KREEP basalts). As discovery and description of igneous rock fragments has increased and understanding has been changed or at least modified, so the affiliations and potential relationships have been re-interpreted. In consequence, the subdivisions used by various authors have changed. In the early days of lunar highlands studies, presumed igneous rocks were subsumed under the single acronym ANT suite (for anorthosite, norite, troctolite) (starting with Keil et al. 1972), but that acronym is now seen as of little relevance. ANT included both igneous and polymict rocks, the latter especially represented by feldspathic granulites. Although the word "suite" is commonly used in various contexts in lunar highlands studies, in this Chapter we avoid the term as being potentially misleading and largely redundant. Among igneous rocks we make the basic distinction of *ferroan anorthosites* from *Mg-rich rocks*. The Mg-rich rocks group comprise subgroups of *Magnesian plutonic rocks*, *Alkali rocks*, *KREEP basalt and Quartz monzodiorite rocks*, and *Granite/Felsite rocks*. At hand-sample level, nomenclature parallels that of terrestrial igneous rocks, hence names tend to reflect whole rock major element chemistry or modes, or both.

Within the igneous group, the distinction of the *Ferroan anorthosites* from the *Mg-rich rocks* corresponds not only with modal plagioclase but with the way that samples fall on a plot of mafic mineral composition against plagioclase composition (Fig. 36) or a corresponding plot of bulk rock chemistry (Fig. 37, which includes polymict rocks).

Table 10. Classification of lunar highlands rocks.

PRISTINE IGNEOUS ROCKS (including monomict breccias)	*POLYMICT BRECCIAS*
1. Ferroan anorthosites	1. Fragmental breccias
2. Mg-rich rocks	2. Glassy melt breccias and impact glasses
2a. Magnesian plutonic rocks	3. Crystalline melt breccias
Troctolites	4. Clast-poor impact melts
Spinel troctolites	5. Granulitic breccias and granulites
Dunites and ultramafic rocks	6. Dimict (two-component) breccias
Norites	7. Regolith breccias
Gabbronorites	
2b. Alkali rocks	
Alkali anorthosites	
Mafic alkali rocks	
2c. KREEP basalts	
Apollo 15	
Apollo 17	
Quartz monzodiorites	

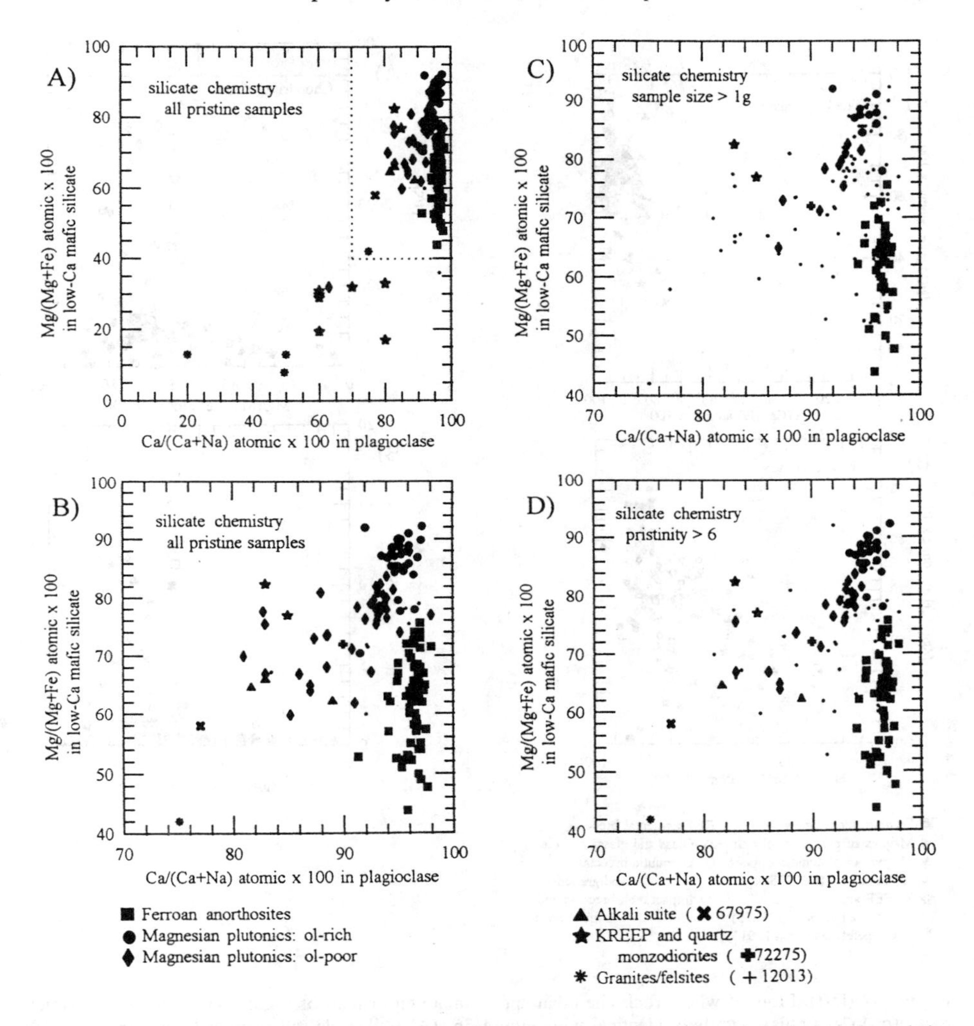

Figure 36. Plots of composition of low-Ca mafic minerals vs. composition of plagioclase for pristine igneous rock samples. (A) Full scale (all compositions possible) for all samples tabulated. (B) Expanded scale of area outlined in (A). (C) Expanded scale; samples larger than 1 g as large symbols, others as dots. (D) Expanded scale; samples with Warren (1993) pristinity index >6 as large symbols, others as dots.

Ferroan anorthosites also have higher Ti/Sm (roughly chondritic) and higher Sc/Sm than most Mg-rich rocks (Fig. 38). The distinction also corresponds with the inference that ferroan anorthosites formed in a reasonably straightforward manner as cumulates from an early magma system of global extent (primary), whereas all other rocks appear to have formed in more complicated processes that include assimilation and remelting (secondary). The few ferroan anorthosites for which reliable crystallization ages have been obtained are ancient (Table 11).

The main groupings among the diverse Mg-rich rocks are somewhat arbitrary and there are no formally agreed-upon criteria. Indeed, there is some disagreement for specific samples about which group is the appropriate one. The groups used do not necessarily

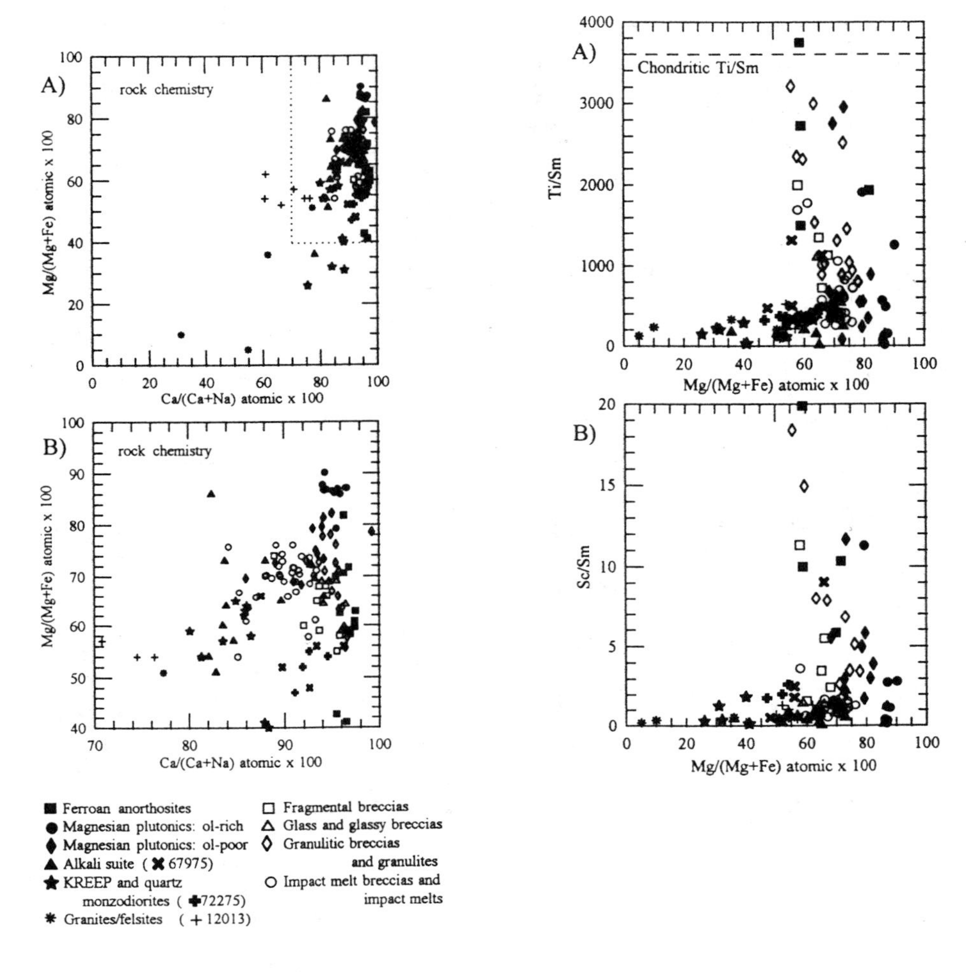

Figure 37 (left). Plots of whole rock chemical approximate equivalent of Figure 36, including polymict rocks (open symbols). Symbols identical with Figure 36. (A) Full scale (all compositions possible). (B) Expanded scale of area outlined in (A). Polymict rocks tend to be less extreme than pristine igneous rocks.

Figure 38 (right). Plots of whole rock chemistry for pristine igneous and polymict highlands rocks. Symbols as in Figure 37. (A) Ti/Sm (mass) vs. mg# × 100. (B) Sc/Sm (mass) vs. mg# × 100.

reflect genetic relationships, or even similar origins, among hand samples within a group, or deny them across groups. The ages and εNd values for KREEP basalts, granites, troctolites, and norites from a variety of lunar sites form a linear array suggesting genetic relationships (Shih et al. 1992). The *Magnesian plutonic rocks* generally have less than 0.1 wt % K_2O and calcic plagioclase (generally An_{90} or greater), as well as the characteristics of plutonic cumulate origin. This distinguishes them from the *Alkali rocks*, which generally have K_2O more than 0.1 wt % and more sodic plagioclases, and tend to have less-well-equilibrated mineral compositions suggestive of shallower origins. There appears to have been little investigation as to whether a 0.1 wt % K_2O distinction has any real significance or is entirely arbitrary. Although alkali rocks have high incompatible element abundances,

Table 11. Radiogenic crystallization ages (Ga) for igneous lunar highlands rocks. Decay constants from from Steiger and Jaeger (1977).

Sample	^{40}Ar-^{39}Ar (*)	Rb-Sr	Sm-Nd	U-Pb, Pb-Pb	Main reference
Ferroan anorthosites					
22013;9002	4.51 +/- ?				Huneke and Wasserburg (1979)
60025			4.44 +/- 0.02		Carlson and Lugmair (1988)
60025				4.51 +/- 0.01	Hanan and Tilton (1987)
67016 cl			4.56 +/- 0.07		Alibert et al. (1994), Alibert (1994)
62236			4.36 +/- 0.03		Borg et al. (1997)
67435;33a cl	4.35 +/- 0.05				Dominik and Jessberger (1978)
67435;33b cl	4.33 +/- 0.04				
Mg-rich plutonic rocks					
Troctolites					
76535		4.51 +/- 0.07			Papanastassiou and Wasserburg (1976)
76535			4.26 +/- 0.06		Lugmair et al. (1976)
76535	4.19 +/- 0.02				Husain and Schaeffer (1975)
76535	4.16 +/- 0.04				Huneke and Wasserburg (1975)
76535	4.27 +/- 0.08				Bogard et al. (1975)
76535				4.27 +/- ?	Hinthorne et al. (1975)
14306,150 (?)				4.245 +/- 0.075	Meyer et al. (1989b)
Dunites					
72417		4.47 +/- 0.10			Papanastassiou and Wasserburg (1975)
Norites					
14305;91 (?)				4.211 +/- 0.005	Meyer et al. (1989b)
15445;17			4.46 +/- 0.07		Shih et al. (1993a)
15445;247			4.28 +/- 0.03		Shih et al. (1993a)
15455;228		4.49 +/- 0.13	4.53 +/- 0.29		Shih et al. (1993a)
72255		4.08 +/- 0.05			Compston et al. (1975)
73215;46,25	4.19 +/- 0.01				Jessberger et al. (1977b)
77215		4.33 +/- 0.04	4.37 +/- 0.07		Nakamura et al. (1976)
78235				4.426 +/- 0.065	Premo and Tatsumoto (1991)
78236	4.39 +/- ?	4.29 +/- 0.02	4.43 +/- 0.05		Nyquist et al. (1981)
78236	4.11 +/- 0.02				Aeschlimann et al. (1982)
78236			4.34 +/- 0.04		Carlson and Lugmair (1981b)
Gabbronorites					
67667			4.18 +/- 0.07		Carlson and Lugmair (1981a)
73255c			4.23 +/- 0.05		Carlson and Lugmair (1981b)
Alkali rocks					
14066;47 (?)				4.141 +/- 0.005	Meyer et al. (1989b)
14304 cl b			4.34 +/- 0.08 (	4.108 +/- 0.053	Snyder et al. (1995b)
14306;60 (?)				4.20 +/- 0.03	Compston et al. (1984)
14321;16 c				4.028 +/- 0.006	Meyer et al. (1989b)
67975;131				4.339 +/- 0.005	Meyer et al. (1989b)

most do not have the KREEP relative abundance pattern. These rocks are not alkaline in classification schemes used for terrestrial rocks; they have far too little Na_2O and K_2O. The *KREEP basalts* are distinguished from the alkali rocks by their having the archetypal KREEP incompatible element pattern and abundances, and thus include *the Quartz monzodiorites*. The latter are the only KREEP rocks that do not have the textures of extrusive rocks, though they are shallow intrusives. Although comparatively enriched in alkalis, even these rocks have too little Na_2O and K_2O to be considered alkalic rocks in terrestrial classifications. The *Granites/Felsites*, are defined by their high silica and potassium contents. They also have lower incompatible element abundances than KREEP and a non-KREEP incompatible element pattern, including a distinctly v-shaped rare earth element pattern and depleted La/Ba. KREEP seems to be ultimately related to the early

Table 11 (continued)

Sample	$^{40}Ar-^{39}Ar$ (*)	Rb-Sr	Sm-Nd	U-Pb, Pb-Pb	Main reference
KREEP basalt and Quartz monzodiorite rocks					
A15 KREEP basalts					
15382	3.84 +/- 0.05				Stettler et al. (1973)
15382	3.85 +/- 0.04				Turner et al. (1973)
15382		3.82 +/- 0.02			Papanastassiou and Wasserburg (1976)
15386		3.86 +/- 0.04			Nyquist et al. (1975)
15386			3.85 +/- 0.08		Carlson and Lugmair (1979)
15434 particle		3.83 +/- 0.05			Nyquist et al. (1975)
A17 KREEP basalts					
72275		3.93 +/- 0.04			Compston et al. (1975)
72275		4.04 +/- 0.08	4.08 +/- 0.07		Shih et al. (1992)
Quartz monzodiorites					
15405,57				4.297 +/- 0.035	Meyer et al. (1996)
15405,145				4.309 +/- 0.120	Meyer et al. (1996)
Granite and felsite rocks					
12013		>4.08			Quick et al. (1981b)
12033,507				3.883 +/- 0.003	Meyer et al. (1996)
12034,106				>3.916 +/- 0.17	Meyer et al. (1996)
14082,49				4.216 +/- 0.007	Meyer et al. (1996)
14303 cl				4.308 +/- 0.003	Meyer et al. (1996)
14311,90				4.250 +/- 0.002	Meyer et al. (1996)
14321 B1(=cl?)				4.010 +/- 0.002	Meyer et al. (1996)
14321 cl	(K-Ca:			3.965 +/- 0.025	Meyer et al. (1996)
14321 cl	4.060 +/- 0.07	4.04 +/- 0.03	4.11 +/- 0.20		Shih et al. (1985, 1993b)
72215 mix		3.95 +/- 0.03			Compston et al. (1975)
73215,43		3.82 +/- 0.05			Compston et al. (1977)
73235,60				4.218 +/- 0.004	Meyer et al. (1996)
73235,63				4.320 +/- 0.002	Meyer et al. (1996)
73235,73				>4.156 +/- 0.00	Meyer et al. (1996)

Notes:
(*) only given if suggestive of original crystallization age
(a) disturbed and suspect

ferroan anorthosite-producing magma system as its residue (as urKREEP; Warren and Wasson, 1979b), but is also inferred to be important in the generation of many other Mg-rich rocks, alkali rocks, and granites/felsites by source contamination, assimilation, silicate liquid immiscibility, or even metasomatism (see e.g. Snyder et al. 1995b). If so, then there is at least one indirect link among all the magma systems. The distinction of all these groups is somewhat ambiguous, arbitrary, and inconsistent, but nonetheless useful. Indeed many authors classify quartz monzodiorites and granites among alkali rocks (e.g. Jolliff 1991), and ultimately petrogenetic relationships may well be proven.

The polymict breccias are divided largely on the basis of matrix texture and the relationship of matrix to the enclosed fragments, and not *a priori* on chemical composition. In practise, there is some correlation of chemistry and textural group. For instance, nearly all granulites (texturally re-equilibrated, metamorphosed breccias) contain between 24 and 28 wt % Al_2O_3 and lack KREEP contamination. To a first order, polymict breccias can be modelled as mixtures of the identified pristine igneous rocks (e.g. Figs. 37, 38, 39, 40), but complications and uncertainties in the nature of both the KREEP and the Mg-rich end-members abound.

The sequence through fragmental breccias, glassy melt breccias, crystalline melt breccias, and clast-poor melt breccias essentially reflects an increasing proportion of melt matrix from almost no melt to almost all melt (Simonds et al. 1976). Granulitic breccias

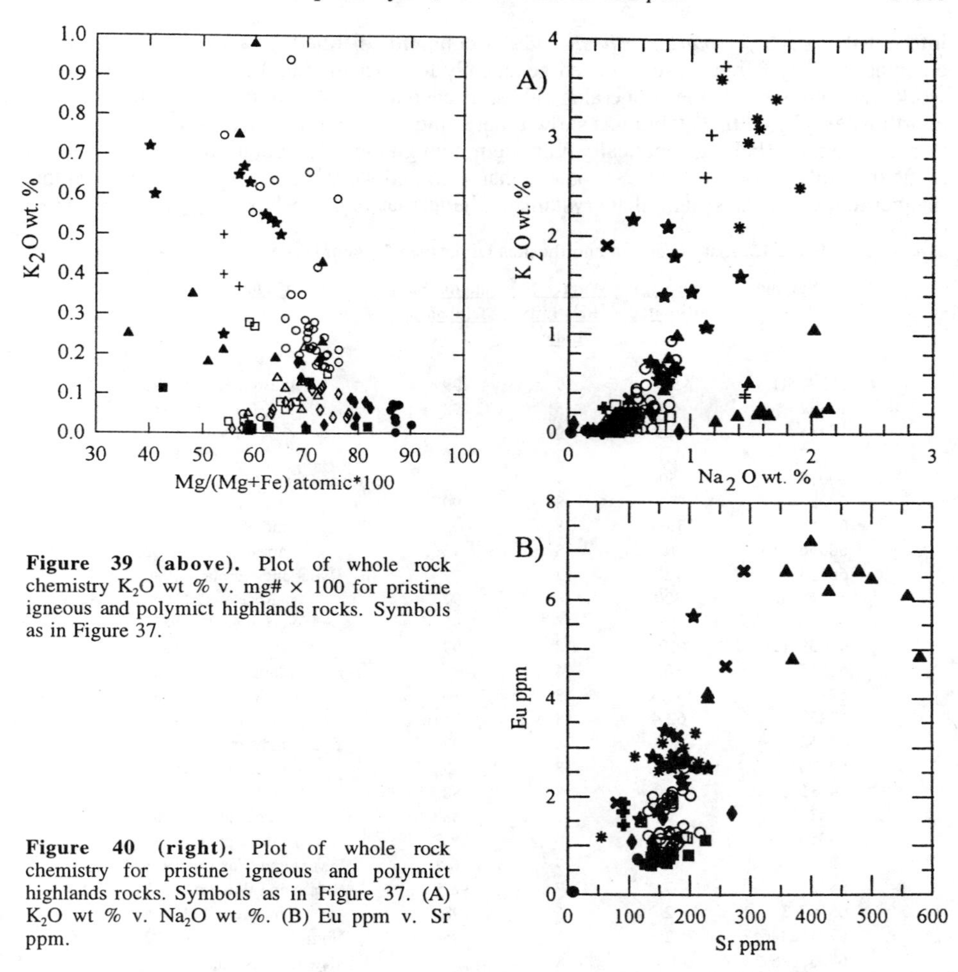

Figure 39 (above). Plot of whole rock chemistry K_2O wt % v. mg# × 100 for pristine igneous and polymict highlands rocks. Symbols as in Figure 37.

Figure 40 (right). Plot of whole rock chemistry for pristine igneous and polymict highlands rocks. Symbols as in Figure 37. (A) K_2O wt % v. Na_2O wt %. (B) Eu ppm v. Sr ppm.

could have had any polymict origin but have been annealed and fairly well equilibrated texturally and in mineral composition by thermal processing. Dimict breccias are recognized at the hand sample scale, and those identified are mainly a crystalline melt breccia or clast-poor impact melt combined with some other form of feldspathic rock such as granulitic breccia or ferroan anorthosite. Regolith breccias are really a subcategory of either fragmental or glassy matrix breccias, but preserve features of the regolith from which they were derived, such as agglutinates or solar wind components.

PRISTINE IGNEOUS ROCKS

Ferroan anorthosites

Definition and overview. Dowty et al. (1974a) recognized that highland samples with more than 90 vol % plagioclase (anorthosites) were characterized by very calcic plagioclase (>An_{96}) and, in comparison with norites, troctolites, and bulk highland compositions, iron-rich pyroxenes and olivines (mg# < 0.70). These samples also had low abundances of incompatible and alkali elements, and originally a coarse grain size. They inferred that such *ferroan anorthosite* was a coherent, distinctive, and widespread lunar cumulate rock type not closely related genetically to other highland rock types. Continued work has confirmed the mineralogical and chemical distinctiveness of the ferroan anorthosites (Figs. 36, 37), but has shown that some hand samples with as much as 15 vol % mafic minerals belong genetically with them, though most do contain only a few percent mafic minerals (Table A5.41). No other lunar highland samples can be related in a simple manner to the magma system that crystallized the ferroan anorthosites.

Table 12. List of ferroan anorthosites larger than 1 gram. [Pristinity index >5]

Sample	mass (g) estimated	Warren '93 Pristinity index	~modal % feldspar	Main reference
15295c41	5.3	9	99	Warren et al. (1990)
15362	4.2	5	98	Ryder (1985)
15415	269	9	99	Ryder (1985)
15437	1.1	7	80	Ryder (1985)
60015	4600	7	99	Ryder and Norman (1979, 1980)
60025	1836	8	90	James et al. (1991)
60055	35.5	8	98	Ryder and Norman (1980)
60056	16	5	95	Warren et al. (1983c)
60135	120	7	77	Ryder and Norman (1980)
60215c30	300	8	97	Rose et al. (1975)
60515	17	8	95	Warren et al. (1983c)
60639c19	10	8	99	Warren and Wasson (1978)
61015	300	6	96	James et al. (1984)
62236	57.3	8	86	Nord and Wandless (1983)
62237	62.4	8	85	Ebihara et al. (1992)
62255	800	8	97	Ryder and Norman (1979)
62275	443	8	93	Warren et al. (1983c)
64435c210A	100	8	98	James et al. (1989)
64435c239	6	6	83	James et al. (1989)
65315	285	8	98.5	Ebihara et al. (1992)
65325	65	7	98.5	Warren and Wasson (1978)
65327	7	7	98.5	Warren and Wasson (1978)
65767c3	2	6	98	Dowty et al. (1974a)
67016c346*?	2	6	95	Norman and Taylor (1992)
67035c26	2.3	8	80?	Ryder and Norman (1978)
67075c17	50	7	96	Haskin et al. (1973)
67455c30	1.7	7	95	Ryder and Norman (1978)
67535	0.99	6	93	Stöffler et al. (1985)
67539c7	1.5	7	95.5	Stöffler et al. (1985)
67635	9.1	8	92	Stöffler et al. (1985)
67636	3.2	7	97	Stöffler et al. (1985)
67637	2.3	7	96	Stöffler et al. (1985)
67915c12-1	50	7	85	Taylor and Mosie (1979)
67915c26	1	7	85	Marti et al. (1983)
73217c35	1.7	6	95	Warren et al. (1983b)
77539c15	6.2	6	99	Warren et al. (1991a)

Ferroan anorthosite samples larger than 1 g are listed with some of their characteristics in Table 12. Most ferroan anorthosite samples are brecciated or at least cataclasized, and their recognition as igneous rocks is based on their low siderophile element abundances, high modal content of plagioclase feldspar, primitive isotopic compositions and lack of a KREEP component, and some relict igneous textures. They evidently formed as coarse-grained igneous rocks that cooled slowly as shown by exsolution in pyroxenes and very

Figure 41. Photomicrographs of ferroan anorthosites, all crossed polarized light, all fields of view ~2 mm. (A) 15415,19: plutonic igneous/recystallized/fractured. (B) 60025,129: coarsely cataclasized (little fine-grained matrix). (C) 61015,161: finely granulated, abundant fine-grained matrix. (D) 60015,128: coarsely cataclasized/shocked (partly maskelynite).

low CaO (<0.08 wt %) in olivines. They are not conceivable products of accumulation from even a large-scale impact melt because their mg# and their incompatible element abundances are so much lower than typical upper crustal compositions. At least some anorthosites appear to be genomict (mixtures from a related sequence) (Ryder 1982, McGee 1993) and suggest that at least early in lunar history impact mixing could occur into a purely anorthositic terrain without addition of KREEP or meteoritic siderophiles.

Ferroan anorthosites are the most abundant type of pristine highland sample, and are most common at the Apollo 16 site in the central highlands. They also form the majority of the tiny clasts identified in those lunar meteorites that sample other (though unidentified) parts of the highlands. Remotely-sensed chemical and mineralogical data for the lunar surface strongly imply that anorthosite, which we can reasonably assume to be *ferroan* anorthosite, is a dominant component of at least the upper part of the highlands crust over most of the Moon (e.g. Lucey et al. 1995,1997, Peterson et al. 1997). Ferroan anorthosites have primitive radiogenic isotopic signatures and ages as great as 4.5 Ga (Table 11). Following the early models of Wood et al. (1970a,b) and Anderson et al. (1970), they are widely inferred to have originated by flotation of plagioclase from—or suspension within—a global magma ocean early in lunar history.

Petrography. Ferroan anorthosites display a range of textures, grain sizes, and shock damage (Fig. 41), as well of mineral chemistry. The vast majority of large samples of ferroan anorthosite contain more than 85 vol % calcic anorthite, most more than 95 vol % (Table A5.41). More mafic members may be more common than large rocks suggest (Jolliff and Haskin 1995). Pyroxene is the second most abundant mineral, in most cases mainly originally a pigeonite that has exsolved augite and inverted to orthopyroxene (Dixon and Papike 1975). In some cases (e.g. 67065) pigeonite is only partly inverted (McCallum et al. 1975). Some orthopyroxene-augite pairs might be the original crystallization products, and in rare cases augite is the dominant pyroxene (e.g. 15415: James 1972). Several ferroan anorthosite samples contain small amounts of olivine, and in some cases olivine is more abundant than pyroxene. The unusually mafic hand samples 62236 and 62237 are also unusually rich in olivine. Traces of ilmenite, aluminous Cr-spinel, FeNi metal, troilite, and a silica phase are also reported for ferroan anorthosites. McGee (1993) reported the rare presence of a (Zn,Fe)S phase. There is an obvious problem with modal representivity, especially for minor and accessory minerals, but also for pyroxene and olivine.

Relict igneous textures are rarely preserved, and most of the ferroan anorthosites are cataclasized to severely brecciated, in some cases melted or with abundant maskelynite or shock-induced twinning. Where original textures are preserved, the pyroxene tends to surround the plagioclases as an adcumulus or orthocumulus phase, but in most cases the mafic minerals form interstitial or fragmental small grains (Fig. 41). Although subdivision on the basis of shock deformation has been made (e.g. Dixon and Papike 1975), a more meaningful subdivision is on relict textures, where they can be discerned. James (1980) identified two pregranulation groups: coarse-grained granular and medium-grained granoblastic. Original plagioclase grain sizes of the order of 2 to 4 mm can be inferred, but the surrounding pyroxene oikocrysts may have been somewhat larger. The famous "Genesis Rock" 15415 has a unique texture, with some extremely coarse plagioclase grains (up to 3 cm, though most are 2 to 4 mm) in polygonal and anhedral grains that attest to brecciation and severe thermal metamorphism (James 1972). Plagioclases are extremely calcic and have a maximum range of compositions of about 3 mol %, from about $An_{95.9\text{-}98.9}$ (Fig. 42) (McGee 1993, Papike et al. 1997b). Plagioclase compositions within some samples have a range almost as great as that of the group as a whole. A few have ranges of less than 1 mol % An (e.g. 15415, which is recrystallized). For most samples there is a reasonable positive correlation of Fe and Mg in plagioclase with the An content, and for those that do not there is evidence for subsolidus re-equilibration (e.g. McGee 1993). Some elemental redistribution, particularly Na, seems to have been in response to shock.While there is a wide total range in low-Ca pyroxene compositions with mg# varying from about 0.72 to about 0.50 for most samples (Fig. 43A), many individual samples show a much more restricted range, and individual grains appear to be homogeneous. Samples with no variation or small but continuous ranges of compositions are interpreted as monomict (McGee 1993); others are considered to be genomict rather than representing an original variation within a hand sample. The pyroxene compositions in all the anorthosites show a regular continuous progression across the quadrilateral, and their Ti, Al, and Cr contents show generally smooth trends (Fig. 44). These minor

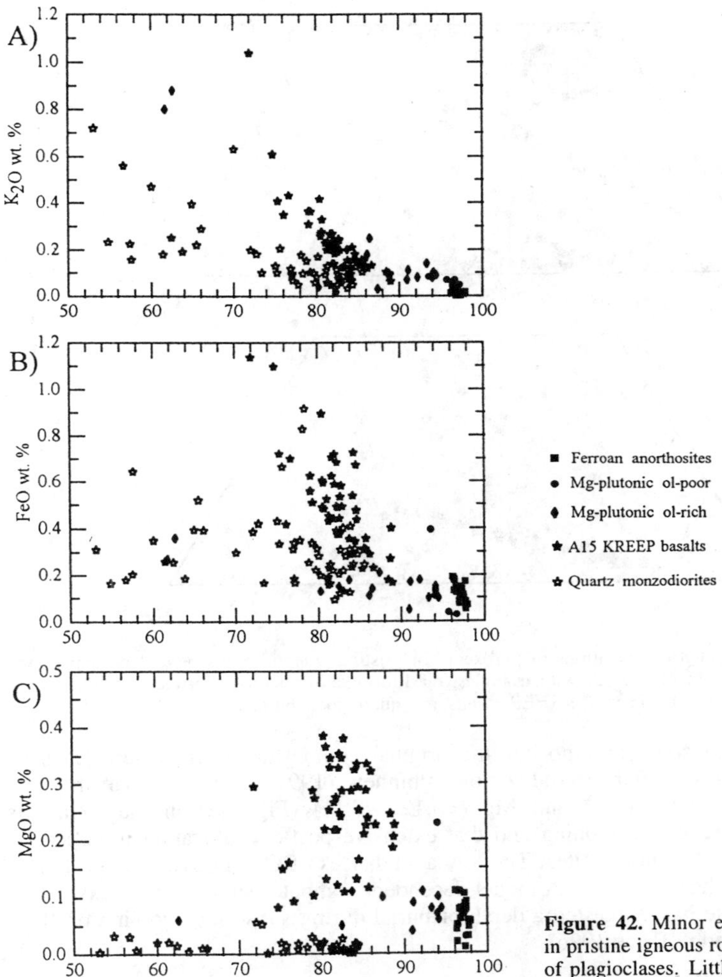

Figure 42. Minor elements in plagioclases in pristine igneous rocks, against An content of plagioclases. Little published data exists for alkali rocks or granites/felsites.

elements in pyroxenes tend to be of lower abundance at a given En than Mg-rich rocks. McGee (1993) found that there was no strong positive or negative correlation of An in plagioclase with mg# in coexisting pyroxene. Olivine grains, which have a total range in mg# similar to that of pyroxenes (Fig. 45), are also homogeneous (e.g. Ryder 1982, James et al. 1991). They also have low Ca contents (mainly <0.08 wt % CaO, Fig. 45A) (James et al. 1991, Bersch et al. 1991) indicative of very slow cooling. They also contain very low Cr_2O_3 (Fig. 45B), and very low Ni contents (<60 ppm Ni; Ryder 1983).

James et al. (1989, 1991) and McGee (1993, 1994) subdivided ferroan anorthosites into four subgroups on the basis of mineral compositions and modes: mafic magnesian, mafic ferroan, anorthositic sodic, and anorthositic ferroan (the most typical). Detailed petrographic studies including precise analyses of minor elements in mineral phases (e.g. Fe, Mg in plagioclases) led these authors to conclude that these anorthosites of all four subgroups are generally consistent with derivation from a common fractionating parent

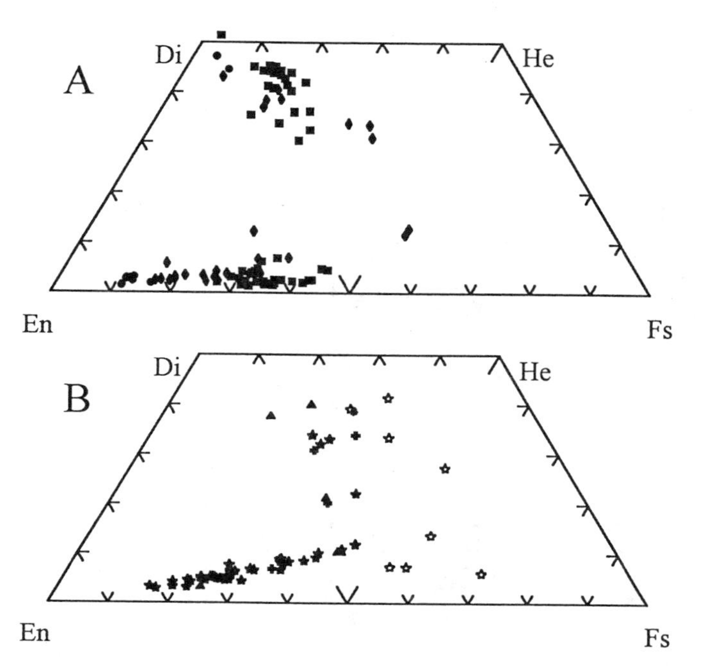

Figure 43. Major element compositions of pyroxenes in pristine igneous rocks plotted onto pyroxene quadrilateral. Symbols as in Figure 42, with triangles added for alkali rocks. (A) Ferroan anorthosites and Mg-rich plutonic rocks. (B) alkali rocks, KREEP basalts, and quartz monzodiorites.

magma. However, in some anorthosites the original compositional characteristics have been altered during or after crystallization. Phinney (1991) suggests that the low abundances of Fe (<0.2 wt %) and Mg (<0.12 wt %) (Fig. 42) in plagioclase is inconsistent with igneous partitioning and that extensive post-crystallization modification has taken place in all the anorthosites. The low abundances of Ca in olivine suggest slow plutonic cooling for these samples. A model dependent computer simulation of exsolution in augite and pigeonite has suggested a depth of burial during subsolidus cooling of 10 to 20 km (McCallum and O'Brien 1996).

Chemistry, radiogenic isotopes, and parent magmas. Chemical analyses for representative ferroan anorthosites are given in Table A5.32. Apart from the obvious high Al_2O_3 and CaO and the low mg#, the ferroan anorthosites are characterized by low incompatible element abundances with trivalent rare earth elements less than chondritic (Fig.46), and much lower than those for most other igneous rock types (Figs. 47 to 50). The rare earth abundance patterns show a large positive Eu anomaly, larger than for terrestrial rocks because the low oxidation state of the Moon allows a much higher fraction of divalent Eu, which is relatively plagiophile. The characteristics, including the low abundance of mafic minerals and virtual absence of incompatible element-rich residual phases, suggest that little trapped magma can be present in these rocks (e.g. Haskin et al. 1981, Ryder 1982). Incompatible element abundances are just as low in those samples (62236, 62237) that do contain significant amounts of mafic minerals, suggesting these minerals are adcumulate. Estimates of parent magma calculated from bulk rock compositions suggest that the parent magmas had refractory incompatible element ratios close to chondritic (e.g. flat rare earth patterns) (e.g. Hubbard et al. 1971b, James et al. 1991). Other element ratios, such as Ti/Sm and Sc/Sm (Figs. 38A,B) are also close to

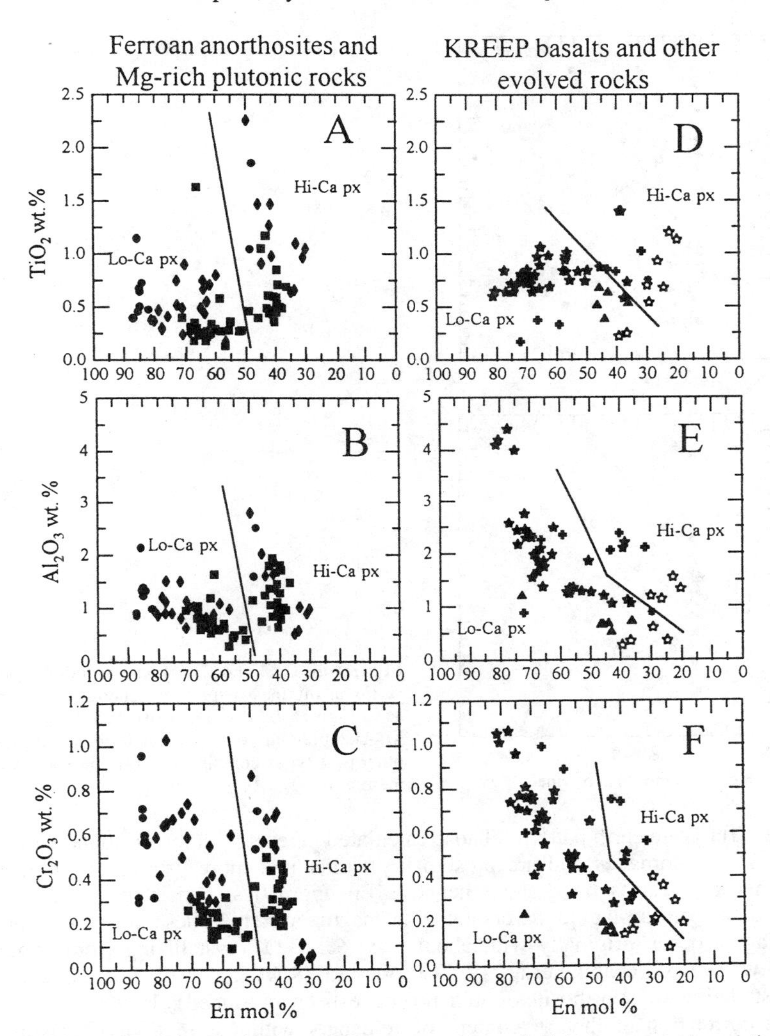

Figure 44. TiO_2, Al_2O_3, and Cr_2O_3 wt % plotted against En mol % in pyroxenes in pristine igneous rocks. Symbols as in Figures 42 and 43. Approximate fields for low-Ca and high-Ca pyroxenes distinguished (there is mild overlap). (A,B,C) for ferroan anorthosites and Mg-rich plutonic rocks, (D,E,F) for KREEP basalts and other evolved rocks.

chondritic and serve to further distinguish ferroan anorthosites from most Mg-rich rocks and KREEP rocks (McKay et al. 1978, Norman and Ryder 1980a,b).

Analyses of trace elements in the cores of plagioclase can be inverted to estimate parental magma compositions if they have not been compromised by post-crystallization effects and the relevant mineral/melt partition coefficients are known. Reliable analyses using ion probe techniques have been made for rare earths and other elements by Floss (1991), Floss et al. (1991, 1998), Jolliff and Hsu (1996), and Papike et al. (1997b) (e.g. Fig. 51A, below). Melts parental to the anorthosite plagioclase are estimated to contain rare earth element concentrations at 10 to 50 times chondritic abundances (Fig. 51B, below) with fairly

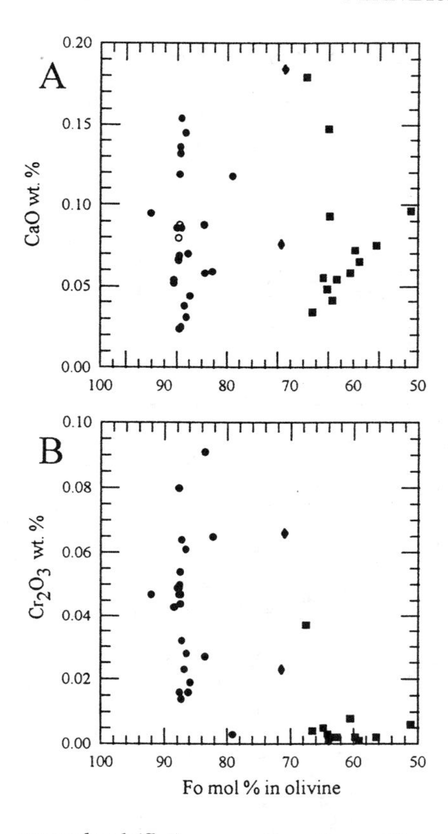

Figure 45. CaO and Cr_2O_3 wt % plotted against mol % Fo in olivines in pristine igneous rocks. Symbols as in Figures 42, 43, 44. Most data is for olivine-rich Mg-rich plutonic rocks and for ferroan anorthosites, as other rock types contain little olivine and few data for them exist.

unevolved (flat) rare earth patterns. Those calculated parents with lower abundances of rare earths lack Eu anomalies and are presumably earlier and more primitive, whereas those with higher rare earths have small negative Eu anomalies, consistent with successive crystallization from a plagioclase-crystallizing magma system. These are consistent with a model magma ocean originally with about 5 wt % Al_2O_3 crystallizing from about 78 to about 90 wt % solid (Papike et al. 1997b) (prior to 78 wt %, no plagioclase would have been crystallizing so anorthosites could not have been produced). However, the general lack of a correlation of the An content of feldspars with the mg# of coexisting mafic minerals (Fig. 36) (unlike terrestrial mafic intrusion sequences) remains without a completely satisfying explanation. In general the chemical data is deemed compatible with complex adcumulus growth (Morse 1982, Ryder 1982), but others suggest more complex processes (e.g. Haskin et al. 1981, Jolliff and Haskin 1995).

Petrographic data suggests that the ferroan anorthosite parent magma crystallized in the sequence olivine (+ Al-chromite?) $\Rightarrow$ low-Ca pyroxene $\Rightarrow$ high-Ca pyroxene $\Rightarrow$ plagioclase, in contrast with Mg-rich rocks in which plagioclase in general seems to have preceeded pyroxene. While the general model of anorthosite genesis has been linked in a complementary manner with mare basalt mantle sources in a magma ocean (with the anorthosites accounting for the low Al, Ca, and negative Eu anomaly of mare sources), this relationship cannot be very direct or simple: Some of the mare sources are too magnesian, contain too much high-Ca pyroxene, and too much Ni to be directly complementary to ferroan anorthosites (e.g. Ryder 1982, 1991), and so modification, possibly by mixing

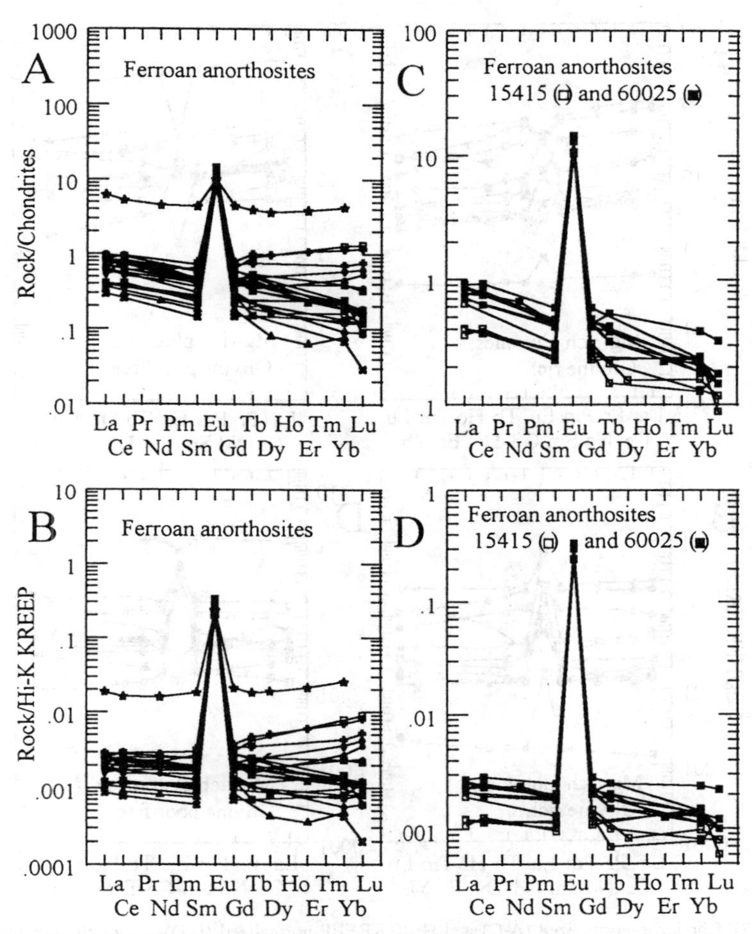

Figure 46. Chondrite-normalized (A,C) and Hi-K KREEP-normalized (B,D) rare earth element plots for ferroan anorthosites. In this figures, in Figures 47–50, and in Figures 56 and 57, the chondrite normalization is that of Nakamura (1974), and the Hi-K KREEP is that of Warren (1989). All chondrite-normalized diagrams use a common scale and all Hi-K KREEP-normalized diagrams use a common scale to allow easy comparison, with the exception of Figures 46C,D, which have expanded scales. These latter compare the within rock and between rock patterns of ferroan anorthosites 60025 and 15415 and show that there is more variation among anorthosites than within one sample, at least for these two rocks.

during solid-state convection of the mantle, must have occurred.

Geochronological data on ferroan anorthosites are sparse (Table 11). Argon isotope systems have all been strongly disturbed by shock-heating events of varied ages and do not reflect original crystallization ages. Zircon is so rare and tiny as to preclude U-Pb ages on that phase. The analytical difficulties of obtaining isochrons on nearly monomineralic rocks, especially those with such low incompatible element abundances, has also precluded much definitive geochronology, especially as shock has at least partly disturbed these systems too. The mafic sample 62237, for instance, failed to produce an isochron for Rb-Sr isotopic analyses, and the data indicate the possibility of both Rb loss by volatilization and Rb redistribution by metamorphism or other processes after crystallization (Snyder

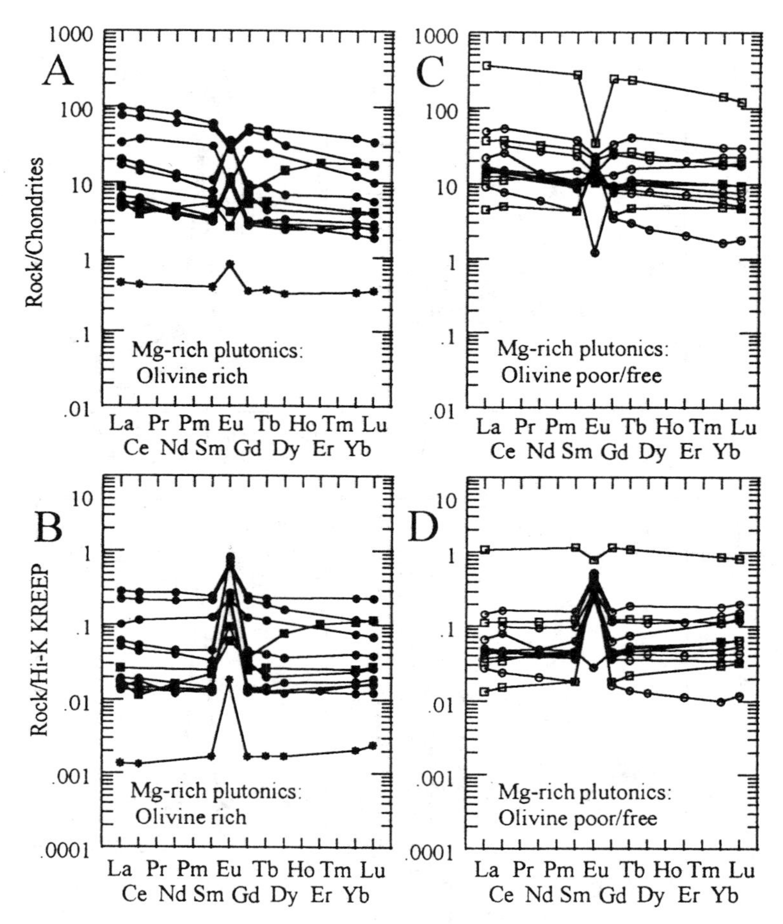

Figure 47. Chondrite-normalized (A,C) and Hi-K KREEP-normalized (B,D) rare earth element plots for Mg-rich plutonic rocks. Among larger rocks the olivine-rich varieties (A,B) tend to have lower rare earth element abundances than olivine-poor varieties; those with high rare earths in (A,B) are some small Apollo 14 samples. The asterix pattern is the dunite 72415-8.

et al. 1994a). Furthermore, any isochron ages might reflect prolonged subsolidus cooling rather than an actual crystallization age. Nonetheless, the low $^{87}Sr/^{86}Sr$ ratios for ferroan anorthosites approach those of primitive meteorites, implying that they have been separated from any significant concentrations of Rb for a length of time comparable with the age of the Moon (Nyquist 1977, Nyquist and Shih 1992). A Sm-Nd internal isochron age of 4.44±0.02 Ga for 60025 (Carlson and Lugmair 1988) is younger than a Pb-Pb model age of 4.51±0.01 Ga for the same sample (Hanan and Tilton 1987). A ferroan anorthosite clast from 67016 yielded a Sm-Nd internal isochron age of 4.56±0.07 Ga (Alibert et al. 1994, Alibert 1994). All these data suggest very early formation, differentiation, and cooling of the Moon. However, Borg et al. (1997) obtained a much younger age of 4.36±0.03 Ga for mafic sample 62236, with an initial εNd of 3.2±0.5 that suggests derivation from a source with depleted light rare earth elements. Premo et al. (1989) analyzed separates from two samples in an attempt to elucidate the lead isotope ratios of the earliest lunar materials. Data do not form linear arrays but are generally consistent with a "cataclysm" array with

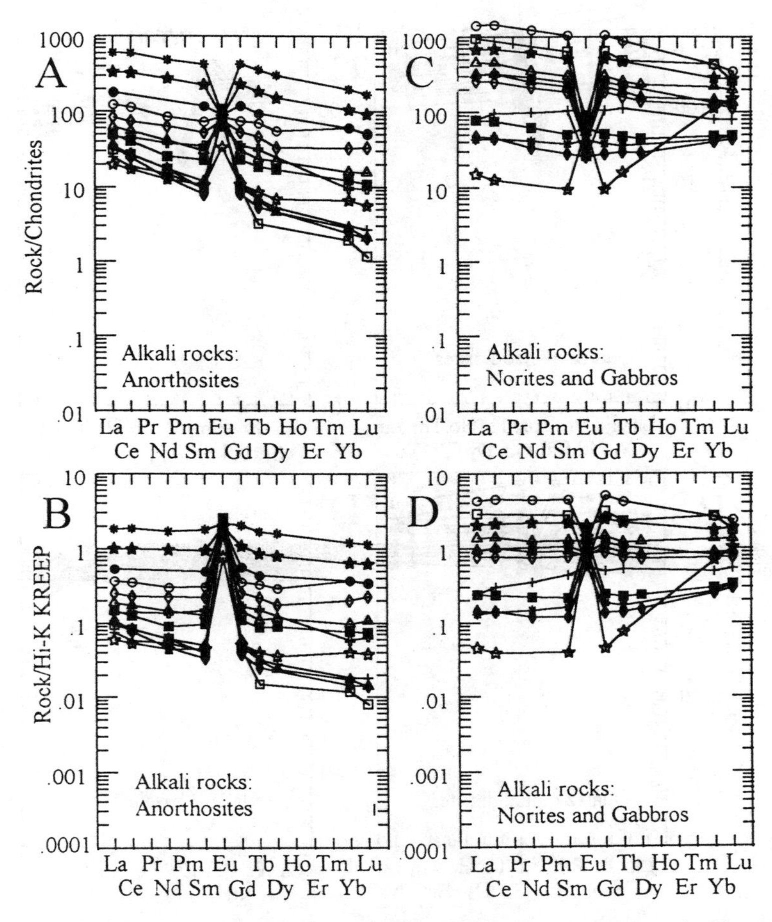

Figure 48. Chondrite-normalized (A,C) and Hi-K KREEP-normalized (B,D) rare earth element plots for alkali rocks. Anorthosites (A,B) tend to have steeper slopes than KREEP (Fig. 49) and norites and gabbros tend to have more varied heavy rare earth slopes than KREEP. Several of these rocks have abundances higher than Hi-K KREEP.

intercepts at ~4.09 and 4.35 Ga. With various assumptions, the data suggest that the Moon had a high μ value ($^{238}U/^{204}Pb$) (>300) from the outset, i.e. it was depleted in volatiles. While all these data are consistent with very early formation, differentiation, and cooling of the Moon and of ferroan anorthosite production, they are not all consistent with fairly simple sequential derivation from a fractionating magma ocean that maintained fairly flat rare earth patterns as is suggested from calculation of parental magmas from bulk rock and mineral data. These inconsistencies remain to be rationalized in some model of early lunar history.

Distribution and relationships. The Apollo samples and lunar meteorites by themselves demonstrate that anorthosite and anorthositic rocks are a major component of the lunar crust. However, while they are abundant at the Apollo 16 site, ferroan anorthosites are rare at the Apollo 15 site, and all but absent from the Apollo 14 and Apollo 17 sites. Some constraints on their global distribution are available from Apollo gamma-ray

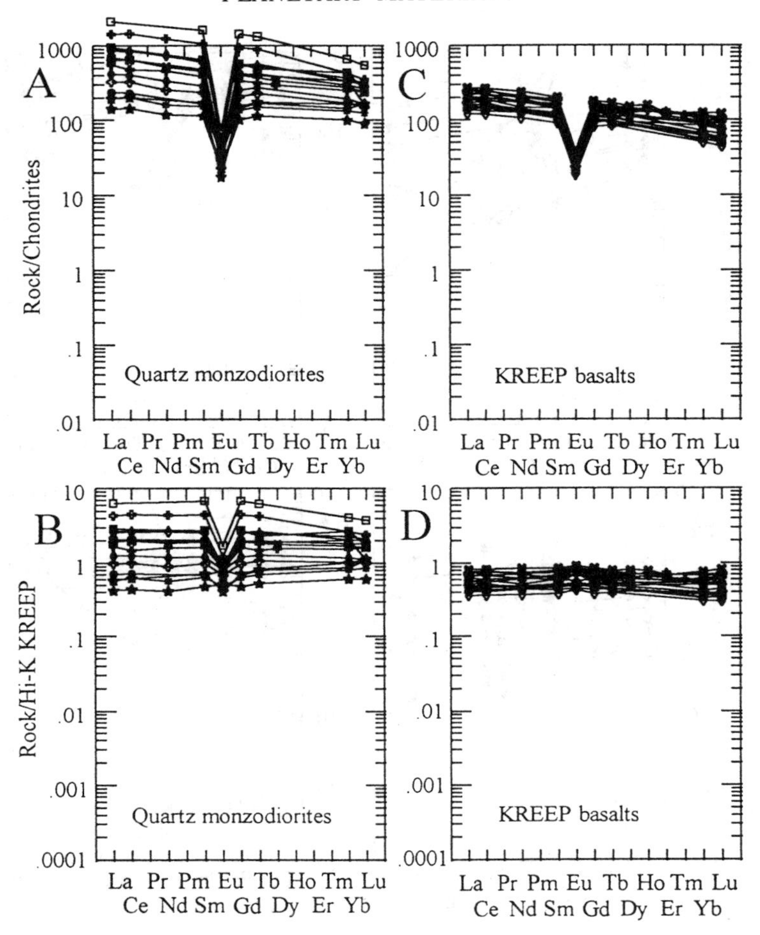

Figure 49. Chondrite-normalized (A,C) and Hi-K KREEP-normalized (B,D) rare earth element plots for quartz monzodiorites and KREEP volcanic samples. The Hi-K KREEP is based largely on Apollo 14 breccias, so its similarity to the volcanic KREEP at Apollo 15 and 17 is noteworthy. The quartz monzodiorites have patterns fairly similar to KREEP, more so than most alkali rocks (Fig. 48).

and x-ray spectroscopy from orbit, and from spectral reflectance studies from the Earth, the Galileo spacecraft, and the Clementine spacecraft. While these data cannot distinguish ferroan anorthosites from any other kind of potential anorthosite, they do confirm the iron-poor and feldspathic nature of the bulk of the highlands. Indeed the Clementine data suggests that vast areas of the lunar highlands surface consist of material with less than ~4 wt % FeO (Lucey et al. 1995, 1997) consistent with ferroan anorthosite, and material with less than 1 wt % FeO is exposed in the inner uplifted rings of some multiring basins (Petersen et al. 1997). However, orbital data suggests that the crust becomes more mafic at depth such that anorthosite does not represent the whole crust (e.g. Spudis and Davis 1986). It may not even have been present in the crustal section that was the target for the Serenitatis basin-forming event (Spudis 1993, Ryder et al. 1997). Considerably more information is needed before the vertical and horizontal distribution of ferroan anorthosite in the lunar crust can be satisfactorily described and used in understanding lunar crustal evolution.

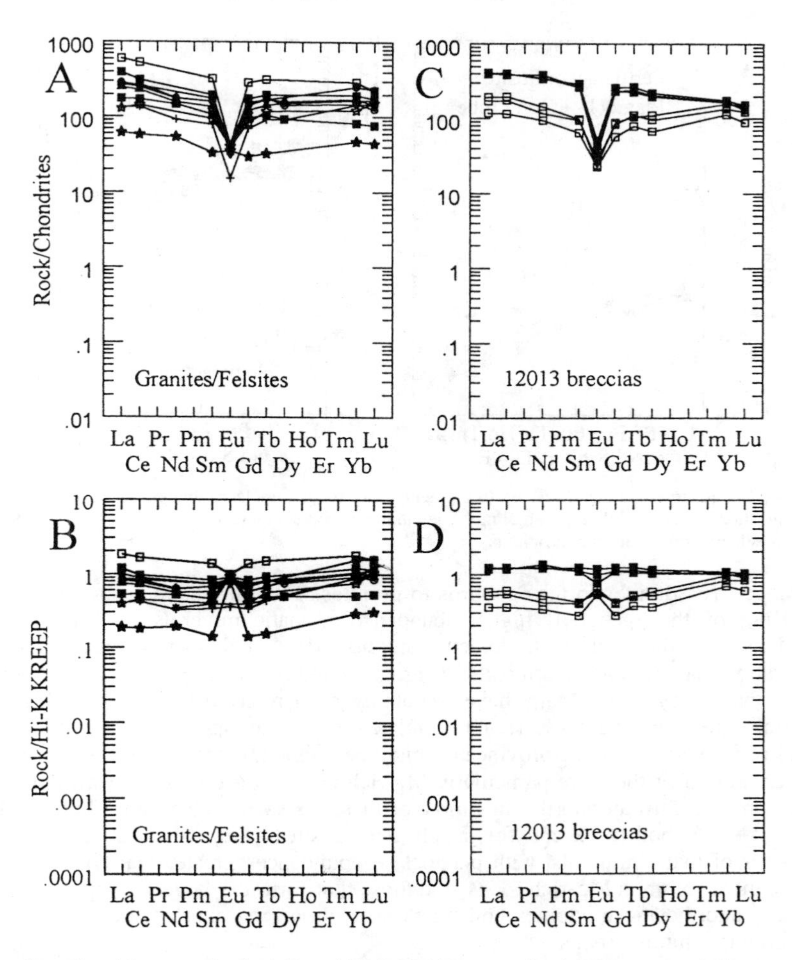

Figure 50. Chondrite-normalized (A,C) and Hi-K KREEP-normalized (B,D) rare earth element plots for granites/felsites (A,B) and the breccias from 12013 (C,D; closed symbols = black breccias, open symbols = gray potassic breccias). The difference between these patterns and KREEP, except for 12013 black breccias, is conspicuous as a V-shape.

A magma that crystallized huge amounts of plagioclase must also have crystallized simultaneously huge amounts of olivine or pyroxene or both. Such iron-rich complementary ultramafic rocks have not been found among lunar crustal samples, and the obvious inference is that they lie buried in the Moon beneath the crust. Some of the characteristics inferred for mare basalt sources are consistent with their being complementary to the ferroan anorthosites and indeed are used to confirm the magma ocean model of lunar evolution and the cumulate remelting hypothesis for mare basalts. Nonetheless, the early lunar evolution was complex and the complementary nature was subsequently modified (Ryder 1991).

Mg-rich rocks

Definition and overview. The compositional range of the Mg-rich rocks is not well-defined, and they are a very diverse group (Norman and Ryder 1979, James 1980).

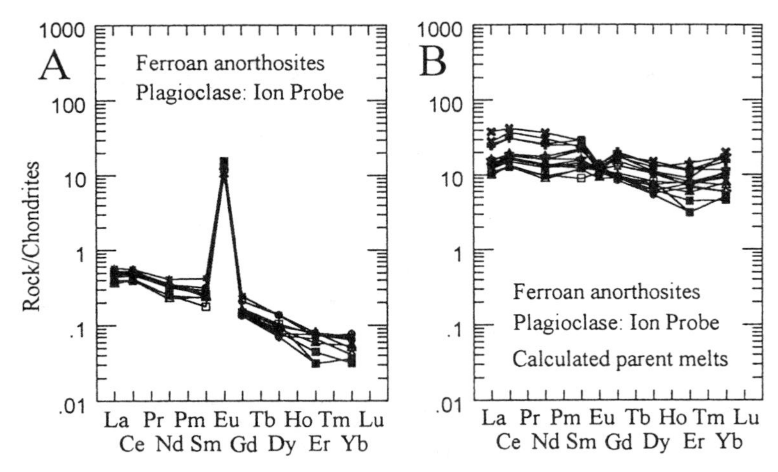

Figure 51. Chondrite-normalized rare earth element plots for plagioclases analyzed by ion microprobe in ferroan anorthosites (A) and their parental melts calculated by inversion (B) (data of Papike et al. 1997b, and using their selected distribution coefficients).

They range from dunites to ferrogabbros to granites. They are distinguished from ferroan anorthosites by their generally higher abundance of mafic minerals, or higher mg#, or elevated incompatible and alkali element contents. Most of them have lower Ti/Sm and Sc/Sm ratios than ferroan anorthosites (Fig. 38), indicating derivation from completely different magma systems. Many have frequently been referred to as Mg-suite rocks, or magnesian suite rocks, but they are not all related in even an approximate way to common magmas or form a geographic province as might be implied by the use of the word "suite". However, not all of them are particularly Mg-rich either, nor do all have a high mg#. In comparison with ferroan anorthosites most of these rocks do have either higher MgO or higher mg# (or both); except for alkali anorthosites, they also all contain greater abundances of mafic minerals, with plagioclase abundances greater than 70 vol % being rare. The most common Mg-rich rocks, and those that form the largest samples, are highly magnesian anorthositic troctolites and magnesian norites or anorthositic norites that are coarse-grained plutonic rocks.

Mg-rich rocks are listed with some of their characteristics in Tables 13-16, and apart from the dominant norites and troctolites, for which only samples larger than 0.5 g are listed, these include all reasonably documented samples. Unlike Heiken et al. (1991*Lunar Sourcebook*), we have included Alkali rocks, KREEP basalts (including quartz monzodiorites), and Granites/Felsites as Mg-rich rocks, mainly because recent interpretations relate some of them in some way to each other and to magnesian plutonic rocks (e.g. Snyder et al. 1995a,b). Modal analyses for rocks within these subgroups are given in Tables A5.42 - A5.45. Compared with ferroan anorthosites, the Mg-rich rocks as a whole contain many more mineral phases, mainly accessory minerals, reflecting higher incompatible element abundances. All but those with mg# > 0.88 have more sodic plagioclase than do the ferroan anorthosites (Fig. 36).

While Mg-rich rocks are common at the Apollo 14, 15, and 17 sites, especially as larger clasts within impact melt samples, they are sparse and small at the Apollo 16 landing site. Remote sensing data such as obtained on the Clementine mission suggest that most of the highlands surficial regions contain only a minor proportion of mafic rock, but they may be more common at depth. Nonetheless, data from Ca in olivines (e.g. Ryder 1992) and

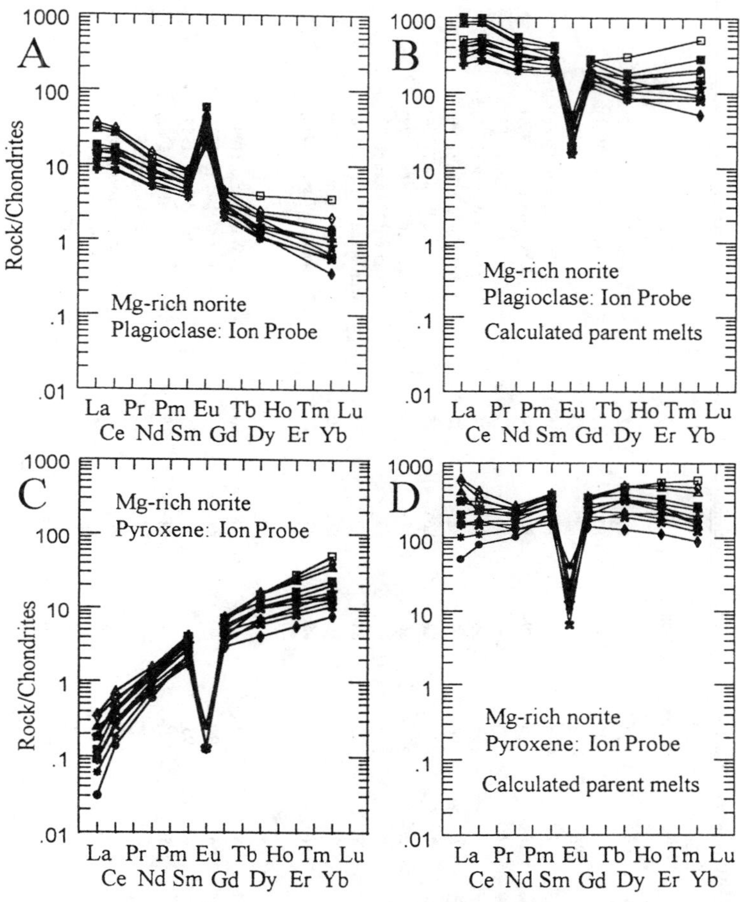

Figure 52. Chondrite-normalized rare earth element plots for plagioclases (A) and pyroxenes (C) in Mg-rich plutonic norites, analyzed by ion microprobe, and parental melts calculated from them by inversion (B,D) (data of Papike 1994, 1996, using their selected distribution coefficients). Open symbols are Apollo 14 samples.

from pyroxene exsolution (e.g. McCallum and O'Brien 1996) suggest that many of the Mg rich rocks are of shallow crustal rather than deep crustal origin. Gamma-ray data from the Apollo orbiting spacecraft show a preponderance for the incompatible-element rich material to be concentrated around the Imbrium region. Given the high incompatible element abundances calculated for the parent magmas even of the more primitive Mg-rich rocks, mass balance shows that they cannot occupy more than about 20% of the global crust. Unlike the ferroan anorthosites, mineral isochrons can be obtained on many of these samples, and some of the more evolved types contain zircons that produce reliable Pb ages. The geochronological data (Table 11) suggest a wide range in ages for Mg-rich rocks, some perhaps as old as ferroan anorthosites, but others much younger, down to the age of the later basin-forming events.

Magnesian plutonic rocks. Most of the larger Mg-rich rocks are plutonic or hypabyssal in origin. Most of these are brecciated or cataclasized and are recognized as

Table 13. List of Mg-rich plutonic rocks

(troctolites and norites, larger than 0.5g; others, all documented)

Sample	mass (g)	Warren '93 Pristinity index	~Plagioclase An mol %	~Low-Ca pxroxene Mg#	~Olivine. Mg#	~Modal % feldspar	Main reference
Troctolites							
14172c11	0.7	7	94	--	87	65	Warren and Wasson (1980)
14179c6	0.7	6	94	--	87	70	Warren et al. (1981)
14303c194	2.0	6	95	--	88	70	Warren and Wasson (1980)
14304c95("a")	0.9	7	94	--	87	55	Goodrich et al. (1986)
14321c1020	9.2	7	95	89	86	70	Lindstrom et al. (1984)
14321c1024	0.7	7	95	--	80	85	Warren et al. (1981)
15455c106	3.0	7	95	85	83	71	Warren and Wasson (1979)
60035c21	0.7	6	96	89	88	57	RD Warner et al. (1980)
73146	3.0	7	95	88	86	85	Warren and Wasson (1979)
73235c127	0.7	5	96	86	83	60	Warren and Wasson (1979)
76255c57("U5B")	2.0	8	96	91	89	77	Ryder and Norman (1979)
76335	465.0	8	96	88	87	77	Warren and Wasson (1978)
76535	155.0	9	96	86	88	50	Ryder and Norman (1979)
76536	10.3	7	--	86	83	70	Warren and Wasson (1979)
Spinel-troctolites							
12071c10	1.3	6	97	--	>78	70	Warren et al. (1990)
14304c109("q")	0.0	6	94	--	87	?	Goodrich et al. (1986)
15295 c	--	--	93	--	91	75	Marvin et al. (1989)
15445c71("A")	1.5	5	92?	--	92	35	Ryder (1985)
15445G (F?)	4.0	--	--	--	90	50	Ryder and Norman (1979)
15455 H	2?	--	96	91	92	25	Baker and Herzberg (1980)
65785c4	0.3	8	96	84	83	65	Dowty et al. (1974b)
67435c77	0.1	8	97	--	92	40	Ma et al. (1981b)
72435	--	--	96	70	73	~80	Dymek et al. (1976)
73263 particles	--	--	96	90	90	~70	Bence et al. (1974)
76503 particles	--	--	96	90	90	~70	Bence et al. (1974)
77517c disagg	--	--	97	90	90	--	RD Warner et al. (1978b)

Table 13. List of Mg-rich plutonic rocks (continued)

Sample	mass (g)	Warren '93 Pristinity index	~Plagioclase An mol %	~Low-Ca pxroxene Mg#	~Olivine. Mg#	~Modal % feldspar	Main reference
Ultramafics							
12033;503Harzb	0.1	8	--	91	89	--	Warren et al. (1990)
14161;212,1Perid	<0.1	7	--	87	85	1	Morris et al. (1990)
14161;212,4Dunite	<0.1	5	--	--	85	--	Morris et al. (1990)
14304c121("d")	0.1	6	--	--	89	--	Warren et al. (1987)
14305c389Pxite	<0.1	7	--	91	90	1	Shervais et al. (1984a)
14321c1141Dunite	0.1	6	--	--	89	2	Lindstrom et al. (1984)
72415-8 Dunite	55.2	9	94	87	87	4	Ryder and Norman (1979)
Norites							
14318c146	1.2	6	87	73	71	55	Warren et al. (1983)
14318c150	0.5	6	83	78	74	65	Warren et al. (1986)
15360;11	0.7	9	93	78	--	65	Warren et al. (1990)
15361	0.9	8	94	84	--	40	Warren et al. (1990)
15445c17("B")	10.0	8	95	82	--	63	Shih et al. (1990)
15455c228	200.0	9	93	83	--	70	Ryder (1985)
72255c42	10.0	8	93	75	--	40	Ryder and Norman (1979)
77035c130	100.0	7	93	79	--	60	Warren and Wasson (1979)
77075/77215	840.0	8	91	71	--	55	Ryder and Norman (1979)
78235/78255	395.0	8	93	81	--	47	Warren and Wasson (1979a)
78527	5.2	4	93	80	77	50	Warren et al. (1983b)
Gabbronorites							
14161;7044	<0.1	6	88	64	--	60	Jolliff et al. (1993)
14161;7350(*)	<0.1	6	96	--	--	90	Jolliff et al. (1993)
14304c114("h")	<0.1	6	89	--	68	40	Goodrich et al. (1986)
14311c220	0.2	5	85	60	--	75	Warren et al. (1983c)
61224;6	0.3	8	83	67	--	34	Marvin and Warren (1980)
67667	7.9	7	91	78	71	24	Warren and Wasson (1979a)
67915c163 (**)	0.2	6	63	--	32	43	Marti et al. (1983)
73255c27,45	0.9	7	89	74	--	53	James and McGee (1979)
76255c72("U5A")	0.1	7	86	67	--	39	Ryder and Norman (1979)
76255c82("U4")	300.0	6	87	65	--	41	Warren et al. (1986)

(*) technically a troctolitic anorthosite; (**) also listed with Alkali rocks (Table 14).

originally igneous and plutonic on relict textural and chemical composition grounds, but a few, such as troctolite 76535, also retain igneous—or simply meta-morphosed—textures. They are distinguished from all other Mg-rich rocks largely on their lower alkali contents (K_2O generally less than 0.1 wt %) including the presence of calcic plagioclase (generally $An_{>90}$; Fig. 42). They tend to have more magnesian pyroxenes than ferroan anorthosites, alkali rocks, quartz monzodiorites, or granites (Fig. 43), but some gabbronorites at least can have fairly iron-rich pyroxenes. It is commonly assumed that these magnesian plutonic rocks intruded existing anorthositic crust, and that they formed layered mafic complexes (similar to the Stillwater Complex) (e.g. James 1980). However, strong evidence for either is not yet available. There is a actually a spatial dissociation between ferroan anorthosites samples (dominating the Apollo 16 landing site) and Mg-rich plutonic rock samples (dominating the Apollo 14, 15, and 17 landing sites) suggesting that lateral and vertical separations dominate the distributions rather than intrusions of one lithology into another. In that many or even most of these samples are plutonic, cumulate, and plagioclases-bearing, they must be from differentiated basaltic bodies, but there is no actual evidence for layering.

Troctolites. Magnesian plutonic igneous rocks consisting mainly of olivine and plagioclase have been collected from all highland sites, but are most common among Apollo 14 and Apollo 17 samples, where ferroan anorthosites are correspondingly rare. Despite their ubiquity, most are rather small samples, with only 4 larger than 5 g, and only another 4 larger than one gram. Even among troctolites there is a diversity of modes (Table A5.42) (some are troctolitic anorthosites) and mg# (Table A5.33) They have been recognized as pristine igneous rocks on the basis of their extreme compositions, phase homogeneity, lack of siderophile contamination, some relict textures, and isotopic data.

Sample 76535 (155 g) is the best preserved of the larger troctolite samples. Its texture is that of an unbrecciated coarse-grained plagioclase-olivine cumulate (Figs. 53A,B) but has generally been interpreted as showing extensive subsolidus annealing and re-equilibration (Gooley et al. 1974, Dymek et al. 1975b). The texture is granular polygonal with smooth curved grain boundaries and abundant 120° triple junctions. The grain size is 2 to 3 mm, rarely larger. A few percent of low-Ca pyroxene and trace amounts of Ca-rich pyroxene, Cr-spinel, Ca-phosphates, baddelyite, metallic iron, and other phases are present, in mesostasis areas and in symplectites. The symplectites of Cr-spinel + augite + ortho-pyroxene occur at olivine-plagioclase boundaries, and are inferred to have formed from that assemblage under a pressure of 0.06 to 0.15 GPa (10 to 30 km depth) (Gooley et al. 1974). The silicate minerals are homogeneous in composition, with plagioclase as calcic as in ferroan anorthosites ($An_{98\text{-}95.5}$) and olivine $Fo_{87\text{-}88}$ (Figs. 36, 42, 45). The other large troctolites 76335 (503 g) and 76536 (10 g) appear to be similar to 76535 in many mineralogical and chemical respects, except that 76536 is more iron rich (Fo_{83}) (Ryder and Norman 1979), but they are more feldspathic and texturally cataclasized (Fig. 53C). 76535 appears to be alone in having a diverse set of trace minerals. Some other troctolites also have more iron-rich olivines e.g. the clast in 15455 (Fo_{83}). Nonetheless most are among the rocks with highest mg# on the Moon other than some spinel-bearing troctolites (below). Along with their higher mg# than ferroan anorthosites, troctolites (and spinel troctolites and dunites) have higher Cr_2O_3 in their olivines (Fig. 45B), but nonetheless Ni is rarely more than 100 ppm (Ryder 1983). The CaO in their olivines, presumably some reflection of cooling rate, varies from as little as 0.02 wt % to as much as 0.15 wt % (Fig. 45A), but clearly far below volcanic olivine abundances.

Several magnesian plutonic troctolites have been found as clasts in the Apollo 14 impact melt breccias (e.g. Warren et al. 1981, Shervais et al. 1983, Lindstrom et al. 1984). Most are magnesian ($Fo_{85\text{-}90}$) but some extend to more iron-rich compositions (Fo_{77}) with

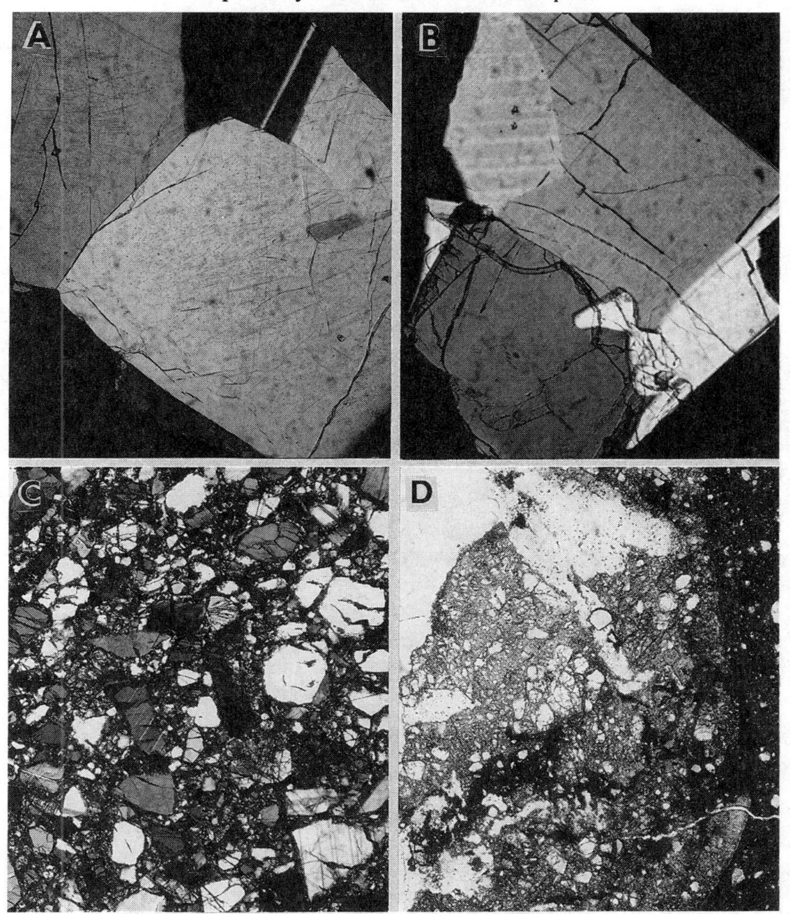

Figure 53. Photomicrographs of Mg-rich plutonic rocks, all fields of view ~2 mm. (A) (B) troctolite 76535,15 (grain mount), crossed polarized light: cumulate or recrystallized, plagioclase and olivine. (C) troctolite 76536,15, crossed polarized light: granulated or cataclasized. (D) spinel-troctolite clast in 15445,133, plane light: crushed and finely granulated. Dark area to right is aphanitic melt matrix.

calcic plagioclase ($An_{94\text{-}97}$). Most of these troctolites have plagioclase abundances greater than mafic mineral abundances, in some cases making them anorthositic troctolites or even troctolitic anorthosites (magnesian-anorthosites of Lindstrom et al. 1984), but most of these are small and thus not necessarily representative. In many of these Apollo 14 fragments the plagioclase was shocked and is maskelynite. The largest sample, an estimated 9 g clast from breccia 14321, is crushed but appears to have had a granoblastic texture with crystals over 2 mm long. Phases are homogeneous plagioclase (An_{95}) and olivine (Fo_{87}), with trace amounts of orthopyroxene, diopside, and Cr-spinel (Warren et al. 1981, Lindstrom et al. 1984). Its mode, while not necessarily representative of a parent rock, clearly has plagioclase in excess of mafic minerals.

Chemical analyses for some magnesian plutonic troctolites are listed in Table A5.33, with rare earth element abundances shown in Figures 47A,B. Large Apollo 17 troctolitic rocks have similar and low rare earth element abundances, with positive Eu anomalies

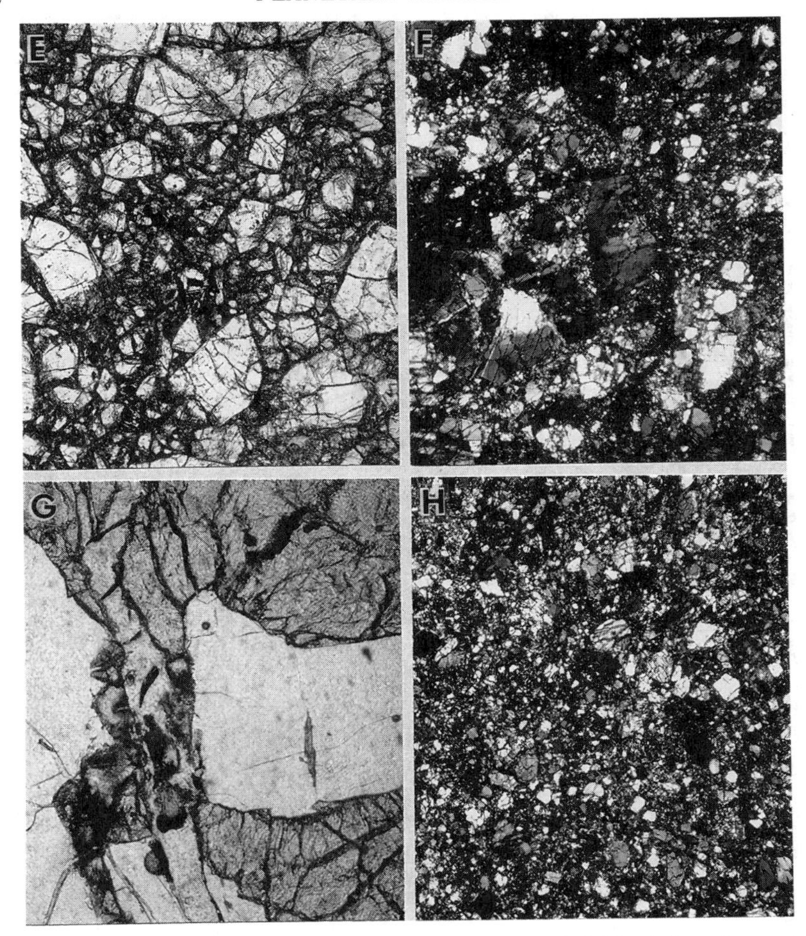

Figure 53 (cont'd). Photomicrographs of Mg-rich plutonic rocks, all fields of view ~2 mm. (E) dunite 72415,54, plane light: cataclasized and recrystallized. (F) norite clast in 15445,133, crossed polarized light; intensely cataclasized, partly maskelynite. (G) norite 78235,48, plane light; shocked but not cataclasized, plagioclase partly maskelynite. (H) gabbronorite ("feldspathic lherzolite") 67667,19, crossed polarized light; extensively granulated, mainly fine matrix.

indicative of plagioclase accumulation. However, the incompatible element abundances are roughly an order of magnitude greater than in typical ferroan anorthosites. Haskin et al. (1974) calculated a probable parent magma for 76535 that had no Eu anomaly and about 30x chondritic light rare earth abundances, with about 16 wt % of the parental magma trapped in 76535. Parent magmas with higher abundances are possible if the trapped magma is less. In contrast, the Apollo 14 troctolites, including the large clast from 14321, show both higher abundances and a greater range of rare earth elements, including some with negative Eu anomalies. Lindstrom et al. (1984) suggest that these Apollo 14 samples represent a magnesian suite that is related to magnesian anorthosites and even dunite from the same breccias. These Apollo 14 troctolites are not members of the alkali rocks, which have more than ~0.1 wt % K_2O (Snyder et al. 1995b) and which generally have even higher rare earth element abundances, more sodic plagioclase, more Fe-rich mafic minerals, and rarely include troctolitic lithologies.

Shervais (1994), Shervais and Stuart (1995), and Shervais and McGee (1997) summarized ion probe analyses of silicate mineral phases in six Apollo 14 magnesian troctolitic rocks. For five of these, parent magmas calculated by assuming that the mineral cores were in equilibrium with that magma were similar to medium- to high-K KREEP, except for the Eu anomalies. The calculated Eu contents are much higher than in KREEP, but the appropriate distribution coefficients are in doubt. The sixth troctolite was somewhat light rare earth element depleted.

Hess (1994) has discussed some of the problems with understanding the petrogenesis of lunar troctolites, particularly the apparent co-saturation of the parent with olivine and plagioclase, but not with pyroxene, and the high-mg# of such a plagioclase-producing magma. Even assimilation of plagioclase, which is an inefficient process, would cause mg# of a magma to decrease as olivine must crystallize to allow assimilation. Possibly at least a part of the interior of the Moon has an mg# higher than the inferred whole-Moon value of 0.80 to 0.84 (e.g. Jones and Delano 1989), as argued by Warren (1986). Indeed, rare relict olivine fragments in breccias have Fo_{93-94} (e.g. Ryder et al. 1997), even higher than the most magnesian found in igneous rocks, attesting to the presence of unsampled magnesian igneous rocks that exacerbate the mg# problem.

Radiogenic isotope data are generally lacking for Mg-troctolites, with the exception of 76535, for which K-Ar, Rb-Sr, Sm-Nd, and U,Th-Pb data have been published. (Table 11). While the Rb-Sr data have been interpreted as suggesting an ancient age of 4.51±0.07 Ga (Papanastassiou and Wasserburg 1976), all other systems point to an age of about 4.25 Ga (see detailed discussion in Premo and Tatsumoto 1992), with some later disturbances apparent. Meyer et al. (1989b) reported a U-Pb zircon age of 4.245±0.075 Ga for a magnesian troctolite (Fo_{88}) from the Apollo 14 landing site, but the presence of zircon and the tiny sample size make its affinity (or even it's igneous origin) obscure. Snyder et al. (1995a) reported whole-rock Nd and Sr isotopic ratios for a sample of troctolitic anorthosite from the Apollo 14 landing site, and the data are compatible with an age of 4.2 Ga and an εNd of -1 at that time, similar to alkali anorthosites.

Spinel troctolites. Troctolites characterized by the presence of a chromian pleonaste ("pink spinel") have so far been found only as small clasts within polymict breccias. Only the clast A in 15445 is known to have a mass larger than 1 g. Most of the spinel troctolites are severely shocked and cataclasized (Fig. 53D) (hence the common term "spinel cataclasite"), and 12071c10 (Warren et al. 1990) is almost all glass. The presence of the spinel is significant: The troctolites completely lack this phase (it is not even a trace phase), whereas the spinel troctolites always contain at least several per cent and thus it is an essential, not just accessory, phase. The modes vary greatly (Table A5.42), from very feldspathic to ultramafic, but because of the small fragment sizes few are reliable. The tiny clasts in 67435 (4 x 4.5 mm and 4 x 5 mm in thin section) described by Prinz et al. (1973) and Ma et al.(1981b) have a preserved cumulus texture with olivine (Fo_{92}) and spinel poikilitically enclosed by calcic plagioclase (An_{97}) (Figs. 36, 42, 45), and about 5 vol % spinel. These clasts also contain rare grains of Ni-rich metal. The grain size is about 1 mm, quite fine by terrestrial plutonic standards. Most other spinel troctolites are also magnesian (e.g. 15445 A Fo_{92}; 15445 H Fo_{91-93}; 73263 fragment Fo_{90}) and typically contain aluminous enstatite (yet pyroxene is absent from the 67435 clasts) (Figs. 43A and 44A,B,C). Mineral grains within a sample tend to be very homogeneous, and the olivines have very low CaO (<0.05 wt %) (Fig. 45). The two clasts in 72435 in contrast have Fo_{72-74} and one contains a grain of cordierite (Dymek et al. 1976). A tiny clast in breccia 15295 also contained 8 vol % cordierite, along with pleonaste spinel, olivine (Fo_{91}) and plagioclase (An_{94}) (Marvin et al. 1989). This assemblage is corundum-normative and suggestive of an origin by metamorphism of a spinel cumulate near the base of the lunar crust. Herzberg (1978) and Herzberg and Baker (1980) calculated minimum depths for equilibrium of 26 km in the Moon for spinel troctolites.

Bulk rock chemical data is hardly likely to be representative on these small fragments, but available data (Table A5.33) show low abundances of rare earth elements similar to those of Apollo 17 troctolites (Figs. 47A,B), except for an analysis of 15445 A that shows an unusual pattern enriched in heavy rare earth elements (Ridley et al. 1973b). A parental magma composition calculated for one of the 67435 clasts (Ma et al. 1981b) has light rare earth element abundances 20 x chondrites and a negative Eu anomaly if about 15 wt % magma is trapped, or more if there is less trapped magma; this is extremely evolved for such a magnesian rock. These spinel troctolites, like other most other Mg-rich rocks, also have subchondritic Ti/Sm and Sc/Sm. It has been shown by Longhi (1978) that a pleonaste-spinel cumulate cannot be derived from a bulk Moon composition with a chondritic Ca/Al ratio: Subchondritic Ca/Al ratios are required. Therefore it seems likely that the origin of spinel troctolites requires a complex set of processes that includes assimilation of plagioclase and contamination with an incompatible element-rich material, presumably related to KREEP (Ma et al. 1981b). The high mg# itself is a problem (Hess 1994), especially given that the bulk Moon is almost certainly more iron-rich than the Earth's mantle; while this applies to troctolites it is even more severe for the more magnesian spinel troctolites. Unfortunately there is no radiogenic isotope data relevant to genesis available for spinel troctolites, mainly because the samples are extremely small, cataclasized, and have little pyroxene or interstitial phases.

Dunites and ultramafic rocks. Only a single large fragment of ultramafic rock exists among the lunar samples: a dunite chipped as several separate fragments (72415-72418) from a 10 cm clast in Boulder 3 at Station 2 on the Apollo 17 mission. This rock is recognized as pristine igneous on account of its mode and chemical composition (which are not plausible as a mixture) and the non-meteoritic ratios of its siderophile elements. While several other ultramafic fragments ranging from dunite to pyroxenite have been reported, these are all so small (<0.2 g or rarely as much as 2 mm in section) as to make their parent rock at the handsample scale unknown.

Dunite 72415-8 is yellowish to greenish gray. It is intensely cataclasized, macroscopically appearing to have 30 vol % yellow-green olivines larger than 1 mm set in a matrix of finer-grained but similarly colored material. The mode has 93 vol % olivine, which has a composition of Fo_{86-89} (Table A5.42; Fig. 45) (Dymek et al. 1975b, Ryder 1992). Besides small amounts of plagioclase, orthopyroxene, and augite, there are trace amounts of Cr-spinel (some in symplectites) and metal, and extremely rare troilite, whitlockite, and Cr-Zr armalcolite. The amount of plagioclase varies significantly among thin sections. There is very little sign of any original igneous texture (Fig. 53E), the sample having been extensively crushed. Many fragments show subgrains and strain bands, and even polygonalized olivines, attesting to a complex history of shock, deformation, and recrystallization. The range in olivine compositions is greater than that among olivines in single hand samples from layered igneous intrusions, and greater than that in troctolite 76535. The range is expressed as zoning of individual crystals, and along with the CaO contents of ~0.1 wt %, indicates that the dunite is a shallow-cooled rock rather than deep-seated in origin (Ryder 1992). The plagioclase is calcic (An_{94-97}) for the most part, though ranges down to An_{89} (Fig. 42). The Fe-metal contains up to 33 wt % Ni.

The dunite has low abundances of rare earth elements and elements compatible with plagioclase (e.g. Al_2O_3 <2 wt %) and is ultrabasic (Table A5.33). Overall the rare-earth pattern is flat at about 0.4 x chondrites, with a positive Eu anomaly (Figs. 47A,B). A modelled parental magma (Laul and Schmitt 1975) has a light-rare earth element enriched pattern about 15 x chondrites, but the data can be modelled in varied ways. McKay et al. (1979), using trapped liquid and updated partition coefficients, suggested substantially lower abundances of rare earths in the parent and a subchondritic Ca/Al ratio. A few subsamples show substantial Ir (2 to 3 ppb) and enrichments of other siderophile and volatile elements, and these have been inferred to be of indigenous rather than meteoritic origin (e.g. Morgan and Wandless 1988).

Papanastassiou and Wasserburg (1975) analyzed subsamples of 72417 for Rb and Sr isotopes, defining a precise age of 4.47±0.10 Ga from a primary isochron. The data define a very ancient crystallization event, and the initial $^{87}Sr/^{86}Sr$ is indistinguishable from BABI. Pb-Pb and U-Pb isotopic data (Premo and Tatsumoto 1992) also suggest an age of 4.52±0.06 Ga, older than but within uncertainty of the Rb-Sr age, and the Pb-Pb age is certainly older than 4.37 Ga. Thus the dunite, however it formed, is among the oldest of lunar rock samples and may be contemporaneous with the ferroan anorthosites (Table 11).

Other small fragments of ultramafic materials tend to be magnesian (mg# >0.85) but in most cases little can be established about their source rock formations because of the small sample sizes. Warren et al. (1990) described a tiny Apollo 12 sample of harzburgite (olivine plus low-Ca pyroxene) with traces of Cr-spinel and Fe-metal. The original grain size was originally 1-2 mm; the olivine ($Fo_{89.5}$) has remained intact but the enstatite has a mottled texture. The enstatite is remarkably Ca-poor ($En_{91}Wo_{0.4}$). Warren et al. (1990) infer that this harzburgite is indeed igneous, despite the extremely high Ir and Au abundances, which they imply are indigneous like those in the dunite 72415-7. The dunite particle in 14321 described by Lindstrom et al. (1984) was inferred by them to have been part of a single plutonic sequence that also produced the Mg-troctolites from the same sample. This fragment has rare earth element abundances more than 20 x higher than in dunite 72415-7. The small, more iron-rich (mg# ~0.80) peridotite and dunite fragments from Apollo 14 described by Morris et al. (1990) do not contain enough Ir to be measured by INAA techniques. The dunite is similar in chemistry to plagioclase-poor splits of 72415-7, with low rare-earth element contents, unlike other Apollo 14 dunites. The four small cataclasized dunite particles described by Snyder et al. (1995a) also have Ir too low to measure by INAA, but these have a range in mg# from 0.76 to 0.88 and are elevated in rare earths up to 30 x chondrites with negative Eu anomalies.

Norites. Magnesian plutonic norites consist almost entirely of cumulus orthopyroxene and plagioclase. In some cases the orthopyroxene is inverted pigeonite. Nearly all are cataclasized (Fig. 53F), with maskelynite common, but there are rare vestiges of igneous texture and grain sizes over 1 mm. The Apollo 17 Station 8 boulder, though shocked and with largely maskelynitised and flowed plagioclase and veinlets of its own meteorite-contaminated shock melt (Jackson et al. 1975) retains its cumulate texture (Fig. 53G) with grains as much as a centimeter across. Nearly all norites are from the Apollo 15 and 17 landing sites, and none have so far been identified among Apollo 16 samples. The largest bodies of igneous rocks actually observed by astronauts in the highlands are cataclasized norites: Two norites of meter-size were sampled on the Apollo 17 mission, one being the discrete boulder at Station 8 referred to above, the other a clast in a larger impact melt breccia boulder (Station 7). No other igneous highlands rock samples, including ferroan anorthosites, were sampled from such large masses. The coarse textures, abundance of pyroxene, and availability of large samples have allowed more mineral isochron ages to be obtained for norites than for ferroan anorthosites or magnesian troctolites.

The Apollo 15 and Apollo 17 norites are clearly distinguished from troctolites by their complete lack of olivine, and there does not seem to be any kind of gradation in mode from these norites to troctolites. Further, the most magnesian norites have an mg# of only 0.83

(their low-Ca pyroxenes would coexist with olivine ~Fo_{86}), in contrast with the most magnesian troctolites (mg# 0.88) and spinel troctolites (mg# 0.92). At one time gabbroic rocks were commonly subsumed under the heading "norites" or described with them (e.g. Norman and Ryder 1979, James 1980). However, James and Flohr (1983) clearly established the distinctions between norites and gabbronorites on several chemical and mineralogical features, and their almost certain derivation from quite different magmas. In particular the norites almost completely lack a discrete high-Ca pyroxene phase (it exists only as exsolved lamellae or trace interstitial phase), and they have a much wider variety of trace phases. (Some of the gabbroic rocks have been further distinguished from magnesian gabbros as being alkalic, but none of the common norites have been distinguished as alkalic).

Modal data for norites are given in Table A5.43. The Apollo 17 norites have subequal amounts of pyroxene and plagioclase, but the two large Apollo 15 clasts have a higher proportion of plagioclase. Within a given sample the major silicate phases are homogeneous, and the samples might well have undergone considerable subsolidus re-equilibration. There is a wide range in pyroxene compositions *among* rocks, with orthopyroxenes ranging from as low as mg# 0.65 in the Apollo 17 Station 7 boulder to about 0.83 in the 15455 clast (Figs. 43, 44); the plagioclase composition varies sympathetically but not greatly (An_{95-88} mainly). The pyroxenes contain higher abundances of the minor elements Ti, Al, Cr at a given En content than do those of ferroan anorthosites (Fig. 44). Only in the most iron-rich norites does exsolution of high-Ca pyroxene exist, e.g. 77215 (Chao et al. 1976) where it is inverted pigeonite. Where high-Ca pyroxene otherwise exists it is an interstitial phase, crystallized late in the petrogenetic sequence. Mineral chemical and textural features suggest that norites are deep plutonic rather than merely hypabyssal rocks, perhaps crystallized at depths of 8 to 30 km in the crust, as suggested for the Apollo 17 Station 8 boulder (Jackson et al. 1975).

The Apollo 15 and Apollo 17 magnesian plutonic norites contain a wide variety of other mineral phases (in contrast with ferroan anorthosites, troctolites, and even gabbronorites which have a much more restricted set), but these are in trace amounts. Almost all show a silica mineral, potash feldspar, Ca-phosphates, Zr-minerals, Fe-Ti-Cr oxides, Fe-metal, troilite, and "granitic" glass as trace interstitial phases (James and Flohr 1983). Niobian rutile is also present in these norites, in contrast with other magnesian plutonic rocks (James and Flohr 1983). These all presumably reflect the presence of small amounts of trapped parental magma.

A few small fragments of possibly magnesian plutonic norites have been identified from other sites, mainly among Apollo 14 samples. These are much less surely pristine than the larger Apollo 15 and Apollo 17 samples, and their significance rather obscure. Each one is somewhat different from the others. Some of them might be better listed as alkali rocks, in that they contain elevated abundances of K_2O (>0.2 wt %), but they have been discussed as magnesian plutonic norites by those who described them (e.g. Warren 1993, Snyder et al. 1995a). While one sample certainly has low K_2O (0.03 wt %) it differs from the larger norites in being much more magnesian (mg# ~0.88) and containing 1 vol % olivine. Several of the other samples also contain olivine,and have mg# generally less than 0.75. The two clasts in 14318 are each about 1 cm across, and have similar though not identical silicate compositions.

Chemical analyses for magnesian plutonic norites are listed in Table A5.33, with rare earth element abundances shown in Figures 47C,D. Incompatible trace element abundances in magnesian plutonic norites are generally higher than in troctolites, except for small troctolite samples from Apollo 14. The abundances of incompatible elements among norites roughly correlate (inversely) with mg#. There is a range in Eu anomalies from positive (samples with lower rare earth abundances) to negative (higher abundances) that appears independent of plagioclase abundance—indeed, it is controlled not by variation in Eu but by variation in the other rare earth elements, and probably by Ca-phosphates. The abundances of incompatible elements are quite evolved in contrast with the major element compositions, as is also the case for KREEP basalts. A relationship between magnesian plutonic norites and KREEP magmas has long been suggested on the basis of Ti/Sm (Fig. 38), rare earth element abundances, and isotopic systematics (e.g. McKay et al. 1978, Lugmair and Carlson 1978, Norman and Ryder 1980a,b).

Papike et al. (1994, 1996) used an ion probe to measure rare earth abundances in the cores of pyroxenes and plagioclases in twelve norites from the Apollo 14, 15, and 17 landing sites, with the objective of calculating the rare earth abundances in the melts parental to these phases. In both plagioclase and pyroxene, the rare earth abundances increase from Apollo 17 to Apollo 15 to Apollo 14 norite by a factor of about two (Fig. 52A,C). The estimated rare earth element concentrations from both phases show some evidence of slight postcrystallization redistribution (or slightly inappropriate distribution coefficients), but the calculated equilibrium melts from each phase for a rock are remarkably consistent. The estimated parents (Figs. 52C,D) are similar to high-K KREEP, especially for the Apollo 15 norites. Snyder et al. (1995a) modelled the fractional crystallization of a KREEP basalt magma, and found that cumulates from the first 55 wt % of crystallization would look chemically like magnesian plutonic norites (subsequent cumulates would be like alkali rocks). Such magnesian plutonic cumulates would contain low amounts of trapped liquid. The residual liquids calculated by Snyder et al. (1995a) are within the range of parental magmas calculated by inversion by Papike et al. (1994, 1996). Thus a strong case can be made on chemical grounds that magnesian plutonic norites were derived as cumulates of KREEP basalt magmas.

More internal isochrons are available for norites than any other lunar highlands rock group, because of the opportunity provided by the availability of samples and their abundance of pyroxene. Some analyses show disturbed isochrons, especially in the Rb-Sr system, and in almost all cases ^{40}Ar-^{39}Ar ages show younger disturbances or ages than the isochrons. Some U-Pb data are also available (Table 11). Many of these data have been discussed by Nyquist and Shih (1992) and Shih et al. (1993a). Some of the norites appear to have formed very early in lunar history, others at later times. The apparent oldest, the anorthositic norite clast in the 15455, has a Rb-Sr isochron that is slightly disturbed, and a Sm-Nd isochron that is more disturbed. The argon was strongly disturbed at around 3.8 Ga. The data suggest a crystallization age of ~4.5 Ga, as old as ferroan anorthosites, and the initial ^{87}Sr/^{86}Sr is within uncertainty of LUNI (Shih et al. 1993a). Two splits only 1 cm apart from the other large Apollo 15 norite, the clast in 15445, give significantly different ages from each other according to Sm-Nd isochrons (the Rb-Sr isochrons for both were too disturbed to extract age information). Shih et al. (1993a) interpret the clast as being composite, though of very similar lithologies. (This seems geologically implausible).

The Apollo 17 Station 8 boulder (78235-8, 78255-6) uniquely gives fairly consistent ages across several systems (Nyquist and Shih 1992) (Table 11). Nyquist et al. (1981) argued for a comparatively old age of 4.43±0.05 Ga as derived from more retentive Sm-Nd system, although another Sm-Nd determination gave 4.34±0.04 Ga. Even ^{40}Ar-^{39}Ar and U-Pb determinations suggest ages in this range, though also show effects of the later shock and intrusion event at ~3.9 to 4.1 Ga. The Station 7 boulder shows a similar age range. Only the 72255 norite clast may have a somewhat younger crystallization age (4.08 Ga). The very oldest ages that suggest contemporaneity with ferroan anorthosites and primitive Sr isotopes are difficult to reconcile with a KREEP magma origin, in that KREEP as an evolved entity appears to require almost complete solidification of the magma ocean. The magmatic and spatial relationships among the oldest norites and the ferroan anorthosites, if they indeed overlap in time, remain obscure. In any case, the range of ages of magnesian norites shows that they were not produced in a single magmatic episode, but in separate intrusions over some period of early lunar history. However, it is possible that in reality all these intrusions were solidified before 4.3 Ga, and represent significant crust-building events.

The larger magnesian plutonic norites, with the exception of the Apollo 17 Station 8 boulder, are all clasts within the low-K KREEP (low-K Fra Mauro) impact melts that have been commonly inferred as being the products of the Imbrium and Serenitatis basins (e.g. Ryder and Wood 1977). Even the Station 8 boulder is likely to have been excavated by the Serenitatis event given its shock age. These melts themselves are noritic in composition, and thus it is likely that the crustal sources of these melts were largely norites similar to those represented by the clasts, with the addition of troctolite and KREEP-norite in some form. These basin-scale impact melts lack ferroan anorthosites either as a chemical component or as clasts, and thus there is no physical evidence that the norite parent magmas actually intruded an anorthositic crust. If the parental magmas of the norites were high-K KREEP, then their proportion of the total lunar crust must be quite small, somewhat less than 20%. If norites dominate the targets in the Imbrium-Serenitatis region, then there must be considerable lateral variability, with other crustal areas such as the farside with far less norite, even beneath the surficial anorthosite.

Gabbronorites. Magnesian gabbronorites are all small samples, and mainly clasts in breccias. Only the clast in 76255 is large (about 300 g, but much of this is impure, with matrix veinlets), and the only other sample larger than a gram is 67667 (8 g). Gabbronorites were recognized as distinct from Mg-norites by James and Flohr (1993). Their most obvious petrographic difference is their greater abundance of high-Ca pyroxene (Table A5.43), not merely as a separate phase but as exsolved from low-Ca pyroxene. James and Flohr (1983) listed 5 other characteristics of gabbronorites in comparison with norites: (1) more sodic plagioclase for a given mg# in low-Ca pyroxene, (2) ilmenite the dominant Ti-bearing mineral, and less variety of minor phases including a lack of Nb- and Zr-rich minerals. (3) Cr-spinel that is richer in Al_2O_3 and poorer in Cr_2O_3. (4) higher Ti/Sm and Sc/Sm at a given mg# (Fig. 38). (5) younger ages. The age distinction is based on very limited data (Table 11). The 6 samples of gabbronorite recognized by James and Flohr (1983) include the sample 67667 previously referred to as "feldspathic lherzolite." Warren (1993) included four tiny Apollo 14 samples with the group. Two of the samples listed by James and Flohr (1983), the 76255 gabbro and the 67915 sodic ferrogabbro clasts (listed in the both Tables 13 and 14), contain K_2O abundances greater than 0.3 wt %, and thus could be considered with alkalic rocks. These observations obviously show the blurred distinctions of lunar rock groups that reflect the inadequate data on their characteristics and understanding of their relationships.

Plagioclase tends to be rather less abundant than in the norites (<50 vol %) and the ratio of high-Ca pyroxene to low-Ca pyroxene is very varied (Table A5.43). The high-Ca pyroxene forms thick rims on the low-Ca pyroxenes, quite unlike those in norites. The gabbronorites tend to be iron-rich in comparison with norites, and to contain greater abundances of ilmenite. The nonetheless small abundance of ilmenite suggests that it is not cumulate, and thus not independently responsible for the higher Ti/Sm. In at least several cases, the minerals in the gabbronorites are zoned (more so than the norites), and these reflect original features suggestive of hypabyssal crystallization, or at least faster cooling than norites. McCallum and O'Brien (1996) studied exsolution of pyroxene in two gabbronorites, the 67915 sodic ferrogabbro clasts, and the 76255 gabbro clast. From modelling of exsolution growth, they concluded that these samples had cooling rates consistent with crystallization in the uppermost crust, not deep crust. At least some of the gabbronorites contain olivine, and 67667 contains 50 vol % olivine with zoned crystals (Fo_{73-68}). Its plagioclase and pyroxene grains are also zoned, consistent with fairly shallow crystallization, but this is a finely granulated sample (Fig. 53H).

The gabbronorites are diverse not only in mineralogy (Figs. 42, 43, 44, 45) but in chemistry (Table A5.33). The incompatible trace element abundances tend to be lower than in norites (as expected from the paucity of trace phases such as zircon and phosphates), but there is overlap. The sodic ferrogabbro fragments in 67915, with a low mg# (0.30 to 0.40), and some of the Apollo 14 gabbronorites have higher incompatible element abundances than most norites, other than those Apollo14, more alkalic fragments. The rare earth element abundance patterns tend to be different from KREEP. The chemical data, especially the distinct Ti/Sm (Fig. 38), strongly suggest that the gabbronorites crystallized from magmas different from those of the norites. These differences might reflect source differences or assimilation differences; at present the data are too scarce for adequate understanding of the origin of these few gabbronorites. Unfortunately, it does not appear that any of the gabbronorites have so far been investigated with the ion microprobe, which could shed considerable light on their parental magmas.

There is little geochronological data for gabbronorites (Table 11). Carlson and Lugmair (1981a) produced an undisturbed and precise Sm-Nd isochron age for 67667 of 4.18±0.07 Ga. The clast in 73255 gave a slightly older age (Carlson and Lugmair 1981b). Two fragments from the Apollo 14 site gave zircon U-Pb ages between 4.1 and 4.2 Ga. These are listed by Nyquist and Shih (1992) as gabbronorites, but the affinity of these samples is uncertain; other gabbronorites do not contain zircon. A third sample listed by Nyquist and Shih (1992) as a gabbronorite, 67975, is also dated by zircon U-Pb as over 4.3 Ga. However, this is an alkali rock, and thus it is probably not to be included with gabbronorites. The suggestion by James and Flohr (1983) that the gabbronorites are younger than the norites remains valid.

The magnesian gabbronorites are only a small part, in number and in size, of the igneous rock collection from the lunar highlands. In contrast with magnesian norites, they have not so far been identified at the Apollo 15 site, but they have been identified at the Apollo 16 site. At the Apollo 17 site, they occur in the same impact melt varieties that contain the norites. Those at the Apollo 14 site, because of their small unrepresentative size and perhaps dubious pristinity, remain obscure in significance. The particular role of gabbronorites in lunar crustal evolution thus remains in doubt.

Alkali rocks. Hubbard et al. (1971b) recognized that a few anorthosite fragments were distinctly higher in potassium than the more common type, but little attention was paid to such rare alkaline fragments until Warren and Wasson (1980) recognized and emphasized their chemical and general geographical distinction. Since then more have been identified, including less-feldspathic fragments, and they have been considered as a significant group, especially among Apollo 12 and Apollo 14 samples. The use of the word *rock* is somewhat of an exaggeration; they are known only as small fragments. Of the 27 samples listed by Warren (1993) only one is larger than 1 g and most are no larger than a few millimeters across (the list actually has 20 samples, but one is really 8 separate clasts of 67955). With that small size comes a corresponding decline in the confidence of any particle actually being of igneous origin, and in representivity of the sample.

There is no formal or agreed definition of an alkali rock. Warren and Wasson (1980) identified only 5 alkali anorthosites as enriched in both alkaline and incompatible elements. Warren (1993) describes the group as apparent intrusives distinguished by relatively sodic plagioclase (generally more sodic than An_{86}) and high concentrations of incompatible trace elements. Snyder et al. (1995b) define alkali rocks as containing in excess of 0.1 wt % K_2O and generally greater than 0.3 wt % Na_2O, and required them to have elevated incompatible element abundances (e.g. La > 30 × chondrites) and evolved mineral compositions (e.g. plagioclase An_{86-76} and pyroxene En_{70-40}). Such a definition would include all KREEP basalts, quartz monzodiorites, and granites/felsites, but alkali rocks rarely have the archetypical KREEP incompatible element abundance pattern (Fig. 48), nor the high abundances of K_2O that these other evolved rocks have (Figs. 39, 40A). They do have a wide range in trace element abundance levels, and typically high Na_2O, Eu, and Sr (Figs. 40A,B).

There are some differences among authors about whether a specific sample is an alkali rock or not. For instance, Warren's (1993) list contains 27 samples, whereas Snyder et al. (1995b) refer to 38 samples (Table 14), and there is disagreement on several samples. Clearly more work on both sample description and classification criteria are required. Many alkali rock fragments are alkali anorthosites, and others are gabbroic or noritic; no dunitic or troctolitic alkali rocks seem to have been so far described. Alkali rock fragments may be related to KREEP basalts as cumulates from evolved KREEP (or quartz monzodiorite) magmas (e.g. Jolliff et al. 1993, Snyder et al. 1994b, 1995b, 1996), but various magmatic origins, assimilations, and metasomatisms have been invoked and their origins remain obscure. Orbital chemistry does not distinguish this rock type from typical KREEP as a component of highland regolith.

Alkali anorthosites. Alkali anorthosites are a common type of alkali rock, and so far appear to be almost restricted to the Apollo 12 and 14 sites, i.e. western. Nearly all are severely cataclasized, but some relict igneous textures have survived. Modes range from nearly pure anorthosite to anorthositic norites with ~84 vol % plagioclase and ~16 vol % pigeonite (Table A5.44). Olivine has not been found in any samples, and accessory minerals include augite, potash feldspar, ilmenite, whitlockite, a silica phase, and Fe-Ni metal (Warren and Wasson 1980, Shervais et al. 1983, 1984b, Warren et al. 1990, and others). The presence of whitlockite has received considerable discussion (e.g. Warren et al. 1983c, Snyder et al. 1992, Jolliff et al. 1993), in particular as to whether it was part of an original assemblage or a later introduction (e.g. by metasomatism, Snyder et al. 1992). Current evidence favors whitlockite as a part of the original assemblage in all rocks, rather than metasomatic (e.g. Jolliff et al. 1993, Snyder et al. 1996). Where relict textures are preserved, it appears that pyroxene and ilmenite are interstitial to the plagioclase and thus might be post-cumulus (Shervais and Taylor 1986). There is a range of compositions of plagioclase and pyroxene among rocks (Figs. 42, 43, 44). In at least some cases the plagioclase is zoned within a sample (e.g. Snyder et al. 1995b), although Shervais and Taylor (1986) refer to plagioclase as unzoned primocrysts. Pyroxene tends to be fairly uniform within a sample, but is not always so (Warren et al. 1990, Snyder et al. 1995b). It rarely shows significant exsolution, and the lack of equilibria suggests rapid cooling and thus a shallow origin rather than a deep plutonic one.

The largest sample, an alkaline noritic anorthosite clast in 14047 described by Warren et al. (1983b), has relict grains showing that both pyroxenes and plagioclases were larger than 1 mm, although the fragment has been severely cataclasized. Its minerals are zoned with plagioclase varying from An_{87-86} and pyroxene from ~$En_{50}Wo_6$ to $En_{40}Wo_{15}$. The pigeonites show little exsolution (maximum 1 μm lamellae). There are also traces of potash feldspar, ilmenite, and Fe-metal.

Chemical analyses for alkali anorthosites are included in Table A5.34. Most of these determinations are by instrumental neutron activation of very small subsamples, most less than 50 mg, and make interpretation poorly constrained (or, in the word of Snyder et al.

1995b, "challenging"). By definition, the major element chemistry of alkali anorthosites has high abundances of aluminum, calcium, and alkalis, and correspondingly low abundances of ferromagnesian elements. They tend to be high in sodium, but not in potassium (Fig. 40A). Snyder et al. (1995b) noted that there was almost a 2-order variation in most incompatible elements, whereas Eu was almost uniform in concentration (mainly 6 to 8 ppm). Thus both strongly negative and strongly positive Eu anomalies exist among alkali anorthosites. The plagiophile elements Eu, Sr, Na (Figs. 40A,B), and Ba are much higher in abundance than they are in ferroan anorthosites. In almost all cases FeO is greater than MgO.

Because of their small sample sizes, little isotopic or geochronological data is available for alkali anorthosites (Table 11). A clast from 14321 described as alkali anorthosite contained zircon dated using U-Pb ion microprobe techniques as 4028±6 Ma (Meyer et al. 1989a,b). Snyder et al. (1995b) reported a Sm-Nd isochron age of 4108±53 Ma for the clast "b" in 14304, with an εNd of -1.0±0.2, which is similar to that of KREEP at around the same time. The Rb-Sr in the same fragment appears to have been disturbed to give scatter such that an isochron "age" of 4336±81 Ma is suspect.

Warren and Wasson (1979a) suggested that the alkali anorthosites (of which few were then recognized and described) were precipitates from either magnesian suite or ferroan anorthosite suite magmas that had assimilated urKREEP. Most suggestions since then have similarly included a KREEP incorporation of some kind. Snyder et al. (1995b) calculated anorthositic and noritic/gabbroic adcumulates and orthocumulates from successive residual liquids of a fractionating KREEP basalt magma. Alkali anorthosites are consistent with being among such cumulates, preceding crystallization of most alkali gabbronorites and quartz monzodiorites. Since the last dregs of the magma ocean crystallized prior to ~4300 Ma, these melts cannot represent direct remnants and must involve remelting. Ion probe analyses of the mineral phases in an alkali anorthosite (14304 "b") show the plagioclase to be varied in rare earth element abundances, but pyroxene more uniform (Snyder et al. 1994b). The data are reasonably consistent with a quartz monzodiorite parent for alkali anorthosites. Further work (Snyder et al. 1996) including other elements appears consistent with this conclusion.

Mafic alkali rocks. The norites, gabbros, and gabbronorites that are alkaline are much less restricted to the western highlands, with a similar number of samples from both east and west. Based largely on the study of the clasts in 67975, James et al. (1987) subdivided them into magnesian and ferroan varieties. The more magnesian samples ($En_{>50}$) contain bytownite, hypersthene, augite, silica, ilmenite, apatite, whitlockite, chrome spinel, zircon, potash feldspar, iron metal, and troilite. The more ferroan samples contain a ternary plagioclase, pigeonite, augite, potash feldspar, silica, and the same other minor phases.

Relict textures are rare because of the common granulation, and appear to be xenomorphic-granular. Plagioclase grains are rarely as large as 750 μm or pyroxene as large as 500 μm across and tend to be anhedral. Exsolution of pyroxene is rare and thin, less than a few microns across, implying fast cooling. There is a wide range of compositions of phases among fragments (Figs. 43, 44), but not within fragments in most cases (e.g. James et al. 1987, Goodrich et al. 1986). The ternary feldspars in the ferroan varieties contain as much of the orthoclase component as of the albite component (>15 mol % of each), although published reliable analyses are rare.

Chemical analyses of mafic alkali rocks are included in Table A5.34. As with the alkali anorthosites, most of these analyses are by instrumental neutron activation of very small subsamples. The analyses show a great diversity of most elements, even the major ones, e.g. Al_2O_3 from 7 to 25 wt %, MgO from 4 to 16 wt %. There is a great variation in

Table 14. List of alkali rocks

Sample	mass (g) (if available)	Warren' 93 Pristinity index	Warren '93 name if not Alk. suite	Snyder et al. '95 name	~Plagioclase An mol %	~Modal % feldspar	Main reference
12003,179/210	0.10	6		A	82	100	Warren et al. (1990)
12033,425/501	0.13	8		A	83	99	Warren et al. (1990)
12033,550/532	0.02	6		A	83	96	Laul (1986), Simon and Papike (1985)
12033,555/534	0.07	3	Mg-norite	N	81	49	Laul (1986), Simon and Papike (1985)
12033,97.7	0.10	5		A	88	100	Hubbard et al. (1971b)
12037,178/177	0.02	6	?Alk suite	A	~92	~97	Laul (1986), Simon and Papike (1985)
12042,280/281			X	N	~85		Laul (1986), Simon and Papike (1985)
12073c120/122	0.08	7		A	79	99	Warren and Wasson (1980)
14047c112/113	1.65	7		A	81	84	Warren et al. (1983b)
14066c49/51	0.01	8		A	81	85	Shervais et al. (1983)
14160,106/105	0.19	7		A	82	100	Warren and Wasson (1980)
14160,197/217			X	A	~70	80	Snyder et al. (1992)
14161,7245	0.04		X	A		~95	Jolliff et al. (1991)
14303,44			X	N	86		Hunter and Taylor (1983)
14304c122"b"	0.49	5		A	82	98	Warren et al. (1987)
14304c86"g"	0.23	7		N	82	14	Goodrich et al. (1986)
14305c283WhtA	0.14	7		A	85	95	Warren et al. (1983c)
14305c400	0.67	6		A	76	99	Shervais et al. (1984a)
14305c91	0.14	7		N	86	90	Hunter and Taylor (1983)
14305,279/301	0.23	8	Mg-troct	A	94	85	Warren et al. (1983c)
14311,96			X	N	85		Hunter and Taylor (1983)
14311,220	0.23	5	Mg-gabnor	N	85	75	Warren et al. (1983c)

Table 14. List of alkali rocks (continued)

Sample	mass (g) (if available)	Warren' 93 Pristinity index	Warren '93 name if not Alk. suite	Snyder et al. '95 name	~Plagioclase An mol %	~Modal % feldspar	Main reference
14313c70WhtA	0.03	5		N	83	50	Warren et al. (1983c)
14316,6/12	0.002	7	?KREEP	N	84	60	Warren et al. (1981)
14318,146/149	1.20	6	Mg-norite	N	87	55	Warren et al. (1983b)
14321c1060WhtA	0.14	6		A	86	96	Warren et al. (1983c)
15405c170	0.04	7		N	89	70	Lindstrom et al. (1988)
15405c181	0.10	5		A	84	99	Lindstrom et al. (1988)
67915,163	0.23	7	Mg-gabnor	N	63	43	Marti et al. (1983)
67975,14			X	N	88	50	James et al. (1987)
67975,44Nm		7	X	N	86	21	James et al. (1987)
67975,44Nf			X	N	~70	63	James et al. (1987)
67975,62			X	N	85	38	James et al. (1987)
67975,86			X	N	82	15	James et al. (1987)
67975,117N		7	X	N		50	James et al. (1987)
67975,131N		7	X	N	85	85	James et al. (1987)
67975,136N		7	X	N	~70	42	James et al. (1987)
67975,42N		7	X	N	85	48	James et al. (1987)
77115c19	0.60	6		-	95	70	Winzer et al. (1974)

X = not included by Warren (1993)
* = Monzogabbro by Jolliff et al. (1993)

A = Alkali anorthosite
N = Alkali norite or gabbronorite

incompatible element abundances of about 2-orders of magnitude, with Eu and Sr lower than in the alkali anorthosites but nonetheless quite constant. Thus both strong positive and negative Eu anomalies are present among these samples. The incompatible element abundance patterns are not those of KREEP but in some cases approximate it. Those that have greater abundances than typical KREEP have a pattern of excess whitlockite (steep heavy rare earth elements with Gd>Lu).

There is little isotopic or geochronological data available for mafic alkali rocks (Table 11). Meyer et al. (1989a,b) reported an age for zircon of 4339±5 Ma in one of the alkali gabbronorites from 67975 using U-Pb ion microprobe techniques, and of 4141±5 Ma in a supposed alkali gabbronorite, zircon-rich, from 14066. Snyder et al. (1995b) reported Sm-Nd whole rock data for two splits of alkali norite clast "g" in 14304, which provide an εNd similar to the alkali anorthosite from the same rock if the same age is assumed. Ar data from sodic ferrogabbro clasts in 67915 showed large (60%) low-temperature ^{40}Ar losses, and all suggestions of ages from it represent thermal events that affected the boulder and not the crystallization of the sodic ferrogabbro (Marti et al. 1983).

Snyder et al. (1995b) included mafic alkali rocks in their modelling of alkali samples as cumulates from the crystallization of KREEP basalts. They found the chemical characteristics to be compatible with an origin as cumulates from successive residual liquids, with most of the gabbronorites crystallizing after the alkali anorthosites, contemporaneously with or after alkali norites, and before quartz monzodiorites. Nonetheless, it seems likely that the alkali anorthosites must have had some complementary mafic rocks. Ion microprobe analyses of rare earth elements in mineral phases in alkali rocks were made by Shervais and McGee (1997). These data suggest most of the samples crystallized from magmas with rare-earth patterns similar to high-K KREEP (though with an apparent higher abundance of Eu), and not significantly different from those parental to non-alkaline magnesian rocks (i,e., Mg-rich plutonic rocks). Snyder et al. (1994b) obtained rare earth element abundances by ion probe in phases in an alkali gabbro in 14318 and deduced a parent magma similar to quartz monzodiorite. Further ion probe work on mineral phases in other samples, and including other elements, suggested evolved magmas with KREEP element abundances as parents (Snyder et al. 1996), and more evolved than those that produced the non-alkaline magnesian rocks.

KREEP basalt and quartz monzodiorite rocks. The two definitive traits shared by rocks characterized as KREEP are highly enriched concentrations of incompatible minor and trace elements and ratios among these elements similar to those characteristic of the "type" KREEP samples, i.e. the soils and typical breccias from Apollo 14 (Meyer 1977, Warren and Wasson 1979b, Warren 1989). The bulk of the incompatible elements in rocks and soils on the surface of the Moon appear to be associated with KREEP, which was discovered in Apollo 12 samples by Meyer and Hubbard (1970) and given its acronymised sobriquet by Hubbard et al. (1971a). The main exceptions of rocks that are rich in incompatible elements but do not have the KREEP relative abundances (or nearly) are the rare granites/felsites and some of the alkali rocks. All these incompatible element-rich rocks are actually alkaline.

While the KREEP component is ubiquitous in lunar samples, it dominantly occurs as a cryptic component of brecciated materials, and igneous rocks carrying the KREEP signature are uncommon, although KREEP contamination of magmas has been inferred. Reviews and summaries of the nature and origin of KREEP have been provided by Meyer (1977), Warren and Wasson (1979b), and Warren (1989), among others. While the history of KREEP is not known in detail, most authors infer its ultimate origin in the residual dregs of a primordial and global magma ocean, because partial melting models do not explain the constancy of relative trace element abundances (partial melting is necessary in *subsequent* development of igneous rocks that contain KREEP). Warren and Wasson (1979b) used the term urKREEP to designate this residual component, whose precise characteristics must necessarily be inferred, not measured. Model ages from KREEP materials for Rb-Sr and Sm-Nd isotopes suggest that the KREEP relative abundance pattern was established at about 4.36 to 4.42 Ga (see Nyquist and Shih 1992).

Igneous rocks with the KREEP abundance pattern are rare and small (Table 15), except for the Apollo 15 KREEP basalts which are not rare but are small. They consist of two groups of basaltic volcanic rocks, and some fragments of more evolved (higher incompatible element abundances) and more slowly cooled rocks that have been designated quartz monzodiorite or quartz monzogabbro. They are all recognized as pristine igneous rocks on the basis of their homogeneous textures and complete lack of any clastic material, as well as their complete lack of meteoritic contamination. Warren (1993) listed a few other particles as possibly pristine KREEP igneous rocks. Because Th is easily sensed using γ-ray techniques, KREEP is detectable from orbit, but is not distinguished from alkali rocks or granites/felsites, nor is its igneous form distinguished from its cryptic polymict form. Nonetheless, the Apennine Bench has been correlated with the Apollo 15 KREEP basalts because of its Th abundance and landforms (Spudis 1978, Hawke and Head 1978).

Apollo 15 KREEP basalts. Igneous-textured KREEP basalts were an unexpected find as numerous small fragments and particles among Apollo 15 samples. They are clearly distinguished mineralogically from mare basalts by their higher plagioclase abundances and the presence of orthopyroxene (Table A5.45). Only two samples, 15382 (3.2 g) and 15386 (7.5 g) were over the 1 cm size to be numbered as individual rocks, and these were part of the rake material, and not individually collected. Smaller fragments are common among the fines and as clasts in breccias (commonly as much as 20 vol %) both on the mare plains and on the Apennine Front. They have been described by Dymek et al. (1974), Basu and Bower (1976), Dowty et al. (1976), Meyer (1977), Dymek (1986), and Ryder (1987), among many others.

A variety of grain sizes and textures characterize the Apollo 15 KREEP basalts, although any given particle is homogeneous (Figs. 54A,B,C,D,E), although the minerals are chemically and thus optically zoned. Intersertal/intergranular to subophitic types are the most common, but fine fasciculate to radiate varieties and glomeroporphyritic varieties occur (Ryder 1987), though most are not porphyritic. Plagioclase forms a reasonably continuous network of interlocking laths with pyroxene in between. Many pyroxenes have cores of orthopyroxene, overgrown by pigeonite and in turn by augite. Olivine (Fo_{70-75}) is an extremely rare early phase, and is absent from most samples. Acicular ilmenite occurs with later pyroxene, and commonly associated with cristobalite and a glassy mesostasis. In some cases the mesostasis is a clear yellow glass, in others it is a more opaque cryptocrystalline assemblage. Immiscibility is common in the mesostasis. The late-stage phases include fayalite, phosphates, Fe-metal, and troilite.

Plagioclase crystals are strongly zoned (An_{90-70}) (Fig. 42), and are distinctly higher in K_2O (Fig. 42A) than plagioclase in other rocks. They are also higher in Fe and Mg than plutonic rocks, including the quartz monzodiorites (Fig. 42B,C). The pyroxenes are strongly zoned (Figs. 43B, 44). Their orthopyroxene cores, as much as En_{85}, include rounded and embayed cores of varied alumina content (1 to 3 wt % Al_2O_3) discontinuously mantled with pigeonite or augite. They might represent partially resorbed grains that had crystallized prior to eruption (Dymek 1986). Ryder (1987) interpreted the presence of orthopyroxene phenocrysts, orthopyroxene-plagioclase glomerocrysts, and abundant clear yellow glass residues in different samples as requiring two-stage cooling histories, more consistent with a dynamic volcanic than an impact origin.

Table 15. List of KREEP basalts and quartz monzodiorites.
[KREEP basalts: coarse fines or larger or equivalent fragments; quartz monzodiorites: most described.]

Sample	mass (g)	Warren '93 Pristinity index	~Plagioclase An mol %	~high-Ca pyroxene Mg#	~Modal % feldspar	Main reference
KREEP basalts						
14161,7048	0.02	5	range	range	55	Jolliff et al. (1991)
15007,290/291	0.06	7	range	range	–	Warren et al. (1983c)
15007,292/293	0.03	7	range	range	–	Warren et al. (1983c)
15007,294	0.01	4	range	range	–	Warren et al. (1983c)
15007,302	0.02	5	range	range	–	Warren et al. (1983c)
15007,304	0.02	6	range	range	–	Warren et al. (1983c)
15024,11	0.21	5	range	range	–	Ryder and Sherman (1989)
15264,4	0.28	6	range	range	–	Ryder and Sherman (1989)
15304,6	0.31	5	range	range	49	Ryder and Sherman (1989)
15314,34	0.11	5	range	range	47	Ryder and Sherman (1989)
15382	3.20	7	range	range	41	Ryder (1985)
15386	7.50	7	range	range	43	Ryder (1985)
15404,5	0.17	5	range	range	–	Ryder and Sherman (1989)
15405c68	0.08	5	range	range	40	Ryder (1985)
15434,16	0.44	5	range	range	45	Ryder and Sherman (1989)
15434,17	0.20	5	range	range	48	Ryder and Sherman (1989)
15434,18	0.74	5	range	range	45	Ryder and Sherman (1989)
15434189	0.41	5	range	range	–	Ryder and Sherman (1989)
15434192	0.10	5	range	range	–	Ryder and Sherman (1989)
15434194	0.05	5	range	range	–	Ryder and Sherman (1989)
15434,21	0.22	5	range	range	32	Ryder and Sherman (1989)
15434,25	0.14	5	range	range	–	Ryder and Sherman (1989)
15434,29	0.30	5	range	range	–	Ryder and Sherman (1989)
15434,8	0.32	5	range	range	42	Ryder and Sherman (1989)
15564,16	0.16	5	range	range	–	Ryder and Sherman (1989)
67015c310	0.04	5	range	range	72	Marvin et al. (1987)
72275c91	2.73	7	range	range	40	Salpas et al. (1987)
Quartz monzodiorites			(~averages)	(~averages)		
14161,7069	0.02	7	70	35	41	Jolliff (1991)
14161,7373	0.02	7	70	49	28	Jolliff (1991)
14161,7264 bx	0.02	–	71	62	50	Jolliff et al. (1991)
15403,23c	–	–	–	–	–	Marvin et al. (1991)
15403,24,7001	0.04	8	70	32	50	Marvin et al. (1991)
15403,7002	0.00	7	60	31	45	Marvin et al. (1991)
15403,71a	0.01	8	60	30	50	Marvin et al. (1991)
15403,71b	0.00	7	–	30	30	Marvin et al. (1991)
15403,71c	0.00	7	20	–	60	Marvin et al. (1991)
15405c56	2.50	8	–	–	46	Ryder (1985)
15434,10	0.12	7	60	–	30	Ryder and Martinez (1991)
15434,12	0.62	7	80	33	40	Ryder and Martinez (1991)
15434,14 bx	0.15	7	80	–	40	Ryder and Martinez (1991)
15459c315	0.10	5	60	29	59	Lindstrom et al. (1988)

bx = brecciated, probably impure (not strictly monomict)

Figure 54. Photomicrographs of evolved rocks, all fields of view ~2 mm. (A) KREEP basalt 15386,5, plane light: coarse-grained. (B) KREEP basalt 15382,7, plane light: finer-grained, mainly crystalline. (C) KREEP basalt 15434,49, plane light: finer-grained, abundant cryptocrystalline mesostasis. (D) KREEP basalt 15404,19, crossed polarized light: glomerocrysts in subophitic groundmass.

Some chemical analyses are included in Table A5.35. As noted by Dowty et al. (1976), these basalts lack meteoritic siderophile contamination. The Apollo 15 KREEP basalts have a small but real range in chemical composition (Ryder 1989, 1994). They string out along the plag-px cotectic in the Ol-Pl-Qz pseudoternary system, and are consistent with being a sequence formed by fractional crystallization of plagioclase and low-Ca pyroxene. There is a negative correlation between MgO (11 to 7 wt %) and Sm (25 to 38 ppm). While this might be construed as resulting from sampling errors for the small fragments, duplicate analyses are consistent, and the MgO also correlates with the composition of the orthopyroxene cores in the samples. Thus a real fractionation requiring 30 vol % plagioclase and pyroxene separation is indicated (Ryder 1989).

The Apollo 15 KREEP basalts are enriched in all incompatible elements (refractory incompatibles ~150 x chondrites), which have relative abundances very similar to those of

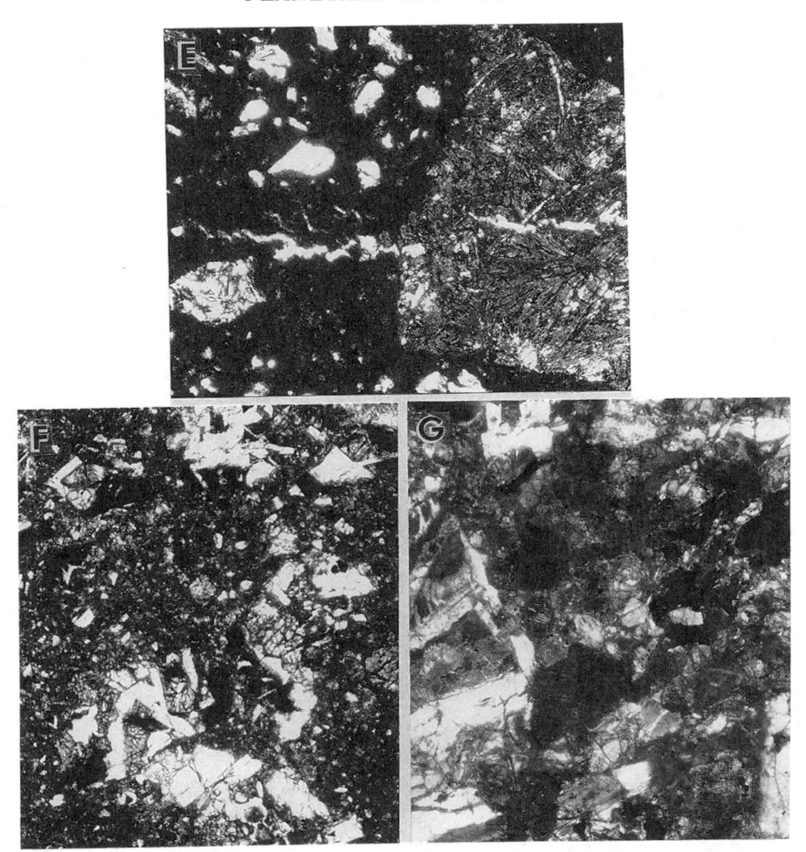

Figure 54 (cont'd). Photomicrographs of evolved rocks, all fields of view ~ 2 mm. (E) KREEP basalt clast (bottom) in KREEP impact melt (dark material), plane light: KREEP clast is fasciculate with interstitial cryptocrystalline groundmass. (F) fragments of Apollo 17 KREEP basalt in 72275,129 matrix, plane light. (G) quartz monzodiorite 15405, crossed polarized light.

Apollo 14 and "average" KREEP materials (Fig. 49). However, the absolute abundances are only about 60 wt % of Apollo 14 KREEP or Warren's (1989) high-K KREEP, and they are thus commonly referred to as "medium-K" KREEP. They have a deep negative Eu anomaly. They also have low abundances of Ni (<50 ppm) and low Ni/Co (~2 or less). For rocks with such evolved rare earths, they have a high mg# (0.55 to 0.65, actually similar to mare basalts), and this appears to be most readily explained as a consequence of mixing between urKREEP and early-post-magmasphere Mg-rich intrusions into the region of the lower crust (Warren 1988, 1989), or sinking of urKREEP into mantle sources. Remelting at a later time is still required.

The Apollo 15 KREEP basalts were extruded at 3.84±0.02 Ga, according to Ar-Ar, Sm-Nd, and Rb-Sr systems, although uncertainties on some determinations are quite large (Table 11; also summaries of data in Shih et al. 1992, Ryder 1994). One of three fragments analyzed by Rb-Sr gives a distinct initial $^{87}Sr/^{86}Sr$, suggesting either assimilation or distinct flows (or both). However, like all KREEP the model ages, and thus the main fractionations from bulk Moon compositions, are much older and indicate that little fractionation among the incompatible elements occurred during the melting events at ~3.84 Ga.

Figure 54 (cont'd). Photomicrographs of evolved rocks. (H) monzogabbro 14161,7373, backscattered electron image. Darkest phases are feldspars and silica-potash feldspar intergrowths, striped gray phases are exsolved pyroxenes, light gray phase is mainly whitlockite, near-white is ilmenite.

Exposed within 50 km of the Apollo 15 landing site, and occupying a stratigraphic position between the formation of the Imbrium basin and the extrusion of mare basalt is the Apennine Bench Formation (Spudis 1978, Hawke and Head 1978). It probably underlies the mare units at the landing site. This unit shows compelling evidence for emplacement as volcanic flows, and orbital data shows that it contains about 11 ppm Th, i.e. the same as Apollo 15 KREEP basalts and can confidently be stated as their source. Thus these basalts represent a post-Imbrium volcanic eruption, possibly induced by the Imbrium event which can be scarcely older (Ryder 1994).

Apollo 17 KREEP basalts. Fragments of this basalt type are far less common than the Apollo 15 KREEP basalts (Table 15). In fact they appear only as several clasts, up to a 1.6 cm across, in one breccia sample, 72275, of Boulder 1 Station 2 at the Apollo 17 landing site (Fig. 54F). Some brecciated clasts are larger. They have been described by Ryder et al. (1975b, 1977) and Salpas et al. (1987). The basalts are subophitic to intersertal, and are mainly pigeonite enclosing plagioclase in roughly equal proportions (Table A5.44). Unlike the Apollo 15 KREEP basalts, they contain no orthopyroxene. Olivine is similarly extremely rare. A complex opaque cryptocrystalline mesostasis of ilmenite, Fe-metal, a silica phase, feldspars, augite, troilite, and phosphate is conspicuous.

The samples show much less textural variation than the Apollo 15 KREEP basalts, and none contain phenocrysts or glomerocrysts, nor clear glass mesostasis. Plagioclases longer than 0.5 mm are rare and are more calcic than the Apollo 15 varieties; Salpas et al. (1987) even reported compositions as calcic as An_{98}. The pigeonite cores are only as magnesian as $En_{73}Wo_6$, and zone erratically towards ferroaugites.

The Apollo 17 KREEP basalts lack meteoritic siderophile contamination. They are more iron-rich and lower in incompatible element abundances than the Apollo 15 KREEP basalts (Table A5.35). Ryder et al. (1977) described them as intermediate between KREEP basalts and mare basalts because of their chemical characteristics compared with Apollo 14 and 15 KREEP. Such characteristics include high Sc (~50 ppm) and low Sr (~90 ppm). Salpas et al. (1987) provided chemical data for a number of fragments, and concluded that there is only a small range of compositions. Although the rare earth abundance pattern is that of KREEP, the Apollo 17 KREEP basalts are slightly more depleted in heavy rare earths (La/Yb of more than 3.5, cf. high-K KREEP of ~3.1). They cannot be related to the Apollo 15 KREEP basalts by any simple process.

The Apollo 17 KREEPy basalts are also distinct in age (Table 11) and isotopic characteristics from the Apollo 15 KREEP basalts. Rb-Sr data for one clast gave an age of ~4.01 Ga and Sm-Nd data for another an age of ~4.08 Ga, the latter consistent with a disturbed Rb-Sr age for the same fragment (summary in Shih et al. 1992). Thus this basalt predates the Serenitatis impact (~3.89 Ga). The two Rb-Sr ages and corresponding initial $^{87}Sr/^{86}Sr$ are distinct and to Shih et al. (1992) suggest derivation from distinct flows several hundred million years apart (as with the two Apollo 15 norites, this does not seem geologically plausible to this author). The initial $^{87}Sr/^{86}Sr$ are lower than those of the Apollo 15 KREEP basalts, and the εNd is not so negative.

Quartz monzodiorites. The name quartz monzodiorite was first applied by Ryder (1976) to a centimeter-sized crystalline clast in sample 15405, an impact melt breccia of KREEP basalt composition. It is used here as a group name for a distinct and fairly coherent set of sodic plagioclase-silica-potash feldspar-bearing fragments of igneous origin that have the KREEP incompatible element pattern, but in higher abundances than even Apollo 14 KREEP (Jolliff 1991; Marvin et al. 1991, Ryder and Martinez 1991). (Quartz monzogabbro is a more precise name for at least some samples, those with plagioclase An > 50 mol %.) All known are from the Apollo 14 and Apollo 15 landing sites, and are the most incompatible-element-enriched samples known from the Moon (light rare earths 700 to 1200 × chondrites). They are distinguished here from granites and felsites which do not have the KREEP pattern and have lower incompatible-element abundances, although Warren (1993) includes most quartz monzodiorites with his granites. Snyder et al. (1995b) included both the quartz monzodiorites and granites with alkali rocks, and indeed clear petrogenetic relationships can be discerned (e.g. Jolliff 1991). Most quartz monzodiorites have SiO_2 less than 60 wt % and K_2O less than ~2 wt %, lower than most granite/felsite fragments. The clast in 15405 (~2.5 g) remains the only one discovered larger than 1 g. Modal data for quartz monzodiorites are included in Table A5.45, but the unrepresentivity of the tiny samples is an obvious impediment to their associations and interpretions. Unrepresentivity is exacerbated by the variety of residual phases such as whitlockite and the grain sizes coarser than KREEP basalts.

The quartz monzodiorites consist mainly of sodic plagioclase (by lunar standards), exsolved iron-rich pyroxene, granophyric intergrowths of silica and potash feldspar, phosphate, ilmenite, and traces of other phases including baddeleyite and zircon. The phosphate is mainly whitlockite with subordinate apatite. The major phases are commonly 1 mm or more across, and original textures have been preserved e.g. in the 15405 clasts, although shock deformation is apparent (Fig. 54G). While the pyroxenes are exsolved, usually finely, generally neither the plagioclases nor the pyroxenes are equilibrated, with a range in any given sample (Figs. 42, 43, 44). Thus petrographically quartz monzodiorites appear to be of hypabyssal origin. However, sample 14161,7373 has equilibrated and coarsely exsolved pyroxene (Fig. 54H) (Jolliff, pers. comm.). Some samples have an abundance of phosphate (up to 18 vol % in the case of 15403,71A, 10 vol % in 14161,7373), in which both apatite and whitlockite are present. In most cases however the phosphate is is less than 1 vol %.

Chemical analyses of quartz monzodiorites are included in Table A5.35. In general they are iron-rich, potassic KREEP compositions. However, because they have Eu abundances similar to average KREEP materials, they have a much deeper negative Eu anomaly (Fig. 49). On the basis of the rare earth patterns the two samples from the Apollo 15 coarse fines, 15434,10 and 15434,12 would not be included with quartz monzodiorites, because they are enriched in heavy rare earths compared with KREEP, but the clear similarity of their mineral phase compositions to those in the 15405 quartz monzodiorite makes it reasonably clear that they are related to it; the small sample sizes have compromised the modes and the chemical compositions.

Meyer et al. (1996) reported U-Pb ion microprobe ages for zircons for the quartz monzodiorite clast in 15405 described by Ryder (1976) and for another similar clast from that sample (Table 11). Combined data give concordia intersections at 4.294±0.026 and 1.320 +240/-280 Ga. The upper age is interpreted as the age of the crystallization of the zircon and its granophyric surrounding material, and the lower age that of a considerable disturbance, which is most likely that of the creation of the 15405 impact melt. This disturbance is seen in other isotopic systems (Rb-Sr, Ar-Ar) and has obliterated most of their record of the crystallization of the quartz monzodiorite (summary and references in Meyer et al. 1996). Radiogenic age information for other quartz monzodiorites is not yet available.

Quartz monzodiorites have frequently been grouped and discussed with granites and felsites (Shervais and Taylor 1986, Snyder et al. 1995b, Jolliff 1991, Ryder 1976, and many others) and there may indeed be some genetic relationship. Most of the data indicates that quartz monzodiorites are fractional crystallization products of a KREEP basalt magma (Jolliff 1991, Marvin et al. 1991, Snyder et al. 1995b, and others). Snyder et al. (1995b) modelled such crystallization and inferred that quartz monzodiorites could be produced by such means and consist of 90 to 60% cumulate crystals and 10 to 40 wt % parent melt. However, there appears to be no reason to suggest that the 15405 quartz monzodiorite is not almost entirely a melt composition. Ryder (1976) inferred that the 15405 quartz monzodiorite fragments were differentiation products of the local Apollo 15 KREEP basalts, as these were essentially the only other clast type recognized in 15405 and the bulk composition of the impact melt was also very similar to those basalts. However, while the petrography and chemistry remain consistent with such an origin, the geochronological data of Meyer et al. (1996) appear to rule out that direct of a relationship. The coincidence of the spatial association of these fragments in 15405 is therefore a wonder. If granites/felsites are related to quartz monzodiorites, it is by continued differentiation and removal of whitlockite and other phases from the magma, and in at least some cases by the onset of immiscibility (e.g. Jolliff 1991, Snyder et al. 1995b). However, Meyer et al. (1996) suggest that granophyres with KREEP rare earth element patterns (i.e. quartz monzodiorites) are older than those without (i.e. granites/felsites).

Granite/felsite rocks. Small fragments consisting largely of silica and potash feldspar, referred to as granites or felsites, have long been recognized in highlands rocks (e.g. Ryder et al. 1975a,b; James and Hammarstrom 1977, Quick et al. 1977). While a felsite should be aphanitic, the choice of the term granite or felsite appears to be largely arbitrary in lunar studies; and all are fine-grained by terrestrial granite standards. Snyder et al. (1995b) distinguish true granites as rare and felsites as fairly common. Their igneous origin is inferred mainly on textural as well as plausibility grounds (because of their extreme compositions) because siderophile element data are rare. Only two fragments larger than 1 g have been recognized (Warren 1993) (Table 16). Many small fragments are brecciated or even glassy and remelted, but the chemistry of even these "rhyolitic" samples has not been severely compromised and provides clues to petrogenesis. In some cases the term "granophyre" has also been used (incorrectly) as a catch-all that commonly includes quartz monzodiorites (e.g. Meyer et al. 1996). Here we distinguish granites/felsites from quartz monzodiorites: granites/felsites have higher modal silica and potash feldspar, and commonly little pyroxene or plagioclase feldspar, and for the most part do not have the KREEP incompatible element pattern. Instead they tend to have a more V-shaped rare earth element pattern (with lower abundances than in quartz monzodiorites), and higher K/La and Ba/La, among other elemental distinctions. Most of the granite/felsite fragments are from the Apollo 14 landing site, where tiny crystalline fragments and glass of granitic composition are ubiquitous; Shervais and Taylor (1986) estimate about 0.5 vol %.

Table 16. List of granites / felsites

Sample	mass (g)	Warren '93 Pristinity index	Plagioclas An mol %	~high-Ca pyroxene Mg#	~Olivine Mg#	~Modal % feldspar	Main reference
12013 bx	--	--	--	--	--	--	Quick et al. (1977)
12033,507	1.20	6	50	--	8	55	Warren et al. (1987)
12070,102-5	<0.01	7	50	30	13		Marvin et al. (1991)
14001,28,2 bx	0.03	7	80	--	--	45	Morris et al. (1990)
14001,28,3 bx	0.02	7	80	--	--	45	Morris et al. (1990)
14001,28,4 bx	0.02	7	80	--	--	45	Morris et al. (1990)
14004,94 glass/bx	0.15	--	--	--	--	40	Snyder et al. (1992)
14004,96 glass/bx	0.10	--	--	--	--	35	Snyder et al. (1992)
14161,7269 glass/bx	0.04	5	67	50	--	50	Jolliff (1991)
14303c204	0.17	7	75	40	42	60	Warren et al. (1983a)
14321c1028	1.80	9	--	5	2	60	Warren et al. (1983a)
15403,71c	--	--	--	45	--		Marvin et al. (1991)
73215c43,3	0.02	7	60	35	19	50	James and Hammarstrom (1977)
73255,27,3	<0.01	--	--	--	--	--	Blanchard and Budahn (1979)
A17 Bl 1 St. 2 many	--	--	--	--	--	--	Ryder et al. (1975a)

bx = brecciated, probably impure (not strictly monomict)

The most common texture in granites/felsites is that of a granophyric (*sensu lato*) intergrowth of a silica phase and potash feldspar. The scale of this intergrowth is diverse, in some cases even within individual samples (Warren et al. 1987). The granophyre exists either almost alone or with sodic plagioclase or with a ternary feldspar (e.g. Ryder et al. 1975a, Shervais and Taylor 1986) that plot in the "forbidden" region of the ternary. Accessory minerals include pigeonite, augite-ferroaugite, fayalite, ilmenite, zircon, and phosphates; silica-potash-rich glass is common (Ryder et al. 1975a, Warren et al. 1983a, 1987; Shervais and Taylor 1986). These are sufficiently disparate to show that there are significant differences among granites. In several cases the mafic minerals are more magnesian rather than more iron-rich than those in quartz monzodiorites. Unlike the quartz monzodiorites, pyroxene exsolution appears to be uncommon or even absent among the granites. Modal data show great differences (Table A5.46), but these are at least partly a result of the unrepresentative nature of the tiny samples.

Lunar granite/felsite compositions exhibit a range of elemental abundances (Table A5.36), an unknown amount of which is a result of the unrepresentative small sample sizes. Nonetheless they have distinct features: characteristic high SiO_2 (~60 to 75 wt %) and K_2O (~2 to 9 wt %) abundances by definition, and correspondingly very low abundances of MgO, FeO, and compatible elements such as Sc, Ni, Co, and Cr, as well as low mg# (0.05 to 0.50). They have low abundances of rare earth elements compared with most quartz monzodiorites, but nonetheless in the 100 to 300 × chondrite range with a strong negative Eu (Fig. 50). Among their most characteristic features are V-shaped chondrite-normalized or KREEP-normalized rare earth abundance patterns. They are also characterized by high Ba and Ba/light rare earth element ratios (reflecting potash feldspar). They are variably enriched in Hf and Zr, and typically strongly enriched in Rb, Cs, Nb, Ta, Th, and U.

Because of their high abundances of K, Rb, rare earth elements, and the presence of zircon, lunar granites/felsites are in some ways attractive targets for radiogenic isotope studies (Table 11), despite their small size. Some indirect results, for instance Rb-Sr analyses on felsite-bearing breccia 12013 (Quick et al. 1981b) suggested that some were old (~4.4 Ga). However, others are clearly younger. The large clast in 14321 was studied by Rb-Sr, Sm-Nd, and K-Ca techniques and despite its partial melting and recrystallization (Warren et al. 1983a), the clast produced concordant ages of ~4.1 Ga (summary in Nyquist and Shih 1992, Shih et al. 1993b). The ^{40}Ar-^{39}Ar system was reset at ~3.9 Ga. Two zircons that are undoubtedly from this clast (Meyer et al. 1996) yield a slightly younger age (ion probe U-Pb) of 3.96 Ga, and this probably reflects disturbance during the remelting episode. Meyer et al. (1996) reported several other precise ion probe U-Pb ages on zircons from granite/felsite lithologies, with a range from 3.8 to 4.32 Ga. The old ages are consistent with other data that suggest that the original magma ocean had crystallized completely within about 200 Ma of the birth of the Moon. However, ages for the "granophyres" suggest continuous granitic magmatism until at least around the time of the formation of the Imbrium basin.

Many have suggested that lunar granites/felsites have formed by silicate liquid immiscibility following production of a moderately Fe-rich melt by fractional crystallization (e.g. GJ Taylor et al. 1980, Warren et al. 1983a, 1987, Jolliff 1991, and many others), despite the rarity of the corresponding high-Fe melt. It seems likely that both fractional crystallization (including phosphate separation) and liquid immiscibility are required in the production of the characteristics of lunar granites/felsites (Neal and Taylor 1989b, Jolliff 1991, Jolliff et al. 1993). Crystallization of whitlockite and zircon deplete the magma in rare earths, zirconium, and hafnium, but enrich it in potassium, rubidium, cesium, and barium, prior to the onset of immiscibility. Given the uncertainties in the characteristics of the granites/felsites imposed by their small sample sizes, and the complexity of the proposed processes, we cannot yet be confident that we understand lunar granite/felsite petrogenesis.

HIGHLAND POLYMICT BRECCIAS

Nearly all of the samples collected in the lunar highlands are polymict breccias produced in impacts. They are mixtures of materials derived from different locations and different kinds of lunar bedrock. They contain varying proportions of broken or clastic rock fragments (*clasts*) and material melted by impacts (*impact melt*). Fragmental breccias and impact-melt breccias are ultimately derived from igneous rocks, plus a little meteoritic debris that is almost entirely cryptic. However, because of the multiple impacts to which these components have been subjected (unlike the terrestrial cases), most of the individual fragments in breccia samples are themselves fragments of older fragmental breccias and impact-produced melt rocks. No polymict samples were collected directly from bedrock.

Polymict rocks show a great diversity of texture, grain size, and chemical composition. These features reflect the original components, the effects of the impact that assembled the components, and metamorphism. Assembly and lithification can result from one impact if there is enough interstitial molten material. Many fragmental breccias are very friable and fall to pieces with little handling.

Three kinds of breccia matrix are distinguished: solidified melt (crystalline or glassy), clastic (fragmental), and metamorphic (recrystallized). A melt matrix consists of material that is entirely distinct from the clasts. In contrast, clastic and metamorphosed matrices are arbitrarily defined on the basis of grain size (usually less than 25 µm). They generally have a wide and seriate range of fragment sizes, and the fragments have populations and histories similar to those of the larger clasts.

For the most part the chemical compositions of polymict breccias fall within extremes demonstrated by pristine igneous rocks (e.g. Figs. 37, 38, 38, 40, and compare rare earth diagrams Figs. 46, 47, 48, 49, 50 with Figs. 56, 57 below). Much of their chemistry can be explained as mixtures of such rocks. However, exact matches are not possible, and especially for the more KREEP-rich rocks the identity of at least some igneous components are not matched exactly by those of any known igneous rocks. This has led to considerable discussion as to what the lunar crust contains, and how representative the known pristine rocks actually are (e.g. Wasson et al. 1977, Ryder 1979, Ringwood et al. 1986).

A single large meteoritic impact can produce a variety of polymict rocks that range from totally molten (and now crystalline) materials to masses of unconsolidated fragments. These materials may then be reworked by later impacts and incorporated into new breccias, many of which are therefore polygenetic. It is almost impossible to trace a breccia sample, or one of its fragments, to a particular impact or a specific parent crater. Age determinations for these rocks can be very complex. The exceptions are some samples collected from rather small craters, e.g. glass-lined craters at the Apollo 15 landing site, and the inference by most authors that impact melts at the Apollo 17 landing site are products of the Serenitatis basin-forming event.

Nomenclature and classification

At the time of the Apollo missions, little work had been done on terrestrial impact craters, and an accepted classification for impact-produced polymict rocks did not exist. As a result, the nomenclature for lunar polymict rocks was built up rapidly during the missions in a piecemeal and uncoordinated fashion. The nomenclature changed as genetic interpretations developed and were modified, and consequently was confusing, duplicative (many different names are applied by different workers to the same samples) and irrational (many terms are neither inclusive nor exclusive). During Apollo mission years, there was a general assumption that the bulk of any ejecta blanket was deposited hot, giving rise to the incorrect inference that many crystalline samples were "recrystallized breccia". We now consider many of these to have been formed by the crystallization of impact melts.

A Nomenclature Committee produced a more consistent, unified classification and nomenclature for lunar highland rocks (Stöffler et al. 1980). This widely adopted classification for polymict rocks leans heavily on concepts from terrestrial impact craters, and is used here with slight modifications (Table 10). Individual terms are based largely on observed and objective features in preference to genetic interpretation. Additional characteristics (composition, detailed texture, grain size, visible shock effects) are used as modifiers. This system still represents some compromise between purely nongenetic and commonly used genetic terms. The composite nature of polymict lunar rocks makes their classification to some extent dependent upon the size of a sample. For instance, small samples of a fragmental breccia might consist of individual pieces of metamorphosed crystalline rock, impact melt rock, or plutonic igneous rock.

A distinction is made between the mechanical assembly of the components of a breccia and their lithification into coherent rocks. In those breccias in which the same impact event produces both assembly and lithification, the latter is promoted mainly by the large amount of melt produced by shock-heating during impact (Simonds et al. 1976). Such a breccia is a mixture of impact-produced melt (hot) and clastic (cooler) material (Fig. 55). These thermally equilibrate in a short time (less than 100 sec; Onorato et al. 1976) and then cool more slowly and uniformly. The temperature produced immediately after mixing is largely a function of the melt/clast ratio; breccias with more hot melt than cold clasts have higher temperatures. This temperature has a dominant effect on the the amount of crystallization, and therefore on classification. Regolith breccias and granulitic breccias, for which assembly of the components and lithification of them are separate and distinct events, do not plot on the diagram. Dimict breccias (with two distinct components) might be considered a special case of impact melt breccias.

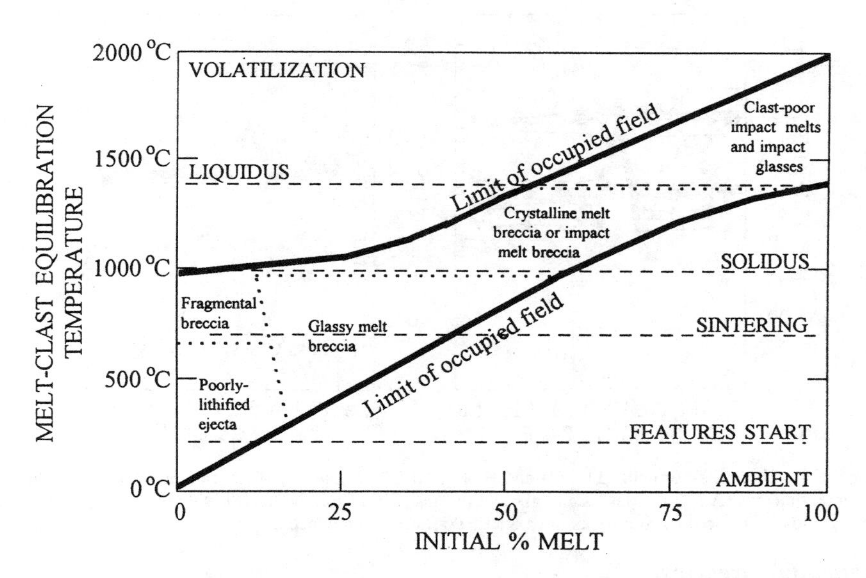

Figure 55. Generalized scheme showing relations among petrographically distinct polymict breccias that were produced and lithified in a single impact event (thus dimict breccias, granulites, and regolith breccias are not shown). The petrography is a function of the percent initial melt (total = melt + clasts), and the equilibrium temperature attained between melt and clasts. Subseqent cooling has some but less effect on final product. Figure modified from Simonds et al. (1976).

Craters in the lunar highlands are formed in target rocks that have previously been multiply impacted, reworked, and mixed together. Because of multiple impacts, the highland polymict rocks consist of fragments and chemical components derived from a broad spectrum of materials. This feature is illustrated in Figures 37, 38, and 40. There is a wide and overlapping range of chemical compositions among fragmental breccias, glassy breccias, and melt rocks (Figs. 56, 57). Nonetheless, this range is far less than that shown by pristine igneous rocks: impact is overall an homogenizing process.

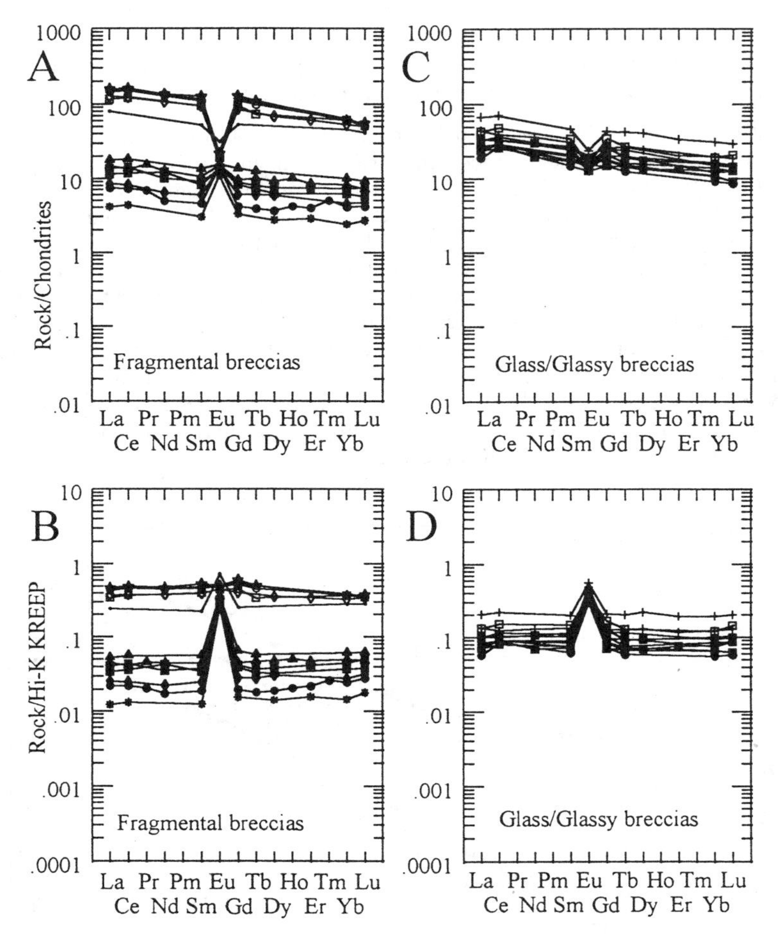

Figure 56. Chondrite-normalized (A,C) and Hi-K KREEP-normalized (B,D) rare earth element plots for fragmental breccias (A,B) and glasses and glassy breccias (C,D) from the lunar highlands. Except for the Eu anomaly, most of these polymict rocks have KREEP relative abundances.

Fragmental breccias

Fragmental breccias consist of a variety of angular discrete fragments of rocks or single mineral crystals embedded in a matrix of finely comminuted similar materials (Figs. 58A,B). Many are porous and friable. The fragments are bonded only at limited points of contact, probably by a very thin layer of glass (Nord et al. 1975), and deformation during lithification has been minor. The matrices, and many individual clasts, are contaminated with meteoritic siderophile elements. Arbitrarily, fragmental breccias do not include regolith breccias that contain solar wind components and agglutinitic glass.

Most known fragmental breccias were collected at the Apollo 16 landing site, and these are highly enriched in plagioclase. Nearly all were collected from the rim of North Ray Crater, and are from a local subsurface unit. Some were chipped from boulders. At the Apollo 14 site, a few fragmental breccias collected near the rim of Cone Crater are less feldspathic. At the Apollo 17 site, a fragmental breccia makes up the matrix of Boulder 1 at

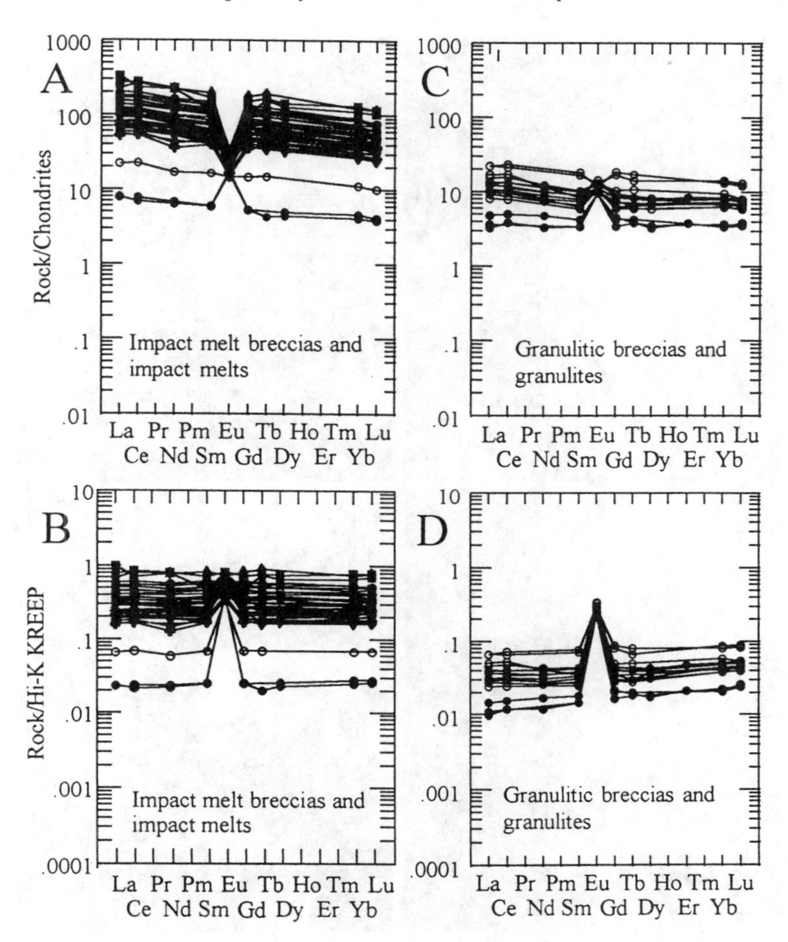

Figure 57. Chondrite-normalized (A,C) and Hi-K KREEP-normalized (B,D) rare earth element plots for impact melt breccias and impact melts (A,B) and granulites and granulitic breccias (C, D). Granulites, with low abundances of rare earth elements, do not have KREEP relative abundances.

Station 2 (Fig. 54F). It seems likely that a large part of the upper highland crust consists of fragmental breccia, varied in composition between different regions.

Petrography and chemistry. Fragmental breccias consist of clasts of other breccias, igneous rocks, and single mineral grains (Figs. 58A,B). Most are glass-poor. Most clasts are angular; from submicroscopic to several centimeters across, and with varyied intensities of shock metamorphism. Stöffler et al. (1980) distinguished two subclasses, according to the presence or absence of glassy melt particles with the same composition as the bulk rock. This distinction has not often been attempted (Norman 1981: 67016; Marvin et al. 1987: 67015).

The Apollo 16 fragmental breccias contain plagioclase and three dominant types of rock fragments: cataclastic anorthosite, granulitic breccia, and fragment-rich feldspathic impact melts (James 1981, Stöffler et al. 1981, 1985, JL Warner et al. 1976, and others). The impact-melt clasts are dark, while the anorthosites and granulites are light colored. The

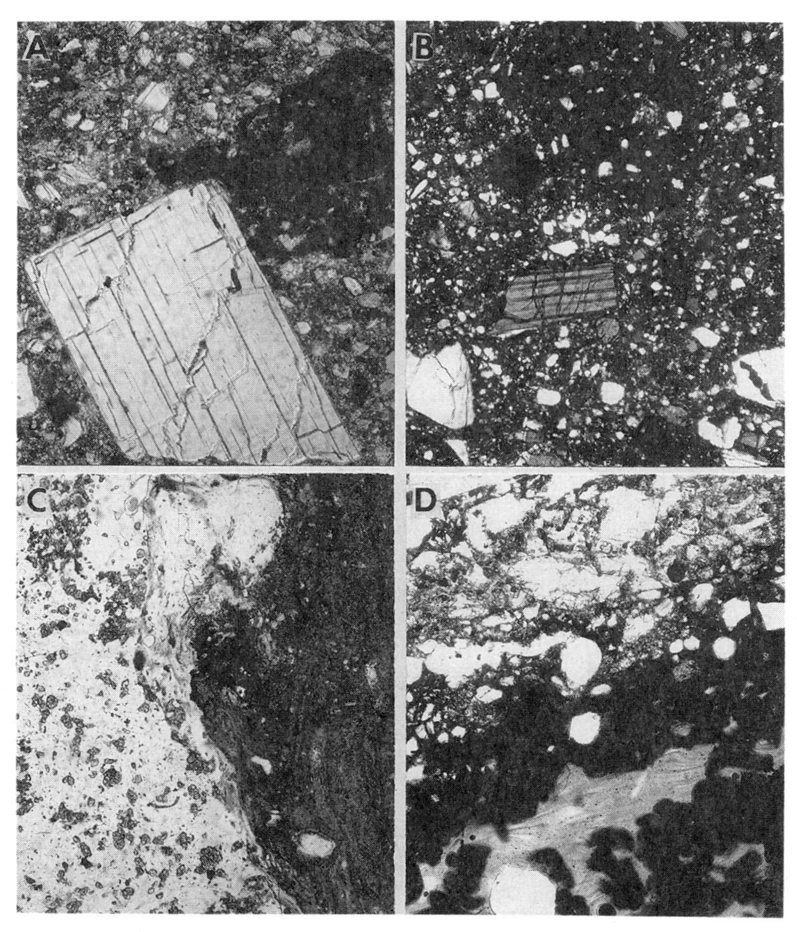

Figure 58. Photomicrographs of feldspathic breccias and glassy breccias; all fields of view ~2mm. (A) feldspathic fragmental breccia 60016,95, plane light. Large pyroxene is angular. (B) feldspathic fragmental breccia 67016,111, crossed polarized light. (C) Flow-glass intruding/coating granulitic impactite in 68815,153, plane light. (D) Clear banded (center) and devitrified/crystallized glass injecting/coating cataclasized anorthosite in 61015,161.

Apollo 16 fragmental breccias are not all the same, and the impact-melt clasts have significant variations in mineral and chemical composition. Stöffler et al. (1985) found that the relative proportions of different rock types in the lithic fragments of the North Ray Crater fragmental breccias (Table A5.47) and the compositions of metal grains in the matrices vary considerably from sample to sample. Their heterogeneity and the coarse fragment size produce variations even among determinations on the same sample. They are all highly aluminous and have similar, but not identical, chemical compositions (Table A5.37, and lower group in Figs. 56A,B). The mg# varies substantially, although all the samples are rather ferroan when compared to the full range of lunar highland samples. They are dominated by ferroan anorthosites and other ferroan plutonic highland rocks (e.g. Stöffler et al. 1985). Most of the incompatible elements (Figs. 56A,B) are from a small component of KREEPy and magnesian impact melts.

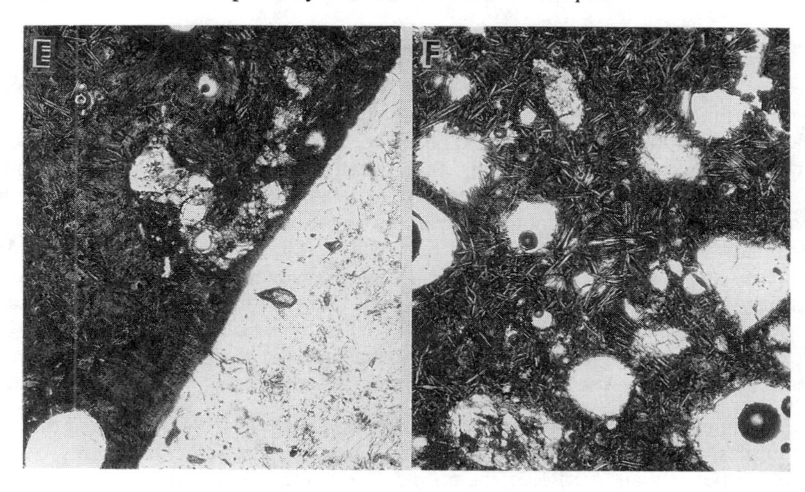

Figure 58 (cont'd). Photomicrographs of feldspathic breccias and glassy breccias, all fields of view ~2mm. (E) devitrified, vesicular, and chilled glass with anorthosite in 60015,30, plane light. (F) coarser-devitrification several millimeters from anorthosite in 60015,30, plane light.

The Apollo 14 samples contain much less plagioclase and hence lower alumina than the Apollo 16 samples (Table A5.37). They contain a wide range of fragment types; fine-grained KREEPy impact melts and troctolitic breccias are prominent. Other fragment types include fine-grained or glassy melt rocks, granulites, norites, anorthosites, and rare mare-like basalts. Fragmental breccia 72275, the matrix of Boulder 1 at Station 2, is a complex rock, a crushed mixture of KREEPy basalt and fine-grained KREEPy impact melts, norites, and granulites (Ryder et al. 1975b). Some fragments have melt rinds, suggesting inflight agglomeration. The KREEPy basalts in this sample are reflected in its chemical composition (Table A5.37).

The fragments of single minerals in all fragmental breccias range widely in composition, and demonstrate both the diverse and polymict nature of the source materials and the lack of thermal metamorphism. The mineral fragments generally resemble those in the coarser-grained rock fragments such as ferroan anorthosites, but in some cases suggest unsampled rock types.

Ages. A fragmental breccia may have two "ages," one for assembly, another for lithification. These may be the same or different. Neither event can generally be dated by radiogenic isotope methods because heating during either event was not enough to reset the isotopic systems. However, Marvin et al. (1987) identified melt clasts in 67015 that had been hot and plastic during assembly. They were dated at ~3.9 Ga (^{40}Ar-^{39}Ar method), an age that is common for melt fragments. Because no fragments younger than ~3.9 Ga have been identified among Apollo 16 fragmental breccias, it has been assumed that these breccias were assembled by a major impact 3.9 Ga. Coatings of glass are observed on some breccia samples. They could provide minimum age estimates for the time of lithification, but there is little reliable information available. In most cases the glass coats appear to be much younger.

The few age data for the KREEP impact melt clasts that dominate the Apollo 14 fragmental breccias show that they are ~3.8 to 3.9 Ga old (Bernatowicz et al. 1978). There is no compelling reason to believe that their chronology is significantly different from that of other highland rocks. The age of the matrix of Boulder 1 at Station 2 at the Apollo 17 site is at present known only to be younger than its clasts, which have been dated at 3.9 Ga (Leich et al. 1975, Compston et al. 1975).

Origins. Fragmental breccias were assembled and lithified while cool, in most cases about 3.9 Ga. Lacking regolith materials, their components must have been derived from depth during the formation of craters that penetrated the regolith, and the breccias themselves assembled as blanket-like deposits around these craters. The Apollo 16 fragmental breccias may represent the Descartes Formation, thought to be a large deposit of ejecta from the Nectaris Basin (e.g. Stöffler et al. 1985, James 1981, Norman 1981), which is consistent with orbital geochemical data. However, KREEP-bearing Apollo 16 breccias may be derived from the younger Cayley Plains, which may contain ejecta from the Imbrium Basin (Stöffler et al. 1985, Korotev 1997). The bulk of ejecta from any impact crater is deposited cold, so it seems quite likely that much of the upper lunar crust consists of material similar to the Apollo 16 feldspathic fragmental breccias.

The origin of the Apollo 14 fragmental breccias is much less clear, because the context of the samples at the site is less well known. They contain some fragments of deeper crustal rock types as well as KREEP melts, and their overall bulk compositions may not be too different from those of the local, near-surface regoliths. The Apollo 17, Boulder 1, Station 2 matrix appears to have been produced by a smaller impact than that which formed the Apollo 16 fragmental breccias, because it is dominated by materials (basalt and impact melt) that originally crystallized at the lunar surface.

Glassy melt breccias and impact glasses

Glassy polymict materials range from extremely clast-rich breccias with glassy matrices (glassy melt breccias) to glass-rich bodies that are clast-poor or clast-free (impact glasses). They can be considered as impact melts that cooled too rapidly to crystallize. Devitrifed glasses, in which microscopic crystals have formed in the solidified glass, are also included in the category of glassy rocks. Glassy melt breccias have a wide range of fragment types and may have more than one generation of glass. Their shapes are generally irregular and some are slaggy . Impact glasses appear to have formed and cooled in single impacts. Many are ovoid, some are hollow, and many have a free melt surface, in contrast with most breccias which have broken surfaces.

All sampled lunar glass particles more than a few millimeters across are impact-produced, not volcanic. In contrast with impact glasses, volcanic glasses contain only rare broken rock and mineral fragments, are homogeneous, and lack the siderophile elements that indicate meteoritic contamination. Examples of impact glasses that form splashed or entrained coatings on chemically unrelated rocks are common.

Petrography and chemistry. The common characteristic of these diverse rocks is a glassy groundmass, which in impact glasses constitutes virtually the entire rock. Glassy melt breccias contain fragmental minerals and rocks, and schlieren and vesicles are common. Some are slaggy or cindery. The groundmass glass may be heterogeneous and may have several generations, but many samples have a dominant "latest" glass groundmass, commonly including veins and protrusions throughout the rock.

Few detailed studies of glassy breccias have been made. 68815, chipped from a boulder, consists of many lobes of glass (Fig. 58C), which vary in composition, color, and banding. Brown et al. (1973) described it as a "fluidized lithic breccia". The clasts are commonly angular and consist dominantly of fine-grained impact melts that are much less aluminous than either the glass or the average composition of the rock. Two large clasts are granulitic breccias (Fig. 58C). The overall chemical composition of this breccia (Table

A5.38) is similar to that of local soils.

Discrete impact glasses are relatively smaller, generally less than a few centimeters. They characteristically have at least one chilled surface. Even impact glasses are rarely homogeneous on scales of more than a few millimeters; heterogeneous and schlieren-rich glasses dominate. Some resemble volcanic bombs., but most impact glasses are coatings or drapings. Those from the Apollo 16 site have been the most studied (See et al. 1986, Morris et al. 1986, Borchardt et al. 1986). The glass drapings have sharp boundaries against the host rock, and differ from the host rock in chemical composition. The presence of distinct crystallization fronts (e.g. Figs. 58D,E,F) in both glass bombs and melt splashes indicates that devitrification has occurred (See et al. 1986). Other samples have partly intersertal textures, indicating crystallization from a melt, yet the outwardly glassy appearances make it more convenient to include them with the impact glasses rather than impact melts. Fragments are common as inclusions in impact glasses. Mineral fragments, particularly plagioclase, are most abundant (>80 vol %), and anorthosites dominate the lithic clasts. Some fragments of older impact-melt rocks and breccias are present.

At the Apollo 15 landing site, a few glass samples were collected from the glassy linings of small craters a meter or two in diameter. These samples have not been studied in detail, but they have characteristics like Apollo 16 glasses. At the Apollo 15 and 17 landing sites, even most highland glasses have an obvious mare basalt component. Large (5 to 10 cm) samples of impact glasses are uncommon from the Apollo 11, 12, and 14 missions, and the few specimens have not been well characterized.

Chemical compositions of some glassy melt breccias and impact glasses are listed in Table A5.38, with rare earth elements shown in Figures 56C,D. The Apollo 16 impact glasses are fairly homogeneous individually, but there is considerable variation among samples (See et al. 1986). Most form a group chemically unlike local regoliths, with higher mg# and lower Al_2O_3 contents. These appear to be mixtures of anorthosite and a low-Fe type of very high-alumina impact melt. A second group is like the local regoliths and is comparatively enriched in a higher-Fe type of impact melt or anorthositic norite. All the glassy specimens are very enriched in meteoritic material, averaging over 7 wt % (Table A5.38), apparently from chondrites (Morris et al. 1986). Impact glasses from the Apollo 15 site are chemically similar to the local regoliths and also have high meteoritic siderophile contamination. Most of the Apollo 15 glasses have a significant mare basalt component.

Ages. Few reliable ages for glassy material have been obtained. In general, the maximum ages of formation of glassy melt breccias can only be constrained loosely from ages of their clasts (Table 17), and the minimum even more loosely from surface exposure ages. Most samples appear to be less than 1 Ga. Most impact glasses at the Apollo 16 landing site appear to have been on the lunar surface for less than 10 Ma, as estimated from micrometeoritic erosion rates and from the preservation of thin glass coatings (Hörz et al. 1975). The dominant group has surface exposure ages of 1.5 to 2 Ma, the other group of about 50 Ma (Morris et al. 1986). Borchardt et al. (1986) dated several samples from North Ray Crater using ^{40}Ar-^{39}Ar techniques, and found them all to be younger than the ~3.9 Ga age estimated for the Cayley Formation. Their solidification ages ranged from ~0.4 to 3.8 Ga, but their surface exposure ages were restricted to 40 to 50 Ma. Data for other glasses are virtually nonexistent. Mare components in the Apollo 15 glass samples limit the glasses to be less than ~3 Ga. The exposure ages are much younger, e.g. the glass coating on sample 15205 is ~1 Ma (see summary in Ryder 1985).

Origins. The production of glass requires rapid cooling, achieved only in small cooling units, i.e. small volumes. Many glasses in the glassy polymict samples are smaller than a few centimeters and could have been produced either in small impacts or in the most

Table 17. Representative radiogenic crystallization ages (Ga) for polymict lunar highlands rocks. Decay constants from Steiger and Jaeger (1977).

Sample (*)	Description	^{40}Ar-^{39}Ar	Rb-Sr	Main reference
Fragmental breccias				
14064,31	KREEP melt clasts	3.81 +/- 0.04		Bernatowicz et al. (1978)
67015,320	feldspathic melt blobs	3.90 +/- 0.01		Marvin et al. (1987)
67015,321	VHA melt blob	3.93 +/- 0.01 (K-Ar)		Marvin et al. (1987)
Glassy breccias and glass				
61015,90	coat	1.00 +/- 0.01		Marvin et al. (1987)
63503 particle lm	glass fragment	2.26 +/- 0.03		Maurer et al. (1978)
67567,4	slaggy bomb	0.84 +/- 0.03		Borchardt et al. (1986)
67627,11	slaggy bomb	0.46 +/- 0.03		Borchardt et al. (1986)
67946,17	slaggy bomb	0.37 +/- 0.04		Borchardt et al. (1986)
Crystalline melt breccias				
14063,215	poikilitic impact melt	3.89 +/- 0.01		Stadermann et al. (1991)
14063,233	aphanitic impact melt	3.87 +/- 0.01		Stadermann et al. (1991)
14167,6,3	melt	3.82 +/- 0.06		Stadermann et al. (1991)
14167,6,7	melt	3.81 +/- 0.01		Stadermann et al. (1991)
15294,6	poikilitic, Gp.Y	3.870 +/- 0.012		Dalrymple and Ryder (1993)
15304,7	ophitic, Gp. B	3.870 +/- 0.012		Dalrymple and Ryder (1993)
15356,9	poikilitic, Gp. C	3.836 +/- 0.011		Dalrymple and Ryder (1993)
15359,12	poikilitic, Gp. C	3.868 +/- 0.011		Dalrymple and Ryder (1993)
60315,6	poikilitic	3.88 +/- 0.05		Husain and Schaeffer (1973)
63503 particle lc	VHA?	3.93 +/- 0.04		Maurer et al. (1978)
65015	poikilitic		3.84 +/- 0.02	Papanastassiou et al. (1972b)
65015	poikilitic	3.87 +/- 0.04		Kirsten et al. (1973)
65785	ophitic	3.91 +/- 0.02		Schaeffer and Schaeffer (1977)
72215, 144	aphanite; felsite melts	3.83 +/- 0.03		Schaeffer et al. (1982)
72255	aphanite; felsite melts	3.85 +/- 0.04		Schaeffer et al. (1982)
72255,238b	aphanite	3.869 +/- 0.016		Dalrymple and Ryder (1996)
73215	aphanite; felsite melt		3.84 +/- 0.05	Compston et al. (1977)
77075,18	veinlet (Serenitatis)	3.93+/- 0.03		Stettler et al. (1974)
72395,96	poikilitic (Serenitatis)	3.893 +/- 0.016		Dalrymple and Ryder (1996)
72535,7	poikilitic (Serenitatis)	3.887 +/- 0.016		Dalrymple and Ryder (1996)
76315,150	poikilitic (Serenitatis)	3.900 +/- 0.016		Dalrymple and Ryder (1996)
76055	magnesian, poikilitic	3.92 +/- 0.05		Turner et al. (1973)
76055,6	magnesian, poikilitic		3.78 +/- 0.04	Tera et al. (1974)
Clast-poor impact melts				
14073	subophitic 14310-group		3.80 +/- 0.04	Papanastassiou and Wasserburg (1971)
14074	subophitic 14310-group	3.80 +/- 0.01		Stadermann et al. (1991)
14276	subophitic 14310-group		3.80 +/- 0.04	Papanastassiou and Wasserburg (1971)
14310	subophitic		3.79 +/- 0.04	Papanastassiou and Wasserburg (1971)
14310	subophitic	3.88 +/- 0.05		York et al. (1972)
14310	subophitic; plag	3.82 +/- 0.04		Turner et al. (1972)
65795	subophitic, very feldspathic		3.81 +/- 0.04	Deutsch and Stoffler (1987)
60635	subophitic 68415-group		3.75 +/- 0.03	Deutsch and Stoffler (1987)
65055	subophitic 68415-group	3.89 +/-0.02		Jessberger et al. (1977)
67559	subophitic 68415-group		3.76 +/- 0.04	Reimold et al. (1985)
68415	subophitic		3.76 +/- 0.01	Papanastassiou and Wasserburg (1972a)
68415	subophitic	3.80 +/- 0.06		Kirsten et al. (1973)
68416	subophitic 68415-group		3.71 +/- 0.02	Compston et al. (1977)
Granulitic breccias and granulites				
14063,207		3.90 +/- 0.02		Stadermann et al. (1991)
14179,11	clast	3.97 +/- 0.01		Stadermann et al. (1991)
15418,50		3.98 +/- 0.06		Stettler et al. (1973)
67215,8		3.75 +/- 0.11		Marvin et al. (1987)
67415		3.96 +/- 0.04		Marvin et al. (1987)
67483,13,8		4.20 +/- 0.05		Schaeffer and Husain (1973)
72255,235b	clast	3.850 +/- 0.020		Dalrymple and Ryder (1996)
77017,46		3.91 +/- 0.02		Phinney et al. (1975)
78155		4.16 +/- 0.04		Turner and Cadogan (1975)
78527		4.146 +/- 0.017		Dalrymple and Ryder (1996)
79215		3.91 +/- ?		Oberli et al. (1979)

(*) including split number if given by authors

distant (distal) or uppermost portions of the ejecta of larger craters. Most samples show chemical similarities to their local regolith, suggesting that they have been produced by fairly small-scale craters that did not penetrate the existing regolith. Several at the Apollo 15 site were collected as linings of small craters, so their regolith origin is not in doubt.

Glass coatings differ in composition from their host rocks, and were not derived from them. Many of the glassy coatings on lunar rocks occur on several sides of the host rock indicating that they were deposited onto their hosts inside growing crater cavities or during ballistic flight. The glass coatings were then "piggy-backed" on the rocks to their collection sites (See et al. 1986).

The surface exposure ages of the Apollo 16 impact glass samples fall into two groups: about 2 Ma and about 50 Ma, which were presumably produced or exposed by the South Ray Crater and the North Ray Crater events, respectively (Arvidson et al. 1975, See et al. 1986, Morris et al. 1986, Borchardt et al. 1986). Not all the glasses were produced by the North Ray Crater event; some are older than 50 Ma. The South Ray Crater samples, however, might have been produced from that event itself (See et al. 1986). Most of these Apollo 16 glasses are not identical with the local regolith. They appear to be dominated by subregolith materials of the Cayley Plains, consistent with the observation that they lack materials (agglutinates, glass droplets, etc.) formed and reworked at the lunar surface. They were still formed in relatively small (<2 km diameter) craters in post-Imbrian times.

Crystalline melt breccias or impact-melt breccias

Crystalline melt breccias are coherent rocks that contain clastic material in a finer-grained groundmass that formed by the crystallization of a silicate melt (Simonds 1975, Simonds et al. 1976, Onorato et al. 1976). The older literature commonly refers to them as recrystallized or metamorphic breccias. Clasts range from a few percent to more than half the rock (Table A5.48). The more the clastic debris, the more glassy the groundmass, grading arbitrarily into glassy melt breccias. Many of the clasts show evidence of shock metamorphism or of partial resorption by the melt. The melt breccias invariably have high siderophile element contents resulting from meteoritic contamination. Some are vesicular. Crystalline melt breccias span virtually the entire range of chemical compositions observed in highland samples (Table A5.39; Figs. 57A,B). They are common at all the lunar highland sites and among the regolith particles of highland origin in the Apollo 11 and 12 samples.

Petrography and chemistry. Many samples of crystalline melt breccias have a grossly homogeneous appearance and a generally even distribution of clasts (Fig. 59). The crystalline groundmass in such samples tends to be poikilitic to subophitic. Only rarely is anything approaching a porphyritic texture observed, and then only in the most feldspathic samples. The groundmass is generally fine- to medium-grained (Fig. 59), with poikilitic pyroxenes typically no larger than a millimeter across. Samples that are more heterogeneous, more clast-rich, and with clasts occurring as schlieren in the groundmass tend to be finer-grained and even partly glassy. Many descriptions of crystalline melt breccias have been published (e.g. Dymek et al. 1976, Simonds 1975, Simonds et al. 1976, James 1976, Ryder and Bower 1976, 1977, with a review by Vaniman and Papike 1980).

The clasts are predominantly of plagioclase; with pyroxene, olivine, and rock fragments present in lesser amounts. Plagioclase clasts are more calcic and the mafic mineral clasts more magnesian than in the groundmass, i.e, more refractory (e.g. Ryder et al. 1997). In samples with a coarser grained groundmass, clastic material tends to be less abundant, even more refractory, and to include fewer rock fragments. For example, the

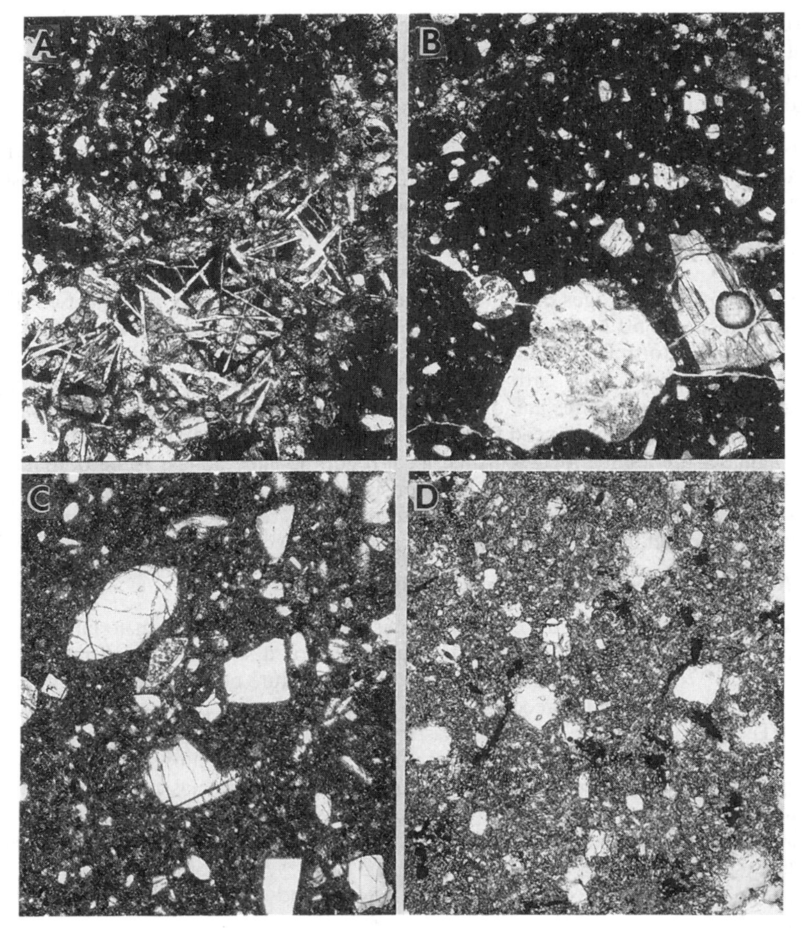

Figure 59. Photomicrographs of impact melt breccias, all fields of view ~2mm. (A) 14321,240, plane light: blobby aphanitic clast-rich melt and clasts of high-alumina mare basalt (bottom). (B) aphanitic melt breccia 72255,89, plane light. (C) micropoikilitic impact melt breccia 77115,57, probably Serenitatis melt, plane light. (D) poikilitic impact melt breccia 76015,98, probably Serenitatis melt, plane light.

aphanitic melt breccias at the Apollo 17 site contain abundant clasts of a diverse rock population. The coarser poikilitic-subophitic melt breccias contain few clasts and, among these, lithic fragments are scarce (e.g. Spudis and Ryder 1981). The crystallized groundmass of these breccias consists dominantly of plagioclase and a low-Ca pyroxene or less commonly olivine (Table A5.48). Ilmenite, ulvöspinel, and a variety of accessory minerals, including globules of Fe-Ni metal and sulfide, are also present. In the more common poikilitic groundmass, plagioclase tends to be small (less than a few tenths of a millimeter) and stubby, and dozens may be embedded in the larger pyroxene (normally pigeonite, less commonly augite) or olivine crystals. In the subophitic groundmass, the plagioclase crystals are more similar in size to the mafic crystals. llmenite tends to be lath-shaped. A glassy mesostasis contains the Fe-rich mafic minerals, a silica phase, and minor minerals such as zircon, zirconolite, baddeleyite, whitlockite, and apatite.

Chemical analyses of representative samples of crystalline impact-melt breccias are

Figure 59 (cont'd). Photomicrographs of impact melt breccias; all fields of view ~2mm. (E) (F) aluminous impact melt 61016,22, plane light and crossed polarized light respectively: carries anorthositic clast material and the plagioclase has been subsequently shocked to maskelynite. (G) clast-poor impact melt 14310,170, plane light: texture is somewhat heterogenous. (H) clast-poor feldspathic impact melt 68415,136, crossed polarized light.

given in Table A5.39. Most have abundances and ratios of incompatible elements that indicate a significant amount of a KREEP component (Figs. 57A,B). In some Apollo 12 and Apollo 14 samples, KREEP is particularly abundant. The KREEP abundance is roughly inversely correlated with alumina, and those with highest KREEP have 16 to 22 wt % Al_2O_3 and 7 to 12 wt % FeO. All samples are contaminated with meteoritic siderophile elements. Crystalline melt breccias and clast-poor impact melts fall into distinct compositional groups (e.g. Korotev 1994, Spudis and Ryder 1987, Vaniman and Papike 1980). A very detailed and comprehensive data set for Apollo 16 impact melts is provided by Korotev (1994). Four main groups have been recognized among the Apollo 16 melts, while most of the Apollo 17 poikilitic melt rocks form a single group. Very tightly-clustered compositional groups may represent samples derived from single impacts.

Ages. A large proportion of the material in crystalline melt breccias was totally melted and degassed during formation. Under such extreme conditions, the radiogenic isotopic systems can be completely reset. However, the clasts are relatively cold when mixed with the melt, and in the clast-rich varieties they cool the melt quickly. As a result, the crystalline groundmass is commonly fine grained, and mineral separations are difficult or impossible. Possible Rb migration also makes Rb-Sr interpretations unreliable (e.g. Reimold et al. 1985). The ^{40}Ar-^{39}Ar dating method is more commonly used than the Rb-Sr method for these rocks, but the clasts tend to be incompletely degassed. In many cases later disturbances preclude defined plateau crystallization ages.

Representative ages for crystalline melt breccias are listed in Table 17. Most samples have ages ~3.8 to 3.9 Ga. No sample is very far out of this range, implying that the collected samples all formed during the period of large impacts that produced at least the younger mare basins (e.g. Imbrium ~3.84 Ga, Serenitatis ~3.89 Ga; Dalrymple and Ryder 1993, 1996).

Origin. Crystalline melt breccias are of impact-melt origin as shown by their clasts, their commonly non-igneous melt chemical compositions (particularly excess alumina), and their meteoritic contamination. A large volume of completely melted rock forms near the point of impact, and moves outward across the developing crater (Simonds et al. 1976). It picks up cold debris, and eventually forms a sheet within the crater or even spills outside. The hot melt and colder clasts come to an equilibrium temperature at which some clasts are preserved and the melt crystallizes (often fine-grained) rather than chilling into a glass.

Some crystalline melt breccias have compositions similar to upper-crustal or local compositions, implying origin from near-surface rocks involved in relatively small impacts. However, several groups, particularly those with approximately basaltic compositions and moderate KREEP contents (e.g. particularly the low-K Fra Mauro groups) do not. Terrestrial crystalline impact-melt rocks can be chemically modelled by mixing compositions corresponding to the observed clasts, but lunar ones commonly cannot. This condition implies that these melts contain a KREEP-rich chemical component not represented by the observed clasts (e.g. Ryder 1979).

A case can be made that at least some of these samples represent melts generated in large basin-forming impacts, incorporating deeper levels of the crust. Possible examples are samples 15445 and 15455 (from Imbrium?; Ryder and Bower 1977); the Apollo 17 poikilitic melts (from Serenitatis?; Ryder and Wood 1977), which definitely represent an important part of the local massifs at the Apollo 17 site; and perhaps the Apollo 16 poikilitic melts. The clasts, both lithic and mineral, are obviously valuable clues to the rocks in the target and near-target areas. Ryder et al. (1997) provide a detailed account of the variety of mineral fragments in the Apollo 17 poikilitic rocks and their implications for the rocks around the target.

Clast-poor impact melts

Clast-poor impact melts are coherent. They are commonly fine-grained and petrographically resemble volcanic rocks. They are much less common than melt breccias, but their chemical compositions span a similar wide range. Clast-poor impact-melt rocks have been interpreted by some workers as internally-generated igneous rocks, with the rare included fragments described as xenocrysts (Kushiro et al. 1972, Crawford and Hollister 1974). A distinction can be made on the basis of siderophile-element abundances that indicate meteoritic contamination in the impact melts, although even that criterion has been questioned (Crawford and Hollister 1974, Delano and Ringwood 1978, Ringwood and Seifert 1986, Ringwood et al. 1986). The microscopic textures in impact melts are more heterogeneous than those of demonstrably igneous rocks. Most lunar impact melts have chemical compositions unlike melts that could be expected from partial melting of the lunar interior. In particular they tend to be more aluminous, a feature not attributable in their case to the accumulation of plagioclase during crystallization.

Petrography and chemistry. Clast-poor impact melts tend to have a range of fine-grained crystalline textures that are generally ophitic, and rarely vitrophyric or intersertal (Fig. 59G,H). Some impact-melt rocks have a slightly porphyritic texture with larger plagioclase crystals. The textures of impact-melt rocks are more heterogeneous than those in normal igneous rocks. In particular, impact-melt rocks have a much wider range in grain size within a single thin section, ranging from ragged plagioclase phenocrysts to irregular granular clots. Small crystal-lined cavities are common.

The minerals in clast-poor impact-melt rocks are virtually the same as those in lunar igneous rocks. Nearly all impact-melt rocks contain plagioclase and pyroxene (normally a low-Ca variety), and olivine is present in some samples. Fe-Ni-Co metal and Fe-sulfide, typically troilite, are ubiquitous as accessory phases. Modal data for selected samples are given in Table A5.48, and reflect the variety of compositions shown by the chemical data (Table A5.39).

The only clast-poor melt composition among the Apollo 14 samples is represented by several specimens, including 14310 and 14276. Sample 14310 was originally thought to be an igneous KREEP basalt; but it is too heterogeneous, not quite clast-free, and contains meteoritic siderophile elements. The incompatible-element abundances are only slightly lower than Apollo 14 regolith samples and the mg# is similar, although they have higher Al_2O_3.

At the Apollo 16 site, several samples from separate locations have compositions similar to samples 68415 and 68416 (Table A5.39), which were chipped from a single boulder. This composition appears to be similar to that of the regolith from the nearby Descartes Highlands. Other Apollo 16 impact-melt samples, such as 64455 and the distinctive 62295, differ from each other, from the 68415 sample group, and from the local regolith compositions.

Ages. During formation, most of the components of clast-poor impact melts were totally molten or extensively degassed, resetting the radiogenic isotopes. Thus there are consistent, meaningful crystallization ages by a variety of systems (Table 17). Most impact melts correspond closely to the ~3.8 to 3.9 Ga ages for other breccias. However, the 68415 sample group is "young", ~3.75 Ga. Relict clasts in the melts, even if not petrographically detectable, still cause problems with the isotopic systems because they may not be completely melted or outgassed. Argon in relict plagioclase fragments, in particular, may preserve ages of older events.

Origin. Clast-poor impact melts crystallized from impact-produced silicate liquids that were once entirely above their liquidus temperatures (~1300°C to 1200°C). If these melts contained clastic debris, then it virtually all dissolved into the melt before the melt had cooled below the liquidus temperature. The slightly heterogeneous textures result mainly from relict, heterogeneously-distributed nucleation sites (Ryder and Bower 1976, Lofgren 1977).

Many impact melts evidently crystallized in units about the size of typical lava flows, according to their crystal sizes (and therefore the inferred cooling rates). Clast-poor impact melt rocks are found in some of the larger terrestrial impact craters, with clast-bearing impact-melt rocks stratigraphically above and below them. An impact-melt sheet more than a few meters thick requires a crater at least a few kilometers in diameter. The chemical compositions suggest that at least some of the lunar impact-melt samples are not derived from basin-scale impact melts, but from melts produced in smaller craters. The Apollo 16 group that includes sample 68415, in particular, appears to mimic upper crustal materials of the Descartes Formation.

Granulitic breccias and granulites

Granulitic breccias and granulites are coherent crystalline rocks that have been recrystallized by heating at temperatures over 1000°C for long periods of time. The textures are characterized by rounded polyhedral grain shapes (i.e. granoblastic texture) or the presence of many small grains within fewer larger grains (poikiloblastic texture) (JL Warner et al. 1977, Lindstrom and Lindstrom 1986).

Granulitic breccias contain relicts of clasts, whereas granulites, which are much less abundant, do not. Both commonly consist of a mosaic of grains whose boundaries meet at angles of about 120°, a texture produced by thorough recrystallization (Stewart 1975). Mineral compositions are uniform and equilibrated. All samples are aluminous (>25 to 29 wt % Al_2O_3), and there is a range of mg#. They have very low abundances of the incompatible elements, i.e. a negligible KREEP component. All granulitic breccias and granulites are contaminated with meteoritic siderophile elements.

Both granulitic breccias and granulites have been found at all the Apollo and Luna landing sites, and in the lunar meteorites. Among the Apollo samples, they are small individual rocks, clasts in other breccias, and isolated regolith particles.

Petrography and chemistry. The textures of granulitic breccias and granulites are extremely varied, ranging from granoblastic to poikiloblastic, from homogeneous to heterogeneous, and from coarse- to fine-grained (Fig. 60). Granoblastic breccias tend to have fine-grained matrices (Figs. 60C,D), whereas poikiloblastic breccias tend to be coarse grained throughout (Figs. 60A,B). The clasts are as recrystallized as the matrices. Several workers have interpreted some of the poikiloblastic samples as igneous, i.e. poikilitic (Hollister 1973: 67955; Ashwal 1975: 77017).

The granoblastic rocks are dominated by anhedral and equant plagioclase (70 to 80 vol %) (Table A5.49). Smaller pyroxenes (15 to 25 vol %) and small anhedral olivines (absent to a few percent) occur within and between the larger plagioclase grains. The poikiloblastic rocks tend to have blocky plagioclase grains, partially enclosed in pyroxene oikocrysts. Ilmenite is rare and generally subhedral. Olivine inclusions commonly form strings or "necklaces" inside large plagioclase grains; the strings possibly indicate the location of pre-metamorphic grain boundaries. In most samples, pyroxene and olivine grains are of nearly constant composition and are unzoned (e.g. McCallum et al. 1974, Ashwal 1975, Bickel 1977, Lindstrom and Lindstrom 1986). In contrast, Norman (1981) found that the granulitic clasts in fragmental breccia 67016 have ranges of mineral compositions within individual clasts. Among granulitic breccias, the pyroxene and olivine grains have mg# ranging from less than 0.50 to more than 0.85 (review in Lindstrom and Lindstrom 1986).

Chemical compositions of the granulitic breccias and granulites are given in Table A5.40. Most are normatively equivalent to those of anorthositic norites (JL Warner et al. 1977, Lindstrom and Lindstrom 1986), with low abundances of incompatible elements (Figs. 57C,D), and contaminated with meteoritic siderophile elements. Lindstrom and Lindstrom (1986) divided the range into two groups: ferroan [mg# < 0.70] and magnesian [mg# > 0.70]. Although the major-element data do not show a clear break, the element Sc appears to define the groups more precisely. Ferroan granulites have about twice as much Sc, and tend to have lower abundances of the incompatible elements.

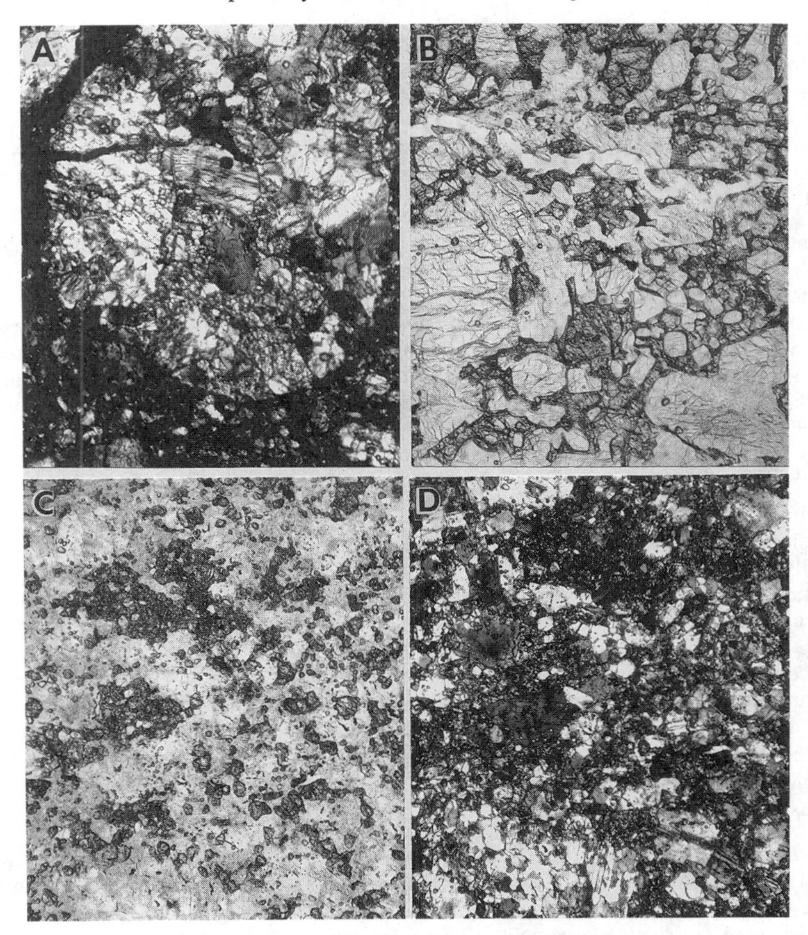

Figure 60. Photomicrographs of granulites and granulitic breccias; all fields of view ~2mm. (A) coarse granulite 67955,53, crossed polarized light: cataclasized subsequent to recrystallization. (B) coarse granulite 77017,85, plane light: poikiloblastic (arguably poikilitic) texture, little deformed subsequent to crystallization. (C) (D) fine granulite 76235,17, plane light and crossed polarized light respectively: brecciated and granulated, recrystallized, subsequently shocked.

Ages. There are few data on the ages of granulitic rocks (Table 17). Most are too fine-grained to permit the mineral separations necessary for Rb-Sr or Sm-Nd internal-isochron age determinations. The available ^{40}Ar-^{39}Ar data suggest that the granulitic rocks formed as much as 4.3 Ga, and that for some samples, the ^{40}Ar-^{39}Ar system was reset by heating at ~3.8 to 3.9 Ga (review in JL Warner et al. 1977). The lack of the KREEP component in the breccias suggests that they were assembled before KREEP became widely dispersed on the lunar surface by ~3.9 Ga.

Origin. The clast-matrix structure of the granulitic breccias and their meteoritic contamination demonstrates that they were derived from even older breccias, metamorphosed and recrystallized by prolonged heating to temperatures of more than 1000°C (Stewart 1975, Bickel, 1977, McCallum et al. 1974). The KREEP-poor precursor rocks were present early in lunar history, and the prolonged heating necessary to produce the textures is different from the effects produced by the intense bombardment about 3.9 Ga. However, Cushing et al. (in press) suggest that contact with an impact melt is sufficient for the observed metamorphism. The textures in the granulitic breccias could be a result of burial at a depth of several kilometers, possibly during an early period with greater heat flow from the lunar interior (JL Warner et al. 1977). Younger ages of ~3.9 Ga might represent isotopic closure by excavation from depth during the ~3.9 Ga bombardment period.

The chemical and textural characteristics suggest that some of the granulites may have been derived from distinctive anorthositic norite precursors; most, however, are clearly polymict (Lindstrom and Lindstrom 1986). The bulk chemical compositions of granulites are similar to the average lunar surface compositions estimated from remotely-sensed data (SR Taylor 1982, Lucey et al. 1995), even more so than the Apollo 16 fragmental breccias. There appears to be no difference in Apollo site distribution among the ferroan and magnesian granulitic rocks, but the available orbital measurements are not adequate to preclude regional differences.

Dimict breccias

Dimict breccias, which are relatively rare, consist of two distinct lithologies combined into a single rock. All are from the Apollo 16 site. In most examples, the two lithologies are mutually intrusive, i.e. neither can be defined as the host. They resemble complex veins of dark and light components. The dark material is a fine-grained crystalline melt breccia or a nearly clast-poor impact melt breccia that is similar in all samples and contains about 21 wt % Al_2O_3 (McKinley et al. 1984). The light-colored material is an anorthositic breccia whose main constituents are crushed and shattered anorthosite fragments or less aluminous breccias (nearly all containing more than 30 wt % Al_2O_3; McKinley et al. 1984, James et al. 1984). Dimict breccias cannot be recognized in samples of less than a few centimeters, because of the scale of mutual intrusion. Available surface exposure ages suggest that all the dimict breccia samples were ejected about 2 Ma by the South Ray Crater event. Several samples have glass coats.

Petrography and chemistry. There are several detailed studies, reviewed by McKinley et al. (1984) and James et al. (1984). The mutually intrusive character is well displayed by sample 61015. Most of the contacts between the two lithologies are smoothly curving surfaces; some are shear planes. Not all samples can be shown to be mutually intrusive, and in some specimens the melt phase is perhaps more clearly the host.

The dark-colored impact melt lithology is a crystalline melt breccia. In most, the melt contains less than 5 vol % of clastic material. The fine-grained melt consists of plagioclase laths, equant olivine, and pyroxene. A glassy mesostasis, producing an intersertal texture, is common. The few clasts are mainly plagioclase, and lithic clasts are very rare. The melt composition is consistent at about 21 wt % Al_2O_3 with incompatible-element contents of about 80 to 90 x that of chondritic meteorites. This composition is remarkably constant and is similar to other impact melts found at the Apollo 16 site. The dimict breccias have mg# of about 0.70, and have meteoritic contamination.

The light-colored lithology is composed dominantly of cataclastic anorthosite, containing more than 95 vol % plagioclase. Low-Ca pyroxene is more common than olivine or high-Ca pyroxene. In their chemistry and mineralogy, these anorthosites are similar to the igneous ferroan anorthosites. In dimict breccias, the compositional range of the mafic minerals (pyroxene, olivine) extends to more magnesian compositions (see data compilations in Ryder and Norman, 1980), and this greater range suggests that these light portions are actually polymict, or that ferroan anorthosites have a wider range of compositions than occurs in discrete anorthosite samples. The anorthosites in dimict rocks are shattered, and the individual crystals are granulated and have undulatory (strain) extinction under polarized light.

Ages. There are as yet no ages available for the dark crystalline melt breccia phase of dimict breccias, the lithology that offers the best possibilities for age dating. However, they are presumably ~3.9 by. old. Two ^{40}Ar-^{39}Ar determinations for the anorthosites give good ages of 3.92±0.02 Ga and 3.91±0.01 Ga (Jessberger et al. 1977). These ages probably date the intense heating and deformation of the anorthosites, i.e. the impact event. The original crystallization of the anorthosites presumably took place much earlier.

Origin. Terrestrial analogs for lunar dimict breccias occur in the basement rocks below large impact craters, where impact-produced melt was injected into hot shocked rock to form complex dike-like bodies (e.g. Stöffler et al. 1979). Such melt-rich dikes have been referred to as pseudotachylite, although that term is more appropriately applied to veins resulting from near-*in situ* friction melting. The lunar dimict breccias were probably created in the same way. The tight chemical clustering of the Apollo 16 dimict breccias suggests that they formed in a single but unknown impact (McKinley et al. 1984). Although the samples appear to be associated with South Ray Crater, this crater is too young and too small to be their origin, and that event merely excavated them. James et al. (1984) suggested that they were produced in a crater of about 50 to 150 km in diameter, but Spudis (1984) noted that their composition is not that of the local crust and advocated an origin in the Nectaris basin-forming event.

Regolith breccias

Regolith breccias form at or very near the lunar surface as the result of impacts into regolith. Their characteristics are covered more completely in the Lunar Regolith section of this Chapter. Regolith breccias contain identifiable regolith components, such as solar wind gases, glass spherules, and agglutinates. They range from friable clods to coherent, glass-bonded breccias. Some may represent ancient regoliths (McKay et al. 1986), rather than regolith developing today. Many of them are not merely the lithified equivalents of ordinary regolith; their agglutinate content is low, and some appear to be coarser-grained than modern lunar soils. They are also less mature than most modern regoliths (McKay et al. 1986).

Most regolith breccias are not strictly highland samples; most were collected from mare surfaces or contain a mare component. These samples are probably younger than 3.8 Ga, the earliest time of widespread mare volcanism. However, some regolith breccias contain only or mostly highland material, including 18 specimens from the Apollo 16 landing site (McKay et al. 1986) and several from the Apollo 14 landing site (e.g. 14318; Kurat et al. 1974). Pure highland regolith breccias are rare at other Apollo landing sites, but examples collected as lunar meteorites in Antarctica contain only minor mare components (Palme et al. 1991, Warren and Kallemeyn 1991).

ACKNOWLEDGMENTS

This chapter benefitted greatly from reviews by Brad Jolliff, Steve Simon and Jeff Taylor. J. J. Papike and C. K. Shearer were supported for this work by NASA Grant #NAG5-4253 and the Institute of Meteoritics. G. Ryder was supported by the Lunar and Planetary Institute which is operated by the Universities Space Research Association under contract NASW-4066 with the National Aeronautics Administration. This paper is Lunar and Planetary Institute Contribution #949.

REFERENCES

Adler I, Trombka JI, Schmadebeck R, Lowman P, Blodget H, Yin L, Eller E, Podwysocki M, Weidner JR, Bickel AL, Lum RKL, Gerard J, Gorenstein P, Bjorkholm P, Harris B (1973) Results of the Apollo 15 and 16 x-ray experiment. Proc Lunar Sci Conf 4:2783-2791

Aeschlimann U, Eberhardt J, Geiss J, Grögler N, Kurtz J, Marti K (1982) On the age of cumulate norite 78236: An ^{39}Ar-^{40}Ar study (abstr). *In:* Lunar Planet Sci XIII:1-2, Lunar and Planetary Institute, Houston

Agrell SO, Peckett A, Boyd FR, Haggerty SE, Bunch TE, Cameron EN, Dence MR, Douglas JAV, Plant AG, Traill RJ, James OB, Keil K, Prinz M (1970) Titanian chromite, aluminian chromite, and chromian ulvöspinel from Apollo 11 rocks. Proc Apollo 11 Lunar Sci Conf, p 81-86

Albee AL, Chodos AA (1970) Microprobe investigations on Apollo 11 samples. Proc Apollo 11 Lunar Sci Conf, p 135-157

Albee AL, Chodos AA, Gancarz AJ, Haines EL, Papanastassiou DA, Ray L, Tera F, Wasserburg GJ, Wen T (1972) Mineralogy, petrology, and chemistry of a Luna 16 basaltic fragment, sample B-1. Earth Planet Sci Lett 13:353-367

Ali MZ, Ehmann WD (1977) Chemical characterization of lunar core 60010. Proc Lunar Sci Conf 8:2967-2981

Alibert C (1994) Erratum to Alibert C, Norman MD, McCulloch MT (1994) An ancient Sm-Nd age for a ferroan noritic anorthosite clast from lunar breccia 67016. Geochim Cosmochim Acta 58:5369

Alibert C, Norman MD, McCulloch MT (1994) An ancient Sm-Nd age for a ferroan noritic anorthosite clast from lunar breccia 67016. Geochim Cosmochim Acta 58:2921-2926

Allton JH, Waltz SR (1980) Depth scales for Apollo 15, 16, and 17 drill cores. Proc Lunar Planet Sci Conf 11:1463-1477

Anders E, Ebihara M (1982) Solar-system abundances of the elements. Geochim Cosmochim Acta 46:2363-2380

Anderson AT Jr, Bunch TE, Cameron EN, Haggerty SE, Boyd FR, Finger LW, James OB, Keil K, Prinz M, Ramdohr P, El Goresy A (1970) Armalcolite: A new mineral from the Apollo 11 samples. Proc Apollo 11 Lunar Sci Conf, p 55-63

Anderson AT Jr (1973) The texture and mineralogy of lunar peridotite, 15445,10. J Geol 81:219-226

Anderson AT Jr, Crewe AV, Goldsmith JR, Moore PB, Newton JC, Olsen EJ, Smith JV, Wyllie PJ (1970) Petrologic history of the Moon suggested by petrography, mineralogy, and crystallography. Science 167:587-590.

Antonenko I, Head JW, Pieters CM (1997) Cryptomare delineations using craters as probes to lunar stratigraphy: Use of multispectral Clementine data as a tool (abstr). *In:* Lunar Planet Sci XXVIII:47-48, Lunar and Planetary Institute, Houston

Arndt J, Flad K, Feth M (1979) Radiative cooling experiments on lunar glass analogues. Proc Lunar Planet Sci Conf 10:355-373

Arndt J, von Engelhardt W, Gonzales-Cabeza I, Meier B (1984) Formation of Apollo 15 green glass beads. Proc Lunar Planet Sci Conf 15th in J Geophys Res 89:C225-C232

Arai T, Warren PH (1996) VLT-mare glasses of probable pyroclastic origin in lunar meteorite breccias Yamato 793274 and QUE94281. Antarctic Meteorites XXI:4-7

Arai T, Warren PH, Papike JJ, Shearer CK, Takeda H (1997) Trace element chemistry of volcanic glasses in lunar meteorites Y 793274 and QUE 94281. NIPR Symposium (in press)

Arvidson R, Crozaz G, Drozd RJ, Hohenberg CM, Morgan CJ (1975) Cosmic ray exposure ages of features and events at the Apollo landing sites. The Moon 13:67-79

Ashwal LD (1975) Petrologic evidence for a plutonic igneous origin of anorthositic norite in 67955 and 77017. Proc Lunar Sci Conf 6:221-230

Axon HJ, Goldstein JI (1973) Metallic particles of high cobalt content in Apollo 15 soil samples. Earth Planet Sci Lett 18:173-180

Baker M, Herzberg CT (1980) Spinel cataclasites in 15445 and 72435: Petrology and criteria for equilibrium. Proc Lunar Planet Sci Conf. 11:535-553.

Baldridge WS, Beaty DW, Hill SMR, Albee AL (1979) The petrology of the Apollo 12 pigeonite basalt suite. Proc Lunar Planet Sci Conf 10:141-179

Baldwin RB (1949) The Face of The Moon. University of Chicago Press, Chicago

Bansal BM, Church SE, Gast PW, Hubbard NJ, Rhodes JM, Wiesmann H (1972) Chemical composition of soil from Apollo-16 and Luna-20 sites. Earth Planet Sci Lett 17:29-35

Basu A, Bower JF (1976) Petrography of KREEP basalt fragments from Apollo 15 soil. Proc Lunar Sci Conf. 7:659-678.

Basu A, McKay DS (1979) Petrography and provenance of Apollo 15 soils. Proc Lunar Planet Sci Conf 10:1413-1424

Basu A, McKay DS (1984) Petrologic comparisons of Cayley and Descartes on the basis of Apollo 16 soils from Stations 4 and 11. Proc Lunar Planet Sci Conf 14th in J Geophys Res 89:B535-B541

Basu A, McKay DS (1985) Composition of agglutinitic glass. Part II: Rationale and interpretation (abstr). *In:* Lunar Planet Sci XVI:41-42. Lunar and Planetary Institute, Houston

Basu A, McKay DS, Fruland RM, (1978) Origin and modal petrography of Luna 24 soils. *In:* RB Merrill, JJ Papike (eds) Mare Crisium: The View From Luna 24, p 321-337, Pergamon, New York

Beaty DW, Albee AL (1978) Comparative petrology and possible petrogenetic relations among the Apollo 11 basalts. Proc Lunar Planet Sci Conf 9:359-463

Beaty DW, Hill SMR, Albee AL, Ma M-S, Schmitt, RA (1979a) The petrology and chemistry of basaltic fragments from the Apollo 11 soil, part I. Proc Lunar Planet Sci Conf 10:41-75

Beaty DW, Hill SMR, Albee AL (1979b) Petrology of a new rock type from Apollo 11: Group D basalts (abstr). *In:* Lunar Planet Sci X:89-91, Lunar and Planetary Institute, Houston

Bell JF, Hawke BR (1984) Lunar dark-haloed impact craters: Origin and implications for early mare volcanism. J Geophys Res 89:6899-6910

Bence AE, Papike JJ (1972) Pyroxenes as recorders of lunar basalt petrogenesis: Chemical trends due to crystal-liquid interaction. Proc Lunar Sci Conf 3:431-469

Bence AE, Papike JJ, Lindsley DH (1971) Crystallization histories of clinopyroxenes in two porphyritic rocks from Oceanus Procellarum. Proc Lunar Sci Conf 2:559-574

Bence AE, Delano J.W, Papike JJ, Cameron KL (1974) Petrology of the highlands massifs at Taurus-Littrow: An analysis of the 2-4 mm soil fraction. Proc Lunar Sci Conf 5:785-827.

Bernatowicz TJ, Hohenberg CM, Hudson B, Kennedy BM, Podosek FA (1978) Argon ages for lunar breccias 14064 and 15405. Proc Lunar Planet Sci Conf 9:905-919

Bersch MG, Taylor GJ, Keil K, Norman MD (1991) Mineral compositions in pristine lunar highlands rocks and the diversity of highlands magmatism. Geophys. Res. Lett 18:2085-2088

Bickel CE (1977) Petrology of 78155: An early, thermally-metamorphosed polymict breccia. Proc Lunar Sci Conf 8:2007-2027

Bickel CE, Warner JL, Phinney WC (1976) Petrology of 79215: Brecciation of a lunar cumulate. Proc Lunar Sci Conf 7:1793-1819

Binder AB (1978) On fission and the devolatilization of a Moon of fission origin. Earth Planet Sci Lett 41:381-385

Binder AB (1982) The mare basalt magma source region and mare basalt magma genesis. Proc Lunar Planet Sci Conf 13:A37-A53

Binder AB (1985) The depths of the mare basalt source region. Proc Lunar Sci Conf 15th in J Geophys Res 89:C396-C404

Blanchard DP, Budahn JR (1979) Remnants from the ancient crust: Clasts from consortium breccia 73255. Proc Lunar Planet Sci Conf 10:803-816

Blanchard DP, McKay GA (1981) Remnants from the ancient lunar crust III: Norite 78236 (abstr). *In:* Lunar Planet Sci XII:83-85, Lunar and Planetary Institute, Houston

Blanchard DP, Haskin LA, Jacobs JW, Brannon JC, Korotev RL (1975) Major and trace element chemistry of Boulder 1 at Station 2, Apollo 17. The Moon 14:359-371

Blanchard DP, Brannon JC, Haskin LA, Jacobs JW (1976a) Sample 15445. Chemistry. *In:* Interdisciplinary Studies by the Imbrium Consortium, Vol 2, p 60, Smithsonian Astrophysical Observatory, Cambridge, MA

Blanchard DP, Jacobs JW, Brannon JC, Haskin LA (1976b) Major and trace element compositions of matrix and aphanitic clasts from consortium breccia 73215. Proc Lunar Sci Conf 7:2179-2187

Blanchard DP, Jacobs JW, Brannon JC (1977) Chemistry of ANT-suite and felsite clasts from consortium breccia 73215 and of gabbroic anorthosite 79215. Proc Lunar Sci Conf 8:2507-2524

Blewett DT, Lucey PG, Hawke BR, Jolliff BL (1997a) Clementine images of the lunar sample-return stations: Improvements to the TiO_2 mapping technique (abstr). *In:* Lunar Planet Sci XXVIII:119-120, Lunar and Planetary Institute, Houston

Blewett DT, Lucey PG, Hawke BR, Jolliff BL (1997b) FeO mapping of the Moon: refinement using images of the sample-return stations (abstr). *In:* Lunar Planet Sci XXVIII:121-122, Lunar and Planetary Institute, Houston

Bogard DD, Nyquist LE, Bansal BM, Wiesmann H, Shih C-Y (1975) 76535: An old lunar rock. Earth Planet Sci Lett 26:69-80

Bogard DD, McKay DS, Morris RV, Wentworth SJ, Johnson P (1985) Regolith breccias from Apollo 15 and 16: Petrology, rare gases, and FMR maturity (abstr). *In:* Lunar Planet Sci XVI:73-74, Lunar and Planetary Institute, Houston

Borchardt R, Stöffler D, Spettel B, Palme H, Wänke H, Wacker K, Jessberger EK (1986) Composition, structure, and age of the Apollo 16 subregolith basement as deduced from the chemistry of post-Imbrium melt bombs. Proc Lunar Planet Sci Conf 17 th in J Geophys Res 91:E43-E54

Borg L, Norman M, Nyquist L, Snyder G, Taylor L, Lindstrom M, Wiesmann H (1997) A relatively young samarium-neodymium age of 4.36 Ga for ferroan anorthosite 62236 (abstr). Met and Planet Sci 32:A18

Braddy D, Hutcheon ID, Price RB (1975) Crystal chemistry of Pu and U and concordant fission track ages of lunar zircons and whitlockites. Proc Lunar Sci Conf 6:3587-3600

Bratt SR, Solomon SC, Head JW, Thurber CH (1985) The deep structure of lunar basins: Implications for basin formation and modification. J Geophys Res 90:3049-3064

Brett R (1975a) Thicknesses of some lunar mare basalt flows and ejecta blankets based on chemical kinetic data. Geochim Cosmochim Acta 39:1135-1143

Brett R (1975b) Reduction of mare basalts by sulfur loss (abstr). *In:* Lunar Science VI:89-91, Lunar Science Institute, Houston

Brett R, Butler P Jr, Meyer C Jr, Reid AM, Takeda H, Williams R (1971) Apollo 12 igneous rocks 12004, 12008, 12009, and 12022: A mineralogical and petrological study. Proc Lunar Sci Conf 2:301-317

Brown GM, Emeleus CH, Holland JG, Peckett A, Phillips R (1971) Picrite basalts, ferrobasalts, feldspathic norites, and rhyolites in a strongly fractionated lunar crust. Proc Lunar Sci Conf 2:583-600

Brown GM, Peckett A, Phillips R, Emeleus CH (1973) Mineral-chemical variations in the Apollo 16 magnesio-feldspathic highland rocks. Proc Lunar Sci Conf 4:505-518

Brown GM, Peckett A, Emeleus CH, Phillips R (1974) Mineral-chemical properties of Apollo 17 mare basalts and terra fragments (abstr). *In:* Lunar Science V:89-91, Lunar Science Institute, Houston

Brown GM, Peckett A, Emeleus CH, Phillips R, Pinsent RH (1975) Petrology and mineralogy of Apollo 17 mare basalts. Proc Lunar Sci Conf 6:1-13

Brunfelt AO, Heier KS, Nilssen B., Steinnes E., Sundvoll B (1972) Elemental composition of Apollo 15 samples (abstr). *In:* JW Chamberlain, C Watkins (eds) The Apollo 15 Lunar Samples, p 195-197, Lunar Science Institute, Houston

Burnham CW (1971) The crystal structure of pyroxferroite from Mare Tranquillitatis. Proc Lunar Sci Conf 2:47-57

Busche FD, Prinz M, Keil K, Bunch TE (1972) Spinels and the petrogenesis of some Apollo 12 igneous rocks. Am Mineral 57:1729-1747

Butler P Jr (1978) Recognition of lunar glass droplets produced directly from endogenous liquids: The evidence from S-Zn coatings. Proc Lunar Planet Sci Conf 9:1459-1471

BVSP (Basaltic Volcanism Study Project) (1981) Basaltic Volcanism on the Terrestrial Planets. Pergamon, New York

Cameron EN (1971) Opaque minerals in certain lunar rocks from Apollo 12. Proc Lunar Sci Conf 2:193-206

Cameron KL, Cameron M (1973) Mineralogy of ultramafic nodules from Knippa Quarry, near Uvalde, Texas (abstr). Geol Soc Am Abstr with Progr 5:566

Campbell HW, Hess PC, Rutherford MJ (1978) Ilmenite crystallization in non-mare basalts (abstr). *In:* Lunar Planet Sci IX:149-151, Lunar and Planetary Institute, Houston

Carlson RW, Lugmair GW (1979) Sm-Nd constraints on early lunar differentiation and the evolution of KREEP. Earth Planet Sci Lett 45:123-132

Carlson RW, Lugmair GW (1981a) Sm-Nd age of lherzolite 67667: Implications for the processes involved in lunar crustal formation. Earth Planet Sci Lett 56:1-8

Carlson RW, Lugmair GW (1981b) Time and duration of lunar highlands crust formation. Earth Planet Sci Lett 52:227-238

Carlson RW, Lugmair GW (1988) The age of ferroan anorthosite 60025: Oldest crust on a young Moon? Earth Planet Sci Lett 90:119-130

Carter JL, Clanton US, Fuhrman R, Laughon RB, McKay DS, Usselman TM (1975) Morphology and composition of chalcopyrite, chromite, Cu, Ni-Fe, pentlandite, and troilite in vugs of 76015 and 76215. Proc Lunar Sci Conf 6:719-728

Champness PE, Dunham AC, Gibb FGF, Giles HN, MacKenzie WS, Stumpfl EF, Zussman J (1971) Mineralogy and petrology of some Apollo 12 samples. Proc Lunar Sci Conf 2:359-376

Chao ECT, Minkin JA, Frondel C, Klein C Jr, Drake JC, Fuchs L, Tani B, Smith JV, Anderson AT, Moore PB, Zechman GR Jr, Trail RJ, Plant AG, Douglas JAV, Dence MR (1970a) Pyroxferroite, a new calcium-bearing iron silicate from Tranquillity Base. Proc Apollo 11 Lunar Sci Conf, p 65-79

Chao ECT, James OB, Minnkin JA, Boreman JA, Jackson ED, and Raleigh CB (1970b) Petrology of unshocked crystalline rocks and evidence of impact metamorphism in Apollo 11 returned lunar sample. Proc Apollo 11 Lunar Sci Conf, p 287-314

Chao ECT, Boreman JA, Desborough GA (1971) Unshocked and shocked Apollo 11 and 12 microbreccias: Characteristics and some geologic implications. Proc Lunar Sci Conf 2:797-816

Chao ECT, Minkin JA, Thompson CL (1976) The petrology of 77215, a noritic impact breccia. Proc Lunar Sci Conf 7:2287-2308

Chen HK, Lindsley DH (1983) Apollo 14 very low titanium glasses: Melting experiments in iron-platinum alloy capsules. J Geophys Res 88:B335-342
Chen HK, Delano JW, Lindsley DH (1982) Chemistry and phase relations of VLT volcanic glasses from Apollo 14 and Apollo 15. J Geophys Res 87:A171-A181
Circone S, Agee CB (1996) Compressibility of molten high-Ti mare glass: Evidence for crystal-liquid density inversions in the lunar mantle. Geochim Cosmochim Acta 60:2709-2720
Clanton US, McKay DS, Laughon RB, Ladle GH (1973) Iron crystals in lunar breccias. Proc Lunar Sci Conf 4:925-931
Clanton US, McKay DS, Waits G, Fuhrman R (1978) Sublimate morphology on 74001 and 74002 orange and black glassy droplets. Proc Lunar Planet Sci Conf 9:1945-1957
Compston W, Foster JJ, and Gray CM (1975) Rb-Sr ages of clasts from within Boulder 1, Station 2, Apollo 17. The Moon 14:445-462
Compston W, Foster JJ, Gray CM (1977) Rb-Sr systematics in clasts and aphanites from consortium breccia 73215. Proc Lunar Sci Conf 8:2525-2549
Compston W, Williams IS, Meyer C Jr (1984) Age and chemistry of zircon from late-stage lunar differentiates (abstr). *In:* Lunar Planet Sci XV:182-183, Lunar and Planetary Institute, Houston
Crawford ML, Hollister LS (1974) KREEP basalt: A possible partial melt from the lunar interior. Proc Lunar Sci Conf 5:399-419
Cushing JA, Taylor G J, Norman MD, Keil K (1998) The granulitic impactite suite: Impact melts and metamorphic breccias of the early lunar crust. Earth Planet Sci Lett (in press)
Dalrymple GB and Ryder G (1993) ^{40}Ar-^{39}Ar spectra of Apollo 15 impact melt rocks by laser step-heating and their bearing on the history of lunar basin formation. J Geophys Res 98:13085-13095
Dalrymple GB and Ryder G (1996) Argon-40/argon-39 age spectra of Apollo 17 highlands breccia samples by laser step heating and the age of the Serenitatis basin. J Geophys Res 101:26069-26084
Dalton J, Hollister LS, Kulick CG, Hargraves RB, (1974) The nature of the chromite to ulvöspinel transition in mare basalt 15555 (abstr). *In:* Lunar Science V:160-162, Lunar Science Institute, Houston
Davis PA, Spudis PD (1985) Petrologic province maps of the lunar highlands derived from orbital geochemical data. Proc Lunar Sci Conf 7:D61-D74
Davis PA, Spudis PD (1987) Global petrologic variations on the Moon: A ternary-diagram approach. Proc Lunar Sci Conf 7th Part 2 in J Geophys Res 92:E387-E395
Delano JW (1979) Apollo 15 green glass: chemistry and possible origin. Proc Lunar Planet Sci Conf 10:275-300
Delano JW (1980) Chemistry and liquidus phase relations of Apollo 15 red glass: Implications for the deep lunar interior. Proc Lunar Planet Sci Conf 11:251-288
Delano JW (1986) Pristine lunar glasses: Criteria, data and implications. Proc Lunar Planet Sci Conf 16:D201-D213
Delano JW (1990) Pristine mare glasses and mare basalts: Evidence for a general dichotomy of source regions (abstr). *In:* Workshop on Lunar Volcanic Glasses: Scientific and Resource Potential; LPI Tech Rept 90-02, p 30-31, Lunar and Planetary Institute, Houston
Delano JW, Livi K (1981) Lunar volcanic glasses and their constraints on mare petrogenesis. Geochim Cosmochim Acta 45:2137-2149
Delano JW and Ringwood AE (1978) Siderophile elements in the lunar highlands: Nature of the indigenous component and implications for the origin of the Moon. Proc Lunar Planet Sci Conf 9:111-159
Delano JW, Hanson BZ, Watson EB (1994) Abundance and diffusivity of sulfur in lunar picritic magmas (abstr). *In:* Lunar Planet Sci XXV:325-326, Lunar and Planetary Institute, Houston
Dence MR, Douglas JAV, Plant AG, Traill RJ (1970) Petrology, mineralogy and deformation of Apollo 11 samples. Proc Apollo 11 Lunar Sci Conf, p 315-340
Deutsch A and Stöffler D. (1987) Rb-Sr-analyses of Apollo 16 melt rocks and a new age estimate for the Imbrium basin: Lunar basin chronology and the early heavy bombardment of the Moon. Geochim Cosmochim Acta 51:1951-1964
Dickinson JE, Hess PC (1983) Role of whitlockite and apatite in lunar felsite (abstr). *In:* Lunar Planet Sci XIV:158-159, Lunar and Planetary Institute, Houston
Dickinson T, Taylor GJ, Keil K, Schmitt RA, Hughes SS, Smith MR (1985) Apollo 14 aluminous basalts and their possible relationship to KREEP. Proc Lunar Planet Sci Conf 15th in J Geophys Res 90:C365-C374
Dixon JR, Papike JJ (1975) Petrology of anorthosites from the Descartes region of the Moon: Apollo 16. Proc Lunar Sci Conf 6:263-291
Dominik B, Jessberger EK (1978) Early lunar differentiation: 4.42-AE old plagioclase clasts in Apollo 16 breccia 67435. Earth Planet Sci Lett 38:407-415

Donaldson CH, Usselman TM, Williams RJ, Lofgren GE (1975) Experimental modeling of the cooling history of Apollo 12 olivine basalts. Proc Lunar Sci Conf 6:843-869

Dowty E, Keil K, Prinz M (1972) Anorthosite in the Apollo 15 rake samples from Spur Crater (abstr). *In:* JW Chamberlain, C Watkins (eds) The Apollo 15 Lunar Samples, p 62-66, Lunar Science Institute, Houston

Dowty E, Prinz M, Keil K (1974a) Ferroan anorthosite: A widespread and distinctive lunar rock type. Earth Planet Sci Lett 24:15-25

Dowty E, Keil K, Prinz M (1974b) Igneous rocks from Apollo 16 rake samples. Proc Lunar Sci Conf 5:431-445

Dowty E, Keil K, Prinz M, Gros J, Takahashi H (1976) Meteorite-free Apollo 15 crystalline KREEP. Proc Lunar Sci Conf 7:1833-1844

Drake MJ, McCallum IS, McKay GA, Weill DF (1970) Mineralogy and petrology of Apollo 12 sample no. 12013. A progress report. Earth Planet Sci Lett 9:103-123

Duke MB, Nagle JS (1974) Lunar Core Catalog. NASA JSC Publication JSC-09252, NASA Johnson Space Center, Houston

Duke MB, Woo CC, Bird ML, Sellers GA, Finkelman RB (1970a) Lunar soil: Size distribution and mineralogical constituents. Science 167:648-650

Duke MB, Woo CC, Sellers GA, Bird ML, Finkelman RB (1970b) Genesis of lunar soil at Tranquillity Base. Proc Apollo 11 Lunar Sci Conf, p 347-361

Duncan AR, Grieve RAF, Weill DF (1975) The life and times of Big Bertha: Lunar breccia 14321. Geochim Cosmochim Acta 39:265-273

Dymek RF (1986) Characterization of the Apollo 15 feldspathic basalt suite (abstr). *In:* Workshop on the Geology and Petrology of the Apollo 15 Landing Site. LPI Tech Rept 86-03, p 52-57, Lunar and Planetary Institute, Houston

Dymek RF, Albee AL, and Chodos AA (1974) Glass-coated soil breccia 15205: Selenologic history and petrologic constraints on the nature of its source region. Proc Lunar Sci Conf 5:235-260

Dymek RF, Albee AL, Chodos AA (1975a) Comparative mineralogy and petrology of Apollo 17 mare basalts: Samples 70215, 71055, 74255, and 75055. Proc Lunar Sci Conf 6:49-77

Dymek RF, Albee AL, and Chodos AA (1975b) Comparative petrology of lunar cumulate rocks of possible primary origin: Dunite 72415, troctolite 76535, norite 78235, and anorthosite 62237. Proc Lunar Sci Conf 6:301-341

Dymek RF, Albee AL, and Chodos AA (1976) Petrology and origin of Boulders #2 and #3, Apollo 17 Station 2. Proc Lunar Sci Conf 7:2335-2378

Ebihara M, Wolf R, Warren PH, Anders E (1992) Trace elements in 59 mostly highland Moon rocks. Proc Lunar Planet Sci 22:417-426

El Goresy A, Ramdohr P (1975) Subsolidus reduction of lunar opaque oxides: Textures, assemblages, geochemistry, and evidence for a late-stage endogenic gaseous mixture. Proc Lunar Sci Conf 6:729-745

El Goresy A, Ramdohr P, Taylor LA (1971a) The geochemistry of the opaque minerals in Apollo 14 crystalline rocks. Earth Planet Sci Lett 13:121-129

El Goresy A, Ramdohr P, Taylor LA (1971b) The opaque minerals in the lunar rocks from Oceanus Procellarum. Proc Lunar Sci Conf 2:219-235

El Goresy A, Taylor LA, Ramdohr P (1972) Fra Mauro crystalline rocks: Mineralogy, geochemistry, and subsolidus reduction of opaque minerals. Proc Lunar Sci Conf 3:333-349

El Goresy A, Ramdohr P, Medenbach O (1973a) Lunar samples from Descartes site: Opaque mineralogy and geochemistry. Proc Lunar Sci Conf 4:733-750

El Goresy A, Ramdohr P, Pavicevic M, Medenback O, Müller O, Gentner W (1973b) Zinc, lead, chlorine, and FeOOH-bearing assemblages in the Apollo 16 sample 66095: Origin by impact of a comet or a carbonaceous chondrite. Earth Planet Sci Lett 18:411-419

El Goresy A, Prinz M, Ramdohr P (1976) Zoning in spinels as an indicator of the crystallization histories of mare basalts. Proc Lunar Sci Conf 7:1261-1279

Epstein S, Taylor HP Jr, (1972) O18/O16, Si30/Si28, C13/C12 and D/H studies of Apollo 14 and 15 samples. Proc Lunar Sci Conf 3:1429-1454

Floss C (1991) Rare earth element and other trace element microdistributions in two unusual extraterrestrial igneous systems: The enstatite achondrite (aubrite) meteorites and the lunar ferroan anorthosites. PhD thesis, Washington University, St Louis

Floss C, James OB, McGee JJ, Crozaz G (1991) Lunar ferroan anorthosites: Rare earth element measurements of individual plagioclase and pyroxene grains (abstr) *In:* Lunar Planet Sci XXII:391-392, Lunar and Planetary Institute, Houston

Floss C, James OB, McGee JJ, Crozaz G (1998) Lunar ferroan anorthosite petrogenesis: Clues from trace element distributions in FAN subgroups. Geochm Cosmochim Acta (in press)

Friel JJ, Goldstein JI (1976) An experimental study of phosphate reduction and phosphorus-bearing lunar metal particles. Proc Lunar Sci Conf 7:791-806

Friel JJ, Goldstein JI (1977) The relationship between lunar metal particles and phosphate minerals. Proc Lunar Sci Conf 8:3955-3965

Frondel JW (1975) Lunar Mineralogy. Wiley, New York

Fuchs LH (1970) Fluorapatite and other accessory minerals in Apollo 11 rocks. Proc Apollo 11 Lunar Sci Conf, p 475-479

Galbreath KC, Shearer CK, Papike JJ. Shimizu N (1990) Inter- and intra-group compositional variations in Apollo 15 pyroclastic green glass: An electron- and ion-microprobe study. Geochim Cosmochim Acta 54:2565-2575

Gancarz AJ, Albee AL, Chodos, AA (1971) Petrologic and mineralogic investigations of some crystalline rocks returned by the Apollo 14 mission. Earth Planet Sci Lett 12:1-18

Gancarz AJ, Albee AL, Chodos, AA (1971) Comparative petrology of Apollo 16 sample 68415 and Apollo 14 samples 14276 and 14310. Earth Planet Sci Lett 16:307-330

Gatehouse BM, Grey IE, Lovering JF, Wark DA (1977) Structural studies on tranquillityite and related synthetic plases. Proc Lunar Sci Conf 8:1831-1838

Gay P, Brown MG, Muir ID (1972) Mineralogical and petrographic features of two Apollo 14 rocks. Proc Lunar Sci Conf 3:351-362

Geiss J, Eberhardt P, Grogler N, Guggisberg S, Maurer P, Stettler A (1977) Absolute time scale of lunar mare formation and filling. Phil Trans Roy Soc London A285:151-158

Goldstein JI, Axon HJ (1973) Composition, structure, and thermal history of metallic particles from 3 Apollo 16 soils, 65701, 68501, and 63501. Proc Lunar Sci Conf 4:751-775

Goldstein JI, Blau PJ (1973) Chemistry and thermal history of metal particles in Luna 20 soils. Geochim Cosmochim Acta 37:847-855

Goodrich CA, Taylor GJ, Keil K, Kallemeyn GW, Warren PH (1986) Alkali norite, troctolites, and VHK mare basalts from breccia 14304. Proc Lunar Planet Sci Conf 16th in J Geophys Res 91:D305-D318

Gooley R, Brett R, Warner J, Smyth JR (1974) A lunar rock of deep crustal origin: Sample 76535. Geochim Cosmochim Acta 38:1329-1339

Green DH, Ringwood AE, Ware NG, Hibberson WO, Major A, Kiss E (1971a) Experimental petrology and petrogenesis of Apollo 12 basalts. Proc Lunar Sci Conf 2:601-615

Green DH, Ware NG, Hibberson, WO, Major A (1971b) Experimental petrology of Apollo 12 mare basalts. Part I, sample 12009. Earth Planet Sci Lett 13:85-96

Green DH, Ringwood AE, Hibberson WO, Ware NG (1975) Experimental petrology of Apollo 17 mare basalts. Proc Lunar Sci Conf 6:871-893

Grieve RAF, Stöffler D, and Deutsch A (1991) The Sudbury structure: Controversial or misunderstood? J Geophys Res 96:22753-22764

Grove TL (1982) Use of lamellae in lunar clinopyroxenes as cooling rate speedometers: An experimental calibration. Am Mineral 67:251-268

Grove TL, Bence AE (1977) Experimental study of pyroxene-liquid interaction in quartz-normative basalt 15597. Proc Lunar Sci Conf 8:1549-1579

Grove TL, Lindsley DH (1978) Compositional variation and origin of lunar ultramafic green glass (abstr). *In:* Lunar Planet Sci IX:430-432, Lunar and Planetary Institute, Houston

Grove TL, Vaniman DT (1978) Experimental petrology of very low-Ti (VLT) basalts. *In:* RB Merrill, JJ Papike (eds) Mare Crisium: The View from Luna 24, p 445-472, Pergamon, New York

Grove TL, Walker D (1977) Cooling histories of Apollo 15 quartz-normative basalts. Proc Lunar Sci Conf 8:1501-1520

Guggisberg S, Eberhardt P, Geiss J, Grögler N, Stettler A, Brown GM, Peckett A (1979) Classification of the Apollo 11 mare basalts according to Ar^{39}-Ar^{40} ages and petrological properties. Proc Lunar Planet Sci Conf 10:1-39

Hafner SS, Niebuhr HH, Zeira S (1973) Ferric iron in plagioclase crystals from anorhtosite 15415 (abstr). *In:* Lunar Science IV:326-328, Lunar Science Institute, Houston

Haggerty SE (1971a) Compositional variations in lunar spinels. Nature Phys Sci 233:156-160

Haggerty SE (1971b) Subsolidus reduction of lunar spinels. Nature Phys Sci 234:113-117

Haggerty SE (1972a) Apollo 14 subsolidus reduction and compositional variations of spinels. Proc Lunar Sci Conf 3:305-333

Haggerty SE (1972b) Chemical characteristics of spinels in some Apollo 15 basalts (abstr). *In:* JW Chamberlain, C Watkins (eds) The Apollo 15 Lunar Samples, p 92-97, Lunar Science Institute, Houston

Haggerty SE (1972c) Luna 16: An opaque mineral study and a systematic examination of compositional variations of spinels from Mare Fecunditatis. Earth Planet Sci Lett 13:328-352

Haggerty SE (1972d) Solid solutions, subsolidus reduction and compositional characteristics of spinels in some Apollo 15 basalts. Meteoritics 7:353-370
Haggerty SE (1973a) Armalcolite and genetically associated opaque minerals in the lunar samples. Proc Lunar Sci Conf 4:777-797
Haggerty SE (1973b) Luna 20: Mineral chemistry of spinel, pleonaste, chromite, ulvöspinel, ilmenite, and rutile. Geochim Cosmochim Acta 37:857-867
Haggerty SE (1977) Apollo 14: Oxide, metal, and olivine mineral chemistries in 14072 with a bearing on the temporal relationships of subsolidus reduction. Proc Lunar Sci Conf 8:1809-1829
Haggerty SE (1978a) Luna 24: Systematics in spinel mineral chemistry in the context of an intrusive petrogenetic grid. *In:* RB Merrill, JJ Papike (eds) Mare Crisium: The View From Luna 24, p 523-536, Pergamon, New York
Haggerty SE (1978b) The redox state of planetary basalts. Geophys Res Lett 5:443-446
Haggerty SE, Boyd FR, Beil PM, Finger LW, Bryan WB (1970) Opaque minerals and olivine in lavas and breccias from Mare Tranquillitatis. Proc Apollo 11 Lunar Sci Conf, p 513-538
Hanan BB, Tilton GR (1987) 60025: Relict of primitive lunar crust? Earth Planet Sci Lett 84:15-21
Hansen EC, Smith JV, Steele IM (1980) Petrology and mineral chemistry of 67667, a unique feldspathic lherzolite. Proc Lunar Planet Sci Conf 11:523-533
Haselton JD, Nash WP (1975) A model for the evolution of opaques in mare lavas. Proc Lunar Sci Conf 6:747-755
Haskin LA, Korotev RL (1977) Test of a model for trace element partition during closed-system solidification of a silicate liquid. Geochim Cosmochim Acta 41:921-939
Haskin LA, Helmke PA, Blanchard DP, Jacobs JW, Telander K (1973) Major and trace element abundances in samples from the lunar highlands. Proc Lunar Sci Conf 3:1275-1296
Haskin LA, Shih C-Y, Bansal BM, Rhodes JM, Wiesmann H, Nyquist LE (1974) Chemical evidence for the origin of 76535 as a cumulate. Proc Lunar Sci Conf 5:1213-1225.
Haskin LA, Jacobs JW, Brannon JC, Haskin MA (1977) Compositional dispersions in lunar and terrestrial basalts. Proc Lunar Sci Conf 8:1731-1750
Haskin LA, Lindstrom MM, Salpas PA, Lindstrom D (1981) On compositional variations among lunar anorthosites. Proc Lunar Planet Sci Conf 12B:41-66
Hawke BR, Head JW (1978) Lunar KREEP volcanism: Geologic evidence for history and mode of emplacement. Proc Lunar Planet Sci Conf 9:3285-3309
Hawke BR, Lucey PG, Bell JF, Spudis PD (1990) Ancient mare volcanism (abstr). *In:* Workshop on Mare Volcanism and Basalt Petrogenesis: Astounding Fundamental Concepts (AFC) Developed Over the Last Fifteen Years, LPI Tech Rept 91-03, p 5-6, Lunar and Planetary Institute, Houston
Head JW (1974) Lunar dark mantle deposits: Possible clues to the distribution of early mare deposits. Proc Lunar Sci. Conf 5:207-222
Head JW (1976) Lunar volcanism in space and time. Rev Geophys Space Phys 14:265-300
Head JW, Wilson L (1992) Lunar mare volcanism: Stratigraphy, eruption conditions, and the evolution of secondary crusts. Geochimica Cosmochim Acta 56:2155-2175
Head JW, Wilson L (1997) Lunar mare basalt volcanism: Early stages of secondary crustal formation and implications for petrogenetic evolution and magma emplacement processes (abstr). *In:* Lunar Planet Sci XXVIII:545-546, Lunar and Planetary Institute, Houston
Heiken G (1975) Petrology of lunar soils. Rev Geophys Space Phys 13:567-587
Heiken G and McKay DS (1977) A model for eruption behavior of a volcanic vent in eastern Mare Serenitatis. Proc Lunar Sci Conf 8:3243-3255
Heiken GH, McKay DS, Fruland RM (1973) Apollo 16 soils: Grain size analysis and petrography. Proc Lunar Sci Conf 4:251-265
Heiken GH, McKay DS, Brown RW (1974) Lunar deposits of possible pyroclastic origin. Geochim Cosmochim Acta 38:1703-1718
Heiken GH, Morris RV, McKay DS, Fruland RM (1976) Petrographic and ferromagnetic resonance studies of the Apollo 15 deep drill core. Proc Lunar Sci Conf 7:93-111
Heiken GH, Vaniman DT, and French BM (eds) (1991) Lunar Sourcebook: A User's Guide to the Moon. Cambridge University Press, Cambridge, UK
Herzberg CT (1978) The bearing of spinel-cataclasites on the crust-mantle structure of the Moon. Proc Lunar Planet Sci Conf 9:319-336
Herzberg CT, Baker MB (1980) The cordierite- to spinel-cataclasite transition: Structure of the lunar crust. *In:* JJ Papike, RB Merrill (eds) Proc Conf Lunar Highlands Crust, p 113-132, Pergamon, New York
Hess PC (1991) Pristine mare glasses: Primary magmas? (abstr). *In:* Workshop on Mare Volcanism and Basalt Petrogenesis: Astounding Fundamental Concepts (AFC) Developed Over the Last Fifteen Years, LPI Tech Rept 91-03, p 17-18, Lunar and Planetary Institute, Houston

Hess PC (1993) Ilmenite liquidus and depths of segregation of high-Ti picritic glasses (abstr). *In:* Lunar Planet Sci XXIV:649-650, Lunar and Planetary Institute, Houston
Hess PC (1994) Petrogenesis of lunar troctolites. J Geophys Res 99:19083-19093
Hess PC, Finnila A (1997) Depths of segregation of hi-TiO_2 picrite mare glasses (abstr). *In:* Lunar Planet Sci XXVIII:559-560, Lunar and Planetary Institute, Houston
Hess PC, Parmentier EM (1995) A model for the thermal and chemical evolution of the Moon's interior: Implications for the onset of mare volcanism. Earth Planet Sci Lett 134:501-514
Hess PC, Rutherford MJ, Guillemette RN, Ryerson FJ, Tuchfeld HA (1975) Residual products of fractional crystallization of lunar magmas: An experimental study. Proc Lunar Sci Conf 6:895-910
Hewins RH, Goldstein JI (1974) Metal-olivine associations and Ni-Co contents in two Apollo 12 mare basalts. Earth Planet Sci Lett 24:59-70
Hewins RH, Goldstein JI (1975) The provenance of metal in anorthositic rocks. Proc Lunar Sci Conf 6:343-362
Higuchi H, Morgan JW (1975) Ancient meteoritic component in Apollo 17 boulders. Proc Lunar Sci Conf 6:1625-1651
Hinthorne JR, Conrad R, Andersen CA (1975) Lead-lead age and trace element abundances in lunar troctolite 76535 (abstr). *In:* Lunar Sci VI:373-375, Lunar Science Institute, Houston
Hlava PF, Prinz M, Keil K (1972) Niobian rutile in an Apollo 14 KREEP fragment. Meteoritics 7:479-485
Hodges EN, Kushiro I (1974) Apollo 17 petrology and experimental determination of differentiation sequences in model Moon compositions. Proc Lunar Sci Conf 5:505-520
Hollister LS (1973) Sample 67955: A description and a problem. Proc Lunar Sci Conf 4:633-641
Hörz F, Cintala M (1997) Impact experiments related to the evolution of planetary regoliths. Met Planet Sci 32:174-209
Hörz F, Gibbons RV, Gault DE, Hartung JB, Brownlee DE (1975) Some correlation of rock exposure ages and regolith dynamics. Proc Lunar Sci Conf 6:3495-3508
Hörz F, Grieve R, Heiken G, Spudis P, Binder A (1991) Lunar Surface Processes. *In:* GH Heiken, DT Vaniman, BM French (eds) Lunar Sourcebook: A User's Guide to the Moon, p 61-120, Cambridge Univ Press, Cambridge, UK
Houck KJ (1982) Modal petrology of six soils from Apollo 16 double drive tube core 64002. Proc Lunar Planet Sci Conf 13th in J Geophys Res 87:A210-A220
Housley RM, Cirlin EH, Grant RW (1973) Characterization of fines from the Apollo 16 site. Proc Lunar Sci Conf 4:2729-2735
Hu H-N, Taylor LA (1977) Lack of chemical fractionation in major and minor elements during agglutinate formation. Proc Lunar Sci Conf 8:3645-3656
Hu H-N, Taylor LA (1978) Soils from Mare Crisium: Agglutinitic glass chemistry and soil development. *In:* RB Merrill, JJ Papike (eds) Mare Crisium: The View from Luna 24, p 291-302, Pergamon, New York
Hubbard NJ, Gast PW (1972) Chemical composition and origin of nonmare lunar basalts. Proc Lunar Sci Conf 2:999-1020
Hubbard NJ, Minear JW (1975) A physical and chemical model of early lunar history. Proc Lunar Planet Sci Conf 6:1057-1085
Hubbard NJ, Meyer C, Gast PW, Wiesmann H (1971a) The composition and derivation of Apollo 12 soils. Earth Planet Sci Lett 10:341-350
Hubbard NJ, Gast PW, Meyer C, Nyquist LE, Shih C-Y, Wiesmann H (1971b) Chemical composition of lunar anorthosites and their parent liquids. Earth Planet Sci Lett 13:71-75
Hubbard NJ, Gast PW, Rhodes JM, Bansal BM, Wiesmann H, Church SE (1972) Nonmare basalts: Part II. Proc Lunar Sci Conf 3:1161-1179
Hubbard NJ, Rhodes JM, Gast PW, Bansal BM, Shih C-Y, Wiesmann H, Nyquist LE (1973) Lunar rock types: The role of plagioclase in non-mare and highland rock types. Proc Lunar Sci Conf 4:1297-1312
Hubbard NJ, Rhodes JM, Wiesmann H, Shih C-Y, Bansal BM (1974) The chemical definition and interpretation of rock types returned from the non-mare regions of the Moon. Proc Lunar Planet Sci Conf 5:1227-1246
Hughes SS, Delano JW, Schmitt RA (1988) Apollo 15 yellow-brown volcanic glass: Chemistry and petrogenetic relations to green volcanic glass and olivine-normative mare basalts. Geochim Cosmochim Acta 52:2379-2391
Hughes SS, Delano JW, Schmitt RA (1989) Petrogenetic modeling of 74220 high-Ti orange volcanic glasses and the Apollo 11 and 17 high-Ti mare basalts. Proc Lunar Planet Sci Conf 19:175-188
Huneke JC, Wasserburg GJ (1979) Sliva iz piroga (plum out of the pie): K/Ar evidence from Luna 20 rocks for lunar differentiation prior to 4.51 AE ago (abstr). *In:* Lunar Planet Sci X:598-600, Lunar and Planetary Institute, Houston.

Hunter RH, Taylor LA (1983) The magma ocean as viewed from the Fra Mauro shoreline: An overview of the Apollo 14 crust. Proc Lunar Planet Sci Conf 13th in J Geophys Res 88:D591-D602

Husain L, Schaeffer OA (1973) ^{40}Ar-^{39}Ar crystallization ages and ^{38}Ar-^{37}Ar cosmic ray exposure ages of samples from the vicinity of the Apollo 16 landing site (abstr). *In:* Lunar Sci IV:406-409, Lunar and Planetary Institute, Houston.

Husain L, Schaeffer OA (1975) Lunar evolution: The first 600 million years. Geophys Res Lett 2:29-32

Ivanov AV, Florensky KP (1975) The role of vaporization process in lunar rock formation. Proc Lunar Sci Conf 6:1341-1350

Ivanov AV, Ilin NP, Loseva LE, Senin VG (1973) Composition of metallic iron and some coexisting phases in samples from highland lunar region returned by automatical station Luna-20. Geochimiya 12:1782-1792

Jackson ED, Wilshire HG (1968) Chemical composition of the lunar surface at the Surveyor landing sites. J Geophys Res 73:7621-7629

Jackson ED, Sutton RL, Wilshire HG (1975) Structure and petrology of a cumulus norite boulder sampled by Apollo 17 in Taurus-Littrow valley, the Moon. Geol Soc Am Bull 86:433-442

James OB (1972) Lunar anorthosite 15415: Texture, mineralogy, and metamorphic history. Science 175:432-436

James OB (1976) Petrology of aphanitic lithologies in consortium breccia 73215. Proc Lunar Sci Conf 7:2145-2178

James OB (1980) Rocks of the early lunar crust. Proc Lunar Planet Sci Conf 11:365-393

James OB (1981) Petrologic and age relations of Apollo 16 rocks: Implications for subsurface geology and the age of the Nectaris basin. Proc Lunar Planet Sci Conf 12B:209-233

James OB, Flohr MK (1983) Subdivision of the Mg-suite noritic rocks into Mg-gabbronorites and Mg-norites. Lunar Planet Sci Conf 13th in J Geophys Res 88:A603-A614

James OB, Hammarstrom JG (1977) Petrology of four clasts from consortium breccia 73215. Proc Lunar Sci Conf 8:2459-2494

James OB, Jackson ED (1970) Petrology of the Apollo 11 ilmenite basalts. J Geophys Res 75:5793-5824

James OB, McGee JJ (1979) Consortium breccia 73255: Genesis and history of two coarse-grained "norite" clasts. Proc Lunar Planet Sci Conf 10:713-743

James OB, McGee JJ (1980) Petrology of mare-type basalt clasts from consortium breccia 73255. Proc Lunar Planet Sci Conf 11:67-86

James OB, Brecher A, Blanchard DP, Jacobs JW, Brannon JC, Korotev RL, Haskin LA, Higuchi H, Morgan JW, Anders E, Silver LT, Marti K, Braddy D, Hutcheon ID, Kirsten T, Kerridge JF, Kaplan IR, Pillinger CT, Gardiner LR (1975) Consortium studies of matrix of light gray breccia 73215. Proc Lunar Sci Conf 6:547-577

James OB, Hedenquist JW, Blanchard DP, Budahn JR, Compston W (1978) Consortium breccia 73255: Petrology, major and trace element chemistry, and Rb-Sr systematics of aphanitic lithologies. Proc Lunar Planet Sci Conf 9:789-819

James OB, Flohr MK, Lindstrom MM (1984) Petrology and geochemistry of lunar dimict breccia 61015. Proc Lunar Planet Sci Conf 15th in J Geophys Res 89:C63-C86

James OB, Lindstrom MM, Flohr MK (1987) Petrology and geochemistry of alkali gabbronorites from lunar breccia 67975. Proc Lunar Planet Sci Conf 17th in J Geophys Res 89:E314-E330

James OB, Lindstrom MM, Flohr MK (1989) Ferroan anorthosite from lunar breccia 64435: Implications for the origin and history of ferroan anorthosites. Proc Lunar Planet Sci Conf 19:219-243

James OB, Lindstrom MM, McGee JJ (1991) Lunar ferroan anorthosite 60025: Petrology and chemistry of mafic lithologies. Proc Lunar Planet Sci 21:63-87

Jessberger EK, Dominic B, Kirsten T, Staudacher T (1977a) New ^{40}Ar-^{39}Ar ages of Apollo 16 breccias and 4.43 AE old anorthosites (abstr). *In:* Lunar Planet Sci VIII:511-513, Lunar and Planetary Institute, Houston

Jessberger EK, Kirsten T, Staudacher T (1977b) One rock and many ages—Further K-Ar data on consortium breccia 73215. Proc Lunar Sci Conf 8:2567-2580.

Johnson TV, Mosher JA, Matson DL (1977) Lunar spectral units: a northern hemispheric mosaic. Proc Lunar Sci Conf 8:1013-1028

Jolliff BL (1991) Fragments of quartz monzodiorite and felsite in Apollo 14 soil particles. Proc Lunar Planet Sci 21:101-118

Jolliff BL (1993) A monazite-bearing clast in Apollo 17 melt breccia (abstr). *In:* Lunar Planet Sci XXIV:725-726, Lunar and Planetary Institute, Houston

Jolliff BL (1997) Clementine UV-VIS multispectral data and the Apollo 17 landing site: What can we tell and how well (abstr). *In:* Lunar Planet Sci XXXVIII:671-672, Lunar and Planetary Institute, Houston

Jolliff BL, Haskin LA (1995) Cogenetic rock fragments from a lunar soil: Evidence of a ferroan noritic-anorthosite pluton on the Moon. Geochim Cosmochim Acta 59:2345-2374

Jolliff BL, Hsu W (1996) Geochemical effects of recrystallization and exsolution of plagioclase of ferroan anorthosite (abstr). *In:* Lunar Planet Sci XXVII:611-612, Lunar and Planetary Institute, Houston

Jolliff BL, Korotev RL, Haskin LA (1991) Geochemistry of 2-4 mm particles from Apollo 14 soil (14161) and implications regarding igneous components and soil-forming processes. Proc Lunar Planet Sci 21:193-219

Jolliff BL, Haskin LA, Colson RO, Wadhwa M (1993) Partitioning in REE-saturated minerals: Theory, experiment, and modelling of whitlockite, apatite, and evolution of lunar residual magmas. Geochim Cosmochim Acta 57:4069-4094

Jolliff BL, Rockow KM, Korotev RL (1996) QUE 94281: Shallow plutonic VLT components and highland components (abstr). *In:* Lunar Planet Sci XXVII:615-616, Lunar and Planetary Science Institute, Houston.

Jones JH, Delano JW (1989) A three component model for the bulk composition of the Moon. Geochim Cosmochim Acta 53:513-527

Kaula WM, Schubert G, Lingenfelter RE, Sjogren WI, Wollenhaupt WR (1972) Analysis and interpretation of lunar laser altimetry. Proc Lunar Sci Conf 3:2189-2204

Kaula WM, Schubert G, Lingenfelter RE, Sjogren WI, Wollenhaupt WR (1974) Apollo laser altimetry and inference as to lunar structure. Proc Lunar Sci Conf 5:3049-3058

Keil K, Bunch TE, Prinz M (1970) Mineralogy and composition of Apollo 11 lunar samples. Proc Apollo 11 Lunar Sci Conf, p 561-598

Keil K, Prinz M, Bunch TE (1971) Mineralogy, petrology and chemistry of some Apollo 12 samples. Proc Lunar Sci Conf 2:319-341

Keil K, Kurat G, Prinz M, Green JA (1972) Lithic fragments, glasses and chondrules from lunar fines. Earth Planet Sci Lett 13:243-256

Keller LP, McKay DS (1992) Micrometer-sized glass spheres in Apollo 16 soil 61181: Implications for impact volatilization and condensation. Proc Lunar Planet Sci 22:137-141

Kesson SE (1975) Mare basalts: Melting experiments and petrogenetic interpretations. Proc Lunar Sci Conf 6:921-944

Kesson SE, Lindsley DH (1974) Mare basalt petrogenesis- A review of experimental studies. Rev Geophys Space Phys 12:361-374

Kesson SE, Lindsley DH (1975) The effect of Al^{3+}, Cr^{3+}, and Ti^{3+} on the stability of armalcolite. Proc Lunar Sci Conf 6:911-920

Kesson SE, Lindsley DH (1976) Mare basalt petrogenesis—A review of experimental studies. Rev Geophys Space Phys 14:361

King EA Jr, Carman MF, Butler JC (1970) Mineralogy and petrology of coarse particulate material from the lunar surface at Tranquillity Base. Proc Apollo 11 Lunar Sci Conf, p 599-606

Kirk RL, Stevenson DJ (1989) The competition between thermal contraction and differentiation in the stress history of the Moon. J Geophys Res 95:12133-12144

Kirsten T, Horn P, Kiko J (1973) ^{39}Ar-^{40}Ar dating and rare gas analysis of Apollo 16 rocks and soils. Proc Lunar Sci Conf 4:175-1784

Klein C Jr, Drake JC, Frondel C (1971) Mineralogical, petrological, and chemical features of four Apollo 12 lunar microgabbros. Proc Lunar Sci Conf 2:265-284

Korotev RL (1976) Geochemistry of grain-size fractions of soils from the Taurus-Littrow valley floor. Proc Lunar Sci Conf 7:695-726

Korotev RL (1985) Geochemical survey of Apollo 15 regolith breccias (abstr). In Lunar Planet Sci XVI:459-460. Lunar and Planetary Institute, Houston

Korotev RL (1991) Geochemical stratigraphy of two regolith cores from the Central Highlands of the Moon. Proc Lunar Planet Sci 21:229-289

Korotev RL (1994) Compositional variation in Apollo 16 impact melt breccias and inferences for the geology and bombardment history of the Central Highlands of the Moon. Geochim Cosmochim Acta 58:3931-3969

Korotev RL (1997) Some things we can infer about the Moon from the composition of the Apollo 16 regolith. Met Planet Sci 32:447-478

Korotev RL and Morris RV (1993) Composition and maturity of Apollo 16 regolith core 600013/14. Geochim Cosmochim Acta 57:4813-4826

Korotev RL, Morris RL, Jolliff BL, Schwarz C (1997) Lithological variation with depth and decoupling of maturity parameters in Apollo 16 regolith core 68001/2. Geochim Cosmochim Acta 61:2989-3002

Krähenbühl U, Grütter A, von Gunten HR, Meyer G, Wegmüller F, Wyttenbach A (1977) Volatile and non-volatile elements in grain size fractions of Apollo 17 soils 75081, 72461, and 72501. Proc Lunar Sci Conf 8:3901-3916

Kridelbaugh SJ, Weill DF (1973) The mineralogy and petrology of ilmenite basalt 75055 (abstr). EOS Trans AGU 54:597

Kurat G, Keil K, Prinz M (1974) Rock 14318: A polymict breccia with chondritic texture. Geochim Cosmochim Acta 38:1133-1146

Kushiro I, Ikeda Y, Nakamura Y (1972) Petrology of Apollo 14 high-alumina basalt. Proc Lunar Sci Conf 3:115-131

Langevin Y (1982) Evolution of an asteroidal regolith: Granulometry, mixing, and maturity (abstr). *In:* Workshop on Lunar Breccias and Soils and their Meteoritic Analogs, LPI Tech Rept 82-02, p 87-93, Lunar and Planetary Institute, Houston

Langevin Y, Arnold JR (1977) The evolution of the lunar regolith. Ann Rev Earth Planet Sci 5:449-489

Laul JC (1986) Chemistry of the Apollo 12 highland component. Proc Lunar Planet Sci Conf 16th in J Geophys Res 91:D251-D261

Laul JC, Schmitt RA (1973) Chemical composition of Apollo 15, 16, and 17 samples. Proc Lunar Sci Conf 4:1349-1367

Laul JC, Schmitt RA (1975) Dunite 72417: A chemical study and interpretation. Proc Lunar Sci Conf 6:1231-1254

Laul JC, Papike JJ (1980a) The lunar regolith: Comparative chemistry of the Apollo sites. Proc Lunar Planet Sci Conf 11:1307-1340

Laul JC, Papike JJ (1980b) The Apollo 17 drill core: Chemistry of size fractions and the nature of the fused soil component. Proc Lunar Planet Sci Conf 11:1395-1413

Laul JC, Wakita H, Showalter DL, Boynton WV, Schmitt RA (1972a) Bulk, rare earth, and other trace elements in Apollo 14 and 15 and Luna 16 samples. Proc Lunar Sci Conf 3:1181-1201

Laul JC, Wakita H, Schmitt RA (1972b) Bulk and REE abundances in anorthosites and noritic fragments (abstr). *In:* JW Chamberlain, C Watkins (eds) The Apollo 15 Lunar Samples, p 221-224, Lunar Science Institute, Houston

Laul JC, Hill DW, Schmitt RA (1974) Chemical studies of Apollo 16 and 17 samples. Proc Lunar Sci Conf 5:1047-1066

Laul JC, Vaniman DT, Papike JJ, Simon SB (1978a) Chemistry and petrology of size fractions of Apollo 17 deep drill core 70009 - 70006. Proc Lunar Planet Sci Conf 9:2065-2097

Laul JC, Vaniman DT, Papike JJ (1978b) Chemistry, mineralogy, and petrology of seven > 1mm fragments from Mare Crisium. *In:* R. Merrill, JJ Papike (eds) Mare Crisium: The View from Luna 24, p 537-568. Pergamon, New York

Laul JC, Papike JJ, Simon SB (1981) The lunar regolith: Comparative studies of the Apollo and Luna sites. Chemistry of soils from Apollo 17, Luna 16, 20, and 24. Proc Lunar Planet Sci Conf 12B:389-407

Lee D-C, Halliday AN, Snyder GA, Taylor LA (1997) Age and origin of the Moon. Science 278:1098-1103

Leich DA, Kahl SB, Kirschbaum AR, Niemeyer S, Phinney D (1975) Rare gas constraints on the history of Boulder 1, Station 2, Apollo 17. The Moon 14:407-444

Lindsley DH (1967) The join hedenbergite-ferrosilite at high pressures and temperatures. Carnegie Inst Wash Yearbk 65:230-232

Lindsley DH, Burnham CW (1970) Pyroxferroite: Stability and x-ray crystallography of synthetic $Ca_{0.15}Fe_{0.85}SiO_3$ pyroxenoid. Science 168:364-367

Lindsley DH, Papike JJ, Bence AE (1972) Pyroxferroite: Breakdown at low pressure and high temperature (abstr). *In:* Lunar Science III:483-485, Lunar Science Institute, Houston

Lindstrom MM, Haskin LA (1978) Causes of compositional variation within mare basalt suites. Proc Lunar Planet Sci Conf 9:465-486

Lindstrom MM, Haskin LA (1981) Compositional inhomogeneities in a single Icelandic theleiite flow. Geochim Cosmochim Acta 45:15-31

Lindstrom MM, Lindstrom D (1986) Lunar granulites and their precursor anorthositic norites of the early lunar crust. Proc Lunar Planet Sci Conf 16th in J Geophys Res 91:D263-D276

Lindstrom MM, Salpas PA (1981) Geochemical studies of rocks from North Ray Crater, Apollo 16. Proc Lunar Planet Sci Conf 12B:305-322

Lindstrom MM, Nava DF, Lindstrom DJ, Winzer SR, Lum RKL, Schuhmann PJ, Schuhmann S, Philpotts JA (1977) Geochemical studies of the white breccia boulders at North Ray Crater, Descartes region of the lunar highlands. Proc Lunar Sci Conf 8:2137-2151

Lindstrom MM, Knapp SA, Shervais JW, Taylor LA (1984) Magnesian anorthosites and associated troctolites and dunite in Apollo 14 breccias. Proc Lunar Planet Sci Conf 15th in J Geophys Res 89:C41-C49

Lindstrom MM, Marvin UB, Vetter SK, Shervais JW (1988) Apennine Front revisited: Diversity of Apollo 15 highland rock types. Proc. Lunar Planet Sci Conf 18:169-185

Lofgren GE (1974) An experimental study of plagioclase crystal morphology: Isothermal crystallization. Amer J Sci 274:243-273

Lofgren GE (1977) Dynamic crystallization experiments bearing on the origin of textures in impact-generated liquids. Proc Lunar Sci Conf 8:2079-2095

Lofgren GE, Lofgren EM (1981) Catalog of lunar mare basalts greater than 40 grams. Part 1. Major and trace elements. LPI Contribution 438, Lunar and Planetary Institute, Houston

Lofgren G, Donaldson CH, Williams RJ, Mullins O (1974) Experimentally reproduced textures and mineral chemistry of A-15 quartz basalts (abstr). *In:* Lunar Science V:458-460. Lunar Science Institute, Houston

Lofgren GE, Grove TL, Brown R, Smith DP (1979) Comparison of dynamic crystallization techniques. Proc Lunar Planet Sci Conf 10:423-438

Lofgren G, Donaldson CH, Usselman TM (1975) Geology, petrology and crystallization of Apollo 15 quartz-normative basalts. Proc Lunar Sci Conf 6:79-99

Longhi J (1978) Pyroxene stability and the composition of the lunar crust. Proc Lunar Planet Sci Conf 9:285-306

Longhi J (1987) On the connection between mare basalts and picritic volcanic glasses. Proc Lunar Planet Sci Conf 17:E349-E360

Longhi J (1990) Silicate liquid immiscibility in isothermal crystallization experiments. Proc Lunar Planet Sci Conf 20:13-24

Longhi J (1992) Experimental petrology and petrogenesis of mare volcanics. Geochimica Cosmochim Acta 56:2235-2251

Longhi J, Walker D, Hays JF (1972) Petrography and cyrstallization history of basalts 14310 and 14072. Proc Lunar Sci Conf 3:131-139

Longhi J, Walker D, Hays JF (1976) Fe, Mg, and silica in lunar plagioclase (abstr). *In:* Lunar Science VII:501-503, Lunar Science Institute, Houston.

Longhi J, Walker D, Grove TL, Stolper EM, Hays JF (1974) The petrology of Apollo 17 mare basalts. Proc Lunar Sci Conf 5:447-469

Lovering JF, Wark DA (1971) Uranium-enriched phases in Apollo 11 and 12 basaltic rocks. Proc Lunar Sci Conf 2:151-158

Lovering JF, Wark DA (1975) The lunar crust: Chemically defined rock groups and their K-U fractionation. Proc Lunar Sci Conf 6:1203-1217

Lovering JF, Wark DA, Reid AF, Ware NG, Keil K, Prinz M, Bunch TE, El Goresy A, Ramdohr P, Brown GM, Peckett A, Phillips R, Cameron EN, Douglas JAV, Plant AG (1971) Tranquillityite: A new silicate mineral from Apollo 11 and Apollo 12 basaltic rocks. Proc Lunar Sci Conf 2:39-45

LSPET (Lunar Sample Preliminary Examination Team) (1970) Preliminary examination of lunar samples from Apollo 12. Science 167:1325-1339

LSPET (Lunar Sample Preliminary Examination Team) (1972) The Apollo 15 lunar samples: A preliminary description. Science 175:363-375

LSPET (Lunar Sample Preliminary Examination Team) (1973a) The Apollo 16 lunar samples: Petrographic and chemical description. Science 179:23-34

LSPET (Lunar Sample Preliminary Examination Team) (1973b) Apollo 17 lunar samples: Chemical and petrographic description. Science 182:659-672

Lucey P G, Taylor GJ, Maleret E (1995) Abundance and distribution of iron on the Moon Science 268:1150-1153

Lucey P G, Taylor GJ, Maleret E (1995) Global abundance of FeO on the Moon: Improved estimates from multispectral imaging and comparisons with the lunar meteorites (abstr). *In:* Lunar Planet Sci XXVIII:849-850, Lunar and Planetary Institute, Houston

Lugmair GW and Carlson RW (1978) The Sm-Nd history of KREEP. Proc Lunar Planet Sci Conf 9:689-704

Lugmair GW and Marti K (1978) Lunar initial Nd^{143}/Nd^{144}: Differential evolution of the lunar crust and mantle. Earth Planet Sci Lett 39:349-357

Lugmair GW, Marti K, Kurtz JP, Scheinin NB (1976) History and genesis of lunar troctolite 76535. Proc Lunar Sci Conf 7:2009-2033

Ma M-S, Schmitt RA, Taylor GJ, Warner RD, Lange DE, Keil K (1978) Chemistry and petrology of Luna 24 lithic fragments and <250 (m soils: Constraints on the origin of VLT mare basalts. *In:* Merrill RB, Papike JJ (eds) Mare Crisium: The View from Luna 24, p 569-592. Pergamon, New York

Ma M-S, Lui Y-G, Schmitt RA (1981a) A chemical study of individual green glasses and brown glasses from 15426: Implications for their petrogenesis. Proc Lunar Planet Sci Conf 12B:915-933

Ma M-S, Schmitt RA, Taylor GJ, Warner RD, Keil K (1981b) Chemical and petrographic study of spinel troctolite in 67435: Implications for the origin of Mg-rich plutonic rocks (abstr). *In:* Lunar Planet Sci XII:640-642, Lunar and Planetary Institute, Houston

Marti K, Aeschlimann U, Eberhardt P, Geiss J, Grogler N, Jost DT, Laul JC, Ma M-S, Schmitt RA, Taylor GJ (1983) Pieces of the ancient lunar crust: Ages and composition of clasts in consortium breccia 67915. Proc Lunar Planet Sci Conf 14th in J Geophys Res 88:B165-B175
Marvin UB (1971) Lunar niobian rutile. Earth Planet Sci Lett 11:7-9
Marvin UB, Warren PH (1980) A pristine eucrite-like gabbro from Descartes and its exotic kindred. Proc Lunar Planet Sci Conf 11:507-521
Marvin UB, Lindstrom MM, Bernatowicz TJ, Podosek FA, Sugiura N (1987) The composition and history of breccia 67015 from North Ray crater. Proc Lunar Planet Sci Conf 17th in J Geophys Res 92:E472-E490
Marvin UB, Carey JW, Lindstrom MM (1989) Cordierite-spinel troctolite, a new magnesium-rich lithology from the lunar highlands. Science 243:925-928
Marvin UB, Lindstrom MM Holmberg BB, Martinez RR (1991) New observations on the quartz monzodiorite-granite suite. Proc Lunar Planet Sci21:119-135
Maurer PPE, Geiss J, Grögler N, Stettler A, Brown GM, Peckett A, Krähenbühl U (1978) Pre-Imbrian craters and basins: Ages, compositions, and excavation depths of Apollo 16 breccias. Geochim Cosmochim Acta 42:1687-1720
McCallister RH, Taylor LA (1973) Kinetics of ulvöspinel reduction—Synthetic study and applications to lunar rocks. Earth Planet Sci Lett 17:357-364
McCallum IS, O'Brien HE (1996) Stratigraphy of the lunar highland crust: Depths of burial of lunar samples from cooling rate studies. Am Mineral 81:1166-1175
McCallum IS, Mathez EA, Okamura FP, Ghose S (1974) Petrology and crystal chemistry of poikilitic anorthositic gabbro 77017. Proc Lunar Sci Conf 5:287-302
McCallum IS, Okamura FP, Ghose S (1975) Mineralogy and petrology of sample 67075 and the origin of lunar anorthosites. Earth Planet Sci Lett 26:36-53
McCord TB, Pieters C, Feierberg MA (1976) Multispectral mapping of the lunar surface using ground-based telescopes. Icarus 29:1-34
McCord TB, Grabow M, Feierberg MA, MacLaskey D, Pieters C (1979) Lunar multispectral maps: Part II of the lunar nearside. Icarus 37:1-28
McGee JJ (1993) Lunar ferroan anorthosites: Mineralogy, compositional variations, and petrogenesis. J Geophys Res 98:9089-9105
McGee JJ (1994) Lunar ferroan anorthosite subgroups (abstr). *In:* Lunar Planet Sci XXV:875-876, Lunar and Planetary Institute, Houston
McKay DS, Greenwood WR, Morrison DA (1970) Origin of small lunar particles and breccia from the Apollo 11 site. Proc Apollo 11 Lunar Sci Conf, p 673-694
McKay DS, Heiken GH, Taylor RM, Clanton US, Morrison DA, Ladle GH (1972) Apollo 14 soils: Size distribution and particle types. Proc Lunar Sci Conf 3:983-995
McKay DS, Clanton US, Ladle G (1973) Scanning electron microscope study of Apollo 15 green glass. Proc Lunar Sci Conf 4:225-238
McKay DS, Fruland RM, Heiken GH (1974) Grain size and evolution of lunar soils. Proc Lunar Sci Conf 5:887-906
McKay DS, Dungan MA, Morris RV, Fruland RM (1977) Grain size, petrographic, and FMR studies of double core 60009/10: A study of soil evolution. Proc Lunar Sci Conf 8:2929-2952
McKay DS, Basu A, Nace G (1980) Lunar core 15010/11: Grain size, petrology, and implications for regolith dynamics. Proc Lunar Planet Sci Conf 11:1531-1550
McKay D S, Bogard DD, Morris RV, Korotev RL, Johnson P, Wentworth SJ (1986) Apollo 16 regolith breccias: Characterization and evidence for early formation in the mega-regolith. Proc Lunar Planet Sci Conf 16th in J Geophys Res 91:D277-D303
McKay GA, Wiesmann H, Nyquist LE, Wooden JL, and Bansal B (1978) Petrology, chemistry, and chronology of 14078: Chemical constraints on the origin of KREEP. Proc Lunar Planet Sci Conf 9:661-687
McKay GA, Wiesmann H, Bansal B (1979) The KREEP-magma ocean connection (abstr). *In:* Lunar Planet Sci X:804-806, Lunar and Planetary Institute, Houston
McKinley JP, Taylor GJ, Keil K (1984) Apollo 16: impact melt sheets, contrasting nature of the Cayley Plains and Descartes Mountains, and geologic history. Proc Lunar Planet Sci Conf 14th in J Geophys Res 89:B513-B524
McKay DS, Williams RJ (1979) A geologic assessment of potential lunar ores. *In:* J Billingham, W Gilbreath, BO'Leary (eds) Space Resources and Space Settlements., NASA SP-428:243-250, NASA, Washington, DC
Mehta S, Goldstein JI (1979) Analytical electron microscopy study of submicroscopic metal particles in glassy constituents of lunar breccias 15015 and 60095. Proc Lunar Planet Sci Conf 10:1507-1521

Mendell WW (ed) (1985) Lunar Bases and Space Activities of the 21st Century. Lunar and Planetary Institute, Houston

Metzger AE, Parker RE (1979) The distribution of titanium on the lunar surface. Earth Planet Sci Lett 45:155-171

Metzger AE, Trombka JI, Peterson LE, Reedy DC, Arnold JR (1973) Lunar surface radioactivity: Preliminary results of the Apollo 15 and Apollo 16 gamma-ray spectrometer experiements. Science 179:800-803

Meyer C Jr (1977) Petrology, mineralogy, and chemistry of KREEP basalt. Phys Chem Earth 10:239-260

Meyer C Jr (1978) Ion microprobe analyses of aluminous lunar glasses: A test of the "rock type" hypothesis. Proc Lunar Planet Sci Conf 9:1551-1570

Meyer C Jr, Hubbard NJ (1970) High potassium, high phosphorous glass as an important rock type in the Apollo 12 soil samples. Meteoritics 5:210-211

Meyer C Jr, McKay DS, Anderson DH, Butler P Jr (1975) The source of sublimates on the Apollo 15 green and Apollo 17 orange glass sample. Proc Lunar Sci Conf 6:1673-1699

Meyer C Jr, Williams IS, Compston W (1989a) Zircon-containing rock fragments within Apollo 14 breccia indicate serial magmatism from 4350 to 4000 million years (abstr). *In:* Workshop on Moon in Transition: Apollo 14, KREEP, and Evolved Lunar Rocks, LPI Tech Rept 89-03, p 75-78, Lunar and Planetary Institute, Houston

Meyer C Jr, Williams IS, Compston W (1989b) 207Pb/206Pb ages of zircon-containing rock fragments indicate continuous magmatism in the lunar crust from 4350 to 3900 million years (abstr). *In:* Lunar Planet Sci XX:691-692, Lunar and Planetary Institute, Houston

Meyer C, Williams IS, Compston W (1996) Uranium-lead ages for lunar zircons: Evidence for prolonged period of granophyre formation from 4.32 to 3.88 Ga. Met Planet Sci 31:370-387

Meyer HOA, Boctor NZ (1974) Opaque mineralogy: Apollo 17, rock 75035. Proc Lunar Sci Conf 5:707-716

Meyer HOA, McCallister RH (1977) The Apollo 16 deep drill core. Proc Lunar Sci Conf 8:2889-2907

Minkin JA, Thompson CL, Chao ECT (1978) The Apollo 17 Station 7 boulder: Summary of study by the International Consortium. Proc Lunar Planet Sci Conf 9:877-903

Misra KC, Taylor LA (1975a) Characteristics of metal particles in Apollo 16 rocks. Proc Lunar Sci Conf 6:615-639

Misra KC, Taylor LA (1975b) Correlation between native metal compositions and the petrology of Apollo 16 rocks (abstr). *In:* Lunar Science VI:566-568, Lunar Science Institute, Houston

Morgan JW, Wandless GA (1984) Surface-correlated trace elements in 15426 lunar glass (abstr). *In:* Lunar Planet Sci XV:562-563, Lunar and Planetary Institute, Houston

Morgan JW, Wandless GA (1988) Lunar dunite 72415-72417: Siderophile and volatile trace elements (abstr). *In:* Lunar Planet Sci XIX:804-805, Lunar and Planetary Institute, Houston

Morris RV (1976) Surface exposure indices of lunar rocks: A comparative FMR study. Proc Lunar Sci Conf 7:315-335

Morris RV (1977) Origin and evolution of the grain-size dependence of the concentration of fine-grained metal in lunar soils: The maturation of lunar soils to a steady-state stage. Proc Lunar Sci Conf 8:3719-3747

Morris RV (1980) Origins and size distribution of metallic iron particles in the lunar regolith. Proc Lunar Planet Sci Conf 11:1697-1712

Morris RV (1985) Determination of optical penetration depths from reflectance and transmittance measurements on albite powders (abstr). *In:* Lunar Planet Sci XVI:581-582, Lunar and Planetary Institute, Houston.

Morris RV, See TH, Hörz F (1986) Composition of the Cayely Formation at Apollo 16 as inferred from impact melt splashes. Proc Lunar Planet Sci Conf 17th in J Geophys Res 91:E21-E42

Morris RV, Lauer HV Jr, Gose WA (1979) Characterization and depositional and evolutionary history of the Apollo 17 deep drill core. Proc Lunar Planet Sci Conf 10:1141-1157.

Morris RW, Taylor GJ, Newsom HE, Keil K, Garcia SR (1990) Highly evolved and ultramafic lithologies from Apollo 14 soils. Proc Lunar Planet Sci Conf 20:61-75

Morse SA (1982) Adcumulus growth of anorthosite at the base of the lunar crust. Proc Lunar Planet Sci Conf 13th in J Geophys Res 87:A10-A18

Murali AV, Ma M-S, Laul JC, Schmitt RA (1977) Chemical composition of breccias, feldspathic basalt, and anorthosites from Apollo 15 (15308, 15359, 15382, and 15362), Apollo 16 (60018 and 65785), Apollo 17 (72435, 72536, 72559, 72735, 72738, 78526, and 78527) and Luna 20 (22012 and 22013) (abstr). *In:* Lunar Sci VIII:700-702, Lunar Science Institute, Houston

Nakamura N (1974) Determination of REE, Ba, Fe, Mg, Na, and K in carbonaceous and ordinary chondrites. Geochim Cosmochim Acta 38:757-775

Nakamura N, Tatsumoto M, Nunes P, Unruh DM, Schwab AP, Wildeman TR (1976) 4.4 b.y.-old clast in Boulder 7, Apollo 17: A comprehensive chronological study by U-Pb, Rb-Sr, and Sm-Nd methods. Proc Lunar Sci Conf 7:2309-2333

Naney MT, Crowl DM, Papike JJ (1976) The Apollo 16 drill core: Statistical analysis of glass chemistry and the characterization of a high alumina-silica poor (HASP) glass. Proc Lunar Sci Conf 7:155-184

Nava DF (1974) Chemical composition of some soils and rock types from the Apollo 15, 16, and 17 lunar sites. Proc Lunar Sci Conf 5:1087-1096

Neal CR, Taylor LA (1992) Petrogenesis of mare basalts: A record of lunar volcanism. Geochim Cosmochim Acta 56:2177-2211

Neal CR, Taylor LA, Lindstrom MM (1988) Apollo 14 mare basalt petrogenesis: Assimilation of KREEP-like components by a fractionating magma. Proc Lunar Planet Sci Conf 18:139-153

Neal CR, Taylor LA, Patchet AD (1989a) High alumina (HA) and very high potassium (VHK) basalt clasts from Apollo 14 breccias, Part 1-Mineralogy and petrology: Evidence of crystallization from evolving magmas. Proc Lunar Planet Sci Conf 19:137-145

Neal CR, Taylor LA, Schmitt RA, Hughes SS, Lindstrom MM (1989b) High alumina (HA) and very high potassium (VHK) basalt clasts from Apollo 14 breccias, Part 2. Whole rock geochemistry: Further evidence of combined assimilation and fractional crystallization within the lunar crust. Proc Lunar Planet Sci Conf 19:147-161

Neal C R, Hacker MD, Taylor LA, Schmitt RA, Lui Y-G (1992) The petrogenesis of Apollo 12 mare basalts, part 1: The "lumpers" versus the "splitters" (abstr). *In:* Lunar Planet Sci XXIII:975-976, Lunar and Planetary Institute, Houston

Nehru CE, Prinz M, Dowty E, Keil K (1974) Spinel-group minerals and ilmenite in Apollo 15 rock samples. Am Mineral 59:1220-1234

Nehru CE, Warner RD, Keil K (1976) Electron Microprobe Analyses of Opaque Mineral Phases from Apollo 11 Basalts. Univ of New Mexico Spec Publ 17, Albuquerque

Nehru CE, Warner RD, Keil K, Taylor GJ (1978) Metamorphism of brecciated ANT rocks: Anorthositic troctolite 72559 and norite 78527. Proc Lunar Planet Sci Conf 9:773-788

Neumann GA, Zuber T, Smith DE, Lemoine FG (1996) The lunar crust: Global structure and signature of major basins. J Geophys Res Planets:16841-16843

Newsom HE (1986) Constraints on the origin of the moon from the abundance of molybdenum and other siderophile elements. *In:* WK Hartmann, RJ. Phillips, GJ Taylor (eds) Origin of the Moon, p 203-230, Lunar and Planetary Institute, Houston

Newton RC, Anderson AT, Smith JV (1971) Accumulation of olivine in rock 12040 and other basaltic fragments in the light of analysis and syntheses. Proc Lunar Sci Conf 2:575-582

Nord Gl Jr, Wandless M-V (1983) Petrology and comparative thermal and mechanical histories of clasts in breccia 62236. Proc Lunar Planet Sci Conf 13th in J Geophys Res 88:A645-A657

Nord GL, Christie JM, Heuer AH, Lally JS (1975) North Ray Crater breccias: An electron petrographic study. Proc Lunar Sci Conf 6:779-797

Norman MD (1981) Petrology of suevitic lunar breccia 67016. Proc Lunar Planet Sci Conf 12B:235-252

Norman MD, Ryder G (1979) A summary of the petrology and geochemistry of pristine highlands rocks. Proc Lunar Planet Sci Conf 10:531-559

Norman MD, Ryder G (1980) Geochemical evidence for the role of ilmenite and clinopyroxene in the early lunar differentiation (abstr). *In:* Lunar Planet Sci XI:821-823, Lunar and Planetary Institute, Houston

Norman MD, Ryder G (1980) Geochemical constraints on the igneous evolution of the lunar crust. Proc Lunar Planet Sci Conf 11:317-331

Norman MD, Taylor SR (1992) Geochemistry of lunar crustal rocks from breccia 67016 and the composition of the Moon. Geochim Cosmochim Acta 56:1013-1-24

Norman MD, Coish RA, Taylor LA (1978) Glasses in the Luna 24 core and petrogenesis of ferrobasalts. *In:* RB Merrill, JJ Papike (eds) Mare Crisium: The View from Luna 24, p 281-290, Pergamon, New York

Norman MD, Keil K, Griffin WL, Ryan CG (1995) Fragments of the ancient lunar crust: Petrology and geochemistry of ferroan noritic anorthosites from the Descartes region of the Moon. Geochim Cosmochim Acta 59:831-847

Nyquist LE, Bansal BM, Wiesmann H (1975) Rb-Sr ages and initial $^{87}Sr/^{86}Sr$ for Apollo 17 basalts and KREEP basalt 15386. Proc Lunar Sci Conf 6:1445-1465

Nyquist LE (1977) Lunar Rb-Sr chronology. Phys Chem Earth 10:103-142

Nyquist LE, Shih C-Y (1992) The isotopic record of lunar volcanism. Geochim Cosmochim Acta 56:2213-2234

Nyquist LE, Hubbard NJ, Gast PW, Bansal BM, Wiesmann H, Jahn B (1973) Rb-Sr systematics for chemically-defined Apollo 15 and 16 materials. Proc Lunar Sci Conf 4:1823-1846

Nyquist LE, Weismann H, Shih C-Y, Bansal BM (1976) 15405 quartz-monzodiorite: SuperKREEP. *In:* Interdisciplinary Studies by the Imbrium Consortium, Vol 2, p 60, Smithsonian Astrophysical Observatory, Cambridge, MA

Nyquist LE, Smith C-Y, Wooden JL, Bansal BM, Wiesmann H (1979) The Sr and Nd isotopic record of Apollo 12 basalts: Implications for lunar geochemical evolution. Proc Lunar Planet Sci Conf 10:77-114

Nyquist LE, Reimold WU, Bogard DD, Wooden JL, Bansal BM, Wiesmann H, Shih C-Y (1981) A comparative Rb-Sr, Sm-Nd, and K-Ar study of shocked norite 78236: Evidence of slow cooling in the lunar crust. Proc Lunar Planet Sci Conf 12B:167-197

Oberbeck VR, Quaide WL (1968) Genetic implications of lunar regolith thickness variations. Icarus 9:446-465

Oberli F, Huneke JC, Wasserburg GJ (1979) U-Pb and K-Ar systematics of cataclysm and precataclysm lunar impactites (abstr). *In:* Lunar Planet Sci X:940-942, Lunar and Planetary Institute, Houston

O'Hara MJ, Biggar GM, Richardson SW, Ford CE, Jamieson BG (1970) The nature of seas, mascons, and the lunar interior in the light of experimental studies. Proc. Apollo 11 Lunar Sci Conf, p 695-710.

Onorato PIK, Uhlmann DR, Simonds CH (1976) Heat flow in impact melts: Apollo 17 Station 6 boulder and some applications to other breccias and xenolith-laden melts. Proc Lunar Sci Conf 7:2449-2467

Onorato PIK, Uhlmann DR, Taylor LA, Coish RA Gamble RP (1978) Olivine cooling speedometers. Proc Lunar Planet Sci Conf 9:613-628

Paces JB, Nakai S, Neal CR, Taylor LA, Halliday AN, Lee D-C (1991) A Sr and Nd isotopic study of Apollo 17 high-Ti mare basalts: Resolution of ages, evolution of magmas, and origins of source heterogeneities. Geochim Cosmochim Acta 55:2025-2043

Palme H, Baddenhausen H, Blum K, Cendales M, Dreibus G, Hofmeister H, Kruse H, Palme C, Spettel B, Vilcsek E, Wänke H, Kurat G (1978) New data on lunar samples and achondrites and a comparison of the least fractionated samples from the Earth, the Moon, and the eucrite parent body. Proc Lunar Planet Sci Conf 9:25-57

Palme H, Spettel B, Jochum KP, Dreibus G, Weber H, Weckwerth G, Wänke H, Bischoff A, Stöffler D (1991) Lunar highland meteorites and the composition of the lunar crust. Geochim Cosmochim Acta 55:3105-3122

Papanastassiou DA, Wasserburg GJ (1971) Rb-Sr ages of igneous rocks from the Apollo 14 mission and the age of the Fra Mauro formation. Earth Planet Sci Lett 12:36-48

Papanastassiou DA, Wasserburg GJ (1972a) The Rb-Sr age of a crystalline rock from Apollo 16. Earth Planet Sci Lett 16:289-298

Papanastassiou DA, Wasserburg GJ (1972b) Rb-Sr systematics of Luna 20 and Apollo 16 samples. Earth Planet Sci Lett 17:52-63

Papanastassiou DA, Wasserburg GJ (1975) Rb-Sr study of a lunar dunite and evidence for early lunar differentiates. Proc Lunar Sci Conf 6:1467-1489

Papanastassiou DA, Wasserburg GJ (1976) Rb-Sr age of troctolite 76535. Proc Lunar Sci Conf 7:2035-2054

Papanastassiou DA, DePaolo DJ, Wasserburg GJ (1977) Rb-Sr and Sm-Nd chronology and genealogy of mare basalts from the Sea of Tranquillity. Proc Lunar Sci Conf 8:1639-1672

Papike JJ (1987) Chemistry of the rock-forming silicates: Ortho, ring, and single-chain structures. Rev Geophys 25:1483-1526

Papike JJ (1988) Chemistry of the rock-forming silicates: Multiple-chain, sheet, and framework silicates. Rev. Geophys 26:407-444

Papike JJ (1996) Pyroxene as a recorder of cumulate formational processes in asteroids, Moon, Mars, Earth: Reading the record with the ion microprobe. Am Mineral 81:525-544

Papike JJ, Cameron M (1976) Crystal chemistry of silicate minerals of geophysical interest. Rev Geophys Space Phys 14:37-80

Papike JJ, Bence AE, Brown GE, Prewitt CT, Wu CH (1971) Apollo 12 clinopyroxenes: Exsolution and epitaxy. Earth Planet Sci Lett 10:307-315

Papike JJ, Bence AE, Lindsley DH (1974) Mare basalts from the Taurus-Littrow region of the Moon. Proc Lunar Sci Conf 5:471-504

Papike JJ, Hodges FN, Bence AE, Cameron M, Rhodes JM (1976) Mare basalts: Crystal chemistry, mineralogy, and petrology. Rev Geophys Space Phys 14:475-540

Papike JJ, Vaniman DT (1978) Luna 24 ferrobasalts and the mare basalt suite: Comparative chemistry, mineralogy, and petrology. *In:* RB Merrill, JJ Papike (eds) Mare Crisium: The View from Luna 24. p 371-401, Pergamon, New York

Papike JJ, Simon SB, White C, Laul JC (1981) The relationship of the lunar regolith <10 mm fraction and agglutinates. Part I: A model for agglutinate formation and some indirect supportive evidence. Proc Lunar Planet Sci12B:409-420

Papike JJ, Simon SB, Laul JC (1982) The lunar regolith: Chemistry, mineralogy, and petrology. Rev Geophys Space Phys 20:761-826

Papike JJ, Fowler GW, Shearer CK (1994) Orthopyroxene as a recorder of lunar crust evolution: An ion microprobe investigation of Mg-suite norites. Am Mineral 79:796-800

Papike JJ, Fowler GW, Layne GD, Shearer CK (1996) Ion microprobe investigation of plagioclase and orthopyroxene from lunar Mg-suite norites: Implications for calculating parental melt REE concentrations and for assessing postcrystallization REE redistribution. Geochim Cosmochim Acta 60:3967-3978

Papike JJ, Spilde MN, Adcock CT, Fowler GW, and Shearer CK (1997a) Trace element fractionation by impact-induced volatilization: SIMS study of lunar HASP Samples. Am Mineral 82:630-634

Papike JJ, Fowler GW, Shearer CK (1997b) Evolution of the lunar crust: SIMS study of plagioclase from ferroan anorthosites. Geochim Cosmochim Acta 61:2343-2350

Peeples WJ, Sill, WR, May TW, Ward SH, Phillips RJ, Jordan RL, Abbott EA, Killpack TJ (1978) Orbital radar evidence for lunar subsurface layering in Maria Serenitatis and Crisium. J Geophys Res 83:3459-3468

Peterson CA, Hawke BR, Lucey PG, Taylor GJ, Blewett DT, Spudis PD (1997) Spacecraft and groundbased identification of lunar anorthosite (abstr). Met and Planet Sci 32:A106

Philpotts JA, Schumann S, Kouns CW, Lum RKL, Bickel AL, Schnetzler CC (1973) Apollo 16 returned samples: lithophile trace-element abundances. Proc Lunar Sci Conf 4:1427-1436

Phinney D, Kahl SB, Reynolds JH (1975) ^{40}Ar-^{39}Ar dating of Apollo 16 and 17 rocks. Proc Lunar Sci Conf 6:1593-1608

Phinney WC (1981) Guidebook for the Boulders at Station 6, Apollo 17. Curatorial Branch Publication 55, Johnson Space Center, Houston

Phinney WC (1991) Lunar anorthosites, their equilibrium melts, and the bulk Moon. Proc Lunar Planet Sci 21:29-49

Pieters CM (1978) Mare basalt types of the front side of the Moon: A summary of spectral reflectance data. Proc Lunar Planet Sci Conf 9:2825-2849

Pieters CM (1983) Strength of mineral absorption features in the transmitted component of near-infrared reflected light: First results from RELAB. J Geophys Res 88:9534-9544

Pieters CM (1990) The probable continuum between emplacement of plutons and mare volcanism in lunar crustal evolution (abstr). *In:* Workshop on Mare Volcanism and Basalt Petrogenesis: Astounding Fundamental Concepts (AFC) Developed Over the Last Fifteen Years, LPI Tech Rept 91-03, p 35-36, Lunar and Planetary Institute, Houston

Pieters C, McCord TB (1975) Classification and distribution of lunar mare basalt types. *In:* Papers Presented to the Conference on Origins of Mare Basalts and Their Implications for Lunar Evolution, p 125-129, Lunar Science Institute, Houston

Pieters CM, Head JW, Adams JB, McCord TB, Zisk SH, Whitford-Stark JL (1980) Late high titanium basalts of the western maria: Geology of the Flamstead region of Oceanus Procellarum. J Geophys Res 85:3513-3938

Powell BN, Aitken FK, Weiblen PW (1973) Classification, distribution, and origin of lithic fragments from the Hadley-Apennine region. Proc Lunar Sci Conf 4:445-460

Premo WR, Tatsumoto M (1992) U-Th-Pb isotopic systematics of lunar norite 78235. Proc Lunar Planet Sci 21:89-100

Premo WR, Tatsumoto M (1992) U-Th-Pb, Rb-Sr, and Sm-Nd isotopic systematics of lunar troctolitic cumulate 76535: Implications on the age and origin of this early lunar, deep-seated cumulate. Proc Lunar Planet Sci 22:381-397

Premo WR, Tatsumoto M, Wang J-W (1989) Pb isotopes in anorthositic breccias 67075 and 62237: A search for primitive lunar lead. Proc Lunar Planet Sci Conf 19:61-71

Prinz M, Dowty E, Keil K, Bunch TE (1973) Spinel troctolite and anorthosite in Apollo 16 samples. Science 179:74-76

Quaide W, Bunch T (1970) Impact metamorphism of lunar surface materials. Proc Apollo 11 Lunar Sci Conf, p 711-729

Quick JE, Albee AL, Ma M-S, Murali AV, and Schmitt RA (1977) Chemical compositions and possible immiscibility of two silicate melts in 12013. Proc Lunar Sci Conf 8:2153-2189

Quick JE, James OB, Albee AL (1981a) Petrology and petrogenesis of lunar breccia 12013. Proc Lunar Planet Sci Conf 12B:117-172

Quick JE, James OB, Albee AL (1981b) A reexamination of the Rb-Sr isotopic systematics of lunar breccia 12013. Proc Lunar Planet Sci Conf 12B:173-184

Ramdohr P (1972) Lunar pentlandite and sulfidization reactions in microbreccia 14315,9. Earth Planet Sci Lett 15:113-115

Ramdohr P, El Goresy A (1970) Opaque minerals of the lunar rocks and dust from Mare Tranquillitatis. Science 167:615-618

Reid AM, Meyer C, Harmon RS, Brett R (1970) Metal grains in Apollo 12 igneous rocks. Earth Planet Sci Lett 9:1-5

Reid AM, Lofgren GE, Heiken GH, Brown RW, Moreland G (1973) Apollo 17 orange glass, Apollo 15 green glass and Hawaiian lava fountain glass. EOS Trans AGU 54:606-607

Reimold WU, Nyquist LE, Bansal BM, Wooden JL, Shih C-Y, Weismann H, McKinnon IDR (1985) Isotope analysis of crystalline impact melt rocks from Apollo 16 Stations 11 and 13, North Ray Crater. Proc Lunar Planet Sci Conf 15th in J Geophys Res 90:C431-C448

Rhodes JM (1973) Major and trace element analyses of Apollo 17 samples (abstr). EOS Trans AGU 54:611-612

Rhodes JM, Hubbard NJ (1973) Chemistry, classification, and petrogenesis of Apollo 15 mare basalts. Proc Lunar Sci Conf 4:1127-1148

Rhodes JM, Blanchard DP (1980) Chemistry of Apollo 11 low-K mare basalts. Proc Lunar Planet Sci Conf 11:49-66

Rhodes JM, Rodgers KV, Shih C-Y, Bansal BM, Nyquist LE, Wiesmann H, Hubbard NJ (1974) The relationship between geology and soil chemistry at the Apollo 17 landing site. Proc Lunar Sci Conf 5:1097-1117

Rhodes JM, Hubbard NJ, Wiesmann H, Rodgers KV, Bannon JC, Bansal BM (1976) Chemistry, classification, and petrogenesis of Apollo 17 mare basalts. Proc Lunar Sci Conf 7:1467-1489

Ridley WI (1975) On high-alumina mare basalts. Proc Lunar Sci Conf 6:131-145

Ridley WI, Brett R (1973) Petrogenesis of basalt 70035: A multistage cooling history (abstr). EOS Trans AGU 54:611-612

Ridley W, Reid AM, Warner JL, Brown RW (1973a) Apollo 15 green glasses. Phys Earth Planet Int 7:133-136

Ridley WI, Hubbard NJ, Rhodes JM, Wiesmann H, and Bansal B (1973b) The petrology of lunar breccia 15445 and petrogenetic implications. J Geol 81:621-631

Ringwood AE, Essene E (1970) Petrogenesis of Apollo 11 basalts, internal constitution, and origin of the Moon. Proc Apollo 11 Lunar Sci Conf, p 769-799

Ringwood AE, Kesson SE (1976) A dynamic model for mare basalt petrogenesis. Proc Lunar Sci Conf 7:1697-1722

Ringwood AE, Seifert S (1986) Nickel-cobalt abundance systematics and their bearing on lunar origin. *In:* WK Hartmann, RJ Phillips, and GJ Taylor (eds) Origin of the Moon, p 249-278, Lunar and Planetary Institute, Houston

Ringwood AE, Seifert S, Wanke H (1986) A komatiite component in Apollo 16 highland breccias: Implications for the nickel-cobalt systematics and bulk composition of the Moon. Earth Planet Sci Lett 81:105-117

Roedder E (1984) Fluid Inclusions. Rev Mineral 12, 644 p

Roedder E, Weiblen PW (1970) Lunar petrology of silicate melt inclusions. Apollo 11 rocks. Proc Apollo 11 Lunar Sci Conf, p 801-837

Roedder E, Weiblen PW (1971) Petrology of silicate melt inclusions, Apollo 11 and Apollo 12 and terrestrial equivalents. Proc Lunar Sci Conf 2:507-528

Rose HJ Jr, Cuttitta F, Annell CS, Carron MK, Christian RP, Dwornik EJ, Greenland L, Dignon GT Jr (1972) Compositional data for twenty-one Fra Mauro lunar materials. Proc Lunar Sci Conf 3:1215-1229

Rose HJ Jr, Cuttitta F, Berman S, Carron MK, Christian RP, Dwornik EJ, Greenland L, Dignon GT Jr (1973) Compositional data for twenty two Apollo 16 samples. Proc Lunar Sci Conf 4:1149-1158

Rose HJ Jr, Baedecker PA, Berman S, Christian RP, Dwornik EJ, Finkelman RB, Schnepfe MM (1975) Chemical composition of rocks and soils returned by the Apollo 15, 16, and 17 missions. Proc Lunar Sci Conf 6:1363-1373

Ross M, Bence AE, Dwornik EJ., Clark JR, Papike JJ (1970) Mineralogy of lunar clinopyroxenes, augite and pigeonite. Proc Apollo 11 Lunar Sci Conf, p 839-848

Ross M, Huebner JS, Dowty E (1973) Delineation of the one atmosphere augite-pigeonite miscibility gap for pyroxenes from lunar basalt 12021. Am Mineral 58:619-635

Rutherford MJ, Vogel RA (1975) A15 Green glass volatiles and oxidation state (abstr). *In:* Lunar Planet Sci XXVI:1205-1206, Lunar and Planetary Institute, Houston

Rutherford MJ, Hess PC, Daniel GH (1974) Experimental liquid line of descent and liquid immiscibility for basalt 70017. Proc Lunar Sci Conf 5:569-583

Rutherford MJ, Dixon S, Hess P (1980) Ilmenite saturation at high pressure in KREEP basalts: Origin of KREEP and Hi-TiO_2 mare basalts (abstr). *In:* Lunar Planet Sci XI:966-967, Lunar and Planetary Institute, Houston

Ryan JG, Langmuir CH (1987) The systematics of lithium abundances in young volcanic rocks. Geochim Cosmochim Acta 51:1727-1741

Ryan JG, Langmuir CH (1988) Beryllium systematics in young volcanic rocks: Implications for 10Be*. Geochim Cosmochim Acta 52:237-244

Ryder G (1976) Lunar sample 15405: Remnant of a KREEP basalt-granite differentiated pluton. Earth Planet Sci Lett 29:255-268

Ryder G (1979) The chemical components of highlands breccias. Proc Lunar Planet Sci Conf 10:561-581

Ryder G (1982) Lunar anorthosite 60025, the petrogenesis of lunar anorthosites, and the bulk composition of the Moon. Geochim Cosmochim Acta 46:1591-1601

Ryder G (1983) Nickel in olivines and parent magmas of lunar pristine rocks (abstr). *In:* Workshop on Pristine Highlands Rocks and the Early History of the Moon, LPI Tech Rept 83-02, p 66-68, Lunar and Planetary Institute, Houston

Ryder G (1985) Catalog of Apollo 15 Rocks. Curatorial Publication 20787, 3 vols, NASA Johnson Space Center, Houston

Ryder G (1987) Petrographic evidence for nonlinear cooling rates and a volcanic origin for Apollo 15 KREEP basalts. Proc Lunar Planet Sci Conf 17th in J Geophys Res 92:E331-E339

Ryder G (1989) Petrogenesis of Apollo 15 KREEP basalts (abstr). *In:* Lunar Planet Sci XX:936-937, Lunar and Planetary Institute, Houston

Ryder G (1991) Lunar ferroan anorthosites and mare basalt sources: The mixed connection. Geophys Res Lett 18:2065-2068

Ryder G (1992) Chemical variation and zoning of olivine in lunar dunite 72415: Near-surface accumulation. Proc Lunar Planet Sci 22:373-380

Ryder G (1994) Coincidence in time of the Imbrium basin impact and Apollo 15 KREEP volcanic flows: The case for impact-induced melting. *In:* BO Dressler, RAF Grieve, VL Sharpton (eds) Large Meteorite Impacts and Planetary Evolution, Geol Soc Am Spec Paper 293:11-18

Ryder G, Bower JF (1976) Poikilitic KREEP impact melts in the Apollo 14 white rocks. Proc Lunar Sci Conf 7:1925-1948

Ryder G, Bower JF (1977) Petrology of Apollo 15 black-and-white rocks 15445 and 15455—Fragments of the Imbrium impact melt sheet? Proc Lunar Sci Conf 8:1895-1923

Ryder G, Martinez RR (1991) Evolved hypabyssal rocks from Station 7, Apennine Front, Apollo 15. Proc Lunar Planet Sci 21:137-150

Ryder G, Marvin UB (1978) On the origin of Luna 24 basalts and soils. *In:* RB Merrill, JJ Papike (eds) Mare Crisium: The View from Luna 24, p 339-356, Pergamon, New York

Ryder G, Norman MD (1979) Catalog of pristine non-mare materials Part 1 Non-anorthosites Revised. NASA Johnson Space Center Curatorial Facility JSC 14565

Ryder G, Norman MD (1980) Catalog of Apollo 16 rocks. JSC 16904, NASA Lyndon Johnson Space Center, Houston

Ryder G, Sherman SB (1989) The Apollo 15 coarse-fines (4-10 mm). JSC # 24035, NASA Johnson Space Center, Houston

Ryder G, Spudis PD (1980) Volcanic rocks in the lunar highlands. *In:* JJ Papike, RB Merrill (eds) Proc Conf Lunar Highlands Crust, p 353-375, Pergamon, New York

Ryder G, Spudis PD (1987) Chemical composition and origin of Apollo 15 impact melts. Proc Lunar Planet Sci Conf 17th in J Geophys Res 92:E432-E446

Ryder G, Wood JA (1977) Serenitatis and Imbrium impact melts: Implications for large-scale layering in the lunar crust. Proc Lunar Planet Sci Conf 8:655-668

Ryder G, Stoeser DB, Marvin UB, Bower JF (1975a) Lunar granites with unique ternary feldspars. Proc Lunar Sci Conf 6:435-449

Ryder G, Stoeser DB, Marvin UB, Bower JF, Wood JA (1975b) Boulder 1, Station 2, Apollo17: Petrology and petrogenesis. The Moon 14:327-357

Ryder G, Stoeser DB, Wood JA (1977) Apollo 17 KREEPy basalt: A rock type intermediate between KREEP and mare basalts. Earth Planet Sci Lett 35:1-13

Ryder G, Norman MD, Taylor GJ (1997) The complex stratigraphy of the highland crust in the Serenitatis region of the Moon inferred from mineral fragment chemistry. Geochim Cosmochim Acta 61:1083-1105

Salpas PA, Taylor LA, Lindstrom MM (1987) Apollo 17 KREEPy basalts: Evidence for the non-uniformity of KREEP. Proc Lunar Planet Sci Conf 17th in J Geophys Res 92:E340-E348

Sato M (1978) Oxygen fugacity of basaltic magmas and the role of gas forming elements. Geophys Res Lett 5:447-449

Sato M, Hickling NL, McLane JE (1973) Oxygen fugacity values of Apollo 12, 14, and 15 lunar samples and reduced state of lunar magmas. Proc Lunar Sci Conf 4:1061-1079

Schaal RB, Hörz F (1977) Shock metamorphism of lunar and terrestrial basalts. Proc Lunar Sci Conf 8:1697-1729

Schaal R, Hörz F (1980) Experimental shock metamorphism of lunar soil. Proc Lunar Planet Sci Conf 11:1679-1695

Schaber GG (1973) Lava flows in Mare Imbrium: Geologic evaluation from Apollo orbital photography (abstr) *In:* Lunar Science IV:653-654, Lunar and Planetary Institute, Houston

Schaeffer OA, Husain L (1973) Early lunar history: Ages of 2 to 4 mm soil fragments from the lunar highlands. Proc Lunar Planet Sci 4:1847-1863

Schaeffer GA, Schaeffer OA (1977) ^{39}Ar-^{40}Ar ages of lunar rocks. Proc Lunar Sci Conf 8:2253-2300

Schaeffer OA, Warasila R, Labotka TC (1982) Ages of Serenitatis breccias (abstr). *In:* Lunar Planet Sci XIII:685-686, Lunar and Planetary Institute, Houston

Schultz PH, Spudis PD (1979) Evidence for ancient mare volcanism. Proc Lunar Planet Sci Conf 10:2899-2918

Schultz PH, Spudis PD (1983) The beginning and end of mare volcanism on the Moon (abstr). *In:* Lunar Planet Sci XIV:676-677, Lunar and Planetary Institute, Houston

Schuraytz BC, Ryder G (1988) A new petrochemical data base of Apollo 15 olivine-normative mare basalts (abstr). *In:* Lunar Planet Sci XIX:1041-1042, Lunar and Planetary Institute, Houston

Schuraytz BC, Ryder G (1991) The contrast of chemical modelling with petrographic reality: Tapping of Apollo 15 olivine-normative mare basalts magma (abstr). *In:* Lunar Planet Sci XXII:1199-1200, Lunar and Planetary Institute, Houston

Schurmann K, Hafner SS (1972) On the amount of ferric iron in plagioclases from lunar igneous rocks. Proc Lunar Sci Conf 3:615-621

Scoon JH (1972) Chemical analyses of lunar samples 14003, 14311, and 14321. Proc Lunar Sci Conf 3:1335-1336

See TH, Hörz F, Morris RV (1986) Apollo 16 impact-melt splashes: Petrography and major-element composition. Proc Lunar Planet Sci Conf 17th in J Geophys Res 91:E3-E20

Shearer CK, Papike JJ (1993) Basaltic magmatism on the Moon: A perspective from volcanic picritic glass beads. Geochimica Cosmochim Acta 57:4785-4812

Shearer CK, Papike JJ, Simon SB, Shimizu N (1989) An ion microprobe study of the intra-crystalline behavior of REE and selected trace elements in pyroxene from mare basalts with different cooling and crystallization histories. Geochim Cosmochim Acta 53:1041-1054

Shearer CK, Papike JJ, Simon SB, Galbreath KC, Shimizu N, Yurimoto Y, Sueno S (1990) Ion microprobe studies of REE and other trace elements in Apollo 14 'volcanic' glass beads and comparison to Apollo 14 mare basalts. Geochim Cosmochim Acta 54:851-867

Shearer CK, Papike JJ, Galbreath KC, Shimizu N (1991) Exploring the lunar mantle with secondary ion mass spectrometry: A comparison of lunar picritic glass beads from the Apollo 14 and Apollo 17 sites. Earth Planet Sci Lett 102:134-147

Shearer, CK, Layne GD, Papike JJ (1994) The systematics of light lithophile elements in lunar picritic glasses: Implications for basaltic magmatism on the Moon and the origin of the Moon. Geochim Cosmochim Acta 58:5349-5362

Shearer CK, Papike JJ, Layne GD (1996a) Deciphering basaltic magmatism on the Moon from the compositional variations in the Apollo 15 very low-Ti picritic magmas. Geochim Cosmochim Acta 60:509-528

Shearer CK, Papike JJ, Layne GD (1996b) The role of ilmenite in the source region for mare basalts: Evidence from niobium, zirconium, and cerium in picritic glasses. Geochim Cosmochim Acta 60:3521-3530

Shearer CK, Weidenbeck MG, Fowler GW, Papike JJ (1997) Volatiles in planetary mantles. The behavior of sulfur in lunar picritic magmas and the Moon's mantle (abstr). Geol Soc Am Abs with Progs 29 #6:A-192

Shervais JW (1994) Ion microprobe studies of lunar highland cumulate rocks: Preliminary results (abstr). *In:* Lunar Planet Sci XXV:1265-1266, Lunar and Planetary Institute, Houston

Shervais JW, McGee JJ (1997) KREEP in the western lunar highlands: An ion microprobe study of alkali and Mg-suite cumulates from the Apollo 12 and 14 sites (abstr). *In:* Lunar Planet Sci XXVII:1301-1302, Lunar and Planetary Institute, Houston

Shervais JW, Stuart JB (1995) Ion microprobe studies of lunar highland cumulate rocks: New results (abstr). *In:* Lunar Planet Sci XXVI:1285-1286, Lunar and Planetary Institute, Houston

Shervais JW, Taylor LA (1986) Petrologic constraints on the origin of the Moon. *In:* WK Hartmann, RJ Phillips, and GJ Taylor (eds) Origin of the Moon, p173-201, Lunar and Planetary Institute, Houston

Shervais JW, Taylor LA, Laul JC (1983) Ancient crustal components in Fra Mauro breccias. Proc Lunar Sci Conf 14th in J Geophys Res 88:B177-B192

Shervais JW, Taylor LA, Laul JC, Smith MR (1984a) Pristine highlands clasts in consortium breccia 14305: Petrology and geochemistry. Proc Lunar Planet Sci Conf 15th in J Geophys Res 89:C25-C40
Shervais JW, Taylor LA, Laul JC (1984b) Very high potassium (VHK) basalt: A new type of aluminous mare basalt from Apollo 14 (abstr). *In:* Lunar Planet Sci XV:768-769, Lunar and Planetary Institute, Houston
Shervais JW, Taylor LA, Lindstrom MM (1985a) Apollo 14 mare basalts: Petrology and geochemistry of clasts from consortium breccia 14321. Proc Lunar Planet Sci Conf 15th in J Geophys Res 89:C375-C395
Shervais JW, Taylor LA, Laul JC, Shih C-Y, Nyquist LE (1985b) Very high potassium (VHK) basalt: Complications in lunar mare basalt petrogenesis. Proc Lunar Planet Sci Conf 16th in J Geophys Res 90:D3-D18
Shih C-Y, Nyquist LE, Bogard DD, Wooden JL, Bansal BM, Wiesmann H (1985) Chronology and petrogenesis of a 1.8g lunar granite clast: 14321,1062. Geochim Cosmochim Acta 49:411-426
Shih C-Y, Nyquist LE, Bansal BM, Wiesmann H (1992) Rb-Sr and Sm-Nd chronology of an Apollo 17 KREEP basalt. Earth Planet Sci Lett 108:203-215
Shih C-Y, Nyquist LE, Dasch EJ, Bogard DD, Bansal BM, Wiesmann H (1993a) Ages of pristine noritic clasts from lunar breccias 15445 and 15455. Geochim Cosmochim Acta 57:915-931
Shih C-Y, Nyquist LE, Wiesmann H (1993b) K-Ca chronology of lunar granites. Geochim Cosmochim Acta 57:48274841.
Shoemaker EM, Morris EC, Batson RM, Holt HE, Larson KB, Montgomery DR, Rennilson JJ, Whitaker EA (1968) Television observations from Surveyor. *In:* Surveyor Project Final Report, Part II, JPL Tech Rept 32-1265, p 21-136, Jet Propulsion Laboratory, Pasadena
Simon SB, Papike JJ (1985) Petrology of the Apollo 12 highland component. Proc Lunar Planet Sci Conf 16th in J Geophys Res 90:D47-D60
Simon SB, Papike JJ, Laul JC (1981) The lunar regolith: Comparative studies of the Apollo and Luna sites. Petrology of soils from Apollo 17, Luna 16, 20, and 24. Proc Lunar Planet Sci Conf 12B:371-388
Simon SB, Papike JJ, Shearer CK (1984) Petrology of Apollo 11 regolith breccias. Proc Lunar Planet Sci Conf 15th in J Geophys Res 89:C109-C132
Simon SB, Papike JJ, Gosselin DC, Laul JC (1985) Petrology and chemistry of Apollo 12 regolith breccias. Proc Lunar Planet Sci Conf 16th in J Geophys Res 90 D75-D86
Simon SB, Papike JJ, Hörz F, See TH (1986a) An experimental investigation of agglutinate melting mechanisms: Shocked mixtures of Apollo 11 and 16 soils. Proc Lunar Planet Sci Conf 17th in J Geophys Res 91:E64-E74
Simon SB, Papike JJ, Gosselin DC, Laul JC (1986b) Petrology, chemistry, and origin of Apollo 15 regolith breccias. Geochim Cosmochim Acta 50:2675-2691
Simon SB, Papike JJ, Laul JC, Hughes SS, Schmitt RA (1988a) Apollo 16 regolith breccias and soils: Recorders of exotic component addition to the Descartes region of the Moon. Earth Planet Sci Lett 89:147-162
Simon SB, Papike JJ, Laul JC (1988b) Chemistry and petrology of the Apennine Front, Apollo 15, Part 1: KREEP basalts and plutonic rocks. Proc Lunar Planet Sci Conf 18:187-201
Simon SB, Papike JJ, Shearer CK, Hughes SS, Schmitt RA (1989) Petrology of Apollo 14 regolith breccias and ion microprobe studies of glass beads. Proc Lunar Planet Sci Conf 19:1-17
Simonds CH (1975) Thermal regimes in impact melts and the petrology of the Apollo 17 Station 6 boulder. Proc Lunar Sci Conf 6:641-672
Simonds CH, Warner JL (1981) Petrochemistry of Apollo 16 and 17 samples (abstr). *In:* Lunar Planet Sci XII:993-995, Lunar and Planetary Institute, Houston
Simonds CH, Warner JL, Phinney WC, McGee PE (1976) Thermal model for impact breccia lithification: Manicouagan and the Moon. Proc Lunar Sci Conf 7:2509-2528
Sippel RF (1971) Luminescence petrography of the Apollo 12 rocks and comparative features in terrestrial rocks and meteorites. Proc Lunar Sci Conf 2:247-263
Skinner BJ (1970) High crystallization temperatures indicated for igneous rocks from Tranquillity Base. Proc Apollo 11 Lunar Sci Conf, p 891-895
Smith JV (1974) Lunar mineralogy: A heavenly detective story. Part I Am Mineral 59:231-243
Smith JV, Steele IM (1974) Intergrowths in lunar and terrestrial anorthosites with implications for lunar differentiation. Am Mineral 59:673-680
Smith JV, Steele IM (1976) Lunar mineralogy: A heavenly detective story. Part II. Am Mineral 61:1059-1116
Smith JV, Anderson AT, Newton RC, Oslen EJ, Wyllie PJ, Crewe AV, Isaacson MS, Johnson D (1970) Petrologic history of the Moon inferred from petrography, mineralogy, and petrogenesis of Apollo 11 rocks. Proc Apollo 11 Lunar Sci Conf, p 897-925

Smyth JR (1974) The crystal chemistry of armalcolites from Apollo 17. Earth Planet Sci Lett 24:262-270

Snyder GA, Taylor LA, Liu Y-G, Schmitt RA (1992) Petrogenesis of the western highlands of the Moon: Evidence from a diverse group of whitlockite-rich rocks from the Fra Mauro Formation. Proc Lunar Planet Sci 22:399-416

Snyder GA, Taylor LA, Halliday AN (1994a) Rb-Sr isotopic systematics of lunar ferroan anorthosite 62237 (abstr) *In:* Lunar Planet Sci XXV:1309-1310, Lunar and Planetary Institute, Houston

Snyder GA, Taylor LA, Jerde E A (1994b) Evolved QMD-melt parentage for lunar highlands alkali suite cumulates: Evidence from ion-probe rare-earth element analyses of indvidual minerals (abstr). *In:* Lunar Planet Sci XXV:1311-1312, Lunar and Planetary Institute, Houston

Snyder GA, Neal CR, Taylor LA, Halliday AN (1995a) Processes involved in the formation of magnesian-suite plutonic rocks from the highlands of Earth's Moon. J Geophys Res 100:9365-9388

Snyder GA, Taylor LA, Halliday AN (1995b) Chronology and petrogenesis of the lunar highlands alkali suite: Cumulates from KREEP basalt crystallization. Geochim Cosmochim Acta 59:1185-1203

Snyder GA, Crozaz G, Taylor LA (1996) Probing the crust of the Earth's Moon: Trace elements in minerals from post-magma ocean highlands rocks (abstr). *In:* Lunar Planet Sci XXVII:1235-1236, Lunar and Planetary Institute, Houston

Snyder GA, Neal CR, Taylor LA, Halliday AN (1997) Anatexis of lunar cumulate mantle in time and space: Clues from trace-element, strontium and neodymium isotopic chemistry of parental Apollo 12 basalts. Geochim Cosmochim Acta 61:2731-2748

Solomon SC (1975) Mare volcanism and lunar crustal structure. Proc Lunar Sci Conf 6:1021-1042

Solomon SC (1977) The relationship between crustal tectonics and internal evolution on the Moon and Mercury. Phys Earth Planet Interiors 15:135-145

Spera FJ (1992) Lunar magma transport phenomena. Geochim Cosmochim Acta 56:2253-2266

Spudis PD (1978) Composition and origin of the Apennine Bench Formation. Proc Lunar Planet Sci Conf 9:3379-3394

Spudis PD (1984) Apollo 16 site geology and impact melts: Implications for the geologic history of the lunar highlands. Proc Lunar Planet Sci Conf 15th in J Geophys Res 89:C95-C107

Spudis PD (1993) The Geology of Multiring Impact Basins. Cambridge Univ Press, Cambridge, UK

Spudis PD, Davis PA (1986) A chemical and petrologic model of the lunar crust and implications for lunar crustal origin. Proc Lunar Planet Sci Conf 17th in J Geophys Res 91:E84-E90

Spudis PD, Ryder G (1981) Apollo 17 impact melts and their relation to the Serenitatis basin. *In:* PH Schultz and RB Merrill (eds) Multi-ring Basins, Proc Lunar Planet Sci Conf 12A:133-148

Spudis PD, Ryder G (1985) Geology and petrology of the Apollo 15 landing site: Past, present, and future understanding. EOS Trans AGU 66:721-726

Stanin FT, Taylor LA (1979a) Armalcolite/ilmenite: Mineral chemistry, paragenesis, and origin of textures. Proc Lunar Planet Sci Conf 10:383-405

Stanin FT, Taylor LA (1979b) Ilmenite/armalcolite: Effects of rock composition, oxygen fugacity and cooling rate (abstr). *In:* Lunar Planet Sci X:1160-1162, Lunar and Planetary Institute, Houston

Stanin FT, Taylor LA (1980) Armalcolite: An oxygen fugacity indicator. Proc. Lunar Planet Sci Conf 11:117-124

Stadermann FJ, Heusser E, Jessberger EK, Lingner S, Söffler D (1991) The case for a younger Imbrium basin: New ^{40}Ar-^{39}Ar ages of Apollo 14 rocks. Geochim Cosmochim Acta 55:2339-2349

Steele AM, Colson RO, Korotev RL, Haskin LA (1992) Apollo 15 green glass: Compositional distribution and petrogenesis. Geochim Cosmochim Acta 56:4075-4090

Steele IM, Smith JV, Grossman L (1972) Mineralogy and petrology of Apollo 15 rake samples: I. Basalts. (abstr). *In:* JW Chamberlain, C Watkins (eds) The Apollo 15 Lunar Samples, p 158-160, Lunar Science Institute, Houston

Steiger RH, Jaeger E (1977) Subcommission on geochronology: Convention on the use of decay constants in geo- and cosmochronology. Earth Planet Sci Lett 36:359-362

Stettler A, Eberhardt P, Geiss J, Grögler N, Maurer P (1973) Ar^{39}-Ar^{40} ages and Ar^{37}-Ar^{38} exposure ages of lunar rocks. Proc Lunar Sci Conf 4:1865

Stettler A, Eberhardt P, Geiss J, Grögler N (1974) ^{39}Ar-^{40}Ar ages of samples from the Apollo 17 Station 7 boulder and implications for its formation. Earth Planet Sci Lett 23:453-461

Stewart DP (1975) Apollonian metamorphic rocks (abstr). *In:* Lunar Sci VI:774-776, Lunar Science Institute, Houston

Stöffler D, Knöll H-D, Maerz U (1979) Terrestrial and lunar impact breccias and the classification of lunar highland rocks. Proc Lunar Planet Sci Conf 10:639-675

Stöffler D, Knöll H-D, Marvin UB, Simonds CH, Warren PH (1980) Recommended classification and nomenclature of lunar highland rocks—a committee report. *In:* JJ Papike, RB Merrill (eds) Proc Conf Lunar Highlands Crust, p 51-70, Pergamon, New York

Stöffler D, Ostertag R, Reimold WU, Borchardt R, Malley J, Rehfeldt A (1981) Distribution and provenance of lunar rock types at North Ray Crater, Apollo 16. Proc Lunar Planet Sci Conf 12B:185-207

Stöffler D, Bischoff A, Borchardt R, Burghele A, Deutsch A, Jessberger EK, Ostertag R, Palme H, Spettel B, Reimold WU, Wacker K, Wänke H (1985) Composition and evolution of the lunar crust in the Descartes Highlands, Apollo 16. Proc Lunar Planet Sci Conf 15th in J Geophys Res 90:C449-C506

Stolper EM (1974) Lunar ultramafic glasses. BA thesis, Harvard Univ, Cambridge, MA

Sung CM, Abu-Eid RM, Burns RG (1974) Ti^{3+}/Ti^{4+} ratios in lunar pyroxenes: Implications to depth of origin of mare basalt magma. Proc Lunar Sci Conf 5:717-726

Takeda H, Miyamoto M, Ishii T, Lofgren GE (1975) Relative cooling rates of mare basalts at the Apollo 12 and 15 sites as estimated from pyroxene exsolution data. Proc Lunar Sci Conf 6th 987-996

Tatsumoto M, Knight RJ, Allégre CJ (1973) Time differences in the formation of meteorites as determined from the ratio of lead-207 to lead-206. Science 180:1279-1283

Tatsumoto M, Premo WR, Unruh DM (1987) Origin of lead from green glass of Apollo 15426: A search for primitive lunar lead. Proc Lunar Planet Sci Conf 17th in J Geophys Res 92:E361-E371

Taylor GJ, Keil K, Warner RD (1977) Very low-Ti mare basalts. Geophys Res Lett 4:207-210

Taylor GJ, Warner RD, Keil K (1978) VLT mare basalts: Impact mixing, parent magma types, and petrogenesis. *In:* RB Merrill, JJ Papike (eds) Mare Crisium: The View from Luna 24, p 357-370, Pergamon, New York

Taylor GJ, Warner RD, Keil K (1979) Stratigraphy and depositional history of the Apollo 17 drill core. Proc Lunar Planet Sci Conf 10:1159-1184

Taylor GJ, Warner RD, Keil K, Ma M-S, Schmitt RA (1980) Silicate liquid immiscibility, evolved lunar rocks and the formation of KREEP. *In:* JJ Papike, RB Merrill (eds) Proc Conf Lunar Highlands Crust, p 339-352, Pergamon, New York

Taylor GJ, Warren PH, Ryder G, Pieters C, Lofgren G (1991) Lunar Rocks. *In:* GH Heiken, DT Vaniman, BV French (eds) Lunar Sourcebook: A User's Guide to the Moon, p 183-284, Cambridge University Press, Cambridge, UK

Taylor LA (1988) Generation of native Fe in lunar soil. *In:* SW Johnson, JP Wetzel (eds) Engineering, Construction, and Operations in Space: Proceedings of Space '88, p 67-77, American Society of Civil Engineers, New York

Taylor LA, Cirlin EH (1986) A review of ESR studies on lunar samples. In ESR Dating and Dosimetry, p 19-29, Ionics, Tokyo

Taylor LA, McCallister RH (1972) Experimental investigation of significance of zirconium partitioning in lunar ilmenite and ulvöspinel. Earth Planet Sci Lett 17:105-111

Taylor LA and Mosie AB (1979) Breccia Guidebook #3, 67915, JSC #16242 Curatorial Branch #50, Lyndon Johnson Space Center, Houston

Taylor LA, Williams KL (1973) Cu-Fe-S phases in lunar rocks. Am Mineral 58:952-954

Taylor LA, Kullerud G, Bryan WB (1971) Opaque mineralogy and textural features of Apollo 12 samples and a comparison with Apollo 11 rocks. Proc Lunar Sci Conf 2:855-871

Taylor LA, Williams RJ, McCallister RH (1972) Stability relations of ilmentite and ulvöspinel in the Fe-Ti-O system and application of these data to lunar mineral assemblages. Earth Planet Sci Lett 16:282-288

Taylor LA, Mao HK, Bell PM (1973a) "Rust" in the Apollo 16 rocks. Proc Lunar Sci Conf 4:829-839

Taylor LA, McCallister RH, Sardi O (1973b) Cooling histories of lunar rocks based on opaque mineral geothermometers. Proc Lunar Sci Conf 4:819-828

Taylor LA, Mao HK, Bell PM (1974) Identification of the hydrated iron oxide mineral akaganeite in Apollo 16 lunar rocks. Geology 2:429-432

Taylor LA, Uhlmann DR, Hopper RW, Misra KC (1975) Absolute cooling rates of lunar rocks: Theory and application. Proc Lunar Sci Conf 6:181-191

Taylor LA, Onorato PIK, Uhlmann DR, Coish RA (1978) Subophitic basalts from Mare Crisium: Cooling rates. *In:* RB Merrill, JJ Papike (eds) Mare Crisium: The View From Luna 24, p 473-482, Pergamon, New York

Taylor LA, Shervais JW, Hunter RH, Shih CY, Bansal BM, Wooden J, Nyquist LE, Laul JC (1983) Pre 4.2 AE mare basalt volcanism in the lunar highlands. Earth Planet Sci Lett 66:33-47

Taylor SR (1982) Planetary Science: A Lunar Perspective. Lunar and Planetary Institute, Houston

Taylor SR, Jakês P (1974) The geochemical evolution of the moon. Proc Lunar Sci Conf 5:1287-1305

Taylor SR, Bence AE (1975a) Trace element characteristics of the mare basalt source region: Implications of the cumulate versus primitive source model, *In:* Papers Presented to the Conference on Origins of Mare Basalts and Their Implications for Lunar Evolution, p 159-163, Lunar Science Institute, Houston

Taylor SR, Bence AE (1975b) Evolution of the lunar highland crust. Proc Lunar Sci Conf 6:1121-1141

Taylor SR, Rudowski R, Muir P, Graham A, Kaye M (1971) Trace element chemistry of lunar samples from the Ocean of Storms. Proc Lunar Sci Conf 2:1083-1099

Taylor SR, Gorton MP, Muir P, Nance WB, Rudowski R, Ware N (1973) Composition of the Descartes region, lunar highlands. Geochim Cosmochim Acta 37:2665-2683

Taylor SR, Gorton MP, Muir P, Nance WB, Rudowski R, Ware N (1974) Lunar highland composition (abstr). *In:* Lunar Sci V:789-791, Lunar Science Institute, Houston

Taylor SR, Norman MD, Esat TM (1993) The Mg-Suite and the highland crust: An unsolved enigma (abstr). Lunar Planet Sci XXIV:1413-1414, Lunar and Planetary Institute, Houston

Tera F, Wasserburg (1974) U-Th-Pb systematics on lunar rocks and inferences about lunar evolution and the age of the moon. Proc Lunar Sci Conf 5:1571-1599

Tera F, Wasserburg GJ (1976) Lunar ball games and other sports (abstr). *In:* Lunar Sci VII:858-860, Lunar Science Institute, Houston

Tera F, Eugster O, Burnett DS, Wasserburg GJ (1970) Comparative study of Li, Na, K, Rb, Cs, Ca, Sr and Ba abundances in achondrites and in Apollo 11 lunar samples. Proc Apollo 11 Lunar Sci Conf, p 1637-1657

Tera F, Papanastassiou DA, Wasserburg GJ (1974) Isotopic evidence for a terminal lunar cataclysm. Earth Planet Sci Lett 22:1-21

Toksoz MN, Dainty AM, Solomon SC, Anderson KR (1974) Structure of the Moon. Rev Geophys Space Phys 12:539-567

Tsay FD, Chan SI, Manatt SL (1971) Ferromagnetic resonance of lunar samples. Geochim Cosmochim Acta 5:865-875

Turner G, Cadogan PH (1975) The history of lunar bombardment inferred from ^{40}Ar-^{39}Ar dating of highland rocks. Proc Lunar Sci Conf 6:1509-1538

Turner G, Huneke JC, Podosek FA, Wasserburg GJ (1972) Ar^{40}-Ar^{39} systematics in rocks and separated minerals from Apollo 14. Proc Lunar Sci Conf 3:1589-1612

Turner G, Cadogan PH, Yonge CJ (1973) Argon selenochronology. Proc Lunar Sci Conf 4:1889-1914

Uhlmann DR, Klein L, Kritchevsky G, Hopper RW (1974) The formation of lunar glasses. Proc Lunar Sci Conf 5:2317-2331

Uhlmann DR, Onorato PIK, Yinnon H, Taylor LA (1979) Partitioning as a cooling rate indicator (abstr). *In:* Lunar Planet Sci X:1253-1255, Lunar and Planetary Institute, Houston

Unruh DM, Stille P, Patchett PJ, Tatsumoto M (1984) Lu-Hf and Sm-Nd evolution in lunar mare basalts. Proc Lunar Planet Sci Conf 14th in J Geophys Res 89:B17-B25

Usselman TM, Lofgren GE (1976) The phase relations, textures, and mineral chemistries of high titanium mare basalts as a function of oxygen fugacity and cooling rate. Proc Lunar Sci Conf 7:1345-1363

Usselman TM, Lofgren GE, Donaldson CH, Williams RJ (1975) Experimentally reproduced textures and mineral chemistries of high-titanium mare basalts. Proc Lunar Sci Conf 6:997-1020

Vaniman DT, Papike JJ (1977a) The Apollo 17 drill core: Characterization of the mineral and lithic component (sections 70007, 70008, 70009). Proc Lunar Sci Conf 8:3123-3159

Vaniman DT, Papike JJ (1977b) The Apollo 17 drill core: Modal petrology and glass chemistry (sections 70007, 70008, 70009). Proc Lunar Sci Conf 8:3161-3193

Vaniman DT, Papike JJ (1977c) Very low-Ti (VLT) basalts: A new mare rock type from the Apollo 17 drill core. Proc Lunar Sci Conf 8:1443-1471

Vaniman DT, Papike JJ (1980) Lunar highland melt rocks: Chemistry, petrology, and silicate mineralogy. *In:* JJ Papike, RB Merrill (eds) Proc Conf Lunar Highlands Crust, p 271-337, New York

Vaniman DT, Labotka TC, Papike JJ, Simon SB, Laul JC (1979) The Apollo 17 drill core: Petrologic systematics and the identification of a possible Tycho component. Proc Lunar Planet Sci Conf 10:1185-1227

Vaniman D, Dietrich J, Taylor GJ, Heiken G (1991) Exploration, samples, and recent concepts of the Moon. *In:* G Heiken, D Vaniman, B French (eds) Lunar Sourcebook: A User's Guide to the Moon, p 5-26, Cambridge Univ Press, Cambridge, UK

Wagner TP, Grove TL (1997) Experimental constraints on the origin of lunar high-Ti ultramafic glasses. Geochimica et Cosmochimica Acta 61:1315-1327.

Walker D, Longhi J, Hays JF (1972) Experimental petrology and origin of Fra Mauro rocks and soil. Proc Lunar Sci Conf 3:797-817

Walker D, Longhi J, Grove TL, Stolper EM, Hays JF (1973) Experimental petrology and origin of rocks from the Descartes highlands. Proc Lunar Sci Conf 4:1013-1032

Walker D, Longhi J, Stolper EM, Grove TL, Hays JF (1975) Origin of titaniferous lunar basalts. Geochim Cosmochim. Acta 39:1219-1235

Walker D, Longhi J, Stolper EM, Grove TL, Hays JF (1976) Differentiation of an Apollo 12 picrite magma. Proc Lunar Sci Conf 7:1365-1389

Walker D, Longhi J, Lasaga AC, Stolper EM, Grove TL, Hays JF (1977) Slowly cooled microgabbros 15555 and 15065. Proc Lunar Sci Conf 8:1521-1547

Walker RJ, Papike JJ (1981) The relationship of the lunar regolith <10 mm fraction and agglutinates. Part II: Chemical composition of agglutinate glass as a test of the "fusion of the finest fraction" (F3) model. Proc Lunar Planet Sci12B:421-431

Wänke H, Baddenhausen H, Balacescu A, Teshke F, Spettel B, Dreibus G, Palme H, Quijano-Rico M, Kruse H, Wlotzka F, Begemann F (1972) Multielement analyses of lunar samples and some implications of the results. Proc Lunar Sci Conf 3:1251-1268

Wänke H, Baddenhausen H, Dreibus G, Jagoutz E, Kruse H, Palme H, Spettel B, Teshke F (1973) Multi-element analyses of Apollo 15, 16, and 17 samples and the bulk composition of the Moon. Proc Lunar Sci Conf 4:1461-1481

Wänke H, Palme H, Baddenhausen H, Dreibus G, Jagoutz E, Kruse H, Spettel B, Teshke F, Thacker R (1974) Chemistry of Apollo 16 and 17 samples: bulk composition, late-stage accumulation, and early differentiation of the Moon. Proc Lunar Sci Conf 5:1307-1335

Wänke H, Palme H, Baddenhausen H, Dreibus G, Jagoutz E, Kruse H, Palme C, Spettel B, Teshke F, Thacker R (1975) New data on the chemistry of lunar samples: Primary matter in the lunar highlands and the bulk composition of the Moon. Proc Lunar Sci Conf 6:1313-1340

Wänke H, Palme H, Kruse H, Baddenhausen H, Cendales M, Dreibus G, Hofmeister H, Jagoutz E, Palme C, Spettel B, Thacker R (1976) Chemistry of lunar highlands rocks: A refined evaluation of thes primary matter. Proc Lunar Sci Conf 7:3479-3499

Ware NG, Lovering JF (1970) Electron microprobe analyses of phases in lunar samples Science 167:517-520

Warner JL, Simonds CH, Phinney WC (1976) Genetic distinction between anorthosites and Mg-rich plutonic rocks: New data from 76255 (abstr). *In:* Lunar Science VII:915-917, Lunar Science Institute, Houston

Warner JL, Phinney WC, Bickel CE, Simonds CH (1977) Feldspathic granulitic impactites and pre-final bombardment lunar evolution. Proc Lunar Sci Conf 8:2051-2066

Warner RD, Keil K, Prinz M, Laul JC, Murali AV, Schmitt RA (1975) Mineralogy, petrology, and chemistry of mare basalts from Apollo 17 rake samples. Proc Lunar Sci Conf 6:193-220

Warner RD, Planner HN, Keil K, Murali AV, Ma M-S, Schmitt RA, Ehmann WD, James WD, Jr, Clayton RN, Mayeda TK (1976) Consortium investigation of breccia 67435. Proc Lunar Planet Sci Conf 7:2379-2402

Warner RD, Taylor GJ, Keil K, Planner HN, Nehru CE, Ma M-S, Schmitt RA (1978a) Green glass vitrophyre 78526: An impact melt of very low-Ti mare basalt composition. Proc Lunar Planet Sci Conf 9:547-564

Warner RD, Taylor GJ, Mansker WL, Keil K (1978b) Clast assemblages of possible deep-seated (77517) and immiscible-melt (77538) origins in Apollo 17 breccias. Proc Lunar Planet Sci Conf 9:941-958

Warner RD, Taylor GJ, Conrad GH, Northrop HR, Barker S, Keil K, Ma M-S, Schmitt RA (1979) Apollo 17 high-Ti mare basalts: New bulk compositional data, magma types, and petrogenesis. Proc Lunar Planet Sci Conf 10:225-247

Warner RD, Taylor GJ, Keil K (1980a) Petrology of 60035: Evolution of a polymict ANT breccia. *In:* JJ Papike, RB Merrill (eds) Proc Conf Lunar Highlands Crust, p 377-394, Pergamon, New York

Warner RD, Taylor GJ, Keil K, Ma M-S, Schmitt RA (1980b) Aluminous mare basalts: New data from Apollo 14 coarse fines. Proc Lunar Planet Sci Conf 11:87-104

Warren PH (1985) The magma ocean concept and lunar evolution. Ann Rev Earth Planet Sci 13:201-240

Warren PH (1986) Anorthosite assimilation and the origin of the Mg/Fe related bimodality of pristine Moon rocks: Support for the magmasphere hypothesis. Proc Lunar Planet Sci Conf 16th in J Geophys Res 91:D331-D343

Warren P H (1988) The origin of pristine KREEP: Effects of mixing between urKREEP and the magmas parental to the Mg-rich cumulates. Proc Lunar Planet Sci Conf 18:233-241

Warren PH (1989) KREEP: Major-element diversity, trace element uniformity (almost) (abstr). *In:* Workshop on Moon in Transition: Apollo 14, KREEP, and Evolved Lunar Rocks, LPI Tech Rept 89-03, p 149-153, Lunar and Planetary Institute, Houston

Warren PH (1993) A concise compilation of petrologic information on possibly pristine nonmare Moon rocks. Am Mineral 78:360-376

Warren PH (1994) Lunar and Martian delivery services. Icarus 111:338-363

Warren PH, Kallemeyn GW (1984) Pristine rocks (8th foray): "Plagiophile" element ratios, crustal genesis, and the bulk composition of the moon. Proc Lunar Planet Sci Conf 15th in J. Geophys Res 89:C16-C24

Warren PH, Kallemeyn GW (1991) The MacAlpine Hills lunar meteorite and implications of the lunar meteorites collectively for the composition and origin of the Moon. Geochim Cosmochim Acta 55:3123-3138

Warren PH, Kallemeyn GW (1993) The ferroan-anorthositic suite, the extent of primordial lunar melting and the bulk composition of the moon. J Geophys Res 98:5445-5455

Warren PH, Wasson JT (1977) Pristine nonmare rocks and the nature of the lunar crust. Proc Lunar Sci Conf 8:2215-2235

Warren PH, Wasson JT (1978) Compositional-petrographic investigation of pristine nonmare rocks. Proc Lunar Planet Sci Conf 9:185-217

Warren PH, Wasson JT (1979a) The compositional-petrographic search for pristine non-mare rocks—third foray. Proc Lunar Planet Sci Conf 10:583-610

Warren PH, Wasson JT (1979b) The origin of KREEP. Rev Geophys Space Phys 17:73-88

Warren PH, Wasson JT (1980) Further foraging for pristine nonmare rocks: Correlations between geochemistry and longitude. Proc Lunar Planet Sci Conf 11:431-470

Warren PH, Taylor GJ, Keil K, Marshall C, Wasson JT (1981) Foraging westward for pristine non-mare rocks: Complications for petrogenetic models. Proc Lunar Planet Sci12B:21-40

Warren PH, Taylor GJ, Keil K, Shirley DN, Wasson JT (1983a) Petrology and chemistry of two "large" granite clasts from the Moon. Earth Planet Sci Lett 64:175-185

Warren PH, Taylor GJ, Keil K, Kallemeyn GW, Rosener PS, Wasson JT (1983b) Sixth foray for pristine nonmare rocks and an assessment of the diversity of lunar anorthosites. Proc Lunar Planet Sci Conf 13th in J Geophys Res 88:A615-A630

Warren PH, Taylor GJ, Keil K, Kallemeyn GW, Shirley DN, Wasson JT (1983c) Seventh foray: Whitlockite-rich lithologies, a diopside-bearing troctolitic anorthosite, ferroan anorthosites, and KREEP. Proc Lunar Planet Sci Conf 14th in J Geophys Res 88:B151-B164

Warren PH, Shirley DN, Kallemeyn GW (1986) A potpourri of pristine Moon rocks, including a VHK mare basalt and a unique, augite-rich Apollo 17 anorthosite. Proc Lunar Planet Sci Conf 16th in J Geophys Res 91:D319-D330

Warren PH, Jerde EA, Kallemeyn GW (1987) Pristine Moon rocks: A "large" felsite and a metal-rich ferroan anorthosite. Proc Lunar Planet Sci Conf 17th in J Geophys Res 92:E303-E313

Warren PH, Jerde E, Kallemeyn GW (1990) Pristine moon rocks: An alkali anorthosite with coarse augite exsolution from plagioclase, a magnesian harzburgite, and other oddities. Proc Lunar Planet Sci Conf 20:31-59

Warren PH, Jerde E, Kallemeyn GW (1991) Pristine moon rocks: Apollo 17 anorthosites. Proc Lunar Planet Sci 21:51-61

Warren PH, Kallemeyn GW, Kyte FT (1997) Siderophile element evience indicates that Apollo 14 high-Al mare basalts are not impact melts (abstr). *In:* Lunar Planet Sci XXVIII:1501-1502, Lunar and Planetary Institute, Houston

Wasson JT, Boynton WV, Kallemeyn GW, Sundberg LL, Wai CM (1976) Volatile compounds released during lunar lava fountaining. Proc Lunar Sci Conf 7:1583-1595

Wasson JT, Warren PH, Kallemeyn GW, McEwing CE, Mittlefehldt DW, Boynton WV (1977) SCCRV, a major component of highlands rocks. Proc Lunar Sci Conf 8:2237-2252

Watson EB, Green TH (1981) Apatite/liquid partition coefficients for the rare earth elements and strontium. Earth Planet Sci Lett 56:405-421

Wechsler BA, Prewitt CT, Papike JJ (1976) Chemistry and structure of lunar synthetic armalcolite. Earth Planet Sci Lett 29:91-103

Weigand PW (1973) Petrology of a coarse-grained Apollo 17 ilmenite basalt (abstr). EOS Trans AGU 54:621-622

Weill DF, Grieve RA, McCallum IS, Bottinga Y (1971) Mineralogy-petrology of lunar samples: Microprobe studies of sample 12021 and 12022; Viscosity of melts of selected lunar compositions. Proc Lunar Sci Conf 2:413-430

Wenk HR, Wilde WR (1973) Chemical anomalies of lunar plagioclase, described by substitution vectors and their relation to optical and structural properties. Contrib Mineral Petrol 41:89-104

Wentworth S, Taylor GJ, Warner RD, Keil K, Ma M-S, Schmitt RA (1979) The unique nature of Apollo 17 VLT mare basalts. Proc Lunar Planet Sci Conf 10:207-223

Wilhelms DE, McCauley J (1971) Geologic map of the near side of the Moon. US Geol Surv Map I-703

Wilhelms DE (1987) The Geologic History of the Moon. US Geol Surv Prof Pap 1348

Williams KL, Taylor LA (1974) Optical properties and chemical compositions of Apollo 17 armalcolites. Geology 2:5-8

Winzer SR, Nava DF, Schuhmann S, Kouns CW, Lum RKL, Philpotts JA (1974) Major, minor, and trace element abundances in samples from the Apollo 17 Station 7 boulder: Implications for the origin of early crustal rocks. Earth Planet Sci Lett 23:439-444

Winzer SR, Nava DF, Schuhmann PJ, Lum RKL, Schuhmann S, Lindstrom MM, Lindstrom DJ, Philpotts JA (1977) The Apollo 17 "melt sheet": Chemistry, age, and Rb/Sr systematics. Earth Planet Sci Lett 33:389-400

Wood JA, Dickey JS, Marvin UB, Powell BN (1970a) Lunar anorthosites. Science 167:602-604

Wood JA, Dickey JS, Marvin UB, Powell BN (1970b) Lunar anorthosites and a geophysical model of the Moon. Proc Apollo 11 Lunar Sci Conf, p 965-988

Wopenka B, Jolliff BL, Zinner E, Kremser DT (1996) Trace element zoning and incipient metamictization in a lunar zircon: Application of three microprobe techniques. Am Mineral 81:902-912

Yoder HS (1976) Generation of Basaltic Magma. National Academy of Science, Washington, DC

Yoder HS, Tilley CE (1962) Origin of basalt magmas: An experimental study of natural and synthetic rock systems. J Petrol 3:342-532

York D, Kenyon WJ, Doyle RJ (1972) ^{40}Ar-^{39}Ar ages of Apollo 14 and 15 samples. Proc Lunar Sci Conf 3:1613-1622

APPENDIX, CHAPTER 5

The following pages contain the Tables labeled A5.1 through A5.49 in the text.

Table A5.1. Listing of rock types and corresponding abbreviations for mineral chemistry tables (A5.2-A5.49).

Rock Type	Table Symbol
Mare Basalts	
High-Ti (>9% TiO_2)	
Apollo 11	
high-K (>0.3 K_2O)	A-11 HK
low-K (<0.11% K_2O)	A-11 LK
Apollo 17	A-17
Low Ti (1.5-9%) TiO_2)	
Apollo 12	
pigeonite (<10% MgO, <5% TiO_2)	A-12 pig
olivine (>10% MgO, <5% TiO_2)	A-12 ol
ilmenite (<10% MgO, >5% TiO_2)	A-12 ilm
Apollo 15	
pigeonite (<10% MgO, <5% TiO_2)	A-15 pig
olivine (>10% MgO, <5% TiO_2)	A-15 ol
Aluminous, Low -Ti (>10% Al_2O_3, 2-5% TiO_2)	
Luna 16	L-16 Al
Apollo 14	A-14 Al
Very-High-K (>0.6% K_2O)	VHK
Very Lo-Ti (<1.5% TiO_2)	
Apollo 17	A-17 VLT
Luna 24	L-24 VLT
Highland Igneous and Monomict Rocks	
KREEP rocks (highK, rare earths, and P)	KREEP
Ferroan Anorthosites (Low-Na plagioclase, low-Mg pyroxene)	Fan
Mg-rich Rocks (variable plagioclase, high-Mg pyroxene)	Mg Rock
Alkali Anorthosites (relatively high-Na plagioclase)	Alk an
Granite (or felsite) (K-feldspar, silica minerals)	Granite
Highland Polymict Rocks	
Fragmental Breccias (mainly fragments)	Frag br
Glassy Melt Breccias (glass with some fragments)	Glassy br
Crystalline Melt Breccias (crystallized melt with some fragments)	Cryst br
Clast-poor Impact Melt Rocks (crystallized melt with rare fragments)	Melt rock
Granulitic Breccias (recrystallized)	Gran br
Dimict Breccias (two rock types joined)	Dimict
Regolith Breccias (compressed soil)	Reg br
Soil (single crystals in soil, from unknown rock types)	Soil

Table A5.2. Pyroxene analyses from mare basalts.

	High-Ti Basalts								
	Apollo 17			Apollo 11 High-K			Apollo 11 Low-K		
	1.	3.	6.	8.	9.	10.	14.	15.	16.
	Chemical Compositon (Weight Percent)								
SiO_2	44.50	52.10	53.80	44.10	49.90	45.20	47.00	49.32	47.09
Al_2O_3	7.70	1.64	0.96	1.43	2.56	0.88	1.85	3.17	0.86
TiO_2	6.00	1.33	0.78	1.06	2.47	0.72	1.41	2.49	0.72
Cr_2O_3	0.98	0.50	0.31	0.06	0.58	0.06	0.29	0.45	0.06
FeO	8.10	16.70	18.40	45.80	11.10	43.90	28.00	10.95	41.07
MnO	0.28	0.30	0.0	0.59	0.18	0.59	0.53	0.08	0.67
MgO	12.00	20.40	22.80	1.72	15.90	1.40	7.20	14.80	2.29
CaO	20.70	6.80	3.68	3.52	16.50	6.12	12.60	17.95	8.04
Na_2O	0.19	0.0	0.0	0.03	0.0	0.0	0.0	0.0	0.0
Total	100.45	99.77	100.73	98.31	99.19	98.87	98.88	99.21	100.80
	Cation Formula Based on 6 Oxygens (Ideal Pyroxene = 4 Cations per 6 Oxygens)								
Si	1.668	1.934	1.967	1.923	1.877	1.952	1.908	1.861	1.965
^{IV}Al	0.332	0.066	0.033	0.074	0.114	0.045	0.089	0.139	0.035
Total tet	2.000	2.000	2.000	1.997	1.991	1.997	1.997	2.000	2.000
Ti	0.169	0.037	0.021	0.035	0.070	0.023	0.043	0.071	0.023
^{VI}Al	0.008	0.006	0.008	0.0	0.0	0.0	0.0	0.002	0.007
Cr	0.029	0.015	0.009	0.002	0.017	0.002	0.009	0.013	0.002
Fe	0.254	0.519	0.562	1.671	0.349	1.585	0.950	0.345	1.433
Mn	0.009	0.009	0.0	0.022	0.006	0.022	0.018	0.003	0.024
Mg	0.670	1.129	1.242	0.112	0.891	0.090	0.436	0.832	0.142
Ca	0.831	0.271	0.144	0.165	0.665	0.283	0.548	0.726	0.359
Na	0.014	0.0	0.0	0.003	0.0	0.0	0.0	0.0	0.0
Total Cations	3.984	3.986	3.986	4.007	3.989	4.002	4.001	3.992	3.990
	Cation Ratios: Ca: Mg: Fe and Fe/(Fe+Mg)								
Ca	47.4	14.1	7.4	8.4	34.9	14.5	28.3	38.1	18.6
Mg	38.2	58.9	63.7	5.7	46.8	4.6	22.5	43.7	7.4
Fe	14.5	27.0	28.9	85.8	18.3	80.9	49.1	18.2	74.1
Fe/(Fe+Mg)	0.27	0.31	0.31	0.94	0.28	0.95	0.69	0.29	0.91

Table A5.2. (continued)

	Low-Ti Basalts									
	Apollo 12 Ilmenite		Apollo 12 Pigeonite			Apollo 12 Olivine			Apollo 15 Pigeonite		
	17.	18.	20.	21.	22.	26.	27.	28.	31.	32.	33.
	Chemical Compositon (Weight Percent)										
SiO_2	49.00	47.32	45.00	53.51	45.23	49.40	45.60	51.77	52.40	43.20	51.50
Al_2O_3	3.46	1.41	2.15	0.66	0.54	0.90	1.27	1.51	1.89	9.66	2.77
TiO_2	2.13	1.14	1.41	0.35	0.73	1.29	0.99	1.08	0.36	3.99	0.63
Cr_2O_3	0.97	0.18	0.09	0.74	0.05	0.13	0.0	0.37	1.04	0.66	1.06
FeO	13.31	30.28	43.10	16.69	46.54	25.70	41.90	16.34	16.90	20.10	14.50
MnO	0.24	0.43	0.51	0.29	0.56	0.44	0.52	0.32	0.0	0.0	0.0
MgO	15.64	2.64	0.41	24.26	0.34	11.10	0.29	15.34	23.90	8.58	14.20
CaO	14.44	16.24	7.69	2.61	6.43	10.50	9.16	13.99	2.62	13.80	14.70
Na_2O	0.05	0.01	0.0	0.0	0.03	0.0	0.0	0.01	0.03	0.05	0.03
Total	99.24	99.65	100.36	99.11	100.45	99.46	99.73	100.73	99.14	100.04	99.39
	Cation Formula Based on 6 Oxygens (Ideal Pyroxene = 4 Cations per 6 Oxygens)										
Si	1.853	1.943	1.910	1.973	1.947	1.948	1.945	1.941	1.935	1.676	1.941
^{IV}Al	0.147	0.057	0.090	0.027	0.027	0.042	0.055	0.059	0.065	0.324	0.059
Total tet	2.000	2.000	2.000	2.000	1.974	1.990	2.000	2.000	2.000	2.000	2.000
Ti	0.061	0.035	0.045	0.010	0.024	0.038	0.032	0.030	0.010	0.116	0.018
^{VI}Al	0.008	0.012	0.018	0.001	0.00	0.00	0.009	0.007	0.017	0.117	0.064
Cr	0.029	0.006	0.003	0.022	0.002	0.004	0.00	0.011	0.030	0.020	0.032
Fe	0.421	1.040	1.530	0.515	1.675	0.848	1.495	0.512	0.522	0.652	0.457
Mn	0.008	0.015	0.018	0.009	0.020	0.015	0.019	0.010	0.00	0.00	0.00
Mg	0.882	0.162	0.026	1.333	0.022	0.652	0.018	0.857	1.315	0.496	0.798
Ca	0.585	0.715	0.350	0.103	0.297	0.444	0.419	0.562	0.104	0.574	0.594
Na	0.004	0.001	0.00	0.00	0.003	0.00	0.00	0.001	0.002	0.004	0.002
Total Cations	3.998	3.986	3.990	3.993	4.017	3.991	3.992	3.990	4.000	3.979	3.965
	Cation Ratios: Ca: Mg: Fe and Fe/(Fe+Mg)										
Ca	31.0	37.3	18.4	5.3	14.9	22.8	21.7	29.1	5.3	33.3	32.1
Mg	46.7	8.4	1.4	68.3	1.1	33.6	1.0	44.4	67.8	28.8	43.2
Fe	22.3	54.3	80.3	26.4	84.0	43.6	77.4	26.5	26.9	37.9	24.7
Fe/(Fe+Mg)	0.32	0.87	0.98	0.28	0.99	0.57	0.99	0.37	0.28	0.57	0.36

Table A5.2. (continued)

	Low-Ti Basalts			Aluminous, Low-Ti Basalts						Very Low-Ti Basalts			
	Apollo 15 Olivine			Luna 16		Apollo 14		VHK		Luna 24		Apollo 17	
	36.	37.	38.	39.	40.	41.	42.	43.	44.	45.	46.	48.	49.
				Chemical Compositon (Weight Percent)									
SiO_2	49.90	51.40	47.00	49.24	46.28	51.30	45.32	51.21	51.90	49.13	44.95	48.28	53.09
Al_2O_3	1.66	2.15	1.16	3.03	0.98	1.73	1.68	1.97	1.64	3.76	0.88	2.91	1.45
TiO_2	1.21	0.83	0.92	2.48	1.16	0.84	1.62	0.83	1.45	0.89	0.76	1.26	0.20
Cr_2O_3	0.22	0.75	0.0	0.59	0.10	0.73	0.02	0.72	0.42	0.89	0.00	0.17	0.73
FeO	25.80	13.60	36.60	13.27	38.67	18.60	30.32	19.07	15.17	16.95	43.35	31.81	17.45
MnO	0.0	0.0	0.0	0.28	0.55	0.0	0.30	0.47	0.37	0.28	0.54	0.43	0.27
MgO	11.40	16.80	5.59	13.22	3.13	21.20	0.57	20.21	11.67	11.68	0.12	11.01	23.60
CaO	10.60	14.40	7.81	18.01	8.75	4.37	18.24	5.37	18.16	16.87	7.88	5.93	3.63
Na_2O	0.0	0.0	0.0	0.09	0.01	0.02	0.07	-	-	0.00	0.00	0.00	0.00
Total	100.79	99.93	99.08	100.21	99.63	98.79	98.14	99.85	100.78	100.45	98.48	101.80	100.42
				Cation Formula Based on 6 Oxygens (Ideal Pyroxene = 4 Cations per 6 Oxygens)									
Si	1.935	1.921	1.948	1.86	1.94	1.928	1.913	1.915	1.956	1.875	1.955	1.887	1.945
^{IV}Al	0.065	0.079	0.052	0.14	0.05	0.072	0.084	0.086	0.044	0.125	0.045	0.113	0.055
Total tet	2.000	2.000	2.000	2.000	1.99	2.000	1.997	2.000	2.000	2.000	2.000	2.000	2.000
Ti	0.035	0.023	0.029	0.07	0.04	0.024	0.051	0.023	0.041	0.026	0.025	0.037	0.006
^{VI}Al	0.011	0.015	0.005	0.0	0.0	0.005	0.0	0.001	0.029	0.044	0.000	0.021	0.007
Cr	0.007	0.022	0.0	0.02	0.003	0.022	0.001	0.021	0.013	0.027	0.000	0.005	0.021
Fe	0.837	0.425	1.269	0.42	1.36	0.585	1.070	0.596	0.478	0.541	1.577	1.040	0.535
Mn	0.0	0.0	0.0	0.009	0.002	0.0	0.011	0.014	0.012	0.009	0.020	0.014	0.008
Mg	0.659	0.936	0.345	0.74	0.20	1.187	0.036	1.126	0.656	0.664	0.008	0.641	1.288
Ca	0.440	0.577	0.347	0.73	0.40	0.176	0.825	0.215	0.733	0.690	0.367	0.248	0.142
Na	0.0	0.0	0.0	0.006	0.001	0.001	0.006	-	-	0.000	0.000	0.000	0.000
Total Cations	3.989	3.998	3.995	3.995	3.996	4.000	3.997	3.996	3.962	4.001	3.997	4.006	4.007
				Cation Ratios Ca: Mg: Fe and Fe/(Fe+Mg)									
Ca	22.8	29.8	17.7	38.6	20.4	9.0	42.7	11.3	39.3	36.2	18.6	12.8	7.2
Mg	34.0	48.3	17.6	39.2	10.2	61.0	1.9	57.5	35.1	34.9	0.4	33.0	65.6
Fe	43.2	21.9	64.7	22.2	69.4	30.0	55.4	31.2	25.6	28.9	81.0	54.2	27.2
Fe/(Fe+Mg)	0.56	0.31	0.79	0.36	0.87	0.33	0.97	0.35	0.42	0.45	0.99	0.62	0.29

Table A5.3. Pyroxene analyses from clast-poor melt rocks, crystalline melt breccias, and KREEP rocks.

	A-16 Melt Rocks			A-16 Cryst Melt Breccias			A-14 Melt Rocks			KREEP		
	2.	3.	4.	8.	9.	10.	13.	14.	15.	25.	26.	27.
Chemical Compositon (Weight Percent)												
SiO_2	52.8	49.8	53.2	53.6	51.8	53.4	53.5	45.2	53.2	51.8	49.5	53.5
Al_2O_3	1.6	0.9	1.1	1.8	2.4	1.7	0.81	1.3	3.7	1.9	1.1	1.6
TiO_2	0.50	1.1	0.23	1.1	1.6	1.2	0.40	1.5	0.7	40.85	0.82	0.51
Cr_2O_3	0.68	0.01	0.63	0.87	0.87	0.54	0.38	0.08	0.6	70.85	0.27	0.79
FeO	14.9	26.4	13.2	10.9	6.4	13.1	18.0	41.2	11.4	19.0	30.8	14.1
MnO	0.21	0.38	0.19	0.21	0.15	0.20	0.30	0.45	0.23	0.25	0.49	0.20
MgO	21.7	9.9	27.3	29.1	17.3	26.2	22.3	4.4	28.2	23.0	11.0	26.9
CaO	8.2	11.6	2.2	2.8	20.4	4.2	5.0	4.6	2.6	2.2	6.2	1.7
Na_2O	0.00	0.00	0.00	0.00	0.10	0.00	0.10	0.10	0.00	0.00	0.00	0.00
Total	100.59	100.09	98.05	100.38	101.02	100.54	100.79	98.83	100.74	99.85	100.18	99.30
Cation Formula Based on 6 Oxygens (Ideal Pyroxene = 4 Cations per 6 Oxygens)												
Si	1.932	1.958	1.952	1.905	1.888	1.919	1.962	1.911	1.881	1.920	1.958	1.941
^{IV}Al	0.068	0.042	0.047	0.075	0.103	0.072	0.035	0.065	0.119	0.081	0.042	0.059
Total tet	2.000	2.000	1.999	1.980	1.991	1.991	1.997	1.976	2.000	2.001	2.000	2.000
Ti	0.014	0.033	0.006	0.030	0.044	0.033	0.014	0.060	0.025	0.024	0.025	0.014
^{VI}Al	0.003	0.000	0.000	0.000	0.000	0.000	0.000	0.000	0.035	0.000	0.009	0.009
Cr	0.020	0.000	0.018	0.024	0.024	0.015	0.011	0.003	0.019	0.025	0.008	0.023
Fe	0.456	0.869	0.403	0.324	0.194	0.394	0.552	1.456	0.337	0.588	1.019	0.429
Mn	0.006	0.013	0.006	0.006	0.005	0.006	0.009	0.016	0.007	0.008	0.016	0.006
Mg	1.186	0.582	1.491	1.541	0.945	1.405	1.219	0.277	1.486	1.269	0.649	1.454
Ca	0.323	0.491	0.084	0.107	0.797	0.160	0.196	0.208	0.099	0.088	0.263	0.065
Na	0.000	0.000	0.000	0.000	0.006	0.000	0.007	0.005	0.000	0.000	0.000	0.000
Total Cations	4.008	3.988	4.007	4.012	4.006	4.004	4.005	4.001	4.008	4.003	3.989	4.000
Cation Ratios: Ca: Mg: Fe and Fe/(Fe+Mg)												
Ca	16.4	25.1	4.3	5.4	41.1	8.2	10.0	10.7	5.1	4.5	13.5	3.3
Mg	60.2	29.8	75.1	77.9	48.6	71.5	62.0	14.3	77.3	65.0	33.3	74.4
Fe	23.4	45.1	20.6	16.7	10.3	20.3	28.0	75.0	17.5	30.5	53.2	22.3
Fe/(Fe+Mg)	0.28	0.60	0.21	0.17	0.17	0.22	0.31	0.84	0.18	0.32	0.61	0.23

Table A5.4. Pyroxene analyses from ferroan anorthosites and Mg-rich rocks.

	Ferroan Anorthosites			Mg-rich Rocks						
				Norite			Troctolite		Dunite	
	2.	3.	5.	12.	13.	14.	16.	17.	18.	19.
Chemical Compositon (Weight Percent)										
SiO_2	53.42	52.71	53.00	54.18	53.31	55.3	55.9	53.48	56.05	54.13
Al_2O_3	0.68	0.28	0.51	1.02	1.10	1.30	1.5	1.00	0.96	1.22
TiO_2	0.36	0.17	0.24	0.29	0.66	0.20	0.4	0.53	0.28	0.11
Cr_2O_3	0.19	0.11	0.09	0.67	0.64	0.51	0.6	0.72	0.26	1.11
FeO	9.70	24.18	22.60	12.89	4.80	12.35	7.4	2.87	6.94	2.71
MnO	0.24	0.54	0.62	0.23	0.15	0.25	0.1	0.06	0.15	0.11
MgO	13.83	20.36	21.90	29.50	16.77	28.7	32.7	18.11	32.29	18.40
CaO	22.14	1.33	0.80	1.19	22.29	1.56	1.5	23.44	2.24	22.50
Na_2O	0.03	-	0.00	-	-	-	-	0.02	0.01	0.05
Total	100.59	99.68	99.76	99.97	99.72	100.17	100.1	100.23	99.18	100.34
Cation Formula Based on 6 Oxygens (Ideal Pyroxene = 4 Cations per 6 Oxygens)										
Si	1.98	1.99	1.984	1.940	1.958	1.967	1.945	1.946	1.967	1.959
^{IV}Al	0.02	0.01	0.016	0.044	0.042	0.033	0.055	0.043	0.033	0.041
Total tet	2.00	2.00	2.000	1.984	2.000	2.000	2.000	1.989	2.000	2.000
Ti	0.01	0.01	0.007	0.008	0.018	0.005	0.010	0.014	0.007	0.003
^{VI}Al	0.01	0.00	0.006	0.000	0.006	0.021	0.006	0.000	0.007	0.011
Cr	0.01	0.00	0.003	0.020	0.018	0.015	0.017	0.021	0.007	0.032
Fe	0.30	0.76	0.707	0.386	0.148	0.366	0.215	0.087	0.203	0.082
Mn	0.01	0.02	0.020	0.008	0.004	0.008	0.002	0.002	0.004	0.003
Mg	0.77	1.15	1.222	1.574	0.918	1.520	1.697	0.981	1.687	0.991
Ca	0.88	0.05	0.032	0.046	0.878	0.060	0.056	0.914	0.084	0.872
Na	-	-	0.000	-	-	-	-	0.002	0.001	0.003
Total Cations	3.99	3.990	3.997	4.026	3.990	3.995	4.003	4.010	4.000	3.997
Cation Ratios: Ca: Mg: Fe and Fe/(Fe+Mg)										
Ca	45.1	2.5	1.6	2.28	45.16	3	2.8	46.1	4.2	44.8
Mg	39.5	58.7	62.3	78.47	47.25	78	86.2	49.5	85.5	50.9
Fe	15.4	38.8	36.1	19.25	7.59	19	10.9	4.4	1.03	4.2
Fe/(Fe+Mg)	0.28	0.40	0.37	0.20	0.14	0.19	0.11	0.08	0.11	0.08

Table A5.5. Plagioclase analyses from mare basalts

	High-Ti Basalts						Low-Ti Basalts			Aluminous, Low-Ti		Very Low-Ti Basalts		
	Apollo 17		Apollo 11 High-K		Apollo Low-K		A-12 Ilmenite	A-12 Pigeonite	A-15 Olivine	L-16 Al	A-14 Al	VHK	Luna 24	A-17 VLT
	1.	2.	3.	4.	5.	6.	7.	8.	9.	10.	11.	12.	13.	15.
Chemical Compositon (Weight Percent)														
SiO_2	46.9	48.6	49.8	48.30	45.72	46.30	48.10	44.50	46.70	45.18	45.10	44.09	45.54	45.03
Al_2O_3	34.5	31.7	30.50	32.40	34.53	34.00	32.00	35.20	31.80	33.87	33.90	35.18	34.31	34.93
FeO	0.28	0.63	0.81	0.51	0.30	0.99	1.17	0.81	2.62	0.54	0.29	0.25	0.92	0.90
MgO	0.26	0.22	0.25	0.22	0.07	0.30	0.08	0.08	1.16	0.16	0.0	-	0.06	0.26
CaO	17.4	17.0	15.80	16.60	18.64	17.10	17.50	19.20	17.30	18.80	18.80	19.54	18.89	19.42
Na_2O	1.34	1.55	1.95	1.84	0.70	1.05	0.40	0.58	0.88	0.86	0.88	0.55	0.65	0.46
K_2O	0.05	0.02	0.25	0.19	0.01	0.09	0.20	0.02	0.06	0.06	0.03	0.03	0.00	0.03
Total	100.73	99.72	99.36	100.06	99.97	99.83	99.45	100.39	100.52	99.47	99.00	99.64	100.37	101.03
Cation Formula Based on 8 Oxygens (Ideal Pyroxene = 5 Cations per 8 Oxygens)														
Si	2.141	2.237	2.295	2.216	2.108	2.136	2.222	2.056	2.160	2.100	2.105	2.048	2.100	2.067
^{iv}Al	1.853	1.721	1.657	1.753	1.877	1.849	1.743	1.917	1.734	1.850	1.865	1.927	1.865	1.891
Total tet	3.994	3.958	3.952	3.969	3.985	3.985	3.965	3.973	3.894	3.950	3.970	3.975	3.965	3.958
Fe	0.011	0.024	0.031	0.020	0.012	0.038	0.045	0.031	0.101	0.020	0.011	0.009	0.036	0.034
Mg	0.018	0.015	0.017	0.015	0.005	0.021	0.006	0.006	0.080	0.010	0.000	-	0.004	0.018
Ca	0.850	0.836	0.780	0.816	0.921	0.845	0.866	0.950	0.857	0.930	0.940	0.972	0.933	0.955
Na	0.119	0.138	0.174	0.164	0.063	0.094	0.036	0.052	0.079	0.080	0.080	0.049	0.058	0.041
K	0.003	0.001	0.015	0.011	0.001	0.005	0.012	0.001	0.004	0.004	0.002	0.001	0.000	0.002
Total Cations	4.995	4.972	4.969	4.995	4.987	4.988	4.930	5.013	5.015	4.994	5.003	5.006	4.996	5.008
Cation Ratio Fe/Fe(Fe+Mg), and Molecular Proportion of Orthoclase (Or), Albite (Ab), and Anorthite (An)														
Fe/(Fe+Mg)	0.38	0.62	0.65	0.57	0.71	0.65	0.89	0.85	0.56	0.67	1.00	1.00	0.90	0.65
Or	0.3	0.1	1.5	1.1	0.1	0.5	1.3	0.1	0.4	0.4	0.2	0.1	0.0	0.2
Ab	12.2	14.1	18.0	16.6	6.4	10.0	3.9	5.2	8.4	7.9	7.8	4.8	5.8	4.1
An	87.5	85.7	80.5	82.3	93.5	89.5	94.8	94.7	91.2	91.7	92.0	95.1	94.2	95.7

Table A5.6. Plagioclase analyses from highland clast-poor melt rocks, crystalline melt breccias, and KREEP rocks.

	A-16 Melt Rocks			A-16 Cryst Melt Breccias			A-14 Melt Rocks			KREEP		
	2.	3.	4.	7.	8.	9.	12.	13.	14.	23.	24.	25.
Chemical Compositon (Weight Percent)												
SiO_2	47.6	43.9	45.4	44.6	45.8	43.8	47.4	43.8	46.4	49.5	47.1	51.3
Al_2O_3	33.3	36.7	35.7	35.9	34.3	36.1	33.4	36.1	34.0	30.8	32.5	30.8
FeO	0.52	0.15	0.46	0.12	0.18	0.15	0.39	0.15	0.06	0.38	0.16	0.48
MgO	0.00	0.05	0.07	0.01	0.07	0.00	0.14	0.19	0.29	0.10	0.17	0.08
CaO	17.3	20.2	18.1	19.3	17.9	19.7	17.2	19.2	17.1	15.9	17.1	14.7
Na_2O	1.6	0.27	0.97	0.40	0.92	0.39	1.6	0.51	1.4	2.2	1.5	2.4
K_2O	0.10	0.01	0.09	0.03	0.06	0.01	0.29	0.05	0.21	0.34	0.11	0.33
Total	100.42	101.28	100.79	100.36	99.23	100.15	100.42	100.00	99.46	99.22	98.64	100.09
Cation Formula Based on 8 Oxygens (Ideal Pyroxene = 5 Cations per 8 Oxygens)												
Si	2.181	2.009	2.078	2.051	2.123	2.024	2.173	2.026	2.143	2.285	2.191	2.334
^{IV}Al	1.796	1.980	1.926	1.948	1.874	1.968	1.804	1.968	1.851	1.676	1.783	1.650
Total tet	3.977	3.989	4.004	3.999	3.997	3.992	3.977	3.994	3.994	3.961	3.974	3.984
Fe	0.020	0.006	0.018	0.005	0.007	0.006	0.015	0.006	0.002	0.015	0.006	0.018
Mg	0.000	0.003	0.005	0.001	0.005	0.000	0.010	0.013	0.020	0.007	0.012	0.006
Ca	0.850	0.991	0.888	0.953	0.888	0.976	0.845	0.952	0.846	0.787	0.855	0.719
Na	0.139	0.024	0.086	0.036	0.084	0.035	0.046	0.125	0.060	0.133	0.210	0.003
K	0.006	0.001	0.005	0.002	0.003	0.000	0.017	0.003	0.012	0.020	0.007	0.019
Total Cations	4.992	5.014	5.006	4.996	4.984	5.009	5.006	5.014	4.999	4.986	4.987	4.956
Fe/(Fe+Mg)	1.00	0.66	0.79	0.83	0.58	1.00	0.61	0.21	0.10	0.68	0.34	0.76
Or	0.60	0.10	0.50	0.10	0.30	0.00	1.70	0.30	1.30	2.00	0.70	2.00
Ab	14.00	2.40	8.80	3.60	8.50	3.50	14.20	4.60	12.70	19.60	13.30	22.10
An	85.40	97.60	90.70	96.20	91.20	96.50	84.10	95.10	86.00	78.40	86.00	75.90

Table A5.7. Plagioclase analyses from ferroan anorthosites and Mg-rich rocks.

				Mg-rich Rocks			
	Ferroan Anorthosites			Norite		Troctolite	
	2.	3.	4.	11.	12.	13.	14.
Chemical Compositon (Weight Percent)							
SiO_2	43.92	44.6	44.6	46.48	45.27	43.40	44.1
Al_2O_3	36.24	35.2	36.0	32.75	35.18	35.41	35.3
FeO	0.09	0.20	0.18	1.21	0.10	0.18	0.04
MgO	-	0.06	0.07	1.51	0.08	0.11	-
CaO	19.49	20.0	19.5	17.77	18.79	19.22	18.7
Na_2O	0.26	0.35	0.38	0.71	0.66	0.58	0.43
K_2O	-	0.01	0.03	0.09	0.14	0.08	0.07
Total	100.00	100.4	100.8	100.52	100.22	98.98	98.64
Cation Formula Based on 8 Oxygens (Ideal P.agioclase = 5 Cations per 8 Oxygens)							
Si	2.03	2.057	2.046	2.137	2.083	2.033	2.063
^{IV}Al	1.97	1.915	1.947	1.775	1.908	1.952	1.943
Total tet	4.00	3.972	3.993	3.912	3.991	3.985	4.006
Fe	0.00	0.008	0.007	0.047	0.004	0.007	0.002
Mg	-	0.004	0.005	0.104	0.005	0.008	-
Ca	0.97	0.987	0.959	0.876	0.927	0.964	0.937
Na	0.02	0.031	0.034	0.063	0.059	0.052	0.038
K	-	0.001	0.002	0.005	0.008	0.004	0.004
Total Cations	4.99	5.003	5.000	5.007	4.994	5.020	4.987
Cation Ratios Fe/(Fe+Mg), and Molecular Proportin of Orthoclase (Or), Albite (Ab), and Anorthite (An)							
Fe/(Fe+Mg)	1.00	0.67	0.58	0.31	0.44	0.47	1.00
Or	-	0.05	0.2	0.6	0.9	0.4	0.4
Ab	2.0	3.0	3.4	6.7	5.9	5.1	3.9
An	98.0	96.9	96.4	92.7	93.2	94.5	95.7

Table A5.8. Olivine analyses from mare basalts

	High-Ti Basalts					Low-Ti Basalts								Very Low-Ti Basalts			
			A-11	A-11		A-12		A-12	A-12		A-15						
	Apollo 17		HK	LK		ilm		pig	ol		ol	VHK		Luna 24		Apollo 17	
	1.	2.	3.	5.	6.	7.	8.	9.	10.	11.	12.	13.	14.	15.	16.	18.	19.
Chemical Compositon (Weight Percent)																	
SiO_2	37.60	37.53	37.50	38.60	29.20	38.08	30.60	34.10	37.80	35.40	29.70	37.46	33.50	33.37	29.93	34.99	37.98
Al_2O_3	0.02	0.04	0.04	0.00	0.30	0.02	0.00	0.32	0.00	0.06	-	-	-	0.20	0.29	0.20	0.18
FeO	28.44	27.44	26.20	25.40	68.70	22.15	65.30	39.60	21.10	33.80	64.80	24.71	49.79	48.61	67.15	39.35	23.06
MgO	34.66	35.93	35.80	34.90	0.50	38.86	3.54	23.70	39.20	30.10	3.80	37.39	15.75	17.88	0.97	26.42	37.94
MnO	0.31	0.28	0.22	0.36	1.10	0.24	0.00	0.52	0.24	0.36	0.66	0.36	0.71	0.45	0.74	0.42	0.29
Cr_2O_3	0.19	0.13	0.21	0.28	0.02	0.35	0.00	0.24	0.44	0.11	0.04	-	-	0.12	0.10	0.15	0.55
CaO	0.33	0.29	0.28	0.31	0.26	0.27	0.65	0.59	0.30	0.36	0.40	0.23	0.32	0.48	1.03	0.29	0.41
Na_2O	0.01	0.00	0.00	0.00	0.00	0.01	0.00	0.00	0.00	0.00	0.00	-	-	-	0.00	-	-
Total	101.56	101.61	100.25	99.85	100.08	99.98	100.09	99.07	99.08	100.13	99.46	100.15	100.07	101.11	100.21	101.82	100.41
Cation Formula Based on 4 Oxygens (Ideal Olivine = 3 Cations per 4 Oxygens)																	
Si	0.993	0.986	0.993	1.020	0.987	0.992	1.005	0.984	0.990	0.978	0.987	0.986	1.006	0.985	0.996	0.975	0.989
Al	0.001	0.000	0.001	0.000	0.012	0.001	0.000	0.011	0.000	0.000	0.002	-	-	0.007	0.011	0.006	0.006
Fe	0.628	0.603	0.580	0.561	1.942	0.483	1.794	0.956	0.462	1.781	1.800	0.544	1.251	1.200	1.870	0.917	0.502
Mg	1.364	1.407	1.413	1.374	0.025	1.509	0.173	1.020	1.530	1.240	0.188	1.468	0.705	0.786	0.048	1.097	1.473
Mn	0.007	0.006	0.005	0.008	0.031	0.005	0.000	0.013	0.005	0.008	0.019	0.007	0.018	0.011	0.021	0.010	0.006
Cr	0.004	0.003	0.004	0.006	0.001	0.007	0.000	0.005	0.009	0.002	0.001	-	-	0.003	0.002	0.003	0.011
Ca	0.009	0.008	0.008	0.009	0.009	0.008	0.023	0.018	0.008	0.011	0.014	0.006	0.009	0.015	0.037	0.009	0.012
Na	0.001	0.000	0.000	0.000	0.000	0.001	0.000	0.000	0.000	0.000	0.000	-	-	-	0.000	-	-
Total Cations	3.007	3.013	3.004	2.978	3.007	3.006	2.995	3.007	3.004	3.020	3.011	3.011	2.989	3.007	2.985	3.017	2.999
Cation Ratios Fe/(Fe+Mg)																	
Fe/(Fe+Mg)	0.32	0.30	0.29	0.29	0.99	0.24	0.91	0.48	0.23	0.39	0.91	0.27	0.64	0.60	0.97	0.46	0.25

Table A5.9. Olivine analyses from highland clast-poor melt rocks and crystalline melt breccias.

	2.	3.	6.	7.	10.	11.	16.	17.
				Chemical Compositon (Weight Percent)				
SiO_2	38.2	37.4	38.9	38.8	39.0	39.1	38.4	38.8
Al_2O_3	0.06	0.02	0.10	0.11	0.16	0.10	0.17	0.14
FeO	25.4	28.8	18.0	20.2	17.4	20.1	22.8	22.2
MgO	37.6	34.5	42.0	41.2	43.0	41.4	39.3	38.3
MnO	0.33	0.28	0.21	0.27	0.14	0.19	0.25	0.15
Cr_2O_3	0.10	0.07	0.22	0.20	0.09	0.08	0.20	0.07
CaO	0.16	0.16	0.13	0.17	0.15	0.16	0.19	0.26
Total	101.85	101.23	99.56	100.95	99.94	101.13	101.31	99.92
				Cation Formula Based on 4 Oxygens (Ideal Olivine = 3 Cations per 4 Oxygens)				
Si	0.989	0.991	0.995	0.990	0.992	0.994	0.988	1.006
Al	0.002	0.001	0.003	0.003	0.005	0.003	0.005	0.004
Fe	0.550	0.638	0.385	0.430	0.369	0.427	0.490	0.481
Mg	1.451	1.363	1.604	1.566	1.630	1.569	1.506	1.480
Mn	0.007	0.006	0.004	0.006	0.003	0.004	0.005	0.003
Cr	0.002	0.001	0.004	0.004	0.002	0.002	0.004	0.001
Ca	0.004	0.005	0.004	0.005	0.004	0.004	0.005	0.007
Total Cations	3.005	3.005	2.999	3.004	3.005	3.003	3.003	2.982
				Cation Ratios Fe/(Fe+Mg)				
Fe/(Fe+Mg)	0.27	0.32	0.20	0.22	0.19	0.22	0.25	0.24

Table 5.10. Olivine analyses from ferroan anorthosites and Mg-rich rocks.

	Ferroan Anorthosites		Mg-rich Rocks: Troctolite			Mg-rich Rocks: Dunite		
	1.	2.	3.	4.	5.	6.	7.	8.
				Chemical Compositon (Weight Percent)				
SiO_2	37.1	35.59	40.3	39.9	40.85	40.24	40.13	39.84
Al_2O_3	0.08	0.0	0.0	-	0.0	<0.01	0.05	0.00
FeO	31.8	34.58	12.3	12.0	11.0	12.29	13.00	13.13
MgO	30.9	30.11	47.96	47.1	48.45	47.65	48.14	48.30
MnO	-	0.42	0.16	0.1	0.11	0.13	0.15	0.17
Cr_2O_3	-	0.05	0.02	<0.02	0.04	0.04	0.05	0.05
CaO	0.06	0.03	0.03	<0.02	0.09	0.13	0.13	0.08
Na_2O	-	-	-	-	0.05	-	0.06	0.02
Total	99.94	100.78	100.77	99.10	100.59	100.48	101.71	101.59
				Cation Formula Based on 4 Oxygens (Ideal Olivine = 3 Cations per 4 Oxygens)				
Si	1.009	0.979	0.991	0.994	1.000	0.993	0.980	0.978
Al	0.003	-	-	-	-	-	0.001	-
Fe	0.723	0.795	0.253	0.250	0.225	0.254	0.265	0.269
Mg	1.253	1.235	1.760	1.750	1.768	1.753	1.752	1.768
Mn	-	0.010	0.003	0.001	0.002	0.003	0.003	0.003
Cr	-	0.001	0.000	0.000	0.001	0.001	0.001	0.001
Ca	0.002	0.001	0.001	0.000	0.002	0.003	0.003	0.002
Na	-	-	-	-	0.002	-	0.003	0.001
Total Cations	2.990	3.014	3.008	2.995	3.000	3.007	3.008	3.022
				Cation Ratios Fe/(Fe+Mg)				
Fe/(Fe+Mg)	0.37	0.39	0.13	0.12	0.11	0.13	0.13	0.13

Table A.511a. Analyses of miscellaneous lunar silicate minerals

	Zircon					Pyroxferroite			K-Feldspar				Silica Polymorphs			
						A-11		A-12			A-12	A-14	A-11	A-11		
	Soil	Reg br		Granite		LK		pig	Granite	Soil	pig	Al	HK	LK		
	1.	2.	3.	4.	5.	6.	7.	8.	9.	10.	11.	12.	13.	14.	15.	16.
	Chemical Compositon (Weight Percent)															
SiO_2	32.2	35.5	32.11	32.41	32.41	46.92	45.86	45.23	60.93	61.0	57.2	63.49	98.0	98.0	97.6	96.9
Al_2O_3	0.09	1.3	-	0.2	-	0.76	0.36	0.54	22.52	20.5	20.1	19.44	1.56	0.92	1.1	1.03
TiO_2	-	0.2	-	0.19	-	0.74	0.37	0.73	-	-	-	-	0.48	0.27	0.2	-
Cr_2O_3	-	-	-	0.03	-	0.12	0.08	0.05	-	-	-	-	-	-	-	-
FeO	0.32	0.8	-	0.02	0.35	42.48	44.33	46.54	0.13	-	0.35	0.03	0.18	0.05	0.3	0.35
MnO	-	-	-	-	-	0.65	0.76	0.56	-	-	-	-	-	-	-	
MgO	0.18	0.1	-	0.19	-	2.41	0.90	0.34	0.0	-	-	-	-	-	-	-
CaO	-	-	-	0.26	-	6.69	6.56	6.43	3.87	2.07	0.55	0.35	0.34	0.16	0.3	0.52
Na_2O	0.09	-	-	-	-	-	-	0.03	3.46	1.24	0.81	0.74	0.19	0.15	0.0	0.05
K_2O	-	-	-	-	-	-	-	-	7.99	10.9	11.1	14.76	-	-	0.0	0.25
ZrO_2	64.7	61.5	67.23	66.93	63.48	-	-	-	-	-	-	-	-	-	-	-
HfO_2	-	0.6	0.88	-	3.01	-	-	-	-	-	-	-	-	-	-	-
BaO	-	-	-	-	-	-	-	-	1.19	2.73	9.3	0.44	-	-	-	-
Total	97.58	100.0	100.22	100.23	99.25	100.77	99.22	100.45	100.09	98.44	99.41	99.24	100.75	99.55	99.5	99.10
	Cation Compositions Based on Relevant Number of Oxygens															
Si	1.004	1.054	0.986	0.987	1.01	1.965	1.978	1.95	2.788	2.88	2.819	2.954	0.979	0.988	0.986	0.985
Al	0.003	0.045	-	0.007	-	0.038	0.018	0.03	1.215	1.14	1.168	1.064	0.018	0.011	0.013	0.012
Ti	-	0.004	-	0.009	-	0.023	0.012	0.02	-	-	-	-	0.004	0.002	0.002	-
Cr	-	-	-	0.001	0.004	0.003	0.00	-	-	-	-	-	-	-	-	-
Fe	0.008	0.020	-	0.001	0.009	1.488	1.599	1.68	0.005	-	0.014	0.000	0.002	0.000	0.003	0.003
Mn	-	-	-	-	-	0.023	0.028	0.02	-	-	-	-	-	-	-	-
Mg	0.008	0.004	-	0.004	-	0.150	0.058	0.02	-	-	-	-	-	-	-	-
Ca	-	-	-	0.008	-	0.300	0.303	0.30	0.190	0.105	0.029	0.016	0.004	0.002	0.003	0.006
Na	0.005	-	-	-	-	-	-	0.00	0.307	0.113	0.077	0.065	0.004	0.003	-	0.001
K	-	-	-	-	-	-	-	-	0.466	0.656	0.698	0.875	-	-	-	0.003
Zr	0.984	0.890	1.007	0.994	0.962	-	-	-	-	-	-	-	-	-	-	-
Hf	-	0.005	0.008	-	0.027	-	-	-	-	-	-	-	-	-	-	-
Ba	-	-	-	-	-	-	-	-	0.021	0.050	0.180	0.006	-	-	-	-
Total	2.012	2.022	2.001	2.011	2.008	3.991	3.999	4.02	4.992	4.944	4.985	4.980	1.011	1.006	1.007	1.010
No. of Oxygens	4	4	4	4	4	6	6	6	8	8	8	8	2	2	2	2

Table A.5.11b. Tranquillityite analyses from mare basalts.

	A-11			A-12	A-12
	LK			ol	pig
	1.	2.	3.	6.	7.
	Chemical Compositon (Weight Percent)				
SiO_2	13.66	13.77	13.98	13.00	14.7
Al_2O_3	0.87	0.90	0.83	0.70	1.71
TiO_2	19.75	20.66	20.01	17.50	19.7
Cr_2O_3	0.06	0.19	0.13	-	-
FeO	43.00	42.37	42.90	41.78	42.3
MnO	0.36	0.30	0.34	0.25	0.22
CaO	1.04	1.11	1.17	1.00	1.53
Zr_2O_3	16.96	16.79	16.35	17.80	17.3
HfO_2	0.05	0.06	0.04	0.60	-
Y_2O_3	2.51	2.73	2.61	4.71	1.34
Total	98.26	98.88	98.36	97.34	98.80
	Cation Formula Based on 24 Oxygens				
Si	2.859	2.849	2.906	2.805	3.009
Al	0.141	0.151	0.094	0.178	0.000
Subtotal	3.000	3.000	3.000	2.983	3.009
Ti	2.916	2.900	2.870	2.839	2.587
Al	0.074	0.069	0.109	0.000	0.413
Cr	0.010	0.031	0.021	0.000	0.000
Subtotal	3.000	3.000	3.000	2.839	3.000
Fe	7.527	7.332	7.458	7.538	7.241
Ti	0.193	0.315	0.258	0.000	0.446
Mn	0.064	0.053	0.060	0.046	0.038
Ca	0.233	0.246	0.261	0.231	0.336
Subtotal	8.017	7.946	8.037	7.815	8.061
Zr	1.731	1.694	1.657	1.873	1.727
Hf	0.003	0.004	0.002	0.037	0.000
Y	0.279	0.300	0.288	0.540	0.146
Subtotal	2.013	1.998	1.947	2.450	1.873
Total cations	16.030	15.944	15.984	16.087	15.943

Table A5.12. Ilmenite analyses from lunar rocks and soils.

	A-11 HK		A-11 LK	A-12 ol	A-11 ol	A-12 pig	A-12 ilm	A-12 pig	A-12 ilm
	3.	4.	6.	7.	8.	9.	10.	11.	12.
	Chemical Compositon (Weight Percent)								
TiO_2	52.22	54.53	53.0	53.8	53.0	52.25	52.34	52.91	53.9
Al_2O_3	<0.03	<0.03	-	0.02	0.03	0.04	0.28	1.13	0.42
Cr_2O_3	0.53	0.53	0.52	0.51	0.75	0.18	0.24	0.19	0.39
V_2O_3	-	-	-	0.07	0.04	-	-	-	0.04
FeO	44.38	44.84	45.1	40.5	44.1	46.74	45.16	46.30	45.4
MgO	1.39	1.14	0.75	5.14	2.28	0.47	0.39	0.01	0.10
MnO	0.15	0.015	0.45	0.42	0.41	0.30	0.34	0.31	0.36
CaO	0.21	0.21	0.10	-	-	-	-	-	-
ZrO_2	-	-	-	-	-	-	-	-	-
SiO_2	0.09	0.09	0.23	-	-	-	-	-	
Total	98.97	101.49	100.15	100.46	100.61	99.98	9875	100.85	100.61
	Cation Formula Based on 3 Oxygens (Ideal Ilmenite = 2 Cations per 3 Oxygens)								
Ti	0.989	1.014	0.996	0.983	0.985	0.991	1.000	0.988	1.007
Al	0.000	0.000	-	0.001	0.001	0.001	0.008	0.033	0.012
Cr	0.010	0.010	0.010	0.010	0.015	0.004	0.005	0.004	0.008
V	-	–	-	0.001	0.001	-	-	-	0.001
Fe	0.935	0.918	0.942	0.822	0.912	0.986	0.959	0.962	0.943
Mg	0.052	0.042	0.028	0.186	0.084	0.018	0.015	0.000	0.004
Mn	0.003	0.003	0.009	0.009	0.009	0.006	0.007	0.007	0.008
Ca	0.006	0.006	0.003	-	-	-	-	-	-
Zr	-	-	-	-	-	-	-	-	-
Si	0.002	0.002	0.006	-	-	-	-	-	-
Total	1.997	1.995	1.994	2.012	2.007	2.006	1.994	1.994	1.983

	A-12 ol	Mg Rock (anorth.)	Glassy br	Melt rock	A-14 Al	A-15 ol	A-17 Basalt	Luna 16 Soil	Luna 20 Soil
	13.	14.	15.	16.	17.	18.	19.	20.	22.
	Chemical Compositon (Weight Percent)								
TiO_2	52.7	51.7	59.35	52.66	53.24	50.70	53.1	52.00	53.7
Al_2O_3	0.12	0.19	0.60	0.10	0.09	0.09	0.15	0.17	-
Cr_2O_3	0.23	0.58	0.85	0.38	0.35	0.23	0.39	0.58	0.39
V_2O_3	0.06	-	-	-	-	-	-	-	-
FeO	46.7	37.7	34.30	44.62	45.30	46.84	45.7	44.70	37.1
MgO	0.09	8.2	3.85	0.83	0.72	0.27	0.72	0.86	8.0
MnO	0.37	0.20	-	0.32	0.47	0.52	0.44	0.50	0.36
CaO	-	0.31	-	0.12	-	0.16	-	0.11	0.58
ZrO_2	-	-	-	-	0.05	-	0.04	-	-
SiO_2	-	0.30	-	0.39	0.16	0.50	-	0.23	0.18
Total	100.27	99.18	98.95	99.42	100.38	99.31	100.54	99.15	100.31
	Cation Formula Based on 3 Oxygens (Ideal Ilmenite = 2 Cations per 3 Oxygens								
Ti	0.998	0.943	1.062	0.994	0.998	0.970	0.996	0.987	0.966
Al	0.004	0.005	0.017	0.003	0.003	0.003	0.004	0.005	-
Cr	0.005	0.011	0.016	0.008	0.007	0.005	0.008	0.012	0.007
V	0.000	-	-	-	-	-	-	-	-
Fe	0.983	0.765	0.682	0.936	0.944	0.997	0.953	0.943	0.742
Mg	0.004	0.297	0.136	0.031	0.027	0.010	0.027	0.032	0.285
Mn	0.008	0.004	-	0.007	0.010	0.011	0.009	0.011	0.007
Ca	-	0.008	-	0.003	-	0.004	-	0.003	0.015
Zr	-	-	-	-	0.001	-	0.000	-	-
Si	-	0.007	-	0.010	0.004	0.013	-	0.006	0.004
Total	2.002	2.040	1.913	1.992	1.994	2.013	1.997	1.999	2.026

*Columns 3 and 4 are averages of multiple analyses.

Table A5.13. Spinel analyses from lunar rocks and soils.

	A-11 LK		A-12 ol	A-12 pig	A-12 ilm	A-12 pig	A-12 ilm		A-12 ol		A-12 ilm		A-12 ol		A-12 pig	
	1.	2.	4.	5.	6.	7.	9.	10.	12.	13.	14.	15.	16.	17.	18.	19.
Chemical Compositon (Weight Percent)																
TiO_2	20.9	21.2	3.8	4.7	6.27	7.90	10.8	20.6	23.7	25.3	24.70	24.8	28.6	29.1	32.0	33.3
Al_2O_3	8.61	4.20	12.2	12.5	11.0	12.17	11.0	8.00	4.03	3.03	5.89	5.18	3.3	2.35	2.3	2.1
Cr_2O_3	23.5	21.9	49.1	48.8	43.8	41.61	34.9	24.1	17.6	16.6	15.59	15.1	11.9	8.71	2.8	1.2
V_2O_3	0.4	-	-	-	0.74	-	0.92	0.4	0.42	0.38	-	0.75	-	0.19	0.3	-
FeO	42.1	46.4	26.8	26.3	33.3	35.98	37.8	44.4	51.7	52.8	49.55	51.6	54.1	57.4	63.3	62.8
MgO	4.23	4.74	7.8	7.6	3.92	1.26	3.94	2.6	0.81	2.01	2.78	2.63	2.7	0.59	0.1	0.1
MnO	0.25	0.50	-	0.4	0.27	-	0.20	0.3	0.53	0.41	0.37	0.39	0.4	1.31	0.3	0.4
CaO	<0.03	0.15	-	-	-	0.15	-	-	-	-	0.34	-	-	-	-	-
SiO_2	-	0.46	-	-	-	0.36	-	-	-	-	0.18	-	-	-	-	-
ZrO_2	<0.10	-	-	-	-	-	-	-	-	-	-	-	-	-	-	-
Total	99.99	99.55	99.70	100.30	99.30	99.43	99.56	100.40	98.79	100.53	99.40	100.45	101.0	99.65	101.1	99.9
Cation Formula Based on 4 Oxygens (Ideal Spinel = 3 Cations per 4 Oxygens)																
Ti	0.541	0.566	0.096	0.117	0.163	0.205	0.281	0.537	0.654	0.685	0.659	0.662	0.764	0.807	0.880	0.924
Al	0.349	0.176	0.481	0.488	0.449	0.493	0.446	0.327	0.174	0.129	0.246	0.217	0.138	0.102	0.100	0.091
Cr	0.638	0.615	1.299	1.279	1.198	1.138	0.953	0.660	0.510	0.473	0.438	0.424	0.334	0.254	0.080	0.035
V	0.011	-	-	-	0.021	-	0.025	0.012	0.012	0.011	-	0.021	-	0.006	0.009	-
Fe	1.213	1.378	0.750	0.729	0.962	1.060	1.093	1.287	1.587	1.589	1.471	1.536	1.608	1.770	1.938	1.938
Mg	0.217	0.251	0.389	0.375	0.203	0.065	0.203	0.132	0.044	0.108	0.147	0.139	0.143	0.032	0.007	0.011
Mn	0.007	0.015	-	0.011	0.007	-	0.006	0.008	0.017	0.013	0.011	0.012	0.012	0.041	0.010	0.013
Ca	0.000	0.006	-	-	-	0.005	-	-	-	-	0.013	-	-	-	-	-
Si	-	0.016	-	-	-	0.012	-	-	-	-	0.006	-	-	-	-	-
Zr	0.000	-	-	-	-	-	-	-	-	-	-	-	-	-	-	-
Total	2.976	3.023	3.015	2.999	3.003	2.978	3.007	2.963	2.998	3.008	2.991	3.011	2.999	3.012	3.024	3.012

	A-12 pig				A-14 Al			Melt Rocks			A-15 pig		A-15 ol
	20.	21.	23. core*	25. rim*	27.	28.	29.	33.	34.	35.	36.	37.	38.
Chemical Compositon (Weight Percent)													
TiO_2	33.6	33.8	4.8	29.9	4.95	3.31	2.79	30.23	32.4	32.3	1.71	2.46	4.10
Al_2O_3	2.3	1.6	12.5	2.5	16.11	20.1	21.74	2.21	2.10	1.33	10.64	10.84	11.76
Cr_2O_3	0.8	0.2	48.6	7.1	39.56	39.1	38.17	4.52	0.97	0.32	54.87	53.30	46.28
V_2O_3	0.3	0.2	-	-	0.70	0.63	0.56	-	0.01	0.04	1.33	-	-
FeO	63.1	62.9	28.0	59.6	33.29	31.7	29.24	60.53	62.5	63.5	24.38	26.38	35.21
MgO	0.1	0.0	6.2	0.3	4.02	4.62	5.98	0.87	0.66	0.26	7.23	6.15	2.35
MnO	0.3	0.3	0.4	0.4	0.25	0.24	0.18	0.40	0.44	0.47	0.27	0.41	0.44
CaO	-	-	-	-	-	-	-	0.07	-	-	0.05	-	0.03
SiO_2-	-	-	-	-	-	-	-	0.30	-	-	0.19	-	0.19
ZrO_2	-	-	-	-	0.00	0.04	0.00	-	0.19	0.08	-	-	-
Total	100.5	99.0	100.5	99.8	98.88	99.74	98.66	99.13	99.27	98.30	100.67	99.54	100.36
Cation Formula Based on 4 Oxygens (Ideal Spinel = 2 Cations per 4 Oxygens)													
Ti	0.926	0.951	0.120	0.829	0.127	0.082	0.069	0.845	0.908	0.923	0.043	0.063	0.107
Al	0.100	0.069	0.491	0.109	0.646	0.781	0.840	0.097	0.092	0.059	0.418	0.434	0.480
Cr	0.024	0.005	1.281	0.207	1.063	1.020	0.989	0.133	0.029	0.010	1.447	1.433	1.266
V	0.009	0.007	-	-	0.019	0.017	0.015	-	0.000	0.001	0.036	-	-
Fe	1.936	1.968	0.781	1.838	0.947	0.875	0.801	1.881	1.948	2.017	0.680	0.750	1.019
Mg	0.007	0.000	0.308	0.017	0.204	0.227	0.292	0.048	0.037	0.015	0.360	0.312	0.121
Mn	0.009	0.009	0.011	0.012	0.008	0.007	0.005	0.013	0.013	0.015	0.008	0.012	0.013
Ca	-	-	-	-	-	-	-	0.003	-	-	0.002	-	0.001
Si	-	-	-	-	-	-	-	0.011	-	-	0.006	-	0.007
Zr	-	-	-	-	0.000	0.001	0.000	-	0.003	0.001	-	-	-
Total	3.011	3.009	2.992	3.012	3.014	3.010	3.011	3.031	3.030	3.041	3.000	3.004	3.014

	A-15 ol				L-16 Soil		L-16 Al	L-20 Soil	
	39.	41.	43.	45.	47.	49.	51.	54.	55.
Chemical Compositon (Weight Percent)									
TiO_2	6.10	13.85	21.10	28.14	0.83	6.93	30.16	15.22	0.54
Al_2O_3	11.35	8.04	5.03	2.82	17.36	23.33	1.90	6.67	45.50
Cr_2O_3	42.79	30.54	19.85	9.16	49.00	30.76	6.68	36.32	18.47
V_2O_3	-	-	-	-	-	-	0.18	-	-
FeO	34.35	42.71	50.11	56.19	25.33	30.11	60.04	35.70	24.18
MgO	3.35	3.23	2.20	1.58	7.05	7.90	0.17	4.89	10.76
MnO	0.46	0.33	0.44	0.45	0.53	0.38	0.31	0.35	0.20
CaO	0.20	0.33	0.29	0.23	0.20	0.01	0.05	0.07	0.16
SiO_2	0.41	0.48	0.31	0.45	0.41	0.13	0.15	0.17	0.43
ZrO_2	-	-	-	-	-	-	-	-	-
Total	0.99.01	99.51	99.33	99.02	100.71	99.55	99.64	99.39	100.24
Cation Formula Based on 4 Oxygens (Ideal Spinel = 2 Cations per 4 Oxygens)									
Ti	0.0159	0.367	0.574	0.776	0.020	0.116	0.840	0.379	0.012
Al	0.465	0.334	0.214	0.122	0.664	0.875	0.083	0.279	1.526
Cr	1.175	0.852	0.567	0.266	1.257	0.774	0.196	0.989	0.415
V	-	-	-	-	-	-	0.005	-	-
Fe	0.998	1.260	1.515	1.724	0.688	0.801	1.861	1.058	0.575
Mg	0.174	0.170	0.119	0.086	0.341	0.375	0.009	0.258	0.456
Mn	0.014	0.010	0.013	0.014	0.015	0.010	0.010	0.011	0.005
Ca	0.007	0.012	0.011	0.009	0.007	0.000	0.002	0.003	0.005
Si	0.014	0.017	0.011	0.017	0.013	0.004	0.006	0.006	0.012
Zr	-	-	-	-	-	-	-	-	-
Total	3.007	3.022	3.024	3.014	3.005	2.955	3.012	2.983	3.006

*Analyses 22 to 26 show the core-to-rim zonation in one spinel crystal

Table A5.14. Armalcolite analyses from lunar rocks and soils.

	Armalcolite Type 1:Fe-Mg Armalcolite												
	A-11 HK		Frag br	Reg br	A-11 Soil	Reg br	A-14 Soil	A-17 Soil		A-17 Basalt			
	1.	2.	3.	4.	5.	6.	7.	8.	9.	11.	12.	14. core*	15. rim*
	Chemical Compositon (Weight Percent)												
TiO_2	70.9	73.4	75.6	72.0	71.9	75.15	69.41	73.4	73.2	72.93	71.61	74.3	72.5
FeO	16.9	15.3	11.9	14.7	11.32	14.30	21.23	16.0	15.7	16.39	16.30	13.5	17.6
MgO	8.6	7.70	8.12	8.7	11.06	6.95	3.59	6.63	6.77	5.77	6.63	7.95	5.32
CaO	-	0.01	-	0.32	-	0.16	0.31	0.28	-	-	0.44	-	-
MnO	0.02	0.08	-	0.07	0.01	0.09	0.10	0.11	0.10	0.08	0.09	0.00	0.02
Al_2O_3	1.8	1.62	1.87	1.48	0.97	1.80	2.60	1.96	2.00	1.87	1.77	1.93	1.91
Cr_2O_3	1.3	2.15	1.81	1.94	1.26	1.88	2.05	1.65	1.78	1.62	1.69	2.17	1.43
V_2O_3	-	<0.5	-	0.07	-	-	-	-	0.71	-	0.27	-	-
SiO_2	-	-	-	-	-	0.40	0.21	-	-	-	0.30	-	-
Nb_2O_5	-	-	-	-	-	-	0.17	-	-	-	-	-	-
Y_2O_3	-	-	-	-	-	0.05	0.05	-	-	-	-	-	-
ZrO_2	-	-	-	-	-	0.05	0.38	-	-	-	-	-	-
Total	99.52	100.26	99.30	99.28	96.52	100.83	100.10	100.03	100.26	98.66	99.10	99.85	98.78
	Cation Formula Based on 5 Oxygens (Ideal Armalcolite = 3 Cations per 5 Oxygens)												
Ti	1.897	1.968	2.012	1.917	1.938	1.926	1.919	1.979	1.967	1.927	1.952	1.984	1.993
Fe	0.506	0.456	0.352	0.438	0.342	0.408	0.653	0.480	0.469	0.482	0.494	0.397	0.538
Mg	0.459	0.409	0.428	0.462	0.595	0.353	0.197	0.354	0.361	0.302	0.357	0.421	0.290
Ca	-	0.000	-	0.012	-	0.006	0.012	0.011	-	-	0.016	-	-
Mn	0.001	0.002	-	0.002	0.000	0.003	0.003	0.003	0.003	0.002	0.002	0.000	0.001
Al	0.076	0.068	0.077	0.062	0.041	0.072	0.113	0.083	0.084	0.045	0.048	0.081	0.082
Cr	0.037	0.061	0.051	0.055	0.036	0.051	0.060	0.047	0.050	0.045	0.048	0.061	0.041
V	-	-	-	0.002	-	-	-	-	0.020	-	0.007	-	-
Si	-	-	-	-	-	0.014	0.008	-	-	-	0.010	-	-
Nb	-	-	-	-	-	-	0.003	-	-	-	-	-	-
Y	-	-	-	-	-	0.001	0.001	-	-	-	-	-	-
Zr	-	-	-	-	-	0.083	0.007	-	-	-	-	-	-
Total	2.976	2.964	2.920	2.950	2.952	2.917	2.976	2.957	2.954	2.803	2.934	2.944	2.945

	Type 2: Cr-Zr-Ca Armalcolite					Type 3: Zr Armalcolite			
	Cryst br	A-15 Soil	Cryst br	A-15 Soil	L-20 Soil	A-15 Soil	A-15 Soil	A-15 Soil	A-15 Soil
	16.	17.	18.	19.	20.	21.	22.	23.	24.
	Chemical Compositon (Weight Percent)								
TiO_2	66.52	68.58	71.2	68.8	65.42	67.94	68.16	71.84	71.72
FeO	9.33	9.78	9.1	13.4	10.66	17.61	17.33	14.08	13.44
MgO	2.31	2.38	1.9	1.7	1.98	7.09	6.78	8.80	9.41
CaO	3.40	3.72	3.1	3.1	3.40	0.35	0.35	0.33	0.35
MnO	0.13	0.21	-	0.2	0.10	0.08	0.02	0.08	0.08
Al_2O_3	1.49	2.12	1.7	0.9	1.48	0.97	0.97	0.94	0.98
Cr_2O_3	10.31	9.67	8.8	4.3	7.67	1.46	1.49	1.49	1.45
V_2O_3	-	-	-	-	-	-	-	-	-
SiO_2	0.23	0.47	0.6	0.2	0.27	0.18	0.23	0.23	0.19
Nb_2O_5	0.37	<0.05	-	-	-	0.65	0.58	0.20	0.44
Y_2O_3	<0.05	<0.05	-	-	-	0.53	0.53	0.01	0.01
ZrO_2	6.01	4.4	4.4	6.1	6.55	3.76	3.92	2.76	2.39
REE	-	-	-	1.3	-	-	-	-	-
Total	100.10	101.33	100.8	100.0	97.53	100.62	100.36	100.76	100.46
	Cation Formula Based on 5 Oxygens (Ideal Armalcolite = 3 Cations per 5 Oxygens)								
Ti	1.834	1.849	1.915	1.921	1.855	1.871	1.880	1.927	1.923
Fe	0.286	0.293	0.272	0.416	0.336	0.539	0.531	0.420	0.401
Mg	0.126	0.127	0.101	0.094	0.111	0.387	0.371	0.468	0.500
Ca	0.134	0.143	0.119	0.123	0.137	0.014	0.014	0.013	0.013
Mn	0.004	0.006	-	0.006	0.003	0.002	0.001	0.002	0.002
Al	0.064	0.090	0.072	0.039	0.066	0.042	0.042	0.040	0.041
Cr	0.299	0.274	0.249	0.126	0.229	0.042	0.043	0.042	0.041
V	-	-	-	-	-	-	-	-	-
Si	0.009	0.017	0.021	0.007	0.010	0.007	0.008	0.008	0.007
Nb	0.008	-	-	-	-	0.011	0.010	0.003	0.007
Y	-	-	-	-	-	0.010	0.010	0.000	0.000
Zr	0.108	0.077	0.077	0.110	0.120	0.067	0.070	0.048	0.042
REE	-	-	-	0.017	-	-	-	-	-
Total	2.872	2.876	2.826	2.859	2.867	2.992	2.980	2.971	2.977

Table A5.15. Analyses of other oxides from lunar rocks and soils.

	Rutile			Baddeleyite			
	A-11 LK	Frag br	L-20 Soil	A-12 ilm	A-12 pig	Melt Rock	L-20 Soil
	1.	3.	5.	6.	8.	9.	10.
	Chemical Compositon (Weight Percent)						
SiO_2	-	0.93	0.13	0.39	<0.01	-	0.18
TiO_2	96.62	87.29	97.23	1.97	2.4	-	1.82
Al_2O_3	1.91	0.13	0.02	-	<0.01	-	0.54
Cr_2O_3	0.30	0.56	0.48	-	-	-	0.13
V_2O_3	-	-	-	-	-	-	0.06
FeO	0.22	7.45	2.34	3.25	7.4	-	0.45
MnO	0.03	0.14	0.01	<0.02	-	-	0.17
MgO	0.05	3.20	0.04	<0.02	-	0.06	0.14
CaO	0.37	0.28	0.10	-	<0.1	-	0.16
Nb_2O_5	-	0.55	-	-	-	-	0.49
ZrO_2	-	0.07	-	91.9	90.3	98.23	94.7
HfO_2	-	-	-	3.23	-	1.70	1.65
REE	-	-	-	-	-	-	-
Total	99.50	100.60	100.35	100.74	100.1	99.99	100.49
	Cation Formula Based on 2 Oxygens (Ideal Rutile or Baddeleyite = 1 Cation per 2 Oxygens)						
Si	-	0.013	0.002	0.008	-	-	0.004
Ti	0.971	0.898	0.980	0.030	0.037	-	0.028
Al	0.030	0.002	0.000	-	-	-	0.013
Cr	0.003	0.006	0.005	-	-	-	0.002
V	-	-	-	-	-	-	0.001
Fe	0.002	0.085	0.026	0.056	0.126	-	0.008
Mn	0.000	0.002	0.000	-	-	-	0.003
Mg	0.001	0.065	0.001	-	-	0.002	0.004
Ca	0.005	0.004	0.001	-	-	-	0.003
Nb	-	0.003	-	-	-	-	0.004
Zr	-	0.000	-	0.915	0.900	0.989	0.933
Hf	-	-	-	0.019	-	0.010	0.010
REE	-	-	-	-	-	-	-
Total	1.012	1.078	1.015	1.028	1.063	1.001	1.013

Table A5.16. Analyses of sulfide minerals from lunar rocks and soils.

	Troilite			Chalcopyrite	Cubanite	Bornite	Sphalerite	
	A-12 ilm		A-12 ol	A-12 pig		A-16 Soil	Cryst br	
	1.	2.	3.	5.	6.	7.	8.	9.
	Chemical Compositon (Weight Percent)							
Ti	0.15	0.05	0.24	-	-	-	-	-
Cr	0.02	0.02	-	-	-	-	-	-
Fe	63.2	63.1	63.4	30.0	40.4	12.7	14.8	17.6
Cu	-	-	-	33.6	22.8	60.7	-	-
Mg	-	-	-	-	-	-	-	-
Zn	-	-	-	-	-	-	51.0	48.2
Ni	0.03	0.03	0.10	-	-	0.07	-	-
Co	0.08	0.06	0.12	0.85	0.87	0.2	-	-
S	36.4	36.0	36.4	35.2	35.7	26.2	33.0	33.7
P	-	-	-	–	-	0.1	-	-
Total	99.88	99.26	100.26	99.65	99.77	99.97	98.8	99.5
	Atomic Formulae: Ideal troilite = FeS, chalcopyrite = $CuFeS_2$, cubanite = $CuFe_2S_3$, bornite = Cu_5FeS_4, sphalerite = (Zn,Fe)S							
Ti	0.003	0.001	0.004	-	-	-	-	-
Cr	0.000	0.000	-	-	-	-	-	0.996
Fe	0.996	1.002	0.996	0.986	1.964	1.133	0.256	0.300
Cu	-	-	-	0.971	0.974	4.758	-	-
Mg	-	-	-	-	-	-	-	-
Zn	-	-	-	-	-	-	0.752	0.701
Ni	0.000	0.000	0.001	-	-	0.006	-	-
Co	0.001	0.001	0.002	0.026	0.040	0.017	-	-
S	0.999	0.996	0.996	2.016	3.022	4.071	0.992	0.999
P	-	-	-	-	-	0.016	-	-
Total	1.999	2.000	1.999	3.999	6.000	10.001	2.000	2.000

TableA5.17. Analyses of native Fe metal from lunar rocks and soils.

	A-11 LK	A-11 Soil	A-12 pig	A-12 ilm	A-12 ol	Melt Rock	A-15 Soil	Fan	Glassy br	Cryst br	Melt Rock	Cryst br	Melt Rock
	1.	3.	4.	5.	6.	7.	9.	11.	12.	13.	14.	15.	17.
	Chemical Compositon (Weight Percent)												
Fe	98.95	85.82	97.54	95.7	67.4	91.71	82.74	91.33	92.66	94.69	90.45	93.31	78.41
Ni	-	13.4	0.61	1.72	26.7	7.45	5.0	6.52	4.74	4.45	0.91	5.46	20.23
Co	0.80	0.48	1.24	2.59	2.37	-	11.8	0.41	0.43	0.30	7.86	0.54	0.78
P	-	-	-	-	-	-	0.02	0.31	1.44	0.29	0.01	0.19	0.14
Si	0.12	0.14	0.16	-	-	-	-	-	-	-	-	-	-
Cr	0.09	0.04	0.05	-	0.03	-	0.46	-	-	-	-	-	-
Mn	-	0.07	0.07	-	-	-	-	-	-	-	-	-	-
S	-	-	-	-	-	-	-	0.06	0.02	0.02	0.05	0.01	0.03
Σ	100.0	99.86	99.67	100.01	96.47	99.16	100.0	98.63	99.29	99.75	99.28	99.51	99.59

Table A5.18a. Whitlockite compositions from Apollo 14 (data from Jolliff et. al. 1993)

	,7044	,7233	,7264	,7350	,7373
P_2O_5 (wt %)	43.17	43.76	43.51	43.08	43.32
SiO_2	0.45	0.34	0.42	0.24	0.13
FeO	0.88	0.44	0.93	0.17	1.97
MnO	0.03	0.03	0.03	0.02	0.01
MgO	3.11	3.47	3.22	3.60	2.60
CaO	40.36	40.36	39.74	39.31	41.27
Na_2O	0.49	0.45	0.30	0.19	0.61
Y_2O_3	3.00	3.32	3.06	3.19	2.42
Sum REE_2O_3	7.29	7.58	8.38	9.10	6.33
Total	98.78	99.75	99.59	98.90	98.66
	Cation Formula based on 56 Oxygens				
P	13.862	13.904	13.910	13.905	13.916
Si	0.171	0.128	0.159	0.092	0.049
Sum (tet)	14.033	14.032	14.069	13.997	13.965
Fe^{2+}	0.279	0.138	0.294	0.054	0.625
Mn	0.010	0.010	0.010	0.006	0.003
Mg	1.758	1.941	1.813	2.046	1.471
Ca	16.401	16.229	16.080	16.058	16.778
Na	0.360	0.327	0.220	0.140	0.449
$Y+REE^{3+}$	1.583	1.669	1.734	1.879	1.339
Sum(other)	20.391	20.314	20.151	20.183	20.665
	Rare Earth Elements (ppm)				
Y	23600	26150	24100	25050	19050
La	7870	8770	9780	10190	7330
Ce	21400	22550	25100	28300	19300
Pr	2730	2650	3170	3670	2410
Nd	13290	13540	14570	17340	11360
Sm	3650	3170	4040	4420	2980
Eu	30	54	31	27	47
Gd	3760	3390	3950	4320	3140
Tb	630	640	650	680	520
Dy	4130	4590	4580	4260	3290
Ho	840	950	920	840	690
Er	2230	2530	2450	2090	1730
Tm	290	310	350	270	210
Yb	1620	1760	2080	1530	1160
Lu	190	220	250	170	150
La/Yb	4.9	5.0	4.7	6.7	6.3
Mg/(Fe+Mg)	0.86	0.93	0.86	0.97	0.70

Table A5.18b. Apatite compositions from Apollo 14 (data from Jolliff et. al. 1993)

	,7044	,7233	,7264	,7350	,7373
P_2O_5 (wt %)	41.22	39.96	40.90	41.47	41.61
SiO_2	0.42	0.76	0.57	0.34	0.30
FeO	0.43	0.29	0.42	0.03	0.50
MnO	0.03	0.04	0.03	<0.01	0.07
MgO	0.14	0.18	0.19	0.09	0.07
CaO	55.56	54.20	54.33	54.85	54.58
Na_2O	0.03	0.03	0.07	0.01	0.03
Y_2O_3	0.04	0.24	0.23	0.14	0.14
REE_2O_3	0.10	0.56	0.58	0.35	0.35
F	3.01	3.15	3.70	3.45	3.00
Cl	1.18	1.00	0.53	0.71	0.72
Sum	102.16	100.41	101.55	101.44	101.37
-O ≡ F	1.27	1.33	1.56	1.45	1.26
-O ≡ Cl	0.27	0.23	0.12	0.16	0.16
New Sum	100.62	98.85	99.87	99.83	99.95
	Cation Formula based on 12.5 Oxygens				
P	2.941	2.912	2.942	2.972	2.974
Si	0.036	0.065	0.048	0.028	0.025
Sum (tet)	2.977	2.977	2.990	3.000	2.999
Fe^{2+}	0.030	0.021	0.030	0.002	0.035
Mn	0.002	0.003	0.003	<0.001	0.005
Mg	0.017	0.023	0.024	0.012	0.009
Ca	5.016	4.998	4.945	4.974	4.937
Na	0.005	0.004	0.011	0.002	0.005
$Y+REE^{3+}$	0.005	0.028	0.028	0.017	0.017
Sum(other)	5.075	5.077	5.041	5.007	5.008
F	0.804	0.859	0.994	0.923	0.801
Cl	0.168	0.147	0.076	0.101	0.103
Sum F, Cl	0.972	1.006	1.070	1.024	0.904
	Rare Earth Elements (ppm)				
Y	300	1860	1780	1120	1090
La	105	580	540	300	380
Ce	300	1530	1610	910	1000
Pr	40	200	225	127	134
Nd	184	950	1160	690	620
Sm	52	320	310	192	176
Eu	2.5	6.6	4.2	2.3	7.0
Gd	50	320	350	250	180
Tb	9.0	55	67	41	30
Dy	55	350	340	224	219
Ho	12	71	69	41	43
Er	30	201	178	104	117
Tm	4.4	30	25	15	15
Yb	27	175	142	80	85
Lu	3.6	20	14	8.6	10
La/Yb	3.9	3.3	3.8	3.8	4.5
Mg/(Fe+Mg)	0.36	0.52	0.44	0.86	0.20

Table A5.19. Modal and major element data for High-Ti Basalts.
Modal data compiled by Papike et al. (1974).
Major element data compiled by BVSP (1981), Papike et al. (1974), Lofgren and Lofgren (1981).

	Apollo 11 High-K Basalts									Apollo 11 Low-K Basalts					
			10017					10072		10020					
	10022	10057	1	2	3	4	10024	1	2	1	2	3	4	10045	10062
Pyroxene	48.9	50.9	47.6	51.0	59.4	49.7	52.2	59.4	52.0	43.3	54.8	52.9	44.0	53.2	52
Olivine	0.6									6.4	4.8	2.9	6.4	3.1	5
Plagioclase	15.6	19.2	26.9	21.5	25.1	18.0	16.4	20.4	18.5	30.6	21.4	28.5	30.7	26.9	24
Opaques	26.3	15.7	15.3	20.2	14.9	23.9	21.8	14.8	22.1	15.4	17.1	13.5	15.3	12.8	18
Silica		<0.1	1.6	1.1		0.7	0.2			3.6	0.9	1.7	2.7	1.8	
Mesotasis	88.6		8.5	6.1	8.3	9.0		7.3		0.6			0.3		1
Vesciles and holes		10.8					1.5				0.9	0.6			
Others		3.3	0.1				3.7								
SiO_2	40.1	40.23	40.64				40.25	41.00		39.79				39.32	39.8
TiO_2	12.2	11.4	11.78				11.9	11.30		10.62				11.21	10.74
Al_2O_3	8.6	9.42	7.98				8.09	9.5		9.93				9.51	10.22
FeO	18.9	19.38	19.65				19.46	18.7		19.21				19.41	19.22
MnO	0.25	0.22	0.24				0.24	0.25		0.26				0.28	0.3
MgO	7.74	7.65	7.68				7.53	7.03		7.93				7.91	7.08
CaO	10.7	10.42	10.65				10.66	11.0		11.32				11.17	11.47
Na_2O	0.46	0.56	0.51				0.52	0.51		0.39				0.36	0.41
K_2O	0.3	0.3	0.29				0.3	0.36		0.04				0.05	0.08
Cr_2O_3	0.37	0.35	0.36				0.38	0.32		0.39				0.41	
Sum	99.62	99.92	99.58				99.93	99.97		99.88				99.63	99.32
Mg/Mg+Fe	.42	.41	.41				.41	.40		.42				.42	.40

	Apollo 11 Low-K Basalts							Apollo 17 Low-K Basalts					
	10003					10044							
	1	2	10050	10058	10047	1	2	74235	70215	74275	70017	70035	75055
Pyroxene	48.7	51.7	50.2	45.7	44.8	46.4	47.3	3.5	42	45.7	50	46	51.4
Olivine	0.5		1.2					7.3	7	10.1	1	2.5	
Plagioclase	34.8	29.0	29.5	37.1	37.8	34.1	33.1		29	13.7	26	26	29.1
Opaques	14.6	18.2	14.8	10.8	10.6	12.3	14.4	34.9	18	30.4	22	22	13.4
Silica	1.0	0.3	2.8	5.1	4.5	6.3	5.2		4		1		3.0
Mesotasis	0.1		0.6		0.6			54.3			2.0	1	
Vesciles and holes		0.5			0.9				1.5				
Others	0.3	0.2	0.9		1.7					0.1			3.0
SiO_2	39.72		40.62	41.63	42.16		42.23	39.02	37.79	38.43	38.54	37.84	40.60
TiO_2	10.50		9.61	9.91	9.43		9.0	12.28	12.97	12.70	12.99	12.97	10.79
Al_2O_3	10.43		10.87	10.98	9.89		10.94	8.91	8.85	8.72	8.65	8.85	9.67
FeO	19.80		16.51	18.06	19.11		18.37	18.93	19.66	18.14	18.25	18.46	18.01
MnO	0.30		0.26	0.27	0.28		0.26	0.27	0.27	0.26	0.25	0.28	0.29
MgO	6.69		7.82	6.17	5.67		6.11	8.90	8.44	10.36	9.98	9.89	7.05
CaO	11.13		12.65	11.81	12.15		12.22	10.78	10.74	10.32	10.28	10.07	12.35
Na_2O	0.40		0.35	0.65	0.45		0.48	0.38	0.36	0.35	0.39	0.35	0.43
K_2O	0.06		0.06	0.08	0.11		0.11	0.08	0.05	0.07	0.05	0.06	0.08
Cr_2O_3	0.25		0.35	0.22	0.18		0.21	0.47	0.41	0.65	0.50	0.61	0.27
Sum	99.32		99.10	99.78	99.43		99.93	100.02	99.54	100.00	99.88	99.38	99.54
Mg/Mg+Fe	.38		.48	.38	.35		.37	.46	.43	.50	.49	.49	.41

Table A5.20 Modal and Major Element Data for the Pigeonite Basalts. Modal data compiled by Papike et al. (1974). Major element data compiled by Papike et al. (1974), BVSP (1981); Lofgren and Lofgren (1981).

	Apollo 12								Apollo 15								
	12052	12053	12065 (1)	12065 (2)	12021 (1)	12021 (2)	12021 (3)	12064	15597	15595	15499	15476 (1)	15476 (2)	15475	15076	15085	15058
Pyroxene	68.1	68.8	67.8	70	66.0	70.5	62.6	55.8	50.1	48.1	41.8	69.7	66.2	64	66.3	46.2	66.3
Olivine	3.9	2	2.8	0.8				1.6			0.8						1.8
Plagioclase	17.2	20.3	17.4	18.8	22.4	22.7	30.7	29.4				24.7	26.3	24	28.5	47.9	27.1
Opaques	10.8	8.3	11.2	10.0	11.6	5.5	5.6	7.1	<0.1	0.3		3.4	4.0	3.5	2.4	3.5	2.8
Silica				0.3		1.3	0.3	5.1				1.7	3.3		2.1	0.7	1.5
Mesostasis		0.6		0.1			0.8										
Others			0.8					0.9	49.9*	51.6*	57.3*	0.5	0.2	8.1	0.6	1.7	0.6
SiO_2	46.40	46.21	46.54		46.68			46.30	47.98	48.07	47.81	48.15		48.15	48.44	47.73	48.14
TiO_2	3.28	3.32	3.28		3.53			3.99	1.80	1.77	1.78	1.79		1.77	1.92	1.96	1.69
Al_2O_3	10.16	10.14	10.45		10.78			10.73	9.44	9.06	9.11	9.78		9.44	8.97	9.92	8.89
FeO	20.15	19.77	19.66		19.31			19.89	20.23	20.23	20.19	20.27		19.98	20.33	19.69	19.86
MnO	0.27	0.28	0.26		0.26			0.27	0.30	0.30	0.28	0.27		0.30	0.29	0.26	0.27
MgO	8.22	8.17	7.97		7.39			6.49	8.74	9.21	9.33	8.52		8.85	8.61	8.84	9.28
CaO	10.80	11.01	10.94		11.38			11.77	10.43	10.52	10.34	10.62		10.58	10.52	10.63	10.27
Na_2O	0.27	0.26	0.29		0.31			.28	0.32	0.35	0.32	0.31		0.27	0.34	0.33	0.28
K_2O	0.07	0.06	0.07		0.07			0.7	0.06	0.05	0.05	0.06		0.06	0.07	0.04	0.03
Cr_2O_3	0.52	0.49	0.48		0.40			0.37	0.48	0.52	0.57	0.45		0.63	0.31	0.52	0.66
Sum	100.14	99.72	99.87		100.11			100.16	99.78	100.08	99.78	100.23		100.03	99.80	99.43	99.37
Mg/Mg+Fe	.42	.42	.42		.41			.37	.44	.45	.45	.43		.44	.43	.45	.45

*very fine grained quench groundmass

Table A5.21 Modal and Major Element Data for Olivine Basalts. Data compiled by Papike et al. (1976); BVSP (1981); Lofgren and Lofgren (1981).

	Apollo 12										Apollo 15				
	12009	12004	12002	12075 (1)	12075 (2)	12018	12020	12040 (1)	12040 (2)	12035	15545 (1)	15545 (2)	15556	15016 (1)	15555 (2)
Pyroxene	9.9	63.6	57.8	61.4	58.0	62.3	61.4	46.0	45	41.6	61.4	67.3	57	63.9	65
Olivine	21.7	12.5	15.4	19.0	20.5	11.4	11.4	29.3	20	32.0	8.6	3.7	0.1	7.5	5
Plagioclase		14.4	17.8	5.6	13.2	19.8	20.7	19.8	20	19.7	23.5	24	38	22.2	25
Opaques	0.6	9.1	8.3	14.0	8.1	6.3	5.6	4.3	13	6.2	6.0	3.8	3.2	5.9	3.5
Silica		0.4	0.1			0.2	0.2				0.5	0.8	0.2	0.3	
Mesostasis	67.8		0.6		0.2		0.7		2	0.5					
Others								0.6			0.3	10.3	0.4		
SiO_2	45.03	44.91	43.56	44.93		43.9	44.57	43.88		43.17	45.21		45.73	44.08	44.57
TiO_2	2.90	2.84	2.6	2.69		2.59	2.76	2.45		2.28	2.41		2.68	2.28	2.10
Al_2O_3	8.59	8.26	7.87	8.39		7.97	7.77	7.27		8.03	8.59		9.62	8.38	8.69
FeO	21.03	21.34	21.66	20.47		20.96	20.98	21.09		22.20	22.15		21.92	22.74	22.53
MnO	0.28	0.29	0.28	0.27		0.27	0.27	0.27		0.29	0.30		0.29	0.32	0.29
MgO	11.55	12.60	14.88	13.86		15.23	14.40	16.45		15.49	10.28		7.95	11.30	11.36
CaO	9.42	9.02	8.26	8.59		8.33	8.60	8.01		8.08	9.82		10.77	9.27	9.40
Na_2O	0.23	0.21	0.23	0.27		0.22	0.22	0.17		0.21	0.31		0.27	0.27	0.27
K_2O	0.06	0.07	0.05	0.06		0.05	0.06	0.05		0.05	0.04		0.05	0.04	0.09
Cr_2O_3	0.55	0.61	0.96	0.60		0.62	0.61	0.63		0.49	0.68		0.70	0.85	0.61
Sum	99.64	100.15	100.35	100.13		100.14	100.24	100.27		100.29	99.79		99.98	99.53	99.91
Mg/Mg+Fe	.50	.51	.55	.55		.56	.55	.58		.55	.45		.39	.47	.47

Table A5.22. Modal and major element data for Apollo 12 Ilmenite Basalts. (Papike et al. 1976)

	12022	12063	12051 1	12051 2	12005
Pyroxene	58.6	63.7	60.8	56.7	56.5
Olivine	16.5	2.8			30.0
Plagioclase	12.0	22.2	21.7	31.2	11.0
Opaques	11.2	8.1	10.7	7.9	2.4
Silica		1.6	2.2	3.4	
Mesotasis		1.6	2.7	0.8	.1
Vesciles and holes			1.9		
SiO_2	42.77	43.48	45.31		41.56
TiO_2	4.85	5.0	4.68		2.76
Al_2O_3	9.08	9.27	9.95		5.30
FeO	21.75	21.26	20.22		22.27
MnO	.25	0.28	.28		.30
MgO	11.01	9.56	7.01		19.97
CaO	9.47	10.49	11.39		6.31
Na_2O	.38	.31	.29		.16
K_2O	.07	.06	.06		.04
Cr_2O_3	.56	.44	.31		.75
Sum	100.19	100.15	99.50		99.42
Mg/Mg+Fe	.47	.45	.38		.62

Table A5.23. Modal and major element data for High Aluminum Basalts (BVSP 1981, Dickenson et al. 1985).
For Very High Potassium Basalts (Shervais et al. 1985b, Warner et al. 1980).

	High Alumina Basalts							Very High Potassium Basalts		
	14321			14053	14072	14702	Luna 16 B-1	14305, 390	14168, 39	14305, 304/370
	average group 1	average group 3	average group 5							
Pyroxene				50		49.9	50			
Olivine						2.5	tr			
Plagioclase				40		38.3	40			
Opaques				3		7.7	7			
Silica				2		1.7				
Mesostasis				tr		tr	tr			
Others				tr		tr	tr			
SiO_2				46.4	45.2		43.80	45.3	47.8	43.8
TiO_2	2.2	2.7	2.6	2.64	2.57		4.90	2.2	1.7	2.4
Al_2O_3	12.7	12.5	11.8	13.6	11.1		13.65	13.0	12.4	9.9
FeO	16.2	16.9	17.5	16.8	17.8		19.35	16.0	15.5	18.1
MnO	0.22	0.24	10.24	0.26	0.27		0.20	0.20	2.2	.23
MgO	7.9	8.2	10.3	8.48	12.2		7.05	9.9	11.0	15.0
CaO	11.2	10.8	10.8	11.2	9.8		10.40	10.6	9.9	9.0
Na_2O	0.60	0.42	0.39	0.44	0.32		0.33	0.41	0.37	0.34
K_2O	0.16	0.009	0.007	0.10	0.08		0.15	0.80	0.57	0.62
Cr_2O_3	0.37	0.34	0.46	0.40	.51		0.28	0.59	0.54	0.58
Sum				100.32			100.11	99.00	100.00	99.97
Mg/Mg+Fe	0.46	0.51	0.46	0.48	0.45		0.39	0.52	0.56	0.60

Trace amount is indicated by 'tr.'

Table A5.24 Modal and Major element data for VLT Basalts. Data from Vaniman and Papike (1977c); BVSP (1981); Laul et al. (1978b); Taylor et al. (1978).

	Apollo 17						Luna 24			
	70008, 356	70008, 370	70007, 328V	78526, 1	average vitrophyric VLT	average microporphyritic VLT	24077, 4	24174, 7	24182 25-100	24109, 52-18
Pyroxene	60.0	61.1					8.0	58.1		
Olivine	5.3	4.3					10.6	5.4		
Plagiocalse	31.1	27.7					2.7	34.2		
Opaques	1.0	0.7					0.1	1.2		
Silica	1.6	0.6						1.1		
Mesotasis		4.2					78.6			
Others	1.0	1.4								
SiO_2	48.1		48.8	46.7	48.0	49.2	46.0	46.0	46.82	45.2
TiO_2	0.36		0.69	0.92	0.68	0.41	0.75	1.1	1.00	0.89
Al_2O_3	11.2		10.0	10.0	11.1	11.51	10.1	12.1	12.58	13.8
FeO	18.2		17.9	18.6	18.0	17.4	22.4	22.1	20.46	20.5
MnO	0.26		0.30	0.24	0.27	0.30	0.26	0.28	0.24	0.27
MgO	11.0		11.8	12.2	11.5	10.5	10.5	6.0	6.65	6.35
CaO	10.2		9.4	10.0	10.1	10.6	10.8	11.6	12.49	12.7
Na_2O	0.15		0.06	0.12	0.14	0.13	0.21	0.27	0.10	0.24
K_2O	0.01		0.02	<0.01	0.02	0.03	0.016	0.022	0.02	0.01
Cr_2O_3	0.60		0.69	0.74	0.71	0.51	0.42	0.3	0.16	0.19
Sum	100.1		99.7	99.52	100.52	100.58			100.52	100.2
Mg/Mg+Fe	.52		.54	.54	.47	.48			.37	.36

Table A5.25. Trace element analyses and selected high-Ti mare basalts. Trace element data compiled by BVSP (1981); Lofgren and Lofgren (1981).

	Apollo 11 High-K basalts						Apollo 11 Low-K basalts				
	10022	10057	10017	10024	10072	10049	10020	10045	10062	10003	10050
La	26.4	28.2	26.6		22.7	28.8	8.11	6.7	14.5	14.7	8.2
Ce	68.5	75	77.3	76.6	69	82.8	25.8	22.5	44.8	45.5	37
Nd	65	69	59.5	66.1	51	62.8	23.9	21.1	37.5	38.3	36
Sm	21.2	20.8	20.9	23.4	17.9	22.3	9.47	8.42	13.7	14	15.1
Eu	2.01	2.18	2.14	2.21	2.07	2.29	1.6	1.54	2.06	1.76	2.15
Gd	23.9	26	27.4	28.6	26	29.3	12.8	13.2	18.2	19	19.9
Dy	30.1	34.7	31.7	33.6	31.2	33.4	15.8	14.5	20.4	21.6	28
Er	15.819	20.0	19.3	16	30.9	10.0	9.7	12.8	13.4		
Yb	17.518.8	19.2	16.6	16.6	20.2	9.87	8.53	12.3	13	12.9	
Lu	2.47	2.66	2.66		2.24		1.43	1.17	1.73	1	1.8
Y		170	159	168	161		84	73	103	112	103
Ba	277	319	309	285	294	330	77.1	10	140	106	92
Rb	5.57		5.63	5.64	5.7	6.2	0.63	0.62	0.81	0.5	.75
Sr	163	173	175	167	168	161	150	138	196	153	171
Zr	130	517	476	375	504		224	194	319	309	265
Nb		28.7	27.4	25	27.8		17.6	14		21	21
Hf	21.516.9	17.9	20	12	17.3		7.7	11.8	11.6	13.5	
Th		3.41	2.96	4.1	3.16	4.03	.678	.765	0.9	.97	.531
U		.87	.78	.67	.86	.81	0.192	0.17	0.27	.254	.156
Li	11.5	14	18.1		14		12	15		9	11
Sc	76	87	86	76.2	77	80.9	85	78	75	74	93
V	70	47	46	37	62		81	98	75	63	88
Co	29	25.4	31	28.4	27.2	24	5.7	16.1	13.8	14.1	14
Ni	10	2	60	20	3		<2	20	15	2.7	2

	All Low-K Basalts			Apollo 17 Low-K Basalts					
	10058	10047	10044	74236	70215	74275	70017	70035	75055
La	11.5	11.3	12	11.4	5.22	6.33		4.79	6.27
Ce	40.2	46	44	22.8	16.5	21.4	10.7	16.4	21.5
Nd	41.2		50	25.3	16.7	22.8	12.1	18.2	23.9
Sm	17.2	18.9	17.9	10.5	6.69	9.19	5.13	7.63	10.05
Eu	2.64	2.71	2.69	2.10	1.37	1.8	1.62	1.82	2.09
Gd	23.6		24	16.6	10.4	14.8		11.0	15.7
Dy	27			18.8	12.2	16.3	10.2	14.1	18.1
Er	16.3			11.1	7.4	9.66	6.31	8.4	10.72
Yb	15.5	18.2	15	9.85	7.04	8.47	6.25	7.79	9.79
Lu	2.14	2.88	1.96		1.03		0.95	1.17	
Y	147	134	147	160		81.5	71.2	75	112
Ba	117	88	95	82.2	56.9	67.3	43	62.1	76.2
Rb	0.98	0.93	1.15	0.612	0.356	1.2	0.28	0.461	0.584
Sr	218	194	224	186	121	153	168	174	191
Zr	376	334	366	263	192	248	218	217	272
Nb	28.4	23	21	10	20.8	22.1	18.5	20	25
Hf	11.2	13.2	12		6.33	8.55	6.4		7.2
Th	1.1	0.724	0.98	0.40	0.34	0.465	0.198		0.447
U	0.2	0.224	0.28	0.12	0.13	0.136	0.060	0.091	0.136
Li	11.4	16.3	11.5	13.3	7.1	9.6	8.57	8.7	9.35
Sc	80.8	92	92	81.4	85.9	75.1	80	82.5	82.7
V	46	13	38.5	61	50	79	146		
Co	14.4	12.2	11	19.1	21.3	23.5	32	20.7	14.5
Ni	2	20	4	1	3	3	<3	2	2

Table A5.26 Trace element analyses of selected Pigeonite Basalts. Trace element data compiled by BVSP (1981); Lofgren and Lofgren (1981). Values in parts per million.

	Apollo 12					Apollo 15							
	12052	12053	12065	12021	12064	15597	15595	15499	15476	15475	15076	15085	15058
La		7.32	6.68		6.76	4.86	5.4		5.9	4.01	7.38		
Ce	18.8	20.9	24	19.8	17.5	13				13.1	15.1		14.5
Nd	14.7	15.3	24	14.4	16.0	9.3				8.87	10.6		10.9
Sm	4.91	5.25	4.5	4.84	5.51	3.09	3.9		4.3	2.93	3.52		3.9
Eu	1.04	1.1	1.06	1.116	1.161	0.84	0.81		1.13	0.48	0.98		0.91
Gd	6.87			6.59	7.2	4.4					4.95		5.0
Dy	7.74	8.01	7.64	7.86	9.03	4.51			3.5	4.59	5.6		5.9
Er	4.55	4.88		4.53	6	1.9				2.7	3.4		3.2
Yb	4.32	4.71	3.78	4.12	4.59	2.13	2.3		3.4	2.35	2.77		2.54
Lu	0.651	0.662	0.59	0.64	0.67	0.301	0.4			0.35	0.33		0.39
Y		52	40	50.5	41				31.1	31.4	29		21.1
Ba	75	84.4	67	71.1	70	52		69	74	45.2	62.7	60	62
Rb	1.26	1.24	1.19	1.14	1.05	1.13	0.90	1.00	1.5	0.70	0.92	1.8	2.0
Sr	116	138	113	129	135	111	99	108	115	111	112	112	99.2
Zr	121	138	140	123	114			109	110	89	97	92	98
Nb	7	10	16	14	7			5.4	5.6	5.9	6.2	6.6	4.9
Hf	3.8	4	3.9	4.1	3.9		2		3.6	2.4	2.1		2.2
Th	1.282	1.34	1.06	0.932	0.842	0.53		0.59	0.733	0.4	0.59		0.52
U	0.365	0.322	0.27	0.261	0.221	0.14		0.16	0.192	0.12	0.153		0.13
Li	8.04	4.8	6	8.37	6.7				8	15.3	5.6		
Sc	50.6	56.4	56.5	49.8	63.1		45		40.6	47.7	47		42
V	149	148	150	160	119			189	160	130	135	165	
Co	38.4	30	38.8	27.7	27.2	39.6	42	44	38	44.6	41	41	42
Ni	5.9	9.9	20	16	6.9	30.0		19	15	8.9	11	23	31

Table A5.27 Trace element analyses of selected Olivine Basalts. Trace element data compiled by BVSP (1981); Lofgren and Lofgren (1981). Values in parts per million.

	Apollo 12								Apollo 15			
	12009	12004	12002	12075	12018	12020	12040	12035	15545	15556	15016	15555
La	6.1	5.43	6.02	6.34					4.93	4.8	5.58	
Ce	16.8	15	17	16.1	15.9	16.1	15.3	11.5	13.9	18	15.6	8.06
Nd	16	12.9	12.30	11.6	11.8	12	12	8.91	9.82		11.4	6.26
Sm	4.53	3.2	4.24	3.94	3.91	4.5	4.03	3.22	3.29	4	4.05	2.09
Eu	0.94	0.82	0.853	0.828	0.834	0.839	0.796	0.751	0.895	1	0.97	0.688
Gd	5.2	4.7	5.65	5.3	5.55	5.43	5.6	4.32	4.48		5.4	2.9
Dy	7.13	5.5	6.34	6.22	6.54	6.13	6.36	5.07	4.68	4.4	5.74	3.27
Er	3.6	3.84	3.89	3.73	3.8	3.75	3.71	3.09	2.67	3.3.	3.1	1.7
Yb	3.74	3.17	3.78	3.71	3.42	3.69	3.38	3.04	2.16	1.59	2.62	1.45
Lu	0.551	0.44		0.508	0.52		0.521	0.423	0.31	0.39	0.32	
Y		36	39	50	30	32	22	29	33	50	21	23
Ba	60	55	67	64	60	64	57	47	46.7	59	61	32.2
Rb	1.05	1.12	1.04	0.99	1.04	1.00	1.0	0.69	0.75	0.84	0.81	0.445
Sr	96	94	101	94	89	94	86	84	104	88	91.4	84.4
Zr	107	110	106	132	89	97	57	81	190	91	86	76
Nb	6	7	8.5	16	5	5	2	4		7	10	4.3
Hf	4	5.1	2.5	2.7	2.5	3.8	2.39	1	2.2	3.1	2.6	
Th	0.881	0.92	0.747	0.62	0.879	0.71	0.47	0.682	0.43	0.56	0.50	0.46
U	0.243	0.30	0.219	0.19	0.248		0.16	0.199	0.13	0.15	0.12	0.13
Li		11		6	7.5	5.7	6.7			9.0	4.6	6.4
Sc	46	43.8	38.3	43	38.9	45.4	37.5	33.7	42	43.1	39.1	
V	153	145	175	180	140	146	153	130	168	266	200	
Co	49	47.9	65.8	61	58	61	61.1	53.2	21	50.3	54	
Ni	51.9	52	63.9	63	55	50	40	32.9	68.9	64.9	85.9	41.9

Table A5.28. Trace elements (ppm) in Apollo 12 Ilmenite Basalts. Data compiled by BVSP (1981); Lofgren and Lofgren (1981).

	Apollo 12			
	12022	12063	12051	12005
La		6.24	6.53	
Ce	17.4	17.8	19.2	10.2
Nd	14.4	16	15.4	
Sm	5.38	6.48	5.68	2.99
Eu	1.26	1.36	1.23	0.62
Gd	7.71	9.4	7.89	
Dy	9.37	11.3	9.05	
Er	5.42	5.3	5.57	
Yb	5.69	5.4	5.46	2.66
Lu		0.79		0.41
Y	68	65	48	28
Ba	55	64	73.6	35
Rb	0.738	0.93	0.909	
Sr	143	150	148	83
Zr	180	133	128	66
Nb	6	7.9	7	4.3
Hf	3.6	4.6	3.1	2.4
Th	0.71	0.679	1.0	
U	0.198	0.191	0.26	
Li	9.51	5.9	7.4	
Sc	53.8	60.8	58	37.1
V	180	135	102	
Co	42.5	43.4	35.1	71
Ni	41.9	20	5.9	90

Table A5.30. Trace elements (ppm) in selected VLT Basalts. From Wentworth et al. (1979), Laul et al. (1978b), and Ma et al. (1978).

	70007, 296	78526, 1	24077, 4	24174, 7	24109, 78
La	.88	1.2	1.80	2.87	2.0
Ce			5.5	8.6	
Nd			5	7.0	
Sm	0.70	1.0	1.39	2.10	1.9
Eu	0.24	0.30	0.58	0.83	0.58
Gd					
Dy	1.5	2.0	1.7	2.9	2.8
Er					
Yb	1.1	1.4	1.22	2.00	1.9
Lu	0.19	0.23	0.18	0.31	0.29
Y					
Ba			40	50	
Rb					
Sr			90	110	
Zr			<40	50	
Nb					
Hf	0.62	0.6	1.1	1.4	1.1
Th			0.20	0.20	
U					
Li					
Sc	54	50	436	57	47
V	202	226	150	140	170
Co			59.3	43.3	
Ni			70	30	80

Table A5.29. Trace elements (ppm) in selected High Alumina Basalts (left) and Very High Potassium Basalts (right) From Dickenson et al. (1985) and Shervais et al. (1985b).

	High Alumina Basalts						Very High Potassium Basalts		
	14321				Luna 16		14305,	14168,	14305
	Group 1	Group3	Group 5	14053	14072	B-1	390	39	304/370
La	25	11.3	3.4	13	6.76		12.0	5.2	4.6
Ce	65	30	8	34.5	17.9		31	14	11.7
Nd	40	21	6.3	21.9	13		20	11	8.5
Sm	12.5	6.6	2.3	6.56	3.93		6.10	3.4	2.6
Eu	1.45	1.24	0.71	1.21	0.88		0.91	0.71	0.75
Gd				8.59	4.2				
Dy	14.9	10.2	0.45	10.5	6				
Er				6.51	3.5				
Yb	8.3	6.0	3.2	6	4.05		5.35	3.6	3.2
Lu	1.21	0.89	0.61		0.61		0.82	0.53	0.47
Y				54.7	38				
Ba	159	112	53	146	128	218	500	100	200
Rb				2.19	1.4	1.58		9.76	14.2
Sr				98	108	437		59	67
Zr	320	170	70	215	166			80	80
Nb				15.7	11.5				
Hf	8.7	4.7	1.9	9.8	6.9		3.7	2.4	2.5
Th	2.3	0.9	0.4	2.101			1.6	0.7	0.61
U				0.592		0.30	<0.5		<0.2
Li									
Sc	59	56	62	55	47.1		55	52	50
V	102	116	121				120	139	120
Co	29	27	29	25	32		34	31	37
Ni				14	31		70	40	60

Table A5.31 Selected analyses of picritic glasses from the Apollo missions. These glasses were selected to show range in glass population and most common composition. 1=A15 Green C, 2=A15 Green A, 3=A16 Green B, 5=A15 Green D, 6=A15 Green E, 7=A14 Green B Type 1, 8=A14 Green Type 2, 9=A14 VLT, 10=Apollo 11 Green, 11=A17 VLT, 12=A14 Green A. 13=LAP, 14=A15 Yellow, 15=A14 Yellow, 16=A17 Yellow, 17=A17 Orange 1, 18= A17 Orange 74220, 19=A15 Orange, 20=A17 Orange 2, 21=A11 Orange, 22=A14 Orange, 23=A15 Red, 24=A14 Red-Black, 25=A12 Red (Shearer and Papike, 1993)

	1a	1b	2a	2b	3	4a	4b	5a	5b	6a	6b	7a	7b	8	9a	9b	10a
SiO_2	48.68	47.55	45.78	46.04	44.10	45.68	44.90	45.01	45.00	45.18	45.61	44.41	44.43	45.71	46.53	47.49	44.5
TiO_2	0.22	0.26	0.51	0.38	0.40	0.43	0.41	0.42	0.40	.54	0.46	0.63	.48	.89	.78	.61	.47
Al_2O_3	7.06	7.84	7.37	7.61	7.79	7.73	7.85	7.45	7.37	7.39	7.74	7.23	7.03	8.50	8.30	10.05	7.65
Cr_2O_3	.60	.60	0.39	.59	.40	.54	.46	.59	.50	.43	.69	.50	.55	.50	.37	.51	.45
FeO	16.30	16.34	19.07	19.69	21.82	18.63	18.91	19.66	20.57	19.00	18.86	19.27	19.37	18.90	18.54	17.64	21.3
MnO	.34	.29	0.32	.22	.25	.26	.34	.27	.27	.22	.30	.28	.38	.20	.32	.10	.32
MgO	18.66	18.59	17.32	17.58	17.00	17.74	17.50	17.89	18.30	18.22	18.38	18.75	19.12	16.16	16.42	14.54	17.6
CaO	8.43	8.02	8.18	8.53	8.55	8.38	8.42	8.19	8.25	7.85	8.14	8.05	7.89	8.74	8.48	9.43	8.37
Na_2O	0.17	0.18	0.05	.19	.10	.17	.20	.13	.20	.18	.09	.25	.13	.34	.14	.34	.28
K_2O	0.00	0.00	0.00	0.00	0.00	0.00	0.00	0.00	n.a.	n.a.	n.a.	n.a.	.01	.02	.08	.06	.00
Total	100.46	99.67	98.99	100.86	100.41	99.56	98.99	99.61	100.86	99.01	100.27	99.37	99.39	99.96	99.96	100.77	100.9
Normative Mineralogy																	
Or	0	0	0	0	0	0	0	0	0	0	0	0	.06	.12	.47	.35	0
Ab	1.44	1.52	.42	1.61	.85	1.44	1.69	1.10	1.69	1.52	.76	2.12	1.10	2.88	1.18	2.88	2.37
An	18.50	20.59	19.89	20.00	20.81	20.33	20.52	19.75	19.21	19.36	20.72	18.61	18.57	21.61	21.79	25.72	19.62
Di	19.02	15.66	17.01	18.35	17.85	17.43	17.45	17.16	17.86	16.07	16.12	17.48	16.86	17.91	16.70	17.34	18.05
Hy	42.29	39.78	35.25	29.48	22.09	30.40	26.60	28.24	23.91	30.08	31.05	23.01	24.94	28.62	36.05	35.40	20.34
Ol	17.91	20.74	24.91	29.86	37.47	28.38	31.29	31.71	36.70	30.36	29.76	36.25	36.17	26.44	21.75	17.17	39.01
Cm	.88	.88	.57	.87	.59	.79	.68	.87	.74	.63	1.02	.70	.81	.70	.54	.75	.66
Il	.42	.49	.97	.72	.76	.82	.78	.80	.76	1.03	.87	1.20	.91	1.69	1.48	1.16	.89
Trace Elements (ppm)																	
Sc	37.2	35.8	38.8	39.9	29.8	37.5	36.8	37.0	37.6	37.7	36.5	28	28	34	42	35.7	43
V	158	155	164	166	110	161	155	159	163	168	161.0	92	121	119	152	160	100
Ni	90	90	170	170		150	150			170	170.0	185	185		125	125	135
Co	47.3	41.4	65.4	70.7	60.9	53.5	63.1	63	69.2	62.2	57.7	54	63	62	94	66.3	72
Sr	12.7	9.7	20.2	23.2	28.9	23.0	20.6	20.2	24.4	27.1	26.1	11	13	34	49	50.7	49.0
Zr	12.0	9.2	18.5	22.7	18.4	25.2	19.7	17.6	18.9	24.5	22.6	42	55	146	170	201	41.7
Ba	7.9	5.5	11.5	14.6	29.0	14.6	13.1	12.1	13.7	17.4	14.9	15	36	81	113	120.2	25.1
La	.76	.60	.88	.92	1.26	1.62	.99	1.11	1.07	1.55	1.82	3.5	4.3	5.4	n.a.	n.a.	1.40
Ce	1.92	1.67	2.32	2.37	3.03	4.00	2.61	2.46	2.72	3.91	4.63	8.5	10.1	14.4	20.0	29.5	3.37
Nd	1.05	.86	1.60	1.43	2.13	2.20	1.38	1.17	1.52	2.23	2.52	5.8	6.2	9.6	12.3	16.8	2.36
Sm	.34	.30	.46	.36	.73	.75	.46	0.71	.46	.85	.96	1.7	1.9	3.0	3.4	4.2	.72
Eu	.06	.09	.19	.13	.36	.20	.13	.12	.17	.14	.22	.30	.10	.3	.5	0.30	0.25
Dy	.61	.58	.94	.73	1.30	1.11	.69	.81	.72	1.05	1.24	2.6	2.8	4.8	4.5	5.3	1.39
Er	.44	.44	.60	.44	.86	.75	.51	.89	.45	.78	.92	1.7	1.7	2.8	2.5	3.2	.94
Yb	.53	.52	.72	.53	.83	.77	.50	.94	.61	.77	.93	1.5	1.5	2.0	n.a.	3.0	1.00

Table A5.31 Continued.

	10b	11a	11b	12a	12b	13	14a	14b	15a	15b	16a	16b	17a	17b	18a	18b	19a
SiO_2	44.3	46.3	46.0	44.81	45.15	41.25	4.375	43.05	41.92	41.14	40.2	39.8	39.30	40.15	38.30	38.50	37.21
TiO_2	.43	.55	.86	.93	1.00	2.76	2.70	3.58	4.82	4.59	6.57	6.84	8.44	8.83	8.85	8.89	9.16
Al_2O_3	7.33	9.82	10.2	6.72	6.87	10.33	8.34	8.58	6.52	6.00	7.82	8.50	6.74	7.21	5.70	5.77	5.67
Cr_2O_3	.50	.46	.54	.51	.64	.28	.55	.52	.40	.29	.69	.54	.63	.60	.81	.63	.54
FeO	21.7	19.5	18.8	22.16	23.00	22.65	21.81	21.22	24.10	24.10	21.6	21.7	21.60	21.21	22.40	22.90	24.17
MnO	.36	.30	.24	.21	.33	.19	.30	.52	.32	.22	.35	.36	.30	.26	.30	.33	.36
MgO	16.6	14.1	13.8	16.01	15.68	13.23	12.63	13.06	13.74	14.62	12.5	11.8	13.80	13.09	15.00	14.30	15.11
CaO	8.16	9.41	9.86	8.16	7.93	10.15	8.62	8.35	7.90	7.49	8.49	8.60	7.46	7.95	7.23	7.24	7.18
Na_2O	0.19	.16	.23	.18	.21	.00	.54	.34	.68	.50	.32	.31	.47	.34	.40	.44	.37
K_2O	.00	.01	.00	.02	.00	.00	.00	n.a.	.14	.07	.02	.00	.04	.05	.01	.03	.05
Total	99.6	100.6	100.4	99.71	100.81	100.84	99.24	99.22	100.54	99.02	98.5	98.5	98.78	99.69	99.00	99.03	100.1
Normative Mineralogy																	
Or	0	0	0	.12	0	0.00	0.0	0.0	0.83	0.41	0.12	0.00	.24	.30	.06	.18	.30
Ab	1.61	1.35	1.95	1.52	1.78	0.0	4.57	2.88	5.75	4.23	2.71	2.62	3.98	2.88	3.38	3.72	3.38
An	19.15	26.08	26.80	17.47	17.80	28.20	20.33	21.89	14.33	13.92	19.84	21.80	16.16	18.00	13.73	13.68	19.40
Di	17.62	17.05	18.28	19.03	17.85	18.50	18.64	16.18	20.62	19.24	18.39	17.25	17.09	17.57	18.11	18.24	16.54
Hy	25.24	33.91	30.67	29.74	31.65	13.46	26.88	30.08	17.62	19.51	24.86	24.54	24.29	29.60	20.92	21.63	26.22
Ol	34.40	20.49	20.41	29.31	28.89	35.05	22.88	20.63	31.65	32.56	19.16	18.45	20.08	13.72	24.82	23.78	15.82
Cm	.74	.68	.79	.75	.94	.41	.81	.77	.59	.43	1.02	0.79	.93	.84	1.19	.93	.79
Il	.82	1.04	1.63	1.77	1.90	5.24	5.13	6.80	9.15	8.72	12.48	12.99	16.03	16.77	16.81	16.88	17.40
Trace Elements (ppm)																	
Sc	38.5	39	31	44	36	44	36.1	41.3	43	43	46	43	47	39	45	48	50
V	110	133	127	175	142	111	137	118	73	88	114	115	137	104	110	134	n.a.
Ni	135	150	150	115	115		85	85	82	82	55	55	46	46	70	70	33
Co	60.1	54	54	111	83	48	49.4	54.6	60	54	48	47	52	51	62	59	48
Sr	83	63	51	67	56	128	139	151	172	160	120	120	224	180	198	213	272
Zr	43.0	35	33	204	160	345	285	166	297	213	118	135	208	167	193	205	248
Ba	40.3	21	31	359	83	240	145	86.9	154	122	34	47	89	73	78	88	115
La	1.77	1.60	2.78	n.a.	n.a.	n.a.	22.1	9.58	10.6	9.6	3.99	7.62	6.73	8.44	5.85	8.70	6.81
Ce	4.63	3.68	6.70	30.0	23.1	39.1	57.8	25.4	30.2	28.3	11.42	22.22	17.5	22.6	16.1	23.7	21.2
Nd	2.52	2.66	4.80	18.5	14.3	28.8	30.9	14.1	23.9	22.7	11.24	21.49	15.8	19.9	15.3	22.0	21.9
Sm	1.06	1.03	1.79	4.7	3.8	8.0	10.2	5.45	8.1	7.1	4.89	8.98	6.73	7.98	6.40	8.76	8.99
Eu	0.13	0.24	0.51	0.5	0.5	.7	1.44	1.14	1.9	0.8	1.69	2.54	1.45	2.02	1.86	2.51	3.02
Dy	1.23	1.53	2.99	5.8	4.8	10.2	9.95	6.13	13.0	11.5	7.23	13.58	7.88	9.90	8.00	11.5	12.6
Er	.88	1.01	2.08	3.2	2.6	5.8	5.48	3.67	7.1	5.9	4.26	8.20	4.43	5.39	4.22	6.18	6.85
Yb	.90	.95	1.72	n.a.	n.a.	5.1	5.56	3.80	6.6	5.6	3.26	7.09	3.03	3.29	2.96	5.43	5.70

Table A5.31 Continued.

	19b	20a	20b	21a	21b	21c	22a	22b	22c	23a	23b	23c	24a	24b	25a	25b	25c
SiO_2	37.48	39.04	39.28	38.14	38.25	38.42	37.88	35.92	37.50	37.02	36.69	37.28	33.55	34.14	34.15	32.97	34.51
TiO_2	9.12	8.96	9.07	9.83	10.17	9.97	12.50	13.35	12.07	11.50	13.69	11.49	16.75	17.01	15.75	16.28	16.70
Al_2O_3	5.70	7.82	7.55	4.56	4.46	4.44	5.73	6.50	6.01	8.65	6.87	8.23	4.03	3.80	5.00	4.99	4.63
Cr_2O_3	.67	.55	.84	.68	.72	.52	.70	.69	.71	.49	.80	.60	.97	.74	.93	1.77	.82
FeO	23.88	23.07	23.35	24.28	24.32	24.47	20.83	21.20	21.12	22.39	20.59	22.47	22.43	23.48	23.44	22.69	21.86
MnO	.20	.32	.27	.26	.26	.48	.31	.21	.42	.27	.23	.22	.26	.35	.31	.39	.30
MgO	15.15	11.32	11.90	14.70	14.84	14.78	13.90	12.46	13.09	10.73	11.84	11.29	15.18	13.60	13.00	13.75	13.12
CaO	7.37	7.94	8.02	7.22	7.23	7.11	6.94	7.66	7.22	8.53	7.68	8.24	6.30	6.39	6.50	6.65	7.59
Na_2O	.44	.40	.35	.21	.29	.27	.48	.69	.74	.62	.69	.56	.20	.45	.00	.12	.21
K_2O	.07	.06	.03	.05	.00	.01	.15	.12	.20	.06	n.a.	.06	.08	.20	.03	.45	.10
Total	100.0	99.46	100.5	99.93	100.5	100.5	99.42	98.80	99.08	100.2	99.08	100.48	99.75	100.16	99.11	100.06	99.84
Normative Mineralogy																	
Or	.41	.35	.18	0.30	0.00	0.06	0.00	0.00	0.00	0.00	0.00	0.00	0.00	0.00	0.16	.00	.20
Ab	2.96	3.55	3.13	1.78	2.45	2.28	4.06	5.84	6.26	5.25	5.84	4.74	1.69	3.81	0.00	1.02	1.78
An	18.82	18.54	13.72	11.35	10.87	10.87	13.48	14.64	13.08	20.82	15.65	19.94	10.10	8.35	13.64	13.08	11.69
Di	17.31	18.04	17.97	20.07	20.49	20.03	16.18	18.32	17.48	17.36	18.19	17.27	16.55	17.79	14.92	13.67	20.35
Hy	25.86	25.14	14.95	24.26	23.79	24.66	30.63	18.89	22.46	16.96	23.30	19.13	25.01	26.00	31.93	27.62	26.59
Ol	16.95	15.77	31.53	22.52	22.57	22.87	9.96	14.48	16.38	17.19	8.93	16.62	12.98	10.36	7.27	10.12	6.29
Cm	.99	.81	1.24	1.00	1.06	.77	1.03	1.02	1.04	.72	1.18	.88	1.43	1.09	1.37	2.60	1.21
Il	17.32	17.02	17.23	18.67	19.31	18.93	23.74	25.35	22.92	21.84	26.00	21.82	31.81	32.30	29.91	30.92	31.71
Trace Elements (ppm)																	
Sc	52	61	50	60	54	61	34	44	55	50	31	54	59	53	54	53	58
V	230	280	139	136	122	141	174	108	154	111	125	113	214	205	208	198	295
Ni	33	33	33	30	30	30	30	30	70	<50	<50	<50	<22	<22	n.a.	n.a.	n.a.
Co	48	47	63	70	59	64	44	48	60	42	23	44	68	76	24	45	38
Sr	270	267	136	172	147	174	239	245	295	263	139	267	284	247	253	210	255
Zr	180	171	156	200	206	206	413	499	522	274	154	306	483	393	380	355	399
Ba	127	131	53	61	48	56	271	320	483	161	149	151	260	395	270	247	270
La	6.80	8.90	7.13	4.48	4.14	4.46	n.a.	n.a.	n.a.	14.4	18.3	14.6	n.a.	n.a.	14.2	18.8	15.4
Ce	21.0	24.9	18.7	13.7	13.0	14.3	65.0	69.6	57.7	41.7	51.1	43.6	35.5	38.2	45.0	65.3	44.9
Nd	21.9	21.1	14.8	15.2	13.3	15.8	48.7	47.3	42.4	26.9	29.5	27.7	33.0	31.5	34.1	58.0	38.9
Sm	8.87	8.38	6.99	6.45	5.85	6.41	13.8	13.8	13.3	8.83	11.0	9.36	11.0	10.4	12.5	15.3	14.3
Eu	2.85	2.69	1.50	1.85	1.49	1.37	2.30	2.70	1.90	2.17	2.05	2.23	3.0	2.6	2.35	2.92	2.35
Dy	12.6	10.3	7.37	10.2	9.9	10.8	16.9	16.1	15.1	11.4	11.5	11.5	12.3	11.8	16.7	20.5	18.3
Er	6.80	5.78	4.72	5.33	5.11	5.52	9.0	8.4	7.5	5.47	6.21	5.86	6.2	6.0	6.78	7.61	8.62
Yb	5.49	6.15	5.44	5.11	4.48	4.96	n.a.	n.a.	n.a.	7.04	7.05	7.25	n.a.	n.a.	6.51	8.09	7.50

Table A5.32. Chemical analyses of ferroan anorthosites.

	15415 ,177	15415 ,123	15415 12	15362 ,7	15363 ,1	60015 ,64	60015 ,63	60025 ,45	60025 ,26	60215 ,30	61016 ,184	62236 ,9	62237 ,26A	62237 ,6	65315 ,32	67016 ,326/8	67075 ,37	67075 ,11	67455 ,30A
wt.%																			
SiO_2		44.1	44.9		45.3	44.2		45.3		44.5	45.0	43.3			44.3	45.3		45.4	
TiO_2		0.02	0.02		0.12		<0.06	<0.02			0.02		<0.1		0.01	0.40		0.10	
Al_2O_3	35.5	35.5	35.7	32.3	28.0	35.7	35.5	34.2		35.5	34.9	29.7	27.8	32.0	34.9	26.2	32.1	31.2	35.5
Cr_2O_3	0.003		0.003	0.004	0.095	0.006	0.007	0.004	0.002	0.050	0.002	0.075	0.054	0.048	0.003	0.083	0.051	0.082	0. 0041
FeO	0.20	0.23	0.21	0.23	4.77	0.26	0.35	0.50	0.32	0.15	<0.05	4.00	6.65	0.44	0.31	6.56	2.94	3.96	0.42
MnO	0.006		0.004	0.004	0.067		0.009	0.008		0.010	<0.005	0.059	0.084	0.054	0.008	0.090	0.046	0.060	0.009
MgO	<0.5	0.09	0.53	0.30	3.85	0.35		0.21		0.14	<0.03	3.80	5.80	0.36	0.25	5.30	2.30	3.14	0.60
CaO	21.0	19.7	20.6	17.0	16.8	19.3	20.0	19.8	18.5	19.3	19.6	17.4	15.9	17.5	19.1	15.8	18.5	17.1	19.8
Na_2O	0.356	0.340	0.380	0.390	0.288	0.360	0.410	0.450	0.445	0.400	0.410	0.213	0.202	0.222	0.303	0.280	0.278	0.260	0.314
K_2O		<0.01	0.017	0.011	0.022	0.130	<0.01	0.113		0.020	0.005	0.015		0.013	0.007	0.020		0.013	
P_2O_5		0.010									0.047				0.008			0.016	
Sum		99.9	102.4		99.3	100.3		100.6		100.1	99.9	98.6			99.1	100.1		101.4	
Mg#		0.41	0.82	0.70	0.59	0.71		0.43		0.62		0.63	0.61	0.60	0.59	0.59	0.58	0.59	0.72
ppm																			
Sc	0.4		0.04	0.7	9.3		0.6		0.5	<2		5.3	3.6	3.5	0.4	17.5	4.7	7.7	0.3
V						4.7	<7			5.8						23.5			
Co	0.2		0.3	0.3	6.0		0.7	0.7	0.6	<2		9.0	14.4	10.8	0.6	7.0	6.4	7.3	0.2
Ni	13.3		3.0	10.0	9.1			1.1	14.7	1.8		3.5		5.8	1.4	37.0			14.0
Rb						0.1		<0.1		<1	0.0				0.2			0.7	
Sr	198	184	173		140				226	121	149		137		167	137	162	127	152
Cs	0.025														0.002			0.030	
Ba	6.0		6.5			10.0	8.0		10.4	<10	6.0		6.0		4.8	29.0	6.0	13.0	9.0
La	.133		.210	.230	.310		.130	.280	.289			.180	.179	.210	.120	2.040	.285	.320	.096
Ce	.330			.720	.520	.630		.650	.670		.253	.680	.490	.440		4.510	.820	.800	.220
Nd						.350		.420			.145			.280		2.830		.120	
Sm	.054		.062	.120	.131	.080	.051	.092	.097		.036	.092	.083	.074	.040	.880	.145	.160	.029
Eu	.805		.820	.800	.880	.810	.810	1.040	1.110		.671	.590	.605	.650	.740	.770	.646	.630	.688
Gd						.060										.970		.300	
Tb	.007				.042		<0.01	.020	.018				.016	.018		.180	.035	.047	.007
Dy			.054				<0.06	.190			.027				.056	1.220		.330	
Er								.050			.014					.850		.240	
Yb	.028		.035	.040	.280		.020	.048	.040		.017	.130	.100	.084	.026	.920	.155	.250	.015
Lu	.006		.004	.006	.045		.003	.006	.005			.021	.018	.012	.004		.026	.040	.001
Zr					<100	1				<10					15	30			
Nb															<0.2	2.4			
Hf	0.014		0.017	0.050	<0.150		0.013	0.020	0.008				0.035		0.490	0.650	0.064	0.120	0.007
Th					<0.100							0.040				0.130			
U			0.002												<0.0006	0.040		0.005	
ppb																			
Ir					1.030							0.008		0.015					
Au			0.770		0.064							0.008		0.146	1			0.660	
	Haskin et al. (1981)	LSPET (1972)	Wanke et al. (1975)	Murali et al. (1978)	Warren et al. (1987)	SR Taylor et al. (1973)	Laul and Schmitt (1973)	Haskin et al. (1973)	Haskin et al. (1981)	Rose et al. (1975)	Philpotts et al. (1973) Nava (1974)	Warren and Wasson (1978)	Haskin et al. (1981)	Warren and Wasson (1978)	Wanke et al. (1976)	Norman and Taylor (1992)	Haskin et al. (1981)	Wanke et al. (1975)	Lindstrom and Salpas (1981)

Table A5.33. Chemical analyses of Mg-rich plutonic rocks.

TRAN troctolitic anorthosite TR troctolite SPTR spinel troctolite D dunite AN anorthositic norite N norite GN gabbro-norite

	12071 c10 TR	14303 c194 TRAN	14305 c264 TRAN	14305 c279 TRAN	14321 c1020 TRAN	76335 ,38 TR	76535 ,21-22 TR	76536 ,16 TR	76536 ,9 TR	15445 ,71-A SPTR	15445 ,103A-G SPTR	72415 D	72417 D
wt.%													
SiO_2	45.6	43.4	43.6	43.4	43.0	43.4	42.9	42.4	43.5		37.5	39.9	
TiO_2	0.22	0.03	0.04	0.05	0.06	0.07	0.05		0.07	0.59	0.25	0.03	
Al_2O_3	24.9	27.0	28.0	28.5	28.7	27.6	20.7	26.2	21.0	7.2	14.7	1.5	1.3
Cr_2O_3	0.103	0.038	0.020	0.029	0.146	0.060	0.110	0.080	0.120			0.340	0.340
FeO	5.02	3.16	2.83	2.25	2.59	2.97	4.99	3.60	4.94		6.40	11.34	11.90
MnO	0.077	0.034	0.031	0.026	0.025	0.037	0.070	0.040				0.130	0.113
MgO	10.80	11.77	11.44	8.29	9.45	10.28	19.10	13.60	17.42	31.10	33.00	43.61	45.40
CaO	14.3	14.4	14.3	15.9	15.1	15.0	11.4	13.3	11.8	1.9	4.8	1.1	1.1
Na_2O	0.325	0.406	0.434	0.469	0.379	0.307	0.200	0.290	0.280	0.100	0.140	<0.02	0.019
K_2O	0.020	0.073	0.073	0.069	0.075		0.030	0.040	0.060		0.022	0.000	0.003
P_2O_5							0.030					0.040	
Sum	101.4	100.3	100.8	99.0	99.5	99.7	99.6	99.6	99.2		96.8	98.1	
Mg#	0.79	0.87	0.88	0.87	0.87	0.86	0.87	0.87	0.86		0.90	0.87	0.87
ppm													
Sc	7.8	3.9	1.8	2.6	1.7			1.8	2.4		3.4		4.3
V													50.0
Co	17.0	23.4	19.0	12.6	9.2	15.6		20.0	25.6		50.4		55.0
Ni	37.0	46.0	19.0	10.0	32.0	<20	25.0	5.0	32.0		820.0	173.0	160.0
Rb							0.2					<0.2	
Sr	164			211			114					11	8
Cs	<0.29												
Ba	27.0	430.0	630.0	450.0	250.0	46.0	33.0	49.0		25.0			4.1
La	1.61	31.70	5.60	6.60	25.00	2.12	1.51	1.90	11.00	1.84	2.86		0.15
Ce	4.90	77.00	12.20	14.80	61.00	5.30	3.80	4.10	31.90	3.20			0.37
Nd	2.20	49.00	7.10	8.00	38.00		2.30	2.50		2.89			
Sm	0.69	12.00	1.56	2.16	10.50	0.70	0.61	0.65	6.03	1.05	1.19		0.08
Eu	0.890	2.320	2.720	2.600	2.070	0.910	0.730	0.780	0.745	0.196	0.310		0.061
Gd							0.73			2.09			
Tb	0.15	2.31	0.20	0.40	1.90	0.13		0.13	1.13		0.26		0.02
Dy	1.10			2.33	10.40		0.80			4.88			0.11
Er							0.53			4.04			
Yb	0.63	8.40	0.83	1.44	4.20	0.56	0.56	0.44	2.67	3.89	0.90		0.07
Lu	0.10	1.15	0.13	0.19	0.56	0.08	0.08	0.06	0.34	0.57	0.14		0.01
Zr	<90	260	120	<50	38		24					3	
Nb							1.2					0.3	
Hf	0.45	4.50	0.25	0.37	0.15	0.45	0.52	0.36	1.04		0.74		0.01
Th	0.16	3.70	0.22	0.46	2.00	0.16		0.20	4.20		0.27		
U	<0.240	0.55	0.17	0.15	<0.400	0.10	0.06						
ppb													
Ir	0.080	0.130	<0.05	0.009	0.053	0.130		0.051					
Au	0.038		0.016	0.190	0.170	0.013		0.020					
	Warren et al. (1990)	Warren and Wasson (1980)	Warren and Wasson (1980)	Warren et al. (1983)	Warren et al. (1981)	Warren and Wasson (1978)	Haskin et al. (1974) Rhodes et al. (1974)	Warren and Wasson (1979)	Simonds and Warner (1981) Blanchard (unpub.)	Ridley et al. (1973b)	Blanchard et al. (1976a) and unpub.	LSPET (1973) Rhodes (1973)	Laul and Schmitt (1975)

Table Table A5.33 continued. Chemical analyses of Mg-rich plutonic rocks.

TRAN tr TRAN troctolitic anorthosite TR troctolite SPTR spinel troctolite D dunite AN anorthositic norite N norite GN gabbro-norite

		15445 ,17 N	15445 ,104 N	15445 ,228 N	15455 ,20 AN	15455 ,9015 AN	72255 c2 N	77035 c130 N	77075 ,27 N	77215 ,45 N	78236 ,3 N	14161 ,7044 GN	61224 ,6 GN	67667 ,3 GN	76255 95 GN
wt.%	wt.%														
SiO_2	SiO_2	48.7	47.7		44.4	47.7	52.0	50.3	51.1	51.3	50.2	46.4	50.7	42.4	48.8
TiO_2	TiO_2	0.15	0.27		<0.07	0.10	0.30	0.20	0.34	0.32	0.18	2.35	0.40	1.03	0.08
Al_2O_3	Al_2O_3	23.8	23.0		26.2	27.0	15.5	19.1	15.0	15.1	17.7	17.8	13.2	7.6	16.8
Cr_2O_3	Cr_2O_3		0.248		0.064	0.172	0.161	0.324	0.380	0.320	0.310	0.172	0.291	0.378	0.192
FeO	FeO	3.88	3.90		4.20	2.80	7.40	5.79	10.67	10.07	6.49	8.99	9.91	17.10	8.12
MnO	MnO	0.080			0.048	0.049	0.122	0.096	0.170	0.160	0.120	0.120	0.159	0.201	0.121
MgO	MgO	9.94	10.20		10.90	6.90	15.90	11.90	12.90	12.56	14.28	8.77	12.77	26.40	12.10
CaO	CaO	12.9	12.8		14.3	14.8	9.1	11.8	8.8	9.0	10.1	12.0	11.6	5.3	11.6
Na_2O	Na_2O		0.320		0.360	0.440	0.330	0.044	0.380	0.430	0.310	0.770	0.910	0.158	0.686
K_2O	K_2O		0.066	0.059	<0.06	0.080	0.080	0.088	0.180	0.140	0.040	0.420	0.017	0.023	0.190
P_2O_5	P_2O_5									0.110	0.080	0.980			
Sum	Sum	99.4	98.5		100.5	100.0	100.9	99.6	99.9	99.4	99.7	99.0	100.0	100.6	98.7
Mg#	Mg#	0.82	0.82		0.82	0.81	0.79	0.79	0.68	0.69	0.80	0.63	0.70	0.73	0.73
ppm	ppm														
Sc	Sc		7.1			5.3	13.2	10.9	16.6		11.2	18.2	20.8	24.4	17.3
V	V				16.0										
Co	Co		10.3		10.0	27.2	29.0	22.0	33.0		28.2	13.7	23.6	26.0	14.3
Ni	Ni				12.0	21.0		9.5	6.1			<90	8.3	4.4	23.0
Rb	Rb			1.1						3.5					4.2
Sr	Sr			138	270					105					156
Cs	Cs											0.570			0.350
Ba	Ba			59.7	42.0	125.0		96.0	160.0	166.0		670.0	32.0	51.0	184.0
La	La		4.02	5.11	3.00	4.80	16.00	5.50	7.20		4.47	119.00	1.47	3.60	12.10
Ce	Ce		10.80	12.60	6.70	11.80	46.00	13.00	22.00	27.20	12.80		4.30	9.60	32.00
Nd	Nd			7.79	3.73	7.40		8.60	8.50	16.80			<9	7.60	20.20
Sm	Sm		1.81	2.13	0.88	1.74	7.60	2.19	3.00	4.68	1.93	55.50	0.87	2.09	5.80
Eu	Eu		0.870	1.070	1.670	1.380	1.750	0.093	0.980	1.080	0.820	2.610	1.430	0.780	1.570
Gd	Gd			2.40	0.95					6.64					
Tb	Tb		0.46		0.14	0.35	1.90	0.49	0.74		0.53	10.90	0.22	0.45	1.23
Dy	Dy			2.69	0.84					7.08					8.00
Er	Er			1.61	0.46					4.51					
Yb	Yb		1.72	1.48	0.36	1.22	6.60	2.20	3.90	4.98	2.12	31.20	1.06	2.20	4.00
Lu	Lu		0.26	0.21	0.06	0.17	1.01	0.32	0.59	0.77	0.32	4.09	0.16	0.32	0.63
Zr	Zr				11	70				171		580	170		120
Nb	Nb				1.0										3.8
Hf	Hf		1.36		0.17	0.67	5.50	1.90			1.70	15.40	0.55	1.40	
Th	Th		0.82		0.23	0.59		1.10			0.60	14.70	0.19		1.40
U	U				0.05	0.18		0.31				3.00			0.38
ppb	ppb														
Ir	Ir					0.020		0.050	0.270				0.148	0.013	0.077
Au	Au					0.023		0.026	0.026				0.079	0.029	0.139
		Ridley et al. (1973)	Blanchard et al. (1976a) and unpub.	Shih et al. (1993)	SR Taylor et al. (1973)	Warren and Wasson (1980)	Blanchard et al. (1975)	Warren and Wasson (1979)	Warren and Wasson (1978)	Winzer et al. (1974)	Blanchard and McKay (1981)	Jolliff et al. (1991, 1993)	Marvin and Warren (1980)	Warren and Wasson (1979)	Warren et al. (1986)

Table A5.34. Chemical analyses of alkali rocks.

AN anorthosite N norite MG monzogabbro GN gabbro-norite

	12003 ,210 AN	12033 ,501 AN	12033 ,532 AN	12073 ,120 AN	14047 ,112 AN	14160 ,106 AN	14160 ,197 AN	14161 ,7245 AN	14304 ,122 AN	14305 ,283 AN	14305 ,279 AN	14321 ,1060 AN
wt.%												
SiO_2		47.3		48.3	48.1	47.9	53.0				43.4	
TiO_2	<2.2	0.06	<0.2	0.13	0.42	0.04	1.99		<2.2	<0.17	0.05	
Al_2O_3	32.1	32.9	33.0	32.3	28.2	33.1	26.8		35.9		28.5	
Cr_2O_3	0.047	0.003	0.002	0.014	0.040	0.007	0.035	0.016	0.007	0.016	0.029	0.009
FeO	3.10	0.30	0.28	1.10	4.25	0.26	2.73	1.87	0.51	2.55	2.25	0.95
MnO	0.045	0.006	0.005	0.018	0.062	0.005	0.029		0.008	0.037	0.026	0.015
MgO	<8.3	0.30	<0.5	0.35	2.77	0.15			<14.3		8.30	
CaO	15.4	16.9	16.4	15.8	14.8	16.5	12.3	17.7	18.2	16.8	15.9	19.6
Na_2O	1.580	1.560	1.600	2.140	1.560	1.650	2.020	1.740	2.040	1.480	0.470	1.390
K_2O	0.250	0.190	0.200	0.250	0.210	0.180	1.050		0.210	0.510	0.069	0.170
P_2O_5												
Sum		99.5		100.4	100.4	99.8	100.0				99.0	
Mg#		0.64		0.36	0.54	0.51					0.87	
ppm												
Sc	6.4	0.4	0.3	2.0	9.2	0.6	4.8	4.6	1.4	8.6	2.6	3.5
V	<61	<55	<10				8.0		<64			
Co	3.6	0.3	0.2	7.1	6.7	6.3	2.6	8.4	1.0	3.6	12.6	1.1
Ni	<64	0.4	<50	71.0	9.5	4.0	57.0	42.0	<40	13.0	10.0	<17
Rb	<15	<5					28.0		<4.2	5.3		
Sr	480	430	500				360		560	400	211	430
Cs	<0.72	49.00			0.22		1.21	0.11	<0.2	0.15		
Ba	460	360	350	620	600	850	960	600	298	1180	450	610
La	20.6	11.6	10.3	16.2	27.7	11.7	41.0	59.6	8.2	201.0	6.6	111.0
Ce	47.8	23.3	23.0	35.0	64.0	21.4	99.0		16.1	520.0	14.8	280.0
Nd	26.0	11.0	11.0	16.0	40.0	8.3	54.0		9.4	305.0	8.0	173.0
Sm	6.1	2.4	1.9	4.6	10.7	1.6	14.9	23.9	2.0	86.0	2.2	46.0
Eu	6.60	6.20	6.45	8.40	5.90	7.70	6.60	6.91	6.10	7.20	2.60	6.60
Gd												
Tb	1.10	0.31	0.15	0.87	2.17	0.25	3.30	4.40	0.32	16.70	0.40	8.60
Dy	7.50	1.56		5.80	11.30		19.00		1.80	103.00	2.33	52.00
Er												
Yb	3.50	0.62	0.42	2.70	7.20	0.52	13.30	13.30	0.66	43.00	1.44	23.20
Lu	0.53	0.07	0.04	0.37	1.13	0.07	1.70	1.74	0.09	5.70	0.19	3.15
Zr	250	50	<50	310	470	140		527	108	196	<50	850
Nb												
Hf	2.89	0.85	<0.3	1.40	6.80	0.33	12.80	12.70	0.27	2.40	0.37	17.50
Th	1.49	0.16	<0.2	1.20	2.85	0.21	12.20	6.50	0.16	21.60	0.46	11.50
U	0.33	<0.08	<0.2	0.90	0.71	<0.3	3.30	1.90	<0.26	3.00	0.15	2.10
ppb												
Ir	<6.7	<0.03		0.11	0.76	<0.1	1.100		<2.5	<2.7	0.009	<5
Au	<4	<0.075		1.71	0.07	0.06	1.800	<5			0.190	
	Warren et al. (1990)	Warren et al. (1990)	Laul (1986)	Warren et al. (1986)	Warren et al. (1983b)	Warren and Wasson (1980)	Snyder et al. (1992)	Jolliff et al. (1991)	Warren et al. (1987)	Warren et al. (1983c)	Warren et al. (1983c)	Warren et al. (1983c)

Table Table A5.34 continued. Chemical analyses of alkali rocks.

AN anort AN anorthosite N norite MG monzogabbro GN gabbro-norite

		15405 ,181 AN	12033 ,534 N	12042 ,280 N	14161 ,7264 MG	14311 ,220 GN	14316 ,6 N	14318 ,156 GN	15405 ,170 N	67975 ,42 GN	67975 ,44 GN	67975 ,117 GN	67975 ,131 GN
wt.%	wt.%												
SiO_2	SiO_2				46.3		46.9	49.4					
TiO_2	TiO_2		2.00	2.30	2.50		1.50	0.62	0.36	4.00	4.60	0.64	0.36
Al_2O_3	Al_2O_3	30.4	16.2	14.8	12.5		10.6	18.7	24.4	15.7	7.2	16.5	28.3
Cr_2O_3	Cr_2O_3	0.001	0.125	0.200	0.369	0.105	0.239	0.158	0.083	0.191	0.393	0.091	0.094
FeO	FeO	0.30	11.20	12.40	13.00	7.10	10.40	8.00	4.30	10.00	17.20	11.80	6.90
MnO	MnO		0.120	0.135	0.210	0.099	0.139	0.106		0.170	0.262	0.179	0.134
MgO	MgO	1.00	9.50	9.30	8.60		15.90	12.30	4.40	5.10	12.40	8.30	7.50
CaO	CaO	17.6	9.4	9.4	10.2	15.4	8.4	10.1	16.4	12.8	15.7	9.1	13.5
Na_2O	Na_2O	1.810	0.890	0.820	1.130	1.210	0.783	0.662	0.910	0.493	0.291	0.310	0.925
K_2O	K_2O		0.980	0.750	1.080	0.110	0.428	0.228		0.350		1.900	
P_2O_5	P_2O_5				2.160								
Sum	Sum				98.1		95.3	100.3					
Mg#	Mg#	0.86	0.60	0.57	0.54		0.73	0.73	0.65	0.48	0.56	0.56	0.66
ppm	ppm												
Sc	Sc	0.6	22.0	24.5		15.1	19.0	12.8	9.3	26.3	53.1	14.1	17.2
V	V		25.0	40.0						41.0	88.0	18.0	30.0
Co	Co	0.4	39.3	40.0		5.2	38.6	20.3	18.0	10.6	19.4	11.0	5.5
Ni	Ni		400.0	410.0		<120	295.0	52.0	<50	<70	<70	<80	
Rb	Rb	<2				<30		6.1	5.0	15.0	14.0	56.0	20.0
Sr	Sr	580	230	170		370			230	175	80	180	260
Cs	Cs	0.19						0.29	0.26	0.37	0.25	0.76	0.47
Ba	Ba	180	1400	1000	4390	750	850	470	890	400	400	5000	2200
La	La	15.1	150.0	105.0	314.0	25.8	85.0	15.6	470.0	101.6	26.5	14.2	5.0
Ce	Ce	39.2	390.0	280.0		65.0	218.0	39.0	1254.0	279.0	83.6	38.5	10.9
Nd	Nd	24.0	220.0	160.0		39.0	134.0	21.1	780.0	181.0	62.0	26.0	
Sm	Sm	7.1	61.0	44.0	131.0	10.2	37.4	5.7	213.0	50.7	21.0	7.7	1.9
Eu	Eu	4.85	4.10	2.65	2.00	4.80	2.90	1.96	4.00	2.86	1.88	3.24	4.67
Gd	Gd												
Tb	Tb	1.48	12.00	8.80	23.00	2.24	7.60	1.40	42.00	10.60	5.25	1.70	0.75
Dy	Dy		80.00	54.00		15.20	48.00	9.90					
Er	Er												
Yb	Yb	2.08	48.00	32.00	96.00	10.60	27.00	8.90	94.00	29.40	17.70	9.37	25.00
Lu	Lu	0.32	6.67	4.55	8.80	1.68	3.90	1.49	11.90	4.34	2.71	1.67	5.19
Zr	Zr	80	1380	960	2080	1240	1430	370	220	1300	350	150	12700
Nb	Nb												
Hf	Hf	2.29	44.00	34.00	53.00	28.30	25.00	8.00	11.00	36.00	11.00	4.22	330.00
Th	Th	0.53	28.10	18.00		3.60	14.20	3.80	39.40	9.50	2.55	1.85	2.18
U	U	0.07	8.20	5.50		1.12	4.00	1.25	1.60	2.07	1.11	0.53	4.20
ppb	ppb												
Ir	Ir	<1				<5	5.300	0.200	<3				
Au	Au	<1						0.270	<6				
		Lindstrom et al. (1988)	Laul (1986)	Laul (1986)	Jolliff et al. (1993)	Warren et al. (1983c)	Warren et al. (1981)	Warren et al. (1983b)	Lindstrom et al. (1988)	James et al. (1987)	James et al. (1987)	James et al. (1987)	James et al. (1987)

Table A5.35. Chemical analyses of quartz monzodiorites and KREEP basalts.

	14161 ,7069 QMD	14161 ,7373 QMD	14161 ,7264 QMD (*)	15403 ,24 QMD	15403 ,23c QMD	15434 ,10/,178a QMD	15434 ,10/,178b QMD	15434 ,12/,179 QMD	15405 ,152 QMD	15405 ,85a QMD	15405 ,85b QMD	15459 c315 QMD
wt.%												
SiO_2	53.6	44.9	46.3			56.9		56.0	55.4			
TiO_2	2.40	1.80	2.49			1.13		1.11	2.60			
Al_2O_3	12.6	8.5	12.5			6.4		9.8	11.9			
Cr_2O_3	0.053	0.143	0.370	0.149	0.133	0.160	0.150	0.260	0.220		0.178	0.172
FeO	13.99	16.05	12.98	15.00	10.80	18.60	26.20	14.60	14.10		15.10	12.90
MnO	0.200	0.200	0.210			0.280		0.270	0.180			
MgO	2.68	6.30	8.64			4.70		5.50	3.80			
CaO	9.0	12.9	10.2	9.7	8.6	8.3		10.5	8.9			11.3
Na_2O	1.41	0.70	1.13	0.79	1.01	0.52	0.46	0.67	0.81		0.87	0.86
K_2O	1.60	0.60	1.08	1.40	1.44	2.17		0.72	2.10	1.71	1.80	
P_2O_5	1.72	4.98	2.16			1.33		0.29				
Sum	99.3	97.1	98.0			100.5		99.7	100.0			
Mg#	0.26	0.41	0.54			0.31		0.40	0.32			
ppm												
Sc	30.2	42.2		34.0	24.4	39.3	59.4	42.9	29.0		30.7	29.6
V												
Co	7.2	15.0		9.9	9.1	23.3	81.4	17.6	8.0		7.8	10.0
Ni	<110	<100			25.0							<40
Rb	52.0	21.0		34.0	28.0	62.0	43.0	18.9		40.6		46.0
Sr	160	207		140	171	118		147		154		190
Cs	1.60	0.36		0.93	0.92	0.48	0.46	0.56			1.10	1.25
Ba	2050	740	4389	1370	1320	2039	1560	745	1900	1490		1100
La	228.0	696.0	314.0	281.0	138.0	65.3	63.7	46.4	183.0	224.0	210.0	108.0
Ce				726.0	350.0	174.2	174.0	123.0	413.0	555.0	560.0	280.0
Nd				470.0	211.0	101.0		74.0	287.0	328.0		160.0
Sm	97.0	326.0	131.0	121.0	58.3	30.8	34.5	22.9	77.4	92.0	93.0	47.2
Eu	3.35	5.68	2.00	2.82	2.61	1.54	1.34	1.70	2.75	2.69	2.52	2.26
Gd												
Tb	18.7	62.2	23.0	24.9	12.8	7.0	8.2	5.3	14.9		19.7	10.7
Dy									101.0	116.0		
Er												
Yb	73.6	146.0	96.0	81.2	42.3	29.3	36.1	21.9	55.2	60.9	65.0	36.9
Lu	10.2	18.7	8.8	10.5	5.6	4.3	5.6	3.0	8.2	8.1	9.0	5.1
Zr	4240	7150	2080	1400	990	1043	727	788	1620			1510
Nb												
Hf	100.0	163.0	53.0	44.9	27.0	30.3	25.3	22.8	44.7		51.0	36.5
Th	44.0	37.0		50.1	28.2	23.0	16.4	10.7	39.4		43.0	22.3
U	12.0	5.4		10.9	7.0	7.4	4.0	3.1	11.1			6.4
ppb												
Ir												<4
Au						<5		18.60				<7
	Jolliff (1991) Jolliff et al. (1993)	Jolliff (1991) Jolliff et al. (1993)	Jolliff et al. (1993)	Marvin et al. (1991)	Marvin et al. (1991)	Ryder and Martinez (1991)	Ryder and Martinez (1991)	Ryder and Martinez (1991)	GJ Taylor et al. (1980)	Nyquist et al. (1976)	Blanchard (unpub)	Lindstrom et al. (1988)

(*) also listed with alkali rocks

Table A5.35 continued. Chemical analyses of quartz monzodiorites and KREEP basalts.

	15382 ,9 A15K	15382 ,14 A15K	15386 ,19 A15K	15386 ,1 A15K	15434 ,189/,191 A15K	15434 ,18/,199-A A15K	15263 ,42 A15K	72275 ,357 A17K	72275 ,415 A17K	72275 ,359 A17K	72275 ,91 A17K
wt.%											
SiO_2				50.8	51.6	52.8					48.0
TiO_2	2.17	2.20	1.90	2.23	1.92	2.14	1.70	1.54	1.48	1.03	1.40
Al_2O_3	14.9	16.4	15.3	14.8	16.4	15.2	15.3	14.5	13.3	13.7	13.5
Cr_2O_3		0.309	0.355	0.310	0.320	0.290	0.310	0.286	0.478	0.477	0.460
FeO	9.20	10.00	10.20	10.55	10.00	10.10	9.90	13.90	15.50	14.00	15.00
MnO		0.143	0.150	0.160	0.140	0.150	0.140	0.172	0.228	0.215	0.156
MgO	7.40	9.80	10.50	8.17	9.20	7.40	9.60	6.80	9.30	9.60	10.00
CaO	7.1	10.4	9.5	9.7	9.8	9.4	9.4	10.8	11.1	10.4	10.5
Na_2O	0.85	0.81	0.81	0.73	0.78	0.89	0.74	0.51	0.47	0.40	0.29
K_2O	0.63	0.53	0.50	0.67	0.55	0.65	0.54				0.25
P_2O_5				0.70	0.46	0.62					
Sum				98.8	101.2	99.6					99.6
Mg#	0.59	0.64	0.65	0.58	0.62	0.57	0.63	0.47	0.52	0.55	0.54
ppm											
Sc		19.0	22.0		19.4	23.2	20.2	51.9	48.9	47.4	61.0
V		60.0	62.0					97.0	118.0	134.0	
Co		17.0	23.0		19.7	18.6	20.4	30.9	33.2	34.5	37.0
Ni		28.0	12.5		<20	25.0			42.0	52.0	
Rb	16.1		14.0	18.5	14.7	23.5		12.0	10.0	15.0	
Sr	195			187	187	170	230	92	90	91	
Cs			0.80		0.62	1.02		0.40	0.46	0.35	
Ba	793	610	650	837	664	871	750	500	430	330	
La	79.5	68.1	58.0	83.5	63.0	83.5	68.0	61.7	51.4	39.7	48.0
Ce	212.0	218.0	147.0	211.0	164.0	216.0	170.0	155.0	139.0	102.0	131.0
Nd	127.0		80.0	131.0	92.0	125.0	110.0	108.0	83.0	66.0	
Sm	35.2	29.7	25.5	37.5	28.8	38.2	31.6	28.9	24.0	18.3	23.0
Eu	2.77	2.72	2.40	2.72	2.39	2.63	2.60	1.87	1.69	1.42	1.58
Gd	42.9			45.4			36.0				
Tb		6.2	5.3		5.3	7.1	6.1	5.8	4.1	3.9	4.5
Dy	45.7	40.0	32.0	46.3			39.0				
Er	28.1			27.3							
Yb	24.0	19.2	18.2	24.4	19.6	25.1	20.4	15.5	14.3	11.0	11.9
Lu	3.4	2.7	2.5		2.7	3.6	3.1	2.2	1.9	1.5	1.8
Zr	1170	966	970		729	995	880	610	640	450	
Nb											
Hf	32.7	27.0	21.0		23.6	31.5	22.7	20.5	18.0	13.5	18.0
Th		10.0	10.0		10.5	14.9	12.0	6.7	5.8	4.5	
U	3.7		2.8		2.9	4.3		2.0	1.6	1.2	
ppb											
Ir			0.06								
Au			0.22		<3	<2		<4	<4	<5	
	Hubbard et al. (1973) Nyquist al. (1973)	Murali et al. (1977)	Warren et al. (1978) Warren and Wasson (1978)	Rhodes and Hubbard (1973) Hubbard et al. (1974)	Ryder and Sherman (1979)	Ryder and Sherman (1979)	Simon et al. (1988b)	Salpas et al. (1987)	Salpas et al. (1987)	Salpas et al. (1987)	Blanchard et al. (1975)

Table A5.36. Chemical analyses of granites/felsites

	12033 ,507/,517 (25% cont.)	14001 ,28/,3	14001 ,28/,4	14001 ,28/,2	14004 ,94	14004 ,96	14161 ,7269	14303 ,204/,206	14321 ,1027/,1028	73215 ,43/,3c	73255 ,27/,3	12013 ,10/,16 Black bx	12013 ,10/,09 Black bx	12013 ,10/,01a Black bx	12013 ,10/,16b Gray Bx	12013 ,10/,28 Gray Bx	12013 ,10/,12a Gray Bx
wt.%																	
SiO_2	65.0				69.0	69.0			74.2								
TiO_2	1.50	1.40	1.00	1.80	1.35	1.14		0.75	0.33		0.26	1.90	4.60	3.30	0.75	0.86	0.30
Al_2O_3	12.9	8.8	11.3	9.6	12.3	13.2		18.5	12.5		12.3	12.5	11.9	14.6	11.1	9.8	10.1
Cr_2O_3	0.070	0.022	0.015	0.014	0.013	0.020	0.099	0.080	0.002	0.015	0.010	0.130	0.144	0.162	0.151	0.086	0.252
FeO	7.60	12.20	11.00	9.10	7.70	6.30	7.40	5.57	2.32	2.98	3.10	14.20	13.30	13.70	9.60	6.00	14.00
MnO	0.040	0.130	0.130	0.110	0.106	0.081		0.060	0.020		0.040	0.158	0.164	0.166	0.122	0.077	0.150
MgO	2.40	<0.83	<0.99	<1.2				3.30	0.07		0.20	10.70	8.60	9.00	6.30	5.50	8.50
CaO	4.9	5.5	5.7	3.9	5.1	5.1	7.5	8.8	1.3		0.5	7.3	8.9	9.7	4.1	3.8	4.6
Na_2O	1.47	1.40	1.90	1.70	1.56	1.54	0.71	1.25	0.52	0.19	0.53	1.45	1.47	1.45	1.28	1.17	1.12
K_2O	2.96	2.10	2.50	3.40	3.10	3.20		3.60	8.60	7.00	7.55	0.37	0.50	0.40	3.74	3.03	2.60
P_2O_5					3.10	3.20											
Sum	98.8				103.3	102.8			99.8								
Mg#	0.36							0.51	0.05		0.10	0.57	0.54	0.54	0.54	0.62	0.52
ppm																	
Sc	14.9	20.0	20.0	15.0	15.8	13.6	15.6	10.7	3.0	4.8	2.3	23.0	30.0	28.0	21.0	17.0	25.0
V								23.0			7.0	33.0	46.0	34.0	47.0	42.0	61.0
Co	8.0	3.8	3.9	2.7	2.4	2.7	15.1	14.1	0.9	2.1	1.5	29.0	24.0	27.0	12.0	8.6	24.0
Ni	22.0						110.0	<60	4.9								
Rb	79	87	89	110	87	92	107	114	210			16	33	10	124	101	87
Sr	156	170	190	170	150	110	190	210	55		215						
Cs	3.0	2.0	2.2	2.3	2.0	2.9	5.0	2.2	5.7			1.8	2.1	1.7	4.0	4.3	3.1
Ba	4540	2300	3100	3600	2610	2810	2290	2100	2160		5740	660	1430	390	4150	4160	3760
La	82.0	200.0	130.0	110.0	91.0	87.0	95.3	58.0	44.3	42.9	20.3	135.0	130.0	135.7	64.0	39.0	56.5
Ce	199.0	460.0	270.0	250.0	218.0	237.0		149.0	117.0	125.0	50.0	333.0	344.0	347.0	170.0	100.0	151.0
Nd	100.0				111.0	113.0		93.0	58.0		34.0	242.0	219.0	215.0	91.0	59.0	74.0
Sm	27.7	66.0	40.0	36.0	32.5	32.5	35.6	22.0	15.9	19.0	6.7	53.0	53.0	59.0	20.0	13.0	19.0
Eu	3.10	3.30	3.00	2.80	2.53	2.82	2.69	3.30	1.17	3.11	2.71	2.52	3.29	3.89	1.85	1.74	2.14
Gd																	
Tb	6.8	15.0	9.3	8.1	8.1	8.4	8.0	4.9	4.3	5.6	1.5	10.6	11.4	12.5	4.9	3.7	5.2
Dy	52.0				55.0	49.0		32.0	31.5			67.0	70.0	75.0	38.0	23.0	33.0
Er																	
Yb	36.7	64.0	43.0	37.0	38.0	33.0	55.3	18.0	32.2	27.2	10.3	36.0	37.0	39.0	35.0	25.0	30.0
Lu	5.8	7.7	5.9	5.0	4.9	4.3	7.9	2.6	5.1	5.3	1.5	4.8	4.5	5.2	5.2	3.0	4.2
Zr	1310						1240	1020	660			940	1140	2070	1130	720	690
Nb																	
Hf	36.8	120.0	43.0	41.0	41.0	51.0	33.0	21.6	13.9	25.6	16.0	32.0	38.0	63.0	23.0	25.0	27.0
Th	40.0	42.0	28.0	35.0	29.0	30.0	66.0			39.9	9.5	17.2	17.0	24.5	44.8	35.6	47.5
U	11.8	13.0	8.2	9.1	8.2	9.9	20.0					5.3	6.2	10.4	12.1	10.4	14.2
ppb																	
Ir	1.27	<5.7	<5.1	<1.6	<1.4	<1.7	<6	<2.8	0.05								
Au	0.51				10.00	<4	<4		0.04								
	Warren et al. (1987)	Morris et al. (1990)	Morris et al. (1990)	Morris et al. (1990)	Snyder et al. (1992)	Snyder et al. (1992)	Jolliff et al. (191)	Warren et al. (1983a)	Warren et al. (1983a)	Blanchard et al. (1977)	Blanchard and Budahn (1979)	Quick et al. (1977)	Quick et al. (1977)	Quick et al. (1977)	Quick et al. (1977)	Quick et al. (1977)	Quick et al. (1977)

Table A5.37 Chemical analyses of fragmental breccias

	14063 ,37/A10	14063 ,46	67015 ,106	67015 avge. mx.	67016 ,173	67016 ,63	67115 ,16	67455 ,17	67455 ,15	72275 ,101	72275 ,2
wt.%											
SiO_2		44.7	46.0		44.8	44.9	44.8	44.9	45.2	46.2	47.5
TiO_2	1.50	1.48	0.48	0.51	0.35	0.22	0.24	0.30	0.20	0.94	0.91
Al_2O_3	22.0	22.3	29.5	29.0	28.2	30.1	31.2	30.4	30.6	18.4	17.0
Cr_2O_3	0.180	0.210	0.060	0.080	0.080			0.080	0.061	0.383	0.360
FeO	7.00	6.71	3.64	4.26		3.45	2.60	3.41	4.37	11.90	11.58
MnO	0.081	0.080	0.050		0.050			0.050	0.054	0.182	0.180
MgO		10.80	3.87	5.10	4.33	3.70	3.03	2.30	3.35	9.90	9.35
CaO	13.1	12.7	15.4	15.7	16.7	16.8	17.8	18.3	18.1	11.7	11.7
Na_2O	0.835	0.760	0.518	0.507	0.491	0.470	0.510	0.410	0.380	0.490	0.380
K_2O	0.170	0.150	0.082		0.019	0.060	0.080	0.030	0.026	0.265	0.280
P_2O_5		0.22					0.02	0.02	0.01		0.35
Sum		100.1	99.6			99.7	100.1	100.2	102.3	100.4	99.7
Mg#		0.74	0.65	0.68		0.66	0.68	0.55	0.58	0.60	0.59
ppm											
Sc	13.6	13.0	7.5	6.7	6.6	10.0	2.3	6.2	6.8	30.6	
V	29.0	33.0	12.0		21.2	23.0	15.0	6.9	17.8		
Co	20.0	16.0	9.7	16.9	10.4	15.0	6.2	4.3	10.0	226.0	
Ni		110.0	110.0	187.0	100.0	110.0	62.0	16.0	28.0	950.0	67.0
Rb		5.0	1.4			0.7	1.2	4.6			8.7
Sr		205	195	189	162		180	145	140		121
Cs			0.05			0.04					
Ba	360	250	86	83	60	69	50	28	40		350
La	26.5	27.0	4.9	5.8	3.8	4.5			1.4	38.0	41.0
Ce			11.6	15.7	10.1	12.1			3.7	104.0	106.0
Nd			7.9		7.6	6.3					67.4
Sm	10.6		2.1	2.7	1.6	1.8			0.6	19.1	18.8
Eu	2.40		1.16	1.18	1.06	1.00			0.84	1.46	1.49
Gd			2.6			2.4			0.9		23.4
Tb			0.5	0.6	0.3	0.4				3.4	
Dy			3.1		2.1	2.5			0.9		23.2
Er			1.9			1.7			0.6		13.7
Yb	10.0	6.8	1.8	2.2	1.4	1.6	0.6		0.5	13.3	11.6
Lu	1.4		0.2	0.3	0.2	0.3			0.1	1.7	1.7
Zr		260	55			48	22	12	17		613
Nb		16.00	4.10			4.12			1.30		32.00
Hf			1.67	2.18	1.16	1.36			0.40	14.10	14.60
Th			0.70	1.07		0.73				5.60	5.29
U	1.10		0.22	0.28		0.17			0.05		1.56
ppb											
Ir					10.0				4.0		
Au			1.1		4.8				1.0		
	Laul et al. (1972a)	Rose et al. (1972)	Wanke et al. (1975)	Marvin et al. (1987)	Wanke et al. (1976)	SR Taylor et al. (1974)	Rose et al. (1973)	Rose et al. (1973)	Wanke et al. (1973)	Blanchard et al. (1975)	Hubbard et al. (1974) LSPET (1973)

Table A5.38 Chemical analyses of glasses and glassy breccias

	60095 ,34 Sphere	60015 ,189 Coat	60015 ,54 Coat	60135 ,8 Coat	60639 ,34 Coat	62255 ,52 Coat	67115 ,52 Coat/Vein	67115 ,17 Coat	61016 ,342 injected	61195 ,49 Gl bx	61195 ,29 Gl bx	68815 ,9 Gl bx	68815 ,130 Gl bx
wt.%													
SiO_2	44.8	44.7		43.7	44.9	45.3		44.2	44.1	43.9	45.4	45.1	46.6
TiO_2	0.48	0.38	0.36	0.17	0.28	0.36		0.64	0.19	0.46	0.50	0.49	0.52
Al_2O_3	25.9	27.3	27.0	30.8	29.2	25.5		27.8	31.7	27.9	26.8	27.2	26.8
Cr_2O_3	0.113	0.098	0.095	0.103	0.080	0.120	0.081	0.100	0.070	0.099	0.100	0.100	0.100
FeO	5.72	5.75	4.80	4.77	3.79	5.53	5.08	5.02	3.20	4.75	5.13	4.75	5.00
MnO			0.064								0.060	0.060	0.060
MgO	7.89	7.17	7.00	3.82	4.72	8.03		5.09	3.22	5.94	5.56	5.88	6.40
CaO	14.6	15.2	14.8	17.2	16.5	14.3		15.7	17.5	16.0	15.4	15.5	15.1
Na_2O	0.310	0.470	0.546	0.350	0.440	0.540	0.560	0.470	0.310	0.360	0.460	0.420	0.530
K_2O	0.120	0.130	0.094	0.050	0.090			0.140		0.060	0.114	0.180	0.217
P_2O_5													
Sum	100.0	101.2		100.9	99.9	99.6		99.2	100.4	99.5	99.5	99.6	101.3
Mg#	0.71	0.69	0.72	0.59	0.69	0.72		0.64	0.64	0.69	0.66	0.69	0.70
ppm													
Sc	7.1	6.1	5.8	5.2	4.4	5.4	8.5		3.8	8.2	8.5		7.2
V			20.0										22.5
Co	40.0	66.0	42.0	43.0	35.0	63.0	20.0		24.0	21.0	27.1		30.2
Ni	412.0	1250.0	900.0	632.0	627.0	1317.0	218.0		280.0	256.0	410.0	206.0	500.0
Rb								1.3			3.9		8.8
Sr											166	175	160
Cs								0.06			0.14		0.46
Ba	200	151	100	87	78	92	120	137		154	152		300
La	14.40	11.06	11.10	7.88	7.90	8.68	9.08	9.00	6.30	10.90	14.60		22.30
Ce	41.70	29.10	28.00	22.40	25.80	24.80	25.20	25.20	23.50	31.80	34.00		61.00
Nd			19.00					12.30			23.00		
Sm	6.98	4.93	5.50	3.61	3.56	4.04	4.29	3.48	3.00	5.26	6.30		9.40
Eu	1.220	0.980	1.010	1.210	1.270	1.040	1.340	1.170	0.970	1.600	1.200		1.840
Gd								4.10			7.23		11.90
Tb	1.27	1.14	0.90	0.72	0.67	0.94	0.76	0.68	0.59	0.92	1.25		2.00
Dy			6.00					4.54			8.30		14.10
Er								3.03			4.60		7.60
Yb	4.27	3.34	3.50	2.39	2.35	2.68	3.04	2.77	2.00	3.58	4.37		6.86
Lu	0.71	0.49	0.49	0.31	0.31	0.43	0.43	0.43	0.29	0.52	0.64		1.00
Zr			70					137			194	266	331
Nb								11.20			11.80	16.00	20.00
Hf	4.90	3.49	3.20	2.30	2.41	3.01	3.02	2.53	2.25	3.82	4.58		7.50
Th	2.60	2.95	1.70	1.64	1.62	1.94	1.35	1.49	0.59	2.26	2.30	3.70	3.74
U	0.99	0.61	0.41	0.46	0.48	0.46	0.56	0.40	0.09	0.47	0.72		1.09
ppb													
Ir			23.0								11.3		11.0
Au			8.0								6.1		15.0
	Morris et al. (1986)	Morris et al. (1986)	Laul et al. (1973)	Morris et al. (1986)	Morris et al. (1986)	Morris et al. (1986)	Morris et al. (1986)	SR Taylor et al. (1973)	Morris et al. (1986)	Morris et al. (1986)	Wanke et al. (1975)	LSPET (1973a)	Wanke et al. (1974)

Table A5.39. Chemical analyses of clast-rich and clast-poor impact melts

	14066 ,31	14068 ,3	14305 ,81	14321 ,286	14321 ,915	14321 ,65	15405 ,112	15357 ,14	15359 ,10	15445 ,25	15445 ,243	15455 ,14	15455 ,257	60315 ,216	60315 ,3	65015 ,60	65015 ,60	61015 ,100
wt.%																		
SiO_2	49.2	47.2	48.4		48.0	47.8	50.7	49.0	47.1	44.6	46.3	47.3	46.0		45.6		47.6	
TiO_2	1.00	1.39	1.52		1.75	2.06	1.79	1.19	1.08	1.47	1.80	1.35	1.56		1.27		1.23	
Al_2O_3	15.9	13.3	16.3		15.4	15.2	16.1	15.2	17.9	16.7	19.0	17.1	17.3		17.2		19.5	
Cr_2O_3	0.174		0.194	0.289	0.201	0.260	0.224	0.165	0.146	0.256	0.168	0.263	0.282					
FeO	10.10	10.00	10.45	13.16	11.42	12.25	11.50	9.90	8.90	9.83	10.40	8.79	10.30	8.36	10.53	7.49	8.52	7.12
MnO	0.120	0.130	0.134		0.150	0.170	0.183		0.133	0.139	0.129				0.120		0.110	
MgO	11.40	17.60	10.28		9.44	10.73	7.60	14.30	12.70	16.00	12.40	13.30	16.00		13.15		9.66	
CaO	10.1	8.3	9.9			9.9	10.4	9.7	11.1	10.0	11.3	10.6	10.2	10.6	10.4	13.2	12.3	13.3
Na_2O	0.84	0.75	0.77	0.67	0.78	0.78	0.88	0.55	0.54	0.55	0.58	0.58		0.66	0.56	0.56	0.57	0.48
K_2O	0.94	0.59	0.64	0.34	0.56	0.62	0.75	0.42	0.17	0.16	0.22	0.17	0.17	0.00	0.35	0.45	0.35	
P_2O_5		0.55			0.68	0.41	0.65		0.44	0.21	0.48				0.45			
Sum	99.8	99.8	98.6			100.2	100.8	100.4	100.2	99.9	102.8	99.5	101.8		99.6		99.8	
Mg#	0.67	0.76	0.64		0.60	0.61	0.54	0.72	0.72	0.74	0.66	0.73	0.73		0.69		0.67	
ppm																		
Sc	20.0		24.0	36.0	23.6		23.3	20.4	16.1		18.2	13.0	18.2	15.0		14.5	14.9	10.3
V												39.0						
Co	30.0		31.0	33.6	42.2		16.9	38.4	26.5		39.3	22.0	41.2	25.3		23.9	27.1	48.8
Ni			200.0		390.0		22.0	231.0	231.0		245.0	184.0	484.0	380.0	191.0	332.0	400.0	800.7
Rb		14.5	25.0		14.8		22.0	11.0	4.7	3.6		2.7	4.0	11.6	9.8	10.4	9.2	
Sr		139	190		188					160		141		170	156	163		
Cs			1.36		0.69		0.82	0.36	0.20		0.15	0.16	0.18	0.59		0.45	0.42	
Ba		780	830	540	940		940	390	314	237	232	370	288	479		606		253
La	75.0		109.0	51.5	88.0		74.5	44.2	35.3	22.1	21.6	32.0	23.7	50.1	45.5	65.9	60.0	26.2
Ce	200.0	157.0	200.0	132.0	237.0		228.0	117.0	95.0	57.6	57.0	81.0	64.0	127.8	113.0	170.0	153.0	69.3
Nd		93.4	140.0	77.0	142.0		139.0	76.0	60.0	35.7	37.0	47.0	43.0	79.0	71.3	102.0	98.0	
Sm		28.1	23.0	25.0	34.3		39.5	21.8	16.8	10.1	10.6	12.8	11.3	22.7	20.1	29.9	26.5	12.4
Eu	2.76	2.01	2.60	2.07	2.69		2.42	1.83	1.65	1.85	1.93	1.82	1.87	2.04	1.89	1.99	2.12	1.49
Gd		29.1	38.0		43.6					11.9		15.5			23.8		33.0	
Tb	7.8		7.4		7.7		8.9	4.4	3.6		2.1	2.4	1.7	4.4		5.8	5.2	2.6
Dy	39.0	35.1	43.0		48.3					13.2		16.0			26.3		38.0	
Er			32.0		29.3							3.8					7.4	
Yb	25.1	20.0	24.2	16.3	28.3		26.5	14.6	10.2	6.9	6.9	9.8	7.0	15.6	14.0	19.3	19.1	8.2
Lu	3.6		3.5	2.4	3.9		3.9	2.1	1.5	1.0	1.1	1.5	1.0	2.1		2.6	2.6	1.1
Zr				375	1210		955	345	385	315	189	480		810	640	924		230
Nb					75.00													
Hf	30.0		26.0	17.3	29.2		29.8	15.5	11.1	10.5	7.6	9.8	8.0	2.0		2.4	19.3	1.0
Th			17.4		13.8		16.6	4.8	7.2		1.9	5.3	3.1	2.1		2.7		1.0
U		3.5	5.2		4.0		4.9	1.9	2.6	0.8	1.1	1.4	1.8		2.1			
ppb																		
Ir			10.0		8.0									9.4		4.3		21.1
Au			6.7		7.8									8.1		10.6		4.1
	Wanke et al. (1972)	Hubbard et al. (1972)	Wanke et al. (1972)	Duncan et al. (1975)	Palme et al. (1978)	Scoon (1972)	Ryder and Spudis (1987)	Ryder and Spudis (1987)	Ryder and Spudis (1987)	Ridley et al. (1973b)	Ryder and Spudis (1987)	SR Taylor et al. (1973)	Ryder and Spudis (1987)	Korotev (1994)	Hubbard et al. (1973) LSPET (1973a)	Korotev (1994)	Haskin et al. (1973)	Korotev (1994)

Table Table A5.39 continued. Chemical analyses of clast-rich and clast-poor impact melts

		61015 ,56	64815 ,9	64815 ,10	61225 ,1	67475 ,10	63335 ,18	72255 ,287A1	72395 ,46	72395 ,51	72435 ,1	76015 ,22	76215 ,27	76215 ,28	76295 ,14	76295 ,35	77135 ,2	77135 ,4	77135 ,91
wt.%	wt.%																		
SiO_2	SiO_2	44.9				44.4		45.1	46.9	46.6	45.8	46.2	46.0	46.1	47.0	47.6	46.1	46.1	46.3
TiO_2	TiO_2	0.62		1.50		0.36	0.34	1.42	1.75	1.36	1.54	1.52	1.52	1.24	1.39	1.64	1.54	1.48	1.31
Al_2O_3	Al_2O_3	24.5		20.0		30.9	31.5	22.3	18.1	18.0	19.2	17.2	17.8	18.7	18.3	17.7	18.0	18.5	19.8
Cr_2O_3	Cr_2O_3							0.201	0.204	0.214	0.200	0.192				0.202	0.172	0.193	0.161
FeO	FeO	6.71	9.37	9.07	7.17	2.91	2.60	6.75	9.29	9.13	8.70	9.81	8.70	8.08	9.09	9.05		8.80	8.28
MnO	MnO	0.080		0.120		0.040	0.035	0.100	0.120	0.125	0.110	0.130					0.130	0.116	0.100
MgO	MgO	8.88		12.60		2.57	2.00	9.35	11.97	11.97	11.63	13.03	12.21	12.43	10.78	9.78	12.63	11.60	11.78
CaO	CaO	13.7	12.0	11.8	14.5	17.2	17.6	12.3	11.3	11.0	11.7	10.8	11.1	11.5	11.5	11.5		10.7	11.7
Na_2O	Na_2O	0.49	0.52	0.53	0.47	0.60	0.69	0.42	0.69	0.62	0.52	0.70				0.82		0.71	0.56
K_2O	K_2O	0.13	0.27	0.26	0.14	0.04	0.05	0.22	0.29	0.24	0.22	0.27	0.28	0.24	0.26	0.29	0.29	0.26	0.21
P_2O_5	P_2O_5	0.20																	
Sum	Sum	100.2		55.9		99.0		98.2	100.6	99.3	99.6	99.7	97.7	98.4	98.3	98.5		98.4	100.3
Mg#	Mg#	0.70		0.71		0.61	0.58	0.71	0.70	0.70	0.70	0.70	0.71	0.73	0.68	0.66		0.70	0.72
ppm	ppm																		
Sc	Sc	8.9	23.1	21.7	9.9		4.4	14.2	18.7	18.8						17.8		15.7	
V	V			51.0															
Co	Co	60.9	57.3	37.3	61.5		5.0	16.0	31.1	33.9						19.9		37.7	
Ni	Ni	1160.0	830.0	460.0	957.0			92.0	260.0	296.0	112.0				203.0	160.0	110.0	425.0	
Rb	Rb		6.4		3.7	0.7		12.2	6.2		3.9	6.4	6.9	6.1	5.4		7.3	7.4	
Sr	Sr	166	132		177	216		151	167	157	172	172					172	201	171
Cs	Cs		0.33		0.18			0.35	0.19	0.21								0.28	
Ba	Ba	280	340	410	189	49	40	221	386	322	334	348	352	294	376		337	367	294
La	La	20.8	33.9	34.3	19.3		2.6	17.3	39.7	26.2	31.7		33.4	27.3	37.8	37.5	32.1	32.6	
Ce	Ce	56.4	95.9	86.0	50.7	6.7	6.0	46.4	95.0	69.7	80.6	83.3	83.6	68.9	95.7	102.0	81.2	86.0	59.2
Nd	Nd	34.0	56.0	55.0	26.0	4.2	4.0	22.6	61.0	43.0	51.3	52.8	52.2	43.7	60.0		51.6	49.0	41.1
Sm	Sm	8.6	15.7	15.7	9.1	1.2	1.2	8.0	16.8	12.4	14.5	14.9	14.9	12.3	16.9	17.0	14.6	14.9	11.2
Eu	Eu	1.30	1.72	1.70	1.28	1.28	1.32	1.27	1.93	1.71	1.88	1.94	1.99	1.70	1.91	2.11	1.99	2.04	1.80
Gd	Gd	11.0				1.4													
Tb	Tb	1.9	3.6	3.4	1.8		0.2	1.7	3.7	2.7						3.9		3.1	
Dy	Dy	11.3		23.0		1.7	1.5												
Er	Er	2.8																	
Yb	Yb	6.5	12.2	11.7	6.3	1.0	0.9	6.5	12.4	9.1	10.1	10.6	10.9	9.0	12.0	12.2	10.5	10.4	8.1
Lu	Lu	0.9	1.7	1.6	0.9	0.1	0.1	0.9	1.9	1.2						1.7		1.4	1.2
Zr	Zr	292	472		272			215	570	329	473	490	495	459	541		494	423	
Nb	Nb																		
Hf	Hf	6.8	1.5	11.1	0.8		0.6	6.6	13.7	9.9	12.7	12.9				13.2		11.8	
Th	Th	2.6	1.2	5.6	0.9		0.3	3.5	6.1	4.7		5.4	5.2	4.6	6.1	5.6		5.4	
U	U	0.7		1.6				0.9	2.1	1.3		1.5	1.5	1.3	1.8			1.5	
ppb	ppb																		
Ir	Ir	29.0	13.7	9.0	17.1		2.0	3.2	11.0	8.1								11.7	
Au	Au	20.0	4.9	8.4	3.1		4.0	4.3	4.7	7.2								9.6	
		Palme et al. (1978)	Korotev (1994)	Wasson et al. (1977)	Korotev (1994)	Lindstrom et al. (1977)	Laul et al. (1974)	Ryder (unpub)	Wanke et al. (1975)	Ryder and Stockstill (unpub)	Hubbard et al. 1974) LSPET (1973b)	Rhodes et al. (1974) Hubbard et al . (1974)	Simonds (1975) Hubbard et al . (1974)	Simonds (1975) Hubbard et al . (1974)	Simonds (1975) Hubbard et al . (1974)	Phinney (1981)	Hubbard et al. 1974) LSPET (1973b)	Ryder and Stockstill (unpub)	Winzer et al. (1977)

Table Table A5.39 continued. Chemical analyses of clast-rich and clast-poor impact melts

		76055 ,5	76055 ,40	72255 ,79	72255 ,44	73215 ,161	73215 av. of 11	73235 ,49	73235 ,91	73255 ,253,13b	14310 ,101/,130	14310 ,245	62295 ,52	62295 ,166	68415 ,027	68415 ,6	68416 ,40
wt.%	wt.%																
SiO_2	SiO_2	44.7	45.6	45.0	46.7	48.1	46.4	46.4	46.7		47.2		45.1			45.4	45.6
TiO_2	TiO_2	1.24	1.28	0.90	0.76	0.80	0.70	0.63	0.65	0.85	1.24		0.72			0.32	0.30
Al_2O_3	Al_2O_3	16.5	15.9	20.7	20.8	19.4	20.6	21.2	20.5	18.0	20.1		20.8			28.6	28.4
Cr_2O_3	Cr_2O_3	0.190	0.200	0.234	0.241	0.168	0.248	0.219	0.197	0.336							
FeO	FeO	9.11	9.21	8.31	8.10	7.63	7.30	7.33	7.40	9.30	8.38	8.07	6.32	7.04	4.19	4.25	4.22
MnO	MnO	0.110	0.116	0.129	0.117	0.123	0.104		0.100	0.130	0.110		0.080			0.060	0.060
MgO	MgO	16.33	16.50	11.30	9.90	10.20	11.60	10.70	11.60	11.60	7.87		14.71			4.38	4.64
CaO	CaO	9.9	9.7	12.0	12.6	11.0	11.6	12.5	11.9	10.3	12.3	12.0	12.1	11.5	15.9	16.4	16.3
Na_2O	Na_2O	0.48	0.57	0.58	0.48	0.62	0.50	0.47	0.46	0.48	0.63	0.72	0.46	0.45	0.47	0.41	0.44
K_2O	K_2O	0.21	0.18	0.21	0.20	0.66	0.20	0.18	0.20		0.49	0.00	0.08	0.11	0.00	0.06	0.08
P_2O_5	P_2O_5		0.20								0.34		0.15			0.07	0.07
Sum	Sum	98.7	99.5	99.4	99.9	98.7	99.3	99.6	99.7		98.7		100.5			100.0	100.1
Mg#	Mg#	0.76	0.76	0.71	0.69	0.70	0.74	0.72	0.74	0.69	0.63		0.81			0.65	0.66
ppm	ppm																
Sc	Sc		14.0	18.2	18.8	14.3	14.7	17.0	13.4	20.8		18.6	10.3	9.5	8.5		9.2
V	V												29.1				
Co	Co		43.1	28.9	24.5	24.5	25.3	33.0	26.3	33.1		21.6	23.0	56.3	11.4		10.0
Ni	Ni	155.0	490.0	260.0	150.0	200.0	165.0	250.0	205.0	236.0	64.0	170.0	330.0	787.0	107.0	49.0	205.0
Rb	Rb	5.2	5.6			5.5		3.1			12.8	13.0	5.6	5.8	2.3	2.1	
Sr	Sr	157	158						150	189	188	171	132	152	175	185	130
Cs	Cs		0.09					0.15				0.57	0.56	0.51	0.09		
Ba	Ba	253	285					315		358	617	644	197	186	73		76
La	La	22.6	25.1	31.0	31.7	41.0	27.0	24.0	24.5	26.9	56.4	56.8	19.2	18.0	7.3		
Ce	Ce	56.3	65.0	79.0	83.3	105.0	70.0	61.0	58.5	72.1	144.0	146.0	52.2	46.5	19.4		
Nd	Nd	35.8	40.0			62.0		33.5		35.4	87.0	84.0	33.0	27.9	10.6		
Sm	Sm	10.1	10.6	15.7	12.9	18.6	12.5	9.0	9.4	12.6	24.0	25.1	8.1	8.3	3.3		
Eu	Eu	1.71	1.73	1.45	1.39	1.63	1.43	1.20	1.25	1.42	2.15	2.11	1.21	1.15	1.12		
Gd	Gd										28.1		10.4				
Tb	Tb		2.4	2.8	2.8	3.6	2.6	1.9	2.2	2.7		5.0	1.8	1.6	0.7		
Dy	Dy										32.7		11.0				
Er	Er												2.4				
Yb	Yb	7.6	8.7	10.5	10.5	13.0	9.1	7.5	7.9	9.0	18.4	18.3	6.3	5.8	2.5		1.8
Lu	Lu		1.2	1.3	1.4	1.9	1.3	1.2	1.1	1.3		2.5	0.9	0.8	0.3		
Zr	Zr	341	345					350	343		842	860	283	249	119	98	80
Nb	Nb										52.00		18.00			5.60	
Hf	Hf		8.8	11.2	10.5	13.7	9.5	6.5	7.9	10.4		2.4	6.6	0.7	0.3		
Th	Th		3.5	5.4	4.3	7.5	4.1	4.3	3.8	4.6	11.0	2.9	2.7	0.8	0.4		
U	U		0.9					1.1	1.1				0.8				
ppb	ppb																
Ir	Ir		13.0									2.1	5.5	16.4	3.1		
Au	Au		7.2						3.1			10.6	7.0	3.0	1.2		
		Hubbard et al. 1974) LSPET (1973h)	Palme et al. (1978)	Blanchard et al. (1975)	Palme et al. (1978)	James et al. (1975)	Blanchard et al. (1976b)	SR Taylor et al. (1974)	Wanke et al. (1974)	James et al. (1978)	Hubbard 3	Korotev (pers comm)	Wanke et al. (1976)	Korotev (1994)	Korotev (1994)	LSPET (1973a)	Rose et al. (1973)

Table A5.40. Chemical analyses of granulitic breccias and granulites.

	15418 ,51a	15418 Sawdust	67215 ,8a	67415 ,3	67415 ,33A	67955 ,56	67955 ,43	72215 ,76	72235 ,36	72559 ,5	76230/5	77017 ,2	77017 ,57	78155 ,2	78155 ,127	79215 ,39/,1
wt.%																
SiO_2	44.2			44.6		45.0		44.7	44.5	42.4	44.5	44.1		45.6	45.4	44.8
TiO_2	0.27	0.37	0.33	0.32	0.42	0.27	0.28	0.50	0.80	0.20	0.20	0.41	0.75	0.27	0.29	0.50
Al_2O_3	26.6	26.4	27.4	26.0	27.0	27.7	27.1	27.3	25.8	28.5	27.0	26.6	26.0	25.9	25.3	27.4
Cr_2O_3	0.127	0.277	0.110		0.103	0.096	0.122	0.126	0.146	0.140	0.110	0.129	0.140	0.140	0.140	0.108
FeO	6.65	7.50	6.39	4.60	4.60	3.84	4.14	4.80	6.19	4.70	5.14	6.19	6.20	5.82	5.63	4.86
MnO	0.100	0.085		0.060		0.050	0.059	0.067	0.080	0.050	0.060	0.080	0.085	0.100	0.085	0.064
MgO	5.08	5.30	5.30	7.77	7.50	7.69	8.13	7.19	8.52	8.41	7.63	6.06	6.00	6.33	6.42	7.40
CaO	16.0	15.8	16.2	15.1	14.6	15.5	15.1	14.9	14.4	15.3	15.2	15.4	14.5	15.2	15.2	14.4
Na_2O	0.270	0.282	0.300	0.520	0.490	0.400	0.455	0.483	0.420	0.350	0.350	0.300	0.310	0.330	0.380	0.580
K_2O	0.013	0.011		0.041		0.050	0.041	0.113	0.110	0.100	0.060	0.060	0.050	0.080	0.073	0.119
P_2O_5											0.05	0.03		0.04		
Sum	99.3			99.0		100.6		100.2	101.0	100.2	100.3	99.4		99.8	98.9	100.2
Mg#	0.58	0.56	0.60	0.75	0.74	0.78	0.78	0.73	0.71	0.76	0.73	0.64	0.63	0.66	0.67	0.73
ppm																
Sc		12.7	12.8		6.1		7.2	7.7	9.8	6.5			12.0		13.3	8.1
V		42.0											40.0			
Co		77.0	12.0		14.9	18.2		11.9	43.7	37.1			24.0		14.3	18.9
Ni			26.0		164.0		170.0	50.0	500.0	494.0	166.0	95.0	290.0	53.0	80.0	126.0
Rb	0.17										0.45	1.31		2.06	2.01	
Sr	140		150	177	183	169	168				146	142		147	141	
Cs			0.07		0.07		0.07								0.11	
Ba	19	70	16	61	63	62	73			70	50	49	30	59	64	
La	1.07	1.20	1.58		5.56	4.45	5.02	7.30	7.20	3.20	3.04	3.48	3.30	4.02	4.28	2.65
Ce	3.31		4.30	9.62	14.70	11.30	13.30	18.30	20.00	8.30	7.54	8.90	9.00	10.20	11.30	6.80
Nd	2.09		3.00	6.23	7.20	7.09	7.70				4.64	5.56	5.00	6.29	7.30	
Sm	0.69	0.69	0.86	1.84	1.73	2.02	2.08	3.36	3.66	1.27	1.34	1.60	1.50	1.81	1.69	1.19
Eu	0.726	0.730	0.726	1.110	1.075	0.973	0.980	1.000	0.920	0.800	0.805	0.794	0.780	0.874	0.862	0.840
Gd	1.25					2.57					1.70	2.01		2.32	2.30	
Tb		0.18	0.20		0.39		0.50	0.66	0.80	0.30			0.30		0.39	0.28
Dy	1.12	1.20		2.75		2.81					2.02	2.35	2.40	2.64	2.63	
Er	0.85														1.90	
Yb	0.74	0.81	0.75	1.75	1.46	1.74	2.08	3.10	2.90	1.58	1.37	1.50	1.60	1.73	1.83	1.37
Lu	0.12	0.12	0.13	0.26	0.23	0.25	0.27	0.44	0.41	0.23	0.20	0.23	0.21	0.26	0.27	0.24
Zr		180	25		55	124	66				42	59			54	
Nb											3.2	4.1		4.8	2.0	
Hf		0.80	0.80		1.70	3.10	1.79	2.40	2.20	1.30		1.50	1.50	4.80	1.49	1.20
Th			0.10		1.11	1.03	0.84	1.30	1.60	0.77	0.72		0.40	1.01	0.84	0.32
U	0.05		0.03		0.19	0.38	0.26		0.43	0.23	0.20	0.22		0.28	0.24	
ppb																
Ir			1.10		4.00		6.80	2.95		13.60		17.00	10.00		3.90	
Au							2.00	0.79		26.70		5.65	3.00		0.68	
	Bansal et al. (1972)	Laul et al. (1972b)	Lindstrom and Lindstrom (1986)	Lindstrom et al. (1977)	Lindstrom and Lindstrom (1986)	Hubbard et al. (1974)	Palme et al. (1978)	Blanchard et al. (1975) Higuchi and Morgan (1976)	Blanchard et al. (1975)	Warren and Wasson (1978)	LSPET (1973b) Hubbard et al. (1974)	LSPET (1973b) Hubbard et al. (1974)	Laul et al. (1974)	LSPET (1973b) Hubbard et al. (1974)	Wanke et al. (1976)	Blanchard et al. (1977)

Table A5.41. Modal mineralogy (vol %) of some ferroan anorthosites of the lunar highlands.

	10056c	14312c	15362	15363	15415	60015	65315	62236	62237	64435c	67016c	74114,5
Plagioclase	~100	~100	97-98	85	99	98.5	98.5	85	82.6	81.1	68-69	96
Olivine	tr	tr?		tr	tr		tr	4	16.1	15.5		1
Opx/Pig	tr	tr?	~0.5	12	tr	1.3	1.4	10	0.8	2.4	~19	0.8
High-Ca px			~1.5	2	tr	tr	tr	1	0.1	0.9	~8	
Mg-spinel												
Cr,Fe-spinel			tr			tr		tr	0.1	0.1		tr
Fe-Ni metal	0.3			0.4	tr	tr				tr		
Troilite			tr	0.1		tr		tr	tr		3-4	tr
Ilmenite			tr		tr	tr		tr	tr		tr	
Phosphates												
Others			tr		tr							

10056c 14312c Warren et al. (1983b); 15362 Dowty et al. (1972); 15363 Warren et al. (1987); 15415 from compilation of Ryder (1995); 60015, 65315 Dixon and Papike (1975); 62236 Warren and Wasson (1978), Nord and Wandless (1983); 62237 Dymek et al. (1975b); 64435c James et al. (1989); 67016c Norman et al. (1995); 74114,5 Warren et al. (1991)

Table A5.42. Modal mineralogy (vol %) of some olivine-rich Mg-rich plutonic rocks of the lunar highlands.

	Troct 76535 (a)	Troct 76535 (b)	Troct 76335	Troct 14321 c1 (a)	Troct 14321 c1 (b)	Troct 14321 c	Sp troct 67435 c1	Sp troct 67435 c2	Sp troct 15445 cA (a)	Sp troct 15445 cA (b)	Sp troct 73263	Sp troct 72435 (a)	Sp troct 72435 (b)	Dunite 72415-8
Plagioclase	58	35	88	60	75	40	26	35	35	5	70	83-89	80	4
Olivine	37	60	12	40	24	60	69	52	50	>40	15-20	2-7	10	93
Opx/Pig	4	5	tr		1					<40	5	2-6	1	2
High-Ca px	tr	tr			0	tr								1
Mg-spinel							5	13	15	15	5	1-11	10	
Cr,Fe-spinel	tr	tr		tr		tr								tr
Fe-Ni metal	tr	tr					tr	tr						tr
Troilite	tr	tr					tr							
Ilmenite						tr					tr			
Phosphates	tr	tr												
Others	tr	tr							tr	tr			tr	tr

76535 (a) Gooley et al. (1974) (b) Dymek et al. (1975); 76335 Warren and Wasson (1978); 14321c1 (a) Lindstrom et al. (1984) (b) Warren et al. (1981);14321c Lindstrom et al. (1984); 67435c1 Prinz et al. (1973); 67435 c2 Ma et al. (1981b); 15445A (a) Ryder and Bower (1977) (b) Anderson (1973);73263,1,11 Bence et al. (1974); 72435 clasts (a) Baker and Herzberg (1980) (b) Dymek et al. (1976); 72415-8 Dymek et al. (1975)

Table A5.43. Modal mineralogy (vol %) of some noritic Mg-rich plutonic rocks of the lunar highlands.

	Norite	Norite	Anor-norite	Norite	Norite	Norite	Norite	Norite	Gab-norite	Gab-norite	Gab-norite	Norite	Gab-norite	Gab-norite
	14318c	15445c	15455c	78235	77035	72255c	73255c	77215	67667	73255c	61224	76255c	76255c	67915c
Plagioclase	~55	60-65	70-75	48	~60	~40	83	52	23	53	34	51	39	42
Olivine	~12								50					
Opx/Pig	~35	35-40	25-30	51	~60	~60	15	45	21	40	43	38	4	5
High-Ca px			tr	0.01	tr	<1	1	1	5	5	22	11	57	23
Mg-spinel														
Cr,Fe-spinel			tr	tr	tr	tr	tr	tr	tr	tr	tr	tr		
Fe-Ni metal	tr	tr	tr	tr		tr	tr	tr	tr	tr	tr		tr	tr
Troilite	tr		tr	tr		tr	tr	1.5	tr	tr	tr	tr	tr	0.4
Ilmenite	tr	tr	tr		tr	tr	tr	0.1	1	0.5		<1	tr	5
Phosphates		tr	tr	0.03		tr	tr	tr		tr				0.2
Others		tr	tr	tr	tr	tr	1	1		tr		tr		25(a)

(a) mainly silica

14318c Warren et al. (1983b) 15445c, 15455c, 78235-6, 72255c, 77215 James and Flohr (1983); 77035 Warren and Wasson (1979); 73255c, 73255c James and McGee (1979); 67667 Hansen et al. (1980); 61224 Marvin and Warren (1980); 76255 JL Warner et al. (1976); 67915c GJ Taylor et al. (1980)

Table A5.44. Modal mineralogy (vol %) of some alkali suite fragments of the lunar highlands.

	12033 ,425	12073 c	14047 c	14066 ,49	14303 ,44	14305 ,91	14305 ,400	14305 c2	14313 (a) c	14321 c5	15362	15405 ,170	15405 ,181	67915 ,163+	67975 ,14	67975 ,136N
Plagioclase	99	99	84	85	~80	~90	99.5	95	~40	~96	97-98	x	~100	40-45	50	50
Olivine					x											
Opx/Pig		1	16	15	~20	~4?		~2	~20	1-2	~0.5	x		5	46	33
High-Ca px	1				x	~1?	0.4	~1	~5	~1	~1.5			20-25	3	2
Mg-spinel																
Cr,Fe-spinel											tr				tr	tr
Fe-Ni metal								tr				x		tr	tr	tr
Troilite			tr					tr			tr	x		tr	tr	tr
Ilmenite	tr		tr	tr	tr	~4	tr	tr		1	tr	x	tr	5-10	0.5	2
Phosphates							tr	~2	~35	1-2		x	tr	~0.25		5
Silica								tr				x		15-25	0.5	3
Potash Feldspar			tr	tr		~1		tr				x		1		5
Others								tr			tr	x				tr

(a) very tiny sample x = reported present

12033,145 Warren et al. (1990); 12073c Warren and Wasson (1980); 14047c Warren et al. (1983b); 14066,49 Shervais et al. (1983); 14303,44 14305,91 Hunter and Taylor (1983); 14305,400 Shervais et al. (1984a); 14305c 14313c 14321c5 Warren et al. (1983c); 15362 Dowty et al. (1972) 15405,170 15405, 181 Lindstrom et al. (1988); 67915,163+ Marti et al. (1983), GJ Taylor et al. (1980); 67975,14 67975,136N James et al. (1987)

Table A5.45. Modal mineralogy (vol %) of KREEP basalt and quartz monzodiorite fragments of the lunar highlands

	Average 6 frags 15K	15382 15K	15386 15K	72275 17K	14161 ,7069 QMD	14161 ,7373 QMD	15403 ,71A QMD	15403 ,71B QMD	15405 ,51 QMD	15434 ,10 QMD	15434 ,12 QMD
Plagioclase	41	49	35	40	30.1	26.3	29		34.8	<1	41
Olivine				tr	tr						
Opx/Pig					19.9	31.1			18		
High-Ca px	35	34	50	40	9.5	15.9	14	42	18	31	53
Mg-spinel											
Cr,Fe-spinel											
Fe-metal	tr		x	tr	tr		tr		0.2		
Troilite	tr		x	tr		0.6	tr	tr	tr	4	tr
Ilmenite			3	x	2.4	2.0	tr	3	1.6	6	2
Phosphates			x	tr	3.7	11.1	18	1	0.8	2	tr
Silica	4	4	10	x	12.2	5.2	5		15	~22	3
Potash Feldspar				tr	11.1	2.0			11	~33	1
Others	20(a)	13(b)		20(c)	11.1(d)	5.8(e)	33(f) +tr	53(f) +tr	0.6 (g)		

15K = Apollo 15 KREEP basalts; 17K = Apollo 17 KREEP basalts; QMD = Quartz monzodiorites

(a) 17% mesostasis and 3 % opaques (b) 8% mesostasis, 5% opaques (c) mesostasis
(d) 10.5 % of ~ 2/3 potash feldspar, 1/3 silica, plus 0.6% zircon (e) 4.8 % of ~ 1/2 potash feldspar, 1/2 silica, plus 1.0 % zircon
(f) potash feldspar-silica intergrowths (g) zircon

Average 6 fragments Powell et al. (1973) 15382 Dowty et al (1972); 15386 Steele et al. (1972); 14162,7069 14161,7373 Jolliff (1991); 15403,71A 15403,71B Marvin et al. (1991); 15405,51 GJ Taylor et al. (1980); 15434,10 15434,12 Ryder and Martinez (1991)

Table A5.46. Modal mineralogy (vol %) of some granite/felsite fragments of the lunar highlands.

	12033 ,507	12070 ,102-5	14303 ,204 (a)	14321 ,1027 (a)	15403 ,71C	73215 c43
Plagioclase	6	16	~20		9	x
Olivine	1	tr	tr	<1		tr
Opx/Pig						
High-Ca px	1	4	4	tr		
Mg-spinel						
Cr,Fe-spinel						
Fe metal			tr	tr	tr	
Troilite	tr		tr		tr	
Ilmenite	1	tr	1	tr	tr	tr
Phosphates	tr	tr	tr		<1	tr
Silica	~33		~35			x
Potash Feldspar	~49		~40			x
Others	9 (b)	80 (c)	tr	~99(d)	90 (e)	x(f)

x = reported present
(a) not including shock glass (b) Si-Fe-rich glass (c) granophyric intergrowth silica, potash feldspar, plagioclase (d) graphic intergrowth potash feldspar ~60% silica ~40%
(e) granophyric intergrowth silica,potash feldspar, and ternary feldspar
(f) granophyric intergrowth potash feldspar ~60%, silica ~40 %

12033,507 Warren et al. (1987); 14303,204 14321,1027 Warren et al. (1983a);
12070,102-5 15403,71C Marvin et al. (1991); 73215 c43 James and Hammarstrom (1977)

Table A5.47. Modal clast populations (vol %) for fragmental breccias from the Apollo 16 landing site.

	63588 ,4	63595 ,4	67455 ,57+,61	67495 ,3	67517 ,3	67526 ,4	67527 ,1
Cataclastic anorthosites	34.6	13.9	49.7	27.4	86.9	87.2	85.0
Granulites and granulitic bx	15.3	22.5	8.1	18.9	5.4	5.2	1.8
Feldspathic microporphyritic bx	1.3	20.8	33.8	44.9	6.3	7.6	13.2
Mafic microporphyritic melt bx		0.6					
Intergranular melt bx	22.7	6.5	2.7	3.5			
Micropoikilitic melt bx		3.8					
Subophitic-intersertal melt bx	3.1	2.8	7.8	1.1			
Glass and glassy bx	22.9	29	3.2	4.2	1.4		
	67528 ,1	**67545 ,4**	**67547 ,1**	**67549 ,5a +,5b**	**67639 ,2**	**67645 ,3**	**67647 ,2**
Cataclastic anorthosites	61.2	90.2	5.5	26.8	15.4	18.1	3.8
Granulites and granulitic bx	21.2		46.0		18.2	20.1	1.3
Feldspathic microporphyritic bx	11.7	9.8	41.5	49.0	62.6	55.8	88.4
Mafic microporphyritic melt bx				2.2			
Intergranular melt bx	1.3			12.4			
Micropoikilitic melt bx							
Subophitic-intersertal melt bx					3.7	2.5	5.7
Glass and glassy bx	4.7		7.0			3.4	0.8

bx = breccias

Stoffler et al. (1985)

Table A5.48. Modal mineralogy (vol %) of clast-rich and clast-poor melt breccias of the lunar highlands.

	14276 CP	14310 CP	62295 CP	62295 CP	64455 CP	65055 CP	67747 CP	67559 CP	68415 CP	68416 CP	60315 CR	60335 CR	62235 CR	66095 CR	72435 CR	77135 CR
Plagioclase	64.6	59.0	35.5	x	59.9	68.5	67.2	83.1	82.2	73	43.2	56.5	34.1	45.0	63.0	41.1
Olivine			28.3	x	6.5		16.3		3.1	4.5	6.1	9.0	6.0	10.2	8.1	15.1
Opx/Pig	25.0	26.8		tr	32.7(*)	27.6(*)	8.0(*)	13.8(*)	7.5	~20(*)	39.0	0.6	36.3	30.6(*)	21.0	30(*)
High-Ca px	4.0	6.1		tr					4.4		4.9	13.4	6.0		3.8	
Mg-spinel			6.5	4-5												
Cr,Fe-spinel																
Fe-Ni metal	0.3	0.1	<0.1	tr	0.2	0.7	0.3	0.1	0.2		0.7	0.1	1.2	1.2	0.1	0.2
Troilite	<0.1	<0.1		tr	0.3	0.4	0.4	0.3				0.8	0.1	0.1		0.1
Ilmenite	1.2	1.8		tr		0.15(c)	0.7(c)	0.1			2.1	0.3	2.8		1.9	1.4
Phosphates	0.5	0.3										<0.1			0.3	
Silica															0.6	
Potash Feldspar	tr															
Clasts											4	15.8	13.5	11.5	(5-10)(e)	12.1
Others	3.5(a)	4.4(a)	30(b)	tr	0.4(a)	2.2(a)	7.0(a)	2.0(a)	2.1(a)	2.0(d)		3.6(a)		1.4(a)	0.9(a)	

CP = Clast-poor CR = clast-rich x = present
(*) includes high-Ca px (a) mainly mesostasis (b) glass (c) includes ulvospinel and armalcolite (d) opaques, mainly ulvospinel, ilmenite, troilite, metal (e) not separated in mode

14276 14310 68415 Gancarz et al. (1972); 62295 Walker et al. (1973); 62295,64 68416 Brown et al. (1973); 64455 BVSP (1981);
65055 60315 60335 62235 66095 77135 Vaniman and Papike (1980); 67747 67559 Reimold et al.(1985); 72435 Dymek et al. (1976)

Table A5.49. Modal mineralogy (vol %) of granulites and granulitic breccias of the lunar highlands.

	67485 ,4	67488 ,3	67746 ,1	67566 ,2	67947 ,3	67955 ,49-1	67955 ,49-2	67955 ,22	72559	73215 29,9	73215 46,25	73215 46,33	76235	77017 ,81	78155 mx avge	79215 avge
Plagioclase	89.9	87.0	74.6	82.2	86.9	81.8	79.3	78.5	74.5	x	x	x	70	~75	73.5	81.3
Olivine	1.6	2.4		7.9	1.3	9.3	8.7	6.0	14.4	x	x	x	10	~5	3.6	9.2
Opx/Pig	3.1	1.9	24.9	4.4	1.3	6.6	11.6	14.5(*)	10.2	x	x	x	20	~10	19.0	4.2
High-Ca px	5.1	8.6		5.2	10.4	2.0	0.3		x	x	x	x		~10	3.1	4.1
Mg-spinel									x					x		
Cr,Fe-spinel									x	tr	tr	tr	x	x		
Fe-Ni metal									x	tr	tr	tr	x	x		
Troilite									x	tr	tr	tr	x	x		
Ilmenite									x	tr	tr	tr	<0.1	x		
Phosphates										tr		tr				0.8
Silica																
Potash Feldspar									x	tr	tr	tr				
Others	0.3(a)	0.15(a)	0.5(a)	0.35(a)		0.3(a)		1.0(a)	0.9(b)	tr	tr	tr			0.8(c)	0.4(d)

mx = matrix x = present
(*) includes high-Ca px (a) opaques (b) (c) 0.4% cr-spinel + ilmenite, 0.4% metal+ sulfide (d) 0.3% oxides, 0.1% metal + sulfide

67485,4 67488,3 67746,1 67566,2 67947,3 67955,49-1 67955-49-2 Stoffler et al. (1985); 67955,22 Hollister (1973); 72559 Nehru et al. (1978);
73215,29,9 73215,46,25 73215,46,33 James and Hammarstrom (1977); 76235 Simonds (1975); 77017,81 McCallum et al. (1974);
78155 Bickel (1977); 79215 Bickel et al. (1976)

Chapter 6

MARTIAN METEORITES

Harry Y. McSween, Jr.

Department of Geological Sciences
University of Tennessee
Knoxville, Tennessee 37996
mcsween@utk.edu

Allan H. Treiman

Lunar and Planetary Institute
3600 Bay Area Boulevard
Houston, Texas 77058
treiman@lpi.jsc.nasa.gov

INTRODUCTION

Mars, as revealed by observations from spacecraft, has had a unique geologic history with many intriguing parallels to that of the Earth. Imaging results from the Mariner 4, 6, and 7 flyby missions in the 1960s were misleading, giving an impression of a lunar-like surface because they returned images of only the ancient, heavily cratered southern highlands. Global mapping by the Mariner 9 orbiter in 1971 revealed the full surface of Mars, which includes vast, younger volcanic plains and huge volcanic constructs in the northern hemisphere, a yawning canyon system, seasonally changing polar caps, and geomorphic evidence of drainage and catastrophic flooding by water early in the planet's history. Viking orbiters and landers in 1976 and the Mars Pathfinder lander in 1997 provided much rich detail about martian geology, and the Mars Global Surveyor mission promises more insights. Comprehensive summaries of martian geophysics, geology, surface properties, and atmosphere have been published previously in *Mars* (Kieffer et al. 1992).

The dozen presently recognized *SNC* (*shergottite-nakhlite-chassignite*) meteorites are generally accepted to be martian igneous rocks. These basaltic and ultramafic rocks define a common oxygen isotope mass fractionation line (Clayton and Mayeda 1996) distinct from those of other solar system bodies and share other geochemical characteristics (McSween 1994) that indicate their formation by partial melting of evolved source regions with broadly similar compositions and redox states (Longhi et al. 1992, Bertka and Fei 1997). Because the assortment of meteorite lithologies has now expanded beyond the bounds of the original SNC classification and is likely to become even broader as more martian meteorites are discovered, the meteoritic (SNC) nomenclature has become somewhat archaic. To minimize confusion, we will describe these meteorites using both the traditional meteorite nomenclature and petrologic classifications in the IUGS-approved terminology for igneous rocks.

These meteorites provide critical constraints on the geochemical and geophysical properties of the martian core and mantle (Dreibus and Wänke 1985, Treiman et al. 1986, Longhi et al. 1992, Collinson 1997, Gaetani and Grove, 1997), the chronology of planetary differentiation, magmatism, and bombardment (Shih et al. 1982, Chen and Wasserburg 1986a, Jagoutz 1991, Ash et al. 1996, Lee and Halliday, 1997), the global volatile inventory and outgassing history (Carr and Wänke 1992, McSween and Harvey

0275-0279/98/0036-0006$05.00

1993), interactions between the atmosphere, hydrosphere, and lithosphere (Wright et al. 1990, Karlsson et al. 1992, Watson et al. 1994), and near-surface weathering processes (Gooding 1992). One meteorite has also been suggested to provide evidence for possible biologic activity on Mars (McKay et al. 1996), a proposal that has sparked great interest and debate.

This review focuses specifically on the mineralogy and petrology of martian meteorites, which relate directly to the planet's magmatism, impact history, and weathering. Although this work touches on some of the implications of these samples for understanding other aspects of martian geology (and possibly biology?), it does not attempt to duplicate other reviews which address that broader subject (e.g. McSween 1985 1994 1997). We will, however, reiterate the evidence that links these meteorites to Mars, as well as describe their removal from Mars and delivery to Earth. We will also compare the mineralogy of these meteorites to that inferred for martian rocks and soils based on interpretations of remote-sensing spectral and chemical data. This paper does not include a discussion of CI carbonaceous chondrites as possible martian sedimentary rocks (Brandenburg 1996), as we consider that hypothesis to be unsubstantiated and implausible (Treiman 1996a). Further information on the SNC meteorites, including sample availability as well as a listing of the minerals occurring in them, can be found in the *Mars Meteorite Compendium* (Meyer 1996). Representative compositions of minerals in SNC meteorites are tabulated in an appendix at the end of this chapter.

SOURCE AND DELIVERY OF SNC METEORITES

Evidence for a Martian origin

Shoemaker et al. (1963) were prescient in suggesting the possibility that impact ejecta might be liberated from Mars and ultimately collide with the Earth, and they wondered whether such materials could ever be recognized. The first suggestions that SNC meteorites were from Mars (McSween et al. 1979a, Walker et al. 1979, Wasson and Wetherill 1979) were based on the relatively young crystallization ages (<1.3 Ga) of the few meteorites then available. The duration of igneous (volcanic) activity on a body is directly related to its size (Fig. 1) because large bodies have low surface-to-volume ratios and so lose internal heat more slowly than small bodies. Bodies of asteroidal size, like 4 Vesta and other achondrite

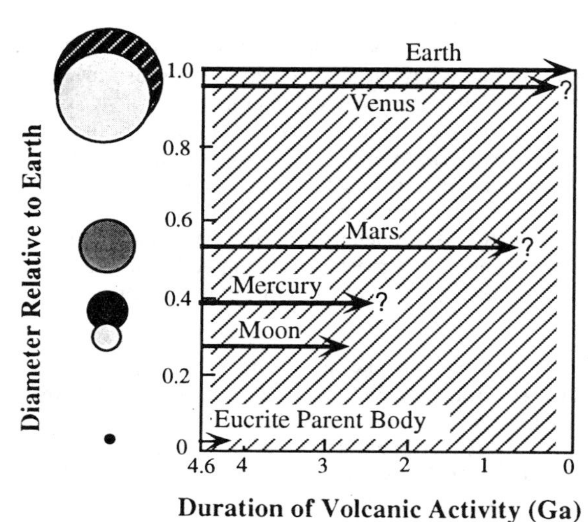

Figure 1. The duration of planetary igneous activity, as judged from crater-counting ages of volcanic surfaces and radiometric ages of samples, is related to planet size. The ages of SNC meteorites, ranging from ~4.5 Ga to as young as 180 Ma (cross-hatched area), suggest their derivation from a planetary body (modified from McSween 1994).

parent bodies (see Chapter IV), experienced melting very early in solar system history but have not apparently produced magmas since ~4.4-4.5 Ga. The ages of currently known martian meteorites extend from ~4.5 Ga to at least 330 Ma, and possibly to 180 Ma (discussed below). Igneous activity on the SNC parent body ranging over most, if not all, of geologic history demands that these meteorites were derived from a large planetary body (Fig. 1), or one like Io with an extrinsic heat source. Although this argument does not link these meteorites specifically to Mars, that planet is an attractive option because of its modest size and proximity to Earth. Another observation made at the time was that chemical analyses of martian duracrust-free soil by Viking landers were remarkably similar to those of shergottites (Walker et al. 1979).

The idea that meteorites could come from Mars remained alive (e.g. Wood and Ashwal 1981, Vickery and Melosh 1983) but highly disputed until Bogard and Johnson (1983) discovered Ar trapped within impact-melted glass in shergottite EETA79001. The isotopic compositions and relative abundances of Ar and other noble gases, N_2, and CO_2 (Becker and Pepin 1984, Swindle et al. 1986), are a remarkable match for martian atmospheric abundances (Fig. 2). The composition of the atmosphere of Mars is unique, so far as we know, and serves as a geochemical fingerprint linking EETA79001 to its parent planet. Moreover, shock experiments have demonstrated that noble gases implanted during impact melting are not isotopically fractionated (Weins and Pepin 1988). Similar trapped gases have now been found in other highly shocked SNC meteorites (e.g. Marti et al. 1995, Turner et al. 1997), convincing most skeptics that these meteorites are from Mars. Although comparison of shock-implanted gases with the *modern* martian atmosphere may be criticized, the gas-implanting shock events are thought to have occurred recently, perhaps as the meteorites were ejected from the planet's surface.

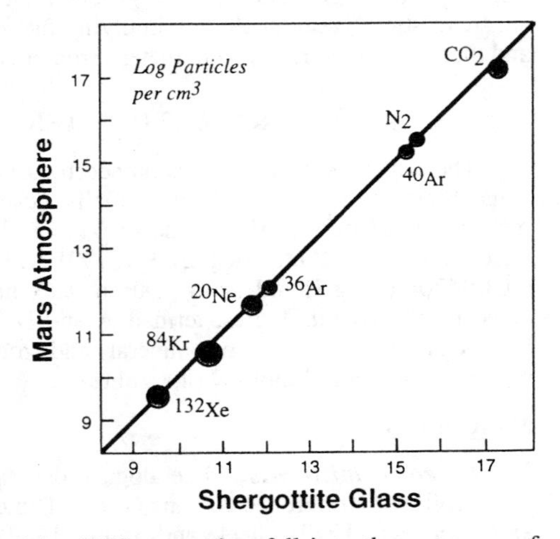

Figure 2. The abundances of N_2, CO_2, and various noble gas isotopes trapped in impact-melted glass from the EETA79001 shergottite match measurements of the martian atmosphere by Viking spacecraft (modified from Pepin and Carr 1992). Uncertainties in the analyses are encompassed within the circles.

Other arguments for a martian origin of the SNC meteorites fall into the category of plausibility tests. For example, the low remanent magnetization recorded in the shergottites (Collison 1986), thought to have formed 330 to 180 Ma ago, is consistent with what is known about the present-day magnetic field of Mars. The highly fractionated rare earth element (REE) patterns of SNC meteorites appear to require garnet in the source region (Nakamura et al. 1982, Longhi 1991), which implies a planet with substantial internal pressures. Also, the presence of pre-terrestrial hydrous minerals and salts in these meteorites (Gooding 1992) is consistent with photogeologic evidence for abundant water on Mars in its distant past.

Removal from Mars and delivery to Earth

A major difficulty initially encountered by the hypothesis that SNC meteorites are from Mars was that no known mechanism could remove rocks from the surface of a planet without totally melting or vaporizing them (Wetherill 1976). This difficulty was underscored by the apparent absence of meteorites from the Moon—if there were no meteorites from the nearest, relatively modest-sized body, how could one expect to find meteorites from a larger, more distant planet? The discovery in Antactica of a lunar highlands breccia, first reported in 1983, demonstrated that meteorites could be ejected from large bodies and opened the way for serious consideration of martian meteorites.

The only plausible event with sufficient energy to launch rocks at martian escape velocity (~5 km/sec) is a large meteor impact. After a number of inventive attempts to model the ejection of modest-sized rocks without completely destroying them (e.g. Wasson and Wetherill 1979, Nyquist 1983), an explanation by Melosh (1984) gained favor. In this model, shock waves from large impacts accelerate fragments at or near the surface to the required velocity. Given the violent process for extracting meteorites from Mars, it is not surprising that some of them have experienced severe shock metamorphism and, in some cases, shock melting. In fact, it is astonishing that a few of them exhibit little or no discernable shock effects.

The fate of rocks escaping Mars orbit has been explored by numerical integration of their orbital histories (Wetherill 1984, Gladman et al. 1996). The efficiency of the delivery of martian ejecta to Earth may be as high as ~7.5%, with about a third of terrestrial encounters occurring within 10 Ma. These results are consistent with the measured cosmic-ray exposure ages for SNC meteorites, which vary from 3 to 16 Ma (Eugster et al. 1997). The exposure ages cluster into groups that are consistent with petrology and other characteristics (Treiman 1995a), implying that cosmic-ray exposure was initiated at launch and that the ejected rocks were meter-sized or smaller.

BASALTIC SHERGOTITES (BASALTS)

The basaltic shergottites are named for Shergotty, a ~5 kg meteorite which fell in the Bihar State of India in 1865 (originally described by Tschermak, 1872). The Zagami meteorite (~18 kg) fell in Katsina Province, Nigeria, in 1962. Other basaltic shergottites were recovered in Antarctica: EETA79001 (7.9 kg found at Elephant Moraine in 1979), and QUE94201 (12 g found in the Queen Alexandria Range in 1994). EETA79001 actually consists of two lithologies, termed A and B, that appear to represent distinct magmas. Lithology A also contains mineral and rock fragments (xenocrysts and xenoliths) representing a third lithology (lherzolite).

Mineralogy

Igneous minerals. Mineralogical descriptions of basaltic shergottites can be found in the following references: Binns (1967), Duke (1968), Smith and Hervig (1979), Stolper and McSween (1979), Steele and Smith (1982), McSween and Jarosewich (1983), Smith et al. (1983), Treiman (1985), Stöffler et al. (1986), McCoy et al. (1992), Treiman and Sutton.(1992), and McSween et al. (1996).

Clinopyroxenes, both *pigeonite* and *augite*, are the dominant minerals in these basalts. Both pyroxene phases are strongly zoned, and in thin section the brown pyroxenes have yellowish-brown rims, reflecting Fe-enrichment at grain margins. Different meteorites exhibit distinct pyroxene zoning patterns. In Shergotty and Zagami, pigeonite and augite tend to have homogeneous magnesian cores and iron-rich rims, although some pyroxene

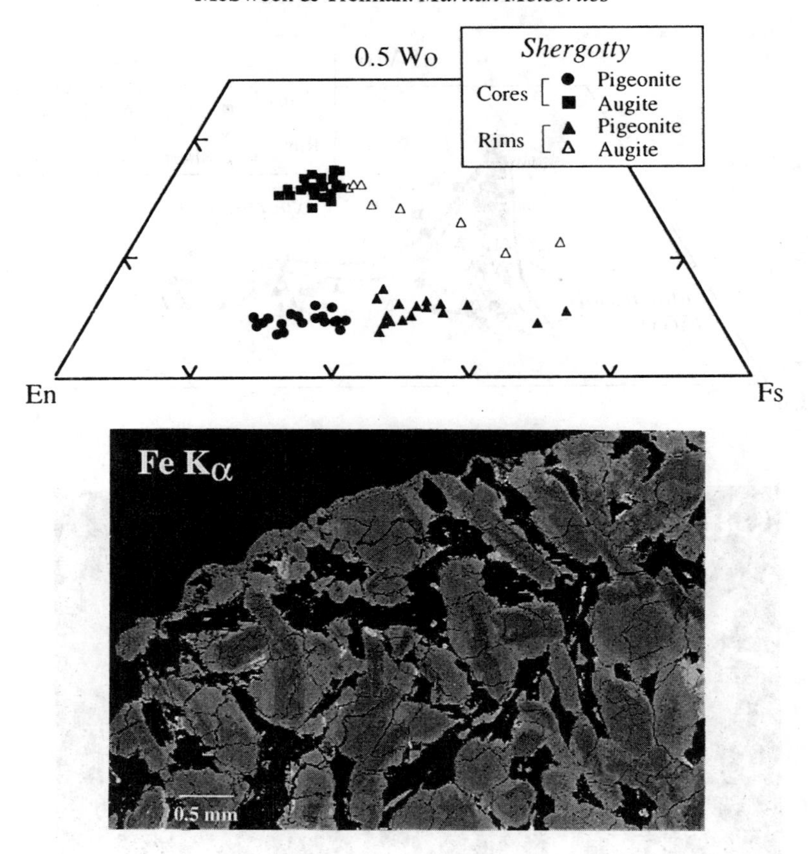

Figure 3. Pyroxenes in Shergotty consist of relatively homogeneous magnesian cores of pigeonite and augite with Fe-rich rims (analyses by Stöffler et al. 1986). The accompanying Fe$K\alpha$ map of Shergotty clearly shows the magnesian cores, interpreted to be cumulus crystals.

grains exhibit more irregular zoning. Typical core and rim compositions are given in Table A1, and zoning trends in Shergotty are illustrated in Figure 3. The homogenous magnesian cores of pigeonite ($En_{57\text{-}60}Wo_{12}$) and augite ($En_{48}Wo_{32}$) have been interpreted as cumulus crystals (Stolper and McSween 1979, McCoy et al. 1992, Treiman and Sutton 1992), possibly representing phenocrysts crystallized at depth and physically concentrated in the magma. In contrast, pyroxenes in QUE94201 and EETA79001 lithology B display complex zoning patterns that are similar to those formed by continuous crystallization of lunar basaltic melts (McSween et al. 1996, Mikouchi et al. 1997). Pyroxene cores have nuclei of magnesian pigeonite mantled by augite, in turn rimmed by strongly zoned ferroan pigeonite and pyroxferroite (Table A1 and Fig. 4). Mantling of pigeonite by augite reflects increasing Ca concentration in residual liquids due to suppression of plagioclase crystallization, and the subsequent replacement of augite by ferroan pigeonite is correlated with the onset of plagioclase crystallization. Thus EETA79001 lithology B and QUE94201 may represent liquid compositions without cumulus pyroxenes. Some portions of the Zagami meteorite, termed the dark mottled lithology by McCoy et al. (1995), may be similar to QUE94201.

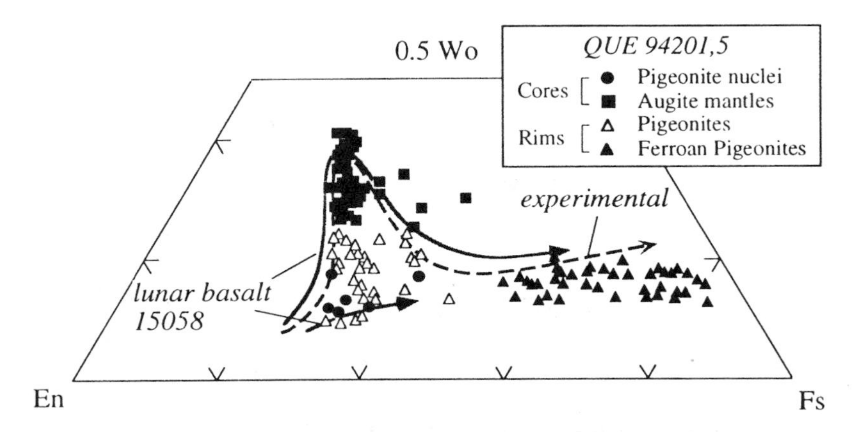

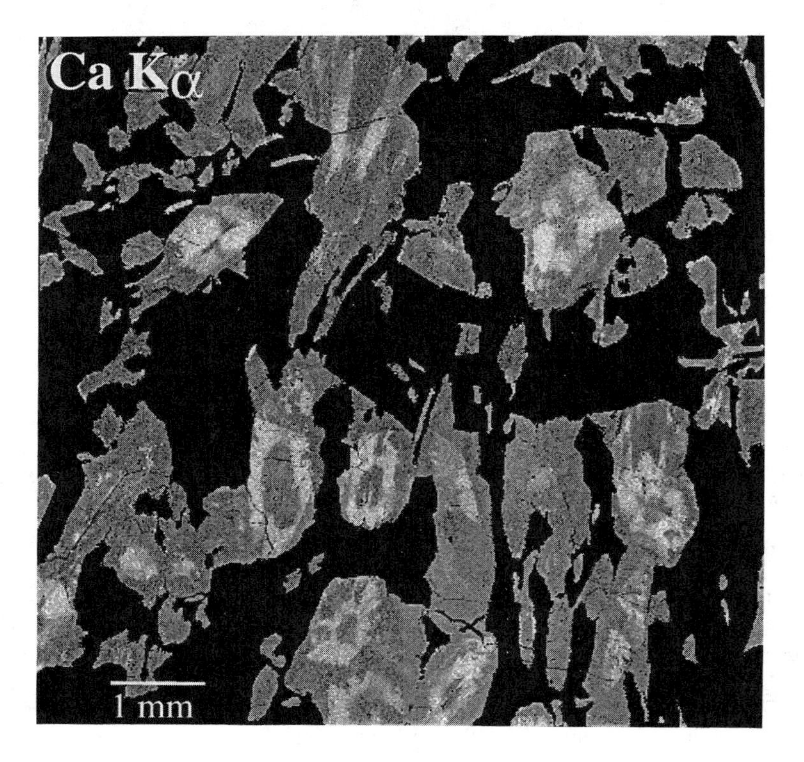

Figure 4. Pyroxenes in the QUE94201 basaltic shergottite show complex zoning, consisting of magnesian pigeonite cores mantled by augite, in turn mantled by ferroan pigeonite and pyroxferroite (after McSween et al. 1996). The accompanying Ca $K\alpha$ map of QUE94201 clearly illustrates the augite mantles. The observed trend mimics that of pyroxenes in lunar basalt 15058 and in continuous crystallization experiments on the same lunar basalt composition.

The Mn/Fe ratios of shergottite pyroxenes (and other SNC pyroxenes) are slightly higher than those in HED achondrites and considerably higher than those in lunar pyroxenes (Stolper and McSween 1979, McSween et al. 1996). This abundance ratio is useful in distinguishing martian meteorites from other samples. Both Fe and Mn have similar volatilities, so it is thought that various bodies accreted with essentially the same (i.e. chondritic) Fe/Mn ratio. These elements also fractionate similarly during melting, so the Fe/Mn ratio of a basalt is nearly identical to that of its source mantle. However, during core separation Fe is fractionated from Mn, so core formation imprints a characteristic Fe/Mn ratio on the complementary silicate portion of the planet. That ratio is readily determined from pyroxenes in basalts.

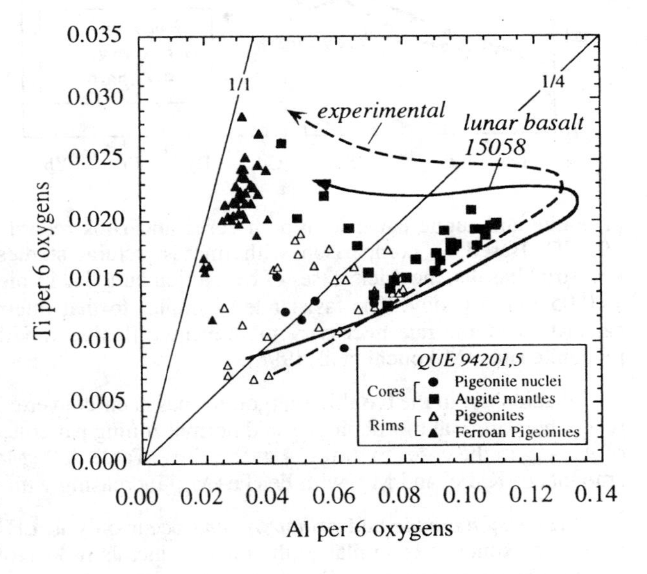

Figure 5. Molar Ti versus Al in pyroxenes of the QUE94201 shergottite (after McSween et al. 1996). With progressive crystallization, Al first increases and then decreases, the latter representing the onset of plagioclase crystallization. This trend, seen in all basaltic shergottites, is similar to that measured in pyroxenes of lunar basalt 15058 and in continuous crystallization experiments.

Minor and trace element zoning patterns in pyroxenes (Stolper and McSween 1979, Treiman and Sutton 1992, Wadhwa et al. 1994, McSween et al. 1996) are consistent with their inferred crystallization histories. An increase, and then decrease, in Al relative to Ti (Fig. 5) reflects delayed crystallization of plagioclase, followed by its onset, in all the basaltic shergottites. Ion microprobe analyses indicate that incompatible elements like Y, Zr, and Ti covary with Fe/(Fe+Mg), whereas Cr concentrations are anticorrelated with Ti and Fe/(Fe+Mg). Sc, which in most terrestrial basalts behaves compatibly in low-Ca pyroxenes, is incompatible in shergottite pigeonites. A few synchrotron X-ray microprobe analyses of Ni, Cu, Zn, and Ga in pyroxenes are also available (Treiman and Sutton 1992). Ion probe measurements of rare earth element abundances in pyroxenes (Lundberg et al. 1988, Wadhwa et al. 1994, McSween et al. 1996) indicate LREE-depleted patterns that are parallel to the whole-rock patterns, suggesting closed-system fractional crystallization. REE abundances in pigeonites are lower than in coexisting augites, and most grains show small negative Eu anomalies (Fig. 6).

The only TEM and electron diffraction studies of shergottite pyroxenes, in Shergotty and Zagami (Brearley 1991, Müller 1993), reported complex exsolution microstructures as well as localized shock effects. Two generations of augite lamellae have exsolved from pigeonite in both the magnesian cores and iron-rich rims. The thicker set, exsolved parallel to (001), generally ranges from 250 to 100 nm in thickness, whereas a thinner set, exsolved parallel to (100), crosscuts the thicker lamellae. The compositions of coexisting

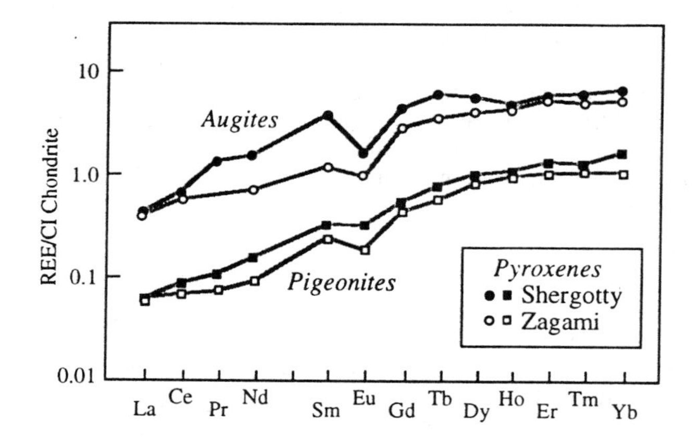

Figure 6. REE abundances in pyroxenes of basaltic shergottites (after Wadhwa et al. 1994).

pigeonite and augite lamellae in both cores and rims record equilibration temperatures of ~950°C. Based on comparison with microstructural studies of pigeonites in lunar and terrestrial basalts, the thicknesses of exsolution lamellae in pigeonites suggest cooling rates of 0.05 to 0.5°C/day. The faster rate is similar to that inferred for QUE94201 based on calculation of the rate necessary to prevent diffusive modification of Mg-Fe zoning in pigeonite rims (Mikouchi et al. 1996).

Plagioclase in the basaltic shergottites has been converted to *maskelynite*. Most grains retain their original morphologies and normal zoning patterns, generally ranging from cores of An_{57-66} in the cores to An_{52-43} in the rims (Table A2). Maskelynites also contain small amounts of K, Fe, and Mg, with Fe/(Fe+Mg) increasing with albite content.

Magnesian olivine and *orthopyroxene* occur only in EETA79001, lithology A. These phases are xenocrysts similar to the major minerals in lherzolitic shergottites, so they will be described in that section. *Fayalitic olivine* (Fa_{90-98}) occurs as an accessory mineral in the late-stage mesostasis of Shergotty and Zagami (Smith and Hervig 1979, Stolper and McSween 1979, McCoy et al. 1993) and QUE94201 (McSween et al. 1996). Olivine was noted in the Mössbauer spectra of Zagami by Vistisen et al. (1992).

The oxide minerals in basaltic shergottites are normally *ilmenite* ($Ilm_{95}Hm_5$) and *titanomagnetite* ($Mt_{37}Usp_{63}$). Ilmenite occurs both as anhedral grains intergrown with titanomagnetite and as thin lamellae within titanomagnetite. The oxide compositions (Table A4) define a temperature of 860°C and an fO_2 of 10^{-14}, corresponding approximately to the quartz-fayalite-magnetite buffer assemblage (Stolper and McSween 1979). However, QUE94201 contains virtually hematite-free ilmenite and and magnetite-free ulvöspinel (Table A4), indicating more reducing conditions near the iron-wüstite buffer (McSween et al. 1996). The primary ulvöspinel grains in QUE94201 now consist of mosaic intergrowths of titanomagnetite and ilmenite, resulting from subsolidus oxidation-exsolution. Lithology A of EETA79001 also contains chromite xenocrysts (described later).

Sulfide minerals are primarily *pyrrhotite* ($Fe_{0.92-0.94}S$) containing minor Ni and Cu (Smith and Hervig 1979). Small grains of pure magnetite associated with pyrrhotite in QUE94201 allow the possibility that the primary sulfide in this meteorite may have been troilite, which underwent secondary oxidation or weathering (McSween et al. 1996). The sulfur isotopic compositions ($\delta^{34}S$ = -1.9 to +0.8‰) of pyrrhotites in basaltic shergottites are unfractionated from the chondritic ratio (Greenwood et al. 1997). *Pentlandite* was also noted in EETA79001 (McSween and Jarosewich 1983).

Phosphates include both *merrillite* and *chlorapatite*, but merrillite is by far the dominant phosphate phase. (The term "whitlockite" is often used interchangeably with "merrillite" in describing meteortic phosphate, and "whitlockite" has been used almost exclusively in literature on martian meteorites. However, Rubin (1997b) noted that structural differences between these phases indicate that "merrillite" is the appropriate name.) REE abundances in phosphates were analyzed by Wadhwa et al. (1994) and McSween et al. (1996). Merrillite is the primary carrier of REE, and mass balance calculations indicate that bulk-rock REE concentrations can be accurately accounted for with 0.5 to 1.5 wt % merrillite, the approximate modal abundance of this phase. Apatite in Zagami has a very high D/H ratio (δD = +3000 to +4300‰) approaching that of the martian atmosphere (Watson et al. 1994).

Glassy mesostasis in the basaltic shergottites commonly contains silica in addition to fayalite. Stöffler et al. (1986) described a 200 µm grain of *α-quartz* in Shergotty. *Baddeleyite* has been described in Shergotty (Smith and Hervig 1979) and QUE94201 (McSween et al. 1996).

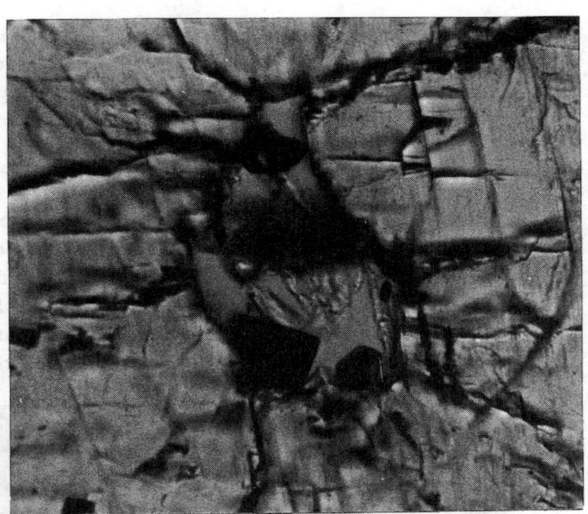

Figure 7. Magmatic inclusion (~0.16 mm across) in the core of a Shergotty pyroxene, viewed in plain light. Hercynitic spinel with magnetite cores (opaque cubes at bottom), kaersutite (dark grain at center), and sulfide (opaque grain at top) are set in glass.

Trapped magmatic inclusions in the cores of pyroxenes in Shergotty, Zagami, and EETA79001 (Fig. 7) commonly contain glass, amphibole, and hercynitic spinel (Treiman 1985 1997). The amphibole is *ferro-kaersutite* or *titano-alumino-ferro-tschermakite* (terminology of Leake et al. 1997), with high concentrations of Ti and Al, very low OH (~1/10 of that expected), and δD = +500 to+1670‰ (Watson et al. 1994). The spinel is approximately $FeAl_2O_4$ and has a distinctive green color. Treiman (1985) suggested that the crystallization of these minerals in trapped melt inclusions required pressures of at least 0.1 GPa, but more recent studies (Popp et al 1995a 1995b) imply that low-OH kaersutite may be stable at lower pressures.

Shock metamorphic minerals and effects. Shergotty is the type specimen for *maskelynite* (Tschermak, 1872), a diaplectic glass formed from plagioclase. The maskelynite generally lacks crystal structure, as indicated by its isotropism. However, powder diffraction of maskelynite separates reveals small amounts of coherently diffracting plagioclase (Hörz et al. 1986), and thermoluminescence studies of Shergotty and Zagami likewise indicate small quantities of crystalline material in maskelynite (Hasan et al. 1986). El Goresy et al. (1997) described two kinds of "maskelynite" in Shergotty: Type 1 grains are smooth and anhedral with no detectable chemical zoning, whereas Type 2 grains have remnant birefringence and twinning and exhibit fractures and zoning. They interpreted the Type 1 grains to be shock melts rather than diaplectic glasses.

The shock pressures required to form maskelynite vary with plagioclase composition. Shergotty and Zagami have experienced equilibrium shock pressures of ~29 GPa, as determined from index of refraction measurements of maskelynite calibrated to shock pressure (Stöffler et al. 1986, Langenhorst et al. 1991), although small pockets and veins of impact melt suggest higher local pressures of 60-80 GPa. EETA79001 experienced a slightly higher equilibrium shock pressure of ~34 GPa (Stöffler et al. 1986), and some maskelynite grains show evidence of flow structures suggesting melting (McSween and Jarosewich 1983). Shock melt pockets and veins are also more prominent in this meteorite. The shock melts consist of brown vesicular glass containing relict crystals and secondary skeletal olivine and pyroxene crystals. The compositions of shock-melted glasses are similar to the bulk composition of the host lithology, but are slightly enriched in the plagioclase component (McSween and Jarosewich 1983). Shock melt has also been reported in QUE94201 (Mikouchi et al. 1996).

Ringwoodite and *majorite*, the high-pressure polymorphs of olivine and pyroxene, respectively, were tentatively reported within shock veins in EETA79001 (Steele and Smith 1982). Stöffler et al. (1986) also reported a possible grain of *stishovite* in Shergotty. Other shock effects in all basaltic shergottites include mosaicism, polysynthetic twinning, and planar features developed in pyroxenes, mechanical twinning in ilmenite, and mosacism and strongly reduced birefringence in quartz (Stöffler et al. 1986, Müller 1993).

Alteration minerals. Small quantities of secondary alteration minerals in SNC meteorites formed by reaction with aqueous fluids, either within the shallow crust or as weathering products on the surface of Mars. Small platy grains of an Fe-rich, Al-poor illite-like *clay* occur in EETA79001, and sulfur pyrolysis experiments suggest the presence of S- and Cl-rich aluminosilicate *mineraloids* (Gooding and Muenow 1986). Tiny veins and disseminated grains of granular *calcite* (Fig. 8), sometimes associated with laths of *gypsum*

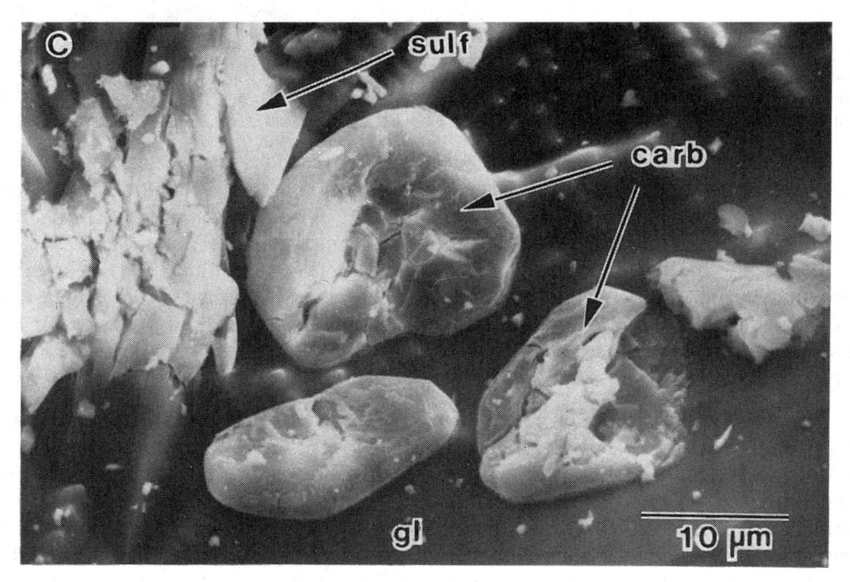

Figure 8. Secondary alteration phases in EETA79001 shergottite glass include carbonate and Ca-sulfate. This SEM photomicrograph was taken by Gooding et al. (1988).

and *Mg-phosphate*, occur within impact-melted glass (Gooding et al. 1988). A portion of the water extracted from basaltic shergottites has O isotopic compositions that are distinct from those of the igneous silicates (Karlsson et al. 1992), implying that the alteration fluid was not in equilibrium with the lithosphere. The isotopic compositions of C and O extracted from the calcite are distinct from terrestrial carbonates (Wright et al. 1988), supporting an extraterrestrial origin. Sulfates, aluminosilicate clays, and secondary silica have been found in QUE94201 (Wentworth and Gooding 1996), but it has not yet been established that these phases are extraterrestrial.

Most of the secondary minerals in shergottites occur as isolated grains either along fractures or partially included in impact glasses. A few vesicles in EETA79001 contain fluffy white deposits of *Ca carbonate*, *Ca sulfate*, and unidentified *phosphates*, the so-called "white druse" (Martinez and Gooding 1986, Gooding et al. 1988, Gooding and Wentworth 1991). The origin of this material is unclear; its O isotopes appear to be martian (Clayton and Mayeda 1988, Wright et al. 1988), but it contains sufficient ^{14}C to suggest a terrestrial origin (Jull et al. 1992).

Figure 9. Backscattered electron image of QUE94201, showing preferred orientation of pyroxene grains with interstitial maskelynite and other phases.

Petrology, geochemistry, and geochronology

Petrology. The basaltic shergottites exhibit foliated textures produced by the partial alignment of pyroxene prisms (Fig. 9), with interstitial maskelynite and other phases. This fabric was initially attributed to crystal accumulation in a shallow intrusion or thick flow (Duke 1968, Stolper and McSween 1979), a conclusion supported by phase equilibrium experiments demonstrating that Shergotty and Zagami are not multiply saturated with pyroxenes and feldspar at low pressure (Stolper and McSween 1979). Although most workers accept a crystal accumulation model for these meteorites, the foliation is now commonly attributed to flow alignment in extruded lavas (e.g. McSween 1994). The confinement of melt inclusions containing amphibole and spinel to the magnesian cores of pyroxene crystals in Shergotty and Zagami supports the idea that the cores are cumulus and crystallized under differing conditions (McCoy et al. 1992), but there is now doubt that the presence of amphibole requires a high confining pressure. Petrographic and experimental estimates of the proportion of cumulus pyroxene cores in these meteorites give conflicting results (Stolper and McSween 1979, McCoy et al. 1992, Treiman and Sutton 1992, McCoy and Lofgren 1996, Hale et al. 1997). As already noted, QUE94201 and EETA79001 lithology B may not contain cumulus pyroxenes.

EETA79001 contains two igneous lithologies joined along a diffuse, planar contact (Fig. 10). McSween and Jarosewich (1983) interpreted these lithologies as flows, but Mittlefehldt et al. (1997) suggested that lithology A might be an impact melt. This scenario is consistent with the partially resorbed clasts (occuring mostly as xenocrysts) of ultramafic lherzolite in this basalt, which are difficult to explain on thermal grounds (Wadhwa et al. 1994). However, it seems unlikely that igneous activity and impact would occur so closely in time that both lithologies would have the same crystallization ages (Wooden et al. 1982).

Zagami contains pockets (up to cm-size) of a distinct, "dark mottled lithology," rich in fayalite, oxides, phosphate, and mesostasis (Vistisen et al. 1992, McCoy et al. 1993). This material represents the last dregs of melt during the fractional crystallization of this magma. This unusual differentiate has not been found as pockets in other basaltic shergottites, but it is petrographically similar to the small QUE94201 shergottite (McSween et al. 1996).

Geochemistry. Complete elemental analyses of all basaltic shergottites except QUE94201 were tabulated by Treiman et al. (1986); an analysis of QUE94201 was published by Warren and Kallemeyn (1997), and analyses from other sources are available in abstract form (see Meyer 1996). All these basalts are Fe-rich and Al-poor compared to terrestrial basaltic counterparts, and their compositions are thought to reflect the composition and redox state of the martian mantle (Longhi et al. 1992). Samples tend to be heterogeneous in trace element abundances, because these elements are distributed among a few rare phases. The abundances of chalcophile elements in shergottites and those inferred for the martian mantle are low, reflecting segregation of sulfide into the core (Treiman et al. 1986). Conversely, moderately volatile element abundances are fairly high (Laul et al. 1986), implying that Mars accreted abundant volatiles. Certain element ratios appear to be uniform in these meteorites, or nearly so, and may serve as geochemical fingerprints to identify members of this group: Fe/Mn and K/U (McSween et al. 1979a); K/La, Al/Ti, and Na/Ti (Treiman et al. 1986); and Ga/Al (Warren and Kallemeyn 1997).

It is difficult to relate Shergotty and Zagami to the Antarctic shergottites based on trace element and radiogenic isotope systematics. The initial Sr, Nd, and Pb isotopic compositions for Shergotty and Zagami are different enough from those of EETA79001 to require that they formed from separate magmas (Jones 1989). Assimilation of an isotopically distinct crustal component rich in incompatible elements by the parent magma of Shergotty and Zagami provides a possible explanation for these differences (Shih et al. 1982, Longhi and Pan 1989). Rare earth element patterns for all basaltic shergottites are LREE-depleted and lack Eu anomalies (e.g. Smith et al. 1984). The LREE depletion in QUE94201 is more extreme than for other basaltic shergottites (McSween et al. 1996), and its initial Nd isotope ratio of $\varepsilon\ ^{143}Nd = +48$ (Borg et al. 1997) implies that the extreme LREE depletion occurred early in Mars' history.

Geochronology. The crystallization ages of most basaltic shergottites have been difficult to interpret. Whole-rock Rb-Sr and U-Th-Pb "isochrons" of ~4.5 Ga and Sm-Nd "isochrons" of ~1.3 Ga are commonly interpreted as mixing lines between different

Figure 10. A slab of the EETA79001 meteorite contains two lithologies, labelled A and B, joined along a diffuse, planar contact. The sketch illustrates the location of a thin section bridging the contact; a photomicrograph of the thin section is shown below (from McSween 1985). Lithology A has a finer-grained groundmass and contains xenocrysts of olivine, orthopyroxene, and chromite. Lithology B is more typical of a basaltic shergottite.

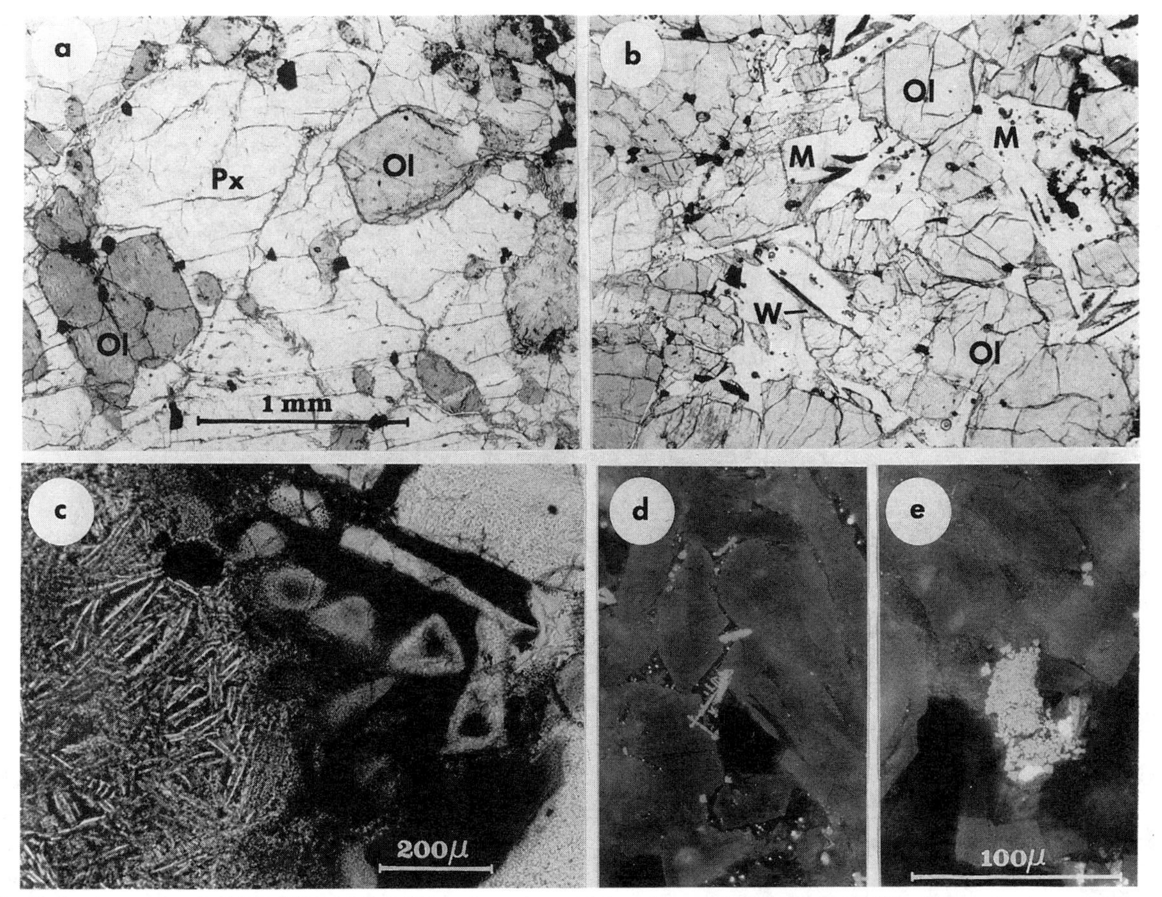

Figure 11. Photomicrographs of ALHA77005. (a) Poikilitic orthopyroxene (Px) encloses subhedral olivines (Ol) and chromite (opaque) in some areas of the meteorite, whereas other areas (b) consist of olivine, maskelynite (M), whitlockite (W), and interstitial pyroxenes. (c) Pockets of impact melt consist of brown glass with varying amounts of vitrophyric or hollow olivine crystallites, (d) dendritic chromite, and (e) small clusters of chromite euhedra.

reservoirs (e.g. Jones 1986), although that is not universally accepted. Rb-Sr and U-Th-Pb mineral isochrons for Shergotty, Zagami, and EETA79001 give virtually identical ages of 180 ± 20 Ma (Shih et al. 1982) and 190 ± 30 Ma (Chen and Wasserburg 1986), which are usually interpreted to indicate the time of crystallization. Various controversies related to attempts to disentangle crystallization and shock ages for these meteorites were summarized by McSween (1994). On the other hand, QUE94201 yields identical Rb-Sr and Sm-Nd crystallization ages of 327 ± 12 Ma and 327 ± 19 Ma respectively (Borg et al. 1997).

LHERZOLITIC SHERGOTTITES (LHERZOLITES)

All known lherzolitic shergottites were recovered in Antarctica. The group consists of ALHA77005 (a 482 g sample found in the Allan Hills in 1977), Y793605 (a 16 g meteorite found in the Yamato Mountains in 1979), and LEW88516 (a 13 g specimen recovered at Lewis Cliff in 1988). Xenocrysts and largely disaggregated xenoliths of lherzolite also occur within lithology A of the EETA79001 basaltic shergottite.

Mineralogy

Igneous minerals. Mineralogical descriptions of the lherzolitic shergottites can be found in McSween et al. (1979b), Smith and Steele (1984), Harvey et al. (1993), Ikeda (1994 1997), Treiman et al. (1994), Mikouchi and Miyamoto (1996 1997), and Gleason et al. (1997).

Olivine varies in composition from Fo_{76-60} (Table A3) within the meteorites of this group, although different meteorites show different mean values. Olivines poikilitically enclosed by pyroxenes are rounded and slightly more magnesian than the euhedral to subhedral grains in non-poikilitic areas (Fig. 11a,b). Most grains contain minor Mn, Ca, and Ni. The olivines in all three lherzolites have a distinctive brown color, apparently caused by oxidation (~4.5% of the total Fe in ALHA77005 olivine is trivalent , and charge transfer produces an absorption band that results in the color—Ostertag et al. 1984). In all three meteorites, the olivines are too Fe-rich to be in equilibrium with coexisting pyroxenes, apparently because subsolidus reequilibration of olivine is faster than pyroxene (Harvey et al. 1993).

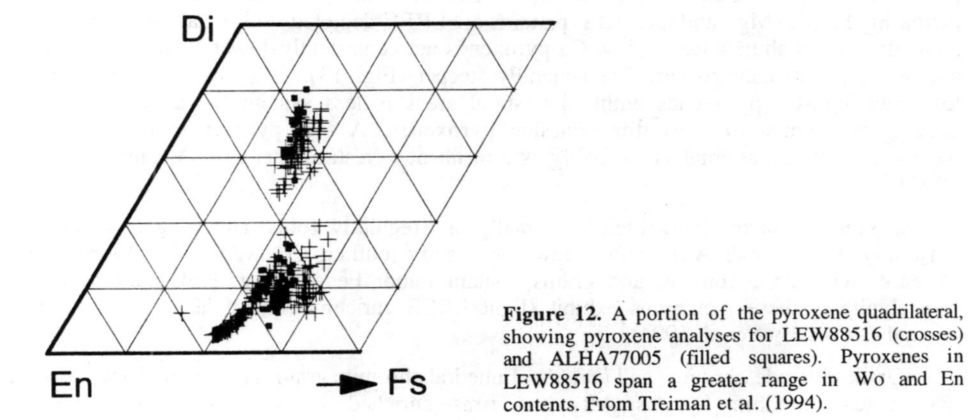

Figure 12. A portion of the pyroxene quadrilateral, showing pyroxene analyses for LEW88516 (crosses) and ALHA77005 (filled squares). Pyroxenes in LEW88516 span a greater range in Wo and En contents. From Treiman et al. (1994).

A variety of pyroxenes occur in the lherzolites. Magnesian low-Ca pyroxene, possibly orthopyroxene (although its structural state is unknown), mostly forms large oikocrysts (Fig. 11a). These crystals are monotonically zoned from $En_{78}Wo_4$ to $En_{65}Wo_{15}$ (Fig. 12).

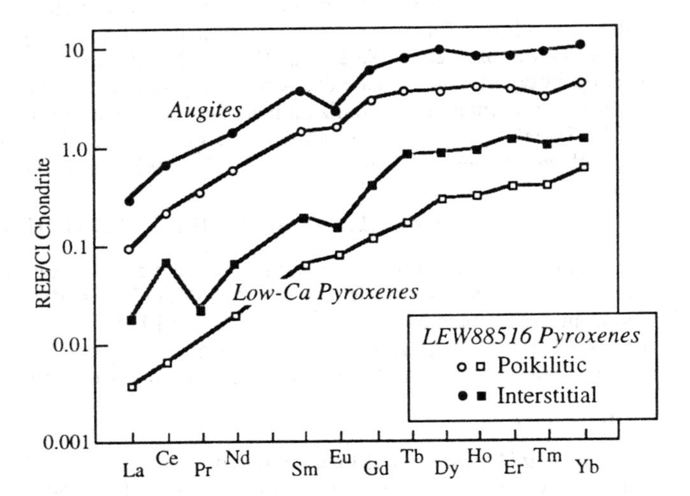

Figure 13. REE abundances in pyroxenes of the LEW88516 lherzolitic shergottite. After Wadhwa et al. 1994.

Magnesian *pigeonite* forms a broad scatter of compositions at the end of this trend (Fig. 12). The pigeonite grains occur as smaller grains in interstitial areas. *Augite* forms exsolution lamellae in low-Ca pyroxenes, as well as small interstitial grains. The augites are apparently zoned from $En_{45}Wo_{40}$ to lower Ca contents (Fig. 12), although this trend may represent overlap of the electron beam on exsolved phases. Representative pyroxene analyses are given in Table A1. Application of two- and three-pyroxene geothermometry to all three meteorites gives equilibration temperatures of ~1150°C (Ishii et al. 1979, Harvey et al. 1993, Mikouchi and Miyamoto 1996). Pyroxene compositions in LEW88516 are slightly more Fe-rich than those in ALHA77005 and Y953605 (Harvey et al. 1993, Treiman et al. 1994, Mikouchi and Miyamoto 1996).

Minor and trace elements in pyroxenes were analyzed by Lundberg et al. (1990), Harvey et al. (1993), and Wadhwa et al. (1994 1997). Poikilitic low- and high-Ca pyroxenes show increases in Ti, Al, Sc, Y, Zr, Cr, and V and decreases in Cr with increasing Fe/(Fe+Mg), and their REE patterns are LREE-depleted with small, negative Eu anomalies. REE abundances in low-Ca pyroxenes are consistently lower than in augites, and the slopes of their patterns are generally steeper (Fig. 13). Trace element zoning in low- and high-Ca pyroxenes within interstitial areas is less systematic, although REE patterns are similar to those for poikilitic pyroxenes. A few pyroxenes in Antarctic meteorites have Ce anomalies, probably as a result of terrestrial alteration (Wadhwa et al. 1994).

Maskelynite in the lherzolites is normally or irregularly zoned and ranges between An_{45} and An_{60} (Table A2), with a few more sodic outliers (Harvey et al. 1993). K increases with albite content, and grains contain minor Fe and Mg. REE patterns for maskelynite in these meteorites exhibit slight LREE enrichment with large postive Eu anomalies (Wadhwa et al. 1994).

Oxides include *chromite* and *ilmenite*. Euhedral chromite grains are zoned (Table A4), with cores of $Chm_{81}Sp_{14}Usp_{2}Mt_{3}$ and rims enriched in Ti towards ulvöspinel compositions where they abut maskelynite, and enriched in Al where they are in contact with pyroxene and olivine (Fig. 14) (McSween et al. 1979b, Ikeda 1994). Some grains appear to be fairly homogeneous *ulvöspinel*, probably as a result of failure to section through the core. As much as ~10 wt % of the Fe in chromites may be Fe_2O_3. Ilmenite is

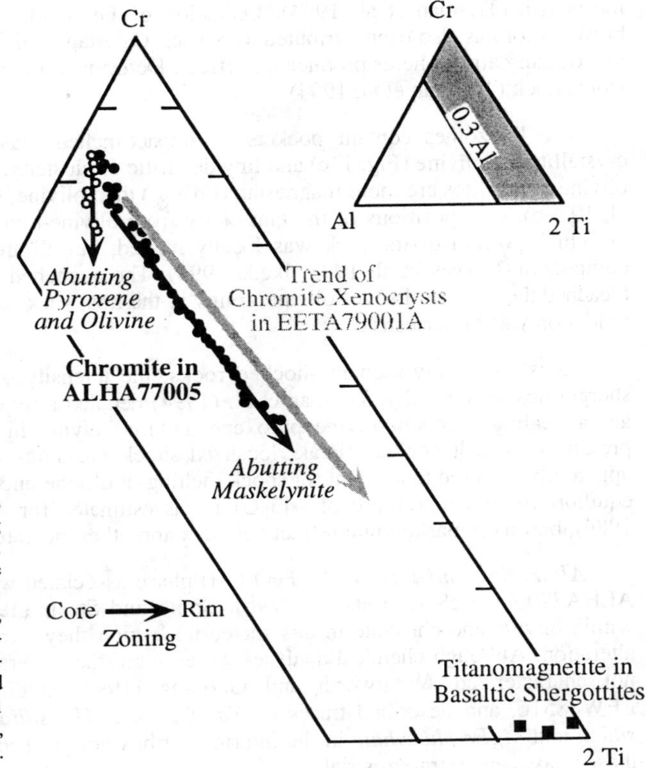

Figure 14. Molar compositions of spinels in ALHA77005 (modified from McSween et al. 1979b, and Ikeda 1994). Chromites in contact with pyroxene and olivine have slightly aluminous rims, whereas those abutting maskelynite have Ti-rich rims. The zoning trend for chromite xenocrysts in EETA79001 (McSween and Jarosewich 1983), and compositions of titanomagnetites in the basaltic shergottites Shergotty (Stolper and McSween 1979), EETA79001 (McSween and Jarosewich 1983), and QUE94201 (McSween et al. 1996) are also illustrated.

virtually hematite-free (Table A4) but contains signficant Mg and minor Mn and Cr (Ikeda 1994).

Other phases include *merrillite* and sulfides, both *pyrrhotite* and *pentlandite*. Pyrrhotite contains minor Ni, Co, Cu, and Zn (Ikeda 1994). Merrillite is the major REE carrier in lherzolites, as it is in basaltic shergottites. The REE patterns for merrillites are LREE-depleted with small negative Eu anomalies (Wadhwa et al. 1994). Pyrrhotite in LEW88516 contains S with similar isotopic composition ($\delta^{34}S$ = -1.9‰) to that in basaltic shergottites (Greenwood et al. 1997).

Magmatic inclusions in the centers of olivine grains contain glass and augite rich in Al, Ti, and Ca compared to other augites in lherzolites. Other accessory minerals in the inclusions are chromite, spinel, low-Ca pyroxene, sulfide, ilmenite, and phosphate, but amphibole has not been reported (Harvey et al. 1993). The inclusions have been analyzed by INAA for trace element abundances (Lindstrom et al. 1993).

Shock metamorphic minerals and effects. Plagioclase has been converted to maskelynite in all the lherzolitic shergottites. In ALHA77005, Ikeda (1994) described some grains of feldspathic glass with thin plagioclase rims, which he attributed to shock melting and subsequent crystallization of plagioclase. Thermoluminescence measurements on this meteorite (Hasan et al. 1986) support the identification of small quantities of plagioclase. Some maskelynite grains show flow lines and vesicles (Treiman et al. 1994).

Olivine shows closely spaced planar elements, and pyroxenes exhibit twinning and mosaicism (Treiman et al. 1994). Oxidation of Fe in olivine to produce its distinctive brown color has also been attributed to impact (Ostertag et al. 1984), as shock experiments in oxidizing atmospheres produce this effect. Deformation twinning in ilmenite also results from shock (Treiman et al. 1994).

The lherzolites contain pockets of impact-melted glass with skeletal and hollow crystallites of olivine (Fig. 11c) and tiny dendritic or clustered chromites (Fig. 11d, e). The olivine crystallites are more magnesian (Fo_{73-86}) than olivine in the host rock (McSween et al. 1979b). Compositions of the glass vary from olivine-rich to pyroxene-rich, depending on which portion of the rock was locally melted, but all are enriched in the plagioclase component (Harvey et al. 1993, Ikeda 1994). The quenched melts have recrystallized and bleached the adjacent brown olivine grains in the host rock, demonstrating that the olivine oxidation was preterrestrial.

As is commonly seen in shocked rocks, the intensity of shock effects in lherzolitic shergottites vary locally. Treiman et al. (1994) detailed a sequence of shock metamorphic and annealing effects in olivine, pyroxene, and maskelynite in LEW88516, as a function of proximity to melt pockets. Peak, localized shock pressures in excess of 60-80 GPa are apparently required to account for shock melting of olivine and pyroxene (Ikeda 1994). An equilibrium shock pressure of ~43 GPa was estimated for ALHA77005 (Stöffler et al. 1986), based on maskelynite refractive index and other indicators.

Alteration minerals. An FeO(OH) phase associated with sulfides was reported in ALHA77005 (McSween et al. 1979b). Smith and Steele (1984) described dark patches within olivine and chromite in this meteorite, which they attributed to terrestrial aqueous alteration. Although chemical analyses were given, the mineralogy of these materials was not characterized. Wentworth and Gooding (1993) studied both ALHA77005 and LEW88516, and described traces of *Na- Ca- and KFe-sulfates*, a low-Al silicate *clay*, *silica*, and *MgFe-phosphate* in the interiors. It has not yet been demonstrated that any of these phases are extraterrestrial.

Petrology, geochemistry, and chronology

Petrology. Hand specimens of the lherzolitic shergottites are heterogeneous on a cm scale, with dark regions composed of large pyroxenes poikilitically enclosing olivine and chromite and light regions composed mostly of olivine and maskelynite with interstitial pyroxenes and other phases (cf. Fig. 11a and b). Olivine and chromite in both regions are thought to be cumulus, and olivine in ALHA77005 has a preferred crystallographic orientation (Berkley and Keil 1981). Virtually identical crystallization sequences were described for LEW88516 (Harvey et al. 1993) and ALHA77005 (Ikeda 1994): olivine and chromite, followed by poikilitic pyroxenes, later joined by interstitial pyroxenes, plagioclase, ilmenite, merrillite, and sulfides. Wadhwa et al. (1994) interpret the post-accumulation histories of these rocks to represent closed-system fractional crystallization, based on trace element zoning patterns in various minerals.

The three meteorites are petrographically very similar. However, they were found at sites in Antarctica in different ice flow drainages, separated by hundreds or thousands of kilometers. They exhibit subtle differences in mineral compositions, and the distinct terrestrial ages for LEW88516 and ALHA77005 support the idea that they are not paired (Treiman et al. 1994). However, the cosmic-ray exposure ages (~3.7 Ma) of all three lherzolites are similar (Eugster and Polnau 1997), suggesting that they were ejected from Mars in the same impact event.

Xenocrysts of low-Ca pyroxene, olivine, and chromite in lithology A of EETA79001 have compositions that are the same as in their lherzolite counterparts, leading to the hypothesis that this basaltic shergottite magma intruded lherzolite and incorporated pieces of it as xenoliths (McSween and Jarosewich 1983, Steele and Smith 1983). Thus, this meteorite is possibly derived from the same geographic area as the lherzolitic shergottites, and plausibly liberated in the same impact event. The younger cosmic-ray exposure age of EETA79001 (~0.5 Ma) may indicate breakup of a larger rock in space at that time.

Geochemistry. Chemical analyses for lherzolitic shergottites were tabulated by Treiman et al. (1986 1994), Meyer (1996), Warren and Kallemeyn (1996 1997), and Gleason et al. (1997). Compared to basaltic shergottites, these meteorites have lower abundances of Si and Al, and higher Fe and Mg, reflecting the abundance of olivine. The lherzolites have similar diagnostic element ratios to those in basaltic shergottites (Treiman et al. 1986), but their absolute trace element abundances are generally lower by an order of magnitude (McSween et al. 1979a).

Accurate REE abundances in whole-rock samples are difficult to obtain, because of the problem in obtaining representative samples of these coarse-grained, heterogeneous rocks. However, a whole-rock REE pattern for ALHA77005 (Smith et al. 1984) is in good agreement with calculated REE abundances using ion microprobe analyses of phases and modal analyses (Lundberg et al. 1990). REE abundances in this and other lherzolites (e.g. Treiman et al. 1994) are much lower than in basaltic shergottites, and their patterns are strongly LREE-depleted with distinctive "humps" at Tb that trail off towards lower HREE abundances.

Chronology. Crystallization ages for the lherzolitic shergottites suffer from the same uncertainties that plague the basaltic shergottites. Rb-Sr ages of 187 ± 12 Ma (Shih et al. 1982) and 154 ± 6 Ma (Jagoutz 1989) have been accepted as the crystallization age for ALHA77005, although the Sm-Nd age is ~325 Ma (Shih et al. 1982). Radiogenic isotope dates for other lherzolites are not available.

NAKHLITES (CLINOPYROXENITES/WEHRLITES)

There are three known nakhlite meteorites. The type meteorite, Nakhla, fell in 1911 at El-Nakhla el-Baharîya, an oasis in northern Egypt. It fell as many individual stones; the known mass is approximately 10 kg, but some reports have put the total mass at ~40 kg. Lafayette is a single stone of 0.8 kg, and was recognized in 1931 at Purdue University. Its pristine fusion crust would seem to imply that it was collected soon after it fell, but ^{14}C measurements give a fall date of ~8,000 years ago (Jull et al. 1997a). Governador Valadares, a single stone of 0.16 kg, was found in Minas Gerias province, Brazil, in 1958. It is unweathered and was probably found soon after it fell. Considering the uncertain histories of Lafayette and Governador Valadares, it remains possible that they are all pieces of the Nakhla fall.

Mineralogy

Igneous minerals. The nakhlites are igneous rocks, and their mineralogy is dominated by phases common in basaltic igneous rocks on Earth and throughout the solar system. Augite is by far the most abundant mineral, olivine is second, and all other minerals are minor. Mineralogical descriptions of nakhlites, and tabulations of mineral compositions, can be found in: Bunch and Reid (1975), Boctor et al. (1976), Berkley et al. (1980), Smith et al. (1983), Treiman (1986 1990 1993), Gooding et al. (1991), Harvey and McSween (1992 a,b), and Treiman et al. (1993).

Sub-calcic *augite* (Table A1 and Fig. 15) is present as euhedra and subhedra of 0.5 mm length on average. The augite is black in hand sample, and very pale green in thin section. Simple twins on {100} are common. In Nakhla and Governador Valadares, the bulk of each augite grain is a nearly homogeneous core of $\sim Wo_{38.5}En_{37.5}$ (Smith et al. 1983, Treiman 1990, Harvey and McSween 1992b). Outside the core is a rim of nearly constant Wo content but monotonically increasing Fe/Mg ratio to $\sim Wo_{33}En_{23}$. The pattern of Fe/Mg change is consistent with igneous fractionation from a reservoir of limited volume, e.g. a magma pocket trapped in a cooling cumulate (Treiman 1990). Ferroan low-Ca pyroxene of variable composition discontinuously surrounds the rim zone, and sometimes replaces portions of both rim and core zones (Figs. 15 and 16). This late low-Ca pyroxene ranges in composition from $\sim Wo_{02}En_{38}$ to $Wo_{11}En_{24}$ to $Wo_{24}En_{16}$ (Berkley et al. 1980, Treiman 1990, Harvey and McSween 1992b). Lafayette has a comparable range of pyroxene compositions, although the distinction between core and rim is weak (Harvey and McSween 1992b).

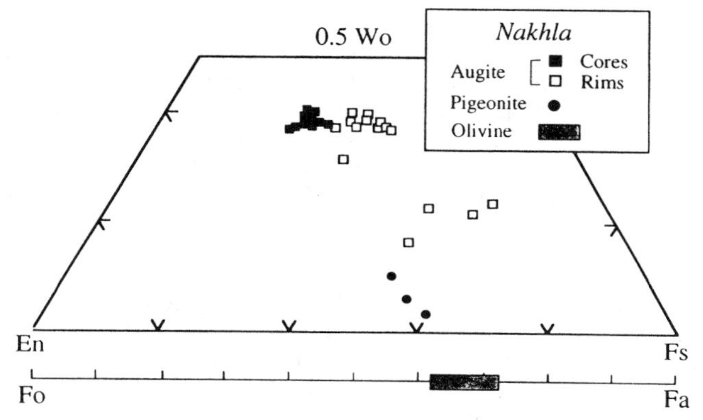

Figure 15. Pyroxene and olivine compositions in Nakhla (modified from Harvey and McSween 1992b). Augite cores have more Fe-rich rims, and pigeonite of variable composition replaces augite. Olivine is too Fe-rich to be in equilibrium with augite and is thought to have suffered subsolidus re-equilibration.

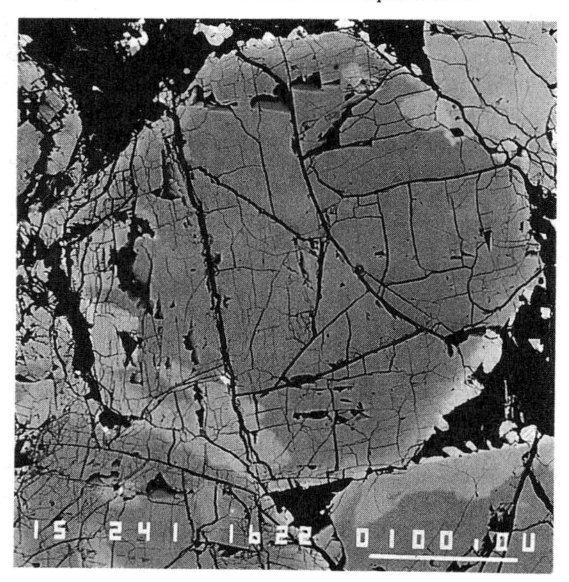

Figure 16. Backscattered electron image of Nakhla augite, showing the homogeneous core and a higher Z (higher Fe) rim and replacement zones. The horizontal bar is 100 μm long.

Minor and trace element zoning patterns in nakhlite pyroxenes are complex. Although the augite cores have nearly constant Ca-Mg-Fe contents, their abundances of trivalent and tetravalent elements can vary by factors of two or three (Treiman 1990, Wadhwa and Crozaz 1995). McKay et al. (1994) found a bimodal distribution of Al contents in the cores and suggested that they might be sector-zoned. Outside the cores, minor and trace element abundances are consistent with fractional crystallization: Al, Ti, Zr, and the REE increase together while Mg and Cr decrease (Treiman 1990, Wadhwa and Crozaz 1995a).

There appear to have been no TEM, electron diffraction, or X-ray diffraction studies of pyroxenes from the nakhlites.

Olivine is the second most abundant mineral in the nakhlites. It is most apparent as postcumulus infillings among euhedral augite grains, but is also prominent in some sections as subhedral grains up to 4 x 3 mm. It is clear and olive-green in hand sample, and colorless in thin section; compared to augite, olivine has higher birefringence, contains rounded magmatic inclusions, and contains feathery chromite inclusions. The chemical compositions of olivine in Nakhla and Governador Valadares (Table A3) are somewhat variable, from Fo_{30} to Fo_{17} and CaO from 0.20 to 0.60% (Berkley et al. 1980, Treiman 1990, Harvey and McSween 1992b). This variability appears as a slight normal zoning in Fe/Mg from core to rim, and as oscillatory zoning in CaO (Treiman 1990, Harvey and McSween 1992b). The zoning in Fe/Mg probably reflects continuous diffusive exchange with late (evolving) magma (Longhi and Pan 1989, Harvey and McSween 1992b); variations in CaO content may reflect original growth zoning. Olivine in Lafayette is effectively of constant composition, Fo_{23}, with CaO contents ranging only from 0.15 to 0.35% (Berkley et al. 1980, Harvey and McSween 1992b). Nakhlite olivines contain significant Ni (~ 200 ppm; Bunch and Reid 1975, Smith et al. 1983), but extremely low abundances of the REE (Wadhwa and Crozaz 1995a).

Olivines in the nakhlites contain numerous inclusions. Most abundant are dark feathery lamellae, 1-2 μm wide and up to 20 μm long, consisting of augite and magnetite. They may be oxidation products of kirschteinite exsolution lamellae (Yamada et al. 1997). Rounded to angular multi-phase inclusions are also common; they have been interpreted as the products of magma trapped in the olivines as they grew, i.e. magmatic inclusions (Treiman 1986, 1990, 1993; Harvey and McSween 1992a). Their mineralogy is described below.

Low-Ca pyroxenes, including pigeonite and orthopyroxene (Wo as low as 3%), occur only as overgrowths on augite and as replacements of olivine and augite (Treiman 1990, Harvey and McSween 1992b). The low-Ca pyroxene in Lafayette is pigeonite according to Berkley et al. (1980) at ~Wo_{10}, but orthopyroxene according to Harvey and McSween (1992b) at Wo_3. The crystal structures of the low-Ca pyroxenes are not known. Early reports of pigeonite exsolution in the augite (Berkley et al. 1980) have not been confirmed.

Plagioclase and *alkali feldspar* occur as radiating sprays of lath-shaped grains in the mesostasis among the larger augite and olivine grains. The plagioclase composition is $An_{23-26}Ab_{60-68}Or_{03-09}$, and the alkali feldspar is $An_{04-06}Ab_{20-42}Or_{52-76}$ (Berkley et al. 1980); concentric and sector zoning are reported. The feldspars retain their normal birefringence. Wadhwa and Crozaz (1995a) have analyzed nakhlite plagioclase for REE.

Titanomagnetite is the principal oxide mineral in the nakhlites, and occurs as ~0.1 mm grains among the augite and olivine (Bunch and Reid 1976, Boctor et al. 1976, Berkley et al. 1980). Most of the titanomagnetite shows abundant exsolution lamellae of *ilmenite* and possibly *ulvöspinel*; discrete grains of ilmenite (partly replaced by *rutile*) and a homogeneous chromian magnetite are also present (Bunch and Reid 1976, Boctor et al.

1976). These oxide minerals (Table A4) are consistent with an oxidation state near the quartz-fayalite-magnetite buffer assemblage, 740°C and fO_2 = 10^{-17} (Reid and Bunch 1985). The intrinsic oxygen fugacity measurements of Delano and Arculus (1980) imply much more reduced conditions, somewhat above the iron-wüstite buffer.

Sulfide minerals in the nakhlites are poorly characterized. *Pyrite* is the apparently the most abundant sulfide in Lafayette, and it contains lamellae of *marcasite* (Bunch and Reid 1975, Boctor et al. 1976, Berkley et al. 1980). Minor *chalcopyrite* is also present (Bunch and Reid 1975, Bunch and Reid 1975, Meyer 1996). Iron monosulfide is common, and has been called *troilite* or FeS (Bunch and Reid 1975, Boctor 1976, Weincke 1978, Berkley et al. 1980). However, it is much more likely to be *pyrrhotite*, given the oxidation state of the nakhlites and the presence of pyrite and marcasite. In fact, Weincke's (1978) chemical analysis of 'troilite' from Nakhla appears to be pyrrhotite of composition $Fe_{0.97}S$.

Other minor minerals in the nakhlites include *chlorapatite*, a *silica* phase, and alkali-silica glass (Bunch and Reid 1975, Berkley et al. 1980). Wadhwa and Crozaz (1995a) have analyzed the chlorapatite for REE, and it is a significant contributor to the rare earth budgets of the nakhlites.

Magmatic inclusions in the nakhlite olivines (Fig. 17) have a slightly different mineralogy than the bulk rock (Harvey et al. 1992a, Treiman 1993). The inclusions consist principally of aluminous augite, silica-rich glass, and Ti-magnetite (ilmenite exsolutions not reported); less abundant minerals include pigeonite, silica, chlorapatite, alkali feldspar, Ti-Al chromite, pyrrhotite, and hercynite spinel (Harvey and McSween 1992a, Treiman 1993). A reported analysis of kaersutite in a Governador Valadares inclusion (Harvey and McSween 1992a) does not have amphibole stoichiometry. The inclusions do not contain plagioclase.

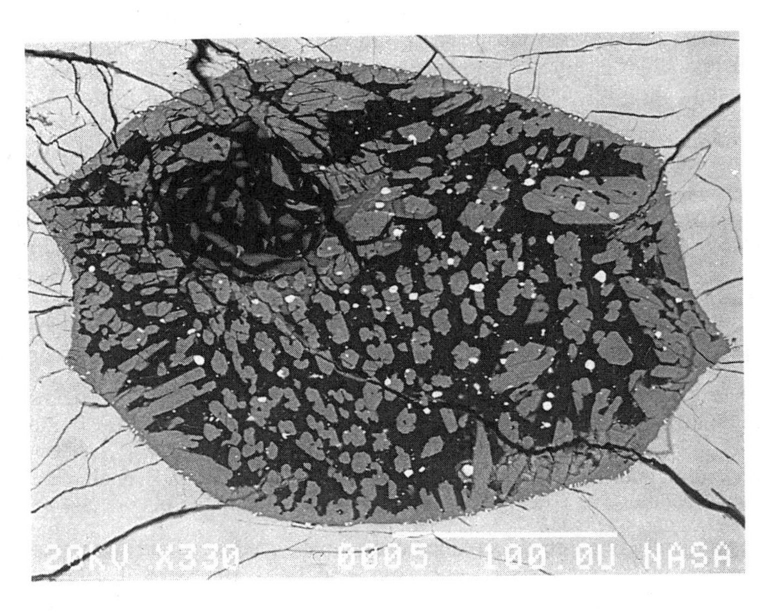

Figure 17. Backscattered electron image of a magmatic inclusion in Nakhla olivine. Aligned pyroxenes are set in dark, silica-rich glass, with small crystals of magnetite (bright dots). The large oval at upper left is a vapor bubble. Scale bar measures 100 microns.

Shock metamorphic minerals and effects. The nakhlites show little evidence of shock. The plagioclase in the nakhlites remains crystalline and has not been converted to maskelynite. To our knowledge, no high-pressure phases, such as ringwoodite or majorite, have been reported from the nakhlites. Shock effects seem to be limited to polysynthetic twinning of augite on {001} (Berkley et al. 1980), minor granulation along fracture ('gouge') zones, and possibly some minor local shock melts.

Alteration minerals. The nakhlites contain a complex assemblage of low-temperature alteration minerals, first recognized and suggested to be extraterrestrial by Bunch and Reid (1975) and Reid and Bunch (1975). Gooding et al. (1991) proved the case by showing that alteration materials in Nakhla were transected by, and decomposed at, the meteorite's fusion crust; Treiman et al. (1993) demonstrated the same relationship for Lafayette. Stable isotope and noble gas studies have confirmed that the hydrous alteration products are martian (Karlsson et al. 1992, Drake et al. 1994, Swindle et al. 1995, Leshin et al. 1996, Romanek et al. 1996).

"Iddingsite' is the prominent alteration material—rusty red veinlets and patches in grain boundaries and replacing olivine (Bunch and Reid 1975, Boctor et al. 1976, Berkley et al. 1980). Nakhlite 'iddingsite' consists principally of ferroan *smectite* and *iron oxides* (Ashworth and Hutchison 1975, Gooding et al. 1991, Treiman et al. 1993). The smectite is poorly crystalline and variable in composition, and the iron oxides include two-ring ferrihydrite and magnetite (Gooding et al. 1991, Treiman et al. 1993). Chemical compositions of the iddingsite and its constituent phases are given by Bunch and Reid (1975), Boctor et al. (1976), Gooding et al. (1991), Treiman et al. (1993), and Treiman and Lindstrom (1997). Swindle et al (1995, 1997) dated its formation by K-Ar methods as a few hundred Ma.

The iddingsite is accompanied by many ionic salt minerals, including *Ca-sulfate* (gypsum?), *Ca-carbonate* (calcite?), *siderite*, *Mg-phosphate* (epsomite? or kieserite?), and *halite* (Chatzitheodoridis and Turner 1990, Gooding et al. 1991, Treiman et al. 1993, Saxton et al. 1997, Vicenzi et al. 1997).

Petrology, geochemistry, and geochronology

Petrology. The nakhlite meteorites (Fig. 18) are considered to be cumulate igneous rocks, enriched in augite and olivine relative to their parental magma (Reid and Bunch 1975, Berkley et al. 1980, Treiman 1986, 1990, 1993; Harvey and McSween 1992b). Berkley et al. (1980) showed that the nakhlite pyroxenes are weakly aligned, consistent with flow or crystal settling. Considering the sharp chemical zoning in the nakhlite pyroxenes and the acicular habit of its plagioclase, Treiman (1986) proposed that the naklites were surface flows or near-surface intrusions.

In fact, rocks of comparable mineralogy, compositions, and textures do occur in thick flows and sills on Earth (Treiman 1987). Friedman et al. (1995) studied the crystal size distributions (CSD) of the augites. They inferred that the augites grew rapidly from magma, and that the nakhlite cumulates formed by settling of crystal clusters or of crystal mush. Following accumulation and crystallization of intercumulus liquids, the nakhlites experienced varying degrees of subsolidus annealing, which caused olivines to reequilibrate (Harvey and McSween 1992b).

Geochemistry. Elemental analyses of the nakhlites were tabulated by Treiman et al. (1986) and Meyer (1996). As befit augite-olivine cumulates, the nakhlites are relatively rich in Ca, Mg, and Fe, and relatively poor in many incompatible elements like Al and Ti. Bulk abundances of incompatible elements are not far different from those of Shergotty and Zagami, but the nakhlites are relatively enriched in highly incompatible elements, e.g.

LREE/HREE = 3 × CI. As with the shergottites, moderately volatile elements are relatively abundant and chalcophile elements are quite depleted; some element abundance ratios appear to be essentially identical in nakhlites and shergottites (McSween et al. 1979a, Treiman et al. 1986, Laul et al. 1986, Warren and Kallemeyn 1997).

Because the nakhlites are cumulates, the compositions of their parent magmas have been difficult to retrieve. Various geochemical and experimental treatments have yielded a range of estimated compositions (Treiman 1986, Longhi and Pan 1989, Harvey and McSween 1992a, Treiman 1993, Kaneda et al. 1997), but all are ferroan basalts with low alumina content — Al_2O_3 between 5 and 9% wt. This low Al_2O_3 is manifested as the late crystallization and low abundance of plagioclase, and early crystallization of augite. Such Al-depleted magmas represent melts from a mantle previously depleted in Al (Longhi and Pan 1989, Treiman et al. 1995).

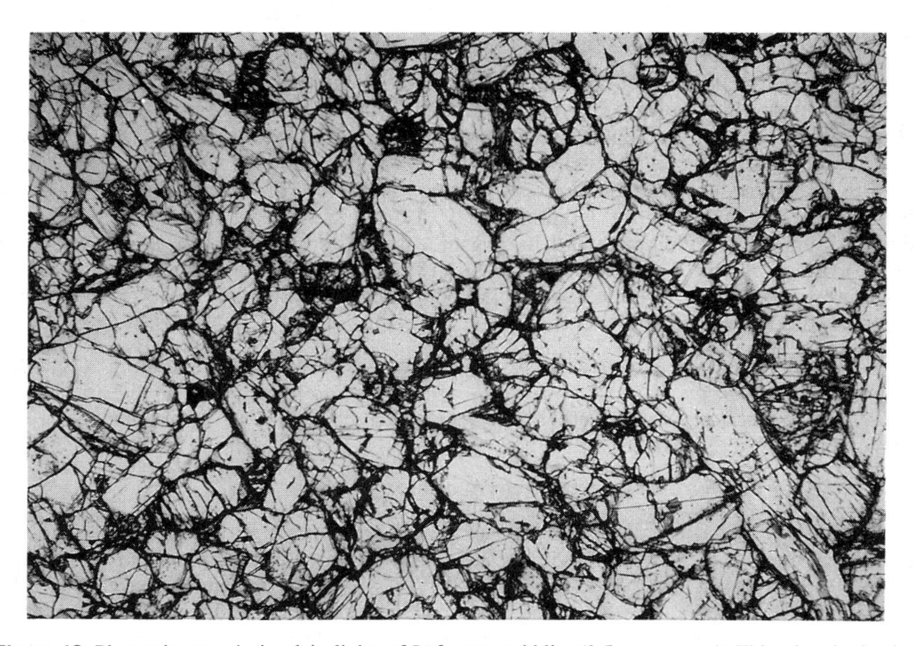

Figure 18. Photomicrograph, in plain light, of Lafayette nakhlite (3.5 mm across). This view is dominated by cumulus augite crystals.

Geochronology. Unlike the shergottites, the geochronology of the nakhlites is simple. All chronometric systems (U-Th-Pb, ^{87}Rb-^{87}Sr, ^{147}Sm-^{143}Nd, K-Ar) yield ages near 1.3 Ga, which are generally interpreted as the times of crystallization (Podosek and Huneke 1973, Papanastassiou and Wasserburg 1974, Gale et al. 1975, Wooden et al. 1979, Nakamura et al. 1982a, Chen and Wasserburg 1986b, Shih et al. 1996). It is possible that the nakhlites crystallized at slightly different times: ages for Governador Valadares and Lafayette are near 1.34 Ga, but ages for Nakhla center on 1.22 Ga. An older age of ~4.4 Ga is recorded in the U-Th-Pb systems (Chen and Wasserburg 1986b), and probably represents mantle differentiation or core formation. This ancient event is also recorded as an excess of ^{142}Nd, decay product of the short-lived isotope ^{146}Sm, in the nakhlites (Harper et al. 1995). The low-temperature alteration material, iddingsite, in the nakhlites formed sometime between 600 and 100 Ma (Swindle et al. 1995, 1997).

CHASSIGNY (DUNITE)

The Chassigny meteorite is unique, the only martian dunite. It was seen to fall in Haute-Marne, France, on October 3, 1815. It is not known if Chassigny fell as a single stone or several, but its total mass was about 4 kg (Meyer 1996).

Mineralogy

Igneous minerals. Chassigny is interpreted as an igneous rock, an olivine-chromite cumulate (Fig. 19), possibly with some cumulus augite and pigeonite (Mason et al. 1975, Floran et al. 1978, Boynton et al. 1976, Wadhwa and Crozaz 1995). Among the cumulus minerals is an intercumulus assemblage of pyroxenes, plagioclase, and minor phases. Modern mineralogical descriptions of Chassigny, and tabulations of its mineral compositions, can be found in Floran et al. (1978), Johnson et al. (1991), and Wadhwa and Crozaz (1995).

Figure 19. Chassigny, viewed in plain light (2 mm across). This view shows euhedral to subhedral ovline, pyroxene, and plagioclase (upper left), opaque chromite, and a magmatic inclusion (lower left).

Olivine is by far the most abundant mineral in Chassigny, comprising approximately 90% of its volume (Prinz et al. 1974, Wadhwa and Crozaz 1995). It occurs as euhedra and subhedra averaging 1.5 mm long among the intercumulus phases, and commonly includes euhedral chromite grains (Floran et al. 1978). The chemical composition of Chassigny olivine is essentially constant, and rather ferroan compared to terrestrial dunites (Table A3). The olivines show no preferred orientation (Floran et al. 1978).

The olivines contain rounded inclusions, as large as 200 μm diameter, with a unique minerals assemblage (at least for a meteorite): silicate glass, augite, low-Ca pyroxene, kaersutite, chorapatite, pyrrhotite, chromite, pentlandite and biotite, but no feldspars (Floran et al. 1978, Johnson et al. 1991). The inclusions have been interpreted as magmatic, partially crystallized droplets of magma trapped in the olivine as it grew. Inclusion pyroxenes are distinctly more aluminous and titanian than those in the bulk rock (e.g. 4.9% Al_2O_3 in inclusion augite vs. 1.2% in intercumulus augite) and slightly more magnesian (Floran et al. 1978). Righter et al. (1997) analyzed the inclusion glass and biotite for trace elements.

Augite and *orthopyroxene* account for ~5% of Chassigny. The pyroxenes form poikilitic and interstitial grains among and around the cumulus olivines; at least some of the pyroxenes may have been cumulus themselves (Boynton, et al. 1976, Floran et al. 1978, Wadhwa and Crozaz 1995). The original igneous minerals appear to have been augite and pigeonite, the latter inferred from regions of orthopyroxene with abundant fine exsolution lamellae of augite (Wadhwa and Crozaz 1995). This exsolved pyroxene is associated with regions of homogeneous, unexsolved orthopyroxene and augite. The pyroxenes all have essentially the same molar Fe/(Fe+Mg) (Table A1); variations in Ca content can be ascribed to mixing of end-member augite and orthopyroxene in the analytical volume. Despite the constancy of their divalent cation composition, pyroxenes in Chassigny preserve significant variations in abundances of tri- and tetravalent cations (Wadhwa and Crozaz 1995). Abundances of Al, Ti, Y, and Ce (and other rare earths) are positively correlated in both augite and orthopyroxene; augites show a 2-fold range in Ti content correlated with and 8-fold range in Ce content (Wadhwa and Crozaz 1995). There appear to have been no TEM, electron diffraction, or X-ray diffraction studies of pyroxenes from Chassigny.

Feldspars are present as intercumulus grains among the olivine and pyroxenes. Individual feldspars are chemically homogeneous, but there is a wide continuous range of compositions. Most of the analyzed grains are *plagioclase* (oligoclase/anorthoclase, ~$An_{10}Ab_{80}Or_{10}$), but individual analyses extend to labradorite, ~$An_{60}Ab_{30}Or_{10}$, and *sanidine*, $An_{02}Ab_{28}Or_{70}$ (Floran et al. 1978; Table A2). Wadhwa and Crozaz (1995) analyzed the plagioclase for rare earth elements.

The Chassigny *kaersutite* was the first find of a hydrous amphibole in a meteorite, and the first find of amphibole in the martian meteorites (Prinz et al. 1974, Floran et al. 1978). In cation content it is a typical kaersutite: rich in Ca, Ti and Al, and with an incompletely filled A site. Analyzed $OH^- + F^- + Cl^-$ sum to ~0.40 anions per formula unit, while their O(3) site must contain 2.000 total anions. The deficit in this site must be filled by ~1.60 O^{2-} anions, making this amphibole an oxy-kaersutite (Hawthorne 1981). Popp et al. (1995a,b) shown that the oxy substitution in kaersutite can be charge balanced by trivalent and tetravalent cations in the C (or 'M') sites.

A single grain of *trioctahedral mica* in Chassigny is the only documented occurrence of this mineral group in the martian meteorites (Johnson et al. 1991, Rubin 1997). The mica was described as biotite, but its molar Mg/Fe ratio of 2.07 places it nearly on the (arbitrary) boundary between biotite and phlogopite. If any of its iron is trivalent, as seems likely, Chassigny's mica should probably be classified as *phlogopite*. The mica is quite rich in titanium, comparable in many respects to those in terrestrial alkaline rocks. The normalized formula has less Si+Al than required for the tetrahedral sites of a biotite; this may suggest Ti^{4+} in tetrahedral sites. Analyzed OH^- + F^- + Cl^- sum to ~1.70 anions per formula unit, while the formula should contain 4.00 total anions. The deficit in this site must be filled by ~2.30 O^{2-} anions, making this mica an *oxy-titan-phlogopite*.

Chromite is the principal oxide mineral in Chassigny (Table A4), and is present as euhedra to 40 µm across in olivine and as larger subhedra associated with feldspar and pyroxene among the olivines (Tschermak, 1885, Floran et al. 1978). Chassigny chromites contain significant ferric iron, constant $Fe^{2+}/(Fe^{2+} + Mg)$, and constant Cr/(Cr + Al). However, Ti and Fe^{3+} contents are somewhat variable, and present an inverse relationship between [2 Ti^{4+} + Fe^{3+}] and [Cr^{3+} + Al^{3+}] content (Floran et al. 1978). Other oxide phases include *ilmenite*, *baddelyite*, and *rutile*, which are always found together associated with intercumulus feldspar (Floran et al. 1978).

Sulfide minerals are reported to include *pyrrhotite*, *marcasite*, and *pentlandite* (Floran et al. 1978). The 'troilite' analyzed by Floran et al. (1978) is apparently pyrrhotite, $(Fe_{0.87}Ni_{0.01})S$. Floran et al. (1978) suggested that the marcasite is a terrestrial weathering product, but the nakhlites and ALH 84001 contain preterrestrial (martian) marcasite and/or pyrite. Pentlandite has only been found in the magmatic inclusions.

The phosphate in Chassigny is *chlorapatite*, which occurs as slender prisms. Floran et al. (1978) and Wadhwa and Crozaz (1995) analyzed the chlorapatite for REE, and it is a significant contributor to Chassigny's rare earth budget.

Shock metamorphic minerals and effects. Chassigny shows variable effects of shock metamorphism, ranging from slightly perceptible to melting. Most of the olivine appears little shocked, with some undulatory extinction and with fractures radiating from magmatic inclusions. Elsewhere, the olivine is cut by planar deformation lamellae, principally oriented near {130}, {100} and {010} (Sclar and Morzenti 1971, Floran et al. 1978, Greshake and Langenhorst 1997). Locally, olivine shows intense mosaicism and reduced birefringence. Olivine-composition melts were produced in pools and along fractures; this material is now glassy or devitrified to fine feathery crystals (Melosh et al. 1983); Tschermak (1885) may have described a similar assemblage. Pyroxenes appear to be little deformed, save fracturing and local recrystallization (Floran et al. 1978, Greshake and Langenhorst 1997). Feldspars show undulatory extinction and reduced birefringence; maskelynite and feldspar-composition melts are rare to absent (Floran et al. 1978, Melosh et al. 1983).

Overall, shock effects in plagioclase and olivine are consistent with a shock stage of S4 (moderately shocked) corresponding to shock pressures of 15 to 35 GPa (Stöffler et al. 1991). However, the presence of planar deformation features suggests a shock stage of S5, and olivine-composition melts suggest S6, consistent with local shock pressures in excess of 55 Gpa (Stöffler et al. 1991).

Alteration minerals. Chassigny contains very little aqueous or low-temperature alteration material. Wentworth and Gooding (1994) found discrete grains of Ca carbonate (calcite), Mg-sulfate (gypsum or bassanite), and Mg-carbonate (magnesite and hydromagnesite?) in veinlets crossing the primary igneous minerals. No clays or ferric oxides were found, but traces of P and Cl are consistent with low-temperature materials in the other martian meteorites (e.g. Treiman et al. 1993). It is not certain that these materials are martian, though Wentworth and Gooding (1994) infer that they probably are. Surprisingly, bulk Chassigny contains a significant amount of water, ~0.1% H_2O (Karlsson et al. 1992, Leshin et al. 1996), far beyond the contributions from its amphibole and biotite (Watson et al. 1994). The D/H ratio of this water is indistinguishable from terrestrial water (Leshin et al. 1996).

Petrology, geochemistry, and geochronology

Petrology. Chassigny is interpreted as a cumulate igneous rock, enriched in olivine, chromite, and pyroxenes relative its parental magma (Prinz et al. 1974, Floran et al. 1978, Longhi and Pan 1981, Wadhwa et al. 1995). Initial crystallization was rapid enough to allow the olivines in Chassigny to entrap droplets of magma, which eventually became the magmatic inclusions. Cooling was, however, slow enough to permit the olivines and pyroxenes to diffusively equilibrate, and for the pigeonite to exsolve to orthopyroxene and augite. The integrated compositions of exsolved pigeonites and the compositions of the most ferromagnesian augites suggest magmatic temperatures around 1200°C (Wadhwa and Crozaz 1995). The compositions of the most calcian augites suggest temperatures as low as ~700°C (Floran et al. 1978).

Among the martian meteorites, Chassigny has been associated with the nakhlites in having the following characteristics: a LREE-enriched parent magma, a parent magma with liquidus olivine and augite, and a comparable crystallization age (see below). However, this association has been questioned by Wadhwa and Crozaz (1995), based in part on the find of possibly cumulus pigeonite. They suggest that Chassigny is more closely related to the lherzolitic shergottites.

Geochemistry. Elemental analyses of Chassigny were tabulated by Treiman et al. (1986) and Meyer (1996). As befits an olivine cumulate, Chassigny is rich in Mg and Fe, and poor incompatible elements like Al and Ti.

Because Chassigny is a cumulate, some effort has been expended to recover the composition of its parent magma. Most workers agree that Chassigny's parent magma was a low-Al ferroan basalt, strongly enriched in incompatible elements (e.g. Longhi and Pan 1991, Wadhwa and Crozaz 1995). The first estimates of the parent magma composition (Floran et al. 1978) were based on broad-beam electron microprobe analyses of the large magmatic inclusions in olivines. However, these compositions are not realistic as they would crystallize plagioclase before augite (Longhi and Pan 1981). Johnson et al. (1991) estimated a parent magma composition by working backward from electron probe analyses of individual phases in magmatic inclusions. Using mineral/melt distribution coefficients and a least-squares minimization approach, they inferred a composition that would crystallize plagioclase after augite and olivine, but before pigeonite. Longhi and Pan (1991) derived a parent magma composition from analyses of cumulus augite and mineral/melt distribution coefficients, and their composition follows a similar crystallization path to that of Johnson et al. (1991). However, recent observations of Wadhwa and Crozaz (1995) seem to show that pigeonite in Chassigny crystallized before augite and plagioclase. Clearly, more work remains.

The water content of Chassigny's parent magma is particularly important for understanding volatiles on Mars. Based on semi-quantitative SIMS analyses for water in its amphiboles, Floran et al. (1978) inferred a relatively high (but not quantified) water content. Johnson et al. (1991) inferred that Chassigny's kaersutites were water-rich, and so inferred that the parental magma (as entrapped in the inclusions) had ~1.5% H_2O. However, quantitative SIMS analyses of Chassigny kaersutite (Watson et al. 1994) gave only ~0.15% H_2O, considerably considerably less than previously estimated. Popp et al. (1995a,b) noted that the low water content of the Chassigny kaersutite could have been controlled by Ti content and oxidation state, and does not necessarily imply a water-poor magma.

Geochronology. Unlike the shergottites, the geochronology of Chassigny appears simple. All chronometric systems (^{87}Rb-^{87}Sr, ^{147}Sm-^{143}Nd, K-Ar) yield ages near 1.3 Ga, which are generally interpreted as its time of crystallization (Lancet and Lancet 1971, Bogard and Nyquist 1979, Nakamura et al. 1982b, Jagoutz 1996). An event near 4.5 Ga, presumably silicate differentiation, is recorded as an excess of ^{142}Nd, the decay product of the short-lived isotope ^{146}Sm (Harper et al. 1995, Jagoutz et al. 1996).

ALH84001 (ORTHOPYROXENITE)

The ALH84001 meteorite (1.9 kg) was found in the Allan Hills, Antarctica, in 1984. It was originally classified as a diogenite. Using a strictly non-genetic classification, this meteorite is a diogenite—a pyroxenite composed mostly of low-Ca pyroxene. Its identification as a martian meteorite was not made until a decade after its recovery (Mittlefehldt 1994). This meteorite does not fit into any of the previously established SNC categories.

Mineralogy

Igneous minerals. Mineralogical descriptions of ALH84001 are given by Berkley and Boynton (1992), Mittlefehldt (1994), Treiman (1995b) and Harvey and McSween (1996).

ALH84001 consists predominantly of coarse *orthopyroxene* crystals of uniform composition ($En_{70}Wo_3$) (Table A1). Contents of Al, Ti, and Cr also show very limited variation (Mittlefehldt 1994). As for other martian meteorites, the REE patterns in orthopyroxene are LREE-depleted with small negative Eu anomalies (Papike et al. 1994), but they do not show the trace element zoning characteristic of other SNC pyroxenes (Wadhwa and Crozaz 1994). No exsolution of clinopyroxene has been observed optically or in microprobe analyses, but minor *augite* ($En_{45}Wo_{43}$) occurs in interstitial regions. Two-pyroxene geothermometry indicates equilibration temperatures of ~875°C (Treiman 1995b).

Harvey and McSween (1996) found a few small patches of embayed *olivine* (Fo_{35}) within some orthopyroxene grains. Like the olivine in lherzolitic shergottites, this olivine is apparently too Fe-rich to be in equilibrium with the host pyroxene. They suggested that the olivines were relics from a reaction with a CO_2-rich fluid to produce orthopyroxene, but they could also be relics from a magmatic reaction to form orthopyroxene.

Maskelynite occurs interstially, with compositions generally varying between An_{31} and An_{37} (Table A2). However, some maskelynites have much higher concentrations of Na and K, varying from $Ab_{66}Or_7$ to $Ab_{67}Or_{17}$. Mittlefehldt (1994) noted that some maskelynite analyses were non-stoichiometric, and argued that they were mixtures of feldspar and silica. REE concentrations in maskelynite are LREE-enriched with pronounced positive Eu anomalies. Wadhwa and Crozaz (1994) attributed differences in the REE patterns of melts in equilibrium with maskelynite and orthopyroxene to metasomatism, but Treiman (1996b) suggested that the difference reflected subsolidus chemical equilibration.

Euhedral *chromite* occurs throughout the rock. These grains contain 5 to 8 wt % Fe_2O_3, as well as minor Al and Ti (Berkley and Boynton 1992, Mittlefehldt 1994) (Table A4). Some chromites show slight zoning, although the variations are not consistent. Another igneous accessory phase is *apatite*, which occurs as small interstial grains. Its REE pattern is LREE-enriched, like the maskelynite (Wadhwa and Crozaz 1994). Although the apatite has the highest REE abundances in the rock, it is not the major REE carrier because of its low modal abundance. Igneous sulfides are very uncommon in ALH84001.

Kirschvink et al. (1996) have interpreted magnetic susceptibility measurements as indicating the presence of *pyrrhotite*, but this phase appears to be quite rare, at least as can be detected using optical and electron microbeam methods.

Alteration and putative biogenic minerals

Fracture zones in ALH84001 contain a rich variety of secondary phases that have been attributed to alteration by hydrothermal or CO_2-rich fluids and to possible biogenic processes. McKay et al. (1996) presented a number of arguments for biologic activity, and we will address only those that are mineralogic in nature. Most prominent are rounded globules of *carbonate* (Fig. 20), consisting of orange cores of ankerite with minor calcite and rims of strongly zoned magnesite and breunnerite (Fig. 21). The magnesite-breunnerite sequence is repeated twice on the rims. Carbonate is also found as irregular patches (Treiman 1995b) in fracture zones and within veins of plagioclase melt glass (Scott et al. 1997). The microstratigraphy in all carbonates is the same, except where it was terminated because the open spaces became occluded. Descriptions of the textural relationships between the carbonate and host rock are conflicting; for example, Treiman (1995b) observed carbonate replacing maskelynite, McKay and Lofgren (1997) and McKay et al. (1997) noted carbonate being intruded by plagioclase glass, and Scott et al. (1997) proposed that carbonate crystallized from impact melts now represented by plagioclase glass. Romanek et al. (1994) argued that these carbonates formed at low temperatures (<80°C) by reaction of an aqueous fluid with the rock. Noting the unusual carbonate compositions and the absence of hydrous silicates, Harvey and McSween (1996) supported the suggestion of Mittlefehldt (1994) that the carbonates formed at high temperatures (>650°C), possibly by reaction of a CO_2-rich fluid with the rock. Treiman (1995b) and Valley et al. (1997) criticized the application of carbonate geothermometry and phase equilibria to these carbonates, instead favoring the formation of metastable carbonates at low temperatures. The oxygen isotopic compositions of carbonates correlate

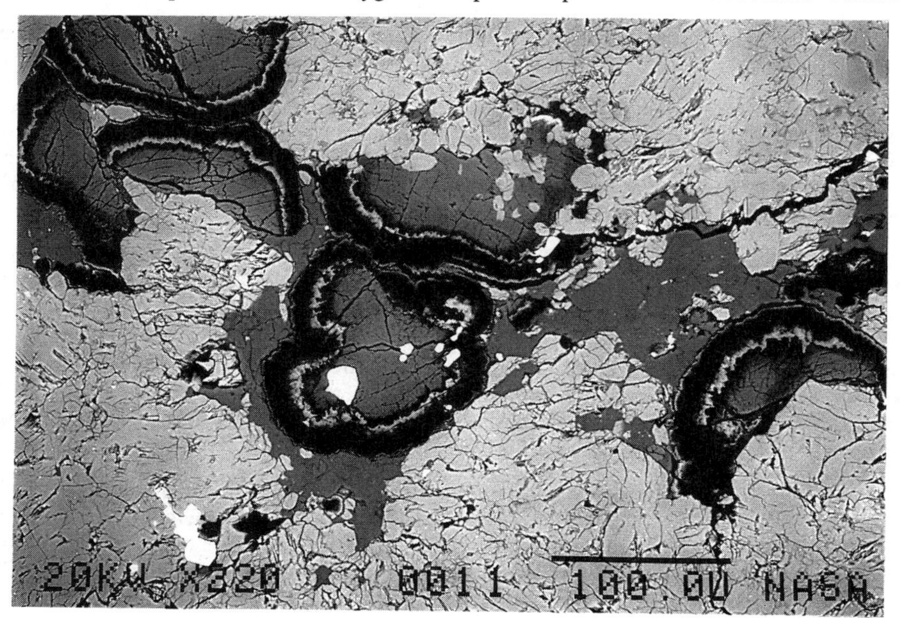

Figure 20. Backscattered electron image of carbonate globules in ALH84001. Ankerite cores are rimmed by multiple zones of magnesite (black) and siderite-rich carbonate (white). Scale bar is 100 microns.

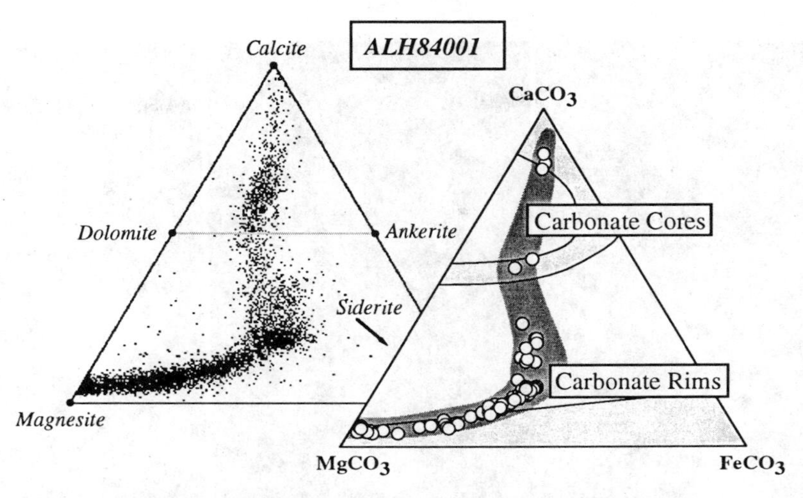

Figure 21. Compositions of zoned carbonate globules in ALH84001 (modifed from Harvey and McSween 1996). Triangle on the left shows more than 6,700 semi-quantitative analyses, illustrating that the carbonates are extremely fine-grained; overlaps suggest that calcite coexists with ankerite, and ankerite abuts the most Fe-rich member of the magnesite-siderite solid solution series. Fully quantitative microprobe analyses are shown in the right triangle, which is the 700°C phase diagram for the Ca-Mg-Fe carbonate system.

with their elemental compositions (Leshin et al. 1997), which may suggest either equilibration with aqueous fluids during temperature excursions of at least 250°C, or Raleigh fractionation of CO_2-rich fluids at an even wider range of temperatures. However, C isotopic compositions may not support these scenarios (Eiler et al. 1997). Carbon isotopic compositions demonstrate that they are extraterrestrial, but imply some exchange with terrestrial carbon (Jull et al. 1997b). REE patterns for carbonates were measured by Wadhwa and Crozaz (1995b) and Shearer et al. (1997), and were attributed to metasomatizing fluids to account for differences in these patterns from those of the primary minerals.

When examined under TEM, the carbonates are seen to contain a variety of *magnetite* and *sulfide* morphologies that were suggested by McKay et al. (1996) to be biogenic minerals. Nanophase magnetites were noted to be similar in size (20-100 nm diameter, magnetically "single domain" crystals) and shape to those produced by terrestrial magnetotactic bacteria, and sulfides (pyrrhotite and greigite) were suggested to have formed by sulfate-respiring organisms. Monoclinic 4C pyrrhotite was identified by its composition and (111) basal spacing of 0.57 nm (although the basal spacing of this phase, corresponding to (002), is 0.53 nm, and the (111) spacing is 0.47 nm), but greigite was identified only by its morphological similarity to terrestrial biogenic greigite. In addition, McKay et al. (1996) described minute "ovoid and elongated forms" seen in SEM as possible nanofossils.

Bradley et al. (1996) attempted to replicate these observations using TEM. They discovered nanophase whiskers and platelets of magnetite (Fig. 22), sometimes with axial screw dislocations, internal structures not characteristic of biogenic magnetite. The unusual assortment of magnetite morphologies and the spiral growth mechanism are consistent with their formation by vapor deposition. Bradley et al. (1996) argued that these grains were deposited at high temperatures, by analogy with magnetite whiskers found in sublimates from volcanic fumaroles. Moreover, they noted that the platelets and whiskers were similar

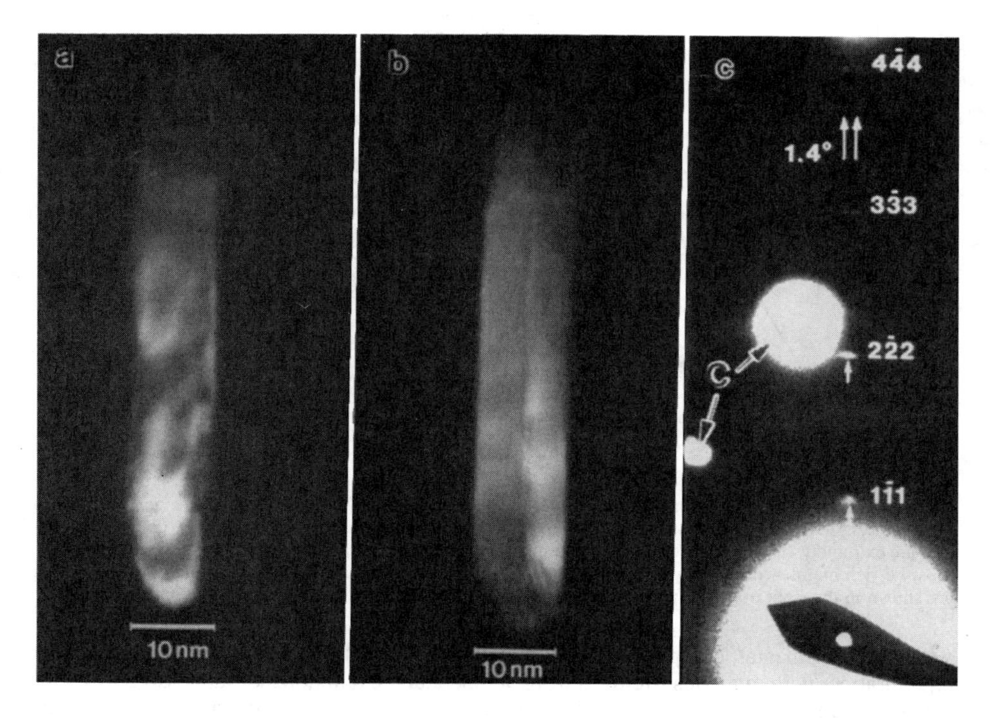

Figure 22. TEM darkfield images of two nanophase magnetite whiskers in ALH84001 carbonate (from Bradley et al. 1996). The whisker in (a) appears twisted and free of internal dislocations, whereas the whisker in (b) has an axial screw dislocation. Splitting of hkl diffractions, seen in the selected area diffraction pattern (c), is caused by helical lattice distortions resulting from the screw axis. Diffractions from the surrounding carbonate are labelled "C."

in size and morphology to the ovoid and elongated forms described by McKay et al. (1996), and the absence of other similar materials in the carbonates suggested that these might be the nanofossils. However, no study has yet identified a nanofossil in the SEM and then sectioned the same particle for TEM analysis. Bradley et al. (1997) found that some magnetite whiskers without screw dislocations had grown epitaxially onto the carbonate substrate and on other magnetites. Epitaxial growth is another mechanism common in vapor deposition.

Other minerals associated with carbonate and believed to be part of the alteration assemblage include *pyrite* and other trace sulfides. The pyrite occurs as small euhedral grains in the fracture zones. Its sulfur is isotopically heavy ($\delta^{34}S$ = +2.0 to +7.3‰; Shearer et al. 1996, Greenwood et al. 1997), which appears to be inconsistent with its formation by reduction of sulfate through chemosynthetic pathways used by known terrestrial organisms. The occurrence of *pentlandite* was reported by Shearer et al. (1997), and Wentworth and Gooding (1995) described feathery grains of ZnS within carbonates.

Silica has been reported at the juncture where some carbonate globules have growth together (Harvey and McSween 1996), in shock-melted veins (Scott et al. 1997, Valley et al. 1997), and also as euhedra within orthopyroxene (Kring and Gleason 1997). As noted above, non-stoichiometric maskelynite analyses have also been attributed to mixture of feldspar and silica (Mittlefehldt 1994).

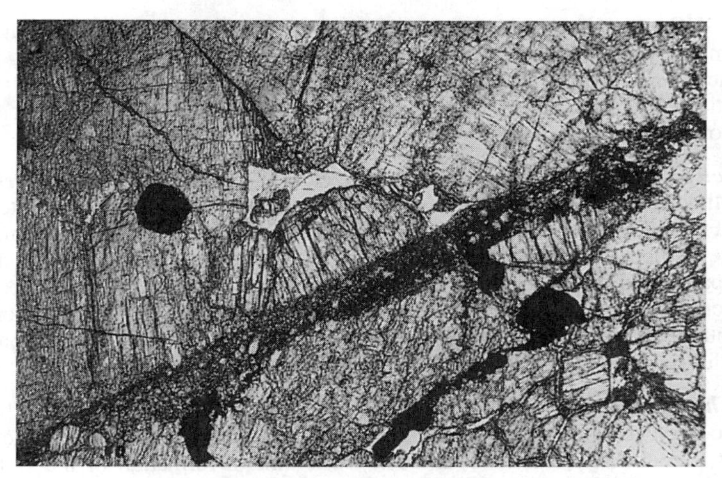

Figure 23. Photomicrograph of ALH84001 showing fracture zones crosscutting orthopyroxenite. The image measures approximately 2 mm (courtesy of D. Mittlefehldt).

Shock metamorphic minerals and effects. ALH84001 is cut by numerous deformation zones a few mm wide (Fig. 23), composed of fine grains of the same minerals that compose the host rock. Mittlefehldt (1994) termed these "crush zones," emphasing their cataclastic origin. Treiman (1995b) used the term "granular bands" to call attention to their granulitic (recrystallized) texture. Some areas of these zones exhibit mortar or augen textures, with swirled stringers of chromite traceable to chromites in the adjacent rock. Treiman (1995b) interpreted these zones as melt-breccia dikelets or crystalline cataclasites.

Maskelynite occurs as small, irregularly shaped grains. Veins of plagioclase glass formed by impact melting also occur (McKay and Lofgren 1997, Scott et al. 1997). Carbonates are commonly cut by microfaults with offsets, and radial fractures occur around chromite and maskelynite. Orthopyroxene exhibits strain birefringence.

Petrology, geochemistry, and geochronology

Petrology. ALH84001 is a coarse-grained orthopyroxenite cumulate with homogeneous pyroxene, suggesting crystallization in a plutonic environment. Interstitial phases apparently represent material crystallized from intercumulus liquid. The meteorite now consists of coherent clasts up to a few cm across bounded by recrystallized fracture zones. The petrologic history of this meteorite, as envisioned by Mittlefehldt (1994), was as follows: Crystallization and accumulation of orthopyroxene and chromite from basaltic magma was followed by crystallization of plagioclase from intercumulus liquid, and then succeeded by formation of carbonate and pyrite. Shock metamorphism then formed fracture zones, and additional carbonate was deposited. Treiman's (1995b) petrogenesis differs in having only one period of carbonate deposition, but two shock events. His order of events after formation of the cumulate rock was as follows: Shock metamorphism produced fracture zones, followed by thermal metamorphism that resulted in textural annealing and equilibration of all minerals. Subsequently, fluids were introduced, which precipitated carbonates and pyrite. Following that, the rock experienced a second shock metamorphism, possibly the ejection event, which converted plagioclase to maskelynite and deformed carbonates.

The oxygen isotopic composition of this stone (Clayton and Mayeda 1996) clearly links it to SNC meteorites, and it has certain mineralogical properties, such as the presence of sodic maskelynite and Fe^{3+}-bearing chromite, that resemble SNC minerals. However, the basaltic parent magma for ALH84001 is probably not related to the parent magmas of other martian meteorites. This meteorite is, after all, 3 to 4 billion years older than other members of this clan (see below).

Geochemistry. Major, minor, and trace elements were analyzed by Dreibus et al. (1994), Mittlefehldt (1994), and Warren and Kallemeyn (1996). The meteorite has high Mg/(Mg+Fe) and relatively high abundances of volatile elements. The abundances of the siderophile elements Ni, Ir, Au, and Os are very low compared to other martian meteorites. The REE pattern exhibits a depletion in LREE and a negative Eu anomaly, as expected for a cumulate orthopyroxenite. The La/Lu ratio is higher than would be predicted for a parent magma with chondritic REE ratios, but can be explained by inclusion of a small amount intercumulus liquid in the rock.

Geochronology. The crystallization age of ALH84001 is 4.5 ± 0.13 Ga, based on a Sm-Nd isochron (Nyquist et al. 1995). Argon isotope dating gives a shock age of 4.0 ± 0.1 Ga (Ash et al. 1996, Turner et al. 1997). The time of formation of the carbonates is unclear and disputed. From $^{40}Ar/^{39}Ar$ measurements, Knott et al. (1995) suggested that the carbonate formed at ~3.6 Ga, which was cited by McKay et al. (1996). On the other hand, the Rb-Sr data of Wadhwa and Lugmair (1996) suggested an age of ~1.4 Ga.

MARTIAN MINERALOGY INFERRED FROM REMOTE SENSING AND SPACECRAFT DATA

Beyond inferences from the SNC meteorites, the mineralogy of Martian surface materials is known only indirectly. Reflection spectra, both telescopic and spacecraft, can be interpreted in terms of mineralogy—absorption bands in the spectra can be assigned to specific minerals or mineral groups. The chemical composition and magnetic properties of the martian soil, or dust, were analyzed in situ by the Viking lander spacecraft, and those analyses provide some indirect constraints on the soil mineralogy. From imagery and reflection spectra, the surface of Mars can be characterized in terms of three spectrally distinct units: low-albedo (dark), gray, Fe^{2+}-rich regions interpreted to be mixtures of dark rocks and residual soils; high-albedo (bright), red, Fe^{3+}-rich regions thought to be covered with aeolian dust; and intermediate albedo regions thought to be indurated soils (Presley and Arvidson 1988, Murchie et al. 1993). Images from the Viking landers (Fig. 24) indicate rock abundances covering 10 to 20% of the surface, and thermal emissivity measurements (Christensen 1986) suggest that some dark areas may contain proportions of rock as high as 35%. In this section we compare what is inferred about the mineralogy of the martian surface, especially the dark rocky regions, from remote sensing spectral measurements and chemical data from the Viking and Mars Pathfinder landers with what has been learned from the study of martian meteorites.

Igneous rocks

Near-infrared reflectance spectra of the dark regions exhibit two crystal-field absorption bands (near 1 and 2 mm) that indicate the presence of pyroxenes with variable Ca and Fe^{2+} contents (Soderblom 1992). As illustrated in Figure 25, the spectra of three low-albedo areas on Mars (Hesperia, Iapygia, and two measurements for Syrtis Major taken at different times—Singer et al. 1980, Mustard et al. 1993) are similar to spectra for basaltic shergottites, but not to lherzolitic shergottites, nakhlites, or ALH84001 (McFadden 1987, Sunshine et al. 1993, Bishop et al. 1994). The 2 μm absorption band is partly masked by martian atmospheric CO_2, as indicated by the widths of the boxes, and both

Figure 24. Photograph of the martian surface at the Viking 2 landing site in Utopia Planitia, taken in 1979. The abundant rocks are partly covered by frost.

absorption bands are probably broadened by compositional scatter in pyroxenes. The absorption bands for Mars and for the SNC meteorites are actually composite spectra of more than one pyroxene phase; however, Sunshine et al. (1993) deconvolved the overlapping spectral bands of pigeonite and augite in EETA79001, as shown by open symbols with tie-lines in Figure 25. High-resolution orbital spectra of Mars obtained by future orbiting spacecraft may be able to similarly deconvolve pyroxene spectral components and thus identify coexisting pyroxenes in surface rocks. As an example, Mustard and Sunshine (1995) determined differences in the modal proportions of pigeonite

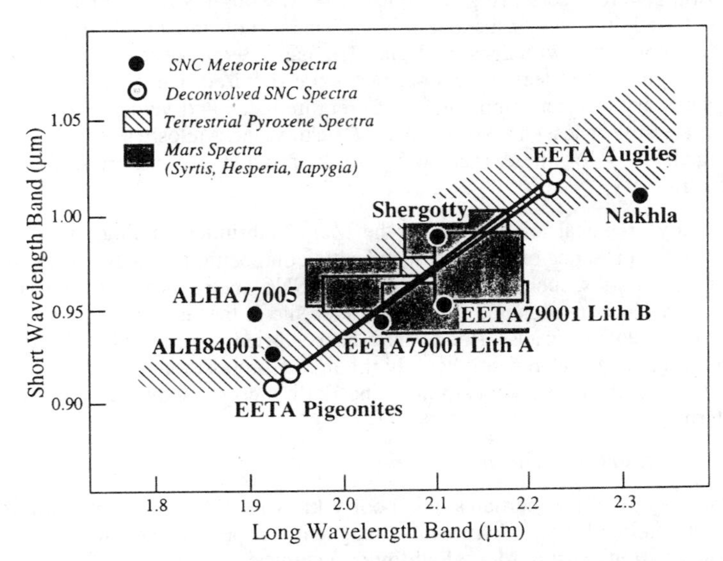

Figure 25. Long- and short-wavelength band positions for SNC meteorites, low-albedo regions of Mars, and terrestrial pyroxenes of varying composition (modifed from McSween 1994). The martian spectra are similar to those of basaltic shergottites, suggesting that these meteorites may represent widespread lavas. The deconvolved spectra for EETA79001 (Sunshine et al. 1993) separate overlapping bands for pigeonite and augite, whose individual spectra are joined by tie-lines.

and augite in comparing high-resolution spectra from Phobos II for Eos Chasma (within Valles Marineris) and Nili Patera (a caldera in Syrtis Major). In any case, the spectral properties of all regions examined so far suggest that basaltic shergottites are probably common lava types on the martian surface (Singer and McSween 1993, Mustard and Sunshine 1995).

The spectral field defined by terrestrial pyroxenes with various compositions and crystal structures is illustrated in Figure 25 by the cross-hatched area (Cloutis and Gaffey 1991). Orthopyroxene plots at the lower left corner of this distribution, and augite plots at the upper right. The spectra of ALHA77005 and ALH84001, both of which contain appreciable orthopyroxene, plot in the appropriate corner, and EETA79001 lithology A, which contains xenocrysts of orthopyroxene, plots closer to this corner than does lithology B. Likewise, the Nakhla spectra plot in the augite field. High-resolution spectra of Mars may thus be able to locate possible source craters for these lithologies. It is noteworthy, however, that the meteorites for which plutonic origins are inferred do not appear to be common surface rocks.

None of the areas surveyed show spectra features indicative of olivine or plagioclase, although abundant pyroxenes could mask limited amounts of these minerals. Magnetite or other opaque phases are inferred to be present in the dark regions, based on their low albedos and other spectral details (Mustard et al. 1993).

The chemical composition of martian duracrust-free soil at the Viking landing sites is similar to that of basaltic shergottites (Baird and Clark 1981, Warren 1987). This may imply derivation from a widespread volcanic protolith, and supports the inference from spectra that shergottites are common surface rocks.

Specific craters on the martian surface have been proposed as the sites from which SNC meteorites were ejected (e.g. Nyquist 1983, Mouginis-Mark et al. 1992). These studies were based on a concensus that only the Tharsis volcanic terrain is young enough to have supplied meteorites with ages <1.3 Ga. Trieman (1995a) used the properties of SNC meteorites to argue that at least three crater sites are required. Consideration of the size of the crater necessary to eject fragments of the required size and velocity appears to restrict choices to craters of at least 12 km diameter (Vickery and Melosh 1987). The location of specific launch sites for these meteorites would provide ground truth for interpreting martian spectra.

Preliminary chemical analyses from the Mars Pathfinder landing site (Rieder et al. 1997) indicate the presence of rocks with andesitic compositions. These rocks share certain chemical characteristics, such as high Fe/Mg, with SNC meteorites, but their compositions are considerably more fractionated. The Pathfinder site at the mouth of an outflow channel was originally selected because floods may have carried samples of the ancient martian crust from the heavily cratered southern highlands. Compositional similarity between the Pathfinder rocks and mean composition of the Earth's crust might imply parallel differentiation patterns.

Soils and weathering products

The mineralogy of the martian soil is poorly known, with limited and ambiguous data coming from Viking lander instruments (magnetic properties, life sciences, and major element composition), from Mars Pathfinder chemical analyses, and from reflectance spectral measurements in the visible and near-infrared. Future spacecraft missions will attempt to fill this void in our knowledge, both with landed mineralogic instruments and with remote sensing in different wavelength ranges, notably the thermal infrared.

Iron oxide minerals are prominent in the martian soil, as suggested by its red-orange color. Reflection spectral studies suggest that the soil contains nanophase hematite and minor crystalline hematite (Morris et al. 1989, 1993; Bell et al. 1990, Bell 1992, Murchie et al. 1993). The soil also contains a few percent of a highly magnetic mineral which is probably titanomagnetite (Hargraves et al. 1977, Baird and Clark 1981). Other iron minerals may also be present, including schwertmannite (a hydroxylated ferric sulfate) and ferrihydrite-intercalated clays (Bishop et al. 1993, Bishop and Murad 1996).

Salt minerals are also important. The Viking lander XRF and Pathfinder APXS analyses of martian soil show high and variable abundances of S which are correlated with those of Mg; both elements are anticorrelated with Si (Clark 1993). Both S and Mg are more abundant in crusted soil samples, which suggests the presence of a water-soluble Mg-sulfate salt; its identity is not known, but possible choices include kieserite $MgSO_4 \cdot H_2O$, bloedite $MgNa_2(SO_4)_2 \cdot 4H_2O$, and loweite $MgNa_2(SO_4)_2 \cdot 2.5H_2O$ (Clark and van Hart 1981). Sulfate might also be present as ferric sulfate intercalations in smectitic clays or as schwertmannite (Bishop et al. 1995, Bishop and Murad 1996). Infrared light absorptions characteristic of sulfate (or bisulfate) minerals are observed in reflection spectra of Mars, but the mineralogic hosts for these anions are not clear (Pollack et al. 1990, Blaney and McCord 1995, Bell et al. 1996). Detectable chorine in the soil suggests the presence of chloride salts, NaCl seems likely (Clark and van Hart 1981), and the positive correlation of Mg and Cl in soil analyses may suggest a Mg-bearing chloride (Clark 1993). The Viking XRF analyses admit the possibility of Mg carbonate minerals in the soil, but probably not Ca carbonates (Clark 1993). Anhydrous carbonates have not been detected spectroscopically, although there is some evidence for hydrous magnesium carbonate minerals (e.g. Calvin et al. 1994). Current interpretations of the Viking life science experiments are not consistent with abundant carbonate minerals (Banin et al. 1992).

The silicate mineralogy of the martian soil is known poorly. Reflection spectra show no absorptions characteristic of olivine or pyroxenes, although the bulk composition of the soil is close to that of basaltic shergottites (Baird and Clark 1981, Banin et al. 1992). The bulk composition and color of the soil have suggested that it is palagonite, or altered basaltic glass (Banin et al. 1992). Palagonite on Earth consists of principally of nanophase phyllosilicates and ferric iron oxides (e.g. Banin et al. 1992), and some palagonites have reflection spectra (visible through near-infrared) that are essentially identical to that of the martian soil (e.g. Morris et al. 1993). This model for the soil silicate is attractive in its simplicity, but cannot explain all available data. Results from the Viking life science and soil reactivity experiments are more consistent with crystalline clays than with palagonite (Banin et al. 1992), and near-infrared absorptions attributable to metal—OH bonds (as in clays) are weak and unusually sharp for known phyllosilicates on Earth (Bell et al. 1994, Bell 1996). The mineral scapolite has been suggested as a significant component of the martian soil, but the spectral absorptions that suggested scapolite are also consistent with other sulfate or carbonate minerals (Clark et al. 1990, Bell 1996).

The nature of the martian soil is one manifestation of chemical weathering on Mars, which has been explored by Gooding (1978) and Siderov and Zolotov (1986) from theoretical thermodynamic bases. The martian meteorites contain some low-temperature alteration minerals of martian origin, as described above, which can be used as constraints on weathering and soil mineralogy (Gooding 1992). A mixture of smectite and illite such as occur in the martian meteorites, plus salt and iron oxide minerals, matches the chemical composition of the martian soil and is perhaps more consistent with the Viking life science experiment results than is palagonite (Gooding 1992). On the other hand, the Viking XRF analyses can be accommodated as a mixture of a single silicate component, a single salt component, and an Fe-Ti oxide (Clark 1993).

ACKNOWLEDGMENTS

We appreciate helpful discussions with J.H. Jones and D.W. Mittlefehldt, the photography assistance of Debra Reub, and constructive reviews by R.P. Harvey and T.J. McCoy. This work was partly supported by NASA grants NAG-54541 to HYM and NAGW-5098 to AHT.

REFERENCES

Ash RD, SF Knott, G Turner (1996) A 4-Gyr shock age for a martian meteorite and implications for the cratering history of Mars. Nature 380:57-59

Ashworth JR, R Hutchison (1975) Water in non-carbonaceous stony meteorites. Nature 256:714-715

Baird AK, BC Clark (1981) On the original igneous source of martian fines. Icarus 45:113-123

Banin A, BC Clark, H Wänke (1992) Surface chemistry and mineralogy. In: Mars. HH Kieffer, BM Jakosky, CW Snyder, MS Matthews (eds) p 594-625. Univ Arizona Press, Tucson, AZ

Becker RH, RO Pepin (1984) The case for a martian origin of the shergottites: Nitrogen and noble gases in EETA79001. Earth Planet Sci Lett 69:225-242

Bell JF III (1992) Charge-coupled device imaging spectroscopy of Mars. 2. Results and implications for martian ferric mineralogy. Icarus 100:575-597

Bell JF III (1996) Iron, sulfate, carbonate, and hydrated minerals on Mars. In: Mineral Spectroscopy: A Tribute to Roger G. Burns. MD Dyar, C McCammon, MW Shaefer (eds) Geochemical Soc Spec Pub 5:359-380

Bell JF III, TB McCord, PD Owensby (1990) Observational evidence for crystalline iron oxides on Mars. J Geophys Res 9:14,447-14,461

Bell JF III, JB Pollack, TR Geballe, DP Cruickshank, R Freedman (1994) Spectroscopy of Mars from 2.04 to 2.44 μm during the 1993 opposition: Absolute calibration and atmospheric vs mineralogic origin of narrow absorption features. Icarus 111:106-123

Berkley JL, NJ Boynton (1992) Minor/major element variation within and among diogenite and howardite orthopyroxenite groups. Meteoritics 27:387-394

Berkley JL, K Keil, M Prinz (1980) Comparative petrology and origin of Governador Valadares and other nakhlites. Proc Lunar Planet Sci Conf 11:1089-1102

Berkley JL, K Keil (1981) Olivine orientation in the ALHA77005 achondrite. Am Mineral 66:1233-1236

Bertka CM, Y Fei (1997) Mineralogy of the martian interior up to core-mantle boundary pressures. J Geophys Res 102:5251-5264

Binns RW (1967) Stony meteorites bearing maskelynite. Nature 211:1111-1112.

Bishop JL, E Murad (1996) Schwertmannite on Mars? Spectroscopic analyses of schwertmannite, its relationship to other ferric minerals, and its possible presence in the surface of Mars. In: Mineral Spectroscopy: A Tribute to Roger G. Burns. MD Dyar, C McCammon, MW Shaefer (eds) Geochemical Soc Spec Pub 5:227-358

Bishop JL, CM Pieters, RG Burns (1993) Reflectance and Mössbauer spectroscopy of ferrihydrate-montmorillonite assemblages as Mars soil analog materials. Geochim Cosmochim Acta 57:4583-4595

Bishop J, C Pieters, J Mustard, S Pratt, T Hiroi (1994) Spectral analyses of ALH84001, a meteorite from Mars (abstr). Meteoritics 29:444-445

Bishop JL, CM Pieters, RG Burns, JO Edwards, RL Mancinelli, H Fröschl (1995) Reflectance spectroscopy of ferric sulfate-bearing montmorillonites as Mars soil analog materials. Icarus 117:101-119

Blaney, D. L., T. B. McCord (1995) Indications of sulfate minerals in the martian soil from Earth-based spectroscopy. J Geophys Res 100:14,433-14,441

Boctor NZ, HOA, G Kullerud (1976) Lafayette meteorite: Petrology and opaque mineralogy. Earth Planet Sci Lett 32:6

Bogard DD, LE Nyquist (1979) ^{39}Ar-^{40}Ar chronology of related achondrites (abstr). Meteoritics 14:356

Bogard DD, P Johnson (1983) Martian gases in an Antarctic meteorite. Science 221:651-654

Boynton WV, PM Starzyk, RA. Schmitt (1976) Chemical evidence for the genesis of the ureilites, the achondrite Chassigny, and the nakhlites. Geochim Cosmochim Acta 40:1439-1477

Borg LE, LE Nyquist, LA Taylor, H Wiesmann, C-Y Shih (1997) Rb-Sr and Sm-Nd isotopic analyses of QUE94201: Constraints on martian differentiation processes (abstr). Lunar Planet Sci XXVIII:133-134

Bradley JP, RP Harvey, HY McSween Jr (1996) Magnetite whiskers and platelets in the ALH84001 martian meteorite: Evidence of vapor phase growth. Geochim Cosmochim Acta 60:5149-5155

Bradley JP, HY McSween Jr, RP Harvey (1997) Epitaxial growth of single-domain magnetite in martian meteorite ALH84001 (abstr). Met Planet Sci 32:A20

Brandenburg JE (1996) Mars as the parent body of the CI carbonaceous chondrites. Geophys Res Lett 23:961-964

Brearley AJ (1991) Subsolidus microstructures and cooling history of pyroxenes in the Zagami shergottite (abstr). Lunar Planet Sci XXII:135-136

Bunch TE, AM Reid (1975) The nakhlites, part 1: Petrography and mineral chemistry. Meteoritics 10:303-315

Calvin WM, TVV King, RN Clark (1994) Hydrous carbonates on Mars? Evidence from Mariner 6/7 infrared spectrometer and ground-based telescopic spectra. J Geophys Res 99:14,659-14,675

Carr MH, H Wänke (1992) Earth and Mars: Water inventories as clues to accretional histories. Icarus 98:61-71

Chatzitheodoridis E, G Turner (1990) Secondary minerals in the Nakhla meteorite (abstr). Meteoritics 25:354

Chen JH, GJ Wasserburg (1986a) Formation ages and evolution of Shergotty and its parent planet from U-Th-Pb systematics. Geochim Cosmochim Acta 50:955-968

Chen JH, GJ Wasserburg (1986b) S =/ N =? C (abstr). Lunar Planet Sci XVII: 113-114

Christensen PR (1986) The spatial distribution of rock on Mars. Icarus 68:217-238

Clark BC (1993) Geochemical components in Martian soil. Geochim Cosmochim Acta 57:4575-4581

Clark BC, DC. van Hart (1981) The salts of Mars. Icarus 45:370-378

Clark RN, GA Swayze, RB Singer, JB Pollack (1990) High-resolution reflectance spectra of Mars in the 2.3-μm region: Evidence for the mineral scapolite. J Geophys Res 95:14,463-14,480

Clayton RN, TK Mayeda (1988) Isotopic composition of carbonate in EETA79001 and its relation to parent body volatiles. Geochim Cosmochim Acta 52:925-927

Clayton RN, TK Mayeda (1996) Oxygen isotope studies of achondrites. Geochim Cosmochim Acta 60:1999-2017

Cloutis EA, MJ Gaffey (1991) Pyroxene spectroscopy revisited: Spectral-compositional correlations and relationships to geothermometry. J Geophys Res 98:10,973-11,016

Collinson DW (1986) Magnetic properties of Antarctic shergottite meteorites EETA79001 and ALHA77005: Possible relevance to a martian magnetic field. Earth Planet Sci Lett 77:159-164

Collinson DW (1997) Magnetic properties of Martian meteorites: Implications for an ancient Martian magnetic field. Met Planet Sci 32:803-811

Delano JW, RJ Arculus (1980) Nakhla: Oxidation state and other constraints (abstr). Lunar Planet Sci XI:219-221

Drake MJ, TD Swindle, T Owen, DS Musselwhite (1995) Fractionated martian atmosphere in the nakhlites? Meteoritics 29:854-859

Dreibus G, H Wänke (1985) Mars, a volatile-rich planet. Meteoritics 20:367-381

Dreibus G, A Burghele, KP Jochum, B Spettel, F Wlotzka, H Wänke (1994) Chemical and mineral composition of ALH84001: A martian orthopyroxenite (abstr). Meteoritics 29:461

Duke MB (1968) The Shergotty meteorite: Magmatic and shock metamorphic features. In: Shock Metamorphism of Natural Materials. BM French, NM Short (eds) Mono Book Corp, Baltimore, MD, p 613-621

Eiler JM, JW Valley, EM Stolper (1997) Stable isotopes in ALH84001: An ion microprobe study (abstr). Met Planet Sci 32:A38

El Goresy A, B Wopenka, M Chen, G Kurat (1997) The saga of maskelynite in Shergotty (abstr). Meteor Planet Sci 32:A38-A39

Eugster O, E Polnau (1997) Mars-Earth transfer time of lherzolite Yamato-793605. Proc NIPR Symp Antarctic Met 10 (in press)

Eugster O, A Weigel, E Polnau (1997) Ejection times of Martian meteorites. Geochim Cosmochim Acta 61:2749-2758

Floran RJ, M Prinz, RF Hlava, K Keil, CE Nehru, JR Hinthorne (1978) The Chassigny meteorite: a cumulate dunite with hydrous amphibole-bearing melt inclusions. Geochim Cosmochim Acta 42:1213-1229

Friedman RC, GJ Taylor, AH Treiman (1995) Processes in thick lava flows: Nakhlites (Mars) and Theo's Flow (Ontario, Earth) (abstr). Lunar Planet Sci XXVI:429-430

Gaetani GA, TL Grove (1997) Partition of moderately siderophile elements among olivine, silicate melt, and sulfide melt: Constraints on core formation in the Earth and Mars. Geochim Cosmochim Acta 61:1829-1846

Gale NH, JW Arden, R Hutchison (1975) The chronology of the Nakhla achondrite meteorite. Earth Planet Sci Lett 26:195-206

Gladman B, JA Burns, M Duncan, P Lee, HF Levinson (1996) The exchange of impact ejecta between terrestrial planets. Science 271:1387-1390

Gleason JD, DA Kring, DH Hill, WV Boynton (1997) Petrography and bulk chemistry of Martian lherzolite LEW 88516. Geochim Cosmochim Acta 61:4007-4014

Gooding JL (1978) Chemical weathering on Mars: Thermodynamic stabilities of primary minerals (and their alteration products) from mafic igneous rocks. Icarus 33:483-513

Gooding JL (1992) Soil mineralogy and chemistry on Mars: Possible clues from salts and clays in SNC meteorites. Icarus 99:28-41

Gooding JL, DW Muenow (1986) Martian volatiles in shergottite EETA79001: New evidence from oxidized sulfur and sulfur-rich aluminosilicates. Geochim Cosmochim Acta 50:1049-1059

Gooding JL, SJ Wentworth (1991) Origin of "white druse" salts in EETA79001 (abstr). Lunar Planet Sci XXII:461-462

Gooding JL, SJ Wentworth, ME Zolensky (1988) Calcium carbonate and sulfate of possible extraterrestrial origin in the EETA79001 meteorite. Geochim Cosmochim Acta 52:909-915

Gooding JL, SJ Wentworth, ME Zolensky (1991) Aqueous alteration of the Nakhla meteorite. Meteoritics 26:135-143

Greenwood JP, LR Riciputi, HY McSween Jr (1997) Sulfur isotopic compositions in shergottites and ALH84001, and possible implications for life on Mars. Geochim Cosmochim Acta 61:4449-4453

Greshake A, F Langenhorst (1997) TEM characterization of shock defects in minerals of the martiain meteorite Chassigny (abstr). Met Planet Sci 32:A52

Hale VPS, HY McSween Jr, GA McKay (1997) Cumulus pyroxene in Shergotty: The discrepancy between experimental and observational studies (abstr). Lunar Planet Sci XXVIII:495-496

Hargraves, R. B., D. W. Collinson, R. E. Arvidson, C. R. Spitzer (1977) The Viking magnetic properties experiment: Primary mission results. J Geophys Res 82:4547-4558

Harper CL Jr, LE Nyquist, B Bansal, H Wiesmann, C-Y Shih (1995) Rapid accretion and early differentiation of Mars indicated by ^{142}Nd/^{144}Nd in SNC meteorites. Science 267:213-217

Harvey RP, HY McSween Jr (1992a) The parent magma of the nakhlite meteorites: Clues from melt inclusions. Earth Planet Sci Lett 111:467-482

Harvey RP, HY McSween Jr (1992b) The petrogenesis of the nakhlites: Evidence from cumulate mineral zoning. Geochim Cosmochim Acta 56:1655-1663

Harvey RP, HY McSween Jr (1996) A possible high-temperature origin for the carbonates in martian meteorite ALH84001. Nature 382:49-51

Harvey RP, M Wadhwa, HY McSween Jr, G Crozaz (1993) Petrography, mineral chemistry, and petrogenesis of Antarctic shergottite LEW88516. Geochim Cosmochim Acta 57:4769-4783

Hasan FA, M Haq, DWG. Sears (1986) Thermoluminescence and the shock and reheating hsitory of meteorites—III. The shergottites. Geochim Cosmochim Acta 50, 1031-1038

Hawthorne FC (1981) Crystal chemistry of the amphiboles. In: Amphiboles and other hydrous pyriboles—Mineralogy. Veblen DR (ed). Rev Mineral 9A:1-102

Hörz F, R Hanss, C Serna (1986) X-ray investigations related to the shock history of the Shergotty achondrite. Geochim Cosmochim Acta 50:905-908

Ikeda Y (1994) Petrography and petrology of ALH-77005 shergottite. Proc NIPR Symp. Antarctic Met 7:9-29

Ikeda Y (1997) Petrology of the Y-793605 lherzolitic shergottite (abstr). Met Planet Sci 32:A64-A65

Ishii T, H Takeda, K Yanai (1979) Pyroxene geothermometry applied to a three-pyroxene achondrite from Allan Hills, Antarctica and ordinary chondrites. Mineral J 9:460-481

Jagoutz E (1989) Sr and Nd isotopic systematics in ALHA77005: Age of shock metamorphism in shergottites and magmatic differentiation of Mars. Geochim Cosmochim Acta 53:2429-2441

Jagoutz E (1991) Chronology of SNC meteorites. Space Sci Rev 56:13-22.

Jagoutz E (1996) Nd isotopic systematics of Chassigny (abstr). Lunar Planet Sci XXVIII:597-598

Jérémine E, J Orcel, A Sandréa (1962) Étude mineralogique et structural de la météorite de Chassigny. Bull Soc franc Mineral Cristal 85:262-266

Johnson MC, MJ Rutherford, PC Hess (1991) Chassigny petrogenesis: Melt compositions, intensive parameters, and water contents of Martian (?) magmas. Geochim Cosmochim Acta 55:349-366

Jones JH (1986) A discussion of isotopic systematics and mineral zoning in the shergottites: Evidence for a 180 m.y. igneous crystallization age. Geochim Cosmochim Acta 50:969-977

Jones JH (1989) Isotope relationships among the shergottites, the nakhlites and Chassigny. Proc Lunar Planet Sci Conf 19:465-474

Jull AJT, DJ Donahue, TD Swindle, MK Burkland, GF Herzog, A Albrecht, J Klein, R Middleton (1992) Isotopic studies relevant to the origin of the "white druse" carbonates on EETA79001 (abstr). Lunar Planet Sci XXIII:641-642

Jull AJT, CJ Eastoe, S Cloudt (1997a) Terrestrial age of the Lafayette meteorite and stable isotopic composition of weathering products (abstr). Lunar Planet Sci XXVIII:685-686

Jull AJT, CJ Eastoe, S Cloudt (1997b) Isotopic composition of carbonates in the SNC meteorites, Allan Hills 84001 and Zagami. J Geophys Res 102:1663-1669

Kaneda K, GA McKay, L Le (1997) Synthetic and natural nakhla pyroxenes: Minor elements composition (abstr). Lunar Planet Sci XXVIII:693-694

Karlsson HR, RN Clayton, EK Gibson Jr, T. K. Mayeda (1992) Water in SNC meteorites: Evidence for a martian hydrosphere. Science 255:1409-1411

Kieffer HH, BM Jakosky, CW Snyder, MW. Matthews, eds (1992) Mars. Univ Arizona Press, Tucson, AZ, 1498 p

Kirschvink JL, A Maine, H Vali (1997) Paleomagnetic evidence of a low-temperature origin of carbonate in the martian meteorite ALH84001. Science 275:1629-1633

Knott SF, RD Ash, G Turner (1995) $^{40}Ar/^{39}Ar$ dating of ALH84001: Evidence for the early bombardment of Mars (abstr). Lunar Planet Sci XXVI:765-766

Kring DA, JD Gleason (1997) Magmatic temperatures and compositions on early Mars as inferred from the orthopyroxene-silica assemblage in Allan Hills 84001 (abstr). Met Planet Sci 32:A74

Lancet MS, K Lancet (1971) Cosmic ray and gas retention ages of the Chassigny meteorite. Meteoritics 6:81-85

Langenhorst F, D Stöffler, D Klein (1991) Shock metamorphism of the Zagami achondrite (abstr). Lunar Planet Sci XXII:779-780

Laul JC, MR Smith, H Wänke, E Jagoutz, G Dreibus, H Palme, B Spettel, A Burghele, ME Lipschutz, RM Verkouteren (1986) Chemical systematics of the Shergotty meteorite and the composition of its parent body (Mars). Geochim Cosmochim Acta 50:909-926

Leake BE and 21 coauthors (1997) Nomenclature of amphiboles: Report of the Subcommittee on Amphiboles of the International Mineralogical Association, Commission on New Minerals and Mineral Names. Am Mineral 82:1019-1037

Lee D-C, AN Halliday (1997) Core formation on Mars and differentiated asteroids. Nature 388:854-857

Leshin LA, S Epstein, EM Stolper (1996) Hydrogen isotope geochemistry of SNC meteorites. Geochim Cosmochim Acta 60:2635-2650

Leshin LA, KD McKeegan, PK Carpenter, RP Harvey (1997) Oxygen isotopic constraints on the genesis of carbonates from martian meteorite ALH84001. Geochim Cosmochim Acta (in press)

Lindstrom DS, AH Treiman, RR Martinez (1993) Trace element analysis of magmatic inclusions in ALHA77005 by micro-INAA (abstr). Met Planet Sci 28:386-387

Longhi J (1991) Complex magmatic processes on Mars: Inferences from the SNC meteorites. Proc Lunar Planet Sci Conf 21:695-709

Longhi J, V. Pan (1989) The parent magmas of the SNC meteorites. Proc Lunar Planet Sci Conf 19:451-464

Longhi J, E Knittle, JR Holloway, H Wänke (1992) The bulk composition, mineralogy and internal structure of Mars. In: Mars. HH Kieffer, BM Jakosky, CW Snyder, MS Matthews (eds) p 185-208. Univ Arizona Press, Tucson, AZ

Lundberg LL, G Crozaz, G McKay, E Zinner (1988) Rare earth element carriers in the Shergotty meteorite and implications for its chronology. Geochim Cosmochim Acta 52:2147-2163

Lundberg LL, G Crozaz, HY McSween Jr. (1990) Rare earth elements in minerals of the ALHA77005 shergottite and implications for its parent magma and crystallization history. Geochim Cosmochim Acta 54:2535-2547

Marti K, JS Kim, AN Thakur, TJ McCoy, K Keil (1995) Signatures of the martian atmosphere in glass of the Zagami meteorite. Science 267:1981-1984

Martinez RR, JL Gooding (1986) New saw-cut surfaces of EETA79001. Antarctic Meteorite Newsletter 9:23

McCoy TJ, GE Lofgren (1996) The crystallization of the Zagami shergottite: A 1 atm experimental study (abstr). Lunar Planet Sci XXVII:839-840

McCoy TJ, GJ Taylor, K Keil (1992) Zagami: Product of a two-stage magmatic history. Geochim Cosmochim Acta 56:3571-3582

McCoy TJ, K Keil, GJ Taylor (1993) The dregs of crystallization in Zagami (abstr). Lunar Planet Sci XXIV:947-948

McCoy TJ, M Wadhwa, K Keil (1995) Zagami: Another new lithology and a complex, near-surface magmatic history (abstr). Lunar Planet Sci XXVI:925-926

McFadden LA (1987) Spectral reflectance of SNC meteorites: Relationships to martian surface composition. Lunar Planet. Inst. Tech. Rept. 88-05:88-90

McKay DS, EK Gibson Jr, KL Thomas-Keprta, H Vali, CS Romanek, SJ Clement, XDF Chillier, CR Maechling, RN Zare (1996) Search for past life on Mars: Possible relic biogenic activity in martian meteorite ALH84001. Science 273:924-930

McKay GA, GE Lofgren (1997) Carbonates in ALH84001: Evidence for kinetically controlled growth (abstr). Lunar Planet Sci XXVIII:921-922

McKay GA, T Mikouchi, G. E. Lofgren (1997) Carbonates and feldspathic glass in ALH84001: Additional complications (abstr). Met Planet Sci 32:A87-A88

McSween HY Jr (1985) SNC meteorites: Clues to martian petrologic evolution? Rev Geophys 23:391-416

McSween HY Jr (1994) What we have learned about Mars from SNC meteorites. Meteoritics 29:757-779

McSween HY Jr (1997) Evidence for life in a martian meteorite? GSA Today 7:1-7

McSween HY Jr and RP Harvey (1993) Outgassed water on Mars: Constraints from melt inclusions in SNC meteorites. Science 259:1890-1892

McSween HY Jr, E. Jarosewich (1983) Petrogenesis of the Elephant Moraine A79001 meteorite: Multiple magma pulses on the shergottite parent body. Geochim Cosmochim Acta 47:1501-1513

McSween HY Jr, EM Stolper, LA Taylor, RA Muntean, GD O'Kelley, JS Eldridge, S Biswas, HT Ngo, ME Lipschutz (1979a) Petrogenetic relationship between Allan Hills 77005 and other achondrites. Earth Planet Sci Lett 45:275-284

McSween HY Jr, LA Taylor, EM Stolper (1979b) Allan Hills 77005: A new meteorite type found in Antarctica. Science 204:1201-1203

McSween HY Jr, DD Eisenhour, LA Taylor, M Wadhwa, G Crozaz (1996) QUE94201 shergottite: Crystallization of a martian basaltic magma. Geochim Cosmochim Acta 60:4563-4569

Melosh HJ (1984) Impact ejection, spallation, and the origin of meteorites. Icarus 59:234-260

Melosh HJ, AH Treiman, RAF. Grieve (1983) Olivine composition glass in the Chassigny meteorite: Implications for shock history. EOS 64:254

Meyer C (1996) Mars Meteorite Compendium—1996. Report 27672, NASA/Johnson Space Center, Houston, TX, 175 p [Available on the Internet at http://www-curator.jsc.nasa.gov/curator/antmet/mmc/mmc.htm]

Mikouchi T, M Miyamoto (1996) A new member of lherzolitic shergottite from Japanese Antarctic meteorite collection: Mineralogy and petrology of Yamato 793605. Proc NIPR Symp Antarctic Met 21:104-106. Nat'l Inst Polar Res, Tokyo

Mikouchi T, M Miyamoto (1997) Major and minor element distributions in pyroxene and maskelynite from martian meteorite Yamato-793605 and other lherzolitic shergottites: Clues to their crystallization histories (abstr). Antarctic Meteorites XXII:109-112

Mikouchi T, M Miyamoto, GA McKay (1996) Mineralogy and petrology of new Antarctic shergottite QUE94201: A coarse-grained basalt with unusual pyroxene zoning (abstr). Lunar Planet Sci XXVII:879-880

Mikouchi T, M Miyamoto, GA McKay (1997) Similarities in zoning of pyroxenes from QUE94201 and EETA79001 martian meteorites (abstr). Lunar Planet Sci XXVIII:955-956

Mittlefehldt DW (1994) ALH84001, a cumulate orthopyroxenite member of the martian meteorite clan. Meteoritics 29:214-221

Mittlefehldt DW, DJ Lindstrom, MM Lindstrom, RR Martinez (1997) Lithology A in EETA79001—Product of impact melting on Mars (abstr). Lunar Planet Sci XXVIII:961-962.

Morris RV, DG Agresti, HV Lauer Jr, JA Newcombe, TD Shelfer, AV Murali (1989) Evidence for pigmentary hematite on Mars based on optical, magnetic, and Mössbauer studies of superparamagnetic (nanocrystalline) hematite. J Geophys Res 94:2760-2778.

Morris RV, DC Golden, JF Bell III, HV Lauer Jr, JB Adams (1993) Pigmenting agents in Martian soils: Inferences from spectral, Mössbauer, and magnetic properties of nanophase and other iron oxides in Hawaiian palagonitic soil PN-9. Geochim Cosmochim Acta 57:4597-4609

Mouginis-Mark PJ, TJ McCoy, GJ Taylor, K Keil (1992) Martian parent craters for the SNC meteorites. J Geophys Res 97:10,213-10,225

Müller WF (1993) Thermal and deformational history of the Shergotty meteorite deduced from clinopyroxene microstructure. Geochim Cosmochim Acta 57:4311-4322

Murchie S, J Mustard, J Bishop, J Head III, C Pieters (1993) Spatial variations in the spectral properties of bright regions on Mars. Icarus 105:454-468

Mustard JF, JM Sunshine (1995) Seeing through the dust: Martian crustal heterogeneity and links to the SNC meteorites. Science 267:1623-1626

Mustard JF, S Erard, J-P Bibring, JW Head, S Hurtrez, Y Langevin, CM Pieters, CJ Sotin (1993) The surface of Syrtis Major: Composition of the volcanic substrate and mixing with altered dust and soil. J Geophys Res 98 (E2):3387-3400

Nakamura N, DM Unruh, M Tatsumoto, R Hutchison (1982a) Origin and evolution of the Nakhla meteorite inferred from the Sm-Nd and U-Pb systematics and REE, Ba, Sr, Rb abundances. Geochim Cosmochim Acta 46:1555-1573

Nakamura N, H Komi, H Kagami (1982b) Rb-Sr isotopic and REE abundances in the Chassigny meteorite (abstr). Meteoritics 17:257-258

Nyquist LE (1983) Do oblique impacts produce martian meteorites? J Geophys Res 88 (suppl):A785-A798

Nyquist LE, B Bansal, H Weissmann, C-Y Shih (1995) "Martians" young and old: Zagami and ALH84001 (abstr). Lunar Planet Sci XXVI:106-1066

Ostertag R, G Amthauer, H Rager, HY McSween Jr (1984) Fe^{3+} in shocked olivine crystals of the ALHA787005 meteorite. Earth Planet Sci Lett 67:162-166

Papanastassiou DA, GJ Wasserburg (1974) Evidence for late formation and young metamorphism in the achondrite Nakhla. Geophys Res Lett 1:23-26

Papike JJ, GW Fowler, GD Layne, MN Spilde, CK Shearer (1994) ALH84001 A "SNC orthopyroxenite": Insights from SIMS analysis of orthopyroxene (abstr). Lunar Planet Sci XXV:1043-1044

Pepin RO, MH Carr (1992) Major issues and outstanding questions. In: Mars. HH Kieffer, BM Jakosky, CW Snyder, MS Matthews (eds) p 120-146. Univ Arizona Press, Tucson, AZ

Podosek, F.A., J. C. Huneke (1973) Argon 40–argon 39 chronology of four calcium-rich achondrites. Geochim Cosmochim Acta 37:667-684

Pollack JB, TL Roush, R Witteborn, J Bregman, D Wooden, C Stoker, OB. Toon, D Rank, B Dalton, R Freedman (1990) Thermal emission spectra of Mars (5.4 - 10.5 μm): Evidence for sulfates, carbonates, and hydrates. J Geophys Res 95:14,595-14,627

Presley MA, RE Arvidson (1988) Nature and origin of materials exposed in the Oxis Palus - Western Arabia - Sinus Meridiani region, Mars. Icarus 75:499-517

Popp RK, D Virgo, MW Phillips (1995a) H deficiency in kaersutitic amphiboles: Experimental verification. Am Mineral 80:1347-1350

Popp RK, D Virgo, HS Yoder Jr, TC Hoering, MW Phillips (1995b) An experimental study of phase equilibria and Fe oxy-component in kaersutitic amphibole: Implications for the f_{H_2} and a_{H_2O} in the upper mantle. Am Mineral 80:534-548

Prinz M, PH Hlava, K Keil (1974) The Chassigny meteorite: A relatively iron-rich cumulate dunite (abstr). Meteoritics 9:393-394

Reid AM, TE Bunch (1975) The nakhlites part II: Where, when, and how. Meteoritics 10:317-324

Reider R, T Economou, H Wänke, A Turkevich, J Crisp, J Brückner, G Dreibus, HY McSween (1997) The chemical composition of martian soil and rocks returned by the mobile alpha proton x-ray spectrometer: Preliminary results in x-ray mode. Science 278:1771-1774

Righter K, RL Hervig, DA Kring (1997) Ion microprobe analyses of SNC meteorite melt inclusions (abstr). Lunar Planet Sci XXVIII:1181-1182

Romanek CS, MM Grady, IP Wright, DW Mittlefehldt, RA Socki, CT Pillinger, EK Gibson Jr (1994) Record of fluid-rock interactions on Mars from the meteorite ALH84001. Nature 372:655-657

Romanek CS, AH Treiman, JH Jones, EK Gibson Jr, R Socki (1996) Oxygen isotope evidence for aqueous activity on Mars: $\delta^{18}O$ of Lafayette iddingsite (abstr). Lunar Planet Sci XXVII:1099-1100

Rubin AE (1997a) Mineralogy of meteorite groups. Met Planet Sci 32:231-247

Rubin AE (1997b) Mineralogy of meteorite groups: An update. Met Planet Sci 32:733-734.

Saxton JM, IC Lyon, E Chatzitheodoridis, G Turner (1997) Oxygen isotopic composition of Nakhla carbonate (abstr). Met Planet Sci 32:A114-A115

Sclar CB, SP Morzenti (1971) Shock-induced planar deformation features in olivine from the Chassigny meteorite. Meteoritics 6:310-311

Scott ERD, A Yamaguchi, AN Krot (1997) Petrological evidence for shock melting of carbonates in the martian meteorite ALH84001. Nature 387:377-379

Shearer CK, GD Layne, JJ Papike, MN Spilde (1996) Sulfur isotopic systematics in alteration assemblages in martian meteorite ALH84001. Geochim Cosmochim Acta 60:2921-2926

Shearer CK, MN Spilde, M Wiedenbeck, JJ Papike (1997) The petrogenetic relationship between carbonates and pyrite in martian meteorite ALH84001 (abstr). Lunar Planet Sci XXVIII:1293-1294

Shih C-Y, LE Nyquist, DD Bogard, GA McKay, JL Wooden, BM Bansal, H Wiesmann (1982) Chronology and petrogenesis of young achondrites, Shergotty, Zagami, and ALHA77005: Late magmatism on a geologically active planet. Geochim Cosmochim Acta 46:2323-2344

Shih C-Y, LE Nyquist, H Wiesmann (1996) Sm-Nd systematics of the nakhlite Governador Valadares (abstr). Lunar Planet Sci XXVII:1197-1198

Sidorov Yu I, MYu Zolotov (1986) Weathering of martian surface rocks. In: Chemistry and Physics of Terrestrial Planets. S Saxena (ed) p 191-223. Springer, New York

Shoemaker EM, RJ Hackman, RE Eggleton (1963) Interplanetary correlation of geologic time. Adv Astronautical Sci 8:70-89

Singer RB, HY McSween Jr (1993) The igneous crust of Mars: Compositional evidence from remote sensing and the SNC meteorites. In: Resources of Near-Earth Space. JS Lewis, MS Matthews, MMI Guerrieri (eds) p 709-736. Univ Arizona Press, Tucson, AZ

Singer RB, RN Clark, PD Owensby (1980) Mars: New regional near-infrared spectrophotometry (0.65-2.50 μm) obtained during the 1980 apparition (abstr). Bull Am Astron Soc 22:680

Smith JV, RL Hervig (1979) Shergotty meteorite: Mineralogy, petrology, and minor elements. Meteoritics 14:121-142

Smith JV, IM Steele (1984) Achondrite ALHA77005: Alteration of chromite and olivine. Meteoritics 19:121-133

Smith JV, IM Steele, CA Leitch (1983) Mineral chemistry of the shergottites, nakhlites, Chassigny, Brachina, pallasites and ureilites. J Geophys Res 88 (suppl): B229-B236

Smith MR, JC Laul, M-S Ma, T Huston, RM Verkouteren, ME Lipschutz, RA Schmidt (1984) Petrogenesis of the SNC (shergottites, nakhlites, chassignites) meteorites: Implications for their origin from a large, dynamic planet, possibly Mars. J Geophys Res 69:B612-B630

Soderblom LA (1992) The composition and mineralogy of the martian surface from spectroscopic observations: 0.3 μm to 50 μm. In: Mars. HH Kieffer, BM Jakosky, CW Snyder, MS Matthews (eds) p 557-593. Univ Arizona Press, Tucson, AZ

Steele IM, JV Smith (1982) Petrography and mineralogy of two basalts and olivine-pyroxene-spinel fragments in achondrite EETA79001. J Geophys Res 87:A375-A384

Stöffler D, R Ostertag, C Jammes, G Pfannschmidt, PR Sen Gupta, SB Simon, JJ Papike, RH Beauchamp (1986) Shock metamorphism and petrography of the Shergotty achondrite. Geochim Cosmochim Acta 50:889-913

Stöffler D, K Keil, ERD Scott (1991) Shock metamorphism of ordinary chondrites. Geochim Cosmochim Acta 55:3845-3867

Stolper EM, HY McSween Jr (1979) Petrology and origin of the shergottite meteorites. Geochim Cosmochim Acta 43:1475-1498

Sunshine JM, LA McFadden, CM Pieters (1993) Reflectance spectra of the Elephant Moraine A79001 meteorite: Implications for remote sensing of planetary bodies. Icarus 105:79-91

Swindle TD, MW Caffee, CM Hohenberg (1986) Xenon and other noble gases in shergottites. Geochim Cosmochim Acta 50:1001-1015

Swindle TD, MK Burkland, JA Grier, DJ Lindstrom, AH Treiman (1995) Noble gas analysis and INAA of aqueous alteration products from the Lafayette meteorite: Liquid water on Mars < 350 Ma ago (abstr). Lunar Planet Sci XXVII:1385-1386

Swindle TD, JA Grier, B Li, E Olson, DJ Lindstrom, AH Treiman (1997) K-Ar ages of Lafayette weathering products: Evidence for near-surface liquid water on Mars in the last hundred million years (abstr). Lunar Planet Sci XXVIII:1403-1404

Treiman AH (1985) Amphibole and hercynite spinel in Shergotty and Zagami: Magmatic water, depth of crystallization, and metasomatism. Meteoritics 20:229-243

Treiman AH (1986) The parental magma of the Nakhla achondrite: Ultrabasic volcanism on the shergottite parent body. Geochim Cosmochim Acta 50:1061-1070

Treiman AH (1987) Geology of the nakhlite meteorites: Cumulate rocks from flows and shallow intrusions (abstr). Lunar Planet Sci XVIII:1022-1023

Treiman AH (1990) Complex petrogenesis of the Nakhla (SNC) meteorite: Evidence from petrography and mineral chemistry. Proc Lunar Planet Sci Conf 20:273-280

Treiman AH (1993) The parent magma of the Nakhla (SNC) meteorite, inferred from magmatic inclusions. Geochim Cosmochim Acta 57:4753-4767

Treiman AH (1995a) S =/ NC: Multiple source areas for martian meteorites. J Geophys Res 100:5329-5340

Treiman AH (1995b) A petrographic history of martian meteorite ALH84001: Two shocks and an ancient age. Meteoritics 30:294-302

Treiman AH (1996a) Comment on "Mars as the parent body of the CI carbonaceous chondrites" by J. E. Brandenburg. Geophys Res Lett 23:3275-3276

Treiman AH (1996b) The perils of partition: Difficulties in retrieving magma compostions from chemically equilibrated basaltic meteorites. Geochim Cosmochim Acta 60:147-155

Treiman AH (1997) Amphibole in martian meteorite EET 79001 (abstr). Met Planet Sci 32:A129-A130

Treiman AH, DJ Lindstrom (1997) Trace element geochemistry of martian iddingsite in the Lafayette meteorite. J Geophys Res102:9,153-9,164

Treiman AH, SR Sutton (1992) Petrogenesis of the Zagami meteorite: Inferences from synchrotron X-ray (SXRF) microprobe and electron microprobe analyses of pyroxenes. Geochim Cosmochim Acta 56:4059-4074

Treiman AH, MJ Drake, J Hertogen, MJ Janssens, R Wolf, E Ebihara (1986) Core formation in the earth and shergottite parent body (SPB): Chemical evidence from basalts. Geochim Cosmochim Acta 50:1071-1091

Treiman AH, RA Barrett, JL Gooding (1993) Preterrestrial aqueous alteration of the Lafayette (SNC) meteorite. Meteoritics 28:86-97

Treiman AH, GA McKay, DD Bogard, DW Mittlefehldt, M-S Wang, L Keller, ME Lipschutz, MM Lindstrom, D Garrison (1994) Comparison of the LEW88516 and ALHA77005 martian meteorites: Similar but distinct. Meteoritics 29:581-592

Treiman AH, GJ Taylor, R Friedman (1995) Nakhla and its look-alikes: Al-depleted magmas and mantle differentiation on Mars and the Earth (abstr). Lunar Planet Sci XXVI:1419-1420

Tschermak G (1872) Die Meteoriten von Shergotty and Gopalpur. Sitzungsber. Akad. Wiss. Wien Math Naturwiss. Kl. 65:122-146

Tschermak G (1885) The Microscopic Properties of Meteorites (translated 1964 by JA Wood, EM Wood). Smithsonian Contrib Astrophys 4:6, Smithsonian Institution, Washington, DC

Turner G, SF Knott, RD Ash, JD Gilmour (1997) Ar-Ar chronology of the Martian meteorite ALH84001: Evidence for the early bombardment of Mars. Geochim Cosmochim Acta 61:3835-3850

Valley JW, JM Eiler, CM Graham, EK Gibson, CS Romanek, EM Stolper (1997) Low-temperature carbonate concretions in the martian meteorite ALH84001: Evidence from stable istopes and mineralogy. Science 275:1633-1638

Vicenzi EP, K Tobin, PJ Heaney, TC Onstott, J Chun (1997) Carbonate in Lafayette meteorite: A detailed microanalytical study (abstr). Met Planet Sci 32:A132-A133

Vickery AM, HJ Melosh (1983) The origin of SNC meteorites: An alternative to Mars. Icarus 56:299-318

Vickery AM, HJ Melosh (1987) The large crater origin for the SNC meteorites. Science 237:738-743

Vistisen L, D Petersen, MB Madsen (1992) Mössbauer spectroscopy showing large scale inhomogeneity in the presumed martian meteorite Zagami. Physica Scripta 46:94-96.

Wadhwa M, G Crozaz (1994) First evidence for infiltration metasomatism in a martian meteorite, ALH84001 (abstr). Meteoritics 29:545

Wadhwa M, G Crozaz (1995a) Trace and minor elements in minerals of nakhlites and Chassigny: Clues to their petrogenesis. Geochim Cosmochim Acta 59:3629-3645

Wadhwa M, G Crozaz (1995b) Constraints on the rare earth element characteristics of metasomatizing fluids in the martian meteorite ALH84001 (abstr). Lunar Planet Sci XXVI:1451-1452

Wadhwa M, GW Lugmair (1996) The formation age of carbonates in ALH84001 (abstr). Met Planet Sci 31:A145

Wadhwa M, HY McSween Jr, G. Crozaz (1994) Petrogenesis of shergottite meteorites inferred from minor and trace element microdistributions. Geochim Cosmochim Acta 58:4213-4229

Wadhwa M, G McKay, G Crozaz (1997) Trace element distributions in Yamato 793605, a chip off the "martian lherzolite" block (abstr). Antarctic Meteorites XXII:197-199

Walker D, EM Stolper, JF Hays (1979) Basaltic volcanism: The importance of planet size. Proc Lunar Planet Sci Conf 10:1995-2015

Warren PH (1987) Mars regolith versus SNC meteorites: Possible evidence for abundant crustal carbonates. Icarus 70:153-161

Warren PH, GW Kallemeyn (1996) Siderophile trace elements in ALH84001, other SNC meteorites and eucrites: Evidence of heterogeneity, possibly time-linked, in the mantle of Mars. Met Planet Sci 31:97-105

Warren PH, GW Kallemeyn (1997) Yamato-793605, EET79001, and other presumed martian meteorites: Compositional clues to their origins. Proc NIPR Symp Antarctic Met (in press)

Wasson JT, GW Wetherill (1979) Dynamical, chemical, and isotopic evidence regarding the formation locations of asteroids and meteorites. In: Asteroids. T Gehrels (ed) p 926-974. Univ Arizona Press, Tucson

Watson LL, ID Hutcheon, S Epstein, EM. Stolper (1994) Water on Mars: Clues from deuterium/hydrogen and water contents of hydrous phases in SNC meteorites. Science 265:86-90

Weincke HH (1978) Chemical and mineralogical examination of the Nakhla achondrite. Meteoritics 13: 660-664

Weins RC, RO Pepin (1988) Laboratory shock emplacement of noble gases, nitrogen, and carbon dioxide into basalt, and implications for trapped gases in shergottite EETA79001. Geochim Cosmochim Acta 52:295-307

Wentworth SJ, JL Gooding (1993) Weathering features and secondary minerals in Antarctic shergottites ALHA77005 and LEW88516 (abstr). Lunar Planet Sci XXIV:1507-1508

Wentworth SJ and JL Gooding (1994) Carbonate and sulfate minerals in the Chassigny meteorite: Further evidence for aqueous activity on the SNC parent body. Meteoritics 29:860-863

Wentworth SJ., JL Gooding (1995) Carbonates in the martian meteorite, ALH84001: Water-borne but not like the SNCs (abstr). Lunar Planet Sci XXVI:1489-1490

Wentworth SJ., JL Gooding (1996) Water-based alteration of the martian meteorite, QUE94201, by sulfate-dominated solutions (abstr). Lunar Planet Sci XXVII:1421-1422

Wetherill G (1976) Where do meteorites come from? Geochim Cosmochim Acta 40:1297-1317

Wetherill G (1984) Orbital evolution of impact ejecta from Mars. Meteoritics 19:1-12.

Wood CA, LD Ashwal (1981) SNC meteorites: Igneous rocks from Mars? Proc Lunar Planet Sci Conf 12:1359-1375

Wooden JL, LE Nyquist, DD Bogard, BM Bansal, H Wiesmann, C-Y Shih, GA McKay (1979) Radiometric ages for the achondrites Chervony Kut, Governador Valadares, Allan Hills 77005 (abstr). Lunar Planet Sci X, 1379-1380

Wooden JL, C-Y Shih, LE Nyquist, BM. Bansal, H Wiesmann, GA McKay (1982) Rb-Sr and Sm-Nd isotopic constraints on the origin of EETA79001: A second Antarctic shergottite (abstr). Lunar Planet Sci XXIII:879-880

Wright IP, MM Grady, CT Pillinger (1988) Carbon, oxygen, and nitrogen isotopic compositions of possible martian weathering products in EETA79001. Geochim Cosmochim Acta 52:917-924

Wright IP, MM Grady, CT Pillinger (1990) The evolution of atmospheric CO_2 on Mars: The perspective from carbon isotope measurements. J Geophys Res 95:14,789-14,794
Yamada I, T Mikouchi, M Miyamoto, T Murakami (1997) Lamellar inclusion in olivine from Nahkla (SNC) meteorite (abstr). Lunar Planet Sci XXVIII:1597-1598

APPENDIX

Representative Mineral Compositions in SNC Meteorites

The Tables on the following pages contain electron microprobe analyses of pyroxenes (A1), feldspars (or maskelynite) (A2), olivines (A3), and oxides (A4) in basaltic and lherzolitic shergottites, nakhlites, Chassigny, and ALH84001.

Table A1. PYROXENES

	Shergotty				Zagami				EETA79001				QUE94201			
	Pig Core	Pig Rim	Aug Core	Aug Rim	Pig Core	Pig Rim	Aug Core	Aug Rim	Pig Core	Pig Rim	Aug Core	Aug Rim	Low-Ca Px	Pig Core	Aug Core	Pyrox Rim
	Smith & Hervig (1979)				T. McCoy (unpub)				McSween & Jarosewich (1983)				McSween et al. (1996)			
Chemical Composition (Weight Percent)																
SiO_2	54.6	49.7	53.6	50.5	51.94	49.29	52.08	50.33	53.8	49.9	52.1	50.7	54.7	51.7	50.8	47.9
Al_2O_3	0.74	0.66	0.96	0.87	0.69	0.53	0.97	0.93	0.70	0.83	1.50	1.21	0.60	1.39	2.33	0.66
TiO_2	0.08	0.40	0.20	0.52	0.13	0.59	0.17	0.77	0.11	0.59	0.26	0.66	0.11	0.52	0.69	0.93
Cr_2O_3	0.43	0.10	0.70	0.17	0.41	0.03	0.72	0.15	0.45	0.27	0.62	0.44	0.49	0.46	0.84	0.06
FeO	16.9	31.1	11.9	25.8	18.90	31.67	12.09	21.18	16.7	27.5	15.0	16.9	15.2	20.5	12.3	34.2
MnO	0.53	0.78	0.41	0.68	0.66	0.91	0.42	0.59	0.63	0.84	0.52	0.59	0.60	0.58	0.43	0.90
MgO	21.8	11.0	16.9	9.62	19.17	11.09	16.00	11.00	23.0	12.6	17.6	13.6	24.5	17.9	14.5	7.48
CaO	6.11	6.64	16.7	12.2	6.35	6.50	16.59	15.46	4.38	7.41	11.8	14.5	3.03	6.98	17.9	7.86
Na_2O	0.07	0.09	0.15	0.11	0.09	0.08	0.18	0.16	0.08	0.11	0.20	0.19	0.04	0.08	0.19	0.05
Total	101.28	100.03	101.42	100.51	98.34	100.69	99.22	100.57	99.85	100.05	99.60	98.79	99.27	100.11	99.98	100.04
Cation Formula Based on 6 Oxygens (Ideal Pyroxene = 4 Cations per 6 Oxygens)																
Si	1.975	1.967	1.972	1.990	1.977	1.955	1.962	1.949	1.979	1.955	1.953	1.958	1.998	1.95	1.91	1.95
^{IV}Al	.025	.031	.028	.010	.023	.025	.038	.042	.021	.039	.047	.042	.002	.05	.09	.03
Total tet	2.000	1.998	2.000	2.000	2.000	1.980	2.000	1.991	2.000	1.94	2.000	2.000	2.000	2.00	2.00	1.98
Ti	.002	.012	.006	.015	.004	.018	.005	.023	.002	.017	.006	.018	.002	.01	.02	.03
^{VI}Al	.007	bd	bd	.030	.008	bd	.005	bd	.009	bd	.019	.012	.023	.01	.01	bd
Cr	.013	.003	.020	.005	.012	.001	.021	.004	.013	.008	.018	.012	.014	.01	.03	.00
Fe	.514	1.030	.366	.846	.602	1.050	.381	.686	.513	.898	.470	.550	.465	.65	.39	1.16
Mn	.016	.026	.013	.023	.021	.031	.013	.019	.019	.027	.015	.018	.018	.02	.01	.03
Mg	1.182	.650	.923	.561	1.088	.656	.899	.635	1.260	.737	1.985	.784	1.333	1.01	.81	.45
Ca	.239	.282	.658	.514	.259	.276	.670	.641	.171	.310	.472	.599	.118	.28	.72	.34
Na	.005	.007	.011	.009	.007	.006	.013	.012	.005	.008	.014	.013	.002	.01	.01	.00
Total Cations	3.978	4.008	4.011	4.003	4.001	4.018	4.007	4.011	3.992	3.945	3.999	4.006	3.975	4.00	4.00	3.99
Cation Ratios: Ca: Mg: Fe and Fe/(Fe+Mg)																
Ca	.124	0.144	0.338	0.268	0.133	0.139	0.344	0.327	0.088	0.088	0.245	0.310	0.062	.34	.38	.17
Mg	.611	.331	.474	.292	.558	.331	.461	.324	.648	.379	.511	.406	.696	.52	.42	.23
Fe	.265	.525	.188	.440	.309	.530	.195	.350	.264	.462	.244	.284	.242	.34	.20	.60
Fe/(Fe + Mg)	.303	.613	.284	.621	.356	.615	.298	.519	.289	.549	.323	.412	.259	.39	.33	.72

Table A1. PYROXENES (Cont'd)

	EETA79001				QUE94201				Nakhla			Lafayette	Gov. Valadares
	Pig Core	Pig Rim	Aug Core	Aug Rim	Pig Core	Pig Rim	Aug Core	Aug Rim	Pig	Aug Core	Aug Rim	Aug Core	Aug Core
	McSween & Jarosewich (1983)				McSween et al. (1996)				Treiman (1989)			Harvey & McSween (1992a)	
Chemical Composition (Weight Percent)													
SiO_2	55.2	53.4	53.6	52.5	55.05	52.17	52.21	52.48	49.64	52.68	49.82	51.36	51.51
Al_2O_3	0.48	0.83	1.29	1.72	0.62	1.34	1.68	1.44	0.69	1.03	1.45	0.85	0.74
TiO_2	0.11	0.47	0.32	0.39	0.07	0.43	0.36	0.28	0.37	0.34	0.44	0.23	0.20
Cr_2O_3	0.52	0.48	0.89	0.84	0.58	0.49	0.77	0.67	0.01	0.47	0.03	0.39	0.37
FeO	14.0	16.3	9.21	9.62	14.37	16.28	10.64	10.07	36.17	14.18	25.57	14.12	13.98
MnO	0.40	0.47	0.22	0.25	0.50	0.59	0.47	0.38	1.03	0.44	0.71	0.41	0.40
MgO	25.9	23.4	17.6	16.9	25.59	20.01	17.60	17.60	7.73	12.97	7.27	12.79	13.16
CaO	3.84	4.07	16.5	17.0	3.02	8.24	15.59	16.40	5.15	18.35	14.86	18.50	18.48
Na_2O					0.06	0.12	0.17	0.18	0.12	0.22	0.23	0.21	0.18
Total	100.45	99.42	99.63	99.22	99.85	99.67	99.48	99.49	100.91	100.67	100.38	98.87	98.91
Cation Formula Based on 6 Oxygens (Ideal Pyroxene = 4 Cations per 6 Oxygens)													
Si	1.985	1.975	1.975	1.953	1.99	1.95	1.94	1.95	1.965	1.969	1.994	1.972	1.97
^{IV}Al	.015	.025	.025	.047	.01	.05	.06	.05	.024	.320	.006	.028	.03
Total tet	2.000	2.000	2.000	2.000	2.00	2.00	2.00	2.00	2.000	2.000	2.000	2.000	2.00
Ti	.003	.013	.009	.011	.00	.01	.01	.01	.010	.013	.011	.007	.01
^{VI}Al	.005	.011	.031	.028	.02	.01	.01	.01	.022	.035	.027	.011	.00
Cr	.015	.014	.026	.025	.02	.01	.02	.02	.014	.001	.000	.012	.01
Fe	.421	.504	.284	.299	.44	.51	.33	.31	.446	.845	1.215	.454	.45
Mn	.012	.015	.007	.008	.02	.02	.02	.01	.014	.024	.350	.013	.01
Mg	1.388	1.290	.967	.937	1.38	1.11	.98	.98	.429	.463	1.153	.732	.75
Ca	.148	.138	.651	.677	.12	.33	.62	.65	.629	.222	.018	.761	.76
Na	nd	nd	nd	nd	.00	.01	.01	.01	.018	.010	.016	.015	.01
Total Cations	3.992	3.985	3.975	3.985	3.99	4.01	4.01	4.00	4.015	3.993	3.982	4.005	4.01
Cation Ratios: Ca: Mg: Fe and Fe/(Fe+Mg)													
Ca	.076	.071	.342	.354	.06	.17	.32	.34	.117	.386	.331	.392	.385
Mg	.709	.668	.508	.354	.71	.57	.51	.51	.244	.382	.227	.376	.381
Fe	.215	.261	.149	.156	.23	.26	.17	.16	.640	.237	.443	.232	.234
Fe/(Fe + Mg)	.233	.281	.227	.242	.24	.32	.25	.24	.724	.378	.661	.382	.380

Table A1. PYROXENES (Cont'd)

	Chassigny			ALH84001	
	Aug	Pig	Opx	Aug	Opx
	Wadhula & Crozaz (1996)			Treiman (1995)	
Chemical Compostion (Weight Percent)					
SiO_2	53.11	53.74	53.50	53.72	54.41
Al_2O_3	1.03	0.61	0.72	0.56	0.56
TiO_2	0.28	0.14	0.19	0.30	0.18
Cr_2O_3	0.77	0.41	0.30	0.49	0.29
FeO	9.33	15.90	17.89	7.48	17.50
MnO	0.34	0.50	0.59	0.27	0.51
MgO	16.71	22.98	24.35	15.94	25.59
CaO	17.87	5.21	1.87	21.01	1.52
Na_2O	0.32	0.11	0.05	0.37	0.02
Total	99.77	99.60	99.47	100.14	100.58
Cation Formula Based on 6 Oxygens (Ideal Pyroxene = 4 Cations per 6 Oxygens)					
Si	1.969	1.979	1.973	1.981	1.975
^{IV}Al	.031	.021	.027	.019	.024
Total IV	2.000	2.000	2.000	2.000	1.999
Ti	.008	.004	.005	.008	.005
^{VI}Al	.014	.005	.003	.005	.300
Cr	.022	.012	.009	.014	.009
Fe	.290	.490	.552	.231	.531
Mn	.010	.015	.019	.009	.016
Mg	.923	1.262	1.339	.876	1.384
Ca	.710	.205	.074	.830	.059
Na	.023	.008	.001	.026	.002
Total Cations	4.000	4.001	4.002	4.000	4.000
Cation Ratios: Ca: Mg: Fe and Fe/(Fe+Mg)					
Ca	.369	.105	.038	.425	.003
Mg	.480	.645	.681	.452	.701
Fe	.151	.250	.281	.119	.269
Fe/(Fe + Mg)	.239	.279	.292	.209	.277

Table A2. MASKELYNITE and FELDSPAR

	EETA 79001		QUE94201		ALHA77005		LEW88516	Nakhla	Chassigny		ALH84001
	Mask Core	Mask Rim	Mask Core	Mask Rim	Mask Core	Mask Rim	Mask	Plag	Plag	Sanid	Mask
	McSween & Jarosewich (1983)		McSween et al. (1996)		McSween et al. (1979a)		Harvey et al. (1993)	Weineke (1978)	Floran et al. (1978)		Mittlefehldt (1994)
Chemical Composition (Weight Percent)											
SiO_2	52.5	55.8	53.0	56.3	54.0	55.8	55.69	60.14	62.6	64.7	61.9
Al_2O_3	29.8	27.5	29.2	27.2	28.4	26.5	27.67	23.78	24.1	20.5	24.7
FeO	0.72	0.61	0.37	0.57	0.26	0.57	0.34	0.92	0.51	0.61	0.29
MgO	nd	nd	nd	nd	nd	nd	nd	0.06	nd	nd	0.05
CaO	12.7	10.5	12.9	10.6	10.7	8.96	10.25	6.44	6.3	0.95	6.29
Na_2O	4.38	5.16	4.11	5.02	5.20	6.00	5.41	7.37	7.0	5.0	7.07
K_2O	0.15	0.32	0.04	0.17	0.20	0.40	0.26	0.91	0.61	7.5	0.97
Total	100.25	99.89	99.62	99.86	98.76	98.23	99.61	99.62	101.1	99.17	101.27
Cation Formula Based on 8 Oxygens (Ideal Plagioclase = 5 Cations per 8 Oxygens)											
Si	2.381	2.519	2.41	2.54	2.467	2.556	2.52	2.716	2.749	2.927	2.721
^{IV}Al	1.594	1.463	1.56	1.45	1.530	1.431	1.47	1.266	1.247	1.093	1.278
Total tet	3.975	3.982	3.97	3.99	3.997	3.987	3.99	3.982	3.996	4.020	3.999
Fe	0.026	0.022	0.01	0.02	.010	.022	.01	.004	.019	.023	.011
Mg	nd	nd	nd	nd	nd	nd	nd	.004	nd	nd	.002
Ca	.616	.504	0.63	0.50	.524	.440	.50	.312	.296	.046	.296
Na	.385	.451	0.37	0.44	.461	.533	.47	.645	.596	.439	.603
K	.008	.018	.00	.01	.012	.023	.01	.052	.034	.433	.054
Total Cations	5.010	4.977	4.97	4.97	5.004	5.005	4.99	4.999	4.941	4.960	4.966
Molecular Proportion of Orthoclase (Or), Albite (Ab), and Anorthite (An)											
Or	.008	.018	.00	.01	.526	.442	.51	.052	.036	.473	.057
Ab	.382	.464	.37	.44	.462	.535	.48	.639	.644	.472	.632
An	.610	.518	.63	.50	.012	.023	.01	.309	.320	.050	.311

Table A3. OLIVINE

	EETA79001	ALHA77005	LEW88516	Nakhla	Lafayette	Gov. Valadares	Chassigny
	McSween & Jarosewich (1983)	McSween et al. (1979a)	Harvey et al. (1993)	Treiman (1993)	Harvey & McSween (1992a)		Floran et al. (1978)
			Chemical Composition (Weight Percent)				
SiO_2	38.3	37.5	37.0	33.46	31.94	32.00	37.1
Al_2O_3	0.02	nd	nd	0.05	nd	nd	0.08
FeO	24.0	24.4	27.47	48.42	52.09	52.61	27.6
MgO	36.8	37.1	34.7	17.36	13.41	13.16	34.0
MnO	0.55	0.67	0.58	0.95	0.97	1.01	0.53
Cr_2O_3	0.07	0.07	nd	0.02	nd	nd	0.04
CaO	0.26	0.25	0.14	0.37	0.19	0.36	0.19
Total	100.00	99.99	99.89	100.63	98.63	99.17	99.54
			Cation Formula Based on 4 Oxygens (Ideal Olivine = 3 Cations per 4 Oxygens)				
Si	1.005	.990	.99	0.994	.99	.99	.997
Al	.000	nd	nd	0.002	nd	nd	.003
Fe	.527	.532	.62	1.203	1.36	1.37	.620
Mg	1.437	1.460	1.39	.769	.62	.61	1.362
Mn	.011	.015	.01	.024	.03	.03	.012
Cr	.001	.001	.00	.001	nd	nd	.001
Ca	.007	.007	.01	.012	.01	.01	.001
Total Cations	2.988	3.005	3.02	3.005	3.01	3.01	2.996
			Cation Ratio Fe/(Fe+Mg)				
Fe/(Fe+Mg)	.268	.267	.308	.610	.69	.70	.313

Table A4. OXIDES

	EET79001			QUE94201		ALH77005			LEW88516		
	Ulvospinel	Ilmenite	Chromite	Ulvospinel	Ilmenite	Chromite Core	Chromite Rim	Illmenite	Chromite Core	Chromite Rim	Ilmenite
	McSween & Jarosewich (1983)			McSween et al. (1996)		McSween et al. (1979a)			Harvey et al. (1993)		
Chemical Composition (Weight Percent)											
TiO_2	30.7	51.1	0.81	30.6	52.6	1.02	11.1	54.3	0.90	8.67	53.71
Al_2O_3	1.72	0.19	7.11	2.07	0.08	6.95	7.14	0.18	5.62	6.39	0.09
Cr_2O_3	0.04	0.34	57.8	0.26	0.04	58.3	38.5	0.62	58.04	40.50	0.76
FeO	67.0	45.4	32.3	64.5	46.4	28.9	37.6	39.4	30.31	38.78	39.79
MgO	0.26	1.76	2.35	0.06	0.14	5.04	5.39	5.52	3.59	3.66	4.56
MnO	0.66	0.74	0.53	0.58	0.73	0.62	0.72	0.7	0.48	0.53	0.78
Total	100.38	99.53	100.90	98.07	99.99	100.83	100.45	100.72	98.94	98.53	99.69
Cation Formula Based on 4 Oxygens (Ideal Spinel = 3 Cations per 4 Oxygens)											
Ti	.869	.907	.021	.87	1.00	.027	.292	.988	0.02	0.23	1.98
Al	.075	.005	.294	.09	.00	.286	.294	.005	0.23	0.27	0.00
Cr	.000	.006	1.612	.01	.00	1.607	1.063	0.012	1.63	1.13	0.03
Fe	2.109	.958	.954	2.01	.98	.842	1.098	.797	0.91	1.16	1.63
Mg	.014	.066	.123	.00	.01	.262	.281	.199	0.19	0.19	0.33
Mn	.021	0.16	.015	.02	.02	.013	.016	.015	0.01	0.02	0.03
Total	3.088	1.895	3.019	3.00	3.01	3.037	3.044	2.016	2.99	3.07	4.00
Fe^{2+}/Fe^{3+}	7.61	bd	13.45	11.5	bd	9.94	11.5	bd	9.85	7.78	bd

Table A4. OXIDES (Cont'd)

	Nakhla		Chassigny		ALH84001
	Magnetite	Ilmenite	Chromite	Ilmenite	Chromite
	Treiman (1993)	Boctor et al. (1978)	Floran et al. (1978)		Mittlefehldt (1994)
	Chemical Composition (Weight Percent)				
TiO_2	14.81	51.3	3.6	51.2	2.23
Al_2O_3	6.83	0.24	6.6	0.70	8.53
Cr_2O_3	0.34	0.12	45.8	0.29	47.7
Fe	75.00	44.56	37.3	39.6	36.0
MgO	0.48	0.86	3.8	5.5	3.85
MnO	0.41	0.62	0.61	1.56	0.39
Total	97.87	97.70	97.71	98.85	98.70
	Cation Formula Based on 4 Oxygens (Ideal Spinel = 3 Cations per 4 Oxygens)				
Ti	0.622	1.974	.150	1.857	.058
Al	0.045	0.015	.417	.038	.350
Cr	0.008	0.005	1.940	.011	1.313
Fe	2.063	1.871	1.592	1.599	1.037
Mg	0.040	0.063	.304	.396	.202
Mn	0.020	0.027	.027	.640	.012
Total	2.798	3.955	4.430	4.541	2.972
Fe^{2+}/Fe^{3+}	1.59	56.6	3.65	7.30	4.21

Chapter 7

COMPARATIVE PLANETARY MINERALOGY: CHEMISTRY OF MELT-DERIVED PYROXENE, FELDSPAR, AND OLIVINE

James J. Papike

Institute of Meteoritics
Department of Earth and Planetary Sciences
University of New Mexico
Albuquerque, New Mexico 87131

INTRODUCTION

The purpose of this chapter is to provide some comparisons of the chemical compositions of pyroxene, feldspar, and olivine that crystallized from melts originating in differing planetary environments (chondrules, Earth, Moon, Mars, and Vesta). Because these melts formed from chemically distinct planetary reservoirs and because they experienced differing intensive thermodynamic parameters, the minerals that crystallized from these melts should reflect these differing conditions. Because of space and time constraints, this comparative mineralogical study will be limited in scope.

The data for terrestrial basalts is presented as averages for seven sample suites: Archean, Columbia Plateau, ocean floor, Hawaiian, island arc, Keweenawan, and Rio Grande. Therefore, although a small number of points will be plotted on the following data displays for terrestrial basalts, they represent 1263 pyroxene, 269 olivine, and 708 feldspar analyses (BVSP 1981). The details concerning these basalt suites are reported in BVSP (1981) while a specific discussion of the silicate mineralogy is reported by Papike (1981). The data for lunar mare basalts and lunar highland melt rocks are tabulated in the appendix to Chapter 5 (lunar samples) in this volume. Data for the silicate minerals in martian rocks are tabulated in the appendix to Chapter 6 (martian meteorites). Chemical data for pyroxene and olivine in chondrules are from the Semarkona ordinary chondrite, which is a good example of an unequilibrated chondrite containing chondrules with phenocrysts that still record melt-mineral partitioning uncompromised by re-equilibration (refer to Chapter 3, this volume). The pyroxene data for 4 Vesta is from six unequilibrated eucrites (Pun and Papike 1996); the feldspar data is from Table HED-4, Chapter 4, this volume (analyses 13-24).

Oxide analysis for these silicate phases were converted to mineral formulae (e.g. Papike 1987, 1988) before preparing the chemical data displays. The index of melt fractionation (the x-axis on some plots) used in this study is X_{Fe} = Fe/(Fe+Mg) atomic.

Chemical data used in this study were obtained by electron microprobe techniques. An important advantage of these techniques is that they involve very good spatial resolution down to ~ one micrometer. However, a serious limitation is that no direct estimate is made of Fe^{3+}/Fe^{2+}, Ti^{4+}/Ti^{3+}, or Cr^{3+}/Cr^{2+}. A general discussion of chemical data for the rock-forming silicates, their crystal structures, and recommended methods of calculated mineral formulae is provided by Papike (1987, 1988).

PLAGIOCLASE FELDSPAR

Plagioclase feldspar contains a small but significant amount of Fe and Mg. Figure 1 plots Fe versus % An and shows that Fe varies from 0 to ~0.05 atoms per formula unit (afu). Although there has been considerable discussion of how Fe is incorporated into the

0275-0279/98/0036-0007$05.00

plagioclase crystal structure, serious questions still remain. Ferric iron can substitute for Al in tetrahedral coordination and limited Fe^{2+} and Mg can substitute for Ca. Jolliff and Hsu (1996) suggest that Fe and Mg in plagioclase from lunar ferroan anorthosites is largely tetrahedral as in $CaSi_3(Fe,Mg)O_8$. They suggest that exsolution of Fe and Mg from plagioclase involves the reaction $CaSi_3(Fe,Mg)O_8 \rightarrow Ca(Mg,Fe)Si_2O_6$ (augite) + SiO_2.

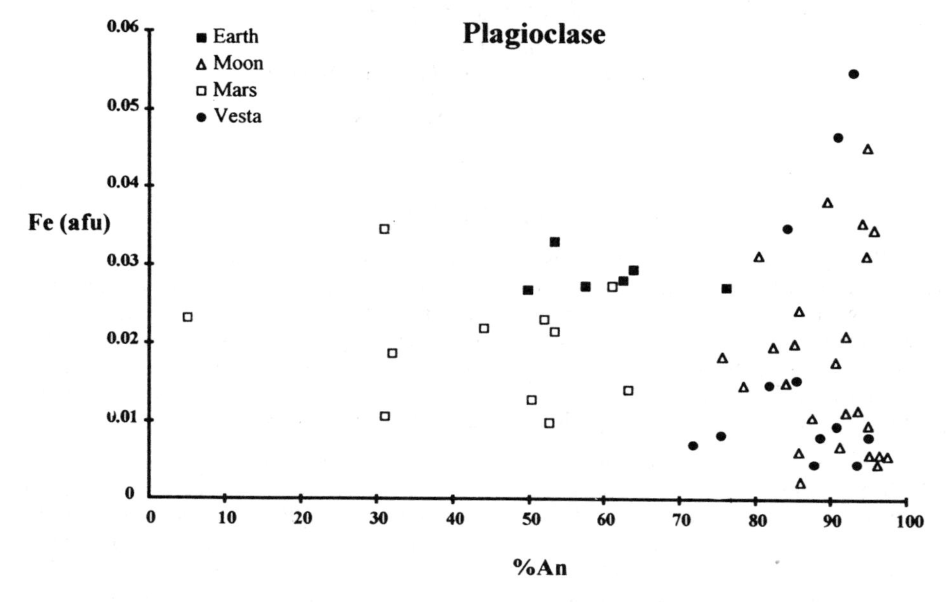

Figure 1. Fe atoms per formula unit (afu) versus % anorthite in plagioclase from planetary melt rocks.

Figure 2 is a plot of K versus An content. The data array shows interesting groupings with martian plagioclase ploting at higher Na and K, terrestrial plagioclase ploting in an intermediate position, and plagioclase from the Moon and 4 Vesta ploting at the lowest Na and K contents. This is an interesting result because it demonstrates the utility of minerals as geochemical recorders, an important attribute when considering planetary regolith samples or small non-representative lithic fragments. In this case, the basaltic melts reflect the relative volatile budget of planetary mantle reservoirs and the plagioclase, which crystallized from these basalts, also records this information. A number of previous studies have discussed the volatile content of the sampled planetary bodies (e.g. Drake et al. 1989) and their results are consistent with these observations.

OLIVINE

Figure 3a is a plot of Al versus X_{Fe}. Basically, it shows that small amounts of Al can substitute into the olivine crystal structure. Perhaps the most obvious feature of this diagram is the very low Al content of chondrule olivine; in most analyses Al is below detection. Figure 3b is a plot of Cr versus X_{Fe} for olivine and appears on the same diagram as Figure 3a for comparison. Whereas the Al content of Semarkona chondrule olivine is very low, Cr is quite high relative to olivine from other planetary basalts. Perhaps a major contributing factor to this systematic is that much of the Cr is in the reduced Cr^{2+} valance state which is compatible with the olivine structure. In a study of chromium partitioning

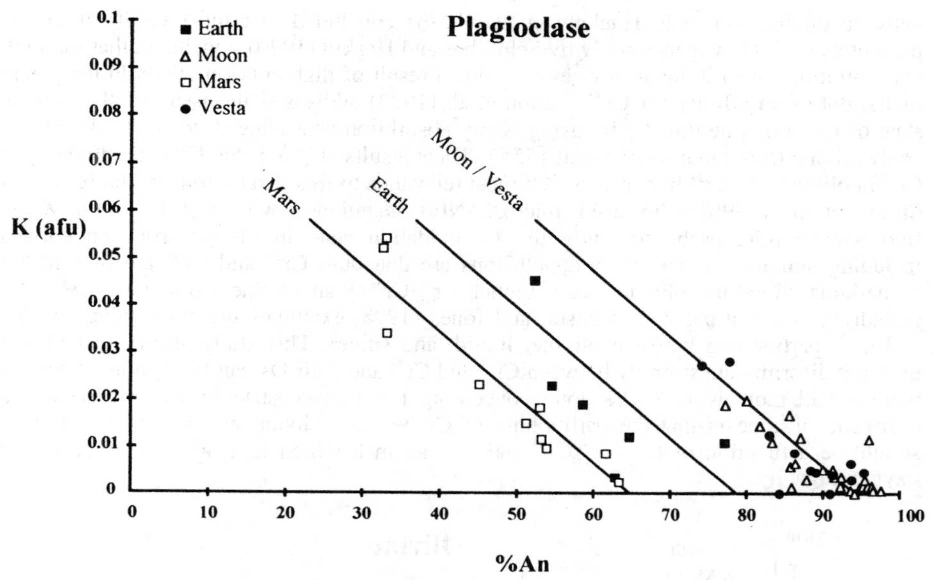

Figure 2. K versus % anorthite in plagioclase from planetary melt rocks. Trend lines are approximate.

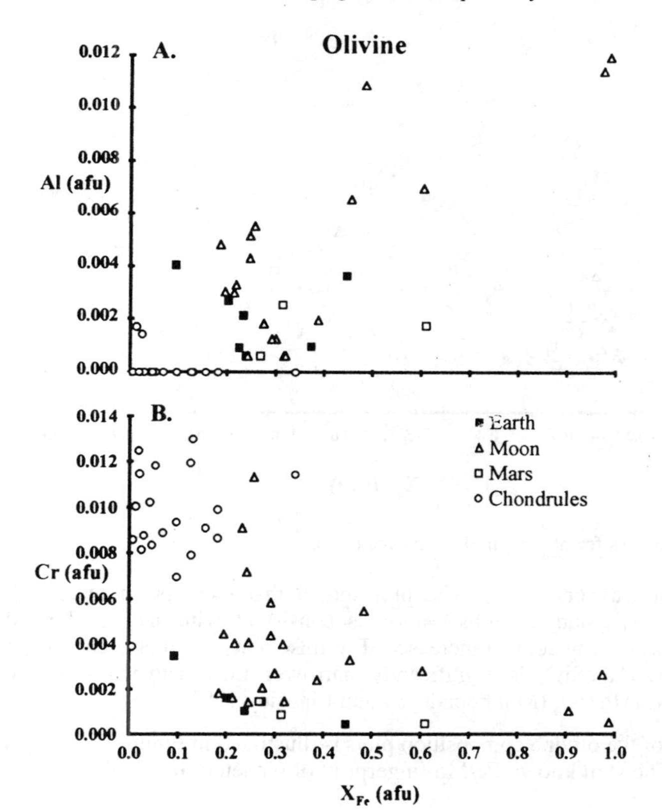

Figure 3. Fe/(Fe+Mg) atomic = X_{Fe} for olivine from planetary melt rocks versus (A) Al, (B) Cr.

between olivine and melt, Huebner et al. (1976) concluded that most Cr in olivine is present as Cr^{2+}. However, a study by Schreiber and Haskin (1976) concluded that the high concentrations of Cr in lunar olivines are a direct result of high concentrations in the parent melts, not of an affinity for Cr^{2+}. Sutton et al. (1993) addressed the issue of the valance state of Cr in olivine directly by using X-ray absorbtion near edge structure (XANES) to study olivine from lunar mare basalt 15555. Their results showed that Cr is predominently Cr^{2+} in olivine and Cr^{3+} in pyroxene. Of most relevance to this discussion are the results of Sutton et al. (1996) who used microXANES techniques with a synchroton X-ray fluorescence mircoprobe to study the Cr oxidation state in olivine from chondrules including Semarkona. The main conclusions are that both Cr^{2+} and Cr^{3+} are present but Semarkona chondrule olivine has a higher Cr^{2+}/Cr^{3+} than olivine from the ALH77307 chondrite. A recent paper by Hanson and Jones (1998) examines the systematics of Cr^{3+} and Cr^{2+} partitioning between olivine, liquid, and spinel. This study shows that olivine does not discriminate strongly between Cr^{3+} and Cr^{2+} and their Ds can be similar. Although there is still more work to be done concerning the valance state of Cr in olivine and pyroxene, it appears that the partitioning of Cr between olivine and pyroxene Cr_{ol}/Cr_{px} should be a function of the Cr^{2+}/Cr^{3+} ratio in the melt which is a direct function of the oxygen fugacity.

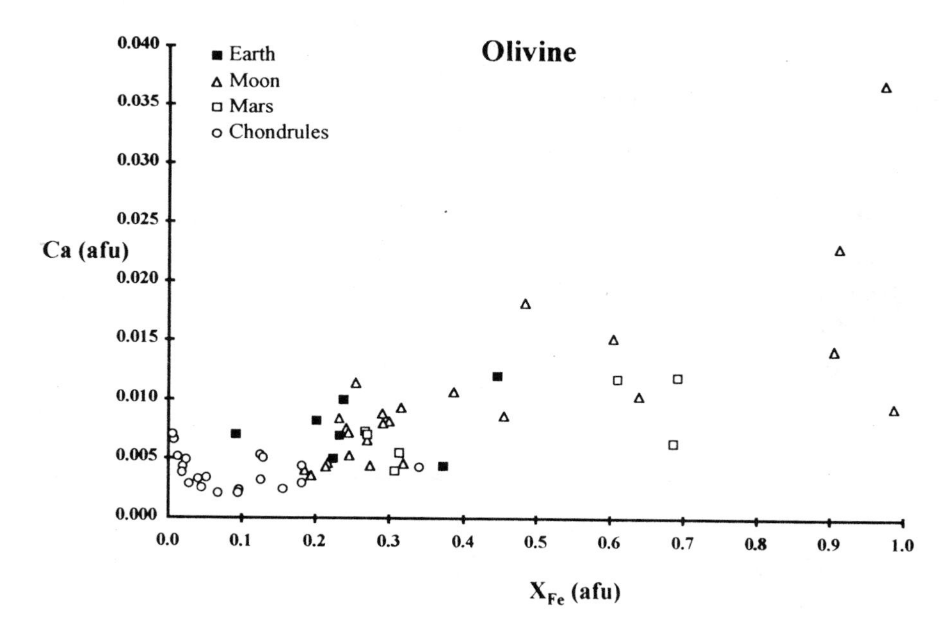

Figure 4. Ca versus X_{Fe} systematics for olivine in planetary melt rocks.

Figure 4 is a plot of Ca versus X_{Fe}. The plot shows that there is an increasing concentration of Ca with X_{Fe}, and this observation is consistent with increased solid solubility of Ca in olivine as its iron content increases. The miscibility gap between fayalite Fe_2SiO_4 and kirschsteinite $CaFeSiO_4$ is significantly narrower then between forsterite Mg_2SiO_4 and monticellite $CaMgSiO_4$ (Mukhopadhyay and Lindsley 1983).

The most interesting of the olivine composition plots is illustrated in Figure 5, which is a plot of Mn against Fe. The well known Fe/Mn fingerprint of planetary reservoirs (e.g.

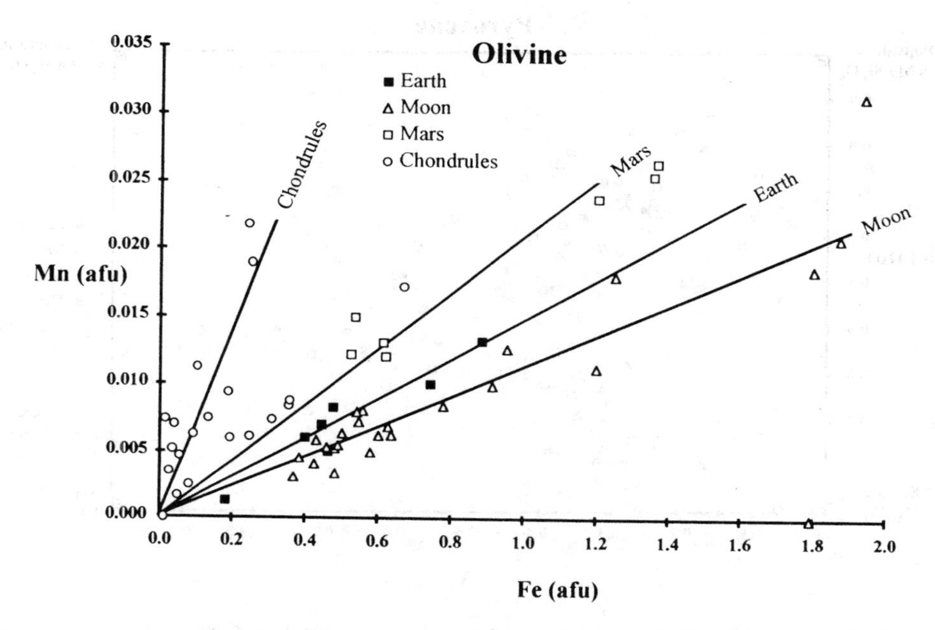

Figure 5. Mn versus Fe systematics for olivine from planetary melt rocks. Trend lines are approximate.

Drake et al. 1989) is clearly indicated in this plot. The Fe and Mn contents of olivine from the Moon fall along a trajectory at low Mn/Fe with the Earth on a trajectory slightly above it. The Mn/Fe of olivine from martian basalts fall on an intermediate trajectory and the Mn/Fe for Semarkona chondrule olivine plot at the highest Mn/Fe. Several factors must affect this Mn/Fe ratio including the composition of the planetary body, whether core formation took place, and the oxygen fugacity that controls the Fe^{2+}/Fe^{o} ratio. In the highly reduced Semarkona chondrules, the low fo_2 leads to low Fe^{2+} concentration and thus high Mn/Fe in olivine. However, oxygen fugacity and metal separation cannot be the only factors because Mn/Fe for olivine from martian meteorites is significantly higher than the Moon trajectory. Also, the Earth, which has a significantly higher fo_2 than the Moon, has basalts whose olivine compositions plot on a higher Mn/Fe trajectory. We will discuss the significance of Mn/Fe systematics more when we discuss pyroxenes.

PYROXENE

Pyroxene is a very effective recorder of mineral-melt partitioning (e.g. Bence and Papike 1972, Papike 1996), and its crystal structure accommodates a wide variety of cations. Figure 6 is a plot of Ca versus X_{Fe} and can be thought of as an alternative data display to the well-known pyroxene quadrilateral. The pyroxene endmembers plot on this diagram as follows: diopside, $CaMgSi_2O_6$ at Ca=1,X_{Fe}=0; hedenbergite, $CaFe^{2+}Si_2O_6$ at Ca = 1, X_{Fe} = 1; enstatite, $MgSiO_3$ at Ca = 0, X_{Fe} = 0; ferrosilite, $Fe^{2+}SiO_3$ at Ca = 0, X_{Fe} = 1. Pyroxene from chondrules comprise both augite and enstatite and reflect the large miscibility gap between these phases. Chondrule pyroxene in Semarkona is very magnesian relative to the other suites. Pyroxenes from lunar mare basalts and highland melt rocks show a large compositional range including compositions near X_{Fe}=1.0. These compositions with Ca < 0.5 afu and X_{Fe}~1 represent metastable phases at low pressures. Some compositions represent pyroxferroite, $Ca_{1/7}Fe_{6/7}SiO_3$ which is a metastable

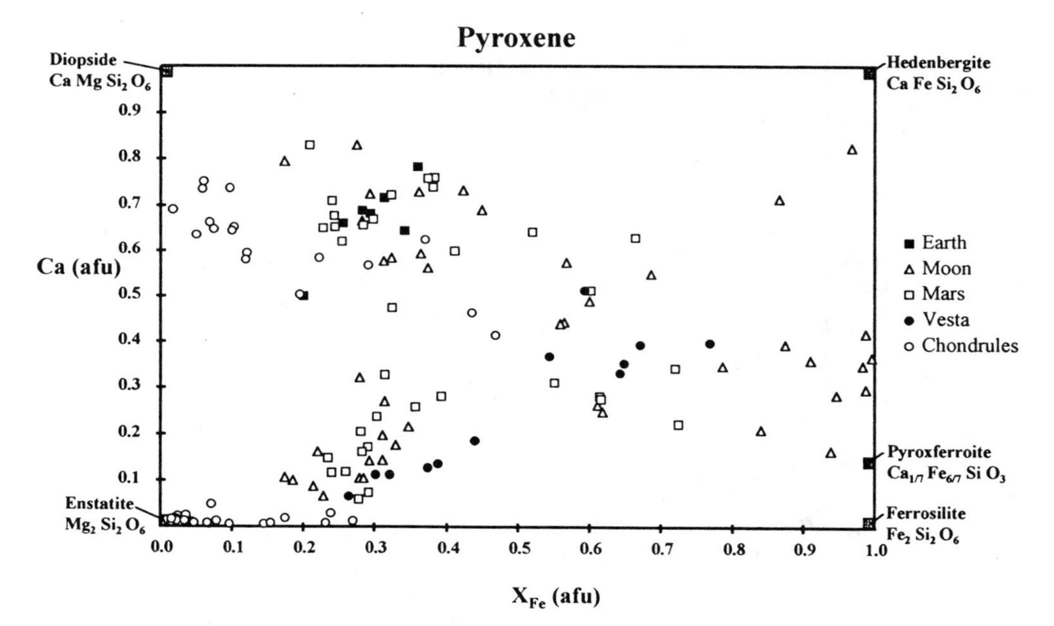

Figure 6. Ca versus X_{Fe} for pyroxene from planetary melt rocks. See text for discussion.

pyroxenoid under the low pressure mare basalt crystallization conditions (Bence and Papike 1972, Lindsley et al. 1972). The points plotting near Ca 0.8-0.9 and X_{Fe} 0.8-1.0 represent hedenbergite, which is a member of the stable (1 atm) 3-phase assemblage hedenbergite–fayalite–SiO_2. The pyroxene compositions for martian basalts cover a similar compositional range as those from lunar mare basalts. The pyroxenes from the terrestrial basalts considered here are augites. Pyroxenes from 4 Vesta basalts originate at pigeonite compositions and zone to iron-enriched varieties.

Figure 7 shows Na versus X_{Fe} systematics in pyroxene. Na contents of pyroxene from Semarkona chondrules show a large range of concentrations (augite has much higher Na concentration than enstatite; see Chapter 3 for a discussion) with many Na concentrations below detection. The Na contents in pyroxene from lunar and 4 Vesta basalts have low concentrations reflecting the volatile-depleted nature of their parent bodies. Pyroxene from martian basalts show higher Na contents consistent with the plagioclase data (Fig. 2).

The systematics of the other than quadrilateral components "other" (Cameron and Papike 1981) Ti, Al and Cr are illustrated in Figures 8a,b,c and 9a,b,c. These cations, which are not included in pyroxene quadrilateral (Ca-Mg-Fe^{2+}-Si-O) compositional space, involve coupled substitutions when entering the pyroxene structure. They involve site change excesses when they substitute into the pyroxene M1 octahedral site for divalent Mg or Fe^{2+}. If Cr is mainly Cr^{3+} and not Cr^{2+} and if Ti is both Ti^{4+} and Ti^{3+} in the pyroxene structure then coupled subsitutions (not involving Na) are: $^{VI}Fe^{3+}$–^{IV}Al, $^{VI}Cr^{3+}$–^{IV}Al, $^{VI}Ti^{3+}$–^{IV}Al, $^{VI}Ti^{4+}$–$2^{IV}Al$, ^{VI}Al–^{IV}Al. Figure 8a show the Ti systematics for pyroxene. The two high Ti points for mare basalts are from the high Ti variety with TiO_2 ~ 10 wt %. Al systematics are illustrated in Figure 8b. Pyroxene from chondrules show a large variation in Al content with Al in augite at much higher concentration then in enstatite. Figure 8c shows Cr systematics. Pyroxene in Semarkona chondrules show the highest Cr contents on this diagram. Figures 9a,b,c show the systematics of Ti, Cr, and Al as a function of Ca

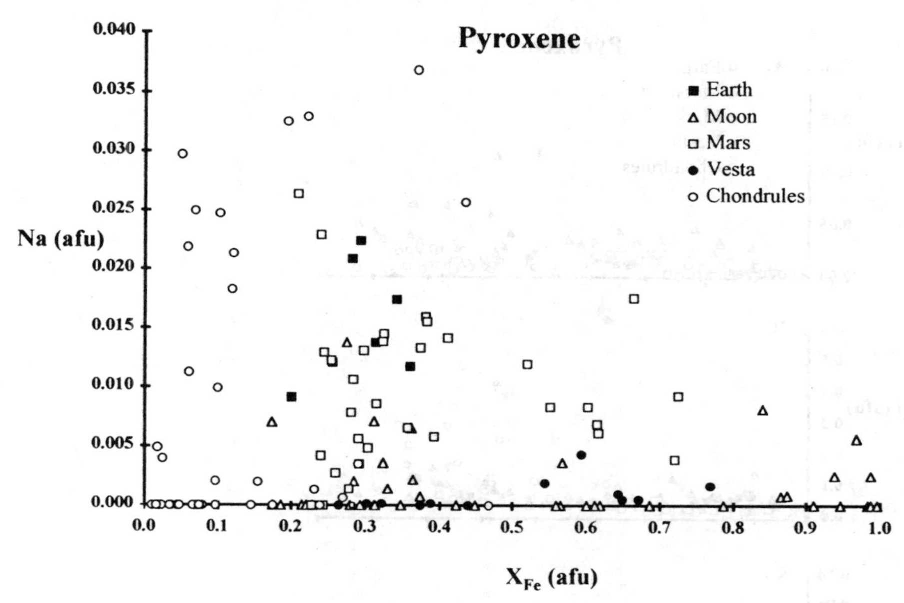

Figure 7 (above). Na versus X_{Fe} systematics for pyroxene from planetary melt rocks.

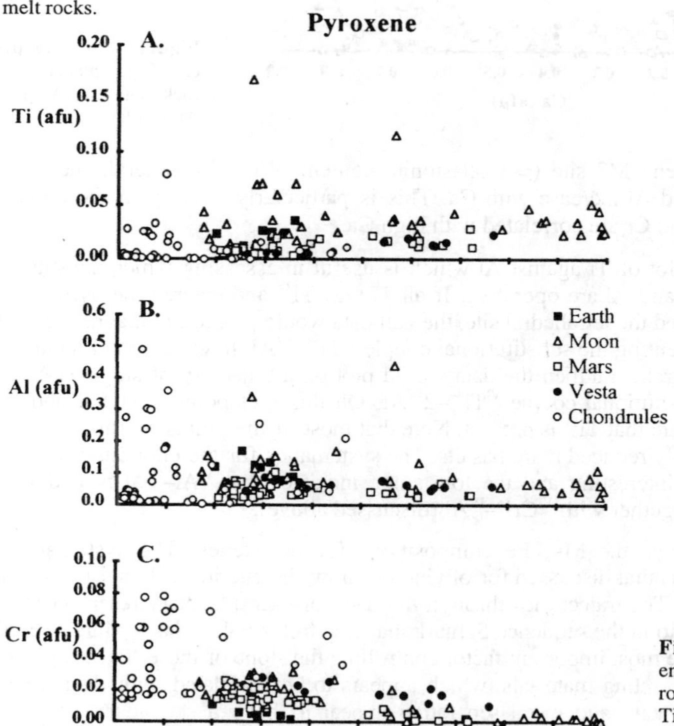

Figure 8. X_{Fe} for pyroxene from planetary melt rocks plotted against (A) Ti, (B) Al, (C) Cr.

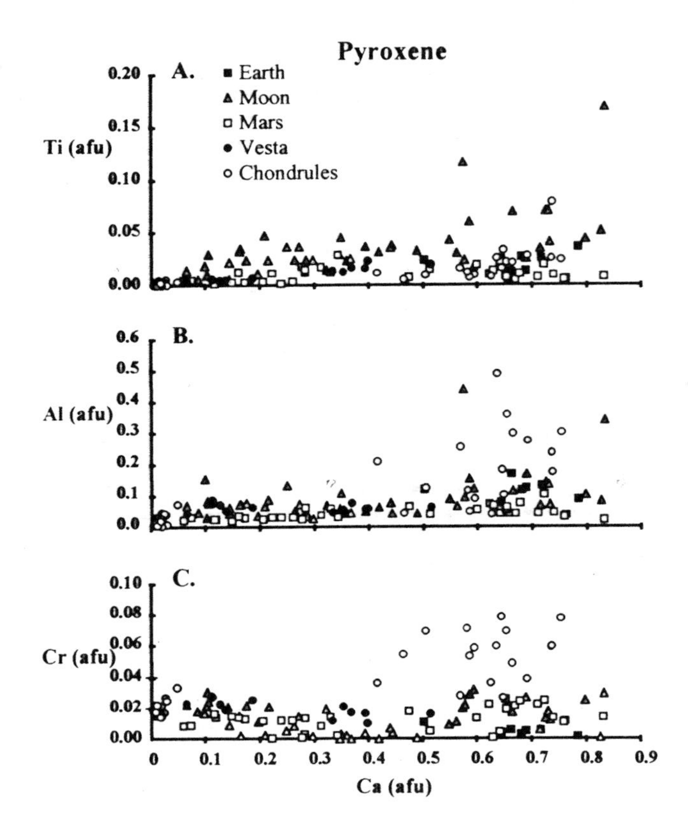

Figure 9. Ca for pyroxene from planetary melt rocks against (A) Ti, (B) Al, (C) Cr.

content of the pyroxene M2 site (= wollastonite content, Wo). In general, the concentrations of Ti, Cr, and Al increase with Ca. This is particularly striking in the chondrule data where high Al and Cr are correlated with high Ca.

Figure 10 is a plot of Ti against Al which is useful in assessing which substitutional couples involving Ti and Al are operative. If all Ti was Ti^{3+} and entered the pyroxene M1 site while all Al entered the tetrahedral site, then all data would plot in the diagram on a line with Ti/Al = 1 representing the substitutional couple $^{VI}Ti^{3+}$–^{IV}Al. If, on the other hand, all Ti is Ti^{4+} and all Al is tetrahedral then the data would plot on a trajectory of slope Ti/Al = 1/2 representing the substitutional couple $^{VI}Ti^{4+}$–$2^{IV}Al$. On this plot points that fall above the Ti/Al = 1/2 line indicate that Ti^{3+} is present. Note that most of the points plotting above this line are from the highly reduced mare basalts. The systematics for the chondrule pryoxenes on this diagram are interesting and the low slope indicates that ^{VI}Al–^{IV}Al is a dominant substitional couple together with $^{VI}Cr^{3+}$–^{IV}Al (discussed above).

Figure 11 displays the Mn, Fe compositions for pyroxenes. The systematics are essentially the same as that discussed for olivine but now the eucrites (olivine free) from 4 Vesta can be plotted. The trajectories through the data for each planetary reservoir show a decreasing Mn/Fe ratio in the sequence Semarkona chondrules > 4 Vesta > Mars > Earth > Moon. Apparently the most important factor controlling the slope of these trajectories is the Mn/Fe ratio of the accreting materials, which appears to be correlated with distance from the sun. Oxygen fugacity and metal separation appear to have a secondary effect. The

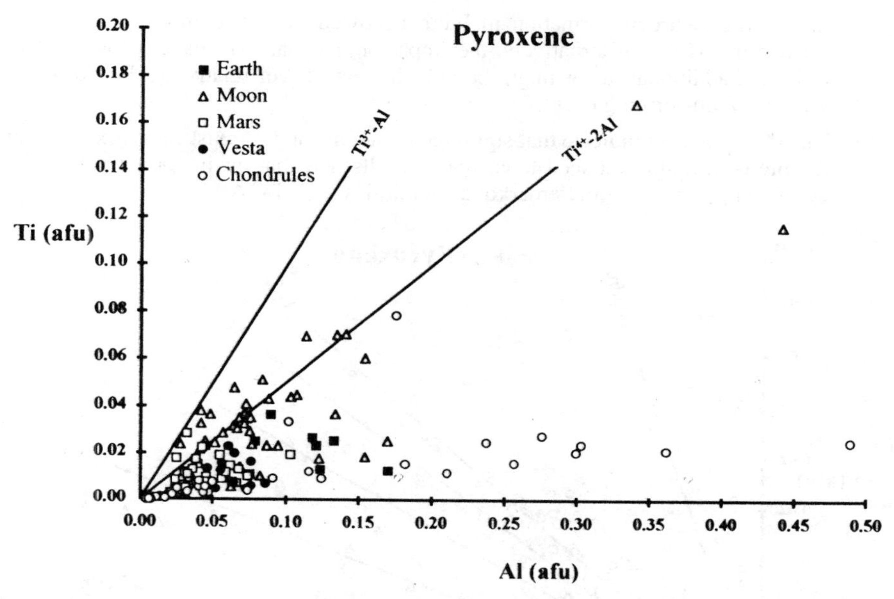

Figure 10. Ti versus Al systematics for pyroxene from planetary melt rocks. See text for discussion.

slightly higher Mn/Fe ratio for terrestrial basalts compared to lunar may reflect core formation on Earth *after* the "giant impact" which allegedly separated terrestrial material to combine with impactor material to form the Moon. Alternatively, the impactor material may have had a lower Mn/Fe ratio than the Earth's mantle. An additional factor might relate to the higher volatility of Mn compared to Fe and thus Mn was depleted relative to Fe during the violent Moon-forming event.

CONCLUSIONS

This brief review of the comparative chemistry of feldspar, olivine, and pyroxene from planetary melts demonstrates that these minerals are powerful recorders of the geochemical and petrologic processes on different planetary bodies. Some specific important information obtained from electron microprobe data on these phases is:

1. The planetary mantle sources have variable Na contents with Mars > Earth > Moon and 4 Vesta.
2. High Cr abundances in olivine phenocrysts from Semarkona chondrules indicate high Cr activity in the melt and high Cr^{2+}/Cr^{3+} with Cr^{2+} favoring olivine and Cr^{3+} favoring pyroxene.
3. Higher Ca contents in Fe-rich olivine than Mg-rich are consistent with a narrower miscibility gap between kirschsteinite, $CaFe^{2+}SiO_4$, and fayalite, $Fe_2^{2+}SiO_4$, than between monticellite, $CaMgSiO_4$, and forsterite, Mg_2SiO_4.
4. Mn/Fe ratios in olivine and pyroxene show decreasing values in the sequence Semarkona chondrules > 4 Vesta > Mars > Earth > Moon. The major factor controling these ratios appears to be the Mn/Fe of the accreting material, with Mn/Fe increasing with distance from the sun. A secondary factor is core formation which perhaps explains the higher Mn/Fe of terrestrial basalts relative to lunar. If

this is true then core formation in Earth followed the giant impact that allegedly formed the Moon. Alternatively, the impactor material may have a lower Mn/Fe ratio. An additional factor might be volatile loss of Mn relative to Fe during the violent Moon-forming event.

5. Ti, Al systematics indicate that significant amounts of Ti^{3+} exist in pyroxenes from the high-Ti, highly-reduced lunar mare basalts and that an important substitional couple in pyroxene from Semarkona chondrules is $^{VI}Al-^{IV}Al$.

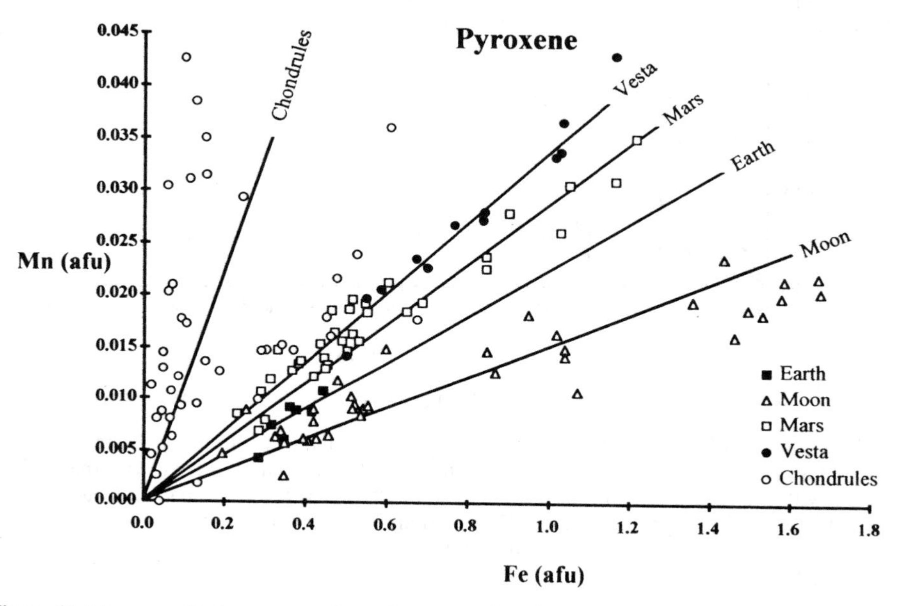

Figure 11. Mn versus Fe for pyroxene from planetary melt rock. Trend lines are approximate.

ACKNOWLEDGMENTS

This research was supported by NASA grant MRA 97-282 and the Institute of Meteoritics. Grant Fowler is thanked for his help with plotting the data and Rhian Jones for help with assembling the chondrule data. I thank John Jones, Rhian Jones, and Mike Spilde for their reviews of this manuscript.

REFERENCES

BVSP (Basaltic Volcanism Study Project) (1981) Basaltic Volcanism on the Terrestrial Planets. Pergamon, New York 1286 p

Bence AE, Papike JJ (1972) Pyroxenes as recorders of lunar basalt petrogenesis: Chemical trends due to crystal-liquid interaction. Proc Lunar Sci Conf 3:431-469

Cameron M and Papike JJ (1981) Structural and chemical variation in pyroxenes. Am Mineral 66:1-50

Drake, MJ, Newsom HE, Capobianco CJ (1989) V, Cr, and Mn in the Earth, Moon, EPB and SPB and the origin of the Moon: Experimental studies. Geochim Cosmochim Acta 53:2101-2111

Hanson B and Jones JH (1998) The systematics of Cr^{3+} and Cr^{2+} partitioning between olivine, liquid, and spinel. Am Mineral (in press)

Huebner JS, Lipin BR, Wiggens LB (1976) Partitioning of chromium between silicate crystals and melts. Proc Lunar Sci Conf 7:1195-1220

Jolliff BL and Hsu W (1996) Geochemical effects of recrystallization and exsolution of plagioclase of ferroan anorthosite. Lunar Planet Sci 27:611-612

Jolliff BL and Hsu W (1996) Geochemical effects of recrystallization and exsolution of plagioclase of ferroan anorthosite. Lunar Planet Sci 27:611-612

Lindsley DH, Papike JJ, Bence AE (1972) Pyroxferroite: Breakdown at low pressure and high temperature. Abstr 3rd Lunar Sci Conf, 483-485

Mukhopadhyay DK, Lindsley DH (1983) Phase relations in the join kirschsteinite ($CaFeSiO_4$)–fayalite (Fe_2SiO_4). Am Mineral 68:1089-1094

Papike JJ (1981) Silicate Mineralogy of Planetary Basalts. In: Basaltic Volcanism on the Terrestrial Planets. Pergamon, New York, 340-363

Papike JJ (1987) Chemistry of the rock-forming silicates: Ortho, ring, and single-chain structures. Rev Geophys 25:1483-1526

Papike JJ (1988) Chemistry of the rock-forming silicates: Multiple-chain, sheet, and framework structures. Rev Geophys 26:407-444

Papike JJ (1996) Pyroxene as a recorder of cumulate formational processes in asteroids, Moon, Mars, Earth: Reading the record with the ion microprobe. Am Mineral 81:525-544

Pun A and Papike JJ (1996) Unequilibrated eucrites and the equilibrated Juvinas eucrite: Pyroxene REE systematics and major, minor, and trace element zoning. Am Mineral 81:1438-1451

Schreiber HD and Haskin LA (1976) Chromium in basalts: Experimental determination of redox states and partitioning among synthetic silicate phases. Proc Lunar Sci Conf 7:1221-1259

Sutton SR, Jones KW, Gordon B, Rivers ML, Bajt S, Smith JV (1993) Reduced chromium in olivine grains from lunar basalt 15555: X-ray absorption near edge structure (XANES) Geochim Cosmochim Acta 57:461-468

Sutton SR, Bajt S, Jones R (1996) In situ determination of chromium oxidation state in olivine from chondrules. Lunar Planet Sci 27:1291-1292

INDEX

(compiled by Rhian H. Jones)

METEORITE INDEX

(Compiled by Adrian J. Brearley)

LUNAR SAMPLE INDEX

(Compiled by Adrian J. Brearley)